FOUNDATION ENGINEERING HANDBOOK

FOUNDATION ENGINEERING HANDBOOK

Second Edition

Edited by

HSAI-YANG FANG Ph.D.

Professor of Civil Engineering and Director, Geotechnical
Engineering Division, Fritz Engineering Laboratory, Lehigh University

VNR VAN NOSTRAND REINHOLD
_____ New York

Van Nostrand Reinhold
115 Fifth Avenue
New York, New York 10003

Chapman and Hall
2–6 Boundary Row
London, SE1 8HN, England

Thomas Nelson Australia
102 Dodds Street
South Melbourne 3205
Victoria, Australia

Nelson Canada
1120 Birchmount Road
Scarborough, Ontario M1K 5G4, Canada

16   15   14   13   12   11   10   9   8   7   6   5   4   3   2   1

Library of Congress Cataloging-in-Publication Data

Foundation engineering handbook / edited by Hsai-Yang Fang. — 2nd ed.
        p.   cm.
     Includes bibliographical references.
     ISBN 0-442-22487-7
     1. Foundations—Handbooks, manuals, etc.   2. Soil mechanics—
—Handbooks, manuals, etc.   I. Fang, Hsai-Yang.
     TA775.F675   1990   89-70490
     624.1'5—dc20   CIP

To

G. M. Cornfield
H. Bolton Seed
Aleksandar S. Vesić
Hans F. Winterkorn

for their pioneering work

CONTENTS

Preface   xv
List of Contributors   xvii

1   SUBSURFACE EXPLORATIONS AND SAMPLING
John Lowe III and Philip F. Zaccheo

1.1   Introduction   1
1.2   Planning an Exploration Program   1
1.3   Reconnaissance Investigations   2
1.4   Explorations for Preliminary Design   6
1.5   Explorations for Detailed Design   8
1.6   Geophysical Explorations   8
1.7   Soundings and Probings   12
1.8   Boring Methods   14
1.9   Excavation Methods for Exploration   29
1.10   Groundwater Investigations   30
1.11   Test Grouting   37
1.12   Representative but Disturbed Samples   37
1.13   Undisturbed Samples   43
1.14   In Situ Soil Testing in Boreholes   51
1.15   Rock Coring   53
1.16   Miscellaneous Exploratory Techniques   59
1.17   Preservation, Shipment, and Storage of Samples   61
1.18   Logs of Subsurface Explorations   64
1.19   Contracting and Supervising Exploratory Programs   68
1.20   Subsurface Explorations Reports   69

2   SAMPLING AND PREPARATION OF MARINE SEDIMENTS
Ronald C. Chaney

2.1   Introduction   72
2.2   Offshore/Nearshore Soil Samples   75
2.3   Handling, Wrapping, and Labeling   80
2.4   Storage   82
2.5   Preparation of Soil Samples   83
2.6   Laboratory versus In-Situ Properties   84

3   SOIL TECHNOLOGY AND ENGINEERING PROPERTIES OF SOILS
Hans F. Winterkorn and Hsai-Yang Fang

3.1   Definition of Soil   88
3.2   Description and Identification of Soils   88
3.3   Rocks and Their Classification   92
3.4   Physical Properties Employed in Engineering Classifications of Soil Materials   99
3.5   Soil Classification Systems   102
3.6   Soil Types by Deposition or Other Special Features   109

3.7 The Structure of Noncohesive Soil Systems 115
3.8 The Structure of Cohesive Soils 119
3.9 Capillarity and Conduction Phenomena 125
3.10 Effect of Temperature on Engineering Properties of Soils 129
3.11 Densification (Compaction) 131
3.12 Consolidation 132
3.13 Shear Strength 134
3.14 In-Situ Measurements of Soil Properties 138

4 BEARING CAPACITY OF SHALLOW FOUNDATIONS
Wai-Fah Chen and William O. McCarron

4.1 Introduction 144
4.2 Methods of Analysis 144
4.3 Soil Governing Parameters 145
4.4 Bearing Capacity by the Upper-Bound Method 145
4.5 Bearing Capacity by the Lower-Bound Method 150
4.6 Footing Depth and Shape and Inclined and Eccentric Loads 152
4.7 Footing Shape, Depth, and Inclination Effects 154
4.8 Nonhomogeneous Foundations and Anisotropic Strength 154
4.9 Influence of Groundwater Table 156
4.10 Comments on Bearing Capacity Solutions 157
4.11 Sliding Stability of Gravity Structures 157
4.12 Choice of Safety Factor 158
4.13 Example Problems 159
4.14 Numerical Evaluation of Bearing Capacity 161

5 STRESS DISTRIBUTION AND SETTLEMENT OF SHALLOW FOUNDATIONS
Robert D. Holtz

5.1 Introduction 166
5.2 Settlement of Shallow Foundations 167
5.3 Applicability of the Theory of Elasticity to Calculation of Stresses and Displacements in Earth Masses 169
5.4 Calculation of Initial Distortion Settlements 170
5.5 Distortion Settlement of Granular Soils 177
5.6 Calculation of Stress Distributions 185
5.7 Consolidation Settlements 192
5.8 Secondary Compression Settlements 209
5.9 Tolerable Criteria 212
5.10 Foundation Treatment Alternates 216

6 EARTH PRESSURES
G. W. Clough and J. M. Duncan

6.1 At-Rest Lateral Pressures 224
6.2 Active and Passive Lateral Earth Pressures 224
6.3 Soil–Structure Interaction for Unmoving Walls 228
6.4 Earth Pressures due to Surface Loads 229
6.5 Earth Pressures due to Compaction 230
6.6 Relation Between Earth Pressures and Wall Movements 232
6.7 Earth Pressures for Design 234

7 DEWATERING AND GROUNDWATER CONTROL
J. Patrick Powers

7.1 Impact of Groundwater on Construction 236
7.2 Design of Structures Below the Water Table 236

7.3   Methods of Groundwater Control 236
7.4   Field Pumping Tests 239
7.5   Design of Predrainage Systems 241
7.6   Managing Groundwater Control 244
7.7   Disputes over Groundwater Control 245
7.8   Cost of Groundwater Control 245
7.9   Undesirable Side-Effects of Dewatering 246
7.10   Permanent Dewatering Systems 246
7.11   Ground Freezing 247

8 COMPACTED FILL
Jack W. Hilf

8.1   Introduction 249
8.2   Soil Compaction 249
8.3   Compaction Equipment 273
8.4   Control of Compaction 294
8.5   Miscellaneous Problems in Compaction 309

9 SOIL STABILIZATION AND GROUTING
Hans F. Winterkorn and Sibel Pamukcu

9.1   Introduction 317
9.2   Principle of Soil Stabilization 317
9.3   Methods of Soil Stabilization 318
9.4   Stabilization of Soils with Granular Skeleton 319
9.5   Chemical Stabilization 324
9.6   Cement Stabilization 328
9.7   Lime Stabilization 337
9.8   Ash and Slag Stabilization 344
9.9   Bituminous Stabilization 346
9.10   Thermal and Electrokinetic Stabilization 349
9.11   Construction 351
9.12   Grouting Principles 353
9.13   Grouting Techniques 354
9.14   Planning of the Grouting Project 358
9.15   The Grouting Process 363
9.16   Control of Grouts and Grouting Operations 368
9.17   Examples of Applications of Grouting 369

10 STABILITY OF EARTH SLOPES
Hsai-Yang Fang and George K. Mikroudis

10.1   Introduction 379
10.2   Factors Affecting Slope Stability Analysis 379
10.3   Factor of Safety 380
10.4   Slope Stability Analysis Procedure: Limit Equilibrium Methods 381
10.5   Slope Stability Analysis Procedure: Limit Analysis Methods 395
10.6   Methods Considering Seismic Effects 399
10.7   Slope Stability in Soils Presenting Special Problems 404
10.8   Computer-Aided and Expert Systems for Slope Stability Analysis 406

11 LANDSLIDES
Bengt B. Broms and Kai S. Wong

11.1   Introduction 410
11.2   Causes of Landslides 411
11.3   Consequences of Landslides 415

11.4   Slope Movements Preceding Landslides   415
11.5   Classification   417
11.6   Investigations for Landslides   425
11.7   Analysis of Landslides   427
11.8   Instrumentation   432
11.9   Methods of Correcting Landslides   435

12   RETAINING STRUCTURES AND EXCAVATIONS
Thomas D. Dismuke

12.1   Introduction   447
12.2   Restrained Retaining Structures   447
12.3   Gravity Structures   478
12.4   Cantilever Retaining Walls   503
12.5   Special Structures   504
12.6   Sheet Piling and H-Piles   507

13   PILE FOUNDATIONS
Bengt H. Fellenius

13.1   Introduction and Background   511
13.2   Aspects for General Consideration   512
13.3   The Shaft Resistance   513
13.4   Toe Resistance   516
13.5   Capacity Determined from In-Situ Field Testing   517
13.6   Installation Considerations   517
13.7   Residual Compression   517
13.8   The Neutral Plane   518
13.9   Capacity of a Pile Group   520
13.10   Summary of Design Procedure for Capacity and Strength   520
13.11   Settlement of Pile Foundations   521
13.12   Static Testing of Piles   524
13.13   Pile Dynamics   529
13.14   Horizontally Loaded Piles   531
13.15   Seismic Design of Lateral Pile Behavior   533
13.16   Design Example   533

14   DRILLED SHAFT FOUNDATIONS
Fred H. Kulhawy

14.1   Introduction   537
14.2   General Behavior Patterns   537
14.3   Axial Compression Capacity   538
14.4   Drained Axial Compression Capacity   539
14.5   Undrained Axial Compression Capacity   545
14.6   Axial Uplift Capacity   548
14.7   Belled Shafts   549
14.8   Compression and Uplift Displacements   549
14.9   Other Site and Loading Conditions   550
14.10   Closing Comments on Design   551
14.11   Pertinent Standards and Specifications   551

15   FOUNDATION VIBRATIONS
George Gazetas

15.1   Introduction   553
15.2   Machine Foundation Vibrations: Statement of the Problem   553

15.3 Soil Moduli and Damping—Field and Laboratory Testing Procedures 555
15.4 Harmonic Vibration of Block Foundations: Definition and Use of Impedances (Dynamic "Springs" and "Dashpots") 564
15.5 Computing Dynamic Impedances: Tables and Charts for Dynamic "Springs" and "Dashpots" 569

16 EARTHQUAKE EFFECTS ON SOIL–FOUNDATION SYSTEMS
Part I: Prior to 1975
H. Bolton Seed

16.1 Introduction 594
16.2 Influence of Soil Conditions on Shaking Intensity and Associated Structural Damage 597
16.3 Influence of Soil Conditions on Ground Settlement 614
16.4 Influence of Soil Conditions on Liquefaction Potential 614
16.5 Influence of Soil Conditions on Landslides 619

Part II: From 1975 to 1989
Ronald C. Chaney and Sibel Pamukcu

16.6   Introduction 623
16.7   Influence of Soil Conditions on Ground Settlement 624
16.8   Influence of Soil Conditions on Liquefaction Potential 628
16.9   Influence of Soil Conditions on Landslides 635
16.10 Behavior of Clays and Silts During Cyclic Loading 653
16.11 Remedy of Earthquake Damage on Soil–Foundation Systems 661

17 FOUNDATION PROBLEMS IN EARTHQUAKE REGIONS
Leonardo Zeevaert

17.1 Introduction 673
17.2 Irrotational Seismic Waves 673
17.3 Shear Seismic Waves 675
17.4 Case History 677

18 OFFSHORE STRUCTURE FOUNDATIONS
Ronald C. Chaney and Kenneth R. Demars

18.1 Introduction 679
18.2 Loading on Foundations 684
18.3 Pile Structures in Marine Environment 684
18.4 Gravity Platforms 702
18.5 Anchor Uplift Capacity 712
18.6 Pipelines 716
18.7 Jack-Up Platforms 723
18.8 Hydraulic Filled Islands 727

19 FOUNDATIONS IN COLD REGIONS
Arvind Phukan

19.1 Introduction 735
19.2 Frozen Soils: Phases and Temperature Profile 735
19.3 Design Approach 737
19.4 Design Considerations 738
19.5 Shallow Foundations 742
19.6 Pile Foundations 745

20 GEOTECHNICS OF HAZARDOUS WASTE CONTROL SYSTEMS
Jeffrey C. Evans

20.1 Introduction 750
20.2 Soil–Waste Interactions 750
20.3 Permeability and Compatibility Testing 752
20.4 Hazardous Waste Control Systems 760
20.5 Covers and Liners 762
20.6 Vertical Barrier Systems and Cutoff Walls 765
20.7 General Guidance 775

21 REINFORCED EARTH
F. Schlosser and M. Bastick

21.1 Introduction 778
21.2 Principle and Advantages of Reinforced Earth 778
21.3 History and Development 779
21.4 Behavior of Reinforced Earth 780
21.5 Design Methods 786
21.6 Material Specifications 790
21.7 Effects of Water and Dynamic Loading 791
21.8 Applications 792

22 GEOSYNTHETICS IN GEOTECHNICAL ENGINEERING
Robert M. Koerner

22.1 Introduction 796
22.2 Geotextiles 796
22.3 Geogrids 800
22.4 Geonets 802
22.5 Geomembranes 804
22.6 Geocomposites 810

23 DEEP COMPACTION OF GRANULAR SOILS
Bengt B. Broms

23.1 Introduction 814
23.2 Vibration Methods 815
23.3 Displacement Methods 821
23.4 Loading Methods 827
23.5 Control Methods 828
23.6 Cost Comparisons 829

24 STABILIZATION OF SOIL WITH LIME COLUMNS
Bengt B. Broms

24.1 Introduction 833
24.2 Principle of the Lime Column Method 837
24.3 Applications of the Lime Column Method 847
24.4 Laboratory and Field Investigations 850
24.5 Control Methods 852
24.6 Schedule for Design and Construction 853

25 DURABILITY AND PROTECTION OF FOUNDATIONS
Thomas D. Dismuke

25.1 Introduction 856
25.2 Facility Function and Failure Criteria 857

25.3   Properties of Construction Materials, Soil, and Water   857
25.4   Site Data   861
25.5   Behavior of Materials in Soil and Water   863
25.6   Evaluation of Site Data   864
25.7   Protection of Foundations   865

26   GROUND ANCHORS AND SOIL NAILS IN RETAINING STRUCTURES
Ilan Juran and Victor Elias

26.1   Introduction   868
26.2   Principles, Historical Development, and Fields of Application   868
26.3   Technology, Construction Process, and Structural Elements   871
26.4   Soil–Inclusion Interaction: Pull-Out Capacity Estimates   877
26.5   Application Criteria: Advantages and Limitations   883
26.6   Feasibility Evaluation   884
26.7   Short- and Long-Term Performance of Anchors and Nails   884
26.8   Durability Considerations   889
26.9   Design of Anchored Walls and Nailed Soil-Retaining Structures   890
26.10   Technological Development and Research Needs   902

APPENDIX   907

INDEX   911

PREFACE

More than ten years have passed since the first edition was published. During that period there have been a substantial number of changes in geotechnical engineering, especially in the applications of foundation engineering.

As the world population increases, more land is needed and many soil deposits previously deemed unsuitable for residential housing or other construction projects are now being used. Such areas include problematic soil regions, mining subsidence areas, and sanitary landfills. To overcome the problems associated with these natural or man-made soil deposits, new and improved methods of analysis, design, and implementation are needed in foundation construction. As society develops and living standards rise, tall buildings, transportation facilities, and industrial complexes are increasingly being built. Because of the heavy design loads and the complicated environments, the traditional design concepts, construction materials, methods, and equipment also need improvement. Further, recent energy and material shortages have caused additional burdens on the engineering profession and brought about the need to seek alternative or cost-saving methods for foundation design and construction.

Rapid and extensive developments in techniques of foundation analysis and design continue to occur. Some of the major developments that have already significantly influenced the instruction and design practice in the foundation engineering profession include the extensive work on computer-aided design techniques for analyzing the stress distribution of soil foundation-structure systems, a new understanding of the behavior and strength of soil under various loads, and improved understanding of the response of various types of superstructures to time-dependent loading and to foundation structures. New knowledge of the interaction between foundations and their environments, such as erosion, subsidence, hazardous and toxic wastes, and earthquakes has also been developed.

The second edition attempts to update the material in view of the developments cited above. The applications will also be expanded to cover such subjects as landfills, cold regions, and offshore foundations, and foundation techniques which include reinforced earth, nails/anchors, and geosynthetics.

In the framework of the second edition, the authors present concisely the basic principles and material parameters pertinent to each subject and illustrate with practical examples the engineering application of these principles within the limitations set by the material parameters. The original style and philosophy of the individual authors has been retained as much as possible.

The primary purpose of this handbook is to serve the practicing engineer. It can also be used as a textbook in applied geotechnical engineering. With the updating of some chapters and the addition of new chapters, this book will also be useful in conjunction with design in other areas, such as offshore/nearshore structures, hazardous waste control systems, transportation facilities, and structural engineering, to name just a few.

The editor expresses his thanks to all those who have generously given advice and encouragement in the preparation of this handbook. Thanks are also due to Ms. Eleanor Nothelfer for her general aid in preparing the manuscript.

Hsai-Yang Fang

LIST OF CONTRIBUTORS

Bastick, M.   Terre Armée International, Paris, France

Broms, Bengt B.   Professor of Civil Engineering, Nanyang Technological Institute, Singapore

Chaney, Ronald C.   Professor and Director, Fred Telonicher Marine Laboratory, Humboldt State University, Trinidad, CA

Chen, Wai-Fah   Professor and Head, Structural Engineering Department, School of Civil Engineering, Purdue University, West Lafayette, IN

Clough, G. W.   Professor and Head, Department of Civil Engineering, Virginia Polytechnic Institute and State University, Blacksburg, VA

Demars, Kenneth R.   Associate Professor of Civil Engineering, University of Connecticut, Storrs, CT

Dismuke, Thomas D.   Senior Consultant, Bethlehem Steel Corporation, Bethlehem, PA

Duncan, J. M.   University Distinguished Professor, Department of Civil Engineering, Virginia Polytechnic Institute and State University, Blacksburg, VA

Elias, Victor   V. Elias and Associates, P. A. Consulting Engineers

Evans, Jeffrey C.   Associate Professor of Civil Engineering, Bucknell University, Lewisburg, PA

Fang, Hsai-Yang   Professor and Director, Geotechnical Engineering Division, Lehigh University, Bethlehem, PA

Fellenius, Bengt H.   Professor of Civil Engineering, University of Ottawa, Ottawa, Canada

Gazetas, George   Professor of Soil Mechanics, National Technical University of Greece, Athens, Greece and State University of New York, Buffalo, NY

Hilf, Jack W.   Consulting Engineer, Aurora, CO

Holtz, Robert D.   Professor of Civil Engineering, University of Washington, Seattle, WA

Juran, Ilan   Professor and Head, Department of Civil and Environmental Engineering, Brooklyn Polytechnic University, NY

Koerner, Robert M.   Bowman Professor of Civil Engineering and Director, Geosynthetic Research Institute, Drexel University, Philadelphia, PA

Kulhawy, Fred H.   Professor of Geotechnical Engineering, Cornell University, Ithaca, NY

Lowe, John III   Consulting Engineer, Yonkers, NY

McCarron, William O.   Senior Research Engineer, Amoco Production Co., Tulsa, OK

Mikroudis, George K.   President, AVANSE Ltd., Athens, Greece

Pamukcu, Sibel   Assistant Professor of Civil Engineering, Lehigh University, Bethlehem, PA

Phukan, Arvind   Professor of Civil Engineering, University of Alaska at Anchorage, Anchorage, AL

Powers, J. Patrick   Consultant, AQUON Ground Water Engineering, Hackettstown, NJ

Schlosser, F.   Ecole Nationale des Ponts et Chaussees, Paris, France

Seed, H. Bolton   Professor of Civil Engineering, University of California, Berkeley, CA

Winterkorn, Hans F.   Professor of Civil Engineering and Geophysics — Emeritus, Princeton University, NJ

Wong, Kai S.   Associate Professor, Nanyang Technological Institute, Singapore

Zaccheo, Philip F.   Consultant, Ebasco Services Inc., New York, NY

Zeevaert, Leonardo   Professor of Civil Engineering, Emeritus, Universidad Nacional Autónoma de México, México, D.F.

1 SUBSURFACE EXPLORATIONS AND SAMPLING

JOHN LOWE III
Consulting Engineer

PHILIP F. ZACCHEO
Staff Consultant, Ebasco Services Incorporated

1.1 INTRODUCTION

The proper design of civil engineering structures requires adequate knowledge of subsurface conditions at the sites of the structures and, when structures are to consist of earth or rockfill materials, of subsurface conditions at possible sources of construction materials. The structures may be divided into three categories.

1. Structures for which the basic problem is the interaction of the structure and the surrounding ground. Such structures include foundations, retaining walls, bulkheads, tunnel linings, and buried pipes. The main point of interest is the load–deflection characteristics of the interface.
2. Structures constructed of earth such as highway fills, earth and rockfill dams, bases and subbases for pavements, and backfill behind retaining walls. Besides the interaction of the earth structure with the adjacent ground, properties of the construction materials are required for determining the action of the earth structure itself.
3. Structures of natural earth and rock as natural slopes and cut slopes. In this case, knowledge of the properties of the natural materials is required.

In order to perform this design work properly, the engineer must have a good understanding of the problems encountered in making subsurface explorations and of the various tools available to make subsurface explorations. Specialists in soil and rock engineering and/or geology are required for planning, conducting, and supervising the programs of subsurface explorations.

The types of subsurface information required for design include, but are not limited to, the following:

1. Areal extent, depth, and thickness of each identifiable soil stratum, within a limited depth dependent on the size and nature of the structure, together with a description of the soil including its degree of density if cohesionless and degree of stiffness if cohesive.
2. Depth to top of rock and the character of the rock, including such items as lithology; areal extent, depth, and thickness of each stratum; strike, dip, and spacing of joints and bedding planes; presence of fault and shear zones; and state of weathering or decomposition.
3. Location of groundwater and the presence and magnitude of artesian pressures.
4. Engineering properties of the soil and/or rock in situ such as permeability, compressibility, and shear strength.

The procedures for obtaining subsurface information may be divided into the two broad categories of indirect and direct methods. Indirect methods include: aerial photography and topographic map interpretation, and the use of existing geological reports, maps, and soil surveys.

Direct methods comprise the following:

1. Geologic field reconnaissance, including the examination of in situ materials in natural and man-made exposures such as river banks, escarpments, highway and railway cuts, quarries, and existing shafts and tunnels.
2. Soundings and probings.
3. Borings, test pits, trenches, shafts, and adits from which representative disturbed and/or undisturbed samples of the in situ materials may be obtained.
4. Simple field tests, such as the Standard Penetration Test (SPT) and the static cone penetration test, whose results have been correlated with engineering properties on a general basis.
5. Field tests such as the vane shear dilatometer and pressure-meter tests, seepage and water-pressure tests, plate bearing tests, the CBR test, and pile load tests, wherein the engineering properties of the in situ materials are measured directly.

It is the purpose of this chapter to present essential information for the complete range of subsurface explorations. Included are descriptions of the planning of an exploration program; indirect methods of exploration; drilling equipment and techniques; sampling equipment and techniques; field test procedures; and suggestions for reporting subsurface exploration information. A list of references is given at the end of the chapter.

1.2 PLANNING AN EXPLORATION PROGRAM

1.2.1 Purpose of Explorations and Phased Execution

The basic purpose of an exploration program is to provide the engineer with a knowledge of the subsurface conditions at the site of an engineering project. Normally, the explorations provide information required for the safe and economical design of a project and inform the construction engineer about the materials and conditions he will encounter in the field. At times, the explorations may be used to obtain information for the analysis of the failure of an engineering structure.

Explorations are normally accomplished in a phased sequence as follows:

1. Reconnaissance investigations
2. Explorations for preliminary design
3. Explorations for detailed design
4. Explorations during construction

Each phase of explorations together with the engineering done in that phase discloses problems that require further investigation in the next phase. Not all phases are required on all projects; the fourth phase generally is not necessary.

The number, type, location, size, and depth of the explorations are dependent upon the nature and size of the project and on the degree of complexity and critical nature of the subsurface conditions. A general rule of thumb is that the cost of the subsurface explorations for design should be in the range of 0.5 to 1.0 percent of the construction cost of the project. The lower percentage is for large projects and for projects with less-critical subsurface conditions; the higher percentage is for smaller projects and for projects with critical subsurface conditions. About half the cost would be expended for explorations for preliminary design and about half for detailed design. A very much smaller amount of money would be expended for explorations in the reconnaissance investigation phase. No rule of thumb can be given for the cost of explorations during construction. Such explorations are used to investigate special problems that may arise during construction or to better delineate the materials in borrow areas or quarries in connection with the contractor's scheduling of his operations. Generally they are not required but, when used, their cost can vary widely from one project to another.

The combined cost of planning the subsurface explorations, supervising the explorations, laboratory testing, and reporting the results usually amounts to about the same cost as the explorations. In general, it is justifiable to spend additional money on explorations and related testing and engineering as long as the savings that can be effected in the project construction cost on the basis of the information obtained are significantly greater than the cost of the explorations plus related engineering work.

Local building codes often specify the minimum number of borings required for a given size and type of structure. In the case of a lightweight structure that is to be founded in an area of relatively uniform subsurface conditions, this minimum number of borings generally will suffice and all borings may be completed during a single exploration program.

At times, because of deadlines set for completion for the engineering work, the argument is given that since the information from a full and proper exploration program will not be available in time for use in the design, the exploration program should be cut. In such cases, the much-preferred procedure is to proceed with the engineering on the basis of the best assumptions that can be made from the available subsurface information, but to continue with the full and proper program of explorations and testing. The information obtained will then either confirm the assumptions made to complete the engineering on time or indicate where changes in the design assumptions have to be made. Frequently, any modifications required by the changes indicated by the full exploration program can be made without undue difficulty and in a timely manner.

The sequence in which the explorations are to be performed is often left to the discretion of the drilling contractor. In such instances, the sequence will be governed by the ease of operation for the driller. Movement of the rigs between borings will be kept to a minimum; all borings in one area of the site may be drilled before those in another area are started. Often it is not only advantageous but essential that the engineer designate the sequence of the explorations. If, for instance, the borings being drilled are intended to fill gaps in a geologic profile based on previous explorations, the information from a given boring could preclude the need for one or more other borings that had been programmed. The sequence may also be dictated by time limitations if the time available for design is short. It is not unusual for project design and laboratory testing to be concurrent with the explorations. Under these conditions, it may be necessary to obtain samples first from specific areas in order that testing may progress in a timely fashion.

1.2.2 Type and Number of Drilling Rigs

The types of rigs used will depend primarily on the type, size, and depth of the explorations; the location of the explorations, that is, whether they are onshore or offshore; the accessibility of the area to be explored; the types of rigs available in the area; and the terrain or sea conditions. The types of rigs used for exploration work and their applicability to various conditions are discussed in a subsequent section. The minimum number of rigs required to perform an exploration program is dependent on the time available for the execution of the program, the rate of advancement of the holes by the selected rigs, and the sequence of explorations. The estimated rate of advancement of the hole should include time allowances for equipment breakdowns, movement of the rigs from one location to another, and standby due to weather.

1.2.3 Types of Sampling Equipment

The sampling equipment to be used will depend on the type of information required and the characteristics of the materials to be sampled. If only classification of the soil strata is required, disturbed samples will suffice and samplers such as the split-tube drive sampler may be used. If, on the other hand, the ultimate goal is the determination of the engineering properties of the soils by laboratory testing, more sophisticated equipment such as the thin-wall tube and double-tube core barrel samplers will be required. The sizes and types of samplers will depend also upon such factors as the presence or absence of gravel; the maximum size of particle to be sampled; the type of material to be sampled, that is, cohesionless or cohesive; the density of cohesionless materials; the consistency of cohesive materials; and the location of the material to be sampled with reference to the groundwater level. The types of samplers available and the specific conditions under which each may be used are discussed in detail in the paragraphs on samplers and sampling techniques.

1.3 RECONNAISSANCE INVESTIGATIONS

1.3.1 Purpose and Scope

The reconnaissance investigations provide information for prefeasibility studies and for planning the explorations for the succeeding phase, explorations for preliminary design. This program, for a localized project such as a building that is to be constructed on a preselected site, will be somewhat limited in scope. However, when a dam or highway project is under consideration, several alternative sites or alignments must be considered. The information obtained in this phase aids in the selection of the alternative sites or alignments for investigation. A large portion of the work during this phase falls into the category of research. Also included would be field reconnaissance by a geologist and a soils engineer plus such geophysical explorations and borings as are deemed essential.

1.3.2 Research

Any investigation begins with a thorough search for all existing information that could shed light on subsurface conditions at the site, including both old and recent topographic maps, geologic maps, aerial photographs, geologic and subsurface exploration reports and records of governmental agencies and private firms, university publications, and articles in engineering and geologic journals. The sources of these items vary from country to country and even between the political subdivisions of a country. One of the major sources of geologic and topographic information within the continental United States is the United States Geologic Survey (USGS). The types of information available from the USGS, as well as from other sources, are discussed below.

1.3.3 Topographic Maps

Various types of maps are useful in the planning of an exploration program. Topographic maps provide information on the accessibility of the site of the work and the terrain, both of which may exert a strong influence on the types of rigs used for the work. Topographic maps may also be used in much the same manner as aerial photographs. A knowledge of geomorphology often will permit a trained observer to surmise much about the geology of a site on the basis of the land forms and drainage patterns shown on the topographic maps. The maps do not have the detail of aerial photographs and, therefore, limit the capabilities of the observer. However, in the complete absence of aerial photographs or as an initial step, the use of the maps is worthwhile. The amount of information that can be derived from such maps also depends on the area involved and the amount of detail shown. General characteristics of the soil and/or rock are commonly revealed by the topography. Geomorphic soil forms such as coastal and flood plains, deltas, alluvial fans, terraces, dunes, eskers, drumlins, and other features are easily recognized. Swamp areas normally are designated directly on the maps and the drainage patterns of soils often give an indication of particle size and the degree of induration. In rock areas, structure is often revealed in such details as the course of a river or the slope of a hill. Under proper conditions, it is possible to determine structural features such as dip and strike, folding, faulting, and relative consistency.

The major source of topographic maps of the United States is the United States Geologic Survey (USGS). The USGS publishes a series of quadrangle maps, known as the National Topographic Map Series, which covers the United States and its territories and possessions. Each map covers a quadrangle area bounded by lines of latitude and longitude. Maps covering areas of 7.5' of latitude by 7.5' of longitude are plotted to scales of 1:24000 and 1:31680 for the continental United States; similar coverage of Puerto Rico is provided at scales of 1:20000 and 1:30000 and of the Virgin Islands at 1:24000. Fifteen-minute maps of the continental United States are at 1:62500; 30' maps are at 1:125000 and one-degree maps at 1:250000. Maps of Alaska covering 15' of latitude by 20' to 30' of longitude are available at 1:63360; other maps of Alaska are also available at 1:250000 and 1:1000000. Hawaii is partially covered by series of maps at 1:62500 and 1:24000. Some shaded-relief maps and metropolitan area maps are also published by the USGS. A complete list of all USGS maps is presented in the U.S. Geological Survey (1965) and the supplements thereto that are published monthly.

Topographic maps of the United States are also produced by the Army Map Service and the United States Coast and Geodetic Survey (USC&GS). The former are based on the United States Military Grid and are at scales of 1:25000 for 7.5' quadrangles; 1:50000 for 15' quadrangles; and 1:250000 for 30' quadrangles. Maps of larger areas are plotted at scales of 1:250000 and 1:500000. The USC&GS publishes aeronautical and coastline charts. The former are small-scale maps which are primarily used for those areas of the United States and its territories that have not been mapped on a large scale by the USGS. The coastline charts, plotted at scales from 1:10000 to 1:80000 are useful for offshore work.

Other sources of topographic information for the United States include the U.S. Army Corps of Engineers, which publishes topographic maps and charts of some rivers and adjacent shores plus the Great Lakes and their connecting waterways; the U.S. Forest Service, which publishes forest reserve maps; and the Hydrographic Office of the Department of the Navy, which publishes nautical and aeronautical charts.

The sources of maps for overseas work are varied. Two excellent sources of general information are the American Geographic Society maps, which cover a large portion of South and Central America, and the British Admiralty charts.

1.3.4 Geologic Maps

The major source of geologic maps and information within the United States is the USGS, which has published books, maps, and charts in various forms since 1879. One of the most useful series of these publications is the "Indexes to Geologic Mapping in the United States." This series comprises a map of each state on which are shown the areas for which geologic maps have been published. A color code is used to give the approximate scale of each map. A text on the sheet also gives the source of publication, scale, date, and author of each geologic map and a complete list of USGS reports on the state. The maps distributed by the USGS also include a geologic map of the United States at a scale of 1:2500000 and several other series of maps, the best known and most widely used of which are (1) Folios of the Geologic Atlas of the United States; (2) Geologic Quadrangle Maps of the United States; and (3) Mineral Resources Maps and Charts. Each of the folios, which were published until 1945, consisted of a text describing the geologic history of the area covered and several quadrangle maps showing the topography, geology, underground structure, and mineral deposits of the area. Since 1945, the folios have been replaced by individual quadrangle maps. These maps often include structure sections, columnar sections, and other graphic geologic data, as well as descriptive material. The mineral resources maps include information on such detailed topics as the sand and gravel deposits of a state, the construction materials of a state, the geology of limited areas, and the location of possible sources of riprap.

The book publications of the USGS may be divided into two general categories. The first consists of those books, papers, etc., variously referred to as bulletins, circulars, mineral resources publications, monographs, and professional papers that are detailed geologic studies of limited scope. The second category includes the water supply papers, which deal with various topics, among which are included detailed studies of surface and subsurface water flow in specific areas. These papers often include detailed geologic descriptions and maps of the area under consideration.

In addition to the above items, the USGS also distributes certain state geologic maps and a quarterly entitled "Abstract of North American Geology," which contains abstracts of the most significant papers published within the quarter.

Geologic information is also available from state and local governmental agencies, the Geological Society of America, and universities. The majority of the states have geologic surveys or an equivalent agency responsible for the gathering and the dissemination of geologic information. The available data may take the form of geologic maps, geologic reports, and records of explorations made in conjunction with state highway construction. State and local authorities also often maintain records of all wells drilled within their jurisdiction. A list of the names and addresses of state geologists is contained in Hunt (1984), Appendix B.

The Geological Society of America publishes geologic maps, as well as a monthly journal and special volumes that treat in detail specific geologic topics on locales. Included among the former are general geologic maps of North and South America, "The Glacial Map of the United States," and "The Loessial Deposits or Windblown Soils in the United States." Detailed geologic maps of limited areas are also often found in the journal articles and special volumes.

The libraries of local colleges and universities frequently contain considerable detailed geologic information in the form of theses. These libraries are likewise often the source of many out-of-print geologic publications.

1.3.5 Soil Surveys

The soil surveys conducted by various governmental agencies are also a useful source of information for the engineer planning a subsurface exploration program. These surveys, which consist of the mapping of surface and near-surface soils over a large expanse of land, are of two types, agricultural and engineering. Since both types usually cover an entire county, the information contained in them is, of necessity, generalized. This information, published in the form of text and maps, is particularly useful for projects such as highways.

Agricultural soil surveys in the United States have been conducted by the Department of Agriculture (USDA) in conjunction with state agencies since the early 1900s. The results of the surveys are presented in the form of reports and maps that commonly, but not always, cover a complete county. The reports, in general, contain a description of the areal extent, physiography, relief, drainage patterns, climate, and vegetation, as well as the soil deposits of the area covered. The soil descriptions are done in accordance with the pedologic method. The maps show the extent and derivation of the various deposits. During the period over which these reports and maps have been published, the criteria for their preparation has varied significantly. The earlier reports were directed entirely toward the use of the soils for agricultural purposes and the accuracy of the mapping varied significantly. In spite of this, the information that was presented has been utilized successfully for engineering purposes by some state highway organizations that have correlated the observed behavior of pavements with the classifications as given in the USDA reports and maps. One of the states that has made extensive use of correlations of engineering characteristics with the pedologic units is Michigan (1960). The more recently published reports now contain engineering-oriented sections prepared by the Bureau of Public Roads in conjunction with the Soil Conservation Service, which is the agency responsible for conducting the surveys and publishing the reports and maps. In addition, in some states, engineering supplements to the agricultural survey reports have been prepared by local authorities. These supplements provide data on the drainage characteristics of the materials and anticipated engineering problems. *Highway Research Board Bulletin* 22-R (Committee on Surveying, Mapping, and Classification of Soils, 1957) contains a list of the soil survey reports published to the time of its printing and rates these surveys on the basis of the adequacy of the soil mapping performed. A current list of published soil surveys may be obtained from the U.S. Department of Agriculture, Soil Conservation Service.

A few engineering soil surveys are available. An excellent example is the Engineering Soil Survey of New Jersey (Rogers, 1950). The survey reports comprise a general volume and an individual volume for each county in the state. The general volume describes the climate, physiography, geology, and soils of the state, the mapping and soil-testing techniques used, and the symbolic notation used for the identification of the various soil types. Each county volume comprises a text and a soils map. Included in the text are general data on the physiography, surface drainage, and geology of the area covered and detailed information concerning the major soil formations found in the area. This information includes the geologic identification and general characteristics of the parent formation; the nature of the underlying formations; the land form; the types of soils included under the general grouping; the engineering classification of the material; the drainage characteristics; and a discussion of the engineering aspects of the material. The soils map that accompanies the report delineates the areal extent of the various materials by means of symbolic notation. This notation is basically a three-part code system, which by a combination of letters and numbers designates the type of geologic formation in which the soil occurs, identifies the soil in accordance with AASHO Designation M 145-49, and indicates the prevailing or average subsurface drainage conditions.

1.3.6 Aerial Photographs

One of the most useful sources of information the planner of a subsurface exploration program can have at his disposal is a series of aerial photographs of the site and surrounding vicinity. This is particularly true in areas where little or no work has previously been done and in the case of projects that have a large areal extent, such as irrigation developments, or projects that have a large longitudinal extent such as highways. The use of aerial photographs predates World War II. However, it was the use of such photographs by the armed forces during the war that gave the impetus to their use for various facets of engineering work in civilian life during the postwar era. Among the major uses for aerial photographs are topographic mapping, geologic mapping, and soils surveys. A brief discussion of their use for the latter two purposes is given below. Innumerable technical papers and texts concerning the subject have been published, among which are American Society of Photogrammetry (1980), Lueder (1959), and Ray (1960), listed at the end of this chapter. The reader is referred to these and similar publications for detailed discussions of the subject.

The aerial photographs used in connection with geologic and soils work are almost exclusively vertical photographs. They may be black-and-white or colored. The black-and-white prints normally used are 9 inches square, although other sizes are used; color transparencies are generally 4.5 or 9 inches square. Color photographs are generally preferred because (a) objects are easier to identify when they appear in their natural color; (b) fine details and small objects can be identified more positively than on black-and-white photographs at the same scale; and (c) the cause of tonal variations is more readily established. However, the high cost of color photography has somewhat limited its use, and black-and-white photographs are still used in most engineering work today.

Aerial photographs are taken from an aircraft flying at a prescribed altitude along preestablished lines. The altitude from which the photographs are taken is a function of the desired scale of the photographs and the equipment used. In photo-geologic work, scales normally range between 1:6000 and 1:40 000 depending upon the degree of accuracy required. The flight lines are so located as to provide for overlapping of photographs from adjacent lines. The photographs along each flight line also are taken so as to provide overlapping coverage. The coverage is such that each photograph includes approximately 60 percent of the area covered by the preceding photograph. This provides the interpreter with pairs of photographs which, when viewed with the aid of a stereoscope, provide the three-dimensional view of the land surface essential to proper interpretation. A large portion of the United States has been covered by photography of this type and some of the photographs are available for purchase from various agencies of the government and from private firms involved in aerial survey work. A map entitled "Status of Aerial Photography in the United States," published by the United States Geological Survey, indicates the areas of the United States that have been mapped by aerial photographs and contains a list of government agencies and private firms that have photographs for purchase. A similar map entitled "Status of Aerial Mosaics in the United States" is also available. In addition, records of all reported aerial photographic coverage including the scales of the photographs and the equipment used are maintained by the USGS Map Information Office.

In the hands of a well-trained and experienced specialist, aerial photographs can be used to reveal subsurface conditions with surprising accuracy. Geologic information is obtained from aerial photographs through a process of deductive and inductive reasoning based upon a thorough understanding of the general geology of the area in which the project is to be located and an extensive knowledge of geology, geomorphology, pedology, groundwater hydrology, and soil engineering. The quantity and quality of the information the interpreter can obtain from any series of photographs depends on several factors, chief among which are the scale and quality of the photographs, the density of the vegetative cover, the degree of relief exhibited by the terrain, and the ability and experience of the interpreter. All other conditions being equal, the maximum amount of information will be obtained when the aerial photographs are used in conjunction with field and laboratory investigations that can be used for verification and correction of interpretations. The information that may be obtained from aerial photographs includes, but is not limited to, the type of bedrock; structural characteristics of the rock such as joint patterns, bedding planes, folds, and faults; the type and thickness of overburden; surface and subsurface drainage characteristics; the depth to groundwater; and the relative percentages of sands and gravels.

In the analysis of the photographs, the interpreter relies on the fact that surface features are controlled by subsurface geologic conditions and the composition of the overburden materials so that, under a given set of conditions, a classic pattern will evolve. The major features that are utilized are landform, drainage, erosion, vegetation, and photographic tone. It is important to note that no single one of these features is sufficient in itself. The combined information from each of the features is essential to proper analysis.

The identification of the landform as a dune, terrace deposit, alluvial fan, esker, moraine, or other type of deposit often permits the type of material to be established within given limits and thus yields an initial appraisal of the situation. This identification is particularly important in those cases where large areas or extended lengths are being mapped; in the case of small sites it is not uncommon to have a single landform present. Drainage patterns aid in the identification of soil type and in the structural characteristics of the underlying rock. Soils of high permeability will be well drained as long as topographic conditions are conducive to drainage. Under such conditions, the drainage channels will normally be widely spaced, that is, the drainage pattern will be coarse-textured; in some cases it may even be absent. In impervious materials the channels will be close, that is, fine-textured. Large numbers of lakes or ponds without streams also are indicative of impervious materials. In bedrock areas where the resistance to erosion is relatively uniform, a dendritic drainage pattern is expected. The presence of trellis, annular, parallel, and other drainage patterns is usually indicative of structural control of flow and, therefore, of the presence of folds and faults. Relatively straight channels in bedrock areas may be indicative of faults or of tilted interbedded sedimentary deposits of varying resistance to erosion.

Gully erosion profiles, both transverse and longitudinal, further define the materials. In general, gullies in granular soils are V-shaped and have steep gradients. Cohesive soils with little or no granular component have gullies that are characterized by a broadly rounded V-shape and a uniform low gradient. Gullies in loess and in cohesive soils having a significant amount of granular particles are U-shaped and have low gradients.

Vegetation, although one of the more difficult features to evaluate, is often indicative of subsurface conditions. Trees growing along the edges of terraces may be indicative of water seeping from the deposits, and orchards are commonly found in well-drained areas. Also, certain types of vegetation are in themselves indicative of conditions. For example, willows and hemlocks require substantial amounts of moisture, whereas poplars and scrub oaks are found in areas of low moisture. Aspen, on the other hand, thrives in a wide variety of soil types and under a wide variety of soil conditions, thereby lending to the difficulty of interpretation. The trained specialist will carefully weigh the relationship between the type of vegetation, the soil type and moisture content, the topography, and other pertinent factors in determining the significance, if any, of the presence of vegetation.

Insofar as photographic tone is concerned, light tones are generally, but not always, indicative of well-drained materials, whereas dark tones indicate poorly drained material. However, as in the case of vegetation, care is required in interpretation since tone is affected by several factors, all of which must be considered. For example, topographic position may preclude drainage of an otherwise free-draining material and give it a dark appearance.

1.3.7 Field Reconnaissance

Subsequent to a review of the available data disclosed by the research described above, and prior to the drilling of any exploratory holes, the proposed site should be thoroughly inspected by a geologist and/or a soils engineer. The primary objective of the reconnaissance is to obtain as much surface and subsurface information as possible without drilling exploratory holes or excavating test pits. The types of information to be obtained include accessibility of the site, topography, soil profile, bedrock lithology and structure, and surface and subsurface drainage. In determining soil and bedrock information, maximum use should be made of exposures occurring both naturally and as a result of construction. River banks, natural escarpments, quarries, and highway and railways cuts are some of the many sources of information concerning the nature and thickness of soil strata and the bedrock lithology and structure. They may also reveal, through the presence of seeps or springs, information on groundwater flow in the area. The geologic information obtained should be entered on sketch maps.

A second source of information that should not be overlooked is adjacent property owners. In the case of structures in urban and suburban areas, adjacent property owners may be able to provide the results of borings and soil tests performed prior to the design of their structures, information on the type of foundation used, and records of the performance of the foundation. In the case of projects located in outlying areas, residents may often provide information on groundwater conditions and sources of construction materials.

1.3.8 Reconnaissance Stage Explorations

One purpose of the reconnaissance survey and exploration phase is to provide information for the preparation of a rough estimate of the cost of site development. Depending on the nature and amount of information obtained through the research portion of this phase, it may be deemed essential to put down a limited number of borings and/or test pits in order to check the findings of the research unless existing borings or test-pit data are already available for this. These explorations may consist of a few small-diameter borings or test pits. The explorations may be located to develop geologic profiles or at the locations of major structures. In this stage, borings that are to penetrate only overburden are seldom larger than $2\frac{1}{2}$ inches in diameter; if bedrock is to be cored, the holes preferably should be 4 inches in diameter in order to permit coring with a NX-size barrel.

1.4 EXPLORATIONS FOR PRELIMINARY DESIGN

1.4.1 Purpose and Scope

The primary objective of these explorations is to obtain sufficient subsurface data to permit the selection of the types, locations, and principal dimensions of all major structures comprising the proposed project and the making of a sound estimate of its cost. The preliminary designs based on these explorations are suitable for economic and technical feasibility reports and project planning reports. The depth, thickness, and areal extent of all major soil and rock strata that will be affected by the construction must be established in reasonable detail. For projects where earth and rock construction materials are required, sources of these materials should be investigated to establish the quantity and quality of the materials available. Disturbed and undisturbed samples of the foundation and borrow materials must be obtained for laboratory testing to provide a basic knowledge of the engineering properties of the various materials. The extent of the program will depend on the nature of both the project and the subsurface conditions. Since it is not unusual for the exploratory program and the design to proceed concurrently, the exploration program must be flexible; it is incumbent upon the planner to constantly review the information obtained so that as the design and explorations proceed problem areas can be defined. Additional explorations can then be added so that the designer can be provided with adequate information for a rational design.

If detailed geologic maps are not available from the reconnaissance stage, field mapping is done to prepare such maps. Geologic sections are prepared also. The maps and sections are adjusted as necessary as information becomes available from the subsurface exploration program.

1.4.2 Type and Spacing of Explorations

The program for preliminary design may consist of any or all of the various types of explorations described in subsequent paragraphs. Regardless of the type of project for which the explorations are to be made, the objectives of this program are to establish in reasonable detail the stratigraphy together with a basic knowledge of the engineering properties of the overburden and bedrock formations that will be affected by or will have an effect upon the new structures, and to locate and determine the quality and approximate quantity of construction materials within an economical haul distance from the site.

Simply stated, the number and spacing of the explorations should be such as to properly achieve these objectives at minimum cost. The number and spacing of the explorations should be consistent with the type and extent of the project and with the nature of the subsurface conditions. Building codes often stipulate the number of borings required for structures, many times on the basis of area per boring. Usually foundation conditions can be adequately investigated using the number of borings required by the code but the designer should not hesitate to make additional borings in critical areas if it is necessary for proper design of the structure. The planner of an exploration program should not be reluctant to space borings closely, to locate them in running water, or to incline them if he deems it essential to the accomplishment of the objectives of the program.

Rigid rules for the number and spacing of explorations cannot and should not be established. In general, however, emphasis should be placed on locating borings to develop typical geologic cross sections; for example, in the case of a river valley, the explorations should be located on lines across the valley and the spacing of the borings on these lines should be much closer than the spacing of the lines, since geologic variations will occur in very much shorter distances across the valley than up and down the valley. Some general guidelines for various types of projects are detailed below.

For localized projects such as office buildings or building complexes that occupy areas of relatively limited extent, an initial program for a single building might consist of a boring at each corner of the structure and one in the middle. Where several structures are involved, as in a housing development, borings might be located and spaced so as to provide geologic sections across the site in mutually perpendicular directions. In either case, additional borings would be put down as the site conditions become apparent.

For larger projects of limited extent, such as dam projects, the planner will seek to establish geologic profiles across the valley at the axis, and at the downstream toe and upstream toe of the dam; at all major appurtenant structures such as the spillway, diversion tunnels or conduits, outlet works, and powerhouse; along access roads; and in quarries and borrow areas. Test pits and/or trenches will also be excavated in the borrow areas to obtain representative samples for testing of compacted specimens. Profile lines for the dam and appurtenant structures will each contain several borings. It is not uncommon for rivers to flow along old fault lines or to have old gorges that have been refilled with alluvial material. Therefore, it is frequently convenient to have some of the borings in the vicinity of the river inclined in order to facilitate mapping of such fault zones and buried valleys. Explorations along the alignment of long tunnels will depend to a large extent on the geology of the area. Borings will be located at portal areas and at points along the alignment to establish the geology.

In the case of quarries and borrow areas, either a series of geologic profiles is established or, where geologic conditions are uniform, explorations are located at the intersections of grid lines. Grids in these areas are often initiated at a spacing of 200 to 400 feet depending on the size of the area. Test pits to obtain samples of proposed borrow materials are located on the basis of the profiles established from borings or auger holes.

For projects such as highways and railroads which have large longitudinal extent compared to their width, explorations generally fall into three categories. The first consists of shallow borings along the alignment. These are generally spaced so as to verify the delineations of large areas of similar materials along the alignment as determined by aerial photographs and geologic mapping. The second category includes borings for major structures such as bridges, retaining walls, and large embankments and borings in areas of cut slopes to permit the determination of safe slopes. These are generally deeper and more closely spaced. The initial borings for bridges are generally spaced so that there is a minimum of one per abutment and one for each pier or for every other pier. Borings for retaining walls and cut slopes are located to give longitudinal and transverse profiles. The exact spacing will depend on the variability of the soil profile. The third category of borings involves those required to establish the quality and quantity of construction material available from quarries and borrow areas. Explorations for these areas should follow the same procedures as for the case of dams described in the preceding paragraph.

1.4.3 Depth of Explorations

Just as rigid rules cannot be established for the spacing of borings, similar types of rules have no place in determining the depth to which explorations are carried. However, general guidelines are available to the program planner. Two major factors will control the depth of exploration, namely, the magnitude and distribution of the load imposed by the structures under consideration and the nature of the subsurface conditions. In all projects, the borings, as a minimum, must extend to a depth sufficient to reveal the nature of all materials that could be significantly affected by the loads imposed by the structure and that, by settlement and/or shear failure, could affect the integrity of the structure. In cases such as the design of dam projects, where control of seepage must be considered, the exploration, as a minimum, should be carried to the nearest relatively impermeable stratum beneath the proposed foundation level. When, based on the above criteria, a boring extends to or into bedrock, it is prudent to carry the boring at least 10 feet and, at times as much as 15 feet or more beneath the top of apparently sound rock so as to establish the fact that bedrock rather than boulder has been encountered. Application of these general criteria to specific projects is outlined in the paragraphs that follow.

For localized structures, such as buildings, it is common practice to carry explorations to a depth beneath the loaded area of 1.5 to 2.0 times the least dimension of the structure. In the case of a building founded on spread footings or on a mat foundation, the level of the loaded area is the base of the footings or mat; for buildings founded on bearing piles or caissons, the level may be taken as the top of the bearing stratum to which the supporting elements are carried. Contours of normal stresses on horizontal planes and of maximum shearing stresses show that at a depth equal to 1.5 to 2.0 times the least dimension of the structure the normal stresses will be approximately 10 percent and the maximum shear stresses approximately 5 percent of the imposed load. Generally, stresses of this magnitude or less are insignificant. Taylor (1948) cautions against the blind use of this rule of thumb for establishing the depth of explorations and cites the possibility of deeply buried strata of soil with poor bearing characteristics causing excessive settlements even when they are below the recommended depth. Where geologic conditions are not well established it is always prudent to carry a minimum of one boring into bedrock to guard against such eventualities. This boring should be performed first and if there is an indication of deeply buried weak strata sufficient number of the remaining borings should be carried deep enough to establish the configuration of such strata.

Tunnels and underground installations such as caverns and chambers require explorations to the elevations of the underground installations. Knowledge of the nature of the material is required with regard to excavation and support of the underground openings. Often, these explorations are extremely deep and expensive and only one or two borings can

be justified. When feasible, it is better to utilize pilot bores or adits to establish the conditions.

The depths of explorations in proposed quarries and borrow areas will be controlled by the anticipated depth and thickness of the material to be excavated and used. The planner must make maximum use of aerial photographs, geologic reconnaissance, and available geologic information to establish the approximate limits of the source. He must first drill a few holes that will verify his assumptions as to the type of material present and the depth and thickness of the deposit. Only then will he be able to establish a proper program including the depth of exploration.

The depths of the explorations along the alignment of a highway or railroad will depend on the knowledge of subsurface conditions as based on geology, soil surveys, and previous explorations, and on the configuration of the highway or railroad at any given point. In areas of light cut and fill with no special problems, explorations should extend a minimum of 5 feet below proposed subgrade. However, where deep cuts are to be made, large embankments or embankments across marshland are to be constructed, or subsurface information indicates the presence of weak layers, the depth will depend primarily on the existing and proposed topography and the nature of the subsoil. Therefore no guidelines can be given except that the borings should be deep enough to provide information on materials that may cause problems with respect to stability, settlement, and drainage. Borings for structures are carried to depths established on the basis of the principles outlined for establishing the depths of borings for buildings. Depths of explorations for quarries and borrow areas would be as described for such areas when used in connection with dams.

1.4.4 Sampling

The majority of the samples taken will be of the representative but disturbed type such as those obtained by split-barrel samplers. This will permit visual identification and classification of the materials encountered as well as identification by means of grain size, water content, and Atterberg limit tests. In order to obtain a basic knowledge of the engineering properties of the materials that will have an effect on the design, a limited number of undisturbed samples such as those obtained with thin-wall tube samplers or double-tube core barrel soil samplers, will be required for possible shear, consolidation, and permeability tests. The number taken should be sufficient to obtain information on the shear strength and consolidation characteristics of each major stratum.

Sampling within the boreholes may be either continuous or intermittent. In the former case, samples are obtained throughout the entire length of the hole; in the latter, samples are taken every 5 feet and at every change in material. Initially it is preferable to have a few holes with continuous sampling so that all major strata present may be identified. The horizontal and vertical extent of these strata may then be established by intermittent sampling in later borings.

1.4.5 Reporting of Results

The results of the explorations and laboratory testing are usually presented in the form of a geology and soils report. The report may be in the form of an individual volume or it may be incorporated as a chapter or an appendix in an economic and technical feasibility report or in a project planning report. A discussion of the contents of a soils and geology report is given in the last section of this chapter. An outline for such a report also is presented.

1.5 EXPLORATIONS FOR DETAILED DESIGN

1.5.1 Type and Number of Explorations

The objective of this phase of explorations is to fill in any gaps in the previous program and to make such additional explorations as are necessary so that subsurface conditions at each structure are well defined. When the subsurface information from the two phases is combined, the resulting information must be adequate for preparation of bidding plans and specifications for construction. In this phase the precise location of each structure is set and additional borings are made so that an adequate number of borings are located at each structure. From the previous program certain problems may have been disclosed and for these more elaborate exploration methods or field tests or larger-diameter samples may be required. Otherwise, the type of explorations made in this phase may be very similar to those made in the previous phase.

1.6 GEOPHYSICAL EXPLORATIONS

1.6.1 Use of Geophysical Methods

Two geophysical methods, seismic and electrical resistivity, have proven useful as rapid means of obtaining subsurface information and as economical supplements to borings in exploratory programs for civil engineering purposes. Such geophysical explorations supply information for bedrock profiling, define the limits of granular borrow areas and large organic deposits, and yield a general definition of subsurface conditions including the depth to groundwater. However, there are numerous limitations to the information obtained by these methods and they should not be expected to give reliable or useful results for all subsurface conditions. To insure the optimum utilization of such exploration techniques, individuals experienced in both soils and geophysical theories should be consulted to determine the applicability of geophysical procedures to the area under investigation; to plan, design, and supervise the geophysical exploration program; and to interpret the data. All geophysical information should be spot-checked by borings and/or other direct methods of exploration.

1.6.2 Theory of Seismic Methods

Seismic methods of exploration are based on the fact that shock waves travel at different velocities through different types of materials. Since the velocity of wave propagation depends on numerous factors such as density, moisture, texture, void space, and elastic constants, the nature and stratification of subsurface materials can be determined. However, most subsurface materials are nonhomogeneous and anisotropic and this makes the analysis of seismic exploration data somewhat complex.

In seismic explorations, artificial impulses are produced either by the detonation of explosives or a mechanical blow (usually with a heavy hammer) at ground surface or at shallow depth within a hole. These artificial shocks generate three types of waves, namely, compression, shear, and surface waves; in general, only compression (longitudinal) waves are observed. These are classified as either direct, reflected, or refracted waves. Direct waves travel in approximately straight lines from the

source of impulse to the surface. Reflected or refracted waves undergo a change in direction of propagation when they encounter a boundary separating media of different seismic velocities. Waves that are turned back when they encounter such a boundary are called reflected waves; those that undergo a change or a bending in the direction of propagation are said to be refracted. It is mainly refraction and reflection seismic methods that are used for subsurface profiling in engineering exploration. Of these, the refraction method is most commonly used for civil engineering purposes, since the reflection methods on land are limited to providing information on subsurface materials at depths greater than approximately 1000 feet below ground surface.

In addition to subsurface profiling, seismic methods have also been used to determine engineering properties of soil and rock. The velocity of compression waves, and sometimes shear waves, determined by subsurface seismic methods such as uphole, downhole, and crosshole surveys are used to compute in situ values of the moduli of rock and soil (Sec. 1.14.4).

1.6.3 Seismic Refraction Method

Seismic refraction methods have been used to investigate subsurface conditions from ground surface to depths of approximately 1000 feet. A typical arrangement of field equipment for the investigation of a two-layer system is shown in the lower part of Figure 1.1. Point A on this diagram is the source of the seismic impulse. Points D_1 through D_{12} represent the locations of the detectors or geophones, whose spacing is dependent on the amount of detail required and the depth to the strata being investigated. In general, the spacing must be such that the length from D_1 to D_{12} is three to four times the depth to be investigated. The geophones are connected by cable to recording devices, which may be truck-mounted or may be portable units placed at ground surface. A high-speed camera is used to record the time at which the seismic impulse is generated and the time of arrival of the wave front at each geophone. A continuous profile along a line may be obtained by moving the geophones along the line, generating a new

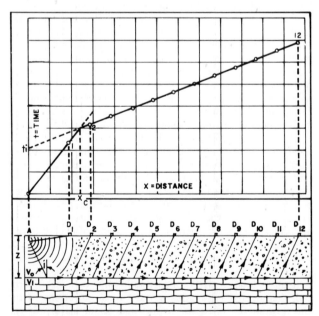

Fig. 1.1 Schematic representation of refraction of seismic energy at a horizontal interface and the resultant time–distance graph. (*Courtesy of Electro-Technical Labs, Inc.*)

impulse from the same source point each time the geophones are moved, and recording, for each shot, the time of initiation of the wave and the "first arrival times."

The data from the film records are plotted in the form of time–distance graphs such as that shown in the upper part of Figure 1.1. The slopes of this plot represent the velocity of the seismic wave as it passes through the various subsurface materials present. These velocities are used in standard formulas to determine the depth to the interface between the layers of material of differing seismic wave velocity. They are also used to obtain a general idea of the nature of the materials present. Typical seismic wave velocities for various materials are given in Table 1.1. Detailed investigation procedures for seismic refraction studies and the formulas used for depth determinations are presented by Jakosky (1950), Mooney (1977), and U.S. Army Corps of Engineers (1979).

The formulas used to determined subsurface strata thicknesses from seismic refraction survey data are based on the following assumptions:

1. Each stratum is homogeneous and isotropic.
2. The boundaries between strata are inclined or horizontal planes.
3. Each stratum is of sufficient thickness to reflect a change in velocity on the time–distance plot.
4. The velocity of wave propagation for each succeeding stratum increases with depth.

Unfortunately, more often than not, the subsurface conditions encountered do not satisfy all of the above assumptions and it is the effect of the variations from the assumed conditions that must be carefully evaluated to provide a proper interpretation of seismic data; for example, a stratum that has a thickness less than one quarter of the depth from the ground surface to the top of the stratum or one in which the velocity of wave propagation is less than that of an overlying stratum will be "masked," that is, its presence will not be reflected in the time–distance plot. In addition, the range of the velocity of wave propagation for a material may overlap that of the material immediately above or below and it may be difficult, if not impossible, to distinguish between the two strata. There also are some strata in which the velocity of wave propagation continuously increases with depth and causes difficulty in the determination of the actual wave velocity. Therefore, borings should always be performed in conjunction with seismic surveys to corroborate the interpretation of the data and to insure maximum reliability.

In situations where the boundaries between strata are inclined, a single time–distance plot will indicate only the average thickness of the strata. It is standard practice to eliminate this problem by "reverse profiling," which consists of obtaining a time–distance plot for shocks initiated at both ends of the line of geophones.

Modified field techniques also are useful in specific situations such as investigations for buried channels. Among these techniques is the "fan shooting" method, in which the geophones are located at approximately the same radial distance from the impulse point. If the time required for a wave to reach any geophone is significantly different from the measured times for the other geophones, the existence of a material of different velocity is indicated.

Seismic refraction surveys have been used successfully on numerous engineering projects. They have proven particularly valuable in the reconnaissance stage, during which they may be used as a rapid, efficient means to obtain information that often permits the elimination of alternate sites or alignments without having to perform a large number of borings. Among the most useful information provided is the depth to bedrock, the thickness and areal extent of gravel borrow areas, the

TABLE 1.1 TYPICAL VALUES OF SEISMIC WAVE VELOCITIES FOR VARIOUS MATERIALS.

Material	Velocity (ft/sec)	
	Compression Waves	Shear Waves
ROCK		
Agglomerate	5 000–6 000	—
Anhydrite, gypsum	11 500–21 400	—
Basalt		
Sound	18 300–21 100	10 500
Weathered and fractured	9 000–14 000	
Chalk		
Above groundwater	6 000–13 000	—
Below groundwater	8 000	—
Conglomerate	7 920–8 000	—
Diabase	19 700	
Diorite	19 100	10 100
Dolomite	10 700–20 200	—
Dunite	24 400–28 400	12 500–14 400
Gabbro	21 300	11 200
Granite		
Sound	13 100–20 000	6 900–10 800
Highly weathered	10 500	—
Decomposed	1 540–2 200	—
Granodiorite	15 000–15 800	10 200
Greenstone	13 300–16 100	—
Limestone		
Hard	16 400–20 200	9 500–10 700
Soft	5 600–13 900	—
Salt	14 400–21 400	—
Phyllite	10 000–11 000	—
Sandstone	4 620–14 200	—
Shale		
Hard	9 000–15 400	—
Soft or weathered	2 600–8 000	—
SOIL		
Alluvium	1 640–6 600	—
Clay	3 000–9 200	—
Glacial deposits		
Moraine		
Dry	2 500–5 000	—
Saturated	5 000–7 000	—
Till	5 600–7 400	—
Gravel	1 500–3 000	—
Sand	4 600–8 400	—
Sand, cemented	2 800–3 200	—
Sand, loose	5 940	1 650
Talus, loose rock	1 250–2 500	—
Top soil		
Dry	600–900	—
Wet	1 000–2 500	—
OTHER		
Ice	12 050	—
Water, fresh	4 700–5 500	—
Water, salt	4 800–5 000	—

Sources: Clark (1966); Dobrin (1976); Jakosky (1950); U.S. Army Corps of Engineers (1979).

location of the groundwater table and the susceptibility of materials to excavation by ripping with a tractor. Alternative dam site locations often may be eliminated in the reconnaissance stage because of a very deep sound rock foundation. Similarly, the bedrock profile information may indicate an excessive amount of rock excavation for one highway alignment as compared to another. Seismic methods may also be used to provide general information during preliminary investigations for bridges, buildings, and other structures. However, the design of the foundations for such structures requires information regarding particular physical properties of the materials present that is not obtainable by seismic methods.

1.6.4 Seismic Reflection Methods

Continuous seismic profiling, a sonar seismic method based on the principles of reflection, is used extensively in studies for deep water projects. The method utilizes sonic waves produced at or near the water surface to provide a continuous profile of the sea bottom and to map the underlying materials. Systems referred to as the Sparker, Boomer, and Pinger have been developed for investigations requiring both shallow and deep penetration of the materials underlying the sea bottom. Detailed descriptions of these systems are presented in an article by Van Reenan (1964).

Continuous seismic profiling has been used for subsurface investigations for pipeline crossings, dam sites, bridges, marine structures, and dredging projects.

1.6.5 Electrical Resistivity Method

The electrical resistivity method of subsurface exploration is based on the fact that different materials offer different resistance to the passage of an electric current. Thus, by the determination of vertical and lateral variations in this resistance it is possible, within certain limitations, to infer the stratification and lateral extent of subsurface deposits. In this method, the resistance to passage of the current is determined by measurement of the specific resistance (resistivity) of the material, which is defined as the resistance in ohms between opposite faces of a unit cube of the material. The common units of resistivity measurement used in subsurface exploratory work are the ohm-foot, ohm-centimeter, and ohm-meter. In soils, the resistivity of the particles is high; similarly the resistivity of groundwater, if pure, is high. Therefore, if there is to be a passage of current through a soil mass it will be almost exclusively through electrolytic action due to the presence of dissolved salts in the groundwater. Consequently, the resistivity of a soil is primarily dependent on the moisture content and the concentration of these dissolved salts. It also is influenced to varying degrees by the void ratio, particle size, stratification, and temperature. In rocks other than mineralized deposits, the resistivity similarly is dependent primarily on the moisture content and the concentration of dissolved salts in the groundwater. It also is affected by porosity, dip and strike, soundness, and temperature.

Several methods involving different electrode arrangements have been developed for making field resistivity measurements. Among these are the Wenner, Schlumberger, and Lee Methods. The major portion of the civil engineering resistivity surveys performed in the United States have been accomplished by the Wenner Method, whereas the Schlumberger Method is utilized for most European surveys.

The Wenner arrangement (Fig. 1.2) consists of four equally spaced electrodes driven approximately 8 inches into the ground. In this method a d.c. or very low-frequency a.c. current of known magnitude is passed between the two outer (current) electrodes, thereby producing within the soil an electric field, whose pattern is determined by the resistivities of the soils present within the field and the boundary conditions. The potential drop for the surface current flow line is measured by means of the inner electrodes and the resistivity is calculated by the formula shown in Figure 1.2. This formula assumes that the material within the limits of the electric field is homogeneous. If it is not, the value obtained represents an average resistivity

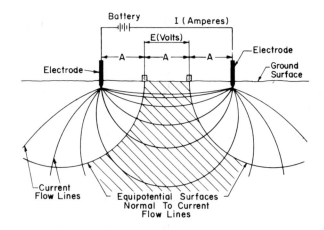

Apparent Resistivity, $\rho = \dfrac{2\pi A E}{I}$

Fig. 1.2 Diagram showing theory of earth resistivity methods. (*After U.S. Army Corps of Engineers.*)

for the material within the shaded zone of Figure 1.2 and is largely dependent on the resistivities of the materials to a depth equal to the electrode spacing A.

Two types of resistivity surveys are used for subsurface explorations, namely, electric profiling and electric sounding. The purpose of electric profiling is to provide information concerning lateral variations in subsurface materials. This is accomplished by maintaining a constant electrode spacing as the electrodes are moved across an area and a resistivity measurement is made for each new location of the electrode spread. A common practice is to advance the electrode spread a distance equal to the electrode spacing by moving the rear electrode to the front position after each reading. Data from a series of such traverses across an area may be presented in the form of a series of contours of equal resistivity. Electric profiling is used to delineate the areal extent of pervious deposits, locate faults, and locate steeply dipping contacts between materials.

The purpose of electric sounding is to provide information on the variation of subsurface materials with depth. This is accomplished by maintaining the center of the electrode spread at a given location and taking a series of resistivity readings as the electrode space A of Figure 1.2 is increased. As the spacing is increased, the depth of material that affects the apparent resistivity increases, and changes in material are reflected in the resistivity values obtained. Electric sounding is used to estimate the depth to bedrock, sand and gravel, or water-bearing strata and to estimate the thicknesses of strata.

Several methods have been developed to interpret the results of electric sounding surveys. These include the Standard Curve Method, the Inflection Method, and the Moore Method. The Standard Curve Method is based on a comparison of the field and theoretical curves of resistivity versus electrode spacing. Wetzel and McMurray (1937) and Roman (1934) have prepared theoretical curves for three- and two-layer systems, respectively. The curves have been developed for specified ratios of the resistivities of each layer and specified ratios of the depths to the layers. If the standard and field curves are plotted on log–log paper and the field curve can be matched to a theoretical curve, the depths of the layers can be ascertained.

The Inflection Method is based on the assumption that the apparent resistivity is a measure of the average resistivity to a depth approximately equal to the electrode spacing and thus changes in the slope of the curve of resistivity versus electrode spacing are indicative of strata boundaries. The peaks and valleys of such curves for three-layer systems have given approximate depths to the top of the third layer. However, in general, the depth to the top of the second layer cannot be determined accurately by this method, nor is the method applicable to cases in which the resistivity continually increases or decreases.

The Moore Method is a modified version of the Inflection Method in which the cumulative resistivity, that is, the sum of the resistivities for a given electrode spacing and all smaller spacings, is plotted versus the given spacing. The spacings at the intersections of tangents drawn to intersect near the points of maximum curvature of the cumulative curve are taken as the depths to strata boundaries.

Success with the resistivity method has been documented for several types of applications. Moore (1961) reports the use of the resistivity method to obtain rock profiles. Experiments conducted at the U.S. Army Corps of Engineers Waterways Experiment Station (1943) indicate good agreement between borings and the results of resistivity surveys to determine the depths of strata boundaries. These experiments also indicated that the method could be used to distinguish between pervious alluvium and clays. Golder and Soderman (1963) describe the successful use of the resistivity method to determine the thickness of peat, soft clay, and sand over a relatively shallow glacial till, and Liesch (1969) reports its use to locate water-bearing aquifers. However, although there are many instances of the successful use of the resistivity method, many factors affect the field measurements and the data can often be misleading or difficult to interpret. Among the factors affecting interpretation are the broad range of resistivity values for a given material and the overlap of ranges of values for different materials, near-surface irregularities, and stray potentials. These are discussed below.

Szechy (1957) records resistivity values of 10^2 to 1.7×10^3 ohm-m for dry granular material. Clays, because of exchangeable ions and double layers even in the partially saturated state, have low resistivities that range from 3 to 50 ohm-m and sandy clays range from 10 to 100 ohm-m. Rocks, in general, have high resistivities; however, weathered rocks and soft rocks can have very low resistivities. Values ranging from 30 to 10^5 ohm-m have been recorded for sandstone and values as high as 10^{15} ohm-m have been measured in porphyry. The values cited indicate the wide range of resistivity that may occur for a given type of material and the overlap that exists for the ranges of resistivity of different materials. Both factors make the determination of the exact nature of the materials present and the location of strata boundaries difficult to determine in many instances. For example, the boundaries between an organic soil and soft clay, between stiff clay and soft clay shale, or between loose sand and coarse sandstone gravel usually cannot be detected. Also, the difference in the resistivity of clay above and below the water table may not be reflected in the data because of the double layers of ions and water that exist around clay particles even above the groundwater table.

The values obtained in resistivity surveys are strongly influenced by near-surface irregularities since the measurements are made at ground surface. The tests conducted by the U.S. Army Corps of Engineers Waterways Experiment Station (1943) indicate that surface anomalies or irregularities such as sloughs and crevasse deposits can affect the overall readings significantly. Other local irregularities such as ditches, road surfaces, and backfilled depressions may also affect readings. The effect of such irregularities is sometimes monitored by overlapping profiles or by smoothing the resistivity curves. On the other hand, resistivity surveys can be used to locate anomalies as shear zones and faults in bedrock that otherwise has relatively uniform resistivity.

Potentials generated by ore bodies, stray potentials, or tramp materials within the ground also can affect the electric field created by resistivity survey equipment. Jakosky (1950) describes the use of self-potential mapping of soils to locate ore bodies beneath the surface and indicates that potentials from such ore bodies may be as great as 500 to 1000 mV. Also, Golder and Soderman (1963) report a case in which resistivity measurements indicated the existence of a channel in the bedrock beneath a river. It was found later that the low resistivity, which had been interpreted as being due to the existence of a channel, was actually due to the presence of a vertical bed of graywacke containing pyrites. Potentials also may arise because of buried electric cables, pipeline galvanic action, subway electric systems, and grounded power currents. These often limit the usefulness of the resistivity method in urban areas.

The above limitations notwithstanding, the electrical resistivity method is a useful exploratory tool when its limitations are recognized and the interpretation of the field data is done by experienced personnel. The method must always be used in conjunction with borings if reliable data are to be obtained. It is, in general, not as accurate or reliable as the seismic method. However, it is a faster and more economical means of exploration. The required equipment has a low initial cost and is completely portable.

1.6.6 Impulse Radar

Probing subsurface conditions by impulse radar (Morey, 1974), referred to as ground penetrating radar (GPR), electromagnetic subsurface profiling (ESP), and subsurface interface radar (SIR), has been successfully used in the field of civil engineering purposes since 1970. The technique has been used for many purposes, among which are the continuous profiling of strata between borings; the profiling of the surface of bedrock and the groundwater table; the detection of voids in soil and rock; the detection of voids within and below pavements; the location of utilities and buried objects; the detection of holes in clay liners; and the location of reinforcing bars in pavements. As with the seismic and electrical resistivity methods described above, it is necessary to have borings for correlation and calibration when the technique is used to delineate subsurface conditions.

This method of exploration consists of the radiation of repetitive electromagnetic impulses into the ground from the surface and the recording of the travel time of the pulses reflected from the ground surface and from discontinuities in the subsurface profile. The travel time of the reflected pulses are used to determine the depth to the discontinuities and to delineate these discontinuities. Reflection of the radar signals is caused by differences in the conductivity of the materials through which the signal passes and the relative dielectric constants of the materials penetrated.

The equipment used for ground penetrating radar surveys includes a radar impulse generator; low- and high-frequency antennas that are used both to transmit the applied radar signal and to recieve the reflected signals; a graphic recorder; a magnetic tape recorder, which is optional; and a converter for d.c. operation. Antennas with high frequencies, that is, 300 to 900 MHz, produce a greater resolution of detail over a shallower depth, whereas antennas with low frequencies (80 to 120 MHz) provide greater penetration but with less resolution. Harding Lawson Associates (1986) indicate that objects less than 1 inch in size can be detected to a depth of 2 feet with the 900-MHz antenna and that penetrations up to 60 feet with a resolution of about 2 feet can be obtained with the 80- to 120-MHz antennas.

During operation, all of the equipment for the radar survey except the antenna is usually mounted in a van or similar vehicle. The sled-mounted antenna is towed over the ground surface behind the vehicle at speeds that, in general, range from 0.5 to 1 mph. Distances along the alignment can be measured by a bicycle-wheel type of measuring device mounted to the vehicle bumper. In those cases where the terrain is not suitable for a vehicle, the antennas can be towed by hand. The remaining equipment, which is connected to the antenna by a cable several hundred feet in length, is moved from one intermediate point to another. The intermediate points are located so that when the antenna is towed for the full length of the cable in both directions from each point, complete coverage of the line is obtained.

As the antenna is towed, either by vehicle or by hand, reflected signals are printed graphically on the strip-chart recorder, thereby permitting preliminary field evaluation. The use of a magnetic tape recorder in conjunction with the strip recorder permits later playback and processing of data.

1.7 SOUNDINGS AND PROBINGS

1.7.1 Description of Equipment and Method

The terms *sounding* and *probing* are used synonymously to represent that method of exploration in which a rod is made to penetrate overburden deposits by means of dynamic or static loading and a continuous or semicontinuous record of the resistance to penetration is obtained. The term *jet probing* is often used to refer to that process in which a rod is jetted through the overburden until it can penetrate no farther and a record is made of the depth of penetration. In the case of a jet probing, the sole purpose is to attempt to determine the depth to the bedrock surface. However, in the case where a rod is used, while the determination of the depth to bedrock may be one of the objectives, the penetration resistance is often used in an attempt to delineate changes in materials, to correlate the resistance to penetration with various soil properties, and to determine the required penetration of piles to obtain a specific capacity. This method of exploration has been used extensively in Europe. In particular, it has found wide application in the Low Countries, where extensive soft deposits overlie the firm bearing strata to which foundations must be carried. Soundings also are particularly useful in areas such as deltaic plains where the soil profile is erratic. In such areas they may be used economically to obtain a more detailed delineation of the soil strata either by measurement of the resistance to penetration between the intermittent samples in individual borings or by measurement of the resistance throughout the entire profile depth at locations between widely spaced borings. Figure 1.3a shows the log of a boring in a delta area in which intermittent samples were taken and a Terzaghi-type penetrometer was used to determine the density of the sands and the consistency of the clays encountered. Figure 1.3b shows the soil profile along a highway as determined from 40 similar borings in which a cone penetrometer was used as a supplement to intermittent sampling.

In its simplest form the apparatus used to make soundings today consists of a conical point attached to a steel rod, which is commonly referred to as a drive point or cone penetrometer. A schematic drawing of a version that incorporates a special coupling and an expendable drive point into the basic apparatus is shown in Figure 1.4a. Soundings performed using this type of penetrometer consist of pushing or driving the point into the ground and recording the pressure required to achieve a specified penetration or the number of blows per foot required to drive the point with a specified hammer weight and drop. Drive point penetrometers, in general, are economical of equipment in that all of the components required except for

special point couplings and drive points are pieces of equipment such as standard hammers, drive heads, driving guides, pull pieces, and drill rods, which would be available as standard equipment with a test boring rig. The drive points used commonly have a cone angle of 60° and range from 1.3 to 2.75 inches in diameter. The special couplings for the apparatus shown in Figure 1.4a are available in sizes that permit the attachment of the point to standard drill rods ranging from E-size to N-size. In practice, the diameter of the point used is always larger than the rods to which it is attached, so that friction on the sounding rods is minimized and retrieval of the rods is thereby facilitated.

Many other variations of the basic penetrometer exist. In general, these variations have been developed in an effort to minimize the effect of side friction or surcharge on the resistances measured in advancing the point. Figures 1.4b and 1.4c illustrate two of the points in use. Figure 1.4b shows a schematic section of the Dutch cone penetrometer (CPT) in which the $\frac{5}{8}$-inch diameter steel rod to which the cone is attached is encased in a $\frac{3}{4}$-inch diameter standard steel pipe. The purpose of the latter is to permit the direct measurement of the point resistance by preventing the development of friction along the rod should the hole tend to close after the cone has passed. It is also possible to obtain an estimate of the side friction by measuring the force required to push the pipe sleeve ahead. This requires the introduction, immediately above the cone, of a short length of sleeve pipe having the same diameter as the cone while using a smaller-diameter pipe for the remainder of the sleeve pipe. The cone and sleeve pipe are then advanced alternately and the force required to move each is measured. The cone used on the Dutch penetrometer generally has a base area of 10 cm^2 when it is used in soft soils; a stronger version of the penetrometer having a cone with a base area of 20 cm^2 is used for deep soundings and dense or stiff materials. The lighter apparatus may be advanced by hand to limited depths; the stronger apparatus requires some mechanical means for achieving penetration.

Standards for this cone penetration test are given in ASTM D-3441, and guidelines for performance and geotechnical design by Schmertmann (1978). In the late 1960s, the Dutch cone penetrometer was improved so that the cone resistance and the friction on the sleeve could be measured electrically and simultaneously, that is, without alternate movement of the cone and sleeve. This arrangement permitted more accurate determination of the friction ratio—frictional resistance/cone resistance. Sandy soils have high cone resistance and low friction ratio, whereas clayey soils have relatively low cone resistance and high friction ratio. For cone resistance and friction ratio to be representative of a particular stratum, the stratum must have some minimum thickness. Jamiolkowski et al. (1985) recommends a minimum thickness of about 70 cm for a thin stiff layer embedded in a soft soil mass, and of 20 to 30 cm for a thin soft layer in a stiff deposit. Schmertmann (1978) gives correlations of CPT values with relative density and effective friction angle of sands, and with undrained shear strength for clays.

In the 1970s the piezocone (CPTU) was developed. This instrument measures the piezometric pressure at the cone during penetration. When the standard penetration rate is about 2 cm/sec, practically undrained penetration conditions obtain in homogeneous cohesive soils, and almost drained penetration conditions obtain in relatively clean sands (fines content less than 10 percent). Jamiolkowski et al. (1985) state that the various CPTU pore pressure coefficients that have been developed are more sensitive than the CPT friction ratio to changes in soil type; that classification charts available in the literature are specific only for a particular piezocone geometry and location of piezometer tip; and that CPTU pore

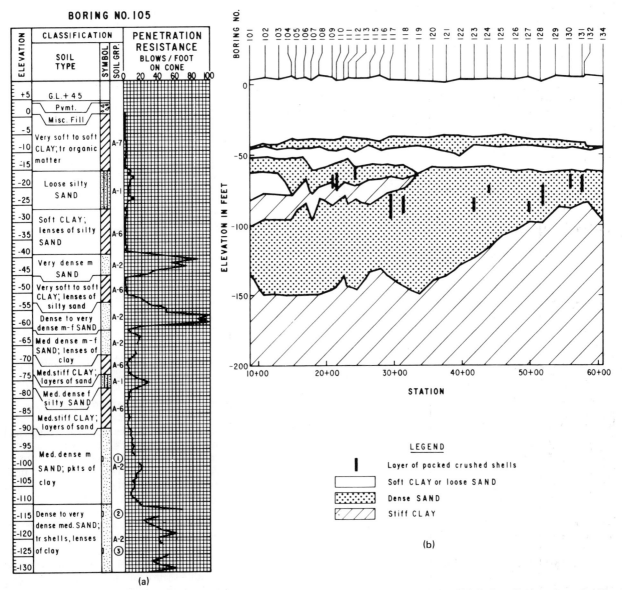

Fig. 1.3 (a) Boring log and plot of resistance to penetration of Terzaghi-type cone penetrometer. (b) Soil profile based on classification of intermittent samples from borings and delineation of strata of differing consistency or density by cone penetrometer resistance.

pressures reflect not only the type of soil, but also its stress history (particularly whether it is normally consolidated or overconsolidated).

Figure 1.4c shows a wash point penetrometer developed by Terzaghi for performing soundings in sand deposits. The penetrometer consists of a 2.75-inch diameter cone attached to a 1.5-inch diameter double extra-strong pipe encased in 3-inch diameter extra-strong pipe. The point is provided with upward-deflected water passages. In operation the point resistance is measured as the point is advanced approximately 10 inches into the soil. Liquefaction of the sand is then accomplished by introducing water, which is pumped into the sounding rod to which the point is attached, flows out of the point through the upward-deflected jets and returns to the surface through the annular space between the drill rod and casing as shown in Figure 1.4c. As the liquefaction occurs, the pipe sleeve is advanced to meet the cone; the flow is then stopped and the procedure is repeated. Alternatively, drilling mud may be used instead of water and then the casing is not required. The purpose of using the casing or drilling mud is to reduce side friction on the sounding rod.

1.7.2 Use and Limitations

The penetrometer is primarily useful in the mapping of soil strata during the early stages of explorations when the number of borings that can be drilled is normally limited. The soundings generally have the advantage of speed and low cost when compared to borings. Some planners of exploratory programs substitute several soundings for a single boring with the intent of obtaining more information. Others prefer to think of soundings as a tool that can be used as required to obtain additional information between borings at minimal cost once it has been ascertained that conditions are erratic. Soundings are particularly useful when performed to obtain information on stratification that normally would not be available until additional borings were performed in a later stage of exploration. Hvorslev (1949) has stated that soundings often reveal the presence of strata that are not recovered or observed in sampling operations. The planner, however, also must be aware of the limitations as well as the advantages of soundings. One major limitation is that either no samples or only wash samples are obtained from soundings; therefore, strata cannot be definitely

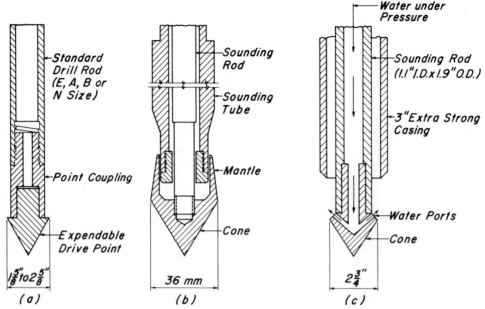

Fig. 1.4 Cone pentrometers: (a) drive point penetrometer; (b) Dutch cone penetrometer; (c) Terzaghi wash-point penetrometer.

identified by soundings alone. In addition, the possibility of obtaining misleading results caused by the presence of gravel, boulder, or wood erratics within the soil strata must be considered. Interpretation of the results obtained from soundings consequently requires considerable experience, particularly in those cases in which correlations between the penetration resistance and engineering properties of the soils penetrated are to be developed.

1.8 BORING METHODS

1.8.1 General Description of Operation

A boring may be defined as any vertical, inclined, or horizontal hole drilled in the ground for the primary purpose of obtaining samples of the overburden or rock materials present and thereby permitting the determination of the stratigraphy and/or the engineering properties of those materials. The hole, in addition, may be utilized for the in situ determination of such engineering properties as permeability and shear strength, the determination of lateral earth pressure, the observation of fluctuations in the groundwater level, the determination of pore water pressures by means of piezometers inserted in the holes, and the measurement of deformations by the installation of devices such as extensometers.

The procedure used to make exploratory borings may be divided into two basic operations: (1) advancing the hole to the depths at which samples are to be obtained, and (2) sampling of the soil or rock. The variability of the materials encountered and the need of samples for various purposes has resulted in the development of many different techniques and types of equipment for both of these operations. Methods for advancing the hole are classified, in general, according to the manner in which materials are removed. Among the techniques commonly employed are displacement, continuous sampling, augering, wash boring, percussion drilling, rotary drilling, and hammer drilling. The techniques for soil sampling are classified, in general, by the mechanical configuration of the sampler and/or the name of its developer. The samplers in common use include spiral, bucket, and hollow-stem augers; solid and split-tube drive samplers; the Shelby tube, stationary piston, retractable plug, and Denison double-tube core barrel samplers; single- and double-tube rock core barrels with diamond or carbide insert bits; shot core drills; and others.

1.8.2 Equipment Used for Making Borings

The machinery used to advance the hole and take samples, commonly referred to as a drill rig, in general consists of: (1) a motor that provides power to operate a hammer to drive casing and to operate a winch to raise and lower the drilling and sampling equipment; provides rotary motion where required to turn augers or coring equipment; and provides downward pressure to push samplers into the ground; (2) a water pump or air compressor that provides water or air under pressure for the removal of the cuttings from the drill hole and for the cooling of rotary bits; (3) a winch to raise and lower drilling tools and casing; and (4) a tripod, four-legged frame mast, or derrick equipped with a sheave for use in conjunction with the winch in raising and lowering drill tools. These rigs are available in a wide variety of sizes and configurations, and are discussed in detail in the paragraphs dealing with methods of advancing drill holes. The unit selected will depend upon (1) the availability of specific rigs; (2) the location of the work; (3) the materials to be penetrated; (4) the type of sampling required; (5) the size and depth of the hole; and (6) the method of penetration. In remote areas or areas of difficult access, lightweight units that can be carried by men or pack animals are commonly used. The other extreme in rig size is represented by the heavy-duty units that are capable of putting down large-diameter holes thousands of feet below ground surface. These are used when explorations for deep underground construction are to be carried out from ground surface.

Drill rigs may be equipped for transport in several ways. In instances where drilling is to be done along highways, streets, or level terrain, the rig is often mounted on the back of a truck (Figs. 1.7, 1.12, 1.13c and d) or on a small trailer (Fig. 1.8). If the terrain is rugged or the area of operation is limited, the rig will normally be mounted on skids (Figs. 1.13a and b) so that it may be dragged along the ground and can be placed in a small area. In such areas, rigs mounted on crawler tractors are also used.

Borings are frequently performed in soils beneath a water surface such as that of an ocean, river, or lake. Special rigs such as the Vibracore (Fig. 1.41) have been developed for sampling

Fig. 1.5 Platform-mounted drilling rig marine borings.

soils beneath the water and these are discussed in a later paragraph. However, the majority of marine work in relatively shallow water still is performed using land-type rigs mounted on barges, floating platforms supported by pontoons or oil drums, and fixed platforms supported by piles or spuds (Fig. 1.5). Lightweight rigs mounted directly on the casing, once the latter has been driven a sufficient distance into overburden, have also been used frequently (Fig. 1.6). The use of casing-mounted drill rigs eliminates difficulties in rotary drilling caused by the up-and-down motion of the drill rods and core barrel that normally occurs with floating equipment.

Fig. 1.6 Casing-mounted drilling rig. (*Courtesy of Acker Drill Co., Inc.*)

1.8.3 Stabilization of Boreholes

Two common problems in exploratory work are caving of the sides of the drill hole and heaving of the bottom of the hole. The latter occurs to some extent in all holes, whether above or below the groundwater table, owing to the stress release caused by removal of material from the hole. However, it is most serious in the case of holes below the groundwater table, since water seeping into the bottom of the hole from the surrounding area can result in considerable disturbance to the soil to be sampled. This disturbance is normally minimized by stipulating in the specifications that the level of the drilling fluid in the hole must at all times be equal to or higher than the elevation of the groundwater. Thus, any seepage will be from the hole to the surrounding area and will stabilize rather than disturb the base of the hole. This procedure should not be used when undisturbed samples are to be taken in the zone above the water table, since it will result in an increase in water content that may destroy the structure of the material to be sampled. Loess and some residual soils are highly susceptible to such structural disturbance.

The susceptibility of the walls of a boring to caving is dependent on the nature of the material into which the hole is drilled, the depth of the hole, and groundwater conditions. In general, uncased holes above the water table will remain stable except in purely cohesionless oils. Boreholes in cohesive soils may remain open for a considerable distance below the groundwater table. The danger of collapse increases, however, once the groundwater has been reached. Closing of a hole below the groundwater level may occur even in rock. Cases have been observed in which the jointing in the rock was such as to form small fragments that, owing to the presence of clay along the joint planes, easily slid from the wall of the hole. This can necessitate redrilling of a portion of the hole as a minimum and has been known to cause the loss of expensive core barrels when it was found impossible to dislodge rock fragments that wedged against the drill rod while the core barrel was still in the hole. Thus, it is common practice to stabilize holes, particularly those extending below the groundwater level, by lining them with drive pipe or casing or by means of drilling fluids, grouting, or freezing.

Drill holes may be lined with either drive pipe or casing. Both are provided in sections, 1 to 10 feet in length, that are coupled together as they are driven, jacked, or drilled into the ground as the hole progresses. The primary differences between drive pipe and casing are in the wall thickness, the type of thread, and the design of the couplings. Drive pipe is the stronger of the two and is therefore preferred for use when conditions are such that the lining will be subjected to high stresses during installation. In general, drive pipe is driven or jacked into the ground, whereas casing is jacked or rotated into place because it is not as rugged as drive pipe. However, under proper conditions, casing is also driven. When the pipe or casing is driven, it is usually protected by a shoe of case-hardened steel at its lower end. The shoe resembles a coupling in size and shape but has a thickened lower edge, the inside of which is beveled to facilitate penetration and to guide displaced material into the pipe or casing as it is driven. When casing is to be drilled into place, a casing shoe or bit having diamonds or carbide inserts is used in lieu of the beveled shoe. Driving of pipe or casing is permissible except when holes are being drilled in soils such as loose sands and silts, which are easily disturbed by vibration. In such instances the casing should be jacked. Drilled casing is sometimes used when it is known that the bedrock is of such poor quality that it will have to be cased.

Drive pipe may be standard or extra-heavy, butt-welded, black steel pipe or extra-heavy seamless steel pipe. Standard-weight pipe may be used when the overburden is relatively shallow, the driving is not expected to be difficult, and the pipe is to be left in place. Extra-heavy pipe is required when the overburden is relatively deep, difficult driving is anticipated, and the pipe is to be reused. Some drillers prefer to use extra-heavy pipe exclusively, because of its greater durability. Drive pipe is available with nominal inside diameters ranging from 2 to 6 inches. The pipe is normally furnished with recessed couplings that have an outside diameter larger than that of the pipe. Flush-jointed pipe has been specially fabricated from extra-heavy drive pipe (Hvorslev, 1949) but it is not as durable as drive pipe with outside couplings. Pertinent data regarding the dimensions and weights of the various pipes and couplings commonly used are presented in Table 1.2.

Casing is fabricated from cold-drawn seamless steel tubing and is supplied with either flush joints or flush couplings. In the flush-joint casing, the tube is so threaded that couplings are not required; in the flush-coupled casing, the couplings are fabricated so that the outside diameter of the coupling is the same as that of the pipe. Flush-joint and flush-coupled casings

are preferred to drive pipe by many drillers because the flush exterior presents less resistance to penetration and pulling. Flush-joint casing also eliminates the danger of samples being caught at the coupling connections and jarred on extraction from the hole.

Casing is manufactured in accordance with standards set forth by the Diamond Core Drill Manufacturers Association (1980). These standards designate sizes not only for casing and the reaming shells, shoes, and bits used with it, but also for drill rods, core barrels, and other exploratory equipment. The standard sizes for casing are designated by double letters such as EX, EW, AX, AW, etc. The standards are such that the casing will nest, that is, casing of a given size will telescope into the next larger casing. Thus, if a length of casing cannot be advanced beyond a given depth and wall support is still required to continue a boring to its planned depth, the next smaller size of casing can be telescoped into the casing in place and the hole may be continued at a minimum reduction in diameter. Also, the standard dimensions are such that the outside diameter of a given casing is small enough to fit into the hole made by the next larger size coring bit and the inside diameter of a given casing will admit a core barrel of the same size designation. Flush-coupled casing is available in sizes from EX to HX and flush-joint casing is available in sizes from RW to ZW. The dimensions and weights of the most commonly used sizes of both types of casing are shown in Table 1.2. The dimensions for other sizes are given in the *Standards Bulletin* of the Diamond Core Drill Manufacturers Association (1980).

Lining a borehole with drive pipe or casing, commonly referred to as "casing the hole," or supporting the walls of the borehole with hollow-stem augers (Section 1.12.6) are the most effective methods of supporting the walls of a boring. They are used in those cases where it is cheaper than using drilling mud or where no other means of support is acceptable. In general, this includes all borings in soft and loose soils and all borings in which field permeability tests and long-term groundwater observations are to be made. Drilling muds and grout are not acceptable in the latter case since they contaminate the soil or rock and reduce permeability. Short sections of casing also are used frequently at the top of uncased borings to maintain the opening, which otherwise tends to be eroded by fluid flowing from the hole.

Drilling fluid in its simplest form is merely water. More commonly the term refers to mixtures of water and various substances that produce suspensions that support the walls of a borehole by means of their high specific gravity and thixotropy and an impervious lining (mud cake) that forms along the walls of the boring. Drilling fluids, when used without casing, serve the dual function of supporting the walls of the

TABLE 1.2 STANDARD SIZES OF DRILL TOOLS.

Drill Rods—Flush Coupled

Size	O.D. (in)	I.D. (in)	Weight (lb/ft)	Coupling I.D. (in)	O.D. (mm)	I.D. (mm)	Weight (kg/m)	Coupling I.D. (mm)
E[a]	1-5/16	7/8	2.7	7/16	33.3	22.2	4.0	11.1
A[a]	1-5/8	1-1/8	3.8	9/16	41.3	28.6	5.7	14.3
B[a]	1-7/8	1-1/4	3.6	5/8	47.6	31.7	5.4	15.9
N[a]	2-3/8	2	5.0	1	60.3	50.8	7.4	25.4
EW[b]	1-3/8	15/16	2.8	7/16	34.9	23.8	4.2	11.1
AW[b]	1-3/4	1-1/4	4.3	5/8	44.5	31.8	6.4	15.8
BW[b]	2-1/8	1-3/4	4.3	3/4	53.9	44.4	6.4	19.0
NW[b]	2-5/8	2-1/4	5.5	1-3/8	66.6	57.1	8.2	34.9
HW[b]	3-1/2	3-1/16	8.8	2-3/8	88.9	77.8	13.1	60.3

[a] Original diamond-core drill tool designations.
[b] Current standards of the Diamond Core Drill Manufacturers Association (DCDMA).

TABLE 1.2 (*Continued*)

Casing—Flush Jointed—DCDMA Standards

Size	O.D. (in)	I.D. (in)	Weight (lb/ft)	O.D. (mm)	I.D. (mm)	Weight (kg/m)
EW	1-13/16	1-1/2	2.8	46.0	38.1	4.2
AW	2-1/4	1-29/32	3.9	57.1	48.4	5.8
BW	2-7/8	2-3/8	7.0	73.0	60.3	10.4
NW	3-1/2	3	8.4	88.9	76.2	12.5
HW	4-1/2	4	11.7	114.3	101.6	17.41

Casing—Flush Coupled—DCDMA Standards

Size	O.D. (in)	I.D. (in)	Weight (lb/ft)	Coupling I.D. (in)	O.D. (mm)	I.D. (mm)	Weight (kg/m)	Coupling I.D. (mm)
EX	1-13/16	1-5/8	1.80	1-1/2	46.0	41.3	2.68	38.1
AX	2-1/4	2	2.90	1-29/32	57.2	50.8	4.31	48.4
BX	2-7/8	2-9/16	5.90	2-3/8	73.0	65.1	8.78	60.3
NX	3-1/2	3-3/16	7.80	3	88.9	81.0	11.60	76.2

Casing—Standard Drive Pipe

Nominal Size (in)	O.D. (in)	I.D. (in)	Weight (lb/ft)	Coupling O.D. (in)	Nominal Size (mm)	O.D. (mm)	I.D. (mm)	Weight (kg/m)	Coupling O.D. (mm)
2	2-3/8	2-1/6	5.5	2-7/8	50.8	60.3	52.4	8.18	73.0
2-1/2	2-7/8	2-15/32	9.0	3-3/8	63.5	73.0	62.7	13.39	85.7
3	3-1/2	3-1/16	11.5	4	76.2	88.9	77.8	17.11	101.6
3-1/2	4	3-9/16	15.5	4-5/8	88.9	101.6	90.5	23.06	117.3
4	4-1/2	4-1/32	18.0	5-3/16	101.6	114.3	102.4	26.78	131.8

Casing—Extra-Heavy Drive Pipe

Nominal Size (in)	O.D. (in)	I.D. (in)	Weight (lb/ft)	Coupling I.D. (in)	Nominal Size (mm)	O.D. (mm)	I.D. (mm)	Weight (kg/m)	Coupling I.D. (mm)
2	2-3/8	1-15/16	5.0	2-7/32	50.8	60.3	49.2	7.44	56.4
2-1/2	2-7/8	2-21/64	7.7	2-5/8	63.5	73.0	59.1	11.46	66.7
3	3-1/2	2-29/32	10.2	3-1/4	76.2	88.9	73.8	15.18	82.5
3-1/2	4	3-23/64	12.5	3-3/4	88.9	101.6	85.3	18.60	95.3
4	4-1/2	3-53/64	15.0	4-1/4	101.6	114.3	97.2	22.32	107.9

Diamond Core Bits

	DCDMA Standards					Wire Line			
Size[a]	Core Diam.[b] (in)	Hole Diam.[c] (in)	Core Diam.[b] (mm)	Hole Diam.[c] (mm)	Size	Core Diam.[b] (in)	Hole Diam.[c] (in)	Core Diam.[b] (mm)	Hole Diam.[c] (mm)
EWG & EWM	0.845	1.485	21.5	37.7	AQ Wire Line	1-1/16	1-57/64	27.0	48.0
AWG & AWM	1.185	1.890	30.0	48.0	BQ Wire Line	1-7/16	2-23/64	36.5	60.0
BWG & BWM	1.655	2.360	42.0	59.9	NQ Wire Line	1-7/8	2-63/64	47.6	75.8
NWG & NWM	2.155	2.980	54.7	75.7	HQ Wire Line	2-1/2	3-25/32	63.5	96.0
2-3/4" × 3-7/8"	2.690	3.875	68.3	98.8	PQ Wire Line	3-11/32	4-53/64	85.0	122.6
4" × 5-1/2"	3.970	5.495	100.8	139.6					
6" × 7-3/4"	5.970	7.750	151.6	196.8					

[a] Size designation: First letter is approximate tool size; second letter designates a given group of tools; third letter designates a particular tool design.
[b] I.D. of core bit set.
[c] O.D. of reaming shell set.

hole and acting as a transportation medium for the cuttings that must be removed from the hole. The use of water is normally restricted to holes in rock and the stronger soils since its low specific gravity compared to that of the drilling muds makes it unsuitable for use in holes through cohesionless soils and the weaker cohesive soils. Drilling suspensions consisting of mixtures of water and locally available fat clays have sometimes been used to support holes through these soils. However, such mixtures are not always completely satisfactory from an economic or technical viewpoint and it is more common to use commercially available products such as Aquagel, Quick-gel, and Volclay. These products consist of highly colloidal, gel-forming thixotropic clays such as bentonite, with various chemical admixtures that provide a more uniform and workable suspension. In U.S. Army Corps of Engineers (1972), it is indicated that one of the most satisfactory suspensions under average conditions is a mixture of approximately 6 percent bentonite by weight with water. Borings in highly pervious soils, loose sands, and soft clays require heavier suspensions whose proportions are normally established as the work progresses or on the basis of previous work in similar soils. The primary advantages of the use of drilling muds are its lower cost as compared to the use of casing and its tendency to minimize stress relief in the soil adjacent to the boring. A major disadvantage, as indicated in the section on casing, is that it cannot be used for borings in which permeability or pressure tests are to be performed.

Drilling foam also has been used as an alternative to the drilling fluids described above. The foam is a low-viscosity, biodegradable material, similar in consistency to shaving cream, that has a high capacity for removal of cuttings from the hole. The low velocity at which the foam is pumped into the hole also provides the advantage of minimum disturbance to the walls of the borehole. Foam-drilling, as the procedure is known, has been found useful in the drilling of holes in existing dams and in holes above the groundwater table.

Grout is often used to stabilize portions of borings that pass through earth or rock materials that are extremely susceptible to caving. These include some deposits of gravel and boulders and rock that is highly fractured or contains cavities, faults, or fissures. The grout may be introduced into the unstable formation either by gravity or under pressure, depending upon the nature of the unstable material. In either case the grout is normally allowed to stand in the lower portion of the hole until it has set. This necessitates redrilling of a portion of the hole which constitutes a disadvantage of this procedure. However, an overriding advantage of this method is that it permits continuation of the boring at the same diameter, whereas telescoping of the casing, the only other alternative in many cases, requires reduction of the size of the hole.

1.8.4 Displacement Method of Advance

The displacement method of advancing the hole consists of driving a plug sampler through the overburden to the depth at which a sample is desired, pushing or driving the sampler to obtain the sample, and withdrawing the sampler to extract the sample. The process is repeated as many times as necessary to obtain samples at varied depths of penetration. In this method, there is no attempt to stabilize the hole or to remove material from above the proposed level of sampling. The method is particularly useful in stiff cohesive soils and cohesionless soils exhibiting some degree of apparent cohesion. In such materials the hole will remain open to a limited depth and the sampling operation is simplified, since it is possible to return the sampler to the last sampling elevation prior to commencing driving. The method is also useful in cases where it is desired to obtain samples of stiff or dense materials located below thick deposits of soft clays and silts. It may be utilized both in reconnaissance and detailed investigations. Retractable plug (Fig. 1.23), silt, and cup samplers ranging from 1 to 3 inches in diameter are commonly used. The 1-inch retractable plug sampler has proven particularly useful in reconnaissance work because it can be utilized with a portable tripod and a hand-lifted hammer and therefore can easily be taken into more inaccessible areas. One disadvantage of the displacement method is that driving or pushing of the sampler with a closed end disturbs the soil beneath the level to which the sampler is driven or pushed with the end closed. However, this depth is normally limited to approximately 3 times the diameter of the sampler.

1.8.5 Continuous-Sample Method of Advance

Erratic foundation conditions, foundations containing thin critical strata, detailed explorations for large structures, and explorations in areas where little foundation information is available often necessitate sampling throughout the entire depth of a boring. Also, explorations in bedrock consist almost exclusively of continuous rock coring. In such borings little, if any, cleaning of the hole is required prior to sampling and the advancement of the hole will be achieved solely through sampling. Samplers producing either disturbed or undisturbed samples of the soils may be utilized. This method, in general, is more expensive than those using intermittent sampling. However, it also has the advantage of providing foundation information on all strata present.

1.8.6 Augering

Drill holes may be advanced by rotating a soil or rock auger while at the same time applying a downward pressure on the auger to assist in obtaining penetration. The augers used may be divided into two broad categories depending upon whether they are advanced manually or by a power rig.

The most commonly used hand augers include the Iwan, ship, closed spiral, and open spiral augers. These are described in Section 1.12.4. In operation, a hand auger is attached to the bottom of a length of pipe that is provided with a crossarm at the top. The hole is advanced by turning this crossarm at the same time as the operator presses the auger into the ground. As the auger becomes filled with soil, it is taken from the hole and the soil is removed as the hole is advanced. Additional lengths of pipe are added as required. The pipe most commonly used is $\frac{3}{4}$-inch diameter standard; however, drill rods having a somewhat larger diameter and a greater wall thickness are sometimes used when the hole is to be deep and the stresses in the pipe will be high.

Power augers vary in size from small-diameter double-tube augers to large-diameter bucket-type augers. These are described in Section 1.12.5. The power units used to drill holes with these augers consist of (1) a motor, which through a system of gears drives the spindle to which the auger is attached, and (2) some means of applying downward pressure to the spindle as it is rotated. The size of the power unit depends on the size of the auger and the nature of the material to be penetrated. A large unit mounted on a flatbed truck is shown in Figure 1.7. In this unit, the auger is attached to a hollow-bore octagonal spindle. Downward pressure on the auger is provided by introducing hydraulic pressure into the bore in the spindle. Figure 1.8 shows a smaller trailer-mounted rig used to obtain samples with a double-tube spiral auger. In this model the downward pressure is applied by means of the lower arm and ratchet shown. The hole is advanced in the same manner with the power auger as with the hand auger. The auger is moved downward until it has a full load of soil on the flight, in the bucket, or in the double tube. The spindle is then retracted and the soil is removed. Deep holes are made possible by the use of extension rods or continuous-flight augers.

Auger borings are used primarily in cases where there is no need for undisturbed samples and where the drilling will be in soils where the borehole will stay open without casing or drilling mud, that is, generally above groundwater table. They are used extensively in surveys to obtain representative samples of the various materials present for general identification and classification purposes; for general soil surveys to map large areas for engineering and agricultural purposes; in borrow areas to establish the quality and quantity of material available; and of the equipment makes them ideally suited to providing adequate coverage for the extremely long, narrow areas covered by these projects.

Borings drilled with augers have the disadvantage that the samples are mixed and, in general, it is difficult if not impossible to locate the exact changes in the soil strata. Two exceptions to this are the holes drilled with hollow-stem augers or with the largest sizes of bucket augers. The former permit the taking of both disturbed and undisturbed samples as described in Section 1.12.6. The latter permit a man to enter the hole to inspect and sample the soil in situ and to obtain large-volume samples from depth for laboratory testing. The auger holes are also limited in the depth to which they can be carried. The exact depth to which any hole can be carried is a function of the types of soil in the profile, the type of auger being used, the amount of power that can be delivered to turn the auger, and the location of the groundwater table. Hand augers, in general, are used above the water table and to a maximum depth of 30 feet. Power augers of the continuous-flight type may be used to depths of 100 feet and sometimes more. Holes drilled with short-flight augers are limited in depth by the travel length of the spindle to which they are attached. The depth is commonly 10 to 20 feet. Also, augers, other than the hollow-stem type in which the auger flight remains in the hole and serves as a casing, are generally restricted to operations above groundwater level unless casing is used.

1.8.7 Wash Boring Method

The term "wash boring" refers to the process in which a hole is advanced by combination of chopping and jetting to break the soil or rock into small fragments called cuttings, and washing to remove the cuttings from the hole. In the past this method also has been used to obtain samples commonly referred to as "wash samples." These consisted of the cuttings that are carried to the surface by the drilling fluid. The samples are obtained

Fig. 1.7 Truck-mounted auger rig. (*Courtesy of Acker Drill Co., Inc.*)

between drive sample borings to fill in details in geologic profiles. They have been found particularly useful in highway, railway, and airfield projects where low cost, the rapidity with which the drilling can be accomplished, and the high mobility

Fig. 1.8 Trailer-mounted auger rig.

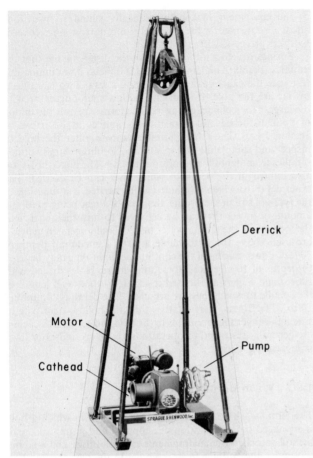

Fig. 1.9 Wash boring rig. (*Courtesy of Sprague & Henwood, Inc.*)

the surface, and the segregation of the particles during the sedimentation process in the sump tank. Thus, this method is no longer acceptable as a means of obtaining samples. Some indication of the changes in strata may be inferred from the reaction of the chopping tools as the hole is advanced or on the basis of the color of the drill water or type of sediment carried by it. However, the wash boring method today is primarily useful as a means of advancing a hole in the interval between samples taken with the various samplers that have been developed to obtain both disturbed and undisturbed intact samples.

The rigs used to advance holes by the wash boring method vary depending upon the types of samples to be obtained. Figure 1.9 shows a wash boring rig suitable for use when only disturbed samples are required. Rotary rigs such as those illustrated in Figures 1.13a–d and described in Section 1.8.9 may also be used to advance holes by the wash boring method. These have hydraulic or screw-feed rotary mechanisms that, in addition, provide jacking force for taking undisturbed soil samples and rotation for taking rock cores. The elements of either a wash boring or rotary rig that are required to advance a hole by the wash boring method are designated in Figure 1.10, which illustrates a rotary rig in the process of advancing a hole. The elements consist basically of (1) a three- or four-legged pipe derrick having a sheave at its top through which the rope or cable from the cathead passes down to the drilling tools; (2) a motor that through a system of gears, drives a drum, commonly referred to as a cathead, which is used to transfer power from the motor to a rope or cable that raises and lowers the chopping bit and which is used for lowering and withdrawing the string of drill tools in the hole; and (3) a water pump to provide the jetting action required to wash material from the hole.

The wash boring method may be used in both cased and uncased holes. Even in the latter it is common practice to install a short section of casing with a wash tee at the top of the hole to stabilize the top of the borehole. Figures 1.10a and b show, respectively, a rig as it appears when casing is being driven and when the hole is being advanced. The basic procedure for a cased hole is as follows: after a sample is taken at ground surface, a length of casing of the type described in Section 1.8.3

by allowing the cuttings to settle out of suspension in a sump tank or a sump hole in the ground adjacent to the drill rig. Samples thus obtained generally are not representative of the material in situ owing to the breakdown of particles by chopping, the loss of fines in transporting the particles to

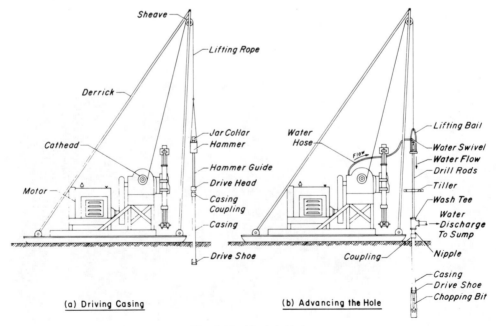

(a) Driving Casing

(b) Advancing the Hole

Fig. 1.10 Wash boring.

is driven into the ground. The soil within the casing and for some distance below it is then removed by chopping, jetting, and washing. The borehole is advanced by the chopping, jetting, and washing action until a change in soil stratum is encountered, as evidenced by a change in the color of the wash water or the behavior or feel of the string of tools, or until a distance equal to the maximum spacing specified for taking samples is reached. The maximum spacing specified for taking samples usually is 5 feet although in deep boreholes 10 feet is occasionally used. Once removal of the material to the prescribed depth is accomplished, a sample is taken, the next section of casing is added, and the casing is driven to the depth of the bottom of the last sample. This process is continued until the required elevation or depth of penetration of the boring is reached. Where the borehole will not stay open for the full distance between samples, as in the case of cohesionless sands, the borehole is advanced some safe distance such as one or two feet and then the casing is driven to the bottom of the hole. Advancing by chopping, jetting and washing, and casing driving is thus performed intermittently until the depth prescribed for the next sample is reached. In uncased holes the procedure is similar except that the installation of casing is omitted completely or just a short length of casing with a wash tee is installed at the top of the hole. Details of the procedures used are given below.

Driving of the casing (Fig. 1.10a) is accomplished by attaching to the coupling at the top of the casing a unit commonly referred to as a guide-and-pull piece assembly, which consists of a metal drive head, a length of double extra-heavy drive pipe (hammer guide), and a drive pipe coupling (jar collar). The hammer, a heavy metal cylinder with a hole along its axis large enough to accommodate the hammer guide, is inserted between the drive head and the jar coupling. It is then raised by means of the rope, which has two or three turns around the cathead, and allowed to fall freely, transmitting its energy to the casing as it strikes the drive head. The weight of the hammer and the height of drop generally are standardized, so that within any group of borings the resistance to penetration, as measured by the number of blows per foot required to drive the casing, may be obtained. It should be noted that, although the casing blow count is sometimes used in an attempt to delineate strata, it is by no means as satisfactory as the blow count on a drive sampler. This results from the difficulty of determining the portions of the blow count on the casing attributable to skin friction along the sides of the casing and to point resistance. The skin friction depends upon whether the casing couplings are flush joint or larger than the casing. The point resistance depends on the size of the hole created below the casing in advancing the hole with the chopping bit. This problem does not exist with the drive sampler. Drive hammers are generally available in weights ranging from 75 to 600 lb. It is common to use a 140-lb or 200-lb hammer for driving samplers and a 200–300-lb hammer for driving casing. The drop height generally ranges between 1 and 3 feet.

To clean a section of casing once it has been driven, the string of tools, consisting of the drill rods with a chopping bit at the bottom and a water swivel and lifting bail at the top, is connected to the pump by a heavy-duty hose attached to the water swivel (Fig. 1.10b). It is also attached to the cathead by means of a rope, which passes through the sheave and is attached to the lifting bail. The tools are then lowered to the level of soil in the casing, and drilling fluid under pressure is introduced at the bottom of the hole by means of water passages in the drill rods and the chopping bit. At the same time, the bit is raised and dropped by means of the rope attached to the lifting bail. Each time the rods are dropped they also are partially rotated manually by means of a wrench placed around the rods or a device called a tiller (Fig. 1.10b) which is attached to the rods. The latter process helps to break up the material at the base of the hole. The resulting cuttings are carried to the surface in the drilling fluid that flows in the annular space between the drill rods and the inside of the casing or the walls of the uncased hole. If the fluid used is drilling mud, it is usually discharged into a sump or tank so that the cuttings will settle out of the suspension and the fluid may be recirculated. The same procedure is often used when the drilling fluid is water. However, if the water supply is abundant the return flow is commonly wasted. When cleaning of the hole has advanced to the level at which a sample is to be taken, the chopping bit is usually raised slightly above the bottom of the hole and circulation of the drilling fluid is continued until the fluid is free of sediment from the hole. Cuttings not removed from the borehole tend to settle to the bottom of the borehole and become the upper part of the next sample, which is undesirable.

Standard diamond core drill rods are the type most commonly used in explorations for civil engineering projects. These are flush-coupled, hollow rods fabricated from cold-drawn seamless steel tubing and provided with square threads for the rapid making and breaking of joints. The rods are available in two series of sizes. Sizes E, A, B, and N constitute an obsolete series that is still available for use with similar equipment still in service. Sizes EW, AW, BW, and NW represent the new "W" series standardized by the Diamond Core Drill Manufacturers Association (DCDMA) (1980). The new series is larger in both inside and outside diameter. This provides for a greater flow of drill water into the hole and a somewhat higher velocity of return flow. The latter, which is due to the decrease in the annular space between the rods and the walls of the hole, tends to improve removal of cuttings from the hole. The dimensions and weights of both series of rods are shown in Table 1.2. Descriptions and sizes of other types of rods used for exploratory drilling are presented by Hvorslev (1949).

Several types of chopping bits are used. Some of the more common are shown in Figure 1.11. The straight and chisel chopping bits are used in sands, silts, clays, and very soft rocks. The cross chopping bits are used when coarse gravel, boulders, or rock are encountered. Bits for exploration work normally range between $1\frac{7}{16}$ and $5\frac{5}{8}$ inches in width depending upon the size of the borehole. Bits may be obtained with either downward- or upward-pointing jet passages for the drilling fluid. The bits with upward-pointing jet passages are safer for cleaning the bottom of the hole since the jet cannot cause gouging of the material to be sampled. With a downward-pointing jet, washing of the hole before sampling must be done gently.

The wash boring method can be used to advance both small- and large-diameter holes. The equipment used for this method is light and relatively inexpensive. The primary disadvantages are that the method is slow in the stiffer and coarser-grained soils and it is not efficient in materials such as

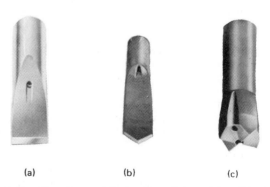

(a)   (b)   (c)

Fig. 1.11 Chopping and jetting bits: (a) straight; (b) chisel; (c) cross. (*Courtesy of Acker Drill Co., Inc.*)

hard or cemented soils, rock, and materials that contain boulders.

1.8.8 Percussion Drilling

The percussion drilling method of advancing a boring is of common use in drilling water wells and is referred to, also, as cable tool drilling or churn drilling. Advance of the hole by this method is accomplished by (1) alternately raising and dropping a combination of heavy drilling tools to break the material to be removed and form a slurry of this material and the groundwater or water introduced into the hole; and (2) periodically removing the slurry thus formed by means of the bailers described below. The amount of water introduced into the hole in this method is kept to the minimum required to form the slurry. In certain materials such as soft soils and cohesionless materials below groundwater table, it is not uncommon for the hole to be made without having to add water.

A typical percussion drilling rig is shown in Figure 1.12a. The combination of heavy tools used to break the material in percussion drilling is commonly referred to as the drilling tool string. The lowest element of this string is a heavy drill bit that has a beveled edge at its base for breaking the material to be penetrated. Several types of bits (Fig. 1.12d) are available. The shape of the bit used depends to a large extent on the nature of the materials to be penetrated. Immediately above the drill bit is a series of elements referred to as drill stems (Fig. 1.12b). The primary function of these drill stems, which are made of high-carbon steel, is to provide a large portion of the weight required to cause the bit to penetrate. They also serve to guide the drill string in the hole. The number of stems used depends on the size of the hole and the material being drilled. If the hole is to be drilled in a formation that has a tendency to cave and cause the tools to be jammed in the hole, the drill stems will be surmounted by a drilling jar (Fig. 1.12c) except when the hole is first started. The drilling jar consists of heavy steel links that can slide within one another in the same manner as the links of a chain and can be used to drive tools out of the hole when they become jammed. The drilling jar in turn is connected to the rope socket or, in some instances, to a sinker bar, which in its turn is connected to the rope socket. The sinker bar is a unit similar to the drill stem but shorter in length. The rope socket is the means by which the cable used to raise and lower the drill string is attached to the string.

Removal of material from the hole in percussion drilling is accomplished by means of bailers and sand pumps. A bailer consists of a pipe having a valve at its lower end and a bail at its upper end. The valve retains material in the bailer as it is lifted to the surface. The bail provides a connection for the sand line, a separate cable line on the percussion rig that is used to operate the bailer. Bailers are classified according to the type of valve used. Two types are common, the flat-valve bailer and the dart-valve bailer. The flat-valve bailer (Fig. 1.12e) has a flat valve that opens to receive material when the bailer is lowered and closes to retain the material when it is lifted. The diameter of the bailer is normally 1 to 2 inches less than the diameter of the hole. In cleaning the hole, the bailer is lowered to the bottom of the hole and then moved a few inches up and down to create a pumping action that causes the slurry in the hole to enter the bailer. When the bailer has been loaded it is lifted to the surface and tipped upside down to remove the material collected.

The dart-valve bailer (Fig. 1.12f) has a valve that consists of a dart-shaped flat steel piece at the top of which is a cone-shaped valve. When the bailer reaches the bottom of the hole the dart strikes the bottom, lifting the cone-shaped valve off of its seat and allowing the slurry and cuttings to enter the bailer. As the bailer is lifted the valve drops onto the seat and retains the collected material. The bailer is emptied by touching the dart against the ground, thereby opening the valve. Should the amount of cuttings collected be great enough to prevent opening of the valve, the bailer must be emptied in the same manner as the flat-valve bailer.

Sand pumps (Fig. 1.12g) consist of a tube inside of which is a plunger. The plunger, when moved up and down, creates a suction that causes the slurry and cuttings to flow into the tube through sidewall openings at the bottom of the pump. The tube contains a valve at its base that retains the material as the pump is lifted to the surface. The pump is emptied by removing the valve. The bottom of the pump may be equipped with a standard bottom or a bit bottom. The latter is used when the material in the hole must be broken prior to removal.

The most commonly used of the three devices for cleaning holes is the dart-valve bailer, primarily because of the ease with which it can be emptied. Both flat-valve bailers and sand pumps provide well-cleaned holes, but the former requires a greater amount of time for emptying than the dart-valve bailer and the sand pump is more expensive than either of the bailers.

Driving of the casing in percussion drilling is accomplished by attaching steel drive clamps to the square section of a drill stem and raising and dropping the stem a short distance so that the drive clamps strike a drive head or coupling affixed to the top of the casing. The drill string provides adequate weight to hammer the casing down.

Changes in the nature of the materials penetrated by this method are noted by observations similar to those used in wash boring. These include rate of progress, behavior of the drilling tools, color of the slurry, and character of the cuttings.

The primary advantage of the percussion method is that it may be used in most materials and is particularly useful for borings that must penetrate coarse granular materials, soils such as glacial till containing boulders, and rock such as limestone which may contain caverns. One of the major disadvantages in using this method for exploratory work is that it generally is not economical for holes less than 4 inches in diameter. An equally important factor is that the heavy blows of the drill tools have been known to disturb the material at the bottom of the hole that is to be sampled. Cohesionless soils can be made denser, cohesive soils can become remolded, and the slurry and cuttings may be driven into the in situ soil. In addition, it is difficult to detect thin layers and slight changes in material when this method is used.

1.8.9 ODEX Drilling

The ODEX drilling method was developed by Atlas Copco-Sandvik for drilling through soils with cobbles and boulders. The name ODEX is an acronym for *Overburden drilling with an eccentric bit*. The bit consists of a central lower portion that chops and rotates and an upper portion, which, during drilling, moves outward to enlarge the borehole so that the casing can readily follow the bit. The casing shoe has a shoulder that is engaged by the bit unit. The casing is thus pulled down by the bit as it advances. A cross section of the bit is shown in Figure 1.13. Because the casing is subjected continuously to the percussion action of the bit, it is made up using welded connections rather than threaded connections. Thus an experienced welder must be available at site to make connections whenever additional sections of casing are to be added. The percussion and rotating action of the bit may be provided from the surface or a "down-the-hole" hammer may be used. The latter is more efficient and is preferred, except for shallow holes. Drilling is usually carried out using air to actuate the down-the-hole hammer and to carry the soil and rock chips

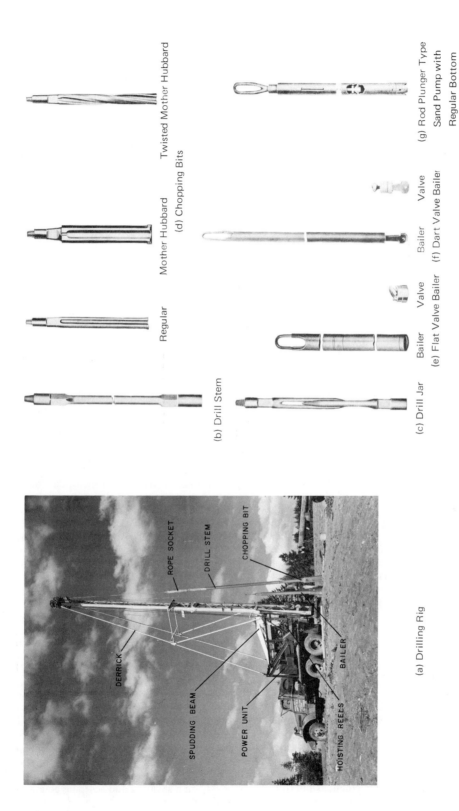

(g) Rod Plunger Type Sand Pump with Regular Bottom

Twisted Mother Hubbard

(d) Chopping Bits

Mother Hubbard

Valve   Bailer   (f) Dart Valve Bailer

Regular

Bailer   Valve   (e) Flat Valve Bailer

(b) Drill Stem

(c) Drill Jar

ROPE SOCKET

DRILL STEM

CHOPPING BIT

DERRICK

SPUDDING BEAM

POWER UNIT

BAILER

HOISTING REELS

(a) Drilling Rig

Fig. 1.12 Percussion drilling equipment. (*Courtesy of Bucyrus–Erie Co.*)

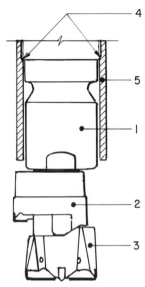

Fig. 1.13 ODEX bit and casing shoe: (1) guide; (2) reamer; (3) pilot bit; (4) shoulder; (5) casing shoe.

to the surface. To facilitate removal of soil and chips from deeper holes, foam may be used. The use of air or foam is advantageous in connection with drilling in earthfill dams in order to preclude the possibility of hydraulic fracturing as might occur if water or drilling mud were used. The ODEX bit can be removed at any time and the casing entered with soil sampling or coring equipment.

1.8.10 Rotary Drilling

In the rotary drilling method, the hole is advanced by rotating a drill string consisting of a series of hollow drill rods to the bottom of which is attached either a cutting bit or a core barrel with a coring bit. Cutting bits shear off chips of the material penetrated and thus are used primarily for penetrating overburden between the levels at which samples are required. Coring bits, on the other hand, cut an annular hole around an intact core that enters the barrel and is retrieved. Thus, the core barrel is used primarily in bedrock, which, under most circumstances, is cored continuously. As the rods with the bit or barrel are rotated, a downward pressure is applied to the drill string to obtain penetration, and drilling fluid under pressure is introduced into the bottom of the hole through the hollow drill rods and passages in the bit or barrel. The drilling fluid serves the dual function of cooling the bit as it enters the hole and removing the cuttings from the bottom of the hole as it returns to the surface in the annular space between the drill rods and the walls of the hole. In an uncased hole, it also serves to support the walls of the hole.

The procedure described above is referred to as straight rotary drilling. A second method of rotary drilling commonly used is referred to as reverse rotary or reverse circulation drilling. The difference in the two methods is primarily in the circulation of the drilling fluid to remove the cuttings and in the equipment used; also, the reverse rotary is limited to use with noncoring bits. In the reverse rotary method, as the rods are rotated, the drilling fluid is introduced under gravity into the annular space between the drill rods and the walls of the hole. The fluid, laden with cuttings from the bottom of the hole, returns to the surface via the hollow drill rods. The return flow is accomplished by (a) application of a head at the top of

the annulus relative to the discharge end of the drill rods; (b) application of suction on the drill rods; (c) introduction into the drill rods of a supply of air that mixes with the slurry and causes it to be removed by air lift. When compared to straight rotary drilling, this method has the dual advantage of (1) minimization of disturbance to the walls of the hole owing to the higher head in the hole and more outward seepage pressure on the hole walls; and (2) more rapid and efficient removal of cuttings from the hole because the area of the drill rods is less than the annulus, thereby giving higher upward velocity. However, it is best suited to holes 12 inches and larger in diameter.

The vast majority of holes in exploratory work are drilled by the straight rotary method. Rigs such as the ones shown in Figures 1.14a and b are commonly used to drill these holes. The basic elements of the rig are similar to those used for wash borings with the exception that a drill head is added. The drill head consists of a rotary drive mechanism and a means of applying downward pressure to the drill rods as they are rotated. In the rotary drive mechanism, spiral bevel gears are used to transmit power from the motor to a drive quill, which commonly has a hexagonal or octagonal bore. The quill in turn drives a strong hollow steel spindle of similar cross section referred to as the drive head, which has a chuck attached to its lower end. Drill rods and flush steel casing are fed into the hole directly through the hollow spindle and are gripped by the chuck so that the rotary motion of the spindle is transferred to the rods or casing. In the model shown in Figure 1.14a the spindle is also connected at its upper end to two hydraulic cylinders by means of a steel crosshead yoke. These cylinders are used to apply a downward pressure, through the yoke, to the drill rods as they are rotated. The same cylinders are also used to provide the downward force necessary to push samplers such as the Shelby tube and piston samplers to obtain undisturbed samples. Some rotary rigs are equipped with a screw-feed mechanism (Fig. 1.14b) in lieu of the hydraulic feed unit. However, the latter is preferred because it provides a uniform push during the sampling process and is thus less likely to cause disturbance to the sample. Drilling fluid is supplied to the drill rods in the same manner as outlined in the discussion of the wash boring method. Another type of rig used to sink holes by the straight rotary method is shown in Figure 1.14c. In this rig rotary power is transmitted to the drill tools through a rotary table. This type of rig is particularly useful in drilling large-diameter holes.

The rotary rigs described above are extremely versatile in that they may be used for wash borings and auger boring as well as rotary work. As a result they are used extensively and are the types of rigs most commonly encountered in exploratory work. Figure 1.14d illustrates still another type of rig. This rig may be used for straight rotary, reverse rotary, and auger boring. The photograph shows the rig as it appears when the reverse rotary method is being used. The small-diameter hose used to supply the air required for removal of the cuttings by air eduction and the large-diameter line through which the cuttings and drilling fluid are discharged may be seen at the drill head. The characteristics of some of the more commonly used rotary rigs, including two portable rigs, are shown in Table 1.3.

Several types of cutting bits are available for use in rotary drilling. Some of these are shown in Figure 1.15. In general, these bits may be divided into two broad categories, drag bits and rock bits. The type of bit used will be determined primarily by the characteristics of the material to be penetrated. Drag bits rely on a shearing and scraping action to remove material and therefore are suitable primarily for use in overburden; some may also be used in soft rock. Included in this category are fishtail, bladed, replaceable blade, and carbide insert bits (Figs. 1.15a–c). The term fishtail is applied by Hvorslev (1949) to a bit resembling a straight chopping bit with a split cutting

Wire Drum Hoist
Cathead
Power Unit
Derrick
Hydraulic Feed
Drill Head
Octagonal Spindle
Drill Chuck

(a)

(b)

(c)   (d)

Fig. 1.14 Rotary drill rigs: (a) hydraulic feed; (b) screw feed; (c) rotary table; (d) reverse circulation. (*Courtesy of Acker Drill Co., Inc., and Calweld.*)

edge, each half of which has been turned slightly in the direction of rotation. The term bladed bit is applied to bits having two, three, or four blades or wings that have been forged to a thin cutting edge and turned slightly in the direction of rotation. The tips of the cutting edges of both types of bit are made of tungsten carbide alloy for wear resistance. The carbide insert bits are similar except that they do not have turned edges and the insert forms the cutting edge. These are commonly available with three or four wings. The Hawthorne replaceable-blade bit and similar bits commonly have three insert blade bits that are themselves individually replaceable. All of the drag bits have passages through which the drilling fluid may be pumped. These jets are directed at the blades for cleaning purposes. The fishtail bit and the two-bladed bit are used in sands, clay, and other soft soils. The three- and four-bladed bits of the fixed blade, carbide insert, and replaceable blade types are suitable for use in firmer soils and in somewhat harder rock than the fishtail.

The rock bits used in rotary drilling are classified as noncoring or coring bits depending upon whether the rock is broken into small fragments and washed out of the hole or is recovered in the form of an intact core. Noncoring bits are used to advance a hole when there is no need for samples to be taken. They are of three types, the cone bit, the roller bit, and the diamond plug bit. The cone and roller bits have teeth milled on the surfaces of cones and rollers which are so mounted that the teeth rotate as the bit is turned. The cone type bits are available with two or three cones, the latter of which (Fig. 1.15d) is commonly referred to as a tricone bit. The roller bit has two rollers on inclined axes and two rollers on a horizontal axis which are mounted perpendicular to the inclined rollers. The spacing and height of the teeth on both the cone and roller bits depend on the material to be penetrated. Long and widely spaced teeth are used for soft materials; short and closely spaced teeth are for hard materials. Diamond plug bits are of three types, namely, concave, pilot, and taper. The concave bit (Fig. 1.15e) is used in relatively soft rock. The pilot bit, which has a lead section of smaller diameter than the remainder of the bit is used in hard rock and in vertical holes in rock

TABLE 1.3 CHARACTERISTICS OF TYPICAL ROTARY DRILLING RIGS.

Make and Model of Rig	Capacity						Rotary Feed							
	Rated Depth Capacity in Feet[a]						Hoist		Engine Power[b] (hp)	Feed Length[c] (in)	Chuck[d]			Weight[e] (lb)
	Hollow-Stem Auger	Rod Size (DCDMA Standards)					Capacity (ft)	Cable Size (in)			No. of Speeds	Range (rpm)	Size (in)	
		EW	AW	BW	NW	HW								
Acker Packsack	—	50	—	—	—	—	—	—	10	—	1	3600	1-3/8	115
Mobile Minuteman	50	50	—	—	—	—	—	—	7	44	3	105–1114	1-3/8	225
Mobile B-24 Surveyor	50	—	100	—	—	—	—	—	25	77-3/4	—	0–155	1-3/4	1465
Smit Winkie	—	400	350	150	—	—	—	—	10	—	2	1200–2800	—	225
Mobile B-31	75	—	—	—	—	—	—	—	—	68	5	56–418	2-5/8	2600
Mobile B-47	100	—	—	—	250	—	—	—	20	68	5	56–418	2-5/8	3650
Longyear L-65	—	600	500	350	300	—	—	—	18	24	—	0–1500	—	290
Longyear 24	—	730	500	—	—	—	65	5/16	22	24	9	203–1964	1-13/16	1110
Sprague & Henwood 37	—	500	450	400	—	—	65	3/8	48	24	3	647–1800	2-1/8	1550
Acker Soil Sentry	70	—	625	—	500	—	—	—	—	72	4	0–600	3-1/2	3230
CME 45C	75	800	—	—	500	—	—	—	37	68	4	35–475	2-5/8	3850
Sprague & Henwood 37H	—	—	725	600	—	—	65	3/8	22	24	3	647–1800	2-1/8	1800
Acker Bush Master	—	—	1050	1125[g]	900[g]	—	365	3/8	48	72	4	0–1200	3-1/2	4490
Longyear Hydra Core 28	—	—	1400[g]	1000[g]	600[g]	—	—	—	46	70	3	480–1250	2-1/4	1463
Longyear 34	—	—	1600	1300	1000	600	190	1/2	31	24	8	14–1850	2-7/8	2895
Mobile B-53	150	—	1600	—	1000	—	—	—	124	78	10	27–716	4-1/2	5076
CME 55	125	—	—	—	1000	—	—	—	120	72	4	100–650	2-5/8	6500
Mobile B-61 HD	200	—	—	—	1000	—	200	7/16	64	—	4	92–583	2-7/8	8000
Mobile B-80	148	—	—	—	1000	—	150	7/16	50	168	10	27–716	4-5/8	—
Acker Teredo Mark II	—	1800	1625	1250	1000	—	150	1/2	46	24	4	141–900	2-7/8	3500
S & H 40-CL	—	1875	1390	1350	1050	—	120	7/16	57	24	8	64–1800	2-5/8	3400
S & H Hycore Jr. 100	—	—	—	—	1050[g]	—	—	—	57	138	4	0–1675	3-3/4	4500
S & H Hycore Jr. 200	125	—	—	—	1050[g]	—	—	—	—	138	3	0–650	3-3/4	4780
Failing 1500-S	140	—	1500	—	—	—	525	1/2	60	30	4	35–220	2-3/8	22000[f]
Acker ADII	150	—	1600	—	—	—	120	1/2	70	72	8	63–969	2-5/8	5400
CME 75	—	—	—	—	1650	—	90	1/2	151	72	5	100–700	2-5/8	7100
Acker Core-Max	—	—	—	2250	1800	1100	—	—	73	132	8	0–754	3-1/2	8500
Longyear 38	—	—	2800	2300	1800	1100	130	9/16	54	24	8	35–1850	2-7/8	3045
Acker Mountaineer	—	—	3250	2500	2000	1300	200	5/8	50	24	8	61–1484	3-1/2	3300
Failing 2000-CF	—	—	—	2500	2000	—	370	5/8	170	—	5	23–163	2-7/8	—
S & H 142-HD	—	2250	3200	3100	2400	—	270	5/8	59	24	4	126–1440	2-7/8	4200
Failing 2500-A	—	—	5000	3500	2500	—	350	3/4	150	—	5	140–1350	3-1/2	42160[f]
Longyear 44	—	—	3100	3200	2500	1600	150	5/8	81	24	4	0–1529	3-3/4	4775
S & H Hycore I	—	—	—	—	2800	—	525	3/4	130	138	—	0–1529	3-3/4	—
Failing 3000-HD	—	—	—	—	3000	3000	—	—	200	—	5	—	4-1/2	—
Acker WAIII-C	—	—	7000[g]	5500[g]	4200[g]	2800[g]	—	—	152	288	4	0–754	3-1/2	16500
Longyear Hydro-44	—	—	4550	4700	3650	2350	150	5/8	148	24	—	20–1250	3-1/2	3200
S & H Hycore II	—	—	—	—	4000	—	—	—	204	138	—	0–1529	3-3/4	—

[a] Rated capacity is greater if wireline equipment is used.
[b] Horsepower ratings vary with rpm. Most models are offered with a choice of horsepower. A choice of gasoline, diesel, air, electric, LPG or power takeoff is also available for many models.
[c] Other lengths are available on some models.
[d] Other chuck sizes and rpm ranges are available on some models.
[e] Weight varies with the options requested.
[f] Includes weight of truck.
[g] Wireline sizes (AQ, BQ, NQ, HQ).

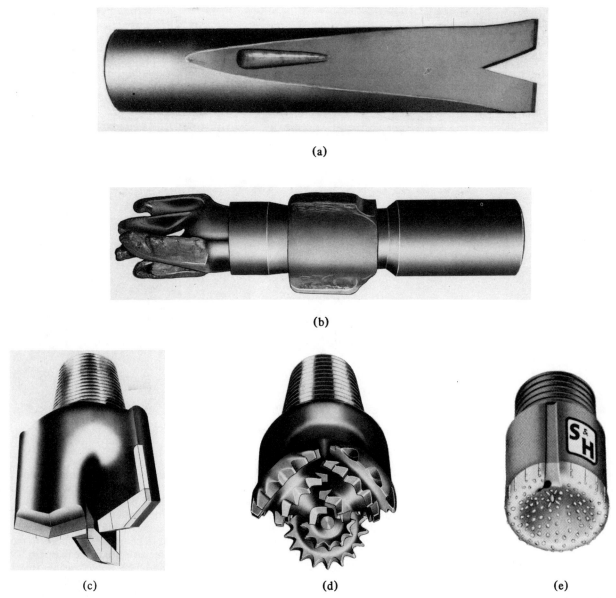

(a)

(b)

(c)   (d)   (e)

Fig. 1.15 Rotary bits: (a) fishtail bit; (b) Hawthorne replaceable-blade drag bit; (c) carbide insert drag bit; (d) tricone bit; (e) diamond plug. (*Courtesy of Sprague & Henwood, Inc.*)

strata of differing hardness. The taper-type is particularly well adapted for use in very hard rock. However, it is common practice to advance holes through bedrock by taking continuous cores with core barrels such as those described in Sections 1.15.7 through 1.15.10. In such cases a coring bit is required that attaches to the bottom of the core barrel and cuts an annular hole rather than a hole covering the entire cross section of the boring. Four basic categories of coring bits are available, namely, diamond, carbide insert, sawtooth, and calyx or shot bits. These are described in Sections 1.15.4 and 1.15.11.

At times exploratory drilling is carried out from lined galleries or tunnels below groundwater table. In such cases it is important that drilling be carried out using a backpressure on the return drilling fluid flow at least equal to the ambient groundwater at the point of drilling. To maintain such a backpressure, drilling may be done through a nipple grouted into the wall of the gallery and the adjacent rock; a stuffing box and valve should be attached to the end of the nipple. The stuffing box and valve can also be attached to the wall of the gallery by a flange connection. A picture of a set-up used by

Soletanch is shown in Figure 1.16. Here, a ball-and-socket connection is attached to the concrete lining of the gallery using bolts and a gasket. The ball-and-socket connection permits orientation of the drilling at various angles besides perpendicular to the surface of the lining. The core barrel or plug bit used should be equipped with a one-way valve, such as a ball valve, which permits drilling fluid to go down the drill string when its pressure exceeds the ambient groundwater pressure, but no flow to occur out of the drill string when its pressure is less than the ambient pressure. The drill string is first inserted through the stuffing box, then the valve is opened and the drill string is extended to the point of drilling. Between the above-mentioned valve and the embedded nipple or bolted flange are a pressure gauge and gated ports to allow release of the return flow. As drilling is carried out, release of return flow is regulated to maintain a backpressure at least equal to the ambient groundwater pressure at the point of drilling.

Rotary drilling for putting down holes in soil may be used for borings 2.5 inches in diameter and larger. However, because of the more expensive equipment required, this method is

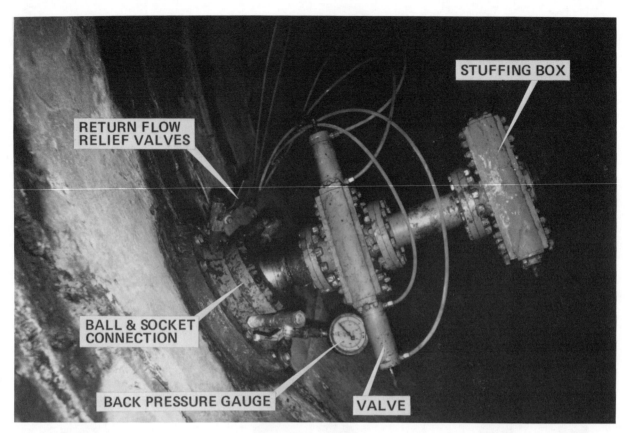

RETURN FLOW RELIEF VALVES

STUFFING BOX

BALL & SOCKET CONNECTION

BACK PRESSURE GAUGE

VALVE

Fig. 1.16 Apparatus to permit drilling under backpressure as developed by Soletanch.

generally used only for larger-diameter holes. The rotary drilling method has several advantages. It is more rapid, in general, than the other methods that have been described and usually results in less disturbance to the material to be sampled. The primary disadvantage of this method is that it is not well adapted for use in materials containing a large percentage of particles of gravel size or larger, since these particles will rotate beneath the drill bits and cannot be easily broken. Thus, a nest of gravel will continually remain at the base of the hole.

1.8.11 Becker Drilling

A highly effective method of penetrating overburden, developed by Becker Drilling Ltd. of Canada, is illustrated schematically in Figure 1.17. The method, devised particularly for use in sand, gravel, and boulders, consists of driving a double-walled casing into the ground with a diesel pile hammer while, at the same time, air or water under pressure is delivered to the bottom of the hole through the annular space between the two pipes. The bottom of the casing is equipped with a specially designed toothed bit, which under the blows of the hammer breaks the material being penetrated into small fragments. These fragments are removed by the air or water, which returns to the surface via the inner casing. Air or water may be introduced interchangeably during drilling so that the removal of all types of materials is facilitated. At the surface, the return flow is ejected through a vent at the top of the casing and into a hose that delivers the material to either a cyclone and then to a collector bucket or directly into collector buckets. The samples thus collected provide a continuous record of the materials penetrated. In addition, since the inner casing is always open

to the bottom of the hole, drilling may be stopped at any point so that samples may be taken with standard sampling devices such as the split barrel sampler or by coring. A hydraulic feed rotary head is usually provided with the rig, permitting the use of core barrels or tricone bit. Drilling of rock may be done either to identify bedrock or to penetrate a boulder. In the case of boulders, blasting may be resorted to to break up the boulder to permit further penetration of the Becker casing. Holes may be drilled at angles up to 45° from the vertical.

A truck-mounted model of the rigs used to advance holes by this method is shown in Figure 1.17a. The basic elements of this rig comprise an air compressor, a pump, a diesel pile hammer, a rotary drive unit, a hydraulic hoist, a casing puller, a mast, and a cyclone for the collection of samples.

The casing used in this method is specially fabricated. It is flush-jointed and is connected by tapered threads to facilitate making and breaking of the casing string. The threaded end sections are separate units welded to each length of pipe so that damaged threads are easily removed and replaced with a welding torch.

The bits used generally contain chisel-type teeth of tempered steel or nickel alloy. The teeth are set slightly ahead of the skirt of the bit wall. Between the teeth the skirt itself is tapered towards the center of the bit so as to force that material not directly beneath the opening away from the bit.

This method can be used to penetrate overburden to depths of 250 feet. Standard casing sizes range from $5\frac{1}{2}$ in o.d. $\times 3\frac{1}{4}$ in i.d. to $6\frac{5}{8}$ in o.d. $\times 3\frac{7}{16}$ in i.d.; however, casings to 18 or 20 inches o.d. have been installed in connection with construction of relief wells. In addition to its use for advancing holes, the method has several additional applications in subsurface exploration programs. These include drilling holes for (1) installation of instrumentation such as piezometers, inclinometers, etc.; (2) test

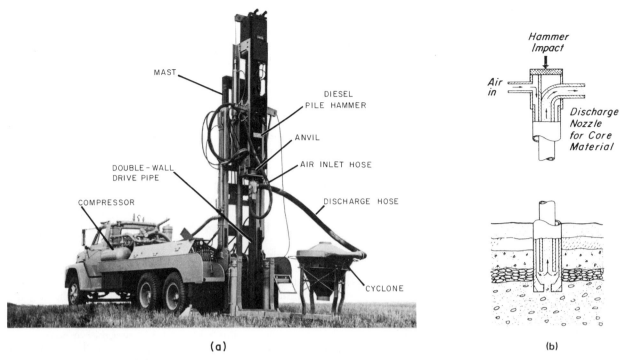

Fig. 1.17 (a) Hammer drill. (b) Schematic diagram showing principle of hammer drill.

grouting; and (3) the installation of pipes to freeze soil prior to sampling or excavating in collapsible granular materials.

The major advantages of the method are that it is rapid and inexpensive and that it is the best known method of penetrating bouldery materials. The rate of progress, even in boulders and cobbles and gravels, is extremely high. On at least one project it has been found that in the time required to drill 20 feet in bouldery materials by rotary methods, it was possible to advance 240 feet by hammer drilling. It is particularly useful in rapid mapping of alluvial materials over large areas. The method also has a number of disadvantages. The greatest of these is that the pressure at the bottom of the casing is reduced far below the hydrostatic pressure from the groundwater table. This is particularly true when compressed air is used for removal of the broken material. It also occurs, although to a lesser extent, when water is used, since the removal of the material depends on the eduction principle. Flow of groundwater to the bottom of the hole under these conditions can result in a great amount of disturbance. In addition, when skip-graded mixtures of sand and boulders are being penetrated, the pressure differential may cause excess sand to be sucked into the casing if penetration of the casing is temporarily halted by a boulder. This would result in unrepresentative samples owing to a recovery greater than 100 percent. Among the other disadvantages of the method is the lack of direct identification of the material. It is not difficult to ascertain changes in strata. However, since the material brought to the surface has been broken, considerable reliance must be placed on judgment and the behavior of the equipment as drilling progresses to establish the exact nature of the material penetrated. Also, in materials consisting of large and small particles, such as in sandy gravels, there will always be a tendency for a larger portion of the fine particles to be brought to the surface. Still another disadvantage is that the material directly beneath the hole is disturbed owing to the action of the pile driver. Thus, it is questionable that undisturbed samples can be obtained. However, regardless of these disadvantages, hammer drilling remains one of the more effective exploratory methods when properly utilized.

1.8.12 Backfilling Boreholes

Open boreholes, as well as other open exploratory excavations, are a hazard and should be backfilled when they are no longer required. In a large number of cases, backfilling boreholes with locally available soil tamped in place will suffice. However, under certain circumstances backfilling with grout is advisable. The U.S. Army Corps of Engineers (1972) recommends the use of grout in cases where it is essential to prevent the movement of water from one stratum to another and to preclude piping of material to the surface through the borehole. Included in this category are all borings on the land side of levees, downstream of dams and proposed embankments, and at the proposed locations of structures. The Corps also recommends that holes upstream of dams and proposed embankments be grouted if they penetrate natural valley blankets of impervious material and extend into pervious strata. The grout should be placed from the bottom of the hole upward by means of a pipe so that all of the drilling fluid will be displaced from the hole. The use of bentonite or a similar material to inhibit shrinkage and insure a good seal is recommended. A mix of 88 to 92 percent of portland cement with 12 to 8 percent of bentonite has been found adequate.

1.9 EXCAVATION METHODS FOR EXPLORATION

1.9.1 Test Pits and Test Trenches

Excavations that are large enough to permit the entrance of one or more persons represent one of the most valuable and dependable means of exploration, since they permit detailed examination of the subsurface materials in situ. They also provide a means to obtain for laboratory testing large-size undisturbed and disturbed samples of the materials encountered.

Test pits and trenches may be excavated by hand or by power equipment such as ditching machines, backhoes, bulldozers, scrapers, draglines, and other types of construction excavators. Special-purpose drilling rigs such as large-diameter bucket augers may also be used. One limitation placed on the use of power equipment is that it should not be used too close to locations where undisturbed samples are to be taken.

The size of the excavation will depend primarily on the space required for efficient excavation and on economic limits. Test pits normally have a cross section 4 to 10 feet square or in diameter; test trenches are usually 3 to 6 feet wide and may be extended for any length required to reveal conditions along a specific line. In general, test trenches are relatively shallow, whereas test pits may be deep. However, it is common to keep even test pit depths to a minimum owing to the expense involved. In the United States, 25 feet is considered by some to be the economic limit. However, larger and deeper excavations have been used when special problems justified the expenses. One such case involved the foundation of a dam that consisted of limestone. An extensive program of borings yielded a confused picture of subsurface conditions and it was not until an excavation 100 feet by 200 feet at the top and approximately 40 feet deep was made that the nature of the in situ material was more clearly understood. Deeper excavations are justifiable in countries where labor is relatively inexpensive. Pits more than 50 feet in depth and shafts up to 200 feet in depth have been excavated in such areas without exceeding economic limits.

Stabilization of the sides of test pits and trenches, when required, is generally achieved by sloping the walls at a sufficiently flat angle or by the use of sheeting. Excavations to depths of approximately 5 feet often do not require support of the walls. However, in those cases where support of these shallow excavations is required it is generally more economical to slope the side walls; deeper excavations are normally more economical if sheeted. Depending on the nature of the soils and the depth of the excavation below groundwater level, dewatering also may be required for stability of the side walls and to prevent the bottom of the pit from heaving as well as to keep the excavation dry. This is an important consideration for excavations in cohesionless material below the groundwater table.

Large-diameter borings deeper than 5 feet should be cased before inspection. The Corps of Engineers recommends the use of large-diameter corrugated steel pipe for holes above the groundwater table and in relatively stable soils. In deep borings in soft, highly unstable materials, including highly fractured rock, corrugated steel pipe may suffice or it may be necessary to use thick-wall steel pipe into which holes have been cut to permit inspection and testing of the side walls as required.

The safety aspects of test pits should not be overlooked in the planning or execution phases of the work. Four major precautions to be taken are: material removed from the excavation should be deposited at least 2 feet and preferably 5 feet or more from the edge of the pit or trench; the side walls should be adequately braced or sloped prior to personnel entering the pit; the cage or bosun's chair used to lower the personnel into the hole for the inspection should have two lines attached to it, the second serving as a safety line; and proper ventilation should be provided, particularly in drilled holes.

1.9.2 Exploratory Adits and Test Shafts

An exploratory adit is a horizontal or near horizontal excavation made by mining methods. In general, these adits are excavated predominantly in rock. The term "test shaft" is used to refer to a vertical excavation, generally in rock, and to very deep test pits. The major purpose of each of these methods of exploration is to permit a detailed in situ examination of the nature of the rock and its structural features such as joints, fractures, faults,

and shear zones. Adits may be utilized also for in situ tests to determine the modulus of deformation of the rock. The adit and shaft are utilized most frequently in the exploration of proposed dam projects. Adits and shafts are among the more costly forms of exploration but also are among the most informative. They are justifiable where large structures are involved and where a clear picture of subsurface conditions cannot be obtained through ordinary means. The dimensions of the adits are 4 ft × 6 ft or 5 ft × 7 ft when hand mucking and hauling equipment are used and 6 ft × 8 ft minimum when power mucking and hauling equipment are used.

1.10 GROUNDWATER INVESTIGATIONS

1.10.1 Purpose of Investigations

Groundwater investigations are of two types, those used to determine groundwater levels and pressures and those used to determine the permeability of the subsurface materials. The former includes measurements to determine the elevation of the groundwater surface or table and its variation with the season of the year; the location of perched water tables; the location of aquifers; and the presence of artesian pressures. Determination of the permeability of soil or rock strata is required in connection with seepage studies for leakage past dams, yield of wells, groundwater lowering, etc. Water levels and pressures may be measured in existing wells, in boreholes, and in specially installed observation wells. Permeability is determined by means of various types of seepage, pressure, and pumping tests.

1.10.2 Observations of Water Levels and Pressures

Existing Wells Many states require the drillers of water wells to file logs of the wells. These are a good source of information of the materials penetrated and the water level at the time of drilling. The owning agency of public or commercial wells may have records of the water levels since installation and these are a source of additional information. In the case of private wells, records are not usually kept. When this is the case, the water level may be measured as described in the paragraph on boreholes. The owner of a private well may also be able to give information as to the fluctuation of the water level and periods when the water has been low or the well has been dry.

Boreholes It is common to establish the water table elevation at a site by measuring the depth to water in the boreholes. The length of time required for water levels in boreholes to stabilize at the groundwater level is a function of the permeability of the soil. There is no doubt that the water should be allowed to stand for a minimum period, preferably 24 hours, following completion of the hole. However, even under these circumstances, the reading may be accurate only if the soil is pervious. Accurate readings can be obtained if readings are taken over a long period of time. In one-shift-a-day drilling the groundwater level is usually observed as the first order of business in the morning. This gives 14–16 hours time for stabilization and this is frequently adequate. Drilling mud obscures observations of the groundwater level owing to filter cake action and its specific gravity being greater than water.

The depth to groundwater is usually measured by means of a chalked tape, a tape with a float, or an electric water-level indicator. In the first method a short section of the lower end of a metal tape is chalked. The tape with a weight attached to its end is then lowered until the chalked section has passed slightly below the water surface. The depth to the water is determined by subtracting the depth of penetration of the line into water, as measured by the water line in the chalked section, from the total depth from the ground surface. In the second method a tape with a float attached to its end is lowered until the float hits the water surface and the tape goes slack. The tape is then lifted until the float is felt to touch the water surface and it is just taut; the depth is then measured. Some floats are equipped with a whistle that sounds when the float hits the water surface. The electric depth indicator consists of a weighted probe attached to the lower end of a length of electrical cable that is marked at intervals to indicate the depth. When the probe reaches the water a circuit is completed and this is registered by a meter mounted on the cable reel. The electric indicator has the advantage that it may be used in extremely small holes.

Observation Wells The term "observation well" is applied to any well or drilled hole used for the purpose of long-term studies of groundwater levels and pressures. Existing wells and boreholes in which casing is left in place are often used to observe groundwater levels. These, however, are not considered to be as satisfactory as wells constructed specifically for the purpose. The latter may consist of a standpipe or a piezometer installed in a previously drilled exploratory hole or a hole drilled solely for use as an observation well. Details of typical installations are shown in Figure 1.18. In cases where pressures in specific zones are required, bentonite or a similar material is used to seal the piezometer within the zone. It is also customary to use a seal at the surface and to slope the fill at the top of the hole away from the pipe in order to prevent the entrance of surface water. The top of the pipe should also be capped to prevent the entrance of foreign material.

The piezometer shown in Figure 1.18a is a wellpoint piezometer; hydraulic and diaphragm type piezometers may be used. The reader is directed to Hvorslev (1949), U.S. Army Corps of Engineers (1969), U.S. Bureau of Reclamation (1968), and Dunnicliff (1988) for detailed discussions of various types of piezometers.

1.10.3 Seepage Tests for Estimation of Field Permeability

General Discussion Seepage tests in boreholes constitute one means of determining the permeability of overburden in situ. They are particularly valuable in the case of materials such as sands or gravels, undisturbed samples of which are difficult or impossible to obtain. Three types of test are in common use; namely the falling, the rising, and the constant water level methods. In general, either the rising or the falling level methods should be used if the permeability is low enough to permit accurate determination of the water level. In the falling level test, the flow is from the hole to the surrounding soil and there is danger of clogging of the soil pores by sediment in the test water used. This danger does not exist in the rising level test, where water flows from the surrounding soil to the hole, but there is the danger of the soil at the bottom of the hole becoming loosened or quick if too great a gradient is imposed at the bottom of the hole. If the rising level is used, the test should be followed by sounding of the base of the hole with drill rods to determine whether heaving of the bottom has occurred. The rising level test is the preferred test. In those cases where the permeability is so high as to preclude accurate measurement of the rising or falling water level, the constant level test is used.

Holes in which seepage tests are to be performed should be drilled using only clear water as the drilling fluid. This precludes the forming of a mud cake on the walls of the hole or clogging of the pores of the soil by drilling mud. The tests are performed intermittently as the borehole is advanced. When the hole reaches the level at which a test is desired, the hole is cleaned and flushed using clear water pumped through a drill tool with shielded or upward-deflected jets. Flushing is continued until a clean surface of undisturbed material exists at the bottom of the hole. The permeability is then determined by one of the procedures given below. Specifications sometimes require a limited advancement of the borehole without casing upon completion of the first test at a given level, followed by cleaning, flushing, and repeat testing. The difficulty of obtaining satisfactory in situ permeability measurements makes this requirement a desirable feature since it permits verification of test results.

Data that must be recorded for each test regardless of the type of test performed include (1) the depth from ground surface to the groundwater surface both before and upon completion of the test; (2) the inside diameter of the casing; (3) the height of the casing above ground surface; (4) the length of the casing during the test; (5) the diameter of the borehole below the casing; (6) the depth to the bottom of the boring from the top of the casing; (7) the depth to the standing water level from the top of the casing; and (8) a description of the material tested.

Falling Water Level Method In this test, the casing is filled with water, which is then allowed to seep into the soil. The rate of drop of the water level in the casing is observed by measuring the depth of the water surface below the top of the casing at 1, 2, and 5 minutes after the start of the test and at 5-minute intervals thereafter. These observations are made until the rate of drop becomes negligible or until sufficient readings have been obtained to satisfactorily determine the permeability. Other required observations are listed in the general paragraph above. Formulas for the computation of the permeability are shown in Figure 1.19.

Rising Water Level Method This method, most commonly referred to as the time lag method (U.S. Army Corps of Engineers, 1951), consists of bailing the water out of the casing and observing the rate of rise of the water level in the casing at intervals until the rise in water level becomes negligible. The rate is observed by measuring the elapsed time and the depth of the water surface below the top of the casing. The intervals at which the readings are required will vary somewhat with the permeability of the soil. The readings should be frequent enough to establish the equalization diagram shown in Figure 1.19. In no case should the total elapsed time for the readings be less than 5 minutes. A plot of the observations such as the diagram shown in Figure 1.19 should be made during the test to insure that sufficient readings have been taken and that the test results are valid. Formulas for the determination of the permeability by this method also are given in Figure 1.19. As noted above, a rising level test should always be followed by sounding of the bottom of the holes to determine whether the test created a quick condition.

Constant Water Level Method In this method water is added to the casing at a rate sufficient to maintain a constant water level at or near the top of the casing for a period of not less than 10 minutes. The water may be added by pouring from calibrated containers or by pumping through a water meter. In addition to the data listed in the above general discussion, the data recorded should consist of the amount of water added to the casing at 1, 2, and 5 minutes after the start of the test and at 5-minute intervals thereafter until an adequate determination of the permeability has been made. Formulas for the determination of the permeability by this method are given in Figure 1.19.

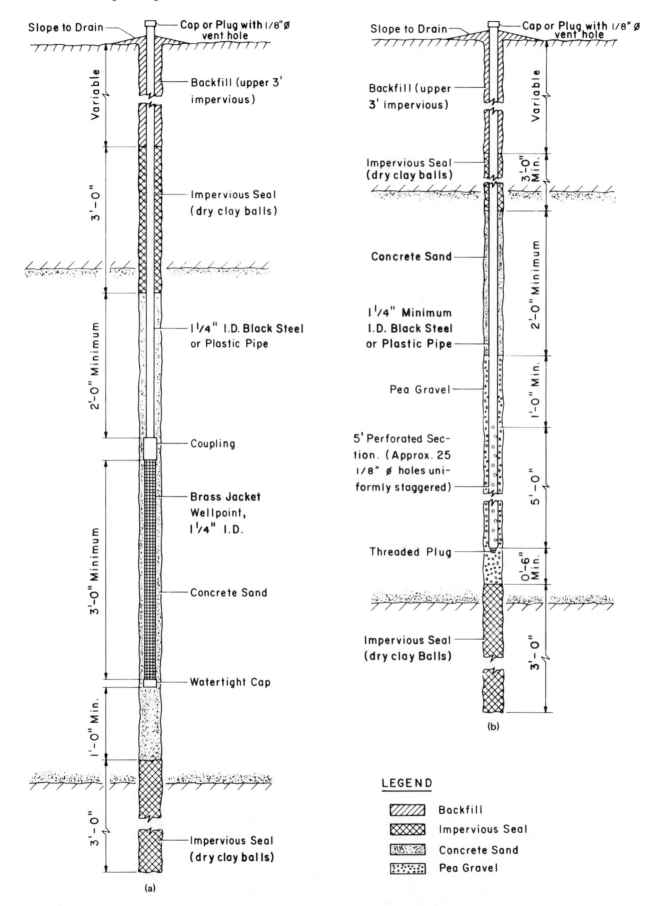

Fig. 1.18 Observation wells: (a) wellpoint piezometer; (b) standpipe.

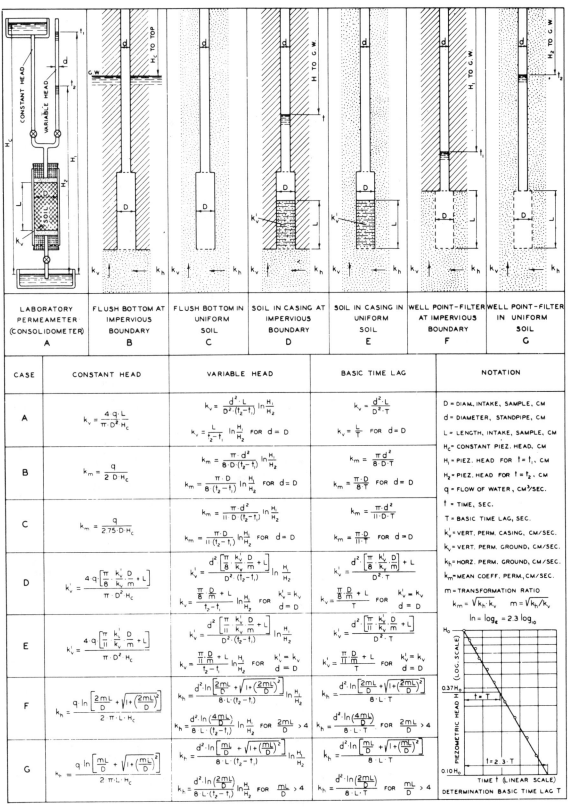

Fig. 1.19 Formulas for determination of permeability from seepage tests. (*After Hvorslev, 1949.*)

1.10.4 Pressure Tests for Estimation of Field Permeability

Tests in which water under pressure is forced into rock in situ through the walls of boreholes provide a means of determining the apparent permeability of the rocks and yield information regarding its soundness. The information thus obtained is used primarily in seepage studies. It is also frequently used as a qualitative measure of the grouting required for impermeabilization or strengthening of the rock. Pressure tests should be performed only in holes that have been drilled with clear water. The reasons for this are detailed in the general paragraph on seepage tests.

The apparatus used for pressure tests in rock is illustrated schematically in Figure 1.20a. It comprises a water pump, a manually-adjusted automatic pressure relief valve, pressure gauges, a water meter, and a packer assembly. The packer assembly, shown in Figure 1.20b, consists of a system of piping to which two expandable cylindrical rubber sleeves, called packers, are attached. The packers, which provide a means of sealing off a limited section of borehole for testing, should have a length five times the diameter of the hole. They may be of the pneumatically or mechanically expandable type. The former are preferred since they adapt to an oversized hole whereas the latter may not. However, when pneumatic packers are used, the test apparatus must also include an air or water supply connected, through a pressure gauge, to the packers by means of a high-pressure hose as shown in Figure 1.20a. The piping of the packer assembly is designed to permit testing of either the portion of the hole between the packers or the portion below the lower packer. Flow to the section below the lower packer is through the interior pipe; flow to the section between

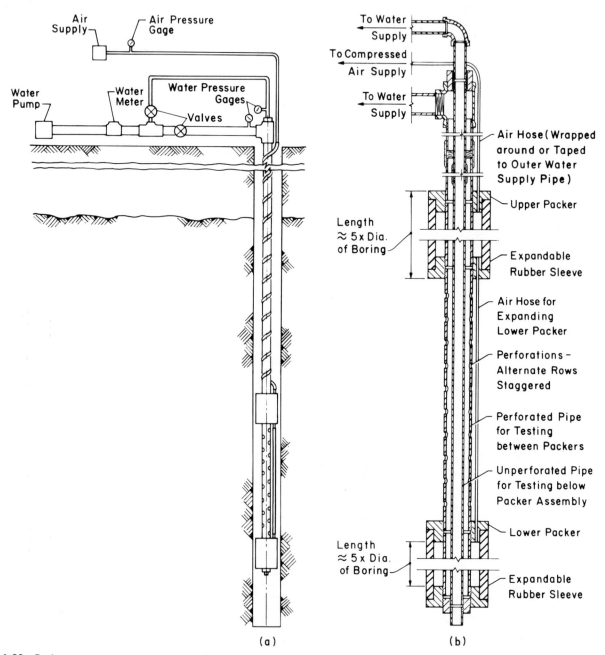

Fig. 1.20 Packer-type pressure-test apparatus for determining the permeability of rock: (a) schematic diagram; (b) detail of packer unit.

the packers is provided by perforations in the outer pipe, which have an outlet area two or more times the cross-sectional area of the pipe. The packers are normally set 2, 5, or 10 feet apart and it is common to provide flexibility in testing by having assemblies with different packer spacings available, thereby permitting the testing of different lengths of the hole. The wider spacings are used for rock that is more uniform; the short spacing is used to test individual joints that may be the cause of high water loss in otherwise tight strata.

The test procedure used depends upon the condition of the rock. In rock that is not subject to cave-in, the following method is in general use. After the borehole has been completed it is filled with clear water, surged, and washed out. The test apparatus is then inserted into the hole until the top packer is at the top of the rock. Both packers are then expanded and water under pressure is introduced into the hole, first between the packers and then below the lower packer. Observations of the elapsed time and the volume of water pumped at different pressures are recorded as detailed in the section on pumping below. Upon completion of the test, the apparatus is lowered a distance equal to the space between the packers and the test is repeated. This procedure is continued until the entire length of the hole has been tested or until there is no measurable loss of water in the hole below the lower packer. If the rock in which the hole is being drilled is subject to cave-in, the pressure test is conducted after each advance of the hole for a length equal to the maximum permissible unsupported length of hole or the distance between the packers, whichever is less. In this case, the test is limited, of course, to the zone between the packers.

Regardless of which procedure is used, a minimum of three pressures should be used for each section tested. The magnitude of these pressures are commonly 15, 30, and 45 psi above the natural piezometric level. However, in no case should the excess pressure above the natural piezometric level be greater than 1 psi per foot of soil and rock overburden above the upper packer. This limitation is imposed to insure against possible heaving and damage to the foundation. In general, each of the above pressures should be maintained for 10 minutes or until a uniform rate of flow is attained, whichever is longer. If a uniform rate of flow is not reached in a reasonable time, the engineer must use his discretion in terminating the test. The quantity of flow for each pressure should be recorded at 1, 2, and 5 minutes and for each 5-minute interval thereafter. Upon completion of the tests at 15, 30, and 45 psi the pressure should be reduced to 30 and 15 psi, respectively, and the rate of low and elapsed time should once more be recorded in a similar manner.

Observation of the water take with increasing and decreasing pressure permits evaluation of the nature of the openings in the rock. For example, a linear variation of flow with pressure indicates an opening that neither increases or decreases in size. If the curve of flow versus pressure is concave upward it indicates that the openings are enlarging; if convex, the openings are becoming plugged. The reader is directed to Cambefort (1964) for a detailed discussion of the interpretation of pressure tests. Additional data required for each test are as follows: (1) depth of hole at time of each test; (2) depth to bottom of top packer; (3) depth to top of bottom packer; (4) depth to water level in borehole at frequent intervals; (5) elevation of piezometric level; (6) length of test section; (7) radius of hole; (8) length of packer; (9) height of pressure gauge above ground surface; (10) height of water swivel above ground surface; and (11) description of material tested. Item (4) is importance since a rise in water level in the borehole may indicate leakage around the top packer. Leakage around the bottom packer would be indicated by water rising in the inner pipe.

The formulas used to compute the permeability from pressure test data are (from U.S. Bureau of Reclamation, 1968):

$$k = \frac{Q}{2\pi LH} \ln \frac{L}{r} \qquad L \geqslant 10r$$

$$k = \frac{Q}{2\pi LH} \sinh^{-1} \frac{L}{2r} \qquad 10r > L \geqslant r$$

where

k = permeability
Q = constant rate of flow into the hole
L = length of the test section
H = differential head on the test section
r = radius of the borehole

These formulas provide only approximate values of k since they are based on several simplifying assumptions and do not take into account the flow of water from the test section back to the borehole. However, they give values of the correct magnitude and are suitable for practical purposes. A graphical solution of the more commonly used upper equation is given in Figure 1.21.

1.10.5 Pumping Tests for Estimation of Field Permeability

Continuous pumping tests are used to determine the water yield of individual wells and the permeability of subsurface materials in situ. The data provided by such tests are used to determine the potential for leakage through the foundations of water-retaining structures and the requirements for construction of dewatering systems during excavation. The test consists of pumping water from a well or borehole and observing the effect on the water table by observations of the water levels in the hole being pumped and in adjacent observation wells. The observation wells should be of the piezometer type. The depth of the test well will depend on the location, in profile, of the strata to be tested. The number, location, and depth of the observation wells will depend on the estimated shape of the groundwater surface after drawdown. The reader is referred to Leonards (1962, chapter 3), for formulas that will permit an estimate of the drawdown curve under various boundary conditions. The number of wells used and their location should be selected so as to provide a clear picture of the drawdown curve in various directions from the test well. As a minimum, observation wells should be located on two perpendicular lines passing through the center of the test well. Along each of the four radial lines thus created there should be a minimum of four wells, the innermost of which should be within 25 feet of the test well. The outermost should be located near the limits of the effect of drawdown and the middle wells should be located to give the best definition of the drawdown curve based on its estimated shape.

The pump used for the test should have a capacity 1.5 to 2 times the maximum anticipated flow and should have a discharge line sufficiently long to obviate the possibility of the discharge water recharging the strata being tested. Auxiliary equipment required includes an air line to measure the water level in the test well, a flow meter, and measuring devices to determine the depth to water in the observation wells. The air line, complete with pressure gauge, hand pump, and check valve, should be securely fastened to the pumping unit with its lower end at the deepest planned pumping level but in no case closer than 2 feet to the end of the suction line. The flow meter should be of the visual type, such as an orifice. The depth-measuring devices for the observation wells may be any of the types described in Section 1.10.2.

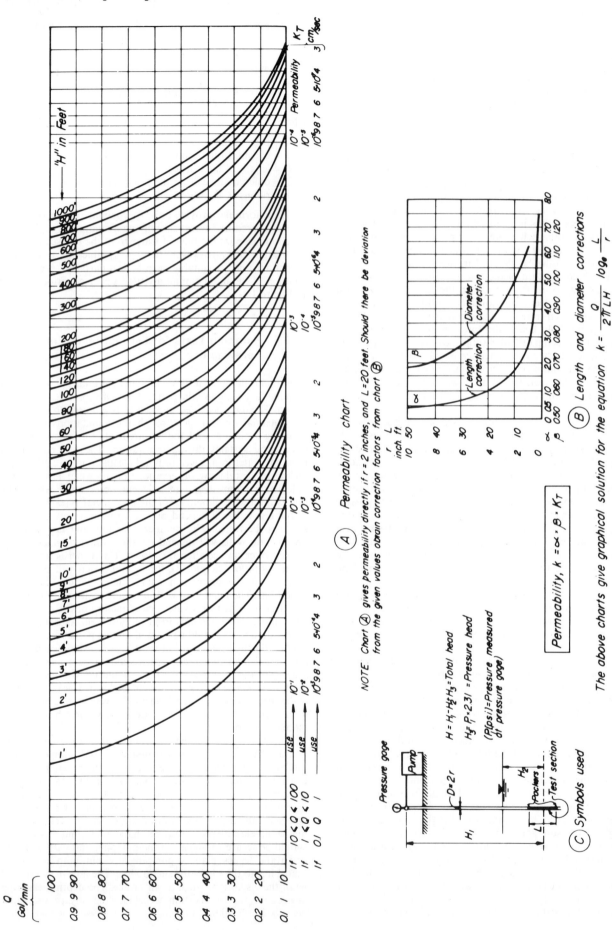

Fig. 1.21 Graphical solution for determining permeability from packer-type test. (*From Handbook of Applied Hydraulics, 3d ed., C. V. Davis and K. E. Sorensen. Copyright 1969 by McGraw-Hill Book Company, New York, N.Y. Used with permission of McGraw-Hill Book Company.*)

The test procedure is as follows. Upon completion of the well or borehole, the hole is cleaned and flushed, the depth of the well is accurately measured, the pump is installed, and the well is developed. The well is then tested at $\frac{1}{3}$, $\frac{2}{3}$, and full capacity. Full capacity is defined as the maximum discharge attainable with the water levels in the test and observation wells stabilized. Each of the discharge rates is maintained for 4 hours after further drawdown in the test and observation wells has ceased or for a maximum of 48 hours, whichever occurs first. The discharge must be maintained constant during each of the three stages of the test and interruptions of pumping are not permitted. If pumping should accidentally be interrupted, the water level should be permitted to return to its full nonpumping level before pumping is resumed. Upon completion of the drawdown test, the pump is shut off and the rate of recovery is observed.

The basic test-well data that must be recorded are: (1) location, top elevation, and depth of well; (2) size and length of all blank casing in the well; (3) diameter, length, and location of all screen casing used; also type and size of screen opening and material of which screen is made; (4) type of filter pack used, if any; (5) water elevation in well prior to testing; and (6) location of bottom of air line. Basic data required for each observation well are: (1) location, top elevation, and depth of well; (2) size and elevation of bottom of casing (after installation of piezometer); (3) location of all blank casing sections; (4) manufacturer, type, and size of piezometer; (5) depth and elevation of piezometer; and (6) water level in well prior to testing. Pump data required are the manufacturer's model designation, pump type, maximum capacity, and capacity at 1800 rpm.

The drawdown test data required for each discharge rate consist of the discharge and drawdown of the test well and the drawdown of each observation well at the time intervals shown below for the various ranges of elapsed time.

Elapsed Time	Time Interval for Readings
0–10 min	0.5 min
10–60 min	2.0 min
1–6 h	15.0 min
6–9 h	30.0 min
9–24 h	1 h
24–48 h	3 h
>48 h	6 h

The required recovery curve data consist of readings of depth to water in the test and observation wells at the same time intervals given above. Readings are continued until the water level returns to the prepumping level or until adequate data has been obtained.

1.11 TEST GROUTING

1.11.1 General Description

This type of field investigation is used (1) to assess the groutability of a given foundation material; (2) to obtain a quantitative measure of the amount of work required to grout a given material; or (3) to prove that the use of a grouting procedure other than that specified will produce an end product equivalent to or better than that obtained by the use of the specified procedure. The latter frequently occurs in connection with the desire to use percussion-drilled holes in lieu of specified rotary-drilled holes. Test grouting consists of injecting grout into a limited area utilizing either the same procedure as will be used in the final construction in the case of (1) and (2) or

a somewhat modified procedure in the case of (3). The records kept are similar to those in a full-scale grouting program and include information pertaining to the drilling and washing of the grout holes and the pressures, grout mixes, and amount of grout used in injecting individual sections of each hole. The reader is referred to U.S. Army Corps of Engineers (1963) for details of these records. The results of these investigations are generally evaluated on the basis of the amount of grout consumed, the extent of grout penetration from a given hole as revealed by core borings and split spacing procedure, the impermeabilizing effect of the grouting as revealed by pressure tests, or a combination of all three.

The cost of this type of investigation is high and it is not used in design stages unless particular grouting problems are suspected. Normally comparison of use of percussion versus rotary grout hole drilling would be done at the beginning of the construction stage.

1.12 REPRESENTATIVE BUT DISTURBED SAMPLES

1.12.1 General Description

Samplers may be divided into two broad categories based on the degree of disturbance suffered by the samples that are retrieved. In general, samples are categorized as disturbed and undisturbed. There is no truly undisturbed sample since the penetration of the sampling tube, no matter how thin the tube or refined the sampling technique, results in some disturbance to the in situ material.

A disturbed sample may be defined as one that contains all of the constituents of the in situ material in proper proportions but that has suffered sufficient disturbance to its structure so that the results of laboratory tests to determine engineering properties such as compressibility, shear strength, and permeability would not be properly representative of the material in situ. Disturbed samples generally are used for identification tests such as visual classification, water content, grain size analyses and Atterberg limit tests, and for specific gravity and compaction tests. Disturbed samples are also used for the preparation of compacted specimens for permeability, shear, and compressibility tests on materials proposed for use in earth structures. Included in this category are samples obtained by driving a sampler into the ground, by auger boring, or by normal test pit excavation.

An undisturbed sample is one obtained with samplers and sampling techniques designed to preserve as closely as possible the natural structure of the material. These samples are suitable for shear, consolidation, and permeability tests of foundation materials. They may be used, also, for all tests for which disturbed samples are used. Undisturbed samples include those obtained by thin-wall tube samplers with and without stationary pistons, Denison-type double-tube core barrel samplers, and careful excavation of soil from test pits.

The characteristics of some of the more popular samplers and sampling techniques in use today are presented in the paragraphs that follow.

1.12.2 Split-Barrel Sampler

This sampler, also referred to as the split-tube or split-spoon sampler, is the most commonly used soil-sampling device. It is a modification of the solid-tube drive sampler, which was one of the first samplers developed. It differs from the solid-tube sampler in that it has a sample retainer located immediately above the barrel shoe and that the barrel in which the sample is retained is split longitudinally, thereby facilitating removal of the sample. As illustrated in Figure 1.22, the split-barrel sampler consists of a barrel shoe, a split barrel or tube, a solid sleeve, and a sampler head. In some models the solid sleeve is omitted and the head is so designed that it threads directly onto the barrel. Other modified versions of the sampler provide for the inclusion of a single long liner or a series of 1-inch high sectional liners that extend for the full length of the sampler. Still another version, referred to as the Lynac sampler, has heavier walls than the regular split-barrel sampler and is better adapted to hard driving conditions.

As indicated by Figure 1.22, when the shoe and the sleeve of this type of sampler are unscrewed from the split barrel, the two halves of the barrel may be separated and the sample may be extracted easily. A ball valve incorporated in the sampler head facilitates the recovery of cohesionless materials. This valve seats when the sampler is being withdrawn from the borehole, thereby preventing water pressure on the top of the sample from pushing it out. If the sample tends to slide out because of its weight, vacuum tends to develop at the top of the sample to retain it. Provisions for retention of the sample also are made at the bottom of the sampler, where a retaining device may be used if necessary. The device shown in Figure 1.22 is a flap type retainer. This retainer permits the soil to enter during driving but upon withdrawal of the sampler from the hole the flap closes and retains the sample. An alternative type of retainer is the basket retainer that consists of a series of fingers of spring steel mounted vertically on the periphery of a metal ring. The tips of the fingers deflect toward the center of the sampler to form a basket shape. As soil enters the sampler it easily pushes the steel fingers back against the walls of the sampler. If upon withdrawal of the sampler the sample tends to slide out, the points of the spring move in and grip the sample. Other types of retainers for use under various conditions also are available. One, referred to as the L.A.D. retainer, consists of spring steel fingers and a plastic sleeve. The latter drops over the fingers if the sample tends to fall out and effectively closes the opening at the base of the sampler.

In general, the split-spoon is available with inside diameters ranging from 1.5 to 4.5 inches in 0.5-inch increments. Barrels are available in standard lengths of 18 and 24 inches and have

a wall thickness of 0.25 inch. The 1.5-inch-diameter sampler is popular because correlations have been developed between the number of blows required for penetration of this sampler and the relative density of cohesionless soils and the shear strength of cohesive soils. The larger-diameter samplers are used when gravel particles are present. An advantage of the larger-size barrels is that they provide more material for classification tests.

Sampling with the split-barrel sampler is accomplished by driving the sampler into the ground with a drive hammer, whose weight will depend on the size of the sampler. Generally the 1.5-inch sampler is driven with a 140-lb hammer dropped 30 inches; the larger samplers are driven with a 300-lb hammer dropped 18 inches. It is common practice to record the number of blows for each 6 inches of the total sampler penetration. The number of blows required to drive the 1.5-inch sampler a distance of 12 inches using the stipulated hammer weight and drop is the Standard Penetration Resistance (SPT values) developed by Terzaghi and Peck (1967). If a sampler is driven 18 inches, the blows for the last 12 inches of penetration are used as the SPT value in order to eliminate the effect of any disturbance or sediment at the bottom of the hole. Similarly, when a drive of 24 inches is made, the blows for the middle 12 inches are used. In order to assure the uniformity and accuracy of SPT values it is essential that the hammer have a free fall. This can be accomplished by using a rope wrapped around a cathead instead of a cable to hoist the hammer and by not restraining the rope during the drop. The energy imparted to the drill rods under good operating conditions amounts to 60–80 percent of that available (weight of hammer times height of drop). Whether the drill rods are size A or N does not appear to make a significant difference. However, in the 1930s to 1950s a $1\frac{1}{2}$-inch standard pipe was frequently used for drill rods. This difference could cause earlier SPT values to be somewhat larger than values that would be obtained currently. Whichever size of drill rods is used, they must be made up tight to prevent excessive loss of energy as the blow of the hammer is transmitted to the sampler.

The relationship between SPT values and compactness of cohesionless sands and silts is given in Figure 1.23, and between SPT values and consistency of cohesive soils in Figure 1.24. Burmister's standard relationships for a heavier hammer and a larger sampler, as well as formulas for correlating the blow counts for other hammers and drop heights, and other size samplers, are also given in Figures 1.23 and 1.24. Several investigators have noted that the SPT value for a cohesionless soil at a particular relative density increases with increase in effective overburden pressure. Peck, Hanson, and Thornburn (1973) give the following equation for this increase:

$$N_1 = C_N N$$

where

N_1 = number of blows at effective overburden pressure of 1 ton/ft^2

$C_N = 0.77 \log_{10} \dfrac{20}{\bar{\sigma}_0}$

$\bar{\sigma}_0$ = actual effective overburden pressure obtaining at depth of SPT

N = SPT measured

A similar type of increase is not expected for a cohesive soil.

Many correlations have been made between SPT values and soils properties. These correlations and their validity are discussed in detail by de Mello (1971), Fletcher (1965), ASCE (1986), and Riggs (1986).

Split-barrel sampling is used primarily in those instances where it is necessary to determine the stratification, identification, consistency, and density of the soils present at a site. Seed (1979)

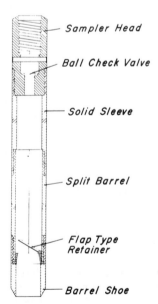

Fig. 1.22 Split-barrel sampler. (*Courtesy of Sprague & Henwood, Inc.*)

— Sampler Head

— Ball Check Valve

— Solid Sleeve

— Split Barrel

— Flap Type Retainer

— Barrel Shoe

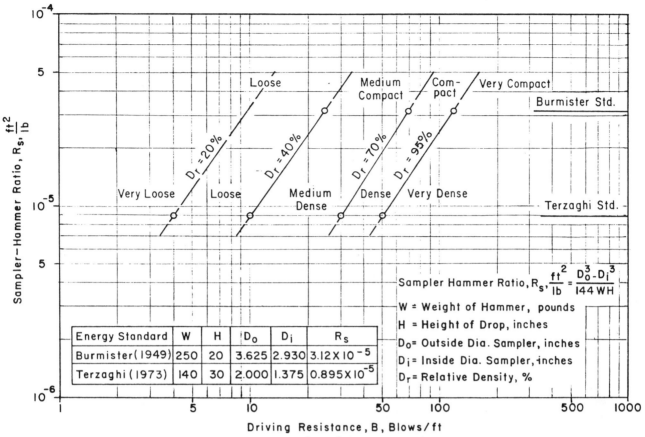

Fig. 1.23 Sampler driving resistance vs. compactness—cohesionless sands and silts.

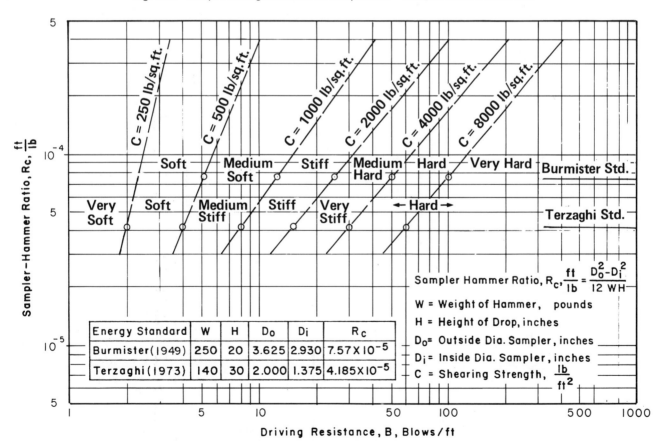

Fig. 1.24 Sampler driving resistance vs. consistency—cohesive soils.

describes the use of blow counts from split-barrel sampling to evaluate the liquefaction potential of soils. Practically all subsurface explorations are initiated with this type of sampling and frequently design can be made on the basis of the split-barrel sample data. This is particularly the case where a minor type structure is involved and where foundation conditions are good. In other instances the split-barrel borings are supplemented with sampling by one or more of the undisturbed type samplers. The primary advantages of split-barrel sampling are that it is simple, quick, and economical.

1.12.3 One-Inch Retractable Plug Sampler

This sampler was developed by O. J. Porter about 1930. As shown by Figure 1.25a, the sampler consists of a retractable piston rod and plug inside a casing and master tube; a cutting shoe; and a driving mechanism. The master tube, which forms the lower portion of the sampler, is usually made to accommodate from four to seven 6-inch-long brass liners into which the soil penetrates as the sampler is driven into the ground. The steps in the operation of the sampler are shown in Figure 1.25. Initially, the sampler, with the piston rod and plug locked to a special coupling between the master tube and the casing (Fig. 1.25a), is driven into the ground by lifting the 30-lb drive weight by hand and letting it fall upon the drive head. When

the sampler reaches the depth at which sampling is desired, the piston rod is rotated to unlock it from the coupling, raised, and rotated once more to lock it to the coupling in an elevated position (Fig. 1.25b). The unit is then driven a distance equal to the length of the brass liners to obtain a sample (Fig. 1.25c). At the end of the sampling operation (Fig. 1.25d) the plug is rotated further to develop a tight seal at the top of the sampler. Then the entire unit is jacked out of the ground or "bumped" out by lifting the drive hammer to strike a pull cap attached to the top of the guide. The next sample is taken by driving the unit down the same hole to the next depth of sampling and repeating the operation. The samples obtained by this procedure are disturbed but are representative. The fact that they are retrieved in brass liners that can be capped, sealed, and shipped to the laboratory makes them convenient for determination of water content.

This sampler is used for both reconnaissance investigations and for mapping strata between borings in which more sophisticated sampling is done. It is used primarily for determining the thickness of surficial deposits of soil silts and clays as occur in swamps and estuaries but it is also useful in the finer-grained cohesionless materials. The unit can be operated to depths of 60 feet or more depending on the nature of the deposits penetrated and the location of the groundwater table. One of its main advantages is that it can be hand-carried into areas difficult for access by conventional equipment. It is

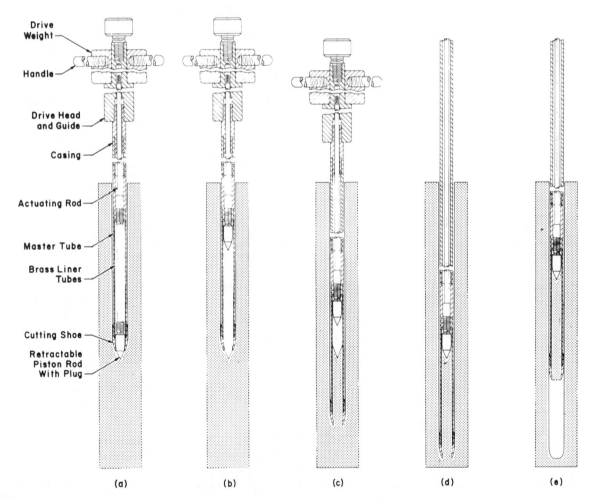

Fig. 1.25 Schematic drawing of 1-inch retractable plug sampler, illustrating principal parts of sampler and sequence of operations in sampling: (a) sampler driven to sampling depth with retractable plug locked in driving position; (b) sampler at sampling depth with plug locked in retracted position; (c) sampling drive in progress; (d) drive completed; (e) sampler and sample being withdrawn from hole.

economical and provides an inexpensive form of obtaining information between larger-diameter borings.

1.12.4 Hand Augers

Manually operated augers may be classified into three general groups as follows: (1) those commonly referred to as "post-hole diggers"; (2) small-diameter helical or screw-type augers; and (3) spiral augers. One of the better known of the first group is the Iwan auger illustrated in Figure 1.26d. Similar augers, which are not illustrated, include the Vicksburg hinged auger and the spoon auger. The Vicksburg auger, which also may be used as a power auger, is hinged so that one side of the auger may be dropped for ease of removal of the material sampled. The post-hole type augers are available in diameters ranging from 3 to 9 inches. Augers of this type, in general, retain the sampled material better than the other hand-operated augers. In general, they are useful for sampling all types of soils except cohesionless materials below the water table and hard or cemented soils. The lack of rigidity of the Vicksburg sampler also makes it unsuited for sampling of soils containing gravel.

The second group of augers includes those consisting of a helical flight on a solid stem and the ship- and worm-type augers, which resmeble wood augers. The ship auger, shown in Figure 1.26a, and the worm augers, are best suited to use in cohesive materials. The helical auger may be used in either cohesive or cohesionless materials above the water table. These augers are available for hand operation in diameters from approximately 2 to 3.5 inches.

Spiral augers were developed for use in those cases where helical and screw augers do not work well. The closed spiral auger (Fig. 1.26b) is used in dry clay and gravelly soils. The Jamaica open spiral auger is most useful in loosely consolidated deposits. These augers are available in the same size range as the preceding group.

For additional information concerning augers, the reader is directed to Section 1.8.6 where the uses, the limitations, the

1.12.5 Power Augers

Motor-driven augers may be classified as flight augers or bucket augers. The former are available in sizes ranging from 2 to 48 inches in diameter; the latter range from 12 to 96 inches in diameter. The flight augers are available in several types depending on their intended use. Figure 1.27a shows a single-flight auger designed for use in clays and other unconsolidated deposits. It is equipped with a pilot bit that makes the initial penetration prior to the ripping of the material by the cutter teeth. Figures 1.27b and c show double-flight augers designed for use in both earth and rock. The former is capable of penetrating most soils including those containing large gravel and boulders and may also be used in soft rock. The latter is capable of penetrating hard rock as well as soil. In operation all of the above augers are attached to a drilling rod referred to as the "Kelly" rod, which is rotated and pressed downward to achieve penetration. The sampling procedure consists of advancing the Kelly with the auger for the height of the flight or until the flight has become filled with soil. The Kelly is then raised until the auger is clear of the hole and the soil is thrown free from the cutter head owing to rotation of the Kelly. The hole is advanced by repeating this process until the required depth is reached. Obviously, the shorter the flight the more time is used in raising and lowering the auger to remove the soil. Therefore, it is common to attach an additional height of spiral to a cutter head so as to form a high spiral auger such as the one shown in Figure 1.27d.

The maximum depth of penetration that can be achieved conveniently with the above augers is limited by the length of the Kelly rod that can be accommodated by the drilling rig used. In general, the depth is limited to 10 to 20 feet. The use of continuous or conveyor flight augers, such as the one shown in Figure 1.28a, overcomes this disadvantage and permits holes

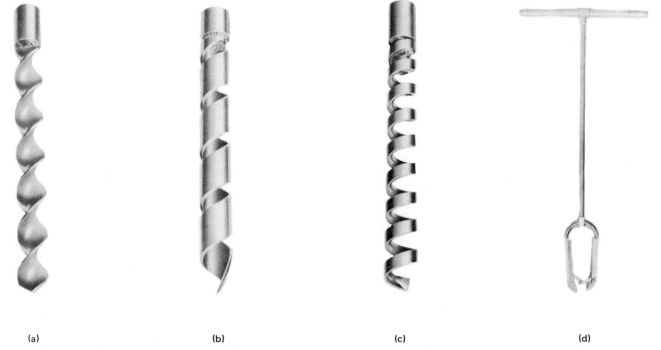

(a)   (b)   (c)   (d)

Fig. 1.26   Hand augers: (a) ship auger; (b) closed spiral auger; (c) open spiral auger; (d) Iwan auger. (*Courtesy of Acker Drill Co., Inc.*)

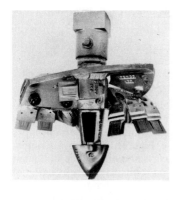 

(a)   (b)

(c)   (d)

Fig. 1.27 Power augers: (a) single-flight earth auger; (b) double-flight earth–rock auger; (c) double-flight rock auger; (d) high spiral auger. (*Courtesy of Mobile Drilling, Inc. and Acker Drill Co., Inc.*)

to depths of 100 feet or more. In this type of auger, the soil rises to the top of the hole on the spiral flight and is sampled as it emerges. As the hole is advanced additional sections of spiral flight are added. Both finger-type cutter heads (Fig. 1.28a) and fishtail bits are used in conjunction with continuous flight augers. Although this type of auger has the advantage of greater maximum depth of hole compared to augers with shorter flights, it has the disadvantage of a greater possibility of the mixing of soil from a given level with soils from strata above. This disadvantage has been overcome by the hollow-stem auger which is shown in Figure 1.28b and is described in Section 1.12.6.

In addition to flight augers, which have been described above, bucket augers similar to the one shown in Figure 1.28b are frequently used for explorations. These buckets are basically an open-top metal cylinder having one or more slots cut in their bases to permit the entrance of soil and rock as the bucket is rotated. At the slots the metal of the base is reinforced and teeth or a sharpened cutting edge are provided to break up the material being sampled. The buckets can be designed to sample any material except solid rock. They are available in nominal diameters ranging from 10 to 96 inches. The larger buckets are used for excavating shafts for large caisson piles. Holes about 30 inches in diameter are sufficiently large to permit the entrance of a man for in situ inspection and sampling of foundation material. When such holes are used, care must be exercised to see that they are well ventilated and that there are no poisonous gases present when personnel enter for inspection or sampling.

In operation the bucket auger is attached to a Kelly rod driven by a rotary table similar to the one shown on the rig in Figure 1.14c. The bucket is rotated and hydraulic pressure is exerted downward to cause penetration of the bucket until it has been filled. Then rotation is stopped, the Kelly is raised, the bucket is tipped to the side of the hole, and a tripping mechanism is activated to open the base of the bucket and drop the retrieved soil. The process is repeated until the desired depth is reached.

The major advantages of the bucket auger are that it permits rapid excavation of deep holes of small to large diameter. A disadvantage is that it cannot be used in cohesionless materials below the water table.

1.12.6 Hollow-Stem Auger

The hollow-stem auger consists of (1) a section of seamless steel tube with a spiral flight to which are attached a finger-type cutter head at the lower end and an adapter cap at the top; and (2) a center drill stem composed of drill rods to which are attached a center plug with a drag bit at the lower end and an adapter at the top. The adapters at the top of the drill stem and auger flight are designed to permit advancement of the auger with the plug in place. As the hole is advanced, additional lengths of hollow-stem flights and center stem are added as required. The center stem and plug may be removed at any time during the drilling to permit disturbed, undisturbed, and core sampling below the bottom of the cutter head by utilizing the hollow-stem flights as casing. This also permits the use of augering in loose deposits below the groundwater table. It has been stated by some that undisturbed samples taken in this manner are better than those taken from a cased hole since the disturbance caused by advancing the auger is much less than that caused by driving the casing. Augers of this type are available with hollow stems having inside diameters from $2\frac{3}{4}$ to 6 inches. Hollow-stem augers are also particularly useful for drilling vertical holes in existing earthfill dams since they do not require the use of drilling fluid. In many instances, if boring methods requiring water or drilling mud were used, hydraulic fracturing of the material surrounding the borehole would occur.

1.12.7 Bulk Sampling in Test Pits

This method of sampling consists merely of hand or machine excavation of soils from the wall or base of a test pit without regard for the disturbance of the soil structure. It may be accomplished by hand shovel, dragline, clamshell bucket, backhoe, or other devices. The method is used primarily when a large volume of soil is required for laboratory testing. A typical example is the case of borrow materials to be used in the construction of a dam or the base course of a highway. If the material is relatively homogeneous, these samples may be taken equally well by hand or by machine. However, in stratified materials, hand excavation may be required. In the sampling of such materials it is necessary to consider the manner in which the material will be excavated for construction. If it is likely that the material will be removed layer by layer through the use of scrapers, samples of each individual material will be required and hand excavation from the base or wall of the pit may be a necessity to prevent unwanted mixing of the soils. If, on the other hand, the material is to be excavated from a vertical face, then the sampling must be done in a manner that will produce a mixture having the same relative amounts of each layer as will be obtained during the borrow area excavation. This can usually be accomplished by hand-excavating a shallow

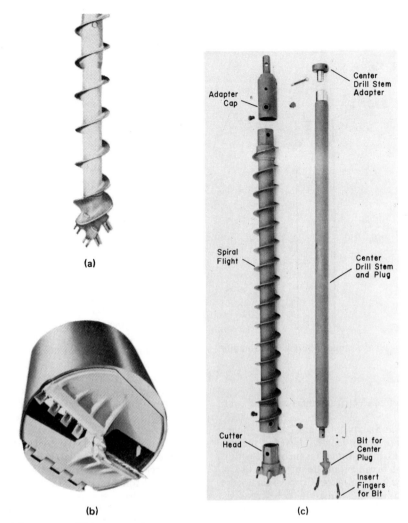

(a)

(b)

(c)

Fig. 1.28 Power augers: (a) conveyor flight auger with finger-type cutter head; (b) bucket auger; (c) hollow-stem auger. (*Courtesy of Mobile Drilling, Inc., Acker Drill Co., Inc., and Hughes Tool Co.*)

trench down the walls of the test pit within the depth range of the materials to be mixed.

1.13 UNDISTURBED SAMPLES

1.13.1 Thin-Wall (Shelby) Tube Sampler

This is the sampler most commonly used in the United States to obtain undisturbed samples. The sampler, as illustrated by Figure 1.29, consists of a thin-wall metal tube connected to a sampler head similar to that used with the split-barrel sampler. The sampler head contains a ball check valve and ports, which permit the easy escape of drilling fluid or air from the sample tube as the sample enters it. The thin-wall tube, which is normally formed from $\frac{1}{16}$- to $\frac{1}{8}$-inch metal, is drawn in at the lower end and is reamed so that the inside diameter of the cutting edge is 0.5 to 1.5 percent less than that of the tube. The exact percentage is governed by the size and wall thickness of the tube, since the sampling tube must conform to the general requirement that the area ratio of undisturbed samplers should be less than 13 percent and preferably less than 10 percent. The area ratio is given by the expression

$$\text{Area ratio (\%)} = \frac{(\text{o.d.})^2 - (\text{i.d.})^2}{(\text{i.d.})^2} \times 100$$

in which o.d. and i.d. are the outside and inside diameters of the sampling tube, respectively.

Sampler tubes usually are fabricated from cold-drawn seamless steel tubing. A major disadvantage of the use of steel for the sampler is the danger of corrosion in cases where the samples must be stored for more than a few days or weeks prior to testing. Corrosion can lead to the development of adhesion between the soil and the tube, making it difficult to remove the sample from the tube without causing disturbance. Also, in certain soils, chemical changes may occur that significantly alter the engineering properties of the soil. As a result, steel tubes are often coated with lacquer or treated with a rust inhibitor prior to use. Sampling tubes of brass and stainless steel have also been used to minimize or avoid corrosion. Stainless steel is corrosion resistant. Brass, although it offers more corrosion resistance than steel, may also corrode if samples must be retained for more than a few months. Therefore, brass tubes also are sometimes coated with lacquer. Since both brass and stainless steel tubes are expensive, it has been general practice to limit their use to special cases in which the cost can be justified.

Directly related to the corrosion problem is also the type of material used for sample tube caps. If two different metals are in contact with the soil in a tube, an electric current may be generated that will aggravate the corrosion problem through electrolytic action. Therefore, tube caps should be made of the

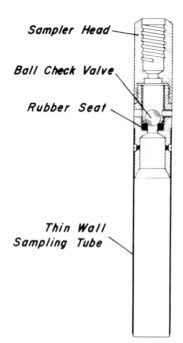

Sampler Head

Ball Check Valve

Rubber Seat

Thin Wall
Sampling Tube

Fig. 1.29 Thin-wall "Shelby tube" sampler. (*Courtesy of Sprague & Henwood, Inc.*)

same metal as the tube or of an electrically inert material such as plastic.

The most commonly used Shelby tubes are 2 and 3 inches in diameter and 24 and 30 inches in length. However, the use of tubes in the 4- to 6-inch-diameter range has increased within the past few years and some organizations require their use as a matter of standard practice. The U.S. Army Corps of Engineers recommends the use of 5-inch i.d. tubes in cohesive soils and shallow deposits of cohesionless material. In the deeper cohesionless deposits, 3-inch i.d. tubes are recommended since the penetration resistance of the larger tubes exceeds the pushing capacity of the more commonly used drilling rigs. The use of the large-diameter tubes has the advantage that it is possible to cut four small-diameter triaxial specimens of fine-grained material from the same level of a 5-inch tube. This is extremely important in soils, such as residual soils, which are not homogeneous and in which the engineering properties may vary with depth.

The thin-wall tube sampler is used primarily for sampling soft to stiff cohesive soils. In such materials sampling is accomplished by pushing the tube into the soil at a rate of $\frac{1}{4}$ to $\frac{1}{2}$ foot per second. At times the sampler is used for obtaining samples of hard cohesive soils or sand. In such instances, the wall thickness of the tube is $\frac{1}{8}$-inch and the sampler is driven into the ground. Depending upon the soil more or less disturbance may result from the driving. Upon completion of the sampling operation, the sampler is withdrawn from the hole and the tube is separated from the sampler head. Approximately 1 inch of soil for identification tests is removed from each end of the tube and the samples are placed in separate jars. Both the tube and the jars are then capped and preserved as described in Section 1.17.2 prior to shipment to the laboratory. Since cuttings from the advancement of the hole often remain at the base of the hole in spite of careful cleaning, caution must be exercised to insure that the jar sample from the top of the tube does not contain sludge, and that any sludge be removed from the top of the tube sample before capping. The jars used should be small enough to permit complete filling by the samples taken. This will prevent local moisture content changes in the samples

due to evaporation of soil moisture and subsequent condensation on the walls of the jars.

1.13.2 TAMS Thin-Wall Tube Sampler with Liners

This sampler is a modification of the Lowe-Acker thin-wall stationary piston sampler with liners (Fig. 1.31) described below (Section 1.13.4), in which the piston and actuating rod are removed and the sampler head is replaced by a ball check-valve head similar to the one used with the Shelby tube sampler. Its principal advantages over the standard thin-wall tube sampler are (1) samples are retrieved in lightweight, chemically and electrolytically inert plastic liners; and (2) the short liners minimize the amount of tube cutting and sample handling required in the laboratory.

1.13.3 Thin-Wall Stationary Piston Sampler

The thin-wall stationary piston sampler (Fig. 1.30) is basically a thin-wall tube sampler with a piston, piston rod, and a modified sampler head. The piston rod is $\frac{1}{2}$ inch in diameter and fits easily inside the hollow drill rod used to lower the sampler into the hole. Joints in the piston rod are displaced about 6 inches from joints in the drill rods. The unit may be lowered on the drill rod since a conical ball bearing catch, termed the piston rod lock on the figure, prevents the piston rod from slipping downward with respect to the head of the sampler. To prevent upward movement of the piston as the sampler is lowered into the borehole, the piston rod has a short section of left-handed threads that engages a matching section of threads in the sampler head. By rotating the piston rod counterclockwise, the rod is threaded into the sampler head and the piston is locked at the bottom of the sampler; to release the piston from the sampler, the piston rod is given several clockwise turns. This method of locking the piston during the lowering of the sampler into the borehole is required for sampling deep below groundwater or in boreholes filled with heavy drilling fluid in order to prevent the piston from rising under the fluid pressure.

The piston sampler is available with tubes having an inside diameter from $1\frac{3}{8}$ to $4\frac{7}{8}$ inches. Tube lengths for all but the largest of these range from $20\frac{3}{4}$ to 30 inches; the largest-diameter tube can be obtained with a length of 54 inches. The most frequently used size is $2\frac{7}{8}$ inches inside diameter, 30 inches long. This retrieves a sample 24 inches long.

The procedure used to obtain a sample is as follows. The sampler, with its piston located at the base of the sampling tube, is lowered into the borehole. When the sampler reaches the bottom of the hole, the piston rod is held fixed relative to the ground surface and the thin-wall tube is pushed into the soil at a rate of $\frac{1}{4}$ to $\frac{1}{2}$ foot per second by hydraulic pressure or mechanical jacking. The sampler is never driven. Upon completion of sampling, the sampler is removed from the borehole and the vacuum between the piston and the top of the sample is broken by means of a vacuum-breaking device provided for this purpose in the piston. The piston head and the piston are then removed from the tube and jar samples are taken from the top and bottom of the sample for identification tests.

The stationary piston sampler is used for sampling soft to stiff cohesive soils. The quality of samples obtained is excellent and the probability of obtaining a satisfactory sample is high. One of its major advantages is that the fixed piston tends to prevent the entrance of excess soil at the beginning of sampling, thus precluding recovery ratios greater than 100 percent. It also

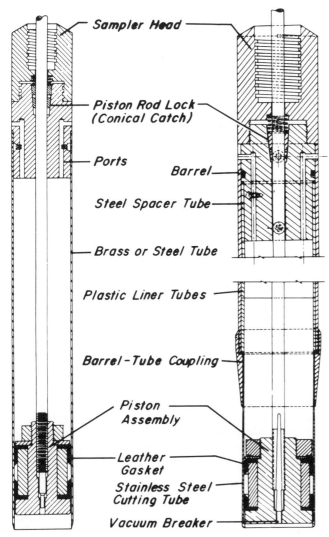

Sampler Head

Piston Rod Lock
(Conical Catch)

Ports

Barrel

Steel Spacer Tube

Brass or Steel Tube

Plastic Liner Tubes

Barrel-Tube Coupling

Piston
Assembly

Leather
Gasket

Stainless Steel
Cutting Tube

Vacuum Breaker

Fig. 1.30 (left) Thin-wall stationary piston sampler. (*Courtesy of Sprague & Henwood, Inc.*)

Fig. 1.31 (right) Lowe–Acker stationary piston sampler with liners. (*Courtesy of Acker Drill Co., Inc.*)

tends to prevent too little soil from entering near the end of sampling. Thus, the opportunity for 100 percent recovery is enhanced. The head used on this sampler also acts more positively to retain the sample than the ball valve of the thin-wall tube samplers.

1.13.4 Lowe-Acker Stationary Piston Sampler with Liners

This sampler, illustrated in Figure 1.31, consists of a sampler head, a piston and piston rod, a barrel with plastic liner tubes, a barrel tube coupling, a stainless steel cutting tube, and a vacuum breaker. The first three components are similar to the comparable parts in the standard stationary piston described in the preceding paragraph. The major differences occur in the barrel that replaces the sample tube of the standard sampler. Starting at the head, the barrel contains a steel spacer tube, a 6-inch and a 9-inch plastic liner and the upper end of a stainless steel cutting tube. These are held within the barrel by a 3-inch tapered barrel tube coupling below the bottom of which the thin-wall cutting shoe with a sharp cutting edge extends for a

length of approximately 6 inches. Because the lower 9 inches of the sampler consists of this thin-walled tube and the tapered coupling, the effect of the thicker-walled barrel section on the sampling is believed to be negligible.

The liner tubes have an inside diameter of 2.8 inches and a wall thickness of $\frac{1}{8}$ inch. They are made of laminated phenolic resin-impregnated paper and are lightweight, durable, watertight, and both chemically and electrolytically inert. Thus, they are ideal for shipment and when provided with plastic end caps may be used for the storage of samples for long periods of time without the danger of corrosion of loss of water. Because of their short length, the tubes do not have to be cut in the laboratory. Therefore, the tubes are reusable and disturbance of the sample due to cutting of the tube is eliminated.

The piston sampler with liners is used primarily to sample soft to stiff cohesive soils. It may also be used in sands and very stiff cohesive soils. However, because of its thick-walled barrel section, its penetration resistance is high and a greater thrust is required for sampling than in the case of the standard thin-wall piston sampler.

The sampling procedure is identical to that used with the standard piston sampler except for removal of the sample. Following the breaking of the vacuum below the piston and removal of the head, the barrel tube coupling is removed. The sample liners and stainless steel cutting tube are then pushed out of the outer barrel without vibration by means of mechanical or hydraulic force in such a manner as to prevent disturbance to the sample. As the cutting tube and liners emerge they are severed from the adjacent liner by means of a fine wire saw. The sample from the cutting tube may be extruded into a glass jar or into a liner tube as desired. If the sample is extruded into a liner, a small jar sample for identification tests is taken from the bottom of the tube.

1.13.5 Osterberg and McClelland Piston Samplers

The Osterberg piston sampler is a modification of the thin-wall stationary piston sampler. A schematic sketch illustrating the sampler and its method of operation is presented in Figure 1.32. The modification consists of the incorporation of a second piston, termed the actuating piston, and a pressure cylinder. Sampling is accomplished by the introduction of fluid pressure on top of the actuating piston. This pressure forces the actuating piston and the sampling tube to move down past the fixed piston until the former comes in contact with the latter. This sampler has been used primarily in the 5-inch-diameter size for the purpose of obtaining samples for consolidation tests. Two of its advantages are: (1) the sampler cannot be over-pushed, since the push stops when the actuating piston contacts the fixed piston; and (2) only one set of drills rods, the set used to hold the stationary piston, is required. The actuating rods used in the standard stationary piston sampler are eliminated by the use of the actuating piston. Pressure to operate this piston is brought from the ground surface to the sampler through the hollow drill rods. A variation of this type of sampler is Acker's GUS sampler (Gregory undisturbed sampler), which can be used with standard thin-wall tubes or with the Lowe-Acker stationary piston sampler with liners. In this case the actuating piston stops its downward motion when it passes parts which release the actuating pressure.

Another variation of this sampler has been developed by McClelland. In the McClelland variation the actuating piston is held in position with a shear pin while pressure is built up above it. When sufficient pressure has been developed, the pin shears and the sampler moves rapidly into the soil. The purpose of this variation is to permit a quick jacking of the sampler into the soil.

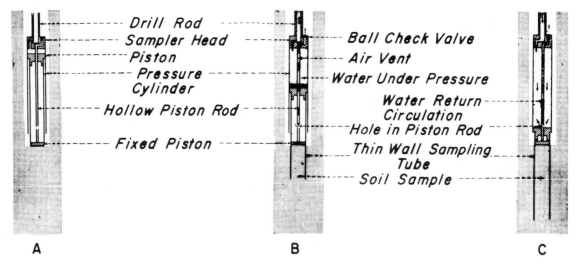

Fig. 1.32 Osterberg piston sampler.

1.13.6 Swedish Foil Sampler

This sampler, developed by W. Kjellman and T. Kallstenius, was designed to increase the length of undisturbed sample that could be obtained in a single sampling operation while still maintaining a quality of sample equivalent to that obtained by thin-wall and piston samplers in use at the time of its development. Sample lengths up to 75 feet are feasible. The ultimate goal was to permit the sampling of overburden to firm ground in a single operation. The principle on which the sampler is based is the complete elimination of sliding resistance between the sample and the inside wall of the sampler. This is accomplished by encasement of the soil in metal foil as it enters the sampler.

The principle of operation and cross sections of the sampler are shown in Figure 1.33. As illustrated, the sampler consists of a lower section, referred to as the sampler head, a series of sample barrels that are attached to the top of the sampler head, and a loosely fitting floating piston that is attached to a rod or chain extending to ground surface. The sampler head consists of a tapered lower section with a sharp cutting edge and a double-walled upper section that houses 16 rolls of metal foils. The metal foil strips run downward from rolls within the thick wall, pass through horizontal slots in the inside wall of the

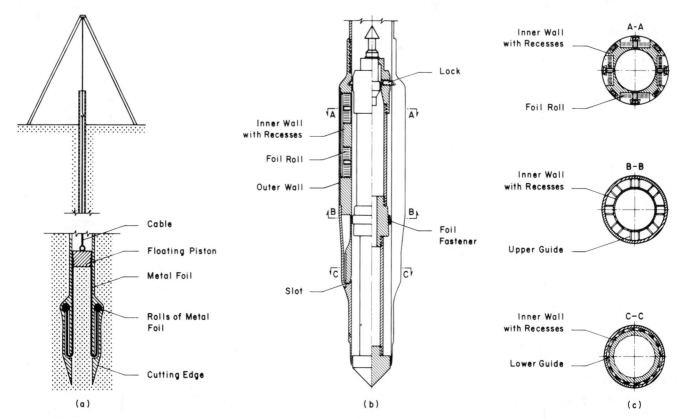

Fig. 1.33 Swedish foil sampler: (a) schematic drawing—principle of operation; (b) sampler head; (c) cross sections through sampler head. (*Courtesy of Sprague & Henwood, Inc.*)

sampler and then extend upward along the inside wall to meet the piston to which they are attached. The foils are generally made from cold-rolled mild steel; foils of tempered steel, which is stronger but more corrosion prone, are also available. The foil is available in thicknesses between 0.0012 and 0.005 inch. The maximum length of foil that can be handled by the sampler depends on the thickness; the rolls are capable of accommodating 130 feet of 0.002-inch-thick foil.

The barrel sections that are attached to the top of the sampler head and extend to ground surface are 8.2 feet long and have an inside diameter of 2.68 inches. They are connected by special couplings designed to reduce disturbance during the removal of the sample. The couplings are split longitudinally into two pieces held together at the top and bottom by conical rings locked in place by screws. This feature permits the barrel sections to be separated from one another without turning.

In general, sampling is accomplished by pushing the sampler into the ground and this is the procedure described below. The reader is referred to Kjellman et al. (1950) for other procedures and a more detailed discussion of the sampler.

Sampling by pushing is accomplished as follows. The sampler, with the piston locked to the barrel at the level of the cutting edge, is lowered to the bottom of the hole. The piston is then freed from the barrel but remains attached to a chain or rod which extends to ground surface and which is then locked and remains locked to the drill rig during the sampling drive. The latter operation forces the piston and the tops of the metal foils that are attached to it to remain at a fixed level as the remainder of the sampler is then pushed past the piston and into the ground. As the sampler penetrates, the strips of foil unwind and encase the soil as it enters the sampler. Additional sections of sampler barrel are added as required until the total push is completed. Upon completion of the push, the sampler is removed from the hole. As each section of sample barrel emerges from the hole, it is uncoupled. This leaves a short section of foil-covered sample accessible. The foils are cut and both ends of the barrel are capped. The section of soil between the barrels is also retrieved and put in a jar and all samples are preserved for shipment to the laboratory.

The Swedish foil sampler is primarily used for sampling soft cohesive soils. It is uniquely adapted for sampling of thinly stratified soils and for sampling of very soft and semiliquid soils, which frequently occur at harbor bottoms. By sampling through the very soft surficial layers and into somewhat stiffer layers below, enough friction can be developed between the foil and the stiffer soil to hold both the lower layers and the overlying very soft layers in the sampler during withdrawal from the ground. The sampler may also be used to sample stiff and dense soils and even soils containing stones. However, this requires special techniques.

1.13.7 Denison Double-Tube Core Barrel Soil Sampler

This sampler was developed by H. L. Johnson about 1939 for use on the Denison Dam Project in Texas. It, as well as the samplers described in Sections 1.13.8, 1.13.9, and 1.13.10, take samples at the bottom of a borehole following the same principles as the taking of tube samples in test pits described in Section 1.13.12. The basic components of the sampler (Fig. 1.34) are an outer rotating core barrel with a bit; an inner stationary sample barrel with a cutting shoe; inner and outer barrel heads; an inner barrel liner; and an optional basket-type core retainer. The coring bit may be either a carbide insert bit or a hardened steel sawtooth bit, depending on the material to be sampled. The shoe of the inner barrel has a sharp cutting edge. The cutting edge may be made to lead the bit by 0.5 to

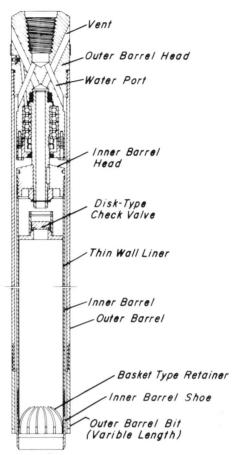

Fig. 1.34 Denison double-tube core barrel soil sampler. (*Courtesy of Sprague & Henwood, Inc.*)

3 inches through the use of coring bits of different lengths. The longest lead is used in soft and loose soils because the shoe can easily penetrate these materials and the longer penetration is required to provide the soil core with the maximum protection against erosion by the drilling fluid used in the coring. The minimum lead is used in hard or dense soils and in soils containing gravel.

The outer and inner barrel heads are connected through thrust bearings that permit the application of downward pressure on the stationary inner barrel while the outer barrel is rotated. A disk valve and a series of ports within the heads provide for venting of the drilling fluid from the inner barrel as the sample enters. The inner barrel liner, in which the sample is preserved, shipped and stored, originally consisted of a 28-gage galvanized steel cylinder with a single $\frac{1}{4}$-inch soldered lap joint flush on the outside. Today, seamless thin brass, stainless steel, and phenolic resin-impregnated paper liners, which are superior to the lap-joint liners, are more commonly used.

The exact operating procedure for the Denison sampler varies with the material being sampled and is determined by trial for each soil condition encountered. In general, the procedure is as follows. The sampler is lowered to the bottom of the hole and a downward pressure from 100 to 250 psi is applied by the hydraulic feed mechanism of a rotary drill. Circulation of the drilling fluid, which cools the coring bit and removes cuttings from the hole, is then initiated. The fluid, under pressure, is introduced to the bottom of the hole via the drill rods and the space between the inner and outer barrels. Return of the fluid to the surface is via the annular space between the wall of the hole or casing and the outside wall of the sampler. The rate of flow should be the minimum consistent with removal of the cuttings. Once this rate has been established, rotation of the outer barrel is started and continued until the sample is obtained. Downward pressure should be maintained during the entire coring operation and rotational speeds should not exceed 100 rpm in order to produce quality samples.

Upon completion of the coring operation, the sampler is removed from the hole and the outer barrel, sampler head, and cutting shoe are removed. The sample liner is then pushed from the inner barrel by hydraulic or mechanical force. Some soil will normally be retrieved in the cutting shoe. This material and approximately 1 inch of undisturbed material removed from the top of the liner are placed in jars. Both these samples, which are used for identification tests, and the liner sample are then preserved as described in Section 1.17.2.

The Denison sampler was originally constructed for taking 6-inch-diameter cores and this is still a frequently used size. Currently it is also available in sizes that produce cores having nominal diameters of $2\frac{3}{8}$, $2\frac{13}{16}$, $4\frac{3}{32}$, and $6\frac{5}{16}$ inches. Standard core lengths are 2 and 5 feet.

The Denison sampler is used primarily in stiff to hard cohesive soils and in sands, which are not easily sampled with thin-wall samplers owing to the large jacking force required for penetration. Samples of clean sands may be recovered by using driller's mud, a vacuum valve, and a basket catch. The sampler is also suitable for sampling soft cohesive soils.

1.13.8   TAMS Double-Tube Core Barrel Soil Sampler

This sampler was developed in the 1950s at a time when the Denison sampler still was available only in the 6-inch-diameter size and thin-gage galvanized steel cylinders were the most commonly used type of liner. The purposes of the new sampler were to permit the retrieval of Denison-type samples from the 4-inch diameter borings commonly used for detailed explorations in the United States and to provide a liner that was more suitable for shipment of samples and was adaptable to the sample ejection procedures used for Shelby and piston-type samples. As illustrated in Figure 1.35, the basic components of the sampler are identical to those of the Denison sampler as described above. The major differences between the samplers, other than size, are as follows: (1) a ball check-valve is used in lieu of the disk valve used in the Denison sampler; (2) the cutting edge of the inner barrel is made to lead the coring bit of the outer barrel by varying amounts through the use of cutting shoes of various lengths rather than the use of coring bits of various lengths as in the case of the Denison sampler; this provides a less expensive way of obtaining a range in the amount of lead; (3) the sampler has a 24-inch long, $\frac{1}{16}$-inch thick liner tube made of phenolic resin-impregnated paper. This tube is chemically inert. It also is easily cut by saw in the laboratory and the $\frac{1}{16}$-inch wall provides a suitable bearing edge for ejection of the specimens by jacking. The galvanized

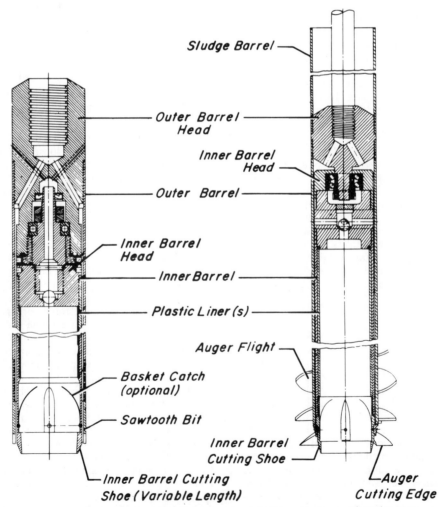

Fig. 1.35   (left) TAMS double-tube core barrel soil sampler. (*Courtesy of Acker Drill Co., Inc.*)

Fig. 1.36   (right) TAMS double-tube auger sampler with liners. (*Courtesy of Acker Drill Co., Inc.*)

liner originally used in the Denison sampler was thin-walled and, therefore, was not well adapted to removal of specimens by jacking.

The sampling technique and the areas of application of this sampler are identical to those described for the Denison sampler. However, because of the small clearance between the inner and outer barrels, clear water is used for drilling. The TAMS sampler has an outside diameter of $3\frac{1}{2}$ inches and produces a sample that is approximately 2.8 inches in diameter and 24 inches long.

1.13.9 TAMS Double-Tube Auger Sampler with Liners

The sampler shown in Figure 1.36 was developed in 1951 for the purpose of obtaining undisturbed samples of silty sand and sandy silt above the groundwater table, where use of drilling fluid would alter the natural water content of the material and perhaps the structure of the material as well. Excellent results have been obtained in the determination of the natural unit weight and water content of silty materials to a depth of 20 feet in the foundation of an airfield pavement.

The auger operates on a principle similar to that of the Denison double-tube core barrel soil sampler described herein, except that a helical conveyor rather than circulating drilling

fluid is used to remove the cuttings. The auger consists of a rotating outer barrel with a helical flight and a stationary inner barrel with liners. The inner barrel has a cutting shoe and basket-type sample retainer at its lower end and is attached, at its upper end, to a head with a ball check-valve. The outer barrel is attached to a second head, which is connected to the inner barrel head through a thrust bearing. The latter permits the inner barrel to be forced into the ground without rotation while the outer barrel is rotated to form the sample by cutting an annular space. The cutting shoe of the inner barrel is fixed and always leads the bit of the outer barrel spiral. The liners in which the sample is obtained, available in brass or phenolic resin-impregnated paper, have an inside diameter of $2\frac{3}{4}$ inches and are 6 inches long. The sampler may be operated from a relatively lightweight rig as demonstrated in Figure 1.8.

1.13.10 Pitcher Sampler

The Pitcher sampler (Figure 1.37) is basically a Denison sampler in which the inner barrel is spring loaded so as to provide for the automatic adjustment of the distance by which the cutting edge of the inner barrel leads the coring bit. The primary components of this sampler as shown in Figure 1.37a are an outer rotating core barrel with a bit and an inner stationary,

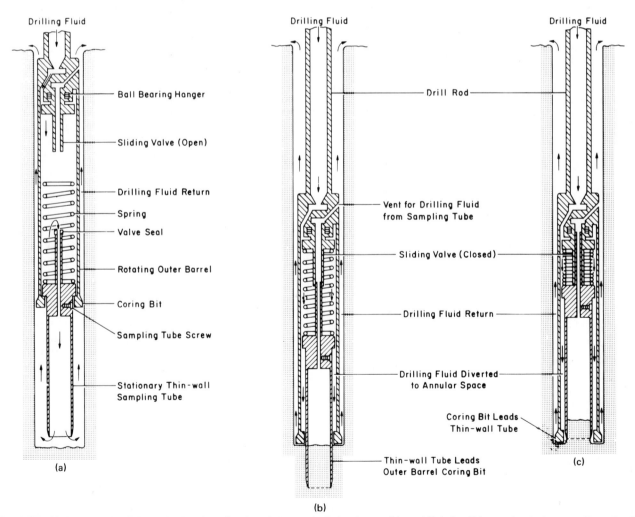

Fig. 1.37 Pitcher sampler. Schematic drawing showing: (a) sampler being lowered into drill hole; (b) sampler during sampling of soft soils; (c) sampler during sampling of stiff or dense soils. (*Courtesy of Mobile Drilling, Inc.*)

spring-loaded, thin-wall sampling tube that leads or trails the outer barrel drilling bit, depending on the hardness of the material being penetrated.

The sampling procedure and principle of operation of the sampler are as follows. After the drill hole has been cleaned, the sampler is lowered to the bottom of the hole (Fig. 1.37a). During this operation, the thin-wall sampling tube is suspended from the outer barrel and the sliding valve at the top of the barrel remains open. In this configuration, drilling fluid may be introduced through the sliding valve and it will be directed through the thin-wall tube to the bottom of the hole. Some drillers take advantage of this to flush cuttings from the bottom of the hole. However, this procedure does not provide as sure a means of removing cuttings as a cleanout auger or chopping bit with upward-deflected jets.

When the sampler reaches the bottom of the hole, the inner tube meets resistance first and the outer barrel slides past the tube until the spring at the top of the tube contacts the top of the outer barrel. At the same time, the sliding valve closes so that the drilling fluid is forced to flow downward in the annular space between the tube and the outer barrel and then upward between the sampler and the wall of the hole. If the soil to be penetrated is soft, the spring will compress slightly (Fig. 1.37b) and the cutting edge of the tube will be forced into the soil as downward pressure is applied. This causes the cutting edge to lead the bit of the outer barrel. If the material is hard, the spring compresses a greater amount and the outer barrel passes the tube so that the bit leads the cutting edge of the tube (Fig. 1.37c). The amount by which the tube or barrel leads is controlled by the hardness of the material being penetrated. The tube may lead the barrel by as much as 6 inches and the barrel may lead the tube by as much as $\frac{1}{2}$ inch. Sampling is accomplished by rotating the outer barrel at 100 to 200 rpm while exerting downward pressure. In soft materials sampling is essentially the same as with a thin-wall sampler and the bit serves merely to remove the material from around the tube. In hard materials the outer barrel cuts a core, which is shaved to the inside diameter of the sample tube by the cutting edge and enters the tube as the sampler penetrates. In either case, the tube protects the sample from the erosive action of the drilling field at the base of the sampler.

Upon completion of the sampling drive, the sampler is removed from the hole, and the inner tube, which is used to ship and store the sample, is removed from the sampler. Samples are preserved in the manner described for thin-wall tube samples in Section 1.17.2.

The Pitcher sampler is manufactured in sizes that will enter holes from $4\frac{3}{8}$ to $7\frac{3}{4}$ inches in diameter; the corresponding range of sample diameters is from $2\frac{1}{2}$ to $5\frac{7}{8}$ inches. Samplers capable of producing samples 3 to 5 feet in length are available.

This sampler is particularly adapted to sampling deposits consisting of alternately hard and soft layers.

1.13.11 Large-Diameter Double Tube Core Barrels

Core barrels of this type are used primarily to obtain cores of bedrock and, therefore, are described in detail in the section on rock coring. They are mentioned here because parts are readily available that permit the rapid conversion of these barrels to soil samplers that have proven useful in obtaining undisturbed samples of materials that could not be adequately sampled by other means. The standard conversion parts include, among others, a sample liner, a stop ring against which the top of the liner bears, and a basket-type core retainer. The bits are usually sawtooth bits faced with hard metal or insert-type bits with hard metal blades. However, in materials such as dense glacial till containing cobble and boulder-size particles and rock

containing soft clay seams, the retrieval of samples is made difficult owing to the erosion of the softer components by the drilling fluid during the slow penetration of the harder components by insert bits. In such instances diamond bits are utilized.

1.13.12 Block and Tube Sampling from Test Pits and Test Trenches

In areas where equipment is not available for undisturbed sampling or where labor is inexpensive, undisturbed samples of both granular and cohesive soils may be taken from the base of test pits or trenches by hand excavation of blocks or by pressing sampling tubes into the ground. In block sampling, successful sampling of granular materials requires sufficient cohesion or apparent cohesion due to capillarity. Excellent samples of both granular and cohesive materials may be obtained by this method if care is exercised in both the excavation and preservation of the samples. The sampling procedure is as follows. The bottom of the test pit or trench is carefully trimmed and leveled with a shovel, trowel, or other excavating tool that will not disturb the material to be sampled. A column of soil, generally square in cross section and having sides 1 inch shorter than those of the box in which the sample is to be shipped, is then formed by excavating the surrounding material to a depth $\frac{1}{2}$ inch less than the depth of the box. The box, minus both top and bottom, is then centered around the specimen and microcrystalline wax, such as that described in Section 1.16.1, is poured into the space between the sides of the sample and box and also across the top of the sample until the level of wax reaches the top of the box. The wax is allowed to congeal and the top of the box is then attached to the sides. This is done with screws to avoid the disturbance that would be caused by driving nails. After attachment of the top, the sample is severed from the ground by a spade or trowel and is turned upside down. The material at the base of the sample is then removed to a depth of $\frac{1}{2}$ inch below the sides of the box and wax is poured across the base of the sample until it fills the box. In the latter operation it is necessary to remove all foreign material from the previously poured wax to insure a good bond between it and the newly poured wax. After the wax has congealed the bottom of the box is attached with screws.

This method of sampling is advantageous in cases where large-sized samples of undisturbed material are required; where the materials to be sampled contain gravel and pebble-sized particles; and where labor is inexpensive. When compared to tube samples or core samples, block samples have the disadvantage of requiring a significantly larger amount of trimming as well as the possibility of disturbance of the material in cutting laboratory test specimens.

Sampling in test pits and trenches may also be accomplished by means of sampling tubes and cylinders. In general, thin-wall tubes are used for soils that do not contain a large percentage of coarse particles; sampling cylinders, such as a CBR mold equipped with a cutting edge, are used for coarse-grained soils. The procedure is as follows. The bottom of the pit or trench is prepared as described above. The sampling tube is aligned vertically and pushed a short distance into the soil. The soil around the tube and for a short distance below the cutting edge is carefully removed by a trowel or other hand-tool so that the cylinder of soil remaining is slightly larger than the sampling tube. Final trimming of the cylinder of soil thus exposed is accomplished by advancing the tube to the depth of excavation. This process, of alternately roughly trimming a short segment of sample by excavation and accomplishing final trimming by advancing the tube, is repeated until the soil sample protrudes $\frac{1}{2}$ to 1 inch above the tube. The sample is severed from the ground by cutting $\frac{1}{2}$ to 1 inch below the cutting edge of the tube. It is then trimmed flush with the bottom and top of the tube and the tube is capped and sealed. During the sampling process, it is essential that the tube be kept in a vertical position both to maintain the diameter of the sample and to preserve its cylindrical shape. In order to accomplish this, it is common practice to utilize a metal plate placed on top of the sampling tube to insure that the force used to push the tube into the ground is as evenly distributed as possible around the periphery of the tube. In the cases of both block and tube sampling, two jar samples also are obtained for identification tests, one each from the trimmings at the top and bottom of the sample. All samples are preserved as described in Section 1.17.2.

1.14 IN SITU SOIL TESTING IN BOREHOLES

1.14.1 Vane Shear Tests

This test is believed to provide one of the best determinations of the in situ shear strength of cohesive soils. A vane, such as one of those shown in Figure 1.38, is attached to the lower end of a drill string and pressed 1 to 2 feet into the undisturbed soil below the bottom of the borehole. Casing should be used for the borehole and should be well cleaned before lowering the drill string with the vane. The drill rods should be equipped with ball-bearing bushings about every 25 feet to eliminate frictional forces on the rods during testing. The test is performed by rotating the drill rods with a torque wrench or a more accurate torque-measuring device. Record is kept of magnitude of torque versus degrees of rotation. Usually, the torque will

reach a peak value and then drop off with continued rotation. Thus, both a peak shear strength and an ultimate and/or remolded shear strength can be obtained from the test. Shearing takes place primarily on the vertical cylindrical surface at the maximum diameter of the vanes. Shearing also takes place on the conical surfaces created by the tapered portions of the vanes at the top and bottom of the device and on the surface of the rod to which the vanes are attached. All of the above-mentioned shear surfaces should be taken into account in computing the shear strength of the soil from the torque measured. Since the shear strength determined applies primarily to shearing on a vertical surface through the soil, if a soil is varved, it would give the average strength of the various layers and not the strength of the weakest layers. Also the test does not give any information on shear strength on other than vertical planes if the soil has anisotropic shear strength properties.

1.14.2 Flat Dilatometer

The flat dilatometer was developed by Marchetti in the late 1970s in Italy. The dilatometer consists of a stainless steel blade 15 mm thick, 96 mm wide, and 150 mm long as illustrated in Figure 1.39. It has a sharp cutting edge. A 60-mm diameter stainless steel membrane is centered on and flush with one side of the blade. The dilatometer is attached to a string of drill rods and pressed, or at times driven, into the ground. A single, combination gas and electrical line extends through the drill rods and down to the blade from a surface control and pressure readout box. The dilatometer is pressed into the ground in 15 to 30 cm increments. The force or blows required to cause

Fig. 1.38 Vane shear equipment: vanes, torque wrenches, bushings for drill rods, and casing cap. (*Courtesy Acker Drill Co., Inc.*)

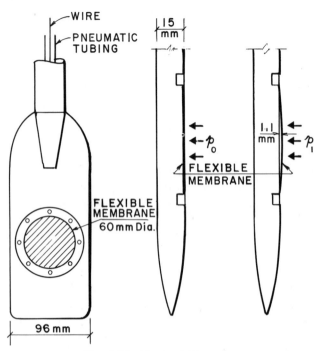

Fig. 1.39   Marchetti dilatometer.

penetration provide information similar to CPT and SPT information. At each increment of penetration, three pressures are measured: the pressure "A" required to cause the diaphragm to just lift off; the pressure "B" to cause 1.1 mm deflection of the diaphragm at its center; and upon the release of pressure, the pressure "C", which obtains when the diaphragm first becomes seated again. The observed pressures are corrected for the stiffness of the diaphragm. The corrected pressure A is designated p_0 and pressure B is p_1. Marchetti (1980) proposed the following three index parameters:

$$I_D = \frac{p_1 - p_0}{p_0 - \mu_0}, \qquad \text{Material or Deposit Index}$$

$$K_D = \frac{p_0 - \mu_0}{\bar{\sigma}_{v0}}, \qquad \text{Lateral Stress Index}$$

$$E_D = 38.2(p_1 - p_0) \qquad \text{Dilatometer Modulus}$$

where

μ_0 = in situ pore pressure
$\bar{\sigma}_{v0}$ = in situ effective overburden stress

From empirical correlations, quantative estimates of the horizontal earth pressure coefficient, K_0, the overconsolidation ratio OCR, and the tangent one-dimensional compressibility modulus $\Delta e_v / \Delta \bar{\sigma}_v$, can be obtained as well as identification of soil types. Schmertmann (1986) illustrates the use of dilatometer data for the rapid computation of settlement.

1.14.3   Self-Boring Pressuremeter

Self-boring pressuremeters were developed independently in France (PAFSOR device: Baguelin et al., 1972) and in England (Camkometer: Wroth and Hughes, 1973). These pressuremeters are an improvement over Ménard's pressuremeter in that

disturbance to the in situ soil caused by predrilling of the borehole and by inserting the pressuremeter in it is, if not eliminated, greatly reduced. The purpose of the pressuremeter test is to determine for the in situ soil the following: the lateral stress; in the case of cohesive soils, the undrained stress–strain properties; in the case of cohesionless material, the peak angle of internal friction under drained conditions. In both the PAFSOR and Camkometer devices the outside diameter of the cutting edge is exactly the same as the outside diameter of the device. The cutting edge is sharp. As the device is pressed into the soil, all material occupying the proposed test location of the device progressively enters the lower end of the device, from where it is removed by a cutting tool located a short distance back from the cutting edge. During installation of the device the inflatable membrane occupies a recess in the device's cylindrical wall. The PAFSOR device is inserted with the membrane slightly inflated so that its outside diameter is the same as the outside diameter of the rest of the device. The PAFSOR device is 215 cm long and 13.2 cm in diameter; the expandable section is 52.8 cm long and is located close to the upper end of the device. The volumetric strain that occurs during inflation is measured. The Camkometer is approximately 1.0 m long and has a diameter of 8 cm. The expandable section is 64 cm long and located midway along the length of the device. The outside diameter of the membrane in its recess is exactly the same as the outside diameter of the rest of the device. Besides measuring volumetric strain, the Camkometer device measures radial strain with three feelers equally spaced around the circumference. The pore water pressure at the face of the membrane is also measured. The self-boring pressuremeter presently offers the best opportunity for measuring in situ horizontal stress and stress history. However, it is a rather elaborate device. Discussion of interpretation of results is given by Jamiolkowski et al. (1985) and Wroth (1984).

1.14.4   Determination of Engineering Properties of Rock and Soil by Seismic Methods

In order to analyze deformations and stresses resulting from dynamic loads on subsurface soils and rocks, four dynamic parameters must be known for the material: Young's modulus (E), the shear modulus (G), Poisson's ratio (v), and the damping factor (h). These values vary primarily with the magnitude of dynamic strain, but also with the frequency and loading condition. The three most common seismic methods for measuring the above-mentioned dynamic parameters are uphole, downhole, and crosshole surveys.

In downhole surveys the energy source, which is located at ground surface, may be a wooden or metal plate that is struck with a sledgehammer or a light, shallow-buried explosive charge. The detectors, which are located in the borehole, are geophones if the hole is dry and hydrophones if the hole contains water or drilling fluid. The geophones are velocity-sensitive devices activated by movement of their supports; the hydrophones are pressure-sensitive instruments that are activated by hydrodynamic pressure changes in the borehole fluid. In uphole surveys the energy source, which is located in the borehole, may be a mechanical device consisting of a hammer and plate, an explosive charge or an air gun that initiates seismic waves by the rapid release of high-pressure compressed air. One or more geophones located at the ground surface serve as detectors. Both the downhole and uphole surveys are used to provide information on the velocities of compression and shear waves, which are determined by measuring the first arrival times of waves at the detector. The parameters E, G, and v can be determined as follows from these velocities if the mass density of the material is known (Ohya, 1986):

$$v = \left(1 - 2\left(\frac{V_s}{V_p}\right)^2\right)\bigg/\left(2 - 2\left(\frac{V_s}{V_p}\right)^2\right)$$

$$G = V_s\rho$$

$$E = 2V_s^2\rho\,\frac{(2 + v)(1 + v)}{v}$$

where

V_p = compression wave velocity
V_s = shear wave velocity and
ρ = mass density

In addition, since these waves must pass through each layer between the energy source and the detectors, the use of the downhole and uphole methods permit the detection of layers that are difficult or impossible to detect by surface refraction surveys (Section 1.6) and thereby supplement profiling information obtained in the refraction surveys. These layers include strata of lower seismic velocity located beneath those of higher velocity and strata that are too thin and/or have seismic velocities similar to those of adjacent strata. Downhole surveys also permit the determination of shear wave velocities for strata at depths that cannot be investigated by surface methods.

In the crosshole method, the energy source is placed at a selected elevation in one borehole and a detector is placed at approximately the same elevation in one or more adjacent boreholes. The energy source may be either mechanical or explosive, as described for the uphole and downhole methods. If shear wave velocities are to be determined, however, a surface vibrator or a vibrator that can be inserted into the borehole is preferred, because these sources minimize the difficulty of identifying the arrivals of the shear waves. The detectors used for crosshole surveys are velocity-type geophones. Since the travel path of the seismic waves in the crosshole survey is essentially horizontal or parallel to strata boundaries, the effect of adjacent layers on the measurements is minimized or eliminated and the seismic velocities measured are essentially for a single stratum. This is in contrast to the uphole and downhole methods wherein the effect of adjacent strata, although reduced when compared to surface refraction methods, is not eliminated to the extent that it is in crosshole surveys. Consequently, although physical properties of the rock and soil penetrated can be determined from the wave velocities measured in uphole and downhole surveys, the crosshole method has become the preferred method for this purpose. It must be recognized, however, that regardless of which method is used, the dynamic properties as determined from seismic velocities are the result of low-strain, short-duration loading and are valid only at small strains. Ambraseys and Hendron (1968) notes that the field seismic velocity is also often used to estimate the in situ modulus of rock masses. He further notes that the modulus as determined from seismic velocity is always higher than the static modulus of deformation determined from plate bearing or chamber tests because of the reasons listed above. Both Deere (1968) and Ambraseys and Hendron (1968) discuss correlation of the ratio between the static and seismic moduli with assessments of rock quality such as RQD. These correlations permit the selection of design values based on the seismic moduli, which are always too high for use in design.

Detailed procedures for the performance of uphole, downhole, and crosshole seismic surveys together with discussions of the equipment used, the interpretation of the data obtained, and the applications of the methods are given in U.S. Army Corps of Engineers (1979) and Ballard (1976).

1.15 ROCK CORING

1.15.1 General

Rock coring is the process in which a sampler, consisting of a tube (core barrel) with a cutting bit at its lower end, cuts an annular hole in a rock mass, thereby creating a cylinder or core of rock that is recovered in the core barrel or within a second or inner tube within the core barrel. Rock cores may be obtained by the use of either percussion or rotary drilling methods. Current practice is to use rotary methods almost exclusively because of the higher quality of samples obtained. Therefore, only rotary core barrels and coring techniques are described here. The reader desiring information concerning percussion coring is directed to Gordon (1958) and Hvorslev (1949).

The primary purpose of core drilling is the same as for undisturbed sampling, that is, to obtain an intact sample truly representative of the in situ material. The behavior of a rock mass is affected not so much by the type and hardness of the material composing the rock itself but more significantly by the nature of fractures in the rock. The size, spacing, and orientation of fractures, degree of weathering of fractures, and the presence of soil within the fractures are critical items. Generally the resistance of a rock mass to sliding depends on the strength of soil within fractures or shear zones. Unfortunately, the obtaining of undisturbed cores with intact fractures and shear zones is so difficult as to be generally impossible. In some instances in better rock and with the use of proper equipment and good drilling technique, close to 100 percent core recovery is achieved. For lesser-quality rock only the recently developed Integral Coring Method appears to give a reasonable chance of recovering fractures and shear zones intact.

1.15.2 Drilling Technique

The basic procedure for coring has been described in Section 1.8.10. Insofar as proper technique is concerned, it is essential that the driller understand that the primary objective is to obtain the maximum percentage core recovery and the maximum amount of information rather than to achieve the maximum rate of production. In all cases the drilling procedure to be followed is the one that brings about the highest percentage recovery and the exact procedure must be determined in the field. Variations in the speed of rotation, the downward pressure on the core barrel, the pressure at which the drilling fluid is introduced into the hole, and the length of hole drilled (run length) prior to removal of the core are major items that must be controlled by the driller. In general, coring should be initiated with short runs, both because the upper portions of the rock mass are commonly highly fractured and also because the elevations of any core losses can be more accurately determined. The length of the runs may be increased to as much as 5 or 10 feet when conditions indicate that this is permissible. However, under no circumstances should coring be continued when it is obvious that the core barrel is blocked. This can only result in a grinding down of the rock and loss of core. In zones that are highly fractured or where the barrel continually becomes blocked it is essential that short runs be used even though this means removal of the entire string of drilling tools every foot or even every few inches unless a wireline core barrel is being utilized.

In general, core barrels are operated at speeds from 50 to 1750 rpm. Essentially, the harder the rock the faster the permissible speed. However, the ultimate factor determining the speed is the amount of vibration encountered as the speed is increased. Operating speeds in good rock are indicated by

Acker (1974) to be 800 to 1200 rpm for EX and AX bits; up to 800 rpm for BX bits; and up to 600 rpm for NX bits.

Downward pressure on the bit is determined on the basis of experience. Rod vibration or "chatter" generally indicates the need for a reduced bit pressure. Coring in soft rock also requires low bit pressure.

The pressure under which the drilling fluid should be introduced into the hole is the minimum consistent with adequate removal of cuttings from the hole and proper cooling of the bit.

1.15.3 Drilling Equipment and Methods

The wide variation in the characteristics of rock materials and the conditions under which they must be sampled has led to the development of a wide variety of core barrel assemblies and special drilling methods in an effort to achieve high-quality sampling and a high percentage recovery. Efforts have been directed toward equipment that is capable of coring, within a single sampling operation, materials ranging from extremely soft to extremely hard, and equipment in which erosion of the core by the drilling fluid is minimized. Devices to improve core retention also have received attention.

The core barrels currently in use are of three basic types, namely, the single tube, the rigid-type double tube, and the swivel-type double tube. A special type of single-tube barrel, which is referred to as a shot core or calyx barrel, is also commonly used to obtain large-size cores. Core barrels are available in a variety of sizes ranging from those that produce cores having a nominal diameter of $\frac{13}{16}$ inch to those producing 48-inch and larger cores. The Diamond Core Drill Manufacturers Association (DCDMA) has established standards for the more commonly used sizes of diamond core drill equipment. These standards cover not only the core barrels and their component parts but also drill rods and casing. A portion of these standards is shown in Table 1.2; a complete description of the standards is given in DCDMA (1980).

Basically, core barrel assemblies comprise, from top to bottom, the following components, which are threaded to one another: a head section; tubular sections variously referred to as outer tubes or core barrels and inner tubes or inner barrels; liners; reaming shells; and coring bits. Core retainers or lifters are devices located at the lower end of the barrel and designed to hold the core in the barrel. The nature of the reaming shells, core retainers, and coring bits is similar for most barrels. It is the arrangement of the core retainer and both the nature and arrangement of the remaining items that distinguish one type of barrel from another. Therefore, in the discussion that follows, coring bits, reaming shells, and core retainers are described separately; the remaining elements are described as they are related to one another in each of the specific types of barrel designated at the start of this section. Also described are wireline core barrels that were developed to increase the speed of drilling; a sampling procedure, the Integral Coring Method; and calyx core drilling.

1.15.4 Coring Bits

The coring bit is the bottommost component of the core barrel assembly. It is the grinding action of this component that cuts the core from the rock mass. Four basic categories of bits are in use: diamond, carbide insert, sawtooth, and calyx or shot bits. Shot bits are described in the paragraph on calyx drilling; the others are described below.

Diamond coring bits may be of the surface set or diamond-impregnated type. In the former (Figs 1.40a and b), the

Fig. 1.40 Coring bits: (a) diamond with conventional waterways; (b) diamond with bottom-discharge waterways; (c) carbide insert, blade type; (d) carbide insert, pyramid type; (e) sawtooth. (*Courtesy of Sprague & Henwood, Inc. and Acker Drill Co., Inc.*)

diamonds are set in the metal matrix at the interior and exterior faces of the bit near its bottom and also on the bottom or cutting face of the bit. The diamond-impregnated bit, on the other hand, has small pieces of diamond embedded throughout the metal matrix of the bit. The diamonds used for both types of bit are commonly West African, processed, or Congo bortz. Surface set bits of standard design, which will provide good performance under average conditions, are readily available from drill equipment manufacturers. However, the wide variation in the hardness, abrasiveness, and degree of fracturing encountered in rock has led to the design of bits to meet specific conditions known to exist or encountered at given sites. Thus, wide variations in the quality, size, and spacing of diamonds, in the composition of the metal matrix, in the face contour, and in the type and number of waterways are found in bits of this type. Similarly, the diamond content and the composition of the metal matrix of impregnated bits also are varied to meet differing rock conditions.

Diamond coring bits are the most versatile of all coring bits since they can produce high-quality cores in rock materials ranging from soft to extremely hard. No other type of bit will produce a satisfactory core in extremely hard rock or in deposits comprising alternating layers of hard and soft rock. Compared to other types, diamond bits in general permit more rapid coring and, as noted by Hvorslev (1949), exert lower torsional stresses on the core. The latter permits the retrieval of longer cores and cores of small diameter. Comparing the two types, the diamond-impregnated type of bit is particularly well adapted for use in drilling extremely abrasive materials that can cause the dislodging of the diamonds in the surface set bit. In such materials the impregnated bit has the advantage that as the bit is worn, new diamonds are exposed to continue the cutting action. On the other hand, the surface set bit can be reset as the diamonds become worn, whereas the impregnated bit is used until it is worn out.

Carbide insert bits are of several types. Two types, the standard and the pyramid, are shown in Figures 1.40c and d, respectively. These bits use tungsten carbide in lieu of diamonds to penetrate the material being cored. Bits of this type are used to core soft to medium-hard rock. They are less expensive than diamond bits. However, the rate of drilling is slower than with diamond bits.

In sawtooth bits, shown in Figure 1.40e, the cutting medium comprises a series of teeth that are commonly cut into the bottom of the bit. The teeth are faced and tipped with a hard metal alloy such as tungsten carbide to provide wear resistance and thereby to increase the life of the bit. These bits have the advantage of being less expensive than diamond bits. However, they do not provide as high a rate of coring and do not have a salvage value. The sawtooth bit is used primarily to core overburden and very soft rock.

An important feature of all bits that should be noted is the type of waterways provided in the bits for the passage of the drilling fluid. Bits are available with so-called "conventional" waterways, which are passages cut on the interior face of the bit (Fig. 1.40a), or with bottom discharge waterways, which are internal and discharge at the bottom face of the bit behind a metal skirt separating the core from the discharge fluid (Fig. 1.40b). Bottom discharge bits should be used when coring soft rock or rock having soil-filled joints to prevent erosion of the core by the drilling fluid before the core enters the core barrel.

1.15.5 Reaming Shells

The reaming shell is a metal sleeve, threaded at both ends, which serves as a coupling between the core barrel and the bit. It is slightly larger in diameter than the core barrel and its surface is set with diamonds, has insert strips with diamonds or has carbide insert strips. These are set to a diameter slightly larger than that of the coring bit. The shell thus reams the hole to enlarge the passageway for return of the drilling fluid to the surface and thereby serves to maintain the gauge of the hole and to reduce friction on the sides of the core barrel as well as wear on the sides of the bit as the barrel is moved in and out of the hole.

1.15.6 Core Lifters or Retainers

There are two devices commonly used to retain the core as the core barrel is removed from the borehole. These are the split-ring core lifter and the basket retainer. The split-ring lifter is a tapered split ring of tempered steel that is fluted on either its interior of exterior surface. It is held in place by the tapered inner face of the coring bit (Fig. 1.41) or the core lifter case (Fig. 1.43). As the core passes through the bit, the core lifter spreads to permit the core to enter the recovery tube. When the barrel is withdrawn from the hole and the core tends to slip down, the tapered shape of the core lifter causes the lifter to jam between the barrel and core and so grip the core. This type of lifter is used primarily to retain cores of sound rock.

The basket retainer, shown as a component of the Denison core barrel in Figure 1.34, comprises a base ring to the periphery of which are welded curved strips or fingers of spring steel. These fingers initially rise vertically and then curve toward the center of the ring. The steel used may be stiff or extremely flexible. Retainers with stiff fingers are used when soft rock and dense or hard soil are being cored; flexible fingers are used when extremely soft or fine-grained soils are cored with barrels such as the Denison. The basket retainer may be held in place by the tapered inner face of a bit or may be set in a recess at the upper end of the bit, which holds it in place against the bottom of the core recovery tube. In operation, the fingers of the basket retainer spread to permit the core to enter the recovery tube as the core barrel is advanced. If the core tends to slip out as the sampler is removed from the hole, the fingers dig into the core and retain it.

1.15.7 Single-Tube Core Barrels

These are the most rudimentary, the least expensive, and the most durable of the core barrels. As shown in Figure 1.41, the barrel from top to bottom consists of a core barrel head, a core barrel, a reaming shell, a core lifter, and a coring bit. Each component, except the core lifter, is threaded to the piece immediately above. The core lifter is held in place by the tapered inner surface of the coring bit and by the bottom of the reaming shell, against which it bears when the core is passing through it.

In operation, the single-tube barrel is rotated as a downward force is applied and drilling fluid is introduced to the hole under pressure. The fluid flows from the drilling rods into the core barrel tube where it passes between the core and the walls of

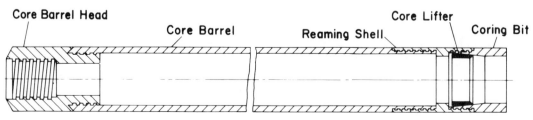

Fig. 1.41 Single-tube core barrel. (*Courtesy of Sprague & Henwood, Inc.*)

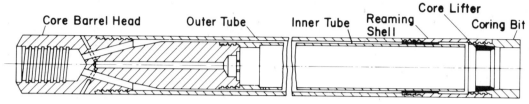

Fig. 1.42 Rigid type double-tube core barrel. (*Courtesy of Sprague & Henwood, Inc.*)

the core barrel to the bottom of the bit. From there it flows upward in the drill hole, between the barrel and the walls of the hole, carrying the cuttings to the surface. In some cases a piece of tube, called a sludge barrel or calyx, is threaded to the top of the core barrel head. This serves to collect the larger particles of the cuttings, which tend to drop from suspension as the drilling fluid flowing from the small annular space around the barrel enters the larger annular space around the drill rods above and has its velocity reduced.

Single-tube core barrels of this type are available in sizes ranging from EWX size (Table 1.2) to 6½ inches o.d. and in lengths of 2, 5, and 10 feet. The smallest of the barrels is referred to as a starting barrel. This barrel is required when a hole is started directly in rock because the clearance between the chuck of the drilling rig and the ground surface normally is not adequate to accommodate a larger barrel.

The main disadvantage of the single-tube core barrel is that the entire core is subject to the erosive action of the drilling fluid as it passes through the core barrel. Because of this, the use of the single-tube barrel should be restricted to hard rock that is unaffected by the flow of the fluid.

1.15.8 Double-Tube Core Barrels

The double-tube core barrel was developed to minimize the erosive action of the drilling fluid on the core and thereby to improve core recovery. Double-tube core barrels fall into two basic categories, the rigid type and the swivel type. A variation of the swivel type known as the Series "M" barrel is extensively used.

The rigid type double-tube barrel (Fig. 1.42) comprises a core barrel head, an outer barrel, an inner core recovery tube, a reaming shell, a core lifter, and a coring bit. In this barrel, water passages in the head are so arranged that the drilling fluid reaches the coring bit via an annular space between the inner and outer tubes, both of which are rigidly attached to the head. In operation, both the inner and outer tubes turn when the rotary force is applied. As the coring bit at the bottom of the outer tube cuts the core, the core passes through the core lifter and into the inner tube, which protects it from erosion by the drilling fluid. Holes at the bottom of the inner tube provide for a small flow of drilling fluid between the core and the wall of the inner tube in order to minimize the possibility of accumulating cuttings that would cause friction between the

core and the tube and thereby cause large torsional forces to be applied to the core.

The rigid type double-tube barrel is available in sizes from EWX to NWX with barrel lengths of 2, 5, 10, 15, and 20 feet. The major advantages of this type of barrel are that, in general, it provides for protection of the core against the erosive action of the drilling fluid and will provide a higher percentage recovery than the single-tube barrel. It is used primarily in medium to hard rock that is not highly fractured.

The swivel type double-tube barrel differs from the rigid type in that the inner tube remains stationary while the outer barrel is rotated. This minimizes the possibility of core disturbance through torsional forces and thereby improves recovery. Two forms of the swivel type barrel are available, the conventional and the Series "M" barrels. The upper portions of both barrels are similar and as illustrated in Figure 1.43 comprise an outer rotating tube, an inner stationary tube, and outer and inner tube heads. As in the rigid type, water passages direct the flow of the drilling fluid into the annular space between the two tubes and vents provide for the exit of water from the barrel. The inner tube assembly is suspended from the outer tube head in such manner that a downward force can be applied to both tubes while only the outer tube is rotated. In the sampler illustrated, this is accomplished by the use of ball bearings; roller bearings also may be used.

The conventional and the Series "M" barrels differ from each other in the arrangement of the lower portion of the barrel. In the conventional barrel the reaming shell, core lifter, and coring bit are as shown for the rigid barrel in Figure 1.42. In this case the inner barrel terminates above the core lifter which has a 5° taper and operates within the bevel section of the bit. The configuration for the Series "M" barrel is as shown in Figure 1.43. Here, the core lifter, which has a 2.5° taper and is thinner than the conventional lifter, operates within a lifter case attached to the bottom of the inner tube. This case has a thin-wall bottom that extends almost to the cutting face of the coring bit. The arrangement used in the Series "M" barrel makes this barrel superior to the conventional barrel in two respects. In the conventional barrel the lifter may tilt and block the entrance to the inner tube or the lifter may rotate with the bit and cause grinding of the core. In addition, the portion of the core from the bottom of the bit to the bottom of the inner tube is exposed to the erosive action of the drilling fluid. In the Series "M" barrel the core is in the inner tube and somewhat aligned before it encounters the lifter, and the lifter remains

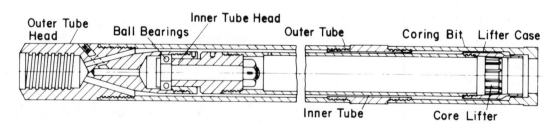

Fig. 1.43 Swivel type double-tube core barrel, series "M" with ball bearings. (*Courtesy of Sprague & Henwood, Inc.*)

oriented because the inner barrel moves little, if at all. Therefore, the possibility for blocking or grinding is minimized. The effective extension of the inner barrel to the cutting face of the bit by the lifter case also minimizes the area of core exposed to the drilling fluid.

Conventional and the Series "M" barrels are available in sizes ranging from EWX and EWM to NWM, respectively (see Table 1.2). They may be obtained in lengths of 2, 5, 10, 15, and 20 feet.

The conventional barrels are used in coring fractured or broken formations that are of average hardness and that are not excessively susceptible to erosion. The Series "M" barrels are particularly adapted to achieving a high recovery in badly fractured or broken strata or in soft and friable formations that are easily eroded.

1.15.9 Large-Diameter Double-Tube Core Barrels

In addition to the double-tube core barrels discussed above, there are available large-diameter barrels that are similar in construction and operation to the Series "M" barrels. The large barrels are available in the following nominal sizes: $2\frac{3}{4}$ inch i.d. $\times$ $3\frac{7}{8}$ inch o.d., 4 inch i.d. $\times$ $5\frac{1}{2}$ inch o.d., and 6 inch i.d. $\times$ $7\frac{3}{4}$ inch o.d. Actual dimensions are given in Table 1.2. The two largest of these are equipped with a sludge barrel attached to the outer barrel head in order to collect large particles that are too heavy to be carried to the surface by the return flow of the drilling fluid. These particles tend to settle out of suspension and often cause barrels to be wedged in boreholes.

In general, the larger the size of a core the better the recovery. Consequently, the large-diameter barrels are frequently used when highly erodible material such as soft or friable rock is to be cored. Conversion units, which were described in Section 1.13.11, are available to permit the use of these barrels to obtain undisturbed soil samples in liners. Therefore, they may be used when it is necessary to obtain cores of overburden containing large particles and when large-diameter samples are required for testing.

1.15.10 Wireline Core Barrels

The use of the core barrels described above requires the removal of the entire string of drill rods from a borehole whenever it is necessary to remove core from the barrel. In the drilling of deep holes, this is an extremely time-consuming operation, which is partially eliminated by the use of the wireline core barrel (Fig. 1.44), introduced in 1954. The wireline core barrel assembly consists of an inner barrel assembly that can be retrieved independently of the outer barrel assembly through the required special drill rod. The outer barrel assembly consists of a tube at whose lower end is a reaming shell that couples the tube to a diamond coring bit. The upper end of the tube is attached to the special large-diameter wireline drill rods. The retrievable inner barrel assembly consists of the core recovery tube with a core lifter at its lower end and a swivel type ball-bearing head at its upper end. Attached to the top of the swivel head is a locking head with a spearhead at its upper end to permit retrieval of the barrel by a device called an overshot. The overshot has at its upper end a socket attaching it to a wireline and at its lower end a lifting dog that grips the spearhead during the retrieval operation. An optional feature of the wireline barrel is a water shut-off valve, which is made a part of the inner barrel assembly. This valve is a rubber

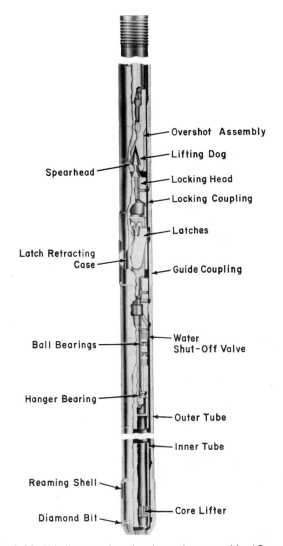

Fig. 1.44 Wireline core barrel and overshot assembly. (*Courtesy of E. J. Longyear.*)

washer which, when a core block occurs, is squeezed out to fill the annular space between the outer and inner barrels. This causes the pump pressure to rise, thereby alerting the driller to the core block and permitting unnecessary grinding of the core to be averted.

In operation, the outer barrel, to which the inner barrel assembly is locked, is rotated to cut the core. When the run has been completed or a block occurs, the hole is flushed and the drill string is broken at the first joint above ground. The overshot is lowered on the wireline and the lifting dog grasps the spearhead at the top of the inner tube. When the wireline is pulled, the latches that couple the inner and outer barrel assemblies are disengaged and the inner barrel assembly may be lifted to the ground surface. The inner tube assembly may be returned to the bottom of the hole in several ways. In a dry hole it must be lowered with the overshot. In holes filled with water the assembly may be dropped to the bottom. However, in deep holes and relatively flat inclined holes water circulation is used to pump the assembly into place. The drill string and outer barrel assembly are removed only when it is necessary to replace a bit.

Wireline core barrels are available in sizes that produce cores $1\frac{1}{16}$, $1\frac{7}{16}$, $1\frac{7}{8}$, $2\frac{1}{2}$, and $3\frac{11}{32}$ inches in diameter (see Table 1.2). Core barrel assemblies are available in 5-, 10-, and 15-foot lengths.

In addition to increasing the rate of progress, the wireline barrel has the following advantages: (1) in formations that are prone to caving, the danger is reduced since the drill string and outer barrel are not removed after each run and the opportunity for loosening of material in the walls of the hole is thereby decreased; (2) bit life is increased by reducing the number of times the bit must core through caved material on reentry into the hole; and (3) if the barrel has the optional water shut-off valve, grinding of the core is decreased and recovery thereby increased.

1.15.11 Calyx or Shot Core Drilling

This method was once used extensively to obtain cores having a wide range of sizes. In current practice it has been replaced to a large extent by the methods previously described and its primary application today is in the drilling of holes having diameters of 30 inches and larger. In this method of drilling, single-tube core barrels are used exclusively. The cutting action is provided by a slotted bit of mild steel and by very hard steel shot, which is fed into the drill hole with the wash water and reaches the bit via the annular space between the core and the wall of the barrel. The single-tube barrel consists of a heavy-walled steel tube with a thick circular steel plate near its upper end. This head plate has a hole at its center for passage of the drilling fluid and several bolt holes to permit the heavy drill pipe to be flange-coupled to it. A thin plate slightly smaller in diameter than the inside diameter of the barrel and located slightly below the head plate is held to the head plate by spacer bolts. This plate acts as a deflector for the entering wash water and causes it to carry the steel shot to the periphery of the core rather than permit it to drop the shot on top of the core. The barrel has a mild-steel bit, $\frac{7}{8}$ to $1\frac{3}{8}$ inches thick, which is formed by welding a second tube to the interior wall of the lower portion of the barrel or by welding a section of heavy-walled pipe to the base of the barrel. Slots cut into the bit at its lower end facilitate the movement of the shot to the bottom and outside of the bit.

In operation, as the barrel is rotated, the shot, which becomes wedged beneath and around the slotted bit, is crushed into abrasive particles. These particles, some of which become embedded in the mild steel bit, provide the cutting action required. Two critical items regarding the cutting are the rate of feed of the shot and the rate of flow of the wash water. Too great an amount of shot will cause the bit to ride on the shot without cutting; too little shot will yield a less than optimum rate of advance. Insofar as the wash water is concerned, too great a rate of flow will cause the shot to be carried away from the bit and too low a rate will hinder removal of the cuttings. In connection with the removal of cuttings there is still another problem immediately above the barrel. In this area, the cross section of the hole becomes so large that the velocity of the water is significantly reduced and the cuttings drop from suspension. Since a velocity adequate to carry the cuttings to the surface would carry the shot away from the bit, the cuttings are provided for by attaching to the top of the barrel a tube called a calyx or sludge barrel, into which the particles fall.

The core barrel used in drilling the large-diameter calyx hole does not have a core lifter. The core may be retrieved in one of two ways. In the first, a special barrel equipped with a core lifter is used. The lifter operates on the same principle as the tapered split-ring core lifter but is not identical to it. The barrel also has, at its lower end, provisions for the placement of powder charges and contains a conduit for carrying blasting wires to the top of the barrel. The charges are used to break the core from the rock mass. In the second method, the core is broken loose by rotating the core barrel after wedging the core to the barrel. The core is removed by means of a lifting hook inserted in a small hole drilled in the center of the core.

The primary advantage of the large-diameter calyx hole is that it provides a hole that will permit the entrance of men for detailed inspection of subsurface material in situ. It also provides a larger core for inspection and, in general, as a result of the larger size, a higher percentage recovery. However, it is an expensive method of exploration and is used only in special circumstances.

1.15.12 Integral Coring Method

This sampling technique was developed by Dr. Manuel Rocha (1971a and 1971b) in response to the need for rock cores truly representative of in situ rock masses, including their discontinuities such as open, tight, or clay-filled joints, shear zones, and cavities. The method consists of taking a core at whose center is a steel rod or pipe previously inserted and bonded to the rock mass to hold the mass together during the coring operation. This technique, when properly used, produces 100 percent recovery and provides oriented cores. The method may be used to obtain cores throughout the entire length of a boring or at selected locations. The procedure used is as follows. A hole with a minimum diameter of 7.5 cm is drilled to the depth at which integral coring is required. A second, small-diameter hole is then drilled coaxially with the first hole and extending from the bottom of the first hole to a depth equal to the length of the required core. The diameter of the second hole must be large enough to accommodate a rod or pipe of adequate stiffness to minimize deformation of the core during the coring but should be as small as feasible to increase the thickness of the annular sample recovered. The majority of integral coring done to date has been accomplished using boreholes 7.5 cm in diameter with reinforcing pipe holes 2.6 cm in diameter. The length of the coring run must be as long as possible to minimize the cost but it must not be so long as to result in difficulties due to hole deviation. Although the maximum core length to date has been 1.5 m, difficulties with appreciably longer lengths are not anticipated.

Upon completion of the inner small-diameter hole, the reinforcing pipe coupled to positioning rods by a special connecting element is lowered into the smaller hole. The pipe, which has been notched at the top and bottom, is carefully positioned so that these notches are in a known direction, thereby providing for core orientation. The pipe used is commonly 5 mm smaller in diameter than the hole. It is bonded to the rock mass by cement grout or epoxy resin grout injected through the drill rods and the pipe itself, which is perforated. The binding agent is often dyed to enable it to be distinguished from the core which it may penetrate. Following injection of the grout, the positioning rods and the special connecting element, which permits the rods to be detached from the reinforcing pipe, are removed as soon as practicable so that they do not become stuck in the hole as the grout sets. Once the grout has set sufficiently, a core of rock with the reinforcing pipe in its center is taken by overcoring with a standard core barrel. Satisfactory results have been achieved with both single- and double-tube core barrels. However, as in the commonly used coring methods, the double tube is preferred when coring poor rock.

This method has proven to be highly successful. Its primary disadvantage is the increased time required for drilling and the related higher costs. In many instances, however, the increased information obtained readily justifies the additional time and cost.

1.15.13 Oriented Core

The solution of problems in rock mechanics, such as the stability analysis of slopes and the design of tunnels in rock, is dependent on proper characterization of the rock mass involved. This requires that the geotechnical engineer plan the subsurface exploration program to determine to the best extent possible, not only the nature, depth and lateral extent of the various rock strata present but also the nature and extent of discontinuities within the mass. Information is required concerning the frequency and orientation of such discontinuities as joints, shear zones, and fault zones, the degree of weathering along the discontinuities, and the nature of materials filling the joints if such materials are present. The Integral Coring Method described above offers one method of determining the orientation, that is, the strike and dip of these discontinuities, in addition to ascertaining the nature of any joint filling present. Other methods of determining the orientation of discontinuities are in use as well. Among these are the use of scribed core as produced with the Christensen–Hugel core barrel, the Craelius orientation device, and marking of the top of the core with chopping bits.

Orientation with the Christensen–Hugel core barrel is accomplished by producing rock core with three continuous, longitudinal grooves together with a multishot film record of the orientation of these grooves as the rock is cored. Scribing of the core is accomplished by tungsten carbide knives spaced 90°, 120°, and 150° apart around the circumference of the shoe of the inner tube of the core barrel. Orientation is provided by a magnetic survey instrument that consists of a compass, a unit to measure the angle of inclination of the hole, a camera, a camera timer, and a battery pack housed in an upper inner barrel section. Both the core barrel and a minimum of 10 feet

Fig. 1.45 Core goniometer used for direct measurement of dip direction, dip angle, and strike from scribed core. (*Courtesy of Boyles Bros. Drilling Company.*)

of drill rod above the barrel are made of nonmagnetic material to preclude interference with the survey instrument. In operation the reference or master groove, which is the groove farthest from the other two grooves, is aligned with the magnetic survey instrument to provide orientation. As coring proceeds, the inclination of the hole and the bearing of the master groove on the core are recorded on film at a preset short time interval. The film record thus obtained provides the basis for a near-continuous log of the inclination of the hole and the strike of the master groove. This log is used to physically orient each core piece so that the strike of the geologic discontinuities present may be determined. The core goniometer (Fig. 1.45), a device that permits a direct, accurate measurement of the dip and strike of joint and bedding planes, shear zones, contacts, and foliation, is often used to simplify these measurements. Each piece of core is clamped in the core holder so that the angle of inclination of the core and the direction of the master groove are as shown on the log. The sighting ring is then aligned with each joint plane, bedding plane, or other feature under consideration. The dip direction and strike of a feature are read from the base ring; the dip angle is read from the protractor.

The Craelius device for core orientation consists of a metal housing containing a spring-loaded conical probe and six metal prongs. The device is clamped to the lower end of the core barrel by spring-actuated teeth. The barrel, with the device attached, is then lowered into the borehole and when the prongs of the device come in contact with the rock at the bottom of the hole, they move to define the profile of the rock surface. The prongs are locked in place when the conical probe, which is spring-loaded, comes in contact with the rock and is released. At the same time, the device is released and moves up in the core barrel as the next core is taken. Upon completion of coring, the device, whose orientation was fixed at the bottom of the hole by a special marking mechanism within the metal housing, is matched to the profile at the top end of the core and the core is thus oriented.

Core orientation in its simplest and least-expensive form consists of marking the rock in the bottom of the borehole immediately prior to each core run. This can be accomplished by using a straight chopping bit and marking the orientation of the blade on the drill rods as they are lowered into the ground. Once the bottom of the hole is reached, the bit is aligned in a known direction using the marks on the drill rods and the chopping bit is hit with a hammer to mark a straight line on the rock. This method is by no means considered the equivalent of the methods described above and it has many limitations. For example, if the rock is too hard it may not be possible to achieve a clearly visible mark, and if it is too soft it may break from the impact of the bit. In addition, if the rock at the bottom of the hole has an inclined surface, there will be only point contact between the bit and the rock surface and it will not be possible to mark a line on the surface. Nevertheless, under those circumstances in which it will work, the method provides a relatively inexpensive procedure for the orientation of rock core.

1.16 MISCELLANEOUS EXPLORATORY TECHNIQUES

1.16.1 Borehole Photography

The NX borehole camera, developed by the U.S. Army Corps of Engineers (1960), is designed to photograph the interior surfaces of boreholes for the purpose of detecting bedrock imperfections that may have a significant bearing on design but may not be revealed by normal coring procedures because of poor core recovery. The camera provides a continuous record of a borehole in the form of undistorted cylindrical photographs that are of such a quality that slight changes in color and fractures as small as 0.01 inch are detectable. These photographs may be obtained in dry or water-filled holes. In the case of water-filled holes, however, the depth is limited to that at which the hydrostatic head is 500 feet.

The camera housing is a stainless steel tube approximately 33 inches long and 2.75 inches in diameter, with a quartz window extending over its entire circumference 5 inches above its lower end. The basic components of the camera, from bottom to top, comprise an oil-damped compass, a hollow truncated conical mirror, a high-voltage circular flash tube, a motor-driven conventional 16-mm motion picture camera with a 15-mm lens, and a power condenser and relay unit. The hollow truncated mirror is supported on the oil-damped compass, which is so positioned that the mirror is directly opposite the quartz window. The mounting of the mirror is such that when viewed from above the compass face is visible in the center of the torus formed by the sloping faces of the mirror. In operation, the camera focuses on the mirror and photographs the face of the compass and the mirror image of the wall of the borehole. Illumination of the wall is provided by the flash tube located midway between the mirror and the camera lens. Both the film advance and the flash are activated by a current-pulsing device on a camera-lowering rig at ground surface. The camera photographs a 1-inch length of borehole with every exposure and makes 16 exposures per foot of borehole. Thus, an overlap is provided from one photo to the next.

A three-conductor armored cable is used to lower and raise the camera. The rate of movement is controlled by a lowering device at ground surface. This device consists of a pay-out reel with a hand crank and gear drive, and a cable guide reel geared to a 16-notch pulsing wheel and depth counter. The pay-out reel has a capacity of 500 feet of camera cable. Electrical power may be provided from available sources or by a portable 117-volt a.c. generator. A small power supply cabinet on the lowering device transforms the 117-volt a.c. circuit to one d.c. circuit of 450 volts and two of 50 volts. The highest voltage is transformed to 15 000 volts in the flash tube circuit. One of the 50-volt circuits operates the camera motor drive and synchronizing relay; the second provides the synchronizing pulse.

Standard practice is to take the photographs as the camera is withdrawn from the hole because the camera is steadier on ascent than on descent. It is also standard practice to lower into the hole a dummy camera of the same dimensions and weight as the actual camera prior to inserting the actual one. This is to insure that the camera will not become stuck as a result of surface irregularities on the wall of the hole.

The photographs obtained show the compass in the center of a torus containing the image of the borehole walls. They are viewed as still photographs on a flat or cylindrical screen. The flat screen is generally used for editing the film; the cylindrical screen is used for viewing the photo as one would view a core. A modified Eastman Recordak Model PM microfilm viewer is used for projection.

The primary advantages of borehole photography are: (1) information can be obtained concerning in situ conditions in zones where there has been no recovery or core recovery is so poor as to preclude proper interpretation of conditions; and (2) the cost of borehole photography is less than the cost of explorations such as large-diameter calyx holes or adits, which, until the development of the Integral Coring Method, would be required to give the same level of reliability to the subsurface information. Disadvantages are the limited depth of focus of the field of view and difficulties with the clarity of water in the drill hole.

1.16.2 Television

Television cameras capable of being lowered into an NX boring have been used to observe in situ rock conditions by display on a closed-circuit TV screen for live viewing and by recording on video tape for replay. It is also possible to obtain photographs of the walls of a boring by this method if a camera is mounted on the receiving monitor. Cameras of various design are available. One camera, manufactured by IBAK of Kiel, West Germany, has an outside diameter of 3 inches and utilizes a 1-inch vidicon tube, which by switching heads can be made to view axially, radially, or with a 180° lens opening. The tube can also be fitted with a 50-mm telescope zoom lens with a powerful miniature floodlight. This feature permits the measurement of underground distances, since the time required for focusing to various distances is known. The size and depth of limestone solution caverns in the left abutment at Gathright Dam, Virginia, were estimated by this method.

1.16.3 Submarine Sampling

Offshore explorations are generally accomplished by drilling rigs mounted on rafts, barges, pile-supported platforms and, occasionally, movable oil-well drilling platforms. In addition, many samplers have been developed for sampling of surface and near-surface sediments beneath the sea (Noorany, 1985). These include free-fall samplers and samplers driven or pushed into the sea bottom by rigs placed on the ocean floor and operated by divers or remotely controlled from the surface.

Samplers operating on the free-fall principle include gravity corers and piston samplers. The gravity corer is basically a sampling tube that is weighted to cause penetration and that has a retainer at its lower end to prevent loss of the sample as the tube returns to the surface. These samplers, lowered from the ship by a wrench, have release mechanisms that are triggered as the sampler approaches the ocean floor and allow the sampler to free-fall into the sediment. The Boomerang Sampler (Noorany and Gizienski, 1970) is a self-contained unit that consists of a core tube inside a coring barrel that has a weight and a sample retainer at its lower end. The barrel is surmounted by a ballast unit and a float unit consisting of two glass spheres and a hollow rubber ball. This sampler is dropped from the ship and allowed to fall freely for the entire water depth. The impact of the sampler triggers a release mechanism that permits the core tube and the float unit to float to the surface. A flashing light in one of the glass spheres marks the location of the sampler.

Another free-fall sampler is the N.G.I. gas-operated piston sampler (Anderson et al., 1965). This is a torpedo-shaped sampler approximately 9.8 feet long and 8.7 inches in diameter at its widest part. The sampler, from top to bottom, consists of a cap with a lifting hook, a propulsion section containing a gas generator, and a main body section which houses a 2.1-inch i.d. sampling tube with a movable piston at its upper end and a piston rod that threads into the propulsion unit at its upper end and has a fixed piston at its lower end. In operation, the sampler is dropped from a ship and falls freely into the sediments at the ocean floor under the influence of gravity. When the penetration has ceased, a time switch, preset to complete a circuit after penetration, causes a charge of rocket propellant to ignite and generate gas, which drives the sampling tube approximately 5.4 feet past the fixed piston. The sampler is then retrieved by means of a line attached to the lifting hook. This sampler is designed to operate in depths of water up to 1150 feet. Its depth of penetration depends, as with all free-fall samplers, on the available amount of kinetic and potential energy and the resistance of the soil penetrated.

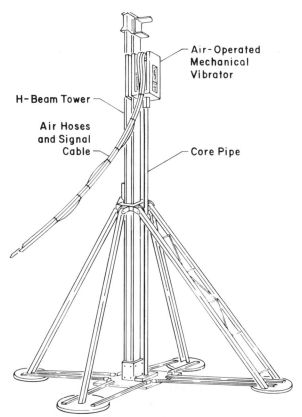

Fig. 1.46 Vibracore drill for underwater sampling. (*Courtesy of Alpine Geophysical Associates, Inc.*)

Among the rigs that are placed on the ocean floor to obtain samples is the Vibracore (Fig. 1.46). The principal elements of this rig are an H-beam tower supported by four legs with foot pads; an air-operated mechanical vibrator; a 4.0-inch i.d. by 4.5-inch o.d. core pipe containing a $\frac{1}{8}$- or $\frac{1}{4}$-inch thick plastic liner; and a portable penetrometer. The rig is powered by compressed air supplied at 100 to 125 psi by a shipboard compressor having a minimum capacity of 300 ft^3/min. The air is supplied to the vibrator via a $1\frac{1}{4}$-inch hose and vented to the atmosphere by two similar hoses.

In operation, the Vibracore is lowered to the ocean or river bottom by means of a crane or davits. Once the rig is in position, air is fed to the vibrator and the core pipe is driven for a length of 20 or 40 feet, depending on the size of the rig being used. During the sampling, the penetrometer on the rig measures the rate and amount of penetration and transmits this information via a signal cable to an onboard strip-chart recorder. Upon completion of the sampling operation, the entire rig is lifted to the deck of the barge for extraction of the plastic liner.

The rig is designed to sample sediments that cannot be penetrated by the more commonly used sampling devices. The time required for a complete cycle of lowering, sampling, lifting, and sample extraction depends on the material sampled. The depth to which the unit may be used is limited by the length of the compressor lines, the size of the compressor, and the penetrability of the soil.

1.17 PRESERVATION, SHIPMENT, AND STORAGE OF SAMPLES

1.17.1 Sealing

Samples, both disturbed and undisturbed, must be carefully sealed and packed prior to shipment to the laboratory. Sealing

is best accomplished with a microcrystalline wax such as Socony Vacuum's Mobil Wax No. 2300 or 2305, Gulf Oil Corporation's Petrowax A, Humble Oil Company's Microvan No. 1650 or Standard Oil Company's Eskar 50. The U.S. Army Corps of Engineers has found that a mixture of 50 percent microcrystalline wax and 50 percent paraffin is also satisfactory. The use of paraffin alone should be avoided because it becomes brittle and cracks easily, thereby breaking the seal. It is best to apply the wax at a temperature just above the congealing point. This is particularly true when sealing undisturbed samples, since the wax, which comes into direct contact with the sample, will harden on contact and thus will not penetrate the sample.

1.17.2 Preservation

Jar Samples Small-volume disturbed but representative samples, such as those obtained from drive samplers and those extracted from the top and bottom of undisturbed samples, should be placed in large-necked, round, screw-top, air-tight glass jars. These jars are commercially available in various sizes. The size used should be such that the sample will fill the jar as nearly as possible. This will prevent loss of moisture from the sample through evaporation and subsequent condensation on the walls of the jar. Samples should be placed in the jars as soon as they are retrieved and the jars should immediately be capped and sealed to preserve the original moisture content as closely as possible. Sealing is accomplished by dipping the cap and threads into wax.

The jars used for preserving the samples are commonly supplied in cardboard boxes with corrugated cardboard inserts to separate one jar from another. If the laboratory is nearby or if the samples are to be transported in a car, it is normally satisfactory to pack the jars in these boxes. However, if the jars are to be shipped by a commercial carrier, it is necessary to pack them in wooden boxes with partitions to insure against breakage. In addition, cushioning material such as vermiculite, styrofoam, or excelsior is sometimes used in the individual jar compartments.

Tube Samples Undisturbed samples are generally preserved in the sample liners or sample tubes in which they were retrieved. A second method commonly used for cohesive soils is to preserve them in cardboard or metal containers. When the samples are preserved in the liner, the procedure is as follows. Approximately 1 inch of material is removed from the top and bottom of the sample and preserved as jar samples. The ends of the sample tube or liner are then filled to the top with wax added in increments to prevent the formation of voids. If the sample is short compared to the length of the tube in which it was recovered, the following procedure may be used. A 2-inch thick seal of wax is placed directly against the sample. A block of wood or similar filler is then used to fill as much of the remaining void as possible. Wax is then poured around and over the top of the block until the void is completely filled. Sealing of the tube is completed by capping both ends with tight-fitting plastic, copper, or galvanized steel caps bound on with friction tape and then dipping the caps and tape in wax.

When the samples are to be preserved in cardboard or metal tubes, the tubes should have an inside diameter approximately 1 inch greater than the sample diameter and a length about 2 inches greater than that of the sample. The tubes commonly used are of multi-ply wax-coated cardboard. They may be open-ended or may be similar to the cans with metal bottoms that are used to form concrete test cylinders. In either case, the sample must be completely enclosed in wax. If open-end tubes are used, the specimen is placed on a spacer block inside of the tube and is centered within the tube. Wax is then poured into the annular space between the tube and the sample until it is filled and then over the top of the sample until the wax is level with the top of the tube. When this wax has congealed the sample and tube are inverted and the spacer block is removed. The remaining void is then filled with wax. If cardboard tubes with metal bottoms are used, a plug of wax approximately 1 inch thick is poured into the bottom of the can. When this has congealed the sample is centered inside the container. Wax is then poured into the annular space between the sample and the tube until it is filled, and then over the top of the sample until the tube is full. In all cases where wax is to be placed against already congealed wax, the congealed wax should be thoroughly cleaned to rid it of foreign material and an initial small pour of hot wax should be made to melt the previously placed wax and form a satisfactory bond between the two pours. Prior to placement in the container, the sample may be wrapped in waxed paper or aluminum foil to prevent penetration of wax into porous portions of the sample and to facilitate removal of the wax at a later time.

In the case of undisturbed samples of cohesionless materials, the U.S. Army Corps of Engineers (1972) generally does not require that samples be sealed. Instead, a perforated expanding packer that will permit drainage of excess water from the sample is installed at each end of the sample to hold it in place. One exception to this procedure is the case in which the samples are not stored in humid rooms over long periods. In such instances the samples are sealed by pouring wax over the packers after they have been installed and the excess water has drained. In cases in which the tests to be performed on samples of cohesionless materials require the preservation of the structure of the soil, the Corps also utilizes freezing techniques. Freezing may be accomplished by the use of insulated containers containing alcohol or dry ice. Excess water is drained from the specimen prior to freezing to prevent destruction of the structure due to the water expanding when it turns to ice.

Undisturbed samples must be packed carefully for shipment to prevent disturbance due to shock, freezing, and vibration. This is particularly true if they are to be transported by a commercial carrier. It is common practice to pack the samples in wooden boxes constructed of plywood having a minimum thickness of $\frac{3}{4}$ inch. Samples may be packed one or more to a box. When more than one sample is packed in a box, interior partitions of plywood should be used to provide individual compartments for the samples. Samples should be packed in a vertical position with the top down and should be completely surrounded on the sides, top, and bottom by approximately 2 to 3 inches of cushioning material. The boxes should be labelled to clearly indicate the top of the box and should contain the notations "This Side Up" and "Fragile—Handle with Care." If freezing temperatures are expected during shipment, an additional note should indicate "Do Not Subject to Freezing Temperatures."

Bulk Samples   Large-volume disturbed but representative samples such as those taken from large-diameter auger holes or test pits and trenches should be preserved in durable waterproof bags of approximately 50-lb capacity that are capable of withstanding shipment without puncture. The bags commonly used are made of heavy-gauge polyethylene sheeting or high-count drill lined with vinyl plastic.

Block Samples   A method for the preservation of undisturbed block samples of both cohesionless and cohesive materials taken from test pits or trenches is described as part of the sampling procedure outlined in Section 1.13.12. An alternative method of preserving block samples of materials with sufficient cohesion to withstand handling without disturbance is as follows. A layer of cheesecloth is wrapped around the entire block and is then covered with a layer of microcrystalline wax. This is repeated until three layers of cheesecloth and wax surround the sample. The block may be shipped in this condition if it is to be transported by one of the field supervisory staff. If it is to be shipped by commercial carrier, the block should be packed in a box made of plywood of $\frac{1}{2}$-inch minimum thickness. There should be a minimum of 2 inches of cushioning material between the sample and the box on all sides.

Cores   Rock cores are stored in partitioned boxes of the type and size shown in Figure 1.47. The cores are boxed in the same sequence in which they were taken from the drill hole. Their arrangement in the boxes is as follows. With the core box opened so that the hinged cover is away from the viewer and the partitioned section is adjacent to him, the core is arranged in order of decreasing elevation starting at the left end of the partition nearest the hinges, proceeding to the right and continuing from left to right in succeeding partitioned areas. Core boxes are numbered in sequence with Box No. 1 containing the core of highest elevation. The cores from each drilling run are separated from the core from adjacent runs by wooden blocks on which the depths of the beginning and end of the run are clearly and permanently marked as indicated in Figure 1.47. Blocks, marked as illustrated in the figure, also are used to indicate core loss. If the loss can be pinpointed, the block is placed at the depth of the loss; otherwise it is placed at the end of the run in which the loss occurred. Labelling of the box is discussed in a subsequent paragraph.

It also is common practice to preserve some cores by sealing them in waxed paper or aluminum foil. This is done to preserve representative samples of the materials encountered, cores of material susceptible to slaking, and cores containing soil-filled fractures of fault-zone gouge. The method used is merely to wrap the waxed paper or foil around the core and seal the ends or the entire core by dipping in microcrystalline wax.

Soil cores that are not required to be preserved in an undisturbed condition may be wrapped and sealed as described above and stored in the same type of box as rock cores. Cores that must be kept in an undisturbed condition should be preserved in the same manner as undisturbed samples preserved outside of their sample tubes or liners.

It is common practice today to obtain color photographs of the cores in each box as soon as practical after completion of a hole. This provides an excellent record of the "as retrieved" condition of the cores, which is particularly important in the case of air slaking materials, and permits the engineer or geologist to review the nature of the rock, as required, at subsequent times. Also, the photographs provide a record of the correct sequence of the core pieces in case the core box is spilled accidentally or cores are not returned to their proper place by persons examining them. The photographs should be taken from directly above the box and should include the inside of the cover of the core box, which contains the project name, boring number, box number, and depth covered by the box. A maximum of two core boxes and preferably one box should be included in each photograph. If the cores have dried out prior to taking the photographs, the cores should be wetted with a light water spray or a damp cloth to accentuate the color of the cores.

1.17.3 Labeling

All samples must be clearly marked so as to leave no doubt as to their exact source. In general, the information shown should include the project name; exploration identifying number such as the boring or test pit number; sample number; top elevation of the hole or pit; depth of the sample below ground surface;

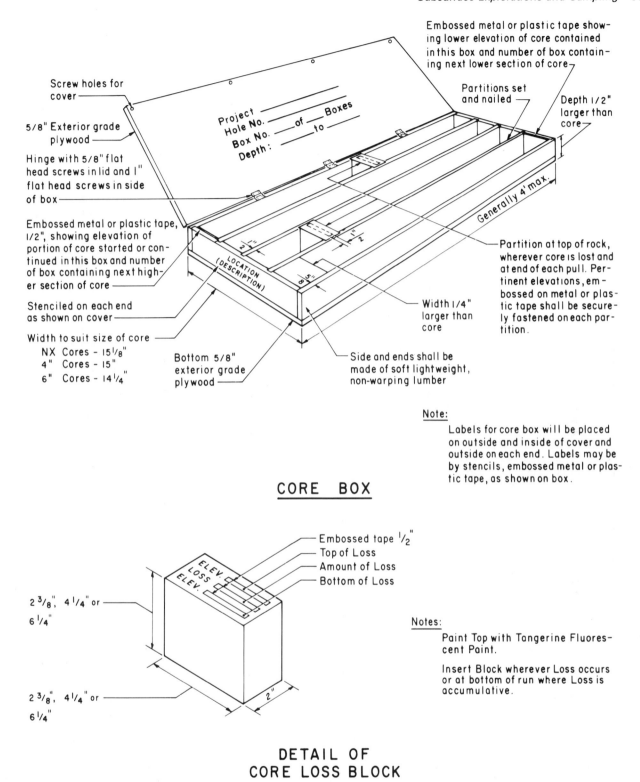

Screw holes for cover

5/8" Exterior grade plywood

Hinge with 5/8" flat head screws in lid and 1" flat head screws in side of box

Embossed metal or plastic tape, 1/2", showing elevation of portion of core started or continued in this box and number of box containing next higher section of core

Stenciled on each end as shown on cover

Width to suit size of core
NX Cores - 15 1/8"
4" Cores - 15"
6" Cores - 14 1/4"

Project _____
Hole No. _____ of _____ Boxes
Box No. _____ to _____
Depth: _____

LOCATION (DESCRIPTION)

Bottom 5/8" exterior grade plywood

Embossed metal or plastic tape showing lower elevation of core contained in this box and number of box containing next lower section of core

Partitions set and nailed

Depth 1/2" larger than core

Generally 4' max.

Partition at top of rock, wherever core is lost and at end of each pull. Pertinent elevations, embossed on metal or plastic tape shall be securely fastened on each partition.

Width 1/4" larger than core

Side and ends shall be made of soft lightweight, non-warping lumber

Note:
Labels for core box will be placed on outside and inside of cover and outside on each end. Labels may be by stencils, embossed metal or plastic tape, as shown on box.

CORE BOX

Embossed tape 1/2"
Top of Loss
Amount of Loss
Bottom of Loss

ELEV.
LOSS
ELEV.

2 3/8", 4 1/4" or 6 1/4"

2 3/8", 4 1/4" or 6 1/4"

2"

Notes:
Paint Top with Tangerine Fluorescent Paint.

Insert Block wherever Loss occurs or at bottom of run where Loss is accumulative.

DETAIL OF CORE LOSS BLOCK

Fig. 1.47 Typical core storage box. (*After U.S. Army Corps of Engineers.*)

description of the material; and, where applicable, the number of blows to drive the sampler. In addition, sampling tubes or liners should be marked to indicate the top of the liner and the level of the top and bottom of the sample within the liner.

All sample markings should be weatherproof and wearproof. Jar samples are commonly marked with gummed labels. These should be overlaid with clear tape. Similar labels are often used on sampling tubes and liners. However, it is preferable to identify such samples by painting directly onto the tubes. Block samples should have an internal and an external label. The internal label may be enclosed in a plastic envelope and sealed in with the last layer of wax. The outer label may be pasted or stapled to the shipping box or, preferably, the information should be painted directly onto the box. Bulk samples should be identified with a plastic enclosed label inside the bag and a mailing tag or similar label attached to the drawstring of the bag. Core boxes should be labeled as shown in Figure 1.47. Note that the project name and the box number should be stenciled on both ends of the box as well as the cover. This makes identification more efficient when the core boxes are stacked.

1.17.4 Shipment

If at all possible, samples should be transported to the laboratory in a motor vehicle by a member of the engineer's field supervisory staff who is familiar with the careful handling required. If the samples are to be shipped by commercial carrier, instructions should be given for special handling and, if possible, the engineer should instruct directly those who will be involved in the actual handling of the samples. In cases where the samples are to be handled by several different companies, the best protection is to be certain that the packing and labeling of the boxes are as described in Sections 1.17.2 and 1.17.3.

It should be noted that special procedures are required in instances where samples are shipped from project sites outside of the continental limits of the United States to laboratories within the country. The entry into the United States of all soil and rock samples is controlled by regulations of the U.S. Department of Agriculture (USDA). These regulations require that, prior to entry, samples must be heat-treated to kill plant pests that may be present. The regulations cover samples from all foreign countries and Hawaii, Guam, Puero Rico, the Virgin Islands, and the Panama Canal Zone. Engineering firms engaged in overseas work may obtain permits from the USDA to have samples passed without heat treatment in order to avoid the alteration of the engineering properties of samples and to preserve the undisturbed state of samples. Permits are valid for a limited time period and must be periodically renewed. Firms acquiring such permits must abide by USDA regulations regarding the shipping, storage, use, and disposal of such samples. Similar regulations are applicable to samples shipped out of USDA quarantined domestic areas to nonquarantined domestic areas.

1.18 LOGS OF SUBSURFACE EXPLORATIONS

Two types of logs are normally prepared in connection with subsurface exploratory work. The first, called a field log, is commonly an extremely detailed chronological record of drilling and sampling operations. The second, called a final log, is basically a graphical log of the lithology on which the pertinent subsurface information and the results of field and laboratory testing are reported. Details of both types of log are described below.

1.18.1 Field Logs of Subsurface Explorations

A legible, concise, and complete record of all significant information pertaining to the drilling and sampling operations within each borehole must be maintained concurrently with the advancement of the hole. Boring specifications should require the maintenance of such records, called field logs, by the driller and the submittal of one or more copies of the log of each hole to the engineer upon completion of the hole. A similar record should also be prepared independently by the engineer or geologist assigned to supervise the operations in the borehole. The purpose of duplicate records is to insure that all observations of the two people most closely connected with the work are recorded and to provide a check in the event of the discovery of some apparent discrepancy after the program has been completed. These records are commonly kept on forms prepared specifically for the purpose or in bound field notebooks of the type used by surveyors. The use of forms is to be preferred, since it eliminates the need for constant repetition of column headings and allows more time to be devoted to observing the exploration work.

The format for field logs varies from one organization to another but, in general, the information recorded is the same. An example of one type of field log is presented in Figure 1.48. Logs such as these are used in the selection of representative samples for testing, in the preparation of geological profiles, and in the preparation of the final logs that are presented in engineering reports and contract drawings. Therefore, they must contain all information necessary to completely define the subsurface profile and groundwater conditions. The log should be a complete chronological record of the drilling and sampling operations within the hole, including delays. Among the information to be recorded is the following:

1. Reference information comprising the project number, title, and location; the exploration number or letter designation; the exploration location by coordinates, station, and offset, or referred to permanent nearby structures; the inclination of the boring and, if inclined, the bearing or azimuth of the dip of the hole; the reference elevation, that is, the elevation from which all depth measurements are made; and the datum of the reference elevation, for example, Mean Sea Level, etc.
2. Personnel information, including the names of the drilling contractor, the driller, and the inspecting engineer or geologist.
3. Equipment data, consisting of the manufacturer's name and model designation for the drill rig, motorized equipment used to excavate test pits and trenches, and seepage- and pressure-testing equipment.
4. Sampling and coring information consisting of the following:
   a. For all sampling and coring operations: the sample type and number; the inside diameter, outside diameter, and length of the sample tube; the depth at the start and at the completion of the sampling drive or push or coring "run"; the length of sample or core recovered; the recovery, which is defined as the ratio, expressed in percent, of the length of sample or core recovered to the length of the sampling drive or push or coring "run"; and a complete visual description of each sample or core including color, type of material, density, or consistency of soil, hardness of rocks, stratification, rock structure, moisture conditions, etc. The description should be made immediately following the retrieval of the sample or core so that it represents the "as retrieved" classification. This is particularly important when sampling materials that tend to break down on exposure. Among these are air-slaking shales, and rocks containing expansive clay minerals.
   b. For drive samplers: the weight and height of drop of the drive hammer and the number of blows required for each 6 inches of penetration of the sampler.
   c. For pressed or pushed samplers: the hydraulic pressure required to push the sampler into the ground and the rate of penetration.
   d. For soil or rock coring: average rotational speed and downward hydraulic pressure of the core barrel and average rate of penetration.
   e. For rock coring: the Rock Quality Designation (RQD) devloped by Deere (1964). The RQD is the ratio, expressed as a percentage, of the aggregate length of core pieces over 4 inches long in a run divided by the length of the run.

BORING NO. *B-9*

SHEET *1* OF *10*

FIELD LOG - SUBSURFACE EXPLORATION

JOB NO. _____*9847*_____ JOB TITLE *Overlook Hydroelectric Project*

LOCATION *Hivolt, R.M.* COORDINATES *N 119,579.05 E 90,848.91*

DRILL *Failing 1500* ANGLE *Vertical* REFERENCE ELEV. *850.15* DATUM *MSL*

DRILLING CONTRACTOR *J. Lynne* DRILLER *W. Edward* INSPECTOR *P. John*

DEPTH IN FEET (ELEVATION)	BLOWS ON SAMPLER FOR 6 INCHES (% RECOVERY)	SYMBOL	SAMPLE NUMBER AND DESCRIPTION OF MATERIAL	SAMPLER AND BIT	CASING TYPE	BLOWS/FOOT ON CASING	DATE	DEPTH IN FEET FROM	TO	DESCRIPTION OF OPERATION AND REMARKS
0	4,7,7,6 (95%)		1 Sample No. 1: Reddish-brown f sandy CLAY; trace roots; stiff. (CL)	2" SS			Apr. 5 1971	0	2	8:00 A.M. - 12:00 Set rig over hole and prepared to drill.
	(100%)		2 Sample No. 2: Reddish-brown f sandy CLAY; stiff. (CL)	6" I.D. DEN.				0	2	2:00 P.M. - 2:30 P.M. Started hole by driving split spoon (SS) from 0 to 2'. Sample length = 1.9'.
5	(93.5%)		3 Sample No. 3 (Upper 15"): Dark reddish-brown c-m-f sandy CLAY; trace f gravel; m.s. 3/8", very stiff. (CL)	6" I.D. DEN.	Uncased as of April 6					2:30 P.M. - 2:50 P.M. Cleaned hole with tricone bit.
	(91.7%)		4	6" I.D. DEN.						2:50 P.M. - 5:00 P.M. Further work delayed due to only Denison starting barrel at site being used at second rig.
10	(95.9%)		5	6" I.D. DT CB			Apr. 6			7:00 A.M. - 9:00 A.M. Prepare equipment to take Denison sample using sawtooth bit.
	(83.4%)		6 Sample No. 3 (Lower 15"): Brown clayey c-m-f SAND and c-m-f GRAVEL; m.s. 3"; dense. (GC)	6" I.D. DT CB				2.0	4.5	9:00 A.M. - 9:20 A.M. Took Sample 2. Rotary speed=120 RPM; downward pressure =100 psi. Sample length = 30"; loss = 0.
15	(100%)		7	6" I.D. DEN.				4.5	7.0	9:30 A.M. - 10:10 A.M. Took Sample 3. 100 RPM; 110 psi. Drilling harder in lower 15"; drill water changed color at ~ 15". C = 28"; L = 2".
	(41.7%)		8 Sample No. 4: Brown clayey c-m-f SAND and c-m-f GRAVEL; m.s. 2"; dense. (GC)	6" I.D. DTCB				7.0	9.5	10:20 A.M. - 10:40 A.M. Took Sample 4. 125 RPM; 120 psi. C = 27.5"; L = 2.5".
20			Sample No. 5: (Upper 11"): Brown clayey c-m-f SAND and c-m-f GRAVEL; m.s. 3"; dense. (GC)					7.0	9.5	10:40 A.M. - 10:55 A.M. Cleaned hole with chopping bit with upward deflected jets.
			Sample No. 5 (Lower 13"): Brown f sandy CLAY; very stiff.					9.5	11.5	11:00 A.M. - 11:25 A.M. Took Sample 5. Bottom-discharge diamond bit; 125 RPM; 120 psi. Drilling easier at 11". C = 23"; L = 1".
			Sample No. 6: Brown f sandy CLAY; very stiff. (CL)					11.5	13.5	11:30 A.M. - 12:00 A.M. Took Sample 6. 100 RPM; 90 psi. C = 20"; L = 4".
			Sample No. 7: Brown f sandy CLAY; very stiff. (CL)					13.5	16.0	1:00 P.M. - 1:15 P.M. Took Sample 7. Sawtooth bit. 100 RPM; 95 psi. C = 30"; L = 0.
			Sample No. 8: Brown c-m-f gravelly c-m-f					16.0	17.0	1:15 P.M. - 1:45 P.M. Took Sample 8. Diamond bit. 130 RPM; 100 psi. Tools pulled from hole when barrel would not advance farther. Possible boulder has been encountered. C = 5"; L = 7".

Fig. 1.48 Typical field boring log.

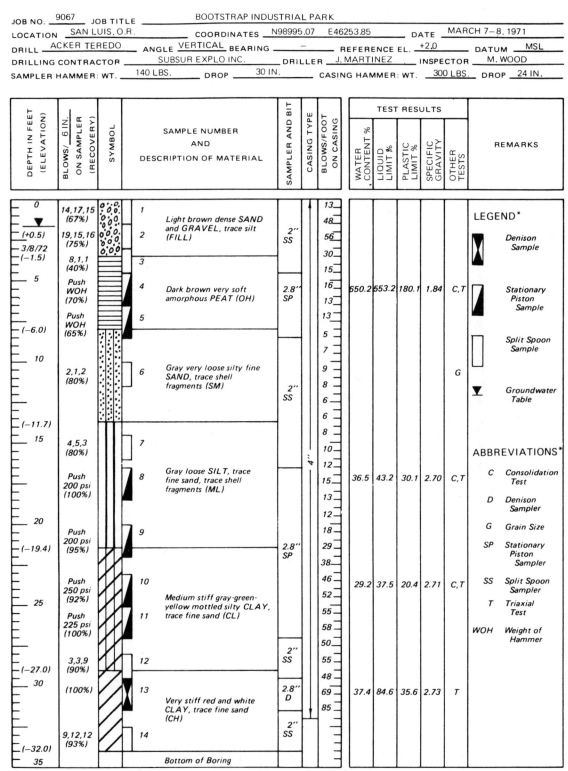

Fig. 1.49 Typical final log for boring in overburden.

BORING NO. _____PH-12_____

SHEET _____1_____ OF _____1_____

SUBSURFACE EXPLORATION LOG

JOB NO. _____9807_____ JOB TITLE _____RUSHING WATER HYDROELECTRIC DEVELOPMENT_____

LOCATION _BASALTVILLE, M.N._ FEATURE _____POWERHOUSE_____ COORDINATES _N10,650.15, E9,865.45_ DATE _____MAY 9 - 12, 1972_____

DRILL _SPRAGUE & HENWOOD 40-C_ ANGLE _____VERTICAL_____ BEARING _____–_____ REFERENCE EL. _____699.2_____ DATUM _MSL_

DRILLING CONTRACTOR _____SUBSUR EXPLO INC._____ DRILLER _____R. ARNOLD_____ INSPECTOR _____M. STEEL_____

SAMPLE HAMMER: WT. _____140 LBS._____ DROP _____30 IN._____ CASING HAMMER: WT. _____300 LBS._____ DROP _____24 IN._____

DEPTH IN FEET (ELEVATION)	BLOWS / 6 IN. ON SAMPLER (% RECOVERY)	CORES RUN NO.	CORES NO. AND SIZE OF CORE PIECES	CORES % RECOVERY	CORES RQD	SOIL DESCRIPTION OR ROCK LITHOLOGY	SYMBOL	ROCK STRUCTURE	SAMPLER AND BIT	CASING TYPE	BLOWS/FOOT ON CASING	PRESSURE TEST RESULTS DEPTH IN FEET FROM	PRESSURE TEST RESULTS TO	PRESSURE PSI	LOSS, GPM	TIME, MIN.	REMARKS
0 ▼ (697.7) 3/8/72 (695.2)	1,1,2 (60%) 2,3,5 (82%)					1 Gray, soft silty CLAY 2 Brown, loose coarse to fine SAND					1 2 4 4 13 15 18 18 20						**LEGEND***
(690.0) 10		1	2 @ 6"	80	77												□ Split Spoon Sample
		2	4 @ 7"	100	100	DENSE BASALT, greenish gray to black, hard, slightly weathered		Closely spaced, open joints at 35°, 45°, 60° and vertical. Joints stained.				9.2	19.2	3 6 12 6 3	2.8 4.6 8.7 4.4 2.9	15 15 15 15 15	
		3	5 @ 7" 1 @ 2"	91	94												■ Double Tube Core Barrel
(681.7) 20		4	2 @ 3" 3 @ 2"	22	0	VESICULAR and AMYGDALOIDAL BASALT, greenish gray, medium hard, highly weathered, many small calcite amygdules.		Closely to moderately jointed; joints clay filled; joints @ 34° and 60°. 1" clay seam at 19.2'.				19.2	29.2	6 12 24 12 6	9.2 8.0 1.3 1.3 0	15 15 15 15 15	▼ Groundwater Table
		5	4 @ 6"	100	91												**ABBREVIATIONS.**
		6	4 @ 7"	99	94												SS Split Spoon Sampler
(669.9) 30		7	3 @ 12" 1 @ 8"	100	99												_Flow contact at 28.5'._
		8	1 @ 48" 1 @ 12"	100	100	DENSE BASALT, reddish gray to black, very hard, slightly weathered.		Moderately jointed; incipient joints 0.04" thick at 30° and 45°; also vertical joints filled with calcite.				29.2	39.2	8 16 32 16 8	0 0 0 0 0	15 15 15 15 15	
		9	1 @ 16" 3 @ 12" 1 @ 4.4"	100	100												
40		10	1 @ 36" 1 @ 24"	100	100							39.2	49.2	10 20 40 20 10	0 0 0.1 0 0	15 15 15 15 15	
		11	2 @ 18" 2 @ 11"	97	96												
(650.9) 50		12	1 @ 36"	100	100	VESICULAR and AMYGDALOIDAL BASALT, greenish gray, medium hard, with calcite amygdules; slightly weathered.		Moderately jointed; joints open; some filled with clay; joints at 30°, 45° and 60°.				49.2	59.2	13 26 52 26 13	3.7 7.0 11.2 6.8 3.1	15 15 15 15 15	
		13	1 @ 9" 6 @ 8" 1 @ 3"	100	95												
60 (638.2)		14	3 @ 9" 4 @ 3"	72	50												_Flow contact at 58.4'._
		15	2 @ 12" 2 @ 5"	100	100	DENSE BASALT, black, very hard.		Moderately jointed; joints at 30°, 45°, 70° and vertical; all joints tight. Vertical joints filled with calcite.				59.2	68.0	16 32 64 32 16	0.3 0.5 0.6 0.3 0.2	15 15 15 15 15	
(631.2) 70		16	4 @ 6" 2 @ 9"	100	100												
						Bottom of Boring											

(Sampler and Bit column: 2" SS, 4"; NX Size Series "M" Core Barrel with Bottom Discharge Diamond Bit. Casing Type: None.)

* The legend and abbreviations shown here in the "Remarks" column are normally presented on a separate page.

Fig. 1.50 Typical final log for boring predominantly in rock.

5. Description of material penetrated but not sampled, as determined from drilling or chopping action or changes in the color of the drill water.

6. Casing information, consisting of the size of the casing; the depth at which casing was added; the length of casing added; the final depth of the bottom of the casing; the weight and height of drop of the hammer and the number of blows for each 12 inches of penetration for driven casing; and, for drilled casing, the average rotational speed and downward pressure on the casing and the average rate of penetration.

7. Seepage- and pressure-test information, comprising the depths at which tests were performed and the time required

for each test. The actual test data is recorded on forms for that purpose.

8. Groundwater information, consisting of the depth to the water surface in the hole, recorded daily, at the start and close of work. These readings should be continued after completion of the hole until the water level in the hole has stabilized.

9. Artesian pressure information, including the depths at which artesian pressures were encountered, the measured heads, and the time at which each measurement was made.

10. Elevation of the top and bottom of the hole and the top of the rock, if encountered.

11. The date and time of all operations and delays, including, but not limited to, drilling, sampling, seepage and pressure testing, artesian pressure measurement, machine breakdown, and injuries.

12. Miscellaneous information that may aid in the interpretation of subsurface conditions. This would include the depth at which drill water is lost or regained, the amount and color of the return water, and the depth at which a change in drilling action occurs. The latter would include the depth at which rod vibration starts or stops, the depth at which the rate of penetration or ease of penetration changes, etc.

13. Any additional information that the driller, engineer, or geologist considers pertinent to the interpretation of subsurface conditions.

1.18.2 Final Logs of Subsurface Explorations

The final log is essentially a condensation of the field log that has been refined on the basis of the results of laboratory tests and often has a limited amount of laboratory test data incorporated into it. The final log should present a clear, concise, accurate picture of subsurface conditions since it is commonly the form in which subsurface information is utilized by the design engineer. Final logs are also used in engineering reports and are provided as a part of the contract documents for construction of the project. Information that should be shown on final logs is illustrated by Figures 1.49 and 1.50. Figure 1.49 shows a convenient format for use when a boring is totally or primarily in overburden. The format shown in Figure 1.50 is for holes primarily in rock. The major advantages of the latter format for rock holes is that separation of the lithologic description from the structure description permits a more rapid visualization of the stratification and provides space for a more convenient and adequate presentation of structural features.

1.19 CONTRACTING AND SUPERVISING EXPLORATORY PROGRAMS

1.19.1 Contract Documents

Obtaining the maximum amount of quality subsurface information at the least cost requires close control of the drilling and sample techniques. Detailed contract documents are thus essential to the success of exploratory work. Many federal and state agencies and many private firms have formulated contract documents based on extensive experience in subsurface work. In general, the formats for these documents are similar. Typical contract documents comprise an invitation to bid, a proposal form, general provisions, technical provisions, and a location plan(s) showing the program.

The proposal form consists of a list of all payment items and the estimated quantity of each plus space for the insertion of unit and total prices for each item. Generally there is a lump-sum bid item for mobilization and demobilization of equipment and men at the site. Other items are generally on a unit-price basis. These items include price per foot of drilling, price per sample, price per test, and hourly rate for standby. Also included is a plant and equipment schedule on which the bidder indicates the number, type, capacity, age, condition, and present location of the principal items of equipment he proposes to use.

The general provisions include, but are not limited to, items that concern commencement and completion dates for the work; liquidated damage payments; method of payment (e.g., monthly); procurement of permits and licenses to drill on public or private lands; protection of existing structures; existing physical information and data; maintenance of traffic; subletting; indemnity; insurance; responsibility for provision of water and power; control of the work by the engineer; conditions under which the contract may be terminated; and posting of a payment and performance bond.

The technical provisions detail the following: the purpose and scope of the work, the quantities and locations of the explorations and the percentage variation in the quantities permitted without a change in unit prices; the work included in the mobilization and demobilization item; the proposed locations and sizes of the exploratory holes, pits, etc.; the number and location of instruments (e.g., piezometers and inclinometers) to be installed, and a schedule of the holes in which seepage and pressure tests are to be performed; the right of the engineer to control the type, number, and location of all sampling and other work; the records to be maintained by the contractor; the type and size of containers used for storage of samples; the labeling and shipping of the samples; type and amount of equipment to be used; the method of advancing the hole and the casing; the samplers and sampling techniques to be used, including a description of the sampler, the rate of pushing in samplers, and the driving energy for split-spoon samplers, etc.; the procedures for rock coring; the methods of excavation and barricading of pits and trenches; the methods of preserving, storing, and shipping samples; seepage and pressure test procedures; the method for measuring artesian pressures; installation procedures for piezometers and observation wells; backfilling of holes; photographing of cores; and the method of measurement and payment for the items listed in the proposal form.

1.19.2 Letting the Contract

Contracts for subsurface explorations may be let on the basis of competitive bidding or through negotiation. Contracts awarded by competitive bidding may be further subdivided into those in which the bidding is open to all interested parties and those in which the bidding is open only to selected and/or prequalified firms. Contracts to which governmental agencies are a party must in almost all cases be awarded through open competitive bidding and the contract must, in general, be awarded to the low bidder. Contracts to which private organizations are a party may be restricted to selected bidders and also may, on the basis of an evaluation of which bid is most advantageous to the owner, be awarded to a bidder other than the low bidder. Such an evaluation would take into consideration the quality of the equipment and capability of the drillers proposed to be assigned to the job and the time schedule proposed.

Negotiated contracts are utilized primarily for small contracts and in cases in which the time available prior to the required commencement of drilling precludes the use of the bidding

Journal of the Soil Mechanics and Foundations Division, American Society of Civil Engineers, **87**, No. SM 3, Proceedings Paper 2838, June, pp. 69–100.

Morey, R. M. (1974), Continuous subsurface profiling by impulse radar, *Subsurface Explorations for Underground Excavation and Heavy Construction*, American Society of Civil Engineers, New York, N.Y.

Noorany, I. (1985), Offshore sampling and in situ testing: 1981 update, *Updating Subsurface Samplings of Soils and Rocks and Their In Situ Testing*, Engineering Foundation, New York, N.Y.

Noorany, I. and Gizienski, S. F. (1970), Engineering properties of submarine soils: State-of-the-art review, *Journal of the Soil Mechanics and Foundations Division, American Society of Civil Engineers*, **96**, No. SM 5, Proceedings Paper 7536, Sept., pp. 1735–1762.

Ohya, S. (1986), In situ P and S wave velocity measurements, *Use of In Situ Tests in Geotechnical Engineering*, Geotechnical Special Publication No. 6, American Society of Civil Engineers, New York, N.Y.

Peck, R. B., Hanson, W. E., and Thornburn, T. H. (1973), *Foundation Engineering*, 2nd ed., John Wiley and Sons, Inc., New York, N.Y.

Ray, R. G. (1960), *Aerial Photographs in Geologic Interpretation and Mapping*, U.S. Department of the Interior, Geological Survey Professional Paper 373, U.S. Government Printing Office, Washington, D.C.

Riggs, C. O. (1986), North American standard pentration test practice: An essay, *Use of In Situ Tests in Geotechnical Engineering*, Geotechnical Special Publication. No. 6, American Society of Civil Engineers, New York, N.Y.

Rocha, M. (1971a), A method of integral sampling of rock masses, *Rock Mechanics and Engineering Geology* III/1, pp. 1–12.

Rocha, M. (1971b), Some applications of the new integral sampling method in rock masses, *Memoria No. 397*, Laboratório Nacional de Engenharia Civil, Lisbon, Portugal.

Rogers, F. C. (1950), *Engineering Soil Survey of New Jersey, Report No. 1*, Engineering Research Bulletin No. 15, College of Engineering, Rutgers University, Edwards Bros., Inc., Ann Arbor, Mich.

Roman, I. (1934), Some interpretations of earth resistivity data, *Geophysical Prospecting* (American Institute of Mining and Metallurgical Engineers), **110**, pp. 183–200.

Schmertmann, J. H. (1978), *Guidelines for CPT Performance and Design*, Report No. FHWA-TS-78-209, U.S. Dept. of Transportation, Federal Highway Administration, Offices of Research and Development, Washington, D.C.

Schmertmann, J. H. (1986), Dilatometer to compute foundation settlement, *Use of In Situ Tests in Geotechnical Engineering*, Geotechnical Special Publication No. 6, American Society of Civil Engineers, New York, N.Y.

Seed, H. B. (1979), Soil liquefaction and cyclic mobility evaluation for level ground during earthquakes, *Journal of the Geotechnical Engineering Division, American Society of Civil Engineers*, **105**, No. GT2, February, pp. 201–255.

Szechy, K. (1957), *Foundations*, Technical Publishing Company, Budapest, Hungary, p. 115.

Taylor, D. W. (1948), *Fundamentals of Soil Mechanics*, John Wiley and Sons, Inc., New York, N.Y.

Terzaghi, K. and Peck, R. B. (1967), *Soil Mechanics in Engineering Practice*, 2nd ed., John Wiley and Sons, Inc., New York, N.Y.

U.S. Army Corps of Engineers (1943), *Seismic and Resistivity Geophysical Explorations Methods*, Waterways Experiment Station Technical Memorandum No. 198–1, Vicksburg, Miss.

U.S. Army Corps of Engineers (1951), *Time Lag and Soil Permeability in Groundwater Observations*, Waterways Experiment Station Bulletin No. 36, Vicksburg, Miss.

U.S. Army Corps of Engineers (1960), *Geological Investigations*, Engineer Manual EM 1110-1-1801, U.S. Government Printing Office, Washington, D.C.

U.S. Army Corps of Engineers (1963), *Foundation Grouting: Field Technique and Inspection*, Engineer Manual EM 1110-2-3503, U.S. Government Printing Office, Washington, D.C.

U.S. Army Corps of Engineers (1969), *Instrumentation of Earth and Rockfill Dams (Groundwater and Pore Pressure Observations)*, Engineer Manual EM 1110-2-1908, U.S. Government Printing Office, Washington, D.C.

U.S. Army Corps of Engineers (1972), *Soil Sampling*, Engineer Manual EM 1110-2-1907, U.S. Government Printing Office, Washington, D.C.

U.S. Army Corps of Engineers (1979), *Geophysical Exploration*, Engineer Manual EM 1110-1-1802, U.S. Government Printing Office, Washington, D.C.

U.S. Burea of Reclamation Department of the Interior (1968), *Earth Manual*, 1st ed., U.S. Government Printing Office, Washington, D.C.

U.S. Geological Survey, Department of the Interior (1965), *Publications of the Geological Survey*, Washington, D.C.

Van Reenan, E. D. (1964), Subsurface exploration by sonar seismic systems, *American Society for Testing and Materials Special Technical Publication No. 351, Symposium on Soil Exploration*, Philadelphia, Pa., pp. 60–73.

Wetzel, W. W. and McMurray, H. V. (1937), A set of curves to assist in the interpretation of the three-layer resistivity problem, *Geophysics*, **2**, No. 4, p. 239.

Wroth, C. P. (1984), The interpretation of in situ soil tests, *Geotechnique*, **34**, No. 4, pp. 449–489.

Wroth, C. P. and Hughes, J. M. O. (1973), An instrument for the in situ measurement of the properties of soft clays, *Proceedings of the VIII ICSMFE*, Vol. 1.2, Moscow.

2 SAMPLING AND PREPARATION OF MARINE SEDIMENTS

RONALD C. CHANEY, Ph.D., P.E.
Professor and Director, Fred Telonicher Marine Laboratory,
Humboldt State University

2.1 INTRODUCTION

2.1.1 Process of Site Evaluation

The recent development in the exploitation of the resources of the ocean floors of the world for petroleum, natural gas, and other minerals, along with waste disposal has resulted in an increased interest in the application of geotechnical sciences to the marine environment (Anderson, 1981; Chaney, 1984; Chaney and Fang, 1985, 1986; Chaney et al., 1986; Fang and Chaney, 1985, 1986; Fang and Owen, 1977; Lee, 1985; Richards and Chaney, 1981, 1982; Winterkorn and Fang, 1971). The selection of an offshore site is essential for most seafloor engineering projects. It is a process strongly influenced by engineering judgment that is often constrained by economic, political, environmental, and societal considerations. Fletcher (1969) considers that the purpose of an offshore site investigation is to "secure such information by procedures appropriate to the project and to report the findings in sufficient technical detail to provide a basis for economic studies and design decisions." The process of site evaluation from the recognition of a siting problem to the development of a problem solution involves a number of discrete steps as shown in Figure 2.1 (Chaney et al., 1985). These steps are: (1) determination of environmental loading, (2) site reconnaissance, (3) development of a stratigraphic model based on both a combination of sampling and in-situ testing, and (4) interpretation of data.

2.1.2 Scope

The purpose of this chapter is to describe geotechnical methods used for sampling offshore sites and transport and preparation of soil samples. The discussion will be limited to obtaining undisturbed samples and will not consider disturbed grab and dredge samples. For further information on grab and dredge samples the reader is referred to *Oceans Magazine* (1969). The chapter is organized into an introductory section to provide background information, followed by five sections that consider (1) offshore soil sampling, (2) handling, wrapping, and labeling, (3) storage, (4) preparation of soil samples, and (5) laboratory versus in-situ properties.

2.1.3 Marine Sedimentary Environment

The ocean floor itself can be divided into two major regions: the continental margins and the ocean basins. The continental

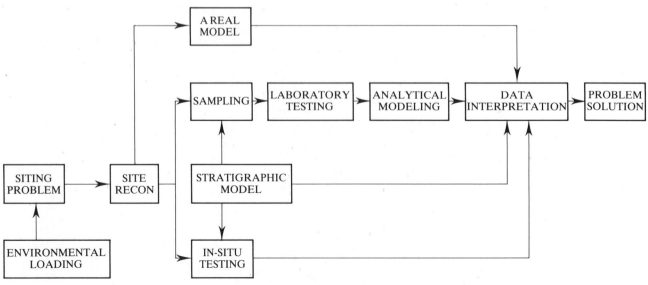

Fig. 2.1 Process of site evaluation. (*After Chaney et al., 1985.*)

margins include the continental slope and continental shelf and form only a small percentage (21 percent) of oceanic area. The remainder of the area is basically the oceanic basins. In addition to the earth's major oceans, there are also a number of much smaller bodies of water that are largely landlocked. These bodies of water include the Arctic Ocean, Mediterranean Sea, Black Sea, Gulf of Mexico, Red Sea, Bering Sea, Baffin Bay, and the North Sea. Each of these bodies of water is distinctive, and each has physical characteristics strongly influenced by the climate and topography of the surrounding continents, river discharge, and a number of other factors. Their sedimentary characteristics thus vary widely.

The behavior of fine-grained sediment is governed to a large extent by its structure of fabric and mineralogy. The development of flocculated clay structures is encouraged in a seawater environment (Casagrande, 1932). The double-layer theory has been generalized to indicate that, for a given mineralogy, the electrochemical environment at the time of deposition is the most important variable in determining clay structure. If there are no environmental changes, the structure of fine-grained sediments acquired at the time of deposition is usually not destroyed or altered. Under such conditions, the strength characterization of most sediments can be defined in terms of original structure and gross geologic events that produced changes in overburden pressure.

The strength of marine sediments is normally dependent on a great variety of factors because of changes in environmental conditions. Diagenesis is the postdepositional process involving physical and chemical changes in sediment, after deposition, that convert it into consolidated rock. Thirty diagenetic processes that are active to varying degrees in the marine environment and that can alter the microstructure of fine-grained sediment structures have been identified by Krombein (1942). For any given seafloor location, the aerial and vertical variability of strength characterization depends on the number, type, and magnitude of the individual diagenetic and biologic processes. Conflicting processes or alternating magnitudes can produce variability both laterally and vertically.

2.1.4 Types and Distribution of Submarine Soils

The seafloor environment is predominantly depositional rather than erosional. As a consequence, marine sediments exhibit more uniformity than normally found on land. Typically, marine sediments are broadly classified on whether the sediments are land derived (terrigenous) or are the result of marine activity (pelagic). The pelagic sediments can be further divided into inorganic or organic materials. Inorganic pelagic materials are typically clay-size material. In contrast, organic materials are primarily the skeletal remains of marine organisms. These materials are either calcium carbonate ($CaCO_3$) or silica (SiO_2). A short summary of marine carbonate materials is presented in Table 2.1 as given by Chaney et al. (1982). The presence of calcium carbonate ($CaCO_3$) in marine sediments is controlled

TABLE 2.1 MARINE CARBONATE MATERIALS[a].

Name	Shape	Smallest Subparticle Shape	Depth Location, m	Environment — High-Energy Area	Environment — Low-Energy Area	Porosity	Overall Material Behavior[b]	Chemical Constituent
Foraminiferan shells (tests)	1 mm		3500	Benthic forams	Planktonic forams / globigerina	Porous (porcellaneous)	Cohesionless (sandy-silt)	Calcite (Mg-calcite in small quantities)
Pteropod shells (tests)	1–2 mm	No discrete subparticles	3000	Tropical and subtropical waters	—	Porous (porcellaneous)	Cohesionless (sand)	Aragonite
Coccolithic plants (nanno-fossils)	0.01 mm		3500–5000	—	Mainly latitudinal variation of species type	No inherent porosity (hyaline)	Cohesionless (silt)	Calcite shells of the algal family— skeletons may be protected by organic or $MgCO_3$ coatings
Corals	20 cm	No discrete subparticles	0–35	Branching, massive and encrusting corals	Foliose and encrusting corals	Porous (porcellaneous)	Cohesionless (sand-gravel)	Mg-calcite, aragonite
Precipitate	0.1–0.6 mm / Highly variable	No discrete subparticles	35–200	Oolites	Pelletoids	No inherent porosity (hyaline)	Cohesionless (sand)	Mg-calcite
Benthic materials	—	—	35–200	Branching coraline algae, forams, planktonic feeders	Mollusks, benthic forams echinoid fragments	—	—	Mg-calcite

[a] Chaney et al. (1982). Copyright ASTM, reprinted with permission.
[b] Qualitative description of material response (that is, sand, sandy silt, sandy-gravel, silt) based on assumed particle size.

TABLE 2.2 TYPICAL GEOTECHNICAL DATA AND THEIR RANGES IN MARINE DEPOSITS[a].

Parameters	Percent	References
Size composition		
Clay 2 micron	35 to 60	
Silt	40 to 60	
Sand	<10	
Moisture content (in place)	60 to 1800	
Activity	0.33 to 1.33	
Clay minerals		
Illite	60 to 75	
Kaolinite	~10	
Montmorillonite	5 to 20	
Void ratio	0.50 to 9.0	Richards and Parker (1968)
Sensitivity	1.6 to 26	
Liquid limit	72 to 121	Fang and Owen (1977)
Plastic limit	34 to 51	
Field moisture equivalent	65 to 78	Winterkorn and Fang (1971)
Centrifuge moisture equivalent	55 to 68	
Shrinkage limit	7 to 10	
Compression index	2.3	

[a] Chaney and Fang (1986). Copyright ASTM, reprinted with permission.

by biologic productivity and the calcium carbonate compensation depth (CCD). The CCD is the depth in the water column at which $CaCO_3$ is dissolved. The presence of $CaCO_3$ material below the CCD is primarily dependent on the amount of biologic material available, although shallow-water materials can be transported into deeper areas by means of turbidity currents and debris flows. Calcareous materials are found in the equatorial regions and the North Atlantic. The equatorial Pacific typically has alternating bands of siliceous and calcareous ooze. Pelagic clay, wind-blown sediments, are found where biogenic (calcareous) oozes are absent. Typical geotechnical data of marine deposits are presented in Table 2.2.

2.1.5 Planning

A typical offshore field program consists of two principal components: the geophysical survey and the geotechnical investigation. These two phases are usually carried out with different vessels and at different times, with the geophysical part preceding the geotechnical. The program is generally defined in terms of whether the investigation is considered site-specific or structure-specific. In the first case, the location of the site is known and the soil conditions govern the foundation design and basic structure type. In the second case, the type of structure has already been selected and some change in the location of the site is allowed to ensure that suitable soil conditions exist.

A complete offshore soils study can be divided in a number of distinct steps. An outline of the steps involved in most site evaluation-selection projects is the following:

1. Define the environmental, oceanographic, geologic, geophysical, geotechnical, marine biota, construction, political, and other factors to be evaluated. The following is a detailed listing of some of the factors that need to be evaluated (adopted from Richards et al., 1976):

   Atmospheric Factors
   a. Winds
   b. Physical properties: temperature, humidity, and pressure
   c. Precipitation: types, quantities, duration, frequency, and pressure

   d. Visibility: types of restrictions, distance, duration, frequency, and recurrence interval
   e. Energy transmissibility: reflection and refraction of low, medium, and high radio frequencies

   Oceanographic Factors
   a. Physical–chemical properties
   b. Sealevel variation: storm surges, tides, tsunamis, and seiches
   c. Waves
   d. Currents
   e. Sea ice

   Geological and Geophysical Factors
   a. Subbottom materials
   b. Topography
   c. Stability of subbottom materials
   d. Rate of erosion and deposition
   e. Geophysical properties of soil and rock
   f. Seismicity

   Geotechnical Factors
   a. Soil and rock classification
   b. Soil areal and vertical variability
   c. Shear strength, deformation characteristics, and bearing capacity
   d. Compressibility characteristics, and permeability
   e. Stability
   f. Liquefaction potential under environmental loads
   g. Breakout resistance

   Marine Biota Factors
   a. Populations
   b. Sensitivity to changes in temperature, turbidity, and pollutants
   c. Effect of structure in providing increased protection and breeding grounds
   d. Effects of biota

   Construction Factors
   a. Availability of construction materials
   b. Availability of construction sites
   c. Availability of labor having appropriate skills
   d. Availability of construction equipment
   e. Weather conditions during working period
   f. Shore-support facilities
   g. Communication and survey facilities
   h. Availability of support industry (i.e., divers)
   i. Facilities for personnel transfer

   Political, Demographic, and Geographic Factors
   a. Basic location in relation to shore, principal cities, markets, and raw materials
   b. Political jurisdictions
   c. Availability of and regulations regarding shore-support facilities and connections
   d. Distance offshore
   e. Aesthetic and acoustical requirements and considerations
   f. Regulations regarding discharge of wastes, heated water, and other materials
   g. Possibility of sabotage

   Other Factors to be Considered in Uses of Adjacent Sea, Land, and Air Areas
   a. Proximity to shipping lanes
   b. Navigational channels, aids, and restrictions
   c. Pleasure and recreational craft
   d. Fishing vessels; fishing grounds; oyster and shrimp beds, etc.; sensitivity to disturbance
   e. Military operations
   f. Proximity to other structures and operations and effect on or by their intakes, discharges, and use of sea
   g. Air traffic
   h. Use of adjacent land areas and up-current areas; their effect on structure; and their effect on their activities

i. Proximity to submarine telephone and power cables, oil and gas pipelines
2. Determine the level of detail required.
3. Define the geographic areas to be evaluated.
4. Conduct a search for any available data.
5. Define necessary additional data requirements based on items 1, 2, and 4.
6. Define an acquisition plan for obtaining additional data.
7. Interface with project planners, designers, installers, and operational personnel for confirmation that the intended solutions will provide environmental data compatible with their requirements within acceptable time and economic limits.
8. Execute the proposed solution.

2.2 OFFSHORE/NEARSHORE SOIL SAMPLES

Methods for obtaining marine soil samples can be subdivided into four operational procedures, which are (1) self-contained sampling tools, (2) fixed seabed structures, (3) sampling tools used in conjunction with a drill string, and (4) bottom platforms (ASTM, 1988; International Society for Soil Mechanics and Foundation Engineering, 1981; Moore and Heath, 1978). The self-contained tools are capable of exploring soil conditions to approximately 40 ft (12.8 m). Deeper investigations will require an operation either through the drill string or a bottom platform.

2.2.1 Self-Contained Samplers

A self-contained sampling tool such as a gravity (or drop) core sampler or a vibratory corer is the most inexpensive method of obtaining soil samples offshore. These tools can generally be operated from small supply vessels (less than 160 ft (51.2 m)) that have a crane or winch and an A-frame. Sampling can be done with fairly loose ship positioning, so anchoring in deep water will generally not be required. One of the shortcomings of these tools is the use of gravity or a mechanical vibrator on the tool that provides the required sampler insertion force. The available penetration force driving into a soil formation is usually depleted at a 40 ft (12.8 m) depth. Sample disturbance of recovered samples generally exceeds realistic limits for structural design purposes, and sample recovery can be low.

Gravity Coring Devices

Deep sea gravity corers are surface-deployed devices capable of recovering bottom samples. They have been constructed in sizes capable of obtaining samples up to 140 ft (44.8 m) in length and 4.3 in (109 mm) in diameter. Their mechanical construction does not limit them to any specific water depth other than mandated by winch capabilities. There are three basic types of gravity coring sampling devices:

a. Free-fall, self-surfacing, ballast release
b. Cable deployed and recovered, valve, and core catcher retention
c. Cable deployed and recovered, piston, and core catcher retention

The maximum sample recovery is typically 3.2 ft (1 m), 20 ft (6.4 m), and 137 ft (43.8 m), respectively. Figure 2.2 shows a simple cable-controlled gravity corer assembly that is designed around a weight stand, pipe, core cutter, and core catcher. At the upper end of the tube is a one-way valve that provides an exit for water.

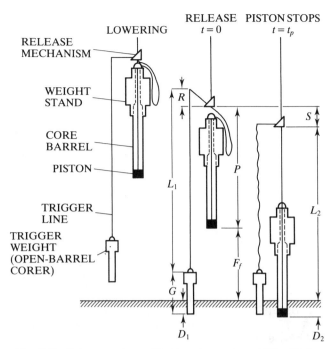

Fig. 2.2 Piston sampler and its triggering system. (*After Preslan, 1969.*)

Bottom Platform

Platform samplers differ from gravity-driven devices in that they are lowered to the seafloor by a cable or driven to a specified location by a submersible. The advantage of this type of sampler is accurate positioning, pistons that are truly fixed relative to the seafloor, and less-destructive sample retainers (Demars and Taylor, 1971). The sample is taken by either pushing or vibrating the sampler into the sediment.

Three diving systems use a modified onshore drilling rig placed on the seabed for operation by divers. The first is lightweight soil or rock-coring rigs or both for shallow waters operated by free swimming divers, while core barrels are individually hoisted to the surface. The second is drilling inside of a dry chamber under ambient air pressure (Hunt, 1984). The third is drilling inside of a dry chamber under atmospheric pressure, a method developed in Holland for seabed cone tests and sampling to shallow depths. The general applicability of these systems is limited by the high costs of diving to relatively shallow depths in friendly waters, or for very specialized applications.

Vibratory corers were developed to recover cores of granular material that could not be recovered by gravity corers. The tool consists of a steel pipe that is driven into the seafloor by applying vibrations at the top. The vibrations are developed by either an eccentric-weight hammer or a piston hammer. An umbilical cable is required for either hammer type to supply their power. Largely because of umbilical capacity of winches and power losses in umbilical lines, vibratory corer operations have been limited to water depths of 200 ft (64 m). Penetrations of the tool rarely exceed 30 ft (9.6 m), although some do have a capability of 60 ft (19.2 m). Soil samples obtained are generally too disturbed for general strength or compressibility soil testing other than classification tests.

A large number of vibratory corers are available from various manufacturers throughout the world. The different types can be classified into three categories based on the power source used for the hammer, which can be (1) electrical, (2) pneumatic,

TABLE 2.3 SUMMARY OF THE GENERAL PROPERTIES OF VIBRATORY CORERS.

Power Source	Ship Positioning	Water Depth, ft	Depth Below Seafloor, ft	Core Diameter, in
Electrical	Loose	1000–2900	3–250	3
Pneumatic	Loose	200– 600	10– 65	2– 6
Hydraulic	Loose	100– 650	10–100	3–12

or (3) hydraulic. A further description of these tools is presented in Table 2.3.

2.2.2 Fixed Seabed Structures

Another series of three systems makes use of fixed seabed structures above the wave level with an onshore drilling system mounted on top of the structure. In water depths of up to 15.6 ft (5 m), a platform can be built on site by scaffolding. To a water depth of about 62.6 ft (20 m), a whole structure can be placed on the seabed, relying on gravity to stand up. In greater water depths of up to about 375.4 ft (120 m), use has been made of towers that are kept upright by wires connected to a moored vessel. Generally the top of the platform is small and because of weight limitations only light drilling rigs can be used. For deeper drilling, a tender barge has to be used for storage. This method is relatively time-consuming (4 to 10 days for a deep borehole) and costly. These systems have been in use in areas with a moderate climate where the standard offshore drilling techniques were not yet known or not available. A third category is the use of a small jack-up rig. In practice, this is limited to coastal waters having depths of about 46.9 ft (15 m) to 62.6 ft (20 m) or less.

2.2.3 Rotary Drilling and Sampling Through a Drill String

Sampling through a drill string which extends below a support ship with a moon pool allows incremental samples to be obtained. The drill string cannot withstand much lateral movement, the ship is typically anchored in shallow water at four points or dynamically positioned if in deeper water. The use of this procedure allows the following: (1) an avenue for retrieving core barrels and returning them to considerable depths below the seafloor, and (2) the drill string keeps the hole open and can be rotated with a drill bit to advance the hole.

The three distinctly different types of drill rigs or vessels commonly used for geotechnical offshore site investigations are shown in Figure 2.3 (Richards and Zuidberg, 1986). The first type, "North Sea Type—Ship-Integral," has a large derrick mounted over a center well, which is typically about 12.5 ft × 12.5 ft (4 m by 4 m) in size. The drill rig with all its components is an integral part of the ship. This vessel may be dynamically positioned over a site or anchored with up to three bow and three stern anchors. The use of a bow thruster enables anchoring without the assistance of another boat. The drill string will always be heave-compensated. Heave compensation is a method employed offshore to minimize the effect of waves on the drilling and sampling process. A more thorough discussion of heave compensation will be presented later in the chapter. The second type, "North Sea Type—Portable," has a portable heave-compensated drill rig that can be readily added to or removed from a vessel. While the type shown in Figure 2.3 is in the preferable location over a relatively small center well located midships for minimum heave in rough seas, the rig can also be located over the stern. Various vessels are available having center wells, with a diameter of about 2.5 ft (750 mm) to 3.0 ft (900 mm). The third type, "Gulf of Mexico—Portable," consists of a portable onshore drill rig that usually is not heave compensated but which may be adapted for offshore use. This type of drill rig is commonly located over a small center well.

The drill ship system is based on the use of straight flush rotary drilling to advance the drill string to the required depth for sampling or in-situ testing below the surface of the seabed. In this method, water (occasionally) or weighted mud (usually) is used to carry the soil cuttings to the seabed surface along the outside of the drill string. The hole is prevented from caving in by the pressure within the mud column. For long-term stability in less-stable soils during drilling for oil, the hole is protected with a casing that is cemented against the borehole wall. In offshore soil borings, the drilling of one hole only takes one or two days, and the long-term stability is usually not a problem. Short-term stability problems are controlled by the proper use of drilling muds; otherwise, the installation of casings and conductors would require more time and money than required for the soil boring itself.

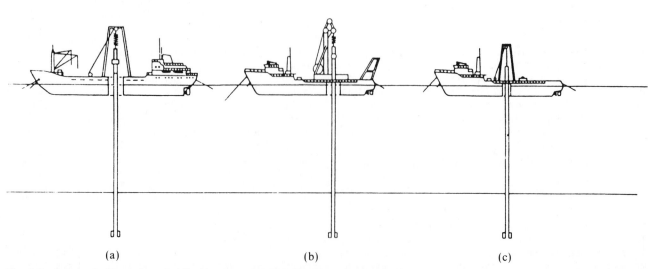

(a)   (b)   (c)

Fig. 2.3 Schematic illustration of drill ship types. (a) Ship-integral. (b) Portable, North Sea type. (c) Portable, Gulf of Mexico type. (*Richards and Zuidberg, 1986.*) Copyright ASTM, reprinted with permission.

Drill Rig and Other Drilling Components

A geotechnical offshore drill rig may include the following components: (1) a derrick; (2) a drilling winch and a drilling line running over a fixed sheave assembly (crown block) mounted on the top of the derrick to a movable sheave assembly (traveling block); (3) a heave compensator usually incorporated either in the crown block or in the traveling block, which keeps the top of the drill string under relatively constant tension on a heaving drill ship; (4) a mud plant, which consists of mud mixing unit, storage tank, pump and piping system that extends to the top of the drill string where the connection is called a mud swivel; (5) either (a) a power swivel, in close proximity to the mud swivel, which rotates the drill string at the top, or (b) a power tong that grips and rotates the drill string about 4.7 ft (1.5 m) above the drill floor; and (6) sections of drill pipe that collectively are called the drill string (Fig. 2.4). A borehole is advanced by rotating the drill string so that the drill bit cuts the soil, which is washed away by a stream of mud flowing through the inside of the drill string and out of the drill bit. At each site, the drilling will normally be stopped every 4.7 ft (1.5 m) to 9.4 ft (3 m) for sampling or testing or both. At such time, the mud flow will be stopped and the top of the mud swivel will be opened to accommodate the carrier tool, which will be lowered down the inside of the drill string by a separate wireline or umbilical. If a power tong is used, the drill string, suspended from an elevator at deck level, can be "broken" at a pipe joint near the drill floor to provide an opening in which the carrier tool is inserted into the drill string. Where required for sampling and in-situ testing, the drill string is then attached again to the heave compensator by slings. The power tong subsystem was common in the early 1970s but has subsequently been phased out in favor of the power swivel.

Drill String and Drill Bit

Drill pipe sections that are coupled into a drill string are typically of the oilfield type, that is, 5 in (127 mm) in diameter, according to the American Petroleum Institute (API) standards. This pipe has a 4 in (100 to 108 mm) inside diameter at the pipe or tool joints. This diameter controls the type of carrier tools, samplers, and sensors that can be used for geotechnical purposes. The bottom few sections of drill pipe typically are composed of heavy thick-walled drill pipes called drill collars, which are designed to increase the drill bit load on the soil and rock, and to help keep the drill string above the drill collars constantly in tension. Major offshore geotechnical companies use lock-in carrier tools, and one section of the drill collar is specially modified to accept these tools.

The outside diameter of the drill bit may vary typically from about 8.3 in (210 mm) to 9.1 in (230 mm). A diameter is usually chosen that is appropriate for the expected soil type and drilling conditions. A diameter larger than the drill collars permits the ready passage of mud and soil cuttings to the surface of the seabed where they accumulate adjacent to the drill string. The drilling bits are open to give the test rods and sensors or samplers access to the soil below. Typical geotechnical drill bits are (1) drag or wing bit, (2) five-winged bit, and (3) Kor King roller bit (Zuidberg et al., 1986). When drilling hard soil, a special bit is available to convert the open bit into a full-face drilling bit (EXXON, 1986).

Vertical Stabilization of the Drill

The drill string must be kept in tension to avoid buckling and subsequent damage to or loss of the string. This requires a bit load to be exerted on the soil. It is desirable for the bit load

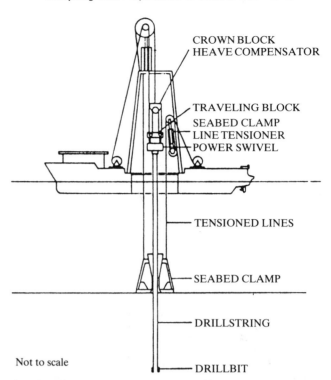

Fig. 2.4 Schematic diagram of the lower part of a geotechnical straight-flush rotary drill system. (*Richards and Zuidberg, 1986.*) Copyright ASTM, reprinted with permission.

to be constant on the soil when drilling. In practice a constant bit load is rarely possible for a variety of reasons. Loads on the underlying soil when drilling or sampling are summarized in Figure 2.5. Heave compensation rarely, if ever, completely eliminates all vertical load variation on the drill bit when drilling. As a consequence, in soft marine soils some degree of compression and tension loading has affected the soil below the drill bit. During the drilling operation, the required load on the drill bit generally ranges from 2 to 25 kN (0.45 to 5.6 kip). In contrast, during push-in or piston sampling, a bit load of 10 to 100 kN (2.2 to 22.5 kip) is usually required. Excessive bit load, as well as drilling mud pressure, may adversely affect the in-situ state of stress in the soil below the drill bit, particularly in soft cohesive soils (Zuidberg, 1979; Richards and Zuidberg, 1985). As a consequence, the control of bit pressure is important when drilling. When performing push or piston sampling or in-situ tests, the drill string must have minimum vertical movement, and preferably none at all. This is particularly important for soft soils.

When drilling from a vessel that is heaving in rough seas, the drill string will heave with the vessel if rigidly attached to it. At the other extreme, if the drill string were to stand alone and if the water depth were more than about 93.8 ft (30 m), the string would buckle and collapse because of its own weight, currents, and so forth. To prevent this, the drill string may be connected to the derrick on the vessel by a combination of springs and dashpots called a heave compensator. Using a heave compensator a specified tensile load can be applied to the top of the drill string. This allows: (1) the drill string to be under tension to prevent buckling, and (2) the load on the bit to be made appropriate to the soil to be drilled (minimum in soft clays and increasing in stiff soils). There are three principal methods that have been used or are used to stabilize the vertical movement of the drill string: (1) vessel heave compensation, (2) seabed clamp, and (3) a downhole anchor. Each will be described in the following sections.

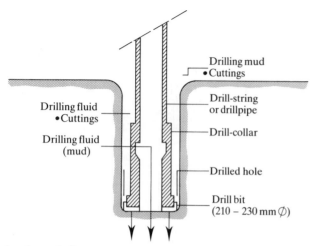

Loads on Soil

Drill Bit Load
- Drilling: 0.2 to 2.5 tonnes, usually
- Testing/sampling: 1 to 10 tonnes, usually
- If poorly heave-compensated, soil suction when bit moves up and "percussion" when bit moves down

Drilling Fluid Load
- Mud pressure normally hydrostatic
- In some clays, pressure may be increased to about 200 kPa
- Seawater density = 1.025 Mg/m³
- Drilling mud density = 1.05 to 1.08 Mg/m³

Fig. 2.5 Schematic diagram of loads on soil due to drilling operation. (*Modified from Richards and Zuidberg, 1985.*)

There are two common methods of heave compensation. The first is the top-mounted crown block heave compensator. The principle of operation is shown in Figure 2.6. The traveling block, power swivel, and mud swivel are attached by a wire to a crown block located on top of the derrick. The crown block in turn is connected to a hydraulic ram that damps the ship movement because of heave. The heave that can be allowed depends on the behavior characteristics of the vessel. It ranges for the "ship integral" system from 9.8 ft (3 m) to 19.7 ft (6 m) for drilling and 6.6 ft (2 m) to 16.4 ft (5 m) for testing. An alternative subsystem is the traveling block heave compensator. In this method, the hydraulic tensioner is placed between the traveling block and the mud swivel and power swivel. Because of this location, the power and mud swivels cannot closely approach the top of the derrick, which for a given height of a derrick results in significantly less working space available for the tool insertion during the drilling sequence. Recently some of these compensators have been offset from the axis of the borehole to overcome this problem. The traveling-block heave compensators used offshore, depending on the type, can accommodate a maximum ship's heave of about 6.6 ft (2 m) to 13 ft (4 m) during drilling or testing. A recent development for heave compensation for soft sediments is the use of a "Hard Tie" as shown in Figure 2.6. A detailed explanation of this procedure is given in Zuidberg et al. (1986).

In addition to the heave compensators that are used during both drilling and sampling the other two methods of drill string stabilization are used only for high-quality sampling or testing or both. The most common method is a heavy (70 to 100 kN; 17.5 to 22.5 kip) seabed clamp that rests on the seabed and can hydraulically clamp the drill string upon actuation from aboard the vessel. There are two advantages to the seabed clamp, which is a positive and reliable method of drill string stabilization: (1) to add reaction to the drill string and (2) to supplement

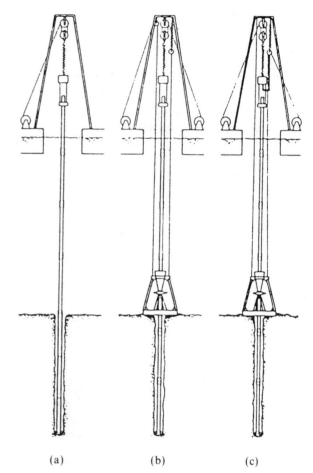

Fig. 2.6 Schematic diagram of three methods of drill-string stabilization against vertical movement. (a) Drilling with heave compensator. (b) Drilling with heave compensator and seabed frame. (c) Drilling with hand-tie.

the role of the shipboard heave compensator when working in very soft soils.

The seabed clamp also can serve as a reentry base (Zuidberg, 1979). In the rare case that a drill bit is worn or tools have been lost downhole and cannot be fished out of the drill string, the drill string can be pulled back for inspection and, subsequently, using the guides of the seabed clamp a new drill string can be inserted into the same hole.

The last method of heave compensation, the downhole anchor (Zuidberg and Windle, 1980) or inflatable packer, is now rarely used to stabilize the drill string since seabed clamps came into general use several years ago.

Typically, 15 to 20 min is required to take a sample or perform a test using modern carrier tool equipment. This time includes both lowering and retrieving the carrier tool.

Samplers Operated in Motion-Uncompensated Drill String

For deep penetration geotechnical investigation, soil conditions will need to be explored with a drill rig. Soil sampling equipment lowered through a motion-uncompensated drill string is the simplest and probably least-expensive procedure for obtaining soil samples at depth. During this type of exploration program, the soil string moves with the vessel. Sampling tools available for this type of drilling program are either isolated or nonisolated from the drill string during the sampling operation. The isolated tools can be used under more severe weather

TABLE 2.4 OFFSHORE SOIL SAMPLERS USED THROUGH THE DRILL STRING.

Name and Owner	Equipment Complexity	Equipment Weight	Operational Method	Ship Positioning	Water Depth, ft	Depth Below Seafloor, ft
Wireline percussion/ several available	Simple	Light	MUDS[a]	Precise	2000	Total
Pressurized core barrel/ Texas A & M	Complicated	Light	MUDS	Precise	450	Total
Hydraulic piston corer (HPC-15)/Scripps	Simple	Light	MUDS/SDS[b]	Precise	No limit	650
Latch-in sampler/ MEI & others	Simple	Light	MUDS/SDS	Precise	No limit	No limit
Swordfish/ McClelland Engineers	Complicated	Light	SDS	Precise	1800	Total
Wipsampler/Fugro	Simple/Average	Light	SDS	Precise	5000	Total
Piston sampler/Fugro	Average	Light	SDS	Precise	5000	Total
Ambient pressure sampler (APS)/Fugro	Complicated	Light	SDS	Precise	5000	Total
Stingray/ McClelland Engineers	Complicated	Heavy	SDS	Precise	3000	Total
Dolphin/ McClelland Engineers	Average	Light	TSP[c]	Precise	No limit	Total

[a] MUDS = Motion-uncompensated drill string.
[b] SDS = Stabilized drill string.
[c] TSP = Tethered seabottom platform.

conditions, while the nonisolated tools require good weather conditions and a long natural ship's heave period for proper operation. Consequently, it is generally recommended that operations undertaken with an uncompensated drill string have both types of samplers on board. During good working conditions the nonisolated tools have the greatest potential for recovering high-quality soil samples. A listing of important properties of each tool discussed in this section is presented in Table 2.4.

Wireline Percussion Sampler A wireline percussion sampler was developed in the 1960s for use in geotechnical investigations from anchored supply vessels. This sampler was considered a major breakthrough because it could be operated with a motion-uncompensated drill string offshore. The tool consists of a sampler head and sliding weight of either 175 or 300 lb (80 or 136 kg). The sampler head is designed with an outside vent and check ball. The sliding weight, which is raised with the wire line, typically has a maximum vertical travel of 10 ft (3.2 m). The sampler is driven by raising and then "free falling" the sliding weight. Emrich (1971) gives a more detailed description of the tool.

The percussion sampler is capable of recovering either granular or cohesive soil samples. Also, the tool can be modified with appropriate head adapters to serve the various types of samplers such as split-barrel and thin-wall tubes. The major shortcoming of this type of sampler is the disturbance caused by percussion sampling (Hvorslev, 1949). In addition, drillers have experienced problems in controlling the downhole performance of the sampler in deep water (DeGroff, 1985). In particular, the weight of the wire line becomes so large in comparison to the sliding weight that the actual sampler penetration and the position of the sliding weight are difficult to distinguish.

Several investigators have quantified the effects offshore wireline percussion sampling has on shear strength and other soil properties. For example, Emrich (1971) compared shear strength obtained from soil samples recovered with different samplers and found the strength of percussion samples to be about 20 to 50 percent lower than push samples. Other investigators (Sullivan, 1978; Semple and Johnston, 1979), have studied the effect percussion sampling has on

North Sea soil deposits. These studies showed the percussion sampler did not have a deleterious effect on shear strength but did adversely affect stress–strain properties.

Latch-in Push Samplers Latch-in push samplers are relatively new to the world of offshore geotechnical investigations. The latch-in sampler consists of a latching mechanism that locks into the drill bit or drill pipe. A thin-wall sampling tube is attached to the latching mechanism and extends beyond the drill bit. Samples are obtained by lowering the drill pipe, which provides the required reaction, to a depth equal to the predetermined sample length. The drill pipe is then raised and the sample tube is extracted from the soil formation. The latch-in mechanism is then released, and the sampler is retrieved on a wire line. This operation is depicted in Figure 2.7.

This type of sampler has been used with a motion-uncompensated drill string. However, the success of recovering a high-quality sample with a moving drill string system is dependent upon the period of motion for the work vessel. If the period is too short, the driller cannot recover the sampler before it is "double pushed" or bent.

The tool has only mechanical parts; therefore, there are no real depth limitations. The latch-in sampler is used to penetrations of 3000 ft (959 m) without difficulty.

Hydraulic Piston Corer (HPC-15) As part of the Deep Sea Drilling Project, a hydraulic piston corer was developed that can be used with motion-uncompensated drill pipe (Driscoll and Hollister, 1969). This is a proven deepwater sampler that has taken high-quality 2.5 in (63.5 mm) diameter soil samples in water depths greater than 6500 ft (2078 m). Sample recovery is generally 90 percent or more, which is outstanding considering the total sample length of 14 ft (4.5 m). In general, the mechanics of the tool are similar to the land-based Osterberg piston sampler.

The two basic assemblies of the HPC-15 are the inner fixed piston assembly and outer sampling barrel. Before sampling, the outer barrel is locked into a position with shear pins so that the core catcher extends just below the stationary piston head. For sampling, the tool is lowered down the drill string until it seats against a stationary sealing top sub. The circulation pump is then started and the drilling fluid begins to build up

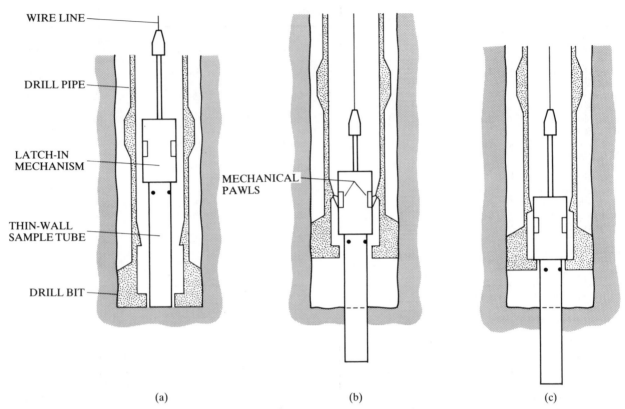

WIRE LINE

DRILL PIPE

LATCH-IN
MECHANISM

MECHANICAL
PAWLS

THIN-WALL
SAMPLE TUBE

DRILL BIT

(a)   (b)   (c)

Fig. 2.7 Operation of a typical latch-in sampler (a) at rest; (b) pushing; (c) pulling. (*Campbell et al., 1982.*) Copyright Offshore Technology Conference.

pressure against the stationary seal and a traveling piston on the core barrel. When the pressure approaches 1500 psi (10 350 kPa), the pins shear and the outer barrel is thrust into the soil formation within a 1 to 2 second period. The sampler is then retrieved and disassembled on the work deck.

Pressurized Core Barrel The pressurized core barrel was developed to sample gas-charged sediments and maintain them at downhole pressure (Denk et al., 1981; Zuidberg et al., 1984). This prevents the free and dissolved gas from expanding within the recovered soil specimen. Expansion of gas can cause alterations to the unit weight and degree of saturation as well as stress–strain and strength characteristics of the soil. Even in non-gas-charged sediments the core barrel could be used to maintain in-situ stress on a sample, thus providing a more "perfect" sample. The typical tool has two sections: (1) a lower part where the core or push sample is stored and (2) an upper section enclosing the hydraulics and control system. During sampling, the sample tube is hydraulically extended beyond the lower section of the tool to take the soil sample and then retracted back inside an outer barrel. The end of the barrel is then sealed with a ball valve. The tool is then wirelined to the surface.

There are some present limitations to this core barrel. For example, the present tools are typically designed for a total working depth of only 1000 ft (320 m), and the testing laboratory is limited to pressures equivalent to 225 ft (72 m) of water. Consequently, the recovered soil samples below a 225 ft (72 m) water depth will experience some stress release during testing.

Samplers Operated in Stabilized Drill String

To obtain high-quality soil samples during adverse weather or sea conditions, the drill string needs to be vertically stabilized

with reference to the soil formation. This can be accomplished by using a motion compensator as discussed earlier. Reaction loading required to push a sampling tube can be provided by adjusting the compensator so that some of the drill string weight is supported by the drill bit bearing against the bottom of the borehole. The "stored" bit load force can then be used to push the sampler. Alternatively, a seafloor jacking unit or drill pipe clamping unit operated in conjunction with a motion compensator can provide the required reaction loading.

2.3 HANDLING, WRAPPING, AND LABELING

The procedure used for handling, storage, and preparation of samples after they have been taken and the general strategy employed in testing (location and time of test) strongly affect the strength and compressibility results that are ultimately obtained. A summary of the typical steps for sampling and preparation of undisturbed marine sediments is presented in Figure 2.8. The methodologies employed by industry and academic/government laboratories are different.

The development of the procedure utilized by industry has been described in a number of publications (Emrich, 1971; Williams and Aurora, 1982; Campbell et al., 1982; Ladd and Azzouz, 1983; Young et al., 1983). Typically the procedure involves extruding the core from an unlined thin metal sampling tube (Shelby Tube). Selected sections of the core are then visually classified and water content, density, and strength index testing performed. Strength index testing is performed using any or all of the following: miniature vane, Torvane, pocket penetrometer, and occasionally the fall cone penetrometer. If a particularly soft material is encountered, all but the fall cone test can be performed directly in the tube. Additional

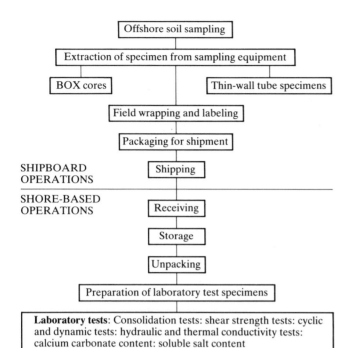

```
                    Offshore soil sampling
                            |
        Extraction of specimen from sampling equipment
              |                               |
        BOX cores                   Thin-wall tube specimens
              |                               |
                 Field wrapping and labeling
                            |
                  Packaging for shipment
                            |
SHIPBOARD                Shipping
OPERATIONS
─────────────────────────────────────────────
SHORE-BASED              Receiving
OPERATIONS
                            |
                         Storage
                            |
                        Unpacking
                            |
            Preparation of laboratory test specimens
                            |
```

Laboratory tests: Consolidation tests: shear strength tests: cyclic and dynamic tests: hydraulic and thermal conductivity tests: calcium carbonate content: soluble salt content

Fig. 2.8 Steps for sampling and preparation of laboratory undisturbed marine soil test specimens.

representative samples are subsampled and wrapped first in plastic wrap followed by aluminum foil prior to being placed into a cylindrical plastic container and sealed with plastic electrical tape.

Samples are extruded while still on shipboard for the following reasons (Young et al., 1983):

1. Stiff clays may "set up" in the tube and require larger extrusion forces if there is a delay.
2. Extruding allows one to test and evaluate the samples and better define the stratigraphy.
3. If samples are extruded, the project engineer can better assess the need to obtain another sample before drilling to the next sampling interval.

These considerations override Hvorslev's (1949) recommendation to preserve the sample "in the tube to avoid disturbance caused by removal and handling of the unprotected sample under adverse conditions in the field." They also override studies that show that using a required extrusion stress exceeding 900 percent of the unconfined compressive strength causes an undrained shear strength reduction of 10 to 20 percent (Arman and McManis, 1976; Broms, 1980). Young et al. (1983) also indicate that disturbance resulting from extrusion can be reduced with increased sample diameter and length, reverse extrusion (back through the opening that the sample originally entered), and trimming of disturbed zones from the sample.

In contrast to industry, most university and government laboratory investigators obtain marine sediment samples in plastic liners within a metal core barrel. The plastic liner is typically cellulose acetate butyrate (CAB). Studies by Keller et al. (1961) have shown that CAB has poor water retention properties and use of a polyester such as polyvinyl chloride is preferred. Samples taken by academic and government laboratories are usually taken with gravity or piston corers, although they can also be obtained with drill string samplers as is the case in the Deep Sea Drilling Project. Upon

removal from the core barrel they are cut transversely into lengths ranging from 1.6 ft (0.5 m) (Almagor and Wiseman, 1977) to 4.8 ft (1.5 m) (Koutsoftas and Fisher, 1976; Bennett et al., 1977). The plastic tube cutting techniques include, among others, a stainless steel rotary-blade hand-held cutter (Dremo Drill or equivalent), a table saw, a hot wire to melt the plastic (Hironaka, 1968; Tirey, 1972), or a hack saw. For techniques where the plastic is cut first, the soil itself is usually cut with a fine wire.

Before the cores are processed in a shore laboratory, X-rays are typically taken to identify disturbed zones, layers of different materials (especially sands), and such items as pebbles, nodules, or shells (Richards and Parker, 1968; Krinitzsky, 1970; Roberts et al., 1976; Allen et al., 1978; Lambert et al., 1981; ASTM D 4452, 1987). An example of X-rays taken on box cores from the prodelta region of the Eel River in Northern California is shown in Figures 2.9 and 2.10.

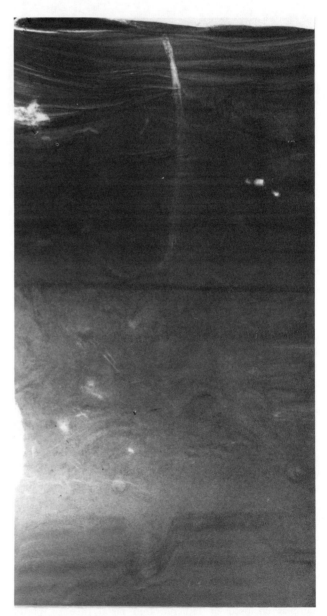

Fig. 2.9 X-ray radiograph of box core from the Eel River prodelta in Northern California (87-6a), showing burrows, laminae, and bioturbation. (*After Borgeld, 1987.*)

Fig. 2.10 X-ray radiograph of box core from the Eel River prodelta in Northern California (87-7), showing laminae and disturbance. (*After Borgeld, 1987.*)

2.4 STORAGE

2.4.1 Prevention of Moisture Loss

Specimens that require more advanced test techniques (i.e., triaxial and consolidation test) are typically sent to shore-based laboratories. Samples for these tests are selected on shipboard. If the sample has already been extruded, it is typically wrapped first in plastic wrap followed by aluminum foil, and then placed in a cylindrical container. The container is either made of cardboard, which is subsequently filled with microcrystalline wax, or plastic, which is capped and then sealed with plastic electrical tape. In contrast, if the sample is still in the sampling tube, then the ends are sealed with rubber caps and plastic electrician's tape and then dipped in wax. Arman and McManis (1976) and Young et al. (1983) warn against using wax that is excessively hot because it might lead to water migration, pore-pressure changes, and drying. Arman and McManis (1976) in addition also recommend the use of polyethylene bags and aluminum foil instead of wax.

2.4.2 Vertical Versus Horizontal Storage

Samples enclosed by either a thin metal tube or plastic liner are generally stored vertically (Bennett et al., 1977; Lambert et al., 1981), although there is at least one report of horizontal storage (Almagor and Wiseman, 1977). The primary advantage of vertical storage is that it preserves the natural stress orientation. The disadvantages of vertical storage are (1) settlement of the sample if there is significant disturbance or water drainage, and (2) cores tend to be unstable unless supported laterally by some type of core holder. In contrast, horizontally stored cores have less vertical effective stress and tend to settle less under their own weight. If they do settle, however, a space is opened along the top. If the core section is placed in a vertical orientation again, soft surface sediment may flow into the void, producing a smear zone (Winters, 1986). In summary, vertical storage is probably best, although the samples should be cut to a fairly short length (approximately 1 m). If samples are stored horizontally, a note to that effect should be made on the core sample information report. In addition, samples should be stored in such a way as to prevent vibrations and shocks from disturbing the sediment (ASTM D4220, 1987).

2.4.3 Storage Time

Most workers agree that testing should be conducted soon after the sample is taken, although there is not full agreement as to how soon it should be defined. Research studies have indicated that, if the samples are not properly sealed, some aging effects may occur and alter the physiochemical and the mechanical properties of the clay, even though there is no apparent loss of water content with time (La Rochelle et al., 1976; Torrance, 1976; Lessard and Mitchell, 1985). More recent work by La Rochelle et al. (1986) has shown that if air is prevented from reaching the surface of the clay during storage, by the use of plastic sheets dipped in wax, no aging effect should occur. A summary of selected sample storage studies is presented in Table 2.5. Based on these studies, several recommendations on sample storage have been made. Young et al. (1983) recommended testing within 7 days if possible, although as long as 30 to 45 days storage may not be harmful. Hironaka and Smith (1968) recommended testing within 1 week. In summary, it is apparent from the most recent work that if samples are wrapped in plastic sheets dipped in wax then they can be stored for many years without changes in properties.

2.4.4 Temperature of Storage

Storage of specimens at near in-place temperatures (2 to 5°C) to prevent the growth of bacteria and chemical changes is

TABLE 2.5 SUMMARY OF SELECTED SAMPLE STORAGE STUDIES.

Storage Time	Result	Reference
10–20 days	Approximately 20% change in strength and maximum past pressure	Arman and McManis (1976)
7.5 months	No change in shear strength for samples stored under water	Hagerty (1974)
1 year	10% loss in strength with 37 mm wax end plug	Broms (1980)
1.5 years	Decreased shear strength	Bozozuk (1971)
3 years	No change in strength and Otterberg limits for samples wrapped in plastic sheets dipped in wax	La Rochelle et al. (1986)

recommended by most workers (Richards and Parker, 1968; Lee and Clausner, 1979; Lambert et al., 1981; Rau and Chaney, 1988). Comparative studies involving strength testing of refrigerated and nonrefrigerated samples show mixed results. Two studies (Hagerty, 1974; Broms, 1980) showed no storage temperature effects on shear strength, while one study by Lee (1973) showed that storage at room temperature for one year reduced the shearing strength by about 25 percent.

In summary, refrigerated storage is probably desirable. Sample freezing must be avoided under all circumstances.

2.5 PREPARATION OF SOIL SAMPLES

The preparation of undisturbed soil samples for laboratory testing can be considered in three separate cases. These cases are (1) the sample has enough strength to support itself during the trimming operation and subsequent transport to the testing device; (2) the sample has some structural integrity but cannot support its own weight either during the trimming process or transport stages; and (3) the sample behaves as a viscous liquid. The sample in case 1 can be treated using standard geotechnical practice as presented in Chapter 1 of this handbook and will not be discussed further. The sample in case 3 cannot be handled using normal procedures. The two alternatives that can be used in case 3 are: (1) removal of the material from the sampling tube and testing using a viscometer, or (2) testing in the sampling tube using a laboratory miniature vane shear device (ASTM D4648, 1987) or other strength measuring device. The remainder of this section will be devoted to handling a sample that fits into case 2.

The two principles of the methodology used in trimming a sample that cannot support itself for testing in a triaxial pressure cell are (1) at no time is the sample handled manually, and (2) it is necessary to maintain lateral stability of the sample at all times. The basic procedure that meets these two criteria is a subsampling procedure using a thin-walled piston sampler that is guided vertically into the sample. A schematic illustration of this apparatus is shown in Figure 2.11. In operation, the thin-walled tube is pushed into the core. Once the desired sample is in the tube a piston is inserted into the top of the sampling tube to provide suction to aid in retention and later extruding of the sample. The sampling tube with the soil is then removed from the core. The sampling tube with the soil is then inserted in an assembled split mold with latex rubber membrane that is attached to the bottom platen of the triaxial cell using O-rings. A vacuum should be applied to the split mold to hold the membrane open. The filter stone and filter paper should already be in place on the bottom platen. The sample is then extruded from the sampling tube, using the piston, into the latex rubber membrane. Once extruded, a filter paper and filter stone are then placed on top of the sample. Then a lightweight plastic top platen is placed on top of the sample and the membrane is removed from the split mold and slipped over the top platen. The vacuum is removed from the split mold. Lateral support for the sample is provided by the latex rubber membrane or, if additional support is needed, a small vacuum is applied to the top of the sample through the top cap. The vacuum that is being applied should always be less than the desired effective stress at which sample is to be tested, to prevent over-consolidation from occurring. This procedure is similar to one

(a)

(b)

(c)

Fig. 2.11 Operation of thin-walled piston sampler. (a) Piston sampler components, sediment core section, and sampler support frame. (b) Piston and tube in contact with sediment surface. (c) Piston still at sediment surface level, tube pushed into sediment. (*After Winters, 1987.*)

developed by Landva et al. (1986) for handling soft organic material in consolidometer testing, and by Winters (1986) for triaxial testing of soft marine clays.

2.6 LABORATORY VERSUS IN-SITU PROPERTIES

The prediction of the in-situ characteristics of marine sediments can be accomplished utilizing a combination of in-situ and laboratory testing and analytical procedures. A summary of the advantages and disadvantages of each of these methodologies is presented in Table 2.6. Utilization of laboratory-generated strength and compressibility information requires consideration of the changes in the sediment stress state and condition during the process of coring from the ground, transportation of the core to the laboratory, and handling and trimming a specimen before testing. The properties that change include:

1. Pore fluid content and effective stress from disturbance or by stress release during perfect sampling. The sample stress state is changing typically from an anisotropic to isotropic stress condition.
2. Bulk expansion and gases come out of solution due to the hydrostatic pressure release.
3. Biologic degradation of organic samples during long-term storage.
4. Fabric degradation and destruction of weak cementation bonds from disturbance.
5. Particle fracture in carbonates during coring releases excess water.

The various disturbance mechanisms have been discussed in detail by Richards and Parker (1968) and Broms (1980). Under no conditions, however, can a sample be considered to be completely undisturbed; the properties are always altered from their in-situ state before testing. The effects of perfect sampling (disturbance caused by stress release only) on the undrained shear strength characteristics of sediments have long been recognized (Skempton and Sowa, 1963; Ladd and Lambe, 1963; Noorany and Seed, 1965; Ladd et al., 1977). These studies indicate that perfect sampling leads to a reduction of undrained shear strength on the order of 8 to 30 percent for normally consolidated clays. The influence of other forms of sample disturbance on the strength characteristics of clays has been clarified by extensive research (Ladd and Lambe, 1963; Noorany and Seed, 1965; Casagrande, 1944). These studies show the effects of sample disturbance on the strength behavior of sediments can be so substantial that results from laboratory tests are vastly different from the strength characteristics of in-situ sediments.

The above studies along with others have developed rational methods of estimating in-situ sediment strength characteristics from specially designed series of laboratory tests on partially disturbed but good-quality samples. The methodologies for correcting laboratory tests to in-situ strength generally depend on evaluating the effects of volume change and stress history on undrained strength. These methods are generally used to extrapolate, in various ways, plots of strength versus void ratio (e) and overconsolidation ratio (OCR). The methods include (1) multiplying an estimated undisturbed lab sample strength by an empirical correction factor (Casagrande, 1944; Casagrande and Rutledge, 1947; Calhoon, 1956; Schmertmann,

TABLE 2.6 SUMMARY OF THE ADVANTAGES AND DISADVANTAGES OF IN-SITU, LABORATORY AND ANALYTICAL APPROACHES TO GEOTECHNICAL ENGINEERING TO PROBLEM SOLVING[a].

In-Situ Studies	Laboratory Studies	Analytical Studies
A. Advantages		
1. Ability to measure existing in-situ soil strength by insertion of probe	1. Ability to stimulate and control a variety of environmental and/or structural loads	1. Ability to perform parametric and sensitivity studies
2. Ability to measure existing in-situ pore pressure by insertion of probe	2. Ability to investigate phenomena occurring over longer time span than normal in-situ test	2. Economic if computer program or theory exists
3. Ability to quantify sands, sensitive or quick clays, and soils containing free gas or significant dissolved gas by use of a probe inserted in the deposit	3. Economic under certain conditions	3. Ability to make predictions on future loading conditions
4. Ability to measure variations in density gradients in sediment using remote sensing techniques	4. Provides fundamental parameters for analytical models	4. Ability to investigate phenomena occurring over long time span
5. Ability to measure surface expression of soil deposits using remote sensing techniques	5. Ability to perform parametric sensitivity studies	
6. Ability to measure initial in-situ stress conditions		
B. Disadvantages		
1. Insertion of probe in soil generates excess pore pressure if soil is saturated	1. Boundary conditions imposed by test apparatus	1. Requires calibration of analytical technique
2. Insertion of probe in soil causes localized disturbance	2. Interpretation of scaled results to prototype application	2. Development of task-specific computer programs expensive and takes time
3. Interpretation difficult without adjacent soil samples	3. Possible sample disturbance due to	3. Availability of input parameters is limited
4. Control of boundary conditions either difficult or impossible	a. fabric changes	4. Utilizes highly idealized models
5. Difficult and expensive to perform long-term studies	b. stress relief	
6. Not possible to measure future environmental loadings	c. desiccation	
7. Typically expensive		

[a] Chaney et al. (1985). Copyright ASTM, reprinted with permission.

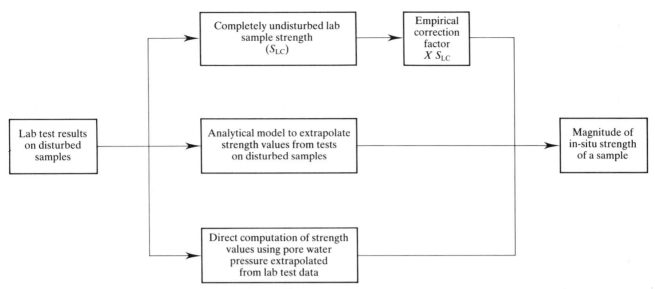

Fig. 2.12 General methodologies for correcting laboratory triaxial and simple shear test results to in-situ strengths. (*After Chaney et al., 1985.*)

1956; Bjerrum, 1972), (2) use of an analytical model to extrapolate strength values from tests on disturbed samples (Ladd and Foote, 1974; Ladd et al., 1977; Ko and Sture, 1981), and (3) direct computation of strength values using pore-water pressure extrapolated from lab test data (Lambe, 1961; Gibbs and Coffey, 1969; Lee, 1973, 1974). A schematic representation of the available correction processes is shown in Figure 2.12.

REFERENCES

Almagor, G. and Wiseman, G. (1977), Analysis of submarine slumping in the continental slope off the coast of Israel, *Marine Geotechnology*, **2**, pp. 349–388.

Allen, R. L., Yen, B. C., and McNeill, R. L. (1978), *Proceedings of the 10th Annual Offshore Technology Conference*, Houston, Texas, Paper No. 3212, pp. 1391–1398.

Anderson, A. (1981), Exploration, sampling, and in-situ testing of soft clay, *Soft Clay Engineering*, E. W. Brand and R. P. Brenner, eds., Elsevier, Amsterdam, pp. 239–308.

Arman, A. and McManis, K. L. (1976), Effects of storage and extrusion on sample properties, *Soil Specimen Preparation for Laboratory Testing, STP 599*, American Society for Testing and Materials, pp. 66–87.

ASTM (1987a), Standard methods for X-ray radiography of soil samples—D4452, *Annual Book of ASTM Standards, Section 4*, American Society for Testing and Materials, Vol. 04.08, pp. 918–933.

ASTM (1987b), Standard practices for preserving and transporting soil samples—D4220, *Annual Book of ASTM Standards, Section 4*, American Society for Testing and Materials, Vol. 04.08, pp. 719–728.

ASTM (1987c), Standard test method for laboratory miniature vane shear test for saturated fine-grained clayey soil—D4648, *Annual Book of ASTM Standards, Section 4*, American Society for Testing and Materials, Vol. 04.08.

ASTM (1988), Standard guide for core sampling submerged, unconsolidated sediments—D4823, *Annual Book of ASTM Standards, Section 4*, American Society for Testing and Materials, Vol. 04.08, pp. 585–597.

Bennett, R. H., Lambert, D. N., and Hulbert, M. H. (1977), Geotechnical properties of a submarine slide area on the U.S. continental slope northeast of Wilmington Canyon, *Marine Geotechnology*, **2**, pp. 245–262.

Bjerrum, L. (1972), Embankments on soft ground, *Proceedings American Society of Civil Engineers, Specialty Conference on Performance of Earth and Earth-Supported Structures*, Purdue University, **2**, pp. 1–54.

Borgeld, J. (1987), Personal communication, Dept. of Oceanography, Humboldt State University.

Bozozuk, M. (1971), Effect of sampling, size and storage in test results for marine clay, *Sampling of Soil and Rock, STP 493*, American Society for Testing and Materials, pp. 121–131.

Broms, B. B. (1980), Soil sampling in Europe: State of the art, *Journal of the Geotechnical Engineering Division, ASCE*, **106**, No. GT1, pp. 65–100.

Calhoon, M. L. (1956), Effect of a sample disturbance on the strength of a clay, *Transactions of the American Society of Civil Engineers*, **121**.

Campbell, K. J., Dobson, B. M., and Ehlers, C. J. (1982), Geotechnical and engineering geological investigations of deep water sites, *Proceedings of the 14th Annual Offshore Technology Conference*, Houston, **1**, Paper No. 4169, pp. 25–37.

Casagrande, A. (1932), The structure of clay and its importance in foundation engineering, *Journal of the Boston Society of Civil Engineers*, **19**, No. 4, p. 168.

Casagrande, A. (1944), U.S.E.D. Triaxial Research Program, *Progress Report No. 7*, Harvard University.

Casagrande, A. and Rutledge, P. C. (1947), *Cooperative Triaxial Shear Research*, Waterways Experiment Station, Vicksburg, Miss.

Chaney, R. C. (1984), Methods of predicting the deformation of the seabed due to cyclic loading, *Proceedings of Symposium on Seabed Mechanics*, International Union of Theoretical and Applied Mechanics, University of Newcastle, Newcastle, pp. 159–167.

Chaney, R. C., Slonim, S. S., and Slonim, S. M. (1982), Determination of calcium content in soils, *Symposium on Performance and Behavior of Calcareous Soils, ASTM STP 777*, K. R. Demars and R. C. Chaney, eds., American Society for Testing and Materials, pp. 3–15.

Chaney, R. C., Demars, K. R., and Fang, H. Y. (1985), Toward a unified approach to soil property characterization, *Symposium On Strength Testing of Marine Sediments: Laboratory and In Situ Measurements, ASTM STP 883*, R. C. Chaney and K. R. Demars, eds., American Society for Testing and Materials, pp. 425–439.

Chaney, R. C. and Fang, H. Y. (1985), Liquefaction in the coastal environment: An analysis of case histories, *Proceedings of the 2nd Shanghai Symposium on Marine Geotechnology and Nearshore/Offshore Structures*, Envo Publishing Co. Inc., pp. 32–64.

Chaney, R. C. and Fang, H. Y. (1986), Static and dynamic properties of marine sediments: A state of the art, *Symposium On Marine Geotechnology and Nearshore/Offshore Structures, ASTM STP 923*, R. C. Chaney and H. Y. Fang, eds., American Society for Testing and Materials, pp. 74–111.

Chaney, R. C., Richards, A. F., and Murray, C. N. (1986), Abyssal plains: Potential sites for nuclear waste disposal, *Proceedings of the International Symposium On Environmental Geotechnology*, Envo Publishing Co. Inc., **1**, pp. 276–284.

DeGroff, W. (1985), Offshore sampling equipment, Unpublished manuscript.

Demars, K. R. and Taylor, R. J. (1971), *Naval Seafloor Sampling and In-Place Test Equipment: A Performance Evaluation*, Naval Civil Engineering Laboratory Technical Report R-730, Port Hueneme, Calif.

Denk, E. W. et al. (1981), A pressurized core barrel for sampling gas-charged marine sediments, *Proceedings, 13th Offshore Technology Conference*, Houston, **4**, pp. 43–52.

Driscoll, A. H. and Hollister, C. D. (1969), The WHOI giant piston core, state-of-the-art, *Proceedings 10th Annual Conference Marine Technology Society*.

Emrich, W. J. (1971), Performance study of soil samplers for deep penetration marine borings, *Symposium On Sampling of Soil and Rock*, *ASTM STP 483*, American Society for Testing and Materials, pp. 30–50.

EXXON (1986), Rotary drill bits, *EXXON USA 2nd Quarter Report*, pp. 16–23.

Fang, H. Y. and Owen, T. D., Jr. (1977), Fracture–swelling–shrinkage behavior of soft marine clays, *Geotechnical Aspects of Soft Clays*, Brenner and Brand, eds., Asian Institute of Technology, Bangkok, Thailand, pp. 15–25.

Fang, H. Y. and Chaney, R. C. (1985), Causes of foundation instability of nearshore/offshore structures and improvement techniques, *Proceedings of Shanghai Symposium on Marine Geotechnology and Nearshore/Offshore Structures*, Tongji University Press, Shanghai, pp. 575–590.

Fang, H. Y. and Chaney, R. C. (1986), Geo-environmental and climatological conditions related to marine structural design along the China coastline, *Proceedings of the Symposium on Marine Geotechnology and Nearshore/Offshore Structures*, *ASTM STP 923*, American Society for Testing and Materials, pp. 149–160.

Fletcher, G. A. (1969), Marine site investigations, *Handbook of Ocean and Underwater Engineering*, J. Meyers, C. H. Holm, and R. F. McAllister, eds., McGraw-Hill Book Co., Inc., New York, N.Y., pp. 8-10 to 8-31.

Gibbs, H. J. and Coffey, C. T. (1969), Application of pore pressure measurements to shear strength of cohesive soils, *U.S. Bureau of Reclamation Report No. EMS-761*.

Hagerty, R. (1974), Usefulness of spade cores for geotechnical studies and some results from the northeast Pacific, *Deep-Sea Sediments, Physical and Mechanical Properties*, A. L. Inderbitzen, ed., Plenum Press, New York, N.Y., pp. 169–186.

Hironaka, M. C. (1968), A remotely controlled incremental seafloor corer, *Technical Note N-1462*, Naval Civil Engineering Laboratory, Port Hueneme, Calif.

Hironaka, M. C. and Smith, R. J. (1968), Foundation study for materials test structures, *Proceedings, Civil Engineering in the Oceans*, American Society of Civil Engineers, New York, N.Y., pp. 489–530.

Hunt, R. E. (1984), *Geotechnical Engineering Investigation Manual*, McGraw-Hill Book Co., Inc., New York, N.Y.

Hvorslev, M. J. (1949), *Subsurface Exploration and Sampling of Soils for Civil Engineering Purposes*, The Engineering Foundation, New York, N.Y.

International Society for Soil Mechanics and Foundation Engineering (1981), *International Manual for the Sampling of Soft Cohesive Soils*, Tokai University Press, Tokyo, 129 pp.

Keller, G. H., Richards, A. F., and Recknagel, J. H. (1961), Prevention of water loss through CAB plastic sediment core liners, *Deep Sea Research*, **8**, No. 2, pp. 148–151.

Ko, H. Y. and Sture, S. (1981), State of the art: Data reduction and application for analytical modeling, *Laboratory Shear Strength of Soil*, *STP 740*, American Society for Testing and Materials, pp. 329–386.

Koutsoftas, D. C. et al. (1976), Evaluation of Vibracorer as a tool for underwater geotechnical explorations, *Proceedings of the 8th Annual Offshore Technology Conference*, Houston, Texas, **3**, pp. 107–122.

Krinitzsky, E. L. (1970), *Radiography in the Earth Sciences and Soil Mechanics*, Plenum Press, New York, N.Y.

Krombein, W. C. (1942), Physical and chemical changes in sediments after deposition, *Journal of Sedimentary Petrology*, **12**, No. 3, pp. 111–117.

La Rochelle, P., Sarraith, J., Roy, M., and Tavenas, F. (1976), Effect of storage and reconsolidation on the properties of champlain clays, *Soil Specimen Preparation For Laboratory Testing*, *ASTM STP 599*, American Society for Testing and Materials, pp. 126–146.

La Rochelle, P., Leroueil, S., and Tavenas, F. (1986), A technique for long term storage of clay samples, *Canadian Geotechnical Journal*, **23**, pp. 602–605.

Ladd, C. C. and Lambe, T. W. (1963), The strength of undisturbed clay determined from undrained tests, *ASTM STP 361*, American Society for Testing and Materials, pp. 342–371.

Ladd, C. C. and Foott, R. (1974), New design procedures for stability of soft clays, *Journal of the Geotechnical Engineering Division, ASCE*, **100**, No. GT7, pp. 763–786.

Ladd, C. C., Foott, R., Ishihara, K., Schlosser, F., and Poulos, H. G. (1977), Stress deformation and strength characteristics, *Proceedings 9th International Conference on Soil Mechanics and Foundation Engineering*, **2**, pp. 421–494.

Ladd, C. C. and Azzouz, A. S. (1983), Stress history and strength of stiff offshore clays, *Geotechnical Practice In Offshore Engineering*, S. G. Wright, ed., American Society of Civil Engineers, New York, N.Y., pp. 65–80.

Lambe, T. W. (1961), Residual pore pressures in compacted clay, *Proceedings of the 5th International Conference on Soil Mechanics and Foundation Engineering*, Paris, **1**, pp. 207–211.

Lambert, D. N., Bennett, R. H., Sawyer, W. B., and Keller, G. H. (1981), Geotechnical properties of continental upper rise sediments—Veatch Canyon to Cape Hatteras, *Marine Geotechnology*, **4**, No. 4, pp. 281–306.

Landva, A. O., Pheeney, P. E., La Rochelle, P., and Briaud, J. L. (1986), Structures on peatland—geotechnical investigations, *Proceedings of the 39th Canadian Geotechnical Conference*, pp. 31–52.

Lee, H. J. (1973), *In Situ Strength of Seafloor Soils Determined From Tests On Partially Disturbed Cores*, Technical Note N-1295, U.S. Naval Civil Engineering Laboratory, Port Hueneme, Calif.

Lee, H. J. (1974), The role of laboratory testing in the determination of deep-sea sediment engineering properties, *Deep-Sea Sediments, Physical and Mechanical Properties*, A. L. Inderbitzen, ed., Plenum Press, New York, N.Y., pp. 111–127.

Lee, H. J. (1985), State of the art: Laboratory determination of the strength of marine soils, *Strength Testing of Marine Sediments: Laboratory and In Situ Measurements*, *ASTM STP 883*, R. C. Chaney and K. R. Demars, eds., American Society for Testing and Materials, pp. 181–250.

Lee, H. J. and Clausner, J. E. (1979), *Seafloor Soil Sampling and Geotechnical Parameter Determination Handbook*, Technical Report R-873, U.S. Naval Civil Engineering Laboratory, Port Hueneme, Calif.

Lessard, G. and Mitchell, J. K. (1985), The causes and effects of aging in quick clays, *Canadian Geotechnical Journal*, **22**, pp. 335–346.

Moore, Jr., T. C. and Heath, G. R. (1978), Sea floor sampling techniques, *Chemical Oceanography*, Vol. 7, 2nd Edn, J. P. Riley and R. Chester, eds., Academic Press, New York, N.Y., pp. 75–126.

Noorany, I. and Seed, H. B. (1965), In situ strength characteristics of soft clays, *Journal of the Soil Mechanics and Foundation Division, ASCE*, **91**, No. SM2, pp. 49–80.

Oceans Magazine (1969), OCMGA, Vol. 3, March.

Preslan, W. L. (1969), Accelerometer-monitored coring, *Proceedings, Civil Engineering in the Oceans II*, ASCE Conference, Miami Beach, Fla., pp. 655–678.

Rau, G. and Chaney, R. (1988), Triaxial testing of marine sediments with high gas contents, *Symposium On Advanced Methods of Triaxial Testing of Soil and Rock*, *ASTM STP 977*, American Society for Testing and Materials, pp. 338–352.

Richards, A. F. and Parker, H. W. (1968), Surface coring for shear strength measurements, *Proceedings, Civil Engineering in the Oceans*, American Society of Civil Engineers, pp. 445–488.

Richards, A. F., Ling, S. C., and Gerwick, Jr., B. C. (1976), Site selection for offshore facilities, *Ocean Engineering*, **3**, Pergamon Press, New York, pp. 189–206.

Richards, A. F. and Chaney, R. C. (1981), Present and future geotechnical research needs in deep ocean mining, *Marine Mining*, **2**, No. 4, pp. 315–337.

Richards, A. F. and Chaney, R. C. (1982), Marine slope stability—A geological approach, *Proceedings of the NATO Conference On Marine Slides and Other Mass Movements*, S. Saxon and J. K. Nieuwenhuis, eds., pp. 163–172.

Richards, A. F. and Zuidberg, H. M. (1985a), In situ testing and sampling offshore in water depths exceeding 300 m, *Offshore Site Investigations*, Graham and Trottman, London, pp. 129–163.

Richards, A. F. and Zuidberg, H. M. (1985b), In-situ determination of the strength of marine soils, *Symposium On Strength Testing of Marine Sediments: Laboratory and In-Situ Measurements, ASTM STP 883*, R. C. Chaney and K. R. Demars, eds., American Society for Testing and Materials, pp. 11–40.

Richards, A. F. and Zuidberg, H. M. (1986), Sampling and in situ geotechnical investigations offshore, *Marine Geotechnology and Nearshore/Offshore Structures, ASTM STP 923*, R. C. Chaney and H. Y. Fang, eds., American Society for Testing and Materials, pp. 51–73.

Roberts, H. H., Cratsley, D. W., and Whelan, T. (1976), Stability of Mississippi River Delta sediments as evaluated by analysis of structural features in sediment borings, *Proceedings of the 8th Annual Offshore Technology Conference*, Houston, Texas, **1**, Paper No. 2425, pp. 10–15.

Schmertmann, J. (1956), Discussion of Paper, "Effect of sample disturbance on the strength of a clay", *Transactions of the American Society of Civil Engineers*, **121**, pp. 940–950.

Semple, R. M. and Johnston, J. W. (1979), Performance of "Stingray" in soil sampling and in situ testing, *Proceedings of the International Conference on Offshore Site Investigation*, London, pp. 169–182.

Skempton, A. W. and Sowa, U. A. (1963), The behavior of saturated clays during sampling and testing, *Geotechnique*, **13**, No. 4, pp. 269–290.

Sullivan, R. A. (1978), Platform site investigation, *Civil Engineering*, London, Feb.

Tirey, G. B. (1972), Recent trends in underwater soil sampling methods, *Underwater Soil Sampling, Testing, and Construction Control, ASTM STP 501*, American Society for Testing and Materials, pp. 42–54.

Torrance, J. K. (1976), Pore water extraction and the effect of sample storage on the pore water chemistry of Leda Clay, *Soil Specimen Preparation for Laboratory Testing, ASTM STP 599*, American Society for Testing and Materials, pp. 147–157.

Williams, J. P. and Aurora, R. P. (1982), Case study of an integrated geophysical and geotechnical site investigation program for a North Sea platform, *Proceedings, 14th Annual Offshore Technology Conference*, Houston, Texas, Paper No. 4168, pp. 11–24.

Winters, W. (1987), (personal communication), U.S.G.S., Woods Hole, Mass.

Winterkorn, H. F. and Fang, H. Y. (1971), Some lessons from other disciplines, *Proc. The Intern. Symposium on the Engineering Properties of Sea-Floor Soils and Their Geophysical Identification*, UNESCO, NSF, and University of Washington, July, pp. 1–10.

Young, A. G., Quiros, G. W., and Ehlers, C. J. (1983), Effects of offshore sampling and testing on undrained soil shear strength, *Proceedings of the 15th Annual Offshore Technology Conference*, Houston, Texas, Paper No. 4465, pp. 193–204.

Zuidberg, H. M. (1979), New system for offshore geotechnical investigations, *New Technologies for Exploration and Exploitation of Oil and Gas Resources*, Vol. 2, Graham and Trotman, London, pp. 995–1009.

Zuidberg, H. M. and Windle, D. (1980), High capacity sampling using a drill string anchor, *Proceedings of the Offshore Site Investigation*, London, pp. 149–158.

Zuidberg, H. M., Richards, A. F., and Geise, J. M. (1986), Soil exploration offshore, *4th International Geotechnical Seminar, Field Instrumentation and In-Situ Measurements*, Nanyang Technological Institute, Singapore.

Zuidberg, H. M., Schrier, W. H., and Pieters, W. H. (1984), Ambient pressure sampler system for deep ocean soil investigations, *Proceedings of the 16th Offshore Technology Conference*, **1**, Paper No. 4679, pp. 283–290.

3 SOIL TECHNOLOGY AND ENGINEERING PROPERTIES OF SOILS

HANS F. WINTERKORN, Dr. phil. nat. (Deceased)
Professor of Civil Engineering and Geophysics,
Princeton University

HSAI-YANG FANG, Ph.D.
Professor of Civil Engineering,
Lehigh University

3.1 DEFINITION OF SOIL

Soil, in the engineering sense, comprises all materials found in the surface layer of the earth's crust that are loose enough to be moved by spade or shovel. Such materials are natural systems that are normally composed of solid, liquid, and gaseous phases. The solid phases are contributed by particulate matter of inorganic or organic character. The liquid phase is usually an aqueous electrolyte solution. The gaseous phase in contact and exchange with the atmosphere may have a different composition from the latter, depending on location and biologic activity within the soil. Since water and air content vary with variation in environmental conditions, soils are normally characterized by their particulate components, while the air and water contents are considered together as porosity. However, in assaying the actual physical properties of a soil system, due consideration must be given to the volume percentages of the component phases as well as to the distribution of the different phases throughout the system.

3.2 DESCRIPTION AND IDENTIFICATION OF SOILS

3.2.1 Size of Soil Particles

Terminology for size fractions as used by various agencies is given in Figure 3.1. Solid soil components may range in size from small boulders (> 30 cm) to colloidally dispersed (< 1 μm)

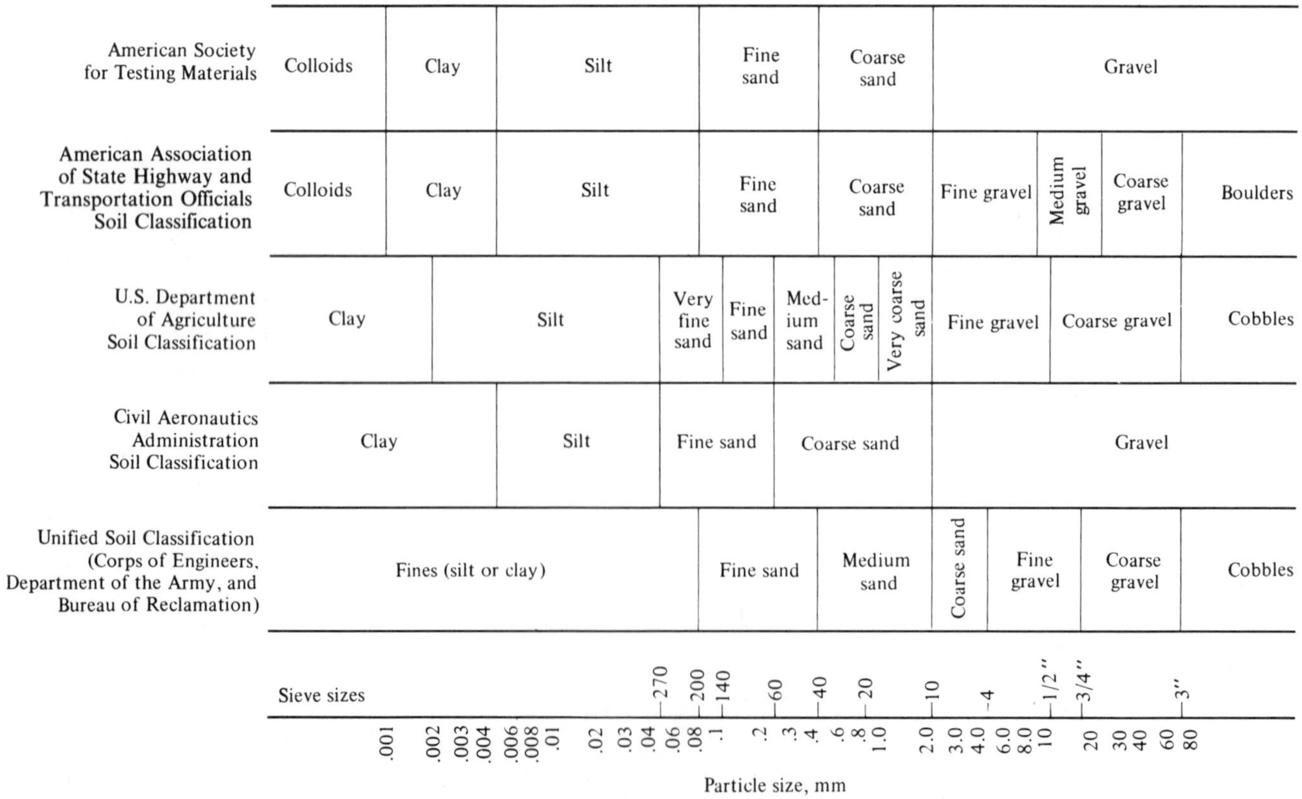

Fig. 3.1 Soil: separate size limits of ASTM, AASHTO, USDA, CAA, U.S. Corps of Engineers, and USBR.

mineral and organic particles, and in mineral character from practically unchanged fragments of igneous, sedimentary, and metamorphic rock through a wide range of weathering products to typical creations (clay minerals, hydrous oxides, organic matter, etc.) of natural soil formation in the biosphere.

3.2.2 Soil Texture

A soil may be composed of only one size fraction of narrow range (beach sand, pebble beaches, loess) or any number of size fractions in continuous or gap grading. The size composition of a soil is called its texture. Figure 3.2 presents a triangular diagram of the textural soil classes as defined by the U.S. Soil Survey. This classification considers only the material that passes the U.S. No. 10 sieve (< 2 mm); the presence of larger particles is denoted by the modifiers gravelly, stony, and cobbly.

If used as components of concretes, mortars, or similar construction materials, gravels are also called coarse aggregate and sands are called fine aggregate. Fractions that pass the U.S. No. 200 sieve ($< 75 \, \mu$m) are called soil fines. In normal soils the very fine sand, silt, and clay constituents do not act

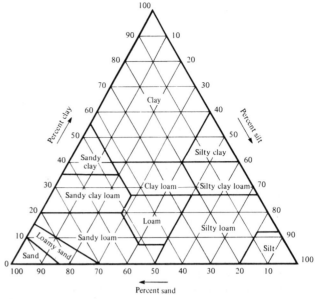

Fig. 3.2 U.S. Department of Agriculture textural classification chart.

TABLE 3.1 FIELD METHOD FOR IDENTIFICATION OF SOIL TEXTURE[a].

Soil Texture	Visual Detection of Particle Size and General Appearance of the Soil	Squeezed in Hand and Pressure Released		Soil Ribboned Between Thumb and Finger When Moist
		When Air Dry	When Moist	
Sand	Soil has a granular appearance in which the individual grain sizes can be detected. It is free-flowing when in a dry condition	Will not form a cast and will fall apart when pressure is released	Forms a cast that will crumble when lightly touched	Cannot be ribboned
Sandy loam	Essentially a granular soil with sufficient silt and clay to make it somewhat coherent. Sand characteristics predominate	Forms a cast that readily falls apart when lightly touched	Forms a cast that will bear careful handling without breaking	Cannot be ribboned
Loam	A uniform mixture of sand, silt and clay. Grading of sand fraction quite uniform from coarse to fine. It is mellow, has somewhat gritty feel, yet is fairly smooth and slightly plastic	Forms a cast that will bear careful handling without breaking	Forms a cast that can be handled freely without breaking	Cannot be ribboned
Silt loam	Contains a moderate amount of the finer grades of sand and only a small amount of clay; over half of the particles are silt. When dry it may appear quite cloddy and can readily be broken and pulverized to a powder	Forms a cast that can be freely handled. Pulverized, it has a soft flourlike feel	Forms a cast that can be handled freely. When wet, soil runs together and puddles	Will not ribbon but it has a broken appearance, feels smooth and may be slightly plastic
Silt	Contains over 80% of silt particles with very little fine sand and clay. When dry, it may be cloddy, readily pulverizes to a powder with a soft flourlike feel	Forms a cast that can be handled without breaking	Forms a cast that can be handled freely. When wet, it readily puddles	It has a tendency to ribbon with a broken appearance, feels smooth
Clay loam	Fine-textured soil breaks into hard lumps when dry. Contains more clay than silt loam. Resembles clay in a dry condition. Identification is made on physical behavior of moist soil	Forms a cast that can be handled freely without breaking	Forms a cast that can be handled freely without breaking. It can be worked into a dense mass	Forms a thin ribbon that readily breaks, barely sustaining its own weight
Clay	Fine-textured soil breaks into very hard lumps when dry. Difficult to pulverize into a soft flourlike powder when dry. Identification based on cohesive properties of the moist soil	Forms a cast that can be handled freely without breaking	Forms a cast that can be handled freely without breaking	Forms long, thin flexible ribbons. Can be worked into a dense, compact mass. Considerable plasticity
Organic soils	Identification based on the high organic content. Muck consists of thoroughly decomposed organic material with considerable amount of mineral soil finely divided with some fibrous remains. When considerable fibrous material is present, it may be classified as peat. The plant remains or sometimes the woody structure can easily be recognized. Soil color ranges from brown to black. These soils occur in lowlands, in swamps or swales. They have high shrinkage upon drying			

[a] After AASHTO (1970, 1986).

independently but as secondary and higher structural units of greater or lesser stability in the presence of water; also the water affinity, the most important property of the clay particles, depends primarily upon their mineralogical character and not on their size. Therefore, except for research purposes, no particle size distribution is determined for soil fines; instead, their interaction with water is determined by the Atterberg limit tests (see Section 3.4) and, if indicated, by additional special procedures. Table 3.1 gives field methods for identifying textural soil types according to AASHTO (1970, 1986).

3.2.3 Soil Horizon and Soil Profile

Soils are three-dimensional systems; they have a two-dimensional areal extent and a third, depth, dimension. Whether they are geologic deposits (unconsolidated sediments) or formed on site by the interaction of geologic parent material, climatic factors, topography, and living organisms, soils show areal limitations and changes with depth. Horizontal as well as vertical transition into another soil type may be gradual or abrupt depending on geologic and soil-forming factors. When vertical changes are due to different geologic deposits, the resulting layers are called strata; when caused by soil-forming factors, they are called horizons. The set of horizons, from the soil surface to the original or physically altered parent rock, is known as the profile. The horizon containing the parent material or substrate is commonly referred to as the C-horizon. The top layer, which spans from the surface deposit of decaying plant litter to a depth at which the organic matter is completely humified, is called the A-horizon. Between the A- and C-horizons lies the B-horizon, which is usually a locus of accumulation of material washed down from the A-horizon in suspension or colloidal solution by percolating precipitation water. Both the A- and B-horizons develop at the expense of the C-horizon or parent

material. If distinct differentiation has taken place within the three primary horizons, they are divided into subhorizons and denoted, respectively, as A_0, A_1, A_2, ..., B_1, B_2, ..., etc. (see Fig. 3.3).

3.2.4 Soil Structure (Pedological)

Soil structure refers to the mutual interaction of soil particles and their arrangement in space. Noncoherent soils (sands and gravels) form single grain structures. Coherent soils (silt, clay) may form massive structures if they suffer no volume change with change in moisture content or if they are so located that no such changes occur except for continued consolidation of water-saturated sediments. Crop, range, and forest soils usually show in their A- and B-horizons aggregation to secondary and higher structural units whose shape may be platy, prismatic, columnar, blocky, nuciform (nut-like), granular, or crumblike. Natural formation of this macroscopic soil structure is always related to water loss and concomitant shrinking of expansive cohesive soils. This water loss may be of general character, occasioned by the moisture and temperature regime in a soil, or it may be localized, as in the vicinity of plant roots or of small growing ice lenses as in winter freezing. These shrinkage structures may become water resistant to a greater or lesser degree by expelling (i.e., moving to the surface of the secondary aggregates during shrinkage) organic matter of lesser water affinity than possessed by the soil minerals or by films of organic or less water-affine mineral matter being deposited there during the normal soil-forming process. Especially water-stable secondary aggregates are the casts of worms and other small animals. The amount of water-stable secondary aggregates is determined by comparing the results of wet sieving of a soil sample with the results of particle size determination on a completely dispersed sample.

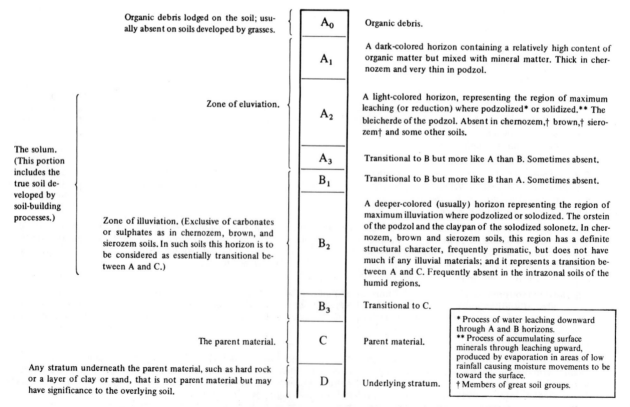

Fig. 3.3 A hypothetical soil profile having all the soil horizons. (*After PCA, 1973.*)

The U.S. Soil Survey differentiates between natural secondary structures or *peds* and aggregations resulting from: (a) plowing or other mechanical working (clods that, when dry, easily slake in water); (b) fragments produced by rupture of a cohesive soil mass; (c) concretions formed by local concentration of compounds that irreversibly cement the soil particles together. The peds are classified according to three criteria: (1) type—shape and arrangement of peds; (2) class—size; (3) grade—distinctness and durability (water resistance). A rather detailed description of types and classes of soil structure recognized by the U.S. Soil Survey is contained in Soil Classification (USDA, 1960).

3.2.5 Soil Fabric

In addition to their macrostructure, natural soils possess fine structures of various kinds that are designated as soil fabric. Their study and elucidation are the objectives of micropedology, which uses hand lenses and the pedologic microscope for field investigations, and thin-section techniques and the petrographic microscope for laboratory work; ultrafine structures are determined with ultrapak techniques and with normal and scanning electron microscopy (McManis et al., 1983; Osipov, 1983).

Soil structure, from macro to ultrafine, influences or even dominates many engineering properties such as permeability, bearing capacity, shear resistance, workability, energy requirements for comminution as in soil stabilization, and thermal and electrical conductivity. In view of the much more inclusive definition of soil used by the engineer, his concept and understanding of structure must be wider than that of the pedologist. It must include, for example, the structure of unconsolidated sediments that may range from disperse or single-grain massive to loose flocculated when precipitated in a marine environment of high salt content. Soil structure in this wider sense will be discussed in Sections 3.7 and 3.8.

3.2.6 Soil Color

Significant differences in color exist not only between different soils but also between different horizons of the same soil. The color may be inherited from the parent material or represent chemical weathering products, whose differential transmission to different horizons produces color variation, or it may be due to organic matter in various amounts and degrees of humification. One differentiates between uniform, spotted, streaked, and mottled colors, which all have physical or chemical significance. For color description, the Munsell notations should be used whenever possible. These take into account (1) hue—dominant spectral (rainbow) color; (2) value—relative lightness of color (approximately the square root of total amount of light); (3) chroma—relative purity of spectral color. Munsell charts for color comparison are commercially available.

Red color due to the presence of nonhydrated Fe_2O_3 (hematite, or bloodstone), and also clear yellow and brown colors, indicate good drainage and aeration; bluish grey colors (reduced or bivalent iron compounds) indicate bad drainage and anaerobic conditions; mottled iron colors show restricted permeability and aeration. Black and dark brown colors are characteristic of organic matter (see Section 3.5) except in rare cases when they are due to manganese, titanium, and other dark-colored minerals, while white colors may be due to preponderance of silica, lime, gypsum, and relatively pure clay deposits.

3.2.7 Visual–Manual Procedure

Visual–manual procedure proposed by ASTM (ASTM D2488) is to standardize the terms to be used when describing a soil for engineering purposes. Descriptive terms used in the procedure include angularity, particle shape, moisture condition, consistency, cementation, dry strength, dilatancy, toughness, plasticity, and others as presented in Table 3.2. The visual procedure is useful in conjunction with Unified Soil Classification System (ASTM D2487-83) as illustrated in Table 3.16 and Figures 3.11 to 3.13. Detailed discussions on this aspect are given by Howard (1984, 1987).

3.3 ROCKS AND THEIR CLASSIFICATION

3.3.1 Origin and Genesis

Rocks are of interest in their own right as foundation support; also they serve as parent material for natural soil formation, and their unconsolidated representatives fall within the engineering definition of soils. In addition, crushed rock serves as an important construction material. Depending upon their formation, rocks are classified as igneous, sedimentary, and metamorphic. Igneous rocks have solidified from a molten or partly molten siliceous solution (magma); sedimentary rocks are naturally consolidated or unconsolidated transported materials. Metamorphic rocks are the result of subjection of igneous or sedimentary rocks to elevated temperatures and/or pressures.

Igneous rocks comprise about 80 percent and metamorphic rocks about 15 percent of the terrestrial and suboceanic earth crust, leaving about 5 percent for the sedimentary rocks; however, about 75 percent of the surface of the continental platforms, and a considerably higher proportion of the ocean floors carry a veneer of sediments. The general characteristics and engineering properties of the more important igneous and sedimentary rocks are shown in Tables 3.3, 3.4, and 3.8. Thermal metamorphosis of sedimentary rocks usually produces harder and tougher materials; limestones crystallize into marbles, and sandstones into quartzites. Metamorphosis by pressure alone is fairly rare, since the differential stresses involved cause shear movement and development of friction heat. This type of metamorphosis results in laminated or banded structures and in mechanical properties that vary with direction. Granite is changed into gneiss, clays and shales into slates.

Igneous rocks are cemented associations of minerals whose size and properties depend not only on the elementary composition of the melt but also on temperature, pressure, and rate of cooling of the system; therefore, as Table 3.3 shows, different types of rock can originate from melts of the same composition. The rate of cooling determines mainly the size of the crystals by its influence on the respective rates of formation of crystal nuclei and of crystal growth. In the same rock formation, the faster-cooling surface layers have smaller crystals than the deeper ones. The higher the silica content of a magma the greater the rate of crystal growth under the same environmental conditions. The size of the individual crystals in rocks, as in metals, determines many important physical properties, with an optimum effect usually at some intermediate size.

Seventy-five to eighty percent of the weight of igneous rocks is contributed by oxygen, silicon, and aluminum; five further elements (iron, calcium, sodium, potassium, and magnesium) make up about 16 percent. These elements form minerals of which more than 90 percent belong to a few groups as shown in Table 3.5. Table 3.6 gives a classification of metamorphic rock

TABLE 3.2 CRITERIA FOR DESCRIPTION AND IDENTIFICATION OF SOILS (ASTM D2488-84).

Description	Criteria	Description	Criteria

a. *Criteria for Describing Angularity of Coarse-Grained Particles*

Angular	Particles have sharp edges and relatively plane sides with unpolished surfaces
Subangular	Particles are similar to angular description but have rounded edges
Subrounded	Particles have nearly plane sides but have well-rounded corners and edges
Rounded	Particles have smoothly curved sides and no edges

b. *Criteria for Describing Particle Shape*

The particle shape shall be described as follows where length, width, and thickness refer to the greatest, intermediate, and least dimensions of a particle, respectively.

Flat	Particles with width/thickness > 3
Elongated	Particles with length/width > 3
Flat and elongated	Particles meet criteria for both flat and elongated

c. *Criteria for Describing Moisture Condition*

Dry	Absence of moisture, dusty, dry to the touch
Moist	Damp but no visible water
Wet	Visible free water, usually soil is below water table

d. *Criteria for Describing the Reaction With HCl*

None	No visible reaction
Weak	Some reaction, with bubbles forming slowly
Strong	Violent reaction, with bubbles forming immediately

e. *Criteria for Describing Consistency*

Very soft	Thumb will penetrate soil more than 1 in (25 mm)
Soft	Thumb will penetrate soil about 1 in (25 mm)
Firm	Thumb will indent soil about $\frac{1}{4}$ in (6 mm)
Hard	Thumb will not indent soil but readily indented with thumbnail
Very hard	Thumbnail will not indent soil

f. *Criteria for Describing Cementation*

Weak	Crumbles or breaks with handling or little finger pressure
Moderate	Crumbles or breaks with considerable finger pressure
Strong	Will not crumble or break with finger pressure

g. *Criteria for Describing Structure*

Stratified	Alternating layers of varying material or color with layers at least 6 mm thick; note thickness
Laminated	Alternating layers of varying material or color with the layers less than 6 mm thick; note thickness
Fissured	Breaks along definite planes of fracture with little resistance to fracturing
Slickensided	Fracture planes appear polished or glossy, sometimes striated
Blocky	Cohesive soil that can be broken down into small angular lumps which resist further breakdown
Lensed	Inclusion of small pockets of different soils, such as small lenses of sand scattered through a mass of clay; note thickness
Homogeneous	Same color and appearance throughout

h. *Criteria for Describing Dry Strength*

None	The dry specimen crumbles into powder with mere pressure of handling
Low	The dry specimen crumbles into powder with some finger pressure
Medium	The dry specimen breaks into pieces or crumbles with considerable finger pressure
High	The dry specimen cannot be broken with finger pressure. Specimen will break into pieces between thumb and a hard surface
Very high	The dry specimen cannot be broken between the thumb and a hard surface

i. *Criteria for Describing Dilatancy*

None	No visible change in the specimen
Slow	Water appears slowly on the surface of the specimen during shaking and does not disappear or disappears slowly upon squeezing
Rapid	Water appears quickly on the surface of the specimen during shaking and disappears quickly upon squeezing

j. *Criteria for Describing Toughness*

Low	Only slight pressure is required to roll the thread near the plastic limit. The thread and the lump are weak and soft
Medium	Medium pressure is required to roll the thread to near the plastic limit. The thread and the lump have medium stiffness
High	Considerable pressure is required to roll the thread to near the plastic limit. The thread and the lump have very high stiffness

k. *Criteria for Describing Plasticity*

Nonplastic	$\frac{1}{8}$ in. (3 mm) thread cannot be rolled at any water content
Low	The thread can barely be rolled and the lump cannot be formed when drier than the plastic limit
Medium	The thread is easy to roll and not much time is required to reach the plastic limit. The thread cannot be rerolled after reaching the plastic limit. The lump crumbles when drier than the plastic limit
High	It takes considerable time rolling and kneading to reach the plastic limit. The thread can be rerolled several times after reaching the plastic limit. The lump can be formed without crumbling when drier than the plastic limit

Identification of Inorganic Fine-Grained Soils from Manual Tests

Soil Symbol	Dry Strength	Dilatancy	Toughness
ML	None to low	Slow to rapid	Low or thread cannot be formed
CL	Medium to high	None to slow	Medium
MH	Low to medium	None to slow	Low to medium
CH	High to very high	None	High

TABLE 3.3 IGNEOUS ROCKS[a].

Acid (Over 66% SiO$_2$)	*Intermediate* (55%–66% SiO$_2$)	*Basic* (Under 55% SiO$_2$)

Coarse-grained plutonic rocks with individual crystallites larger than 1/20 inch. Tend to be brittle due to presence of large crystals, hence coarsest-grained representatives unsuitable for road construction
Examples:

Granite	Syenit	Gabbro
Granodiorite	Diorite	Norite

Medium-grained intrusive (hypabyssal) rocks with individual crystallites between 1/20 and 1/200 inch. Often possess intergrown structure which together with small crystal sizes produces toughness. This group includes the best road construction aggregate
Examples:

Microgranite	Porphyry	Dolerite
Granoporphyre	Porphyrite	Diabase

Fine-grained extrusive (volcanic) rocks with individual crystallites below 1/200 inch (no longer visible with the unaided eye). Similar to medium-grained rock but tend to be glassy, brittle and splintery with increasing SiO$_2$ content
Examples:

Rhyolite	Trachyte	Basalt
Felsite	Andesite	Spilite

Decreasing SiO$_2$ content ———————————➤

Increasing toughness ———————————➤

Increasing specific gravity ———————————➤

Increasing quality as an aggregate for road construction ———————➤

[a] After *Soil Mechanics for Road Engineers*, Department of Scientific and Industrial Research, Her Majesty's Stationary Office, London, 1952 (Crown copyright).

TABLE 3.4 SEDIMENTARY ROCKS[a].

Calcareous. Predominant Mineral CaCO$_3$ (calcite)

Limestone (CaCO$_3$):	Specific gravity 2.65–2.75.
Dolomite (CaMg(CO$_3$)$_2$):	Half of the CaCO$_3$ replaced by MgCO$_3$. Specific gravity 2.7–2.8.
Dolomitic limestones:	Part of the CaCO$_3$ replaced by MgCO$_3$. Softer than sound igneous rock. Tend to become slippery if used for surface dressing owing to physical and chemical polishing and to form rock powder if used in stone bases subject to heavy and frequent traffic loads. This powder makes such bases frost susceptible. Become harder with increasing MgCO$_3$ content.

Siliceous. Predominant Mineral Quartz or Chalcedony (SiO$_2$)

Sandstone:	Specific gravity 2.60–2.75. Strength depends on density and type of natural cementing substance that binds sand grains together. Usually not suitable for road construction.
Quartzite:	Specific gravity 2.55–2.65; tends to be brittle.

Argillaceous. Predominant Mineral Clay

Clay, shale:	Very fine-grained and soft, unsuitable for road aggregate.
Mudstone:	Very fine-grained, often laminated and splintery; when metamorphosed, sometimes useful as a road aggregate.

[a] After *Soil Mechanics for Road Engineers*, Department of Scientific and Industrial Research, Her Majesty's Stationary Office, London, 1952 (Crown copyright).

according to Farmer (1968). Table 3.7 presents the Mohs Hardness Scale for minerals together with simple field tests.

3.3.2 Engineering Properties and Classification of Rocks

The properties of rocks vary significantly within the same rock type and even within the same formation. Hence, tabulated data may serve only as general indicators of the expected range of properties. The actual properties of rock in situ must be determined by appropriate tests for any major projects. For purposes of general orientation, elasticity and strength properties are given in Table 3.8 for various rock types.

Other important properties of rocks and natural rock bodies are:

1. Permeability to water and effect of water on elastic and strength properties
2. Creep of rocks under high stresses, and underlying rheologic properties

TABLE 3.5 AVERAGE MINERALOGICAL COMPOSITION OF IGNEOUS ROCKS, IN PERCENT BY WEIGHT[a].

Constituent	Norm Calculated from Average Composition of Igneous Rocks Given by— Vogt	Norm Calculated from Average Composition of Igneous Rocks Given by— Clarke and Washington	Mineral	Mode According to Clarke
Orthoclase	21.6 ⎫	18.8 ⎫	Feldspars	59.5
Albite	29.2 ⎬ 67.0	33.0 ⎬ 67.5	Quartz	12.0
Anorthite	16.2 ⎭	15.7 ⎭	Hornblende and pyroxene	16.8
Quartz	17.6	9.0	Biotite	3.8
CaSiO$_3$	2.0 ⎫	3.3 ⎫	Titanium minerals	1.5
MgSiO$_3$	6.1 ⎬ 11.8	8.8 ⎬ 17.7	Apatite	0.6
FeSiO$_3$	3.7 ⎭	5.4 ⎭	Other rock-making minerals	5.8
MnSiO$_3$		0.2		
Ilmenite	1.0	2.0		
Hematite	2.2	3.1		
Apatite	0.4	0.7		
Total	100.0	100.0	Total	100.0

[a] After Rankama and Sahama (1950). Reproduced with permission of the University of Chicago Press.

3. Dynamic properties including acceptance, transmission and dispersion of seismic energy
4. Thermal and electric capacities and transmission properties
5. Response upon exposure to environmental conditions that differ physically and chemically from those of the original rock environment

For engineering purposes, rock classification consists of two basic assessments (AASHTO, 1988): that for intact character, such as a hand specimen or small fragment; and in-situ character, or engineering features of rock masses. Figure 3.4 presents various intact-rock classification systems based on the strength of the rock material (Bieniawski, 1980). The other physical parameter frequently used for identification and classification of rock mass is the rock-quality designation (RQD) as proposed by Deere (1963). This parameter is a quantitative index based on a core-recovery procedure that incorporates only those pieces of core 100 mm (4 in) or more in length. The RQD is a measure of drill-core quality, and it disregards the influence of orientation, continuity, joint tightness and gauge (infilling). Therefore, the RQD cannot serve as the only parameter for the full description of a rock mass. However, this parameter is easy to use and simple to understand, and practical engineers have used this index widely for the preliminary identification and classification of rock mass.

The Unified Rock Classification System (URCS) is used commonly in the Forest Service of the U.S. Department of Agriculture (Williamson, 1980). The URCS was originally conceived in 1959, and it has been extended and refined since then. The basic elements include four major physical rock properties: (1) degree of weathering; (2) strength; (3) discontinuity or directional weakness; and (4) gravity or unit weight. By establishing limiting values of these elements by using field tests and observations combined with other geotechnical information, URCS permits a rough estimate of rock performance such as foundation and excavation suitability, slope stability, material use, blasting characters, and hydraulic conductivity.

3.3.3 Properties and Engineering Classification of Shales

Shales predominate among the sedimentary rocks in the earth's crust; their properties vary from those of "solid" rock that must be blasted for excavation to those of soil-like materials that fall within the engineering definition of soil. Obviously, their properties, identification, and classification deserve special attention. Figure 3.5 shows the place of shales within the scheme of sedimentary rocks while Figure 3.6 gives a geological classification of shales (Mead, 1936). We differentiate between compaction of soil-like shales in which the bonding material is essentially the water substance even if present only in very small amounts, and cemented, rock-like shales in which the particulate components may be cemented by calcareous, siliceous, ferruginous, gypsiferous, phosphatic, or other bonding agents, or may be welded together by recrystallization. The chemical nature of bonding agents can usually be determined by simple field or laboratory tests.

Rock-like shales normally preserve their strength and integrity even during repeated exposure to wetting and drying cycles, while soil-like shales slake under these conditions. The time required for the slaking of standardized specimens, the size and character of the slaked particles, as well as the general slaking picture represent valuable clues with regard to the engineering behavior of shales, as for soils.

The bearing on the engineering properties of soil-like shales of such physical, physicochemical, and chemical factors as grain size composition, mineral associations, types and amounts of exchangeable ions, chemical character of electrolytes in the pore water, etc., is of the same nature as their effect on normal soils. As this will be treated in a later section, consideration here will be concentrated on those factors that give shales their distinguishing characteristics. One of these is the mode of breaking and the degree of fissility shown. Ingram (1953) recognizes three dominant types of breaking characteristics: massive, flaggy, and flaky. In the case of intermediate types the dominant feature is given.

Massive structure and resulting breakage mode can be expected from sediments predominantly of silt or larger-sized particles within a limited size range that have not undergone marked shear displacement in their development and consolidation.

Fissility or cleavage characteristics increase with increasing content of plate-shaped minerals (clays and mica) in parallel arrangement because of the manner of sedimentation and overburden pressures. Tectonic disturbances may decrease the parallel order and hence the degree of fissility. The effect of organic matter and of other factors depends on the degree to which they favor or impede the parallel arrangement of the plate-like mineral constituents. The higher the degree of fissility, the greater is the variation as a function of direction of such physical properties as modulus of elasticity, tensile, compressive, and shear strength, thermal conductivity, and water permeability.

The porosity, the void ratio, and the degree of packing of a shale depend on its mineral and granulometric composition, its mode of sedimentation, its stress and deformation history, its chemical history, and the duration of exposure to different

TABLE 3.6 CLASSIFICATION OF METAMORPHIC ROCKS[a].

Classification	Rock	Description	Major Mineral Constituents
Massive	Hornfels	Microfine grained	Quartz
	Quartzite	Fine grained	Quartz
	Marble	Fine to coarse grained	Calcite or dolomite
Foliated	Slate	Microfine grained, laminated	Clay minerals, mica
	Phyllite	Soft, laminated	Mica, clay minerals
	Schist	Altered, hypabyssal rocks, coarse grained	Feldspars, quartz, mica
	Gneiss	Altered granite	Hornblende

[a] After Farmer (1968).

TABLE 3.7 HARDNESS OF MINERALS.

Mohs' Scale	Standard Mineral	Chemical Composition	Field Test can be scratched with—
1	Talc	$Mg_3Si_4O_{10}(OH)_2$	Finger nails—easily
2	Gypsum	$CaSO_4 \cdot 2H_2O$	—with difficulty
3	Calcite	$CaCO_3$	Knife—easily
4	Fluorite	CaF_2	—with moderate pressure
5	Apatite	$Ca_5(PO_4)_3(OH, F, Cl)$	—with difficulty
6	Orthoclase	$KAlSi_3O_8$	—no longer
7	Quartz	SiO_2	Gives sparks with steel
8	Topaz	$Al_2SiO_4(OH, F)_2$	—
9	Corundum	Al_2O_3	—
10	Diamond	C	—

TABLE 3.8 ENGINEERING PROPERTIES OF VARIOUS ROCKS.

Rock Types	Bulk Density, g/cm³	Porosity, %	Poisson Ratio	Compressive Strength, kg/cm²	Tensile Strength, kg/cm²	E^b, 10^5 kg/cm²	$\phi_a{}^c$, degree	$q_a{}^d$, kg/cm²	Dynamic Shear Modulus, 10^5 kg/cm²	Wave Velocity, m/sec	Rate of Water Absorption, %	Coefficient of Saturation
Basalt	2.5–3.3	0.1–1.0	0.02–0.20	1 500–3 000	34–300	6–10	45–60	50–60	7.1–11.4	4 500–6 500	0.29–0.31	0.69
Coal	0.7–2.0			50– 500	20–50	1–2						
Diorite	2.5–3.3	0.1–0.5	0.02–0.25	1 800–3 000	150–300	7–10	40–60	40–60	7.1–11.4	3 000–6 800	0.30–0.38	0.59
Dolerite	2.2–3.0	0.1–0.5		2 000–3 500	150–350	8–11	55–60					
Dolomite	2.2–3.0	0.4–5.0	0.16–0.36	800–2 500	150–250	4–8.4		30–50	3.3–7.8	3 000–6 800		0.80
Gabbro	2.7–3.1	0.1–0.2	0.02–0.16	1 000–3 000	150–300	7–11		40–60	8.6–11.4	5 200–6 800		
Gneiss	2.5–3.0	0.5–1.5	0.05–0.30	500–2 000	50–200	1.5–7		30–50	5.0–9.1	3 700–6 500	0.10–0.70	
Granite[a]	2.6–3.3	0.5–1.5	0.02–0.36	750–2 500	21–250	2–6	45–60	30–60	5.0–9.4	3 000–6 800	0.10–0.70	0.55
Limestone	1.7–3.1	0.4–43	0.04–0.50	100–3 500	6–250	1–8	27–50	12–40	1.0–9.4	2 500–6 000	0.10–4.50	0.35
Marble	2.5–3.3	0.2–2	0.16–0.36	700–2 500	20–200	1–3.4	35–50	40–50	5.0–8.2	3 500–6 000	0.10–0.80	
Mudstone	2.0–2.5	22–32	0.20–0.40	35– 600	3–42	2–5	9–60	12–40	0.5–4.4	1 800–5 250	2.14–8.20	
Quartzite	2.65	0.1–0.5	0.10–0.20	1 500–3 600	57–300	4.5–14.2	50–60	30–60	5.6–14.2	5 000–6 500		
Sandstone	1.2–3.0	0.7–34	0.05–0.25	100–1 800	2–250	0.5–8	27–50	12–40	0.5–9.1	1 400–4 000	0.20–7.00	0.69
Shale	1.6–2.7	1.6–33	0.05–0.30	100–1 000	14–100	1.3–3.5	15–30	12–40	1.9–3.3	1 400–3 000	0.10–0.30	
Slate	2.5–3.3	0.1–0.5	0.16	1 000–2 000	70–200	2.2–3.4	40–50	40–50	7.1–7.8	3 000–6 500	0.10–0.30	0.82

Data obtained from Farmer (1968), Winterkorn and Fang (1975), Tianjin University et al. (1979), and Goodman (1980).
[a] Weathered granite ranges from 1 to 5 percent, and decomposed granite can reach as high as 20 percent.
[b] At zero load.
[c] Apparent friction angle.
[d] Allowable bearing pressure.

$$\text{Coefficient of saturation} = \frac{\text{rate of water absorption}}{\text{degree of saturation}}$$

$$\text{Degree of saturation} = \frac{\text{amount of water absorbed by rock under 150 atmospheres}}{\text{dry weight of rock}}$$

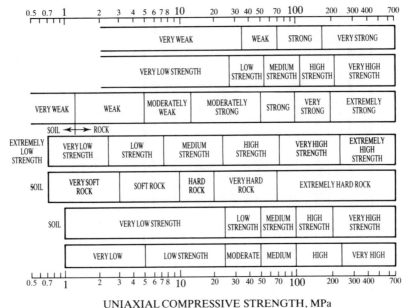

UNIAXIAL COMPRESSIVE STRENGTH, MPa

Note: 1 MPa = 145 lbf/in².

Fig. 3.4 Comparisons of various existing intact-rock classification systems. (*After Bieniawski, 1980.*)

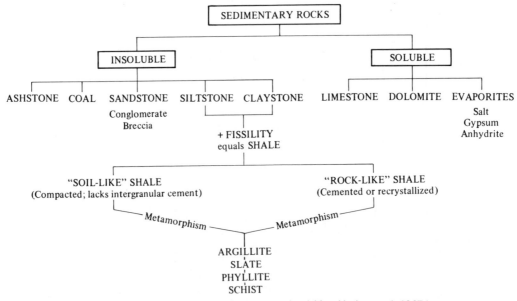

Fig. 3.5 Classification of sedimentary rocks. (*After Underwood, 1967.*)

environmental conditions. Under otherwise comparable conditions, the void ratio decreases with increasing range in particle size. Porosities and corresponding void ratios of shales may range from about 3 percent to more than 52 percent and from 0.03 to more than 1.07, respectively; moisture contents may range from less than 5 percent to as high as 35 percent for some clay shales. Shales are called saturated if their pores are completely filled with water even though they may possess additional water intake and swelling capacity. The mechanical strength of shales decreases exponentially with increasing void ratio and water content. With cemented shales it is a function of the strength of the cementing material and the ratio of its volume to the pore volume. The compressive strength of shale may range from less than 25 psi for the weaker-compaction

shales to more than 15 000 psi for well-cemented shales. Young's moduli have been observed of less than 20 000 psi to over 2×10^6 psi.

With respect to permeability, swelling, shrinkage, consistency, and related properties, shales obey the same basic laws as soils (TRB, 1981, 1982).

Great differences are often observed between the in situ strength and elastic properties of shales and the results obtained in laboratory tests. Contributory factors are (1) improper or inadequate sampling, (2) disturbance of shale structure, (3) difference in stress conditions, and especially (4) the rebound in overconsolidated sediments where the release of strain energy may break the weaker bonds and permit the entrance of air into the expanded pore space. Lowering of strength occurs even

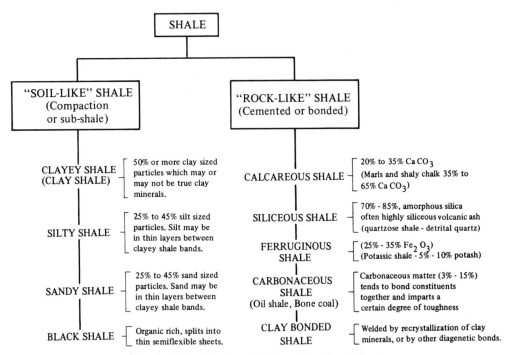

Fig. 3.6 Classification of shale. (*After Mead, 1936.*)

TABLE 3.9 AN ENGINEERING EVALUATION OF SHALES[a].

Laboratory Tests and In Situ Observations (1)	Unfavorable (2)	Favorable (3)	High Pore Pressure (4)	Low Bearing Capacity (5)	Tendency to Rebound (6)	Slope Stability Problems (7)	Rapid Sinking (8)	Rapid Erosion (9)	Tunnel Support Problems (10)
	Physical Properties — Average Range of Values		*Probable In Situ Behavior*						
Compressive strength (psi)	50 to 300	300 to 500	✓	✓					
Modulus of elasticity (psi)	20 000 to 200 000	200 000 to 2×10^6		✓					✓
Cohesive strength (psi)	5 to 100	100 to >1500			✓	✓			✓
Angle of internal friction (degrees)	10 to 20	20 to 65			✓	✓			✓
Dry density (pcf)	70 to 110	110 to 160	✓					✓(?)	
Potential swell (%)	3 to 15	1 to 3			✓	✓		✓	✓
Natural moisture content (%)	20 to 35	5 to 15	✓			✓			
Coefficient of permeability (cm/sec)	10^{-5} to 10^{-10}	$>10^{-5}$	✓			✓	✓		
Predominant clay minerals	Montmorillonite or illite	Kaolinite and chlorite	✓			✓			
Activity ratio[b]	0.75 to >2.0	0.35 to 0.75				✓			
Wetting and drying cycles	Reduces to grain sizes	Reduces to flakes					✓	✓	
Spacing of rock defects	Closely spaced	Widely spaced		✓		✓		✓(?)	✓
Orientation of rock defects	Adversely oriented	Favorably oriented		✓		✓			✓
State of stress	> Existing overbruden load	≅ Overburden load			✓	✓			✓

[a] After Underwood (1967).

[b] Activity ratio = $\dfrac{\text{plasticity index}}{\text{clay content}}$.

TABLE 3.10 SOME PHYSICAL PROPERTIES OF TYPICAL SHALES[a].

Name of Shale Formation, Age, Locality (1)	Compressive Strength, psi (2)	Modulus of Elasticity, psi (3)	Cohesion, psi (4)	Angle of Internal Friction, ϕ, deg (5)	Dry Density, pcf (6)	Potential Swell, % (7)	Natural Moisture, % (8)	Predominant Clay Minerals (9)	Activity Ratio (10)	Source of Data (11)
Bearpaw, Cret., Canada — Weathered	7–84	7500	3 to 6	6 to 20	85–95	0.5–2	29–36	Illite, Montmorillonite	0.30	Ringheim (1964), Morton (1965), Peterson (1954, 1958), laboratory and field slope analyses
Unweathered	154–406	18 000	22	30	95–108	5–20	19–27	Mixed-layer	>1.5	
Pierre, Cret., S. Dakota	70–1 400	20 000–140 000	2 to 30	8–25	95–110	3–5	18–27	Do	0.3 to >2	Corps of Engineers, Missouri River Division
Ft. Union, Tert., N. Dakota	70–1050	11 200–56 000	10	20 / 16	95–115	2 (?)	16–24	Illite	—	Corps of Engineers, South West Division
Pepper, Cret., Texas	28–70	—	2–6	7–14	110	—	20	Illite Mont. Kaolinite	1.2	
Del Rio, Cret., Texas	56–154		1–8	19–28	119	—	17	Illite, Kaolinite, Mont.	1.0	
Trinity, Cret., Texas	30–170	2400–33000	0–7	26	115–133	—	11–17	—	—	
Taylor, Cret., Texas	250–550	6000–20000	1.5–25	8–30	112–118	—	15–18	—	—	
Composite Cyclothem of Pennsylvanian Shales (Eastern Ohio and Western Penn.) — Silty Clayey Carbonaceous	210 / 4165	1 000 000	56 / 1 562	23	138	—	9.1	—	—	Corps of Engineers, Pittsburgh District
Clay Bonded	2084	—	931	7	—	—	—	—	—	
Clayey Ferruginous	1661	—	488	29	—	—	—	—	—	
Sandy	3674	486 500	1 600	9	—	—	—	—	—	
Niobrara, Calc. Sh., Cret., Colo.								Illite, Beid.	—	U.S. Geological Survey, Denver, Colorado
Mowry, Cret., Colo.								Kaolinite, Chlorite, Illite	—	
Graneros, Cret., Colo.								Kaol., Ill., Mix-layer Mont., Illite, Beid.	—	
Morrison, Jura., Colo.									—	
Laramie, Cret., Colo.									—	
Mauv, Calc. Shale, Cam., Utah	5 220	2.3×10^6	1 160	64	164	—	2	—	—	U.S. Bureau of Reclamation, Denver (1953)
Quartzose Sh., Cam., Utah	17 770	2.3×10^6	3 390	45	165	—	4	—	—	

[a] After Underwood (1967).

when the rebound takes place in water under prevention of air entrance. Vees and Winterkorn (1967) found great differences in the shear resistance of kaolinite and attapulgite clays at the same saturated void content and the same normal pressure between virgin specimens and others that had been previously overconsolidated with subsequent pressure release and reconsolidation under water.

Table 3.9 taken from Underwood (1967) lists physical properties considered important for the engineering evaluation of shales and shows their correlation with probable in situ behavior. Table 3.10 gives physical properties of typical shales.

3.4 PHYSICAL PROPERTIES EMPLOYED IN ENGINEERING CLASSIFICATIONS OF SOIL MATERIALS

3.4.1 Particle Size

The engineering classifications in common use are based on the size composition of the solid constituents and on their interaction with the water substance as evidenced by volume and consistency changes. Since water interaction is dominated by total amount of surface present in a sample and since the ratio of surface to volume (specific surface) increases with decreasing particle size, this interaction and concomitant consistency changes are normally determined on the fraction passing the No. 40 sieve (< 0.042 cm). The specific surface (S) for cubic and spherical particles can be calculated from the equation $S = 6/d$, where d is the diameter of the sphere or the side length of the cube. Accordingly, the specific surfaces for particles corresponding to the openings of the 1.5-inch, No. 4, No. 10, No. 40, and No. 200 sieve, and to colloidal size (10^{-4} cm), are 1.6, 13, 30, 143, 800, and 60 000 cm^2/cm^3, respectively. The use of the -40 sieve fraction instead of even smaller ones takes into account that in natural soils the smallest-sized particles are usually aggregated to larger effective units that may even include fine sand particles. Even with the thorough disturbance normally involved in soil used for highway construction, these effective units are not broken down into their primary particulate constituents.

For determination of particle size distribution (ASTM D422) in the coarse range, standardized sieves with openings from 125 mm to 75 μm (U.S. No. 200 sieve) are available. (For international nomenclature, see Table 3.11.) Except where a fraction of specified size must be separated for special tests, any sieve set that gives a fair coverage of the size range of the soil particles may be employed, especially since the data are normally plotted to yield continuous curves from which the percentage smaller than a certain size may easily be picked. Noncohesive soils are sieved dry. Cohesive soils are sieved wet with flowing water with the No. 200 sieve at the bottom of the set. Sedimentation methods for particle size determination are based on the law of Stokes for equilibrium fall velocity of particles in viscous (liquid or gas) media where the gravitational force F_g acting on a particle is equal to the viscous resistance F_v of the liquid.

$$F_g = \frac{4\pi r^3}{3}(\gamma_s - \gamma_l)g = 6\pi r\eta \frac{dl}{dt} = F_v \qquad (3.1)$$

$$v = \frac{dl}{dt} = \frac{2}{9}\frac{r^2(\gamma_s - \gamma_l)g}{\eta} = cd^2 \qquad (3.2)$$

$$\text{and} \quad d = \sqrt{\frac{v}{c}}$$

in which

dl = distance traveled by particle in time dt

$\dfrac{dl}{dt} = v$ (velocity of fall)

r = radius of particle

γ_s = density of particle

γ_l = density of liquid or gaseous medium

η = viscosity of medium

$c = \dfrac{1}{18}\left(\dfrac{\gamma_s - \gamma_l}{\eta}\right)g$ is a constant as long as γ_s, γ_l, g, and η are constant, implying that the temperature also must be constant. When this is not true, pertinent corrections must be made.

Stokes' law holds true only for nonhydrated spherical particles that are not so large that steady-fall conditions are not attained during the time and distance of fall available and not so small that counterdisplacement by Brownian movement equals or exceeds the displacement due to gravity. Despite its limitations, the law is very useful for mechanical analysis and for fractionation of soil particles within the range between 0.1 and 0.001 mm.

3.4.2 Specific Gravity

Knowledge of the specific gravity (ASTM D854) of soil particles is necessary not only for sedimentation analysis but also for assessing the volumetric contributions of the different size fractions to a soil system. For the majority of mechanical analyses, the generally accepted value of 2.65 for the specific gravity of soil fines is sufficiently accurate. If more exact data are desired, the pycnometer method is used for the smaller soil fractions and Archimedes' principle for large gravel and cobbles. The specific gravity of the larger-sized soil constituents is influenced by the presence of inaccessible pores and that of the smallest particles by their interaction with water or with other liquids used as immersion medium (Waidelich, 1958; Andrews et al., 1967). Typical value of specific gravity of some clay minerals is given in Table 3.12.

3.4.3 Consistency of Soil–Water Systems

The original soil consistency tests of Atterberg (1911) have been further developed and standardized to the present methodology, which is widely used by soil engineers. They are normally made on the fraction that passes the U.S. No. 40 (425 μm) sieve; it is often advantageous to make them also on the No. 200 (75 μm) sieve fraction.

The *liquid limit* (LL; W_L) is the water content at which two halves of a soil cake prepared in a standardized manner in the cup of a standardized device will flow together for a distance of $\frac{1}{2}$ inch (1.25 cm) along the bottom of the groove when the cup is dropped 25 times for a distance of 1 cm at the rate of 2 drops per second onto a hard rubber base (ASTM D423).

A simple procedure for determining the liquid limit of soils has been developed by Fang (1960). The method is derived from the definition of the flow index as follows:

$$I_f = \frac{LL - W_n}{\log N - \log 25} \qquad (3.3)$$

or

$$LL = W_n + I_f \log \frac{N}{25} \qquad (3.4)$$

TABLE 3.11 COMPARISON OF VARIOUS STANDARD SIEVE SERIES[a].

U.S.A. (1)		Tyler (2)	Canadian (3)		British (4)		French (5)		German (6)
Standard[b]	Alternate	Mesh Designation	Standard	Alternate	Nominal Aperture	Nominal Mesh. No.	Opg. M.M.	No.	Opg.
125 mm	5″								
106 mm	4.24″								
100 mm	4″								
90 mm	3½″								
75 mm	3″								
63 mm	2½″								
53 mm	2.12″								
50 mm	2″								
45 mm	1¾″								
37.5 mm	1½″								
31.5 mm	1¼″								
26.5 mm	1.06″	1.05″	26.9 mm	1.06″					
25.0 mm	1″								25.0 mm
22.4 mm	7/8″	0.883″	22.6 mm	7/8″					
19.0 mm	3/4″	0.742″	19.0 mm	3/4″					20.0 mm
16.0 mm	5/8″	0.624″	16.0 mm	7/8″					18.0 mm
13.2 mm	0.530″	0.525″	13.5 mm	0.530″					16.0 mm
12.5 mm	1/2″								12.5 mm
11.2 mm	7/16″	0.441″	11.2 mm	7/16″					
9.5 mm	3/8″	0.371″	9.51 mm	3/8″					10.0 mm
8.0 mm	5/16″	2½″	8.00 mm	5/16″					8.0 mm
6.7 mm	0.265″	3	6.73 mm	0.265″					
6.3 mm	1/4″								6.3 mm
5.6 mm	No. 3½	3½	5.66 mm	No. 3½					
4.75 mm							5.000	38	5.0 mm
	4	4	4.76 mm	4					
4.00 mm	5	5	4.00 mm	5			4.000	37	4.0 mm
3.35 mm	6	6	3.36 mm	6	3.35 mm	5			
							3.150	36	3.15 mm
2.80 mm	7	7	2.83 mm	7	2.80 mm	6			
2.36 mm	8	8	2.38 mm	8	2.40 mm	7	2.500	35	2.5 mm
2.00 mm	10	9	2.00 mm	10	2.00 mm	8	2.000	34	2.0 mm
1.70 mm	12	10	1.68 mm	12	1.68 mm	10	1.600	33	1.6 mm
1.40 mm	14	12	1.41 mm	14	1.40 mm	12			
1.18 mm							1.250	32	1.25 mm
1.00 mm	16	14	1.19 mm	16	1.20 mm	14			
	18	16	1.00 mm	18	1.00 mm	16	1.000	31	1.0 mm
850 μm	20	20	841 μm	20	850 μm	18			
							0.800	30	800 μm
710 μm	25	24	707 μm	25	710 μm	22			
							0.630	29	630 μm
600 μm	30	28	595 μm	30	600 μm	25			
500 μm	35	32	500 μm	35	500 μm	30	0.500	28	500 μm
425 μm	40	35	420 μm	40	420 μm	36			
							0.400	27	400 μm
355 μm	45	42	354 μm	45	355 μm	44			
							0.315	26	315 μm
300 μm	50	48	297 μm	50	300 μm	52			
250 μm	60	60	250 μm	60	250 μm	60	0.250	25	250 μm
212 μm	70	65	210 μm	70	210 μm	72			
180 μm	80	80	177 μm	80	180 μm	85	0.200	24	200 μm
							0.160	23	160 μm
150 μm	100	100	149 μm	100	150 μm	100			
125 μm	120	115	125 μm	120	125 μm	120	0.125	22	125 μm
106 μm	140	150	105 μm	140	105 μm	150			
							0.100	21	100 μm
90 μm	170	170	88 μm	170	90 μm	170			90 μm
							0.080	20	80 μm
75 μm	200	200	74 μm	200	75 μm	200			
									71 μm
63 μm	230	250	63 μm	230	63 μm	240	0.063	19	63 μm
									56 μm

TABLE 3.11 (*Continued*)

U.S.A. (1)		Tyler (2)	Canadian (3)		British (4)		French (5)		German (6)
Standard[b]	Alternate	Mesh Designation	Standard	Alternate	Nominal Aperture	Nominal Mesh. No.	Opg. M.M.	No.	Opg.
53 μm	270	270	53 μm	270	53 μm	300			50 μm
							0.050	18	
45 μm	325	325	44 μm	325	45 μm	350			45 μm
							0.040	17	40 μm
38 μm	400	400	37 μm	400					

[a] After Tyler W. S., Inc. (1970).
[b] These sieves correspond to those recommended by ISO (International Standards Organization) as an International Standard and this designation should be used when reporting sieve analysis intended for international publication.
(1) U.S.A. Sieve Series—ASTM Specification E-11-70.
(2) Tyler Standard Screen Scale Sieve Series.
(3) Canadian Standard Sieve Series 8-GP-1b.
(4) British Standards Institution, London BS-410-62.
(5) French Standard Specifications, AFNOR X-11-501.
(6) German Standard Specification DIN 4188.

where

I_f = flow index = slope of flow curve
LL = liquid limit
N = number of blows ($17 < N < 36$)
W_n = moisture content at N blows

The term $I_f \log N/25$ is called the *moisture correction factor* and is a function of number of blows and soil type (as reflected in the flow index). These factors have been prepared in the form of a simple chart and table. For accurate results, the flow index can also be determined from the following equation:

$$I_f = 0.36 W_n - 3 \qquad (3.5)$$

The *plastic limit* ($PL; W_p$) is the lowest moisture content expressed as a percentage of the oven-dry weight of a soil at which it can be rolled into threads of $\frac{1}{8}$-inch diameter without the threads breaking into pieces (ASTM D424).

The *plasticity index* ($PI; I_p$) is the numerical difference between the liquid and plastic limits: $PI = LL - PL$ and represents the moisture range in which plastic properties dominate soil behavior, though plastic properties may also be present below the PL and above the LL (ASTM D424). Typical relationships between the PL and PI for various soils are shown in Figure 3.7.

The *field moisture equivalent* (FME) is the minimum water content at which the smooth surface of a soil pat will not absorb within 30 seconds a drop of water placed on it. It is a measure of the water absorption power of the sample (ASTM D426).

The *centrifuge moisture equivalent* (CME) is the water content of a previously saturated soil sample after being centrifuged for 1 h under a force equal to 1000 times that of gravity (ASTM D425).

The relative positions of the PL, LL, FME, and CME are important in the assessment of soil behavior toward water. An $FME > LL$ indicates the danger of autogenous liquefaction of a soil in the presence of free water. If both FME and CME exceed 30 and if $FME > CME$ the soil probably expands after release of load and is classified as *expansive soil*. The engineering properties of expansive soils are discussed in Section 3.6.9.

The *shrinkage limit* (SL) according to ASTM D427 is that moisture content, expressed as a percentage of the weight of the oven-dried soil, at which reduction in moisture content will not cause a decrease in volume of the soil mass, but at which an increase in moisture content will cause an increase in the volume of the soil mass.

The shrinkage, and its counterpart the swelling, of a cohesive soil is greatly influenced by the mutual arrangement of its constituent particles, that is, its structure. Consequently, the degree of disturbance of this structure which depends on the method of sample preparation and testing is reflected in the numerical values obtained for the SL. For thoroughly disturbed clay pastes, low SL values indicate a dispersed structure, while high values indicate a flocculated structure.

Activity has been defined by Skempton (1953) as the ratio of the plasticity index and the clay fraction. Typical values for various clay minerals are given as follows:

Clay Mineral	Activity
Muscovite	0.23
Kaolinite	0.40
Illite	0.90
Ca-Montmorillonite	1.5
Na-Montmorillonite	6.0

TABLE 3.12 TYPICAL VALUE OF SPECIFIC GRAVITY OF SOME CLAY MINERALS.

Clay Mineral	Specific Gravity	Clay Mineral	Specific Gravity
Aragonite	2.94	Illite	2.60
Attapulgite	2.61	Kaolinite	2.50–2.61
Augite	3.20–3.40	Limonite	3.40–4.00
Biotite	3.00–3.10	Magnetite	5.17
Calcite	2.70–2.72	Mica	2.80–3.20
Chlorite	2.60–3.00	Montmorillonite	2.40–2.51
Dolomite	2.85–2.90	Muscovite	2.80–2.90
Gypsum	2.30–2.34	Orthoclase	2.56–2.60
Hematite	5.10–5.20	Quartz	2.50–2.80
Hematite (hydrous)	4.20–4.40		

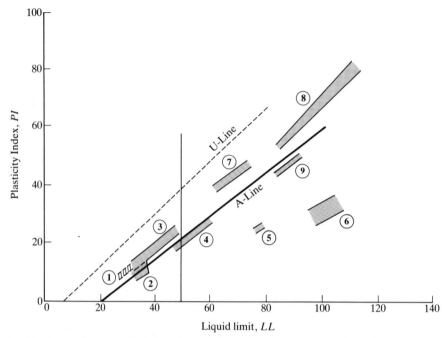

Fig. 3.7 Typical relationships between the liquid limit and the plasticity index for various soils. 1, Silty clay, Pennsylvania; 2, silty loess, Kansas-Nebraska; 3, alluvial deposits, Shanghai; 4, decomposed granite, Hong Kong; 5, halloysite clay, Guam; 6, lake sediments (diatomaceous and pumice), N. California; 7, glacial lake deposits, N. Dakota; 8, expansive clay (sodium-montmorillonite). (*Data from USBR, 1973, and others.*)

Other indices commonly used for engineering purposes are as follows:

$$Toughness\ Index\ (TI, I_t) = \frac{plasticity\ index}{flow\ index}$$

$$Liquidity\ Index\ (LI, I_l) = \frac{water\ content - plastic\ limit}{plasticity\ index}$$

$$Consistency\ Index\ (CI, I_c) = \frac{liquid\ limit - water\ content}{plasticity\ index}$$

3.5 SOIL CLASSIFICATION SYSTEMS

A classification is the simplest method of discovering order in the bewildering multiplicity of nature. It is the process of recognizing and identifying classes that possess significant characteristics in common. The members may themselves be classes or individuals; the latter represent the lowest class while the highest is the most inclusive category. Classification is the first step of a science, but as a science itself develops this is normally reflected in the classification system, giving it better definition and refining its logical structure. If possible, the properties according to which objects are classified should themselves be evident causes of many other properties. Rigorous pursuit of this principle leads to a philosophical or natural classification (Encyclopedia Britannica, 1950). An example is the present formulation of the periodic chart of the elements from which theoretically the entire chemical science may be derived. This example also illustrates the problem inherent in rigorously scientific or natural classifications: they are either too abstract to permit direct use in practical problems or they become too cumbersome. Practically useful classifications are commonly the technical ones that are centered on the same level of abstraction as that of the practical problems or subjects for which the classification is to serve a useful purpose. Only

as many levels above and below this center should be included as serve a practical purpose. This is the case with the several soil classifications now employed by the engineer. A general discussion of soil classification for engineering purposes is given by Liu (1967).

3.5.1 The Pedologic Soil Classification System

Pedology is concerned with the study of soil for its own sake independently of the application of the information and concepts developed by it. Its realm is the pedosphere that lies beside or between the atmosphere, the lithosphere, and the hydrosphere. To the pedologist, soil is a unique natural body formed from mineral parent material by interacting meteorologic, climatic, and biologic factors. Accordingly, pedologic soil includes only the A-, B-, and part of the C-horizon of a soil or that part of an engineering soil that has been formed by the above-named factors. If undisturbed by extraneous agents for a sufficiently long time, pedologic soil is expected to attain a kind of dynamic equilibrium with its natural environment.

Pedology may be divided into three phases:

1. The study of soil genesis or soil formation as resulting from the soil forming factors: parent material, climate, plant and animal organisms, topography, time, and man's intervention
2. The classification of pedologic soils within a general climatic frame but on the pragmatic basis of actually observable soil types characterized by their profile which is the result of the soil forming factors
3. Mapping of soil areas in accordance with this classification, grouping together soils of similar profile characteristics and therefore of similar overall character and engineering behavior

Different systems of pedologic classification have been developed and are being employed at the present time. Most useful for the engineer and others dealing with shallow foundations in the U.S. and cooperating countries is the Thorp and Smith (1949) formulation of the original Marbut system (USDA, 1938). Marbut lists the following features as essential for the definition of a soil unit: number, color, texture, structure, thickness, chemical and mineral composition, relative arrangement of the various horizons, and the geology of the parent material. An individual soil unit, the soil type has at least two names, a *series* or family name and a *class* (texture) name; for example, Sassafras loam. A *soil series* comprises all soils that have the same:

1. Parent material: (a) solid rock (igneous, sedimentary, metamorphic), (b) loose rock (gravels, sands, clays, other sediments)
2. Special features of parent material (residual or transported by wind, water, ice, or combinations)
3. Topographic position (rugged to depressed)
4. Natural drainage (excessive to poor)
5. Profile characteristics

The different series usually have geographic names indicative of the location where they were first recognized and described (e.g., Sassafras, Putnam, Cecil). The soil series are the most important to the soil engineer and are described in detail and mapped by the U.S. Soil Survey.

Identification of a soil in the field as belonging to a certain series makes automatically available the information already gathered regarding this soil. To this knowledge the results of additional engineering tests can be added and be available for future use.

The higher categories of the pedologic classification of 1949 are shown in Table 3.13. The higher the category, the less the number of actual properties that are indicated by the place of a soil in it. However, the indices employed for placement in the higher categories such as salinity, alkalinity, chemical and mineral composition of the clay fraction, presence or absence of a lime or other concentration zone, etc., are also indices of troubles to be expected or opportunities offered and are therefore important warning signs or guide posts for country-wide and continent-wide planning.

3.5.2 The AASHTO Soil Classification System

The classification system of the American Association of State Highway and Transportation Officials evolved from the U.S. Bureau of Public Roads system of classifying soils in accordance with their performance as subgrades underneath highway pavements. There are seven basic groups, A-1 to A-7; the members of each group have similar load bearing values and engineering characteristics under normal service conditions. The overall quality as a subgrade material decreases with increasing classification number; however, this is not true for all service conditions. Groups A-1 to A-3 soils possess in the densified state an effective granular skeleton formed of sand-size and larger grains. Groups A-4 to A-7 soils possess no such bearing skeleton and their engineering behavior is governed essentially by the amount and water affinity of its silt-clay components (-200 sieve fraction). The A-2 group is subdivided into A-2-4 to A-2-7 subgroups; the last number identifies the type of -200 sieve fraction present.

The classification is based on the results of sieve tests employing sieves Nos. 200, 40, 10 and/or larger openings where indicated, and of consistency tests (liquid limit and plasticity index) performed on the fraction passing the No. 40 sieve. Differentiation between the quality within a certain group is made by the group index (*GI*), which is calculated as follows:

$$GI = (F - 35)[0.2 + 0.005(LL - 40)]$$
$$+ 0.01(F - 15)(PI - 10) \qquad (3.6)$$

TABLE 3.13 SOIL CLASSIFICATION IN THE HIGHER CATEGORIES[a].

Order	Suborder	Great Soil Groups
Zonal soils	1. Soils of the cold zone	Tundra soils
	2. Light-colored soils of arid regions	Desert soils Red desert soils Sierozem Brown soils Reddish-brown soils
	3. Dark-colored soils of semiarid, subhumid, and humid grasslands	Chestnut soils Reddish chestnut soils Chernozem soils Prairie soils Reddish prairie soils
	4. Soils of the forest-grassland transition	Degraded chernozem Noncalcic brown or Shantung brown soils
	5. Light-colored podzolized soils of the timbered regions	Podzol soils Gray wooded or Gray podzolic soils Brown podzolic soils Gray-brown podzolic soils Red-yellow podzolic soils
	6. Lateritic soils of forested warm-temperature and tropical regions	Reddish-brown lateritic Yellowish-brown lateritic soils Laterite soils
Intrazonal soils	1. Halomorphic (saline and alkali) soils of imperfectly drained arid regions and littoral deposits	Solonchak or Saline soils Solonetz soils Soloth soils
	2. Hydromorphic soils of marshes, swamps, seep areas, and flats	Humic-glei soils (includes wiesenboden) Alpine meadow soils Bog soils Half-bog soils Low-humic-glei soils Planosols Groundwater podzol soils Groundwater laterite soils
	3. Calcimorphic soils	Brown forest soils (braunerde) Rendzina soils
Azonal soils		Lithosols Regosols (includes dry sands) Alluvial soils

[a] After Thorp and Smith (1949).

where *F* is the percent passing the No. 200 sieve, *LL* is the liquid limit, in percent, and *PI* is the plasticity index, also in percent. The group index is given in parentheses after the soil group, for example A-6(7).

General quality of subgrade soil as indicated by the group index:

Excellent	A-1-a (0) soils
Good	(0–1)
Fair	(2–4)
Poor	(5–9)
Very poor	(10–20)

TABLE 3.14 CLASSIFICATION OF SOILS AND SOIL-AGGREGATE MIXTURES[a].

General Classification	Granular Materials (35% or Less Passing No. 200)							Silt-Clay Materials (More than 35% Passing No. 200)			
	A-1		A-3	A-2				A-4	A-5	A-6	A-7
Group classification	A-1-a	A-1-b		A-2-4	A-2-5	A-2-6	A-2-7				A-7-5, A-7-6
Sieve analysis, percent passing:											
No. 10	50 max.	—	—	—	—	—	—	—	—	—	—
No. 40	30 max.	50 max.	51 min.	—	—	—	—	—	—	—	—
No. 200	15 max.	25 max.	10 max.	35 max.	35 max.	35 max.	35 max.	36 min.	36 min.	36 min.	36 min.
Characteristics of fraction passing No. 40:											
Liquid limit	—		—	40 max.	41 min.	40 max.	41 min.	40 max.	41 min.	40 max.	41 min.
Plasticity index	6 max.		N.P.	10 max.	10 max.	11 min.	11 min.	10 max.	10 max.	11 min.	11 min.[b]
Usual types of significant constituent materials	Stone fragments, gravel and sand		Fine sand	Silty or clayey gravel and sand				Silty soils		Clayey soils	
General rating as subgrade	Excellent to good							Fair to poor			

[a] AASHTO (1986, 1988).
[b] Plasticity index of A-7-5 subgroup is equal or less than LL minus 30. Plasticity index of A-7-6 subgroup is greater than LL minus 30.

The AASHTO subgrade soil classification is shown in Table 3.14; Figures 3.8 and 3.9 give charts for graphical determination of the group index.

3.5.3 Federal Aviation Agency (FAA) Classification

This classification was originally based on mechanical analysis, plasticity characteristics, expansive qualities, and California bearing ratio. It included evaluation of the quality of the soils as supports for flexible and rigid pavements, respectively, and for climatic conditions of different severity. This was too much for any system to carry and has been shed in the present FAA system, which is based on granulometric composition and the plasticity characteristics of soils.

The textural classification is based on the grain size composition of the fraction that passes the No. 10 sieve, and utilizes the biaxial chart shown in Figure 3.10, which also contains definitions for sand, silt, and clay. Table 3.15 shows the FAA soil classification with soil groups E-1 to E-13. If test results on the fine-grained soils E-6 to E-12 place them in more than one group, Figure 3.11 can be used to decide the proper group. Upgrading of the soil by one or two classes is allowed when the fraction retained on the No. 10 sieve exceeds 45 percent in the case of groups E-1 to E-5 and 55 percent for the remaining groups. The amount of upgrading depends additionally on granulometric and mineral composition of the +10 sieve fraction.

3.5.4 Unified Soil Classification System

This system grew out of the soil classification and identification system developed by Casagrande in 1948. The system was significantly revised in 1983 (Howard, 1984). The essence of the system and its nomenclature is shown in Table 3.16 and Figures 3.11 to 3.13.

The significant changes and revisions adopted (ASTM D2487-83) include the following:

1. Soil classification consists of both a name and a symbol such as: CL-lean clay, or sand lean clay, and gravelly lean clay with sand.

2. The names (Figures 3.11, 3.12) were standardized. These names have a single unique name for each symbol (except for organic silts and clays). The corresponding symbols are: GW—well graded gravel; GP—poorly graded gravel; GM—silty gravel; GC—clayey gravel; SW—well-graded sand; SP—poorly graded sand; SM—silty sand; SC—

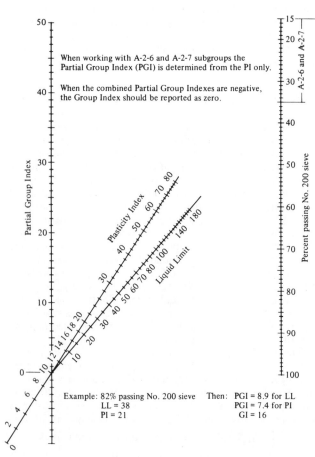

When working with A-2-6 and A-2-7 subgroups the Partial Group Index (PGI) is determined from the PI only.

When the combined Partial Group Indexes are negative, the Group Index should be reported as zero.

Example: 82% passing No. 200 sieve Then: PGI = 8.9 for LL
LL = 38   PGI = 7.4 for PI
PI = 21   GI = 16

Fig. 3.8 Group index chart. (*AASHTO, 1986, 1988.*)

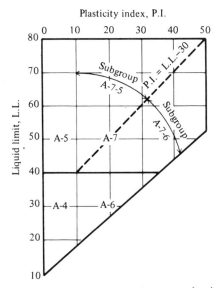

Plasticity index, P.I.

Fig. 3.9 Liquid limit and plasticity index ranges for the A-4, A-5, A-6, and A-7 subgrade groups.

clayey sand; CL—lean clay; ML—silt; OL—organic silt or organic clay; CH—fat clay; MH—elastic silt; OH—organic silt or organic clay; and PT—peat (Table 3.16).
3. Organic silts and clays were redefined as shown in Figure 3.11(b).
4. In Figure 3.13, the upper limit or "U" line was added to the plasticity chart to aid in the evaluation of test data. This line was recommended by Casagrande as an empirical boundary for natural soils.

Details of the field and laboratory procedures employed in the use of this classification and on the engineering characteristics of the various soil classes are found in the *Earth Manual* of the USBR (1973), and Howard (1984, 1987). Ueshita and Nonogaki (1971) have proposed a modification of the classification system for coarse soils based on gradation, maximum dry density, optimum moisture content, California bearing ratio value, and coefficient of permeability.

TABLE 3.15 FAA CLASSIFICATION OF SOILS FOR AIRPORT CONSTRUCTION.

| | *Mechanical Analysis* | | | | | |
| | *Material Finer than No. 10 Sieve* | | | | | |
Soil Group	Retained on No. 10 Sieve[a], Percent	Coarse Sand Passing No. 10, Retained on No. 60, Percent	Fine Sand Passing No. 60, Retained on No. 270, Percent	Combined Silt and Clay Passing No. 270, Percent	LL	PI
E-1	0–45	40+	60–	15–	25–	6–
E-2	0–45	15+	85–	25–	25–	6–
E-3	0–45	—	—	25–	25–	6–
E-4	0–45	—	—	35–	35–	10–
E-5	0–45	—	—	45–	40–	15–
E-6	0–55	—	—	45+	40–	10–
E-7	0–55	—	—	45+	50–	10–30
E-8	0–55	—	—	45+	60–	15–40
E-9	0–55	—	—	45+	40+	30–
E-10	0–55	—	—	45+	70–	20–50
E-11	0–55	—	—	45+	80–	30+
E-12	0–55	—	—	45+	80+	—
E-13	Muck and peat—field examination					

[a] Classification is based on sieve analysis of the portion of the sample passing the No. 10 sieve. When a sample contains material coarser than the No. 10 sieve in amounts equal to or greater than the maximum limit shown in the table, a raise in classification may be allowed provided the coarse material is reasonably sound and fairly well graded.

3.5.5 Organic Soils and their Engineering Classification

Organic soils are those whose solid constituents consist predominantly of vegetable matter in various stages of decomposition or preservation. They are commonly designated as bog, muskeg, and moor soils with differentiation between peat and muck soils on the one hand, and coastal marshland soils on the other. Muck indicates a higher degree of decomposition of the vegetable

Textural class	Percent sand	Percent silt	Percent clay
Sand	80-100	0-20	0-20
Sandy loam	50-80	0-50	0-20
Loam	30-50	30-50	0-20
Silty loam	0-50	50-80	0-20
Silt	0-20	80-100	0-20
Sandy clay	50-80	0-30	20-30
Clay loam	20-50	20-50	20-30
Silty clay loam	0-30	50-80	20-30
Sandy clay	50-70	0-20	30-50
Silty clay	0-20	50-70	30-50
Clay	0-50	0-50	30-100

Fraction	Sieve size	Grain size, mm.
Coarse sand	#10–#60	2.0-0.25
Fine sand	#60–#270	0.25-0.05
Silt	<#270	0.05-0.005
Clay	–	<0.005

Fig. 3.10 FAA textural classification of soils.

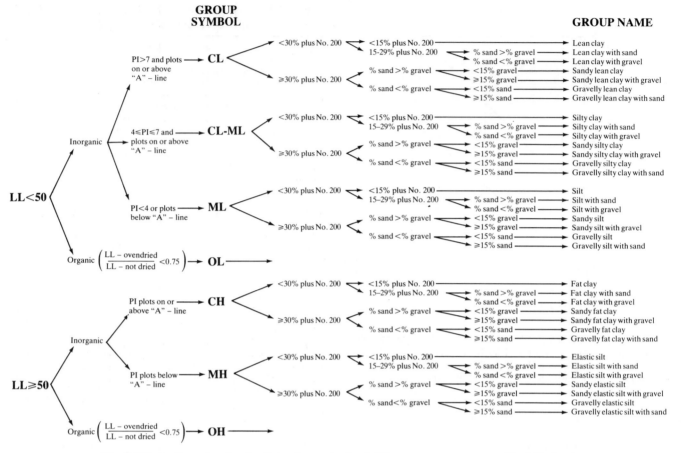

Fig. 3.11(a) Flow chart for classifying fine-grained soil (50 percent or more passes No. 200 sieve).

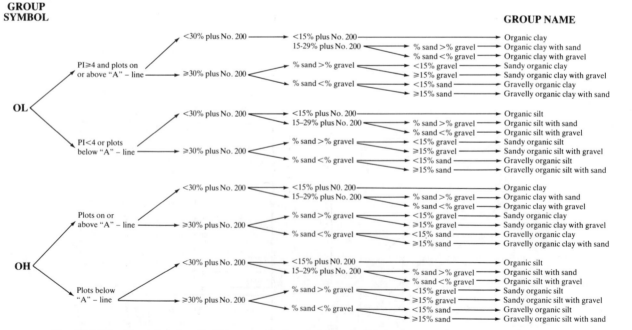

Fig. 3.11(b) Flow chart for classifying organic fine-grained soil (50 percent or more passes No. 200 sieve).

TABLE 3.16 UNIFIED SOIL CLASSIFICATION SYSTEM (ASTM D2487-83).

Criteria for Assigning Group Symbols and Group Names Using Laboratory Tests[a]			Soil Classification	
			Group Symbol	Group Name[b]
Coarse-grained soils More than 50% retained on No. 200 sieve	Gravels More than 50% of coarse fraction retained on No. 4 sieve	Clean gravels Less than 5% fines[c]	$Cu \geq 4$ and $1 \leq Cc \leq 3^{e}$ → GW	Well-graded gravel[f]
			$Cu < 4$ and/or $1 > Cc > 3^{e}$ → GP	Poorly graded gravel[f]
		Gravels with fines More than 12% fines[c]	Fines classify as ML or MH → GM	Silty gravel[f,g,h]
			Fines classify as CL or CH → GC	Clayey gravel[f,g,h]
	Sands 50% or more of coarse fraction passes No. 4 sieve	Clean sands Less than 5% fines[d]	$Cu \geq 6$ and $1 \leq Cc \leq 3^{e}$ → SW	Well-graded sand[i]
			$Cu < 6$ and/or $1 > Cc > 3^{e}$ → SP	Poorly graded sand[i]
		Sands with fines More than 12% fines[d]	Fines classify as ML or MH → SM	Silty sand[g,h,i]
			Fines classify as CL or CH → SC	Clayey sand[g,h,i]
Fine-grained soils 50% or more passes the No. 200 sieve	Silts and clays Liquid limit less than 50	Inorganic	$PI > 7$ and plots on or above "A" line[j] → CL	Lean clay[k,l,m]
			$PI < 4$ or plots below "A" line[j] → ML	Silt[k,l,m]
		Organic	$\dfrac{\text{Liquid limit} - \text{oven dried}}{\text{Liquid limit} - \text{not dried}} < 0.75$ → OL	Organic clay[k,l,m,n] / Organic silt[k,l,m,o]
	Silts and clays Liquid limit 50 or more	Inorganic	PI plots on or above "A" line → CH	Fat clay[k,l,m]
			PI plots below "A" line → MH	Elastic silt[k,l,m]
		Organic	$\dfrac{\text{Liquid limit} - \text{oven dried}}{\text{Liquid limit} - \text{not dried}} < 0.75$ → OH	Organic clay[k,l,m,p] / Organic silt[k,l,m,q]
Highly organic soils	Primarily organic matter, dark in color, and organic odor		PT	Peat

[a] Based on the material passing the 3-in. (75-mm) sieve.
[b] If field sample contained cobbles or boulders, or both, add "with cobbles or boulders, or both" to group name.
[c] Gravels with 5 to 12% fines require dual symbols:
GW-GM well-graded gravel with silt
GW-GC well-graded gravel with clay
GP-GM poorly graded gravel with silt
GP-GC poorly graded gravel with clay
[d] Sands with 5 to 12% fines require dual symbols:
SW-SM well-graded sand with silt
SW-SC well-graded sand with clay
SP-SM poorly graded sand with silt
SP-SC poorly graded sand with clay

[e] $Cu = D_{60}/D_{10}$ $Cc = \dfrac{(D_{30})^2}{D_{10} \times D_{60}}$
[f] If soil contains $\geq$ 15% sand, add "with sand" to group name.
[g] If fines classify as CL-ML, use dual symbol GC-GM, or SC-SM.
[h] If fines are organic, add "with organic fines" to group name.
[i] If soil contains $\geq$ 15% gravel, add "with gravel" to group name.
[j] If Atterberg limits plot in hatched area, soils is a CL-ML, silty clay.
[k] If soil contains 15 to 29% plus No. 200, add "with sand" or "with gravel," whichever is predominant.
[l] If soil contains $\geq$ 30% plus No. 200, predominantly sand, add "sandy" to group name.

[m] If soil contains $\geq$ 30% plus No. 200, predominantly gravel, add "gravelly" to group name.
[n] $PI \geq 4$ and plots on or above "A" line.
[o] $PI < 4$ or plots below "A" line.
[p] PI plots on or above "A" line.
[q] PI plots below "A" line.

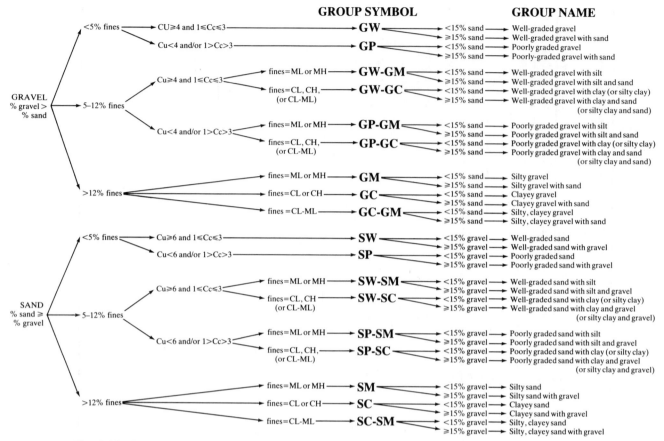

Fig. 3.12 Flow chart for classifying coarse-grained soils (more than 50 percent retained on No. 200 sieve).

matter or intermixing with mineral soil consituents in contrast to the purely vegetable peats that have well-preserved plant remains. Several types of moors are recognized depending on the source of water supply (high moors—water mainly from precipitation; low moors—drainage from surrounding areas), topographic characteristics, and types of underlying soil or rock. Pedologically, organic soils are intrazonal hydromorphous soils, and may occur within any macroclimatic zone as long as hydrologic and topographic conditions provide basins of standing water or land areas with a rising water table (e.g., consequence of beaver dam). Because of the influence of

environmental and plant ecologic factors, these soils are best considered as organic terrain (muskeg) and classified in accordance with genetic principles as has been done by the Muskeg Subcommittee of the National Research Council of Canada. The subsequent treatment is based mainly on the work of this committee (MacFarlane, 1969).

The parent material of organic soils is the native vegetation that may cover a wide range of aquatic, marsh, and swamp forest plants. Each visible forest or other vegetation above a land or water surface has a counterpart, a sort of distorted mirror image, below this surface. This counterpart is formed by the root systems, which are functionally and geometrically related to the normally visible part. Actually, the subterranean or subaqueous part may contribute more to peat and muck formation that the more rapidly decomposed organic matter from the surface vegetation. However, the functional and geometric relationship between the visible and invisible portions of the vegetation proves the importance of the surface cover for purposes of classification of organic terrain.

Table 3.17 presents nine pure coverage classes designated A to I. Pure classes seldom exist by themselves but are usually combined with others. Complete description of coverage may require two or three letters arranged in order of their contribution. No letter is awarded for a contribution of less than 25 percent. Table 3.18 lists the occurrence of prevailing conditions for the various cover classes. Table 3.19 gives topographic features pertinent to muskeg formation and its field description. The characteristics of the subsurface material, the organic soil that is the product of the interaction of topography, vegetation, high water table and decomposition and preservation processes are shown in Table 3.20. Sixteen categories of organic soil are recognized, based on the extent

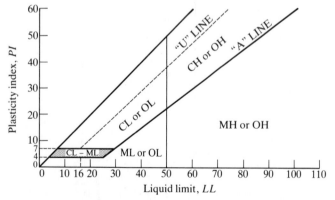

Fig. 3.13 Plasticity chart for classification of fine-grained soils and fine-grained fraction of coarse-grained soils. Equation of "A" line: horizontal at $PI = 4$ to $LL = 25.5$, then $PI = 0.73(LL - 20)$. Equation of "U" line: vertical at $LL = 16$ to $PI = 7$, then $PI = 0.9(LL - 8)$.

TABLE 3.17 PROPERTIES DESIGNATING NINE PURE COVERAGE CLASSES[a].

Coverage Type (Class)	Woodiness vs. Nonwoodiness	Stature (Approximate Height)	Texture (Where Required)	Growth Habit
A	Woody	15 ft or over	—	Tree form
B	Woody	5–15 ft	—	Young or dwarfed trees or bush
C	Nonwoody	2–5 ft	—	Tall, grasslike
D	Woody	2–5 ft	—	Tall shrub or very dwarfed tree
E	Woody	Up to 2 ft	—	Low shrub
F	Nonwoody	Up to 2 ft	—	Mats, clumps, or patches, sometimes touching
G	Nonwoody	Up to 2 ft	—	Singly or loose association
H	Nonwoody	Up to 4 in	Leathery to crisp	Mostly continuous mats
I	Nonwoody	Up to 4 in	Soft or velvety	Often continuous mats, sometimes in hummocks

[a] After MacFarlane (1969).

to which the following types of structural components are present: (1) amorphous and granular, (2) woody or nonwoody fine fibers, and (3) wood particles and coarse woody fibers. The relationship between peat structure and important engineering characteristics under in situ conditions can be gleaned from Table 3.21.

The properties of organic soils must be determined in situ, on location, and in the laboratory in as close a state of disturbance or nondisturbance as the respective engineering use may require (Arman, 1970; Yamanouchi, 1977; Jarrett, 1983). While the discussion of pertinent tests and of the physical meaning of the data obtained can be found in Sections 3.7–3.14, and while certain muskeg properties may vary within a wide range, a few of the more important characteristics are given.

Peats, especially those of predominantly fibrous constitution, have a spongelike nature that accounts for their *high natural water content* (50 to 2000 percent, which is 10 to 100 times greater than in mineral soils), *high void ratio* (normally 5 to 15 but may be as high as 25), *high drying shrinkage* (up to 50 percent), *large compressibility*, and *low bearing capacity*.

The permeability of natural peat deposits varies widely despite their high void ratios or porosities depending on the effective size of the voids, on the portion of the water held physicochemically at the external surfaces of the particulate constituents and on that which produces internal swelling or turgidity; also, permeability is often much greater in the horizontal than in the vertical direction.

The specific gravity of peats ranges from 1.1 to 2.5; values above 2 indicate marked contribution by mineral matter, which may be checked by the determination of the ash content. The low specific gravity of both organic matter and water leads to low unit weights for the natural peat. Tensile and shear strength of peats even in their natural wet state is provided by the felt-like interweaving of their fibrous constituents. This is the reason why tensile and shear strengths do not always increase with decreasing water content. The aqueous phase of most peats is acid with pH values ranging from 4 to 7, but values as low as 2 and as high as 8 have been encountered.

After drying out, peats upon rewetting do not regain their original high water content. Disturbance of the natural peat structure decreases its strength properties. The sensitivity of peats ranges from 1.5 to 10. Discussions and comparisons of

various testing methods and classification systems (Figure 3.14) for peats and organic soils are given by Al-Khafaji and Andersland (1981), Landva et al. (1983), and Burwash and Wiesner (1984).

3.6 SOIL TYPES BY DEPOSITION OR OTHER SPECIAL FEATURES

Engineering soils include unconsolidated sediments transported to their present place by glaciers, water, and air, as well as pedological (or residual) soils formed in place from these materials and from local bedrock. The different transporting agents have different carrying capacities and also affect the properties of their loads in different ways. It is therefore important to recognize and name such soils in accordance with the means of their transportation and manner of deposition (Winterkorn and Fang, 1976).

3.6.1 Alluvial Sediments

Alluvial sediments have been transported by running water and have settled out when the speed of water flow was no longer sufficient to carry them. For this reason they possess a relatively narrow range of particle size irrespective of whether they are cobbles and gravels from rushing rivers and creeks, sands from moderately moving rivers, or clays from sluggish rivers or from precipitation water moving in sheets down the sides of gentle slopes. Alluvial soils are those that are developing from recently deposited alluvium and do not exhibit horizon development.

3.6.2 Glacial Deposits

Glacial deposits have been transported by glaciers whose action may be likened to that of giant bulldozers that push all sorts of materials ahead with droppings on the side and grinding underneath. Spring or general melting of the glaciers stops their forward movement and permits settling out or further

TABLE 3.18 OCCURRENCE OF PREVAILING CONDITIONS FOR THE VARIOUS COVER CLASSES[a].

Predominant Class	Engineering Significance	Predominant Class	Engineering Significance
A	(1) Presence of large woody erratics in the peat (2) The position of relatively shallow depths of peat for the landscape as a whole (3) Location of best drained peat (4) Location of best drained mineral soil sublayer (5) Presence of highly permeable peat (6) Vicinity of lowest summer temperatures in the peat (7) Location of the coarsest, most durable peat (8) Best conditions for static load and dynamic loading		(3) Conditions accommodating to certain articulated wheeled vehicles (4) Good cohesion and tensility, moderate elasticity even when water shows at the surface in the field (5) Easily drainable conditions (for free water)
B	Same as for A above, but less intensively represented	F	Presence of highly critical conditions when prominent in the formula (1) Low points on drainage gradients (2) Muskeg with centers of extremely low bearing potential whether wet or relatively dry (3) Peat of low tensile strength and showing little elasticity unless the local water table is consistently high (small open pools the year round) (4) Sites where shear strength is lowest in muskeg at frequent intervals; water is not excessive
C	Predominance rare, except in tropical and subtropical locations (for example, Guyana, Brazil, Paraguay and Uruguay, and possibly Southern Rhodesia, Nigeria, Israel, Malaysia, etc.)	G	Rarely predominates in the formula, is indicative of a highly fluctuating water table
D	(1) Linear drainage, often an open water course (2) Lagg condition around a confined muskeg (bog) (3) Traps present (4) Good, but highly elastic, bearing conditions; difficult to consolidate and with marked patterned local differentials as to rate of consolidation (5) Features highly conducive to spring flooding (6) Silt in the mineral soil sublayers with highly mixed aggregate from outwash (7) Features conducive to differential settlement (often abrupt) under load	H	When predominant indicates presence of: (1) Permafrost and late seasonal subsurface ice conditions of uneven contour (2) Maximum range of microtopographic amplitude (often abrupt) for all muskeg (3) Local imponding and highly irregular, dissected drainage gradients (4) Relatively locally degraded peat (structurally and mechanically disrupted)
E	Equally important as D above and is very common in temperate, arctic, and arctic zones (1) High order of homogeneity in peat, even in relation to microtopography in which mounds, ridges, and ice knolls are important (2) Peat difficult to re-wet once drained of gravitational water	I	Unless Class I is the only component comprising the cover formula (which is rare), it lacks prominence. When it is a single contributing factor in cover it is very local, usually no more than 4 or 5 m in area of coverage, and the following occurs: (1) Vehicle immobilization on the second pass for amphibious vehicles (2) The base of minor or major drainage gradients

[a] After MacFarlane (1969). Reproduced with permission of the University of Toronto Press.

movement of suspended rock particles by flowing water. Glacial deposits may vary in size composition from boulders to clays. Glaciers produce a disordered landscape, often with inhibited drainage and development of bogs. Differentiation is made between:

a. *Glacial drift*—rock debris that has been transported by glaciers and deposited either directly from the ice or from the meltwater. It may or may not be heterogeneous.
b. *Till*—unstratified glacial drift deposited by the ice and consisting of clay, silt, sand-gravel, and boulders, intermingled in any proportion.
c. *Glaciofluvial deposits*—material moved by glaciers and subsequently sorted and deposited by streams flowing from the melted ice. The deposits are stratified and may occur in

the form of outwash plains, deltas, kames, eskers, and kame terraces.

3.6.3 Fluvial and Lacustrine Deposits

Fluvial or *river deposits* may range from boulders to colloidal clay depending on the speed of river flow and the consequent carrying capacity. At a particular location the particle size range is usually relatively narrow.

Lacustrine deposits are the usually fine-grained materials deposited on lake bottoms. They may contain appreciable amounts of organic matter and also of fragments of shells and skeletons from aquatic animals. Such deposits are called *marl* if they are rich in calcium carbonate.

TABLE 3.19 TOPOGRAPHIC FEATURES[a].

Contour Type	Feature	Description
a	Hummock	Includes "tussock" and "niggerhead", has tufted top usually vertical sides, occurring in patches, several to numerous
b	Mound	Rounded top, often elliptic or crescent-shaped in plane view
c	Ridge	Similar to Mound but extended, often irregular and numerous; vegetation often coarser on one side
d	Rock gravel plain	Extensive exposed areas
e	Gravel bar	Eskers and old beaches (elevated)
f	Rock enclosure	Grouped boulders overgrown with organic deposit
g	Exposed boulder	Visible boulder interrupting organic deposit
h	Hidden boulder	Single boulder overgrown with organic deposit
i	Peat plateau (even)	Usually extensive and involving sudden elevation
j	Peat plateau (irregular)	Often wooded, localized and much contorted
k	Closed pond	Filled with organic debris, often with living coverage
l	Open pond	Water rises above organic debris
m	Pond or lake margin	Abrupt
n	Pond or lake margin	Sloped
o	Free polygon	Forming a rimmed depression
p	Joined polygon	Formed by a system of banked clefts in the organic deposit

[a] After MacFarlane (1958). Reproduced with permission of the National Research Council, Canada.

3.6.4 Aeolian Deposits

Aeolian deposits range from *sand dunes* to *loess* deposits whose particles are predominantly of silt size but with a certain amount of fine sand and aggregated clay particles present. The valley loesses (Missouri, Mississippi, Rhine) are typically developed in areas peripheral to those covered by the last ice sheets. Other important loess deposits are Argentina and China (ACTA, 1961). Typical loess has a calcium carbonate content that acts as a bonding agent, which, though weak, allows the loess to form vertical or even overhanging walls on the banks of streams. Slopes at angles less than 90° are easily eroded. Summarization and comparison of engineering properties of loess in the United States is given by Sheeler (1968).

3.6.5 Laterite Soils

Laterite soils (SSSA, 1970) are extreme types of Latosols (see pedologic soil classification) that form a suborder of zonal (climatic) soils and include soils formed under forested tropical and humid conditions. Their clay fractions have low silica–sesquioxide ratios (< 2), low activity and base-exchange capacity. Laterites have low contents of soluble constituents and of most primary minerals. Their secondary aggregates are usually highly stable and water-resistant, and in extreme forms (pisolith) may serve as aggregate in soil stabilization. Lesser degrees of laterization are found in laterite and lateritic soils formed in warm, temperate, and tropical regions and include the yellow podzolic, red podzolic, yellowish-brown lateritic, and lateritic great soil groups. For an outline of the weathering of the tropical

TABLE 3.20 CLASSIFICATION OF PEAT STRUCTURE[a].

Predominant Characteristic	Category	Name
Amorphous-granular	1	Amorphous-granular peat
	2	Nonwoody, fine-fibrous peat
	3	Amorphous-granular peat containing nonwoody fine fibers
	4	Amorphous-granular peat containing woody fine fibers
	5	Peat, predominantly amorphous-granular, containing nonwoody fine fibers, held in a woody, fine-fibrous framework
	6	Peat, predominantly amorphous-granular containing woody fine fibers, held in a woody, coarse-fibrous framework
	7	Alternate layering of nonwoody, fine-fibrous peat and amorphous-granular peat containing nonwoody fine fibers
Fine-fibrous	8	Nonwoody, fine-fibrous peat containing a mound of coarse fibers
	9	Woody, fine-fibrous peat held in a woody, coarse-fibrous framework
	10	Woody particles held in nonwoody, fine-fibrous peat
	11	Woody and nonwoody particles held in fine-fibrous peat
Coarse-fibrous	12	Woody, coarse-fibrous peat
	13	Coarse fibers criss-crossing fine-fibrous peat
	14	Nonwoody and woody fine-fibrous peat held in a coarse-fibrous framework
	15	Woody mesh of fibers and particles enclosing amorphous-granular peat containing fine fibers
	16	Woody, coarse-fibrous peat containing scattered woody chunks
	17	Mesh of closely applied logs and roots enclosing woody coarse-fibrous peat with woody chunks

[a] After MacFarlane (1969). Reproduced with permission of the University of Toronto Press.

soils of the Hawaiian Islands, which is also pertinent with respect to clay formation under different climates, see Figures 3.15a and b (Sherman, 1952; Lyon Associates, 1971).

As the process of laterization progresses, these soils are usually stabilized more easily by cement, bitumen, and other agents, except when they contain certain organic impurities as may accumulate under extensive cultivation of earth nuts and of sugar-producing cane and beet crops (Winterkorn and Chandrasekharan, 1951). For treatment of lateritic soils from an engineering point of view see M.O.P. (1959), Vallerga and VanTil (1970), and Gidigasu (1975).

TABLE 3.21 RELATIONSHIP BETWEEN PEAT STRUCTURE AND IMPORTANT ENGINEERING CHARACTERISTICS IN NATURAL STATE[a].

Predominant Structural Characteristics	Properties in Which		
	Highest	Intermediate	Lowest
(1) Amorphous-granular	Unit weight	Void ratio Compressibility	Water content Permeability Shear strength Tensile strength
(2) Fine fibrous	Water content Void ratio Compressibility	Permeability Shear strength Tensile strength	Unit weight
(3) Coarse fibrous (woody)	Permeability Shear strength Tensile strength	Water content Unit weight	Void ratio Compressibility

[a] After MacFarlane (1969). Reproduced with permission of the University of Toronto Press.

3.6.6 Residual Soil

Residual soil is produced by the in-situ decomposition of the underlying rock and the action of the pertinent soil forming factors such as microclimate, flora, fauna, and geometric features. Chemical breakdown is particularly active in hot humid regions with production and decomposition of large amounts of organic materials. The physical structure and engineering properties of residual soils are unique. The texture and minerology may still reflect the original rock structure with the added complication of decreased weathering with increased depth below the ground surface. Detailed discussions on engineering properties of residual soils and applications are given by Deere and Patton (1971), de Mello (1972), Martin (1977), and Townsend (1985).

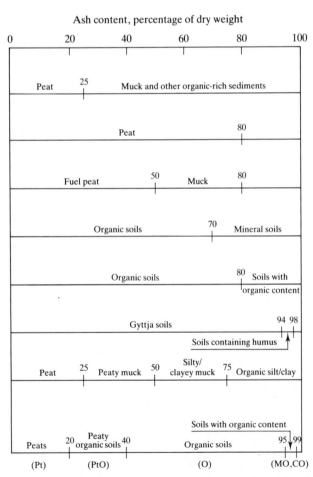

Fig. 3.14 Comparisons of various existing peats and organic soil classification systems. (*After Landva et al., 1983.*)

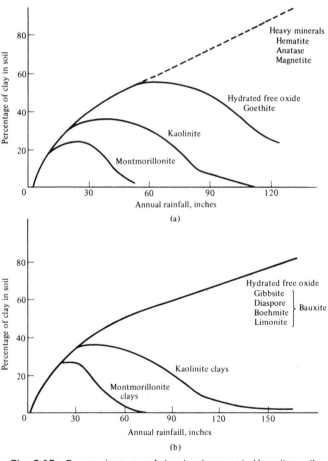

Fig. 3.15 Progressive types of clay development in Hawaiian soils under (a) a climate having alternating wet and dry seasons; (b) a continuously wet climate. (*After Sherman, 1952; Lyon Associates, 1971.*)

3.6.7 Dispersive Clay

Some fine-grained soils, called dispersive soils, with a higher content of dissolved pore-water sodium than ordinary soils, rapidly erode, forming tunnels and deep gullies by a process in which the individual clay particles go into suspension in slow-moving water (colloidal erosion), damaging earth dams, canals, and other structures. Dispersive soils cannot be differentiated from ordinary soils by conventional geotechnical tests. In general, dispersive clays have a preponderance of dissolved sodium cations in the pore water, whereas ordinary, erosion-resistant clays have a preponderance of dissolved calcium and magnesium cations. The first clear description of this phenomenon was provided in the early 1960s by Australian engineers, following investigations of piping failures of many small farm dams. Detailed discussions of identification and applications are given by Perry (1975) and Sherard and Decker (1977), and in Chapter 8.

3.6.8 Saline-alkali Soils

Saline-alkali soils have more than 15 percent of their base exchange capacity saturated with Na^+ ions and contain appreciable quantities of soluble salts. The electric conductivity of their saturation extract is greater than 4 millimhos per cm (at $25°C$) and the pH in the saturated soil solution is usually 8.5 or less. *Saline soils* are nonalkali soils that contain appreciable amounts of soluble salts, which impair crop production. Both saline-alkali and saline soils may be deleterious to contacting concrete structures (Winterkorn, 1948) especially when they contain appreciable amounts of Na^+ and Mg^{2+} sulfates. Also, these salts may cause detrimental soil expansion occasioned by drop in ambient temperature, which resembles frost heave although it takes place above the freezing temperature of water (Blaser and Scherer, 1969).

3.6.9 Expansive Soils (Including African and Asian Black-cotton Soils)

The problems and economic losses caused by expansive clay soils and the underlying soil-physical causes have been of considerable interest ever since low-cost road construction and soil stabilization began transformation from an art to an applied science. The greatly increased interest in this subject since the work of Wooltorton (1950) coincided with the expansion of settlements into arid and semiarid areas, and those with tropical climates and alternating wet and dry seasons and explains the important recent contributions by Indian, African, Australian, Chinese and Israeli researchers (Aitchison, 1965; Kassiff et al., 1969; Tianjin University et al., 1978; and Chen, 1979).

Soils that exhibit greatest volume changes from dry to wet state usually possess a considerable percentage of montmorillonite clay or of related three-layer clay minerals.

Swelling (and shrinking) soils do not have a continuous granular skeleton with sufficient interstitial porosity to accommodate the volume changes of the silt-clay fractions due to increase (or decrease) in moisture content. In disturbed compacted soils the presence or absence of an effective skeleton may be deduced from the size composition of the soil and the Atterberg limits of the soil fraction (ASTM D4318). There exist, however, in arid regions, natural soils of high sand and low silt-clay content with such structure that the silt-clay fraction forms bridges between loosely packed sand grains. If such soils are exposed to water, the cohesive bridges are weakened and the soils may collapse under their own weight or super-imposed loads (Gibbs, 1965). This illustrates the importance of structural features, in addition to granulometric and volumetric relationships in the interaction of soil solids with water.

The total amount of swell of a soil body, usually expressed as a percentage of its original volume, represents the difference in the sum of the phase volumes of solid, liquid, and gaseous soil constituents in the initial and final states. Assuming the volume of the solid constituents as essentially constant, the difference is due to changes of the volumes of the liquid and gaseous constituents. Factors that influence the volume changes of swelling clays are:

a. General compositional
   (1) The grain size composition of the whole soil
   (2) The amount, mineral character and exchange ion coverage of the clay fraction and, to a lesser but often significant degree, of the other size fractions
   (3) The types and concentration of electrolytes in the soil solution and in the source of swelling water
b. Initial condition—phase composition and structural arrangement as defined or influenced by
   (1) dry density
   (2) amount and electrolyte content of soil-water
   (3) soil structure as influenced by previous chemical and stress and strain history (including static, impact, or kneading densification)
   (4) environmental temperature and pressure conditions including the effect of surcharges

For the identification and characterization of expansive soils the following criteria are used:

(1) The USBR uses the colloid content (percent < 0.001 mm), *PI*, and *SL* as criteria for identification of expansive clays as shown in Table 3.22. Since the respective Atterberg limits are routinely determined on soils and the colloid content is also easily determined, this method should be preferred to the single index methods such as the *PI* criterion suggested by Seed et al. (1962), and the *SL* criterion suggested by Altmeyer (1956) and Ring (1966), except where specific relationships have been established for a certain soil region. A Na-montmorillonite soil may show a lower *SL* and a higher free swell value than its Ca-equivalent but may present less of an expansion danger because of the more energetic water absorption and higher swelling pressure developed by the latter. The same criticism holds true for the free swell test (Holtz and Gibbs, 1956) in

TABLE 3.22 RELATION OF SOIL INDEX PROPERTIES AND PROBABLE VOLUME CHANGES FOR HIGHLY PLASTIC SOILS[a].

Data from Index Tests[b]			Estimation of Probable Expansion[c]	
Colloid Content (Percent minus 0.001 min)	Plasticity Index	Shrinkage Limit (Percent)	Percent Total Volume Change (Dry to Saturated Condition)	Degree of Expansion
>28	>35	<11	>30	Very high
20–31	25–41	7–12	20–30	High
13–23	15–28	10–16	10–20	Medium
<15	<18	>15	<10	Low

[a] USBR, *Earth Manual* (1973).
[b] All three index tests should be considered in estimating expansive properties.
[c] Based on a vertical loading of 1.0 psi as for concrete canal lining. For higher loadings the amount of expansion is reduced, depending on the load and on the clay characteristics.

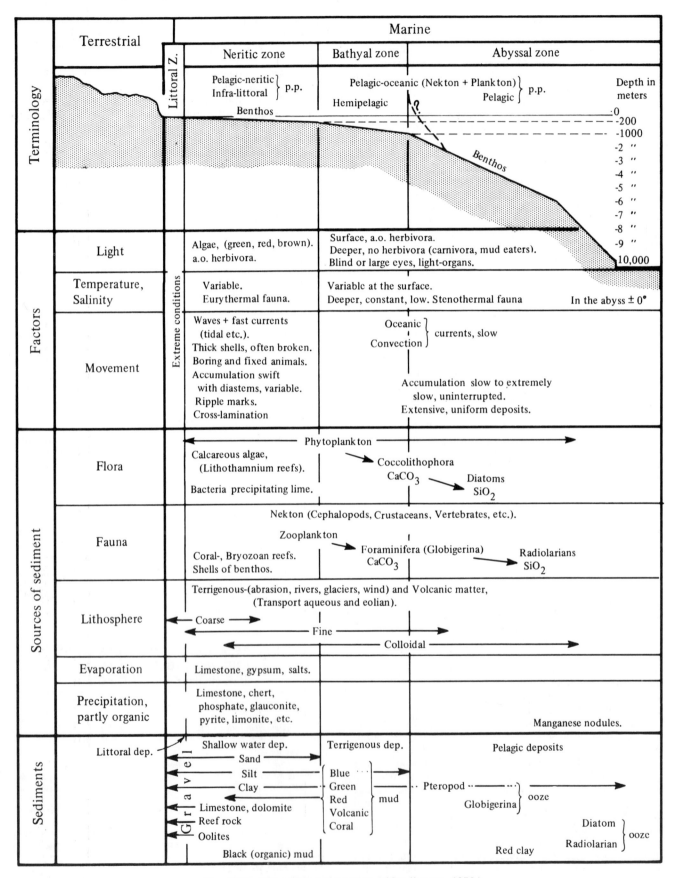

Fig. 3.16 The marine environment. (*After Kuenen, 1950.*)

TABLE 3.23 DISTRIBUTION OF CLAY MINERALS IN THE OCEAN[a].

Location	Number of Samples	Chlorite (Percent)	Montmorillonite (Percent)	Illite (Percent)	Kaolinite (Percent)
North Atlantic	202–206	10	16	55	20
South Atlantic	196–214	11	26	47	17
North Pacific	170	18	35	40	8
South Pacific	140–151	13	53	26	8
Indian	127–129	12	41	33	17

[a] After Griffin et al. (1968).

which a certain bulk volume of dry soil is placed into a water-filled graduated cylinder and the volume change noted.

(2) Other indicative tests requiring more equipment include the moisture density relationships as determined by the Standard and Modified AASHTO and Dietert compaction tests, the CBR and swell tests, the AASHTO T-190 expansion pressure test with or without modifications and the PVC (potential volume change) test (AASHTO, 1988).

(3) The most useful and reliable quantitative methods are of the conventional consolidometer (oedometer) type, employing an oedometer and a loading frame of some type and soil samples of from 2 to 4 inches in diameter and $\frac{1}{2}$ to 1 inch in height.

General discussions relating to the engineering properties, measuring techniques, evaluation and control of expansive soils are given by Gromko (1974), Chen (1979), O'Neill and Poormoayed (1980), Anderson and Lade (1981), and TRB (1985).

3.6.10 Marine Soils

About 71 percent of the earth's surface is covered by oceans. The loose sediments on the ocean floor qualify as engineering soils, and the interaction of the ocean waters with coast materials puts its mark on the coastal soil while elutriating its finer constituents for marine deposition. As shown in Figure 3.16, marine sediments are made up of terrestrial and marine contributions. The former represent the particulate material eroded from the shore or carried as bottom load or in suspension by rivers and creeks, by iceberg, tree root, and kelp rafting, as well as mineral matter in true or colloidal solution, for example, bicarbonates of bivalent metal ions, and silica sols and gels from which the marine flora and fauna form internal and external skeletons. The marine contribution is represented by the organic and inorganic (skeletal) remnants of dead marine life and by precipitation sediments from oversaturated solutions. The terrestrial contribution normally decreases, in both proportion and particle size, with increasing distance from the shore line, while the marine contribution increases. Local deviations from this rule are normally attributable to slumping, mud slides, and turbidity (gravity) currents (Kuenen, 1950).

Because of the classifying action of wave attack on the shore material and of the seaward transportation, most marine deposits have a relatively narrow particle size range (Winterkorn and Fang, 1970; Fang and Chaney, 1986). The surface sediments on the continental shelf may be sands, silts, and clays depending on location, geology, and proximity to rivers. There are three general types of deep sea deposits; viz., brown (red) clays, calcareous ooze, and siliceous ooze. The term ooze denotes that more than 30 percent of the sediments are of biotic origin. Equatorial and polar regions are especially high in organic productivity. These sediments are calcareous at depths of less than 1400 feet and siliceous in deeper waters. In areas of lesser biotic activity, calcareous ooze is found in shallow and brown (red) clay in deeper waters.

(a) *Calcareous ooze.* Calcareous ooze is nonplastic, creamy to white in color with particles of sand to silt size that crush easily. Some oozes containing 50 to 100 percent water show undrained shear strengths of 70 to 220 psf.

(b) *Siliceous ooze.* Siliceous ooze consists mainly of the remains of siliceous plant skeletons which are found in a large belt around the Antarctic and also northeast of the Japanese Islands; in a few areas of the equatorial Pacific ooze from radiolaria is predominant.

(c) *Brown clay.* Brown clay of terrestrial origin transported by wind and/or water is found in most of the deeper portions of the oceans. Table 3.23 gives the distribution of clay minerals in the oceans according to Griffin et al. (1968). Only montmorillonite (smectite) is believed to be formed in the marine environment. In marine clays normally about 60 percent of particles are less than 2 μm in size; also they are very low in $CaCO_3$ content.

A state-of-the-art review of the engineering properties of marine soils are given by Noorany and Gizienski (1970), Chaney and Fang (1986), and in Chapters 2 and 18.

3.7 THE STRUCTURE OF NONCOHESIVE SOIL SYSTEMS

Soil structure may be defined as the geometric arrangement of its identifiable particulate constituents and their physical and physicochemical interaction, which may or may not influence the structural morphology and stability. While the particulate components of most engineering materials vary only within a narrow range of size and shape, the more or less independently acting soil constituents may range from the size of boulders (> 30 cm) to that of the proton ($< 10^{-8}$ cm) and in shape from spherical through regular and distorted geometric figures to plate, needle, fiber, and ribbon-like extremes. In addition, the interparticle forces may range from practically nonexistent to strong electrostatic bonds.

To bring some order into the practically infinite number of structural possibilities in soils, we start with geometric relationships that can easily be visualized on granular systems composed of macroscopic particles. The resulting laws and concepts are, because of their geometric nature, not bound to any particular size or size range and these laws are, of course, directly applicable to coarse noncohesive soil systems. However, the knowledge and the feeling for the importance of shape and geometry developed in the contemplation of assemblies of particles of visible size will aid our understanding of the importance of geometric factors in systems with molecular and colloidal constituent sizes.

Primary structure refers to the natural arrangement of the constituent particles of a soil in what may be regarded as a continuous system such as a body of sand or gravel or a natural secondary unit in a cohesive soil such as the "peds." The structure of noncohesive granular soils is an important part of the discipline of granulometry.

This discipline is concerned with the physical properties of systems composed of grainlike particles such as sand, gravel, crushed rock, bird shot, seeds, breakfast cereals, powders, and others. It embraces the study and measurement of the size, shape, and surface features of individual particles as well as the influence of these properties and of the gradation of the particles on the packing characteristics, mechanical resistance properties, and permeability of multiparticle systems. While granulometry deals mainly with the type of materials listed above, many of its laws are of a geometric nature and are equally true for particles of atomic size as they are for gravels and boulders. Thus, knowledge originally obtained on assemblies of atoms and molecules can be utilized for sand and gravel systems, and vice versa.

3.7.1 Special Granulometric Terms

1. *Effective size*—the size of the screen opening permitting 10 percent of the granular material to pass and retaining 90 percent.
2. *Uniformity coefficient*—the ratio of the size of screen openings passing 60 percent to that passing 10 percent. The smaller this ratio, the more uniform the sand.
3. *The fineness modulus*—a measure of gradation developed by Abrams (1918) and widely used in concrete technology. It is defined as one-hundredth of the sum of the cumulative percentages retained in the sieve analysis when using the U.S. sieve series: $1\frac{1}{2}$ in, $\frac{3}{4}$ in, $\frac{3}{8}$ in, No. 4, No. 8, No. 16, No. 30, No. 50, and No. 100.
4. *The Santos constant*—a measure of influence of grading on consistency properties of soils. It is defined by the formula

$$a = \frac{\Sigma y}{100n} \qquad (3.7)$$

in which Σy = sum of percentages passing each of a set of n sieves, that with the smallest openings being a No. 200. The set usually consists of the following size openings: Nos. 8, 16, 30, 50, 100, 200.

3.7.2 Volumetric Terms

Volumetric relationships in granular systems are usually expressed by such terms as:

1. *Phase volumes*—the portions of the total volume contributed by the various solid, liquid, and gaseous phases of the system, expressed as fractions or percentages of the bulk volume.
2. *Porosity (n)*—the fraction or percentage of the bulk volume not occupied by solid matter.
3. *Void ratio (e)*—the ratio of the volume fraction not occupied by solid matter to that occupied by solid matter.
4. *Absolute volumes*—the actual volumes occupied by the various phases, usually expressed in volumetric units employed in engineering or fractions thereof such as cubic foot, cubic yard, or cubic feet per cubic yard. The sum of the absolute phase volumes must give the total or bulk volume.

3.7.3 Size and Shape Measurement

I. Direct Methods
1. Slide caliper (large pieces).
2. Ring (particle must be able to pass through ring in any orientation).
3. Ordinary light microscope (down to 5×10^{-4} cm).
4. Ordinary and scanning electron microscope.

5. X-ray diffraction for particles of atomic dimensions.

II. Sieving and Screening
Size is defined by the size of the openings of a sieve through which the particle just passes and that on which it is retained. Standard sieve sets are used for determining size ranges and separating materials into fractions of predetermined size ranges.

The sample used for sieve analysis must be representative and must have been reduced from a larger representative sample by quartering or by means of a sample splitter. The proper weight of the sample to be tested depends on the largest pieces. The Standard U.S. Sieve Series ranges from an opening size of 10.78 cm to 3.7×10^{-3} cm.

III. Filtration as a Form of Sieving
1. Filter papers of varous pore sizes down to 10^{-4} cm.
2. Nitrocellulose films to 10^{-7} cm.
3. Molecular filters, natural and artificial zeolites, and ultramarines down to molecular (H_2O) and ionic (Ba^{++}) dimensions.

Note: Filtering action depends not only on the dimensions of the respective particles and holes but also on the electric charges of the particles and the pore walls.

IV. Sedimentation and Elutriation Methods
Methods in this section are based on Stokes' law
1. Standardized sedimentation methods. Pertinent sedimentation procedures for soils are the pipette method of the International Soil Science Society and the hydrometer methods of Bauyoucos and Casagrande, in which dispersed soil is permitted to fall in a column of water. Air as a settling medium is used in one of several methods for the determination of particle size of portland cement.
2. Elutriation methods. A battery of cylindrical vessels of different cross sections is so arranged and interconnected that the same stream of water flows upward in each vessel but with a velocity that is greatest in the first vessel and stepwise smaller in each succeeding vessel. At the end of the test, the minimum size of the particles remaining in each vessel is that whose rate of fall is equal to the upward velocity of the flowing water.

V. Indirect Methods Based on Surface Area
The surface area per unit volume of particles is $3/r$ for spheres and $6/d$ for cubes. This relationship permits the calculation of an average or effective particle size from the surface area S per unit volume of a material.

$$S = 3/r \qquad \text{or} \qquad d = 6/S \qquad (3.8)$$

where

r = radius of sphere
d = side length of cube

Several methods are available for determination of surface area per unit volume of solid particles utilizing the absorption of liquids and gases, respectively, on mineral surfaces.

VI. Shape Measurement and Characterization
1. General
   The shape of particles that compose a granular system influences the degree of density and the internal friction and workability of the system, as well as its permeability to water, air, and other fluids.

   Often particle shape is given in general terms such as spherical, rounded, angular, irregular, or simply by differentiation between crushed stone and natural gravels and sands. Shapes may also be characterized by the closest geometric form such as spherical, cylindrical, elliptic, cubic, prismatic, plate- and needle-shaped. For visual identification of degree of angularity or sphericity, charts like those in Figure 3.17 are available.

For large particles, the general shape may be identified by reference to such charts while the length, width, and thickness dimensions can be measured by use of slide caliper, micrometer, or other means. For small particles, photographic methods may be employed. However, all these direct methods are rather tedious and for practical purposes particle shape is usually defined by indirect methods. Such methods utilize the fact that particle shape greatly influences the degree of density achieved in an assembly of particles by means of standardized methods of densification. The more the particle shape deviates from the spherical, the greater is the difference between the maximum and minimum densities of particle assemblies achieved by appropriate densifications methods.

2. Numerical Expression of Particle Shape
   (a) Sphericity
      Sphericity was defined by Wadell (1932) as the ratio of the surface area of a sphere of the same volume as the particle to the surface area of the particle. The usefulness of this method depends mainly on the ease with which surface areas can be measured.
   (b) Volumetric Coefficient
      Defined by Joisel (1948) as the ratio of the actual volume of a particle to that of a sphere in which it can just be enclosed. Accordingly:

$$V.C. = \frac{6v}{\pi a^3} \qquad (3.9)$$

where

v = volume of particle
a = largest dimension

Shape of Particle	V.C.
Sphere	1
Cube	0.37
Tetrahedron	0.22
Round gravel	0.34
Angular stone	0.22
Plates	0.07
Needles	0.01

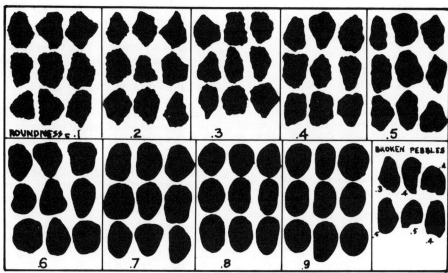

(a) Pebble images for visual roundness.

(b) Chart for determining the degree of angularity of particles.

Fig. 3.17 Angularity charts. (*After Krumbein, 1941.*)

(c) Elongation Ratio, Flatness Ratio, Shape Factor

If a is the largest, b the intermediate, and c the smallest dimension of a particle, then

$$b/a = q = \text{elongation ratio}$$

$$c/b = p = \text{flatness ratio}$$

$$ca/b^2 = p/q = F = \text{shape factor}$$

By plotting p against q, one obtains the shape factor as shown in Figure 3.18.

The subdivision into the four shape groups is in accordance with Zingg (1935). Aschenbrenner (1956) developed a chart by means of which p and q values can be converted into sphericity ψ in accordance with the equation

$$\psi = \frac{12.8(p^2 q)^{1/3}}{1 + p(1 + q) + 6[1 + p^2(1 + q^2)]^{1/2}} \quad (3.10)$$

(d) Indirect Methods for Evaluating Particle Shape

These are based on the loosest states that particles of a narrow size range assume in standardized dumping procedures. They can be refined by reference also to the densest state obtainable by an appropriate method. Representative are the procedures developed by the National Crushed Stone Association (Gray, 1964).

3.7.4 The Laws of Granulometry

Just as any other physical law, the laws of granulometry were derived from observations and experimentation on actual physical systems. However, in their most encompassing form they reflect abstractions necessary to obtain the simplest possible mathematical expression. Therefore, in the application of these laws to actual physical systems, the abstractions that have been made in their derivation must be properly corrected for. Also, these laws may be expected to best express the behavior of actual physical systems if their constituent particles approach closely the ideal, that is, spherical shape.

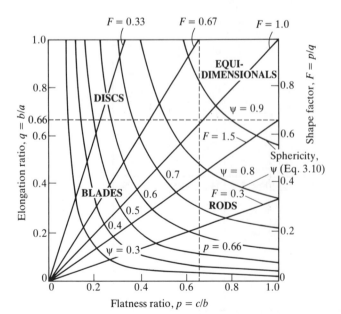

Fig. 3.18 Classification of particle shapes. (*Based on Zingg, 1935; Lees, 1964.*)

3.7.5 Packings of Particles and Their Primary Structure

The theoretically possible types of "continuous, incompressible, uniform packings" of *identical spheres* are given by Kézdi (1964). Their porosities range from 72 to 26 percent. With sand-sized spheres and rounded sand particles, it is very difficult in practice to get uniform packings with porosites higher than 50 percent or less than 36 percent, nor are the packings between these limits of a uniform character. Rather, photographs by Kolbuszewski (1965) and by Mogami (1967) show that in a sand mass of a certain bulk porosity there are domains of regularly packed particles (cubic, $e = 0.91$; orthorhombic, $e = 0.65$; etc.) in various orientations, separated by interphases of more loosely and irregularly packed particles. The lower the void ratio, the less the volume proportion of the interphases and the greater that of the more orderly and densely packed domains. In contrast to their geometric abstractions, material spheres in terrestrial environments possess packing characteristics that are influenced by sphere size, surface/volume ratio, adsorbed gas and water layers, and interparticle friction. In general, packing density decreases with decreasing particle size. For a thorough treatment of systematic packing of spheres with particular relation to porosity and permeability, see Graton and Fraser (1935).

For *simple polygonal shapes* that can be fitted together without intervening cavities, the minimum theoretical porosity approaches zero. However, normal handling and compaction methods usually result in higher maximum and minimum porosities than obtained for spheres of equivalent size, which is especially true for crushed rock particles having sharp corners and edges and rough surfaces. If the edges and corners have been rounded off by natural transportation or by milling, such particles pack easily to greater densities than spherical ones.

Binary Systems Theoretically, the densest system obtainable with spheres of two different sizes is one on which the larger spheres are packed to lowest porosity (26 percent) and the interstices filled in densest packing by spheres so small that the extent and curvature of the pore walls do not affect their packing. The minimum theoretical porosity would then be 6.75 percent. A practical though very imperfect approach to this type of packing is the macadam pavement, in which first a layer of uniformly sized broken stone is laid with as much interlocking as possible and the interstices and filled with successively smaller-sized stone and finally sand, and the process is repeated for the next layer.

When mixtures of larger and smaller particles are used, the smaller always interfere with the packing of the larger (Furnas, 1931).

Ternary Systems Packing interference is also very evident in ternary systems. According to Feret (1892) and subsequent workers, greatest density is achieved by proportions of about 1 part of the smallest and 2 parts of the largest-size materials. Systems composed of about equal parts of all three components exhibit the smallest difference between practically obtainable maximum and minimum porosities.

Continuous Gradings Continuous gradings yield low-porosity mixtures requiring little compactive effort and are therefore of great practical importance in soil stabilization and in the making of concretes with hydraulic and bituminous cements (Holl, 1971). The greater the range from the maximum to the minimum particle size, the less the porosity of the system. Representative values for various ratios of D_{max}/d_{min} can be calculated with the equations:

(a)
$$n(\%) = 38.5 - 8 \log_{10} \frac{D_{max}}{d_{min}}$$
for rounded gravel and sand (3.11)

(b)
$$n(\%) = 47.5 - 8 \log_{10} \frac{D_{max}}{d_{min}}$$
for crushed stone (3.12)

where D_{max} and d_{min} are the respective diameters of the largest and smallest particles present. The calculated values are for rodded mixtures and lie between those for maximum and minimum obtainable densities. For most natural materials of relatively narrow gradation, the range of easily obtainable and reproducible porosity lies between 36 percent and 46 percent. For a particular sand, the loosest practically obtainable state is easily determined by the method of Kolbuszewski (1965). The porosity of the practically densest state is usually 9.5 percent lower than that of the loosest.

3.8 THE STRUCTURE OF COHESIVE SOILS

The structure of cohesive soils with regard to both its geometric features and its resistance to deformation and destruction is intimately related to the interaction of the surfaces of the solid soil particles with the molecular and ionic constituents of the soil solution or aqueous phase of the soil. Because of the molecular nature of this interaction its understanding presupposes knowledge of the molecular structure of the particle surfaces, which itself is a function of that of the particles. Restricting our considerations to mineral soil particles we assume for good and sufficient reasons that they are primarily composed of ions and held together by electrostatic forces. The consequences of this assumption with regard to the contributions of the various ions to the "solid" volume of the minerals in the earth's crust are shown in Table 3.24. Thus, we conclude that the minerals in the earth's crust, and especially those in the soil layers, are essentially oxygen ion structures held together by smaller cations. However, the actual contributions of these cations to the total volumes of the respective minerals can often be neglected vis-à-vis the larger differences caused by the type of packing they impose on the oxygen ions. This is illustrated

by the data in Table 3.25 for minerals often found in the silt and sand fractions of soils. This table also shows that water and ice are also essentially very loosely packed oxygen ion structures.

3.8.1 Ionic Structure of Soil Minerals

Since solid particles can react only on their surfaces (external and internal), special interest devolves on the crystalline and amorphous particles of clay size and on the clay minerals themselves, which because of their smallness and their usually platy, fibrous, and ribbon-like shapes contribute most to the total internal surface of a cohesive soil system. There are two primary units in the formation of clay structures: the oxygen tetrahedron formed around a silicon ion, and the oxygen octahedron formed around an aluminum or magnesium ion. The tetrahedra form sheets such as also occur in the mineral cristobalite and the octahedra form sheets such as occur in the minerals gibbsite and brucite. The simplest clay mineral structures are two-layer units formed of alternating tetrahedral and octahedral sheets. The three-layer structures have two tetrahedral sheets with an octahedral one sandwiched in between. The structural arrangement of the oxygen ions is the predominant feature into which electrical neutralization by the available cations must fit as well as possible. The structural predominance of the O^{2-} ions can be easily visualized from the volumes of the Si^{4+}, Al^{3+}, and O^{2-} ions, which are 0.256, 0.796, and 9.65 cubic angstroms, respectively. If the cationic glue does not provide enough positive charges for neutralization within the mineral structure, as in replacement of Si^{4+} by Al^{3+} or of Al^{3+} by Mg^{2+}, then the excess negative charges of the oxygen structures must be neutralized on the mineral surfaces by adsorption of positive ions including H^+ from the environment. These cations are exchangeable for others. This property is called the base-exchange capacity and is usually expressed in milliequivalents per 100 grams of dry clay (or soil). The great variety of possible substitutions of cations within the clay mineral structures and intergrowth of various structural layers leads to a great diversity of actual clay minerals. A classification scheme for phyllo (layer) silicates related to clay minerals is shown in Table 3.26 (SSSA, 1970).

3.8.2 The Electrical Structure of the Mineral Surfaces

From the evidence so far presented we can make several important qualitative statements regarding the electrical structure of the mineral surface, assuming for the sake of simplicity absence of adsorbed water molecules and of exchange ions:

1. Even if the number of positive charges within a soil mineral structure equals that of the negative ones assuming overall neutralization, the mineral surfaces possess a slightly negative character because of the larger size and easier accessibility of the O^{2-} and OH^- building stones. This can be pictured as a diffuse negative electric surface field.
2. If in the gibbsite sheet of a three-layer clay mineral Al^{3+} ions are replaced by an equal number of Mg^{2+} ions, the resulting electrical unbalance is transmitted to the mineral surface with attendant spreading or diffusion of the negative charge.
3. If in the surface (cristobalite) sheet of a three-layer clay mineral Si^{4+} ions are replaced by Al^{3+} ions, the resulting negative charges originating closer to the surface produce a more patchy surface electric field of lesser uniformity than in case (2).

TABLE 3.24 AVERAGE COMPOSITION OF THE EARTH CRUST AND OF THE A AND B HORIZONS OF SOILS.

| Element | Earth Crust (Percent) | | | Soil Horizon | |
	Weight	Mole	Ionic Volume	A (Percent by Weight)	B
O^a	48.18	63.40	94.00	52.78	49.73
Si	28.60	21.50	0.83	35.70	35.70
Al	7.13	5.62	0.69	5.30	6.24
Fe	4.38	1.64	0.44	2.35	4.05
Ca	3.52	1.85	1.42	0.74	0.71
Mg	2.35	2.04	0.62	0.36	0.51
K	2.54	1.37	2.07	1.43	1.71
Na	2.52	2.32	1.40	0.61	0.61
Mn	0.06	0.023	0.01	0.09	0.09
Ti	0.47	0.206	0.03	0.54	0.54
P	0.13	0.088	0.001	0.06	0.06
S	0.13	0.086	0.002	0.05	0.05

^{a} Oxygen percentage does not include contribution from free water content.

TABLE 3.25 VOLUMETRIC RELATIONSHIPS IN SOIL MINERALS.

Ions	Si^{4+}	Al^{3+}	Mg^{2+}	Li$^+$	Na$^+$	K$^+$	O^{2-}	OH$^-$	F$^-$
Volume (Å^3)	0.25	0.79	1.98	1.98	3.94	9.85	9.65		9.85

Volumes of Component Ions in Å^3

Volume percentages and mole volumes of oxygen ions in various minerals if volumes of cations are assumed to be negligible

Mineral	Formula	Vol. % O^{2-} Ions	Mole Weight	Density (g/cm^3)	Volume per Gram Mole of O^{2-} (cm^3)
Ice	H_2O	100	18	0.92	19.65
Water	H_2O	100	18	1.00	18.00
Orthoclase	$KAlSi_3O_8$	87	278	2.53–2.6	13.4–13.7
Muscovite	$KH_2Al_3Si_3O_{12}$	90	398	2.76–3.1	10.7–12
Quartz	SiO_2	99	60	2.65	11.35
Corundum	Al_2O_3	95	102	4.0	8.5

4. If the neutralization of negative charges originating in the oxygen structures involves large positive ions within the surface layer, such as the K^+ ion, which is of about the same size as the O^{2-} ion and fits into structural surface cavities, then a number of positive patches are found in the surface electric field that serve as sinks for the Faraday lines issuing from the negatively charged areas. Such positive patches also occur at broken corners and edges. Electric surface fields of nonuniform character exhibit much greater dispersion with increasing distance from the particle surfaces than do uniform fields.

5. The top and bottom surfaces of a kaolinite crystal having different compositions also differ in their physicochemical properties, including their interaction with water.

6. The dispersion of a uniform electric field with distance from its surface or origin increases with decreasing radius of curvature of the surface. Hence the effective depth of the field is largest for plane surfaces and smaller for spheres, edges, and corners.

3.8.3 Structural Properties of the Water Substance

The peculiar properties of the water substance and the various modes of its interaction with mineral surfaces and dissolved ions derive from the electric and geometric structure of the H_2O molecule. According to the simple, but for our purpose sufficiently accurate, electrostatic model of Born et al. (1924), the HOH molecule is approximately spherical with a radius of about 1.38 Å while that of the O^{2-} ion is about 1.32 Å. The H^+ ions are so located on the surface of the O^{2-} ion that they form an angle of about 105° with its center. The ensuing lack of coincidence of the centers of positive and negative charges within the water molecule gives the latter a permanent dipole of 1.87×10^{-18} electrostatic unit (esu). The strength of this dipole combined with the geometry of the water molecule results in a strong directed interaction between adjoining molecules, which because of its directional characteristics prevents their close packing. This is illustrated by the following data giving the volumes occupied by one mole (6.02×10^{23}) of spheres of the same size as the water molecule in various symmetrical packings.

Absolute volume of solids	6.62 cm^3
Hexagonal rhombohedral	8.94 cm^3
Tetragonal spheroidal	9.49 cm^3
Orthorhombic	10.92 cm^3
Cubic	12.69 cm^3
Normal water	18.00 cm^3
Ice	19.65 cm^3

Accordingly, the packing of the H_2O molecules in normal water and ice is so loose that the "solid" molecular matter occupies only 36.8 and 33.7 percent, respectively, of a given volume. The packing may become even looser if the water molecules are vicinal to a plane surface possessing a uniform negative charge field that attracts their positively charged portions, which are mutually repellant. On the other hand, the strong electric fields around dissolved cations may break the weaker H_2O–H_2O dipole bonds and produce a denser packing within their sphere of influence. This gives rise to apparent negative volumes of ions in aqueous solutions, as illustrated by the data below:

Ion	H^+	Li^+	Na^+	K^+	NH_4^+	Mg^{2+}	Ca^{2+}	Ba^{2+}	Al^{3+}	OH^-
Apparent volume in solution (Å^3)	−8	−8	−8.5	+8	+12	−48	−45	−33	−87	−2

The data tabulated below may serve for general and specific orientation with regard to the energies involved in molecular interactions that are basic for the understanding of soil–water relationships:

Interacting pair	Type of attraction	Equilibrium separation (Å)	Energy of interaction (kcal/mole)
Na^+ F^-	Ion–Ion	1.88	157
Na^+ OH_2	Ion–Dipole	2.14	21.6
OH_2 OH_2	Dipole–Dipole	2.37	4.84
Ne Ne	Coupled electronic oscillation	3.30	0.0613

3.8.4 The Clay–Water Micelle and Its Structure

The clay micelle is an important structural unit in aqueous clay suspensions and comprises the solid clay particle itself as well as its sphere of influence in the surrounding water or aqueous solution. The sphere of influence corresponds to the distance in which exchange ions originating from the particle surface may be found in a dynamic equilibrium between the electric attractive forces from the charged particles (attenuated by the dielectric properties of the water) and the inherent kinetic energy of the ions, which themselves may be surrounded by strongly held water molecules. The total volume of this sphere of influence within the aqueous phase minus the amount of bound or restrained water of ionic hydration or particle surface hydration represents "osmotic" swelling and is for geometric reasons especially large on flat particle surfaces (Winterkorn,

TABLE 3.26 TERMINOLOGY IN CLAY MINERALOGY AND CLASSIFICATION SCHEME[a].

Type	Group (x = Charge per Formula Unit)	Subgroup	Species[b]
1:1	Kaolinite- serpentine	Kaolinites	Kaolinite, halloysite
	x ~ 0	Serpentines	Chrysotile, lizardite, antigorite
2:1	Pyrophyllite-talc	Pyrophyllites	Pyrophyllite
	x ~ 0	Talcs	Talc
	Smectite *or* montmorillonite- saponite	Dioctahedral smectites *or* montmorillonites	Montmorillonite, beidellite, nontronite
	x ~ 0.24–0.6	Triocathedral smectites *or* saponites	Saponite, hectorite, sauconite
	Vermiculite	Dioctahedral vermiculite	Dioctahedral vermiculite
	x ~ 0.6–0.9	Trioctahedral vermiculite	Trioctahedral vermiculite
	Mica[c]	Dioctahedral micas	Muscovite, paragonite
	x ~ 1	Trioctahedral micas	Biotite, phlogopite
	Brittle mica	Dioctahedral brittle micas	Margarite
	x ~ 2	Trioctahedral brittle micas	Clintonite
2:1:1	Chlorite	Dioctahedral chlorites (4–5 octahedral cations per formula unit)	
	x variable	Trioctahedral chlorites (5–6 octahedral cations per formula unit)	Pennine, clinochlore, prochlorite

[a] After Soil Science Society of America (1970).
[b] Only a few examples are given.
[c] For further information see Soil Science Society of America July 1987 manual.

1936); also, because of its osmotic character, its volume is reduced when the dispersion medium contains dissolved ions or molecules that compete for the available solvent to a degree expressed by their osmotic pressure.

In the general case, an aqueous clay suspension with an open surface or air–water interface may contain the water substance in the following simultaneously existing states:

1. In solid solution in the surface layer of the clay (or other mineral particles) similarly as water is soluble in silver nitrate and other solids; the extent of this solution is governed by the chemical character and lattice structure of the mineral.
2. In oriented (constrained) condition on the solid surfaces resulting from the superposition of a purely geometric "wall" effect and the pattern and intensity of the electric field

characterizing the particle surface; this includes the tendency of the packing pattern of the O^{2-} and OH^- ions of the solid surface to impress itself on the vicinal layers of water molecules. In the case of montmorillonite, the density of this layer appears to be below that of normal liquid water (Low, 1961).

3. In an oriented and densified condition in the strong electric fields issuing from small cations in solution and from highly charged corners and edges of the solid particles. The strong interaction between water and cations (also called hydratation) is associated with the heat of wetting of the dry particles and the heat of solution of the respective ions, which result from the difference in the heat capacity of water molecules in the free and constrained states, respectively.
4. In a relatively free state with the normal water structure located between the hydrated cations and the hydrated particle surface from which they issue and in whose attraction sphere they are held by electrostatic forces; it represents the osmotic swelling water.
5. Free water or aqueous solutions between the different micelles within the dispersion.
6. Oriented water at the air–water interface, which according to Henniker and McBain (1948) is several hundred molecules thick at normal temperature.

Actually, the above presentation of different structural states of the water substance in a clay–water system represents only a very limited picture of the practically unlimited possibilities and facets of the interaction of the solid soil components with the water substance that occur under natural conditions. Even the next logical step from the structure of a single micelle to the interaction of two or more micelles has been handled theoretically only for the simplest possible cases and neglecting very important available information, as a result of which the end product of such present-day theoretical development is critically inadequate for engineering purposes. A general discussion on soil–water interaction is given by Fang (1989).

3.8.5 Other Consequences of the Micelle Structure

The solid portions of clay and similarly behaving humus micelles may be considered as large polyvalent anions. If these are surrounded and neutralized only by H^+ ions, they are in effect molecules of polybasic acids; if completely neutralized by other cations, they form salts of these acids. These salts may give an alkaline reaction in water if the base strength of the cations is greater than the strength of the polyvalent acid. Partial neutralization may result in "neutral" or in acid salts. In the case of too great an acidity of the system, the solid particle itself will be attacked, with concomitant freeing of Al^{3+} and/or Fe^{3+} or other structural cations that may take exchange-ion positions. The types and amounts of exchange ions on clay particles affect the structure of the clay micelle and through this the engineering behavior of clays and clay soils.

3.8.6 Soil Acidity

All soils contain H ions in their aqueous phase since water itself is dissociated into H^+ and OH^- ions as shown below:

Temperature in °C	0	12	25	35	50
Mole/liter of H^+ or $OH^- \times 10^7$	0.28	0.73	1.00	1.45	2.3

At 25°C the concentration product $[H^+] \times [OH^-]$ is 10^{-14} and the H^+ and OH^- concentration each 10^{-7} in the case of neutrality. Higher H^+ concentrations are due to exchangeable H ions on the soil particles and/or the presence of mineral or organic acids, e.g., H_2SO_4 from oxidation of pyrites, H_2CO_3, and other organic acids from the decomposition of vegetation. Soil acidity varies with the season. It is normally expressed as the negative logarithm of the H^+ ion concentration, which is called the pH. Accordingly, a pH of 7 indicates neutrality and a value less than 7 denotes acidity, which increases with decreasing pH value. A solution of 1 mole of H^+ ions per liter has a pH of zero. Soil acidity is important to the soil engineer because of its corrosive effect on metals and other construction materials; on the other hand, acidity is sometimes desirable because of its catalytic effect on certain reactions employed in soil stabilization.

One differentiates between actual acidity (free H^+ ions) and exchange and hydrolytic acidity (loosely or strongly adsorbed H^+ ions), the latter representing potential acidity. The actual acidity is determined on a soil-water dispersion which must be kept in suspension because of the "suspension effect" which is illustrated below after Pallmann (1930):

Relative amount of clay in suspension	100	50	33.3	18.2	9.7	0 (ultra-filtrate)
pH			4.47	4.72	4.93	5.20	5.31	5.64

The pH determination by potentiometric method (electric pH meter) gives the degree of acidity while the total amount of potential acid is determined by titration. It is expressed by the amount of base required for the neutralization of 100 g of soil. Titrations are performed on filtered soil extracts. These are made with water for "actual" acidity, with KCl solutions for exchange acidity and with Ca-acetate solutions for hydrolytic acidity.

3.8.7 Buffer Action of Soil

Buffering is the capacity of soil or similar systems to resist a change in pH by addition of acids or bases. It is characterized by a buffer curve obtained by plotting pH values against increments of added base or acid. The "buffer area" is the area enclosed by the buffer curve and a curve obtained in a nonbuffering material such as purified fine sand.

3.8.8 The Clay Fraction and the Behavior of Whole Soils

The actual contribution of the clay fraction to the engineering behavior of whole soils depends not only on its quantity and

the physicochemical properties of its micelles but also on the relative amounts and characteristics of the other soil constituents, including the aqueous and gaseous phases with which the clay particles interact. While especially in the case of disturbed soils the granulometric considerations (treated in Section 3.7) are of primary importance, decisive influence may often be exerted by:

1. The state of aggregation of the clay particles. Are they dispersed and independently acting, dispersed but aggregated to dense secondary structures or peds with external protective films of mineral or organic character, or do they form loosely aggregated flocculant structures?
2. The location of the clay particles: do they form films around the coarser constituents that, while interfering to a certain extent with their packing, endow the system with cohesive properties, or do they act as fillers in the pore space of a granular skeleton formed by the coarse constituents?
3. The degree and strength of physicochemical interaction with the coarser-sized constituents that may be gauged by slaking experiments.

The types and water resistance of the various secondary structures formed may be gauged from the results of shrinkage limit and slaking tests made, respectively, on the clay and No. 40 sieve fractions and on the whole soil under consideration. Important in the evaluation of slaking test data is the heat of wetting produced by mixing the soil or soil fraction (after oven-drying and cooling in a vacuum desiccator over P_2O_5) with an excess of water. The heat of wetting is closely related to the hygroscopicity W_H, which is defined as the percentage of water based on the weight of the dry soil that satisfies the adsorption capacity of the particle surfaces such that further water addition does not release additional heat of wetting. The hygroscopicity, which corresponds approximately to the percent of water absorbed from the atmosphere over a 10 percent aqueous solution of H_2SO_4 is, like the heat of wetting, mainly a function of the amounts and types of exchangeable cations and of the extent and character of the particle surfaces. The heat of wetting is determined in a calorimeter and expressed in calories per gram of dry soil. Table 3.27 gives H_w values for typical soils and clays including homoionic variants.

A high shrinkage limit combined with a large heat of wetting indicates soils that are especially vulnerable to slaking and loss of stability when placed in contact with water while in the dry state. The heat of wetting in calories per gram of dry soil is numerically equal to one-half of the hygroscopicity over

TABLE 3.27 HEATS OF WETTING OF DIFFERENT CLAYS AND WHOLE SOILS INCLUDING EFFECT OF DIFFERENT EXCHANGE CATIONS.

	(a) H-Clays					
Clay Type	Bentonite	Lufkin	Wabash	Putnam	Susquehanna	Cecil
$Si_2/(Al_2O_3 + Fe_2O_3)$	5.0	3.8	3.2	3.2	2.3	1.3
H_w, cal/g	20	15	13.9	13.8	11.7	5.9

	(b) Typical data for soil of different size composition (Landolt-Bornstein, 1952)		
Soil Type	Sand	Sandy Loam	Loam	Clay
H_w (cal/g)	0.54	1.65	2.41	7.85

	(c) Effect of exchange cation					
	H	Li	Na	K	Ca	Ba
Putnam Clay	13.6	12.0	12.0	9.5	15.0	13.9
Wabash Clay	13.9	11.4	11.6	8.9	15.3	14.9
Bentonite	20	—	16	—	22	—

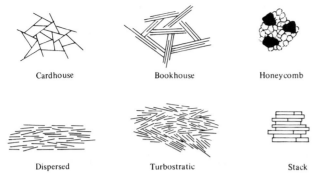

Cardhouse   Bookhouse   Honeycomb

Dispersed   Turbostratic   Stack

Fig. 3.19 Idealized clay structures. (*After Barden and Sides, 1971.*)

10 percent H_2SO_4 and about three-fourths of the hygroscopicity over 30 percent H_2SO_4.

3.8.9 Micro- and Macrostructures

The concepts of pedologic soil structures from macroscopic to ultrafine were introduced and briefly defined in Section 3.7. Section 3.7 presented the general granulometric and geometric aspects of the structure of cohesionless soil materials, and the preceding portions of Section 3.8 dealt with structures on the molecular and micellar (colloidal) level that are basic for the understanding of structure formation on levels of larger dimensions. While the structure of micelle associations is being revealed with increasing clarity by the use of the scanning electron microscope, it may suffice here to refer to the simplified presentation in Figure 3.19 (Barden and Sides, 1971). Of more immediate importance to the engineer is the understanding of the origin and influence of soil properties of micro- and macrostructures. Since structures on higher dimensional levels owe their forms and characteristics largely to those of their smaller building blocks, a meaningful discussion of larger structures must make frequent reference to their smaller constituents.

A classification of natural soil fabric structures, as revealed by the pedologic microscope (Kubiena, 1938) in the field and by the petrologic microscope in the laboratory, is given in Table 3.28, which also shows their relationship to the great pedologic soil groups. The microstructure of a soil, just as that of any other construction material, has a significant influence on its engineering properties and behavior.

Sedimentary soils often show structural characteristics acquired at the time of deposition, especially if they have been protected from the effect of normal climatic factors by sufficiently thick superimposed soil layers. Thus, the flocculated structure described by Casagrande (1932) is typical for clay deposits in salt water. Such flocculated structures may persist even after the causative element—the salt content of the aqueous medium—has been removed by percolating fresh water until the activation energy required for structural change is supplied from the outside in the form of mechanical energy by shock or soil working or by drying, freezing, and electroosmotic or other processes. An excellent example for this is the sensitivity of the leached marine clays of Norway (Bjerrum, 1954). On the other hand, the dispersed or single-particle structure of Na-clays may be retained even after leaching with solutions containing bivalent cations as long as the system is kept saturated with water.

Macroscopic soil structure in the form of secondary aggregates presupposes the properties of cohesion and swelling on the part of the soil material. Structure of this type is caused by forces acting from the outside on the soil system. Such forces are associated with soil cultivation, boring and digging by land animals, root activity of plants, meteorologic cycles that result in freezing and thawing and wetting and drying, and electroosmosis. Natural formation of macroscopic soil structure is always related to water loss and concomitant shrinkage of swelling cohesive soils. The basic phenomenon can be observed easily on drying mud flats. As water is lost from the surface, tension forces are established in the drying surface layer, which, because of the water loss, loses its ability to relieve these tension forces by plastic flow. These stresses are finally relieved by the formation of shrinkage cracks that break up the surface layer into pieces of more or less distinct geometric shape. If the soil material is homogeneous, the ideal pattern of the shrinkage cracks is hexagonal in accordance with the law of least energy. This cracking pattern produces the greatest stress relief with the least amount of work involved. If the soil system is nonhomogeneous, as in the presence of organic matter of different water affinity and greater mobility than the mineral soil components, the hexagons tend to become rounded as the organic matter becomes concentrated at the surfaces of the fissures. The drying of a film made with a solution of shellac in dry acetone will result in the same type of shrinkage pattern, while disturbing the homogeneity of the solution by addition of a small amount of water will tend to round the corners of the polygons.

The actual size and shape of shrinkage structures formed in soils depends on a number of factors of soil composition and condition as well as on the rate of drying and the geometry of the primary and secondary drying surfaces. However, the basic principle remains that soil structure formation is caused by

TABLE 3.28 KUBIENA CLASSIFICATION OF SOIL FABRIC.

Coating of Mineral Grains	Arrangement of Fabric	Occurrence
Grains not coated	Grains embedded loosely in a dense ground mass	Lateritic soils
	Grains united by intergranular braces	Chernozems, brown earths, lateritic soils
	Intergranular spaces containing loose deposits of flocculent material	Sandy prairie soils
Grains coated	Grains cemented in a dense ground mass	Desert crusts, podsol B-horizons
	Grains united by intergranular braces	B-horizons of podsolized brown forest soils
	Intergranular spaces empty	B-horizons of iron and humus podsols

water loss and shrinkage even if the drying is not caused by evaporation from a surface, but by other processes such as water absorption by plant roots or water migration within the soil to crystallization nuclei and growing ice lenses as in the freezing of moist silty and clayey soils. The great importance and widespread occurrence in nature of polygonal (especially hexagonal) patterns has been thoroughly discussed by Thompson (1942). For the effect of clay mineral and exchange-ion type on the structural patterns developed by freeze–thaw and wet–dry cycles, see Czeratzki and Frese (1958).

Where structure formation is due to freezing, the latter does not proceed uniformly throughout the soil mass, but occurs at locations where the specific water condition is most suitable for nucleus formation; from there, ice formation proceeds along preferred directions. It is often easier for water to move to the growing ice crystals than for the freezing front to advance into the soil. The actual ice structure formed may range from *Kammeis* to rhythmic ice-banding and lensing. Ice structure formation and its effect on soil structure are different for different textural, mineral, and chemical types. In addition, they are influenced by pore geometry, moisture content, and other environmental factors, and, perhaps decisively, by the physico-chemical phenomenon that water located at the basal planes of the clay particles resists the formation of normal ice and tends to migrate to places where ice is already being formed under more favorable conditions.

3.8.10 Soil Macrostructure and the Great Soil Groups

The relationship between macroscopic soil structure and the great pedologic soil groups is shown in Baver's classification of soil structure (Fig. 3.20). The existence of such a relationship, which was originally established on a purely observational basis, points to underlying physical and physicochemical causes.

The structural pattern of a soil may be considered as a permanent inherent soil property; the individual soil structural units, however, may be ephemeral, lasting only until the next rain, or they may be possessed of various degrees of slaking resistance. The water resistance of secondary soil structural

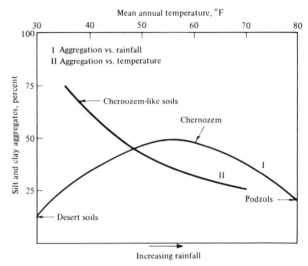

Fig. 3.21 The effect of climate on soil aggregation. (*After Baver, 1968.*)

units is usually determined by wet sieving, elutriation, or sedimentation analysis without use of dispersion agents. Extensive investigations on formation of water-resistant soil aggregates have led to correlations with climatic soil-forming factors. Correlations of the percentage of water-stable soil aggregates with the temperature and moisture regimes are shown in Figure 3.21, which illustrates the following important facts:

(1) In soils formed under humid conditions, the percentage of water-stable aggregates varies from about 25 for podzolic soils (low temperatures) to approximately 95 for lateritic soils (high temperatures).

(2) For temperate climates, the percentage of water-stable aggregates decreases with increasing amount of precipitation; however, the aggregation of desert soils tends to be of a different (flocculant) character as compared with that of soils developed under intermediate and high moisture conditions.

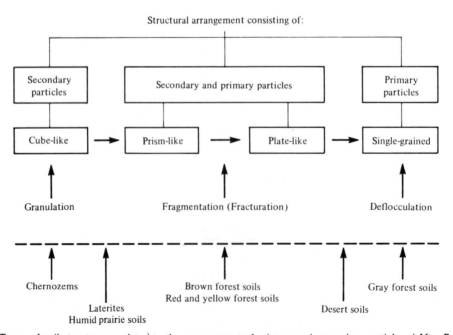

Fig. 3.20 Types of soil structure as related to the arrangement of primary and secondary particles. (*After Baver, 1940*).

(3) In semiarid (chernozem) soils, the percentage of stable aggregates decreases with increasing temperature as does the content in organic matter (Baver, 1968). It should be noted in this connection that the organic matter in semiarid soils differs in chemical character from that in podzolic (moist, cold) soils (Waksman and Hutchings, 1935; Winterkorn and Eckert, 1940).

The general pedologic evidence points to the following more specific physicochemical conclusions.

(1) Whether a dry cohesive soil possesses a monolithic structure of a granular or fragmented structure depends upon its swelling and shrinkage properties. A nonswelling soil will tend to possess a monolithic structure.

(2) The water resistance of fragmented and granular soil aggregates is a function of the amount and type of organic matter concentrated at the outside of the aggregates and of the accessibility and water affinity of the internal surface of the secondary aggregates.

(3) The internal cohesion of soil aggregates may be: (a) ephemeral if due only to surface tension of moisture films; (b) reversible if due to reversible hydrophilic clays and mineral and organic colloids; (c) practically irreversible if due to dehydrated irreversible hydrous iron and aluminum oxides— such cohesion can properly be called cementation (Winterkorn and Tschebotarioff, 1947).

(4) If the term cohesive soil is employed in a general engineering sense rather than in its restricted pedologic meaning, one may include among the more or less irreversible cementing agents such materials as colloidal silica, complex phosphates, calcium carbonate, and powdered igneous rock, reduced to very fine size by either glacial action or volcanic activity, which may possess hydraulic cementing properties to a greater or lesser extent.

In accordance with the preceding discussion, the water resistance of soil aggregates is due mainly to either or both cementation and coating by water-resistant substances. In the stabilization of cohesive soils for use as bases underneath pavements, these secondary aggregates, rather than the primary soil particles, are waterproofed and/or bonded to produce a better engineering material. This illustrates the importance of knowledge on the degree of soil aggregation and the stability of natural soil aggregates. The latter also have a determinant influence on the mechanical resistance properties of the natural soil as well as on its ability to transmit thermal, electrical, and mechanical energy.

3.9 CAPILLARITY AND CONDUCTION PHENOMENA

3.9.1 Capillarity

All surfaces and interfaces of solids and liquids are loci of free energy whose universal tendency to decrease (by the second law of thermodynamics) leads to the phenomenon of surface and interfacial tension. One consequence of this is the rise of a liquid in a capillary whose walls are wetted by the liquid and its depression in a capillary whose walls have no affinity for the liquid. The basic equation for capillary rise or depression is

$$h = \pm \frac{2T \cos \alpha}{r \gamma g} \qquad (3.13)$$

where

h = height of capillary rise in cm
T = surface tension in dyne/cm
α = angle of wetting (zero for most soil mineral/water systems; hence $\cos \alpha = 1$)
r = radius of capillary in cm
γ = density of liquid in $g \cdot cm^{-3}$
g = gravity acceleration in $cm \cdot sec^{-2}$

Since $2/r$ equals the ratio of wetted perimeter to cross section of liquid column and also that of wetted area to liquid volume per unit length of liquid column, the equation may be written

$$h = \frac{S\,T}{n\,\gamma g} \qquad (3.14)$$

for a homogeneous capillary system whose solid surfaces are completely wettable by the liquid, where S is the internal or capillary surface per unit volume and n is the pore space per unit volume. This equation holds reasonably well for nonswelling soil systems in which the thickness of the strongly adsorbed water films is negligible as compared with the pore dimensions. Where it is of special importance, the potential capillary rise or "suction" in a soil should be determined experimentally by any of the pertinent methods listed in Table 3.29, taken from Marshall (1959).

Often the rate of capillary rise is more important than the ultimate height. In such cases one employs the equation

$$\frac{dz}{dt} = \frac{k}{n}\left(\frac{h-z}{z}\right) \qquad (3.15)$$

in which

dz = distance differential in vertical direction
dt = time differential
k = coefficient of permeability of the soil
h = capillary potential, height of capillary rise
z = distance of capillary meniscus from groundwater level

By integration we obtain

$$t = \frac{nh}{k}\left(\ln\frac{h}{h-z} - \frac{z}{h}\right) \qquad (3.16)$$

where t is the time required for the meniscus of the capillary water to rise to the height z above the free water level.

3.9.2 Conduction Phenomena

There exist so many analogies in the conduction of various types of energy and materials through continuous conducting systems that it is advantageous to treat conduction phenomena on the same phenomenological basis. The scientific justification for this is that all conduction is a consequence of the second law of thermodynamics, which relates spontaneous phenomena to degradation of free energy and increase in entropy and the fact that all forms of energy can be expressed as products of capacity and intensity factors. If the intensity factors such as temperature, electric and chemical potentials, elevation in a gravity field, etc., vary from one location to another in a conducting system, then conduction will ensue spontaneously. All conduction equations are based on the fact that in first approximation the rate of conduction is directly proportional to the gradient of the intensity factor ($\Delta P/\Delta L$) and to a coefficient of conductivity k, which is a parameter of the physical system. In place of the conductivity coefficient, its inverse, the resistivity $1/k = R$ may be used. Use of the latter is advantageous in the case of composite systems in which

TABLE 3.29 RELATION BETWEEN THE VARIOUS UNITS USED FOR EXPRESSING RETENTION OF WATER[a].

	Suction		Capillary Potential or Free Energy	pF	Equivalent Radius of Largest Pores Which Could be Full of Water	
Soil Moisture Condition	*Length h, in cm, of Vertical Column of Water or Millibars*	*(Atmospheres or Bars)*	*(ergs/g or ergs/cm³)*	*log h*	*(10⁻⁴ cm or μm)*	*Instruments Suitable for Measuring the Suction*
At suction of 1 cm (soil saturated	1	0.001	-9.8×10^2	0	1 500	Tensiometer for suction or pressure plate)
At suction of 100 cm (corresponding uncritically to field capacity)	100	0.1	-9.8×10^4	2.0	15	Tensiometer (or suction or pressure plate)
At suction of 15 atmospheres (corresponding to permanent wilting point)	15 000	15	-1.5×10^7	4.2	0.1	Pressure membrane apparatus
At relative vapor pressure of 0.85 (soil feels dry)	220 000	220	-2.2×10^8	5.4	0.007	Vapor adsorption methods
Conversion from suction in cm (at 20°C when appropriate)	*h* cm or *h*/1.022 mb	*h*/1035 atm or *h*/1022 bars	$-980\,h$	log *h*	0.15/*h*	

[a] After Marshall (1959). Reproduced with permission of the Commonwealth Agriculture Bureau.

components are in series arrangement, resulting in additivity of resistances, while the former is indicated for parallel arrangement, in which conductances are additive. The fundamental one-dimensional-flow or conduction equation is

$$\frac{dV}{dt} = ki = \frac{1}{R}i \qquad (3.17)$$

in which

> dV/dt = volume of flow (yield) per unit cross section of system in unit time
> k = coefficient of conductivity
> R = coefficient of resistivity
> i = gradient = $\Delta P/\Delta L$ = ratio of driving potential difference to length over which difference occurs

3.9.3 Hydraulic Conductivity

Equation 3.17 becomes Darcy's law for hydraulic conductivity in soils if k is the coefficient of hydraulic conductivity or permeability. Because of the laminar nature of the flow, theoretical calculation of k was attempted by a number of workers on the basis of the fundamental Poiseuille–Hagen equation for flow through capillaries:

$$\frac{dV}{dt} = \frac{\pi r^4}{8\eta}i \qquad (3.18)$$

in which

> r = radius of capillary
> η = coefficient of viscosity of liquid

Best known among the attempts of theoretical derivation of k is perhaps the Kozeny–Carman equation

$$k = \left(\frac{1}{\eta}\right)\frac{1}{k_p t^2 S^2}\frac{n^3}{(1-n)^2} \qquad (3.19)$$

in which

> k_p = pore shape factor
> t = tortuosity
> S = specific surface per unit volume
> n = porosity
> η = viscosity

A simpler equation that gives essentially the same results for noncohesive granular soil systems is the following from Winterkorn (1942):

$$k = \frac{nr^2}{8\eta} \qquad (3.20)$$

Since $2/r = S/n$ we obtain

$$k = \frac{n}{2\eta}\frac{n^2}{S^2} \qquad (3.21)$$

which has the advantage that n/S can be calculated from the height of capillary rise and n can be determined independently. Winterkorn (1963) presents the results of actual permeability measurements together with coefficients calculated by the Winterkorn method (Eq. 3.21). Because of interaction of the mineral surfaces with the water substance and also of secondary structure formation in clay-containing soils, no such theoretical equation holds strictly for natural clay-containing soils. Hence, meaningful hydraulic conductivities must be determined experimentally. Detailed discussions on the factors affecting test results are given by Fang (1985), Fang and Evans (1988), Evans and Fang (1988), and Chapter 20.

Equations (3.19) to (3.21) are derived for saturated laminar flow and do not hold for other types, for example, flow or creep along pore walls or flow through a porous system containing air bubbles. Because of the expansion of the latter with increasing temperature, a warm rain may percolate through a soil more slowly than a cold rain although the viscosity of water decreases with increasing temperature.

The coefficient of permeability is determined in the laboratory on representative soil specimens either by means of the constant-head permeameter for relatively high k-values or by the falling

TABLE 3.30 RANGES OF HYDRAULIC CONDUCTIVITY *k* OF SOME TYPICAL SOILS, ROCKS, AND OTHER CONSTRUCTION MATERIALS[a].

Type of Materials	Hydraulic Conductivity (cm/sec)
Soils	
Gravel	$>10^{-1}$
Sand gravel	10^{-1}–10^{-3}
Sand	10^{-3}–10^{-5}
Silty clay	10^{-5}–10^{-7}
Clay	$<10^{-7}$
Rocks	
Basalt	10^{-4}–10^{-5}
Calcite	0.70–93×10^{-9}
Dolomite	4.6–12×10^{-9}
Granite	0.05–0.20×10^{-9}
Mudstone	600–$2\,000 \times 10^{-9}$
Limestone	0.70–120×10^{-9}
Marble	10^{-4}–10^{-5}
Shale	10^{-3}–10^{-4}
Slate	0.07–0.16×10^{-9}
Sandstone	160–$12\,000 \times 10^{-9}$
Construction Materials	
Asphalt pavement	10^{-6}–10^{-8}
Concrete pavement	10^{-5}–10^{-7}
Gravel base (compacted)	10^{-4}–10^{-6}
Stone base (compacted)	10^{-3}–10^{-5}
Fly ash (loose)	10^{-4}–10^{-6}
Fly ash (dense)	10^{-5}–10^{-8}

[a] Data from Serafim and Del Campo (1965), Farmer (1968), Winterkorn and Fang (1975), and others.

head permeameter for low *k*-values, or it is calculated from the results of consolidation tests. It may also be calculated from seepage, pumping, and pressure tests in the field (see Chapters 1 and 7). The pertinent equation for the constant-head permeameter is

$$k = \frac{V}{Ait} \qquad (3.22)$$

where V is the water yield in time t, A is the cross sectional area of the specimen and $i = \Delta P/L$, i.e., the difference in pressure at the two ends of the specimen. For the falling head permeameter

$$k = 2.3 \frac{La}{At} \log \frac{h_1}{h_2} \qquad (3.23)$$

where a is the cross-sectional area of the standpipe in which the hydraulic head falls from h_1 to h_2 in time t while the water volume $a(h_1 - h_2)$ flows through the soil specimen of thickness L. For procedural details and discussion of the permeability test, see Bowles (1986). Table 3.30 summarizes pertinent information on coefficients of permeability values for soils, rocks, and other construction materials.

The previously mentioned relationship between all types of conduction phenomena is especially close for water conduction in soils under pressure gradients, on one hand, and under electric, thermal, and concentration gradients on the other.

3.9.4 Electroosmosis

Water possesses an electric structure that causes its interaction with the electrically charged soil minerals and the exchangeable ions that cover or surround them as well as with the ions in the aqueous soil solutions. It is this property of water that causes it to move in the capillary pore space of soils upon application of an electric potential (Reuss, 1809). For a development of theoretical concepts and practical applications of electroosmosis, see Gray and Mitchell (1967), Segall et al. (1980), and Khan et al. (1989), as well as the pertinent sections in Chapters 9 and 11 of this book. For practical use of this phenomenon, the electroosmosis transmission coefficient k_e (volume of water transmitted through a unit cross section in unit time by application of a potential of 1 volt/cm normal to the cross section) is determined in the laboratory or field and used in a Darcy-type equation.

3.9.5 Thermoosmosis

Because of the change with temperature of the electric soil mineral/water interaction structure, water also moves in clay soils upon application of thermal gradients. In saturated low-porosity systems the flow is of a film type, while in unsaturated systems possessing a significant air phase, thermal gradients also produce water movement by diffusion and microdistillation (Winterkorn, 1947, 1955; HRB, 1958, 1969).

3.9.6 Heat Transmission in Soils

Heat is a form of energy. A given amount of heat energy can be expressed as the product of a capacity and an intensity factor. The heat capacity of a substance is the quantity of heat necessary to produce unit change of temperature in a unit mass. If it is expressed in "normal" calories per degree Celsius per gram, the heat capacity is numerically equivalent to the specific heat, since one normal calorie is the amount of heat required to raise the temperature of one gram of water from 3.5°C to 4.5°C. The heat capacity of a gram-atom of an element is its atom heat and that of a gram mole its mole heat. An older measure of heat quantity is the British Thermal Unit (BTU), which is the heat required to raise the temperature of one pound of water at its greatest density by 1°F. The heat capacity C_s of a system is the sum of the capacities of its components:

$$C_s = g_1 c_1 + g_2 c_2 + g_3 c_3 + \cdots + g_i c_i \qquad (3.24)$$

where g_1 to g_i represent the masses of the various components and c_1 to c_i the respective capacities per unit mass. Heat capacities are temperature dependent. The total heat content U of a system having a capacity C_s and the absolute temperature of T K is

$$U = \int_0^T C_s \, dt \qquad (3.25)$$

According to Kersten (1949) the specific heats of common soil minerals differ only slightly, being 0.189 for Ottawa sand and 0.193 for silt loam at 140°F (60°C) with a sufficiently accurate value for common soil solids of 0.19 at this temperature and of 0.16 at 0°F (-17.8°C) and using linear interpolation for intermediate temperatures. The specific heat of soil–water mixtures is calculated from their weight proportions and specific heats. To obtain the practically more significant heat capacity per unit volume, that per unit mass is multiplied by the density of the substance; taking for soil minerals an average density value in the CGS system of 2.65 yields 0.505 and 0.425 cal/cm^3 at 60°C and at -17.8°C, respectively.

Heat moves spontaneously from higher to lower temperatures because of the resulting increase in entropy (Q/T). While heat transmission utilizes all available means and mechanisms (radiation, convection, macro- and microdistillation, conduction), in soils it can be ascribed essentially to conduction (a molecular phenomenon) employing effective experimentally determined coefficients of conductivity or of resistivity, although these may actually include contributions from microdistillative and other mechanisms.

Heat conduction follows the law

$$Q = Akit \qquad (3.26)$$

in which

Q = amount of heat flowing through a cross section A, under a thermal gradient i, during time t

k = coefficient of conductivity, that is, amount of heat flowing in unit time through unit cross section under a unit gradient

By definition $k = 1/R$ and $R = 1/k$, where R is the coefficient of resistivity. Since $k = Q/Ait$, its dimensions and numerical value depend on those chosen for Q, A, i, and t. Most generally used is the calorie-CGS system in which $[k] = \mathrm{cal/cm^2 \cdot sec \cdot {}^\circ C \cdot cm^{-1}} = \mathrm{cal/cm \cdot sec \cdot {}^\circ C}$ and the corresponding $[R] = \mathrm{cm \cdot sec \cdot {}^\circ C/cal}$. Table 3.31 summarizes pertinent information on parameters and units for determination of thermal properties of soil.

Older work employs the British engineering units BTU, foot and/or inch, and degrees Fahrenheit. For interrelationships between various common units see Table 3.32.

Instead of the calorie the joule may be used (1 cal = 4.185 joule = 4.185×10^7 erg); since a joule/sec is a watt, $[k] = \mathrm{watt/cm \cdot {}^\circ C}$, and $[R] = \mathrm{cm \cdot {}^\circ C/watt}$. For this system, the unit of R is also called a thermal ohm.

The results of an extensive study of the thermal properties of soils by Kersten (1949) may be summarized as follows.

The thermal conductivity for soil above the freezing point of water increases slightly with increase in mean temperature.

At the optimum moisture content for compaction the conductivity below freezing averages about 17 percent greater than above freezing.

TABLE 3.32 CONVERSION TABLE FOR THE COEFFICIENT k OF THERMAL CONDUCTIVITY.

Multiply	By	To Obtain
$\mathrm{BTU/(h)(ft^2)(deg\,F/ft)}$	4.134×10^{-3}	$\mathrm{cal/(sec)(cm^2)(deg\,C/cm)}$
$\mathrm{Watts/(cm^2)(deg\,C/cm)}$	57.780	$\mathrm{BTU/(h)(ft^2)(deg\,F/ft)}$
$\mathrm{Watts/(cm^2)(deg\,C/cm)}$	0.239	$\mathrm{cal/(sec)(cm^2)(deg\,C/cm)}$

At constant moisture content, an increase in density results in an increase in conductivity. The rate of increase is about the same at all moisture contents.

At constant density, an increase in moisture content causes an increase in conductivity, which is true up to the point of saturation and holds for frozen as well as unfrozen soils.

For saturated, unfrozen soils, the conductivity decreases with decrease in density. For saturated, frozen soils, no well-defined relationship between density and conductivity was found.

The thermal conductivity varies, in general, with the texture of soils. At a given density and moisture content, the conductivity is relatively high in coarse-textured soils such as gravel or sand, somewhat lower in sandy loam soils, and lowest in fine-textured soils such as silt loam or clay.

The thermal conductivity of a soil is dependent upon its mineral composition. Sands with high quartz contents have greater conductivities than sands with high contents of such minerals as plagioclase, feldspar, and pyroxene. Soils with relatively high contents of kaolinite have relatively low conductivities.

To predict thermal conductivity, the following four empirical formulas were suggested by Kersten (1949):

Silt and clay soils

Unfrozen: $\quad k = [0.9 \log \omega - 0.2]10^{0.01\gamma} \qquad (3.27)$

Frozen: $\qquad k = 0.01(10)^{0.022\gamma} + 0.085(10)^{0.008\gamma}(\omega) \qquad (3.28)$

TABLE 3.31 PARAMETERS AND UNITS FOR DETERMINATION OF THERMAL PROPERTIES OF SOIL.

Aspect	Quantity	Equation	Typical Unit SI	Typical Unit fps
Heat retained by soil (heat retention)	Mass heat capacity[a] (C_m)	$C_m = \dfrac{\text{Heat flow}}{(\text{Mass})(\text{Temperature change})}$	$\dfrac{\mathrm{J}}{(\mathrm{kg})({}^\circ\mathrm{C})}$	$\dfrac{\mathrm{BTU}}{(\mathrm{lb})({}^\circ\mathrm{F})}$
	Volumetric heat capacity[b] (C_v)	$C_v = C_m\gamma \quad$ (see note c)	$\dfrac{\mathrm{J}}{(\mathrm{cm^3})({}^\circ\mathrm{C})}$	$\dfrac{\mathrm{BTU}}{(\mathrm{ft^3})({}^\circ\mathrm{F})}$
	Specific heat[a,b] (C_{sp})	$C_{sp} = \dfrac{(C_{ms})\,\text{substance}}{(c_{mw})\,\text{water}}$	Dimensionless	
Dissipation of heat (heat transmission)	Thermal conductivity[a] (k)	$k = \dfrac{Q}{iAt}$	$\dfrac{\mathrm{W}}{(\mathrm{cm})({}^\circ\mathrm{C})}$	$\dfrac{\mathrm{BTU}}{(\mathrm{ft})(\mathrm{s})({}^\circ\mathrm{F})}$
	Thermal resistivity[a,b] (R)	$R = \dfrac{1}{k}$	$\dfrac{(\mathrm{cm})({}^\circ\mathrm{C})}{\mathrm{W}}$	$\dfrac{(\mathrm{ft})(\mathrm{s})({}^\circ\mathrm{F})}{\mathrm{BTU}}$
	Thermal diffusivity[b] (α)	$\alpha = \dfrac{k}{C_v}$	$\dfrac{\mathrm{cm^2}}{\mathrm{s}}$	$\dfrac{\mathrm{ft^2}}{\mathrm{h}}$

[a] Experimentally determined.
[b] Calculated quantity.
[c] γ = unit weight of soil = ρg, where ρ = density and g = gravitational acceleration.

Sandy soils

Unfrozen: $k = [0.7 \log \omega + 0.4] 10^{0.01\gamma}$ (3.29)

Frozen: $k = 0.076(10)^{0.013\gamma} + 0.032(10)^{0.0146\gamma}(\omega)$ (3.30)

In these formulas, the thermal conductivity k is in British thermal units per square foot per inch per hour per degree Fahrenheit, the moisture content ω is in percent of the dry soil weight, and γ is the dry density in pounds per cubic foot. The formulas for the silt and clay soils apply for moisture contents of 7 percent or more; those for the sandy soils, of one percent or more. The formulas for sandy soils are based largely on tests on fairly clean sands. For sandy soils with a relatively high silt and clay content, conductivity values intermediate between those calculated by the two formulas might be a reasonable prediction. Maximum error for use of these formulas is 25 percent.

3.9.7 Diffusivity

The diffusivity is the quotient of the thermal conductivity and the heat capacity per unit volume; its units in the CGS system are cm^2/sec. The diffusivity value may be determined by calculation if the thermal conductivity, specific heat, and density of a soil are known. Based on the work done by Kersten (1949), the following general conclusions regarding the diffusivity of soils may be drawn:

1. Changes in temperature do not cause significant changes in diffusivity unless they pass through the freezing point. In the latter case the increase is the greater the higher the moisture content.
2. At low or moderate moisture content, an increase in moisture content of either a frozen or an unfrozen soil causes an increase in diffusivity.
3. An increase in density of a soil causes a slight increase in diffusivity.
4. Mineral composition affects diffusivity to about the same degree that it affects conductivity.

Table 3.33 presents average resistivity values in thermal ohms (Rho) for some soil constituents and allied materials.

TABLE 3.33 AVERAGE RESISTIVITY VALUES (RHO) FOR SOME SOIL CONSTITUENTS AND ALLIED MATERIALS[a].

Material	Rho
Quartz $\parallel$	7.9
Quartz $\perp$	14.9
Quartz, random orientation	11.0
Quartz glass	79.0
Granite	26–58
$CaCO_3$ $\perp$	26.3
Marble	34–48
Limestone, dense	45
Ice	45
Sandstone	50
Dolomite	58
Slate	67
Water	165
Mica $\perp$	170
Pine wood $\parallel$	265
Pine wood $\perp$	608
Organic material wet	400
Organic material dry	700
Air	4000

[a] After Winterkorn (1961).

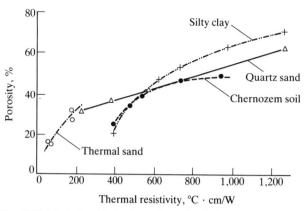

Fig. 3.22 Relationship between thermal resistivity and porosity of sand and clay. (*After Winterkorn and Fang, 1975.*)

TABLE 3.34 THERMOOSMOTIC MOISTURE TRANSMISSION COEFFICIENTS FOR SEVERAL HOMOIONIC MODIFICATIONS OF NEW JERSEY SOIL[a].

Soil Material		Percent H_2O per Dry Weight of Soil	Thermoosmotic Transmission Coefficient (cm^3/cm^2 sec per $°C/cm \times 10^6$)
H-kaolinite		30.6	3.4
H-illite (grundite)		23.5	1.6
H-montmorillonite		61.8	1.9
Natural		23.1	3.7
K		21.9	2.5
Mg	Hagerstown soil	22.7	2.2
H		22.5	1.9
Cu		22.7	1.6
Al		23.1	1.5
Na		21.5	0.9

[a] After Winterkorn (1961).

Figure 3.22 gives thermal resistivities for different types of dry soils as a function of their porosity. The curves indicate that below a porosity of about 50 percent the thermal conductivities tend to be additive, while above that porosity the resistivities tend to be so. Thermoosmotic transmission coefficients for various soil materials from Hagerstown, New Jersey, are given in Table 3.34. For a presentation and discussion of various physical phenomena that contribute to the overall thermal transmission in soils, see Winterkorn (1961) and Farouki (1966). For a comprehensive presentation of theoretical and practical aspects of thermal conductivity in soils, see Van Rooyen and Winterkorn (1957, 1959), Salomone and Kovacs (1984), and Radhakrishna et al. (1984).

3.10 EFFECT OF TEMPERATURE ON ENGINEERING PROPERTIES OF SOILS

The effect of temperature on the soil–water interaction could be expected from pertinent physicochemical considerations and was studied by Baver and Winterkorn (1935) on extracted natural clays over a wide moisture range and a temperature range from 30°C to 99°C. They separated the total interaction (swelling) water at elevated moisture contents into (1) hydration water whose binding is associated with a considerable amount of heat of wetting and (2) osmotic water, the relatively unrestrained water between the hydrated cations in what many call the diffuse double layer. Both types of water are intrinsically related to the dielectric constant of water. Since the latter decreases with increasing temperature, and more rapidly than the kinetic energy of the exchange ions increases with temperature, a decrease in both the hydration water and the osmotic water in a saturated soil system may be expected with increasing temperature. This means that at constant total water content, increase in temperature decreases the portion that is bound or restrained and increases the portion that is free while also decreasing the viscosity of the latter.

While this temperature effect is quite straightforward in relatively pure clay–water systems within a limited range of temperature, it becomes more complicated even in these systems when high temperatures may cause marked dispersion or flocculation effects depending on clay mineral and exchange ion type. Even greater complexity obtains in the presence of nonclay minerals of silt and larger size. The basic phenomenon, however, is a decrease in interaction water with increasing temperature. This decrease will normally result in an increase in the coefficient of permeability with increasing temperature. However, in the presence of a strong dispersion effect, the permeability may actually decrease, especially if the clay is contained within the interstices of a coarse-granular skeleton.

Further discussions on effect of temperature on engineering properties of soils are given by HRB (1969), Mitchell and Kao (1978), Chaney et al. (1983), and Salomone and Kovacs (1984).

3.10.1 Frost Action in Soils

A heat conduction problem of great importance to the foundation engineer is the penetration of freezing temperatures into soils and the accompanying physical phenomena. The forceful expansion of confined water upon freezing can destroy porous rock and loosen soil. Commonly of greater significance is the accumulation of the water substance in the form of ice lenses, resulting in frost heave in winter and morass formation during thawing. A freezing front penetrating into moist soil starts only a limited number of crystallization centers. Water moves to these from the surrounding soil, especially from lower depths that are at higher temperatures. As it freezes water gives off about 80 calories per gram of ice formed. If the heat released by the freezing water just balances the heat lost by conduction to the earth surface, the freezing front becomes stationary and thick ice lenses may be formed, especially if the groundwater level or other water reservoir is close by. Without such close supply of free water, the frost may advance until a new moist layer is encountered. Here nuclei and ice lens formation begin anew. Daily and other short-period temperature fluctuations make the larger ice lenses grow at the expense of the smaller ones.

The following criteria for tendency of ice lens formation and frost damage are used:

(1) Casagrande (Taber, 1929, 1930) considers the < 0.02 mm fraction as determinant for frost susceptibility with a modifying influence of the size composition of the whole soil as expressed by its uniformity coefficient C_u. At $C_u = 15$, a soil is frost-susceptible if the < 0.02 mm fraction exceeds 3 percent; at $C_u = 5$, a soil is frost-susceptible if the < 0.02 mm fraction exceeds 10 percent.

According to Ducker (see Ruckli, 1950) all soils having more than 3 percent of < 0.02 mm fraction irrespective of their uniformity coefficient are frost-susceptible. A few very uniform soils with at most 10 percent > 0.1 mm and at least 25 percent

grains between 0.05 and 0.02 mm are frost-susceptible even if they do not have a < 0.02 mm fraction.

(2) Beskow (1935) considers both the granulometry of the soil and the height of capillary rise. He established four classes for presence or lack of frost susceptibility:

0. No danger under any circumstances
1a. Danger only at very high groundwater elevation (frost heave and creep in subgrades and embankment)
1b. As above, also frost damage in pavements at very high groundwater levels and slow freezing
2. Normally frost heave and ice-lens formation at groundwater depths of 1.5 m in the case of sediments (homogeneous) and of 1 m in the case of moraines (heterogeneous)
3. Frost heave but no ice lens formation—loamy soils
4. Neither frost heave nor ice lens formation—stiff loams and clays

(3) The Freiberg criterion considers in addition to the granulometry and the height of capillary rise also the rate of water supply by capillarity to the frost zone and the frost duration during which this supply takes place. The assumption made in the derivation of this criterion that the capillarity acts as a driving force is very doubtful (Ruckli, 1950; Winterkorn, 1955).

(4) The geologic petrographic criterion (Keil, 1937; Ruckli, 1950). All mechanically consolidated easily weathered, especially layered, sedimentary rock may be considered as dangerously frost susceptible even in the nonweathered condition. This includes all such rock that in crushed condition would not be acceptable for a Telford or macadam base (Packlage). Regarding the frost susceptibility and other pertinent characteristics of a crushed stone base, see Figure 3.23.

(5) Description of frozen soils based on visual examination and simple manual tests was proposed by ASTM (ASTM D4083-83). This practice is intended primarily for use by soils

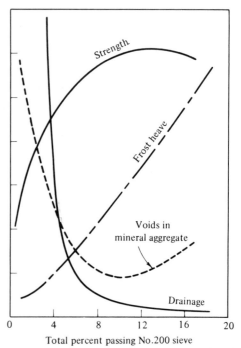

Fig. 3.23 Characteristics of graded crushed stone base as affected by the quantity of material passing No. 200 sieve. (*After National Crushed Stone Association, 1972.*)

engineers and technicians in the field, where the soil profile or samples from it may be observed in a relatively undisturbed (frozen) state, or it may be used in the laboratory to describe the condition of relatively undisturbed soil samples that have been maintained in a frozen condition following their acquisition in the field.

The system for describing and classifying frozen soil is based on an identification procedure that involves three steps: step (1) consists of a description of the soil phase; step (2) consists of the addition of soil characteristics resulting from the frozen state; and step (3) consists of a description of the important ice strata associated with the soil. In addition to the description of the soil profile at a given site it is normally advantageous to describe the local environmental and geologic conditions such as terrain features, vegetation cover, depth and type of snow cover, local relief and drainage conditions, and depth of thaw (Shockley, 1978).

The frost problem in foundation engineering has been treated comprehensively by Ruckli (1950) and Washburn (1969). The effect of soil freezing on the formation of soil structure has been well presented by Czeratzki and Frese (1958), and the influence of meteorological factors on the extent of road damage caused by frost has been thoroughly treated by Kübler (1963). The effect of frost penetration and temperature related to the performance of soil-pavement systems has been evaluated by Fang (1969). The experimental data of this study were obtained from AASHO Road Test.

In large areas of Siberia and also Canada and Alaska, permanently frozen soil (permafrost) is encountered at depths that depend on the climate and thermal conductivity of the surface soil. The frozen subsoil prevents drainage of the water from spring thaw. Structures founded on permanently frozen soil must be separated from it by insulating material or must counteract normal heat conduction by refrigeration (Muller, 1947). Some of the permafrost is of fissile character and, not being in equilibrium with the present climate, it will not re-form once melted.

Further discussions on various aspects of geotechnical problems on frozen ground are given by Andersland and Anderson (1978), Phukan (1985), Michalowski (1989), and Chapter 19.

3.11 DENSIFICATION (COMPACTION)

Theoretical and practical aspects of the densification of cohesive and noncohesive soils are treated extensively in Chapter 8 (Compacted Fill) and Chapter 23 (Deep Compaction of Granular Soils). Granulometric factors that influence the susceptibility to densification of purely granular soils and of cohesive soils with a granular bearing skeleton are treated in Section 3.7 and Chapter 9 (Soil Stabilization and Grouting), while Sections 3.9 and 3.10 contain data and concepts that are pertinent for the understanding of the role of soil–water interaction and of temperature in the densification of cohesive soils. Personal integration of this information by the handbook user should serve his needs for understanding the many facets of soil densification and provide a basis for successful application of the available methodologies and tools. However, we should point out here a few items of considerable engineering importance whose treatment in other places may not have been as explicit as desirable.

(1) To avoid irregular and meaningless moisture density curves in laboratory and field tests on highly cohesive soils it is imperative that the moisture be evenly distributed throughout the secondary soil aggregates. This may take from 1 to 7 days

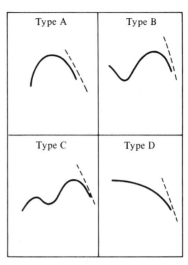

Fig. 3.24 Four types of compaction curves found from laboratory investigation. (*After Lee and Suedkamp, 1972.*)

in the case of highly cohesive soils to which water is added in the dry condition. This time may be reduced by the use of live steam.

(2) Even properly performed tests may not yield the generally expected paraboloid curves. Andrew et al. (1967) found two optimum moisture contents for the compaction of Na-bentonite, one at 100 percent H_2O yielding 73 lb/ft^3 and one at 50 percent H_2O yielding 65 lb/ft^3. Lee and Suedkamp (1972) reported the existence of four types of moisture–density curves as shown in Figure 3.24. They are (A) single-peak, (B) $1\frac{1}{2}$ peak, (C) double peak, and (D) oddly shaped with no distinct optimum moisture content. They relate the different types to the liquid limit ranges as follows: $LL < 30$—double and $1\frac{1}{2}$ peaks; LL from 30 to 70—typical single peak; $LL > 70$—peaked and oddly shaped curves.

(3) Because the viscosity of water and the degree and intensity of the interaction of mineral surfaces with water are functions of temperature, the moisture–density relationships are also temperature dependent. In general, the temperature effect increases with increasing specific surface (surface per unit volume) and therefore with clay content, and is also a function of the clay mineral and exchange-ion type and the electrolytes in the aqueous phase. An important modifying factor is the presence or absence of a granular bearing skeleton in the compacted soil. If it is present, the predominant effect of temperature increase may be a lower optimum moisture content; if it is absent, the increase in compaction temperature generally increases the maximum density and decreases the optimum moisture content. If, as is usually done, densification specifications are translated into a number of roller passes at a certain moisture content, serious undercompaction may result if the work is performed at lower than assumed temperature. A considerable number of cases are on record of sheep foot rollers sinking in daytime to the drum surface into "compacted" soil from which they had "walked out" with practically no impression on the surface during night operation.

(4) If cohesive soils, densified to maximum density at optimum moisture content are subject in service to wetting–drying and to freezing–thawing cycles, they may incur dangerous loss in bearing capacity and proper provisions should be made against this eventuality. Especially detrimental is the sudden flooding of cohesive soils when very dry. The factors governing the attack of water on dry cohesive soil systems and the various types of response have been studied by Winterkorn (1942) and Winterkorn and Choudhury (1949).

3.12 CONSOLIDATION

When a load is applied to a saturated compressible soil mass, the load is usually carried initially by the water in the pores, because the water is relatively incompressible when compared with the soil structure. The pressure that results in the water because of the load increment is called *hydrostatic excess pressure* because it is in excess of that pressure due to the weight of water. If the water drains from the soil pores, the hydrostatic excess pressure and its gradient gradually decrease and the load increment is shifted to the soil structure. In other words, the transfer of load is accompanied by a change in volume of the soil mass equal to the volume of water drained. This process is known in soil mechanics as *consolidation*.

Consolidation is one of the most important processes in soil engineering. It accounts for the settlement of structures founded on compressible soils, and its understanding together with pertinent laboratory tests permits the calculation of total settlement as well as of its time rate. Consolidation (as well as its inverse: expansion) is affected by and itself affects such physical parameters as permeability, compressibility, and shear strength.

A theory relating pressure, time, and volume change was proposed by Karl Terzaghi in the early 1920s and has become known as the Terzaghi theory of consolidation. One of the major assumptions in the Terzaghi theory is that volume change and the outflow of pore water occur in one direction only. For this reason it is sometimes referred to as the one-dimensional consolidation theory; its theoretical derivation may be found in standard textbooks (Taylor, 1948; Lambe and Whitman, 1979; Holtz and Kovacs, 1981).

If the present effective overburden pressures are the maximum pressures to which the soil has ever been subjected at any time in its history, the deposit is referred to as *normally consolidated*. A soil deposit that has been fully consolidated under a pressure larger than that of the present overburden is called *overconsolidated*. If soil deposits are not fully consolidated under the present overburden pressure, as is the case for recent fills, they are called *underconsolidated*. The settlement behavior of such deposits is discussed in detail in Chapter 5.

Laboratory consolidation tests are usually performed in floating or fixed-ring type oedometers employing cylindrical soil specimens of from about 2 to $4\frac{7}{16}$ inches in diameter and from about $\frac{3}{4}$ to $1\frac{1}{2}$ inches high, with the load being applied in increments. Detailed tests procedures are given by Bowles (1986). Factors influencing the test results include sample size, pressure increment, loading duration, type of pore fluid, initial water content, degree of sample disturbance, etc. Some of the more important factors are discussed in Chapter 5.

The test data are commonly plotted as deformation (dial reading) versus logarithm of time or square root of time, as illustrated in Figures 3.25 to 3.27, for the purpose of estimating 100 percent of primary consolidation and to compute the consolidation coefficient c_v and other parameters, by means of the equations presented in the following paragraphs.

Casagrande Logarithm-of-Time Method

Make a plot of dial reading versus log of time (see Fig. 3.25), extend the two straight portions of the curve to intersect at 100 percent primary consolidation. The D_0 on the semi-logarithmic plot (the theoretical dial reading, $t = 0$) can be obtained by selecting any two points on the early part of the curve whose times are in the ratio of 1 to 4. The D_{50} point is

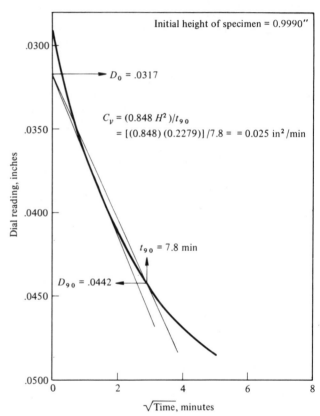

Initial height of specimen = 0.9990″

$D_0 = .0317$

$$C_v = (0.848\,H^2)/t_{90}$$
$$= [(0.848)(0.2279)]/7.8 = = 0.025 \text{ in}^2/\text{min}$$

$t_{90} = 7.8$ min

$D_{90} = .0442$

Fig. 3.26 Square root of time method.

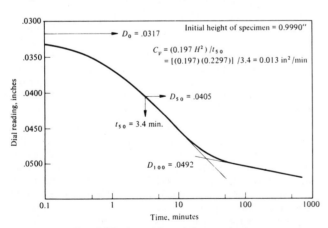

$D_0 = .0317$
Initial height of specimen = 0.9990″

$$C_v = (0.197\,H^2)/t_{50}$$
$$= [(0.197)(0.2297)]/3.4 = 0.013 \text{ in}^2/\text{min}$$

$D_{50} = .0405$

$t_{50} = 3.4$ min.

$D_{100} = .0492$

Fig. 3.25 Logarithm of time method

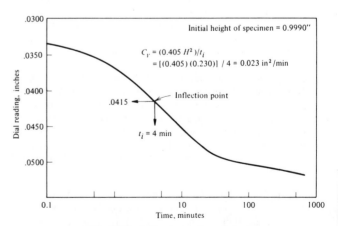

Initial height of specimen = 0.9990″

$$C_v = (0.405\,H^2)/t_i$$
$$= [(0.405)(0.230)]/4 = 0.023 \text{ in}^2/\text{min}$$

Inflection point

.0415

$t_i = 4$ min

Fig. 3.27 Inflection point method.

halfway between the D_0 and D_{100} points. The consolidation coefficient c_v can be computed by

$$c_v = \frac{0.197H^2}{t_{50}} \qquad (3.31)$$

where H is the average thickness or the longest drainage path during the given load increment and t_{50} is the time at 50 percent consolidation.

Taylor Square-Root-of-Time Method

Make a plot of dial reading versus square root of time. Extending the straight-line portion back to $t = 0$, we obtain the corrected zero reading D_0. Through D_0 draw a straight line having an inverse slope 1.15 times the tangent line, and intersect the laboratory curve at D_{90} (see Fig. 3.26). The quantity c_v can be computed by the square-root-of-time method as

$$c_v = \frac{0.848H^2}{t_{90}} \qquad (3.32)$$

where t_{90} is the time at 90 percent consolidation.

Inflection-Point Method

The inflection-point method (Cour, 1971) is based on the fact that time factor T corresponding to the inflection point of a semilogarithmic consolidation curve is fixed and equal to $T = 0.405$. Thus, the coefficient of consolidation c_v can be readily computed from

$$c_v = 0.405H^2/t_i \qquad (3.33)$$

in which t_i is the instant at which the inflection point occurs in a plot of dial reading versus the logarithm of time (Fig. 3.27). The following method for locating the inflection point is recommended. If the time curve is plotted, t_i can easily be recognized with a reasonable degree of accuracy. If the time curve is not plotted, or if a more precise location of the inflection point is desired, it can be defined as the point at which the absolute value of the tangent to the time curve on a semilogarithmic plot reaches a maximum.

Other methods for determination of the coefficient of consolidation include that of Naylor-Doran (1948) and the balanced area method (Teves and Moh, 1968). These two methods both employ successive approximation. If time and effort can be expended for improvement of degree of accuracy of the test results these two methods may be valuable.

The consolidation results may also be plotted as void ratio e versus logarithm of pressure (see Fig. 3.28). The slope of this curve is called the compression index, C_c or

$$C_c = \frac{\Delta e}{\log(p_2/p_1)} \qquad (3.34)$$

The compression index can be estimated roughly from an empirical relationship with the liquid limit proposed by Terzaghi and Peck (1967) for normally consolidated clay with low to moderate sensitivity:

$$C_c = 0.009(LL - 10) \qquad (3.35)$$

For inorganic silty clay, Hough (1957) proposed an approximate expression for the compression index that uses the initial void ratio e_0:

$$C_c = 0.30(e_0 - 0.27) \qquad (3.36)$$

Based on the initial moisture content ω_0, Koppula (1981) recommended another approximate expression, which can be

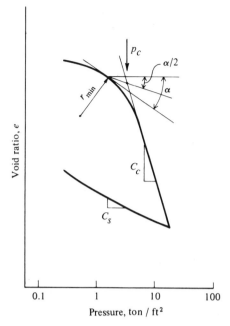

Fig. 3.28 Maximum preconsolidation pressure. (*After Casagrande, 1936.*)

written as:

$$C_c = 0.01\omega_0 \qquad (3.37)$$

Other approximate expressions for compression index C_c are given by Azzouz et al. (1976), Goldberg et al. (1979), and Rendon-Herrero (1980).

Typical values for the compression index C_c of various soils are given in Table 3.35. To express the expansion which may occur upon unloading of a soil sample, one commonly uses the swell index C_s, which is defined by:

$$C_s = \frac{\Delta e}{\log(p_2/p_1)} \qquad (3.38)$$

The C_s values are always much smaller than the C_c values for the virgin compression.

The *coefficient of compressibility* a_v is the slope of the e versus p curve and can also be found from the e versus $\log p$ curve as

$$a_v = \frac{0.435C_c}{p} \qquad (3.39)$$

in which p is the average pressure for the increment.

The *coefficient of permeability* k can be calculated from consolidation test results as

$$k = \frac{C_v a_v \gamma_w}{1 + e} \qquad (3.40)$$

where γ_w is the unit weight of water.

3.12.1 Preconsolidation Pressure

The maximum intergranular pressure p_c larger than the present overburden pressure is called preconsolidation pressure or past pressure, and is usually determined by the Casagrande (1936) graphical technique.

The Casagrande method involves selecting the point corresponding to the minimum radius of curvature on the e versus

TABLE 3.35 TYPICAL VALUES FOR COMPRESSION INDEX, C_c.

Soil	Liquid Limit	Plastic Limit	Virgin Compression Index		References
			Undisturbed	Remolded	
Boston blue clay	41	20	0.35	0.21	
Chicago clay	58	21	0.42	0.22	
Louisiana clay	74	26	0.33	0.29	Lambe and Whitman (1979)
New Orleans clay	79	26	0.29	0.26	
Fort Union clay	89	20	0.26		
Mississippi loess	23–43	17–29	0.09–0.23		Sheeler (1968)
Delaware organic silty clay	84	46	0.95		Schmidt and Gould (1968)
Indiana silty clay	36	20	0.21	0.12	
Marine sediment, B.C. Canada	130	74	2.3		Finn et al. (1971)
Shanghai soft clay	43	23	0.42		Fang (1980); Gao et al. (1986)

log p curve (see Fig. 3.28). At this point horizontal and tangent lines are drawn and the angle between them is bisected, then the straight-line portion of the curve is projected back to intersect the bisector of the angle. The pressure corresponding to this point of intersection is equal to the maximum past preconsolidation pressure. Other methods for determining the preconsolidation pressure include those of Burmister (1951), Schmertmann (1955), and Brumund et al. (1976). These methods are discussed by Holtz and Kovacs (1981) and Yong and Townsend (1986) and in Chapter 5 of this book.

3.13 SHEAR STRENGTH

The shear strength of soil systems is usually expressed by the Coulomb equation

$$S = c + \sigma \tan \phi \qquad (3.41)$$

where

S = shear strength
c = cohesion
ϕ = angle of internal friction
σ = normal stress on the shear plane

Shear parameters c and ϕ in Equation 3.41 can be determined in the laboratory or in the field (see Section 3.14). In the laboratory test, the direct shear and triaxial tests are commonly used. Brief discussion of these two tests are presented as follows.

3.13.1 Laboratory Shear Test

1. Direct Shear Test

The direct shear test, one of the earliest developed, provides a measure of the shearing resistance of a cohesive or cohesionless soil across a predetermined failure plane. It has been criticized because the failure plane is determined by the test method and not by the soil properties, and because of difficulties associated with controlling the sample volume. However, it is a simple test and should not be overlooked.

The soil specimen is enclosed in a box consisting of an upper and a lower half. The lower half can slide underneath the upper half of the box, which is free to move vertically. The normal load is applied on the top of the upper half of the box. A horizontal force is applied to the lower half of the box. In the case of wet cohesive soils, stones are used to permit drainage of water from the specimens. A stress–strain curve is obtained by plotting the shear stress versus shear displacement. To obtain

the failure envelope, at least three tests using various normal stresses are performed on specimens of the same soil. The cohesion and friction can also be obtained graphically from this plot.

2. Triaxial Test

A cylindrical soil specimen is encased in a thin rubber membrane with rigid caps and pistons on both ends and placed inside a triaxial cell. The cell is then filled with a fluid and, by application of pressure to the fluid, the specimen may be subjected to hydrostatic compressive stress. Shear stresses in the specimen are created by applying additional vertical stress. This additional vertical stress is called *deviator stress* $\Delta \sigma$. The deviator stress is steadily increased until failure of the specimen occurs. Drainage of water from the specimen is measured by a burette. Details of triaxial test and pore-water pressure measurements are described by Bishop and Henkel (1962) and Bowles (1986). To obtain Mohr's envelope, several triaxial tests should be performed on specimens of the same soil using various confining (cell) pressures σ_3 (see Fig. 3.29). From Figure 3.29 the stresses τ and σ can be obtained either graphically or by means of the formulas

$$\tau = \frac{\sigma_1 - \sigma_3}{2} \sin 2\alpha \qquad (3.42)$$

$$\sigma = \frac{\sigma_1 + \sigma_3}{2} + \frac{\sigma_1 - \sigma_3}{2} \cos 2\alpha \qquad (3.43)$$

From a simultaneous solution of the above equations, we have

$$\sigma_1 = \sigma_3 \tan^2 \left(45° + \frac{\phi}{2} \right) + 2c \tan \left(45° + \frac{\phi}{2} \right) \qquad (3.44)$$

$$\sigma_3 = \sigma_1 \tan^2 \left(45° - \frac{\phi}{2} \right) - 2c \tan \left(45° - \frac{\phi}{2} \right) \qquad (3.45)$$

where

τ = shear stress
σ_1 = major principal stress ($= \sigma_3 + \Delta \sigma$)
σ_3 = minor principal stress (confined pressure)
α = angle between normal stress and major principal stress

Point A in Figure 3.29 has the coordinates

$$x = \frac{\sigma_1 + \sigma_3}{2}, \quad y = \frac{\sigma_1 - \sigma_3}{2} \qquad (3.46)$$

The line or curve connecting these points is called a *stress path*. The stress path is just like a Mohr circle construction as it represents states of stress.

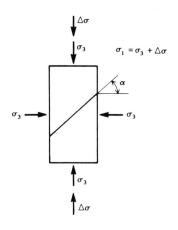

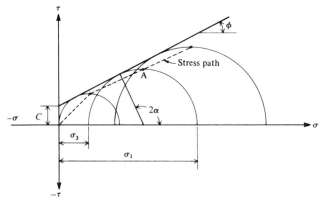

Fig. 3.29 Triaxial test.

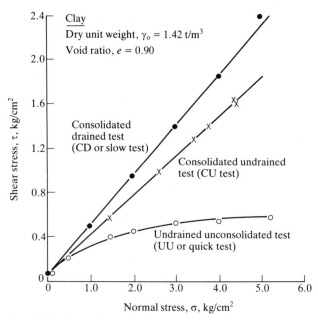

Fig. 3.30 Comparison of test results from laboratory shear tests under various loading and drainage conditions. (*After Tianjin University et al., 1978.*)

There are three conditions under which shear tests may be performed, viz., undrained, drained, and consolidated-undrained. These conditions are used to evaluate the effect of pore water pressure.

(a) *Unconsolidated-Undrained Test* (UU or Quick Test). No drainage is allowed during application of confining pressure σ_3 or normal load in direct shear. In the case of an embankment constructed rapidly over a soft clay deposit or a strip loading placed rapidly on a clay deposit, the UU analysis should be performed. The unconfined compression test is considered as a special case of the unconsolidated-undrained test with confining pressure equal to zero. The deviator stress at failure is called unconfined compressive strength q_u.

(b) *Consolidated-Undrained Test* (CU Test). Drainage is allowed during application of the confining or normal load. The sample is consolidated with respect to the applied pressure as observed via drainage (or vertical deformation in the direct shear test). No drainage is allowed during the shear test. In the case of rapid drawdown behind an earth dam or rapid construction of an embankment on a natural slope, the CU test should be performed.

(c) *Consolidated-Drained Test* (CD or Slow Test). The difference between the CD test and the CU test is that drainage takes place during the test and the test is slow enough so that pore pressures do not build up. In the case of an embankment constructed very slowly in layers over a soft clay deposit, an earth dam with steady-state seepage, and a strip footing on a clay deposit a long time after construction, the CD test should be performed.

The same soil under different loading and drainage conditions presents different stress–strain relationships, as shown in

Figure 3.30. In general, the CD or slow test gives the greater internal friction angle ϕ, while the UU or quick test provides the minimum value of ϕ.

3.13.2 Effective Stress

The effective stress σ' is equal to the total stress σ minus the pore water pressure u, that is:

$$\sigma' = \sigma - u \tag{3.47}$$

Skempton (1954) proposed a method for evaluating the pore pressure for both saturated and partially saturated soil as

$$\Delta u = B[\Delta\sigma_3 + A(\Delta\sigma_1 - \Delta\sigma_3)] \tag{3.48}$$

in which

$\Delta u =$ change in pore pressure due to increased stresses

$A, B =$ pore-pressure parameters

$\Delta\sigma_1, \Delta\sigma_3 =$ change in major and minor principal stresses

$\Delta\sigma_1 - \Delta\sigma_3 =$ deviator stress

The parameter B ranges from 0 for dry soil to 1.0 for saturated soil. The parameter A is mainly dependent on the type of soil and its previous stress history; typical values are given in Table 3.36.

TABLE 3.36 TYPICAL VALUES OF PARAMETERS A[a].

Soil Type	Parameter A (at failure)
Sensitive clay	1.2–2.5
Normal consolidated clays	0.7–1.3
Overconsolidated clays	0.3–0.7
Heavily overconsolidated sandy clays	−0.5–0.0
Very loose fine sand	2.0–3.0
Medium fine sand	0.0
Dense fine sand	−0.3
Loess	−0.2

[a] From Bjerrum (Lambe, 1982) and Wu (1966).

The values of A and B can be measured experimentally in both laboratory and field. (For more discussion, see Lambe (1962) and Lambe and Whitman (1979)).

3.13.3 Sensitivity

Most clays lose a portion of their strength and stiffness when remolded. The main cause of this phenomenon may be reorientation of particles into less favorable positions. If the clay regains a portion of its strength with elapsed time, the phenomenon is referred to as *thixotropy*.

The sensitivity of a soil to remolding is measured by the ratio of undisturbed strength to remolded strength as defined by Terzaghi (1944). Commonly, the unconfined compression test is used. From the point of view of their sensitivity to remolding, clays may conveniently be classified as follows (Skempton and Northey, 1952; Bjerrum, 1954):

Sensitivity	Classification
<2	insensitive
2–4	moderately sensitive
4–8	sensitive
8–16	very sensitive
16–32	slightly quick
32–64	medium quick
>64	quick

The sensitivity of most clays ranges from 2 to 4, that of peat from 1.5 to 10, and of marine clays from 1.6 to 26.

3.13.4 Tensile Strength

Tensile strength data are useful for predicting the cracking behavior of earth dams, highway pavement, and stabilized soil structures. For its determination, a simple test method called the double-punch test or unconfined-penetration (UP) test may be used (Fang and Chen, 1971; Fang and Fernandez, 1981). This test uses two steel disks centered on both top and bottom surfaces of a cylindrical soil specimen; load is then applied on the disks until the specimen reaches failure. The tensile strength is computed by the following formula:

$$\sigma_t = \frac{P}{\pi(KbH - a^2)} \qquad (3.49)$$

in which

σ_t = tensile strength
P = load
b = radius of the specimen
a = radius of disk
H = height of specimen
K = constant = 1.0

A height–diameter ratio of the specimen varying from 0.8 to 1.2 and a ratio of the diameter of the specimen to the diameter of the disk varying from 0.2 to 0.3 are suitable for the test.

Comparison of tensile strength of various construction materials determined by both UP test (Eq. 3.49) and the conventional split-tension test are presented in Figure 3.31. Good agreement between both tensile strength results has been observed. The UP test has the advantage that the test can be conveniently performed in conjunction with routine California

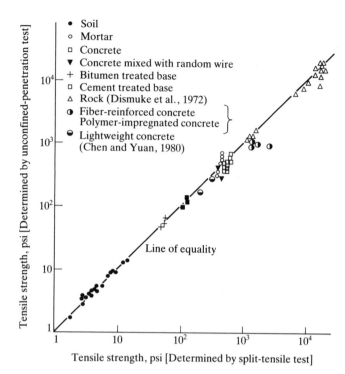

Fig. 3.31 Comparison of tensile strength of various construction materials determined by unconfined-penetration and split-tensile tests. [1 psi = 6.9 kPa (kN/m^2).]

Bearing Ratio (CBR) and compaction tests. The split-tension test and other tensile strength test methods measure the tensile strength across a predetermined failure plane, whereas the UP test always causes failure on the weakest plane, which results in the measurement of the true tensile strength.

3.13.5 Residual Shear Strength

For analysis of shear characteristics of overconsolidated clays, ordinary shear tests are not suitable because they give too high a shear value. Skempton (1964) showed that the strength remaining in laboratory samples after large shearing displacement corresponded closely with the computed strength from actual landslides (Chapters 10 and 11); therefore, he proposed a residual strength concept as shown in Figure 3.32 and Equation 3.50:

$$S_r = \sigma' \tan \phi \qquad (3.50)$$

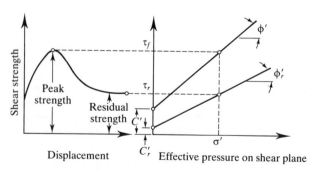

Fig. 3.32 Residual shear strength.

where

S_r = residual shear strength
σ' = effective pressure on shear plane
ϕ'_r = residual friction angle

In examining Figure 3.32, the peak strength τ_f, the corresponding effective friction angle ϕ', and the effective cohesion c' are used for conventional slope stability analysis. However, for overconsolidated clays, ϕ'_r and c'_r are suggested. The c'_r value is usually very small or zero. The residual shear parameters presented in Equation 3.50 can be obtained from slow-drained shear tests (Kenney, 1967; Webb, 1969; Bishop et al., 1971; CRRI, 1976, 1979). Correlations between these shear parameters with conventional soil constants such as liquid limit (LL),

plasticity index (PI), and liquidity index (LI) are shown in Figures 3.33 to 3.35. Further discussions on test methods, data interpretation and applications are given by Ramiah et al. (1970), Voight (1973), Townsend and Gilbert (1973), and Cancelli (1977).

As indicated in Figure 3.30, the same soil under different loading and drainage conditions presents a different stress–strain relationship. Therefore, careful examination for use of the proper test method and proper selection of strength parameters is important. Table 3.37 provides some guidelines

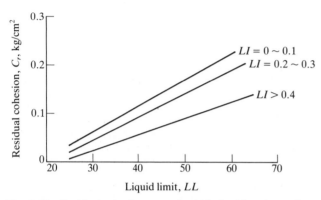

Fig. 3.33 Residual cohesion versus liquid limit with various values of Liquidity Index, LI = (water content − plastic limit)/plasticity index. (*Data from CRRI, 1979.*)

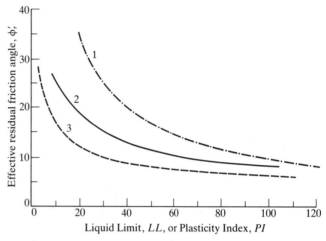

Fig. 3.35 Effective residual friction angle versus Liquid Limit or Plasticity Index. 1, ϕ'_r vs. LL (Jamiolkski and Pasqualini, cited by CRRI, 1979); 2, ϕ'_r vs. PI (De Beer, 1969); 3, ϕ'_r vs. *PI* (Russian, 1965 cited by CRRI, 1979).

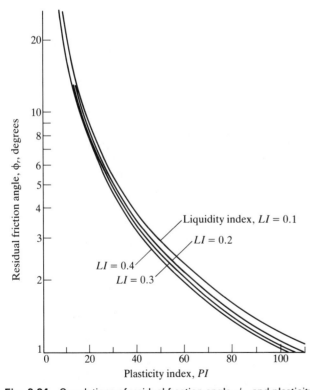

Fig. 3.34 Correlations of residual fraction angle, ϕ_r, and plasticity index, PI, with various values of Liquidity Index, LI = (water content − plastic limit)/(plasticity index). (*Data from CRRI, 1979.*)

TABLE 3.37 SELECTION OF STRENGTH PARAMETERS AND THEIR APPLICATIONS.

Strength Parameters and Type of Shear Tests	Applications
Unconsolidated-undrained test (UU or quick test)	Embankment constructed rapidly over a soft clay deposit
	A strip loading placed rapidly on a clay deposit
	Short-term slope stability analysis
Consolidated-undrained test (CU test)	Rapid drawdown behind an earth dam
	Rapid construction of an embankment on a natural slope
Consolidated-drained test (CD or slow test)	Embankment constructed very slowly in layers over a soft clay deposit
	An earth dam with steady-state seepage
	A strip footing on a clay deposit a long time after construction
Residual strength, S_r	Slope stability analysis on overconsolidated clay
Tension test, σ_t	Short-term stability analysis
	Analysis of progressive failure
	Field control of stabilized materials
Unconfined compression test, q_u	Short-term stability analysis
	Field control of stabilized materials

TABLE 3.38 RELATIONSHIP BETWEEN SPT (N), RELATIVE DENSITY (D_r) AND ANGLE OF INTERNAL FRICTION (ϕ) OF COHESIONLESS SOILS.

Type of Soil	SPT, N	Relative Density, D_r	Angle of Internal Friction, ϕ	
			Peck et al. (1974)	Meyerhof (1956)
Very loose sand	<4	<0.2	<29	<30
Loose sand	4–10	0.2–0.4	29–30	30–35
Medium sand	10–30	0.4–0.6	30–36	35–40
Dense sand	30–50	0.6–0.8	36–41	40–45
Very dense sand	>50	>0.8	>41	>45

to assist selecting the proper shear test method for preliminary analysis and design of foundation engineering projects. Further discussions on the mechanism of shear strength, test methods, data analysis, and interpretations are given by Yong and Townsend (1981), Saada and Townsend (1981), Donaghe et al. (1988), and Tuncan et al. (1989).

With regard to cohesionless soils, Table 3.38 shows general relationships between relative density (D_r), standard penetration test (SPT), and angle of internal friction (Meyerhof, 1956; Peck et al., 1974). Figure 3.36 shows the relationship between void ratio (e) and coefficient of internal friction of various rock materials (Mikuni, 1980). These data can be used for preliminary design. Fundamental theoretical treatment are

given by Farouki and Winterkorn (1964), Winterkorn (1971), and Fang (1989).

3.14 IN-SITU MEASUREMENT OF SOIL PROPERTIES

Field tests are conducted to determine the in-situ strength, deformation, hydraulic conductivity, and commonly used methods are listed in Table 3.39. Methods relating to the shallow foundations, landslides, earthquake regions, and off-shore foundations are discussed in Chapters 1, 5, 9, 11, 16, and 18. Hydraulic conductivity and groundwater measurements are presented in Chapters 1 and 7. Water pressure tests (AASHTO, 1988) are performed in situ, within test borings, to measure the permeability of a soil or rock mass. Pressure testing helps in locating zones of leakage and measuring the capacity of such zones for transmitting water, and is useful in estimating grouting (Chapter 9) and dewatering (Chapter 7) requirements for construction purposes.

Figure 3.37 presents an interrelationships of soil classification and some selected in-situ measurement methods including California Bearing Ratio (CBR), cone penetration test (CPT), standard penetration test (SPT) and self-boring pressuremeter test (SBPMT). This chart can be used to make preliminary estimations of bearing values of subgrade soil as well as its classification with given in-situ parameters. It can also be used to derive approximate correlations between different in-situ parameters, verify test results or identify areas where more and detailed data base is needed (Pamukcu and Fang, 1989).

General discussions on in-situ measurement of soil properties, including test procedures, data acquisition, data interpretation, and practical applications are given by ASCE (1976), Campanella and Robertson (1985), and Clemence (1986). Problems relating to the contaminated soil–water systems and hazardous and toxic waste sites are given by Johnson et al. (1985), Fang (1986, 1987), Woods (1987), and Chapter 20.

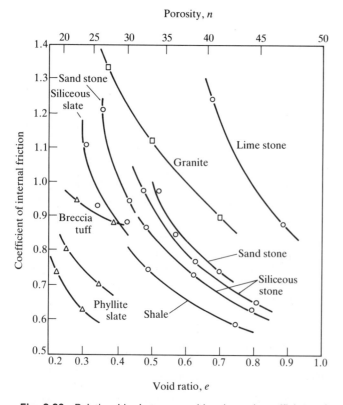

Fig. 3.36 Relationship between void ratio and coefficient of internal friction of various rock materials. □, Miboro dam; ▲, Kuzuryu dam; ○, Misakubo dam. Max size 200 mm. (*After Mikuni, 1980.*)

TABLE 3.39 COMMONLY USED METHODS FOR IN-SITU MEASUREMENT OF SOIL PROPERTIES.

Type of Test Methods	References
A. Strength/Deformation Measurements	
California Bearing Ratio (CBR)	ASTM D1883, D4429, AASHTO T193
Bearing value	ASTM D1195, D1196, AASHTO T221, T222
Vane shear test	ASTM D2573; Richards (1988)
Standard penetration test (STP)	ASTM D1586; AASHTO T206
Cone penetration test (CPT)	ASTM D3441; Schmertmann (1978)
Pressuremeter test (PMT)	Sanglerat (1979); Briaud and Audibert (1986)
Self-boring pressuremeter test (SBPMT)	Jezequel and Le Mahute (1979); Benoit and Clough (1986)
Dilatometer (DMT)	Schmertmann (1986)
Other shear tests	ASCE (1976); Campanella and Robertson (1985); Clemence (1986)
B. Hydraulic Conductivity Measurements	
Pumping test and groundwater monitoring techniques	Freeze and Cherry (1979); Powers (1981); Collins and Johnson (1988)
Water pressure test	AASHTO (1988)

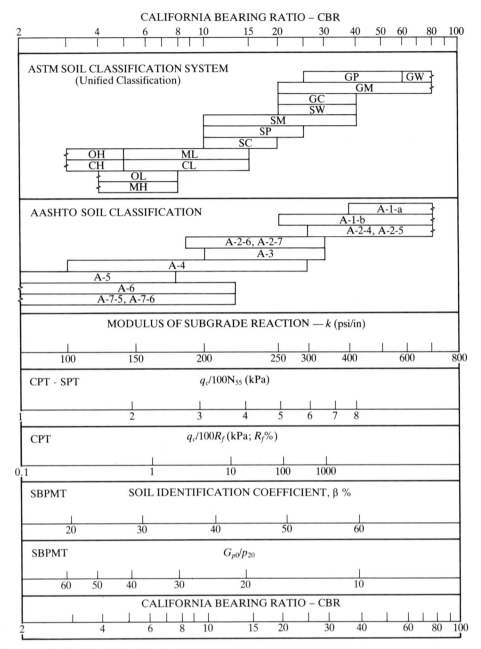

Fig. 3.37 Chart for approximate interrelationships between soil classification, bearing values, and some in-situ parameters: q_c, cone tip bearing; N_{55}, SPT blow count/ft; R_f, friction ratio (percent); G_{p0}, shear modulus at 0 percent strain; p_{20}, pressure at 20 percent strain; CBR, California Bearing Ratio. (*After Pamukcu and Fang, 1989.*)

REFERENCES

AASHTO is new acronym for AASHO = American Association of State Highway & Transportation Officials.

AASHO Road Test (1962), *Pavement Research*, Report No. 5, Highway Research Board Special Report 61E.

AASHTO (1970, 1986), *Standard Specifications for Highway Materials and Methods of Sampling and Testing*, part 1.

AASHTO (1988), *Manual on Subsurface Investigations*.

AASHTO (1987), *Maintenance Manual*. Amer. Association of State Highway & Transportation Officials, Washington, D.C.

ACTA (1961), *Research on Loess*, ACTA Instituti Constructions et Architecturae Academiae Sinicae (in Chinese).

Abrams, Duff A. (1918), *Design of Concrete Mixtures*, Bulletin, Structural Materials Laboratory, Lewis Institute, Chicago, Illinois.

Aitchison, G. D. (ed.) (1965), *Moisture Equilibria and Moisture Changes in Soils Beneath Covered Areas*, Butterworth, Australia.

Al-Khafaji, A. W. N. and Andersland, O. B. (1981), Ignition test for soil organic content measurement, *Journal of the Geotechnical Engineering Division, ASCE*, **107**, No. GT-4, pp. 465–479.

Altmeyer, W. T. (1956), Discussion of paper by Holtz and Gibbs on engineering properties of expansive clays, *Trans. ASCE*, **121**, pp. 666–669.

Andersland, O. B. and Anderson, D. M. (1978), *Geotechnical Engineering for Cold Regions*, McGraw-Hill Book Co., New York, N.Y.

Anderson, J. N. and Lade, P. V. (1981), The expansion index test, *ASTM Geotechnical Testing Journal*, **4**, No. 2, pp. 58–67.

Andrews, R. E., Gawarkiewicz, J. J., and Winterkorn, H. F. (1967), Comparison of the interaction of three clay minerals with water, dimethyl sulfoxide, and dimethyl formamide, *Highway Research Record*, No. 209, pp. 66–78.

Arman, A. (1970), Engineering classification of organic soils, *Highway Research Record*, No. 310, pp. 75–89.

ASCE (1976), *In Situ Measurement of Soil Properties*, American Society of Civil Engineers, New York, N.Y., Vols. 1 and 2.

ASTM (1985), *1985 Annual Books of ASTM Standards*, Section 4, Construction, Volume 04.08 Soil and Rock; Building Stones.

Aschenbrenner, B. C. (1956), A new method of expressing particle sphericity, *Journal of Sedimentology and Petrology*, **26**, pp. 5–31.

Atterberg, A. (1911), Die Plastizität der Tone, *Int. Mitt. für Bodenkunde*, **I**, pp. 10–43.

Azzouz, A. S., Krizek, R. J., and Corotis, R. B. (1976), Regression analysis of soil compressibility, *Soils and Foundations*, **16**, No. 2, pp. 19–29.

Barden, L. and Sides, G. (1971), Sample disturbance in the investigation of clay structure, *Géotechnique*, **21**, No. 3, pp. 211–222.

Baver, L. D. (1940), *Soil Physics*, John Wiley & Sons, Inc., New York, N.Y., p. 224.

Baver, L. D. (1968), The effect of organic matter on soil structure, *Pontificiae Academiae Scientiarum*, Scripta, Varia.

Baver, L. D. and Winterkorn, H. F. (1935), Sorption of liquids by soil colloids, II: Surface behavior in the hydration of clays, *Soil Science*, **40**, No. 5, pp. 403–419.

Benoit, J. and Clough, G. W. (1986), Self-boring pressuremeter tests in soft clay, *Journal of Geotechnical Engineering, ASCE*, **112**, No. 1, pp. 60–78.

Beskow, G. (1935), Soil freezing and frost heaving (in Swedish), *Sveriges Geol. Undersökning*, No. 375, translated into English by J. O. Osterberg, Northwestern University, Evanston, Ill., 1947, p. 145.

Bieniawski, Z. T. (1980), Rock classifications: state of the art and prospects for standardization, *Transportation Research Record*, 783, pp. 2–9.

Bishop, A. W. and Henkel, D. J. (1962), *The Measurement of Soil Properties in Triaxial Test*, Edward Arnold, Ltd., London.

Bishop, A. W. et al. (1971), A new ring shear apparatus and its application to measurement of residual strength, *Géotechnique*, **21**, No. 4, pp. 273–328.

Bjerrum, L. (1954), Geotechnical properties of Norwegian Marine clays, *Géotechnique*, **4**, No. 2, p. 49.

Blaser, H. D. and Scherer, O. J. (1969), Expansion of soils containing sodium sulfate caused by drop in ambient temperatures, *Highway Research Board*, Special Report 103.

Born, M. Heisenberg, and Hund (1924), as quoted by Eucken, A. in *Lehrbuch der Chemischen Physik* (1930), Akademische Verlagsgesellschaft, Leipzig.

Bowles, J. E. (1986), *Engineering Properties of Soils and Their Measurement*, McGraw-Hill Book Co., New York, N.Y.

Briaud, Jean-Louis and Audibert, Jean M. E. (1986), The pressuremeter and its name applications, *ASTM STP 950*.

Brumund, W. F., Jonas, E., and Ladd, C. C. (1976), Estimating in-situ maximum past preconsolidation pressure of saturated clays from results of laboratory consolidometer tests, *Transportation Research Board, Special Report 163*, pp. 4–12.

Burmister, D. M. (1951), The application of controlled test methods in consolidation testing, *ASTM STP 126*, p. 83.

Burwash, A. L. and Wiesner, W. R. (1984), Classification of peats for geotechnical engineering purposes, *Proceedings of the 3rd International Specialty Conference on Cold Regions Engineering*, **2**, pp. 979–998.

Campanella, R. G. and Robertson, P. K. (1985), Recent developments in in-situ testing of soils, *Proceedings of the 11th ICSMFE*, **2**, pp. 849–854.

Cancelli, A. (1977), Residual shear strength and stability analysis of a landslide in fissured overconsolidated clays, *Bulletin of the International Association of Engineering Geology*, No. 16.

Casagrande, A. (1932), The structure of clay and its importance in foundation engineering, *Journal Boston Society of Civil Engineering*, **19**, No. 4, p. 168.

Casagrande, A. (1936), The determination of the pre-consolidation load and its practical significance, *Proc. 1st International Conf. on Soil Mechanics and Foundation Engineering*, **3**, p. 60.

Casagrande, A. (1948), Classification and identification of soils, *Trans. ASCE*, 1948, pp. 901–992.

Chaney, R. C. and Fang, H. Y. (1986), Static and dynamic properties of marine sediments: a state of the art, *ASTM STP 923*, pp. 74–111.

Chaney, R. C. et al. (1983), Suggested test method for determination of thermal conductivity of soil by thermal-needle procedure, *ASTM*

Geotechnical Testing Journal, **6**, No. 4, pp. 220–225.

Chen, F. H. (1979), *Foundations on Expansive Soils*, Elsevier Science Publishing Co., Ltd., New York, N.Y.

Chen, W. F. and Yuan, R. L. (1980), Tensile strength of concrete: double-punch test, *Journal of the Structural Division, ASCE*, **106**, No. ST8, pp. 1673–1693.

Clemence, S. P. (ed.) (1986), *Use of In Situ Tests in Geotechnical Engineering*, Geotechnical Special Publication No. 6. American Society of Civil Engineers, New York, N. Y.

Collins, A. G. and Johnson, A. I. (eds.) (1988), Ground-water contamination, Field methods, *ASTM STP 963*.

Cour, F. R. (1971), Inflection point method for computing C_v, *Journal of the Soil Mechanics and Foundations Division, Proc. ASCE*, **97**, No. SM-5, pp. 827–831.

CRRI (1976), *Collected Papers on Landslides*, Chinese Railway Research Institute, Lanzhou, China (in Chinese).

CRRI (1979), *Landslides*, Vol. 2, Chinese Railway Research Institute, Lanzhou, China (in Chinese).

Czeratzki, W. and Frese, H. (1958), Importance of water in the formation of soil structure, *Highway Research Board*, Special Report 40.

De Beer, E. E. (1969), Experimental data concerning clay slopes, *Proceedings of the 7th International Conf. on Soil Mechanics and Foundation Engineering*, **2**.

Deere, D. U. (1963), Technical description of rock cores for engineering purposes, *Rock Mechanics and Engineering Geology*, **1**, p. 18.

Deere, D. U. and Patton, F. D. (1971), Slope stability in residual soils, *Proceedings of the 4th Panamerican Conference of Soil Mechanics and Foundation Engineering*, **1**, pp. 87–170.

De Mello, V. F. B. (1972), Thoughts on soil engineering applicable to residual soils, *Proceedings of the 3rd Southeast Asian Conference on Soil Engineering*, Hong Kong, pp. 5–34.

Dismuke, T. D., Chen, W. F., and Fang, H. Y. (1972), Tensile strength of rock by the double-punch method, *Rock Mechanics*, Springer-Verlag, Vol. 4, pp. 79–87.

Donaghe, R. T., Chaney, R. C., and Silver, M. L. (1988), Advanced Triaxial Testing of Soil and Rock, *ASTM STP 977*.

Encyclopedia Britannica (1950), Classification.

Evans, J. C. and Fang, H. Y. (1988), Triaxial permeability and strength testing of contaminated soils, R. T. Donaghe, R. C. Chaney, and M. L. Silver (eds.), *ASTM STP 977*, pp. 387–404.

Fang, H. Y. (1960), Rapid determination of liquid limit of soils by flow index method, *Highway Research Board Bulletin 254*, pp. 30–35.

Fang, H. Y. (1969), Influence of temperature and other climatic factors on the performance of soil-pavement systems, *Highway Research Board*, Special Report 103, pp. 173–185.

Fang, H. Y. (1980), Geotechnical properties and foundation problems of Shanghai soft clays, *ASCE Convention and Exposition*, Portland, Preprint 80-176, April.

Fang, H. Y. (1985), Soil–pollutant interaction effects on the soil behavior and the stability of foundation structures, *Environmental Geotechnics*, A. A. Balkema Publisher, Rotterdam, pp. 155–163.

Fang, H. Y. (ed.) (1986, 1987), *Environmental Geotechnology*, Proceedings of the International Symposium, Vol. 1, 1986, Vol. 2, 1987, Envo Publishing Co., Inc., Bethlehem, Pa.

Fang, H. Y. (1989), Particle theory: a unified approach for analyzing soil behavior, *Proceedings of the 2nd International Symposium on Environmental Geotechnology*, **1**, Envo Publishing Co., Inc., Bethlehem, Pa., pp. 167–194.

Fang, H. Y. and Chen, W. F. (1971), New method for determination of tensile strength of soils, *Highway Research Record*, No. 345, pp. 62–68.

Fang, H. Y. and Fernandez, J. (1981), Determination of tensile strength of soils by unconfined-penetration test, *ASTM STP 740*, pp. 130–144.

Fang, H. Y. and Chaney, R. C. (1986), Geo-environmental and climatological conditions related to coastal structural design along the China coastline, *ASTM STP 923*, pp. 149–160.

Fang, H. Y. and Evans, J. C. (1988), Long-term permeability tests using leachate on a compacted clayey liner material, *ASTM STP 963*, pp. 397–404.

FAO (1968), Nomenclature used in World Soils Map, *Food and Agricultural Organization UNESCO*, Rome Italy.

Farmer, I. W. (1968), *Engineering Properties of Rocks*, E. and F. N. Spon Ltd., London, p. 180.

Farouki, O. T. (1966), Physical properties of granular materials with reference to thermal resistivity, *Highway Research Record*, No. 128, pp. 25–44.

Farouki, O. T. and Winterkorn, H. F. (1964), Mechanical properties of granular systems, *Highway Research Record*, No. 52, pp. 10–42.

Feret, R. (1892), Sur la compacite des mortiers hydrauliques, *Annales des Ponts et Chaussées*, Memoires et Documents, 7e Series, Tome IV, Paris, pp. 5–164.

Finn, L. W. D., Byrne, P. M., and Emery, J. J. (1971), Engineering properties of a marine sediment, *Proc. International Symposium on the Engineering Properties of Sea-Floor Soils and Their Geophysical Identification*, Univ. of Washington, Seattle, Washington, July, pp. 110–120.

Freeze, R. A. and Cherry, J. A. (1979), *Groundwater*, Prentice-Hall, Inc., Englewood Cliffs, N.J.

Furnas, C. C. (1931), Grading aggregates; I, Mathematical relations for beds of broken solids of maximum density, *Industrial and Engineering Chemistry*, 23, pp. 1052–1058.

Gao, D. Z., Wei, D. D., and Hu, Z. X. (1986), Geotechnical properties of Shanghai soils and engineering applications, *ASTM STP 923*, pp. 161–177.

Gibbs, H. J. (1965), Paper on collapsible soils, presented at the *1965 Texas A&M International Symposium on Expansive Soils*.

Gidigasu, M. D. (1975), *Laterite Soil Engineering*, Elsevier Science Publishing Co. Ltd., New York, N.Y.

Goldberg, G. D., Lovell, Jr., C. W., and Miles, R. D. (1979), Use the geotechnical data bank, *Transportation Research Record 702*, pp. 140–146.

Goodman, R. E. (1980), *Introduction to Rock Mechanics*, John Wiley and Sons, Inc., New York, N.Y.

Graton, L. C. and Fraser, H. J. (1935), Systematic packing of spheres with particular relation to porosity and permeability, *Journal of Geology*, 43, pp. 785–909.

Gray, J. E. (1964), Method for determining particle shape of sands, *National Crushed Stone Association*, Washington, D.C.

Gray, D. H. and Mitchell, J. K. (1967), Fundamental aspects of electroosmosis in soils, *Journal of the Soil Mechanics and Foundations Division, Proc. ASCE*, 93, No. SM-6, pp. 209–236.

Griffin, J. J., Windom, H., and Goldberg, E. D. (1968), The distribution of clay minerals in the world ocean, *Deep-Sea Research*, 15, pp. 433–459.

Gromko, G. J. (1974), Review of expansive soils, *Journal of the Geotechnical Engineering Division, ASCE*, 100, No. GT-6, pp. 667–687.

Henniker, J. C. and McBain, J. W. (1948), The depth of a surface zone of a liquid, *Stanford Research Institute*, Stanford, California.

Holl, A. (1971), *Bituminöse Strassen*, Bauverlag Gmbh, Wiesbaden and Berlin.

Holtz, W. G. and Gibbs, H. J. (1956), Engineering properties of expansive clays, *Trans. ASCE*, 121, pp. 641–677.

Holtz, R. D. and Kovacs, W. D. (1981), *An Introduction to Geotechnical Engineering*, Prentice-Hall, Inc., Englewood Cliffs, N.J.

Hough, B. K. (1957), *Basic Soil Engineering*, The Ronald Press Co., New York, N.Y., pp. 114–115.

Howard, A. K. (1984), The revised ASTM standard on the unified classification system, *ASTM Geotechnical Testing Journal*, 7, No. 4, pp. 216–222.

Howard, A. K. (1987), The revised ASTM standard on the description and identification of soils (visual–manual procedure), *ASTM Geotechnical Testing Journal*, 10, No. 4, pp. 229–234.

HRB (1958), *Water and Its Conduction in Soils*, International Symposium, Highway Research Board, Special Report 40.

HRB (1969), *Effect of Temperature and Heat on Engineering Behavior of Soils*, Proceedings of an International Conference, Highway Research Board, Special Report 103.

Ingram, R. L. (1953), Fissility of mudrocks, *Bulletin, Geological Society of America*, 64, August.

Jarrett, P. M. (ed.) (1983), *Testing of Peats and Organic Soils*, ASTM STP 820.

Jezequel, J. F. and Le Mahute, A. (1979), *The Self-boring Pressuremeter Model 76 (PAF 76). User's manual* (English translation by J. Canou and M. T. Tumay, Louisiana State University).

Johnson, A. I., Frobel, A. K., Cavalli, N. J., and Bernt, A. C. (eds.) (1985), *Hydraulic Barriers in Soil and Rock*, ASTM STP 874.

Joisel, A. (1948), Crushing and fragmentation of rocks, *Annales Institut Technique du Batiment et des Travaux Publics*, no. 26, Paris, France.

Kassiff, G., Livneh, M., and Wiseman, G. (1969), *Pavements on Expansive Clays*, Jerusalem Academic Press, Israel.

Keil, K. (1937), see Ruckli, R. (1950), *loc. cit.*

Kenney, T. C. (1967), Influence of mineral composition on the residual strength of natural soils, *Proc. Geotechnical Conf. Oslo*, 1, pp. 123–129.

Kersten, M. S. (1949), Thermal properties of soils, *University of Minnesota Institute of Technology Bulletin*, No. 28, Minneapolis.

Kézdi, Árpád (1964), Discussion of paper by Farouki and Winterkorn, *Highway Research Record*, 52, pp. 42–59.

Khan, L. I., Pamukcu, S., and Kugelman, I. J. (1989), Electro-osmosis in fine-grained soil, *Proceedings of the 2nd International Symposium on Environmental Geotechnology*, 1, Envo Publishing Co. Inc., Bethlehem, Pa., pp. 39–47.

Kolbuszewski, J. (1965), *Sand Particles and Their Density*, Lecture given to the Materials Science Club Symposium on Densification of Particulate Materials, London, February 26.

Koppula, S. D. (1981), Statistical estimation of compression index, *ASTM Geotechnical Testing Journal*, 4, No. 2, pp. 68–73.

Krumbein, W. C. (1941), Measurement and geological significance of shape and roundness of sedimentary particles, *Journal of Sedimentology and Petrology*, 11, No. 2, pp. 64–72.

Krumbein, W. C. and Pettijohn, F. G. (1938), *Manual of Sedimentary Petrography*, D. Appleton-Century Co., New York, N.Y.

Kubiena, W. L. (1938), *Micropedology*, Collegiate Press, Ames, Iowa.

Kubiena, W. L. (1970), *Micromorphological Features of Soil Geography*, Rutgers University Press, New Brunswick, N.J.

Kübler, G. (1963), Influence of meteorological factors on the extent of road damage caused by frost, Highway Research Board Frost Damage Symposium, *Highway Report Record*, No. 33.

Kuenen, P. H. (1950), *Marine Geology*, John Wiley and Sons, Inc., New York, N.Y.; Chapman & Hall, London.

Lambe, T. W. (1962), Pore pressures in a foundation clay, *Journal of the Soil Mechanics and Foundations Division, Proc. ASCE*, 88, No. SM-2, pp. 19–47.

Lambe, T. W. and Whitman, R. V. (1979), *Soil Mechanics*, John Wiley and Sons, Inc., New York, N.Y.

Landolt-Börnstein (1952), Section 3237-Bodenkunde, *Astronomy & Geophysics*, Springer Verlag, 3, pp. 358–368.

Landva, A. O., Korpijaakko, E. O., and Pheeney, P. E. (1983), Geotechnical classification of peats and organic soils, *ASTM STP 820*, pp. 37–51.

Lee, P. Y. and Suedkamp, R. J. (1972), Characteristics of irregularly shaped compaction curves of soils, *Highway Research Record*, No. 381, pp. 1–9.

Lees, G. (1964), The measurement of particle shape and its influence in engineering materials, *Journal, British Granite and Whinstone Federation*, 4, No. 2.

Liu, T. K. (1967), A review of engineering soil classification systems, *Highway Research Record*, No. 156, pp. 1–22.

Low, P. F. (1958), Condition of water in soil systems and its response to applied force fields, *Highway Research Board*, Special Report 40, pp. 55–64.

Low, P. F. (1961), Physical chemistry of clay–water interaction, *American Society of Agronomy*, 13, pp. 279–327.

Low, P. F. (1968), Mineralogical data requirements in soil physical investigations, *Mineralogy in Soil Science and Engineering*, SSSA Spec. Pub. No. 3, Soil Science Soc. of America.

Lyon Associates, Inc. (1971), *Laterite and Lateritic Soils and Other Problem Soils of Africa*, an engineering study for Agency for International Development AID/CSD-2164.

MacFarlane, I. C. (1958), Guide to a field description of Muskeg, *Technical Memorandum 44, Associate Committee on Soil and Snow Mechanics*, National Research Council of Canada, Ottawa.

MacFarlane, I. C. (ed.) (1969), *Muskeg Engineering Handbook*, University of Toronto Press.

McManis, K. L., Ferrell, Jr., R. E., and Arman, A. (1983), Interpreting the physical properties of a clay using microanalysis techniques, *ASTM Geotechnical Testing Journal*, 6, No. 2, pp. 87–92.

Marshall, C. E. (1930), A new method of determining the distribution curve of polydisperse colloidal systems, *Proc. Roy. Soc.*, 26A.

Marshall, T. J. (1959), Relations between water and soil, *Technical Communication No. 50, Commonwealth Bureau of Soils*, Harpenden, England.

Martin, R. E. (1977), Estimating foundation settlements in residual soils, *Journal of the Geotechnical Engineering Division, ASCE*, 103, No. GT-3, pp. 197–212.

Mead, W. J. (1936), Engineering geology of damsites, *Transactions, 2nd International Congress on Large Dams*, Washington, D.C., **4**, p. 183.

Meyerhof, G. G. (1956), Penetration tests and bearing capacity of cohesionless soils, *Journal of the Soil Mechanics and Foundations Division, Proc. ASCE*, **82**, No. SM-1, pp. 866-1 to 866-19.

Michalowski, R. L. (ed.) (1989), *Cold Regions Engineering*, ASCE Publication No. 680.

Mikuni, E. (1980), Rockfill dams in Japan, *Geotechnical Engineering*, **11**, No. 2, pp. 93–133.

Mitchell, J. K. and Kao, T. C. (1978), Measurement of soil thermal resistivity, *Journal of the Geotechnical Engineering Division, ASCE*, **104**, No. GT-10, pp. 1307–1320.

M.O.P. Ministerios das Obras Publicas e do Ultramar (1959), As laterites do Ultramar Portugues, *Memoria No. 141, Laboratorio Nacional de Engenharia Civil*, Lisboa.

Mogami, T. (1967), Mechanics of granular material composed of particles of various sizes, *Japanese Society of Civil Engineers*, No. 137.

Mohr, E. C. J. and Van Baren, F. A. (1954), *Tropical Soils*, Interscience Publishers, New York, N.Y.

Muller, S. W. (1947), *Permafrost and Related Engineering Problems*, Edwards Bros. Inc., Ann Arbor, Michigan.

Nascimento, U., De Castro, E., and Rodrigues, M. (1964), Swelling and Petrifaction of Laterite Soils, Technical Paper no. 215, *Laboratorio Nacional de Engenharia Civil Ministerio das Obras Publicas*, Lisbon, Portugal.

Naylor, A. H. and Doran, I. G. (1948), Precise determination of primary consolidation, *Proc. 2nd International Conf. on Soil Mechanics and Foundation Engineering*, **1**, p. 34.

NCSA (1972), National Crushed Stone Assoc., *RETS Digest*, April.

Noorany, I. and Gizienski, S. F. (1970), Engineering properties of submarine soils, State-of-Art Review, *Journal of Soil Mechanics and Foundation Division, Proc. ASCE*, Sept., pp. 1735–1762.

O'Neill, M. W. and Poormoayed, N. (1980), Methodology for foundations on expansive clays, *Journal of the Geotechnical Engineering Division, ASCE*, **106**, No. GT-12, pp. 1345–1367.

Osipov, V. I. (1983), Methods of studying clay microstructure, *ASTM Geotechnical Testing Journal*, **6**, No. 1, pp. 10–17.

Pallmann, H. (1930), Doctoral dissertation ETH Zürich, *Kolloid Beihefte*, **30**, p. 344.

Pamukcu, S. and Fang, H. Y. (1989), Development of a chart for preliminary assessments in pavement design using some in situ soil parameters, *Transportation Research Record*, No. 1235, pp. 38–44.

PCA (1973), *Soil Primer*, Portland Cement Association, Skokie, Ill.

Peck, R. B., Hanson, W. E., and Thornburn, T. H. (1974), *Foundation Engineering*, John Wiley and Sons, Inc., New York, N.Y.

Pereira Dos Santos, M. P. (1955), Prediction of consistency limits of soils and soil mixtures, *Highway Research Board, Bulletin 108*, pp. 67–74.

Perry, E. B. (1975), Piping in earth dams constructed of dispersive clay: Literature review and design of laboratory tests, *Technical Report S-75-15*, US Army Engineer Waterways Experiment Station.

Phukan, A. (1985), *Frozen Ground Engineering*, Prentice-Hall, Inc., Englewood Cliffs, N.J.

Powers, J. P. (1981), *Construction Dewatering—A Guide to Theory and Practice*, John Wiley and Sons, Inc., New York, N.Y.

Radhakrishna, H. S., Lau, K. C., and Crawford, A. M. (1984), Coupled heat and moisture flow through soils, *Journal of Geotechnical Engineering, ASCE*, **110**, No. 12, pp. 1766–1784.

Ramiah, B. K., Dayalu, N. K., and Purushothamaraj, P. (1970), Influence of chemicals on residual strength of silty clay, *Soils and Foundations*, **X**, No. 1, March, pp. 25–36.

Rankama, K. and Sahama, Th. G. (1950), *Geochemistry*, The University of Chicago Press, Chicago, Ill., p. 130.

Rendon-Herrero, O. (1980), Universal compression index equation, *Journal of the Geotechnical Engineering Division, ASCE*, **106**, No. GT-11, pp. 1179–1200.

Reno, W. H. and Winterkorn, H. F. (1967), The thermal conductivity of kaolinite clay as a function of type of exchange ions, density and moisture content, *Highway Research Record*, No. 209, pp. 79–85.

Reuss, F. F. (1809), Sur un nouvel effet de l'électricité galvanique, *Proc. of the Imperial Russian Naturalist Society*, Moscow, **2**, pp. 327–337.

Richards, A. F. (1988), Vane shear strength testing in soils: Field and laboratory studies, *ASTM STP 1014*.

Richards, L. A. (ed.) (1947), *The Diagnosis and Improvement of Saline and Alkali Soils*, U.S. Regional Salinity Laboratory, Riverside, Calif., U.S. Department of Agriculture.

Ring, G. W. III (1966), Shrink swell potential of soils, *Highway Research Record*, No. 119.

Rittinger, V. (1867), *Testing Sieves and Their Uses*, Handbook 53 (1967), W. S. Tyler Co., Mentor, Ohio.

Rodenbush, W. H. and Buswell, A. M. (1958), Properties of water substance, *Highway Research Board*, Special Report 40.

Rothfuchs, G. (1935), How to obtain densest possible asphaltic and bituminous mixtures (in German), *Bitumen*, **3**, March.

Ruckli, R. (1950), *Der Frost im Baugrund*, Springer Verlag, Wien, Austria.

Ruiz, C. L. (1962), Osmotic interpretation of the swelling of expansive soils, *Highway Research Board Bulletin 313* (also: Publication no. 24 Direccion de Vialidad, Provincia de Buenos Aires).

Russell, E. W. (1934), The interaction of clay with water and organic liquids as measured by specific volume changes and its relation to the phenomena of crumb formation in soils, *Philosophical Transactions, Royal Society of London*, Series A, **233**, pp. 361–389.

Saada, A. S. and Townsend, F. C. (1981), State of the art: laboratory strength testing of soils, *ASTM STP 740*, pp. 7–77.

Salomone, L. A. and Kovacs, W. D. (1984), Thermal resistivity of soils, *Journal of the Geotechnical Engineering Division, ASCE*, **110**, No. 3, pp. 375–389.

Sanglerat, G. (1972), *The Penetrometer and Soil Exploration*, Elsevier Science Publishers, Amsterdam, The Netherlands.

Schmertmann, J. M. (1955), The undisturbed consolidation of clay, *Trans. ASCE*, **120**, p. 1201.

Schmertmann, J. H. (1978), Guidelines for cone penetration test, performance and design, US Department of Transportation, *Federal Highway Administration Report No. FHWA-TS-78-209*.

Schmertmann, J. H. (1986), Suggested method for performing the flat dilatometer test, *ASTM Geotechnical Testing Journal*, **9**, No. 2, pp. 93–101.

Schmidt, T. J. and Gould, J. P. (1968), Consolidation properties of an organic clay determined from field observations, *Highway Research Record*, No. 243, pp. 38–48.

Seed, H. B., Woodward, R. J., and Lundgren, R. (1962), Prediction of swelling potential for compacted clays, *Journal Soil Mechanics and Foundations Division, Proc. ASCE*, **88**, No. SM-3, pp. 53–87.

Seed, H. B., Mitchell, J. K., and Chan, C. K. (1962), Studies of swell and swell pressure characteristics of compacted clays, *Highway Research Board Bulletin 313*, pp. 12–39.

Segall, B. A., O'Bannon, C. E., and Jubson, A. M. (1980), Electro-osmosis chemistry and water quality, *Journal of Geotechnical Engineering Division, ASCE*, **106**, No. GT-10, pp. 1148–1152.

Serafim, J. L. and Del Campo, A. (1965), Interstitial pressures on rock foundations of dams, *Journal of the Soil Mechanics and Foundations Division, Proc. ASCE*, **91**, No. SM-5, September, pp. 66.

Sheeler, J. B. (1968), Summarization and comparison of engineering properties of loess in the United States, *Highway Research Record*, No. 212, pp. 1–9.

Sherard, J. L. and Decker, R. S. (eds.) (1977), *Dispersive Clays, Related Piping and Erosion in Geotechnical Projects*, ASTM STP 623.

Sherman, G. D. (1952), The genesis and morphology of the alumina-rich laterite clays, *Am. Inst. Min., Met. and Pet. Eng.*, N.Y., pp. 154–161.

Shockley, W. G. (1978), Suggested practice for description of frozen soils (visual-manual procedure), *ASTM Geotechnical Testing Journal*, **1**, No. 4, pp. 228–233.

Skempton, A. W. (1953), Soil mechanics in relation to geology, *Proc. of the Yorkshire Geological Society*, **29**, Part 1, No. 3, pp. 33–62.

Skempton, A. W. (1954), The pore pressure coefficients *A* and *B*, *Géotechnique*, **4**, No. 4, pp. 143–147.

Skempton, A. W. (1964), Long-term stability of clay slopes, *Géotechnique*, **XIV**, No. 2, pp. 77–101.

Skempton, A. W. and Northey, R. D. (1952), The sensitivity of clays, *Géotechnique*, **3**, No. 1, pp. 30–53.

Soil Mechanics for Road Engineers (1952), Road Research Laboratory, Her Majesty's Stationary Office, London.

SSSA, Soil Science Society of America (1970), *Glossary of Soil Science Terms*, SSSA, 677 South Segoe Road, Madison, Wis.

SSSA (1972), *Soil Water*, Soil Science Society of America, Madison, Wis.

Taber, S. (1929), Frost heaving, *Journal of Geology*, **38**, pp. 429–461.

Taber, S. (1930), Freezing and thawing of soils as factors in the destruction of road pavements, *Public Roads*, **11**, pp. 113–132.

Taylor, D. W. (1948), *Fundamentals of Soil Mechanics*, John Wiley and Sons, Inc., New York, N.Y.

Terzaghi, K. (1944), Ends and means in soil mechanics, *Engineering Journal (Canada)*, **27**, p. 608.

Terzaghi, K. and Peck, R. B. (1967), *Soil Mechanics in Engineering Practice*, John Wiley and Sons, Inc., New York, N.Y.

Teves, A. S. and Moh, Z. C. (1968), Compressibility of soft and medium Bangkok clays, *Research Report No. 4, Asian Institute of Technology*, Thailand, p. 117.

Thompson, D'A. W. (1942), *Growth and Form*, Cambridge University Press, Cambridge, England.

Thorp, J. and Smith, G. D. (1949), Higher categories of soil classification, Order, suborder, and great soil groups, *Soil Science*, **67**, pp. 117–126.

Tianjin University, Harbin Architectural Engineering Institute, Xian Institute of Metallurgy and Construction and Chongqing Architectural Engineering Institute (1978), *Soil Mechanics and Foundation Engineering*, Chinese Construction Publishing Co., Beijing, China (in Chinese).

Townsend, F. C. (1985), Geotechnical characteristics of residual soils, *Journal of Geotechnical Engineering, ASCE*, **111**, No. 1, pp. 77–94.

Townsend, F. C. and Gilbert, P. A. (1973), Tests to measure residual strength of some clay shales, *Géotechnique*, **23**, No. 2, pp. 267–271.

TRB (1981), *Shales and Swelling Soils*, Transportation Research Record 790.

TRB (1982), *Overconsolidated Clays: Shales*, Transportation Research Record 873.

TRB (1985), *Evaluation and Control of Expansive Soils*, Transportation Research Record 1032.

Tuncan, M., Pamukcu, S., and Hu, Z. X. (1989), Development of multipurpose triaxial apparatus for testing of soils undercoupled influence of thermal-chemical-hydraulic and electrical potential, *Proceedings of the 2nd International Symposium on Environmental Geotechnology*, Envo Publishing Co., Inc., Bethlehem, Pa., pp. 111–123.

Tyler, W. S., Inc. (1970), *Testing Sieves and Their Use*, Handbook 53, Mentor, Ohio.

U.S. Bureau of Reclamation (1973), *Earth Manual*, Federal Center, Denver, Colo.

USDA (1938), *Soils and Man*, Yearbook of Agriculture, U.S. Department of Agriculture, Washington, D.C.

USDA (1960), *Soil Classification (a Comprehensive System)*, 7th Approximation, U.S. Department of Agriculture, Washington, D.C.

Ueshita, K. and Nonogaki, K. (1971), Classification of coarse soils based on engineering properties, *Soils and Foundations*, **11**, No. 3, pp. 91–111, Japan.

Underwood, L. B. (1967), Classification and identification of shales, *Journal of the Soil Mechanics and Foundations Division, Proc. ASCE*, **93**, No. SM-6, pp. 97–116.

Vallerga, B. A. and Van Til, C. J. (1970), Classification and engineering properties of lateritic materials, *Highway Research Record*, No. 310, pp. 52–67.

Van Rooyen, M. and Winterkorn, H. F. (1957), Theoretical and practical aspects of the thermal conductivity of soils, *Highway Research Board Bulletin 168*, pp. 143–205.

Van Rooyen, M. and Winterkorn, H. F. (1959), Structural and textural influences on the thermal conductivity of soils, *Proc. Highway Research Board*, **38**, pp. 576–621.

Vees, E. and Winterkorn, H. F. (1967), Engineering properties of several pure clays as functions of mineral type, exchange ions and phase composition, *Highway Research Record*, No. 209, pp. 55–65.

Voight, B. (1973), Correlation between Atterberg plasticity limits and residual shear strength of natural soils, *Géotechnique*, **23**, No. 2, pp. 265–267.

Wadell, H. (1932), Volume, shape and roundness of rock particles, *Journal of Geology*, **40**, pp. 443–451.

Wadell, H. (1935), Volume, shape and roundness of quartz pebbles, *Journal of Geology*, **43**, pp. 250–280.

Waidelich, W. C. (1958), Influence of liquid and clay mineral type on consolidation of clay-liquid systems, *Highway Research Board*, Special Report 40, pp. 24–42.

Waksman, S. A. and Hutchings, I. J. (1935), Chemical nature of organic matter in different soil types, *Soil Science*, **40**, pp. 347–363.

Washburn, A. L. (1969), *Weathering, Frost Action, and Patterned Ground in the Mesters Vig District, Northeast Greenland*, Kobenhaven, C.A., Reitzels Forlag.

White, H. E. and Walton, S. F. (1937), Particle packing and particle shape, *Journal American Ceramic Society*, **20**, pp. 155–166.

Webb, D. L. (1969), Residual strength in conventional triaxial tests, *Proceedings of the 7th International Conference on Soil Mechanics and Foundation Engineering*, **1**.

Williamson, D. A. (1980), Uniform rock classification for geotechnical engineering purposes, *Transportation Research Record*, 783, pp. 9–14.

Winterkorn, H. F. (1936), Studies on the surface behavior of bentonites and clays, *Soil Science*, **41**, No. 1, pp. 25–32,

Winterkorn, H. F. (1942), Mechanism of water attack on dry cohesive soil systems, *Soil Science*, **54**, pp. 259–273.

Winterkorn, H. F. (1947), Fundamental similarities between electro-osmosis and thermoosmosis, *Proc. Highway Research Board*, **27**, pp. 443–455.

Winterkorn, H. F. (1948), Engineering uses and limitations of pedology for regional exploration of soils, *Proc. 2nd Int. Conf. Soil Mechanics and Foundations Engineering*, **1**, Rotterdam, Netherlands.

Winterkorn, H. F. (1953), *Macromeritic Liquids*, Symposium on Dynamic Testing of Soils, ASTM Special Technical Publication No. 156, pp. 77–89.

Winterkorn, H. F. (1955), Water movement through porous hydrophilic systems under capillary, electric and thermal potentials, *ASTM STP* 163, pp. 27–35.

Winterkorn, H. F. (1961), The behavior of moist clay soil in a thermal energy field, *Proc. 9th National Clay Symposium, Clay and Clay Minerals*, Pergamon Press, New York, N.Y.

Winterkorn, H. F. (1967), Application of granulometric principles for optimization of strength and permeability of granular drainage structures, *Highway Research Record*, No. 203, pp. 1–7.

Winterkorn, H. F. (1971), Analogies between macrometric and molecular liquids, and the mechanical properties of sand and gravel assemblies, *Chemical Dynamics* (papers in honor of Henry Eyring), Wiley-Interscience, New York, N.Y.

Winterkorn, H. F. and Baver, L. D. (1934), Sorption of liquids by soil colloids, I, *Soil Science*, **38**, No. 4.

Winterkorn, H. F. and Eckert, G. W. (1940), Consistency and physico-chemical data of a loess pampeano soil, I & II, *Soil Science*, **49**, pp. 73–82, and pp. 479–488.

Winterkorn, H. F. and Tschebotarioff, G. P. (1947), Sensitivity of clay to remolding and its possible causes, *Proc. Highway Research Board*, **27**, pp. 432–435.

Winterkorn, H. F. and Choudhury, A. N. D. (1949), Importance of volume relationships in soil stabilization, *Proc. Highway Research Board*, **29**, pp. 553–560.

Winterkorn, H. F. and Chandrasekharan, E. C. (1951), Laterite soils and their stabilization, *Highway Research Board Bulletin 44*, pp. 10–29.

Winterkorn, H. F. and Fang, H. Y. (1970), Mechanical resistance properties of ocean floors and beaches in light of the theory of macromeritic liquids, *Proc. Inter Ocean 70*, Düsseldorf, **2**, pp. 43–46.

Winterkorn, H. F. and Fang, H. Y. (1975), Soil technology and engineering properties of soils, *Foundation Engineering Handbook*, Chapter 2, Van Nostrand Reinhold Co., Inc., New York, N. Y., pp. 67–120.

Winterkorn, H. F. and Fang, H. Y. (1976), Engineering properties of some problematic soils and rocks, *Analysis and Design of Building Foundations*, Chapter 2, Envo Publishing Co., Inc., Bethlehem, Pa., pp. 17–36.

Woods, R. D. (ed.) (1987), *Geotechnical Practice for Waste Disposal*, ASCE Geotechnical Special Publication No. 13.

Woodward-Clyde & Associates (1967), *Expansive Clay Soils*, prepared for Portland Cement Association, Los Angeles, Calif.

Wooltorton, F. L. D. (1950), Movements in the desiccated alkaline soils of Burma, *Proc. ASCE*, **116**, January.

Wu, T. H. (1966), *Soil Mechanics*, Allyn and Bacon, Inc., Boston, p. 431.

Yamanouchi, T. (ed.) (1977), *Engineering Problems of Organic Soils in Japan*, Research Committee on Organic Soils, Japanese Society of Soil Mechanics and Foundation Engineering.

Yong, R. N. and Townsend, F. C. (1981), *Symposium on Laboratory Shear Strength of Soil*, ASTM STP 740.

Zingg, Th. (1935), Beitrag zur Schotteranalyse, *Schweizer. min. pet. Mitt.*, **15**, p. 39–140.

4 BEARING CAPACITY OF SHALLOW FOUNDATIONS

WAI-FAH CHEN, Ph.D.
Professor and Head
Structural Engineering Department
School of Civil Engineering
Purdue University

WILLIAM O. McCARRON, Ph.D.
Senior Research Engineer
Amoco Production Company

4.1 INTRODUCTION

Foundations, like the structures or equipment they support, are usually designed to meet certain serviceability and strength criteria. Serviceability conditions dictate that the foundation should perform such that under normal operating loads the structure or equipment it supports may fulfill its design purpose. These serviceability limitations are typically described by settlement or other motion limitations. The strength criteria have the purpose of insuring that the foundation has sufficient reserve strength to resist the occasionally large load that may be experienced due to extreme environmental forces or other sources. In most, but not all cases, the serviceability or settlement criteria and the strength criteria may be treated as unrelated design tasks. Serviceability is typically a long-term consideration for the foundation that may depend on time-dependent consolidation characteristics. Foundation strength, or bearing capacity, may be a short-term problem such as an embankment construction on an undrained clay foundation or a long-term problem in which the maximum foundation load may appear at some unknown time.

This chapter will consider only the strength or bearing capacity of shallow foundations. A shallow foundation may be defined as one in which the embedment depth of the foundation is less than its least characteristic dimension. Usually, the bearing capacity of a foundation is determined by limit equilibrium, limit analysis, or slip-line solutions. The variety of solutions available for a particular problem may lead to some uncertainty about which is the more appropriate procedure. In the following, the basis of these solution procedures will be summarized and methods for their use presented.

4.2 METHODS OF ANALYSIS

At the present time, the analysis of foundations can be made by employing one of the following four widely used methods:

1. Slip-line method
2. Limit equilibrium methods
3. Limit analysis methods
4. Finite-element methods

The first three methods are used in association with stability problems where the bearing capacity is sought. If, instead, the foundation settlement or stress distribution within the soil mass are of prime interest, then the fourth method must be used. Brief descriptions of the first three procedures are given here.

The slip-line method involves construction of a family of shear or slip-lines in the vicinity of the footing loads. These slip-lines, which represent the directions of the maximum shear stresses, form a network known as slip-line fields. The plastic slip-line field is bounded by regions that are rigid. For plane strain problems, there are two differential equations of plastic equilibrium and one equation for the yield condition available for solving for the three unknown stresses. These equations are written with respect to curvilinear coordinates that coincide with slip-lines. If the foundation boundary conditions are given only in terms of stresses, these equations are sufficient to give the stress distribution without any reference to the stress–strain relationship. However, if displacements or velocities are specified over part of the boundary, then the constitutive relation must be used to relate the stresses to the strains and the problem becomes much more complicated. Although solutions may be obtained analytically, numerical and graphical methods are often found necessary (see Sokolovskii, 1965; Brinch Hansen, 1961, 1970).

The methods described in the well-known textbooks by Terzaghi (1943) and by Taylor (1948), or the methods developed by Meyerhof (1951) are all classified here as methods of limit equilibrium. They can best be described as approximate approaches to constructing the slip-line fields. The solution requires that assumptions be made regarding the shape of the failure surface and the normal stress distribution along such a surface. The stress distribution usually satisfies the yield condition and the equations of static equilibrium in an overall sense. By trial and error, it is possible to find a most critical location of the assumed slip surface from which the capacity of the footing can be calculated.

In addition to the yield condition, the limit analysis methods consider the soil stress–strain relationship in an idealized manner. This idealization, termed normality or the flow rule, establishes the limit theorems on which limit analysis is based. The methods offer an upper and a lower bound to the true solution. The upper-bound solution is calculated from a kinematically admissible velocity field that satisfies the velocity boundary conditions and is continuous except at certain discontinuity surfaces where the normal velocity must be continuous, but the tangential velocity may undergo a jump on crossing a boundary. Similarly, the lower-bound solution is determined from a statically admissible stress field that satisfies the stress boundary conditions, is in equilibrium, and nowhere violates the failure condition. If the two solutions coincide, then the methods give the true answer for the problem considered. A good treatment of the subject is given by Chen (1975) and Chen and Liu (1990).

The methods described above are related in a manner. Most of the slip-line solutions give kinematically admissible velocity fields and thus can be considered as an upper-bound solution provided that the velocity boundary conditions are satisfied. If the stress field within the plastic zone can be extended into the rigid region so that the equilibrium and yield conditions are satisfied, then the solution constitutes a lower bound. Thus, slip-line solutions may be exact solutions. Shield (1955) has shown this for many cases. The extensive work that has been done on the stability analysis, including using the slip-line method, is summarized in the book by Sokolovskii (1965).

Limit equilibrium methods utilize the basic philosophy of the upper-bound rule, that is, a failure surface is assumed and the least answer is sought. However, it gives no consideration to soil kinematics and the equilibrium conditions are satisfied only in a limited sense. Therefore, limit equilibrium solutions are not necessarily an upper bound or a lower bound. However, any upper-bound solution from limit analysis will obviously be a limit equilibrium solution. Nevertheless, the method has been the most widely used owing to its simplicity and reasonably good accuracy.

The limit analysis method itself has many striking features that should appeal to researchers, as well as engineers. The problem formulation is generally simple and an analytical solution is always assured. In simple problems, it has been shown to yield reasonable answers when compared to limit equilibrium solutions. Its capability of providing a means for bounding the true solution is noteworthy. Finally, the method is efficient and can be extended to solve more difficult footing problems for which other methods have so far failed.

4.3 SOIL GOVERNING PARAMETERS

The bearing capacity of footings depends not only on the mechanical properties of the soil (cohesion c and friction angle ϕ), but also on the physical characteristics of the footing (width B, depth D, length L, and roughness δ). For a Coulomb material, Cox (1962) has shown that for a smooth surface footing bearing on a soil subjected to no surcharge, the fundamental dimensionless parameters associated with the stress characteristic equations are ϕ and $G = \gamma B/2c$, where γ is the unit weight of the soil. When G is small, the soil behaves essentially as a cohesive weightless medium. If G is large, soil weight rather than cohesion is a principal source of bearing strength. For most practical cases, one can expect that ϕ lies in the range of $0°$ to $40°$ and G will range from 0.1 to 1.0. These limits assume that c ranges from 500 to 1000 psf, and that the footing width ranges from 3 to 10 ft. The dimensionless bearing capacity q_0/c depends only on the angle of internal friction of the soil ϕ, the dimensionless soil weight parameter G, footing base friction angle δ, surcharge depth ratio D/B, and the base dimensions B and L.

For the most part, the bearing capacity of footings on soils have in the past been calculated by a superposition method, suggested by Terzaghi (1943), in which the contributions to the bearing capacity from different soil and loading parameters are summed. These contributions are represented by the expression

$$q_0 = cN_c + qN_q + \frac{\gamma B}{2} N_\gamma \qquad (4.1)$$

where q_0 is the average pressure over the footing contact area A, q is the overburden or surcharge pressure at the foundation

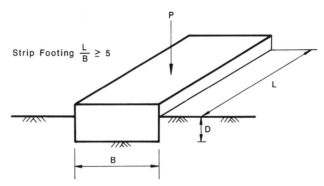

Fig. 4.1 Footing geometry.

base and the bearing capacity factors N_c, N_q, and N_γ represent the effects due to soil cohesion, surface loading, and soil unit weight, respectively. Equation 4.1 is valid for strip footings subjected to vertical center loads. However, other geometries are more common (Fig. 4.1). The parameters N are all functions of the angle of internal friction ϕ. Terzaghi's quasiempirical method assumed that these effects are directly superposable, whereas the soil behavior in the plastic region is nonlinear and thus superposition does not hold for general soil bearing capacities. The reason for using the simplified (superposition) method is largely the mathematical difficulties encountered when using conventional equilibrium methods.

4.4 BEARING CAPACITY BY THE UPPER-BOUND METHOD

Three basic conditions must be satisfied in the solution of a solid mechanics problem, namely, the stress equilibrium equations, the stress–strain (constitutive) relations, and the compatibility equations. In an elastic–plastic material, however, there is as a rule a three-stage development in a solution when the applied loads are gradually increased in magnitude from zero (the initial elastic response, the intermediate contained plastic flow, and the unrestricted plastic flow). The complete solution by this approach is cumbersome for all but the simplest problems, and methods are needed to furnish the load-carrying capacity in a more direct manner. Limit analysis is a method that enables a definite statement to be made about the collapse load without carrying out the step-by-step elastic–plastic analysis.

In contrast to slip-line and limit equilibrium procedures, the limit analysis method considers the stress–strain relationship of a soil, but in an idealized manner. This idealization, termed *normality*, establishes limit theorems on which limit analysis is based. Within the framework of this assumption, the approach is rigorous and the techniques are competitive with those of limit equilibrium, and in some instances are simpler. The plastic limit theorems of Drucker et al. (1952) may conveniently be employed to obtain upper and lower bounds of the collapse load for stability problems, such as the critical heights of unsupported vertical cuts or the bearing capacity of non-homogeneous soils.

The criteria that upper- and lower-bound theorems satisfy are shown in Figure 4.2 in relation to the solution requirements for general boundary-value problems. The two main limit theorems for a body or an assemblage of bodies of an elastic–perfectly plastic material may be restated as follows:

1. *Lower-bound.* The collapse load, calculated from a statically admissible stress field that satisfies all stress boundary conditions, is in equilibrium, nowhere violates the failure

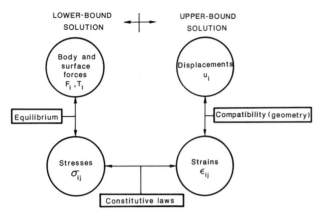

Fig. 4.2 Relationship between equilibrium and compatibility and solution type.

criterion, and is always lower, or at most, equal to the actual collapse load.

2. *Upper-bound.* The collapse load, calculated from a kinematically admissible velocity field of which the rate of external work done exceeds the rate of internal dissipation is always greater than the actual collapse load.

The upper-bound technique thus considers only velocity or failure modes and energy dissipation. The stress distribution need not be in equilibrium and is only defined in the deforming regions of the mode. The lower-bound technique, on the other hand, considers only equilibrium and yield conditions. It gives no consideration to material kinematics. The effect of the changes in geometry on equilibrium conditions is also neglected. The theorems do not require the stress or velocity fields to be continuous. In fact, discontinuous velocity fields not only prove convenient, but often resemble the actual collapse mechanism. This is in marked contrast to discontinuous stress fields, which rarely resemble the actual state.

The solution to upper-bound limit problems is obtained by writing the internal and external plastic work components, equating them, and then minimizing the solution to obtain the least upper bound. The internal work components are due to differential motion between adjacent bodies and plastic deformations of deforming regions. Figures 4.3 and 4.4 show work expressions for Tresca and Coulomb bodies (Chen, 1975). The Tresca and Coulomb criteria are given by

$$c = \frac{\sigma_1 - \sigma_3}{2} \qquad \text{Tresca} \qquad (4.2a)$$

$$\tau = c + \sigma_n \tan \phi \qquad \text{Coulomb} \qquad (4.2b)$$

or

$$\sigma_1 - \sigma_3 = 2c \cos \phi + (\sigma_1 + \sigma_3) \sin \phi \qquad \text{Coulomb} \qquad (4.2c)$$

where σ_1, σ_3 and σ_n are the maximum principal stress, minimum principal stress and the normal stress (compression positive) on the failure plane, respectively. The intermediate principal stress has no influence on the Tresca and Coulomb failure conditions. For the special condition $\phi = 0$, the Coulomb criterion reduces to the Tresca criterion.

Figure 4.5 shows three assumed failure mechanisms for determining the bearing capacity of a strip footing on a Tresca material. The solutions are outlined below.

Figure 4.5a illustrates a simple failure mechanism. The solution is found by evaluating the internal and external rates of work and then equating them. The external work sources are the footing load and surcharge. The internal work sources are due to discontinuous movement along the surfaces AB, BC,

and OB. The internal and external rates of work dissipation are

$$W_{\text{EXT}} = Pv - \gamma DBv$$

$$W_{\text{INT}} = W_{\text{AB}} + W_{\text{OB}} + W_{\text{BC}}$$

$$= \sqrt{2}B\sqrt{2}vc + 2Bvc + \sqrt{2}B\sqrt{2}vc = 6Bvc \qquad (4.3)$$

The bearing capacity is found by setting $W_{\text{EXT}} = W_{\text{INT}}$ and solving for q_0, which gives

$$q_0 = \frac{P}{B} = 6c + \gamma D \qquad (4.4)$$

The mechanism in Figure 4.5b involves a surcharge being pushed upward by the rigid region OBC and the region OAB under the footing, which is made up of many rigid triangles. The rate of external work due to the footing load, soil weight, and surcharge is

$$W_{\text{EXT}} = Pv - \frac{B^2}{2}\gamma v - BD\gamma v + \frac{B^2}{2}\gamma v \qquad (4.5)$$

The first term is due to the footing load; the second term is due to the soil weight of OBC; the third term is due to the surcharge; and the final term is due to the soil weight of region OAB. The rate of internal work dissipation is

$$W_{\text{INT}} = W_{\text{OAB}} + W_{\text{AB}} + W_{\text{BC}} + W_{\text{OB}}$$

$$= 2c\pi \frac{vB}{2} + \sqrt{2}v\sqrt{2}Bc + cvB \qquad (4.6)$$

Equating the two terms and solving for q_0 gives

$$q_0 = \frac{P}{B} = c(\pi + 3) + \gamma D \qquad (4.7)$$

The solution for the mechanism shown in Figure 4.5c is determined by equating the external work done by the footing load, foundation weight, and surcharge to the internal work dissipated by the deforming region ABDE and the relative motion along the surfaces AE, ED, DB, AB, FE, and DC.

$$W_{\text{EXT}} = Pv - \frac{\gamma B^2}{2}v + \frac{\gamma B^2}{2}v - \gamma DBv$$

$$W_{\text{INT}} = W_{\text{ABDE}} + W_{\text{AE}} + W_{\text{ED}} + W_{\text{DB}} + W_{\text{AB}} + W_{\text{FE}} + W_{\text{DC}}$$

$$W_{\text{ABDE}} = 2c\dot{\varepsilon}(\text{vol}) = 2cB^2\frac{v}{B}$$

$$W_{\text{AE}} = c\left(\frac{v}{2} + \frac{v}{2}\right)L_{\text{AE}} = cvB = W_{\text{BD}}$$

$$W_{\text{FE}} = c\left(\sqrt{2}\frac{v}{2}\right)L_{\text{FE}} = c\frac{v}{\sqrt{2}}\sqrt{2}B = W_{\text{DC}}$$

$$W_{\text{ED}} = c\left(\frac{v}{4} + \frac{v}{4}\right)\frac{L_{\text{ED}}}{2} = \frac{cBv}{4}$$

$$W_{\text{AB}} = 0 \qquad \text{(smooth footing)}$$

$$W_{\text{AB}} = W_{\text{ED}} \qquad \text{(rough footing)} \qquad (4.8)$$

The first term of W_{EXT} is due to the footing load; the second term represents the soil weight effects of AFE and BDC; the third term is the weight contribution of ABDE; and the final term is due to the surcharge. Setting $W_{\text{EXT}} = W_{\text{INT}}$ and solving, the bearing capacity is then

$$q_0 = \frac{P}{B} = 6.25c + \gamma D \qquad \text{(smooth footing)}$$

$$q_0 = \frac{P}{B} = 6.50c + \gamma D \qquad \text{(rough footing)} \qquad (4.9)$$

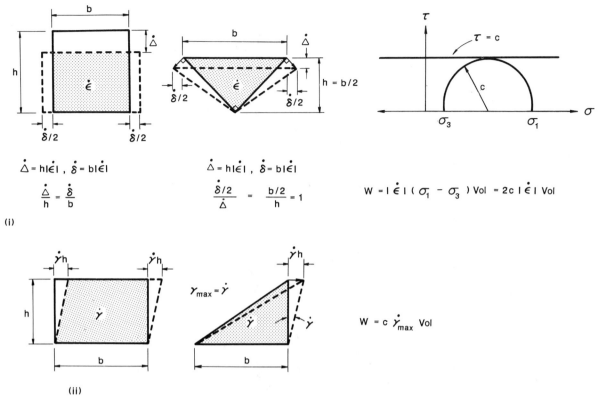

$$\dot\Delta = h|\dot\epsilon|, \quad \dot\delta = b|\dot\epsilon|$$

$$\frac{\dot\Delta}{h} = \frac{\dot\delta}{b}$$

(i)

$$\dot\Delta = h|\dot\epsilon|, \quad \dot\delta = b|\dot\epsilon|$$

$$\frac{\dot\delta/2}{\dot\Delta} = \frac{b/2}{h} = 1$$

$$W = |\dot\epsilon|(\sigma_1 - \sigma_3)\,\mathrm{Vol} = 2c\,|\dot\epsilon|\,\mathrm{Vol}$$

$$\gamma_{max} = \dot\gamma$$

$$W = c\,\dot\gamma_{max}\,\mathrm{Vol}$$

(ii)

Fig. 4.3a Energy dissipation and failure mechanisms for homogeneously deforming regions of Tresca material. (i) Simple vertical compression and lateral expansion deformation. (ii) Simple shear deformation.

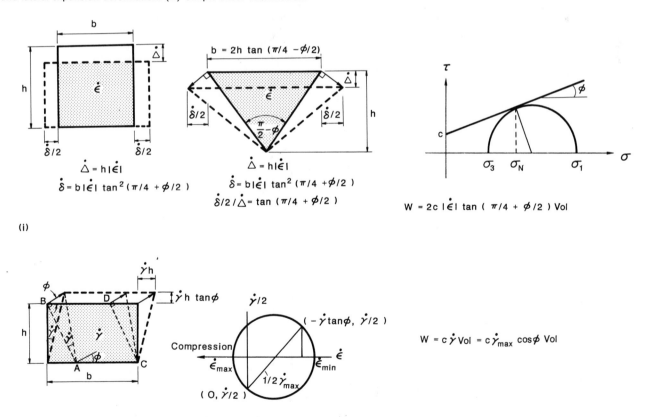

$$b = 2h\,\tan(\pi/4 - \phi/2)$$

$$\dot\Delta = h|\dot\epsilon|$$

$$\dot\delta = b|\dot\epsilon|\tan^2(\pi/4 + \phi/2)$$

$$\dot\Delta = h|\dot\epsilon|$$

$$\dot\delta = b|\dot\epsilon|\tan^2(\pi/4 + \phi/2)$$

$$\dot\delta/2/\dot\Delta = \tan(\pi/4 + \phi/2)$$

$$W = 2c\,|\dot\epsilon|\tan(\pi/4 + \phi/2)\,\mathrm{Vol}$$

(i)

$$\dot\gamma h\tan\phi$$

$$(-\dot\gamma\tan\phi, \dot\gamma/2)$$

Compression

$$\dot\epsilon_{max}$$

$$\dot\epsilon_{min}$$

$$1/2\,\dot\gamma_{max}$$

$$(0, \dot\gamma/2)$$

$$W = c\,\dot\gamma\,\mathrm{Vol} = c\,\dot\gamma_{max}\cos\phi\,\mathrm{Vol}$$

(ii)   (iii)

Fig. 4.3b Energy dissipation and failure mechanisms for homogeneously deforming regions of Coulomb material. (i) Simple vertical compression and lateral expansion deformation. (ii) Simple shear deformation. (iii) Mohr circle.

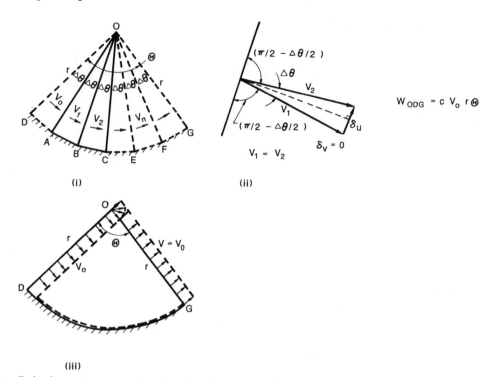

Fig. 4.4a Energy dissipation and failure mechanisms for radial regions of pure cohesive soils. (i) Rigid triangles. (ii) Velocity relation. (iii) Displaced pattern.

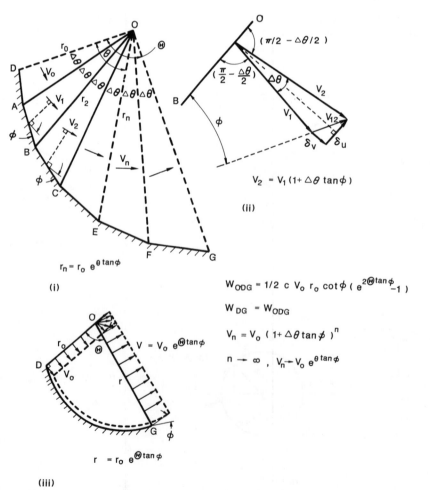

Fig. 4.4b Energy dissipation and failure mechanisms for logspiral regions of c–ϕ soils. (i) Rigid triangles. (ii) Velocity relation. (iii) Displaced pattern.

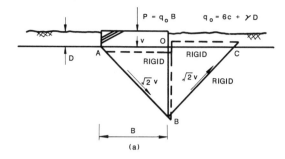

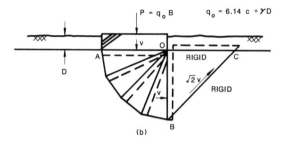

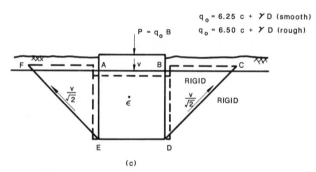

Fig. 4.5 Bearing capacity on Tresca material.

The failure mechanism shown in Figure 4.6 for a footing on a Coulomb material is that determined by Hill (1950). Since the mechanism is symmetrical, we may consider only the left side. Again we determine the solution by equating the internal and external rates of energy dissipation. The external work is due to the footing load. The internal work is due to the deforming region ADC and the differential slip along the surfaces OC, CD, and DE. No work is dissipated along the surfaces AC and AD owing to the nature of the assumed deformation in ADC. The work components are

$$W_{\text{EXT}} = \frac{P v_p}{2}$$

$$W_{\text{ACD}} = \frac{c v_0 B \cot \phi}{8 \cos \left(\frac{1}{4}\pi + \frac{1}{2}\phi\right)} \left(e^{\pi \tan \phi} - 1\right) = W_{\text{CD}}$$

$$W_{\text{OC}} = c v_0 \cos \phi \left[\frac{B}{4 \cos \left(\frac{1}{4}\pi + \frac{1}{2}\phi\right)}\right]$$

$$W_{\text{ED}} = c \left[v_0 \cos \phi \, e^{\left(\frac{1}{2}\pi\right) \tan \phi}\right] \left[\frac{B e^{\left(\frac{1}{2}\pi\right) \tan \phi}}{4 \cos \left(\frac{1}{4}\pi + \frac{1}{2}\phi\right)}\right] \quad (4.10)$$

These expressions are developed by considering the work expressions in Figure 4.4b and the kinematic relations of Figure 4.6. Collecting terms,

$$P = cB \cot \phi \left[e^{\pi \tan \phi} \tan^2 \left(\tfrac{1}{4}\pi + \tfrac{1}{2}\phi\right) - 1\right] \quad (4.11)$$

The limit of Equation 4.11 as ϕ approaches zero is $P = (2 + \pi)Bc$, which is the well-known exact solution for a footing on a Tresca material.

Using limit analysis, Chen (1975) has evaluated the bearing capacity factors N of Equation 4.1. These values, summarized in Table 4.1, are given by the following relationships.

$$N_q = e^{\pi \tan \phi} \tan^2 \left(\tfrac{1}{4}\pi + \tfrac{1}{2}\phi\right) \quad (4.12)$$

$$N_c = \cot \phi \, (N_q - 1) \quad (4.13)$$

$$N_\gamma = 2(1 + N_q) \tan \phi \tan \left(\tfrac{1}{4}\pi + \tfrac{1}{5}\phi\right) \quad (4.14)$$

It is worth pointing out that while the correct N_γ expression has for some time remained unsettled, the above values of N_c and N_q are the same as those given by Meyerhof and Brinch Hansen and are generally accepted as the correct or exact values. The differences in the reported N_γ are substantial, ranging from about one-third to double the values reported here. Andersen (1972) and Georgiadis and Michalopoulos (1985) have determined that the expression for N_γ provided by Brinch Hansen (1970) provides a lower bound to values determined experimentally. The data presented by Georgiadis and Michalopoulos indicate that Equation 4.14 provides a very good representation of the data for $\phi < 40°$ from a design standpoint. Brinch Hansen's expression has the form

$$N_\gamma = 1.5(N_q - 1) \tan \phi \quad (4.15)$$

On the basis of a statistical analysis of test results, Ingra and Baecher (1983) have suggested that for strip footings

$$N_\gamma = e^{(0.173\phi - 1.646)} \quad (4.16)$$

which gives N_γ values greater than Chen's expression (Eq. 4.14) for $\phi < 50°$.

Numerous authors have noted that the correct value of ϕ to be used in the bearing capacity equations may be different from that value measured by conventional triaxial tests. The value of ϕ measured in plane strain tests may be 10% greater

TABLE 4.1 BEARING CAPACITY FACTORS[a].

ϕ	N_q	N_c	N_γ	N_q/N_c	$\tan \phi$
0	1.000	5.142	0.000	0.194	0.000
2	1.197	5.632	0.156	0.212	0.035
4	1.432	6.185	0.350	0.232	0.070
6	1.716	6.813	0.595	0.252	0.105
8	2.058	7.527	0.909	0.273	0.141
10	2.471	8.345	1.313	0.296	0.176
12	2.973	9.285	1.837	0.320	0.213
14	3.586	10.370	2.522	0.346	0.249
16	4.335	11.631	3.422	0.373	0.287
18	5.258	13.104	4.612	0.401	0.325
20	6.399	14.835	6.196	0.431	0.364
22	7.821	16.883	8.316	0.463	0.404
24	9.603	19.323	11.173	0.497	0.445
26	11.854	22.254	15.049	0.533	0.488
28	14.720	25.803	20.351	0.570	0.532
30	18.401	30.139	27.665	0.611	0.577
32	23.177	35.490	37.849	0.653	0.625
34	29.440	42.163	52.182	0.698	0.675
36	37.752	50.585	72.594	0.746	0.727
38	48.933	61.351	102.050	0.798	0.781
40	64.195	75.312	145.191	0.852	0.839
42	85.373	93.706	209.435	0.911	0.900
44	115.307	118.368	306.920	0.974	0.966
46	158.500	152.096	458.018	1.042	1.036
48	222.297	199.257	697.926	1.116	1.111
50	319.053	266.878	1089.456	1.195	1.192

[a] Chen (1975).

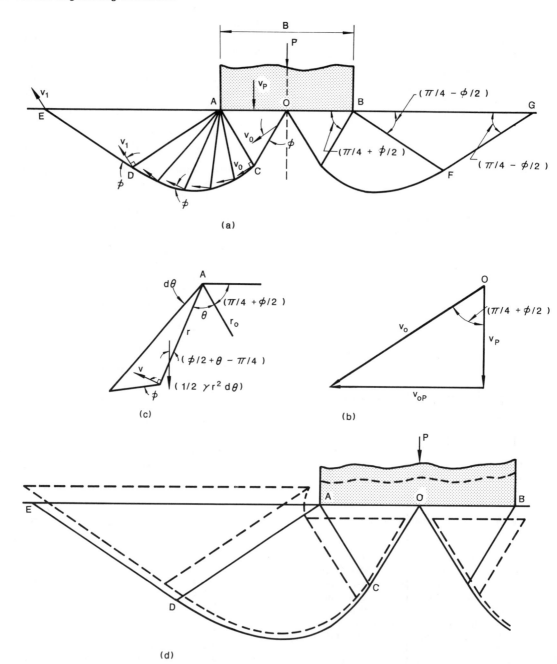

Fig. 4.6 Bearing capacity on Coulomb material. (a) Hill mechanism. (b) Velocity diagram. (c) Differential element. (d) Resulting deformation pattern.

than the triaxial test value. Meyerhof (1963) has suggested the following modification for values measured in triaxial tests:

$$\phi = \left(1.1 - 0.1 \frac{B}{L}\right)\phi_{\text{triaxial}} \qquad (4.17)$$

4.5 BEARING CAPACITY BY THE LOWER-BOUND METHOD

To obtain a lower bound of the bearing capacity problem, the simple stress field of Figure 4.7a was selected, which gives $q_0 = 2c$ as a lower bound. A better approximation is obtained by considering the weight of the soil mass (Fig. 4.7b) and adding

a hydrostatic stress field, giving $q_0 = (4c + \gamma D)B$ (Fig. 4.7c). Superposition of a hydrostatic stress field has no effect on the failure condition for $\phi = 0$ materials.

A better solution may be obtained using physical intuition and noting that the load in the soil "spreads" as distance from the footing increases. Therefore one might select the stress field in Figure 4.8b, which has a vertical stress component of $3c$ below the footing. By superimposing the stress field of Figure 4.7b, the lower-bound solution is increased to $q_0 = 5c + \gamma D$, which is very close to the exact solution.

Using lower-bound methods, Chen (1975) and Shield (1955) have developed solutions to the problem of a rigid footing over a foundation of finite thickness. Figure 4.9 shows the discontinuous stress field for a strip footing. The stress fields are in equilibrium and nowhere violate the failure condition for the

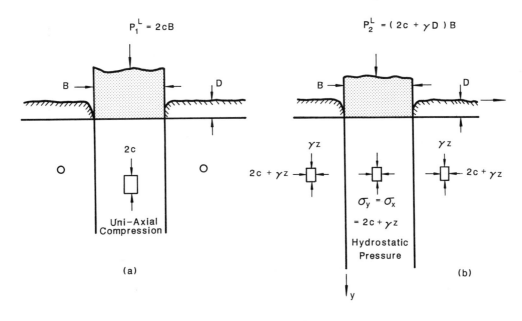

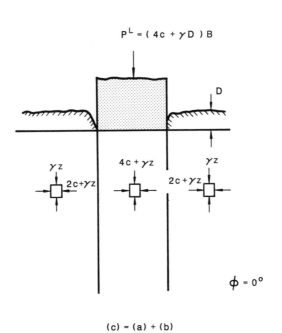

Fig. 4.7 Lower-bound bearing capacity example for Tresca material (simple stress field).

Tresca material. The average pressure over Oa is found to be

$$q_0 = c\left[5 + 4\cos 2\beta\left(1 - \frac{H}{B}(1.154 + 2\cot\beta)\right)\right] \quad (4.18)$$

for $B/H > 3.154$ and $\beta < 45°$. The maximum q_0 is found by maximizing Equation 4.18 with respect to β. For $B/H < 3.154$, q_0 should be taken as $5c$ (lower bound).

Using similar stress fields, Shield (1955) has developed solution for strip, square, and circular footings on Tresca foundations of finite depth. Those solutions are:

$$q_0 = c\left(3.42 + 0.48\frac{H}{B} + \frac{B}{4H}\right) \quad \text{strip footing} \quad (4.19)$$

where $B/H = 5.982 + 5.656n$ and n is an integer depending on the number of discontinuous stress fields;

$$q_0 = c\left(3.42 + \frac{B}{6H} + 3.64\frac{H}{B} - 1.04\frac{H^2}{B^2}\right) \quad \text{square footing}$$

$$(4.20)$$

where $B/H = 5.982 + 5.656n$; and

$$q_0 = c\left(2.72 + \frac{B}{6H} + 6.1\frac{H}{B} - 1.76\frac{H^2}{B^2}\right) \quad \text{circular footing}$$

$$(4.21)$$

where $B/H = 3.154 + 2.828n$.

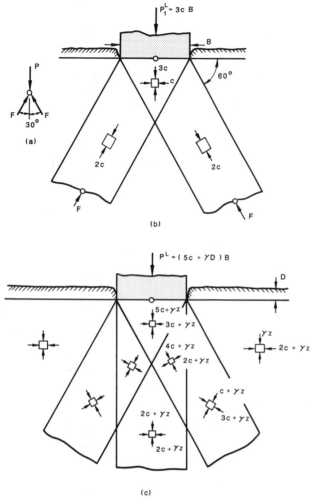

Fig. 4.8 Lower-bound bearing capacity example for Tresca material (complex stress field).

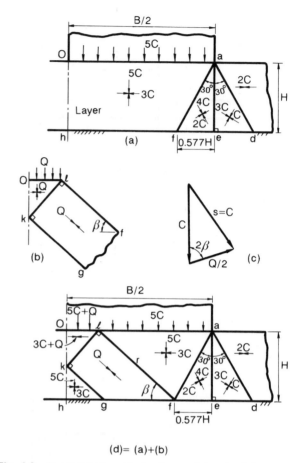

Fig. 4.9 Stress field for strip footing on finite-depth foundation.

4.6 FOOTING DEPTH AND SHAPE AND INCLINED AND ECCENTRIC LOADS

The prior discussion has been concerned mainly with strip footings; however, the majority of foundation design is concerned with footings of finite size. Chen (1975), Meyerhof (1963), and Brinch Hansen (1970) have developed solutions for footings of various shapes. Chen used limit analysis, while Meyerhof and Brinch Hansen used quasiempirical modifications of limit equilibrium and slip-line solutions, respectively. Brinch Hansen, Meyerhof, Vesić (1975), and Murff and Miller (1977a) have addressed the problem of eccentric and inclined loads on foundations. The methods of Meyerhof and Brinch Hansen (as modified by Vesić) are presented here because they are the most general and the most commonly used.

The bearing capacity of foundations subject to eccentric loads is commonly determined by using a reduced footing area located below the eccentric load (Fig. 4.10). The procedures outlined in Figure 4.10 are empirical and based on observations of footing tests. For shapes other than rectangular, the effective foundation area may be determined as that of the equivalent rectangle constructed so that the geometric center coincides with the load center and follows as closely as possible the adjacent contour of the actual base. With reference to Figure 4.10, the equivalent base dimensions for a rectangular footing with eccentric loads are

$$L' = L - 2e_1 \qquad B' = B - 2e_2 \qquad (4.22)$$

Equations 4.22 are empirical relations based on test observations (Meyerhof, 1953). The equivalent area for a circular footing may be evaluated as (see Fig. 4.10)

$$A' = 2S = B'L'$$

$$L' = \left[2S \left(\frac{R+e}{R-e} \right)^{0.5} \right]^{0.5}$$

$$B' = L' \left(\frac{R+e}{R-e} \right)^{0.5} \qquad (4.23)$$

where

$$S = \frac{\pi R^2}{2} - \left[e\sqrt{R^2 - e^2} + R^2 \arcsin (e/R) \right]$$

These area reduction factors for circular and rectangular sections are plotted in Figure 4.11.

Highter and Anders (1985) have provided additional graphical solutions to determine the equivalent areas of rectangular and circular footings subjected to eccentric loads.

The bearing capacity equation used for inclined and eccentric loads has the following form:

$$q_0 = N_c s_c i_c d_c c + N_q s_q i_q d_q q + \frac{\gamma B}{2} N_\gamma s_\gamma i_\gamma d_\gamma \qquad (4.24)$$

where q_0 is the vertical component of the footing load and the factors s, i, and d are the footing shape, load inclination, and

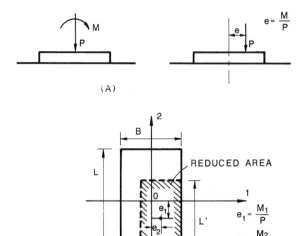

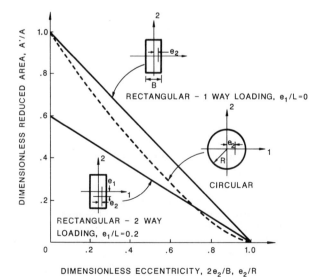

Fig. 4.10 Reduced footing area for eccentric loads. (*API, 1987.*) (A) Equivalent loadings. (B) Reduced area—rectangular footing. (C) Reduced area—circular footing.

Fig. 4.11 Area reduction factors for eccentrically loaded footings. (*API, 1987.*)

footing depth factors, respectively, given in Tables 4.2 and 4.3. An iterative solution is required for inclined loads because the footing load components H and V appear in the factors i. Figure 4.12 shows the geometry for an inclined footing load. The equivalent dimensions B' and L' should be used, in the case of eccentric loads, in the expression of Tables 4.2 and 4.3. The ultimate footing load should be based on the reduced footing dimensions.

TABLE 4.2 MEYERHOF FOOTING DEPTH AND LOAD INCLINATION BEARING CAPACITY MODIFIERS[a]**.**

$$q_0 = N_c s_c i_c d_c c + N_q s_q i_q d_q q + \frac{\gamma B}{2} N_\gamma s_\gamma i_\gamma d_\gamma$$

For $D < B$:

$$d_c = 1 + 0.2 \sqrt{N_\phi}\, \frac{D}{B}$$

$$d_q = d_\gamma = 1 \qquad (\phi = 0°)$$

$$d_q = d_\gamma = 1 + 0.1 \sqrt{N_\phi}\, \frac{D}{B} \quad (\phi > 10°)$$

$$i_c = i_q = (1 - \alpha/90°)^2$$
$$i_\gamma = (1 - \alpha/\phi)^2$$
$$N_\phi = \tan^2(\tfrac{1}{4}\pi + \tfrac{1}{2}\phi)$$

[a] Meyerhof (1963).

TABLE 4.3 BRINCH HANSEN FOOTING DEPTH AND LOAD INCLINATION BEARING CAPACITY MODIFIERS[a]**.**

$$q_0 = N_c s_c i_c d_c c + N_q s_q i_q d_q q + \frac{\gamma B}{2} N_\gamma s_\gamma i_\gamma d_\gamma$$

For $D < B$:

$$d_c = 1 + 0.4 \frac{D}{B} \qquad (\phi = 0°)$$

$$d_c = d_q - \frac{1 - d_q}{N_c \tan \phi} \qquad (\phi > 0°)$$

$$d_q = 1 + 2 \tan \phi (1 - \sin \phi)^2 \frac{D}{B}$$

$$d_\gamma = 1$$

$$i_c = 1 - \frac{mH}{Ac(\pi + 2)} \qquad (\phi = 0°)$$

$$i_c = i_q - \frac{1 - i_q}{N_c \tan \phi} \qquad (\phi > 0°)$$

$$i_q = \left[1 - \frac{H}{V + Ac \cot \phi}\right]^m$$

$$i_\gamma = \left[1 - \frac{H}{V + Ac \cot \phi}\right]^{m+1}$$

$$m = m_L \cos^2 \theta_n + m_B \sin^2 \theta_n$$

$$m_B = \frac{2 + B/L}{1 + B/L} \qquad m_L = \frac{2 + L/B}{1 + L/B}$$

θ_n is the projected direction of load in the plane of the footing, measured from the side of length L

[a] As modified by Vesić (1975).

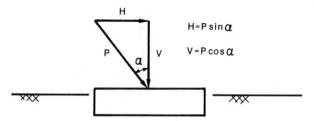

Fig. 4.12 Inclined footing load.

$H = P \sin \alpha$

$V = P \cos \alpha$

Bowles (1982) has suggested that the Brinch Hansen inclination factors *i* should not be used in conjunction with the shape factors *s*. The use of Meyerhof's expressions with Equation 4.24 has been verified by comparison with laboratory experiments (Hanna and Meyerhof, 1981).

4.7 FOOTING SHAPE, DEPTH, AND INCLINATION EFFECTS

The footing size, orientation and position relative to the ground surface have an influence on the bearing capacity. Brinch Hansen (1970) and Vesić (1975) have addressed these problems. The bearing capacity may be determined as

$$q_0 = N_c s_c d_c g_c b_c c + N_q s_q d_q g_q b_q q + \frac{\gamma B}{2} N_\gamma s_\gamma d_\gamma g_\gamma b_\gamma \quad (4.25)$$

where q_0 is the vertical component of the footing load and the factors *s*, *d*, *g*, and *b* are the footing shape, footing depth, ground inclination, and base inclination factors, respectively. The expressions for the factors *s*, *b*, and *g* are given in Tables 4.4 and 4.5. The geometric parameters for inclined footings and ground surfaces are given in Figure 4.13. The equivalent dimensions B' and L' should be used, in the case of eccentric loads, in the expression of Tables 4.4 and 4.5. The ultimate

TABLE 4.4 MEYERHOF AND BRINCH HANSEN FOOTING SHAPE BEARING CAPACITY MODIFIERS.

$$q_0 = N_c s_c i_c d_c c + N_q s_q i_q d_q q + \frac{\gamma B}{2} N_\gamma s_\gamma i_\gamma d_\gamma$$

Meyerhof (Meyerhof, 1963)

$$s_c = 1 + 0.2 N_\phi \frac{B}{L}$$

$$s_q = s_\gamma = 1.0 \qquad (\phi = 0°)$$

$$s_q = s_\gamma = 1 + 0.1 N_\phi \frac{B}{L} \qquad (\phi > 10°)$$

$$N_\phi = \tan^2 (\tfrac{1}{4}\pi + \tfrac{1}{2}\phi)$$

Brinch Hansen (After Vesić, 1975)

$$s_c = 1 + \frac{B}{L} \frac{N_q}{N_c}$$

$$s_q = 1 + \frac{B}{L} \tan \phi$$

$$s_\gamma = 1 - 0.4 \frac{B}{L}$$

For circular footing use $B/L = 1$.

TABLE 4.5 BRINCH HANSEN FOOTING AND GROUND INCLINATION BEARING CAPACITY MODIFIERS[a].

$$q_0 = N_c s_c d_c g_c b_a c + N_q s_q d_q g_q b_q q + \frac{\gamma B}{2} N_\gamma s_\gamma d_\gamma g_\gamma b_\gamma$$

Footing Inclination Factors

$$b_q = b_\gamma = (1 - \alpha \tan \phi)^2$$

$$b_c = 1 - \frac{2\alpha}{\pi + 2} \qquad (\phi = 0°, \ \alpha \text{ in radians})$$

$$b_c = b_q - \frac{1 - b_q}{N_c \tan \phi} \qquad (\phi > 0°)$$

Ground Inclination Factors

$$g_q = g_\gamma = (1 - \tan \omega)^2 \qquad (\phi > 0°)$$

$$g_c = 1 - \frac{2\omega}{\pi + 2} \qquad (\phi = 0°, \ \omega \text{ in radians})$$

$$g_c = g_q - \frac{1 - g_q}{N_c \tan \phi} \qquad (\phi > 0°)$$

Restrictions: $\alpha < 45°$, $\omega < 45°$, $\omega < \phi$.
For ground inclination use $N_\gamma = -2 \sin \omega$.

[a] As modified by Vesić (1975).

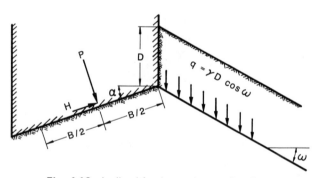

Fig. 4.13 Inclined footing and ground surface.

footing load should be based on the reduced footing dimensions in the case of an eccentric load.

Bowles (1982) has suggested the factors *s*, *d*, *g*, and *b* should not be used in conjunction with the load inclination factors *i*. Vesić (1975) has suggested that for the sloping ground condition and $\phi = 0$ (Tresca material) soils,

$$N_\gamma = -2 \sin \omega \quad (4.26)$$

4.8 NONHOMOGENEOUS FOUNDATIONS AND ANISOTROPIC STRENGTH

The conditions of nonhomogeneous foundations and anisotropic soil strength have been treated by Reddy and Srinivasan (1967) using limit equilibrium procedures and by Chen (1975) using the upper-bound method. These authors examined the common condition of a strip footing supported by a two-layer undrained clay (Tresca material, $\phi = 0$) foundation. Chen also considered the case of a linearly increasing strength with depth. The results of Chen agree well with those of Redding and Srinivasan and are shown in Figure 4.14.

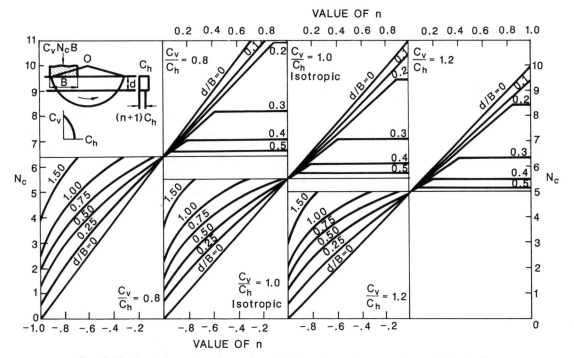

Fig. 4.14 Bearing-capacity factors for cohesive soil on a layered anisotropic foundation.

The bearing-capacity parameter N_c for the case of linearly increasing strength with depth is shown in Figure 4.15. Figures 4.14 and 4.15 also include the effects of anisotropic strength for a Tresca material. The parameters c_v and c_h represent the vertical and horizontal strengths, respectively. The vertical strength c_v, for example, is determined by taking a vertical sample and testing it with the major principal stress in the vertical direction.

Duncan and D'Orazio (1984) have discussed the development of simplified bearing-capacity analyses for layered foundations. Comparisons between experimental and analytical investigations of layered foundations have been reported by Hanna (1981a, 1981b) and Satyanarayana and Garg (1980).

Meyerhof and Hanna (1978) and Hanna and Meyerhof (1980) have presented solutions for layered foundations consisting of a sand overlying a weak clay (Fig. 4.16). The procedure developed by those authors is summarized by the following bearing capacity equation for strip footings:

$$q_{uv} = q_{bv} + \gamma_1 H^2 [1 + 2D(\cos \alpha)/H] K_s i_s \frac{\tan \phi}{B} - \gamma_1 H \leqslant q_{tv}$$

(4.27)

where q_{uv} is the vertical component of the ultimate bearing capacity q_u in the direction of the load; q_{bv} and q_{tv} are the vertical components of the limit loads q_b and q_t on thick beds of the lower and upper soil, respectively; α is the load inclination with respect to vertical; and i_s is the load inclination factor obtained from Figure 4.17. The value of the parameter K_s is obtained by entering the graphs in Figure 4.17 with the appropriate values of ϕ and q_2/q_1, where $q_1 = 0.5\gamma_1 B N_\gamma$ (for homogenous upper sand) and $q_2 = cN_c$ (for homogenous lower clay).

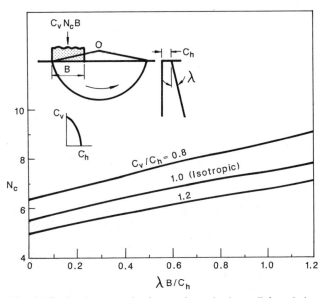

Fig. 4.15 Bearing-capacity factors for cohesive soil foundation with linearly increasing strength.

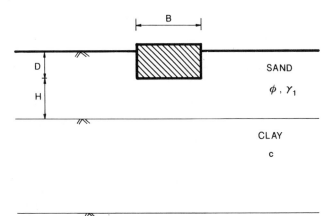

Fig. 4.16 Layered foundation for Meyerhof and Hanna solution.

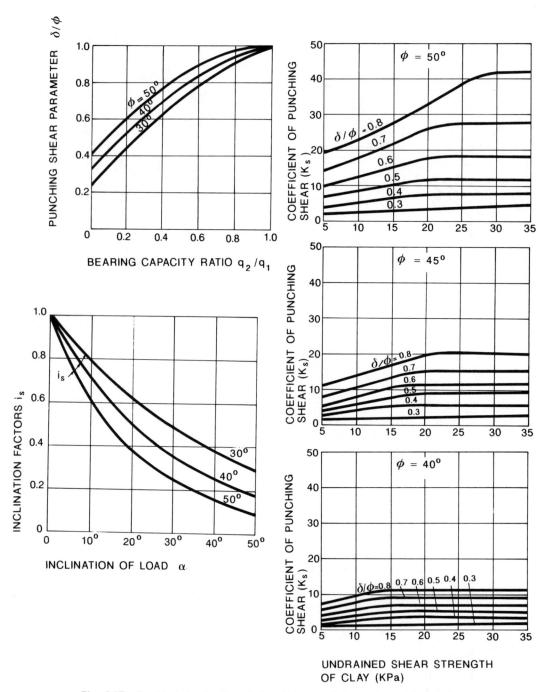

Fig. 4.17 Graphical aids for Meyerhof and Hanna solution to layered foundation.

4.9 INFLUENCE OF GROUNDWATER TABLE

The position of the groundwater table may have a significant effect on the bearing capacity of shallow foundations. Generally the submergence of soils will cause loss of the apparent cohesion due to capillary stresses or weak cementation bonds. At the same time the effective unit weight of submerged soils will be reduced to about one-half the weight of the same soils above water. Thus, through submergence, all three terms involved in the bearing capacity equation are reduced. For this reason, it is essential that the bearing capacity analysis be made assuming the highest possible groundwater level at the particular level for the expected lifetime of the structure. The assessment of this

level must be made taking into consideration the probability of temporary high levels that could be expected in some locations during heavy rainstorms or floods, although they may not appear in the official records.

If the highest groundwater level is within the depth $z_w < B$ below the foundation level (Fig. 4.18) the effective weight γ of the soil below the foundation base should be taken as

$$\gamma = \gamma' + \frac{z_w}{B}(\gamma_m - \gamma') \tag{4.28}$$

where γ' is the submerged unit weight and γ_m is the moist unit weight of soil, corresponding to the minimum moisture content of the soil above the water table (Meyerhof, 1953). If the water

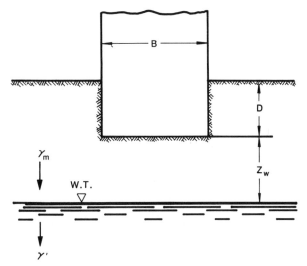

Fig. 4.18 Influence of groundwater table.

table is permanently below depth $z_w = B$, γ should be taken equal to γ_m. For a water table at or above the level of the foundation base, the submerged unit weight γ' should be used.

4.10 COMMENTS ON BEARING CAPACITY SOLUTIONS

It is worthwhile to review again the assumptions involved in developing the bearing capacity solutions presented here. With respect to Figure 4.2, it is seen that the bearing capacity solutions are not complete solutions, in that they do not satisfy all the requirements for the solution of a boundary-value problem. The solutions usually satisfy either equilibrium in a limited form (e.g., limit equilibrium, lower-bound, or slip-line methods) or kinematic compatibility (e.g., upper-bound or slip-line methods). The solutions do not satisfy the constitutive relations and, therefore, provide no information on deformations. The solutions do not consider the effects of changing geometry, boundary conditions, or load direction.

The bearing capacity equations (Eqs. 4.24 and 4.25) are modifications of Equation 4.1, which was formulated for strip footings subjected to vertical center loads. The expressions for the bearing capacity modifiers i, s, d, g, and b are largely empirical in nature. Therefore, it is expected that these factors may not satisfactorily account for all foundation and soil behavior. For instance, the behavior of a footing under an inclined load having a positive eccentricity should be expected to be different than the same footing subjected to an inclined load with a negative eccentricity (Vivatrat and Watt, 1983). However, the modified bearing capacity equations (Eqs. 4.24 and 4.25) recognize no such differences.

The performance of actual foundations may be influenced by the foundation size. This may be true for several reasons. First, as the foundation size increases, so does the possibility that nonhomogeneous soils or weak seams may occur within the zone of influence of the foundation. From an analysis standpoint, the Coulomb failure criterion (from which the bearing capacity factors N are derived) is fairly simple. While it performs remarkably well, it does ignore the influence of the intermediate principal stress, which is known to affect the actual soil failure stress state. This may explain the frequent use of a modified friction angle ϕ (Eq. 4.17) in bearing-capacity calculations.

With respect to horizontal footings on the surface of Tresca foundations ($\phi = 0$) subjected to vertical center loads, Chen

(1975) has noted that the bearing capacity of square and circular footings is always greater than that of a strip footing.

The accurate determination of bearing capacity involves minimizing upper-bound solutions, maximizing lower-bound solutions, or minimizing limit equilibrium solutions. The effort of minimizing or maximizing the solution involves evaluating different kinematic or equilibrium conditions. Zhang and Chen (1987) have used variational calculus to evaluate the stability of slopes using the upper-bound method. Garber and Baker (1977) have used the variational method to determine bearing capacity. The variational procedure generally allows one to determine the best failure mechanism and the associated normal stress distribution along the slip surface.

The effects of foundation roughness (footing–soil interface friction) may increase the bearing capacity. The kinematic mechanism of Figure 4.5b allows no relative motion between the footing and soil and thus may be considered an upper-bound solution for the finite-friction case. The stress field shown in Figure 4.8c has no base-shear components and thus the solution may be considered as a lower bound for the case of finite friction. Apparently rather small values of the interface friction δ (10° to 15°) are required to obtain rough footing effects. The Hill mechanism shown in Figure 4.6 is representative of the deformation mode that has been observed for smooth footings (Ko and Davidson, 1973) and thus may provide conservative bearing capacity solutions.

4.11 SLIDING STABILITY OF GRAVITY STRUCTURES

Offshore production of oil reserves has resulted in the construction and installation of approximately 20 gravity structures in ocean water depths ranging from 100 to 500 ft (30 to 150 m). Permanent structures have been installed in the North Sea, while mobile structures are in use in the U.S. and Canadian Beaufort Sea. These structures are unique in that the transverse loads control the foundation design. The design wave for the Ekofisk oil-storage tank, the first gravity structure in the North Sea, results in a transverse load of 173 300 kips (770 MN) compared to its effective weight of 419 000 kips (1860 MN) (Bjerrum, 1973).

One method of insuring adequate sliding resistance is through the use of "shear skirts" on a structure base, which penetrate the foundation surface. Shear skirts are often proposed because of the likelihood of nonuniform foundation contact or the existence of weak surface sediments. Murff and Miller (1977b) have developed an upper-bound solution for evaluating the capacity of such a system. Figure 4.19 shows the configuration

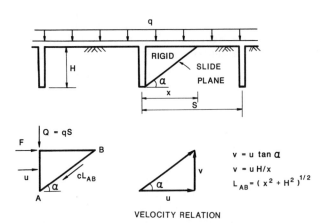

VELOCITY RELATION

Fig. 4.19 Shear skirt geometry.

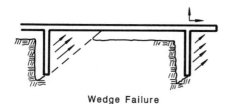

Wedge Failure

Flow-Under Failure

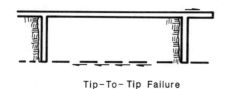

Tip-To-Tip Failure

Fig. 4.20 Shear skirt resistance modes.

of one such system. The sliding resistance is generated through passive wedge bearing on the shear skirt. The amount of resistance is determined by the structure surcharge q, shear skirt height H, and shear skirt spacing S.

The upper-bound solution is obtained by equating the internal and external rates of energy dissipation and minimizing the relationship to obtain the least upper bound. The work components are

$$W_{EXT} = Fu - Qv = Fu - Q\frac{H}{x}u$$

$$W_{INT} = cL_{AB}\frac{u}{\cos\alpha} = c(H^2 + x^2)\frac{u}{x} \qquad (4.29)$$

Solving for the skirt load F (force/unit length) and minimizing with respect to x gives

$$F = 2c\left(H^2 + \frac{QH}{c}\right)^{1/2} \qquad (4.30)$$

Other modes of failure are also possible, as shown in Figure 4.20. In general the three modes that must be considered are, in order of increasing resistance: Murff wedge, flow-under, and tip-to-tip. The resistance for the latter two modes are (Fig. 4.20):

$$F = 8Hc \quad \text{flow-under} \qquad (4.31)$$

$$F = Sc \quad \text{tip-to-tip} \qquad (4.32)$$

4.12 CHOICE OF SAFETY FACTOR

The analyses described in this chapter are all made with the purpose of assessing the magnitude of the ultimate load P or the ultimate pressure q_0 at which the foundation may experience a bearing capacity failure. As mentioned in the introduction, foundations are designed so as to possess an adequate margin of safety against this type of failure.

TABLE 4.6 MINIMUM SAFETY FACTORS FOR DESIGN OF SHALLOW FOUNDATIONS[a].

Preliminary note. The selection of safety factors for design cannot be made properly without assessing the degree of reliability of all other parameters that enter into design, such as design loads, strength and deformation characteristics of the soil mass, etc. In view of this, each case is to be considered separately by the designer. The following table may be used as a guide for permanent structures in reasonably homogeneous soil conditions.

Category	Typical Structures	Characteristics of the Category	Soil Exploration	
			Complete	Limited
A	Railway bridges Warehouses Blast furnaces Hydraulic Retaining walls Silos	Maximum design load likely to occur often; consequences of failure disastrous	3.0	4.0
B	Highway bridges Light industrial and public buildings	Maximum design load may occur occasionally; consequences of failure serious	2.5	3.5
C	Apartment and office buildings	Maximum design load unlikely to occur	2.0	3.0

[a] Vesić (1975).

Remarks:
1. For temporary structures, these factors can be reduced to 75 percent of the above values. However, in no case should safety factors lower than 2.0 be used.
2. For exceptionally tall buildings, such as chimneys and towers, or generally whenever progressive bearing capacity failure may be feared, these factors should be increased by 20 to 50 percent.
3. The possibility of flooding of foundation soil and/or removal of existing overburden by scour or excavation should be given adequate consideration.
4. It is advisable to check both the short-term (end-of-construction) and long-term stability, unless one of the two conditions is clearly less favorable.
5. It is understood that all foundations will be analyzed also with respect to maximum tolerable total and differential settlement. If settlement governs the design, higher safety factors must be used.

Vesić (1975) has suggested the total factors of safety F_s in Table 4.6 on the basis of classification of structure, knowledge of foundation conditions, and the consequences of failure. Lower factors of safety may be applied to temporary structures, while higher factors of safety may apply to structures that will regularly experience their maximum design load. Meyerhof (1970, 1984) discussed the total factors of safety F_s given in Table 4.7 and the use of the load and resistance factors (partial

TABLE 4.7 VALUES OF MINIMUM TOTAL SAFETY FACTORS[a].

Failure Type	Item	Safety Factor, F_s
Shearing	Earthworks	1.3–1.5
	Earth-retaining structures, excavations	1.5–2
	Foundations	2–3
Seepage	Uplift, heave	1.5–2
	Exit gradient, piping	2–3

[a] Meyerhof (1984).

TABLE 4.8 VALUES OF MINIMUM PARTIAL FACTORS[a].

Category	Item	Load Factor	Resistance Factor
Loads	Dead loads	(f_d) 1.25 (0.85)	
	Live loads, wind or earthquake	(f_1) 1.5	
	Water pressures	(f_u) 1.25 (0.85)	
Shear strength	Cohesion (c) (stability; earth pressures)		(f_c) 0.65
	Cohesion (c) (foundations)		(f_c) 0.5
	Friction $(\tan \phi)$		(f_ϕ) 0.8

Note: Load factors given in parentheses apply to dead loads and water pressures when their effects are beneficial, as for dead loads resisting instability by sliding, overturning or uplift.

[a] Meyerhof (1984).

factors) given in Table 4.8. The higher values in Table 4.7 are applied to the normal loads and service conditions, while the lower values are applied to the maximum loads and worst environmental conditions.

The basic philosophy using total factors of safety is that the foundation should be capable of resisting a load F_s times greater than the design load. The load and resistance factor design (LRFD) method applies separate or partial factors to the loads and soil resistance. The load factors are provided mainly for variability and pattern of loading, which differ for dead loads, live loads, environmental loads, and water pressures. The resistance factors consider the variability and uncertainty of assessment of soil resistance, which differ for the cohesive and friction components. Thus, the factored shear strength of soil at the ultimate limit state may be expressed as

$$\tau = f_c c + \sigma_n f_\phi \tan \phi \qquad (4.33)$$

for the Coulomb criterion. The factors f_c and f_ϕ are the resistance factors for the cohesive and friction components, respectively. It is evident from Equation 4.33 that the total factor of safety obtained will depend on the relative contributions of the cohesive and friction components.

Whitman (1984) has recently reviewed the application of the related topic of risk analysis to geotechnical engineering.

4.13 EXAMPLE PROBLEMS

EXAMPLE 4.1

A rectangular footing (Fig. 4.21) 28 ft wide and 84 ft long is to be placed at a depth of 10 ft in a deep stratum of soft, saturated clay (bulk unit weight 105 lb/ft³). The water table is at 8 ft below ground surface. Find the ultimate bearing capacity under the following two conditions:

a. assuming that the rate of application of dead and live loads is fast in comparison with the rate of dissipation of excess pore-water pressures caused by loads, so that undrained conditions prevail at failure;

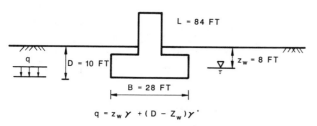

$$q = z_w \gamma + (D - z_w)\gamma'$$

Fig. 4.21 Footing geometry.

b. assuming, as the other extreme, that the rate of loading is slow enough that no excess pore-water pressures are introduced in the foundation soil.

The strength parameters of the soil, obtained from unconsolidated, undrained tests are $c_u = 0.22$ ton/ft², $\phi_u = 0$. Consolidated, drained tests give $c_d = 0.04$ ton/ft², $\phi_d = 23°$.

CONDITION (a)

Submerged unit weight of soil: $\gamma' = 105 - 62 = 43$ lb/ft³.
Overburden stress: $q = [(8)(105) + (2)(43)]/(2000)$
$\qquad\qquad = 0.463$ ton/ft².
Bearing capacity factors (Table 4.1): $N_c = 5.14$; $N_q = 1$;
$\qquad\qquad\qquad\qquad N_\gamma = 0$.
Shape factors (Table 4.4, Brinch Hansen):

$$s_c = 1 + \frac{B}{L}\frac{N_q}{N_c} = 1 + (1/3)(0.19) = 1.065$$

$$s_q = 1.00$$

Ultimate bearing pressure (Eq. 4.24):

$$q_0 = cN_c s_c + qN_q s_q$$

$$q_0 = (0.22)(5.14)(1.065) + (0.463)(1)(1.00)$$

$$= 1.21 + 0.46 = 1.67 \text{ ton/ft}^2$$

CONDITION (b)

Bearing capacity factors: $N_c = 18.05$; $N_q = 8.66$; $N_\gamma = 9.70$.
Shape factors:

$$s_c = 1 + \frac{B}{L}\frac{N_q}{N_c} = 1 + (1/3)(0.48) = 1.16$$

$$s_q = 1 + \frac{B}{L}\tan\phi = 1 + (1/3)(0.42) = 1.14$$

$$s_\gamma = 1 - 0.4\frac{B}{L} = 1 - (0.4)(1/3) = 0.87$$

Ultimate bearing pressure:

$$q_0 = cN_c s_c + qN_q s_q + \frac{\gamma B}{2}N_\gamma s_\gamma$$

$$q_0 = (0.04)(18.05)(1.16) + (0.463)(8.66)(1.14)$$

$$+ (1/2)(43)(28)(9.7)(0.87)/(2000)$$

$$= 0.72 + 4.57 + 2.54 = 7.83 \text{ ton/ft}^2$$

EXAMPLE 4.2

Solve the problem described in Example 4.1 if the footing is placed at the same depth (10 ft) in a deep stratum of medium dense sand. Assume for sand a saturated unit weight of 118 lb/ft^3 and an average moist unit weight above the water table of 100 lb/ft^3. Drained triaxial tests on sand samples show that the angle ϕ of shearing resistance of sand varies with mean normal stress σ_0 according to the equation

$$\phi = \phi_1 - (5.5°) \log_{10}(\sigma_0/\sigma_1)$$

where $\phi_1 = 38°$ is the angle of shearing resistance at a mean normal stress $\sigma_1 = 1$ ton/ft^2.

Submerged unit weight of sand: $\gamma' = 118 - 62 = 56$ lb/ft^3. Overburden stress: $q = [(8)(100) + (2)(56)]/(2000)$
$$= 0.456 \text{ ton/ft}^2.$$

To find the mean normal stress a preliminary estimate of bearing capacity is needed. It is assumed for this preliminary analysis that $\phi = 34°$.

Bearing capacity factors: $N_q = 29.44$; $N_\gamma = 52.18$.
Shape factors:

$$s_q = 1 + \frac{B}{L} \tan \phi = 1 + (1/3)(0.67) = 1.22$$

$$s_\gamma = 1 - 0.4 \frac{B}{L} = 1 - 0.4(1/3) = 0.87$$

Ultimate bearing pressure (Eq. 4.24):

$$q_0 = qN_qs_q + \frac{\gamma B}{2} N_\gamma s_\gamma$$

$$q_0 = (0.456)(29.44)(1.22)$$
$$+ (1/2)(56)(28)(52.18)(0.87)/(2000)$$
$$= 16.4 + 17.8 = 34.2 \text{ ton/ft}^2$$

Mean normal stress along the slip surface (De Beer, 1965):

$$\sigma_0 = \tfrac{1}{4}(q_0 + 3q)(1 - \sin \phi)$$
$$\sigma_0 = \tfrac{1}{4}[34.2 + (3)(0.456)](1 - 0.559) = 3.92 \text{ ton/ft}^2$$

Representative angle of shearing resistance:

$$\phi = 38° - (5.5°)(0.593) = 34.7 \cong 35°$$

The analysis is now repeated with $\phi = 35°$:

$$N_q = 33.3 \quad N_\gamma = 62.3$$

$$s_q = 1 + \frac{B}{L} \tan \phi = 1 + (1/3)(0.7) = 1.23$$

$$s_\gamma = 0.87$$

$$q_0 = qN_qs_q + \frac{\gamma B}{2} N_\gamma s_\gamma$$

$$q_0 = (0.456)(33.3)(1.23)$$
$$+ (1/2)(56)(28)(62.3)(0.87)/(2000)$$
$$= 18.7 + 21.2 = 39.9 \text{ ton/ft}^2$$

In view of small change in mean normal stress from the previously found value, this answer is retained.

EXAMPLE 4.3

For the footing discussed in Example 4.1, find the ultimate bearing capacity in conditions (a) and (b) if the footing reaction acts 3 ft off-center in the direction of the short side B ($e_B = 3$ ft) and if the inclination of the reaction is in the same direction. Assume that the horizontal component of the reaction is equal to half of the ultimate value given by

$$H = V \tan \phi + Ac$$

CONDITION (a)

Effective width of the footing: $B' = 28 - (2)(3) = 22$ ft.
Horizontal reaction: $H = 0.5H_{max} = (0.5)(22)(84)(0.22)$
$$= 203.3 \text{ ton.}$$
Exponent m_B (Table 4.3): $m_B = [2 + (1/3)]/[1 + (1/3)]$
$$= 1.75.$$

Inclination factors (Table 4.3):

$$i_q = 1$$

$$i_c = 1 - [(1.75)(203.3)/(406.6)(5.14)] = 0.83$$

Ultimate bearing pressure (see calculations from Example 4.1):

$$q_0 = cN_cs_ci_c + qN_qs_qi_q$$

$$q_0 = (1.21)(0.83) + (0.46)(1)$$

$$= 1.00 + 0.46 = 1.46 \text{ ton/ft}^2$$

CONDITION (b)

Assume $\tan \delta = \tan \phi_d = 0.42$, $c_a = 0$.

$$H = (0.5)(0.42)V = 0.21V$$
$$i_q = (1 - 0.21)^{1.75} = 0.66 \quad \text{(Table 4.3)}$$
$$i_c = 0.66 - [(1 - 0.66)/(18.05)(0.42)] = 0.62$$
$$i_\gamma = (1 - 0.21)^{2.75} = 0.52$$

$$q_0 = cN_cs_ci_c + qN_qs_qi_q + \frac{\gamma B}{2} N_\gamma s_\gamma i_\gamma$$

$$q_0 = (0.72)(0.62) + (4.57)(0.66) + (2.54)(0.52)$$
$$= 0.45 + 3.02 + 1.32 = 4.79 \text{ ton/ft}^2$$

EXAMPLE 4.4

For the same footing find the ultimate bearing capacity if the reaction acts 6.5 ft off-center in the direction of the long side, and if the inclination is in the same direction. Assume that the horizontal component is equal to the ultimate value given by

$$H = V \tan \phi + Ac$$

CONDITION (a)

Effective length of the footing: $L' = 84 - (2)(6.5) = 71$ ft.
$H = H_{max} = Ac = (28)(71)(0.22) = 437.4$ ton.
Exponent m_L (Table 4.3): $m_L = (2 + 3)/(1 + 3) = 1.25$.

$$i_c = 1 - [(1.25)(437.4)/(437.4)(5.14)] = 0.76$$

$$i_q = 1$$

$$q_0 = cN_cs_ci_c + qN_qs_qi_q$$

$$q_0 = (1.21)(0.76) + (0.46)(1) = 1.38 \text{ ton/ft}^2$$

CONDITION (b)

$$H = 0.42V$$
$$i_q = (1 - 0.42)^{1.25} = 0.51$$
$$i_c = 0.51 - [(1 - 0.51)/(18.05)(0.42)] = 0.45$$
$$i_\gamma = (1 - 0.42)^{2.25} = 0.29$$

$$q_0 = cN_cs_ci_c + qN_qs_qi_q + \frac{\gamma B}{2}N_\gamma s_\gamma i_\gamma$$
$$q_0 = (0.72)(0.45) + (4.57)(0.51) + (2.54)(0.29)$$
$$= 0.32 + 2.33 + 0.74 = 3.39 \text{ ton/ft}^2$$

EXAMPLE 4.5

For the footing discussed in preceding examples, find the ultimate bearing capacity if the footing base is tilted 1 (vertical) to 4 (horizontal). Assume, as in Example 4.3, that the reaction is 3 ft off-center in the direction of the short side B, inclined in the same direction, with a horizontal reaction equal to one-half the ultimate value from

$$H = V \tan \phi + Ac$$

CONDITION (a)

Angle of base tilt: $\alpha = \tan^{-1}(1/4) = 0.245$.
Base tilt factors (Table 4.5):

$$b_c = 1 - (2)(0.245)/(3.14 + 2) = 0.90$$
$$b_q = 1$$

Ultimate bearing pressure (see calculations from Example 4.3):

$$q_0 = cN_cs_cb_c + qN_qs_qb_q$$
$$q_0 = (1.00)(0.90) + 0.46 = 0.90 + 0.46 = 1.36 \text{ ton/ft}^2$$

CONDITION (b)

Base tilt factors (Table 4.5)

$$b_q = b_\gamma = [1 - (0.245)(0.42)]^2 = 0.80$$
$$b_c = 0.80 - [(1 - 0.80)/(18.05)(0.42)] = 0.77$$

$$q_0 = cN_cs_cb_c + qN_qs_qb_q + \frac{\gamma B}{2}N_\gamma s_\gamma b_\gamma$$
$$q_0 = (0.45)(0.77) + (3.02)(0.80) + (1.32)(0.80)$$
$$= 0.35 + 2.42 + 1.06 = 3.83 \text{ ton/ft}^2$$

EXAMPLE 4.6

For the footing discussed in Example 4.1, find the ultimate bearing capacity if the ground slopes 5 (horizontal) to 1 (vertical). The load is assumed to remain central and vertical.

CONDITION (a)

Angle of ground slope:

$$\omega = \tan^{-1}(1/5) = 0.20 = 11.3°$$
$$\sin \omega = 0.19 \quad \cos \omega = 0.98$$

Ground slope factor (Table 4.5):

$$g_c = 1 - [(2)(0.20)/(3.14 + 2)] = 0.92$$

Bearing capacity factor (Eq. 4.26):

$$N_\gamma = -2 \sin \omega = -0.38$$

Ultimate bearing pressure (see calculations from Example 4.1):

$$q_0 = cN_cs_cg_c + qN_qs_q \cos \omega + \frac{\gamma B}{2}N_\gamma s_\gamma$$
$$q_0 = (1.21)(0.92) + (0.46)(0.98)$$
$$- [(1/2)(43)(28)(0.38)(0.87)]/(2000)$$
$$= 1.02 + 0.45 - 0.10 = 1.37 \text{ ton/ft}^2$$

CONDITION (b)

Ground slope factors (Table 4.5):

$$g_q = g_\gamma = (1 - 0.20)^2 = 0.64$$
$$g_c = 0.64 - (1 - 0.64)/[(18.05)(0.42)] = 0.59$$

$$q_0 = cN_cs_cg_c + qN_qs_qg_q + \frac{\gamma B}{2}N_\gamma s_\gamma g_\gamma$$
$$q_0 = (0.72)(0.59) + (4.57)(0.98)(0.64) + (2.54)(0.64)$$
$$= 0.42 + 2.87 + 1.62 = 4.91 \text{ ton/ft}^2$$

EXAMPLE 4.7

For a layered footing consisting of a sand layer ($\phi = 45°$) over a weak clay ($c = 20$ kPa), compute the bearing capacity if the footing is loaded vertically. Use the following dimensions for the strip footing on a saturated footing (see Fig. 4.16):

$$B = \tfrac{1}{3}\text{ m} \quad D = \tfrac{1}{3}\text{ m} \quad H = \tfrac{2}{3}\text{ m}$$

and

$$\gamma_1 = 1.9 \text{ g/cm}^3$$

Using Meyerhof and Hanna's procedure,

$$q_1 = 0.5\gamma BN_\gamma$$
$$= 0.5(1.9 - 1)(9.8)(0.33)(382) = 556$$
$$q_2 = cN_c = 20(5.14) = 103 \text{ kPa}$$
$$\frac{q_2}{q_1} = 0.185$$

From Figure 4.17, $\delta/\phi = 0.55$ and $K_s = 10$. Then

$$q_u = q_b + \gamma H^2\left(1 + \frac{2D}{H}\right)K_s\frac{\tan \phi}{B} - \gamma H$$
$$= 103 + (1.9 - 1)(9.8)(0.66)^2[1 + 2(0.33)/0.66]$$
$$\times (10)(\tan 45°)/0.33 - (1.9 - 1)(9.8)(0.66)$$
$$= 103 + 231 - 6 = 328 \text{ kPa}$$

4.14 NUMERICAL EVALUATION OF BEARING CAPACITY

The classical methods of determining bearing capacity have proven very useful. However, under some circumstances it is necessary to resort to alternative procedures for analysis. This may be true when a knowledge of the foundation deformations is required, or when classical limit solutions do not exist for the problem at hand. At the present time, sufficient experience exists with the constitutive modeling of soils that accurate predictions of foundation performance may be obtained for a wide variety of problems. This experience indicates that the greatest confidence is in the solutions of problems involving monotonic loading. Some problems arise when attempting to analyze cyclic loading problems involving plastic action in both loading and reversed loading.

If we limit ourselves to trying to determine accurately the bearing capacity of footings and do not necessarily require that accurate deformation predictions be determined, a number of simple plasticity models exist that are very useful for finite-element applications. It should also be understood that a number of models exist (plasticity-based as well as other formulations) that will give acceptable limit loads as well as deformation predictions. However, these models generally require a greater effort (being computationally expensive and having more model parameters to determine) to use and are not widely available. In the following, we will discuss some computational requirements for finite-element solutions and display the use of two material models in obtaining collapse loads for foundation problems.

The correct solution of boundary-value problems involving the prediction of footing bearing capacity requires that accurate numerical procedures be adopted in discretizing the foundation, solving the equilibrium equations, and implementing the soil constitutive model. Detailed discussions on the first two topics are given by Bathe (1982) and Kardestuncer and Norrie (1987). Chen (1982) and Chen and Baladi (1985) have discussed the latter topic. Boundary-value problems involving large deformation inelastic analysis are discussed in a recent book by Chen and Mizuno (1990). Despite the relative complexities of these topics, there are some simple rules of thumb that may simplify the analysis process.

1. With regard to finite-element type selection, plasticity analyses generally require the use of higher-order elements. The 8-node element (Fig. 4.22) with reduced integration (2×2) has been shown to be very efficient for plane-strain problems.
2. Nonlinear problems require solution iterations to satisfy equilibrium and convergence requirements. One method of assessing the acceptability of a solution is by comparing the norm of the incremental displacement vector R to the norm of the total displacement vector D in the manner

$$\frac{[R^T R]^{0.5}}{[D^T D]^{0.5}} \leqq \text{TOL} \qquad (4.34)$$

where TOL is in some tolerance. The incremental displacement R is that due to the equilibrium iteration.

Experience has shown that satisfactory results are obtained for $0.001 < \text{TOL} < 0.01$.

3. The selection of a material model should be made in recognition of the problem to be solved, the material (sand or clay), and the mode of soil response (drained or undrained). The simplest model to satisfy the selection criteria should be adopted.

Chen and McCarron (1983) presented a brief comparison of the effects of element selection, equilibrium tolerance, and material model selection on bearing-capacity solutions; Figure 4.23 presents the results. The finite-element model in Figure 4.22 was used. That study concluded that computationally efficient and accurate solutions were obtained using the 8-node (parabolic) element with reduced integration (2×2). It was also observed that the 4-node (linear) element results in poor solutions for plasticity problems. Constant-strain triangle (CST) elements also provide good solutions. These conclusions are in agreement with the general observations of many researchers.

The two material models to be discussed here are the Drucker–Prager model (Chen, 1982) and the CAP model (Chen, 1982; Chen and Baladi, 1985). The present version of the CAP model involves simply the addition of an elliptical cap to the Drucker–Prager model. Figure 4.24 shows the cap model in the stress space.

The Drucker–Prager model represents a cone in the stress base with a circular trace in the deviatoric plane. Mathematically, it may be described as

$$f = \alpha I_1 - J_2^{1/2} + \kappa = 0 \qquad (4.35)$$

where α and κ are material parameters. The stress invariants I_1 and J_2 are given by

$$I_1 = \sigma_1 + \sigma_2 + \sigma_3 \qquad (4.36)$$

$$J_2 = \tfrac{1}{6}[(\sigma_1 - \sigma_2)^2 + (\sigma_2 - \sigma_3)^2 + (\sigma_3 - \sigma_1)^2] \qquad (4.37)$$

J_2 is related to the octahedral shear stress in the form

$$\tau_{\text{oct}} = (\tfrac{2}{3}J_2)^{1/2} \qquad (4.38)$$

Thus, Equation 4.35 represents the relationship between shear strength and effective mean pressure.

The cross sections of the Drucker–Prager and Coulomb models in the deviatoric plane are shown in Figure 4.25. Since the Drucker–Prager model has a circular cross section in the deviatoric plane, it cannot accurately predict both triaxial compression and extension strengths of soils with a single set of material parameters. Matching conditions (Chen, 1982) for the Drucker–Prager model with the Coulomb condition are given in Table 4.9.

The CAP model used here is a simple extension of the Drucker–Prager model obtained by adding a work-hardening elliptical cap. The elliptical cap is described mathematically as

$$f = (I_1 - l)^2 + R^2 J_2 - (X - l)^2 = 0 \qquad (4.39)$$

where l is the position of the intersection of the cap and the Drucker–Prager surface, R is the ratio of the horizontal to the vertical axis of the elliptical cap, and X is the position of the cap intersection with the hydrostatic axes.

The cap position X is related to the amount of plastic volumetric strain (compaction) in the form

$$\varepsilon_{\text{vol}}^p = W(e^{-DX} - 1) \qquad (4.40)$$

where W and D are material parameters.

The bearing-capacity solutions for a strip footing are presented in Figure 4.23 for the Drucker–Prager and CAP models. The Drucker–Prager model parameters were obtained by using the plane-strain matching conditions with the Coulomb criteria (Table 4.9). Both of the models give the same failure

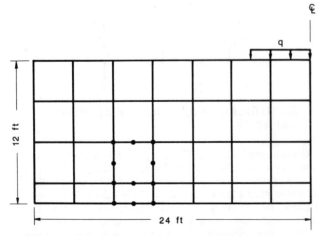

Fig. 4.22 Foundation model for footing problem.

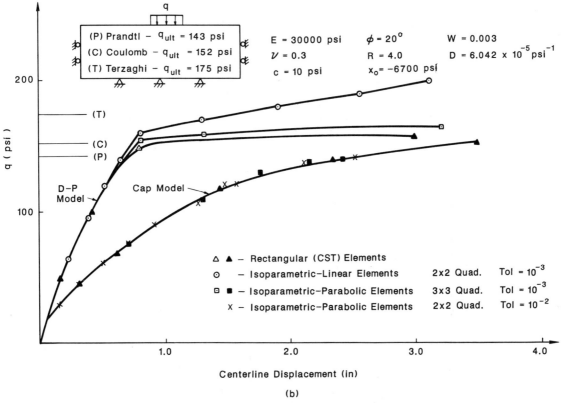

Fig. 4.23 Footing response for Drucker–Prager and Cap models.

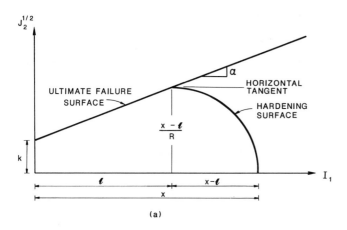

(a)

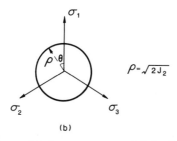

Fig. 4.24 Plasticity models in stress space: (a) $I_1 – J_2^{1/2}$ plane; (b) deviatoric plane.

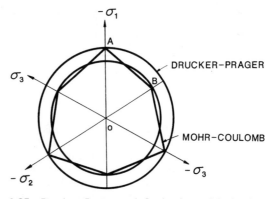

Fig. 4.25 Drucker–Prager and Coulomb models in deviatoric plane.

TABLE 4.9 MATCHING CONDITIONS FOR DRUCKER–PRAGER AND COULOMB MODELS[a].

Condition	α	κ
Triaxial compression	$\dfrac{2\sin\phi}{\sqrt{3}\,(3-\sin\phi)}$	$\dfrac{6c\cos\phi}{\sqrt{3}\,(3-\sin\phi)}$
Triaxial extension	$\dfrac{2\sin\phi}{\sqrt{3}\,(3+\sin\phi)}$	$\dfrac{6c\cos\phi}{\sqrt{3}\,(3+\sin\phi)}$
Plane strain	$\dfrac{2\sin\phi}{[9+12\tan^2\phi]^{0.5}}$	$\dfrac{6c\cos\phi}{[9+12\tan^2\phi]^{0.5}}$

[a] Chen (1982).

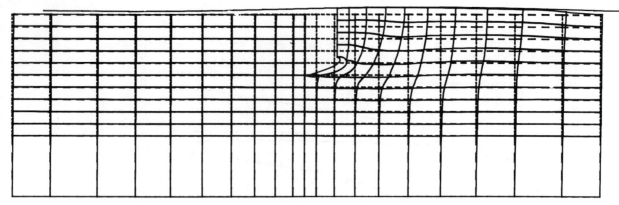

Fig. 4.26 Deformed finite-element model for shear skirt: $H = 5$ ft; $S = 50$ ft; $c = 1$ ksf; $\phi = 0$.

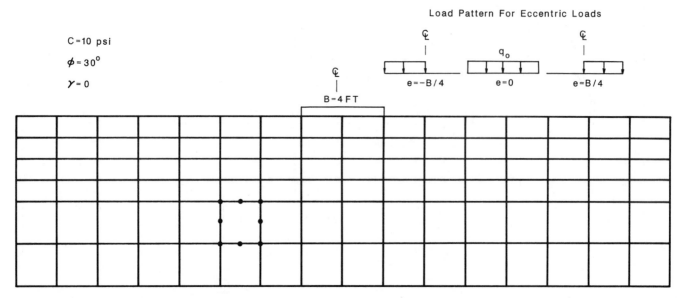

Fig. 4.27 Foundation model for eccentric and inclined loads.

load, as they should since the Drucker–Prager surface controls the shear strength for both models. The CAP model gives a softer response owing to the action of plastic strains as the stress paths strike the cap surface. The cap surface attempts to simulate the behavior of a normally consolidated specimen under virgin loading. The analyses in Figure 4.23 represent drained loading conditions. McCarron and Chen (1987) have simulated undrained conditions with the CAP model.

Figure 4.26 shows the deformed mesh for a shear skirt analysis (Section 4.11). The soil was modeled using the Drucker–Prager model with $\alpha = 0$ to simulate undrained ($\phi = 0$) conditions using a total stress analysis. The finite-element solution for the transverse sliding resistance was 29 kip/ft (39 kN/m) of shear skirt length. The Murff solution (Eq. 4.29) gives a sliding resistance of 32 kip/ft (43 kN/m).

The final example of a material model application is again with the Drucker–Prager model. Bearing-capacity solutions are obtained for the footing shown in Figure 4.27. The effects of eccentric and inclined loads are investigated. The results are compared in Table 4.10 with the limit solutions. The finite-element solutions compare quite well with the limit solutions. The largest difference in the solutions is for the inclined load with $e = B/4$. The finite-element solution gives a much larger bearing capacity, since a larger contact area was maintained

TABLE 4.10 COMPARISON OF FINITE-ELEMENT AND LIMIT SOLUTIONS.

Methods	q_0 (psf)				
	Center Vertical	$e = B/4$ Vertical	$e = -B/4$ Inclined	$e = 0$ Inclined	$e = B/4$ Inclined
Limit	49 490	46 440	—	34 060	32 230
Finite-element	51 580	49 680	50 330	32 830	37 370

Notes: Inclined load limit solution with Brinch Hansen procedure ($H = 0.2V$). Limited solutions not applicable to negative eccentricity with inclined load.

in the model ($3B/4$) than is accounted for in the limit solution ($B/2$).

REFERENCES

API (1987), API-RP-2A, *Recommended Practice for Planning, Design and Constructing Fixed Offshore Platforms*, American Petroleum Institute, Washington, D.C.

Andersen, K. H. (1972), *Bearing Capacity of Shallow Foundations on*

Cohesionless Soils, Internal Report 51404-1, Norwegian Geotechnical Institute.

Bathe, K.-J. (1982), *Finite Element Procedures in Engineering Analysis*, Prentice-Hall, Inc., Englewood Cliffs, N.J.

Bjerrum, L. (1973), Geotechnical problems involved in foundations of structures in the North Sea, *Geotechnique*, **23**, No. 3, pp. 319–358.

Bowles, J. E. (1982), *Foundation Analysis and Design*, McGraw-Hill Book Co., Inc., New York, N.Y.

Brinch Hansen, J. (1961), *A General Formula for Bearing Capacity*, Bulletin No. 11, Danish Geotechnical Institute, Copenhagen.

Brinch Hansen, J. (1970), *A Revised and Extended Formula for Bearing Capacity*, Bulletin No. 28, Danish Geotechnical Institute, Copenhagen.

Chen, W. F. (1975), *Limit Analysis and Soil Plasticity*, Elsevier, Amsterdam.

Chen, W. F. (1982), *Plasticity in Reinforced Concrete*, McGraw-Hill Book Co., Inc., New York, N.Y.

Chen, W. F. and McCarron, W. O. (1983), Modeling of soils and rocks based on concepts of plasticity, *Proceedings of the International Symposium on Recent Developments in Laboratory and Field Tests and Analysis of Geotechnical Problems*, Bangkok, Balkema, pp. 467–510.

Chen, W. F. and Baladi, G. Y. (1985), *Soil Plasticity: Theory and Implementation*, Elsevier, Amsterdam.

Chen, W. F. and Liu, X. L. (1990), *Limit Analysis in Soil Mechanics*, Elsevier, Amsterdam.

Chen, W. F. and Mizuno, E. (1990), *Nonlinear Analysis in Soil Mechanics*, Elsevier, Amsterdam.

Cox, A. D. (1962), Axially symmetric plastic deformations in soils—II—Indentation of ponderable soils, *International Journal of Mechanical Science*, **4**, pp. 371–380.

De Beer, E. E. (1965), Bearing capacity and settlement of shallow foundations, *Proceedings of the Symposium on Bearing Capacity and Settlement of Foundations*, Duke University, pp. 15–34.

Drucker, D. C., Greensberg, H. J., and Prager, W. (1952), Extended limit design theorems for continuous media, *Quarterly of Applied Mathematics*, **9**, pp. 381–389.

Duncan, J. M. and D'Orazio, T. B. (1984), Stability of steel storage tanks, *ASCE Journal of Geotechnical Engineering*, **110**, No. 9, pp. 1219–1238.

Garber, M. and Baker, R. (1977), Bearing capacity by variational method, *ASCE Journal of Geotechnical Engineering Division*, **103**, No. GT11, pp. 1209–1225.

Georgiadis, M. and Michalopoulus, A. P. (1985), Bearing capacity of gravity bases on a layered soil, *ASCE Journal of Geotechnical Engineering*, **111**, No. 6, pp. 712–729.

Hanna, A. M. (1981a), Foundations on strong sand overlying weak sand, *ASCE Journal of the Geotechnical Engineering Division*, **107**, No. GT7, pp. 915–927.

Hanna, A. M. (1981b), Experimental study on footings in layered soil. *ASCE Journal of the Geotechnical Engineering Division*, **107**, No. GT8, pp. 1113–1127.

Hanna, A. M. and Meyerhof, G. G. (1980), Design charts for ultimate bearing capacity of foundations on sand overlying soft clay, *Canadian Geotechnical Journal*, **17**, pp. 300–303.

Hanna, A. M. and Meyerhof, G. G. (1981), Experimental evaluation of bearing capacity of footings subjected to inclined loads, *Canadian Geotechnical Journal*, **18**, pp. 599–603.

Higher, W. H. and Anders, J. C. (1985), Dimensioning footings subjected to eccentric loads, *ASCE Journal of Geotechnical Engineering*, **111**, No. 5, pp. 659–665.

Higher, W. H. and Anders, J. C. (1987), Closure and errata, dimensioning footings subjected to eccentric loads, *ASCE Journal*

of Geotechnical Engineering, **113**, No. 2, pp. 186–187.

Hill, R. (1950), *The Mathematical Theory of Plasticity*, Clarendon Press, Oxford.

Ingra, T. S. and Beacher, G. B. (1983), Uncertainty in bearing capacity of sands, *ASCE Journal of Geotechnical Engineering*, **109**, No. 7, pp. 899–914.

Kardestuncer, H. and Norrie, D. H. (1987), *Finite Element Handbook*, H. Kardestuncer and D. H. Norrie (eds.), McGraw-Hill Book Co., Inc., New York, N.Y.

Ko, H.-Y. and Davidson, L. W. (1973), Bearing capacity of footings in plane strain, *ASCE Journal of the Soil Mechanics and Foundation Division*, **99**, No. SM1, pp. 1–23.

McCarron, W. O. and Chen, W. F. (1987), A capped plasticity model applied to Boston Blue Clay, *Canadian Geotechnical Journal*, **24**, No. 4, pp. 630–644.

Meyerhoff, G. G. (1951), The ultimate bearing capacity of foundations, *Geotechnique*, **2**, No. 4, pp. 301–332.

Meyerhof, G. G. (1953), The bearing capacity of foundations under eccentric and inclined loads, *Proceedings of the Third International Conference on Soil Mechanics and Foundation Engineering*, **1**, pp. 440–445.

Meyerhof, G. G. (1963), Some recent research on the bearing capacity of foundations, *Canadian Geotechnical Journal*, **1**, No. 1, pp. 16–26.

Meyerhof, G. G. (1970), Safety factors in soil mechanics, *Canadian Geotechnical Journal*, **7**, No. 4, pp. 349–355.

Meyerhof, G. G. (1984), Safety factors and limit states analysis in geotechnical engineering, *Geotechnique*, **21**, pp. 1–7.

Meyerhof, G. G. and Hanna, A. M. (1978), Ultimate bearing capacity of foundations on layered soils under inclined load, *Canadian Geotechnical Journal*, **15**, pp. 565–572.

Murff, J. D. and Miller, T. W. (1977a), Foundation stability on nonhomogeneous clays, *ASCE Journal of the Geotechnical Engineering Division*, **103**, No. GT10, pp. 1083–1094.

Murff, J. D. and Miller, T. W. (1977b), Stability of offshore gravity structure foundations by the upper-bound method, *Proceedings of the Offshore Technology Conference*, **3**, pp. 147–154.

Reddy, A. S. and Srinivasan, R. J. (1967), Bearing capacity of footings on layered clays, *ASCE Journal of the Soil Mechanics and Foundation Division*, **93**, No. SM2, pp. 83–99.

Satyanarayana, B. and Garg, R. K. (1980), Bearing capacity of footings on layered c–ϕ soils, *ASCE Journal of Geotechnical Engineering Division*, **106**, No. GT7, pp. 819–825.

Shield, R. T. (1955), The plastic indentation of a layer by a flat punch, *Quarterly of Applied Mathematics*, **13**, No. 1, pp. 27–46.

Sokolovskii, V. V. (1965), *Statics of Granular Material*, Pergamon Press, New York, N.Y.

Taylor, D. W. (1948), *Fundamentals of Soil Mechanics*, John Wiley and Sons, Inc., New York, N.Y.

Terzaghi, K. (1943), *Theoretical Soil Mechanics*, John Wiley and Sons, Inc., New York, N.Y.

Vesić, A. S. (1975), Bearing capacity of shallow foundations, *Foundation Engineering Handbook*, 1st edn., H. F. Winterkorn and H. Y. Fang (eds.), Chapter 3, Van Nostrand Reinhold Company, Inc., New York, N.Y.

Vivatrat, V. and Watt, B. J. (1983), Stability of Arctic gravity structures, *Proceedings of the Conference on Geotechnical Practice in Offshore Engineering*, ASCE, pp. 267–287.

Whitman, R. V. (1984), Evaluating calculated risk in geotechnical engineering, *ASCE Journal of Geotechnical Engineering*, **110**, No. 2, pp. 145–188.

Zhang, X. J. and Chen, W. F. (1987), Stability analysis of slopes with general nonlinear failure criterion, *International Journal for Numerical and Analytical Methods in Geomechanics*, **11**, pp. 33–50.

5 STRESS DISTRIBUTION AND SETTLEMENT OF SHALLOW FOUNDATIONS

ROBERT D. HOLTZ, Ph.D., P.E.
Professor of Civil Engineering
University of Washington

5.1 INTRODUCTION

5.1.1 Basic Requirements for a Good Foundation

The basic requirements for a good foundation are that (1) it is safe against complete collapse or failure of the soils upon which it is founded; (2) it experiences no excessive or damaging settlements or movements; (3) environmental and other factors (see below) are properly considered; and (4) the foundation is economically feasible in relation to the function and cost of the overall structure.

Environmental factors and other considerations that may adversely affect the construction and performance of the foundation include:

1. Frost action
2. Shrinking or swelling soils
3. Earthquakes and vibrations
4. Groundwater
5. Underground defects
6. Adjacent structures, excavations, and property lines
7. Scour and wave action

The influence of environmental and other factors has been discussed by Sowers (1962), Perloff and Baron (1976), among others; Sowers (1974) describes some dramatic examples. For a discussion of the effects of frost action in cold regions, see Chapter 19. Shrinking and swelling soils are discussed by Gromko (1974), Chen (1975), Snethen (1979, 1980), and Meyerhof and Fellenius (1985).

A discussion of the effects of foundation vibrations and earthquakes can be found in Chapters 15, 16, and 17. Scour and wave action are discussed in Chapter 18 for offshore structures.

5.1.2 Steps in Ordinary Foundation Design

In order to develop the most economical foundation for a particular structure, the geotechnical engineer ordinarily goes through the following steps (Perloff and Baron, 1976):

1. Establish the scope of the problem.
2. Investigate the conditions at the site.
3. Formulate a trial design.
4. Establish a model of the subsurface to be analyzed.
5. Determine the loads and soil parameters.
6. Perform the analyses.
7. Compare results with other models and experience.
8. Modify the design.
9. Observe the construction.

These steps are followed no matter what is the purpose of the foundation. Analyses are carried out for stability (bearing capacity) and settlement.

The primary requirement is that the foundation must be safe against possible instability; that is, the foundation soils must have adequate bearing capacity to support the loads of the structure. Thus, a bearing capacity analysis is normally done first to insure that the factor of safety is adequate. Bearing capacity calculations are discussed in standard foundation engineering textbooks such as those of Vesić (1975) and Perloff and Baron (1976), and Chapter 4 of this book.

If the bearing capacity is satisfactory, settlement analyses are carried out to estimate both the immediate and long-term settlements. If the bearing capacity is inadequate and/or settlements are too large, the soils are a good candidate for soil improvement or a deep foundation must be used. These points will be discussed later in this chapter.

5.1.3 Shallow Foundations

G. A. Leonards (personal communication, 1973) defines a shallow foundation as one in which the structural loads are transmitted to the soil at an elevation required for the function of the structure itself. Thus, a shallow foundation is not necessarily one that is near the ground surface, but one that is "shallow" in relation to the structure it is supporting. For example, a high-rise building with a five-story underground basement may still be supported on shallow foundations (e.g., spread footings or a mat foundation), even though the foundation elevation might be 20 or 25 m below street level.

Generally, the most economical shallow foundation is isolated spread footings, that is, footings whose area is less than 40% of the total area of the structure. If the footing area is larger than that or combined or strap footings are required, then a mat foundation may be more economical. The various types of shallow foundations and their structural design are discussed by, for example, Peck et al. (1974) and Bowles (1975a, b).

Because most shallow foundations have adequate bearing capacity, or since at least an adequate bearing capacity can relatively easily be achieved, the performance of most shallow foundations is controlled by their settlements. Bearing capacity rarely controls design. With the high factors of safety ordinarily used, especially with building foundations, there is a very low probability of failure. In considering settlements of shallow foundations, three questions must be answered:

1. How do we estimate the movements of the foundation for any given design?
2. What are the tolerable movements?
3. If the estimated movements are greater than the tolerable movements, then what do we do?

In this chapter, only the vertical movements or settlements of shallow foundations are considered. This does not mean that horizontal movements are not important—they are; however, a different approach must be used to make estimates of these movements, as discussed in Chapter 6.

5.2 SETTLEMENT OF SHALLOW FOUNDATIONS

5.2.1 Components of Settlement

The time–settlement history of a shallow foundation or earth structure is illustrated schematically in Figure 5.1. The total settlement s is the sum of the three components as shown in the figure:

$$s = s_i + s_c + s_s \qquad (5.1)$$

in which s_i is the *immediate* or *distortion* settlement, s_c is the *consolidation* settlement, and s_s is the *secondary compression* settlement. The immediate component is that portion of the settlement that occurs essentially with load application, primarily as a consequence of distortion (change of shape not change of volume) in the foundation soils. The distortion settlement is generally not elastic, although it is often calculated using elastic theory, especially when cohesive compressible materials are involved.

The other two components of settlement result from the gradual expulsion of water from the voids and the concurrent compression of the soil skeleton. The distinction between consolidation and secondary compression settlements is made on the basis of the physical processes that control the time rate of settlement. Consolidation settlement refers to settlements due to *primary consolidation*, in which the time rate of settlement is controlled by the rate at which water can be expelled from the void spaces in the soil. During *secondary compression*, the rate of settlement is controlled largely by the rate at which the soil skeleton itself yields, compresses, and creeps after the excess hydrostatic pressure is zero; that is, at a constant effective stress. The transition time between these two processes is arbitrarily identified as that time at which excess pore water pressure Δu becomes essentially zero. This time, denoted t_p, is shown in Figure 5.1.

Because the response of soils to applied loads is not linear, the superposition implied by Equation 5.1 is not strictly valid.

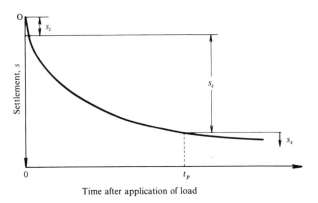

Fig. 5.1 Schematic time–settlement history of the settlement of a shallow foundation. (*Perloff, 1975.*)

However, a consistent and practical alternate approach has not yet been developed, and experience indicates that this approach yields reasonable predictions of settlements for many soil types.

The time–settlement relationship shown in Figure 5.1 is applicable to all soils. However, it should be recognized that the time scale and relative magnitudes of the three components may differ by orders of magnitude for different soil types. For example, water flows so readily through most clean granular soils that the expulsion of water from the pores is, for all practical purposes, instantaneous. Thus, foundations and earth structures on clean granular soils settle essentially simultaneously with the application of load.

The relative importance of each of these components for settlement analyses depends on the soil type, as shown in Table 5.1. As mentioned above, for granular materials, only the immediate settlement is of concern. For clay soils, consolidation settlement is the major concern, but immediate and secondary settlements must also be checked. For organic soils, especially fibrous peats, most of the settlement is secondary compression. Because of their high permeability, consolidation settlement occurs so rapidly as to almost be "immediate", and both components are usually combined for analysis purposes.

5.2.2 Causes of Settlement

It is instructive to look at the causes of settlement that may affect the performance of a structure or foundation. Sowers and Sowers (1970) have listed these as in Table 5.2. It is interesting that only the settlements due to structural loading and groundwater lowering are readily computed. In the cases of environmental loads or load-independent settlements, only the general susceptibility of a particular soil type to such settlements can be stated. Estimates of magnitudes and rates of settlement are virtually impossible in these cases.

5.2.3 Steps in Settlement Analyses

1. Establish the soil profile including the location of the groundwater table. Determine which layer or layers are

TABLE 5.1 RELATIVE IMPORTANCE OF IMMEDIATE, CONSOLIDATION, AND SECONDARY SETTLEMENT FOR DIFFERENT SOIL TYPES.

Soil Type	Immediate Settlement	Consolidation Settlement	Secondary Compression
Sands	Yes	No	No
Clays	Possibly	Yes	Possibly
Organic soils	Possibly (Yes)	Possibly (No)	Yes

TABLE 5.2 CAUSES OF SETTLEMENT[a]

Cause	Form of Mechanism		Amount of Settlement	Rate of Settlement
Structural load	Distortion (change in shape of soil mass)		Compute by elastic theory or empirical rules and procedures	Instantaneous
	Consolidation: change in void ratio under stress	Immediate	Stress-void-ratio curve	From time curve
		Primary consolidation	Stress-void-ratio curve	Compute from Terzaghi theory
		Secondary	Compute from log time-settlement	Compute from log time—settlement
Environmental load	Shrinkage due to drying		Estimate from stress–void ratio or moisture–void ratio and moisture loss limit-shrinkage limit	Equal to rate of drying, seldom can be estimated
	Consolidation due to water table lowering		Compute from stress–void ratio and stress change	Compute from Terzaghi theory
Load-independent (but may be aggravated by load) often environment-related, but not dependent	Reorientation of grains—shock and vibration		Estimate limit from relative density (up to 60–70%)	Erratic, depends on shock, relative density
	Structural collapse—loss of bonding (saturation, thawing, etc.)		Estimate susceptibility and possibly limiting amount	Begins with environmental change, rate erratic
	Raveling, erosion into openings, cavities		Estimate susceptibility but not amount	Erratic, gradual or catastropic, often increasing
	Biochemical decay		Estimate susceptibility	Erratic, often decreases with time
	Chemical attack		Estimate susceptibility	Erratic
	Mass collapse—collapse of sewer, mine, cave		Estimate susceptibility	Likely to be catastrophic
	Mass distortion, shear-creep or landslide in slope		Compute susceptibility from stability analysis	Erratic, catastrophic to slow
	Expansion—frost, clay expansion, chemical attack (resembles settlement)		Estimate susceptibility, sometimes limiting amount	Erratic, increases with wet weather

[a] After Sowers and Sowers (1970). Reproduced with permission of Macmillan Publishing Co.

compressible. Compute the total, neutral, and effective stress profile with depth.

2. Estimate the magnitude and rate of application of the loads applied to the foundation, both during construction and during the estimated economic and service life of the structure. In some structures, the loads applied to the foundation are provided by the structural engineer or architect. In other situations, for example, embankments and tanks, the foundation engineer may estimate the loads.

3. Estimate the change in stress with depth. If the loading is one-dimensional in nature (that is, if the width of the loaded area is significantly greater than the thickness of the compressible layer), then one-dimensional loading and compression may be assumed. In such a case, the change in stress at depth is equal to the stress applied at the surface. If, on the other hand, the width of the loaded area is equal to or less than the thickness of the compressible layer, three-dimensional loading occurs and the applied surface stresses dissipate with depth. Elastic theory or empirical methods are commonly used to estimate the change in stress with depth, but probabilistic methods may also be used.

4. Estimate the preconsolidation pressure. Compare with the effective stress profile computed in (1) above. Determine whether the soil is normally consolidated or overconsolidated.

5. Calculate the consolidation settlements.

6. Estimate the time rate of consolidation settlements.

7. Estimate the rate of secondary compression.

8. See Table 5.1. If necessary, estimate the immediate or distortion settlement. If the foundation soils are cohesive, use elastic theory. If granular, use empirical methods.

5.2.4 Scope and Organization of Chapter

The chapter is arranged in a sequence approximately corresponding to the computation process for the individual components of settlement described above. Following a discussion of the applicability of the theory of elasticity to the calculation of displacements and stresses in earth masses, methods for estimating the immediate settlements of subsurface materials that are primarily cohesive or cohesionless are described. Next is a discussion of how stresses at depth may be estimated for different loading geometries and boundary conditions. Then the calculation of consolidation settlement including time rate and secondary compression are described. To judge whether the calculated settlements can be tolerated, a discussion of tolerable settlements of buildings and other structures is presented. Finally, the chapter discusses what to do if the estimated settlements are not tolerable. Emphasis is placed on foundation treatment methods not covered elsewhere in this handbook.

Much of this chapter is based on Perloff (1975) and Perloff and Baron (1976). In addition, my colleague G. A. Leonards of Purdue University provided many helpful insights into the fine art of applied foundation engineering.

5.3 APPLICABILITY OF THE THEORY OF ELASTICITY TO CALCULATION OF STRESSES AND DISPLACEMENTS IN EARTH MASSES (After 65 years of soil mechanics, don't we have anything better?)

5.3.1 Rationale for Use of Elastic Theory

The distribution of stresses in earth masses is often estimated using the corresponding distribution in a linear elastic medium with boundary conditions approximating those in the problem of interest. In some cases, elastic theory is also used to estimate displacements as well. Although soils do not behave as linear elastic materials, the rationale for this practice has been the availability of solutions to problems for which the boundary conditions corresponded reasonably well to the boundary conditions for foundation engineering problems, as well as the lack of generally accepted alternatives. Experimental and analytical studies have been carried out to investigate the degree to which the results of elastic theory are applicable to earth masses. Perloff (1975) and Harr (1977) have summarized the conclusions of these investigations.

5.3.2 Applicability for Stress Calculations

Homogeneous Masses When the boundary conditions of the linear elastic analytical model approximate the in-situ boundary conditions, the stress distribution interpreted from the field measurements corresponds reasonably well to that predicted by linear elastic theory, probably because of small deformations and a high factor of safety against collapse. Another reason they probably also work is that specific material constants are not required in vertical stress distributions as predicted by the Boussinesq solutions. Thus, as long as the loading is well below yield with a high factor of safety and only vertical stresses are desired, linear elastic solutions give reasonable results.

Layered Systems Data concerning stress distributions within layered systems is very limited. Only a very few studies are reported in the literature, and those that are do not agree very well with the predictions of multilayer elastic theory. One of the difficulties is the nature of the interlayer shear stress. Solutions are only possible if interlayer shear is assumed to either be zero (frictionless) or with perfect fixity. Neither case is likely to be true in actuality.

In summary, it seems that elastic theory is satisfactory for homogeneous masses but not very good for layered systems. However, it is still used by pavement engineers (e.g., Yoder and Witczak, 1975) for estimating stresses in layered systems, even though it is recognized that the predictions are not very accurate.

5.3.3 Applicability of Elastic Theory to Displacement Calculations

Because displacements depend directly on the nature of the assumed constitutive law, the ability of elastic theory to predict displacements depends more on in-situ nonlinearity and material inhomogeneity than it does on the stress calculations themselves. Settlements due to initial undrained distortion of saturated or nearly saturated cohesive soils, which are subject to moderate increments of stress and where the elastic parameters can be assumed to be approximately constant throughout the mass, may be estimated reasonably well by elastic theory. Again, small strains and a high factor of safety are necessary.

On the other hand, when the in-situ soil conditions are markedly different from the assumptions of linear elastic theory, its use is inappropriate. Examples include the case of cohesionless soil deposits in which the equivalent elastic modulus depends significantly on confinement and where the stress increment due to loading varies significantly throughout the strata; thus, a constant equivalent modulus is not appropriate. It is possible that techniques such as the finite element method (Duncan, 1972; Desai and Christian, 1977) may be appropriate in this situation. Although it may be too cumbersome for routine applications, the finite element method is recommended for major projects, especially when the modulus cannot be assumed to be constant. For routine calculations, semiempirical methods that have been developed are appropriate.

5.3.4 Alternative Approach Using Probabilistic Theory

M. E. Harr (1977) developed an alternate approach to stress distribution problems based on the theory of probability. In particulate media such as soils, the requirements of a continuum theory such as the theory of elasticity are so markedly different from reality that a probabilistic approach has great merit. In contrast to the Boussinesq theory, the properties of the soil are incorporated in the probabilistic stress distribution approach through a parameter v, which is related to the in-situ coefficient of lateral stress. Harr (1977) shows that the distribution value of the expected normal stress often agrees better with the few published field measurements. Experiments with embedded pressure cells have invariably demonstrated that the theory of elasticity predicts too large vertical stresses and too small lateral stresses in the region of the load. These results more closely correspond to predictions by the probabilistic approach.

In the case of compacted soils, where the coefficient of lateral stress is likely to be rather large, serious disagreements are observed with elastic theory. However, this can be accounted for quite nicely by an increase in the parameter v, as shown by Harr (1977). Finally, another great advantage of probabilistic theory is that multilayer systems can readily be treated, and no assumptions regarding interlayer stress conditions are required.

Equations for calculating stress distributions using the probabilistic approach will be presented in Section 5.6.5.

5.4 CALCULATION OF INITIAL DISTORTION SETTLEMENTS

5.4.1 Distortion Settlement and Contact Stress

As mentioned earlier, distortion settlement occurs because of a change in shape of the soil mass rather than because of a change in volume. The shape of the deflected soil profile depends on whether the soil is predominantly cohesive or granular and whether the loaded area is rigid or flexible. The possibilities are shown in Figure 5.2. In the case of rigid foundations, the settlements produced are of course uniform, whereas the contact stress distributions under the foundations are very nonuniform. In the case of cohesive soils, at the outer edges of a rigid foundation on a perfectly elastic soil, the stress is infinite. In actuality, as shown in Figure 5.2a, it is limited by the shear strength of the soil. With rigid footings on granular materials, because the confinement is less at the outer edges, the stress is also less. For a very wide footing on granular material (e.g., a stiff mat foundation), settlement would be fairly uniform; near the middle of the mat, the contact stress also would be quite uniform.

As expected, the contact stress distribution for a flexible loaded area is also uniform, but the settlement profiles are quite different, depending on whether the soil is cohesive or granular. These cases are shown in Figure 5.2b. In the case of cohesive soils, which include saturated clays and many rocks, the surface will deform in a shape that is concave upward. The shape of the settlement profile on a granular material is exactly the opposite, concave downward, again because the confining stress near the edges of the footing is lower than in the center. If the sand is confined, it has a higher modulus than at the edges, which means that there is less settlement in the center than at the edges. If the flexible loaded area is very large, then the settlements near the center of the area are relatively uniform

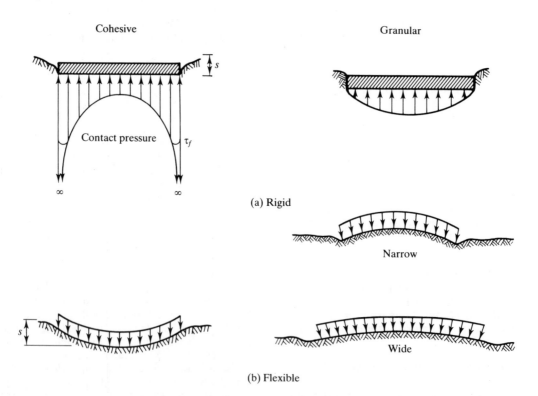

(a) Rigid

(b) Flexible

Fig. 5.2 Distribution of settlement and contact stresses for rigid and flexible loaded areas on cohesive and cohesionless soils.

and less than at the edges (Fig. 5.2b). Contact stress distributions are important for the design of foundations and footings (Bowles, 1975a). For the structural design of footings, a linear contact stress distribution is often assumed although this is obviously incorrect from a soil mechanics point of view.

5.4.2 Immediate Settlement of Cohesive Foundations

For soils that are predominately cohesive, linear theory of elasticity is used to estimate the magnitude of initial settlements. Soil profiles are typically simplified, although some solutions involving multiple layer theory are available. Homogeneity and isotropy are implicitly assumed so that only two elastic parameters, the modulus of elasticity E and Poisson's ratio v are needed. This approach works reasonably well on clay soils if the applied stress level is low; that is, if the factor of safety is large and we do not have plastic yielding in the foundation. If foundation yielding is likely to occur, another approach is recommended, which will be described below.

In many foundations on cohesive soils, the immediate or distortion settlement is a relatively small part of the total vertical movement and, thus, rough estimates are acceptable. A discussion of relative importance of immediate and consolidation settlement will be given later in this section.

A Distributed Load at or Near the Surface of a Deep Layer When the foundation problem can be approximated as one or more uniformly distributed loads acting on circular or rectangular areas near the surface of a relatively deep stratum, the vertical settlement can be estimated by

$$s_i = C_s q B \left(\frac{1 - v^2}{E_u} \right) \qquad (5.2)$$

where

s_i = settlement of a point on the surface
C_s = shape and rigidity factor
q = magnitude of the uniformly distributed load
B = characteristic dimension of the loaded area as shown in Figure 5.3
E_u = Young's modulus (undrained)
v = Poisson's ratio

The coefficient C_s accounts for the shape and rigidity of the loaded area and for the position of the point for which the settlement is being calculated. Values of C_s are given in Table 5.3.

EXAMPLE 5.1

A structure is to be supported on a stiff reinforced concrete mat foundation whose dimensions are 20 m by 50 m. The load on the mat is to be uniformly distributed; its magnitude is 65 kPa. The mat rests on a deep saturated deposit of saturated clay for which the average undrained Young's modulus is approximately 40 MPa. Estimate the immediate settlement at the center and corner of the mat.

Solution

Since the mat is stiff, use the rigid factors from Table 5.3a. With $L/B = 50/20 = 2.5$, the shape factors for both the center and corner are determined by interpolation to be $C_s = 1.20$. Thus from Equation 5.2 the immediate surface settlement at both the center and corner of the mat is $s_i = 1.20(65)(20) [(1 - 0.5^2)]$ divided by $[(40 \times 10^3)] = 0.029$ m $= 29$ mm.

For comparison, the shape factors for a flexible mat would be determined by interpolation to be

At the center $C_s = 1.63$
At the corner $C_s = 0.81$

Thus, the immediate surface settlements are

At the center $s_i = 40$ mm
At the corner $s_i = 20$ mm

A mat foundation is usually neither completely flexible nor completely rigid, depending on its size and thickness and how heavily reinforced it is. If it is large, the distribution of contact pressure may be nearly uniform over its center portion. At the corners and edges, however, the rigidity of the mat may be significant (owing to its thickness and the amount of reinforcing), and settlements are likely to be less than predicted. In a saturated clay, because of settlement in the middle portion of the foundation, some heave may occur in the outer portions because of undrained (constant volume) loading and shear.

Effect of Layered Systems In actuality, most soil profiles are not homogeneous and deep. If the thickness of the top layer is large relative to the dimensions of the loaded area, immediate surface settlement may be calculated as if the soil were a homogeneous layer of infinite depth. However, if the upper stratum is relatively thin, the effect of layering must be taken into consideration. This is likely to be especially important when a soft compressible stratum is underlain by rock or very hard or dense soils. This special case may be approximated by a layer of elastic material of finite thickness underlain by a rigid base. Settlements for this case may be determined by Equation 5.2, but using a shape factor C_s that accounts for the presence of the rigid base. Values for these shape factors C_s are tabulated in Table 5.3b for the settlement under the center of a rigid circular area and under the corner of flexible rectangular areas. These shape factors depend upon both the shape of the loaded area and the thickness of the compressible stratum relative to the width of the loaded area, as illustrated in Figure 5.3.

Examination of Table 5.3 indicates the importance of the presence of a rigid boundary. When $H/B = 0.5$, the reduction in surface displacements of the center of the loaded area relative to that for the halfspace is greater than 50 percent.

EXAMPLE 5.2

Compute the immediate settlement at the center of the uniformly loaded (flexible) area measuring 6 m × 6 m. The applied surface stress is 200 kPa and the depth to firm bottom is 3 m. Assume the undrained elastic modulus is 10 000 kPa and $v = 0.5$.

Solution

Use the C_s values for the corners of four equally sided rectangles 3 m × 3 m. In this case, $H/B = 1$, $L/B = 1$, and,

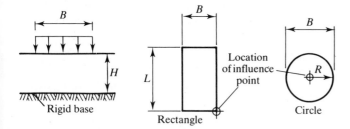

Fig. 5.3 Notation for loaded areas, shown in plan view. (*U.S. Navy, 1982.*)

TABLE 5.3 SHAPE AND RIGIDITY FACTORS, C_s, FOR CALCULATING SETTLEMENTS OF POINTS ON LOADED AREAS AT THE SURFACE OF AN ELASTIC HALFSPACE[a]

a. Infinite Depth

Shape and Rigidity	Center	Corner	Edge/Middle of Long Side	Average
Circle (flexible)	1.00		0.64	0.85
Circle (rigid)	0.79		0.79	0.79
Square (flexible)	1.12	0.56	0.76	0.95
Square (rigid)	0.82	0.82	0.82	0.82
Rectangle: (flexible) length/width				
2	1.53	0.76	1.12	1.30
5	2.10	1.05	1.68	1.82
10	2.56	1.28	2.10	2.24
Rectangle: (rigid) length/width				
2	1.12	1.12	1.12	1.12
5	1.6	1.6	1.6	1.6
10	2.0	2.0	2.0	2.0

b. Limited Depth Over a Rigid Base

H/B	Center of Rigid Circular Area Diameter = B	Corner of Flexible Rectangular Area				
		L/B = 1	L/B = 2	L/B = 5	L/B = 10	L/B = ∞ (strip)
			$v = 0.50$			
0	0.00	0.00	0.00	0.00	0.00	0.00
0.5	0.14	0.05	0.04	0.04	0.04	0.04
1.0	0.35	0.15	0.12	0.10	0.10	0.10
1.5	0.48	0.23	0.22	0.18	0.18	0.18
2.0	0.54	0.29	0.29	0.27	0.26	0.26
3.0	0.62	0.36	0.40	0.39	0.38	0.37
5.0	0.69	0.44	0.52	0.55	0.54	0.52
10.0	0.74	0.48	0.64	0.76	0.77	0.73
			$v = 0.33$			
0	0.00	0.00	0.00	0.00	0.00	0.00
0.5	0.20	0.09	0.08	0.08	0.08	0.08
1.0	0.40	0.19	0.18	0.16	0.16	0.16
1.5	0.51	0.27	0.28	0.25	0.25	0.25
2.0	0.57	0.32	0.34	0.34	0.34	0.34
3.0	0.64	0.38	0.44	0.46	0.45	0.45
5.0	0.70	0.46	0.56	0.60	0.61	0.61
10.0	0.74	0.49	0.66	0.80	0.82	0.81

[a] After U.S. Navy (1982).

from Table 5.3, $C_s = 0.15$:

$$S_i = 0.15(200)(3)\left(\frac{1 - 0.5^2}{10\,000}\right) \times 4 = 27 \text{ mm}$$

If the soil profile consists of a relatively thin stiffer layer underlain by a less stiff layer of greater depth, then the stresses from the surface load must be distributed to the top of the less compressible layer. Use the stress distribution techniques discussed in Section 5.6.

Analytical and/or numerical methods for the determination of displacements in multilayered systems are available for cases other than those in Table 5.3 (see Poulos and Davis, 1974). A number of multilayer solutions are now available in computer codes. Except for pavements and special foundations, however, the use of multilayered computer analyses is generally not justified, because the material parameters are not accurately determined, the boundary conditions and interface conditions between the strata are not that well known, and, finally, approximations may be required to fit the geometry of the real problem to that for which the solution is available. In many situations an approximate analysis of the intermediate settlement is sufficient, as illustrated in the following two examples.

EXAMPLE 5.3

The mat foundation of Example 5.1, 20 m × 50 m supporting a uniform normal load of 65 kPa, is founded on a soil profile shown in Figure 5.4. The profile indicates a layer of stiff clay over a more compressible layer that is in turn underlain by shale. Assume these conditions are representative of the entire site. Estimate the immediate surface settlement at the center of the mat.

Solution

Assume the shale acts as a rigid base and above it a *single* stratum of thickness $H = 15$ m. Thus,

$$\frac{H}{B} = \frac{15}{10} = 1.5 \qquad \frac{L}{B} = \frac{25}{10} = 2.5$$

The shape factor C_s obtained from Table 5.3b by linear

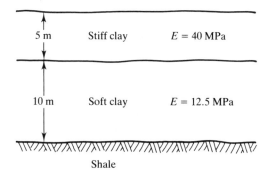

Fig. 5.4 Soil profile for Example 5.3.

interpolation is 0.21. Substituting this value into Equation 5.2 leads to a calculated surface settlement (assuming $v = 0.5$) of

$$s_i = (0.21)(65)(10)\left(\frac{1 - 0.5^2}{E}\right) \times 4 = 410 \times \frac{10^3}{E} \quad (\text{kPa} \cdot \text{m})$$

The potential immediate surface displacement can be bounded by estimating the settlement using the moduli of the two compressible strata, or $10 \text{ mm} < s_i < 33 \text{ mm}$. A better estimate can be obtained by using an *equivalent* Young's modulus weighted by the relative thicknesses of the two strata, or

$$E_{eq} = \frac{5(40 \times 10^3) + 10(12.5 \times 10^3)}{15} = 21.7 \times 10^3 \text{ kPa}$$

Thus, the immediate settlement is 19 mm.

The settlement predicted by this approach could exceed that determined by multilayer solutions because the load distribution effect of the upper stiffer layer is not accounted for in the weighted average E_{eq}.

Another way to obtain an estimate of s_i, especially when the stiffness of the upper layer is much greater than that of the lower layer, is to assume that the immediate settlement results primarily from distortions within the less stiff layer. This layer is extended all the way from the ground surface to the top of the underlying shale and the elastic settlement is calculated as before using Table 5.3 and Equation 5.2. To account for the actual depth of the less-stiff material, the settlement that would occur in the thinner layer of the material overlying the rigid base is subtracted from the settlement determined assuming the entire depth was soft. See Perloff (1975) for an example.

Correction for Low Factor of Safety and Large Undrained Shear Deformations If the factor of safety against bearing capacity failure (see Chapter 4) is less than about 3, the immediate settlement should be modified to take into account undrained plastic yielding that occurs in the foundation. A semiempirical procedure that takes this into account was developed by D'Appolonia et al. (1971). See also U.S. Navy (1982) and Foott and Ladd (1981).

Heave of Excavations It may occur that a significant portion of the heave of excavations above compressible strata arises from undrained distortions within the strata. This heave, and the subsequent settlement resulting when the structural load within the excavation is applied to the foundation within the excavation, may be significant, particularly in the case of a partially or fully compensated mat foundation. For major projects, use of the finite element method is recommended. See Clough and Schmidt (1977) for discussions and examples.

For preliminary estimation purposes, a common approach is to use the loading approximation in Figure 5.5a, in which the upper boundary of the elastic medium is presumed to be at the base of the excavation, subjected to an upward uniform strip load of magnitude $-\gamma D^2$, where D is the depth of the excavation. Such an analysis fails to account for the influence of the material surrounding the excavation on the distribution of stresses and therefore displacements within the soil below this excavation.

An alternate approach has been developed by Perloff (1975) based on the work of Baladi (1968), who obtained solutions for the heave at the base of a strip excavation in a linear elastic medium. By determining the heave at various depths within the elastic medium, an approximation of the expected heave when only a limited depth of deformable material overlies a rigid boundary (Fig. 5.5b) can be made. Such an analysis may be useful for preliminary estimation purposes in lieu of a finite element analysis. The magnitude of heave or rebound at the base of a strip excavation of rectangular cross section is determined by this method from

$$r_d(\text{strip}) = \Delta_{\text{strip}} \frac{\gamma D^2}{E} \quad (5.3)$$

in which Δ_{strip} is a dimensionless heave factor whose magnitude is determined by the geometry of the excavation and the position for which the rebound r_d is being calculated. The heave factor for the centerline and the edge of the excavation are given in Figure 5.6, for a variety of excavation geometries and depths to a rigid boundary.

When the length-to-width ratio of the excavation (L/B) is less than about 5, the limited length of the excavation must be accounted for. This is done by assuming that the relative effect of excavation shape on heave of a rectangular excavation will be similar to that for the settlement of a uniformly loaded area. Thus, the heave of a rectangular excavation can be calculated by

$$r_d = C_r'' \Delta_{\text{strip}} \frac{\gamma D^2}{E} \quad (5.4)$$

in which C_r'' is determined from shape and rigidity factors given by Perloff (1975) and plotted in Figure 5.7 for base heave at the center and midpoint of the long side of the excavation. The use of these figures is illustrated in the following example.

EXAMPLE 5.4

A foundation excavation, $20 \text{ m} \times 30 \text{ m}$ in plan, is to be carried out to a depth of 10 m in the soil profile shown in Figure 5.8. Estimate the heave at the center of the excavation due to undrained distortion of the silty clay layer.

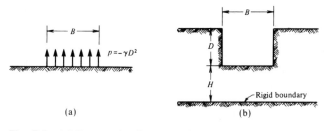

Fig. 5.5 (a) Common loading approximation for heave. (b) Linear elastic excavation analysis. (*Perloff, 1975.*)

(a)

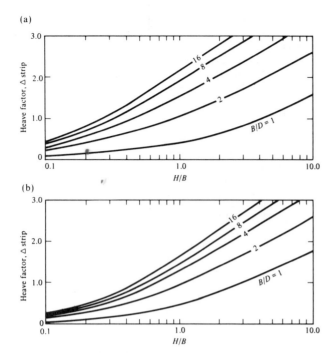

(b)

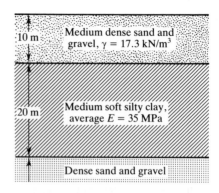

Fig. 5.8 Soil profile for Example 5.4.

and from Figure 5.6a, $\Delta_{\text{strip}} = 1.08$. For

$$\frac{L}{B} = \frac{30}{20} = 1.5$$

the shape factor is, from Figure 5.7a, $C_r'' = 0.97$. Thus, the heave can be calculated from Equation 5.4 as

$$r_d = (0.97)(1.08)\frac{(17.3 \text{ kN/m}^2)(10)^2}{35 \times 10^3 \text{ kPa}}$$

$$= 0.05 \text{ m} = 52 \text{ mm}$$

In the *Canadian Foundation Engineering Manual* (Meyerhof and Fellenius, 1985), a method is shown for calculating settlement in compressible soils using the net stress at a characteristic point. The net stress includes the decrease in applied foundation stress due to the excavated soil. Lambe and Whitman (1969) used the stress path method (Lambe, 1967) to predict heave of excavations.

Fig. 5.6 Heave at base of strip excavation in linearly elastic medium of limited thickness. (a) Heave at center line. (b) Heave at edge. (*Based on analysis of Baladi, 1968.*)

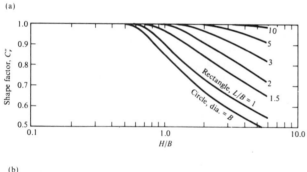

(a)

(b)

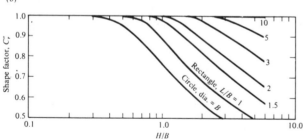

Fig. 5.7 Factors for correcting heave of strip excavation for shape of excavation. (a) Shape factor for heave at center. (b) Shape factor for heave at midpoint of long side. (*Data from Egorov, 1958, as cited by Harr, 1966.*)

5.4.3 Evaluation of Elastic Parameters

The magnitude of the calculated immediate distortion settlements (and heave) depends directly on the values of the elastic parameters (Young's modulus and Poisson's ratio) used in the calculations. Because cohesive soils are not linear elastic materials, these "elastic parameters" must be properly evaluated so that when they are substituted into the appropriate equations, correct estimates of the initial distortion settlement will be obtained.

For saturated clay soils, which deform at constant volume during the limited time required to develop the elastic distortion settlement, a Poisson's ratio of $v = 0.5$, corresponding to an incompressible medium, is usually assumed. Although this assumption may not be strictly correct, the magnitude of the computed settlement is not especially sensitive to small changes in Poisson's ratio.

Laboratory Tests Determination of the appropriate value of the equivalent Young's modulus E_u, an undrained modulus, is much more difficult. The ideal way would be to use the initial slope or tangent modulus of the stress–strain curve, as determined from triaxial compression or unconfined compression tests (Fig. 5.9). Alternatively, a secant modulus could be used, determined for the stress level estimated to occur in the field, for example, 25 or 50 percent of σ_{max}. There is, however, ample laboratory and field evidence to indicate that the values so obtained are too low, often only a small percentage of the field value. There are two primary reasons for this discrepancy. (1) Sample disturbance during sampling and preparation of the

Solution

Assuming that the underlying dense sand and gravel acts as a rigid boundary, we have

$$\frac{B}{D} = \frac{20}{10} = 2 \qquad \frac{H}{B} = \frac{20}{20} = 1$$

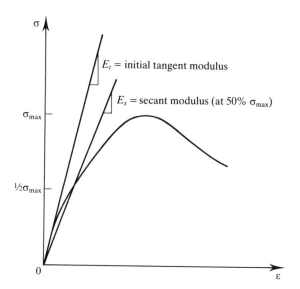

Fig. 5.9 Definitions of the initial tangent and secant moduli, for example, at 50 percent σ_{max}.

specimen for laboratory testing leads to a reduced undrained stiffness; the modulus is one of the properties most sensitive to disturbance effects. (2) Defects such as fissures are common in a great many sedimentary soil deposits. These inhomogeneities are usually unimportant to the settlement of a structure because they are small relative to the dimensions of the foundation. However, such defects may strongly influence a small laboratory test specimen and produce spuriously low measured modulus values in a laboratory test, especially unconfined compression tests. Because these two factors reduce the E_u, there is a tendency to overpredict the immediate settlements in the field.

On the basis of limited laboratory evidence, Perloff (1975) proposes a procedure for obtaining a suitable value of an equivalent field modulus. Ladd et al. (1977) and Foott and Ladd (1981) recommend the use of the direct sample shear test on high-quality undisturbed samples.

Field Plate Load Tests Because of the problems that affect the determination of the undrained modulus in laboratory tests, field plate bearing tests (ASTM D1194) are sometimes conducted for important projects. In these tests, all the parameters in Equation 5.2 are known except the factor $(1 - v^2)/E_u$, which can then be determined by back calculation. However, it should be kept in mind that if the seat of the settlement of the plate is different from that of the foundation, load tests on small plates, typically 30 cm to 75 cm in diameter, cannot be simply extrapolated to predict the settlements of prototype foundations; that is, the settlement of the foundation may be influenced by the presence of compressible strata that are far below the zone of influence of a small test plate. The best approach is to have access to the compressible layers during the exploration program to conduct the plate load tests. Disturbance of the surface can be partially overcome by cycling the load at least five times. This test should be carried out at the expected foundation elevation. Because of the relatively shallow influence of the loaded plates, it may be advisable to use two different size plates and test at two different depths, and scale up the modulus to the prototype foundation. A further complicating factor in the plate tests is that once the load on the plate exceeds about half the ultimate or failure load, settlements start to accelerate as the load is increased. Thus, back-calculated E_u values are very dependent upon the level of shear stress imposed by the plate.

Empirical Relations Because of all the problems with laboratory determination of E_u and because large scale field loading tests are expensive, it is common to assume that E_u is somehow related to the undrained shear strength. A common approximation (Bjerrum, 1963, 1972) is to use the ratio E_u/τ_f ranging between 500 and 1500, with τ_f determined either by the field vane shear or the undrained triaxial compression test. The lowest value is for highly plastic clays where the applied load is large compared to the value of $\sigma'_p - \sigma'_{v0}$; (that is, the stress added to the foundation is relatively large). The higher value is for clays of low plasticity, where the added load is relatively small. D'Appolonia et al. (1971) reported an average E_u/τ_f of 1200 for load tests at ten sites, but for the clays of higher plasticity the range was 80–400. Values have been found ranging from 40 to 3000 (Simons, 1974). These cases plus a few others taken from the literature are plotted versus PI in Figure 5.10. Stiff fissured soils and glacial tills are not included. There is much scatter for $PI < 50$ and not much data for $PI > 50$. It seems reasonable to simply use Bjerrum's recommendation (E_u/τ_f of 500 and 1500), modified as required by procedures developed by D'Appolonia et al. (1971) for estimating immediate settlements of soft clays.

Another factor that strongly affects the undrained shear strength of clays is stress history, and stress history also affects Young's modulus. Information from Ladd et al. (1977) is shown in Figure 5.11. The relationship is not so simple because E_u/τ_f depends strongly on the level of shear stress. In general, however, it decreases with increasing overconsolidation ratio for a given stress level, as shown in Figure 5.11. Duncan and Buchignani (1976) also present a relationship between undrained modulus and OCR (Table 5.4) that may be used.

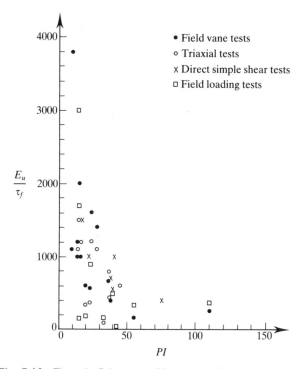

Fig. 5.10 The ratio E_u/τ_f versus PI as reported by various authors (see Holtz and Kovács, 1981, for references cited). Sources: Bozozuk (1963); Bozozuk and Leonards (1972); D'Appolonia et al. (1971); DiBaglio and Stenhamar (1975); Hansbo (1960); Holtz and Holm (1979); Ladd and Edgers (1972); LaRochelle and Lefebvre (1971); Raymond et al. (1971); Simons (1974); Tavenas et al. (1974).

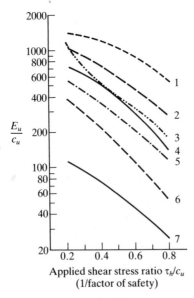

No.	Description	c_u/σ'_{vc}
$1^{(1)}$	Portsmouth sensitive CL clay $S_f \geq 10$, LL = 35, I_p = 15	0.20
$2^{(1)}$	Boston CL clay LL = 41, I_p = 22	0.20
$3^{(1)}$	Bangkok CH clay LL = 65, I_p = 41	0.27
$4^{(1)}$	Maine organic CH – OH clay LL = 65, I_p = 38	0.285
$5^{(2)}$	AGS CH clay LL = 71, I_p = 40	0.255
$6^{(1)}$	Atchafalaya CH clay LL = 95, I_p = 75	0.24
$7^{(3)}$	Taylor River peat w_n = 500%	0.46

(1) From Ladd and Edgers (1972)
(2) MIT for Dames and Moore
(3) MIT for Haley and Aldrich

$E_u = 3\tau_h/\gamma$ τ_h = applied horizontal shear stress

$c_u = (\tau_h)_{max}$ γ = shear strain

(a)

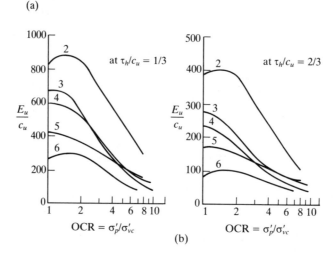

(b)

Fig. 5.11 Effect of OCR and shear stress level on E_u/τ_f from direct simple shear tests. (a) Normalized secant modulus vs. stress level for normally consolidated soils. (b) Normalized secant modulus vs. overconsolidation ratio. (*Data from Ladd et al., 1977, and Foott and Ladd, 1981.*)

TABLE 5.4 RELATIONSHIP BETWEEN UNDRAINED MODULUS AND OVERCONSOLIDATION RATIO[a].

	E_u/τ_f		
OCR	PI < 30	30 < PI < 50	PI > 50
< 3	600	300	125
3–5	400	200	75
> 5	150	75	50

[a] After Duncan and Buchignani (1976) and U.S. Navy (1982).

In-Situ Tests Besides using a plate bearing test to determine the in-situ elastic modulus, it would be possible to use it directly for estimating the immediate settlements. As before, the primary requirement is that the soil volume under the foundation stressed by the plate should have some relation to the volume stressed by the proposed foundation. Included in this possibility

are large-scale loading tests utilizing, for example, an embankment fill or a large tank of water. In these latter cases, geotechnical instrumentation is required for measurement of settlements and excess pore pressures.

Another approach is to use the pressuremeter test (Baguelin et al., 1978; Meyerhof and Fellenius, 1985). Through selection of appropriate empirical coefficients, the immediate settlement can be determined from the pressuremeter modulus. Other candidate in-situ tests for this purpose are the screw plate compressometer and the Marchetti dilatometer.

5.4.4 Importance of Immediate Settlement Calculations

As mentioned earlier, the immediate or distortion settlement is often not a significant portion of the total foundation settlement, so only rough estimates are ordinarily required for estimating this component of the total settlement. However, if as occurs in certain circumstances, the immediate settlement is an important part of the total settlement, then it is worth the effort to obtain a good estimate of the undrained elastic modulus. Recommendations in Section 5.4.3 above are appropriate in this case.

From Equation 5.2, immediate settlement is directly proportional to the applied load. Sometimes, to provide an adequate factor of safety in bearing capacity, the footing size is increased or a large mat foundation is chosen. Designers should realize that the immediate settlement also increases as the foundation size increases.

Large footings and heavy loads cause large immediate settlements, especially in weaker soils, and a correction for low factor of safety and large undrained shear deformations has already been mentioned. As noted by Foott and Ladd (1981), these occur especially in soils that have a high plasticity index and also contain organic matter. They recommend estimating the soil moduli using K_0 consolidated direct simple shear tests and using the empirical observations by D'Appolonia et al. (1971) for lateral yield and low factors of safety. With appropriate consideration of soil type and available consolidation time, their method should indicate conditions where initial and creep-induced settlements could be a significant design problem.

Burland et al. (1977) make some empirical observations to help designers decide when immediate settlements are likely to be an important part of the total settlement of a foundation. For soft soils in which the applied stress will exceed the preconsolidation pressure, typically the immediate settlement component is about 10 percent of the consolidation component. Therefore, it is relatively unimportant, although even this magnitude may be a problem if the structure is sensitive to rapid settlements. On the other hand, for stiff soils in which the applied stress does not exceed the preconsolidation pressure, then the immediate settlement may be as much as 50 to 60 percent of the total settlement. It also decreases as the depth of the compressible layer decreases. Even for deep layers of overconsolidated or stiff clay, the immediate settlement is unlikely to exceed 70 percent of the total settlement, and it may in an extreme case be as low as 25 percent for nonhomogeneous and anisotropic soils. Average values of the ratio s_i/s appear to range from 0.5 to 0.6. The total settlement s, because the secondary component is so small in stiff soils, can be estimated quite adequately from oedometer tests.

Effect of Footing Size on Immediate Settlement Perloff and Baron (1976) give three instructive examples on what happens to immediate settlements as the size of footings is changed. Increasing the size of a square footing to support a given total load leads to a reduction in elastic settlement in proportion to the increased footing dimension. In other words, with the same modulus, if the applied stress q is reduced, one should obtain less settlement.

Is this true for a strip footing with a constant load per unit length? Increasing the width to reduce the unit pressure does *not* reduce the elastic settlement, as long as a constant total load per unit length is applied.

Finally, they consider two square footings of width B and nB, respectively, which are subjected to the same unit pressure. So the total load carried by each footing is not the same. Their analysis shows that the larger footing will settle n times as much as the smaller one. Thus, different sized footings subjected to the same unit pressure will not have equal immediate settlements.

5.5 DISTORTION SETTLEMENT OF GRANULAR SOILS

Virtually all settlement of granular soils can be considered to be immediate (Table 5.1). This is because even if the sands are

below the groundwater table, and completely saturated, excess pore pressures dissipate rapidly during loading. Although the magnitude of settlement may be significantly less than might be obtained with similar foundations on cohesive soils, the settlements of structures on sand must be considered and accurately estimated because most structures are more sensitive to distortion settlement with rapid loadings than they are if the distortion occurs slowly. Further, granular soils are more likely to be heterogeneous than many sedimentary clay deposits. Because we have no rational theory for prediction of the settlement of shallow foundations on granular soils, we use empirical procedures in engineering practice, and these procedures will be the subject of this section.

We begin by a discussion of the sources of settlement and the factors that cause shear strains in granular soils when they are loaded. Then we describe a number of methods that have been developed for estimating immediate settlements on cohesionless soils, concentrating especially on the Schmertmann (1970) method and other procedures using in-situ tests.

The sources of settlement in granular soils are: (1) shear strains, which result in changes in shape after loading, and (2) changes in volume, which can be either positive or negative (dilation or compression). Both factors are functions of the initial void ratio and confining pressure, and both occur in granular soils and result in surface settlements.

5.5.1 Factors Affecting Sand Compressibility (and How to Determine Them)

Although granular materials are not elastic, we can consider the elastic constants in Equation 5.2 to be equivalent approximations. If so, then the magnitude of load (the contact pressure and size of the footing) directly affects the settlement of a granular layer. The thickness of the granular layer must also influence the settlement. Thickness is determined by the site exploration program and the geology of the site.

The magnitude of the load, contact pressure, and size of footing are determined from the preliminary design. Sometimes presumptive bearing values are used to size the foundation elements or, if a bearing capacity analysis has been carried out previously, estimated working footing stresses may be used.

G. A. Leonards (1987, personal communication) has indicated that the primary factors that influence the compressibility of granular materials are:

1. Soil characteristics
2. State of stress in the ground
3. State of compaction
4. Stress history

Soil Characteristics Characteristics of granular soils such as gradation, grain size, angularity, roughness, and mineral hardness affect compressibility. For the same packing, relative density, stress history, and stress level, a better gradation decreases compressibility, while increasing angularity results in an increase in compressibility. On the other hand, increasing the roughness of the grains and their size decreases the compressibility. The propensity for grain crushing depends on the type of mineral, particle shape, and stress level. Some minerals are more susceptible to crushing than others. Everything else being equal, the compressibility increases as hardness decreases. A number of other crushing factors are discussed by Hardin (1985). At ordinary footing loads, crushing of mineral grains does not significantly contribute to settlement. However, grain crushing could be important in micaceous sands and silts.

Soil characteristics are determined through a program of site investigation and laboratory testing.

State of Stress The second important factor is the state of stress in the ground. If the sand is subjected to a large horizontal stress, there is less tendency towards volume change and less settlement. Large horizontal stresses result from prestressing, discussed below, either due to geologic factors or to construction loadings. Even if some of the vertical overburden stress is removed, a certain percentage of the horizontal stress remains. K_0, the coefficient of lateral earth pressure at rest, may increase from approximately 0.5 up to 1.0 or greater. Therefore, if the soil is again loaded, less settlement will occur. Other factors related to the state of stress include the location of groundwater table, depth of foundation, void ratio, and possible prestressing and prestraining. These last two factors are discussed below.

How is the state of stress determined? A quantitative determination is not easy. From the geology and site investigation, certain inferences can be made about the possibility of prestraining and therefore the presence of large residual lateral stresses in the ground. Plate load tests may be useful in this regard as are other in-situ measurements such as those from earth pressure cells and pressuremeter and dilatometer tests.

State of Compaction State of compaction includes the packing, density, and orientation of the sand particles. It is very important to know the initial state of compaction and its variability. For example, if the deposit is loose and variable, large and possibly detrimental settlements will occur in the foundation. If the deposit is dense, then the sand tends to expand when sheared and small, often negligible, settlements result.

How is the state of compaction determined? Sands are very difficult to sample undisturbed; therefore, in-situ tests, the SPT and Dutch cone penetrometer, for example, are used to correlate with in-situ density.

Most compressibility obviously will occur with looser deposits. In terms of relative density or density index, it has been estimated that greater than 90 percent of the compression occurs between $D_r = 0.10$ and 0.70. Very little settlement occurs as the density index exceeds 0.70. In terms of packing at a given relative density, the compressibility of the least favorable packing is probably 2 to 4 times the compressibility of most favorable packing. It has a relatively minor importance for footing settlements.

Stress History The most important factor influencing the compressibility of granular soils is the stress history, or more precisely, the strain history of the deposit. If the sand deposit has been previously loaded or strained, a large decrease in compressibility (increase in equivalent modulus), and therefore a large decrease in settlement, results. The normally consolidated compressibility is at least five times greater, with typical values between 8 and 16 times, and it may even approach 30 times greater than the overconsolidated or prestrained compressibility. This occurs because if a sand is loaded in compression, for example in a triaxial test, and somewhere before failure the sample is unloaded and then reloaded, the unload/reload modulus E_r is much steeper than the initial tangent modulus E_t (Fig. 5.12). This effect is greater in a loose sand. The unload/reload modulus in very loose sand can easily be 5 to 30 times less than the initial tangent modulus. If the designer overestimates settlements by a factor of 5 to 30, then expensive and unnecessary foundation treatment or deep foundations may be selected. Therefore, it is very important to determine if possible whether a sand deposit has previously been loaded.

Prestraining and its effects have been discussed by Dahlberg (1975), Lambrechts and Leonards (1978), and Jamiolkowski et al. (1985).

Because sands are very easily disturbed during sampling, it is almost impossible to determine in a laboratory test whether

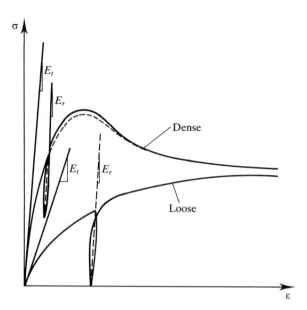

Fig. 5.12 Stress–strain curve for a typical sand in loose and dense states.

a sample has been prestrained. Therefore, indirect means are used. Examination of the geology and geologic history of the deposit can be very instructive. For example, if construction excavation and replacement has taken place, or if sand dunes have moved over the area, as often happens in lake shore and sea coastal regions, then the sand has almost certainly been preloaded. In-situ tests used to measure soil variability such as the SPT or the Dutch cone penetrometer (CPT) are quite insensitive to changes in modulus due to prestraining because they measure the ultimate or failure strength of the material. Penetration resistance is increased very slightly, probably less than 10 percent, owing to prestraining, whereas an increase in compressibility from 5 to 30 times can occur. Therefore, correlations between, for example, SPT and/or CPT and modulus can be in error by a factor of 5 to 30.

Plate load tests and other in-situ tests also may be used to determine whether a sand deposit has been prestressed. Good references on prestraining and in-situ tests include Dahlberg (1975) and Jamiolkowski et al. (1985). The prerequisites for a successsful plate load test on sand are discussed in some detail by Terzaghi and Peck (1967).

Plate load tests may be somewhat difficult to interpret. A small prestress effect may occur if an excavation is made to enable the plate test to be performed near the proposed foundation elevation. The location of the groundwater table may also affect the results, as will soil variability. If the deposit has a highly variable in-situ density, the plate load test will give inappropriate results. The most important point is to use a correct factor to scale from the plate load test up to the footing size, and the Terzaghi and Peck (1967) chart and scaling factor have been slightly modified by G. A. Leonards (personal communication, 1987). The Terzaghi and Peck charts were developed for normally consolidated sands above the groundwater table and relative density measured by the SPT. If the results of a plate load test indicate a significant difference from the charts, the deposit is most likely prestressed.

Use of correlations developed for normally consolidated sands may yield very misleading results if they are used for sands that have been previously prestrained. Therefore, it is best to try to measure in-situ compressibility directly by the use of plate load and other in-situ tests such as the pressuremeter, dilatometer, or screwplate, as discussed near the end of this section.

5.5.2 Procedures to Estimate the Settlements of Foundations on Granular Materials

Elastic Theory and Recent Modifications According to Harr (1966), elastic theory is utilized for estimating the settlements of shallow foundations on granular materials in the U.S.S.R. This process must involve some empirical corrections, because of the great influence confining pressure has on the modulus and compressibility. A related development is the Oweis (1979) method for predicting settlements based on an equivalent linear model using the deformation that would be caused by plate load tests at depth. The model was calibrated by use of actual plate load test data. Recently Bowles (1987) proposed that settlements of shallow foundations on sands could be estimated using Steinbrenner's (1936) modification of the Boussinesq equations by adjusting appropriately the strain influence factor. For the cases he investigated, good results were obtained. In another recent development, Hardin (1987) proposed a model for one-dimensional strain that appears to represent very adequately the shear stress–strain behavior for normally consolidated cohesionless soils over a wide range of stresses. The model is potentially useful for estimating settlements of structures that can be approximated by one-dimensional strain.

Empirical Procedures In view of the difficulties with (1) elastic theory and (2) plate load and other in-situ tests, a number of empirical procedures have been developed. These include the procedures by Terzaghi and Peck (1967) and by Peck et al. (1974) based on the results of the standard penetration test (SPT). Schmertmann (1970) developed a procedure using cone penetration tests to determine in-situ compressibility, and this method and its later modifications will be described in some detail in this chapter. Other empirical in-situ procedures will also be described briefly. Finally, Burland and Burbidge (1985) have conducted an extensive study of some 200 case histories of settlements of shallow foundations on granular materials and expanded on a similar study reported by Burland et al. (1977). We will use the Burland procedures to check results from the other empirical procedures.

5.5.3 Schmertmann (1970) Procedure

In 1970, J. H. Schmertmann proposed a new procedure for estimating the settlement of shallow foundations on granular soils. Although empirical, the procedure has a rational basis in the theory of elasticity, finite element analyses, and observations from field measurements and laboratory model studies. From the theory of elasticity, the distribution of vertical strain ε_z within the linear elastic halfspace subjected to a uniformly distributed load over an area at the surface can be determined by

$$\varepsilon_z = \frac{\Delta q}{E} I_z \qquad (5.5)$$

where

Δq = the intensity of the uniformly distributed load
E = Young's modulus of the elastic medium
I_z = a strain influence factor, which depends only upon the Poisson's ratio and the location of the point for which the strain is being evaluated

Based on the results of displacement measurements within sand masses loaded by model footings, as well as finite element analyses and deformations of materials with nonlinear stress–strain behavior, the distribution of strain within loaded granular masses is very similar in form to that for a linear elastic medium.

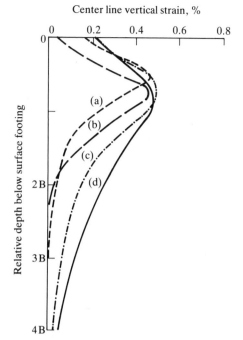

Center line vertical strain, %

Fig. 5.13 Comparisons of vertical strain distributions from FEM studies and from rigid model tests. (a) Hartman FEM axisymmetric, $B = 6.1$ m, $\Delta p = 287$ kN/m². (b) Brown model, $B = 150$ mm, $L/B = 1$, $\Delta p = 10$ kN/m². (c) Hartman FEM plane strain, $B = 6.1$ m, $\Delta p = 383$ kN/m². (d) Brown model, $B = 150$ mm, $L/B = 4$, $\Delta p = 10$ kN/m². Both (a) and (c) are with Duncan and Chang $K = 60\,000$, $\phi = 42°$, $v = 0.4$, $K_0 = 0.50$. Both (b) and (d) are averages of 3 tests. (*Schmertmann et al., 1978.*)

Some typical results of model tests and finite element analyses reported by Schmertmann (1978) are shown in Figure 5.13. Other results are shown by Schmertmann (1970) and Perloff (1975). On the basis of these observations, Schmertmann (1970) suggested that for practical purposes the distribution of vertical strain within a granular mass could be expressed by Equation 5.5 in which the Young's modulus might vary from point to point. The strain influence factor could be approximated by a triangle with a maximum value of 0.6 at $z/B = 0.5$ and $I_z = 0$ at a depth of $z/B = 2$. Schmertmann (1970) refers to this as a "2B–0.6 distribution."

The surface settlement s_i is the integration of the strains:

$$s_i = \int_{z=0}^{\infty} \varepsilon_z \, dz \qquad (5.6a)$$

which can be approximated by

$$s_i = \Delta q \int_0^{2B} \frac{I_z}{E} \, dz \qquad (5.6b)$$

This relationship can be approximated further as a summation of settlements of convenient approximately homogeneous layers, or

$$s_i = C_1 C_2 \Delta q \sum_{i=1}^{n} \left(\frac{I_z}{E}\right)_i \Delta z_i \qquad (5.6c)$$

in which

Δq = net load intensity at the foundation depth
I_z = strain influence factor from the 2B–0.6 distribution
E = appropriate Young's modulus at the middle of the ith layer of thickness Δz_i
C_1, C_2 = correction factors as described below

To incorporate the effect of strain relief due to embedment and yet retain simplicity for practical design purposes, the method assumes that the 2B–0.6 distribution of the strain influence factor is unchanged but its maximum value is modified. The suggested factor is

$$C_1 = 1 - 0.5\left(\frac{\sigma'_{v0}}{\Delta q}\right) \geqslant 0.5 \qquad (5.7)$$

where

σ'_{v0} = effective in-situ overburden stress at the foundation depth

Δq = net foundation pressure

In all cases, however, it is suggested that this correction factor not be less than 0.5.

Schmertmann (1970) also included a second correction factor, C_2, to account for some time-independent increase in settlement that was observed even for foundations on presumably cohesionless soils. In the cases studied by Schmertmann, time-dependent settlements probably occurred as a result of the consolidation of thin strata of silts and clays within the sands. Consequently, because the elastic distribution is inappropriate for cohesive soils and the method uses the Dutch cone penetration test (CPT) to estimate modulus, which is questionable for cohesive soils, the use of the correction factor C_2 is not recommended; therefore, use C_2 equal to 1.0 in Equation 5.6c.

No account was taken in the original procedure of the influence of foundation shape on the strain distribution, because as a foundation shape changes from approximately axisymmetric to approximately plane strain conditions, the angle of shearing resistance increases and the stresses at a given depth also increase. These two effects were thought to cancel each other, giving a strain distribution that is, perhaps, not very different for a wide range of length-to-width ratios.

Model test results suggest that when a rigid boundary lies within the 2B–0.6 distribution, the distribution of the strain influence factor will be simply truncated at the depth, with the slopes of the distribution remaining as for the homogeneous case.

Modifications of 1978 A number of modifications have been made by Schmertmann et al. (1978) and Schmertmann (1978). The strain influence diagram was modified slightly on the basis of extensive analytical studies, and axisymmetric and plane strain loadings are now considered separately. The modified strain influence diagram is shown in Figure 5.14. Note that the depth of the strain influence factor goes to 2B for the axisymmetric case and to 4B for plane strain conditions. The maximum value of the influence factor is at least 0.5 plus an incremental increase relative to the effective vertical overburden pressure at the depth of the maximum value. An explanation of the pressure terms in I_{zp} is shown in Figure 5.14b. Schmertmann (1978) recommends that if L/B is greater than 1 and less than 10, both the axisymmetric and plane strain cases can be calculated and interpolated for the actual L/B ratio.

As before, this method is only appropriate for normally loaded sands where the bearing capacity of the sand is adequate. If the sand has been prestrained by previous loading, then the real settlements will, as explained earlier, be greatly overpredicted by this method. Schmertmann (1978) recommends that a tentative reduction in settlement after preloading or other means of compaction of half the predicted settlement be used, and this is probably still conservative. There may also be some additional settlement effect due to dynamic, cyclic, or vibratory loads. This of course is a very serious potential problem for loose sands below the water table. Some type of densification or prestressing is an easy and effective way of reducing the potential for liquefaction or other undesirable behavior.

The correction factors C_1 (Eq. 5.7) and C_2 are unchanged. Also, as before, the correction factor C_2 is subject to question.

The use of this method to estimate the settlement of a shallow foundation on sand is illustrated by an example later in this section.

$C_2 = 1 + 0.2 \log\left(\frac{T}{0.1}\right)$ T=in years

v. conservative (39)back

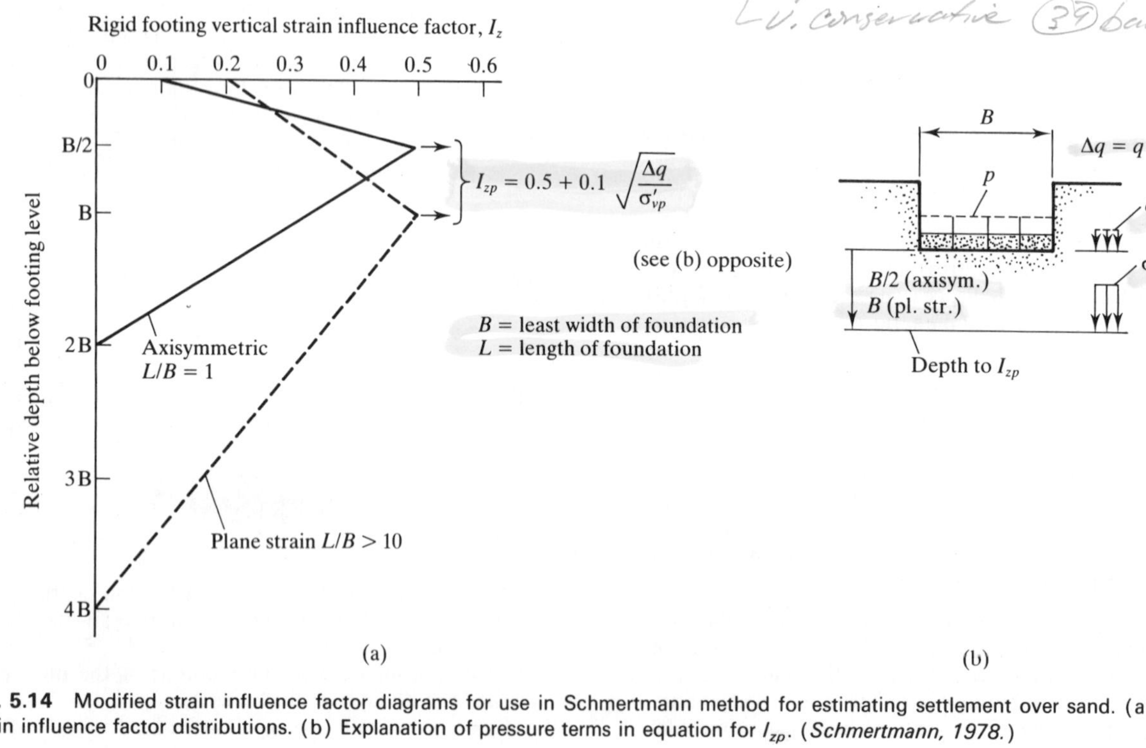

Fig. 5.14 Modified strain influence factor diagrams for use in Schmertmann method for estimating settlement over sand. (a) Simplified strain influence factor distributions. (b) Explanation of pressure terms in equation for I_{zp}. (*Schmertmann, 1978.*)

Determination of Equivalent Young's Modulus Using the CPT
In order to use the Schmertmann (1970, 1978) method it is necessary to estimate the stiffness of the soil in terms of an equivalent Young's modulus at various depths. Schmertmann (1970) suggested that this could be done using the Dutch cone penetration resistance of the soils at the site. The Dutch cone is a quasistatic cone penetrometer (CPT) that was developed in Holland in the early 1930s. The CPT provides a convenient and rapid way of measuring the bearing capacity and providing soil properties through correlations at various depths. In the case of normally loaded cohesionless materials (not prestressed significantly to pressures above the present in-situ overburden pressure), the CPT bearing capacity q_c has been correlated with Young's modulus E_s by DeBeer (1965) and Webb (1969). The relationship suggested by Schmertmann (1970) is

$$E_s = 2q_c \tag{5.8}$$

where q_c = CPT bearing capacity (in kgf/cm² or tonf/ft²). The *Canadian Foundation Engineering Manual* (CFEM) (Meyerhof and Fellenius, 1985) suggests the following:

$$E_s = kq_c \tag{5.9}$$

where

  $k = 1.5$ for silts and sand
  $ = 2$ for compact sand
  $ = 3$ for dense sand
  $ = 4$ for sand and gravel

Schmertmann (1978) and Schmertmann et al. (1978) recommend that for axisymmetric conditions,

$$E_s = 2.5q_c \tag{5.10}$$

and for plane strain conditions,

$$E_s = 3.5q_c \tag{5.11}$$

Presumably these values should be used with essentially free-draining sands and silty sands. The E_s is the equivalent Young's modulus for sand for footing-type loadings. The CFEM (1985) calls E_s the "apparent modulus of elasticity." It is related to the constrained modulus M by the relationship

$$M = \frac{E(1-v)}{(1+v)(1-2v)} \tag{5.12}$$

For drained granular materials, Poisson's ratio v is probably around $\frac{1}{4}$ to $\frac{1}{3}$, which means that $M = 1.2E_s$ to $1.5E_s$.

As described previously, because the stiffness of the cohesionless material is more strongly influenced than its strength by prestress effects, the correlations given in Equations 5.8 to 5.11 are likely to underestimate the equivalent Young's modulus for overconsolidated cohesionless soils. Prestress will also influence the strain distribution; however, the extent is not known. How to account for the influence of prestress was discussed earlier in this section, and some additional remarks will now be made.

In fact, Bellotti et al. (1986) state that there is no unique relationship between the modulus and q_c because the ratio is highly dependent on the soil and the strain and stress history of the deposit. In addition, with prestressed sands, this relationship also depends on the relative density of the sand, a further complicating factor. Jamiolkowski et al. (1985) present a complete review of the problems of obtaining deformation moduli from CPT tests. The relationships mentioned above between modulus and q_c may be determined through calibration chamber tests for specific types of sands, but unfortunately such tests are expensive and the equipment is not generally available.

There seem to be two contributors to the observed increase in modulus due to overconsolidation. The first is the strain

hardening produced by accumulated plastic strains. The second corresponds to an increase in the σ'_{h0} for a given level of σ'_{v0}. Some recent research by Jamiolkowski et al. (1988) is shown in Figures 5.15 and 5.16 as determined from calibration chamber tests. They note that the ratios M/q_c and E/q_c are much higher

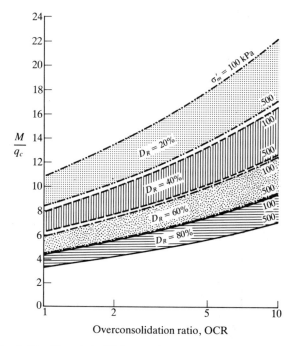

Fig. 5.15 *M* vs. q_c for Ticino sand.

$$\frac{M}{q_c} = C_0 P_a \left(\frac{\sigma'_m}{p_a}\right)^{C_1} \text{OCR}^{C_2} \exp(C_3 D_R)$$

M = tangent constrained modulus
σ'_m = mean effective stress
$C_0 = 14.48$; $C_1 = -0.116$; $C_2 = 0.313$; $C_3 = -1.123$; $R = 0.95$;
$p_a = 1$ bar $= 98.1$ kPa. (*Jamiolkowski et al., 1988.*)

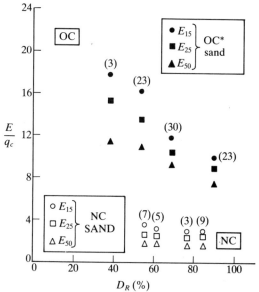

( ) Number of CK_oD triaxial compression
tests considered

(*) $2 \leqslant \text{OCR} \leqslant 8$

Fig. 5.16 *E* vs. q_c for Ticino sand. (*Jamiolkowski et al., 1988.*)

for the mechanically overconsolidated sands than for the normally consolidated sands. Therefore, without an a priori knowledge or assumption of the stress history of the deposit, it is impossible to select a reliable value of the design modulus versus q_c ratio. Note also from Figure 5.15 that the mean effective stress also influences the M/q_c ratio.

Meigh (1987) gives a good summary of recent work using CPT tests for estimating sand deformability.

In estimating the settlements of a number of footings at a granular site, especially if the CPT values are highly erratic, the approach proposed by Peck et al. (1974) for the SPT is recommended. The heaviest column load is placed over the loosest SPT area or profile and the lightest column load is placed over the densest SPT area or profile at the site to obtain an estimate of the maximum probable differential settlement.

EXAMPLE 5.5

The bridge pier shown schematically in Figure 5.17a is to be constructed at the groundwater table, which is 2 m below the surface of a medium loose to dense sand stratum of considerable depth. The CPT profile is given in Figure 5.17b. Determine the settlement of the pier.

Solution

1. Determine the L/B ratio. For this example, $L/B = 8.8 \approx$ 10. Therefore, use the plane strain influence diagram.
2. Plot the strain influence factor distribution as shown in Figure 5.17c. The maximum strain influence value is determined from Figure 5.14 as $I_{zp} = 0.5 + 0.1\sqrt{\Delta q/\sigma'_{vp}}$

$$\Delta q = 178 - 15.7(2\,\text{m}) = 147\,\text{kPa}$$
$$\sigma'_{vp} = 15.7(2\,\text{m}) + (15.7 - 9.8)(2.6\,\text{m}) = 47.6\,\text{kPa}$$

Thus, $I_{zp} = 0.68$, and this maximum value occurs at a depth of B or 2.6 m below the foundation. The triangle extends to a depth of $4B$ or 10.4 m.

3. Based on the q_c profile and the strain influence factor diagram, divide the $4B$ depth into a convenient number of layers, as shown in Figure 5.17c and columns 1 and 2 of Table 5.5.
4. Determine the average q_c for each layer, and then calculate the corresponding E_s from Equation 5.11, as indicated in columns 5 and 6 of Table 5.5.
5. Locate the depth of the middle of each layer (column 3 of Table 5.5) and determine the I_z value for the depth from Figure 5.17c. The values are given in column 4 of Table 5.6.
6. Calculate $(I_z/E_s)\Delta z$ for each layer and sum the results in column 7, Table 5.5.
7. Determine C_1 from Equation 5.7. Initial overburden pressure at the foundation depth, $\sigma'_{v0} = 15.7 \times 2\,\text{m} = 31\,\text{kPa}$. The *net* foundation pressure, Δq is 178 kPa − 31 kPa = 147 kPa. Thus, $C_1 = 1 - 0.5(\sigma'_{z0}/\Delta q) = 1 - 0.5(31/147) = 0.89$.
8. Calculate settlement from Equation 5.6c: $s_i = C_1\Delta q \Sigma (I_z/E_s)_i \Delta z_i = 0.89(147)(0.237) = 31\,\text{mm}$.

Use of Other In Situ Tests to Estimate E The static CPT bearing capacity is more reliable than the standard penetration

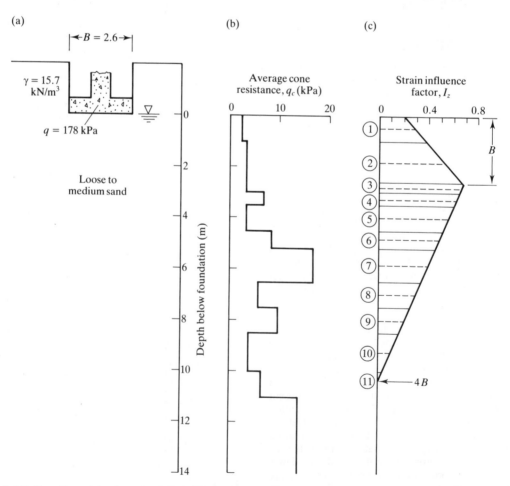

Fig. 5.17 (a) Soil profile and footing; (b) CPT profile; (c) strain influence diagram for Example 5.5. (*After Schmertmann, 1978.*)

TABLE 5.5 DATA FOR EXAMPLE 5.5[a].

Layer, i (1)	Δz (m) (2)	Distance z_i to center of Δz_i (m) (3)	I_z (4)	Average q (MPa) (5)	Average E_s (MPa) (6)	$I_z \Delta z / E_s$ (m/MPa) (7)
1	1.0	0.5	0.28	2.5	8.75	0.032
2	1.6	1.8	0.52	3.5	12.25	0.068
3	0.4	2.8	0.66	3.5	12.25	0.022
4	0.5	3.25	0.62	7.0	24.5	0.013
5	1.0	4.0	0.55	3.0	10.5	0.052
6	0.7	4.85	0.48	8.5	29.75	0.011
7	1.3	5.85	0.39	17.0	59.5	0.009
8	1.0	7.0	0.29	6.0	21.0	0.014
9	1.0	8.0	0.20	10.0	35.0	0.006
10	1.5	9.25	0.09	4.0	14.0	0.010
11	0.4	10.2	0.02	6.5	22.75	0.0003
	$\Sigma = 10.4 = 4B$					$\Sigma = 0.237$

[a] Modified from Schmertmann (1978).

TABLE 5.6 CORRELATION BETWEEN CPT RESISTANCE q_c AND SPT RESISTANCE N[a].

Soil Type	q_c / N[b]
Silts, sandy silts, slightly cohesive silt–sand mixtures	2–4
Clean, fine to medium sands and slightly silty sands	3–5
Coarse sands and sands with little gravel	4–5
Sandy gravel and gravel	6–8

[a] After Schmertmann (1970, 1978).
[b] Units of q_c are kgf/cm^2 or tons/ft^2.
Units of N are blows/ft or blows/30 cm.

test (SPT) for granular deposits without substantial gravel, because the SPT is a dynamic test and has considerable experimental difficulties and uncertainties. However, the SPT is widely used throughout the world, and it is recognized that SPT data may be available for a number of sites where CPT data are not. To permit the use of SPT data as a "temporary expedient", Schmertmann (1970) suggests that a conservative correlation between CPT data and SPT resistance can be used, such as is given in Table 5.6. It is of course preferable to obtain CPT resistance directly when using the Schmertmann (1970, 1978) method.

On the basis of a number of studies, Robertson et al. (1983) found the q_c / N ratio to vary with grain size of the penetrated sands (Fig. 5.18). This chart can be used to convert N_{SPT} values for 55 percent efficiency to CPT data. If other SPT equipment that produces a higher than average energy is used, then the q_c / N values will be slightly higher than those shown in Figure 5.18. Kasim et al. (1986) report additional q_c / N values that confirm the relationship in Figure 5.18. Additional aspects of SPT–CPT correlations are discussed by Robertson et al. (1983).

A similar relationship between q_c / N and grain size as determined by Burland and Burbidge (1985) from a large number of case histories in the United Kingdom is given in Figure 5.19. Note the different units in the ordinates, and that Figure 5.19 includes fine to medium gravels.

Other correlations of the form $E_s = a + bN$ relating the SPT N value and average settlement modulus have been reported by Anagnostopoulos (1990). From laboratory tests,

$$E_s = 7.5 + 0.8N \qquad (5.13a)$$

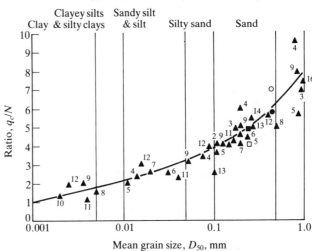

q_c, bars; N, blows/ft (1 bar = 100 kPa)

Fig. 5.18 Variation of q_c / N ratio with mean grain size. (*After Robertson et al., 1983.*)

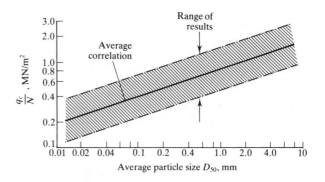

Fig. 5.19 Relationship between cone penetration test and standard penetration test. (*From Burland and Burbidge, 1985.*)

while back analysis from 30 case histories of foundations with a width of 1 to 4 m gives

$$E_s = 4.8 + 1.25N \qquad (5.13b)$$

In both these equations, E_s is in megapascals.

Skempton (1986) presents an excellent review of SPT procedures and problems.

Other in-situ methods such as the plate load tests, screw plate load tests, pressuremeter, dilatometer, and seismic velocity, may be used to obtain the settlement or compression modulus directly.

The plate load test (ASTM D1194) is one of the older in-situ tests and, as discussed earlier in this section, it can be used to determine whether a deposit has been prestressed. It can also provide reliable estimates of the deformation moduli and therefore of settlements. Performing the PLT at depth is difficult and expensive and this led to the development of the screw plate load test (SPLT). For SPLT procedures, see Schmertmann (1970a). Dahlberg (1975), Jamiolkowski et al. (1985), and Robertson (1986) present good reviews of both the PLT and SPLT.

These same authors, plus Mair and Wood (1987), discuss the pressuremeter test (PMT) and self-boring pressuremeter (SBP). Dahlberg (1975) also used the conventional PMT in an investigation of overconsolidated sands.

The dilatometer test (DMT) is a relatively recent addition to the family of in-situ tests and it is gaining acceptance in North America and Europe. A good review of the DMT is provided by Jamiolkowski et al. (1985) and Robertson (1986). The use of this test for predicting foundation settlements will be described below.

Another in-situ method uses the seismic velocity determined from downhole and crosshole techniques to obtain the elastic moduli for settlement analyses. This approach was proposed by Swiger (1974) and successfully applied by Konstantinidis et al. (1986) and by Martin et al. (1986), among others.

5.5.4 Other Methods for Estimating Settlement of Shallow Foundations on Sand

Dilatometer Methods Schmertmann (1986) describes a method of computing foundation settlement based on the results of dilatometer (DMT) tests. The procedure is for one-dimensional compression and uses correlations of DMT-determined parameters to obtain the constrained (one-dimensional) modulus M. Comparison with the results of a number of full-scale settlement measurements indicates quite good predictions. Leonards and Frost (1988) take a somewhat different approach to the prediction of settlements of shallow foundations utilizing the DMT. The advantages of their method are that the dilatometer modulus E_D is used directly and possible prestress is taken care of by considering the E_D/q_c ratios for the deposit. Thus, it is possible to avoid large overpredictions when the possibility of prestress is not appropriately considered.

SPT Methods As noted by Bellotti et al. (1986), there are significant disadvantages to using the SPT, CPT, or DMT because of the differences in moduli as determined for normally consolidated versus overconsolidated sands. As these devices only are modestly sensitive to stress and strain history of the sands, it may be better to refer to the correlations between some average values of the penetration resistance and the settlement of actual foundations rather than to use direct single correlations between individual values of N_{SPT} and the moduli in cohesionless soils.

This was the approach taken by Burland and Burbidge (1985), who developed an indirect method based on an extensive

review of more than 200 case histories of the settlement of shallow foundations, tanks, and embankments on sands and gravels. Most of the subsurface information was average SPT blow counts, but some CPT results were available. The following is a condensation of the Burland and Burbidge (1985) method, as provided by Meigh (1985).

The average settlement s (mm) of a foundation on sand or gravel may be expressed as

$$s = f_s \times f_l f_t [(q' - \tfrac{2}{3}\sigma'_{v0})B^{0.7}I_c] \qquad (5.14)$$

where

q' = average gross effective applied pressure (kN/m²)
σ'_{v0} = maximum previous effective overburden pressure
B = width (m)
I_c = compressibility index
f_s = shape correction factor
f_l = correction factor for thickness of sand layer
f_t = time factor

Note that the maximum previous overburden pressure is σ'_{v0} for both an overconsolidated sand or for loading at the base of an excavation.

The compressibility index I_c is read from the solid line on Figure 5.20, or it can be computed from

$$I_c = \frac{1.71}{\bar{N}^{1.4}} \qquad (5.15)$$

where $\bar{N}$ = average SPT blow count over the depth of influence of the foundation, Z_I. This depth of influence Z_I can be determined from Figure 5.21. Note that the N values are not corrected for overburden pressure but the Terzaghi correction for sands below the water table is applied:

$$N_{corrected} = 15 + 0.5(N - 15) \qquad (5.16)$$

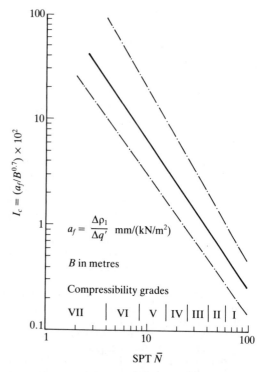

Fig. 5.20 Relationship between compressibility index (I_c) and mean SPT blow count ($\bar{N}$) over depth of influence. Chain dotted lines show upper and lower limits. (*Burland and Burbidge, 1985.*)

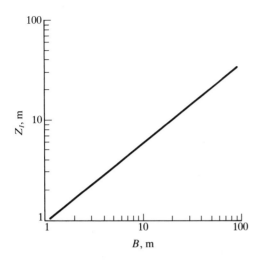

Fig. 5.21 Relationship between width of loaded area B and depth of influence z_1 (within which 75 percent of the settlement takes place). (*Burland and Burbidge, 1985.*)

for $N > 15$. If $N < 15$, no correction for water table is recommended. When the material consists of gravel or sandy gravel, a correction such as

$$N_{\text{corrected}} = 1.25N \qquad (5.17)$$

is appropriate. The shape correction factor f_s is given by

$$f_s = \left(\frac{1.25L/B}{L/B + 0.25} \right)^2$$

where the thickness of the granular layer below the foundation, H_s, is less than the depth of influence Z_I. The correction factor f_l for the thickness of the layers is given by

$$f_l = \frac{H_s}{Z_I} \left(2 - \frac{H_s}{Z_I} \right) \qquad (5.18)$$

where, as before, Z_I is taken from Figure 5.21. The time correction factor f_t is given by

$$f_t = \frac{s_t}{s_i} = \left(1 + R_3 + R_t \log \frac{t}{3} \right) \qquad (5.19)$$

where

- t = time in years ($t \geqslant 3$ years)
- R_3 = time-dependent settlement (expressed as a ratio of the immediate settlement s_i) that takes place during the first 3 years after construction
- R_t = time-dependent settlement (again expressed as a ratio of s_i) that takes place each log cycle of time after 3 years.

For static loads, conservative values of R_3 and R_t are 0.3 and 0.2, respectively. Thus, at $t = 30$ years, $f_t = 1.5$. For fluctuating loads, conservative values of R_3 and R_t are 0.7 and 0.8, respectively, so that at $t = 30$ years, $f_t = 2.5$.

A few explanatory points follow. For a normally consolidated sand, the immediate settlement s_i in mm, corresponding to the average effective foundation pressure q' (kN/m^2), is

$$s_i = q'B^{0.7}I_c \qquad (5.20a)$$

The dimension of B is meters, and values of I_c are obtained from Figure 5.20 or Equation 5.15. For an overconsolidated sand, or for loading at the base of an excavation for which the previous effective overburden pressure is σ'_{v0}, the average settlement s_i (mm) corresponding to the average gross effective

pressure q' (where $q' > \sigma'_{v0}$) has two components:

$$s_i = \sigma'_{v0} B^{0.7} \frac{I_c}{3} + (q' - \sigma'_{v0}) B^{0.7} I_c$$

$$= (q' - \tfrac{2}{3}\sigma'_{v0})(B^{0.7} I_c) \qquad (5.20b)$$

When $q' < \sigma'_{v0}$

$$s_i = q' B^{0.7} \frac{I_c}{3} \qquad (5.20c)$$

The case records examined by Burland and Burbidge (1985) indicated that foundations on sands and gravels exhibited time-dependent settlement. However, they were not able to find a distinct pattern. Foundations with fluctuating loads such as tall chimneys, bridges, silos, and turbines exhibited much larger time-dependent settlements than those subjected only to static loads. The time correction factor f_t given above reflects these observations.

Finally, it should be noted that the probable limits of accuracy of Equations 5.14 and 5.15 can be assessed from the upper and lower limits of I_c shown as the chain-dotted lines in Figure 5.20, which are approximately two standard deviations above and below the mean line. If the factor of safety against bearing capacity is less than about 3, the pressure–settlement curve may be nonlinear and the Burland–Burbidge method will probably underestimate the settlements. The case histories they studied involved quartzitic sand and gravel deposits. Sites with carbonate and other mineralogically different sands and gravel deposits should not be analyzed by this method unless the deformation properties of these deposits are similar to those of quartzitic deposits. Finally, for a given project it may be useful to refer to the case histories given by Burland and Burbidge (1985), with similar structures or ground conditions.

Burland et al. (1977) suggested a method that provides a useful quick check using SPT results to indicate whether a settlement problem exists. The method uses the data collected in Figure 5.22, in which the settlement per unit pressure (mm/(kN/m^2)) is plotted against foundation width for loose, medium, and dense sands. The data points are taken from the literature and referenced in Burland et al. (1977). "Upper limit" lines for the three relative densities are indicated in the figure. In each case, Burland et al. (1977) suggest that the probable settlement can be taken as equal to half the upper limit values indicated in the figure and that maximum settlement usually does not exceed about 1.5 times the probable value. Soils in the upper zone, which tend to be loose, slightly silty and organic sands, probably should not be used for shallow building foundations; however, the line could provide a useful preliminary assessment of settlement of structures such as storage tanks on loose sands. Burland et al. (1977) also comment that loose silts are a very difficult foundation material and may be even less suitable for supporting footings than soft normally consolidated clay. As Terzaghi and Peck (1967) recommend, medium and dense silts should be treated as nonplastic materials the same as for sands, and silts with some plasticity should be considered as clay soils.

5.6 CALCULATION OF STRESS DISTRIBUTIONS

As discussed in the following section, determination of the distribution of stresses within a compressible soil mass is an important part of the process of estimating settlements of cohesive soil strata. If a very large area is loaded—an area in which the thickness of the compressible layer is significantly less than the minimum dimension of the loaded area—then the

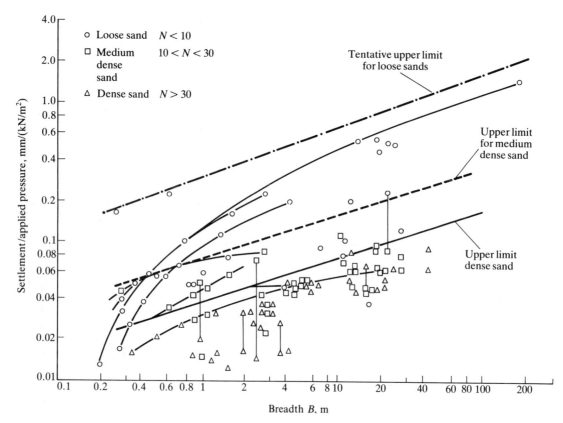

Fig. 5.22 Observed settlement of footings on sand of various relative densities. (*After Burland et al., 1977.*)

loading can be considered to be one-dimensional, and the stress increase with depth is equal to the applied stress at the surface. However, near the edge or end of the area it might be expected that a certain amount of attenuation of stress with depth would occur, because no stress is applied beyond the edge. Similarly, with a loaded area of limited size, it might be expected that the applied stress at the surface would dissipate rather rapidly with depth.

For shallow foundation problems, solutions from the theory of elasticity or approximate methods such as the "2:1 method" are most commonly used to evaluate stresses with depth under areas of limited extent. The justification for the use of the theory of elasticity was presented in Section 5.3. Frequently used solutions from the theory of elasticity, usually in the form of convenient graphical charts, can be found in Scott (1963), Harr (1966), Poulos and Davis (1974), Perloff (1975), Perloff and Baron (1976), Holtz and Kovacs (1981), and U.S. Navy (1982), among others. In some cases, the equations are given and solutions to often-used problems can be readily programmed on a microcomputer or a programmable calculator. As these solutions are available in most design offices, they will not be presented here. Instead, the probabilistic approach to stress distribution developed by Harr (1977) will be presented because it is not generally available.

First, a few remarks about the conventional approach to calculating stress distributions. For shallow foundation problems, the most important original solution was given by Boussinesq (1885) for the distribution of stresses within a linear elastic halfspace due to a point load applied perpendicularly to the surface. For the vertical stress, the solution is solely in terms of geometry, that is, no material properties are involved. Only the ratio of stress to strain should be a constant; small strains are assumed so that higher-order terms may be neglected. For horizontal and shear stresses, Poisson's ratio is required.

As most foundation loadings do not act at a point but are distributed over an area, vertical stresses at depth are found by integrating the Boussinesq equations over areas of different geometric configurations. As linear elasticity is assumed, superposition holds. Thus, the effects of different magnitudes and shapes of loaded areas and different distances from the point under consideration can be obtained. A simple but useful graphical integration procedure in the form of influence diagrams was developed by Newmark (1942).

The magnitude of stress that is determined from stress distribution charts or solutions to Boussinesq equations is that due only to the applied loading being considered. Preexisting in-situ stresses must be added to those determined by the use of these diagrams.

5.6.1 Distribution of Stresses

After the stress increments due to surface loadings at a variety of points within a soil mass have been determined, it may be useful to depict these stresses graphically relative to their points of action. Several methods for doing this are commonly used. One method is to plot a profile of the stress increments as a function of depth below selected points under the loaded area. Figure 5.23a shows the distribution of vertical stress beneath the centerline of a uniformly loaded circular area, and beneath a point a distance of two radii from the centerline. The stress is plotted so that the ordinate is the depth expressed as a ratio of the radius, and the abscissa is the magnitude of the stress normalized by the load p. Note that the difference in the attenuation of stress with depth depends on the location of the vertical section. At the centerline, the stress due to the loaded area is a maximum immediately below the loaded area, and attenuates to less than 10 percent of the surface stress at a depth

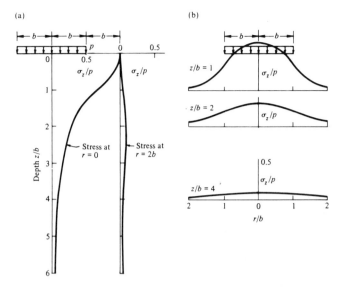

Fig. 5.23 Distribution of vertical stress due to a loaded circular area on a linear elastic halfspace (a) along vertical lines, (b) along horizontal lines. (*Perloff, 1975.*)

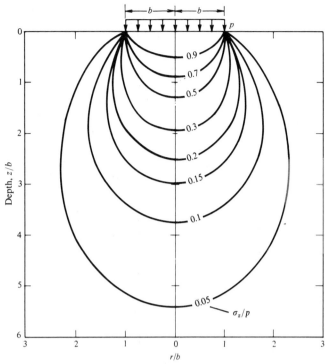

Fig. 5.24 Contours of constant vertical stress (isobars) beneath a uniformly loaded circular area on a linear elastic halfspace. (*Perloff, 1975.*)

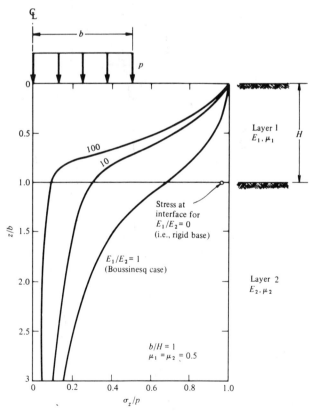

Fig. 5.25 Vertical normal stress beneath the center of a uniformly loaded circular area at the surface of a two-layer elastic system. (*Modified from Burmister, 1958, by Perloff, 1975.*)

equal to twice the width of the loaded area ($z/b = 4$ in Figure 5.23a). Beneath the point outside the loaded area, the stress is zero at the ground surface and increases to a maximum at a depth of approximately 1.2 times the diameter of the loaded area ($z/b = 2.4$). However, the magnitude of this stress changes only slightly with depth below this point.

The distribution of vertical stress along selected horizontal lines beneath the center of the loaded area is shown in Figure 5.23b. The stress is more concentrated beneath the area at shallow depths, but tends to spread more at increasing depths.

Another useful way to view the distribution of stresses is as contours of equal vertical stress called isobars (or arctic taverns), as shown in Figure 5.24 for a circular loaded area. The area contained within a given stress contour experiences stresses larger than the stress level indicated by that contour. For example, the zone within the stress contour for which $\sigma_z = 0.05p$ contains all of the material subjected to stresses (resulting from the loaded area) of that magnitude or greater. Because of the shape of this zone, it is often referred to as the bulb of pressure. Note that for a loaded area of a given shape on the surface of a linear elastic halfspace, the size of the pressure bulb is proportional to the size of the loaded area. Thus, when considering the settlement potential of a large structure, it should be remembered that the stresses increase with depth in direct proportion to the size of the loaded area.

5.6.2 Effects of Layered Systems

It is frequently necessary to determine the stresses in a compressible layer that underlies one or more layers of different mechanical properties. This problem has been analyzed by considering layered systems consisting of different elastic properties. Solutions are discussed by Poulos and Davis (1974), and Perloff (1975) gives a number of references. The most extensive use of layered elastic theory has been by pavement engineers.

Typical results of one such analysis are given in Figure 5.25, in which a uniformly distributed load is shown acting on a circular area on the surface of a two-layer elastic system. In this case, the thickness of the upper layer has been chosen equal to the radius of the loaded area. The vertical stress distribution under the centerline of the loaded area is shown as a function of depth for ratios of Young's moduli $E_1/E_2 = 100$, 10, and 1. When the upper layer is significantly stiffer than the lower layer, the stress in the lower layer is greatly reduced. Conversely, the stress in a layer underlain by a very stiff layer is markedly increased, as shown by the stress at the interface between the layers when the lower one is rigid. Thus, it might be expected that the stresses in a relatively thin compressible layer underlain by a stiff granular material or rock would be higher than those predicted by Boussinesq.

5.6.3 Approximate 2:1 or 60° Stress Distribution Methods

One of the simplest means for computing the distribution of stress beneath a loaded area with depth is to use the 2 to 1 (2:1) method. This is a very popular empirical approach based on the assumption that the area over which the load acts increases in a systematic way with depth. Sometimes a 60° distribution angle is used ("the 1.73 to 1 method"). Since the same vertical force is spread over an increasingly larger area, the unit stress decreases with depth as shown in Figure 5.26. In Figure 5.26a, a continuous footing is seen in elevation view. At depth z, the enlarged area of the footing increases by $z/2$ on each side. The width at depth z then is $B + z$, and the stress σ_z at that depth is $P/(B + z)$.

By analogy, a rectangular footing of width B and length L would have an area of $(B + z)(L + z)$ at a depth z, as shown in Figure 5.26b. The corresponding stress at depth z is also shown in the figure.

The relationship between the approximate distribution of stress determined by the 2:1 method and the exact distribution

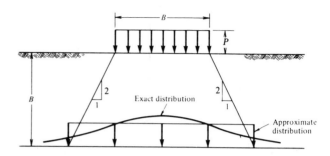

Fig. 5.27 Relationship between vertical stress below a square uniformly loaded area as determined by approximate and exact methods. (*Perloff, 1975.*)

is illustrated in Figure 5.27. Here, the vertical stress distribution at a depth $z = B$ beneath the uniformly loaded square area of width B, along a line that passes beneath the centerline of the area, is shown. Also shown is the assumed uniform distribution at depth B determined by the 2:1 method. The discrepancy between these two methods decreases as the ratio of depth to the size of the loaded area increases.

5.6.4 Effects of Gravity Structures

Embankments In most situations in foundation engineering, the loads imposed on a soil by the foundation can be simply represented by a system of boundary stresses without significant error in the calculated stress distributions. However, for stresses due to earth embankments, such an approximation may lead to important differences in the calculated stresses. Perloff (1975) suggests that a more reasonable approach is to consider the embankment and foundation as a single body loaded only by its own weight. As noted by Poulos and Davis (1974), there may be some inaccuracy in the results given by Perloff (1975) and Perloff and Baron (1976). The results are based on original studies by Baladi (1968), who used conformal mapping to solve the differential equations. The way self-weight is considered causes some inconsistencies, as the width to height ratio of the embankment becomes large. Thus, in using the Perloff (1975) charts, care should be exercised for cases where $L/H > 1$.

For important projects in which an accurate estimation of stresses is important, a finite element analysis (e.g., Duncan, 1972) is recommended.

Effect of Soft Foundations As shown by finite element analyses of a stiff elastic embankment overlying a less-stiff elastic foundation of limited depth (Perloff, 1975), the distribution of stresses arising from the weight of an embankment is in general affected by the magnitude of the relative stiffness of the embankment and its foundation. As the relative stiffness of the embankment increases, the vertical stress beneath the center decreases, especially as the ratio of the moduli becomes large. Shear stress at the base of the embankment is significantly decreased, and only a minor increase in horizontal stresses is observed. However, for modular ratios greater than 10, the lower half of the embankment exhibits rather large tensile stresses. Whether these actually occurred in an embankment would depend on the stress–strain characteristics of the embankment material and possible cracking and stress redistribution.

Stress Relief Due to Excavations Another circumstance in which stresses due to gravity forces are frequently of interest is that of unloading due to an excavation. Evaluation of the heave (rebound) that occurs undrained and could be analyzed by

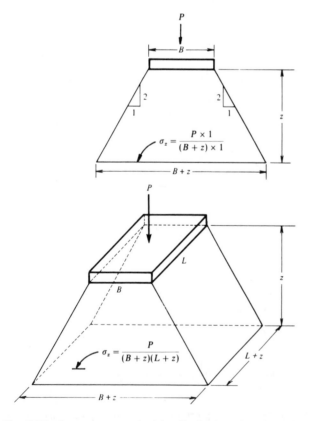

$$\sigma_z = \frac{P \times 1}{(B + z) \times 1}$$

$$\sigma_z = \frac{P}{(B + z)(L + z)}$$

Fig. 5.26 Approximate method for distribution of vertical stress due to surface load.

elastic theory was discussed previously in Section 5.4. As with the case of embankments, for important projects, detailed finite element analyses (e.g., Duncan, 1972) are appropriate. Design and performance of excavations in soft clays are discussed by Clough and Schmitt (1981).

5.6.5 The Probabilistic Approach (Harr, 1977)

Harr (1977), using the theory of probability, developed procedures for estimating the distribution with depth of stress at a point, as well as distributed loads, on the surface of a particulate medium. Assuming only that the medium is homogeneous, the distribution of vertical normal stress at a point depends only on the porosity of the medium and the expected value of the vertical normal stress at that point. Harr (1977) presents the theoretical background and derives the probabilistic stress distribution equations. These equations for several practical situations are given in this section, and examples to illustrate the use of the probabilistic approach are also presented.

Point Load The expected value for the vertical stress $\bar{\sigma}_z$ for a concentrated point load of intensity P acting at the origin of the coordinate system shown in Figure 5.28 is

$$\bar{\sigma}_z = \frac{P}{2\pi vz^2} \exp\left(-\frac{x^2 + y^2}{2vz^2}\right) = \frac{P}{2\pi vz^2} \exp\left(-\frac{r^2}{2vz^2}\right) \tag{5.21}$$

The v is the new coefficient of lateral earth pressure, and its physical meaning will be discussed later in this section. Figure 5.28b solves Equation 5.21 for a range of v values.

EXAMPLE 5.6

If $v = 1/5$, find the expected value of the vertical normal stress due to a concentrated vertical force of 100 kN at the point $r = 2$ m and $z = 3$ m.

Solution

The arrow in Figure 5.28 indicates that for the given conditions $vz^2\bar{\sigma}_z/P = 0.055$. Hence, the expected value of the vertical normal stress is

$$\bar{\sigma}_z = \frac{0.055(100)}{(1/5)(9)} = 3.1 \text{ kN/m}^2$$

For comparison, Boussinesq gives for this case $\sigma_z = 2.2 \text{ kN/m}^2$.

Line Load The expected value of the vertical normal stress under a line load of intensity P acting normal to the surface at the origin of coordinates (Fig. 5.29) is

$$\bar{\sigma}_z = \frac{P}{z\sqrt{2v\pi}} \exp\left(-\frac{x^2}{2vz^2}\right) \tag{5.22}$$

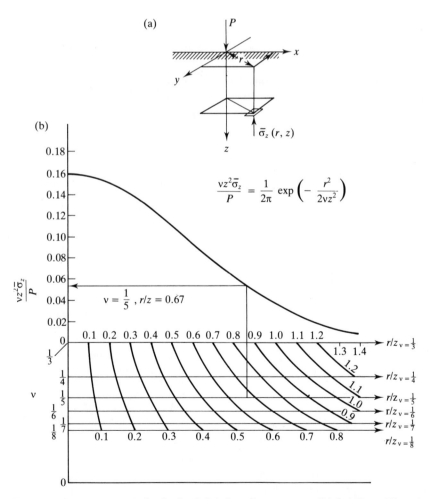

Fig. 5.28 Stress distribution under a concentrated point load. (a) Coordinate system. (b) Solution of Equation 5.21. (*Harr, 1977.*)

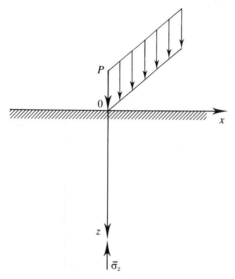

Fig. 5.29 Line load geometry.

Strip Load For a uniform normal load q acting on a strip of width $2a$ (Fig. 5.30b), the expected vertical normal stress under the centerline ($x = 0$) is

$$\bar{\sigma}_z = 2q\psi\left(\frac{a}{z\sqrt{v}}\right) \tag{5.23}$$

In general, this value is

$$\bar{\sigma}_z = q\left[\psi\left(\frac{x+a}{z\sqrt{v}}\right) - \psi\left(\frac{x-a}{z\sqrt{v}}\right)\right] \tag{5.24}$$

Note that values of the function $\psi(\ )$ are given in Table 5.7 and are tabulated values of the standardized normal distribution curve for a normally distributed random variable. Figure 5.30a is a plot of $\bar{\sigma}_z/q$ for a range of values of a and $z\sqrt{2v}$.

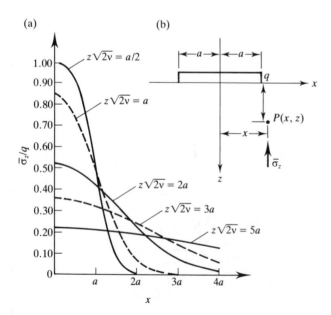

Fig. 5.30 Uniform normal load over strip. (a) Normalized vertical stress versus width of strip for several values of $z\sqrt{2v}$. (b) Geometry. (*After Harr, 1977.*)

EXAMPLE 5.7

Find the expected value of the vertical normal stress at a point $x = 2$ m and $z = 4$ m for a uniformly distributed load $q = 100 \, \text{kN/m}^2$ acting over a strip 8 m wide. Assume $v = \pi/8$.

Solution

Equation 5.24 becomes

$$\bar{\sigma}_z = 100\left[\psi\left(\frac{(2+4)}{4(0.63)}\right) - \psi\left(\frac{(2-4)}{4(0.63)}\right)\right]$$

$$= 100[\psi(2.39) - \psi(-0.80)]$$

From Table 5.7,

$$\bar{\sigma}_z = 100(0.4916 + 0.2881) = 78.0 \, \text{kN/m}^2$$

The theory of elasticity gives in this case $\sigma_z = 73.5 \, \text{kN/m}^2$.

Harr (1977) also gives equations for the expected value for the shear stress and horizontal normal stress under a uniform strip load.

Multilayer Systems The expected stress distribution at a depth z due to a line load of intensity P acting normal to the surface of a particulate medium at the origin of coordinates was given by Equation 5.22. Harr (1977) has extended the analysis to consider multiple layers of thickness h and with different values of the parameter v (Fig. 5.31). The equivalent thickness of the upper $N - 1$ layers is

$$H_{N-1} = h_1\sqrt{v_1/v_N} + h_2\sqrt{v_2/v_N} + \cdots + h_{N-1}\sqrt{(v_{N-1})/v_N} \tag{5.25}$$

and the expected vertical stress in the Nth layer (where z_N is the vertical distance into the Nth layer as measured from its upper boundary) is

$$\bar{\sigma}_z = \frac{P}{H_{N-1} + z_N}\sqrt{\frac{1}{2\pi v_N}}\exp\left(-\frac{x^2}{2v_N(H_{N-1} + z_N)^2}\right) \tag{5.26}$$

EXAMPLE 5.8

A three-layer system such as is shown in Figure 5.31 is subjected to a line load of $P = 900 \, \text{kN/m}$. The system geometries and properties are $h_1 = 1$ m, $v_1 = 0.4$; $h_2 = 2$ m, $v_2 = 0.3$; $v_3 = 0.2$ and h_3 is very deep. Find the expected value for the vertical normal stress 3 m into the third layer immediately under the line load (or 6 m below the ground surface).

Solution

From Equation 5.25, the equivalent thickness is

$$H = 1\sqrt{0.4/0.2} + 2\sqrt{0.3/0.2} = 3.86 \, \text{m}$$

Thus, from Equation 5.26, we have for the expected vertical normal stress at a depth of 3 m in the third layer immediately under the line load ($x = 0$),

$$\bar{\sigma}_z = \frac{900}{3.86 + 3}\sqrt{\frac{1}{2\pi(0.2)}} = 117 \, \text{kN/m}^2$$

Uniform Vertical Load Over a Rectangular Area The expected vertical stress under a corner of a rectangular area of size $a \times b$ is

$$\bar{\sigma}_z = q\psi\left(\frac{a}{z\sqrt{v}}\right)\psi\left(\frac{b}{z\sqrt{v}}\right) \tag{5.27}$$

TABLE 5.7 VALUES OF THE FUNCTION ψ^a.

$$\psi(z) = \frac{1}{\sqrt{2\pi}} \int_0^z \exp\left[-\frac{U^2}{2} \right] dU$$

for $z > 2.2$, $\psi(z) \approx \frac{1}{2} - \frac{1}{2}(2\pi)^{-1/2} \exp\left[-\frac{z^2}{2} \right]$

Note: $\operatorname{erf} z = 2\psi(z\sqrt{2})$

z	0	1	2	3	4	5	6	7	8	9
0	0	.003 969	.007 978	.011 966	.015 953	.019 939	.023 922	.027 903	.031 881	.035 856
.1	.039 828	.043 795	.047 758	.051 717	.055 670	.059 618	.063 559	.067 495	.071 424	.075 345
.2	.079 260	.083 166	.087 064	.090 954	.094 835	.098 706	.102 568	.106 420	.110 251	.114 092
.3	.117 911	.121 720	.125 516	.129 300	.133 072	.136 831	.140 576	.144 309	.148 027	.151 732
.4	.155 422	.159 097	.162 757	.166 402	.170 031	.173 645	.177 242	.180 822	.184 386	.187 933
.5	.191 462	.194 974	.198 466	.201 944	.205 401	.208 840	.212 260	.215 661	.219 043	.222 405
.6	.225 747	.229 069	.232 371	.235 653	.234 914	.242 154	.245 373	.248 571	.251 748	.254 903
.7	.258 036	.261 148	.264 238	.257 305	.270 350	.273 373	.276 373	.279 350	.282 305	.285 236
.8	.288 145	.291 030	.293 892	.296 731	.299 546	.302 337	.305 105	.307 850	.310 570	.313 267
.9	.315 940	.318 589	.321 214	.323 814	.326 391	.328 944	.331 472	.333 977	.336 457	.338 913
1.0	.341 345	.343 752	.346 136	.348 495	.350 830	.353 141	.355 428	.357 690	.359 929	.362 143
1.1	.364 334	.366 500	.368 643	.370 762	.372 857	.374 928	.376 976	.379 000	.381 000	.382 977
1.2	.384 930	.386 861	.388 768	.390 651	.392 512	.394 350	.396 165	.397 958	.399 727	.401 475
1.3	.403 200	.404 902	.406 582	.408 241	.409 877	.411 492	.413 085	.414 657	.416 207	.417 736
1.4	.419 243	.420 730	.422 196	.423 641	.425 066	.426 471	.427 855	.429 219	.430 563	.431 888
1.5	.433 193	.434 476	.435 745	.436 992	.438 220	.439 429	.440 620	.441 792	.442 947	.444 083
1.6	.445 201	.446 301	.447 384	.448 449	.449 497	.450 529	.451 543	.452 540	.453 521	.454 486
1.7	.455 435	.456 367	.457 284	.458 185	.459 070	.459 941	.460 796	.461 636	.462 462	.463 273
1.8	.464 070	.464 852	.465 620	.466 375	.467 116	.467 843	.468 557	.469 258	.469 946	.470 621
1.9	.471 283	.471 933	.472 571	.473 197	.473 610	.474 412	.475 002	.475 581	.476 148	.476 705
2.0	.477 250	.477 784	.478 308	.478 822	.479 325	.479 818	.480 301	.480 774	.481 237	.481 691
2.1	.482 136	.482 571	.482 997	.483 414	.483 823	.484 222	.484 614	.484 997	.485 371	.485 738
2.2	.486 097	.486 447	.486 791	.487 126	.487 455	.487 776	.488 089	.488 396	.488 696	.488 989
2.3	.489 276	.489 556	.489 830	.490 097	.490 358	.490 613	.490 863	.491 106	.491 344	.491 576
2.4	.491 802	.492 024	.492 240	.492 451	.492 656	.492 857	.493 053	.493 244	.493 431	.493 613
2.5	.493 790	.493 963	.494 132	.494 297	.494 457	.494 614	.494 766	.494 915	.495 060	.495 201
2.6	.495 339	.495 473	.495 604	.495 731	.495 855	.495 975	.496 093	.496 207	.496 319	.496 427
2.7	.496 533	.496 636	.496 736	.496 833	.496 928	.497 020	.497 110	.497 197	.497 282	.497 365
2.8	.497 445	.497 523	.497 599	.497 673	.497 744	.497 814	.497 882	.497 948	.498 012	.498 074
2.9	.498 134	.498 193	.498 250	.498 305	.498 359	.498 411	.498 462	.498 511	.498 559	.498 605
3.0	.498 650	.498 694	.498 736	.498 777	.498 817	.498 856	.498 893	.498 930	.498 965	.498 999
3.1	.499 032	.499 065	.499 096	.499 126	.499 155	.499 184	.499 211	.499 238	.499 264	.499 289
3.2	.499 313	.499 336	.499 359	.499 381	.499 402	.499 423	.499 443	.499 462	.499 481	.499 499
3.3	.499 517	.499 534	.499 550	.499 566	.499 581	.499 596	.499 610	.499 624	.499 638	.499 651
3.4	.499 663	.499 675	.499 687	.499 698	.499 709	.499 720	.499 730	.499 740	.499 749	.499 758
3.5	.499 767	.499 776	.499 784	.499 792	.499 800	.499 807	.499 815	.499 822	.499 828	.499 835
3.6	.499 841	.499 847	.499 853	.499 858	.499 864	.499 869	.499 874	.499 879	.499 883	.499 888
3.7	.499 892	.499 896	.499 900	.499 904	.499 908	.499 912	.499 915	.499 918	.499 922	.499 925
3.8	.499 928	.499 931	.499 933	.499 936	.499 938	.499 941	.499 943	.499 946	.499 948	.499 950
3.9	.499 952	.499 954	.499 956	.499 958	.499 959	.499 961	.499 963	.499 964	.499 966	.499 967

[a] After Harr (1977).

EXAMPLE 5.9

A uniformly distributed vertical load of 25 kN/m² acts over an area of 4 m by 8 m at the surface of a particulate medium. Find the expected vertical normal stress 6 m below the center of the area. Assume $v = 1/3$.

Solution

Divide the 4 m by 8 m area into four areas, each with sides 2 m × 4 m, and use Equation 5.27. Thus,

$$\bar{\sigma}_z = 4(25)\psi\left(\frac{2}{6\sqrt{1/3}}\right)\psi\left(\frac{4}{6\sqrt{1/3}}\right)$$

$$= 100\psi(0.58)\psi(1.15)$$

$$= 100(0.219)(0.375) = 8.2 \text{ kN/m}^2$$

The theory of elasticity gives for this case $\sigma_z = 7.3 \text{ kN/m}^2$.

Uniform Vertical Load Over a Circular Area The expected value of the vertical stress under the center of a uniformly loaded

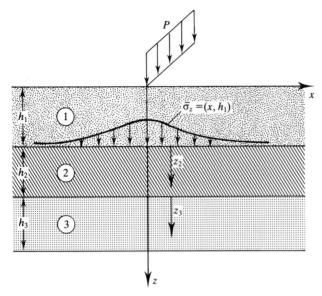

Fig. 5.31 Multilayer line load. (*Harr, 1977.*)

(intensity q) circular area of radius a is

$$\bar{\sigma}_z = q\left[1 - \exp\left(-\frac{a^2}{2vz^2}\right)\right] \qquad (5.28)$$

EXAMPLE 5.10

Find the expected vertical stress under the center of a tank 10 m in diameter at a depth of 10 m if the tank exerts a surface stress of 100 kPa. Assume $v = 1/3$.

$$\bar{\sigma}_z = 100\left[1 - \exp\left(-\frac{10^2}{2(1/3)(10^2)}\right)\right] = 78\,\text{kPa}$$

The theory of elasticity for this case would give a value of 65 kPa.

Harr (1977) also gives solutions for the cases of parabolic loading over a circular area and for tangential or horizontal loads acting on the surface of particulate media. Other solutions are possible for oddly shaped loaded areas by developing probabilistic Newmark type charts.

The v Parameter   Not much has been said about the coefficient v that we have used throughout this section as an indicator of the soil characteristics important in the transmission of surface stress throughout a soil body. As discussed by Harr (1977), the v parameter is somehow related to the various states of lateral stress in the soil mass. In a granular medium, it is probably best represented physically by the coefficient of earth pressure at rest K_0, although under some circumstances, a state of stress closer to either active or passive earth pressure states might be more appropriate. For the traditional means of calculating these pressures, see Chapter 6; Harr (1977) and Bourdeau (1986) discuss some experiments to determine K.

For distribution of foundation stresses, the use of a reasonable lateral stress coefficient K for the parameter v would be appropriate. For example, if the soils at the site were highly overconsolidated, then a $K > 1$ should be used. The probabilistic approach to stress distribution allows the geotechnical engineer to explicitly consider the probable stress state at the site.

The assumptions involved in the probabilistic approach are more realistic than in the traditional Boussinesq approach to stress distribution. Harr (1977) describes a series of experiments

in which stresses were measured in sand and clay deposits. In general, a higher peak stress was observed, that is, higher than would be predicted by the theory of elasticity, but the stress attenuated faster than that theory would indicate, both with depth and with distance from the loaded area. This behavior is exactly predicted by the probabilistic stress distributions.

5.7 CONSOLIDATION SETTLEMENTS

As mentioned in Section 6.1, when foundation loads are transmitted to cohesive subsoil, a tendency for volumetric strains occurs. In the case of saturated materials, an increase in pore water pressure (in excess above the static groundwater conditions) results immediately upon load application. Consolidation then is the process in which there is a reduction in volume due to the expulsion of water from the pores of the soil. The dissipation of excess pore water pressure in the soil is accompanied by a concurrent increase in effective stress. The analysis of the volumetric strains that result, and the settlements accompanying them, is greatly simplified if it is assumed that such strains occur only vertically. Such an assumption is reasonable when the geometric and boundary conditions in the field are such that vertical strains dominate, the conditions of one-dimensional loading and strain. One-dimensional loading and strains are likely to occur, for example, when the dimensions of the loaded area are large with respect to the thickness of the compressible stratum and/or when the compressible material lies between two stiffer soil strata whose presence tends to reduce the magnitude of horizontal strains. Using the assumption of one-dimensional loading and compression, consolidation settlements of cohesive soils are usually calculated using the following steps.

1. Determine the initial conditions and effective stress profile with depth. From the site investigation, determine thickness of strata, location of groundwater table, soil density, etc. See Chapter 1.
2. Determine the magnitude and geometry of the loads from nature of the problem and project requirements.
3. Estimate the change in stress with depth. If one-dimensional loading exists, then the stress at depth is equal to the surface loading. If not, then use approximate methods, theory of elasticity, or probabilistic methods. See Section 5.6.
4. Determine the preconsolidation stress (σ_p') profile with depth. Is the deposit normally consolidated or overconsolidated? Use laboratory oedometer tests on undisturbed samples (Chapter 3) or use in-situ tests (Chapter 3 and later in this section).
5. Calculate the consolidation settlement, described later in this section.
6. Determine the time rate for primary consolidation settlement and secondary compression (discussed later in this chapter).

5.7.1 Ultimate One-Dimensional Consolidation Settlement

In order to predict the final or ultimate consolidation settlement of compressible strata, it is necessary to determine the relationship between in-situ void ratio and effective vertical stress for the materials involved. It should be noted that the final or ultimate consolidation settlement corresponds to the situation wherein essentially complete dissipation of excess pore water pressure in the compressible strata has occurred. For almost all naturally occurring cohesive soil deposits, mechanical disturbance of the soil sample will yield a laboratory relationship that is different from that in the field.

The influence of rebound during sampling and disturbance on the observed compressibility of cohesive soils is illustrated in Figure 5.32. The loading history for a point in a normally consolidated saturated deposit is shown on a void ratio versus log effective consolidation pressure diagram (Fig. 5.32). The in-situ virgin compression curve is indicated as a solid line down to the point that represents the in-situ conditions at which the overburden pressure σ'_{v0} is equal to σ'_p, the preconsolidation pressure. An additional increment of load on this deposit will produce a void ratio change as a continuation of the in-situ virgin compression curve (shown dashed). Owing to rebound and sample disturbance effects, however, the effective consolidation pressure for this specimen brought into the laboratory is reduced, as shown in the figure, even though the void ratio and water content remain essentially constant. When the specimen is reloaded in the laboratory oedometer or consolidation apparatus, a void ratio decrease occurs, and the solid laboratory curve shown in the figure results. The more disturbance there is, the more rounded are the laboratory reconsolidation curves.

In the case of an overconsolidated clay (Fig. 5.32b), the in-situ stress history is represented by the solid in-situ virgin compression curve to the point at which the maximum past pressure σ'_p is reached, after which the stress is reduced to the existing overburden pressure σ'_{v0} (shown by a solid rebound curve). Subsequent reloading in the field would produce the dashed in-situ recompression curve, which would tend to rejoin the field virgin compression curve when the preconsolidation stress was exceeded. Again, the effect of stress release and sample disturbance is to reduce the effective consolidation stress at approximately a constant water content and void ratio. This leads to the laboratory reconsolidation curve (shown as a solid line in the figure). Increased compressibility and a decrease in the observed preconsolidation pressure result with increasing sample disturbance.

The stress history of a deposit or sample is generally expressed by the overconsolidation ratio OCR,

$$\text{OCR} = \frac{\sigma'_p}{\sigma'_{v0}} \qquad (5.29)$$

where

σ'_p = preconsolidation pressure
σ'_{v0} = the vertical effective overburden pressure

Normally consolidated soils have an OCR = 1, while soils with an OCR > 1 are preconsolidated or overconsolidated. It is possible for soils to be underconsolidated, and this eventuality will be discussed later in this section.

Preconsolidation Stress   It is evident from the shapes of the compressibility curves in Figure 5.32 that if a foundation applies a stress increment such that the final stress is less that the preconsolidation pressure, the settlement of the clay layer will be relatively small because the decrease in void ratio will be small. However, if the stress increment is such that the final stress is greater than the preconsolidation pressure, much larger settlements will occur. In the case of foundations supporting buildings, settlements in this latter case may be unacceptably large and alternate foundation systems will probably be necessary. On the other hand, for a number of important structures such as highway embankments and tanks, large total settlements may be tolerable, and both cases may need to be evaluated. Thus, the determination of the preconsolidation stress or OCR profile is often the most important step in settlement analyses for foundations.

Based on the behavior in consolidation tests (e.g. Fig. 5.32), it was previously thought that the preconsolidation pressure was primarily due to previous loading, usually of a geologic nature. However, it has become evident in recent years that the profile of preconsolidation stress observed in some deposits is greater than the maximum past pressure that could have existed during its geologic history. Thus, preconsolidation stress can apparently arise from phenomena other than previous geologic loading, phenomena such as desiccation, secondary compression, thixotropy, weathering, etc., as indicated in Table 5.8. These phenomena have been discussed in general by Brumund et al. (1976); see also Holtz and Kovacs (1981), Jamiolkowski et al. (1985), and Dyer et al. (1986), among others (see references in Table 5.8). A comparison of the compression curves having different preconsolidation pressure mechanisms is shown in Figure 5.33. Apparent preconsolidation or quasi-preconsolidation due to secondary compression is discussed in detail by Perloff (1975).

How is the preconsolidation pressure determined? Several procedures have been proposed for determining σ'_p, and Leonards (1962) and Crawford (1986) give a good review of the methods. The most popular method by far is the Casagrande (1936) construction. Another common method is to simply draw tangents to the virgin compression curve and the recompression curve; the intersection is called the preconsolidation stress. Burmister (1951) proposed another method for one-dimensional compression tests, as did Schmertmann (1955). This latter

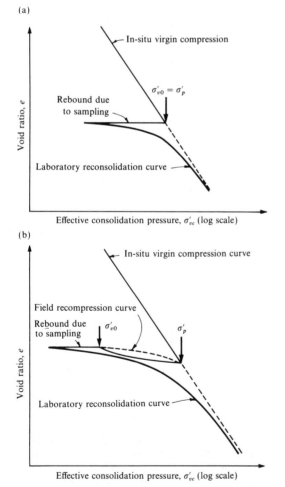

(a)

In-situ virgin compression

$\sigma'_{v0} = \sigma'_p$

Rebound due
to sampling

Void ratio, e

Laboratory reconsolidation curve

Effective consolidation pressure, σ'_{vc} (log scale)

(b)

In-situ virgin compression curve

Field recompression curve

Rebound due
to sampling σ'_{v0} σ'_p

Void ratio, e

Laboratory reconsolidation curve

Effective consolidation pressure, σ'_{vc} (log scale)

Fig. 5.32   Effect of sample disturbance on compressibility. (a) Normally consolidated clay ($\sigma'_{v0} = \sigma'_p$). (b) Overconsolidated clay ($\sigma'_{v0} < \sigma'_p$). (*Perloff, 1975.*)

TABLE 5.8   PRECONSOLIDATION PRESSURE MECHANISMS (FOR HORIZONTAL DEPOSITS WITH GEOSTATIC STRESSES)[a].

Category	Description	Stress History Profile	In-Situ Stress Condition	Remarks / Reference
A. Mechanical one-dimensional	1. Changes in total vertical stress (overburden, glaciers, etc.) 2. Changes in pore pressure (water table, seepage conditions, etc.)	Uniform with constant $\sigma'_p - \sigma'_{v0}$ (except with seepage)	K_0, but value at given OCR varies for reload vs. unload	Most obvious and easiest to identify
B. Desiccation	1. Drying due to evaporation, vegetation, etc. 2. Drying due to freezing	Often highly erratic	Can deviate from K_0, e.g. isotropic capillary stresses	Drying crusts found at surface of most deposits; can be at depth within deltaic deposits
C. Drained creep (aging)	1. Long-term secondary compression	Uniform with constant σ'_p / σ'_{v0}	K_0, but not necessarily normally consolidated value	Leonards and Altschaefl (1964); Bjerrum (1967, 1972, 1973)
D. Physicochemical	1. Natural cementation due to carbonates, silica, etc. 2. Other causes of bonding due to ion exchange, thixotropy, "weathering", etc.	Not uniform	No information	Poorly understood and often difficult to prove. Very pronounced in eastern Canadian clays, e.g. Sangrey (1972), Bjerrum (1967, 1973), Quigley (1980)

[a] Jamiolkowski et al. (1985) and Dyer et al. (1986).

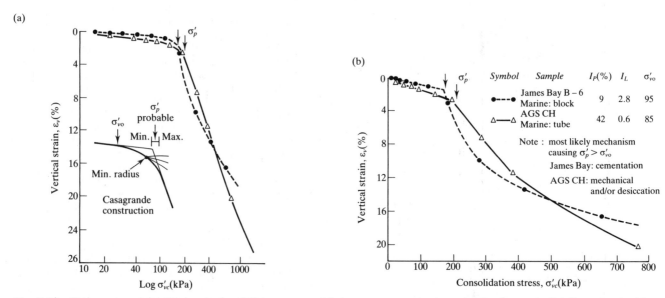

Fig. 5.33 Compression curves for clays having differing preconsolidation pressure mechanisms. (a) Semilog scale. (b) Natural scale. (*From Jamiolkowski et al., 1985.*)

procedure is discussed later in this section in connection with his methods for predicting the field compressibility curves.

Leonards (1962), Bjerrum (1973) and Brumund et al. (1976) suggest that the load increment ratio be reduced around the assumed value of the preconsolidation pressure to provide a more precise definition of the "break" in the consolidation curve. This may require testing two specimens, one using LIR = 1 to get a rough idea of the σ'_p. Then the procedures as described above may be used to determine the most probable value of σ'_p. Bjerrum (1973) recommends the following. (1) Load the sample to the σ'_{v0} in three steps. (2) Estimate the preconsolidation pressure using Casagrande or some other convenient method. (3) Continue loading with reduced load increments, for example (estimated $\sigma'_p - \sigma'_{v0}$)/3. The load duration should be equal to the time of primary consolidation, t_p, and the Taylor $\sqrt{\text{time}}$ construction is appropriate. (4)

When the load is greater than σ'_p, load increment ratios of 0.5 or greater should be used.

In all cases, a range of maximum and minimum values should be given, as noted by Brumund et al. (1976), Holtz and Kovacs (1981), and others (Fig. 5.33a).

Continuous loading tests such as the constant rate of strain (CRS), constant rate of load (CRL), and constant gradient test (CGT) have all been used with success to determine σ'_p because they provide a continuous $\varepsilon_v - \sigma'_v$ curve. This makes determination of the "break" in the curve generally obvious, provided that the samples are of high quality (Holtz et al., 1986). Sällfors (1975) developed a simple method to evaluate σ'_p from the CRS test. He showed from field loading tests on sensitive Swedish clays that the method predicts the in-situ preconsolidation pressure very well.

Another approach is to use the compression modulus method developed by Janbu (1963) and his coworkers. The Janbu modulus method will be described later in this section.

A number of empirical procedures for determining the σ'_p have also been proposed. Skempton (1954b) in a discussion of Lambe (1953) gave the well-known relation

$$\frac{\tau_f}{\sigma'_{v0}} = 0.11 + 0.0037(PI) \qquad (5.30)$$

for normally consolidated clays. Leonards (1962) extended the relationship to higher-*PI* soils, and since then a number of authors have provided correlations between τ_f/σ'_{v0} and the plasticity index, *PI*. Figure 5.34 is a compilation of some of these relationships. If your clay does not fall within correlations such as Equation 5.30, then it is probably overconsolidated. This relationship was also the basis of a proposal by Anderson and Lukas (1981) for determining σ'_p.

A relationship commonly used in Sweden was suggested by Hansbo (1957) for normally consolidated clays:

$$\frac{\tau_f}{\sigma'_{v0}} = 0.45 LL_{\text{cone}} \qquad (5.31)$$

Karlsson and Viberg (1967) note that σ'_p is commonly used in place of σ'_{v0}.

Leroueil et al. (1980) proposed an intriguing method using an oedometer, in which the pore pressure induced by a large single increment of load could be measured at the bottom of the specimen. Their initial data looked very good, but in a later paper (Leroueil et al., 1983), they showed that the results unfortunately did not predict as well as originally anticipated. Morin et al. (1983) also compared laboratory with in-situ preconsolidation pressures at five sites. They concluded that the σ'_p in-situ value could be estimated satisfactorily using conventional oedometer tests on high-quality samples. This same conclusion was also reached by Jamiolkowski et al. (1985), Holtz et al. (1986), and Dyer et al. (1986).

A method developed by Becker et al. (1987) utilizes the strain energy determined in a one-dimensional consolidation test to determine the preconsolidation pressure (Fig. 5.35).

Factors that influence the measured value of σ'_p are discussed in detail by Jamiolkowski et al. (1985) and Dyer et al. (1986). The most influential is the effect of sample disturbance and methods of assessing it on the shape of the compression curve (e.g. Brumund et al., 1976; Holtz et al., 1986). Test equipment and procedures (Olson, 1986) and environmental factors may also affect the σ'_p and compressibility of clay samples.

The preconsolidation pressure may also be estimated more or less accurately from certain in-situ tests. In-situ tests are attractive because of the difficulties associated with sample disturbance and different laboratory testing procedures, although in-situ tests are not without problems, too. Use of the piezocone (CPTU) to profile stress history is described by Jamiolkowski et al. (1985) and Dyer et al. (1986). Additional work using the piezocone is reported by Mayne and Holtz (1989) and Mayne and Bachus (1988). Determination of the OCR by the use of the dilatometer test (DMT) has been reported by Marchetti (1980), Lacasse and Lunne (1983), Sonnenfield et al. (1985), and Mayne (1987). Mayne and Mitchell (1988) suggest that the field vane shear test may be used for a similar purpose. Much additional research is needed to verify the validity of OCR and σ'_p profiles determined by the various in-situ tests in a variety of deposits with different test histories and preconsolidation mechanisms (Table 5.8).

Reconstruction of Field Compressibility Curves In order to accurately predict settlements in the field, it is necessary to obtain the field compressibility curve from the one observed in the laboratory. The most common procedure for doing this is the method developed by Schmertmann (1955). First, the preconsolidation pressure σ'_p is determined by using any of the procedures discussed above. If the deposit is normally consolidated at the depth of the sample (i.e., if $\sigma'_p \approx \sigma'_{v0}$), then the field virgin compression curve can be reconstructed as shown in Figure 5.36. The steps are:

1. Plot point *B*. This point represents the in-situ effective stress and void ratio at the depth of the specimen. With high-quality undisturbed samples and good laboratory technique, the initial void ratio e_0 of the consolidation specimen should be equal to the field void ratio.

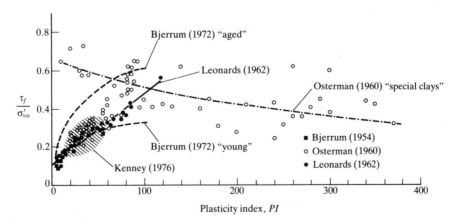

Fig. 5.34 Relationship between the ratio τ_f/σ'_{v0} and plasticity index for normally consolidated clays. (*See Holtz and Kovács, 1981 for references cited.*)

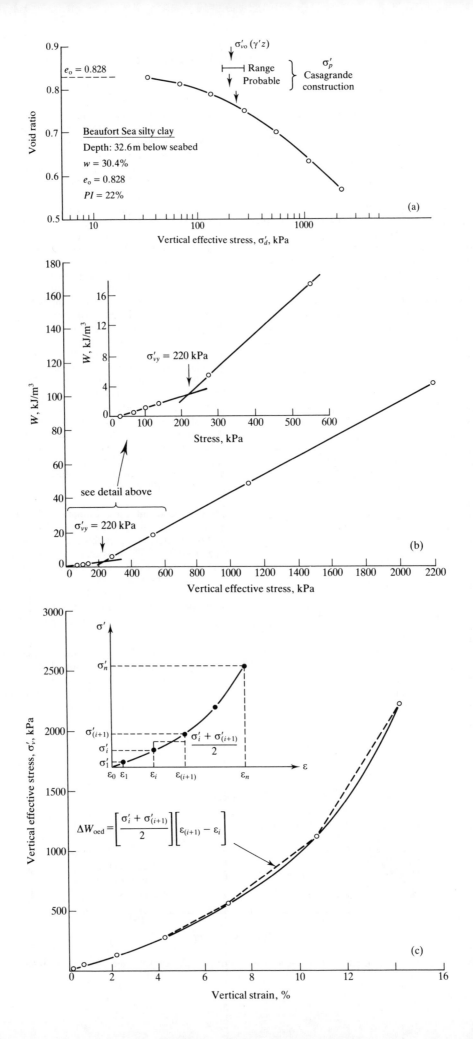

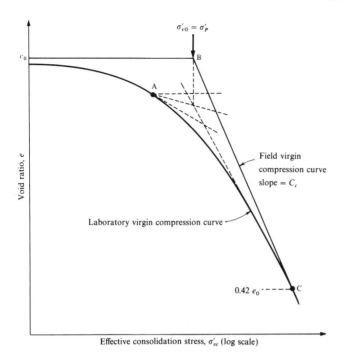

Fig. **5.36** Construction of the virgin field compression curve for normally consolidated cohesive soil. (*After Schmertmann, 1955.*)

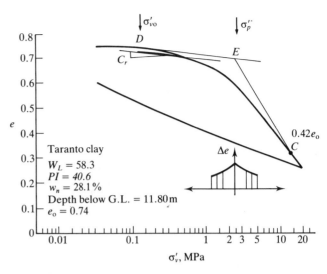

Fig. **5.37** Reconstruction of the field virgin compression curve and determination of the σ'_p by the Schmertmann (1955) method for a heavily overconsolidated clay. (*Dyer et al., 1986.*)

2. Extend the laboratory virgin compression curve to point C, which is $0.42e_0$. On the basis of many laboratory tests, Schmertmann (1955) found that the laboratory compression curve for varying degrees of disturbance intersected the field virgin compression curve at approximately $0.42e_0$.
3. Construct line BC. Draw a line between the in-situ σ'_p and $0.42e_0$. This line is the estimated field compression curve.

If the preconsolidation pressure σ'_p is found to be larger than the present overburden pressure σ'_{v0}, then by definition the sample is overconsolidated. In this case, the steps required to reconstruct the in-situ compressibility relation are as illustrated in Figure 5.37:

1. Plot point D. This point represents the in-situ effective stress and void ratio at the depth of the specimen. Draw a line through point D parallel to the mean slope of the rebound–reconsolidation curve. Thus, some time during the test the specimen will have to be partially unloaded and then reloaded to obtain the rebound–reconsolidation loop shown in Figure 5.37. As discussed below, it is preferable to perform this procedure prior to reaching the preconsolidation pressure; however, it may also be done after the σ'_p, although the accuracy may be somewhat less.
2. Extend the line through D until it intersects the σ'_p (which was previously estimated, as before). The straight line DE is assumed to be the field recompression curve with slope C_r.

It is important to determine C_r from the rebound–recompression cycle rather than from the initial recompression curve of the laboratory specimen, because sample disturbance usually causes this initial slope to be much steeper and more rounded than in situ. Research (Leonards, 1976) summarized later, has found that the value determined from a rebound–recompression cycle is more likely to approximate the field results.

3. Establish point C. This is the point at which the laboratory virgin compression curve intersects the void ratio equal to $0.42e_0$.
4. Construct line EC. This line is the estimated field virgin compression curve.

Research to verify the determination of the field virgin compression curve using the Schmertmann (1955) procedure is quite limited. According to Ladd (1971), the slope of the virgin compression curve is about 8 to 10 percent greater with the Schmertmann procedure.

Schmertmann (1955) also proposed that the in-situ preconsolidation pressure may be estimated from the shape of the Δe versus $\log \sigma'_{v0}$ plot, where Δe is the difference in the ordinate between the laboratory curve and the estimated in-situ compression curve (see small inset in Fig. 5.37). The field compression curve producing the most symmetrical Δe–$\log \sigma'$ relationship is the "correct" one. This procedure is illustrated by an interesting case history reported by Peck (1974), in which the Schmertmann procedure was used to avoid the use of deep foundations at a nuclear power plant. By utilizing a better prediction of the in-situ compressibility relation, it was determined that the likely in-situ preconsolidation pressure was about 650 kPa rather than 280 kPa, a value predicted from rather poor compressibility curves on disturbed samples (Fig. 5.38).

A recent paper by Leroueil (1988) summarized important work on consolidation of natural clays, and applicable to this section is his discussion of in-situ compressibility. It appears that the "end of primary" consolidation curve, as determined by the common oedometer test (Brumund et al., 1976; Leonards, 1976), is not unique and not necessarily representative of in-situ behavior. This means that Hypothesis A (Ladd et al., 1977; and Jamiolkowski et al., 1985) is not correct, at least on the basis of evidence Leroueil (1988) has collected to date. He discussed a number of case histories in which field in-situ curves determined from loading tests were compared with different types of laboratory compressibility curves. He concluded that secondary compression or creep must occur during primary consolidation.

Fig. **5.35** (opposite) Oedometer test on normally consolidated Beaufort Sea clay. (a) Void ratio–log stress relationship. (b) Work per unit volume interpretation of test data. (c) Stress–strain relationship. (*Becker et al., 1987.*)

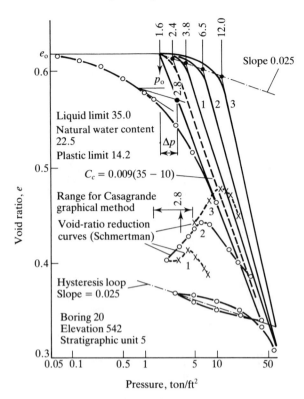

Fig. 5.38 Use of the Schmertmann (1955) procedure for obtaining the in-situ σ'_p (1 ton/ft^2 = 95.8 kN/m^2). (*Peck, 1974.*)

That is, observed in-situ compressibility relations cannot be explained by primary consolidation alone; some creep or "secondary" effects must be going on at the same time in order to obtain the observed field compressibility curves. The best evidence indicates that the in-situ preconsolidation pressure σ'_p is approximately equal to that determined in conventional 24-hr incremental load tests, which is the standard test (ASTM D2435). It is possible to obtain this value also from CRS tests if appropriate strain rates are chosen. Leroueil (1988) also discusses effects of stress path and sample disturbance, and the importance of high-quality sampling are again emphasized.

Calculation of Settlement—Normally Consolidated Clay The basic equation for calculating the settlements s in terms of the change in void ratio Δe of the soil layer is

$$s = \frac{\Delta e}{1 + e_0} H_0 \qquad (5.32)$$

where

e_0 = initial void ratio
H_0 = initial height of the compressible layer

The problem in any settlement calculation is to determine Δe for a given load increment applied to the compressible layer. When test results are plotted in terms of the void ratio versus the logarithm of effective stress (Figs. 5.36 and 5.37), then the slope of the compression curve is called the compression index C_c. In practice, it is quite common to plot the consolidation test results as percent consolidation (or vertical strain) versus the logarithm of effective stress (Brumund et al., 1976; Holtz and Kovacs, 1981). In this case, the slope of the virgin curve is called the modified compression index C_{ce}. Sometimes C_{ce} is called the compression ratio because it is numerically equal to the $C_c/(1 + e_0)$.

To calculate the consolidation settlement s_c, Equation 5.32 is used together with an appropriate formulation for Δe, or

$$s_c = C_c \frac{H_0}{1 + e_0} \log \frac{\sigma'_2}{\sigma'_1} = C_{ce} H_0 \log \frac{\sigma'_2}{\sigma'_1} \qquad (5.33a)$$

If the soil is normally consolidated, σ'_1 equals the existing vertical overburden stress σ'_{v0}, and σ'_2 includes the additional stress $\Delta\sigma_v$ applied by the structure, or

$$s_c = C_c \frac{H_0}{1 + e_0} \log \frac{\sigma'_{v0} + \Delta\sigma_v}{\sigma'_{v0}} = C_{ce} H_0 \log \frac{\sigma'_{v0} + \Delta\sigma_v}{\sigma'_{v0}}$$

$$(5.33b)$$

For calculation purposes, it is usually more accurate to divide compressible layers into strata that are sufficiently thin that the error in assuming an average σ'_{v0} and $\Delta\sigma$ throughout a stratum is small. Usually the basis for this division is the soil description and soil properties, which may be considered uniform over a particular stratum. Sometimes, if the compressible layer is thick and the properties are reasonably uniform throughout it, subdivision is made based on depth alone, for example, into layers of a convenient thickness for calculation. The settlement Δs_{ci} of the ith layer can be assumed to result from the void ratio change Δe_i at the middle of the ith layer, or

$$\Delta s_{ci} = \frac{\Delta e_i}{1 + e_{0i}} H_i \qquad (5.34)$$

where e_{0i} is the initial void ratio at the middle of the ith layer and H_i is the thickness of that layer. The total settlement then is the sum of the settlements of the n individual layers, or

$$s_c = \sum_{i=1}^{n} \Delta s_{ci} = \sum_{i=1}^{n} \frac{\Delta e_i}{1 + e_{0i}} H_i \qquad (5.35)$$

With increasing depth, the error introduced by assuming an average σ_{v0} and $\Delta\sigma'$ throughout the depth of the layer decreases, because the rate of change of the applied pressure increment with respect to depth decreases as the depth increases. Thus, the thickness of the strata subdividing an approximately homogeneous clay may be increased conveniently as the depth increases.

If the layer is normally consolidated and compression occurs entirely along the virgin compression curve, the void ratio change for each layer is

$$\Delta e_i = C_{ci} \log \left(\frac{\sigma'_{v0} + \Delta\sigma}{\sigma'_{v0}} \right)_i \qquad (5.36)$$

in which σ'_{v0} is the initial effective vertical overburden stress at the middle of the layer, $\Delta\sigma$ is the increment in vertical stress at the middle of the layer, and C_{ci} is the estimated in-situ value of the compression index (Fig. 5.36). Equation 5.36 can also be written in terms of the modified compression index C_{ce}.

A settlement calculation for a normally consolidated clay stratum is illustrated in the following example.

EXAMPLE 5.11

A large tank 50 m in diameter and 10 m high is to be constructed at a site near San Francisco Bay. The soil profile at the site is shown in Figure 5.39. The tank will be supported by a mat that is assumed to distribute the load uniformly to the soil 1 m below the ground surface. Determine the ultimate settlement at the center of the tank due to consolidation of the Bay mud.

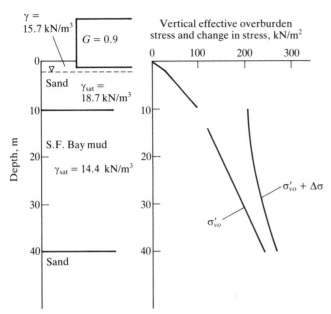

Fig. 5.39 Soil profile and stress distribution with depth for Example 5.11.

Solution

The initial effective overburden pressure and the increase in vertical effective stress within the clay stratum under the middle of the tank are shown in Figure 5.39. The increase in stress due to the tank was calculated by the probabilistic method, specifically Equation 5.28, using $v = 0.6$, which is approximately equal to K_0 for San Francisco Bay mud.

The stress applied by the tank was determined from the specific gravity of the soil (0.9) or $q = 0.9(9.81)(10) = 88 \text{ kN/m}^2$.

The void ratio versus pressure relation from a typical consolidation test from a sample taken from a nearby site is given in Figure 5.40. Assuming $G_s = 2.7$, the in-situ void ratio is 2.60. The preconsolidation stress is about 7.5 kPa for the sample and the in-situ modified compression index $C_{c\varepsilon} = 0.35$. First, as an exercise, determine the settlement for the entire 30-m layer.

$$s_c = 0.35(30 \text{ m}) \log \frac{172 + 52}{172} = 1.20 \text{ m}$$

To improve accuracy, divide the clay into three 10-m layers. Determine the average value of σ'_{v0} and $\Delta\sigma$ for each layer from Figure 5.39. Assume that the value of $C_{c\varepsilon}$ obtained from the laboratory sample (Fig. 5.40) is representative for the entire stratum. Use Equation 5.35. In this case $s_c = 0.774 + 0.402 + 0.208 = 1.38 \text{ m}$, or a value about 15 % greater than if the entire layer is used.

Note that in calculating q or $\Delta\sigma$ at depth, z = the depth of the middle of the layer minus 1 m, because the tank is founded 1 m below the ground surface.

Calculation of Settlement—Overconsolidated Clay Since over-consolidated soils are encountered more often in engineering practice than are normally consolidated soils, it is important to know how to make settlement calculations for these soil deposits.

First check whether the soil is preconsolidated, as discussed previously. If the soil layer is definitely overconsolidated, then check to see whether the stress added by the engineering

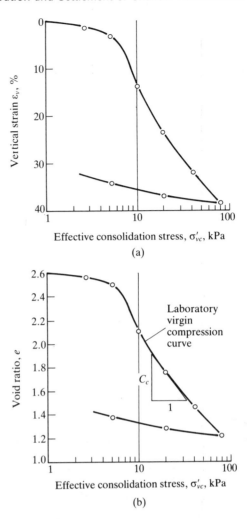

Fig. 5.40 Consolidation test data presented as (a) percent consolidation (or strain) versus log effective stress, and (b) void ratio versus log effective stress. (*Holtz and Kovács, 1981.*)

structure $\Delta\sigma_v$ plus the σ'_{v0} exceeds the preconsolidation pressure σ'_p. Whether or not it does can make a large difference in the amount of settlement calculated, as is shown in Figure 5.41.

If $\sigma'_{v0} + \Delta\sigma_v \leqslant \sigma'_p$ (Fig. 5.41a), then use Equation 5.33b, but with the recompression indices C_r or $C_{r\varepsilon}$ in place of C_c and $C_{c\varepsilon}$, respectively. The recompression index C_r is defined just like C_c but using the average slope of the recompression part of the e versus log σ'_{vc} curve (Fig. 5.37). If the data are plotted in terms of ε_r versus log σ'_{vc}, then the slope of the recompression curve is called the modified recompression index $C_{r\varepsilon}$ (sometimes called the recompression ratio). C_r and $C_{r\varepsilon}$ are related just like C_c and $C_{c\varepsilon}$, or $C_{r\varepsilon} = C_r/(1 + e_0)$.

To calculate settlement of overconsolidated clays, Equation 5.33b becomes

$$s_c = C_r \frac{H_0}{1 + e_0} \log \frac{\sigma'_{v0} + \Delta\sigma_v}{\sigma'_{v0}} = C_{r\varepsilon} H_0 \log \frac{\sigma'_{v0} + \Delta\sigma_0}{\sigma'_{v0}}$$

(5.37)

when $\sigma'_{v0} + \Delta\sigma_v \leqslant \sigma'_p$. Since C_r is usually much less than C_c, the settlements occurring when $\sigma'_{v0} + \Delta\sigma_v \leqslant \sigma'_p$ are much less than if the soil were normally consolidated (Fig. 5.41a).

If the added stress caused by the structure exceeds the preconsolidation stress, then much larger settlements may be expected. This is because the compressibility of the soil is

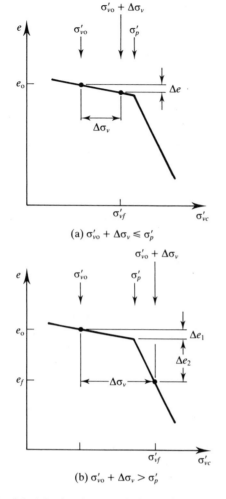

(a) $\sigma'_{vo} + \Delta\sigma_v \lessgtr \sigma'_p$

(b) $\sigma'_{vo} + \Delta\sigma_v > \sigma'_p$

Fig. 5.41 Principle of settlement calculations for overconsolidated soils. (*Perloff and Baron, 1976; Holtz and Kovács, 1981.*)

much greater on the virgin compression curve than on the recompression curve (Fig. 5.32). For the case where $\sigma'_{vo} + \Delta\sigma_v > \sigma'_p$, the settlement equation consists of two parts: (1) the change in void ratio or strain on the recompression curve from the original in-situ conditions (e_0, σ'_{vo}) or (ε_{vo}, σ'_{vo}) to σ'_p; and (2) the change in void ratio or strain on the virgin compression curve from σ'_p to the final conditions of (e_f, σ'_{vf}) or (ε_{vf}, σ'_{vf}). Note that $\sigma'_{vf} = \sigma'_{vo} + \Delta\sigma_v$. These two parts are shown graphically in Figure 5.41b. The complete settlement equation then becomes

$$s_c = C_r \frac{H_0}{1 + e_0} \log \frac{\sigma'_{vo} + (\sigma'_p - \sigma'_{vo})}{\sigma'_{vo}}$$
$$+ C_c \frac{H_0}{1 + e_0} \log \frac{\sigma'_p + (\sigma'_{vo} + \Delta\sigma_v - \sigma'_p)}{\sigma'_p} \quad (5.38a)$$

This equation reduces to

$$s_c = C_r \frac{H_0}{1 + e_0} \log \frac{\sigma'_p}{\sigma'_{vo}} + C_c \frac{H_0}{1 + e_0} \log \frac{\sigma'_{vo} + \Delta\sigma_v}{\sigma'_p} \quad (5.38b)$$

In terms of the modified indices, we have

$$s_c = C_{r\varepsilon} H_0 \log \frac{\sigma'_p}{\sigma'_{vo}} + C_{c\varepsilon} H_0 \log \frac{\sigma'_{vo} + \Delta\sigma_v}{\sigma'_p} \quad (5.38c)$$

Both Equations 5.38b and 5.38c give the same results. It could

be argued that in the right-hand term of Equation 5.38b the void ratio corresponding to the preconsolidation pressure on the field virgin compression curve should be used. Although technically correct, it does not make a significant difference in the answer.

If the degree of overconsolidation varies throughout the compressible layer, simply divide the entire stratum into several layers and apply the appropriate equation to calculate the average settlement for each layer; then sum up the settlements by Equation 5.35.

Determination of C_r and $C_{r\varepsilon}$ As mentioned, because of sample disturbance, the slope of the initial recompression portion of the laboratory consolidation curve (Fig. 5.32) is too steep and yields values of C_r and $C_{r\varepsilon}$ that are too large. Leonards (1976) discussed why in-situ values are generally smaller than those obtained from laboratory measurements: (1) disturbance during sampling, storage, and preparation of test specimens; (2) recompression of gas bubbles in the voids; and (3) errors in test procedures and methods of interpreting test results. This latter item includes the problem of reproducing the in-situ state of stress in the specimen.

Leonards recommends that the σ'_{v0} be applied to the specimen and that it be innundated and allowed to come to equilibrium for at least 24 hr before starting the incremental loading. Any tendency to swell should be controlled. Then the consolidation test is continued with relatively large load increments, e.g. LIR = 1. To reproduce as closely as possible the in-situ stress state, Leonards recommends that the sample be consolidated to slightly less than σ'_p and then be allowed to rebound. This is the first cycle shown in Figure 5.42. If σ'_p is not known, then consolidate initially to $\sigma'_{v0} + \Delta\sigma$ only, which is presumably less than σ'_p. The determination of C_r or $C_{r\varepsilon}$ is over the range of $\sigma'_{v0} + \Delta\sigma_v$, as shown in Figure 5.42. It is common practice to take the average slope of the two curves. The actual values of the recompression index depend on the stress at which the rebound–reload cycle starts, especially whether it starts at a stress less than or greater than the σ'_p. See the difference in slopes of the rebound curves shown in the figure. The C_r also depends on the OCR to which rebounding and reloading take place, for example, the ratio of σ'/σ'_r in Figure 5.42. The final consideration affecting the value of C_r is the presence of gas bubbles in the pores of the soil. Use of back pressure can sometimes take care of this problem.

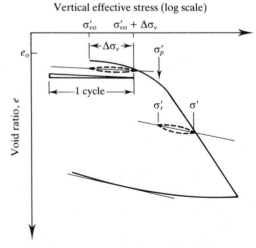

Fig. 5.42 Typical consolidation curve showing the recommended procedure for determining the C_r. (*After Leonards, 1976.*)

EXAMPLE 5.12

The problem of Example 5.11 is reconsidered for the case in which the clay stratum is overconsolidated. The settlements calculated assuming normally consolidated clay were rather large, but as the consolidation data used in the calculations (Fig. 5.40) were not from the site of the tank, high-quality undisturbed samples were taken at the exact site of the proposed tank and oedometer tests were conducted. It was found that the stratum was actually partially overconsolidated, as shown in the effective stress and preconsolidation stress profile in Figure 5.43.

The procedure suggested by Leonards (1976), described above, was used to determine the average $C_{r\varepsilon}$ to be 0.007 at the site. $C_{c\varepsilon}$ was 0.35 as in Example 5.11. The soil profile is again conveniently divided into three layers. Layer I, the upper layer, is entirely overconsolidated. Use Equation 5.37 or $s_{cI} = 0.007 (10 \, \text{m}) \log(208/126) = 0.015 \, \text{m}$.

Layer II from 20 to 30 m depth is both normally consolidated and overconsolidated. Use Equation 5.38 to include the settlement contributions of both the settlement on the recompression curve and that occurring on the virgin curve:

$$s_c = 0.007(10 \, \text{m}) \log(200/172)$$
$$+ 0.35(10 \, \text{m}) \log(172 + 52)/200$$
$$= 0.005 + 0.172 = 0.177 \, \text{m}$$

From Figure 5.43, the average σ'_p for this layer is about 200 kPa.

The third layer, III, is all normally consolidated and calculations are exactly the same as for that layer in Example 5.11; $s_c = 0.208 \, \text{m}$.

Total settlement is 0.40 m, which as expected is considerably less than when the entire stratum was assumed to be normally consolidated.

Examples 5.11 and 5.12 emphasize the importance of a careful site investigation and laboratory testing program in foundation engineering. Depending on what settlements a given structure can tolerate (Section 5.9), overestimation of the consolidation settlements may cause the selection of an expensive deep foundation or foundation treatment alternate that is entirely unnecessary and a waste of money.

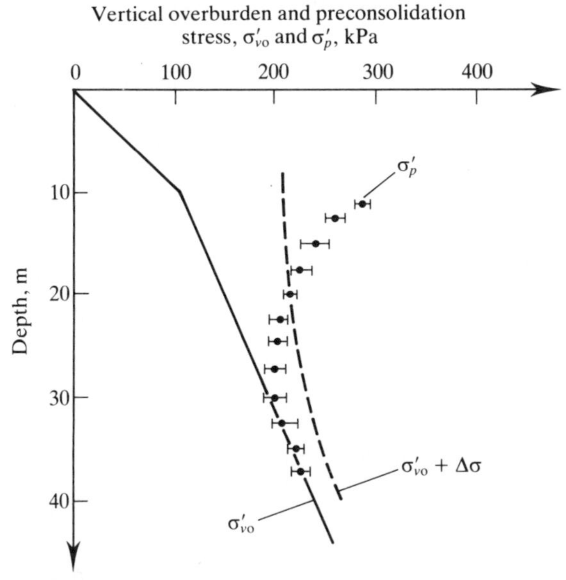

Janbu's Modulus Method   As shown in Figure 5.33b, the load–deformation relationship in one-dimensional compression is generally not linear. As the stress increases, the strain increases also, but at a decreasing rate. The slope of the curve at any point is called the tangent modulus M_t, and it may be expressed by

$$M_t = \frac{d\sigma'}{d\varepsilon} \tag{5.39}$$

where

$d\sigma' =$ increment of effective stress
$d\varepsilon =$ increment of strain

Janbu (1963, 1965, 1967) has shown that the tangent modulus can be determined by the following empirical relation:

$$M_t = m\sigma_a \left(\frac{\sigma'}{\sigma_a}\right)^{1-a} \tag{5.40}$$

where

$m =$ modulus number (dimensionless)
$a =$ stress exponent
$\sigma' =$ effective stress in the ε direction
$\sigma_a =$ reference stress $= 100 \, \text{kN/m}^2 \approx 1$ atmosphere

The reference stress is included to make Equation 5.40 dimensionally correct.

For settlement calculations, the strain at a typical element can be expressed as

$$\varepsilon = \int_{\sigma'_{v0}}^{\sigma'} \frac{1}{M_t} \, d\sigma' \tag{5.41}$$

Combining Equations 5.40 and 5.41 and solving for ε gives the following equations for calculating the strain induced in the soil layer from an increase in effective stress to σ'_f, the final effective stress:

$$\varepsilon = \frac{1}{ma}\left[\left(\frac{\sigma'_f}{\sigma_a}\right)^a - \left(\frac{\sigma'_{v0}}{\sigma_a}\right)^a\right] \quad \text{if} \quad a \neq 0 \tag{5.42a}$$

and

$$\varepsilon = \frac{1}{m} \ln\left(\frac{\sigma'_f}{\sigma'_{v0}}\right) \quad \text{if} \quad a = 0 \tag{5.42b}$$

The values of the modulus number m and stress exponent a to be used in a particular case can be determined by conventional testing in the laboratory (oedometer, triaxial, etc.) or in the field (in-situ tests, field observations, etc.).

The slope of the stress–strain curve in, for example, Figure 5.33b, is M_t. If the values of the tangent modulus are plotted versus effective stress, characteristic curves are obtained for different material types, and a few of these are shown in Figure 5.44. The case in Figure 5.44c is typical for a normally consolidated clay. Note that the minimum point in the curve is very close to the preconsolidation pressure σ'_p, and it is often assumed that the modulus method is an alternative way of determining the preconsolidation pressure

In terms of conventional oedometer data:

$$m = 2.3 \frac{1 + e_0}{C_c} \tag{5.43a}$$

Fig. 5.43   Vertical effective overburden stress, change in stress, and preconsolidation profile for Example 5.12.

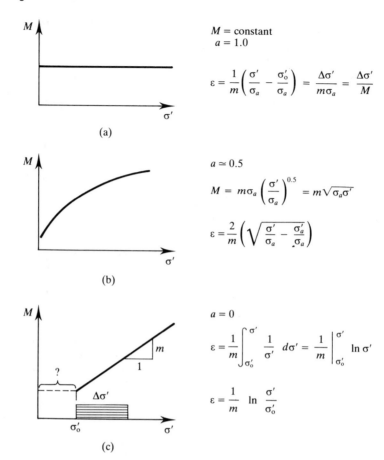

$$M = \text{constant}$$
$$a = 1.0$$

$$\varepsilon = \frac{1}{m}\left(\frac{\sigma'}{\sigma_a} - \frac{\sigma'_o}{\sigma_a}\right) = \frac{\Delta\sigma'}{m\sigma_a} = \frac{\Delta\sigma'}{M}$$

(a)

$$a \simeq 0.5$$

$$M = m\sigma_a\left(\frac{\sigma'}{\sigma_a}\right)^{0.5} = m\sqrt{\sigma_a\sigma'}$$

$$\varepsilon = \frac{2}{m}\left(\sqrt{\frac{\sigma'}{\sigma_a}} - \frac{\sigma'_a}{\sigma_a}\right)$$

(b)

$$a = 0$$

$$\varepsilon = \frac{1}{m}\int_{\sigma'_o}^{\sigma'}\frac{1}{\sigma'}\,d\sigma' = \frac{1}{m}\bigg|_{\sigma'_o}^{\sigma'}\ln\sigma'$$

$$\varepsilon = \frac{1}{m}\ \ln\frac{\sigma'}{\sigma'_o}$$

(c)

Fig. 5.44 Moduli for (a) equivalent elastic, (b) elastoplastic, and (c) plastic materials.

For the case of the rebound/reconsolidation modulus:

$$m_r = 2.3\frac{1 + e_0}{C_r} \qquad (5.43b)$$

To calculate settlement in an overconsolidated clay using the tangent modulus approach, an equation analogous to Equation 5.38c as follows is used:

$$\varepsilon = \frac{1}{m_r}\ln\left(\frac{\sigma'_p}{\sigma'_{v0}}\right) + \frac{1}{m}\ln\left(\frac{\sigma'_{v0} + \Delta\sigma}{\sigma'_p}\right) \qquad (5.44)$$

Values of the stress exponents and modulus numbers vary widely with soil type, density, and strength; typical values are given in Table 5.9 from Janbu (1963, 1967). Additional results are given in Janbu (1985). The variation of the stress exponent and the modulus number depend on the density of the soil expressed in terms of porosity, as shown in Figure 5.45.

As noted by Janbu (1963, 1985), the tangent modulus approach permits the use of a single method of analysis for a wide variety of geologic materials. Variation in deformation modulus with stress level can be considered and potential sources of error due to variation in C_c and e_0 are eliminated. As with all other prediction methods, high-quality samples and laboratory tests are essential for accurate predictions of settlement using the tangent modulus method.

Effect of Strain Rate It has been known since the late 1950s that rate of loading strongly influences the compressibility of clays (Hamilton and Crawford, 1959; Crawford, 1964). Figure 5.46 indicates the effect for constant rate of strain (CRS) and incremental load (IL) testing (see also Holtz et al., 1986).

TABLE 5.9 TYPICAL VALUES OF THE STRESS EXPONENT a AND THE MODULUS NUMBER m[a].

Soil or Rock Type	Stress Exponent, a	Modulus Number, m
Rock		
High strength	1	1 000 000–1000
Low strength	1	1 000–300
Tills: very dense to dense	1	1 000–300
Gravel	0.5	400–40
Sand		
Dense	0.5	400–250
Compact	0.5	250–150
Loose	0.5	150–100
Silt		
Dense	0.5	200–80
Compact	0.5	80–60
Loose	0.5	60–40
Clays		
Silty clay and clayey silt		
hard	0	60–20
stiff	0	20–10
soft	0	10–5
Soft marine clays and		
organic clays	0	20–5
Peat	0	5–1

[a] Meyerhoff and Fellenius (1985).

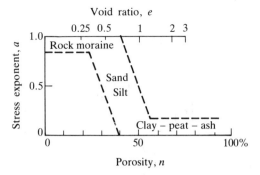

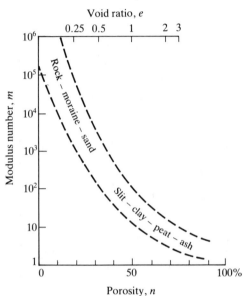

Fig. 5.45 Variation of m and a with porosity and void ratio. (*Meyerhoff and Fellenius, 1985.*)

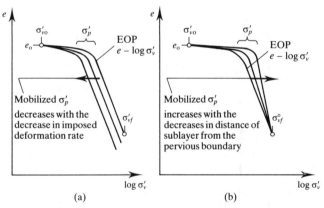

Fig. 5.46 Compression behavior of soft clays observed in laboratory consolidation tests. (a) CRS testing. (b) IL testing. (*Mesri and Choi, 1987.*)

Crawford (1988) and Leroueil (1988) present comprehensive summaries and discussions of the effect of strain rate on compressibility. Practical implications for incremental loading and CRS tests are given, and the recommendations are that the strain rate for CRS tests should be about 10^{-7}/sec or less. Leonards (1985) suggests testing samples thick enough that the strain rates are less than 10^{-8}/sec for the more plastic clays.

Leonards (1985) also observed that if the yield locus is plotted in terms of changes in volumetric and shear strain, then yielding can be considered to be independent of strain rate. He also noted that in a large number of oedometer tests the preconsolidation pressure occurs at about the same strain but not at the same stress, even as the strain rate during the test is varied (Fig. 5.46). Thus, he recommends that CRS testing should be at a rate slow enough that the strain rate effects are not important.

Leroueil (1988) states that at large strains and for strain rates less than 10^{-7}/sec, the strain rate dependence of the effective stress at a given strain can be described by a linear $\log \sigma'_v$–$\log \dot{\varepsilon}'_v$ relation; that is, $\Delta \log \dot{\varepsilon}'_v = m \, \Delta \log \sigma'_v$, with experimental data indicating that $m = 32$.

Mesri and Feng (1986) recommend the following equation for strain rate $\dot{\varepsilon}_v$ in CRS tests for determining the end of primary e–$\log \sigma'_v$ curves:

$$\dot{\varepsilon}_v = \frac{k_{v0}}{2^{C_c/C_k} H_d^2} \frac{\sigma'_p}{\gamma_w} \frac{C_\alpha}{C_c} \tag{5.45}$$

where

k_{v0} = initial vertical coefficient of permeability
$C_k = \Delta e/(\Delta \log k_v)$
H_d = maximum drainage distance
σ'_p = preconsolidation pressure corresponding to the end of primary e–$\log \sigma'_v$
$C_\alpha = \Delta e/(\Delta \log t)$

Underconsolidated Soils Occasionally a stratum of cohesive soil indicates a preconsolidation pressure less than the existing overburden pressure. This occurs where a deposit has not yet reached equilibrium under the applied overburden stresses. Examples include areas of recent landfilling or where the groundwater table has recently been lowered. The classic case is the Flushing Meadows Park in New York, site of the 1939 World's Fair (Burmister, 1942). The area was converted to the LaGuardia Field Airport, and considerable problems resulted because of underconsolidation. When an additional load was applied to the soil, such as from a fill or a structure, settlements occurred as expected in response to that load. In addition, however, settlements occurred due to the preexisting load under which equilibrium had not yet been reached.

Perloff (1975) gives a procedure for calculating the settlement due to both components.

Although underconsolidated soils occur relatively infrequently, it is important that they be recognized as such when they are encountered. Failure to do so may lead to settlements far in excess of those predicted on the basis of the assumption of a normally consolidated deposit. If underconsolidation is suspected, in-situ piezometric measurements are required.

Effect of Footing on Consolidation Settlements Perloff and Baron (1976), using simplified analyses, show the effect of footing size on consolidation settlements. The first case they consider is a square footing, $B \times B$, supporting a total load q. What is the effect on the consolidation settlement of increasing the footing width by a factor n? They show that the relative settlement of the two footings depends not only on the relative sizes but also on the magnitude of the applied stress in relation to the existing overburden stress; this is the load–increment ratio (LIR). An increase in footing size supporting a constant total load results in a marked reduction in consolidation settlement. It will be recalled (Section 5.4.4) that there was a similar effect for immediate settlements on elastic foundations. In this case, the effect is somewhat more pronounced for small LIR.

The next condition Perloff and Baron (1976) investigated was a strip footing with a constant load Q per unit length of the footing. They found that the effect of increasing the footing width was to reduce the consolidation settlement, and the effect was more pronounced for smaller LIR. This is not the same effect as with a strip footing on an elastic foundation (Section 5.4.4).

The third case they investigated was the consolidation settlement of two square footings of width B and nB, respectively, which are subjected to the same unit pressure q. Note that in this case the total load carried by each footing is not the same. They found that an increase in footing size produced an increase in settlement due to one-dimensional compression and that the effect was more important as the LIR increases. Therefore, both elastic and consolidation settlements will increase as footing size increases, even if the unit stress stays the same.

Approximate Methods for Estimating Compressibility Because of the time and expense involved in consolidation testing, it is sometimes desirable to be able to relate compression indices to classification of soils. Such relationships are useful for preliminary design estimates and for checking the validity of consolidation test results. Azzouz et al. (1976) list a number of equations for this prediction (see also Holtz and Kovacs, 1981).

It should be noted that typical values of C_r range from 0.015 to 0.035 (Leonards, 1976). The lower values are for clays of lower plasticity and low OCR. Values of C_r outside the range of 0.005 to 0.05 should be questioned.

5.7.2 Three-Dimensional Effects

As noted at the beginning of Section 5.7.1, all of the discussions of settlements have been restricted to one-dimensional compression. In those cases in which the thickness of the compressible strata is large relative to the dimensions of the loaded area, the three-dimensional nature of the problem may influence the magnitude and rate of settlement. The best approach for problems of this nature is a three-dimensional numerical analysis, but these have not yet become generally accepted in practice. Semiempirical approaches such as that of Skempton and Bjerrum (1957) and the stress path method (Lambe, 1967) are more commonly used.

Skempton and Bjerrum Correction The Skempton–Bjerrum (1957) approach assumes one-dimensional consolidation, as usual, and corrects for three-dimensional effects using the Skempton (1954a) pore pressure coefficient A. The values of A chosen were generally those at failure, or A_f. As pore pressures at failure are usually closest to the maximum value, use of the A_f values may result in an overprediction of the three-dimensional effect. In any event, Leonards (1976) recommends a modification using the equation

$$s_c = \alpha s \qquad (5.46)$$

The coefficient α depends on the overconsolidation ratio (OCR) and the ratio of the width of the loaded area to the depth or thickness of the clay stratum. If the width of the loaded area exceeds four times the thickness of the stratum, or if the depth to the top of the stratum is twice the width of the loaded area, α is assumed to equal 1. The values of α may be obtained from Figure 5.47.

Lambe and Whitman (1969) point out that the stress path assumed in the Skempton–Bjerrum method is discontinuous and therefore cannot be strictly correct. They recommend that on important projects the actual effective stress path should be estimated and testing techniques developed to provide a more realistic stress path.

Stress Path Method An alternate approach described by Lambe (1964, 1967) and Lambe and Whitman (1969) is the

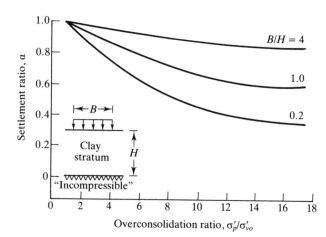

Fig. 5.47 Relation between settlement ratio and overconsolidation ratio. (*Leonards, 1976, and U.S. Navy, 1982.*)

stress path method. A specimen of soil is presumed to represent an average element or elements underneath the centerline or in other key areas under the structure. The specimen is subjected in the laboratory to the same effective stress as estimated to occur in the field. Vertical and horizontal deformations of this specimen are measured and summed up over the thickness of the compressible stratum to provide estimates of field vertical and horizontal movements.

Lambe (1967) and Lambe and Whitman (1969) present examples of the successful use of the stress path method. Estimates of stresses to be applied in the laboratory tests are made using elastic theory or other approximate methods. High-quality undisturbed samples and good laboratory techniques are requisite for the success of this method.

5.7.3 Time Rate of Consolidation Settlement

The preceding discussion has been concerned with determination of the final or ultimate consolidation settlement, that is, that which has occurred after all excess pore water pressure has dissipated and the soil is in equilibrium under an effective stress state. The process of consolidation involves expulsion of water from the pores of the soil being compressed, in response to an excess pore pressure gradient. The amount of settlement in a clay layer is directly related to how much water has been squeezed out of the voids in the clay. This amount of water and thus the change in void ratio is in turn directly related to the amount of excess pore water pressure that has dissipated. Therefore, the rate of settlement is directly related to the rate of excess pore pressure dissipation. In order to predict the rate of settlement of a foundation, the theory of consolidation is used, which predicts the pore pressure and void ratio at any point in time and space in the consolidating clay layer. The change in thickness (settlement) of the layer after the beginning of loading can be determined by integration of the consolidation equation over the thickness of the layer.

The theory of consolidation that is most commonly used in soil mechanics is a one-dimensional theory developed by Terzaghi in the 1920s. Its derivation and solution are given in detail by Taylor (1948), Leonards (1962), Perloff and Baron (1976), and Holtz and Kovacs (1981), among others.

Terzaghi's One-Dimensional Consolidation Theory The assumptions of the Terzaghi theory are:

1. Drainage and compression are one-dimensional.
2. The compressible soil layer is homogeneous and completely saturated with water.

3. Mineral grains in the soil and the water in the pores are incompressible.
4. Darcy's law governing the egress of water from the soil pores is valid.
5. The applied load increment produces only small strains; therefore, both the compressibility and permeability remain essentially constant during the consolidation process.
6. There is a unique relationship between the change in void ratio Δe and the change in effective stress $\Delta\sigma'$, that is, there is no secondary compression.

The basic one-dimensional consolidation equation is a special case of the diffusion equation from mathematical physics. It can be written as:

$$c_v \frac{\partial^2 u}{\partial z^2} = \frac{\partial u}{\partial t} \qquad (5.47)$$

where

$$c_v = \frac{k}{\rho_w g} \frac{1 + e_0}{a_v} \qquad (5.48)$$

where

k = Darcy coefficient of permeability
ρ_w = density of water
g = gravitational constant
e_0 = initial or in-situ void ratio
a_v = coefficient of compressibility (the slope of the void ratio–σ' relationship for the soil)

The coefficient c_v is called the coefficient of consolidation and it is analogous to the diffusion coefficient. It contains the material properties that govern the process of consolidation, and it has dimensions of $L^2 t^{-1}$ or m^2/s. The Terzaghi one-dimensional consolidation equation could just as well be written in three dimensions but most of the time one-dimensional consolidation is assumed. In general, the c_v is not a constant but it is assumed to be so, in order to make the equation linear and more easily solvable. Solutions of this equation using Fourier series are given by Taylor (1948), Leonards (1962) and Holtz and Kovacs (1981). Approximate solutions using a finite difference approach are given by Harr (1966) and Perloff and Baron (1976).

The Fourier series solution is of the form:

$$u = (\sigma'_2 - \sigma'_1) \sum_{n=0}^{\infty} f_1(Z) f_2(T) \qquad (5.49)$$

where Z and T are dimensionless parameters. The first term, Z, is a geometry parameter, and it is equal to z/H. The second term, T, is known as the time factor, and it is related to the coefficient of consolidation c_v by

$$T = c_v \frac{t}{H_{dr}^2} \qquad (5.50)$$

where

t = time
H_{dr} = length of the longest drainage path

From Equation 5.48, the time factor can also be written as

$$T = \frac{k(1 + e_0)}{a_v \rho_w g} \frac{t}{H_{dr}^2} \qquad (5.51)$$

The drainage path for double drainage would be equal to half the thickness H of the clay layer, or $2H/2 = H_{dr}$. If we had only a singly drained layer, the drainage path would still be H_{dr}, but then it would be equal to the thickness H of the layer.

The progress of consolidation after some time t and at any depth z in the consolidating layer can be related to the

void ratio at that time and the final change in void ratio. This relationship is called the consolidation ratio, and it is expressed as

$$U_z = \frac{e_1 - e}{e_1 - e_2} \qquad (5.52)$$

where e is some intermediate void ratio, as shown on Figure 5.48. This equation is the ratio of ordinates corresponding to AB and AC. In terms of stresses and pore pressures, Equation 5.52 becomes

$$U_z = \frac{\sigma' - \sigma'_1}{\sigma'_2 - \sigma'_1} = \frac{\sigma' - \sigma'_1}{\Delta\sigma'} = \frac{u_i - u}{u_i} = 1 - \frac{u}{u_i} \qquad (5.53)$$

where σ' and u are intermediate values corresponding to e in Equation 5.52 and u_i is the initial excess pore pressure induced by the applied stress $\Delta\sigma'$. Note that $\Delta\sigma' = -\Delta u$.

U_z is zero at the start of loading, and it gradually increases to 1 (or 100 percent) as the void ratio decreases from e_1 to e_2. At the same time, of course, as long as the total stress remains constant, the effective stress increases from σ'_1 to σ'_2 as the excess hydrostatic stress (pore water pressure) dissipates from u_i to zero. The consolidation ratio U_z is sometimes called the degree of consolidation of percent consolidation, and it represents conditions at a point in the consolidating layer. It is now possible to put our solution for u in Equation 5.49 in term of the consolidation ratio (Eq. 5.52), or

$$U_z = 1 - \sum_{n=0}^{\infty} f_1(Z) f_2(T) \qquad (5.54)$$

The solution to this equation is shown graphically in Figure 5.49 in terms of the dimensionless parameters already defined. From Figure 5.49 it is possible to find the degree of consolidation (and therefore u and σ') for any real time after the start of loading and at any point in the consolidating layer. The time factor T can be calculated from Equation 5.50, given the c_v for the particular soil deposit, the total thickness of the layer, and boundary drainage conditions. It is applicable to any one-dimensional loading situation where the soil properties can be assumed to be the same throughout the compressible layer.

Figure 5.49 is also a picture of the progress of consolidation. The isochrones (lines of constant T) in Figure 5.49 represent the degree or percent consolidation for a given time factor throughout the compressible layer. For example, the percent

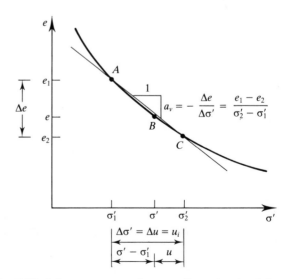

Fig. 5.48 Laboratory compression curve. Note: $\sigma' - \sigma'_1 = (\sigma'_2 - \sigma'_1)$ $- u = u_i - u$.

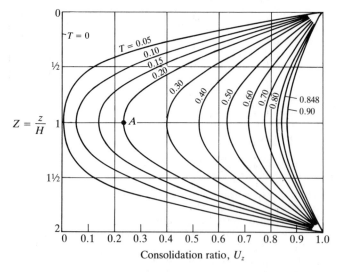

Fig. 5.49 Consolidation for any location and time factor in a doubly drained layer. (*After Taylor, 1948.*)

TABLE 5.10 RESULTS OF TERZAGHI THEORY FOR A LINEAR DISTRIBUTION OF EXCESS PORE PRESSURE.

U_{avg}	T
0.1	0.008
0.2	0.031
0.3	0.071
0.4	0.126
0.5	0.197
0.6	0.287
0.7	0.403
0.8	0.567
0.9	0.848
0.95	1.163
1.0	∞

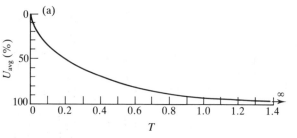

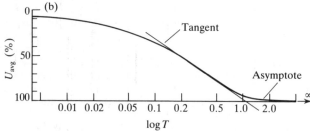

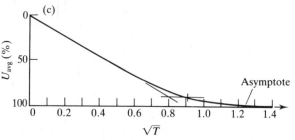

Fig. 5.50 U_{avg} versus T. (a) Arithmetic scale. (b) Log scale. (c) Square root scale. (*Holtz and Kovać s, 1981.*)

consolidation at midheight of a doubly drained layer (total thickness $= 2H$) for a time factor equal to 0.2 is approximately 23 percent (see point A in Fig. 5.49). However, at the same time (and time factor) at other locations within the soil layer, the degree of consolidation is different. At 25 percent of the depth, for example, $z/H = \frac{1}{2}$ and $U_z = 44$ percent. Similarly, near the drainage surfaces at $z/H = 0.1$, for the same time factor, because the gradients are much higher, the clay is already 86 percent consolidated, which means that at that depth and time, 86 percent of the original excess pore pressure has dissipated and the effective stress has increased by a correspondent amount.

In most cases, we are not interested in how much a given point in a layer has consolidated. Of more practical interest is the average degree of percent consolidation of the entire layer. This value, denoted by U or U_{avg}, is a measure of how much the entire layer has consolidated and thus it can be directly related to the total settlement of the layer at a given time after loading. Note that U can be expressed as either a decimal fraction or a percentage.

To obtain the average degree of consolidation over the entire layer corresponding to a given time factor, the area under the T curve of Figure 5.49 is obtained. How the integration is done mathematically is shown in Taylor (1948) and Holtz and Kovacs (1981). Table 5.10 presents the results of the integration for the case where a linear distribution of excess pore water pressure is assumed.

The results in Table 5.10 are shown graphically in Figure 5.50, arithmetically, semilogarithmically, and versus $\sqrt{T}$. Note that as T becomes very large, U asymptotically approaches 100 percent. This means that, theoretically, consolidation never stops but continues indefinitely. It should also be pointed out that the solution for U versus T is dimensionless and applies to all types of problems where $\Delta\sigma = \Delta u$ varies linearly with depth. Solutions for cases where the initial pore pressure distribution is sinusoidal, half sine, and triangular are presented by Leonards (1962) and Taylor (1948).

When the initial hydrostatic excess u_i is not constant, the consolidation of the strata is more complex. Three examples of variable u_i are represented in Figure 5.51 by cases IB, II, and III. In the upper portion of the figure, u_i diagrams for these cases are shown, as described by Taylor (1948).

Actual cases may be approximated by combinations of cases. For example, diagram III represents a typical case and its u_i

relationship may be expressed as a linear relationship minus a sinusoidal relationship. Taylor (1948) shows that the solution for case III is obtained by weighting U values from consolidation curves I and II in proportion to the areas of the respective u_i diagrams. Because consolidation curve III is not appreciably different from curve I, the latter is assumed to be an adequate representation of typical cases in nature.

Casagrande (1938) and Taylor (1948) provide the following use approximations:

For $U < 60$ percent,

$$T = \frac{\pi}{4} U^2 = \frac{\pi}{4}\left(\frac{U\%}{100}\right)^2 \tag{5.55}$$

For $U > 60$ percent,

$$T = 1.781 - 0.933 \log(100 - U\%) \tag{5.56}$$

Cases: I*A* I*B* II III

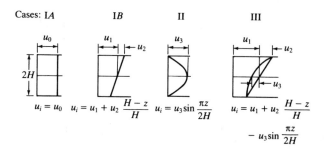

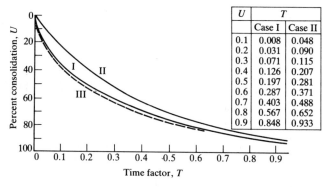

Fig. 5.51 Consolidation curves according to the Terzaghi theory for different distributions of initial excess pore water pressure. (*Taylor, 1948.*)

U	T	
	Case I	Case II
0.1	0.008	0.048
0.2	0.031	0.090
0.3	0.071	0.115
0.4	0.126	0.207
0.5	0.197	0.281
0.6	0.287	0.371
0.7	0.403	0.488
0.8	0.567	0.652
0.9	0.848	0.933

At any time between the application of the load producing consolidation and the time at which essentially 100 percent consolidation has occurred, the progress of consolidation can be expressed as

$$U_{avg} = \frac{s(t)}{s_c} \qquad (5.57)$$

where s_t = settlement at any time and s_c = final or ultimate consolidation (primary) settlement at $t = \infty$. When s_c has been estimated, Equation 5.57 can be used to predict the time–settlement history of interest, provided that U_{avg} can be determined.

EXAMPLE 5.13

A doubly drained 12-m thick clay layer has $c_v = 0.7 \, \mathrm{m^2/yr}$. It is estimated that there will be 75 mm of total settlement due to consolidation of the clay· layer. Determine the consolidation settlement that will have occurred after 3 years.

Solution

The time factor (Eq. 5.50) for $t = 3$ yr is

$$T = \frac{(0.7)(3)}{(6)^2} = 0.058$$

From Figure 5.50, $U = 27$ percent. Therefore, the settlement at the end of 3 years is

$$s_c = 0.27(0.075 \, \mathrm{m}) = 0.02 \, \mathrm{m} = 20 \, \mathrm{mm}$$

EXAMPLE 5.14

A 6-m thick clay layer drained at both boundaries is consolidating. If $c_v = 0.7 \, \mathrm{m^2/yr}$, how long will it take to achieve 50 percent of the ultimate settlement?

Solution

From Figure 5.50 or Table 5.10, the T corresponding to $U_{avg} = 50$ percent is 0.197. Using Equation 5.50,

$$t = \frac{0.197(3)^2}{0.7} = 2.5 \, \mathrm{yr}$$

EXAMPLE 5.15

If the clay stratum described in Example 5.14 were 12 m thick instead of 6 m, how long would it take to achieve 50 percent of the ultimate consolidation settlement?

Solution

Using Equation 5.50,

$$\frac{t_2}{t_1} = \frac{T_2(H_2)^2/c_{v2}}{T_1(H_1)^2/c_{v1}}$$

Since $T_1 = T_2$ and $c_{v1} = c_{v2}$,

$$\frac{t_2}{t_1} = \left(\frac{H_2}{H_1}\right)^2$$

$$t_2 = \frac{6^2}{3^2}(2.5) = 4(2.5) = 10 \, \mathrm{yr}$$

Effect of Boundary Drainage Conditions When the initial excess pore pressure distribution is symmetrical about the mid-height of a layer drained at both boundaries, the distribution of excess pore water pressure will be symmetrical about this plane at all times, as shown in Figure 5.49. Thus, the distribution of excess pore pressure in one-half of the stratum is the same as that in a stratum one-half the thickness but drained at only one boundary. Thus, the results in Figure 5.49 and Table 5.10 can also be used just as well for a singly drained stratum, as shown in the following example.

EXAMPLE 5.16

A clay stratum 1.5 m thick is drained only at its upper boundary. It is estimated that there will be a total of 50 mm of consolidation settlement at the point in question if the coefficient of consolidation is $2 \, \mathrm{m^2/yr}$. Determine the settlement due to consolidation at the end of 1 year.

Solution

The time factor corresponding to 1 yr is

$$T = \frac{(2)(1)}{(1.5)^2} = 0.89$$

From Table 5.10 and Figure 5.50, $U = 84$ percent. Thus, at the end of 1 year

$$s_c(t) = 0.84(0.050) = 0.042 \, \mathrm{m} = 42 \, \mathrm{mm}$$

Superposition of Results Because the differential equation of one-dimensional consolidation is assumed to be linear, solutions to the equation may be superimposed, provided the appropriate distribution of initial excess pore pressures is used. Thus, if an additional excess pore pressure distribution can be represented as the sum of two or more of the distributions, then the average degree of consolidation for such a case can be computed from the average degrees of consolidation corresponding to the appropriate initial excess pore pressure. This is done by summing the degrees of consolidation weighted in proportion to the relative areas of the initial distributions. Initial distributions are given by Taylor (1948), Leonards (1962), and Perloff (1975). Perloff (1975) and Perloff and Baron (1976) give an example of how to carry out this superposition.

Internal Drainage Layers In cases in which a compressible layer contains one or more seams of pervious material that act as drainage layers, the average degree of consolidation may also be found by taking advantage of the linearity of the mathematical procedure:

$$U = \frac{1}{s_c}(U_1 s_{c1} + U_2 s_{c2} + \cdots + U_N s_{cN}) \quad (5.58)$$

where

U = average degree of consolidation

s_c = ultimate consolidation of settlement, both for the entire compressible layer

U_i = average degree of consolidation

s_{ci} = ultimate consolidation settlement, both for the ith portion of the stratum, where $i = 1$ to N

EXAMPLE 5.17

A 6-m thick stratum drained at the top boundary is consolidating due to a uniform initial excess pore pressure distribution. A thin horizontal seam of sand acts as a drain at a depth of 3 m from the upper boundary. The bottom of the stratum is undrained. It has been estimated that the ultimate settlement of the upper 3 m is 75 mm, and that the lower 3 m is 50 mm. For the entire stratum, $c_v = 0.85\,\text{m}^2/\text{yr}$.

a. Determine the time at which the average degree of consolidation for the entire stratum is 50 percent.
b. If the sand seam were not present, how long would it take for 50 percent of the ultimate settlement to occur?

Solution

a. Consider the stratum to consist of two layers: (1) the upper layer 3 m thick and doubly drained, and (2) the lower layer 3 m thick and singly drained. Thus, $H_{dr1} = 1.5\,\text{m}$ and $H_{dr2} = 3\,\text{m}$.

$$s_{c1} = 75\,\text{mm} \quad s_{c2} = 50\,\text{mm} \quad s_c = 125\,\text{mm}$$

From Equation 5.58,

$$U = U_1(75/125) + U_2(50/125) = 50 \text{ percent} \quad (a)$$

and

$$T_1 = \frac{c_v t_1}{H_{dr1}^2} \quad T_2 = \frac{c_v t_2}{H_{dr2}^2}$$

But $t_1 = t_2$ and $c_{v1} = c_{v2}$. So $T_1/T_2 = H_{dr2}^2/H_{dr1}^2 = (3)^2/(1.5)^2$ or $T_2 = 0.25 T_1$. Assume $U < 60$ percent; therefore, from Equation 5.55, $U = \sqrt{4T/\pi}$. So,

$$U_1 = \sqrt{4/\pi}\sqrt{T_1} \quad \text{and} \quad U_2 = \sqrt{4/\pi}\sqrt{T_2} \quad (b)$$

Inserting Equation (b) into Equation (a), we obtain

$$\sqrt{4/\pi}\sqrt{T_1}(75/125) + \sqrt{4/\pi}\sqrt{0.25\,T_1}(50/125) = 0.5$$

Solving for T_1, we obtain $T_1 = 0.30$. And,

$$t = \frac{0.3(1.5)^2}{0.85} = 0.79\,\text{yr} \quad \text{or } 9.5 \text{ months}$$

Note that if $U > 60$ percent, then Equation 5.55 does not hold and the values in Figure 5.50 or Equation 5.56 must

be used. Perloff (1975) uses a trial-and-error solution to obtain approximately the same answer.

b. With no internal drainage layer, $H = 6\,\text{m}$ (only one drainage boundary). Then from Table 5.10 or Figure 5.50, for $U = 50$ percent, $T = 0.197$ and

$$t = \frac{0.197(6)^2}{0.85} = 8\,\text{yr}$$

The effect of the sand seam acting as a drainage layer within the clay stratum is to reduce the time to $U_{\text{avg}} = 50$ percent to about 1/10 the magnitude of that without the sand seam.

Settlements generally occur much more rapidly than predicted using c_v determined in one-dimensional consolidation tests. The presence of thin undetected drainage seams and layers is often an important factor in this observation, as shown in Example 5.17. Therefore, a critical aspect of the exploration of cohesive strata of which the time rate of settlement is to be estimated is the location of possible drainage layers. As noted by Jamiolkowski et al. (1985), the piezocone is an excellent instrument for this purpose. Other possibilities include continuous undisturbed sampling, but generally piezocone exploration is significantly cheaper.

Contiguous Compressible Strata Contiguous compressible strata with different consolidation characteristics pose a formidable analytical challenge. Until the general availability of high-speed digital computers, it was conventional practice to simplify the problem so that available analytical solutions could be used for estimating the time rate of settlements and/or changes in effective stress. Such situations are most easily investigated by numerical methods such as the finite-difference procedure (Harr, 1966; Perloff and Baron, 1976), because it permits ready incorporation of multiple layer systems, changes in permeability and compressibility with changing effective stress, and even variable boundary conditions. With the development of versatile spreadsheet software for personal computers, the use of finite-difference type solutions for contiguous compressible strata is made even simpler.

Examples of analytical approximations can be found in Lambe and Whitman (1969) and U.S. Navy (1982).

Three-Dimensional Consolidation As summarized by Schiffman et al. (1969), several attempts have been made to analyse theoretically the time rate of consolidation in three dimensions. When the three-dimensional nature of the deformations and dissipation of excess pore pressure are considered, the total stresses as well as the effective stresses are a function of time. Using the example of the strip load applied normal to the surface of a saturated compressible halfspace, Schiffman et al. (1969) showed that the components of total stress changed sufficiently that the maximum shear stress beneath the strip load increased during the consolidation process to a maximum magnitude nearly twice its initial value. As a further consequence of this effect, the change in total stresses during consolidation led to an increase in excess pore pressure in a zone beneath the strip load. This is called the Mandel–Cryer effect.

Large-Strain Consolidation As mentioned at the start of this section, the conventional Terzaghi consolidation theory is a small-strain theory. That is, the strains produced by the loads are assumed to be small relative to the thickness of the consolidating layer. This condition is not always the case in a number of important practical problems, especially those involving sedimentation as well as consolidation, such as in marine geotechnology and mine-waste disposal. For a good review of the theory of one-dimensional consolidation including nonlinear finite-strain theory, see Schiffman et al. (1984) and Schiffman et al. (1988).

The basic formulation for large-strain nonlinear finite-strain consolidation was first made by Gibson et al. (1967).

5.8 SECONDARY COMPRESSION SETTLEMENTS

As mentioned in Section 5.2.1, secondary compression is that portion of the time-dependent settlement that occurs at essentially constant effective stress. During secondary compression the rate of volume change is not controlled by the rate at which pore water can flow from the soil, and therefore it does not depend upon the thickness of the compressible layer. Field rates of secondary consolidation are estimated from laboratory tests or empirical relationships.

The following discussion is from Holtz and Kovacs (1981).

The secondary compression component of settlement results from compression of bonds between individual clay particles and domains, as well as other effects on the microscale that are not yet clearly understood. A complicating factor is that in the field it is difficult to separate secondary compression from consolidation settlement, especially if the consolidating clay layer is relatively thick. Parts of the layer near the drainage surfaces may be fully consolidated, and therefore undergoing "secondary" compression, while portions near the center of the layer are still in "primary." Both types of settlements contribute to the total surface settlement, and separating the effects in order to predict the final surface settlement is not easy.

In this section a practical working hypothesis, acceptable for engineering practice, for estimating secondary compression is described, and two methods for making estimates of secondary settlement are presented.

5.8.1 Working Hypothesis

Raymond and Wahls (1976) and Mesri and Godlewski (1977) define the *secondary compression index* C_α as

$$C_\alpha = \frac{\Delta e}{\Delta \log t} \qquad (5.59)$$

where

Δe = the change in void ratio along a part of the void ratio versus the logarithm of time curve between times t_1 and t_2

Δt = the time between t_2 and t_1

This definition is analogous to the primary compression index C_c, defined as $\Delta e/\Delta \log \sigma'$. Sometimes it is given the symbol $C_{\alpha e}$ to distinguish it from the modified secondary compression index $C_{\alpha \varepsilon}$, or

$$C_{\alpha \varepsilon} = \frac{C_{\alpha e}}{1 + e_p} \qquad (5.60)$$

where

$C_{\alpha e}$ = the secondary compression index, Equation 5.59

e_p = the void ratio at the start of the linear portion of the e versus log t curve. (One could also use e_0, the in-situ void ratio, with no appreciable loss of accuracy.)

Sometimes $C_{\alpha \varepsilon}$ is called the secondary compression ratio, or the rate of secondary consolidation. As Ladd et al. (1977) note, $C_{\alpha \varepsilon} = \Delta \varepsilon/\Delta \log t$.

The secondary compression index, $C_{\alpha e}$, are the modified secondary compression index, $C_{\alpha \varepsilon}$, can be determined from the slope of the straight line portion of the dial reading versus log time curve that occurs after primary consolidation is complete. Usually the ΔDR is determined over one log cycle of time. The corresponding change in void ratio is calculated from the settlement equation (Eq. 5.32) because the height of specimen for that increment and e_0 is known.

To provide a working hypothesis for estimating secondary settlements, the following assumptions are made, based on the work of Ladd (1971) and others and summarized by Raymond and Wahls (1976), about the behavior of fine-grained soils in secondary compression.

1. C_α is independent of time (at least during the time span of interest).
2. C_α is independent of the thickness of the soil layer.
3. C_α is independent of the load increment ratio (LIR), as long as some primary consolidation occurs.
4. The ratio C_α/C_c is approximately constant for a wide range of geotechnical materials over the normal range of engineering stresses.

Typical dial reading versus log time behavior curves illustrating these assumptions for a normally consolidated clay are shown in Figure 5.52. The rate of secondary compression expressed in terms of settlement (ΔR) per log cycle is assumed to be independent of the thickness of the specimen as well as of the load increment. There is some effect, however, of the consolidation stress and, as Mesri and Godlewski (1977) point out, C_α is strongly dependent on the final effective stress.

Note that just because $C_{\alpha e}$ may be a constant, the rate of secondary compression is not a constant, but decreases as t

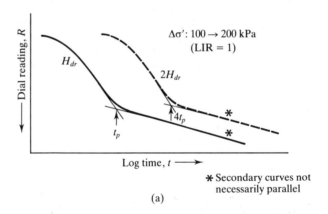

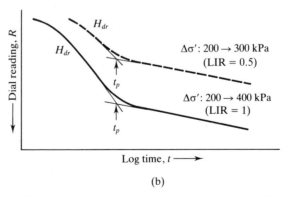

Fig. 5.52 Typical secondary compression behavior from the working hypothesis by Raymond and Wahls (1976). (a) Effect of drainage distance. (b) Effect of load increment ratio and consolidation stress. (*From Holtz and Kovacs, 1981.*)

increases (Leonards, personal communication, 1988). Equation 5.59 can be written (Mesri and Castro, 1987):

$$C_{\alpha e} = \frac{\partial e}{\partial \log t} = 2.3t \frac{\partial e}{\partial t}$$

and $\partial e / \partial t$ is, by definition, the rate of secondary compression. Therefore, $\partial e / \partial t = C_{\alpha e}/2.3t$, which means that even if $C_{\alpha e}$ is constant, $\partial e / \partial t$ is not but is inversely proportional to t.

The working hypothesis is useful as a first approximation for estimating secondary settlements. However, some aberrations in the actual long-term settlement response of a foundation should be expected, because the assumptions are an over-simplification of real behavior. For example, the secondary compression curves of Figure 5.52 may not actually be parallel or may even have a constant slope. There is evidence that C_α may change with time, both in the laboratory (Mesri and Godlewski, 1977) and in the field (Leonards, 1973). Also, the duration and therefore the magnitude of secondary settlement is a function of the time required for completion of primary consolidation (t_p), and the thicker the consolidating layer, the longer the time required for primary consolidation. Even if the strain at the end of primary consolidation for both thin and thick layers is about the same, which is questionable (Leroueil, 1988), there is limited evidence (Aboshi, 1973) that the slopes may not be parallel and that C_α may decrease as the thickness of the soil layer increases.

Assumptions (3) and (4) are approximately correct. Assumption (3) was verified by Leonards and Girault (1961) and Mesri and Godlewski (1977), except that the load increment must be sufficient to go well beyond the preconsolidation stress. The fourth assumption, that the ratio $C_{\alpha e}/C_c$ is approximately a constant, has also been verified by Mesri and Godlewski (1977) and Mesri and Choi (1984) for a wide variety of natural soils. As indicated in Table 5.11, the average value of $C_{\alpha e}/C_c$ is about 0.04 ± 0.01 for a majority of soft inorganic clays and 0.05 ± 0.01 for the highly organic plastic clays; in no case did a value exceed 0.1. They also showed that this ratio holds at any time, effective stress, and void ratio during secondary compression. The only exception, as shown by Leonards and Girault (1961, fig. 3), seems to be the load increment that spans the preconsolidation stress, σ'_p. However, as noted by Mesri and Castro (1987), this discrepancy results from using an average C_c over an increment, rather than an instantaneous C_c, to determine the $C_{\alpha e}/C_c$ ratio. The correct graphical procedure is shown in Figure 5.53 (Mesri and Castro, 1987).

5.8.2 Empirical Estimates

If, for some reason, determination of $C_{\alpha e}$ from laboratory test data is not practical, the $C_{\alpha e}/C_c$ data of Table 5.11 for similar

TABLE 5.11 VALUES OF $C_{\alpha e}/C_c$ FOR NATURAL SOILS[a].

Organic silts	0.035–0.06
Amorphous and fibrous peat	0.035–0.085
Canadian muskeg	0.09–0.10
Leda clay (Canada)	0.03–0.06
Postglacial Swedish clay	0.05–0.07
Soft blue clay (Victoria, B.C.)	0.026
Organic clays and silts	0.04–0.06
Sensitive clay, Portland, Maine	0.025–0.055
San Francisco Bay mud	0.04–0.06
New Liskeard (Canada) varved clay	0.03–0.06
Mexico City clay	0.03–0.035
Hudson River silt	0.03–0.06
New Haven organic clay silt	0.04–0.075

[a] Modified from Mesri and Godlewski (1977).

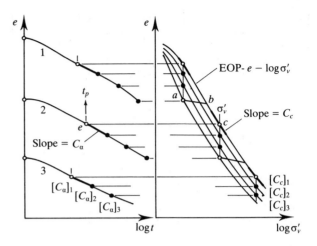

Fig. 5.53 Corresponding values of C_a and C_c at any instant (e, σ'_p, t) during secondary compression. (*Mesri and Castro, 1987.*)

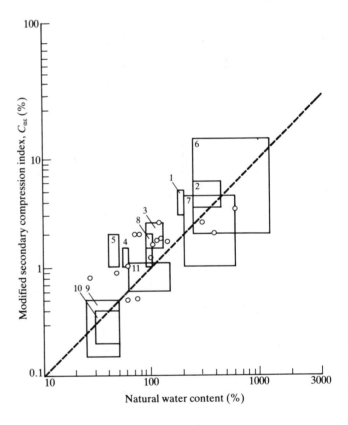

Fig. 5.54 Modified secondary compression index versus natural water content. (*After Mesri, 1973, which should be consulted for details of the references cited.*) 1, Whangamarino clay (Newland and Allely, 1960). 2, Mexico City clay (Leonards and Girault, 1961). 3, Calcareous organic silt (Wahls, 1962). 4, Leda clay (Crawford, 1965). 5, Norwegian plastic clay (Bjerrum, 1967). 6, Amorphous and fibrous peat (Lea and Brawner, 1963). 7, Canadian muskeg (Adams, 1965). 8, Organic marine deposits (Keane, 1965). 9, Boston blue clay (Horn and Lambe, 1965). 10, Chicago blue clay (Peck, personal files). 11, Organic silty clay (Jonas, 1965). ○, Organic silt, etc. (Moran et al., 1958).

soils may be used, or an average $C_{\alpha e}/C_c$ value of 0.04 or 0.05 may be acceptable for preliminary calculations. Another method to estimate the modified secondary compression index from the natural water content of the soil is shown in Figure 5.54.

How to estimate secondary settlements is shown in the following Examples.

EXAMPLE 5.18

An oedometer test on a sample of San Francisco Bay mud gave the following time rate of consolidation data for the load increment of 40 to 80 kPa. This load increment represents the anticipated load in the field. The initial void ratio e_0 is 2.855, $w_n = 105.7$ percent, $LL = 88$, $PL = 43$, and $\rho_s = 2.7 \, \text{Mg/cm}^3$. The initial height of the test specimen is 25.4 mm, and the initial dial reading is 12.700 mm. Using the basic specimen data, the void ratio was determined for any height or thickness of the test specimen, and these results are shown in Column 3 of Table 5.12.

Assume the consolidation settlement s_c is 30 cm and that it occurs after 25 years. The thickness of the compressible layer is 10 m. Compute the amount of secondary compression that would occur from 25 to 50 years after construction. Assume the time rate of deformation for the load range in the test approximates that occurring in the field.

Solution

To evaluate C_α, use Equation 5.59. First plot a void ratio versus $\log t$ curve from the given data. This is shown in Figure 5.55. C_α is then found to be 0.052. Note that $C_\alpha = \Delta e$ when $\Delta \log t$ covers one full log cycle. The corresponding modified secondary compression index $C_{\alpha\varepsilon}$ (Eq. 5.60) is $0.052/(1 + e_p) = 0.052/(1 + 2.372) = 0.0154$; e_p is obtained from Figure 5.55 at the end of primary consolidation.

To calculate secondary settlement s_s, use the basic settlement equation (Eq. 5.32). Now, however, Δe is a function of time and not stress. Substituting Δe from

Equation 5.59 into Equation 5.32 and using e_p for e_0, we obtain

$$s_s = \frac{C_\alpha}{1 + e_p}(H_0)(\Delta \log t)$$

$$= \frac{0.052}{1 + 2.372}(10 \, \text{m}) \log \frac{50}{25}$$

$$= 0.046 \, \text{m} = 4.6 \, \text{cm} \tag{5.61}$$

Thus, $s = s_c + s_s = 30 + 4.6 = 34.6 \, \text{cm}$ in 50 years. This excludes any immediate settlement s_i that may also have occurred.

TABLE 5.12   TEST DATA.

(1) Dial Reading, mm	(2) Elapsed Time, min	(3) Void Ratio
11.224	0	2.631
11.151	0.1	2.620
11.123	0.25	2.616
11.082	0.5	2.609
11.019	1.0	2.600
10.942	1.8	2.588
10.859	3.0	2.576
10.711	6	2.553
10.566	10	2.531
10.401	16	2.506
10.180	30	2.473
9.919	60	2.433
9.769	100	2.410
9.614	180	2.387
9.489	300	2.368
9.373	520	2.350
9.223	1350	2.327
9.172	1800	2.320
9.116	2850	2.311
9.053	4290	2.301

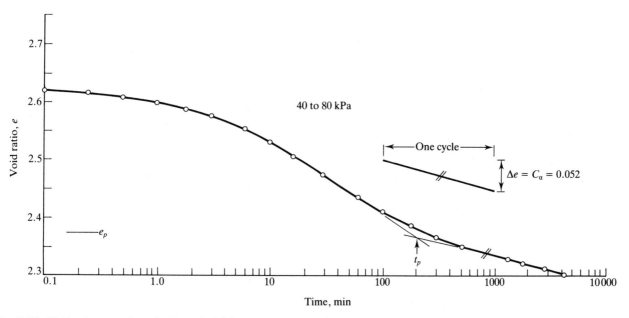

Fig. 5.55 Void ratio versus log t for Example 5.18.

The secondary settlement may also be computed by means of Equations 5.32 and 5.60, where

$$s_s = C_{\alpha\varepsilon} H_0 (\Delta \log t)$$

$$= 0.0154 \, (10 \, \text{m}) \log \frac{50}{25}$$

$$= 0.046 \, \text{m, as before} \qquad (5.62)$$

EXAMPLE 5.19

Given the data given in Example 5.18 and the knowledge that the average C_c is 1.23; the initial water content of the specimen is 105.7 percent. From the data in Table 5.11 and Figure 5.54, estimate $C_{\alpha e}$ and $C_{\alpha\varepsilon}$, and compare with the values calculated in Example 5.18.

Solution

From Table 5.11, for San Francisco Bay mud, use an average value of C_α/C_c of 0.05. Therefore,

$$C_{\alpha e} = 0.05(C_c) = 0.05(1.23) = 0.062$$

From Equation 5.60, $C_{\alpha\varepsilon} = C_{\alpha e}/1 + e_p$. From Figure 5.54, $e_p = 2.372$. Therefore,

$$C_{\alpha\varepsilon} = \frac{0.062}{1 + 2.372} = 0.018$$

A second way to estimate the modified secondary compression index is to use Figure 5.54 where $C_{\alpha\varepsilon}$ is plotted versus water content. For this example, the initial water content is 105.7 percent. If the dashed line is used in Figure 5.54, a value of $C_{\alpha\varepsilon}$ of about 0.01 (or higher) is obtained. From Example 5.18, $C_{\alpha e} = 0.052$ and $C_{\alpha\varepsilon} = 0.015$. The agreement using the approximate values is acceptable for preliminary design estimates.

5.8.3 Relative Importance

As stated in Section 5.2.1 and Table 5.1, secondary settlements are important for foundations constructed on organic soils, especially fibrous peats, and they may be important for foundations on clay soils. Whether secondary settlements need to be considered in detail depends on (1) the magnitude of the secondary compression relative to the primary consolidation settlement, and (2) the rate of consolidation settlement. When the relative magnitude of the secondary to primary is large, i.e., when $s_s/s_c > 1$, then good estimates of the secondary settlement should be made. When the rate of consolidation settlement is rapid, i.e., when t_p occurs within a few weeks to months, then the magnitude of secondary settlement will be relatively important, and again, good estimates of secondary are advisable. These two conditions tend to occur in soils with active clay minerals (montmorillonites, illites, etc.), small load increment ratios, and short drainage distances (e.g., deposits with intermediate drainage layers; fibrous peats with a high permeability).

If Δs_s is the secondary compression per log cycle, and s_c is the consolidation settlement, then a consistent relation between s_s/s_c and the load increment ratio (LIR) has been demonstrated by Leonards and Girault (1961) and is shown in Figure 5.56. Below a certain value of the LIR, the secondary compression

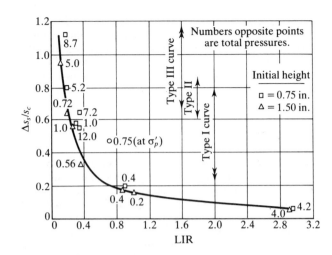

Fig. 5.56 Effect of load increment ratio on rate of secondary compression for undisturbed Mexico City clay. (*After Leonards and Girault, 1961, in Leonards, 1968.*)

per unit of consolidation settlement increases rapidly as the LIR decreases. When the load increment spans the preconsolidation pressure, secondary compressions are anomalously high. (This observation is also useful as a check on the value of σ'_p interpreted from the consolidation curve.)

If a relationship such as the one shown in Figure 5.56 is obtained, it can be used to estimate secondary compressions in the field as follows (Leonards, 1968).

1. The compressible strata are subdivided into layers and the consolidation settlement (s_c) is computed for each layer using the procedure described in Section 5.7.
2. Using the known values of LIR, values of $\Delta s_s/s_c$ are obtained for each layer from the relation shown in Figure 5.56.
3. By multiplying values of s_c obtained in step (1) by values of $\Delta s_s/s_c$ obtained in step (2), values of Δs_s for each layer can be obtained.
4. The sum of the values of Δs_s for each layer in step (3) is the total secondary compression per log cycle.

This procedure was used by Leonards (1968) to predict the secondary settlements of buildings in Drammen, as shown in Figure 5.57. The range in discrepancies is from 5 to 75 percent.

It should be kept in mind that secondary settlements occur relatively slowly, during one or more log cycles of time after the end of primary. Thus, depending on the design life of the structure, construction costs, maintenance costs, etc., the impact of secondary settlement on the actual performance of a structure may be minimal, even though its magnitude is relatively large.

5.9 TOLERABLE CRITERIA

For important structures or structures thought to be particularly sensitive to small differential settlements, it is recommended

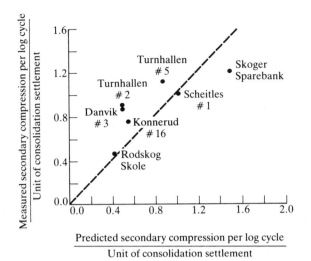

Fig. 5.57 Secondary compression for seven buildings in Drammen, Norway. (*Leonards, 1968.*)

that total settlements be computed or estimated at a sufficient number of points or typical footings to establish the likely overall settlement pattern for the structure. From this pattern, it is possible to determine the overall settlement profile and the greatest differential settlement between adjacent foundation units, columns, along bearing walls around the periphery of tanks, etc. Determination of the settlement that will cause significant architectural or structural damage is an extremely complex indeterminant analytical problem. Factors affecting such an analysis include the variability in the soil properties, uncertainty of the structural materials and rigidity, construction sequence, time rate and uniformity of settlement, contact pressures, stiffness and rigidity of the footing connections, and the nature of the actual loads transmitted to the foundation units. Consequently, the analysis of tolerable settlements and development of criteria for tolerable settlements have been established almost entirely empirically on the basis of observations of settlement and damage in actual buildings.

If an attempt is made to model analytically the structure and calculate the effect of differential settlements, one obtains ridiculously low allowable differential settlements because of the large bending moments that will be calculated in the frame. Some yielding in the structure actually occurs, but how much is unknown; therefore, empirical evidence from the performance of actual buildings is used to establish tolerable settlement criteria. It is important to be able to predict reasonably well the actual settlements of a structure so that proper preparation may be made for tolerating those settlements either in the foundation, structure, or perhaps in some sort of soil improvement or alteration of the structure's geometry, configuration, or in some rare cases, even its stiffness. Types of settlement include total, tilting, and distortion, or differential movement. A common empirical rule is that the differential settlement is equal to about one-half of the calculated total settlement. Terzaghi and Peck (1967) suggest that for footings on sand, the differential settlement is unlikely to exceed 75 percent of the estimated total settlement. For clays, however, differential settlement may be greater than this value and may even approach the total settlement. References on tolerable and differential settlements include Skempton and MacDonald

(1956), Polshin and Tokar (1957), Bjerrum (1963), Burland and Wroth (1974). Wahls (1981) provides an excellent review of the tolerable settlement of buildings. Summary information is also given in U.S. Navy (1982) and *Canadian Foundation Engineering Manual* (Meyerhof and Fellenius, 1985).

5.9.1 Definitions

Figure 5.58 indicates the definitions of settlement terminology for (a) settlement without tilt and (b) settlement with tilt. The differential settlement, δ_{AB} results in an angular distortion of β_{AB} between points A and B. Distortion can be expressed in terms of either δ/l or a distortion angle β. Both are used.

Both architectural and structural damage may be observed. Architectural damage includes (1) cracking of plaster and masonry brick walls, which may be unsightly, (2) damage that may be unpleasant (in the case of broken sewer lines and connections to loading docks), and (3) damage that may even be dangerous (as in the case of gas lines breaking, or tilting damage to adjacent buildings). Architectural damage can occur from causes other than foundation settlement. Examples include changes in temperature, moisture (swelling clays; shrinkage due to trees), vibrations due to traffic, wind, blasting, etc.

Structural damage is any damage that reduces the ultimate ability of the structure to carry and resist the load imposed on it. Structural damage in any situation is unacceptable, and the foundation design must be such that structural damage is avoided at all costs. In foundation design, there is no point in trying to estimate the severity of structural damage. It must be simply avoided. The propensity for structural damage is influenced greatly by the rate at which the settlements occur and the rigidity and type of framing system. Settlements that occur very slowly, over periods of decades or more, may be tolerable by masonry or reinforced concrete frame structures. These same settlements, if they occurred within a period of a few months or few years, would result in severe structural damage.

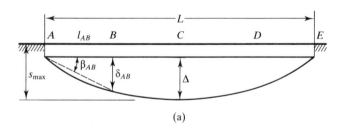

(a)

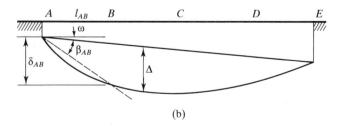

(b)

Fig. 5.58 Definitions of settlement terminology. (*Wahls, 1981.*)

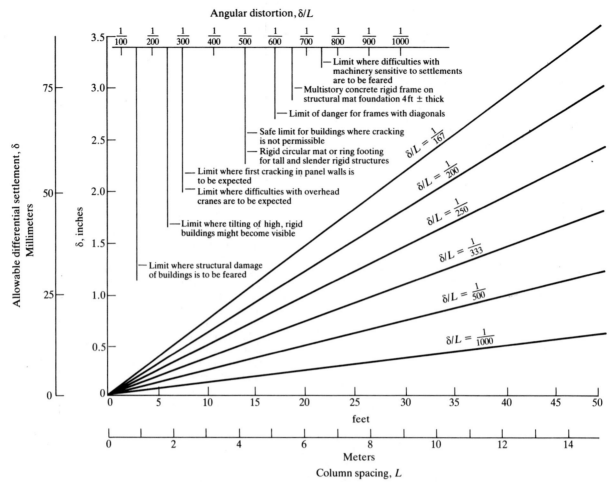

Fig. 5.59 Tolerable settlements for buildings. (*Bjerrum, 1963; U.S. Navy, 1982.*)

TABLE 5.13 LIMITING SETTLEMENTS FOR STRUCTURES[a].

Type of Movement		Limiting Factor	Maximum Allowable Settlement or Differential Movement
Total settlement		Drainage	6 to 12 in
		Access	12 to 24 in
		Probability of nonuniform settlement	
		Masonry-walled structure	1 to 2 in
		Framed structures	2 to 4 in
		Smokestacks, silos, mats	3 to 12 in
Tilting	Most maintain B.C. safety	Stability against overturning	Depends on height and weight
		Tilting of smokestacks, towers	$0.004h$
		Rolling of trucks, etc.	$0.01L$
		Stacking of goods	$0.01L$ ($\sim 1/100$)
		Machine operation—cotton loom	$0.003L$
		Machine operation—turbogenerator	$0.0002L$
		Crane rails	$0.003L$
		Drainage of floors	$0.01L$ to $0.02L$
Differential movement		High continuous brick walls	$0.0005L$ to $0.001L$ ($\sim 1/300$)
		One-story brick mill building, wall cracking	$0.001L$ to $0.02L$
		Plaster cracking (gypsum)	$0.001L$ ($1/600$)
		Reinforced-concrete building frame	$0.0025L$ to $0.004L$ ($\sim 1/150$ to $1/170$)
		Reinforced-concrete building curtain walls	$0.003L$
		Steel frame, continuous	$0.002L$
		Simple steel frame	$0.005L$

[a] After Sowers (1962), Perloff and Baron (1976, p. 517).
Note: L = distance between adjacent columns that settle different amounts, or between any two points that settle differently. Higher values of allowable settlement are for regular settlements and more tolerant structures. Lower values are for irregular settlements and critical structures.

TABLE 5.14 LIMITING ANGULAR DISTORTION[a].

Category of Potential Damage	Δ/l (2)
Danger to machinery sensitive to settlement	1/750
Danger to frames with diagonals	1/600
Safe limit for no cracking of buildings[b]	1/500 (1/600 plaster cracking)
First cracking of panel walls	1/300
Difficulties with overhead cranes	1/300
Tilting of high rigid buildings becomes visible	1/250
Considerable cracking of panel and brick walls	1/150
Danger of structural damage to general buildings	1/150
Safe limit for flexible brick walls, $L/H > 4$[b]	1/150

[a] Wahls (1981).
[b] Safe limits including factor of safety.

STRUCTURE		TOLERABLE DISTORTION $\dfrac{\Delta_{max}}{L}$ or β
A. Unreinforced load-bearing walls (L and H are respectively length and height of the wall from top of footing)	Sagging for $L/H < 3$	$\dfrac{\Delta_{max}}{L} = 1/3500$ to $1/2500$
	for $L/H > 5$	$\dfrac{\Delta_{max}}{L} = 1/2000$ to $1/1250$
	Hogging for $L/H = 1$	$\dfrac{\Delta_{max}}{L} = 1/5000$
	for $L/H = 5$	$\dfrac{\Delta_{max}}{L} = 1/2500$
B. Jointed rigid concrete pressure conduits (Maximum angle change at joint 2 to 4 times average slope of settlement profile. Longitudinal extension affects damage.)		1/65
C. Circular steel petroleum or fluid storage tanks.		$\beta < 1/300$ $\beta' = 1/500$ to $1/300$

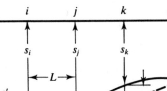

Points on tank perimeter

$\beta = \dfrac{s_i - s_j}{L}$

$\beta' = \left(s'_j - \dfrac{s'_i + s'_k}{2}\right)\dfrac{1}{L}$

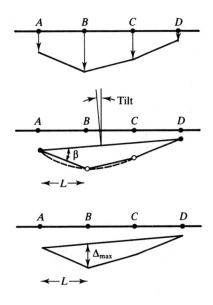

Fig. 5.60 Tolerable differential settlement for miscellaneous structures. (*After U.S. Navy, 1982.*)

It may be possible to alter the foundation or framing system to avoid structural damage, although this is rarely done today. Altering the foundation system is more common. The use of various types of deep foundations, mats, or raft foundations, or foundation soil improvement are the common foundation alternates.

5.9.2 Empirical Criteria*

Commonly used empirical criteria for estimating tolerable differential settlements are given in Figure 5.59. Tables 5.13 and 5.14 give some additional information including limiting total settlements and tilting for various structures. Figure 5.60 gives tolerable differential settlements for miscellaneous structures such as tanks, pressure conduits, etc.

Even if there is no structural damage, total settlements can present problems, especially with utilities, loading docks, drainage, etc. Total settlements can also affect adjacent buildings by causing them either to tilt or to settle. In some process industries and sewage treatment plants, gravity flow may be compromised. In some cases, it may be more economical to change the process or install pumps than to reduce total and differential settlements.

Magnitudes of tolerable total settlements are indicated in Table 5.13. If total settlements in a structure exceed 150 mm, utility connections should be flexible. This may be more economical than attempting to reduce the total settlement. Loading docks, for example, cannot operate if the total settlement exceeds about 150 mm but it may be more economical to initially raise the dock by about that amount if one would expect, for example, a total settlement of 300 mm. A similar approach may be appropriate with bridge and bridge-approach spans. Criteria in this case would be the maximum allowable change in grade.

Tilting of a structure may also be a problem. The classic example is the Leaning Tower of Pisa in Italy, where the tilt is on the order of 1 in 10, or considerably greater than the values given in Table 5.13. If excessive tilt occurs, eccentric loads on foundations can result, causing redistribution of stress with time and additional settlements. Progressive tilting may also impair the function of the structure. Typically, the function limits tilt before a bearing capacity failure occurs.

Limits on tilt are indicated in Table 5.13. It is interesting that a number of linear accelerators and turbine generators are operating today without apparent distress with tilts much greater than the limits indicated in the table.

Tilting of smoke stacks and towers is often limited more by their proximity to adjacent structures than some sort of impairment in actual safety or operating conditions. The eye is extremely sensitive to differences in verticality.

In general, tilt rarely governs the design.

Distortion settlements usually expresses maximum angular distortion in Figure 5.59 and Table 5.14 or δ/l. Limits are indicated for cracking plaster, damage to masonry walls, and structural damage to reinforced concrete frames. In the latter case, if the settlement occurs in less than 10 years, then structural damage is likely. As mentioned previously, the differential or distortion settlement that a structure is able to tolerate depends greatly on the rate of settlement. It is very difficult to predict this value with certainty; it is much easier to predict the range of total and ultimate settlements and apply a rule of thumb,

such as that the differential settlements equal about half the total settlements.

In summary, it is very important to be able to predict the magnitude and rate of settlements. In doing this, the function of the structure must be taken into account. Settlements must not impair the function of the structure. Finally, avoid structural damage. Keep in mind that what is tolerable is empirical. Many structures are OK with settlements greatly exceeding the allowables as indicated in the tables. Others, however, have experienced much less settlement and exhibit undesirable architectural or in some cases structural damage. The establishment of tolerable criteria is one of the most difficult problems facing the foundation engineer.

5.9.3 Effect of Structural Rigidity

The effect of structural rigidity is discussed for a number of cases by Burland and Wroth (1974), Wahls (1981), and U.S. Navy (1982). Estimates of differential settlements are normally less accurate than total or average settlement computations, because the soil–structure interaction is so difficult to predict. As noted in Figure 5.2, if the structure is completely rigid, uniform total settlement and thus no differential settlement would occur. On the other hand, complete flexibility implies uniform contact pressures between the foundation and the soil and differential settlements. Actual conditions are probably always between the two extremes. However, depending on the magnitude of the relative stiffness, mats can be defined as either rigid or flexible for practical purposes, and this is discussed in U.S. Navy (1982).

5.10 FOUNDATION TREATMENT ALTERNATES

In this section, we briefly discuss what to do if the estimates of settlements exceed the tolerable limits as discussed above. The alternates are:

1. Go to deep foundations.
2. Alter the structure.
3. Alter the foundation.
4. Improve the properties of the foundation soil.

Each of these items will be discussed in this section. Table 5.15 lists some methods for reducing or accelerating settlements.

In selecting one or more of these alternates, the foundation designer must consider not only the short-term or initial construction cost but also the long-term operation and maintenance costs. Time required to construct an alternate must be taken into consideration and, most important, the certainty involved in the construction and installation of the alternate must be taken into account. Even if a given alternate may cost more initially, if it is more certain in its execution, construction, and performance, it may be a better choice. Certainty of construction and installation depends on the soil and site conditions, the nature of the design and construction, type of design and method of contracting for the construction and design, and who has responsibility for inspection and what type of control there exists over the quality of inspection during construction. All these factors influence the final selection of foundation and alternates.

---

*I am indebted to my colleague, G. A. Leonards, Purdue University, for many of the following concepts.

TABLE 5.15 METHODS OF REDUCING OR ACCELERATING SETTLEMENT OR COPING WITH SETTLEMENT[a].

Method	Comment
Procedures for linear fills on swamps or compressible surface stratum	
Excavation of soft material	When compressible foundation soils extend to depth of about 10 to 15 ft, it may be practicable to remove entirely. Partial removal is combined with various methods of displacing remaining soft material
Displacement by weight of fill	Complete displacement is obtained only when compressible foundation is thin and very soft. Weight displacement is combined with excavation of shallow material
Jetting to facilitate displacement	For a sand or gravel fill, jetting within the fill reduces its rigidity and promotes shear failure to displace soft foundation. Jetting within soft foundation weakens it to assist in displacement
Blasting by trench or shooting methods	Charge is placed directly in front of advancing fill to blast out a trench into which the fill is forced by the weight of surcharge built up at its point. Limited to depths not exceeding about 20 ft
Blasting by relief method	Used for building up fill on an old roadway or for fills of plastic soil. Trenches are blasted at both toes of the fill slopes, relieving confining pressure and allowing fill to settle and diplace underlying soft materials
Blasting by underfill method	Charge is placed in soft soil underlying fill by jetting through the fill at a preliminary stage of its buildup. Blasting loosens compressible material, accelerating settlement and facilitating displacement to both sides. In some cases, relief ditches are cut or blasted at toe of the fill slopes. Procedure is used to swamp deposits up to 30 ft thick
Procedures for preconsolidation of soft foundations	
Surcharge fill	Used where compressible stratum is relatively thin and sufficient time is available for consolidation under surcharge load. Surcharge material may be placed as a stockpile for use later in permanent construction. Soft foundation must be stable against shear failure under surcharge load
Accelerating consolidation by vertical drains	Used where tolerable settlement of the completed structure is small, where time available for preconsolidation is limited, and surcharge fill is reasonably economical. Soft foundation must be stable against shear failure under surcharge load
Vertical drains with or without surcharge fill	Used to accelerate the time for consolidation by providing shorter drainage paths
Wellpoints placed in vertical sand drains	Used to accelerate consolidation by reducing the water head, thereby permitting increased flow into the sand drains. Particularly useful where potential instability of soft foundation restricts placing of surcharge or where surcharge is not economic
Vacuum method	Variation of wellpoint in vertical sand drain but with a positive seal at the top of the sand drain surrounding the wellpoint pipe. Atmospheric pressure replaces surcharge in consolidating soft foundations
Balancing load of structure by excavation	Utilized in connection with mat or raft foundations on compressible material or where separate spread footings are founded in suitable bearing material overlying compressible stratum. Use of this method may eliminate deep foundations, but it requires very thorough analysis of soil compressibility and heave

[a] U.S. Navy (1982).

5.10.1 Deep Foundations

Deep foundations are probably the most common alternate selected by foundation engineers when settlements exceed tolerable limits. Deep foundations are discussed in Chapters 13 and 14.

5.10.2 Alteration of the Structure

A number of possibilities exist in this case. It may be possible to move or reorient the structure so as to obtain better foundation and settlement conditions, depending on the site, its location, and intended function. In urban areas, where land values are high, moving the structure is often very difficult, if not impossible. In rural areas, there may be more freedom to move the structure.

The loads may be lightened or redistributed in an attempt to reduce the weight of various components of the structure. This alternate may be difficult if the structure is already designed, as usually happens in most building and bridge foundations. Typically, the foundations are the last component of the system to be designed, which is unfortunate in some situations, as a different or alternate type of framing system might result in a more economical foundation.

In some cases, it may be possible by altering the framing system to make a structure more rigid. Alternatively, the structure may be made more flexible. Simple supports may be used instead of continuous spans, which may permit larger distortion. Judicious use of construction joints or hinges, to allow portions of a building to settle with respect to one another, are often quite economical.

In increasing the rigidity of the structure, it should be kept in mind that the stiffness of the structure depends upon the geometry and framing system chosen and not necessarily on the stiffness of the individual structural elements. The classical example is of course a truss, which is inherently stiff because of geometry and not necessarily because of the stiffness of the individual members. Depending on rate of settlements anticipated, stiffening the structure may be more detrimental in terms of architectural damage and potential structural damage. Unless the foundation is also stiffened, making a stiffer structure is not necessarily a wise approach.

Reducing or redistributing loads is another possibility. An example would be to construct fewer stories than originally anticipated. It may be possible to consider moving or changing operations within a structure so that the heaviest loads are not necessarily over the poorest soils. Another example is a building with a central tower with wings, which may exacerbate differential settlement and distortion in the structure. However, use of construction joints may take care of some of these problems.

Finally, it may be possible to use jacks under key columns in the structure to adjust for differential settlements. This alternative does require measurements to be made periodically and maintenance personnel to adjust the jacks to compensate for differential movements. This procedure will require great care in execution, especially when it must be carried out over a very long time. However, a number of examples of the use of jacks exist, the most notable being in Mexico City where a number of high-rise buildings are adjusted periodically to maintain their total settlements consistent with the general regional settlement in the city.

5.10.3 Alteration of the Foundation

A number of possibilities exist in this category. Technically, deep foundations should be included under this heading, but they have already been mentioned separately.

Perhaps the most obvious course is to use a fully or partially compensated foundation. In this case, excavation reduces the net load applied to the soil. Sometimes this is called a "floating" foundation. Over large areas, control of a compensating foundation to reduce differential settlements is not easy and, consequently, a large mat or raft type foundation must often be used.

For typical soil conditions and weights of buildings, excavation of one basement level is approximately equivalent to 10 stories of average building height. Deep excavations, of course, can present construction problems. Thus, depending on the specific and geotechnical conditions at the site, there are limits as to how many stories' compensation can be achieved.

Another alternate is to use combined footings or even a raft or mat foundation completely covering the area under the structure. Raft and mat foundations have their own problems, but are a possibility that should be considered. It may also be possible to change the elevation of the bearing stratum, for example, by excavating softer or looser upper-lying materials.

Another possibility is to vary the contact pressure beneath the footings or adjust the initial elevations of the footings. In order for these alternates to be successful, a complete picture of soil variability at the site must be obtained.

Another possibility that has been used in rare cases is to underpin a shallow foundation during construction. By starting with an economical shallow foundation, immediate use of the structure is possible. As construction progresses, underpinning of the structure can take place. This alternate is less attractive today because advancements in geotechnical and foundation engineering enable us to make better predictions and foundation designs than in the past.

5.10.4 Alteration of Soil Properties

In this category are included what are commonly called soil improvement or foundation treatment methods. A number of possibilities exist and many depend on the type of soils involved, whether granular or cohesive, and the nature of the structure involved. Some methods are more economical than others, and local experience is very important in this regard.

Many problem foundation soils can be improved by:

1. Reducing the load.
2. Replacing the problem materials with more competent materials.
3. Increasing the shear strength and reducing the compressibility of the problem materials.
4. Transferring loads to more competent layers.
5. For embankments and other similar foundations, reinforcement is also possible.

Methods include removal or excavation of the problem materials and replacement by suitable fill. For buildings, this process is called a structural fill. Soft clay foundations can be improved by consolidation using surcharge fills, often combined with vertical drains to accelerate consolidation. Chemical alteration and stabilization is also possible. Techniques include lime and cement columns, grouting and other injections, electroosmosis, and thermal (freezing) techniques.

Physical stabilization or densification is appropriate in some cases, especially with loose granular materials. Appropriate methods include dynamic compaction, blasting, vibroreplacement and vibroflotation techniques, sand compaction piles, stone columns, and even water. Reinforcement techniques include geotextiles and geogrids, fascines, Wager short sheet piles, anchors, root piles, and reinforced walls.

To help assess which method of ground improvement may be the most important, Mitchell (1981) suggests that the following factors be considered.

1. The operating criteria for the structure; for example, stability requirements, allowable total and rate of settlement, maintenance needs, etc. This will establish the level of improvement required in terms of properties such as strength, modulus, compressibility, etc.
2. The area, depth, and total volume of the soil to be treated.
3. Soil type and its initial properties.
4. Availability of materials, for example, sand, gravel, water, admixtures, etc.
5. Availability of equipment and required skills.
6. Environmental factors such as waste disposal, erosion, water pollution, and effects on adjacent facilities or structures.
7. Local experience and preference.
8. Time available.
9. Cost.

General references on foundation and soil improvement methods include Mitchell (1970, 1981), Broms (1979), Eggestad (1983), Broms and Anttikoski (1983), Welsh (1987), and Holtz (1989). These references include detailed discussions on many of the specific treatment methods and their design, as well as specific references on design, construction, and a number of useful case histories.

Foundation treatment methods discussed in this handbook include a number of chapters, specifically Chapter 23 on dynamic compaction, Chapter 24 on lime and stone columns, etc., Chapter 9 on grouting and stabilization, Chapter 8 on excavation and replacement, Chapter 21 on reinforcement.

REFERENCES

Aboshi, H. (1973), An experimental investigation on the similitude in the consolidation of a soft clay, including the secondary creep settlement, *Proceedings of the Eighth International Conference on Soil Mechanics and Foundation Engineering*, Moscow, **4.3**, p. 88.

Anagnostopoulos, A. G. (1990), The compressibility of cohesionless soils, *Geotechnik*, No. 3.

Anderson, T. C. and Lukas, R. G. (1981), Preconsolidation pressure predicted using $s_u/\bar{p}$ ratio, *Laboratory Shear Strength of Soil*, *ASTM, STP* 740, pp. 502–515.

Azzouz, A. S., Krizek, R. J., and Corotis, R. B. (1976), Regression analysis of soil compressibility, *Soils and Foundations*, **16**, No. 2, pp. 19–29.

Baguelin, Jezequel, and Shields (1978), *The Pressuremeter and Foundation Engineering*, Trans Tech Publications, Clausthal, Germany.

Baladi, G. Y. (1968), Distribution of stresses and displacements within and under long elastic and viscoelastic embankments, Ph.D. Thesis, Purdue University.

Becker, D. E., Crooks, J. H. A., Been, K., and Jefferies, M. G. (1987), Work as a criterion for determining in situ and yield stresses in clays, *Canadian Geotechnical Journal*, **24**, No. 4, pp. 549–564.

Bellotti, R., Ghionna, V., Jamiolkowski, M., Lancellotta, R., and Manfredini, G. (1986), Deformation characteristics of cohesionless soils from in situ tests, *Proceedings of In Situ '86, ASCE Specialty Conference on Use of In Situ Tests in Geotechnical Engineering*, Geotechnical Special Publication No. 6, ASCE, pp. 47–73.

Bjerrum, L. (1963), Allowable settlement of structures, *Proceedings of the 3d European Conference on Soil Mechanics and Foundation Engineering*, Wiesbaden, **2**, pp. 135–137.

Bjerrum, L. (1967), Engineering geology of Norwegian normally consolidated marine clays as related to settlements of buildings. 7th Rankine Lecture, *Geotechnique*, **XVII**, No. 2, pp. 81–118.

Bjerrum, L. (1972), Embankments on soft ground, *Proceedings of the ASCE Specialty Conference on Performance of Earth and Earth Supported Structures*, Purdue University, **II**, pp. 1–54.

Bjerrum, L. (1973), Problems of soil mechanics and construction on soft clays, State of the art report to Session IV, *Proceedings of the Eighth International Conference on Soil Mechanics and Foundation Engineering*, Moscow, **3**, pp. 111–159.

Bourdeau, P. L. (1986), Analyse probabiliste des tassements d'un massif de sol granulaire, Doctor of Technical Science Thesis No. 628, Federal Technical University of Lausanne, 349 pp. Translated by B. Corey for USAE Waterways Experiment Station, Vicksburg, Miss.

Boussinesq, J. (1885), *Application des Potentiels à L'Étude de L'Équilibre et due Mouvement des Solides Élastiques*, Gauthier-Villars, Paris.

Bowles, J. E. (1975a), Spread footings, *Foundation Engineering Handbook*, H. F. Winterkorn and H. Y. Fang, eds., Chapter 15, Van Nostrand Reinhold, New York, N.Y. pp. 481–503.

Bowles, J. E. (1975b), Combined and special footings, *Foundation Engineering Handbook*, H. F. Winterkorn and H. Y. Fang, eds., Chapter 16, Van Nostrand Reinhold, New York, N.Y. pp. 504–527.

Bowles, J. E. (1987), Elastic foundation settlements on sand deposits, *Journal of Geotechnical Engineering, ASCE*, **113**, No. 8, pp. 846–860.

Broms, B. B. (1979), Problems and solutions to construction in soft clay, *Proceedings of the 6th Asian Regional Conference on Soil Mechanics and Foundation Engineering*, Singapore, Guest Lecture, **II**, pp. 3–38.

Broms, B. B. and Anttikoski, U. (1983), Soil stabilization, General Report, Specialty Session 9, *Proceedings of the 8th European Conference on Soil Mechanics and Foundation Engineering*, Helsinki, **3**, pp. 1298–1301.

Brumund, W. F., Jonas, E., and Ladd, C. C. (1976), Estimating in situ maximum past (preconsolidation) pressure of saturated clays from results of laboratory consolidometer tests, *Estimation of Consolidation Settlement*, Manual of Practice, Transportation Research Board,

Special Report 163, pp. 4–12.

Burland, J. B. and Wroth, C. P. (1974), Allowable and differential settlement of structures including damage and soil–structure interaction, *Proceedings of the Conference on Settlement of Structures*, Cambridge, England, pp. 611–654.

Burland, J. B., Broms, B. B., and De Mello, V. F. B. (1977), Behavior of foundations and structures, State of the Art Report, Session II, *Proceedings of the 9th International Conference on Soil Mechanics and Foundation Engineering*, Tokyo, **2**, p. 517.

Burland, J. B. and Burbidge, M. C. (1985), Settlement of foundations on sand and gravel, *Proceedings of the Institution of Civil Engineers*, **78**, Part 1, pp. 1325–1381.

Burmister, D. M. (1942), Laboratory investigation of soils in Flushing Meadows Park, *Transactions, ASCE*, **68**, pp. 187–200.

Burmister, D. M. (1951), The application of controlled test methods in consolidation testing, *Symposium on Consolidation Testing of Soils*, *ASTM STP* 126, p. 83.

Casagrande, A. (1936), The determination of the preconsolidation load and its practical significance, Discussion 34, *Proceedings of the First International Conference on Soil Mechanics and Foundation Engineering*, Cambridge, **III**, pp. 60–64.

Casagrande, A. (1938), Notes on Soil Mechanics—First Semester, Harvard University (unpublished), 129 pp.

Chen, F. H. (1975), *Foundations on Expansive Soils*, Elsevier, Amsterdam.

Clough, G. W. and Schmidt, B. (1981), Design and performance of excavations and tunnels in soft clay, *Soft Clay Engineering*, E. W. Brand and R. P. Brenner, eds., Chapter 8, Elsevier, Amsterdam, pp. 567–634.

Crawford, C. B. (1964), Interpretation of the consolidation test, *Journal of the Soil Mechanics and Foundations Division, ASCE*, **90**, No. SM5, pp. 87–102.

Crawford, C. B. (1986), State of the art: Evaluation and interpretation of soil consolidation tests, *Consolidation of Soils: Testing and Evaluation, ASTM STP* 892, pp. 71–103.

Crawford, C. B. (1988), On the importance of rate of strain in the consolidation test, *Geotechnical Testing Journal, ASTM*, **11**, No. 1, pp. 60–62.

Dahlberg, R. (1975), Settlement characteristics of preconsolidated natural sands, Swedish Council for Building Research, Document D1:1975.

D'Appolonia, D. J., Poulos, H. G., and Ladd, C. C. (1971), Initial settlement of structures on clay, *Journal of the Soil Mechanics and Foundations Division, ASCE*, **96**, No. SM10, pp. 1359–1377.

DeBeer, E. E. (1965), Bearing capacity and settlement of shallow foundations on sand, Lecture No. 3, *Proceedings of the Symposium on Bearing Capacity and Settlement of Foundations*, Duke University, pp. 15–33.

Desai, C. S. and Christian, J. T. (eds.) (1977), *Numerical Methods in Geotechnical Engineering*, McGraw-Hill Book Co., Inc., New York, N.Y.

Duncan, J. M. (1972), Finite-element analyses of stresses and movements in dams, excavations, and slopes: State of the art, *Proceedings of the Symposium on Applications of the Finite Element Method in Geotechnical Engineering*, **1**, USAE Waterways Experiment Station, Vicksburg, Miss., pp. 267–326.

Duncan, J. M. and Buchignani, A. L. (1976), *An Engineering Manual for Settlement Studies*, University of California, Berkeley.

Dyer, M., Jamiolkowski, M., and Lancellotta, R. (1986), Experimental soil engineering and models for geomechanics, *Proceedings of NUMOG II*, Ghent (preprint), pp. 1–34.

Eggestad, A. (1983), Improvement of cohesive soils, State-of-the-Art Report, *Proceedings of the 8th European Conference on Soil Mechanics and Foundation Engineering*, Helsinki, **3**, pp. 991–1007.

Egorov, K. E. (1958), Concerning the question of the deformation of bases of finite thickness, *Mekhanika Gruntov*, Sb. Tr. No. 34, Gosstroiizdat, Moscow.

Foott, R. and Ladd, C. C. (1981), Undrained settlement of plastic and organic clays, *Journal of the Geotechnical Engineering Division, ASCE*, **107**, No. GT8, pp. 1079–1094.

Gibson, R. E., England, G. L., and Hussey, M. J. L. (1967), The theory of one-dimensional consolidation of saturated clays, I. Finite non-linear consolidation of thin homogeneous layers, *Geotechnique*, **17**, pp. 261–273.

Gromko, G. J. (1974), Review of expansive soils, *Journal of the*

Geotechnical Engineering Division, ASCE, **100**, No. GT6, pp. 667–687.

Hamilton, J. J. and Crawford, C. B. (1959), Improved determination of preconsolidation pressure of a sensitive clay, *Papers on Soils, ASTM STP 254,* pp. 254–271.

Hansbo, S. (1957), A new approach to the determination of the shear strength of clay by the fall-cone test, *Proceedings No. 14,* Swedish Geotechnical Institute, Stockholm.

Hardin, B. O. (1985), Crushing of soil particles, *Journal of Geotechnical Engineering, ASCE,* **111**, No. 10, pp. 1177–1192.

Hardin, B. O. (1987), 1-D strain in normally consolidated cohesionless soils, *Journal of Geotechnical Engineering, ASCE,* **113**, No. 12, pp. 1449–1467.

Harr, M. E. (1966), *Foundations of Theoretical Soil Mechanics,* McGraw-Hill Book Co., Inc., New York, N.Y.

Harr, M. E. (1977), *Mechanics of Particulate Media—A Probabilistic Approach,* McGraw-Hill Book Co., Inc., New York, N.Y.

Harr, M. E. (1987), *Reliability Based Design in Civil Engineering,* McGraw-Hill Book Co., Inc., New York, N.Y.

Holtz, R. D. (1989), Treatment of problem foundations for highway embankments, *NCHRP Synthesis of Highway Practice 147,* Transportation Research Board.

Holtz, R. D. and Kovacs, W. D. (1981), *An Introduction to Geotechnical Engineering,* Prentice-Hall, Inc., Englewood Cliffs, N.J.

Holtz, R. D., Jamiolkowski, M. B., and Lancellotta, R. (1986), Lessons from oedometer tests on high quality samples, *Journal of Geotechnical Engineering, ASCE,* **112**, No. 8, pp. 768–776.

Jamiolkowski, M., Ladd, C. C., Germaine, J. T., and Lancellotta, R. (1985), New developments in field and laboratory testing of soils, Theme Lecture No. 2, *Proceedings of the XI International Conference on Soil Mechanics and Foundation Engineering,* San Francisco, **1**, pp. 57–154.

Jamiolkowski, M., Ghionna, V. N., Lancellotta, R., and Pasqualini, E. (1988), New applications of penetration tests in design practice, *Proceedings of the First International Symposium on Penetration Testing (ISOPT 1),* Orlando, Fla. (preprint).

Janbu, N. (1963), Soil compressibility as determined by oedometer and triaxial tests, *Proceedings of the 3d European Conference on Soil Mechanics and Foundation Engineering,* Wiesbaden, **1**, pp. 19–25; **2**, pp. 17–21.

Janbu, N. (1965), Consolidation of clay layers based on non-linear stress–strain, *Proceedings of the 6th International Conference on Soil Mechanics and Foundation Engineering,* Montreal, **2**, pp. 83–87.

Janbu, N. (1967), Settlement calculations based on the tangent modulus concept, three guest lectures at Moscow State University, Bulletin No. 2, Soil Mechanics Department, Norwegian Institute of Technology.

Janbu, N. (1985), Soil models in offshore engineering, *Geotechnique,* **XV**, No. 3, pp. 241–281.

Karlsson, R. and Viberg, L. (1967), Ratio c/p' in relation to liquid limit and plasticity index, with special reference to Swedish clays, *Proceedings of the Geotechnical Conference,* Oslo, **1**, pp. 43–47.

Kasim, A. G., Chu, M. Y., and Jensen, C. N. (1986), Field correlation of cone and standard penetration tests, *Journal of Geotechnical Engineering, ASCE,* **112**, No. 3, pp. 368–372.

Konstantinidis, B., Van Riessen, G., and Schneider, J. P. (1986), Structural settlements at a major power plant, *Settlement of Shallow Foundations on Cohesionless Soils: Design and Performance,* Geotechnical Special Publication No. 5, ASCE, pp. 54–73.

Lacasse, S. and Lunne, T. (1983), DMT in two soft marine clays, Norwegian Geotechnical Institute Publication 146, Oslo, pp. 1–8.

Ladd, C. C. (1971), Settlement analyses for cohesive soils, Research Report R71-2, Soils Publication 272, Department of Civil Engineering, Massachusetts Institute of Technology.

Ladd, C. C., Foott, R., Ishihara, K., Schlosser, F., and Poulos, H. G. (1977), Stress–deformation and strength characteristics: State of the art report, *Proceedings of the 9th International Conference on Soil Mechanics and Foundation Engineering,* Tokyo, **2**, pp. 421–494.

Lambe, T. W. (1953), The structure of inorganic soil, *Proceedings, ASCE,* **79**, Separate No. 315.

Lambe, T. W. (1964), Methods of estimating settlement, *Journal of the Soil Mechanics and Foundations Division, ASCE,* **90**, No. SM5, pp. 43–67; also in *Design of Foundations for Control of Settlement,* ASCE, pp. 47–72.

Lambe, T. W. (1967), Stress path method, *Journal of the Soil Mechanics and Foundations Division, ASCE,* **93**, No. SM6, pp. 309–331.

Lambe, T. W. and Whitman, R. V. (1969), *Soil Mechanics,* John Wiley and Sons, Inc., New York, N.Y.

Lambrechts, J. R. and Leonards, G. A. (1978), Effects of stress history on deformation of sands, *Journal of the Geotechnical Engineering Division, ASCE,* **104**, No. GT11, pp. 1371–1387.

Leonards, G. A. (1962), Engineering properties of soils, *Foundation Engineering,* G. A. Leonards, ed., Chapter 2, McGraw-Hill, Inc., New York, N.Y., pp. 66–240.

Leonards, G. A. (1968), Predicting settlement of buildings on clay soils, *Foundation Engineering,* Chicago Soil Mechanics Lecture Series, Northwestern University, Evanston, Ill., pp. 41–50.

Leonards, G. A. (1973), Discussion of "The Empress Hotel, Victoria, British Columbia: Sixty-five Years of Foundation Settlements", *Canadian Geotechnical Journal,* **10**, No. 1, pp. 120–122.

Leonards, G. A. (1976), Estimating consolidation settlements of shallow foundations on overconsolidated clays, *Special Report 163,* Transportation Research Board, pp. 13–16.

Leonards, G. A. (1985), Discussion of Theme Lecture No. 2 by Jamiolkowski et al., *Proceedings of the 11th International Conference on Soil Mechanics and Foundation Engineering,* San Francisco, **5**, pp. 2674–2675.

Leonards, G. A. and Girault, P. (1961), A study of the one-dimensional consolidation test, *Proceedings of the 5th International Conference on Soil Mechanics and Foundation Engineering,* Paris, **1**, pp. 116–130.

Leonards, G. A. and Altschaeffl, A. G. (1964), Compressibility of clay, *Journal of the Soil Mechanics and Foundations Division, ASCE,* **90**, No. SM5, pp. 133–156.

Leonards, G. A. and Frost, J. D. (1988), Settlement of shallow foundations on granular soils, *Journal of Geotechnical Engineering, ASCE,* **114**, No. 7, pp. 791–809.

Leroueil, F. (1988), Tenth Canadian Geotechnical Colloquium: Recent developments in consolidation of natural clays, *Canadian Geotechnical Journal,* **25**, No. 1, pp. 85–107.

Leroueil, S., LeBihan, J. P., and Tavenas, F. (1980), An approach for the determination of the preconsolidation pressure in sensitive clays, *Canadian Geotechnical Journal,* **17**, No. 3, pp. 446–453.

Leroueil, S., Tavenas, F., Samson, L., and Morin, P. (1983), Preconsolidation pressure of Champlain clays, Part II, Laboratory determination, *Canadian Geotechnical Journal,* **20**, No. 4, pp. 803–816.

Mair, R. J. and Wood, D. M. (1987), *Pressuremeter Testing, Methods and Interpretation,* CIRIA Ground Engineering Report on In Situ Testing, Butterworths, London.

Marchetti, S. (1980), In situ tests by flat dilatometer, *Journal of the Geotechnical Engineering Division, ASCE,* **106**, No. GT3, pp. 299–321.

Martin, W. O., McCoy, J. W., and Hunt, D. D. (1986), Settlement of a reactor containment on sand, *Settlement of Shallow Foundations on Cohesionless Soils: Design and Performance,* Geotechnical Special Publication No. 5, ASCE, pp. 74–90.

Mayne, P. W. (1987), Determining preconsolidation stress and penetration pore pressures from DMT contact pressures, *Geotechnical Testing Journal, ASTM,* **10**, No. 3, pp. 146–150.

Mayne, P. W. and Bachus, R. C. (1988), Profiling OCR in clays by piezocone soundings, *Proceedings of the First International Symposium on Penetration Testing (ISOPT-I),* Orlando, Florida, pp. 857–864.

Mayne, P. W. and Holtz, R. D. (1988), Profiling stress history from piezocone soundings, *Soils and Foundations,* **28**, No. 1, pp. 16–28.

Mayne, P. W. and Mitchell, J. K. (1988), Profiling of overconsolidation ratio in clays by field vane, *Canadian Geotechnical Journal,* **25**, No. 1, pp. 150–157.

Meigh, A. C. (1985), In situ testing using the static cone penetrometer, Draft Report to CIRIA, London, Appendix G.

Meigh, A. C. (1987), *Cone Penetration Testing, Methods and Interpretation,* CIRIA Ground Engineering Report: In Situ Testing, Butterworths, London.

Mesri, G. (1973), Coefficient of secondary compression, *Journal of the Soil Mechanics and Foundations Division, ASCE,* **99**, No. SM1, pp. 123–137.

Mesri, G. and Godlewski, P. M. (1977), Time- and stress-compressibility interrelationship, *Journal of the Geotechnical Engineering Division, ASCE,* **103**, No. GT5, pp. 417–430.

Mesri, G. and Choi, Y. K. (1984), Discussion of "Time effects on the stress–strain behaviour of natural soft clays", *Geotechnique*, **XXXIV**, No. 3, pp. 439–442.

Mesri, G. and Feng, T. W. (1986), Discussion of "Stress–strain–strain rate relation for the compressibility of sensitive natural clays", by Leroueil et al., *Geotechnique*, **36**, No. 2, pp. 283–287.

Mesri, G. and Castro, A. (1987), C_a/C_c concept and K_0 during secondary compression, *Journal of Geotechnical Engineering*, ASCE, **113**, No. 3, pp. 230–247.

Mesri, G. and Choi, Y. K. (1987), Closure to discussion on "Settlement analysis of embankments on soft clays", by Mesri and Choi, *Journal of Geotechnical Engineering*, ASCE, **113**, No. 9, p. 1078.

Meyerhof, G. G. and Fellenius, B. H. (eds.) (1985), *Canadian Foundation Engineering Manual*, 2d ed., Canadian Geotechnical Society.

Mitchell, J. K. (1970), In-place treatment of foundation soils, *Journal of the Soil Mechanics and Foundations Division*, ASCE, **96**, No. SM1, pp. 73–110.

Mitchell, J. K. (1981), Soil improvement—State-of-the-art report, Session 12, *Proceedings of the 10th International Conference on Soil Mechanics and Foundation Engineering*, Stockholm, **4**, pp. 506–565.

Morin, P., Leroueil, S., and Samson, L. (1983), Preconsolidation pressure of Champlain clays, Part I: In situ determination, *Canadian Geotechnical Journal*, **20**, No. 4, pp. 782–802.

Newmark, N. M. (1942), Influence charts for computation of stresses in elastic foundations, *University of Illinois Engineering Experiment Station Bulletin*, Series No. 338, **61**, No. 92, Urbana, Illinois, reprinted 1964, 28 pp.

Olson, R. E. (1986), State of the art: Consolidation testing, *Consolidation of Soils: Testing and Evaluation*, ASTM STP 892, pp. 7–70.

Oweis, I. S. (1979), Equivalent linear model for predicting settlement of sand bases, *Journal of the Geotechnical Engineering Division*, ASCE, **105**, No. GT12, pp. 1525–1544.

Peck, R. B. (1974), The selection of soil parameters for the design of foundations, 2d Nabor Carrillo Lecture, *7th National Meeting of the Mexican Society for Soil Mechanics*, Guadalajara, Mexico, pp. 9–49.

Peck, R. B., Hanson, W. E., and Thornburn, T. H. (1974), *Foundation Engineering*, 2d ed., John Wiley and Sons, Inc., New York, N.Y.

Perloff, W. H. (1975), Pressure distribution and settlement, *Foundation Engineering Handbook*, H. F. Winterkorn and H. Y. Fang, eds., Chapter 4, Van Nostrand Reinhold, New York, N.Y., pp. 148–196.

Perloff, W. H. and Baron, W. (1976), *Soil Mechanics, Principles and Applications*, Ronald Press, New York, N.Y.

Polshin, D. E. and Tokar, R. A. (1957), Maximum allowable nonuniform settlement of structures, *Proceedings of the 4th International Conference on Soil Mechanics and Foundation Engineering*, London, **I**, pp. 402–406.

Poulos, H. G. and Davis, E. H. (1974), *Elastic Solutions for Soil and Rock Mechanics*, John Wiley and Sons, Inc., New York, N.Y.

Quigley, R. M. (1980), Geology, mineralogy and geochemistry of Canadian soft soils: a geotechnical perspective, *Canadian Geotechnical Journal*, **17**, No. 2, pp. 261–285.

Raymond, G. P. and Wahls, H. E. (1976), Estimating 1-dimensional consolidation, including secondary compression of clay loaded from overconsolidated to normally consolidated state, *Special Report 163*, Transportation Research Board, pp. 17–23.

Robertson, P. K. (1986), In situ testing and its application to foundation engineering, *Canadian Geotechnical Journal*, **23**, No. 4, pp. 573–594.

Robertson, P. K., Campanella, R. G., and Wightman, A. (1983), SPT–CPT correlations, *Journal of Geotechnical Engineering*, ASCE, **109**, No. 11, pp. 1449–1459.

Sällfors, G. (1975), Preconsolidation pressure of soft, high-plastic clays, Technical Doctor's Thesis, Chalmers University of Technology, Gothenburg, Sweden.

Sangrey, D. A. (1972), Naturally cemented sensitive soils, *Geotechnique*, **22**, No. 1, pp. 139–152.

Schiffman, R. L., Chen, A. T. F., and Jordan, J. C. (1969), An analysis of consolidation theory, *Journal of the Soil Mechanics and Foundations Division*, ASCE, **95**, No. SM1, pp. 285–312.

Schiffman, R. L., Pane, V., and Gibson, R. E. (1984), The theory of one-dimensional consolidation of saturated clays, IV. An overview of nonlinear finite strain sedimentation and consolidation, *Sedimentation/Consolidation Models*, R. N. Yong and F. C. Townsend, eds., ASCE, pp. 1–29.

Schiffman, R. L., Vick, S. G., and Gibson, R. E. (1988), Behavior and properties of hydraulic fills, *Proceedings of the ASCE Specialty Conference on Hydraulic Fill Structures*, Fort Collins, Colorado.

Schmertmann, J. H. (1955), The undisturbed consolidation behavior of clay, *Transactions*, ASCE, **120**, pp. 1201–1233.

Schmertmann, J. H. (1970), Static cone to compute static settlement over sand, *Journal of the Soil Mechanics and Foundation Division*, ASCE, **96**, No. SM3, pp. 1011–1043.

Schmertmann, J. H. (1970a), Suggested method for screw-plate load test, *Special Procedures for Testing Soil and Rock for Engineering Purposes*, ASTM STP 479, pp.81–85.

Schmertmann, J. H. (1978), Guidelines for cone penetration test performance, and design, Federal Highway Administration, Report FHWA-TS-78-209.

Schmertmann, J. H. (1986), Dilatometer to compute foundation settlement, *Proceedings of In Situ '86, A Specialty Conference on Use of In Situ Tests in Geotechnical Engineering*, Geotechnical Special Publication No. 6, ASCE, pp. 303–321.

Schmertmann, J. H., Hartman, J. P., and Brown, P. R. (1978), Improved strain influence factor diagrams, *Journal of the Geotechnical Engineering Division*, ASCE, **104**, No. GT8, pp. 1131–1135.

Scott, R. F. (1963), *Principles of Soil Mechanics*, Addison Wesley, Reading, Mass.

Simons, N. E. (1974), Normally consolidated and lightly over-consolidated cohesive materials, Review Papers, Session II, *Proceedings of the BGS Conference on Settlement of Structures*, Cambridge, pp. 500–530.

Skempton, A. W. (1954a), The pore-pressure coefficients A and B, *Geotechnique*, **IV**, pp. 143–147.

Skempton, A. W. (1954b), Discussion on "The structure of inorganic soil" by Lambe, ASCE, Separate No. 478, pp. 19–22.

Skempton, A. W. (1986), Standard penetration test procedures and the effects in sands of overburden pressure, relative density, particle size, aging, and overconsolidation, *Geotechnique*, **XXXVI**, No. 3, pp. 425–447.

Skempton, A. W. and MacDonald, D. H. (1956), Allowable settlement of buildings, *Proceedings of the Institution of Civil Engineers*, **5**, Part III, pp. 727–768.

Skempton, A. W. and Bjerrum, L. (1957), A contribution to the settlement analysis of foundations on clay, *Geotechnique*, **7**, No. 4, pp. 168–178.

Snethen, D. R. (1979), Technical guidelines for expansive soils in highway subgrades, Federal Highway Administration, Report No. FHWA-RD-79-51.

Snethen, D. R. (1980), Expansive soils in highway subgrades: Summary, Federal Highway Administration, Report No. FHWA-TS-80-236.

Sonnenfeld, S., Schmertmann, J. H., and Williams, R. (1985), A bridge site investigation, *Strength Testing of Marine Sediments*, ASTM STP 883, pp. 515–535.

Sowers, G. F. (1962), Shallow foundations, *Foundation Engineering*, G. A. Leonards, ed., Chapter 6, McGraw-Hill, Inc., New York, N.Y.

Sowers, G. F. (1974), Analysis and design of lightly-loaded foundations, State of the Art Presentation in the *Proceedings of the ASCE Specialty Conference on Analysis and Design in Geotechnical Engineering*, Austin, Texas, **II**, pp. 49–78.

Sowers, G. B. and Sowers, G. F. (1970), *Introductory Soil Mechanics and Foundations*, 3d ed., Macmillan Company, New York, N.Y.

Steinbrenner, W. (1936), A rational method for determination of vertical normal stress under foundations, *Proceedings of the 1st International Conference on Soil Mechanics and Foundation Engineering*, Cambridge, **2**, pp. 142–143.

Swiger, W. F. (1974), Evaluation of soil moduli, State of the Art Presentation in the *Proceedings of the ASCE Specialty Conference on Analysis and Design in Geotechnical Engineering*, Austin, Texas, **II**, pp. 79–92.

Taylor, D. W. (1948), *Fundamentals of Soil Mechanics*, John Wiley and Sons, Inc., New York, N.Y.

Teng, W. C. (1975), Mat foundations, *Foundation Engineering Handbook*, H. F. Winterkorn and H. Y. Fang, eds., Chapter 17, Von Nostrand Reinhold, New York, N.Y., pp. 528–536.

Terzaghi, K. and Peck, R. B. (1967), *Soil Mechanics in Engineering Practice*, 2d ed., John Wiley and Sons, Inc., New York, N.Y.

U.S. Navy (1982), *Soil Mechanics*, Design Manual 7.1, Department of the Navy, Navy Facilities Engineering Command (NAVFAC).

Vesić, A. S. (1975), Bearing capacity of shallow foundations, *Foundation Engineering Handbook*, H. F. Winterkorn and H. Y. Fang, eds., Chapter 3, Von Nostrand Reinhold, New York, N.Y., pp. 121–147.

Wahls, H. E. (1981), Tolerable settlement of buildings, *Journal of the Geotechnical Engineering Division*, ASCE, **107**, No. GT11, pp. 1489–1504.

Webb, D. L. (1969), Settlement of structures on deep alluvial sandy sediments in Durban, South Africa, *Proceedings of the Conference on In Situ Investigations in Soils and Rock*, London, B.G.S., pp. 181–188.

Welsh, J. P. (ed.) (1987), Soil improvement—A ten year update, *Proceedings of a Symposium by the Committee on Placement and Improvement of Soils*, Geotechnical Engineering Division, ASCE, Atlantic City, Geotechnical Special Publication No. 12.

Yoder, E. J. and Witczak, M. W. (1975), *Principles of Pavement Design*, 2d ed., John Wiley and Sons, Inc., New York, N.Y.

6 EARTH PRESSURES

G. W. CLOUGH, Ph.D., P.E.
Professor and Head of Department
of Civil Engineering
Virginia Polytechnic Institute
and State University

J. M. DUNCAN, Ph.D., P.E.
University Distinguished Professor
Department of Civil Engineering
Virginia Polytechnic Institute
and State University

Design of earth-retaining structures requires knowledge of the earth and water loads that will be exerted on them. The first methods for determination of earth loads acting on retaining structures were developed in the eighteenth and nineteenth centuries by Coulomb and Rankine. These were based on idealized concepts where the retaining structure is rigid and moves as a unit. Also, the soil that loads the wall is assumed to be "wished in place," and to undergo systematic, prescribed failure patterns as the wall displaces. These assumptions ignore the true effects of soil–structure interaction, and the processes of construction of the system. Nonetheless, the Coulomb and Rankine methods provide simple and reasonably accurate means for estimating earth loads, and remain useful tools today.

During this century, new, and often complex retaining structures have been developed. These structures lead to questions about the effects of variable system flexibility, and alternative forms of rigid body rotation relative to simpler structures. The result is a redistribution of earth loads from the more flexible to the stiffer portions of the system, and other forms of stress transfer. In spite of the complexity of the actual problems, design solutions to predict earth pressures for the new structures have tended to be based on empirical modifications of the Rankine and Coulomb theories.

Examples of common earth-retaining structures are shown in Figure 6.1. The simplest form is the gravity wall, which has enough rigidity to avoid bending deformations but which can, and typically does, move as a unit. Flexible walls such as the bulkhead or the excavation support wall undergo bending deformations such that the earth pressures are described according to the relative flexibilities within the system. Buried structures such as the U-frame lock or the culvert bring into play vertical earth loads as well as lateral earth pressures. Internal earth pressures are important to cofferdams and silos, systems that confine earth within the structure. Finally, there are the "self-contained" earth-retention structures of the reinforced-earth type, which are acted upon by internal and external earth loads.

Most conventional methods for predicting earth pressures produce a diagram with a linear or some slightly more complicated regular geometric shape. In actuality, we know from physical measurements and recent numerical computations that even in the case of the simplest structure, earth pressure distributions are not linear (Handy, 1985; Clough and Duncan, 1971). However, for design purposes, geometrically uncomplicated shapes are preferred in the interest of ease of calculations. The development of such diagrams usually incorporates an attempt to build in a degree of conservatism, and leads to the neglect of certain parts of the problem in the

interest of simplicity. Over time, the use of such diagrams often leads to a lack of understanding of the principles used to develop them. Further, the overlapping of factors of safety used on the geotechnical side, combined with those used in the structural side, blurs the specific role of important parameters. In many cases our present practice leads to designs that are conservative, but there are instances where they can be unconservative. Thus, we arrive at the incongruous situation where we have earth-retaining structures standing that in theory should fail, or we have structures failing, usually owing to excessive deformation, that in theory should not fail. In this chapter we will take the approach that we are concerned primarily with practical design, but that we also need to understand the fundamentals of the earth pressure problem for improved design. In the period since the first edition of this handbook was published, new developments have been made, and we will attempt to integrate these into the basic approach where appropriate.

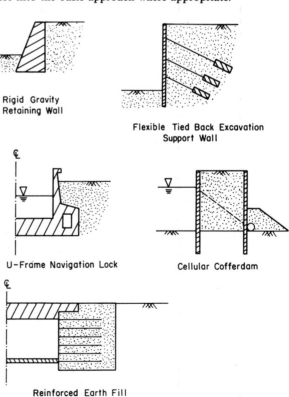

Fig. 6.1 Typical structures acted upon by lateral earth pressures.

6.1 AT-REST LATERAL PRESSURES

At-rest pressures exist in level ground, and develop under long-term conditions as the soil is deposited and acted upon by changes in the loading environment as caused by erosion, glaciers, and physicochemical processes. At-rest pressures rigorously only apply for walls that are placed into the ground with a minimum of disturbance and that remain unmoved during loading, or for unmoving, frictionless walls with a backfill placed with a minimum of compactive effort. In practice such conditions are rarely achieved. However, at-rest pressures are still useful in design as either a baseline against which other pressure states can be judged or as an assumed conservative choice for the design loading.

At-rest effective lateral pressures are often assumed to follow a linear distribution (Fig. 6.2), with the effective lateral pressure σ'_x taken as a simple multiple of the vertical effective pressure σ'_z:

$$\sigma'_x = K_0(\sigma'_z) \tag{6.1}$$

In homogeneous, dry soil with a constant K_0 and unit weight, both the vertical and lateral pressures are linearly distributed. With the presence of a water table, the at-rest pressure distribution exhibits a break in slope at the water table, reflecting the use of submerged unit weights to determine vertical effective stresses (Fig. 6.2).

Our early concepts of the parameter K_0 were formed on the basis of normally consolidated soils. Jaky (1944) proposed a relationship between K_0 and the drained friction angle ϕ' for normally consolidated soils:

$$K_0 = 1 - \sin \phi' \tag{6.2}$$

Numerous studies have confirmed the general validity of this empirical equation (Brooker and Ireland, 1965; Mayne and Kulhawy, 1982). However, results from laboratory experiments and in-situ tests have shown that the K_0 value also varies as a function of overconsolidation ratio (OCR) and stress history. For the case of a soil that has been subjected to one or more cycles of unloading, Schmidt (1966) proposed that K_0 can be determined as a function of its value in the normally consolidated state using the relationship

$$K_{0u} = K_{0nc}(\text{OCR})^\alpha \tag{6.3}$$

in which K_{0u} is the coefficient for unloading, K_{0nc} is the coefficient for the normally consolidated soil, and α is a dimensionless coefficient. Experimental data have confirmed this relationship, and Mayne and Kulhawy (1982) showed that, for most soils, α can be taken as $\sin \phi'$.

Soils that are overconsolidated and are in the process of being reloaded pose a difficulty in that Equation 6.3 does not apply. For this condition, a more complex equation is needed as well as a full knowledge of the stress history of the soil (Mayne and Kulhawy, 1982). For practical purposes, it may

TABLE 6.1 TYPICAL COEFFICIENTS OF LATERAL EARTH PRESSURE AT REST.

Soil type	Coefficient of Lateral Earth Pressure			
	OCR = 1	OCR = 2[a]	OCR = 5[a]	OCR = 10[a]
Loose sand	0.45	0.65	1.10	1.50
Medium sand	0.40	0.60	1.05	1.55
Dense sand	0.35	0.55	1.00	1.50
Silt	0.50	0.70	1.10	1.60
Lean clay, CL	0.60	0.80	1.20	1.65
Highly plastic clay, CH	0.65	0.80	1.10	1.40

[a] Unloading cycle.

be enough to know that the K_0 during reloading falls about halfway between that for unloading and normally consolidated conditions. Also, K_0 might be directly determined through in-situ testing methods.

Table 6.1 presents typical values for K_0 for a subset of soils. For other conditions, K_0 values can be determined directly from Equations 6.2 and 6.3, and/or using in-situ testing techniques.

Because the K_0 value in a given soil often varies with depth, and the soil types themselves may change with depth, the at-rest lateral pressure distribution is typically not linear as shown in Figure 6.2. Self-boring pressuremeter tests in clays with overconsolidated profiles induced by desiccation have demonstrated that the K_0 under such conditions decreases with depth in the soil deposit and reaches a steady state where the desiccation effects are no longer present (Clough and Denby, 1980).

6.2 ACTIVE AND PASSIVE LATERAL EARTH PRESSURES

Most walls move, either by global shifting or by local deformations. These movements cause adjustments to occur in the earth loads and the pressure distributions. Conventional means for assessing the effects of system movements are to set them into the context of extreme conditions. These are referred to as the active and passive earth pressure loadings.

6.2.1 Active Pressure

Assuming that a gravity wall with no friction on its face is translated away from a soil mass that is initially at the at-rest condition, then the soil mass adjacent to the wall will pass into a failure state as shown in Figure 6.3. At this stage, the

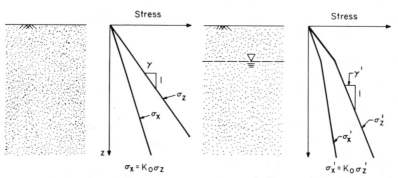

Fig. 6.2 At-rest earth pressure distribution—homogeneous soil.

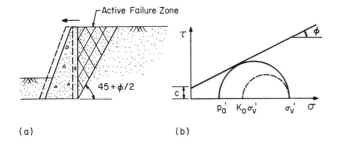

(a)   (b)

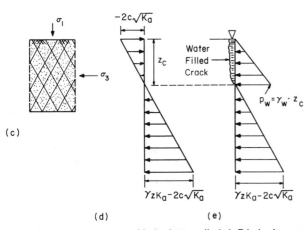

(c)

(d)   (e)

Fig. 6.3 Active pressure—frictionless wall. (a) Frictionless wall moves away from backfill. (b) Stress state in active failure. (c) Active failure zone. (d) Theoretical active pressure distribution. (e) Water-filled crack in tension zone.

soil fails with the vertical stress unchanged from its original value, but with the lateral pressure decreased to a minimum value that can be defined using the Mohr–Coulomb failure criterion. The minimum lateral pressure is known as the active pressure, and denoted by the symbol p_a. It is desirable to reach this condition if possible, since it reduces the amount of load that the wall will have to carry while allowing the soil to share in the load-bearing process.

For the frictionless wall with a level backfill, the active pressure can be calculated from the geometry of the Mohr diagram in Figure 6.3 by the equation

$$p_a = k_a \gamma z - 2c\sqrt{k_a} \qquad (6.4)$$

where $k_a = \tan^2(45° - \phi/2)$, and is referred to as the active pressure coefficient. Other terms are γ, the unit weight; ϕ, the friction angle; c, the cohesion; and z, the depth below the ground surface. The distribution of active pressure as shown in Figure 6.3 is linear. If the soil has a cohesion component, the soil is in a state of tension of a depth of $2c/\gamma\sqrt{k_a}$. Ordinarily, it should not be assumed that this portion of the diagram will act on a wall, but rather that a tension crack will form to this depth, and fill with water, which then exerts a positive pressure on the wall.

Equivalent Fluid Unit Weight If the backfill is composed of cohesionless soil, as is often the case, then the active earth pressure equation reduces to

$$p_a = k_a \gamma z \qquad (6.5)$$

This can also be written as:

$$p_a = \gamma_{eq} z \qquad (6.6)$$

where the term γ_{eq} is known as the equivalent fluid unit weight for active pressure loading, and equals $k_a\gamma$. This term is often used in design, and it should be realized in using it that the simplifying assumptions used in the derivations of this point are also incorporated in the equivalent fluid unit weight concept.

Surcharge and Nonhomogeneous Conditions Design conditions often call for incorporation of a surcharge on the ground surface adjacent to the wall. In the case of a frictionless wall, the active pressure due to soil weight and surcharge, as shown in Figure 6.4, can be calculated using the equation

$$p_a = k_a(\gamma z + q_s) \qquad (6.7)$$

where q_s is the surcharge pressure.

Where a water table is situated above the bottom of the wall, or the soil involved is nonhomogeneous, Equations 6.4 and 6.7 can be used if the proper allowance is made for the submergence effect and the changing properties for the soil layers. Figure 6.5 illustrates these considerations for cohesionless soil.

Force Polygon Solution for Active Loadings The equations presented to this point are limited to consideration of relatively simple conditions. More complex conditions can be included using a force polygon analysis based on assumed kinematic failure mechanisms developing in the soil. One of the more important conditions that can be considered in this way is the case of friction developing between the wall and the soil as a result of relative movements between them. Figure 6.6 illustrates this situation for the case of a wall translating away from a homogeneous soil.

Assuming a straight-line failure surface in the backfill as the wall moves away from the soil, the equilibrium of the soil wedge bounded by the wall and the backfill failure surface can be examined in the force polygon in Figure 6.6. The force E required to maintain equilibrium is exerted by the wall. In the most general situation, the critical value of the force between the wall and the soil is found by working with trial slopes of failure wedge until the maximum value of the stabilizing force E is obtained.

For relatively simple conditions where the soil backfill is level, and the wall face is vertical, the inclination of the failure surface in the soil that yields the minimum earth loading is $45° + \phi/2$ to the horizontal. Under these conditions, if wall friction is zero, then the kinematic force polygon procedure yields the same answer for the active load as Equation 6.4. If the wall friction is positive in the sense shown in Figure 6.6, then the active loading for most cases is slightly reduced from the case of no friction. More importantly, the vertical shear

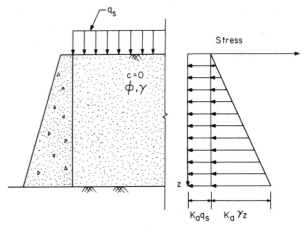

Fig. 6.4 Frictionless wall with surcharge.

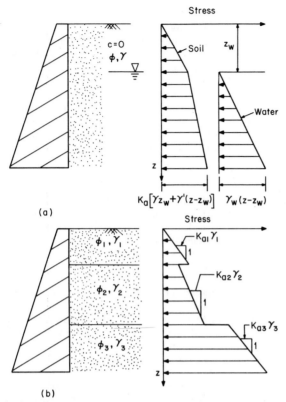

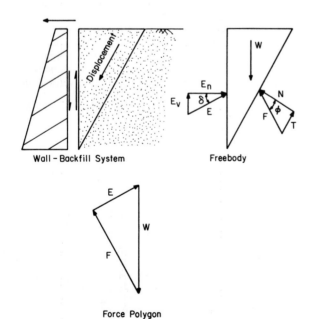

Fig. 6.5 Active pressures for frictionless wall in presence of groundwater table and nonhomogeneous soil conditions. (a) Groundwater table. (b) Nonhomogeneous cohesionless soil.

force that is generated helps to combat overturning and increases the resistance against sliding of the wall.

A general formula can be developed for active earth load acting on a wall for the case of a homogeneous soil backfill with arbitrary degrees of wall friction, wall slope, and backfill surface slope. Assuming that the failure surface in the backfill is a straight line, the formula is as shown in Figure 6.7. In the

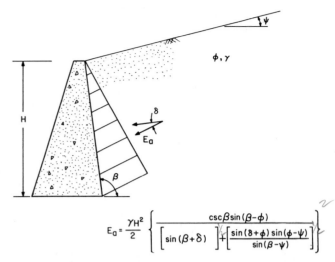

$$E_a = \frac{\gamma H^2}{2} \left\{ \frac{\csc\beta\sin(\beta-\phi)}{\left[\sin(\beta+\delta)\right] + \left[\frac{\sin(\delta+\phi)\sin(\phi-\psi)}{\sin(\beta-\psi)}\right]} \right\}$$

Fig. 6.7 Closed-form solution for active earth loading, rough wall, sloping wall face, and backfill.

event of relatively complex backfill or wall geometries or surcharge conditions, then the exact failure surface that yields the minimum earth load can be found by a trial procedure. A number of references describe this process, and examples can be found in the original edition to this handbook.

Further Comments on Active Load Determinations The kinematic analysis in Figure 6.6 assumes that the failure surface is a straight line. In fact, in the most general case of a soil whose failure is governed by a Mohr–Coulomb criterion, and which has a friction component, the correct failure surface under active conditions consists of a log spiral, as shown in Figure 6.8. However, in the active state, the log-spiral shape is reasonably approximated by a straight line, and the resultant load predicted using the simple straight-line failure mechanism is within 10 percent of that obtained with the more exact log-spiral mechanism.

Table 6.2 presents values for the active pressure coefficient that allow calculation of the active loading resultant as shown for conditions where wall friction, sloping backfill, and a sloping wall face exist. These coefficients are based on the log-spiral failure surface assumption. A graphical format for the active pressure coefficient from the log-spiral analysis that is useful for many practical problems is given in Figure 6.9. It assumes a vertical wall face and horizontal backfill. For conditions encountered that deviate from those described in Table 6.2 or in Figure 6.9, the trial procedure can be used assuming straight-line failure surfaces in the soil.

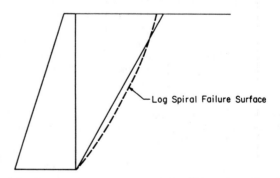

Fig. 6.6 Force polygon solution for active loading.

Fig. 6.8 Comparison of log-spiral and straight-line failure surfaces for active conditions.

TABLE 6.2 VALUES OF k_a FOR LOG SPIRAL FAILURE SURFACE.

δ, deg	ψ, deg	β, deg	ϕ, deg 20	25	30	35	40	45
		−10	0.37	0.30	0.24	0.19	0.14	0.11
	−15	0	0.42	0.35	0.29	0.24	0.19	0.16
		10	0.45	0.39	0.34	0.29	0.24	0.21
		−10	0.42	0.34	0.27	0.21	0.16	0.12
0	0	0	0.49	0.41	0.33	0.27	0.22	0.17
		10	0.55	0.47	0.40	0.34	0.28	0.24
		−10	0.55	0.41	0.32	0.23	0.17	0.13
	15	0	0.65	0.51	0.41	0.32	0.25	0.20
		10	0.75	0.60	0.49	0.41	0.34	0.28
		−10	0.31	0.26	0.21	0.17	0.14	0.11
	−15	0	0.37	0.31	0.26	0.23	0.19	0.17
		10	0.41	0.36	0.31	0.27	0.25	0.23
		−10	0.37	0.30	0.24	0.19	0.15	0.12
$\phi°$	0	0	0.44	0.37	0.30	0.26	0.22	0.19
		10	0.50	0.43	0.38	0.33	0.30	0.26
		−10	0.50	0.37	0.29	0.22	0.17	0.14
	15	0	0.61	0.48	0.37	0.32	0.25	0.21
		10	0.72	0.58	0.46	0.42	0.35	0.31

[a] After Caquot and Kerisel (1948).

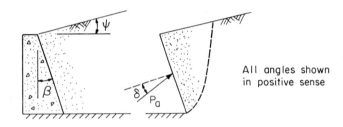

All angles shown in positive sense

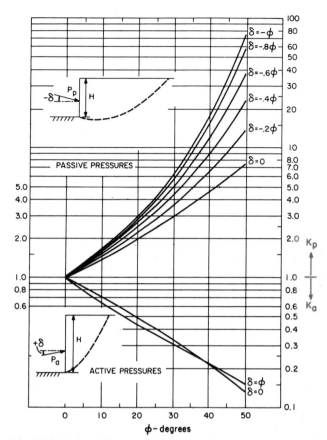

Fig. 6.9 Active and passive pressure coefficients for vertical wall and horizontal backfill, based on log-spiral failure surfaces. (*After Caquot and Kerisel, 1948.*)

6.2.2 Passive Pressures

Passive pressure conditions develop where a structure is forced into a soil mass. This situation is most commonly associated with the soil located on the opposite side of the wall from the backfill (Fig. 6.10). Assuming that a frictionless wall is forced into a soil mass that is originally at-rest, the end result will be that a portion of the soil mass will pass into a passive failure condition as shown in Figure 6.11. The soil fails with the vertical stress unchanged from its original value, but with the horizontal stress increased to a maximum value as defined by the Mohr–Coulomb failure criterion. The maximum pressure is denoted by the symbol p_p, and it is defined from the geometry of the Mohr diagram in Figure 6.10 by the equation

$$p_p = \gamma z k_p + 2c\sqrt{k_p} \qquad (6.8)$$

where k_p is the passive pressure coefficient, and can be expressed as follows:

$$k_p = \tan^2\left(45° + \frac{\phi}{2}\right) \qquad (6.9)$$

In Figure 6.10, the passive pressure distribution defined by Equation 6.8 is shown to be linear, and in compression throughout.

A uniform surcharge for cohesionless soils can be incorporated into Equation 6.8 in the form

$$p_p = k_p(\gamma z + q_s) \qquad (6.10)$$

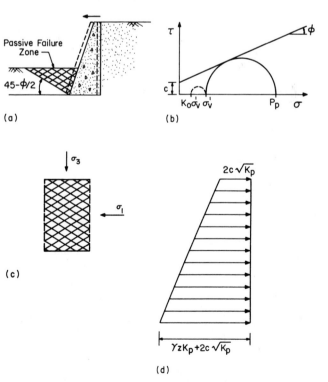

Fig. 6.10 Passive pressure—active wall. (a) Frictionless wall moves into soil. (b) Stress state in passive failure. (c) Passive failure zone. (d) Theoretical pressure distribution.

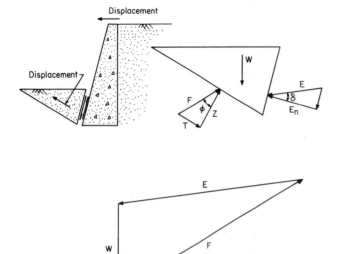

Fig. 6.11 Force polygon solution for passive conditions.

Force Polygon Solution For Passive Loads The flexibility of the force polygon solution for the active loading cases was seen earlier. This solution allows accommodation of the effects of wall friction, sloping wall faces, sloping soil surfaces, and other factors. A similar approach can be used for the passive case, as illustrated in Figure 6.11, using the assumption that the failure surfaces in the soil for the passive state are straight lines. However, except for the case of a frictionless wall, the actual failure surface for passive failure is markedly nonlinear, and is displaced well below the most critical plane failure surface (Fig. 6.12). As a result, the passive resistances calculated using the straight line can be much higher than those calculated using log-spiral surfaces, and should not be used for values of wall friction angle (δ) greater than half of ϕ.

General Comments on Passive Load Determinations Table 6.3 presents passive pressure coefficients that are derived from analyses with log-spiral surfaces. The table allows passive loads to be determined for a range of wall friction angles, wall slopes, and backfill slopes. Figure 6.9 also includes passive pressure coefficients based on the log-spiral theory in the simplified condition of vertical wall face and horizontal backfill.

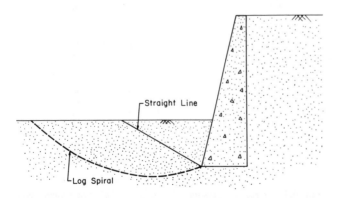

Fig. 6.12 Comparison of straight-line and log-spiral failure surfaces for passive conditions.

TABLE 6.3   VALUES OF k_p FOR LOG-SPIRAL FAILURE SURFACE.

δ, deg	ψ, deg	β, deg	ϕ, deg 20	25	30	35	40	45
		−10	1.32	1.66	2.05	2.52	3.09	3.95
	−15	0	1.09	1.33	1.56	1.82	2.09	2.48
		10	0.87	1.03	1.17	1.30	1.33	1.54
		−10	2.33	2.96	3.82	5.00	6.68	9.20
0	0	0	2.04	2.46	3.00	3.69	4.59	5.83
		10	1.74	1.89	2.33	2.70	3.14	3.69
		−10	3.36	4.56	6.30	8.98	12.2	20.0
	15	0	2.99	3.86	5.04	6.72	10.4	12.8
		10	2.63	3.23	3.97	4.98	6.37	8.2
		−10	1.95	2.90	4.39	6.97	11.8	22.7
	−15	0	1.62	2.31	3.35	5.04	7.99	14.3
		10	1.29	1.79	2.50	3.58	5.09	8.86
		−10	3.45	5.17	8.17	13.8	25.5	52.9
$-\phi°$	0	0	3.01	4.29	6.42	10.2	17.5	33.5
		−10	2.57	3.50	4.98	7.47	12.0	21.2
		−10	4.95	7.95	13.5	24.8	50.4	11.5
	15	0	4.42	6.72	10.8	18.6	39.6	73.6
		10	3.88	5.62	8.51	13.8	24.3	46.9

[a] After Caquot and Kerisel (1948).

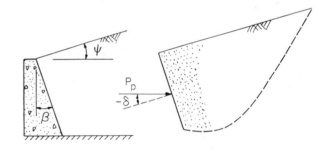

6.3   SOIL–STRUCTURE INTERACTION FOR UNMOVING WALLS

Conventional thought has it that when at-rest earth pressures are assumed to act on a wall, then there is no need to consider the possibility of wall-to-soil shear, or downdrag. In fact, downdrag will naturally develop in certain situations, one of the more prominent being when backfill is placed in layers behind a wall. During placement, the soil will settle relative to the wall under its own weight, and mobilize a downdrag force. This force will act to stabilize the wall in that it resists overturning, and it adds to the normal force acting on the base of the wall, helping to prevent sliding of the wall. The role of the downdrag force is important because, for many problems, the designer conservatively assumes at-rest pressures as the lateral loading, and then neglects the downdrag force. Thus, a double factor of safety is added into the design, leading to excessive conservatism. This helps explain why many existing walls stand when theoretical analyses suggest they should fail.

The amount of the downdrag force that actually develops is a function of the friction between the wall and the soil. Typically, the friction force is mobilized with very small movements. Consideration should be given to the effects that such a force will have on wall safety, when assessment is made of the degree of conservatism to be used in design of new structures or evaluation of the safety of existing structures.

6.4 EARTH PRESSURES DUE TO SURFACE LOADS

Vertical loads on the surface of the ground increase both the vertical and lateral pressures in the ground. Loads on the backfill surface near an earth-retaining structure cause increased earth pressures on the structure.

6.4.1 Uniform Surcharge Loads

A uniform surcharge pressure applied to the ground surface over a large area causes a uniform increase in vertical pressure of the same amount,

$$\Delta p_v = q_s \tag{6.11}$$

in which Δp_v = increase in vertical pressure due to surcharge, and q_s = surcharge pressure. The surcharge pressure also causes an increase in lateral pressure,

$$\Delta p_h = k q_s \tag{6.12}$$

in which Δp_h = increase in horizontal pressure due to surcharge, and k is an earth pressure coefficient. For active earth pressure conditions, $k = k_a$; for at-rest conditions, $k = k_0$; and for passive earth pressure conditions, $k = k_p$.

Owing to the fact that the surcharge loading is applied over a large area (theoretically, an infinitely large area) both the vertical pressure due to the surcharge (Eq. 6.11) and the horizontal pressure due to the surcharge (Eq. 6.12) are constant at all depths.

6.4.2 Point Loads, Line Loads, and Strip Loads

When the surface loading is not uniform, or does not act over a large area, more complex calculations are needed to estimate the magnitude of the induced lateral stresses. As shown in Figure 6.13, the horizontal pressure induced by a vertical point load varies with depth and distance along the wall.

Although exact solutions to the problem shown in Figure 6.13 have not been developed, a simple approximation has been found that is accurate enough for practical purposes. Boussinesq developed expressions for the stresses induced within an elastic mass by a point load acting on the surface. According to this

solution, the horizontal stress can be expressed as

$$\Delta p_h = \frac{Q}{2\pi R^2} \frac{3zr^2}{R^3} - \frac{R(1 - 2v)}{R + z} \tag{6.13}$$

in which Q = the magnitude of the point load, expressed in units of force; $R^2 = x^2 + y^2 + z^2$; $r^2 = x^2 + y^2$; x and y are horizontal distances from the load to the stress point; z = depth of stress point below surface; and v = Poisson's ratio.

Boussinesq's solution can be used to develop an expression for the horizontal stress on a wall due to point load on the surface if two simplifying assumptions are made: (1) the wall does not move, and (2) the wall is perfectly smooth (there is no shear stress between the wall and the soil). Under these conditions the stress induced on the wall would be the same as the stress induced in an elastic half-space by two loads of equal magnitude situated as shown in Figure 6.14. The second load (called the image or imaginary load) would cause equal and opposite normal displacements on a plane midway between it and the real load, thus enforcing the zero-horizontal-displacement boundary condition at the wall. Thus, the horizontal pressures on the wall are twice as large as the horizontal stress induced in an elastic half-space, and can be calculated from the expression

$$\Delta p_h = \frac{Q}{\pi R} \frac{3zx^2}{R^3} \frac{R(1 - 2v)}{R + z} \tag{6.14}$$

in which x = horizontal distance from load to wall, $y = 0$, and the other terms are as defined for Equation 6.13.

Spangler (1938) and Terzaghi (1954) performed experiments to compare measured and calculated pressures on walls due to point loads. These experiments confirmed the fact that doubling the free-field stress (i.e., using the stress calculated from Equation 6.14), provides a good approximation to measured values of earth pressures on walls.

The same procedure has been used to develop expressions for stresses due to line loads and strip loads. For an infinitely

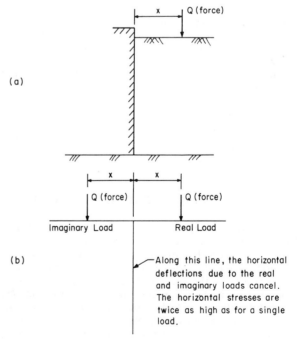

Fig. 6.14 Use of an imaginary load to enforce a zero-displacement condition at a wall. (a) A point load near a wall. (b) Two point loads on an elastic half-space.

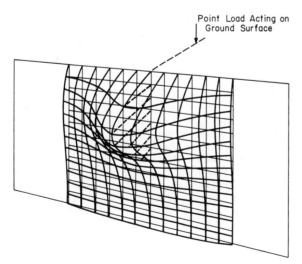

Fig. 6.13 Earth pressure data due to a point load. (*After Spangler, 1938.*)

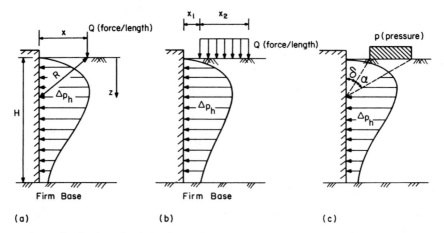

Fig. 6.15 Earth pressures due to line loads and strip loads. (a) Infinitely long line load parallel to wall. (b) Finite line load perpendicular to wall. (c) Uniformly loaded strip parallel to wall.

long line load parallel to the wall, as shown in Figure 6.15a, Scott (1963) developed the expression

$$\Delta p_h = \frac{4p}{\pi} \frac{x^2 z}{R^4} \qquad (6.15)$$

in which p = magnitude of line load, x = distance from line load to wall, z = depth below surface, and $R^4 = x^4 + z^4$.

For a line load of finite length oriented perpendicular to a wall, as shown in Figure 6.15b, Peck and Mesri (1987) have derived the expression

$$\Delta p_h = \frac{Q}{\pi z} \left(\frac{1}{\left[1 + \left(\frac{z}{x_2} \right)^2 \right]^{3/2}} - \frac{1 - 2v}{\left[1 + \left(\frac{z}{x_2} \right)^2 \right]^{1/2} + \frac{z}{x_2}} \right.$$
$$\left. - \frac{1}{\left[1 + \left(\frac{z}{x_1} \right)^2 \right]^{3/2}} + \frac{1 - 2v}{\left[1 + \left(\frac{z}{x_1} \right)^2 \right]^{1/2} + \frac{z}{x_1}} \right) \quad (6.16)$$

in which x_1 = distance from near end of line load to wall, x_2 = distance from far end of line load to wall, and the other terms are as defined previously.

Scott (1963) developed the following expression for the stress on a wall due to a vertically loaded strip of infinite length oriented parallel to a wall, as shown in Figure 6.15c:

$$\Delta p_h = \frac{2p}{\pi} [\alpha - \sin \alpha \cos (\alpha + 2\delta)] \qquad (6.17)$$

in which α and δ are the angles shown in Figure 6.14c.

6.5 EARTH PRESSURES DUE TO COMPACTION

When compaction equipment moves across the backfill adjacent to a wall, it induces added earth pressures on the wall. These added pressures can be estimated using procedures described in the previous section. When the compaction equipment moves away, a portion of the added earth pressure continues to act on the wall owing to the inelastic behavior of the soil. The magnitudes of these residual horizontal earth pressures have been studied by Broms (1971), Broms and Ingleson (1971), Rehnman and Broms (1972), Coyle et al. (1974), Coyle et al. (1976), and Carder et al. (1977, 1980).

Duncan and Seed (1986) developed a procedure for estimating the magnitudes of residual earth pressures due to compaction.

A typical distribution of these pressures with depth is shown in Figure 6.16. It can be seen that the residual earth pressure increases rapidly with depth in the upper 5 ft, and less rapidly at greater depths. At depths below about 25 ft in this particular example, there is no residual earth pressure due to compaction. Below about 25 ft the earth pressure is equal to the normal earth pressure at rest.

Williams et al. (1987) have used the analytical procedures developed by Duncan and Seed (1986) to develop earth pressure charts and tables of adjustment factors that can be used to make estimates of residual earth pressures due to compaction. These charts and tables make the computations easier, and they provide insight into the importance of the various factors that govern the magnitudes of compaction-induced earth pressures.

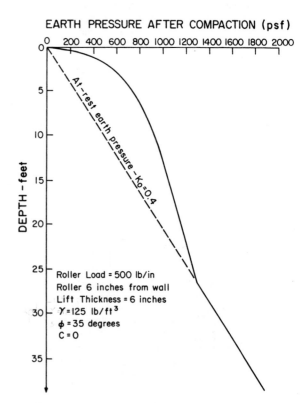

Fig. 6.16 Residual earth pressure after compaction of backfill behind an unyielding wall.

Charts for earth pressures due to compaction by rollers, vibratory plates, and rammers are shown in Figures 6.17, 6.18, and 6.19, respectively. The pressures in these figures were calculated using the procedures developed by Duncan and Seed (1986). The linear pressure variations in the lower parts of the diagrams correspond to various values of earth pressure at rest. To estimate the distribution of residual earth pressures following compaction using these charts, select the appropriate curve in the upper portion of the figure, and continue it until it meets the appropriate K_0 line. The resulting distribution has the form shown in Figure 6.16.

It may be seen that a number of parameter values were held constant in developing the design charts shown in Figures 6.17, 6.18, and 6.19. Variations in the values of these parameters have some influence on the calculated values of compaction-induced earth pressures. The effects of deviations from the standard values of these parameters can be taken into account through the adjustment factors in Tables 6.4 and 6.5. In cases where the actual conditions differ from those considered in developing the design charts, the earth pressures obtained from the charts are multiplied by correction factors from Table 6.4 or 6.5.

To illustrate the use of these earth pressure charts and correction factors, consider this example. Estimate the horizontal earth pressure at a depth of 5 ft below the surface after compaction in 6-in lifts by multiple passes of a Bomag BW 35 walk-behind vibratory roller. The static weight on one drum is 628 lb, and the centrifugal force on one drum is 2000 lb. The length of the drum is 15.4 inches. Thus $q = 2628/15.4 = 171$ lb/in.

From Figure 6.17, at a depth of 5.0 ft, find $p_h = 340$ psf. Adjustments must be made to this value, however, to account for the facts that: (1) the ϕ for the soil is 40° rather than the standard 35°, (2) the length of the roller is 15.4 inches rather than the standard 84 inches, and (3) the roller approaches within 0.2 ft of the wall rather than the standard 0.5 ft. The adjustment

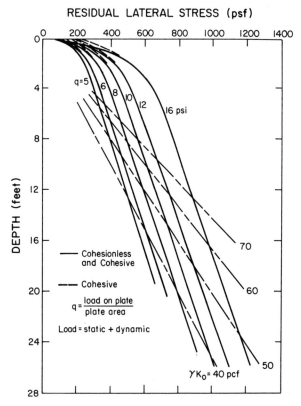

Fig. 6.18 Earth pressures due to compaction by vibratory plates.

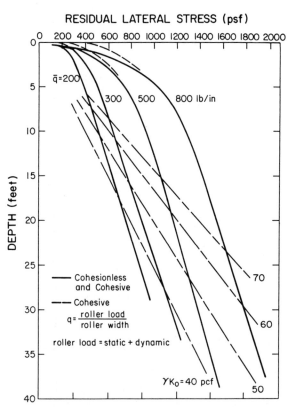

Fig. 6.17 Earth pressures due to compaction by rollers.

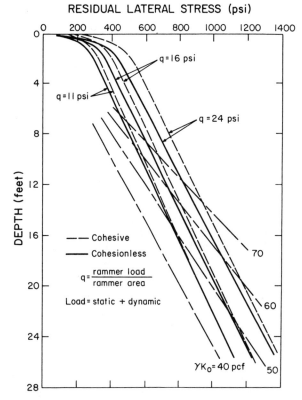

Fig. 6.19 Earth pressures due to compaction by rammer plates.

TABLE 6.4 ADJUSTMENT FACTORS FOR EARTH PRESSURES INDUCED BY COMPACTION WITH ROLLERS.

Variables				Multiplier Factors for z =			
				2 ft	4 ft	8 ft	16 ft
Lift thickness and distance from wall (*x*) (adjustments for these two factors are combined)	6-in lifts		$x = 0$	1.70	2.00	1.90	1.85
			$x = 0.2$ ft	1.50	1.85	1.70	1.65
			$x = 0.5$ ft	1.00	1.00	1.00	1.00
			$x = 1.0$ ft	0.85	0.86	0.87	0.88
	12-in lifts		$x = 0$	1.05	1.10	1.15	1.20
			$x = 0.2$ ft	1.00	1.05	1.10	1.10
			$x = 0.5$ ft	0.90	0.94	0.98	1.00
			$x = 1.0$ ft	0.70	0.70	0.70	0.70
Roller width (*w*)			$w = 15$ in	0.90	0.85	0.85	0.90
			$w = 42$ in	0.95	0.95	0.95	0.95
			$w = 84$ in	1.00	1.00	1.00	1.00
			$w = 120$ in	1.00	1.00	1.00	1.00
Friction angle (*φ*)			$\phi = 25°$	0.70	0.80	0.90	1.10
			$\phi = 30°$	0.85	0.90	0.95	1.05
			$\phi = 35°$	1.00	1.00	1.00	1.00
			$\phi = 40°$	1.25	1.15	1.10	1.00

TABLE 6.5 ADJUSTMENT FACTORS FOR EARTH PRESSURES INDUCED BY COMPACTION WITH VIBRATING PLATES AND RAMMERS.

Variables				Multiplier Factors for z =			
				2 ft	4 ft	8 ft	16 ft
Lift thickness and distance from wall (*x*) (adjustments for these two factors are combined)	4-in lifts		$x = 0$	1.00	1.00	1.00	1.00
			$x = 0.5$ ft	0.79	0.81	0.82	0.83
	6-in lifts		$x = 0$	0.83	0.85	0.87	0.90
			$x = 0.5$ ft	0.66	0.69	0.71	0.75
Vibrating plate area			240 in²	0.85	0.85	0.90	0.95
			480 in²	1.00	1.00	1.00	1.00
			960 in²	1.15	1.15	1.15	1.10
Rammer plate area			72 in²	0.85	0.85	0.90	0.95
			144 in²	1.00	1.00	1.00	1.00
			288 in²	1.15	1.15	1.15	1.10
Friction angle (*φ*)			$\phi = 25°$	0.80	0.90	1.05	1.25
			$\phi = 30°$	0.85	0.95	1.00	1.10
			$\phi = 35°$	1.00	1.00	1.00	1.00
			$\phi = 40°$	1.15	1.10	1.00	0.90

factors for these non-standard values are estimated using the values summarized in Table 6.4. The values of the adjustment factors (called R) are: $R_x = 1.8$, $R_w = 0.85$, $R_\phi = 1.14$.

Using this information from Figure 6.17 and Table 6.4, it is estimated that the postcompaction lateral earth pressure is equal to $p_h = (340 \text{ psf})(1.8)(0.85)(1.14) = 590$ psf. This value compares to a value of 570 psf calculated by means of detailed computer analyses performed using the methods developed by Duncan and Seed (1986).

By using the same procedure to estimate pressures at other depths, the distribution of earth pressures after compaction can be estimated. At the depth where these become smaller than the estimated at-rest pressures, the lateral pressures are equal to the at-rest values, as shown in Figure 6.16.

Postcompaction earth pressures estimated using Figures 6.17, 6.18, and 6.19 and Tables 6.4 and 6.5 apply to conditions where

the wall is stiff and nonyielding. These pressures would provide a conservative (high) estimate of pressures on flexible walls or massive walls whose foundation support conditions allow them to shift laterally or tilt away from the backfill during compaction. Such movements would reduce the earth pressures. The reduction would be expected to be less near the surface, where the compaction-induced loads would tend to "follow" the wall as it deflected or yielded.

6.6 RELATION BETWEEN EARTH PRESSURES AND WALL MOVEMENTS

As a wall moves toward the backfill, the earth pressures increase; as it moves away from the backfill, the earth pressures decrease. Ultimately, after sufficiently large movements, the limiting

TABLE 6.6 APPROXIMATE MAGNITUDES OF MOVEMENTS REQUIRED TO REACH MINIMUM ACTIVE AND MAXIMUM PASSIVE EARTH PRESSURE CONDITIONS.

| Type of Backfill | Values of Δ/H^a | |
	Active	Passive
Dense sand	0.001	0.01
Medium-dense sand	0.002	0.02
Loose sand	0.004	0.04
Compacted silt	0.002	0.02
Compacted lean clay	0.01^b	0.05^b
Compacted fat clay	0.01^b	0.05^b

[a] Δ = movement of top of wall required to reach minimum active or maximum passive pressure, by tilting or lateral translation. H = height of wall.
[b] Under stress conditions close to the minimum active or maximum passive earth pressures, cohesive soils creep continually. The movements shown would produce active or passive pressures only temporarily. With time, the movements will continue if pressures remain constant. If movement remains constant, active pressures will increase with time and passive pressures will decrease with time.

conditions of maximum passive and minimum active earth pressures are reached. If the movements continue after the maximum passive or minimum active pressures are reached, the earth pressures remain constant. Eventually, with sufficiently large movements, the pressure would change further as a result of the altered geometric conditions. The movements required to reach the minimum active or maximum passive pressures, however, do not result in appreciable changes in geometry.

The amount of movement required to reach the limiting conditions has been investigated experimentally, and by means of the finite-element method. A number of these investigations are summarized in Table 6.6. The results in Table 6.6 show:

- The movements required to reach the extreme pressures are proportional to the height of the wall, at least to a first approximation.
- The movement required to reach the maximum passive earth

pressure is of the order of ten times as large as the movement required to reach the minimum active earth pressure.

- The movements required to reach the extreme pressures are larger for loose, compressible soils than for dense, incompressible soils. For any cohesionless backfill the movement required to reach the minimum active condition is no more than about 1 inch in 20 feet ($\Delta/H = 0.004$). The movement required to reach the maximum passive conditions is no more than about 1 inch in 2 feet ($\Delta/H = 0.04$). These criteria (1 inch in 20 feet and 1 inch in 2 feet) provide simple, easy-to-remember guidelines for the amounts of movement required to reach the pressure extremes, and in most cases they are conservative.

Variations of the value of the earth pressure coefficient k with wall movement are shown in Figure 6.20 for dense and loose sand. The figure is drawn for the ideal condition where the backfill begins from at-rest pressures, with $K_0 = 1 - \sin \phi'$. This would be the case for a wall or a backfill that was "wished" into place.

Beginning from the at-rest condition, with $K_0 = 0.5$ for the loose sand and $K_0 = 0.29$ for the dense sand, the pressures increase as the wall moves toward the backfill and decrease as the wall moves away. Because the dense sand is stiffer than the loose sand, the pressures change more rapidly with wall movement for the dense sand.

A similar diagram is shown in Figure 6.21 for a compacted-sand backfill behind a wall. The figure applies to a backfill compacted to a medium-dense condition with no movement of the wall. The average value of K_0 after compaction, which would vary with compaction procedure and wall height, has been assumed to be 1.00 in Figure 6.21. Because the value of K_0 is increased as compared to the conditions shown in Figure 6.20, the movement required to reach the minimum active earth pressure condition is increased, and the movement required to reach the maximum passive pressure is decreased. Even though compaction has an effect on the amount of movement required to reach the extremes, the rules of thumb of 1 inch in 20 feet for active and 1 inch in 2 feet for passive still provide reasonable estimates of the movements required to reach the extreme pressure conditions.

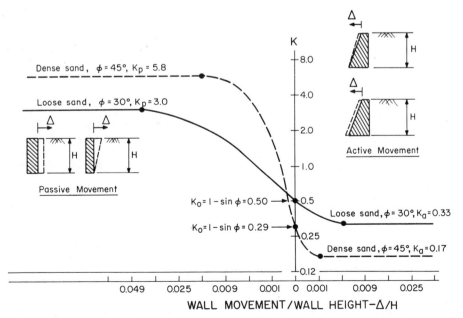

Fig. 6.20 Relationship between wall movement and earth pressure for ideal cases of walls that are "wished" into place.

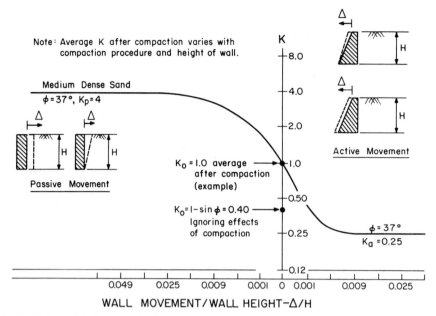

Fig. 6.21 Relationship between wall movement and earth pressure for a wall with compacted backfill.

6.7 EARTH PRESSURES FOR DESIGN

Although the earth pressures exerted on structures and walls by the adjacent backfill depend on soil–structure interaction, and are thus influenced by a number of different factors, it is nevertheless possible to estimate earth pressures with sufficient accuracy for most practical purposes without resorting to elaborate or complex analyses. To estimate earth pressures for design, the questions that must be answered are the following.

(1) Will the backfill be drained? If not, the major design consideration will be water pressure. If water can pond behind a wall, the wall must be designed to resist the hydrostatic water pressure plus the earth pressure exerted by the buoyant backfill. Usually in this condition the water pressures exceed the earth pressures, often by a considerable margin. If the backfill will be drained throughout its life, the wall need only be designed for earth pressures.

(2) What kind of backfill will be used? Free-draining cohesionless backfills are easy to compact, and exert relatively low pressures on walls. Walls backfilled with cohesionless backfills can be designed for minimum active earth pressures provided they can yield as much as 1 inch in 20 feet. Cohesive backfills are harder to compact and have higher at-rest pressures. Walls backfilled with cohesive soils cannot be designed for active earth pressures even if movements as large as 1 inch in 20 feet are acceptable, because cohesive soils creep. Walls with cohesive backfills that are designed for active earth pressures will continue to move gradually throughout their lives, usually with another episode of movement each time the backfill is thoroughly soaked by rainfall infiltration or rising groundwater levels. Even if wall movements as large as 1 inch in 20 feet are tolerable, walls backfilled with cohesive soils must be designed for pressures between active and at-rest.

(3) How much movement of the wall can be tolerated? Walls that can tolerate very little or no movement should be designed for at-rest pressures, including the effects of compaction-induced pressures. Earth pressures due to compaction can be estimated using Figures 6.17, 6.18, and 6.19, and Tables 6.4 and 6.5. Walls that can tolerate movements as large as 1 inch in 20 feet can be designed for minimum active earth pressures if the backfill

is free-draining and cohesionless. If the backfill is cohesive, pressures intermediate between active and at-rest must be used for design. Wall movements after compaction relieve compaction-induced earth pressures. Therefore, when movements as large as 1 inch in 20 feet after compaction are tolerable, earth pressures due to compaction need not be considered for design. Excessive compaction of the backfill, however, can induce large movements of walls. Especially during later stages of backfilling, heavy equipment should be kept away from the wall.

(4) Will the surface of the backfill support surcharge loads? Earth pressures due to uniform surcharge loads can be estimated using Equation 6.12. The value of k used in the equation should be equal to k_a, K_0, or an intermediate value, as appropriate for the type of backfill and the amount of wall movement that can be tolerated. Earth pressures due to point loads, live loads, or strip loads can be estimated by means of the theory of elasticity, using Equations 6.14 through 6.18. Although these equations were developed for nonyielding walls, it is appropriate to use them for all walls. To use lower pressures might lead to a condition where the wall would move an additional amount each time a load was applied to the backfill.

In many cases earth pressures for design can be based on answers to these questions, combined with judgment and experience, without laboratory tests on the soils or extensive theoretical analyses of the earth pressures. Table 6.7 gives values of earth pressure coefficients and equivalent fluid unit weights that can be used for design of walls of moderate height, up to about 20 feet.

For higher walls, consideration should be given to use of detailed design procedures, and determination of backfill properties through laboratory tests on samples of the backfill soil compacted to the expected field conditions. If cohesive backfills are used behind high walls, however, an appropriate empirical adjustment should be made to account for the effects of creep. Earth pressures for cohesive backfills estimated using methods that do not allow for creep effects should be increased, using judgment, to allow for long-term increases in earth pressures as a consequence of the tendency for these soils to creep.

TABLE 6.7   EARTH PRESSURES FOR DESIGN.

	Equivalent Fluid Unit Weights and Pressure Coefficients							
	Level Backfill				Backfill 2(H) on 1(V)			
	At-Rest		$\Delta/H = 1/240$		At-Rest		$\Delta/H = 1/240$	
Type of Soil	γ_{eq} (lb/ft³)	k	γ_{eq} (lb/ft³)	k	γ_{eq} (lb/ft³)	k	γ_{eq} (lb/ft³)	k
Loose sand or gravel	55	0.45	40	0.35	65	0.55	50	0.45
Medium-dense sand or gravel	50	0.40	35	0.25	60	0.50	45	0.35
Dense sand or gravel	45	0.35	30	0.20	55	0.45	40	0.30
Compacted silt (ML)	60	0.50	40	0.35	70	0.60	50	0.45
Compacted lean clay (CL)	70	0.60	45	0.40	80	0.70	55	0.50
Compacted fat clay (CH)	80	0.65	55	0.50	90	0.75	65	0.60

Note: $p_h = \gamma_{eq}z + kq_s$.

c_{eq} = equivalent fluid unit weight, z = depth below ground surface, k = horizontal earth pressure coefficient, q_s = uniform surcharge pressure.

REFERENCES

Broms, B. (1971), Lateral earth pressures due to compaction of cohesionless soils, *Proceedings of the 4th Budapest Conference on Soil Mechanics and Foundation Engineering*, pp. 373–384.

Broms, B. and Ingleson, I. (1971), Earth pressures against abutment of rigid frame bridge, *Geotechnique*, 21, No. 1, pp. 15–28.

Brooker, E. W. and Ireland, H. O. (1965), Earth pressure at rest related to stress history, *Canadian Geotechnical Journal*, 2, No. 1, pp. 1–15.

Carder, D. R., Pocock, R. G., and Murray, R. T. (1977), Experimental retaining wall facility-lateral stress measurements with sand backfill, *Transport and Road Research Laboratory Report*, No. LR 766.

Carder, D. R., Murray, R. T., and Krawczyk, J. V. (1980), Earth pressures against an experimental retaining wall backfilled with silty clay, *Transport and Road Research Laboratory Report*, No. LR 946.

Caquot, A. and Kerisel, J. (1948), *Tables for the Calculation of Passive Pressure, Active Pressure and Bearing Capacity of Foundations*, Gauthier-Villars, Imprimeur-Libraire, Libraire du Bureau des Longitudes, de L'Ecole Polytechnique, Paris.

Clough, G. W. and Duncan, J. M. (1971), Finite element analyses of retaining wall behavior, *Journal of the Soil Mechanics and Foundations Division, ASCE*, 97, No. SM12, pp. 1657–1674.

Clough, G. W. and Denby, G. M. (1980), Self boring pressuremeter study of San Francisco bay mud, *Journal of the Geotechnical Division, ASCE*, 106, No. GT1, pp. 45–63.

Coyle, H. M., Bartoskewitz, R. E., Milberger, L. J., and Butler, H. D. (1974), Field measurement of lateral earth pressures on a cantilever retaining wall, *Transportation Research Record*, No. 517, pp. 16–29.

Coyle, H. M. and Bartoskewitz, R. E. (1976), Earth pressures on precast panel retaining wall, *Journal of the Geotechnical Engineering Division, ASCE*, 102, No. GT5, pp. 441–456.

Duncan, J. M. and Seed, R. B. (1986), Compaction-induced earth pressure under K_0-conditions, *Journal of the Geotechnical Engineering Division, ASCE*, 112, No. 1, pp. 1–22.

Handy, R. L. (1985), The arch in soil arching, *Journal of the Geotechnical Engineering Division, ASCE*, 111, No. 3, pp. 302–318.

Jaky, J. (1944), The coefficient of earth pressure at-rest, *Journal for Society of Hungarian Architects and Engineers*, Budapest, Hungary, pp. 355–358.

Mayne, P. W. and Kulhawy, F. H. (1982), K_0–OCR relationships in soil, *Journal of the Geotechnical Engineering Division, ASCE*, 108, No. GT6, pp. 851–872.

Peck, R. B. and Mesri, G. (1987), Discussion of "Compaction-induced earth pressures under K_0-conditions" by Duncan J. M. and Seed, R. B., *Journal of Geotechnical Engineering, ASCE*, 113, No. 11, pp. 1406–1410.

Rehnman, S. E. and Broms, B. B. (1972), Lateral pressures on basement wall. Results from full-scale tests, *Proceedings of the 5th European Conference on Soil Mechanics and Foundation Engineering*, 1, pp. 189–197.

Schmidt, B. (1966), Discussion of "Earth pressures at-rest related to stress history," *Canadian Geotechnical Journal*, 3, No. 4, pp. 239–242.

Scott, R. F. (1963), *Principles of Soil Mechanics*, Addison-Wesley, Reading, Mass.

Spangler, M. G. (1938), Lateral pressures on retaining walls caused by superimposed loads, *Proceedings of the 18th Annual Meeting of The Highway Research Board*, Part II, pp. 57–65.

Terzaghi, K. (1954), Anchored bulkheads, *Transactions, ASCE*, 119, pp. 1954.

Williams, G. W., Duncan, J. M., and Sehn, A. L. (1987), Simplified chart solution of compaction-induced earth pressures on rigid structures, *Geotechnical Engineering Report*, Virginia Polytechnic Institute and State University, Blacksburg, Va.

7 DEWATERING AND GROUNDWATER CONTROL

J. PATRICK POWERS, P.E.
Consultant, AQUON Ground Water Engineering

7.1 IMPACT OF GROUNDWATER ON CONSTRUCTION

Whenever excavation must take place below the water table, groundwater affects the project. It affects the function and design of the facility, and the cost of its construction. Groundwater is a frequent cause of disputes between the owner and the contractor. Dewatering by unsuitable methods can under some conditions cause damage to adjacent properties, and result in third party litigation. Under some conditions dewatering may be harmful to the environment. Activities involving groundwater are closely regulated in many areas. The process of obtaining permits is often tedious, and sometimes authorities require special procedures that can be expensive.

Engineers responsible for all the phases of a project, from initial planning and budgeting through final construction need to be aware of the potential impact of groundwater so that their decisions will be effective. This chapter describes problems that groundwater has created, and presents methods that have proven effective in alleviating those problems. References are provided to more detailed treatment of the subject.

7.2 DESIGN OF STRUCTURES BELOW THE WATER TABLE

Factors that planners and designers should consider include the following.

Waterproofing More elaborate methods are necessary, particularly with very deep structures, to insure watertight integrity. Flexible membranes are recommended so that minor structural cracking does not cause leaks. There are many cases where expensive underground space has been rendered unusable by seepage. Postconstruction corrective measures such as grouting or permanent dewatering may be costly.

Hydrostatic Pressure Slabs and walls must be designed to carry both soil loads and water pressure. Where column spacing is wide, slab strengthening can be a major cost.

Uplift The weight of the structure must be sufficient to resist buoyancy. If the calculations credit the weight of the upper floors of a building for example, or the equipment in a lift station, then temporary dewatering must be continued until the necessary weight is in place. There have been instances where structures have heaved owing to premature release of groundwater control.

Relieved Structures Drydocks, sewage treatment tanks, and lift stations are often provided with drainage systems to lower the hydrostatic pressure when they are emptied. Building basements have been designed with dewatering systems that permanently depress the water table. A relieved design can sharply reduce first cost, but the drainage systems demand special methods and annual costs must be considered (Section 7.10).

Depth versus Cost Often a deep excavation is an option. The planner may be considering underground parking in an urban area, or the designer may be comparing the cost of piling to a deep floating foundation. For such decisions a reliable estimate of dewatering cost is essential. There have been cases where a major design change became necessary after construction began, because of unexpectedly difficult dewatering.

Permanent Effects on the Groundwater Regime Structures below the water table can affect the groundwater regime, sometimes undesirably. A cut-and-cover subway may create a dam in the path of natural groundwater flow, raising water tables upgradient and depressing them downgradient. Relieved structures can permanently depress the water table. Sewers placed in gravel bedding can modify groundwater flow patterns. If these factors cause undesirable side-effects (Section 7.9) corrective measures can be incorporated during construction. Drainage may be advisable around or beneath cut-and-cover subways. Relieved structures can be surrounded with cutoffs. Gravel bedding around pipes can be plugged at intervals with clay or concrete. Such measures are more expensive after construction.

7.3 METHODS OF GROUNDWATER CONTROL

The many methods that have been developed for the control of groundwater fall into three basic categories.

Open Pumping is the process in which as the excavation is advanced the water is allowed to flow in and is collected in ditches and sumps, then pumped away. The open pumping method is lowest in direct dewatering cost, and is viable under certain conditions. Under other conditions it can seriously hamper other operations, and can result in catastrophic failure.

Conditions favorable for open pumping include shallow subgrades, no more than a few feet below water table, and stable soils of low to moderate permeability such as dense, well-graded tills, clayey sands, firm clays, or fissured rock. There must be room for relatively flat slopes, or the excavation must be supported.

Open pumping is not effective in nonplastic sands and silts, since boiling can loosen the soil at subgrade, harming its bearing capacity. When excavating into an aquifer of high transmissibility, a large initial rate of pumping will occur during storage depletion. Open pumping will be cumbersome, unless the excavation is carried out very slowly. Where there are existing structures nearby, loss of soils due to open pumping may cause damage. If artesian pressure exists in a confined aquifer below subgrade, the condition cannot be relieved by open pumping. Conditions that favor or preclude open pumping are more fully discussed in Powers (1981).

During project planning, open pumping should not be budgeted for unless there is reasonable probability that conditions favorable to the method exist. When applying the method on a project underway, continuous observation is recommended to determine whether an undesirable situation is developing. A contingency plan incorporating one of the more positive methods of groundwater control is advisable.

Predrainage is the process of lowering the water table prior to excavation, using one of several methods available.

Deep Wells with individual pumps are most suitable for free-draining soils, and where the bottom of the aquifer is well below the subgrade of the excavation (Fig. 7.1). In this situation the yield per well in the dewatered condition is high, and relatively few wells are required. Since the cost per well is high, if the wells must be spaced very closely to accomplish the desired result the method may not be cost-effective.

Wellpoint Systems utilize groups of closely spaced wellpoints connected to a common suction header and pump. The cost per wellpoint is low, and in stratified soils where the spacing must be close the installation cost is lower. However, wellpoints depend on atmospheric pressure to lift the water to the pump, and the amount the water can be lowered with the method is limited to about 15 to 18 ft (5 to 6 m) at sea level, less at higher elevations. For greater drawdowns, multistages of wellpoints are required.

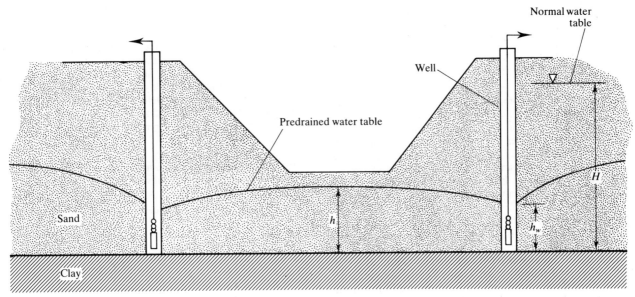

Fig. 7.1 Predrainage with deep wells.

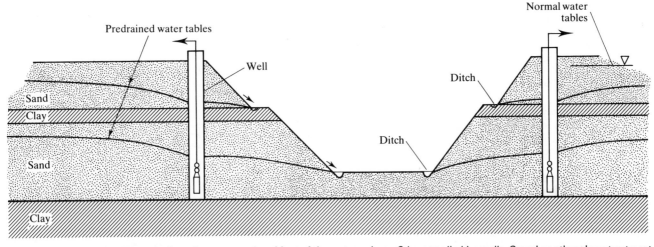

Fig. 7.2 Combination of deep wells and open pumping. Most of the water volume Q is controlled by wells. Gravel or other slope treatment may be necessary where residual seepage escapes.

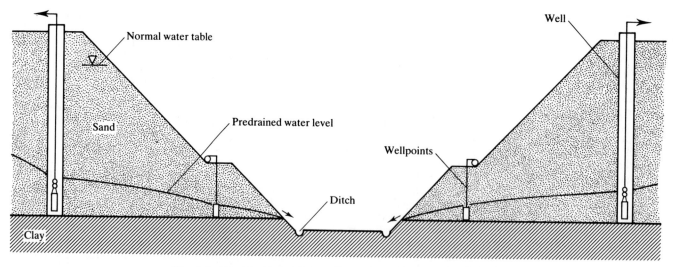

Fig. 7.3 Combination of deep wells, wellpoints, and open pumping.

Ejector Wellpoints incorporate a nozzle and venturi in the riser pipe. This method of pumping, similar to the domestic jet pump, eliminates the suction lift limitation of conventional wellpoints. Water tables have been lowered 100 ft or more using a single stage with this method. The pumping efficiency is poor, and for volumes in excess of 1000 gpm ejectors are rarely cost-effective.

Cutoff and Exclusion encompass a variety of methods that have been developed to obstruct entry of water to the excavated area.

Vertical cutoffs include driven *steel sheet piling* (NAVFAC DM-7, 1971, and Chapter 12) and concrete *diaphragm walls* (Xanthakos, 1979) built in panels in a bentonite slurry. The *slurry trench* (EPA, 1979, and Chapter 20) normally involves a continuous trench under bentonite slurry, backfilled with a mixture of soil and bentonite, or with bentonite/cement grout.

Permeation Grouting has been used to provide partially effective cutoffs and the newer *jet grouting* process has produced effective cutoffs in some soils (Baker, 1982; Karol, 1983). *Ground freezing* (Maishman, 1975; Sanger, 1968) has been successful as water cutoff and ground support under a wide variety of conditions.

Horizontal water cutoffs have been constructed by the *tremie seal* method (NAVFAC DM-7, 1971) and by *ground freezing*.

Water can be excluded from tunnels and shafts during excavation by *compressed air*, the *slurry shield*, and the *earth-pressure shield* (Bickel and Keusel, 1982; Pequignot, 1963; Richardson and Mayo, 1975).

Each of the methods of cutoff and exclusion has characteristics that make it more or less suitable under various ground conditions and project requirements (Powers, 1981).

Methods in Combination

Given the wide variety of methods for groundwater control and the many variations in project requirements, it is not surprising that frequently the most cost-effective solution is one that combines two or more of the methods on a single project.

Deep Wells and Open Pumping Figure 7.2 shows an excavation in stratified soil where open pumping alone is not a viable method. To fully predrain the water would require a great many wells. However, a well array on reasonable spacing, say 20 to 40 ft, will drain the more permeable strata to the extent that residual seepage can be controlled safely with sumps. This combination is most effective where recharge is remote and

storage can be depleted by lengthy pumping in advance of excavation.

Deep Wells, Wellpoints and Open Pumping Figure 7.3 shows a deep excavation that penetrates a bed of impermeable clay. A large number of wells would be required to predrain the water close enough to the clay that open pumping would be viable. A lesser number of wells have been installed, to lower the water table to within suction lift of a single-stage wellpoint system. The residual seepage past the wellpoints is handled with open pumping.

Steel Sheet Piling and Wellpoints Figure 7.4 shows an intake structure in open water. It may be difficult and costly to drive

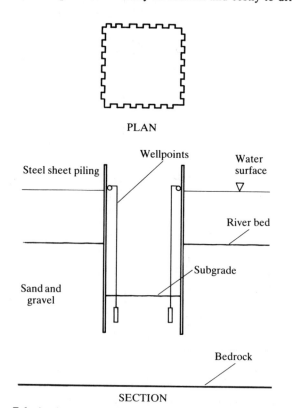

Fig. 7.4 Intake structures, combination of cutoff and predrainage.

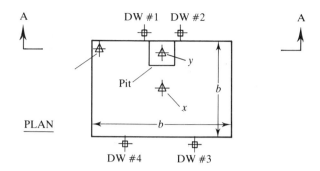

PLAN

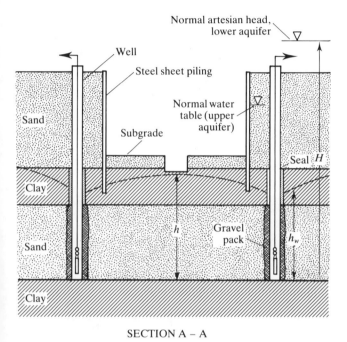

SECTION A – A

Fig. 7.5 Cutoff and pressure relief.

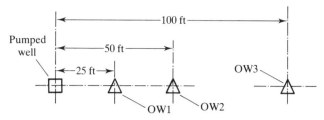

Fig. 7.6 Typical pumping test array. If recharge or barrier boundaries are suspected, lines of observation wells may be arrayed in several directions.

A typical array is shown in Figure 7.6. The test well is designed to stress the aquifer sufficiently to produce meaningful drawdowns in the observation wells (Powers, 1981). The analysis techniques assume that the aquifer is homogeneous and isotropic within the zone of pumping influence, that all water pumped is from aquifer storage, and that it is instantaneously released. Natural aquifers almost never meet the requirements, so the analyst's major challenge is to interpret from anomalies in the data the characteristics of this aquifer and how it departs from the ideal. For this purpose the semilog plots of the Jacob modification are most suitable, and will be discussed here. More elaborate methods are available, and are useful in special conditions. Certain criteria on length of the test must be fulfilled before the Jacob modification is valid (Driscoll, 1986; Powers, 1981).

The test must be operated long enough to identify recharge or barrier boundaries, or delayed storage release. In a water table aquifer, 7 days is recommended. In an artesian aquifer, 24 hours is often sufficient. The pumping should be continuous at a constant rate.

Water levels are recorded periodically prior to pumping to observe any natural fluctuations (tides, river stages, etc.) When pumping begins, early water levels are recorded on a frequent schedule, typically every few minutes for an hour. Readings can then be spaced out logarithmically with time. The drawdowns are plotted against the log of time for each observation well, and versus log of distance for all observation wells at various times. Figure 7.7 shows typical plots in an ideal aquifer. It is good practice to make the plots as the test proceeds, so that it can be decided whether enough information has been gathered before terminating the test. Recovery readings are taken when pumping stops, on a frequency similar to pumpdown.

At some point into the test, samples should be recovered for chemical and bacteriological analysis.

The plots of Figure 7.7 can be used to calculate the key parameters of the aquifer, *transmissibility T* and *storage coefficient C_s*. Transmissibility is defined as the ability of the aquifer to transmit water, per unit width normal to flow direction, under unit hydraulic gradient. In US units, T is measured in gallons per day per foot (gpd/ft). The storage coefficient is the yield of the aquifer in unit volume of water per unit volume of soil dewatered or pressure relieved. It is unitless. Referring to Figure 7.7:

Q = constant flow rate, gpm
δ = observed drawdown from static, feet
$\Delta\delta$ = drawdown difference per log cycle, feet
t = time since pumping started, minutes
t' = time since pumping stopped, minutes
δ' = residual drawdown during recovery period
t_0 = zero-drawdown intercept on log-time plot
r = radius from pumped well to observation well, feet
R_0 = radius of influence, zero-drawdown intercept on log distance plot

---

the sheeting to rock to provide cutoff. A wellpoint system is used to control the seepage under the toe of the sheets, providing a dry and stable subgrade, and assuring passive strength of the sand to support the toe of the sheeting. In some such cases, the most cost-effective combination may be to extend the toe of the sheeting deeper and use wells rather than wellpoints.

Cutoff and Pressure Relief Figure 7.5 shows steel sheeting extending to a clay layer. Artesian pressure beneath the clay presents a risk of heave. Deep wells are used to relieve the pressure.

7.4 FIELD PUMPING TESTS

Every geotechnical investigation should include the basic observations of groundwater conditions described in Section 1.10. If a potential groundwater problem has been identified, more elaborate procedures are advisable, such as additional borings completed as observation wells, sieve analysis of samples, and research on dewatering and water supply history in the area. When it appears that groundwater may significantly affect the design of the structure or the cost of the project, or result in disputes or third-party claims, then a field pumping test should be considered.

From the drawdown versus log time plot (Fig. 7.7a),

$$T = \frac{264Q}{\Delta\delta} \quad (7.1)$$

$$C_s = \frac{Tt_0}{4790r^2} \quad (7.2)$$

From the drawdown versus log distance plot (Fig. 7.7b),

$$T = \frac{528Q}{\Delta\delta} \quad (7.3)$$

$$C_s = \frac{Tt}{4790R_0^2} \quad (7.4)$$

From the recovery plot of residual drawdown versus $\log t/t'$ (Fig. 7.7c),

$$T = \frac{264Q}{\Delta\delta} \quad (7.5)$$

In a uniform aquifer, transmissibility can be defined as the product of the *permeability* k in gpd/ft and the aquifer thickness B in feet:

$$T = kB \quad (7.6)$$

Conversely, the permeability can be calculated from the transmissibility determined as above from a pump test, by dividing by the aquifer thickness as determined from boring logs. But this is only for a uniform aquifer. Most aquifers have seams with quite variable k, and calculated k is some sort of average value. It can distort the dewatering analysis (Section 7.5), and must be used with judgment.

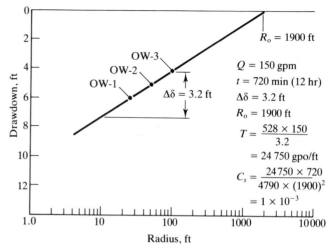

$R_0 = 1900$ ft
$Q = 150$ gpm
$t = 720$ min (12 hr)
$\Delta\delta = 3.2$ ft
$R_0 = 1900$ ft

$$T = \frac{528 \times 150}{3.2}$$
$$= 24\,750 \text{ gpo/ft}$$

$$C_s = \frac{24750 \times 720}{4790 \times (1900)^2}$$
$$= 1 \times 10^{-3}$$

Fig. 7.7b Drawdown versus log r.

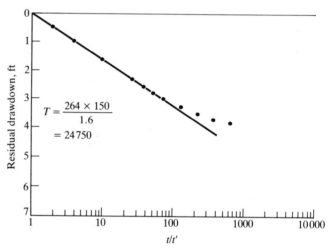

$$T = \frac{264 \times 150}{1.6}$$
$$= 24750$$

Fig. 7.7c Residual drawdown versus $\log \dfrac{t}{t'}$.

An advantage of the Jacob plots is that any departure from the ideal straight line is readily apparent, and the analyst is forewarned. He then must compare the values of T and C_s obtained from the various plots and attempt to judge from the shape and offset of the curves the nature of the anomalies in the natural aquifer with which he is dealing. Note that in an ideal aquifer that fulfills all the assumptions, the values of T and C_s as calculated from all three plots of Figure 7.7 are the same. When they are not, there is a problem. The shape and offset of the curves provide clues. Figure 7.8 illustrates some distortions commonly encountered.

Figure 7.8a shows the effect on time plots of recharge and barrier boundaries. Figure 7.8b shows the effect of delayed storage release. If T is calculated from the portion of the curve that has been flattened by delayed release, the value can be grossly in error. Figure 7.8b also illustrates the effect of a pumping interruption. Note the difficulty if no observations of drawdown were being made during the interruption. Figure 7.8c illustrates the effect of tides.

Figure 7.8d shows the effect on distance plots of recharge and barrier boundaries, and Figure 7.8e shows the effect of partial penetration. The partial penetration effect becomes insignificant at a radius r roughly equal to 1.5 times the saturated thickness of the aquifer, B. A quick scan of the variations in shape of these distorted curves and in their

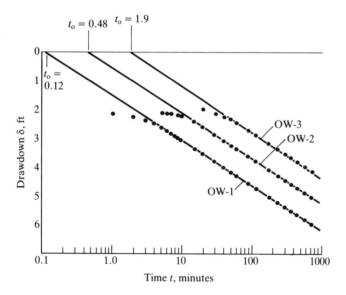

	OW-1	OW-2	OW-3
Radius r	25 ft	50 ft	100 ft
$\Delta\delta$	1.6	1.6	1.6
t_0	0.12	0.48	1.9
$T = \dfrac{264Q}{\Delta\delta}$	24 750	24 750	24 750
$C_s = \dfrac{Tt}{4790r^2}$	9.9×10^{-4}	9.9×10^{-4}	9.8×10^{-4}

Fig. 7.7a Drawdown versus log time t.

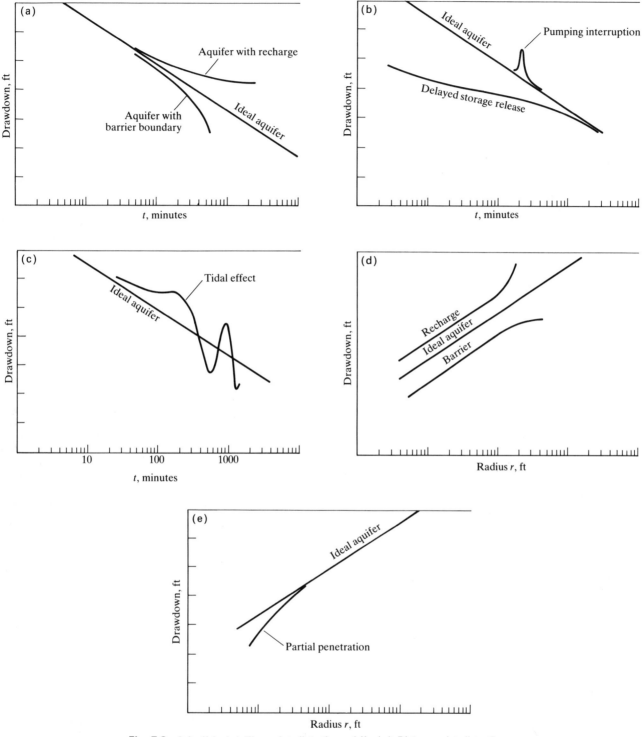

Fig. 7.8 (a), (b), (c) Time plot distortions. (d), (e) Distance plot distortions.

zero-drawdown intercepts will demonstrate how far calculated values can be in error.

7.5 DESIGN OF PREDRAINAGE SYSTEMS

Well Systems For the design and cost projection of a typical well system, it is necessary to estimate the total Q to achieve the required drawdown, and the yield Q_w of each well in the dewatered condition. These determine the number N of wells

required. Large total Q does not necessarily mean high cost. A free-draining aquifer may be dewatered at reasonable cost with a few large wells, whereas a project in stratified soils may require a great many wells, perhaps in combination with wellpoints, and although the total Q is less the cost may be greater.

In addition to the number and capacity of wells, it is necessary to estimate the area within which significant drawdown will occur, since this determines whether side-effects will have an impact on adjacent properties. It is also advisable to make a judgment on what conditions will exist within the excavation, based on soil descriptions and laboratory testing. If an impermeable stratum exists above subgrade, will perched water seep in over it, eroding the slopes? (Fig. 7.2). If the impermeable material is clay, the problem may be manageable; but if it is nonplastic silt, stability problems may be severe. Such considerations may affect the choice among water-control methods discussed in Section 7.3.

Mathematical Models

Confined Aquifer In simple aquifer situations, a suitable design approach is to model the excavation as a single well, with effective radius equivalent to the well system, to estimate the total Q. The problem illustrated in Figure 7.5 will serve as an example of the design procedures.

A rectangular structure of dimensions a and b is to penetrate through an upper aquifer of sand to a foundation in clay. Steel sheeting will be used to cut off the water in the upper aquifer, and to support the sides of the excavation. However, the clay is thin, and another aquifer beneath it must be pressure-relieved to protect the excavation from bottom heave.

It is proposed to surround the cofferdam with a system of pumped wells to relieve the lower aquifer. If we consider the well system as a single well of equivalent radius r_w, then

$$r_w = \sqrt{\frac{ab}{\pi}}$$ (7.7)

and

$$Q = \frac{kB(H - h_w)}{229 \ln R_0/r_w}$$ (7.8)

where

Q = total yield of the system, gpm
k = permeability, gpd/ft^2
B = thickness of aquifer, feet
R_0 = radius of influence, feet
H = initial head in the aquifer, feet
h_w = final head at the equivalent well

Selection of values to use in Equation 7.8 requires judgment, and is based on information from many sources, including a field pumping test, the boring logs, data on surface hydrology (to evaluate recharge from infiltration, and from surface water bodies), and from data on groundwater hydrology (to evaluate recharge from other aquifers). Inexperienced judgment in the selection of values for Equation 7.8 has resulted in gross error in analyses of this type.

The product kB is of course the transmissibility T as determined from the field pumping test. B can be estimated from boring logs, and an average k can be calculated from Equation 7.6. Experience shows that values of T and k are frequently misinterpreted (Powers, 1986). In most pumping tests, the indicated values are for horizontal permeability k_h. This value is only appropriate when flow to the dewatering system will be essentially horizontal. In sedimentary soils, the ratio of horizontal to vertical permeability, k_h/k_v, is rarely less than 3, and can be 10, 100, or larger. If flow to the dewatering system is other than horizontal, use of k_h in Equation 7.8 introduces error that can be significant. For example, if the upper water table aquifer in Figure 7.5 were to be dewatered, the flow would be both horizontal and vertical. A procedure that has given reasonable estimates in such a situation is as follows: knowing k_h from a pumping test, then k_v is estimated from a study of the boring logs, and the general geology.

The value used in the mathematical mode is the isotropic permeability k_i.

$$k_i = \sqrt{k_h k_v}$$ (7.9)

The radius of influence R_0 can be defined as the horizontal limit of the zone that will be affected by pumping. It is the zero-drawdown intercept on a Jacob distance plot (Fig. 7.7b). If all water pumped was from aquifer storage, then R_0 could be estimated by rearranging Equation 7.4:

$$R_0 = \sqrt{\frac{Tt}{4790C_s}}$$ (7.10)

Thus, R_0 without recharge or barrier boundaries is a function of the transmissibility and storage coefficient, and expands with the square root of time. In many dewatering situations, R_0 expands to a value less than indicated by Equation 7.10, because of recharge of surface water or leakage from other aquifers. Judgment must be made on any potential sources of recharge so that an equilibrium R_0 can be estimated. Since R_0 is a log function, a precise estimate is not necessary.

Within the operating period of a pumping project R_0 can vary. During heavy rains it can contract. Along volatile rivers when spring floods cause broad inundation of the flood plain, R_0 can shrink dramatically, and Q increases. Unless provision has been made in the design of the dewatering system, it may be overpowered. The U.S. Army Corps of Engineers monitors the stages of many rivers. Hydrographs can be researched, and topographic maps studied to evaluate the possibility of major inundation. In many areas, flood insurance maps are available to show the inundation during storms of various frequencies.

Initial head H in the aquifer can vary seasonally with precipitation or the stage of an adjacent river, and can be affected by other pumping in the vicinity. Appropriate safety factors should be included.

Final head h_w at the radius of the equivalent well must be lower than final head h that is desired under the excavation. The differential $h - h_w$ is a function of k, B, Q_w, and the geometry of the system.

When appropriate values have been estimated for the variables, total Q can be estimated from Equation 7.8. An estimate is then made of Q_w, the yield of an individual well in the dewatered condition, and the required number of wells N can be calculated:

$$N = \frac{Q}{Q_w}$$ (7.11)

Well yield Q_w is best evaluated from pump test data. In extrapolating the test results, adjustments must be made for conditions at the well that are different from those during the test. A factor of major significance is l_w, the saturated thickness of aquifer in contact with the wall of the borehole. In uniform aquifers, Q_w is directly proportional to l_w.

The phenomenon of *well loss* has been studied by Walton (1970) and others. No well can provide frictionless entry, so some well loss always occurs. Poorly constructed wells can have very high well loss. Unsatisfactory practices include the use of bentonite as a drilling fluid, inadequate flushing of the borehole to remove drilling detritus before completion, low-quality wellscreen, filter packs that are inappropriate to the aquifer or that have been ineffectively placed, and insufficient well development (Powers, 1981; Driscoll, 1986). Well loss during a pump test can be evaluated by estimating the drawdown at the radius of the borehole from the Jacob drawdown–distance plot, and comparing it with the pumping level in the well. It can also be estimated from *step drawdown tests* conducted before or after the constant rate test (Walton, 1970).

In the absence of a pumping test, Q_w can be estimated crudely with a modified version of the Sichart formula (Powers, 1981),

$$Q_w = (0.04)l_w r_w \sqrt{k} \qquad (7.12)$$

where

r_w = the borehole radius in inches
l_w = the saturated aquifer thickness in contact with the borehole
k = estimated permeability in gpd/ft^2

Sichart estimates are sometimes reasonably close, but can be far from actual, in either direction.

Water Table Aquifer Analysis of dewatering in a water table aquifer is quite different from pressure relief in a confined aquifer. The water table aquifer shown in Figure 7.1 has a variable phreatic surface, and its saturated thickness H changes during pumping. Q can be calculated from the relationship

$$Q = \frac{k(H^2 - h_w^2)}{458 \ln R_0/r_w} \qquad (7.13)$$

Values for the variables are estimated as in the discussion above. However, in water table aquifers an average k calculated from Equation 7.6 can introduce significant error. The aquifer will be partially dewatered. If the remaining saturated thickness h has a higher k than the average, higher Q must be expected. It is of note that when the deeper k is lower than average, the dewatering is more difficult. Q_w is lower, more wells are required, and although the total Q is less the cost may be significantly higher.

Other mathematical models that have proven useful in analyzing various aquifer situations are given in Figure 7.9.

The method of *cumulative drawdowns* can provide more accurate analysis of the confined aquifer in Figure 7.5 than the simple mathematical model. Accuracy is not the principal advantage however, since with the variables imprecisely known, precision in analysis is of limited value. The chief advantage of cumulative drawdowns is the method's ability to analyze many points within the flow regime.

It is assumed in Figure 7.5 that

$a = 60$ ft
$b = 40$ ft
$H - h = 27$ ft under general subgrade
$H - h = 30$ ft under the pit

From the pump test of Figures 7.6 and 7.7, values are estimated as

$kB = 25\,000$ gpd/ft
$R_0 = 2500$ ft at pumping time t when pressure relief must be accomplished.
$Q_w = 175$ gpm in the desired condition

A plot is constructed showing the effect of each well (Fig. 7.10). The slope of the curve can be calculated by rearranging Equation 7.3:

$$\Delta\delta = \frac{528Q}{T} \qquad (7.14)$$

The drawdown at any point of interest is the summation of the drawdowns due to each well at its radius from the point, as read from Figure 7.10. The cumulative drawdowns are shown in Table 7.1. Note that the well array is asymmetric, to provide greater drawdown at point y under the pit. The method is useful in checking the predicted value of Q_w. Aquifer drawdown at

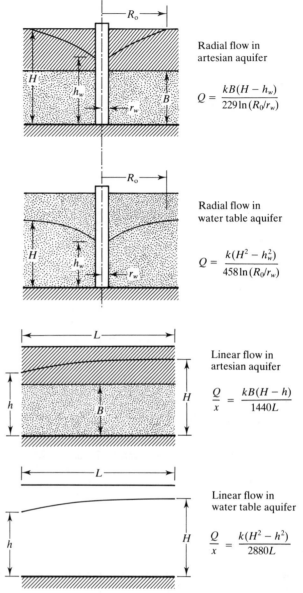

Radial flow in artesian aquifer

$$Q = \frac{kB(H - h_w)}{229\ln(R_0/r_w)}$$

Radial flow in water table aquifer

$$Q = \frac{k(H^2 - h_w^2)}{458\ln(R_0/r_w)}$$

Linear flow in artesian aquifer

$$\frac{Q}{x} = \frac{kB(H - h)}{1440L}$$

Linear flow in water table aquifer

$$\frac{Q}{x} = \frac{k(H^2 - h^2)}{2880L}$$

Fig. 7.9 Useful mathematical models: L = distance to line source (feet); x = unit length of drainage trench.

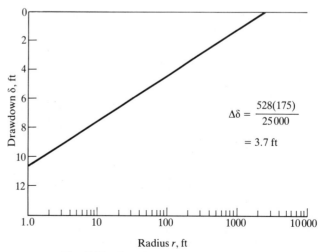

$$\Delta\delta = \frac{528(175)}{25\,000}$$

$$= 3.7 \text{ ft}$$

Fig. 7.10 Cumulative drawdown analysis.

TABLE 7.1   CUMULATIVE DRAWDOWN ANALYSIS.

Well No.	Point x		Point y		Point z		DW # 1	
	r	δ	r	δ	r	δ	r	δ
DW-1	25	7.4	10	8.8	20	7.7	1.0	12.5
DW-2	25	7.4	10	8.8	35	6.9	15	8.2
DW-3	27.6	7.2	41	6.6	58	6.0	51	6.3
DW-4	27.6	7.2	41	6.6	43	6.5	46	6.4
Total δ		29.2		30.8		27.1		33.4

well # 1 has been calculated in Table 7.1, using an effective radius of the well as 1 ft. If the remaining head h_w in the aquifer is less than the aquifer thickness B, the well may not yield what was estimated.

The cumulative drawdown method is suitable for confined aquifers, and gives reasonably reliable results in water table aquifers where the initial saturated thickness is not reduced by pumping by more than 20 percent.

Flow Net Analysis is useful when the geometry of the problem is too complex to apply a mathematical model or the cumulative drawdown method. Cedergren (1968) discusses the construction and application of flow nets. Complex situations in either plan or section can be analyzed. Difficulty occurs when the problem is complex in three dimensions. Perhaps some future geohydrologist whose hobby is three-dimensional chess will develop a tool for such problems.

Fragment Analysis developed by Harr (1977) is a mathematical approach similar to flow nets. Harr derived a group of relationships that estimate conditions in various fragments of a flow regime. He made several simplifying assumptions, while pointing out that the error thus introduced is well below the errors caused by variation in permeability and other factors. Fragment analysis is most useful when multiple iterations are required to find an optimum solution, and repeated drawing of flow nets is cumbersome. For example, if in Figure 7.3 it is desired to examine the cost-effectiveness of various penetrations of the steel sheet piling, fragment analysis can be useful. The method shares with flow nets the difficulty presented by a problem in three dimensions.

Computer Analysis has proven useful when aquifer situations are complex. Modeling techniques have been developed by a number of investigators (Akin, 1982; Driscoll, 1986). They can be particularly effective in saving time when multiple iterations are required. The game of "what if" is helpful in estimating the potential impact of conditions other than those assumed. What if another nearby aquifer is recharging the one under study? What if k is different from that assumed? What if the adjacent river experiences a storm of hundred-year frequency during the dewatering, and inundates the flood plain? Computer models with their remarkable speed can save considerable time in evaluating such possibilities. In practice, without the computer the analyses frequently just do not get accomplished. Experience demonstrates, however, that there is an occupational hazard with computer analysis. One becomes so fascinated with the method that not enough time and effort is devoted to making the judgments on aquifer parameters discussed above. If those judgments are poor, the result may be in serious error, no matter how sophisticated the analysis method.

7.6   MANAGING GROUNDWATER CONTROL

Where a significant groundwater problem exists, its successful solution depends on each participant in the project carrying out his functions in a responsible manner. This section offers suggestions on effective performance of the various functions, based on observations of past projects where groundwater created unnecessary difficulty, and observations of those projects where skilled management kept the groundwater problem within control.

The *owner* has perceived the need for a facility and has arranged its financing. Working through his consultants he seeks a contractor who will build a structure of the desired quality on schedule and within budget, and without disputes.

The *geotechnical engineer* must be alert for groundwater indications during the initial investigation (Section 1.10). Where a problem with significant potential appears, he should recommend a budget for closer study (Section 7.4).

The *planner* faces chronic budget restraints in the initial stage. But if his decisions are to be sound, he must have a reliable evaluation of the cost and impact of the water problem. Some decisions can be tentative until more data is available. Underground parking may or not be provided; piling or a deep mat foundation can be options; a lift station and force main may be substituted for a deep gravity interceptor. But when planning an underground metro, the options are limited. The planner's decision when to expand the groundwater investigation is based on many factors, and must frequently be a compromise.

When the role of the *designer* becomes dominant, tentative decisions must be made firm. The groundwater study must be essentially complete so that choices between options can be made, based on reliable estimates of cost and other factors. A key decision is how to allocate responsibility for controlling the water. If such decisions must subsequently be reversed because they were based on inadequate information, costs escalate.

Permits   Groundwater activity in many jurisdictions is regulated by the authorities. Permits may be required for drilling wells, for extracting groundwater and for releasing it to the environment. Sometimes hearings are required, which can be time-consuming. The permit requirements should be investigated well in advance of advertising for bids, to avoid unexpected delay. If the contractor is required to obtain some of the permits, it is good practice to alert bidders of the fact, and discuss the proposed plan with the regulating authorities ahead of time.

Contractor-designed Systems for groundwater control are normal practice, for good reasons. Water control affects other operations, such as excavation, ground support, and scheduling. Given latitude, the bidder can use his ingenuity to be more competitive. The contractor's operations must be regulated, however, to insure that the foundation properties of the subsoil will not be impaired, that the structure can be built safely and on schedule, and that adjacent structures will not be endangered. A recommended *technical specification* is:

> Control of groundwater shall be accomplished in a manner that will preserve the strength of the foundation soils, will not cause instability of the excavated slopes, and will not result in damage to existing structures. Where necessary to this purpose, the water will be lowered in advance of excavation by wells, wellpoints, or similar methods. Open pumping will not be permitted if it results in boils, loss of fines, softening of the subgrade, or slope instability. Wells and wellpoints will be installed with suitable screen and filters so that pumping of fines does not occur. Discharge will be arranged to facilitate sampling by the engineer.

When a potential problem has been identified it is good practice to draw bidders' attention to it; for example, if the soil at subgrade is sensitive to upward seepage, if there is deep artesian pressure, or if the aquifer is sensitive to pumping interruptions.

The contractor should be required to provide a *submittal* of his plan for groundwater control, prior to beginning its installation. The engineer reviews this to assure it is in accordance with the specifications and with good practice. The review and approval does not release the contractor from his responsibility for the plan's adequacy. The contractor when preparing his submittal usually has available only the prebid information. On complex projects, the specifications sometimes require a *second submittal*. After the system has been installed, the contractor must demonstrate by test operation that the system will accomplish its purpose.

Owner-designed Dewatering Systems are sometimes effective, but more often they create problems. The designer has only the prebid information at hand, and tends to be overly conservative. Where the contractor is responsible he can develop more data as the installation proceeds. Each well becomes a test boring, each pumping test adds insight. The contractor has incentive to locate his wells skillfully and build them of good quality to minimize his costs. If he is paid for each well the incentive is lost. With owner-designed systems skilled inspectors are necessary, and such specialists may not be readily available. Perhaps the most pernicious problem with owner design is mixed responsibility. If the control is a combination of predrainage and open pumping (Fig. 7.2) disputes arise over whether sufficient wells were provided to make the open pumping manageable.

The *minimum system* concept has been applied in some cases with satisfactory results. The specifications require, for example, that a minimum number of wells be installed, but the contractor is made responsible for the adequacy of the system, including additional wells that may be necessary. The owner is assured against overoptimism by the contractor, which could result in schedule delays or other problems. The contractor has incentive to build high-quality wells, to avoid extra expense. Any open pumping is clearly his responsibility.

The *contractor's options* in executing the dewatering are several. He may buy or rent the dewatering equipment and install it with his own forces, a frequent practice on routine projects. Where the dewatering is complex he can engage a *specialty subcontractor* to execute the work. He may invite proposals on a turnkey subcontract, where the specialist undertakes to install and remove a guaranteed system for a lump sum, and operate and maintain it for a price per unit time. Or the contractor may elect to subcontract part of the work, such as well drilling, on a unit price basis, and retain the risk.

7.7 DISPUTES OVER GROUNDWATER CONTROL

Groundwater is involved in a high percentage of the disputes that occur in underground construction, whether between the contracting parties or with third parties. Disputes can escalate the cost of a project, and delay its completion. Some number of disputes is inevitable, given the risks of the underground, but with preparation many disputes can be equitably settled in timely fashion, avoiding costly litigation. This section suggests procedures to reduce the probability of disputes, and to expedite the settlement of those that occur. It is based on observations of disputes that have occurred.

The primary preventive is an appropriate *geotechnical investigation* (Sections 1.10 and 7.4). When a problem has been identified and evaluated it is less likely to result in a dispute. The data should be provided accurately and completely to potential bidders.

The *differing site condition clause* used on federally funded projects and many others accomplishes two purposes. It avoids contingencies in the bids, since the contractor is assured that in the event of unexpected conditions he will receive an equitable adjustment. It also defines clearly what an unexpected condition is. The Federal clause (DOT, 1976; ASCE, 1988) has a large body of legal precedent that enables the parties to understand what has been agreed to. To qualify for an adjustment under this clause the condition must be different from that portrayed in the contract documents, and also different from what an experienced contractor would expect. The procedure for reporting the condition is detailed.

Substantive progress has been made in developing procedures to avoid disputes among the contracting parties (ASCE, 1988). These include the issuance of a *Geotechnical Design Summary Report*, which provides the bidders with not only the data from the investigation but the opinions and conclusions of the owner's consultants, on which the design was based. A *Dispute Review Board* is recommended to facilitate settlement of disagreements as early as possible.

If the geotechnical investigation identifies conditions where the groundwater control may affect third parties, preventive measures should be considered (Section 7.9). It has been common practice for owners to place the responsibility for the side-effects of dewatering on their contractors. If the side-effects are beyond the contractor's control, the practice is not recommended. Some courts have held for example, that if in the planning of the project the owner and his engineers contemplated lowering the water table, and the process subsequently causes injury to third parties, the owner is at least partly responsible.

7.8 COST OF GROUNDWATER CONTROL

Groundwater does not lend itself to rule-of-thumb estimating. Costs per gallon of water pumped, per cubic yard of excavation below the water table, or per lineal foot of trench dewatered, can all vary by an order of magnitude or more. Even the cost per well varies surprisingly. On a rapidly advancing trench excavation where the well is in operation for less than a week and the pump, screen, and casing are repeatedly reused, the cost per well installed is quite low. A well for long-term operation in an urban street drilled with difficulty and requiring sophisticated surface accessories can cost 50 times as much. Given these variations it is advisable to cost each project on an individual basis, perhaps with the assistance of an experienced specialist.

For the combined deep well and open pumping system illustrated in Figure 7.1, the elements of cost will include some or perhaps all of the following elements:

1. *Mobilization* (lump sum, per unit time on rentals)
   (a) Permits and submittal preparation
   (b) Rental or purchase of dewatering equipment
       (i)   Wellscreen and casing
       (ii) Well pumps, controls, wiring, discharge column
       (iii) Sump pumps, controls, wiring, hose
       (iv) Discharge manifold
       (v)   Standby generators
       (vi) Electric distribution cable and switchgear
       (vii) Flow meters
   (c) Transportation to and from jobsite
       (i)   Dewatering equipment
       (ii) Drill rig

    (iii)   Hydraulic crane
    (iv)   Loader/backhoe
(d) Electric power drop

2. *Installation and Removal* (lump sum)
   (a) Well drilling and development
   (b) Well filter gravel
   (c) Drilling fluid additive
   (d) Observation wells
   (e) Gravel and geotextiles for sumps and ditches
   (f) Cold weather protection
   (g) Installation labor (with hydraulic crane, truck, loader, and small backhoe as required)
      (i)   Well pumps
      (ii) Discharge lines
      (iii) Construction of sumps and ditches
      (iv) Sump pumps
      (v) Electric distribution system
      (vi) Electric pump connections
      (vii) Electric substation
     (viii) Standby generators
   (h) Removal labor (with hydraulic crane as required)
      (i)   Pumps, discharge lines
      (ii) Electric equipment
      (iii) Standby generators
      (iv) Well abandonment
      (v) Grouting of sumps and ditches

3. *Operation and Maintenance* (per unit time)
   (a) Electric power
   (b) Generator fuel
   (c) Maintenance labor (with crane and backhoe as required)
      (i)   Maintenance of sumps and ditches
      (ii) Removal and replacement of pumps
      (iii) Pump repair
      (iv) Periodic readings of observation wells, and preparation of reports
   (d) Maintenance material
      (i)   Replacement pumps and motors.
      (ii) Repair parts for pumps, controls, generators
      (iii) Lube supplies for generators
   (e) Specialist superintendent

On certain projects *extraordinary costs* are incurred that can be significant. They have included:

• Treatment of dewatering effluent
• Periodic chemical analysis of effluent
• Special reports to the engineer or to regulating authorities
• Artificial recharge

7.9 UNDESIRABLE SIDE-EFFECTS OF DEWATERING

Under certain conditions, the process of dewatering can have side-effects that are harmful to the project under construction, to other facilities nearby, or to the environment. It is important to identify such conditions during the planning stage, so that procedures can be instituted to control any damage. If a problem emerges unexpectedly after construction has begun, the costs of dealing with it escalate, and the project may be delayed. Powers (1985) presents procedures for identifying and evaluating problems of side-effects, and methods for dealing with them.

Improper dewatering such as open pumping under the wrong conditions, can cause damage to the structure being built or to adjacent structures. The principles in Section 7.3 can prevent such avoidable occurrences.

Ground settlement can occasionally be a problem. Lowering the water table increases the effective stress in the soil. The stress increase is usually modest, and most soils are not affected significantly. But if there are compressible soils in the vicinity, such as peat, organic silt, or soft clays, settlement may occur. Whether the settlement causes significant damage depends on the thickness and consolidation characteristics of the compressible deposit, the depth of drawdown and the duration of pumping, the foundations of the structures within the zone affected, and the type of their construction.

Untreated timber piles or other wooden structures below the water table can be damaged if the dewatering process exposes them to oxygen.

Groundwater supplies in the vicinity may be affected, by temporary reduction in the yield of supply wells, by salt water intrusion, or by the expansion of contaminant plumes.

The delicate ecological balance of *wetlands* can be upset by dewatering, particularly if pumping continues for an extended period. Trees or other plantings in *urban parks* may be affected.

Dewatering discharge sometimes contains substances that make treatment necessary before the water can be released into the surface environment. Natural substances include hydrogen sulfide and methane. Commonly encountered man-made contaminants include volatile organics, petroleum products, acid wastes, and sanitary sewage.

When problems from side-effects such as the above have been identified, their potential severity should be evaluated and procedures planned to deal with them. It may be necessary to control the water without dewatering, using one of the cutoff or exclusion methods (Section 7.3). Sometimes it is cost-effective to accept minor building damage from settlement and arrange for its repair. Temporary replacement water supplies can be provided to groundwater users. Artificial recharge can be employed to restrict the influence zone from the pumping (Powers, 1981, 1985).

7.10 PERMANENT DEWATERING SYSTEMS

Permanent pumping systems are employed for many purposes, including relieved structures such as drydocks, building foundations, and sewage treatment tanks; to prevent seepage into leaking structures; and to contain or recover pollutants in groundwater. The design of permanent systems is similar to that for temporary construction dewatering systems but there are significant differences. Comparison of cost of construction versus operation and maintenance is necessary to determine the optimum balance. Typically, better-quality construction can be justified by long-term savings.

Materials of Construction are selected for long life and low maintenance. A clear understanding of the groundwater chemistry, including potential variations with time is essential. Samples should be recovered during the pumping test, and analyzed at a laboratory experienced in groundwater problems. Piping and wellscreens are frequently made from PVC or ABS plastics, but if the system is to recover solvents from the ground such materials are unsuitable. Fittings, couplings, and valves should be selected on the basis of the chemistry as well as their function.

Pumps are designed for efficient operation, to reduce power costs, and also to avoid surging and cavitation that might cause maintenance and repair problems. Pumping test data must be extrapolated to conditions that can be expected during extended pumping. The impeller, volute, and other parts in contact with the water must be of suitable materials. Pumps are available in cast iron, plastic, fiber glass, bronze of various alloys suitable to different environments, and stainless steel. If solvents are present in the water, the elastomers in the mechanical seal must be of viton or some material resistant to the particular substance.

Automatic Controls for permanent systems can be complicated if the purpose is to be accomplished effectively. Many systems operate intermittently to control the water table within design limits. Water-level probes installed in an observation well initiate the starting and stopping of the pump. With systems of several wells, there may be multiple probes that start one pump, then start another if the water level continues to rise. Often a high-level alarm probe is used with a remote signal to alert maintenance personnel to a developing problem.

Encrustation can be a problem with permanent systems. It is a function of the groundwater chemistry and bacteriology, and should be investigated during the pumping test. Agents that have created problems include high hardness, which can result in precipitation of calcium carbonate; iron and manganese, particularly in the presence of *Crenothrix* or other iron-fixing bacteria; and organic slimes, sometimes encountered near sewage treatment plants.

Where there is encrustation potential, special features are advisable. Wells or wellpoints are designed for low entrance velocity, to retard the encrustation. Wellheads are arranged for easy access for acid treatment, cleaning, and redevelopment. Provision is made for safe handling and disposal of the acid. Experience shows that encrustation can be more readily removed if the work is done on a timely, periodic basis. If the condition advances to the stage where the encrustation is developing in the soil outside the wellscreen, acidization may be ineffective and the well has to be replaced. Similarly, if encrustation buildup in a pump advances too far it can cause mechanical damage.

The tops of the wells or wellpoints and the connecting piping are typically buried, to protect against weather and vandalism or for aesthetic reasons. Provision is necessary for access for well inspection, adjustment, and maintenance. Manholes, curb boxes, or pitless adaptors have been employed.

7.11 GROUND FREEZING

In the ground-freezing process (Sanger, 1968; Maishman, 1975) heat is pumped from the ground through freeze pipes until the pore water freezes, forming a cutoff and ground support around an excavation. A great variety of frozen structures can be formed (Maishman and Powers, 1982). Since freezing performs a dual function, it is cost-effective on many projects, particularly small, deep excavations and tunnels.

The shaft in Figure 7.11 is a typical application. Refrigerated brine is circulated through the freeze pipes (Fig. 7.12), which are typically arranged in series/parallel loops. As heat is removed, isotherms move outward in a manner very similar to drawdown contours around a water well. Heat pumping is analogous to pumping groundwater, and the mathematical relationships used for the analysis are essentially the same. When the 32°F isotherms intersect, the frozen diaphragm is complete. As the process continues the frozen structure increases in thickness. An observation well in the center will exhibit measurable rises in water level as the freeze moves inward. Temperature monitor holes enable observation with thermocouples to confirm that the wall has reached its design strength before excavation begins. The freezing is continued until the excavation is complete and the shaft has been lined. If the inner surface of the frozen wall is to be exposed to ambient conditions for an extended period, an insulating blanket is installed.

The brine is chilled by portable refrigeration plants, typically of 100 to 300 horsepower. On larger projects a group of such

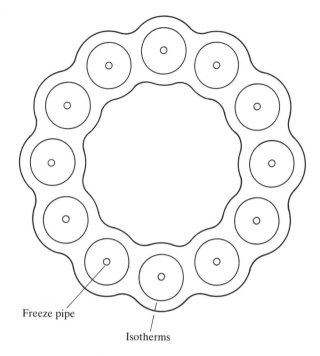

PLAN

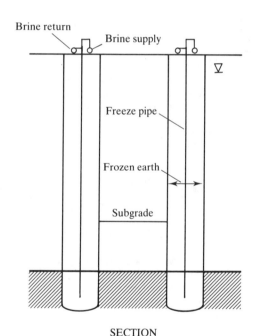

SECTION

Fig. 7.11 Ground freezing.

plants may be employed, with a central brine-circulating system. The heat is released to the atmosphere through an evaporative condenser, which requires a fresh water supply of about 5 to 10 gpm.

Quality Control is essential to accomplishing the desired result on schedule. Freeze pipes are spaced closely, typically 4 to 6 ft on centers, in order to achieve the design thickness within a reasonable formation period. If a borehole drifts out of plumb during drilling, the design spacing may not be achieved at depth. It is good practice to survey each hole with a directional inclinometer. If spacing between two pipes exceeds the allowable,

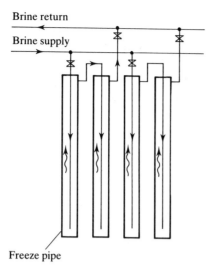

Brine return

Brine supply

Freeze pipe

Fig. 7.12 Brine circulation.

an additional pipe is installed. Each freeze pipe must be hydrostatically tested, since any brine leakage would impair the freeze. Appropriate instrumentation includes an array of temperature-monitoring holes and temperature-monitoring points in the brine system, as well as the instruments on the refrigeration plant.

Groundwater Movement in the vicinity of the freeze can create problems if the velocity exceeds 3 or 4 ft per day. The movement presents an additional heat load on the freezing system. Natural velocities over 3 ft per day are unusual, but transients can occur, for example, in the flood plain of a volatile river. When ground freezing is being considered, it is advisable to investigate whether any groundwater is being pumped nearby.

Liquid Nitrogen is another process that has been employed for ground freezing. The cost per unit of heat extracted is much higher than with refrigerated brine, but nitrogen can be cost-effective on small projects of short duration that are located near a manufacturing plant producing the fluid. The nitrogen is stored in a vacuum-insulated vessel on the site and is expended in the process. Special trucks refill the vessel periodically. The gas is nontoxic, but precautions are necessary to prevent its accumulation in enclosed areas. On a liquid-nitrogen installation, freeze pipes are typically smaller in diameter than with brine, and more closely spaced. Metallic hoses and other

special devices and materials are used to withstand the extremely low temperatures, on the order of $-320°F$.

REFERENCES

ASCE (1988), *Avoiding and Resolving Disputes in Underground Construction*, ASCE, New York, N.Y.

Akin, J. E. (1982), *Application and Implementation of Finite Element Methods*, Academic Press, New York, N.Y.

Baker, W. H. (ed.) (1982), *Grouting in Geotechnical Engineering*, ASCE, New York, N.Y.

DOT (1976), *Better Contracting for Underground Construction*, U.S. National Committee on Tunneling Technology, DOT-TST-76-48, U.S. Department of Transportation, Washington, D.C.

Bickel, J. and Keusel, T. R. (1982), *Tunnel Engineering Handbook*, Van Nostrand Reinhold Co., New York, N.Y.

Cedergren, H. (1968), *Seepage, Drainage and Flow Nets*, John Wiley and Sons, Inc., New York, N.Y.

Driscoll, F. (ed.) (1986), *Groundwater and Wells*, 2d ed., Johnson Division, St. Paul, Minn.

EPA (1979), *Slurry Trench Construction*, EPA-540/2-84-001, U.S. Environmental Protection Agency, Cincinnati, Ohio.

Harr, M. (1977), *Mechanics of Particulate Media*, McGraw-Hill Book Co., Inc., New York, N.Y.

Karol, R. (1983), *Chemical Grouting*, Marcel Dekker, New York, N.Y.

Krynine, D. and Judd, W. (1957), *Principles of Engineering Geology and Geotechnics*, McGraw-Hill Book Co., Inc., New York, N.Y.

Maishman, D. (1975), Ground freezing, *Methods of Treatment of Unstable Ground*, F. G. Bell, ed., Newnes-Butterworths, London.

Maishman, D. and Powers, J. (1982), Ground freezing in tunnels—three unusual applications, *Third International Symposium on Ground Freezing*, ISGF, Hanover, N.H.

NAVFAC DM-7 (1971), Department of the Navy, Washington, D.C.

Pequignot, C. A. (1963), *Tunnels and Tunnelling*, Hutchinson and Co., London.

Powers, J. P. (1981), *Construction Dewatering—A Guide to Theory and Practice*, John Wiley and Sons, Inc., New York, N.Y.

Powers, J. P. (ed.) (1985), *Dewatering—Avoiding Its Unwanted Side Effects*, Underground Technology Research Council, ASCE, New York, N.Y.

Powers, J. P. and Burnett, R. G. (1986), Permeability and the field pumping test, *In Situ '86 Specialty Conference*, ASCE, Blacksburg, Va.

Richardson, H. W. and Mayo, R. (1975), *Practical Tunnel Driving*, McGraw-Hill Book Co., Inc., New York, N.Y.

Sanger, F. J. (1968), Ground freezing in construction, *Journal of the Soil Mechanics and Foundations Division*, ASCE, **94**, No. SM-1, Proc. Paper 5743, pp. 131–158.

Walton, W. (1970), *Ground Water Resource Evaluation*, McGraw-Hill Book Co., Inc., New York, N.Y.

Walton, W. (1987), *Ground Water Pumping Tests*, Lewis Publishers, Inc., Chelsea, Michigan.

Xanthakos, P. (1979), *Slurry Walls*, McGraw-Hill Book Co., Inc., New York, N.Y.

8 COMPACTED FILL

JACK W. HILF, Ph.D, P.E.
Consulting Engineer
Aurora, Colorado

8.1 INTRODUCTION

As a construction material, soil has been used since antiquity with both success and failure. The widespread availability and relative economy of earth material continue to make it attractive for use in foundations, embankments, and as backfill. It has long been recognized, first empirically and then scientifically, that compaction changes the physical properties of soils—in some cases tremendously. For example, a properly compacted, well-graded gravel may be 15 times as resistant to deformation under a bearing load as the same material in the loose state.

To be used effectively, compaction must be tailored to the soil type, moisture condition, and subsequent environment of the compacted product. Thus, the ability of the engineer or job superintendent to identify the soil type accurately takes on prime importance. Wasted effort, such as that in attempting to compact clean sands with sheepsfoot rollers, can result from inattention to the recognition of soil type. The fallacy of making twice as many passes of the roller in an attempt to compensate for overly thick layers or overly wet soil can be avoided by an understanding of the compaction process in cohesive soils.

In Section 8.2 the elements of soil compaction for cohesive and noncohesive soils are discussed, their respective compaction characteristics are described, and the effect of compaction on soil physical properties important in engineered construction is explained. In Section 8.3 we discuss the various types of compaction equipment and indicate what to expect from their use with different soil types. The procedures by which quality control of compaction can be achieved are covered in Section 8.4. Miscellaneous problems of compacted fills such as expansive clays, dispersive clays, frost action, and slopes are treated in Section 8.5.

8.2 SOIL COMPACTION

Compaction is the process by which a mass of soil consisting of solid soil particles, air, and water is reduced in volume by the momentary application of loads, such as rolling, tamping, or vibration. Compaction involves an expulsion of air without significantly changing the amount of water in the soil mass. Thus, the moisture content of the soil, which is defined as the ratio of weight of water to weight of dry soil particles, is normally the same for a loose, uncompacted soil as for the same soil after compaction to a denser state. Since the amount of air is reduced without change in the amount of water in the soil mass, the degree of saturation increases. In most soils, however, the expulsion of all of the air cannot be achieved by compaction; hence 100 percent saturation does not occur. When it is used as a construction material the significant engineering properties of soil are its shear strength, its compressibility, and its permeability. Compaction of the soil generally increases its shear strength, decreases its compressibility, and decreases its permeability.

When considering the compaction of soils, two broad classifications of soils can be considered separately: (1) cohesive soils, and (2) cohesionless soils. Cohesive soils are those that contain sufficient quantities of silt or clay to render the soil mass virtually impermeable when properly compacted. Such soils are all varieties of clays, silts, and silty or clayey sands and gravels, which include those that fall into the Unified Soil Classification Groups CH, CL, MH, ML, SC, SM, GC, GM, and boundary groups of any two of these (Bureau of Reclamation *Earth Manual*, 1968a). On the other hand, cohesionless soils are the relatively clean sands and gravels that remain pervious even when well compacted. Soil Groups SW, SP, GW, and GP and boundary groups of any two of these represent such soils (see Chapter 3).

8.2.1 Cohesive Soils

An important characteristic of cohesive soils is the fact that compaction improves their engineering properties of shear strength and compressibility. Compaction of cohesive soils has been proved to follow the principles stated by Proctor (1933). Although there are several laboratory compaction standards and many different types of compactive efforts used in construction of compacted fills of cohesive soils, the effect of the water content of the soil on the resulting dry density (dry unit weight) is similar for all methods. For each compaction procedure there is an "optimum" moisture content that results in the greatest dry density or state of compactness. At every other moisture content, both dry and wet of the "optimum," the resulting dry density is less than this maximum. Figure 8.1 shows two moisture–density curves (Proctor curves) for different amounts of compactive effort on the same soil. Note that a different Proctor curve is obtained for each compactive effort, but each curve has the characteristic peaked shape.

The Standard Proctor Compaction Test (ASTM Test for Moisture-Density Relations of Soils Using 5.5-lb Rammer and 12-inch Drop D698, A; also AASHTO T-99) is made in a $\frac{1}{30}$-ft^3 cylinder mold using three layers each compacted by 25 blows of the rammer dropped 12 inches for an energy input of 12 375 foot-pounds per cubic foot. There are several variations of this standard, one using the same energy input but compacting the soil in a $\frac{1}{20}$-ft^3 cylinder in three layers with the same weight of rammer dropped 18 inches (Bureau of Reclamation *Earth Manual*, 1968) and others such as Test for Moisture–Density Relations of Soils Using 10-lb Rammer and 18-inch Drop,

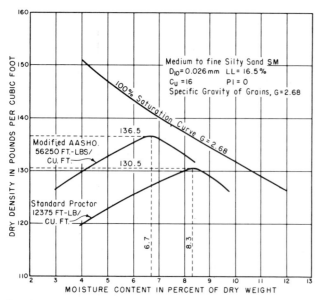

Fig. 8.1 Moisture–density curves of a cohesive soil for different compactive efforts.

ASTM Methods D 1557, A, B, C, D, also modified AASHTO T-180 using 56 250 ft-lb per ft³ of energy. In Figure 8.1, the soil is the same and the compactive effort varies. Figure 8.2 shows Proctor curves for three different cohesive soils using the Standard Proctor Compactive Effort (12 375 ft-lb per ft³). This illustrates the fact that each cohesive soil has its own characteristic moisture–density curve for a given compactive effort. The compactive effort used to obtain the curve in Figure 8.2 has been found to approximate the compaction achieved in the field by 12 passes of a 20-ton dual-drum sheepsfoot roller on 8- to 9-inch loose layers of cohesive soils. Figure 8.3 shows the average field compaction curves for three different cohesive soils used in earth dams, together with the corresponding standard Proctor laboratory curves.

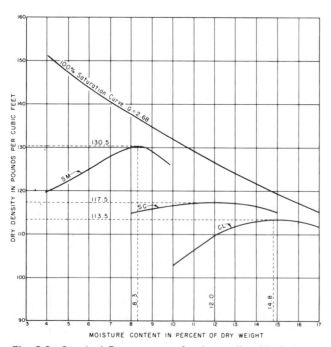

Fig. 8.2 Standard Proctor curves for three soils with the same specific gravity.

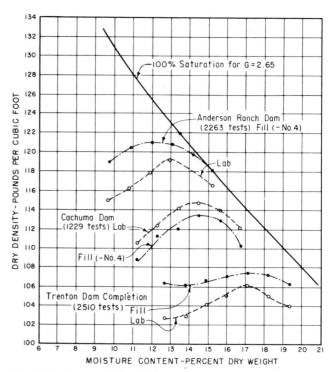

Fig. 8.3 Average field and laboratory compaction curves for three dam embankment soils.

Although most cohesive soils used in compacted fills have their own characteristic compaction curves for a given compactive effort, in some soils formed by the weathering of rocks in place (residual soils) the compaction curve is not unique but changes in ways depending on the moisture content at the start of the compaction test. The properties of many of these residual soils change irreversibly on drying, so that if a moist sample from a borrow area is predried and its compaction curve is determined by adding water (the normal procedure), the result is usually a higher dry density and lower optimum moisture content than if the Proctor test were performed starting at the natural moisture content and adding water or subtracting water (by drying if necessary) to obtain points on the compaction curve. Moreover, the Atterberg limits and engineering properties of strength, compressibility, and permeability are affected by predrying of moisture-sensitive soils; hence, for proper design and construction of compacted fills it is essential to identify these soils and to take into account their special characteristics.

This phenomenon of irreversible change of properties of these residual soils on drying is attributed to the presence of the clay minerals halloysite and/or allophane in the soils. Lambe and Martin (1955, 1956) and Mitchell (1976) provided data on soil composition and engineering properties for several soils containing appreciable percentages of halloysite.

Halloysite is related to the clay mineral kaolinite but consists of crystals of hollow cylinders typically 1 μm in length and of diameter about $\frac{1}{5}$ to $\frac{1}{10}$ of the length; kaolinite consists of crystals of pseudohexagonal plates of comparable size. Hydrated halloysite ($4H_2O$) contains a monomolecular layer of water within its mineral layers; when this water is completely removed by oven drying the result is dehydrated halloysite ($2H_2O$) (Nemecz, 1981; Bohn et al., 1985). The Atterberg limits of the two halloysites are different, but both plot below the A-line of Casagrande's plasticity chart and fall in the MH group of the Unified Soil Classification System.

Walker (1960) reported on a deeply weathered soil containing 25 percent of partially hydrated halloysite that had a natural moisture content of 28 percent and an optimum moisture content of 33.6 percent at standard Proctor dry density of 85.9 pounds per cubic foot. When this soil was air-dried prior to compaction, the corresponding optimum and maximum standard compaction values were 29.6 percent and 88.4 pcf, respectively. This indicates that an irreverisble change in the soil had occurred during air drying. The effects of predrying depend on the amount of air drying or oven drying and are accentuated when the percentage of halloysite and the natural moisture content of the soil are high.

Wesley (1973) discussed basic engineering properties of halloysite and allophane clays in Indonesia. He described allophane as consisting of gel-like fragments of random alluminosilicate held together by random links at a relatively small number of points. Much water is enclosed in the very open structure. Drying of this material causes an irreversible formation of hard grains. Changes in Atterberg limits and compaction characteristics of allophanes on drying are generally greater than those of halloysite. He concluded that both allophane and halloysite are good engineering materials despite their high natural moisture contents. Their location below the A-line of the plasticity chart and the changes in properties on drying are indicative of these soils.

The satisfactory engineering properties of these high-liquid-limit soils are evidenced by the performance of an appreciable number of earth dams constructed of soils containing these clay minerals. Terzaghi (1958), in reporting on the Sasumua Dam (halloysite content about 59 percent) stated that the high shear strength and low compressibility of halloysite soils have been found comparable to those of clays having much lower liquid limits, although dry densities are low (typically 60 pounds per cubic foot) and optimum moisture contents are high (50 percent or more) for pure halloysite. The 81-ft high Fena River Dam in Guam (Bureau of Reclamation, 1950; *Engineering News-Record*, 1950), constructed of soils containing about 85 percent halloysite is one of the dams that has given very satisfactory performance.

Walker (1960) pointed out that a weathered sandstone soil, when placed at laboratory optimum moisture content and compacted by heavy rollers, incurred particle breakdown to the extent that its moisture content was found to be 4 percent dry of the laboratory optimum determined after field compaction. These instances emphasize the importance of special care in using residual soils for compacted fills.

Hilf (1956) provided an explanation of the compaction characteristics of a cohesive soil as well as of its compressibility and shear strength properties by considering the soil mass as a three-phase system of solid particles, air, and water. The solid particles generally are mineral grains of various sizes and shapes occurring in every conceivable arrangement. Depending on the geologic processes that determined its occurrence, a soil in its natural state in the ground may have single-grained structure or compound structure. In the former type, each particle is supported by contact with several other grains. In the latter type, large voids are enclosed in a skeleton or arches of individual fine grains (honeycomb structure) or of aggregations of colloidal-size particles into chains or rings (flocculent structure). Compound structure is the result of sedimentation of particles that are small enough to exhibit appreciable surface activity. Soils with compound structure are usually of low density (large void volumes) but may have developed considerable strength owing to the compression of the arches of the soil skeleton. When these soils are remolded, their structure is changed and it approaches the single-grained structure, depending on the thoroughness of remolding. Remolding tends to densify the soil by expulsion of air. The processes of excavating, placing, and compacting cohesive soils in modern fills constitute a high degree of remoldings, so that

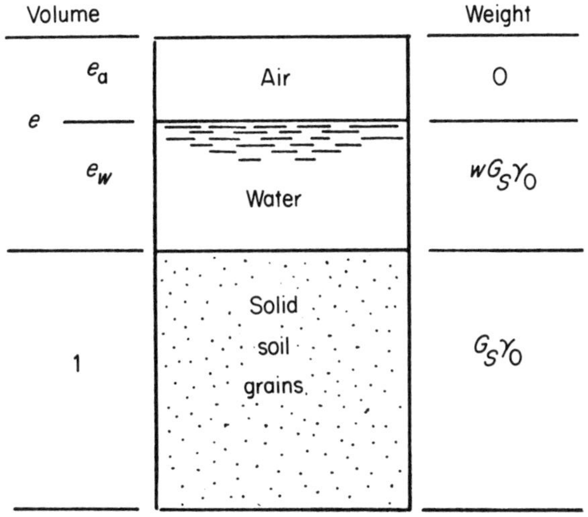

Fig. 8.4 Weight and volume relationships for soils.

$$n = \frac{e}{1 + e}$$

$$\gamma_w = \frac{G_s(1 + w)}{1 + e} \gamma_0 \qquad s = \frac{e_w}{e}$$

n = porosity; e = void ratio; e_w = water void ratio; e_a = air void ratio; γ_d = dry unit weight (dry density); γ_w = wet unit weight (wet density); γ_s = saturated unit weight (saturated density); γ_b = buoyant unit weight (buoyant density); w = water content in percent of dry weight; G_s = specific gravity of grains; S = degree of saturation; γ_0 = unit weight of water.

the structure of the final product bears little resemblance to that of the source deposit. For example, the characteristics of loessial deposits that stand in high vertical cuts as a result of natural structure are absent from the compacted soil made of the same material.

Soil is porous—that is, it contains interconnected void spaces between the grains, thus permitting the flow of fluids through the soil mass. It is a well-known fact that the volume of voids in a soil mass is less important, from the standpoint of permeability, than the size of the pores. Thus, a clay soil with an average grain size of 0.002 mm containing 50 percent voids by volume may be 1000 times less permeable than a sand of average grain size 0.5 mm containing 30 percent voids by volume. The amount of voids in a soil mass may be expressed as its porosity n, the volume of voids per unit volume of soil mass, usually expressed as a percentage; the void ratio e is the volume of voids per unit volume of solid soil particles, usually expressed as a decimal. The relation between these values, as well as the nomenclature and the weight–volume relationships for partially saturated soils, is shown in Figure 8.4.

Stresses on the Soil Skeleton

The solid particles in a compacted soil mass of single-grained structure can be considered to be a skeleton through which forces may be transmitted by grain-to-grain contact. The percentage of the surface area of a particle that is in contact with other particles in the mass is known to be small; for granular soils it is less than 1 percent. Terzaghi (1936) has shown that even for clays it is small enough to be neglected in soil mechanics computations. Although a soil mass is far from being an ideal, homogeneous, isotropic material, the concept of stress on the soil skeleton is the same as for other engineering materials; that is, normal and shear stresses are considered to be acting on the soil skeleton. These stresses are the so-called *effective stresses* acting at a point on a plane passing through the soil mass without cutting through any soil particles, but passing through the void spaces and the many points of contact between soil particles.

Under the action of effective stresses, the soil skeleton generally undergoes both elastic deformation and alteration of its structure by particle rearrangement. The relationship of the volume of the soil mass to the effective normal stresses applied to the soil skeleton is known as the *compressibility*. Similarly, the *shear strength* of a soil mass depends on the ability of the soil skeleton to resist shear stresses. Hence, the mechanical properties of soils are controlled entirely by the stresses on the soil skeleton—the effective stresses.

All compacted cohesive soils are a three-phase system, since perfectly dry soils are extremely rare in nature and it is impossible to expel all air from a wet soil by compaction. The soil skeleton is surrounded by a pore fluid consisting of water, water vapor, and air. The liquid phase, or soil-water, will be considered first.

It has been demonstrated (Briggs, 1897) that in a mass of moist soil the phenomenon of surface tension results in a negative (less than atmospheric) pressure in the water. This has been called *soil pull, capillary potential, capillary pressure, suction pressure, pressure deficiency,* and *capillary tension.* The soil-water is considered to exist in the form of a continuous film covering the grains and menisci near the contacts between grains (see Fig. 8.5). The menisci have a particular curvature that corresponds to a particular capillary pressure u_c for a given temperature, water content, and state of packing for a given soil. The remainder of the void space surrounding the wetted soil skeleton is occupied by air. It has been shown (Hilf, 1956) that the curvature of the menisci, which determines the capillary pressure in the soil-water, is unaffected by a change in air pressure.

The pressure in the air u_a of the voids of the soil mass that has been compressed without permitting escape of the pore fluids can be calculated by combining Boyle's law of compressibility of air with Henry's law of solubility of air in water (Eq. 8.1). The final pore water pressure—that is, the pressure in the water surrounding the soil skeleton—will be the algebraic sum of the pore air pressure and the capillary pressure, that is, $u = u_a + u_c$:

$$u_a = \frac{P_a \Delta e}{e_{a_1} + h e_w} \qquad (8.1)$$

where

P_a = atmospheric pressure
Δe = change in void ratio during compression without drainage
e_{a_1} = air void ratio after compression
e_w = water void ratio
h = coefficient of solubility of air in water by volume

Theories of Compaction

Several investigators—Proctor (1933), Hogentogler (1936), Hilf (1956), Lambe (1960), Olson (1963), and Barden and Sides (1970)—have attempted to explain the shape of the moisture–density curve for cohesive soils. All of these explanations must be considered tentative since they are based almost entirely on what appeared to be logical for the state of knowledge at the time rather than on reproducible measurements. It now appears that the process that results in the familiar peaked compaction curve is quite complex, involving capillary pressures, hysteresis, pore air pressure, pore water pressure, permeability, surface phenomena, osmotic pressures, and the concepts of effective stress, shear strength, and compressibility.

Proctor (1933) believed that the moisture in a relatively dry soil created capillary effects that held particles together, resulting in high frictional resistance that opposed the compaction forces. Thus, the compaction of a soil at a very low moisture content results in a hard and firm fill due to capillary moisture. His reasoning as to the effect of additional moisture on compaction is essentially as follows.

Compacting the soil by the same method but with a higher moisture content causes a greater rearrangement of the variously sized soil particles owing to the increased lubrication furnished by the additional water. The result is a soil of greater density but one that is less firm.

By compacting the soil with increasing amounts of water, this effect continues until the point at which the moisture content, combined with a small amount of contained air that the compaction process cannot remove, becomes just sufficient to fill the voids when the compaction process is completed. The soil now has the greatest density (least voids) that this method of compaction can obtain.

A still higher moisture content limits the compaction to a point at which the voids equal the volume of the contained air and water, resulting in a compacted soil with more voids, less density, and increased plasticity (softness). This effect continues with the addition of more water until the soil becomes too soft to sustain compacting equipment.

Proctor's explanation provided him with an excellent method of controlling the construction of cohesive fills, considering the limited knowledge then existing of pore pressures and shear strength.

Hogentogler (1936) considered that the shape of the compaction curve reflects four stages of wetting of the soil—hydration, lubrication, swelling, and saturation. Figure 8.6 illustrates Hogentogler's explanation. In this figure, the moisture–density curve has the same ordinate but the abscissa is not moisture content in percent of dry weight of soil, but a moisture content in percent of volume of solids plus volume of water. He believed that this plot provided four straight lines corresponding to his idea of stages of wetting. *Hydration* was believed to involve both absorption of water within the solid

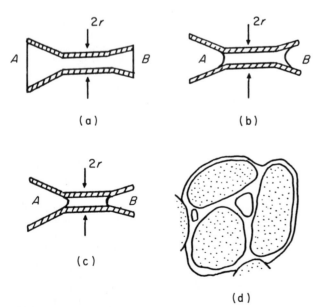

Fig. 8.5 Capillary pressure in soils. (a) Water surfaces are flat. (b) Menisci are curved. (c) Maximum curvature of menisci. (d) Unsaturated soil.

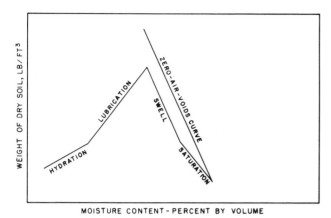

MOISTURE CONTENT-PERCENT BY VOLUME

Fig. 8.6 Hogentogler's explanation of the compaction curve.

particles and attachment of water as thin films on their surfaces. This water was considered to be highly viscous and gluelike. *Lubrication* was the process whereby as the water films thicken part of the water acts as a lubricant to facilitate the rearrangement of particles being compacted into closer association without, however, excluding all air. The maximum moisture content of this stage of wetting may be termed the lubrication limit. It is the optimum moisture content at which maximum density is attained. This process is similar to Proctor's concept of lubrication.

Water in excess of the lubrication limit was believed by Hogentogler to cause *swelling* of the soil mass without changing the amount of air from that existing at the lubrication limit. Hence, the additional water merely acts to displace the soil particles.

During the *saturation* stage, virtually all the air is displaced and the soil becomes truly saturated and merges with the zero-air-voids curve, as shown in Figure 8.6.

It is now known that the compaction process does not result in complete saturation and that the compaction curve on the wet side of optimum tends to parallel rather than intersect the zero-air-voids curve. Also, more recent study of adsorbed water indicates that the viscous water is limited to a thickness of only a few molecules and that moisture contents used in field compaction are far in excess of the amounts in which the viscous water concept is significant.

Hilf (1956) used the theory of pore water pressures in unsaturated soils to provide the basis for an understanding of the process of compaction and of the importance of moisture and density control of embankments. Instead of the familiar moisture–density relation shown in Figure 8.1, it is desirable to use a method of plotting based on void ratio and water–void ratio rather as shown in Figure 8.7. The latter compaction curve is similar in shape to the usual Proctor curve. The minimum void ratio point corresponds to the maximum density–optimum water content point of that curve. The corresponding dry density and water content scales can be plotted alongside the e and e_w scales as shown. In Figure 8.7 the curves representing degrees

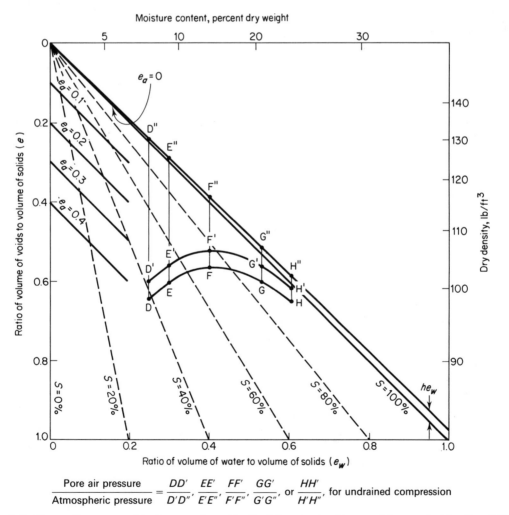

$$\frac{\text{Pore air pressure}}{\text{Atmospheric pressure}} = \frac{DD'}{D'D''}, \frac{EE'}{E'E''}, \frac{FF'}{F'F''}, \frac{GG'}{G'G''}, \text{ or } \frac{HH'}{H'H''}, \text{ for undrained compression}$$

Fig. 8.7 Compaction curve void ratio vs. water void ratio. (Also dry density vs. moisture content for $G_s = 2.60$)

of saturation(s) are straight lines, since $s = e_w/e$. The loci of points representing constant air content are also linear and are parallel to the 100-percent-saturation line (zero-air-voids line), since $e_a = e - e_w$. The addition of the line whose vertical distance above the $s = 100$-percent line represents he_w permits the magnitude of air pressure u_a of a sealed and loaded sample of compacted cohesive soil to be visualized. Equation 8.1 can be written for placement condition D:

$$\frac{u_a}{P_a} = \frac{\Delta e}{e_{a_1} + he_w} = \frac{DD'}{D'D''} \tag{8.2}$$

The corresponding values of u_a/P_a for Points E, F, G, and H are

$$\frac{EE'}{E'E''}, \quad \frac{FF'}{F'F''}, \quad \frac{GG'}{G'G''}, \quad \frac{HH'}{H'H''}$$

respectively.

The shape of the compaction curve is explained by the concepts of capillary pressure and pore air pressure. Dry soils are difficult to compact because the appreciable friction force caused by high curvature of menisci resists the compactive effort. Air, however, is expelled quickly because the air voids are relatively large. When the soil is compacted at an increased water content, the menisci flatten and cannot resist compactive effort as well as do the drier soils; hence, density increases until a maximum point is reached. The decrease of density with increasing water content beyond the optimum can be attributed to the trapping of air and building up of pore air pressure, which reduces the effectiveness of compaction.

It is now known that in soils of low degree of saturation the air voids are interconnected. Gilbert (1959) and Langfelder et al. (1968) reported that at about optimum water content the air voids in a compacted cohesive soil are no longer interconnected and the air permeability is essentially zero. This tends to support Hilf's pore pressure explanation of the shape of the compaction curve. However, Hilf assumed that the negative pressures in the moisture films surrounding the soil grains were interconnected for water contents used in field compaction and that these negative pressures acted to result in an all-around effective compressive stress on the soil skeleton equal in magnitude to the negative pressure. Subsequent investigations (Bishop, 1959; Bishop et al., 1960) indicate that the capillary pressure may not act fully as an effective stress but as a lesser effective stress by a factor χ that is said to vary with the degree of saturation (0 for completely dry soil and 100 percent for fully saturated soil). Measurement of this χ factor cannot be done directly and it has been found to be difficult to determine. Hence, this subject is still controversial.

Lambe (1960) attempted to explain the shape of the compactive curve in terms of surface chemical theories. At low water contents, flocculation of particles is caused by high electrolyte concentration, which reduces osmotic repulsion. Flocculation causes low densities. More water decreases electrolyte concentration, allowing double layers to develop more fully and producing a more dispersed soil structure. The increased dry density was assumed to result from such dispersion that "permits the particles to slide past each other into a more oriented and denser bed." Lambe considered the term "lubrication" to describe properly the effect of adding more water to the soil.

Olson (1963) concluded that the existing physicochemical theory of compaction does not have general applicability, and that advances in knowledge in supporting fields of science have applied limitations to Proctor's and Hogentogler's theories that were not apparent when those theories were being developed. Rather, Olson explained the shape of the compaction curve as follows using the effective stress concept.

For compaction at relatively low moisture contents, increases in moisture increase the degree of saturation, which results in higher pore air pressures and pore water pressures. This weakens the soil by reducing the effective stresses between the particles. The soil particles slide over one another until sufficiently large lateral stresses and horizontal shearing stresses with the layer previously compacted have developed to give the soil the requisite effective stress. As the soil is subjected to further blows, the effective stresses increase owing to three factors: increases in the residual lateral total stresses, the increasing negative residual pore water pressures, and the fact that the shear-induced increases in pore water pressure become progressively smaller. Small increases in dry density continue to occur as more and more blows are applied, since local concentration of shear stress will cause localized densification.

The same line of reasoning applies as the water content is increased further, except that the decreasing air permeability may result in the development of significant pore air pressures. Eventually, enough water may be added to the soil so that air channels become discontinuous, and the air is trapped. When the air voids become completely discontinuous, the air permeability of the soil drops to zero (Fig. 8.8), and no further densification is possible. The soil has reached the so-called *optimum moisture content*.

The problem remains of explaining how the soil can develop sufficient strength to resist the foot pressure when the water content is increased above the optimum point because, at these water contents, increases in water content cause decreases in the compacted dry density. The solution to this problem may consist of two parts, both related to the large foot penetrations encountered on the wet side of optimum.

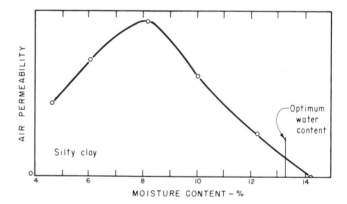

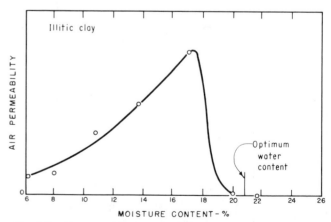

Fig. 8.8 Air permeability curves of compacted soil. (*After Olson, 1963.*)

First, for a dense soil, pore water pressure peaks at small strains and then decreases continuously with strain, becoming negative for highly overconsolidated soils at high strains. It is possible that the pore water pressure increases slightly as the foot pressure is applied and then decreases again, becoming sufficiently negative at large strains to allow the soil to resist the foot pressure.

Second, for a slight penetration, the compaction foot is somewhat similar to a shallow footing. The bearing capacity of a shallow footing is approximately six times the shearing strength of the soil (assuming a shear type failure). When the foot penetrates deeply in the soil, it simulates the action of a deep footing where the bearing capacity is perhaps 10 times the shearing strength of the soil. Hence, the soil can resist the same foot pressure with a smaller shearing strength, provided the foot penetration increases considerably. This is precisely what happens when the water content is increased beyond that at the optimum point. If the water content is increased too much, the foot penetrates all the way to the base of the mold and the test is concluded. The same process occurs in the field when sheepsfoot rollers sink into the soil until the weight is carried by the drum.

Barden and Sides (1970) made an experimental investigation of the engineering behavior of a compacted, partly saturated lean clay and attempted to relate this to direct microscopic observations of the clay structure. Their results indicate that the soil structure is markedly affected by the compaction moisture content. The shape of the compaction curve was attributed to the fact that at low compaction moisture contents, the low dry density is caused by the presence of large air-filled macropores between $\frac{1}{8}$- to $\frac{1}{4}$-inch-diameter pellet-like macropeds that have high strength and are able to resist the compaction pressures without much distortion. As the compaction moisture content is increased, the macropeds become wetter and consequently weaker. Thus, during compaction they are more easily distorted and the size of the voids is reduced. Further increase in the compaction moisture content causes the macropeds to become even weaker and during compaction they are distorted to fill the macropores, which tend to disappear. At this compaction moisture content the dry density is a maximum, and an increase in compaction moisture content causes the dry density to fall as the water layers between the soil particles increase in thickness. Barden and Sides show that the air voids become occluded (noncontinuous) near the optimum moisture content.

Figure 8.9a is a photograph of the structure (magnified 110 times) of a lean clay compacted 2.8 percent dry of optimum. Figure 8.9b is the same soil compacted 5.2 percent wet of optimum (magnified 110 times).

In summary, the effective stress explanation of the shape of the compaction curve (Hilf, 1956; Olson, 1963; Barden and Sides, 1970) appears more likely than explanations involving concepts of lubrication and viscous water. It is reasonable to assume that loose moist soil, whether prepared for compaction in the laboratory or in a fill, consists of lumps of particles that are held together by effective stress caused by capillarity. The drier the soil, the harder will be the lumps. The compaction process attempts to deform these lumps and make them coalesce. A given compactive effort will be more successful in doing this if the lumps are softer, as when additional water is added, than when the moisture content is low and the lumps are hard. As Barden and Sides have shown, the occlusion of the air paths in the compacted soil mass at the optimum moisture content provides a reasonable explanation of the limit of effectiveness of a given compactive effort. With additional water the compaction process can no longer expel air efficiently and transient pore air pressures can develop that resist the compactive effort. This explanation appears to agree with

Fig. 8.9 Compacted clay air-dried and viewed in scanning electron microscope at magnification of 110 ×. (a) Compacted 2.8 percent dry of optimum. (b) Compacted 5.2 percent wet of optimum. (*After Barden and Sides, 1970.*)

known facts and does not preclude the existence of osmotic pressures, surface phenomena, or χ factors of effective stress. Further research based on new measuring techniques will be needed to verify or disprove it.

Shear Strength of Compacted Cohesive Soils

The shear strength of a given compacted cohesive soil depends on the density and the moisture content at the time of shear. Although the effective strength parameters C' and ϕ' of such soils may be determined in the laboratory by "slow" shear tests where the small sample is permitted to drain during shear, these soils are so impermeable that under field conditions their strength for a considerable length of time after placement depends on undrained or partially drained conditions. The pore water pressures developed while the soil is being subjected to shear are of great importance in determining the strength of such soils. Before the theories of pore water pressure were well understood, the resistance of a compacted cohesive soil to penetration provided evidence of the phenomenon, which now can be explained by pore pressure theory. Figure 8.10 shows the resistance to penetration of a compacted silty clay for different conditions of placement under Proctor standard compaction effort. Note that the soil placed at 10 percent moisture content (3.3 percent dry of optimum) at 104 lb/ft³ resists penetration (2900 psi) very well, that is, it has a high shear strength. This is attributable to the high negative pore water pressures (capillary pressure) in the voids.

Penetration resistance decreases as the optimum water content is approached even as the density increases, and penetration resistance continues to decrease for wet of optimum placement conditions. This relationship, first developed by Proctor (1933), was used by him in determining how much drier than optimum he should place soil in embankments. He assumed that once placed the soil could become saturated by rainfall or by reservoir water, and that he could determine how soft it would become by the penetration resistance corresponding to the density ordinate on this figure. For example, the value of density corresponding to 300-psi penetration resistance when saturated could be set as a control standard.

Pore pressures produced by volume changes coincident with the shearing process act to reduce the apparent strength of a compacted cohesive soil. As can be seen in Figure 8.7, these pore pressures increase rapidly with increases in water content in the vicinity of the peak of the compaction curve. Compaction of the soil at water contents slightly less than optimum often results in a net increase in strength because the slight reduction in friction value (which accompanies the reduction in density) is more than compensated for by the comparatively large reduction in pore pressure that is thereby obtained.

The subject of shear strength of compacted cohesive soils was considered in detail at the American Society of Civil Engineers Research Conference on Shear Strength of Cohesive

Soils in June 1960. The Proceedings of the conference should be consulted for further study. For example, Gibbs et al. (1960) discussed shear strength of compacted soils used in fills.

During construction of a rolled fill, the objective is for each layer of soil to be identical and to be compacted at the same moisture content and to the same density. Immediately following compaction, the soil is assumed to be virtually unstressed externally. However, capillary pressures (negative pore water pressures) exist in the soil. These stresses are accompanied by normal effective stresses that are equal in all directions within the soil layer (gravity forces on the capillary water are neglected within the layer). Hence, no shear stresses exist in the soil layer considered. As construction proceeds, the load of superimposed layers of fill simultaneously applies normal and shear stresses to the soil below, causes it to change in volume, and induces pore air pressures and changes in capillary pressures.

It is apparent then that in the fill the shear process for each layer of soil starts at the same void ratio; hence, the usual laboratory process of applying different compressive loads to speciments of soil prior to shear does not follow the stress history of the fill. It appears desirable to conduct the laboratory triaxial test on compacted cohesive soils in such a manner that each specimen whose strength will determine the Mohr envelope is subjected to shear stresses while at the anticipated density and moisture content.

One procedure used by the Bureau of Reclamation to accomplish this is to prepare specimens at the same moisture content but at lower densities than the contemplated placement condition, with each specimen at a different density. A specimen is then placed in the triaxial machine and compressed with an all-around pressure without drainage, to the desired density. By measuring volume change, accurate control of density can be accomplished. In the application of chamber pressures to bring all sealed specimens to the same density, pore water pressures will be developed. If the initial densities of the specimens are properly chosen, the effective lateral stress will be different for each specimen. In this manner, a strength envelope on the basis of effective stresses can be obtained for an unsaturated soil that is comparable to fill conditions to the extent that shear started at placement density and moisture content.

Also, to avoid prestressing the specimens as the volume of the soil decreases during shear, a test method of constant effective lateral stress has been developed (Gibbs and Hilf, 1957). In this method, the failure criteria of maximum deviator stress and maximum principal effective stress ratio coincide, thus avoiding ambiguity.

The combination procedure of starting the shear process on sealed, compacted soil specimens all at the same moisture content and density, conducting the test so that no prestress occurs, and measuring pore water pressures, results in the strength envelopes shown in Figure 8.11. Two envelopes are shown for the same data: (1) the solid circles are effective stress circles based on pore pressures measured at the top of the specimens using perforated end plates ($\frac{1}{32}$-inch-diameter holes) to contact the soil; and (2) the dashed circles are effective stress circles based on pore pressures measured at the bottom of the specimens using a P-3 filter disk (23-psi bubbling pressure or air entry value). The two different values of pore pressure, effective stress, C intercept, and tan ϕ are shown paired in the figure. The upper value corresponds to the top measurement.

Using a commercial kaolinite ($LL = 57$ percent, $PL = 37$ percent, $G_S = 2.67$) and a real silty clay (Higgins clay) ($LL = 38$ percent, $PL = 21$ percent, $G_S = 2.79$, 100 percent < 0.1 mm), Lee and Haley (1968) investigated their strengths at confining pressures up to 1000 psi. The tests included samples compacted dry of optimum and wet of optimum to the same

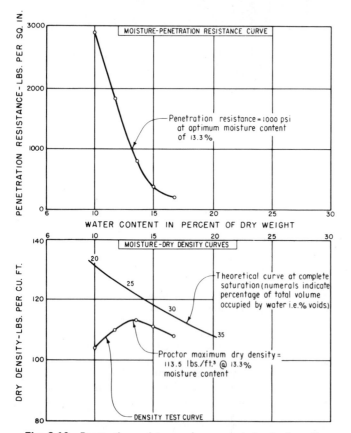

Fig. 8.10 Penetration resistance of a compacted cohesive soil.

SPECIMEN NO.	DRY DENSITY (PCF)			MOISTURE CONTENT (%)		DEGREE OF SATURATION (%)		TEST VALUES AT FAILURE					SHEAR VALUES CORRECTED FOR PORE PRESSURE	
	PLACE-MENT	WETTED	CONSOLI-DATED	PLACE-MENT	WETTED	PLACE-MENT	WETTED	PORE PRESSURE (PSI)	EFFECTIVE LATERAL PRESSURE (PSI)	VOLUME CHANGE (% OF INITIAL)	AXIAL STRAIN (%)	DEVIATOR STRESS (PSI)	TAN φ	COHESION (PSI)
4	116.8	—	118.2	13.6	—	81.5	—	$\left(\begin{smallmatrix}12.0\\6.5\end{smallmatrix}\right)$	$\left(\begin{smallmatrix}12.5\\18.0\end{smallmatrix}\right)$	−3.00	17.73	61.4		
5	116.0	—	118.3	13.5	—	79.4	—	$\left(\begin{smallmatrix}24.0\\17.9\end{smallmatrix}\right)$	$\left(\begin{smallmatrix}24.5\\30.6\end{smallmatrix}\right)$	−4.81	19.72	76.9		
6	115.2	—	118.7	13.1	—	75.4	—	$\left(\begin{smallmatrix}61.9\\57.6\end{smallmatrix}\right)$	$\left(\begin{smallmatrix}50.9\\55.2\end{smallmatrix}\right)$	−7.14	23.09	115.1	$\left(\begin{smallmatrix}0.47\\0.58\end{smallmatrix}\right)$	$\left(\begin{smallmatrix}16.0\\10.0\end{smallmatrix}\right)$

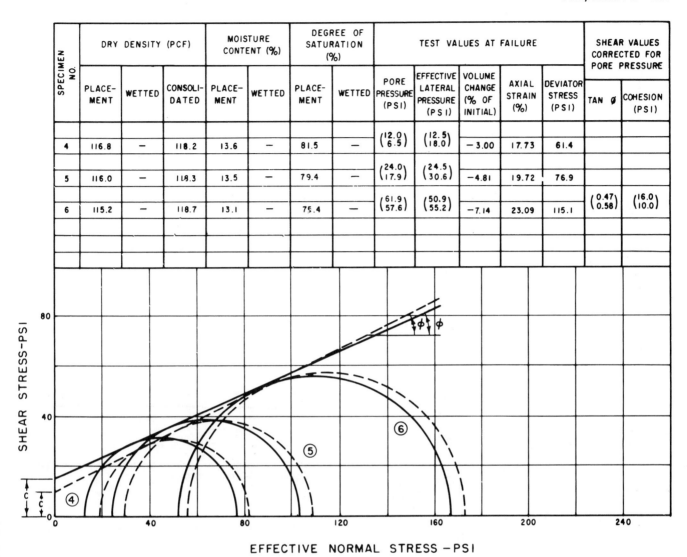

Fig. 8.11 Shear strength of a compacted cohesive soil.

dry density. Trends were similar, but only the results of the real clay (Higgins) will be discussed here. Samples prepared dry of optimum (flocculated structure) exhibit very brittle stress–strain curves in an unconfined compression test and were up to three times stronger than samples of the same density placed wet of optimum by kneading compaction (dispersed structure) or by static compaction (flocculated structure). In an unconsolidated undrained (UU) test, however (i.e., with lateral support), the samples prepared dry of optimum were strong in comparison to the samples prepared wet of optimum, but the stress–strain data indicate that the samples no longer exhibit extremely brittle characteristics. In fact, all of the stress–strain curves for consolidated–undrained (CU) tests on Higgins clay were approximately the same shape, and none exhibited a particularly brittle loss of strength at high strains. In consolidated–undrained tests (wherein the samples were fully saturated after consolidation) at the same consolidation pressure, samples prepared dry of optimum were found to have higher moisture contents and similar stress–strain curves, and approximately equal to slightly lower undrained strengths than otherwise identical samples prepared wet of optimum by kneading compaction. For these CU conditions, the Mohr envelopes of failure based on either total or effective stresses appeared to pass through the origin and become progressively

flatter with increased confining pressure. On an effective stress basis, the CU (saturated) angle of friction reduced from 30° at 100-psi confining pressure to 21° at 500-psi confining pressure. Comparable values for total stresses, CU (saturated) were 20° to 11°, respectively.

Figures 8.12 to 8.15 show the compaction characteristics, unconfined compression test data, UU test data, CU test data, and Mohr envelopes. These data appear to favor placement on the dry side of optimum rather than the wet side of optimum for the same density so long as an appreciable confining pressure exists. The undrained strength of the dry soil is much greater as shown in Figures 8.12 and 8.13, and even under the severe conditions of full saturation at the consolidated density the soil originally placed dry of optimum has no significantly different brittleness (Fig. 8.14) or strength (Fig. 8.15) than the consolidated soil originally placed wet of optimum. Since stability of slopes during construction is a UU condition (pore pressure develops with little opportunity for drainage) the advantage of placing cohesive soils dry of optimum becomes apparent so far as shear strength is concerned. This laboratory evidence is substantiated by considerable field data as well as theory.

In a comprehensive review of theoretical methods for predicting pore pressures developed during construction of embankments and summary of observed pore-pressure data

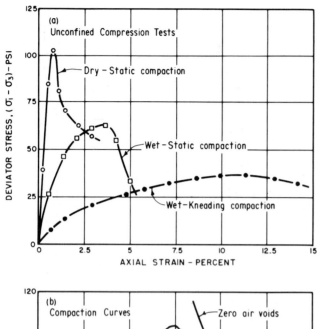

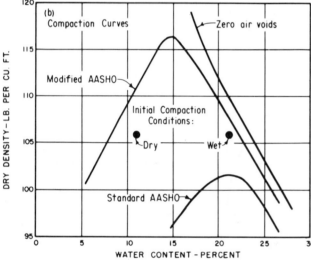

Fig. 8.12 Compaction and unconfined compression characteristics of Higgins clay. (*After Lee and Haley, 1968.*)

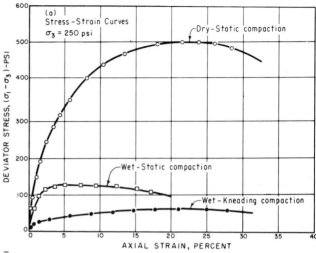

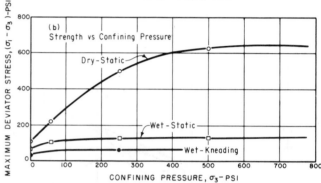

Fig. 8.13 Unconsolidated–undrained tests on compacted Higgins clay. (*After Lee and Haley, 1968.*)

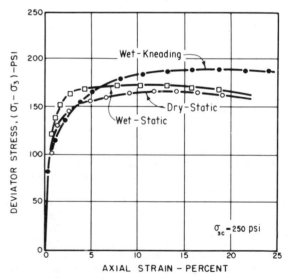

Fig. 8.14 Stress–strain curves for CU tests on compacted Higgins clay. (*After Lee and Haley, 1968.*)

from the files of the Corps of Engineers (CE) offices and published data from other agencies, Sherman and Clough (1968) reported that "Low dams (less than 100 ft high) as well as high dams are subject to development of significant pore pressures when the placement-water content is above optimum."

Nevertheless, there is an increasing tendency to compact the cores of earth dams on the wet side of optimum moisture content because of the belief that by so doing the fill will be made more flexible and have less tendency to develop cracks. There may be some validity in this procedure for very thin core dams where drainage of the high pore pressures that develop can occur, but there appears to be no good reason to intentionally make a compacted cohesive fill unstable by placing it wetter than necessary for good compaction.

Compressibility of Compacted Cohesive Soils

The compressibility of a soil is the relation between effective stress on the soil skeleton and the volume change. Cohesive soils vary in compressibility depending on the amount and character of the fines and the amount and gradation of the coarse particles they contain. The compressibility of a compacted cohesive soil depends on its density and its moisture content at the time of loading. In a series of tests made at the Bureau of Reclamation (1952) in the one-dimensional consolidometer (confined compression tests) a fine-grained soil, silt (ML) (LL 27.1 percent, PI 4.8 percent, $G_S = 2.68$, 97.2 percent passing No. 200 sieve), was compacted at various

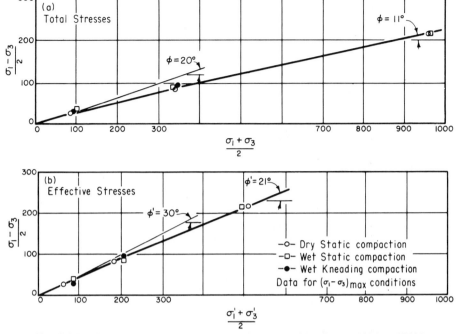

Fig. 8.15 Modified Mohr diagrams for Higgins clay. (*After Lee and Haley, 1968.*)

water contents to represent 4 percent dry of optimum, 2 percent dry of optimum, optimum, and 2 percent wet of optimum. Then each sample was subjected to loads of 37.5 psi, 150 psi, and 300 psi, with drainage permitted. While under the 300-psi load, the soil was saturated. The volume changes measured in these tests are given in Table 8.1.

Note that the placement moisture affected the compressibility more than did the dry density (compactness) of the soil. The significant increase in volume change that occurred when the 4 percent dry-of-optimum sample was saturated while under a 300-psi load is attributable to softening of the soil. This phenomenon of additional settlement on saturation (collapse) under load is characteristic of soils that are compacted considerably dry of optimum. Note that the sample compacted 2 percent dry of optimum did not compress additionally under the 300-psi load. However, other tests on this soil where saturation was done under loads of 150 psi and 75 psi showed a small amount of additional settlement on saturation (0.03 percent and 0.04 percent, respectively). This phenomenon does not occur with soils compacted at optimum water content and with soils compacted wet of optimum.

It has long been known that a compacted cohesive soil that had been placed at too dry a moisture content would undergo collapse when saturated under load. The lower limit of moisture control used by the Bureau of Reclamation was determined by a series of confined compression tests in which soil specimens were placed at various moisture contents and densities corre-

sponding to the laboratory compaction curve, and were subjected to various axial loads and saturated while under load. The criterion of no additional settlement on saturation for a given load (Holtz, 1948) was used to establish the lower moisture limit, although no satisfactory theory had been advanced to explain the collapse phenomenon or to provide a means of extrapolating laboratory results.

In Figure 8.16 (Holtz, 1948) are presented the data for determining the limiting moisture contents in terms of dry density, placement moisture, and applied pressures. The soil was a SC-CL with 50 percent clayey fines. The initial placement condition is represented by a 33-blow Proctor compaction curve (16 335 ft-lb/ft^3) which was estimated to match field compactive effort. The four solid lines above the compaction curve, labeled 25 psi, 100 psi, 175 psi, and 250 psi, represent consolidated densities obtained by drained confined compression tests versus placement moisture for those respective loads *before saturation*. The four dashed lines designated 25 psi, 100 psi, 175 psi, and 250 psi represent consolidated densities *after saturation*, without change of load. The vertical difference between the solid and dashed lines shows the magnitude of settlement on saturation or "collapse." The heavy solid line plotted through the intersection of the solid and dashed load lines represents the lower limit of placement moisture for the respective fill pressures if no additional settlement on saturation is to be permitted. The plot showed that soil to be subjected to overburden load of 100 psi should not be placed drier than 1.5 percent dry of optimum to avoid any tendency to collapse on saturation. The heavy dashed lines marked 10, 20, 30, 40, 50, and 60 represent pore pressure in percentages of total vertical pressure calculated from Boyles' and Henry's laws and Terzaghi's effective stress equation. In 1948 when this paper was presented, the effect of capillarity was neglected in the pore pressure calculations. The figure shows that if soil is placed 1.5 percent wet of optimum and subjected to a total load of 100 psi, it could develop pore pressure of 30 percent of the applied load if drainage did not occur during the construction period.

The compressibilities of the compacted soil used in Figure 8.16 for three placement conditions—1 percent dry of optimum moisture content (where there is no danger of collapse),

TABLE 8.1 CONFINED COMPRESSION OF A SILT AT VARIOUS PLACEMENT CONDITIONS.

| Placement Objective | γ_d | w | $S\%$ | Volume Change in Percent Under psi Loads | | | | |
				37.5	75	150	300	Sat. 300
4 percent dry	105.5	13.0	59.6	1.2	1.7	2.6	5.5	8.0
2 percent dry	108.3	15.2	74.8	1.4	2.1	3.4	5.6	5.6
Optimum	109.7	17.4	88.4	3.6	4.5	5.9	7.9	7.9
2 percent wet	107.4	19.4	93.4	3.8	5.0	6.5	8.5	8.6

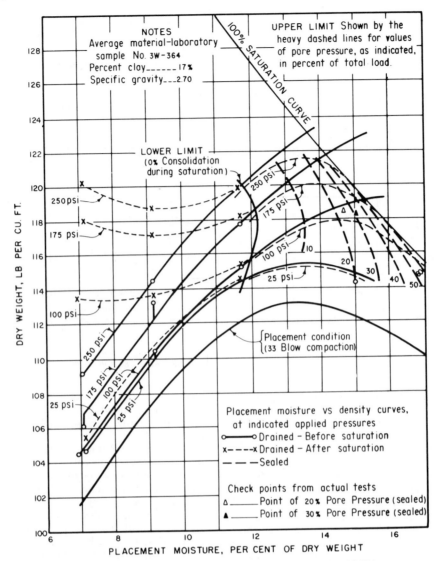

Fig. 8.16 Placement moisture control. (*After Holtz, 1948.*)

optimum, and 1 percent wet of optimum—are given in Table 8.2. Note that there is a significant increase in compressibility of the soil placed wet of optimum over that of the soil placed dry of optimum although the dry density is the same.

Hilf (1956) attempted an explanation of the collapse phenomenon by recognizing that the confined compression test actually subjects the soil specimen to shear stresses. In Figure 8.17a, circle 1 represents the effective stresses at a point in the soil at equilibrium under a given applied load with

drainage permitted. Point O represents the stresses in the soil after compaction but prior to loading. While the soil is changing volume under the applied load, its safety factor is unity. At equilibrium under an applied axial load its safety factor against collapse is indicated by the ratio of the ordinate of point M, on the potential failure plane, to the ordinate of a point vertically above M on the Mohr envelope. When the soil is saturated under constant applied axial load, the capillary pressure u_c, which is always negative while the soil is not fully saturated,

TABLE 8.2 EFFECT OF PLACEMENT CONDITIONS ON VOLUME CHANGE IN CONFINED COMPRESSION FOR A SAND CLAY.

Effective Axial Load, psi	Placement Condition		
	1% dry of optimum (113.0 pcf)	Optimum (113.5 pcf)	1% wet of optimum (113.0 pcf)
25	1.9	1.9	2.3
100	2.9	3.7	4.8
175	5.2	5.9	7.6
250	6.8	7.8	9.0

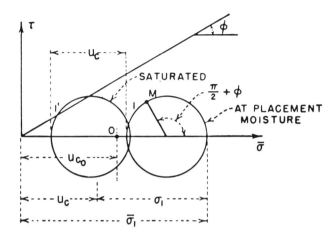

a. SOIL IS PLACED DRY OF OPTIMUM

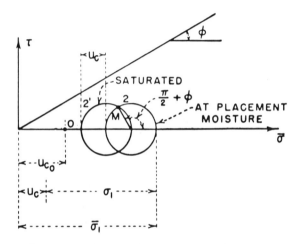

b. SOIL IS PLACED WET OF OPTIMUM

Fig. 8.17 Settlement on saturation in confined compression tests.

increases to zero. This is illustrated in Figure 8.17a by circle 1 being translated towards the origin by an amount equal to u_c, to give circle 1'. Additional settlement will occur in this case since circle 1' intersects the Mohr envelope. (Note that while Hilf equated the numerical value of u_c to effective stress, the same principle would hold if the effective stress equivalent of u_c and u_{c_0} in Figure 8.17 were reduced by Bishop's χ factor.)

A specimen placed wetter than optimum and compacted to the same density as the soil represented by circle 1 may be represented by circle 2 in Figure 8.17b. Its diameter will in general be less than that of circle 1 since σ_1 is the same and the absolute value of u_c is smaller than the corresponding value for the specimen compacted at a moisture content dry of optimum. Saturation of this soil will result in translation of circle 2 to circle 2', resulting in no failure. At higher densities near the peak of the compaction curve, the value of ϕ will increase, thereby providing a greater safety margin against settlement when the soil is saturated under applied load.

Barden et al. (1969) reported on laboratory tests made in a consolidation cell in which total stress, pore water pressure, and pore air pressure were measured on three soils varying in clay content from 10 to 15 percent. The test procedure permitted varying the stress path with change of volume. In lieu of using Bishop's effective stress parameter χ, an equation for one-dimensional compression proposed by Coleman (1962) was

considered:

$$\frac{dv}{v} = C_1 \, d(\sigma - u_a) + C_2 \, d(u_a - u_w)$$

where $(u_a - u_w) = -u_c$.

They concluded that the "general transition from a flocculent structure on the dry side of optimum to a dispersed structure on the wet side of optimum results in a very different compression behavior for clays compacted dry and wet of optimum." Coefficients C_1 and C_2 appeared to be "functions only of the current stresses." The main cause of stress path dependency appeared to be "a reversal in the direction of saturation, probably due to hysteresis between the saturation and desaturation processes." (Note: This hysteresis phenomenon was first proposed by Haines (1930), whose work on capillarity in soils was referred to by Hilf (1956)).

According to Barden et al. (1969), "The collapse mechanism is controlled by three factors: (a) a potentially unstable structure, such as a flocculent type associated with soils compacted dry of optimum, or with loess soils; (b) a high applied stress which further increases the instability; and (c) a high suction which provides the structure with a temporary rigidity and whose removal on wetting leads to collapse. The absence of any one of these three factors removes the possibility of significant collapse."

Bureau of Reclamation experience indicates that collapse can occur with condition (b) absent, as when dry, low-density fine-grained soils are wetted in a cut section of a canal that has not been preponded.

8.2.2 Cohesionless Soils

Because these soils are relatively pervious even when compacted, they are not affected significantly by their water content during the compaction process. Consequently, the peaked curved relationship between dry density and water content (Proctor curve) that is characteristic of all cohesive soils is ill defined or nonexistent for clean sands and gravels. For a given compactive effort on the latter soils, the dry density obtained is high when the soil is completely dry and high when the soil is completely saturated, with somewhat lower densities occurring when the soil has intermediate amounts of water. The explanation for this involves the phenomenon of bulking in sands where small capillary stresses in the partly saturated soil tend to resist the compactive effort. This bulking phenomenon is not present in completely dry sand and disappears when the moist sand is saturated.

For these soils, where the Proctor curve concept is not applicable, the normally used compaction criterion is *relative density*, introduced by Terzaghi (1925), defined by

$$D_r = \frac{e_{max} - e}{e_{max} - e_{min}} \tag{8.3}$$

where

e_{max} = void ratio of the soil in its loosest state
e = void ratio of the soil being tested
e_{min} = void ratio of the soil in its densest state
D_r = relative density, usually expressed as a percentage

Since void ratio is related to dry density for a given specific gravity, Equation 8.3 can be written

$$D_r = \frac{\gamma d_{max}(\gamma d - \gamma d_{min})}{\gamma d(\gamma d_{max} - \gamma d_{min})} \tag{8.4}$$

where

γd_{max} = dry density of the soil in its densest state
γd_{min} = dry density of the soil in its loosest state
γd = dry density of the soil being tested

Relative density can also be expressed in terms of porosity:

$$D_r = \frac{(n_{max} - n)(1 - n_{min})}{(n_{max} - n_{min})(1 - n)} \qquad (8.5)$$

where

n_{max} = maximum porosity (loosest state)
n_{min} = minimum porosity (densest state)
n = porosity of the soil being tested

Terzaghi (1925) and Bjerrum et al. (1960) considered it possible to judge whether a sand is deposited in a loose or dense state only on the basis of its relative density and its compactibility. Terzaghi defined the ranges of relative denseness

TABLE 8.3 COMPACTIBILITY (F) OF COHESIONLESS SOILS
(where $F = (e_{max} - e_{min})/e_{min}$).

Unified Classification	γ_{min}	γ_{max}	e_{min}	e_{max}	Max. size	D_{10}	C_u	C_c	F
SP-SM	90	108	0.54	0.84	#16	0.058	6.0	2.2	0.555
SM	75	97	0.83	1.36	$\frac{3}{4}''$	0.0065	31	5.5	0.638
SP	92	112	0.48	0.80	#4	0.15	3.0	0.93	0.667
SP	93	113	0.46	0.77	$1\frac{1}{2}''$	0.16	2.4	0.92	0.674
SP	95	116	0.43	0.74	#4	0.30	3.7	1.0	0.721
SP-SM	92	113	0.46	0.80	$\frac{3}{4}''$	0.08	3.0	0.88	0.739
SP	85	107	0.54	0.94	#30	0.10	2.3	1.3	0.740
SP	97	118	0.40	0.70	$1\frac{1}{2}''$	0.11	3.2	1.2	0.750
SP	99	120	0.38	0.67	$1\frac{1}{2}''$	1.8	4.4	0.76	0.763
SM-ML	83	108	0.62	1.11	#4	0.012	8.3	1.5	0.790
SP-SM	79	103	0.60	1.08	#30	0.09	2.4	1.5	0.800
SP	103	124	0.33	0.60	$\frac{3}{8}''$	0.17	5.0	0.75	0.818
SM	105	126	0.31	0.57	5″	0.02	350	0.30	0.838
SP-SM	87	112	0.48	0.90	#4	0.08	3.0	1.3	0.875
SM	82	108	0.54	1.02	#16	0.023	6.5	1.4	0.889
SW-SM	95	119	0.39	0.74	3″	0.05	10	1.4	0.897
SP	98	122	0.36	0.69	#4	0.37	5.1	1.2	0.917
SW-SM	98	125	0.34	0.71	3″	0.07	6.8	1.0	1.088
SP-SM	97	124	0.33	0.70	$\frac{3}{4}''$	0.10	5.0	1.4	1.121
SP-SM	84	115	0.44	0.97	$1\frac{1}{2}''$	0.085	4.7	1.4	1.205
SP-SM	94	123	0.34	0.76	$1\frac{1}{2}''$	0.12	4.4	1.3	1.235
SM	99	128	0.31	0.70	3″	0.02	240	1.8	1.258
SP-SM	80	114	0.44	1.06	#16	0.07	3.7	1.6	1.409
SW-SM	80	116	0.42	1.07	$1\frac{1}{2}''$	0.074	6.6	2.4	1.547
SM	83	120	0.38	0.99	#4	0.015	26	6.1	1.605
SM	102	134	0.23	0.62	$\frac{3}{4}''$	0.01	120	1.9	1.695
GN-GM	113	127	0.31	0.47	3″	0.14	86	1.2	0.517
GP-GM	112	129	0.32	0.52	3″	0.03	200	0.50	0.625
GW-GM	116	133	0.26	0.44	5″	0.17	171	2.2	0.692
GP-GM	110	128	0.30	0.51	3″	0.11	191	15	0.700
GP-GM	117	133	0.24	0.41	5″	0.125	160	4.0	0.708
GW-GP	111	130	0.27	0.49	3″	0.20	105	7.5	0.815
GP	116	134	0.23	0.43	5″	0.27	111	6.2	0.870
GW	119	139	0.24	0.45	3″	0.51	45	2.2	0.875
GW	120	139	0.20	0.39	3″	0.45	51	1.6	0.950
GW	119	139	0.21	0.41	3″	0.18	94	1.1	0.952
GW	111	132	0.25	0.49	3″	2.9	9.7	1.8	0.960
GP	115	136	0.22	0.44	5″	0.38	29	0.61	1.000
GP	114	135	0.22	0.45	3″	2.0	11	0.77	1.045
GW-GM	121	141	0.19	0.39	3″	0.30	77	2.3	1.052
GM	122	141	0.17	0.36	$1\frac{1}{2}''$	0.025	381	3.0	1.118
GW-GM	114	137	0.21	0.45	3″	0.60	16	1.2	1.143
GW	112	138	0.20	0.48	3″	2.0	12	1.3	1.400
GW	109	137	0.21	0.52	3″	2.0	14	2.6	1.476
GP	114	140	0.18	0.45	3″	1.7	10	0.76	1.500
GM	101	132	0.25	0.64	$1\frac{1}{2}''$	0.03	260	12	1.560
GW-GM	111	139	0.19	0.49	3″	1.8	13	2.3	1.578
GP	115	142	0.17	0.44	3″	0.31	87	8.2	1.588
GW	123	146	0.13	0.34	3″	0.21	124	1.1	1.615
GW-GM	110	139	0.19	0.50	5″	0.42	43	2.1	1.631
GW-GM	115	142	0.17	0.45	3″	0.15	133	1.1	1.647
GP-GM	112	140	0.18	0.48	3″	0.42	26	4.2	1.667
GW-GM	112	140	0.18	0.48	5″	0.25	56	1.0	1.667
GW-GM	114	142	0.16	0.45	3″	1.2	15	1.7	1.812
GP	112	141	0.17	0.48	3″	1.4	7.1	0.73	1.823
GW-GM	118	147	0.12	0.40	3″	1.3	19	1.1	2.333

of sand as follows

$0 < D_r < \frac{1}{3}$ Loose sand
$\frac{1}{3} < D_r < \frac{2}{3}$ Medium compact sand
$\frac{2}{3} < D_r < 1$ Dense sand

He defined compactibility as

$$F = \frac{e_{max} - e_{min}}{e_{min}} \qquad (8.6)$$

In well-graded cohesionless soils such as SW or GW, $e_{max} - e_{min}$ is large and e_{min} is small; hence F is large. These soils are easily compacted.

In uniform soils such as certain types of SP and GP, $e_{max} - e_{min}$ is small and e_{min} is large; hence F is small and the soil more difficult to compact. Table 8.3 lists compactibilities for a variety of soils.

Burmister (1948) showed that the relative density of noncohesive soils was a more significant parameter than dry density alone insofar as engineering properties of the soil are concerned. His work has been verified and extended by other investigators (D'Appolonia, 1970).

As indicated in Equations 8.3 and 8.4, determination of the relative density of a soil requires measuring its dry density in place, its dry density in the loosest state, and its dry density in the densest state (or the three corresponding void ratios). The density in place and minimum density (loosest state) present no particular difficulty, but a generally accepted method of determining the maximum density (densest state) of all cohesionless soils has not yet been found (see Section 8.4).

Zolkov and Wiseman (1965) studied uniform, fine subangular quartz beach sands. They used ASTM D 1557-58T (10-lb hammer) to obtain maximum density, frequently at 0 percent compaction moisture. They point out that a percentage of maximum density can be very misleading, for if $\gamma_{max}/\gamma_{min} = 1.25$, 80 percent $\gamma_{max} = \gamma_{min}$. The quantity $\gamma_{max}/\gamma_{min}$ varied from 1.17 to 1.35. For uniform spheres, $\gamma_{max}/\gamma_{min} = 1.4$. Figure 8.18 shows the relationship between percentage of maximum density to relative density D_r for various ratios of maximum to minimum densities.

Figure 8.19 from the Bureau of Reclamation *Earth Manual* (1968b) shows maximum and minimum densities of typical sand and gravel soils.

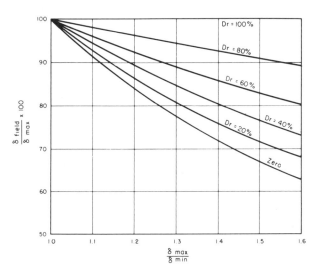

Fig. 8.18 Relative density, maximum and minimum density relationships. (*After Zolkov and Wiseman, 1965.*)

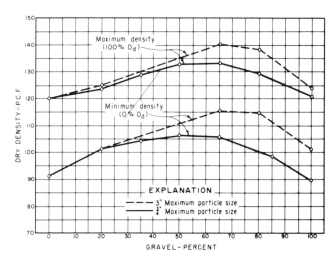

Fig. 8.19 Maximum and minimum densities of typical sands and gravel soils (101-D-173). (*After Bureau of Reclamation* Earth Manual, *1968b.*)

Compressibility of Compacted Sands and Gravels Under Static Loads

The relationship between relative density and volume change under load is illustrated in Figure 8.20 for a Platte River medium to coarse clean sand (SW). The sand had a maximum density of 124.0 pcf, a minimum density of 92.5 pcf, and a specific gravity of 2.63.

Tests were made in 1956 by the Bureau of Reclamation in the one-dimensional consolidation apparatus with the sand in a wet condition. These confined compression tests show that this sand is more than twice as compressible at a relative density of about 40 percent than it is at a relative density of 70 percent.

The effect of relative density on compressibility is accentuated at higher loads; for example, under a 25-psi load the consolidation (volume change) is 0.86 percent for a relative density of 73.1 percent and 1.86 percent for a relative density of 39.4 percent, but under a load of 200 psi, the consolidations are 1.7 and 5.1, respectively, for those relative densities.

Terzaghi and Peck (1967) reported similar results for compressibility of confined layers of loose and dense sand and showed that the shape of the particles affect volume change under load; sands with flat particles (sand–mica mixtures) are more compressible than sand alone. Crushing of the sand grains appeared to occur at pressures of about 100 kg/cm^2. Roberts and DeSouza (1959) made high-pressure (up to 14 000 psi) confined compression tests on well-rounded uniform quartz sand. They concluded that at sufficiently high pressures, sand may be more compressible than clay, primarily owing to crushing and fracturing of individual sand grains. The pressure at which breakdown occurs depends on the angularity and initial void ratio of the sand.

The foregoing results were for confined compression in which volume change is due to vertical movement only. The results of these studies show that sand is relatively incompressible at low pressures, that at high pressures there can be considerable volume change due to crushing of grains, and that compression can continue for a considerable period of time.

Lee and Seed (1967a) reported on tests on saturated washed Sacramento River sand (0.297 in to 0.149 in, subangular to subrounded, $G = 2.68$, $e_{min} = 0.61$, $e_{max} = 1.03$) in triaxial compression. Samples (1.4-in diameter, 3.4-in high) were prepared at different initial void ratios and confined under a seating load of 0.7 psi, after which volume changes were

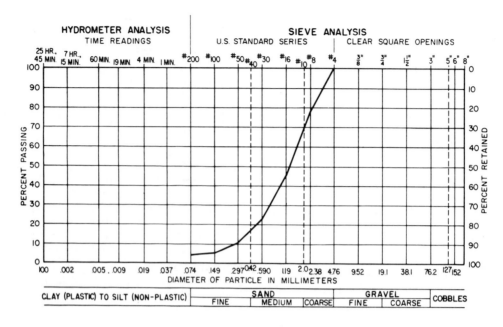

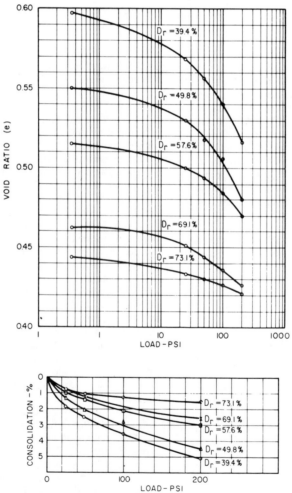

Fig. 8.20 Compressibility of a clean medium sand at various relative densities.

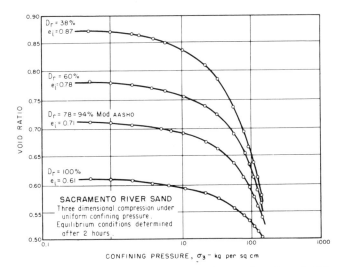

Fig. 8.21 Pressure–void ratio curves for sand at four initial densities. (*After Lee and Seed, 1967a.*)

observed for all pressures above this load. In all cases the samples drained immediately on application of load (excess pore pressure dissipated) but continued to change volume significantly with time for about 2 hr. Figure 8.21 shows the results of these tests at four initial densities. Even the dense sand compressed significantly under static all-around pressure.

The compression characteristics of cohesionless materials of large size, including rockfills, were studied by Marachi et al. (1969). The effect of particle size refers to the possible differences in deformation characteristics of soils similar in every respect except particle size. Similarity or modeling requires that the gradation curves be parallel and the particle shape be the same. Triaxial compression tests were made on specimens up to 36 inches in diameter using field samples up to 6 inches in size. Comparisons were made with modeled specimens 12 inches in

diameter and 2.8 inches in diameter. Figures 8.22 to 8.25 show the gradations and isotropic compression results for three materials tested. It was concluded that modeling of rockfill materials does not materially affect their isotropic consolidation characteristics.

Figure 8.25 shows that the Oroville Dam material, consisting of rounded to subrounded particles, was less compressible than the Pyramid Dam material and crushed basalt, which were angular in shape. This appears to substantiate Casagrande's (1965) conclusion that rockfill materials composed of well-graded and well-rounded particles are superior in their mechanical properties to uniform, angular rockfill materials and thus are more suitable for use in high dams. However, the Oroville gravel was denser than the angular materials tested, which undoubtedly contributed to the smaller compressibility of the gravels.

These authors also reviewed previous work on particle breakdown during compression and shear and tested the effect of models of grain-size distribution on crushing of rockfill materials. They concluded that all materials tested underwent some degree of breakdown and that shear caused more break-down than isotropic compression. The amount of particle breakage increased as the maximum particle size increased, as would be expected from Griffith's crack theory (1921, 1924). This theory indicates that all materials that undergo brittle-type fracture under the action of stress contain innumerable cracks and flaws, and that the probability of occurrence of inherent flaws and cracks increases with increase in size, making the tensile strength of the material decrease with increasing size.

Kjaernsli and Sande (1966) made confined compression tests on various materials in a 500-cm diameter, 25-cm high compressometer up to 500 metric tons per square meter. Crushed angular syenite, gneiss, and limestone, rounded gravel, and subrounded gravel were tested. They concluded that compressibility is decreased by increased hardness of rock and is smaller for rounded smooth-surfaced aggregates than for angular rough-surfaced aggregates. For a given type of aggregate, the compressibility is decreased by increasing relative

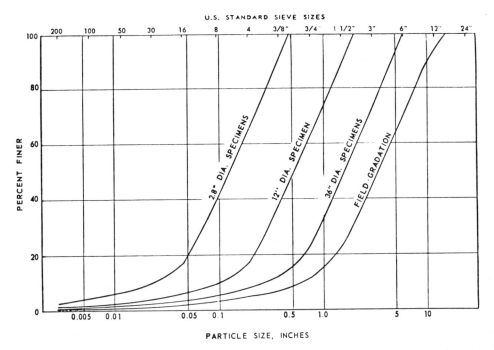

Fig. 8.22 Grain-size distribution for the modeled Pyramid Dam material. (*After Marachi et al., 1969.*)

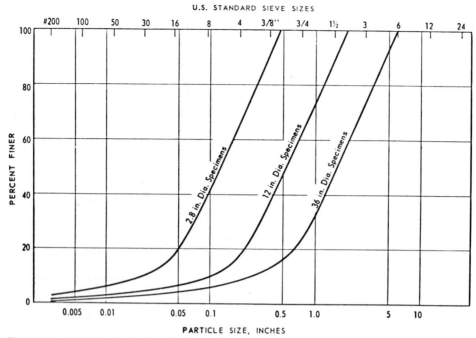

Fig. 8.23 Grain-size distribution for the modeled crushed basalt rock. (*After Marachi et al., 1969.*)

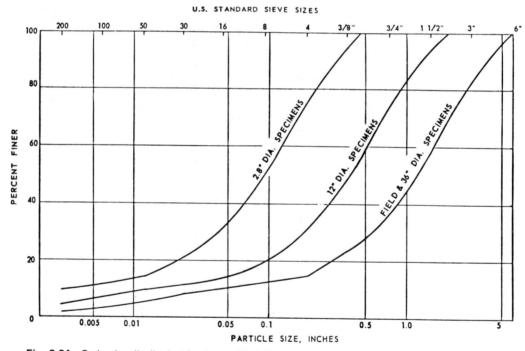

Fig. 8.24 Grain-size distribution for the modeled Oroville Dam material. (*After Marachi et al., 1969.*)

density, and is smaller for well-graded than for uniform-sized particles. The following practical conclusions are given:

Volume compression tests on specimens of aggregates have shown that the compressibility of aggregates of even hard rocks can be rather high and higher than for a morainic material with water content and density corresponding to approximately Proctor Optimum conditions. Minimum compressibility might be obtained by placing natural gravel, generally rounded and smooth-surfaced and broadly graded, in sufficiently thin layers to avoid segregation which would lead to a narrow grading, and by moistening the fill during compaction. By the same method of placing, tunnelspoil of good rock, which usually is relatively broadly graded, should also give a relatively good fill from the point of view of small settlements and deformations. Broadly graded material such as gravel and tunnelspoil will, when placed in high lifts, lead to segregation and low density, and therefore to rather large settlements and deformations. Following these thoughts, material placed in high lifts should possibly be of big size and rather narrowly graded, and should also be wetted or sluiced during dumping. The settlements and deformations of fills placed in high lifts must, however, be expected to be relatively larger anyhow.

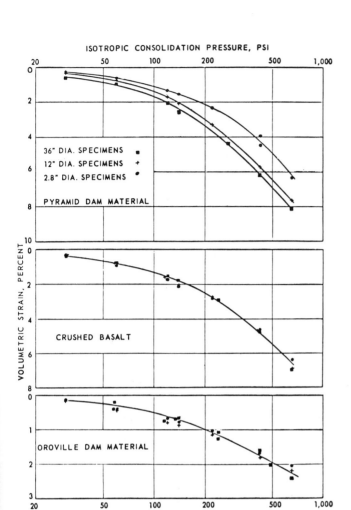

Fig. 8.25 Isotropic consolidation of molded rockfill materials. (*After Marachi et al., 1969.*)

Shear Strength of Compacted Sands and Gravels

Loose sands and gravels are known to have less resistance to shear than the same soils in a dense state. The relationship between relative density and the angle of internal friction (ϕ) is illustrated by Figure 8.26.

The values of ϕ in this figure represent those corresponding to the peak strength, which is considerably higher than the value calculated from ultimate strength of compacted sands and gravels, which expand during shear. Loose sands do not exhibit this peak during shear. Burmister (1948) pointed out that because of the possibility of progressive failure in a backfill or embankment, use of ϕ corresponding to the peak value is not warranted. He suggested that only 50 to 75 percent of the increase of the value of ϕ over its value at $D_r = 0$ be used depending on the amount of restraint against progressive failure. For example, in Figure 8.26, the coarse to fine sand with $\phi = 34.5°$ at $D_r = 0$ shows an increase of 11.5° to $\phi = 46°$ at $D_r = 100$; hence the range to be used would be 40° to 43° for $D_r = 100$ where progressive failure is possible.

Zolkov and Wiseman (1965) reported that for beach and dune sands measured angles of internal friction from triaxial shear tests varied only slightly (from 30° to 32°) at relative densities close to zero, but increased to a range of from 36° to 45° at relative densities close to 100 percent. These authors found that a plot of tan ϕ versus void ratio as in Figure 8.27 had less scatter. Such a plot may be found useful for particular

sands, but cannot be generalized, since Sand A, for example, in Figure 8.26 had values of $e = 1.23$ for $D_r = 100$ percent, $\phi = 38°$ and $e = 1.88$ for $D_r = 0$, $\phi > 27.5°$.

On the other hand, the values of 36° to 45° at $D_r = 100$ mentioned by Zolkov and Wiseman, although somewhat greater than the general trend of results, are not inconsistent with other data. Considering the nonuniformity of compaction procedures used to obtain $D_r = 100$ percent and differences in test procedures in obtaining values of ϕ, it appears that the relative density concept is a reasonable approach to obtaining desired shear strength of sands.

The stress–strain–volume characteristics of dense (Fig. 8.28) and loose (Fig. 8.29) sand were investigated by Lee and Seed (1967a). As illustrated in Figure 8.28, at low confining pressures the sand expands during shear and it exhibits a peak strength at low values of axial strain. Increasing the confining pressure has three effects: it reduces the brittle characteristics of the stress–strain curve, it increases the strain to failure, and it decreases the tendency to dilate. In fact, if the confining pressure is high enough, then even though the sand may have been prepared to 100 percent relative density, and further densified by the effect of the all-around pressure, the volume changes during shear decrease continuously. The stress–strain–volume change characteristics of dense sand at high confining pressures are not unlike those of loose sand at low pressures.

The results of a similar series of drained triaxial tests on loose sand are shown in Figure 8.29. The pattern is similar to that for dense sands, except that at low pressures the tendency for dilation is not so strong as for dense sands, while at high pressures the tendency for compression is greater.

These authors concluded that dense sands dilate when sheared at low confining pressures and exhibit a brittle-type stress–strain curve. When sheared at high confining pressures, the same dense sand becomes compressible and exhibits a more plastic stress–strain relationship with high strains at failure.

The drained shearing resistance of sand appears to be governed by three components: sliding friction, dilatancy and particle crushing, and rearranging. Friction may be assumed to be essentially constant, though it may vary slightly with changes in confining pressure and crushing of particles. At low pressures dilatancy causes a significant increase in angle of friction and accounts for the steep failure envelopes commonly observed for dense sands. Crushing becomes progressively more effective with increasing confining pressure. At medium pressures it partly offsets the reduced effects of dilatancy, but not sufficiently to prevent a flattening of the failure envelope due to the reduction in dilatancy effects. At high pressures, particle crushing and rearranging requires considerable energy, and, together with a possible small increase in friction, causes the failure envelope to cease flattening and to continue to rise at a constant or perhaps a slightly increasing slope.

According to Lee and Seed (1967a), sand at any initial density can be made to shear with no volume changes at failure by testing at the appropriate critical confining pressure. For Sacramento River sand, even samples compacted to the maximum laboratory densities could be prevented from dilating by confining pressures of about 16 kg/cm². For Ottawa standard sand, however, the samples still dilated in tests at confining pressures as high as 100 kg/cm². The crushing strength of the individual grains of a cohesionless material thus has a large influence on the strength mobilized at constant volume.

For Sacramento River sand, the angle of friction determined by drained tests showing no volume change at failure was found to be essentially independent of confining pressure.

The effect of method of test on the shear strength of sands has been reported by Marachi et al. (1969), who conducted plane strain shear tests as well as triaxial shear tests on sands. They found that the angle of internal friction of the sand is not

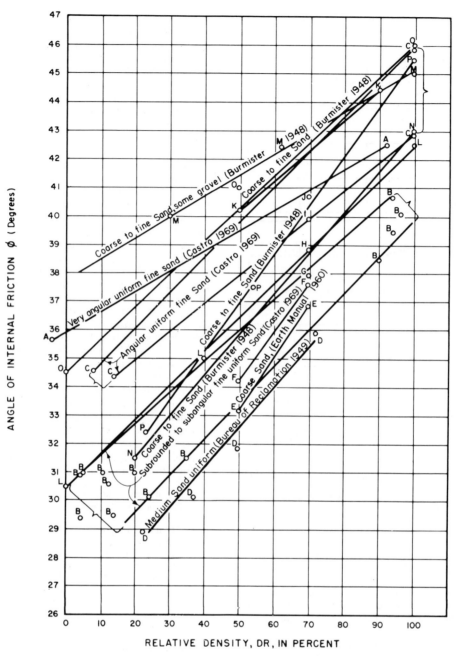

Fig. 8.26 Relative density vs. friction angle for cohesionless soils.

affected by specimen shape or size under plane strain conditions. For very dense sand, ϕ was 7° greater for plane strain shear than for triaxial shear. This reduced to about 3° for loose sand specimens and indicated that at $D_r = 0$ there would be no difference in ϕ.

These authors made tests of large rockfill specimens (36-in diameter), intermediate-sized rockfill specimens (12-in diameter), and small specimens (2.8-in diameter), all with parallel grain size curves and the same material (same shape of grains). Their results indicated that the angle of internal friction of the smallest particles (2.8-in diameter specimen) was 3 to 4 degrees higher than that of the large-sized particles (up to 6 in) in the 36-in diameter specimen, regardless of confining pressure or material type. (Three quite different types of materials were tested.) The intermediate-sized material (12-in diameter specimen) had

angles of internal friction 1 to 1½ degrees lower than the smallest material.

Pertinent data from the large rockfill tests including tests reported by Marsal (1965, 1967a, b) are compiled in Table 8.4. The table lists values of angle of internal friction, axial strain at failure, and volumetric strain at failure for those tests conducted at a confining pressure of 350 psi for which information was available. It may be observed that for the dredger tailings, which contain rounded particles, the values of axial and volumetric strains at failure are 6.5 and 1.5 percent, respectively, while for all of the angular materials these values are greater than 13 and 5.5 percent, respectively. The values of the volumetric strain at failure for all of the angular materials for which data were available ranged between 5.5 and 13 percent. Among these materials, granitic gneiss and shale showed the

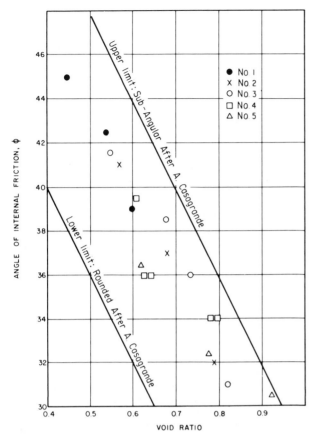

Fig. 8.27 Angle of internal friction vs. void ratio. (*After Zolkov and Wiseman, 1965.*)

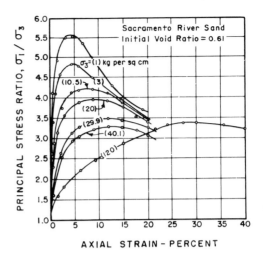

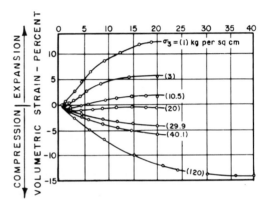

Fig. 8.28 Stress–strain–volume change data for dense sand. (*After Lee and Seed, 1967a.*)

greatest volumetric compression at failure. The volumetric compression at failure for the rest of the angular materials, however, ranged between 5.5 and 6.5 percent.

Seed and Lee (1966) demonstrated that the critical confining pressure (the value of confining pressure for which there is no volume change at failure) is a significant characteristic of a soil and that the undrained strength of noncavitating samples (where the negative pressure does not exceed one atmosphere) can be determined with a satisfactory degree of accuracy from a knowledge of the critical confining pressure of the soil and the angle of friction corresponding to constant volume conditions. Values of the critical confining pressure for the rockfill materials are estimated and listed in Table 8.4. As can be seen, the critical confining pressure for a rockfill material composed of rounded particles, that is, dredger tailings, is substantially greater than that for the rockfill materials composed of angular particles. Among the angular materials the well-graded basalt (Marsal, 1967a) had the highest critical confining pressure. The granitic gneiss and shale material appeared to have the lowest critical confining pressure.

The pertinent data for the high-pressure tests made by Marachi et al. (1969) with the intermediate-sized specimens are compiled in Table 8.5. The conclusions to be drawn from this table are very similar to those from Table 8.4. The values of the angle of internal friction range between 37° to 41°, if data for the uniform 1-in to $\frac{3}{4}$-in quartzite are excluded. The values of the axial and volumetric strains at failure are lower for the better-graded materials composed of rounded and hard rocks than for the uniformly graded or angular materials. The values of the critical confining pressures are also highest for the dense sand and gravel and the dredger tailings materials that are composed of rounded particles.

From the above discussions the authors concluded that in general the variations in the angles of internal friction for the rockfill materials investigated to date are not more than about 8 degrees. Within this range, the well-graded materials containing rounded particles show the highest angles of internal friction. These materials also show the lowest values of axial and volumetric strains at failure and the highest critical confining pressure. These observations are in agreement with Casagrande's (1965) conclusion that well-graded materials are more suitable for high dams than narrowly graded materials.

"Liquefaction" of Saturated Sands and Gravels

One type of landslide is the "flowing" of natural slopes of saturated sand, which, when triggered by relatively minor forces, resembles a heavy liquid and from initial slopes of about 20° may come to rest on flat slopes of 4° or less. Hundreds of such slides have occurred in the last 200 years in Zeeland, Holland (Koppejan et al., 1948). Similar "flows" of natural sand banks have been reported along the Mississippi River (Waterways Experiment Station, 1967) and in hydraulic fills (Middlebrooks, 1942; Hazen, 1920). Common characteristics of the sands involved in flow slides are uniform gradation, rounded grains, and loose state of compaction. Seed (1968) in the ASCE 1967 Terzaghi lecture listed slides associated with 37 earthquakes in which liquefaction has played a part. Saturated sandy soils in a loose to medium dense condition were liquefied during earthquakes varying in magnitude from 5.5 to 8.5 (Richter scale) and at epicentral distances varying from several miles to

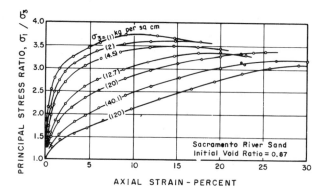

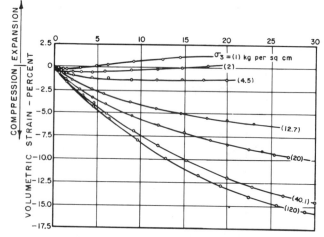

Fig. 8.29 Stress–strain–volume change data for loose sand. (*After Lee and Seed, 1967a.*)

hundreds of miles. Flow slides have occurred in large masses of fairly dry granular soils and in loess (Terzaghi, 1950). In all of these events, the phenomenon of pore pressure, either pore water pressure in the saturated sands or pore air pressure in the dry soils, appears to play a prominent role.

In 1935 Casagrande (1936) introduced the hypothesis of critical density or critical void ratio of sand with relation to its volume change characteristics during shear. His initial concept was that at this critical density, a cohesionless soil could undergo any amount of deformation or actual flow without volume change. This critical void ratio could be reached either from a loose state wherein the density increases (the soil compresses) during shear or from a dense state wherein the density decreases (the soil dilatates). Subsequently Casagrande (1938) realized that the critical void ratio was not a property of a given sand but that it decreases as the confining pressure applied to it increases. Roscoe and Schofield (1958) were able to reach a critical void ratio starting from either a loose or dense state using drained tests in a simple shear device. They also found that the critical void ratio decreased with increasing confining pressure.

Casagrande's and Roscoe's work considered shear deformations of a static nature. Since flow slides are known to have occurred during earthquakes, several investigators, starting with Maslov (1957), have studied the effect of vibrations on a saturated sand. Seed and his collaborators made a comprehensive study of controlled cyclic loadings on deformation and pore pressures developed in saturated sands. They tested at various void ratios a clean, uniform, subangular to subrounded quartz and feldspar sand of specific gravity 2.68 known as Sacramento River sand of grain size between 0.297 mm and 0.149 mm with

maximum and minimum void ratios of 1.03 and 0.61, respectively. Lee and Seed (1967b) considered four degrees of failure for these cyclic tests: (1) initial liquefaction, (2) partial liquefaction, (3) complete liquefaction, and (4) failure. They determined quantitative effects of initial void ratio, confining pressure, magnitude of cyclic stress, number of stress cycles, and failure criterion on the susceptibility of a sand to liquefaction and the development of large shear strains. Test conditions were designed to represent earthquake ground motions on soil below level ground when no significant drainage could occur during shaking. They concluded that a complex relationship among the variables determined the susceptibility of a sand to liquefaction.

At relative densities below about 50 percent, initial liquefaction and complete liquefaction occurred together; however, at relative densities above about 70 percent, a considerable number of stress cycles was required after initial liquefaction to develop large strain amplitudes.

An approximate linear relationship between the relative density and the stress was required to cause initial liquefaction in a given number of stress cycles. At relative densities below about 50 percent, the same relationship was applicable to the stress required to cause failure. However, small increases in relative density above 60 percent resulted in substantial increases in the cyclic stress required to cause failure in a given number of cycles, and the magnitude of this effect increased with the relative density.

At all densities, the cyclic stress required to cause initial liquefaction in a given number of stress cycles increased almost linearly with increase in confining pressure; the relationship between the cyclic stress required to cause 20 percent strain amplitude in 100 stress cycles and the confining pressure was also essentially linear for samples with relative densities less than about 80 percent.

For all densities and confining pressures, the relationship between the cyclic deviator stress and the logarithm of the number of stress cycles required to cause failure is approximately linear over a range from 5 to 200 stress cycles, with the stress required to cause failure decreasing as the number of stress cycles increases.

At all densities, the cyclic stresses required to induce liquefaction or failure are much less than those required to cause failure in static drained or static undrained tests conducted at the same confining pressure.

Casagrande (Castro, 1969) considered that the terms for different kinds of liquefaction used by Maslov (1957), Seed and Lee (1966), and Lee and Seed (1967b) were descriptive of a different phenomenon from the liquefaction that produces flow slides. Casagrande suggested that the term "cyclic mobility" be used for the behavior reported by Maslov and by Seed and collaborators. Under Casagrande's guidance, Castro (1969) investigated the phenomenon of liquefaction in the laboratory using triaxial $\bar{R}$ (consolidated–undrained) tests on three sands at various void ratios.

Liquefaction failure was taken as the sudden and continuous increase in axial deformation (say from 1 percent to 25 percent) in a fraction of a second caused by a sudden and substantial increase in pore pressure, which indicates a substantial change in the arrangement of the sand grains. This continuous yielding at constant void ratio and constant resistance corresponds to the concept of critical void ratio, as originally defined by Casagrande.

The concept of "limited liquefaction" was defined by the occurrence of liquefaction except that the rapid deformation stopped by itself at about 19 percent axial strain. In Figure 8.30 are plotted the combinations of void ratio e_c (or relative density D_{rc}) and effective confining pressure $\bar{\sigma}_c$ for all $\bar{R}$ (consolidated–undrained) tests on one of the sands tested, using different

TABLE 8.4 STRENGTH AND DEFORMATION CHARACTERISTICS OF LARGE ROCKFILL SPECIMENS.
   Confining Pressure = 350 psi.

Dam or Place	Material	Gradation	Particle Shape	ϕ (deg)	ε_{lf}, percent	ε_{vf}, percent	σ_{3crit}, psi	Reference
Oroville	Dredger tailings	Well-graded 6"–fines	Rounded	43	6.5	1.5	120	(1)
Pinzandaran	Sand and gravel (dry)	Well-graded 8"–fines	Rounded	39	8	4.7	60	(3)
San Francisco	Basalt	Well-graded 8"–$\frac{1}{4}$"	Angular	39	15	6	60	(2)
San Francisco	Basalt	Well-graded 3"–$\frac{1}{4}$"	Angular	38				(3)
San Francisco	Basalt	Poorly-graded 6"–fines	Angular	37	20	6.5	40	(1)
Malpaso	Conglomerate (dry)	Well-graded 8"–fines	Angular	37	13	4.5	20	(3)
El Infiernillo	Silicified conglomerate (dry)	Poorly-graded 8"–fines	Angular	36.5	14	5.5	30	(4)
Pyramid	Argillite	Poorly-graded 6"–fines	Angular	36.5	20	5.5	25	(1)
El Infiernillo	Diorite (dry)	Poorly-graded 8"–fines	Angular	35	15	10	25	(3)
El Granero	Shale[a]	Well-graded 8"–$\frac{1}{4}$"	Angular	35	>14	>10	10	(3)
El Granero	Shale[a]	Poorly-graded 8"–$\frac{3}{4}$"	Angular	33	>14	>10	5	(3)
Mica	Granitic gneiss[a]	Well-graded 8"–fines	Angular	32	>14	6	20	(2)
Mica	Granitic gneiss[a]	Poorly-graded 8"–$1\frac{1}{3}$"	Angular	25	14	10	5	(2)

[a] Tests were not carried to failure.
(1) Marachi et al. (1969).
(2) Marsal (1967a).
(3) Marsal (1967b).
(4) Marsal (1965).

TABLE 8.5 STRENGTH AND DEFORMATION CHARACTERISTICS OF INTERMEDIATE-SIZE ROCKFILL SPECIMENS.

Dam or Place	Material	Gradation	Particle Shape	σ_3, psi	ϕ (deg)	ε_{lf}, percent	ε_{vf}, percent	σ_{3crit}, psi	Reference
Furnas Dam, Trans'n.	Quartzite	Well-graded 1"–fines		510	42	7	2	80	(1)
Oroville	Dredger tailings	Well-graded 2"–fines	Rounded	500	40	8	1.5	110	(2)
El Infiernillo	Silicified conglomerate (dry)	Well-graded 1$\frac{1}{2}$"–$\frac{1}{6}$"	Angular	350	40.8	14.5	5.6	50	(3)
Mica	Sandy gravel ($D_r = 87$ percent)	Well-graded 3"–fine sand	Rounded	450	39.5	7	2	140	(1)
Furnas Dam, Trans'n.	Quartzite	Well-graded $\frac{3}{8}$"–fines		510	39	7	2	80	(1)
Furnas	Hard quartzite	Well-graded $\frac{3}{8}$"–fines		510	39	12	3	40	(1)
Mica	Granitic gneiss	Well-graded 1$\frac{1}{2}$"–$\frac{1}{6}$"	Angular	350	39.3	14.5	5.3	20	(3)
San Francisco	Basalt	Poorly-graded 2"–fines	Angular	500	38	19	8	30	(2)
Pyramid	Argillite	Poorly-graded 2"–fines	Angular	500	37	19	8	25	(2)
Mica	Sand gravel ($D_r = 50$ percent)	Well-graded 3"–fines	Rounded	450	37	18	5	25	(1)
Furnas	Hard quartzite	Uniformly-graded 1"–$\frac{3}{8}$"	Angular	510	34	>20	13	25	(1)

(1) Casagrande (1965).
(2) Marachi et al. (1969).
(3) Marsal (1965).

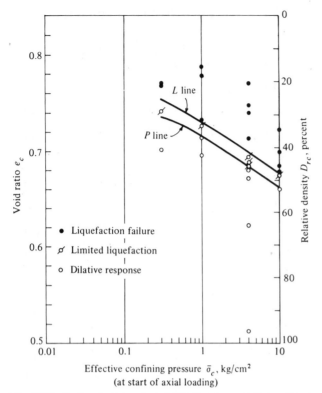

Fig. 8.30 Types of stress–strain behavior observed during *R* tests on sand B. (*After Castro, 1969.*)

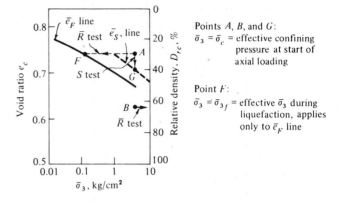

Points *A*, *B*, and *G*:
$\bar{\sigma}_3 = \bar{\sigma}_c$ = effective confining pressure at start of axial loading

Point *F*:
$\bar{\sigma}_3 = \bar{\sigma}_{3f}$ = effective $\bar{\sigma}_3$ during liquefaction, applies only to $\bar{e}_F$ line

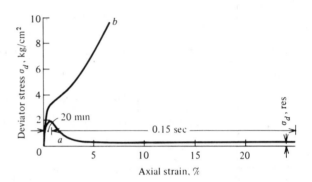

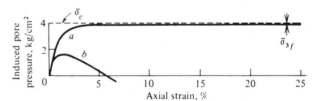

Fig. 8.31 Summary of liquefaction of sands. (*After Castro, 1969.*)

symbols for different types of stress–strain behavior. The borderline *L* separates the stress–strain behavior of liquefaction (above the line) from the "limited liquefaction" stress–strain behavior. The borderline *P* separates the latter from the cases where the stress–strain behavior is dilative and the resistance to shear increases. The plot shows that the type of stress–strain behavior, that is, liquefaction, limited liquefaction, or dilative, is a function of both the void ratio e_c and the effective confining pressure $\bar{\sigma}_c$. For example, a specimen of sand B with a void ratio of 0.70 ($D_r = 42$ percent) would have a dilative behavior in an $\bar{R}$ (CU) test with $\bar{\sigma}_c = 2$ kg/cm², and would develop liquefaction failure with $\bar{\sigma}_c = 4$ kg/cm².

The principal findings of Castro's (1969) work are summarized in Figure 8.31. The ordinates are void ratio e_c and relative density D_{rc}, after consolidation under the effective confining pressure $\bar{\sigma}_c$, prior to axial loading. The abscissae are $\bar{\sigma}_c$ and effective minor principal stress $\bar{\sigma}_{3f}$ during liquefaction. The results of all tests on this sand in which liquefaction occurred are located on or close to the critical void ratio line, $\bar{e}_F$.

When a denser specimen is subjected to an $\bar{R}$ (consolidated–undrained) test with the same confining pressure (e.g., starting at point *B* below the $\bar{e}_F$ line in Figure 8.31), the resulting stress–strain curves are dilative, pore pressures become negative, and the strength increases to values even higher than those obtained in an *S* (consolidated–drained) test.

The $\bar{e}_F$ line was found to control the stress condition during liquefaction for anisotropically consolidated specimens as well and for specimens in which liquefaction was induced by cyclic loading.

Some *S* tests were performed and critical void ratios from these tests are plotted as an $\bar{e}_S$ line that is not coincident with but falls above the $\bar{e}_F$ line. Casagrande explains this by assuming that during a liquefaction failure the structure of a saturated sand changes rather abruptly into a "minimum resistance structure" or "flow structure," which differs in the arrangement

of the grains from the "normal structure" of a sand that governs the stress–strain behavior during *S* tests.

The results of the $\bar{R}$ (CU) tests with cyclic loading in which liquefaction developed (always starting at small strains) were found to be governed also by the $\bar{e}_F$ line. During cyclic loading in which liquefaction, as defined by Casagrande, did not occur, pore pressure and cyclic strains increased with each cycle, eventually resulting in momentary zero effective stresses when the deviator stress passed through the zero value. Castro (1969) attributes this to radical redistribution of moisture content within the specimen from the lower to upper portion of the specimen, resulting in a nonhomogeneous specimen that can no longer be considered representative of in-situ elements that remain homogeneous.

Although Seed and collaborators' and Casagrande's concepts of "liquefaction" are somewhat different, both schools of thought emphasize the importance of providing adequate relative density in cohesionless soils that may become saturated and subjected to static or especially to dynamic shear stresses. Castro (1969) concluded that of the three sands tested, the one with the subrounded to subangular grains was more susceptible to liquefaction than the sands with angular grains. A critical review of liquefaction and cyclic deformation of sands is given by Casagrande (1979) and in Chapter 16.

Kishida (1969) reported on field observations made where eruption of water and soil had been observed during three earthquakes of intensity V to VI (Japanese Meteorological Agency scale). Most of the liquefaction of sands occurred under these conditions, for sand in which $D_{50} < 2$ mm and $C_u < 10$. The effective overburden pressure was less than 2.0 kg/cm^2; $D_r < 75$ percent and no fine-grained soil strata lay above the saturated sands.

D'Appolonia (1970) reported that the general conditions for liquefaction qualitatively documented by the Japanese based on the Niigata earthquake of 1964 were: the soils contained less than 10 percent fines (silt and clay sizes), D_{60} was between 0.2 mm and 1.0 mm, C_u was between 2 and 5, and the blow count per foot in standard penetration tests was less than 15. He suggested that liquefaction might occur in cohesionless soils of relative density less than 50 percent during ground accelerations of approximately $0.1g$; while for relative densities greater than 75 percent, liquefaction is unlikely for most earthquake loadings. For foundation soils that are susceptible to liquefaction, D'Appolonia (1970) suggests the following procedure: (1) excavate the liquefiable material and replace it in a dense state by compaction, (2) bypass the questionable material by using piles to support the structure, and (3) densify the soil in situ by vibratory compactors, vibroflotation, stone columns, or other suitable means (see Section 8.3). For machinery foundations the granular soils should be compacted to a depth (significant depth) of 1.5 times the average width of the foundation and for an annular distance around the building equal to at least one-half the average width of the foundation.

Where estimates of ground motion can be made using, for example, the work of Whitman and Richart (1967), D'Appolonia (1970) suggested relative densities of foundations subjected to dynamic loads and soil strain caused by the ground motion as follows:

Soil Strain Caused by Ground Motion	Suggested Minimum Relative Density
Small (10^{-5} to 10^{-3} in/in)	70 percent
Intermediate (10^{-3} to 10^{-2} in/in)	80 percent
Large (10^{-2} to 10^0 in/in)	90 percent

To reduce the risk of liquefaction the granular soil should be densified to a minimum relative density of 85 percent in the upper portions and at least 70 percent within the zone of influence (significant depth and area) of the foundation.

8.3 COMPACTION EQUIPMENT*

A large variety of mechanical equipment is available for compaction of soils, but soil type and moisture condition will often dictate the type of equipment and methods of use. The choice of compaction equipment depends also on the intended function of the compacted fill.

As was discussed in Section 8.2, desirable characteristics of compacted fill are low compressibility and high shear strength. In addition, low permeability is essential for compacted fill in water-retaining structures. In cohesive soils, low compressibility and high shear strength can be correlated with high densities. The requirement for low permeability in water-retaining structures precludes the use of equipment or construction methods that will produce layering or laminations in compacted fill. Substantial variation in density (density gradients) in individual lifts are also to be avoided if a homogeneous fill is desired.

* Edward W. Gray, Jr., Civil Engineer, U.S. Bureau of Reclamation, prepared this section under the direction of the author.

Types of Compaction Equipment

Compaction equipment produces compaction by the amount and type of energy it applies to the soil. Table 8.6 shows types and typical uses of compaction equipment (NAVDOCKS DM-7, 1962). Comparative studies of different types of compaction equipment using large-scale field tests have been made by the U.S. Corps of Engineers, by the British Road Research Laboratory, and by the U.S. Bureau of Reclamation. The following maximum dry densities, expressed as a percentage of modified AASHTO density (56 250 ft-lb/ft^3), were obtained in silty clay fills built in 6-inch lifts, with six passes of the compaction equipment (Waterways Experiment Station, 1949):

10 000-lb wheel load rubber tires	92–94 percent
20 000-lb wheel load rubber tires	92–93 percent
40 000-lb wheel load rubber tires	93–94 percent
250-psi sheepsfoot roller	92 percent
500-psi sheepsfoot roller	91–92 percent
750-psi sheepsfoot roller	91–92 percent

These data indicate that increasing wheel loads or contact pressures did not result in significant density increases. It should be noted, however, that the wheel pressures of the rubber-tired equipment remained constant and that the drums of the sheepsfoot roller remained in contact with the soil throughout the series of tests (i.e., the rollers did not walk out).

The British Road Research Laboratory (Lewis, 1960) compacted four different soils with five different types of compacting equipment: three-wheel smooth-wheel rollers, sheepsfoot rollers, a pneumatic-tire roller, vibrating baseplate compactor, and a vibrating roller. The high-pressure pneumatic-tire roller produced the highest compaction in a heavy clay, but a club-footed sheepsfoot roller produced almost the same compaction. The heavy pneumatic-tire roller and the vibrating roller produced the highest compaction in a sandy clay. Vibratory rollers produced the highest compaction in a well-graded sand and in a clay, sand, and gravel mixture.

A large-scaled field test using a vibratory sheepsfoot roller on a cohesive soil was conducted by the Bureau of Reclamation (Tiedemann and Fink, 1969). The soil was a lean clay ($LL = 34$ percent, $PI = 15$ percent) obtained from borrow pits adjacent to the test section. Data from the field indicated the following:

The maximum degree of compaction obtained with the vibratory roller was approximately the same to 1.5 percent lower than for Bureau specified tamping rollers. However, the shapes of the relative compaction curves obtained with the tamping rollers were flatter than those obtained with the vibratory roller. Also, the optimum for the tamping rollers tended to be dry of the laboratory optimum, whereas the vibratory roller optimum was wet of the laboratory optimum. Therefore, the tamping roller can place material dry of the laboratory optimum without any loss in density. Also, the tamping roller would tend to produce a more uniform embankment.

Since the effect of the number of passes on the density was not determined for the tamping rollers, the relative efficiency of the two types of rollers could not be definitely determined. However, based on the results obtained from this and other investigations, it would appear that the vibratory roller was more efficient because fewer roller passes appeared necessary.

There appears to be no type of compaction device completely suitable for all soils and situations. Each of the major types of compaction equipment has its range of applicability with respect to type of soil, density desired, and cost.

Sheepsfoot (Tamping) Rollers It has been reported (*Southwest Builder and Contractor*, 1936) that the sheepsfoot roller originated as a result of a flock of sheep crossing a scarified, oil-treated road surface in Southern California in 1906. A patent

TABLE 8.6 COMPACTION EQUIPMENT AND METHODS (NAVDOCKS DM-7).

Equipment Type	Applicability	Requirements for Compaction of 95 to 100 Percent Standard Proctor Maximum Density			Possible Variations in Equipment
		Compacted Lift Thickness, in	Passes or Coverages	Dimensions and Weight of Equipment	
Sheepsfoot rollers	For fine-grained soils or dirty coarse-grained soils with more than 20 percent passing the No. 200 sieve. Not suitable for clean coarse-grained soils. Particularly appropriate for compaction of impervious zone for earth dam or linings where bonding of lifts is important	6	4 to 6 passes for fine-grained soil; 6 to 8 passes for coarse-grained soil	Soil type / Foot contact area, in^2 / Foot contact pressures, psi: Fine-grained soil $PI > 30$ — 5 to 12 — 250 to 500; Fine-grained soil $PI < 30$ — 7 to 14 — 200 to 400; Coarse-grained soil — 10 to 14 — 150 to 250. Efficient compaction of soils wet of optimum requires less contact pressures than the same soils at lower moisture contents	For earth dam, highway and airfield work, drum of 60-in dia., loaded to 1.5 to 3 tons per lineal foot of drum is generally utilized. For smaller projects 40-in dia. drum, loaded to 0.75 to 1.75 tons per lineal foot of drum is used. Foot contact pressure should be regulated so as to avoid shearing the soil on the third or fourth pass.
Rubber tire rollers	For clean, coarse-grained soils with 4 to 8 percent passing the No. 200 sieve	10	3 to 5 coverages	Tire inflation pressures of 60 to 80 psi for clean granular material or base course and subgrade compaction. Wheel load 18 000 to 25 000 lb	Wide variety of rubber tire compaction equipment is available. For cohesive soils, light wheel loads, such as provided by wobble-wheel equipment, may be substituted for heavy-wheel load if lift thickness is decreased. For cohesionless soils, large-size tires are desirable to avoid shear and rutting.
	For fine-grained soils or well-graded, dirty coarse-grained soils with more than 8 percent passing the No. 200 sieve	6 to 8	4 to 6 coverages	Tire inflation pressures in excess of 65 psi for fine-grained soils of high plasticity. For uniform clean sands or silty fine sands, use large size tires with pressures of 40 to 50 psi	

on the "Petrolithic" roller was issued in that year. Rollers with feet 7 inches long, 4 in^2 in area, and exerting unit pressures of 75 psi were used to compact earth storage reservoirs in California in 1912. This type of roller was reported to be "the only one which would compact a fill in layers without producing lamination, giving a uniform degree of compaction and density."

Sheepsfoot rollers are found in many combinations of empty and loaded weights, drum lengths and diameters, and tamping foot lengths and end areas. Loaded weights vary from about 6000 pounds for some of the smaller towed units to more than 80 000 pounds for larger, multidrummed self-propelled units. Drum lengths range from 48 to 72 inches, and drum diameters range from 40 to 72 inches. Foot lengths range from 7 to 10 inches with end areas usually in the range of $5\frac{1}{2}$ to 10 in^2. Figure 8.32 shows a 20-ton towed sheepsfoot roller and Figure 8.33 a self-propelled sheepsfoot roller weighing about 40 tons.

Variations of sheepsfoot rollers include tractors with wheels of flat steel rims to which tamping feet are fastened, compactors that substitute steel rings for drums, and compactors that substitute discs for drums. Table 8.7 lists specifications data for 90 sheepsfoot rollers (Gray, 1968).

The U.S. Bureau of Reclamation has built over 100 earth dams for which density and moisture data are available for the rolled earthfill portions. The soil types represented by these data cover the entire range of impervious soils from silty and clayey gravels to silts and heavy clays. The rolled earthfill has been compacted, in almost all cases, by 12 passes of a heavy sheepsfoot (tamping) roller in 6-inch compacted lifts.

Hilf (1957) gave results of compacting 39 of these dams and provided a statistical method of obtaining roller curves from control tests made during construction. Figures 8.34 through 8.37 are statistical curves of more recent data which compare field compaction with laboratory compaction for a wide range of soils compacted by the Bureau roller. The field compaction curve appears to be slightly above the laboratory curve. For these curves, the ordinate D is the ratio of fill dry density to Proctor maximum dry density, and the ordinate D/C is the

TABLE 8.6 (*Continued*)

Equipment Type	Applicability	Requirements for Compaction of 95 to 100 Percent Standard Proctor Maximum Density			Possible Variations in Equipment
		Compacted Lift Thickness, in	Passes or Coverages	Dimensions and Weight of Equipment	
Smooth wheel rollers	Appropriate for subgrade or base course compaction of well-graded sand-gravel mixtures	8 to 12	4 coverages	Tandem type rollers for base course or subgrade compaction, 10 to 15 ton weight, 300 to 500 lb per lineal inch of width of rear roller	3-Wheel rollers obtainable in wide range of sizes. 2-Wheel tandem rollers are available in the range of 1 to 20 ton weight. 3-Axle tandem rollers are generally used in the range of 10 to 20 ton weight. Very heavy rollers are used for proof rolling of subgrade or base course.
	May be used for fine-grained soils other than in earth dams. Not suitable for clean well-graded sands or silty uniform sands	6 to 8	6 coverages	3-Wheel roller for compaction of fine-grained soil; weights from 5 to 6 tons for materials of low plasticity to 10 tons for materials of high plasticity	
Vibrating baseplate compactors	For coarse-grained soils with less than about 12 percent passing No. 200 sieve. Best suited for materials with 4 to 8 percent passing No. 200, placed thoroughly wet	8 to 10	3 coverages	Single pads or plates should weigh no less than 200 lb. May be used in tandem where working space is available. For clean coarse-grained soil, vibration frequency should be no less than 1600 cycles per minute	Vibrating pads or plates are available, hand-propelled or self-propelled, single or in gangs, with width of coverage from $1\frac{1}{2}$ to 15 ft. Various types of vibrating-drum equipment should be considered for compaction in large areas.
Crawler tractor	Best suited for coarse-grained soils with less than 4 to 8 percent passing No. 200 sieve, placed thoroughly wet	10 to 12	3 to 4 coverages	No smaller than D8 tractor with blade, 34 500 lb weight, for high compaction	Tractor weights up to 60 000 lb.
Power tamper or rammer	For difficult acces, trench backfill. Suitable for all inorganic soils	4 to 6 in for silt or clay, 6 in for coarse-grained soils	2 coverages	30-lb minimum weight. Considerable range is tolerable, depending on materials and conditions	Weights up to 250 lb; foor diameter 4 to 10 in

Fig. 8.32 A 20-ton towed sheepsfoot roller.

Fig. 8.33 A self-propelled sheepsfoot roller weighing about 40 tons.

ratio of the Proctor density at fill moisture content to Proctor maximum dry density. (The value C is the ratio of fill dry density to the dry density obtained in laboratory compaction cylinder using standard Proctor effort at fill moisture content.)

The Bureau of Reclamation 1970 Standard Paragraphs for sheepsfoot rollers used in earth dam construction are as follows:

Rollers—Tamping rollers shall be used for compacting the earthfill. The rollers shall meet the following requirements:

(1) Roller drums—Tamping rollers shall consist of two or more roller drums mounted side by side in a suitable frame. Each drum of a roller shall have an outside diameter of not less than 5 feet and shall be not less than 5 feet nor more than 6 feet in length. The space between two adjacent drums, when on a level surface, shall be not less than 12 inches nor more than 15 inches. Each drum shall be free to pivot about an axis parallel to the direction of travel. Each drum ballasted with fluid shall be equipped with at least one pressure-relief valve and with at least one safety head as shown on Drawing No. (40-D-6001) or with approved equivalent types. The safety head shall be equal to union type safety heads as manufactured by Black, Sivals and Bryson, Inc., Kansas City, Missouri, with rupture discs suitable for between 50- and 75-psi rupturing pressure.

The pressure-relief valve shown is a manually operated valve and shall be opened periodically. Personnel responsible for opening pressure-relief valves shall be instructed to ascertain that valve openings are free from plugging to assure that any pressure developed in roller drums is released at each inspection.

(2) Tamping feet—At least one tamping foot shall be provided for each 100 square inches of drum surface. The space measured on the surface of the drum, between the centers of any two adjacent tamping feet, shall be not less than 9 inches. The length of each tamping foot from the outside surface of the drum shall be not more than 11 inches and shall be maintained at not less than 9 inches. The cross-sectional area of each tamping foot shall be not more than 10 square inches at a plane normal to the axis of the shank 6 inches from the drum surface, and shall be maintained at not less than 7 square inches nor more than 10 square inches at a plane normal to the axis of the shank 8 inches from the drum surface.

(3) Roller weight—The weight of a roller when fully loaded shall be not less than 4000 pounds per foot of length of drum.

The loading used in the roller drums and operation of the rollers shall be as required to obtain the specified compaction. If more than one roller is used on any one layer or fill, all rollers so used shall be of the same type and essentially the same dimensions.

Rollers operated in tandem sets shall be towed in a manner such that the prints of the tamping feet produced by the tandem units do not overlap. The design and operation of the tamping roller shall be subject to the approval of the contracting officer who shall have the right at any time during the prosecution of the work to direct such repairs to the tamping feet, minor alterations in the roller, and variations in the weight as may be found necessary to secure optimum compaction of the earthfill materials. Rollers shall be drawn by crawler-type or rubber-tired tractors. The use of rubber-tired tractors shall be discontinued if the tires leave ruts that prevent uniform compaction by the tamping roller. Tractors used for pulling rollers shall have sufficient power to pull the rollers satisfactorily when drums are fully loaded with sand and water.

At the option of the contractor, self-propelled tamping rollers conforming with the above requirements may be used in lieu of tractor-drawn tamping rollers. For self-propelled rollers, in which steering is accomplished through the use of rubber-tired wheels, the tire pressure shall not exceed 40 psi. During the operation of rolling, the spaces between the tamping feet shall be maintained clear of materials which would impair the effectiveness of the tamping rollers.

Table 8.8 gives the pertinent data for heavy tamping rollers as agreed to in 1970 by the Corps of Engineers and the Bureau of Reclamation. Hilf (1957) in discussing sheepsfoot rollers used in earth dam construction said:

The ability of heavy sheepsfoot rollers to add appreciably to the amount of mixing obtained by normal earthwork operations is an outstanding characteristic of this type of equipment. The use of special equipment for thorough mixing of soil to obtain maximum homogeneity is usually prohibitive from the standpoint of cost on a large earthwork project. Good construction requires, however, that the excavation and placement methods obtain as much mixing as practicable. Of special importance is the problem of uniformity of moisture content when water is added on the spread lift. Because of the mixing action of the feet and the appreciable number of passes used, compaction by the sheepsfoot roller contributes significantly to uniformity of soil and of moisture, hence, to homogeneity of the fill. The inability of a heavy tamping roller to walk out completely is a built-in safety feature of such rollers. Although good construction practice requires that hard, smooth surfaces left by travel or rubber-tired hauling equipment should be reworked before the succeeding loose lift is placed, it should be recognized that penetration by the feet of heavy tamping rollers will diminish the hazard of such surfaces remaining in the fill.

Rubber-tired Rollers Rubber-tired rollers range from small wobble-wheel rollers to heavy rollers weighing over 100 tons. For smaller rollers, tires may be arranged so that complete coverage of a lift is achieved with a single pass of the roller. The axles of the rubber-tired rollers may be set to allow the wheels to wobble or may be set for straight rolling. Wheel loads for heavy rubber-tired rollers may be as much as 50 tons with tire inflation pressures up to 150 psi. Large four-wheeled rollers in the 50- to 100-ton class usually carry 90-psi tire pressure, but they may be equipped to carry 150 psi. These large rollers are usually towed by heavy crawler-type tractors. The body of a rubber-tired roller may be segmented in such a way that each wheel can be ballasted individually. Self-propelled rubber-tired rollers may weigh as much as 35 tons gross weight. Figure 8.38 shows a heavy-duty multiple box pneumatic-tired compaction roller.

The results of many full-scale field compaction tests (Johnson and Sallberg, 1960) using rubber-tired rollers indicate that 100 percent standard Proctor maximum dry density can be attained by compacting 5- to 12-inch loose lifts with a nominal number of coverages (4 to 8) with a wide range in wheel loads and tire inflation pressures.

Rubber-tired rollers can usually compact in less time and at lower cost than a sheepsfoot roller. In geographic areas where rainfall may occur frequently, construction of earth embankments is expedited by the use of rubber-tired rollers that seal the compacted surface and retard infiltration. The entire surface of a layer to be compacted is subjected to the compaction force of a rubber-tired roller, and 100 percent coverage can be obtained in one or two roller passes.

The field optimum moisture content for a rubber-tired roller occurs at a higher degree of saturation than that for a sheepsfoot roller. This may be detrimental in embankments where high construction pore pressures cannot be tolerated.

The rubber-tired roller leaves a smooth compacted surface and subsequently does not produce significant bonding and blending of successive lifts. Also, compaction by rubber-tired roller is sensitive to the moisture content of the soil. A moisture content wet of laboratory moisture content decreases the bearing capacity of the soil and increases the number of passes of a rubber-tired roller needed to produce a given soil density. The relatively smooth surface of a rubber tire can neither aerate a wet soil nor mix water in a dry soil. A heavy roller with high-pressure tires may require extra tractive effort and possibly precompaction with lighter rollers when (a) the soil moisture content is wet of laboratory optimum moisture content, or (b) the uncompacted lift is thick and the roller expends energy pushing and shoving the loose lift.

Heavy rubber-tired rollers are not recommended for initial rolling of heavy (highly plastic) clay soils but are effective and economical in a wide range of soils from clean sand to silty clay.

The pressure exerted on fill material by rubber-tired rollers is roughly equal to the tire inflation pressures. The sidewalls of the rubber tires will cause some unequal pressure distribution depending on sidewall stiffness.

Figure 8.39 shows field compaction curves for rubber-tired rollers related to laboratory compaction (Johnson and Sallberg, 1960). In summarizing the effect of wheel load and tire inflation pressure, they found:

a. The contact area and contact pressure under the tires, both of which affect the state of compaction produced, are functions of the wheel load and the tire inflation pressure.
b. An increase in the wheel load or in the tire inflation pressure produces an increase in the roller maximum dry unit weight with a corresponding decrease in optimum moisture content.
c. The greater the wheel load and the tire inflation pressure, the greater the unit weight at any depth. However, increasing the tire inflation pressure without proportionately increasing wheel load tends to produce greater compaction near the surface.

Other sources indicate that:

a. Increasing the tire inflation pressure is more effective than increasing the wheel load in producing higher densities.
b. Increasing the tire size while maintaining the same tire inflation pressure will increase the compaction at greater depths.

Increasing the number of coverages of a rubber-tired roller will result in higher maximum densities and lower field optimum water contents, although increasing the tire pressure is probably more effective when the number of passes is held constant.

According to Turnbull et al. (1949):

> The effect of number of coverages of rubber-tired rollers on soil density is of importance and has a relationship similar to that given for sheepsfoot rollers. The effect of thickness of lift is particularly pronounced, however; a deficiency in contact pressure can never be fully overcome by a thinner lift, and decreasing the thickness of lift is an expedient to be used only when adequate contact pressure cannot be attained.

The Corps of Engineers 1967 Guide Specifications Paragraph for rubber-tired rollers used in earth dam construction is as follows:

> *Rubber-tired Rollers.* Rubber-tired rollers shall have a minimum of four wheels equipped with pneumatic tires. The tires shall be of such size and ply as can be maintained at tire pressures between 80 and 100 pounds per square inch for a 25 000 pound wheel load during rolling operations. The roller wheels shall be located abreast and be so designed that each wheel will carry approximately equal load in traversing uneven ground. The spacing of the wheels will be such that the distance between the nearest edges of adjacent tires will not be greater than 50 percent of the tire width of a single tire at the operating pressure for a 25 000 pound wheel load. The roller shall be provided with a body suitable for ballast loading such that the load per wheel may be varied, as directed by the Contracting Officer, from 18 000 to 25 000 pounds. The roller shall be towed at speeds not to exceed five miles per hour. The character and efficiency of this equipment shall be subject to the approval of the Contracting Officer.

Vibratory Rollers Large tractor-towed or self-propelled vibratory rollers are used to compact granular fills and embankments as well as rockfill material for earth dams. Adequate vibration must meet the requirements of having sufficient force (dead weight plus dynamic force) acting through the required distance (amplitude) and giving sufficient time for movement of soil grains (frequency) to take place (Johnson and Sallberg, 1960).

Most vibratory rollers are designed to produce vertical forces. Vibratory rollers may be equipped with rubber tires, smooth wheel drums, or tamping feet, and have been used successfully in compacting sand, gravel, rockfill, and some cohesive soils. Another form of vibratory compaction equipment is the vibrating base-plate compactor, which produces results similar to that of vibratory rollers. The frequency and dead-weight of vibratory rollers must be matched to the material being compacted: heavyweight rollers with low-frequency vibrations for gravel or rockfill, light- to medium-weight rollers with high-frequency vibrations for sands, and heavyweight rollers with low-frequency vibrations for clays. Tests indicate that vibratory rollers compact best when the soil is at or slightly wetter than optimum moisture content (Hall, 1968). A Bureau of Reclamation study (Tiedemann and Fink, 1969) indicated that moisture control in cohesive soils could be critical because of the peaked nature of the field compaction curves for vibratory

TABLE 8.7 SPECIFICATIONS FOR SHEEPSFOOT ROLLERS.
(Unpublished Memo, Gray, 1968.)

Brand Name and Model Number		Date of Source	Number of Drums	Drum Length, in	Drum Diameter, in	Oscillation	Space Between, in
Browning	HD-114	SI-6635P	2	60	60	Yes	?
Browning	HD-112	SI-6635P	2	48	40	Yes	?
Browning	HD-120	SI-6635P	2	60	40	Yes	?
Bros	SP-255D	5-65	2	60	60	Yes	14
Bros	SP-3DT	8-60	3	60	60	Yes	69
Bros	SP-446T	3-67	4	Rear 50 Front 40	60	Yes	—
Bros	SP-446P	12-66	4	?	72	Yes	—
Bros	GR2-9¼	1-63	2	60	60	Yes	12
Bros	M2-5½	1-63	1	48	40	Yes	12
Bros	G2-8	1-63	1	60	60	Yes	12
Bros	G26-8	1-63	2	72	60	Yes	12
Caterpillar	824B	5-66	4	44½	51	Front No Rear Yes	52 est
Caterpillar	824B	5-66	4	44½	51	Front No Rear Yes	52 est
Caterpillar	834	5-66	4	48	60	Front No Rear Yes	54 est
Caterpillar	834	5-66	4	48	60	Front No Rear Yes	54 est
Ferguson	SP-22	12-59	2	72	60	Yes	12
Ferguson	SP-120	9-60	2	60	60	Yes	16
Ferguson	SP-120B	7-66	2	60	60	Yes	14
Ferguson	SP-120D	10-65	2	60	60	Yes	12
Ferguson	SP-112	7-66	2	48	48	Yes	12
Ferguson	SP-112W	9-60	2	48	48	Yes	14
Ferguson	SP-84	5-61	2	36	?	No	3
Ferguson	SP-96M	12-59	4	72 or 84	60 or 72	Yes	?
Ferguson-Gebhard	22	4-66	2	72	60	Yes	16
Ferguson-Gebhard	22T	4-66	1	72	60	—	—
Ferguson-Gebhard	22 Mod	4-66	2	72	60	Yes	16
Ferguson-Gebhard	120	4-66	2	60	60	Yes	15
Ferguson-Gebhard	120-T	4-66	1	60	60	—	—
Ferguson-Gebhard	120 Mod	4-66	2	60	60	Yes	15
Ferguson Tamping Reclamation	USBR-2	4-66	2	60	60	Yes	14
Ferguson	A-32	4-66	2	72	72	Yes	?
Ferguson	112	4-66	2	48	40	Yes	6
Ferguson	112T	4-66	1	48	40	—	—
Ferguson	112W	4-66	2	48	40	Yes	6
Ferguson	112W-T	4-66	1	48	40	—	—
Ferguson	112W-48	4-66	2	48	48	Yes	6
Ferguson	112W-48T	4-66	1	48	48	—	—
Grace	E230	8-66	2	36	54	No	c
Grace	E236	8-66	2	36	54	No	c
Grace	P-104	8-66	2	48	40	Yes	12
Grace	R-112	8-66	2	48	40	Yes	14
Grace	4 × 5-95	8-66	2	48	60	Yes	18
Grace	5Y5-120	8-66	2	60	60	Yes	14
Grace	5Z6-138	8-66	2	60	72	Yes	14
Holt	DW-20	3-62	2	53	60	No	49½
Hyster	C450A	10-67	2	80	66½	Yes	—
Hyster	C455A	11-67	2	72	60	—	—
Hyster	DW20A	1-62	4	32	53	No	6
Hyster	C400B Tractor	9-61	2	34.6	61	No	60½
Hyster	C400B Compactor	9-61	2	32	53	No	6
Hyster	Model D	9-63	2	32	53	No	6
Hyster	C150C	3-64	2	60	60	Yes	15
Hyster	C410A	1-63	2	32	53	Yes	12.4
R. G. LeTourneau	M50-55	5-62	4	60	60	Yes	15
LeTourneau-Westinghouse	120	1-65	2	60	60	Yes	15
LeTourneau-Westinghouse	W-2	11-65	2	48	42	Yes	7
Littleford Chester	6060	11-47	2	60	60	Yes	18
Littleford Chester	4840	11-46	2	48	40	Yes	11
McCoy	USHD-55	1-60	2	60	60	Yes	10½
McCoy	USHD-65	1-60	2	60	72	Yes	?
Pactor	3-60	9-66	3	Front 30 Rear 46.5	73	No	20
Pactor	3-40	3-66	3	Front 27 Rear 43	53	No	10½

TABLE 8.7 *(Continued)*

No. Feet Per Ft²	Foot Length, in	Foot Area, in²	Foot Spacing, in	Foot % of Periph.	Empty Wt. Per l.f.	Loaded Wt. Per l.f.	Pressure Re. Valves	Maximum Speed	Motivation
1.46	9	7	11.2	5.4	1 550	3 550	—	—	Towed
2.69	7	6¼	6.7	8.7	750	1 700	—	—	Towed
2.31	7	6¼	7.2	7.4	700	1 650	—	—	Towed
1.54	8	7	8	5.9	2 300	3 100	—	8.9	S.P.[a]
1.43	9¼	9½	10.2	7.2	5 400	5 400	—	7.0	S.P.
1.65	9¼	9½	10	8.3	4 100	4 100	—	18.0	S.P.
—	—	55	—	—	—	62 000	—	18.0	S.P.
1.53	9¼	7	11	5.7	2 080	4 140[b]	—	—	Towed
2.69	7	5½	7	10.2	760	1 750	—	—	Towed
1.43	8	7	11	5.5	1 620	3 740	—	—	Towed
1.53	8	7	11	5.9	1 720	3 790	—	—	Towed
1.69	8	12	9	10.9	3 640	4 460	—	17.4	S.P.
1.31	7½	29¾	11.2	20.6	3 910	4 730	—	17.0	S.P.
1.53	8	12	9.6	10.1	4 120	4 980	—	20.4	S.P.
1.43	7½	29¾	10.6	23.7	4 440	5 290	—	20.0	S.P.
1.53	9¼	7½	11.6	6.0	3 280	4 280	—	8.0	S.P.
1.53	8	7½	10.6	6.3	2 230	3 300	—	8.0	S.P.
1.53	9	9½	10.6	7.8	2 820	3 840	—	7.5	S.P.
1.53	7½ or 9½	7.1 or 9.5	10.6	6.0 to 7.8	3 290	4 310	—	10.0	S.P.
2.22	8	7.1	8.6	8.2	2 260	2 890	—	6.3	S.P.
2.22	8	5.6	8.6	6.5	1 480	2 160	—	6.3	S.P.
—	8	5.6	—	—	2 150	2 720	—	8.0	S.P.
?	?	?	?	?	?	?	—	5.2	S.P.
1.53	8	7½	11.6	6.3	1 790	3 750	—	—	Towed
1.53	8	7½	11.6	6.3	1 710	3 630	—	—	Towed
1.53	9½	7½	11.6	6.0	2 240	4 200	—	—	Towed
1.53	8	7½	10.6	6.0	1 560	3 680	—	—	Towed
1.53	8	7½	10.6	6.3	1 660	3 720	—	—	Towed
1.53	9½	7½	10.6	6.0	2 300	4 412	—	—	Towed
1.53	10	7.1	9.4	5.6	2 850	4 740	—	—	Towed
1.48	10	9½	11.6	7.7	2 790	4 140	—	—	Towed
2.66	7	5½	8.2	7.6	810	1 710	—	—	Towed
2.66	7	5½	8.2	7.6	760	1 660	—	—	Towed
2.66	8	5½	8.2	7.6	1 140	2 020	—	—	Towed
2.66	8	5½	8.2	7.6	1 090	1 980	—	—	Towed
2.22	8	5½	8.6	6.4	1 310	2 610	—	—	Towed
2.22	8	5½	8.6	6.4	1 260	2 560	—	—	Towed
?	8	7	—	—	c	c	—	—	S.P.
?	9	9½	—	—	c	c	—	—	S.P.
2.48	8	7	7.9	8.9	900	1 750	—	—	Towed
2.67	7¼	5½	9.1	7.5	790	1 670	—	—	Towed
1.51	8	7	12	5.8	1 690	4 550	—	—	Towed
1.54	8	7	12	5.8	1 700	3 810	—	—	Towed
1.46	8	7	12	5.9	1 800	4 930	—	—	Towed
1.72	9¼	9½	9.4	8.7	4 180	4 760	Yes	15.3	S.P.
2.84	6½	21½	10.5	17.8	3 880	3 880	—	17.0	S.P.
1.53	9¾	7.9	10.0	6.3	4 180	4 180	Yes	11.5	S.P.
2.43	7	21	9.8	28.0	4 360 Tractor 2 470 Towed Unit	8 100 Tractor 6 850 Towed Unit	—	17.0	S.P.
2.22	7	21	9.2	26.3	5 630	7 610	—	—	S.P.
2.43	7	21	9.8	28.0	2 310	6 880	—	—	Towed
2.43	7	21	9.8	28.0	2 260	6 300	—	15.0	Towed
1.53	7½	7½	10.6	6.4	1 440	3 630	—	—	Towed
2.03	7	21	9.7	23.4	2 660	4 530	—	—	Towed
1.53	9	10	9.5	8.2	4 000	4 000	—	5.0	S.P.
1.54	8¼	7.1	9.5	5.9	1 770	6 936	—	—	Towed
2.00	8	5.1	8.5	5.1	750	1 680	—	—	Towed
1.54	8	7.1	10.5	6.0	1 900	4 100	—	—	Towed
2.52	8	7.1	8	8.8	1 020	1 970	—	—	Towed
1.53	8	7.1	10.6	6.0	1 500	3 530	—	—	Towed
1.46	8¼ or 9¼	6 to 9	12	6.0	2 350	5 050	—	—	Towed
1.44	5½	9.6 to 13.4	8.5	—	5 960	5 960	—	14.0	S.P.
2.18	10½	9.6 to 13.4	8	14.5	4 800	4 800	—	15.0	S.P.

TABLE 8.7 *(Continued)*

Brand Name and Model Number		Date of Source	Number of Drums	Drum Length, in	Drum Diameter, in	Oscillation	Space Between, in
Pactor	3-40	3-66	3	Front 27 Rear 43	53	No	10½
Southwest	454	7-66	3	Front 48 Rear 60	72	Yes	12½
Southwest	55HS	7-66	2	60	60	Yes	12
Southwest	355RR	7-66	3	60	60	Yes	14
Southwest	2DL-96R	7-66	2	48	40	Yes	8
Southwest	2DM-120R	7-66	2	60	60	Yes	8
Southwest	2DH-RR	7-66	2	60	60	Yes	12
Southwest	55RR	7-66	2	60	60	Yes	12
Southwest	1DH-RR	CM-44	1	60	60	—	—
Southwest	SD-56	4-66	1	72	60	—	—
Southwest	BR-R	CM-44	2	60	60	Yes	?
Tampo	SP-500	7-66	2	60	60	Yes	15
Tampo	H-1	7-66	1	48	40	—	—
Tampo	H-2	7-66	2	48	40	Yes	12
Tampo	H-3	7-66	3	48	40	Yes	12
Tampo	H2R	7-64	2	48	40	Yes	12
Tampo	H2WL	7-66	2	48	40	Yes	12
Tampo	H2WH	7-66	2	48	40	Yes	12
Tampo	H21	7-66	2	48	40	Yes	12
Tampo	H20	7-66	2	60	40	Yes	12
Tampo	501	7-66	1	60	60	—	—
Tampo	502	7-66	2	60	60	Yes	14
Tampo	501R	7-66	1	60	60	—	—
Tampo	502R	7-66	2	60	60	Yes	14
Tampo	502WL	7-66	2	60	60	Yes	14
Tampo	S-1	7-66	1	48	40	—	—
Tampo	S-2	7-66	2	48	40	Yes	12
Tampo	502-X	7-66	2	60	60	Yes	14
Wagner	WC-24 Front	8-61	2	48	60	No	72
	Rear		1	60	60	—	—

[a] S.P. = self-propelled.
[b] Total weight of machine.
[c] Depends upon tractor on which tamping wheels are mounted.

sheepsfoot rollers. The same Bureau study indicated that maximum compaction was obtained at moisture contents slightly wet of optimum moisture content. Figure 8.40 shows the vibratory sheepsfoot roller used in the Bureau study.

In order for a vibratory roller to secure bond between lifts, the feet of a vibratory sheepsfoot roller should penetrate the entire lift thickness, as with a static roller.

The speed and number of passes of a vibratory roller are critical, inasmuch as these factors determine the number of dynamic load applications available for each point of the compacted fill. Increasing the number of passes increases the compactive effort and also increases the effective depth of compaction. However, the curve of compaction vs. number of passes begins to level out after a few passes, and beyond that point an increase in depth of compaction requires many passes.

Thickness of compacted lifts is controlled by the weight and vibration frequency of a vibratory roller: lightweight, high-frequency rollers obtain satisfactory densities in thin lifts and heavyweight, low-frequency rollers obtain satisfactory densities in thick lifts.

D'Appolonia et al. (1969) described field studies of the effects of several variables in vibratory compaction, including roller weight and size, roller operating frequency and forward speed, lift height, number of roller coverages, sand particle size and gradation, and soil-water content. The sand at the site tested is a poorly graded dune sand, the mean particle size is 0.18 mm, and the uniformity coefficient is about 1.5. Particles are angular to subangular, and are predominantly coarse. Using procedures suggested by Burmister (1964), the dune sand has a minimum density of 88.5 lb/ft³ and a maximum density of 110 lb/ft³.

Their conclusions include:

Surface compaction of in situ granular deposits with light vibratory rollers is not effective for achieving high densities at depths greater than 4 feet or 5 feet. The compacted density in any depth increases with the number of roller coverages, but after about five passes a large increase in the number of passes is required to achieve a significant increase in density.

When sand fill is placed in lifts, the lift height selected should be small enough so that a loose layer is not trapped near the interface between lifts. This can be achieved by choosing a lift height that is not significantly greater than the depth at which the maximum amount of compaction is achieved for a single lift height. Moreover, the lift height should not be significantly less than the depth at which maximum compaction occurs for a single lift or else much of the compactive effort will be lost through repeated overvibration of near-surface layers.

Roller operating frequency has a large effect on compacted density. Compacted density as well as ground acceleration and dynamic stress increased with operating frequency. Because vibratory rollers used in this study had a maximum operating frequency of about 30 cycles per second, it was not possible to establish the full amplitude response spectra for the rollers. However, for operating frequencies of less than 30 cycles per second, the roller-soil system appears to be highly damped, indicating that a clearly defined resonant frequency does not exist. For the highly damped roller–soil system with the dynamic forces proportional to the square of the operating frequency, ground acceleration and dynamic stress, and thus compacted density, will not decrease appreciably if frequencies greater than the resonant frequency are used. Therefore, the operating frequency should be at least as large as the resonant frequency to obtain the most efficient use of the vibratory roller.

Measurements of horizontal stress in vibratory compacted sand fill indicate that the lateral stresses are significantly greater than

TABLE 8.7 *(Continued)*

No. Feet Per Ft²	Foot Length, in	Foot Area, in²	Foot Spacing, in	Foot % of Periph.	Empty Wt. Per l.f.	Loaded Wt. Per l.f.	Pressure Re. Valves	Maximum Speed	Motivation
1.44	5½	9.6 to 13.4	8	—	4 360	4 360	—	12.0	S.P.
1.51	9¼	9	9.8	7.6	5 000	8 380	—	10.0	S.P.
1.30	8	14	10.5	10.0	2 910	4 860	—	—	Towed
1.52	9¼	7	9.5	5.7	2 450	4 170	Yes	—	Towed
2.29	7	6	7.7	7.1	790	1 725	—	—	Towed
1.52	8	6	9.7	5.9	1 440	3 475	—	—	Towed
1.52	9¼	7	9.5	5.7	2 080	4 045	Yes	—	Towed
1.52	9¼	7	9.5	5.7	2 340	4 315	Yes	—	Towed
1.52	9¼	7	9.5	5.7	1 840	3 740	—	—	Towed
1.53	9¼	7	10.5	5.7	1 970	3 830	—	—	Towed
?	10	7	?	?	3 000	4 720	—	—	Towed
1.53	8	7	10.6	5.9	2 770	3 590	—	14.0	S.P.
2.67	7	6	8.2	8.2	800	1 280	—	—	Towed
2.67	7	6	8.2	8.2	790	1 270	—	—	Towed
2.67	7	6	8.2	8.2	780	1 270	—	—	Towed
2.29	7	7	8.6	8.2	890	1 820	—	—	Towed
2.29	7	7	8.6	8.2	820	1 740	—	—	Towed
2.48	7	5½	8.4	7.0	840	1 810	—	—	Towed
2.10	8	5	9.0	5.2	780	1 260	—	—	Towed
2.14	7	6	9.7	6.6	660	1 590	—	—	Towed
1.53	7	6	10.6	5.2	1 440	3 410	—	—	Towed
1.53	7	6	10.6	5.2	1 440	3 410	—	—	Towed
1.53	8	7	10.6	5.9	1 680	3 660	—	—	Towed
1.53	8	7	10.6	5.9	1 680	3 660	—	—	Towed
1.53	7	7	10.6	6.0	1 490	3 450	—	—	Towed
2.67	7	6	8.2	8.2	750	1 230	—	—	Towed
2.67	7	6	8.2	8.2	760	1 240	—	—	Towed
1.53	9¼	6	11.2	4.8	2 160	4 100	—	—	Towed
1.59	9	7.1	9.4	6.0	4 810	4 810	—	24.0	S.P.
1.53	9	7.1	9.4	5.8	5 300	6 520	—	—	S.P.

the at-rest earth pressure. Horizontal stresses were found to increase with the number of roller passes and also with roller operating frequency.

Morehouse and Baker (1969) reported the compaction of a clean, poorly graded sand by a smooth-drum vibrating roller. They found that for the roller used, eight coverages achieved relative densities in excess of 70 percent to a depth of 5½ ft and relative densities in excess of 50 percent to a depth of 10 ft. Using settlement data obtained from base plates installed in the sand fill, it was found that additional coverages of the roller continued to densify the sand above the 6 to 7 ft depth, but that for more than eight coverages the effect per coverage was substantially less than that for coverages in the one to eight range. The upper 1½ ft of the test fill tended to loosen under the action of the compactor.

Hall (1968) reported on the evaluation of vibratory rollers on three types of soil at the U.S. Army Engineers Waterways Experiment Station. The compactive effort of three smooth-drum vibratory rollers was compared to that of a 50-ton rubber-tired roller. Field tests were conducted under shelter to provide better control over all phases of the testing program. The smallest of the rollers weighed 3150 lb, had a 30- by 50-inch drum, and operated at a fixed 3600 rpm. The heaviest roller weighed 7000 lb, had a 48- by 60-inch drum, and operated over a range of 800 to 1600 rpm. The medium roller weighed 5270 lb, had a 38- by 84-inch drum, and operated over a range of 600 to 1400 rpm. Operating frequencies of the three rollers encompassed the range over which vibratory rollers then operated. Each of the three rollers was used to compact three soil types: lean clay,

crushed limestone, and clean sand. Each soil was compacted wet of, dry of, and at optimum moisture content. Density in sand compacted with vibratory rollers appears to be a cyclic function of coverages. The lowest-frequency middle-weight compactor performed the best in sand. Densities in crushed limestones and lean clay generally increased in direct proportion to the dead weight of vibratory rollers. Except in crushed limestone, vibratory rollers will not produce densities to significantly greater depths than rubber-tired rollers. For comparable width thickness of compaction much lighter vibratory rollers may be substituted for heavy rubber-tired rollers; however, there is a limit to the amount of weight reduction that can be achieved using vibratory rollers.

Schmertmann (1970) and Chapter 23 discussed deep sand densification by surface vibratory rollers, stating that it appeared that sand thicknesses of 5 to 7 ft and perhaps more can be densified from the surface in a manner adequate for many ordinary foundation designs. This fact has major economic importance in Florida, where a 5- to 10-ft layer of loose surface sand is a common situation.

The 1967 Corps of Engineers Guide Specifications Paragraph for vibratory rollers used in earth dam construction is as follows:

Vibratory Rollers. Vibratory rollers shall have a total static weight of not less than 20 000 lb, with at least 90 percent of the weight transmitted to the ground through a single smooth drum when the roller is standing in a level position hitched to a towing vehicle. The diameter of the drum shall be between 5 and 5.5 feet and the width between 6 and 6.5 feet. The unsprung weight of drum, shaft, and internal mechanism shall not be less than 12 000 lb. The frequency of vibration during operation shall be between 1000 and

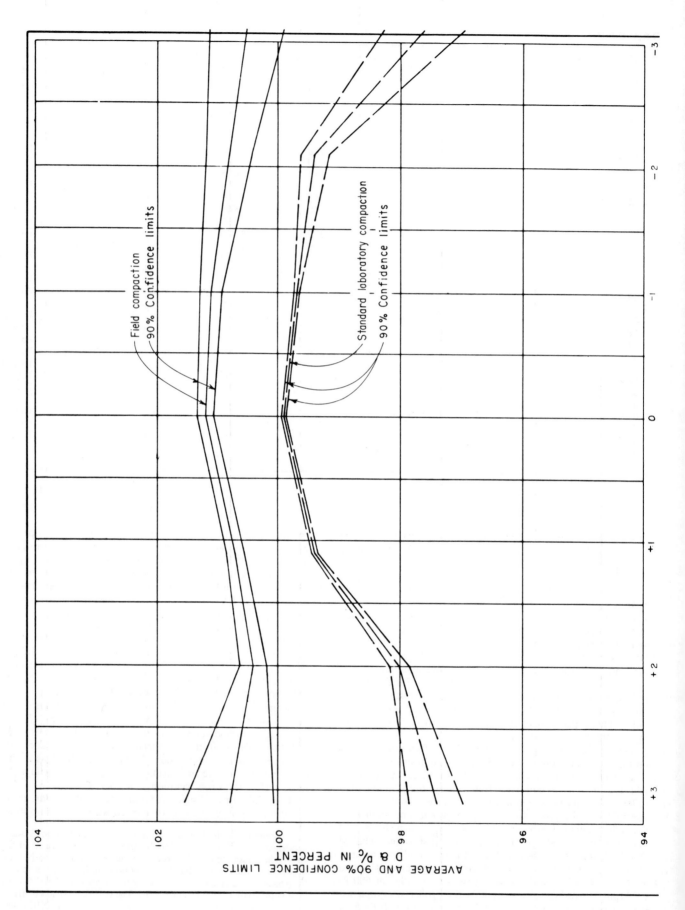

AVERAGES

General soil classification	Liquid limit	Plastic index	Percentage plus #4	Water content (minus #4)(wf) % by dry weight	Dry density compaction cyl. (minus #4) p.c.f.	Dry density fill (minus #4) p.c.f.	Proctor max. dry density (minus #4) p.c.f.	Water content opt. (wo) (minus #4) % by dry weight	wo − wf	C in percent	D in percent
ML	25.0	3.0	0	17.3	105.2	106.8	106.0	17.7	0.4	101.5	100.8

DISTRIBUTION -- PERCENTAGE OF TESTS

MOISTURE CONTROL			DRY UNIT WEIGHT CONTROL				STD. DEVIATIONS		
Drier than wo − 3.2	Between wo − 3.2 to wo + 0.2	Wetter than wo + 0.2	Less than .95 γd max.	Between .95 γd max. .98 γd max.	Between .98 γd max. γd max.	Greater than γd max.	wo − wf	C	D
0.8	66.9	32.3	0.0	1.6	14.8	83.6	±1.3	±2.2	±2.0

VARIATION OF FILL WATER CONTENT FROM LAB. OPTIMUM PERCENT BY DRY WEIGHT (wo−wf)

RATIO OF FILL DRY DENSITY TO LAB MAXIMUM DRY DENSITY (D IN PERCENT)

RATIO OF CYLINDER DRY DENSITY AT FILL WATER CONTENT TO LAB. MAX. DRY DENSITY (% IN PERCENT)

RANGE wo − wf	AVERAGE wo − wf	NO OF TESTS n	AVERAGE x̄	STANDARD DEVIATION σ	90% CONFIDENCE LIMITS	AVERAGE x̄	STANDARD DEVIATION σ	90% CONFIDENCE LIMITS
−3.8 to −2.8	−3.1	22	100.5	±1.65	±0.620	97.6	±1.77	±0.666
−2.7 to −1.7	−2.1	78	100.8	±2.00	±0.379	98.4	±1.22	±0.232
−1.6 to −0.6	−1.0	278	101.1	±1.90	±0.188	99.7	±0.51	±0.051
−0.5 to +0.5	0	609	101.2	±2.05	±0.137	99.9	±0.60	±0.041
0.6 to 1.6	+1.1	530	100.7	±2.04	±0.147	99.4	±0.69	±0.050
1.7 to 2.7	+2.0	236	100.4	±2.06	±0.223	98.0	±1.49	±0.161
2.8 to 3.8	+3.1	37	100.8	±2.63	±0.741	96.4	±1.58	±0.446

Fig. 8.34 Relation between field and laboratory compaction (minus No. 4 material). Zone 1 —borrow area—A, C, N and required excavation—Red Willow Dam.

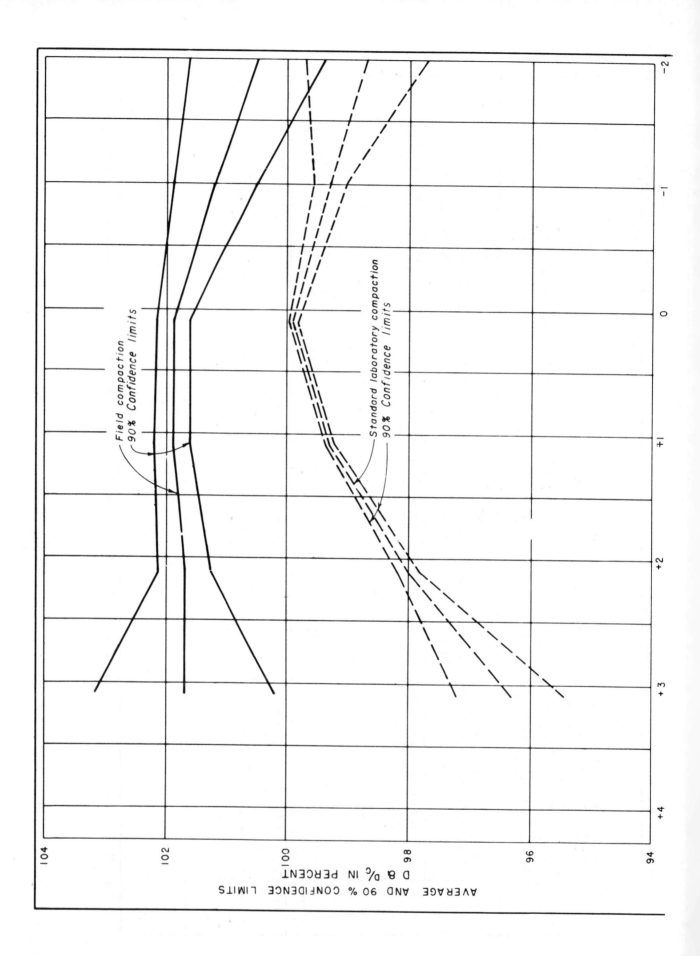

General soil classification	Liquid limit	Plastic index	Percentage plus #4	Water content (minus #4) (w_f) % by dry weight	Dry density compaction cyl. (minus #4) p.c.f.	Dry density fill (minus #4) p.c.f.	Proctor max. dry density (minus #4) p.c.f.	Water content opt. (w_o) (minus #4) % by dry weight	w_o − w_f	C in percent	D in percent	Drier than w_o − 3.2	Between w_o + 0.2 / w_o − 3.2	Wetter than w_o + 0.2	Less than 95 γd max.	Between 95 γd max. / 98 γd max.	Between 98 γd max. / γd max.	Greater than γd max.	w_o − w_f	C	D
				AVERAGES								**DISTRIBUTION - PERCENTAGE OF TESTS** (MOISTURE CONTROL / DRY UNIT WEIGHT CONTROL)							**STD. DEVIATIONS**		
CL	27.0	13.0	6.4	13.2	112.9	115.7	114.0	14.1	0.9	102.5	101.5	1.5	86.4	13.1	0.5	4.1	16.2	79.2	±1.1	±2.3	±2.1

VARIATION OF FILL WATER CONTENT FROM LAB. OPTIMUM PERCENT BY DRY WEIGHT (w_o−w_f)		NO OF TESTS	RATIO OF FILL DRY DENSITY TO LAB MAXIMUM DRY DENSITY (D IN PERCENT)			RATIO OF CYLINDER DRY DENSITY AT FILL WATER CONTENT TO LAB. MAX. DRY DENSITY (% IN PERCENT)		
RANGE w_o−w_f	AVERAGE w_o−w_f	n	AVERAGE $\bar{X}$	STANDARD DEVIATION σ	90% CONFIDENCE LIMITS	AVERAGE $\bar{X}$	STANDARD DEVIATION σ	90% CONFIDENCE LIMITS
2.8 to 3.8	3.1	10	101.7	±2.47	±1.508	96.3	±1.45	±0.883
1.7 to 2.7	2.1	82	101.7	±2.34	±0.433	98.0	±0.94	±0.174
0.6 to 1.6	1.1	118	101.9	±1.97	±0.302	99.3	±0.45	±0.069
0.5 to −0.5	0.1	102	101.9	±1.65	±0.272	99.9	±0.51	±0.084
0.6 to −1.6	−1.0	26	101.2	±2.04	±0.701	99.3	±0.72	±0.248
−1.7 to −2.7	−2.0	6	100.5	±1.25	±1.124	98.7	±1.11	±0.996

Fig. 8.35 Relation between field and laboratory compaction (minus No. 4 material). Zone 1—Paonia Dam.

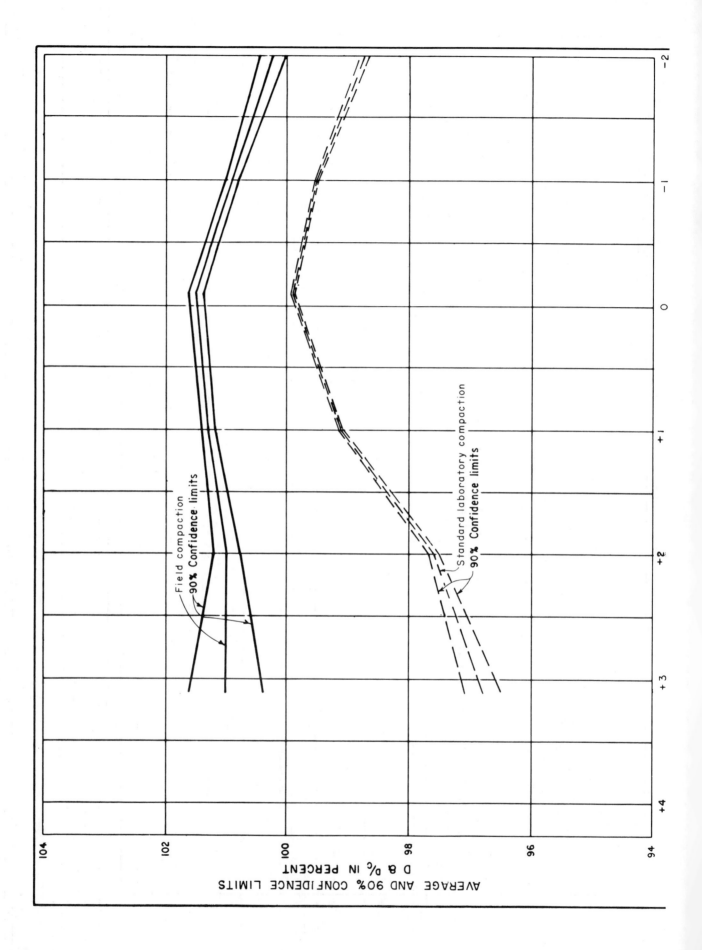

AVERAGES

General soil classification	Liquid limit	Plastic index	Percentage plus #4	Water content (minus #4) (w_f) % by dry weight	Dry density compaction cyl. (minus #4) p.c.f.	Dry density fill (minus #4) p.c.f.	Proctor max. dry density (minus #4) p.c.f.	Water content opt. (w_o) (minus #4) % by dry weight	$w_o - w_f$	C in percent	D in percent
SM	*	*	0.0	15.2	111.3	113.5	112.3	16.5	0.3	102.0	101.0

| DISTRIBUTION — PERCENTAGE OF TESTS | | | | | | | STD. DEVIATIONS | | |
MOISTURE CONTROL			DRY UNIT WEIGHT CONTROL				$w_o - w_f$	C	D
Drier than $w_o - 3.2$	Between $w_o - 3.2$ $w_o + 0.2$	Wetter than $w_o + 0.2$	Less than .95 γ_d max.	Between .95 γ_d max. .98 γ_d max.	Between .98 γ_d max. γ_d max.	Greater than γ_d max.			
0.9	62.2	36.5	0.1	1.4	30.7	67.8	±1.3	±2.1	±1.9

* Data not available

| VARIATION OF FILL WATER CONTENT FROM LAB. OPTIMUM PERCENT BY DRY WEIGHT ($w_o - w_f$) | | NO OF TESTS | RATIO OF FILL DRY DENSITY TO LAB MAXIMUM DRY DENSITY (D IN PERCENT) | | | RATIO OF CYLINDER DRY DENSITY AT FILL WATER CONTENT TO LAB. MAX. DRY DENSITY (% IN PERCENT) | | |
RANGE $w_o - w_f$	AVERAGE $w_o - w_f$	n	AVERAGE $\bar{X}$	STANDARD DEVIATION σ	90% CONFIDENCE LIMITS	AVERAGE $\bar{X}$	STANDARD DEVIATION σ	90% CONFIDENCE LIMITS
-2.7 to -1.7	-2.1	142	100.2	±1.55	±0.217	98.7	±0.59	±0.082
-1.6 to -0.6	-1.0	538	100.9	±1.62	±0.115	99.5	±0.26	±0.018
-0.5 to +0.5	-0.1	534	101.5	±1.86	±0.132	99.9	±0.50	±0.035
0.6 to 1.6	1.0	830	101.3	±2.04	±0.116	99.1	±0.56	±0.032
1.7 to 2.7	2.0	309	101.0	±2.26	±0.212	97.6	±0.94	±0.088
2.8 to 3.8	3.1	43	101.0	±2.37	±0.617	96.8	±1.18	±0.307

Fig. 8.36 Relation between field and laboratory compaction (minus No. 4 material). Zone 1—borrow area A—Foss Dam.

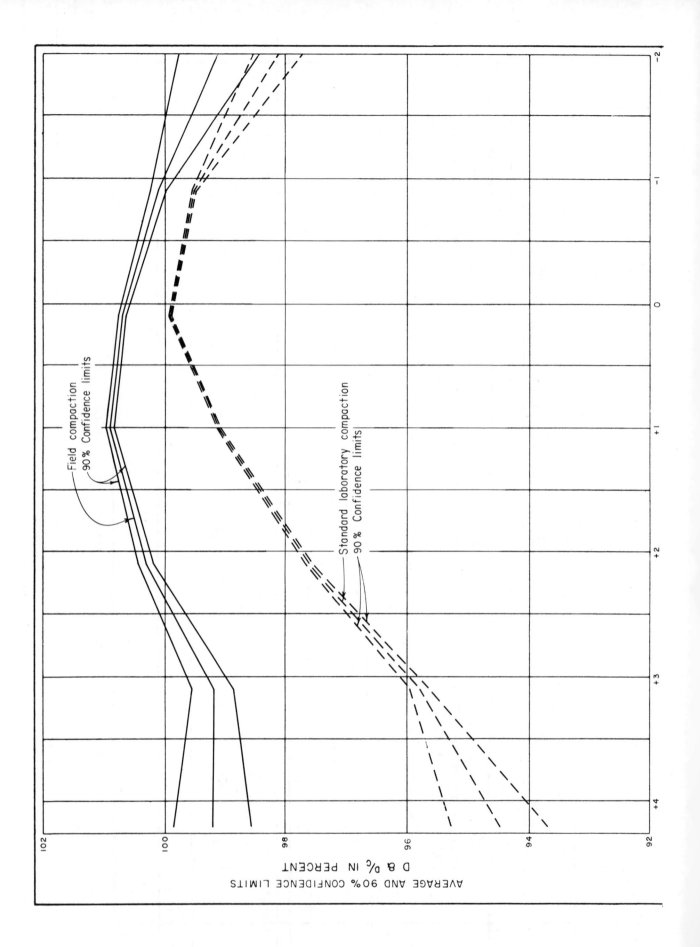

General soil classification	Liquid limit	Plastic index	Percentage plus #4	Water content (minus #4)(w_f) % by dry weight	Dry density compaction cyl. (minus #4) p.c.f.	Dry density fill (minus #4) p.c.f.	Proctor max. dry density (minus #4) p.c.f.	Water content opt. (w_o)(minus #4) % by dry weight	$w_o - w_f$	C in percent	D in percent	DISTRIBUTION — PERCENTAGE OF TESTS — MOISTURE CONTROL — Drier than $w_o - 3.2$	Between $w_o - 3.2$ $w_o + 0.2$	Wetter than $w_o + 0.2$	DRY UNIT WEIGHT CONTROL — Less than .95 yd max.	Between .95 yd max .98 yd max.	Between .98 yd max. yd max.	Greater than yd max.	STD DEVIATIONS — $w_o - w_f$	C	D
CL-SM	28.0 20.0	10.0 3.0	1.2	13.7	114.6	116.3	115.7	14.5	0.8	101.5	100.5	1.0	88.1	10.9	0.2	5.6	31.1	63.1	±0.9	±2.0	±1.8

VARIATION OF FILL WATER CONTENT FROM LAB. OPTIMUM PERCENT BY DRY WEIGHT ($w_o - w_f$)

RANGE $w_o - w_f$	AVERAGE $w_o - w_f$	NO OF TESTS n	RATIO OF FILL DRY DENSITY TO LAB MAXIMUM DRY DENSITY (D IN PERCENT) AVERAGE $\bar{X}$	STANDARD DEVIATION σ	90% CONFIDENCE LIMITS	RATIO OF CYLINDER DRY DENSITY AT FILL WATER CONTENT TO LAB. MAX. DRY DENSITY (% IN PERCENT) AVERAGE $\bar{X}$	STANDARD DEVIATION σ	90% CONFIDENCE LIMITS
-1.7 to 2.7	-2.0	15	99.1	±1.40	±0.659	98.1	±0.88	±0.414
-0.6 to -1.6	-0.9	297	100.1	±1.35	±0.130	99.5	±0.40	±0.038
+0.5 to -0.5	0.1	1735	100.7	±1.51	±0.060	99.9	±0.50	±0.020
0.6 to 1.6	1.0	2286	100.9	±1.91	±0.065	99.1	±0.61	±0.021
1.7 to 2.7	2.1	805	100.3	±2.30	±0.134	97.6	±0.94	±0.055
2.8 to 3.8	3.1	137	99.2	±2.46	±0.350	95.8	±1.10	±0.157
3.8 to 4.8	4.2	11	99.2	±2.27	±1.299	94.5	±1.38	±0.790

Fig. 8.37 Relation between field and laboratory compaction (minus No. 4 material). Zone 1—Navajo Dam.

289

TABLE 8.8 CHARACTERISTICS OF SHEEPSFOOT ROLLERS SPECIFIED BY U.S. BUREAU OF RECLAMATION AND CORPS OF ENGINEERS FOR COMPACTING EARTHFILL PORTIONS OF EARTH DAMS.

Characteristic	Requirement
Number of drums	Two or more side by side
Drum length	Minimum—60 inches
	Maximum—72 inches
Drum diameter	Minimum—60 inches
Oscillation between drums	Required to prevent dual or triple drum units from acting as a single drum on uneven surfaces or rocky soils
Number of tamping feet	Minimum of one per each 100 in² of drum surface
Spacing of tamping feet measured on drum surface	Not less than 9 inches between centers
Space between drums	Minimum—12 inches
	Maximum—15 inches
Length of tamping feet	Minimum—9 inches
	Maximum—11 inches
Tamping foot area	Not more than 10 in² at 6 inches from the drum surface. Not less than 7 in² nor more than 10 in² at 8 inches from the drum surface
Pressure relief valves	Required to prevent bursting of drums ballasted with sand and water due to air pressure built up by friction
Weight of roller (empty)	Maximum—Corps of Engineers uses 2500 lb/ft of drum length. Bureau of Reclamation does not have this requirement
Weight of roller (ballasted)	Minimum—4000 lb/ft of drum length
Speed	Maximum—Corps of Engineers limits speed to 5 mph for either towed or self-propelled rollers. Bureau of Reclamation does not have this requirement

1500 fpm, and the dynamic force shall be not less than 40 000 lb at 1400 fpm. The roller shall be towed at speeds not to exceed 1.5 miles per hour by a crawler tractor with a minimum drawbar rating of 50 horsepower. The equipment manufacturer shall furnish sufficient data, drawings, and computations for verification of the above specifications, and the character and efficiency of this equipment shall be subject to the approval of the Contracting Officer.

Figure 8.41 shows a smooth-drum vibratory roller.

Smooth-drum Rollers Smooth-drum rollers may have steel drums, front and rear, or one drum may be replaced by smooth steel wheels. Smooth-drum rollers may be combined with accessories to produce vibrations for compacting sand, gravel, and rockfill. Smooth-drum rollers are not recommended for compacting cohesive soils in earth dams because of their low unit pressures and the smooth embankment surface that they leave.

Pad-type Tamping Rollers Tamping foot end areas of pad-type tamping rollers range from 20 to 30 in², and the total end areas

may occupy up to 25 percent of the surface of an imaginary cylinder constructed through the ends of the tamping feet. The tamping foot end areas of conventional sheepsfoot rollers occupy about 5 to 10 percent of the same imaginary surface. Large end areas in combination with high unit pressures produce excellent densities in a wide range of soils, but tamping feet with large end areas do not blend and mix embankment materials as effectively as conventional sheepsfoot rollers with tamping foot end areas in the 5 to 10 in² range.

Modern self-propelled, pad-type tamping rollers are capable of high speeds while compacting and are highly maneuverable because of articulated frames. The speed of these rollers, however, may be limited by the criteria for achieving proper soil compaction. The high maneuverability makes pad-type tamping rollers especially effective in compacting materials adjacent to rough or irregular surfaces.

Track-type Tractors In the absence of suitable vibrating equipment, granular soils may be compacted by track-type tractors. The U.S. Bureau of Reclamation usually specifies that

Fig. 8.38 A heavy-duty multiple-box pneumatic-tired compaction roller.

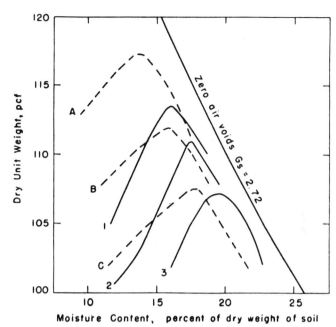

Fig. 8.39 Comparison of laboratory compaction curves (dashed lines) and pneumatic-tired roller compaction curves (solid lines) for a lean clay soil (*LL* = 36, *PI* = 15). (*After Johnson and Sallberg, 1960.*)

Compaction Effort

A—Modified AASHO,[a] 5 layers, 55 blows per layer, 10-lb hammer, 18-in. drop, 56,022 ft lb/cu ft

B—Intermediate,[b] 5 layers, 26 blows per layer, 10-lb hammer, 18-in. drop, 26,483 ft lb/cu ft

C—Equal to AASHO,[a] 5 layers, 12 blows per layer, 10-lb hammer, 18-in. drop, 12,223 ft lb/cu ft

1—Four coverages,[b] 31,250-lb wheel load, 16.00 × 21-in. tire, inflation pressure 150 psi

2—Four coverages,[b] 25,000-lb wheel load, 18.00 × 24-in. tire, inflation pressure 90 psi

3—Four coverages,[b] 15,875-lb wheel load, 18.00 × 24-in. tire, inflation pressure 50 psi

[a] 6-in. diameter × 45-in. high mold.

[b] Four coverages require 8 passes of roller.

compaction of clean sands, or sands and gravels may be performed by four passes (full coverages) of the treads of a 40 000-pound crawler-type tractor.

Construction Equipment Routing of construction equipment can be directed so as to provide compaction of embankment materials. Their use should be undertaken with the understanding that a uniform, high-density fill will probably not be produced.

Vibroflotation Vibroflotation is a patented method for obtaining high densities in sandy soils. Vibroflotation is carried out by jetting a special probe down into the sand deposit to be densified. By a combination of vibration and saturation, a quick condition is created in the sand adjacent to the probe. The sand at a short distance from the probe is densified and water jets on the probe are then used to carry additional sand to the bottom of the probe. The additional sand is added from the surface as the probe is slowly withdrawn.

Mitchell (1970) and Chapter 23 described the method of vibroflotation as a compaction process for cohesionless soils, and gave 18 examples of vibroflotation applications. He concluded that loose cohesionless soils can be densified effectively using vibroflotation. However, the presence of fines in excess of about 20 percent may seriously impair the effectiveness of this method. Vibroflotation can be used without regard to the location of the water table. Figure 8.42 shows the range of the soil grain sizes suitable for compaction by vibroflotation. Mitchell indicated that a review of the published records revealed that in soils suitable for vibroflotation treatment relative densities of at least 70 percent can usually be obtained at points midway between compaction.

Allowable bearing pressures subsequent to vibroflotation treatment are commonly in the range of 2 to 3 tons per square foot. While costs will vary with local conditions and the final soil condition specified, they may be on the order of $2.50 to $5.00 per cubic yard.

A study by D'Appolonia et al. (1955) found that (1) the relative density obtained by vibroflotation is not increased above 70 percent at points more than 3 ft from a single vibroflot compaction; (2) the overlapping effect for spacings greater than 8 ft is small; (3) spacings less than 6 ft should give relative densities greater than 70 percent within the compacted area; (4) the effect of adjacent compactions can be superimposed;

Fig. 8.40 Vibratory sheepsfoot roller.

Fig. 8.41 Smooth-drum vibratory roller.

and (5) square and triangular patterns give about the same result, but the triangular pattern is preferred because it gives the greatest compaction effort overlap.

A spacing of 7 to 8 ft is often chosen. Clean coarse sand can probably be compacted satisfactorily with spacings of 8 ft or more, whereas finer material containing clay inclusions may require much closer spacings.

Basore and Boitano (1969) reported on an opportunity to evaluate the effectiveness of both vibroflotation and compaction piles in densifying 30 ft of loose to medium dense sand fill, in order to improve the safety of the site against liquefaction, since a 3-story office building was to be placed on it. The vibroflotation method densified the granular material by vibration. The compaction piles densified primarily by displacement although some vibration occurred during driving. Sand compaction piles were installed by driving a hollow steel mandrel with a false bottom to the required depth, filling the mandrel with sand, applying air pressure to the top of the sand column, and withdrawing the mandrel. They concluded that vibroflotation and compaction piles were effective methods of

densifying the sand fill at Treasure Island if proper spacing between compaction points was used. Experience gained on this project indicated that vibroflotation produced a more uniform, dense fill. Compaction piles must be placed closer together than vibroflotation compaction points to densify sand to the same relative density. The cost of installing a compaction pile is much less than that of performing a single vibroflot compaction point. However, based upon a limited amount of cost information it appeared that vibroflotation was more economical than compaction piles in terms of cost per cubic yard.

Densification of the building site by compaction piles spaced 4 ft and 5 ft on centers produce a fill with an average relative density in excess of 75 percent. The densification process was not completely uniform and several tests fell below the specified minimum relative densities of 75 percent beneath footings and 65 percent below floor areas.

In 1948 the Bureau of Reclamation conducted field tests of vibroflotation equipment at Enders Dam in Nebraska. Attempts were made to densify a clean, free-draining sand and a sandy

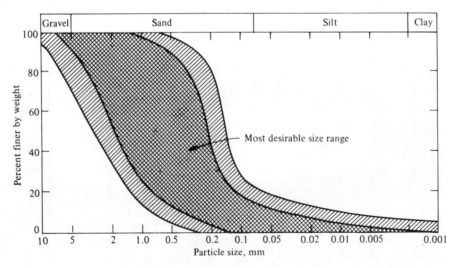

Fig. 8.42 Range of soil grain sizes suitable for compaction by vibroflotation. (*After Mitchell, 1970.*)

Fig. 8.43 Vibroflot used at Enders Dam, Nebraska.

Fig. 8.44 Gasoline-powered backfill tamper in use at Fort Cobb Dam, Oklahoma.

silt containing approximately 60 percent by weight of particle sizes smaller than a No. 200 sieve. The vibroflotation process was effective in increasing the density of the sand, but did not produce an increase in density of the silt. The tests conducted in the silt resulted in a column of sand (introduced from the top of the hole) replacing silt which washed out. There was no intermixing of the added sand and the in-place silt. Figure 8.43 shows the vibroflot used at Enders Dam, Nebraska.

Backfill Tampers Hand-held, air- or gasoline-powered backfill tampers are used to compact soils in areas where larger equipment cannot be used. Criteria for tampers are that they should be able to compact soil to the required unit dry weight at the required moisture content and achieve a relatively uniform density from top to bottom of the lift.

Backfill tampers can effectively compact soils to lift thicknesses comparable to those compacted by sheepsfoot and rubber-tired rollers. Tests performed by the British Road Research Laboratory (Johnson and Sallberg, 1960) using a heavy (1350-lb) frog rammer and four different 250-pound tampers indicated that maximum densities greater than British Standard Maximum (comparable to Proctor Standard Effort) could be obtained at moisture contents slightly dry of laboratory optimum moisture content. Other results of these tests showed that increasing the size of the baseplate decreased the maximum unit weight and increased the optimum moisture content. Figure 8.44 shows a gasoline-powered backfill tamper in use at Fort Cobb Dam, Oklahoma.

Baseplate Vibratory Compactors The distinguishing characteristic of compactors in this category is a vibrating rectangular or circular baseplate. The baseplates may be mounted individually for one-man operation, or they may be gang-mounted on machines. Tests have shown baseplate vibrators to be capable of producing adequate compaction in all types of soils. Figure

8.45 shows a rectangular baseplate vibrator in use at Glen Elder Dam spillway, Kansas.

Converse (1957), in discussing compaction of cohesive soils by baseplate vibratory compactors, summarized basic rules for such compaction as follows:

1. Operate at or near resonant frequency of the soil oscillator mass.
2. Supply a dynamic force approximately equal to the dead weight of the oscillator.
3. Have the moisture content of the soil equal to or slightly higher than the optimum for maximum density as determined in the laboratory.

Fig. 8.45 A rectangular baseplate vibrator in use at Glen Elder Dam spillway, Kansas.

4. Provide sufficient dead weight to give a unit dead weight bearing pressure of the oscillator on the soil of about 10 to 12 psi.

The British Road Research Laboratory (Johnson and Sallberg, 1960) in tests of the five sizes of baseplate compactors on four different types of soils found that:

1. Two of the compactors were able to attain unit weights in excess of 100 percent of standard maximum unit weight for a heavy clay soil at 3 to 9 percent dry of standard laboratory optimum moisture content.
2. Dry unit weights ranging from 101.7 to 106.4 percent of standard maximum unit weight were attained on a sandy clay at 1 percent dry of standard laboratory optimum moisture content.
3. Dry unit weights ranging from 105.8 to 111.6 percent of standard maximum unit weight were attained on a well-graded sand at approximately a standard laboratory optimum moisture content.
4. Dry unit weights ranging from 98.4 to 109.3 percent of standard maximum unit weight were attained on a gravel–sand–clay mixture at approximately modified (AASHTO) laboratory optimum moisture content.

Field tests by various organizations have indicated that baseplate vibratory compactors attain their maximum unit weights in a relatively small number of passes, depending on the thickness of lift being compacted. As with vibrating rollers, however, the slower the speed of travel, the more vibrations that will occur at a given point, and the fewer the passes required to attain a given soil unit weight. Also the soil unit weight is greater for a given number of passes.

Comparison of Performance of Sheepsfoot and Rubber-tired Rollers in Cohesive Soils The density produced in a given layer of earthfill, assuming the moisture content is proper, is a function of the level of compactive effort, which depends upon the roller weight and the number of passes. The question concerning compactive effort is how much of it goes toward densifying the soil. The energy expended in the continual shearing of soils wet of optimum moisture content by a heavy sheepsfoot roller and the pushing and shoving of earth by a heavy rubber-tired roller do not accomplish any significant densification. The compaction method and compaction equipment must be selected to be appropriate to the soil bearing capacity.

The compacted earthfill constructed by a sheepsfoot roller is more uniform with regard to density, moisture content, and material than that compacted by a rubber-tired roller used alone. The tamping feet, by penetrating the full depth of an uncompacted layer, will blend and bond successive layers. Sheepsfoot rollers can compact effectively over a wider range of moisture contents than can a rubber-tired roller.

The surface left by a rubber-tired roller is smooth and hard and must be scarified to obtain bond with the succeeding layer. Consequently, in earthfills that are not designed to store water—and hence where horizontal permeability is not of importance—the rubber-tired roller could be an appropriate choice. A rubber-tired roller can compact faster and with fewer passes than a sheepsfoot roller, and fewer pieces of compaction equipment may be needed for a large job. This frees the working surface of several pieces of equipment and the need for scarifying the layer surfaces may thus be outweighed by the speed and efficiency of the rubber-tired rollers. The rubber-tired roller is more effective in compacting granular soils where the great weight of such a roller can be used to advantage. In areas where a rainy season occurs, the sealing of an embankment surface by the normal operation of a rubber-tired roller can be a distinct advantage.

The field optimum for sheepsfoot rollers occurs at a degree of saturation of about 80 percent, but the field optimum moisture content for rubber-tired rollers may occur at about 90 percent. This places a restriction on the use of rubber-tired rollers for high embankments, where construction pore pressures may be detrimental to stability.

Field tests have shown that rubber-tired rollers produced laminations (shear surfaces or slickensides) in the soil at moisture content wet of optimum that were more pronounced than those produced with sheepsfoot rollers under the same conditions. Hilf (1957) reported that these laminations could be produced in the laboratory using a sheepsfoot compactor machine on a clayey loess (CL). The tests made with various compactive efforts showed that shear surfaces occurred only on samples compacted wet of optimum moisture content that corresponded to an air content of less than 6 percent of the volume of the soil mass for each of the four compactive efforts used.

The heavy sheepsfoot roller (4000 lb per linear foot of drum) and medium weight (15 000 to 30 000 lb wheel weights) rubber-tired rollers will usually compact soils to densities in the vicinity of the Proctor maximum dry unit weight. Heavier rubber-tired rollers are used to compact subgrades and base courses for heavy-duty roads and airfields.

8.4 CONTROL OF COMPACTION

Control Criteria for Cohesive Soils

The importance of control of compaction of embankments and backfill on earth dams, canals, highways, and airfields is widely recognized. The relationships between the state of compactness of a soil (its dry density or void ratio) and the engineering properties of compressibility and shear strength have been discussed in Section 8.2. For cohesive soils, it was shown that the moisture content during compaction and the resulting dry density or void ratio have significant effects on these properties. Since the measurement of compressibility and shear strength requires time-consuming and expensive laboratory tests conducted by highly skilled personnel, the use of the simpler density and moisture measurements has been generally accepted for field control of cohesive soils.

Most specifications for compaction of such soils require the dry density of the compacted soil mass to be a stated percentage of some laboratory standard. There are two major standards in use: (1) Standard Proctor Compaction (ASTM Designation D-698 or AASHTO Designation T-99, both with compactive effort of 12 375 ft-lb/ft^3; this standard is generally used for embankments and backfills of cohesive soils); and (2) modified AASHTO T-180 or ASTM D-1557, both of which are 56 250 ft-lb/ft^3; this standard is generally used for base courses under flexible or rigid highway and airfield pavements. For the construction of hydraulic structures, moisture control may be specified in addition to a density requirement. Although the requirements for density and moisture control differ, depending on the purpose of the earthfill, virtually all authorities use these criteria.

The use of a percentage of a standard laboratory maximum dry density (for example, 98 percent), rather than a stated value for dry density (for example, 115 lb/ft^3), stems from the inherent variability of soils. Small changes in size and gradation from place to place within a borrow area will result in significant changes in absolute values of maximum dry density and optimum moisture content. Table 8.9 illustrates the variability of numerical values of density and moisture in consecutive control tests on soils used in two earth dams. Both soils were gravel-free (100 percent $<\frac{1}{4}$-in size). The Trenton Dam soils

TABLE 8.9 STANDARD PROCTOR COMPACTION VALUES OF SOILS FROM CONSECUTIVE CONTROL TESTS.

	Trenton Dam, 1952			*Cheney Dam, 1964*	
Control Test No.	*Proctor Optimum Moisture Content, Percent*	*Maximum Proctor Dry Density, pcf*	*Control Test No.*	*Proctor Optimum Moisture Content, Percent*	*Maximum Proctor Dry Density, pcf*
6-5-A1-R	17.0	102.8	4-1-A3-R	20.3	106.1
6-5-A2-R	18.8	102.5	4-1-B1-R	22.0	102.6
6-5-A3-R	17.4	103.3	4-1-B2-R	19.7	103.2
6-5-A4-R	15.4	104.3	4-1-B3-R	17.9	109.5
6-5-A5-R	18.0	105.5	4-2-A1-R	16.8	110.9
6-6-A2-R	18.2	103.7	4-2-A2-R	20.4	104.3
6-6-A3-R	17.5	104.5	4-2-A3-R	21.4	101.0
6-6-A4-D	17.8	104.8	4-2-B1-R	20.5	104.3
6-6-A5-R	18.7	104.1	4-2-B2-R	19.7	104.8
6-7-A1-R	17.9	104.2	4-2-B3-R	20.7	103.1
6-7-A2-R	17.5	104.8	4-3-A1-R	17.9	108.3
6-7-A3-R	18.0	103.0	4-3-A2-R	19.6	105.2
6-7-A4-R	16.6	104.0	4-3-A3-R	22.1	101.9
6-7-A5-R	17.2	104.7	4-6-B1-R	19.0	107.7
6-9-A1-R	17.7	104.0	4-6-B2-R	23.7	99.7
6-9-A2-R	17.4	106.9	4-6-B3-R	20.0	105.6
6-9-A3-D	18.2	104.9	4-7-A1-R	21.1	105.6
6-9-A4-R	17.3	105.3	4-7-A2-R	20.6	103.6
6-9-A5-R	17.4	102.9	4-7-A3-R	21.3	103.1
6-10-A1-R	17.0	105.4	4-7-B1-R	22.3	99.9
6-10-A2-R	17.4	104.6	4-7-B2-R	19.8	104.5
6-10-A3-R	16.4	107.2	4-7-B3-R	22.0	101.6
6-10-A4-R	17.3	103.8	4-8-A1-R	20.1	105.5
6-11-A1-R	17.8	103.4	4-8-A2-R	20.1	104.9
6-11-A5-D	17.8	105.6	4-8-A3-R	21.2	103.1
6-12-A1-R	17.3	104.8	4-8-B1-R	24.6	97.8
6-12-A2-R	17.8	104.4	4-8-B2-R	19.6	105.9
6-12-A4-R	16.4	103.6	4-8-B3-R	17.2	109.0
6-12-A5-D	17.6	103.9	4-9-A1-R	20.7	106.2
6-13-A1-R	16.3	108.1	4-9-A2-R	22.3	100.5
6-13-A2-R	17.8	104.9	4-9-A3-R	20.9	103.6
6-13-A3-R	17.5	112.1	4-9-B1-R	20.1	104.0
6-13-A4-R	17.7	106.3	4-9-B2-R	19.2	106.8
6-13-A5-R	17.7	103.5	4-9-B3-R	21.2	102.8
6-14-A1-R	18.1	102.6	4-10-A1-R	22.4	102.5
6-14-A2-R	17.5	101.8	4-10-B1-R	22.5	101.9
Minimum value	15.4	102.5		16.8	97.8
Maximum value	18.7	112.1		24.6	110.9
Mean value	17.5	104.4		20.7	104.2
Std. deviation	0.64	1.42		1.72	2.86

were obtained from what would be considered a very uniform aeolian deposit (loess) while the Cheney Dam soils were obtained from a more variable alluvial deposit.

It is more desirable, therefore, to require a fill density equal to the maximum density (or 98 percent of it) of a laboratory standard than to specify a numerical value of dry density. This procedure ensures that a compactive effort comparable to that of the laboratory standard is achieved in the fill. Since the designs are usually based on values of compressibility and shear strength obtained from laboratory tests on samples compacted at the standard density, this procedure is valid. Also, use of a percentage of standard dry density eliminates the need to consider variations in average specific gravity of the soil that often occur with gradation changes, especially changes in percentage of gravel and sand sizes.

Specifications for control of compaction may include the statement "Each layer shall be compacted to 100 percent (or 98 percent) of the dry density obtained in a standard laboratory compaction test. The moisture content of the material prior to

and during compaction shall be controlled to fall within the range of optimum moisture content of this standard to 2 percent less than the optimum moisture content." Some specifications do not state a percentage of standard density, but, instead, attempt to achieve the comparable compactive effort by specifying the type of equipment to be used, the thickness of layer to be compacted, and the number of coverages or "passes" of the equipment. Usually the moisture requirement, with relation to optimum, is stated, to ensure that the field compactive effort is used effectively in this type of specifications.

The basis for control of compaction is a hole in the fill from which compacted soil is extracted and weighed. The weight of soil can be divided by the measured volume of the hole to obtain the wet density, γ_{wet}, of the fill. This value can be converted to dry density of fill, γ_{df}, by the relation:

$$\gamma_{df} = \frac{\gamma_{\text{wet}}}{1 + w_f} \qquad (8.7)$$

where w_f is moisture content of fill in percentage of dry weight of soil.

The ratio $\gamma_{df}/\gamma_{dm} = D$ (usually expressed as a percentage) is the criterion of compaction, where γ_{dm} is the maximum dry density of the soil for a given laboratory compaction standard. This density is the ordinate of the peak point of the Proctor dry density curve shown in Figure 8.46a; the abscissa of the peak point is the optimum moisture content w_o. Also shown in this figure are the points of an example fill wet density and its corresponding dry density, designated $\gamma_{df}(1 + w_f)$ and γ_{df}, respectively.

In the past the value D was obtained by the following steps:

1. Obtain fill wet density.
2. Determine fill moisture content.
3. Obtain fill dry density by Equation 8.7.
4. Add (or remove) water from several specimens of the sample removed from the fill and compact them to get ordinates of curve A in Figure 8.46a.
5. Determine moisture contents for each specimen to obtain abscissas of curve A.
6. Obtain curve B from curve A by Equation 8.7.
7. Divide fill dry density by the ordinate of the peak point of curve B.

This procedure is relatively simple. However, it takes time, mainly for drying specimens to obtain the moisture contents (steps (2) and (5) above). Standards of the American Society for Testing Materials require at least 16 hours of oven drying at 110°C for proper moisture content determination of clayey soils.

On most construction jobs, this amount of time is unacceptable and would make the control engineer merely a job historian; hence, many short-cut, approximate methods have been used.

The Proctor penetration needle has been used to obtain the approximate difference between optimum moisture content and fill moisture content, as prescribed in Designation E-22, Bureau of Reclamation *Earth Manual* (1968c). Rapid methods of determining moisture content values include: the alcohol-burning method (Highway Research Board, 1952a), the Buoyoucos alcohol method using a hydrometer (Bouyoucos, 1931), and a moisture meter using calcium carbide to generate acetylene in a closed container connected to a pressure gauge (Reinhold, 1955). All of these rapid moisture methods are either only approximate or give correct values of moisture content only for certain kinds of cohesive soils. Hence, the standard oven method must be used in addition, or at least as a periodic check.

The use of radioactive materials for determination of moisture content and density has been under investigation since 1949. Results reported by Horonjeff and Javette (1956) showed some promise, but difficulties in calibration and in adapting laboratory-type instruments for field use were reported. Dunn and McDougall (1970) reported that since about 1959 nuclear density devices have been available commercially for use by state highway departments and by others interested in measuring soil densities. In 1965 a conference was held to correlate these devices for variations in measurements among different gauges. For backscatter-type gauges the average standard error was reported as ± 11.0 pcf, and for transmission-type gauges the average standard error reported was ± 7.53 pcf. These errors were unacceptable. In 1969 the North Carolina State University Nuclear Soil Gauge Workshop-Symposium was held to present methods of improving nuclear-gauge calibration techniques. The average standard error or average difference between measured and actual densities for backscatter gauges was ± 3.81 pcf and for direct transmission gauges was ± 2.60 pcf. This would appear to be a significant improvement. However, the average standard deviations, or average differences in readings among individual gauges from their average readings, remained essentially the same for the 1969 workshop as for the 1965 conference. Dunn and McDougall concluded that the backscatter gauges, already the most convenient density gauges available, can also be made the most accurate by the

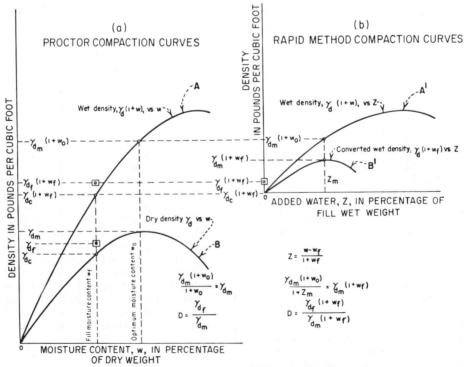

Fig. 8.46 Comparison of compaction curves—Proctor and Rapid Methods.

optimum design of a dual-gauge system. They predicted that dual-gauge design with "errors of 1 pcf or less should soon be available." It must be remembered that the nuclear method is designed to measure numerical values of density and moisture only. The need to compare these values with the laboratory standard remains.

If a dry density curve is assumed for the soil being tested, then an approximate dry density of the fill can be obtained from a graph by compacting the soil in the laboratory cylinder at fill moisture content and determining the wet density (Highway Research Board, 1952b). Another method uses the wet density at fill moisture content and the Proctor needle reading of the soil in the cylinder to choose one of a set of typical curves from which the approximate maximum dry density and optimum moisture content can be found (Highway Research Board, 1952c).

Where the major portion of fill material is a fine-grained cohesive soil with liquid limit greater than 40 percent, the U.S. Corps of Engineers (1970a) uses a correlation between liquid limit and Proctor maximum dry density and optimum moisture content as the basis for an approximate density control method. The liquid limit is determined for material taken from the field moisture–density test and the corresponding Proctor maximum dry density and optimum moisture content are obtained from the correlation curves. To facilitate results, a one-point liquid limit procedure may be used (see U.S. Corps of Engineers, 1949; ASTM, 1970a), although one of every five liquid limit tests is performed by the standard four-point procedure. This method requires drying of the soil in the liquid limit test procedure and requires conversion of field wet density to dry density. Errors in correlation, one-point liquid limit, and rapid drying are likely to make this approximate method susceptible to important inaccuracies.

Prior to 1957 the only correct method was to obtain the value γ_{dm} (maximum dry density for the standard used) for the material extracted from the hole. This was done for many years by the Bureau of Reclamation and by many other construction organizations. In 1957 Hilf (1959a, b) developed an exact method of dry density control using wet densities only.

Hilf's Rapid Method (Hilf, 1959a, b)

If the Proctor curve is obtained for each sample extracted from the fill, it is possible to calculate the ratio of fill dry density to maximum Proctor dry density D precisely from wet densities, thus saving the time required for moisture content determinations. This is done by dividing the fill wet density $\gamma_{df}(1 + w_f)$ by the value $\gamma_{dm}(1 + w_f)$, which is obtained as shown in Figure 8.46b. In this plot, the ordinate of curve A′ is wet density (exactly the same as the ordinate of curve A in Figure 8.46a). The abscissas, however, are not moisture content w, but the added water Z in percentage of fill wet weight. For the soil at fill moisture content, whatever that moisture content may be, $Z = 0$. The relation between Z and w is

$$Z = \frac{wW_s - w_f W_s}{W_s(1 + w_f)} = \frac{w - w_f}{1 + w_f} \qquad (8.8)$$

where W_s is dry weight of soil. The value of Z can be negative.

Curve A′ can be plotted by keeping track of the amount of water mixed with a given weight of moist soil prior to compaction. If the ordinate of every point of curve A′ is divided by the quantity $(1 + Z)$, a new curve B′ is obtained whose ordinates are

$$\frac{\gamma_d(1 + w)}{1 + Z} = \gamma_d(1 + w_f) \qquad (8.9)$$

Since w_f is a constant for every point on the curve, the maximum ordinate of curve B′ must have the value $\gamma_{dm}(1 + w_f)$. Dividing this value into the fill wet density $\gamma_{df}(1 + w_f)$ gives the desired value D.

The location of the peak point of the curve B′ shows whether the soil is at optimum moisture content (w_o) or is less than or greater than optimum. However, the exact magnitude of the difference between optimum moisture content and fill moisture content is unknown. If Z_m is the abscissa of the peak point of curve B′, it can be shown that

$$w_o - w_f = \frac{Z_m(1 + w_o)}{1 + Z_m} \qquad (8.10)$$

Since w_o is unknown, it must be estimated in order to obtain the magnitude $w_o - w_f$. However, an error in estimating w_o results in a much smaller error in the value $w_o - w_f$ for small values of Z_m. For example, for $Z_m = +0.02$, an error of 0.05 in estimating w_o is reduced to $(0.02/1.02)(0.05) = 0.00098$ for $w_o - w_f$. This small error is acceptable for control purposes.

To avoid the necessity of estimating w_o for each density test, a set of curves was prepared that automatically estimates w_o for the coordinates of the peak point of curve B′. Figure 8.46b shows the relationship between the peak point of curve B′ and the wet density at optimum moisture content. The approximate relation between the latter value and the optimum moisture content was determined for 1300 soils compacted by standard Proctor compaction and is shown in Figure 8.47. The relationships of Figures 8.46 and 8.47 were combined to obtain the curved lines shown in Figure 8.48. In this figure, the diagonal lines are a nomograph to convert wet densities (ordinates of curve A′ in Fig. 8.46b) to converted wet densities (ordinates of curve B′) by dividing by $(1 + Z)$.

Figure 8.48 is an example of Hilf's method, using the form designed for the Bureau of Reclamation's version of Proctor laboratory compaction standard which used the $\frac{1}{20}$-ft^3 compaction cylinder. By changing the weight of soil sample and of water added in proportion to the size of cylinder used, this form is suitable for density control for any compaction standard. (For a $\frac{1}{30}$-ft^3 compaction cylinder 5.0 lb of moist soil is used.) To determine whether the correction lines for moisture control are applicable for a given soil and compaction standard, the wet density at optimum moisture content versus the optimum moisture content for that soil should be plotted on Figure 8.47. If the point falls within the area of points that determined the line, no changes need be made; this is usually the case. If the correction curves of Figure 8.48 are not available, the procedure is (1) multiply the maximum ordinate of curve B′ by $(1 + Z_m)$ to find $\gamma_{dm}(1 + w_o)$; (2) estimate w_o from the curve shown in Figure 8.47, or a similar curve for the soils and compactive efforts used; and (3) find $w_o - w_f$ from Equation 8.10.

Figure 8.48 is an example in which the fill moisture content is less than optimum. The test data are:

Fill wet density = 127.5 lb per cubic foot = $\gamma_{df}(1 + w_f)$

Point	Wet Density in Pounds per Cubic Foot	Added Water in Percent of Fill Wet Weight Z	Converted Wet Density in Pounds per Cubic Foot
(1)	123.4	0	123.4
(2)	128.6	2	126.1
(3)	124.6	4	119.8
By the parabola method (Fig. 7.49)			
0		$Z_m = 1.6$	126.3

$(\gamma_{dm}(1 + w_f))$

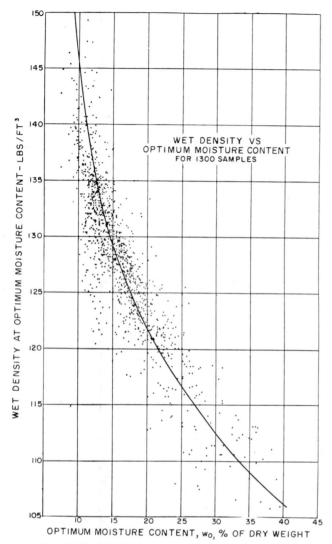

WET DENSITY VS OPTIMUM MOISTURE CONTENT FOR 1300 SAMPLES

Fig. 8.47 Wet density vs. optimum moisture content.

Then,

$$D = \frac{\gamma_{df}(1 + w_f)}{\gamma_{dm}(1 + w_f)} = \frac{127.5}{126.3} = 101.0 \text{ percent}$$

$$w_0 - w_f = Z_m + \text{correction value} = +1.6 + 0.2 = 1.8 \text{ percent}$$

(dry of optimum)

These values are sufficient for use in accepting or rejecting the compacted fill according to criteria established for the work.

After the fill moisture content has been determined by drying a sample to constant weight at 110°C, the field density test is completed for record purposes as follows:

$$w_f = 15.0\%$$

$$\gamma_{df} = \frac{\gamma_{df}(1 + w_f)}{1 + w_f} = \frac{127.5}{1.15} = 110.9 \text{ lb/ft}^3$$

$$\gamma_{dm} = \frac{\gamma_{dm}(1 + w_f)}{1 + w_f} = \frac{126.3}{1.15} = 109.8 \text{ lb/ft}^3$$

$$w_o = w_f + (1 + w_f)Z_m = 0.15 + (1.15)(0.016)$$

$$= 0.168 \text{ or } 16.8\%$$

If the ordinate of point (2) on the converted wet density curve (curve B′) is smaller than the ordinate of point (1), then point (3)

is obtained by drying a weighed specimen and reweighing to determine Z_3, which will be negative, and then compacting after thorough mixing. When the ordinate of point (2) on curve B′ is only slightly smaller than that of point (1), $Z_3 = +1$ percent is often used. With these modifications, the procedure shown in Figure 8.48 is applicable for soils at optimum moisture content and for soils wet of optimum. For soils that have obviously been compacted several percent wet of optimum, points (2) and (3) can be obtained by drying, mixing, weighing, and compacting. The determination of the peak point of curve B′ in Figure 8.46b can be made by obtaining a sufficient number of points, closely spaced, to establish it precisely. However, to avoid the necessity for more than three points, a parabola with vertical axis can be fitted to the points. Figure 8.49 describes the graphical solution for Z_m and the ordinate of the peak point.

If points (1), (2), and (3) are equally spaced horizontally in 2-percent increments, the values of y_m and Z_m can be tabulated for various values of the difference in ordinate between points (1) and (2) and (1) and (3). Such tabulations are used as a check and in some instances as a substitute for the graphical parabola method by the Bureau of Reclamation *Earth Manual* (1968d).

Discussion of Hilf's Rapid Method The Rapid Method of Control described here has been used satisfactorily on about 150 earth dams since 1957 and has been adopted for control of all compacted cohesive soils on Bureau of Reclamation projects. It has been used extensively in Brazil (de Mello et al., 1960a) and is an Australian Standard (Hosking, 1973). It is included as an alternate method in the Department of the Navy *Design Manual* (NAVDOCKS DM-7, 1962), in the U.S. Corps of Engineers (1970b) *Construction Quality Control Guide*, and as an ASTM suggested method of test for Rapid Compaction Control (ASTM, 1970b).

The time required for performing the test (excluding the time required to dig and measure the density hole) has varied from as little as 6 minutes to $1\frac{1}{2}$ hours depending on the amount of gravel-sized particles that must be removed from the soil. The average time for gravel-free soil is 25 minutes; 1 hour for gravelly soils. On large earth dams, mechanical mixers for adding water uniformly with the soil, mechanical screeners to remove the gravel-sized particles, and mechanical compactors are used to facilitate the work and achieve greater uniformity. For the control of highway and canal embankments, portable or trailer-mounted equipment is used, and the operations of compaction, mixing, and screening are done manually. The test requires a scale of at least 25 lb capacity sensitive to 0.01 lb.

Experience has shown that to obtain maximum accuracy, the field density hole (or sample) should be large enough for at least three laboratory compactions without reuse. Also it is desirable to prevent moisture loss by using a covered watertight container when transferring soil from the field to the laboratory and to keep the material covered in the laboratory when not using it.

The primary purpose of this control method is to provide an irrefutable basis for acceptance or rejection of the compacted fill in the shortest possible time. As explained, this is done without requiring the determination of the values of dry density or of water content. There appears to be little justification for using approximate quick moisture methods or relying on assumed compaction curves to compare with the fill density when a method that gives the *exact* ratio of fill dry density to laboratory maximum dry density is available.

Moisture control is considered essential in the construction of high earth dams, and it is the key to effective use of compaction equipment in all earth embankments. The method described provides moisture control information in terms of the difference between optimum moisture content and fill

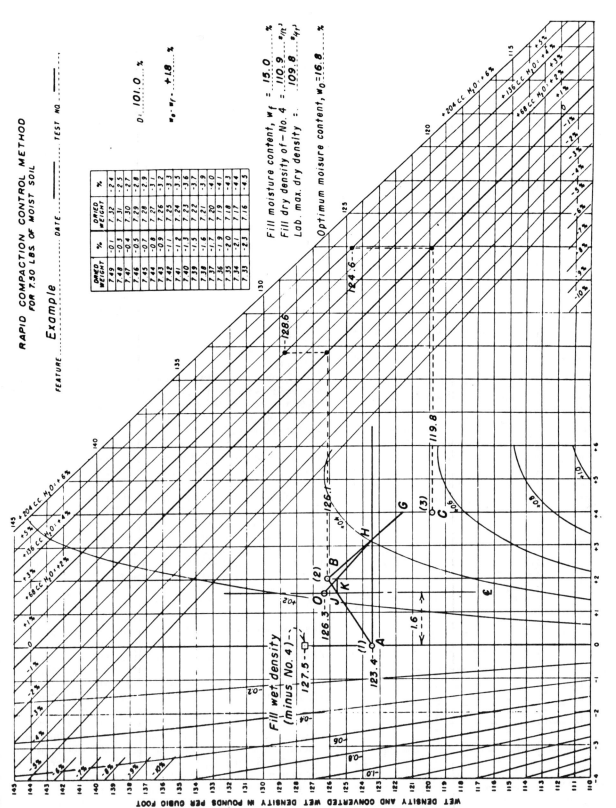

ADDED WATER IN PERCENT OF FILL WET WEIGHT
FOR PEAK POINT OF CURVE WHEN CORRECTED AS SHOWN BY CURVED LINES

ALSO Wo − Wf (%)

Fig. 8.48 Example of rapid method in which fill moisture content is less than optimum.

Parabola Method

Graphical solution for vertex, O, of a parabola whose axis is vertical, given three points A, B, and C. If more than three points are available, use the three closest to optimum.

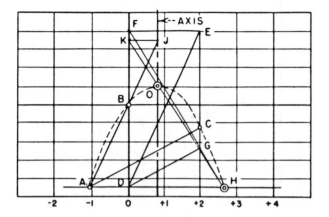

1. Draw horizontal base line through the left point, A, and draw vertical lines through points B and C.
2. Draw line DE parallel to AB, point E lies on the vertical line through point C, project E horizontally to establish point F on the vertical line through B.
3. Draw line DG parallel to AC, point G lies on the vertical line through point C.
4. Line FG intersects the base line at H. Axis of parabola bisects AH, draw the axis.
5. Intersection of line AB with the axis is at J, project J horizontally to K, which lies on the vertical line through point B.
6. Line KH intersects the axis at 0, the vertex.

NOTE: If points A, B, and C are equally spaced horizontally (this is true when 2 points are obtained by adding water or when soil is dried exactly 2 percent) steps 2 and 3 above are eliminated. Point F coincides with point B and point G is halfway between the base line and point C. Hence, point H is obtained by drawing BG and point O is obtained by steps 5 and 6 as usual. See graph below.

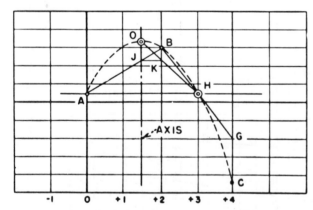

Fig. 8.49 Graphic solution for peak point of parabola.

moisture content. Although the value is not precise, experience has shown that the error is negligible for control purposes.

An additional advantage of this method over the approximate methods is that the data obtained for control purposes in the former can later be used to obtain the correct values of dry density and water content for record purposes. This is accomplished by obtaining only one moisture content—the fill moisture content—by drying a sample in the oven.

The method described is believed to be particularly applicable for control of compaction highways and canal embankments where the compaction characteristics of the soil are likely to vary along their routes, making the assumption of a dry density curve highly speculative.

In 1980, a programmable hand calculator was used to perform all the calculations to obtain D, C, and $w_o - w_f$ of the Rapid Method for input values of field wet density, cylinder wet density at field moisture, and two additional cylinder wet densities each converted to field moisture conditions (divided by $1 + Z$). The analytical solution of a parabola with vertical axis, given three points (Hilf, 1959b), is used to determine the peak ordinate of the converted wet density curve and its abscissa, Z_m. From the relation shown in Figure 8.46(b), the ordinate of the peak point of the parabola multiplied by $(1 + Z_m)$ gives the value of an ordinate in Figure 8.47 for which the corresponding abscissa from the graph or its analytical representation is w_o. The program then obtains the approximate value of $w_o - w_f$ from Equation 8.10. If a printer is coupled to the calculator, all input and output values can be displayed and identified. Figure 8.50 is an example showing the relation between the Rapid Method and Standard Proctor Compaction (without predrying) using the results of the computer program.

Compaction Control for Cohesionless Soils

The engineering characteristics of cohesionless soils, such as compressibility and shear strength, have been shown in Section 8.2 to be related to their relative density, that is, their rate of compactness between the loosest and densest states. Specifications for control of compaction of cohesionless soils either require a stated percentage of relative density to be achieved or specify type of equipment, thickness of layer, and number of passes of equipment. The latter, so-called equipment specification, is also based on relative density using experience or a previously constructed test section to determine what equipment and procedure are likely to obtain the desired relative density.

To obtain the density in place, the field density test procedure (Bureau of Reclamation *Earth Manual*, 1968e; Bowles, 1986) can be used. Size of density hole or pit is selected according to the maximum size of particle. For fine sands, a 6-in diameter 9-in deep, or 8-in diameter 12- to 14-in deep hole can be used. For soils containing appreciable gravel, 10- to 12-in diameter 12- to 14-in deep conical holes are often used. For cohesionless soils containing appreciable cobbles, a small test pit at least 2 by 3 ft in size and $1\frac{1}{2}$ to 2 ft deep can be used. The usual procedure requires weighing all the material extracted from the hole and measuring the volume of the hole that it came from.

A relative density test is made on the materials extracted from the fill. To determine the minimum density or 0-percent relative density, oven-dried material is used and the soil is carefully poured into a container of known volume with a free-fall of about 1 inch, then carefully struck off without jarring or vibrating. For the maximum density (100 percent relative density determination), the soil is placed either in the oven-dry or saturated condition in a container of known volume and vibrated on a vibratory table at up to 3600 vibrations per minute for 8 min under a surcharge of 2 psi, after which the volume and dry weight are measured. See Part B, Designation E-12, Bureau of Reclamation *Earth Manual* (1968f). The relative density is then obtained by Equation 8.4.

For sands and gravels containing a significant amount of fines, the maximum density obtained by vibration may be somewhat less than can be obtained by extremely heavy impact compaction such as the modified AASHTO procedure.

Before w_f is known			After w_f is known = 17.4%[c]	
Wet density, lb/ft³	Added water Z, %	Converted wet density CWD, lb/ft³	Dry density = CWD/$(1 + w_f)$	Moisture content $w = w_f + (1 + w_f)Z$
Fill 125.8[a]			107.2	w_f = 17.4%
Pt.0 128.4	$Z_0 = 0$[a]	128.4[a] CWD_0	109.4	w_f = 17.4%
Pt.1 124.2	$Z_1 = 2$[a]	121.8[a] CWD_1	103.8[d]	$w_1 = 19.7\%$[d]
Pt.2 123.7	$Z_2 = -2.3$[a]	126.6[a] CWD_2	107.8[d]	$w_2 = 14.7\%$[d]
Peak point 127.9	$Z_M = -0.74$[b]	128.9[b] CWD_M	109.8	$w_{OPT} = 0.174 + (1.174)(-0.0074) = 16.5\%$

(Z_M and CWD_M are coordinates of the peak point of the parabola)

(Exact) $D = 125.8/128.9 = 97.6\%$[b]   (Exact) $D = 107.2/109.8 = 97.6\%$

(Exact) $C = 125.8/128.4 = 98.0\%$[b]   (Exact) $C = 107.2/109.4 = 98.0\%$

(Approx.) $w_{OPT} - w_f = -0.87\%$, wet[b]   (Exact) $w_{OPT} - w_f = 16.5 - 17.4 = -0.9\%$, wet

[a] input to computer
[b] calculated by computer program

[c] completion of test for records
[d] essential only if Proctor curve is to be plotted

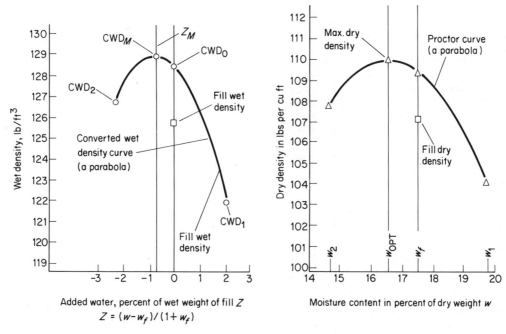

Fig. 8.50 Relation of rapid compaction control method to standard Proctor compaction (without predrying). Example using English units.

When relatively deep foundation fills are compacted by surface vibrators, vibroflotation, or other devices, it is often inconvenient to obtain samples of the compacted material and run relative density tests in the laboratory. Hence, the field penetration test with the split-spoon sampler (Designation E-21, Bureau of Reclamation *Earth Manual*, 1968g) provides a rapid approximate method for determining the relative density of the soil. Figure 8.51a shows a relationship of blow count to relative density based on research of the Bureau of Reclamation (Holtz and Gibbs, 1957). Figure 8.51b shows the same data plotted in a manner suggested by Coffman (1960).

Schmertmann (1970, 1978) suggested that the static cone penetration test (Schmertmann, 1967) is superior to the standard penetration test in evaluating the condition of the sand before and after compaction. He cautioned about converting standard penetration (N) values as well as cone bearing values to relative density for the after-compaction condition. He stated that the work of D'Appolonia, Whitman, and D'Appolonia (1969) showed that surface compaction can improve the soil in two ways, that is, it increases both density and lateral stress. Such lateral stress increase will likely increase N values and cone bearing as well. This suggests that direct sampling and testing

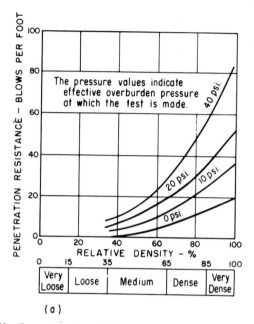

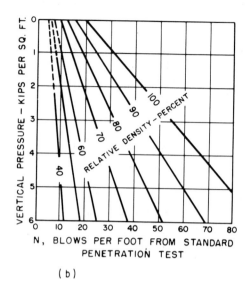

(a)

(b)

Fig. 8.51 General relationship between penetration resistance and relative density for dry and saturated cohesionless sands. ((a) *After Bureau of Reclamation* Earth Manual, *1968*; (b) *After Coffman, 1960.*)

should be used to determine relative density after compaction by surface vibration rather than to rely on indirect, sounding measurements.

Caution must be exercised in using the standard penetration test as a universal method of determining relative density of sands in situ. This is evidenced by the results reported by Zolkov and Wiseman (1965). More than one thousand standard penetration tests were performed on a natural deposit of very uniform fine dune sand ($D_{10} = 0.15$ mm, $D_{50} = 0.18$ mm, $D_{60}/D_{10} = 1.3$).

The results of tests in the upper 2 meters where field densities were also determined are shown in Figure 8.52. It appears that the relative density in situ was substantially underestimated by published penetration resistance values. These authors also reported that unloaded sand foundations exhibited higher than expected blow counts probably owing to prestressing and the residual horizontal stresses remaining after removal of the surcharge.

Because of the time required for preparation of cohesionless soil for the relative density test, which requires drying the soil in an oven for the minimum density procedure (ASTM D 2049, 1964), the development of a more rapid relative density test procedure has been attempted in the field. At the Bureau of Reclamation's Navajo Dam the time required for a relative density test for sand and gravel material was 4 hours. This could be reduced to about 30 minutes by comparing the volume of material extracted from the hole with the minimum volume of the same material, obtained by vibrating the material after saturation in a large cylindrical container. This gave a percentage of maximum density that the field forces were attempting to correlate with relative density. This was unsuccessful, but the idea of comparative volume measurements suggests a possible rapid relative density test procedure.

It should be recognized that the relative density given by Equation 8.3,

$$D_r = \frac{e_{max} - e}{e_{max} - e_{min}}$$

can be determined precisely by comparing the volumes of the material extracted from the density hole with the maximum and minimum volumes of all the material extracted as follows:

$$D_r = \frac{V_{T\,max} - V_T}{V_{T\,max} - V_{T\,min}} \tag{8.11}$$

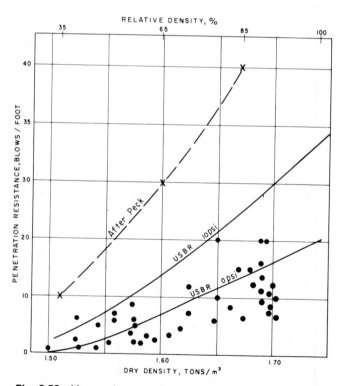

Fig. 8.52 Measured penetration resistance vs. dry density for depths from zero to 2 meters for a dune sand.

where

V_T = volume of density hole
$V_{T\,max}$ = loosest volume of material extracted from the hole
$V_{T\,min}$ = densest volume of material extracted from the hole

The loosest and densest volumes are obtained in the laboratory by using containers large enough to hold the loosest (maximum) volume of the material extracted from the hole and performing the minimum and maximum density procedure. Volumes of soil are obtained by subtraction from the known volume of a container using the sand or water methods for measuring the unoccupied volume of the container. The possibility of using the soil at fill moisture to determine maximum volume by careful placing in the container and then transferring the loose soil to a container equipped with a vibrator and water to determine minimum volume should also be carefully examined as a rapid field relative density procedure that is particularly applicable to coarse sands and gravels where the extra handling of materials for drying and weighing take considerable time and effort. The bulking properties of moist fine to medium sands may require them to be dried for accurate determination of the maximum volume; hence the error due to bulking should be determined prior to use of the proposed method on these sands.

Statistical Control of Compaction

Statistical evaluation of results of control tests was first reported by Davis (1953), who called attention to errors and misconceptions arising from arithmetical averaging of the results of control tests and proposed statistical methods of evaluation, similar to those commonly used for quality control in industry. He proposed use of cumulative frequency plots for establishment of allowable limits of variation and emphasized that the use of statistical methods requires standardized sampling and testing procedures for a particular project, separate analysis for each borrow area and for each compaction method or compactive effort, and elimination of nonrepresentative samples and tests.

Data on compaction of cohesive soils from 72 earth dams constructed by the Bureau of Reclamation were summarized by Davis (1966) and are given in Table 8.10. The control values of difference between optimum and fill moisture contents, $w_o - w_f$; ratio of fill to cylinder dry density at fill moisture content, $C = \gamma_{df}/\gamma_{dc}$; and ratio of fill dry density to maximum laboratory dry density, $D = \gamma_{df}/\gamma_{dm}$, are given together with the frequency distributions and statistical measures of dispersion. Using the normal distribution curve (Fig. 8.53) as the statistical model for control of density and moisture content, Davis (1966) used Work Sheets (Figs. 8.54 and 8.55) and cumulative frequency plots (Fig. 8.56) to evaluate control achieved. He concluded that moisture control $(w_o - w_f)$ can be based on a standard deviation of less than about 1.5 percent and density control (D) can be based on a standard deviation of less than about 3.0 percent.

Turnbull et al. (1966) reported on the variation of density and moisture parameters on several Corps of Engineers projects. They concluded that rather substantial and consistent variations occur that are larger than earthwork designers generally expect. They anticipated that substantial advantages would result from the adaption of statistical methods to soil compaction control. However, they pointed out that entirely satisfactory structures had been built in the past without the use of statistical quality control methods, and stated that current methods would continue to be useful in evaluating construction quality.

In a discussion of Turnbull's paper, Smith and Prysock (1966) reported on a research project that involved randomly sampling accepted embankment material at 50 locations from each of three highway projects in the State of California. Smith and Prysock agreed that variation in earthwork compaction should be less on well-controlled jobs and greater on poorly controlled jobs. However, they emphasized that compaction distribution curves for a particular job should not be used to compare or evaluate the degree of control employed on another job unless field conditions, including embankment materials, were similar.

In another discussion of Turnbull's paper, Abdun-Nur (1966) pointed out that the term "quality control" is not applicable for acceptance testing by the owner's personnel, which sould be termed "engineering control" or "acceptance control." He stated that quality control (as developed by industry) is possible only through the use of probability and statistical principles, and that it should be left in the hands of the contractor or producer, who is the only one who can control his own work.

The practicability of specifying compaction control using statistical concepts was discussed by de Mello et al. (1960b), who used frequency curves for analyzing compaction data from Tres Marias and Santa Branca Dams in Brazil. They considered that there were three basic methods of reducing the percentage of density test values below a given specifications limit: (1) shift the entire field compaction curve by increasing the compactive effort; (2) decrease the standard deviation of the lot (and of the fill) by greater uniformity of construction; and (3) increase the number of tests per lot on which the decision to recompact is based. They recognized that all these methods involve additional cost, which can be compared with the alternative of not attempting to improve the quality of the fill, but making due compensation in the design by using flatter slopes, etc.

Specifications for the Bureau of Reclamation Cutter Dam, issued in 1970, contain the following provisions for statistical control of compaction including definite rejection criteria and allowance for variations based on frequency distribution concepts:

Moisture and density control.

(1) General. Each layer of the material on the embankment shall be compacted by 12 passes of the tamping roller as provided in Subparagraph g. below, which shall be the minimum compacting effort to be performed by the contractor. During compaction, the placement moisture content and dry density of the earthfill shall be maintained within the control limits specified below.

To determine that the moisture content and dry density requirements of the compacted earthfill are being met, field and laboratory tests will be made at frequent intervals on samples taken at embankment locations determined by the contracting officer. Field and laboratory tests will be made by the contracting officer in accordance with Designations E-11, E-24, and E-25, of the Bureau of Reclamation *Earth Manual.*

Materials not meeting the specified moisture content and dry density requirements, as determined by the tests, shall be reworked until approved results are obtained. Reworking may include removal, rehandling, reconditioning, rerolling or combinations of these procedures. The contractor shall be entitled to no additional allowance above the prices bid in the schedule by reason of any work required to achieve the placement moisture content and density specified in this paragraph.

(2) Moisture control. The standard optimum moisture content is defined as, "That moisture content which will result in a maximum dry unit weight of the soil when subjected to the Bureau of Reclamation Proctor Compaction Test." The maximum dry weight, in pounds per cubic foot, obtained by the above procedure is the Proctor maximum dry density. The Bureau of Reclamation Proctor Compaction Test (Designation E-11 of *Earth Manual*) is the same as ASTM Designation: D 698, Method A, except that a 1/20-cubic-foot compaction mold is used and the tamping rod is dropped from a height of 18 inches.

TABLE 8.10 SUMMARY OF EARTHWORK CONTROL STATISTICS.

Column groups: **Averages** — Water Content, Dry Densities, Proctor max., C & D Percent. **Distribution — Percentage of Tests** — Moisture Control and Dry Unit Weight Control. **Standard Deviations** — C and D.

DAM	Dam Identification Number	Year Completed	Number of Tests	General Soil Classification	Liquid Limit	Plasticity Index	Percentage Plus #4	Water Content (minus #4) (w_f) % by Dry Weight	Dry Density Compaction cyl pcf	Dry Density Fill (minus #4) pcf	Dry Density (minus #4) pcf	Proctor max. Dry Density (minus #4) pcf	Water Content opt (w_o) (minus #4) % by Dry Weight	$w_o - w_f$	C, Percent	D, Percent	Drier than $w_o-3.2$	Between $w_o-3.2$ $w_o+0.2$	Wetter than $w_o+0.2$	Less than 0.95 γ_d max.	Between 0.95 / 0.98 γ_d max.	Between 0.98 γ_d max.	Greater than γ_d max.	$w_o - w_f$ (Std Dev)	Std Dev C	Std Dev D
Green Mountain	1	1943	1066	SC	21.3	7.6	35.2	8.4	131.9	131.8	131.9	131.8	8.8	0.4	99.9	99.9	0.2	87.1	12.7	5.5	15.5	35.0	44.0	±0.6	n.a.	±2.2
Deerfield	2	1946	60	n.a.	n.a.	n.a.	19.1	13.1	119.2	118.1	119.2	118.1	13.1	0.0	99.1	99.1	0.0	61.7	38.3	6.5	14.5	56.0	23.0	±1.4	n.a.	±1.6
Long Lake	3	1948	396	SM	17.8	0.2	0.0	14.9	106.3	108.5	108.7	108.7	17.0	2.1	102.1	99.8	27.3	65.1	7.6	6.5	21.0	24.0	48.5	±1.9	n.a.	±3.1
Angostura	4	1949	117	SC	22.0	7.0	20.0	11.2	120.9	123.3	122.2	122.2	11.7	0.5	102.0	100.9	2.6	80.3	17.1	3.5	14.5	16.5	65.5	±1.0	n.a.	±3.0
Heart Butte	5	1949	365	SM-SC / CL	27.6 / 24.5	6.0 / 8.0	2.7	14.5	111.8	113.2	113.3	113.3	15.4	0.9	101.3	99.9	8.8	70.1	21.1	5.5	15.5	25.5	53.5	±1.6	n.a.	±2.8
Medicine Creek	6	1949	1347	ML	19.5	1.0	0.0	16.3	106.3	105.0	107.7	107.7	17.3	1.0	98.8	97.5	6.0	88.4	5.6	18.0	37.0	26.0	19.0	±1.3	n.a.	±2.8
O'Sullivan	7	1949	1739	SM-ML	22.0	NP	0.0	16.3	103.6	105.9	105.2	105.2	17.7	1.4	102.2	100.7	14.1	67.9	18.0	8.0	10.0	20.0	62.0	±1.9	n.a.	±3.3
South Coulee	8	1949	410	SM	n.a.	n.a.	6.5	16.9	102.6	103.5	105.2	105.2	19.1	2.2	100.9	98.4	27.5	61.0	11.5	18.0	26.0	22.0	34.0	±1.9	n.a.	±3.6
Horsetooth	9	1949	285	CL	26.0	11.6	2.2	11.7	108.1	112.6	114.9	114.9	13.9	2.2	104.2	98.0	28.4	64.9	6.7	23.5	25.5	18.0	33.0	±1.7	n.a.	±4.0
Soldier Canyon	10	1949	294	CL	27.0	12.8	1.9	11.6	108.3	112.1	114.1	114.1	14.2	2.6	103.5	98.2	33.3	60.6	6.1	20.0	25.0	21.0	34.0	±1.8	n.a.	±4.1
Jackson Gulch	11	1949	203	CL	40.3	21.3	5.2	18.9	97.4	100.7	100.4	100.4	21.4	2.5	103.4	100.3	35.0	53.7	11.3	12.0	15.5	20.0	52.5	±3.1	n.a.	±4.1
Dixon Canyon	12	1949	240	CL	27.0	12.8	1.0	12.1	107.7	111.2	112.9	112.9	14.9	2.8	103.2	98.5	41.3	55.0	3.7	17.0	27.0	21.5	34.5	±1.7	n.a.	±3.7
Spring Canyon	13	1949	221	CL	28.8	14.0	3.2	12.0	106.6	111.3	112.6	112.6	14.9	2.9	104.4	98.8	40.3	58.3	1.4	14.0	21.0	20.0	45.0	±1.5	n.a.	±3.7
Olympus	14	1949	79	SM	17.6	NP	0.9	10.2	117.5	116.5	120.1	120.1	11.6	1.4	99.1	97.0	13.9	81.0	5.1	31.0	23.0	16.0	30.0	±1.8	n.a.	±5.0
Granby	15	1950	550	SC	21.6	7.6	27.2	9.6	126.0	126.8	127.0	127.0	10.0	0.4	100.6	99.8	1.1	77.1	21.8	3.5	18.0	26.5	52.0	±1.1	n.a.	±2.4
Anderson Ranch	16	1950	2139	SC	28.2	12.2	11.0	12.3	117.8	120.3	118.9	118.9	12.8	0.5	102.1	101.2	1.4	76.0	22.6	0.0	4.5	24.0	71.5	±1.0	n.a.	±2.1
Davis	17	1950	492	CL / SM	31.4 / 21.3	12.5 / 1.0	15.1	10.8	119.2	120.7	120.7	120.7	12.0	1.2	101.3	100.0	10.8	71.5	17.7	6.0	19.0	23.5	51.5	±1.7	n.a.	±2.9
Dickinson	18	1950	279	SM	n.a.	n.a.	1.2	13.7	111.1	112.3	113.2	113.2	15.3	1.6	101.1	99.2	14.7	71.7	13.6	5.0	17.5	42.5	35.0	±1.6	n.a.	±2.0
Platoro	19	1951	304	GC-SC	30.0	12.7	39.3	11.8	122.5	118.8	123.3	123.3	12.2	0.4	97.0	96.4	0.4	80.9	18.7	27.5	38.0	18.0	16.5	±0.8	n.a.	±3.7
Bonny	20	1951	2100	ML-SM	21.0	1.5	0.0	14.2	106.9	109.7	108.5	108.5	15.2	1.0	102.6	101.1	8.7	71.5	19.8	3.0	7.0	20.5	69.5	±1.6	n.a.	±2.3
Cedar Bluff	21	1951	2692	CL	25.8	12.0	0.0	13.7	109.1	110.1	111.0	111.0	14.8	1.1	100.9	99.2	11.1	67.8	21.1	9.5	22.5	25.0	43.0	±1.8	n.a.	±2.6
Enders	22	1951	888	ML	20.2	1.0	0.0	13.2	109.4	111.6	111.5	111.5	14.5	1.3	102.0	100.1	10.2	75.8	14.0	5.5	15.5	24.0	55.0	±1.7	n.a.	±2.7
North Coulee	23	1951	291	ML	21.6	1.4	0.0	15.7	100.8	101.5	103.0	103.0	18.2	2.5	100.7	98.5	33.7	52.9	13.4	12.5	28.0	23.5	36.0	±2.2	n.a.	±3.0
Shadehill	24	1951	1042	SC	25.3	10.2	26.1	11.8	119.3	119.7	121.2	121.2	13.0	1.2	100.3	99.8	6.0	84.4	9.6	11.5	23.0	33.0	32.5	±1.4	n.a.	±3.1
Shadehill Dike	25	1951	85	CL	31.0	11.0	0.0	17.3	101.8	102.0	103.0	103.0	18.9	1.6	100.2	99.0	15.3	75.3	9.4	1.5	20.0	41.0	37.5	±1.7	n.a.	±1.8
Boysen	26	1952	863	SM-SC / GM-GP	20.7 / 18.5	4.7 / 2.0	35.3	11.1	120.5	122.9	122.0	122.0	11.4	0.3	102.0	100.7	3.2	62.5	34.3	12.0	14.0	23.0	51.0	±1.5	n.a.	±4.6
Big Sandy	27	1952	262	SM	n.a.	n.a.	2.0	16.6	106.6	111.0	107.7	107.7	17.5	0.9	104.1	103.1	8.4	72.9	18.7	0.0	4.5	16.0	79.5	±1.6	n.a.	±3.3
Carter Lake	28	1952	839	CL	28.0	13.8	6.0	13.4	111.2	113.7	112.9	112.9	14.8	1.4	102.2	100.7	7.2	85.2	7.6	2.5	13.0	26.5	58.0	±1.3	n.a.	±4.5
Keyhole	29	1952	486	ML-CL	n.a.	n.a.	Neg.	13.8	111.9	112.5	113.2	113.2	14.5	0.7	100.5	99.4	2.9	75.4	21.7	4.0	16.5	42.5	37.0	±1.4	n.a.	±1.7
Lauro	30	1952	249	SC	n.a.	n.a.	16.1	11.4	113.0	116.3	116.0	116.0	13.3	1.9	104.6	101.9	17.7	76.7	5.6	1.5	8.0	14.0	76.5	±1.4	n.a.	±3.0
Rattlesnake	31	1952	157	SM	23.3	3.8	5.3	12.1	114.4	116.3	115.5	115.5	13.5	1.4	101.7	100.7	10.2	80.2	9.6	5.0	23.0	25.0	47.0	±1.1	n.a.	±2.8
Cachuma	32	1953	1324	SC-SM	25.5	6.3	36.3	13.8	113.8	112.3	114.8	114.8	14.6	0.8	98.7	97.8	1.6	85.2	13.2	17.0	31.5	25.0	26.5	±1.1	n.a.	±3.1
Falcon (Schedule #1)	33	1953	385	ML-CL	23.0	9.0	6.2	13.3	109.2	111.5	111.5	111.5	15.0	1.7	102.0	99.9	11.9	83.2	4.9	5.0	17.0	26.0	52.0	±1.4	n.a.	±2.5
Falcon (Schedule #2)	34	1953	2354	ML-CL	22.0	6.0	2.2	11.6	117.1	116.8	118.8	118.8	12.7	1.1	99.7	98.3	8.5	78.0	13.5	9.0	31.5	34.0	25.5	±1.4	n.a.	±2.3

Jamestown	35	1953	381	CL	28.9	11.6	5.0	18.5	99.4	103.0	101.4	20.1	1.6	103.5	101.6	7.1	85.0	7.9	2.5	7.0	21.0	69.5	±1.2	n.a.	±3.2
Ortega	36	1953	70	CL-CH	40.0	24.0	0.9	13.5	114.7	115.6	115.5	14.3	0.8	100.8	100.1	0.0	87.1	12.9	0.0	7.0	41.0	52.0	±0.9	n.a.	±1.4
Trenton (Foundation)	37	1953	1448	ML-CL	27.0	6.0	0.0	16.5	105.1	106.1	106.5	17.0	0.5	101.0	99.6	6.0	61.5	32.5	4.0	17.0	34.0	45.0	±1.8	n.a.	±2.1
Trenton	38	1953	3185	ML-CL	27.0	6.0	0.0	16.5	104.5	106.5	105.8	17.2	0.7	101.9	100.7	4.5	73.2	22.3	0.5	7.5	31.0	61.0	±1.5	n.a.	±2.3
Willow Creek	39	1953	203	SC	41.5	21.0	24.6	15.7	108.0	110.3	109.5	16.7	1.0	102.1	100.7	1.0	86.7	12.3	4.5	7.5	28.5	59.5	±1.1	n.a.	±2.4
Flatiron	40	1953	222	SC-CL	26.8	8.4	4.2	13.0	114.5	116.4	115.5	13.7	0.7	101.7	100.8	1.8	75.7	22.5	0.0	10.0	30.0	60.0	±1.2	n.a.	±2.1
Vermejo #1	41	1954	381	CL	34.7	18.4	2.0	17.2	108.4	110.0	109.8	17.9	0.7	101.5	100.2	2.4	75.0	22.6	6.5	18.5	26.5	48.5	±1.1	n.a.	±3.6
Kirwin	42	1955	3803	CL	42.0	15.0	0.0	18.8	104.9	103.6	102.7	19.2	0.4	101.7	100.9	2.3	71.1	26.6	0.5	6.0	29.0	64.5	±1.0	±2.2	±2.0
Sly Park	43	1955	226	CL	36.0	14.0	11.1	15.2	111.1	110.6	111.8	15.0	0.2	99.5	98.9	0.0	51.3	48.7	7.0	23.0	29.0	41.0	±1.2	±2.6	±2.6
Weber Aqueduct	44	1955	50	CL	29.0	13.0	0.0	15.7	111.2	109.0	111.9	15.9	0.2	98.0	97.4	2.0	62.0	36.0	17.5	23.0	39.5	20.0	±1.3	n.a.	±2.8
Alamogordo (Enlarg.)	45	1956	92	SC	23.0	12.0	5.9	11.6	115.0	120.4	116.3	12.5	0.9	104.7	103.5	0.0	95.7	4.3	0.0	4.5	7.0	88.5	±0.7	n.a.	±2.4
Pactola	46	1956	1037	CL	30.0	12.0	12.0	15.5	111.4	111.6	112.4	16.0	0.5	100.2	99.3	3.6	71.7	24.7	2.0	24.0	34.0	40.0	±1.4	n.a.	±2.3
				CL	38.0	19.0																			
Tiber (Main Embankment)	47	1956	853	ML	23.0	2.0	5.8	14.2	112.3	114.7	113.6	14.8	0.6	102.1	101.0	4.9	72.1	23.0	2.0	14.0	18.0	66.0	±1.6	n.a.	±3.2
				CL	19.0	10.0																			
Tiber (Dike)	48	1956	280	CL-CH	48.0	28.0	0.0	19.7	102.0	104.5	103.4	19.9	0.2	102.5	101.1	5.7	55.0	39.3	3.0	12.0	19.0	66.0	±1.8	n.a.	±3.0
Webster	49	1956	3042	CL	37.0	19.0	0.0	17.6	104.5	106.6	105.2	18.0	0.4	102.0	101.3	2.6	69.0	28.4	1.5	9.0	21.0	68.5	±1.2	±2.8	±2.8
Palisades	50	1957	2108	ML	26.0	4.0	3.3	15.7	107.4	107.4	108.7	16.5	0.8	100.0	98.8	6.0	68.1	25.9	11.0	24.0	26.0	39.0	±1.7	n.a.	±3.5
Haystack	51	1957	110	SM-ML	n.a.	NP.	5.9	20.3	98.8	98.2	100.5	21.2	0.9	99.4	97.7	3.6	81.9	14.5	15.0	35.0	23.0	27.0	±1.4	n.a.	±2.8
Lovewell	52	1957	1330	CL	44.0	26.0	0.0	18.2	104.8	105.8	105.4	18.5	0.3	101.0	100.4	0.6	52.5	46.9	0.0	6.0	38.0	56.0	±1.3	n.a.	±1.7
Pineview (Enlarg.)	53	1957	51	GM	n.a.	n.a.	20.3	14.3	117.9	116.4	119.6	13.7	0.6	98.7	97.3	0.0	39.2	60.8	11.5	34.0	32.0	22.5	±1.5	n.a.	±2.3
Wanship	54	1957	845	CL	26.0	12.0	2.6	13.9	112.8	115.2	114.0	14.5	0.6	102.1	101.1	2.7	74.5	22.8	2.0	11.5	19.5	67.0	±1.3	n.a.	±2.6
Glenco	55	1958	716	ML	n.a.	n.a.	2.6	16.4	108.0	109.0	108.2	16.8	0.4	100.9	100.7	1.4	75.0	23.6	0.0	9.0	34.0	57.0	±1.1	n.a.	±1.9
Helena Valley	56	1958	81	SC-CL	29.0	12.0	23.2	14.9	111.8	111.1	113.0	15.3	0.4	99.4	98.3	2.5	61.7	35.8	12.0	32.0	24.0	32.0	±1.3	n.a.	±2.9
Twitchell (Vaquero)	57	1958	1022	CL	41.0	24.0	19.4	20.1	101.0	101.2	101.7	20.5	0.4	100.2	99.5	2.1	74.8	23.1	4.5	19.0	34.0	42.5	±1.2	±2.7	±2.2
				GM-ML	30.0	5.0																			
Casitas	58	1959	1621	CL	28.0	11.0	2.5	13.8	114.2	115.3	114.9	14.9	1.1	101.0	100.3	4.2	85.4	10.4	1.5	14.0	27.5	57.0	±1.2	n.a.	±2.4
Fort Cobb	59	1959	2415	SM	n.a.	NP	0.0	11.9	112.2	114.8	113.3	12.9	1.0	102.3	101.3	4.3	76.8	18.9	0.2	2.0	22.6	75.2	±1.3	±2.6	±2.6
Vega	60	1959	412	SC-CL	40.0	17.0	18.8	17.8	106.8	106.6	107.3	18.2	0.4	99.8	99.3	0.7	76.5	22.8	3.6	32.6	36.6	27.2	±1.0	n.a.	±2.2
Keene Creek	61	1959	53	SC	49.0	15.0	7.5	41.5	78.3	76.8	78.7	40.7	0.8	98.1	97.6	1.9	20.7	77.4	18.9	37.7	22.1	21.3	±1.2	n.a.	±3.3
Little Wood River	62	1960	263	GC-CL	45.0	19.0	28.5	19.7	98.3	99.0	99.8	21.2	1.5	100.7	99.2	7.2	91.6	1.2	1.9	29.7	42.9	25.5	±1.0	±3.1	±2.6
Foss	63	1961	3893	SM	n.a.	NP	0.0	15.3	111.1	113.2	112.2	15.6	0.3	101.9	101.0	0.9	62.8	36.3	0.1	1.4	32.2	66.3	±1.3	±2.1	±2.0
				CL	38.7	19.3																			
Steinaker	64	1961	533	CL	32.0	15.0	0.0	14.1	112.8	114.6	114.1	15.1	1.0	101.6	100.5	0.9	90.8	8.3	1.7	8.4	31.5	58.4	±1.0	±2.5	±2.3
Emigrant (except Area J)	65	1961	251	CL	32.0	16.0	25.9	13.4	109.1	11.0	111.1	15.1	1.7	101.7	99.9	9.9	81.6	8.5	7.2	23.9	21.3	47.6	±1.3	±4.2	±3.8
Emigrant (Area J)	66	1961	44	CH	69.0	48.0	33.8	22.5	90.0	92.5	92.2	24.8	2.3	102.8	100.4	27.3	70.5	2.2	15.9	25.0	9.1	50.0	±1.7	±6.3	±6.5
Prineville	67	1961	560	SM-ML	34.0	4.0	35.9	17.7	107.9	109.9	108.6	18.0	0.3	101.9	101.2	0.0	77.1	22.9	0.4	11.8	20.2	67.6	±1.0	±2.6	±2.6
Trinity	68	1982	4841	SM-SC	28.5	6.3	32.0	16.0	112.8	112.6	113.8	17.2	1.2	99.8	98.9	0.6	93.1	6.3	7.0	28.8	27.8	36.4	±0.8	±3.1	±3.0
Trinity (Zone 2)	69	1962	958	GM	n.a.	n.a.	42.5	15.6	113.3	108.0	114.6	16.5	0.9	95.3	94.2	4.1	81.0	14.9	50.5	20.6	13.5	15.4	±1.3	±5.2	±5.0
Navajo	70	1962	5541	CL	28.0	10.0	1.2	13.7	114.6	116.3	115.7	14.5	0.8	101.5	100.5	1.0	88.1	10.9	0.2	5.6	31.1	63.1	±0.9	±2.0	±1.8
				SM	20.0	3.0																			
Twin Buttes	71	1962	7249	CL	29.0	17.0	7.5	16.4	108.9	111.4	109.7	16.9	0.6	102.3	101.5	0.5	84.1	15.4	0.2	1.8	24.1	73.9	±1.0	±2.7	±2.3
Red Willow	72	1962	1914	ML	25.0	3.0	0.0	17.3	105.2	106.8	106.0	17.7	0.4	101.5	100.8	0.8	67.2	32.0	0.1	1.9	36.0	62.0	±1.3	±2.2	±2.0
Prosser Creek	73	1962	481	GM-GC	32.2	12.1	31.3	18.7	105.6	105.1	107.2	19.1	0.4	99.5	98.1	1.7	70.0	28.3	15.8	29.1	25.7	29.4	±1.6	±2.9	±2.8
Willard (2nd Stage)	74	1962	1992	CL-ML	25.0	5.0	1.4	16.4	110.1	109.9	111.9	16.6	0.2	99.8	98.2	5.1	58.1	36.8	7.6	37.9	30.1	24.4	±1.7	±2.4	±2.4
Crawford	75	1962	334	GC	36.0	17.0	21.3	17.0	106.7	108.8	107.7	17.9	0.9	102.0	101.0	3.9	75.5	20.6	0.9	6.9	29.0	63.2	±1.4	±2.9	±2.6
Sherman	76	1962	1475	CL	32.0	16.0	0.0	18.5	104.5	106.4	105.6	17.7	0.8	101.9	100.8	1.3	29.5	69.2	0.7	1.0	34.2	64.1	±1.3	±2.1	±2.0
Twin Buttes (Zone 3)	77	1962	570	CL	27.0	14.0	17.8	14.6	113.4	116.5	114.3	15.4	0.8	102.8	101.9	2.6	80.2	17.2	0.9	4.0	21.1	74.0	±1.3	±3.1	±2.9
Twin Buttes (Zone 4)	78	1962	1164	CL	39.0	18.0	9.3	17.6	105.3	107.1	106.1	18.5	0.9	101.7	101.0	1.8	85.7	12.5	0.2	8.7	26.1	65.0	±1.2	±2.6	±2.4
Paonia	79	1962	419	CL	27.0	13.0	6.4	13.2	112.9	115.7	114.0	14.1	0.9	102.5	101.5	1.6	85.4	13.1	0.5	4.1	16.2	79.2	±1.1	±2.3	±2.1
Bully Creek	80	1963	318	ML	34.0	7.0	7.7	23.7	92.7	92.9	93.9	24.9	1.2	100.1	98.9	6.9	78.6	14.5	2.8	32.7	33.1	31.4	±1.4	±2.6	±2.4

Note: All materials are Zone 1 except as noted. Data designated n.a. not available. For Willard Dam (2nd Stage), compaction was primarily by equipment travel.

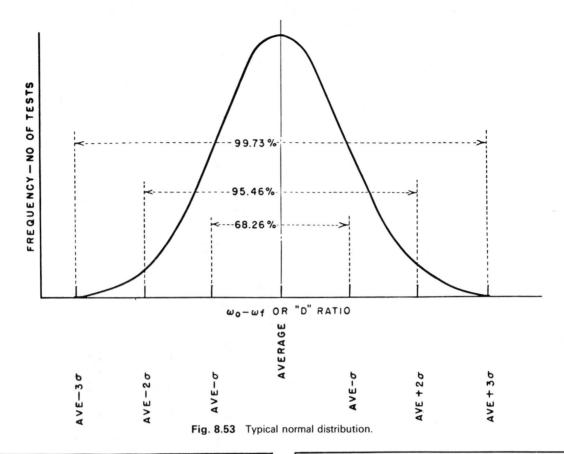

Fig. 8.53 Typical normal distribution.

(Curve labels: 99.73%, 95.46%, 68.26%; axis $\omega_o - \omega_f$ OR "D" RATIO; FREQUENCY—NO OF TESTS; AVE-3σ, AVE-2σ, AVE$-\sigma$, AVERAGE, AVE$+\sigma$, AVE$+2\sigma$, AVE$+3\sigma$)

EARTHWORK CONTROL WORK SHEET — MOISTURE CONTROL

Example   DAM   BORROW AREA "B"   ZONE ____I____

$\omega_o-\omega_f$ = VARIATION OF FILL MOISTURE OF MINUS NO.4 MATERIAL FROM LAB OPTIMUM

	F (PREV)	THIS PERIOD — FREQUENCY OF OCCURRENCE	F	CUM F	CUM %	F	CUM F	CUM %
> 5.7								
5.3-5.7								
4.8-5.2								
4.3-4.7								
3.8-4.2	1					1	1	0.3
3.3-3.7	.							
2.8-3.2	1	I	1	1	1.2	2	3	1.0
2.3-2.7	3	ʅʅ I	6	7	8.2	9	12	4.1
1.8-2.2	14	ʅʅ III	8	15	17.6	22	34	11.6
1.3-1.7	23	ʅʅ ʅʅ	10	25	29.4	33	67	22.9
0.8-1.2	32	ʅʅ ʅʅ ʅʅ ʅʅ	20	45	52.9	52	119	40.6
0.3-0.7	23	ʅʅ IIII	9	54	63.5	32	151	51.5
+0.2 TO -0.2	28	ʅʅ IIII	9	63	74.1	37	188	64.2
0.3-0.7	37	ʅʅ ʅʅ I	11	74	87.1	48	236	80.5
0.8-1.2	21	ʅʅ	5	79	92.9	26	262	89.4
1.3-1.7	15	IIII	4	83	97.6	19	281	95.9
1.8-2.2	8	II	2	85	100	10	291	99.3
2.3-2.7	1					1	292	99.7
2.8-3.2	1					1	293	100
3.3-3.7								
3.8-4.2								
4.3-4.7								
4.8-5.2								
5.2-5.7								
> 5.7								
TOTALS	208		85			293		

% BELOW / % ABOVE

	PREV	THIS PERIOD	TO DATE
Average optimum moisture	14.3	15.4	14.6
Average fill moisture	14.1	14.8	14.3
Mean variation from opt. moisture ($\omega_o-\omega_f$)	0.2	0.6	0.3

σ (This period, 85 tests) = ±1.16
σ (Cum., 293 tests) = ±1.20
PERIOD OF REPORT ___10-26-59___ TO ___11-25-59___
TESTS ___11-2-A1-R___ TO ___11-24-A4-R___

Fig. 8.54

EARTHWORK CONTROL WORK SHEET — DRY UNIT WEIGHT CONTROL

Example   DAM   BORROW AREA "B"   ZONE ____I____

FILL DRY DENSITY / PROCTOR MAXIMUM DRY DENSITY (MINUS NO.4 MATERIAL) ×100 $D =$

	F (PREV)	THIS PERIOD — FREQUENCY OF OCCURRENCE	F	CUM F	CUM %	F	CUM F	CUM %
86.0 - 86.9								
87.0 - 87.9								
88.0 - 88.9								
89.0 - 89.9								
90.0 - 90.9	1					1	1	0.3
91.0 - 91.9								
92.0 - 92.9								
93.0 - 93.9		I	1	1	1.2	1	2	0.7
94.0 - 94.9								
95.0 - 95.9								
96.0 - 96.9								
97.0 - 97.9	8					8	10	3.4
98.0 - 98.9	26	ʅʅ ʅʅ I	11	12	14.1	37	47	16.0
99.0 - 99.9	43	ʅʅ ʅʅ II	12	24	28.2	55	102	34.8
100.0 - 100.9	45	ʅʅ ʅʅ I	11	35	41.2	56	158	53.9
101.0 - 101.9	41	ʅʅ ʅʅ ʅʅ IIII	19	54	63.5	60	218	74.4
102.0 - 102.9	22	ʅʅ ʅʅ I	11	65	76.5	33	251	85.7
103.0 - 103.9	13	ʅʅ ʅʅ III	13	78	91.8	26	277	94.5
104.0 - 104.9	2	II	2	80	94.1	4	281	95.9
105.0 - 105.9	5	II	2	82	96.5	7	288	98.3
106.0 - 106.9		II	2	84	98.8	2	290	99.0
107.0 - 107.9	2	I	1	85	100	3	293	100
108.0 - 108.9								
109.0 - 109.9								
110.0 - 110.9								
TOTALS	208		85			293		

	PREV	THIS PERIOD	TO DATE
Average Proctor max. γ_D (PCF)	114.5	113.2	114.1
Average fill γ_D (PCF)	115.3	114.7	114.8
Mean percentage of Proctor max. γ_D (PCF)	100.7	101.3	100.6
Average rock content (% of plus No.4 by dry weight)	0.0	0.0	0.0

σ (This period, 85 test) = ±2.09
σ (Cum., 293 test) = ±2.07
PERIOD OF REPORT ___10-26-59___ TO ___11-25-59___
TESTS ___11-2-A1-R___ TO ___11-24-A4-R___

Fig. 8.55

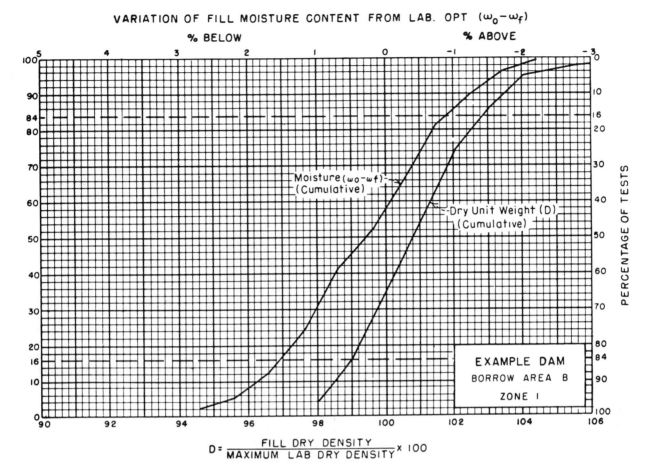

Fig. 8.56 Cumulative frequency of control test results.

The moisture content of the earthfill material prior to and during compaction shall be distributed uniformly throughout each layer of the material. The allowable ranges of placement moisture content are based on design considerations. The moisture control shall be such that the moisture content of compacted earthfill, as determined by tests performed by the contraction officer shall be within the following limits:

a. Material represented by the samples tested having a placement moisture content more than 3.5 percent dry of the standard optimum condition, or more than 1.0 percent wet of the standard optimum condition will be rejected and shall be removed or reworked until the moisture content is between these limits.

b. Within the above limits, and based on a continuous record of tests made by the Government on previously placed and accepted material, the uniformity of placement moisture content shall be such that:

   (aa) No more than 20 percent of the samples of accepted embankment material will be drier than 3.0 percent dry of the standard optimum moisture content, and no more than 20 percent will be wetter than 0.5 percent wet of the standard optimum moisture content.

   (bb) The average moisture content of all accepted embankment material shall be between 0.5 and 1.0 percent dry of the standard optimum moisture content.

The results of all completed earthwork tests will be available to the contractor in the Government laboratory. The Government will inform the contractor when the placement moisture content approaches or exceeds the limits of uniformity specified above, and the contractor shall immediately make adjustments in procedures as necessary to maintain the placement moisture content within the specified limits.

As far as practicable, the material shall be brought to the proper moisture content in the borrow pit before excavation, as provided in Paragraph 66. Supplementary water, if required, shall be added to the material by sprinkling on the earthfill on the embankment and each layer of the earthfill shall be conditioned so that the moisture is uniform throughout the layer.

(3) Density control. Density control of compacted earthfill shall be such that the dry density of the compacted material, as determined by tests performed by the contracting officer, shall conform to the following limits:

a. Material represented by samples having a dry density less than 96.0 percent of its Proctor maximum dry density will be rejected. Such rejected material shall be rerolled until a dry density equal to or greater than 96.0 percent of its Proctor maximum dry density is obtained.

b. Within the above limit and based on a continuous record of tests made by the Government on previously placed and accepted embankment, the uniformity of dry density shall be such that:

   (aa) No more than 20 percent of the material represented by the samples tested shall be at dry densities less than 97.0 percent of Proctor maximum dry density.

   (bb) The average dry density of all accepted embankment material shall be not less than 100.0 percent of the average Proctor maximum dry density.

The results of all completed earthwork tests will be available to the contractor in the Government laboratory.

The Government will inform the contractor when the dry density approaches or exceeds the limits of uniformity specified above, and the contractor shall immediately make adjustments in procedures as necessary to maintain the dry density within the specified limits.

TABLE 8.11 RANDOM NUMBERS[a].

1		2		3		4		5	
A	B	A	B	A	B	A	B	A	B
.576	.730	.430	.754	.271	.870	.732	.721	.998	.239
.892	.948	.858	.025	.935	.114	.153	.508	.749	.291
.669	.726	.501	.402	.231	.505	.009	.420	.517	.858
.609	.482	.809	.140	.396	.025	.937	.310	.253	.761
.971	.824	.902	.470	.997	.392	.892	.957	.640	.463
0.63	.899	.554	.627	.427	.760	.470	.040	.904	.993
.810	.159	.225	.163	.549	.405	.285	.542	.231	.919
.081	.277	.035	.039	.860	.507	.081	.538	.986	.501
.982	.468	.334	.921	.690	.806	.879	.414	.106	.031
.095	.801	.576	.417	.251	.884	.522	.235	.398	.222
.509	.025	.794	.850	.917	.887	.751	.608	.698	.683
.371	.059	.164	.838	.289	.169	.569	.977	.796	.996
.165	.996	.356	.375	.654	.979	.815	.592	.348	.743
.477	.535	.137	.155	.767	.187	.579	.787	.358	.595
.788	.101	.434	.638	.021	.894	.324	.871	.698	.539
.566	.815	.622	.548	.947	.169	.817	.472	.864	.466
.901	.342	.873	.964	.942	.985	.123	.086	.335	.212
.470	.682	.412	.064	.150	.962	.925	.355	.909	.019
.068	.242	.667	.356	.195	.313	.396	.460	.740	.247
.874	.420	.127	.284	.448	.215	.833	.652	.601	.326
.897	.877	.209	.862	.428	.117	.100	.259	.425	.284
.875	.969	.109	.843	.759	.239	.890	.317	.428	.802
.190	.696	.757	.283	.666	.491	.523	.665	.919	.146
.341	.688	.587	.908	.865	.333	.928	.404	.892	.696
.846	.355	.831	.218	.945	.364	.673	.305	.195	.887
.882	.227	.552	.077	.454	.731	.716	.265	.058	.075
.464	.658	.629	.269	.069	.998	.917	.217	.220	.659
.123	.791	.503	.447	.659	.463	.994	.307	.631	.422
.116	.120	.721	.137	.263	.176	.798	.879	.432	.391
.836	.206	.914	.574	.870	.390	.104	.755	.082	.939
.636	.195	.614	.486	.629	.663	.619	.007	.296	.456
.630	.673	.665	.666	.399	.592	.441	.649	.270	.612
.804	.112	.331	.606	.551	.928	.830	.841	.602	.183
.360	.193	.181	.399	.564	.772	.890	.062	.919	.875
.183	.651	.157	.150	.800	.875	.205	.446	.648	.685

[a] After Sherman, Watkins, and Prysock (1967).

For a test sample to be representative of a section of compacted fill, its location must be chosen at random. This means that every point in the section has an equal chance of being included in the sample.

To determine the coordinates of a random sample in an area ready for acceptance testing, let X be the length of the section and Y the width. Using Table 8.11 (Sherman et al., 1967), start at any location and read pairs of numbers A and B up or down but not skipping any numbers. For example, in Column 4 reading down: 0.732, 0.721, 0.153, 0.508, and so on. Now multiply A by X and B by Y to obtain coordinates of the random samples. For $X = 1000$ ft (Sta 4 + 28 to 14 + 28) and $Y = 50$ ft (offset 26 ft right to 76 ft right of centerline), the coordinates of five random samples are computed as shown in Table 8.12.

The number of tests used by various organizations to determine acceptance of compaction varies somewhat among organizations specifying compaction. Table 8.13 shows the frequency of tests used by Government organizations.

More frequent tests than are tabulated are used during the early stages of fill construction until routine procedures are established.

Various statistical approaches (Wald, 1947; Wilks, 1948; Lieberman and Resnikoff, 1955) to determining the number of tests all require many more tests than have been used to control compaction of fills (Wahls et al., 1966). When the equipment and procedural type of specification is used, visual and other sensory inspection of the materials and processes is commonly used to decrease the risk of improper decision on acceptance or rejection.

The availability of rapid methods of testing now permits a sequential inspection plan to be used, reducing the number of tests required for decision making. A sequential inspection plan consists of examining, in sequence, single items that are obtained at random from a lot and, at each step, making one of three possible decisions:

1. The lot is acceptable.
2. The lost is unacceptable.
3. The evidence is not sufficient for either decision without too great a risk of error.

When the third decision is reached, an additional item is inspected (at random) and the same three decisions are reconsidered in the light of this additional information.

A possible application of this sequential method is illustrated by Figure 8.57, which relates to a specification that requires limits of acceptable moisture and density such as for the Bureau of Reclamation Cutter Dam (U.S. Bureau of Reclamation, 1970). This is based on the use of a compaction control method that requires comparison of the density and moisture of each sample of fill with the Proctor values for that identical sample, such as Hilf's rapid method for cohesive soils or a complete relative density test on a sample of compacted fill. Other construction control methods for highway embankment developed at AASHO Road Test are given by Shook (1959), Highway Research Board (1962), and Fang (1975).

TABLE 8.12 COMPUTATION OF RANDOM SAMPLE LOCATION COORDINATES.

Coordinate Along Axis of Fill (X Direction) 0 Corresponds to 4 + 28		Coordinate Transverse to Axis of Fill (Y Direction) 0 Corresponds to Offset 26 ft rt	
Random numbers A (Top of Col 4A down)	Station to be sampled	Random numbers B (Top of Col 4B down)	Offset to be sampled
0.732	11 + 60	0.721	62 ft rt
0.153	5 + 81	0.508	51 ft rt
0.009	4 + 37	0.420	47 ft rt
0.937	13 + 65	0.310	42 ft rt
0.892	13 + 20	0.957	74 ft rt

TABLE 8.13 MINIMUM FREQUENCY OF DENSITY TESTS.

	USBR[a]	*U.S. Navy*[b]	*U.S. Corps of Engineers*[c]	*California Department of Water Resources (Oroville Dam)*[d]
Mass earthwork (embankment)	1/2000 cy	1/2000 cy	1/1000 cy to 3000 cy	1/3000 cy to 4000 cy or 1 every 24 hr
Relatively thin sections; canals and reservoir linings	1/1000 cy	1/1000 cy	—	—
Backfill in trenches or around structures	1/200 cy	1/200 cy to 500 cy	—	—
Minimum per shift on mass earthwork	1	1	—	—
Doubtful areas	1	1	—	—
Pervious materials	1/1000 cy to 1/10 000 cy	—	—	1/50 000 cy to 1/100 000 cy or 1 every 24 hr

[a] Bureau of Reclamation *Earth Manual* (1968), pp. 249, 300.
[b] NAVDOCKS DM-7 (1962).
[c] Corps of Engineers EM 1110-2-2300 (1 Apr 1959), p. 24.
[d] "Control of Earth and Rockfill for the Oroville Dam Embankment" (Gordon and Miller, 1966).

SPECIFICATIONS REQUIREMENTS

Moisture control:

$$w_0 - 3.5 \leq w_f \leq w_0 + 1$$
$$w_0 - 3.0 \leq w_f \text{ For not more than 20\% of samples}$$
$$w_0 + 0.5 \geq w_f \text{ For not more than 20\% of samples}$$
$$w_0 - 1.0 \leq w_f \text{ average} \leq w_0 - 0.5$$

Density control:

$$D \geq 96\%$$
$$D \geq 97\% \text{ For not less than 80\% of the samples}$$
$$D_{average} = 100\%$$

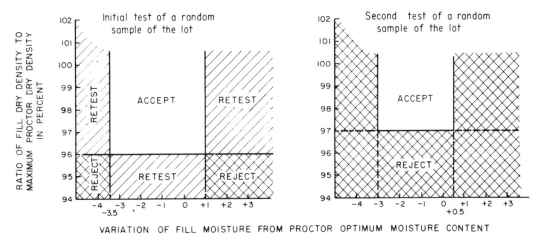

Fig. 8.57 Sequential control tests for acceptance or rejection of a section of compacted fill.

8.5 MISCELLANEOUS PROBLEMS IN COMPACTION

Compaction of Expansive Clays

The volume change characteristics of clays vary with the type of clay mineral present in the soil and the percentage of that clay mineral. Chemically, clay minerals are crystalline hydrous aluminosilicates, often containing small amounts of potassium, sodium, magnesium, and iron. Clay minerals can be classified roughly into two groups: (1) the kaolin group, and (2) the montmorillonite group. The kaolin minerals have fixed crystal lattices or layered structure and exhibit only a small degree of hydration and adsorptive properties. On the other hand, the montmorillonite minerals have expanding lattices and exhibit a high degree of hydration and cation adsorption. A third common clay mineral, illite, is sometimes given as a third classification. It has an expanding lattice structure similar to that of montmorillonites but has much less expansion potential.

The swelling characteristics of compacted cohesive materials have been investigated by many researchers, including Ladd (1960), Seed et al. (1962), Holtz and Gibbs (1956), Holtz (1959), Gizienski and Lee (1965), Sallberg and Smith (1965), and

Lambe (1958). Factors determined by these investigators to influence the swelling characteristics of compacted cohesive soils include the type and amount of clay minerals, the compaction condition, the chemical properties of the pore fluid, the confining pressure applied during the swelling, the time allowed for swelling, the osmotic repulsive pressure, the compression of air in the voids as water permeates the soil mass, the stress history of the material, and alternate wetting and drying cycles. Several attempts have been made to use reliable index tests to predict the probable degree of potential swell for a given soil. Holtz and Gibbs (1956) based their method on plasticity index, shrinkage limit, and colloid content. Ladd and Lambe (1961) based their method on plasticity index, water content at 100 percent relative humidity, percentage of swell under 200-lb/ft^2 load, and percentage volume change. A method suggested by Seed et al. (1962) was based on activity and percentage finer than the 0.002-mm size. A method proposed by Kantey and Brink (1962) was based on the percentage of linear shrinkage, liquid limit, and plasticity index of the soil.

Holtz and Gibbs (1956) reported results of research by the Burea of Reclamation on a quantitative measure of swelling pressure and free swell. They attempted to correlate data from index tests to estimate the probable volume change for expansive soils. Their work resulted in Table 8.14 which shows the relation of soil index properties of expansion potential of high-plasticity clay soils. The colloid content, the plasticity index, and the shrinkage limit are used in this table to estimate the degree of expansion potential.

In a review of existing methods of classification and identification of swelling clays, Kassiff et al. (1969) concluded that there is no one classification method that satisfies all requirements. They indicated, however, that troublesome soils can be identified by use of the shrinkage limit, liquid limit, plasticity index, and free swell (USBR method).

Two properties of expansive soils can be considered separately—swelling percentage and swelling pressure. *Swelling percentage* is the amount of vertical swell or heave expressed as a percentage of an initial sample thickness, obtained when the soil is wetted under a nominal (1 psi) load. This property is determined in the confined compression test. For a particular compacted expansive clay, the relationship of initial moisture and density to its swelling percentage is shown in Figure 8.58. This figure shows that this soil, when compacted to the optimum moisture content with standard compactive effort, will expand about 3 percent of its volume when saturated under 1-psi restraining load. This expansion reduces to 0 at about 3 percent

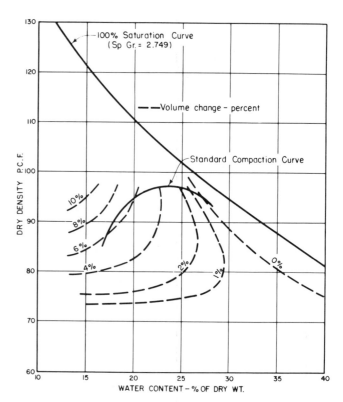

Fig. 8.58 Effect of initial moisture and density on the expansion properties of a compacted expansive clay soil when wetted.

wet of optimum and increases to 6 percent at 3 percent dry of optimum.

Bara (1969) reports on the expansive properties of a high-liquid-limit soil from San Luis Drain, California (liquid limit, 89 percent; *PI*, 65 percent; *SL*, 8 percent). The swelling percentage of this soil was 12 percent at optimum moisture content at standard compactive effort with about 8 percent expansion at placement moisture 5 percent wet of optimum.

When a compacted expansive clay soil has access to water it swells until its internal forces are in equilibrium with its surroundings. If the soil is confined, a *swelling pressure* develops.

Holtz and Gibbs (1956) reported on swelling pressure tests made on Porterville clays from the Delta-Mendota Canal, California, for canal embankment studies. Figure 8.59 shows the pressure necessary to hold the volume expansion of compacted clay at zero in the confined compression apparatus. The heavy dashed lines are loci of placement conditions for equal values of swelling pressure. This figure indicates that a decrease in density is more effective in reducing swelling pressure than increase in moisture content, for this compacted clay.

When the magnitude of restraining load is less than the swelling pressure, some expansion of the compacted soil occurs. Hence a load–expansion curve can be obtained for a given compacted soil by preparing several specimens at the same moisture content and dry density in a confined compression apparatus and allowing each specimen access to water while under different restraining loads. Barber (1956) provided some data on this relationship for a marine (Tuxedo) clay near the District of Columbia, which are given in Table 8.15. The soil was compacted to 118 pcf at 10 percent moisture content.

However, the effect of sequence of loading and exposure to water affects the magnitude of expansion. Holtz and Gibbs (1956) reported that a compacted expansive clay expanded considerably more under a light load (such as 1 psi) than an identical specimen that was first loaded sufficiently to prevent

TABLE 8.14 RELATION OF SOIL INDEX PROPERTIES AND PROBABLE VOLUME CHANGES FOR HIGHLY PLASTIC SOILS.

Data from Index Tests[a]			Estimation of Probable Expansion[b] Percent Total Volume Change (Dry to Saturated Condition)	Degree of Expansion
Colloid Content (Percent minus 0.001 mm)	Plasticity Index	Shrinkage Limit, Percent		
>28	>35	<11	>30	Very high
20–31	25–41	7–12	20–30	High
13–23	15–28	10–16	10–20	Medium
<15	<10	>15	<10	Low

[a] All three index tests should be considered in estimating expansive properties.
[b] Based on a vertical loading of 1.0 psi as for concrete canal lining. For higher loadings the amount of expansion is reduced, depending on the load and on the clay characteristics.

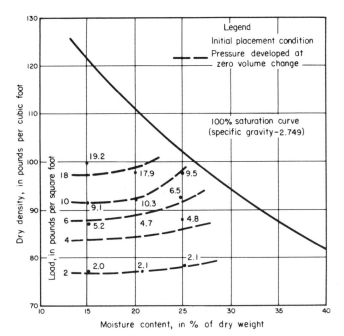

Fig. 8.59 Swelling pressure caused by wetting compacted Porterville clay at various placement conditions.

expansion and then allowed to expand by reducing the load in increments (Chen, 1975).

Problems with expansive clays occur during the construction of fills for light buildings or compacted fills for canal linings where the restraining loads are small. In these cases, the rise of groundwater, seepage, leakage, or elimination of surface evaporation may increase the degree of saturation of the compacted soil, which leads to the tendency to expand. If the swelling pressure developed is greater than the restraining pressure, heave will occur, which may cause structural distress. When such soils must be used, the treatment recommended (NAVDOCKS DM-7, 1962) is to compact the clays as wet as practicable consistent with compressibility requirements. Avoid overcompaction of the general fill and avoid undercompaction of backfill at column footings or in utility trenches, which would accentuate differential movements. Consideration should be given to adding various salts to reduce swelling potential by increasing the ion concentration in the pore water.

Bara (1969) reported on a study of control requirements for embankments built of expansive clay. Since nonexpansive soils were unavailable for constructing low embankments needed to support a concrete lining of a drain, the highly expansive materials from required excavation were used. Laboratory investigations were performed to determine the most desirable placement moisture and densities for these soils. Based on tests made on various expansive clays which had liquid limits of 89 percent and shrinkage limit of only 8 percent, it was

TABLE 8.15 THE EFFECT OF EXPANSION ON THE SWELLING PRESSURE OF A COMPACTED CLAY.

% Displacement Permitted	Pressure Developed, psi
0	200
0.6	151
1.2	119
1.8	105
2.4	75

determined to limit the small zone of compacted embankment materials that would support the concrete lining to soils with a liquid limit of less than 70 percent and also to specify that this clay be placed wet of optimum moisture content to attain compacted dry densities of 89 to 94 percent of standard Proctor maximum.

From the reported studies of compaction of expansive clays, it appeared that compaction on the wet side of optimum moisture content will produce lower magnitudes of swelling and swell pressure. However, it was pointed out in Section 8.2 that compacting on the wet side of the optimum moisture content produces materials with lower strength and high compressibilities. Therefore, the choice of a desired moisture content and dry density for a given method of compaction must be made not from the single criterion of lower swelling characteristics. The final decision should be based on criteria that will ensure a workable range for volume expansion characteristics, strength characteristics, and compressibility characteristics. Moreover, there is evidence (Lambe, 1958; Seed and Chan, 1959) that the effect of soil structure on shrinkage is opposite to the effect of soil structure on swelling. It appears that a dispersed type of soil structure, which is characteristic of compaction wet of optimum, will exhibit lower swell pressure and magnitude of swell but will exhibit more significant shrinkage characteristics.

Dispersive Clays

Severe erosion of slopes of dams and canals as well as complete failures of homogeneous clay dams have resulted from the presence of certain clay soils that are highly erodible in the presence of water of low salt content. Even though the clays are compacted the particles of these dispersive soils go into suspension in flowing water or even in still water.

These soils are not rare and can be found almost anywhere. They often contain the clay minerals montmorillonite or illite and have a higher percentage of dissolved sodium cation in their pore water than do ordinary clays. Dispersive clays have been found in flood-plain deposits, slope wash, lake beds, and loess. Rarely have in-situ weathering products of igneous and metamorphic rocks been found to be dispersive and none have been found in soils derived from limestone (Sherard et al., 1977).

Since the standard tests for classifying soils for engineering purposes do not take into account the sodium content, dispersive clays cannot be identified by Atterberg limits or the Unified Soil Classification System. Most of the dispersive clays are cohesive but a probable lower limit for these soils is 12 percent of the particles finer than 0.005 mm and a plasticity index of 4 percent.

The deflocculation or dispersion of individual colloidal clay particles which repel each other in the presence of relatively pure water distinguishes these soils from non-cohesive silts and very fine sands which also are erodible in water.

Identification of dispersive clays has been made by indirect tests such as the double hydrometer test (ASTM test method D 4221), the crumb test (Holmgram and Flanagan, 1977) and chemical tests that relate the percentage of sodium to total soluble salt content in the pore water. However, the Standard Test Method for Identification and Classification of Dispersive Clay Soils by the Pinhole Test (ASTM Designation 4647-87) has been shown to have the best correlation with field experience.

In the pinhole test method, distilled water flows horizontally initially under a hydraulic head of 50 mm through a 1.0 mm diameter hole punched in the soil specimen. Flow emerging from a dispersive clay will be dark and the hole will enlarge rapidly increasing the flow. Flow from slightly to moderately dispersive clays will be slightly dark with constant hole size.

Flow from non-dispersive clays will be completely clear with no increase in hole size.

Methods A and C of ASTM 4647-87 use the evaluation of cloudiness of effluent, final size of pinhole, and computation of flow to classify the soil. D1 and D2 are dispersive clays that fail rapidly under a 50 mm head; ND4 and ND3 are slightly to moderately dispersive clays that erode slowly under a 50 mm or 180 mm head; ND2 and ND1 are non-dispersive clays with very slight to no colloidal erosion under a 380 mm or 1020 mm head. Method B of ASTM 4647-87 classifies D1, D2, and ND4 soils as D (dispersive); ND3 soils as SD (slightly dispersive); and ND2 and ND1 soils as ND (non-dispersive).

Where protected by adequate sand filters containing fine sand, dispersive clays have not caused problems in many existing dams, but where these clays are in contact with foundation rock or coarse sand and gravel and are subjected to seepage forces they can cause failure by piping through even minute cracks in the rock or through the voids of the gravel. Slopes of dispersive clays are susceptible to severe erosion by rainfall and homogeneous dams of dispersive clays will fail if a crack occurs allowing relatively pure reservoir water to emerge on the downstream slope.

Chemical additives have been successful in controlling or minimizing the erosive behavior of compacted dispersive clays used for foundations of structures or for embankments. McElroy (1987) described more than sixteen years of experience by the U.S. Soil Conservation Service in the use of alum (aluminum sulfate), gypsum (calcium sulfate), hydrated lime (calcium hydroxide), and mixtures of hydrated lime and agricultural lime (calcium carbonate and/or magnesium carbonate).

In designing major dam or canal embankments, if dispersive clays cannot be avoided, chemical treatment should be used in those portions of the structures that are susceptible to erosion, such as the 30 to 60 cm of soil in contact with the foundation of the dam or the outer 60 cm of canal or embankment slope. Compaction of the chemically treated soil should be made at Standard Proctor optimum moisture and density. For the dams, adequate filters both upstream and downstream from the dispersive clay core are essential.

Frost Action in Compacted Soils (Chapters 3 and 19)

The heaving and thawing of ground surfaces and pavements, the deterioration of backslopes, and even the overturning of retaining walls have been attributed to frost action. The actual amount of frost heaving is generally much more than the increase in volume due to the expansion of water within the soil itself, which is about 9 percent of the volume of voids. The additional change in volume is caused by the formation of horizontal layers or lenses of ice distributed throughout the depth of frost penetration. For example, a saturated soil having a void ratio of 0.70 and a depth of 3 ft would heave approximately $1\frac{1}{4}$ inches if suddenly frozen. The same soil, if it is of the proper type, under conditions favorable for frost action might heave as much as 10 inches. Frost heave occurs under these favorable conditions unless the pressures developed by the tendency to heave are counteracted by superimposed loads.

Frost action includes thawing as well as heaving. When the ground thaws, the water released from the ice lenses causes localized oversaturation until the excess water can leave the thawed zone. Since thawing commences from the surface and progresses downward, the frozen zone underlying the initially thawed zone acts as an impervious diaphragm tending to trap the excess water in this upper layer (Osler, 1966).

Soils susceptible to frost action are those with pores of certain sizes. Coarse soils, with large pore sizes, are not affected. Soils with very small voids (clays) can develop high frost heaving

pressures but, because of low permeability, the volume of water available to form ice lenses is limited and frost heaving in clays is not a very serious problem. Casagrande (1931) and Riis (1948) provided the following criteria for soils susceptible to frost action: the fraction smaller than 0.02 mm > 3 percent for nonuniform soils ($C_u = D_{60}/D_{10} > 5$), or > 10 percent for uniform soils.

Increasing the density of the compacted material will increase the frost susceptibility of a coarse-grained soil that is near the 3-percent criterion because compaction decreases the size of the voids. On the other hand, increasing the density of a fine-grained soil that is frost susceptible may decrease the permeability sufficiently to minimize frost heave. Also, compaction on the dry side of optimum reduces frost heave because less water is available to freeze and the corresponding capillary pressures make it difficult for ice crystals to form in the soil.

Even with susceptible soils, a water supply must exist in order for frost action to occur. Ordinarily the amount of water contained within a soil is not enough to cause serious trouble. There must be an additional supply to feed the ice lenses. This water supply is often in the form of a high water table in contact with a silty soil in which the water is carried continually by capillary action into the frost zone to feed the growing ice crystals.

The rate of freezing affects frost action. A slow freeze causes the severest frost action because as the frost penetrates very slowly, there is ample time for ice crystals to form and for the underlying soil to bring up additional water by capillarity. On the other hand, in a quick freeze the water contained within the soil will freeze solid before the ice lenses have time to form.

Designs to prevent the effects of frost action include use of soils that are not susceptible to frost action, controlling the water supply and height of capillary rise, or insulating against frost penetration. Excavating silty soils to the depth of frost penetration and replacing them with clean granular material such as sand or sandy gravel is a common method of avoiding frost action. Another method is to provide adequate surface

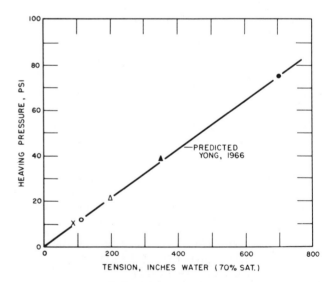

EXPERIMENTAL RESULTS

X AUGREY SAND – HOEKSTRA ET AL, 1965
▲ N.H. SILT   "
● RICHFIELD SILT   "
○ INORGANIC SILT–MCKYES, 1966
△ POTTERS FLINT – PENNER, 1959

Fig. 8.60 Comparison between predicted and experimental results, moisture tension at 70 percent saturation vs. frost heaving pressure.

TABLE 8.16 RECOMMENDED REQUIREMENTS FOR COMPACTION AND SLOPES OF HIGHWAY EMBANKMENTS.

		Condition of Exposure					
		Condition 1 (Not Subject to Inundation)			Condition 2 (Subject to Inundation)		
Revised Public Roads System	Approximate Equivalent, Unified System	Height of Fill, feet	Side Slope	Desired Compaction, % AASHO Maximum Density	Height of Fill, feet	Side Slope	Desired Compaction, % AASHO Maximum Density
A-1	GW, GP, SW, some GM or SM	Not critical	1½ to 1	95+	Not critical	2 to 1	95
A-3	SP	Not critical	1½ to 1	100+	Not critical	2 to 1	100+
A-2-4	Most GM and SM	Less than 50	2 to 1	95+	Less than 10	3 to 1	95
A-2-5					10 to 50		95–100
A-2-6 or 7	GC or SC	Less than 50	2 to 1	95+	Less than 50	3 to 1	95–100
A-4, A-5	ML, MH	Less than 50	2 to 1	95+	Less than 50	3 to 1	95–100
A-6, A-7	CL, CH	Less than 50	2 to 1	95–100	Less than 50	3 to 1	95–100

Notes:
(1) Under Condition 2, higher fills on the order of 35 to 50 ft should be compacted to 100 percent at least for portions subject to inundation. Major fills composed of unusual materials which have low shearing resistance should be analyzed by soil mechanics methods.
(2) For soils of the A-6 or A-7 groups, the lower compaction requirements shown obtain only for low fills (10 to 15 ft or less) not subject to inundation and not carrying large volumes of heavy traffic.
(3) Highly organic soils are not generally suitable for fill construction.

drainage and lowering of the water table so that the water level or zone of capillary rise will not fall within the zone of frost penetration. The use of intercepting ditches to collect ground water and elimination of perched water conditions will help. Frost action can be minimized by using materials within the zone of frost penetration that have insulating value, such as clean sand and sandy gravel provided they are drained.

Studies by Penner (1959) concerning pressures due to frost heave were reviewed by Osler (1966), who reported that measured frost heaving pressures in a uniform silt agreed exactly with predicted values of heaving pressures using Penner's equation. For well-graded soils, the effective size (D_{10}) also correlates well with the predicted pressures.

Using the relation between size of pore and soil suction necessary to remove water from a pore of such size, Yong (1966) determined a relationship between the effective pore size of the silt found from frost heaving experiment and the moisture tension. From the moisture tension–saturation data available for this silt, it was observed that the moisture tension associated with the effective pore size was that obtained at 70 percent saturation in the silt. Yong then checked the relationship between heaving pressures and moisture tension at 70 percent saturation; the results plotted in Figure 8.60 (Osler, 1966), show excellent agreement between theoretical and experimental values.

The relationship between particle and pore sizes and heaving pressures suggests an interesting relationship between frost susceptibility criteria and depth of frost penetration. Osler (1966) shows that a well-graded soil with 3 percent finer than 0.02 mm, which would be classified on the borderline of frost susceptibility, would have an effective size D_{10} of about 0.04 to 0.05 mm, which would correspond to a heaving pressure of 2.5 to 3.0 psi, equal to an overburden pressure of 3.5 ft. If the frost penetration in an area were 3.5 ft, the overburden pressure at the freezing front in this particular soil would be sufficient to prevent ice segregation. Since many of Casagrande's field observations were made in New Hampshire, where a frost penetration of about 3 ft is probably representative of average conditions, the correlation between effective grain size and heaving pressure is striking. Osler points out, however, that in

more northern latitudes where frost penetrations of 5 to 7 ft are common, the frost heaving theory predicts that a soil could have a larger quantity of fines, since the resulting larger heaving pressure would be counteracted by the greater overburden pressure above the frost front. This suggests that a higher quantity of fines would be acceptable under typical winter conditions in Ontario, Canada, than is given by the Casagrande criteria.

Slopes of Compacted Fills

Embankments of compacted fills are used in the construction of highways, railroads, airfields, canals, levees, and earth dams. The last three of these are hydraulic structures, whose design is beyond the scope of this chapter. For other embankments generally 50 ft or less in height that will not be subjected to differential water levels, a general guide to embankment slopes is given in Table 8.16 (U.S. Air Force and U.S. Army, 1954).

REFERENCES

AASHTO (1970, 1986), *Standard Specifications for Highway Materials and Methods of Sampling and Testing*, part 1.

Abdun-Nur, E. A. (1966), Discussion of "Quality control of compacted earthwork," *Journal of the Soil Mechanics and Foundations Division, ASCE*, **92**, No. SM4, p. 116.

ASTM (1970a), Suggested methods of test for securing the liquid limit of soils using one-point data, *Special Procedures for Testing Soil and Rock for Engineering Purposes*, ASTM Special Technical Publication 479, pp. 94–96.

ASTM (1970b), Suggested method of test for rapid compaction control, *Special Procedures for Testing Soil and Rock for Engineering Purposes*, ASTM Special Technical Publication 479, pp. 502–515.

ASTM (1985), *1985 Annual Books of ASTM Standards*, Section 4, Construction, Volume 04.08, Soil and Rock; Building Stones.

Bara, J. P. (1969), Controlling the expansion of desiccated clays during construction, Paper presented to the Second International Research and Engineering Conference on Expansive Clay Soils at Texas A&M University, College Station, Texas.

Barber, E. S. (1956), Discussion on "Engineering properties of expansive clays" by Holtz and Gibbs, *Transactions ASCE*, **121**, p. 672.

Barden, L., Madedor, A. O., and Sides, G. R. (1969), Volume change characteristics of unsaturated clay, *Journal of the Soil Mechanics and Foundations Division, ASCE*, **95**, No. SM1, p. 33.

Barden, L. and Sides, G. R. (1970), Engineering behavior and structure of compacted clay, *Journal of the Soil Mechanics and Foundations Division, ASCE*, **96**, No. SM4, p. 1171.

Basore, C. E. and Boitano, J. D. (1969), Sand densification by piles and vibroflotation, *Journal of the Soil Mechanics and Foundations Division, ASCE*, **95**, No. SM6, pp. 1303–1323.

Bishop, A. W. (1959), The principle of effective stress, *Teknisk Ukeblad*, **39**, pp. 859–863.

Bishop, A. W., Alpan, I., Blight, G. E., and Donald, I. B. (1960), Factors controlling the strength of partly saturated cohesive soils, Research Conference on Shear Strength of Cohesive Soils, Soil Mechanics and Foundations Division, ASCE, Boulder, Colorado, pp. 503–532.

Bjerrum, L., Casagrande, A., Peck, R. B., and Skempton, A. W. (1960), *From Theory to Practice in Soil Mechanics, Selections from the Writings of Karl Terzaghi*, New York, John Wiley and Sons, Inc., New York, N.Y.

Bohn, H. L., McNeal, B. L., and O'Connor, G. A. (1985), *Soil Chemistry*, 2d ed., John Wiley & Sons, Inc., New York, N.Y.

Bouyoucos, G. J. (1931), The alcohol method for determining the water content for soils, *Soil Scientist*, **32**, pp. 173–179.

Bowles, J. E. (1986), *Engineering Properties of Soils and Their Measurement*, McGraw-Hill Book Co., Inc., New York, N.Y.

Briggs, L. J. (1897), The mechanics of soil moisture, Bulletin No. 10, U.S. Department of Agriculture, Division of Soils, Washington D.C., Government Printing Office.

Bureau of Reclamation (1950), *Laboratory Tests on Foundation and Embankment Materials—Fena River Dam—Guam, Marianas Islands*. For the Bureau of Yards and Docks, United States Department of the Navy, Earth Materials Laboratory Report No. EM-230, Bureau of Reclamation, March 21.

Bureau of Reclamation (1952), Placement moisture control and shearing strength studies—Palisades Dam borrow materials—Palisades Project, Idaho, EM-321, Denver, Colorado.

Bureau of Reclamation *Earth Manual* (1968), U.S. Bureau of Reclamation, Denver, Colorado, First Edition, Revised, pp. 455–467.

Bureau of Reclamation *Earth Manual* (1968a), U.S. Bureau of Reclamation, Denver, Colorado, First Edition, Revised, pp. 1–18.

Bureau of Reclamation *Earth Manual* (1968b), U.S. Bureau of Reclamation, Denver, Colorado, First Edition, Revised, p. 42.

Bureau of Reclamation *Earth Manual* (1968c), U.S. Bureau of Reclamation, Denver, Colorado, First Edition, Revised, p. 579.

Bureau of Reclamation *Earth Manual* (1968d), U.S. Bureau of Reclamation, Denver, Colorado, First Edition, Revised, pp. 591–612.

Bureau of Reclamation *Earth Manual* (1968e), U.S. Bureau of Reclamation, Denver, Colorado, First Edition, Revised, pp. 582–591.

Bureau of Reclamation *Earth Manual* (1968f), U.S. Bureau of Reclamation, Denver, Colorado, First Edition, Revised, pp. 467–474j.

Bureau of Reclamation *Earth Manual* (1968g), U.S. Bureau of Reclamation, Denver, Colorado, First Edition, Revised, pp. 574–578.

Burmister, D. M. (1948), The importance and practical use of relative density in soil mechanics, *Proceedings ASTM*, **48**, Philadelphia, Pennsylvania.

Burmister, D. M. (1964), Suggested methods for test for maximum and minimum densities of granular soils, Procedures for Testing Soils, ASTM, Philadelphia, Pennsylvania.

Casagrande, A. (1931), Discussion on "A new theory of frost heaving" by Benkelman and Olmstead, *Proceedings of the Highway Research Board*, **11**, Part 1, pp. 168–172.

Casagrande, A. (1936), Characteristics of cohesionless soils affecting the stability of slopes and earthfills, *Journal of the Boston Society of Civil Engineers*, January.

Casagrande, A. (1938), The shearing resistance of soils and its relation to the stability of earth dams, *Proceedings of the Soils and Foundation Conference of the U.S. Engineer Department*, June.

Casagrande, A. (1965), *Hohe Staudamme*, Communication No. 6, Institute for Foundation Engineering and Soil Mechanics, Techniche Hochscule, Vienna, December.

Casagrande, A. (1979), *Liquefaction and Cyclic Deformation of Sands— A Critical Review*, Harvard Soil Mechanics Series No. 88, Harvard University.

Castro, Gonzalo (1969), Liquefaction of sands (Thesis), Harvard

University Soil Mechanics Series 81, Cambridge, Massachusetts, January.

Chen, F. H. (1975), *Foundations on Expansive Soils*, Elsevier Scientific Publishing Co., Amsterdam.

Coffman, B. S. (1960), Estimating the relative density of sands, *Civil Engineering*, **30**, No. 10, October, pp. 78–79.

Coleman, J. D. (1962), Correspondence to *Geotechnique*, **12**, No. 4.

Converse, F. J. (1957), Compaction of cohesive soil by low-frequency vibration, *Papers on Soils*, ASTM Special Technical Publication No. 206, pp. 73–82.

D'Appolonia, D. J., Whitman, R. V., and D'Appolonia, E. (1969), Sand compaction with vibratory rollers, *Journal of the Soil Mechanics and Foundations Division, ASCE*, **95**, No. SM1, pp. 263–284.

D'Appolonia, E. (1970), Dynamic loadings, *Journal of the Soil Mechanics and Foundations Division, ASCE*, **96**, No. SM1, p. 49.

D'Appolonia, E., Miller, C. E., and Ware, T. M. (1955), Sand compaction by vibroflotation, *Transactions ASCE*, **120**, Paper No. 2730, p. 154.

Davis, F. J. (1953), Quality control of earth embankments, *Third International Conference on Soil Mechanics and Foundation Engineering*, Switzerland, **1**, 1953, p. 218.

Davis, F. J. (1966), Summary of Bureau of Reclamation Experience in Statistical Control of Earth Dam Embankment Construction, Paper presented at the National Conference on Statistical Quality Control Methodology in Highway and Airfield Construction, University of Virginia.

de Mello, V. F. B., Souto Silveira, E. B., and Silveira, A. (1960a), Geotecnica's experience in compaction control of earth dams, *First Pan-American Conference on Soil Mechanics and Foundation Engineering*, Mexico D. F., **II**, p. 637.

de Mello, V. F. B., Souto Silveira, E. B., and Silveira, A. (1960b), True representation of the quality of a compacted embankment, *First Pan-American Conference on Soil Mechanics and Foundation Engineering*, Mexico D. F., **II**, p. 657.

Dunn, W. L. and McDougall, F. H. (1970), Minimizing errors in gamma-ray surface-type density gages: Existing gages and new design concepts, *Highway Research Record* No. 301, Highway Research Board, Washington, D.C.

Engineering News-Record (1950), Navy pushes Guam Dam in race against rainy season, September 21, p. 47.

Fang, H. Y. (1975), Sampling Plans and Construction Control, *Proceedings of the 2nd International Conference on Applications of Statistics and Probability in Soil and Structural Engineering*, **2**, pp. 323–338.

Gibbs, H. J. and Hilf, J. W. (1957), Triaxial shear tests holding effective lateral stress constant, *Proceedings of the Fourth International Conference on Soil Mechanics and Foundation Engineering*, London, **1**, pp. 156–159.

Gibbs, H. J., Hilf, J. W., Holtz, W. G., and Walker, F. C. (1960), Shear strength of cohesive soils, *Research Conference on Shear Strength of Cohesive Soils, Soil Mechanics and Foundations Division, ASCE*, Boulder, Colorado, pp. 33–162.

Gizienski, S. F. and Lee, L. J. (1965), Comparison of laboratory swell tests to small-scale field tests, *Engineering Effects of Moisture Changes in Soils*, Concluding Proceedings, International Research and Engineering Conference on Expansive Clay Soils, Texas A & M Press, College Station, Texas.

Gilbert, O. H. (1959), The influence of negative pore water pressures on the strength of compacted clays, SM Thesis, MIT.

Gordon, B. B. and Miller, R. K. (1966), Control of earth and rockfill for Oroville Dam, *Journal of the Soil Mechanics and Foundations Division, ASCE*, **92**, No. SM3, pp. 1–23.

Gray, E. W., Jr. (1968), Evaluation of manufacturer, Bureau of Reclamation, and Corps of Engineers specifications for tamping rollers, Unpublished memorandum to Chief Designing Engineer, Office of Chief Engineer, U.S. Bureau of Reclamation, Denver, Colorado.

Griffith, A. A. (1921), The phenomena of rupture and flow in solids, *Philosophical Transactions Royal Society of London*, Series A, **221**, pp. 163–198.

Griffith, A. A. (1924), The theory of rupture, *First International Conference on Applied Mechanics*, pp. 55–63.

Haines, William B. (1930), Studies in the physical properties of soil V: The hysteresis effect in capillary properties and the modes of moisture distribution associated therewith, *Journal of Agricultural Science*, London, **20**, pp. 97–115.

Hall, J. W. (1968), Soil Compaction Investigation Report No. 10,

Evaluation of Vibratory Rollers on Three Types of Soils, Technical Memorandum No. 3-271, Corps of Engineers, U.S. Army Engineers, Waterways Experiment Station, Vicksburg, Miss.

Hazen, A. (1920), Hydraulic fill dams, *Transactions ASCE*, **83**, pp. 1713–1745.

Highway Research Board (1952a), *Compaction of Embankments, Subgrades, and Bases*, Bulletin No. 58, Washington, D.C., p. 41.

Highway Research Board (1952b), *Compaction of Embankments, Subgrades, and Bases*, Bulletin No. 58, Washington, D.C., p. 40.

Highway Research Board (1952c), *Compaction of Embankments, Subgrades, and Bases*, Bulletin No. 58, Washington, D.C., p. 42.

Highway Research Board (1962), AASHO Road Test Report 2, Materials and Construction, *Highway Research Board Special Report 61B*, pp. 151–154.

Hilf, J. W. (1956), *An Investigation of Pore-Water Pressure in Compacted Cohesive Soils*, Technical Memorandum 654, U.S. Department of the Interior, Bureau of Reclamation, Denver, Colorado.

Hilf, J. W. (1957), Compacting earth dams with heavy tamping rollers, *Journal of the Soil Mechanics and Foundations Division, ASCE*, **83**, No. SM2, Paper No. 1205.

Hilf, J. W. (1959a), *A Rapid Method of Construction Control for Embankments of Cohesive Soil*, Engineering Monograph No. 26, Bureau of Reclamation, p. 7.

Hilf, J. W. (1959b), *A Rapid Method of Construction Control for Embankments of Cohesive Soil*, Engineering Monograph No. 26, Bureau of Reclamation, p. 23.

Hogentogler, C. A. (1936), Essentials of soil compaction, *Proceedings of the Highway Research Board*, National Research Council, Washington, D.C., pp. 309–316.

Holmgram, G. C. S. and Flanagan, C. P. (1977), Factors affecting spontaneous dispersion of soil materials as evidenced by the crumb test, *Symposium on Dispersive Clays, Related Piping, and Erosion in Geotechnical Projects*, ASTM STP 623, ASTM, pp. 218–239.

Holtz, W. G. (1948), The determination of limits for the control of placement moisture in high rolled earth dams, *Proceedings ASTM*, pp. 1240–1248.

Holtz, W. G. (1959), Expansive clays—properties and problems, *Colorado School of Mines Quarterly*, **54**, No. 4, pp. 89–125.

Holtz, W. G. and Gibbs, H. J. (1956), Engineering properties of expansive clays, *Transactions ASCE*, **121**, pp. 641–663.

Holtz, W. G. and Gibbs, H. J. (1957), Research on determining density of sands by spoon penetration testing, *Proceedings, Fourth International Conference on Soil Mechanics and Foundation Engineering*, **1**, pp. 35–39.

Horonjeff, R. and Javette, D. F. (1956), Neutron and gamma-ray methods for measuring moisture content and density to control field compaction, *Soil Testing Methods*, Bulletin No. 122, Highway Research Board, Washington, D.C., pp. 23–34.

Hosking, A. (1973), Snowy Mountains Hydro-Electric Authority, Cooma North, Australia, private correspondence (Australian Standard A89.21-1973, Test 21—Compaction Control Test (Hilf Method)).

Johnson, A. W. and Sallberg, J. R. (1960), *Factors that Influence Field Compaction of Soils: Compaction Characteristics of Field Equipment*, Bulletin No. 272, Highway Research Board, National Research Council, Washington, D.C.

Kantey, C. A. and Brink, A. B. A. (1962), *Laboratory Criteria for the Recognition of Expansive Soils*, National Building Research Institute, South African Council for Scientific and Industrial Research, Bulletin No. 9, p. 25.

Kassiff, G., Livneh, M., and Wiseman, G. (1969), *Pavements on Expansive Clays*, Jerusalem Academic Press, Jerusalem, Israel.

Kishida, H. (1969), Characteristics of liquefied sands during Mino Owhri, Tohnankai, and Fukui earthquakes, Japanese Society of Soil Mechanics and Foundation Engineering, *Soils and Foundations*, **9**, No. 1, Tokyo, Japan, pp. 75–91.

Kjaernsli, B. and Sande, A. (1966), Compressibility of some coarse-grained materials, Norwegian Geotechnical Institute Publication 66, Oslo.

Koppejan, A. W., Wamelen, B. M., and Weinberg, L. J. (1948), Coastal flow slides in the Dutch province of Zeeland, *Proceedings, Second International Conference on Soil Mechanics and Foundation Engineering*, Rotterdam, **5**, pp. 89–96.

Ladd, C. C. (1960), Mechanisms of swelling by compacted clay, *Highway Research Board Bulletin* No. 245, Washington, D.C.

Ladd, C. C. and Lambe, T. W. (1961), The identification and behavior of expansive clays, *Proceedings, Fifth International Conference on Soil Mechanics and Foundation Engineering*, Paris, France, pp. 201–205.

Lambe, T. W. (1958), The engineering behavior of compacted clay, *Journal of the Soil Mechanics and Foundations Division, ASCE*, **84**, No. SM2, pp. 1655-1 to -35.

Lambe, T. W. (1960), Structure of compacted clay, *Transactions ASCE*, **125**, pp. 682–705.

Lambe, T. W. and Martin, R. T. (1955), Composition and engineering properties of soil (III), *Proceedings of the 34th Annual Meeting, Highway Research Board*, Washington, D.C., pp. 566–582.

Lambe, T. W. and Martin, R. T. (1956), Composition and engineering properties of soils (IV), *Proceedings of the 35th Annual Meeting, Highway Research Board*, Washington, D.C., pp. 661–677.

Langfelder, L. J., Chen, C. F., and Justice, J. A. (1968), Air permeability of compacted cohesive soils, *Journal of the Soil Mechanics and Foundations Division, ASCE*, **94**, No. SM4, p. 999.

Lee, K. L. and Seed, H. B. (1967a), Drained strength characteristics of sands, *Journal of the Soil Mechanics and Foundations Division, ASCE*, **93**, No. SM6, pp. 117–141.

Lee, K. L. and Seed, H. B. (1967b), Cyclic stress conditions causing liquefaction of sand, *Journal of the Soil Mechanics and Foundations Division, ASCE*, **93**, No. SM1, p. 47.

Lee, K. L. and Haley, S. C. (1968), Strength of compacted clay at high pressure, *Journal of the Soil Mechanics and Foundations Division, ASCE*, **94**, No. SM6, p. 1303.

Lewis, W. A. (1960), Full scale compaction studies at the British Road Research Laboratory, *Highway Research Board Bulletin* No. 254, National Research Council, Washington, D.C., pp. 1–12.

Lieberman, G. J. and Resnikoff, G. J. (1955), Sampling plan for inspection by variables, *Journal of the American Statistical Association*, **50**, 457.

Marachi, N. D., Chan, C. K., Seed, H. B., and Duncan, J. M. (1969), *Strength and Deformation Characteristics of Rockfill Materials*, Report No. TE-69-5 to State of California Department of Water Resources, University of California, Department of Civil Engineering, Institute of Transportation and Traffic Engineering.

Marsal, R. J. (1965), Discussion, *Proceedings, Sixth International Conference, Soil Mechanics and Foundation Engineering*, **3**, pp. 310–316.

Marsal, R. J. (1967a), Large-scale testing of rockfill materials, *Journal of the Soil Mechanics and Foundations Division, ASCE*, **93**, No. SM2, pp. 27–43.

Marsal, R. J. (1967b), *Behavior of Granular Soils*, Publication of the Soil Engineering Department of the Universidad Catolica Andres Bello, Caracas, Venezuela.

Maslov, N. N. (1957), Questions of seismic stability of submerged sandy foundations and structures, *Proceedings, 4th International Conference on Soil Mechanics and Foundation Engineering*, London, **1**, pp. 368–371.

McElroy, C. H. (1987), *Engineering Aspects of Soil Erosion, Dispersive Clays, and Loess*, Geotechnical Special Publication No. 10, ASCE, pp. 1–29.

Middlebrooks, T. A. (1942), Fort Peck slide, *Transactions ASCE*, **107**, pp. 723–764.

Mitchell, J. K. (1970), In-place treatment of foundation soils, *Journal of the Soil Mechanics and Foundations Division, ASCE*, **96**, No. SM1, p. 73.

Mitchell, J. K. (1976), *Fundamentals of Soil Behavior*, John Wiley & Sons, Inc., New York, N.Y., 422p.

Morehouse, D. C. and Baker, G. L. (1969), Sand densification by heavy vibratory compactor, *Journal of the Soil Mechanics and Foundations Division, ASCE*, **95**, No. SM4, p. 985.

NAVDOCKS DM-7 (1962), Bureau of Yards and Docks, Department of the Navy, *Design Manual: Soil mechanics, foundations, and earth structures*, Washington, D.C., p. 7-9-8.

Nemecz, E. (1981), *Clay Minerals*, Akademiai Kiado, Budapest, 547p.

Olson, R. E. (1963), Effective stress theory of soil compaction, *Journal of the Soil Mechanics and Foundations Division, ASCE*, **89**, No. SM2, pp. 27–45.

Osler, J. C. (1966), Studies of the engineering properties of frozen soils, *Soil Mechanics Series No. 18*, McGill University, Montreal, Canada.

Penner, E. (1959), The mechanism of frost heaving in soils, *Highway Research Board Bulletin* No. 225.

Proctor, R. R. (1933), The design and construction of rolled earth dams, *Engineering News-Record* **III**, August 31, September 7, 21, and 28.

Reinhold, F. (1955), Soil-cement methods for residential streets in western Germany, *World Construction*, March–April, p. 41.

Riis, J. (1948), Frost damage to roads in Denmark, *Proceedings, Second International Conference on Soil Mechanics and Foundation Engineering*, Rotterdam, **2**, p. 287.

Roberts, J. E. and De Souza, J. M. (1959), The compressibility of sands, *Proceedings, ASTM*, **58**, p. 1269.

Roscoe, K. H. and Schofield, A. N. (1958), On the yielding of soils, *Geotechnique*, **7**, pp. 25–53.

Sallberg, J. R. and Smith, P. C. (1965), Pavement design over expansive clays: Current practices in the United States, Engineering Effects of Moisture Changes in Soils, *Concluding Proceedings, International Research and Engineering Conference on Expansive Clay Soils*, Texas A & M Press, College Station, Texas.

Schmertmann, J. H. (1967), Static cone penetrometers for soil exploration, *Civil Engineering*, **37**, No. 6, June, pp. 71–73.

Schmertmann, J. H. (1970), Sand densification by heavy vibratory compactor, Discussion of Proceedings, Paper 6656, *Journal of the Soil Mechanics and Foundation Division, ASCE*, **96**, No. SM1, p. 363.

Schmertmann, J. H. (1978), *Guidelines for Cone Penetration Test*, Federal Highway Administration Report FHWA-78-209, U.S. Department of Transportation, Washington, D.C.

Seed, H. B. (1968), Landslides during earthquakes due to soil liquefaction, *Journal of the Soil Mechanics and Foundations Division, ASCE*, **94**, No. SM5, pp. 1055–1122.

Seed, H. B. and Chan, C. K. (1959), Structure and strength characteristics of compacted clays, *Journal of the Soil Mechanics and Foundations Division, ASCE*, **85**, No. SM5, pp. 87–128.

Seed, H. B., Mitchell, J. K., and Chan, C. K. (1962), Studies of swell and swell pressure characteristics of compacted clay, *Highway Research Board Bulletin* No. 313, Highway Research Board, Washington, D.C., pp. 12–38.

Seed, H. B., Woodward, R. J., and Lundgren, R. (1962), Prediction of swelling potential for compacted clays, *Journal of the Soil Mechanics and Foundations Division, ASCE*, **88**, No. SM3, pp. 53–58.

Seed, H. B. and Lee, K. L. (1966), Liquefaction of saturated sands during cyclic loading, *Journal of the Soil Mechanics and Foundations Division, ASCE*, **92**, No. SM6, p. 105.

Sherard, J. L., Dunnigan, L. P., and Decker, R. S. (1977), Some engineering problems with dispersive clays, *Symposium on Dispersive Clays, Related Piping, and Erosion in Geotechnical Projects*, ASTM STP 623, ASTM, p. 9.

Sherman, G. B., Watkins, R. O., and Prysock, R. H. (1967), A statistical analysis of embankment compaction, *Highway Research Record* No. 177, Symposium on Compaction of Earthwork and Granular Bases, Highway Research Board, National Research Council, Washington, D.C., pp. 157–185.

Sherman, W. C. and Clough, G. W. (1968), Embankment pore pressures during construction, *Journal of the Soil Mechanics and Foundations Division, ASCE*, **94**, No. SM2, p. 527.

Shook, J. F. (1959), Construction materials control for the AASHO road test, *Journal of the Soil Mechanics and Foundations Division, ASCE*, **85**, No. SM5, pp. 15–29.

Smith, T. W. and Prysock, R. H. (1966), Discussion of "Quality control of compacted earthwork," *Journal of the Soil Mechanics and Foundations Division, ASCE*, **92**, No. SM5, p. 142.

Southwest Builder and Contractor (1936), Tamping feet of a flock of sheep gave idea for sheepsfoot roller, August 7, p. 13.

Terzaghi, K. (1925), *Erdbaumechanik auf bodenphysikalischer*, Grundlage, Vienna, Deuticke.

Terzaghi, K. (1936), Simple tests determine hydrostatic uplift, *Engineering News-Record*, June 18, pp. 872–875.

Terzaghi, K. (1950), Mechanism of landslides, *Application of Geology to Engineering Practice*, Berkey Volume, Geological Society of America, pp. 181–194.

Terzaghi, K. (1958), Design and performance of Sasumua Dam, Paper No. 6252, *Proceedings of the Institution of Civil Engineers*, **9**, pp. 369–388.

Terzaghi, K. and Peck, R. B. (1948, 1967), *Soil Mechanics in Engineering Practice*, John Wiley and Sons, Inc., New York, N.Y., pp. 57–60.

Tiedemann, D. A. and Fink, R. E. (1969), *Compaction of Cohesive Soil with a Vibratory Roller—Cawker City Test Embankment*, Report No. EM-766, Bureau of Reclamation, Denver, Colorado.

Turnbull, W. J., Johnson, S. J., and Maxwell, A. A. (1949), *Factors Influencing Compaction of Soils*, Bulletin No. 23, Highway Research Board, National Research Council, Washington, D.C., p. 18.

Turnbull, W. J., Compton, J. R., and Ahlvin, R. G. (1966), Quality control of compacted earthwork, *Journal of the Soil Mechanics and Foundations Division, ASCE*, **95**, No. SM1, pp. 93–103.

U.S. Air Force and U.S. Army (1954), Control of soils in military construction, AFM 88-52, TM 5-541, Departments of the Air Force and the Army, p. 281.

U.S. Bureau of Reclamation (1970), Bureau of Reclamation Construction Specifications for Cutter Dam, D6845-DC, U.S. Bureau of Reclamation, Denver, Colorado.

U.S. Corps of Engineers (1949), Simplification of the liquid limit procedure, *Technical Memorandum No. 3-286*, Waterways Experiment Station, Vicksburg, Miss.

U.S. Corps of Engineers (1970a), Liquid limit correlation method; embankments, earth, Appendix 2B-AA, Construction, Draft Quality Control Guide for Testing of Materials and Equipment.

U.S. Corps of Engineers (1970b), *Construction Quality Control Guide*.

Wahls, H. E., Fisher, C. P., and Langfelder, L. J. (1966), *The Compaction of Soil and Rock Materials for Highway Purposes*, Department of Civil Engineering, North Carolina State University at Raleigh, North Carolina, p. 274. (Report prepared for Bureau of Public Roads, Washington, D.C.)

Wald, A. (1947), *Sequential Analysis*, John Wiley and Sons, Inc., New York, N.Y.

Walker, F. C. (1960), The use of residual soils in earth and rock-fill dams, *Proceedings, First Panamerican Conference on Soil Mechanics and Foundation Engineering*, Mexico D.F., **2**, pp. 589–602.

Waterways Experiment Station (1949), Soil compaction investigation: Compaction studies on silty clay, Report No. 2, Technical Memorandum No. 3-271, U.S. Corps of Engineers, Vicksburg, Miss.

Waterways Experiment Station (1967), U.S. Corps of Engineers, Potamology Investigations Report 12-18 Verification of empirical method for determining riverbank stability, 1965 data, Vicksburg, Miss.

Wesley, L. D. (1973), Some basic engineering properties of halloysite and allophane clays in Java, Indonesia, *Geotechnique*, **23**, No. 4, pp. 471–494.

Whitman, R. V. and Richart, F. E. (1967), Design procedures for dynamically loaded foundations, *Journal of the Soil Mechanics and Foundations Division, ASCE*, **93**, No. SM6, pp. 169–193.

Wilks, S. S. (1948), Sampling and its uncertainties, *Proceedings ASTM*, **48**, p. 859.

Yong, R. N. (1966), On the relationship between partial soil freezing and surface forces, Paper presented to International Conference on Low Temperature Science, Sapporo, Japan, Report No. d, Phys R (G) Misc 24, Director of Physical Research, Defence Research Board, Canada.

Zolkov, E. and Wiseman, G. (1965), Engineering properties of dune and beach sands and the influence of stress history, *Proceedings, Sixth International Conference on Soil Mechanics and Foundation Engineering*, Montreal, **1**, pp. 134–138.

9 SOIL STABILIZATION AND GROUTING

HANS F. WINTERKORN, Dr. phil. nat. (Deceased)
Professor of Civil Engineering and Geophysics
Princeton University

SIBEL PAMUKCU, Ph.D.
Assistant Professor of Civil Engineering
Lehigh University

9.1 INTRODUCTION

Soil stabilization and grouting are methods of soil improvement. Soil improvement is a combination of physical and chemical methods for regional or mass densification, reinforcement, cementation, and control of drainage and volume stability of soil when it is used as a construction material.

Soil stabilization is manipulation of foundation or base soils with or without admixtures, to increase their load-carrying capacity and resistance to physical and chemical stress of the environment over the service life of the engineered facility. Properties of soil such as strength, stiffness, compressibility, permeability, workability, swelling potential, frost susceptibility, water sensitivity, and volume change tendency may be altered by various methods of soil stabilization. These methods can range from preparation of soil-aggregate and simple compaction, to application of admixtures, to thermal and electrokinetic methods. Among the newer techniques of stabilization are vitrification and electroosmosis, which are emerging technologies used to decontaminate and stabilize contaminated soils.

Grouting is a ground improvement method used both in rock and soil. Rock grouting is essentially waterproofing a rock mass where there is a network of conduits and cracks and excessive fragmentation. Soil grouting is employed in various capacities. Most common applications are: in-situ reinforcement, stabilization and densification of deep foundations, barrier systems to control water flow and corrective measure in postconstruction problems in soft or loose deposits. Soil grouting is a method often used to stabilize deep foundation soils in which the desired improvement of the physical and mechanical properties is achieved through in-situ injection of a stabilizer. The stabilizers are generally mixtures of soil additives or chemicals that work to densify and/or water-proof the soil deposit. This chapter will mainly consider the stabilization aspect of grouting. The reader is referred to other chapters for detailed coverage of application of grouting in other capacities, such as soil reinforcement and barrier systems (Chapters 21 and 25). Although related, for the purpose of continuity, grouting will be discussed under a separate heading from the concepts of soil stabilization in this chapter.

9.2 PRINCIPLE OF SOIL STABILIZATION

Soil stabilization is the collective term for any physical, chemical, or biological method, or any combination of such methods, employed to improve certain properties of a natural soil to make it serve adequately an intended engineering purpose. The different uses of soil pose different requirements of mechanical strength and of resistance to environmental forces. Stabilizing the slope of a cut or a fill against erosion is a vastly different matter from stabilizing a soil to carry a heavy bomber without excessive deformation. Consequently, the term soil stabilization possesses a definite meaning only if the nature of the soil, the engineering requirements, and the environmental conditions are properly identified.

As understood and practiced by the geotechnical engineers, soil stabilization is divided into two parts:

1. The improvement of native soils for construction of shallow foundations, especially for highways, airfields, parking lots, and similar facilities.
2. The improvement of deep foundation soils or large soil masses used for engineering purposes (i.e., in construction of dams and dikes) by way of injection treatment, owing to the difficulty of access, size, or location of the soil mass.

In the applications above, the goals are often to increase strength, bearing, and resistance to deteriorative forces of natural and/or man-made environments; to decrease permeability, volume change tendency and settlement; and to improve stability thus ensuring long-term performance of the foundation soils or earth structures. The first section of this chapter is devoted to principles and methods of improvement of native soils for shallow foundations, using granular, chemical, thermal, and electrokinetic methods of stabilization. Matters falling within stabilization of deep foundation soils or large soil masses are covered in the next section of this chapter, that on grouting.

The Art and Science of Soil Stabilization

In the field of soil stabilization an important aspect that requires attention is the dynamic interaction of the soil foundation with the environment. Soil stabilization involves more than a mere increase in compressive strength and shear resistance or the improvement of other physical properties of soils. It should supply a defense mechanism against the continually acting dynamic forces associated with the subjection of a soil system to daily and seasonal temperature and moisture changes, to microbial and other biological activity, and to chemical instability due to natural or man-made causes, all of which tend to destroy the integrity of the system. The environmental conditions at the time of construction, the random or periodic variation of these conditions over time, and the relative amplitude of such variations may influence the integrity of the system significantly. Some of these conditions are: temperature extremes (including cyclical), submersion, sunlight exposure, cyclic freezing/thawing, degree of saturation, type of contact water (fresh or salt), cyclic wetting/drying, leaching or draining, chemical intrusion, preparation of material, and construction

317

TABLE 9.1 THE THREE PILLARS OF THE SCIENCE OF SOIL STABILIZATION[a].

Pedology	Physical and Chemical Sciences	Body of Knowledge on Traditional Construction Materials
The science of soil formation and of the physical and chemical soil characteristics as function of 1. Parent material 2. Climate 3. Topography 4. Organisms 5. Human activity 6. Time	These allow judgment of soil deficiencies and indicate supplementation in physical and chemical terms that can be translated into economically available materials such as: 1. Inorganic cements 2. Organic cements 3. Waterproofing agents with or without improvement of soil granulometry, etc.	A stabilized soil, being a construction material, must resemble traditional construction materials in its essential mechanical and chemical resistance properties and structure. Especially important are granulometric considerations. Most non-metallic construction materials and most stabilized soils fall into the category of collameritic (cemented-particle) systems.

[a] After Winterkorn (1955a).

sequencing. For instance, stabilization requirements of a soil may increase with increasing proximity of the soil structure to the pavement surface, where the normal temperature and moisture variations exhibit greatest amplitude. Similarly, a highway base course of the same soil type and built for the same traffic loads and frequencies has considerably higher stabilization requirements in the more severe climate of New Brunswick, Canada, than in the mild climate of Florida. Hence, numerically expressed requirements of mechanical strength and durability may possess real meaning only for the specific conditions and within the climatic region for which they have been established.

A distinction must be made between the art and the science of soil stabilization. The latter represents the logical scientific structure, whose primary building stones are empirical facts, collected by observation and experimentation. The essence of this empirical information has been coordinated in accordance with established scientific principles and laws. This structure not only provides a logical scheme for the proper placement and retrieval of experienced knowledge but also permits interpolation into as yet unknown territory, thus serving as a valuable heuristic tool. Consequently, the science of soil stabilization is most important for extension of the art into environments, soil, and climatic provinces where actual experience is lacking.

The art of soil stabilization ranges from local use of the empirical knowledge to practice of methods for such extra-ordinary uses as the stabilization of seafloor soils or lunar and planetary surface materials. The three pillars of science of soil stabilization are shown in Table 9.1.

Historically, major uses of soil stabilization have been: (1) improving adverse ground conditions to facilitate economic development of an area, that is, lifting a region out of mud or sand; (2) providing bases and surfaces for secondary and rural roads, where primary roads are in existence; (3) providing bases for paved surfaces where crushed gravel and rock normally employed are not available economically; (4) for city and suburban streets where the noise-absorbing and elastic properties of certain stabilized soil systems are a definite advantage over other construction methods; (5) for military and other emergencies where an area must be made trafficable within a short period of time.

Soil stabilization involves: (1) diagnosis of the resistance properties of a given soil and required modification of these properties for the intended use in physical, physicochemical, or chemical terms; (2) translation of the modification requirements into available materials and processes, and decision on the use of a specific method or methods on the basis of economy, feasibility, or special (military or other emergency) considerations; (3) construction, generally consisting of preparation of material (in-situ or offsite), application, and densification; and (4) economic considerations relating to the cost of materials, construction, and maintenance for the service life of the structure.

9.3 METHODS OF SOIL STABILIZATION

Among many theoretical and practical possibilities of stabilizing soils, the following have been identified as practical and economical solutions.

(1) *Granular stabilization* is a combination of physical and chemical stabilization methods in which granular bearing skeleton is modified with pore-filling and/or cementing natural and extraneous materials (clay and other concretes and mortars).

(2) *Chemical stabilization* is a general term for all those methods in which chemical rather than physicochemical and physical interactions play a predominant role. It covers the methods in which one or more chemical compounds are added to the soil for treatment. A series of chemical reactions, which can be described by a few simple chemical equations, take place between soil constituents and the additives. Some of the basic reactions that may occur between soil and the additives are hydration, ion-exchange, pozzolanic reaction (cementation), flocculation, precipitation, polymerization, oxidation, and carbonation. There are numerous chemicals and agents used for various stabilization purposes and new ones are being introduced as cheaper alternatives. Vegetation and biotechnical stabilization to protect slopes and prevent erosion of the land can be considered within the subject matter of chemical stabilization since chemical as well as biological processes are involved in such stabilization practices. Utilization of a number of the chemical stabilization methods may be subject to constraints with respect to economy and environmental safety. Some of the widely used and emerging systems are summarized below.

a. *Soil–cement* (*portland cement*). Compacted soil-cement; cement modified soil; plastic soil-cement (PCA, 1971, 1978; Melancon and Shah, 1973).

b. *Soil–lime*. Stabilization of clay or clay fraction of soil using slaked lime (calcium hydroxide), quick lime (calcium oxide), lime–fly-ash (LFA), lime–gypsum (Eggestad, 1983; NLA, 1985; TRB, 1987).

c. *Soil–bitumen*. Soil-bitumen (modified granular bearing skeleton); sand–bitumen (added cohesion); waterproofed granular stabilization; oiled earth surface (Asphalt Institute, 1974, 1982; Baker et al., 1975; Johnson, 1976).

d. *Soil–resin*. Waterproofing cohesive soils with small amounts of resin (less than 2% by dry weight of soil); waterproofing and cementing cohesive or noncohesive soils by means of artificial or natural resins such as cement–resin mixes (polymer cements); organic resins (epoxy, acrylic, poly-acrylate, polyurethane, solvinated resins) (Bell, 1975; WES, 1977; Kézdi, 1979; Mitchell, 1981).

e. *Soil–powerplant wastes.* Stabilization of cohesive soils by adding pozzolanic mineral coal ash, which can be in the form of fly-ash, bottom ash, or FGD (flue gas desulfurization) sludge. Properties of these siliceous residues are dependent on the type of coal used as a fuel—anthracite, bituminous, subbituminous, or lignite—and may or may not require addition of lime for pozzolanic reaction to take place (Hecht and Duvall, 1975; Meyers et al., 1976; Covey, 1980; Usmen et al., 1983; Pamukcu et al., 1987; ASCE, 1987).

f. *Others.* Salts (chlorides); acids (phosphoric acid), lignin, thermoplastic materials for encapsulation (tar, asphalt, poly-ethylene), sodium and calcium silicates, calcium aluminates, sulfur, sulfates, potassium, iron oxide, hydroxy-aluminum, cement basillius (lime–gypsum–alumina), pozzolanic ash of peat and agricultural waste (rice, peanut, and castor bean husk; sugar-cane bagasse), hydromulching materials (wood fiber, waste paper, cellulose pulp), ceramic waste, mining wastes (WES, 1977; Pojasek, 1982; Eggestad, 1983; Cincotto, 1984; Cook, 1984; Ghavami and Fang, 1984; ASCE, 1987).

(3) *Thermal stabilization* is a physical method of stabilization that utilizes heating or freezing for long- or short-term improvement of properties of difficult soil.

a. *Heating* of soils to high temperatures (typically above 300°C) by methods that may involve combustion, electricity, microwave, or laser beam application, to cause permanent changes in their physical properties (Jurdanov, 1978; Eggestad, 1983). Heating methods may be divided up into two groups: (1) in-situ treatment by use of road burners, or heating elements placed in the ground; (2) treatment of the backfill soil in stationary kilns to produce artificial material. Fine-grained soils, especially partially saturated silts, clays, and loesses are most suited for stabilization using heating techniques. The benefits of the treatment are increase in bearing capacity; decrease in water sensitivity, compressibility, and swelling; reduction in lateral pressure; and elimination of collapse properties. Another application of the technique is vitrification of hazardous waste containment sites or contaminated soils to create an inert glass block for control of contaminant migration or encapsulation of the site (Timmerman, 1984).

b. *Freezing* (refrigeration) is used to achieve temporary ground stability or control of groundwater in soft grounds or excavations below the groundwater table; or to achieve continuous stability in permafrost regions where thawing is to be prevented (Shuster, 1972; Jessberger, 1982; and Chapter 19). The method is applicable to all soils and the benefits include reduction of permeability and increase in compressive strength. The temporary improvement of these properties is dependent on time, temperature, and the grain-size distribution of the soil.

(4) *Electrokinetic stabilization.* Stabilization of soils by the influence of an electric field has been used to facilitate consolidation of fine-grained soils and movement of stabilizing agent through dense soils, and to detour natural seepage by providing flow barriers (Winterkorn, 1947b; WES, 1961; Bjerrum et al., 1967; Gray and Mitchell, 1967; Karpoff, 1976; Esrig and Henkel, 1968; Chappel and Burton, 1975; Olsen, 1984). Currently, electrokinetic phenomena are under investigation to develop efficient techniques of in-situ decontamination of soil and groundwater in waste disposal areas and contaminated soils (Segall et al., 1980; Probstein and Renauld,

1986; Ferguson and Nelson, 1986; Mitchell, 1986; Acar et al., 1989; Khan et al., 1989).

9.4 STABILIZATION OF SOILS WITH GRANULAR SKELETON

Soils with particle sizes greater than 0.075 mm are designated as medium to coarse-grained soils. These soils, when compacted, form a granular bearing skeleton through a network of grain-to-grain contact points that is able to transfer load without permanent deformation, and provide frictional resistance and volume stability. They may also contain material with particle sizes less than 0.075 mm without violating the requirements given above if: (1) the volume of the silt-clay size (< 0.075 mm) fraction plus that of the water, normally required to satisfy the capillary and physicochemical sorption capacity, does not exceed the volume of the pore space left by the stable continuous granular skeleton; and (2) the ratio of the size of the smallest bearing grain to that of the largest silt-clay particle is such as to cause no detrimental interference of grain–grain contact of the granular skeleton.

Stabilization of this class of soils is designated "granular stabilization." It involves preparation of mixture of soil-aggregate consisting of stone, gravel, and sand and containing silt-clay, compacted to maximum density to obtain high strength, stability, and durability in all weather conditions. Granular stabilization is used in construction of base, subbase, and surface courses of paved facilities. The primary objective is to obtain a well-proportioned mixture of particles with continuous gradation (well graded) and the desired plasticity. The requirements for composition of mixtures intended for use as bases generally differ from those for use as wearing surfaces. For example, the compositions for base and subbase courses are required to have high stability to transfer load and low capillarity, to resist softening with accumulation of moisture. The compositions for wearing surfaces, on the other hand, need to satisfy the conditions of resisting abrasion and penetration of water, and of capillarity to replace moisture lost by surface evaporation. Therefore, the composition of base and subbase soil-aggregate requires less fine-soil fraction than the composition for wearing surface.

9.4.1 Granulometry and Porosity

The pore volume and the size of the pores formed by the granular skeleton determine the transition of a particular soil to one with or without a bearing skeleton. Fundamentals of granulometry are applied to establish quantitative definitions of granular skeleton with effective compactness (Winterkorn, 1967a). Grain-size distributions that yield minimal porosity values with small densification effort are best presented by the Talbot formula:

$$p = \left(\frac{d}{d_{max}}\right)^m \qquad (9.1)$$

where

p = weight percent of the particles with diameter less than d
d_{max} = maximum particle diameter in the mixture
m = exponent determined empirically

The factor m varies between 0.11 and 0.66. Figure 9.1 presents the relation given by Equation 9.1 for various m values. U.S. Bureau of Public Roads (USBPR, 1962) recommends 0.45 as the best overall value for m. The grain-size distribution of

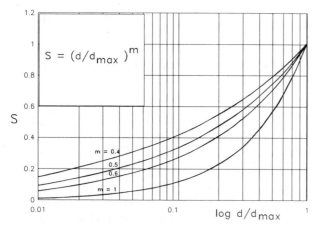

Fig. 9.1 Curve satisfying the Talbot formula

satisfactory mixtures is usually plotted as a parabolic curve following the Rothfuchs formula:

$$p = \frac{d^m - d_{min}^m}{d_{max} - d_{min}^m} \quad (9.2)$$

where d_{min} = minimum particle diameter in the mixture. When the maximum and the minimum particle diameters in a mixture is known, than the composition of a continuous grain distribution can be obtained by using the Rothfuchs formula (Eq. 9.2). The greater the range from maximum to minimum size of the skeleton-forming particles in a continuous gradation, the less the porosity of the system. The average porosity n of continuous grain distribution curves can be expressed as a function of the ratio d_{max}/d_{min} by the following equation:

$$n = A - 8 \log_{10} (d_{max}/d_{min}) \quad (9.3)$$

where

$A = 38.5$ for rounded gravel and sand
$A = 47.5$ for crushed stone

For most natural materials of relatively narrow gradation, the range of easily obtainable and reproducible porosity lies between 36 and 46 percent. Gap-graded mixtures may also be used successfully in granular stabilization. In many cases, economical constraints and limited availability of certain materials necessitate design of a mix which could be best suited for intended stabilization.

9.4.2 Soil Binder

Soils with granular bearing skeleton in the densified state possess volume stability and frictional resistance. They may require bonding or cementation, increase in cohesion, and also decrease in permeability or water storage capacity if deficient in fines. Such stabilized granular soils belong to the class of *collameritic* (colla = glue, meros = particle) systems. In the terminology of materials science, such bonded soils belong to the class of "concretes" if the maximum particle size is larger than the openings of No. 4 sieve (4.76 mm), and to the class of "mortars" if the largest particles are of fine sand size or the size of the openings of No. 40 sieve (0.425 mm). The latter type of soils are also called "soil binder." Table 9.2 summarizes the properties of various collameritic systems.

Complete replacement of natural soil binder in a clay-bonded stabilized gravel (clay concrete) by portland cement produces *portland cement concrete*. Partial replacement leads to systems that possess properties intermediate between those of clay concrete and a portland cement concrete. Similarly, partial replacement of the soil binder by asphalt leads to waterproofed granular soil stabilization, and complete replacement by bitumen and filler leads to bituminous concrete. In a like manner, there exist a continuous series of concretes which include clay concrete, lime concrete, resin concrete, sorel-cement concrete, gypsum plaster concrete, and others. Potentially all known and conceivable inorganic and organic cementing materials and combinations of these can be used to create concrete products as long as they adhere to or can be made to adhere to the soil mineral and particle surfaces. Partial or complete substitution of clay binder with other cementing agents can be done in sand–clay and clay–mortar systems also. The use of such cementing materials is limited by availability; cost; and susceptibility to local climatic conditions, mixing, placing, and densification with the available implements at the site of the construction.

The soil binder or the cementing materials tend to surround the coarse-grain particles and/or form bonding bridges between particles such that the granular system attains rigidity and stability. The strength of such a system is dependent on the strength of the cement and on the shear resistance at the cement–particle interface, as well as on the strength of the granular network.

TABLE 9.2 COLLAMERITICS—THE SCIENCE OF COMPOSITION AND PROPERTIES OF NONMETALLIC CONSTRUCTION MATERIALS[a].

Properties of the Particles	Properties of the Cementing Agents	Examples of Cemented Systems

A. Physical
  I. Granulometry
    Laws of arrangement and packing
    as functions of size, gradation
    and shape factors
  II. Mechanical
    Strength, toughness abrasion
    resistance
B. Physicochemical and chemical
  I. Interaction and bonding with
    cementing agents
  II. Reactivity with deleterious
    substances in environment
	A. Inorganic
	  I. Simple
	    Gypsum and lime plasters
	  II. Complex
	    Sorel-, hydraulic and other
	    cements
	  III. Clay and binder soil
	B. Organic
	  I. Bituminous
	    Asphalts, pitches, tars
	  II. Natural and synthetic resins and
	    other polymers
	  III. Gums, glues of various types, etc.
		1. Mortars with inorganic and organic
		   cements including natural and
		   artificial sand stones
		2. Concretes
		   Portland cement, bituminous,
		   resinous, clay, etc., including
		   naturally cemented conglomerates
		3. Plastics
		   Powder, paper-, cloth-, and fiber-
		   filled; also natural wood in which
		   cellulose fibers are bonded together
		   by lignin

[a] After Winterkorn (1955a).

9.4.3 Specifications on Gradation and Selection of Soil Elements

Soils with enough granular constituents, or aggregates, to form a bearing skeleton, and enough silt–clay, or fines, to provide effective cohesion and cementation are often found in nature and require densification to yield structures of good mechanical stability. However, more frequently, the natural soil will lack some constituents needed to form a continuous bearing skeleton or to provide the necessary cohesion and cementation. In these cases, the desired mixture can be compounded by addition of proper proportions of the aggregates or fines. Sometimes, the natural soil may have such amount and activity of silt–clay constituents that the tolerances for a stable bearing skeleton are exceeded. In such cases, treatment with waterproofing or cementing material, such as lime, may decrease the swelling of the fines and inhibit volume instability under wet conditions. On the other hand, if there is a small fraction of fines or binder in the natural soil, the desired permeability and cohesion may not be achieved. This situation can be remedied by addition of silt–clay or other binder material. The properties of the final mixture are generally controlled and judged by gradation, the liquid limit, and the plasticity index. Standard methods of test for aggregate gradation include ASTM C136 (AASHTO T27) for coarse and fine aggregates; and ASTM C117 (AASHTO T11) for materials finer than the No. 200 obtained by washing. For mixtures expected to include appreciable percentages of fines (passing No. 200 sieve), the standard gradation test follows the ASTM D422 (AASHTO T88) method.

A granular bearing skeleton may be established by several different methods. The choice depends on the soil and other materials available, intended use and special properties desired in the stabilized system, and time constraints for planning and construction. Specifications with respect to gradation are usually quite tolerant, with the emphasis on making the best possible and most economical use of locally available material. For road bases, subbases, and bearing courses the specifications generally set limitations on the percentage and plasticity characteristics of the material passing the No. 40 sieve. This represents an indirect limitation on the water affinity of the soil binder, which when exceeded may alter the continuity of the granular bearing skeleton and introduce excessive volume change with changing water content. The specifications on the amount of soil binder and its plasticity is different for stabilized systems used as wearing surfaces. Such surfaces require a higher percentage of binder material than bases to withstand abrasive effects, inhibit permeation of water to layers below, and replace evaporating water by capillarity.

The theoretical gradations, plasticity properties and suggested treatment suitable for various purposes are given in Figure 9.2. ASTM D1241 (AASHTO M147) gradation requirements for soil-aggregate materials used in construction of subbase, base, and surface courses are given in Table 9.3. Gradings A and B of Table 9.3 are not recommended for wearing surfaces. The specifications for wearing surfaces require a maximum liquid limit of 35, a plasticity index range from 4 to 9, and a minimum of 8 percent passing No. 200 sieve, and maximum size of aggregate is recommended as 1 inch. For bases and subbases the maximum liquid limit is 25, the maximum plasticity index is 6. This indicates a lower percentage of clay than that recommended for surface courses. However, the amount of fines passing No. 200 can be adjusted according to local climatic conditions or the service characteristics of the facility constructed. For example, the percentage of fines may be decreased to prevent frost damage in cold climates, or increased if the base course is designed to perform under an open pavement through which water may penetrate freely. Physical tests, other than plasticity indices, can be used to control the properties and

amount of soil binder. For example, one specification for base courses is that the CBR (California Bearing Ratio) be not less than 80 and the expansion of a compacted sample during soaking not exceed 1 percent. The quality of the coarse aggregate in wearing surfaces is controlled by the Los Angeles abrasion test (ASTM C131, C535; AASHTO T96), which specifies that the percentage of wear shall not exceed 50 percent. The types of natural aggregates and various methods of tests for aggregate quality are given in Baker et al. (1975). The ASTM specification D2940, allows a wide variety of gradations to be used satisfactorily for bases or subbases for highways or airports provided that the variations are kept to a minimum. The requirements of ASTM D2940 are given in Table 9.4. ASTM also designates standard size of coarse aggregate for highway construction in specification D448.

In recent years the economic constraints on construction aggregates with depletion of local supplies and shortages of high-quality natural aggregates have prompted investigators to find ways to extend resources. Successful applications of "synthetic" aggregate products in road-base constructions have been reported (ASTM STP 774, 1980). Among such materials are residues from fluidized-bed combustion boilers, iron and steel slags, and ceramic aggregates produced from mining and industrial wastes (ASTM STP 774, 1980; Fang, 1986; Fang and Pamukcu, 1989; Pamukcu et al., 1989). Use of air-cooled blast furnace slag has had the most common utilization in highway construction. The ASTM definition of blast furnace slag is given in specification C125. Properly produced slags have been observed to perform satisfactorily in base and subbase systems. Such utilization has the added benefit of resource recovery as well as effective disposal of an industrial waste product.

9.4.4 Proportioning and Mixing Soil Elements

Producing a mixture of preconceived continuous gradation, especially in unfamiliar territory and where certain aggregate sizes are not economically available may not only prove expensive but may also result in stabilized systems of lesser quality than possible with the use of available materials. In such cases, it is advisable to determine experimentally the proportions of granular and cohesive soil elements, that when compacted in earth-moist conditions, produce the desired dry density with optimized values of strength and permeability.

The dense system possessing the bearing granular skeleton can be obtained by both continuous (well-graded) or discontinuous (gap-graded) gradation. Continuous graded materials are preferred, often owing to the workability and ease of placement of such materials. However, the choice is generally dictated by economic constraints. The type of natural materials that meet the requirements range from silty or slightly clayey gravels, coarse silty sands, and mixed-grain-size crushed limestone, to certain demolition or industrial by-products. The artificially produced stabilized systems are commonly called "sand–clay" or "stabilized gravel." Sand–clay is a favorable mixture of clay, silt, fine and coarse sand and some fine gravel. Recommended grain-size distribution of sand–clay is grading F of Table 9.3. The stabilized gravel consists of coarse aggregates, sand, and binder soil. These systems prove to be most economical when suitable sources of coarse aggregates, sand, or clay are available locally.

In many cases, it is necessary to mix two or three different types of soils to obtain the desired gradation. The admixture ratio for two soils, whose gradation curves are given, can be determined by the following equation:

$$p = \frac{p_1 + np_2}{1 + n} \tag{9.4}$$

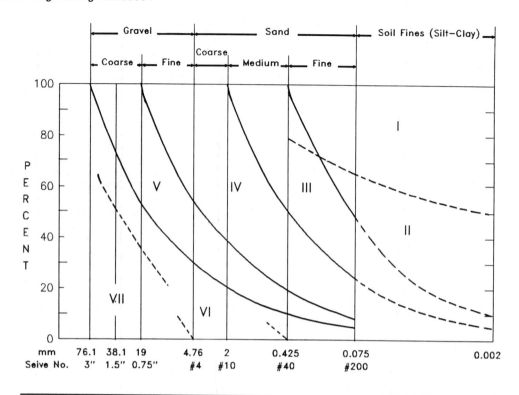

Fig. 9.2 Grain-size distribution and plasticity properties as related to various methods of stabilization.

Area	Properties and Suggested Treatment
I	Heavy clay soils of high plasticity; treatment with quick or hydrated lime by itself or for rendering soil susceptible to stabilization with hydraulic cements or bituminous materials (asphalts, tars, etc.).
II	Intermediate clay contents; hydraulic cements if LL of $-$ No. 40 sieve fraction <40 and $PI < 15$, and bituminous binders if $PI < 10$; in borderline cases toward (I) combined with use of hydrated lime.
III	Well-graded sandy soil, easily stabilized with hydraulic cement or bitumen.
IV	Well-graded soil in the sand–clay (clay–mortar) range easily shaped and compacted at optimum moisture content to form light traffic surfaces if $PI = 4–12$ and bases if $PI = 0–6$. Higher type mortars are obtained by partial or complete substitution of soil binder by inorganic or organic cementing agents.
V	Stabilized gravel or soil concrete suitable for heavy traffic surfaces if $PI = 4–9$ and bases if $PI = 0–6$. Higher type concretes are obtained by partial or complete substitution of soil binder by inorganic or organic cementing agents.
VI	If minimum sizes fall between No. 40 and No. 4 sieve, material has open texture and is subject to raveling under traffic.
VII	Harsh mixtures, difficult to compact and shape during construction.

TABLE 9.3 GRADATION REQUIREMENTS FOR SOIL AGGREGATE MATERIALS [ASTM D1241 (AASHTO M147)].

Sieve Size (Square Openings)	Weight Percent Passing Square Mesh Sieves					
	Type I				Type II	
	Gradation A	Gradation B	Gradation C	Gradation D	Gradation E	Gradation F
2 in (50 mm)	100	100	—	—	—	—
1 in (25 mm)	—	75 to 95	100	100	100	100
⅜ in (9.5 mm)	30 to 65	40 to 75	50 to 85	60 to 100	—	—
No. 4 (4.75 mm)	25 to 55	30 to 60	35 to 65	50 to 85	55 to 100	70 to 100
No. 10 (2.00 mm)	15 to 40	20 to 45	25 to 50	40 to 70	40 to 100	55 to 100
No. 40 (425 μm)	8 to 20	15 to 30	15 to 30	25 to 45	20 to 50	30 to 70
No. 200 (75 μm)	2 to 8	5 to 15	5 to 15	8 to 15	6 to 15	8 to 15

TABLE 9.4 GRADING REQUIREMENTS FOR FINAL MIXTURES FOR BASES OR SUBBASES (ASTM D2940).

Sieve Size (Square Openings)	Design Range[a] (Weight Percentages Passing)		Job Mix Tolerances (Weight Percentages Passing)	
	Bases	Sub-bases	Bases	Sub-bases
2 in (50 mm)	100	100	−2	−3
1½ in (37.5 mm)	95 to 100	90 to 100	±5	+5
¾ in (19.0 mm)	70 to 92	—	±8	—
⅜ in (9.5 mm)	50 to 70	—	±8	—
No. 4 (4.75 mm)	35 to 55	30 to 60	±8	±10
No. 30 (600 μm)	12 to 25	—	±5	—
No. 200 (75 μm)	0 to 8[b]	0 to 12[b]	±3	±5

[a] Job mix formula should be selected with due regard to availability of materials and service requirements of project. Job mix tolerances may permit acceptance test results outside the design range.

[b] Determine by wet sieving. Where local environmental conditions (temperature and availability of free moisture) indicate that in order to prevent damage by frost action it is necessary to have lower percentages passing the No. 200 (75-μm) sieve than permitted in Table 9.4, appropriate lower percentages shall be specified. As a further precaution, a maximum limit of 3 weight percent is recommended for that portion having a diameter smaller than 20 μm.

where

n = admixture ratio

p_1, p_2 = weight percent of particles of soils #1 and #2 with diameter less than a particular size d

p = weight percent of particles of the resulting soil mixture with diameter less than d

Admixture ratio can be determined on the basis of the same equation if the resultant gradation is specified. The resultant curve may be specified to lie between two limit curves. Figure 9.3 illustrates two sets of limit curves for gradation of unsurfaced road material. In such cases, graphical methods can be used to determine the desired admixture ratio (Kézdi, 1979).

A stable soil mixture must also satisfy the plasticity requirements as discussed in the previous section. The strength of the stabilized soil is dependent on the compactness and the strength of the bonds developed between particles. Soil binder and water are the two elements that create the adhesion and bonding between the coarse grains and provide the continuity of the structure by filling in the voids of the bearing skeleton. The continuous granular skeleton is strengthened and stabilized by the added cohesion. During dry weather, shrinkage of soil binder develops tensile forces on the surfaces of the course grains, which has the desirable effect of increased compression on the granular skeleton. On the other hand, in wet weather, swelling of the soil binder might be desirable, as it would reduce the permeability and retard penetration of water. However, introduction of excessive volume change to the system might be detrimental to functioning of the bearing skeleton. Therefore the amount and the properties of the soil binder should be controlled for optimum results.

The recommendations of ASTM and AASHTO for percentage and criteria on plasticity of soil binder for various mixtures of aggregates are given in Tables 9.3 and 9.4 and also Figure 9.2. The following equation can be used to determine the resultant plasticity index, PI, of the mixture when the percentages of the component soils are known:

$$X = \frac{S_1 a_1 X_1 + S_2 a_2 X_2}{S_1 a_1 + S_2 a_2} \qquad (9.5)$$

where

S_1, S_2 = percentage of soils #1 and #2 in the resulting mixture

a_1, a_2 = percent passing sieve No. 40 in soils #1 and #2, respectively

X_1, X_2 = PI of soils #1 and #2, respectively

Equation 9.5 can also be used to determine percentages of soil components necessary to achieve a desired value of PI in the resultant mixture. A similar equation, in which the a values represent the *grain size coefficients* and the X values represent the particular consistency limit, has been used and verified in estimations of Atterberg limits of mixtures. However, it has been pointed out that a modified version of the equation, simply omitting the a values, delivers close approximations also. It should be noted that Equation 9.5 and its modified versions may be less reliable if the component soils are dissimilar with respect to chemical and geological origin and grain size distribution.

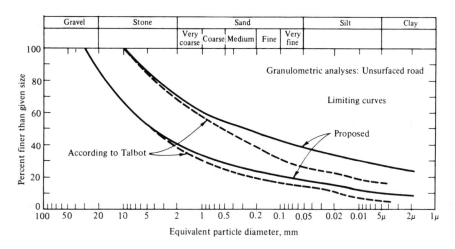

Fig. 9.3 Clay–concrete for laterite soil binder.

9.5 CHEMICAL STABILIZATION

Soil stabilization using chemical admixtures is the oldest and most widespread method of ground improvement. Known applications date back as far as ancient Greek, Egyptian, and Roman times. Chemical stabilization is mixing of soil with one of or a combination of admixtures of powder, slurry, or liquid for the general objectives of improving or controlling its volume stability, strength and stress–strain behavior, permeability, and durability. The fundamental processes that take place in a chemically stabilized soil system are cementation and ion-exchange reactions, alteration of soil surface properties, plugging of the voids, and coating and thus binding of the soil particles. Lime, cement, asphalt, silicates, resins, acids, lignins, metal oxides, siliceous residues of coal, incinerator residue, and some caustic compounds are stabilizing agents that have been studied and used over the years (WES, 1977; Ingles and Lim, 1980; Sutton et al., 1985). Application of these agents depends on the physical and chemical conditions of the natural soil, workability of agent, economic and safety constraints, and specific conditions of the construction.

The most commonly used stabilizing agents are portland cement, lime, fly-ash, and bitumen. Portland cement, lime, and fly-ash fall into the category of active agents, whereas bitumen is an inert agent. The active agents are those that react chemically with soil constituents and/or other admixtures. Inert agents, such as bitumen or resin, provide a stabilized matrix by coating and cementing the soil particles and aggregates. Stabilization with inert agents can also be considered as part of granular stabilization, where a cementing agent or fines are used to bind the noncohesive soil particles together. In the case of cohesive soils, inert agents are used to waterproof the soil in order to maintain low water contents for high strength and bearing capacity. Effectiveness of chemical stabilization depends partly on the interaction between soil particles and the stabilizing agent, and partly on the properties of the agent. Water affinity and water-retention capacity, ion-exchange capacity, clay content, grain-size distribution, and porosity of the soil are some of the factors that affect the interactions between soil particles and the admixtures.

9.5.1 Principles of Solid–Water–Electrolyte Interaction in Cohesive Soils

Cohesive soils usually take any load imposed on them if they are in a coherent, dense state and within a certain intermediate moisture range. In this condition the clay particles "stick" to each other and to the silt and sand particles that may be present. The water molecules involved may be constrained or organized in various ways that may be quite different from those in normal water. As a result, a heavy clay soil containing 15 percent or more water may appear dry and dusty initially. With increasing water content its consistency may change from solid to plastic and then finally to liquid, accompanied by swelling and loss of strength of the soil. The physicochemical interactions between soil constituents (solid–water–electrolyte) determine the extent of such behavior. For example, the cohesion at low moisture contents and the swelling at high moisture contents of some clays can be due to the affinity of the mineral surfaces to water and their interaction with it. Eliminating this affinity may result in destruction of the desirable cohesion and need for addition of cementing materials. The diffuse double-layer thickness of the clay particles, the concentration, size, and valence of the ions near particle surface, and the orientation and structure of water molecules in the vicinity of solid surfaces are some of the principal factors that affect the behavior of cohesive soils. For

better understanding of physicochemical processes involved in chemical stabilization, and for design of effective stabilization systems, it is beneficial to review some of the basic principles of solid–water–electrolyte interaction. An outline of basic principles and some descriptive material is presented below. For detailed information the reader is refered to Chapters 3 and 20, and to Mitchell (1976), Bohn et al. (1985), Sparks (1986), and Bennett and Hulbert (1986).

Diffuse Double Layer

Most solid surfaces exhibit charge development due to internal structural defects (isomorphic substitution), ionic dissolution (adsorption of H^+ or OH^- on the surface; dissociation of surface sites or broken bonds), and ionization of organic groups. For electrical neutrality, an equal amount of charge of the opposite sign accumulates near the charged surface under electrostatic attraction. Clays typically possess negative surface charge and attract positively charged cations, the concentration of which increase towards the surface. Owing to diffusion forces, the cations are also drawn back into the equilibriating solution. The anions are repelled by the surface, and diffusion forces act in the opposite direction. Such a distribution of cations and anions in front of the negatively charged surface of a clay particle is termed *diffuse double layer*. There are a number of models that relate surface charge density to the distribution of ions in the diffuse layer. Among these models are: (1) Gouy–Chapman theory for a single flat double layer; (2) Stern theory of a compact double layer (Stern-layer; Stern, 1924) near the surface, extending into the diffuse double layer away from the surface; (3) theory of interacting flat double layer. The thickness of the diffuse double layer determines much of the interaction between clay particles and thus the range of repulsive and attractive forces. These forces influence the fabric and structure development that account for a number of physical properties and mechanical behavior of soils. The thickness of diffuse double layer is affected by electrolyte concentration, the valence and size of the cation, the dielectric constant of the electrolyte pore fluid, temperature, and pH. Diffuse double-layer theory can be used to describe various properties of clay soils and suspensions, including flocculation and dispersion, osmotic swelling (interparticle swelling), and orientation of particles (with respect to changes in double-layer repulsion forces).

Adsorbed Water

The water molecule (H_2O) has a "V" shaped arrangement of its atoms, with H–O–H angle slightly less than 105°, such that the directionality of the bond creates a permanent dipole. In such a configuration, water can interact with both uncharged surfaces and charged surfaces. Interaction between water molecules and uncharged surfaces arise from van der Waals attraction and forces arising from surface tension at solid–water interfaces. One type of van der Waals attraction is caused by hydrogen bonding. In this bonding the negative dipoles of hydroxyl ions and oxygen atoms cause the orientation of the positive dipole of the hydrogen ends of water molecules. Relatively strong associations might occur with surface oxygens and hydroxyls of clay and water molecules. Interactions between charged surfaces and water arise from Coulombic forces ion-dipole and ion-induced dipole interactions with water molecules. Ion-dipole attraction results in hydration of ions, which is the adsorbtion of water molecule onto the ion surface at those corners of the water molecule of opposite electrical charge. The hydration of cations in the diffuse double layer provides a mechanism for the water molecules to be adsorbed onto clay surfaces. Diffusion of water molecules towards clay surface to equalize the high concentration of cations is another possible mechanism of water adsorbtion onto clay. The structure and properties of the adsorbed water may be quite different from those of free water. Results of a number of studies have indicated that a few layers of adsorbed water may be subject to rotational and translational constraints that diminish with increasing water content. A number of macroscopic property measurements, which include density, specific volume, viscosity, and heat capacity, have been used to draw inferences about the structure and properties of the adsorbed water.

Ion Exchange

The excess of ions of opposite charge (to that of the surface) over those of like charges present in the diffuse double layer are called exchangeable ions. These ions can be replaced by a group of different ions having the same total charge by altering the chemical composition of the equilibrium electrolyte solution. The maximum number of counterions (cations for a negatively charged surface and anions for a positively charged surface) present per unit exchanger under a given set of environmental conditions (temperature, pressure, pH, electrolyte concentration) is the *cation exchange capacity* of the soil. Cation exchange and cation exchange selectivity are probably the most important aspects of surface chemistry of soils. Exchange selectivity depends on the valence, size, and concentration of different ion types, properties of the ions, and exchange sites. In soils the most commonly found cations are calcium (Ca^{2+}), magnesium (Mg^{2+}), sodium (Na^+), and potassium (K^+); the anions are sulfate (SO_4^{2-}), chloride (Cl^-), phosphate (PO_4^{3-}) and nitrate (NO_3^-). Ordinarily, small size cations or cations with higher valence have more replacing power; however, increased concentration of one ion often produces the effect of increased replacing power of that ion regardless of size and valance constraints. Clays also exhibit anion exchange at the edge sites where the electric charge is positive. Organic matter in soil (acidic and carbonyl groups and alcoholic hydroxyls) are a major source of pH-dependent charges. They form electrostatic bonds with cations and contribute to the cation exchange capacity of the soil (Kelley, 1948; White and Zelazny, 1986).

Ion-exchange reactions, double-layer interactions, water adsorption onto clay surfaces, and hydration of counterions in clay soils are the major processes that determine the soil structure and physical properties of a stabilized system in chemical stabilization. The exchange capacity, chemical composition of the pore fluid, and the ionic and organic constituents of the soil influence the selection and long-term performance of an additive used to achieve particular improvements in the soil.

9.5.2 Relationship Between Granulometry of Soil and Additives

Addition of smaller-sized particles to a system of larger-sized particles, in amounts less than necessary to fill the void space, decreases the packing density of larger particles. The interference can be considered negligible if the size ratio of the particles is 100:1. Application of this criterion to clay particles of size 10^{-5} cm (1 μm), limits the size of the particulate additive, for cementation or waterproofing purposes, to 10^{-7} cm (0.01 μm) or about 10 atoms thickness (molecular size). A portland cement, with a surface area of 1700 cm^2/g, or 5355 cm^2/cm^3, and average particle size of 1.8×10^{-3} cm, would cause interference to the packing density of natural soils of grain sizes less than 2×10^{-3} cm, or those passing No. 10 sieve. Thus, the compatibility of the particle size of the additive with the soil

aggregate is an important factor in achieving effective soil stabilization.

The extent to which the pore space is filled with additives is another factor that needs to be considered for effective and economic stabilization. The dry weight of a highly cohesive soil, compacted at its optimum moisture content and maximum density, is usually below 100 lb/ft^3. Assuming an average specific gravity of 2.65 for the soil solids, the porosity can be estimated to be at around 40 percent or more. Then the amount of a molecular additive needed to fill all the available pore space can be determined by taking the product of the density and the porosity of the soil. Establishment of such a nonporous system for stabilization purposes would be unrealistic as well as uneconomic because of the large quantity of the additive required. Consequently, any stabilized system that can be established economically with such cohesive soils will be of a certain porosity. Figure 9.4 shows a typical gradation band for additive-stabilized systems; the bold lines represent the range for 1 to 2 percent bitumen added mixture.

Most construction specifications for stabilization of cohesive soils require that, at the time of compaction, 100 percent of the mixture pass the 2.54-cm screen and a minimum of 80 percent pass the No. 4 sieve, exclusive of any stone or gravel that might be present. The reason for this specification is to detect large-sized "clay lumps" formed by aggregation of cohesive particles (silt < 0.075 mm and clay < 0.002 mm), and held together by cohesion and natural cementation in presence of water (Winterkorn and Tschebotarioff, 1947). The actual stabilization of such soils involves two steps. (1) The stabilizer coats the lumps and penetrates a certain distance; some stabilizers react to break down the lumps by pulling water from the clay. (2) The surface-coated lumps are cemented with the rest of the particles by means of the remaining stabilizer in the mixture of stabilizer added in excess. The extent and the rate of penetration of the stabilizing agents into the soil lumps depend on the size and moisture content of the lump, the type of the stabilizer, and the mechanism by which it reacts with soil constituents, as well as the environmental conditions such as temperature and pressure.

In addition to the natural clay lumps there may be secondary aggregations produced by working of the soil, such as breakdown of large lumps, mixing, and compaction processes. The average size of the lumps in the field is usually considerably larger than that of those in laboratory specimens. This needs to be taken into consideration when practical conclusions are to be drawn from results of laboratory experiments on small-scale specimens. However, this does not negate the fact that valuable theoretical and ultimately practical conclusions can be drawn from such experimentation (Herzog and Mitchell, 1963; Noble, 1967). In addition to the size effects, the environmental conditions during curing and testing, the sequence of admixtures and the time allowed between mixing and compaction must be considered when evaluating laboratory-generated data to design for field stabilization.

9.5.3 Use of Chemicals in Granular Stabilization

Salts are used sometimes as stabilizing agents in construction of wearing surfaces. Calcium chloride ($CaCl_2$) and sodium chloride (NaCl) are the two commonly used salts to control dust, prevent raveling of the surface, maintain moisture and thus decrease volume change, and add to density of the stabilized system. Calcium chloride is a white salt with high affinity for water. It tends to retain water in dry periods and to absorb water to the extent of four to ten times its own weight in wet periods. The net effect of the treatment is reduction in

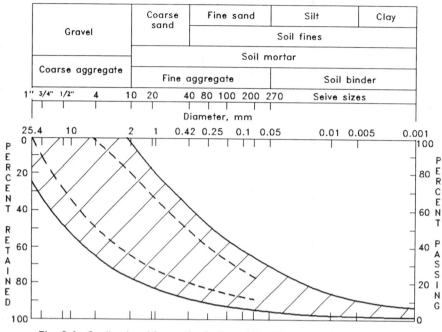

Fig. 9.4 Grading band for mechanically stabilized base courses. (*HRB, 1946.*)

the repulsive forces between fine particles and strengthening of binding, thus preventing dusting or raveling of the surface material. In addition, calcium chloride is an effective stabilizing agent in producing increased densities for given compactive efforts. Another dust-preventive agent used successfully is lignin sulfonate, a by-product of sulfite wood-pulping process, which is either mixed with the surface material or directly sprayed onto the surface. Petroleum resins have been used for dust prevention in similar manner.

Sodium chloride is also used to control moisture and increase density in granular stabilization. It tends to form a barrier to the movement of water and thus decrease volume change tendency. The crystalline nature of sodium chloride facilitates formation of a hard crust on the surface of the stabilized system, thereby reducing surface wear. It should be noted here that use of both salts might have important environmental consequences with respect to surface and groundwater contamination and corrosion effects. Consideration should be given to these issues where necessary.

9.5.4 Other Chemical Stabilization Methods

Numerous chemicals have been used for soil stabilization purposes. A study conducted at the U.S. Army Engineer Waterway Experiment Station (WES, 1977), gives a comprehensive report of materials evaluated as potential stabilizers. The document lists seven major categories of materials, which are acids, asphalts, cements, limes, resins, salts, and others. Only a few of these will be discussed here.

1. Sodium Silicates   are a group of chemicals used in industry as adhesives, cements, cleaning compounds, deflocculants, protective coatings, etc. They are produced at various ratios of $Na_2O:Si_2$. Use of sodium silicate in soil treatment was mentioned as early as 1910, when it was used as a dust palliative and in construction of macadam roads. In soil stabilization, sodium silicate reacts with calcium salts in water solutions and forms an insoluble gel of calcium silicates. The high pH resulting from the presence of sodium silicate might also promote

dissolution of silicates from soil particle surfaces, thereby contributing to the reaction of cementation. Sodium silicate stabilization seems to work well with silica sands, and low-activity clays such as kaolinite, but not, however, with high-activity clays such as montmorillonite.

The amount of sodium silicate used in typical stabilization procedures ranges from 2 to 10 percent. Addition of bentonite, usually on the order of 1 percent has been observed to increase the strength of the mixture substantially. Other additives such as calcium chloride, sodium aluminate, and magnesium carbonate have been shown to result in higher strength and improved durability of sodium silicate–soil mixtures. Unconfined compressive strengths of 1500 psi have been reported with 2.5 percent sodium silicate addition to sandy soils. The treatment, however, has not been always successful, as in the case of a beach sand (Winterkorn, 1949). A comprehensive review of sodium silicate and its use as a soil stabilizer is given by Hurley and Thornburn (1972). Use of silicates as stabilizer is also discussed by Clough et al. (1979).

2. Iron Oxides   have been used in base stabilization in Sweden and Australia (Emery, 1980). Merolit, developed in Sweden is typically made up of 70 percent air-cooled blast furnace slag, 25 percent granulated blast furnace slag, 3 percent iron oxides, and 2 percent hydrated lime plus water. Ingles and Lim (1980) described a process in which they combined iron oxide and sodium silicate solution to treat soil after it had been heated to a high temperature to destroy its water sensitivity.

3. Chlorides   ($CaCl_2$, $NaCl_2$, and $MgCl_2$) have been used for dust control in road construction. Owing to their effect on reduction of water evaporation, they have been used as stabilizing agents as well. Chloride solutions have reduced freezing points and therefore are used to thaw frozen soils and roads. Chlorides also effect the plasticity of soils, resulting in reduced plastic and liquid limits. This is due to the adhesion and flocculation of the particles, which increases the apparent grain size. Same effect increases the permeability of the soil, thus making the soil less susceptible to capillarity and therefore to frost hazards. Calcium chloride reduces soil strength and may have substantial effects in loose soils. Since chlorides are soluble in water they will be leached out of the system by rain if not protected. Subsequent treatments can be made in such cases. The grain-size distribution and plasticity of soils suitable for calcium chloride stabilization are given by the Calcium Chloride Institute (1953).

4. Phosphoric Acid The idea of stabilizing with phosphoric acid (H_3PO_4) was first introduced by Winterkorn (1940). The addition of phosphoric acid and other phosphoric compounds has been shown to increase strength and water resistance of soils. It has been successful in stabilizing loess soil slopes, resulting in increased strength and erosion durability (Evans and Bell, 1981). Phosphoric acid reacts with clay minerals, forming water-insoluble aluminum and iron compounds, such as aluminum metaphosphate. Aluminum metaphosphate is a hard substance soluble in the pH range 2 to 4. As it dissolves in the residual acid it precipitates as a gel, producing a cementing action.

Strength gains of 20 to 50 percent of that of the untreated soil have been achieved by phosphoric acid treatment. It reduces the dry density slightly and the optimum water content changes on the order of 1 percent. Addition of sodium fluorosilicate to phosphoric acid mixed soil has resulted in significant strength increase for the mixtures. Soils containing calcium may not be treated adequately with phosphoric acid since most of the acid will be taken up to neutralize calcium carbonate ($CaCO_3$). The type and quantity of clay minerals, and the preparation, feeding, and admixture types are important features in phosphoric-acid stabilization. Therefore, a laboratory test program using specimens of site soil is recommended prior to field application of the phosphoric-acid stabilization.

5. Aniline–Furfural is a polymer first suggested as a soil stabilizer by Winterkorn. Aniline is an aromatic amine and furfural is a primary aldehyde. Aniline is produced by substituting a hydrogen atom of benzene by an NH_4 group, and furfural is produced by distilling agricultural products like corn cobs in the presence of sulfuric acid. In soil stabilization they are sprinkled on the ground successively in aqueous emulsion form. The stabilization effect of the material is to render the soil water-repellent and therefore decrease water absorbtion capacity and expansion. When aniline and furfural react they form a resin or polymer that provides cementation between soil particles. Through the ion-exchange reaction (organic cations replacing the metallic cations in clay fraction) and coating of the particles, the soil is made water-repellent. Table 9.5 gives a comparison of the strength, water absorbtion, and durability of various soil types stabilized with aniline–furfural, portland cement, and liquid asphalt. Although the results show substantial

TABLE 9.5 COMPARISON OF THE SOIL STABILIZATION EFFECTIVENESS OF ANILINE–FURFURAL RESIN (70:30), PORTLAND CEMENT AND LIQUID ASPHALT.

Soil Type	Type of Percentage of Admixture	10 Days Capillary Absorption[a]		12 Cycles Freeze–Thaw[a]	
		Load Wet, lbs	Absorption, %	Load Wet, lbs	Absorption, %
Miami Gravelly Sandy Loam	None	47	9.8	No test	No test
	10% Portland Cement	1793	9.1	1900	10.7
	6% Liquid Asphalt	328	1.1	185	1.6
	6% Aniline–Furfural	1316	0.6	1015	0.7
Warsaw Sandy Loam	None	131	19.6	No test	No test
	8% Portland Cement	288	19.8	No test	No test
	10% Portland Cement	358	19.2	185	17.5
	12% Portland Cement	715	17.2	No test	No test
	6% Liquid Asphalt	158	2.6	238	2.9
	6% Aniline–Furfural	625	1.2	595	0.9
Vigo Silt Loam	None	32	21.7	No test	No test
	11% Portland Cement	793	18.8	110[b]	21.0
	13% Portland Cement	900	18.7	No test	No test
	6% Liquid Asphalt	277	2.4	Failed	c
	6% Aniline–Furfural	587	0.4	475	0.5
Crosby Sand– Clay Loam	None	Failed	23.0	No test	No test
	10% Portland Cement	1468	11.0	67	21.9
	12% Portland Cement	1573	17.3	No test	No test
	6% Liquid Asphalt	235	1.6	88	2.0
	6% Aniline–Furfural	1065	1.7	No test	No test
Brookston Silty Clay	None	Failed	32.1	No test	No test
	10% Portland Cement	403	18.1	125	18.0
	6% Liquid Asphalt	236	4.8	Failed	c
	4% Aniline–Furfural	572	3.9	No test	No test
Clyde Clay	None	10	32.7	No test	No test
	10% Portland Cement	610	16.7	110	22.3
	6% Liquid Asphalt	87	1.4	45[d]	8.0
	6% Aniline–Furfural	477	1.8	280	2.9

[a] Test procedure described in Winterkorn (1947a).
[b] 10% Portland cement.
[c] No determination.
[d] Only 1 specimen out of 6 completed 12 cycles.

improvement over the others, one of the disadvantages of aniline–furfural treatment is the cost of production of the material. In addition aniline is environmentally hazardous.

6. Other chemical stabilizers that are currently being used or have potential for usage include: phosphorous pentoxide, AM9 (water-soluble acrylamide and diacrylamide), Arapol 7110, Arothane 170, Bisphenol A (Epon 828), calcium acrylate, Epon VIII, Epon 562, Epon 828, Epon 834, Arquad 2HT (dialkyl dimethylammonium chloride), chrome lignin, lignin, sandcrete, sodium methylethyl propyl siliconate (WES, 1977), hydroxyaluminum (Bryhn et al., 1983), gypsum, dihydrated gypsum (Kujala, 1983), rice-husk ash (Cook, 1984), Vinsol, and sulfite liquor (Kézdi, 1979).

9.6 CEMENT STABILIZATION

Cement stabilization of soils involves mixing of pulverized soil, cement, and water, and compacting this mix to a high density, which renders the material resistant to various physical, thermal, and chemical stresses. Depending on the type of soil and size of aggregates used, there are various different products of cement stabilization. As the cement hydrates, a gel is formed that upon hardening forms a cellular matrix that encapsulates the soil particles or forms strong bridges ("spot welding") between the aggretates, thus producing a hard, durable structural material. If size of the soil particles is smaller than that of the cement, the soil particles surround the cement particles and weaker bonds are formed. Nevertheless, when properly mixed and constructed, a cement-stabilized soil system generally performs well for the intended purpose, even when exposed to wetting–drying or freezing–thawing cycles.

Compositions of various types of cements and cementitious materials used in soil stabilization are given in Figure 9.5. These are called hydraulic cements because they react with water to form strongly bonded systems. The common hydraulic cements are mixtures of calcium silicates and aluminates and include the portland, natural, slag, and alumina cements. Also shown in Figure 9.5 are compositional ranges of quicklime, hydraulic lime, and puzzolan cement. Among these hydraulic cements, the portland cement compositional range is stipulated to give the most desirable end-product in reaction with water.

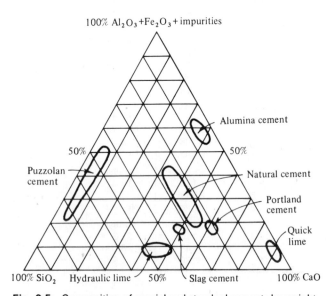

Fig. 9.5 Composition of special and standard cements by weight. (*After Murphy, 1957.*)

TABLE 9.6 APPROXIMATE CHEMICAL COMPOSITION OF PORTLAND CEMENT.

Oxides	Amount, %	Abbreviation
Calcium oxide (CaO)	60–65	C
Magnesium oxide (MgO)	0–5	M
Aluminum oxide (Al_2O_3)	4–8	A
Ferric oxide (Fe_2O_3)	2–5	F
Silicon dioxide (SiO_2)	20–24	S
Sulfur trioxide (SO_3)	1–3	S
Loss on ignition	0.5–3	

Portland Cement

The most commonly used cement in stabilization is portland cement, which is a finely powdered hydraulic cement, essentially consisting of hydraulic calcium silicates (specifications in AASHTO M85 and ASTM C150). The approximate chemical composition of portland cement is given in Table 9.6. The particle size of portland cement ranges from 0.5 to 80 μm, with major part of it passing No. 200 sieve, and the specific gravity of the particles ranges from 3.12 to 3.20. The major compounds in portland cement are (with abbreviated nomenclature also given): tricalcium silicate ($3CaO \cdot SiO_2$; C_3S), bicalcium silicate ($2CaO \cdot SiO_2$; C_2S), tricalcium aluminate ($3CaO \cdot Al_2O_3$; C_3A), and tetracalcium aluminoferrite ($4CaO \cdot Al_2O_3 \cdot Fe_2O_3$; C_4AF). These compounds react with water or water and available lime to form very stable hydrated silicates and aluminates and also calcium hydroxide ($Ca(OH)_2$). One such stable product is called *tobermorite gel* ($3CaO \cdot 2SiO_2 \cdot 3H_2O$), a calcium silicate hydrate (CSH) that resembles the natural tobermorite mineral. Calcium hydroxide, or lime, produced during the hydration of portland cement can further react with the silicates and aluminates of the clay soil and produce more cementing material or tobermorite gel (Herzog and Mitchell, 1963). Presence of organic matter and organic acids (i.e., lignosulfonic acids, carboxylic acids, carbohydrates, phenolic compounds), sulfates (SO_3), and low-pH environment hinder formation of tobermorite gel by immobilizing calcium (Ca^{2+}). As the calcium-adsorbtion capacity of the soil increases, the Ca^{2+} ions freed by the hydration of portland cement become unavailable for the gel formation (Clare and Sherwood, 1954).

The ASTM C150 specification describes five types of portland cements, in which Type I is for standard use. Type II cement contains reduced quantities of C_3A and C_3S to provide resistance for sulfate attack and lower heat of hydration. Type III is high early strength cement with increased quantities of C_3A and C_3S, which also produces high heat of hydration and thus greater drying shrinkage. Type IV contains severely limited quantities of C_3A and C_3S, which renders a low heat of hydration cement. Type V is designed for maximum sulfate attack resistance with strict limitations on C_3A content. Table 9.7 summarizes the chemical requirements for the five types of portland cements discussed above.

Types of Cement Stabilization

Basically there are three types of cement-stabilized systems.

1. Soil–Cement contains sufficient cement to produce a hard and durable construction material and only enough moisture to satisfy the hydration requirements of the cement and the soil, and also to provide sufficient lubrication for the compaction of the mixture to a high density. The resulting material has a well-defined resistance to weathering and mechanical forces. Standard laboratory tests have been developed to predict

TABLE 9.7 CHEMICAL REQUIREMENTS FOR PORTLAND CEMENTS[a].

Chemical Requirement	Cement				
	Type I	Type II	Type III	Type IV	Type V
Silicon dioxide (SiO_2), min, %		21.0			
Aluminum oxide (Al_2O_3), max, %		6.0			
Ferric oxide (Fe_2O_3), max, %		6.0		6.5	
Magnesium oxide (MgO), max, %	5.0	5.0	5.0	5.0	5.0
Sulfur trioxide (SO_3), max, %:					
When $3CaO \cdot Al_2O_3$ is 8% or less	2.5	2.5	3.0	2.3	2.3
When $3CaO \cdot Al_2O_3$ is more than 8%	3.0		4.0		
Loss on ignition, max, %	3.0	3.0	3.0	2.5	3.0
Insoluble residue, max, %	0.75	0.75	0.75	0.75	0.75
Tricalcium silicate ($3CaO \cdot SiO_2$), max, %				35	
Dicalcium silicate ($2CaO \cdot SiO_2$), min, %				40	
Tricalcium aluminate ($3CaO \cdot Al_2O_3$), max, %		8	15	7	5
$3CaO \cdot SiO_2 + 3CaO \cdot Al_2O_3$, max, %		58			
Tetracalcium aluminoferrite ($4CaO \cdot Al_2O_3 \cdot Fe_2O_3$) $+ 2X(3CaO \cdot Al_2O_3)$, max, %					20

[a] After ASTM.

performance of soil–cement, such as strength, durability, water susceptibility, and frost resistance. Soil–cement is commonly used for stabilization of road bases of flexible and rigid pavements, subbases, embankment slopes, earth dam cores, reservoir linings, building foundations, trenches, and for frost protection and reinforcement of load-bearing layers.

2. Cement-modified Soil is an unhardened or semihardened mixture of soil and cement. Relatively small quantities of portland cement are added to a granular or silty clay soil to improve certain physical and chemical properties of the soil. The intended improvements are reduction of volume change tendency and plasticity, and increasing load-bearing capacity of the soil. There is sufficient cement to interact with the silt and clay fractions and to deprive them of their water affinity, but not enough to bond all of the soil particles into a coherent system. The result is an improved soil rather than a new building material with standardized properties such as soil–cement. Cement improvement of soils is often used for erosion and frost protection, and to reduce shrinkage and expansion of foundation and base layers.

3. Plastic Soil–Cement results in a hardened product but contains, at the time of placement, sufficient water to produce a consistency similar to that of a plastering mortar. This allows it to be placed on steep or irregular areas where access of construction equipment is difficult or not possible. This material compares with soil–cement, and like soil–cement it is required to satisfy strict strength and durability conditions. It is most commonly used for lining of ditches, irrigation canals, and trenches, and for protection of such surfaces against erosion.

9.6.1 Application of Soil–Cement

Soils for soil–cement construction can be divided into three groups: sandy and gravelly soils with 10 to 35 percent silt and clay content; sand soils deficient in fines; and silty and clayey soils. To achieve maximum strength and durability in soil–cement mixtures, it is necessary that soils be pulverized and intimately mixed with cement and water. Treatment (curing and compaction), and the amount of cement and water used vary with these soil groups.

In the first group are soils with granular bearing skeleton and sufficient fines to fill partly the voids between the aggregates without interfering with the grain-to-grain contact. These soils require the least amount of cement for adequate hardening. Included are water-deposited sands and gravels, caliche, limerock, and almost all granular materials of which 55 percent pass the No. 4 sieve. If the soil is well-graded, it may contain up to 65 percent gravel retained on No. 4 sieve and still have sufficient fines for effective binding. This group of soils is usually easily pulverized and mixed, and can be used under a wide range of weather conditions.

Sandy soils that are deficient in fines, such as some beach sands, glacial sands, and wind-blown sands, require slightly more cement than the first group for comparable results in hardening. It is often difficult to move construction equipment over these soils. It is therefore recommended that the sand be kept wet to obtain better traction, and that track-type equipment be used. The soft nature of these soils in the initial placement requires that care be taken in the final packing and finishing to insure a smooth, dense surface.

Silty and clayey soils can produce a satisfactory soil–cement, but those with high clay content may be difficult to pulverize and may result in systems with excessive shrinkage properties. The required cement content increases with the amount and water affinity of the clay in the soil. In general, weather conditions affect this group of soil cement more than the previous two groups.

9.6.2 Setting of Soil-Cement Mixtures

Some soils may contain organic substances that prevent or greatly retard setting and hardening of soil–cement. Early studies have shown that the most active organic compounds that retard the setting of cement mixtures are carboxylic acids, carbohydrates, and humus. The possible mechanisms of reduction of hardening can be active compounds interfering

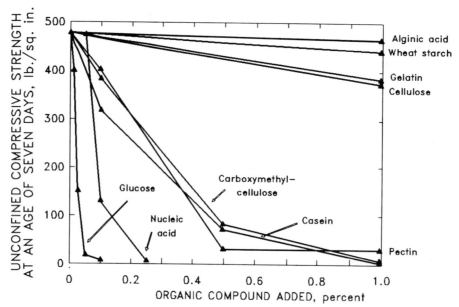

Fig. 9.6 Effects of various organic compounds on the unconfined compressive strength of soil–cement mixtures. (*After Clare and Sherwood, 1954.*)

with the hydration of the cement, change in the crystal structure of the cement matrix, or interruption of bonding between the matrix and the soil particles owing to coating of the organic material around the soil particles. Figure 9.6 illustrates the effects of various organic compounds on the unconfined compressive strength of sand–cement mixtures (Clare and Sherwood, 1954). In this study the organic materials were classified into three groups, in which (1) cellulose, wheat starch, alginic acid, wood, straw, and esparto-grass lignin were designated as inactive or slightly active; (2) carboxymethyl-cellulose, pectin, and casein were designated as active; (3) glucose and nucleic acid were designated as very active.

The retardation of setting of cement is more commonly related to presence of some active compounds than the total organic content. The activity of organic matter in natural soil can be related to the capacity of soil to adsorb calcium ions. The organic matter retains the calcium ions liberated during the hydrolysis of the calcium silicates and aluminates in the cement, limiting the amount of calcium available to form the bonding compounds. Addition of calcium chloride ($CaCl_2$) can remedy this condition by satisfying the calcium-adsorbtive capacity of the organic matter before the hydration of cement has begun. The calcium chloride present in rapid-hardening cement may be sufficient in some cases, and in others an increased proportion of the salt may be needed. Addition of silty-clay to the soil sometimes aids the cement reaction.

Carboxylic and phenolic acidic groups in soil also react with the available calcium ions to give insoluble calcium salts and hence hinder the setting process. The pH of the soil may be a useful indicator for detecting soils unsuitable for cement stabilization. Figure 9.7 illustrates this effect. In some cases free iron (ferric oxide) content may indicate poor reactivity of the soil. The soluble organoiron complexes may become adsorbed onto the silica surfaces and decompose by addition of cement, which raises the pH of the environment. This in turn leaves the organic component free to interfere with the hydration of the cement.

Organic additives are sometimes added to the soil or the cement to aid in dispersion, partial waterproofing, or water retainment, as the case may be. The commercially available admixtures, hydroxylated carboxylic acid and calcium ligno-sulfonate are used to retard the cement setting time and reduce the thickness of the cement gel. These additives have also been shown to increase the strength and durability of soil–cement layers, and also considerably increase the shear strength of a layered system by promoting the bond formation at the interfaces (Arman and Dantin, 1969). Proper bondage at the interface of multiple-layered soil-pavement structures is essential to overcome slipping tendencies between the layers induced by tangential components of normal forces of failure due to tensile cracking. It was demonstrated that retardation of the hydration process of the soil–cement mixture within the first 2 to 7 days of curing was instrumental in the development of bonds between the layers and subsequent increase in the shear strength of the layered system. The amount of additives used in the mixture need to be controlled so as not to allow strength loss below design requirements. The recommendations of the study by Arman and Dantin are: (1) additive concentration of 0.25 lb per bag of portland cement; (2) maximum of 7 hours

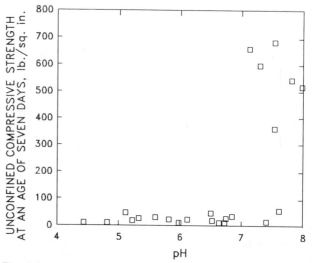

Fig. 9.7 Effect of pH on unconfined compressive strength of soil–cement mixtures. (*After Clare and Sherwood, 1954.*)

of time lapse between mixing of the first layer and placement of the second layer; (3) premixing of additive with water prior to mixing with soil–cement mixture; (4) no reduction in the optimum water content when admixture is used.

9.6.3 Water and Cement Requirements

Practical limitations on susceptibility of soils to cement stabilization derive from the water requirements during the compaction and hardening period. The system must contain enough water for hydration of cement and silt–clay constituents and for workability of the soil. In the working of the soil, water acts as an interparticle lubricant. Water used in soil–cement should be relatively clean and free of harmful amounts of salts, alkalies, acids, or organic matter. Water fit to drink is satisfactory. However, seawater has been used satisfactorily when fresh water was unobtainable.

A well-graded soil containing gravel, coarse sand, and fine sand with or without small amounts of silt or clay requires 5 percent or less cement by weight. The remaining sandy soils generally require 7 percent. Nonplastic or moderately plastic silty soils generally require about 10 percent, and plastic clay soil requires 13 percent or more. The usual ranges of cement requirements for subsurface soils of various AASHTO soil groups are given in Tables 9.8a and 9.8b. These tables present the general relationship between soil character and cement percentage required to produce soil–cement of standard quality.

9.6.4 Shrinkage Cracking and Curing

As the hydration of cement proceeds, drying and evaporation of the surplus water results in shrinkage of the soil–cement system. Since the base of the system is generally restrained by friction, this results in build up of tensile stresses. When the level of stress exceeds the tensile strength of the system, cracking occurs. The severity of the phenomenon increases with increasing water affinity of the soil. This sets a natural limit to the types of soil that can be practically stabilized with portland cement. Limiting the plasticity of soil used, thorough pulverization of soil, and thorough mixing with cement have been demonstrated to produce good results with respect to control of shrinkage cracking. In inhomogeneous mixtures the shrinkage process is nonuniform and therefore results in separation of strata or formation of fissures. However in some

cases a certain amount of strength nonuniformity may be desirable, as early development of cracks can alleviate strain build-up that is likely to lead to large crack widths. Effective load transfer has been shown to occur when the crack openings are less than 0.03 in to 0.04 in (0.077 cm to 0.10 cm). Therefore, controlling the size of crack openings so that they do not exceed the limiting values is as important as minimizing shrinkage in soil–cement systems. This also involves control of crack patterns. A pattern of frequent small cracks is generally preferred to wide cracks at large intervals, owing to the relative ease of remediation of the former. Rough subgrade surfaces induce higher friction resistance, which results in reduced crack spacing and reduced crack widths. For cohesive soils the required cement content increases with the water affinity of the soil. A higher cement content reduces volume change tendency; however, it increases the tensile strength of the soil–cement giving the undesirable pattern of cracks with lower frequency and larger width.

Shrinkage has been shown to be dependent upon some viscoelastic properties and creep behavior of soil–cement (George, 1969). Creep, which is generally defined as the total time-dependent deformation of the material due to load, increases with decreasing relative humidity. In field applications, maintaining the saturation at or below 80 percent has been shown to result in significant minimization of shrinkage cracking. Below 80 percent saturation the continuity of the liquid phase is no longer evident and shrinkage therefore causes only local microfissures confined to these zones of saturation. In general, experience shows that the requirement of the mean value of saturation to be 70 percent, with coefficient of variation of 15 percent, results in cement-stabilized layers with minimum shrinkage cracking.

The density and cement content also influences the amount of shrinkage. Increased cement content and density increases the thermal coefficient of expansion of a soil–cement system, which renders it more susceptible to temperature variations. However, increased cement content reduces volume change tendencies and thus shrinkage. It also increases water absorbtion capacity and reduces the overall permeability of the system. Mixing portland cement with hydrated lime, surface treatments with lime containing a trace of sugar, expansive cement additives (ettingrite), and sodium silicate (George, 1971), and use of expanding cements have been shown to be effective in reducing cracks. It should be noted that the use of "expanding" cements whose expansion is due to sulfate action may not be viable for long-term performance, since some organic substances or sulfur

TABLE 9.8a TYPICAL CEMENT REQUIREMENTS FOR VARIOUS SOIL TYPES[a].

AASHTO Soil Classification	Unified Soil Classification	Normal Range of Cement Requirements[b]		Typical Cement Content for Moisture–Density Test (ASTM D558), percent by weight	Typical Cement Contents for Wet–Dry (ASTM D559) and Freeze–Thaw Tests (ASTM D560), percent by weight
		percent by vol.	percent by wt.		
A-1-a	GW, GP, GM, SW, SP, SM	5–7	3–5	5	3– 5– 7
A-1-b	GM, GP, SM, SP	7–9	5–8	6	4– 6– 8
A-2	GM, GC, SM, SC	7–10	5–9	7	5– 7– 9
A-3	SP	8–12	7–11	9	7– 9–11
A-4	CL, ML	8–12	7–12	10	8–10–12
A-5	ML, MH, CH	8–12	8–13	10	8–10–12
A-6	CL, CH	10–14	9–15	12	10–12–14
A-7	MH, CH	10–14	10–16	13	11–13–15

[a] PCA (1978).
[b] Does not include organic or poorly reacting soils. Also, additional cement may be required for severe exposure conditions such as slope protection.

TABLE 9.8b AVERAGE CEMENT REQUIREMENT FOR B AND C HORIZON SILTY AND CLAYEY; AND SANDY SOILS[a].

(i) Silty and Clayey Soils

AASHTO Group Index	Material Between 0.05 mm, and 0.005 mm, Percent	Cement Content, Percent by Weight — Maximum Density, lb/ft³						
		90–94	95–99	100–104	105–109	110–114	115–119	120 or More
0–3	0–19	12	11	10	8	8	7	7
	20–39	12	11	10	9	8	8	7
	40–59	13	12	11	9	9	8	8
	60 or more	—	—	—	—	—	—	—
4–7	0–19	13	12	11	9	8	7	7
	20–39	13	12	11	10	9	8	8
	40–59	14	13	12	10	10	9	8
	60 or more	15	14	12	11	10	9	9
8–11	0–19	14	13	11	10	9	8	8
	20–39	15	14	11	10	9	9	9
	40–59	16	14	12	11	10	10	9
	60 or more	17	15	13	11	10	10	10
12–15	0–19	15	14	13	12	11	9	9
	20–39	16	15	13	12	11	10	10
	40–59	17	16	14	12	12	11	10
	60 or more	18	16	14	13	12	11	11
16–20	0–19	17	16	14	13	12	11	10
	20–39	18	17	15	14	13	11	11
	40–59	19	18	15	14	14	12	12
	60 or more	20	19	16	15	14	13	12

(ii) Sandy Soils

Material Retained on No. 4 Sieve, Percent	Material Smaller than 0.05 mm, Percent	Cement Content, Percent by Weight — Maximum Density, lb/ft³					
		105–109	110–114	115–119	120–124	125–129	130 or More
0–14	0–19	10	9	8	7	6	5
	20–39	9	8	7	7	5	5
	40–50	11	10	9	8	6	5
15–29	0–19	10	9	8	6	5	5
	20–39	9	8	7	6	6	5
	40–50	12	10	9	8	7	6
30–45	0–19	10	8	7	6	5	5
	20–39	11	9	8	7	6	5
	40–50	12	11	10	9	8	6

[a] PCA (1956).

bacteria may eventually react with the sulfate-containing system. In addition, it should also be taken into consideration that products of some additives, may not be chemically stable under changing environmental conditions. For instance, it has been shown that ettingrites will start to dissappear in acidic conditions with pH as high as 10.8 (Kujala, 1983).

Preventing the loss of excessive amounts of water during curing aids in minimizing the shrinkage cracking. This is often done by covering the stabilized system with various water-proofing agents, water, or emulsified asphalts. Other materials such as waterproof paper, moist straw, or soil can be satisfactory also. The types of bituminous materials most commonly used are RC-250, MC-250, RT-5, and emulsified asphalt (RS-2). Rate of application varies from 0.15 to 0.3 gallons per square yard. When the surface is to be used immediately, the bituminous material should be blotted with sand. Soil–cement mixtures have autogenous healing characteristics, particularly during the initial weeks. When a curing agent such as emulsified asphalt is used over a soil–cement system that may have cracks due to early shrinkage or heavy equipment usage, the emulsion will have a tendency to seep into these cracks and prevent the autogenous healing by isolating the walls of the cracks. Therefore, water, rather than bituminous substances, is specified for curing in many instances (Arman, 1988).

Curing periods vary from 7 to 14 days, extending to 28 days in some cases. Some field studies have shown that light traffic during curing may improve the strength and durability of the soil–cement layer; others recommend keeping off these layers for a period of 7 to 14 days after compaction. In resolving the question of whether or not to allow traffic on the stabilized system, factors such as thickness of the stabilized layer, expected weight and repetition of wheel loads, and expected short-term strength should be considered by the design engineer for each case.

9.6.5 Mixing and Compaction

Depending on the size and conditions of the project, soil stabilization with portland cement, or any other stabilizer, may be achieved with means that range from the most primitive handtools to sophisticated single-pass machines. The basic steps in the construction of a soil–cement base are: (1) pulverizing the soil; (2) spreading the cement and mixing; (3) addition of water; (4) compaction, rolling, and finishing; (5) curing of the completed system; (6) quality control. Pulverization of the soil and thorough mixing are essential factors in achieving satisfactory soil–cement systems.

Pulverization of the soil should be achieved such that, with the exception of stones and gravel, 100 percent of the soil must pass through 1 in sieve and 80 percent of it must pass through No. 4 sieve. The moisture content should be at or near optimum during pulverization. Therefore, aeration or wetting of the soil may be necessary, depending on the moisture content of the soil. Pulverization can be done with various handtools or rotary speed mixers. In-situ mixing equipment may present problems when working to pulverize higher plasticity soils (generally $PI = 18$ or higher). In addition, higher-plasticity clays require larger amounts of cement which may not be economically viable. Therefore, cement stabilization of clays is generally limited to clays of liquid limit below 40, plasticity index not greater than 18, and the percentage passing the 0.002 mm sieve less than 35 percent. Pulverization and uniform mixing of such soils with cement are best accomplished in central mixing plants, generally used on large projects to produce uniform quality soil–cement and fast delivery of the product to the site. For reasons of economy, use of such plants in small projects may not be viable. However, increased installation of central mixing plants in urban areas may provide economic solutions for projects of all sizes in the future. Traveling plants are also used extensively, with a train of equipment added to the mixing equipment to supply water and complete the mixing process.

Various types of rollers can be employed for compaction depending on the type of the soil and the nature of the project. Compaction is done at or near optimum moisture content. For soils susceptible to shrinkage cracking, maintaining the moisture content below optimum may be a beneficial practice. In general, however, because some loss of water is expected during the mixing operations, addition of water slightly in excess of the optimum (1 or 2 percent more than the optimum) can be expected to result in a moisture content close to the optimum. It has been demonstrated through laboratory and field studies that a delay of two or more hours in compaction of the soil–cement after mixing results in reduced durability, compressive strength, and density of the soil–cement systems (Arman and Saifan, 1967). It is recommended that compaction of soil–cement mixtures not be delayed beyond 0.80 times the initial setting time of the cement gel, as specified by the ASTM designation C266 for plant-mixed soil–cement. The effects of delayed compaction may be reduced by increasing the compactive effort, using set-retarding agents such as calcium lignosulfate and hydroxylate carboxylic acid in trace amounts, or in some cases, by increasing the thickness of the base. In general, a maximum of 3 hours is specified as the limiting delay time between mixing the cement and completion of compaction in most projects. It should be noted that delays beyond 3 hours may increase the required compactive effort to a level that may be beyond the capabilities of ordinary field compaction equipment (Cowell and Irwin, 1979).

The most commonly used test for quality control of soil–cement mixtures is the Standard Test for Cement Content of

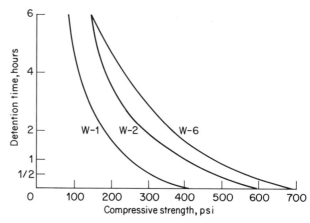

Fig. 9.8 Effect of delayed compaction on compressive strength of soil–cement mixtures. (*After Arman and Saifan, 1967.*)

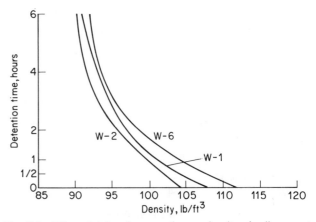

Fig. 9.9 Effect of delayed compaction on density of soil–cement mixtures. (*After Arman and Saifan, 1967.*)

Freshly Mixed Soil–Cement by ASTM D2901. Some advanced methods include onsite determination of cement content to an accuracy of as much as +5 percent. One transportable self-contained piece of equipment makes use of ^{252}Cf neutron source to produce a short-lived ^{49}Ca radioactive isotope followed by measurement of the ^{49}Ca by gamma spectroscopy.

Best usage of the various types of construction equipment and additional information on the construction of soil–cement can be found in the Portland Cement Association publication *Soil–Cement Construction Handbook* (PCA, 1978). Table 9.9 summarizes a number of characteristics of soil–cement application in four different countries (Jonker, 1982).

9.6.6 Laboratory Testing

Because of the varied reactions encountered with different soils, it is necessary to conduct preliminary tests on soil samples to determine the proper cement and moisture contents, any additives that may be required, and degree of compaction to be attained. Detailed laboratory testing procedures have been developed by the Portland Cement Association and adopted by AASHTO and ASTM (PCA, 1971). In design of cement stabilized systems the general considerations are compressive strength, durability, and density. The durability criteria, wet–dry and freeze–thaw resistance, are established by the Portland

TABLE 9.9 CHARACTERISTICS OF SOIL–CEMENT APPLICATION IN FOUR DIFFERENT COUNTRIES[a].

	The Netherlands	*United States*		*Switzerland*	*France*
Type of soil	Sand	Fine-grained	Granular	Gravel-sand type with substantial quantity of fines	Sand
Soil specifications	7-day compression strength with 10% cement and 11% water at least 2.0 MN/m²	—	Often grading specified	Clay-content max. 10%, plasticity-index of cohesive part max. 8%	Max. dimension 6 mm, bearing capacity of sand–cement before hardening is a measure for the extent of application
Mix-design conception and specification	Strength 28-day compression strength 5.0 MN/m² for medium- and high-traffic facilities and 4.0 MN/m² for low-traffic facilities	Durability based on wet–dry and freeze–thaw tests		Durability based on wet–dry and freeze–thaw tests	Strength 90-day direct tensile strength (R_t) $0.2 < R_t < 0.35$ MN/m² class A $0.35 < R_t < 0.5$ MN/m² class B $0.5 < R_t < 0.75$ MN/m² class C $R_t > 0.75$ MN/m² class D
Type of cement	Blast furnace cement quantity of slags up to 70%	Portland cement		Portland cement	Blast furnace cement quantity of slags up to 85%
Quantity of cement	8–12% (140–190 kg/m³)	4–12% (80–200 kg/m³)		3–5% (70–100 kg/m³)	About 7%
Construction method	Mix-in-place in general	Mostly mix-in-place, also mix-in-plant	Mostly mix-in-plant is specified	Mix-in-place only permitted for subgrade Mix-in-plant specified for road-base	Mix-in-plant is specified
Application field Subgrade	Incidental	Application only to a limited extent		Incidental	Sand–cement with $R_t < 0.2$
Flexible pavement Subbase	Medium, high[a]	Medium		Medium, high	Medium, high (all classes permitted)
Base	Low (no subbase)[b]	Low (no subbase)		Low, medium, high	Low (class C and D permitted), medium (class D permitted)
Rigid pavement subbase	High	—	High	High	—
Thickness and thickness-design conception	0.15–0.40 m Equivalence value to asphalt ranging from 0.4 to 0.5 Linear elastic multilayer theory	0.15–0.20 m Equivalence value in respect to unbound high-quality crushed stone base ranging from 1.1 to about 1.6		0.20–0.70 m built up in different layers with thicknesses of 0.20–0.30 m Equivalence value according to U.S.A.	0.15–0.25 m Based on linear elastic multilayer theory

[a] After Jonker (1982).
[b] Low, Medium, and High relate to the traffic loading intensity. Low and Medium (secondary and rural roads, residential streets, parking areas, airport paving for light traffic, container ports). High (major urban and urban highways, airport pavements for heavy aircraft).

Cement Association. The basic tests to determine the above properties are:

1. Test for Compressive Strength of Molded Soil–Cement Cylinders, ASTM D1633
2. Making and Curing Soil–Cement Compression and Flexure Test Specimens in the Laboratory, ASTM D1632
3. Test for Moisture–Density Relations of Soil–Cement Mixtures, ASTM D558 (AASHTO T134)
4. Wetting and Drying Tests of Compacted Soil–Cement Mixtures, ASTM D559 (AASHTO T135)
5. Freezing and Thawing Tests of Compacted Soil–Cement Mixtures, ASTM D560 (AASHTO T136)

The preparation of test specimens in the laboratory should represent on a small scale the steps and processes actually employed in construction. Pulverization or comminution of cohesive soils and clay clods, or homogenization of noncohesive soils by removing the oversize particles, is the initial step in specimen preparation for laboratory testing. Determination and addition of the proper amounts of cement and water, thorough mixing, compaction at or around maximum density and optimum moisture content according to the requirements, and finally proper curing with prevention of water loss are the steps followed in specimen preparation. The optimum moisture content and maximum dry density of the soil–cement mixtures are determined using Standard Proctor method (ASTM D698) with maximum feasible aggregate size $\frac{3}{4}$ in. Specimens, however, are restricted to a maximum content of 45 percent of the particles in the No. 4 to $\frac{3}{4}$ in range. If the soil possesses aggregates larger than $\frac{3}{4}$ in, the oversize is crushed to the No. 4 to $\frac{3}{4}$ in range and added to the No. 4 fraction to a maximum total of 45 percent of all the soil.

The steps involved in the mix design and preparation of test specimens are: (1) preparing soil sample; (2) determining moisture density relations for the soil and soil–cement mixtures (Tables 9.8a and 9.8b); (3) molding Proctor specimens with the selected cement contents at the optimum moisture contents to the maximum dry densities previously determined. Procedures used in design of soil–cement systems vary around the world. The general methods used for design include freeze–thaw and wet–dry tests, unconfined compressive strength, CBR, and in-situ deflection. The wetting and drying tests (W–D) and the freezing and thawing tests (F–T) are conducted over 12 cycles on two sets of 7-day moist-cured specimens. After each cycle of wet–dry or freeze–thaw the specimens are brushed with a steel wire brush and the loss in the material is recorded. The amount of loss permitted after 12 cycles depends on the size and composition of the soil and becomes less with increasing silt–clay content. Table 9.10 gives the total percentage losses by weight allowed for AASHTO soil groups after 12 cycles of durability tests. The soaked unconfined compressive strengths normally obtainable for soil–cement mixtures vary from 200 psi to 1200 psi. This variation depends on the soil type, amount of cement used, and curing periods. Table 9.11 gives the ranges of unconfined compressive strengths for 7-day and 28-day moist-cured and soaked specimens of three different soil groups. Dry densities of soil–cement range from 135 lb/ft³ for well-graded gravel, down to 85 lb/ft³ for silty or clayey soils.

These tests may take up to a month to conduct but are essential for a successful application in which the resistance of the soil–cement to abrasion and disintegration can be ascertained. Because of the time requirement, shorter tests have been developed. These are the "short-cut" tests developed by

the Portland Cement Association based on the statistical evaluation of an extensive number of experimental results involving as many as 2500 soil types. The short-cut method is especially dependable for sandy soils. It should be noted that the method cannot be applied to organic soils or if percentage by weight of particles less than 0.05 mm exceeds 50 and/or percentage by weight of particles smaller than 0.005 mm is less than 20. The "short-cut" tests can be performed within a day or within 7 days if the compressive strength tests are included. The latter are usually performed on 2-in diameter × 2-in high (Dietert) specimens compacted at optimum moisture content to maximum dry density determined by standard Proctor method, and they are moist-cured for 7 days. The short-cut method follows one procedure if the soil–cement mixtures contain no particles larger than 4.75 mm (No. 4 sieve), and another if the mixture contains particles larger than 4.75 mm. Figures 9.10, 9.11, and 9.12 pertain to the former case, whereas

TABLE 9.10 SOIL–CEMENT LOSSES ALLOWED BY AASHTO SOIL GROUPS.

Soil Type	Soil–Cement Losses During 12 Cycles of the Wet–Dry Test
AASHTO groups: A-1, A-2-4, A-2-5, A-3	14%
AASHTO groups: A-2-6, A-2-7, A-4, A-5	10%
AASHTO groups: A-6, A-7	7%

TABLE 9.11 UNCONFINED COMPRESSIVE STRENGTHS OF SOIL–CEMENT MIXTURES OF DIFFERENT SOIL TYPES.

Soil Type	Compressive Strength[a] (psi) 7-day	28-day
Sandy and gravelly soils AASHTO groups A-1, A-2, A-3 Unified groups GW, GC, GP, GM, SW, SC, SP, SM	300–600	400–1200
Silty soils AASHTO groups A-4 and A-5 Unified groups ML and CL	250–500	300–900
Clayey soils AASHTO groups A-6 and A-7 Unified groups MH and CH	200–400	250–600

[a] Specimens moist-cured 7 or 28 days, then soaked in water prior to strength testing.

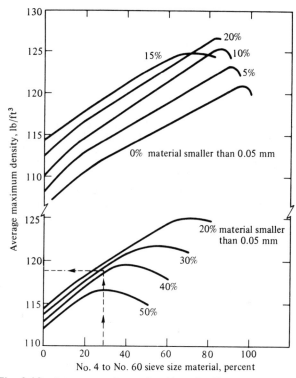

Fig. 9.10 Average maximum densities of soil–cement mixtures having no material retained on the No. 4 sieve.

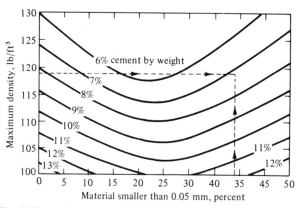

Fig. 9.11 Indicated cement content of soil–cement mixtures having no material retained on the No. 4 sieve.

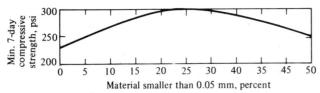

Fig. 9.12 Minimum 7-day compressive strength required for soil–cement mixtures having no material retained on the No. 4 sieve.

Figures 9.13, 9.14, and 9.15 pertain to the latter case. These charts are used through the short-cut test procedure as follows.

1. Performing a sieve analysis of the soil
2. Performing a moisture–density test on soil–cement mixture
3. Determining the indicated cement requirement by the use of charts
4. Verifying the cement requirement by compressive strength tests.

Details and interpretation of the results of the short-cut method are given in the *PCA Soil–Cement Laboratory Handbook* (PCA, 1971).

Tests on existing cement–soil bases have shown that these systems perform well over time, resisting structural damage even under heavy loading (Tayabji et al., 1982). Studies show cement to be working effectively in reducing the plasticity of treated soils over the long term also (Robbins and Packard, 1979; Roberts, 1986). Results of 60-cycle durability tests on laboratory soil–cement samples have also demonstrated the permanency of the CBR value for these systems (Robbins and Packard, 1979).

The durability and compression test standards have been developed on the basis of experience in the successful stabilization of natural soils. The time period allowed in various phases of sample preparation, curing, and testing are related to physical and chemical rate-determining factors such as types of energies of chemical interactions, temperature, thermal conductivity, and

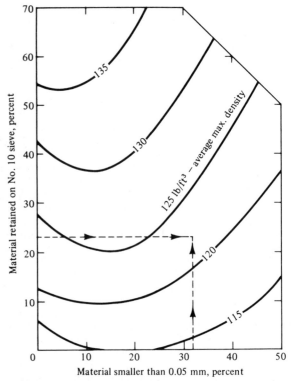

Fig. 9.13 Average maximum densities of soil–cement mixtures having material retained on the No. 4 sieve.

permeability to water and air. Any significant change in one of these factors, for example, a change in the dispersivity of the soil–cement system by means of a chemical additive, will upset the validity of the testing procedure. For instance, if a chemically caused decrease in soil permeability does not allow entrance of water within the time allowed for normal water permeability

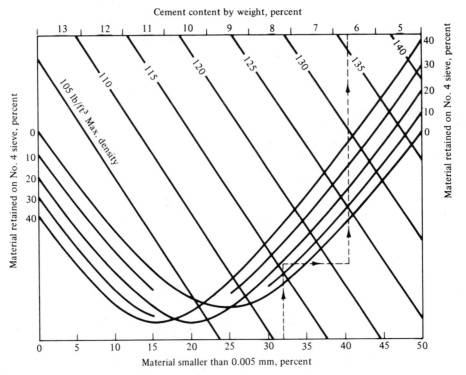

Fig. 9.14 Indicated cement content of soil–cement mixtures having material retained on the No. 4 sieve.

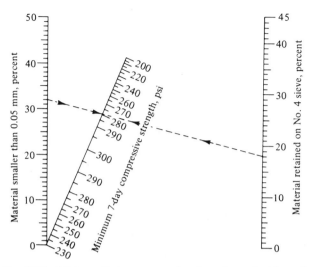

Fig. 9.15 Minimum 7-day compressive strength required for soil–cement mixtures having material retained on the No. 4 sieve.

under a given head of water, then the test results may be deceptive and longer times must be allowed for the water sorption phase. The same holds true when specimens of larger than standard size are used (Winterkorn and Reich, 1962).

Water may penetrate into a soil–cement system not only in the liquid (capillary) phase but also in the vapor phase. Some soil–cement systems, especially those containing certain types of organic matter, are not easily wetted and penetrated by liquid water. They can, however, be penetrated by water vapor. This takes place slowly in actual field installations. It will happen more rapidly, however, in the thawing phase of the freeze–thaw test when water vapor from the surrounding warmer atmosphere will move into and condense in the cold core of the test specimen. It is important that freeze–thaw tests be conducted on such types of soils even when constructions are located in areas where freezing does not normally occur. In these cases, the real function of the test is to "pump" water vapor into the specimen in a similar manner to what would occur over an extended period under natural conditions beneath a pavement.

In order to judge realistically from laboratory tests the beneficial or adverse effects of a given additive on soil properties, a number of factors need to be considered. Among these considerations are specimen size, rate and type of loading (transportation), understanding of standard reactions when common soil types are used and the need to vary test standards with respect to changes that may be induced on certain soil properties by the additives.

9.7 LIME STABILIZATION

Lime is another commonly used additive for soil stabilization or for improving soil properties. It is the oldest known method of chemical stabilization, used by Romans to construct the Appian Way.

Examining Figure 9.5, it can be concluded that the hydration products of portland cement can be duplicated by combining two or more of the primary components, that is, CaO, SiO_2, and Al_2O_3, Fe_2O_3 in the right proportions in presence of water. Since most soils contain silica and aluminosilicates, simple addition of quicklime (CaO) or hydrated lime ($Ca(OH)_2$) and water may suffice to establish the desired composition. Experience has shown that mixtures of most clay soils with quick or hydrated lime and water will form cementatious products in reasonable period of time. Certain soils, however,

may necessitate addition of siliceous compounds such as pozzolans, infusorial earth, or fly-ash to facilitate the reactions.

Three basic chemical reactions take place when pozzolanic clays and lime are mixed in presence of water: (1) cation exchange and flocculation–agglomeration; (2) cementation (pozzolanic reaction); (3) carbonation. Cation exchange and flocculation–agglomeration are the initial reactions between soil and lime and cause immediate improvement of soil plasticity, workability, uncured strength, and load deformation properties. Pozzolanic or cementation reactions are time and temperature dependent. These reactions may continue over several years, with temperatures less than 55°C retarding the reactions and temperatures over 55°C accelerating them. Carbonation is an undesirable reaction and results in relatively insoluble carbonate that will not react with pozzolanic soils and render soils more plastic.

9.7.1 Soil–Lime Reactions

Cation Exchange and Flocculation–Agglomeration Reactions

These reactions result from the replacement of univalent sodium (Na^+) and hydrogen (H^+) ions of soil with divalent (Ca^{2+}) calcium ions of lime. This reaction binds clay, resulting in reduction of clay content and thus plasticity. Lime contents of 3 to 18 percent by volume are used to reduce plasticity of clays and render them manageable for compaction, manipulation, or stabilization with other chemicals. This is referred to as "lime modification" of soils. It is not unusual to reduce plasticity of a 50 *PI* soil to 15 by lime modification. The reaction usually takes place within 96 hours. Saturated soils often require only crude mixing with light equipment (i.e., farm discs pulled by light tractors) or by hand, since lime has a strong affinity for water and the resulting reaction is a harsh one. The reaction breaks down clay lumps by drawing water from the clay, thus creating a drier material.

Agglomeration reaction of lime and soil is used to destroy collapsible characteristics of some silts. Approximately 6 to 9 percent by volume of lime mixed with collapsible silts destroys clay buttresses between silt particles and renders soils stable after 72 hours. Swell potential of clays can be reduced significantly by modifying them with 6 to 12 percent lime by volume. Swell potentials of 7 to 8 percent can be reduced to 0.1 to 0.2 percent by lime modification (Thompson, 1969).

Cementation or Pozzolanic Reactions

When SiO_2 and Al_2O_3, Fe_2O_3 of soil react with available calcium in lime, they produce very stable calcium silicates and aluminates that act as natural cement (tobermorite gel) similar to portland cement. Soils with plasticity indices as high as 37 can be stabilized with single application of 9 to 24 percent by volume of hydrated lime ($Ca(OH)_2$, $Ca(OH)_2 \cdot MgO$) or quicklime (CaO, $CaO \cdot MgO$). Soils with higher plasticity can be stabilized with double application of lime. In this case part of the lime is applied to reduce plasticity and the other part is applied thoroughly and uniformly to obtain cementation and thus stabilization. Cementation is a time-dependent slow reaction. Production of natural cements depends on the presence of pozzolanic clays. It produces lower strengths (80 to 1100 psi) depending on soil type, amount of lime, and curing period, compared to the strength of soil–cement. Soil–lime mixtures also possess autogenous healing properties. As the pozzolanic reaction continues, the healing effect remains effective for many years.

Recent laboratory and field studies have shown that lime can also be used to stabilize some organic soils. Acidic organic soils found in low-lying or formerly low-lying areas composed of mixtures of mineral soils and decayed vegetation, with pH factor as low as 5.4 need approximately 6 percent by volume of lime to neutralize organic acids initially. If pozzolanic clays are available in the organic soil, then additional lime reacts with it to produce a stabilized mixture. Figure 9.16 illustrates the manner in which lime modification improves the strength of organic soils with 22 percent organic content by volume (Arman and Munfakh, 1972).

Carbonation

Carbonation is an undesirable reaction that occurs when the lime added to soil does not react with soil, but draws CO_2 from air or soil to form $CaCO_3$. This occurs when the soil does not contain adequate amount of pozzolanic clay or because excessive amount of lime has been added. $CaCO_3$ is a plastic material. It increases the plasticity of the soil and binds lime so that it cannot react with pozzolanic materials. Therefore, excessive lime addition to soil does not produce beneficial results.

Figure 9.17 shows a flow diagram of various reactions that take place in lime–clay–water systems and the end-products.

9.7.2 Types and Properties of Lime and Lime–Clay–Water System

The lime materials used in construction are as follows:

Type	Formula
Calcia (high-calcium quicklime)	CaO
Hydrated high-calcium lime	Ca(OH)$_2$
Dolomitic lime	CaO + MgO
Normal hydrated or monohydrated dolomitic lime	Ca(OH)$_2$ + MgO
Pressure-hydrated or dihydrated dolomitic lime	Ca(OH)$_2$ + Mg(OH)$_2$

The higher the magnesium content of the quick or hydrated lime, the less is the water affinity and the heat developed in mixing with water. Depending on the raw material, limes may differ in their properties that may effect their reaction with different soils. Therefore, it may be critical to evaluate the

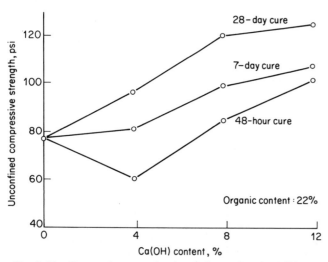

Fig. 9.16 Change in strength of soil–lime mixtures (different curing periods). (*After Arman and Munfakh, 1972.*)

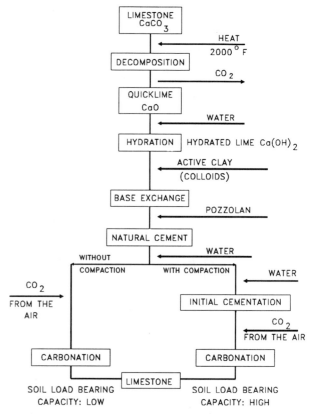

Fig. 9.17 Flow diagram of reactions that take place in the lime–clay–water system. (*After Arman, 1988.*)

properties of lime as well as soil. Some of the laboratory tests for characterization of lime are chemical constituents, pH value, specific gravity, viscosity, color, and odor. Likewise, it is important to characterize the soil intended for the mixture by all or a number of the following tests: particle-size distribution, consistency (Atterberg) limits, pH value, classification, mineralogy, organic content, and moisture–density relationship.

Various theories have been suggested by investigators in explanation of cementation reaction between lime and soil. One theory is that lime is adsorbed onto clay surfaces and reacts with other surfaces to precipitate cementatious products (Diamond et al., 1964). Another theory postulates that silica and alumina are released from clay lattices under high pH of up to 12.4, and they combine with the calcium of lime in aqueous medium (Eades, 1962). Still another theory suggests that lime reacts directly with clay edges, forming a network of cementatious materials (Stocker, 1972). Optical studies show evidence that all of these processes may occur in clay–lime–water systems (Ormsby and Bolz, 1966; Ford et al., 1982). Some of the major factors that influence soil's ability to react with lime are soil pH; organic carbon content; natural drainage; presence of excessive quantities of exchangeable sodium; clay mineralogy; degree of weathering; presence of carbonates, sulfates, or both; extractable iron; sesquioxide ratio; and silica–alumina ratio. Other factors that effect the reaction are particle size and surface-to-volume ratio; solubility of Ca(OH)$_2$; and temperature. Soils with low pH factor require lime in excess of cementation to neutralize the acid. Sulfates have a detrimental effect on the cementation reaction. Sulfates react with calcium and carbonate to form ettringite and thaumasite, which are expansive and may produce swelling characteristics in excess of that of untreated soil (Mitchell, 1986). They also reduce pH and therefore inhibit formation of pozzolanic products.

Solid particles tend to react on their surfaces; therefore, their respective reaction rates are proportional to their surface/volume ratios. This ratio is inversely proportional to the linear particle size. On this basis, the amount of $CaSiO_3$ formed per unit time per unit volume, or rate of reaction, will be proportional to the percentage of the smallest-sized siliceous components, that is, the clay content of the soil. Also, the higher the proportion of silica and the smaller that of aluminum and iron sesquioxides in the smallest-sized fractions, the greater is their reactivity with $Ca(OH)_2$. Reaction between soil and lime occurs in aqueous medium and the rate of reaction is highly dependent on the concentration of the components. Therefore, the solubility of $Ca(OH)_2$ becomes an important factor. This solubility is relatively small: about 1.65 grams of $Ca(OH)_2$ per liter at normal temperature. In a normal clay soil this amount of $Ca(OH)_2$ in solution would very soon be exhausted in the satisfaction of the cation exchange capacity of the clay fraction if there were not some excess solid $Ca(OH)_2$ stored in the system. To insure the greatest possible rate of solution of the solid $Ca(OH)_2$, it should be in a colloidally dispersed state.

The rate of reaction depends not only on the effective concentration of the reactants but also on the temperature of the system. The solubility of $Ca(OH)_2$ decreases with increasing temperature while that of silicon dioxide increases. The overall rate of reaction is more influenced by the effective concentration of the silica than by that of the $Ca(OH)_2$. This is exemplified in the related field of manufacture of sand–lime bricks, in which 4 to 10 percent of $Ca(OH)_2$ is mixed with quartz sand and reacted at a steam pressure of 150 psi, which corresponds to a temperature of about 366°F (185.5°C). The process produces bricks of strength of the order of 4000 psi. Mateos (1964) showed that specimens of lime, fly-ash, and sand soil (6 percent lime, 17.5 percent fly-ash) acquired a 28-day strength of 2400 psi when cured at 120°C. Some specimens failed upon immersion in water when they were cured at 10°C. These results show the significance of curing periods at elevated temperatures and performing lime stabilization in the warm season to insure sufficient strength development before the start of the cold weather.

If a soil does not possess fine-grained siliceous material necessary to react with lime, then such material can be added in the form of volcanic ashes (pozzolan, santorin), defatted diatomaceous earth, or siliceous fly-ashes. Assuming that the end-product of the reaction is tobermorite ($3Ca(OH)_2 \cdot 2SiO_2$), the ratio of $Ca(OH)_2$ to reactive silica is 1.82:1. In practice ratios of fly-ashes to lime vary from 3:1 to 5:1, indicating that the major part of the fly-ash acts as an inert filler. Taking into account the differences in composition of locally available fly-ashes and limes, it is best that the optimum ratios as well as absolute quantities of these additives be determined by test with the soil to be treated.

9.7.3 Effectiveness of Lime Treatment

The changes induced by lime stabilization on physical properties of soils are summarized by National Lime Association given in the *Lime Stabilization Construction Manual* (NLA, 1985) as follows.

1. The plasticity index of the soil drops sharply by a factor of 4 or more in some instances. This is generally due to liquid limit decreasing and plastic limit increasing.
2. The soil is agglomerated, decreasing the soil binder or clay content (minus No. 40 sieve mesh size) substantially.
3. Lime, by pulling water, accelerates breaking up of clay clods during mixing. As a result of agglomeration and

disintegration, the soil becomes more friable and can be worked readily.
4. Lime aids in drying out wet soils quickly, thus speeding up compaction.
5. The shrinkage and swell characteristics of soil is reduced markedly.
6. After curing, unconfined compressive strength increases considerably. This increase can be as much as 40-fold.
7. Load-bearing values, as measured by various tests (CBR, *R*-value, Texas triaxial, plate bearing or *k*-value, etc.) increase substantially.
8. The tensile or flexural strength, as measured by various tests (cohesiometer, split tensile, etc.) increases markedly. Thus, the stabilized layer develops beam strength.
9. The lime-stabilized layer forms a water-resistant barrier by impeding penetration of gravity water from above and capillary moisture from below. Thus, the layer becomes a firm "working table", shedding rain water readily and remaining stable.

According to U.S. experience, treatment with lime is most effective for: (1) stabilization of clay gravel materials to serve as bases for pavements where 2 to 4 percent of $Ca(OH)_2$ by weight of soil is used; (2) stabilization of heavy clay soils to serve as bases (5 to 10 percent by weight lime) or subbases (1 to 3 percent by weight lime) for pavements. Lime treatment has been found less effective for silty-loam soils. It is not recommended for sandy soils except with addition of clay, fly-ash, or other pozzolanic constituents, which serve as both hydraulically reactive ingredients and as filler to improve the reactivity and gradation of the soil. Lime treatment (especially by quicklime) may serve as an important construction aid for treatment of access roads to construction sites and of construction sites themselves if they have become impassible owing to excess precipitation. Soil stabilization with lime should be considered:

1. As a preparative measure for subsequent stabilization of clay soils with cement or bituminous binders and waterproofing agents.
2. As an additional improving measure in granular soil stabilization by controlling the plasticity of the binder material and increasing its cementing power.
3. For improvement of subbases. In addition to increasing the bearing capacity, lime treatment inhibits capillary water movement, thereby preventing water movement beneath the pavement.
4. For stabilized bases underneath all paved surfaces.
5. For independent pavements for secondary and tertiary roads. However, surface treatment with bituminous materials is recommended in such cases.

Examples of field projects where lime stabilization has been successfully used are (ASCE, 1978):

I-10 project, Baton Rouge, Louisiana, where the top 6 inches of a 6-ft high embankment was treated with 4 percent lime
Quicklime treatment of pavements and embankments in France, Belgium and Germany
Stabilization of loessial soil with 3 percent hydrated lime and 2 to 5 percent fly-ash for foundation of a coal power plant in Canton, Illinois
Five percent calcitic lime stabilized clay foundation for a large storage building in a potash plant in Saskatoon, Canada

Lime has also been used successfully in treating swelling soils, an example of which is given by Mitchell and Raad (1973). Other applications include erosion control (Diamond, 1975), and dispersive soil treatment (Hayden and Haliburton, 1976).

In addition to the above, lime stabilization has been shown to work well under hydraulic conditions such as for irrigation canals, levees, and earth dams (Gutschick, 1978). Examples of these applications are: stabilization of earthen slopes of the Friant–Kern canal near Fresno, California; levees on the Mississippi River in the states of Arkansas and Illinois; remedial treatment of existing earth dams and construction of new dams in a number of states in the United States and other countries. Other applications, such as stabilization of soil for lining of stock tanks, earth reservoirs for storage of light fuel oil, hydraulic embankments, and a lake bottom have been reported in the literature (Gutschick, 1978).

Lime-treated soils show improvement in the following properties and characteristics (TRB, 1987):

(1) *Compaction characteristics.* As curing continues, the properties of the treated material continue to change. The maximum density of the treated material decreases with curing time and lime content. The optimum moisture content, however, increases with curing time and lime content.

(2) *Plasticity and workability.* Substantial reduction in plasticity index and increase in shrinkage limit occur. Soils with high *PI* initially require greater quantities of lime addition. First increments of lime addition are often more beneficial than subsequent additions in reducing the plasticity. Lime-treated soil attains a silty and friable texture that causes significant improvement in workability.

(3) *Volume change.* Swell potential and swell pressures are reduced substantially. This is attributed to the reduced affinity of the calcium-saturated clay for water and the formation of a cementitious matrix that resists volumetric expansion. Field moisture contents of lime-treated soils usually stabilize at around optimum and the moisture content variation is small. Therefore shrinkage cracking due to moisture loss is minimized. Laboratory and field data indicate that for typical field conditions, shrinkage of lime-treated soil is not extensive.

(4) *Strength.* Strength of lime-treated soils depend on a number of variables including type of soil, lime, percentage of lime, curing period, and temperature. The popular ways of measuring strength of lime-treated soils are unconfined compression test, stabilometer, and CBR tests. The initial strength or uncured strength is a result of the immediate reactions between lime and soil (cation exchange, agglomeration, and flocculation) and the improvement may be up to 100 percent. The cured strength comes about with cementation and pozzolanic reactions that occur over time. 28-Day unconfined compressive strengths of lime-treated soils cured at 73°F (22.8°C) often show increases of more than 100 psi. With longer curing periods larger increases have been reported. The unconfined strength of 5 percent lime-treated AASHTO Road Test embankment soil cured 75 days at 120°F (48.9°C) was 1580 psi. Lime increases the cohesion of soil substantially and produces minor increase in internal friction angle. Cohesion increases with the unconfined compressive strength and large shear strength gains develop in cured soil–lime mixtures. Tensile strength is an important parameter of soil–lime systems, especially when used in pavement design. Indirect tensile or flexural tests have been used to measure tensile strength of soil–lime mixtures. The ratio of indirect tensile strength to unconfined compressive strength was found to be 0.13 in one study. Based on this, flexural strength is taken as 25 percent of the unconfined compressive strength. It has been observed that lime-treated fine-grained soils often display increased CBR values irrespective of the length of curing and lime-reactivity of soil. Therefore, CBR values may serve as a general measure of strength; however,

use of CBR values is not recommended if results of other strength tests are available (TRB, 1987).

(5) *Stress–strain characteristics.* In soil–lime mixtures the failure stress increases and the ultimate strain decreases compared to natural soils. A generalized stress–strain relationship developed for soil–lime mixtures shows maximum compressive stress to develop at approximately 1 percent strain, regardless of soil type or curing period. Linear relationships of compressive strength versus compressive elastic modulus and flexural strength and flexural modulus were found for cured soil–lime mixtures. Field measurement of elastic moduli of the lime-treated base of an airfield showed a substantial increase from a value of 165 000 psi to 568 000 psi over approximately 2 to 2.5 years following construction. The Poisson's ratios of soil–lime mixtures have been reported to be around 0.31 between 50 to 75 percent of the ultimate compressive strength.

(6) *Fatigue and durability.* Studies have shown that fatigue strengths at 5 million stress repetitions were 54 percent of the ultimate flexural strength on the average. However, as the ultimate strength increases with curing, the fatigue stress, as a percentage of ultimate strength, decreases, and thus the fatigue life of the mixture increases. Durable material can be obtained when lime is mixed with reactive soil. Although some strength reduction and volume change occur, the residual strength is often adequate to meet field requirements. Durability tests on typical soil–lime mixtures showed that the average rate of decrease in unconfined compressive strength is 9 psi per cycle and 18 psi per cycle of freeze–thaw (ASTM D560) for 48 and 96 h curing at 120°F, respectively (Dempsey and Thompson, 1968). Soil–lime mixtures also possess autogenous healing capability. Therefore, the strength reduction caused by freeze–thaw is not cumulative and with curing it would be restored. Figure 9.18 illustrates the influence of cyclic freeze–thaw and immediate curing on unconfined compressive strength.

Table 9.12 shows the general stabilizing effect of lime on different soil types.

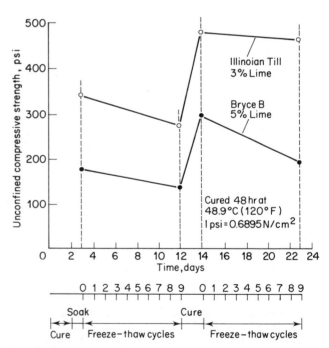

Fig. 9.18 Influence of cyclic freeze–thaw and immediate curing on unconfined strength of lime-stabilized soil. (*After Dempsey and Thompson, 1968.*)

TABLE 9.12 GENERAL STABILIZING EFFECTS OF LIME ON DIFFERENT SOIL TYPES.

Type of Soil	Untreated					Lime Treated[a]				
	Triaxial	CBR	R-Value	k-Value	Cohesiometer	Triaxial	CBR	R-Value	k-Value	Cohesiometer
Heavy clay	5.5	2	20	100	—	3.2–3.5	15–30	55–69	250–350	350–850
Light clay	4.5	5	35	150	—	2.9–3.4	20–40	60–75	300–400	450–700
Sandy clay	3.7	12	50	200	—	2.4–3.0	35–60	65–80	400–500	550–850
Granular soil PI = 8+	3.2	30	65	250	—	1.5–2.7	50–75	70–80+	450+	650+
Clay gravel PI = 6 to 10	2.6	50	75	400	—	1.0–1.6	70–100+	80+	500+	800+

[a] Based on use of 4–6 percent lime for clay soils and 2–4 percent for granular and clay-gravel types. Triaxial and cohesiometer values are based on approximately 18 days of laboratory curing, CBR on 4 days curing (soaked), and R-value on about 2 days curing. The stability values of lime-treated specimens increase markedly with longer or accelerated curing: e.g., curing CBR specimens for 2 days at 120°F prior to soaking will nearly double the CBR values. This accelerated curing would correspond approximately to 30 to 45 days of summer field curing.

9.7.4 Soil–Lime Mixture Design

The main objective in soil–lime mix design is to establish the appropriate lime content. Depending on the stabilization objectives, a number of engineering properties of the soil–lime mixture must be considered. Included in these are consistency (Atterberg) limits, CBR (California Bearing Ratio), swell potential, R-value (Hveem stabiliometer test) (ASTM 2844) and unconfined compression. The strength-test specimens (cylindrical) should be prepared with a length-to-diameter ratio of 2 and their density and moisture must be carefully controlled. Specifying the compactive effort is essential to compare results. In most procedures, mixtures are prepared at or near optimum moisture content as determined by AASHTO. Various test

procedures specify different curing conditions; therefore, the results of one set of tests should not be arbitrarily adopted for comparison without checking the curing procedure. Some of the curing procedures used are given in ASTM D1632 and AASHTO T-220. Studies have shown that elevated curing temperatures in excess of 120°F in the laboratory should be avoided, since the pozzolanic compounds formed at these temperatures can be significantly different from those formed in the field at lower temperatures. In addition, laboratory testing of representative samples is essential when considering use of new stabilizers and additives or improvement of old ones. Otherwise, it may be a dangerous practice if the new stabilizer radically changes the properties of the stabilized system unfavorably in an unexpected manner. Such has been the case

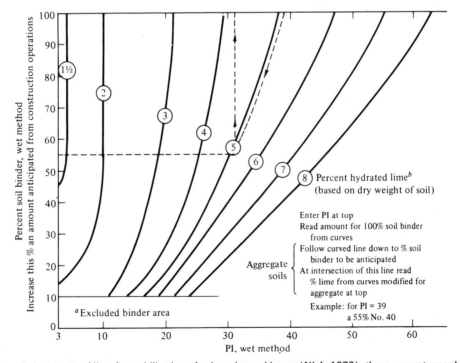

Fig. 9.19 Recommended amounts of lime for stabilization of subgrades and bases (NLA, 1972); these percentages should be substantiated by approved testing methods on any particular soil material. *Notes:* [a]Exclude use of chart for materials with less than 10 percent No. 40 and cohesionless materials (*PI* less than 3). [b]Percentage of relatively pure lime usually 90 percent or more Ca and/or Mg hydroxides and 85 percent or more passing the No. 200 sieve. Percentages shown are for stabilizing subgrades and base courses where lasting effects are desired. Satisfactory temporary results are sometimes obtained by the use of as little as half of the above percentages. Reference to cementing strength is implied when such terms as "lasting effects" and "temporary results" are used. (*Reproduced with permission of the National Lime Association.*)

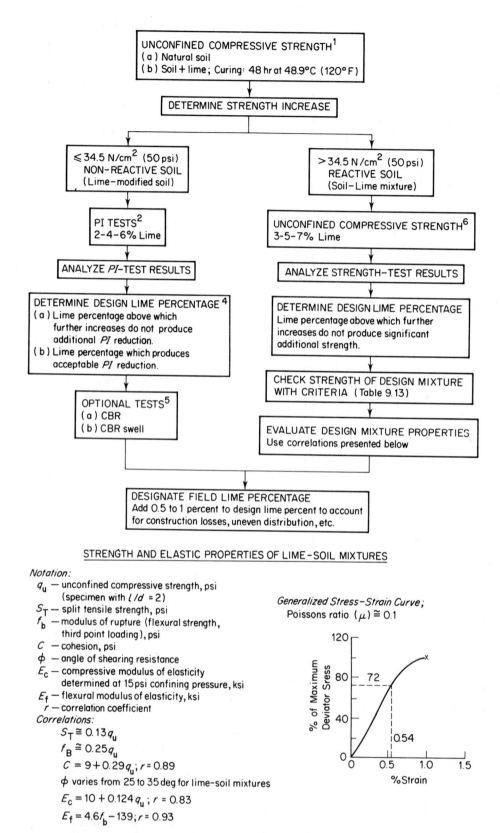

STRENGTH AND ELASTIC PROPERTIES OF LIME-SOIL MIXTURES

Notation:

q_u — unconfined compressive strength, psi
(specimen with $l/d = 2$)
S_T — split tensile strength, psi
f_b — modulus of rupture (flexural strength, third point loading), psi
C — cohesion, psi
ϕ — angle of shearing resistance
E_c — compressive modulus of elasticity determined at 15 psi confining pressure, ksi
E_f — flexural modulus of elasticity, ksi
r — correlation coefficient

Correlations:

$S_T \cong 0.13 q_u$

$f_B \cong 0.25 q_u$

$C = 9 + 0.29 q_u; r = 0.89$

ϕ varies from 25 to 35 deg for lime-soil mixtures

$E_c = 10 + 0.124 q_u; r = 0.83$

$E_f = 4.6 f_b - 139; r = 0.93$

Generalized Stress-Strain Curve;
Poissons ratio $(\mu) \cong 0.1$

Fig. 9.20 Mixture design process for lime-treated soils according to Thompson procedure. [1] All specimens are compacted at optimum water content to maximum dry density. Lime treatment level for (b) may be 5 percent or as determined by the pH procedure (see Note 6). [2] PI tests are conducted 1 h after mixing the lime–soil–water mixture. Mixture is not cured before testing. [3] In some cases more closely spaced treatment levels may be appropriate. [4] Criteria (a) or (b) may be applied depending on the stabilization objective. [5] Conduct tests on design lime content. Curing of CBR specimens before soaking is optional depending on stabilization objective. If swell is not reduced to a satisfactory level, additional CBR tests may be conducted at higher lime contents. Design lime content may be increased if further swell reduction is obtained. Swell considerations are of great importance for lime-modified subgrades. [6] Specimens are compacted at optimum moisture content to maximum dry density. Additional or different (or both) lime percentages may be required for some soils. An estimate of approximate optimum lime content may be obtained by applying the pH test developed by Eades and Grim (1966). (*After Thompson, 1970.*)

when saline and alkaline additives were used as reaction accelerators in soil–cement and soil–lime mixtures (Winterkorn, 1964).

Mixture design criteria can be specified either for stabilization objectives of soil modification (*PI* reduction and improvement of workability), or soil cementation (strength improvement). In the first case the criteria would include: no further decrease in *PI* with addition of lime, acceptable *PI* reduction for the particular stabilization objective, acceptable swell potential reduction, and CBR and *R*-value increase sufficient for anticipated uses. In the second case, the cured mixture must satisfy the minimum strength and durability requirements and design lime content should be the percentage that produces the maximum strength for given curing conditions. Some of the current mixture design procedures in the United States are (TRB, 1987):

1. California procedure
2. Eades and Grim (1966) procedure
3. Illinois procedure
4. Oklohoma procedure
5. South Dakota procedure
6. Texas procedure
7. Thompson procedure
8. Virginia procedure

Figure 9.19 shows the recommended amounts of lime for stabilization of subgrades and bases according to the Texas procedure. Figure 9.20 shows the proposed mixture design process for lime-treated soils according to the Thompson procedure (Thompson, 1970). The Texas procedure is AASHTO T-220, which suggests strength criteria of 100 psi for base construction and 50 psi for subbase construction. A general outline of the procedure is as follows.

1. Lime percentage is selected based on grain size and *PI* data (Fig. 9.19).

2. Optimum moisture and maximum dry density of the mixture are determined in accordance with AASHTO.
3. Test specimens (6-in diameter and 8-in height) are compacted at optimum moisture content.
4. The specimens are placed in a triaxial cell (AASHTO T-212 or Tex-121-E) and cured in the following manner:
   a. Cool to room temperature.
   b. Dry at temperature not exceeding 140°F for about 6 h until one-third to one-half of the molding moisture is removed.
   c. Cool for at least 8 h.
   d. Subject specimens to capillarity for 10 days (AASHTO T-212).
5. Test cured specimens in unconfined compression.

Lime contents are generally specified as percentage of dry weight of soil, or sometimes as volume percentage. Recommended percentages of lime for laboratory testing and for construction vary from 2 to 10 percent by weight. They are 2, 3, and 5 percent for coarse soils (clay–gravels, caliche, sandy soils) having less than 50 percent silt–clay fraction. The lime percentages increase to 5, 7, and 10 percent for coarse soils with silt–clay fraction more than 50 percent. For intermediate soils 3, 5, and 7 percent of lime are indicated. Where severe freezing and thawing conditions prevail, lime percentages of 8 to 12 percent are recommended. In combination with additives, such as fly-ash, 3, 5, and 7 percent of lime are used with fly-ash contents that normally range from 10 to 20 percent.

Lime is commonly used in slaked, or hydrated, form, either as powder or slurry. In rural areas where dust may not present an environmental problem, lime is applied in dry form either from bags or trucks. Otherwise, lime is added to the soil in a slurry formed in an approximate mixture of 70 percent water and 30 percent lime (CaO). When quicklime (CaO) or unslaked lime is used, care should be taken to insure that soil has sufficient water to allow the lime to slake, or hydrate, before final mixing

TABLE 9.13 QUALITY CRITERIA FOR SOIL–LIME MIXTURES[a].

		Strength Requirements for Various Anticipated Service Conditions[c], psi				
	Residual Strength Requirement[b] psi	Extended Soaking for 8 days, psi	Cyclic Freeze–Thaw[d]			
Anticipated Use			3 Cycles	7 Cycles	10 Cycles	
Modified subgrade	20	50	50	90 50	120	
Subbase						
Rigid pavement	20	50	50	90 50[e]	120	
Flexible pavement by cover thickness						
10 in	30	60	60	100 60[e]	130	
8 in	40	70	70	100 75[e]	140	
5 in	60	90	90	130 100[e]	160	
Base	100[g]	130	130	170 150[e]	200	

[a] After Thompson (1970).
[b] Minimum anticipated strength following first winter exposure.
[c] Strength required at termination of field curing following construction to provide adequate residual strength.
[d] Number of freeze–thaw cycles expected in the soil–lime layer during the first winter of service.
[e] Freeze–thaw strength losses based on 10 psi/cycle except for 7-cycle values indicated by superscript *e*, which were based on previously established regression equation.
[f] Total pavement thickness overlying the subbase. The requirements are based on the Boussinesq stress distribution. Rigid pavement requirements apply if cemented materials are used as base courses.
[g] Flexural strength should be considered in thickness design.

and compaction. This slaking, or *mellowing*, period varies from 1 to 4 hours depending on the amount of lime and water content of the soil. Detailed presentation of lime stabilization construction can be found in *Lime Stabilization Construction Manual* (NLA, 1985) and also *Lime Stabilization* (TRB, 1987). The quality criteria of soil–lime mixtures are often based on pavement structural behavior and durability. Table 9.13 (Thompson, 1970) summarizes these criteria for various uses of lime-stabilized soil. The quality control of stabilized systems requires field tests, some of which are: density, moisture penetration resistance (ASTM D1558), load/deflection, erosion and leaching rate (ASTM D3385), and dust analysis (ASTM D1739).

9.7.5 Stabilization With Lime and Pozzolanic Additives

When the natural soil lacks pozzolanic constituents to react with lime, natural or synthetic substances can be added to obtain cementation reactions. Natural pozzolanic materials generally occur in the form of volcanic ash, tuff, and trass. They are not crystalline but ionic in structure. Pozzolana and pozzolanic rocks have been used as cement additives in a number of countries. Crushed basalt rock or quarry waste possess sufficient cementatious properties to be used as additives also. Other than naturally occurring pozzolans such as these, by-products of combustion of coal, colliery shales, and municipal waste possess pozzolanic properties that have been exploited as additives in soil stabilization. Among these materials, fly-ash, bottom ash, boiler slag, and granulated blast-furnace slag, which are all by-products of coal combustion, have been used most extensively. A variety of other additives have been tested to improve soil–lime mixture properties. Good results were obtained with a number of sodium compounds, including sodium metasilicate, sodium sulfate and sodium hydroxide. A comprehensive review of coal-associated by-products used in soil stabilization is presented by Usmen et al. (1983).

9.8 ASH AND SLAG STABILIZATION

9.8.1 Fly-ash

Fly-ash is a by-product of power plants fuelled by pulverized coal. It is recovered from flue gases. Approximately 70 percent of the chemical composition of fly-ash is alumina and silica. It reacts with lime in the presence of water, setting and hardening similarly to hydraulic binders. The chemical composition and physical properties of fly-ashes depend on the coal source, degree of coal pulverization, design of boiler unit, loading and firing conditions, and handling and storage methods. There are four main classes of coal: anthracite, bituminous, subbituminous, and lignite. The fly-ashes from lignite coal (sulfocalcic or class C ash (ASTM C618)) contain higher percentages of calcium and sulfates, and less carbon, silica, and alumina than bituminous fly-ashes (aluminosiliceous or class F ash (ASTM C618)). Table 9.14 shows weight percentages of constituents of typical fly-ashes (Styron, 1980). Figure 9.21 shows the composition of these fly-ashes along with slag and portland cement (OECD, 1984). Fly-ashes with large amount of free lime tend to be very reactive and show hydraulic cementation properties. It should be noted here that the CaO content does not necessarily indicate amount of free lime, since almost all of the lime tends to be chemically combined with silicates and aluminates at the high temperatures at which the ash is generated. As an example, a

TABLE 9.14 WEIGHT PERCENTAGE OF VARIOUS COMPONENTS OF TYPICAL FLY-ASHES[a].

	Weight Percentage	
	Bituminous	Lignite–Subbituminous
Silica (SiO$_2$)	40–55	20–40
Alumina (Al$_2$O$_3$)	25–35	10–30
Ferric oxide (Fe$_2$O$_3$)	5–24	3–10
Magnesium oxide (MgO)	$\frac{1}{2}$–5	$\frac{1}{2}$–8
Sulfur trioxide (SO$_3$)	$\frac{1}{2}$–5	1–8
Titanium oxide (TiO$_2$)	$\frac{1}{3}$–2	$\frac{1}{2}$–2
Carbon (C)	$\frac{1}{2}$–12	$\frac{1}{2}$–2
Moisture	0–3	$\frac{1}{3}$–3
Calcium dioxide (CaO)	$\frac{1}{2}$–4	10–32
Potassium oxide (K$_2$O)	$\frac{1}{2}$–3	$\frac{1}{2}$–4
Sodium oxide (Na$_2$O) (available alkali)	$\frac{1}{2}$–1$\frac{1}{2}$	$\frac{1}{2}$–6

[a] After Styron (1980).

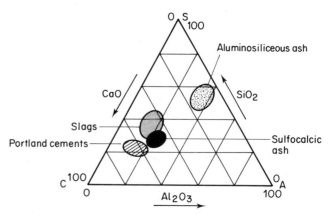

Fig. 9.21 Composition of fly-ashes along with portland cement and slag. (*After OECD, 1984.*)

lignite source fly-ash with 30 percent CaO could have free lime as low as 2 percent. A study shows that ammonium phosphate appears to attack and break down the glassy phase of a calcium-rich fly-ash, liberating the calcium, aluminum, and silicon to form neopozzolanic cementatious reaction products (Bergeson et al., 1984). High carbon content (7 to 10 percent) can inhibit pozzolanic reaction severely.

The color of fly-ashes varies from cream to dark brown or gray. A cream color is indicative of high calcium content, while gray to black color indicates high carbon content. The specific gravity of fly-ashes ranges from 2.1 to 2.6. They are usually composed of silt-size particles of glassy spheres, some crystalline matter, and carbon. The diameter of the glassy spheres ranges from 1 μm to 100 μm with a mean value of 7 μm. The angular carbon particles range from 10 to 300 μm in size. The pozzolanic reaction involves the silica and alumina compounds as well as the free lime. A number of variables control the extent and rate of reaction between soil–lime–fly-ash. They are: (1) quantity of free lime or cement, (2) amount of silica (SiO$_2$) or alumina (Al$_2$O$_3$), (3) adequate water, (4) compacted density, (5) presence of carbon in the fly-ash, (6) fineness of fly-ash, (7) temperature, and (8) age of mixture. Extent and rate of pozzolanic reaction is directly proportional to all but two of these factors, which are numbers (3) and (5) (Meyers et al., 1976). Pozzolanic reactivity of fly-ashes can be tested in accordance with ASTM C593.

Lime-to-fly-ash ratios of 1:2 to 1:7 have been tested and accepted. However, ratios of 1:3 and 1:4 are used typically for purposes of economy and quality. Typical proportions of lime and fly-ash range from 2.5 to 4 percent for lime, and 10 to 15 percent for fly-ash. Laboratory testing of lime–fly-ash (LFA) (or lime–cement–fly-ash (LCFA)) and soil mixtures are conducted on trial specimens cured 7 days at 100°F. Laboratory testing includes moisture–density relationship, compressive strength, and durability. Mixtures cured for 7 days at 100°F normally develop compressive strength that ranges from 500 to 1000 psi. A class C fly-ash produced in Wyoming (Powder River basin coals) contains 30 percent lime, which makes it a good hydraulic binder exhibiting 7-day cured strength of up to 4500 psi. The ratio of flexural to compressive strength of most LFA mixtures ranges between 0.18 and 0.25. Table 9.15 shows the commonly used strength and durability criteria for LFA–soil mixtures as specified by various agencies (Meyers et al., 1976). A comprehensive study on strength development and shrinkage of cement–fly-ash (CFA), cement–lime–fly-ash (CLFA) and lime–fly-ash (LFA) showed that shrinkage of these mixtures is less than for soil–cement mixtures (Natt and Joshi, 1984). Shrinkage was observed to be greater in LFA mixtures than CFA mixtures. All of the mixtures showed strength regain after repeated cycles of freezing and thawing followed by a curing period. A study by Pamukcu et al. (1987) showed that increased addition of LFA to kaolinite clay beyond 10 percent

TABLE 9.15 STRENGTH AND DURABILITY CRITERIA FOR LFA–SOIL MIXTURES[a].

Agency	Test	Criteria
	(a) Strength	
ASTM	ASTM C 593 Unconfined compression test 7-day cure at 100°F (38°C)	Min. 400 psi (2760 kPa)
British Road Research Laboratory	Unconfined compression test 28-day cure	Min. 25 psi (1720 kPa), except 400–500 psi (2760–3450 kPa) for clay soils and severe climatic conditions
	California Bearing Ratio	80% immediately beneath surface, and decreasing with depth
	(b) Durability[b]	
ASTM	ASTM C 593 Vacuum saturation method[c]	Min. 400 psi (2760 kPa)
Portland Cement Association	AASHTO T135-70 and T136-70 wet–dry and freeze–thaw brushing tests	7–14% allowable weight loss, exact value dependent upon soil grain size
British Road Research Laboratory	Durability ratio (ratio of weathered strength to unweathered strength)	Min. 80%
	Iowa freeze–thaw test; index of resistance (ratio of weathered strength to unweathered strength)	Min. 80%

[a] After Meyers et al. (1976).
[b] Applicable in regions where climatic conditions are a factor in pavement performance.
[c] Approved revision; replaces freeze–thaw brushing test.

by weight increased the sensitivity (undrained/remolded strength) of the mixture on the wet side of the optimum water content and decreased it on the dry side. This indicates that fly-ash particles may be electrically charged and influence soil structure and thus mechanical behavior of the mixtures.

The reader is referred to a publication by Meyers et al. (1976) for detailed mixing design procedures of LFA and LCFA. Different mixing design procedures are utilized in countries that produce and use fly-ash. As an example, gravels and sands are treated with 8 to 12 percent damp fly-ash and 2 to 3 percent lime in construction of road bases in France. Again, in northern France road bases of heavy traffic are generally constructed with 91 percent fly-ash, 4 percent lime, and 5 percent phosphogypsum. In such cases, caution must be exercised to avoid swelling problems. Fly-ash, similarly to cement, is not recommended for use with clayey soils alone. This is because the rapid cementing action does not allow time for intimate mixing, thus resulting in an inhomogeneous system. Retarders such as commercial-grade borax, used at 0.45 percent by weight of fly-ash, or a mixture of powdered borax ($\frac{2}{3}$) and boric acid ($\frac{1}{3}$) have been found to work well, especially with fast-setting class C type ashes. Examples of usage of class C type fly-ash in stabilization of various types of soils are stabilization of base rock materials; structural fill and backfill; fine dune sands; and expansive clays later used as a backfill (Lamb, 1985).

Class F fly-ashes require addition of lime for setting. A number of setting agents other than lime have been suggested and used with fly-ash in different countries. These are ash or clinker from iron works and steelworks, sludges from papermills, sulfocalcic fly-ash, alkaline silicates, and natural or waste sulfates (phosphogypsum). In the case of phosphogypsum, mixtures of fly-ash–phosphogypsum and portland cement–phosphogypsum have shown potential as base stabilizers (Gregory et al., 1984). Strengths of mixtures containing less-acidic (pH around 5) phosphogypsum were significantly higher than those containing highly acidic phosphogypsum (pH < 3). Waste calcium sulfate has also been tested as an additive in LFA mixtures in roadway base construction (Usmen and Moulton, 1984). The results of a large-scale field test and laboratory evaluation of strength and durability indicates that large percentages of calcium sulfate can be used successfully in pozzolanic stabilization of base or subbase courses. Another potential composition to be used in soil stabilization is fly-ash and coal refuse material (Head et al., 1982), in which coal refuse is a mixture of earthern materials discarded from coal processing. In a study of characterization of these materials and various mixtures of them, it was concluded that coal refuse and fly-ash mixtures may be beneficial in stabilization when used with one or more cementing agents, such as cement, lime, or emulsified asphalt.

Studies have been conducted to investigate use of FDG (flue gas desulfurization) waste sludge (slurried fly-ash) in soil stabilization also. The American Coal Association is a good source of information on reuse of coal ashes and also information on further advancement in use of fly-ash and its mixtures in soil stabilization.

9.8.2 Bottom Ash, Boiler Slag, and Blast-Furnace Slag

Bottom ash is the solidified ash in the blast furnace. It is porous and brittle and has pozzolanic properties. Bottom ash, when crushed, can be used as an additive to cement or lime. It has been shown to have high water adsorption capacity (1 to 25 percent). Typical values for maximum dry density and optimum moisture content range from 75 to 115 lb/ft^3 and

15 to 30 percent, respectively. Bottom ash has been used in mixture with fly-ash in structural backfills (Leonards and Bailey, 1982).

Boiler slag, originally in molten state, quenches and forms a glassy texture. When crushed, it too can be used as a pozzolanic additive in cements. Typical ranges of maximum dry density and optimum moisture content range from 90 to 100 lb/ft^3 and 14 to 22 percent, respectively (Usmen et al., 1982). Use of bottom ash and boiler slag as a soil-stabilizing additive has not been exploited fully. However, experience in the United States has shown good performance of a LFA–slag roadway base in 5 years of service (Barenberg and Thompson, 1982).

Sudden cooling of blast furnace slag (tailings, flux, and coke-ash) through granulation or pelletization imparts hydraulic properties to the material. In presence of water and an activating agent, such as lime, certain constituents of the slag pass into solution and form insoluble compounds. Blast-furnace slag may be cooled slowly, resulting in a crystalline rock that has no cementatious properties. The granulated form is used to treat coarse-grained soils (namely, sands and gravels) in typical proportions of 10 to 20 percent slag mixed with 1 percent lime. This technique was originally developed in France and has been widely used in Europe. Studies have been conducted in the United States for potential use of blast-furnace slag in stabilized base construction (Emery, 1980).

9.9 BITUMINOUS STABILIZATION

Bituminous stabilization refers to those methods in which bituminous materials are incorporated into a soil or soil–aggregate mixture to construct base courses, and occasionally surface courses, capable of carrying traffic loads under normal conditions of moisture and traffic. Cohesive soils usually have satisfactory bearing capacity at low moisture contents. The purpose of incorporating bitumen into such soils is to water-proof them as a means of maintaining them at low moisture contents and thus high bearing capacities. In the case of noncohesive granular soil, bitumen serves as a bonding or cementing agent. Depending on the granulometric composition and the physical properties of the soil, there are four types of bitumen-stabilized products, as follows.

(1) *Soil bitumen.* Waterproofed cohesive soil. Best results are obtained with soils with grain sizes such that (a) maximum size is not greater than approximately one-third of the compacted thickness or the same as the thickness of a compacted lift if the latter is a fraction of the thickness of the base course; (b) 50 percent passes No. 4 sieve; 35 to 100 percent passes No. 40 sieve; 10 to 50 percent passes No. 200 sieve; (c) the liquid limit does not exceed 40 percent and the plasticity does not exceed 18 percent. Bitumen requirements generally range from 4 to 7 percent of dry weight of the soil.

(2) *Sand bitumen.* Loose sand particles cemented with bitumen. The sand can be beach, river, or wind-blown sand. Existing roadway material, substantially free from organic matter, lumps, or adherent fills of clay are also included in this category. The sand may require admixture of filler material to meet mechanical stability requirements. The natural sand or sand–filler mixture should normally contain less than 12 percent of passing No. 200 sieve fraction. However, in the case of wind-blown or dune sands, up to 25 percent of passing No. 200 fraction may be allowed. This is provided that the portion of sand or sand–filler mixture passing No. 40 sieve has a failed moisture equivalent less than 20 percent and linear shrinkage less than 5 percent. The amount of bituminous binder ranges from 4 to 10 percent. The optimum amount of binder should be determined by density, strength, and water-resistance tests on trial specimens. It should not exceed the pore space of the compacted mineral mix.

(3) *Waterproofed granular stabilization.* A good gradation of soil particles, from coarse to fine and having a high potential density, is waterproofed by addition of 1 to 2 percent bitumen. For recommended gradations of the soil aggregate materials, see Table 9.16 and Figure 9.4.

(4) *Oiled earth.* A soil surface consisting of silt–clay material is stabilized by spraying slow- or medium-curing bitumen cutbacks or emulsions. The coating is made by two or three applications of the oil at a rate of 1 gallon per square yard of soil surface (Winterkorn, 1934). The main purpose of oiled earth is to produce a water- and abrasion-resistant surface.

TABLE 9.16 GRADATION OF SOIL AGGREGATE MATERIALS FOR BITUMINOUS STABILIZATION.

Sieve Analysis	Soil–Bitumen,[a] %	Sand–Bitumen, %	Waterproofed Granular Stabilization, %		
			A	B	C
Passing;					
1½ inch			100		
1 inch	b		80–100	100	
¾ inch			65–85	80–100	100
No. 4	>50	100	40–65	50–75	80–100
No. 10			25–50	40–60	60–80
No. 40	35–100		15–30	20–35	30–50
No. 100			10–20	13–23	20–35
No. 200	10–50	<12; <25[c,d]	8–12	10–16	13–30
Characteristics of Fraction Passing No. 40 Sieve					
Liquid limit	<40				
Plasticity index	<18		<10; <15	<10; <15	<10; <15[e]
Field moisture equivalent		<20[c]			
Linear shrinkage		<5[c]			

[a] Proper or general.
[b] Maximum size not larger than ⅓ of layer thickness; if compacted in several layers, not larger than thickness of one layer.
[c] Lower values for wide and higher values for narrow gradation band of sand. If more than 12 percent passes, restrictions are placed as indicated on field moisture equivalent and linear shrinkage.
[d] A certain percentage of −200 or filler material is indirectly required to pass supplementary stability tests.
[e] Values between 10 and 15 are permitted in certain cases.

Bituminous Materials

Bituminous substances consist almost entirely of carbon and hydrogen with very little oxygen, nitrogen, and sulfur. They vary in consistency from that of a light crude oil to the solid nature of an asphaltite. *Bitumen* is nonaqueous elastic or solid hydrocarbon either naturally occurring (asphaltic) or obtained from destructive distillation (pyrogenic) of organic substances such as tars and pitches. It is soluble in carbon disulfide and not soluble in water, diluted acids or alcohol. *Liquid bitumen* is a combination of bitumen with lightweight volatile solvents, petroleum distillates (gasoline, kerosene, diesel oil), or black coal tar oil. *Tar* is a liquid or semisolid hydrocarbon produced by dry distillation of coal, oil, lignite, or wood.

Most widely used in the United States are the asphaltic bituminous materials, which are obtained from the refining of asphaltic-base crude oils. They are produced within various consistency ranges, measured by standard viscosity and/or penetration tests. These materials are called *asphaltic cements* (AC). The AC materials are semisolid and must be heated before use. For working at lower temperatures, they are diluted with solvents (cutbacks) or emulsified in water. The higher the proportion of solvent, the lower its viscosity, and the greater its volatility, then the shorter is the curing time or the time during which the solvent escapes. The thinned bitumen that enters the soil pore space uniformly coats and cements together the soil particles when the solvent evaporates. Depending on the rate of evaporation of the solvent, the AC materials are divided into three categories: (1) rapid curing (RC)—naphtha; (2) medium curing (MC)—kerosene; (3) slow curing (SC)—heavy diesel oil. These are further divided into grades with respect to viscosity (ASTM D2026). In the case of emulsions, the time required for their setting or breaking determines the curing grade. The typing and grading of liquid tar products is similar to that of asphalt cutbacks and emulsions. Evaluation of 10-year performance of an MC-asphalt cement stabilized sand road in Minnesota gave good results with respect to servicability index, deformations, cracking, and need for maintenance (Skok et al., 1983). Foamed asphalt, formed by mixing asphalt cement and cold water through a foam nozzle, has also been used successfully in construction of low-volume roads (Acott and Myburgh, 1983; Castedo and Wood, 1983).

9.9.1 Classification of Soil–Bitumen Types

A system of scientifically discernible possible types of soil–bitumen is given below. Most of these systems are currently being used in practice.

1. Systems with granular skeleton as main mechanical strength factor, either open (without the −No. 200 material) or impervious (with the −No. 200 material)
   (a) Bearing skeleton: coarse aggregate with or without sand
      (i) Graded course and fine aggregate, the pores of which are filled with silt–clay–bitumen. Example: waterproofed granular soil stabilization.
      (ii) Bitumen-cemented river and bank gravels with or without small proportions of silt–clay. Example: base construction in West Germany or open type American road mix.
      (iii) Bitumen-cemented coarse aggregate (gravel or crushed stone) of narrow size range. Example: construction of layered drains for roadbeds.
   (b) Bearing skeleton: sand
      (i) Sand–bitumen with pore-filling silt–clay or similar −No. 200 material; utilized in Florida and California.

      (ii) Bitumen- (or bitumen resin)-cemented sand without pore-filling fines. Example: military beach stabilization.
2. Systems without gravel or sand skeleton: waterproofed silt–clay materials
   (a) Stabilization of natural material with little or no cohesion, is predominantly silty and has, at most, 20 percent of clay. An example is loess soil stabilization, which renders the material resistant to swelling and shrinkage even after long exposures of wet and dry cycles.
   (b) Stabilization of soils with predominant clay fraction, which occurs in secondary aggregations or crumbs even after comminution and mixing with the bitumen. Examples of such cases are heavy clay soils. It is recommended that such soils be pretreated with 2 percent or more of hydrated lime.

9.9.2 The Mechanical and Physicochemical Aspects of Bituminous Soil Stabilization

The quality of a soil–bitumen system depends upon the quantities and characteristics of its components. These are basically soil, bitumen, and water. The objective is to introduce a sufficient amount of bituminous material to form a thin coating over the surfaces of particles without reducing the intergranular friction and yet allow for adhesion and thus cementation between the particles. In systems without granular skeleton, the amount of bitumen required for satisfactory waterproofing will, in general, increase with increasing clay and colloidal content. However, soils with similar clay contents may require different amounts of bitumen depending on the water affinity of the soil fines. The optimum amounts of bitumen and water should be determined by laboratory testing. Mixing on the wet side of the optimum, though more troublesome, produces better soil–bitumen systems after compaction and drying than does mixing on the dry side.

As in the case of soil–cement and other methods, at the time of compaction, soil–bituminous mixtures containing appreciable amounts of silt–clay must have sufficient moisture to satisfy their water affinity (Winterkorn, 1967b). This value is usually in the vicinity of the optimum moisture content for Proctor compaction. The water plus the liquid bitumen admixture together make up the optimum liquid content for compaction. If this renders the system too sticky to be mixed properly with the available equipment, then pretreatment of soil with quicklime (CaO) or hydrated lime (Ca(OH)$_2$) is recommended. Clays in overly wet state should also undergo similar pretreatment prior to bituminous stabilization. Combinations of lime and bituminous stabilization are often beneficial. Lime pretreatment is also recommended for acidic soils (pH < 6). Soils with high pH and containing high levels of dissolved salts are often not suitable for bituminous stabilization.

As a result of a wide-ranging investigation of the physicochemical factors of importance in bituminous stabilization of silt–clay soils, the following conclusions were drawn (Winterkorn and Eckert, 1940).

(1) *Influence of type of clay.* The larger the ratio of $SiO_2/(Al_2O_3 + Fe_2O_3)$ in the clay minerals, the greater the bitumen requirements for satisfactory stabilization.

(2) *Influence of exchange cations on clays*

a. Bitumen requirements increase with increasing base exchange capacity of the clay and soil.

b. The higher the valence of the exchange cations on the clay (Na^+, Ca^{2+}, Al^{3+}, and others), the higher the efficiency of bituminous stabilization.

c. Among the monovalent cations, Na^+ exerts a detrimental effect consistently, while K^+ may be beneficial in case of illitic clays. It should be noted here that essentially the same effects of clay mineral composition and type of predominant exchange cations hold true in resinous soil stabilization (Winterkorn and Reich, 1962).

(3) Influence of soil organic matter. The effects of soil organic constituents depend on the conditions under which they were formed and accumulated. Acid organic matter from leached forest and river-bottom spoils appears to be detrimental, while neutral and basic organic matter from arid and semiarid regions does not seem to have a detrimental effect, other than that produced at times by the presence of monovalent cations.

(4) Influence of bitumen type and composition. Among the materials tested, the waterproofing effectiveness on silt–clay materials decreased in the following order: normal (high temperature) coal tars > cutbacks from cracked asphalts > cutbacks from noncracked asphalts > low-temperature tars. Separating the asphaltic materials into their group components—asphaltenes, asphaltic resins, and oils—showed a strong correlation of increase in stabilizing power with increasing content in asphaltenes (Winterkorn and Eckert, 1940).

Use of secondary additives in bituminous stabilization can be beneficial if the appropriate additive can be selected for a given soil type. Use of lime as a pretreatment of wet clay soils prior to application of bituminous material was discussed earlier. Additives are used to enhance the adhesion of the bitumen to the soil particles. Phosphorus pentoxide (P_2O_5) (Michaels and Puzinauskas, 1956, 1958) and condensation products of aromatic amines and furfural (Winterkorn, 1938) have shown good results as admixtures. The reader is cautioned with regard to substances whose beneficial effects may be short-term. For a number of soils, admixture of soaps has shown good initial results. However, these are subject to bacterial and fungal decomposition and therefore must be accompanied by bactericidal substances when used as an admixture in soil stabilization (Jones, 1948).

Increasing bitumen content decreases the maximum dry density and increases the optimum moisture content. The strength increase is observed up to a certain percent of bitumen addition and decreases from there on, even falling below the strength of unstabilized soil. For sandy clay the maximum strength is often observed at around 4 percent of bitumen addition. The water absorbtion may also show an increase at lower percentages of bitumen, but will definitely decrease after a threshold value (i.e., 4 percent). Strength generally increases with mixing time. Extended mixing increases the total surface of particles coated with bitumen.

9.9.3 Laboratory Testing of Bitumen-Stabilized Soils

In the absence of a universally accepted method of preparing, curing and testing bitumen-stabilized soil and soil–aggregate specimens, a number of procedures have been developed and are in present use. The more significant ones have been developed from methodologies already established for the testing of soils (CBR, triaxial compression) and for traditional bituminous paving mixtures (Hubbard–Fields, Hveem stabilometer, Marshall method). For fine-grained soils, the convenient and material-saving 2-in × 2-in cylindrical specimens of the Dietert device, originally developed for foundry sand and introduced to soil stabilization by the PCA for testing for comprehensive strength of soil–cement, are used. The general factors involved in the design of bituminous mixtures are stability, durability, flexibility, fatigue resistance, skid resistance, permeability, fracture or tensile strength, workability, and thermal properties. The Asphalt Institute (1982) *Soils Manual* ES-11 lists a number of ASTM and AASHTO specifications for bituminous test methods. The most widely used methods for bituminous mix design are the Marshall method and the Hveem stabilometer method, which are also given in the Asphalt Institute Manual.

The choice of a particular method of testing, including the size of the test specimen, should depend on the largest particle size in the mixture, in addition to convenience and dependability. A direct connection between test results and pavement design, as in the CBR, triaxial compression, and Hveem stabilometer tests is a very desirable feature. All tests methods for bituminous stabilization, as well as other pertinent types of stabilization, have in common the following features.

1. Determining gradation of whole soil sample and Atterberg limits on the − No. 40 sieve fraction.
2. Determining density–moisture relation of the whole sample for a selected standardized compaction method.
3. Selecting from previous experience or theory the type and indicated percentage of bitumen required, and determining the density–moisture relation for the soil–water–bitumen mixture. It is a beneficial practice to record the ease or difficulty of mixing and the homogeneity of the resultant mixture during this test.
4. Designing mix and making sets of test specimens at three different bitumen contents and optimum water contents found in (3) above.
5. Curing the specimens for 7 days in accordance with conditions encountered on the job (no water loss, partial water loss, air drying) and subjecting them to the temperature and moisture conditions simulating severe cases of local conditions (capillary water absorbtion, water immersion, wet–dry and freeze–thaw cycles).
6. Mechanical testing of exposed specimens by a selected method.

For testing the susceptibility of fine-grained soils or of the mortar fraction of coarse-grained soils to stabilization, use of a Dietert-size specimen can be advantageous. Its volume (6.28 in^3 or 103 cm^3) is less than one-ninth that of the Proctor specimen, which allows the making of a relatively large number of specimens. For each composition to be tested, it is recommended that 14 specimens be made, numbered, and tested in unconfined compression in accordance with the following scheme:

Nos. 1 and 14	Tested immediately after molding without loss of water and other volatiles
Nos. 2 and 13	Tested after air drying for 7 days
Nos. 3 and 12	Cured for 7 days under appropriately selected moisture and temperature conditions, and then immersed in water for 7 days
Nos. 4 and 11	Subjected to one cycle of wetting and drying
Nos. 5 and 10	Subjected to 4 cycles of wetting and drying
Nos. 6 and 9	Subjected to 1 cycle of freezing and thawing
Nos. 7 and 8	Subjected to 4 cycles of freezing and thawing

After the cycle exposures, the specimens are immersed in water for 24 hours, their condition is noted, dimensions are checked, and weight is determined. Then these samples, along with the 7-day immersion specimens are tested for unconfined compressive strength. In many cases the strength after 7 days' immersion alone is a sufficiently good indicator and it may not be necessary to conduct the cycle tests.

This testing scheme can be applied to all types of admixtures employed for the stabilization of fine-grained soils for base-course construction. Standard test methods and specifications have been developed by AASHTO, ASTM, RRL, and other agencies, and should be checked in their most recent publications.

9.10 THERMAL AND ELECTROKINETIC STABILIZATION

9.10.1 Thermal Stabilization

Heating and freezing have been used for soil-improvement purposes. Thermal stabilization involves the resolution of following issues: thermal evaluation of heat flow, design of heating or refrigeration systems, performance analysis depending on the strength and stress–strain–time properties of the stabilized soil. The ultimate compressive strength of three soils as a function of temperature is given in Figure 9.22.

Heat-flow analysis can be done similarly to seepage or consolidation. Heat transfer in soils occurs by conduction, convection (free, forced, by thawing), and radiation. The predominant mechanism is conduction, which takes place in three constituents of soil, namely, soil solids, water (liquid, ice or vapor), and pore air. Heat conduction is influenced by thermal properties of soil, mainly thermal conductivity and latent heat of fusion and heat of vaporization of soil water, and heat capacity of soil constituents.

Heat flow behavior may be dominated by the latent heat of fusion of water on freezing and the heat of vaporization of water on heating above 100°C (Mitchell, 1981). Latent heat of fusion is heat that must be added to unit mass of a substance to change it from liquid to solid or solid to liquid without changing temperature. Heat of vaporization is the heat needed to change the substance from liquid to vapor. The latent heat of fusion, L, of water to ice is 333×10^3 J/kg (143.3 BTU/lb), and the heat of vaporization, V, of water is 2260×10^3 J/kg. Freezing or thawing latent heat for soil water, L_s, and heat of

vaporization of soil water, V_s, are given by:

$$L_s = L\left(\frac{w}{100}\right)\gamma_d \quad (\text{J/m}^3) \qquad (9.6)$$

$$V_s = V\left(\frac{w}{100}\right)\gamma_d \quad (\text{J/m}^3) \qquad (9.7)$$

respectively, where w is the water content and γ_d is the dry density (kg/m^3). The heat capacities of ice, water, and soil minerals are given as $C_i = 2098$ (J/kg)/K, $C_w = 4286$ (J/kg)/K, $C_m = 710$ (J/kg)/K, respectively. The heat capacities of unfrozen (C_u) and frozen (C_f) soils are then:

$$C_u = \gamma_d\left(C_m + \frac{C_w w}{100}\right) \qquad (9.8)$$

$$C_f = \gamma_d\left(C_m + \frac{C_i w}{100}\right) \qquad (9.9)$$

The thermal conductivity of soils can be evaluated by a number of methods, some of which are Kernsten's empirical equations (Kernsten, 1949), Mickley's method (1951), Gemant's method (1952), De Vries' method (1952), the empirical equations of Van Rooyen and Winterkorn (1959), and Johansen's method (1975). Detailed discussions of these methods can be found in Farouki (1986). According to Farouki, who conducted a comparative analysis of various methods, Johansen's method seems to best describe measured data. The thermal conductivity of most soils ranges between 0.35 and 3.5 (W/m)/K. Thermal conductivity charts for various soils can also be found in Andersland and Anderson (1978).

9.10.2 Heating

Heating causes permanent changes in soil properties, often rendering the material hard and durable. Treatment is reported to result in significant decrease in compressibility and increase in cohesion, internal friction angle, and modulus of elasticity (Litvinov, 1960; Kujala et al., 1985). Temperatures needed to produce such effects usually range from 300°C to over 1000°C. Soils melt at 1250 to 1750°C. The melting point can be lowered by addition of fluxing agents such as Na_2CO_3. Thermal treatment increases the strength and reduces the water content, compressibility, and swelling potential. The amount of fuel or energy needed to obtain the high temperatures can be costly. Ingles and Metcalf (1973) proposed an equation for estimated cost of thermal stabilization by heating:

$$F = 100\,\frac{(6.4w\gamma_d + 0.25T\gamma_d]}{EC_f} \qquad (9.10)$$

in which w = moisture content (%), γ_d = soil dry density, T = burning temperature, C_f = unit heat capacity of fuel (35 percent open burning, 70 percent closed burning), and F = fuel use per unit volume of soil.

Stabilization by heating has mostly been applied in the U.S.S.R. and Eastern Europe. It has been used successfully to stabilize landslides, improve collapsing soils under existing structures, construct mat foundation, form vitrified piles in place, and reduce lateral stresses on retaining walls and embedded structures. Methods range from combustion to electrical, and investigations of microwave drying and soil fusion by laser beam are being conducted.

In heat treatment, good permeability of the material is desirable to allow the escape of water vapor and injection of stabilizing agent if used. Typical field methods are discussed by Mitchell (1981). Table 9.17 gives the minimum heat requirements

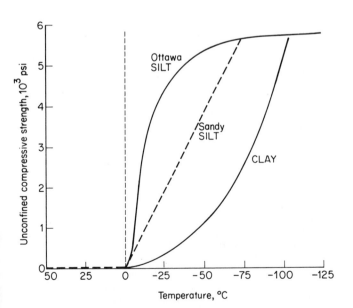

Fig. 9.22 Compressive strength of three soils as a function of temperature. (*After Sayles, 1966.*)

TABLE 9.17 MINIMUM HEAT REQUIREMENT FOR DIFFERENT APPLICATIONS OF HEATING STABILIZATION.

Purpose of Heating	Minimum Treatment Temperature, °C
Reduction of lateral pressure	300–500
Elimination of collapse properties (loess)	350–400
Control of frost heave	500
Massive column construction below frost depth	600
Manufacture of building materials	900–1000

for different applications of the treatment. In the United States, vitrification has been developed for encapsulation of hazardous waste sites (Timmerman, 1984). This is the conversion of contaminated soil into a durable and crystalline monolith through melting by electric heat. The method has been used for immobilization of nuclear waste in the ground.

9.10.3 Freezing

Ground freezing in construction is a relatively old method of stabilization. The first recorded application was in Wales in 1892 to support an excavation. Although an expensive process, artificial ground freezing has been shown to be beneficial and versatile in temporary ground support in excavations or tunneling (Jones, 1982; Sadovsky and Dorman, 1982) or in control of groundwater, especially in soft soils. The freezing of pore water acts as a cementing agent between the particles, causing significant increase in shear resistance and permeability. The method is carried out by injecting a coolant through freezing pipes placed in the ground. The coolant can be liquid nitrogen (boiling point $-196°C$) or brine. When brine is used, longer duration of application may be needed (3–6 months) to adequately freeze the ground. Some of the current applications of ground freezing are temporary underpinning, temporary support for excavations, prevention of groundwater flow into excavated areas, temporary slope stabilization, and temporary containment of toxic/hazardous waste contamination.

Temperature and freezing pressure are two important design parameters in artificial freezing. For a given temperature, the free moisture content of the soil depends on the specific surface of the soil particles. Below $0°C$ the percentage of free water drops quickly over a few degrees, resulting in considerable increase in strength and stiffness of the soil. Pressure due to increased volume of water during freezing develops if moisture content is sufficient and the soil is prevented from expanding. This pressure is much lower in sands than it is in clays and silts. The rate of expansion is maximum in silts. When the mean effective stress is greater than the freezing pressure, then there is no expansion of the frozen soil. Expansion is an important issue in freezing stabilization since it may involve the stability of nearby structures.

The technique of freezing requires study and determination of a number of critical parameters to ensure success and safety. These are the freezing parameters such as energy, freezing time, accurate positioning of freezing elements; groundwater flow and quality; potential ground movements accompanying freezing; and long-term strength and stress–strain properties of the frozen ground (Schuster, 1972). Two basic types of freezing systems are available: (1) expendable refrigerant systems using liquid nitrogen or carbon dioxide, in which the refrigerant is lost to the air; (2) primary plant with circulating coolant systems. The first is fast (hours), but costly and also difficult to control in the field. The latter requires a longer time (days or weeks), but is more economical and easier to control. Details of techniques of ground freezing are available in a number of publications (Schuster, 1972; Braun et al., 1978; Sanger and Sayles, 1979; and Mitchell, 1981).

Other design concerns in ground freezing are consolidation due to thawing of ice lenses that form with 9 percent volume expansion of water, and large creep strength losses over time and creep rupture. Thawing consolidation may be a potential problem in silts and clays owing to the high pore water pressures generated in these soils during melting. Deformation of frozen soil is viscoplastic and it is temperature- as well as stress-dependent. Results of a number of studies conducted to evaluate creep behavior of frozen soils are available in the literature (ISGF, 1985, 1988; and Chapter 19).

9.10.4 Electrokinetic Methods in Soil Stabilization

Electroosmosis, an electrokinetic phenomenon, has been used and can be used in the following methods of ground improvement:

1. Consolidation
2. Dewatering of dredged soil
3. Injection of grouts
4. Decontamination of soil and groundwater

Electroosmosis has been applied successfully over a number of years as a ground-improvement method to dewater, consolidate soft soil deposits, and desalinate arable land (Casagrande, 1949). Gray and Mitchell (1967) studied the fundamental aspects of electroosmosis in three different types of clay (kaolinite, illitic clay, silty clay).

An electrokinetic process occurs when two phases move with respect to each other while the interface is the location of an electric double layer. In this motion, for example, between a solid and a liquid phase, a thin layer of liquid adheres to the solid surface, and the shearing plane between liquid and solid is located in the liquid at some distance from the solid surface. Part of the counterion atmosphere therefore stays with the solid, and part moves with the liquid. The electric double-layer potential at this shearing plane is called the electrokinetic potential or zeta-potential. This potential was shown in a formula by Helmholtz in 1879 as a function of the dielectric constant of the liquid medium, the temperature, the thickness of the electric double layer, and the electric charge per unit area of particles. The electroosmotic flow rate is directly proportional to zeta-potential according to Helmholtz-Smoluchowski theory developed in 1914. The zeta-potential has been shown to be directly related to the specific surface of the solid. In clays, total wetted particle surface is very large. Therefore, substantial electroosmosis should occur in clay soils.

Electroosmotic flow of water occurs when an electric potential is applied across two electrodes (cathode and anode) placed in soil. This flow is also explained as a secondary effect of moving ions (electrolysis) that drag layers of water molecules toward the electrodes. If the pores of the soil are relatively large, the galvanic effect causing electrolysis becomes predominant, and the electric current fails to produce movement of fluid. Absence of surface conductance with respect to the conductance of the pore liquid also causes failure of electroosmotic effects. In clay soil systems, there is an excess of cations in the double layer because the clay particles tend to develop negative surface charge. Owing to the excess cations in the double layer, there is a net transport of water with the cations as they move toward the cathode. The greater the difference between the concentrations of cations and anions, the greater the net drag on the water in the direction of cathode during electroosmosis.

Although the process has been applied in the field to consolidate or dewater soft clay deposits in the past, there seem to be a number of fundamental factors that need to be studied for better understanding of the phenomena and a more generalized application in ground engineering (Olsen, 1984). Some of the important factors that have been identified as influencing electroosmosis are:

1. Electric potential
2. Current density
3. Solvation of ions
4. Change in the viscosity and dielectric constant of the pore fluid
5. pH of the system and pH gradients
6. Temperature of the system and temperature gradients
7. Surface conductance
8. Cation exchange capacity of the soil

An analytical derivation involving all these variables is desirable. However, the interaction between the actual processes are often so complex that a generalized equation is difficult to derive. The electric double layer theory, although lacking the desired generality, provides a simple approach. With more systematic information on the influence of combined factors, the theory can be modified to describe the phenomena more closely.

Large-Scale Application

Electroosmosis has been successfully applied in large-scale projects to improve geotechnical properties of fine-grained soils by dewatering and consolidation (Casagrande, 1959; Gray and Mitchell, 1967; Esrig, 1968; Mitchell, 1970; Chappel and Burton, 1975; Vitayasupakorn, 1986). In considering field applications of electroosmosis, there can be several complicating factors that need to be evaluated. Electrolysis of water, desiccation of soil due to heat generation at the electrodes, mineral precipitation and decomposition, and pH gradients can impair the efficiency of electroosmotic water transport. These effects can be controlled by applying an effective voltage so that most of the electrical energy applied is consumed in moving water. The alkaline environment created by electrolysis of water can, however, be beneficial in attempts at groundwater decontamination. Segall et al. (1980) reported that significant release of organics from clay was observed due to dissolution under highly alkaline conditions.

Electroosmosis is a method that works in low-permeability soils to mobilize water efficiently and rapidly. The coefficient of electroosmotic permeability of typical clays is approximately 100 times larger than their coefficient of hydraulic permeability (Mitchell, 1976). The effectiveness of electroosmosis in drawing chemicals from anode to cathode has been applied to grouting fine-grained soils when chemical grouts cannot be transported through the soils by hydraulic pressure. Use of electroosmosis in void-filling techniques with colloidal chemicals or gels, for example, silicates, bentonite and aluminum hydroxide, has been reported (Mitchell, 1981). The method is also potentially applicable to separation and movement of both adsorbed and nonadsorbed contaminants from clay surfaces (Mitchell, 1986). Evaluations of past experience suggest that if the method can be applied effectively, it should offer significant economy over the other in-situ decontamination techniques. However, there is not sufficient data, nor is there complete enough understanding of the conditions under which the method could be used. This is because the physical and chemical aspects of electrokinetic

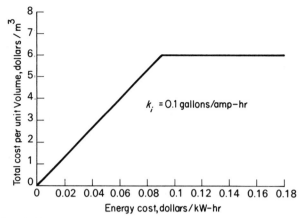

Fig. 9.23 Estimated variation of energy expenditure for a constant electroosmotic transport. (*Acar, 1988.*)

phenomena on different chemical species in the pore fluid of soils are not yet well understood. In addition, the adsorbtion of contaminants on soil particles and the conditions under which they are released are also poorly understood. However, investigations are being conducted to exploit the possibility of the use of electroosmosis in decontamination of polluted soils (Olsen, 1984; Ferguson and Nelson, 1986; Probstein and Renauld, 1986; Khan and Pamukcu, 1989; Khan et al., 1989; Acar et al., 1989).

Energy requirements for electroosmosis can be estimated from

$$\varepsilon = \frac{\Delta E}{k_i} \times 10^{-3} \qquad (9.11)$$

where

ε = energy required per gallon discharged, kilowatt-hours
ΔE = applied voltage, volts
k_i = electroosmotic transport, gallons/ampere-hour

Figure 9.23 (Acar, 1988) shows the estimated variation of energy expenditure for a constant electroosmotic transport of 0.1 gallons/ampere-hour. When the transport is increased 10-fold, the total cost is expected to increase by the same factor. Considering the upper limits, for electroosmotic transport of 0.1 to 1.0 gallons/ampere-hour, the energy cost to move one pore volume of electroosmotic water through 1 m³ of soil may range from $6 to $60. Taking an estimated value of $0.05 energy cost per kilowatt-hour, and application of three pore volumes of throughput for a decontamination process, for instance, the total cost would range from $9 to $90 per cubic meter of soil.

9.11 CONSTRUCTION

Construction of stabilized soil layers may be achieved with the help of power sources that range from simple agricultural tools to high-power single-pass stabilizers. The choice depends on size and location of the job, cost and quality of available labor, state of technical development, and other economic factors.

For methods in present use, excellent construction guides are found in the pertinent bulletins and other publications of the Highway Research Board, Portland Cement Association, Asphalt Institute, National Lime Association, Calcium Chloride Institute, and other agencies and trade organizations, and of the manufacturers of construction equipment. An up-to-date file of the latter and their products is available from the American Roadbuilders Association. While it would exceed the scope of this chapter to discuss the various methods in present construction use, it is important to understand the essence of the unit processes involved, namely, (1) comminution and/or homogenization, (2) mixing, (3) compaction, and (4) curing.

9.11.1 Comminution

After the soil surface is leveled to the desired grade (or the soil to be treated placed into one or more windrows after removal of oversized aggregate), the first step in the case of a cohesive soil is the comminution (also incorrectly called pulverization) of the larger clods. This is done to the extent that 100 percent passes the 1-inch screen and at least 85 percent the No. 4 sieve. This is most easily achieved if the soil has a moisture content that places it in the friable state. The state of optimum friability is slightly below the optimum moisture content for Proctor compaction.

Energy of Comminution The minimum theoretical energy requirement for a specified degree of comminution of a cohesive soil can be calculated from the increase in particle surface and, therefore, surface energy or from tensile strength data (Winterkorn and Schmid, 1971). Comparing such theoretical minimum energies with those actually used in construction reveals a tremendous waste of energy, particularly in the case of rotary comminution machines, compared with which a common plow is 50 times more effective. Among the reasons for the very low power efficiency of the rotary comminutor and the better but still low one for the plow are the following.

1. The high-speed rotary blades, required for the comminution of moist soils, are not as efficient as the slowly moving plow.
2. Considerable energy is wasted in the churning action of the rotary blades.
3. The rotary comminutor is designed for operation in both cohesive and noncohesive soils, not for maximum efficiency in a particular type of soil.

Other important facts are: the plow only cuts one surface, forming the others by bending the soil over sufficiently small radii to produce local tension and shear failures; the rotary blades cut numerous thin slices, losing energy in soil-to-metal friction and adhesion on each cut and on impact of the blades on the soil and on any pebbles or stones encountered; the relative thickness of the blades increases the friction and that of the cutting edges causes considerable compression in the soil before it is cut.

The resulting energy losses per cubic foot of comminuted soil are considerably larger for the fast-moving rotor-cutter than for the slow-moving plow, especially if the latter is designed for use in the particular soil type. Still, there is much room for improvement in both implements.

9.11.2 The Mixing Phase

Scientific approaches have been used to some extent in the design of mixers for portland cement concrete and for asphalt paving mixtures of predominantly coarse granular materials (sand, gravel, crushed stone) and liquids of relatively low viscosity (air, water, low viscosity bitumen). For such systems, optimum mixing times have been found, for which greater time leads to undesirable segregation. Soil-stabilizer systems of predominantly granular nature and comparable consistency can be mixed satisfactorily by available road construction equipment or even by some agricultural implements.

A real problem arises in the case of clay soils of high plasticity and at moisture contents considerably above the plastic limit, especially if the employed stabilizer is itself highly viscous or

forms highly viscous compounds with the soil-water present. Examples of such stabilizers are bituminous materials, resin monomers, polymer solutions and emulsions, and similar organic materials. When the soils are really heavy and moist or wet, the difficulties being to arise in the comminution phase; in order to facilitate comminution and start the stabilization, it is necessary to plow into the soil 3 percent or more of hydrated lime, letting this act for a day or two before final comminution and stabilization with materials that are compatible with the lime. Such materials are portland and other hydraulic cements, road oils, asphaltic cutbacks, and tars.

It should be remembered that the range of friability that is most suitable for both comminution and mixing lies within the plastic range of a cohesive soil and moves toward lower moisture contents with increasing compressive stress in a soil mass. Hence, comminution of a cohesive soil even at optimum moisture content should be either by gentle action, such as that of the plow, which permits the sward to break into natural structural units, or by rapid cutting of thin slices, in which the cutting proceeds at such a fast rate that the soil has no time for plastic deformation. This is attempted by the various types of rototillers. If a machine achieves this type of comminution, the problem of mixing with a stabilizer—even a bituminous one—is greatly facilitated by spraying the stabilizer liquid on the comminuted particles while they are in air suspension in the mixing compartment. This is the principle on which the design of single-pass stabilizers is based. Coal tars often give a less-plastic or "shorter" mixture with most clay soils than asphaltic bitumens. Also, combinations of tar, pitch, and aniline–furfural have made possible the stabilization of wet clay soils that could not be handled at all with Ca-acrylate or liquid bitumens alone.

With regard to comminution and mixing of soils and soil–stabilizer combinations, the following conclusions can be drawn:

1. The moisture content of the soil is the principal controlling factor in the comminution and workability of cohesive soils, soil–asphalt, and similar mixtures.
2. Dry soils may easily be watered to increase their moisture content prior to comminution; sometimes, in desert areas when the groundwater level is not too distant and time not critical, covering of the planed surface by impervious membranes will condense sufficient water in the top layer of the soil to bring it to optimum moisture content within a reasonable period of time.
3. Wet sand and gravel soils usually do not present major problems. Wet clay soils should in general be "dried" by plowing in of quick or hydrated lime for lime stabilization or for subsequent stabilization with asphalt or tar.
4. Effective machine design, with proper regard for cutting edges, tool shapes, and mounting arrangements, can save considerable power and improve the degree of comminution, while lessening its costs.
5. Operation at the lowest possible blade speed consistent with the properties and moisture content of the soil can minimize the power consumption of comminution.
6. The farm plow is still the most efficient comminution device for heavy clay soils.
7. The minimum amount of mixing necessary to obtain complete coverage of the soil aggregates with the stabilizer generally produces the most satisfactory stabilization.
8. Effective spray coverage of the soil aggregates with the stabilizer should reduce mixing time, if not eliminate the need for mixing altogether.
9. None of the commercially available stabilization or mixing machines is ideally suited to the comminution or mixing of cohesive soils, soil–asphalt, and similar mixtures.

10. For heavy clay soils, lime and combined lime–portland cement stabilization offer the least difficulties in comminution and mixing.

9.11.3 Compaction

Compaction or densification is a very important stabilizing procedure for natural soils and soil mixtures as well as for soil systems provided with admixtures of waterproofing and/or cementing agents. There is sufficient information on both the engineering and scientific aspects of compaction in the literature to render unnecessary a detailed treatment at this place. For an excellent discussion see Chapter 8.

The improvement of the quality of stabilized soils with increasing dry density for the same material composition is in accordance with:

1. The concept of the solid and liquid state of macromeritic systems (Winterkorn, 1971)
2. The law of Feret (1892) which relates the strength of a cemented granular system to the ratio of the volume of cement to the volume of the pore space, that is,

$$S = k\left(\frac{c}{1-s}\right)^n \qquad (9.12)$$

where

S = compressive strength of the cemented system
k = essentially the strength of the cement but is influenced by granulometric and material factors
c = absolute volume of the cement
s = absolute volume of the inert granular components
n = constant depending on material and geometric factors

3. The lower porosity, which reduces the amount of free pore water that may accumulate and make the system "quick" under sudden excess loading
4. The smaller effective pore size, which decreases the rate of moisture movement into the stabilized system under adverse environmental conditions, especially if the latter are of a periodic nature.

For normal soils it has been shown that the difference between Proctor and modified AASHO dry density at the same water content is worth about 2 percent of portland cement. However, one must be very careful that the system contains sufficient water to satisfy the needs of both the soil minerals and the portland cement or other stabilizers that require water as a reagent or as a solvent. Care must also be taken that the system is not overcompacted.

9.11.4 Curing

Where water is an important reaction agent—with lime, hydraulic cement, etc.—and the reactions are relatively slow, loss of water must be prevented during the reaction period. This is done most conveniently by spraying a film of bituminous material on the finished surface. This film must be thick enough to effectively prevent evaporation. If this is achieved, the water vapor usually supplied by the lower soil layers during the cooler nighttime is normally sufficient to continue good curing conditions even if some moisture is lost by diffusion at the edges of the stabilized base. Provided that a sufficient amount of moisture is thus retained in the system, the higher the environmental temperatures the greater the rates of reaction.

For the traditional stabilizing agents, construction and curing methods are described in PCA (1956), ARBA (1971),

NLA (1985), TRB (1987) and more recent publications of these organizations. These methods or modifications thereof are applicable in principle also for more recently developed stabilizing agents. A guide to the selection of an appropriate method of stabilization for any set of specific conditions was developed by the U.S. Naval Air Engineering Center (US NAEC, 1969).

9.12 GROUTING PRINCIPLES

Grouting is the injection of appropriate materials under pressure into rock or soil through drilled holes to change the physical characteristics of the formation. The results are sealing of voids, cracks, seams, and fissures in the existing rock or soil, and rendering of them less permeable and stronger. Grouting is often viewed as a versatile method of ground improvement for application in difficult soil and rock conditions. A large portion of information on grouting evolves from experiences in situ. Conferences at national and international level have been important sources of information and also milestones in development of grouting practice (British National Society of the International Society of Soil Mechanics and Foundation Engineering, 1963; Baker, 1982; Jessberger, 1982). This chapter will cover only soil grouting and will exclude the treatment of rock.

Soil grouting has been developed considerably over the past two decades. It has been used to alleviate difficult foundation problems, for remediation or maintenance of existing foundations and earth structures as well as for routine construction cases. Grouting proves especially effective in the following cases.

1. When the foundation has to be constructed below the groundwater table. The deeper the foundation, the longer the time needed for construction; therefore, there is more benefit gained from grouting as compared with dewatering.
2. When there is difficult access to the foundation level. This is very often the case in city work, in tunnel shafts, sewers, and subway construction.
3. When the geometric dimensions of the foundation are complicated and involve many boundaries and contact zones.
4. When the adjacent structure requires that the soil of the foundation strata should not be excavated (extension of existing foundations into deeper layers).

Grouting is a process in which grout in liquid form is pumped into the voids of the soil and then hardens. As a result, the soil is densified and its permeability is decreased or it is waterproofed. Some of the applications of grouting are (Mitchell, 1981):

1. Waterproofing a certain volume of ground below or around a structure in order to achieve permanent or temporary cutoff zone
2. Densification of foundation soils to increase shear strength and reduce compressibility
3. Void filling to prevent excessive settlement
4. Ground strengthening under existing structures to prevent movement during adjacent excavation or pile driving
5. Ground movement control during tunneling
6. Soil strengthening to reduce lateral support requirements
7. Soil strengthening to increase lateral load resistance of piles
8. Stabilization of loose sands against liquefaction
9. Foundation underpinning
10. Slope stabilization
11. Volume change control of expansive soils

9.13 GROUTING TECHNIQUES

Currently there are four types of grouting methods used in practice: compaction (displacement), chemical (permeation), slurry (intrusion), and jet (replacement) grouting (Welsh, 1986). Figure 9.24 illustrates these types of grouting techniques. Each one serves a different purpose and uses different equipment. Chemical and slurry grouting are low-pressure methods with $22.6 \, kN/m^2$ of pressure per meter depth, while jet and compaction grouting are high-pressure methods. Jet grouting uses grouting pressures up to $69\,000 \, kN/m^2$ per meter depth and compaction grouting uses $2700 \, kN/m^2$ pressure per meter. Each method can cause hydraulic fracturing in soils with permeability less than $10^{-1} \, cm/sec$. Methods such as acoustic emission can be used to detect these fractures (Koerner et al., 1985). Bruce (1988) summarizes these grouting techniques, and types and properties of various grouts and discusses some case studies.

Compaction Grouting was developed in the United States. It has been used in projects ranging from remedial type to densification of foundation soils prior to construction and to prevention of settlement in tunneling through soft ground. The control of grout mix consistency is essential for successful operations. A low slump (0 to 2 inch), very stiff soil–cement mortar is injected to displace and compact the soil. Too liquid a mortar would behave similar to slurry grout and fracture through the formation. Compaction grout constituents are silty sand (10 to 30 percent passing No. 200 sieve), cement (fly-ash), additives (fluidifiers, accelerators), and water. A typical application of compaction grouting was in the Bolton Hill subway of Baltimore, in which the settlement of the tubes was prevented during excavation (Baker et al., 1983). Figure 9.25 illustrates a typical section through the subway and approximate zones of grout injection. Another example of compaction grouting is densification of a liquefiable stratum under West Pinpolis Dam in South Carolina (Salley et al., 1987).

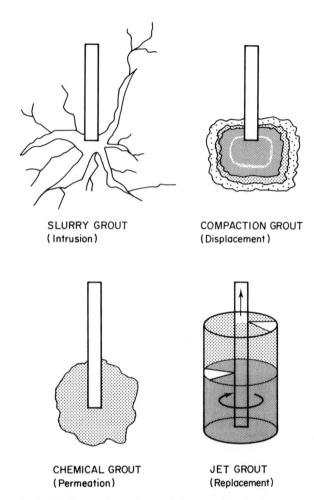

SLURRY GROUT
(Intrusion)

COMPACTION GROUT
(Displacement)

CHEMICAL GROUT
(Permeation)

JET GROUT
(Replacement)

Fig. 9.24 Types of grouting techniques. (*After Welsh, 1986.*)

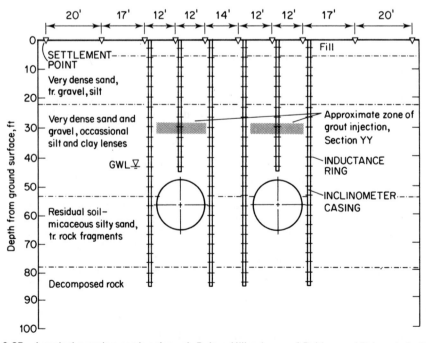

Fig. 9.25 A typical grouting section through Bolton Hill subway of Baltimore. (*Baker et al., 1983.*)

Jet Grouting was introduced in Japan in the early 1970s. The method excavates a cavity by the action of a horizontal high-pressure air/water jet, from a grout pipe that is rotated at a controlled rate. The cavity is then filled with the appropriate grout. The advantage of the method is that it can be applied to most types of soil. The primary use appears to be for underpinning and cutoff walls. The disadvantage of the method is that many variables need to be carefully controlled, namely, grout mix, jet nozzle diameter, jet nozzle pressure, grout flow rate, pipe withdrawal rate, pipe rotation rate. Materials used in this type of grouting are cement slurry, cement/sand mortars, chemical grouts, and cement/bentonite slurry. Figure 9.26 illustrates some of the potential applications of the method. Figure 9.27 shows a section of pile-underpining project using the jet grouting method in Japan (Yahiro et al., 1980). Figure 9.28 shows some options in jet grouting (Coomber, 1985). One of the newest methods of jet grouting was developed in Japan and is called the *Super Soil Stabilization Management*

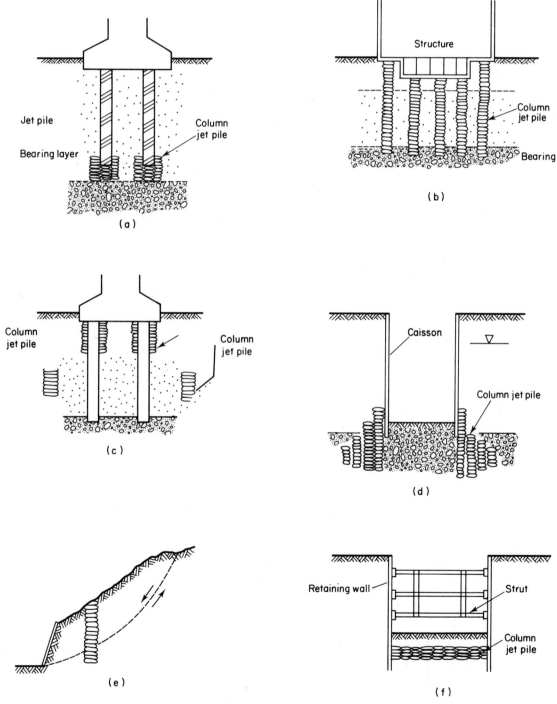

Fig. 9.26 Potential applications of jet-grouting. (a) Improvement of bearing layer for piers or existing piles. (b) Improvement of ground under the existing structure. (c) Improvement of resistance capacity of piers and existing piles against horizontal force. (d) Reduction of uplift. (e) Prevention of slope failure. (f) Soil improvement below excavated bottom. (*Yahiro et al., 1980.*)

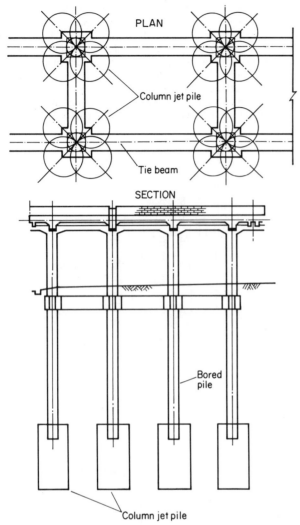

Fig. 9.27 A section of a pile underpinning project using jet grouting in Japan. (*After Yahiro et al., 1980.*)

(SSS-Man) system. Figure 9.29 shows the typical working procedure of the method, in which total and verifiable excavation of the soil prior to grouting is possible.

Chemical Grouting is the injection of properly formulated chemicals into sandy soils. The passing No. 200 fraction of the soil must be less than 20 percent. The resulting product is often sandstone-like material with unconfined compressive strength over $4100 \, kN/m^2$. Chemical grouts are often used for water control purposes owing to their favorable characteristics such as low viscosity and good control of set time. In addition, chemical grouts fill the voids in sandy soils, making them waterproof. Chemical grouts are generally based on sodium silicate solutions and inorganic reagents (e.g., sodium aluminate and sodium bicarbonate). Acrylates (AC-400) and sodium silicate (GEOLOC-4) are commonly used as grout materials. Others are polyurethanes and MS-silicates. Among the newer products is a chemical grout named *Silascol* composed of activated silica liquor and an inorganic reagent based on calcium. It forms stable calcium hydrosilicates with a crystal structure similar to cement. Medium and fine sands have been treated effectively in Italy using this product (Tornaghi et al., 1988). The product may also provide higher safety against pollution by dissolution, as might be the case in other chemical grouts. Figure 9.30 shows soil particle sizes for different chemical grouts along with other grouting and soil-improvement techniques. A typical application includes the Woods Street Station of the Pittsburg subway in which 3 785 000 liters of sodium silicate was used for underpinning purposes (Parish et al., 1983).

Reactivity of various chemical grouts with hazardous wastes and leachates has been investigated (Bodocsi et al., 1988). Of the seven grouts evaluated, acrylate, cement–bentonite, and urethane grouts had the lowest base permeabilities with water. Acrylate performed well with low (25 percent) concentrations of certain compounds, paint and refinery wastes, but was seriously damaged with high (100 percent) concentrations of aniline, acetone, hydrochloric acid, and methanol. Cement–bentonite mix exhibited a decrease in permeability with a number of compounds tested. The urethane grout maintained its low permeability for most of the compounds permeated.

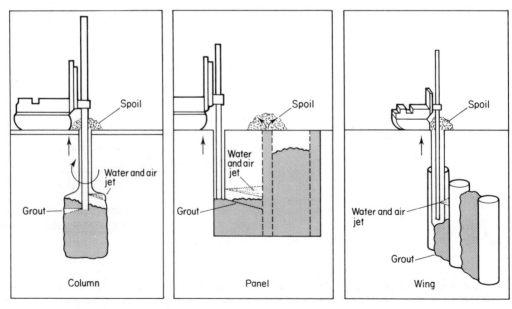

Fig. 9.28 Jet grouting options using the three-fluid system. (*After Coomber, 1985.*)

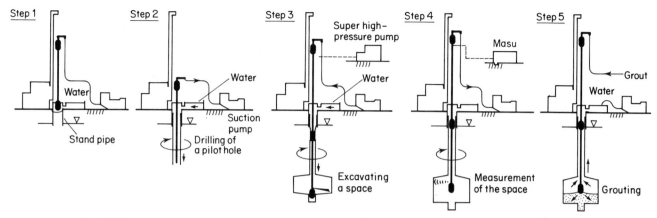

Fig. 9.29 Working procedure of the SSS-Man method of jet grouting. (*After Miki and Nakanishi, 1984.*)

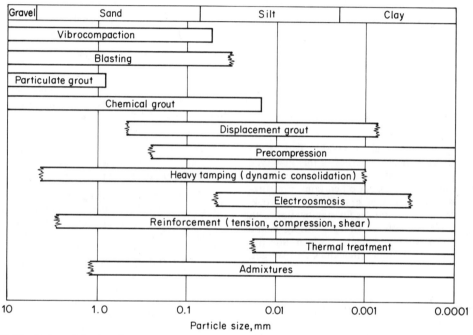

Fig. 9.30 Soil particle sizes for grouting and other ground improvement techniques. (*After Mitchell, 1981.*)

Slurry Grouting is the intrusion of flowable particulate grouts into voids and cracks underground. Slurry or particulate grouting is frequently utilized in the United States primarily to reduce permeability of rock beneath new dams. The grout materials are cement, clay (bentonite), sand, additives, microfine cement, fly-ash, lime, and water. These grouts cannot be injected into soils finer than medium to coarse sands. The "groutability ratios" have become useful in identifying soils and rock that can be injected with particulate grouts (Mitchell, 1981):

FOR SOIL: $N = \dfrac{(D_{15})_{\text{soil}}}{(D_{85})_{\text{grout}}}$ (9.13)

$N > 24$: grouting consistently possible

$N < 11$: grouting not possible

$N = \dfrac{(D_{10})_{\text{soil}}}{(D_{95})_{\text{grout}}}$ (9.14)

$N > 11$: grouting consistently possible

$N < 6$: grouting not possible

FOR ROCK: $N_R = \dfrac{\text{width of fissure}}{(D_{95})_{\text{grout}}}$ (9.15)

$N > 5$: grouting consistently possible

$N < 2$: grouting not possible

where D is grain size

Microfine cement (MC-500) has proven to be beneficial in achieving greater permeation of the grout into shear zones than type III portland cement (Moller et al., 1984). Figure 9.31 shows the rheological properties for various types of particulate grouts (Tornaghi et al., 1988). In recent years a new class of particulate grouts have evolved that have the following properties: (1) improved penetrability under low pressure in sandy-gravelly soils; (2) lower water loss and, therefore, greater volume of voids filled with the same volume of grout; and (3) the possibility of filling all the voids, permeating medium-coarse sands with refined products and minimizing any hydrofracturing effects. One such type of grout is a class of cement–bentonite grout named *Mistra*. It has been used successfully in the construction

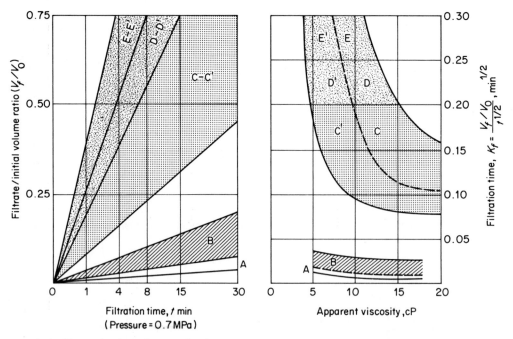

Fig. 9.31 Ranges of significant rheological properties for various types of particulate suspensions. A = bentonite muds; B = cement grouts with special colloidal additive; C = cement–bentonite grouts; D = poorly stabilized cement grouts; E = unstable cement grouts; C', D', E' = types C, D, E with fluidifiers and dispersing agents. (*After Tornaghi et al., 1988.*)

of the Milan subway (Tornaghi et al., 1988). Type B in Figure 9.31 shows the rheological properties of this new type of grout. Typical applications of slurry grouting are rock foundation treatment for dams, rock cut-off curtains, pressure-injected anchors, and stabilization of gravel and shot rock. Expansive soils are also good candidates for stabilization with lime slurry grouting. An example of slurry grouting is the restoration of failed highway embankments in which lime/fly-ash was injected with successful results (Blacklock and Wright, 1986).

9.14 PLANNING OF THE GROUTING PROJECT

A design engineer in charge of a grouting project must decide on the following items:

1. Grout hole locations and geometry, and length and inclination of the boreholes
2. Injection rates and pressures
3. Grout properties (liquid, transition, set), or grouts to be used
4. The program of work in stages

For this the engineer requires the results and data of a complete and reliable soil investigation, a sufficient knowledge of the geometric dimensions of the volume to be grouted, and the degree of improvement aniticipated in soil physical characteristics. Pregrouting investigations often involve the following (Tornaghi, 1983):

1. Hydrogeological and geotechnical investigation, by drilling, sampling, piezometric survey and in-situ testing (for permeability particularly)
2. Laboratory tests on soil samples with reference to grout-ability: grain size, porosity, permeability and chemical–mineralogical composition in some cases
3. Laboratory tests for selection of grout types: rheological and long-term properties of pure grouts and treated soil samples

4. In-situ grouting tests, if necessary, to collect more detailed and reliable information about the performance of the grout in the particular application.

The selection of grouting type and method is depended upon a number of soil properties, an important one being grain-size distribution. Proper sampling is often the key to production of representative grain-size distribution curves. Improper curves may result in selection of expensive or less safe grouting materials and methods. Experience have shown that use of large-size core-barrels (diameter 200 to 300 mm) with a liner, rotated with circulation of drilling fluid provides better samples and thus better analysis of grain-size distribution. Measurement of site permeability is also an important pregrouting investigation. Various methods can be used, the most reliable of which are borehole sections equipped for constant-head pumping tests and piezometers in the surrounding ground. Field tests such as the Lugeon (water pressure) test and Lefranc (open-end pipe) test, or indirect tests such as geophysical soundings and data from logging of fissures may also be used to estimate permeability. Microcurrent meter measurements (Tornaghi, 1983), which are based on the measurement of vertical flow velocity in a well where the water is pumped in and out at constant discharge, help to obtain continuous permeability profiles. The significance of obtaining a permeability profile is that the lower limit of the permeabilities corresponds to the practical groutability limit by permeation of chemical solutions.

Laboratory soil testing involves grain-size distribution, porosity, and permeability. The effective grain size D_{10}, the uniformity coefficient ($U = D_{60}/D_{10}$), and the specific surface are the important parameters that should be evaluated from grain-size analysis. Figure 9.32 shows the groutability of soils in relation to grout and soil permeability and effective grain size (Coomber, 1985). Porosity is a parameter that influences the permeability to some extent and it may be useful to estimate the required amount of grout also.

Laboratory grout testing involves evaluation of their short-term rheological properties and long-term properties and interaction with soil and other surrounding materials that may

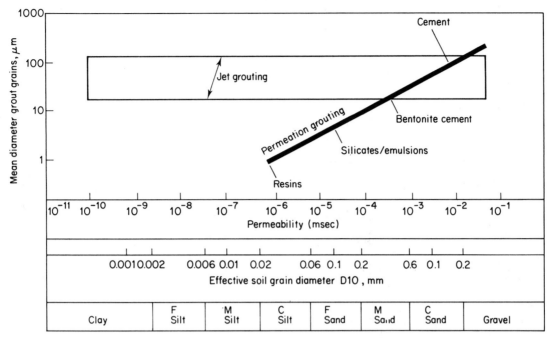

Fig. 9.32 Groutability of soils in relation to grout and soil properties. (*After Coomber, 1985.*)

be present. The properties of grout that need investigation are viscosity during grouting, penetration ability, strength and deformability after setting, resistivity to corrosion, durability, and adherence to soil particles, ground crack, or cavity walls. Some of the laboratory tests that help to evaluate the above are flow cone tests, bleeding and consistency tests, time of set, compressive strength, density and pH tests. From a rheological point of view, there are three types of grouts: (1) particulate suspensions (Binghamian fluids); (2) colloidal solutions (evolutive Newtonian fluids); (3) pure solutions (nonevolutive Newtonian fluids). The particulate suspension grouts are cement grouts that are termed either *stable* or *unstable* depending on the amount of water loss. Pure cement suspensions that exhibit significant water loss by sedimentation can be made *stable* by addition of clay or bentonite, which renders them cohesive and viscous. Colloidal solutions generally consist of diluted sodium silicate with organic or inorganic reagents. Their viscosity increases with time before setting (evolutive). The pure solutions are based on acrylic, phenolic, and amino resins. Their viscosity

may be kept constant until a given setting time. Figure 9.33 shows the rheological behavior of these grouts. Strength tests on pure grouts may be useful to estimate the cohesion of treated corase-grained soils. Tests on treated material, however, may be required for certain types of grouts. Creep and durability are some of the long-term properties that may need to be evaluated for some grouts such as silicate-based mixes.

In-situ grouting tests may be required by some complex designs or to test new methods. If such tests are undertaken, they should be conducted on large scale so that actual design conditions are simulated as closely as possible, otherwise the results may be misleading. Quality control of grouted soils can be accomplished by field permeability tests, laboratory tests on retrieved samples, and in-situ mechanical tests. Comparison of field permeability values before and after grouting should give a good indication of the performance of the grouted system. Laboratory tests should be conducted on undisturbed specimens. Field tests such as SPT, cone penetration, plate-bearing at shallow depths, and pressuremeter tests are suitable for

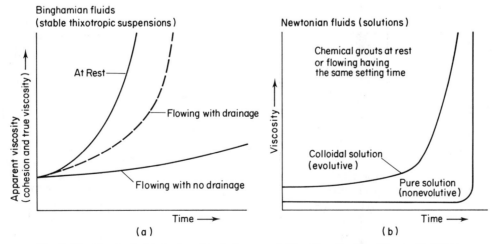

Fig. 9.33 Rheological behavior of typical grouts. (*After Mongilardi and Tornaghi, 1986.*)

gathering field strength and performance. Nondestructive tests, such as crosshole shear wave velocity, crosshole radar, and acoustic emission monitoring are available and have been used in large projects (*Engineering News Record*, 1982) to evaluate some geotechnical properties of grouted soils.

9.14.1 The Study of Soil Investigation Results

Soil investigation must include a series of boreholes providing information about the position of the bedrock, the characteristics and nature of each stratum or soil layer encountered, and a satisfactory knowledge of the hydrogeological data of the ground in which the injection is planned. Such exploration should not be limited to the volume to be treated but should also include the adjacent zones, which may influence the stability of the structure (e.g., excessive settlements) or influence the injection (e.g., important losses of grout through open channels).

Knowledge of the hydrogeological characteristics of the subsoil may be obtained by the following tests or operations.

1. Pumping tests carried out in one or several pumping wells combined with a series of properly located piezometers. Such a test will provide satisfactory information on the average permeability of the soil to be treated.
2. Lefranc type water injection or pumping tests carried out during the exploratory borings (Cambefort, 1964) will provide an idea of the permeability of each type of layer encountered when compared with the geological results of the borings. (One should remember that such tests are only of strictly local significance.)
3. "Micromoulinet" (hydraulic propeller) water tests, which are carried out in a borehole equipped with a strainer pipe, provide a good picture of the relative permeability of each layer as compared to the other layers.

The hydrogeological characteristics gathered from these tests should be compared with the results obtained from tests carried out on disturbed or undisturbed samples by a soil laboratory (grain-size distribution, permeability, etc.). In addition, the chemical properties of the groundwater must be known because

they may have an important influence on the chemical stability of the grout once it is placed into the ground.

A good knowledge of the ground conditions can be achieved by gathering and comparing the results of the various water tests, pumping tests, laboratory tests, and borelogs.

9.14.2 Determination of the Geometric Dimensions of the Zone to Be Tested

The geometric dimensions of the zone to be treated depends on several factors, including the geometric dimensions of the proposed structure (tunnel, galeries, shafts, foundation excavations, cofferdams), and the purpose of the grouting operation, either for waterproofing (total or partial) or for densification, in relation to the nature of the overall structure (dam, tunnel) or to the phase of the work being considered (permanent waterproofing or temporary protection).

Applications of grouting are manifold. Therefore, rather than offering very general advice, some specific examples will be given for which the grouting design was well suited to the overall civil engineering project.

Figure 9.34 illustrates several cases in which a grout curtain was used to cut off seepage through pervious alluvial soils underneath earth and rock fill dams. The curtain extended down to the rock level (a boundary that in some cases may be replaced by an impervious layer, provided such a layer extends sufficiently upstream and has a proper bond with lateral bedrock). Note the increasing thickness of the grout curtain near the surface. At that point the hydraulic gradient is highest and good contact has to be achieved with the clay core. Grouting is not as efficient at the surface because of the lack of overburden dead weight (high grouting pressure may not be employed because of possible surface leaks), and water seepage control, although very difficult, is very important.

As an example of a tunnel excavation, Figures 9.35a and 9.35b show part of the Paris Express Underground Railway, where the grouting treatment had to fit the construction needs and achieve two goals. First, as seen in Figure 9.35a, from a pilot grouting gallery close to the groundwater table it was

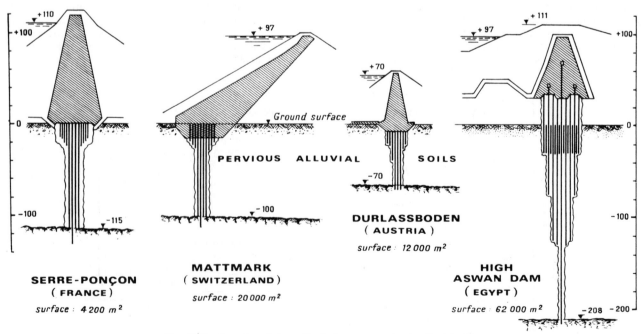

Fig. 9.34 Dam foundations. Typical grout curtains in alluvial soils.

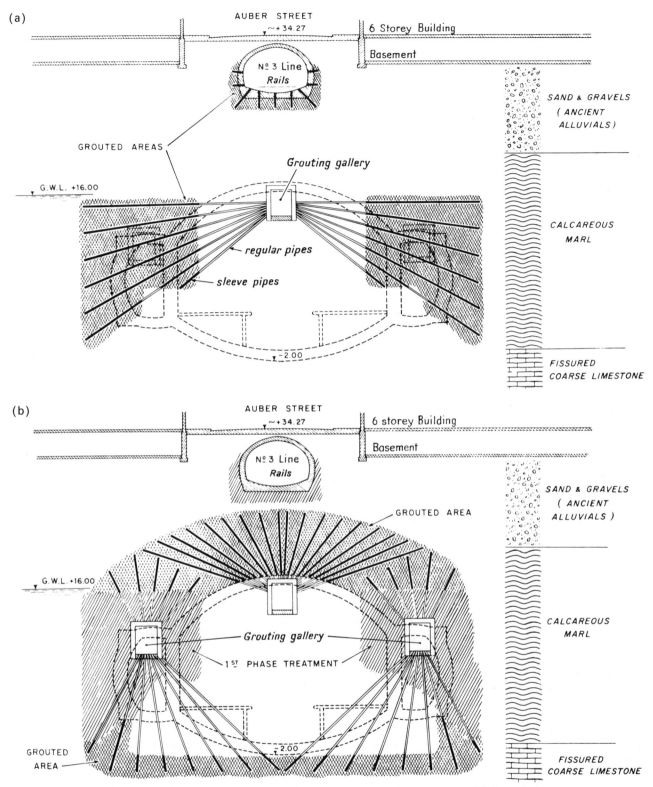

(a)

AUBER STREET
~+34.27

6 Storey Building

Basement

Nº 3 Line
Rails

GROUTED AREAS

Grouting gallery

G.W.L. +16.00

regular pipes

sleeve pipes

-2.00

SAND & GRAVELS
(ANCIENT
ALLUVIALS)

CALCAREOUS
MARL

FISSURED
COARSE LIMESTONE

(b)

AUBER STREET
~+34.27

6 storey Building

Basement

Nº 3 Line
Rails

GROUTED AREA

G.W.L. +16.00

Grouting gallery

1ST PHASE TREATMENT

-2.00

GROUTED AREA

SAND & GRAVELS
(ANCIENT
ALLUVIALS)

CALCAREOUS
MARL

FISSURED
COARSE LIMESTONE

Fig. 9.35 Grout hole layout, Auber Station (Paris subway). (a) First phase treatment. (b) Second phase treatment.

necessary to treat a zone located well below the groundwater table, where two other grouting galleries were to be constructed. Secondly, Figure 9.35b shows how from these two grouting galleries the major protection treatment for the main subway gallery was carried out (Janin and LeSciellour, 1970).

The dimensions of the treated zone should be in proportion to the dimensions of the whole structure and fit in with the construction project so as to resist the applied stresses and to seal the ground. Moreover, they should be such that the grouted zone is thick enough to resist dangers of erosion and of uplifts and to provide proper connection at contact zones, which usually are the weak zones of a grouting treatment.

By grouting we try to fill as thoroughly as possible, or up to a desired degree, all voids within the boundaries of the volume of soil to be treated without wasting any grout. Moreover, such filling should be carried out without disturbing beyond certain accepted limits the existing structure on the ground through bursting or uplift.

The degree to which the voids will be filled depends on the ability of the chosen grout to penetrate into the voids, which grow smaller and smaller during the injection. By knowing the ability of each grout to penetrate soils of given characteristics, one can adjust the sequence of grouts to be used, taking into account the grain-size distribution curve of the virgin ground.

9.14.3 Grout Hole Spacing

The location of the grout holes is determined by the following principal factors: the geometric form of the volume to be treated, the purpose of each line of grout holes, and the nature and radius of action of each grout to be used. It is obvious that the spacing of the holes will be smaller for finer grouts (gels or resins) and larger for coarser grouts (e.g., clay–cement grouts) because of the application of fine grouts for less pervious soils.

However, hole spacing may depend on other considerations, such as the special treatment of a contact zone, the limitations due to work platform dimensions related to drilling machine size, and pretreatment of a zone around a subway, sewers, or basements of existing buildings in order to prevent grout leaking into such structures during the main injection. In no case should these factors affect the spacing of the holes in the main injection zone.

9.14.4 Injectible Soils

The ground may be roughly divided into two groups: fissured rocks and granular soils.

In the first case, impermeabilization is generally obtained by grouting a diluted cement mixture under high pressure. The fissures are opened under the effect of pressure and the grout may progress. Of course, the finer the fissures, the more fluid the grout must be (low cement/water ratio). Thus, it will be possible to treat under good conditions nearly all fissured rock and cohesive soils, taking into consideration both the dilution of the grout and the pressure under which it is used.

Conditions are quite different for granular soils. If the soil contains large gravel or cobbles, one may also grout with a cement mixture. But in case of a finer grain size (sand type), dilution of the grout and increase of the grouting pressure will have little influence. Indeed, the diluted cement-grout will lose its water while in contact with the sand and become a thick mixture, and a pressure increase would only provoke burstings. For the treatment of such soils it is therefore necessary to take into consideration the nature of the grout.

It is absolutely necessary to adapt the nature of the grout to the soils to be treated. If this fundamental rule is neglected, the treatment is likely to fail or superfluous expenses may be incurred.

Figure 9.36 shows the typical ranges of gradation for different grouting techniques discussed here. To each grain size class corresponds one or several appropriate products allowing the injection under the best technical and economic conditions. Progress made in the last ten years has allowed us to fill in all the gaps. These new grouts deal mainly with the treatment of soils of low permeability.

A typical example is the execution of the impervious curtain of the Mission Dam on the Bridge River (Terzaghi and Lacroix,

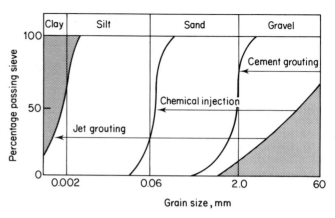

Fig. 9.36 Typical ranges of gradation for different grouting techniques. (*After Welsh et al., 1986.*)

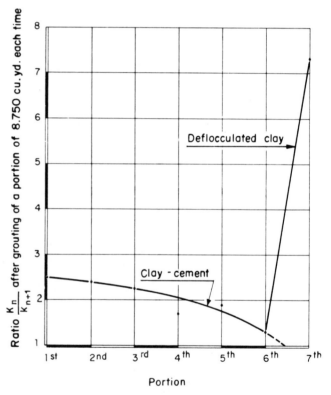

Fig. 9.37 Relationship between the ratio K_n/K_{n+1} and the number of portions of grout injected in the curtain.

1964) in Canada. This curtain rendered it possible to reduce the permeability by 500 times. Thanks to a series of piezometers placed at both sides of the curtain, the permeability was observed in relation to the portions of grout injected. Thus the ratio K_n/K_{n+1} was determined; where K_n is the permeability of the soil after the grouting of portion n and K_{n+1} is the permeability after injecting portion $n + 1$, each portion corresponding to 8750 yd^3 of grout.

The results given in Figure 9.37 show that after every injection the decrease of permeability becomes more and more insignificant. By replacing the clay–cement grout by a more liquid grout (deflocculated clay), a further increase of impermeabilization was achieved.

9.15 THE GROUTING PROCESS

The grouting (injection) process comprises two main sections: knowledge about the grouts and the technique of injecting the grout into the soil.

9.15.1 The Grouts

The finer the soil to be injected, the more liquid the grout has to be; for example, Binghamian suspensions (on the basis of cement or clay) in coarse soils; colloidal solutions more or less viscous (silicate or lignochrom gel, bitumen emulsion, organic colloids) in soils with medium grain size; and pure solutions (organic monomers in water solution) in very fine soils.

Under these conditions the fundamental difference between the grouts is of rheological origin, and classification can be made into nongranular Newtonian grout and granular Binghamian grout.

Nongranular Newtonian Grout The simplest grout corresponds to a liquid without rigidity, whose viscosity—independent of the speed gradient—is invariant with time.

Such types of grout are prepared with organic monomers (acrylamide, phenoplast, aminoplast) diluted in water. The viscosity of these noncolloidal mixtures does not vary before mass polymerization takes place (Fig. 9.38, curve I_1). The discharges are, therefore, at any moment, strictly proportional to the pressures and do not depend on the age of the grout. As the viscosity of these grouts is similar to that of water, the grouts can be used in very fine soils (silt, clayey sand, etc.).

Chemically they belong to three different classes:

1. The acrylamides modified by reticulation. After polymerization, this resin is very elastic but has little strength even at high concentrations. Therefore the acrylamides are

generally used for impermeabilization work rather than for densification (increase in strength or bearing power).

2. The urea–formols (formaldehydes). This resin can be used for both impermeabilization and densification. As its polymerization takes place only in acid media, it is incompatible with calcareous soil. In this case one may carry out a preinjection of acid, which destroys the carbonate, but the void index (as well as the expenses) is also increased.

3. The resorcinol–formols are also convenient for impermeabilization and densification. As the catalysis takes place in a basic medium, the compatibility is perfect in any soil.

It is not always necessary to have available such highly liquid grouts. In many cases, such as for fine or medium sands, the grouting can be accomplished with viscous nongranular mixtures.

Silica gel is a type of grout used very frequently and has been known for more than half a century. According to the concentration of sodium silicate, the initial viscosity varies between 4 and 20 centipoises. These mixtures, purely Newtonian before mass gelification, have a viscosity that increases with time (Fig. 9.38, curve I_2). The setting time is characterized by the appearance of rigidity with simultaneous transformation of the Newtonian liquid into a Binghamian liquid.

As these mixtures have no rigidity, there is a perfect linear relation between rates of discharge and pressure. However, as the viscosity increases with time, a decrease in discharge can be observed at constant pressure. This decrease, reproduced in units of permeability k leads to the curves of Figure 9.39, obtained with gels of variable setting time. If, on the other hand,

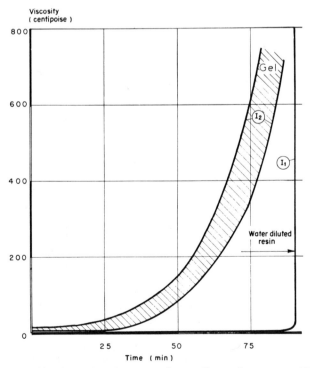

Fig. 9.38 Time–viscosity curve of some Newtonian grouts with equal setting times.

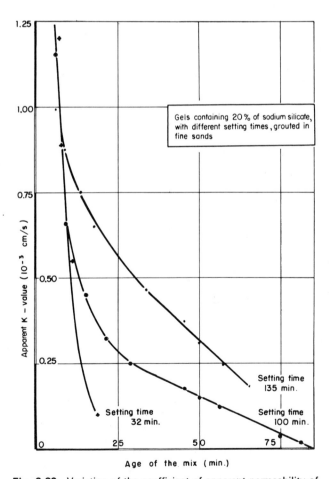

Fig. 9.39 Variation of the coefficient of apparent permeability of a gel during setting.

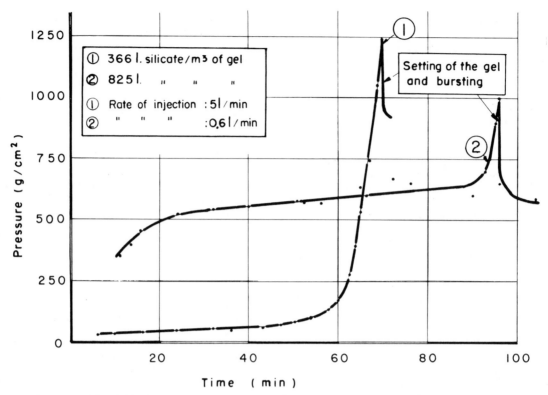

Fig. 9.40 Evolution of injection pressure of gels at constant rate of discharge.

a constant discharge is required, the grouting pressure has to be increased (Fig. 9.40).

The simplest and most frequently used grout corresponding to these characteristics is silica gel. This grout can be used both for impermeabilization and densification.

For the purpose of impermeabilization a very diluted silicate is used (e.g., 1 part of silicate for 9 parts of water). By adding an acid agent or a polyvalent salt, this solution is slowly transformed into a soft gel.

For the purpose of densification, one has to use concentrated silicate solutions (e.g., 7 parts of silicate for 3 parts of water). This solution is immediately transformed into a hard gel by adding an acid agent or polyvalent salt. This reaction of transformation must, therefore, take place in the ground. For this reason the silicate and the gel-forming agent (calcium chloride in the Joosten method, U.S. patent 2081541) are brought down separately through two neighboring boreholes.

This method has two disadvantages:

1. The output of the operation is far from being optimal, for the mixture of the silicate and its agent cannot be perfect. There remains, therefore, silicate that has not set.
2. The interpenetration of the silicate and its agent requires very high grouting pressures.

This method fell out of use some time ago, as it is now possible to obtain hard gels with slow setting time. Some organic substances, very limited in number, are capable of setting the silicate under the same conditions as the different mineral agents although they have no acid function or polyvalent ion. The setting is then slow even with concentrated silicates. Research on these organic gel-forming products dates back to 1958 and 1959: Gandon, Lehmann, Marcheguet, and Tarbouriech (French patent 1,166,581); Caron (French patent 1,164,835); Pecler (U.S. patent 1,230,836); Schmidt and Gendenk (German patent 3914).

In the first method, a dialdehyde (glyoxal) is used, which provokes the gelification of the silicate when it is transformed into glycolic acid by an internal Cannizzaro reaction.

In the second case, the gelification is obtained by the transformation of ethyl acetate into acetic acid and ethyl alcohol.

In the third case, the hydrolysis of an amide forms an ammonium salt, which reacts with the sodium silicate under release of ammonia.

In the fourth case, it is the hydrolysis of an acid chloride that ensures the gelification when the corresponding acid is liberated.

With these organic agents the neutralization of the silicate does not occur immediately, as in the case of mineral agents, but only after the acid has been liberated by the glyoxal, ester, amide, or acid chloride. The polymerization and setting can only begin after the chemical transformation of the agent.

Experience gained in the laboratory showed that one could not fully rely upon silicate gels: indeed, the silica in contact with water disappeared to a great extent. After its gelification the system continues to develop with time by polycondensation of the molecules of SiO_2. This polycondensation is connected with an expulsion of water (syneresis). This explains the very undesirable behavior of pure gels, but when the gel is connected with a skeleton of fine sand, the exudation of the syneresis water becomes practically zero and the system remains, therefore, impervious. The practical conclusion is that a soil injected by means of gel will practically be stable in time, whether the application is for impermeabilization or for densification, provided that the soil is fine, that is, corresponds well to the appropriate size class.

Several other products have been proposed for half a century for the treatment of this class of fine sand: bitumen emulsion, organic colloids, lignosulfites, derivatives of lignosulfites, derivatives of tannins, phenolic precondensates or polyurethanes, etc.

In the literature, these products often appear under code names given by the producer so that one does not know their exact composition and their mode of action.

No matter whether they are on an organic or mineral basis, they correspond to the same rheologic behavior as the silica gels, that is, they lie almost in the same range of curve I_2 in Figure 9.38.

Granular Binghamian Grouts   For the injection of coarse soils, suspensions of cement or clay or their mixtures are used. These loaded grouts are not Newtonian. They are, therefore, different from the preceding products for the double reason that they are Binghamian and that they contain grains.

A Binghamian grout offers at the same time a rigidity f_0 and a viscosity U. The effort F that is necessary to produce a displacement corresponding to a speed gradient dv/dx is

$$F = f_0 + U \frac{dv}{dx} \qquad (9.16)$$

where f_0 and U are constant for a given liquid.

Figure 9.41 gives an example of this relation, that is, shearing speed versus pressure.

This Binghamian relation can also be interpreted in a way that approaches the law of Poisseuille, $F = U' \, dv/dx$, by giving the Binghamian liquid a simple viscosity term. In this case the viscosity term is no longer a constant and decreases as the rate of flow increases (Figure 9.41).

Finally, these grouts are evolutive: their viscosity and particularly their rigidity increase with time.

The discharge versus pressure relation of these Binghamian liquids is quite different from that observed for Newtonian liquids (evolutive or not).

Indeed, according to the above Bingham formula, the displacement of a Binghamian grout can only begin beyond a certain pressure. Cambefort showed that the minimum pressure gradient P/L was a function of both the grout (characterized by its cohesion f_0) and of the soil (characterized by its grain size or its specific surface S and its porosity n): $P/L = f_0 S(1 - n)$.

Figure 9.42 illustrates another example of this initial rigidity.

This brings us to the consequences of the granular nature of these grouts. It is logical that the clay or cement particles of the grout fill the voids of the soil, thus diminishing the permeability during grouting operations. Also, at constant pressure, the discharge decreases even when the apparent viscosity of the grout remains unchanged.

In a soil with a given permeability, this decrease of discharge by sealing the voids will be a function of three parameters: the grain size of solid elements of the grout, the percentage of dry materials, and the state of flocculation.

The first parameter, that of the grain size of solid elements of the grout, is obvious: a grout containing cement grains of $50 \, \mu m$ will not penetrate a soil as easily as a bentonite grout, whose grains do not exceed $1 \, \mu m$.

The second parameter is the percentage of dry material. It was demonstrated in the laboratory and checked on site that slightly loaded grouts would more easily penetrate a soil than a highly loaded grout. This is the reason why the cement quantity is reduced to the strict minimum necessary to obtain the desired resistance (the resistance has nothing to do with the ratio of dry materials but is mainly a function of the cement/water ratio—Fig. 9.43) and the stability of the grout is guaranteed by very low percentages of ultracolloidal clay (i.e., bentonite). Thus, cement–bentonite grout for impermeabilization contains hardly more than $170 \, kg$ of dry materials per $1 \, m^3$ grout and for moderate densification ($20 \, kg/cm^2$ resistance) contains $300 \, kg$ of dry materials.

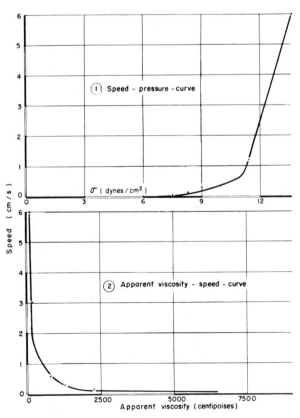

Fig. 9.41   Speed vs. pressure curve and apparent viscosity of a Bingham liquid.

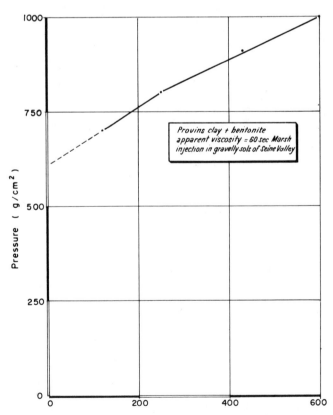

Fig. 9.42   Rate of discharge vs. pressure for a Bingham grout.

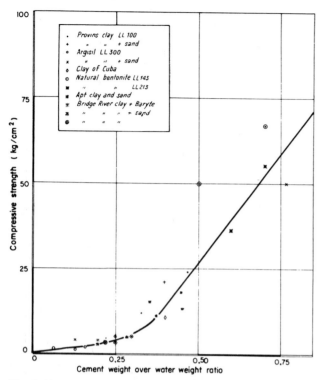

Fig. 9.43 Variation of the compressive strength of clay–cement grouts fabricated with different clay types.

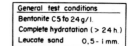

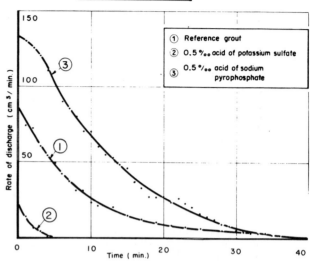

Fig. 9.44 Influence of the degree of flocculation on the rate of discharge. General test conditions: bentonite C5 to 24 g/liter, complete hydration (> 24 hr), leucate sand (0.5–1 nm).

Going further in this direction, it was proposed—and applied in several cases—to use grouts containing about 50 percent air (Cambefort and Caron, 1954). This grout has a high penetration capacity at low pressure. Thus, on a site near Paris where the pressure limit was fixed at 10 kg/cm², the injection rate could be increased by 30 percent by this method.

The third parameter that we consider is the state of flocculation. A stable Binghamian suspension penetrates the soil more easily when it contains few grains and when the diameter of these grains is small. This means that slightly loaded grouts without any cement, that is, clay and bentonite grout, are used for impermeabilization work. To be more successful the clay or bentonite should be dispersed in the grout as elementary grains and not in flocculated form (Fig. 9.44). One of the applications of this principle consists of producing a very penetrating grout including clay or bentonite, a rigidifying agent, and a peptizing agent (Caron, 1964).

9.15.2 Injection Technique

Once the different grouts have been selected, the method of injection has to be determined.

The Injection Method By general definition, injection is the process in which, by means of an applied pressure, a grout is forced into the soil through pipes that have been strategically placed to limit the zone of soil to be treated.

Different injection methods have been developed.

If a single-phase injection is sufficient, there are two processes in which drilling and injection are combined. A very pervious soil may be injected during rotary drilling. During the drilling of the borehole, each time a predetermined distance has been reached the drill rod is withdrawn a certain length and the grout is injected through the drill rod into the soil. During each

injection the top of the grout hole is sealed by a collar (Fig. 9.45c).

Another process, more frequently employed for single-phase injection, is to drive a casing to full depth, withdraw the casing a predetermined length, and inject through it. This method is effective only if the grout does not remount outside the casing (Fig. 9.45a).

The sealed-in sleeve pipe injection (with *tube à manchettes* sleeves) is a multiple-phase process that allows several successive injections in the same zone (Fig. 9.45b). This involves placing a sleeve pipe into a grout hole, which is kept open by casing or by mud. This pipe is permanently sealed in with a sleeve grout composed of a clay–cement mixture. The sleeve grout seals the borehole between the pipe and the soil to prevent the injection grout from channeling along the borehole. This means that, under pressure, the injection grout will break through in radial directions and penetrate into the soil.

The sleeve pipe consists of a steel or plastic tube with a diameter of 1 to 2 inches. Plastic tubes are used to facilitate excavation work afterwards.

At 1-ft intervals small holes are drilled in the pipe to serve as outlets for the grout. The holes are tightly covered by rubber sleeves (*manchettes*) that open only under pressure. The holes and sleeves work as one-way valves. The sleeve pipes are used only in the grouting zone; regular pipes are used for the rest of the grout hole.

In order to inject through a sleeve, a double packer fixed at the end of a smaller-diameter injection pipe is inserted into the sleeve pipe and centered around the sleeve to form a closed chamber with one-way valve outlets (Fig. 9.46).

This method presents numerous advantages. First, there is the possibility of repeating injection several times, which permits the use of grouts with decreasing viscosities. This permits better penetration of the fine voids after the big ones have been closed. More pervious soil layers may be sealed first, regardless of the order of injection level, which prevents loss of high-cost, low-viscosity grouts. Also, the grouting operations are carried out completely independently of drilling and this is very convenient for job organization and the use of the grouting plant.

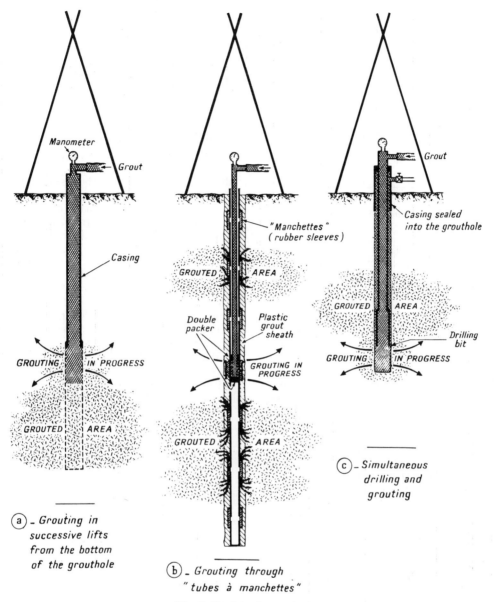

(a) _ Grouting in successive lifts from the bottom of the grouthole

(b) _ Grouting through "tubes à manchettes"

(c) _ Simultaneous drilling and grouting

Fig. 9.45 Injection methods.

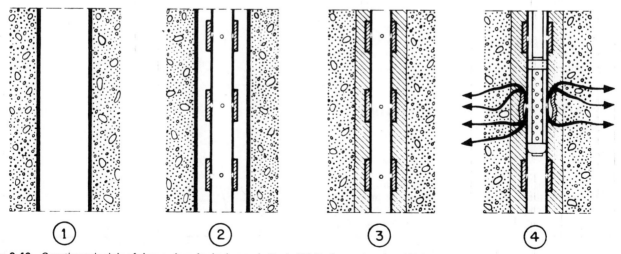

Fig. 9.46 Grouting principle of sleeve pipes (*tube à manchettes*). (1) Boring and casing. (2) Inserting the *tube à manchettes*. (3) Sealing-in of *tube à manchettes* and withdrawal of casing. (4) Injection by means of a double packer.

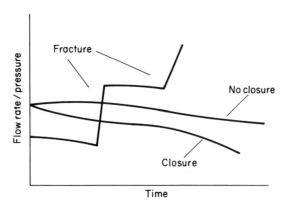

Fig. 9.47 Typical flow rate–pressure ratio curves in grouting. (*After GKN Hayward Baker, 1988.*)

Injection Pressure Discussion of injection pressure usually leads to divided opinion because theoretical considerations do not always agree with practical experience.

The pressure that is measured at the entry of a grout hole is always higher than the overburden stress at the level of injection. Otherwise, it would not be possible to inject a soil, say, 5 m underneath ground surface, with 5 kg/cm² pressure without encountering considerable uplifts.

If the rate of injection is kept constant, the pressure measured at the entry of a grout hole depends on the size of the voids in the soil (i.e., the permeability), on the viscosity of the grout, which may change with time, and on the obstacles through which the grout passes to reach the soil (sleeves, grout sheath).

These three factors acting as resistances determine the relationship between pressure and rate of injection. Economic factors and the action radius of the grout require a high rate of discharge. However, this rate is limited by the pressure that the soil can support in order to avoid phenomena such as hydraulic fracture and fissuring with surface leaks, deformations, and uplifts.

Figure 9.47 shows typical flow rate–pressure ratio curves. Hydraulic fracture is indicated by large increases in flow rate with small increase in pressure.

The Quantity of Injected Grout The estimate of the total grout volume necessary is based on the void volume of the soil; it is necessary to fill to a maximum degree all the voids of the soil volume to be grouted, even those filled with water.

However, the radius of grout flow is very irregular and usually involves a great loss of grout into the neighboring zones. This occurs when one tries to fill the voids to the maximum degree possible with a one-phase injection. The technique is first to seal the boundaries of the soil volume by injecting limited volumes of grout in several steps.

9.16 CONTROL OF GROUTS AND GROUTING OPERATIONS

9.16.1 Control of Grouts

Automatic weighing and volumetric systems insure the exact proportioning of the various constituents of the grout. The working of the batchers and the stability of the basic products are periodically checked by means of grout samplings. These controls concern mainly: viscosity, setting time, decantation, and resistance.

In the case of an anomaly of one of these measures, the working of the batchers will first be checked and if nothing can

be found, one proceeds with more profound controls of the different basic products.

Out of the four suggested controls, the first three are immediate and make it possible to react at once in case of an anomaly. If no divergences are observed, the control of the resistances will certainly be satisfactory. Unfortunately, this important value may only be found 48 hours after the sampling. In this respect one has to distinguish between grouts on the basis of cement and those of the gel or resin type.

With the cement-based grouts the medium—the soil into which the grout is injected—may not be of great importance. The gel or resin-type grouts, however, are essentially glues and therefore the resistance of the treated soil will be higher the smaller the glue joints, that is, the finer the soil.

Controls are therefore made on sand–gel or sand–resin mixtures, the sand being normalized (generally 100–200 μm) or being the sand of the treated ground.

9.16.2 Control of Grouting Operations

The measurement of the rate of discharge and of the pressure gives a good picture of the grouting operations. The efficiency of the treatment can also be checked by observing the variation of permeability with the help of a series of piezometers.

The study of uplift gauges makes it possible to modify the nature of the grout if it no longer penetrates the soil, but provokes hydraulic fracture.

Figure 9.48, for example, resulting from investigations on the Thulagain Dam in Scotland, shows that all voids of the soil could be entirely injected without provoking important uplifts

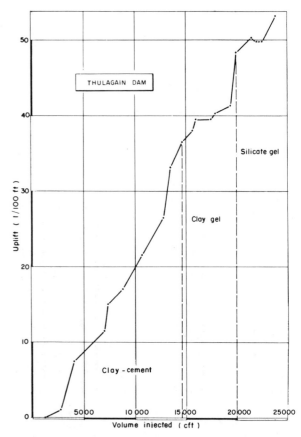

Fig. 9.48 Average uplifts in the center of the test area related to the injected quantities.

by means of successive change from a coarse clay–cement grout to a more elaborate clay gel and finally to a liquid silicate grout.

The main object of Figure 9.48 is to show that it is quite possible to control a job and limit uplifts to a few millimeters by checking the uplift gauges. This, by the way, was not necessary in the above-mentioned case.

9.17 EXAMPLES OF APPLICATIONS OF GROUTING

9.17.1 The Hydroelectric Power Plant of Fessenheim

The power plant is situated at the Grand Alsace Channel near the Rhine River. It rests on alluvial sand and gravel layers more than 200 m thick with permeability values as high as $k = 10^{-2}$ m/sec. Its construction required an excavation of 175×89 m surface and 27 m depth, 23 m of which was below the groundwater table.

Preliminary studies showed that sealing the subsoil would cost less than pumping and groundwater lowering. Sheet pile driving down to 20 m was not feasible in this type of soil. It was decided to inject vertical impermeable diaphragms surrounding the site before excavation was started.

These vertical diaphragms are connected to a horizontal diaphragm injected at a very deep level, thus forming a watertight box big enough to contain the power station. The horizontal diaphragm is situated at such a depth that the weight of the overlying soil body remaining after excavation is large enough to balance the hydrostatic uplift pressure exerted on its undersurface (Fig. 9.49).

The vertical diaphragms were injected through two rows of grout holes that were 3 m apart. The grout holes themselves were also spaced 3 m apart.

Where sand layers were found, a third, intermediate row was drilled. The grout holes that were equipped with sleeve pipes were drilled from an excavation level equal to the highest seasonal groundwater level.

The horizontal diaphragm was injected from grout holes laid out in a 9-m-square grill pattern (Fig. 9.50). This large distance between holes was possible because the horizontal permeability of the alluvials was much higher than the vertical one, and furthermore because at this depth injection pressures as high as 60 to 70 kg/cm² could be applied. Sleeve pipes were placed only in the 8-m-thick injection zone; the upper part of the grout hole was equipped with regular pipes.

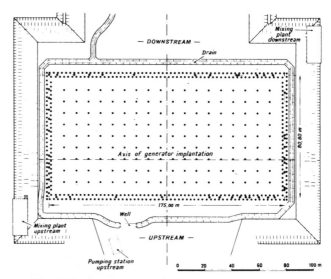

Fig. 9.50 Grout hole layout at Fessenheim power station.

The grout properties varied with the zone of injection. For the vertical diaphragms a clay–cement mixture with a compressive strength of 2 kg/cm² was felt necessary in order to provide more security against sliding of slopes. The horizontal diaphragm was injected with a low-viscosity grout with a base of pure or treated clays. The sand lenses were grouted with silicate gel.

For the injection of a box surface of 30 000 m², a total combined length of 22 500 m of grout holes were drilled and 42 500 tons of dry products were consumed. This corresponds to a rate of 1.4 tons/m² surface. After excavation had been completed, a residual water flow of only 115 liters/sec was left for pumping out, which proved that the grouting was effective, and the Darcy's coefficient of permeability decreased to $k = 3.10^{-6}$ m/sec (Fig. 9.51).

The slopes of excavation are comparable with the downstream slopes of a dam with an injected grout–soil core. Their inclination of 1:1 with some berms is extremely high when compared with the slope of constructed earth dams.

The technique of sealed-in underground boxes has been carried out on a large scale for the first time at Fessenheim and it has proved to be very successful. However, it has to be emphasized that such work always requires a sufficient number of controls during and after construction.

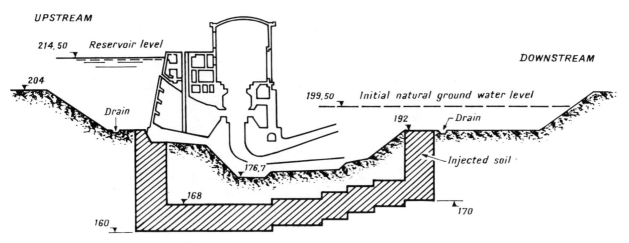

Fig. 9.49 Cross section of injected box at Fessenheim power station.

Fig. 9.51 Clay–cement grouted alluvials at Fessenheim construction site.

9.17.2 Durlassboden Dam

An example of a permanent impervious curtain is found in the work carried out for the Durlassboden Dam in Austria (Kropatschek and Rienossl, 1967). The dam is nearly 70 m high, made of earth with 2.5 million m³ material and 52.5 million m³ useable storage for a 25 MW power plant with a level difference of 120 m downstream. It is founded on an alluvial mass constituting a backfilling of the eroded bedrock (Fig. 9.52). The sedimentation of the deposits is irregular. The river sediments

such as sand, gravel, and pebbles alternate in height and width with sands or silty sands and partially with regular thick silty layers.

Between 30 and 50 m depth an impervious silt or silty sand level was revealed upon which the impervious curtain under the dam could be founded. The average horizontal permeability of the sandy-gravelly alluvials that exist between the bottom of the dam and the silt horizon is $k = 10^{-4}$ m/sec and the horizontal permeability of the silty sand varies, according to the silt content, between 10^{-5} and 10^{-6} m/sec.

A large-scale grouting test composed of seven boreholes that met the silty level at about 50 m depth was carried out in order to prove the injectibility of the alluvials. The permeability decreased from 3×10^{-4} to 4×10^{-7} m/sec.

Impermeabilization by means of injection was therefore possible. This method was the most suitable to comply with the following conditions:

a. Reduction of the permeability of the alluvials to 1×10^{-6} m/sec
b. Avoidance of any damage to the curtain by soil settlements
c. Permitting the curtain to be completed after construction of the earth dam and at the end of the settlement

The grout curtain was composed of eight borehole lines, five of which were 15 to 21 m deep, and the three median lines were drilled down to about 60 m, 5 m of which lay in the silt (Fig. 9.53). The distance between the borehole lines was 3 m and between the median lines 2.5 m. The boreholes themselves lay 3 m apart.

The deep boreholes were drilled with slurry by means of rotary rigs equipped with tricones and by rapid driving in shallow depths. The boreholes were fitted out with *tubes à manchettes* $1\frac{1}{2}$ inches in diameter and in two median lines with tubes 2 inches in diameter for the installation of clinometers.

The rocky abutments were grouted partially according to the conventional method and partially according to the method described above.

Grouting took place in four different phases, the first and second phase by means of a cement/clay mix with a minimum

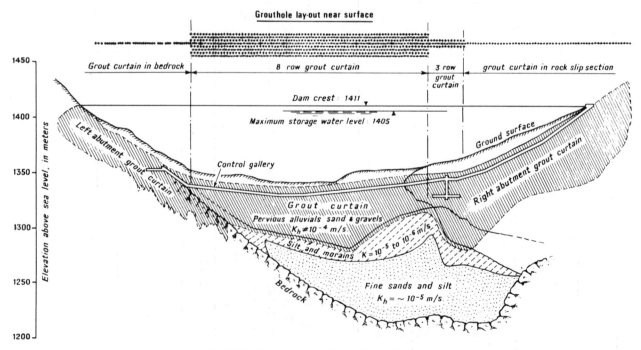

Fig. 9.52 Section along Durlassboden Dam axis.

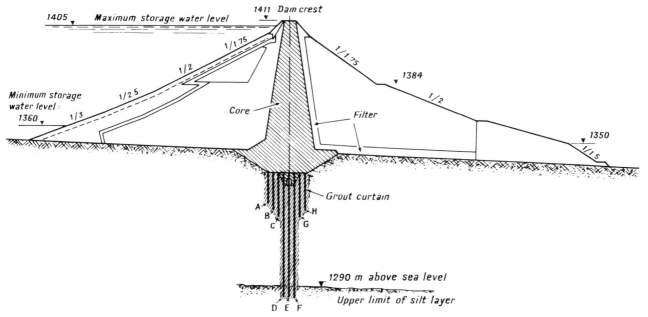

Fig. 9.53 Typical cross section of Durlassboden Dam.

viscosity of 38 Marsh seconds and composed of 500 kg clay with its natural water content and of 150 kg cement and 1000 liters water.

The third phase was carried out with deflocculated bentonite (bentonite gel) composed of bentonite, sodium monophosphate, sodium silicate, and water (Solétanche patent). Its viscosity was not to exceed 38 Marsh seconds.

The fourth phase was carried out with a chemical gel (called Algonite gel) mainly composed of silicate aluminate. Its viscosity amounted to nearly 32 Marsh seconds.

The final grouting pressures were 50 to 60 kg/cm^2 in depth.

For the sealing of 10 579 m^2 cohesionless soil, a total combined length of 20 587 m of borings were necessary. Into the boreholes a total combined length of 15 577 m of *tubes à manchettes* were placed; the quantity of grout necessary for the different phases was as follows:

Phase 1 and 2   22 540 m^3 of clay/cement grout
Phase 3   25 132 m^3 of bentonite gel
Phase 4   6724 m^3 of Algonite gel

Supposing that a zone of 1.5 m around the outer lines was treated by the grouting, the absorption per cubic meter of theoretical soil is equal to 45 percent of this volume for the entire grouted mass.

Besides current controls of the grout, the permeability of the curtain was determined by 290 Lefranc tests after completion of the third phase. The Darcy's coefficient of permeability decreased to 3×10^{-5} m/sec. After the last phase and before the construction of the dam, the arithmetical mean of the Darcy value was 0.8×10^{-6} m/sec.

This result entailed a depression of the water table behind the impervious curtain of 50 m for the highest level in the storage basin despite the considerable water flow of the slopes.

9.17.3 Singapore Mass Transit Railway

An example of application of jet grouting in "block" consolidation from the surface is in construction of the Singapore Mass Transit Railway in 1983–84. The case was first described by Tornaghi and Cippo (1985) and Mongilardi and Tornaghi

(1986) and later discussed by Bruce (1988). The geology of the site consisted of beach sand and fill (3 to 5 m) that overlaid very soft peaty clay, marine clay, and fluvial soils to a depth of about 15 m. The base was hard-to-stiff cohesive soil or soft rock. Groundwater table was 1 to 2 m below surface. Under these circumstances, soil improvement was necessary even for shield excavation. Soil was too fine for permeation grouting. Jet grouting was decided upon in which the grouted section would extend to the full excavation section above the base material and would be limited in the upper part to create an arch of consolidated soil about 1.5 m thick. Figure 9.54 shows sections of the grouted area near Dhoby Ghaut Station. A full-scale field test was conducted in which two different layouts of jet-grouted columns were tested. Two different quantities of grout (600 and 800 liters/m^3 of soil) were injected in each layout, which resulted in four different schemes of grouting. Figure 9.55 illustrates the layout of these test sections. The following general procedure was applied: drilling to 10.5 m depth; treatment from the bottom to 0.5 m of depth; injection of grout with water/cement ratio of 1.6 and grouting pressure of 40 MPa. Instrumentation consisted of two inclinometers to check horizontal soil displacements, two piezometers to record pore pressure build-up, and nine datum points to check vertical soil displacements with reference to a fixed point 20 m distant from the perimeter of the test area. The maximum horizontal displacement was measured to be 23 cm at a depth of 6 m, at 1 m from the axis of the external row of Scheme IV. At a distance of 6 m, the displacements were less than 5 cm. The excess pore pressures were low throughout (0.02 and 0.04 MPa at 3 and 6 m from the perimeter of the treated area). The mean heave values were 30 cm at 1 m distance, 17 cm at 3 m, and 1 cm at 10 m distance.

A total of 190 m^3 of grout was injected, 70 m^3 of which was ejected during grouting. Out of the remaining, 60 m^3 was estimated to correspond to the upward displacement of the soil and the other 60 m^3 caused the radial displacement and compression. Laboratory tests were conducted on samples retrieved from 8 boreholes (S1–S8) approximately 2 weeks after the completion of the test site. Table 9.18 summarizes the composition and strength data of test-section specimens and actual treatment specimens. The strength achieved was well

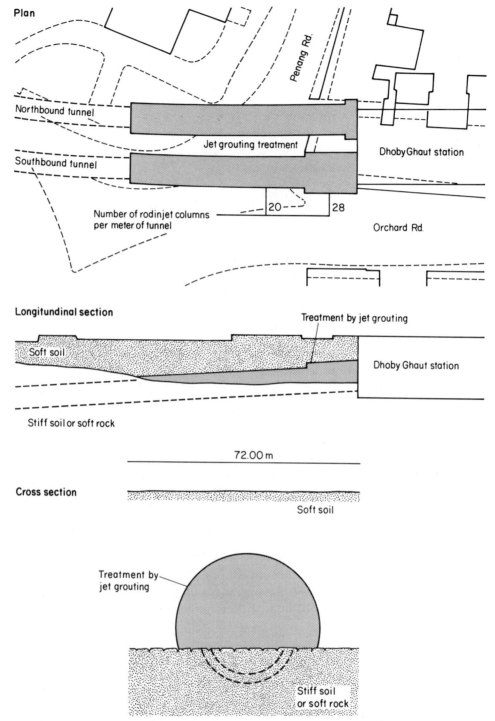

Fig. 9.54 General arrangement of jet grouting MRT 106, Singapore. (*After Tornaghi and Cippo, 1985.*)

over the specified minimum of 0.3 MPa and the water/cement ratio of the treated soil was twice that of the pure grout. A test pit was excavated inside the test area 15 days after the end of the treatment. Overlapping of columns with significant discontinuities was found only in Scheme II and therefore a spacing of 0.7 m was selected for the final design. Overall, 9400 m³ soil was treated with more than 10 000 columns. Ground movements were kept within the safe limit of about 2 cm by adjusting the sequence of grouting. Quality control was accomplished by sampling, laboratory testing, and cone

penetration tests during the project. Excellent performance was recorded during subsequent shield tunneling.

9.17.4 Chemical Grouting in Michigan

An example of chemical grouting was described by Welsh (1983). Chemical grouting was used at an industrial site in Michigan to simultaneously underpin spread footings, act as an excavation support system, and provide a watertight barrier

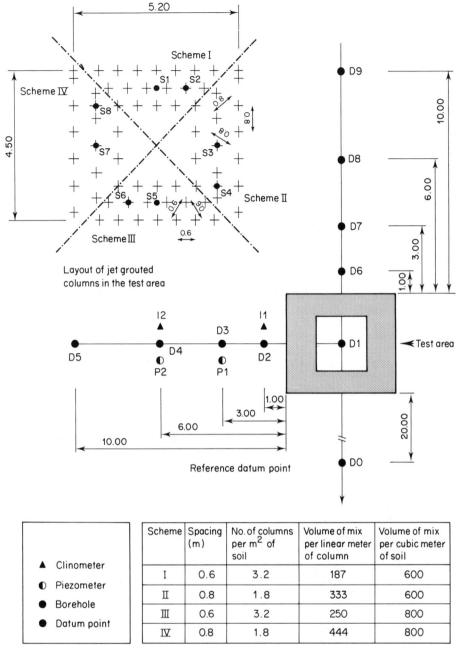

Fig. 9.55 Layout of test section MRT 106, Singapore. (*After Tornaghi and Cippo, 1985.*)

Scheme	Spacing (m)	No. of columns per m² of soil	Volume of mix per linear meter of column	Volume of mix per cubic meter of soil
I	0.6	3.2	187	600
II	0.8	1.8	333	600
III	0.6	3.2	250	800
IV	0.8	1.8	444	800

▲ Clinometer
◐ Piezometer
● Borehole
● Datum point

against very large groundwater flows. The geology of the site showed fine and medium sands with several thin strata of cohesive soil. The sand strata extended well below proposed excavation bottom and the groundwater table was 12 m above. Sheeting and dewatering was ruled out by the hydrogeological conditions and the inability to accept any ground movement. The sand contained less than 20 percent fines and was therefore considered to be chemically groutable.

Three rows of sleeved grout pipes on 0.9-m centers were situated around the perimeter of the pit in order to insure that a minimum of a 1.6-m thick grout wall would be created. The outer rows were injected initially and then the inner row to fill any gaps in the curtain. Grout pipes for the bottom of the pit were also placed on 0.9-m centers from which a 1.9-m thick floor curtain was created. A total of 140 grout pipes were

used to form a grouted soil "bathtub" around the proposed excavation. Figure 9.56 shows the details of the grout layout plan and cross section of the area.

Sodium silicate grout was used in this project. GEOLOC-4 was selected because of its high strength and nonodorous and nontoxic characteristics. GEOLOC-4 consists of 40 percent Grade N sodium silicate, 6 percent organic ester, water, and trace accelerators. As formulated, it produced average strengths of 1035 N/m², and had a viscosity of 4 cP with a gel time of 15 min at 20°C. The components were stored in compartmentized tanker and stream-proportioned and mixed by a series of four parallel pumps with variable-speed transmissions. The liquid volume of the grout was estimated using the following equation:

$$\text{Liquid grout volume} = V(n \times F)(1 + L) \qquad (9.17)$$

TABLE 9.18 COMPOSITION AND STRENGTH DATA OF TEST SECTION AND ACTUAL TREATMENT SPECIMENS, MRT CONTRACT 106, SINGAPORE[a].

		Unit Weight, kN/m^3	U.C. Strength, MPa	Composition, kN/m^3					
				C/W	C	dS	W	W_s (%)	V_g (l/m^3)
Field Trial									
Jet-grouted soil	Center columns	15.91	0.604	0.294	1.93	7.41	6.57	45.3	386
	Between columns	16.47	0.477	0.261	1.62	8.64	6.21	40.6	324
	Average	16.10	0.559	0.285	1.82	7.83	6.45	43.6	365
Ejected mixture		14.91	0.641	0.303	2.18	5.54	7.19	(64.3)	435
Actual Treatment									
Jet-grouted soil		15.31	1.090	0.395	2.76	5.56	6.99	43.0	550
Ejected mixture		16.14	0.519	0.272	1.75	7.97	6.42	44.0	350

[a] After Mongilardi and Tornaghi (1986).

C = cement; W = water; W_s = water content of soil; V_g = volume of grout per unit volume of treated soil.
General assumptions: Fully saturated soil and no drainage.

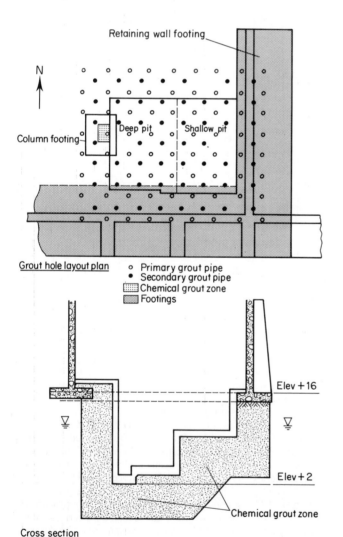

Fig. 9.56 Grout hole layout plan and grout cross section. (*After Welsh, 1983.*)

where

V = total volume of treatment zone
n = soil porosity
F = void filling factor
L = loss factor for grout placed outside the treatment zone

In practice, it is found that typical sodium silicate-based grouts fill from a minimum of about 85 percent up to nearly 100 percent of the void volume. The lower percentage occurs for well-graded sands with fines where only a single stage or limited second stage of grouting is performed under low pressures. The grout loss factor may vary from 5 percent to 15 percent depending upon the shape of the grouting zone, the frequency of injection points per unit volume, and the presence of highly porous layers in the groutable soils. On this specific project, 158 970 liters of chemical was used to treat approximately 1672 m^3 of sand.

Quality control and evaluation of the performance was done via monitoring wells. A wellpoint was installed at elevation 5 m. Highly concentrated dyes were injected outside the grouted zone beyond each wall and floor of the proposed excavation. A subsequent 24-hr pump-down test in the wellpoint showed steadily decreasing water level with no color traces. This verified the good degree of watertightness of the grouted zone. The subsequent excavation revealed sandstone-like material and appeared to have high integrity. Absence of bracing, however, caused creep and about 0.1 m^3 of material to spall off the face of the wall. This was due to a heavy point load nearby the excavation. Subsequently, the walls were partially braced and the excavation remained open 4 days prior to pouring the reinforced concrete. The net groundwater intrusion was estimated to be on the order of 0.3 liters/minute.

Other pertinent examples and case histories of various grouting projects can be found in WES (1984), Parish et al. (1983), Baker et al. (1984), Brand et al. (1988), and in ASCE (1987).

ACKNOWLEDGMENTS

Sibel Pamukcu would like to extend her thanks to Ara Arman for his input and help in revision of this chapter.

REFERENCES

Acar, Y. (1988), Communication, Department of Civil Engineering, Louisiana State University, Baton Rouge, La.

Acar, Y. B., Gale, R. W., Putnam, G., and Hamed, J. (1989), Electrochemical processing of soils: Its potential use in environmental geotechnology and significance of pH gradients, *Proceedings of the 2nd International Symposium on Environmental Geotechnology*, Shanghai, China, **1**, Envo Publishing Co., Inc., Bethlehem, Pa., pp. 25–38.

Acott, S. M. and Myburgh, P. A. (1983), Design and performance study of sand bases treated with foamed asphalt, *Transportation Research Board* No. 898, TRB, National Research Council, Washington, D.C., pp. 232–241.

Andersland, O. B. and Anderson, D. M. (1978), *Geotechnical Engineering for Cold Regions*, McGraw-Hill Book Co., Inc., New York, N.Y.

ARBA (1971), *Materials for Soil Stabilization*, Education and Information Guide, American Roadbuilders Association, Washington, D.C.

Arman, A. (1988), Short Course Notes on *Soil-Cement and Soil-Lime Stabilization*, Louisiana State University, Baton Rouge, La.

Arman, A. and Saifan, F. (1967), The effect of delayed compaction on stabilized soil–cement, *Highway Research Board* No. 198, Washington, D.C., pp. 30–38.

Arman, Ara and Dantin, T. J. (1969), The effect of admixtures on layered systems constructed with soil–cement, *Highway Research Record* No. 263, Washington, D.C., pp. 69–80.

Arman, Ara and Munfakh, G. A. (1972), Lime stabilization of organic soils, *Highway Research Record* No. 381, Washington, D.C., pp. 37–45.

ASCE (1978), *Soil Improvement: History, Capabilities and Outlook*, Report by the Committee on Placement and Improvement of Soils of the Geotechnical Engineering Division of the American Society of Civil Engineering, ASCE, New York, N.Y., 326 p.

ASCE (1987), *Soil Improvement—A Ten Year Update*, ASCE Geotechnical Special Publication No. 12, ed. J. P. Welsh, ASCE, New York, N.Y.

Asphalt Institute (1974), *Mix Design Methods for Liquid Asphalt Mixtures*, Supplement to MS-2 (Misc. 74-2), The Asphalt Institute, College Park, Md.

Asphalt Institute (1982a), *Soils Manual*, MS-10, The Asphalt Institute, College Park, Md.

Asphalt Institute (1982b), *Asphalt Surface Treatments—Specifications*, ES-11, The Asphalt Institute, College Park, Md.

ASTM (1980), *Extending Aggregate Resources*, ASTM STP 774, Philadelphia, Pa.

Baker, R. F., Byrd, L. G., and Mickle, D. G. (eds.) (1975), *Handbook of Highway Engineering*, Van Nostrand Reinhold Co., New York, N.Y.

Baker, W. H. (ed.) (1982), *Proceedings of the Conference on Grouting in Geotechnical Engineering*, ASCE, New Orleans, La.

Baker, W. H., Cording, E. J., and MacPherson, H. H. (1983), Compaction grouting to control ground movements during tunneling, *Underground Space*, **7**, pp. 205–212.

Baker, W. H., Gazaway, H. N., and Kautzmann, G. (1984), Grouting rehabs earth dam, ASCE, *Civil Engineering*, September.

Barenberg, E. J. and Thompson, M. R. (1982), Design, construction and performance of lime, flyash and slag pavement, *Transportation Research Record* No. 839, TRB, National Research Council, Washington, D.C., pp. 1–6.

Bell, F. G. (ed.) (1975), *Methods of Treatment of Unstable Ground*, Butterworth & Co. Ltd., London.

Bennett, R. H. and Hulbert, M. H. (1986), *Clay Microstructure*, IHRDC Publishers, Boston, Mass.

Bergeson, K. L., Pitt, J. M., and Demirel, T. (1984), Increasing cementitious products of a class C fly ash, *Transportation Research Record* No. 998, TRB, National Research Council, Washington, D.C., pp. 41–46.

Bjerrum, L., Moum, J., and Eide, O. (1967), Application of electroosmosis on a foundation problem in Norwegian quick clay, *Geotechnique*, **17**, No. 3, pp. 214–235.

Blacklock, J. R. and Wright, P. J. (1986), Injection stabilization of failed highway embankments, *Transportation Research Board* No. 1104, TRB, National Research Council, Washington, D.C., pp. 7–18.

Bodocsi, A., Bowders, M. T., and Sherer, R. (1988), Reactivity of various grouts to hazardous wastes and leachates, *Project Summary*, EPA/600/S2-88/021, Hazardous Waste Engineering Research Laboratory, Cincinnati, Ohio.

Bohn, L. H., McNeal, B. L., and O'Connor, G. A. (1985), *Soil Chemistry*, John Wiley & Sons, Inc., New York, N.Y.

Brand, A., Blakita, P., and Clarke, W. (1988), Grout supports Brooklyn tunneling, ASCE, *Civil Engineering*, January.

Braun, B., Schuster, J. A., and Burnham, E. W. (1978), Ground freezing for support of open excavations, *Proceedings of the 1st International Symposium on Ground Freezing*, Bochum, Germany, pp. 429–453.

British National Society of the International Society of Soil Mechanics and Foundation Engineering (1963), *Grouts and Drilling Muds in Engineering Practice*, Butterworth, London.

Bruce, D. A. (1988), Developments in geotechnical construction process for urban engineering, *Civil Engineering Practice*, Spring, pp. 49–97.

Bryhn, O., Loken, T., and Aas, G. (1983), Stabilization of sensitive clays with hydroxy-aluminum compared with unslaked lime, *Proceedings of the 8th ECSMFE*, Helsinki.

Calcium Chloride Institute (1953), *Calcium Chloride for Stabilization of Bases and Wearing Courses*, Washington, D.C.

Cambefort, H. (1964), *Injection des Sols*, Eyrolles, Paris, 567 p.

Cambefort, H. and Caron, C. (1954), Method for impermeabilization or consolidation of permeable soils and other porous masses, French patent 1,148,413 (June 26).

Caron, C. (1964), New procedures in the use of clay grouts, and material utilized in their practical application, French patent 1,237,3111.

Casagrande, L. (1949), Electro-osmosis in soils, *Geotechnique*, **1**, pp. 1959–1977.

Casagrande, L. (1959), *Review of Past and Current Work on Electro-Osmotic Stabilization of Soils*, Harvard Soil Mechanics Series No. 45, Harvard University, Cambridge, Mass. (reprinted November 1959, contains supplement of June 1957).

Castedo, L. H. and Wood, L. E. (1983), Stabilization with foamed asphalt of aggregates commonly used in low-volume roads, *Transportation Research Record* No. 898, TRB, National Research Council, Washington, D.C., pp. 232–241.

Chappel, B. A. and Burton, P. L. (1975), Electro-osmosis applied to unstable embankment, *Journal of the Geotechnical Engineering Division*, ASCE, **101**, No. GT8, pp. 733–740.

Clare, K. E. and Sherwood, P. T. (1954), The effect of organic matter on the setting of soil–cement mixtures, *Journal of Applied Chemistry*, No. 4, November, pp. 625–630.

Clough, G. W., Kuck, W. M., and Kasali, G. (1979), Silicate stabilized sands, *Journal of the Geotechnical Engineering Division*, ASCE, **105**, No. GT1, pp. 65–82.

Cincotto, M. A. (1984), Selection of waste materials with pozzolanic activity, *Low Cost and Energy Saving Construction Materials*, **1**, Envo Publishing Co., Inc., Bethlehem, Pa., pp. 287–297.

Cook, D. J. (1984), Production of cements based on rice husk ash, *Low Cost and Energy Saving Construction Materials*, **1**, Envo Publishing Co., Inc., Bethlehem, Pa., pp. 1–20.

Coomber, D. B. (1985), Groundwater control by jet grouting, *Proceedings of the 21st Regional Conference of the Engineering Group of the Geological Society*, Sheffield, England, pp. 485–498.

Covey, J. N. (1980), An overview of ash utilization in United States, *Proceedings of the Fly Ash Applications in 1980 Conference*, Texas A & M University, Texas.

Cowell, M. J. and Irwin, L. H. (1979), Effects of compaction delays and multiple treatments on the strength of cement stabilized soil, *Transportation Research Board* No. 702, TRB, National Research Council, Washington, D.C., pp. 191–198.

Dempsey, B. J. and Thompson, M. R. (1968), Durability properties of lime soil mixtures, *Highway Research Record* No. 235, HRB, National Research Council, Washington, D.C., pp. 61–75.

De Vries, D. A. (1952), Thermal conductivity of soil, *Mededelingen van deLandbouwhogeschool te Wageningen*, Vol. 52, No. 1, pp. 1–73 (translated by Building Research Station, Library Communication No. 759, England).

Diamond, S. (1975), Methods of soil stabilization for erosion control, *Final Report*, Joint Highway Research Project, Purdue University, Indiana.

Diamond, S., White, J. L., and Dolch, W. L. (1964), Transformation of clay minerals by calcium hydroxide attack, *Proceedings of the 12th National Conference on Clays and Clay Minerals*, Pergamon Press, New York, N.Y., pp. 359–379.

Eades, J. L. (1962), Reactions of Ca(OH)$_2$ with Clay Minerals in Soil Stabilization, Ph.D. Thesis, Geology Department, University of Illinois, Urbana.

Eades, J. L. and Grim, R. E. (1966), A quick test to determine lime requirements for lime stabilization, *Highway Research Record* No. 139, HRB, National Research Council, Washington, D.C., pp. 61–72.

Eggestad, A. (1983), Improvement of cohesive soils, *Proceedings of VIII ECSMFE*, Helsinki, Finland, **3**, pp. 991–1007.

Emery, J. J. (1980), Slag utilization in pavement construction, *ASTM STP 774*, pp. 95–118.

Engineering News Record (1982), Shoehorning Pittsburg Subway, July 22, pp. 30–34.

Esrig, M. I. (1968), Pore pressures, consolidation and electrokinetics, *Journal of the Soil Mechanics and Foundation Engineering Division*, ASCE, **94**, No. SM4, pp. 899–921.

Esrig, M. I. and Henkel, D. J. (1968), The use of elektrokinetics in the raising of submerged partially buried metallic objects, *Soil Engineering Series Research Report*, No. 7, Cornell University, Ithaca, N.Y.

Evans, G. L. and Bell, D. H. (1981), Chemical stabilization of loess, New Zealand, *Proceedings of the 10th ICSMFE*, Stockholm, **3**, pp. 649–658.

Fang, H. Y. (ed.) (1986), Environmental geotechnology, *Proceedings of the 1st International Symposium on Environmental Geotechnology*, Allentown, Pa., **1**, Envo Publishing Co., Inc., Bethlehem, Pa., 685 p.

Fang, H. Y. and Pamukcu, S. (eds.) (1989), Environmental geotechnology, *Proceedings of the 2nd International Symposium on Environmental Geotechnology*, Shanghai, China, **1**, Envo Publishing Co., Inc., Bethlehem, Pa., 550 p.

Farouki, O. T. (1986), *Thermal Properties of Soils*, Trans Tech Publications, Germany.

Ferguson, J. F. and Nelson, P. (1986), Migration of inorganic contaminants in groundwater under the influence of an electric field, Position Paper presented at the USEPA—University of Washington Workshop on Elektrokinetic Treatment for Hazardous Waste Remediation, Seattle, Wash.

Ford, C. M., Moore, R. K., and Hajek, B. F. (1982), Reaction products of lime treated southeastern soils, *Transportation Research Record* No. 839, TRB, National Research Council, Washington, D.C., pp. 38–40.

Gemant, A. (1952), How to compute thermal soil conductivities, *Heating, Piping and Air-Conditioning*, **24**, No. 1, pp. 122–123.

George, K. P. (1969), Cracking in pavements influenced by viscoelastic properties of soil-cement, *Highway Research Record* No. 263, Washington, D.C., pp. 47–59.

George, K. P. (1971), Soil-cement base craking: Theoretical and model studies, *Highway Research Board*, presented at 50th Annual Meeting.

Ghavami, K. and Fang, H. Y. (eds.) (1984), *Low Cost and Energy Saving Construction Materials*, **1**, Envo Publishing Co., Inc., Bethlehem, Pa.

GNK Hayward Baker (1988), *Ground Modification*, Seminar Notes, GNK Hayward Baker Inc., Odenton, Md.

Gray, D. H. and Mitchell, J. K. (1967), Fundamental aspects of electroosmosis in soils, *Journal of the Soil Mechanics and Foundation Engineering Division*, ASCE, **93**, No. SM6, pp. 209–236.

Gregory, C. A., Saylak, D., and Ledbetter, W. B. (1984), The use of by-product phosphogypsum for road bases and subbases, *Transportation Research Record* No. 998, TRB, National Research Council, Washington, D.C., pp. 47–52.

Gutschick, K. A. (1978), Lime stabilization under hydraulic conditions, *4th International Lime Congress*, Hershey, Pa.

Hayden, M. L. and Haliburton, T. A. (1976), Improvement of dispersive clay erosion resistance by chemical treatment, paper presented at the 1976 Annual Meeting of the Transportation Research Board, Washington, D.C.

Head, W. J., McQuade, P. V., and Anderson, R. B. (1982), Coal refuse and fly ash compositions: Potential highway base course materials, *Transportation Research Record* No. 839, TRB, National Research Council, Washington, D.C., pp. 11–19.

Hecht, N. L. and Duvall, D. S. (1975), *Characterization and Utilization of Municipal and Utility Sludges and Ashes: Vol III—Utility Coal Ash*, Report prepared for the National Environmental Research Center, U.S.E.P.A.

Herzog, A. and Mitchell, J. K. (1963), Reactions accompanying stabilization of clay with cement, *Highway Research Record* No. 36, Washington, D.C., pp. 146–171.

HRB (1946), Soil bituminous roads, *Current Road Problems Bulletin* No. 12, Highway Research Board, Washington, D.C.

Hurley, C. H. and Thornburn, T. H. (1972), Sodium silicate stabilization of soils: A review of the literature, *Highway Research Record* No. 381, Washington, D.C., pp. 46–79.

Ingles, O. G. and Metcalf, J. B. (1973), *Soil Stabilization Principles and Practice*, John Wiley & Sons, Inc., New York, N.Y.

Ingles, O. G. and Lim, N. (1980), Accelerated laterization. A new method for clay stabilization, *Proceedings of the 7th Conference, Australian Clay Minerals Society*.

ISGF (1985), *Proceedings of the 4th International Symposium on Ground Freezing*, A. A. Balkema, Boston, Mass.

ISGF (1988), *Proceedings of the 5th International Symposium on Ground Freezing*, A. A. Balkema, Rotterdam, Netherlands.

Janin, J. and Le Sciellour, G. F. (1970), Chemical grouting for Paris rapid transit tunnels, *Journal of the Construction Division*, ASCE, No. CO1, pp. 61–74.

Jessberger, H. L. (1982), State of the art report—Ground freezing: Mechanical properties, processes, and design, *Ground Freezing 1980*, Developments in Geotechnical Engineering, **28**, Elsevier Scientific Publishing Co., New York, N.Y., pp. 5–53.

Jessberger, H. L. (1983), Soil Grouting General Report, Specialty Session No. 2, *Proceedings of the 8th ECSMFE*, Helsinki, pp. 1069–1088.

Johansen, O. (1975), Thermal conductivity of soils, Ph.D. Thesis, Trondheim, Norway (CRREL Draft Translation, 637, 1977). ADA 044002.

Johnson, J. C. (1976), Asphalt stabilization, *Materials for Stabilization. Education and Information Guide HC-100A*, American Road Builders Association, Washington, D.C.

Jones, J. S. (1982), State of the art report: Engineering practice in artificial ground freezing, *Ground Freezing 1980*, Developments in Geotechnical Engineering, **28**, Elsevier Scientific Publishing Co., New York, N.Y., pp. 313–326.

Jones, P. C. T. (1948), The microbial decomposition of resinous stabilizing agents in soil, *Proceedings of the 2nd International Conference on Soil Mechanics and Foundation Engineering*, Rotterdam, **4**, pp. 280–284.

Jonker, C. (1982), Sub-grade improvement and soil cement, *Proceedings of the International Symposium on Concrete Roads*, London.

Jurdanov, A. (1978), Special function of deep thermal treatment of soils and its development, *Osnov. Fund. Mech. Grunt.*, **6**, No. 20, pp. 14–16.

Karpoff, K. P. (1976), Stabilization of fine-grained soils by electro-osmotic and electro-chemical methods, *New Horizons in Construction Materials*, ed. H. Y. Fang, Envo Publishing Co., Inc., Bethlehem, Pa., pp. 265–272.

Kelley, W. P. (1948), *Cation Exchange in Soils*, Reinhold Publishing Co., New York, N.Y., 144 p.

Kernsten, M. S. (1949), The thermal properties of soils, *Bulletin 28*, Engineering Experiment Station, University of Minnesota, Minneapolis.

Kezdi, A. (1979), *Stabilized Earth Roads*, Developments in Geotechnical Engineering 19, Elsevier Scientific Publishing Co., New York, N.Y.

Khan, L. I. and Pamukcu, S. (1989), Validity of electro-osmosis for groundwater decontamination, *Environmental Engineering*, Proceedings of the 1989 Specialty Conference, ASCE, Austin, Texas, pp. 563–570.

Khan, L. I., Pamukcu, S., and Kugelman, I. J. (1989), Electro-osmosis in fine grained soil, *Proceedings of the 2nd International Symposium on Environmental Geotechnology*, Shanghai, China, **1**, Envo Publishing Co., Inc., Bethlehem, Pa., pp. 39–47.

Koerner, R. M., Sands, R. N., and Leaird, J. D. (1985), Acoustic emission monitoring of grout movement, *Issues in Dam Grouting*, ASCE, Denver, Colo., pp. 149–155.

Kropatschek, H. and Rienossl, K. (1967), Travaux d'étanchement du sous sol du barrage de Durlassboden, *Transactions 9th International Congress on Large Dams*, Istanbul, **1**, pp. 695–714.

Kujala, K. (1983a), The use of gypsum in deep stabilization, *Proceedings of the 8th ECSMFE*, Helsinki.

Kujala, K. (1983b), The use of x-ray diffraction and scanning electron microscope in stabilization research, *Proceedings of the 8th ECSMFE*, Specialty Session 4, Helsinki, pp. 1145–1146.

Kujala, K., Halkola, H., and Lahtinen, P. (1985), Design parameters for deep stabilized soil evaluated from in-situ and laboratory tests, *Proceedings of the 11th ICSMFE*, San Fransisco, California, **3**, pp. 1717–1720.

Lamb, J. H. (1985), Type C flyash and clay stabilization, *Developments of New and Existing Materials*, Proceedings of a Session sponsored by Materials Engineering Division of ASCE, ASCE Convention, Detroit, Michigan.

Leonards, G. A. and Bailey, B. (1982), Pulverized coal ash as structural fill, *Journal of Geotechnical Engineering, ASCE*, **108**, No. GT4, pp. 517–531.

Litvinov, I. M. (1960), Stabilization of settling and weak clayey soils by thermal treatment, *Highway Research Board Special Report* No. 60, Washington, D.C., pp. 94–112.

Mateos, M. (1964), Heat curing of sand–lime–flyash mixtures, *ASTM Materials Research and Standards*, **4**, No. 5, pp. 212–217.

Melancon, J. L. and Shah, S. C. (1973), *Soil-Cement Study, Final Report*, No. 72, Project No. 68-9S, Louisiana Department of Highways, U.S. D.O.T. Federal Highway Administration.

Meyers, J. F., Pichumani, R., and Kapples, B. S. (1976), Fly-ash as a construction material for highways: A manual, *Report No. FHWA-IP-76-16*, Federal Highway Administration, U.S. D.O.T., Washington, D.C.

Micheals, A. S. and Puzinauskas, V. (1956), Additives as aids to asphalt stabilization of fine grained soils, *Highway Research Board Bulletin* No. 129, National Research Council, Washington, D.C.

Micheals, A. S. and Puzinauskas, V. (1958), Improvement of asphalt stabilized fine grained soils with chemical additives, *Highway Research Board Bulletin* No. 204, National Research Council, Washington, D.C.

Mickley, A. S. (1951), The thermal conductivity of moist soil, *American Institute of Electrical Engineers Transactions*, **70**, pp. 1789–1797.

Miki, G. and Nakanishi, W. (1984), Technical progress of the jet grouting method and its newest type, *Proceedings of the International Conference on In-situ Soil and Rock Reinforcement*, Paris, pp. 195–200.

Mitchell, J. K. (1970), In-place treatment of foundation soils, *Journal of the Soil Mechanics and Foundation Engineering Division, ASCE*, **96**, No. SM1, pp. 73–109.

Mitchell, J. K. (1976), *Fundamentals of Soil Behavior*, John Wiley and Sons, Inc., New York, N.Y.

Mitchell, J. K. (1981), Soil improvement—State of the art report, *Proceedings of the 10th ICSMFE*, Stockholm, **4**, pp. 509–565.

Mitchell, J. K. (1986a), Potential uses of electro-kinetics for hazardous waste site remediation, Position Paper presented at the *USEPA–University of Washington Workshop on Electrokinetic Treatment for Hazardous Waste Remediation*, Seattle, Wash.

Mitchell, J. K. (1986b), Practical problems from surprising soil behavior, *Journal of the Geotechnical Engineering Division, ASCE*, **112**, No. 3, pp. 259–289.

Mitchell, J. K. and Raad, L. (1973), Control of volume changes in expansive earth materials, *Proceedings of Workshop on Expansive Clays and Shales in Highway Design and Construction*, Vol. 2, Federal Highway Administration, Washington, D.C.

Moller, D. W., Minch, H. L., and Welsh, J. P. (1984), Ultrafine cement pressure grouting to control groundwater fractured granite block, *Innovative Cement Grouting*, SP-83, ACI, Detroit, Michigan.

Mongilardi, E. and Tornaghi, R. (1986), Construction of large underground openings and use of grouts, *Proceedings of the International Conference on Deep Foundations*, Beijing, China.

Murphy, G. (1957), *Properties of Engineering Materials*, Int. Text Book Co., Scranton, Pa.

Natt, G. S. and Joshi, R. C. (1984), Properties of cement and lime-flyash stabilized aggregate, *Transportation Research Record* No. 998, TRB, National Research Council, Washington, D.C., pp. 32–40.

NLA (1985), *Lime Stabilization Construction Manual*, Bulletin 326, National Lime Association, Arlington, Va.

Noble, D. F. (1967), Reactions and strength development in portland cement–clay mixtures, *Highway Research Record* No. 198, Washington, D.C., pp. 39–56.

OECD (1984), Use as binder of by-products or natural materials with hydraulic and pozzolanic properties, *Road Binders and Energy Savings*, Road Transport Research, Organization for Economic Co-operation and Development, Paris, France.

Olsen, H. W. (1984), Osmosis and geomechanical processes, Program with Abstracts, *Clay Minerals Society Annual Meeting*, September 30–October 4, 1984, Baton Rouge, La., p. 93.

Ormsby, W. C. and Bolz, L. H. (1966), Microtexture and composition of reaction products in the system kaolin–lime–water, *Journal of the American Ceramic Society*, **49**, No. 7, pp. 364–366.

Pamukcu, S., Kavulich, M. A., and Fang, H. Y. (1987), A parametric sensitivity study on mechanical performance of fly-ash mixed soil, *Proceedings of the 19th Mid-Atlantic Industrial Waste Conference*, Technomic Publishing Co., Inc., Lancaster, Pa., pp. 589–599.

Pamukcu, S., Kugelman, I. J., and Lynn, J. D. (1989), Solidification and reuse of steel industry sludge waste, *Proceedings of 21st Mid-Atlantic Industrial Waste Conference*, Technomic Publishing Co., Lancaster, Pa., pp. 3–15.

Parish, P. W. C., Baker, W. H., and Rubright, R. M. (1983), Underpinning with Chemical Grout, ASCE, *Civil Engineering*, August.

PCA (1956), *Soil-Cement Construction Handbook*, Portland Cement Association, Stokie, Ill.

PCA (1971), *Soil-Cement Laboratory Handbook*, Portland Cement Association, Stokie, Ill.

PCA (1978), *Soil-Cement Construction Handbook*, Portland Cement Association, Stokie, Ill.

Pojasek, R. B. (ed.) (1982), *Toxic and Hazardous Waste Disposal*, Vol. 1, Ann Arbor Science Publishers, Ann Arbor, Mich.

Probstein, R. F. and Renauld, P. C. (1986), Quantification of fluid and chemical flow in electrokinetics, Position Paper presented at the *USEPA–University of Washington Workshop on Electrokinetic Treatment for Hazardous Waste Remediation*, Seattle, Wash.

Robbins, G. E. and Packard, R. G. (1979), Soil cement—A construction material, *Transportation Research Record* No. 702, TRB, National Research Council, Washington, D.C., pp. 173–181.

Roberts, J. D. (1986), Performance of cement-modified soils: A follow up report, *Transportation Research Record* No. 1089, TRB, National Research Council, Washington, D.C., pp. 81–86.

Sadovsky, A. V. and Dorman, Y. A. (1982), The artificial freezing and cooling of soil at construction sites, *Ground Freezing 1980*, Developments in Geotechnical Engineering **28**, Elsevier Scientific Publishing Co., New York, N.Y., pp. 327–331.

Salley, R. J., Foreman, B., Baker, W. H., and Henry, J. F. (1987), Compaction grouting test program: Pinopolis West Dam, *Proceedings of Soil Improvement—A 10 Year Update*, Geotechnical Division, ASCE, Atlantic City Convention, pp. 245–269.

Sanger, F. J. and Sayles, F. H. (1979), Thermal and rheological computations for artificially frozen ground constructions, *Engineering Geology*, No. 13, pp. 311–337.

Sayles, F. H. (1966), *Low Temperature Soil Mechanics*, Technical Note, U.S. Army Cold Regions Research and Engineering Laboratory.

Segall, B. A., O'Bannon, C. E., and Matthias, J. A. (1980), Electro-osmosis chemistry and water quality, Technical Note, *Journal of Geotechnical Engineering, ASCE*, **106**, No. GT10, pp. 1148–1152.

Schuster, J. A. (1972), Controlled freezing for temporary ground support, *Proceedings of the 1st North American Rapid Excavation and Tunneling Conference*, Chicago, **2**, pp. 863–894.

Skok, E. L., Jr., Mathur, T. S., Wenck, N. G., and Ramsey, N. (1983), Ten-year performance report on asphalt stabilized sand road with instrumentation, *Transportation Research Record* No. 898, TRB, National Research Council, Washington, D.C., pp. 232–241.

Sparks, L. D. (ed.) (1986), *Soil Physical Chemistry*, CRC Press Inc., Boca Raton, Fla.

Stern, O. (1924), Zur Theorie der Elektrolytischen Doppelschriht, *Z. Electrochem.*, **30**, pp. 508–516.

Stocker, P. T. (1972), *Diffusion and Diffuse Cementation in Lime and Cement Stabilized Clayey Soils*, Special Report 8, Australian Road Research Board, Victoria, Australia.

Styron, R. (1980), Fly ash specifications and quality control, *Proceedings of the Fly Ash Applications in 1980 Conference*, Texas A&M University, Texas.

Sutton, J. R., Myers, D. A., and Jensen, W. H. (1985), Soil stabilization with cementitious fly ash, Materials and Engineering Testing Session, *ASCE Spring Convention*, Denver, Colorado.

Terzaghi, K. and Lacroix, Y. (1964), The Mission Dam, *Proceedings of the Institution of Civil Engineers*, London.

Tayabji, S. D., Nussbaum, P. J., and Ciolko, A. T. (1982), Evaluation of heavily loaded cement stabilized bases, *Transportation Research Board* No. 839, TRB, National Research Council, Washington, D.C., pp. 6–11.

Thompson, M. R. (1969), Engineering properties of lime–soil mixtures, *Journal of Materials, ASTM*, **4**, No. 4, pp. 968–969.

Thompson, M. R. (1970), Suggested method of mixture design for lime treated soils, *ASTM STP* 479, ASTM, Philadelphia, Pa.

Timmermann, C. L. (1984), Stabilization of contaminated soils by in-situ vitrification, *Proceedings, U.S. Department of Energy, Annual Environmental Systems Symposium*, Bethesda, Maryland, DOE Report No. PNL-SA-11638; Conf-8403105-1.

Tornaghi, R. (1983), Soil Grouting Co-Report, Specialty Session 2, *Proceedings of the 8th ECSMFE*, Helsinki, pp. 1089–1101.

Tornaghi, R. and Cippo, A. P. (1985), Soil improvement by jet grouting for the solution of tunneling problems, *Proceedings of the 4th International Symposium Tunnelling '85*, Brighton, England, Institution of Mining and Metallurgy, British Tunnelling Society, and the Transport and Road Research Laboratory, Dept. of Transport, pp. 265–276.

Tornaghi, R., Bosco, B., and DePaoli, B. (1988), Application of recently developed grouting procedures for tunneling in Milan urban area, *Proceedings of the 5th International Symposium Tunnelling '88*, London.

TRB (1987), *Lime Stabilization: Reactions, Properties, Design and Construction*, State of Art Report 5, Transportation Research Board, Washington, D.C.

Usmen, M. A., Head, W. J., and Moulton, L. K. (1983), Use of coal associated wastes, in low volume roads, *Transportation Research Record* No. 898, National Research Council, Washington, D.C., pp. 268–277.

Usmen, M. A. and Moulton, L. K. (1984), Construction and performance of experimental base course test sections built with waste calcium sulfate, lime and fly-ash, *Transportation Research Record* No. 998, TRB, National Research Council, Washington, D.C., pp. 52–62.

U.S. BPR (1962), *Aggregate Gradation for Highways*, Bureau of Public Roads, Washington, D.C.

U.S. NAEC (1969), *Soil Stabilization—State of the Art Survey*, Vol. II, NAEC, Eng-7469 Code Ident. No. 80020, U.S. Naval Air Engineering Center, Philadelphia, Pa.

Van Rooyen, M. and Winterkorn, H. F. (1959), Structural and textural influences on thermal conductivity of soils, *Highway Research Board Bulletin* No. 168, National Research Council, Washington, D.C., pp. 143–205.

Vitayasupakorn, V. (1986), Development of an electro-osmotic field test for evaluation of consolidation parameters of soils, Ph.D. Dissertation, University of Washington, Seattle, Wash.

Welsh, J. P. (1983), Chemical grouting utilized for underpinning and water control, *Proceedings of the 8th ECSMFE*, Helsinki, pp. 117–180.

Welsh, J. P. (1986), Construction consideration for ground modification projects, *Proceedings of the International Conference on Deep Foundations*, Beijing, China.

Welsh, J. P., Rubright, R. M., and Coomber, D. B. (1986), Jet grouting for support of structures, *Grouting for Support of Structures*, ASCE, Seattle, Wash.

WES (1961), *Electrical Stabilization of Fine-Grained Soils*, Mis. Pap. No. 3-122, Report 7, U.S. Army Engineers Waterways Experiment Station, Vicksburg, Miss.

WES (1965), Soil stabilization investigations of chemically modified quicklime as stabilizing material, Tech. Rep. 3-455 Report 6, U.S. Army Engineers Waterways Experiment Station, Vicksburg, Miss.

WES (1977), Oldham, J.C., Eaves, R.C., and White, D. W., *Materials Evaluated as Potential Soil Stabilizers*, Misc. Pap. S-77-15, U.S. Army Engineers Waterways Experiment Station, Vicksburg, Miss.

WES (1984), *Foundation Grouting Practices at Corps of Engineers Dams*, Tech. Rep. GL-84-13, U.S. Army Engineers Waterways Experiment Station, Vicksburg, Miss.

White, G. N. and Zelazny, L. W. (1986), Charge properties of soil colloids, *Soil Physical Chemistry*, ed. D. L. Sparks, Chapter 2, CRC Press, Boca Raton, Fla.

Winterkorn, H. F. (1934), Oiling earth roads. Application of surface chemistry, *Industrial and Engineering Chemistry*, 26, pp. 815–819.

Winterkorn, H. F. (1938), Affinity of hydrophilic aggregate for asphaltic bitumen. Use of furfural and its resinous derivatives for improving affinity, *Industrial and Engineering Chemistry*, 30, pp. 362–368.

Winterkorn, H. F. (1940), Physicochemical testing of soils and application of the results in practice, *Proceedings of the Highway Research Board*, 20, pp. 798–806.

Winterkorn, H. F. and Tschebotarioff, G. P. (1947a), Sensitivity of clay to remolding and its possible causes, *Proceedings of the Highway Research Board*, 27, pp. 435–442.

Winterkorn, H. F. (1947b), Fundamental similarities between electro-osmosis and thermoosmosis, *Proceedings of the Highway Research Board*, 27, pp. 443–455.

Winterkorn, H. F. (1949), *Final Report on Beach Sand Stabilization Research*, Bureau of Yards and Docks, Department of Navy No. 15087, 1949.

Winterkorn, H. F. (1955a), The science of soil stabilization, *Highway Research Board Bulletin*, 108, pp. 1–24.

Winterkorn, H. F. (1955b), Potentials in moisture migration, *Proceedings of Conference on Building Materials*, Bulletin 1, Division of Building Research, National Research Council of Canada, pp. 86–101.

Winterkorn, H. F. (1964), Critical consideration of the influence of chemical additives on the properties of soil cement (in German), *Proceedings of the International Workshop on Soil Mechanics in Road Construction*, Vienna, Austria.

Winterkorn, H. F. (1967a), Application of granulometric principles for optimization of strength and permeability of granular drainage structures, *Highway Research Record* No. 203, Washington, D.C., pp. 1–7.

Winterkorn, H. F. (1967b), Soil bituminous stabilization—Discussion, *Highway Research Record* No. 198, Washington, D.C., pp. 68–70.

Winterkorn, H. F. (1971), Analogies between macromeritic and molecular liquids and the mechanical resistance properties of sand and gravel assemblies, *Chemical Dynamics*, Henry Eyring Presentation Book, Wiley-Interscience, New York, N.Y., pp. 504–513.

Winterkorn, H. F. and Eckert, G. W. (1940), Physicochemical factors of importance in bituminous soil stabilization, *Proceedings of the Association of Asphalt Paving Technologists*, 11, pp. 204–257.

Winterkorn, H. F. and Reich, T. H. (1962), Effectiveness of certain derivatives of furfural as admixtures to bituminous soil stabilization, *Highway Research Board Bulletin* No. 357, Washington, D.C., pp. 79–94.

Winterkorn, H. F. and Schmid, W. E. (1971), *Soil Stabilization—Basic Parameters*, AFWL-TR-70-3, Air Force Weapons Laboratory, Kirtland Air Force Base, New Mexico.

Yahiro, T., Yoshida, H., and Nishi, K. (1980), *Soil Improvement Method Utilizing A High Speed Water and Air Jet*, KICT Report No. 33, Kajima Institute of Construction Technology, Tokyo, Japan.

10 STABILITY OF EARTH SLOPES

HSAI-YANG FANG, Ph.D.
Professor of Civil Engineering
Lehigh University

GEORGE K. MIKROUDIS, Ph.D.
President
AVANSE, Ltd.

10.1 INTRODUCTION

The failure of a mass of soil in a downward and outward movement of a slope is called a slide or slope failure. Slides occur in almost every conceivable manner, slowly or suddenly, and with or without any apparent provocation. They are usually caused by excavation, by undercutting the foot of an existing slope, by a gradual disintegration of the structure of the soil, by an increase of the pore water pressure in a few exceptionally permeable layers, or by a shock that liquefies the soil.

Two types of slope stability problems occur in clays; short-term stability (end-of-construction case) and long-term stability (steady seepage case). The short-term case applies after a cut is made in a slope. In excavating for a cut, shear stresses are induced that may cause failure in the undrained state. Theoretically it is possible to analyze the stability of a newly cut slope on the basis of either total or effective stresses; however, since it is difficult to ascertain the distribution of pore pressures under these conditions, the $\phi = 0$ method of analysis (total stress method) has proved more successful.

The long-term case is also encountered in natural slopes and should also be considered in analyzing the stability of embankments. In this case, pore pressures may be assumed to be in equilibrium and are determined from considerations of steady seepage; thus, no excess pore pressures are included. This case is analogous to that of the drained shear test, and effective stress parameters should be used.

On the basis of field observations and laboratory analyses it is concluded that for short-term stability analysis of non-fissured clay, the $\phi = 0$ total stress method is satisfactory. For fissured, overconsolidated clays, the $\phi = 0$ analysis is not safe unless account is taken of reduced strength due to the degree of fissures. The effective stress method of analysis should be used for long-term stability analysis (Bjerrum and Kjaernsli, 1957; Kjaernsli and Simons, 1962; Skempton, 1964), of both nonfissured and overconsolidated fissured (discontinuities) clay.

For overconsolidated clays, Skempton (1964) suggested use of the residual shear strength concept for long-term slope analysis. The residual shear strength can be obtained from slow-drained shear tests and its behavior is discussed in Section 3.13 (Chapter 3).

Detailed discussions on the method of selection of strength parameters and failure surfaces for stability investigation in cohesive soils are given by Lowe (1967), Schuster (1968), Cancelli (1977), and Nguyen (1985).

Stability analysis determines whether the given or proposed slope meets the safety requirements: soil mass under given loads should have an adequate safety factor with respect to shear failure and the deformation of the soil mass under the given loads should not exceed certain tolerable limits. The analysis must be made for the worst conditions, which seldom occur at the time of investigation. Not only is knowledge of analytical methods required, but experience and judgment are necessary to predict probable changes in conditions.

10.2 FACTORS AFFECTING SLOPE STABILITY ANALYSIS

There are numerous factors that affect slope stability analysis. These factors include: failure plane geometry, nonhomogeneity of soil layers, tension cracks, dynamic loading or earthquakes, and seepage flow. Brief discussions of each of these factors are presented in the following.

The uses of the circular arc and logarithmic spiral failure planes for stability analysis have been discussed by Spencer (1969) and Chen (1970). On the basis of numerical analysis, Spencer suggests that the circular arc is more critical than the logarithmic spiral arc for the cross section of the rupture surface. However, Chen points out that the shape of the failure plane is not sensitive in the analysis of the stability problem. The advantage in assuming that the logarithmic spiral angle equals the value of the friction angle of the soil is that all intergranular forces acting on the spiral slip surface are directed toward the center of the spiral. Because of this condition, the analysis becomes statically determinate, and it permits the stability problem to be solved in a relatively simple mathematical form.

Soils may be isotropic depending upon the environmental condition of deposition and subsequent stress changes during geological history. Lo (1965), based on Taylor's work, developed a general method of stability analysis for anisotropic soils subject to undrained conditions. A mathematical expression has been obtained for the cases in which the vertical strength is constant with depth and in which it increases linearly with depth. It was found that for steep slopes, the effect of anisotropy is small. However, for flatter slopes, the influence of anisotropy on the stability condition is significant. The effect of anisotropy on the solution can also be found by limit analysis techniques, which are discussed in Section 10.4.

Studying the effect of tension cracks on slope stability, Spencer (1968) found that the overall factor of safety decreased as the depth of the tension crack increased. The reduction in the factor of safety was, however, very small. The effect of water pressure in a tension crack on the position of the critical circle was also found to be rather small.

The effect of dynamic loading including earthquakes on slope stability should be considered. On the basis of laboratory tests, Ellis and Hartman (1967) reported that the dynamic strength of soil may be less or greater than soil strength under static loadings. Therefore, the dynamic soil strength should be determined and then applied to the stability factor of safety, which may be done by expressing the dynamic soil strength as a ratio or percentage of the normal static strength.

The analysis of the behavior of a slope during an earthquake involves the determination of the yield acceleration and the distribution of velocity and displacement that occur at accelerations in excess of yield. The yield acceleration of slopes in cohesionless soils has been studied by Seed and Goodman (1964). Analysis of slopes in cohesive soils and the distribution of velocity and shear strain that occurs due to the effect of shear strain rate on shearing resistance have been reported by Finn (1966). Suggested methods for evaluating slope response to earthquakes and design procedures related to earthquakes are given by Seed (1966, 1967), Sherard (1967), Majumdar (1971), and in Chapter 16 of this book.

If the soil mass is located below the groundwater table or under steady seepage, it is necessary to consider the effect of pore water pressures, and the effective stress should be used instead of the total stress. Pore water pressure can be determined from the groundwater flow net. The Bishop, Morgenstern, and Spencer methods, illustrated later in this chapter, demonstrate the effect of pore water pressure on the stability analysis. Further discussion of this subject may be found in standard textbooks.

10.3 FACTOR OF SAFETY

The factor of safety or degree of safety is used by engineers to indicate whether or not a foundation soil or earth work will fail under the worst service conditions for which it was designed. The present concept for determining the factor of safety for a slope is based on Coulomb's law.

Fellenius (1927) has used the ratio of actual shear strength to critical shear strength, or the ratio of required shear strength (τ) to the available shear strength (S)

$$F_s = \frac{S}{\tau} \tag{10.1}$$

Other factors, suggested by Rendulic (1935) and Jáky (1936), are

$$F_c = \frac{C_c}{C} \tag{10.2}$$

in which F_c, the average factor of safety for the cohesion component of strength, is the ratio between the actual cohesion (C) and the cohesion (C_c) required for stability with full friction mobilized. This ratio is sometimes called the factor of safety with respect to cohesion. Other parameters suggested by Taylor (1937, 1948) are friction angle and critical height. These may be expressed as

$$F_\phi = \frac{\tan \phi_c}{\tan \phi} \quad \text{and} \quad F_H = \frac{H_c}{H} \tag{10.3}$$

in which F_ϕ is the average factor of safety for the frictional component of strength and F_H is called the factor of safety with respect to height. The critical height (H_c) is the maximum height at which it is possible for the slope to be stable and H is the actual height. The factor of safety may also be calculated as the sum of the resisting moments (M_r) divided by the sum of the moments tending to cause failure (M) as (Fellenius, 1927)

$$F = \frac{M_r}{M} \tag{10.4}$$

The application of the methods for computing various factors of safety will be illustrated in later sections. The concept of factor safety as applied to soil and foundation engineering is discussed by Taylor (1948), Fröhlich (1953), Kézdi (1957), and Jumikis (1967).

Wu and Kraft (1967) proposed the use of probability analysis to calculate the safety of foundations for various loads and strength distributions. Applied load and soil strength are considered to be random variables. The appropriate probability function is determined empirically by fitting to the experimental data. The application of reliability theory for analysis of slope failure are given by Alfaro and Harr (1981), Asaoka and

TABLE 10.1 RECOMMENDED FACTORS OF SAFETY FOR SLOPE STABILITY ANALYSIS IN RESIDUAL SOIL REGION[a].

Class	Cutting Type	Factor of Safety	
		(A) Comprehensive Site Investigation[b]	(B) Cursory Site Investigation[c]
1	Road cutting or cutting in remote area where probability of life at risk, owing to failure, is small	1.1	1.2
2	Road cutting on main arterial route where main line communications can be cut and risk to life is possible	1.2	1.3
3	Areas adjacent to buildings where failure would affect stability of building, e.g. car park. Risk to life significant	1.2	1.4
4	Cuts adjacent to buildings where failure could result in collapse of building. Risk to life very great	1.4	Not applicable

[a] After Binnie and Partners (1971); Chiang (1979).
[b] Such a site investigation would, in addition to normal boring and drilling, include a program of laboratory testing to determine shear strength parameters for both soils and rock failures. Joint system surveys would be carried out and likely effects of heavy rainfall on the slopes would also be considered. These effects would be included in the soils and rock stability analyses.
[c] Site investigation under such a classification would be limited to determination of the boundaries of the various grades of material, the type of rock, and also predominant joint patterns in the case of rock stability problems. Shear strength parameters would be derived from back-analysis of failures.

Athanasiou-Grivas (1981), and Wong (1985). Several European countries have adopted such reliability approaches in the development of their standards and codes.

The factor of safety also can be obtained from practical experience, as illustrated in Table 10.1, which provides some guidance in selecting the appropriate factor of safety for slope stability analysis.

10.4 SLOPE STABILITY ANALYSIS PROCEDURE: LIMIT EQUILIBRIUM METHODS

There are numerous methods currently available for performing slope stability analyses. The majority of these may be categorized as limit equilibrium methods. The basic assumption on the limit equilibrium approach is that Coulomb's failure criterion is satisfied along the assumed failure surface, which may be a straight line, circular arc, logarithmic spiral, or other irregular surface. A free body is taken from the slope and starting from known or assumed values of the forces acting upon the free body, the shear resistance of the soil necessary for equilibrium is calculated. This calculated shear resistance is then compared to the estimated or available shear strength of the soil to give an indication of the factor of safety.

Methods that consider only the whole free body include the Culmann method (Taylor, 1948) (Fig. 10.1a) and the friction circle method (Taylor, 1948) (Fig. 10.3a). Another approach is to divide the free body into many vertical slices and to consider the equilibrium of each slice. There are several versions of the

slice method available; the best known are the Swedish circle method (Fellenius, 1927) (Fig. 10.1b) and the Bishop (1955) method (Fig. 10.1c). The details of Bishop's method will be discussed in Section 10.4.2.

If the failure surface can be approximated satisfactorily by two or three straight lines, this type of failure analysis is known as a wedge method. Detailed discussions of the wedge method with numerical examples are given by Perloff and Baron (1976) and Lambe and Whitman (1979).

Extensive refinements of the limit equilibrium method have been undertaken by many investigators. These refinements have generally been concerned with defining a more acceptable failure surface or with modifying the method by which the forces acting on the failure surface are handled. The forces assumed to be acting on a slice in the various methods are summarized in Figures 10.1 and 10.2. Some currently used limit equilibrium methods including those of Taylor (1937, 1948), Bishop (1955), Bishop and Morgenstern (1960), Morgenstern (1963), Spencer (1967), Hunter and Schuster (1968, 1971), Huang (1975, 1980), and Koppula (1984) are presented in Sections 10.4 and 10.6. The majority of these methods are in chart form and cover a wide variety of conditions. Numerical examples are given for each method and are easily used by practical engineers.

The limit equilibrium method for analysis is widely used for slope stability problems; however, it has several inherent weaknesses, the most important of which is that it neglects the soil stress–strain relationship. An alternative method for slope stability analysis called the *limit analysis* method has been introduced (Drucker and Prager, 1952) and has been refined and extended for further applications (Chen, 1975; Snitbhan

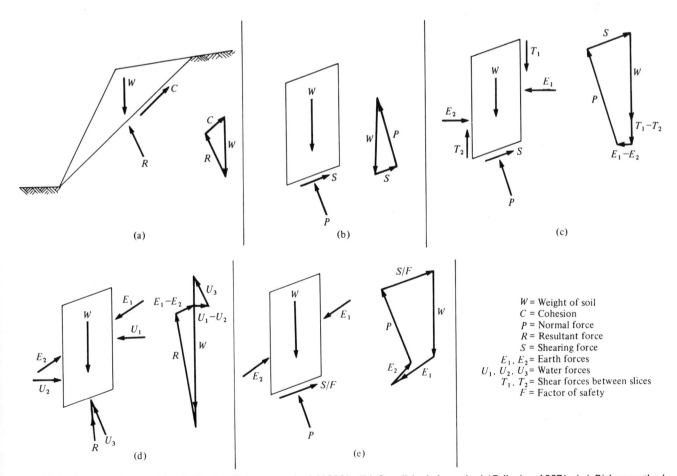

Fig. 10.1 Forces acting on a free body. (a) Culmann method (1866); (b) Swedish circle method (Fellenius, 1927); (c) Bishop method (1955); (d) Lowe and Karafiath method (1960); (e) Spencer method (1967).

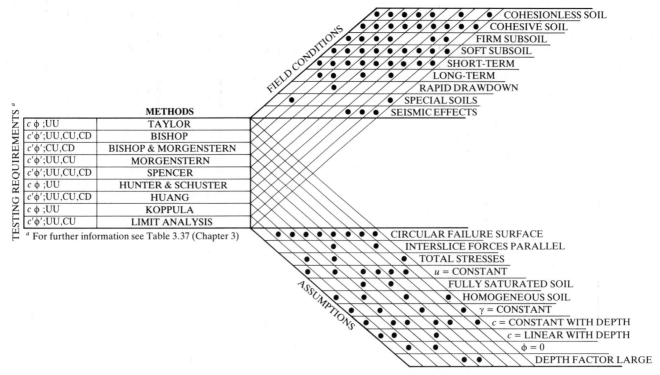

Fig. 10.2 Classification of slope stability analysis.

et al., 1975; Mizuno and Chen, 1984; Zhang and Chen, 1987; Huang and Chen, 1990).

10.4.1 Taylor Method

The Taylor method is based on the friction circle method (Taylor, 1937, 1948), which is illustrated by the diagram shown in Figure 10.3a. The radius of the circular failure surface is designated by R. The radius of the friction circle is equal to $R \sin \phi$. Any line tangent to the friction circle must intersect the circular failure arc at an oblique angle ϕ. Therefore, any vector representing an intergranular pressure at oblique ϕ to an element of the failure surface must be tangent to the friction circle. The analysis is based on total stresses and assumes that the cohesion c is constant with depth. For a given value of ϕ the critical height of a slope is given by the equation

$$H_c = N_s \frac{c}{\gamma} \qquad (10.5)$$

where H_c = critical height, c = cohesion, γ = unit weight of soil, and N_s = stability factor.

The stability factor N_s is a pure number, depending only on the slope angle β and friction angle ϕ. The relationships between N_s, β, and ϕ are shown in Figure 10.3b. In the chart the β value varies from 0° to 90° with ϕ varying from 0° to 25°. The depth factor D, shown in Figure 10.4 is defined as the depth to the hard stratum divided by the height of the slope. In the chart, the value of D varies from 1.0 to ∞ with β varying from 0° to 90°.

EXAMPLE 10.1

A cut is to be excavated in a soil that has a cohesion c of 240 psf, a unit weight of 120 pcf, and an angle of internal friction ϕ of 10°. The design calls for a slope angle β of 60°. What is the maximum depth of cut that can be made and still maintain a factor of safety of 1.5 with respect to the height of the slope?

Solution

From Figure 10.3b, for $\phi = 10°$ and $\beta = 60°$, N_s may be read as 7.15; from Equation 10.5 we obtain

$$H_c = (c)(N_s)/\gamma = (240)(7.15)/120 = 14.3 \text{ ft}$$

and from Equation 10.3

$$H = H_c/1.5 = 14.3/1.5 = 9.5 \text{ ft}$$

EXAMPLE 10.2

A cut is to be excavated in soft clay to a depth of 30 ft. The soil has a unit weight of 120 pcf and a cohesion of 600 psf. Hard stratum underlies a soft layer at a depth of 45 ft below the ground surface. What is the slope angle β, if any, at which failure is likely to occur?

Solution

For a soft clay, ϕ is assumed to be zero. Therefore, Figure 10.4 is applicable, and the depth factor

$$D = 45/30 = 1.5$$

If failure is to occur, the critical height H_c is 30 ft:

$$N_s = \gamma H_c/c = (120)(30)/600 = 6.0$$

From Figure 10.4, for $D = 1.50$ and $N_s = 6.0$, β may be read as 32°.

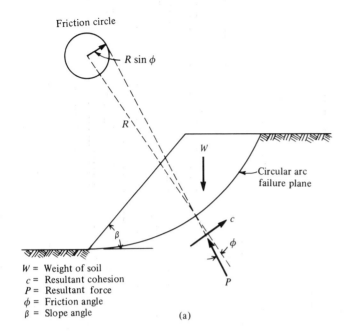

W = Weight of soil
c = Resultant cohesion
P = Resultant force
ϕ = Friction angle
β = Slope angle

(a)

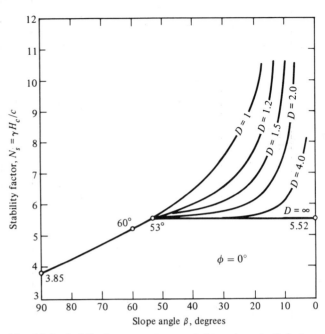

Fig. 10.4 Stability factor vs. slope angle for various depth factors D. (*After Taylor, 1937; and Terzaghi and Peck, 1967.*)

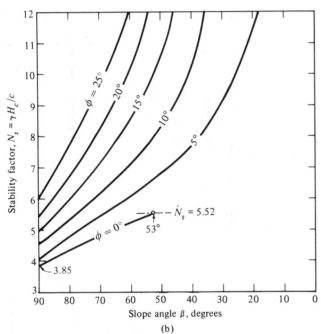

(b)

Fig. 10.3 Taylor method. (a) Friction circle method. (b) Stability factor vs. slope angle with various friction angles ϕ. (*After Taylor, 1937; and Terzaghi and Peck, 1967.*)

10.4.2 Bishop Method

If a slope consists of several types of material with different values of c and ϕ, and if the pore pressures u in the slope are known or can be estimated, the Bishop method of slices (Bishop, 1955) is useful.

In this method the mass of soil *acdbfe* (Fig. 10.5) is divided into many vertical slices. The forces acting on each slice are evaluated from the limit equilibrium of the slices. The equilibrium of the entire mass is determined by summation of the forces on all the slices.

Consider the forces on an individual slice *cdef*. They consist of the weight of the slice W, the surface load acting on the slice Q,

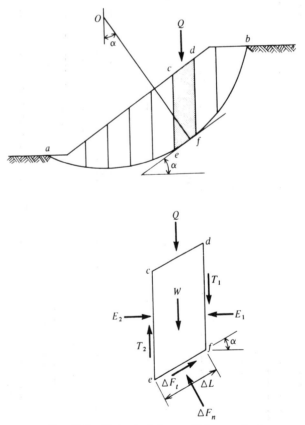

Fig. 10.5 Method of slices—Bishop method.

the normal and shear forces ΔF_n and ΔF_t acting on the failure surface *ef*, and the normal and shear forces E_1, T_1, E_2, and T_2 on the vertical faces *cedf*. The system is statically indeterminate, and to arrive at a solution it is necessary to make certain assumptions concerning the magnitudes and points of application of the forces E and T.

An approximate solution may be obtained by assuming that the resultants of E_1 and T_1 are equal to that of E_2 and T_2 and that their lines of action coincide. This assumption greatly simplifies the calculation. By applying the condition of equilibrium to the slice, we obtain

$$\Delta F_n = (W + Q) \cos \alpha$$
$$\Delta F_t = (W + Q) \sin \alpha \tag{10.6}$$

The unit pressure normal to and shear stresses along ΔL are equal to

$$\sigma_n = \frac{1}{\Delta L}(W + Q) \cos \alpha$$
$$\tau_n = \frac{1}{\Delta L}(W + Q) \sin \alpha \tag{10.7}$$

The shearing strength s is determined by

$$s = c' + \sigma' \tan \phi'$$

where

σ' = effective normal stress
c' = effective cohesion
ϕ' = effective angle of internal friction

The total shear force over the entire arc is the sum of the shear forces on the slices and is equal to $\Sigma(W + Q) \sin \alpha$. The shearing resistance on a slice of the arc is

$$s\Delta L = c' \Delta L + [(W + Q) \cos \alpha - u \Delta L] \tan \phi'$$

The factor of safety is then given by the ratio (Eq. 10.1)

$$F = \frac{s}{\tau}$$

or

$$F = \frac{\Sigma\{c' \Delta L + [(W + Q) \cos \alpha - u \Delta L] \tan \phi'\}}{\Sigma(W + Q) \sin \alpha} \tag{10.8}$$

The accuracy of the analysis may be improved by taking forces E and T into consideration. For the element in Figure 10.5, summation of forces in the vertical direction gives

$$\Delta \bar{F}_n \cos \alpha = (W + Q) + (T_1 - T_2) - u \Delta L \cos \alpha - \Delta F_t \sin \alpha \tag{10.9}$$

If the slope is not on the verge of failure ($F > 1$) the tangential force ΔF_t is equal to the shearing resistance on *ef* divided by F:

$$\Delta F_t = \frac{c' \Delta L}{F} + \Delta \bar{F}_n \frac{\tan \phi'}{F}$$

Substituting this into Equation 10.9 and solving for $\bar{F}_n$, we obtain

$$\Delta \bar{F}_n = \left[(W + Q) + (T_1 - T_2) - u \Delta L \cos \alpha - \frac{c'}{F} \Delta L \sin \alpha\right]$$
$$\times \frac{1}{\cos \alpha + (\tan \phi' \sin \alpha / F)} \tag{10.10}$$

The factor of safey is

$$F = \frac{\Sigma(c' \Delta L + \Delta \bar{F}_n \tan \phi')}{\Sigma(W + Q) \sin \alpha} \tag{10.11}$$

Substituting Equation 10.10 in 10.11 we get

$$F = \frac{\Sigma\{c' \Delta L \cos \alpha + [(W + Q - u \Delta L \cos \alpha) + (T_1 - T_2)] \tan \phi'\}[\cos \alpha + (\tan \phi' \sin \alpha / F)]^{-1}}{\Sigma(W + Q) \sin \alpha} \tag{10.12}$$

In order to obtain the factor of safety F from Equation 10.12 the quantity $T_1 - T_2$ must be evaluated by means of successive approximation. Trial values of E_1 and T_1 that satisfy the equilibrium of each slice, and the conditions

$$\Sigma(E_1 - E_2) = 0 \qquad \Sigma(T_1 - T_2) = 0$$

are used. The calculations can be simplified if the term $\Sigma(T_1 - T_2) \tan \phi'$ is assumed to be 0. The value of F may then be computed by first assuming an arbitrary value for F. This assumed value, together with the soil properties c', ϕ', u, and the slope geometry α are substituted into Equation 10.12 and F is then calculated. If the calculated value differs appreciably from the assumed value, a second approximation is made and the computation is repeated. Figure 10.6 developed by Janbu et al. (1956) helps to simplify the computation procedure.

The above approximation taking $\Sigma(T_1 - T_2) \tan \phi'$ as 0 results in an error of only about 1 percent. The error introduced by using Equation 10.8 is about 15 percent.

For reducing the error introduced by Equation 10.8, Ting (1983) recommended that the area for each slice (Fig. 10.5) be exactly computed and the base inclination angle be obtained from the straight line at the slice bottom. This method yields consistently accurate estimates of the resisting moments, overturning moments, and factor of safety using a relatively small number of slices.

The calculations outlined above refer to only one trial circle. Several circles must be analyzed until the minimum value of factor of safety is determined. Hand calculations, graphical methods and computer programs (Section 10.8) may be used. In many cases, the slope consists of several types of materials with different values of c' and ϕ', or with irregular failure surface. For these conditions, the methods of slices have been extended (Janbu, 1954a; Morgenstern and Price, 1965; Nonveiller, 1965); also see Examples 10.11 and 10.12.

10.4.3 Bishop and Morgenstern Method

The Bishop and Morgenstern (1960) method is based on the Bishop method of slices and considers the pore water pressure by means of the pore pressure ratio r_u, which is defined as

$$r_u = \frac{u}{\gamma h} \tag{10.13}$$

where u = pore water pressure; γ = unit weight of the soil; and h is the depth of the point in the soil mass below the ground surface. The pore pressure ratio is assumed to be constant throughout the cross section, which is then called a homogeneous pore pressure distribution. If there are minor variations in r_u throughout the soil layer, an average value of r_u is used. For steady-state seepage a weighted average of r_u over the layer is satisfactory.

The factor of safety F is defined as

$$F = m - (n)(r_u) \tag{10.14}$$

where m and n, called *stability coefficients*, are determined by use of charts in Figures 10.7 to 10.9. Depth factors D of 1.0, 1.25, and 1.5 are used in this solution, where the depth factor is defined by Taylor as the depth to a hard stratum divided by the height of slope. The cohesion c is constant with depth.

The averaging technique for estimating the pore pressure ratio tends to give an overestimation of the factor of safety that, in an extreme case, will be on the order of 7 percent.

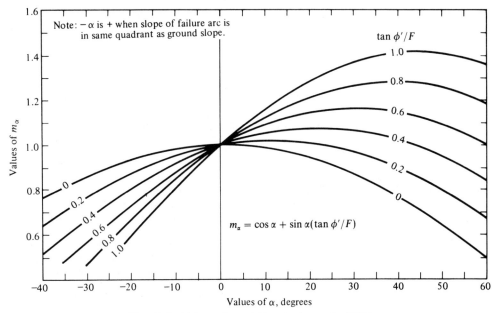

Fig. 10.6 Values of m_α. (*After Janbu et al., 1956.*)

In using this chart solution it is convenient to select the critical depth factor by use of the lines of equal pore pressure ratio, r_{ue}, on the charts of Figures 10.7 and 10.8. The r_{ue} is defined as

$$r_{ue} = \frac{m_2 - m_1}{n_2 - n_1} \qquad (10.15)$$

where n_2 and m_2 are values for a higher depth factor D_2, and n_1 and m_1 correspond to a lower value of the depth factor D_1.

In case the design value of pore pressure ratio is higher than r_{ue} for the given section and strength parameters, then the factor of safety determined with the higher depth factor D_2 has a lower value than the factor of safety determined with the lower depth factor D_1. This is useful to know when no hard stratum exists, or when checking to see whether a more critical circle exists not in contact with a hard stratum.

To determine the minimum factor of safety for sections not located directly on a hard stratum, use the appropriate chart for the given $c'/\gamma H$, and initially, for $D = 1.00$. The values of β and ϕ' indicate a point on the curves of n versus β which is associated with a value of r_{ue} given by the dashed lines. If that value is less than the design value of r_u, the next depth factor, $D = 1.25$, will yield a more critical value of the factor of safety. If, from the chart for $D = 1.25$, the values are checked and r_{ue} is still less than the design value for r_u, move to the chart for $D = 1.50$ with the same value of $c'/\gamma H$.

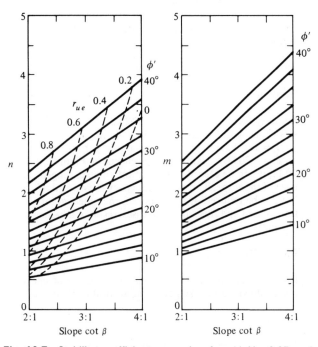

Fig. 10.7 Stability coefficients m and n for $c'/\gamma H = 0.05$ and $D = 1.00$. (*After Bishop and Morgenstern, 1960.*)

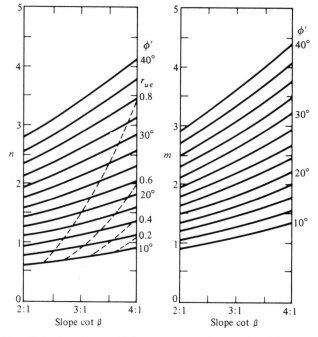

Fig. 10.8 Stability coefficients m and n for $c'/\gamma H = 0.05$ and $D = 1.25$. (*After Bishop and Morgenstern, 1960.*)

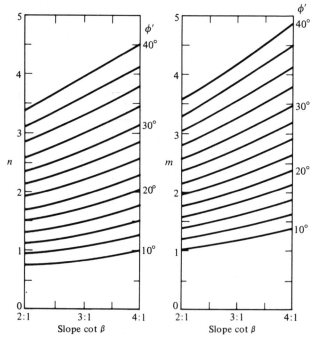

Fig. 10.9 Stability coefficients m and n for $c'/\gamma H = 0.05$ and $D = 1.50$. (*After Bishop and Morgenstern, 1960.*)

Modification of the Bishop and Morgenstern method has been developed by O'Connor and Mitchell (1977). A set of design charts have been produced for slope stability analysis in natural earth slope where $c'/\gamma H = 0.075$ and 0.100. This permits use of the Bishop and Morgenstern stability method in sensitive clay soils such as Canadian Lada clay.

EXAMPLE 10.3 (after Hunter and Schuster, 1971)

A slope is cut so that the slope angle β is 4:1. The cut is 140 ft deep, and hard stratum exists at a depth of 60 ft below the bottom of the cut. The soil has an effective angle of shear resistance ϕ' of 30°. The effective cohesion c' is 770 psf. The unit weight of soil is 110 pcf and it is estimated that the pore pressure ratio r_u is 0.50 for the slope. Determine the minimum factor of safety for the section.

Solution

From the given data we obtain

$$\frac{c'}{\gamma h} = \frac{770}{(110)(140)} = 0.05$$

From Figure 10.7, for $D = 1.0$, $c'/\gamma h = 0.05$, $\phi' = 30°$, $\beta = 4:1$, it is seen that $r_{ue} < 0.5$; therefore, $D = 1.25$ is the more critical depth factor. Then, from Figure 10.8 for $D = 1.25$ and $c'/\gamma h = 0.05$ it is found that $r_{ue} > 0.5$. In this case the maximum value that D could have is $(140 + 60)/140 = 1.43$, and within the limitations of the charts, $D = 1.25$ is thus the critical depth factor. From Figure 10.8 it is seen that

$$m = 3.2 \quad n = 2.8$$

From Equation 10.14 with $r_u = 0.50$, it follows that the minimum factor of safety for the section is

$$F = m - (n)(r_u) = 3.2 - 2.8(0.50) = 1.8$$

Further discussion as well as additional design charts and tables for various depth factors D and $c'/\gamma h$ values are contained in Bishop and Morgenstern's paper.

10.4.4 Morgenstern Method

The Morgenstern (1963) method is based on the Bishop and Morgenstern method of slices. The primary use for this method is to compute the factor of safety of earth slopes during rapid drawdown. As a reservoir level is lowered, the factor of safety decreases if it is assumed that no dissipation of pore pressure occurs during drawdown. Morgenstern assumed that the slope is a simple slope of homogeneous material resting on a rigid impermeable layer at the toe of the slope. The soil comprising the slope has effective stress parameters c' (cohesion) and ϕ' (angle of shear resistance), both of which remain constant with depth. The slope is completely flooded prior to drawdown, that is, a full submergence condition exists.

The pore pressure parameter $\bar{B} = \Delta u/\Delta \sigma_1$ (Skempton, 1954; Bishop, 1954) is assumed to be unity during drawdown. The unit weight γ of the soil is assumed to be constant at twice the unit weight of water or 124.8 pcf. In addition, it is assumed that the pore pressure can be approximated by the product of the height of soil above a given point times the unit weight of water.

The drawdown ratio is defined as L/H, where L is the amount of drawdown and H is the original height of the slope (see Fig. 10.10).

All assumed potential sliding circles must be tangent to the base of the section. This means that the value of H in the stability factor $c'/\gamma H$ and in L/H must be adjusted for intermediate levels of tangency.

Morgenstern's charts cover a range of stability factor $c'/\gamma H$ from 0.125 to 0.050 for slopes of 2:1 to 5:1. The maximum value of ϕ' is 40° (see Figs. 10.10 to 10.12).

EXAMPLE 10.4 (after Morgenstern, 1963)

An embankment has a height H of 100 ft. It is composed of a soil with an effective cohesion c' of 312 psf and an effective angle of shear resistance ϕ' of 30°. The unit weight of the soil is assumed to be 124.8 pcf ($\gamma = 2\gamma_\omega$). The embankment is to have a slope, $\beta = 3:1$. What is the minimum factor of safety for the complete drawdown condition?

Solution

Stability factor $= c'/\gamma H = 312/(124.8)(100) = 0.025$. Using this value, with $\beta = 3:1$, $\phi' = 30°$, and the drawdown ratio $L/H = 1.0$, we obtain the factor of safety from Figure 10.11 as $F = 1.20$. By examining the charts in Figures 10.10 to 10.12, it can be seen that the critical circle is tangent to the base of the slope; if any other tangency is assumed, H would have to be reduced. If H is reduced, then the stability factor is increased and this will, in all cases, result in a higher factor of safety.

EXAMPLE 10.5

It is now required to find the minimum factor of safety for a drawdown to mid-height of the section in Example 10.4.

Solution

Considering slip circles tangential to the base of the slope, the effective height of the section H_e is equal to its actual

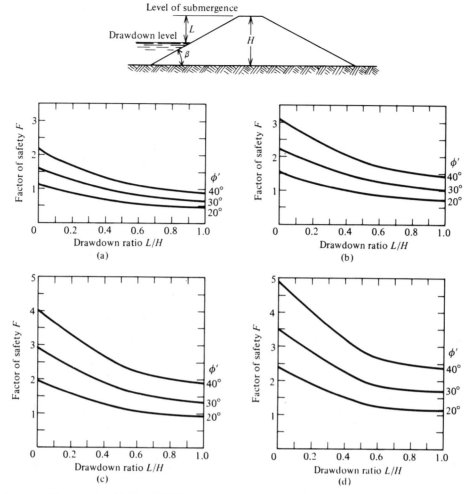

Fig. 10.10 Drawdown stability chart for $c'/\gamma H = 0.0125$. (a) $\beta = 2:1$; (b) $\beta = 3:1$; (c) $\beta = 4:1$; (d) $\beta = 5:1$. (*After Morgenstern, 1963.*)

height and the stability factor remains unchanged as 0.025. With this value of stability factor and $L/H_e = 0.50$, and with other conditions remaining the same, the factor of safety may be read from Figure 10.11 as $F = 1.52$.

Considering slip circles tangential to mid-height of the slope, the effective height is equal to one-half the actual height so that $H_e = H/2 = 100/2 = 50$ ft. Thus $c'/\gamma H_e = 0.05$, and $L/H_e = 1.00$. The minimum factor of safety can be read directly from Figure 10.12 as $F = 1.48$.

Considering slip circles tangential to a level $H/4$ above the base of the slope H_e becomes $3H/4 = 75$ ft; thus the stability factor $c'/\gamma H_e = 0.033$, and $L/H_e = 0.67$. Here the minimum factor of safety can be obtained by interpolation. From Figure 10.11 with $c'/\gamma H_e = 0.025$, the factor of safety is 1.31, and from Figure 10.12 with $c'/\gamma H_e = 0.05$, the factor of safety is 1.61. By linear interpolation, the minimum factor of safety is 1.41, for $c'/\gamma H_e = 0.033$.

These illustrations demonstrate that, for partial drawdown, the critical circle may often lie above the base of the slope, and it is important to investigate several levels of tangency for the maximum drawdown level. In the case of complete drawdown the minimum factor of safety is always associated with circles tangent to the base of the slope and the factor of safety at intermediate levels of drawdown need not be investigated. This may not be the case if the pore pressure distribution during drawdown differs significantly from that assumed by Morgenstern.

10.4.5 Spencer Method

Spencer's (1967) method is based on the work of Fellenius (1927) and Bishop (1955). A cylindrical slip surface is assumed and the earth mass within the surface is divided into small vertical slices. The analysis is in terms of effective stress and satisfies two equations of equilibrium, the first with respect to forces and the second with respect to moments. The interslice forces are assumed to be parallel. The forces acting on a typical slice are shown in Figure 10.1(e). The factor of safety F is defined as the quotient of shear strength available divided by the shear strength mobilized (Eq. 10.1). The mobilized angle of shear resistance ϕ'_m is the angle whose tangent is $\tan \phi'/F$. The depth factor D is assumed to be large. A homogeneous pore pressure distribution is assumed and the pore pressure ratio r_u proposed by Bishop and Morgenstern (1960) is used (as shown in Eq. 10.13). The stability factor N_s is defined as

$$N_s = \frac{c'}{(F)\gamma H} \qquad (10.16)$$

where

c' = cohesion with respect to effective stress
F = factor of safety
γ = unit weight of soil
H = mean height of slice

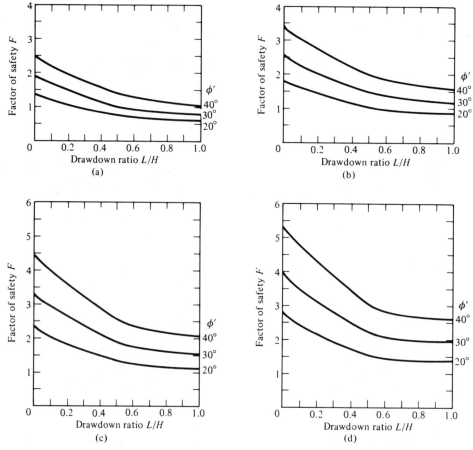

Fig. 10.11 Drawdown stability chart for $c'/\gamma H = 0.025$. (a) $\beta = 2:1$; (b) $\beta = 3:1$; (c) $\beta = 4:1$; (d) $\beta = 5:1$. (*After Morgenstern, 1963.*)

Spencer provides charts for a range of stability factors N_s from 0 to 0.12 with mobilized friction angle ϕ'_m varying from $10°$ to $40°$ and slope angle β up to $34°$. Three pore pressure ratios r_u with values of 0, 0.25, and 0.5 are provided. These charts are shown in Figure 10.13. Values of r_u falling between the charts can be obtained sufficiently accurate for practical purposes by linear interpolation.

EXAMPLE 10.6 (after Spencer, 1967)

An embankment has a height of 100 feet. The soil properties are as follows: $c' = 870$ psf; $\gamma = 120$ pcf; $\phi' = 26°$, and $r_u = 0.5$. Find the slope angle that corresponds to a factor of safety of 1.5.

Solution

The stability factor

$$N_s = c'/(F)\gamma H$$
$$= 870/(1.5)(120)(100)$$
$$= 0.048$$

and

$$\tan \phi'_m = \tan \phi'/F$$
$$= \tan 26°/1.5 = 0.488/1.5 = 0.325$$
$$\phi'_m = 18°$$

Referring to Figure 10.13 for a pore pressure ratio $r_u = 0.5$, the slope corresponding to a stability factor of 0.048 and $\phi'_m = 18°$ is $18.4°$, or 3:1 slope.

10.4.6 Hunter and Schuster Method

The Hunter and Schuster (1968, 1971) method assumes that the potential sliding surface is a circular arc. The soil is saturated to the surface through capillarity and is assumed to be a normally consolidated unfissured clay when $\phi = 0$. The cohesion c varies linearly with depth.

The method considers the depth of the water table by use of the water table ratio M, defined as

$$M = \left(\frac{h}{H}\right)\left(\frac{\gamma_w}{\gamma'}\right) \tag{10.17}$$

where

$h =$ depth from top of slope to the water table during consolidation
$H =$ height of cut
$\gamma_w =$ unit weight of water
$\gamma' =$ submerged or buoyant unit weight of soil

The factor of safety F is defined as:

$$F = \left(\frac{c}{p'}\right)\left(\frac{\gamma'}{\gamma}\right)N_s \tag{10.18}$$

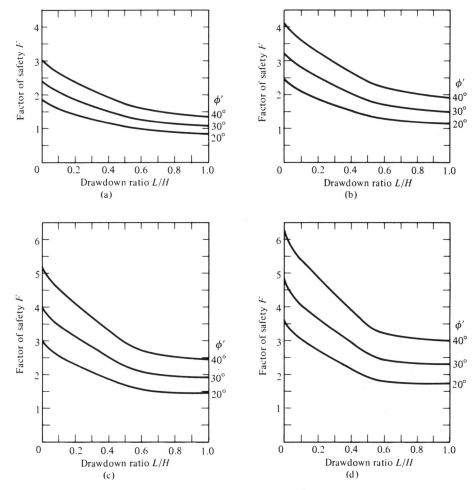

Fig. 10.12 Drawdown stability chart for $c'/\gamma H = 0.05$. (a) $\beta = 2:1$; (b) $\beta = 3:1$; (c) $\beta = 4:1$; (d) $\beta = 5:1$. (*After Mogenstern, 1963.*)

where c = cohesion and p' = effective vertical stress. The c/p' ratio can be estimated from Skempton's (1957) formula $c/p' = 0.11 + 0.0037(PI)$, where PI = plasticity index of the soil in percent and N_s = stability factor, determined from Figure 10.14.

The method also considers the effect of deep, hard strata by using the concept of the depth factor D developed by Taylor. The charts furnished by Hunter and Schuster show the water table ratio varying from 0 to 2 and the depth factor varying from 0 to 4. Following are some example problems using Hunter and Schuster's method.

EXAMPLE 10.7 (after Hunter and Schuster, 1971)

A cut has a height of 15 ft with a slope angle β of 30°. The water table is 5 ft below the ground surface. The unit weight of the soil is 104 pcf and the c/p' ratio for the soil is 0.24. What is the factor of safety for this proposed cut?

Solution

The water table ratio

$$M = \left(\frac{h}{H}\right)\left(\frac{\gamma_w}{\gamma'}\right) = \left(\frac{5}{15}\right)\left(\frac{62.4}{41.6}\right) = 0.50$$

In Figure 10.14 with $M = 0.50$ and $\beta = 30°$, $N_s = 8.9$ (a possible failure).

The factory of safety

$$F = \left(\frac{c}{p'}\right)\left(\frac{\gamma'}{\gamma}\right) \cdot N_s = (0.24)\left(\frac{41.6}{104}\right)(8.9) = 0.85 < 1.00$$

EXAMPLE 10.8 (after Hunter and Schuster, 1971)

A cut is to be made 15 ft deep with slope angle $\beta = 10°$; the water table is 15 ft from the ground surface. Underneath the clay at a depth of 30 ft is a harder, stronger stratum. When tested, the soil showed $\phi = 0°$ on a total stress basis. The c/p' ratio for this soil is 0.24, and its unit weight is 104 pcf. What is the factor of safety for this proposed cut?

Solution

The water table ratio

$$M = \left(\frac{h}{H}\right)\left(\frac{\gamma_w}{\gamma'}\right) = \left(\frac{15}{15}\right)\left(\frac{62.4}{41.6}\right) = 1.50$$

From Figure 10.14, with $\beta = 10°$ and $M = 1.50$, the value of $N_s = 23.9$. Because it is in the deep failure zone, $D = 30/15 = 2.0$ may be important.

From Figure 10.15 for $M = 1.50$, $D = 2.0$, and $\beta = 10°$, it is seen that N_s reduces slightly to 23.2.

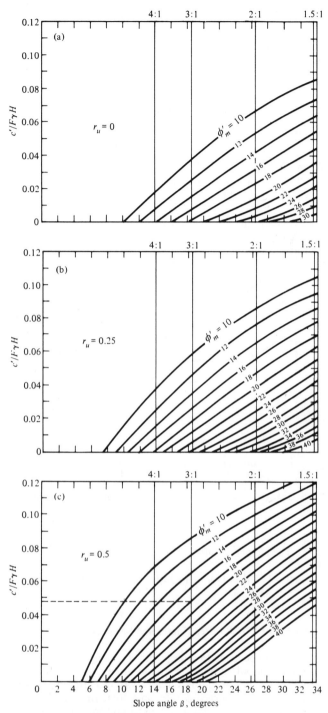

Fig. 10.13 Stability charts. (*After Spencer, 1967.*)

Thus the factor of safety is

$$F = \left(\frac{c}{p'}\right)\left(\frac{\gamma'}{\gamma}\right) \cdot N_s = (0.24)\left(\frac{41.6}{104}\right)(23.2) = 2.23$$

Note that the depth factor, in general, has only a negligible or quite small effect on the factor of safety. Additional design charts for various depth factors D and water table ratios M are contained in Hunter and Schuster's paper.

The solutions developed by Taylor, Bishop, Bishop and Morgenstern, Morgenstern, Spencer, and Hunter and Schuster

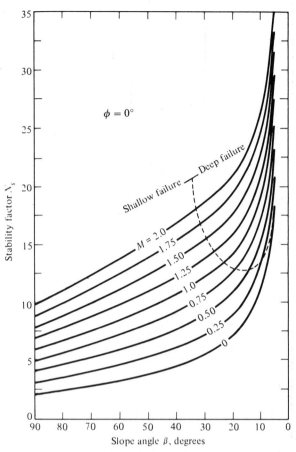

Fig. 10.14 Relationship between slope angle β and stability factor N_s for various water table ratios M. (*After Hunter and Schuster, 1968.*)

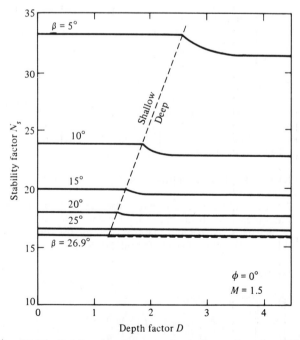

Fig. 10.15 Relationship between depth factor D and stability factor N_s for various slope angles. (*After Hunter and Schuster, 1968.*)

can be applied to a number of slope stability analyses. Of the solutions introduced, those of Taylor and of Hunter and Schuster are best suited to the short-term cases where pore water pressures are not known. The other methods are intended for use in long-term stability cases with known effective stress parameters.

10.4.7 Huang Method

Huang's method (1975, 1983) assumes that the potential sliding surface is a circular arc. Two design charts are developed; one for short-term stability analysis of embankments, and the other for long-term stability analysis. The short-term analysis is employed to insure the stability during or immediately after construction, and the long-term analysis to insure the stability long after construction. The design of earth embankments is generally governed by short-term stability. However, long-term stability should also be considered if the embankments are subjected to steady-state seepage or rapid drawdown.

Short-Term Stability Analysis

The total stress analysis is used to evaluate the short-term stability of embankment, or the stability at the end of construction. It is assumed that soil is completely saturated, and $\phi = 0$ analysis is applied. The stability chart presented in Figure 10.16 is based on a homogeneous simple slope and a circular failure surface. The embankment has a height H, and a slope $S = 1$. A ledge is located at a depth DH below the toe, where D is the depth factor (Fig. 10.4). It can be seen from the figure that the deeper the circle, the smaller the stability factor, so the critical circle is always tangent to the ledge.

The center of the circle is at a horizontal distance XH and a vertical distance YH from the edge of the embankment. By equating the driving moment about the center of the circle due to weight to the resisting moment due to the cohesion of the soil along the failure arc, the developed cohesion c_d is obtained. The stability factor N_s is defined as

$$N_s = \frac{\gamma H}{c_d} \qquad (10.19)$$

in which γ = unit weight of soil. For a section of homogeneous soil with the critical circle passing below the toe, it can be easily proved that the center of the critical circle lies on a vertical line intersecting the slope at mid-height, or $X = 0.5S$. This type of failure surface is called a *midpoint circle*, the results of which are presented by the solid curves in Figure 10.16. If the depth factor D is small, the failure surface may intersect the slope at or above the toe. This type of failure surface is called a *toe or slope circle*, the results of which are presented by the dashed curves in Figure 10.15. For $D = 0$ and $S = 2.5$, the value of X for the center of the critical circle may be different from $0.5S$, as noted in the figure. Following are example problems using Huang's method.

EXAMPLE 10.8 (after Haung, 1975)

An embankment is 20 ft high with a slope of 3:1 and a cohesion of 1500 psf. The foundation consists of two soil layers as shown in Figure 10.17a. The assumption of a cohesion as low as 100 psf is not realistic but is used merely to indicate the contrast in shear strength among different layers. Although the unit weights for different soils are generally not the same, an average unit weight of 130 pcf is assumed. What is the factor of safety for this slope with multiple soil layers?

Solution

Because the weakest layer lies directly above the ledge, the critical circle will be tangent to the ledge, and the depth factor is

$$D = 60/20 = 3$$

When the slope $S = 3$ and depth factor $D = 3$ taken from Figure 10.16, the most critical circle is a midpoint circle with a center located at $YH = (2)(20) = 40$ ft above the top of embankment. Also from Figure 10.16, with $S = 3$ and $D = 3$, the value of the stability factor is 5.7. The developed cohesion c_d, or the cohesion that is actually mobilized, can be determined by Equation 10.19.

$$c_d = \frac{\gamma H}{N_s} = (130)(20)/5.7 = 456 \text{ psf}$$

Because the slope is not homogeneous, the center of the most critical circle may be different from that of a homogeneous slope. However, it was found that, unless D and S are small, the difference between the two centers is generally small, so the critical center for homogeneous slopes can be used for nonhomogeneous slopes as well.

The average effective cohesion c_e, or the maximum cohesion that can be mobilized along the failure arc, can be determined by measuring the length of arc through each soil and taking the weighted average (Fig. 10.17b).

$$c_e = \frac{(24)(1500) + (2)(57)(800) + (142)(100)}{24 + (2)(57) + 142}$$

$$= 505 \text{ psf} \quad (24.2 \text{ kPa})$$

By definition (Section 10.3), the factor of safety

$$F = \frac{c_e}{c_d} = \frac{505}{456} = 1.11$$

If a circle of larger radius with $Y = 6$, or $YH = (6)(20) = 120$ ft is used, from Figure 10.16, $N_s = 6.05$, or

$$c_d = \frac{(130)(20)}{6.05} = 430 \text{ psf} \quad (20.6 \text{ kPa})$$

By measuring the length of arc through each soil layer,

$$c_e = \frac{(28)(1500) + (2)(70)(800) + (180)(100)}{28 + (2)(70) + 180}$$

$$= 494 \text{ psf} \quad (23.7 \text{ kPa})$$

so that, factor of safety

$$F = 494/430 = 1.15$$

It can be seen that the use of the critical center based on a homogeneous slope still yields a smaller factor of safety. When the critical failure surface is tangent to the bottom of a weak stratum and the center is moved upward, the decrease in effective cohesion due to larger radius is not as significant as the decrease

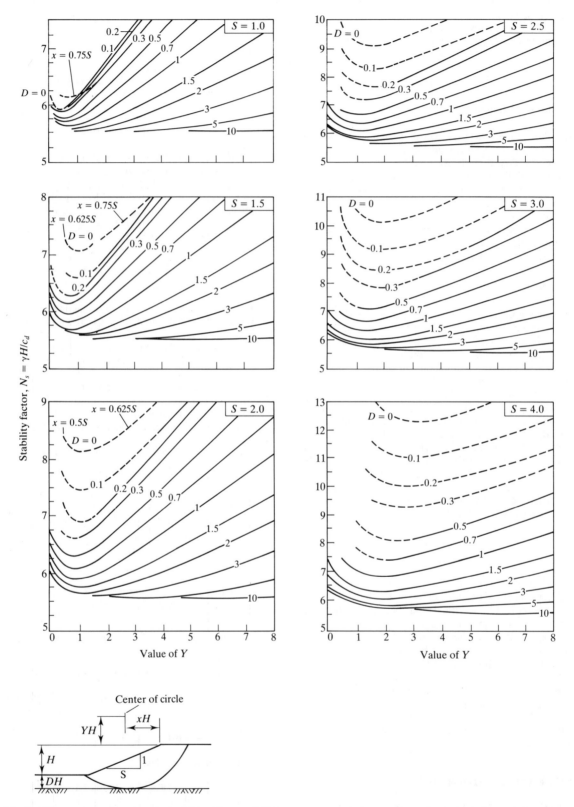

Fig. 10.16 Stability curve for $\phi = 0$ analysis. (*After Huang, 1975.*)

Notes: (1) Solid curves represent failure by midpoint circles and dashed curves represent failure by slope or toe circles. (2) $X = 0.5S$ unless indicated otherwise. (3) Numerals on curves indicate values of D. (4) γ = unit weight and C_d = developed cohesion.

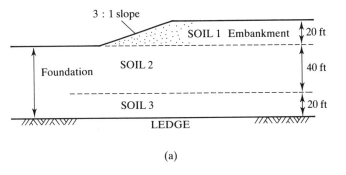

(a)

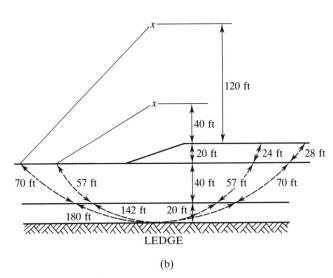

(b)

Fig. 10.17 (a) A nonhomogeneous slope with multiple soil layers —Example 10.8. *Soil data*: Soil 1, c = 1500 psf; soil 2, c = 800 psf; soil 3, c = 100 psf. (b) Failure circles in a nonhomogeneous slope.

in developed cohesion, and thus a greater factor of safety is obtained.

Long-Term Stability Analysis

The effective stress analysis is used to predict the long-term stability of embankment. A set of design charts was developed as shown in Figure 10.18. Two extreme cases are presented: one for zero pore pressure as indicated by the solid curves, and the other for a pore pressure ratio of 0.5 as indicated by the dashed curves. The pore pressure ratio used in the charts is the same as described in the Bishop and Morgenstern method (Eq. 10.13). The factor of safety presented in Figure 10.18 is based on the analysis of a homogeneous slope by the simplified Bishop (1955) method. Part of the data was obtained from the charts and tables by Bishop and Morgenstern (1960). However, the charts shown in Figure 10.18 are different from Bishop and Morgenstern's (Figs. 10.7 to 10.9) in that the ledge is assumed to be at a depth greater than 0.5H from the toe, so its presence has no effect on the factor of safety. If the ledge is located at or close to the toe, a slope circle will result and the factor of safety may be slightly greater. Therefore, the use of Figure 10.18 is on the safe side. The method also considers the cohesion factor C.F., defined as:

$$C.F. = \frac{100c'}{\gamma H} \qquad (10.20)$$

where c' = effective cohesion, γ = unit weight of soil, and H = height of embankment. The major advantage of the chart shown in the figure is its simplicity as illustrated in the following example.

EXAMPLE 10.9 (after Huang, 1975)

An embankment has a height H of 20 ft, slope S of 3:1, effective cohesion c' of 150 psf, effective friction angle ϕ of 30°, a pore pressure ratio of 0.5, and the unit weight of soil is $\gamma = 120$ pcf. What is the factor of safety for the embankment?

Solution

The cohesion factor C.F. is

$$C.F. = \frac{100c'}{\gamma H} = \frac{(100)(150)}{(120)(20)} = 6.25$$

When C.F. = 6.25, $\phi' = 30°$, and $S = 3$, then, the factor of safety F can be read directly from Figure 10.18 as 1.55.

If the embankment is directly placed on a ledge, the factor of safety obtained from the Bishop and Morgenstern chart is 1.64, which is only slightly greater than the 1.55 obtained from Figure 10.18.

10.4.8 Other Methods

Other limit equilibrium methods discussed here include those introduced by Lowe and Karafiath (1960) and Janbu (1954). The Lowe and Karafiath method was developed primarily for analyzing sloping core type earth and rockfill dams under the condition of rapid drawdown. It is used in the following manner. The first step is to determine the stability of the dam for the condition of equilibrium under the high reservoir level presumed to prevail immediately before drawdown. The resultant stresses along the assumed failure circle are determined for this high reservoir condition using the method of slices modified to include consideration of the earth forces as well as the water forces on the sides of the slices (see Fig. 10.1d). The principal stress ratios along the failure circle are estimated by assuming that the direction of the principal planes before drawdown is the same as the direction at failure. The second step is to determine, by laboratory test, the undrained shear strength of the soil for the condition of anisotropic consolidation to the major principal stresses determined in the first step. The third step is to make the stability analysis for the situation immediately after drawdown using the shear strength as determined in the first and second steps. The stability analysis again follows the slices method with earth and water forces considered on the sides of the slices; graphical, analytical and computer solutions are available (Table 10.3).

The Janbu method is also useful in analyzing the influence of partial submergence and drawdown conditions and the effects of tension cracks and surcharge. The method assumes the potential sliding surface to be a cylinder. The shear strength is constant along the entire sliding surface ($\phi = 0$) or, in cases where the soil is layered, the shear strength is constant within each layer. At the instant of failure the shear strength is completely mobilized at every point along the sliding surface. In zones containing tension cracks, the shear strength is neglected.

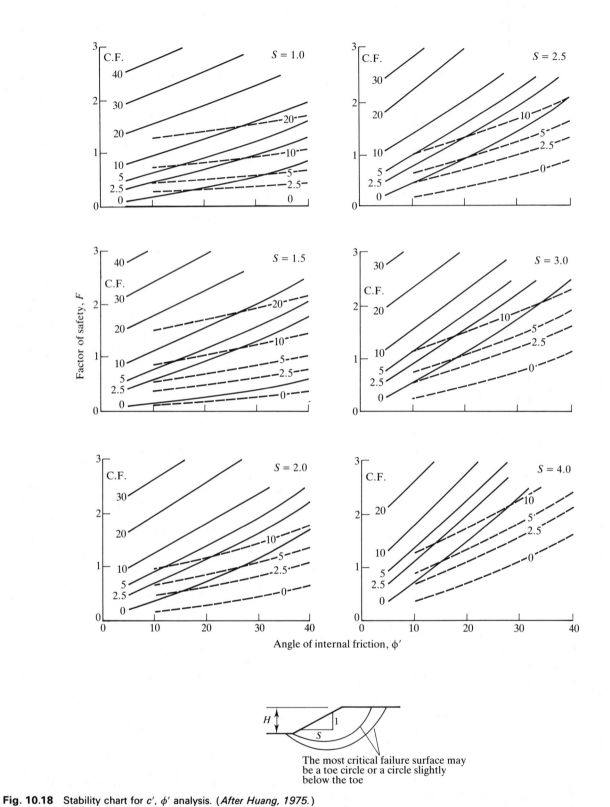

Fig. 10.18 Stability chart for *c'*, *φ'* analysis. (*After Huang, 1975.*)
Notes: (1) Solid curves indicate zero pore pressure and dashed curves indicate a pore pressure ratio of 0.5. The factor of safety for other pore pressure ratios can be obtained by a straight-line interpolation between the solid and the dashed curves. (2) Numerals on curves indicate cohesion factor, C.F. = $100c'/\gamma H$. (3) γ = unit weight and c' = effective cohesion.

10.5 SLOPE STABILITY ANALYSIS PROCEDURE: LIMIT ANALYSIS METHODS

10.5.1 General Discussion

As mentioned in Section 10.4, the limit analysis method is a relatively new technique compared with the conventional limit equilibrium methods. The method uses the concept of a yield criterion and its associated flow rule, which considers the stress–strain relationship. In addition, the limit analysis approach is simple to apply and, in many cases, provides a closed-form solution to a problem. It also provides engineers with the possibility of a clear physical picture and should help them to visualize the mechanics of the problem.

The method is based on two theorems for any body or assemblage of bodies of elastic-perfectly plastic material.

Lower-bound Theorem If an equilibrium distribution of stress can be found that balances the applied load and nowhere violates the yield criterion that includes c, the cohesion, and ϕ, the angle of internal friction, the soil mass will not fail or will be just at the point of failure.

Upper-bound Theorem The soil mass will collapse if there is any compatible pattern of plastic deformation for which the rate of work of the external loads exceeds the rate of internal energy dissipation.

According to the upper-bound theorem, it is necessary to find a compatible failure mechanism (velocity field) in order to obtain an upper-bound solution. A stress field satisfying all conditions of the lower-bound theorem is required for a lower-bound solution. If the upper bounds provided by the velocity field and stress field coincide, the exact value of the collapse load is determined.

The fundamental concept of limit analysis in relation to various problems in soil mechanics and foundation engineering has been described in various references.

The upper-bound theorem solution has been shown to yield reasonable answers when compared with existing limit equilibrium solutions whose validity has been established on the basis of practical experience (Chen and Scawthorn, 1970; Fang and Hirst, 1970). The following section gives examples

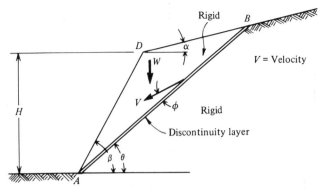

Fig. 10.19 Straight-line plasticity failure mechanism—velocity field (upper-bound solution).

demonstrating the application of limit analysis to problems of earth slope stability. Some derivation steps are included for illustration purposes.

10.5.2 Vertical Cut

A limit analysis of the slope stability problem was first performed by Drucker and Prager (1952). They assumed a straight-line failure plane and suggested the upper-bound failure mechanism shown in Figure 10.19. As the wedge formed by the shear plane slides downward along the discontinuity surface, the rate of work done by the external forces is equal to the vertical component of the velocity multiplied by the weight of the soil wedge. For a vertical cut ($\beta = 90°$) and horizontal backfill ($\alpha = 0°$), the rate of external work is

$$\text{Rate of work (ext.)} = (1/2)\gamma H^2 V \cot \theta \sin (\theta - \phi) \quad (10.21)$$

where V equals the velocity of the wedge and is inclined at angle ϕ to the discontinuity surface. The rate of internal energy dissipation along the discontinuity surface is

$$\text{Rate of work (int.)} = \frac{cH}{\sin \theta} V \cos \phi \quad (10.22)$$

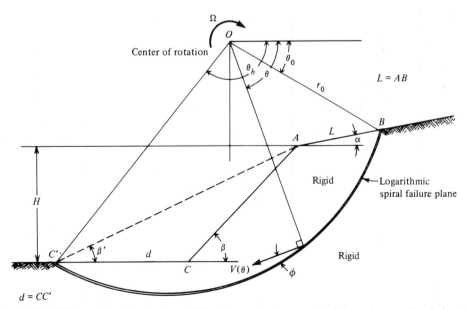

Fig. 10.20 Failure mechanism for the stability of an embankment with failure plane passing below toe.

TABLE 10.2 STABILITY FACTOR, $N_s = H_c\gamma/c$ (LIMIT ANALYSIS SOLUTION).

Friction Angle φ, Degrees	Slope Angle α, Degress	Slope Angle β, Degrees															
		90	85	80	75	70	65	60	55	50	45	40	35	30	25	20	15
0	0	3.83	4.081	4.325	4.57	4.789	5.026	5.25	5.462	5.760	5.86	6.063	6.249	6.51	6.602	6.787	7.35
5	0	4.19	4.502	4.818	5.14	5.469	5.807	6.17	6.526	6.920	7.33	7.839	8.414	9.17	10.130	11.668	14.80
	5	4.14	4.436	4.740	5.05	5.366	5.691	6.03	6.384	6.764	7.18	7.645	8.194	8.93	9.821	11.271	14.62
10	0	4.59	4.971	5.375	5.80	6.249	6.732	7.26	7.844	8.515	9.32	10.298	11.606	13.53	16.636	23.137	45.53
	5	4.53	4.907	5.300	5.72	6.153	6.625	7.14	7.717	8.375	9.14	10.129	11.416	13.26	16.368	22.785	45.15
	10	4.47	4.829	5.207	5.61	6.301	6.487	6.98	7.543	8.180	8.93	9.872	11.109	12.97	15.839	21.957	44.56
15	0	5.02	5.498	6.012	6.57	7.176	7.854	8.64	9.537	10.642	12.05	13.972	16.829	21.71	32.108	69.404	
	5	4.97	5.437	5.940	6.49	7.084	7.754	8.52	9.418	10.513	11.91	13.816	16.652	21.50	31.850	69.047	
	10	4.90	5.363	5.853	6.39	6.971	7.628	8.38	9.262	10.339	11.73	13.591	16.383	21.14	31.378	68.256	
	15	4.83	5.270	5.743	6.28	6.825	7.460	8.18	9.045	10.088	11.42	13.228	15.916	20.59	30.254	65.173	
20	0	5.51	6.099	6.751	7.48	8.299	9.253	10.39	11.799	13.628	16.18	19.998	26.655	41.27	94.632		
	5	5.46	6.040	6.681	7.40	8.212	9.157	10.30	11.687	13.506	16.04	19.850	26.485	41.06	94.377		
	10	5.40	5.969	6.598	7.31	8.105	9.038	10.15	11.542	13.346	15.87	19.641	26.232	40.73	93.776		
	15	5.33	5.882	6.496	7.20	7.970	8.886	9.98	11.347	13.122	15.59	19.322	25.818	40.16	92.898		
	20	5.25	5.773	6.366	7.04	7.793	8.681	9.78	11.066	12.785	15.17	18.770	25.011	39.19	88.632		
25	0	6.06	6.793	7.624	8.59	9.696	11.048	12.75	14.972	18.098	22.92	31.333	50.059	120.0			
	5	6.01	6.735	7.556	8.52	9.611	10.955	12.65	14.864	17.981	22.78	31.188	49.887	119.8			
	10	5.96	6.666	7.475	8.41	9.508	10.842	12.54	14.727	17.829	22.60	30.986	49.635	119.5			
	15	5.89	6.584	7.378	8.30	9.382	10.700	12.40	14.547	17.623	22.37	30.687	49.234	118.7			
	20	5.81	6.483	7.258	8.16	9.220	10.514	12.17	14.297	17.325	21.98	30.198	48.503	117.4			
	25	5.71	6.354	7.104	7.97	9.003	10.257	11.80	13.922	16.851	21.35	29.245	46.759	115.5			
30	0	6.69	7.607	8.675	9.96	11.485	13.439	16.11	19.712	25.413	35.63	58.274	144.199				
	5	6.63	7.550	8.607	9.87	11.400	13.348	16.00	19.607	25.298	35.44	58.127	144.011				
	10	6.58	7.483	8.529	9.79	11.301	13.239	15.87	19.475	25.151	35.25	57.924	143.738				
	15	6.53	7.404	8.436	9.67	11.180	13.104	15.69	19.305	24.956	34.99	57.629	143.307				
	20	6.44	7.309	8.323	9.54	11.029	12.931	15.48	19.076	24.682	34.64	57.159	142.538				
	25	6.34	7.190	8.181	9.37	10.833	12.700	15.21	18.744	24.265	34.12	56.302	140.842				
	30	6.22	7.038	7.995	9.15	10.561	12.369	14.81	18.216	23.544	33.08	54.252	134.524				
35	0	7.43	8.581	9.969	11.68	13.857	16.774	20.94	27.448	39.109	65.53	166.378					
	5	7.38	8.524	9.902	11.60	13.774	16.685	20.84	27.344	38.995	65.39	166.220					
	10	7.32	8.458	9.825	11.51	13.676	16.578	20.71	27.216	38.851	65.22	166.003					
	15	7.26	8.382	9.735	11.41	13.560	16.448	20.55	27.053	38.662	65.03	165.720					
	20	7.18	8.291	9.627	11.28	13.417	16.285	20.36	26.836	38.401	64.74	165.188					
	25	7.11	8.180	9.494	11.12	13.234	16.072	20.07	26.533	38.015	64.18	164.298					
	30	6.99	8.041	9.325	10.93	12.990	15.778	19.73	26.071	37.384	63.00	162.333					
	35	6.84	7.858	9.098	10.66	12.641	15.337	19.21	25.271	36.150	60.80	154.978					
40	0	8.30	9.771	11.608	14.00	17.152	21.724	28.99	41.887	71.485	185.6						
	5	8.26	9.713	11.541	13.94	17.069	21.635	28.84	41.784	71.370	185.5						
	10	8.21	9.649	11.465	13.85	16.974	21.530	28.69	41.657	71.226	185.3						
	15	8.15	9.574	11.377	13.72	16.860	21.405	28.54	41.498	71.038	185.0						
	20	8.06	9.487	11.273	13.57	16.723	21.249	28.39	41.290	70.780	184.6						
	25	7.98	9.382	11.147	13.42	16.551	21.049	28.16	41.002	70.406	184.0						
	30	7.87	9.252	10.989	13.21	16.326	20.779	27.88	40.578	69.812	183.2						
	35	7.76	9.086	10.784	12.95	16.016	20.391	27.49	39.885	68.728	182.3						
	40	7.61	8.863	10.501	12.63	15.551	19.773	26.91	38.525	66.119	181.1						

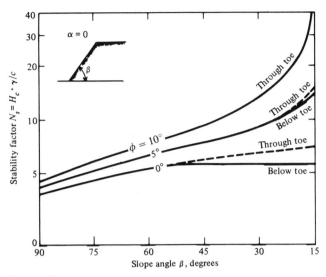

Fig. 10.21 Comparisons of stability factors between, through, and below toe failure conditions. (*After Chen and Giger, 1971.*)

Equating the rate of external work (Eq. 10.21) to the rate of internal energy dissipation (Eq. 10.22), we obtain

$$H \leqslant \frac{2c}{\gamma} \frac{\cos \theta}{\sin (\theta - \phi) \cos \theta} \qquad (10.23)$$

Minimizing the right-hand side of Equation 10.23, we have

$$\theta = \frac{\pi}{4} + \frac{\phi}{2} \qquad (10.24)$$

and

$$H_c \leqslant \frac{4c}{\gamma} \tan \left(\frac{\pi}{4} + \frac{\phi}{2} \right) \qquad (10.25)$$

Thus, it may be seen that the upper-bound plasticity solution for the critical height H_c of a vertical cut yields the same value as the Culmann limit equilibrium solution.

10.5.3 Logarithmic-Spiral Failure Plane

A closed-form mathematical solution of the upper-bound problem for both through toe and below toe have been developed by Chen et al. (1969) and Chen and Giger (1971). The assumed failure mechanism is shown in Figure 10.20. The region $ABC'CA$ rotates as a rigid body about the center of rotation O, with the materials below the logarithmic surface AB remaining at rest. Thus, the surface $C'B$ is a surface of velocity discontinuity. The mechanism can be specified by three variables, θ_0, θ_h, and H, where θ_0, θ_h, and H are the slope angles of the chords OB and OC' and the height of the embankment, respectively. From the geometrical relations it may be shown that the ratios, H/r_0 and L/r_0, can be expressed in terms of

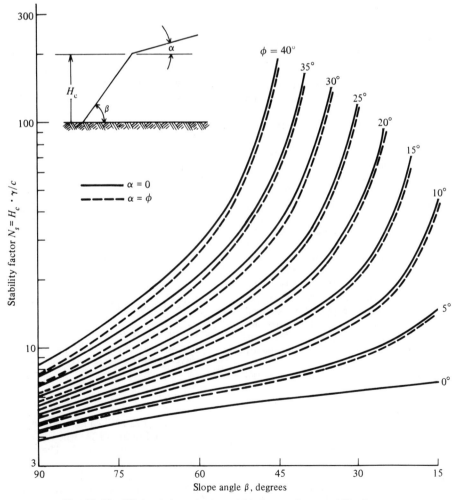

Fig. 10.22 Effect of slope angle and friction angle on stability factor.

the angles θ_0 and θ_h. The rate of internal and external work done by the region can be obtained.

The internal dissipation of energy occurs along the discontinuity surface BC'. The differential rate of dissipation of energy along the surface may be found by multiplying the differential area, $r\,d\theta/\cos\phi$, of this surface by the cohesion c times the tangential discontinuity in velocity $V\cos\phi$ across the surface. The total internal dissipation of energy is found by integration over the whole surface. Further mathematical manipulation results in an expression for the critical height of the form

$$H_c \leqslant \frac{c}{\gamma}\, N_s \qquad (10.26)$$

The term $\gamma H_c/c$ is a dimensionless expression called the stability factor, N_s. A numerical solution of Equation 10.26 in terms of stability factor has been obtained by computer and tabulated in Table 10.2.

The stability factor depends on the slope angles α, β, and the angle of internal friction ϕ. Comparison of stability factor values between, through, and below the toe surfaces are shown in Figure 10.21. There are slight differences when a slope angle β and friction angle ϕ are small. Figure 10.22 shows the stability factor versus slope angle β with α varying from 0 to ϕ. It is indicated that angles β and ϕ significantly affect the stability factor values, while the angle α has little effect.

For computing the factor of safety, the ratio of the height of strength parameters may be used (Eq. 10.3). For considering the pore water effects, use of the effective stress c' and ϕ' is recommended.

10.5.4 Slope in Anisotropic and Nonhomogeneous Soils

As mentioned in Section 10.2, most soils in their natural states exhibit some anisotropy with respect to shear strength and some nonhomogeneity with respect to depth. Lo (1965) developed a method of stability analysis for anisotropic soils in undrained shear using limit equilibrium techniques. Special cases for the friction angle $\phi = 0$ were computed. Based on limit analysis techniques, a more general closed-form solution when friction angles are constant and the soil exhibits both anisotropy and nonhomogeneity has been developed (Chen et al., 1975; Snitbhan et al., 1975). The logarithmic-spiral failure plane is used for the analysis. The techniques used for developing the solution are similar to those described in previous sections. The term *nonhomogeneous* refers to the undrained shear strength, which is assumed to vary linearly with depth. The term *anisotropy* describes the variation of undrained shear strength with direction at a particular point. The directional variation of other soil properties, such as compressibility and permeability, is not considered. The anisotropy factor is defined as the ratio of principal cohesion stress in the horizontal direction to the principal cohesion stress in the vertical direction. The effect of degree of anisotropy on the slope stability factor is shown in Figures 10.23 and 10.24.

For a given slope the stability factor decreases when the anisotropy factor decreases and the friction angle increases. The effect of anisotropy is more significant for flatter slopes than for steep slopes.

EXAMPLE 10.10 (after Fang et al., 1975)

Given a homogeneous slope, where $\beta = 30°$, $\phi = 0$, cohesion $c = 550$ psf, and the unit weight of soil $\gamma = 120$ pcf, find the critical height H_c.

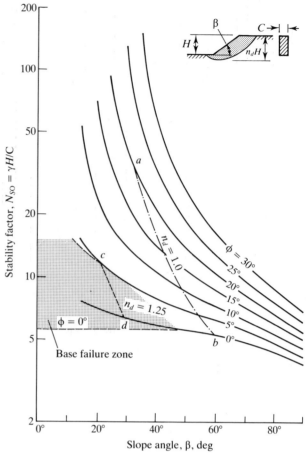

Fig. 10.23 Stability number versus slope angle for isotropic homogeneous slopes (limit analysis solution). Solid curves, through toe; broken curves, below toe; n_d = depth factor. (*After Snitbhan et al., 1975.*)

Solution

From Figure 10.23,

$$N_{s0} = \frac{\gamma H_c}{c} = 5.5 \quad \text{(interpolated)}$$

Therefore,

$$H_c = (5.5)(550)/120 = 25.3 \text{ ft} \quad (7.7 \text{ m})$$

EXAMPLE 10.11

Given a layered slope as shown in Figure 10.24, where $\beta = 30°$, $\phi = 0$, $\gamma = 120$ pcf, and cohesion $c_1 = 550$ psf (layer 1) and $c_2 = 275$ psf (layer 2), find the critical height H_c.

Solution

$c_2 = (1 + n)c_1$. When $c_1 = 550$, and $c_2 = 275$, $n = -0.5$.

From Figure 10.24, when $\beta = 30°$, and $n = -0.5$, the correction factor is 0.72. Therefore, $H_c = (0.72)(25.3) = 18.2$ ft (5.5 m).

For homogeneous slope, $H_c = 25.3$ ft (7.7 m); however, for a layered slope, the critical height H_c for the same slope is reduced from 25.3 ft to 18.2 ft (5.5 m).

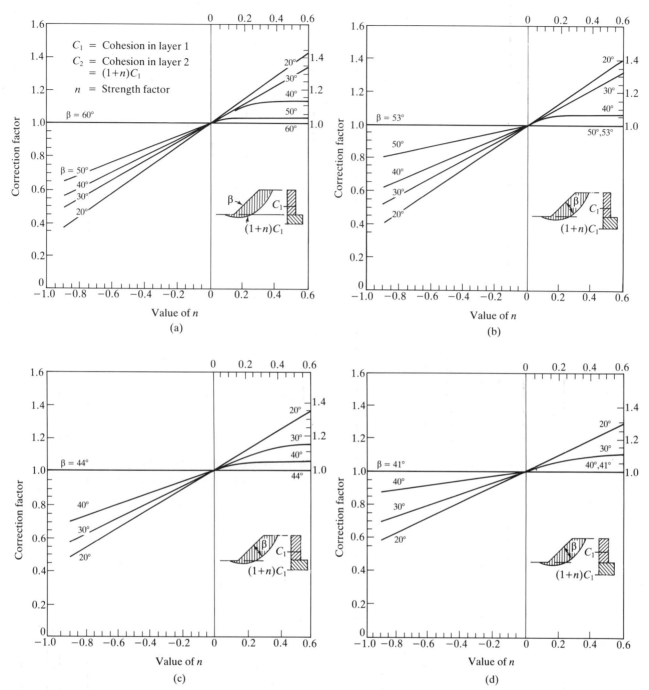

Fig. 10.24 Correction factor versus soil strength factor for $\phi = 0$, 5°, 10°, and 15° (limit analysis solution). (a) $\phi = 0$; (b) $\phi = 5°$; (c) $\phi = 10°$; (d) $\phi = 15°$ (*After Snitbhan et al., 1975.*)

10.6 METHODS CONSIDERING SEISMIC EFFECTS

The seismic stability analysis of slopes requires consideration of lateral shear forces that develop in a soil mass in direct response to the earthquake-induced relative displacements. These displacements are a function of the earthquake intensity of ground motion and the generated spectral motions. At present, seismic analysis and design of slopes utilize computer programs that incorporate a stress-path type method into an elastic dynamic analysis, or elastic-plastic analysis. For preliminary design, or when complicated analyses are not warranted, an alternative to these methods is the simple pseudostatic approach (equivalent static force, ESF methods). In these methods, the earthquake effect is considered in terms of an equivalent horizontal force acting at the center of gravity of the soil mass, and equal to the total soil weight times a seismic coefficient. Two such methods are presented as follows.

10.6.1 Huang Method

The stability charts of this method were developed on the basis of a homogeneous dam constructed with the same material as that of its foundation and since that material has a small effective cohesion, the critical failure surface is then considered as a shallow surface (Huang, 1980, 1983).

Figure 10.25 shows a slope with a height H and a slope $S:1$ (horizontal:vertical). The procedure for locating the most dangerous failure surface is described in Figure 10.25. When a failure circle is determined, the average shear stress developed along the failure can be determined by equating the moment at the center of a circle due to both the weight of the sliding mass and the corresponding seismic force to that due to the average shear stress distributed uniformly over the failure arc. This developed shear stress is proportional to the unit weight of the soil and the height of the slope. The average shear strength along the failure surface also varies with unit weight and height of slope, according to the Mohr–Coulomb theory. The factor of safety is a ratio between the shear strength and the shear stress and can be expressed as:

$$F = \frac{\dfrac{c'}{\gamma H} + \dfrac{(1 - r_u)\tan\phi'}{N_f}}{(1/N_s) + (C_s/N_e)} \qquad (10.27)$$

where

F = factor of safety
c' = effective cohesion
γ = total unit weight of soil
H = height of slope
r_u = pore pressure ratio (ratio between the porewater pressure and the overburden pressure)
ϕ' = effective angle of internal friction
N_f = friction number
N_s = stability factor
C_s = seismic coefficient (the ratio between seismic force and weight)
N_e = earthquake number

Procedures

Given: S = slope; H = height of slope

Let $ab = a'b' = 0.1SH$ (the assumption of $ab = 0.1SH$ is arbitrary but yields good results)

Point O is the middle point between bb' so that $bO = Ob'$. The center of potential failure circle must be along OO' and must pass through points a and a'. By trial, a most dangerous failure surface is drawn. The vertical distance YH can be measured. The value of Y is needed for computing other parameters as shown in the figure.

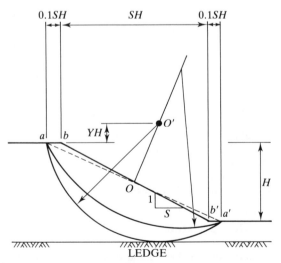

Fig. 10.25 Procedures for locating the potential failure circle in a typical slope. (*After Huang, 1980*).

Equation 10.27 shows that the factor of safety depends on four geometric parameters (H, N_s, N_f, and N_e) and four soil parameters (r_u, γ, c', and ϕ'). Values of N_s, N_f, and N_e can be obtained from Figure 10.26, and that of r_u from the location of the phreatic surface with respect to the failure circle. If the slope is homogeneous, γ, c', and ϕ' are given directly. If the slope is nonhomogeneous, average values of γ, c', and ϕ' must be determined (see Example 10.12).

EXAMPLE 10.12 (after Huang, 1980)

Figure 10.27a shows a 2.5:1 slope, 20 m high, composed of three soil types. The soil data, including c', ϕ', γ, are given. The location of the phreatic surface is also shown in the figure. Assuming a seismic coefficient of 0.1, determine both the static and the seismic factors of safety.

Solution

1. Location of the most dangerous failure surface (see Fig. 10.25); results are given in Figure 10.27b.
2. YH distance measured from Figure 10.27c:

$$YH = 5.5 \text{ m}$$
$$Y = 5.5 \text{ m}/20 \text{ m} \quad (H = 20 \text{ m})$$
$$= 0.275$$

3. Determination of N_s, N_f, and N_e from Figure 10.26. For $S = 2.5$ (given), $Y = 0.275$ (from Step 2). Then from Figure 10.26,

$$N_s = 7.0 \quad N_f = 2.0 \quad N_e = 2.8$$

4. Average unit weight of soil, γ:

$$\gamma = \frac{(131)(18) + (221)(19) + (534)(20)}{131 + 221 + 534} = 19.5 \text{ kN/m}^3$$

5. The length of the failure arc through soils 1, 2, and 3 was measured and found to be 40, 17.6, and 24 m, respectively.
6. The average effective cohesion, c':

$$c' = \frac{(40)(5) + (17.6)(7.5) + (24)(10)}{40 + 17.6 + 24} = 7.0 \text{ kPa}$$

7. The average coefficient of friction, $\tan\phi'$: Since only the component of weight normal to the failure surface is effective in producing friction, to determine the average coefficient of friction, $\tan\phi'$, the weight above the failure surface must be multiplied by $\cos\theta$, where θ is the angle of inclination of the chord as shown by θ_1, θ_2, and θ_3 in Figure 10.27b.

The weight normal to the failure arc in soil 1 is

$$[(131 \times 18) + (187 \times 19) + (293 \times 20)](0.95)$$
$$= 11\,182 \text{ kN/m}$$

For soil 2 the weight normal to the failure arc is

$$[(2 \times 17 \times 19) + (110 \times 20)](0.75) = 2135 \text{ kN/m}$$

For soil 3 the weight normal to the failure arc is

$$(131 \times 20)(0.46) = 1205 \text{ kN/m}$$

Then, the average coefficient of friction, $\tan\phi'$, is

$$\frac{11\,182\tan 25° + 2135\tan 30° + 1205\tan 35°}{11\,559 + 2135 + 1205} = 0.502$$

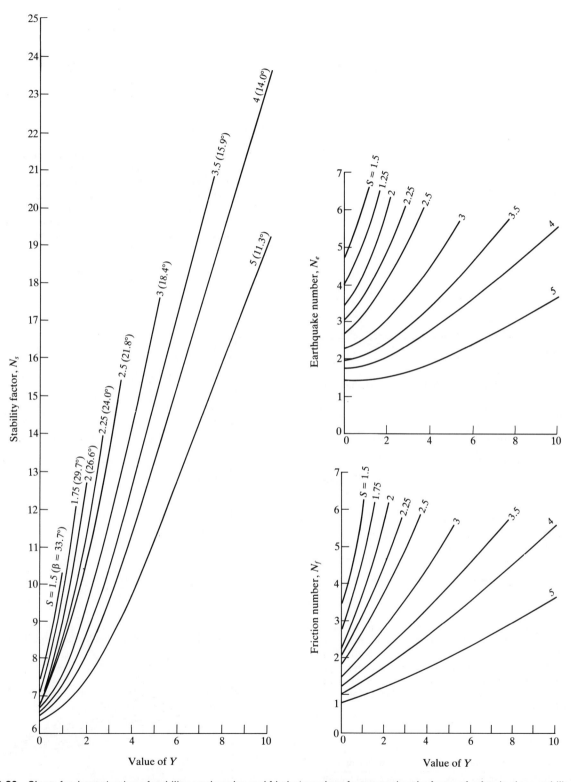

Fig. 10.26 Charts for determination of stability, earthquake, and friction numbers for computing the factor of safety in slope stability analysis. Parameter *Y* used in these charts is determined from Figure 10.25. (*After Huang, 1980.*)

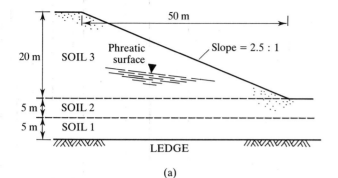

(a)

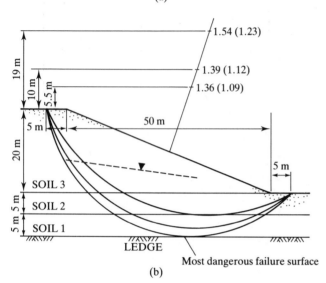

(b)

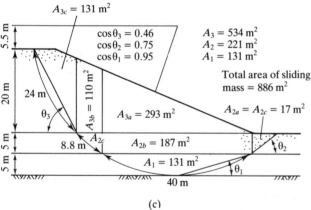

(c)

Fig. 10.27 (a) Example 10.12. A nonhomogeneous slope. *Soil data*: Soil 1, $\gamma = 18$ kN/m³, $c' = 5.0$ kPa, $\phi' = 25°$. Soil 2, $\gamma = 19$ kN/m³, $c' = 7.5$ kPa, $\phi' = 30°$. Soil 3, $\gamma = 20$ kN/m³, $c' = 10.0$ kPa, $\phi' = 35°$. (b) Analysis of a nonhomogeneous slope—locating the potential failure circle. (c) Analysis of a nonhomogeneous slope computation of area, arc length, and chord inclination of critical circle.

8. The average pore pressure, r_u:

$$r_u = \frac{\text{area of sliding mass under water} \times \text{unit weight of water}}{\text{total area of sliding mass} \times \text{average unit weight of soil}}$$

The area of sliding mass under water was measured as 527 m². The average pore pressure ratio r_u is

$$r_u = \frac{(527 \times 9.8)}{(886 \times 19.5)} = 0.299$$

9. Factor of safety, F, computed from Equation 10.27:

Static Factor of Safety

$$F = \frac{7.0/[(19.5)(20)] + (1 - 0.299)(0.502)/2.0}{1/7.0} = 1.36$$

Seismic Factor of Safety

$$F = \frac{7.0/[(19.5)(20)] + (1 - 0.299)(0.502)/2.0}{(1/7.0) + (0.1/2.8)} = 1.09$$

10.6.2 Koppula Method

Koppula's (1984) method applies to the stability analysis of slopes in cohesive soils exhibiting a linear variation in shear strength with depth and a strength at the surface greater than zero. The effect of seismicity is analyzed by circular failure mechanism with earthquake loading included as an equivalent horizontal force. The factor of safety F is defined as the ratio of the restoring moment to the overturning moment and is given by

$$F = N_1\left(\frac{a_0}{\gamma}\right) + N_2\left(\frac{c_0}{\gamma H}\right) \qquad (10.28)$$

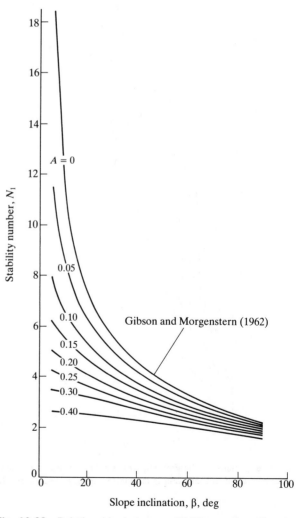

Fig. 10.28 Relationship between stability number N_1, slope inclination β, and seismic coefficient A. (*After Koppula, 1984.*)

in which

N_1 and N_2 = stability factors
γ = unit weight of soil
H = height of the slope
a_0, c_0 = constants used to express the relationship between the strength of the soil with depth

$$C = c_0 + a_0 z \qquad (10.29)$$

where

C = shear strength of soil at depth z below the ground surface
c_0 = shear strength of soil at the ground surface
a_0 = gradient at which the soil strength varies with depth

The use of the stability factors N_1 and N_2 for different values of the seismic coefficient A from charts (Figs. 10.28 to 10.30) is illustrated in the following examples.

EXAMPLE 10.13 (after Koppula, 1984)

Consider a cohesive slope of height H inclined at 60° to the horizontal. Let the shear strength of the soil be given by $a_0/\gamma = 0.01$ and $c_0/\gamma H = 0.2$. Determine the factor of safety.

Solution

When $\beta = 60°$ and the seismic coefficient $A = 0$, from Figure 10.28 we obtain the stability factor $N_1 = 3.23$; and from Figure 10.30 we obtain the stability factor $N_2 = 5.25$. The factor of safety F can be computed from Equation 10.28 as

$$F = (3.23)(0.01) + (5.25)(0.2) = 1.082$$

EXAMPLE 10.14 (after Koppula, 1984)

Consider a cohesive slope with $a_0/\gamma = 0.1$ and $c_0/\gamma H = 0.1$ as soil strength parameters, situated in a seismic zone with seismic coefficient, $A = 0.1g$ intensity. Determine the factor of safety.

Solution

When $\beta = 60°$ and the seismic coefficient $A = 0.1$, from Figure 10.28 we obtain the stability factor $N_1 = 3.0$; and from Figure 10.30 we obtain the stability factor $N_2 = 4.5$.

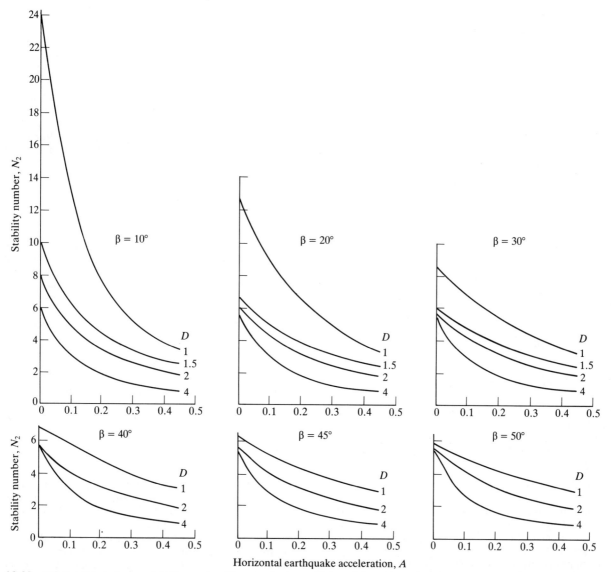

Fig. 10.29 Relationship between stability number N_2 and seismic coefficient A for various slope inclinations β. (*After Koppula, 1984*).

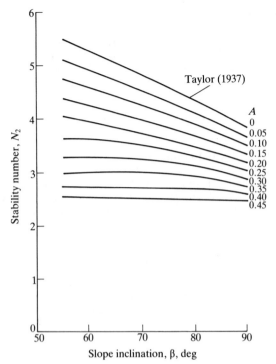

Fig. 10.30 Relationship between stability number N_2, slope inclination β ($\geqslant 55°$), and seismic coefficient A. (*After Koppula, 1984.*)

The factor of safety F can be computed from Equation 10.28 as

$$F = (3.0)(0.1) + (4.5)(0.1) = 0.75$$

These factors of safety calculated from the stability numbers N_1 and N_2 are within 1 to 2 percent of those computed directly from the equation used to derive Figures 10.28 to 10.30 and may be used with confidence to calculate the factor of safety for all practical design purposes.

10.7 SLOPE STABILITY IN SOILS PRESENTING SPECIAL PROBLEMS

The stability of natural or excavated slopes presents unusual problems when dealing with a number of special field conditions. These include special soil types, marine slopes, effects of vegetation on slope stability, and stability of landfills. A brief discussion on these problems is presented as follows.

10.7.1 Difficult Soil Conditions

These include residual soil regions, loess deposits, stiff fissured clays, dispersive clays, and overconsolidated soils. For the residual soil regions, the following design procedures are proposed by Binnie and Partners (1971), Chiang (1979), and MAA Group (1980).

1. The effective shear strength parameters c' and ϕ' are used in design (Section 3.13).
2. Negative pore pressures are normally assumed to be totally eliminated by infiltration in adverse site conditions with regard to rainfall (Lumb, 1962; Chiang, 1979).
3. In practical design, the choice of the cohesion intercept c' is complicated by the sensitiveness of the parameter to sampling and handling (Chapters 1 and 2). A reduction of

40 to 100 percent of laboratory tested results or projected values from back-calculation are usually adopted for analysis-design.

4. The thickness of the wetting band can be estimated (Beattie and Chau, 1976) for a particular slope site and rainstorm together with records of the groundwater level obtained by monitoring standpipes or piezometers for the estimation of the probable rise in phreatic surface.
5. The depth of tension crack in unsaturated slopes and the development of hydrostatic pressure in cracks are also taken into consideration in stability analysis (Spencer, 1969; Ajaz, 1978; Fang et al., 1989).
6. For weathering profiles or relict joints, the methods of analysis assuming circular failure surface are not considered suitable. Alternative methods assuming generalized slip surfaces proposed by Janbu (1954a, b) and Morgenstern and Price (1965) are more applicable.
7. The factor of safety for slope stability analysis for various slopes or cutting types are summarized in Table 10.1.

Loess deposits have high permeability in the vertical direction because they contain networks of interconnected channels that were formed from decayed roots. Protecting such slopes from infiltration and erosion prevents vertical percolation and water destroying the particle bonds and causing slope failure.

Slope failures in stiff-fissured clays and shales (Section 3.3) may occur progressively over a long period of time as they are subjected to very large shearing displacements (Skempton, 1977). When these displacements exceed the shear displacement corresponding to peak strength, the shearing resistance is reduced to the residual value (Fig. 3.32), although, in some cases, these soils may show stable slopes at steeper angles. Further discussions on the stability of slopes for problematic soils and environmental conditions are given by Varnes (1978), Chowdhury (1978, 1980), Leonards (1979), Fukuoka (1980), Gray and Leiser (1982), Imaizumi et al. (1989), and in Chapter 11.

10.7.2 Marine Slope Stability

Marine slope stability analysis is still more of an art, mainly because field confirmation of slope stability studies is minimal. The measurement of in-situ pore pressures is now possible, which allows the application of effective stress analysis methods (Richards, 1978). Consideration of dynamic loading effects is also needed since the driving force for many types of marine slope instability is considered to be wave action. Other major differences in the marine environment include gas in sediments (Rau and Chaney, 1988), soil sampling (Richards and Zuidberg, 1986), and in-situ measurements (Chapter 2). General discussions on these various aspects of marine slope stability are given by Inderbitzen (1965), Winterkorn and Fang (1970), Simpson and Inderbitzen (1971), Richards and Chaney (1982), and Chang et al. (1983).

10.7.3 Effects of Vegetation

Vegetation and tree roots can affect the soil stability in many ways (Gray and Leiser, 1982), such as changes in the soil moisture regime and contribution of roots to the soil strength. For stability analysis of slopes, important data on infiltration, evapotranspiration, and root geometry are needed. Such data are usually scarce. Analysis and design of slopes, considering the effects of vegetation, should be supported by extensive research and field data. A method that considers the effects of

TABLE 10.3 MAIN FEATURES OF SOME COMMERCIALLY AVAILABLE COMPUTER SOFTWARE PACKAGES FOR SLOPE STABILITY ANALYSIS[a].

Program	Originating Program	Methods of Analysis	Failure Surface	Seismic Coefficient	Graphics Support	Loading Types	Vendor Address	Price
GEOSLOPE	STABL4 Purdue Univ.	Simplified Bishop Janbu	Circular Noncircular	—	Output	Surface loadings	GEOCOMP Corp., Concord, MA	$750
GEOSYSTEM		Simplified Bishop	Circular	Yes	Interactive input and output	Distributed	Van Gunten Engineering Software, Fort Collins, CO	$595
PC-SLOPE		Simplified Bishop Ordinary Fellenius Janbu Simplified Spencer's Morgenstern–Price GLE method Corps of Engineers Lowe and Karafiath	Circular Composite Any shape	Yes	Output	—	GEO-SLOPE Programming Ltd, Calgary, Alberta, Canada	$1975
SLOPE8R	U.C. Berkeley	Spencer's	Noncircular	—	—	—	Virginia Polytechnic Institute, Blacksburg, VA	$100
STABGM	U.C. Berkeley	Ordinary method of slices	Circular	—	—	—		$100
STAB8R/G	U.C. Berkeley	Spencer's	Noncircular	—	Output	—	GEOSOFT, Orange, CA	$390
STABR/G	U.C. Berkeley	Ordinary method of slices	Circular	—				$390
Z SLOPE.PC		Nonlinear finite-element Von Mises Drucker–Prager plasticity	Failure surface and factor of safety found in solution	—	Interactive input and output	Combinations of gravity and soil stresses	ZEi/ZACE Ltd., Falls Church, VA	$5000

[a] For additional information, see TRB (1990).

root reinforcement on slope stability is given by Wu (1984). According to this method, the soil containing roots is considered as a special problem of reinforced earth (Chapter 21).

10.7.4 Landfills

The stability analysis of slopes in refuse dams or hazardous waste landfills may require the use of a large computer program. Chen (1986) reports an analysis of a waste disposal site for the Taipei metropolitan area by the computer program STABLE II. A stability failure on a sanitary landfill was analyzed by Dvirnoff and Munion (1986) using the computer program SLOPE. For preliminary analysis and design, however, the use of manual methods of slope stability analysis may be justified. A landfill is composed of different materials that result in a very small effective cohesion and is subjected to seepage (Fang et al., 1976, 1977). Thus, methods based on effective stress analysis could be used. The complications that arise are due mainly to complex geometry and material conditions (Oweis and Khera, 1986, and Chapter 20).

10.8 COMPUTER-AIDED AND EXPERT SYSTEMS FOR SLOPE STABILITY ANALYSIS

10.8.1 Computer-Aided Systems for Slope Stability Analysis

The introduction of personal computers in engineering practice has affected the way engineers may solve geotechnical problems. Microcomputer programs allow the geotechnical engineer to analyze the stability of most soil structures. Such programs can simulate efficiently different slope geometries, stress conditions, water levels, and types of loadings. Thus, they allow the engineer to examine various instability scenarios within hours, in a cost-effective manner.

Several slope stability analyses programs are available to the geotechnical engineer. Many of these programs have been adapted from other programs originally developed on mainframe computers. Most of them are based on a single method of analysis, usually a limit equilibrium method. Others allow the user to select from two or more methods. The results can usually be stored in a file, or sent to the printer or plotter. Some of the programs use computer graphics to display the results and/or interactively change the geometry of the slope or the location of the failure surface.

Table 10.3 summarizes the main features of six slope stability analysis programs, from a survey of commercially available software packages conducted in August, 1988. The first two columns show the name of the program and, if applicable, the original program it was adapted from. The next three columns show the main assumptions of the program, that is, the method of analysis, the type of failure surface, and use of seismic coefficient. The following two columns show additional features of the programs, that is, what kinds of graphics the program supports and types of loading. Finally, the last two columns give the vendor's name and address, and the price of the package at the time of the survey.

10.8.2 Use of Expert Systems for Slope Stability Analysis

Slope stability and landslides assessment involves multiple characteristics and requires interdisciplinary knowledge; utilization of knowledge-based expert systems may provide effective results. Expert systems are intelligent computer programs that are able to perform an intellectual task in a specific field as a human expert would. Systems are being applied to classification problems such as interpretation and diagnosis, as well as to generative problems such as planning and design. At present, most available systems are the single-domain type. A single-domain system consists of the following basic parts: the knowledge/data base, and the shell, which, in general, consists of artificial intelligence techniques, graphics, and risk analysis. Each shell is designed for a particular application. Some shells give more information than others. There are several commercially available shells on the market. GEOTOX, developed by the Envirotronics Corporation, is one such shell and useful for planning, analysis, design, and management (Fang and Mikroudis, 1987; Mikroudis and Fang, 1988; Muhiddin et al., 1989). Multi-Domain System (MDS) is an extension of the GEOTOX model that covers more features and widens the applications. Both GEOTOX and MDS have been used successfully in various environmental and geotechnical problems as well as large structural systems and construction projects. For assessment and analysis of slope stability problems, the information can be obtained from the expert system outlined as follows (Fang and Mikroudis, 1990):

1. *Data Banks.* Storage of vast amounts of data; classification of data; evaluation of factors affecting slope stability analysis; rating the causes of slope failures
2. *Correlation Studies.* Correlation with local environmental and geological conditions such as rainfall, seismic effects; cost/benefit ratio for controlling landslides
3. *Risk Analysis.* Uncertainty for predicting the slope failures; simple statistical analysis; fuzzy processes; stochastic analysis; decision analysis
4. *Computer-aided Design.* Computer control for selecting most critical failure surface; selection of strength parameters for stability analysis; computing factor of safety; control for selecting most economical and efficient remedial action details for landslides.

All information produced with these expert systems includes colorful pictorial displays and/or tabular results at any given stage of interaction. Also, the user can trace back and forth to see what has been done, or may interactively alter technical and/or financial criteria and constraints. The advantage of the computer-integrated systems is that they can lead to a greater degree of unification in the planning, analysis, design, and management processes across many disciplines. There can be an updating of information and an expansion of capacities within the human–machine interface as well as in the subsystems to maintain the up-to-dateness of the overall system at any given time.

NOMENCLATURE

A, A_1, A_2	Area (s)
A	Seismic coefficient (Figs. 10.28 to 10.30)
a_0, c_0	Constants (Eq. 10.28)
c_1, c_2	Cohesion in layer 1, layer 2 (Fig. 10.24)
c, C	Cohesion
c'	Effective cohesion (Fig. 10.2)
c_d	Developed cohesion (Eq. 10.19)
c_e	Average effective cohesion, or maximum cohesion
C_c	Cohesion with full friction mobilized (Eq. 10.2)
C_s	Seismic coefficient (Eq. 10.27)
CD	Consolidated–drained test (Fig. 10.2)
C.F.	Cohesion factor (Eq. 10.20)
CU	Consolidated–undrained test (Fig. 10.2)
d, D	Depth; depth factor (Fig. 10.4)

E_1, E_2	Earth forces (Fig. 10.1)
F	Factor of safety; force
F_c	Factor of safety respect to cohesion
F_H	Factor of safety respect to height
F_s	Factor of safety respect to shear strength
F_ϕ	Factor of safety respect to friction angle
h, H	Height; depth of water table (Eq. 10.17)
h_c, H_c	Critical height (Eqs. 10.3 and 10.5)
L	Length; width of slice; depth of drawdown (Fig. 10.10)
M, M_r	Moment; resisting moment (Eq. 10.4)
M	Water table ratio (Eq. 10.17)
m	Stability coefficient (Eq. 10.14)
n	Stability coefficient (Eq. 10.14)
n	Soil strength factor (Fig. 10.24)
n_d	Depth factor (Fig. 10.23)
N_1, N_2	Stability factor(s) (Eq. 10.28)
N_e	Earthquake number (Eq. 10.27)
N_f	Friction number (Eq. 10.27)
N_s	Stability factor
N_{s0}	Stability factor for homogeneous and isotropic slope (Fig. 10.23)
p'	Effective vertical stress (Eq. 10.18)
P	Normal force
Q	Surcharge load
r_0	Radius
r_u	Pore pressure ratio (Eq. 10.13)
r_{ue}	Equal pore pressure ratio (Eq. 10.15)
R	Resultant
s, S	Shear strength
T_1, T_2	Shear force between slices (Fig. 10.1)
u	Pore water pressure
U_1, U_2	Water forces (Fig. 10.1)
UU	Unconsolidated–undrained test (Fig. 10.2)
V	Velocity of the wedge (Eq. 10.21)
$V(\theta)$	Discontinuous velocity across the failure plane (Fig. 10.19)
W	Weight of soil
Y	Parameter (Figs. 10.16; 10.25 to 10.26)
z	Depth
α, β	Slope angles
θ_1, θ_2	Angles
θ_0, θ_h	Angular variables of a log-spiral curve (Fig. 10.20)
Ω	Angular velocity (Fig. 10.20)
ϕ	Friction angle
ϕ'	Effective friction angle
ϕ'_m	Mobilized friction angle (Fig. 10.13)
γ	Unit weight of soil
γ'	Submerged or buoyant unit weight of soil
γ_w	Unit weight of water
σ_n	Unit normal pressure (Eq. 10.7)
τ_n	Unit shear stress

REFERENCES

Ajaz, A. (1978), Detection and prevention of cracking of clay cores in dams, *Geotechnical Engineering*, **IX**, No. 1, pp. 39–62.

Alfaro, L. and Harr, M. E. (1981), Reliability of soil slopes, *Transportation Research Record*, **809**, pp. 78–82.

Asaoka, A. and Athanasiou-Grivas, D. (1981), Short-term reliability of slopes under static and seismic conditions, *Transportation Research Record*, **809**, pp. 64–70.

Beattie, A. A. and Chau, E. P. Y. (1976), The assessment of landslides potential with recommendations for future research, *Journal of the Hong Kong Institution of Engineering*, February.

Binnie and Partners (1977), *Construction on Slopes Manual*, Hong Kong.

Bishop, A. W. (1954), The use of pore pressure coefficients in practice, *Geotechnique*, **IV**, No. 4, pp. 148–152.

Bishop, A. W. (1955), The use of the slip circle in the stability analysis of slopes, *Geotechnique*, **V**, No. 1, pp. 7–17.

Bishop, A. W. and Morgenstern, N. R. (1960), Stability coefficients for earth slopes, *Geotechnique*, **X**, No. 4, pp. 129–150.

Bjerrum, L. and Kjaernsli, B. (1957), Analysis of the stability of some Norwegian natural clay slopes, *Geotechnique*, **VII**, No. 1, pp. 1–16.

Cancelli, A. (1977), Residual shear strength and stability analysis of a landslide in fissured overconsolidated clays, *Bulletin of the International Association of Engineering Geology*, No. 16.

Chang, C. J., Yao, J. R. P., and Chen, W. F. (1983), Evaluation of seismic factor of safety of a submarine slope by limit analysis, *Proc. Shanghai Symposium on Marine Geotechnology and Nearshore/Offshore Structures*, Tongji University Press/Envo Publishing Co., Inc., Bethlehem, Pa., pp. 262–295.

Chen, W. F. (1970), Discussion of "Circular and logarithmic spiral slip surfaces," by E. Spencer, *Journal of the Soil Mechanics and Foundations Division*, ASCE, **96**, No. SM1, pp. 324–326.

Chen, W. F. (1975), *Limit Analysis and Soil Plasticity*, Elsevier Scientific Publishing Co., Amsterdam.

Chen, R. H. (1986), Slope stability analysis of a waste landfill, *Proc. International Symposium on Environmental Geotechnology*, **1**, Envo Publishing Co., Inc., Bethlehem, Pa., pp. 37–42.

Chen, W. F., Giger, M. W., and Fang, H. Y. (1969), On the limit analysis of stability of slopes, *Soils and Foundations*, **IX**, No. 4, pp. 23–32.

Chen, W. F. and Scawthorn, C. R. (1970), Limit analysis and limit equilibrium solutions in soil mechanics, *Soils and Foundations*, **X**, No. 3, pp. 13–49.

Chen, W. F. and Giger, M. W. (1971), Limit analysis of stability of slopes, *Journal of the Soil Mechanics and Foundations Division*, ASCE, **97**, No. SM1, pp. 19–26.

Chen, W. F., Snitbhan, N., and Fang, H. Y. (1975), Stability of slopes in anisotropic, nonhomogeneous soils, *Canadian Geotechnical Journal*, **12**, No. 1, pp. 146–152.

Chiang, Y. C. (1979), Design and construction practice of slopes in Hong Kong, *Proc. Seminar on Slope Stability and Landslides*, Chinese Institute of Engineers, Taipei, pp. 55–82.

Chowdhury, R. N. (1980), Landslides as natural hazards—mechanisms and uncertainties, *Geotechnical Engineering*, **11**, No. 2, pp. 135–180.

Chowdhury, R. N. (1978), *Slope Analysis*, Elsevier Publishing Co., Amsterdam.

Culmann, C. (1866), *Die graphische Statik*, Meyer and Zeller, Zürich.

Drucker, D. C. and Prager, W. (1952), Soil mechanics and plastic analysis or limit design, *Quarterly of Applied Mathematics*, **10**, pp. 157–165.

Dvirnoff, A. H. and Munion, D. W. (1986), Stability failure of a sanitary landfill, *Proc. International Symposium on Environmental Geotechnology*, **1**, Envo Publishing Co., Inc., Bethlehem, Pa., pp. 25–35.

Ellis, W. and Hartman, V. B. (1967), Dynamic soil strength and slope stability, *Journal of the Soil Mechanics and Foundations Division*, ASCE, **93**, No. SM4, pp. 355–375.

Fang, H. Y. and Hirst, T. J. (1970), Application of Plasticity theory to slope stability problems, *Highway Research Record* No. 323, pp. 26–38.

Fang, H. Y., Snitbhan, N., and Chen, W. F. (1975), Discussion on stability for earth embankments, *Transportation Research Record*, **548**, pp. 12–15.

Fang, H. Y., Slutter, R. G., and Stuebben, G. A. (1976), Stress–strain characteristics of compacted waste disposal material, *New Horizons in Construction Materials*, Envo Publishing Co., Inc., Bethlehem, Pa., pp. 127–138.

Fang, H. Y., Slutter, R. G., and Koerner, R. M. (1977), Load bearing capacity of compacted waste disposal materials, *Proc. Specialty Session on Geotechnical Engineering and Environmental Control*, 9th International Conference on Soil Mechanics and Foundation Engineering, Tokyo, July, pp. 265–278.

Fang, H. Y. and Mikroudis, G. K. (1987), Multi-domains and multi-experts in knowledge-based expert systems, *Proc. International Symposium on Environmental Geotechnology*, **2**, Envo Publishing Co., Inc., Bethlehem, Pa., pp. 355–361.

Fang, H. Y., Mikroudis, G. K., and Pamukcu, S. (1989), Fracture behavior of compacted fine-grained soils, *Fracture Mechanics: Perspectives and Directions* (Twentieth Symposium), *ASTM STP* 1020, pp. 659–667.

Fang, H. Y. and Mikroudis, G. K. (1990), Use of multi-domain knowledge-based expert systems for environmental assessment of slope and landslides, *Proc. 2nd International Symposium on Environmental Geotechnology*, **2**, Envo Publishing Co., Inc., Bethlehem, Pa. (in press).

Fellenius, W. (1927), *Erdstatische Berechnungen* (calculation of stability of slopes), W. Ernst und Sohn, Berlin. (Revised edition, 1939.)

Finn, W. D. L. (1966), Earthquake stability of cohesive slopes, *Journal of the Soil Mechanics and Foundations Division, ASCE*, **92**, No. SM1, pp. 1–11.

Fröhlich, O. K. (1953), A factor of safety with respect to sliding of a mass of soil along the arc of logarithmic spiral, *Proc. Third International Conference on Soil Mechanics and Foundation Engineering*, Zurich, **II**, pp. 230–233.

Fukuoka, M. (1980), Landslides associated with rainfall, *Geotechnical Engineering*, **11**, No. 1, pp. 1–29.

Gibson, R. E. and Morgenstern, N. (1962), A note on the stability of cuttings in normally consolidated clays, *Geotechnique*, **12**, No. 3, pp. 212–216.

Gray, D. H. and Leiser, A. T. (1982), *Biotechnical Slope Protection and Erosion Control*, Van Nostrand Reinhold Co., New York, N.Y.

Huang, Y. H. (1975), Stability charts for earth embankments, *Transportation Research Record*, **548**, pp. 1–12.

Huang, Y. H. (1980), Stability charts for effective stress analysis of nonhomogeneous embankments, *Transportation Research Record*, **749**, pp. 72–74.

Huang, Y. H. (1983), *Stability Analysis of Earth Slopes*, Van Nostrand Reinhold Co., New York, N.Y.

Huang, T. K. and Chen, W. F. (1990), CAP plasticity model for embankment: from theory to practice, *Proc. 2nd International Symposium on Environmental Geotechnology*, **2**, Envo Publishing Co., Inc., Bethlehem, Pa. (in press).

Hunter, J. H. and Schuster, R. L. (1968), Stability of simple cuttings in normally consolidated clays, *Geotechnique*, **XVIII**, No. 3, pp. 372–378.

Hunter, J. H. and Schuster, R. L. (1971), Chart solutions for analysis of earth slopes, *Highway Research Record* No. 345, pp. 77–89.

Imaizumi, S., Nakayama, H., Nakajima, S., and Tajiri, K. (1989), Analysis of susceptibility to slope failure from heavy rainfall using a geomorphic and geological data based system, *Proc. 2nd International Symposium on Environmental Geotechnology*, **1**, Envo Publishing Co., Inc., Bethlehem, Pa., pp. 481–492.

Inderbitzen, A. L. (1965), An investigation of underwater slope stability, *Ocean Science and Ocean Engineering, MTS/ASLO Joint Conference Transactions*, **2**, pp. 1309–1343.

Jáky, J. (1936), The stability of earth slopes, *Proc. First International Conference on Soil Mechanics and Foundation Engineering*, **II**, Harvard University Press, Cambridge, Mass., pp. 125–129.

Janbu, N. (1954a), Application of composite slip surfaces for stability analysis, *Proc. European Conference on Stability of Earth Slopes*, Sweden, **3**, pp. 43–49.

Janbu, N. (1954b), Stability analysis of slopes with dimensionless parameters, *Harvard Soil Mechanics Series* No. 46, Harvard University Press, Cambridge, Mass.

Janbu, N., Bjerrum, L., and Kjaernsli, B. (1956), Veiledning ved losning av fundamenteringsoppgaver, *Norwegian Geotechnical Institute, Publication* No. 16, Oslo.

Jumikis, A. R. (1967), The factor of safety in foundation engineering, *Highway Research Record* No. 156, pp. 23–32.

Kézdi, A. (1957), On the factor of safety, *Proc. 4th International Conference on Soil Mechanics and Foundation Engineering*, **III**, pp. 253–254.

Kjaernsli, B. and Simons, N. (1962), Stability investigations of the north bank of the Drammen River, *Geotechnique*, **XII**, No. 2, pp. 147–167.

Koppula, S. D. (1984), Pseudo-static analysis of clay slopes subjected to earthquakes, *Geotechnique*, **34**, No. 1, pp. 70–79.

Lambe, T. W. and Whitman, R. V. (1979), *Soil Mechanics*, John Wiley and Sons, Inc., New York, N.Y., pp. 353–373.

Leonards, G. A. (1979), Stability of slopes in soft clays, *Proc. 6th Pan-American Conference on Soil Mechanics and Foundation Engineering*, **1**, pp. 223–274.

Lo, K. Y. (1965), Stability of slopes in anisotropic soils, *Journal of the Soil Mechanics and Foundations Division, ASCE*, **91**, No. SM4, pp. 85–106.

Lowe, J. III (1967), Stability analysis of embankments, *Journal of the Soil Mechanics and Foundations Division, ASCE*, **93**, No. SM4, pp. 1–33.

Lowe, J. III and Karafiath, L. (1960), Stability of earth dams upon drawdown, *Proc. First Panamerican Conference on Soil Mechanics and Foundation Engineering*, Mexico City, **2**, pp. 537–552.

Lumb, P. (1962), Effect of rain storms on slope stability, *Proc. Symposium on Hong Kong Soils*, Hong Kong Institution of Engineering, pp. 73–87.

MAA Group (1980), Collection of Technical Papers Published in 1976–1980, MAA Group Consulting Engineers, Singapore.

Majumdar, D. K. (1971), Stability of soil slopes under horizontal earthquake force, *Geotechnique*, **XXI**, No. 1, pp. 84–89.

Mikroudis, G. K. and Fang, H. Y. (1988), GEOTOX-PC: A new hazardous waste management tool, *Microcomputer Knowledge-based Expert Systems in Civil Engineering*, ed. H. Adeli, ASCE, New York, N.Y., pp. 102–117.

Mizuno, E. and Chen, W. F. (1984), Plasticity models for seismic analyses of slopes, *Soil Dynamics and Earthquake Engineering*, **3**, No. 1, pp. 2–7.

Morgenstern, N. (1963), Stability charts for earth slopes during rapid drawdown, *Geotechnique*, **XIII**, No. 2, pp. 121–131.

Morgenstern, N. R. and Price, V. E. (1965), The analysis of the stability of general slip surfaces, *Geotechnique*, **XV**, No. 1, pp. 79–93.

Muhiddin, A. B., Pamukcu, S., and Fang, H. Y. (1989), Use of knowledge-based expert systems for controlling landslides in tropical-urban environment, *Proc. 1st Caribbean Conference on Artificial Intelligence*, University of the West Indies, St. Augustine, Trinidad, West Indies, pp. 63–74.

Nguyen, V. U. (1985), Determination of critical slope failure surfaces, *Journal of Geotechnical Engineering, ASCE*, **111**, No. 2, pp. 238–250.

Nonveiller, E. (1965), The stability analysis of slopes with a slip surface of general shape, *Proc. 6th International Conference on Soil Mechanics and Foundation Engineering*, Montreal, **2**, pp. 522–525.

O'Connor, M. J. and Mitchell, R. J. (1977), An extension of the Bishop and Morgenstern slope stability charts, *Canadian Geotechnical Journal*, **14**, No. 1, pp. 144–151.

Oweis, I. S. and Khara, R. (1986), Criteria for geotechnical construction on sanitary landfills, *Proc. International Symposium on Environmental Geotechnology*, **1**, Envo Publishing Co., Inc., Bethlehem, Pa., pp. 205–222.

Perloff, W. H. and Baron, W. (1976), *Soil Mechanics*, The Ronald Press Co., New York, N.Y., pp. 528–586.

Rau, G. and Chaney, R. C. (1988), Triaxial testing of marine sediments with high gas contents, *Advanced Triaxial Testing of Soil and Rock*, eds R. T. Donaghe, R. C. Chaney, and M. L. Silver, ASTM STP 977, American Society for Testing and Materials, Pa., pp. 338–352.

Rendulic, L. (1935), Ein Beitrag Zur Bestimmung der Gleitsicherheit, *Der Bauingenieur*, No. 19/20.

Richards, A. F. (1978), Marine slope stability: an introduction, *Marine Geotechnology*, **2**, pp. 1–7.

Richards, A. F. and Chaney, R. C. (1982), Marine slope stability— A geological approach, *Proc. NATO Conference on Marine Slides and Other Mass Movements*, pp. 163–172.

Richards, A. F. and Zuiderg, H. M. (1986), Sampling and in-situ geotechnical investigations offshore, *ASTM STP 923*, pp. 51–73.

Schuster, R. L. (1968), Selection of analytical methods and strength parameters for slope stability investigations in cohesive soils, *Highway Research Record* No. 223, pp. 1–8.

Seed, H. B. (1966), A method for earthquake resistant design of earth dams, *Journal of the Soil Mechanics and Foundations Division, ASCE*, **92**, No. SM1, pp. 13–41.

Seed, H. B. (1967), Slope stability during earthquakes, *Journal of the Soil Mechanics and Foundations Division, ASCE*, **93**, No. SM4, pp. 299–323.

Seed, H. B. and Goodman, R. E. (1964), Earthquake stability of slopes of cohesionless soils, *Journal of the Soil Mechanics and Foundations Division, ASCE*, **90**, No. SM6, pp. 43–74.

Sherard, J. L. (1967), Some considerations in earth dam design, *Journal of the Soil Mechanics and Foundations Division, ASCE*, **93**, No. SM4, pp. 377–401.

Simpson, F. and Inderbitzen, A. L. (1971), Shear strength and slope stability of marine sediments in a gullied area, *Proc. International Symposium on the Engineering Properties of Sea-Floor Soil and Their Geophysical Identification*, UNESCO/NSF/University of Washington, pp. 95–109.

Skempton, A. W. (1954), The pore-pressure coefficients A and B, *Geotechnique*, **IV**, No. 4, pp. 143–147.

Skempton, A. W. (1957), Discussion of the planning and design of the new Hong Kong airport, *Proc. of the Institution of Civil Engineers*, **7**, London, pp. 305–307.

Skempton, A. W. (1964), Long-term stability of clay slopes, *Geotechnique*, **XIV**, No. 2, pp. 77–101.

Skempton, A. W. (1977), Slope stability of cuttings in brown London clay, Special Lecture Volume, *Proc. 9th International Conference on Soil Mechanics and Foundation Engineering*, pp. 25–33.

Skempton, A. W. and Golder, H. Q. (1948), Practical examples of the $\phi = 0$ analysis of stability of clays, *Proc. Second International Conference on Soil Mechanics and Foundation Engineering*, **2**, pp. 63–70.

Snitbhan, N., Chen, W. F., and Fang, H. Y. (1975), Slope stability of layered soils, *Proc. 4th Southeast Asia Conference on Soil Engineering*, Kuala Lumpur, Malaysia, pp. 5-26 to 5-29.

Spencer, E. (1967), A method of analysis of the stability of embankments assuming parallel inter-slice forces, *Geotechnique*, **XVII**, No. 1, pp. 11–26.

Spencer, E. (1968), Effect of tension on stability of embankments, *Journal of the Soil Mechanics and Foundations Division*, *ASCE*, **94**, No. SM5, pp. 1159–1173.

Spencer, E. (1969), Circular and logarithmic spiral slip surfaces, *Journal of the Soil Mechanics and Foundations Division*, *ASCE*, **95**, No. SM1, pp. 227–234.

Taylor, D. W. (1937), Stability of earth slopes, *Journal of the Boston Society of Civil Engineers*, **24**, pp. 197–246.

Taylor, D. W. (1948), *Fundamentals of Soil Mechanics*, John Wiley and Sons, Inc., New York, N.Y., pp. 406–479.

Terzaghi, K. and Peck, R. B. (1967), *Soil Mechanics in Engineering Practice*, John Wiley and Sons, Inc., New York, N.Y., pp. 232–255.

Ting, J. M. (1983), Geometric concerns in slope stability analyses, *Journal of Geotechnical Engineering*, *ASCE*, **109**, No. 11, pp. 1487–1491.

TRB (1990), Microcomputer software for geotechnical engineering, *Transportation Research Circular* No. 356, Transportation Research Board/National Research Council.

Varnes, D. J. (1978), Slope movement types and processes, *Landslides: Analysis and Control*, National Academy of Sciences, *Special Report* 176.

Winterkorn, H. F. and Fang, H. Y. (1970), Mechanical resistance properties of ocean floors and beaches in light of the theory of macromeritic liquids, *Proc. Inter Ocean 70*, Dusseldorf, **2**, pp. 43–46.

Wong, F. S. (1985), Slope reliability and response surface method, *Journal of Geotechnical Engineering*, *ASCE*, **111**, No. 1, pp. 32–53.

Wu, T. H. (1984), Effect of vegetation on slope stability, *Transportation Research Record*, **965**, pp. 37–46.

Wu, T. H. and Kraft, L. M. (1967), Probability of foundation safety, *Journal of the Soil Mechanics and Foundations Division*, *ASCE*, **93**, No. SM5, pp. 213–231.

Zhang, X. J. and Chen, W. F. (1987), Stability analysis of slopes with general nonlinear failure criterion, *International Journal for Numerical and Analytical Methods in Geomechanics*, **11**, pp. 33–50.

11 LANDSLIDES

BENGT B. BROMS, Ph.D.
Professor of Civil Engineering
Nanyang Technological Institute
Singapore

KAI S. WONG, Ph.D.
Associate Professor
Nanyang Technological Institute
Singapore

11.1 INTRODUCTION

In many parts of the world, especially in mountainous countries like Chile, Czechoslovakia, Iran, Italy, Japan, Mexico, Norway, Switzerland, and Yugoslavia, landslides are very common and have serious consequences for almost all construction activities in these countries. For example, over 9000 landslides were registered in Czechoslovakia during 1961–1962 (Zaruba and Mencl, 1969). In Japan over 2000 embankment failures occur on the average each year along the lines of the Japanese National Railways alone (Saito and Uezawa, 1969).

Even relatively small changes of the stability may trigger landslides, especially in areas where slides have previously taken place. Several landslides occurred, for example, during the construction of the Panama Canal, threatening to close the canal permanently during World War I. More than 45 million m³ of material had to be removed before the canal could again be opened for traffic. Terzaghi wrote in 1936 that "the catastrophic descent of the slope of the deepest cut on the Panama Canal issued a warning that we are overstepping the limits of our ability to predict the consequences of our actions." This statement is still true today in spite of the considerable advances made during the last ten years in the analysis and the understanding of the mechanisms of landslides. A Japanese soils engineer recently summarized the present state of the art when he wrote that "a landslide devil seems to laugh at human incompetence" (Bjerrum, 1967).

The first sign of an imminent landslide is the appearance of surface cracks in the upper part of the slope, perpendicular to the direction of the movement. These cracks may gradually fill with water, which weakens the soil further and increases the horizontal force that initiates the slide. Frequently, inclined shear cracks and lateral ridges can also be observed on both sides of the slide, as well as a slight bulge at the toe of the slope, as illustrated in Figure 11.1.

Landslides are primarily caused by gravitational forces, but occasionally seismic forces can be a contributing factor. A landslide is primarily the result of a shear failure along the boundary of the moving mass of soil or rock. Failure is generally assumed to occur when the average shear stress along the sliding or slip surface is equal to the shear strength of the soil or rock as evaluated by field or laboratory tests. However, owing to progressive failure, landslides can occur at an average shear stress considerably less than the peak strength of the soil or of the rock as measured by conventional tests such as triaxial or direct shear tests. Progressive failure is generally associated with nonuniform stress distribution along the failure surface and with stratified soils and rocks when the failure surface cuts through materials with different stress–strain properties. Local failure can occur when the maximum shear stress at a point corresponds to the peak shear strength of the soil or of the rock. Calculations indicate that the maximum shear stress occurs at or close to the toe of a slope, that the shear strength of the soil is first exceeded at this point and that the failure then spreads up the slope. Also, the fact that the strain that corresponds to the peak stress increases with increasing normal pressure can contribute to the development of progressive failure.

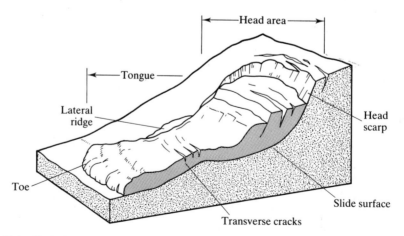

Fig. 11.1 Typical slide in cohesive material. (*After Zaruba and Mencl, 1982; Varnes, 1978.*)

Engineers and geologists often look at landslides from different points of view. The geologist regards landslides as one of the many natural processes that act on the surface of the earth as part of the general geologic cycle. He studies primarily the mechanisms of landslides with respect to geologic and hydrologic features and the resulting landforms, but he rarely makes any detailed measurements. The engineer, on the other hand, tries to determine the maximum angle at which a slope is stable and studies the stability of a slope in terms of a global factor of safety. He often has a very vague understanding of the geologic and hydrologic conditions within the area that is susceptible to landslides.

The following summarizes the present state of the art with respect to landslides. However, it should be borne in mind that "On hardly any other section of the science of foundation engineering does there exist such an extensive literature as on slides of loose masses and their calculation. This fact alone shows how difficult it is to grasp the subject and how little satisfaction exists with the possibilities obtained so far" (Müller, 1968).

11.2 CAUSES OF LANDSLIDES

A large number of factors such as geologic and hydrologic conditions, topography, climate, and weathering affect the stability of a slope and can initiate a landslide. The cause of a landslide can seldom be attributed to only a single factor. Terzaghi (1950) points out that all natural slopes have probably been exposed to more unfavourable conditions than those that occur during the lifetime of a man. If a natural slope fails, it is thus much more probable that the failure has been caused by a gradual decrease of the shear strength than by extreme conditions at the time of failure. The areas affected by landslides are increasing owing to "the engineering achievements that spread a growing population to the hitherto relatively inaccessible slopes. Highways that cut into the toe of landslides, and lubrication from the gardener's hose conspired with nature to increase or regenerate natural slope instability" (Cloud, 1969).

Terzaghi (1950) makes a distinction between external and internal factors. External factors are those that cause an increase of the average shear stress along potential failure planes or surfaces in rock or soil, such as steepening of a slope through excavation or by erosion at the toe of a slope, while internal factors are those that cause a decrease of the average shear strength. Table 11.1 summarizes the factors that may cause a landslide and Figure 11.2 the corresponding variations of the factor of safety.

11.2.1 Construction Operations

Many landslides have occurred during the excavation for highways, railways or canals, and slides are also common in quarries and pits. One spectacular example is the Rissa landslide in Norway, which was triggered by a small excavation for a farm house at the toe of a slope (Gregerson, 1981). The landslide extended over an area of 330 000 m² and involved a volume of about 5 to 6 million m³.

Frequently, heavy buildings can contribute appreciably to a landslide if they are located close to the edge of a slope. Peck (1967) has cited several examples where a very small change of the stability conditions in an old slide area have initiated further sliding. Landslides can also be caused by an increase of the load, from a fill, for example, at the head of the slope, as was the case at Portuguese Bend in California (Merriam, 1960). These slides are caused by an increase of the average

shear stress along potential failure surfaces in the soil mass as a result of construction.

Landslides can also be triggered by vibrations from pile driving or blasting. In Figure 11.3 is shown a slope failure in a soft, sensitive clay that was caused by the driving of piles at the top of a slope. The first three rows of piles were driven successfully. The failure occurred just after the first few piles in the fourth row had been driven (Broms and Bennermark, 1968). One case has occurred recently in Canada (Carson, 1979), where a large landslide in sensitive clay can be attributed to liquefaction caused by pile driving of a silty layer at some depth below the toe. A more detailed discussion of landslide induced by pile driving has been presented by Massarsch and Broms (1981). They pointed out that the increase in pore pressure does not directly affect the stability, as the undrained shear strength of a cohesive soil is independent of a change in total stress. However, the swelling that takes place can result in a gradual decrease of the shear strength. One of the most problematic situations is loose saturated sand or silt seams, where vibration from pile driving can produce an increase in pore pressure that immediately reduces the shear strength or can cause spontaneous liquefaction.

Mining may also initiate landslides. One example is the landslides and the resulting subsidence caused by the coal mining close to Stadice in Czechoslovakia (Păsek and Demek, 1969). Another example is the large landslide that occurred at Turtle Mountain in Alberta, Canada. This landslide has partly been attributed by Terzaghi (1950) to mining operations in the area. Turtle Mountain consists mainly of jointed limestone that overlays a weaker layer where coal was mined at the time of the slide. Creep movement caused the limestone blocks to break. As a result, the shear strength of the rock mass was reduced.

11.2.2 Erosion

Erosion from streams and rivers, glaciers, waves, currents, and wind is responsible for the formation as well as the destruction of many slopes. These agents undercut or oversteepen the slope, which may eventually lead to failure of the slope.

Many slides in quick clay in the Scandinavian countries are triggered by stream erosion (Bjerrum, 1969). The recession of the river banks (Williams et al., 1979) and the regression of the coast line due to current and wave actions can also cause major landslides. Erosion can play an active role in areas with clay shales, which may lead to large retrogressive landslides as has been observed in the Saskatoon area (Hang et al., 1977).

11.2.3 Tectonic Movement

Tectonic movement in the earth's crust can also steepen the slopes and thus trigger landslides. For example, a large part of the Raukumara Peninsula in New Zealand is being uplifted at a rate of 10 mm/year (Gage and Black, 1979). The soft, clay-rich rocks have been severely deformed and crushed. The slopes in this region are continually regraded by flow slides.

Faulting can also change abruptly the slope gradient. Fault zones often contain fractured and crushed rocks. These zones can be weakened further by the percolation of water into the bedrock and by the subsequent chemical weathering of the crushed rocks.

11.2.4 Earthquakes and Vibrations

Landslides can also be triggered by earthquakes or by vibrations from pile driving or blasting. The vibrations can cause

TABLE 11.1 PROCESS LEADING TO LANDSLIDES[a].

Name of Agent	Event or Process That Brings Agent into Action	Mode of Action of Agent	Slope Materials Most Sensitive to Action	Physical Nature of Significant Actions of Agent	Effects on Equilibrium Conditions of Slope
Transporting agent	Construction operations or erosion	1. Increase of height or rise of slope	Every material	Changes state of stress in slope-forming material	Increases shearing stresses
			Stiff, fissured clay, shale	Changes state of stress and causes opening of joints	Increases shearing stresses and initiates process (8)
Tectonic stresses	Tectonic movements	2. Large-scale deformations of earth crust	Every material	Increases slope angle	Increase of shearing stresses
	Earthquakes	3. High-frequency vibrations	Every material	Produces transitory change of stress	
			Loess, slightly cemented sand, and gravel	Damages intergranular bonds	Decrease of cohesion and increase of shearing stresses
			Medium or fine loose sand in saturated state	Initiates rearrangement of grains	Spontaneous liquefaction
Weight of slope-forming material	Process that created the slope	4. Creep on slope	Stiff, fissured clay, shale remnants of old slides	Opens up closed joints, produces new ones	Reduces cohesion, accelerates process (8)
		5. Creep in weak stratum below foot of slope	Rigid materials resting on plastic ones		
Water	Rains or melting snow	6. Displacement of air in voids	Mois sand	Increases pore water pressure	Decrease of frictional resistance
		7. Displacement of air in open joints	Jointed rock, shale		Decrease of cohesion
		8. Reduction of capillary pressure associated with swelling	Stiff, fissured clay and some shales	Causes swelling	
		9. Chemical weathering	Rock of any kind	Weakens intergranular bonds (chemical weathering)	
	Frost	10. Expansion of water due to freezing	Jointed rock	Widens existing joints, produces new ones	
		11. Formation and subsequent melting of ice layers	Silt and silty sand	Increases water content of soil in frozen top layer	Decrease of frictional resistance
	Dry spell	12. Shrinkage	Clay	Produces shrinkage cracks	Decrease of cohesion
	Rapid drawdown	13. Produces seepage toward foot of slope	Fine sand, silt, previously drained	Produces excess pore water pressure	Decrease of frictional resistance
	Rapid change of elevation of water table	14. Initiates rearrangement of grains	Medium or fine loose sand in saturated state	Spontaneous increase of pore water pressure	Spontaneous liquefaction
	Rise of water table in distant aquifer	15. Causes a rise of piezometric surface in slope-forming material	Silt or sand layers between or below clay layers	Increases pore water pressure	Decrease of frictional resistance
	Seepage from artifical source of water (reservoir or canal)	16. Seepage toward slope	Saturated silt	Increases pore water pressure	Decrease of frictional resistance
		17. Displaces air in the voids	Moist, fine sand	Eliminates surface tension	Decrease of cohesion
		18. Removes soluble binder	Loess	Destroys intergranular bond	Decrease of frictional resistance
		19. Subsurface erosion	Fine sand or silt	Undermines the slope	Increase of shearing stresses

[a] After Terzaghi (1950).

412

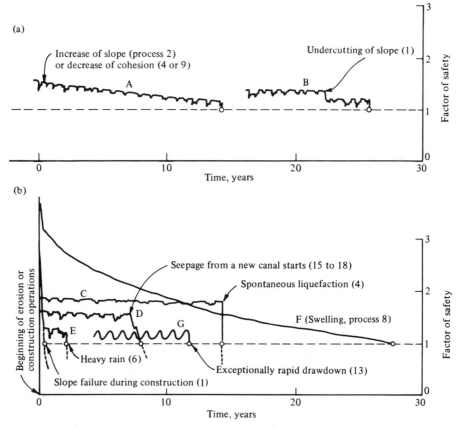

Fig. 11.2 Diagram illustrating the variations of the factor of safety of various slopes prior to a landslide. (a) Old slopes; (b) slopes of recent origin. (*After Terzaghi, 1950.*)

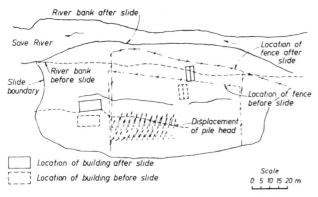

Fig. 11.3 Failure of slope caused by pile driving. (*After Broms and Bennermark, 1968.*)

spontaneous liquefaction of loose sand or of silt, or of loess below the groundwater table, and a reduction of the shear strength of some sensitive clays. The structure of these soils may collapse when the relative density is low. As a result, the pore pressure will increase. When the pore pressure approaches the total overburden pressure, the shear strength of the soil is drastically reduced and the soil becomes essentially a heavy liquid. The landslide at Turnagain Heights in Alaska in 1964 is an example of this type (Seed and Wilson, 1967).

Numerous landslides were triggered by the 1960 earthquakes in Chile. One of the largest known landslides, in loess in the Kansu province in China in 1920, was triggered by an earthquake (Close and McCormick, 1922). The soil became a suspension of silt-sized particles in air with the same properties

as a dense fluid, and the suspension flowed with an extremely high velocity (greater than 3 m/sec) into the valleys and buried entire villages. Seed (1968) has reviewed a number of landslides in loess that have been triggered by earthquakes. The 1949 earthquake at Chait in the U.S.S.R. caused flow slides that buried 21 villages in the Yasman River valley. The debris completely covered the bottom of the valley, forming a 20-km long and over 1-km wide obstruction that was several tens of meters deep. The landslide at Surte, Sweden, in 1950, which occurred in a soft, sensitive clay, was probably triggered by vibrations from pile driving or from a passing train (Jakobson, 1952).

Soils that usually are not affected by vibrations are clays with low sensitivity and dense sand either above or below the groundwater table. Loose sand or silts as a rule are not affected above the groundwater table.

11.2.5 Rains or Melting Snow

Most slope failures occur after a heavy rain or during the spring when the snow melts and water penetrates into cracks and fissures. The addition of water to a slope increases the load owing to the added weight of the water. The shear strength is reduced owing to the increase of the pore water pressures. Water, in fact, has been implicated as the main controlling factor in most slides.

Heavy rain often increases markedly the frequency of landslides. The initiating factor is on many occasions a high-intensity storm following a period of continuous rainfall. Such correlations have been documented by a number of investigators (Onodera et al., 1974; Radbruch-Hall and Varnes, 1976).

Brand et al. (1984) have observed that the landslides in Hong Kong depend on the intensity of the short-period rainfalls with a threshold value of about 70 mm/hr. Antecedent rainfall is not a major factor in landslide occurrence.

An increase of the water content due to snowmelt can also cause slope failures. A number of such failures in sensitive soils in Canada have been reported by Eden and Mitchell (1970) and by Eden (1971).

In Sweden, Norway (Jörstad, 1970), and the eastern part of Canada (Lebuis and Rissmann, 1979), it has been observed that the frequency of landslides is high during the early spring when the snowmelt is combined with rain.

For residual soils, Brand (1981) has pointed out that the balance between infiltration and pore pressure suction plays a major role with respect to the frequency of landslides in Hong Kong. The infiltration reduces the pore suction and thus the effective stress and the shear strength of the soil.

11.2.6 Frost

Frost is another factor of degradation and weathering that gradually reduces the shear strength of both soils and rocks. The formation of ice laminae or lenses in fine-grained soils may cause high pore water pressures during the thawing. The increase of the volume as the water in fissures and cracks freezes may in time cause rockfall. In Figure 11.4 is shown the monthly distribution of rockfalls in Norway in 1951–1955. The largest number of rockfalls occurred during the spring and in the fall when the temperature fluctuates around the freezing point (Bjerrum and Jörstad, 1968).

11.2.7 Dry Spells

The shear strength can also be reduced by cracks and fissures that open up when the overburden pressure is reduced or when the soil is dried out. Shrinkage cracks may develop in shales or clays if the surface is not protected by a layer of sand or by sod. Weathering can also appreciably reduce the shear strength of rocks and may cause failure of a slope.

11.2.8 Rapid Drawdown

Slope failure often occurs, particularly for silty soils, when the water level in, for example, a reservoir is lowered suddenly.

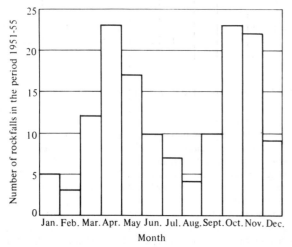

Fig. 11.4 Distribution of rockfalls in Norway, 1951–1955. (*After Bjerrum and Jörstad, 1968.*)

Legget and Bartley (1953) have described a number of slides that occurred during the draining of the steep Rock Lake in Canada. Pope and Anderson (1960) have described a slide at Lookout Point Reservoir in Oregon in stiff clay that was caused by drawdown. The movement accelerated during 1958 until it was about 75 mm/day. The total movement that year was about 7 m. This type of landslide is caused by high pore water pressures along potential failure surfaces in the soil. Methods of analyzing these types of landslides have been developed by Reinius (1948).

11.2.9 Seepage from Artificial Sources of Water

The shear strength of partially saturated soils is reduced when the soil becomes saturated, especially if the soil is fine-grained. The shear strength of such soils comes at least partly from the surface tension in the pore water at the contact points between the individual mineral particles, which is lost when the soil becomes saturated. The resulting false cohesion can prevent failures of vertical cuts in moist, fine sand or silt even if the cut is 3 to 6 m deep. The false cohesion is generally not destroyed above the groundwater table even by heavy rainfalls. However, failure will occur when the slope is submerged for the first time. Also, cemented soils such as loess lose part of their shear strength when they become saturated. The binder material is dissolved by the percolating water and the shear strength of the soil is reduced.

The filling of a reservoir may trigger large landslides, as was the case at Vaiont (Müller, 1964; Kiersch, 1964). Breth (1967a, b) and Lauffer et al. (1967) have described mass movements in moraine and talus material that occurred during the filling of a reservoir at Gapatsch, Austria. The rate increased with increasing reservoir level and the maximum vertical movement was 150 mm/day. The rate decreased rapidly after the filling was temporarily stopped. Over 500 landslides occurred during the filling of the Franklin D. Roosevelt Lake in Washington (Jones et al., 1961).

Leakage from reservoirs or unlined canals may cause failure by retrogressive internal erosion or by piping. As erosion develops, the flow of water increases, and the roof of the conduit thus created may collapse and trigger a slide.

The leaching of salt due to groundwater flow or from infiltration of rain water that takes place in many clays and infiltration of waste water can cause a reduction of the shear strength. This reduction of the shear strength has contributed to many landslides in Scandinavia and Canada (Söderblom, 1973).

11.2.10 Vegetation

Vegetation also affects the stability. The roots of trees and bushes growing on a slope can contribute appreciably to the overall strength of the soil mass. The water content of the soil is reduced by evapotranspiration. As a result, the overall stability is increased (Gray and Leiser, 1982; Sidle et al., 1985). On the other hand, the gradual decay of the root system when the trees are cut or die may gradually decrease the stability (Brown and Shen, 1975). Also, the added weight of the trees and wind may also affect the stability. However, Gray (1970) pointed out that for a given tree load the increase of the shear strength along a potential slip surface from the increased normal stress will be larger than the downslope or driving component when the inclination of the slope is less than the effective angle of internal friction of the soil. The effect of wind can generally be neglected. According to Wu (1976), the increase of the average shear stress from an 80 km/hr wind is only 0.1 kPa.

The effect of clear cutting of a forest on the stability has been investigated by many. Bishop and Stevens (1964) report that the number of slides and the area affected by slides in southeastern Alaska have increased as much as fourfold over a 10-year period after the logging. Wu (1976) found for the same area that the frequency of the landslides after clear cutting was even greater than that reported by Bishop and Stevens.

11.3 CONSEQUENCES OF LANDSLIDES

A major landform-shaping process in mountainous areas throughout the world is landslides. Only when landslides cause death or injury, or structural damage do they receive much attention. The worldwide loss of human life, buildings, and other structures and land is almost impossible to quantify.

The greatest impact is the loss of human life. More than 100 000 were killed by landslides that took place during a series of earthquakes in 1920 in the Kansu province in China. "Landslides that eddied like waterfalls, crevasses that swallowed houses and camel trains, and villages that were swept away under a rising sea of loose earth were a few of the subsidiary occurrences that made the earthquake in Kansu one of the most appalling catastrophes in history. Loose earth cascaded down the valleys and buried every object in its path" (Close and McCormick, 1922).

Another disastrous landslide occurred at Vaiont where about 240 million m³ of rock from Mount Toc slid into the reservoir, creating a wave over 100 m high. At a distance of about 2 km from the slide the wave height was over 70 m. Rock and water climbed 260 m above the reservoir level on the opposite side of the reservoir (Kiersch, 1964). The wave overtopped the dam and washed away the town of Longarone. About 2000 people lost their lives during the flood.

In May 1970, an earthquake-triggered debris flow destroyed the city of Yungay in Peru and killed 18 000 people (Plafker et al., 1971). Incomplete records compiled by UNESCO from 1971 to 1974 showed that 2378 people were killed worldwide (Sidle et al., 1985). In Japan, 1344 have been killed by debris flows and landslides from 1967 to 1972 (Ikeya, 1976).

Aside from the losses of human life, the financial losses from landslides can be large. The cost can be divided into direct and indirect costs. Direct costs include, for example, actual damage to homes and commercial buildings, damage to public utilities, loss of agricultural crops, etc. Examples of indirect costs are preventive measures to reduce future damage, reduced value of real estate in the vicinity of the landslide, and loss of tax revenues. Indirect costs are difficult to evaluate and they can be much greater than the direct costs.

Krohn and Slosson (1976) estimated that the annual damage (direct cost) to buildings and their sites by landslides in the United States alone is around US$400 million. The California Division of Mines and Geology (Alfors et al., 1973) has estimated that the costs of slope movements throughout the United States from 1970 to 2000 will be nearly US$10 billion, or an average of more than US$300 million a year. Schuster (1978) has estimated that the total costs including the indirect costs may even exceed US$1 billion per year.

Relatively accurate cost estimates can be made for individual landslides or for landslides within a relatively small geographic area. For example, the damage caused by landslides, slumps, and debris flows in Los Angeles during the winter of 1951/1952 have been estimated at US$7.5 million (Cloud, 1969). Damage from landslides caused by the Alaska earthquake of 1964 exceeded US$100 million (Coates, 1977).

Often the consequence of a landslide may not be known until many months afterwards. Slides sometimes close mountain valleys and create temporary lakes. When the "dam" is overtopped, disastrous floods occur. In 1912, a 150 million m³ rock slide blocked the Brenno Valley in northern Italy and created a lake with a depth exceeding 50 m. The impounded water broke through two years later and over 60 people lost their lives in the resulting flood. One of the largest landslides in the history of man took place in 1893 at Upper Garhwal in India. The landslide created a natural dam that was over 3 km long and 1.5 km wide. The height of the "dam" was 295 m. When it was overtopped approximately one year later, over 1 billion m³ of water was released and entire towns and villages were swept away by the flood (Legget, 1962).

In addition to loss of lives and property, landslides can also reduce the productivity of agricultural, forest, and range lands. The sediments in streams may destroy breeding grounds of fish and affect the quality of the water downstream.

11.4 SLOPE MOVEMENTS PRECEDING LANDSLIDES

11.4.1 Creep

Some creep takes place in almost all steep earth and rock slopes. The creep movements may be concentrated along a preexisting or potential sliding surface or be distributed evenly throughout the moving soil mass as shown in Figure 11.5. Old landslide areas are particularly affected by creep.

There are many factors that probably contribute to the creep in soils and rocks, but little is known directly about its nature. Terzaghi (1950) distinguishes between seasonal and continuous creep. Seasonal creep, which is caused by temperature and moisture variations in the soil or rock within the surface layer, takes place primarily in clayey or silty soils. The depth of the surface layer that is affected by seasonal creep is often equal to or less than the depth of seasonal temperature and moisture variations. In temperate climates, seasonal creep is caused

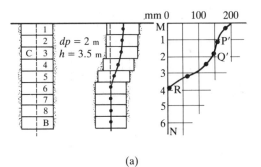

(a)

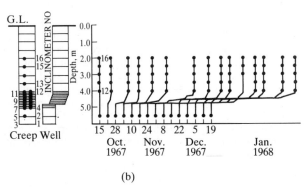

(b)

Fig. 11.5 Creep well observations. (a) Ter-Stepanian (1965); (b) Watari et al. (1977).

primarily by freezing and thawing (solifluction). High pore water pressures develop in the spring, thereby decreasing the shear strength and causing the soil to move downhill. Creep also takes place in tropical areas in many residual soils within the zone of seasonal variations, but the creep rate is generally very low.

A soil or rock that undergoes creep will not necessarily fail but can often sustain a much higher shear stress without failure than that causing creep. However, the creep gradually decreases the shear strength. Analysis by Suklje and Vidmar (1961) of a landslide through a heterogeneous cohesionless soil and an overconsolidated clay indicated that the shear strength of the materials was reduced by creep. The effects of creep on the stability of slopes have been discussed by Goldstein et al. (1965).

Movements caused by surface creep are generally less than a few millimeters per year but they can cause trees to lean, fences and roads to be displaced, and the gradual bending of bedding planes within the soil or rock. Haefeli (1953) has found that seasonal creep is highest at the surface and that it decreases with depth.

Continuous creep is caused by gravitational forces and occurs below the surface zone of seasonal variations. Bjerrum (1967, 1969) has pointed out that continuous creep primarily takes place in shales. Creep appears to become more pronounced with increasing initial strength of the soil or the rock. For example, creep is common in poorly cemented argillaceous schists in the eastern part of Switzerland. These rocks swell and deteriorate when they come into contact with water and air (Haefeli et al., 1953; Moos, 1953). Creep is generally very small in normally consolidated clays owing to the increase of the shear strength that takes place with time.

Continuous creep in shales appears to occur in relatively thick zones. Ter-Stepanian (1965) measured creep movements within a 2.0-m thick zone of fissured stiff clay where the creep velocity increased approximately linearly with depth within the zone. Creep movements have also been measured by Yen (1969) in a 1.0-m thick zone of fissured shale along the coast of southern California.

Ter-Stepanian (1966) distinguishes between planar, rotational, and general creep deformations. Planar or translational creep occurs along a plane that is approximately parallel to the ground surface. This type of creep takes place in relatively long slopes. Rotational creep, which occurs primarily in homogeneous rock masses, causes a rotation of the moving rock or soil mass. The creep movements that are not classified as planar or rotational are called general creep.

11.4.2 Creep Rates

The creep movement can be divided into three stages (Mitchell, 1976): primary, secondary, and tertiary. In the primary stage, the rate of movement decreases with time. It is a transient stage that takes place just after a change of the conditions in the slope and that may last only for a relatively short period of time. In the secondary stage, the rate of movement is relatively constant and may last for tens of years. Haefeli et al. (1953) report that for an old slide area at Klosters in Switzerland the creep rate over a 13-year period was practically constant, around 37 mm/year. At the site of the old West Culebra slide in the Panama Canal, Hirschfeld et al. (1965) observed a steady creep rate of 9 mm/year. In the tertiary stage, the creep rate increases rapidly and eventually leads to failure.

Saito and Uezawa (1961) and Saito (1965, 1969) have suggested that there exists a relationship between time to rupture and strain rate. The time to rupture was found to be inversely proportional to the strain rate and independent of the

soil type. When failure is imminent, the creep rate may be as high as a few centimeters per day. For example, the creep rate at Vaiont was 200 mm/year about two years before the slide occurred and reached about 60 mm/day the week before the slide.

Creep rates between 100 and 260 mm per year were measured for a slope in argillaceous schists at Heizenberg, Switzerland. The inclination of the slope was 14° and the bedding planes were more or less parallel with the ground surface. The total estimated movement was 50 to 150 m (Haefeli, 1953).

The creep rate for a talus slope at Stoss, Switzerland, was 1 to 16 m/year. The talus consisted primarily of marl debris but also contained some sandstone and glacial till. The inclinations of the slope varied from 17° in the upper parts to 8° in the lower parts of the slope. The total volume of material involved was about 2.9 million m³ (Moos, 1953).

Gould (1960) measured creep rates for a number of slopes in heavily overconsolidated clays along the coast of California as shown in Figure 11.6. The average creep rate for a 50-m high slope with an average inclination of 10° was about 0.3 m/year. The movement took place within a narrow zone at a depth of 15 m below the ground surface. The average shear stress along the failure surface corresponded approximately to the residual shear strength of the clay. The geologic history of the area indicates that all slopes will eventually fail when the average shear strength required for stability is larger than the residual shear strength of the soil.

The average creep rate at Gradot Ridge, Yugoslavia, was smaller than 15 mm/year for about 100 years, but the rate increased gradually as failure was approached (Suklje and Vidmar, 1961). Creep in the clay layer caused tensile cracks and a reduction of the shear strength in the overlying rigid tuff cap as well as in the clay.

11.4.3 Total Displacement Before Failure

The total displacement just before failure is primarily dependent on the strain that corresponds to the peak strength and on the thickness of the failure zone. The total movement appears to be smaller for normally consolidated clay than for stiff overconsolidated clay. For example, Zaruba and Mencl (1969) have estimated from field observations that stiff clayey soils can undergo a shear displacement of up to 2.5 percent of the thickness of the failure zone. For stiff shales the corresponding displacement is about 0.8 percent.

The total movement before failure exceeded 1.3 m for the large landslide at Gradot Ridge in Yugoslavia (Suklje and Vidmar, 1961). At Vaiont the total movement before failure was 2.5 m. The total movement for the landslide at Dosan (Saito, 1965) exceeded 0.4 m before failure finally occurred, and the rate of the movement at the day of the slide was 0.3 m.

If the failure zone is relatively thin, the total movement before failure will be small (a few millimeters), while the total movement can be one meter or more if the failure zone is relatively thick.

11.4.4 Thickness of Failure Zone

At the landslides at Folkstone Warren in England, the thickness of the failure zone in stiff fissured clays was a few centimeters (Muir Wood, 1955). The thickness of the shear zone in the stiff fissured clay at Walton's Wood in England was about 20 mm. Orientation of clay particles was observed within a 20- to 30-μm wide band (Morgenstern and Tchalenko, 1967). Also, Gould (1960) has observed that the failure in stiff fissured clays takes place within a narrow failure zone.

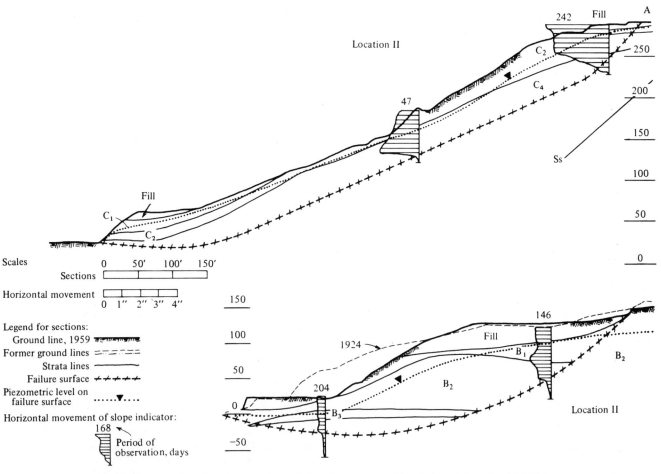

Fig. 11.6 Creep movements for slopes in heavily overconsolidated clays. (*After Gould, 1960.*)

11.5 CLASSIFICATION

A classification system can be of considerable help to the engineer or the geologist when he tries to determine the causes of a landslide and the measures that can be taken to prevent or to correct landslides. Terzaghi (1950) has pointed out that a large number of factors can contribute to a landslide and that landslides can occur in almost all materials ranging from rock to very soft clays, as indicated in Table 11.1. Terzaghi remarked that "a phenomenon involving such a multitude of combinations between materials and disturbing agents opens unlimited vistas for the classification enthusiast."

Several different classification systems have been proposed. Many of these systems have been developed for a particular purpose and for a particular area and cannot be used elsewhere. Thus, Heim (1932) proposed a classification system that is applicable only to rock slides in the Alps, and he recognized over 20 different types.

The classification system proposed by Sverenskii (Zaruba and Mencl, 1969) is used extensively in eastern Europe. Landslides are divided with respect to the shape of the sliding surface into asequent, consequent, and insequent landslides. An asequent slide occurs along curved, mainly cylindrical failure surfaces. This type of slide takes place primarily in intact cohesive soils. The failure surface of consequent slides primarily follows bedding planes, fissures, joints, or other planes of weakness. The failure surface of insequent slides cuts across bedding planes, fissures, and joints. This type of slide generally extends deep into a slope.

Landslides can also be classified with respect to their maximum depth as follows:

Type	Maximum depth D, m
Surface slides	< 1.5
Shallow slides	1.5 to 5.0
Deep slides	5.0 to 20.0
Very deep slides	> 20.0

Skempton (1953) has extended this system to classify landslides with respect to their D/L ratio, where D is the maximum depth and L is the maximum length of the landslide.

Varnes (1978) has proposed the following classification with respect to the slide velocity.

Term	Velocity
Extremely rapid	> 3 m/sec
Very rapid	0.3 m/min to 3 m/sec
Rapid	1.5 m/day to 0.3 m/min
Moderate	1.5 m/month to 1.5 m/day
Slow	1.5 m/year to 1.5 m/month
Very slow	0.06 m/year to 1.5 m/year
Extremely slow	< 0.06 m/year

Landslides can be classified according to their state of activity into active, dormant, and stabilized slides (Zaruba and Mencl, 1969). Active landslides are generally easy to recognize. Trees within an active slide area are not vertical; roads, fences, and telephone poles have been displaced; buildings are damaged. The scarp at the head of an active slide is steep and free of vegetation. Numerous open cracks are often observed.

TABLE 11.2 ABBREVIATED CLASSIFICATION OF SLOPE MOVEMENTS[a].

	Type of Material		
		Engineering Soils	
Type of Movement	Bedrock	Predominantly Coarse	Predominantly Fine
Falls	Rock fall	Debris fall	Earth fall
Topples	Rock topple	Debris topple	Earth topple
Slides			
Rotational	Rock slump	Debris slump	Earth slump
Translational	Rock block slide	Debris block slide	Earth block slide
	Rock slide	Debris slide	Earth slide
Lateral spreads	Rock spread	Debris spread	Earth spread
Flows	Rock flow (deep crack)	Debris flow (soil creep)	Earth flow (soil creep)
Complex	Combination of two or more principal types of movement		

[a] After Varnes (1978).

Dormant landslides are often difficult to recognize. They are generally covered by vegetation. Old scarps, which developed when the landslide was active, have been eroded. Dormant landslides are characterized by trees that lean or are bent and cracks that have been filled by debris. Dormant landslides can generally be activated by even relatively small changes of the stability conditions, for example, by excavation at the toe of the slope. The age of a dormant landslide can often be determined from the age of the oldest tree that is growing straight within the landslide area, from the stage of the erosion, and from the development of recent soil profiles. Stabilized slides that have been stabilized by chemical or mechanical means can be either active or dormant.

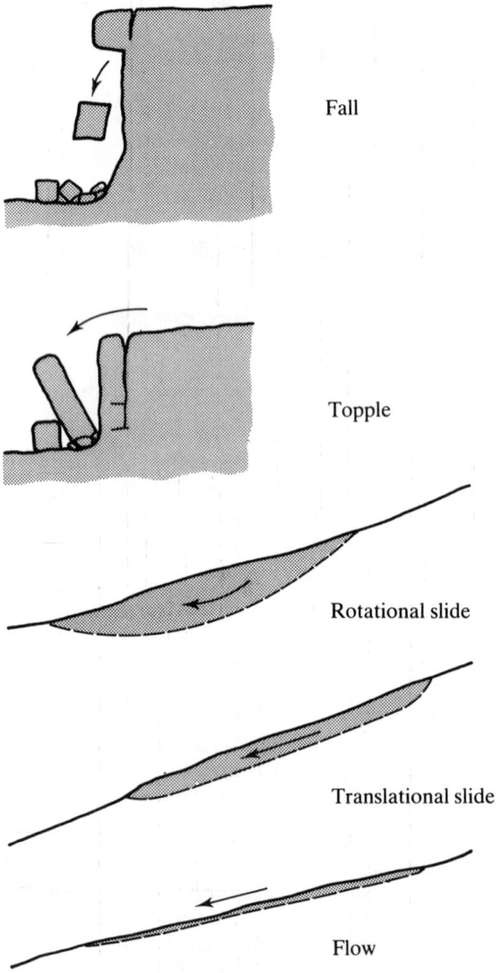

The scheme that enjoys the widest acceptance is that of the Highway Research Board Committee on Landslide Investigation (Varnes, 1958), which has subsequently been updated by Varnes (1978). An abbreviated form of this scheme is shown in Table 11.2. The five principal types of mass movement are: falls, topples, slides, lateral spreads and flows, as illustrated in Figure 11.7. A combination of two or more of the five principal types, whether within various parts of the moving soil mass or at different stages, is referred to as complex slope movement.

11.5.1 Falls

Falls involve the descent of soil and rock masses mostly by free falling, bouncing, and rolling. Movements are very rapid to extremely rapid. For rock falls, velocities of 40 to 70 m/sec have been observed.

Rockfalls can occur in almost all types of rock and are generally caused by frost wedging, weathering, temperature variations, high cleft water pressures, or undercutting. Examples of rockfalls are shown in Figure 11.8. In temperate climates, frost wedging is the most important of these factors. Falls generally occur along bedding planes, joints or local fault zones, or fault planes. More than 200 large rockfalls have been recorded in Norway since 1940. These have generally been caused by freezing or by excess cleft water pressures (Bjerrum and Jörstad, 1968). There are no reliable methods available for calculating the stability of a slope with respect to falls.

Soil falls often occur when an easily erodible material underlies a more erosion-resistant material, for example, where a layer of clean sand or silt underlies a layer of overconsolidated clay (Barzett et al., 1961; Skempton and LaRochelle, 1965). The mechanism of such falls has been discussed by Henkel (1967). The falls of loess along the bluffs of the lower Mississippi River valley is another example of soil fall caused by erosion. Falls along preexisting joints and fissures in residual soils have been described by Lumb (1962).

Falls are one of the main erosion mechanisms in heavily overconsolidated clays. In such clays they generally occur when

Fig. 11.7 Main slide types.

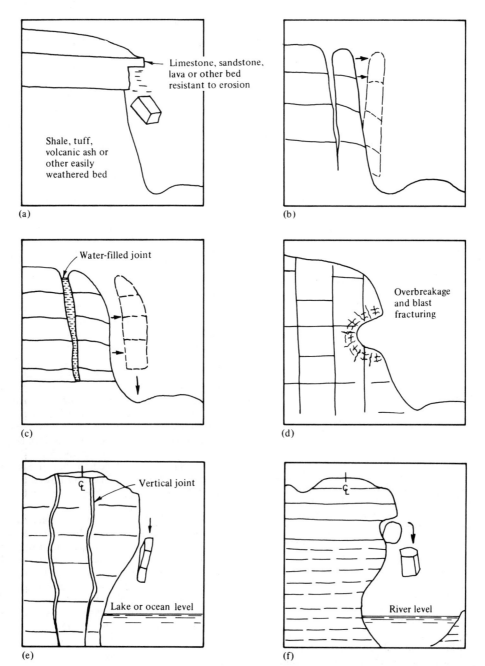

Fig. 11.8 Examples of rockfalls. (a) Differential weathering. (b) Frost wedging in jointed homogeneous rock. (c) Jointed homogeneous rock. Hydrostatic pressure acting on loosened blocks. (d) Homogeneous jointed rock. Blocks left unsupported or loosened by overbreakage and blast fracture. (e) Either homogeneous jointed rock or resistant bed underlain by easily eroded rock. Wave-cut cliff. (f) Either homogeneous jointed rock or resistant bed underlain by easily eroded rock. Stream-cut cliff. (*After Varnes, 1958.*)

rainwater fills the tensile cracks of fissures at the crest of a steep slope. The location of the fissures has a considerable influence on the extent of a fall. Falls that are caused by deep-seated joints or fissures generally fall backward, while those caused by cracks close to the crest often fall forward when the slope is sufficiently steep.

11.5.2 Topples

Topple is tilting without collapse, as shown in Figure 11.7. It involves forward rotation of a unit (or units) about some pivot point caused by gravity forces from adjacent units or by water

pressures in cracks. Most topples have been observed in rocks (deFreitas and Watters, 1973). Toppling may or may not culminate in either falling or sliding. The failure mechanism depends on the geometry of the failing mass and on the orientation and extent of the discontinuities.

11.5.3 Rotational Slides

Different types of rotational slides can be recognized, namely slips, multiple rotational slides, and successive slips, as illustrated in Figure 11.9. The failure surface can be circular or noncircular.

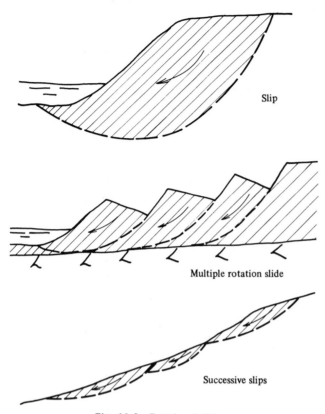

Fig. 11.9 Rotational slides.

Slips or slumps occur generally along concave failure surfaces. The failure surface is approximately circular and the failing soil mass moves as an unit along a relatively thin failure or slip surface (Hultin, 1916; Petterson, 1916). The slump mass is often waterlogged and the lower part of the landslide frequently develops into an earth flow.

The cross section of a typical rotational slip is illustrated in Figure 11.1. The scarp at the head of a rotational failure is generally almost vertical, especially for fine-grained stratified soils. Usually it cannot stay unsupported very long or a new failure will occur. The water that is trapped by the backward tilting of the sliding soil mass may trigger additional slides as the stability of the slope is decreased. This type of movement is most prevalent in uniform deposit of clay shale and compacted clayey fill. The slip surface is relatively deep-seated. Slide of this type can be initiated by erosion at the toe, by steepening, or by surcharging of the slope.

The landslide at Lodalen, Norway, which occurred in a slightly overconsolidated clay, is an example of this type of failure (Sevaldson, 1956). Rotational slides in normally consolidated clays have also been discussed by Ireland (1954), Skempton (1945), and Kankare (1968). Rotational slides in shales have been described by Crandell (1952). These slides occurred in the Pierre shale along the Missouri River in South Dakota.

The maximum velocity in homogeneous clays with a low sensitivity or in residual soils is low, generally about 0.1 to 0.3 m/min. For a rotational slide, the velocity seldom exceeds 0.3 m/min (Terzaghi, 1950).

Multiple Rotational Slides are triggered by an initial, often local slip. This type develops gradually and spreads backward along a common basal failure surface. The multiple rotational slide that took place at Folkstone Warren in Kent, England, has been described by Toms (1953), Muir Wood (1955), and

Hutchinson (1969). The landslides at Meikle River in Alberta (Nasmith, 1964) and at Sandnes in Norway (Bjerrum, 1967) also appear to be of this type.

According to Skempton and Hutchinson (1969), multiple rotational slides occur most frequently in actively eroding slopes of overconsolidated clays, fissured clays, or clay shales that are overlain by a layer of more competent soil or rock.

The so-called "bottleneck" slides in Scandinavia and Canada that take place in quick clays are generally multiple rotational. This type of slide is often triggered by a small local slip at an eroding river bank. When the soil is remolded during the slide, its consistency is changed to that of a heavy liquid and the soil flows out of the slide area. The remolded soil often flows rapidly both upstream and downstream. The failure progresses retrogressively at the unsupported rear scarp as the soil flows out. Slabs of soil from the overlying stiff dry crust can often be seen floating in the moving soil. Even waves and backwater phenomena have been observed (Holmsen, 1953). The width of the slide area frequently increases away from the river bank, which gives the slide area its characteristic "bottleneck" shape.

Bottleneck slides are often difficult to analyze. They occur primarily in leached glacial and postglacial marine clays (quick clays). Bjerrum (1954) has described a bottleneck slide that took place at Ullensåker, Norway (Fig. 11.10), while Crawford and Eden (1967) discuss a similar slide that occurred at Nicolet, Canada. The slide at Ullensåker, Norway, developed in a slope with an initial average inclination of only 3 to 4°. A few hours before the slide occurred, fissures were observed in the lower part of the slope. The slide progressed backward from a 20-m wide initial slide so that a bowl-shaped depression with a diameter of 200 to 300 m was formed. The depth of the depression varied between 5 and 8 m. The consistency of the sliding clay mass changed to that of castor oil owing to the remolding of the soil. The clay flowed about 1.5 km down a brook (Bjerrum, 1954).

Successive Slip (Fig. 11.9) is characterized by a number of shallow rotational slides on slopes of overconsolidated fissured clays. For London clay, successive slips generally occur where the inclination of the slope varies between 9.5° and 12°. The angle of ultimate stability is about 9°, and this angle seems to be related to the residual strength of the clay. Observations indicate that this failure type initiates at the toe of the slope and spreads upward. Examples of successive slips in Japan have been described by Fukuoka (1953).

11.5.4 Translational Slides

Translational slides can be divided into block slides, slab slides, and multiple translational slides, as illustrated in Figure 11.11.

Translational slides take place along bedding planes, faults, cracks, or fissures that are approximately parallel to the ground surface. The failure surface is essentially planar. There is a fundamental difference between rotational and translational failures. The force system that initiates a rotational slide or slump decreases with increasing deformation owing to the backward tilting of the moving soil or rock mass. For a translational slide, on the other hand, the force system that causes the failure remains constant.

Translational failures in clays containing sand or silt seams can be caused by high excess pore water pressures in the seams, especially if the seams dip toward an excavation. Sloughing may occur where the seams are exposed at the toe of a slope. Also, frost action can cause some local sloughing. In many translational slides, the slide mass is greatly deformed or breaks

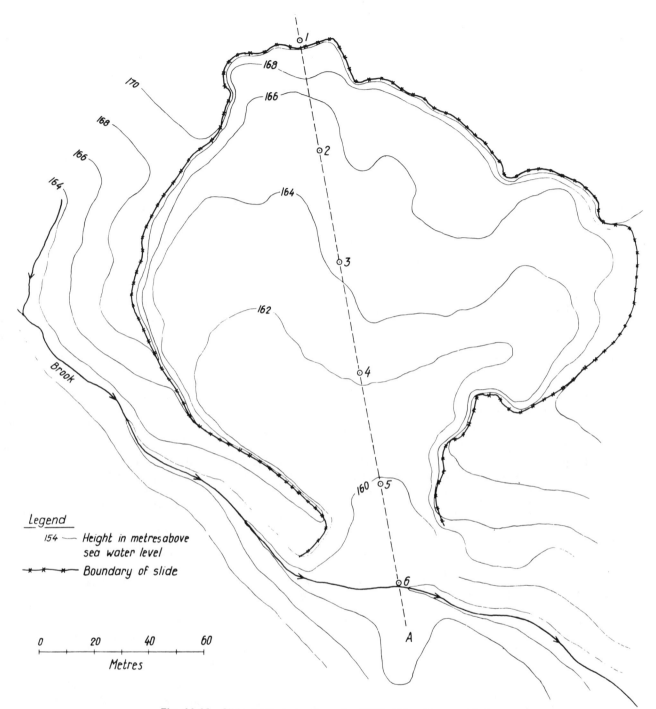

Fig. 11.10 Slide at Ullensakr, December 1953. (*After Bjerrum, 1954.*)

up into many more or less independent units. As deformation and disintegration continues, and especially as the water content is increased, the slide mass can turn into a flow.

Translational Block Slides take place in hard jointed materials along steeply sloping joints, bedding plants, or fault zones. Translational slides are relatively common in rock and occur along predetermined failure surfaces. Such slides are often triggered by the undercutting of a slope, by erosion, or by excavation and occur when the inclination of the slope exceeds the angle of internal friction of the rock mass along the bedding plane. The friction angle, which increases with increasing roughness of the bedding planes, can be reduced by weathering

and by freezing and thawing. The stability of the slopes can also be reduced by excess cleft water pressures in joints or bedding planes if the free flow of water is prevented.

The maximum velocity of rockslides can also be very high. A review by Müller (1964) of published data indicated that the sliding velocity has varied between 0.8 and 10 to 20 m/sec. Analysis of the waves generated by the rockslide at Vaiont indicates that the average velocity of the rock mass must have been at least 25 m/sec (Müller, 1964; Jaeger, 1969). Habib (1967) has shown that during a rockslide sufficient energy can be generated in the form of heat to vaporize the pore water, and that the rock under certain conditions can be sliding on a cushion of vaporized water that carries the weight of the sliding

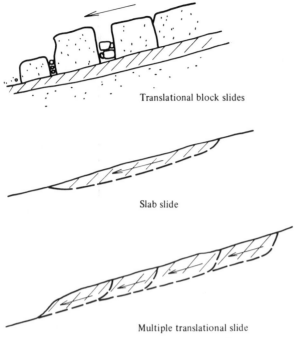

Translational block slides

Slab slide

Multiple translational slide

Fig. 11.11 Translational slides.

rock mass. It is possible that vaporization of the pore water can explain the high velocity that was observed for the Vaiont slide.

Block slides in overconsolidated jointed clays in Italy have been described by Esu (1966). The dimensions of the blocks were primarily governed by the spacing of the cracks. Failure occurred mainly in zones where the clay was closely jointed and where the direction of the joints favored sliding along nearly plane surfaces, as illustrated in Figure 11.12.

A number of translational slides in the form of block slides along faults, bedding planes, or joints at the Mangla, Sukian, and Jari Dams in Pakistan have been described by Fookes and Wilson (1966). Several examples of block slides have been reported by Zaruba and Mencl (1969) where layers of clay, marl, and claystone underlay jointed rocks (Fig. 11.11). Translational block slides that have taken place in Pine

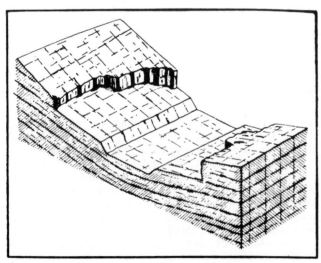

Fig. 11.12 Block diagram of a cut for a canal in the Valdarno clay. (*After Esu, 1966.*)

Mountain, Kentucky, have been described by Froelich (1970). The slides occurred in interbedded siltstone, sandstone, and clay shale.

Block slides along the coast of New Zealand have been described by Benson (1946) where blocks of basalt slid on an underlying sandy claystone. New slides developed as the broken material at the toe of the slope was washed away by waves and coastal currents.

Block slides in jointed limestone interbedded within thin layers of hard clay have been discussed by Henkel (1961). Henkel (1967) also investigated a number of block slides that have taken place in weak sandstone and clay shales. The strata dipped 5° to 11° within the slide area. Water that flowed through the sandstone layers filled the tension joints or cracks along the strike line of the strata. The resulting pore water pressures caused high shear stresses in the underlying clay shale along the base of the sandstone layer.

Slab Slides occur primarily in weathered clay or in shallow-slope debris on bedrock. The failure surface is approximately parallel to the ground surface, and the failing soil mass moves as a unit without distortion. The D/L ratio is rarely larger than 0.1 (Skempton, 1953). Slab slides have been described by Skempton and Delory (1957), Hutchinson (1967) and Vargas and Pichler (1957). The slab slide described by Vargas and Pichler occurred in a steep slope of residual soil after heavy rainfall. The slab slide at Furre, Norway, described by Hutchinson (1961) took place along a thin layer of quick clay.

One of the largest known translational rock slides occurred near Flims in Switzerland during the Pleistocene period. Approximately 12 million m³ of marly limestone slipped along a bed of marly limestone that dipped 7° to 12° toward the Rhine River. The velocity of the slide was so high that the rock mass climbed to a height of 150 m above the valley floor on the other side of the valley (Heim, 1932).

During the 1806 landslide at Mount Rossberg in Switzerland, about 40 million m³ of conglomerates slid along a bed of marl that dipped 19° to 21°. The slide buried the town of Geldau and was attributed to an increase of the average inclination of the slope by tectonic movements and to weathering (Terzaghi, 1950). The landslide at Gros Ventre, Wyoming, in 1925 involved more than 20 million m³ of shale, sandstone and limestone, which slid on a bedding plane with a dip of 18° to 21° (Legget, 1962).

Multiple Translational Slides are generally triggered by an initial slab slide. Skempton and Hutchinson (1969) have pointed out that the size of the individual slides probably decreases and the number of slides increases with increasing cohesion of the soil. Such slides spread gradually upslope as the soil at the rear scarp is gradually weakened by rainwater that fills the cracks above the scarp. Additional slides generally occur after heavy rainfalls or during the spring when the soil thaws. Multiple translational slides have occurred at Jackfield in England (Henkel and Skempton, 1954) and at Portuguese Bend, California (Merriam, 1960).

11.5.5 Lateral Spreads

Lateral spreading is basically a type of a retrogressive translational slide. It involves the fracturing and extension of the underlying weak material, whether soil or bedrock. It can also be caused by liquefaction or plastic flow. The resulting lateral movement breaks up the overlying material into more or less independent units or lumps that may subside, translate, rotate, or disintegrate. This type of slide, which develops rapidly,

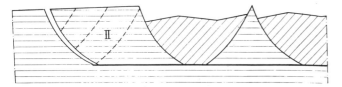

Fig. 11.13 Mechanism of failure by spreading. (*After Odenstad, 1951.*)

frequently occurs where the slope of the ground surface is relatively flat. Failure by spreading is common in varved clay where high pore water pressures occur seasonally in sand or silt seams, as was the case at Sköttorp, Sweden (Odenstad, 1951). The slide mechanism is illustrated in Figure 11.13. The landslide at Turnagain Heights, Alaska, is also of this type. The high pore water pressure that developed during the 1964 earthquake in interbedded layers of loose sand probably caused the failure. The slide apparently initiated behind the crest of the slope, and numerous sharp clay ridges formed within the slide area perpendicular to the direction of the movement. The height of these ridges was 3 to 5 m. The slide spread backward along a common basal failure surface, which was weakened by liquefaction of sand lenses in the clay. Several 30-m long sand ridges with a height of 0.50 to 1 m, which were caused by sand boils, were found within the slide area (Seed and Wilson, 1967). A failure reported by Ward (1948) appears also to be of this type. This failure was probably caused by liquefaction and internal erosion of a layer of fine sand located between two clay strata.

Terzaghi (1950) has described a failure by spreading that occurred in 1930 during a heavy rainstorm when the Swir III hydroelectric plant in the U.S.S.R. was under construction. The failure was attributed to lateral expansion of a stiff greenish clay that underlay a well-compacted and slightly cohesive till. The horizontal expansion of the clay layer caused the overlying till to crack, and the failure occurred when rainwater filled the tension cracks in the till.

Failure by lateral spreading has also taken place at Northampton in England during excavation through layers of limestone, sandstone, and shale down to the underlying Lias clay. The clay at the bottom of the excavation was squeezed out under the weight of the overlying rocks, and cracks developed in the limestone and sandstone beds parallel to the excavation (Hollingworth and Taylor, 1944).

At the slides that were caused by lateral spreading at Algiers in Algeria during heavy rainfalls, blocks of jointed limestone sank into the underlying marl. The marl had been weakened by weathering and leaching (Drouhin et al., 1948). Similar slides (Fig. 11.14) that have taken place at Handlova, Czechoslovakia, have been described by Pǎsek (1967).

11.5.6 Flows

A characteristic feature of a flow is that the material breaks up as it moves down a slope and flows as a viscous fluid. Several different types of flows can be recognized: rock flows, earthflows, debris flows, mudflows, spontaneous liquefaction, and solifluction, as illustrated in Figure 11.15. The rate of movement can range from very slow to extremely rapid.

Rock Flows in general, involve gradual deformation of the rock mass along joints and fractures. There is no distinct failure plane. The movements are generally extremely slow and fairly constant with time. Flow movements may result in folding, bending, bulging, or other manifestations of plastic behavior (Varnes, 1978) as shown in Figure 11.16. This type of movement is often referred to as creep. Radbruch-Hall (1978) has described rock flows in the United States and in other countries. On the other extreme, rock flow can take the form of an avalanche where the rate of movement is extremely high. The rock fall avalanche of Elm, Switzerland, appears to have been of this type. The broken rock mass flowed about 1.5 km down the main valley before it stopped. The high velocity of the slide may be explained by high air pressures that were generated in the broken rock mass.

Earthflows resemble viscous fluid flow and the flow mass can be wet or dry. Slip surfaces within the moving mass are usually not visible. The boundary between the moving mass and the material in place may be a distinct surface or it may be a relatively thick zone. Many earthflows continue to move for many years until the inclination becomes small enough to stop the movement or the increase of the shear strength by, for example, drainage or drying is sufficient.

Skempton and Hutchinson (1969) distinguish between earthflows and mudflows. An earthflow represents a transition between a slide and a mudflow. The breakdown of the moving soil mass is less for an earthflow than for a mudflow and a considerable part of the original vegetation cover is retained. Varnes (1978) uses the term earthflow to describe both mudflow and debris flow. Mudflow applies to slides involving predominantly fine-grained soils with more than 50 percent silt- and clay-size particles, whereas debris flow involves coarse-grained soils.

Debris Flows often develop in arid and semiarid regions where the ground is not covered by vegetation or in talus slopes or gorges that are filled with debris and where a vegetation cover

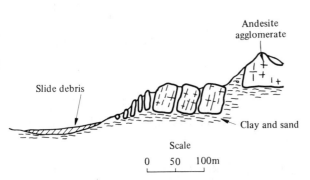

Fig. 11.14 Failure of lateral spreading at Handlova, Czechoslovakia. (*After Pašek, 1967.*)

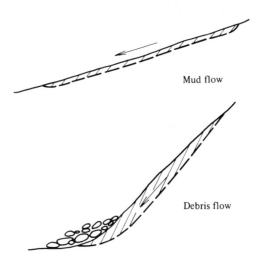

Fig. 11.15 Mud flow and debris flow.

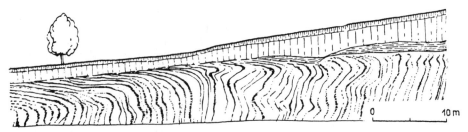

Fig. 11.16 Surface bending of beds in a loam-pit in Prague. (*After Zaruba and Mencl, 1982.*)

is lacking or has been removed. In the Alps, debris flows (called "muren") generally begin above the timber line. Debris flows often start during a heavy rainstorm or sudden floods. Because they often follow old stream beds, debris flows are generally long and narrow. The canyon can be deeply eroded by the flowing material because of its very high density, and the flow may continue for several kilometers. The flow rate ranges from very slow to very rapid, and the mass usually breaks up as it moves down the slope. The water content of the moving mass is usually about 100 percent. Zaruba and Mencl (1982) have described numerous debris flows that have occurred in the High Tetra Mountain in Czechoslovakia during an unusually heavy rainstorm (26 mm/hr) in 1933. A detailed description of debris flows has also been given by Johnson (1970).

Mudflows often take place on mountain slopes where slope debris and weathered material have accumulated. Heavy rainfall increases the weight of the detritus and reduces its shear strength, causing the saturated soil mass to flow down the slope. A number of earthflows in loose slope debris have been described by Zaruba and Mencl (1982). In 1960, an earthflow of this kind occurred near Handlova in Czechoslovakia, which involved 14.5 million m³ of material. The thickness of the moving mass was 10 to 18 m and reached 20 to 25 m in the tongue area. During the first few days of the slide, the flow rate was as high as 6.3 m/day, but the rate decreased gradually. The slope in the root area was 7.3°. The movement involved debris of volcanic rocks and clayey silty sediments accumulated from previous slides.

Mudflows occur frequently in both fissured and intact clay that is interbedded with layers of fine sand. The flows in these soils are caused by erosion due to high pore water pressures in the sand layers or sand seams. Mudflows that have occurred along the Norfolk coast in England have been described by Henkel (1967). Legget and Bartley (1953) have described similar flows that occurred during the draining of the Steep Rock Lake in Canada. The flows took place in varved clays owing to the loosening of the underlaying varves by the percolating groundwater.

Mudflows are also common in regions of former glaciation, where the clays often contain sand or silt pockets or lenses. High pore water pressures can develop in these pockets during heavy rainfalls, which may disintegrate the soil mass into a mixture of sand, silt, and chunks of clay and flow down the slope. Broscoe and Thomson (1969) have described a mudflow at Steel Creek, Yukon, Canada, that occurred in a moraine after a heavy rainfall. The mudflow carried boulders of up to 4 m in diameter, and the material in the mudflow remained fluid for some time after it came to rest. Two weeks later, a 0.15-m thick crust had developed on the 0.6-m thick deposit.

Hutchinson (1970) has described a mudflow in the stiff fissured London clay along the 30-m high coastal cliff at Beltinge in North Kent, England. The material consisted of hard clay fragments in a soft matrix with an undrained shear strength of 5 to 15 kPa as measured by vane tests. The mudflow consisted

of an upper and a lower part called feeder flow and accumulation flow, respectively. The inclination of the feeder flow was about 13° to 18° and its thickness was about 1.0 to 1.5 m. The average inclination of the accumulation flow with a thickness of 5.0 to 5.5 m was 7°. The water content of the matrix was 1 to 9 percent higher than that of the hard clay fragments. The velocity of the mudflow was up to 0.26 m per day during the spring. The movement stopped during the fall. The variation of the flow rate corresponded to the variations of the pore water pressure in the moving material.

The rate of mudflow varies from extremely slow (soil creep) to very rapid (flow avalanche) depending on the inclination of the slope and on the water content of the soil. For clay slopes the creep rate can be less than 1 mm/year. The rate of mudflow is commonly between 5 and 25 m/year. The average rate of the mudflow at, for example, Mount Chasu, Japan, was 25 m/year (Fukuoka, 1953). Moos (1953) reports that the rate of the mudflow at Stoss varied between 1 and 15 m/year. The mudflow rate at Slumgullion, Colorado, has been measured by Crandell and Varnes (1961). The total movement was 58 m between 1939 and 1952, which corresponds to an average rate of 4.5 m/year. The maximum flow rate in 1959 was about 6 m/year in the middle of the mudflow and about 0.75 m/year at the toe.

A special type of earthflow takes place in quick clays, where the flow rate can be very high. For quick clays the residual strength is much less than the peak strength. The shear strength of this material can thus be reduced to a fraction of its initial value because of the remolding. Earthflows in quick clays are common in Scandinavia and in Canada. Failures that have taken place in the sensitive Leda clay in Canada have been described by Crawford and Eden (1967). The high sensitivity of these clays is partly caused by the gradual decrease of the salt content by groundwater flow and by infiltration of rainwater and partly by infiltration of humus or other organic substances carried by the groundwater into the clay. Slides in quick clays can occur even on slopes as small as 3° to 4°. The slides often develop rapidly and without warning.

The slide velocity in quick clay can be as high as 0.5 to 1.5 m/sec, as pointed out by Skempton and Hutchinson (1969). The landslide that took place in quick clay at Furre, Norway, occurred within a few seconds. The central part of the slide came to rest after about 1.0 to 1.5 minutes (Hutchinson, 1961). The velocity of the slide in quick clay at Vaerdalen in Norway in 1893 has been estimated to be 17 m/sec. The remolded material flowed a distance of about 11 km. In Kenogami, Canada, a maximum velocity of 3.5 to 4.5 m/sec was observed. The slope of the valley was about 9° (Terzaghi, 1950; Brzezinski, 1971). The maximum velocity of the quick clay slide at Surte, Sweden, was less than 0.9 m/sec. The total duration of slide was about 3 minutes (Jakobson, 1952).

Spontaneous Liquefaction takes place in loose saturated sand and silt and is initiated by vibrations caused by earthquakes, blast loading, or pile driving. Strong vibrations cause the soil structure to collapse and decrease in volume. This reduction of the volume increases the pore water pressure, and when the pore pressure is equal to the overburden pressure the effective stress between the soil particles is eliminated. Then the soil loses its strength completely and is liquefied. Skaven-Haug (1955) and Andresen and Bjerrum (1968) have described several flow slides that have taken place in a silty sand in Trondheim Harbour, Norway. The inclination of the slope varied between 8° to 11°. Peck and Kaun (1948) have described a flow slide caused by liquefaction in a very loose, fine sand that had been placed by hydraulic dredging. The landslide in 1964 at Valdez, Alaska, was caused by liquefaction of silt, fine sand, and gravel. The standard penetration resistance of the materials varied between 7 and 25 blows/0.3 m. About 90 million m³ was involved (Seed, 1968).

Liquefaction under water can spread rapidly over large areas at a rate of 10 to 100 km/hr. The failure that occurred at Grand Banks outside of Newfoundland in 1929 progressed at a rate of 20 to 110 km/hr for a distance of 550 km (Terzaghi, 1956). The slope gradient was less than 1 percent. The slide in Orkdalsfjord in Norway in 1930, which occurred in a very loose silt, propagated at a rate of 10 to 26 km/hr (Andresen and Bjerrum, 1968).

Solifluction refers to the slow surface movements that take place in primarily silty or clayey soils. When the soil thaws in the spring, the surface material becomes waterlogged and moves slowly down the slope. The rate has been measured by Rapp (1960) in Northern Norway. At slopes inclined at 15° to 25°, the rate was 80 to 300 mm/year. Solifluction is common in mountainous areas in Europe, Canada, and the United States.

11.5.7 Complex Slope Movements

It is quite common for a landslide to involve a combination of one or more of the five principal types of slope movement, either within various parts of the moving mass or during the different stages in the development of a slide. The term *complex slope movement* is used to describe this type of slide.

For example, the lower part of a rotational slide often develops into a flow because of the accumulation of water at the toe. The soil mass has usually been broken up in that region during the slide. When the displacement of a soil or rock mass is controlled by a plane of weakness such as bedding planes, joints, faults, and other discontinuites, the movement can be a combination of rotational and translational slides.

The disaster at Elm, Switzerland, is an example. This landslide started as small rock slides and falls and eventually ended as a flow avalanche. The mass rushed up the opposite side of a small valley, turned, and streamed into the main valley. It flowed for nearly 1.5 km at a high velocity before stopping (Heim, 1932; Varnes, 1978).

11.6 INVESTIGATIONS FOR LANDSLIDES

A landslide investigation may involve:

- Assessment of the risk of landslides in areas where future construction is being planned
- Identification of old landslides and an analysis of the factors responsible for the failure.

Unstable areas usually shows symptoms of past movements and can usually be detected in a detailed field investigation. Proper preventive measures can then be taken to reduce or to eliminate future movements, or an alternate location can be chosen. A field investigation can help to determine the size of old landslides to identify the mode of the movement and to determine the factors responsible for the failure. The investigation should be comprehensive enough that appropriate corrective measures can be taken.

The extent of an investigation is governed by how likely landslides are within an area and by the consequences of a failure. The stability of natural or man-made slopes is generally evaluated from field and laboratory studies and from stability calculations. These studies are as a rule carried out in stages.

The first stage is to identify the factors that may affect the stability. Data have to be collected that are required for the detailed investigation of critical areas (Banks and Meade, 1982) from:

a. Topographic maps that show relief, drainage, vegetation, and cultural use of the area
b. Geologic maps that indicate the surface and the bedrock materials; geologic structures such as faults, joints, and intrusions; water resources and groundwater conditions
c. Hazardous zoning maps showing the location of landslide-prone areas
d. Geologic reports and special reports describing features and problems related to the area
e. Agricultural maps and reports that illustrate soil distribution and origin as well as soil profiles

Some information on the engineering properties of the soil and the rock can also be obtained from aerial photographs and other remote sensing methods such as infrared microwave radiometry and thermal sensing (Sowers and Royster, 1978).

The second stage involves site reconnaissance, where the area is mapped in detail, the extent of the slide is measured, and features such as scarps, cracks, bulges, springs, leaning trees, and displaced fences are identified. In an old landslide area the vegetation cover often differs from that of the surrounding stable slopes. Zaruba and Mencl (1982) have pointed out that certain plants are frequently found in unstable areas. A hydrologic investigation should be carried out in the field to identify the location of surface streams, water-filled depressions, springs, and other related features so that a detailed surveying and subsurface exploration program can be prepared.

The third stage usually involves geodetic survey and subsurface exploration at the site. At the geodetic survey the location and the dimensions of the landslides are determined so that the volume of the moving mass can be estimated and possibly the mode of the movement. Ground controls should be established for photogrammetric mapping and for monitoring of future movements.

Subsurface exploration involves boring, sampling, in-situ testing, test trenches and test pits, as well as geophysical measurements. The objectives are to identify the local geologic conditions, to determine the depth and thickness of the different strata, to obtain quantative data on their physical properties for stability analysis, and to locate the sliding surface.

The location and the variation of the groundwater table and of the pore water pressures in permeable strata are generally determined with open standpipes. In clay and silt, pore pressure gauges with closed systems and short time lags are required. Pore pressures usually vary appreciably during the year and it is often difficult to estimate the maximum pore pressure that can occur, for example, in a thin sand or silt seam.

Geophysical methods such as resistivity and seismic refraction methods as well as sonic logging are often used to determine the depth of the overburden and the quality of the underlying rock.

The fourth stage involves evaluation of the collected data, stability analysis of the affected slope, and design of corrective measures. Often, additional investigations (for example, borings) may be needed in critical areas to supplement the data collected from the initial investigation. Sowers and Royster (1978) point out that this additional work can frequently exceed that originally considered to be adequate by 30 to 50 percent of more.

During the fifth stage the movements of the slopes are monitored in areas where further activities are planned and where corrective measures have been taken. The surveillance should continue long enough for the full range of environment conditions that can reasonably be expected to occur.

11.6.1 Aerial Surveys

From aerial surveys or photogrammetry, an overview of the site conditions can be obtained. Until recently, black-and-white photographs were common. Their accuracy is high and the cost is low. When observed through a stereoscope, the aerial photographs provide an exaggerated three-dimensional model of the area, from which the geomorphology (landforms) can be identified. Colour photographs are particularly valuable for assessing differences in vegetation cover, groundwater conditions, and the stratification of the exposed soil and rock. More sophisticated remote sensing methods such as satellite and infrared photographs can also be helpful for identification of landslides (Rib and Liang, 1978).

11.6.2 Boring Layout

In apparently stable slopes, the borings can be uniformly distributed. In areas where slides have occurred, Philbrick and Cleaves (1958) suggest that at least three borings should be located, along the centerline of the slide just below the main scarp, near the center of slide, and in the bulge area. Four additional boreholes should be drilled next to the slide in areas that have not experienced any significant movement. The foot of the bulge zone is especially important for the design of retaining structures to reduce or eliminate future movements.

11.6.3 Detection of Slide Surface

The failure surface of a moving soil or rock mass can often be located by borings, soundings, test pits, trenches, or tunnels. The use of test pits and tunnels is generally expensive and dangerous because the strength of the disturbed soil or rock within the slide area is often low. The failure surface can in some cases be determined from borings if they are carefully done, since the water content of the soil or of the rock often changes suddenly at the slide surface.

Inclinometers are frequently used to locate the slide surfaces. A less expensive alternative is to use a steel guide rod, which is lowered into a tight-fitting steel or plastic tube (15 to 40 mm in diameter) placed in a vertical borehole. The borehole should be deep enough that it extends through the sliding zone. Usually the guide rod cannot pass through the tube at the level where the slip has occurred owing to creep along this zone.

The sliding surface can often be located by placing in a borehole a plastic pipe where strain gauges have been attached to the inside wall of the pipe. Any additional movement in the soil or in the rock can be determined from the strain guage readings.

In Sweden the slip surface in soft clay is often located by in-situ resistivity measurements, since the salt content and thus the resistivity of the soil often changes suddenly at the slip surface, as shown in Figure 11.17 (Söderblom, 1958, 1969; Hutchinson, 1961).

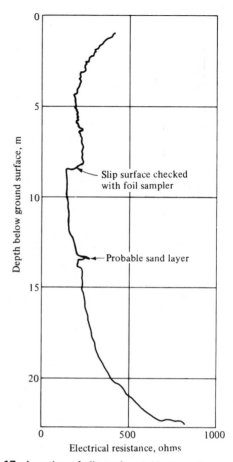

Fig. 11.17 Location of slip surface with the salt sounding tool. (*After Söderblom, 1969.*)

Hutchinson and Highes (1968) were able to locate failure surfaces in the stiff fissured London clay from a micro-paleontological investigation by studying the changes of the fossils and the ancient life forms. Cotecchia and Federico (1984) used a nuclear moisture-density probe to determine the location of the slip surface in a marly gray blue clay in Southern Italy.

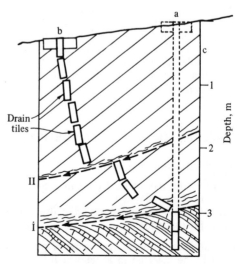

Fig. 11.18 Location of failure surface with drain tiles. (*After Zaruba and Mencl, 1969.*)

With this probe any changes of the water content and of the density in a borehole can be measured.

A shear-strip indicator can also be used to locate slip surfaces. This device consists of a row of electrical resistors that are wired in parallel. The resistors, which are spaced 150 mm apart, are mounted on a Bakelite strip. The strip is placed in a borehole and grouted. An electrical circuit is broken when the shear displacement at any point exceeds 2 to 3 mm. The location of the slip can be determined from the resistance of the intact resistors in the circuit.

Surface movements or the displacements of shallow slides can, for example, be measured by placing concrete blocks or drain tiles at different depths in boreholes or in test pits. The lateral displacement can then be determined by excavating the blocks or tiles after a certain time period, as illustrated in Figure 11.18. Ter-Stepanian (1974) used a stack of concrete rings placed inside a large-diameter borehole to measure slow creep movements. The center of each ring was marked by two crossed wires. The displacement of the upper ring relative to the lower ring was determined with a pendulum.

11.7 ANALYSIS OF LANDSLIDES

Even today it is impossible to make a reliable assessment of the stability of many, if not most, natural slopes. There are few slopes in homogeneous or in simple materials where the shear strength can be expressed in terms of one or two parameters. Calculations are therefore often based on a very rough estimate of the shear strength of the soil or the rock mass, since most soils and rocks are heterogeneous and anisotropic. An assessment can be broadly divided into deformation analysis, liquefaction analysis, and limit equilibrium analysis.

The *deformation analysis*, either static or dynamic, is usually done with the finite-element method. The obvious application of this method is engineered embankments and fill slopes, where the soil mass is relatively uniform. In most natural slopes, the soil properties and the groundwater conditions are complex and uncertain; the relevance of a deformation analysis of these slopes can be questioned.

A *liquefaction analysis* is mainly limited to saturated loose sands, silts, and sensitive clays subjected to strong shaking from earthquake, blasting, or pile driving. However, the dynamic behavior of the soil as well as the characteristics of the vibrations have to be known or estimated. Further details of this method can be found in Chapter 25.

The *limit equilibrium method* is perhaps the most widely used method in practice, where the factor of safety against total failure is determined. Various limit equilibrium methods are described in Chapter 10. The prerequisites of a successful limit equilibrium analysis are a clear understanding of:

1. The mechanism and the mode of failure, whether it is observed or predicted
2. The geometry of the failure plane
3. The hydrologic conditions
4. The shear strength of the soil or rock along the failure plane

The geometry of the slip surface is generally governed by the geologic conditions. In a uniform deposit of clay and shale, the slip surface is usually circular. When there are bedding planes, weak seams, or old sliding surfaces, the slip plane may be circular, spiral-shaped, straight, or a combination of various shapes. The failure surface for an existing landslide can be located in the field by borings, soundings, test pits, trenches, or tunnels. In a stability analysis, the failure surface must be assumed.

The hydrologic condition at the time of failure is often unknown. A realistic assessment of the groundwater level and seepage condition must be made in a back-analysis so that the shear strengths of the soil can be evaluated. If the purpose of the analysis is to obtain design parameters, a conservative assumption on the hydrologic condition is frequently made.

11.7.1 Falls

Falls generally occur in rock and soil along a near-vertical plane that is often controlled by joints, cracks, and fissures, which is rather difficult to quantify in terms of a factor of safety. It is common to assume that falls will occur and to take appropriate measures to control the falls. Piteau and Clayton (1977) suggest that a rock-fall model can be used to determine the final resting position of the debris.

11.7.2 Topples

Topples have been investigated in several recent publications. Methods of analysis have been discussed by Goodman and Bray (1976) and by Evans (1981).

11.7.3 Slides

Slides can be rotational or translational. For rotational earth slides, the simplified Bishop's method of analysis is widely used. If the slip surface is noncircular, other methods such as those proposed by Morgenstern and Price (1965) and by Janbu (1973) are commonly applied. The stability of translational slides is frequently analyzed with the wedge or the infinite-slope method.

Rotational slides in rock are rarely encountered because of the strong influence of structural discontinuities within the rock mass. Vector analysis or a graphical method by means of stereonets (Hendron et al., 1971; Hoek and Bray, 1974) are used to analyze translational slides.

11.7.4 Lateral Spreads

Lateral spreading is caused by the fracturing and extension of a cohesive soil or of a rock mass due to liquefaction or plastic flow of the underlying weak soils or the development of high pore pressure in thin silt or sand seams. This type of failure can be analyzed, as it is basically a type of retrogressive translational slide. The mechanism of the failure has to be understood and the shear strength of the underlying weak soils must be determined.

11.7.5 Flows

A flow is, in general, the postfailure movement of the broken and softened material after a translational slide. A stability analysis should involve the initial failure stage. The analysis of the streaming stage has been discussed by Vallejo (1981). The same methods as for translational slides can be used (Karlsrud et al., 1984).

Flows can also be initiated by spontaneous liquefaction, which is difficult to quantify, as well as flows initiated by solifluction and those caused by the sudden access of water to a debris-mantled slope.

11.7.6 Soft Intact Clay

Normally consolidated clays contain generally few discontinuities. The failure surface is usually circular. The short- and long-term stability is generally analyzed using the total and the effective stress methods, respectively. In the total stress method, the greatest uncertainty is the evaluation of the undrained shear strength. This strength is normally determined in the field by vane tests or in the laboratory by unconsolidated undrained (UU) triaxial or direct shear tests or by fall cone tests. In order to obtain satisfactory results, the measured values have to be adjusted to take into account anisotropy and loading rate, for example. The correction factor proposed by Bjerrum (1972) for field vane tests is widely used. There is no published correction for unconfined compression or for undrained triaxial (UU) tests.

LaRochelle et al. (1974) suggest that the undrained strength at large strains (USALS) should be used instead of the peak strength from UU tests. Trak et al. (1980) found that for Canadian sensitive clays, this strength is approximately equal to $0.22\sigma_c'$, as suggested by Mesri (1975), where σ_c' is the preconsolidation pressure. Similar observations have been made by Tavenas et al. (1980) and by Aas (1981). Larsson (1980) has shown that a normalized undrained strength of about $0.22\sigma_c'$ is also applicable to soft inorganic Scandinavian clays. Tavenas and Leroueil (1981) have pointed out that this relationship may be too low for very young and nearly normally consolidated organic clays.

Even with various correction factors, the results of the analyses are sometimes erratic, since the undrained shear strength as determined from field vane, triaxial, or fall cone tests is no more than an index property. The reliability can be improved if local experience is considered.

Considerable progress has been made during the last few years to obtain more realistic values of the undrained strength. In his extensive state-of-the-art paper, Bjerrum (1973) concluded that "the most realistic value of the undrained shear strength can be obtained by triaxial compression test, direct shear test, and triaxial extension test." It is known as the ADP method (for Active, Direct, and Passive, respectively) or the advanced undrained method (Karlsrud et al., 1984). A similar approach was developed independently at MIT by Ladd and Foott (1974) at about the same time. The method is called SHANSEP, which stands for Stress History and Normalized Soil Engineering Properties. Both methods have been verified by case studies of embankment constructed on soft clay. However, experience with the method for natural slopes in soft clay is still lacking.

Alternatively, the effective stress method can be used where the shear strength parameters c' and ϕ', as well as the pore pressure along the failure plane, have to be determined from, for example, consolidated undrained triaxial compression tests. The pore pressure can be estimated or measured in the field. Janbu (1977) points out that the effective stress method is appropriate for excavated slopes because the change of the pore water pressures to that which corresponds to the long-term steady-state seepage often takes place in a relatively short period of time, as indicated by the failure at, for example, the Kimola (Kenney and Uddin, 1974). Janbu (1977) points out that the effective stress method is in principle more correct and should be used for all natural slopes, including quick clay slopes, as shown by a number of case studies.

The stability of slopes in soft intact clay can be analyzed by either the total or the effective stress methods. To provide a safe slope in soft clays, the safety factor, as calculated by either method, should be sufficient. The total stress method indicates the failure risk with respect to sudden changes of the loading conditions, while the effective stress method using an estimated pore pressure from, for example, a flow net indicates the risk of a failure with respect to gradual changes.

11.7.7 Stiff Intact Clay

Stiff intact clays such as boulder clays or clay tills have generally been heavily overconsolidated by the weight of the overlying thick ice sheet during the last glaciation. These soils, which are relatively rare, contain few or no structural discontinuities because of the remolding when these soils were formed. Boulder clays or clay tills often have a low clay content.

The effective stress method is often used to analyze the stability of natural slopes. Skempton (1977) recommended that for first-time slide in stiff intact clay the peak strength parameters c' and ϕ' should be used. Skempton and Brown (1961) have found for a natural slope of a boulder clay at Selset, Yorkshire, England, that a close correlation exists between the shear strength parameters c' and ϕ' as determined from triaxial drained tests and those back-calculated from stability analyses. Similar results have been reported by Sevaldson (1956) for a slide in an intact overconsolidated clay.

DeBeer and Goelen (1977) report for cut slopes in stiff intact clay in Belgium that the peak shear strength parameters should be used. The cohesion is, however, reduced where appropriate. The worst assumptions are made with respect to the pore water pressure in the slope. The calculated factor of safety should be at least 1.3.

The peak strength parameters can be determined from drained triaxial or direct shear tests. The value of c' is often relatively high, generally larger than 10 kPa. The peak friction angle ϕ' is typically 20° to 30°.

For reactivated slides along old slip surfaces or shear planes, the residual strength parameters c_r' and ϕ_r' are recommended. Bjerrum (1967) has indicated that the stability of natural slopes in stiff intact clays and clay shales can also correspond to the residual shear strength if sufficient time is allowed for the development of progressive failure.

The residual strength parameters can be determined by ring shear tests on undisturbed samples or from reverse direct shear tests. Chandler (1984) reported that the residual strength in the field ϕ_r' generally exceeds that determined by ring-shear tests by 2° to 3° at $c_r' = 0$. A number of correlations of ϕ_r' with liquid limit have been published (Mitchell, 1976; Mesri and Capeda-Diaz, 1986). A comparison of the results is shown in Figure 11.19.

11.7.8 Stiff Fissured Clay

Stiff fissured clays, which are common in temperate climates, are often formed by weathering of argillaceous rock. These clays frequently contain slickensides or fissures that are caused by uneven volume changes during the weathering. Clays in the tropics that have been formed by weathering of rock in situ are

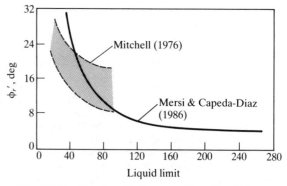

Fig. 11.19 Relationship between ϕ_r' and liquid limit.

generally stiff and fissured. These residual clays often contain the structure of the parent rock, such as joints and bedding planes, when the weathering has not been complete. The residual clays, which have a high plasticity, may contain slickensides and drying cracks close to the ground surface, where the soil is affected by seasonal variations.

The degree of fissuring depends on the properties of the clay and on its stress history. For example, fissuring often increases with increasing clay fraction, with increasing liquid limit, and with increasing degree of overconsolidation.

The clay in the stiff dry crust with a thickness of 1 to 5 m, which in Scandinavia and Canada overlies layers of soft clay, is also as a rule slickensided and fissured. The cracks are caused by the drying of the soil and by weathering. The shear strength of these stiff fissured clays can be appreciably reduced close to an open cut, partly because rainwater fills the cracks that open up during the excavation of the cut. Analysis of several slope failures in test trenches in the weathered fissured crust has indicated that failures can occur at a shear stress that is considerably less than the undrained shear strength of the soil. These failures occurred at an average shear stress that was about 75 percent of the shear strength as determined by vane tests (Barzett et al., 1961).

Slope failures in the stiff fissured London clay often occur in the form of successive slips or translational slab slides along old failure surfaces. Barton (1984) found that the most prevalent mode of failure in overconsolidated clays and soft rocks with flat-lying bedding is the compound slide in which there is some rotation as well as translation. Rotational or compound slides have been observed (Hutchinson, 1967) for deep to moderately deep sliding surfaces in the stiff fissured London clay where the foot of the slope has been eroded severely. Rotational failures in overconsolidated clays have been reported by Hutchinson (1969) and by Skempton and Brown (1961). Slips occur along shallow slip surfaces or in the form of mudflows when the erosion is moderate, while shallow rotational slides will occur if the erosion protection is not supplemented by drainage or by flattening of the slope.

The short-term stability of a cut slope in stiff fissured clay can be analyzed using the total stress method. However, the results are only reliable for as long as the undrained condition prevails. With time, the negative pore pressure generated during excavation will dissipate and the soil will swell. As a result, the shear strength will decrease and the factor of safety will be reduced. It is difficult to estimate the duration of this transient period because of the difficulty of assessing the permeability of the fissured clay and the drainage conditions. If the clay is highly fissured and there are frequent drainage layers, then it may take only a few days to reach equilibrium. In a uniformly fissured clay this transient period may last 50 years or more (Skempton, 1977). Local experience is essential when the time required for the equalization of the pore water pressures is estimated.

Another problem with the total stress method is the difficulty to determine the undrained shear strength in the laboratory because of the fissures in the specimen. The measured strength tends to decrease as the size increases (Ward et al., 1965; Bishop, 1966; Lo, 1970). The scatter of the test results is large for small specimens of fissured clay. The shear strength can vary from that of the unfissured material to a relatively low value for the samples that fail along unfavorably inclined fissures. Skempton and LaRochelle (1965) have reported that fissures can reduce the undrained strength by as much as 30 percent. The extrapolation of the results from the laboratory tests to the shear strength of the soil in situ is also uncertain.

Terzaghi (1936) points out that it is by no means uncommon that the shear strength of a stiff fissured clay may be reduced from 1.0 or 0.2 MPa to 0.03 MPa or less due to infiltration of water into the cracks. Chandler (1984) has documented six short-term failures in stiff fissured clay. The time to failure was in each case relatively short, up to approximately 50 days. Results of back-analyses indicated that the undrained shear strengths in the field were only about 50 to 60 percent of those measured in the laboratory. Analysis of slides in stiff fissured clays by Skempton (1977) indicated that for 6-m to 17-m high slopes in London clay with an inclination ranging from $1\frac{1}{2}$:1 to $2\frac{3}{4}$:1 the time to failure varied between 3 and 35 years. For inclinations ranging from 3:1 to $3\frac{2}{3}$:1 the time to failure ranged from 46 to 65 years when the height was between 6.0 and 12.2 m.

For a long-term stability analysis, it is necessary to estimate the steady-state pore pressures. It is also important to differentiate between first-time slides and reactivated slides. For a first-time slide, Skempton (1977) recommends that the "critical state" or the "fully-softened" strength parameters (c_s' and ϕ_s') should be used. These parameters correspond to the shear strength at about 5 to 15 percent axial strain past the peak value when the measured shear resistance appears to be stabilized with increasing deformation. These parameters can also be determined from normally consolidated remolded clay samples. There are a large number of case records worldwide that support Skempton's recommendation (Eigenbrod and Morgenstern, 1972; Lefebvre and LaRochelle, 1974; Marsal, 1979; Cancelli, 1981).

The "fully softened" strength parameters can be determined by triaxial compression tests. Again the size of specimen is important. Marsland and Butler (1967) report for Barton clay that c_s' was reduced from 22 kPa to 5 kPa when the diameter of the samples was increased from 38 mm to 130 mm, respectively, when ϕ_s' was assumed to be constant (20°) irrespective of sample size. Chandler (1984) reported similar results for the London and the Lias clays for sample diameters ranging from 38 mm to 260 mm.

A reasonable value of ϕ_s' can be determined from small specimens but the value of c' will be overestimated. Chandler (1984) has reported that the measured values from large vertical (260 mm) specimens were $c_s' = 10$ kPa and $\phi_s' = 20°$, while the back-calculated values from failed slopes (Chandler and Skempton, 1974) were $c_s' = 1$ kPa and $\phi_s' = 20°$. For Lias clay, the measured values were $c_s' = 9$ kPa and $\phi_s' = 23°$, while values of $c_s' = 1.5$ kPa and $\phi_s' = 23°$ were obtained from a back-analysis. Therefore, the measured value of c_s' should be reduced before it is used in slope stability analysis.

For reactivated slides, Skempton (1977) states that the residual shear strength ($c_r' = 0$ and ϕ_r') should be used. This strength appears to represent the lower limit with respect to the stability of natural slopes in stiff fissured clays as indicated by Skempton (1964). These results have later been confirmed for clay shales by Sinclair et al. (1966). Bjerrum (1967) draws the same conclusion from a review of published data.

11.7.9 Residual Soils

Residual soils are formed by the in-situ weathering of rocks. Because the weathering proceeds downward from the surface, the ground conditions range from the fully decomposed material at the ground surface (Grade VI) to the intact rock (Grade I). At shallow depth, the soil is relatively uniform without any relict structure. Further down, the weathered mass may have many of the physical features of the parent rock such as joints, faults, and folds. Relict joint surfaces are often stained red or black by a claylike material. They may be slickensided owing to swelling and shrinkage (St. John et al., 1969). The strength of the soil generally increases with depth. The permeability may either increase or decrease with depth depending on the minerals present in the profile and the degree of leaching. It is difficult to carry out a reliable stability analysis of natural or cut slopes in residual soils because of the structural discontinuities in the soil. Joint surfaces, slickensides, etc., are difficult to consider because these are generally not visible in borehole specimens. They are difficult to locate even in test pits. Even if these features are identified during the field exploration, it is practically impossible to assess their size, extent, orientation and frequency or other factors that are required in the analysis and design of cut slopes.

Translational slides are common but rotational slides also occur. The slides are generally relatively shallow. The maximum depth is usually a few meters.

Residual soils, which are relatively uniform and free of unfavorable structural discontinuities, are commonly analyzed using the effective strength parameters (c' and ϕ') as determined by direct shear or triaxial compression tests. The samples are usually saturated by increasing back-pressure. The measured effective cohesion c' is generally small, and a stability analysis is frequently carried out in terms of ϕ' only (Brand, 1984). However, the use of these parameters together with a conservative assumption on the pore pressure distribution is often over-conservative. A back-analysis of stable slopes using this approach often indicates a factor of safety that is less than unity (Ching et al., 1984). This discrepancy has been attributed by many investigators (Morgenstern and de Matos, 1975; Sweeney and Robertson, 1979) to soil suction.

Fredlund et al. (1978) have proposed the following equation for the shear strength of partially saturated soils, which includes the effect of soil suction,

$$s = c' + (\sigma - u_a) \tan \phi' + (u_a - u_w) \tan \phi^b$$

where

$$u_a = \text{pore air pressure in the soil}$$
$$(u_a - u_w) = \text{matrix suction}$$
$$\phi^b = \text{angle of internal friction with respect to a change of } (u_a - u_w)$$

In order to use this equation, ϕ^b and the matrix suction had to be known at various depths. The value of ϕ^b can be determined in the laboratory (Fredlund et al., 1978; Ho and Fredlund, 1982), whereas the determination of the soil suction requires in-situ measurements. The relationship indicated above has been used by Ching et al. (1984) to analyze the stability of slopes in Hong Kong.

Although the contribution of soil suction to the stability of residual soil slopes is uncertain, Brand (1984) has reported that the soil suction for relatively shallow potential slip surfaces in Hong Kong decreases to approximately zero during the wet season. Similar observations have also been made in Singapore. Hence, the contribution of the soil suction to the stability at least at shallow depth would be small. However, measurements made at some depth suggests that not all the suction is destroyed (Sweeney and Robertson, 1979; Sweeney, 1982). The contribution of suction depends largely on the pattern of rainfall and the infiltration of the rainwater.

Because of the difficulties in evaluating the shear strength parameters and the soil suction, attempts have been made to formulate guidelines based on local experience. Brand and Hudson (1982) have published a stability chart for cut slopes in granitic soil in Hong Kong based on a statistical study of 117 stable and failed slopes.

11.7.10 Loess

Loess consists of angular to subangular silt-sized particles that have been cemented together. The soil has a tendency to split more or less vertically along the root holes that are common in loess. Consequently, the permeability in the vertical direction is large in comparison with the horizontal direction. Loess can easily be eroded close to the groundwater table. To decrease the erosion, cuts are generally made vertically. Such cuts can be stable for many years and as a rule do not collapse even during heavy rainstorms. When the soil is flooded, large settlements can occur. It is therefore important to protect the foot of a cut against saturation during heavy rainstorms. Cuts in loess are often made much wider than the roadway to prevent damage or interruption of traffic due to local failure.

The mode of failure in loess can be by earth falls due to erosion at the base. An example is the falls of loess along the bluffs of the lower Mississippi River valley (Varnes, 1978). A number of slides in loess has taken place in Hungary along the Danube River (Andai, 1970). These landslides have generally been triggered by erosion at the toe of the slope.

Translational slides along weak planes have also been observed. Wang and Xu (1984) have reported a loess slide that occurred on March 7, 1983 in Gansu Province, China. A block of loess 200 m thick moved suddenly without previous warning. The landslide destroyed four villages and killed 220 people.

Another mode of failure is earthflow. One example is the loess flow caused by the 1920 earthquake in Kansu Province in China (Close and McCormick, 1922). Another loess flow occurred in Tadzhikistan, south-central Asia during the 1949 Chiat earthquake (Varnes, 1978). The flow destroyed 33 villages and covered the bottom of the valleys to depths of several tens of meters for many kilometers. Casagrande and Shannon (1947) suggest that liquefaction of loess is caused by an increase of the air pressure in the voids during the collapse of the soil structure caused by the strong shaking.

11.7.11 Sand and Gravel

In sand and gravel slopes the mode of failure is usually planar and translational. The stability of these materials can usually be analyzed with the infinite-slope method using the effective strength parameters c' and ϕ' and estimated pore water pressures.

The shear strength of sand and gravel is generally evaluated by triaxial or direct shear tests. The angle of internal friction as determined by drained triaxial tests is frequently larger than that from direct shear tests. Plain strain tests generally give an angle of internal friction that is 3° to 4° higher than that from triaxial tests when the relative density of the sand is high. When the relative density is low, the difference is small (Cornforth, 1961). In the analysis of the slope stability the results from triaxial tests are generally used.

Cuts in sand and gravel above the groundwater table can generally be made at a standard slope of 1.5 : 1. Sand and gravel with high or medium relative density are also stable at the standard slope below the groundwater table. However, flow slides may occur in loose, fine, uniform sand below the groundwater table owing to liquefaction. They can be triggered by pile driving, blasting, earthquakes, or a sudden change of the groundwater table. When a flowslide has been initiated in sand it may not stop until the slope is less than about 10°.

11.7.12 Rocks

The average strength and deformation properties of rock are generally difficult to evaluate from laboratory tests. It is often possible to get only a rough estimate of the shear strength and a reliable stability analysis can only be made when the geologic conditions are very simple.

The stability of a rock slope is rarely governed by the strength of the intact rock. The joint pattern generally has a decisive influence on the stability of unweathered rock. The angle of shearing resistance can often be estimated with sufficient accuracy from a joint survey while the effective cohesion cannot be determined by presently available methods (Terzaghi, 1962). Fortunately, the effective cohesion is typically much less important for the stability of slopes than the angle of shearing resistance. It is frequently difficult to estimate or to measure accurately the cleft water pressures that develop along potential sliding surfaces in rock. Terzaghi (1962) has pointed out that practically nothing is known about the mechanics of deep-seated large-scale rock slides. This statement is still true today.

Slides often occur in rocks that are composed of soft plate-like particles such as mica, schist, talc, or serpentine. Also, anhydrite can cause serious movements, since its volume increases by 33 percent when it is exposed to the atmosphere and is changed into gypsum.

Slides often take place in massive rocks along joints or fault planes. The joints in such massive rocks as granites, some limestones, and dolomites are generally almost randomly oriented. They may, however, contain several sets of continuous joints. The angle of internal friction of jointed rocks can be expected to be at least 65° (Terzaghi, 1962). The critical slope angle for a slope underlain by hard massive rocks with random joint patterns is about 70° provided there are no seepage pressures along the boundary between weathered and unweathered rock. The transition zone between completely weathered material and the underlying unweathered rock is generally relatively thick for granite, gneiss, and other massive rocks. Terzaghi and Peck (1967) have pointed out that in most instances an economical design is to cut the slopes as seems appropriate for the intact material and to provide adequate width at the base of the cut for material from occasional falls to accumulate and to be cleaned out from time to time.

The existing state of stress in the earth's crust will also affect the stability of steep slopes in massive rocks. The overburden pressure increases approximately linearly with depth but local variations may occur. The lateral pressure, which acts parallel with the ground surface, is frequently large, in some cases much larger than the overburden pressure. For example, Hast (1965) measured horizontal stresses in Scandinavian rocks that at a depth of 100 m were more than ten times larger than the total overburden pressure. These high horizontal stresses can contribute to progressive failure in rock and to the gradual development of slip surfaces. The stress relief due to excavation or erosion may cause joints in the rock parallel with the surface. These sheet joints frequently cause rock falls.

Slides are also common in stratified sedimentary rocks. The layer thickness of such rocks can vary from a few centimeters to several meters. The effective cohesion c' is generally low along the bedding planes and is dependent on the joint pattern and the orientation of the joints. The slope may vary between 30° and 90°. Deep-seated slides frequently take place in shales, which constitute about 50 percent of all exposed rocks in the world. Shales have been subjected to moderate to high temperatures and pressures and the particles are bound together. The strength and permanence of the bonds are dependent on the mineralogical composition of the shale, on the maximum temperature, and on the overburden pressure. The properties of a shale may vary from those of a soil to those of a rock. Most shales contain numerous cracks and joints that were caused by expansion when the overburden pressure was reduced, although they are generally closed when the depth exceeds 30 m. Poorly bonded clay shales usually contain thin layers of silt, sandstone, or siltstone. Sometimes they contain layers of bentonite, which considerably affect the slope stability. Well-bonded shales disintegrate upon unloading into angular, hard fragments with a relatively high permeability, while poorly-bonded shales turn into stiff fissured clays with an unconfined compressive strength that is normally larger than 0.5 MPa. The disintegration process is very slow and can only be predicted from field observations. When the shear strength of such shales has been decreased by remolding due to a slide, the shear strength along the failure surfaces will normally not increase with time above the residual value. The loading rate frequently has a large influence on the measured undrained shear strength of clay shales (Casagrande and Wilson, 1951). The undrained shear strength decreased to 35 percent of the initial strength when the duration of the test was increased to one month. It is not possible to predict the slope stability of shales from the results of laboratory tests alone.

The Culebra slides that occurred in 1912 during the excavation of the Panama Canal have been attributed to progressive failure in a montmorillonitic shale of the Cacaraucha formation. Such shales are generally much more troublesome than other types of shale. The slides, which raised the bottom of the canal between 10 and 15 m, were initiated by local failures in the unweathered shale below the canal. The local failures were followed by horizontal expansion of the clay and an approximately horizontal slip plane developed progressively (Bjerrum, 1967). Analysis of old slides indicated that the average shear strength of the clay shales in the east bank had been reduced to 78 percent of the initial value between 1912 and 1915. For the west bank, the shear strength decreased to 65 percent during the same period. Between 1915 and 1948 the average shear strength at the east bank decreased to 20 percent of its initial value and the corresponding decrease for the west bank was up to 24 percent (Binger, 1948). The effective angle of internal friction ϕ' required for stability of the banks agreed quite well with the residual angle of internal friction as determined from laboratory tests (Hirschfeld et al., 1965).

Numerous slides occurred in the Pierre shale during the construction of the Oahe Dam in South Dakota (Knight, 1963). The excavated shale deteriorated rapidly when it was exposed. According to Crandell (1952), slides are so common in the Pierre shale along the Missouri River in South Dakota that "at least 75 percent of the material exposed in the walls of the Missouri River trench and its tributaries has been moved to some extent by landsliding."

Slope failures in weathered shales are generally shallow and extend to the lower part of the weathered zone. The sliding surface is as a rule more or less parallel with the ground surface.

A number of shallow slope failures occurred in the Bearpaw shale during the construction of the South Saskatchewan Dam in Canada (Ringheim, 1964). The failures developed along an almost horizontal sliding surface within the weathered zone. The horizontal expansion of the weathered shale that took place during the excavation owing to high lateral pressures in the rock probably contributed to the progressive development of the slip surface.

The slide that took place during the construction of the Waco Dam in Texas has also been attributed to progressive failure in the underlying fissured clay shale (Beene, 1967). A stability analysis indicated an average drained angle of internal friction between 8° and 15° when the cohesion was neglected. The average angle of internal friction after the slide was 6° compared with a friction angle of 8° to 10° as determined from direct shear test on presplit samples.

Failures in poorly bonded shale during or immediately after the excavation of a cut take place under undrained conditions ($\phi = 0$ analysis) at an average shear stress considerably less than the peak strength as determined from undrained triaxial tests (Peterson et al., 1960).

Shales located close to river valleys are subjected to high shear stresses. These shales have as a consequence been much more weakened due to expansion and an increase of the water content than shales located at some distance from the valleys where the shear stresses are relatively low.

11.8 INSTRUMENTATION

The stability of a slope can be checked by measuring the downslope displacements within the potential landslide area. A potential landslide is characterized by relatively large displacements and a high displacement rate, which increases with time. Various methods have been developed to measure lateral displacements at the surface and at some depth.

Many types and models of instruments are available with different levels of sophistication. The selection of instruments requires a good understanding of failure mechanism, the local geology, and the subsurface conditions if meaningful data are to be obtained. It is preferable to use simple and rugged instruments that will function properly for a long time rather than sophisticated instruments that may give very accurate readings for only a limited period (Wilson, 1967).

11.8.1 Surveying

In conventional surveying, the slope is staked with survey markers such as wooden or metal pegs and mortar patches. Bench marks and transit stations should be located on stable ground. A theodolite with an accuracy of approximately ± 0.2 mm for a 100-m sight distance can be used. The accuracy is dependent on the triangulation layout, the stability of the fixed points, the accuracy of the mounting and of the readings. It is often difficult to establish fixed reference lines because of the settlements of the surrounding ground. The movement should be checked at several sections with measuring chains and sight lines.

It is also possible to measure surface movements by electrooptical distance-measuring instruments. With this method the phase change of a molulated light beam reflected from a reflector placed on the slope is measured. This type of instrument is particularly useful in rugged terrain. It is more accurate, faster, and requires less personnel than ordinary surveying techniques. A portable unit can cover distances from 20 m to 3000 m under ideal conditions with an accuracy of ± 3 mm. More powerful units using light or microwaves can be used for longer distances. However, the accuracy of these instruments is influenced by the weather and the atmospheric conditions. Wilson and Mikkelsen (1978) used this method to monitor the slope movement of an ancient landslide in Washington along the Columbia River, where the active slide was more than 600 m wide and 4800 m long. The yearly movements varied from 1 m up to 9 m. The error was about 100 mm.

11.8.2 Photogrammetry

Photogrammetry can also be used. By comparing stereoscopic pairs taken at different times, the direction and the magnitude of movements can be defined (Shields and Harrington, 1981; Pilot, 1984). The error of this method is about one order of magnitude greater than that of a precise ground survey.

For a slope of an open-pit coal mine in Indonesia, Shields and Harrington (1981) report that the error of the measurement ranged from 0.44 to 3.87 percent. The same technique was used in monitoring the Frasses landslide in Switzerland where Bonnard (1983a, b) reports an accuracy of ± 200 mm for photographs at 1 : 24 000 scale and ± 400 mm when the scale

is 1 : 40 000. The fundamental principle and practical application of this method has been discussed by Bernini et al. (1979).

11.8.3 Tape Extensometers

Surface extensometers can also be used. Pilot (1984) has described a number of different extensometers that cover a distance from less than a meter to several tens of meters. At this method an Invar wire that is subjected to a constant tension between two marks is generally used. For measured distance up to 50 m, the accuracy is typically 10^{-5} to 10^{-6} (Kovari, 1974; Rochet, 1983).

Suemine (1983) has described several landslides in Japan that have been monitored using an Invar-wire extensometer. Several extensometers can be connected in series in order to cover a longer distance. Sassa et al. (1980) used a 1200-m long line of extensometers at the centerline of the Zenktou landslide in Japan.

11.8.4 Tiltmeters

Tiltmeters can also be useful to monitor landslides that have a large rotational component. They are of little use for translational landslides. The older designs are based on the principle of a water level. Sherwood and Currey (1973) have described an electrolytic tiltmeter in which a spirit level containing an electrolyte and three electrodes is used. The change of the resistance between the electrodes when the instrument is tilted is measured by a Wheatstone bridge. A whole range of tiltmeters is available measuring from 0.5° to 5°. The accuracy is 1 to 10 seconds of arc, respectively. Servoaccelerometers have also been used (Wilson and Mikkelsen, 1978). The precision is high and the instrument is light in weight and easy to operate. A number of other high-precision tiltmeters have been described by Pilot (1984).

11.8.5 Crack Gauges

A warning of a forthcoming slope failure can be obtained by monitoring the width of the cracks in the slope. When the movements are large, a simple surveying method is generally sufficient. When movements are small, special electrical displacement gauges are required. Bhandari (1984) has described a simple device incorporating a shock absorber from a motorbike. The accuracy is 0.001 mm.

11.8.6 Inclinometers

Inclinometers are commonly used to measure horizontal movements in areas that are susceptible to landslides. An inclinometer consists in principle of a pendulum that is lowered into a cased borehole. The inclination of the borehole from the vertical is measured at successive levels by strain gauges or other transducers that are attached to the pendulum. Changes of the inclination of the casing and any horizontal movements can be determined by repeating the measurements at different times.

Several types of inclinometers are available as described by Wilson and Mikkelson (1978). Most inclinometers require a slotted casing so that the orientation can be determined. The inclinometer developed at the Swedish Geotechnical Institute (Kallstenius and Bergau, 1961) does not require a slotted casing. The inclination is measured with strain gauges that are attached

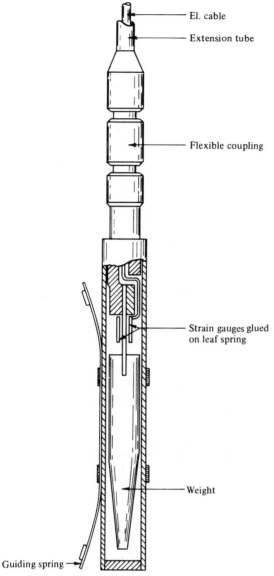

El. cable

Extension tube

Flexible coupling

Strain gauges glued
on leaf spring

Weight

Guiding spring

Fig. 11.20 Inclinometer type SGI. (*After Kallstenius and Bergau, 1961.*)

to a high-strength steel blade, as shown in Figure 11.20. The pendulum is rotated at each measuring level to determine the direction of maximum and minimum inclination of the borehole. A similar inclinometer has also been described by Hutchinson (1970).

Another technique is the use of a series of sensors that are permanently attached to the casing. The main advantage with an in-place inclinometer is that errors connected with the repeated positioning of the inclinometer in the casing are eliminated. The system can also be used for continuous monitoring. The main disadvantage is that only a limited number of sensors can be used in one borehole because of the large number of cables required. Only eight sensors plus cables can be accommodated in a 59-mm diameter casing (Haugen and Isaac, 1984).

11.8.7 Borehole Extensometers

The movements within a soil or rock mass can be determined by borehole extensometers. The movements are then measured by wires or rods that are inserted into a borehole. The lower end of the wires or rods are anchored to the sides or to the bottom of the borehole and the upper end to a dial indicator or an electrical transducer. With a one-wire extensometer the movements of up to four points at different depths can be measured. The maximum length of the wires is about 100 m. The movements can normally be measured with an accuracy of 0.5 mm for a gauge length of 50 m. The accuracy is affected by friction along adjacent wires and the sides of the borehole and by temperature variations in the borehole. The friction can, however, be decreased with spacers and the diameter of the wires should be as large as possible. The temperature effects can be compensated by using two wires with different coefficient of linear expansion that are anchored at the same depth.

There are two types of extensometers available, namely constant-tension and variable-tension extensometers. The wire tension is adjusted in the constant-tension extensometer to a constant value when the extensometer is read. Its function can be checked by measuring the extension of the wire when the tension in the wires is varied. If the wires move freely in the borehole, the extension of the wires will be the same for the same stress change. Constant-tension extensometers are therefore as a rule more accurate and reliable than variable-tension extensometers.

Another method is to measure the relative displacement with a series of metal rings. A moving probe with inductive transducers is used to locate the rings (Bernede, 1977). The accuracy is about 0.02 mm.

11.8.8 Piezometers

The long-term stability of a slope depends to a large extent on the pore or cleft water pressures that develop along potential sliding surfaces. The pore pressure gauges that are used to investigate this factor should be accurate and durable and the time lag should be small. Frequently it is difficult to detect long-term trends owing to temperature effects and changes in the atmospheric pressure. Open standpipes, which in principle are cased observation wells, are used to measure the pore pressures in soils with a high permeability such as sand and gravel.

In soils with medium to low permeability, open standpipes cannot be used because the time lag is too large. Hvorslev (1951) and Penman (1961) have discussed the time lag effects for different types of piezometers. In such soils, pneumatic, hydraulic, or electrical piezometers are used.

Pneumatic type piezometers utilize a sealed porous tip which contains a gas- or fluid-operated diaphragm or valve. The diaphragm or valve is connected to one or two lines that extend through the borehole to the ground surface. When the applied pressure in the connecting line is equal to the fluid pressure at the porous tip, the diaphragm or valve opens. This pressure is assumed to be equal to the pore or cleft water pressure in the surrounding soil or rock. The Warlam piezometer (Warlam and Thomas, 1965) is operated with air while a fluid is used to operate the Glötzl piezometer (Lauffer and Schober, 1964). Wilson and Squier (1969) have pointed out that pneumatic type piezometers are simple to operate and have small time lag and long-term stability, and that the instrumentation terminal can be located at any reasonable elevation with respect to the piezometer.

Hydraulic piezometers have been used extensively to measure pore water pressures over long time periods. This type of gauge consists basically of a porous tip that is directly connected to a pressure gauge. Many types of hydraulic gauges are available. In soil with low to medium permeability a Casagrande-type piezometer is often used. This piezometer consists of a porous tip that is connected to a 10-mm plastic tube. The porous tip is then placed in a borehole, which is sealed in order to decrease the time lag. Hydraulic pore pressure gauges with a single line have the disadvantage that air may collect in the line. Some gauges are provided with double lines so that water can be circulated in order to remove any air that has been trapped in the system.

Electrical piezometers are often unreliable over long times. However, they have in general a very short time lag and can thus be used to measure rapid changes of the pore or the cleft water pressures such as those caused by earthquakes, blast loading, or pile driving. The pore water pressures are determined by measuring the deflection of a thin diaphragm with an electrical transducer. Electrical piezometers have been described by Penman (1961), Cooling (1962), Shannon et al. (1962), Bishop et al. (1964), Brooker and Lindberg (1965), Wissa et al. (1975), and by Wilson and Mikkelson (1978). There are also pore pressure gauges available that record automatically any changes of the pore water pressure.

11.8.9 Tensiometers

It may also be of interest to monitor the soil suction in partially saturated soils, for example, in Hong Kong (Brand, 1982; Sweeney, 1982). Tensiometers consisting of a small porous ceramic tip that is connected to a perspex tube and a vacuum gage have been used. A fully automated system has been installed in Hong Kong employing sensitive differential tensiometers that can measure both positive and negative pore pressures (Brand, 1984).

11.8.10 Measurements of Acoustic Emission

The emission of noise associated with movements in a soil or rock mass can be measured with accelerometers or electrodynamic geophones. The signals are amplified and filtered. The signals exceeding a certain threshold value are selected and the number of impulses during a given interval of time is counted. By installing several microseismic monitors at different levels, it is possible to locate unstable areas.

The method has been applied very successfully in underground mines to locate rockbursts and other instabilities. It is doubtful whether the method can be used to detect potential landslides in soil because of the high attenuation, which dampens the noise very rapidly. Nevertheless, a number of successful applications of this method in soils have been described by Pilot (1984). Cadman and Goodman (1967) successfully located in the laboratory the rupture surface at an embankment failure. Favourable results between movements determined from inclinometer measurements and topographic survey and acoustic emission have been reported by McCauley (1976) for a slide at Thornton Bluffs. Novosad et al. (1977) have described several successful applications in Czechoslovakia, for example, at the slide at Turaneck. Piteau et al. (1978) have measured the acoustic emission down to 150 m depth for the Downies slide and have correlated the emission peaks with the failure level. Koerner et al. (1981) used acoustic emission to monitor the failure of an experimental sand embankment as shown in Figure 11.21.

11.8.11 Slide Warning System

Different slide warning systems are available. They consist of a slide fence as shown in Figure 11.22 or a series of in-place

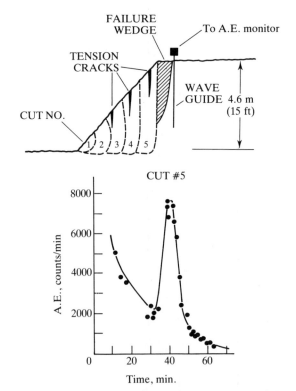

Fig. 11.21 Recording of acoustic emission during the failure of the embankment at Philadelphia, stage 5. (*Koerner et al., 1981.*)

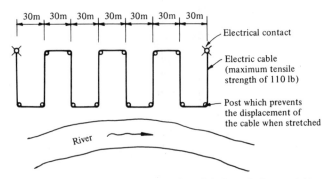

Fig. 11.22 Slide warning system, Swedish State Railways. (*After Broms, 1969.*)

inclinometers, deflectometers, extensometers, and piezometers that are connected with an alarm system. One problem is to determine the threshold value at which the alarm should be triggered. Another difficulty is the planning and execution of the preventive measures to be taken in case of an alarm.

Human factors play an important role even for fully automated systems, since failures are normally caused by long-term rather than by short-term changes. The data must be reviewed periodically by an experienced engineer. False alarms caused by poor judgment can have an adverse effect on faith in the system. Successive false alarms may generate an atmosphere of disbelief and real alarms may be disregarded.

A successful application of a slide warning system at Aurland, Norway, for a road crossing a large unstable avalanche has been reported by Grimstad (1983). The slope is instrumented with extensometers coupled to LVDTs that are connected to traffic lights. When the movement exceeds 20 mm the traffic lights turn red until the situation has been rectified.

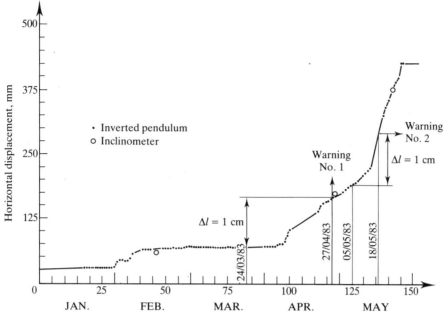

Fig. 11.23 Champ-la-Croix monitoring. Displacement record and triggering of warning (*Delmas and Rodriguez, 1984.*)

Mitchell and Williams (1981) monitored the movements of a slope in the Champlain clay at Ottawa, Canada, that was brought to failure by increasing the hydraulic pressure in recharge wells. The slope was heavily instrumented. It was observed that the monitoring of the cracks, the deformations at the toe of the slope, the piezometric, the microseismic, or the acoustic emission level alone does not give a clear indication of the stability of natural slopes in overconsolidated, fissured, and very plastic clays. Surface extensometers with adjustable threshold limits are preferred.

Haugen and Isaac (1984) designed a continuous monitoring and warning system for a tar sand mine in Alberta, Canada. A surface monitoring system was not sensitive enough to indicate an impending failure. Surface monitoring is mainly useful as a secondary system. It is preferable to use borehole deflection sensors or in-place inclinometers so that any deep-seated movements can be measured directly near the rupture surface. Microseismic monitoring and fiber optics can also be used but further developments and refinements are required.

Delmas and Rodriguez (1984) have described a warning system that has been installed in an unstable natural slope at Champ-la-Crois in France. The warning system consists of servoaccelerometer tiltmeters and an inverted pendulum with potentiometers so that the movement near the rupture surface can be monitored continuously. Measurements are taken every 5 seconds and compared to the mean value of previous measurements. When the total movement exceeds 10 mm, a warning is transmitted to the nearest station, 12 km from the site. Figure 11.23 shows the results for a 5-month period. The 10-mm threshold value was exceeded twice during this period.

11.9 METHODS OF CORRECTING LANDSLIDES

Many methods can be used to correct a landslide, such as flattening of the slope, pressure berms, erosion protection, electroosmosis, drying, lowering of the groundwater table, etc. Some of the methods that have been used to stabilize the slopes along the Göta River in Sweden are shown in Figure 11.24.

Different stabilization methods have been reviewed by Hutchinson (1977). A number of books have also been published (Veder, 1981; Zaruba and Mencl, 1982; Reeves, 1982; Bromhead, 1986). Close cooperation between the engineer and the geologist is usually required in order to reach an economic and safe solution to a complicated landslide problem.

Commonly used correction methods can be classified as follows:

1. *Geometric methods*, in which the geometry of the slope is changed
2. *Hydrological methods*, in which the groundwater table is lowered or the water content of the soil is reduced
3. *Chemical and mechanical methods*, in which the shear strength of the sliding mass is increased or the external force causing the landslide is reduced

11.9.1 Geometric Methods

The stability of a slope can be increased by flattening the slope, by removing part of the soil or the load at the top of the slope, or by constructing pressure berms at the toe. A combination of methods can also be used. For small and shallow slides, the removal of the unstable material is another alternative.

The flattening of a slope can be done either as a uniform regrading from the ditch line to the top of the slope or as several straight slopes separated by benches. For slopes higher than 5 m, benches are generally preferred over a uniform straight slope. The benches will reduce the velocity of the surface runoff and hence the surface erosion. The benches should be provided with paved gutters and catch basins in order to divert the surface water away from the slope. The flattening and the benching of a slope has to be done judiciously to prevent local failures or slumps.

Removal of soil or loads at the top of the slope is only effective for rotational slides. This method is not suited for flows or translational slides parallel to the slope.

Toe berms are effective for many types of landslides. For rotational slides, the berms act as counterweights that reduce the overturning moment. For translational slides, they act as buttresses. In constructing a toe berm, it is often necessary to excavate down to a firm layer or to bedrock. The excavation

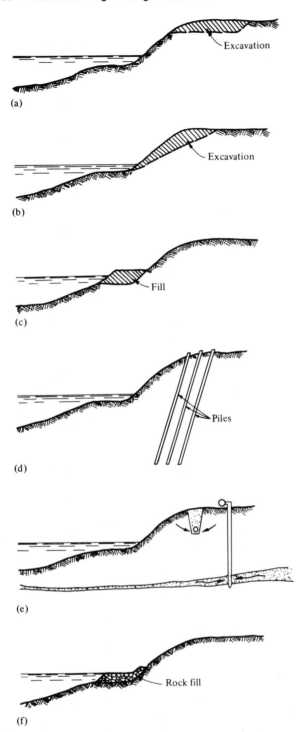

(a)

(b)

(c)

(d)

(e)

(f)

Fig. 11.24 Stabilization of slopes along the Gota River in Sweden. (a) Excavation at top of slope. (b) Flattening of slope. (c) Fill at toe of slope. (d) Driving of piles. (e) Lowering of groundwater level. (f) Erosion protection. (*After Broms, 1969.*)

and the backfilling should be done in narrow strips in order to prevent excessive movements of the landslide debris (Wahlstrom and Nichols, 1969).

A toe berm was used successfully to stabilize a landslide at Sandnes, Norway, where about 5000 m³ of sand was required to stop the movement. The size of the slide was about 150 m by 250 m. The rate of the movement decreased gradually to 2.5 mm/day as more sand was brought in (Bjerrum, 1967).

11.9.2 Hydrologic Methods

Because saturation and an increase of the pore pressures are often the major factors causing slope failures, a properly designed drainage system to intercept the flow is essential in most stabilization work.

Surface Drains

The surface of a landslide mass is generally uneven, hummocky, and deeply fissured. Springs and ponds are common within the slide area. Water accumulating in cracks and fissures can aggravate the slope movement further. One of the first measures to be taken after a landslide is to control the surface water with surface drains. Depressions should be filled and all fissures and cracks should be sealed. Trenches should be dug above the head scarp and paved to divert the surface water from the threatened area. Springs should be contained and diverted away from the slide. Vegetation within a slide area should be retained as much as possible as protection against further erosion. Gunite can be used to reduce the infiltration and to increase the runoff.

Drainage trenches are useful at depths less than 5 m. A typical configuration is shown in Figure 11.25, where a perforated plastic pipe has been embedded in a gravel layer at the bottom of the trench. The gravel is protected by filter fabric. The trench is backfilled with an impervious material.

Weeks (1969) has described a case where shallow rotational slips have been stabilized by a system of longitudinal drains. The 5-m deep and 1-m wide drains were filled with stones.

The softening process of stiff fissured clays can be delayed by draining the area above the slope to a distance that corresponds to about the depth of the cut. The surface of a cut should be sealed in order to reduce the infiltration. Debris flows can often be prevented by the installation of drains along the upper part of the slope and by protecting the surface of the slope by a layer of soil of low permeability. Slides caused by leakage from irrigation canals and from reservoirs may be prevented by watertight linings or cut-off walls. Slides caused by solifluction can be prevented by a blanket of pervious material (Lane, 1948).

Surface drains alone are seldom sufficient to stabilize a slope. This method is often combined with other corrective measures such as a subsurface drainage system.

Horizontal Drains

High pore water pressures in a slope can be decreased with horizontal or inclined 50-mm diameter perforated drains. The required inclination is approximately 5 to 10 percent. The boreholes for the drains can be drilled by a helical hollow-stem auger where the drain pipes are inserted through the stem or

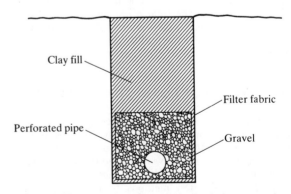

Fig. 11.25 Typical configuration of a drainage trench.

by a rotary drilling method. The pipes can in some cases be jacked into place. The length of the drains may be up to 60 to 70 m. They are generally spaced 5 to 15 m apart or spread out like a fan.

There are many successful applications of horizontal drains. LaRochelle et al. (1977), for example, have used horizontal drains to stabilize a landslide in a stratified sensitive clay. The drains were pushed into soil at the toe of the slope. Horizontal drains can increase the factor of safety of a slope by about 30 percent. Gedney and Weber (1978) have described two cases where horizontal drains were used to stabilize highway slopes in California.

The design of horizontal drains has been studied by Kenney et al. (1977). They have developed a series of design charts. Nonveiller (1981) has investigated the pore pressure dissipation in the zone of influence based on consolidation theory. Satisfactory stabilization can often be achieved in sandy and silty soils within 1 month while 6 months may be required in clays.

Vertical Drains

Vertical sand drains are well suited to correct landslides that contain permeable sand and silt layers. The drains will connect the permeable strata. Horizontal drains can then be installed to intercept the vertical drains near the bottom.

Vertical sand drains with 200 mm diameter and spaced 5 m apart have been used to stabilize a slope in a quick clay (Holm, 1961). Measurements indicated that pore pressures decreased appreciably during the installation of the drains, which gave the slope a sufficient stability.

Drainage Galleries

Drainge galleries are sometimes used to increase the stability of a slope. They are often difficult to construct and are therefore expensive. They are generally very effective owing to their large perimeter area. They should preferably be constructed below a possible sliding or failure surface. The water from the moving soil or rock mass can then be drained into the drainage gallery by vertical boreholes. Drainage galleries have, for example, been used to stabilize an ancient landslide area in Weirton, West Virginia (D'Appolonia et al., 1967). In weak rock the drainage galleries should be backfilled with stones or gravel (Zaruba and Mencl, 1969).

The steep slopes in loess along the Danube River in Hungary have been stabilized by a system of vertical and horizontal drains and deep drainage galleries (Andai, 1970).

Electroosmosis

The shear strength of primarily silty soils can be increased by electroosmosis. With this method an electric current is passed through the soil between two rows of electrodes. The current causes the water in the soil to migrate towards the cathodes. The water content and the pore water pressures in the soil are then reduced. Also, a base exchange takes place during electroosmosis that increases further the shear strength of the soil.

Casagrande et al. (1961) describe a case in which electro-osmosis was used to stabilize a slope of mainly loose saturated silt at Little Pic River in Canada. The slope was treated to a minimum depth of 12 m and the maximum spacing of the electrodes was 3 m. The potential was 100 volts, which corresponded to a gradient of 0.3 V/cm. After 3 months the average water content of the soil had decreased by about 4 percent. The groundwater level had been lowered by 10 m at

the top and 15 m at the toe of the slope. After the treatment it was possible to excavate to 16 m depth at a 1 : 1 slope. Before the treatment, the slope was unstable at a slope of 2.5 : 1. Other successful applications have been discussed by Casagrande (1952).

Fetzer (1967) has described the successful use of electro-osmosis to stabilize a slide in the West Branch Dam, Ohio. High excess pore pressure developed in a 20-m layer of clay that was located below a 24-m high embankment. The natural water content, which varied between 32 and 40 percent, was several percent below the liquid limit. The spacing of the electrodes was 6 m and they were installed by jetting down to a depth of 40 m below the crest of the dam. The potential between the electrodes varied between 50 and 70 V. The application of electroosmosis decreased the piezometric level by 8 m in 30 days at the downstream berm, while at the upstream berm the piezometric level decreased as much as 6.5 m in a month. The average decrease during a 10-month period was 2.0 m/month.

Griffin (1972) reports a case where electroosmosis was used to stabilize a 30-m high slope of very loose silt upstream from the Kootenay Channel power plant in British Columbia. Five rows of electrodes, 100 to 300 m in length, were installed parallel to the slope. The electrodes were installed in pairs. Each pair consisted of a cathode and an anode spaced 3 m apart. The cathodes were wellpoints and the anodes were 50-mm diameter steel tubes jetted in place down to 30 m depth. The potential was 100 to 150 V. About 160 liters/min were pumped from the uppermost row of 30 wellpoints. After a week, the slope could be steepened from 5 : 1 (horizontal to vertical) to 2 : 1. Six months later, the entire slope was stabilized.

In Singapore, an 8-m high earth embankment used as a temporary cofferdam during the construction of a dry dock was stabilized by electroosmosis (Chappel and Burton, 1975). The lateral displacement decreased from up to 1 m/day to less than 10 mm/day after 9 days. The shear strength of the embankment material increased more than 100 percent after 1 month and a significant decrease of the moisture content was also recorded. Other examples where electroosmosis has been used successfully have been described by Henke (1968) and by Pilot (1981).

Short-Circuit Conductors

This method is based on the hypothesis that there exists an electropotential difference between soil layers of different water content. A steel rod or a water pipe driven into these layers creates a short circuit that eliminates the potential differences between the layers. Veder (1981) reports that this method not only stops the migration of water to the slip zone, it also reduces the water content of the soil along the slip surface. In addition, the monovalent ions in the soil are replaced by higher-valency ions from the conductors, which increases the shear strength of the soil. Veder reports several successful applications of this method in Austria and Japan. There are, however, some questions about the efficiency of the method.

Vegetation

Vegetation provides an effective protection against surface erosion and shallow slides. It dries out the soil at the surface, thereby increasing the shear strength. The root system provides a mechanical reinforcement of the soil. A number of vegetation schemes have been used from the planting of grass and shrubs to the more elaborate methods such as contour wattling, contour brush-layering, reed-trench terracing, brush matting, and live staking. These methods have been reviewed by Gray and Leiser (1982).

Consolidation

The possibility of stabilizing the sensitive clays at Turnagain Heights by reconsolidation after the soil has been remolded using a series of small charges has been investigated by Long and George (1967b). Experiments indicated that the shear strength of the clay could be increased significantly if at least 50 percent of the soil were remolded and the overburden pressure increased to approximately 300 kPa.

Thermal Methods

Beles (1957) has described several cases where thermal stabilization has been used to increase the stability of slopes. Bricklike columns were formed in the soil when hot gas with a temperature of at least 800°C was forced through two interconnected vertical 200- to 400-mm diameter holes. The holes were spaced 0.8 to 1.2 m apart. The diameter of the bricklike columns was approximately 1.5 m after 6 to 8 days of treatment.

The stability of a hillside in California was increased by circulating hot air through a series of tunnels, which dried out the surrounding soil. The tunnels were subsequently filled with gravel (Hill, 1934).

Inverted Filters

Slides caused by piping can be prevented by an inverted filter that covers the lower part of the slope. Finzi (1961) describes a case where the slopes of a reservoir were protected against erosion and the pore water pressures were reduced by an inverted filter. The gradation of the transition material satisfied the filter criteria proposed by Terzaghi. The minimum size of the rockfill was 100 mm; 40 percent had a particle size larger than 300 mm.

11.9.3 Chemical and Mechanical Methods

Grouting

Grouting can be used to increase the shear strength and to lower the permeability of coarse-grained soils. Both chemical and cement grouts can be used. Gedney and Weber (1978) report an interesting application of this method where a 90-m deep cut slope in North Carolina was stabilized by injecting cement grouts into large voids in the rock. Zaruba and Mencl (1982) have described another application of this method in Czechoslovakia using aerated cement suspension and mortar. The stabilization of about 8000 m³ of unstable material required 88 tons of cement. 107 boreholes were drilled with a total length of 540 m. The grouting pressure varied from 0.2 to 0.6 MPa.

Lime and Cement Columns

The shear strength of soft clay can be increased with quicklime or cement. The strength increases in general with increasing lime content up to about 10 to 12 percent with respect to the dry weight of the soil. For soft clays with a shear strength of 10 to 15 kPa the relative increase is normally 10 to 50 times the initial shear strength. Broms and Boman (1976) have described a case where the shear strength of the improved soil increased 50 times in 1 year after the mixing, half of which was reached within the first 2 months.

Lime is generally more effective when the plasticity index of the clay is high, while it is advantageous to use cement when the soil is sandy or silty with a low plasticity index. Unslaked lime has the advantage that it does not require the same thorough mixing with the clay compared with cement. The permeability of soil stabilized with lime will increase and the lime columns will function as vertical drains in the soil. The permeability of the soil is normally reduced when cement is used.

The columns are usually 500 mm in diameter with a maximum length of about 15 m. They are manufactured by mixing in situ the soft clay with quicklime (CaO) or with cement using a tool shaped like a giant dough mixer as described by Broms and Boman (1976, 1977) and by Broms (1983).

Lime columns have, for example, been used to increase the stability of a slope close to a bridge abutment, as described by Ekström and Tränk (1979). The lime columns were installed between the bridge abutment and an adjacent stream. The column spacing of 0.5 m was equal to the diameter of the columns. The average shear strength of the clay increased from 10 kPa to over 45 kPa. It was possible to increase the factor of safety using this method to 1.5 with respect to a circular failure.

Another type of lime column has been used that does not involve mixing the lime with the soil. One successful treatment has been reported by Handy and Williams (1967). Approximately 45 000 kg of quicklime was placed in 200-mm diameter predrilled holes at 1.5-m centers throughout the slide area. The lime migrated a distance of 300 mm from the drilled holes in 1 year. Other cases have been described by Brandl (1973, 1976).

Ion Exchange

With this technique (Smith and Forsyth, 1971) the clay is treated along the slip plane with a concentrated chemical solution. Some of the cations in the clay are then replaced. The shear strength of the soil may be increased by as much as 200 to 300 percent. The successful application of this method in Northern California has been described by Mearns et al. (1973).

Freezing

It has been reported that freezing was used to stabilize the right abutment of the Grad Coulee Dam in the Columbia River in Washington during the construction of the dam. A 13-m high dam of frozen soil with a width of 6 m and a length of 30 m was created with 377 special freezing points (Gordon, 1937).

Compaction

Landslides caused by spontaneous liquefaction can be prevented by compacting the soil. Clean sands can be compacted by pile driving or vibroflotation, or by exploding a large number of small charges in the soil.

Geofabric

Geofabric can be used to increase the drainage and to prevent shallow slips. An example of this method has been reported by Broms and Wong (1985a) as shown in Figure 11.26. The height of the slope is 7 m and the inclination was 37° prior to the failure. Geofabric-wrapped drains were constructed in the slope. The spacing of the drains was 3 m. The drains extended beyond the estimated slip surface to provide the needed reinforcement. They were connected to the crib wall at the toe of the slope. The crib wall was filled with crushed rock to provide drainage of water from the transverse drains. Horizontal layers of fabric were also placed just below the ground surface to increase the stability of the slope with respect to shallow slides above the drain as shown in Figure 11.27.

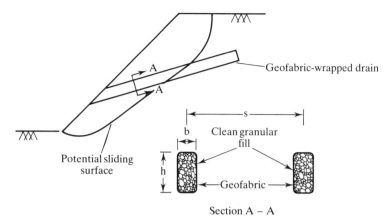

Fig. 11.26 Schematics of slope stabilization using geofabric-wrapped drains. (*After Broms and Wong, 1985a.*)

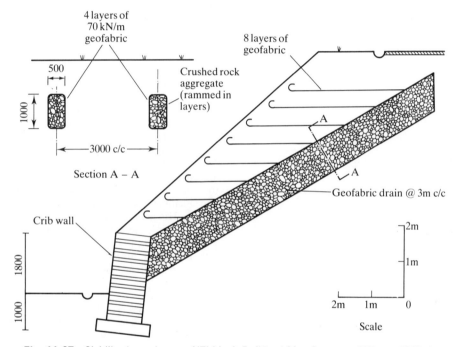

Fig. 11.27 Slabilization scheme—NTI block E slide. (*After Broms and Wong, 1985a.*)

Rock Bolts

Rock bolts have been used primarily to stabilize slab slides in rock by increasing the normal force acting on a potential sliding surface and by dowel action. It is important to reduce the cleft water pressures when rock bolts are used and to prevent the water from freezing in the joints or fissures; otherwise the slope may fail by progressive failure. The disadvantage with rock bolts is that they can corrode if they are not coated and grouted. Some controversy exists, however, about the function of rock bolts in general.

Grouted rock bolts with lengths of up to 13 m were used at the Howard A. Hanson Dam on the Green River in Washington (Christman, 1960).

Soil Nailing and Root Piles

Soil nailing is a form of soil reinforcement that is particularly useful to prevent relatively shallow slides. With this method, steel reinforcement bars, 15 to 30 mm in diameter, are grouted in holes drilled into the slope. The length of the bars is 3 to 10 m depending on the location of the estimated slip surface. The spacing is typically 2 m. Reinforced-concrete beams are used to connect the nails at the ground surface. Recent work on soil nailing has been summarized by Schlosser and Juran (1979). Applications of the method to stabilizing slopes in residual soils have been described by Tan et al. (1985).

Use of root piles is similar to soil nailing in many respects. Steel bars with a diameter of 25 to 75 mm are used, which are driven in clusters in order to form a reinforced soil mass. Several cases have been reported by, for example, Lizzi (1977), Schlosser and Juran (1979), Lizzi and Carnevale (1979), Floss (1979), Dash and Jovino (1980), and Murray (1980) in which this method has been used successfully.

Embankment Piles

Driven piles have also been used to stabilize landslides when the affected area is small. For translational slides, the piles function as dowels. For rotational slides, the piles function either as end-bearing or as friction piles. The efficiency of the small-diameter piles is high owing to the large surface area. The piles are generally driven at the top of the slope. They are normally inclined in order to increase their efficiency.

For rotational slides in soft clay, the piles contribute to the resisting moment through the skin friction along the upper part of the piles, which is located above a potential slip surface in the soil as shown in Figure 11.28. This is generally the case for end-bearing piles. For friction piles, the resistance is governed by the part of the pile that offers the least resistance. The dowel action of the piles is generally not taken into account in a rotational slide because relatively large movements are required to mobilize the dowel action.

The vibrations and the remolding of the soil caused by the pile driving may temporarily reduce the shear strength of the

clay, especially if the sensitivity is high. A reduction of up to 50 percent of the shear strength has been observed (Broms and Wong, 1985b). Broms and Bennermark (1968) have described a landslide in a very sensitive clay that has been initiated by pile driving at the top of a slope.

The sequence of the driving can also affect the stability of a slope in soft clay. The piles located closest to the slope should be driven first. Subsequent piles should be driven away from the slope in order to reduce the lateral displacements of the soil.

Cast-in-place Piles

Cast-in-place piles have a number of advantages over driven piles and conventional retaining walls. Problems associated with the vibration and the remolding of the soil from driven piles are then greatly reduced. The need for excavation and backfilling is eliminated. The piles are typically 300 mm to 1500 mm in diameter. The spacing varies from zero to three times the pile diameter. Lagging can be used between piles to prevent raveling. Anchors or tiebacks are sometimes useful in order to increase the lateral resistance. The design and analysis of cast-in-place piles have been discussed by Nethero (1982) and Nakamura (1984).

Numerous successful applications of this type of retaining structure have been described in the literature (Andrews and Klasell, 1964; Gould, 1970; De Beer and Wallays, 1970; Palladino and Peck, 1972; Nethero, 1982; and Rodriguez Ortiz et al., 1984). However, there are also some unsuccessful cases. An example is the slide at Portuguese Bend in Los Angeles (Merriam, 1960). Heavily reinforced cast-in-place concrete piles, 1200 mm in diameter and 18 m in length, were constructed in an attempt to stop the slide. The landslide continued to move for more than 20 years after the installation of the piles (Schuster and Fleming, 1982). Another example is a highway cut at Potrero Hill in San Francisco (Smith and Forsyth, 1971), where an anchored concrete pile wall was constructed. The wall was 224 m long and 6 m to 18 m high. The grouted anchors extended 9 to 14 m into interbedded sandstone and shale. Parts of the wall failed after a few months. The failure surface passed beneath the anchors. The stability was later restored with longer and larger piles.

Broms and Bjerke (1973) have pointed out the danger in using large-diameter bored pile to retain soft clay. An 11-m high cast-in-place reinforced-concrete pile wall was built in central Sweden to retain the clays. The stabilization was unsuccessful. The soft clay was squeezed like toothpaste into the excavation between the piles.

Retaining Walls

Crib walls, gabion walls, reinforced earth, welded-wire walls, and reinforced-concrete walls have been used successfully to stabilize landslides by increasing the resisting force at the toe. These walls must be designed to provide adequate resistance against sliding, overturning, and general shear failure. Also, anchored sheet-pile walls and anchored bored-pile walls are common. A comprehensive review of the various retaining structures has been made by Gray and Leiser (1982). Details of tieback walls have been reported by Weatherby and Nicholson (1982).

The stability of a slope is temporarily decreased during the construction of a gravity wall when the soil at the toe of the slope is removed. This decrease of the stability can be sufficient to trigger a landslide, especially within an old slide area. It is important that the retaining walls are provided with drains and weep holes to relieve the excess pore water pressures that otherwise can develop behind the walls.

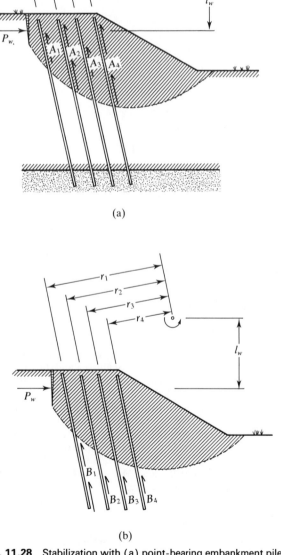

Fig. 11.28 Stabilization with (a) point-bearing embankment piles, (b) floating embankment piles.

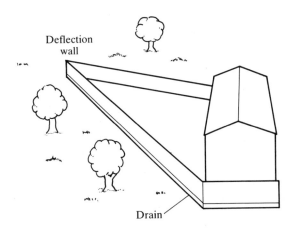

Fig. 11.29 Typical deflection wall layout. (*After Hollingsworth and Kovács, 1981.*)

In England, wide dry-stone toe walls have been used successfully to stabilize cuts in overconsolidated clays (Cassel, 1948). Gravel buttresses were used to stabilize the landslide area along Fourth Avenue in Anchorage, Alaska, after the 1964 earthquake (Long and George, 1967a). A similar procedure was used to stabilize the landslide at Dillon Dam in Colorado (Wahlstrom and Nichols, 1969).

D'Appolonia et al. (1967) used a flexible sheet-pile wall to stabilize an ancient slide area. The sheet-pile wall was anchored with pretensioned ties to prevent local failure at the toe of the slope.

Deflection Walls

Deflection walls can be used to protect buildings in the path of a flow slide. A typical layout is shown in Figure 11.29. A number of successful applications have been summarized by Veder (1981). He reports one interesting case in Austria where a bridge pier was constructed in an area susceptible to creep movements of 10 mm per year. The pier was constructed on bedrock below the creep zone and was protected by an elliptical deflection wall. The clearance between the pier and the wall was about 1.5 m. The wall appears to have functioned effectively.

REFERENCES

Aas, G. (1981), Stability of natural slopes in quick clay. *Proc. 10th Int. Conf. on Soil Mechanics and Foundation Engineering*, Stockholm, **3**, pp. 333–338.

Alfors, J. T., Burnett, J. L., and Gay, T. E. Jr. (1973), *Urban Geology: Master Plan for California*, California Division of Mines & Geology, Bulletin 198.

Andai, P. (1970), Sicherung von Gelanderutschungen am Losshang an der Donau, *Der bauingenieur*, **45**, No. 2, pp. 58–64.

Andresen, A. and Bjerrum, L. (1968), *Slides in Subaqueous Slopes in Loose Sand and Silt*, Norwegian Geotechnical Institute, Oslo, Publication No. 81, pp. 1–9.

Andres, G. H. and Klasell, J. A. (1964), Cylinder pile retaining wall, *Highway Research Record* No. 56, pp. 83–97.

D'Appolonia, E. D., Alperstein, R., and D'Appolonia, D. J. (1967), Behaviour of a colluvial slope, *Journal of the Soil Mechanics and Foundations Division*, ASCE, **93**, No. SM4, pp. 447–473.

Banks, D. C. and Meade, R. B. (1982), Slope stability, control and remedial measures; and reservoir induced seismicity, *Proc. 4th Cong. Int. Ass. of Engrg. Geology*, New Delhi IX, pp. 119–142.

Barton, M. E. (1984), The preferred path of landslide shear surfaces in overconsolidated clays and soft rocks, *Proc. 4th Int. Symp. on Landslides*, Toronto, **III**, pp. 75–80.

Barzett, D. J., Adams, J. L., and Matyas, E. L. (1961), An investigation of a slide in a test trench excavated in fissured sensitive clay, *Proc. 5th Int. Conf. on Soil Mechanics and Foundation Engineering*, **1**, pp. 431–435.

Beene, R. R. W. (1967), Waco Dam slide, *Journal of the Soil Mechanics and Foundation Engineering Division*, ASCE, **93**, No. SM4, pp. 35–44.

De Beer, E. and Geolen, E. (1977), Stability problems of slopes in overconsolidated clays, *Proc. 9th Int. Conf. on Soil Mechanics and Foundation Engineering*, Tokyo, **III**, pp. 31–39.

De Beer, E. E. and Wallays, M. (1970), Stabilization of a slope in schists by means of bored piles reinforced with steel beams, *Proc. 2nd Congr. on Rock Mechanics*, **3**, pp. 7–13.

Beles, A. A. (1957), Le traitment thermique du sol, *Proc. 4th Int. Conf. on Soil Mechanics and Foundation Engineering*, **3**, pp. 266–267.

Benson, W. N. (1946), Landslides and their relation to engineering in the Dunedin district, New Zealand, *Economic Geology*, **41**, pp. 328–347.

Bernede, J. (1977), Appareils de conception récente utilisés actuellement au contrôle des mouvements de terrain. Télémesure associée. *Rev. Française de Geotechnique*, Paris, No. 1, pp. 76–85.

Bernini, F., Inghilleri, G., Mandolesi, P., and Vianelli, G. (1979), Photogrammetric method applied to survey and control of landslides (in Italian. I metodi fotogrammetrici applicati al relievo e controllo delle frane). *Symp. Movimenti Franosi Edinamica dei Versenti*, Salice Terme. Pavie.

Bhandari, R. K. (1984), Simple and economical instrumentation and warning systems for landslides and other mass movements, *Proc. 4th Int. Symp. on Landslides*, Toronto, **I**, pp. 251–274.

Binger, W. V. (1948), Analytical studies of Panama Canal slides, *Proc. 2nd Int. Conf. on Soil Mechanics and Foundation Engineering*, **2**, pp. 54–60.

Bishop, A. W. (1966), The strength of soils as engineering materials, *Geotechnique*, **16**, pp. 91–128.

Bishop, A. W., Kennard, M. F., and Vaughan, P. R. (1964), Developments in the measurement and interpretation of pore pressure in earth dams, *Trans. Congr. Large Dams*, pp. 47–72.

Bishop, D. M. and Stevens, M. E. (1964), Landslides on logged areas in southeast Alaska, *Research Paper NOR-1*, Forestry Service, U.S. Dept. of Agriculture, Juneau, Alaska.

Bjerrum, L. (1954), Stability of natural slopes in quick clay, *Proc. European Conf. on Stability of Earth Slopes*, Stockholm, **2**, pp. 16–40.

Bjerrum, L. (1967), Progressive failure in slopes of overconsolidated plastic clay and clay shales, *Journal of Soil Mechanics and Foundations Division*, ASCE, **93**, No. SM5, pp. 3–49.

Bjerrum, L. (1969), Stability of natural slopes and embankment foundations. *Proc. 7th Int. Conf. on Soil Mechanics and Foundation Engineering*, Mexico, **3**, p. 411.

Bjerrum, L. (1972), Embankment on soft ground, *Proc. Specialty Conf., Performance of Earth and Earth Supported Structures*, ASCE, **2**, pp. 1–54.

Bjerrum, L. (1973), Problems of soil mechanics and construction on soft clays and structurally unstable soils, *Proc. 8th Int. Conf. on Soil Mechanics and Foundation Engineering*, Moscow, **3**, pp. 1111–1159.

Bjerrum, L. and Jörstad, F. (1968), *Stability of Rock Slopes in Norway*, Norwegian Geotechnical Institute, Oslo, Publication No. 79, pp. 1–11.

Bonnard, C. (1983a), Nouvelles techniques de mesures au glissement de La Frasse, *Route et raffic*, No. 1, pp. 19–25.

Bonnard, C. (1983b), Determination of slow landslide activity by multidisciplinary measurement techniques. *Symposium on Field Measurement in Geomechanics*, Zurich (preprint).

Brand, E. W. (1981), Some thoughts on rain-induced slope failures, *Proc. 10th Int. Conf. on Soil Mechanics and Foundation Engineering*, Stockholm, **3**, pp. 373–376.

Brand, E. W. (1982), Analysis and design in residual soils. *Proc. ASCE Spec. Conf. on Engineering and Construction in Tropical and Residual Soils*, Honolulu, pp. 89–143.

Brand, E. W. (1984), State-of-the-art report on landslides in Southeast Asia, *Proc. 4th Int. Symp. on Landslides*, Toronto, **I**, pp. 17–60.

Brand, E. W. and Hudson, R. R. (1982), CHASE—An empirical approach to the design of cut slopes in Hong Kong soils. *Proc. 7th S.E. Asian Geotech. Conf.*, **1**, 1–16. (Discussion, **2**, pp. 61–72 and pp. 77–79.)

Brand, E. W., Premchitt, J., and Phillipson, H. B. (1984), Relationship between rainfall and landslides in Hong Kong, *Proc. 4th Int. Symp. on Landslides*, Toronto, I, pp. 377–384.

Brandl, H. (1973), Stabilization of slippage-prone slopes by lime injections, *Proc. 8th Int. Conf. on Soil Mechanics and Foundation Engineering*, Moscow, **4.3**, pp. 300–301.

Brandl, H. (1976), Die Sicherung von hohen Abschnitten in rutschgefahrdeten Verwitterungsboden. *Proc. 6th European Conf. on Soil Mechanics and Foundation Engineering*, **1.1**, pp. 19–28.

Breth, H. (1967a), The dynamics of a landslide produced by the filling of a reservoir, *Trans. 9th Int. Congr. on Large Dams*, Istanbul, I, pp. 37–45.

Breth, H. (1967b), Calculation of the shearing strength of a moraine subjected to landsliding due to reservoir filling in Kauner Valley, Austria, *Proc. Geotechnical Conf.*, Oslo, **1**, pp. 171–177.

Bromhead, E. N. (1986), *The Stability of Slopes*, Surrey University Press, London.

Broms, B. B. (1969), Stability of natural slopes and embankment foundations, *Proc. 7th International Conference on Soil Mechanics and Foundation Engineering*, Mexico, **3**, pp. 385–394.

Broms, B. B. (1983), Stabilization of soft clay with lime columns, *Int. Seminar on Construction Problems in Soft Soils*, Singapore, pp. BB1–BB30.

Broms, B. B. and Bennermark, H. (1968), Shear strength of soft clay, *Discussion, Proc. Geotechn. Conf.*, Oslo, **2**, pp. 118–120.

Broms, B. B. and Bjerke, H. (1973), Extrusion of soft clay through a retaining wall, *Canadian Geotechnical Journal*, **10**, No. 1, pp. 103–109.

Broms, B. B. and Boman, P. (1976), Stabilization of deep cuts with lime columns, *Proc. 5th European Conf. on Soil Mechanics and Foundation Engineering*, Tokyo, **1**, pp. 207–210.

Broms, B. B. and Boman, P. (1977), Stabilization of soil with lime, *Ground Engineering*, **12**, No. 4, pp. 23–32.

Broms, B. B. and Wong, I. H. (1985a), Stabilization of slopes with geofabric, *3rd Int. Geol. Seminar*, Nanyang Tech. Institute, Singapore, pp. 75–83.

Broms, B. B. and Wong, I. H. (1985b), Embankment piles, *3rd Int. Geol. Seminar*, Nanyang Tech. Institute, Singapore, pp. 167–178.

Brooker, E. W. and Lindberg, D. A. (1965), Field measurement of pore pressure in high plasticity soils, *Engineering Effects of Moisture Changes in Soils: Proc. Int. Res. Engrg. Conf. Expansive Clay Soils*, Texas A & M University, College Station, Texas, pp. 57–68.

Broscoe, A. J. and Thomson, A. (1969), Observations on an Alpine mudflow, Steel Crack, Yukon, *Canadian Journal of Earth Science*, **6**, No. 2, pp. 219–220.

Brown, C. and Shen, M. S. (1975), Effect of deforestation on slopes, *Journal of the Geotechnical Engineering Division, ASCE*, **101**, GT2, pp. 147–165.

Brzezinski, L. S. (1971), A review of the 1924 Kenogami landslide, *Canadian Geotechnical Journal*, **8**, No. 1, pp. 1–6.

Cadman, J. D. and Goodman, R. E. (1967), Landslide noise. *Science*, **158**, No. 3805, pp. 1182–1184.

Cancelli, A. (1981), Evolution of slopes in over-consolidated clays, *Proc. 10th Int. Conf. on Soil Mechanics and Foundation Engineering*, Stockholm, **3**, pp. 377–380.

Carson, M. A. (1979), Le glissement de Rigaud du 3 mai 1978, *Geographic Physique Quaternaire*, **33**, No. 1, pp. 63–92.

Casagrande, A. and Shannon, W. L. (1947), Research on stress deformation and strength characteristics of soils and soft rocks under transient loading, *Harvard Soil Mechanics Series*, No. 31.

Casagrande, A. and Wilson, S. D. (1951), Effect of rate of loading on the strength of clays and shales at constant water content, *Geotechnique*, **2**, No. 3, pp. 251–263.

Casagrande, L. (1952), Electro-osmotic stabilization of soils, *Journal of the Boston Society of Civil Engineers*, **9**, pp. 51–83.

Casagrande, L., Loughney, R. W., and Matich, M. A. J. (1961), Electroosmotic stabilization of a high slope in loose saturated silt, *Proc. 5th Int. Conf. on Soil Mechanics and Foundation Engineering*, **2**, pp. 555–561.

Cassel, F. L. (1948), Slips in fissured clay, *Proc. 2nd Int. Conf. on Soil Mechanics and Foundation Engineering*, **2**, pp. 46–50.

Chandler, R. J. (1984), Delayed failure and observed strengths of first time slides in stiff clay: a review, *Proc. 4th Int. Symp. on Landslides*, Toronto, **II**, pp. 19–25.

Chandler, R. J. and Skempton, A. W. (1974), The design of permanent cutting slopes in stiff fissured clays, *Geotechnique*, **24**, No. 4, pp. 457–466.

Chappell, B. A. and Burton, P. L. (1975), Electro-osmosis applied to unstable embankment, *Journal of the Geotechnical Engineering Division, ASCE*, **101**, GT8, pp. 733–740.

Ching, R. K. H., Sweeney, D. J., and Fredland, D. G. (1984), Increase in factor of safety due to soil suction for two Hong Kong slopes, *Proc. 4th Int. Symp. on Landslides*, Toronto, **I**, pp. 617–624.

Christman, H. E. (1960), Bolts stabilize high rock slopes, *Civil Engineering*, **30**, pp. 98–99.

Close, U. and McCormick, E. (1922), Where the mountains walked. An account of the recent earthquake in Kansu Province, China, which destroyed 100,000 lives, *National Geographic Magazine*, **41**, No. 5, pp. 445–464.

Cloud, P. E. (1969), Geology, cities and surface movement, *Highway Research Record*, No. 271, Washington, D.C., pp. 1–9.

Coates, D. R. (1977), Landslide perspectives, *Review in Engineering Geology*, **3**, pp. 3–28.

Cooling, L. F. (1962), Field measurements in soil mechanics, *Geotechnique*, **12**, No. 2, pp. 77–104.

Cornforth, D. H. (1961), Plane strain failure characteristics of a saturated sand, Ph.D. Thesis, University of London.

Cotecchia, V. and Federico, A. (1984), On the determination of slip surfaces, *Proc. 4th Int. Symp. on Landslides*, Toronto, **III**, pp. 109–110.

Crandell, D. R. (1952), Landslides and rapid flowage phenomena near Pierre, South Dakota, *Economic Geology*, **47**, pp. 548–568.

Crandell, D. R. and Varnes, D. J. (1961), Movement of the Slum-gullion earth flow near Lake City, Colorado, *U.S. Geological Survey Professional Paper* 424 B, pp. 136–139.

Crawford, C. B. and Eden, W. J. (1967), Stability of natural slopes in sensitive clays, *Journal of the Soil Mechanics and Foundations Division, ASCE*, **93**, No. SM4, pp. 419–436.

Dash, U. and Jovino, P. L. (1980), Construction of a root-pile wall at Monessen, Pennsylvania, *Transportation Research Record* No. 749, pp. 13–21.

Delmas, Ph. and Rodriguez, J. P. (1984), Surveillance d'un site de glissement—systeme d'alarme a distance, *Proc. 4th Int. Symp. on Landslides*, Toronto, **II**, pp. 523–528.

Drouhin, G., Gautier, M., and Dervieux, F. (1948), Slide and subsidence of the hills of St Rafael-Télemly, *Proc. 2nd Int. Conf. on Soil Mechanics and Foundation Engineering*, **5**, pp. 104–106.

Eden, W. J. (1971), *Landslides in Clays*, National Research Council of Canada, Canadian Building Digest, CBD 143.

Eden, W. J. and Mitchel, R. J. (1970), The mechanics of landslides in leda clay, *Canadian Geotechnical Journal*, **7**, No. 3, pp. 285–296.

Eigenbrod, K. D. and Morgenstern, N. R. (1972), A slide in Cretaceous bedrock, Devon, Alberta, *Geotechnical Practice for Stability in Open Pit Mining*, ed. C. O. Brawner and V. Milligan, Society of Mining Engineers of the American Institute of Mining, Metallurgical, and Petroleum Engineers, New York.

Ekström, A. and Tränk, R. (1979), Kalkpelarmetoden. Tillämpningar for stabilisering av brostöd och rörgrav. (The lime column method. Stabilization of bridge abutments and trenches), *Proc. Nordic Geotechnical Meeting*, Helsinki, pp. 258–268. (In Swedish).

Esu, F. (1966), Short-term stability of slopes in unweathered jointed clays, *Geotechnique*, **16**, pp. 321–328.

Evans, R. S. (1981), An analysis of secondary toppling rock failures—the stress redistribution method, *Quarterly Journal of Engineering Geology*, **14**, pp. 77–86.

Fetzer, C. A. (1967), Electro-osmotic stabilization of West Branch Dam, *Journal of the Soil Mechanics and Foundations Division, ASCE*, **93**, No. SM4, pp. 85–106.

Finzi, D. (1961), Slope consolidation of the banks of the Monquelfo Reservoir, Italy, *Proc. 5th Int. Conf. on Soil Mechanics and Foundation Engineering*, **2**, pp. 591–594.

Floss, R. (1979), Discussion, design parameters for artificially improved soils, *Proc. 7th European Conf. on Soil Mechanics and Foundation Engineering*, Brighton, **4**, pp. 279–281.

Fookes, P. G. and Wilson, D. D. (1966), The geometry of discontinuities and slope failures in Siwalik clay, *Geotechnique*, **16**, No. 4, pp. 305–320.

Fredlund, D. G. (1978), Usage, requirements and features of slope stability computer software (Canada 1977), *Canadian Geotechnical Journal*, **15**, pp. 83–95.

Fredlund, D. G., Morgenstern, N. R., and Widger, R. A. (1978), The shear strength of unsaturated soils. *Canadian Geotechnical Journal*, **15**, pp. 313–321.

de Freitas, M. H. and Watters, R. J. (1973), Some field examples of toppling failure, *Geotechnique*, **23**, No. 4, pp. 495–514.

Froelich, A. J. (1970), Geologic setting of landslides along south slope of Pine Mountain, Kentucky, *Highway Research Record* No. 323, pp. 1–5.

Fukuoka, M. (1953), Landslides in Japan, *Proc. 3rd Int. Conf. on Soil Mechanics and Foundation Engineering*, **2**, pp. 234–238.

Gage, M. and Black, R. D. (1979), Slope stability and geological investigations at Mangatu State Forest, *Technical Paper 66*, N.Z. Forestry Service, Forestry Research Institute, Wellington.

Gedney, D. S. and Weber, W. G. Jr. (1978), *Design and Construction of Soil Slopes. Landslides: Analysis and Control*, National Academy of Sciences, Washington, D.C., Special Report 176, pp. 172–191.

Goldstein, M. L., Lapidus, L., and Misumsky, V. (1965), Rheological investigation of clays and slope stability, *Proc. 6th Int. Conf. on Soil Mechanics and Foundation Engineering*, **2**, pp. 482–485.

Goodman, R. E. and Bray, J. W. (1976), Toppling of road slopes, *Proc. Specialty Conf. on Rock Engineering for Foundations and Slopes*, Boulder, Colorado.

Gordon, G. (1937), Arc dam of ice stops slide, *Engineering News-Record*, **118**, pp. 211–215.

Gould, J. P. (1960), A study of shear failure in certain tertiary, marine sediments, *ASCE Research Conf. on Shear Strength of Cohesive Soils*, Boulder, Colorado, pp. 615–641.

Gould, J. P. (1970), Lateral earth pressures on rigid permanent structures, *ASCE Conference: Lateral Stresses in the Ground and Design of Earth Retaining Structures*, Ithaca, N.Y., pp. 219–269.

Gray, D. H. (1970), Effects of forest clear-cutting on the stability of natural slopes. *Bulletin, Association of Engineering Geologists*, **7**, No. 1–2, pp. 45–66.

Gray, D. H. and Leiser, A. T. (1982), *Biotechnical Slope Protection and Erosion Control*, Van Nostrand Reinhold Co., New York, N.Y.

Gregerson, O. (1981), *The Quick Clay Landslide in Reese, Norway*, Norwegian Geotechnical Institute, Oslo, Report No. 135.

Griffin, F. (1972), Power house slope stabilized by electroosmosis. *Heavy Construction News*, Wellpoint Corp. of New York.

Grimstad, E. (1983), Assessment of the risks of instability, including seismic risks, *Proc. XVII World Road Congress*, Sydney, pp. 122–127.

Habib, P. (1967), Sur, un mode de glissement des massifs rocheux, *Comptes Rendus des Séances de l'Academie des Sciences, Paris*, Serie A, **264**, pp. 151–152.

Haefeli, R. (1953), Creep problems in soils, snow and ice, *Proc. 3rd Int. Conf. on Soil Mechanics and Foundation Engineering*, **3**, pp. 238–251.

Haefeli, R., Schaerer, Ch., and Amberg, G. (1953), The behaviour under the influence of soil creep pressure of the concrete bridge built at Klosters by the Rhaetian Railway Company, Switzerland, *Proc. 3rd Int. Conf. on Soil Mechanics and Foundation Engineering*, **2**, pp. 175–179.

Handy, R. L. and Williams, W. W. (1967), Chemical stabilization of an active landslide, *Civil Engineering*, **37**, No. 8, pp. 62–65.

Hang, M. D., Sauer, K., and Fredlund, D. G. (1977), Retrogressive slope failure at Beaver Creek, South of Saskatoon, Saskatchewan, Canada, *Canadian Geotechnical Journal*, **14**, No. 3, pp. 288–301.

Hast, N. (1965), Spanningstillståndet i den fasta jordskorpans övre del, Bergmekanik, *Ingeniörsvetenskapsakademiens Meddelande*, **142**, pp. 13–23.

Haugen, M. A. and Isaac, B. A. (1984), Continuous slope monitoring at an oil sand mine, *Proc. 4th Int. Symp. on Landslides*, Toronto, **II**, pp. 529–534.

Heim, A. (1932), Bergsturz und Menschenleben, *Naturförschug Gesselleschäft, Zurich*, **77**.

Hendron, A. J. Jr., Cording, E. J., and Aiyer, A. K. (1971), *Analytical and Graphical Methods for the Analysis of Slopes in Rock Masses*, U.S. Army Engineer Nuclear Cratering Group, Technical Report 36.

Henke, K. F. (1968), Böschungssicherung durch horizontale Drainagebohrungen, *Strassen u. Tiefb.*, **22**, No. 2, pp. 74–80.

Henkel, D. J. (1961), Slide movements on an inclined clay layer in the Avon Gorge in Bristol, *Proc. 5th Int. Conf. on Soil Mechanics and Foundation Engineering*, **2**, pp. 619–624.

Henkel, D. J. (1967), Local geology and the stability of natural slopes, *Journal of the Soil Mechanics and Foundations Division, ASCE*, **93**, No. SM4, pp. 437–446.

Henkel, D. J. and Skempton, A. W. (1954), A landslide at Jackfield, *Proc. European Conf. on Stability of Earth Slopes*, **1**, pp. 90–101.

Hill, R. A. (1934), Clay stratum dried out to prevent landslips, *Civil Engineering*, No. 4, pp. 403–407.

Hirschfeld, R. C., Whitman, R. V., and Wolfskill, L. A. (1965), Engineering properties of nuclear craters: Review and analysis of available information on slopes excavated in weak shales, *Technical Report No. 3-699*, U.S. Army Corps of Engineers, Waterways Experiment Station, Vicksburg, Miss.

Ho, D. Y. F. and Fredlund, D. G. (1982), Increase in strength due to suction for two Hong Kong soils, *Proc. ASCE Spec. Conf. on Engineering Construction in Tropical and Residual Soils*, Honolulu, pp. 263–295.

Hoek, E. and Bray, J. W. (1974), *Rock Slope Engineering*, Institute of Mining and Metallurgy.

Hollingsworth, R. and Kovaćs, G. S. (1981), Soil slumps and debris flows: prediction and protection, *Bulletin of the Association of Engineering Geologists*, **18**, No. 1, pp. 17–28.

Hollingworth, S. E. and Taylor, J. H. (1944), Large-scale superficial structures in the Northampton Ironstone Field, *Quarterly Journal of the Geological Society of London*, **100**, pp. 1–44.

Holm, O. S. (1961), Stabilization of a quick clay slope by vertical sand drains, *Proc. 5th Int. Conf. on Soil Mechanics and Foundation Engineering*, **2**, pp. 625–627.

Holmsen, P. (1953), Landslips in Norwegian quick clays, *Geotechnique*, **2**, No. 5, pp. 187–194.

Hultin, S. (1916), Grusfyliningar för Kajbyggnader, Bidrag till frågen on deras stabilitet, *Tekn Tidskr V. U.*, **46**, H31, pp. 292–294.

Hutchinson, J. N. (1961), A landslide on a thin layer of quick clay at Furre, central Norway, *Geotechnique*, **11**, No. 2, pp. 69–94.

Hutchinson, J. N. (1967), The free degradation of London clay cliffs, *Proc. Geotechnical Conf.*, Oslo, **1**, pp. 113–118.

Hutchinson, J. N. (1969), A reconsideration of the coastal landslides at Folkstone Warren, Kent, *Geotechnique*, **19**, No. 1, pp. 6–38.

Hutchinson, J. N. (1970), Coastal mudflow on London clay cliffs at Bettinge, *Geotechnique*, **20**, No. 4, pp. 412–438.

Hutchinson, J. N. (1977), Assessment of the effectiveness of corrective measures in relation to geological conditions and types of slope movement, *Bulletin of the International Association of Engineering Geologists*, **16**, pp. 131–155.

Hutchinson, J. N. and Highes, M. J. (1968), The application of micropaleontology to the location of a deep-seated slip surface in the London clay, *Geotechnique*, **18**, No. 4, pp. 508–510.

Hvorslev, M. (1951), *Time Lag and Soil Permeability in Groundwater Observations*, Bulletin No. 36, U.S. Army Corps of Engineers, Waterways Experiment Station, Vicksburg, Miss.

Ikeya, H. (1976), *Introduction to Sabo Works—The Preservation of Land Against Sediment Disaster*, The Japan Sabo Association, Tokyo.

Ireland, H. O. (1954), Stability analysis of the Congress Street open cut in Chicago, *Geotechnique*, **4**, No. 4, pp. 163–168.

Jaeger, C. (1969), The stability of partly immerged fissured rock masses, and the Vajont rock slide, *Civil Engineering and Public Works Review*, **64**, No. 761, pp. 1204–1207.

Jakobson, B. (1952), The landslide at Surte on the Gota River, September 29, 1950, *Proceedings of the Royal Swedish Geotechnical Institute*, No. 5.

Janbu, N. (1973), *Slope Stability Computations. The Embankment Dam Engineering, Casagrande Volume*, John Wiley & Sons, Inc., New York, N.Y., pp. 47–86.

Janbu, N. (1977), State-of-the-art report: Slopes and excavations. *Proc. 8th Int. Conf. on Soil Mechanics and Foundation Engineering*, Tokyo, **2**, pp. 549–566.

Johnson, A. M. (1970), *Physical Processes in Geology*, Freeman, Cooper and Co., San Francisco.

Jones, F. O., Embody, D. R., and Peterson, W. L. (1961), *Landslides Along the Columbia River Valley, Northeastern Washington*, U.S. Geological Survey Professional Paper 367.

Jörstad, F. A. (1970), *Leirskred i Norge*, Norwegian Geotechnical Institute, Oslo, Publication No. 83, pp. 1–6.

Kallstenius, T. and Bergau, W. (1961), In situ determination of horizontal ground movements, *Proc. 5th Int. Conf. on Soil Mechanics and Foundation Engineering*, **1**, pp. 481–485.

Kankare, E. (1968), Skredet vid Kimola flottningskanal i södra Finland, *Väg-o Vattenb*, No. 8, Stockholm, pp. 154–161.

Karlsrud, K., Aas, G., and Gregersea, O. (1984), Can we predict landslides hazards in soft sensitive clays?, *Proc. 5th Int. Symp. on Landslides*, Toronto, I, pp. 107–130.

Kenney, T. C., Pazin, M., and Choi, W. S. (1977), Design of horizontal drains for soil slopes, *Journal of the Geotechnical Engineering Division, ASCE*, **103**, GT11, Proc. Paper 13366, pp. 1311–1323.

Kenney, T. C. and Uddin, S. (1974), Critical period for stability of an excavated slope in clay soil, *Canadian Geotechnical Journal*, **11**, No. 4, pp. 620–623.

Kiersch, G. A. (1964), Vaiont Reservoir disaster, *Civil Engineering, ASCE*, **34**, No. 3, pp. 32–39.

Knight, D. K. (1963), Oahe Dam: Geology embankment and cut slopes, *Journal of the Soil Mechanics and Foundations Division, ASCE*, **89**, No. SM2, pp. 99–125.

Koerner, R. M., McCabe, W. M., and Lord, A. E. Jr. (1981), Acoustic emission behavior and monitoring of soils, *Proc. Symp. ASTM*, Detroit, ASTM Special Technical Publication 750, pp. 93–141.

Kovari, K. (1974), Measures de deformation avec le distancemetre ISETH, *CR Journees d'etude, Les procedes modernes de constr des tunnels*, Nice, pp. 206–209.

Krohn, J. P. and Slosson, J. E. (1976), Landslide potential in the United States, *California Geology*, **29**, No. 10, pp. 224–231.

Ladd, C. C. and Foott, R. (1974), New design procedure for stability of soft clays, *Journal of the Geotechnical Engineering Division, ASCE*, **100**, GT7, pp. 763–786.

Lane, K. S. (1948), Treatment of frost sloughing slopes, *Proc. 2nd Int. Conf. on Soil Mechanics and Foundation Engineering*, 3, pp. 281–283.

LaRochelle, P., Trak, B., Tavenas, F., and Roy, M. (1974), Failure of a test embankment on a sensitive Champlain clay deposit, *Canadian Geotechnical Journal*, **11**, No. 1, pp. 142–164.

LaRochelle, P., Lefebvre, G., and Bilodeau, P. M. (1977), Stabilization of a slide in Saint-Jerome, Lac Saint-Jean, *Canadian Geotechnical Journal*, **14**, No. 3, pp. 340–356.

Larsson, R. (1980), Undrained shear strength in stability calculation of embankments and foundations on soft clays, *Canadian Geotechnical Journal*, **17**, No. 4, pp. 591–602.

Lauffer, H., Neuhauser, E., and Schober, W. (1967), Uplift responsible for slope movement during the filling of the Gepatsch Reservoir, *Trans. 9th Int. Cong. on Large Dams*, 1, pp. 669–693.

Lauffer, H. and Schober, W. (1964), The Gepatsch Rockfill Dam in the Kauner Valley, *Trans. 8th Int. Cong. on Large Dams*, Edinburgh, 3, pp. 635–660.

Lebuis, J. and Rissman, P. (1979), Earthflows in the Quebec and Shawington areas, Geological Association of Canada Congress, Quebec, Brochure for field trip B-11.

Lefebvre, G. and LaRochelle, P. (1974), The analysis of two slope failures in cemented Champlain clays, *Canadian Geotechnical Journal*, **11**, No. 1, pp. 89–108.

Legget, R. F. (1962), *Geology and Engineering*, 2nd ed., McGraw-Hill Book Co., Inc., New York, N.Y.

Legget, R. F. and Bartley, M. V. (1953), An engineering study of glacial deposits at Steep Rock Lake, Ontario, Canada, *Economic Geology*, **48**, pp. 513–540.

Lizzi, F. (1977), Practical engrg in structurally complex formations (The "in-situ reinforced earth"), *Proc. Int. Symp. on the Geotechnics of Structurally Complex Formations*, Capri, pp. 327–333.

Lizzi, F. and Carnevale, G. (1979), Les reseaux de pieux racines pour la consolidation des sols. Aspects theoriques et essais sur modiles, *Proc. ENPC-LCPC CR Coll. Int. Reinforcement des Sols*, Paris, pp. 317–324.

Lo, K. Y. (1970), The operational strength of fissured clays, *Geotechnique*, **20**, pp. 57–74.

Long, E. and George, W. (1967a), Buttress design earthquake-induced slides, *Journal of the Soil Mechanics and Foundations Division, ASCE*, **93**, No. SM4, pp. 595–609.

Long, E. and George, W. (1967b), Turnagain slide stabilization, Anchorage, Alaska, *Journal of the Soil Mechanics and Foundations Division, ASCE*, **93**, No. SM4, pp. 611–627.

Lumb, D. (1962), The properties of decomposed granite, *Geotechnique*, **12**, No. 3, pp. 226–243.

Marsal, R. J. (1979), Stability investigations related to clay shales, *Int. Symp. on Soil Mechanics*, Laxaca, **1**, pp. 51–74.

Marsland, A. and Butler, M. E. (1967), Strength measurements on stiff fissured Barton Clay from Fawley, Hampshire, *Proc. Geotech. Conf.*, Oslo, **1**, pp. 139–145.

Massarsch, K. R. and Broms, B. B. (1981), Pile driving in clay slopes, *Proc. 10th Int. Conf. Soil Mechanics and Foundation Engineering*, Stockholm, 3, pp. 469–474.

McCauley, M. L. (1976), Microsonic detection of landslides, *Transportation Research Record* No. 581, pp. 25–30.

Mearns, R., Carney, R., and Forsyth, R. A. (1973), *Evaluation of Ion Exchange Landslide Correction Technique*, Materials and Research Department, California Division of Highways, Report CA-HY-MR-2116-1-72-39.

Merriam, R. (1960), Portuguese Bend landslide, Palos Verdes Hills, California, *Journal of Geology*, **68**, pp. 140–153.

Mesri, G. (1975), Discussion on new design procedure for stability of soft clays, *Journal of the Geotechnical Engineering Division, ASCE*, **101**, No. GT4, pp. 409–412.

Mesri, G. and Capeda-Diaz, A. F. (1986), Residual shear strength of clays and shales, *Geotechnique*, **36**, No. 2, pp. 269–274.

Mitchell, J. K. (1976), *Fundamentals of Soil Behaviour*, John Wiley and Sons, Inc., New York, N.Y.

Mitchell, R. J. and Williams, D. R. (1981), Induced failure of an instrumented clay slope, *Proc. 10th Int. Conf. on Soil Mechanics and Foundation Engineering*, Stockholm, 3, pp. 479–484.

Moos, A. (1953), The subsoil of Switzerland, *Proc. 3rd Int. Conf. on Soil Mechanics and Foundation Engineering*, 3, pp. 252–264.

Morgenstern, N. R. and de Matos, M. (1975), Stability of slopes in residual soils, *Proc. 5th Panam. Conf. on Soil Mechanics and Foundation Engineering*, Buenos Aires, 3, pp. 369–384.

Morgenstern, N. R. and Price, V. E. (1965), The analysis of stability of general slip surfaces, *Geotechnique*, **15**, No. 1, pp. 79–93.

Morgenstern, N. R. and Tchalenko, J. S. (1967), Microstructural observations on shear zones from slips in natural clays, *Proc. Geotechnical Conf.*, Oslo, **1**, pp. 147–152.

Muir Wood, A. M. (1955), Folkstone Warren landslips: Investigations, 1948–1950, *Proceedings of the Institution of Civil Engineers*, **4**, Part 2, pp. 410–428.

Müller, L. (1964), The rock slide in the Vaiont Valley, *Felsmechanik u Ing. geol.*, **2**, pp. 148–212.

Müller, L. (1968), New considerations on the Vaiont slide, *Felsmechanik u Ing. geol.*, **6**, No. 1–2, pp. 1–91.

Murray, R. P. (1980), In-place roadway foundation stabilization, *Transportation Research Record* No. 749, pp. 1–6.

Nakamura, H. (1984), A landslide. Its movement mechanism and control works, *Proc. 4th Int. Symp. on Landslides*, Toronto, **II**, pp. 155–160.

Nasmith, H. (1964), Landslides and Pleistocene deposits in the Meikle River valley of northern Alberta, *Canadian Geotechnical Journal*, **1**, No. 3, pp. 155–156.

Nethero, M. F. (1982), Slide control by drilled pier walls, *Proc. of Application of Walls to Landslide Control Problems*, Las Vegas, ed. R. B. Reeves, pp. 61–76.

Nonveiller, E. (1981), Efficiency of horizontal drains on slope stability, *Proc. 10th Int. Conf. on Soil Mechanics and Foundation Engineering*, Stockholm, 3, pp. 495–500.

Novosad, S., Blaha, P., and Kneijzlik, J. (1977), Geoacoustic methods in the slope stability investigation, *Proc. Symp. Landslides and Other Mass Movements*, Prague, Bull. Int. Assoc. Eng. Geol. No. 16, pp. 229–231.

Odenstad (1951), The landslide at Skottorp on the Lidan River, February 2, 1946, *Proceedings of the Royal Swedish Geotech. Institute*, No. 4.

Onodera, T., Yoshinaka, R., and Kazama, H. (1974), Slope failure caused by heavy rainfall in Japan, *Proc. 2nd Int. Cong. 2nd IAEG*, Sao Paulo, Brazil, **2**, pp. 11.1–11.10.

Palladino, D. J. and Peck, R. B. (1972), Slope failures in an over-consolidated clay, Seattle, Washington, *Geotechnique*, **22**, No. 4, pp. 563–595.

Pásek, J. (1967), Schollenartige Hangbewegungen, *Mitt. Ges. Geol. Bergbaustud*, **18**, pp. 367–378.

Pásek, J. and Demek, J. (1969), Mass movements near the community of Stadice in northwestern Bohemia, *Czechoslovak Academy of Sciences, Inst. of Geography, Brno, Studia Geographica*, **3**.

Peck, R. B. (1967), Stability of natural slopes, *Journal of the Soil Mechanics and Foundations Division, ASCE*, **93**, No. SM4, pp. 403–417.

Peck, R. B. and Kaun, M. V. (1948), Description of a flow slide in loose sand, *Proc. 2nd Int. Conf. on Soil Mechanics and Foundation Engineering*, **2**, pp. 31–33.

Penman, A. D. M. (1961), A study of the response time of various types of piezometer, *Proc. Conf. on Pore Pressure and Suction in Soils*, London, pp. 53–58.

Peterson, R., Jasper, J. L., Rivard, P. J., and Iverson, N. L. (1960), Limitations of laboratory shear strength in evaluating stability of highly plastic clays, *Proc. Res. Conf. Shear Strength of Cohesive Soils*, Boulder, Colorado, pp. 765–791.

Petterson, K. E. (1916), Kajraset i Göteborg den, 5 mars, 1916, *Tekn Tidskr, V U* **46**, H30, pp. 281–287; **46**, H31, pp. 289–291.

Philbrick, S. S. and Cleaves, A. B. (1958), Field and laboratory investigations, *Landslides and Engineering Practice*, ed. E. B. Eckel, Highway Research Board, Special Report 29, pp. 93–111.

Pilot, G. (1981), Gissements de terrain liés directement à des travaux, *Revue Francaise Geotechnique*, No. 17, pp. 55–70.

Pilot, G. (1984), Instrumentation and warning system for research and complex slope stability problems, *Proc. 4th Int. Symp. on Landslides*, Toronto, **I**, pp. 275–306.

Piteau, D. R. and Clayton, R. (1977), Discussion of Paper "Computerized design of Rock Slopes using interactive graphics for the input and output of geometrical date", by A. Cundall, M. D. Voegde, and C. Fairhurst, *Proc. 16th Symp. on Rock Mechanics*, Minnesota, ASCE.

Piteau, D. R., Mylrea, F. H., and Blown, I. G. (1978), Downie slide, Columbia River, British Columbia, Canada, *Rockslides and Avalanches, 1. Natural Phenomena*, ed. B. Voight, Elsevier, Amsterdam, pp. 365–390.

Plafker, G., Eriksen, G. E., and Ferandez Concha, J. (1971), Geological aspects of the May 31, 1970, Peru earthquake, *Bull. Seismol. Soc. Am.*, **61**, No. 3, pp. 543–578.

Pope, R. J. and Anderson, M. W. (1960), The strength properties of clays derived from volcanic rocks, *Proc. Res. Conf. on Shear Strength of Cohesive Soils*, Boulder, Colorado, pp. 315–340.

Radbruch-Hall, D. (1978), Gravitational creep of rock masses on slopes, *Rockslides and Avalanches, 1. Natural Phenomea*, ed. B. Voight, Elsevier, Amsterdam, pp. 607–657.

Radbruch-Hall, D. and Varnes, D. J. (1976), Landslides—cause and effect, *Bulletin of the International Association of Engineering Geologists*, **14**, pp. 205–216.

Rapp, A. (1960), Recent development of mountain slopes in Karkevagge and surroundings, Northern Scandinavia, *Geografiska Annafer*, **XL.II**, pp. 73–200, Uppsala.

Reeves, R. B. (1982), Application of walls to landslide control problem, *ASCE, National Convention*, Las Vegas.

Reinius, E. (1948), *On the Stability of the Upstream Slope of Earth Dams*, Swedish State Committee for Building Research Bulletin No. 12.

Rib, H. T. and Liang, T. (1978), Recognition and identification, *Landslides: Analysis and Control*, National Academy of Sciences, Washington, D.C., Special Report No. 176, pp. 34–80.

Ringheim, A. S. (1964), Experiences with bearpaw shales at the South Saskatchewan River Dam, *Trans. 8th Int. Cong. on Large Dams*, **1**, pp. 529–550.

Rochet, L. (1983), Le distancemetre orientable a fil d'invar DO1, *Notice Laboratoire Regional des Ponts et Chaussees de Lyon*.

Rodriguez Ortiz, J. M., Hernandez Del Pozo, J. C., and Castanedo, F. J. (1984), Slope stabilization in weathered schists and shales by excavation, piles and drainage, *Proc. 4th Int. Symp. on Landslides*, Toronto, **II**, pp. 173–178.

Saito, M. (1965), Forecasting the time of occurrence of a slope failure, *Proc. 6th Int. Conf. on Soil Mechanics and Foundation Engineering*, **2**, pp. 537–541.

Saito, M. (1969), Forecasting time of slope failure by tertiary creep, *Proc. 7th Int. Conf. on Soil Mechanics and Foundation Engineering*, **2**, pp. 677–683.

Saito, M. and Uezawa, H. (1961), Failure of soil due to creep, *Proc. 5th Int. Conf. on Soil Mechanics and Foundation Engineering*, **1**, pp. 315–318.

Saito, M. and Uezawa, H. (1969), Experiments on slope failure and its prevention by drainage for sandy embankments under artificial rainfall, *Quarterly Reports, Railway Technical Research Institute, Japanese National Railways*, **10**, No. 3, pp. 142–148.

Sassa, K., Takei, A., and Kobashi, S. (1980), Landslides triggered by vertical subsidences, *Proc. Int. Symp. on Landslides*, New Delhi, **1**, pp. 49–54.

Schlosser, F. and Juran, I. (1979), Parameters de calcul des sols artificiellement ameliores: Rapport general Seance 8, *Proc. 8th European Conf. on Soil Mechanics and Foundation Engineering*, **8**, pp. 1–29.

Schuster, R. L. (1978), Introduction, *Landslides: Analysis and Control*, National Academy of Sciences, Washington, D.C., Special Report No. 176, pp. 1–10.

Schuster, R. L. and Fleming, R. L. (1982), Geologic aspects of landslide control using walls, *Proc. Application of Walls to Landslide Control Problems*, Las Vegas, ed. R. B. Reeves, pp. 1–18.

Seed, H. B. (1968), Landslides during earthquakes due to soil liquefaction, *Journal of the Soil Mechanics and Foundations Division, ASCE*, **94**, No. SM5, pp. 1055–1122.

Seed, H. B. and Wilson, S. D. (1967), The Turnagain Heights landslide, Anchorage, Alaska, *Journal of the Soil Mechanics and Foundations Division, ASCE*, **93**, No. SM4, pp. 325–353.

Sevaldson, R. A. (1956), The slide in Lodalen, October 6, 1954, *Geotechnique*, **6**, No. 4, pp. 167–182.

Shannon, W. L., Wilson, S. D., and Meese, R. H. (1962), *Field Problems: Field Measurements, Foundation Engineering*, ed. G. A. Leonards, McGraw-Hill Book Co., Inc., New York, N.Y., pp. 1025–1080.

Sherwood, D. E. and Currey, B. (1973), Experience in using electrical tiltmeters, *Proc. Field Inst. in Geotechn. Eng.*, London, pp. 382–395.

Shields, D. H. and Harrington, E. J. (1981), Measurements of slope movements with a simple camera, *Proc. 10th Int. Conf. on Soil Mechanics and Foundation Engineering*, Stockholm, **3**, pp. 521–525.

Sidle, R. C., Pearce, A. J., and O'Loughlin, C. L. (1985), *Hillslope Stability and Land Use*, American Geophysical Union, Water Resources Monograph No. 11.

Sinclair, S. R., Brooker, E. W., and Thomson, S. (1966), Stability of clay shale slopes, *ASCE Conf. on Slope Stability*, Berkeley, California, Discussion to Session 4.

Skaven-Haug, S. (1955), *Undervannsskred i Trondheim havneområde*, Norwegian Geotechnical Institute, Oslo, Publication No. 7, pp. 1–12.

Skempton, A. W. (1945), A slip in the west bank of the Eau Brink Cut, *Journal of the Institute of Civil Engineers*, **24**, No. 7, pp. 267–287.

Skempton, A. W. (1953), Soil mechanics in relation to geology, *Proceedings of the Yorkshire Geological Society*, **29**, Part 1, No. 3, pp. 33–62.

Skempton, A. W. (1964), Long-term stability of clay slopes, *Geotechnique*, **14**, No. 2, pp. 77–101.

Skempton, A. W. (1977), Slope stability of cuttings in brown London clay, *Proc. 9th Int. Conf. on Soil Mechanics and Foundation Engineering*, Tokyo, **3**, pp. 26–270.

Skempton, A. W. and Brown, J. D. (1961), A landslide in boulder clay at Selset, Yorkshire, *Geotechnique*, **11**, No. 4, pp. 280–293.

Skempton, A. W. and Delory, F. A. (1957), Stability of natural slopes in London clay, *Proc. 4th Int. Conf. on Soil Mechanics and Foundation Engineering*, **2**, pp. 378–381.

Skempton, A. W. and Hutchinson, J. (1969), Stability of natural slopes and embankment founds, State-of-the-Art Volume, *Proc. 7th Int. Conf. on Soil Mechanics and Foundation Engineering*, Mexico City, pp. 291–340.

Skempton, A. W. and LaRochelle, P. (1965), The Broadwell slip: A short-term failure in London clay, *Geotechnique*, **15**, No. 3, pp. 221–242.

Smith, T. and Forsyth, R. (1971), Potrero hill slide and correction, *Journal of the Soil Mechanics and Foundations Division, ASCE*, **97**, No. SM3, pp. 541–564.

Söderblom, R. (1958), Saltsonden och dess användning vid bestämning av skredbottnen vid Göta, *Geol Fören. Stockholm Förh.*, **80**, H.1, pp. 87–96.

Söderblom, R. (1969), Salt in Swedish clays and its importance for quick clay formation, Results from some field and laboratory studies, *Proceedings of the Swedish Geotech. Institute*, No. 22.

Söderblom, R. (1973), Nagra nya synpunkter på begreppet kvicklera och kvicklerans betydelse för spridandet av skred, *Nordisk Geo-Technikermpte i Trondheim 24–26 August, 1972. Foredrag*, Norsk Geoteknisk Forening/Norges Geotekniske Institutt, Oslo, pp. 87–94.

Sowers, G. F. and Royster, D. L. (1978), Field investigations, *Landslides: Analysis and Control*, National Academy of Sciences, Washington, D.C., Special Report 176, pp. 81–111.

St. John, B. J., Sowers, G. F., and Weaver, C. E. (1969), Slickensides in residual soils and their engineering significance, *Proc. 7th Int. Conf. on Soil Mechanics and Foundation Engineering*, Mexico City, **2**, pp. 591–597.

Suemine, A. (1983), Observational study on landslide mechanism in the area of crystalline schist (Part 1), *Bull. Dis. Prev. Res. Inst. Kyoto*, **33**, Part 3, pp. 105–128.

Suklje, L. and Vidmar, S. (1961), A landslide due to long-term creep, *Proc. 5th Int. Conf. on Soil Mechanics and Foundation Engineering*, **2**, pp. 727–735.

Sweeney, D. J. (1982), Some in-situ soil suction measurements in Hong Kong's residual soil slopes, *Proc. 7th S.E. Asian Geotechnical Conf.*, Hong Kong, pp. 91–106.

Sweeney, D. J. and Robertson, P. K. (1979), A fundamental approach to slope stability problems in Hong Kong, *Hong Kong Engineer*, **10**, No. 7, pp. 35–44.

Tan, S. B., Tan, S. L., Yang, K. S., and Chin, Y. K. (1985), Soil improvement methods in Singapore, *3rd Int. Geotechnical Seminar*, Nanyang Tech. Inst., Singapore, pp. 249–272.

Tavenas, F. and Leroueil, S. (1981), Creep and failure of slopes in clays, *Canadian Geotechnical Journal*, **18**, pp. 106–120.

Tavenas, F., Trak, B., and Leroueil, S. (1980), Remarks on the validity of stability analyses, *Canadian Geotechnical Journal*, **17**, No. 1, pp. 61–73.

Ter-Stepanian, G. (1965), In-situ determination of the rheological characteristics of soil on slopes, *Proc. 6th Int. Conf. on Soil Mechanics and Foundation Engineering*, **2**, pp. 575–577.

Ter-Stepanian, G. (1966), Types of depth creep of slopes in rock masses, *Proc. 1st Congr. Int. Soc. of Rock Mech.*, **2**, pp. 157–160.

Ter-Stepanian, G. (1974), Depth creep of slopes, *Bulletin of the International Society of Engineering Geologists*, No. 9, pp. 97–102.

Terzaghi, K. (1936), Stability of slopes of natural clay, *Proc. Int. Conf. on Soil Mechanics and Foundation Engineering*, **1**, pp. 161–165.

Terzaghi, K. (1950), Mechanisms of landslides, *Geological Society of America, Engineering Geology (Berkey) Volume*, pp. 83–123.

Terzaghi, K. (1956), Varieties of submarine slope failures, *Proc. 8th Texas Conf. on Soil Mechanics and Foundation Engineering*, Austin, Texas.

Terzaghi, K. (1962), Stability of steep slopes on hard unweathered rock, *Geotechnique*, **12**, No. 4, pp. 251–270.

Terzaghi, K. and Peck, R. B. (1967), *Soil Mechanics in Engineering Practice*, 2nd ed., John Wiley & Sons, Inc., New York, N.Y.

Toms, A. H. (1953), Recent research into coastal landslides at Folkstone Warren, Kent, England, *Proc. 3rd Int. Conf. on Soil Mechanics and Foundation Engineering*, **2**, pp. 288–293.

Trak, B., LaRochelle, P., Tavenas, F., Leroueil, S., and Roy, M. (1980), A new approach to the stability analysis of embankments on sensitive clays, *Canadian Geotechnical Journal*, **17**, No. 4, pp. 526–544.

Vallejo, L. E. (1981), Stability analysis of mudflows on natural slopes, *Proc. 10th Conf. on Soil Mechanics and Foundation Engineering*, Stockholm, **3**, pp. 541–544.

Vargas, M. and Pichler, E. (1957), Residual soil and rock slides in Santos, Brazil, *Proc. 4th Int. Conf. on Soil Mechanics and Foundation Engineering*, **2**, pp. 394–398.

Varnes, D. J. (1958), Landslide types and processes, *Highway Research Board Special Report* No. 29, Landslides and Engineering Practice, pp. 20–47.

Varnes, D. J. (1978), Slope movement types and processes, *Landslides: Analysis and Control*, National Academy of Sciences, Washington, D.C., Special Report 176, pp. 11–33.

Veder, C. (1981), *Landslides and Their Stabilization*, Springer-Verlag, New York, N.Y.

Wahlstrom, E. E. and Nichols, T. C. (1969), The morphology and chronology of a landslide near Dillon Dam, Dillon, Colorado, *Engineering Geology*, **3**, No. 2, pp. 149–174.

Wang, Gong-xian and Xu, Bang-dong (1984), Brief introduction of landslides in loess in China, *Proc. 4th Int. Symp. on Landslides*, Toronto, **I**, pp. 197–208.

Ward, W. H. (1948), A coastal landslip, *Proc. 2nd Int. Conf. on Soil Mechanics and Foundation Engineering*, **2**, pp. 33–38.

Ward, W. H., Marsland, A., and Samuels, S. G. (1965), Properties of the London Clay at the Ashford Common shaft in-situ and undrained strength tests, *Geotechnique*, **15**, pp. 321–344.

Warlam, A. A. and Thomas, E. W. (1965), Measurement of hydrostatic uplift pressure on spillway weir with air piezometers, *Instruments and Apparatus for Soil and Rock Mechanics, ASTM STP 392*, pp. 143–151.

Watari, M. et al. (1977), Mechanism of neocene mudstone landslide-Sarukuyogi landslide, *Proc. 8th Int. Conf. on Soil Mechanics and Foundation Engineering*, Tokyo, **1**, pp. 963–1012.

Weatherby, D. E. and Nicholson, D. J. (1982), Tiebacks used for landslide stabilization, *Proc. Application of Walls to Landslide Control Problems*, Las Vegas, ed. R. B. Reeves, pp. 44–60.

Weeks, A. G. (1969), Effects of counterfort drains on the Seven-oaks Bypass, *Civil Engineering and Public Works Review*, **64**, No. 759, pp. 991–995.

Williams, D. R., Romeril, P. M., and Mitchell, R. J. (1979), Riverbank erosion and recession in the Ottawa area, *Canadian Geotechnical Journal*, **16**, No. 4, pp. 641–650.

Wilson, S. D. (1962), The use of slope measuring devices to determine movements in earth masses, *Field Testing of Soils, ASTM STP 322*, pp. 187–197.

Wilson, S. D. (1967), Investigation of embankment performance, *Journal of the Soil Mechanics and Foundations Division, ASCE*, **93**, No. SM4, pp. 135–156.

Wilson, S. D. and Mikkelsen, P. E. (1978), Field instrumentation, *Landslides: Analysis and Control*, National Academy of Sciences, Washington, D.C., Special Report 176, pp. 112–138.

Wilson, S. D. and Squier, R. (1969), Earth and rockfill dams, *Proc. 7th Int. Conf. on Soil Mechanics and Foundation Engineering*, Mexico, State-of-the-Art Volume, pp. 137–223.

Wissa, A. E. Z., Martin, R. T., and Garlanger, J. E. (1975), The piezometer probe, *ASCE Specialty Conf. on In-situ Measurement of Soil Properties*, North Carolina State University, Raleigh, **1**, pp. 536–545.

Wu, T. H. (1976), *Investigation of Landslides on Prince of Wales Island, Alaska*, Ohio State University, Geotechnical Engineering Report No. 5.

Yen, B. C. (1969), Stability of slopes undergoing creep deformation, *Journal of the Soil Mechanics and Foundations Division, ASCE*, **95**, No. SM4, pp. 1075–1096.

Zaruba, Q. and Mencl, V. (1969), *Landslides and their Control*, Elsevier, Amsterdam.

Zaruba, Q. and Mencl, V. (1982), *Landslides and their Control*, Elsevier, Amsterdam.

12 RETAINING STRUCTURES AND EXCAVATIONS*

THOMAS D. DISMUKE, M.S.C.E., P.E.
Consulting Engineer
Bethlehem Steel Corporation

12.1 INTRODUCTION

Structures that retain lateral forces from soil and/or water are composed of various materials, constructed by numerous procedures and methods, and are used for many purposes. In this chapter the manner of support is used to classify the different types of structures. Reinforced earth structures are not included, as that type of structure is covered in Chapter 21. The three means of support consist of restrained, gravity, and cantilever systems. Each support system has variations—the restrained system having the greatest number and cantilever the least. Gravity structures are far more numerous than the other two since they are used for the construction of many small structures such as soil-retaining walls. The largest retaining structures are of the gravity type. Restrained structures are prominent in the construction of waterfront facilities and excavations. Cantilever structures are the least used because they allow large deflections to occur. Combinations of support types are sometimes used.

Medium and large temporary retaining structures are usually constructed of steel elements because they can be withdrawn and reused a number of times. In many cases wood is used for constructing small temporary structures.

Geostructures and particularly retaining structures that are in place many years are often subjected to severe environments that cause material deterioration. As a result, material durability and/or protection system considerations weigh heavily in the selection of materials and type of structure to be constructed. Chapter 26 addresses the durability of materials in some detail.

The most important structures in each support system classification will be emphasized in this chapter as there is limited space. Some examples of computations are included. Gravity structures such as concrete ("cantilever" and counterfort) retaining walls will not be covered in this chapter. The first edition of this Handbook contains coverage of the subject, as do numerous concrete-design books.

12.2 RESTRAINED RETAINING STRUCTURES

The various uses of these structures includes waterfront facilities, excavation retention structures and single-wall cofferdams. The lateral restraints used include deadmen, walls utilizing passive pressure, batter piles, and driven or drilled-in (pressure-grouted or belled) anchors. The anchors are located behind the retaining

element and are connected to it by tension ties (bars, rods, cable, etc.) or are directly attached. Compression members such as horizontal braces, rakers, circumferential rings, etc., are used to support the retaining element in front (on the unloaded side) of the retaining wall element.

12.2.1 Waterfront Structures

A distinction should be made between the types of anchorages according to yielding capability. This distinction is important because rigid supports tend to attract greater load than does a more flexible or slightly yielding support. The use of tension restraints usually insures that some slight yielding can occur without structural distress and that compression (braced) restraints do not allow yielding.

A. ANCHOR TYPE RESTRAINING SYSTEMS

Anchored structures are usually constructed of steel sheet piling walls; however, aluminum, concrete, and wood sheet piling are used. Small retaining structures (maximum height 10 ft) are generally constructed of wood or aluminum piling. The design procedures are, for the most part, similar for all materials. The stiffness of concrete sheet piling affects the design.

The use of anchored sheet piling structures and the various types of structures are briefly described herein.

The most commonly encountered case is that in which there is only a very small differential water level on both sides of the piling (Fig. 12.1a). The uses include marine and harbor applications, as in wharves, bulkheads, quays, piers, docks, jetties, breakwaters, slipways, river and canal walls, waterfront retaining walls, boat basins, flood walls, levees, and reclamation walls. Permanent sheet piling is also used for coast and river protection works, and for various forms of retaining walls on land, for buildings, and for highways. While in many of the applications a single line of piling is employed, in others such as piers or breakwaters, two parallel lines of sheet piling are anchored to each other and form in effect a double-wall structure.

This less usual but important condition covers cases in which there is a high differential water level across the two sides of the sheet piling, which occurs, for example, where the piling is used to form the walls of a drydock (Fig. 12.1b), a lock, basements of buildings, or other underground structures such as pumping chambers. The sheet piling serves both as temporary support during construction and as the permanent walls of the structure. While the usual range of anchorage methods, referred

*G. M. Cornfield (deceased) authored most of Section 12.2.1A and Section 12.4.

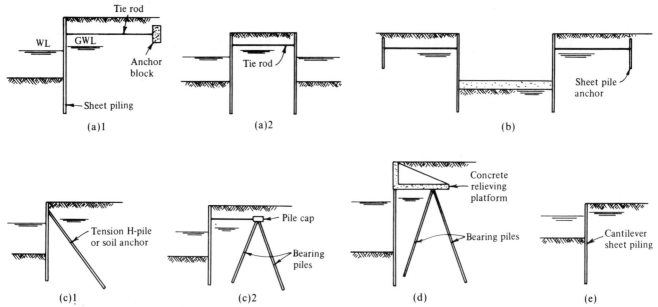

Fig. 12.1 Types of permanent sheet-pile structures. (a) 1, Anchored retaining wall; (a) 2, jetty. (b) Sheet-pile walls dry dock. (c) 1, Raking anchor; (c) 2, A-frame anchor. (d) Relieving-platform wall. (e) Cantilever retaining wall.

to below, are applicable to structures such as drydock walls, in the case of building basements and the like it is sometimes practical to use the ground floor or lower floors to strut the walls apart. For a circular underground chamber, ring walings in compression are the ideal form of support. In all structures in this category it is important to take into account the effect of uplift water pressure on the base.

Anchored walls in the average range of heights, say from 15 ft up to 40 ft or so depending on the soil and site conditions, form the greatest proportion of the permanent use of steel sheet piling. For this reason walls in this particular category will be given most attention in subsequent sections of this chapter dealing with design and details.

Most retaining walls in this category are supported by a simple system of horizontal steel walings along the piling with steel tie rods to transfer the waling loads to isolated anchorages or deadmen. These may be plain concrete blocks, vertical reinforced concrete slabs, or a group of 2 to 6 short sheet piles. In some cases it is worth considering long sheet piles as anchors with the portion below the tie rods being of greater length, which act as cantilever anchorages. In other cases anchorages are constructed as continuous vertical walls of concrete or sheet piling.

The tied rods may be raked steeply—say, at 45° to the horizontal—and connected to soil anchors. Alternatively the anchorage may take the form of steel H tension piles raking back from the piling at about 45° and connected directly to the walings (Fig. 12.1, c.1).

Horizontal tie rods may also be connected to A-frame anchorages formed of forward- and backward-raking bearing and tension piles with or without a substantial block of concrete at the heads of these piles (Fig. 12.1, c.2).

Relieving-platform walls are applicable chiefly to high retaining walls. With the availability of high-modulus sections there may not be the same necessity to resort to a relieving platform as there was in the past when the strongest available section of piling may have been inadequate for a particularly high wall. However, both alternatives should be examined to determine which produces the lowest-cost wall in given circumstances.

A typical relieving-platform wall is illustrated in Fig. 12.1d. It will be seen that the soil pressure on the wall, and consequently the bending movement on the sheet piling, will be reduced by the provision of the reinforced concrete platform behind the piling some distance below ground level. In general, the platform, which is supported on a system of suitable raking piles, also serves as an anchorage for the sheet piling. Where a conventional anchored retaining wall can be constructed without special difficulty, a relieving platform wall would be more costly. Walls of this kind are therefore used where the construction of a conventional wall is not possible or would involve special problems. There are thus several circumstances in which relieving platforms should be considered: (1) where the existence of unsuitable soil prevents the use of conventional anchorages; (2) where there is insufficient space for ordinary tie rods and anchorages; (3) where there are large superloads or loads from crane or rail tracks, requiring bearing piles to support them.

Cantilever-type retaining walls (Fig. 12.1e) are probably least commonly used. In the first place they are only economic for walls having a height no greater than about 10 to 14 ft. Apart from this restriction, cantilever walls are more sensitive than anchored walls to variations in soil conditions and are also subject to larger horizontal deflections. They are therefore more appropriate for temporary structures.

Additional information on cantilever wall systems is given in Section 12.4.

1. Principles of Design of Anchored Walls

Conventional design assumptions and procedures for anchored sheet-pile walls may be divided into two broad groups, depending on the relative amount of penetration of the piling below the final dredge line. The first type is referred to as *free-earth support* (Fig. 12.2a) and the second *fixed-earth support* (Fig. 12.2b).

In the free-earth support type of wall, which has less penetration than the second type, the piling is assumed to be acting as a vertical beam spanning two supports, these being

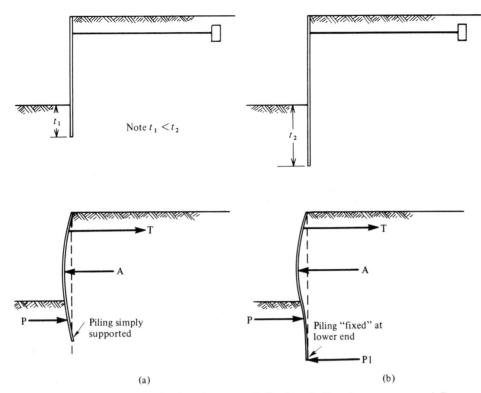

Fig. 12.2 Comparison of (a) free-earth and (b) fixed-earth support. Deflection of piling shown exaggerated. T = support from anchorage; A = loading from soil pressure behind wall; P = support from soil in front of wall; P_1 = support from soil at bottom of piling.

the anchorage system and the soil in front of the piling below the dredge line.

In the case of fixed-earth support the piling has greater penetration below the dredge line and as a result it is assumed that the piling is fixed in direction at the bottom. The wall is then effectively equivalent to a vertical propped beam fixed at the lower end.

(a) Soil Pressure, Bending Moment, and Deflection

Conventional diagrams for soil pressure, bending moment, and deflection are shown in Figure 12.3a for fixed-earth support conditions. The net soil pressure diagram is the difference between the total active and total passive diagrams (dashed line). The quantity T represents the anchorage resistance, A is the total net active pressure force acting on the wall, and P is the total net passive force available below dredge line, while P_1 is a reverse passive force at the bottom of the piling required for obtaining "fixity." The corresponding bending moment diagram shows that the bending moment is a maximum (positive) somewhere between tie rod level and dredge line, with a reverse maximum moment (negative) some distance below dredge line. A point of zero bending moment occurs a short distance below the dredge line. The deflected form of the piling shows maximum deflection occurring close to the position of maximum positive moment and zero slope at the bottom of the piling corresponding to "fixity." Note the position of the point of contraflexure CP corresponding to the position of zero bending moment.

The equivalent pressure, bending moment, and deflection diagrams for free-earth support are given in Figure 12.3b. Here the only forces are T, A, and P, as referred to above, while the bending moment diagram indicates a maximum positive moment only. This moment is significantly greater than that for fixed-earth support for a wall of the same height in the same

soil conditions. The deflection diagram in this case has no point of contraflexure.

(b) Fixed-Earth Support

While a wall in given site and soil conditions can generally be designed either for free- or for fixed-earth support, experience has shown that overall economy results from fixed-earth design. Though longer piles are needed, the required section modulus or bending strength of the piles is less, and the anchor loads tend to be lower. Fixed-earth design automatically provides sufficient penetration to give an adequate factor of safety against outward movement of the piles below dredge line. However, in the case of free-earth support, an extra penetration, over and above that just needed for stability, is required to provide a factor of safety against outward movement of the bottom of the piles.

Accordingly, the later section on detail design will deal essentially with fixed-earth support only. Free-earth support design might, however, be applicable in cases such as (a) when clays exist below dredge line or (b) when medium rock exists at, or near, dredge line so that the piles cannot be driven far enough to provide fixity.

(c) Modes of Failure of Anchored Sheet-Pile Walls

The following is a summary of the main types of failure that can occur, and which proper methods of design and the use of adequate factors of safety are intended to prevent.

Failure of the Anchorage System Wall failure can occur owing to the inadequacy of the bolts attaching the walings to the piling, the walings, the tie rods, and their end fixings, or the anchorages themselves. Failure can also occur if the components of the anchorage system are adequate individually but if the

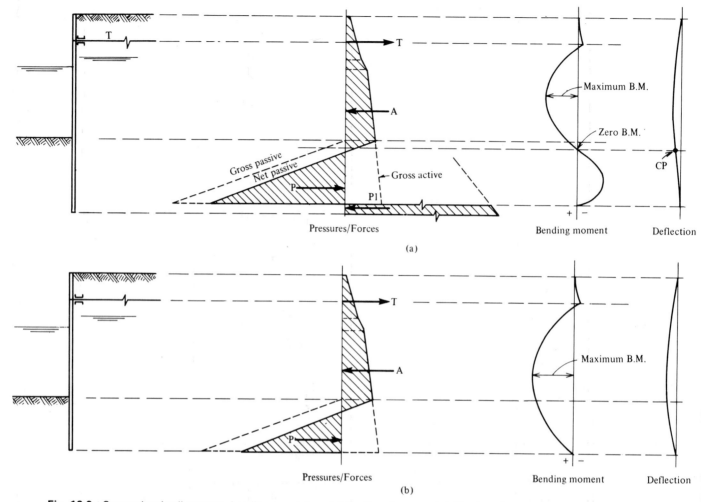

Fig. 12.3 Conventional soil pressure, bending moment, and deflection diagrams. (a) Fixed-earth support; (b) free-earth support.

anchorage is located too close to the sheet-pile wall; that is, if the tie rods are not long enough. Another possibility is anchorage failure due to excessive unintentional surcharge loading *behind* the anchorages. Also to be guarded against is failure of the anchorage system where a horizontal stratum of soft clay exists below the anchors (see Fig. 12.4a).

Movement of the Toes of the Piling Outward at and Below Dredge Level Failure of this type may occur if there is inadequate penetration of the piling. It can also occur if there is unintentional reduction of the penetration, for example, by scour action of ships' propellers or scour due to the action of water currents or waves, or by overdredging (Fig. 12.4b).

Failure of the Sheet Piling in Bending This type of failure may arise if an inadequate section of piling has been used, if the soil pressures have been incorrectly estimated, if unsuitable filling material has been used, if unintentionally large surcharge loading occurs behind the wall, or if the dredge line is unintentionally lowered by scour or overdredging (Fig. 12.4c).

Overall Circular Slip Failure Failure of the wall system as a whole can occur owing to circular slip in soft clay (Fig. 12.4d).

Settlement of Filling Excessive settlement of the filling behind the sheet piling may also be regarded as failure. Such settlement may drag down the tie rods and cause excessive stresses in

them. The settlement may be caused either by consolidation of the filling itself or by settlement of an existing stratum of soft clay due to the extra weight of the filling (Fig. 12.4e).

It is interesting to note that of the failures which have occurred from time to time, most are due to failure of part of the anchorage system or by inadequate penetration below the nominal dredge line. Very few cases of failure have been due solely to the piling failing in bending.

(d) Economics of Alternative Designs

The principal objective of design is to select and detail a safe and adequate structure having a minimum cost. Since the cost of the sheet piling forms the largest single item in a sheet-pile wall, it is worth examining the factors that affect the section of piling used. Apart from driving considerations, which will be referred to later, the strength of the piling will be governed by soil pressures, differential water pressures, surcharge loading, and the effective span of the piling between tie rods and point of contraflexure. Many alternative possibilities can therefore arise.

Another aspect that may have to be considered is the case in which the site and soil conditions require the use of a section of piling heavier than any available. This is another reason for studying means of reducing pressures on the piling. Thus, where soft clay exists below dredge line, it has sometimes been found

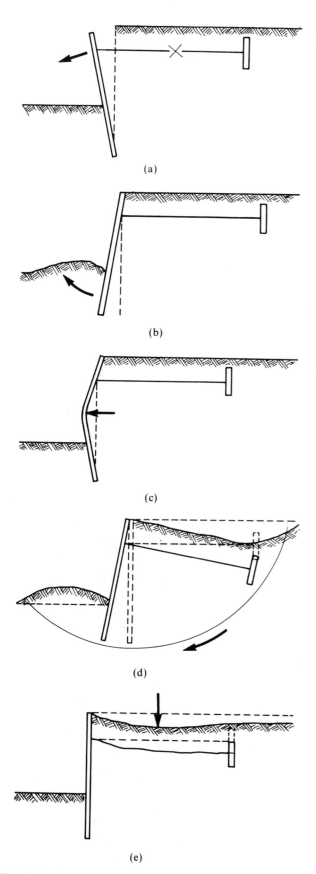

Fig. 12.4 Modes of failure of anchored sheet-pile walls. (a) Failure of anchorage system. (b) Bottom of piles move outward. (c) Failure of piling in bending. (d) Circular slip in soft clay. (e) Settlement behind wall.

economical to dredge out all soft soil beforehand and to replace it with sand. If only clay is cheaply available for filling and reclamation behind a sheet-pile wall, economics dictate that suitable cohesionless filling *must* nevertheless be used immediately behind the wall, while the clay fill can be used for the main part of the reclamation. The advantages of relieving-platform walls in some circumstances have already been referred to. When a great depth of soft clay exists at a site, it may be impossible to construct economically a sheet-pile wall, and the correct solution will then be a concrete deck supported on vertical bearing piles with a soil slope extending under the deck from dredge line to the back of the deck.

2. Soil and Water Pressures

Before proceeding to the actual detail design of a sheet-pile wall, it is necessary to derive a diagram of the pressures and resistances acting on the wall. These will be determined essentially by the height of wall, the soil properties, water levels, and surcharge loadings. Other forces which may have to be considered where applicable are those due to wave action, mooring loads, impact of vessels, and earthquakes.

In Figure 12.5 we present typical information required to design a sheet-pile wall. All the points referred to will influence the final design, and therefore the preliminary investigation of all factors needs to be considered with some care.

The number and siting of boreholes for the soil investigation will depend on the size and character of the project. As a general guide, boreholes might be sunk at 100 to 200-ft centers on the proposed line of the wall, or a short distance in front of it, together with further boreholes immediately in front of the anticipated line of the anchorages. The boreholes should be taken well below the anticipated depth of penetration of the sheet piling. Soil samples should be taken at all changes of strata and at intervals of 5 to 10 ft of depth in each borehole. In the case of cohesionless soils such as sands and gravels, the essential information for design will be derived from standard penetration tests carried out during boring and from mechanical analyses of representative samples. For cohesive soils, that is, clays and silts, undisturbed soil samples should be subjected to unconfined compression tests or preferably to undrained quick triaxial tests, and for particularly soft clay and silts, in-situ vane tests should also be carried out. The unit weight of undisturbed soil samples should be determined.

The properties of a soil that are significant in relation to sheet-pile wall design are its unit weight in its natural (γ) and submerged (γ') states, the angle of internal friction ϕ for a cohesionless soil, and the undrained shear strength or cohesion of a clay or silt (c).

(a) Cohesionless Soils

Typical values of γ, γ', and ϕ for cohesionless soils are given in Table 12.1 but actual values for final design should be determined on the basis of tests. Any filling used behind the piling should be clean, hard, and free-draining, such as sand, gravel, broken stone, or rock fill. Clays and silts should not be used as a filling material as these will produce high lateral pressures.

Active Pressure For *cohesionless soils* where the wall is vertical, and for a horizontal ground surface behind the wall, the horizontal component of intensity of active earth pressure p_a at a depth h is given by

$$p_a = K_a \gamma h$$

where γh is the total intensity of effective vertical pressure at depth h. Thus, the appropriate value of γ should be used for

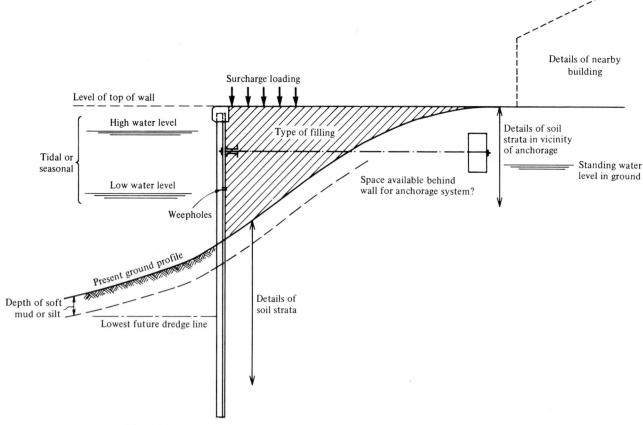

Fig. 12.5 Typical information required for the design of a sheet-pile retaining wall.

both drained and submerged unit weights. Values of K_a for different values of ϕ and δ, the angle of wall friction between the soil and the piling, are given in Table 12.2.

For calculating active pressure, δ may be taken as one-half the value of ϕ. However, δ is sometimes neglected, that is, it is taken as zero when there is some doubt about the actual soil properties at the site; the error is on the safe side.

Silty sands containing more than 5 percent passing a 200-mesh sieve should be the subject of special tests to determine K_a.

Passive Pressure For a vertical wall with horizontal cohesionless soil in front of it, the horizontal component of intensity of passive earth pressure p_p at a depth h below the soil level in front of the wall is given by

$$p_p = K_p \gamma h$$

where γh is the total intensity of effective vertical pressure at the depth h, taking into account both drained and submerged unit weights as may be applicable. Values of K_p are given in Table 12.3.

In the calculation of passive pressure it is usual to take $\delta = \frac{2}{3}\phi$ for anchored sheet-pile walls. However, there may not be sufficient downward resistance to mobilize such a value of wall friction in the case of anchorages unless they have sufficient dead weight, and it is therefore usual to take $\delta = 0$ for the design of anchorages. The possibility of a similar situation occurring in a cantilever retaining wall should also be considered.

The value of K_p for silty sands containing more than 5 percent passing a 200-mesh sieve should be determined on the basis of special soil tests.

TABLE 12.1 COHESIONLESS SOILS—TYPICAL VALUES.

Sands and Gravels	Natural Unit Weight γ, lb/ft³	Submerged Unit Weight γ', lb/ft³	Angle of Internal Friction, ϕ
Loose	90 to 125	55 to 65	30°
Medium	110 to 130	60 to 70	35°
Dense	110 to 140	65 to 80	40°

TABLE 12.2 COHESIONLESS SOILS— ACTIVE PRESSURE COEFFICIENT K_a.

	$\phi = 30°$	$\phi = 35°$	$\phi = 40°$
$\delta = 0°$	0.33	0.27	0.22
$\delta = 10°$	0.31	0.25	0.20
$\delta = 20°$	0.28	0.23	0.19

TABLE 12.3 COHESIONLESS SOILS— PASSIVE PRESSURE COEFFICIENT K_p.

	$\phi = 30°$	$\phi = 35°$	$\phi = 40°$
$\delta = 0°$	3.0	3.7	4.6
$\delta = 10°$	4.0	4.8	6.5
$\delta = 20°$	4.9	6.0	8.8

TABLE 12.4 COHESIVE SOILS—TYPICAL VALUES.

Consistency	Cohesion, c, lb/ft²	Saturated Unit Weight γ, lb/ft³	Submerged Unit Weight γ', lb/ft³
Very stiff	Over 3000	120 to 140	60 to 80
Stiff	1500 to 3000	115 to 135	55 to 75
Firm	750 to 1500	105 to 125	45 to 65
Soft	375 to 750	90 to 110	30 to 50
Very soft	Under 375	90 to 100	30 to 40

(b) Cohesive Soils

Typical values for cohesive soils, such as saturated clays and silts, are given in Table 12.4, but the properties to be used in design should be based on tests of actual undisturbed samples.

Active Pressure For vertical walls with horizontal ground surfaces, the horizontal intensity of active earth pressure p_a at a depth h is given by

$$p_a = \gamma h - 2c$$

Passive Pressure Again for vertical walls with horizontal ground surfaces, the horizontal intensity of passive earth pressure p_p at a depth h is given by

$$p_p = \gamma h + 2c$$

In both the above expressions for active and passive pressure in cohesive soils, it has been assumed that the value of wall adhesion, corresponding to wall friction in cohesionless soils, is zero, which is normal practice in the design of sheet-pile walls.

In the case of the expression for active pressure, it will be observed that when the value of γh is only slightly greater than $2c$, the active pressure p_a becomes very small. As we are then concerned with a small difference between two relatively large quantities γh and $2c$, it will be seen that p_a in these circumstances is very sensitive to the values of both γ and c. In such a case accurate values of γ and c should be obtained from soil tests together with the range of scatter of their values.

When in the case of active pressure, the value of p_a is apparently negative, it is common practice to assume an arbitrary minimum positive value of active pressure equal to about $\frac{1}{3}\gamma h$.

It should be noted that this section refers to in-situ cohesive soils, and not clays or silts, which may be redeposited by man. In any case, clays or silts should not be used as a *filling* behind or in front of sheet-pile walls.

The normal methods of calculation of active and passive pressures in cohesive soils as given above are based on the immediate shear strength determined in undrained triaxial or unconfined compression tests, and for such saturated cohesive soils $\phi = 0$. With the passage of time and the subsequent alteration of pore pressures, it is sometimes considered that changes in the design properties will occur as determined by drained triaxial tests, which reveal a value of c tending towards zero and a value of ϕ' that may for some clays be in the region of 15° to 25°. However, many sheet-pile walls have been built on the usual $\phi = 0$ basis by using the conventional procedures to be described in the next section, and it is thought that it would be unduly conservative to design on the long-term ϕ' basis.

(c) Pressures Due to Water

Cohesionless Soils In cohesionless soils, full water pressure $\gamma_w h$ will act on the piling below water level or groundwater level.

As has already been mentioned, the soil pressure $K_a\gamma'h$ in the same region will be based on the submerged (or buoyant) unit weight of the soil γ'.

Cohesive Soils In cohesive soils the same total pressure will result either if the water pressure is taken separately together with soil pressure based on γ', or if the water is not assumed "free" and only soil pressure acts, being based on γ. This is clear from the expressions for active and passive pressure; thus, for active pressure

$$p_a = \gamma h - 2c$$

but as

$$\gamma = \gamma_w + \gamma'$$

hence

$$p_a = \gamma_w h + (\gamma'h - 2c)$$

Differential Water Levels Where different water levels exist on either side of the piling, as for a quay wall in tidal waters, and where the piling penetrates to a relatively impermeable clay stratum, the net water pressure is simply the difference between the water pressures on both sides, corresponding to the shaded part of the diagram in Figure 12.6.

Where the piles do not penetrate to an impermeable stratum, seepage flow will take place beneath the toes of the piles, and the net water pressure diagram will be different from that shown in Figure 12.6. Additionally, the seepage flow will slightly increase the effective unit weight of soil on the active side and thus increase active pressures, and will decrease the effective unit weight of soil on the passive side, possibly giving rise to a significant decrease in passive pressure. However, in the case of quay walls of the type that form the majority of sheet-pile retaining walls, the water level differential is unlikely to exceed about 3 or 4 ft on a falling tide, and for these conditions it is customary to adopt the net water pressure diagram given in Figure 12.6. Where the differential is likely to be significantly greater—for example, owing to rapid rise in the groundwater table after rainstorms or where the piling forms the wall in a dry dock—it is vital that seepage effects be taken into account (Terzaghi, 1954).

To restrict the differential water wall it is customary to provide anchored sheet-pile walls with a drainage system, which may simply take the form of weepholes in the piling with graded filters behind. High water differentials give rise to high pressures as well as seepage effects where applicable, and can thereby increase the wall cost significantly.

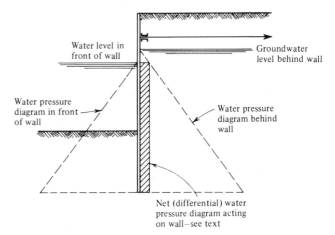

Fig. 12.6 Differential water pressure.

(d) Active Pressure from Surcharge Loading

Pressure Due to Uniformly Distributed Surcharge (q) The active pressure due to a uniform surcharge q is equal to $K_a q$ at any depth. It is usual to regard the surcharge as equivalent to an increase in the wall height, or in the level of ground behind the wall of q/γ ft as this gives the same result.

Pressure Due to Line or Point Surcharge Loads The effect of these can often be dealt with by estimating an equivalent uniform surcharge loading assuming the line or point loads to spread at about 60° to the vertical. For a more precise treatment, reference should be made to Terzaghi (1954). For heavy line and point loads such as those due to wharf cranes, it will normally be found more economic to support these independently on vertical bearing piles and they will consequently not produce any lateral pressure on the wall.

(e) Deriving the Net Pressure Diagram for a Sheet-Pile Wall

The following procedure for obtaining the net pressure diagram is recommended, as it is simple to carry out and avoids the need for much written computation. The procedure is a tabular one and is easy to check.

The method can best be described by reference to an actual example as given in Figure 12.7. The diagram includes the whole of the calculations necessary to produce the final net pressure diagram shown, the following remarks being given only in explanation. The basis of the procedure is the separate tabulation of values of (1) γh on the active side, (2) γh on the passive side, (3) differential water pressures, (4) active soil pressures, and (5) passive soil pressures, as shown in the columns numbered similarly.

In column (1) total γh values on the active side of the wall are set down cumulatively as measured from the top of the wall, γh being the total vertical effective soil pressure at the depth h. Thus, starting with the surcharge 220 lb/ft² the γh value at the top of wall is then 220, all units being in lb/ft². Proceeding to the next lower level, that is, the groundwater level, which is 10 ft lower, simply add $\gamma h = 110 \times 10 = 1100$ to the first figure of 220 to give a total of 1320 at this 10-ft level. Continue adding successive values of γh using γ or γ' as is appropriate. In column (2) the procedure is repeated for γh on the passive side, in this case commencing at the dredge line. Water pressures are dealt with in column (3) using only the net differential water pressure. Thus, at every successive elevation below low-water level the differential water pressure remains constant.

In column (4) p_a values were calculated by multiplying γh, column (1), for each level by the appropriate value of K_a for the particular stratum at that level. Passive pressures are dealt with in column (5) in a similar manner, by multiplying γh, column (2), by appropriate K_p values.

The active pressure diagram can then be plotted using column (4) figures, and differential water pressures from column (3) are added as shown. The passive pressure diagram is plotted to the left of the vertical datum line. Finally the net pressure diagram is obtained simply by "subtraction" as indicated by the area shown shaded.

Cohesive soils can be dealt with in the same way as described above for cohesionless soils. At the top and bottom of a cohesive stratum, γh of column (1) should have subtracted from it the value of $2c$ to obtain p_a of column (4). Passive cohesive soil pressure will be obtained by adding $2c$ to γh of column (3) to set down p_p in column (5).

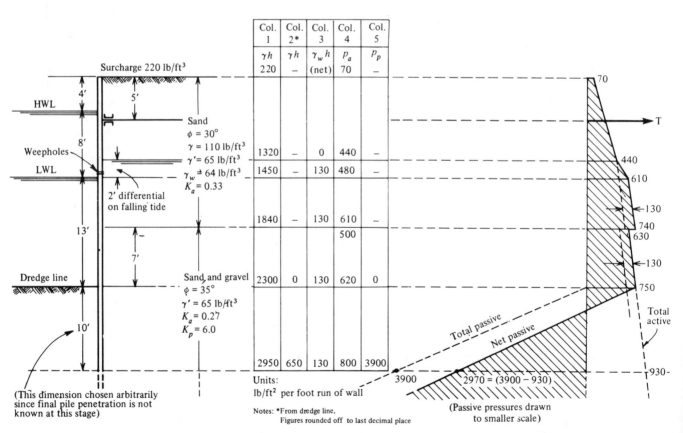

Fig. 12.7 Derivation of pressure diagram.

3. Conventional Design Methods

(a) Anchored Retaining Walls in Cohesionless Soils

Reference will be made to three different methods for the conventional design of sheet-pile walls in *cohesionless* soils. The deflection line procedure, the simpler equivalent beam method, and the nomogram method all refer to the case of fixed-earth support only.

Deflection Line Method The conventional fixed-earth support diagram for soil pressure, bending moment and deflection were shown in Figure 12.3a. After the net soil pressure diagram has been obtained as already described, the bending moment and the deflection diagrams are determined by methods of graphic statics or analytically. A detailed description of the procedure is given in Terzaghi (1943, p. 224).

It is not proposed to describe the graphical deflection line procedure as it is laborious and not often used. The method was, however, the basis of the much-simplified equivalent-beam procedure.

Equivalent-Beam Method This method can best be explained by referring to an actual example, and for this purpose the retaining wall example and pressure diagram already given in Figure 12.7 are reproduced in Figure 12.8 with additional symbols noted thereon.

The design is carried out in the following manner.

1. Determine the distance x, which depends on the value of ϕ below the dredge line, thus:

   when $\phi = 30°$ $x = 0.08H'$
   $\phi = 35°$ $x = 0.03H'$
   $\phi = 40°$ $x = 0$

2. Determine the span L of the equivalent beam, that is, the distance from tie rod level to point of contraflexure in the bending moment diagram:

$$L = H_T + x$$

3. Calculate the total load W on span L, which is the shaded area shown in the pressure diagram.
4. Calculate the maximum bending moment M given by the expression $M = WL/8$. While the derivation of M by this expression is obviously approximate, the error is usually negligible. If desired, a precise value for M can be determined by graphical or analytical methods.
5. Calculate the shear force B at the point of contraflexure by taking moments of the shaded area about the tie rod level. If the unshaded portion of the pressure diagram above tie rod level is neglected, the small resulting error is on the safe side.
6. Calculate the actual total penetration of the piling t below the dredge line from the expression

$$t = x + 1.2\sqrt{\frac{6B}{(K_p - K_a)\gamma'}}$$

where K_p, K_a, and γ' are the applicable values below the dredge line.

7. Calculate T, the anchor pull or tie rod load per unit length of wall, either accurately by taking moments about B or approximately by deducting the value of B from the load W, and adding the load represented by the unshaded part of the pressure diagram above tie rod level.

It will be seen that this procedure results in values for total pile length, maximum bending moment and anchor pull, from which pile section and anchorage design can be carried out.

For a description of the derivation of the equivalent-beam method see Terzaghi (1943, p. 225).

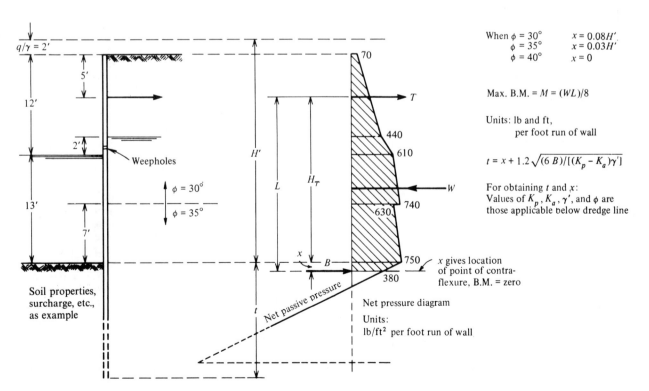

Fig. 12.8 Equivalent-beam method.

Limitations of the equivalent-beam method. The method is applicable for cohesionless soils only and for a small water differential. It is not applicable for an anchored or braced structure or excavation where a significant differential water level occurs, and the method should not be applied where the superload is very large compared with the wall height, nor if the tie rods are located at an exceptionally low level.

Example of Design by the Equivalent-Beam Method The basic data for this example and the pressure diagram are given in Figures 12.7 and 12.8. All the calculations refer to unit length, that is one foot run of wall. The parts of the calculation given correspond to the sections of the description above.

$$H = 25.0 \text{ ft}$$
$$H' = 27.0 \text{ ft}$$
$$H_T = 20.0 \text{ ft}$$
$$\phi = 35° \text{ below dredge line}$$
$$K_p = 6.0 \text{ below dredge line}$$
$$K_a = 0.27 \text{ below dredge line}$$
$$\gamma' = 65 \text{ lb/ft}^3 \text{ below dredge line}$$

1. Since $\phi = 35°$: $x = 0.03 \times 27 = 1.0$ ft (approx.).
2. $L = H_T + x = 20.0 + 1.0 = 21.0$ ft.
3. W is determined by summing the areas of the trapeziums forming the shaded area in Figure 12.8. Thus

$$W = \left(\frac{260 + 440}{2}\right)5 + \left(\frac{440 + 610}{2}\right)2 + \left(\frac{610 + 740}{2}\right)6$$
$$+ \left(\frac{630 + 750}{2}\right)7 + \left(\frac{750 + 380}{2}\right)1$$
$$= 12\,250 \text{ lb per foot run of wall.}$$

4. $M = WL/8$
   $$= 12\,250 \times 21.0/8$$
   $$= 32\,200 \text{ ft-lb per foot run of wall.}$$

5. Taking moments about tie rod level.

   $B \times 21.0$ = sum of moments of area of trapeziums of shaded pressure diagram
   $$= 144\,700 \text{ ft-lb per foot run of wall.}$$
   Hence $B = 6890$ lb per foot run of wall.

6. $t = 1.0 + 1.2\sqrt{\dfrac{6 \times 6890}{(6.0 - 0.27)65}} = 1.0 + 12.6 = 13.6$ ft

   The total pile length is then $25.0 + 13.6 = 38.6$ ft (say, 39 ft).

7. $T = 12\,250 - 6890 + \left(\dfrac{70 + 260}{2}\right)5$
   $$= 6190 \text{ lb per foot run of wall.}$$

Use of Nomogram for Design The nomogram (Cornfield, 1969) shown is intended for the direct design of anchored sheet-pile retaining walls in cohesionless soils. It applies to soil having $\phi = 30°$ (Fig. 12.9) throughout.

The applicable symbols are given in the cross section sketches:

H (ft) = height of wall to dredge level including the extra height equivalent to surcharge loading
q (lb/ft²) = surcharge loading behind the wall
h_e (ft) = extra height equivalent to surcharge loading
$\qquad = q/\gamma = q/110$ ft
γ (lb/ft³) = unit weight of cohesionless soil above water level = 110 lb/ft³
γ' (lb/ft³) = unit weight of submerged cohesionless soil = 65 lb/ft³
γ_w (lb/ft³) = unit weight of water = 64 lb/ft³
H' (ft) = actual height of wall to dredge level
$\qquad = H - h_e$

D (ft) = depth to tie rod level, measured from a level equivalent to top of wall plus h_e
S (ft) = depth to water level in front of wall, measured from a level equivalent to top of wall plus h_e
$H/12$ (ft) = assumed differential water level, front/back of wall, on a falling tide
t (ft) = penetration of piling below dredge level
T (lb/ft) = tie rod load per ft run of wall
M (lb/ft) = maximum bending moment in piling per ft run of wall

The method of use of the nomogram is best illustrated by considering a typical example of a retaining wall design and by following the arrows on the nomogram for $\phi = 30°$ (Fig. 12.9), referring to the following example:
Assume the given data for the retaining wall is:

$H' = 28$ ft
$q = 224$ lb/ft²
$S = 12$ ft at low water level (including h_e)
$D = 6$ ft (including h_e)
$\phi = 30°$
Hence $\quad h_e = 224/110 = 2$ ft
Hence $\quad H = H' + h_e = 28 + 2 = 30$ ft
Hence $S/H = 12/30 = 0.4$
Hence $D/H = 6/30 = 0.2$

Commence at the bottom left of the nomogram with this value of S/H and follow the arrowed lines in order to determine the penetration (t), the maximum bending moment (M), and the tie rod load (T), thus:

1. The penetration of piling (t): the value of t/H is read off as 0.64. Hence $t = 0.64 \times H = 0.64 \times 30 = 19$ ft. The pile length is then $19 + H' = 19 + 28 = 47$ ft.
2. The maximum bending moment (M) is read off directly as 57×1000 ft-lb = 57 000 ft-lb per foot run of wall.
3. The tie rod load (T) is read off directly as 8.3×1000 lb = 8300 lb per foot run of wall.

The nomogram is based on the equivalent-beam method already described and is applicable only to soils that are cohesionless and homogeneous throughout the whole depth of the piling. It is therefore not applicable for cases in which clay strata exist, nor may it be used for the design of cofferdams or strutted excavations, where the differential water level is usually substantial. Weep holes must in any case be provided at lowest water level in anchored retaining walls as shown. The nomogram should not be used for cases in which the surcharge loading is very large or when the design involves tie rods at an unusually great depth, for example when D/H is greater than 0.3.

It may be necessary to use the nomogram twice in order to find the highest values of M, T, and t, which will vary as S varies. The maximum values of M and t usually occur when S/H is between 0.6 and 0.8, approximately. This will be obvious from the families of curves on the left of the nomogram. On the other hand, T is found to have a maximum value when S/H is about 0.4 to 0.5. The value of S/H corresponding to low-water level should be used first, as in the example above. If in this example the tidal variation has been large, so that S/H varies from, say, 0.2 to 0.75, then $S/H = 0.75$ would be used initially, but a check should be made with a lower value of S/H (i.e., a higher water level) to find the maximum value of T.

Selection of Sheet-Pile Section Having obtained the maximum bending moment as described, the next step is to divide this by the allowable working stress for the particular steel to be used in order to obtain the required section modulus of the sheet piling. The procedure is not, however, as straightforward as

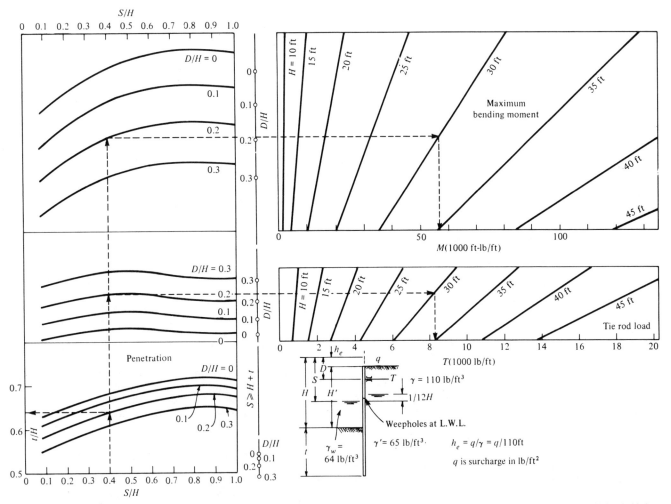

Fig. 12.9 Nomogram, $\phi = 30°$. Anchored steel sheet-pile retaining walls in cohesionless soil. Based on equivalent-beam method. Units: lb and ft, all per foot run of wall. (*After Cornfield, 1969. Acknowledgement is made to the Editor,* Civil Engineering and Public Works Review, *London, for permission to reproduce the nomogram.*)

might be assumed, because there is no generally accepted agreement on what the allowable working stress should be, and some designers also introduce reduction factors into the derived bending moment. This will be covered in greater detail below.

However, the following are some of the alternative procedures that are used to determine the section modulus from the bending moment as calculated by the conventional methods already described.

1. Use a working stress of from 0.6 to 0.7 of the minimum yield stress of steel, the higher value being used when the soil properties are known with confidence.
2. Reduce the bending moment by 25 percent provided the maximum wall deflection is *greater* than 0.5 percent of the wall height, and then use a steel stress equal to 0.6 of the yield.
3. Reduce the bending moment by 33 percent, excluding that part of the bending moment due to differential water pressure, and use a steel stress equal to 0.65 of the yield. No moment reduction is to be used if the wall is essentially a filled type without significant dredging in front.

The use of the above procedures applied to the example illustrated in Figure 12.8, where the maximum bending moment was 32 300 ft-lb per foot run of wall, results in section moduli ranging from 14.5 to 20.1 in³ per foot run of wall if the steel

to be used is ASTM-A 328 having a minimum yield stress of 39 000 lb/in².

Steel sheet piling is now available in steels having a higher yield stress than that referred to above, and it will often be found economic to use such higher grades, as the percentage increase in cost per ton is usually less than the percentage increase in yield stress.

Finally, the section of piling chosen should be checked to ensure that the piles can in fact be driven to the required penetration, which will depend on the soil conditions. Cases can arise in which it may be impossible to reach the necessary penetration if there is insufficient stiffness in the cross section of the piles to withstand the driving forces needed to get the piles down, even though the bending strength of the piling is adequate for the final condition. In such cases it will be necessary either to select the next heavier section of piling or to use a higher-quality steel purely to enable the piles to be driven to the designed penetration.

Tie Rod Loading It is common practice in conventional design to increase the calculated tie rod loading by approximately 20 percent in order to determine the working load for design of the whole anchorage system. One reason for doing this is that the extent of tie rod load increase caused by tightening of the end nuts cannot be determined precisely.

(b) Anchored Retaining Walls in Cohesive Soils

As the equivalent-beam method is not applicable for clay or silt soils it becomes necessary to calculate the maximum bending moment, tie rod load, and penetration from the basic net pressure diagram—for example, that shown in Figure 12.3a. This can be done by a graphical method such as the deflection line procedure already mentioned, or analytically.

For cohesionless soils the designer has a choice of deciding whether to design on free-earth support, though it is normally more economical to choose the fixed-earth support procedure. In the case of cohesive soils this choice is available only when the soil below dredge line is stiff to hard, depending on the wall height. Otherwise it will be found that the free-earth support method will become necessary, unless excessively long piles are used, thus leading to an uneconomical solution. Some engineers feel that, owing to the gradual creep of clay when it is stressed, fixed-earth support cannot be relied on for the whole life of the structure; however, over a period of many years, many walls have been thus designed and no failures have resulted.

When design is done on the free-earth support basis, the actual calculation from the net pressure diagram becomes simpler. Moments are taken about tie rod level and a factor of safety of 2 against failure by outward movement of the piles must be provided; thus, the penetration is determined by trial and error so that the total passive pressure moment about tie rod level is twice the total active pressure moment. The tie rod load and maximum bending moment can then be calculated. The selection of a piling section is done using a working stress from 0.6 to 0.7 times the yield stress of the steel, depending on the accuracy of information on the soil properties.

4. Relieving-Platform Retaining Walls

Reference has been made earlier to relieving-platform sheet-pile retaining walls, and the circumstances in which such walls would be used were given (Fig. 12.1d).

The reinforced-concrete platform is situated at a depth generally not less than one-fourth the wall height. It should be of sufficient width to provide total or almost total shielding of the sheet piling from the effect of the soil above the platform

level. To achieve this it must cover the wedge of potentially unstable soil that starts from a point some distance below dredge line.

The whole wall *structure* has to support the horizontal pressures due to the soil below the platform and the soil above it.

The platform may be supported wholly on bearing piles, or partly on bearing piles and partly on the steel sheet piling. Anchorage is provided by the inclusion of raking piles. Some of the bearing piles in the system may be subject to tension loads, and it is important in analyzing the bearing pile system to apply a load factor, usually 2.0, to the total horizontal thrust on the platform in order to determine the ultimate loads in the bearing piles and tension piles.

A soil pressure diagram is determined as already described and analyzed as for an anchored retaining wall, to determine the penetration of the piling below dredge line, the maximum bending in the piling, and the magnitude of the horizontal force on the platform due to the soil below it.

Two typical examples of relieving-platform walls are shown in Figure 12.10a and b. Wall (a) shows a platform supported partly on the sheet piling and on forward- and backward-raking bearing piles. The existing ground profile is at about the same level as the platform, and dredging is carried out after completion of the structure. Relieving-platform walls can readily be constructed in this condition and also when the existing ground is at a still higher level. Where the existing ground profile is at a low level, the platform would have to be constructed on shuttering and there would be difficulty in introducing the filling under it afterwards. In wall (b) only forward raking piles are used, together with vertical bearing piles; under conditions of horizontal loading these may be subject to tension loading.

5. Anchorages for Sheet-Pile Retaining Walls

The three main types of anchorages were described in Section 12.2.1 (Fig. 12.1), that is, anchorages that derive their stability from the passive resistance of the soil in front of them, steeply raking deep soil anchors or tension piles deriving their resistance from soil at a much lower level, and the A-frame type, again obtaining stability by transference of load to soil

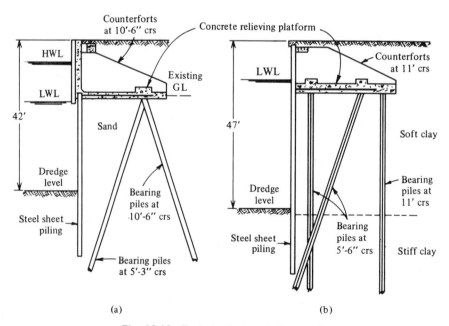

(a)   (b)

Fig. 12.10 Typical relieving-platform walls.

at a much greater depth. Relieving platforms also come within this category.

It is proposed to deal in detail with anchorages utilizing passive resistance since no special design problems arise with the remaining types.

(a) Anchorages Relying on Passive Soil Resistance

The three principal types of anchorage in this category are illustrated in Figure 12.11a, b, and c, namely (a) either mass concrete or reinforced concrete, (b) sheet piling with tie rods located centrally, and (c) cantilever sheet-pile anchors with tie rods located higher than the center of the piles. In each case the anchorages may be of isolated or separate concrete blocks or groups of piles, or alternatively they may be in the form of a continuous wall of concrete or sheet piling.

Because the values of passive soil pressure used in the design of anchorages are ultimate values, it is necessary to apply a load factor, usually not less than 2.0, to the working anchorage load. In an earlier section it was suggested that the working anchor load should be 20 percent greater than the value obtained by conventional design procedures, and this should then be multiplied by 2.0 to determine the required size of anchorage based on the passive and active soil pressure coefficients given in the section on soil pressures. The structural design of the anchorage, tie rod, and connections can be carried out on the working load using allowable working load stresses.

Concrete and sheet-pile anchorages of the type shown in Figure 12.11a and b are designed by equating the net (passive minus active) soil pressure to the ultimate anchor loading. For these anchorages, if the overall height is not less than about 0.6 times the depth from ground surface to bottom of anchorage, the anchorage behaves as if it extended right to ground surface.

Cantilever anchorages (Fig. 12.11c) are designed using the ultimate tie rod load by first determining the level and magnitude of the maximum bending moment in order to select a suitable piling section (based on yield stress), and then finding the level of zero bending moment to calculate the pile length after making a sufficient addition to the penetration to provide balance of all horizontal forces.

Analysis has shown that the loading on the soil from a series of separate anchorages, as distinct from a continuous one, spreads horizontally through the soil so that an isolated anchorage has an effective width of about 1.6 times its actual width; this applies provided that the centers of adjacent isolated anchorages are not less than 1.6 times their width. At this spacing, the sum of the resistance of a number of isolated anchorages is approximately equal to the resistance of a continuous wall of the same height.

The full angle of wall friction δ used for determining K_p can only be mobilized if the anchorage has sufficient dead weight or is otherwise restrained from moving upwards. As it is normally only possible to mobilize a part of δ it is often assumed that $\delta = 0$ for anchorage design, which is on the safe side.

The overall design should be based on the least-favorable combination of conditions applicable to the retaining wall structure as a whole. Thus, the anchor pull of a wharf wall may be smallest at high water, but the anchorage resistance may be a minimum for this condition if it is then partially below groundwater level—the anchor pull may be a maximum at low water, but the anchorage resistance may also be at its maximum value at this time. Similarly the design should consider the possibility of a very high surcharge load behind the anchorage with none in front of it; for example, anchorages should not be located in front of a loaded footing of a building unless the effect of this is taken into account.

The distance of the anchorage from the retaining wall can be critical and is dealt with later in this section.

(b) Example of Anchorage Design

The problem is to find the safe working load that can be applied to the continuous reinforced concrete anchorage illustrated in Figure 12.12 based on the following data:

Water level at ground surface.
Cohesionless soil throughout.

$$\phi = 35°$$
$$\gamma = 110 \text{ lb/ft}^3$$
$$\gamma' = 65 \text{ lb/ft}^3$$
$$\delta = 0$$
$$K_a = 0.27$$
$$K_p = 3.7$$

At 9 ft depth:

$$(\gamma'h) = 9.0 \times 65 = 585 \text{ lb/ft}^2$$
$$p_p = 3.7 \times 585 = 2160 \text{ lb/ft}^2 \ (ac)$$
$$p_a = 0.27 \times 585 = 160 \text{ lb/ft}^2 \ (cd)$$
$$\text{Net } (p_p - p_a) = 2000 \text{ lb/ft}^2 \ (bc)$$

$$\text{Total ultimate resistance} = (p_p - p_a) = \text{Area } (bce)$$
$$= 2000 \times 9/2 = 9000 \text{ lb/ft run}$$
$$\text{of anchorage}$$

For a load factor of 2.0, the allowable working load that can be applied to the anchorage is then 9000/2.0 or 4500 lb/ft run.

Note that if instead of being continuous the anchorage consisted of separate RC slabs 6 × 6 ft in size and spaced at not less than 1.6 × 6 ft (9.6 ft) center to center, the safe working resistance of each slab would be 9.6 × 4500 = 43 200 lb.

(c) Table of Resistances of Block Anchorages or Deadmen

Table 12.5 gives values of the safe resistance of isolated mass concrete block anchorages of the proportions indicated in the

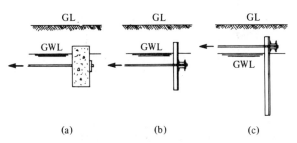

(a)　　　　(b)　　　　(c)

Fig. 12.11 Types of anchorages.

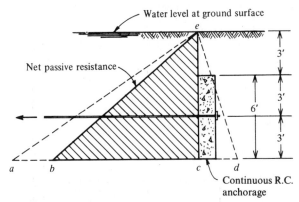

Fig. 12.12 Concrete anchorage.

TABLE 12.5 SAFE RESISTANCE OF ISOLATED SQUARE MASS CONCRETE ANCHORAGE BLOCKS IN UNITS OF 1000 lb.

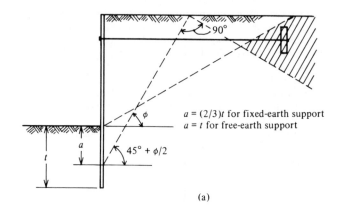

Size of Square Block, B		3 ft	4 ft	5 ft	6 ft	7 ft
Cohesionless	$S = 0$	5.6	13.4	24.6	42.5	66.2
soil	$S = 3$ ft	9.0	17.9	35.8	58.2	89.6
$\phi = 30°$	$S = 6$ ft	9.0	20.2	40.3	67.2	103.0
	$S = 9$ ft	9.0	20.2	40.3	71.7	112.0
Cohesionless	$S = 0$	6.7	15.7	31.4	53.7	85.2
soil	$S = 3$ ft	11.2	24.6	44.8	76.2	114.2
$\phi = 35°$	$S = 6$ ft	11.2	26.9	51.5	87.4	134.4
	$S = 9$ ft	11.2	26.9	51.5	91.8	143.4
Cohesive	lb/ft²					
soil	$c = 750$	11.2	20.2	31.4	47.0	64.9
	$c = 1000$	15.7	26.9	44.8	62.7	87.4
	$c = 1250$	20.2	35.8	56.0	82.8	112.0
	$c = 1500$	24.6	44.8	71.7	103.0	143.4

The values in this table only apply if the anchor blocks are at centers not less than 1.6B.
The values include a factor of safety of approximately 2.0.
Values of resistance are in units of 1000 lb.

figure attached to the table, and are based on the following:

> Uniform soil throughout
> Horizontal soil surface
> Surcharge not exceeding 250 lb/ft²
> Horizontal tie rods at vertical centers of square anchorages
> $\gamma = 110$ lb/ft³
> $\gamma' = 65$ lb/ft³
> $B =$ size of square block, feet
> Factor of safety = 2.0

For cohesionless soils the top of the anchorage is located at a depth $B/2$ below the ground surface. For undisturbed cohesive soils the values given are for the top of the block being located at a constant depth of 3 ft below the surface, and this top 3 ft of cohesive soil has been assumed to provide negligible resistance.

As an example, if $\phi = 35°$, $B = 5$ ft and $S = 5$ ft, the safe resistance is 44.8 units of 1000 lb, that is 44 800 lb. As a factor of safety of 2.0 has been used, the ultimate resistance is $2.0 \times 44\,800 = 89\,600$ lb.

A value of δ of about $\phi/3$ has been used for the tabular values and is applicable for the mass concrete blocks shown. However, if reinforced-concrete slabs or isolated groups of sheet piles are to be used, δ should be taken as zero as their reduced dead weight does not permit mobilization of much wall friction. The values of total resistance given will then be reduced by about 20 percent.

(d) Location of Anchorages

To utilize the full allowable resistance of anchorages designed as already described it is necessary that the tie rods be of a certain minimum length.

In cohesionless soils (Fig. 12.13a) the anchorages should therefore be located in the zone shown shaded. The principle here is to avoid interaction between the passive soil wedge of the anchorages and the active wedge of the retaining wall.

For cohesive soils, two criteria are suggested for the tie rod lengths. First, it is suggested that they should not be shorter than the length of the sheet piles forming the wall, and secondly, a check should be carried out as indicated in Figure 12.13b. A similar check should be done if there is a horizontal stratum of cohesive soil between layers of cohesionless soil.

If, for site reasons, it is not possible to locate the anchorages as recommended above, the allowable resistance of the anchorages is reduced as described by Terzaghi (1943). Alternatively it will then be necessary to use an A-frame anchorage or a relieving platform.

(e) Effect of Wall Friction on Types of Anchorages

Wall friction and its effect on anchor design has already been referred to for the types of anchorages dealt with earlier in this section, where it was assumed that the tie rods were horizontal.

If the tie rods slope so that they are significantly lower at the anchorage end, less wall friction can be mobilized. On the other hand, if steeply sloping tie rods are used with the anchorage end at the higher level there will be no problem about mobilization of wall friction at the anchorage, but a check must then be made regarding wall friction mobilization at the main piling wall on the passive side. This same situation arises

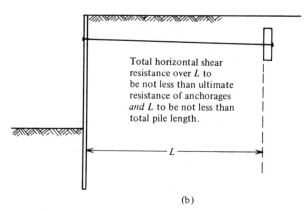

(a)

(b)

Fig. 12.13 Location of anchorage. (a) Cohesionless soil throughout. (b) Cohesive soil throughout or stratum of cohesive soil with cohesionless soil.

in a wall that is supported on the outside face by steeply raking bearing piles. Where steel H tension piles or soil anchors at 45° provide the anchorages, the only problem is the downward component on the piling wall, and the additional vertical stress on the piling should then be checked.

6. Steel Walings and Tie Rods for Anchored Retaining Walls

The walings, or wales, for anchored sheet-pile retaining walls commonly consist of a pair of steel channels with the webs back-to-back and horizontal. It is usual to place the channels behind the sheet piling as it improves the appearance of the wall, makes the channels less subject to corrosion, and eliminates any projection that might interfere with the use of the wall. *Back-bolts* are required to transmit the loading from the piling to the waling.

Steel tie rods having diameters in the range of about $1\frac{1}{2}$ to 4 inches are used to transfer the load from the walings to the anchorages. For high loadings the use of tie rods with upset ends may be justified where the area at the bottom of the threads is at least equal to the rod cross-sectional area, but for the smaller diameters plain threaded rods will usually be more economical. Where the tie rods exceed 30 to 40 ft in overall length, hexagonal couplings or turnbuckles are incoporated to make handling and erection easier on site.

If there is any possibility of settlement of the soil under the tie rods, they should either be supported separately—for example, on light vertical bearing piles at suitable centers depending on tie rod diameter—or they should be installed within a large pipe with the tie rod lying in the bottom initially.

The tie rods and back-bolts have to be provided with washer and bearing plates. These should be of sufficient size and thickness to provide adequate bearing on the sheet piling and walings or on concrete anchor blocks. Sheet-pile anchorages require walings similar to those on the main wall, but as the walings may be placed behind the anchor piles, no back-bolts are required other than some assembly bolts.

Bolted fishplate joints or welded joints are used for connection of long lengths of walings.

(a) Design

The working anchor load per foot run of wall increased by 20 percent, as referred to in the section on design, forms the basis of design of the walings, tie rods, back-bolts, washers and bearing plates in the usual manner for structural steelwork.

The tie rods form a particularly important part of an anchored retaining wall, as failure of any one of them could lead to progressive failure of others and then to extensive collapse of the wall. It is advisable to use a modest tie rod working stress of the order of 0.45 or 0.50 of the minimum yield. Very high-yield steels should be avoided, as the higher elongation at yield associated with steels of low yield strength is a useful safety factor: thus, if an exceptionally high load occurs in one tie rod, and it is able to stretch without failure, its load can then be shared to some extent, via the walings, with adjacent tie rods. The working stress is of course applicable to the cross-sectional area at the bottom of the threads for tie rods that do not have upset ends.

The walings are designed as continuous over several spans, these being the distance between tie rods, often in the range of about 10 to 20 ft depending on the overall size of the wall. Joints in the walings can conveniently be situated opposite the troughs formed by the piles and as close as possible to the "fifth-points" of the span, where the bending moment will be low. The joints in upper and lower channels may be staggered.

7. Other Design Methods for Anchored Sheet-Pile Walls

(a) Introduction

The conventional design procedure described in a previous section is the oldest method and probably still the most widely used, though certain modifications of it have been introduced in some cases. There are, however, several other design methods that have been introduced in the last thirty years or so.

If design is intended to cover the economic selection of sheet piles of appropriate bending strength and length, and of an adequate anchorage system on the basis of a given factor of safety, one would expect a unique answer for a given wall height in a given set of site and soil conditions. But if the various design methods are applied it will be found that a variety of answers will result, and these differ by more than just a few percent (Lamboj and Fang, 1970). Bending moments, pile lengths, and tie rod loads may vary over a wide range, where the highest figures may in some cases be up to 50 percent or more greater than the lowest figures in the range of answers obtained.

This situation prompts two immediate questions. First, *why* are there so many different methods, and second, *which* of these should be used?

There are probably many answers to the first question. The whole problem of anchored sheet-pile wall design is considerably more complex than is at first apparent, and originators of design methods have adopted differing simplifying assumptions and procedures to arrive at their end result. Though thousands of such walls are built throughout the world each year, very few failures occur, which may suggest that factors of safety are unnecessarily high. Attempts have also been made to explain why some very early steel sheet-pile and timber sheet-pile walls still stand although conventional theory indicates a factor of safety of less than 1. Arching of the soil between tie rod level and dredge line was considered by some to produce a reduction in bending moment on the sheet piling. Various soil pressure theories have also helped to confuse the situation. The development of ultimate design methods for steel structures led naturally to its application to anchored retaining walls in order to achieve a given factor against failure rather than a factor of safety on steel stress.

There is no simple answer to the question of which one of the various designs should be used. A number of investigators have compared the results of applying several methods as mentioned above, and in some cases there appears to be an implied suggestion that the method that produces the cheapest final result should be selected. This would be right only if it is known that every one of the methods gives a satisfactory factor against failure in all conditions. However, very few full-size tests have been carried out on anchored walls owing to the expense and difficulty involved, and probably none have been done on full-size walls taken to failure; true factors against failure of actual walls are therefore not known. Most of the design methods are theoretical in that there is little experimental evidence to support them, apart from the apparent lack of failures of walls built according to these procedures, with one exception now to be mentioned.

(b) Rowe's Work on Anchored Sheet-Pile Walls

The one particular exception to the foregoing comments is the series of experimental investigations which were carried out by Professor P. W. Rowe on model anchored walls. This remarkable and important work continued over a period of several years (Rowe, 1952, 1955, 1956, 1957) and led to the first clear understanding of how these walls really functioned under both *working* and *failure* conditions.

It is not proposed to describe this work in any detail, but a careful study of all the relevant papers and ensuing discussions is recommended.

Rowe's initial work (Rowe, 1952, 1955) dealt with walls under working stress conditions and clearly showed that the flexibility of the piling has a vital influence on design, particularly on the bending moment in the piling. Whereas previously it was known that piling flexibility influenced the working condition bending moment, it was believed that the reduction was due to soil arching. Rowe showed that arching did occur initially when there was no forward movement of the wall at tie rod level, but that only a small amount of tie rod and anchor yield as occurs in practice was sufficient to eliminate arching and to produce a triangular distribution of soil pressure behind the wall. Nevertheless, the bending moments on the piling were less with a flexible wall compared with a stiffer wall under *working* stress conditions. For a given tie rod level, the moment reduction is due chiefly to the raising of the point of contraflexure in the piling under the influence of piling flexibility and relative soil compressibility.

The effect can be illustrated by the crude simplified analogy shown in Figure 12.14a. A very flexible vertical beam spans from a point support *A* to a curved rigid support *B* and carries a load *W*. Owing to deflection of the beam, contact with the lower support is at point *C* above the center of the curved support. Now consider Figure 12.14b, where the only difference is that the beam, though of equal strength, is very much stiffer. Its deflection under the same load *W* is very much less and contact with the curved support *B* is at a point *D* only just above its center. The effective span of the beam in (a) is significantly less than in (b), and the maximum moment in (a) is correspondingly less than in (b) for the same load *W*; that is, in these artificial conditions the more flexible beam can be designed for a lower bending moment than the stiffer beam. This is the nature of the effect producing moment reduction in anchored walls, and where the loading is distributed and triangular, in such walls, the maximum bending moment is proportional to the cube of the effective span between tie rod and point of contraflexure. A change of "span length" of only 5 percent would effect a bending moment change of about 16 percent.

Rowe (1956) next proceeded to study experimentally the anchored wall taken to failure and the results are of vital significance in the understanding of the design problem. In the usual practical range of wall heights and site conditions for

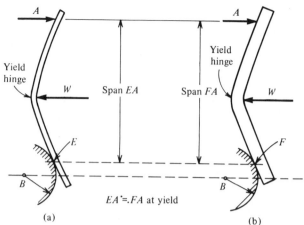

Fig. 12.15 Analogy to show effect of flexibility at failure condition. (a) Flexible beam. (b) Stiff beam.

steel sheet-pile walls, it was found that both piling flexibility and tie rod depth affected bending moment reduction at failure, but variation of flexibility was a less significant factor at yield than under working stress conditions. The rough analogy in Figure 12.14 can be extended to illustrate this in Figure 12.15. The vertical beams are of different stiffness but equal bending strength. When load *W* is increased so that steel yield occurs in both beams, they become approximately equally flexible, the effective spans become equal, and the bending moments are therefore equal. However, at working load conditions where *W* is, say, half the value at yield, the bending moment and hence the working stress in the flexible beam is lower than in the stiffer beam.

Rowe finally recommends design on the basis of conditions at failure, that is, at initial yield of the steel, rather than on the basis of working stress conditions.

(c) An Alternative Suggested Design Procedure Based on Rowe

As an alternative, the author suggests the following simple design procedure based on a study of Rowe's experimental work. It does not go quite as far as Rowe considers may be possible in practice, but even Rowe suggests some caution until full-size trials lead to confirmation of all the model results. The chief advantage of this suggested method is that it is based on the conventional method already described.

A series of examinations of typical designs (Rowe, 1956, and discussion), both by Rowe's failure method and by the conventional method, tended to show that (with some simplifying assumptions) both gave similar final results provided increased *apparent* working stresses were used with the conventional method.

The suggested method for steel sheet-pile anchored walls is as follows:

1. Carry out the initial design exactly as in the section on conventional design for *fixed-earth support* and find the calculated bending moment *M* and calculated tie rod load *T*.
2. Use the graph in Figure 12.16 to get the apparent working steel stress *s* for selection of the section modulus of the piling. In Figure 12.16, *D'* is the tie rod depth below actual top of wall, excluding surcharge, and *H'* the actual height of wall. As an example: If $D'/H' = 0.2$, if the minimum yield stress of the steel is 39 000 lb/in^2, and if the calculated bending moment *M* is 65 000 ft-lb per foot run of wall, then the stress *s* from Figure 12.16 is $0.69 \times 39\,000 = 26\,900$ lb/in^2. The required section modulus is $(65\,000 \times 12)/(26\,900) = 29.0$ in^3 per foot run of wall.

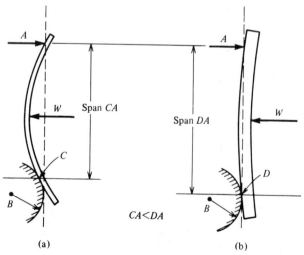

Fig. 12.14 Analogy to show effect of flexibility under working conditions. (a) Flexible beam. (b) Stiff beam.

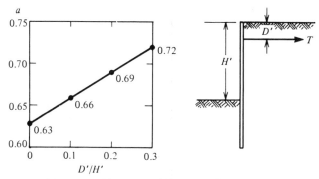

a = Apparent working stress(s)/ Min. yield of steel

Fig. 12.16 Alternate design method—apparent working stress.

3. Increase the calculated tie rod load T by not less than 25 percent and up to 33 percent when D'/H' approaches 0.3. Use this final figure to design the tie rods and walings with the working stresses given in the section on steelwork details, and to design the anchorage with a load factor of not less than 2.0.
4. The penetration of the piling below dredge level is that determined by the conventional method.

Note that this suggested method is not applicable to free-earth support nor to cohesive soils. The procedure should not be applied to relieving-platform walls owing to the relative rigidity of the anchorage system compared with that of normally anchored walls.

While high stresses relative to the yield stress of the steel are apparently suggested above, they are not the actual stresses that occur in the piling under working conditions and which depend significantly on the flexibility of the piling. However, the load factor against yield of the piling will not be less than 2.0.

(d) Other Design Methods

Space permits only a brief mention of some of the other design methods that have been used for anchored sheet-pile walls:

- Tschebotarioff's method (Tschebotarioff, 1973).
- Terzaghi's method (Terzaghi, 1954, with particular reference to the discussion). Permits flexibility reduction according to Rowe.
- Code of Practice, Earth Retaining Structures. A 25-percent reduction of conventional bending moment is permitted.
- German Committee for Waterfront Structures (Committee for Waterfront Structures, 1980). Permits a reduction of up to 33 percent of conventional bending moment under certain conditions only. Also permits design based on Rowe's work or alternatively by Hansen's ultimate-load method (below).
- Hansen's ultimate-load method (Hansen, 1953). This is a failure or ultimate method based on the formation of yield hinges at failure. Flexibility of the piling does not enter into the method.
- Danish rules (Tschebotarioff, 1973). A large moment reduction is permitted based on arching.
- Haliburton (1968) developed a numerical method of analyzing flexible retaining structures. It is based on the familiar beam column on elastic foundation differential equation with a method of handling nonlinear soil response.

(e) Comment on Factors that Affect Bulkheads and Results of Failure Investigations

Casagrande (1973) commented on the conventional design of retaining structures. His comments were based on conclusions drawn from performances of retaining structures and observations of instrumented structures. These comments included:

- Satisfactory performance of retaining structures is dependent on a number of considerations and not just on whether or not the structure can yield sufficiently to reduce the actual earth pressure to the active state.
- As the magnitude of predicted earth pressure can be on the unsafe side by a factor of 1.5 or greater, the safety factor should be 2.5 for instances where uncertainties exist.
- The use of coarse granular materials (gravel, crushed stone, or slag) or backfilling insures continued minimum earth pressure if not compacted. Even light compaction may permanently increase the earth pressure into the passive range.
- The finer the backfill material, the less assurance that active earth pressure will be effective for an extended period. Plastic deformations, vibrations, water level fluctuations, temperature changes and other effects may cause such materials to develop at-rest earth pressure.
- Anchored bulkheads designed in accordance with accepted practice may, with time, yield excessively at the anchor level because of inadequate anchorage and/or because the yield strength of the tie rods has been exceeded. To prevent excessive yielding tie rods and anchorage should be designed for not less than twice the force used in conventional design.

In discussion of Casagrande's paper, Lum (1973) added the comment that a wall founded on plastic clay can fail by sliding at the base although the retaining wall is satisfactorily designed otherwise. Also, Lum indicated that compaction of granular backfill behind a retaining wall on bedrock could develop pressures much greater than at-rest pressure.

Tsinker (1983) reviewed current design practices for anchored bulkheads and presented results of Soviet investigations. He concluded that:

- Anchored bulkheads should be considered as flexible structures regardless of the material of which they are composed and the type of anchorage.
- Total active pressure against bulkheads usually does not exceed that determined by Coulomb's theory.
- Depending on the characteristic of bulkhead deflection, active pressure usually redistributes vertically along a sheet-piling wall, concentrating at the anchors and in the area of the passive pressure.
- Tie rods should be placed with negative sag or inside a pipe.
- Joints at both ends of the tie rod should be designed so they can rotate if the tie rod moves downward.

A bulkhead failure on the Houston ship channel was reported by Daniel and Olson (1982). The bulkhead was designed in accordance with several published recommendations. Less than a month after construction was complete, the backfill began developing cracks and within $3\frac{1}{2}$ months portions had moved outward nearly 10 feet. A general shear failure, including the anchorage, had occurred and in addition a toe failure may have occurred. The authors assigned the cause of the failure to a lack of understanding of soil behavior under project circumstances. A large amount of dredging took place in front of the bulkhead, which lowered the soil strength and governed the design. It is another example of the consequences that can occur if the characteristics of plastic clays are not recognized.

Most of the several design methods used and illustrated have been compared with or developed from model studies. The results have also been correlated with the performances of full-sized structures. The models cannot provide data associated with long-term performance or the effect of construction methods. Correlation of the performance characteristics of actual structures with results developed from theory on models is usually independent of time and very few quantitative values can be obtained. Full-scale instrumented structures provide reliable data because not only are the measurements obtained but the design and service conditions will most likely be documented. Finite-element studies are difficult to develop to adequately represent soil characteristics and results still must be correlated with actual behavior.

(f) Accuracy of Design

The accuracy of any final design rests on four main factors:

1. The accuracy of the information regarding the soil conditions, water levels, and other site factors.
2. The accuracy of assumptions concerning the surcharge, overdredging, scour, and so on.
3. The accuracy of the particular design method or theory used.
4. Constructional aspects—for example, possible overtightening or unequal tightening of tie rods, type of filling and degree of compaction not in accordance with that specified, weepholes clogging if not provided with satisfactory graded filters, and so on.

It can be deduced that factor (3), to which most attention has been given by many investigators, is not the only important one. Considerable judgment is therefore required to assess the effects of the other factors before arriving at the final section modulus, pile length, and anchorage details.

8. Constructional Aspects

It is not proposed to deal in this final section with all details of sheet-pile driving and construction, but it is worth noting some points that should be borne in mind in the early stages of design and which may have some effect on the selection and design of the structure, as well as on the engineer's specification documents.

Sheet piles are usually driven in pairs, using hammers or the double-acting type of diesel hammers. To avoid leaning of the piling and an irregular line it is nearly always preferable to erect a panel of several pairs of piles before starting to drive. The panel should be supported vertically by an arrangement of temporary horizontal walings at two levels, and the end pairs of piles of the panel should be driven first. With such a panel-driving procedure it is more convenient and cheaper to suspend the hammer from a crane and thus avoid the use of hanging leads or a pile frame. The hammer would be provided with purpose-made downward projecting legs to enable it to sit securely on the piles (BSP Pocket Book, 1963).

As the design engineer will be concerned that his retaining wall will be of good appearance, the above procedure for driving is strongly recommended. The actual width of a wall of piling may differ from the theoretical width based on the nominal width of a single pile. A consequence of this is alteration in tie rod spacing, which can be significant over a very long wall. However, correct panel driving can help prevent this effect, as it is possible to check and slightly alter the width of a panel before it is driven, especially with Z-type piles.

The engineer will have already selected a pile section and steel quality for the required penetration and soil conditions so that the piles can in fact be driven to the desired elevation. He should also ensure that the correct hammer is used. One that is too light will tend to damage the pile heads, and an excessively heavy hammer can have the same effect (Cornfield, 1968, 1970).

Sheet piles can be driven in most rocks to penetrations sufficient to provide lateral support to the toes of the piles, provided the correct pile section, steel quality, and hammer are used.

Other constructional points to be considered and which have already been referred to earlier are the possibility of overtightening of tie rods, the type of fill behind the wall and its method of placing, weepholes and also graded filters behind them where appropriate, and support of tie rods during the filling process. During the construction of anchorages, disturbance of the existing soil should be kept to a minimum and correct fill material placed around the anchors should be well compacted.

B. BRACED WATERFRONT RETAINING STRUCTURES

A typical braced structure is shown in Figure 12.17. The batter piles are in compression and located in front of the wall. These structures are stiff in the horizontal plane and unyielding at the restraining supports. The cost of this type of structure is usually greater than that of the anchor type, but is used because of site conditions or the construction method, or because of the completed structure's operational requirements.

The relieving type of structures shown in Figures 12.1d and 12.10a and b are also very stiff structures.

C. WALL SYSTEMS

There are a number of wall systems used for retaining soil in anchored and braced waterfront structures. Figure 12.18 shows the most commonly used walls.

Most walls are constructed of moment-resisting steel sheet piling that, in the United States, is produced with section moduli as high as 60.7 in³/linear foot of wall.

A combination of sections is often used when savings are indicated. Examples of steel section combinations are shown in Figures 12.18a and b. Very large moment-resisting walls can be constructed by combining sections. Cover plates can be welded to the flanges of sheet piling to greatly increase the section modulus. The cover plates are welded to the piling only where the bending strength of the piling section is exceeded.

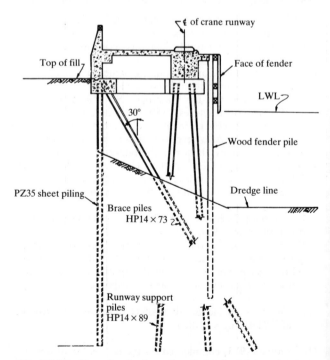

Fig. 12.17 Example of a braced waterfront retaining structure.

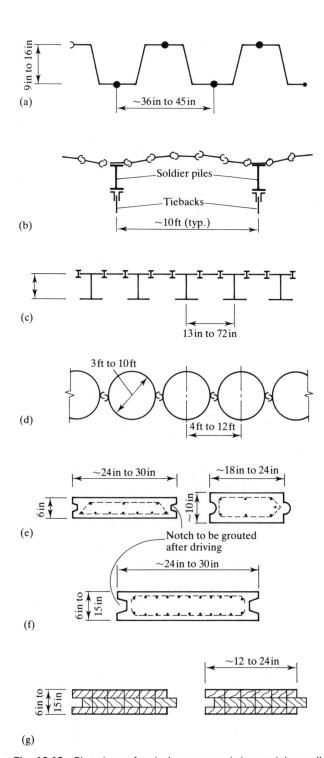

(a)

(b)

(c)

(d)

(e)

(f)

(g)

Fig. 12.18 Plan views of typical moment-resisting retaining wall systems composed of sheet piling or sheet piling and structural sections. (Piling shown in (e), (f), and (g) have beveled tips so that the piling being driven is forced against the previously driven section.) (a) Z-shaped hot and cold rolled steel and aluminum sheet piling. Other shapes include those that interlock at mid-depth of two interlocked sections. (b) Combination of soldier piles (structural sections) and flat web sheet piling. (c) H-shaped piling with or without intervening sheet piling. Flat web or moment-resisting sheet piling can be used between H shapes. (d) Interlocked large-diameter pipe piles. Sheet piling can be placed between the pipe piles. (e) Reinforced-concrete sheet piling. (f) Prestressed concrete sheet piling. (g) Wood sheet piling. Wakefield type shown, but tongue and groove and butted sections are used.

Aluminum sheet piling is rarely combined with other sections and is used for small waterfront structures—particularly in saltwater applications. The maximum section modulus of aluminum sheet piling that the author is aware of is 9.0 in^3/linear foot of wall.

Concrete sheet piling can be produced in a wide range of bending strengths. There are no universally produced standard sheet piling sections.

Wood sheet piling is produced from various species and can be treated with several different preservatives. Wood sheet piling is used for small waterfront structures.

Any type of anchorage or bracing is used with the various wall systems. The bending strength, stiffness, drivability, durability, and maintenance or repair characteristics of a wall system should be included in the considerations during design. The wall material and installation may be the largest expenditure for a given structure, so care is required when choosing the material and type of wall.

12.2.2 Excavation Retention Structures

A. WALL SYSTEMS

The walls of land coffers have more varied components than those of waterfront structures. Figure 12.19 shows most of the commonly used wall systems in the United States.

Performance requirements of soil-retention walls can vary considerably more than those used in water cofferdams or waterfront structures. For some excavation retention walls, movement is of little concern except for the stability of the wall itself. No wall movement is allowed for some excavations where adjacent structures or utilities are located. Wall movement is a function of soil properties, the restraining system, and the installation method. Some of the characteristics of the various wall systems, as shown in Figure 12.20, are listed below.

a. *Sheet piling*
   - Site dewatering not required for installation
   - Installed by vibratory or impact hammers
   - Occasionally used as permanent retaining structure
   - Used with bracing and tieback restraints
b. *Soldier piles and wood lagging*
   - Wall cannot be installed below water without dewatering
   - Soldier piles can be installed by hammers or predrilling
   - Rarely used as permanent retaining structure
   - Normally used with tieback restraints
c. *Soldier piles and concrete diaphragms*
   - Wall rarely installed below water table without dewatering
   - Frequently used as permanent retaining structures above waterline
d. *Cast-in-place walls (slurry walls)*
   - Site dewatering not required for installation
   - Usually restrained by tiebacks
   - Frequently used as a bearing foundation and always left in situ
   - Walls are very stiff
e. *Cast-in-place contiguous piles*
   - Site dewatering not required for installation
   - Rarely used as permanent structure
f. *Prestressed concrete or steel cylinders*
   - Piles are driven into place
   - Site dewatering not required for installation
   - Normally restrained by tiebacks
   - Wall is very stiff
   - Used as permanent retaining structures

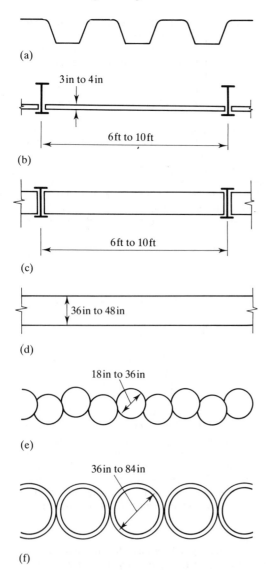

(a)

3 in to 4 in

6 ft to 10 ft

(b)

6 ft to 10 ft

(c)

36 in to 48 in

(d)

18 in to 36 in

(e)

36 in to 84 in

(f)

Fig. 12.19 Plan views of typical excavation retention wall systems. (Wood retention systems are often used for trench retention, but not for larger excavations.) (a) Z-shaped moment-resisting sheet piling. (See Fig. 12.18a.) (b) H sections or wide-flange structurals used as soldier piles with horizontal wood sheeting. Wood sheeting can also be placed in front of the flanges using clips. (c) H sections, wide-flange structurals or reinforced concrete piles used as soldier piles with precast or cast-in-place concrete diaphragms. (d) Cast-in-place reinforced walls. Narrow trenches are dug and filled with slurry to prevent sidewall collapse, then concrete is placed in the trench (which displaces the slurry) and prefabricated rebar cages are set into the concrete. (e) Cast-in-place contiguous concrete piles (staggered or in-line). (f) Drive, prestressed concrete, or steel cylinders.

B. ANCHORED AND BRACED EXCAVATIONS

Single-wall structures are used extensively for land cofferdams, as well as for water cofferdams. Some land excavations are so wide and construction methods are such that bulkhead design methods are applicable. Cells are used occasionally as land cofferdams when large areas and/or deep excavations are required.

Terzaghi and Peck (1967) distinguish between shallow and deep excavations. Excavations of more than 20 ft are classified

as deep, and the data presented is primarily applicable for deep excavations.

Figure 12.20 shows typical configurations of single-wall, retained excavations. The strut-braced excavation has been the most commonly used type. When the width of the excavation is large the raker-braced type may be used. In recent years, tiebacks and anchor systems have been used with increasing frequency and appear to be supplanting the more conventional types. This type of support is generally more costly than either the strut or raker bracing; however, since the excavation is free from bracing, considerable economy may be obtained by not having to work around the bracing. In addition, the chance of accidentally dislodging a support is minimized. Circular and semicircular excavations with diameters ranging from a few feet to several hundred feet have been constructed. The walls of these excavations are usually braced by wide-flange or reinforced-concrete ring wales. Rectangular excavations have also been constructed using ring wales.

Failure modes of retained excavations are shown in Figure 12.21. There are at least four general modes of failure: (1) excessive movement of the wall; (2) support yielding or collapse; (3) bottom heave in cohesive soils; and (4) piping in granular soils. As with other structures, a given deformation may be suitable for one site condition but constitute failure in another. If the excavation is adjacent to buildings, subways, pipelines, or other structures that will be damaged by slight movement, the retaining walls and supports must be designed with consideration for these restrictions. There are few civil structures more difficult to construct successfully than a retained excavation when buildings or other structures are adjacent to the excavation. Employment of proper dewatering and drainage techniques is just as necessary as good structural design. Details of dewatering and drainage techniques are covered in Chapter 7.

Gould (1980) listed the advantages of tiebacks in limiting cofferdam movements compared to internal bracing as follows:

a. In granular soils whose modulus increases with stress level, prestressing by tiebacks makes the soil mass more rigid.
b. Tiebacks are typically prestressed to about 120 percent of design load and locked off between 75 and 100 percent of design, taking up the slack in the cofferdam system. Internal bracing is prestressed (if at all) to not more than about 50 percent of design; loading increases as excavation deepens, causing compression in the cofferdam system.
c. Temperature strains are more important with internal bracing than tiebacks and can cause inward movement at times of falling temperature that is not fully recovered at times of rising temperature.
d. Frequently, bracing is removed and then rebraced to facilitate interior construction, and flexure of the wall occurring with strut removal contributes to movements. Tiebacks do not have to be removed and their unloading can often be staged to allow the take-up of support by the permanent construction.
e. Overexcavation below the future position of internal braces frequently occurs to facilitate soil removal. This encourages deep-seated movements, especially in weak soils. Using tiebacks, the excavation level can be maintained at or slightly above the tieback elevation.

1 Lateral Pressure

With few exceptions (such as relatively large land excavations without rigid boundary requirements), the pressure distribution as given by Coulomb is not used in the design of supporting members for cross-braced excavations. Principal reasons for this are the boundary requirement of limiting wall movement and the sequence of construction.

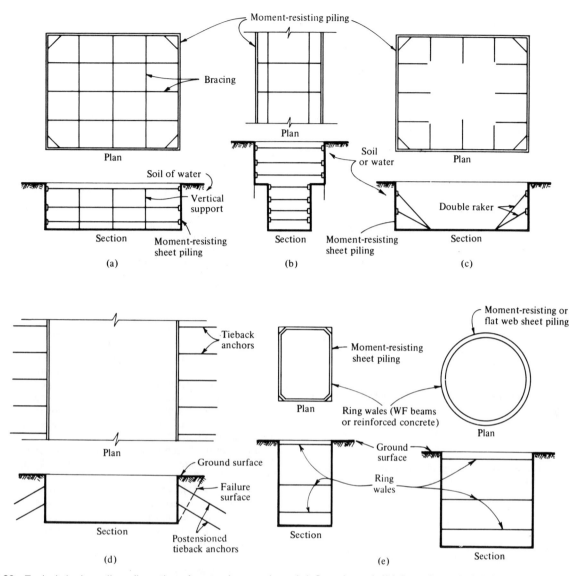

Fig. 12.20 Typical single-wall configuration of retained excavations. (a) Cross braced. (b) Braced trench. (c) Raker braced. (d) Tieback (post-tensioned). (e) Rectangular and circular ring wales.

When an area is to be excavated the usual procedure is to first install the walls (steel sheet piling, H-piles and wood lagging, concrete cylinders, or sheeting) to a given elevation; then make a cut to the first level of support, which is usually at about one-half the height (H_c) the soil would stand without support. The supports (bracing or tiebacks) are loaded to a desired value before excavating to the next support level and repeating the cycle. The wall tends to move toward the excavation with increasing depth, and the final wall profile is similar to that shown in Figure 12.22a.

Raker-braced walls usually yield more at the top than walls supported by other methods because additional earth in front of the wall is removed prior to positioning the bottom of the raker. As a result, the Coulomb pressure distribution is applicable.

Sheeting is usually driven at least several feet below the excavation to prevent raveling of the soil.

When soldier pile and lagging construction is used, the soldier piles are usually driven deeper than the excavation so that they can provide sufficient support for the lagging, which is carried only to the excavation bottom.

Figure 12.22 shows the commonly used various lateral pressure distributions on braced or tieback walls. The pressure variation shown for land cofferdams has had some verification from field studies and can vary widely with field installation practice and soil characteristics.

Peck (1969) has indicated that when tieback and anchor systems are used there is some evidence that the pattern of deformations is different from that when cross-bracing is used. The deformations decrease from top to bottom, and the triangular distribution of pressure may be correct. Until more information is available, the appropriate rectangular or trapezoidal distribution given in Figure 12.22 should be used.

The Terzaghi and Peck method is based on the following conditions:

1. Applies to excavations $\geqslant 20$ feet deep.
2. An artificial loading diagram, designated "apparent" pressure, is used for determining strut loads.
3. Water table is below bottom of excavation. Shear strength of clay is taken in the undrained state. Pore pressures are not considered.

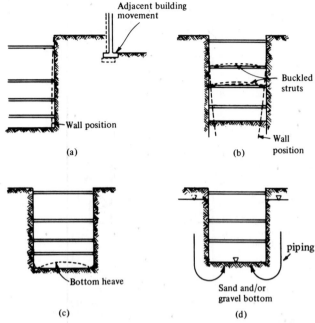

Fig. 12.21 Failure modes of braced excavations. (a) Excessive movement of sheeting. (b) Strut, wale, or ringwale collapse. (c) Bottom heave (plastic clay). (d) Boil (piping).

4. A stability expression has been developed to indicate the performance of an excavation in clay. The expression is

$$N = \frac{\gamma H}{c}$$

Ground movement becomes significant with N values above 3 or 4, and base failure is likely with values $\geqslant 6$.

The Terzaghi–Peck (1967) unit pressure for sand is given as

$$p = 0.65 K_a \gamma H$$

where $K_a = \tan^2(45° - \phi/2)$.

Tschebotarioff gives the unit pressure in sand as

$$p = 0.8 K_a \gamma H \cos \delta$$

where K_a is the same as above.

The following are unit pressures for clays.
TERZAGHI–PECK (SOFT-TO-MEDIUM CLAYS):

$$p = 1.0 K_a \gamma H$$

where

$$K_a = 1 - m \frac{2q_u}{\gamma H}$$

and m is a reduction factor depending on N values. If values of $N > 3$–4 are encountered where full plastic zones can develop and the clay has been preloaded, the value of $m = 1.0$ is appropriate. If the clay is normally loaded use a value $m < 1.0$.
TERZAGHI–PECK (STIFF FISSURED CLAY):

$$p = (0.2\text{–}0.4)\gamma H$$

The lower values to be used when movement of walls is minimal and there is a short construction period.
TSCHEBOTARIOFF (STIFF CLAY):

$$p = 0.3\gamma H$$

(medium clay):

$$p = 0.375\gamma H$$

Peck (1943) proposed the following unit pressure for excavations in layered soils (sand and clay), with sand overlaying clay:

$$p'_m = K'_a \gamma' H$$

where

$$K'_a = 1 - \frac{2q'_u}{\gamma' H}$$

The average strength of the sand and clay was put in terms of equivalent unconfined compression strength:

$$q'_u = \frac{1}{H} [\gamma_s H_s^2 K_s \tan \phi + (H - H_s)n q_u]$$

where the s subscript refers to the sand stratum and n = ratio of field unconfined compressive strength to that determined in the laboratory.

The average unit weight is

$$\gamma' = \frac{1}{H} (\gamma_s H_s + \gamma_c H_c)$$

where the c subscripts refer to the clay stratum.

The accuracy with which pressure can be determined for mixed soils which exhibit cohesion and friction is less than for granular soils which do not have cohesion. In order to determine the pressures we decide which property is dominant for the time the excavation is open, the restrictions on wall movement, permeability and the location of the water table. The distribution of the pressures could be similar to those presented in Figure 12.22.

Conventional techniques may be employed for design of bracing elements in shallow excavations. Usually the size of the bracing normally used is sufficient.

Lambe (1970) has reviewed various methods of determining pressures in excavations and soil movements and has compared the results with data from actual excavations. It was shown that movements of the soil outside of the excavation and strut loads cannot be adequately predicted in most field conditions.

2. Heave and Piping

The common failure modes of retained excavations were shown in Figure 12.21. Those in Figure 12.21c and 12.21d are due to bottom instability problems.

Bottom heave in excavations in clay soils is influenced by shear strength and loading history of the clay as mentioned in the previous section. Figure 12.23 gives a method for determining the factor of safety against heave. Figure 12.23a is for the case in which the full plastic zone can develop under the excavation; Figure 12.23b is for the case in which a hard bottom restricts the development of the plastic zone. The latter case provides a great improvement in the tendency to heave instability. A factor of safety of 1.5 is recommended for determining the resistance to heave. If the safety factor is lower than required, the sheeting is driven deeper. The required penetration of the sheeting may be estimated or arbitrarily set at one-half the width of the excavation. However, Peck (1969) observed that when a deep soft-to-medium clay stratum extends below the excavation bottom little is gained by driving sheeting much below the excavation.

Although rare, heave may occur in cohesionless soils. The factor of safety against heave occurring is

$$F.S. = 2N_\gamma \left(\frac{\gamma_2}{\gamma_1}\right) K_a \tan \phi$$

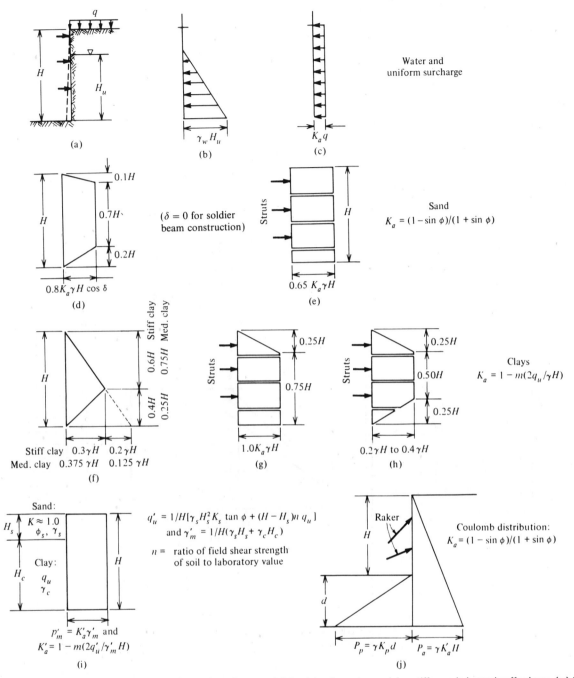

Fig. 12.22 Lateral pressure distribution for computation of strut and tieback loads, wales, and ring stiffeners in braced cofferdams. (a) Section. (b) Water. (c) Surcharge. (d) Tschebotarioff, 1951. (e) Peck, 1969. (f) Tschebotarioff, 1951. (g) Soft-to-medium clay (Terzaghi and Peck, 1967). (h) Stiff fissured clay (Terzaghi and Peck, 1967). (i) Mixed soils. (j) Raker braced.

where

N_γ = bearing capacity factor of the soil below the excavation (Fig. 12.48)
γ_1 = unit weight of soil above the bottom of the excavation
γ_2 = unit weight of soil below the excavation

Piping occurs if the water head is sufficient to produce critical velocities in cohesionless soils. This results in a "quick" condition at the bottom. Figure 12.24 illustrates one method for determining whether a quick condition may occur (NAVFAC DM-7, 1971).

McLean and Krizek (1971) studied the seepage of imperfect cutoffs. They concluded that when sheet piling is driven below an excavation for seepage control the penetration depth is the most important factor in controlling seepage flow.

Wellpoints and other dewatering methods may be used to lower the water table around the excavation. Although the water table is usually lowered, it may cause settlement problems where structures are located adjacent to the excavation. Details of methods for controlling piping and dewatering methods are given in Chapter 7.

The primary functions of the portion of the sheet piling driven more than several feet below the floor of the excavation

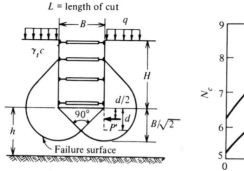

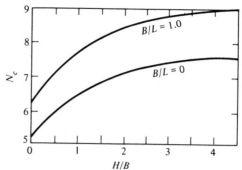

Load on buried length of sheeting:

If $d > \dfrac{2B}{3\sqrt{2}}$

$$P' = .7\,(\gamma HB - 1.4cH - \pi cB)$$

If $d < \dfrac{2B}{3\sqrt{2}}$

$$P' = 1.5dB\,(\gamma HB - 1.4cH - \pi cB)$$

$$\text{F.S.} = N_c \frac{c}{\gamma H + q} \geqslant 1.5$$

where

N_c = bearing capacity factor
c = unit strength of clay
H = excavation depth
γ = unit weight of soil
q = surcharge

If F.S. $<$ 1.5 drive sheeting below bottom of excavation

(a)

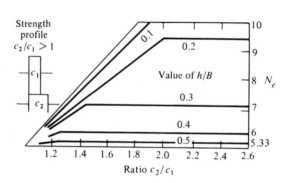

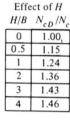

Effect of H	
H/B	N_{cD}/N_c
0	1.00
0.5	1.15
1	1.24
2	1.36
3	1.43
4	1.46

For $B/L = 0$

$$\text{F.S.} = N_{cD} \frac{c_1}{\gamma H + q} \geqslant 1.5$$

where

N_{cD} = bearing capacity factor
c_1 = unit strength of soft stratum

(b)

Fig. 12.23 Stability against heave in cohesive soils. (a) Stability when $h > 0.7B$. (b) Stability when $h < 0.7B$. (*After NAVFAC DM-7, 1971*).

are to prevent heave, increase the water flow path, or restrict water flow by embedment in an impermeable stratum. The depth of the walls is usually a function of the stability of the excavation.

3. Design of Wall and Wall Supports

The type and sizing of bracing elements are mainly dependent on the boundary requirements and soil characteristics. When limitation of movement is required to protect against structures, incremental excavation, numerous supports, high values of strut, raker or anchor tieback preloading, and stiff walls may

be used. Peck (1969) pointed out that the use of extremely stiff walls does not insure against movement. Support spacing is more critical in the excavation area.

If the water table is not lowered below the excavation, care must be taken to construct a reasonably impermeable wall to prevent soil flow, and bottom stability problems may be increased. On the other hand, if the wall is not designed for water loading, drainage from behind the wall must be assured.

At this time, previous experience is the best guide for estimating movement of structures outside of the excavation.

The estimate of wall pressure was covered in the section on lateral pressure, however, the load shown in Figure 12.22e

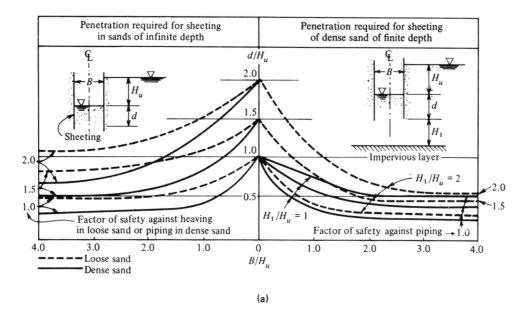

(a)

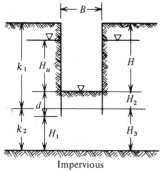

Impervious

Let $K_1 < K_2$ If $H_1 < H_3$ there generally is more flow than given in graph (a) (infinite) above.

If $(H_1 - H_3) > B$ use graph (a) (infinite).

If $(H_1 - H_3) < B$ there is more flow than given in graph (a) (infinite). If $K_2 > 10K_1$ failure head H_u is equal to H_2.

Let $K_1 > K_2$ If $H_1 < H_3$ safety factors are intermediate between those for graph (a) (finite).

If $H_1 > H_3$ graph (a) (finite) is conservative.

Let $K_1 = K_3$
and
$K_1 \gg K_2$ If $(H_2 - d) > B$ use graph (a) (finite) above.

If $(H_2 - d) < B$ pressure relief required so that unbalanced head on fine layer does not exceed weight of H_2.

If fine layer is higher than bottom of excavation the completed excavation is safe, but during construction a blow in may occur—pressure relief then required.

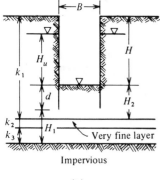

Impervious

(b)

Fig. 12.24 Stability against "piping" in noncohesive soils. (a) Graphs for determining sheeting penetration in granular soils. (b) Piling penetration to prevent piping. (*After NAVFAC DM-7, 1971.*)

should be altered when sizing the wall. A triangular distribution at the top is more appropriate.

When the water table is not lowered and surcharge loads are present, the water load and lateral component of the surcharge load may be added to the pressures given in Figure 12.22. The above procedure may not be the best method for including the effects of the additional loads but, most likely, is a conservative approach.

Figures 12.25 and 12.26 show conventional methods for sizing the bracing elements of a vertical-sided excavation. The struts, rakers, and tieback anchors should have as large—or

larger—a factor of safety as the other bracing elements. The same holds true for ring wales. When determining the buckling load of a ring wale the designer must be aware of the sequence, location, pressure, and materials used for blocking. In addition, stiffeners may be required at the blocking points and at strut or tieback locations in straight wales.

Current ACI and AISC Specifications are used as reference specifications for design of concrete and steel bracing elements.

Figure 12.27 is a design example for sizing bracing elements of a narrow excavation in stratified soil with a high water table.

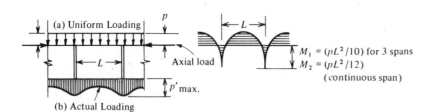

Moment diagram

(a) Continuous beam

(b) Simply supported

Wall

(Loads due to water table elevation and surcharge must be added to soil load if appropriate.)

(a) For continuous spans the maximum moment (3 or more spans):

$$M_{max} = p \frac{H_{max}^2}{10}$$

(b) For simple spans

$$M_{max} = \frac{p \, H_{max}^2}{8}$$

Required section modulus $S = \dfrac{M_{max}}{F_b}$

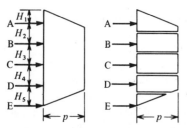

(a) Uniform Loading

Axial load

(b) Actual Loading (not used)

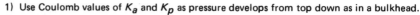
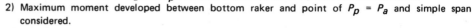

$M_1 = (pL^2/10)$ for 3 spans
$M_2 = (pL^2/12)$
(continuous span)

Wales

1) Compute bending moment
2) Pick trial section and compute bending stress
3) Determine axial stress and KL/r for both axes
4) Select F_a from AISC Column Tables
5) Check effect of combined stress [Sect. 1.6.1– AISC, 8th Ed.)
6) Repeat steps 2 thru 5 if necessary

Wales may be axially loaded due to deflection of end retaining walls

Struts

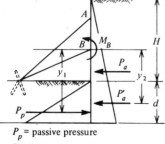

(a) Continuous beam   (b) Terzaghi-Peck, 1967

(a) and (b): Compute Reactions A thru D for (b) using Strut C as typical

1) $R_C = \dfrac{P}{2} (H_3 + H_4)$

2) Try a WF section and determine L/r ratio where γ = radius of gyration
   L = unbraced strut length

3) Determine F_a from current AISC Column Tables

4) Repeat steps 2 and 3 until strut capacity $\geq$ strut load

Rakers

1) Use Coulomb values of K_a and K_p as pressure develops from top down as in a bulkhead.
2) Maximum moment developed between bottom raker and point of $P_p = P_a$ and simple span considered.
3) Piling penetration computed by taking moments about lower brace:

$$P_a' Y_2 - P_p/Y_1 (F.S.) - M_B = 0$$

where M_B = allowable moment at B.

4) Upper raker sized for F.S. of 1.5.

P_p = passive pressure
P_a' = active pressure below point B

Fig. 12.25 Methods of loading and sizing excavation walls, struts, and rakers.

4. General Comments on Anchored and Braced Excavations

The failure of the excavation owing to a general slide should be considered early in the design procedure. This problem is usually confined to large excavations.

There are various procedures that have been used to minimize soil stress changes (and deformations) adjacent to excavations. Some have been used for many years and others have been introduced recently. Narrow cuts made at the periphery of a large excavation allow wales and tiebacks to be installed before the remainder of the excavation is made. Bottom slabs and walls have been constructed in small sections and the soil replaced until most of the slab and wall structure were in place before final removal of soil. Gould (1980) states that at a distance back of the cofferdam equivalent to the depth of the excavated cofferdam the settlement is one-third to two-thirds of the maximum settlement. Also, damage to existing adjacent structures typically results from approximately equal magnitudes of angular distortion from settlement and lateral extension strains of the order of 0.002 or 0.0013.

EARTH ANCHORS

Wall pressure diagram

Soldier pile at 6' to 8' spacing, sheet piling or slurry wall

Anchors
Wire tendons
Alloy bars
H piles

Failure plane

Grouted

$(45° - \phi/2)$

Pressure behind anchors (surcharge effects to be included if present)

Toe anchors may be used to elevate failure plane 1–3

ROCK ANCHORS

Soldier pile, sheet piling or slurry wall

Anchors
Wire tendons
Alloy bars

Failure plane

Soil

Rock

Grout

H

DESIGN AND PRESTRESSING PROCEDURES

Soil Pressure

Upper elastic limit (establishes maximum grout pressure)

Lower elastic limit

Borehole Volume

1. For earth anchor using Bauer (pressure grouted) anchor, develop soil pressure vs. volume change graph by using a Me'nard pressuremeter in borehole.
2. Compute lateral pressure on wall.
3. Assume required number and length of anchors:
   a) Bauer type (press. grouted) ≈ 12 k/ft (sand and gravel)
   b) Grouted type ≈ 6 k/ft (sand and gravel) ≈ 4 k/ft (clays)
   c) Short, end-belled anchors ≈ 100 k (clays)
   d) Grouted rock anchor ≈ 8 k/ft
4. Compute safety factor against area 1, 2, 4, 5 moving into excavation:
   ≥ 1.2 temporary (3 mos.) excavation
   ≥ 1.5 long-term excavation
5. Determine soldier pile size (include vertical component load of anchors)
6. Tension 5% to 10% of anchors to 150% of design load (no movement for 1/2 hour after cycling 150% of load 4 times) and back off to 80% of design load. Tension remainder of anchors to 100% of design load and back off to 80% of design load.
7. *Do not allow settlement* of soldier piles or sheet piling.
8. Typical Wall Movement
   Vertical – .1% H in stiff (compact sand, hard clays) soils to 1.0% H in weak soils
   Lateral – .2% to .4% H in stiff soils to > 1.0% H in weak soils

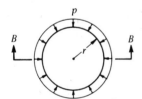

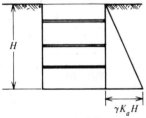

Section *B-B*

$\gamma K_a H$

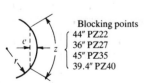

Blocking points
44" PZ22
36" PZ27
45" PZ35
39.4" PZ40

RING WALES

Required Wale Section Modulus $S = \dfrac{MA}{F'A - T}$

S = section modulus (in.^{3})

M = Moment at block point (in.-lbs)

$M = .86 \times eT$ (For wales with pin ended connections. Use .67 in lieu of .66 of moment connection.)

F' = Allowable stress in wale (psi) (based on buckling strength)

T = Thrust; $T = pr$

A = Cross section area (in.^{2})

e = Rise between block points (in.)

$$e = r - \sqrt{r^2 - (g/2)^2}$$

r = Radius of N.A. of wale (in.)

g = Chord length (in.)

Fig. 12.26 Methods for loading and sizing post-tensioned tiebacks and ring wales.

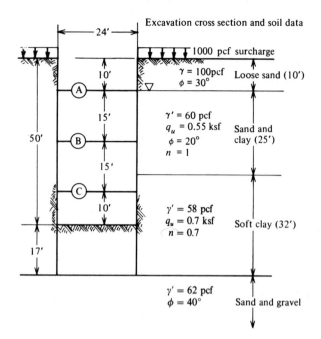

Excavation cross section and soil data

Length of excavation = 144'
Avg. unit soil weight:

$$\gamma' = \frac{1}{50} \left[(.1)(10) + (.06)(25) + (.058)(15) \right] = .067 \text{ ksf}$$

Avg. unconfined compressive strength of all soil:

$$q'_u = \frac{1}{50} \left\{ (.1)(.33)(10)^2(.577) + (50 - 10)(.7) \left[\frac{1}{40} (25 \times .55 + 15 \times .7) \right] \right\} = .38 \text{ ksf}$$

Stability number:

$$N = \frac{\gamma H}{c} = \frac{(.067)(50)}{.7/2} = 9.5 > 4$$

but this soft clay is not normally loaded; ∴ $m = 1.0$.

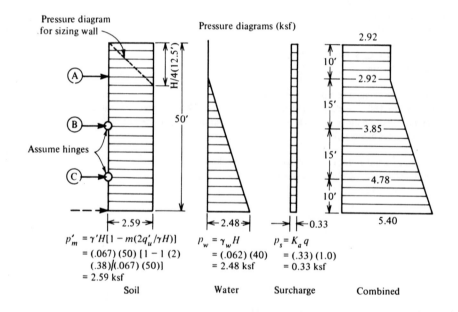

Pressure diagram for sizing wall

Assume hinges

Pressure diagrams (ksf)

$$p'_m = \gamma'H[1 - m(2q'_u/\gamma H)]$$
$$= (.067)(50)[1 - 1(2)$$
$$(.38)/(.067)(50)]$$
$$= 2.59 \text{ ksf}$$
Soil

$$p_w = \gamma_w H$$
$$= (.062)(40)$$
$$= 2.48 \text{ ksf}$$
Water

$$p_s = K_a q$$
$$= (.33)(1.0)$$
$$= 0.33 \text{ ksf}$$
Surcharge

Combined

Strut Loads per Foot of Excavation Length

From Fig. 12.22 determine the apparent strut loads:

Strut A:

$$(10)(2.92) = 29.2 \text{ kips}$$

$$\left(\frac{15}{2}\right)\left[\frac{2.92}{2}\left(2 + \frac{3.85 - 2.92}{2}\right)\right] = \frac{27.0 \text{ kips}}{56.2 \text{ kips}}$$

Strut B:

$$(15)(3.85) = 57.8 \text{ kips}$$

Strut C:

$$\left(\frac{15}{2} + \frac{10}{2}\right)\left[\frac{1}{2}\left(\frac{4.78 + 3.85}{2}\right) + \frac{1}{2}\left(\frac{4.78 + 5.4}{2}\right)\right] = 58.8 \text{ kips}$$

Bottom:

$$\left(\frac{10}{2}\right)\left[\left(\frac{4.78 + 5.4}{4}\right) + \frac{5.4}{2}\right] = 26.2 \text{ kips}$$

Size Struts

Strut spacing = 12.0'
Unbraced strut length = 12.0'
Note: No overstress allowed because of existence of high static water table.

Strut A

Total load: $12 \times 56.2 = 674$ kips
First try, assume $F_a = 19$ ksi (A36 steel)

$$\text{Required area} = \frac{674}{19} = 35.5 \text{ in}^2$$

Try W14 × 120 (Area = 35.3 in²)
$KL = (1.0)(12) = 12$

From AISC Table 1 the allowable axial load is 676 kips > 674 kips

Check: $KL/r = \dfrac{(1.0)(12)^2}{3.75} = 38.4$

From AISC Table 3-36, $F_a = 19.31$ ksi (O.K. to use)

Remainder of struts are sized in the same manner.

Size Sheet Piling ($F_b = 25$ ksi for A328 steel)

Assume overstress factor of 1.5

Find maximum moment per foot of excavation

Moment between B and C, assuming hinges at B and C:

Approx. uniform load $\dfrac{3.85 + 4.78}{2} = 4.32$ kips per foot

Moment $= 432\dfrac{(15)^2}{8} = 121.5$ ft-kips

Required section modulus: $S = \dfrac{(121.5)(12)}{(25)(1.5)} = 38.9$ in^3/ft of wall

Use PZ35 at $S = 48.5$ in^3/ft

Note: If tieback anchors were used, the piling would have to be sized for the downward force component of the anchors.

Size Wales (A36 steel)

Assumptions and conditions:

1) Wales assumed to be axially loaded
2) Flanges of wales supported by sheet piling
3) Apparent strut loads used
4) Load reduction of .67 used (Teng, 1962)

Level A:

Axial load $= (.67)(56.2)\left(\dfrac{12}{2}\right) = 226$ kips

Moment $= (.67)(56.2)\dfrac{(12)^2}{10} = 542$ ft-kips

Try W30 × 173

(Area $= 50.8$ in^2, $r_x = 12.7$ in, $r_y = 3.43$ in, $s = 5.39$ in^3)

Axial stress $f_a = \dfrac{226}{50.8} = 4.45$ ksi

$K_y\dfrac{L_y}{r_y} = (1.0)\dfrac{(12)(12)}{3.43} = 42$

$K_x\dfrac{L_x}{r_x} = (1.0)\dfrac{(12)(12)}{12.5} = 11$

From AISC Column Table 3-36, $F_a = 19.03$ ksi

x–x axis bending stress $f_b = \dfrac{542 \times 12}{539} = 12.07$ ksi

Check for compact section:

$\dfrac{b_f}{2t_f} = 7.04 < 10.8$ (O.K.)

$\therefore F_b = .66\,F_y = (.66)(36) = 24.0$ ksi

Let $c_m = 1.0$ and see if

$$\dfrac{f_a}{F_a} + \dfrac{c_m f_b}{(1 - f_a/F'_e)\,F_b} \leqslant 1.0$$

and

$$\dfrac{f_a}{0.6\,F_y} + \dfrac{f_b}{F_b} \leqslant 1.0$$

Thus

$$\dfrac{4.46}{19.03} + \dfrac{(1.0)(12.07)}{\left(1 - \dfrac{4.46}{\dfrac{12\pi^2(29\,000)}{23(11)^2}}\right)(24.0)} = .74 < 1.0 \text{ (O.K.)}$$

$$\dfrac{4.46}{(0.6)(36.0)} + \dfrac{12.27}{24.0} = .72 < 1.0 \text{ (O.K.)}$$

This wale is larger than necessary, and a smaller member may be tried. Remainder of wales on Levels B and C are similarly sized.

Bottom Stability

From Fig. 12.23a

$B/\sqrt{2} = 24/\sqrt{2} = 16.9' \approx h\ (17.0')$. Therefore, the hard sand and gravel stratum is below the excavation sufficiently far for a full plastic zone to develop.

$\text{F.S.} = N_c\dfrac{c}{\gamma H + q}$

$H/B = \dfrac{50}{24} = 2.08$ and $B/L = \dfrac{24}{144} = .17$

$\therefore N_c = 7.4$

$\text{F.S.} = 7.4\dfrac{.7/2}{(.067)(50) + 1.0} = .60 < 1.5$. Drive sheeting several feet into hard stratum. Maximum moment occurs between struts B and C.

For computation purposes, the water table level was assumed to be static; however, there is always some flow as the "permeability" of sheet piling with ball and socket interlocks has been assumed to be as high as 0.1×10^{-6} cm/sec.

Normally, the water table would be lowered by well points prior to making the excavation. The support elements are large and could be reduced if even partial dewatering were done.

Fig. 12.27 Design of sheet piling and bracing in a retained excavation.

According to the results of a number of strain measurements made on walls and supports of retained structures, the wall bending moments are very different from those initially calculated. This is particularly true for stiff walls. Tieback loads vary greatly from tieback to tieback, but the aggregate tieback loads are generally close to the computed loads. Hansmire et al. (1989) reported on wall and tieback measurements on a slurry wall constructed in Boston. The wall was made part of the permanent structure and the influence of the tieback loads on the wall bending movements was strong enough (from wall bending moments) to predict that long-term design loading would not occur.

Innovations are numerous and will most likely continue to multiply as the job end result requirements can be rigorous, but restrictions on excavation design and construction are not.

It is regrettable that the designer of retaining structures generally has so little control over the execution of his work. Soil pressure is affected by many conditions, not the least of which are the method of construction and the workmanship involved. If specified retaining structure deformations are not adhered to, large pressure changes may occur. Lack of adequate inspection procedures and manpower may result in the use of slugged or other nonconforming welds, misaligned or misplaced support members, and improper sequencing of work.

12.2.3 Single-wall Cofferdams in Water

This type of cofferdam is usually used to provide a coffer for the construction of structures such as bridge piers in rivers, lakes, and estuarial and shallow ocean areas. Almost all such cofferdams are temporary. While adequate design and construction methods are utilized the effects of long-term exposure to the environment, such as corrosion, do not have to be considered. Much ingenuity is used in the design and construction of water cofferdams. Few structures can be so quickly destroyed through carelessness in design, construction, or subsequent construction operations.

Figure 12.28 shows some configurations of single-wall cofferdams and Figure 12.29 is a photograph of a constructed open-water cofferdam. Considerable variety in cofferdam construction has been used to accommodate site conditions and construction requirements. Large, shallow coffers have been constructed with exterior supports as shown in Figure 12.28d. Two single walls of the cofferdam shown in Figure 12.28b were used to reduce the pressure on the cofferdam surrounding the work area. Cantilever cofferdams have been used in shallow water when applicable.

Recognition of conditions and actions that could prevent or assist the successful completion of a coffer requires experience. Some things to consider are:

- If outward movement of the lower portion of the sheet piling occurs during tremie placement, the coffer may collapse or dewatering may be difficult.
- Piling interlocks tend to close during dewatering. It is usually advantageous to quickly dewater the top portion of the coffer to effect a seal in the lower portion of the cofferdam.
- Plastic sheets can be placed on the outside of the sheet piling to reduce inflow during and after dewatering. Cinders, sawdust and other materials have been placed near the interlocks and carried to the interlock by water flow to effect a seal.
- Soil debris that can collect in the trough portion of the sheet piling should be cleaned away before a tremie seal is placed. Otherwise water will enter the coffer at that point during dewatering.
- When bracing is at or below the exterior soil line, it must be tight against the sheet piling to prevent soil movement.
- If the sheet piling bears on rock, it may not be seated adequately to develop a seal and grouting of the toes may be necessary.

A. DESIGN

Most temporary structures are the responsibility of the contractor, but the design and construction are usually overseen by the owner or owner's agent.

The cofferdam must be braced to resist load in all three planes. It must resist hydrostatic and soil pressures. Additional forces from waves and barge mooring and berthing must be considered. The design should accommodate the structure requirements within the coffer and recovery of the cofferdam after the interior construction is completed.

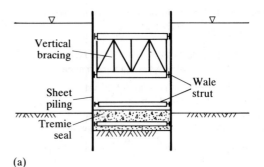

(a)

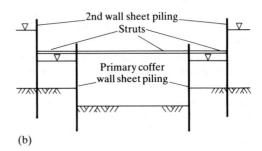

(b)

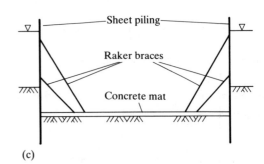

(c)

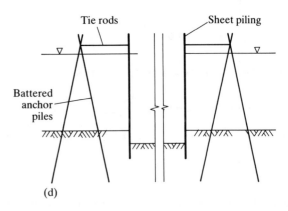

(d)

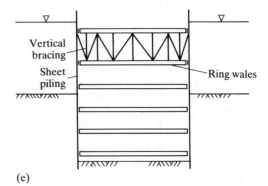

(e)

Fig. 12.28 Cross sections of single-wall cofferdams. (a) Rectangular, braced cofferdam with bottom concrete tremie seal in place. (b) Single-wall cofferdam with intermediate second wall used for reducing load on the primary coffer wall. (c) Raker-braced cofferdam with well-embedded sheet piling to prevent pullout. (d) Exterior supported cofferdam. (e) Circular cofferdam with compression rings and vertical plane bracing. If the ring wales are adequately attached to the sheet piling, the vertical bracing may not be required.

Fig. 12.29 Photograph of an open-water single-wall cofferdam. Foundation piles have been driven and cut off in one section. Note the vertical bracing in the coffer.

The steps outlined below should be included before and during the design of an open water cofferdam.

1. Obtain site data—water depth (minimum and maximum), currents, wind intensity, frequency, duration, soil properties, water-borne traffic, wave type and size.
2. Develop cofferdam and construction data—coffer size, internal structure foundation requirements (piling, soil or rock bearing), open areas required in bracing, mooring facilities and access, construction equipment to be used, sheet piling penetration, dewatering method, construction sequence, and cofferdam removal.
3. Member sizing—develop loading diagrams in horizontal and both vertical planes (Fig. 12.30a), size members for handling underwater or assembling while floating if method of construction requires it.

B. CONSTRUCTION

The sequence of placing cofferdam components is determined by the contractor's experience, internal structure construction, and site conditions. Contractor skill and planning are of utmost importance. Numerous methods of cofferdam construction have been used. Two typical methods of construction are outlined below (see Figure 12.30b):

SHALLOW COFFERDAMS (≤ 40 feet)

1. Place one or more temporary templates at the site and at the proper elevation through the use of spud piles.
2. Place and drive the sheet piling around the template.
3. Remove the template(s) and place the required prefabricated bracing. Bracing can be supported from the sheet piling or by spud piles.
4. Place the vertical bracing prior to or during dewatering.

Depending on the condition of the bottom, a tremie seal may have to be placed before dewatering and if piles are to be driven the tremie seal cannot be placed until the pile driving is completed.

DEEP COFFERDAMS

1. Prefabricate bracing frames and move them to the site by barge or flotation equipment.
2. If multiple frames are to be used, drive spud piles through openings in the stacked frames.
3. Lower each frame to its designated elevation.
4. Place and drive sheet piling around the frames and secure the sheet piling to the frames by the use of divers.
5. Place the vertical bracing prior to or during dewatering.

(See note about the seal and bearing piles above. Almost all deep cofferdams require tremie seals.)

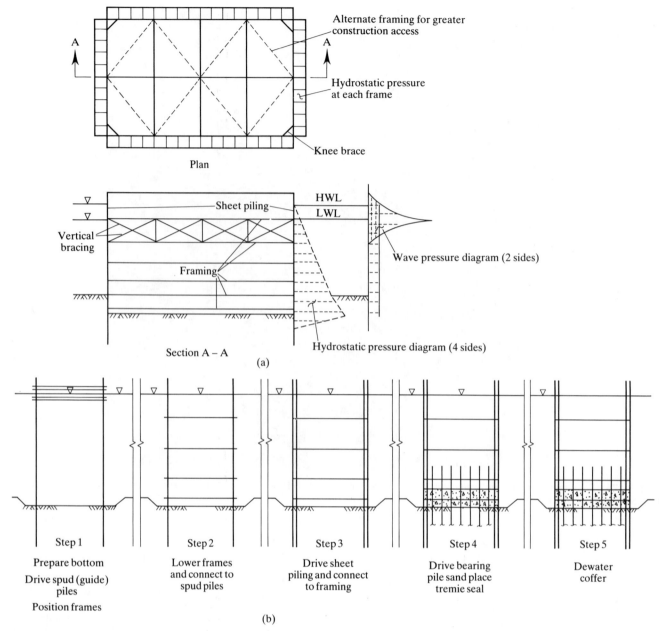

Fig. 12.30 Loads on an open-water single-wall cofferdam and the construction sequence. (a) Typical layout and loading. (b) Typical construction sequence.

12.3 GRAVITY STRUCTURES

This section covers gravity structures such as filled cells made of steel sheet piling, filled cells made of precast concrete, and bin and crib structures.

Crib and bin structures have been used for a far longer time than other structures that utilize materials other than wood. One of the characteristics of bin and crib structures is that they can be constructed without the aid of large handling equipment.

12.3.1 Filled Steel Sheet Piling Cellular Structures

Filled cells are a combination of earth and piling that, individually, are unstable. When soil is placed inside the sheet

piling the cells stabilize, enabling it to resist lateral and vertical forces.

Cellular structures are used for the construction of piers, marginal wharves, breakwaters, cofferdams, graving docks, sand islands, breasting, and mooring dolphins. In addition, they have been used for stabilizing loaded areas behind bulkheads. Figure 12.31 shows most of the commonly used filled cellular configurations.

Cofferdams—composed of any material—represent a somewhat enterprising type of structure owing to the fact that they are usually temporary and have a low safety factor; they can resist large lateral loads, which may cause noticeable cell movement, and are generally installed where difficult site conditions exist. In contrast, where cells are used for the construction of marginal wharves, piers, etc., the lateral loads

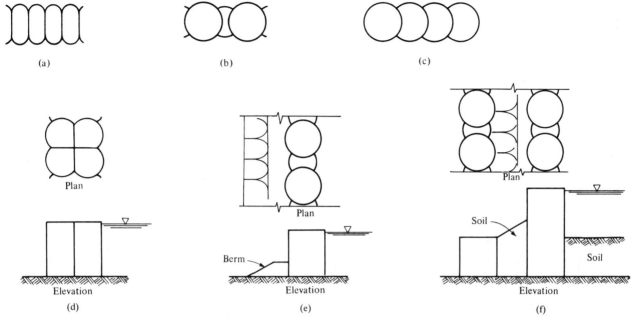

Fig. 12.31 Cellular configurations. Single-cell configurations used for most installations: (a) diaphragm; (b) circular; (c) semicircular. Cellular arrangements to resist high lateral loads: (d) cloverleaf; (e) circular with berm; (f) double circular cells.

are usually smaller, site conditions are likely to be somewhat better and, although simpler configurations are used, greater restrictions as to lateral and vertical movement of the cells are necessary.

Flat web piling is widely used for the larger cells as shallow arch piling tends to flatten out above interlock (hoop) loads of 3 kips per inch. Various cell configurations (such as double cells and cloverleafs), lightweight cell fill, and stepped and berm methods of construction have been used to obtain greater cell heights. Sheet piling with published ultimate interlock strength of up to 28 kips per inch will allow greater cell heights than before.

Generally, the ratio of average width to height of diaphragm and circular cells used in water-retaining cofferdams on rock is 0.85, although much lower ratios (less than 0.6) have been used successfully. On this basis, cells may be classified as being small (with heights up to 40 ft), medium (up to 70 ft), and large (over 70 ft).

A. REQUIRED DATA FOR CELLULAR STRUCTURE DESIGN AND INSTALLATION

Depending on the location and function of the cells, various data concerning the site and soil conditions must be developed for a satisfactory installation.

If cells are to be constructed in a river, the necessary information about the site would include the following:

1. Low- and high-water elevations
2. Probable water flow rate and flow rate changes during the life of the cell structure
3. Occurrence and quantity of ice
4. Scour patterns
5. Cross section and ground line profiles
6. Soil type and thickness
7. Soil permeability, shear strength, and consolidation data
8. Granular borrow and spoil areas
9. Water quality (particularly for permanent installations)
10. Deterioration potential
11. River traffic and site obstructions

Where coastal installations are to be constructed, some of the data (such as flow information as outlined above) may not be useful; however, additional data as listed below is necessary:

1. Wave height and period
2. Tidal range and times of tide change
3. Rapidity of storm buildup
4. Weather-warning station locations
5. Ship impact, bollard loading

The site conditions and structure requirements are then used to establish the design criteria. For instance, the following specific items must be established in the design of a cofferdam:

1. External loading (see examples of external loads in Figure 12.32)
   (a) High- and low-water levels and rate of change
   (b) Wave forces
   (c) Impact from floating objects
2. Seepage
   (a) Quantity of water entering coffer under various heads
   (b) Water collection system in coffer and pump capacity
   (c) Seepage forces
3. Dewatering and flooding
   (a) Size of coffer intake structure
   (b) Pump capacity for dewatering coffer
4. Erosion control
   (a) Size of required riprap, cap stones, etc.
5. Foundation
   (a) Maximum base loading
   (b) Settlement
6. Cofferdam dimension
7. Allowable cell movement
8. Structure life

Some of the listed items may not apply to river and/or open water sites or all site conditions; however, the list was meant to be general in nature. Items 1, 4, 5, 7, and 8 are applicable for the design of piers, wharves, breakwaters, and other structures.

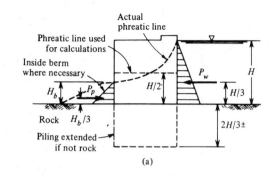

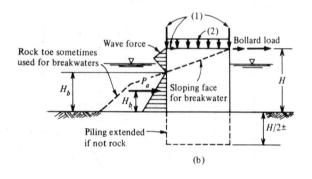

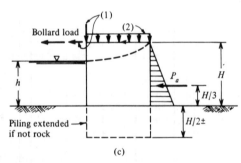

Fig. 12.32 Examples of external loads on cellular structures. (a) Cofferdam. (b) Pier or breakwater: 1, unloading equipment live loads; 2, operations surcharge loading (1000 psf). (c) Marginal wharf (no surcharge behind cell): 1, unloading equipment live loads; 2, operations surcharge loading.

B. DESIGN PROCEDURE

After obtaining the applicable general information outlined in the previous section and making decisions such as whether or not overtopping may be allowed, the engineer may set tentatively the cell location and height; the next step is to size the cells.

Prior to further discussion of the cell sizing procedure, it is necessary to point out that all design methods are based on a rectangular cell configuration. This reduces the computation work in determining cell properties and is a simplification no less rigorous than other assumptions made concerning cell design. Rossow, Demsky, and Mosher (1987) point out that owing to wall curvature there is a component of interlock tension normal to the longitudinal axis of the cofferdam. They conclude that a three-dimensional finite-element analysis is necessary for computations using the cellular cofferdam configuration with simplification, and that approach is not feasible for design offices. Figure 12.33 shows the commonly used relationships for determining cell dimensions.

When sizing the cell various failure modes are considered. These are shown in Figure 12.34 and listed below in the general order of importance:

1. Excessive tilting or base rotation of cell
2. Interlock and connection failure
3. Stability of base and sliding
4. Loss of cell fill due to piling rise
5. Overturning

The order of the above listing is primarily suited for cofferdam application.

Outboard piling pullout is not included as a failure mode because no evidence exists to support the inclusion. Inboard piling plunging is not included for the same reason. Some cellular bulkhead wharves have been constructed with a retaining wall and slab supported by the waterside sheet piling. The slabs have the normal wharf unloading equipment and loads acting on it with no distress to the cells or sheet piling.

C. INTERNAL STABILITY

There are two methods that are widely used to determine the stability of the cells: the horizontal shear method (Cummings, 1957) and the vertical shear method developed by Terzaghi

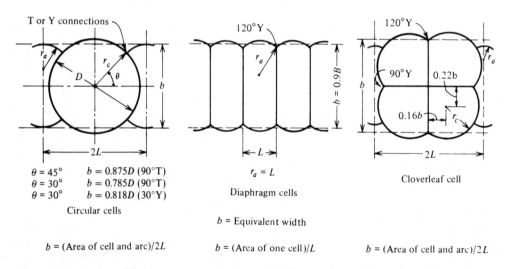

$\theta = 45°$ $b = 0.875D$ (90°T)
$\theta = 30°$ $b = 0.785D$ (90°T)
$\theta = 30°$ $b = 0.818D$ (30°Y)

Circular cells

$b = $ (Area of cell and arc)$/2L$

$r_a = L$

Diaphragm cells

$b = $ Equivalent width

$b = $ (Area of one cell)$/L$

Cloverleaf cell

$b = $ (Area of cell and arc)$/2L$

Section modulus/linear ft $= (1/6) \times (b^2/2L)$ in ft^3 for b and L given in feet

Fig. 12.33 Nomenclature and design properties for common cellular configurations.

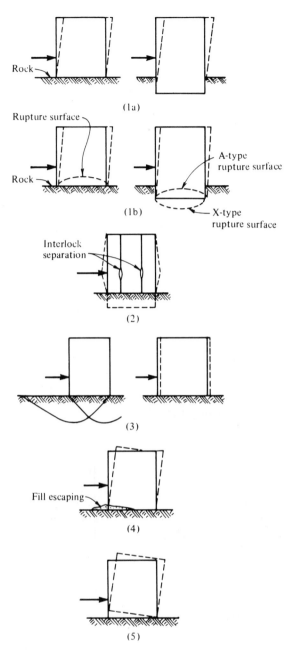

Fig. 12.34 Failure modes of filled cells. (1a) Internal shear (tilting), vertical shear (Terzaghi, 1945; TVA, 1957) and horizontal shear (Cummings, 1957); (1b) rotational shear surface (rotation) (Hansen, 1953; Ovesen, 1962). (2) Interlock separation (TVA, 1957). (3) Base failure and sliding. (4) Piling rise on loaded side (TVA, 1957). (5) Overturning.

(1945) and TVA engineers (1957). Belz (1970) presented an excellent review of both methods. A third but little-used method developed by Hansen (1953) and elaborated by Ovesen (1962) is of interest. The first two methods are based on the assumption that failure of the cells is an internal shear failure. Hansen's concept of failure is not based on internal failure but, rather, on a base failure—assuming the cell fill acts as a unit. Several other stability methods have been proposed but, to date, are only of academic interest.

The vertical shear method was developed in the late 1930s and early 1940s. Since details of this method have been covered elsewhere only the outlines of the method are given in

Figure 12.35. This method was developed to explain the resistance of filled cells to tilting. Terzaghi (1945) assumed a K factor for lateral pressure at the center of the cell that indicates that it was a principal plane. This assumption was corrected because shear is assumed to act in the vertical plane. Ever since there has been controversy as to the correct value of K. Currently used values of K range between $1.2K_a$ and $1.6K_a$.

As in the case of the vertical shear method, the horizontal shear method has been covered in detail elsewhere, so only the basic design method is given in Figure 12.36. Schroeder and Maitland (1979) proposed that a plane of fixity exists for the piling driven into soil below the dredge line or coffer bottom and that pressure calculations should be based on the effective height H_e, the distance from the plane of fixity to the top of the cell. The plane of fixity is defined to be where plastic hinges are located in the sheet piling. For clay foundations the distance from the plane of fixity to the coffer floor is (Hetenyi, 1946)

$$d' = 3.3\left(\frac{EI}{E_s}\right)^{1/4}$$

where

E_s = horizontal spring modulus
I = moment of inertia of piling section
E = modulus of elasticity of the piling material

The equation is valid if the piling is driven into the clay a depth of

$$d' \geqslant 4\left(\frac{EI}{E_s}\right)^{1/4}$$

For granular foundations d' is located by (Hetenyi, 1946)

$$d' = 3.1\left(\frac{EI}{n_h}\right)$$

where n_h = constant of horizontal subgrade and is calculated by

$$n_h = \frac{b_s}{d'} l_n$$

where

b_s = width of single-sheet piling
l_n = constant of horizontal subgrade reaction for anchored bulkheads with free earth support

The horizontal spring modulus E_s is calculated by

$$E_s = b_s K_{se} \frac{l}{d'}$$

where K_{se} is the coefficient of vertical subgrade reaction, whose values are given in Tables 2 and 4 of Terzaghi (1955).

The equation for finding the value of d' in granular soils is valid if the piling is driven into the soil a depth of

$$d' \geqslant 5\left(\frac{EI}{n_h}\right)$$

An alternative for establishing the plane of fixity is done by locating the point where internal (assumed active) and external (assumed passive) pressures are equal.

The assumption that horizontal shear planes develop, as given in the horizontal shear method, could lead a designer to assume that fill in the unloaded side of the cell could be reduced without significantly affecting the results. Although this procedure was inadvertently actually followed in some of the earliest cofferdams, it is not recommended.

The rotation method was conceived by Hansen (1953) after an observation that a circular rupture surface developed at the

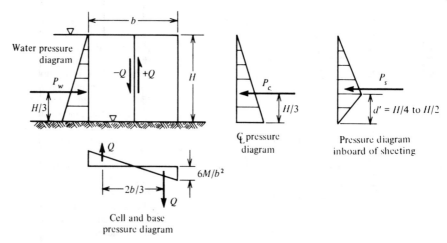

Total shearing force/Unit length: $Q = 1/2 \times (b/2)\dfrac{6M}{b^2} = \dfrac{3}{2}\dfrac{M}{b}$

where $M = P_W H/3$

Total centerline shear resistance: $S_c = P_c \tan \phi$

where $P_c = (1/2)K\gamma H^2$

and $K = .4$ to $.6$

Total interlock friction resistance: $F_T = fP_s = f(1/2)\gamma HK\,(H - d')$

where $f_{ss} = .3$ frictional resistance of steel on steel

Factor of safety: $F.S. = \dfrac{S_c + F_T}{Q} \geqslant 1.25$ (temporary)

$\geqslant 1.50$ (permanent)

Fig. 12.35 Internal stability of filled cell by the vertical shear method.

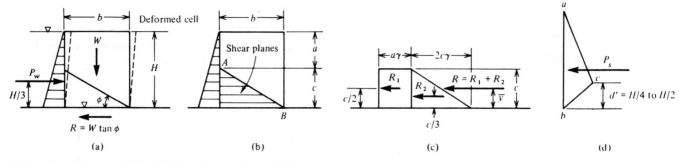

(a) Wedge resistance inside cell. The resistance due to sliding on planes below AB in (b) is

$$R = W \tan \phi = \gamma bH \tan \phi$$

Let $H = a + c$ and $b = c/\tan \phi$. Then

$$R = ac\gamma + c^2\gamma$$

The concept of shear resistance is given in (c). The total moment of resistance per foot of wall about cofferdam base is

$$M_r = R_1 (c/2) + R_2 (c/3)$$

Let $R_1 = ac\gamma$ and $R_2 = c^2\gamma$.

The interlock friction force is given by

$$F_T = P_s L f_{ss}$$

where P_s is the area of diagram abc in (d) (pressure on the interlock) and L equals one-half the distance from center to center of cells.

The total interlock resisting moment about the cofferdam base is

$$M_T = F_T b/L = P_s f_{ss} b$$

and the factor of safety is

$$F.S. = \dfrac{M_r + M_T}{(1/3)P_W H}$$

The minimum recommended factor of safety is 1.25 (temporary) and 1.50 (permanent).

Fig. 12.36 Internal stability of filled cell by the horizontal shear method.

base of a model double-wall cofferdam that was loaded laterally to failure. Two variations of analysis, based on the above observation, were developed. These were the equilibrium and extreme methods. Figure 12.37 illustrates some details of the extreme method. The cell, assumed to be filled with granular material, is founded on a rock or granular base. The equilibrium method is not presented here but may develop into a very useful analysis and design method if, as claimed by Ovesen, deflection of the cell may be determined. The extreme method is easier to use and gives only the ultimate resistance to rotation. The extreme method is used to simplify calculations by utilizing a log spiral as the rupture surface. The log spiral approximates a circular rupture surface, and the force resultant along the rupture surface passes through the rotation point, thereby eliminating those forces from the computations. Figure 12.38 shows the geometrical difference between two spirals with angles of internal friction ϕ equal to 30° and 38°, respectively.

Figure 12.39 is a graph of the internal resistance of cells (as given by the horizontal and vertical shear methods) versus cell height. Since the resistance expression for each method is different, the shape of the curves is of interest. The internal resistance of the cell, as computed by the horizontal shear method, is affected by the cell height more than that of the vertical shear method for cell heights slightly less than 100 feet.

For varying b/H ratios and the conditions given in Figure 12.40, the ratio of the resistance to tilting contributed by interlock friction to that contributed by fill shear is constant (0.19) when computed by the vertical shear method. This ratio changes (reduces from 0.33 to 0.25) with increasing b/H ratios (0.7 to 1.0) for the horizontal shear method. The influence of interlock friction in the rotation method is smaller than in the other two methods.

A comparison of safety factors, as calculated by the three methods, against internal instability for given site conditions appears in Figure 12.40.

The ratio K of lateral to vertical stress in the vertical shear method is taken as 0.6.

D. INTERLOCK AND CONNECTION FORCES

1. Interlock

Terzaghi (1945) concluded that there was no significant change in hoop force due to deformation of the cell caused by lateral loads. His reasoning was based on the fact that cells had not burst after application of lateral loads. Since the publication of Terzaghi's paper, a number of field tests of circular cells have been reported. Results of one test reported by White, Cheney, and Duke (1961) included hoop stress during cell filling and backfilling behind the cell. There were hoop stress changes due to backfilling. In general, the hoop stress increased on the water side of the cell and decreased on the backfilled side; however, the stress values were low and the changes were relatively small.

Other field and model tests have been reported: Schroeder et al. (1977); Schroeder and Maitland (1979); Sorota et al. (1981); Sorota and Kinner (1981); and Schroeder (1987). From these results and results from the Lock and Dam 26 cofferdam tests Rossow et al. (1987) concluded that the crosswall interlock loading near the 30° Y connection may be as much as 20 percent higher than the main cell tension even though the test data is not completely consistent.

The hoop or interlock forces are dependent on the type of soil with which the cell is filled, the diameter or radius of curvature and height of the cell, construction procedure, and surcharge loads. For cellular cofferdams the dewatering and internal drainage conditions affect hoop forces.

The cell hoop force outside the connecting arcs and the arc hoop stress may be calculated by the equation

$$t_a = (\text{or } t_{c1}) = (1/12)p \times r_a \text{ (or } r_c)$$

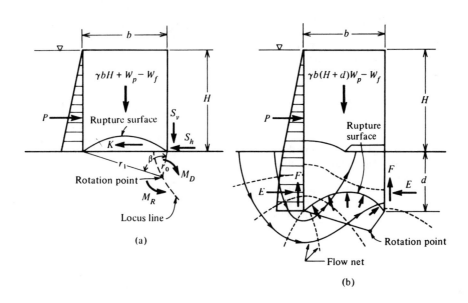

Stability calculations (extreme method):

(1) Generate log spiral locus line $(r = r_0 e^{\alpha \tan \phi})$ for given angle of internal friction ϕ (spiral intersects feet of piling).

(2) Compute loads $(P, \gamma bH, \gamma b(H+d), W_p, W_f, E, \text{ and } F)$ and reactions (S_h, S_v) (K acts for diaphragm cells but not circular cells).

(3) Take moments about points on the locus line. Plot safety factor $\dfrac{M \text{ (Driving)}}{M \text{ (Resisting)}}$ vs. r_0 to obtain minimum safety factor.

(4) Recommended safety factors: 1.15 (temporary), 1.35 (permanent).

Fig. 12.37 Stability of filled cell on rock and soil by the rotation (extreme) method. (a) Rock surface. (b) Sand foundation. (*After Ovesen, 1962*).

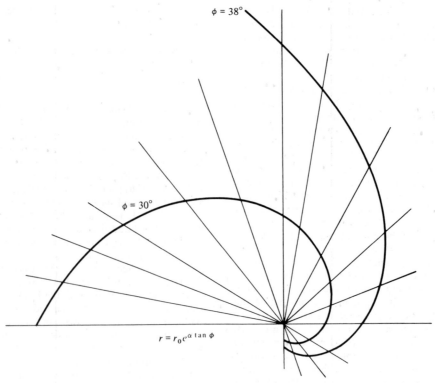

$$r = r_0 e^{\alpha \tan \phi}$$

Fig. 12.38 Log spiral curves.

where

t_a = hoop or interlock force for connecting arcs (lb/linear inch)

t_{c1} = hoop or interlock force for cell outside of arcs (lb/linear inch)

p = lateral unit pressure (lb/ft²) (p is taken as maximum $\frac{1}{4}$ to $\frac{1}{2}$ the distance H from the mudline to the top of the wall in the TVA Method. However, Schroeder and Maitland (1971) suggest that the pressure should be determined at the plane of fixity)

r_a = arc radius (ft)

r_c = cell radius (ft)

The above formula may be used to determine the hoop force in the connecting arc, except that r_c must be replaced by the radius r_a of the connecting arc.

The greatest interlock force, located just inside the arc connection, is often overlooked (TVA, 1957). The hoop cell force at the arc connection is calculated by the TVA formula

$$t_{c2} = p \times (L/12) \times \sec \theta$$

where

t_{c2} = circular cell hoop or interlock force between arcs (lb/linear inch)

$L = \frac{1}{2}$ center-to-center distance between cells (ft)

θ = angle (degrees) between centerline of cells and a line from center of cell to point on cell periphery where connecting arc is attached.

Swatek was the first to use the formula $t_{c2} = pL$ to calculate the maximum crosswall tension (t_{c2}) values. This formula is based on the assumption that the average load being restrained by the crosswall is acting on the cofferdam with a length of L

for a given value of p. The t_{c2} value calculated by the TVA secant formula can vary from 15 to 41 percent higher than Swatek's formula depending on the location of the connection on the main cell.

The relative hoop forces at any level in the cell are shown in Figure 12.41 for circular cells.

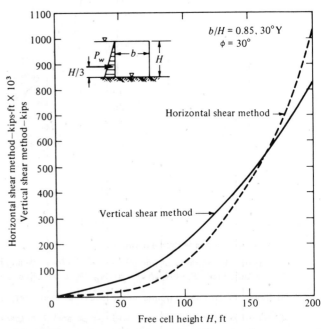

Fig. 12.39 Comparison of internal stability of cell as computed by the vertical and horizontal shear methods

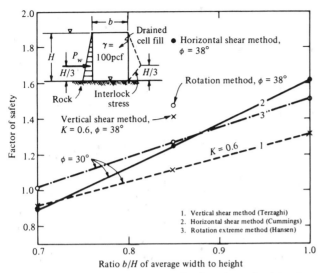

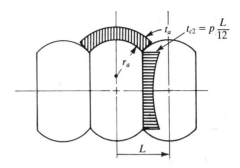

Fig. 12.42 Relative hoop force in diaphragm cell due to cell fill.

Fig. 12.40 Comparison of factor of safety determined by three internal stability calculation methods.

approximately equal loads on all three legs. Hoop force on the front and rear arcs of the cells is calculated using the radius of the arc as r_a. As in the case of that portion of the circular cell between the interlocks, the diaphragm interlock load is reduced by soil friction.

Frictional forces due to soil pressure on both sides of the piling behind or between the connecting arc(s) slightly reduce the hoop or interlock force. In the case of the single-arc bulkhead type, it has been reported that the hoop force is reduced to less than the prestress force caused by filling the cell (White et al., 1961).

The relative hoop and diaphragm forces for diaphragm cells are given in Figure 12.42. The 120° Y connection has

2. T and Y Connections

The most commonly used connections in cellular construction are the T and the Y. These are used primarily to connect the arc to the circular cell, or to connect the diaphragm to the arcs of diaphragm and cloverleaf cells. The connections are the most highly stressed part of the cell, and if failure occurs it usually is within the connection or the connection interlock.

Figure 12.43 shows the theoretical directions of loads acting on T and Y connections. The loading is similar to that given in TVA (1957). Various values of the interlock force between arcs can be evaluated by the method of summation of forces on the T or 30° or 60° Y connections and compared with the values found by using the TVA method. The values of t_{c2}, as calculated by either method, give consistent results. It is interesting to note that the reactive force P' as used in the force-summation method must have unrealistic values to achieve equilibrium. A P' force equal to 40 times the active pressure in T-connection calculations is necessary for equilibrium in some configurations. Actually, the connections undergo some deformation, causing a redistribution of forces.

Figure 12.43a shows a common field position of the arc piling adjacent to the stem. When the arc piling is in that position, considerable additional lateral stem loading occurs. The T can only resist the arc loading by bending of the cell piling and passive soil resistance in the arc area. If riveted T construction is used, the angles that connect the stem to the cell piling web distribute the bending loads through the rivets to the cell piling web. Distribution of load by the rivets to the cell piling web provides a somewhat better loading condition than that of a welded T.

Figure 12.44 is a photograph of a distorted T connection with the cell and arc areas filled. Riveted connections can undergo extensive distortion without failure, and this photograph illustrates the general shape that T connections can assume under high loading.

Using the TVA method, the arc load varies from 30 to 38 percent—average 35 percent—of the maximum load in the cell piling for commonly used cell geometry. As an example, if the temporary maximum hoop load (during cofferdam filling, dewatering, etc.) is 21 kips per inch, the arc load will be approximately 0.35×21, or 7.4 kips per inch. The ultimate stem capacity of some connections is greater than 15 kips per inch (Dismuke, 1970).

Considering "overstressing" of the steel piling in cellular construction, it is well to recall a portion of Terzaghi's

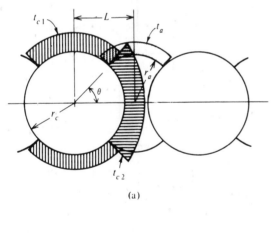

(a)

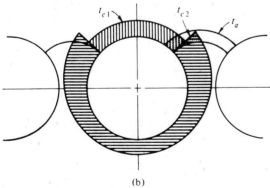

(b)

Fig. 12.41 Relative hoop forces in circular cells. (a) Typical circular cell cofferdam or pier. (b) Typical circular cell bulkhead.

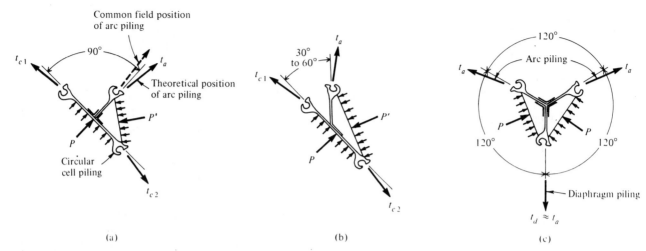

Fig. 12.43 Loads on T and Y connections. (a) T connection (after TVA, 1957). (b) 30° and 60° Y (TVA, 1957). (c) 120° Y for diaphragm cells.

Fig. 12.44 Distorted T connection.

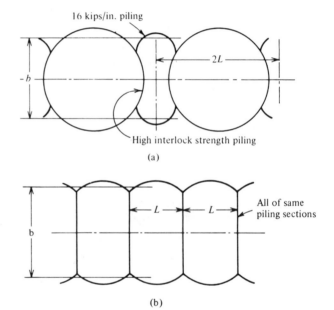

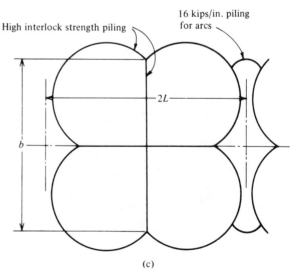

Fig. 12.45 Possible combinations of piling for cellular construction. (a) Circular cells. (b) Diaphragm cells. (c) Cloverleaf cells.

conclusion to his classic paper in which he stated, "In cofferdams, local stressing of the steel beyond the yield point can scarcely be considered a failure...."

The use of combinations of sections with different interlock strengths may, in some cases, be advantageous.

Figure 12.45 shows possible combinations for use in circular cells. The circular configuration causes greater stress variations than the diaphragm type; therefore, it is more applicable to the circular cells.

The design values for interlock strength, as currently used, are

AISI Section	Design Interlock Strength
PSA23	3 kips/inch
PS27.5	8 to 10 kips/inch*
PS31	8 to 12 kips/inch*

*Depending on grade of steel

Figures 12.46 and 12.47 show the variation of the factor of safety with cell height for interlock loads and three different cell diameters. The horizontal shear method gives larger safety factors than the vertical shear method.

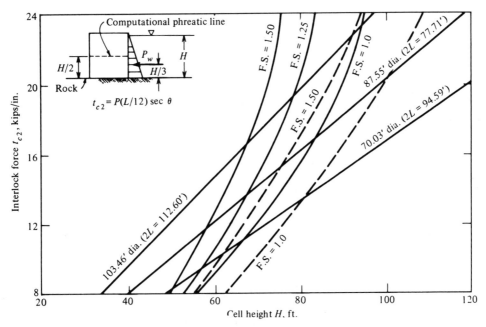

Fig. 12.46 Interlock force vs. cell height of circular cellular cofferdam for various factors of safety. The solid lines represent the stability safety factor by the TVA method and the dashed lines represent that from the Cummings method. In the figure, $\theta = 45°$; $b = 0.875D$; the weight of unsaturated cell fill is 110 pcf; weight of submerged cell fill is 65 pcf; T connections.

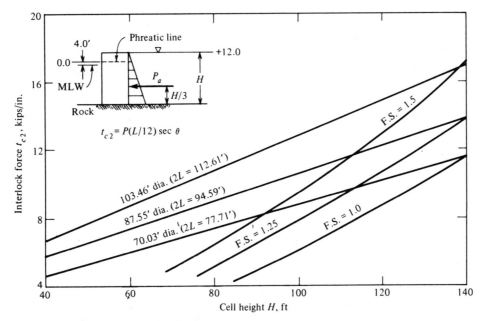

Fig. 12.47 Interlock force vs. cell height of circular cellular bulkhead for various factors of safety. TVA stability method used. In the figure, $\theta = 45°$; $b = 0.875D$; weight of unsaturated soils is 110 pcf; weight of submerged soils is 65 pcf; T connections.

E. BASE STABILITY AND SLIDING

1. Base Stability

Cells not founded on rock must be investigated for base stability. If the base soil is a deep sand structure, the additional cell weight may cause shear failure. If the cells are laterally loaded as in a cofferdam, pile settlement on the inboard side and excessive underseepage may occur. If the base soil is clay, shear failure is possible. Where the clay is soft-to-medium (compressible) the settlement of the soil in and below the cell may cause slippage to occur in the sheet-piling interlocks, thereby overcoming the frictional resistance caused by hoop tension. When a lateral load is applied, the inboard sheeting is highly loaded owing to the soil settlement, and failure could result.

(a) bearing capacity of base

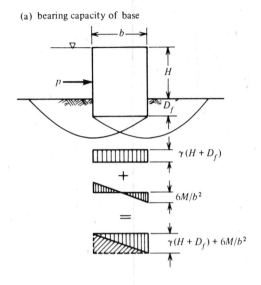

Note: When $D_f = 0$, the critical condition exists.

Bearing Capacity Factors (Terzaghi and Peck, 1967)

Ultimate Bearing Capacity

$$q_f = 1/2\, b\gamma N\gamma + CN_c + \gamma D_f N_q$$
for strip loaded area

$$q_f = 0.6\, \gamma b N\gamma + 1.3\, N_c + \gamma D_f N_q$$
for circular load areas

where

γ = Unit weight of soil around cell (pcf)
b = Equivalent cell width (ft)
N_γ = Bearing capacity factor
C = Cohesion (psf)
N_c = Cohesive factor
D_f = Ground surface to toe of cell (ft)
N_q = Surcharge factor

$$\text{F.S.} = \frac{q_f}{\gamma(H + D_f) + \dfrac{6M}{b^2}} \quad \begin{array}{l} \geqslant 2.0 \text{ for sand} \\ \geqslant 2.5 \text{ for clay} \end{array}$$

(b) Internal instability due to settlement of compressible base (for soft and medium clays, q_u = 400 psf to 1000 psf)

$$\text{F.S.} = \frac{(P_p - P_s)\,(D/2)f_{ss}\,(b/L)\left(\dfrac{L + .25b}{L + .5b}\right)}{M} \quad \begin{array}{l} \geqslant 1.25 \text{ (temporary)} \\ \geqslant 1.5 \text{ (permanent)} \end{array}$$

where

P_s = Inboard pressure
P_p = Passive pressure of berm and/or overburden on inside of cofferdam
f_{ss} = Coefficient of friction steel on steel
M = Overturning moment (see Fig. 12.35)
D = Diameter
b = Equivalent width

Fig. 12.48 Stability of base soils.

The friction force that causes penetration of the inboard sheeting into the sand–base soil when the cell is laterally loaded is

$$F = P_s \tan\delta \geqslant 1.25 \text{ (temporary and permanent)}$$

where δ is the angle of wall friction and P_s is given in Figure 12.36.

Figure 12.48a gives the method of computing the bearing capacity of the base soil. The soil may be sand, clay, or a mixture. The method is based on treating the cell as a rigid structure and, as the base pressure diagram shows, a uniform base pressure. Both assumptions are conservative under most conditions.

Figure 12.48b gives the method for computing the factor of safety of the cell under lateral loading when the base soil is a compressible clay. A heavy berm is required in most installations when soft clays are the base soils.

Seepage under the cell may cause piping to occur on the inboard side. Sufficient buoyancy of the soil causes loss of bearing capacity and passive pressure. There are several ways of effectively dealing with this problem: driving the piling deeper; providing drain filters on the inside of the cofferdam; and placing a berm or increasing the berm height. If the piling is driven deeper, a longer drainage path develops, which reduces the flow rate. Drain filters reduce the buoyancy by permitting greater flow rates. Berms are usually placed over the filters.

When piling is driven to a penetration of $0.67H$ in deep sand, the drainage path is usually sufficiently long to prevent piping. If the water table inside the coffer is lowered $0.17H$ below the ground surface, the penetration may be reduced to $0.33H$.

Chapter 7 covers dewatering, drainage and filter design.

2. SLIDING

The factor of safety against lateral cell movement on rock is computed by

$$F.S. = \frac{Wf_{gr} + P_p}{P}$$

$$F.S. \geqslant 1.25 \text{ (temporary)}$$
$$\geqslant 1.5 \text{ (permanent)}$$

where

W = weight of fill and piling (kips)
f_{gr} = friction of cell fill on rock
P_p = passive pressure on inboard side (kips)
P = lateral pressure from water and/or soil (kips)

When cells are resting on concrete or a wood surface, the value of f_{gr} is reduced and may reach values as low as 0.2. This condition may occur adjacent to a lock, spillway, or other similar structure. A thin stratum of low- or medium-strength clay below the toe of the piling may allow excessive cell movement.

For cells founded on soil, the resistance to sliding may be computed by the method of wedges. The failure plane for all trial failure paths should pass through the toes of the sheet piling. The factor of safety against sliding in this case should be the same as for cells founded on rock.

F. PILING RISE OR PULLOUT

On the loaded side of the cell there is a tendency for the piling to rise due to lateral loading. If this occurs in a cell founded on rock, cell fill is lost and the cell may fail. The factor of safety against piling rise is

$$F.S. = \frac{b(P_w + P_a)f_{gs} + P_pH_b/3}{P_wH/3 + P_aH_s/3} \geqslant 1.25 \text{ (temporary)}$$
$$\geqslant 1.5 \text{ (permanent)}$$

where

b = equivalent cell width
f_{gs} = coefficient of friction between cell fill and the sheet piling
H_b = berm height

The factor of safety against pullout of the outboard piling on sand or clay foundation is

$$F.S. = \frac{\text{pullout capacity/unit length of wall}}{\text{pullout force/unit length of wall}}$$

The pullout force is

$$F_p = \frac{P_wH + P_wd - P_pH_b}{3b(1 + b/4L)}$$

The pullout capacity for cells on sand bases is

$$C_p = \tfrac{1}{2}K_a\gamma d^2 \tan \delta \, 2\pi r_c$$

Pullout capacity for piles driven into clays is

$$C_p = C_a 2\pi rd$$

where C_a = adhesion of clay (psf) (450 to 700 psf for medium stiff clay and 700 to 720 psf for stiff clay).

The factor of safety is then

$$F.S. = \frac{C_p}{F_p}$$

G. OVERTURNING

The cell may be assumed to rotate about its toe. The resultant of the cell weight and lateral forces is restricted to the middle one-third of the cell. The factor of safety is computed by

$$F.S. = \frac{M_o}{M_r} = \frac{Wb/2}{PH/3}$$

where M_r = resisting moment and M_o = overturning moment.

According to Rossow et al. (1987), the eccentricity at the base of the cell that is laterally loaded reduces the effective width of the cell. Therefore, the effective width is $b' = b - 2e$, where e is the distance from the cell center line to the force resultant at the base of the cell.

The overturning concept as a failure mode is used by few designers. The computation is still made, however, in order to determine whether the base reaction falls in the middle one-third of the cell width. If the resultant falls outside the middle one-third, the internal shear stability methods are not applicable, nor are the base stability concepts as indicated by Rossow et al. (1987).

Figure 12.49 gives a comparison of the importance of the various failure modes for a cellular cofferdam on rock. The vertical shear or TVA method of computation was used.

H. EXAMPLES OF CELLULAR DESIGN

Examples of cellular structure designs are given in Figures 12.50 through 12.52. The three methods of internal stability design previously discussed are used in several different foundation conditions.

In the preceding sections recent results and conclusions regarding the design of cellular structures have been presented. This work has provided a basis for further advances in design, but is not conclusive enough to supplant traditional design methods. The traditional designs have worked reasonably well for many years, so the design examples herein represent the traditional methods.

I. FIELD PROCEDURES AND PROBLEMS

Splicing of the sheeting is required for many of the cells that are constructed. During the splicing operation, proper alignment of the interlocks is of utmost importance. In addition, piling of equal width should be used, because a loading discontinuity occurs at the splice since the wider sheet is not loaded. This condition has resulted in tearing of the web of the narrower sheeting (Grayman, 1970).

When driven to an irregular or sloping rock surface, each properly placed piling will be seated on the rock surface, preventing possible loss of cell fill and discontinuities in the hoop forces. However, if the bottom sheeting elevation varies too much from sheet to sheet (1 to 2 ft) or from one side of the cell to the other ($\approx D/4$), a load discontinuity may develop that would result in torn sheets and excessive cell tilt.

A common occurrence is the driving out of interlock. Generally, this occurs under difficult driving conditions or when inexperienced crews are setting and driving. Patterson (1970) has given an excellent review of the proper installation procedure. Piles should not be driven any deeper than absolutely necessary as the chance of driving out interlock increases with driving depth.

Sometimes the specials (connections) are cambered, particularly after having been handled incorrectly and/or subjected to a rough unloading procedure. The cambering causes the special to bow in or out—generally in—thereby creating subsequent pile driving problems. Straightening of the specials may be

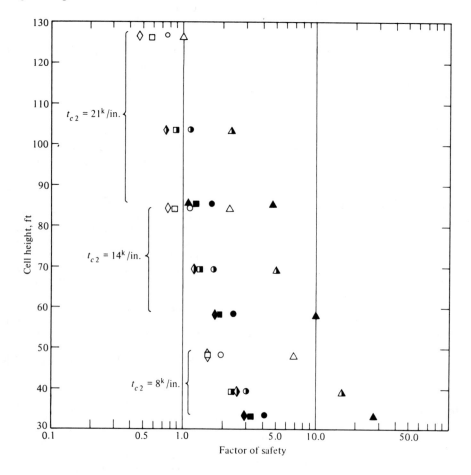

$\theta = 45°, 90°$ $T_s, b = .875D, \phi = 29°$

	70' Dia. Cell	87.55' Dia. Cell	103.40' Dia. Cell
Resistance to Sliding	○	◑	●
Resistance to Piling Rise	□	◨	■
Resistance to Overturning	△	◭	▲
Resistance to Vertical Shear	◇	◈	◆

Fig. 12.49 Cell height vs. factor of safety for various interlock loads and cell diameters (TVA method).

accomplished by mild heating of portions of the special (supported at the ends or at other locations, depending on the cambered portion).

When setting or racking the piling, the specials should be set, plumbed, and tacked in position. As noted above, if cambered specials are used the cell goes out of round and driving may become difficult. A similar condition occurs if the specials are not plumbed in both the radial and tangential axes for circular structures. This has occurred when only one template was used and/or in flowing water. A horizontal rotating guide, which rolls on curved rails mounted close to the template edge, has been used as a guide for setting piling. With this system, two templates are still desirable.

After the specials or four quadrant (corner) pilings are set, the piling between the specials is installed. The setting should proceed from the set piling toward the midpoint between the specials. When the closures are made, two or more sections may have to be lifted and shaken to allow the closure piling to slide into position.

Piling may be driven in increments up to approximately 5 feet at a time all around the perimeter. The incremental driving is repeated until the proper toe elevation is reached.

When circular cells with arcs are filled, the cell is filled first, followed by filling of the connecting arc area between the two filled, stable cells. Occasionally, the cell is partially filled and then the arc area fill is started. This practice tends to distort the piling more than the conventional method and should be avoided. Hydraulic filling of the cell is very desirable at many sites, but the discharge end of the pipe should be moved continuously or at frequent intervals to prevent localized fill build-up, which causes cell distortion.

When filling diaphragm cells, the fill is not placed more than 4 to 5 ft higher in one cell than in the adjacent cell in order to prevent the diaphragms from arching or moving out of a straight line between Y's.

Since the interlocks are tight due to hoop tension, the rate of water draining from the cell can be much slower than the coffer dewatering rate. Therefore, approximately 2-inch-diameter holes are burned in the piling webs as the coffer water recedes. The cells can be designed to hold the full head of water plus the fill, but when the coffer is fully dewatered the hoop tension for most current installations is too high or requires the use of high interlock strength piling. Usually, the cell water is drained at the burned holes. The holes are rodded at frequent intervals to prevent clogging.

When the coffer and seepage water is pumped out, care must be exercised to prevent leakage or discharge from the discharge line from entering the cells. Sometimes an adequate discharge line is disturbed by waves, debris, or careless equipment movement. Leaks then occur at the joints, and the water in the cell can rise rapidly and may cause drastic results (Grayman, 1970).

As the water level on the inside of the coffer is lowered, the interlocks are inspected for evidence of separation. If a separation is evident, straps are sometimes placed across the separation and welded to both sheets. The decision of what to do about a separation depends on its location and importance to the structure as a unit.

Many coffers are designed so that they may be overtopped during high-water periods. As a result, the gate must be sized to allow water to flow into the coffer at a rate that will minimize the increasing lateral pressure on the outside of the cell. The gate may be placed between the circular cells in the connecting arc area. When diaphragm cells are used, a circular arc is sometimes substituted for two diaphragms and the gate is then placed between the circular arcs. Stop logs are generally used for the gate closure.

Land assembly of cells for use in cofferdams or waterfront structures has been used where there is an advantage in cost and/or time compared to conventional construction.

Swatek (1970) has summarized most of the field and design problems associated with cellular construction.

J. DETERIORATION PREVENTION

When cellular structures are used in permanent construction (piers, bulkheads, breakwaters, etc.), consideration must be given to the effect of the environment on the structure during the desired useful life. If the structure is located in fresh water there will be less deterioration of the steel than under brackish or seawater conditions.

Severe corrosion is usually confined to an area from just below mean low water to the top of the splash zone area. At some locations less severe corrosion will occur on the immersed piling below mean low water, but this is primarily attributable to a function of the ratio of the immersed zone to the tidal zone.

Deterioration prevention of the piling may be accomplished through the use of coatings, corrosion-resistant low-alloy steels, and/or cathodic protection.

Stray currents present a slightly different problem in that an impressed current from grounded motors, welding machines, etc., travels through the best conductor in the soil or water on its way back to the generating source. A common method for combating this is to bond (weld) all piling together by using a reinforcing bar and providing a low-potential drain point away from the steel structure. If arc welding is being done adjacent to the cell, the return leads should not be dropped into the water adjacent to the piling.

K. GENERAL COMMENTS ON CELLULAR STRUCTURES

At times, the available cell fill is not of the desired quality. Where this situation occurs the sheet piling is required to

provide a larger share of the stability resistance. This is usually done by placing structural members diagonally across the diaphragms. Cables may be used for the same purpose. The diaphragm type of cell is usually better suited for reinforcement because of the flat diaphragm. Many ways exist to reinforce the basic cell configuration.

In the past, progressive yielding of cell fills has occurred, allowing tilting to take place. The fill may be strengthened by sand piles, combined with dewatering.

Vertical load bearing capacity of flat or shallow arch sheet piling used in curved configurations which resist hoop tension (i.e., have interlock tension) can be high. The piling load capacity is usually sufficient to support much, and possibly all, of the normal unloading equipment utilized for permanent waterfront structures. To utilize this load bearing capacity, the sheets should be continuously interlocked to below the dredge line and the bearing capacity of the underlying soil be adequate.

In cellular construction some designers, in order to reduce costs, step up the toe of the sheet piling from the connecting arcs to the centerline of the line of cells. Depending on the cell and coffer configurations and soil characteristics, there may be as much as 15 ft difference in toe elevation. The practice has been satisfactory for many installations, but caution is advised when considering that configuration. The outboard piling must be driven to a sufficient depth for protection against scour, etc., and bearing capacity should not be computed for an embedment depth greater than the minimum piling embedment. The steps from toe to toe of the piling must be small enough to minimize loading discontinuities.

12.3.2 Double Wall Structures

Figure 12.53 shows a double wall (using moment resisting piling) structure tied together by tie rods. This type of filled structure is used for many of the same purposes as filled cellular structures. Structure stability considerations are similar. Generally, the height to which these structures are constructed is less than cellular structures because bending strength of sheet piling is limited and the structure is not as stable during construction as filled cells unless an extensive support system is employed. The depth of the structure can be increased by integrating large structural members in the walls.

A method of analysis for double wall structures has been advanced by Ohori et al. (1988). A calculation model was based on solutions considering the double wall structure as a composite consisting of piling granular fill, tie rod and the base soil. The solutions included the deformation at the granular fill and resistance of the base soil. The results of the model were compared to the results of fairly large model tests.

To date, the stability design of double wall structures has been similar to that of cellular cofferdam and low bulkheads. This includes all three methods previously discussed.

12.3.3 Filled Concrete Cellular Structures

Very large concrete structures are precast on land or in basins and launched or floated, then towed to the construction site. Having been properly ballasted prior to towing they are then sunk onto a prepared leveling base. Each cell in the module is then filled with granular material. The next module is similarly placed and filled and two aligned slots between the modules are filled with gravel. Concrete is placed in the middle slot to form

Saturation line location

Cell Fill	Slope
Coarse grained sand and gravel	1:1
Gravel, sand silt and clay mixtures	2:1
Fine sands with silts and clays	3:1

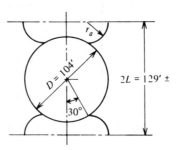

Given:

$H = 100.0'$
$H_s = 30.0'$
$\theta = 30°$
$\gamma = 105$ pcf
$\gamma' = 62$ pcf
f_{gr} (gravel on rock) = .5
f_{gs} (gravel on steel) = .36
f_{ss} (steel on steel) = .3
$\phi = 30°$, $\quad \delta = 20°$ (sand)

Note:
1. Temporary factor of safety classification
2. PS31 Sheet piling section
3. ASTM A572 steel grade

$b = .818D$

El. 100.0 — High water

El. 65.0 — Low water

Saturation line

El. 30.0 — Overburden H_s

P_w

P_a

$H/3$

L_B

P_p

L_B'

Rock el. 0.00

$H_s/3$

Note: All computations are based on per foot of cell width

COMPUTATION BY TVA METHOD

Cell Dimensions

Assume: Average equivalent width $b = .85H = .85(100) = 85'$

$$\text{Diameter } D = \frac{85}{.818} = 103.9 \approx 104'$$

Lateral Pressure

Water: $P_w = .5(.0625)(100)^2 = 313.0$ kips/ft

Soil: $P_a = .5 K_a \gamma' H_s^2 = (.5) \left\{ \dfrac{(\cos 30°)^2}{\cos 20° \left[1 + \sqrt{\dfrac{\sin(30° + 20°)\sin 30°}{\cos 20°}}\right]^2} \right\} (.062)(30)^2$

$\qquad = (.5)(.297)(.062)(30)^2 = 8.3$ kips/ft

F.S. Against Sliding

$$F.S. = \frac{\text{resisting force}}{\text{lateral pressure}} = \frac{b\left[\gamma(H - H') + \gamma'H'\right] f_{gr}}{P_w + P_a}$$

$$= \frac{76.9\left[.105(100 - 75.5) + .062(75.5)\right].5}{313.0 + 8.3}$$

$$= \frac{279.5}{321.5} = .87 < 1.25 \therefore \text{Berm required}$$

Berm (Assume dimensions in fully saturated condition)

Let $L_b = 50'$

Saturation curve in cell is on slope of 2:1

$\qquad \therefore H_b = 100 - 85/2 = 57.5'$

and $L_b' = 50 + 1.75 \times 57.5 = 150.6'$

Berm weight $W_b = 62 [(57.5)(50) + (150.6 - 50)(57.5) 1/2]$

$\qquad\qquad = 336,554$ lb $= 336.6$ kips

Fig. 12.50 Design of circular cell structure used as cofferdam.

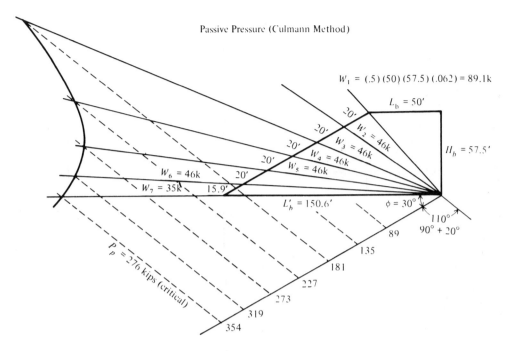

Passive Pressure (Culmann Method)

$W_1 = (.5)(50)(57.5)(.062) = 89.1k$

Horizontal component of passive pressure $= P_p \cos \delta = (276)(\cos 20°) = 259.4$ kips

Horizontal component of passive pressure $= P_p \cos \delta$
$$= (276)(\cos 20°)$$
$$= 259.4 \text{ kips}$$

Force required for berm to slide on rock:

Berm weight $\times f_{gr} = (354)(.5) = 177$ kips $<$ 259 kips

∴ Berm will slide before shear failure occurs.

F.S. Against Sliding with Berm Resistance

$$F.S. = \frac{279.5 + 177}{321.5} = 1.42 > 1.25 \quad O.K.$$

Internal Stability (Vertical Shear Method)

Total shearing force: $Q = \dfrac{3}{2}\dfrac{M}{b}$
$$= \frac{3}{2(85)}\left[313\left(\frac{100}{3}\right) + 8.3\left(\frac{30}{3}\right) - 177\left(\frac{57.5}{3}\right)\right]$$
$$= 139 \text{ kips}$$

Ratio of horizontal earth pressure to vertical pressure at rest:

$$K = \frac{\cos^2 \phi}{2 - \cos^2 \phi} = \frac{(.866)^2}{2 - (.866)^2} = .6$$

(The quantity K should be given values based on experience as the above formula gives larger ratios with better soils—0.6 is O.K. for this cell fill)

Fig. 12.50 *(Continued)*

Total centerline shear resistance $S_c = P_c \tan \phi$

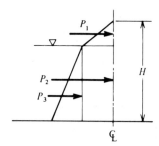

$$P_c = P_1 + P_2 + P_3$$
$$= \tfrac{1}{2}(.6)(.105)(85/4)^2 + (.6)(.105)(85/4)(100 - 85/4) + \tfrac{1}{2}(.6)(.062)(100 - 85/4)^2$$
$$= 14.2 + 105.4 + 115.3 = 235 \text{ kips}$$
$$S_c = 235 \tan 30° = 135.7 \text{ kips}$$

Total interlock friction resistance $F_T = f_{ss} P_s$

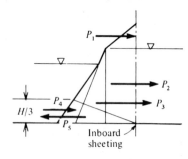

$$P_s = P_1 + P_2 + P_3 + P_4 - P_5$$
$$= 235 + \tfrac{1}{2}(.062)(100 - 85/2)^2$$
$$\quad - \tfrac{1}{2}(100/4)\,[(.6)(.105)(85/4) + (.6)(.062)(100 - 85/4) + (.062)(100 - 85/2)]$$
$$= 235 + 102 - 12.5\,[1.3 + 2.9 + 3.6]$$
$$= 235 + 102 - 98 = 239 \text{ kips}$$

Note: The value of K for computing P_s was equal to that used for computing P_c; however, $K = .4$ (minimum) was recommended by Terzaghi. The Coulomb value of K_a has been used by others.

$$F_T = .3(239) = 71.7 \approx 72 \text{ kips}$$

$$\text{F.S. of internal stability} = \frac{S_c + F_T}{Q}$$

$$= \frac{136 + 72}{139} = 1.5 > 1.25$$

F.S. for Piling Rise

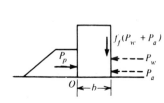

$$\text{F.S.} = \frac{f_{gs}(P_w + P_a)b + P_p H/3}{P_w H/3 + P_a H_s/3}$$

$$= \frac{.36(313 + 8.3)85 + 259(57.5/3)}{313(100/3) + 8.3(30/3)}$$

$$= \frac{14795}{11263} = 1.3 > 1.25$$

Interlock (Hoop) Tension Force

Condition A—Hydraulically filled and no cell drainage:

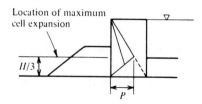

Fig. 12.50 (*Continued*)

Condition B—Cell draining to top of berm:

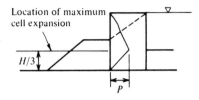

For Condition A:

Use active pressure coefficient $K_a = \tan^2 \left(45° - \dfrac{30°}{2}\right) = .3$

Unit pressure at Elev. 33.0 ($\frac{2}{3}H$). The plane of maximum cell expansion is higher than on bare rock because of the overburden and berm.

$$P = \frac{2}{3}H \left(K_a \gamma' + \gamma_w\right)$$
$$= 67 \left[(.3)(.062) + .062\right]$$
$$= 5.38 \text{ kips/ft}^2$$

Maximum interlock force (at 30° Y)

Note: By Swatek's formula

$$t_{c2} = \frac{PL \sec \theta}{12}$$

$$= \frac{(5.38)(129) \sec 30°}{(12)(2)}$$

$$= 33.4 \text{ kips/linear inch} > t_{ult} (24.0 \text{ kips/linear inch})$$

$$t_{c2} = \frac{PL}{12}$$

$$= \frac{5.38(129)}{(12)(2)} = 28.9 \text{ kips/linear inch}$$

$$= 28.9 > t_{ult} (24.0 \text{ kips/linear inch})$$

For Condition B (partially drained cell):

$$P = \gamma_w \left(\frac{2}{3}H - \frac{b}{2}\right) + K_a \gamma' \left(\frac{2}{3}H - \frac{b}{4}\right) + K_a \gamma \left(\frac{85}{4}\right)$$

$$= .062 (24.5) + .3 (.062)(45.75) + .3 (.105)(21.25)$$

$$= 3.04 \text{ kips/ft}^2$$

Maximum interlock force (at 30° Y)

Note: By Swatek's formula

$$t_{c2} = \frac{(3.04)(129) \sec 30°}{(12)(2)}$$

$$= 18.9 \text{ kips/linear inch}$$

$$\text{F.S.} = \frac{24.0}{18.9} = 1.27 < 2.0$$

$$t_{c2} = \frac{(34.0)(129)}{(12)(2)}$$

$$= 16.3 \text{ kips/linear inch}$$

$$\text{F.S.} = \frac{26.0}{16.2} = 1.49 < 2.0$$

Condition A, which is a fully saturated condition, would exceed the ultimate published interlock strength; therefore, the contractor must avoid this situation during the filling, when the river level may be low, or after coffer dewatering. Condition B, which is a partially drained condition, is not suitable as the factor of safety is 1.27 (< 2.0) for the static condition. Use of a lightweight cell fill such as slag or shells would bring the interlock force within an acceptable value. The sliding and vertical shear values would have to be recomputed using the lower value of γ.

Fig. 12.50 (*Continued*)

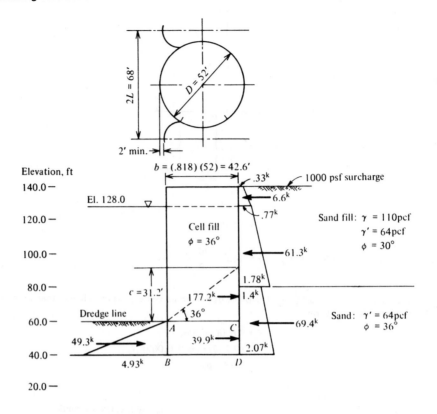

Computation by Horizontal Shear (Cummings) Method

External pressures and forces:

$$K_a = \tan^2 (45° - 30°/2) = .33$$

Active pressure: Elev. 140.0 $P_a = (.33)(1.0) = .33$ ksf

Elev. 128.0 $P_a = .33 + (12)(.11)(.33) = .77$ ksf

Elev. 80.0 $P_a = .77 + (48)(.064)(.33) = 1.78$ ksf

Internal resistance:

Locate ϕ line from dredge line on outboard side (Point A).
Use $\phi = 36°$ all the way as cell is filled with medium sand.
$c = b \tan \phi = 42.6 \tan 36° = 31.2'$
Average unit weight above plane AC:

$$\gamma' = \frac{1}{80} [(.11)(12) + (.064)(68)] = .071 \text{ kcf}$$

$R_1 = ac\gamma' = (80.0 - 31.2)(31.2) .071 = 108.1^k$
$R_2 = c^2 \gamma' = (31.2)^2 (.071) = 69.1^k$
$R = R_1 + R_2 = 108.1 + 69.1 = 177.2^k$

Location of R: $(108.1) \dfrac{31.2}{2} = 1686'^k$

$(69.1) \dfrac{31.2}{3} = \dfrac{719'^k}{2405'^k}$

$\dfrac{2405}{177.2} = 13.6'$ up from plane AC

Resistance of soil in cell below plane AC:
$(42.6)(20)(.064) \tan 36° = 39.9^k$
Point of resistance taken at one-half of distance CD.

$$K_a = \tan^2 (45° - 36°/2) = .26$$

Fig. 12.51 Stability analysis of cellular bulkhead on sand.

Active pressure: Elev. 80.0 (lower stratum)
$$P_a = [(1.0) + (12)(.11) + (48)(.064)] .26 = 1.40 \text{ ksf}$$
Elev. 40.0
$$P_a = 1.40 + (40)(.064)(.26) = 2.07 \text{ ksf}$$

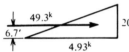

(.33)(12) = 3.96	(3.96)(12/2) = 23.76
(.77 − .33)(12/2) = 2.62	(2.62)(12/3) = 10.48
6.58^k	34.24$^{'k}$
	$\frac{34.24}{6.58} = 5.2'$

(.77)(48) = 37.0	(37.0)(24) = 888
(1.78 − .77)(48/2) = 24.3	(24.3)(48/2) = 389
61.3^k	1277$^{'k}$
	$\frac{1277}{61.3} = 20.8'$

(1.40)(40) = 56.0	(56.0)(40/2) = 1120
(2.07 − 1.40)(40/2) = 13.4	(13.4)(40/3) = 179
69.4^k	1299$^{'k}$
	$\frac{1299}{69.4} = 18.7'$

Passive pressure: $K_p = \tan^2(45° + 36°/2) = \tan^2 63° = 3.85$
at Elev. 40.0 $P_p = (40)(.064).385 = 4.93 \text{ ksf}$

$$(4.93)\left(\frac{20}{2}\right) = 49.3^k$$

Interlock Resistance

Use active pressure inside of cell down to dredge line as cell expansion reduces from dredge line down.

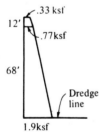

Total interlock force P_s:

$$P_s = 12\left(\frac{.33 + .77}{2}\right) + 68\left(\frac{.77 + 1.9}{2}\right) = 97.4^k$$

.77 + (68)(.64)(.26)
= 1.9 ksf

Driving or Overturning Moment

$$M_D = (6.58)(88 + 5.2) + (61.3)(40 + 20.8) + (69.4)(18.7) = 5638^{'k}$$

Resisting Moment

$$\begin{aligned}
M_R &= M_r + M_T + \text{external passive moment} \\
&= \text{Arm} \times R + \text{Arm} \times R' + P_f f_{ss} b + \text{Arm} \times P_p \\
&= (20 + 13.6)(177.2) + (10)(39.9) + (97.4)(.3)(42.6) + (6.7)(49.3) \\
&= 5954 + 399 + 1245 + 330 \\
&= 7928^{'k}
\end{aligned}$$

Factor of safety against tilting: $F.S. = \dfrac{7928}{5638} = 1.41 < 1.5$, but O.K.

Note: The remaining failure modes are checked in a similar manner, as given in Fig. 12.50.

Fig. 12.51 *(Continued)*

Computation by Rotation (Extreme) Method

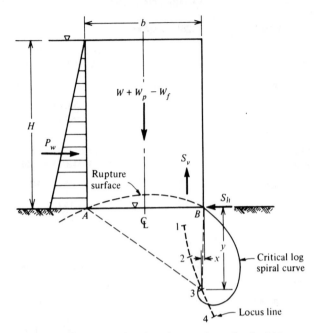

Given:

Circular cell and double arcs

$\theta = 30°$, $30°$ Y and $2L = 115'$

$b/H = .7$ and $H = 100.0'$

$\gamma = 100$ pcf

$\gamma_w = 62.4$ pcf

$\phi = 30°$, $\delta = 20°$

$f_{gr} = .5$

$f_{gs} = .36$

$f_{ss} = .3$

Equivalent Width and Cell Diameter

$b = (.7)(100) = 70'$

$D = \dfrac{70}{818} = 85.6 \approx 86'$

Total Center-to-Center Distance

$2L = 115'$ (from layouts)

Note: All computations based on per foot of cell width.

Cell Fill Weight

$W = b\gamma H = 70(.1)(100) = 700$ kips

Sheet Piling Weight

$W_p = 10.2$ kips (from layout)

Vertical Piling Reactions (acting on unloaded side of cell)

Friction from main cell:

$S_{V1} = \pi r_c K_a \gamma \dfrac{H^2}{2} \tan \delta$

$= \pi(86/2)(.33)(.1)[(100)^2/2] \tan 20° = 8110$ kips

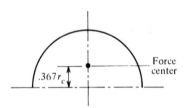

Friction from arc cell:

$S_{V2} = 2(L_1 + L_2) K_a \gamma \dfrac{H^2}{2} \tan \delta$

$= 2(22 + 22)(.33)(.1)[(100)^2/2] \tan 20° = 5290$ kips

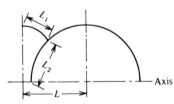

Weight of piling:

$S_{V3} = (\pi r_c + 2L_2) H A_p$

$= [(3.14)(43) + 2(22)](100)(.035)$

$= 627$ kips

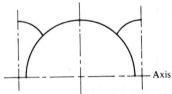

Fig. 12.52 Stability analysis of circular cell cofferdam on rock with drained fill.

Interlock friction:

$$S_{V4} = 2K_a\gamma \frac{H^2}{2} r_c (f_{ss})$$

$$= 2(.33)(.1)(100)^2/2(43)(.3)$$

$$= 4257 \text{ kips}$$

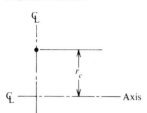

Total vertical reaction:

$$S_V = S_{V1} + S_{V2} + S_{V3} + S_{V4}$$
$$= 8113 + 5280 + 627 \div 4257$$
$$= 18277 \text{ kips}$$

Distance S_V acts from centerline axis:

$$d_1 = \frac{(8110)(.367)(86/2) + (5290)(24.8) + (627)(18.5) + (4257)(86/2)}{18277}$$

$$= 24.8 \text{ ft}$$

Total vertical reaction per foot of cell:

$$S_V = \frac{18277}{115} = 159 \text{ kips}$$

Horizontal Piling Reaction

$$S_h = S_V f_d = (159)(.6) = 95 \text{ kips}$$

(f_d is used here as the piling is usually driven into the rock. The only failure could be bending of the toe of the piling, and to consider friction acting on rock is unusual.)

Weight of Cell Fill Below Failure Surface

As failure surface has not been established, the log spiral locus line is drawn and at least three points are picked so that factors of safety may be computed at each point and then plotted against r_0 to determine the minimum factor of safety. After picking the three points on the locus line the distances r_1 and r_0 are scaled. The location of four points with reference to Pt. B is given under the X and Y Columns.

Point	r_0 (ft)	r_1 (ft)	X (ft)	Y (ft)
1	16	58	+13	9
2	26	68	+ 8	29
3	49	84	+ 2	49
4	65	100	− 6	15

Cell fill below failure surface:

$$W_f = \gamma \left(\frac{r_1^2 - r_0^2}{4\tan\phi} - b\frac{Y}{2} \right)$$

Point 1 $\quad W_{f1} = (.1)\left[\frac{(58)^2 - (16)^2}{4\tan 30°} - (70)\left(\frac{9}{2}\right) \right] = 103 \text{ kips}$

Point 2 $\quad W_{f2} = (.1)\left[\frac{(68)^2 - (26)^2}{4\tan 30°} - (70)\left(\frac{29}{2}\right) \right] = 69 \text{ kips}$

Point 3 $\quad W_{f3} = (.1)\left[\frac{(84)^2 - (49)^2}{4\tan 30°} - (70)\left(\frac{49}{2}\right) \right] = 30 \text{ kips}$

Point 4 $\quad W_{f4} = (.1)\left[\frac{(100)^2 - (65)^2}{4\tan 30°} - (70)\left(\frac{65}{2}\right) \right] = 23 \text{ kips}$

Fig. 12.52 (*Continued*)

$$F.S. = \frac{\text{resisting moment}}{\text{driving moment}}$$

Trial Points	Load	Value, kips	Arm, ft	Moments, kip-ft	
				Driving	Resisting
1	P_w	312	42.3	13198	
	$W + W_p - W_{f1}$	607	22.0		13354
	S_v	159	2.8		445
	S_h	95	9.0		855
				13198	14651

$$F.S. = \frac{14651}{13198} = 1.11$$

2	P_w	312	62.3	19438	
	$W + W_p - W_{f2}$	641	27.0		17307
	S_v	159	2.2		350
	S_h	95	29.0		2755
				19438	20412

$$F.S. = \frac{20412}{19438} = 1.05$$

3	P_w	312	82.3	25678	
	$W + W_p - W_{f3}$	680	33.0		22440
	S_v	159	8.2	1304	
	S_h	95	49.0		4655
				26982	27095

$$F.S. = \frac{27095}{26982} = 1.00$$

4	P_w	312	98.3	30670	
	$W + W_p - W_{f4}$	687	41.0		28167
	S_v	159	16.2	2576	
	S_h	95	65.0		6175
				33246	34342

$$F.S. = \frac{34342}{33246} = 1.032$$

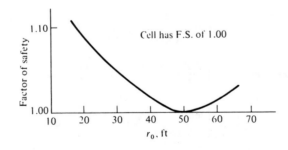

Since this cell has a minimum factor of safety of 1.00, it is close to an unstable condition. The other two methods of computing stability would give lower factors of safety.

The methods of computing interlock forces are the same as given in Fig. 12.50.

Fig. 12.52 *(Continued)*

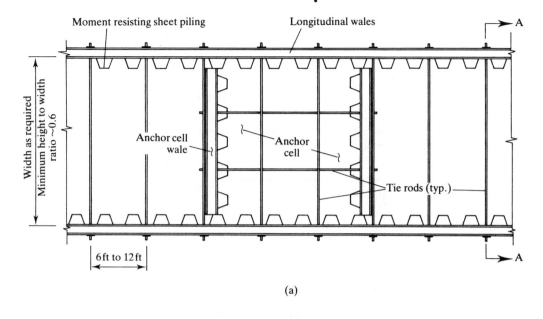

(a)

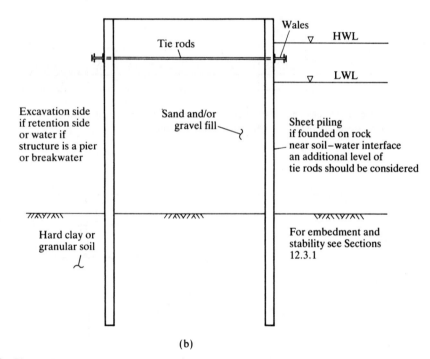

(b)

Fig. 12.53 Plan and section of a double-wall cofferdam. (a) Plan view. (b) Sectioned view.

a watertight seal. Select fill is placed behind the structure and the reinforced retaining wall is constructed, after which the remainder of the fill is placed.

The structures are heavy and the foundation must be adequate to support the large weight. More often than not the site has to be dredged and an adequate depth of granular soil placed before module placement begins, as shown in Figure 12.54.

The design of the modules must take into account launching and towing loadings and stability of the module in water (proper ballasting), as well as the in-situ loads.

12.3.4 Block Concrete, Crib and Bin Structures

Figures 12.55 through 12.57 show typical gravity type retaining structures. Concrete, rock, and soil (granular) are used singly or in combination with other materials to construct retaining structures. Concrete retaining structures are common and are termed "cantilever" although they are gravity walls. Crib type walls such as rock-filled timber frames have in the past been used extensively, but are labor-intensive and of limited strength. Metal and concrete bin type walls are routinely used in highway and railroad retention structures.

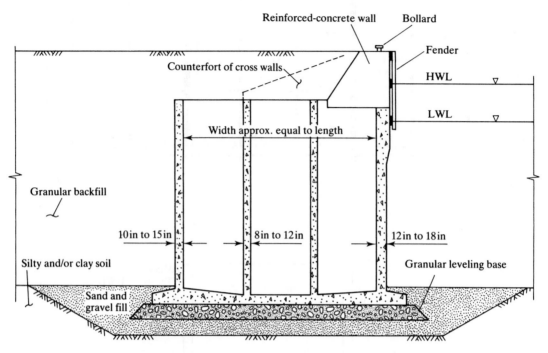

Fig. 12.54 Cross section of concrete cellular wharf. *Notes.* 1: Concrete seal to be placed between modules. 2: Cells and modules are approximately square. 3: Weep holes are necessary between cells of the bottom.

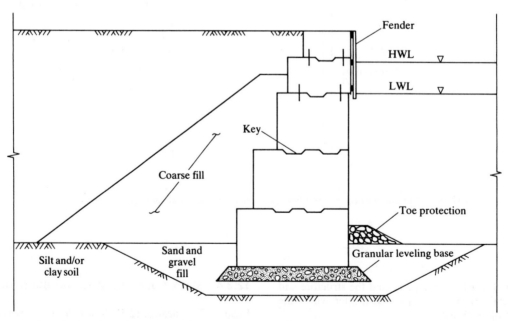

Fig. 12.55 Cross section of a concrete block gravity wharf. *Notes.* 1: Typical maximum block weight = 150 kips. 2: Blocks set on 10° to 15° backslope. 3: Keying of blocks to be done carefully to prevent seating difficulty.

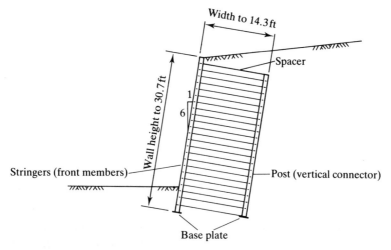

Fig. 12.56 Metal bin type retaining wall. *Notes.* 1: Front of wall is continuous. 2: Side and rear framing can be intermittent. 3: Typical installation is on a 1 : 6 slope but can be installed vertically. 4: Embedment is approximately 3 to 4 ft. 5: Bin backfill must be granular soil, properly compacted.

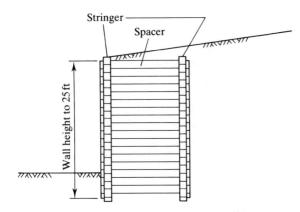

Fig. 12.57 Concrete and wood crib type retaining wall. *Notes.* 1: Front of wall is continuous. 2: Side and rear framing is continuous. 3: Typical installation is vertical but can be erected on slopes of up to 1 : 8. 4: Typical embedment is 2 to 4 ft. 5: Interior fill must be granular soil, properly compacted.

12.4 CANTILEVER RETAINING WALLS

Cantilever sheet-pile retaining walls are particularly sensitive to variations in height, loading, and soil properties. Because of this and also because these walls are subject to larger deflections than anchored walls, they are preferably used in temporary rather than permanent applications. As already mentioned, they are not often used for retained heights of more than about 10 ft and seldom for heights of over 15 ft. Commencing in this range, it will be found that anchored retaining walls will be more economical and do not possess the disadvantages of cantilevered walls.

As much larger penetrations below the dredge line are needed for cantilever walls compared with anchored walls, it becomes even more important to check that the section of piling finally calculated on bending moment considerations will be large enough to enable it to withstand the vertical driving stresses during installation.

12.4.1 Conventional Design of Cantilever Walls

For a cantilever wall of height H, Figure 12.58 shows the final net pressure diagram (shaded) derived as explained earlier. The following are the steps to be taken in determining the required penetration of the piles and the maximum bending moment, using a conventional simplified procedure.

1. By trial find a penetration t_1 such that when moments are taken about e, the moment of the net active pressure load acd is 2.0 times the resisting moment due to the net passive pressure def. This ensures a factor of safety of 2.0 against overturning.

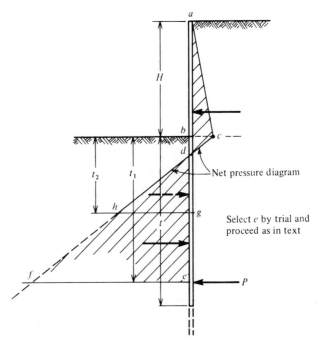

Fig. 12.58 Cantilever wall design.

2. The actual required penetration t is then determined by increasing t_1 by 15 percent, to allow for mobilization of the back passive force P required to ensure stability.
3. Next find by trial the point at penetration t_2 at which the shear force is zero, yielding the position of maximum bending moment. Thus, at point g the net active load acd will equal the net passive load dgh.
4. Take moments about g to find the maximum bending moment. This will be the difference between the moments of acd and dgh about g.
5. Use the bending moment thus found to select a section of sheet piling, based on a working steel stress of 0.5 times the minimum yield stress of the steel.

12.4.2 Cantilever Walls—Standard Design Graphs

In Figure 12.59 graphs are given to enable pile selection and penetration to be quickly determined for certain standard cases of cantilever retaining walls. Two different cases A and B are given for uniform cohesionless soils with $\phi = 30°$ and $35°$ respectively, while case C is for cohesive soils having c values of 500, 1000, and 1500 lb/ft², respectively.

It is recommended, as indicated in Figure 12.58, that t/H be taken as not less than 1.0 in practice, even if theory indicates a lesser penetration for some cases. While the case for a cohesive soil with $c = 500$ lb/ft² is given for comparison, it is not recommended except perhaps for temporary walls under 10 ft in height. Standard case C for cohesive soils applies only for undisturbed soil throughout, that is, where dredging or excavation is carried out in front and the values given are not applicable if clay is used as filling behind the wall.

12.5 SPECIAL STRUCTURES

There are many structures constructed primarily of steel sheet piling that are used for retaining soil and water but are not in the common category of retaining walls. These structures range from silos (for internal storage of coal and other bulk materials) to dolphins. Most of the special applications of sheet piling involve loading that is different from that received in retention of the soil. The techniques used in the determination of structure resistance are similar for all applications. Several of these structures are described below.

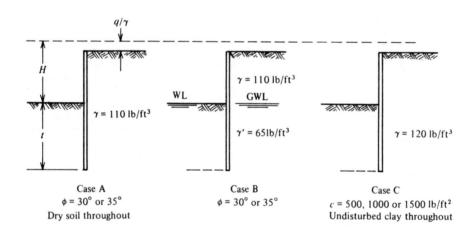

	Case A $\phi = 30°$ or $35°$ Dry soil throughout	Case B $\phi = 30°$ or $35°$	Case C $c = 500, 1000$ or 1500 lb/ft² Undisturbed clay throughout

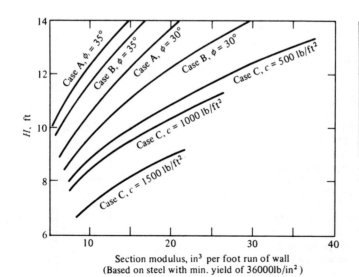

q is surcharge in lb/ft²

	Values of t/H	
Case A	$\phi = 30°$	1.0
	$\phi = 35°$	1.0
Case B	$\phi = 30°$	1.35
	$\phi = 35°$	1.15
Case C	$c = 500$ lb/ft²	*
	$c = 1000$ lb/ft²	1.0
	$c = 1500$ lb/ft²	1.0

*$t/H = 1.5$ when $H = 7.5'$
$= 2.3$ when $H = 9.5'$

Section modulus, in³ per foot run of wall
(Based on steel with min. yield of 36000lb/in²)

Fig. 12.59 Cantilever retaining walls—three cases.

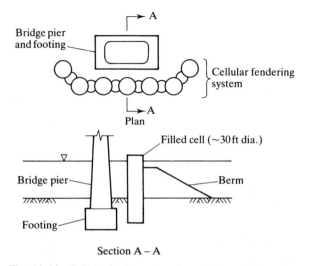

Section A – A

Fig. 12.60 Cellular fender system for a bridge pier in water.

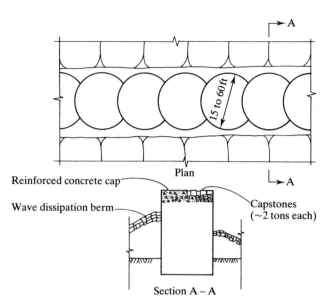

Section A – A

Fig. 12.61 Steel sheet piling cellular breakwater.

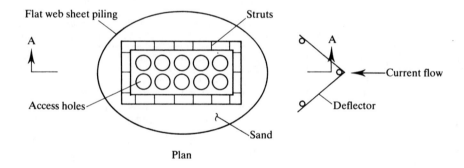

Plan

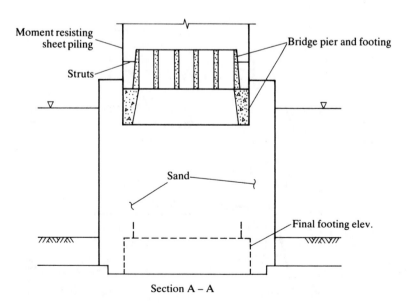

Section A – A

Fig. 12.62 Sand island for bridge pier construction.

12.5.1 Drydocks and Basins

Figure 12.1b shows the cross section of a shipbuilding basin whose walls are constructed of sheet-piling cells. A number of basins have been constructed in this manner. Another common retention system is composed of large concrete walls with counterforts. Both retention systems move when the basin is flooded or emptied.

12.5.2 Turning, Mooring, and Breasting Dolphins and Fendering

These structures have been used as waterfront facilities for many years and are constructed of wood piling, steel sheet piling and bearing piles, and concrete cylinders. Single cells of steel sheet piling filled with granular material are used to resist large vertical and lateral loads. Filled cells are used for protection of piers of major structures such as bridges. Figure 12.60 shows a typical cellular fender system.

12.5.3 Piers and Breakwaters

Filled sheet-piling cells, filled precast concrete cellular modules, and concrete blockwork are used for these structures. Filled double-wall sheet-piling structures with one or more levels of tie rods are also used. Small breakwaters are sometimes constructed of cribbed stone with or without reinforced concrete caps. Figure 12.61 shows a typical breakwater composed of filled sheet-piling cells. Figure 12.54 shows a concrete cellular structure that can be adapted for breakwaters.

Given a suitable foundation and fill material that has been properly placed, cellular structures can resist large lateral forces in addition to vertical loads. The commonly specified pier deck load of 1 kip per square foot and crane (mobile and rail) wheel loads of 100 kips or more are readily supported. Designers often use bearing piles to support crane rail foundations and the retaining and fender support system at the face of a pier. The load capacity of the fill and the sheet piling is much greater than generally thought and some of the bearing piles that are specified are not necessary.

TABLE 12.6 PROPERTIES AND DIMENSIONS OF HOT ROLLED STEEL SHEET PILING. (Courtesy of Bethlehem Steel Corporation.)

Properties

Section Designation	Area, in²	Nominal Width, in	Weight in Pounds Per lin ft of bar	Weight in Pounds Per ft² of wall	Moment of Inertia, in⁴	Section Modulus, in³ Single Section	Section Modulus, in³ Per lin ft of wall	Surface Area, ft² per lin ft of bar Total Area	Surface Area, ft² per lin ft of bar Nominal Coating Area[a]
PLZ23	13.28	24	45.2	22.6	407.5	60.4	30.2	5.98	5.52
PLZ25	14.60	24	49.6	24.8	446.5	65.7	32.8	5.98	5.52
PZ22	11.86	22	40.3	22.0	154.7	33.1	18.1	4.94	4.48
PZ27	11.91	18	40.5	27.0	276.3	45.3	30.2	4.94	4.48
PZ35	19.41	22.64	66.0	35.0	681.5	91.4	48.5	5.83	5.37
PZ40	19.30	19.69	65.6	40.0	805.4	99.6	60.7	5.83	5.37
PSA23	8.99	16	30.7	23.0	5.5	3.2	2.4	3.76	3.08
PS27.5	13.27	19.69	45.1	27.5	5.3	3.3	2.0	4.48	3.65
PS31	14.96	19.69	50.9	31.0	5.3	3.3	2.0	4.48	3.65

[a] Excludes socket interior and ball of interlock.

Dimensions

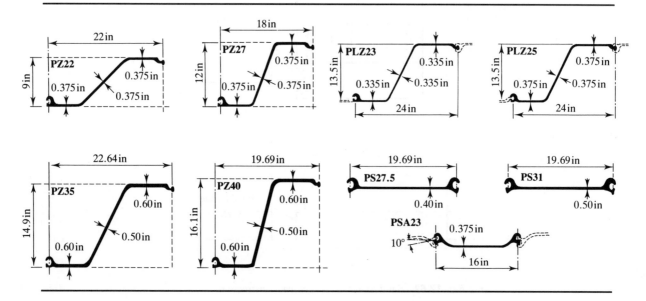

Breakwaters frequently have capstone movement within cells during storms when wave loading occurs. This movement causes localized loading on the top portion of the sheet piling. The most resistant portion of the sheet piling is the fabricated T or Y connection. As a result, the connection attracts loads and the concentration will sometimes cause deformation and/or fracture of the connection or its adjacent piling. Movement of exterior armor against the sheet piling can also cause distress. In those cases where repairs to the piling are required, top-of-cell reinforcement can be accomplished by installation of steel or reinforced-concrete wales or by concrete caps.

12.5.4 Sand Islands

Construction of bridge piers in water is usually done in the dry by erecting a single-wall cofferdam as previously mentioned. However, under suitable conditions sand islands are sometimes constructed (as a single cell) with an elliptical shape to fit, for example, the rectangular shape of a bridge pier. The bridge pier with interior access holes is formed in sections on the sand surface and sunk as the sand is dug out from under the formed sections. Bracing between the bridge pier and the sheeting is placed as necessary. Since the sand fill tends to expand the cell when it is filled, the interlocks tighten and this restricts water movement into the cell. Figure 12.62 shows the general configuration of a sand island.

12.6 SHEET PILING AND H-PILES

Tables 12.6 and 12.7 give the dimensions and properties of sheet piling and H-piles which are produced domestically. These sections are used in all types of retaining and support structures.

NOMENCLATURE

A	Cross-sectional area
B	Width of excavation
C_m	Bending term coefficient in interaction formula (AISC)
C_a	Adhesion of clay to piling
C_p	Piling pullout capacity
D	Cell diameter
D_f	Depth of cell below soil surface in bearing capacity formula
E	Normal component of total earth pressure on cell wall
F	Tangential component of total earth pressure on cell wall

TABLE 12.7 PROPERTIES AND DIMENSIONS OF H-PILES. (Courtesy of the American Institute of Steel Construction, Inc.)

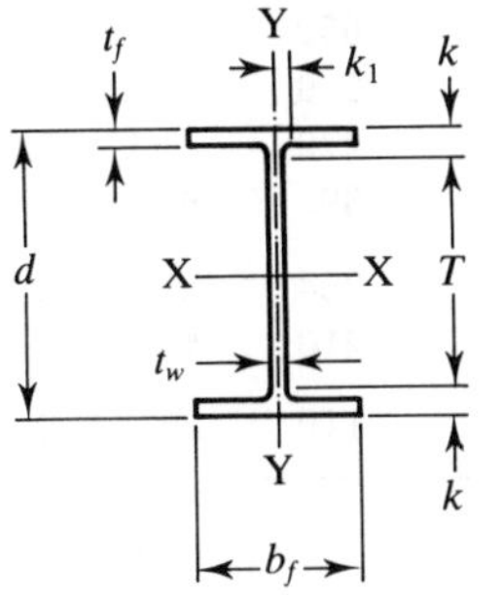

HP Shapes: Dimensions

Designation	Area A, in^2	Depth d, in		Web: Thickness t_w, in		Web: $\frac{t_w}{2}$, in	Flange: Width b_f, in		Flange: Thickness t_f, in		Distance: T, in	Distance: k, in	Distance: k_1, in
HP 14 × 117	34.4	14.21	$14\frac{1}{4}$	0.805	$\frac{13}{16}$	$\frac{7}{16}$	14.885	$14\frac{7}{8}$	0.805	$\frac{13}{16}$	$11\frac{1}{4}$	$1\frac{1}{2}$	$1\frac{1}{16}$
× 102	30.0	14.01	14	0.705	$\frac{11}{16}$	$\frac{3}{8}$	14.785	$14\frac{3}{4}$	0.705	$\frac{11}{16}$	$11\frac{1}{4}$	$1\frac{3}{8}$	1
× 89	26.1	13.83	$13\frac{7}{8}$	0.615	$\frac{5}{8}$	$\frac{5}{16}$	14.695	$14\frac{3}{4}$	0.615	$\frac{5}{8}$	$11\frac{1}{4}$	$1\frac{5}{16}$	$\frac{15}{16}$
× 73	21.4	13.61	$13\frac{5}{8}$	0.505	$\frac{1}{2}$	$\frac{1}{4}$	14.585	$14\frac{5}{8}$	0.505	$\frac{1}{2}$	$11\frac{1}{4}$	$1\frac{3}{16}$	$\frac{7}{8}$
HP 13 × 100	29.4	13.15	$13\frac{1}{8}$	0.765	$\frac{3}{4}$	$\frac{3}{8}$	13.205	$13\frac{1}{4}$	0.765	$\frac{3}{4}$	$10\frac{1}{4}$	$1\frac{7}{16}$	1
× 87	25.5	12.95	13	0.665	$\frac{11}{16}$	$\frac{3}{8}$	13.105	$13\frac{1}{8}$	0.665	$\frac{11}{16}$	$10\frac{1}{4}$	$1\frac{3}{8}$	$\frac{15}{16}$
× 73	21.6	12.75	$12\frac{3}{4}$	0.565	$\frac{9}{16}$	$\frac{5}{16}$	13.005	13	0.565	$\frac{9}{16}$	$10\frac{1}{4}$	$1\frac{1}{4}$	$\frac{15}{16}$
× 60	17.5	12.54	$12\frac{1}{2}$	0.460	$\frac{7}{16}$	$\frac{1}{4}$	12.900	$12\frac{7}{8}$	0.460	$\frac{7}{16}$	$10\frac{1}{4}$	$1\frac{1}{8}$	$\frac{7}{8}$
HP 12 × 84	24.6	12.28	$12\frac{1}{4}$	0.685	$\frac{11}{16}$	$\frac{3}{8}$	12.295	$12\frac{1}{4}$	0.685	$\frac{11}{16}$	$9\frac{1}{2}$	$1\frac{3}{8}$	1
× 74	21.8	12.13	$12\frac{1}{8}$	0.605	$\frac{5}{8}$	$\frac{5}{16}$	12.215	$12\frac{1}{4}$	0.610	$\frac{5}{8}$	$9\frac{1}{2}$	$1\frac{5}{16}$	$\frac{15}{16}$
× 63	18.4	11.94	12	0.515	$\frac{1}{2}$	$\frac{1}{4}$	12.125	$12\frac{1}{8}$	0.515	$\frac{1}{2}$	$9\frac{1}{2}$	$1\frac{1}{4}$	$\frac{7}{8}$
× 53	15.5	11.78	$11\frac{3}{4}$	0.435	$\frac{7}{16}$	$\frac{1}{4}$	12.045	12	0.435	$\frac{7}{16}$	$9\frac{1}{2}$	$1\frac{1}{8}$	$\frac{7}{8}$
HP 10 × 57	16.8	9.99	10	0.565	$\frac{9}{16}$	$\frac{5}{16}$	10.225	$10\frac{1}{4}$	0.565	$\frac{9}{16}$	$7\frac{5}{8}$	$1\frac{3}{16}$	$\frac{13}{16}$
× 42	12.4	9.70	$9\frac{3}{4}$	0.415	$\frac{7}{16}$	$\frac{1}{4}$	10.075	$10\frac{1}{8}$	0.420	$\frac{7}{16}$	$7\frac{5}{8}$	$1\frac{1}{16}$	$\frac{3}{4}$
HP 8 × 36	10.6	8.02	8	0.445	$\frac{7}{16}$	$\frac{1}{4}$	8.155	$8\frac{1}{8}$	0.445	$\frac{7}{16}$	$6\frac{1}{8}$	$\frac{15}{16}$	$\frac{5}{8}$

TABLE 12.7 (*Continued*)

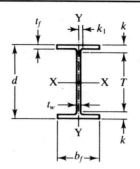

HP Shapes: Properties

Nom-inal Wt. per Ft, lb	Compact Section Criteria					Elastic Properties						Plastic Modulus	
						Axis X–X			Axis Y–Y				
	$\frac{b_f}{2t_f}$	$\frac{h_c}{t_w}$	F_y''', ksi	X_1, ksi	$X_2 \times 10^6$, $(1/ksi)^2$	I, in⁴	S, in³	r, in	I, in⁴	S, in³	r, in	Z_x, in³	Z_y, in³
117	9.2	14.2	—	3870	659	1220	172	5.96	443	59.5	3.59	194	91.4
102	10.5	16.2	—	3400	1090	1050	150	5.92	360	51.4	3.56	169	78.8
89	11.9	18.5	—	2960	1840	904	131	5.88	326	44.3	3.53	146	67.7
73	14.4	22.6	—	2450	3880	729	107	5.84	261	35.8	3.49	118	54.6
100	8.6	13.6	—	4020	571	886	135	5.49	294	44.5	3.16	153	66.6
87	9.9	15.7	—	3510	970	755	117	5.45	250	38.1	3.13	131	58.5
73	11.5	18.4	—	3000	1790	630	98.8	5.40	207	31.9	3.10	110	48.8
60	14.0	22.7	—	2460	3880	503	80.3	5.36	165	25.5	3.07	89.0	39.0
84	9.0	14.2	—	3860	670	650	106	5.14	213	34.6	2.94	120	53.2
74	10.0	16.0	—	3440	1050	569	93.8	5.11	186	30.4	2.92	105	46.6
63	11.8	18.9	—	2940	1940	472	79.1	5.06	153	25.3	2.88	88.3	38.7
53	13.8	22.3	—	2500	3650	393	66.8	5.03	127	21.1	2.86	74.0	32.2
57	9.0	13.9	—	3920	631	294	58.8	4.18	101	19.7	2.45	66.5	30.3
42	12.0	18.9	—	2920	1970	210	43.4	4.13	71.7	14.2	2.41	48.3	21.8
36	9.2	14.2	—	3840	685	119	29.8	3.36	40.3	9.88	1.95	33.6	15.2

F'	Allowable stress in ring wale based on buckling strength	K	Ratio of lateral force to horizontal force and force on cellular cofferdams transverse walls and permeability and effective length factor for axially loaded members
$F.S.$	Factor of safety		
F_a	Axial stress permitted in absence of bending moment		
F_b	Bending stress permitted in absence of axial stress	K_a	Ratio of lateral force to horizontal force, active state
F_p	Piling pullout force	K_0	Ratio of lateral force to horizontal force, at rest state
F_T	Total interlock friction resistance		
F_e'	Euler stress divided by safety factor	K_p	Ratio of lateral force to horizontal force, passive state
F_y	Specified minimum yield stress		
G	Specific gravity	L	One-half center-to-center cell distance from circular cells; distance between diaphragms on diaphragm cells; length of excavation; distance between struts or unbraced column length
H	Height of cell or depth of excavation		
H_b	Height of berm		
H_c	Critical height of excavation or thickness of clay stratum		
		L_b	Width of top of berm
		L_b'	Width of bottom of berm
H_s	Thickness of sand stratum or soil	L_1	Length of arc wall or length of tieback not grouted
H_u	Water head		
$H_{1,2\ldots}$	Distance between struts; stratum thickness	L_2	Length of cell wall or length of tieback that is grouted
H_1	Distance from bottom of piling to impervious stratum		
		M	Moment
H'	Distance from phreatic line to bottom of cell at centerline	M_r (or M_R)	Total soil resisting moment in cell
		M_T	Total interlock resisting moment in cell

M_O (or M_o)	Total overturning moment
N	Stability number for excavations in clay
N_c	Bjerrum or Terzaghi bearing capacity factor
N_{cD}	Bearing capacity factor
N_q	Surcharge bearing capacity factor
N_γ	Bearing capacity factor
P	Lateral pressure
P'	Reactive force at connections in cells or force on sheeting below excavation
P_a	Total active pressure
P_c	Total pressure at centerline of cell
P_s	Total pressure inboard of piling on cell
P_p	Total passive pressure
P_w	Total water pressure
Q	Total centerline shearing force
R	Total resistance due to sliding on planes or strut reaction
R_1	Portion of resistance due to sliding on planes in cells (Cummings' Method)
R_2	Portion of resistance due to sliding on planes in cells (Cummings' Method)
S	Section modulus
S_c	Total centerline shear resistance
S_h	Total horizontal piling reaction
S_V	Total vertical piling reaction
S_{V1}	Vertical piling reaction
S_{V2}	Vertical friction from arc piling
S_{V3}	Vertical reaction from piling weight in one-half of cell
S_{V4}	Vertical reaction from interlock friction
T	Thrust
W	Weight
W_f	Weight of soil below rupture surface
W_p	Weight of piling
Y	Distance of log spiral locus below (or above) foot of cell
Z	Distance between blocking points on ring wale
a	Distance from intersection of centerline to top of cell
b	Equivalent width of filled cells and distance from bottom of cell to centerline intersection
c	Cohesive strength or chord length
c'	Average cohesive strength
d	Penetration of sheeting below ground surface of excavation
d_1	Distance S_V acts from centerline cell axis
d'	Distance from ground line to maximum bulge of cell
e	Void ratio or maximum distance between the chord and rise of a curve
f_a	Computed axial stress
f_b	Computed bending stress
f_d	Coefficient of friction and resistance between piling foot and rock
f_{gr}	Coefficient of friction between gravel and rock
f_{gs}	Coefficient of friction between gravel and steel
f_{ss}	Coefficient of friction between steel and steel
g	Chord length between blocking points on ring wale
h	Distance from bottom of excavation to hard stratum or water depth
k	Permeability
m	Reduction factor
n	Ratio of field q_u to laboratory q_u
p	Unit pressure
p'_a	Average active unit pressure
p'_m	Average unit pressure of mixed soils
p_s	Unit pressure from surcharge

p_w	Unit pressure from water
q	Surcharge or unit load
q_f	Bearing capacity
q_u	Unconfined compressive strength
q'_u	Average unconfined compressive strength
r_a	Arc radius
r_c	Cell radius
r_g	Radius of gyration
r_1	Farthest distance from pole to foot of cell (log spiral curve)
r_0	Closest distance from pole to foot of cell (log spiral curve)
t_a	Arc interlock force
t_{c1}	Outboard interlock force (circular cells)
t_{c2}	Inboard interlock force (circular cells)
t_d	Diaphragm interlock force
t_{ult}	Ultimate interlock strength
α	Variable in polar coordinate system (log spiral curve)
β	Angle between r_1 and r_0 at pole of log spiral curve
γ	Unit weight of soil
γ_1	Unit weight of soil above excavation bottom
γ_2	Unit weight of soil below excavation
γ'	Buoyant unit weight of soil
γ'_m	Average unit weight of soil
γ_w	Unit weight of water
δ	Angle of wall friction
ϕ	Angle of internal friction
ksf	Kips per square foot (1 kip = 1000 pounds)
ksi	Kips per square inch
pcf	Pounds per cubic foot

REFERENCES

Belz, C. A. (1970), Cellular structure design methods, *Proceedings, Conference on Design and Installation of Pile Foundations and Cellular Structures*, ed. H. Y. Fang and T. D. Dismuke, Envo Publishing Co., Lehigh Valley, pp. 319–388.

British Standards Institution (1972), *Foundations, Civil Engineering Code of Practice*, CP 2004, London.

B.S.P. Pocket Book (1963), B.S.P. International Foundations Ltd., Claydon, Suffolk, U.K.

Casagrande, L. (1973), Comments on conventional design of retaining structures, *Journal of the Soil Mechanics and Foundations Division, ASCE*, **99**, No. SM-2, pp. 181–198.

Committee for Waterfront Structures (1980), *Recommendations of the Committee for Waterfront Structures* (English translation), W. Ernst and Son, Berlin.

Cornfield, G. M. (1968), Driven piles: Method and equipment, *Construction News*, London, May 9. [Reprinted in *Impact* (No. 15), Journal of the British Steel Piling Co. Ltd., London.]

Cornfield, G. M. (1969), Direct-reading nomograms for design of anchored sheet pile retaining walls, *Civil Engineering and Public Works Review*, August, London.

Cummings, E. M. (1957), Cellular cofferdams and docks, *Transactions, ASCE*, **125**, pp. 13–34; Discussion, pp. 34–45.

Daniel, D. E. and Olson, R. E. (1982), Failure of an anchored bulkhead, *Journal of the Geotechnical Engineering Division, ASCE*, **108**, No. GT-10, pp. 1318–1327.

Dismuke, T. D. (1970), Stress analysis of sheet piling in cellular structures, *Proceedings, Conference on Design and Installation of Pile Foundations and Cellular Structures*, ed. H. Y. Fang and T. D. Dismuke, Envo Publishing Co., Lehigh Valley, pp. 339–365.

Gould, J. P. (1980), A summary of some performance records, *New Developments in Earth Support—Case Histories*, Metropolitan Section ASCE, Foundations and Soil Mechanics Group Seminar, December.

Grayman, R. (1970), Cellular structures failures, *Proceedings, Conference on Design and Installation of Pile Foundations and Cellular Structures,* ed. H. Y. Fang and T. D. Dismuke, Envo Publishing Co., Lehigh Valley, pp. 383–391.

Haliburton, T. A. (1968), Numerical analysis of flexible retaining structures, *Journal of the Soil Mechanics and Foundations Division, ASCE,* **94,** No. SM-6, pp. 1233–1251.

Hansen, J. B. (1953), *Earth Pressure Calculations,* The Danish Technical Press, The Institution of Danish Civil Engineers, Copenhagen.

Hansmire, W. H., Russell, H. A., Rawnsley, R. P., and Abbott, E. L. (1989), Field performance of structural slurry wall, *Journal of Geotechnical Engineering, ASCE,* **115,** No. GT-2, pp. 141–156.

Hetenyi, M. (1946), *Beams on Elastic Foundation,* University of Michigan Press.

Lambe, T. W. (1970), Braced excavations, *ASCE Specialty Conference,* Cornell University, pp. 149–328.

Lamboj, L. and Fang, H. Y. (1970), Comparison of maximum moment, tie rod force and embedment depth of anchored sheet piles, *Fritz Engineering Laboratory Report* No. 365.1, Lehigh University, Bethlehem, Pa.

Lum, W. (1973), Discussion of "Comments on conventional design of retaining structures," by Leo Casagrande (Proceedings Paper 9568), *Journal of the Soil Mechanics and Foundations Division, ASCE,* **99,** No. SM-8, p. 648.

McLean, F. G. and Krizek, R. J. (1971), Seepage characteristics of imperfect cutoffs, *Journal of the Soil Mechanics and Foundations Division, ASCE,* **105,** No. SM-1, pp. 305–312.

NAVFAC DM-7 (1971), Department of the Navy, Bureau of Yards and Docks Design Manual, Washington, D.C.

Ohori, K., Takahashi, K., Kawai, Y., and Shiota, K. (1988), Static analysis model for double pile wall structures, *Journal of Geotechnical Engineering, ASCE,* **114,** No. GT-7, pp. 810–823.

Ovesen, N. K. (1962), *Cellular Cofferdams, Calculation Methods And Model Tests,* Bulletin No. 14, The Danish Geotechnical Institute.

Patterson, J. H. (1970), Installation techniques for cellular structures, *Proceedings, Conference on Design and Installation of Pile Foundations and Cellular Structures,* ed. H. Y. Fang and T. D. Dismuke, Envo Publishing Co., Lehigh Valley, pp. 393–412.

Peck, R. B. (1943), Earth pressure measurements in open cuts, Chicago, Ill. Subway, *Transactions, ASCE,* **108.**

Peck, R. B. (1969), Deep excavations and tunneling in soft ground, State of the Art Volume, *Seventh International Conference on Soil Mechanics and Foundations,* Mexico, pp. 225–290.

Rossow, M., Demsky, E., and Mosher, R. (1987), *Theoretical Manual For Design Of Cellular Sheet Pile Structures (Cofferdams And Retaining Structures),* Technical Report ITL-87-5, Waterways Experiment Station, Corps of Engineers, U.S. Dept of the Army.

Rowe, P. W. (1952), Anchored sheet pile walls, *Proceedings, Institution of Civil Engineers (London),* **1,** pp. 27–70; Discussion, **1,** pp. 616–647.

Rowe, P. W. (1955), A theoretical and experimental analysis of sheet pile walls, *Proceedings, Institution of Civil Engineers (London),* **4,** pp. 32–87; Discussion, **4,** pp. 828–855.

Rowe, P. W. (1956), Sheet pile walls at failure, *Proceedings, Institution of Civil Engineers (London),* **5,** pp. 276–315; Discussion, **6,** pp. 347–361.

Rowe, P. W. (1957), Limit design of flexible walls, *Proceedings, Midland Soil Mechanics and Foundation Engineering Society,* **1,** pp. 29–40.

Schroeder, W. L. (1987), Wharf bulkhead behavior at Fulton Terminal 6, *Journal of Geotechnical Engineering, ASCE,* **113,** No. GT-6, pp. 600–615.

Schroeder, W. L., Marker, D. K., and Khuayjarernpanishk, Th. (1977), Performance of a cellular wharf, *Journal of the Geotechnical Enginering Division, ASCE,* **103,** No. GT-3, pp. 153–168.

Schroeder, W. L. and Maitland, J. K. (1979), Cellular bulkheads and cofferdams, *Journal of the Geotechnical Engineering Division, ASCE,* **105,** No. GT-7, pp. 823–837.

Sorota, M. D. and Kinner, E. B. (1981), Cellular cofferdam for Trident drydock: Design, *Journal of the Geotechnical Engineering Division, ASCE,* **107,** No. GT-12, pp. 1643–1655.

Sorota, M. D., Kinner, E. B., and Haley, M. X. (1981), Cellular cofferdam for trident drydock: Performance, *Journal of the Geotechnical Engineering Division, ASCE,* **107,** No. GT-12, pp. 1657–1676.

Swatek, E. P., Jr. (1970), Summary—Cellular structure design and installation, *Proceedings, Conference on Design and Installation of Pile Foundations and Cellular Structures,* ed. H. Y. Fang and T. D. Dismuke, Envo Publishing Co., Lehigh Valley, pp. 413–423.

Terzaghi, K. (1943), *Theoretical Soil Mechanics,* John Wiley & Sons, Inc., New York, N.Y.

Terzaghi, K. (1945), Stability and stiffness of cellular cofferdams, *Transactions, ASCE,* **110,** pp. 1083–1119; Discussion, pp. 1120–1202.

Terzaghi, K. (1954), Anchored bulkheads, *Transactions, ASCE,* **119,** pp. 1243–1280; Discussion, pp. 1281–1324.

Terzaghi, K. (1955), Evaluation of coefficients of subgrade reaction, *Geotechnique,* **5,** December.

Terzaghi, K. and Peck, R. (1967), *Soil Mechanics In Engineering Practice,* John Wiley & Sons, Inc., New York, N.Y., pp. 260–266, 394–413.

Tschebotarioff, G. P. (1951), *Soil Mechanics, Foundations and Earth Structures,* McGraw-Hill Book Co., Inc., New York, N.Y.

Tschebotarioff, G. P. (1973), *Foundations, Retaining And Earth Structures,* McGraw-Hill Book Co., Inc., New York, N.Y.

Tsinker, G. P. (1983), Anchored sheet pile bulkheads: Design practice, *Journal of Geotechnical Engineering, ASCE,* **109,** No. GT-8, pp. 1021–1038.

TVA (1957), *Sheet Steel Piling Cellular Cofferdams on Rock,* Technical Monograph No. 75, Vol. 1, U.S. Tennessee Valley Authority.

White, A., Cheney, J. A., and Duke, C. M. (1961), Field study of a cellular bulkhead, *Journal of the Soil Mechanics and Foundations Division, ASCE,* **88,** No. SM-4, Proceedings Paper 2902.

13 PILE FOUNDATIONS

BENGT H. FELLENIUS, P.Eng., Dr.Tech.
Professor in Civil Engineering
University of Ottawa, Ottawa, Canada

13.1 INTRODUCTION AND BACKGROUND

Piles are vertical or slightly inclined, relatively slender structural foundation members. They transmit loads from the superstructure to competent soil layers. Length, method of installation, and way of transferring the load to the soil can vary greatly.

Piles are used for a variety of reasons, as follows:

- A competent soil layer can only be found at depth.
- The soil layers immediately below the structure, while competent, are subject to scour.
- The structure transmits large concentrated loads to the soil that cannot be spread out horizontally by means of a wide, shallow foundation.
- The structure is very sensitive to differential settlement.
- The site has a very high water table or artesian water conditions and the soil is sensitive to the construction of even shallow excavations required for mat or footing foundations.

In some cases, the piles serve only to improve the bearing capacity, density, or stiffness of the surrounding soil without directly carrying the load of the structure. Figure 13.1 gives a few examples of the use of piles, including a schematic display of group arrangements of piles.

The connection between the superstructure and the pile is called the *pile cap*. The upper end of the pile (the end connected to the pile cap) is called the *pile head*, and the bottom end is called the *pile toe*. What lies between the pile head and the pile toe is called the *pile shaft*. In older terminology, the term "skin" was used to refer to the surface of the pile shaft.

Piles can be cylindrical or conical. Conical piles can be smooth-tapered or step-tapered. The cross section of a pile can be circular, octagonal, hexagonal, and even triangular; it can be H-shaped, solid, or hollow. The surface of the shaft can be smooth or grooved. The pile toe can be blunt or pointed, or equipped with a shoe with a blunt or pointed end. The shoe can be of the same diameter as the pile shaft or enlarged; it can be equipped with a separate dowel of hardened steel, a rock point. Figure 13.2 provides a small selection of various cross sections and shapes of piles.

The pile material can be wood, concrete, or steel, or any combination thereof. Wood is naturally a very variable material. The proportioning of concrete and the amount and type of reinforcement differs with intended use and with geographic location (mostly because engineering tradition differs both between and within countries). For steel piles, the yield strength of the steel varies considerably between different countries and different sources.

Piles can be used singly or in groups. Mostly, piles are placed in groups. The behavior of a single pile is different from that of an individual pile in a group. A pile group can consist

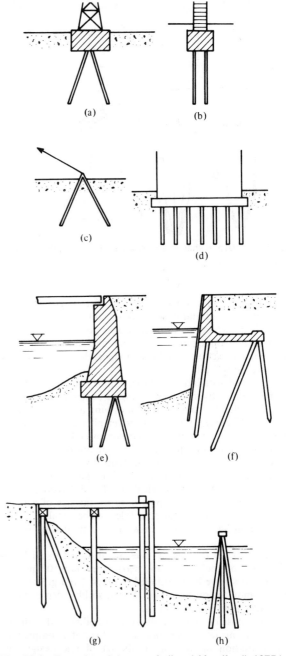

(a)

(b)

(c)

(d)

(e)

(f)

(g)

(h)

Fig. 13.1 Examples of the use of piles. (*After Kezdi, 1975.*)

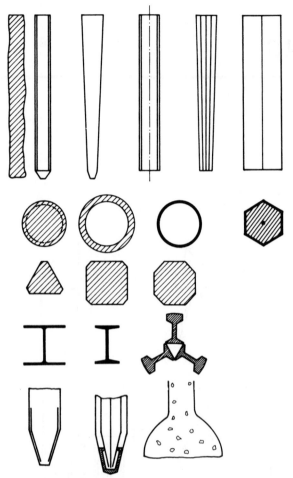

Fig. 13.2 Varous shapes and cross sections of piles. (*Modified after Kezdi, 1975.*)

of a cluster of piles, where the group effect is governing in all directions of load and movement, or consist of a row of piles (a pile bent) where the pile behavior is governed by the group effect in one direction of load and movement, while in the orthogonal direction the piles are independent of the group and behave as single piles.

The list of installation methods can be made very long: Piles can be installed by means of driving or be made in-situ (bored), or be installed by combination of driving and in-situ methods (concrete-filled steel-pipe piles). Driven piles can be head-driven, that is, forced into the ground by means of a pile driving hammer creating a compression force by impacting on the pile head. They can be toe-driven, that is, installed by means of a hammer impacting the pile toe from inside of the pile creating a tensile force in the pile to pull the pile down. Head-driving techniques can be used to toe-drive pipe piles by the use of a mandrel acting on the bottom of the pipe. Piles can be installed by means of vibration hammers with or without combination of jetting techniques.

Last, but far from least, piles are installed in all kinds of soils, even soft rock. For many years, more than 100 metre long slender piles have been installed in clays. Even greater depths are common for offshore piles. Short, stubby, driven piles are frequently used in ablation tills.

It is impossible to cover all aspects of piling within the space allotted to this text. Therefore, this text is not a comprehensive documentation of the state-of-the-art of pile design, but a brief presentation of basics of pile analysis and a few simple

approaches that can assist the practising foundation engineer in routine work.

13.2 ASPECTS FOR GENERAL CONSIDERATION

All of the variations mentioned in the introduction will to a greater or smaller degree influence how piles support a structure. For single piles, the dependency on the more important factors is reasonably well known. About group piles, on the other hand, very little is known.

According to Vesic (1977), there were in 1977 extremely few papers reporting full-scale tests with pile groups: Five pile groups were tested in clay and six in sand. The words "full-scale" are used generously in this context. In all but one case, the full-scale was limited to a diameter of 100 mm and a length of 2 m and the maximum number of piles in a group was nine. Despite some recent full-scale tests (O'Neill et al., 1982), so little is known and less verified of pile group behavior that every general method of analysis and statement of group effect should be considered hypothetical and unproven. Therefore, the practising foundation engineer, working under time constraint and having limited data for input into an analysis, is well advised to consider the quantitative result of all analysis to be uncertain, whether the analysis is based on a simplified theoretical approach, or performed according to the latest advancement in theory and laboratory testing. Also, in the light of the ever increasing liability of the designer, verification in the field is imperative in all pile design.

Verification of the analysis can consist of simply relating the design to the old design and past performance of piles as well as of pile-supported structures in the same general area. Naturally, when such experience is lacking or inadequate, various methods of field testing and observations are required. Then, it is necessary to first know the degree of error involved in the data and second to minimize this degree as much as technically and economically possible. The data should be subjected to a rational analysis and related to a theoretical model of the behavior of the pile(s). Otherwise, conclusions drawn and recommendations made may be erroneous.

No design work is complete unless combined with an educated quality inspection and control program aimed to verify that the construction procedures and revelations agree with the design assumptions.

The bearing capacity of a pile consists of a combination of fully developed shaft and toe capacities. For shaft capacity, a distinction is made between, on the one hand, the mobilization of shear caused by the transfer to the soil of load applied to the pile [positive shaft resistance is mobilized for compression loading (the pile is pushed into the ground) and negative shaft resistance for tension loading (uplift; pull)] and, on the other hand, shear mobilized by the soil moving relative to the pile [negative skin friction when the soil settles relative to the pile and positive skin friction when the soil expands (piles in swelling soil), respectively].

The capacity (the ultimate resistance; the ultimate load) is often difficult to assess even by means of a static loading test. The oldest approach is simply to state that the ultimate load in a test is equal to the applied load when the movement of the pile head is 10 percent of the pile toe diameter. Vesic (1977) listed this definition and some others based on the movement of the pile head. Fellenius (1975b, 1980) presented a comparison of nine methods based on the shape of the load–movement curve (see below).

In the simplest principle, design for capacity consists of determining the allowable load (the service load; the upper limit of the applied load) on the pile by dividing capacity with a factor of safety.

When the service load on the piles stresses the soil, the soil will consolidate and compress and the pile will settle. Usually, the methods of settlement calculation are very simple. In estimating the settlement of piles in sand, for instance, a common approach is to consider the settlement to be equal to 1 percent of the pile head diameter plus the "elastic" compression of the pile under the load.

For calculating the settlement of an essentially shaft-bearing pile group in homogeneous clay soil, Terzaghi and Peck (1967) recommended taking the settlement of the group as equal to that calculated for an equivalent footing located at the lower third point of the pile embedment length and loaded to the same stress and over the same area as the pile group plan area (see Fig. 13.3).

More sophisticated methods for calculation of settlement may be used that employ elastic half-sphere analysis and/or finite-element techniques. Vesic (1977) and Poulos and Davis (1980) presented several such analytical approaches toward calculating settlement on single piles and pile groups. Also in the sophisticated methods, the overly simplified assumptions are made that no stress is present in the pile or piles before load is applied to the pile foundation, the piles are of equal length and have equal capacities, and the soil surrounding the piles is assumed to be homogeneous with depth as well as across the group.

The obvious discrepancy between the conditions in the field on the one hand and the too general and/or ideal conditions (assumed in both the simple and the sophisticated approaches) is very frustrating to the foundation engineer. As a consequence, prediction and analysis of settlement of pile groups in engineering design practice leaves much to be desired.

For the case of piles installed through a multilayered soil deposit, where upper layers settle owing to, for instance, a surcharge on the ground surface or to a general groundwater lowering, a settlement calculation of the pile group is often not performed. The design practice seems to trust that the settlement will somehow be taken care of by deducting dragload from the allowable load. (Dragload is the accumulation of negative skin friction in the settling layers.) When a settlement calculation is carried out, sometimes the load generating settlement is taken

to consist of service load and dragload combined! This practice is very unsatisfactory, because capacity and settlement interact and they cannot be treated separately and independently of each other. However, methods do exist by which capacity and settlement can be calculated. One of the most straightforward methods will be described in the following sections.

13.3 THE SHAFT RESISTANCE

For the analysis of shaft resistance, Johannessen and Bjerrum (1965) and Burland (1973) established that the unit resistance is proportional to the effective overburden stress in the soil surrounding the pile. The constant of proportionality is called beta-coefficient, β, and is assumed to be a function of the earth pressure coefficient, K_s, in the soil, times the soil internal friction, $\tan \phi'$, and times the quotient of the wall friction, M (Bozozuk, 1972). Thus, the unit shaft resistance, r_s, follows the following relations:

$$r_s = \beta\sigma'_z \qquad (13.1)$$

$$\beta = MK_s \tan \phi' \qquad (13.2)$$

where

r_s = unit shaft resistance at depth z
β = Bjerrum–Burland beta-coefficient
σ'_z = effective overburden stress at depth z
M = quotient of wall friction = $\tan \delta'/\tan \phi'$
δ' = effective soil–pile friction angle
ϕ' = effective soil friction angle
K_s = earth pressure coefficient

The terms and symbols are also explained in Figure 13.4.

One can develop a wide range of beta-coefficients from a combination of possible earth pressure coefficients, friction angles, and wall friction quotients. However, it appears that the variation of the beta-coefficient is smaller than the variation of its parts would suggest.

In analyzing measurements on piles subjected to downdrag, Bjerrum et al. (1969) found that the beta-coefficient in a soft silty clay ranged between 0.20 and 0.30. This range can be considered the low boundary of the beta-coefficient. While the theoretical upper boundary obviously can be very large, there is a practical limit governed by the density and strength of the soil in which the pile is driven or otherwise installed. For piles in very dense soil, the upper boundary can approach and exceed a value of 1.0, but usually an upper limit of 0.8 is assumed. Table 13.1 suggests a relative range of beta-values. The ranges shown are very wide and very approximate.

When shaft resistance is mobilized by a compression load (push load) applied to a pile, it is defined as "positive shaft resistance". When mobilized by uplift (pull or tension load), the term is "negative shaft resistance". For reasons of reduction of diameter (Poisson's ratio effect) and unloading of the effective overburden stress when loading in pull as opposed to loading in push, the negative shaft resistance is smaller than the positive shaft resistance. Ratios of 0.50 have been proposed.

It has been suggested that the above effective stress relation (Eq. 13.1) ceases to be valid at a certain critical depth equal to about 10 to 20 pile diameters. Below the critical depth, the unit shaft resistance would be constant and equal to the value at the critical depth. However, the concept of critical depth is unproven and in question. It should therefore be applied with caution, if at all.

Sometimes, particularly in cemented soils and more so for cast in-situ piles than for driven piles, the shaft resistance is a

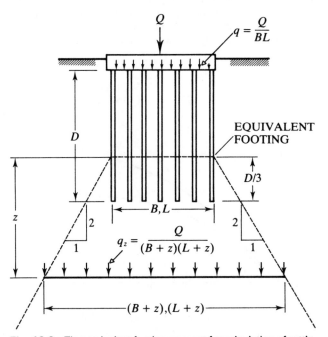

Fig. 13.3 The equivalent footing concept for calculation of settlement of a pile group. (*After Terzaghi and Peck, 1967.*)

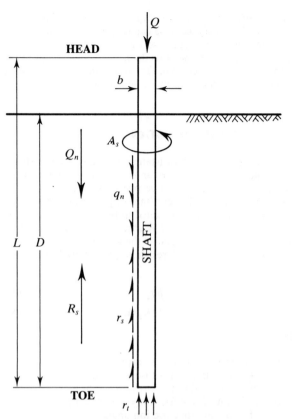

Fig. 13.4 Terms and symbols for pile analysis. Q_d = dead load; Q_l = live load; Q_n = drag load; Q_u = ultimate load (= capacity); R_s = shaft resistance; R_t = toe resistance; R_u = ultimate resistance (= capacity); L = pile length; D = embedment depth; b = pile diameter; A_s = circumferential area; A_t = toe cross-sectional area; N_t = toe bearing capacity coefficient.

$$R_t = A_t r_t - A_t \sigma'_{z=0} N_t$$

$$R_s = \Sigma A_s r_s = \Sigma A_s \beta \sigma'_z \quad \text{or} \quad \Sigma A_s (c' + \beta \sigma'_z)$$

function of both friction and cohesion. Equation 13.1 then changes to:

$$r_s = c' + \beta \sigma'_z \tag{13.3}$$

where c' = effective cohesion intercept.

Although it has been proven conclusively that the transfer of load from a pile to the soil by means of shaft resistance is governed by the effective stress, for piles in clay, a total stress analysis can be useful in site-specific instances. Also, enough information is often not available to support a reliable design based on effective stress analysis. A total stress analysis may then be used, which means that the shaft resistance is equal to the undrained shear strength of the soil and independent of the overburden stress:

$$r_s = \alpha \tau_u \tag{13.4}$$

TABLE 13.1 RANGES OF BETA-COEFFICIENTS.

Soil Type	phi	beta
Clay	25–30	0.23–0.4
Silt	28–34	0.27–0.5
Sand	32–40	0.30–0.8
Gravel	35–45	0.35–0.8

where

τ_u = undrained shear strength
α = proportionality coefficient

However, the total stress analysis can only lead so far and effective stress analysis according to Equations 13.1 and 13.3 provides the better means for analysis of test data and for putting experience to use in a design. Of course, more sophisticated effective stress theories for unit shaft resistance exist. However, in contrast to most of these, the effective stress approach according to Equations 13.1 and 13.3 is not restricted to homogeneous soils, but applies equally well to piles in layered soils and it can easily accommodate non-hydrostatic pore pressures.

The proportionality coefficient is equal to unity in soft and firm clays, but smaller than unity in stiff and hard clays, especially if they are overconsolidated. A useful qualitative reference is illustrated in Figure 13.5, showing that for wood and concrete piles the proportionality coefficient is equal to unity up to a shear strength of about 30 kPa, whereupon it becomes progressively smaller. For steel piles, the coefficient is indicated as smaller than unity even for soft clays.

Equation 13.5 gives the total shaft resistance as the integral of the unit shaft resistance over the embedment depth:

$$R_s = \int_0^D r_s \, dz = \int_0^D A_s (c' + \beta \sigma'_z) \, dz \tag{13.5}$$

where

R_s = total shaft resistance (fully mobilized)
A_s = pile unit circumferential area
D = pile embedment depth

It is important to realize that even simple axial loading of a single pile can be made in several different ways. Figure 13.6 illustrates six cases, A through F, of axial loading. Case A shows a pile loaded with a compression load (push load) at the pile head. The transfer of load to the soil increases the effective stress in the soil and produces compression stress in the pile. The increased stress in the pile causes an increase of pile diameter (Poisson's ratio effect; a minimal increase, of course). These aspects are symbolically indicated in the figure.

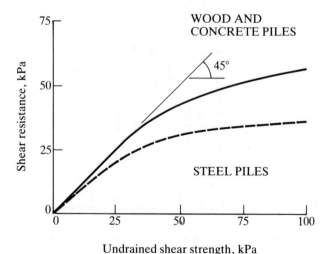

Fig. 13.5 Shaft resistance in clay as a function of undrained shear strength. (*After Tomlinson, 1957.*)

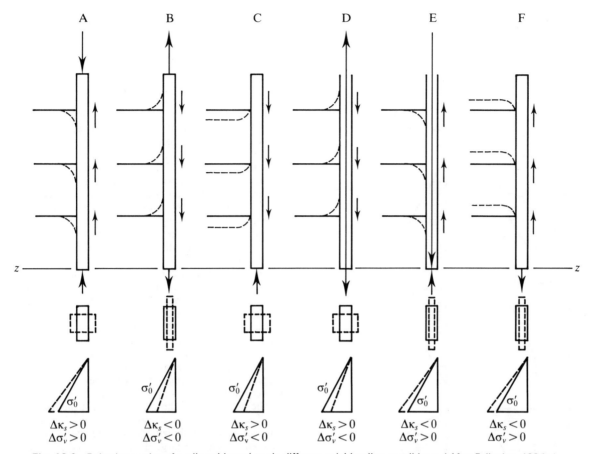

Fig. 13.6 Behavior modes of a pile subjected to six different axial loading conditions. (*After Fellenius, 1984a.*)

Case B shows the effect of an axial tension load (uplift load), when the effective stress in the soil is relieved, the pile is in tension, and the diameter reduces. Case C shows the effect of loading the pile by negative skin friction, when the pile effective stress is relieved, but, in contrast to Case B, the stress in the pile is increased and the diameter is increased. To test for the effect of negative skin friction, the pile would have to be pulled from the toe, as is illustrated in Case D.

Case E shows a pile tested by applying a push load at the toe, which simulates the effect of a pile in swelling soil as modeled in Case F.

The differences between the loading cases shown in Figure 13.6 are slight and the relevance of the distinctions can be questioned. However, it has been observed that the shaft resistance in pull (Case B) is smaller than the shaft resistance in push (Case A).

In contrast to most of the more sophisticated theories for unit shaft resistance in existence, the effective stress approach according to Equation 13.1 and Figure 13.6 is not restricted to homogeneous soils, but equally well applicable to piles in layered soils.

The analysis of load in a pile makes use of mathematical expressions called "transfer functions". Thus, for the case of fully mobilized shaft resistance, the following simple relation is obtained for the load in pile loaded to its full capacity:

$$Q_u = R_s + R_t \qquad (13.6)$$

where

R_s = shaft resistance
R_t = toe resistance

and

$$Q_z = Q_u - \int_0^z r_s dz = Q_u - \int_0^z A_s(c' + \beta \sigma_z') dz \qquad (13.7)$$

where

Q_z = the load in the pile at depth z
Q_u = the ultimate load

Equation 13.7 is the equation for a curved line which curvature increases progressively with increasing effective overburden stress, that is, increasing depth. Notice that a transfer function in a homogeneous soil resulting in a load distribution at failure that does not decrease progressively with depth is not correct.

Figure 13.7 indicates a few commonly suggested transfer functions (Q) with the corresponding distributions of unit shaft resistance (r_s). Of the five examples shown, only the first two, Cases (a) and (b), are reasonable in a homogeneous soil. The others are merely examples of little interest to the practitioner. In the literature, sometimes, evaluations of field test data are presented that show load distributions similar to Cases (c) and (d). Such evaluations should be considered with considerable scepticism. Case (e) does not exist in the real world.

The movement of the pile surface relative to the soil required to mobilize the ultimate shaft resistance is very small. Even at failure, the movement is not a slip, but occurs as a shear deformation in a zone at and extending out from the pile surface. Long-term measurements of load transfer and movements in piles in a slowly settling clay, where movements at depths were measured about one pile diameter away from the pile surface, showed that relative movements of about 1 mm were all that

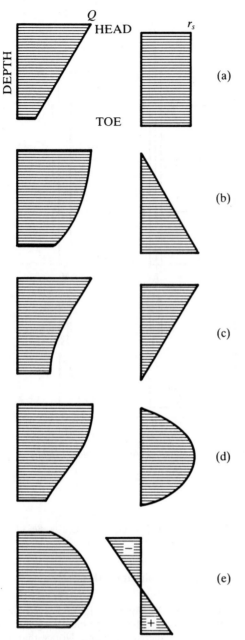

Fig. 13.7 Load transfer functions for five distributions of shaft resistance. (*After Vesic, 1970.*)

was necessary to mobilize the ultimate shaft resistance (Fellenius, 1972; Bjerin, 1977).

13.4 TOE RESISTANCE

The toe resistance is governed by the effective overburden stress according to the following simple relation:

$$R_t = A_t N_t \sigma'_{z=D} \qquad (13.8)$$

where

R_t = the toe resistance = $r_t A_t$
A_t = the pile toe cross-sectional area
N_t = toe bearing capacity coefficient
D = embedment depth
$\sigma'_{z=D}$ = effective overburden stress at the toe

As in the case for the shaft resistance, the unit toe resistance dependency on the effective overburden stress has been claimed to cease at a critical depth—conveniently close to the critical depth of the shaft resistance. For reasons similar to those given for the critical depth with regard to shaft resistance, one can question the existence of the critical depth also in the case of mobilizing toe resistance.

The toe bearing capacity coefficient, N_t, is considered to be related to the N_q factor in the bearing capacity formula for a conventional shallow foundation, that is, it is a function of the effective internal friction angle. Meyerhof (1976) and the *Canadian Foundation Engineering Manual* (1985) suggest that N_t be taken as approximately equal to $3N_q$. Given that, in engineering practice, concern for the toe capacity of a pile only occurs for pile toes placed in dense soils, where the internal friction angle is relatively high, at least about 38°, and that the uncertainty of the N_q relation for a friction angle in the range above 38° is considerable, the suggested relation to N_q is very approximate.

The N_t coefficient depends on soil composition in terms of grain size distribution, angularity and mineralogical origin of the grains, original soil density, density changes due to installation techniques, and other factors. For sedimentary cohesionless deposits, N_t ranges from a low of about 30 to a high of about 120. In very dense, non-sedimentary soils, such as glacial base tills, the N_t coefficient can be considerably higher, but also approach the lower boundary given above. Table 13.2 suggests a relative range of N_t coefficients. The ranges shown are very wide and very approximate.

The movement of the pile toe against the soil necessary to mobilize the ultimate toe resistance is considerably larger than that necessary for mobilizing the shaft resistance. The magnitude required for piles of large diameter is greater than for small-diameter piles. Driven piles, having densified the natural soil below the pile toe and/or preloaded it, require smaller movement as opposed to bored and cast-in-situ piles, not having densified the soil below the pile toe, but, potentially, having disturbed and loosened the soil instead.

For a driven pile, the necessary movement lies in the range of about 3 to 10 percent of the pile toe diameter. However, the load–movement relation is not a straight line. For instance, at a movement of half of that necessary to mobilize the ultimate toe resistance, more than half the toe resistance may be developed. For bored piles, the magnitude is more variable and less predictable.

Although the effective overburden stress governs the toe resistance also in clay, a total stress analysis is sometimes (traditionally) employed wherein the unit toe resistance is set to a factor times the undrained shear strength of the soil at the pile toe. In overconsolidated clays, the factor is sometimes set to a value of 3. Mostly, however, the value is 9. Considering that the undrained shear strength often is in the range of 20 to 30 percent of the effective overburden stress, the N_t factor in these clays becomes equal to about 3, which is the lower boundary of the range given in Table 13.2 for clays.

TABLE 13.2 RANGES OF N_t COEFFICIENTS.

Soil Type	phi	N_t
Clay	25–30	3–30
Silt	28–34	20–40
Sand	32–40	30–150
Gravel	35–45	60–300

13.5 CAPACITY DETERMINED FROM IN-SITU FIELD TESTING

Before the capacity of a pile and/or its required embedment depth can be determined, site information obtained by means of a site exploration program must be obtained. The site investigation includes identification of soil layers and classification of soil properties and strength parameters by sampling and testing of samples. In addition, and most important for determining the soil profile, in-situ testing is performed. Usually, the in-situ testing consists of all or at least one of standard penetration testing, vane shear testing, and cone penetrometer testing. More recently, pressuremeter testing and dilatometer testing are included in site exploration programs.

For many years, the N-index of standard penetration test has been used to calculate capacity of piles. Meyerhof (1976) compiled and rationalized some of the wealth of experience available and recommended that the capacity be a function of the N-index, as follows:

$$R = R_t + R_s = mNA_t + nNA_sD \qquad (13.9)$$

where

  m = a toe coefficient
  n = a shaft coefficient
  N = N-index at the pile toe
  N = average N-index along the pile shaft
  A_t = pile toe area
  A_s = unit shaft area; circumferential area
  D = embedment depth

For values inserted into Equation 13.9 using base SI-units, that is, R in newton, D in meters, and A in square meters, the toe and shaft coefficients, m and n, become:

$m = 400 \times 10^3$ for driven piles and 120×10^3 for bored piles (N/m^2)

$n = 2 \times 10^3$ for driven piles and 1×10^3 for bored piles (N/m^3)

The standard penetration test (SPT) is a subjective and highly variable test. The test and the N-index have substantial qualitative value, but should be used only very cautiously for quantitative analysis. The *Canadian Foundation Engineering Manual* (1985) includes a listing of the numerous irrational factors influencing the N-index. However, when the use of the N-index is considered with the sample of the soil obtained and related to a site and area-specific experience, prediction by the crude and decried SPT test does not come out worse than predictions by other methods of analysis.

The vane shear test provides a value of undrained shear strength that may be applied to Equation 13.4, above. While the vane shear test appears to be useful in soft clays, it is less reliable when used in silts. It should be recognized that no sample is obtained in the test and that not all vanes are alike. Again, when applied with knowledge of the soil layer tested and related to local experience of its prior use, the vane shear strength can be useful for judging pile capacity.

The static cone penetrometer resembles a pile. There is shaft resistance in the form of so-called local friction measured immediately above the cone point, and there is toe resistance in the form of the directly applied and measured cone-point pressure.

When applying cone penetrometer data to a pile analysis, both the local friction and the point pressure may be used as direct measures of shaft and toe resistances, respectively. However, both values can show a considerable scatter. Furthermore, the cone-point resistance, the cone-point being small compared to a pile toe, may be misleadingly high in gravel and layered soils. Schmertmann (1978) has indicated an averaging procedure to be used for offsetting scatter, whether caused by natural (real) variation in the soil or inherent in the test.

The piezocone, which is a cone penetrometer equipped with pore pressure measurement devices at the point, is a considerable advancement on the static cone. By means of the piezocone, the cone information can be related more dependably to soil parameters and a more detailed analysis can be performed. Soil is variable, however, and the increased and more representative information obtained also means that a certain digestive judgment can and must be exercised to filter the data for computation of pile capacity. In other words, the designer is back to square one; more thoroughly informed and less liable to jump to false conclusions, but certainly not independent of site-specific experience.

The pressuremeter and dilatometer have yet to provide the field verification necessary as reference before the devices become useful to the foundation engineer dealing with pile design.

13.6 INSTALLATION CONSIDERATIONS

It is difficult to determine the magnitude of the shaft and toe resistances before the disturbance from the pile installation has subsided. For instance, presence of dissipating excess pore pressures causes uncertainty in the magnitude of the effective stress in the soil and on-going strength gain due to reconsolidation is hard to estimate. Such installation effects can take a long time to disappear, especially in clays. Fellenius and Samson (1976) and Bozozuk et al. (1978) reported observations inside pile groups installed in silty clay where the reconsolidation period was greater than 6 months. For single piles in soft clay, Fellenius (1972) reported a reconsolidation period of 5 months.

There are indications that piles subjected to static loading tests in a soil before it has fully recovered from the installation effect require considerably large movements before the full ultimate resistance is mobilized (Fellenius et al., 1983, 1989) as opposed to piles tested after full recovery.

13.7 RESIDUAL COMPRESSION

Reconsolidation after installation of a pile imposes compression loads in the pile as a result of negative skin friction developing in the upper part of the pile. The induced load, called *residual load* or *residual compression*, is resisted by positive shaft resistance in the lower part of the pile and some toe resistance. If the residual load is not recognized in the evaluation of results from a static loading test, totally erroneous conclusions will be drawn from the test.

Presenting results from laboratory testing of slender model piles in sand, Hanna and Tan (1973) demonstrated in a pioneering paper the effect of not-considering versus considering residual compression when evaluating data from pile testing. Because the sand was deposited around the pile after the pile had been placed in the testing apparatus, it was ensured that the pile was subjected to a compression load before the start of the static loading test. Figure 13.8, modified from Hanna and Tan (1973), shows load-transfer functions evaluated when not considering the residual compression with load distribution curves shown for five applied loads, the largest being the load at induced soil failure, that is, with the ultimate shaft and toe resistances mobilized. Studying the slope of the load distribution curve at failure, it would appear from the upper diagram in

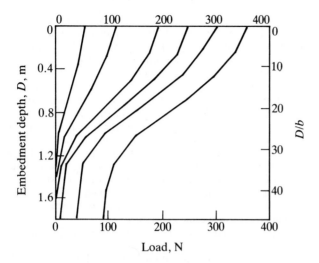

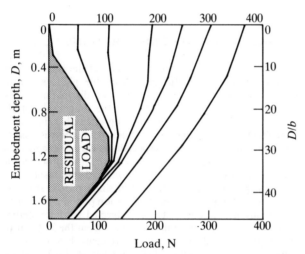

Fig. 13.8 Load distribution in a pile without and with consideration of residual load. (*After Hanna and Tan, 1973.*)

Figure 13.8 that the shaft resistance was the largest along the upper portion of the pile and that it reduced in magnitude with increasing depth to become first constant (straight-line load distribution) and then almost zero near the pile toe. This is seemingly in support of the critical depth concept. However, when, as is shown in the lower diagram in Figure 13.8, the residual compression evaluated to load was included in the analysis, the appearance changed dramatically. The load-distribution line curves progressively in accordance with Equation 13.7 and there is no critical depth or other anomaly shown. It should be noticed that the quantitative effect of including, as opposed to omitting, the residual load in the analysis is that the shaft resistance becomes smaller and the toe resistance becomes larger.

For a discussion in principle of the influence of residual loads on the analysis of piles in compression and tension testing, see Holloway et al. (1978) and Briaud and Tucker (1985).

Residual load in a pile is caused during the installation of a pile, by reconsolidation of the soil after the installation, and by a previous loading cycle. It manifests itself as a compression in the pile caused by negative skin friction in the upper portion of the pile, which is balanced by positive shaft resistance in the lower portion plus some toe resistance.

While the effect of residual loads on the analysis of results from a static loading test have now been realized by the

profession, the fact that every pile, whether loaded by a structure or not, is subjected to a similar interaction with the soil is not, as yet, readily recognized.

13.8 THE NEUTRAL PLANE

As stated earlier, only extremely small relative movements between the pile and the soil are required to mobilize a shaft resistance (in positive as well as negative direction). In combination with the additional fact that the difference in stiffness between the pile material and the soil is considerable and that there are movements and strains in a natural soil, make it clear that, for every pile, a stress transfer exists between the pile and the soil. In other words, there are movements in any and every soil, which are restrained by the pile. The restraint builds up force in the pile, and as there is force but no accelerating movement of the pile, the forces must be balanced—must be in equilibrium. Thus, negative skin friction is induced in the upper portion of the pile, resulting in a load in the pile that increases from zero at the pile head to a maximum at the depth of equilibrium. Below the equilibrium depth, the load decreases by being transferred to the soil by means of positive shaft resistance in combination with the small toe resistance.

Every pile develops an equilibrium of forces between the sum of the dead load (the sustained load) applied to the pile head and the dragload, and the sum of the positive shaft resistance and the toe resistance. The location of the equilibrium is called the *neutral plane* and it is the depth at which the shear stress along the pile changes from negative skin friction to positive shaft resistance. This depth is also where no relative displacement occurs between the pile and the soil.

The key aspect of the foregoing is that the development of a neutral plane and negative skin friction always occurs in piles without any appreciable settlement of the soil around the piles.

Normally, the neutral plane lies below the mid-point of a pile. The extreme case is for a pile on rock, where the location of the neutral plane is at the bedrock elevation. For a dominantly shaft-bearing pile "floating" in a homogeneous soil with linearly increasing shear resistance, the neutral plane lies at a depth which is about equal to the lower third point of the pile embedment length.

The larger the toe resistance, the deeper the elevation of the neutral plane, and, the larger the dead load, the shallower the elevation of the neutral plane.

The load transfer in a pile above the neutral plane during long-term conditions is expressed by Equation 13.10 (as opposed to the load in the pile during a static loading test to failure, where the load transfer follows Eq. 13.7):

$$Q_z = Q_d + \int_0^z r_s\,dz = Q_d + \int_0^z A_s(c' + \beta\sigma_z')dz \quad (13.10)$$

where

z = depth above the neutral plane
Q_z = the load in the pile at depth z
Q_d = the dead load applied to the pile head

Below the neutral plane, the load is expressed by Equation 13.7 (provided that the toe resistance is fully mobilized).

It is usually assumed that the unit skin friction, q_n, is equal to the unit positive shaft resistance, r_s, an assumption that is debatable, but any error results in an overestimation of the dragload and places the neutral plane higher than if q_n is smaller than r_s. That is, the assumption gives results on the conservative side.

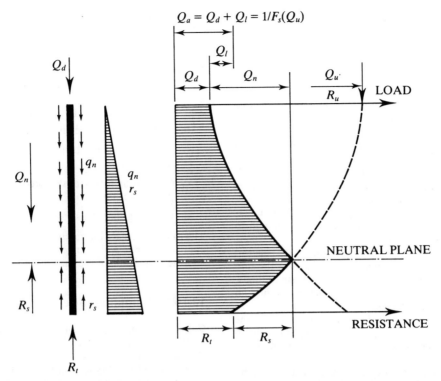

Fig. 13.9 Construing the neutral plane. For terms and symbols, see Figure 13.4. (*After Fellenius, 1989a.*)

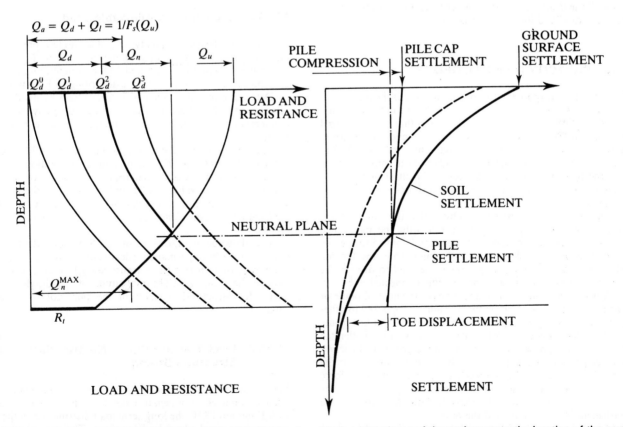

Fig. 13.10 Diagrams of load and resistance and of settlement showing the dependence of the settlement on the location of the neutral plane. The unified design for capacity, negative skin friction, and settlement. (*After Fellenius, 1989a.*)

With larger toe resistance, the elevation of the neutral plane lies deeper into the soil. If an increased dead load is applied to the pile head, the elevation of the neutral plane moves upward.

Figure 13.9 illustrates how to construe the location of the neutral plane. The figure shows the distribution of load in a pile subjected to a service load, Q_d, and installed in a relatively homogeneous soil deposit, where the shear stress along the pile is proportional to the effective overburden stress (for explanations of terms and symbols, see Fig. 13.4).

For reasons of clarity, several simplifying assumptions lie behind Figure 13.9: (a) that any excess pore pressure in the soil caused by the pile installation has dissipated and the pore pressure is hydrostatically distributed; (b) that the shear stress along the pile is independent of the direction of the relative movement, that is, the magnitude of the negative skin friction, q_n, is equal to the magnitude of the unit positive shaft resistance, r_s; and (c) that the toe movement induced is large enough to mobilize some toe resistance, R_t.

As shown, a dragload, Q_n, develops above the neutral plane. The magnitude of the dragload is calculated as the sum (the integral) of the unit negative skin friction. Correspondingly, the total shaft resistance below the neutral plane, R_s, is the sum of the unit positive shaft resistance.

In Figure 13.10, the left-hand diagram illustrates how the elevation of the neutral plane changes with a change in the load, Q_d, applied to the pile head. Notice also that the magnitude of the dragload changes when Q_d changes. The right-hand diagram illustrates the distribution of settlement in the soil as caused by a surcharge on the ground, and/or lowering of the groundwater table, etc., and by the dead load on the pile(s).

Figure 13.10 indicates that the settlement of the pile and the settlement of the soil are equal at the neutral plane. The "kink" in the curve at the neutral plane represents the influence of the dead load on the pile that starts to stress the soil at the neutral plane. If the dead load is zero, the settlement distribution curve has no "kink" and follows the dashed line.

13.9 CAPACITY OF A PILE GROUP

In extending the approach to a pile group, it must be recognized that a pile group is made up of a number of individual piles that have different embedment lengths and that have mobilzed the toe resistance to a different degree. The piles in the group have two things in common, however. They are connected to the same stiff pile cap and, therefore, all pile heads move equally, and the piles must all have developed a neutral plane at the same depth somewhere down in the soil (long-term condition, of course).

Therefore, it is impossible to achieve that the neutral plane is common for the piles in the group, with the mentioned variation of length, etc., unless the dead load applied to the pile head from the cap differs between the piles. Thus, the Unified Method extended to a pile group can be used to discuss the variation of load within a group of stiffly connected piles.

A pile with a longer embedment below the neutral plane or one having mobilized a larger toe resistance as opposed to other piles will carry a greater portion of the dead load on the group. On the other hand, a shorter pile, or one with a smaller toe resistance, as opposed to other piles in the group, will carry a smaller portion of the dead load. If a pile is damaged at the toe, it is possible that the pile exerts a negative—pulling—force at the cap and thus increases the total load on the pile cap. Remember, a dragload will occur without any appreciable settlement in the soil around the piles.

An obvious result of the development of the neutral plane is that no portion of the dead load is transferred to the soil via the pile cap. Unless, of course, the neutral plane lies right at the pile cap and the entire pile group is failing.

13.10 SUMMARY OF DESIGN PROCEDURE FOR CAPACITY AND STRENGTH

The design of a pile or a pile group follows four steps:

   a. Compiling and assessing all site and soil information
   b. Calculating capacity and distribution of shaft and toe resistances
   c. Calculating load-transfer curves determining the neutral plane location
   d. Checking that the structural strength is adequate

The calculations are interactive inasmuch as change of the load applied to a pile will change the location of the neutral plane and the magnitude of the maximum load in the pile.

13.10.1 Compiling Site and Soil Information

Compile first into a table all available data useful for reference when determining shaft and toe resistances, while noting the elevation of the groundwater table and the distribution of pore pressures and identifying soil layers of similar properties and expected behavior. Values, such as unit weights, water contents, shear strengths, N-values, etc., should be tabulated.

Then, use the tabulated data to estimate the beta-coefficients, cohesion intercepts (or undrained shear strength values), and N_t factors, as well as appropriate ranges of such values.

13.10.2 Capacity and Allowable Load

Calculate the bearing capacity, Q_u, of a single representative pile as a sum of the shaft and toe resistances, R_s and R_t, according to Equations 13.5 and 13.8 and determine the load distribution curve for a single pile according to Equation 13.7.

Determine the allowable (or factored) load by dividing the pile capacity with a Factor of Safety, F_s, governed by the degree of uncertainty in the given case, or use the applicable Resistance Factor.

In the beginning of a design process, a range of 2.5 through 3.0 is usually chosen for F_s. Later, as more information becomes available, such as capacity determined by means of static or dynamic tests, the value of F_s may be reduced to the range of 1.8 through 2.0. For a discussion on the factor of safety, see Chapter 23 in the *Canadian Foundation Engineering Manual* (1985).

The allowable load, or—in the ULS design—the factored load, includes both permanent (dead; sustained; Q_d) loads and temporary or transient (live; transient; Q_l) loads. It does not include the dragload. (The magnitude of the dragload only affects the structural strength of the pile, not the bearing capacity.)

13.10.3 Load-Transfer Curve, Neutral Plane, and Structural Strength

Starting with the dead load, Q_d, and increasing the load in the pile by adding effect of negative skin friction, q_n, in accordance with Equation 13.10, the long-term load distribution in the pile above the neutral plane is determined. The neutral plane is where the transfer curve according to Equation 13.7 intersects the curve determined according to Equation 13.10. The construction of the neutral plane is illustrated in Figure 13.9.

The maximum load in the pile is the dead load plus the dragload and it occurs at the neutral plane. The maximum load must not be larger than a certain portion of the structural strength of the pile. The limit is governed by considerations different to those applied to the structural strength at the pile cap. It is recommended that for *straight and undamaged piles*, the allowable maximum load at the neutral plane be limited to 70 percent of the pile strength. For composite piles, such as concrete-filled pipe piles, the load should be limited to a value that induces a maximum of 1.0 millistrain into the pile with no material becoming stressed beyond 70 percent of its strength.

13.11   SETTLEMENT OF PILE FOUNDATIONS

13.11.1   Introduction

Settlement occurs as a consequence of a stress increase causing a volume reduction of the subsoil. It consists of the sum of "elastic" compression of the soil skeleton and free gas present in the voids, which occurs quickly and is normally small, and of consolidation, that is, volume change due to the expulsion of water, which occurs quickly in coarse-grained soils, but slowly in fine-grained soils.

Consolidation settlement is due to the fact that the imposed stress, initially carried by the pore water, is transferred to the soil skeleton, which compresses in the process until all the imposed stress is carried by effective stress. In some soils, creep adds to the compression of the soil skeleton. Creep is compression occurring without an increase of effective stress.

Soil materials do not show a linear relation between stress and strain, and settlement is a function of the relative stress increase. The larger the existing stress before an additional stress is applied, the smaller the induced settlement. Cohesive soils, in particular, have a distinct non-linearity. Of course, these statements are given with due consideration to any pre-consolidation pressure.

When analyzing piles, it is important that settlement is not confused with the movement occurring as a result of the transfer of load to the soil, that is, the movement necessary to build up the resistance to the load. In the case of shaft resistance, this movement is small, but substantial movement of the pile toe into the soil may occur before full toe resistance is mobilized.

13.11.2   Conventional Approach

Settlement is calculated as compression due to increase in stress—that is, the difference between the original and the final effective stresses. The increase is normally not constant throughout the soil volume, but a function of the vertical distribution (spreading) of stress. In engineering practice, the distribution under the mid-point of a footing is usually calculated by the 2:1 method according to Equation 13.11:

$$q_z = q_0 \frac{BL}{(B+z)(L+z)} \qquad (13.11)$$

where

B = footing width (breadth)
L = footing length
q_0 = applied stress (beneath the footing; at the pile cap)
q_z = applied vertical stress at depth z

The settlement is calculated by dividing the soil profile into layers, calculating for each layer the compression caused by the stress increase. The settlement is then equal to the sum of the compressions of the individual layers. Traditionally, the settlement calculation is treated differently in cohesionless and cohesive soils, as follows.

Cohesionless Soil

In cohesionless soil, the calculation of the settlement is carried out according to Hooke's law, as follows:

$$\varepsilon = \frac{1}{E} q_z \qquad (13.12)$$

and

$$S = \Sigma s = \Sigma(\varepsilon h) \qquad (13.13)$$

where

ε = strain induced in a soil layer
E = modulus of elasticity
h = thickness of soil layer
s = compression of soil layer
S = settlement for the footing as a sum of the compressions of the soil layers

The "elastic" modulus method for settlement calculation is an over-simplification and results in a highly inaccurate settlement value and use of the method is discouraged. The tangent modulus method described below is a considerably better approach.

Cohesive Soil

For settlement calculation in cohesive soils, it is generally realized that the elastic modulus approach cannot be used. Instead, conventional calculation makes use of a compression index, C_c, and the original void ratio, e_0, to determine the strain, ε, induced in a layer.

Cohesive soils, however, may be consolidated for a higher stress than the actual effective stress. This higher stress is called the preconsolidation stress, σ'_p. The compression of such soils is much smaller for stresses below the preconsolidation stress and it can be calculated using a compression index, C_{cr}. When in overconsolidated soil and with the final stress larger than the preconsolidation stress, strain, ε, is calculated according to Equation 13.14:

$$\varepsilon = \frac{1}{1+e_0} \left[C_{cr} \ln \frac{\sigma'_p}{\sigma'_0} + C_c \ln \frac{\sigma'_t}{\sigma'_p} \right] \qquad (13.14)$$

A weakness of Equation 13.14 is that the calculation requires three parameters, C_c, C_{cr}, and e_0, and too often in a project design the compression indices and the void ratio value are incompatible. Again, the tangent modulus method described below is a considerably better approach.

13.11.3   The Janbu Tangent Modulus Approach

Stress–strain relation in a soil is non-linear. For a stress increase from where the original stress in the soil is small, the corresponding increase of strain is larger than where the original stress was larger. That is, the slope of the line, the tangent modulus, M_t, increases with increasing original stress. According to a tangent modulus approach proposed by Janbu (1963, 1965), as referenced by the *Canadian Foundation Engineering Manual* (1985), the relation between stress and strain depends on two non-dimensional parameters that are unique for a soil: a stress exponent, j, and a modulus number, m. For most cases, the stress exponent can be assumed to be either 0, which is representative of cohesive soils, or 0.5, which is representative of cohesionless soils.

In cohesionless soils, $j > 0$, the following simple formula governs:

$$\varepsilon = \frac{1}{mj}\left[\left(\frac{\sigma'_1}{\sigma_r}\right)^j - \left(\frac{\sigma'_0}{\sigma_r}\right)^j\right] \quad (13.15)$$

where

ε = the strain induced by the increase of effective stress
σ'_0 = the original effective stress
σ'_1 = the new effective stress
j = the stress exponent
m = the modulus number, which is determined from testing in the laboratory and/or in the field
σ_r = a reference stress, a constant, = 100 kPa (1 tsf)

In an essentially cohesionless, sandy, silty soil, the stress exponent is close to a value of 0.5. By inserting this value and considering that the reference stress is equal to 100 kPa, the formula is simplified to:

$$\varepsilon = \frac{1}{5m}(\sqrt{\sigma'_1} - \sqrt{\sigma'_0}) \quad (13.16)$$

Notice, Equation 13.16 is not independent of the choice of units. The stress values must be inserted in kPa. That is, a value of 2 MPa is to be inserted as the number 2000 and a value of 300 Pa as the number 0.3. In English units Equation 13.16 becomes:

$$\varepsilon = \frac{2}{m}(\sqrt{\sigma'_1} - \sqrt{\sigma'_0}) \quad (13.16a)$$

Again, the equation is not independent of units. Because the reference stress is 1.0 tsf, Equation 13.16a requires that the stress values are inserted in units of tsf.

If the soil is overconsolidated and the final stress exceeds the preconsolidation stress, Equations 13.16 and 13.16a change to:

$$\varepsilon = \frac{1}{5m_r}(\sqrt{\sigma'_p} - \sqrt{\sigma'_0}) + \frac{1}{5m}(\sqrt{\sigma'_1} - \sqrt{\sigma'_p}) \quad (13.17)$$

$$\varepsilon = \frac{2}{m_r}(\sqrt{\sigma'_p} - \sqrt{\sigma'_0}) + \frac{2}{m}(\sqrt{\sigma'_1} - \sqrt{\sigma'_p}) \quad (13.17a)$$

where

σ'_0 = the original effective stress (kPa or tsf)
σ_p = the preconsolidation stress (kPa or tsf)
σ'_1 = the new effective stress (kPa or tsf)
m = the modulus number (dimensionless)
m_r = the recompression modulus number (dimensionless)

Equation 13.17 requires stress units in kPa and Equation 13.17a in tsf.

In cohesive soils, the stress exponent is zero, $j = 0$. Then, in a normally consolidated cohesive soil:

$$\varepsilon = \frac{1}{m}\ln\left(\frac{\sigma'_1}{\sigma'_0}\right) \quad (13.18)$$

and in an overconsolidated soil:

$$\varepsilon = \frac{1}{m_r}\ln\left(\frac{\sigma'_p}{\sigma'_0}\right) + \frac{1}{m}\ln\left(\frac{\sigma'_1}{\sigma'_p}\right) \quad (13.19)$$

By means of Equations 13.15 through 13.19, settlement calculations can be performed without resorting to simplifications such as that of a constant elastic modulus. Apart from knowing the original effective stress and the increase of stress plus the type of soil involved, without which knowledge no settlement analysis can ever be made, the only soil parameter required is the modulus number. The modulus numbers to use in a particular case can be determined from conventional laboratory testing, as well as in-situ tests. As a reference, Table 13.3 shows a range of conservative values typical for various soil types, which is quoted from the *Canadian Foundation Engineering Manual* (1985).

In a cohesionless soil, where previous experience exists from settlement analysis using the elastic modulus approach (Eq. 13.12), a direct conversion can be made between E and m, which results in Equation 13.20 when using SI-units—stress and E-modulus in kPa:

$$m = \frac{E}{5(\sqrt{\sigma'_1} + \sqrt{\sigma'_0})} = \frac{E}{10\sqrt{\sigma'}} \quad (13.20)$$

When using English units and stress and E-modulus in tsf, Equation 13.20a applies:

$$m = \frac{2E}{(\sqrt{\sigma'_1} + \sqrt{\sigma'_0})} = \frac{E}{\sqrt{\sigma'}} \quad (13.20a)$$

Notice, most natural soils have aged and become overconsolidated with an overconsolidation ratio, OCR, that often exceeds a value of 2. The recompression modulus, m_r, is often five to ten times greater than the virgin modulus, m, listed in the table.

In a cohesive soil, unlike the case for a cohesionless soil, no conversion is required as the traditional and the tangent modulus approaches are identical, although the symbols differ. Thus, values from the C_c and e_0 approach are immediately transferable via Equation 13.21:

$$m = \ln 10\left(\frac{1 + e_0}{C_c}\right) = 2.30\left(\frac{1 + e_0}{C_c}\right) \quad (13.21)$$

In cohesive soils, the Janbu tangent modulus approach is much preferred to the C_c and e_0 approach because, when m is determined directly from the testing, the commonly experienced difficulty is eliminated of finding out what C_c value goes with what e_0 value.

TABLE 13.3 TYPICAL AND NORMALLY CONSERVATIVE MODULUS NUMBERS.

Soil Type	Modulus Number	Stress Exponent, j
Till, very dense to dense	1000–300	1
Gravel	400–40	0.5
Sand		
Dense	400–250	0.5
Compact	250–150	0.5
Loose	150–100	0.5
Silt		
Dense	200–80	0.5
Compact	80–60	0.5
Loose	60–40	0.5
Clays		
Silty clay and clayey silt		
Hard, stiff	60–20	0.5
Stiff, firm	20–10	0.5
Soft	10–5	0.5
Soft marine clays and organic clays	20–5	0
Peat	5–1	0

13.11.4 Calculation of Pile Group Settlement

The neutral plane is, as mentioned, the location where there is no relative movement between the pile and the soil. Consequently, whatever the settlement in the soil is in terms of magnitude and vertical distribution, the settlement of the pile head is equal to the settlement of the soil at the neutral plane plus the compression of the pile caused by the applied dead load and the dragload combined.

The simplest method for calculating the settlement of the pile group at the location of the neutral plane is by calculating the settlement for a footing equal in size to the pile cap, placed at the location of the neutral plane, and imposing a stress distribution equal to the permanent (dead) load on the pile cap divided by the footing area. The settlement calculation must include the effect of all changes of effective stress in the soil, not just the load on the pile cap. Notice that the load giving the settlement is the permanent load acting on the pile cap and that neither the live load nor the dragload are included in the settlement calculation.

For a dominantly shaft-bearing pile "floating" in a homogeneous soil with linearly increasing shear resistance, the neutral point lies at a depth which is about equal to the lower third point of the pile embedment length. It is interesting to note that this location is also the location of the equivalent footing according to the Terzaghi–Peck approach illustrated in Figure 13.3. (The assumptions behind the third-point location are that the unit negative skin friction is equal to the positive shaft resistance, that the toe resistance is small, and that the load applied to the pile head is about a third of the bearing capacity of the pile.)

Assume that the distribution of settlement in the soil around the pile is known and follows the "settlement" diagram in Figure 13.10. As illustrated in the diagram for the case of the middle service load, by drawing a horizontal line from the neutral plane to intersection with the settlement curve, the settlement of the pile at the neutral plane can be determined and, thus, the settlement of the pile head. The construction in the figure is valid both for a small settlement that diminishes quickly with depth and for a large settlement that continues to be appreciable well below the pile toe.

The construction in Figure 13.10 has assumed that the induced toe movement (toe displacement) is sufficiently large to fully mobilize the toe resistance. As stated, the movement between the shaft and the soil is always large enough to mobilize the shaft shear—negative skin friction or positive shaft resistance—but if the soil settlement is small, it is possible that the toe movement is not large enough to mobilize the full toe resistance. In such a case, the neutral plane moves to a higher location as determined by the particular equilibrium condition.

In a pile group connected with a stiff cap, all piles must settle an equal amount and the elevation of the neutral plane must be equal for the piles in the group. (The individual capacities may vary, and, therefore, the permanent load actually acting on an individual pile will vary correspondingly.) Then, according to Fellenius (1984, 1989), the settlement of the group is determined as the settlement of an equivalent footing located at the elevation of the neutral plane with the load spreading below the equivalent footing calculated by the 2:1 method.

13.11.5 Summary of Settlement Calculation

Step 1. The soil profile is assessed and divided into layers for calculation, which requires that pertinent soil parameters are assigned to each layer.

Step 2. Calculation of settlement of a pile group requires the prior calculation of bearing capacity including the distribution of load and resistance along the piles, which determines the location of the neutral plane.

Step 3. The pile group is replaced by an equivalent footing at the neutral plane and the increase of stress below the equivalent footing caused by the dead load on the pile group is calculated using the 2:1 method. This stress is added to the change of effective stress caused by other influences, such as fill, excavation, and groundwater lowering.

Step 4. The settlement of each soil layer below the neutral plane as caused by the change of effective stress is determined using the tangent modulus approach and the values are summed to give the soil settlement at the neutral plane. The settlement of the pile group is this value plus the compression of the pile for the dead load and the dragload.

Step 5. Inasmuch as the determination of the neutral plane made use of a fully developed toe resistance, a check is made of the magnitude of settlement calculated below the pile toe. If this value is smaller than about 5 percent of the pile diameter, Step 2 is repeated using an appropriately smaller toe resistance to arrive at a new location of the neutral plane (higher up) and followed by a repeat of Steps 3 through 5, as required.

13.11.6 Special Aspects

The dragload must not be included when considering bearing capacity, that is, the analysis of soil bearing failure. Consequently, for bearing capacity consideration, it is incorrect to reduce the dead load by any portion of the dragload.

The dead load should only be reduced owing to insufficient structural strength of the pile at the location of the neutral plane, or by a necessity to lower the location of the neutral plane in order to reduce the amount of settlement.

Normally, when the pile capacity is reliable, that is, it has been determined from results of a static loading test or analysis of data from dynamic monitoring, a factor of safety of 2 ensures that the neutral plane is located below the mid-point of the pile.

In the design of a pile foundation, provided that the neutral plane is located deep enough in the soil to eliminate settlement concerns for the piles, the settlement of the surrounding soils (and the negative skin friction) are of no concern directly for the pile group. However, where large settlement is expected, it is advisable to avoid inclined piles in the foundation, because piles are not able to withstand lateral or sideways movement and the settlement will bend an inclined pile.

Piles that are bent, doglegged, or damaged during the installation will have a reduced ability to support the service load in a downdrag condition. Therefore, the unified design approach postulates that the pile installation is subjected to stringent quality control directed toward ensuring that the installed piles are sound and that bending, cracking, and local buckling do not occur.

When the design calculations indicate that the settlement could be excessive, increasing the pile length or decreasing the pile diameter could improve the situation. When the calculations indicate that the pile structural capacity is insufficient, increasing the pile section, or increasing the strength of the pile material could improve the situation. When such methods are not practical or economical, the negative skin friction can be reduced by the application of bituminous coating or other viscous coatings to the pile surfaces before the installation, as demonstrated by Bjerrum et al. (1969). (See also Fellenius, 1975a, 1979; Clemente, 1981.) For cast-in-place piles, floating sleeves have been used successfully. It should be recognized, however, that measures such as bitumen coating and sleeves are very expensive, and they should only be considered when other measures for lowering the neutral plane have been shown to be impractical.

The unified design approach shares one difficulty with all other approaches to pile group design, viz., that there is a lack of thorough and representative full-scale observations of load distribution in piles and of settlement of pile foundations. For settlement observations, the lack is almost total with respect to observations of settlement of both the piles and the soil adjacent to pile foundations.

In a typical design case, the shaft and toe resistance for a pile can only be estimated within a margin. To provide the profession with reference cases for aid in design, it is very desirable that sturdy and accurate load cells be developed and installed in piles to register the load distribution in the pile during, immediately after, and with time following the installation. Naturally, such cells should be placed in piles subjected to static loading tests, but not exclusively in these piles (see Dunnicliff, 1982, 1988).

The greatest perceived need lies in the area of settlement observations. It is paradoxical that although pile foundations are normally resorted to for reasons of excessive settlement, the design is almost always based on a capacity rationale, with disregard of settlement. To improve this situation, full-scale and long-term field observation cases are needed. Actual pile foundations should be instrumented to determine both the settlement of the piles and the distribution of settlement in the soil near the piles. No instrumentation for study of settlement should be contemplated without the inclusion of piezometers.

13.12 STATIC TESTING OF PILES

The axial compression testing of a vertical single pile is the most common test performed. However, despite the numerous tests that have been carried out and the many papers that have reported on such tests and their analyses, the understanding of static pile testing in current engineering practice leaves much to be desired. The reason is that engineers have concerned themselves with mainly one question, only—"does the pile have a certain least capacity?"—finding little of practical value in analyzing the pile–soil interaction. However, considerable engineering value can be gained from routine elaboration on a pile test, during the actual testing in the field, as well as in the analysis of the results.

13.12.1 Testing Methods

A static loading test is performed by loading a pile with a gradually or stepwise increasing force, while monitoring the movement of the pile head. The force is obtained by means of a hydraulic jack reacting against a loaded platform or anchors.

The American Society for Testing and Materials, ASTM, publishes three standards, D-1143, D-3689, and D-3966 for static testing of a single pile in axial compression, axial uplift, and lateral loading, respectively.

The ASTM standards detail how to arrange and perform the pile test. Wisely, they do not include how to interpret the tests, because this is the responsibility of the engineer in charge, who is the only one with all the site- and project-specific information necessary for the interpretation.

The most common test procedure is the slow maintained load method referred to as the "standard loading procedure" in the ASTM Designation D-1143 and D-3689 in which the pile is loaded in eight equal increments up to a maximum load, usually twice a predetermined allowable load. Each load level is maintained until zero movement is reached, defined as 0.25 mm/h (0.01 in/h). The final load, the 200 percent load, is maintained for a duration of 24 h. The "standard method" is very time consuming, requiring from 30 to 70 h to complete. It should be realized that the words "zero movement" are very misleading: the "zero" movement rate is equal to a movement of more than 2 m (7 ft) per year!

Each of the eight load increments is placed onto the pile very rapidly; as fast as the pump can raise the load, which usually takes about 20 seconds to 2 minutes. The size of the load increment in the "standard procedure", 12.5 percent of the maximum load, means that each such increase of load is a shock to the pile and the soil. Smaller increments that are placed more frequently disturb the pile less, and the average increase of load on the pile during the test is about the same. Such loading methods provide more consistent, reliable, and representative data for analysis.

Tests that consist of load increments applied at constant time intervals of 5, 10, or 15 minutes are called Quick Maintained-Load Tests or just "Quick Test". In a Quick Test, the maximum load is not normally kept on the pile longer than any other load before the pile is unloaded. Unloading is done in about five steps of no longer duration than about 1 minute. The Quick Test allows for attempting to apply one or more load increments beyond the minimum number that the particular test is designed for, that is, making use of the margin built into the test. In short, the Quick Test is from technical, practical, and economical points of view superior to the "standard loading procedure".

A Quick Test should aim for 25 to 40 increments with the maximum load determined by the amount of reaction load available or by the capacity of the pile. For routine cases, it may be preferable to stay at a maximum load of 200 percent of the intended allowable load. For ordinary test arrangements, where only the load and the pile head movement are monitored, time intervals of 10 minutes are suitable and allow for the taking of two to four readings for each increment. When testing instrumented piles, where the instruments take a while to read (scan), the time interval may have to be increased. To go beyond 20 minutes, however, should not be necessary. Nor is it advisable, because of the potential risk of the influence of time-dependent movements, which may impair the test results. Usually, a Quick Test is completed within 3 to 6 h.

For a description of constant-rate-of-penetration and cyclic methods, see Fellenius (1975b, 1980) and references contained therein.

In routine tests, cyclic loading, or even single unloading and loading phases must be avoided, as they do little more than destroy the possibility of a meaningful analysis of the test results. There is absolutely no logic in believing that anything of value on load distribution and toe resistance can be obtained from an occasional unloading or from one or a few "resting periods" at certain load levels, when considering that we are testing a unit that is subjected to the influence of several soil types, is subjected to residual stress of unknown magnitude, exhibits progressive failure, etc., and when all we know is what is applied and measured at the pile head.

13.12.2 Interpretation of Failure Load

For a pile that is stronger than the soil, the failure load is reached when rapid movement occurs under sustained or slightly increased load (the pile plunges). However, this definition is inadequate, because plunging requires very large movements and it is often less a function of the capacity of the pile–soil system and more a function of the man–pump system.

To be useful, a definition of failure load must be based on some mathematical rule and generate a repeatable value that is independent of scale relations and the opinions of the individual interpreter. Furthermore, it has to consider the shape of the load–movement curve or, if not, it must consider the length of the pile (which the shape of the curve indirectly does).

Davisson (1972) proposed a load limit defined as the load corresponding to the movement that exceeds the elastic compression of the pile by an offset of 4 mm (0.15 in) plus a value equal to the diameter of the pile divided by 120. For example, the offset value for a pile toe diameter of 300 mm (12 in) is 6 mm (0.25 in).

Brinch-Hansen (1963) proposed an 80 percent criterion defining the ultimate load as the load that gives four times the movement of the pile head as obtained for 80 percent of that load. Usually, the 80 percent criterion agrees well with the perceived "plunging failure."

In applying the general work by Kondner (1963), Chin (1970, 1971) proposed a method in which each applied load is divided by its corresponding movement and the resulting number is plotted against the movement. After some initial variation, the plotted values fall on a straight line. The inverse slope of this line is the Chin failure load.

The three methods mentioned for determining failure load are included in the *Canadian Foundation Engineering Manual* (1985). Details on the application of the methods in engineering practice have been presented by Fellenius and Rasch (1990). Figure 13.11 illustrates the variation of the three and six other methods (Fellenius, 1980), when applied to the results of a static loading test on a 40 m (130 ft) long 300 mm (12 in) diameter pile in clay and silt. As shown, the Davisson limit of 2130 kN

(240 tons) is lower than all the others and the Chin value of 2930 kN (330 tons) is the highest. The other seven values are grouped more or less together around an average of 2400 kN (270 tons).

It is difficult to make a rational choice of the best criterion to use, because the preferred criterion depends strongly on a person's past experience. In the case of an engineering report, the preference and experience of the receiver of the report may also influence what criterion to choose.

The Davisson limit has the merit of allowing the engineer, when proof testing a pile for a certain allowable load, to determine *in advance* the maximum allowable movement for this load with consideration of the length and size of the pile. Thus, as proposed by Fellenius (1975b), contract specifications can be drawn up including an acceptance criterion for piles proof tested according to quick testing methods. The specifications can simply call for a test to at least twice the design load, as usual, and declare that at a test load equal to a factor F times the design load the movement shall be smaller than the Davisson offset from the elastic column compression of the pile. Normally, F would be chosen within a range of 1.8 to 2.0. The acceptance criterion could be supplemented with the requirement that the safety factor should also be smaller than a certain minimum value calculated on pile bearing failure defined according to the 80 percent criterion or other preferred criterion.

The Brinch-Hansen 80 percent criterion usually gives a Q_u value that is close to what one subjectively accepts as the true

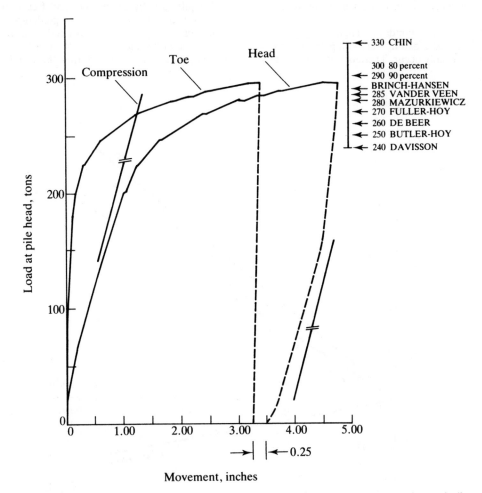

Fig. 13.11 Load–movement diagram from a quick maintained-load static loading test with measurement of pile compression and toe movement. (*After Fellenius, 1980.*)

ultimate failure value. However, the criterion is very sensitive to inaccuracies of the test data.

The Chin method allows a continuous check on the test, if a plot is made as the test proceeds, and a prediction of the maximum load that it will be possible to apply during the test. Sudden kinks or slope changes in the Chin line indicate—give an early warning—that something is amiss with either the pile or with the test arrangement (Chin, 1978).

13.12.3 Influence of Errors

A static loading test is usually considered a reliable method for determining the capacity of a pile. However, even when using new manometers and jacks that have been calibrated together, the applied loads are usually substantially overestimated. The error is usually about 10 to 15 percent of the applied load. Errors as large as 30 to 40 percent are not uncommon. The diagram in Figure 13.12 is from an actual field test and it is representative of the error commonly encountered in a routine static loading test.

The reason for the error is that the jacking system is required to do two things at the same time, that is, both to provide the load and to measure it, and load cells with moving parts are considerably less accurate than those without moving parts. For instance, when calibrating testing equipment in the laboratory, it is ensured that no eccentric loading, bending moments, or temperature variations influence the calibration. In contrast, all of these adverse factors are at hand in the field and influence the test results to an unknown extent, unless a load cell is used.

It is inconceivable that the foundation engineering practice can continue to perform static loading tests with potential errors as large as those that usually occur. Therefore, it is absolutely essential that a load cell be used in all tests. Of course, the jack pressure should still be measured and be used as a back-up.

The above deals with the error of the applied load. The error in movement measurement can also be critical. Such errors do not originate in the precision of the reading (the usual precision is more than adequate) but in undesirable influences, such as heave or settlement of the reference beam during unloading the ground when loading the pile. For instance, one of the greatest villains known to spoil a loading test is the sun: the reference beam must be shielded from sunshine at all times.

It must be remembered that the minimum distances from the supports of the reference beam to the pile and the platform, etc., as recommended in the ASTM standards, really are minimum values, which most often do not give errors of much concern for ordinary testing but which are too short for research or investigative testing purposes.

13.12.4 The Analysis of Results Using Telltale Data

In the routine static loading test, measurements are taken at the pile head only. Yet, in the interpretation of the test results, what is of interest is the distribution of load in the pile and especially at the pile toe. It is impossible to estimate with any worthwhile accuracy the mobilized toe resistance from load–movement data obtained at the pile head.

The test can be substantially enhanced by placing telltales in the pile. A telltale is a rod with its lower end connected to the pile, usually at the toe, and free from the pile along its overall length by means of a guidepipe arrangement. By attaching a dial gage at the upper end of the rod and measuring the change of distance between the rod top and the pile head, the shortening of the pile during the test is monitored. The

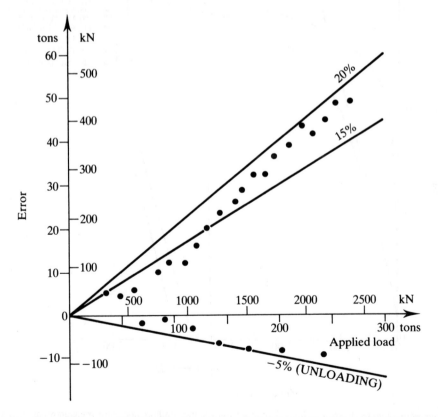

Fig. 13.12 Example of error of load encountered in a routine static loading test. (*After Fellenius, 1984b.*)

movement of the pile toe is obtained as the measured pile shortening subtracted from the movement of the pile head.

In the static loading test on a precast concrete pile, for which results are presented in Figure 13.11, a guidepipe for a telltale had been cast in the pile, allowing a telltale to be inserted to the pile toe after the driving to monitor the compression of the pile and the pile toe movement. With use of some foresight and planning, telltales can be installed rather easily and cheaply in all types of piles. The mentioned ASTM standards include suggestions for telltale arrangements.

The measured compression divided by the telltale length gives the strain of the pile for the applied load, which when combined with the Young's modulus of the pile material results in the average load in the pile over the telltale length. In the case of a constant unit shaft resistance, the average load is equal to the load in the middle of the pile, or the middle of the telltale length. In the case of a linearly increasing unit shaft resistance, the average load is equal to the load at a level lying somewhere between the mid-point and the upper third point. Obviously, knowledge of the distribution of the shaft resistance is essential for the evaluation of the load distribution.

However, an estimation of the toe resistance can be made from the measured shortening of a telltale to the pile toe. The following relations were given by Fellenius (1980) and build on the assumption of constant unit shaft resistance acting along the full length of the pile (the telltale length):

$$Q_{ave} = AE\frac{\Delta L}{L} \tag{13.22a}$$

$$R_t = 2Q_{ave} - Q_{head} \tag{13.22b}$$

$$R_s = Q_h - R_t \tag{13.22c}$$

where

Q_{ave} = average load in the pile
A = cross-sectional area of the pile
ΔL = measured shortening of the pile
L = pile or telltale length
R_t = toe resistance
Q_{head} = load applied to the pile head
R_s = shaft resistance

For linearly increasing unit shaft resistance, the relation for the toe resistance becomes:

$$R_t = 3Q_{ave} - 2Q_{head} \tag{13.23}$$

By means of Equations 13.22 and 13.23, a range of toe resistance values can be bounded by the two extremes of constant and linearly increasing unit shaft resistance. Furthermore, by means of adding a small value to the measured shortening of the pile, one can include an analysis of the effect on the calculated toe and shaft resistances of a residual load, acting on the pile before the static loading test was started.

Having several telltales in a pile results in several values of average load, because three average values of load result from measurement taken by two telltales; the third load value is obtained from the difference in compression measured over the distance between the two telltale ends connected to the pile. Correspondingly, having three telltales results in six load values. There is a practical limit, because from primarily practical considerations of accuracy, it is not worthwhile having telltale lengths and distances shorter than about 10 m (30 ft).

Leonards and Lovell (1978) presented an analysis method for determining the load distribution in a pile instrumented with one telltale, where only the relative distribution of unit shaft resistance needs to be known, or the upper and lower ranges of it in the case of a boundary-type analysis. The shaft resistance does not need to be uniform, but can be of any irregular distribution. Lee and Fellenius (1990) developed the Leonard–Lovell method to include a simultaneous analysis of two or three telltales, which allows a computation of residual load in the pile and its influence on the test results.

When using telltales, the accuracy of the compression measurements must be several times better than the accuracy usually accepted for movement measurements. The nominal precision of measurements of movement using dial gages is usually only 0.025 mm (0.001 in). On special occasions for recording compression using telltales, dial gages with a ten times finer reading precision are used. The actual accuracy of the values is, of course, smaller than the precision, even when neglecting influences on the measuring beam. At best, when using mechanical gages, the error is about 0.1 mm (0.005 in) or larger. The ten times finer gages will have a smaller error, but not ten times smaller.

It is necessary to have dial gages with stems that are long enough to allow the telltale records to be taken during the entire test without having to reset the gages or to shim them, otherwise, errors are introduced that will destroy the value of the records.

Even if movement and compression measurements are complementary readings from the view of mathematics in an analysis, when using telltales, the telltales should always measure compression directly, not movement, because obtaining compression as difference between two measurements of movement introduces large errors.

Apart from the obvious fact that results of an analysis of telltale measurements depend primarily on the accuracy of the measurements, the analysis introduces the modulus of the pile material and the results depend also on how accurately the modulus is known. Steel has a constant modulus and steel piles are very suitable for telltale instrumentation. Concrete, however, does not have a constant modulus over the stress range considered in a static loading test. Therefore, telltale measurements in concrete piles and concreted steel pipe piles are difficult to analyze.

In the analysis of telltale compression measurements from a pile having a stress (or strain) dependent modulus, a diagram should be made showing increment of load over increment of strain plotted against strain, that is, observed tangent (chord) modulus versus strain (Fellenius, 1989b). In Figure 13.13 is shown such a plot, which is from the test of a long prestressed concrete pile tested in a quick maintained-load test to a load exceeding 4500 kN (500 tons). The upper diagram shows load–strain as measured in two telltales and the lower diagram shows the modulus versus strain determined from the data.

As can be seen, after an applied load of about 2000 kN (200 tons), the tangent modulus plots become approximately linear and sloping toward reducing values. The tangent modulus in the beginning of the test is about twice that at the end of the test (from about 38 GPa to about 16 GPa; 5500 ksi to 3900 ksi). The tangent modulus can be obtained from linear regression of the straight line, which solves the coefficients a and b in the following equation:

$$E_t = A\frac{d\sigma}{d\varepsilon} = a\varepsilon + b \tag{13.24}$$

where

E_t = tangent modulus
A = cross-sectional area of the pile
σ = stress in the pile
ε = strain in the pile
a = slope of the tangent modulus line
b = initial tangent modulus; y intercept

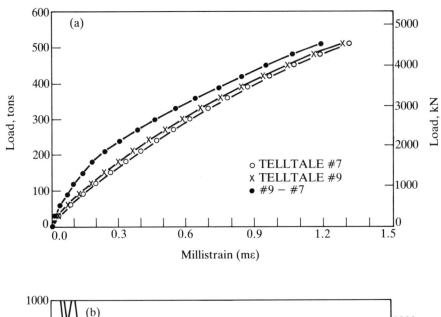

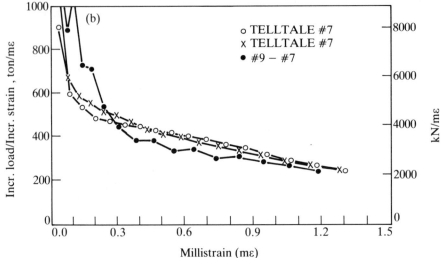

Fig. 13.13 Diagrams of (a) load versus strain and (b) tangent modulus versus strain. (*After Fellenius, 1989b.*)

The initial tangent modulus, term *b* in Equation 13.24, is best determined from the strain measured over an upper portion of the pile, where the load is affected the least by shaft resistance. Alternatively, and better still, strain should be measured over a free-standing (sleeved off) portion of the pile, because the smaller the shaft resistance, the smaller the error in term *b*.

On integrating the line, an equation for the stress–strain curve of the pile as a free-standing column is obtained:

$$\sigma = \frac{1}{2} a\varepsilon^2 + b\varepsilon \qquad (13.25)$$

Equation 13.25 gives the stress–strain relation for the pile as a column, that is, without the influence of shaft resistance, and can be used for evaluating the average load over the telltale length directly from the measured strain. Alternatively, the secant modulus is determined, as follows:

$$E_s = \frac{1}{2} a\varepsilon + b \qquad (13.26)$$

and the load in the pile for a certain induced strain is:

$$Q = AE_s\varepsilon \qquad (13.27)$$

The tangent modulus plot can also be used to evaluate the shaft resistance acting on the pile by making use of the fact that the plot becomes linear when all shaft resistance has been mobilized.

It will be obvious for anyone trying the tangent modulus plot in an analysis that only eight load levels to calculate from are too few and the values are too far apart; the "standard procedure" is not suitable for analysis.

The primary value of telltales is for measuring the toe movement. For evaluating the load in the tested pile, the accuracy of the compression measurements must be very high. This means that mechanical type dial gages, even those with high-precision gradation, are not suitable. The use of linear voltage displacement transducers, LVDTs, is to be preferred.

The tangent modulus analysis shown above is equally suitable to direct measurement of strain, of course. Using strain gages in a test requires more knowhow than using telltales. Therefore, an experienced instrumentation specialist should be included in the project team early in the planning of the test.

When planning a static loading test and considering the inclusion of telltales to provide data for use in the analysis of load distribution, it is recommended that the telltales be limited to one to the toe and one back-up placed, say 10 m (30 ft) above the toe. The rest of the instrumentation for measuring strain should be electrical strain gages. Furthermore, telltale data intended for load distribution analysis must be obtained with an accuracy much greater than normally used for measurements during a static loading test. The minimum precision of the dial gages is 0.01 mm (0.0004 in).

13.13 PILE DYNAMICS

13.13.1 Wave Mechanics

The penetration resistance of driven piles provides a direct means of determining bearing capacity of a pile. In impacting a pile, a short-duration force wave is induced in the pile, giving the pile a downward velocity and resulting in a small penetration of the pile. Obviously, the larger the number of blows necessary to achieve a certain penetration, the stronger the soil. Using this basic principle, a large number of so-called pile driving formulae have been developed for determining pile bearing capacity. All these formulae are based on equalizing potential energy available for driving in terms of weight of hammer times its height of fall (stroke) with the capacity times penetration (set) for the blow. The set often includes a loss term.

The principle of the dynamic formulae is fundamentally wrong as wave action is neglected along with a number of other aspects influencing the penetration resistance of the pile. Nevertheless, pile driving formulae have been used for many years and with some degree of success. However, success has been due less to the theoretical correctness of the particular formulae used and more to the fact that the users possessed adequate practical experience to go by. It is mainly when applied to single-acting hammers that use of a dynamic formula may have some justification. Dynamic formulae are the epitome of an outmoded level of technology and they have been replaced by modern methods, such as the wave equation analysis and dynamic measurements, which are described below.

Pile-driving formulae or any other formula applied to vibratory hammers are based on a misconception. Vibratory driving works by eliminating resistance to penetration, not overcoming it. Therefore, records of penetration combined with frequency, energy, amplitudes, etc., can only relate to the resistance not eliminated, not to the static pile capacity after the end of driving.

Pile-driving hammers are rated by the maximum potential energy determined as the ram weight times the maximum ram travel. However, diesel hammers and double-acting air/steam hammers, but also single-acting air/steam hammers, develop their maximum potential energy only during favorable combinations with the pile and the soil. Then, again, the energy actually transferred to the pile may vary due to variation in cushion properties, pile length, toe conditions, etc. Therefore, a relation between the hammer rated energy and measured transferred energy provides only very little information on the hammer. More information is obtained by relating the energy ratio to the actual potential energy, either measured directly (Likins and Rausche, 1988; Hannigan and Webster, 1988) or—for single-acting diesel hammers—determined from the blow rate. Figure 13.14 illustrates the relation between the blow rate and the hammer stroke. With the stroke known, the potential energy of the hammer in actual operation is obtained by multiplying the stroke by the weight of the ram.

About ten years ago, the wave equation analysis had developed sufficiently that it became a tool for general use by the profession. This constituted a big leap in the understanding of pile driving, because, in contrast to the pile-driving formula, the wave equation is theoretically correct and an analysis includes all aspects influencing the pile driving and penetration

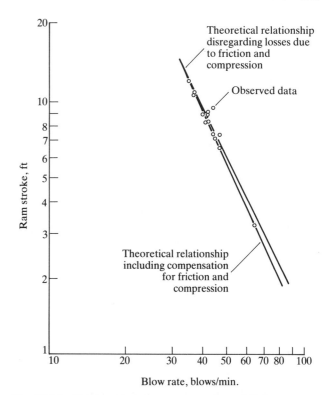

Fig. 13.14 Relationship between ram stroke and blow rate for a single-acting diesel hammer. (*After Fellenius et al., 1978.*)

resistance: hammer mass and travel, combustion in a diesel hammer, helmet mass, cushion stiffness, hammer efficiency, soil strength, viscous behavior of the soil, elastic properties of the pile, to mention some. The commercially available wave equation programs are simple and fast to use and there is, therefore, not even a justification of convenience in continued use of dynamic formulae.

However, the parameters used as input into a wave equation program are really variables with certain ranges of values and the number of parameters included in the analysis is large. Therefore, the result of an analysis is only qualitatively correct, and not necessarily quantitatively correct, unless it is correlated to observations.

The full power of the wave equation analysis is only realized when combined with dynamic measurements during pile driving by means of transducers attached to the pile head. The impact by the pile-driving hammer produces strain and acceleration in the pile which are picked up by the transducers and transmitted via a cable to a data acquisition unit (the Pile Driving Analyzer), which is placed in a nearby monitoring station. The acquisition translates strain and acceleration measurements into force and velocity, displaying these graphically on an oscilloscope. Simultaneously and blow for blow, the energy transferred to the pile is calculated and an estimation is obtained of the bearing capacity of the pile. Impact force, maximum compression force, and maximum tensile force are determined and printed out on a strip chart. The complete procedure is described in the American Society for Testing and Materials Standard for Dynamic Measurements, ASTM D-4945.

The immediate results are usually presented in the form of a "wave trace", which shows the measured force and velocity drawn against time as illustrated in Figure 13.15. As is the convention, the time unit is given in units of L/c, that is, the time it takes for the wave to travel the length of the pile. At time $2L/c$, therefore, the traces show the reflections originating

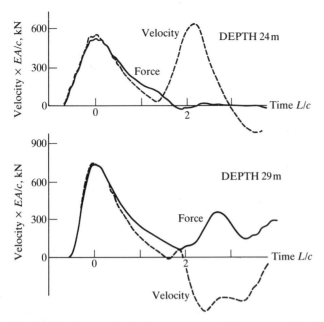

Fig. 13.15 Force and velocity wave traces during easy and hard driving. (*After Authier and Fellenius, 1983.*)

from the pile toe. The peak at zero time (zero L/c) is defined as the point of impact. Because of the wave action, hammer impact force is transmitted to the pile before this time, as well as a considerable time thereafter.

At first, force and velocity are proportional by the so-called acoustic impedance, a material constant (equal to AE/c; the product of the cross-sectional area and the elastic modulus over the wave propagation velocity). Therefore, when velocity and force are plotted to scale of the ratio of impedance, the force and velocity traces at first plot on top of each other. When the impact wave (stress-wave or strain-wave) traveling down the pile meets soil resistance, a reflection of the wave occurs that travels back up the pile. This reflected wave will superimpose on the downward wave, which has the effect of increasing stress at the location of the monitoring transducers and decreasing the velocity. Thus, the two traces will separate with the amount of separation proportional to the soil resistance encountered: at first shaft resistance and, finally, at time $2L/c$, also toe resistance. If there is no or only little resistance at the pile toe, the reflected wave will be in tension and have the effect that the measured velocity increases and the force decreases.

The wave traces shown in Figure 13.15 are taken from the driving of a 43 m long, 460 mm diameter, octagonal shape, concrete pile. The upper set of traces, depth 24 m, is from relatively easy driving. The lower set, depth 29 m, is from when the pile toe entered dense soils. Before time $2L/c$, both sets of traces show only little separation between the force and velocity traces, which is indicative of small shaft resistance acting on the pile. For the upper, "easy-driving" diagram, the traces at time $2L/c$ indicate a velocity peak and essentially zero force, that is, a tension reflection from the pile toe. The lower, "hard-driving" diagram shows a force reflection at time $2L/c$ and a negative velocity, the pile "bounces up" which is indicative of toe resistance. Thus, the traces provide valuable qualitative information on the distribution and magnitude of the soil resistance.

The dynamic measurements can also provide quantitative information of pile static capacity. In the mid-1960s, Dr. G. G. Goble and coworkers of Case Western Reserve University derived a simple relation for calculating capacity from the values

of force and velocity at times $0L/c$ and $2L/c$. In words, the resistance to penetration is equal to the mean of the forces at the two times plus the velocity change between the two times multiplied by the impedance (EA/c) of the pile (Rausche et al., 1985). The static capacity is obtained by subtracting a velocity-dependent portion calculated using a damping factor. The static capacity value is called the Case Method Estimate (of capacity). The method has of course been substantially developed since its first derivation and it is today the mainstay of the capacity determination of dynamically monitored piles.

The dynamic measurements can also be used to investigate damage and defects in the pile, such as voids, cracks, spalling, local buckling, etc. (Rausche and Goble, 1978; Rausche et al., 1988; Middendorp and Reiding, 1988).

13.13.2 CAPWAP Analysis

Dynamic records are routinely stored using a tape recorder (or similar unit) or digitized to a computer for renewed analysis. This enables a more time-consuming and detailed analysis to be performed called the CAPWAP signal matching analysis (Rausche et al., 1972). The CAPWAP analysis provides, first of all, a calculated static capacity and the distribution of resistance along the pile. However, it also provides several additional data, for example, the movement necessary to mobilize the full shear resistance in the soil (the quake) and damping values for input in a wave equation analysis. The principle and the procedure of the CAPWAP analysis signal matching are as follows.

As mentioned, the force and velocity induced by the hammer are proportional via the pile impedance (EA/c) and force and velocity react differently to the reflected wave: force increases and velocity decreases. The resulting separation of the two plots is, therefore, an indication of the size of the resistance: dynamic and static together.

The two measurements, force and velocity, are independent of each other. However, they are caused by the same impact from the hammer and affected by the same soil resistance and they have to follow the same physical laws of wave propagation. The CAPWAP analysis makes use of this situation by taking the input from one measurement, usually the velocity, moderating it by reflections computed from an assumed distribution of damping, quake, and soil resistance, and transferring it to force by means of wave mechanics computations. In a trial and error procedure, the input data are adjusted until the computed force plots on top of the measured force throughout the impact event. The signals have been matched and the CAPWAP analysis has then calibrated the site conditions and provided the static bearing capacity of the pile as well as indicated dynamic parameters.

13.13.3 Pile Integrity Tester

Low-strain integrity testing is used on all types of piles, but in particular on bored piles, caissons, piers, or piles that cannot be subjected to driving. It is performed using a high-sensitivity accelerometer placed on the pile head, an amplifier–receiver, a special small-impact device, and a portable computer with digitizing and graphics capability.

In testing for integrity, a "low-strain" compressive impact wave is generated and the acceleration and velocity records (traces on the screen) of the impact are studied. Damage or defects in the pile will show up on the acceleration and velocity traces, which are graphically displayed and stored on disk for later reprocessing. A special computer program processes the records and includes special effects such as averaging of records

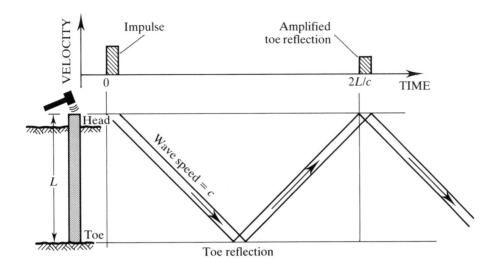

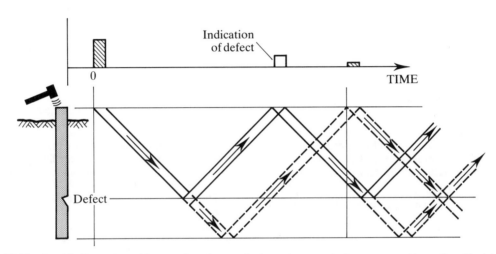

Fig. 13.16 Graphic illustration of how an impulse travels through a sound pile as opposed to a pile with a defect.

from several blows, gradual or exponential amplification of the reflections from down the pile, etc., to enable separation of random reflections from important reflections such as those from cracks, discontinuities, voids, etc. A signal matching procedure similar to the CAPWAP method is also available.

In testing a pile, a slight blow is delivered to the pile head by means of a hand-held hammer. The impact of the hand-held hammer initiates a small strain wave that travels down the pile at the speed of sound. A highly sensitive accelerometer is used to pick up the impact (sonic impulse) and the faint reflection (echo) of the impact from below the pile.

The principles of low-strain testing are illustrated in Figure 13.16, which shows two vertical piles, one having a defect and one being sound. To the right of each pile is shown an upper diagram of velocity integrated from the acceleration measured at the pile head and plotted versus time, and a lower diagram indicating the downward travel of the impulse in the pile. When the strain wave reaches a crack, or a void, in the pile, a reflection in the form of a tensile wave is sent back up to the pile head. If there is no such defect in the pile, the wave travels unimpeded to the pile toe (bottom end of the pile) and reflects from there. The acceleration measurement is picked up by a signal amplifier and sent to a computer, which digitizes, integrates, and processes the record for display on a screen.

The existence of soil shear resistance along the pile shaft dampens the strain wave. Without amplification, the reflected velocity trace would lie very close to the zero axis (Fig. 13.16). The computer processing allows an amplification of the signal that progressively offsets the shear resistance. At times, however, the faint signal can be overshadowed by randomly varying impulses, that is, electronic noise. Then, to filter such random impulses, several contiguous blows may be processed and averaged, which eliminates the noise and allows an indication of damage or defect to stand out in the record.

13.14 HORIZONTALLY LOADED PILES

Because foundation loads act in many different directions, depending on the load combination, piles are rarely loaded in true axial direction only. Therefore, a more or less significant lateral component of the total pile load always acts in combination with an axial load. The imposed lateral component is resisted by the bending stiffness of the pile and the shear resistance mobilized in the soil surrounding the pile.

An imposed horizontal load can also be carried by means of inclined piles, if the horizontal component of the axial pile load is at least equal to and acting in the opposite direction to the imposed horizontal load. Obviously, this approach has its limits as the inclination cannot be impractically large. It should, preferably, not be greater than 4(vertical) to 1(horizontal). Also, only one load combination can provide the optimal lateral resistance.

In general, it is not correct to resist lateral loads by means of combining the soil resistance for the piles (inclined as well as vertical) with the lateral component of the vertical load for the inclined piles. The reason is that resisting an imposed lateral load by means of soil shear requires the pile to move against the soil. An inclined pile will rotate owing to such movement and either push against or pull away from the pile cap, which will substantially change the axial load in the pile. Such combination requires sophisticated computer analysis by means of a suitable program, such as GROUP1 developed by Reese et al. (1990).

In design of vertical piles installed in a homogeneous soil and subjected to horizontal loads, an approximate and usually conservative approach is to assume that each pile can sustain a horizontal load equal to the passive earth pressure acting on an equivalent wall with depth $6b$ and width $3b$, where b is the pile diameter, or face-to-face distance (*Canadian Foundation Engineering Manual*, 1985).

Similarly, the lateral resistance of a pile group may be approximated by the soil resistance on the group calculated as the passive earth pressure over an equivalent wall with depth equal to $6b$ and width equal to:

$$L_e = L + 2B \qquad (13.28)$$

where

L = the length, center-to-center, of the pile group in plan perpendicular to the direction of the imposed loads

B = the width of the equivalent area of the group in plan parallel to the direction of the imposed loads

The lateral resistance calculated according to Equation 13.28 must not exceed the sum of the lateral resistance of the individual piles in the group. That is, for a group of n piles, the equivalent width of the group, L_e, must be smaller than n times the equivalent width of the individual pile, $6b$. For an imposed load not parallel to a side of the group, calculate for two cases, applying the components of the imposed load that are parallel to the sides.

The very simplified approach expressed in Equation 13.28 does not give any indication of movement. Nor does it differentiate between piles with fixed heads and those with heads free to rotate, that is, no consideration is given to the influence of pile bending stiffness. As the governing design aspect with regard to lateral behavior of piles is lateral displacement and the lateral capacity or ultimate resistance is of secondary importance, the usefulness of the simplified approach is very limited in engineering practice.

The analysis of lateral behavior of piles must consider two aspects.

- *The pile response.* The bending stiffness of the pile, how the head is connected (free head, or fully or partially fixed head).
- *The soil response.* The input in the analysis must include the soil resistance as a function of the magnitude of lateral movement.

The first aspect is modeled by treating the pile as a beam on an "elastic" foundation, which is done by solving a fourth-degree differential equation with input of axial load on the pile, material properties of the pile, and the soil resistance as a nonlinear function of the pile displacement.

The derivation of lateral stress may make use of a simple concept called "coefficient of subgrade reaction" having the dimension of force per volume (Terzaghi, 1955). The coefficient is a function of the soil density or strength, the depth below the ground surface, and the diameter (side) of the pile. In cohesionless soils, the following relation is used:

$$k_s = n_h \frac{z}{b} \qquad (13.29)$$

where

k_s = coefficient of horizontal subgrade reaction
n_h = coefficient related to soil density
z = depth
b = pile diameter

The intensity of the lateral stress, p_z, mobilized on the pile at depth z is then as follows:

$$p_z = k_s y_z b \qquad (13.30)$$

where y_z = the horizontal displacement of the pile at depth z.

Combining Equations 13.29 and 13.30:

$$p_z = n_h y_z z \qquad (13.31)$$

The relation governing the behavior of a laterally loaded pile is then as follows (Reese and Wang, 1985):

$$Q_h = EI \frac{d^4 y}{dx^4} + Q_v \frac{d^2 y}{dx^2} - p \qquad (13.32)$$

where

Q_h = lateral load on the pile
EI = bending stiffness (flexural rigidity)
Q_v = axial load on the pile

Design charts have been developed that, for an input of imposed load, basic pile data, and soil coefficients, provide values of displacement and bending moment. See, for instance, the *Canadian Foundation Engineering Manual* (1985).

The design charts cannot consider all the many variations possible in an actual case. For instance, the $p-y$ curve can be a smooth rising curve, can have an ideal elastic–plastic shape, or can be decaying after a peak value. As an analysis without simplifying shortcuts is very tedious and time-consuming, resort to charts has been necessary in the past. However, with the advent of the personal computer, special software has been developed that makes the calculations easy and fast. In fact, as in the case of pile driving analysis and wave equation programs, engineering design today has no need for computational simplifications. Exact solutions can be obtained as easily as approximate ones. Several proprietary and public-domain programs are available for analysis of laterally loaded piles. One of the most widely used and accepted is produced by Reese and Wang (1985).

One must not be led to believe that, because an analysis is theoretically correct, the results also predict to the true behavior of the pile or pile group. The results must be correlated to pertinent experience, and, lacking this, to a full-scale test at the site. If the experience is limited and funds are lacking for a full-scale correlation test, then a prudent choice is necessary of input data, as well as of margins and factors of safety.

Designing and analyzing a lateral test is much more complex than for the case of axial behavior of piles. In service, a laterally loaded pile almost always has a fixed-head condition. However, a fixed-head test is more difficult and costly to perform than a free-head test. A lateral test without inclusion of measurement of lateral deflection down the pile (bending) is of limited value. While an axial test should not include unloading cycles, a lateral test should be a cyclic test and include a large number of cycles at different load levels. The laterally tested pile is much more sensitive to the influence of neighboring piles than is the axially tested pile. Finally, the analysis of the test results is much more complex and requires the use of a computer and appropriate software.

13.15 SEISMIC DESIGN OF LATERAL PILE BEHAVIOR

Seismic design of lateral pile behavior is often taken as being the same as the conventional lateral design. A common approach is to assume that the induced lateral force to be resisted by piles is static and equal to a proportion, usually 10 percent, of the vertical force acting on the foundation. If all the horizontal force is designed to be resisted by inclined piles and all piles, including the vertical ones, are designed to resist significant bending at the pile cap, this approach is normally safe, albeit costly. It cannot be used for lateral design of vertical piles, however.

The seismic wave appears to the pile foundation as a soil movement forcing the piles to move with the soil. The movement is resisted by the pile cap, bending and shear are induced in the piles, and a horizontal force develops in the foundation, starting it to move in the direction of the wave. A half period later, the soil swings back, but the foundation is still moving in the first direction, and, therefore, the forces increase. This situation is not the same as the one originated by a static force.

Seismic lateral pile design consists of determining the probable amplitude and frequency of the seismic wave as well as the natural frequency of the foundation and structure supported by the piles. The first requirement is, as in all seismic design, that the natural frequency must not be the same as that of the seismic wave. Then, the probable maximum displacement, bending, and shear induced at the pile cap are estimated. Finally, the pile connection and the pile cap are designed to resist the induced forces.

There is at present a rapid development of computer software for use in detailed seismic design.

13.16 DESIGN EXAMPLE

A heavy column foundation, which is to support a vertical load of 12 MN (dead-load portion is 9.6 MN), will be placed at a site where the soils consist of an upper 6 m thick layer of organic, compressible clay followed by a 4 m thick sand layer below which a 5 m thick normally consolidated clay layer is deposited. Under the silty clay, depth 15 m, lies a 25 m thick layer of silty sand deposited on bedrock at a depth of 40 m. The groundwater table is located at the ground surface and the pore water pressure in the soil is hydrostatically distributed. The site has been thoroughly investigated and the geotechnical parameters of the soil layers have been determined. The column foundation is to be placed level with the ground surface in the center of a 100 m by 100 m area over which a 2 m thick fill will be placed. Table 13.4 presents geotechnical values for use in this example.

The foundation must be pile-supported and it has been decided to use 12 driven, 300 mm square, precast, prestressed, concrete piles and to install them by means of a Vulcan 010 single-acting air hammer. The piles will have to go well into the silty sand layer, but they will not reach bedrock. The 12 piles will each require a capacity of 1000 kN. The concrete 28-day strength is 50 MPa and the prestressing is by eight 11-mm (7/17-in) strands of yield 1860 MPa (270 ksi). They will be placed at the minimum spacing recommended by the *Canadian Foundation Engineering Manual* (1985), that is, 2 percent of the embedment depth plus 2.5 diameters, which results in a spacing center-to-center of 1.2 m. The piles will be

TABLE 13.4 GEOTECHNICAL PARAMETERS.

Parameter	Units	Fill	Clay	Silty Sand	Silty Clay	Sand
Thickness	m	2.0	6.0	4.0	5.0	25.0
Density	kg/m^3	1800	1600	1900	1850	2000
w	%	—	50	20	36	15
β	—	—	0.25	0.45	0.35	0.50
c'	kPa	—	12.0	0	0	0
N_t	—	—	—	—	—	60
m	—	—	15	200	30	300
m_r	—	—	120	500	450	900
j	—	—	0	0.5	0	0.5
OCR	—	—	1.8	2.0	—	2.0
$\sigma'_1 - \sigma'_p$	kPa	—	—	—	40	—

placed in four rows of three piles and the foundation size is, therefore, 4.0 m by 3.0 m.

Given the particular conditions imposed by the site conditions, the pile, and the hammer to use, the design effort includes calculations of what embedment length of the pile to consider by means of a static analysis of axial capacity, analyzing the drivability of the pile, and checking that the calculated settlement of the pile foundation is acceptable.

Embedment Length

A static analysis begins by determining the distribution of unit shaft resistance in the soil and the toe resistance according to Equations 13.3 and 13.5. When applying the values given in Table 13.4 and stipulating a factor of safety of 3.0 on the total load of 1000 kN, which requires a total pile capacity of 3000 kN, an embedment depth of 23 m is obtained. For the design example, the analysis is carried out by means of the UNIPILE program (Goudreault and Fellenius, 1990) that also yields the depth to the neutral plane and the load in the pile at the neutral plane, 19 m and 1900 kN, respectively. The 1900 kN load translates to a stress of about 20 MPa in the concrete pile, which is much smaller than 70 percent of the 50-MPa concrete cylinder strength and acceptable.

During initial driving, excess pore pressures will develop in the clay strata and, to a degree, also in the silty sand stratum, reducing the effective stress. When excess pore pressures are introduced in the UNIPILE program, the computed static resistance at depth 24 m becomes about 1800 kN, which is the static resistance that the hammer has to overcome in the initial driving. The balance of about 1200 kN will be derived from soil set-up occurring during the reconsolidation after driving and from consolidation of the soils under the weight of the fill.

Drivability Analysis

Drivability analysis combines analyses by the wave equation and static methods. The most common approach is to perform wave equation analysis to produce a so-called bearing graph, which shows the capacity against the penetration resistance. To account for variability of input values, such as hammer efficiency, soil quakes, cushion properties, etc., it is prudent to present a range of curves within an envelope or band. Figure 13.17 presents the band applicable to the design example and calculated by means of the GRLWEAP program (Goble Rausche Likins and Associates, 1988). The bearing graph shows that the required static capacity of about 1800 kN at the end-of-initial-driving (EOID) will be reached at a penetration resistance of about 400 ± 100 blows/m, which corresponds to

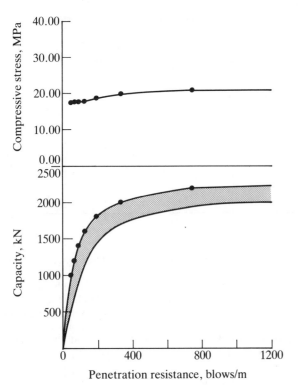

Fig. 13.17 Example design case. Bearing graph from a range wave equation analysis.

about 10 ± 2 blows/25 mm. Continuing to drive beyond a resistance at EOID exceeding 10 blows/25 mm will not be productive. The analysis also shows that the maximum compressive stress induced by the driving is about 22 MPa, which is smaller than two-thirds of the concrete strength of 50 MPa and therefore acceptable.

The bearing graph only presents the conditions toward the end of the initial driving. In a drivability study, one must answer many additional questions, such as what is the accumulated penetration resistance and the maximum tensile stress, which usually occurs before the full embedment depth has been reached. This study can be made by means of GRLWEAP's option of "blow count versus depth", the results of which are shown in Figure 13.18.

Figure 13.18 shows that although the static resistance (the capacity) increases linearly with depth, the penetration resistance (the blow count) increases progressively and the maximum depth to which one can reasonably expect to drive the pile with the assigned hammer is about 26 m. The analysis also suggests that the time for driving the pile will be 45 minutes, excluding splicing, but full-strength mechanical splices are completed in a few minutes and the splicing will add less than 5 minutes to the installation time.

Furthermore, the analysis and Figure 13.18 show that the maximum tensile stress is about 3.5 MPa, or 315 kN. The yield strength of the eight 11-mm strands of 1850 MPa results in an ultimate tensile strength of the pile of 1400 kN. The acceptable tensile stress is 70 percent of this value, that is, about 1000 kN. Hence, the analysis suggests that tensile stress will not become a problem in the driving.

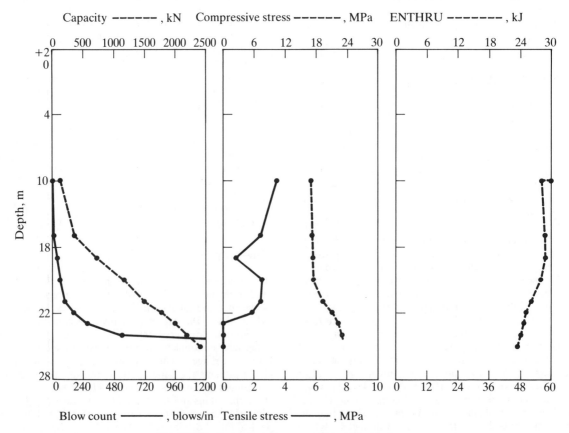

Fig. 13.18 Example design case. Drivability Study.

Settlement Analysis

The static analysis indicated a neutral plane located at an embedment depth of 19 m. By placing an equivalent footing at this depth having the same size as the pile cap ($12 \, m^2$), loading it by the dead load on the pile cap (9600 kN), and distributing this load by means of the 2:1 method, the settlement can be calculated from the parameters given in Table 13.4. The UNIPILE program calculates the settlement for the equivalent footing to be 30 mm. The movement of the pile toe into the sand is 20 mm, which is sufficient to activate the full toe resistance for the driven pile.

Acceptance of the Design

The bearing capacity as well as the structural capacity are within acceptable limits. The assigned hammer, although it is a light one, appears to be sufficient for installing the piles. Acceptance of the design now depends on whether the 30-mm settlement can be accepted. It should be recognized that even though piles have been installed underneath a foundation, there will always be settlement. However, in many instances a limit of 25 mm is acceptable. Most structures can actually tolerate considerably higher values.

If found unacceptable, however, the settlement can be reduced by taking the piles deeper into the sand. Changing to a larger size pile or to another type of pile and installing them to the same 23-m depth will not change the settlement appreciably. In the example design case, there is obviously a margin for using a larger hammer that can take the piles deeper. Perhaps, in the extreme, even all the way to the bedrock, which would be a costly proposition, however.

It is difficult to verify settlement calculations in advance. However, to verify pile capacity is not difficult. For a case similar to the example design case, where no large margin of capacity has been demonstrated, it will be wise to verify the design assumptions by means of dynamic monitoring of the initial driving and CAPWAP analysis of the dynamic data.

The bearing graph in Figure 13.17 shows that the hammer will be able to move the pile against a capacity of up to about 2200 kN. Because of soil set-up, the pile capacity will probably increase to about 3000 kN, which will be evidenced by a penetration restrike resistance in excess of about 15 blows/ 25 mm. To verify the capacity, it will be necessary to bring in a heavier hammer capable of moving the pile against a 3000-kN capacity. Alternatively, other means of capacity verification can be employed, for example, a static loading test or special dynamic methods, such as the STATNAMIC method and device (Bermingham and Janes, 1989).

REFERENCES

Authier, J. and Fellenius, B. H. (1983), Wave equation analysis and dynamic monitoring of pile driving, *Civil Engineering for Practicing and Design Engineers*, **2**, No. 1, pp. 1–20.

Bermingham, P. and Janes, M. (1989), An innovative approach to load testing of high capacity piles, *Proceedings of the International Seminar on Piling and Deep Foundations*, London, May 1989, Vol. 1, ed. J. B. Burland and J. M. Mitchell, A. A. Balkema Publishers, Rotterdam, pp. 409–413.

Bjerin, L. (1977), Pahangskrafter pa langa betongpalar, *Swedish Geotechnical Institute, Report No. 2* (in Swedish).

Bjerrum, L., Johannessen, I. J., and Eide, O. (1969), Reduction of negative skin friction on steel piles to rock, *Proceedings of the 7th International Conference on Soil Mechanics and Foundation Engineering*, Mexico City, **2**, pp. 27–34.

Bozozuk, M. (1972), Downdrag measurement on a 160-ft floating test pile in marine clay, *Canadian Geotechnical Journal*, **9**, No. 2, pp. 127–136.

Bozozuk, M. (1981), Bearing capacity of a pile preloaded by downdrag, *Proceedings of the 10th International Conference on Soil Mechanics and Foundation Engineering*, Stockholm, **2**, pp. 631–636.

Bozozuk, M., Fellenius, B. H., and Samson, L. (1978), Soil disturbance from pile driving in sensitive clay, *Canadian Geotechnical Journal*, **15**, No. 3, pp. 346–361.

Briaud, J.-L. and Tucker, L. (1985), Piles in sand—A method including residual stresses, *Journal of Geotechnical Engineering, ASCE*, **110**, No. 11, pp. 1666–1679.

Brinch-Hansen, J. (1963), Hyperbolic stress–strain response of cohesive soils. Discussion. *Journal of Soil Mechanics and Foundation Engineering, ASCE*, **89**, No. SM-4, pp. 241–242.

Burland, J. B. (1973), Shaft friction of piles in clay, A simple fundamental approach, *Ground Engineering*, London, **6**, No. 1, pp. 30–42.

Canadian Foundation Engineering Manual (1985), Second edition. Part 1: Fundamentals; Part 2: Shallow Foundations; Part 3: Deep Foundations; Part 4: Excavations and Retaining Structures. Canadian Geotechnical Society, Technical Committee on Foundations, BiTech Publishers, Vancouver.

Chin, F. K. (1970), Estimation of the ultimate load of piles not carried to failure, *Proceedings of the 2nd Southeast Asian Conference on Soil Engineering*, Southeast Asian Geotechnical Society, pp. 81–90.

Chin, F. K. (1971), Discussion, Pile test, Arkansas River Project, American Society of Civil Engineers, *Journal of Soil Mechanics and Foundation Engineering, ASCE*, **97**, No. SM-6, pp. 930–932.

Chin, F. K. (1978), Diagnosis of pile condition, Lecture at the 6th Southeast Asian Conference on Soil Engineering, Bangkok, 1977, *Geotechnical Engineering*, **9**, pp. 85–104.

Clemente, F. M. (1981), Downdrag on bitumen coated piles in a warm climate, *Proceedings of the 10th International Conference on Soil Mechanics and Foundation Engineering*, Stockholm, **2**, pp. 673–676.

Davisson, M. T. (1972), High capacity piles, *Proceedings of Lecture Series on Innovations in Foundation Construction*, American Society of Civil Engineers, Illinois Section, Chicago, March 22, pp. 81–112.

Dunnicliff, C. J. (1982), *Geotechnical Instrumentation for Monitoring Field Performance*, National Cooperative Highway Research Program, Synthesis of Highway Practice No. 89, Transportation Research Council, Washington, D.C.

Dunnicliff, C. J. (1988), *Geotechnical Instrumentation for Monitoring Field Performance*, John Wiley & Sons, Inc., New York, N.Y.

Fellenius, B. H. (1972), Downdrag on piles due to negative skin friction, *Canadian Geotechnical Journal*, **9**, No. 4, pp. 323–337.

Fellenius, B. H. (1975a), Reduction of negative skin friction with bitumen coated slip layers, Discussion, *Journal of Geotechnical Engineering, ASCE*, **101**, No. GT-4, pp. 412–414.

Fellenius, B. H. (1975b), Test loading of piles. Methods, interpretation and new proof testing procedure, *Journal of Geotechnical Engineering, ASCE*, **101**, No. GT-9, pp. 855–869.

Fellenius, B. H. (1979), Downdrag on bitumen coated piles. Discussion, *Journal of Geotechnical Engineering, ASCE*, **105**, No. GT-10, pp. 1262–1265.

Fellenius, B. H. (1980), The analysis of results from routine pile loading tests, *Ground Engineering*, London, **13**, No. 6, pp. 19–31.

Fellenius, B. H. (1984a), Negative skin friction and settlement of piles, *Proceedings of Second International Geotechnical Seminar, Pile Foundations*, Nanyang Technological Institute, Singapore.

Fellenius, B. H. (1984b), Ignorance is bliss—And that is why we sleep so well, *Geotechnical News*, Canadian Geotechnical Society and the United States National Society of the International Society of Soil Mechanics and Foundation Engineering, **2**, No. 4, pp. 14–15.

Fellenius, B. H. (1989a), Unified design of piles and pile groups. *TRB Record 1169*, pp. 75–82. Transportation Research Board, Washington, D.C.

Fellenius, B. H. (1989b), Tangent modulus of piles determined from strain data, *ASCE Geotechnical Engineering Division, 1989 Foundation Congress*, ed. F. H. Kulhawy, Vol. 1, pp. 500–510.

Fellenius, B. H. and Samson, L. (1976), Testing of drivability of concrete piles and disturbance to sensitive clay. *Canadian Geotechnical Journal*, **13**, No. 2, pp. 139–160.

Fellenius, B. H., Samson, L., Thompson, D. E., and Trow, W. (1978), Dynamic behaviour of foundation piles and driving equipment, Terratech Ltd. and the Trow Group Ltd., Final Report, Department of Supply and Services, Canada, Research Project, Vols. I and II.

Fellenius, B. H., O'Brien, A. J., Riker, R. E., and Tracy, G. R. (1983), Dynamic monitoring and conventional pile testing procedures, ASCE, *Proceedings of Symposium on Dynamic Measurement of Piles and Piers*, ed. G. G. Goble.

Fellenius, B. H., Riker, R. E., O'Brien, A. J., and Tracy, G. R. (1989), Dynamic and static testing in a soil exhibiting setup, *Journal of Geotechnical Engineering, ASCE*, **115**, No. 7, pp. 984–1001.

Fellenius, B. H. and Rasch, N. C. (1990), *FAILPILE Program for Analysis of Failure Loads in the Static Pile Loading Test*. User Manual, Bengt Fellenius Consultants Inc., Ottawa.

Goble Rausche Likins and Associates (1988), *GRLWEAP Program for Wave Equation Analysis of Pile Driving*, User Manual. Cleveland, Ohio.

Goudreault, P. and Fellenius, B. H. (1990), *UNIPILE Program for Unified Analysis of Piles and Pile Groups Considering Capacity, Negative Skin Friction, and Settlement*. User Manual, Bengt Fellenius Consultants, Inc., Ottawa.

Hanna, T. H. and Tan, R. H. S. (1973), The behaviour of long piles under compressive loads in sand, *Canadian Geotechnical Journal*, **10**, No. 3, pp. 311–340.

Hannigan, P. J. and Webster, S. D. (1988), Evaluation of drive system performance and hammer cushion parameters, *Proceedings of the Third International Conference on the Application of Stress-Wave Theory to Piles*, ed. B. H. Fellenius, BiTech Publishers, Vancouver, pp. 869–878.

Holloway, M., Clough, G. W., and Vesic, A. S. (1978), A rational procedure for evaluating the behavior of impact-driven piles, *ASTM Symposium on Behavior of Deep Foundations*, ed. R. Lundgren, Special Technical Publication STP 670, pp. 335–357.

Janbu, N. (1963), Soil compressibility as determined by oedometer and triaxial tests, *European Conference on Soil Mechanics and Foundation Engineering*, Wiesbaden, **1**, pp. 19–25; **2**, pp. 17–21.

Janbu, N. (1965), Consolidation of clay layers based on nonlinear stress–strain, *Proceedings of the 6th International Conference on Soil Mechanics and Foundation Engineering*, Montreal, **2**, pp. 83–87.

Johannessen, I. J. and Bjerrum, L. (1965), Measurement of the compression of a steel pile to rock due to settlement of the surrounding clay, *Proceedings of the 6th International Conference on Soil Mechanics and Foundation Engineering*, Montreal, **2**, pp. 261–264.

Kezdi, A. (1975), Pile foundations. *Foundation Engineering Handbook*, 1st ed., ed. H. F. Winterkorn and H. Y. Fang, Van Nostrand Reinhold, New York, N.Y., pp. 556–600.

Kondner, R. L. (1963), Hyperbolic stress–strain response. Cohesive soils. *Journal of Soil Mechanics and Foundation Engineering, ASCE*, **89**, No. SM-1, pp. 115–143.

Lee, S. Q. S. and Fellenius, B. H. (1990), *TELLPILE Program for Analysis of Telltale Data from a Static Loading Test*. User Manual, Bengt Fellenius Consultants Inc., Ottawa.

Leonards, G. A. and Lovell, D. (1978), Interpretation of load test on high capacity driven piles, *ASTM Symposium on Behavior of Deep Foundations*, ed. R. Lundgren, Special Technical Publication STP 670, pp. 388–415.

Likins, G. and Rausche, F. (1988), Hammer inspection tools, *Proceedings of the Third International Conference on the Application of Stress-Wave Theory to Piles*, ed. B. H. Fellenius, BiTech Publishers, Vancouver, pp. 659–667.

Meyerhof, G. G. (1976), Bearing capacity and settlement of pile foundations, The Eleventh Terzaghi Lecture, November 5, 1975, *Journal of Geotechnical Engineering, ASCE*, **102**, No. GT-3, pp. 195–228.

Middendorp, P. and Reiding, F. (1988), Determination of discontinuities in piles by low and high strain impacts, *Proceedings of the Third International Conference on the Application of Stress-Wave Theory to Piles*, ed. B. H. Fellenius, BiTech Publishers, Vancouver, pp. 33–43.

O'Neill, M. W., Hawkins, R. A., and Mahar, L. J. (1982), Load transfer mechanism in piles and pile groups. *Journal of Geotechnical Engineering, ASCE*, **108**, No. GT-12, pp. 1605–1623.

Poulos, H. G. and Davis, E. H. (1980), *Pile Foundation Analysis and Design*, Series in Geotechnical Engineering, John Wiley and Sons, Inc., New York, N.Y.

Rausche, F., Moses, F., and Goble, G. G. (1972), Soil resistance predictions from pile dynamics, *Journal of Soil Mechanics and Foundation Engineering, ASCE*, **98**, No. SM-9, pp. 917–937.

Rausche, F. and Goble, G. G. (1978), Determination of pile damage by top measurements, *ASTM Symposium on Behavior of Deep Foundations*, ed. R. Lundgren, Special Technical Publication STP 670, pp. 500–506.

Rausche, F., Goble, G. G., and Likins, G. E. (1985), Dynamic determination of pile capacity, *Journal of Geotechnical Engineering, ASCE*, **111**, No. GT-3, pp. 367–383.

Rausche, F., Likins, G. E., and Hussein, M. (1988), Pile integrity by low and high strain impacts, *Proceedings of the Third International Conference on the Application of Stress-Wave Theory to Piles*, ed. B. H. Fellenius, BiTech Publishers, Vancouver, pp. 44–55.

Reese, L. C. and Wang, S. T. (1985), *Documentation of Computer Program LPILE1*, Ensoft Inc., Austin, Texas.

Reese, L. C., Awoshika, K., Lam, P. H. F., and Wang, S. T. (1990), *Documentation of Computer Program GROUP1*, Ensoft Inc., Austin, Texas.

Schmertmann, J. H. (1978), Guidelines for cone penetration test, performance and design, U.S. Federal Highway Administration, Washington, D.C., Report FHWA-TS-78-209.

Terzaghi, K. (1955), Evaluation of coefficients of subgrade reaction, *Geotechnique*, **5**, No. 4, pp. 297–326.

Terzaghi, K. and Peck, R. B. (1967), *Soil Mechanics in Engineering Practice*, 2nd ed., John Wiley and Sons, Inc., New York, N.Y.

Tomlinson, M. J. (1957), The adhesion of piles driven in clay soils, *Proceedings of the 4th International Conference on Soil Mechanics and Foundation Engineering*, London, **2**, pp. 66–71.

Vesic, A. S. (1970), Load transfer in pile–soil systems, *Proceedings of the Conference on Design and Installation of Pile Foundations and Cellular Structures*, ed. H. Y. Fang, Envo Publishing Co., Bethlehem, Pa., pp. 47–73.

Vesic, A. S. (1977), *Design of Pile Foundations*, National Cooperative Highway Research Program, Transportation Research Board, National Research Council, National Academy of Sciences, Washington, Synthesis of Highway Practice No. 42.

14   DRILLED SHAFT FOUNDATIONS

FRED H. KULHAWY, Ph.D., P.E.
Professor of Civil/Geotechnical Engineering
Cornell University

14.1 INTRODUCTION

A drilled shaft, also known as drilled pier, drilled caisson, caisson, bored pile, etc., is a versatile foundation system that is used extensively on a worldwide basis. In its simplest form, a drilled shaft is constructed by making a cylindrical excavation, placing a reinforcing cage (when necessary), and then concreting the excavation. With available drilling equipment, shaft diameters up to 20 ft (6 m) and depths exceeding 250 ft (76 m) are possible. However, for most normal applications, diameters in the range of 3 to 10 ft (1 to 3 m) are typical. This size versatility allows a single drilled shaft to be used in place of a driven pile group and eliminates the need for a pile cap. In addition, normal construction practices for drilled shafts effectively eliminate the noise and strong ground vibrations that develop during pile driving operations. For these and other secondary reasons, drilled shafts have become both the technical and economic foundation of choice for many design applications. In fact, they have become the dominant foundation type in many geologic settings around the world.

During the past 25 years, major advances have been made in understanding the behavior of drilled shaft foundations. These advances have stemmed largely from extensive field load testing, controlled laboratory testing, and sophisticated numerical simulations. From these studies, consistent patterns of behavior have emerged, leading to the realistic analysis and design equations that are presented in this chapter.

The performance of drilled shaft foundations, as with any other foundation type, is a function of the construction method used. Therefore, construction issues have to be incorporated into the analysis and design process. The design engineer must understand thoroughly the available construction methods and equipment, as well as their applicability for different subsurface conditions. Useful references on these subjects are the recent works by Greer and Gardner (1986) and Reese and O'Neill (1988).

Within this chapter, it is impossible to cover all aspects of drilled shaft foundations because of space limitations. Therefore, attention is focused on the behavior of shafts in soil under axial compression and uplift loading. The only construction issues included are those that impact on the analysis and design process directly. Other loading modes, shafts in rock, and related issues are discussed briefly, and pertinent references are cited.

14.2 GENERAL BEHAVIOR PATTERNS

When a drilled shaft is loaded in compression, the general form of load–displacement curve shown in Figure 14.1 results. The upper curve represents the total load applied to the

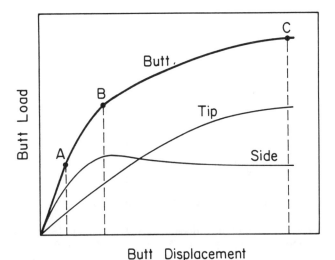

Fig. 14.1 Generalized load–displacement behavior for drilled shaft in compression.

butt (top or head) of the shaft, while the other two curves differentiate the load into its side and tip (base or toe) resistance components. To visualize the reasons for this behavior, consider the force equilibrium diagram in Figure 14.2. The notation used in Figure 14.2a defines the shaft depth (D), diameter (B), and weight (W); the compression capacity of the shaft (Q_c); and the tip (Q_{tc}) and side (Q_{sc}) resistances in compression. Figure 14.2b shows an illustrative distribution of available soil shearing resistance (τ) acting along the tip and side of the shaft. The evaluation of τ will be described later.

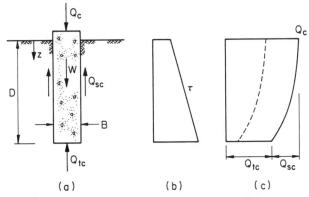

Fig. 14.2 Idealized force equilibrium diagram for drilled shaft in compression.

When a compressive load is applied to the butt of the shaft, downward displacement occurs, which begins to mobilize the soil shearing resistance. This process transfers the load to the supporting soil and results in progressively smaller load in the shaft with depth, as shown by the dashed curve in Figure 14.2c. For this curve, the load transferred to the tip is quite small, corresponding to point A in Figure 14.1. As the butt load is increased further to point B in Figure 14.1, all of the available soil shearing resistance is mobilized along the shaft side and any further load transfer then must develop at the shaft tip. When the butt load is increased further to its maximum value, the full tip resistance is mobilized (point C in Figure 14.1), and the load transfer pattern is given by the solid curve in Figure 14.2c. During the loading process from B to C, the side resistance may increase, decrease, or remain constant, depending on the stress–strain characteristics of the soil–shaft interface. For soils that exhibit significant strain-softening, this factor must be evaluated carefully.

Figures 14.1 and 14.2 together illustrate several very important behavioral issues for drilled shafts in compression. First, the overall load–displacement response is nonlinear. Second, the full side resistance develops at relatively small butt displacements. Third, the full tip resistance develops at large butt displacements. And fourth, the load transfer between the side and tip will be primarily a function of: (1) the available shearing resistance along the side and below the tip, (2) the geometry of the shaft, (3) the load level, and (4) the relative stiffnesses of the shaft and soil. All of these factors must be considered when proportioning a drilled shaft for both capacity requirements and displacement limits.

14.3 AXIAL COMPRESSION CAPACITY

From the equilibrium diagram given in Figure 14.2a, the compression capacity (Q_c) of a drilled shaft is given by

$$Q_c = Q_{sc} + Q_{tc} - W \tag{14.1}$$

in which Q_{sc} = side resistance in compression, Q_{tc} = tip resistance in compression, and W = weight of shaft. As noted in Section 14.2, both Q_{sc} and Q_{tc} are displacement-dependent and develop their limiting values at significantly different displacements.

The side resistance represents the interface shearing resistance (τ) available along the shaft surface and is given by

$$Q_{sc} = \int_{\text{surface}} \tau(z)\,dz \tag{14.2}$$

in which z is the depth shown in Figure 14.2a. For a shaft with circular cross section, Equation 14.2 becomes

$$Q_{sc} = \pi B \int_0^D \tau(z)\,dz \tag{14.3}$$

The value of τ will vary as a function of soil type and loading conditions, as discussed later.

The tip resistance is evaluated as a bearing capacity problem and is given by

$$Q_{tc} = q_{\text{ult}} A_{\text{tip}} = q_{\text{ult}} \pi B^2/4 \tag{14.4}$$

in which q_{ult} = ultimate bearing capacity and A_{tip} = shaft tip area. The general solution for q_{ult} (e.g., Vesić, 1975) is the Terzaghi–Buisman equation given below:

$$q_{\text{ult}} = c N_c + 0.5 B \gamma N_\gamma + q N_q \tag{14.5}$$

in which c = soil cohesion, γ = soil unit weight, q = vertical stress at shaft tip (γD), and N_c, N_γ, N_q = dimensionless bearing capacity factors. This equation includes the Prandtl and

Reissner solutions for a load on a weightless medium, resulting in

$$N_q = \tan^2(45° + \phi/2) \exp(\pi \tan \phi) \tag{14.6}$$

$$N_c = (N_q - 1) \cot \phi \quad \text{(note: as } \phi \to 0, N_c \to 5.14) \tag{14.7}$$

in which ϕ = soil friction angle. The N_γ term is given as

$$N_\gamma \approx 2(N_q + 1) \tan \phi \tag{14.8}$$

which is Vesić's (1975) approximation of the numerical solution by Caquot and Kerisel (1953). These bearing capacity factors are shown in Figure 14.3.

Equation 14.5 was developed for the idealized conditions of general soil shear failure for an infinitely long strip foundation at shallow depth. To extend this equation to actual field conditions, modifiers have been developed by a number of authors. Those presented below are based primarily upon the consistent interpretations of the available data by Vesić (1975) and Hansen (1970), with minor modifications by Kulhawy et al. (1983).

In its general form, the bearing capacity equation is given by

$$q_{\text{ult}} = c N_c \zeta_{cs} \zeta_{cd} \zeta_{cr} + 0.5 B \gamma N_\gamma \zeta_{\gamma s} \zeta_{\gamma d} \zeta_{\gamma r} + q N_q \zeta_{qs} \zeta_{qd} \zeta_{qr} \tag{14.9}$$

The ζ modifiers extend the theory to field conditions and are doubly subscripted to indicate which term they apply to (N_c, N_γ, N_q) and which phenomena they describe (s for foundation shape, d for foundation depth, and r for soil rigidity). For foundations of circular cross section, the modifiers are given in Table 14.1. The shape modifiers depend only on the soil friction angle, ϕ. The depth modifiers vary with ϕ and the shaft depth-to-diameter ratio, D/B. With these depth modifiers, no arbitrary definitions for "shallow" or "deep" are needed, because D/B is accounted for explicitly as a continuous function. Figure 14.4 illustrates a typical variation of ζ_{qd} and shows that depth variations become minor after only a modest foundation embedment.

While the shape and depth modifers are familiar to all designers, the rigidity modifiers, which include ϕ and the rigidity index, I_{rr}, may not be so familiar. The rigidity index (soil

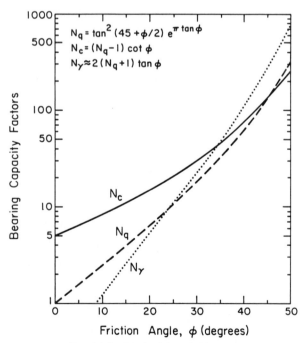

Fig. 14.3 Bearing capacity factors.

TABLE 14.1 GENERAL BEARING CAPACITY MODIFIERS FOR CIRCULAR FOUNDATION.

Modification	Symbol	Value
Shape	ζ_{cs}	$1 + (N_q/N_c)$
	$\zeta_{\gamma s}$	0.6
	ζ_{qs}	$1 + \tan\phi$
Depth	ζ_{cd}	$\zeta_{qd} - [(1-\zeta_{qd})/(N_c\tan\phi)]$
	$\zeta_{\gamma d}$	1
	ζ_{qd}	$1 + 2\tan\phi(1-\sin\phi)^2[(\pi/180)\tan^{-1}(D/B)]$
Rigidity	ζ_{cr}	$\zeta_{qr} - [(1-\zeta_{qr})/(N_c\tan\phi)] \quad \leqslant 1$
	$\zeta_{\gamma r}$	ζ_{qr}
	ζ_{qr}	$\exp\{[-3.8\tan\phi]+[(3.07\sin\phi)(\log_{10}2I_{rr})/(1+\sin\phi)]\} \quad \leqslant 1$

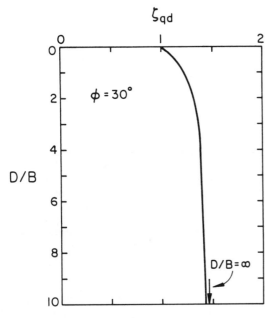

Fig. 14.4 Variation of ζ_{qd} with D/B.

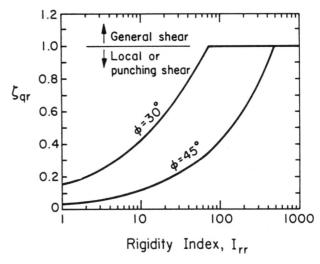

Fig. 14.5 Variation of ζ_{qr} with rigidity index and friction angle.

modulus/soil strength) has not been a traditional geotechnical design parameter, although it is common in structural design. However, it is well known that foundations fail in modes ranging from general to local to punching shear as a function of the soil stiffness (e.g., Vesić, 1975). Some references have suggested somewhat arbitrary criteria to distinguish these different modes and make modifications to the bearing capacity terms as a function of relative density or other index parameters. While these criteria may be convenient and simple, they do not incorporate the soil stiffness directly. The rigidity index is a more rational way of incorporating this influence directly. Fortunately, small variations in I_{rr} do not result in significant variations in the rigidity modifiers. For example, Figure 14.5 shows clearly that a tenfold increase in I_{rr} results in only a two- to threefold increase in the ζ_{qr} for local or punching shear cases. This result is quite important for design, because it means that only the correct "order-of-magnitude" is necessary for rigidity index evaluation. For stiffer soils that fail in general shear, the modifiers all equal 1.

The equations given in this section are used most commonly in either of two alternative forms, based on the soil type and rate of loading. The first is for drained loading, described in

Section 14.4, which develops under most loading conditions in coarse-grained soils such as sands and for long-term sustained loading of fine-grained soils such as clays. The second is for undrained loading, described in Section 14.5, which develops for short-term loading of fine-grained soils.

14.4 DRAINED AXIAL COMPRESSION CAPACITY

When the soil is sufficiently coarse-grained that no excess pore water stresses develop during loading, or for long-term loading in fine-grained soils, drained conditions prevail. The soil then is assumed to behave as a purely frictional material with no cohesion, and effective stress soil parameters are used for design. For these conditions, the axial capacity can be evaluated as follows. In all cases, the seasonal high water table should be used for design.

14.4.1 Drained Bearing Capacity

The ultimate bearing capacity for drained loading, in which the effective stress cohesion ($\bar{c}$) is zero, can be evaluated from Equation 14.9 as follows:

$$q_{ult} = 0.5\,B\,\gamma\,N_\gamma\zeta_{\gamma s}\zeta_{\gamma d}\zeta_{\gamma r} + q\,N_q\zeta_{qs}\zeta_{qd}\zeta_{qr} \quad (14.10)$$

The appropriate values of the ζ modifiers are given in Table 14.1. With $\zeta_{ys} = 0.6$ and $\zeta_{yd} = 1$, Equation 14.10 becomes

$$q_{\text{ult}} = 0.3 \, B \, \bar{\gamma} \, N_\gamma \zeta_{\gamma r} + \bar{q} \, N_q \zeta_{qs} \zeta_{qd} \zeta_{qr} \qquad (14.11)$$

In this equation, $\bar{\gamma}$ is the average effective unit weight from depths D to $D + B$, $\bar{q}$ is the vertical effective stress at depth D, and both ϕ and I_{rr} are the average values from depths D to $D + B$. The depths D and $D + B$ are chosen to reflect the approximate zone within which most of the shearing occurs in bearing capacity. The appropriate value of ϕ to use for drained loading is $\bar{\phi}$, the effective stress friction angle.

As described previously, the rigidity index was introduced by Vesić (1975) to allow a consistent and rational means for analyzing general, local, and punching shear. The basic rigidity index (I_r) can be defined as

$$I_r = \frac{G_d}{\bar{c} + \bar{q}_a \tan \bar{\phi}} \qquad (14.12)$$

in which G_d = drained shear modulus and $\bar{q}_a$ = average vertical effective stress from depths D to $D + B$. For drained loading with $\bar{c} = 0$, Equation 14.12 becomes

$$I_r = \frac{G_d}{\bar{q}_a \tan \bar{\phi}} \qquad (14.13)$$

From elasticity theory, $G = \frac{1}{2}E/(1 + v)$, resulting in

$$I_r = \frac{E_d}{2(1 + v_d)\bar{q}_a \tan \bar{\phi}} \qquad (14.14)$$

in which E_d = drained elastic (Young's) modulus and v_d = drained Poisson's ratio.

Young's modulus can be evaluated directly from a number of different laboratory shear tests or in-situ field tests such as the pressuremeter. Typical ranges in this parameter within working stress ranges are given in Table 14.2. In addition, a large number of empirical correlations have been suggested to determine the modulus from the results of standard penetration, cone penetration, and dilatometer tests. Very useful summaries of these tests and the available correlations are given by Mitchell (1986), Robertson (1986), Skempton (1986), Lutenegger (1987), Mair and Wood (1987), Meigh (1987), and Kulhawy and Mayne (1990). The Skempton, and Kulhawy and Mayne papers are particularly recommended if use of the standard penetration test is contemplated, because they very clearly illustrate the many in-situ variables that must be addressed in properly interpreting this test.

Poisson's ratio in most granular soils ranges from about 0.1 to about 0.4. Based on existing data, Trautmann and Kulhawy (1987) suggested the following approximate relationship for estimating the drained Poisson's ratio

$$v_d = 0.1 + 0.3 \, \phi_{\text{rel}} \qquad (14.15)$$

in which ϕ_{rel} is defined as a relative friction angle, given by

$$\phi_{\text{rel}} = \frac{\bar{\phi} - 25°}{45° - 25°} \qquad (14.16)$$

TABLE 14.2 TYPICAL RANGES IN DRAINED YOUNG'S MODULUS.

Sand Consistency	Drained Young's Modulus, E_d	
	tons/ft²	MN/m²
Loose	50–200	5–20
Medium	200–500	20–50
Dense	500–1000	50–100

with limits of 0 and 1. This approximation is not valid for cemented or calcareous soils.

The basic rigidity index was developed for the condition of no volumetric strain. However, volumetric strain does occur in drained loading, so Vesić (1975) introduced the reduced rigidity index (I_{rr}) as follows:

$$I_{rr} = \frac{I_r}{1 + I_r \Delta} \qquad (14.17)$$

in which Δ = volumetric strain. Based on Vesić's (1975) guidelines, Trautmann and Kulhawy (1987) showed that Δ can be estimated conveniently by

$$\Delta \approx 0.005 \, (1 - \phi_{\text{rel}}) \left(\frac{\bar{q}_a}{p_a} \right) \qquad (14.18)$$

in which p_a = atmospheric pressure in the desired stress units (1 atm = 1.058 tsf = 101.3 kN/m² = 14.7 psi), up to a limit of $(\bar{q}_a/p_a) = 10$. Again, this approximation is not valid for cemented or calcareous soils.

After I_{rr} has been computed, it is compared with the critical rigidity index (I_{rc}), which was developed from cavity expansion theory (Vesić, 1975). For a circular foundation, I_{rc} is given by

$$I_{rc} = 0.5 \exp[2.85 \cot(45° - \bar{\phi}/2)] \qquad (14.19)$$

If $I_{rr} > I_{rc}$, the soil behaves as a rigid-plastic material, general shear behavior results, and $\zeta_{cr} = \zeta_{\gamma r} = \zeta_{qr} = 1$. If $I_{rr} < I_{rc}$, the relative soil stiffness is low, local or punching shear behavior results, and ζ_{cr}, $\zeta_{\gamma r}$, and ζ_{qr} will be less than 1 and must be computed to reduce the ultimate bearing capacity.

To illustrate the above principles, Kulhawy (1984) presented the simplified example in Figure 14.6 for uniform sand deposits. This figure incorporates a number of assumptions regarding material properties and is intended only for illustration. However, this example shows clearly that the tip capacity increases with depth at a continually decreasing rate, increases with soil density, and increases as the water table drops from the ground surface (saturated) to below the depth shown (dry). For simplicity, the first term in Equation 14.11 has not been included. This term often is small compared to the second term, and several authors have suggested that it can be deleted conservatively. However, for larger-diameter and smaller-depth shafts, this term can be significant, and therefore it should always be included in the calculations until it is shown to be a minor component.

Figure 14.6 also demonstrates that rather large values of the tip capacity may be computed. Whether these large values actually can be achieved will ultimately depend on the soil type

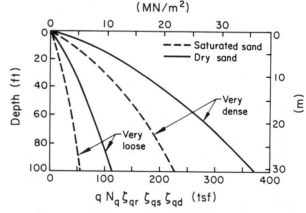

Fig. 14.6 Approximate tip ultimate bearing capacity for deep foundations in sand. (*Kulhawy, 1984, p. 97.*)

and structure and the construction method. For soil deposits that may be structurally sensitive (e.g., lightly cemented sands, calcareous sediments, collapsible soils, etc.), higher imposed foundation stresses may actually "crush" or "collapse" the soil structure in situ and lead to dramatically reduced bearing capacity. Load test evidence in soils of similar geologic environment provides the best guide for assessing this phenomenon. As a secondary approach, the cone penetration test tip resistance may provide an indirect measurement of this soil structure "collapse." If the cone tip resistance achieves an upper limit and remains more or less constant with depth, then the cone resistance is an approximate indicator of the maximum bearing capacity controlled by the soil structure.

A second issue to consider is the construction method and its influence on the soil properties and stresses beneath the shaft tip. The process of augering a cylindrical excavation and removing the soil releases some stresses from the soil mass beneath the excavation. This stress release is minimized with slurry excavation. For shallower excavations of smaller diameter, say to 20 ft (6 m) depth and less than 5 ft (1.5 m) diameter or so, the excavation-induced changes are minor, particularly if the shaft is concreted promptly. However, for much deeper or larger-diameter excavations, or when the concreting is delayed, heave and lateral flow may significantly reduce the strength and deformation properties of the soil. In the limit, this would result in using the fully-softened or critical-state soil friction angle (25° to 35° for many granular soils) and a modulus corresponding to this fully-softened state (typically comparable to a loose consistency). Although these reductions may appear to be dramatic, they are consistent with observations by Vesić (1977) and others that the tip resistance of drilled shafts may be significantly less than that of driven piles, where no excavation is made and driving may densify the soil beneath the tip.

The above arguments also relate to the issue of cleanout at the bottom of the excavation. If the cleanout is less than satisfactory, a significant amount of loose, excavated soil will be present that can result in reduced soil properties at the tip. Extreme cleanout measures are not warranted, however, because the soil properties already are reduced because of stress relaxation.

One final issue is that, in the majority of straight-sided drilled shaft designs in soil, the load transfer is such that only a small percentage of the applied butt load actually is supported by tip resistance. This percentage often is in the range of 5 to 20 percent in uniform soil deposits at working load levels. With this small percentage, the issues of soil structure "collapse", stress relaxation, and excavation cleanout become of relatively small importance in the overall design process. However, these same issues may become important with shorter belled shafts or with shafts penetrating looser soils and tip bearing on denser soils. In these cases, more load is supported by tip resistance, and therefore tip issues are more important.

14.4.2 Drained Side Resistance

As given by Equation 14.3, the drained side resistance is simply a summation of the available soil shearing resistance over the side area of the shaft. This equation can be expanded for drained conditions to give

$$Q_{sc} = \pi B \int_0^D \bar{\sigma}_h(z) \tan \delta \, dz \qquad (14.20)$$

in which $\bar{\sigma}_h$ = horizontal effective stress, which acts as a normal stress on the soil–shaft interface, and δ = effective stress friction

angle for the soil–shaft interface. Expressing the horizontal stress in terms of the vertical stress yields

$$Q_{sc} = \pi B \int_0^D \bar{\gamma} z \, K(z) \tan \delta \, dz \qquad (14.21)$$

in which $\bar{\gamma}$ = effective soil unit weight (submerged below water table; dry or moist above), K = coefficient of horizontal soil stress ($\bar{\sigma}_h / \bar{\sigma}_v$), and $\bar{\sigma}_v$ = vertical effective stress. These terms often are grouped to yield two common alternative forms:

$$Q_{sc} = \pi B \int_0^D \bar{\gamma} z \, \beta(z) \, dz \qquad (14.22)$$

$$Q_{sc} = \pi B \int_0^D f(z) \, dz \qquad (14.23)$$

in which $\beta = K \tan \delta$ and $f = \beta \bar{\gamma} z$ = unit side resistance. Much effort has been expended in the literature to correlate f and β with a variety of soil characteristics, but these efforts have not produced generally useful guidelines because they attempt to generalize too many separate variables into "lumped-together" parameters. Instead, the general form of Equation 14.21 should be used with the key variables δ and K. Use of this equation also allows for a better understanding of the controlling design parameters.

The interface friction angle, δ, controls the available shearing resistance of the soil–shaft interface. Kulhawy et al. (1983) summarized available data and showed that δ can be expressed as a function of $\bar{\phi}$, and that $\delta/\bar{\phi}$ can vary from 1.0 to less than 0.5 as a function of the roughness of the soil–structural material interface. For cast-in-place concrete and good construction techniques, a rough interface develops, giving $\delta/\bar{\phi} = 1$. However, poor slurry construction techniques that leave a thick side slurry cake could reduce this value to 0.8 or lower. This possibility must be considered in design, although it should not develop in well-constructed foundations. The use of permanent casing also will result in a smoother interface, giving $\delta/\bar{\phi}$ on the order of 0.7.

Perhaps the most difficult and important term to evaluate is K, the coefficient of horizontal soil stress. This term is a function of the original in-situ horizontal stress coefficient, K_0, and the stress changes caused by construction, loading, and time. Taking these factors into account may result in K ranging from less than to greater than K_0. Values of K from analysis of field tests range from about 0.1 to over 5. These bounds correspond roughly to the range from minimum active to maximum passive stress states.

To illustrate the importance of K_0, consider the idealized uniform soil profile in Figure 14.7, which shows an overconsolidated "crust" overlying normally consolidated soil. For simplicity in this example, assume $K = K_0$ and $\delta = \bar{\phi}$, with $\bar{\phi}$ being constant with depth. Assume also that the water table is at ground surface, the submerged unit weight = 50 pcf (7.9 kN/m³), and $\bar{\phi} = 37°$. If the deposit was considered to be normally consolidated (NC), the overconsolidation ratio (OCR), K_0, and β would be constant with depth as shown, and the unit side resistance would increase linearly with depth. However, when the preconsolidation is taken into account, the OCR ($= \bar{\sigma}_p / \bar{\sigma}_v$), K_0 (described below), and β all vary as shown, resulting in a nonlinear variation of f with depth within the overconsolidated crust.

Of particular interest is the shape of the unit side resistance curve within the overconsolidated crust. The value of f increases quickly with depth, and then it appears to change little with depth within the top 50 ft (15 m). This shape might suggest that a "limiting" side resistance has been reached, but this is not the case. The shape is only a coincidence that results from the decreasing K_0 and increasing $\bar{\gamma} z$ tending to cancel each other

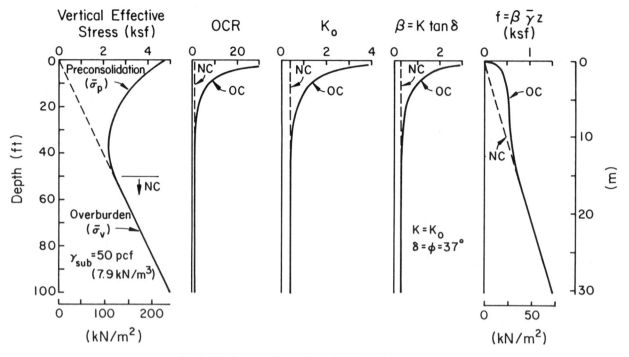

Fig. 14.7 Example illustrating principles of side resistance.

over a particular depth range. Once below the overconsolidated crust, K_0 is a constant, and therefore f increases linearly with depth.

Table 14.3 illustrates the importance of including this stress history effect. In this table, the side resistance was computed for three depths, assuming both the normally consolidated and overconsolidated unit side resistance variations shown in Figure 14.7. The ratio of the computed side resistances shows rather dramatically that, for shorter-shafts that are well within the overconsolidated crust, the capacity is increased greatly by incorporating the stress history effect (K_0) in the soil. However, as the depth increases and penetration into the underlying normally consolidated soil become significant, the stress history effect becomes a minor influence in capacity evaluation. The main lesson from this example is that the soil stress history must be evaluated carefully when a shaft is to be placed in moderately to heavily overconsolidated soil. If the actual high K_0 is disregarded, a very conservative prediction will be made for the side resistance.

Unfortunately, K_0 is not a parameter that traditionally has been determined during site investigations, although it is very important in many geotechnical problems (see, e.g., Schmertmann, 1985). However, a simple perusal of the recent geotechnical literature shows very clearly that increasing attention is being paid to the evaluation of K_0. At the present time, K_0 can be estimated by three procedures. The first

is by interpretation of direct measurements in situ using the pressuremeter, dilatometer, or other in-situ devices. While some problems may exist in test conduct and interpretation, these devices can be used to provide a field estimate of the in-situ K_0.

The second approach is to estimate K_0 from the stress history of the soil deposit. This process involves determining the OCR from direct measurements, such as from consolidation tests, or by reconstructing the geologic history of the soil deposit. While geologic reconstruction may, at first, seem to be a formidable task to the average engineer, it is not to a skilled engineering geologist or geomorphologist, particularly since only the limiting events are of concern. Examples of immediate interest include the maximum height of ice cover in glaciated regions, the maximum extent of denudation or erosion, maximum and minimum water levels and sea levels, and other significant geologic events. This type of information is accessible in the geologic literature and can be used effectively in estimating the range of OCR for the soil deposit. Recent events also must be considered, such as previous structure or fill loading of the site and groundwater fluctuations.

Particular concern should be paid to capillary effects and desiccation (e.g., Kulhawy and Beech, 1987), as well as to other phenomena that may influence the groundwater level, ranging from desiccation beneath boilers to tree roots. For example, Blight (1987) recently showed that clear-cutting a wooded

TABLE 14.3 COMPUTED SIDE RESISTANCE (Q_{sc}) FOR SOIL PROFILE IN FIGURE 14.7.

Shaft Depth		$Q_{sc}(NC)$		$Q_{sc}(OC)$		$Q_{sc}(OC)$
ft	m	kip	kN	kip	kN	$Q_{sc}(NC)$
25	7.6	44	196	107	476	2.43
50	15.2	177	787	258	1148	1.46
100	30.5	707	3145	792	3523	1.12

Note: shaft diameter = 3 ft (0.9 m).

region prior to construction allowed the groundwater to rise some 62 ft (19 m). Changes in this magnitude alter the effective stresses and K_0 significantly.

Once the maximum and current OCR values have been determined, the value of K_0 can be computed (Mayne and Kulhawy, 1982) from

$$K_0 = \frac{\bar{\sigma}_h}{\bar{\sigma}_v} = (1 - \sin \bar{\phi}) \left[\frac{OCR}{OCR_{max}^{(1 - \sin \bar{\phi})}} + \frac{3}{4} \left(1 - \frac{OCR}{OCR_{max}} \right) \right] \tag{14.24}$$

Figure 14.8 illustrates the range of applicability of this equation. Stress path O–A–B describes virgin loading (OCR = OCR_{max} = 1), stress path B–C–D describes primary unloading (OCR = OCR_{max}), and stress path D–E describes primary reloading (OCR_{max} = OCR at point D, and OCR = OCR at point E). The dotted stress path loop above point E indicates typical annual groundwater level fluctuations. Note that points A, C, and E all have the same vertical effective stress but significantly different horizontal stresses and K_0 values.

In geologic settings where it is difficult to evaluate OCR_{max}, it may be necessary to assume that OCR_{max} = OCR, resulting in

$$K_0 = (1 - \sin \bar{\phi}) OCR^{\sin \bar{\phi}} \tag{14.25}$$

With this simplification, the computed K_0 is likely to be on the high side. Judgment then must be exercised in selecting an appropriate design value, considering the overall behavior shown in Figure 14.8.

Empirical methods represent the third approach to evaluating K_0 from the results of both standard penetration and cone penetration tests (Kulhawy and Mayne, 1990). However, as with all empirical correlations, caution and discretion must be exercised in their use. Space limitations preclude a detailed discussion of these methods.

After the variation of the in-situ K_0 with depth is known, the effects of construction, loading, and time can be introduced. Based upon examination of several hundred load tests, Kulhawy et al. (1983) found that K/K_0 normally varies between $\frac{2}{3}$ and 1 and is a function of the construction method and its influence on the in-situ stress. For dry construction, minimal sidewall disturbance, and prompt concreting, the soil disturbance is minimized and K/K_0 approaches 1. For slurry or wet-hole construction, using proper construction techniques, the soil disturbance also is minimized and K/K_0 approaches 1. However, when slurry or wet-hole procedures are not applied properly, the soil stresses may relax significantly, and therefore K/K_0 may reduce to $\frac{2}{3}$. Casing construction under water typically represents an intermediate case. Obviously, when caving and other extreme ground-loss phenomena occur, the soil stresses may be reduced dramatically, to minimum active stresses (K_a) or lower if voids occur. Reese et al. (1981) presented case histories that illustrated this problem rather dramatically.

Careful attention to the construction details is very important to ensure a well-built foundation.

Loading and time also can influence the in-situ K_0. If the shaft excavation is not concreted promptly, soil stress relaxation can occur and therefore reduce the shaft side resistance. To avoid these problems, the shaft should be concreted as soon as possible, preferably within an hour or two of completion. Concreting the following day or later should not be allowed.

From a loading standpoint, much depends on the ratio of the transient live loading to the dead plus sustained live loading. If this ratio is small, as is typical in most building and bridge-type structures, the loading influence is minimal. However, when this ratio is large, as with transmission line, antenna-type, and other similar classes of structures, stress degradation may occur. Research studies on large-scale laboratory models (Turner and Kulhawy, 1987, 1990) have shown that when the peak-to-peak cyclic displacement in two-way repeated loading exceeds about 3 mm (0.1 in), significant capacity reductions can occur, ranging from $\frac{1}{3}$ at low relative densities to as much as 80 percent at high relative densities. These reductions must be evaluated carefully, and conservatism is warranted in these cases. The limited available data on this subject should be reviewed carefully when assessing this loading mode (Turner et al., 1987).

Taking the above into account yields the following general equation for evaluating the side resistance in drained loading:

$$Q_{sc} = \pi B \frac{K}{K_0} \int_0^D \bar{\gamma} z \, K_0(z) \tan [\bar{\phi}(\delta/\bar{\phi})] \, dz \tag{14.26}$$

which can be expressed in incremental layer form as

$$Q_{sc} = \pi B \frac{K}{K_0} \sum_{i=1}^N (\bar{\gamma} z)_i K_{0_i} \tan [\bar{\phi}_i(\delta/\bar{\phi})] \, t_i \tag{14.27}$$

in which t_i = thickness of layer i and all other parameters are evaluated at the layer mid-height. Equation 14.27 is particularly useful for capacity estimates in stratified or highly variable soil profiles.

14.4.3 Friction Angle Evaluation

Although it is very well established in the literature (e.g., Bishop, 1966; Bolton, 1986) that the failure envelope for granular soils commonly is nonlinear, particularly at higher relative densities, this nonlinearity often is ignored in conventional foundation engineering practice. However, disregard of this influence can lead to significant differences in both the tip and side resistances.

Consider the laboratory shear test data shown in Figure 14.9, given by the three strength measurements at three different

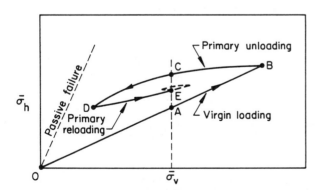

Fig. 14.8 Stress paths for simple stress histories.

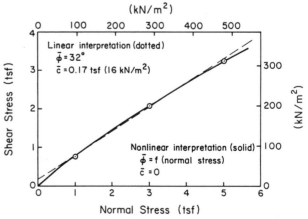

Fig. 14.9 Laboratory shear test data and interpretations.

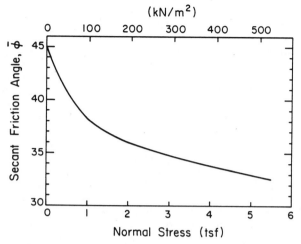

Fig. 14.10 Secant friction angle versus normal stress.

normal stresses. A "traditional" approach would be to fit a straight line to these three points, resulting in $\bar{\phi} = 32°$ and $\bar{c} = 0.17\,\text{tsf}$ ($16\,\text{kN/m}^2$). However, the true failure envelope is given by the curved solid line shown. By taking successive secants from the origin to the curved failure envelope, the values of $\bar{\phi}$ can be evaluated as a function of the normal stresses, as given in Figure 14.10. This latter figure should be used for both tip and side resistance calculations. As noted by Bolton (1986), secant angles "are required in a rational approach" to the strength of coarse-grained soils.

To illustrate the importance of these differences, sample calculations were done for a drilled shaft installed in the soil profile shown in Figure 14.7, with the strength parameters as given in Figures 14.9 and 14.10. The tip resistance is evaluated below for a 3-ft (0.9-m) diameter shaft that is 25 ft (7.6 m)

deep. The water table is assumed at the ground surface, and the submerged unit weight is 50 pcf ($7.9\,\text{kN/m}^3$). Assuming general shear only ($\zeta_{cr} = \zeta_{\gamma r} = \zeta_{qr} = 1$) for simplicity, the bearing capacity is given by

$$q_{\text{ult}} = \bar{c}\,N_c\zeta_{cs}\zeta_{cd} + 0.3\,B\,\bar{\gamma}\,N_\gamma + \bar{q}\,N_q\zeta_{qs}\zeta_{qd}$$

The tip resistance then is computed for the two evaluation methods.

a. Linear failure envelope with $\bar{c} = 340\,\text{psf}$ ($16\,\text{kN/m}^2$) and $\bar{\phi} = 32°$.

$$N_c = 35.49, \qquad N_\gamma = 30.22, \qquad N_q = 23.18$$

$$\zeta_{cs} = 1.653, \qquad \zeta_{qs} = 1.625, \qquad \zeta_{qd} = 1.401, \qquad \zeta_{cd} = 1.418$$

$$q_{\text{ult}} = (340)(35.49)(1.653)(1.418) + 0.3(3)(50)(30.22)$$
$$+ (25 \times 50)(23.18)(1.625)(1.401)$$
$$= 28\,234 + 1360 + 65\,965 = 95\,559\,\text{psf}$$
$$= 47.8\,\text{tsf} \quad (4.58\,\text{MN/m}^2)$$

$$Q_{tc} = (47.8)(\pi)(3 \times 3)/4 = 338\,\text{tons} \quad (3.01\,\text{MN})$$

b. Nonlinear failure envelope with $\bar{c} = 0$ and $\bar{\phi}$ from Figure 14.10 at a stress corresponding to a depth of $D + B/2$ (i.e., $\bar{\sigma}_v = 25 \times 50 + 0.5 \times 3 \times 50 = 1325\,\text{psf}$).

$$\bar{\phi} = 39.7°, \qquad N_\gamma = 103.9, \qquad N_q = 61.58$$

$$\zeta_{qs} = 1.830, \qquad \zeta_{qd} = 1.314$$

$$q_{\text{ult}} = 0.3(3)(50)(103.9) + (25 \times 50)(61.58)(1.830)(1.314)$$
$$= 4676 + 185\,096 = 189\,772\,\text{psf}$$
$$= 94.9\,\text{tsf} \quad (9.09\,\text{MN/m}^2)$$

$$Q_{tc} = 94.9(\pi)(3 \times 3)/4 = 671\,\text{tons} \quad (5.97\,\text{MN})$$

Table 14.4 outlines the side resistance calculations in incremental form (Eq. 14.27), assuming K_0 for simple unloading

TABLE 14.4 EXAMPLE SIDE RESISTANCE EVALUATION.

a. Linear failure envelope with $\bar{c} = 340\,\text{psf}$ ($16\,\text{kN/m}^2$) and $\bar{\phi} = 32°$

	$\bar{\phi}$, deg	K_0	$\bar{c}$, psf	Layer Q_{sc} (tons)		
				Friction	Cohesion	Combined
Same	32.0	4.12	340	2.7	4.8	7.5
layering,	32.0	2.17	340	4.3	4.8	9.1
$\bar{\sigma}_v$, and	32.0	1.53	340	7.2	6.4	13.6
OCR	32.0	1.13	340	10.4	8.0	18.4
	32.0	0.89	340	11.5	8.0	19.5
	32.0	0.73	340	12.1	8.0	20.1

$$Q_{sc} = 88.2\,\text{tons}$$
$$(785\,\text{kN})$$

b. Nonlinear failure envelope with $\bar{c} = 0$ and $\bar{\phi}$ from Figure 14.10 as a function of $\bar{\sigma}_v$

Layer Depth, ft	Layer Mid-depth, ft	$\bar{\sigma}_v$, psf	OCR	$\bar{\phi}$, deg	K_0	Layer Height, ft	Layer Q_{sc}, tons
0–3	1.5	75	60.0	44.7	5.28	3	5.5
3–6	4.5	225	18.0	43.7	2.28	3	6.9
6–10	8.0	400	9.25	42.9	1.45	4	10.2
10–15	12.5	625	5.20	42.1	1.00	5	13.3
15–20	17.5	875	3.31	41.3	0.75	5	13.6
20–25	22.5	1125	2.27	40.3	0.60	5	13.5

$$Q_{sc} = 63.0\,\text{tons}$$
$$(560\,\text{kN})$$

1 ft = 0.305 m; 1 psf = 47.9 N/m²; 1 ton = 8.9 kN.

TABLE 14.5 EXAMPLE CAPACITIES EVALUATED USING LINEAR AND NONLINEAR FAILURE ENVELOPES.

	Linear Interpretation of Figure 14.9	Nonlinear Interpretation of Figure 14.9	Linear / Nonlinear
Q_{tc}	338 tons (3.01 MN)	671 tons (5.97 MN)	0.50
Q_{sc}	88.2 tons (785 kN)	63.0 tons (560 kN)	1.40

(OCR = OCR$_{max}$). The "cohesion" component of Q_{sc} was computed from $(\pi B t_i \bar{c})$, with t_i as the layer height.

Table 14.5 summarizes the results of these calculations and shows rather dramatic differences for this example. The linear interpretation underestimates the tip resistance by a factor of 2, while overestimating the side resistance by a factor of 1.4. Although these differences will vary with the problem geometry, soil stress history and mineralogy, and the shape of the failure envelope, they may be large and unconservative (side resistance in this case). For uplift loading in this example, where the side resistance dominates, this unconservatism could lead to problems.

14.4.4 Foundation Weight

For drained loading conditions, the effective weight of the foundation is given by

$$W = \frac{\pi B^2}{4} [\gamma_c D_w + (\gamma_c - \gamma_w)(D - D_w)] \qquad (14.28)$$

in which γ_c = unit weight of shaft concrete, γ_w = unit weight of water, and D_w = depth to water table.

14.4.5 Extension to True $\bar{c}$, $\bar{\phi}$ Soils

Some natural soil deposits may have a true effective stress cohesion ($\bar{c}$), as well as an effective stress friction angle ($\bar{\phi}$). Among these deposits are cemented sands and gravels, partially saturated soils, and heavily overconsolidated clays. If a geotechnical specialist has determined that a particular soil actually has a true cohesion, then the cohesion term is added to the bearing capacity and rigidity index equations, and a cohesion term is added to the side resistance equation. The procedures for doing this were illustrated for the $\bar{c}$, $\bar{\phi}$ soil in Section 14.4.3 and Table 14.5.

When considering use of the effective stress cohesion, several precautions should be taken. With cemented soils, consideration must be given to whether the construction process or the design stresses will alter or destroy the cementation. If so, the cohesion could be lost permanently. Partially saturated soils normally derive their cohesion from capillary tension in the soil pores. If the water table rises, this tension decreases, eventually to zero when the soil becomes saturated. Careful consideration must be given to the maximum water table during the structure design life when analyzing foundations in partially saturated soils. Finally, the effective stress cohesion in heavily overconsolidated clays decays with time, particularly if the clays are fissured. This process has been described well by Skempton (1964) and others. This decay of cohesion with time must be considered carefully in the context of the structure design life. For all of these soils, the assistance of a geotechnical specialist is warranted.

14.5 UNDRAINED AXIAL COMPRESSION CAPACITY

For undrained conditions, which develop when loads are applied relatively rapidly to saturated fine-grained soils, pore water stresses develop in the soil at constant effective stress and lead to the analysis procedure known commonly as the total stress or $\phi = 0$ method. With this method, the soil strength is described exclusively by c or, more appropriately, s_u, which is defined as the undrained shearing resistance or undrained shear strength. For undrained conditions, the axial capacity can be evaluated as follows.

14.5.1 Undrained Bearing Capacity

For $\phi = 0$ conditions, the bearing capacity factors are defined uniquely as $N_c = 5.14$, $N_\gamma = 0$, and $N_q = 1$, and therefore Equation 14.9 reduces to

$$q_{ult} = 5.14 s_u \zeta_{cs} \zeta_{cd} \zeta_{cr} + q \zeta_{qs} \zeta_{qd} \zeta_{qr} \qquad (14.29)$$

The ζ modifiers are given in Table 14.6, and the introduction of the uniquely defined values into Equation 14.29 results in

$$q_{ult} = 6.17 s_u \zeta_{cd} \zeta_{cr} + q \qquad (14.30)$$

In this equation, q is the soil total vertical stress at the shaft tip ($= \gamma D$), in which γ = soil total unit weight. The value of ζ_{cd} given in Table 14.6 increases rapidly with depth, in much the same way as ζ_{qd} discussed earlier (Fig. 14.4). At $D/B = 5$, ζ_{cd} is 96 percent of the value for $D/B = \infty$.

The rigidity modifier (ζ_{cr}) is again evaluated as an average value from depth D to depth $D + B$, and therefore it is computed for the conditions at $D + B/2$. However, the rigidity index is less complex for undrained loading. In this case, the rigidity index (I_r) is given by

$$I_r = \frac{G_u}{s_u} = \frac{E_u}{2(1 + v_u)s_u} = \frac{E_u}{3 s_u} = I_{rr} \qquad (14.31)$$

in which G_u = undrained shear modulus, E_u = undrained Young's modulus, and v_u = undrained Poisson's ratio, which

TABLE 14.6 UNDRAINED ($\phi = 0$) BEARING CAPACITY MODIFIERS FOR CIRCULAR FOUNDATION.

Modification	Symbol	Value
Shape	ζ_{cs}	1.2
	ζ_{qs}	1
Depth	ζ_{cd}	$1 + 0.33 [(\pi/180) \tan^{-1}(D/B)]$
	ζ_{qd}	1
Rigidity	ζ_{cr}	$0.44 + 0.60 \log_{10} I_{rr} \quad \leqslant 1$
	ζ_{qr}	1

is equal to 0.5 for saturated cohesive soil in undrained loading. Since $v_u = 0.5$, no volumetric strains occur, and therefore the reduced rigidity index (I_{rr}) is equal to I_r.

The undrained shear strength (s_u) is a function of the soil effective stress strength parameters ($\bar{\phi}$), the developed pore water stress during loading (Δu), and the in-situ state of stress in the soil. Accordingly, there is likely to be more inherent variability in s_u than in $\bar{\phi}$. In addition, the measured value of s_u is highly dependent on the test type and boundary conditions employed. Details of the different test types and their inter-relationships are given by Wroth (1984), Jamiolkowski et al. (1985), and Kulhawy and Mayne (1990). As a first-order approximation, Jamiolkowski et al. (1985) suggested the following for evaluating s_u:

$$\frac{s_u}{\bar{\sigma}_v} \approx (0.23 \pm 0.04)\, OCR^{0.8} \qquad (14.32)$$

in which $\bar{\sigma}_v$ = vertical effective stress. Since s_u varies with stress and OCR, both of which vary with depth, the variation of s_u with depth should be made for the highest groundwater conditions likely.

Young's modulus again can be evaluated from a number of different laboratory shear tests or in-situ field tests. Typical ranges of this parameter within working stress ranges are given in Table 14.7 and Figure 14.11. Other correlations of this parameter are given in the references cited earlier in Section 14.4.1, as well as by Wroth (1984) and Kulhawy and Mayne (1990).

After the soil rigidity index has been computed, it is compared to the theoretically based critical rigidity index, I_{rc} (Vesić, 1975). For a circular foundation, $I_{rc} = 8.64$. If $I_{rr} > I_{rc}$,

the soil behaves as a rigid-plastic material, general shear failure will occur, and $\zeta_{cr} = 1$. If $I_{rr} < I_{rc}$, the soil stiffness is low, local or punching shear will occur, and ζ_{cr} will be less than 1 and therefore must be computed. Figure 14.12 illustrates how the undrained bearing capacity varies with the depth-to-diameter ratio and rigidity index. The depth influence becomes very small after only a modest foundation embedment.

The rather low value of the critical rigidity index, and the relatively high values of E_u and E_u/s_u given in Table 14.7 and Figure 14.11, suggest that local or punching shear would be unlikely for drilled shafts in undrained loading. Callanan and Kulhawy (1985) analyzed the results from 18 compression and 38 uplift load tests, using the Mattes and Poulos (1969) elastic model, to back-calculate the "average" E_u corresponding to the load and displacement at one-half the ultimate load in the load test. The results of this study are given in Figure 14.13, which shows that 54 of the 56 load tests had an "average" E_u/s_u of 200 or more, corresponding to $I_{rr} = 67$. The other two cases gave E_u/s_u values of 30 and 150, corresponding to $I_{rr} = 10$ and 50. Higher stress levels would give lower I_{rr} values. These data show that general shear would be the norm, and only occasionally would local or punching shear develop.

In addition, the issues of soil structure "collapse," stress relaxation, and excavation cleanout must be addressed in a similar manner as described previously in Section 14.4.1. The only significant difference is that the depth of stress relaxation influence can be estimated better by the stability number (Peck et al., 1974), N_s, given below for $\phi = 0$ conditions:

$$N_s = 3.85 = \frac{\gamma D_c}{s_u} \qquad (14.33)$$

If the critical depth, D_c, approaches or exceeds $3.85\, s_u/\gamma$, soil plastic yield may develop at the bottom of the shaft excavation. In the extreme, s_u may be reduced to its fully remolded value, while E_u will be reduced by at least the same amount. Incomplete cleanout will complicate this issue even further. This approach neglects arching and geometric constraints, and therefore represents a lower bound on the critical depth.

TABLE 14.7 TYPICAL RANGES IN UNDRAINED YOUNG'S MODULUS.

Clay Consistency	Undrained Young's Modulus, E_u	
	tons/ft²	MN/m²
Soft	25–150	2.5–15
Medium to stiff	150–500	15–50
Very stiff to hard	500–2000	50–200

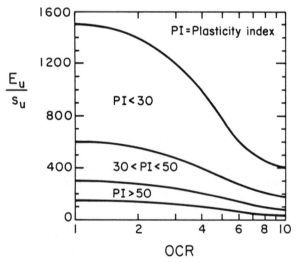

Fig. 14.11 Normalized undrained modulus versus OCR. (*Based on Duncan and Buchignani, 1976.*)

$$N_c\, \zeta_{cs}\, \zeta_{cd}\, \zeta_{cr} = \frac{q_{ult} - q}{s_u}$$

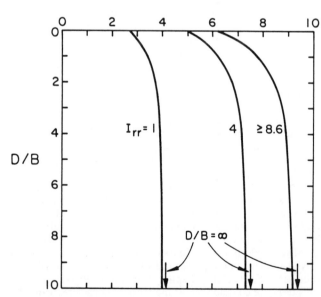

Fig. 14.12 Variation of undrained bearing capacity with normalized depth and rigidity index.

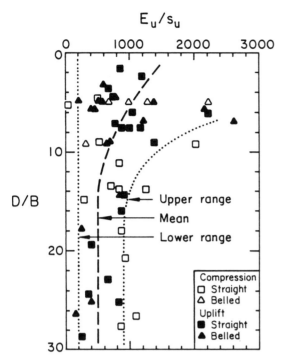

Fig. 14.13 Normalized undrained modulus versus depth for drilled shafts, computed at one-half the ultimate load. (*Callanan and Kulhawy, 1985, p. 3-28.*)

14.5.2 Undrained Side Resistance

As given by Equation 14.3, the undrained side resistance is simply a summation of the available soil shearing resistance over the side area of the shaft. In principle, either total-stress ($\phi = 0$) or effective-stress methods can be used for undrained analysis. However, to use the effective-stress methods given in Section 14.4.2, the excess pore water stress (Δu) developed along the shear surface also must be evaluated. This evaluation is difficult at present, and consequently effective-stress methods have not yet evolved for conventional design practice. However, research is progressing in this direction (e.g., Beech and Kulhawy, 1987), and it is likely that general effective-stress methods will become available for design in the future.

For total stress analysis, the procedures are deceptively simple, with the following design equation:

$$Q_{sc} = \pi B \alpha \int_0^D s_u(z)\,dz \qquad (14.34)$$

in which s_u = undrained shear strength and α = empirical "adhesion" factor. In incremental form, this equation becomes

$$Q_{sc} = \pi B \alpha \sum_{i=1}^{N} s_{ui} t_i \qquad (14.35)$$

in which t_i = thickness of layer i and s_u is evaluated at the layer mid-height. These equations are analogous forms of Equations 14.26 and 14.27. Therefore, α and s_u collectively represent K/K_0, $\bar{\gamma}$, z, K_0, $\bar{\phi}$, $\delta/\bar{\phi}$, and Δu. This parametric "lumping" inevitably leads to more uncertainty for a total stress analysis, unless local calibrations are available in the form of instrumented load tests of comparable geometry in the same soil types.

The α concept was introduced by Tomlinson (1957) as a way to examine the average percentage of the average s_u over the foundation depth that was mobilized during a load test. Subsequent work on α for driven piles (e.g., Randolph and Wroth, 1982; Semple and Rigden, 1984) showed that α can be

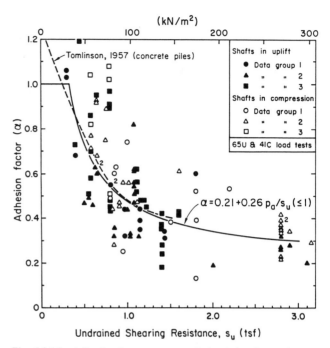

Fig. 14.14 Adhesion factor versus undrained shearing resistance for drilled shafts. (*Stas and Kulhawy, 1984, p. 7-9.*)

correlated to the soil stress history and driven pile geometry. Comparable work on α for drilled shafts has not been done until recently because of an inadequate database on well-documented field load tests. Stas and Kulhawy (1984) evaluated an existing database on 106 drilled shafts and developed the α correlation shown in Figure 14.14. Included in this figure are both compression and uplift tests on drilled shafts with diameters ranging from 0.5 to 5.9 ft (0.15 to 1.80 m) and depths ranging from 4.0 to 66.7 ft (1.22 to 20.3 m). Group 1 tests include good documentation, while Group 3 tests are marginally documented. No particular trends were evident with stress history, shaft geometry, or other variables.

Figure 14.14 shows a significant data scatter, as would be expected when all geotechnical and construction parameters are defined within two terms. However, the values show a consistent trend, and the mean α is given by

$$\alpha = 0.21 + 0.26\left(\frac{p_a}{s_u}\right) \qquad \leqslant 1.0 \qquad (14.36)$$

in which p_a = atmospheric pressure in the same units as s_u. When used for design, these α values must include a sufficient factor of safety that represents the uncertainty in both α and s_u.

14.5.3 Foundation Weight

For undrained loading conditions, the total weight of the foundation is given by

$$W = \frac{\pi B^2}{4} \gamma_c D \qquad (14.37)$$

in which γ_c = unit weight of shaft concrete. However, if the site is under water, the buoyant weight should be used for that portion of the shaft above the soil surface, as given below:

$$W = \frac{\pi B^2}{4} \left[(\gamma_c - \gamma_w) D_w + \gamma_c (D - D_w) \right] \qquad (14.38)$$

in which D = total depth of shaft, D_w = depth of water to top of soil, and γ_w = unit weight of water.

14.6 AXIAL UPLIFT CAPACITY

When a drilled shaft is loaded in uplift, the same general shape of load–displacement curve develops as for compression (Figure 14.1). However, the force equilibrium diagram is different, as shown in Figure 14.15. The notation and forces are given in Figure 14.15a, while Figure 14.15b shows an illustrative distribution of available soil shearing resistance (τ) that can act along the side of the shaft. The load transfer to the soil is shown in Figure 14.15c.

From this equilibrium diagram, the uplift capacity (Q_u) of a drilled shaft is given by

$$Q_u = Q_{su} + Q_{tu} + W \qquad (14.39)$$

in which Q_{su} = side resistance in uplift, Q_{tu} = tip resistance in uplift, and W = foundation weight. The weight term is evaluated in the same way as described previously for compression loading.

14.6.1 Side Resistance

For side resistance, there has been speculation in the literature that the side resistance in uplift would be less than that in compression. However, examination of possible Poisson effects for shafts in soil has shown that these effects are negligible (Kulhawy et al., 1983). Similarly, examination of available load test data has not shown any discernible difference (Stas and Kulhawy, 1984). Some of these data are given in Figures 14.13 and 14.14, which show that the uplift and compression data are uniformly distributed and that the same conclusions would be reached for either data set.

There is one important difference that does occur between shafts loaded in compression or uplift. While both cases develop their side resistance in a cylindrical shear mode, in uplift a cone of soil may develop occasionally that effectively acts as a part of the foundation. This cone is illustrated in Figure 14.15a by a dashed line and occurs because of the failure kinematics of short shafts in soil with high K_0 (Stewart and Kulhawy, 1981; Kulhawy, 1985). Figure 14.16 shows the initial data exhibiting cone breakout, with the numbers at each data point representing the z/D ratio for the normalized cone depth. Subsequent data from Tucker (1987) substantiate the tentative limits given in the figure.

From these data, a reduction of the computed Q_{su} from a cylindrical shear mode is appropriate when D/B is less than 6

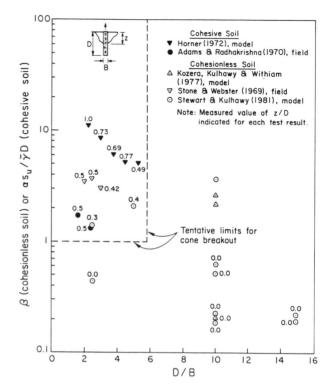

Fig. 14.16 Cone breakout conditions for drilled shafts in uplift. (*Trautmann and Kulhawy, 1987, p. 2-6.*)

and either β or $\alpha s_u/\bar{\gamma}D$ is greater than 1. A suggested relationship to make this reduction is given below (Stas and Kulhawy, 1984):

$$Q_{sur} = \left(\frac{2+\chi}{3\chi}\right)Q_{su} \qquad (14.40)$$

in which Q_{sur} = reduced side resistance in uplift and χ = average value over the full shaft depth of β for drained loading or $\alpha s_u/\bar{\gamma}D$ for undrained loading.

14.6.2 Tip Resistance

The tip resistance in uplift is dramatically different from that in compression. In compression, a rather large resistance can develop from bearing capacity. In uplift, a much smaller resistance can develop from tension and suction. These terms commonly are disregarded, although they can be of importance in some cases.

Tip tension results when good bonding develops between the shaft concrete and the soil at the shaft tip. The resulting tip resistance is given as

$$Q_{tu} = A_{tip}\,s_t \qquad (14.41)$$

in which A_{tip} = tip area and s_t = tip tension. The value of s_t is the minimum tensile strength of the soil or the concrete. However, the tensile strength of soil commonly is low, and normal construction practices usually result in a thin zone of very low-strength soil at the tip. These two points lead to the prudent conclusion of assuming zero tip tension in most cases. However, when very careful construction practices are employed, and when the soil at the tip has significant tensile strength, tip tension may be considered.

Tip suction also can develop in saturated fine-grained soils below the water table during undrained loading. The suction

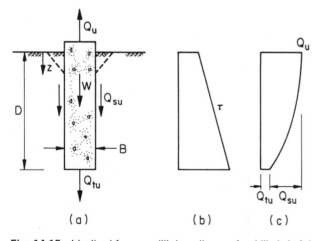

Fig. 14.15 Idealized force equilibrium diagram for drilled shaft in uplift.

stress (s_s) occurs because of a reduction in applied stress at the tip and can be estimated from

$$Q_{tu} = A_{tip} s_s \qquad (14.42)$$

Load tests in the literature have shown that suction stresses may be as large as 1 atmosphere. To estimate this stress, an approximate expression has been developed (Stas and Kulhawy, 1984):

$$s_s = \frac{W}{A_{tip}} - u_i \qquad (14.43)$$

in which u_i = initial pore water stress before loading, given by

$$u_i = \gamma_w (D - h) \qquad (14.44)$$

with h = depth to water table. The suction stress is limited to 1 atmosphere.

Suction develops only in fine-grained soils during undrained loading and dissipates rapidly with time, measured in days to weeks. Therefore, it should only be considered for transient live loading design cases.

14.7 BELLED SHAFTS

Drilled shafts also may be constructed by underreaming the shaft to construct a bell, as shown in Figure 14.17. Bells with 60° angles are most common, but 45° bells also are feasible (Sheikh et al., 1983). Although this type of shaft may appear to have decided advantages over a straight shaft because of the enlarged tip area, it is being used less frequently. Instead, it is often more cost-effective to install a deeper straight shaft, which will minimize any potential cleanout problems. However, there still are many geologic situations where a belled shaft is appropriate.

For the design of a belled shaft, all of the design equations presented previously are still valid. However, some modifications are necessary because of the different geometry. For all tip resistance calculations, the bell diameter (B_b) should be used. For side resistance in compression, the shaft diameter (B_s) should be used and the height of the bell should be disregarded because downward movement can separate the bell from the soil. For weight calculations, the additional bell concrete should be added.

For the side resistance in uplift, no general theory yet exists to evaluate the capacity increase from the bell. However, O'Neill (1987a) has suggested a tentative model, and Kulhawy (1985) has suggested simple empirical guidelines to approximate this behavior. The empirical guidelines are given below. For shafts with D/B_s greater than 10, there is little apparent increase in

uplift side resistance because of the bell, and therefore the design B can be approximated with B_s. For shafts with D/B_s less than 5, the bell influence is significant. A convenient assumption is to use a design B equal to $[B_s + (B_b - B_s)/3]$. For D/B_s from 5 to 10, a linear interpolation between the above limits can be used. Load test analyses by Kulhawy (1985) and Stas and Kulhawy (1984) suggest that these approximations provide a reasonable interim design approach until a general theory is developed.

14.8 COMPRESSION AND UPLIFT DISPLACEMENTS

14.8.1 Compression Displacement

Several methods have been proposed to compute the compression displacement (settlement) of drilled shafts. Many of these have been reviewed by Poulos and Davis (1980), who also recommend a detailed procedure based on elastic theory. This procedure is too lengthy to review here, although it includes many of the significant variables and may be the most comprehensive to date. However, it does not include construction effects, and there are some difficulties in determining the appropriate elastic parameters for input. Alternatively, Vesić (1977) suggested a simple, approximate solution that is based upon elastic theory and empirical correlations with the soil elastic parameters and limited field test data. This simplified method is described below.

The total settlement, ρ_t, of a drilled shaft may be approximated by

$$\rho_t = \rho_f + \rho_{tt} + \rho_{ts} \qquad (14.45)$$

in which ρ_f = settlement from axial deformation of the shaft concrete, ρ_{tt} = settlement of the tip from the tip load, and ρ_{ts} = settlement of the tip from the shaft side load. Both ρ_{tt} and ρ_{ts} include elastic, consolidation, and secondary settlement components. The elastic settlement normally constitutes at least 80 percent of the total (Poulos and Davis, 1980; Vesić, 1977), and secondary effects are usually minimal. However, if compressible soil underlies the foundation, the situation is different. For this case, a conventional settlement analysis is necessary to evaluate the settlement of the underlying stratum caused by the entire drilled shaft foundation (Peck et al., 1974).

The value of ρ_f is determined from conventional formulas for the deformation of a structural material and is given by

$$\rho_f = (Q_t + \alpha_s Q_s) \frac{D}{A E_c} \qquad (14.46)$$

in which Q_t = load transmitted at the foundation tip, Q_s = load transmitted along the shaft side, D = depth, A = cross-sectional area, E_c = concrete Young's modulus, and α_s = coefficient that depends on the distribution of side resistance. For a linearly increasing side resistance, as illustrated in Figure 14.2, $\alpha_s = \frac{2}{3}$. Other possible distributions are described by Vesić (1977). For the distribution of Q_t and Q_s as a function of the total applied load, Q_c, the elastic load transfer analysis of Poulos and Davis (1980), given in Figure 14.18, provides a reasonable estimate for the working stress range. As can be seen, Q_t approaches 30 percent of Q_c for small D/B ratios, but it decreases rapidly as D/B increases. The foundation stiffness factor, K_{FS}, is defined as

$$K_{FS} = \frac{E_c}{E_s} \qquad (14.47)$$

in which E_s = soil elastic modulus (E_u or E_d as given previously).

The tip settlement equations are developed from elastic theory (Poulos and Davis, 1974, 1980). The simplified Vesić

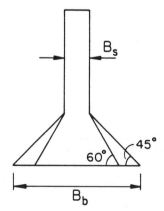

Fig. 14.17 Geometry of belled shaft.

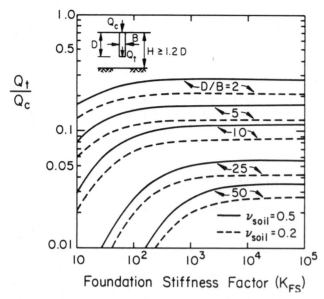

Fig. 14.18 Elastic load transfer to shaft tip. (*Based on Poulos and Davis, 1980, pp. 85, 86.*)

TABLE 14.8 TYPICAL VALUES OF C_t.

Soil Type	C_t for Drilled Shafts
Sand (dense to loose)	0.09–0.18
Clay (stiff to soft)	0.03–0.06
Silt (dense to loose)	0.09–0.12

Source: Excerpted from Vesić (1977), p. 33.

solutions employ elastic theory and empirical correlations, resulting in

$$\rho_{tt} = \frac{C_t Q_t}{B\, q_{ult}} \qquad (14.48)$$

$$\rho_{ts} = \frac{C_s Q_s}{D\, q_{ult}} \qquad (14.49)$$

in which q_{ult} = ultimate bearing capacity at the tip, and C_t and C_s = empirical coefficients. Values of C_t are given in Table 14.8, and C_s is related to C_t by

$$C_s = (0.93 + 0.16\sqrt{D/B})C_t \qquad (14.50)$$

These values of C_t estimate the total, long-term settlement when a firm underlying stratum is at least $10B$ below the foundation tip. At $5B$, the settlement is 88 percent of that from Equation 14.48. At B, the settlement is 51 percent of that from Equation 14.48.

14.8.2 Uplift Displacement

Witham and Kulhawy (1981) demonstrated by numerical modeling that the uplift displacement could be estimated with appropriate modification of the Mattes and Poulos (1969) elastic solution for compression loading. Therefore, the butt movement in uplift, ρ_u, can be computed as follows:

$$\rho_u = \frac{(Q_u - W)I_\rho}{D\, E_s} \qquad (14.51)$$

in which Q_u = uplift load at butt, W = foundation weight, and I_ρ = displacement influence coefficient, given in Figure 14.19.

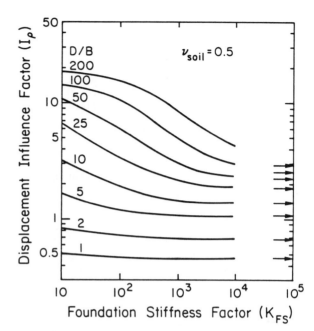

Fig. 14.19 Uplift displacement influence factor. (*Based on Mattes and Poulos, 1969, p. 198.*)

14.8.3 Load Test Observations

When computing the uplift or compression displacements of drilled shafts, the designer should always consider the typical ranges in these parameters that have been observed in full-scale field load tests. For the side resistance in either uplift or compression, numerous studies (e.g., Vesić, 1977; Hirany and Kulhawy, 1988) have shown that the displacement to mobilize the maximum side resistance is typically in the range of 0.4 to 0.6 inches (10 to 15 mm). Similarly, the butt displacement to mobilize the maximum tip resistance in compression is in the range of 4 to 10 percent of the shaft diameter. When tip resistance in uplift is considered, the displacement is comparable to that for side resistance mobilization.

14.9 OTHER SITE AND LOADING CONDITIONS

Because of space limitations, it is impossible to do justice to all design situations. To alleviate this problem to some degree, some general observations and pertinent references are noted below.

For design in expansive clays, consideration must be given to the seasonal variation in water table and its influence on the side resistance, which can change from uplifting to downdragging within the seasonally affected soil depth. This problem requires careful consideration when present. O'Neill (1987b) has given a particularly useful overview of this design case.

Downdrag also can occur when surrounding clay soils settle sufficiently under dead loads to induce downward shearing forces along the shaft sides. This problem is covered well in traditional texts (e.g., Peck et al., 1974) and must be examined carefully because of the potential adverse consequences of this phenomenon.

When lateral loads are imposed on drilled shafts, a highly nonlinear load–displacement behavior results. This behavior usually requires a computer-based iterative solution to address the nonlinearities. Useful approaches to this problem are given by Davidson et al. (1982), Evans and Duncan (1982), and Reese (1984).

For drilled shafts bearing on, or socketed into, rock masses, the behavioral patterns change significantly because of the different characteristics of rock masses. Rock mass evaluation and axial shaft design are described by Kulhawy and Goodman (1987) and Carter and Kulhawy (1988), while lateral and other loading modes are addressed by Carter and Kulhawy (1988).

Design data also are obtained from full-scale load tests, which have to be conducted and interpreted properly to provide meaningful design data. Recent work by Hirany and Kulhawy (1988) has focused exclusively on drilled shafts, and this study has developed consistent criteria for both the conduct and interpretation of drilled shaft load tests under axial, lateral, and other loading modes.

14.10 CLOSING COMMENTS ON DESIGN

As described earlier in this chapter, the axial capacity of a drilled shaft is governed by the weight, tip resistance, and side resistance components. These components also develop their maximum values at dramatically different displacements (weight at initial loading, side resistance and tip resistance in uplift at small displacements, tip resistance in compression at large displacements). Therefore, Equations 14.1 and 14.39 must be evaluated on the basis of the displacement dependency of each term in the equations. Simple addition of the terms and application of a "global" factor of safety are inappropriate unless the component displacements have been addressed.

Therefore, the issues facing the designer include the individual capacity components, their respective displacements, the structure type, the allowable structure displacements, the acceptable factor of safety as a function of loading condition (dead load, sustained live load, transient live load), and the uncertainty in each of these issues and the site data. No single global criterion for the factor of safety can be applied uniformly because of the many variables involved. Each factor has to be evaluated in the context of a particular site, structure, and design to achieve an acceptable solution for the particular project.

14.11 PERTINENT STANDARDS AND SPECIFICATIONS

When addressing overall analysis, design, and construction issues, attention must be paid to pertinent consensus standards and specifications. These documents are particularly useful for construction standards and materials specifications, and they represent typical overall practice. Those which should be reviewed for drilled shafts include the work of ACI Committee 336 (1980, 1985) and the ADSC (1987) for standards and specifications, and the work by Greer and Gardner (1986) and Reese and O'Neill (1988) for construction methods and specifications.

ACKNOWLEDGMENTS

The writer wishes to thank the following individuals for their review comments on the first draft of this chapter: C. N. Baker, Jr., R. G. Bea, S. P. Clemence, R. D. Darragh, J. A. Focht, Jr., C. P. Gupton, R. D. Holtz, T. W. Klym, H. S. Lacy, A. Macnab, M. W. O'Neill, R. B. Peck, H. G. Poulos, H. S. Radhakrishna, L. C. Reese, and J. L. Witham, in addition to my Cornell staff (S. W. Agaiby, P. W. Mayne, K. J. McManus, C. H. Trautmann, and P. C.-W. Wong). The review comments were very helpful in preparing the final version of this chapter.

REFERENCES

ACI Committee 336 (1980), Suggested design and construction procedures for pier foundations, *ACI 336.3R-72 (Reaf. 1980)*, American Concrete Institute, Detroit [being extensively revised at time of writing].

ACI Committee 336 (1985), Standard specification for the construction of drilled piers, *ACI 336.1-79 (Rev. 1985)*, American Concrete Institute, Detroit [being extensively revised at time of writing].

ADSC: The International Association of Foundation Drilling (1987), Standards and Specifications for the Foundation Drilling Industry, ADSC, Dallas, Texas.

Beech, J. F. and Kulhawy, F. H. (1987), Experimental study of the undrained uplift behavior of drilled shaft foundations, *Report EL-5323*, Electric Power Research Institute, Palo Alto, Calif.

Bishop, A. W. (1966), The strength of soils as engineering materials, *Geotechnique*, **16**, No. 2, pp. 91–130.

Blight, G. E. (1987), Lowering of the groundwater table by deep-rooted vegetation—The geotechnical effects of water table recovery, *Proceedings, 9th European Conference on Soil Mechanics and Foundation Engineering*, Dublin, **1**, pp. 285–288.

Bolton, M. D. (1986), The strength and dilatancy of sands, *Geotechnique*, **36**, No. 1, pp. 65–78.

Callanan, J. F. and Kulhawy, F. H. (1985), Evaluation of procedures for predicting foundation uplift movements, *Report EL-4107*, Electric Power Research Institute, Palo Alto, Calif.

Caquot, A. and Kerisel, J. (1953), Sur le Terme de Surface Dans le Calcul des Fondations en Milieu Pulvérulent, *Proceedings, 3rd International Conference on Soil Mechanics and Foundation Engineering*, Zurich, **1**, pp. 336–337.

Carter, J. P. and Kulhawy, F. H. (1988), Analysis and design of drilled shaft foundations socketed into rock, *Report EL-5918*, Electric Power Research Institute, Palo Alto, Calif.

Davidson, H. L., Cass, P. G., Khilji, K. H., and McQuade, P. V. (1982), Laterally loaded drilled pier research, *Report EL-2197*, Electric Power Research Institute, Palo Alto, Calif.

Duncan, J. M. and Buchignani, A. L. (1976), *An Engineering Manual for Settlement Studies*, Dept. of Civil Engineering, University of California, Berkeley.

Evans, L. T., Jr. and Duncan, J. M. (1982), Simplified analysis of laterally loaded piles, *Report UCB/GT/82-04*, University of California, Berkeley.

Greer, D. M. and Gardner, W. S. (1986), *Construction of Drilled Pier Foundations*, John Wiley and Sons, Inc., New York, N.Y.

Hansen, J. B. (1970), A revised and extended formula for bearing capacity, *Bulletin 28*, Danish Geotechnical Institute, Copenhagen, pp. 5–11.

Hirany, A. and Kulhawy, F. H. (1988), Conduct and interpretation of load tests on drilled shaft foundations, *Report EL-5915*, Electric Power Research Institute, Palo Alto, Calif.

Jamiolkowski, M., Ladd, C. C., Germaine, J. T., and Lancellotta, R. (1985), New developments in field and laboratory testing of soils, *Proceedings, 11th International Conference on Soil Mechanics and Foundation Engineering*, San Francisco, **1**, pp. 57–153.

Kulhawy, F. H. (1984), Limiting tip and side resistance—Fact or fallacy, *Analysis and Design of Pile Foundations*, ed. J. R. Meyer, ASCE, New York, N.Y., pp. 80–98.

Kulhawy, F. H. (1985), Drained uplift capacity of drilled shafts, *Proceedings, 11th International Conference on Soil Mechanics and Foundation Engineering*, San Francisco, **3**, pp. 1549–1552.

Kulhawy, F. H., Trautmann, C. H., Beech, J. F., O'Rourke, T. D., McGuire, W., Wood, W. A., and Capano, C. (1983), Transmission line structure foundations for uplift-compression loading, *Report EL-2870*, Electric Power Research Institute, Palo Alto, Calif.

Kulhawy, F. H. and Beech, J. F. (1987), Ground water influences on foundation side resistance. *Proceedings, 9th European Conference on Soil Mechanics and Foundation Engineering*, Dublin, **2**, pp. 707–710.

Kulhawy, F. H. and Goodman, R. E. (1987), Foundations in rock, Chapter 55 in *Ground Engineer's Reference Book*, ed. F. G. Bell, Butterworths, London, pp. 55/1–55/13.

Kulhawy, F. H. and Mayne, P. W. (1990), Manual on estimating soil properties for foundation design, *Report EL-6800*, Electric Power Research Institute, Palo Alto, Calif.

Lutenegger, A. J. (1987), Use of in-situ tests to determine design parameters for drilled shaft foundations, *Proceedings, Short Course* on *Drilled Shafts for Engineering Faculty*, ADSC, Dallas, Texas. (Also *Report 87-4*, Civil Engineering, Clarkson University, Potsdam, N.Y.)

Mair, R. J. and Wood, D. M. (1987), *Pressuremeter Testing*, Butterworths, London.

Mattes, N. S. and Poulos, H. G. (1969), Settlement of single compressible pile, *Journal of the Soil Mechanics and Foundations Division, ASCE*, **95**, No. SM-1, pp. 189–207.

Mayne, P. W. and Kulhawy, F. H. (1982), K_0–OCR relationships in soil, *Journal of the Geotechnical Engineering Division, ASCE*, **108**, No. GT-6, pp. 851–872.

Meigh, A. C. (1987), *Cone Penetration Testing*, Butterworths, London.

Mitchell, J. K. (1986), Settlement analysis and volume change potential assessment using in-situ tests, *Proceedings, Symposium on Interpretation of Field Testing for Design Parameters*, Adelaide, **2**, pp. 45–60.

O'Neill, M. W. (1987a), Use of underreams in drilled shafts, *Proceedings, Short Course on Drilled Shafts for Engineering Faculty*, ADSC, Dallas, Texas.

O'Neill, M. W. (1987b), Drilled shafts in expansive clays—Design and analysis concepts, *Proceedings, Short Course on Drilled Shafts for Engineering Faculty*, ADSC, Dallas, Texas.

Peck, R. B., Hanson, W. E., and Thornburn, T. H. (1974), *Foundation Engineering*, 2nd ed., John Wiley and Sons, Inc., New York, N.Y.

Poulos, H. G. and Davis, E. H. (1974), *Elastic Solutions for Soil and Rock Mechanics*, John Wiley and Sons, Inc., New York, N.Y.

Poulos, H. G. and Davis, E. H. (1980), *Pile Foundation Analysis and Design*, John Wiley and Sons, Inc., New York, N.Y.

Randolph, M. F. and Wroth, C. P. (1982), Recent developments in understanding the axial capacity of piles in clay, *Ground Engineering*, **15**, No. 7, pp. 17–25, 32.

Reese, L. C. (1984), Handbook on design of piles and drilled shafts under lateral load, *Report FHWA-IP-84-11*, Federal Highway Administration, McLean, Virginia.

Reese, L. C., Owens, M., and Hoy, H. (1981), Effects of construction methods on drilled shafts, *Drilled Piers and Caissons*, ed. M. W. O'Neill, ASCE, New York, N.Y., pp. 1–18.

Reese, L. C. and O'Neill, M. W. (1988), Drilled shafts: Construction procedures and design methods, *Report FHWA-HI-88-042*, Federal Highway Administration, McLean, Virginia.

Robertson, P. K. (1986), In-situ testing and its application to foundation engineering, *Canadian Geotechnical Journal*, **23**, No. 4, pp. 573–594.

Schmertmann, J. F. (1985), Measure and use of the in-situ lateral stress, *The Practice of Foundation Engineering (Osterberg Volume)*, ed. R. J. Krizek, C. H. Dowding, and F. Somogyi, Northwestern University, Evanston, Ill., pp. 189–213.

Semple, R. M. and Rigden, W. J. (1984), Shaft capacity of driven pipe piles in clay, *Analysis and Design of Pile Foundations*, ed. J. R. Meyer, ASCE, New York, N.Y., pp. 59–79.

Sheikh, S. A., O'Neill, M. W., and Venkatesan, N. (1983), Behavior of 45 degree underreamed footings, *Report UHCE 83-18*, University of Houston, Houston, Texas.

Skempton, A. W. (1964), Long-term stability of clay slopes, *Geotechnique*, **14**, No. 2, pp. 75–102.

Skempton, A. W. (1986), Standard penetration test procedures and the effects in sands of overburden pressure, relative density, particle size, ageing and overconsolidation, *Geotechnique*, **36**, No. 3, pp. 425–447.

Stas, C. V. and Kulhawy, F. H. (1984), Critical evaluation of design methods for foundations under axial uplift and compression loading, *Report EL-3771*, Electric Power Research Institute, Palo Alto, Calif.

Stewart, J. P. and Kulhawy, F. H. (1981), Experimental investigation of the uplift capacity of drilled shaft foundations in cohesionless soil, *Contract Report B-49(6)* to Niagara Mohawk Power Corporation, Syracuse, N.Y. by Cornell University, Ithaca, N.Y.

Tomlinson, M. J. (1957), The adhesion of piles driven in clay soils, *Proceedings, 4th International Conference on Soil Mechanics and Foundation Engineering*, London, **2**, pp. 66–71.

Trautmann, C. H. and Kulhawy, F. H. (1987), CUFAD-A computer program for compression and uplift foundation analysis and design, *Report EL-4540-CCM, Vol. 16*, Electric Power Research Institute, Palo Alto, Calif.

Tucker, K. D. (1987), Uplift capacity of drilled shafts and driven piles in granular materials, *Foundations for Transmission Line Towers*, ed. J.-L. Briaud, ASCE, New York, N.Y., pp. 142–159.

Turner, J. P. and Kulhawy, F. H. (1987), Prediction of drilled shaft displacements under repeated axial loads, *Proceedings, International Symposium on Prediction and Performance in Geotechnical Engineering*, Calgary, pp. 105–112.

Turner, J. P., Kulhawy, F. H., and Charlie, W. A. (1987), Review of load tests on deep foundations subjected to repeated loading, *Report EL-5375*, Electric Power Research Institute, Palo Alto, Calif.

Turner, J. P. and Kulhawy, F. H. (1990), Drained uplift capacity of drilled shafts under repeated axial loading, *Journal of Geotechnical Engineering, ASCE*, **116**, No. 3, pp. 470–491.

Vesić, A. S. (1975), Bearing capacity of shallow foundations, *Foundation Engineering Handbook*, 1st ed., ed. H. F. Winterkorn and H. Y. Fang, Van Nostrand Reinhold, New York, N.Y., pp. 121–147.

Vesić, A. S. (1977), Design of pile foundations, *Synthesis of Highway Practice 42*, Transportation Research Board, Washington, D.C.

Withiam, J. L. and Kulhawy, F. H. (1981), Analysis procedure for drilled shaft uplift capacity, *Drilled Piers and Caissons*, ed. M. W. O'Neill, ASCE, New York, N.Y., pp. 82–97.

Wroth, C. P. (1984), The interpretation of in-situ soil tests, *Geotechnique*, **34**, No. 4, pp. 449–489.

15 FOUNDATION VIBRATIONS

GEORGE GAZETAS, Ph.D., P.E.
Professor of Soil Mechanics
National Technical University
Athens, Greece
and
State University of New York
Buffalo

15.1 INTRODUCTION

When subjected to dynamic loads, foundations oscillate in a way that depends on the nature and deformability of the supporting ground, the geometry and inertia of the foundation and superstructure, and the nature of the dynamic excitation. Such an excitation may be in the form of support motion due to waves arriving through the ground during an earthquake, an adjacent explosion, or the passage of a train; or it may result from the dynamic forces imposed directly or indirectly on the foundation from operating machines, ocean waves, and vehicles moving on the top of the structure.

Since the very important subject of foundation response during earthquake shaking is treated in the next chapter, attention herein will be focused on determining the vibratory response of foundations to applied loads such as those produced by a machine. A key step in such response analyses (and hence the main thrust of this chapter) is to estimate the dynamic "*spring*" and "*dashpot*" coefficients of flexibly-supported foundations. To this end, an engineering procedure is developed, based on simple algebraic formulae and dimensionless charts, for surface and shallow foundations, embedded foundations, and piles. Note that, in addition to being directly applicable to machine-loaded foundations, much of the information presented could also be used in assessing the dynamic soil–foundation–structure interaction during seismic (or any other ground) shaking. Of course, in such cases the loading arises from inertial (D'Alembert) forces developing in the oscillating superstructure.

This chapter also presents information on the pertinent dynamic soil parameters, and outlines current methods of measuring them in the laboratory and in the field. Some useful results and concepts from dynamics and wave propagation theory are also presented and elucidated when the need arises, throughout the chapter; they provide background information and help in developing a better understanding of the methods presented. The chapter concludes with a number of illustrative realistic examples.

15.2 MACHINE FOUNDATION VIBRATIONS: STATEMENT OF THE PROBLEM

A sketch of a typical rigid block foundation carrying rotatory machinery and supported on a layered soil profile is shown in Figure 15.1. The dynamic loading arises from an unbalanced mass m_0 rotating with an eccentricity r_0 at the operational circular frequency $\omega = 2\pi f$, where f = frequency in cycles per second (Hz). The forces and moments acting on the soil–foundation interface and transmitted into the ground are of the form $m_0 r_0 \omega^2 \cos \omega t$ or, using complex notation, $m_0 r_0 \omega^2 \exp(i\omega t)$;* that is, they vary harmonically with time. Waves are emitted from the interface and propagate in all directions within the deposit. In the presence of the free ground surface and of soil layers with differing stiffnesses these waves undergo numerous reflections and refractions, as well as transformations into surface waves. Much of the energy imparted onto the foundation is diffused by such outward- and downward-spreading waves, while a small portion is dissipated by inelastic action in the soil.

As a result, the soil–foundation interface, and with it the foundation block, undergoes harmonic oscillations of the form $u_0 \cos(\omega t + \varphi)$ or, using complex notation, $u_0 \exp[i(\omega t + \varphi)]$, with frequency-dependent amplitude and phase lag, $u_0 = u_0(\omega)$ and $\varphi = \varphi(\omega)$. The basic goal of the geotechnical design is to limit the amplitudes of all possible modes of oscillation to small enough levels that will neither endanger the satisfactory operation of the machine nor disturb the people working in the immediate vicinity. Charts like the one depicted in Figure 15.1b (based on information from Richart, 1975) may guide the selection of an appropriate upper limit for a satisfactory foundation performance.

Notice that these limiting displacement amplitudes are typically of the order of a hundredth of a centimeter—compared to the several centimeters that is the usual restriction for foundation settlement under static load. A direct consequence is that soil deformations would in the majority of cases by quasielastic, involving negligible nonlinearities and no permanent deformations. Among the possible exceptions are a laterally oscillating piled foundation working at low frequencies, which may induce strains of the order of 0.02 percent in soft clayey layers; and a rocking shallow foundation may induce large strains directly under its edges. Thus, analyses to predict vibration amplitudes assume linear viscoelastic soil behavior, with hysteretic soil damping to model energy losses at these

---

* It has become traditional in dynamics to introduce complex-number notation, which significantly simplifies the computations. The understanding, of course, is that at the end the absolute value (amplitude) and phase angle can be recovered from a complex response $u_1 + iu_2$; the former being equal to $\sqrt{u_1^2 + u_2^2}$ and the latter to $\tan^{-1} u_2/u_1$.

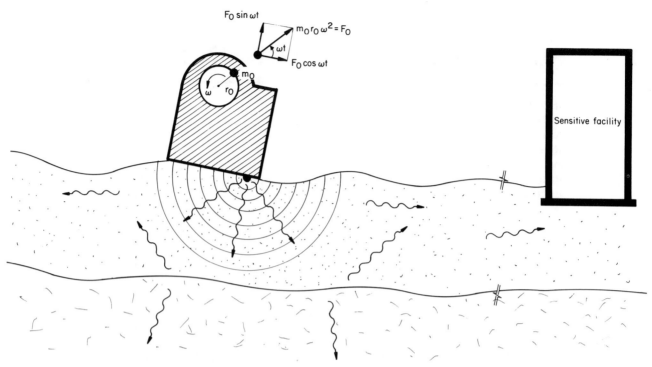

Fig. 15.1(a) The machine foundation problem.

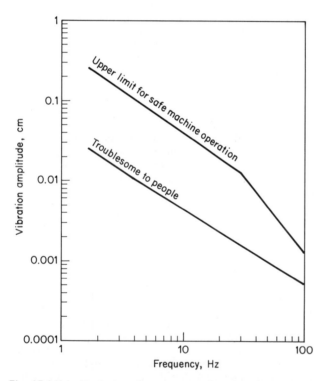

Fig. 15.1(b) Typical performance requirements for machine foundations.

(a) Estimate magnitude and characteristics of the dynamic loads. The most common types of machines include:

- Rotating machinery, which produces sinusoidaly varying forces as already explained (examples: turbines, compressors, pumps, fans)
- Reciprocating machinery, which generates biharmonic loads of the form $F \simeq m_0 r_0 \omega^2 [\exp(i\omega t) + \alpha \exp(2i\omega t)]$, where α is a geometric constant (examples: steam engines, internal-combustion engines, piston-type compressors and pumps)
- Impact producing machines, involving intermittent impulsive loading with a nearly triangular variation of applied force versus time (examples: forging hammers, stamping machines, presses)
- Machines with simultaneous impulsive and rotatory forces, in which the former are due to the main function of the machine (hammering) while the latter generate parasitically from unbalanced wear of the hammers (solid-waste shredders, car-shredders, rotatory rock crushers, all kinds of hammer-mills)

This crucial task will not be further addressed herein, since it has been treated in detail in the first edition of the *Foundation Engineering Handbook* (Richart, 1975). Additional information may be found in Barkan (1962), Richart et al. (1970), Arya et al. (1979), Major (1980), and Prakash and Puri (1988).

(b) Establish the soil profile and determine the appropriate shear modulus and damping, G and β, for each soil layer. In addition to standard geotechnical soil investigation techniques, special dynamic procedures are used today to assess these soil parameters in the field and the laboratory. Section 15.4 presents up-to-date information on this subject.

(c) Guided by experience, select the type and trial dimensions on the foundation, and in cooperation with the client establish performance criteria such as those of Figure 15.1b.

(d) Estimate the dynamic response of this trial foundation, subjected to the load of step (a) and supported by the soil

small strain amplitudes. The low-strain value of the shear modulus (denoted by G_0 or G_{max} in the literature) is the key soil parameter that must be assessed for each layer.

The design of a machine foundation is a trial-and-error process involving the following main steps (engineering tasks).

deposit established in step (b). This key step of the design process usually starts with simplifying and idealizing soil profile and foundation geometry, and involves selecting the most suitable method of dynamic soil–foundation interaction analysis. To this end, several formulations and computer programs have been developed in recent years. Moreover, for the two key parameters, the dynamic stiffness and damping, numerous solutions have been published in the form of parametric dimensionless graphs, applicable to a variety of idealized situations. The main contribution of this chapter is to present in a concise and comprehensive way a complete set of ready-to-use results for the stiffness and damping ("spring" and "dashpot") of foundations on and in several characteristic soil profiles.

(e) Check whether the estimated response amplitude of step (d) at the particular operation frequency conforms with the performance criteria established in step (c). Repeat steps (c), (d), and (e) until a (theoretically) satisfactory design is established. At this stage, two additional checks may be necessary: First, to ensure that the motions transmitted to nearby structures and underground facilities are within safe levels for their uninterrupted functioning—a task usually accomplished with the help of semiempirical energy-attenuation relationships, and guided by experience. Second, if the subsoil contains soft clays and/or loose sands, to investigate the potential for accumulation of large permanent deformations— an unlikely event, requiring shear strain amplitudes well in excess of 0.01 percent.

The design process frequently stops here. However, in case of important projects one or two additional postconstruction steps are necessary:

(f) Monitor the actual motion of the completed foundation and compare with the theoretical predictions of step (d). The necessity of this task arises from the several simplifying assumptions that are unavoidably introduced with even the most sophisticated analyses. Furthermore, experience, and confidence in the advantages of advanced methods of analysis can only be gained through such comparisons of theoretical predictions with reality. Reference is made to Richart et al. (1970), Gazetas and Selig (1985), and Hall (1985) for information on instrumentation and field measurements related to machine foundations and to man-induced vibrations.

(g) Finally, if the actual performance of the constructed foundation does not meet the aforesaid design criteria (step (c)), remedial measures must be devised. These may be, repair of the worn-out parts to minimize unbalanced masses; change of the mass of the foundation or the location of the machinery; stiffening of the subsoil through, for example, grouting; increasing the soil–foundation contact surface; construction of piles through the existing foundation mat; and so on. Steps (d), (e), and (f) must be repeated until a satisfactory design is finally achieved.

This chapter addresses in detail tasks (b) and (d).

15.3 SOIL MODULI AND DAMPING—FIELD AND LABORATORY TESTING PROCEDURES

A vibrating foundation emits shear and dilatational waves into the supporting ground. The former, denoted as S waves, propagate with a velocity V_s that is controlled by the shearing stiffness G and the mass density ρ of the soil:

$$V_s = \sqrt{\frac{G}{\rho}} \qquad (15.1)$$

Dilatational waves, denoted as P waves, propagate with a velocity V_p related to the constrained modulus M_c:

$$V_p = \sqrt{\frac{M_c}{\rho}} \qquad (15.2)$$

For an elastic material, M_c depends on the shear modulus G and the Poisson's ratio v of the soil so that

$$V_p = V_s \sqrt{\frac{2(1 - v)}{1 - 2v}} \qquad (15.3)$$

The relationship V_p/V_s versus v from Equation (15.3) is plotted in Figure 15.2.

Therefore, V_s and V_p, or G and M_c, or G and v, are the equivalent pairs of soil parameters relevant to wave propagation phenomena. Note that waves other than S and P also arise in the ground under an oscillating foundation, most notably Rayleigh and Love waves. All these other waves, however, also relate to G and v, as they are the outcome of combinations ("interferences") of S and P waves.

For the small strains (less than about 0.005 percent) usually induced in the soil by a properly designed machine foundation, shear deformations are the result of particle distortion rather than sliding and rolling between particles. Such deformation is *almost* linearly elastic: the hysteresis loops that *do* develop upon unloading and reloading are very, very narrow. The actual behavior can be simulated quite accurately as that of a *linear hysteretic solid* described through the "tangent-at-the-origin" shear modulus G_0 and a damping ratio β_0.

In fact, the approximation as a linear hysteretic solid is also employed to describe dynamic behavior at large strains. However, as illustrated in Figure 15.3, the appropriate ("equivalent linear") modulus G is the secant modulus, that is, the slope of the line connecting the origin with the tip of the hysteresis loop. G is smaller than G_0 (hence the familiar notation of the latter as $G_{\max}$). At the same time, the area of the hysteresis loop has expanded owing to increased dissipation of energy resulting from sliding at particle contacts. The equivalent linear hysteretic damping ratio β is larger than β_0.

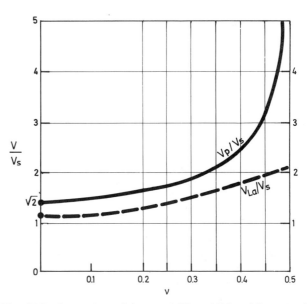

Fig. 15.2 Comparison of the actual (V_p and V_s) and "apparent" (V_{La}) wave velocities used in foundation vibration analyses.

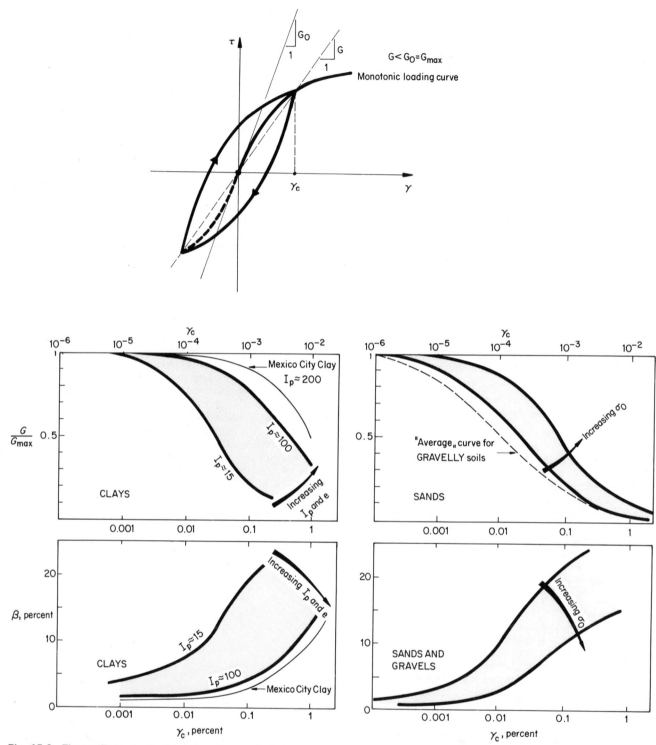

Fig. 15.3 The nonlinear-hysteretic cyclic stress–strain behavior of soils is conveniently represented in terms of modulus decreasing and damping ratio increasing with shear strain amplitude.

Apparently, the bigger the cyclic shear strain, the smaller the "equivalent" modulus G and the larger the "equivalent" damping β. Plots of modulus ratio G/G_{max} and damping ratio β as functions of cyclic strain γ_c have become the traditional way of depicting cyclic stress–strain behavior, following the pioneering work by Seed and Idriss (1970). Figure 15.3 summarizes published data for clays, sands, and gravels, encompassing some recently published information.

15.3.1 Shear Modulus G_{max} and S-Wave Velocity $V_{s,max}$

Factors affecting G_{max} and $V_{s,max}$ From the foregoing discussion it is clear that the low-strain shear modulus, G_{max}, or the corresponding S-wave velocity $V_{s,max} = \sqrt{(G_{max}/\rho)}$, is the single most important soil parameter influencing the response of machine foundations. Laboratory and field tests have revealed a number of factors on which G_{max} and $V_{s,max}$ depend. The following discussion summarizes the most significant findings of these tests.

(1) The two most important parameters influencing G_{max} of all types of soils (granular *and* cohesive) are the mean confining effective stress $\bar{\sigma}_0$ and the void ratio e. From the published results it appears that G_{max} is proportional to $\bar{\sigma}_0^n$, where typically $n \simeq 0.3$ to 0.6 for granular and $n \simeq 0.5$ to 0.9 for silty and clayey soils. Experimental tests with large cubic samples of dry sand at the University of Texas (Knox et al., 1982) have revealed that $V_{s,max}$ (and G_{max}) depend *only* on the stresses $\bar{\sigma}_a$ and $\bar{\sigma}_b$ in the directions of wave propagation and particle motion, respectively; they are independent of the stress $\bar{\sigma}_c$ in the out-of-plane direction.

(2) The static-stress prehistory, expressed for instance through the overconsolidation ratio, OCR, influences mainly the modulus G_{max} of clays. The granular material changes in OCR are adequately accounted by the present void ratio. On the other hand, cyclic prestraining, that is, application of moderately large shear strains for a large number of cycles, tends to increase the modulus of granular soils beyond what is anticipated with the increased void ratio. With cohesive soils the effect of prestraining is not clear.

(3) For cohesive soils, geologic age seems to be of great importance, as it perhaps controls the creation of "bonds" between the clay platelets or clay clusters. In the laboratory, attempts to simulate the natural process of aging are being made by increasing the duration of the initial confining state of stress to several days, before applying the cyclic loading. Increases in G_{max} of the order of 100 percent have often been reported. Aging may also be important for fine-grained cohesionless soils that are partly saturated.

(4) For partially saturated ($S_r \simeq 10$ to 50 percent) fine granular soils (silty sands) capillary stresses may increase G_{max} by 50 to 100 per cent over the value of G_{max} measured in the laboratory on completely dry or on fully saturated samples.

(5) For all soils, cohesionless and cohesive, the frequency, or the rate of loading, has no practical effect on G_{max} (at least within the range of parameters applicable to machine foundations). This means that soil is basically *not* a viscous, but rather a hysteretic, material.

Empirical correlations for G_{max} Several expressions relating G_{max} to other soil parameters have been devised on the basis of laboratory test results.

For granular and cohesive soils Hardin (1978) proposed that

$$G_{max} \simeq 625 \frac{(\text{OCR})^\mu}{0.3 + 0.7e^2} \sqrt{p_a \bar{\sigma}_0} \qquad (15.4)$$

where p_a = the atmospheric pressure in the same units as $\bar{\sigma}_0$ and G_{max}, and μ is a function of the plasticity index I_p plotted in Figure 15.4.

On the other hand, the aforementioned experimental work at the University of Texas (Knox et al., 1982) has concluded that

$$\bar{\sigma}_m = \tfrac{1}{2}(\bar{\sigma}_a + \bar{\sigma}_b) \qquad (15.5)$$

should be used in place of $\bar{\sigma}_0$ in Equation 15.4. ($\bar{\sigma}_a$ and $\bar{\sigma}_b$ are the effective stresses in the directions of wave propagation and particle motion, respectively). Alternatively, the following expression can be used for clean sands:

$$G_{max} \simeq \frac{180}{0.3 + 0.7e^2}(\bar{\sigma}_a \bar{\sigma}_b)^{0.20} p_a^{0.60} \qquad (15.6)$$

Suggested Values of $K_{2,max}$ for Equations 15.7 (Seed and Idriss)

Soil Type	$K_{2,max}$ for stress units of	
	kPa	psf
Loose sand	8	35
Dense sand	12	50
Very dense sand	16	65
Very dense sand and gravel	25 to 40	100 to 150

(a)

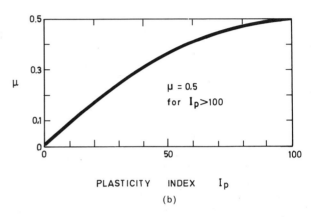

(b)

Fig. 15.4 Suggested values for the coefficients $K_{2,max}$ and μ in Equations 15.4 and 15.7.

Note, however, that in many actual situations S waves will propagate in all directions away from the foundation, and it will not be readily evident which are the directions a and b. Hence it may be as advantageous to use Equation 15.4.

For granular soils Seed and Idriss (1970) developed the simpler expression

$$G_{max} \simeq 1000 K_{2,max} \sqrt{\bar{\sigma}_0} \qquad (15.7)$$

in which the dimensional empirical coefficient $K_{2,max}$ is a function of the (relative) density of the material (dimension: square root of stress) given in Figure 15.4, for both SI and English units.

For saturated clays, G_{max} relates to undrained shear strength S_u:

$$\frac{G_{max}}{S_u} \simeq 1000 \text{ to } 2500 \qquad (15.8)$$

(The geotechnical engineer should not be surprised at such high G_{max}/S_u values. The value $G \simeq 100 S_u$ reported in soil mechanics literature refers to near-failure conditions, that is, at strains in excess of 1 percent.)

Use of empirical expressions such as those of Equations 15.4 to 15.6 may be recommended in practice in several cases: (1) in feasibility studies and preliminary design calculations, before any direct measurements have been performed in the field or laboratory; (2) for final design calculations in small projects, where the cost of proper testing for G_{max} cannot be justified—unless parameter studies reveal a high sensitivity of the response to the "exact" value of modulus; (3) to provide an order-of-magnitude check against the experimentally determined values.

Another empirical correlation of interest is between G_0 and the Standard Penetration Test (SPT) resistance N (blows/ft). Using mostly Japanese data, Seed et al. (1986) have proposed that

$$G_{max} \simeq 20\,000(N_1)_{60}^{1/3} \sqrt{\bar{\sigma}_0} \quad \text{(psf)} \qquad (15.9a)$$

or

$$G_{max} \simeq 4500(N_1)_{60}^{1/3} \sqrt{\bar{\sigma}_0} \quad \text{(kPa)} \tag{15.9b}$$

in which the corrected resistance is given by

$$(N_1)_{60} = N\left(\frac{p_a}{\bar{\sigma}_{v0}}\right)^{1/2}\left(\frac{ER}{0.60}\right) \tag{15.10}$$

where $\bar{\sigma}_{v0}$ = vertical effective overburden stress, and ER = ratio of the energy actually transmitted to the rod of the SPT, divided by the theoretical free-fall energy. Several other empirical correlations between G_{max} and N values have also been proposed in the literature. One that has been frequently quoted in the literature has been proposed by Ohsaki and Iwasaki (1973):

$$G_{max} = 12\,000N^{0.8} \quad \text{(kPa)} \tag{15.11a}$$

$$G_{max} = 240N^{0.8} \quad \text{(ksf)} \tag{15.11b}$$

However, the reliability of such relations is very low, and they should only be used, if necessary, for crude preliminary estimates of soil stiffness.

15.3.2 Constrained Modulus and P-Wave Velocity

Whereas shear (S) waves can propagate only through the mineral skeleton of a soil (fluids offer no shear resistance), dilatational (P) waves can propagate through both the mineral skeleton and the pore water. Since water is far less compressible than any soil skeleton, P-waves in fully saturated soils are essentially transmitted solely through the water phase with a velocity V_p that is of the order of, or somewhat larger than, $V_w \simeq 1500$ m/sec (or 4900 ft/sec)—the velocity of sound waves in water.

On the other hand, the presence of even small amounts of air in the pores might dramatically increase the compressibility of the water–air phase; only the soil skeleton would then resist the induced dilatation: V_p would be essentially the same as the P-wave velocity of a dry, but otherwise identical, soil sample

For a clean sand, Figure 15.5a portrays the sensitivity of V_p to variations in the degree of saturation S_r. As long as saturation remains below about 99 percent V_p is nearly independent of S_r, being a measure of the incompressibility of the soil skeleton. (The small decline from the dry velocity $V_{p,dry}$ at large values of S_r is the consequence of increasing mass density, rather than of decreasing constrained modulus in Equation 15.2.) As S_r approaches 100 percent, V_p jumps to a very high value, $V_{p,sat}$, that is controlled by the pressure wave velocity in water, V_w. For practical purposes, the velocity $V_{p,sat}$ is independent of the type of soil, is similar for clays and sands, and shows only a slight dependence on $\bar{\sigma}_0$ and e, as visualized in Figure 15.4b. Hence, *measuring the P-wave velocity of saturated soils is of little if any value in assessing the actual soil stiffness.*

The foregoing experimental findings can be qualitatively explained with elastic theory. Saturated soil is a practically incompressible material with Poisson's ratio v approaching 0.50. Equation 15.3 would then predict that V_p is far greater than V_s, and, in the limit, $v = 0.5$ and $V_p = \infty$ regardless of V_s—that is, regardless of soil stiffness.

15.3.3 Poisson's Ratio v

For soils that are not close to saturation, v can be obtained from Equation 15.3 once V_s and V_p have been measured:

$$v = \frac{1 - a^2/2}{1 - a^2} \qquad a = V_p/V_s \tag{15.12}$$

This expression, however, is rather unreliable: small errors in the values of V_p or V_s will lead to substantial errors in v.

On the other hand, v shows little sensitivity to soil type, confining pressure, and void ratio, but depends very much on the degree of saturation and the drainage conditions. Consequently, it is not very difficult to make a reasonably good prediction of v if saturation and drainage conditions are known. As an example, the following values are given as a guide in selecting v in practical cases.

Saturated clays and sands, beneath the water table	$v \simeq 0.50^-$
Nearly saturated clays, above the water table	$v \simeq 0.40$
Wet silty sands ($S_r = 50$ to 90 percent)	$v \simeq 0.35$
Nearly dry sands, stiff clays, and rocks	$v \simeq 0.25$

Once v has been estimated, Equation 15.3 is used to determine V_p, unless of course $v \simeq 0.50$ so that, as previously explained, Equation 15.3 is meaningless. An interesting conclusion drawn from studies of foundation vibrations is that the influence of v is not of great significance in most cases; an exception is vertical and rocking oscillations in soils with v approaching 0.50. Hence, small errors in assessing the value of v would likely be of no practical consequence.

15.3.4 Damping Ratio β_0

The low-strain value of material damping, β_0, depends only marginally on such variables as the confining stress and the void ratio. For most soils it ranges between 2 and 6 percent. Since oscillating foundations generate "radiation" damping that may be substantially higher than β_0, the precise value of the latter is usually rather insignificant. (Exceptions are rotational oscillations at low frequencies, and translational oscillations on a shallow soil stratum, again at low frequencies.)

15.3.5 Measurement of Low-Strain Moduli

For satisfactory design of a machine foundation the geotechnical engineer must:

- *Establish* the soil profile, including layering and depth to bedrock, physical characterization and classification of each layer, elevation of water table and groundwater conditions, and extent of lateral homogeneity
- *Determine* with in-situ or laboratory tests the low-strain value of shear modulus G_{max} and *select* proper values for Poisson's ratio v and damping ratio β_0.

Standard subsurface exploration techniques and field and laboratory testing required for static design may provide a complete answer to the first of the foregoing tasks. But, with few exceptions, estimation of soil parameters for dynamic analyses is presently being done increasingly frequently with the help of special "dynamic" procedures in the field and the laboratory. Only a summary of the best techniques for determining G_{max} is offered herein. More detailed information may be found in Richart (1975), Woods (1978, 1985), Stokoe (1980), and Drnevich (1985). Note: most dynamic tests provide an indirect evaluation of G_{max} through measurements of the S-wave velocity $V_{s,max}$

$$G_{max} = \rho V_{s,max}^2 \tag{15.13}$$

in which ρ is the known total mass density of the soil.

In-situ testing procedures have some distinct advantages over laboratory techniques. Sample disturbance, for example, may be more deleterious for determining low-strain soil stiffness (which reflects the exact particle arrangement—"fabric"), than behavior at large strains and failure (after a rearrangement of

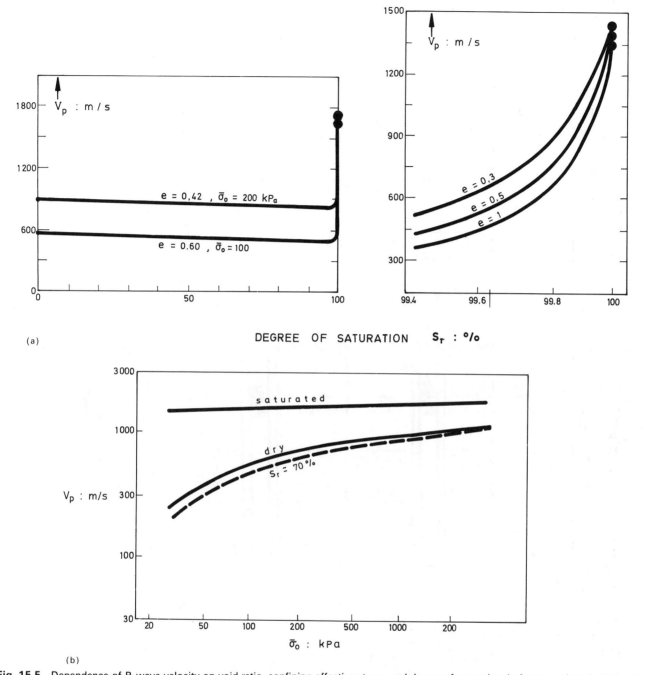

Fig. 15.5 Dependence of P-wave velocity on void ratio, confining effective stress, and degree of saturation (references given in the text).

particles has occurred—"destruction" of the initial "fabric"). Moreover, simulating in the laboratory the effects of stress prehistory, aging, and capillary stresses is not a routine task. In fact, with granular soils even reproducing the in-situ void ratio and geostatic stresses (which control G_{max} according to Equation 15.3), may prove a rather difficult task. With coarse sand and gravel, things get even more complicated.

As a result, in-situ measured moduli are almost invariably found to exceed those measured in the laboratory—sometimes by more than 100 percent. However, when the effects of all the important factors (Figure 15.4) are properly reproduced, laboratory test results can closely match the field test data. Moreover, laboratory tests are valuable for studying the effect of various variables on G_{max}, for determining the damping ratio, and for obtaining V_s and G at moderate and large strains.

15.3.6 Field Procedures

Dynamic in-situ tests induce strains smaller than 10^{-5} and thereby measure $V_{s,max}$ and G_{max}. The list of in-situ testing procedures includes the following.

1. The Crosshole Seismic Survey (or simply crosshole method) This is probably the best geotechnical method for determining the variation with depth of in-situ low-strain S-wave velocity, $V_{s,max}$. Illustrated by a sketch in Figure 15.6, the crosshole method is based on a very simple concept: it generates S waves in a borehole and measures their arrival times at the same elevation in neighboring boreholes. The wave velocity is computed from the travel times and the spacing between the boreholes. For the success, however, of a crosshole test there

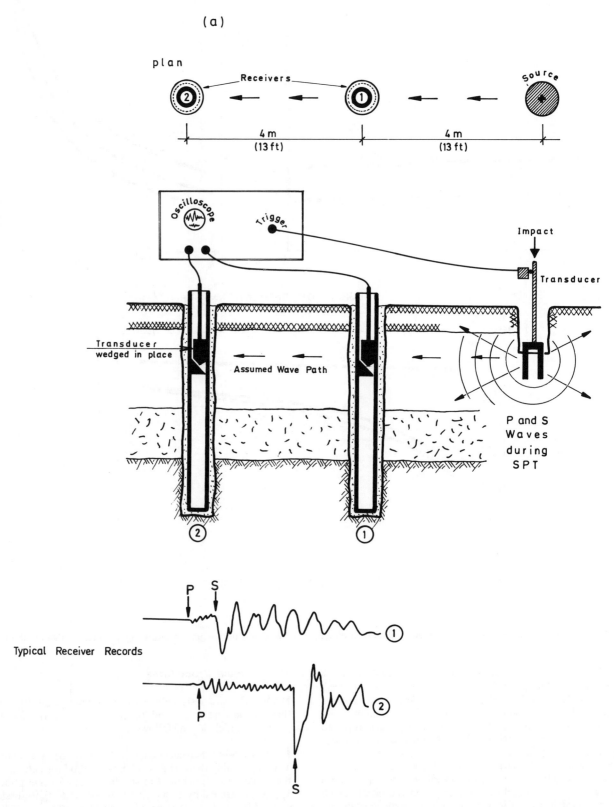

Fig. 15.6 Sketches of (a) the crosshole, (b) the downhole, and (c) the seismic cone penetration tests (references in the text).

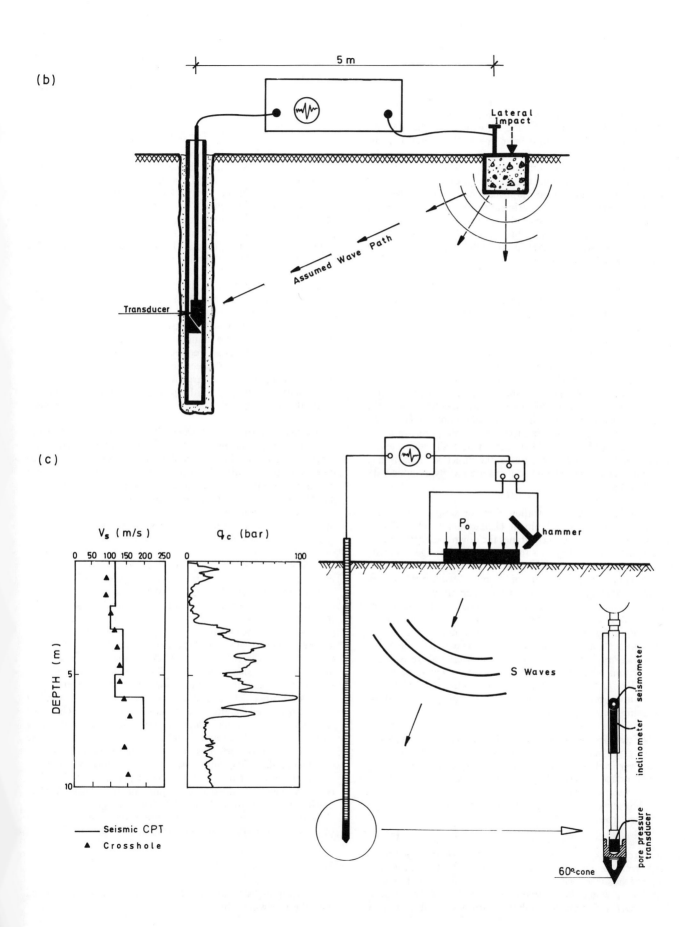

are several requirements. (a) There should be at least two and preferably three boreholes, which are spaced about 3 to 5 m (10 to 15 ft) apart, the verticality of which is instrumentally secured. (b) The source must be rich in shear wave generation and poor in P-wave generation, so that detection of S-wave arrivals is unambiguous (torsional sources are the best in this sense, but the SPT offers a good inexpensive solution). (c) The receivers (geophones) must have a proper frequency response and should be oriented in the direction of the particle motion. Moreover, they must be in "perfect" contact with the surrounding soil, either directly (in case of stiff cohesive soils) or through properly grouted casing (in case of granular and soft cohesive soils). Coupling between geophone transducer and vertical wall should be accomplished with use of specially designed packers. (d) The triggering and recording systems must be accurate. Evidently, "crosshole" would not classify among the most economic in-situ tests, but it *is* one of the most reliable. See Woods (1978), Hoar and Stokoe (1978), and Woods and Stokoe (1985) for more details.

2. The Seismic Downhole Survey (or simply the downhole method) This is the *economic* alternative to crosshole testing. It is explained with the help of Figure 15.6. It needs only one borehole inside which the receiver(s) is (are) placed at various depths while the source is at the surface, 2 to 5 m (6 to 15 ft) away. Travel times of body waves (S or P) between surface and receiver(s) are recorded, and then travel-time versus depth plots are constructed from which $V_{s,max}$ or $V_{p,max}$ of all the layers can be determined. An effective and economic S-wave source consists of a steel-jacketed rigid beam weighted down the ground and struck horizontally with a sledge hammer. However, if the source is placed too close to the borehole, parasitic waves are created and S-wave arrivals cannot be easily identified; if it is placed too far from the source, the direct wave path may not be a straight line. These problems are largely avoided with crosshole testing.

3. The Seismic Cone Penetration Test (or simply the seismic cone) This recent development (Robertson et al., 1985) is sketched in Figure 15.6. It combines the downhole method with cone penetration testing. To this end, a small rugged velocity seismometer is incorporated inside the electronic penetrometer and downhole measurements of seismic S-wave velocity are performed during brief pauses in cone penetration testing. In addition to its speed, a significant advantage of the seismic cone is that with a single sounding test one obtains information for the stratigraphy of the site, the low-strain moduli of the various layers, as well as the (static) strength-related parameters q_c (point bearing stress) and f_s (sleeve frictional resistance). Comparisons with the "crosshole" are very encouraging, as seen in Figure 15.6. A limitation of the method is that it may not be appropriate for some types of soils (such as those containing coarse gravel).

4. The Steady-State Vibration of the Free Surface This method, requiring no boreholes, is based on the fact that a circular footing vertically oscillating with frequency f generates along the free surface primarily Rayleigh (R) waves. Their wavelength λ_R is the distance between any two successive crests (or troughs) of the vibrating surface (Figure 15.7a), and their velocity C_R is calculated from

$$C_R = f\lambda_R \qquad (15.14)$$

Measurement of λ_R is made by moving a seismic geophone away from the vibrator and locating points that are moving in phase. If the subsoil were very deep and homogeneous its S-wave velocity V_s would have been unique and roughly equal to 1.06 times C_R. With real-life inhomogeneous deposits $V_s = V_s(z)$

and $C_R = C_R(z)$, and the value from Equation 15.13 would correspond to a depth of about $\frac{1}{3}$ of the wavelength (the "center" of the R-wave displacement profile). By progressively decreasing the frequency f of vibrations, the wavelength λ_R would increase and the R-wave would affect soil at greater depths, having different properties. Equation 15.14 would at every frequency give a different value of C_R. From these values the velocity profile is constructed as

$$V_s \text{ (at depth } z = \tfrac{1}{3}\lambda_R) \simeq 1.06 f \lambda_R \qquad (15.15)$$

As an example, Figure 15.7b (adapted from Gazetas, 1982) plots the theoretical variation of R-wave velocity versus frequency for a deposit consisting of an inhomogeneous layer over bedrock. For the layer $V_s(z) = V_s(0)(1 + 10z/H)^{1/2}$, where H is its thickness. The bedrock velocity V_{rock} is 8 times $V_s(0)$. We denote by f_s the fundamental frequency of the stratum in shear; $f_s \simeq 0.66V_s(0)/H$. Notice that at frequencies f_r exceeding $15f_s$ the R-wave velocity approaches $V_s(0)$, while at f less than $0.5f_s$, C_R approaches V_{rock}. (Plots like that of Figure 15.7b are called "dispersion" relations.)

Clearly, this method cannot even in theory produce the accurate and detailed (layer-by-layer) information of the three borehole methods. However, it *can* provide: (a) the near-surface wave velocity $V_s(0)$, which controls the radiation damping of high-frequency machine foundations, as well as the response in rocking and torsion at all frequencies; (b) an average (over the horizontal and vertical direction) wave velocity of a stratum over bedrock, covering a large area; and (c) with high-power equipment operating at low frequencies, the velocity of deeper strata that could not be reached inexpensively with a borehole.

5. The Spectral Analysis of Surface Waves This recent development is a very promising evolution of the foregoing steady-state vibration method (Nazarian and Stokoe, 1983). Its goal is to determine the detailed V_{s0} profile, as with "crosshole", but working entirely from the surface. A vertical impact at the surface generates transient Rayleigh (R) waves, which are recorded by vibration transducers located a known distance apart. If the subsoil were very deep and homogeneous (halfspace) the signals of the two transducers would have the same shape. However, in nonhomogeneous or layered deposits the various frequency components generated by the impact propagate at different speeds (recall Figure 15.7), thereby arriving at different relative times at the two locations; hence, the two signals have different shapes. Through a fast Fourier transform spectral analysis of the two signals the "dispersion" relation (that is, the variation of C_R with frequency) is computed for the particular site. The thicknesses and S-wave velocities of each and every layer are then back-calculated by use of an analytical "inversion" procedure. The results of the method seem to be in excellent agreement with crosshole measurements.

Several other field tests are available to the profession but are not discussed herein for various reasons. They include the *seismic refraction survey*, which is good mainly for preliminary surveys covering large areas, and for determining the P-wave velocities of near surface soft layers and the depth to rock (Richart et al., 1970); the *resonant footing method*, in which the resonant frequency of a concrete block placed on the surface is determined and utilized in conjunction with homogeneous halfspace theories to back-calculate the (average) soil modulus (Moore, 1985); and the *standard penetration test*, which may provide indirect crude estimates of moduli (e.g. Eq. 15.9).

15.3.7 Laboratory Procedures

Low-strain values of moduli and wave velocities can be obtained with the following laboratory tests.

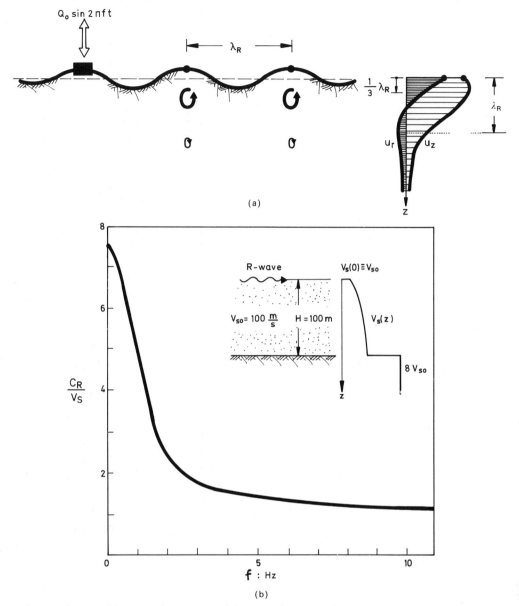

Fig. 15.7 (a) A harmonically-oscillating footing generates Rayleigh (R) waves propagating along the surface of a soil deposit, and "reaching" to a depth of about one wavelength. (b) The R-wave velocity in a nonhomogeneous two-layer stratum decreases with frequency because of the decreasing wavelength.

1. The Resonant Column Test This truly dynamic test is undoubtedly the best widely available today for determining $V_{s,max}$ in the laboratory. It uses solid or hollow cylindrical samples and subjects them to torsional or axial steady-state harmonic excitation with the help of an electromagnetic device (see sketch in Figure 15.8). The frequency of the input vibration is slowly changed until the fundamental resonant condition is determined. The resonant frequency is a function of soil stiffness, sample geometry, and boundary conditions of the apparatus employed. For the case of fixed base and free top sketched in Figure 15.8a the frequency at first resonance is either

$$f_s = \frac{V_s}{4H} = \frac{\sqrt{G/\rho}}{4H} \qquad (15.16)$$

in the torsional mode, or

$$f_L = \frac{V_L}{4H} \simeq \frac{\sqrt{E/\rho}}{4H} \qquad (15.17)$$

in the axial mode. H = the height of the sample, and E = the Young's modulus of the soil, $E = 2(1 + \nu)G$. Equations 15.16 and 15.17 provide G and E, respectively. Material damping ratio can also be estimated either from the free-vibration logarithmic decrement or from the half-power bandwidth of the steady-state response curve. Figure 15.8b plots in dimensionless form the (theoretically determined) response curve. The distribution of shear (or normal) strains along the sample during resonance follows a sinusoidal law:

$$\frac{\gamma(z)}{\gamma(H)} = \sin\left(\frac{\pi}{2}\frac{z}{H}\right) \qquad (15.18)$$

To achieve the development of an almost uniform distribution of strains in the sample, Drnevich (1977) adds a mass at the top as shown in Figure 15.8a. Such a uniformity is highly desirable when V_s at strains exceeding 10^{-5} is needed. The hollow cylinder is also a necessity in such a case, since the distribution of shear strains across the thickness of a solid

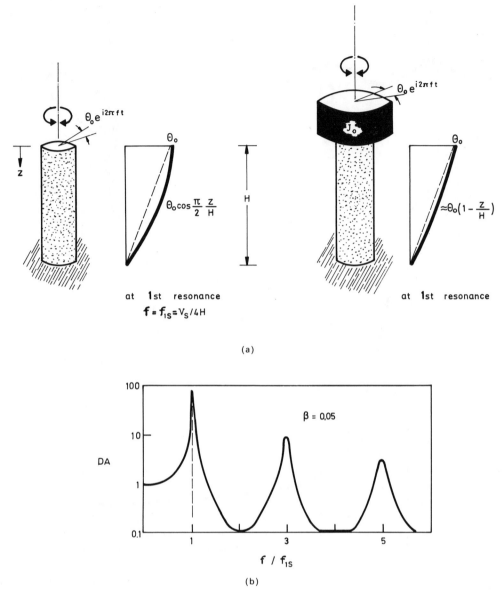

Fig. 15.8 Resonant column test. Distribution of rotation amplitude along sample length in two variants of the test, and dynamic amplification of the top motion versus imposed frequency. (*Based on Woods, 1978; Drnevich, 1985.*)

cylindrical sample in torsion is nonuniform, varying from 0 at the center to a maximum at the periphery. For more details, see Woods (1978) and Drnevich (1985).

2. The Ultrasonic Pulse Test Piezoelectric crystals at one end of the soil sample generate dilatational or shear waves, and at the other end record their arrival. From the travel time and the known sample thickness, the appropriate velocity, V_p or V_s, is calculated. The identification and recognition of the exact wave arrival, requiring the use of an oscilloscope, is by no means a routine operation. The results of this method are in good accord with resonant column data. An advantage of the method is that it can use the same sample to determine both V_p and V_s (and hence Poisson's ratio, or the condition of saturation). Moreover, it can be performed on very soft clays while still retained in the Shelby tube—thus minimizing disturbance.

3. Cyclic Load–Deformation Tests In their standard form, they are appropriate *only* for medium and large strains

$(10^{-4} < \gamma_c < 10^{-1})$ and are used to determine stress–strain hysteresis loops (from which "effective" moduli and damping ratios are deduced, and degradation characteristics are studied). However, in recent years special cyclic triaxial apparatuses have been designed capable of determining moduli at $\gamma_c \geqslant 5 \times 10^{-6}$ (Ladd and Dutko, 1985).

15.4 HARMONIC VIBRATION OF BLOCK FOUNDATIONS: DEFINITION AND USE OF IMPEDANCES (DYNAMIC "SPRINGS" AND "DASHPOTS")

Frequently, machine foundations are constructed as rigid reinforced-concrete blocks, whose response to dynamic loads arises solely from the deformation of the supporting ground. Like any rigid body, such foundations possess six degrees of freedom, three translational and three rotational: (dynamic) displacements along the axes x, y, and z, and (dynamic) rotation

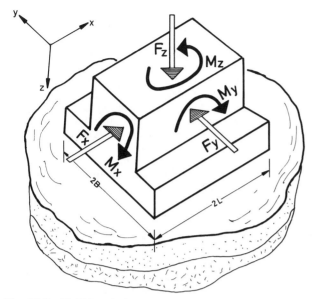

Fig. 15.9 Rigid foundation block with its six degrees of freedom.

around the same axes (Figure 15.9). In this section a general method is presented for computing each of these six dynamic displacements and rotations due to steady-state harmonic excitation (forces and moments). The choice of *harmonic* oscillations is made not only because many machines usually produce unbalanced forces that indeed vary harmonically with time (rototary or reciprocating engines), but also because nonharmonic forces (such as those produced by punch presses and forging hammers) can be decomposed into a (large) number of sinusoids through Fourier analysis.

15.4.1 Vertical Oscillation

Let us explain the method with the help of the easy-to-visualize case of vertical vibrations. Figure 15.10 portrays a rigid foundation block of total mass m, assumed to have a vertical axis of symmetry z passing through the centroid of the soil–foundation contact surface. The foundation is underlain by a deposit consisting of horizontal linearly deforming soil layers. Subjected to a vertical harmonic force $F_z(t)$ along the z axis, this foundation will experience only a vertical harmonic displacement $u_z(t)$. The question is to determine $u_z(t)$ given $F_z(t)$.

To this end, we consider separately the motion of each "body": the foundation block and the supporting ground (Figure 15.10). The two free-body diagrams are sketched in the figure and include the inertial (D'Alembert) forces. The foundation "actions" on the soil generate equal and opposite "reactions", distributed in some unknown way across the interface and having an unknown resultant $P_z(t)$. Furthermore, since in reality the two bodies remain always in contact, their displacements are identical and equal to the rigid body displacement $u_z(t)$. Thus, the dynamic equilibrium of the block takes the form

$$P_z(t) + m\ddot{u}_z(t) = F_z(t) \tag{15.19}$$

and that of the linearly deforming multilayered ground can be "summarized" as

$$P_z(t) = \mathscr{K}_z u_z(t) \tag{15.20}$$

in which $\mathscr{K}_z$ is called the dynamic vertical "impedance", determined for this particular system with one of the methods described in the sequel.

Combining Equations 15.19 and 15.20 leads to

$$m\ddot{u}_z(t) + \mathscr{K}_z u_z(t) + F_z(t) = 0 \tag{15.21}$$

from which it is evident that the key to solving the problem is the determination of the impedance $\mathscr{K}_z$, that is, of the dynamic force-over-displacement ratio according to Equation 15.20. Note also that, as it is well known from structural dynamics, the steady-state solution $u_z(t)$ to Equation 15.21 for a harmonic excitation $F_z(t) = F_z \cos \omega t$ is also harmonic with the same frequency ω.

Theoretical and experimental results reveal that, in Equation 15.20, a harmonic action P_z applied on to the ground and the resulting harmonic displacement u_z have the same frequency ω but are *out of phase*. That is, if

$$P_z(t) = P_z \cos(\omega t + \alpha) \tag{15.22}$$

then u_z can be expressed in the following two equivalent ways:

$$u_z(t) = u_z \cos(\omega t + \alpha + \varphi) \tag{15.23a}$$

$$= u_1 \cos(\omega t + \alpha) + u_2 \sin(\omega t + \alpha) \tag{15.23b}$$

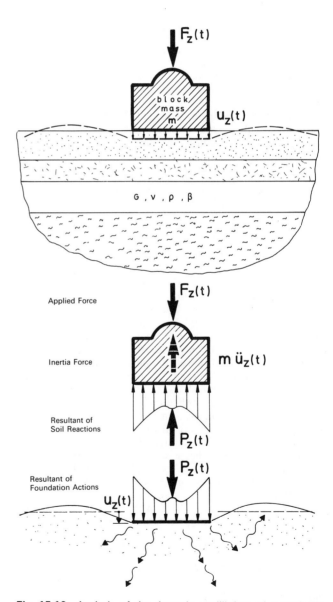

Fig. 15.10 Analysis of the dynamic equilibrium of a vertically vibrating foundation block.

where the amplitude u_z and phase angle φ are related to the inphase, u_1, and the 90°-out-of-phase, u_2, components according to

$$u_z = \sqrt{(u_1^2 + u_2^2)} \qquad (15.24a)$$

$$\tan \varphi = \frac{u_2}{u_1} \qquad (15.24b)$$

We can rewrite the foregoing expressions in an equivalent but far more elegant way using complex number notation:

$$P_z(t) = \bar{P}_z \exp(i\omega t) \qquad (15.25)$$

$$u_z(t) = \bar{u}_z \exp(i\omega t) \qquad (15.26)$$

where now $\bar{P}_z$ and $\bar{u}_z$ are complex quantities:

$$\bar{P}_z = P_{z1} + iP_{z2} \qquad (15.27)$$

$$\bar{u}_z = u_{z1} + iu_{z2} \qquad (15.28)$$

Equations 15.25 to 15.28 are equivalent to Equations 15.22 to 15.24, with the following relations being valid for the amplitudes:

$$P_z = |\bar{P}_z| = \sqrt{(P_{z1}^2 + P_{z2}^2)} \qquad (15.29)$$

$$u_z = |\bar{u}_z| = \sqrt{(u_{z1}^2 + u_{z2}^2)} \qquad (15.30)$$

while the two phase angles, α and φ, are properly "hidden" in the complex forms.

In addition to elegance, it is computational ease that motivates the adoption of complex notation, as will become apparent later on.

With P_z and u_z being out of phase (Eqs. 15.22 to 15.23) or, alternatively, with $\bar{P}_z$ and $\bar{u}_z$ being complex numbers (Eqs. 15.25 to 15.28), the dynamic vertical "impedance" (force–displacement ratio) becomes

$$\boxed{\mathcal{K}_z = \frac{P_z(t)}{u_z(t)} = \frac{\bar{P}_z}{\bar{u}_z} = \text{complex number}} \qquad (15.31)$$

which may be put in the form:

$$\boxed{\mathcal{K}_z = \bar{K}_z + i\omega C_z} \qquad (15.32)$$

in which both $\bar{K}_z$ and C_z are functions of the frequency ω. They can be interpreted as follows. The real component, $\bar{K}_z$, termed "*dynamic stiffness*", reflects the stiffness and inertia of the supporting soil; its dependence on frequency is attributed solely to the influence that frequency exerts on inertia, since soil properties are to a good approximation frequency-independent. The imaginary component, ωC_z, is the product of (circular) frequency times the "*dashpot coefficient*" C_z, which reflects the two types of damping—radiation and material damping—generated in the system, the former due to energy carried by waves spreading away from the foundation, and the latter due to energy dissipated in the soil due to hysteretic action.

Equation 15.32 is a (theoretical *and* experimental) fact for *all* foundation–soil systems. However, the interpretation of $\bar{K}$ and C as dynamic stiffness and dashpot coefficients must be justified. This is easy if we substitute Equation 15.32 into Equation 15.21. We are looking for the harmonic response $\bar{u}_z \exp(i\omega t)$ to the harmonic excitation $F_z \exp(i\omega t)$. Straightforward operations lead to

$$\boxed{m\ddot{u}_z(t) + C_z\dot{u}_z(t) + \bar{K}_z u_z(t) = F_z(t)} \qquad (15.33)$$

and to

$$[(\bar{K}_z - m\omega^2) + i\omega C_z]\bar{u}_z = F_z \qquad (15.34)$$

Equation 15.33 is the equation of motion of a simple oscillator with mass m, spring "constant" $\bar{K}_z$, and dashpot "constant" C_z—justifying our previous interpretation. The quotation marks around the word constant are placed deliberately: in fact, $\bar{K}_z$ and C_z are *not* constant but vary with the frequency ω of oscillation. Nonetheless, Equation 15.33 suggests for the vertical mode of oscillation an analogy between the actual foundation–soil system and the system depicted in Figure 15.11 and consisting of the same foundation but supported on a "*spring*" and "*dashpot*" with characteristic moduli equal to $\bar{K}_z$ and C_z, respectively.

Once these moduli have been established for a particular excitation frequency, $\bar{u}_z$ is obtained from Equation 15.34:

$$\bar{u}_z = \frac{F_z}{(\bar{K}_z - m\omega^2) + i\omega C_z} \qquad (15.35a)$$

and thereby the *amplitude* of oscillation that is of interest is simply

$$u_z = |\bar{u}_z| = \frac{F_z}{\sqrt{(\bar{K}_z - m\omega^2)^2 + \omega^2 C_z^2}} \qquad (15.35b)$$

Conclusion: The soil reaction against a vertically oscillating foundation is fully described with the complex frequency-dependent dynamic vertical *impedance* $\mathcal{K}_z(\omega)$ or, equivalently, the frequency-dependent "*spring*" (stiffness) and "*dashpot*" (damping) coefficients, $\bar{K}_z(\omega)$ and $C_z(\omega)$. Once these parameters have been obtained for the particular frequency (or frequencies) of interest, solving the equation of motion yields the desired amplitude of the harmonic vertical displacement.

15.4.2 Generalization to All Modes of Oscillation

The definition of dynamic impedance given in Equation 15.31 for vertical excitation–response is also applicable to each of the other five modes of vibration. Thus, we define as lateral (swaying) impedance $\mathcal{K}_y$ the ratio of the horizontal harmonic force, $P_y(t)$, imposed in the short direction at the base of the foundation over the resulting harmonic displacement, $u_y(t)$, in the same direction:

$$\mathcal{K}_y = \frac{P_y(t)}{u_y(t)} = \frac{\bar{P}_y}{\bar{u}_y} \qquad (15.36)$$

Similarly,

$\mathcal{K}_x$ = the longitudinal (swaying) impedance (force–displacement ratio), for horizontal motion in the long direction

$\mathcal{K}_{rx}$ = the rocking impedance (moment–rotation ratio), for rotational motion about the long axis of the foundation basemat

$\mathcal{K}_{ry}$ = the rocking impedance (moment–rotation ratio), for rotational motion about the short axis of the foundation basemat

$\mathcal{K}_t$ = the torsional impedance (moment–rotation ratio), for rotational oscillation about the vertical axis

Moreover, in embedded foundations and piles, horizontal forces along principal axes induce rotational in addition to translational oscillations; hence, two more "cross-coupling" horizontal-rocking impedance exist: $\mathcal{K}_{xry}$ and $\mathcal{K}_{yrx}$. They are usually negligibly small in shallow foundations, but their effects may become appreciable for greater depths of embedment, owing to the moments about the base axes produced by horizontal soil reactions against the sidewalls. In piles the "cross-coupling" impedances are as important as the "direct" impedances.

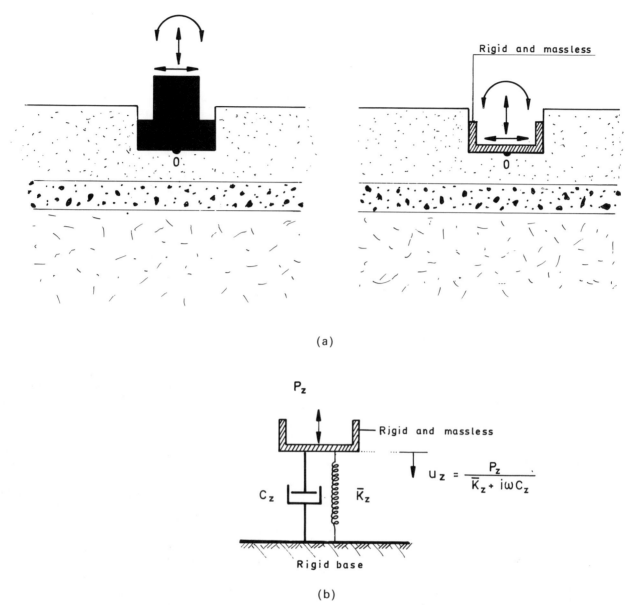

(a)

(b)

Fig. 15.11 (a) A foundation–structure system and the associated rigid and massless foundation. (b) Physical interpretation of the dynamic stiffness ($\bar{K}_z$) and dashpot (C_z) coefficients for a vertically vibrating footing.

Note that throughout this chapter (as in most of the literature) all impedances refer to axes passing through the foundation basemat–soil interface.

The eight impedances turn out to be complex numbers and functions of frequency that can be written in the form of Equation 15.32. Thus, in general, for each mode

$$\mathcal{K}(\omega) = \bar{K}(\omega) + i\omega C(\omega) \qquad (15.37)$$

and the analogy suggested in Figure 15.11 extends to all modes.

Once, for a particular excitation frequency ω, the eight dynamic impedances (or the eight pairs of dynamic stiffness or "spring" and "dashpot" coefficients) have been determined by following the procedures to be presented in this chapter, by recourse to the published literature, or by using available numerical formulations and computer codes, the steady-state response of a *rigid* foundation block to arbitrary harmonic external forces can be computed analytically by application of Newton's laws. Also analytically, one can derive the steady-state response of a flexible structure possessing natural modes and

subjected to harmonic external forces and to harmonic base motion.

15.4.3 Coupled Swaying–Rocking Oscillation

Figure 15.12 portrays a typical rigid block foundation: it has equal depth of embedment along all the sides and possesses two orthogonal vertical planes of symmetry, xz and yz, the intersection of which defines the vertical axis of symmetry, z. The foundation plan also has two axes of symmetry, x and y. For such a foundation the vertical and torsional modes of oscillation along and around the z axis can be treated separately, as was previously illustrated for vertical oscillation. In other words, each of these two modes is uncoupled from all the others.

On the other hand, swaying oscillation in the y direction cannot be realized without simultaneous rocking oscillation about x. This coupling of these two modes is a consequence of the *inertia* of the block and the fact that its center of gravity

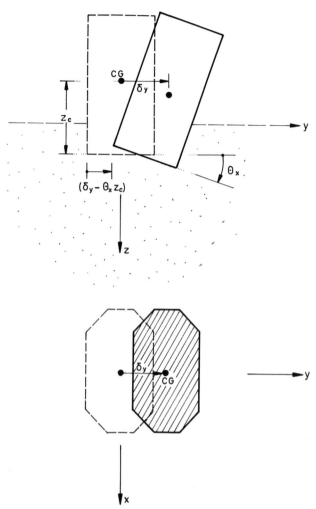

Fig. 15.12 Coupled swaying–rocking oscillations: definition of displacement variables. Top: section. Bottom: plan.

is above the center of pressure of the soil reactions. Thus, if the block is initially being displaced only horizontally, an inertial force arises at the center of gravity and produces a net moment at the foundation base—hence rocking is born. Similarly coupled are swaying in the x direction and rocking around y.

To study the coupled swaying–rocking oscillations of the block in the zy plane, we call δ_y and ϑ_x the horizontal displacement at the foundation center of gravity and the angle of rotation of the rigid block, respectively. Referring to Figure 15.12 and calling $F_y(t)$ and $M_x(t)$ the excitation force and moment at the block center of gravity, one can write the translational force and rotational moment dynamic equilibrium as follows:

$$P_y(t) + m\ddot{\delta}_y(t) = F_y(t) \tag{15.38}$$

$$T_x(t) - P_y(t) \cdot z_c + I_{0x}\ddot{\vartheta}_x(t) = M_x(t) \tag{15.39}$$

where

m = total foundation mass

I_{0x} = mass moment of inertia about a principal horizontal axis, parallel to x and passing through the block center of gravity

P_y and T_x = net horizontal force and rocking moment reactions, acting from the soil against the foundation during swaying and rocking, and referring to the centroid of the foundation basemat

For a harmonic excitation:

$$F_y(t) = F_y \exp(i\omega t) \tag{15.40}$$

$$M_x(t) = M_x \exp(i\omega t) \tag{15.41}$$

in which the amplitudes F_y and M_x may be either constant, or (more typically) proportional to the square of the operational frequency $\omega = 2\pi f$. F_y and M_x result from the operation of the machine.

The steady-state harmonic response can be written in the form:

$$\delta_y(t) = \bar{\delta}_y \exp(i\omega t) \quad \bar{\delta}_y = \delta_{y1} + i\delta_{y2} \tag{15.42}$$

$$\vartheta_x(t) = \bar{\vartheta}_x \exp(i\omega t) \quad \bar{\vartheta}_x = \vartheta_{x1} + i\vartheta_{x2} \tag{15.43}$$

in which $\bar{\delta}_y$ and $\bar{\vartheta}_x$ are complex frequency-dependent displacement and rotation amplitudes at the center of gravity. Note that Equations 15.40 to 15.43 do *not* by any means imply that the two components of motion and the two components of excitation are all in phase. Instead, the true phase angles are "hidden" in the complex form of each displacement component.

Using similar arguments with regard to the soil reactions, one may, without loss of generality, set

$$P_y(t) = \bar{P}_y \exp(i\omega t) \tag{15.44}$$

$$T_x(t) = \bar{T}_x \exp(i\omega t) \tag{15.45}$$

The complex amplitudes $\bar{P}_y$ and $\bar{T}_x$ are related to the complex displacement and rotation amplitudes through the corresponding dynamic impedances. Recalling that the latter are referred to the center of the foundation base, rather than the block center of gravity, one can immediately write

$$\bar{P}_y = \mathscr{K}_y(\bar{\delta}_y - z_c\bar{\vartheta}_x) + \mathscr{K}_{yrx}\bar{\vartheta}_x \tag{15.46}$$

$$\bar{T}_x = \mathscr{K}_{rx}\bar{\vartheta}_x + \mathscr{K}_{yrx}(\bar{\delta}_y - z_c\bar{\vartheta}_x) \tag{15.47}$$

Substituting Equations 15.40 to 15.47 into the governing Equations 15.37 to 15.38 leads to a system of two (coupled) algebraic equations with two unknowns $\bar{\delta}_y$ and $\bar{\vartheta}_x$. The solution is obtained using Kramer's rule:

$$\boxed{\bar{\delta}_y = (B_{22}F_y - B_{12}M_x)N^{-1}} \tag{15.48}$$

$$\boxed{\bar{\vartheta}_y = (B_{11}M_x - B_{12}F_y)N^{-1}} \tag{15.49}$$

in which the following substitutions have been made

$$B_{11} = \mathscr{K}_y(\omega) - m\omega^2 \tag{15.50a}$$

$$B_{12} = \mathscr{K}_{yrx}(\omega) - \mathscr{K}_y(\omega)z_c \tag{15.50b}$$

$$B_{22} = \mathscr{K}_{rx}(\omega) - I_{0x}\omega^2 + \mathscr{K}_y(\omega)z_c^2 - 2\mathscr{K}_{yrx}z_c \tag{15.50c}$$

and

$$N = B_{11}B_{22} - B_{12}^2 \tag{15.51}$$

Notice that, for a particular frequency ω, determination of the motions from Equations 15.48 to 15.51 is a straightforward operation once the *dynamic impedances* $\mathscr{K}_{ij}$ (or the corresponding "spring" and "dashpot" coefficients, $\bar{K}_{ij}$ and C_{ij}) are known. Of course, the computations are somewhat tedious if performed by hand, since complex numbers are involved; but with even small microcomputers the calculations can be done routinely, at minimal cost.

Therefore, it is proposed that this procedure, in connection with an appropriate evaluation of impedances at the frequency (or frequencies) of interest, should be used in analysis of machine foundations vibrating in swaying–rocking.

Vibrations in the vertical and torsional mode (each of which is practically uncoupled from all the other modes in the usual case of nearly symmetric foundations), can be respectively analysed with Equation 15.35 and its torsional counterpart:

$$\vartheta_z = |\bar{\vartheta}_z| = \frac{M_z}{\sqrt{(\bar{K}_z - J_z\omega^2)^2 + \omega^2 C_t^2}} \quad (15.52)$$

in which $\bar{K}_t$ = the dynamic "spring" coefficient for torsion, C_t = the "dashpot" coefficient for torsion, J_z = the moment of inertia of the whole foundation (including the machine) about the vertical z axis, and $M_z \exp(i\omega t)$ = the harmonic external moment around z.

15.5 COMPUTING DYNAMIC IMPEDANCES: TABLES AND CHARTS FOR DYNAMIC "SPRINGS" AND "DASHPOTS"

Several alternative computational procedures and computer codes are in principle available to the engineer wishing to obtain dynamic impedance functions ("springs" and "dashpots") for each specific machine–foundation problem. The choice among these methods depends to a large extent on the required accuracy, which in turn is primarily dictated by the size and importance of the particular project. Furthermore, the method to be selected must reflect the key characteristics of the foundation and the supporting soil. Specifically, one may broadly classify soil–foundation systems according to the following material and geometric characteristics:

- The shape of the foundation (circular, strip, rectangular, arbitrary)
- The type of soil profile (deep uniform deposit, deep multi-layered deposit, shallow stratum on rock)
- The amount of embedment (surface foundation, embedded foundation, piled foundation)

Broadly speaking, the various computational methods can be grouped into four categories, each with its own merits and limitations:

- *Analytical and semi-analytical* methods that can handle multi-layered soil deposits and rectangular surface foundations, but cannot treat embedment (e.g., Luco, 1976; Gazetas and Roesset, 1976, 1979).
- *Dynamic finite-element* methods that can treat surface, embedded, and piled foundations on or in layered soil profiles. Most of these methods are limited to axisymmetric (circular) or plane-strain (strip) situations, that is, they cannot study rectangles and arbitrary shapes; and usually they require the presence of a rigid bottom boundary (bedrock) at relatively shallow depths (Waas, 1972; Kausel, 1974; Lysmer et al., 1975).
- *Combined analytical–numerical* methods that try to take advantage of the capabilities of analytical and numerical approaches. Included in this category are recently developed boundary element methods (Kausel, 1981; Lysmer et al., 1981; Tassoulas, 1981).
- *Approximate techniques* that simplify the physics of the problem and can provide engineering solutions to some very complicated situations (e.g., separation between foundation sidewalls and backfill) that cannot be treated rigorously (Beredugo and Novak, 1972; Meek and Veletsos, 1973; Novak et al., 1978; Nogami, 1979; Gazetas and Dobry, 1984; Wolf, 1985, 1988; Gazetas and Tassoulas, 1987).

Application of most of the rigorous methods and solutions to a specific engineering problem usually involves using a specialized computer code, which may or may not be available. Developing "tailor-made" codes is a very impracticable undertaking, in view of the mathematical complexity of the problem.

Moreover, even when the appropriate sophisticated code *is* available, the effort involved in getting one or two sets of usable results may be such that no time/budget is left for the necessary parametric studies. Such studies are of course critical for exploring various design options and for evaluating the effects of uncertainties in poorly known parameters (e.g., soil properties, or quality of soil–foundation contact).

An alternative engineering approach has been the development of easy-to-use closed-form expressions and graphs, based on the results of rigorous and approximate formulations. This is the approach taken in this chapter.

15.5.1 Presentation of Tables and Graphs

Six large tables (15.1 through 15.6) present comprehensive and easy-to-use information for dynamic "spring" and "dashpot" coefficients. The information is in the form of simple algebraic formulas and dimensionless graphs pertaining to all possible (translational and rotational) modes of oscillation and covering a wide range of idealized soil profiles and foundation geometries. The engineer should be able, by using the tables, to approximate with sufficient accuracy the actual problem in many cases.

Figure 15.13 sketches the soil–foundation systems covered in each table. Specifically:

1. Table 15.1 and the accompanying set of graphs refer to foundations of *any solid shape* resting on the *surface of a homogeneous halfspace*.
2. Table 15.2 and the related graphs are for foundations with *any solid basemat shape partially or fully embedded in a homogeneous halfspace*.
3. Table 15.3 refers mainly to *circular and strip* foundations on the *surface* of a homogeneous soil *stratum* underlain by bedrock (some results are also given for rectangular foundations).
4. Table 15.4 refers to *circular and strip* foundations partially or fully *embedded* in a homogeneous *stratum* underlain by bedrock.
5. Table 15.5 pertains to *square and strip* foundations on the *surface* of some inhomogeneous profiles, in which the modulus increases smoothly with depth according to

$$G = G_0(1 + \alpha\zeta)^n \quad \zeta = z/B \quad (15.53)$$

6. Table 15.6 is mainly for laterally oscillating single *floating piles in two inhomogeneous and a homogeneous stratum or halfspace*; some information is also given for vertical oscillations, and for pile–soil–pile dynamic interaction factors.

Simplicity without any serious compromise in accuracy has been the prime goal when developing these tables. It is believed that, in general, the errors that may result from their use will be well within an acceptable 15 percent. (Use of the approximation symbol, however, implies a slightly inferior accuracy.)

The formulas and graphs given in Tables 15.1 and 15.2 for arbitrarily shaped surface and embedded foundations on or in a homogeneous halfspace have been compiled from some recent publications by the author and his coworkers (Dobry and Gazetas, 1985; Gazetas et al., 1985, 1987; Fotopoulou et al., 1989). They are based on (a) some simple physical models calibrated with results of rigorous boundary-element formulations and (b) data from the literature (most notably from the work of Lysmer, Veletsos, Luco, and Roesset and Kausel—see references). On the other hand, Tables 15.3 and 15.4 pertaining to surface and embedded foundation on a homogeneous stratum over bedrock are based on results by Kausel (1974), Johnson et al. (1975), Gazetas and Roesset (1976, 1979), Elsabee and

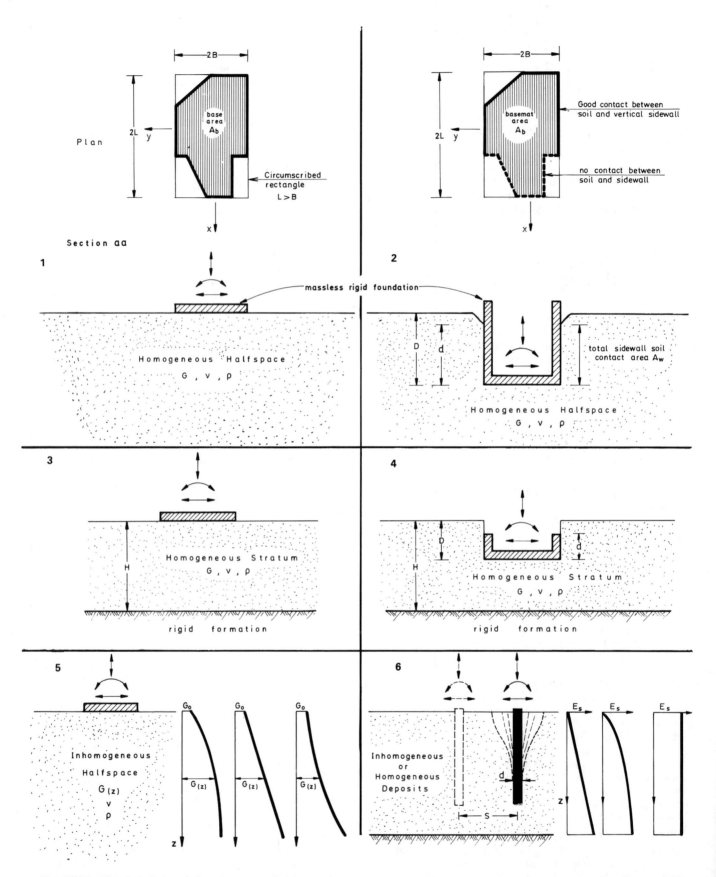

Fig. 15.13 The six soil–foundation systems studied in this chapter. Numbers 1 through 6 refer to the corresponding tables 15.1 to 15.6 and the associated graphs.

Morray (1977), Jakub and Roesset (1977), Kausel and Ushijima (1979), and Chow (1987). The sources of Table 15.5 for surface foundation on a number of inhomogeneous deposits include Hadjan and Luco (1977), Gazetas (1983), Booker et al. (1985), Wong and Luco (1985), Werkle and Waas (1986), and Novak (1987). Finally, the information of Table 15.6 on single piles is based on Roesset (1980a, b), Sanchez-Salinero (1982), Velez et al. (1983), and Gazetas (1984). The pile-group interaction factors are from Dobry and Gazetas (1988) and Gazetas and Makris (1990), calibrated with results from Nogami (1979), Kaynia and Kausel (1982), and Davies et al. (1985).

The reader will certainly find useful detailed information in these original sources. However, Tables 15.1 to 15.6 and the accompanying graphs are sufficient for complete dynamic analyses.

15.5.2 Use of Tables and Graphs; Illustrative examples

1. Surface Foundation on Halfspace

For an arbitrarily shaped foundation mat, the engineer must first determine a circumscribed restangle $2B$ by $2L$ $(L > B)$ using common sense as explained in Figures 15.13 and 15.14. Then, to compute the impedances in the six modes of vibration, from Table 15.1, all he needs is the values of:

• A_b, I_{bx}, I_{by} and J_b = area, area moments of inertia about x, y, and polar moment of inertia about z, of the *actual* soil–foundation contact surface. If loss of contact under part of the foundation (e.g., along the edges of a rocking

$$L/B = 3.5 \ , \ A_b = 57.6 \ m^2 \ (614 \ ft^2) \ , \ I_{bx} \approx 82 \ m^4 \ (9,309 \ ft^4) \ ,$$
$$I_{by} \approx 904 \ m^4 (105,000 \ ft^4) \ , \ \omega = 2\pi (30) = 188.5 \ s^{-1} \ , \ \alpha_o = 1.3$$

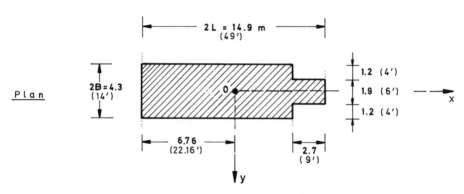

FIRST NUMERICAL EXAMPLE

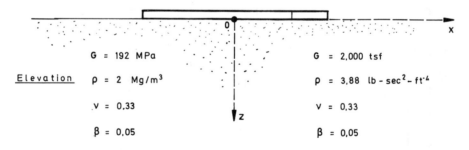

SECOND NUMERICAL EXAMPLE

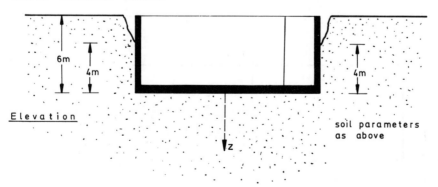

Fig. 15.14 Geometry and material parameters of the two illustrative numerical examples. (Note also that $V_s \approx 310$ m/s, $V_{La} \approx 500$ m/s; $\chi \approx 0.26$.)

TABLE 15.1 DYNAMIC STIFFNESSES AND DASHPOT COEFFICIENTS FOR ARBITRARILY SHAPED FOUNDATIONS ON THE SURFACE OF A HOMOGENEOUS HALFSPACE.

Vibration Mode	Static Stiffness K — Dynamic Stiffness $\bar{K} = K \cdot k(\omega)$ — General Shape (foundation–soil contact surface is of area A_b and has a circumscribed rectangle $2L$ by $2B$; $L > B$)*	Static Stiffness K — Square $L = B$	Dynamic Stiffness Coefficient k (General shape; $0 \leq a_0 \leq 2$)†	Radiation Dashpot Coefficient C (General Shapes)
Vertical, z	$K_z = \dfrac{2GL}{1-v}\,(0.73 + 1.54\chi^{0.75})$ with $\chi = \dfrac{A_b}{4L^2}$	$K_z = \dfrac{4.54GB}{1-v}$	$k_z = k_z\!\left(\dfrac{L}{B}, v; a_0\right)$ is plotted in Graph a	$C_z = (\rho V_{La}A_b)\cdot \tilde{c}_z$ $\tilde{c}_z(L/B, v; a_0)$ is plotted in Graph c
Horizontal, y (in the lateral direction)	$K_y = \dfrac{2GL}{2-v}\,(2 + 2.50\chi^{0.85})$	$K_y = \dfrac{9GB}{2-v}$	$k_y = k_y\!\left(\dfrac{L}{B}; a_0\right)$ is plotted in Graph b	$C_y = (\rho V_s A_b)\cdot \tilde{c}_y$ $\tilde{c}_y(L/B; a_0)$ is plotted in Graph d
Horizontal, x (in the longitudinal direction)	$K_x = K_y - \dfrac{0.2}{0.75-v}\,GL\!\left(1 - \dfrac{B}{L}\right)$	$K_x = K_y$	$k_x \simeq 1$	$C_x \simeq \rho V_s A_b$
Rocking, rx (around longitudinal x axis)	$K_{rx} = \dfrac{G}{1-v}\,I_{bx}^{0.75}\left(\dfrac{L}{B}\right)^{0.25}\!\left(2.4 + 0.5\,\dfrac{B}{L}\right)$ with I_{bx} (I_{by}) area moment of inertia of the foundation–soil contact surface around the $x(y)$ axis	$K_{rx} = \dfrac{3.6GB^3}{1-v}$	$k_{rx} \simeq 1 - 0.20a_0$	$C_{rx} = (\rho V_{La}I_{bx})\cdot \tilde{c}_{rx}$ $\tilde{c}_{rx}(L/B; a_0)$ is plotted in Graphs e and f
Rocking, ry (around lateral axis)	$K_{ry} = \dfrac{G}{1-v}\,I_{by}^{0.75}\left[3\left(\dfrac{L}{B}\right)^{0.15}\right]$	$K_{ry} = K_{rx}$	$\begin{cases} v < 0.45:\\ \quad k_{ry} \simeq 1 - 0.30a_0 \\ v \simeq 0.50:\\ \quad k_{ry} \simeq 1 - 0.25a_0\!\left(\dfrac{L}{B}\right)^{0.30} \end{cases}$	$C_{ry} = (\rho V_{La}I_{by})\cdot \tilde{c}_{ry}$ $\tilde{c}_{ry}(L/B; a_0)$ is plotted in Graph g
Torsional	$K_t = GJ_b^{0.75}\left[4 + 11\left(1 - \dfrac{B}{L}\right)^{10}\right]$ with $J_b = I_{bx} + I_{by}$ being the polar moment of the soil–foundation contact surface	$K_t = 8.3GB^3$	$k_t \simeq 1 - 0.14a_0$	$C_t = (\rho V_s J_b)\cdot \tilde{c}_t$ $\tilde{c}_t(L/B; a_0)$

* Note that as $L/B \to \infty$ (strip footing) the theoretical values of K_z and $K_y \to 0$; the values computed from the two given formulas correspond to a footing with $L/B \approx 20$.

† $a_0 = \omega B/V_s$.

GRAPHS ACCOMPANYING TABLE 15.1

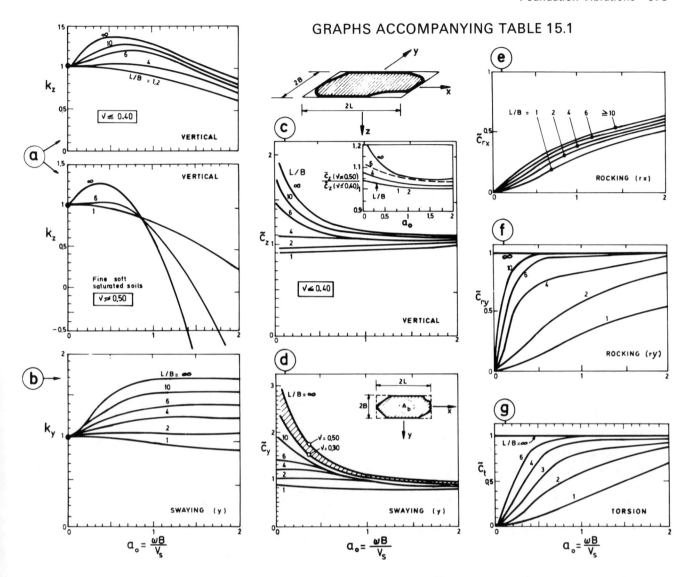

foundation) is likely, the engineer may use his judgment to discount the contribution of this part.

- B and L = semi-width and semi-length of the circumscribed rectangle.
- G and v, or V_s and V_{La} = shear modulus, Poisson's ratio, shear-wave velocity, and "Lysmer's analog" wave velocity. The last is the apparent propagation velocity of compression–extension waves under a foundation and is related to V_s according to

$$V_{La} = \frac{3.4}{\pi(1-v)} V_s \qquad (15.54)$$

- $\omega = 2\pi f$ = the circular frequency (in radians/second) of the applied force (e.g., frequency of operation of the machine).

This table as well as all the other tables gives:

- The dynamic stiffnesses ("springs"), $\bar{K} = \bar{K}(\omega)$, as a product of the static stiffness, K, times the dynamic stiffness coefficient $k = k(\omega)$:

$$\bar{K}(\omega) = K \cdot k(\omega) \qquad (15.55)$$

- The radiation damping ("dashpot") coefficients $C = C(\omega)$. These coefficients do *not* include the soil hysteretic damping, β_0 (the only exception is with Table 15.6 (for piles), where

the ξ_{ij} values combine hysteretic and radiation damping). To incorporate such damping, simply add to the foregoing C value the corresponding material dashpot coefficient $2\bar{K}\beta_0/\omega$:

$$\text{Total } C = \text{radiation } C + \frac{2\bar{K}}{\omega}\beta_0 \qquad (15.56)$$

Numerical Example A numerical example illustrates the use of Table 15.1 and the attached graphs in computing the dynamic stiffnesses ("springs") and damping coefficients ("dashpots"), for four modes of vibration. A sketch of the foundation and lists of all pertinent geometric, material, and load parameters are included in Figure 15.14. The computations follow in SI units, but the results are also given in English units (in parentheses). The excitation frequency is $f = 30$ Hz.

VERTICAL MODE

Static stiffness:

$$K_z = \frac{2.0 \times 192\,000 \times 7.45}{1 - 0.33}\left[0.73 + 1.54(0.26)^{0.75}\right]$$

$$\cong 5500 \times 10^3 \text{ kN/m} \quad (3.8 \times 10^8 \text{ lb/ft})$$

Dynamic stiffness coefficient: $k_z \cong 0.93$

Dynamic stiffness ("spring"):

$\bar{K}_z = 5500 \times 10^3 \times 0.93 = 5.1 \times 10^6$ kN/m   (3.5×10^8 lb/ft)

Radiation dashpot coefficient:

$$C_z = 2.0 \times 500 \times 57.6 \times 1$$
$$\cong 5.8 \times 10^4 \text{ kN} \cdot \text{s} \cdot \text{m}^{-1} \quad (3.8 \times 10^6 \text{ lb-sec-ft}^{-1})$$

Total dashpot coefficient:

$$\text{Total } C_z = 5.8 \times 10^4 + \frac{2 \times 5.1 \times 10^6}{188.5} \times 0.05$$
$$\cong 6.0 \times 10^4 \text{ kN} \cdot \text{s} \cdot \text{m}^{-1} \quad (4.0 \times 10^6 \text{ lb-sec-ft}^{-1})$$

LATERAL HORIZONTAL MODE (y)

Static stiffness:

$$K_y = \frac{2.0 \times 192\,000 \times 7.45}{2 - 0.33} [2 + 2.5(0.26)^{0.85}]$$
$$\cong 4790 \times 10^3 \text{ kN/m} \quad (3.2 \times 10^8 \text{ lb/ft})$$

Dynamic stiffness coefficient: $k_y \approx 1.2$

Dynamic stiffness ("spring"):

$\bar{K}_y = 4790 \times 10^3 \times 1.2 \cong 5.8 \times 10^6$ kN/m   (3.9×10^8 lb/ft)

Radiation dashpot coefficient:

$$C_y = 2.0 \times 310 \times 57.6 \times 1.0$$
$$\cong 3.6 \times 10^4 \text{ kN} \cdot \text{s} \cdot \text{m}^{-1} \quad (2.4 \times 10^6 \text{ lb-sec-ft}^{-1})$$

Total dashpot coefficient:

$$\text{Total } C_y = 3.6 \times 10^4 + \frac{2 \times 5.8 \times 10^6}{188.5} \times 0.05$$
$$\cong 3.9 \times 10^4 \text{ kN} \cdot \text{s} \cdot \text{m}^{-1} \quad (2.7 \times 10^6 \text{ lb-sec-ft}^{-1})$$

ROCKING MODE rx (AROUND THE LONGITUDINAL AXIS)

Static stiffness:

$$K_{rx} = \frac{192\,000}{1 - 0.33} \times (82.0)^{0.75} \times (3.5)^{0.25} \times \left(2.4 + \frac{0.5}{3.5}\right)$$
$$\cong 2.7 \times 10^7 \text{ kN} \cdot \text{m} \quad (1.4 \times 10^{10} \text{ lb-ft})$$

Dynamic stiffness coefficient:

$$k_{rx} \cong 1 - (0.2 \times 1.3) = 0.74$$

Dynamic stiffness ("spring"):

$$\bar{K}_{rx} = 2.7 \times 10^7 \times 0.74$$
$$\cong 2.0 \times 10^7 \text{ kN} \cdot \text{m} \quad (1.07 \times 10^{10} \text{ lb-ft})$$

Radiation dashpot coefficient:

$$C_{rx} = 2.0 \times 500 \times 82.0 \times 0.5$$
$$\cong 4.1 \times 10^4 \text{ kN} \cdot \text{s} \cdot \text{m} \quad (3.7 \times 10^7 \text{ lb-sec-ft})$$

Total dashpot coefficient:

$$\text{Total } C_{rx} = 4.1 \times 10^4 + \frac{2 \times 2.0 \times 10^7}{188.5} \times 0.05$$
$$\cong 5.2 \times 10^4 \text{ kN} \cdot \text{s} \cdot \text{m} \quad (4.5 \times 10^7 \text{ lb-sec-ft})$$

TORSIONAL MODE

Static stiffness

$$K_t = 192\,000 \times (985)^{0.75} \times \left[4 + 11\left(1 - \frac{1}{3.5}\right)^{10}\right]$$
$$\cong 1.5 \times 10^8 \text{ kN} \cdot \text{m} \quad (10.3 \times 10^{10} \text{ lb-ft})$$

Dynamic stiffness coefficient:

$$k_t \cong 1 - (0.14 \times 1.3) \cong 0.82$$

Dynamic stiffness ("spring"):

$$\bar{K}_t = 1.5 \times 10^8 \times 0.82$$
$$\cong 1.2 \times 10^8 \text{ kN} \cdot \text{m} \quad (8.5 \times 10^{10} \text{ lb-ft})$$

Radiation dashpot coefficient:

$$C_t = 2.0 \times 310 \times 985 \times 0.90$$
$$\cong 5.5 \times 10^5 \text{ kN} \cdot \text{s} \cdot \text{m} \quad (4.05 \times 10^8 \text{ lb-sec-ft})$$

Total dashpot coefficient:

$$\text{Total } C_t = 5.5 \times 10^5 + \frac{2 \times 1.2 \times 10^8}{188.5} \times 0.05$$
$$\cong 6.1 \times 10^5 \text{ kN} \cdot \text{s} \cdot \text{m} \quad (4.5 \times 10^8 \text{ lb-sec-ft})$$

This concludes the example computations. Having obtained these "spring" ($\bar{K}$) and "dashpot" (total C), the engineer can use

- Equation 15.35 to obtain the amplitude of vertical oscillation.
- Equation 15.52 to obtain the amplitude of torsional oscillation.
- Equations 15.48 to 15.51 to obtain the amplitudes δ_y and ϑ_y of coupled swaying–rocking oscillation in the yz plane (lateral direction).

To get an insight into the meaning of the computed C_j values ($j = z, y, rx, t$), let us consider above the foundation base of Figure 15.14 a rigid block having a (total) mass $m = 1200$ mg and (total) mass moments of inertia $I_x = 13\,000$ mg $\cdot$ m^2 and $I_z = 38\,000$ mg $\cdot$ m^2, about the base x axis and the vertical z axis, respectively. For each mode an "*effective*" or "*equivalent*" *damping ratio*, ξ_j, can be defined as follows:

Translational modes $\xi_j = \dfrac{C_j}{2\sqrt{\bar{K}_j m}}$ $(j = z, y)$ (15.57a)

Rotational modes $\xi_l = \dfrac{C_l}{2\sqrt{\bar{K}_l I_p}}$ $(l = rx, t; p = x, z)$ (15.57b)

For the four modes considered we obtain:

- Vertical $\xi_z \cong 0.38$ or 38 percent
- Lateral $\xi_y \cong 0.23$ or 23 percent
- Rocking $\xi_{rx} \cong 0.05$ or 5 percent
- Torsion $\xi_t \cong 0.15$ or 15 percent

From these values it is evident that for rotational modes and especially for rocking the effective damping ratio is quite low, while for translational modes and especially vertical oscillation it can be very high. These differences are a direct consequence of different amounts of "*radiation*" *damping* which results from geometric spreading of waves generated at the foundation–soil interface. When the foundation undergoes a vertical oscillation such waves are emitted in phase and "reach" long distances away from the foundation; hence, relatively large amounts of wave energy are "lost" for the foundation—high radiation damping. In contrast, two "points" symmetrically located on the opposite sides of a rocking foundation send off waves that are 180° out of phase and tend to cancel each other out when they meet at distant locations along the centerline (dynamic equivalent of St. Venant's Principle); hence, they cannot "reach" long distances and effectively dissipate little energy from that imparted onto the foundation—low radiation damping.

Such differences in effective damping ratios are typical for surface foundation *on a homogeneous halfspace*. Figure 15.15, adopted from the first edition of this handbook (Richart, 1975), refers to a circular foundation of radius R. Each of the four damping ratios, ξ_z, ξ_x, ξ_{rx}, and ξ_t, is portrayed as a decreasing function of a corresponding mass ratio:

$$\bar{m}_z = \frac{m(1-v)}{4\rho R^3} \qquad \bar{m}_x = \frac{m(2-v)}{8\rho R^3} \qquad (15.58a)$$

$$\bar{m}_{rx} = \frac{3I_x(1-v)}{8\rho R^5} \qquad \bar{m}_t = \frac{I_z}{\rho R^5} \qquad (15.58b)$$

The reader should check the overall similarity between the damping ratios computed in this example and those anticipated from Figure 15.15.

Two important conclusions emerge from Figure 15.15.

1. The consequences of resonance (when the operational frequency coincides with the natural frequency of the foundation–soil system) are far more secious for rocking than for translational vibrations. "Avoiding resonance" in rocking is a prudent consideration. On the other hand, relatively light foundations vibrating vertically on homogeneous halfspace may experience damping, ξ_z, in excess of 50 percent; hence, occurrence of resonance would hardly spell disaster, and avoiding it at any cost (as some older procedures were recommending) might mislead the designer. Recall that the "dynamic amplification" or "dynamic load factor" at resonance, DA_{max}, is almost inversely proportional to ξ_z:

$$DA_{max} \equiv \frac{\max u_z}{\text{static } u_z} = \frac{1}{2\xi_z\sqrt{1-\xi_z^2}} \qquad (15.59)$$

2. As the inertia of a foundation block increases relative to its base dimensions, the effective damping ratios decrease (a consequence of m and I_p being in the denominator in Equation 15.57). Therefore, older design practices of keeping

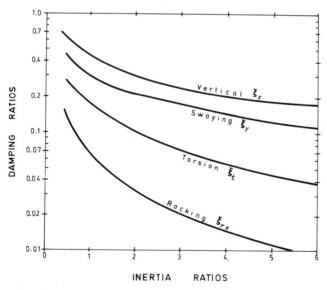

Fig. 15.15 Radiation damping ratios of a circular foundation on a homogeneous half-space. Notice the very small amount of damping generated in rocking. (*Adapted from Richart et al., 1970.*)

the foundation mass quite large are rather unfortunate. Another conclusion from Figure 15.16 is that rigid block foundations may not necessarily be sensitive to high-frequency excitation, because, even in the unlikely event of resonance, peak response may not be excessive. High *natural* frequencies require relatively small masses or moments of inertia and are thereby associated with high damping ratios. As a general rule, the frequency factors $a_0 = \omega B/V_s$ of interest do not exceed 2.

The foregoing conclusions are valid for surface foundations on very deep homogeneous deposits (halfspace). Important factors that may modify (quantitatively *and* qualitatively) these conclusions include foundation embedment, the presence of bedrock at shallow depths, and soil inhomogeneity.

2. Foundation Embedded in Halfspace

With the formulas and graphs of Table 15.2 one can assess the effects of embedment in a variety of realistic situations. The additional parameters that must be known or computed before using this table are:

- D = the elevation below the ground surface of the foundation basemat.
- A_w or d = the total area of the *actual* sidewall–soil contact surface, or the (average) height of the sidewall that is in "good" contact with the surrounding soil. A_w should, in general, be smaller than the nominal area of contact to account for such phenomena as slippage and separation that may occur near the ground surface. The engineer should refer to published results of large-scale and small-scale experiments for a guidance in selecting a suitable value for A_w or d (e.g., Stokoe and Richart, 1974; Erden, 1974; Novak, 1985). Note that A_w or d will not necessarily attain a single value for all modes of oscillation.
- A_{w_s} and $A_{w_{ce}}$, which refer to horizontal oscillations and represent the sum of the projections of all the sidewall areas in directions parallel (A_{w_s}) and perpendicular ($A_{w_{ce}}$) to loading. Again A_{w_s} and $A_{w_{ce}}$ should be smaller than the nominal areas shearing and compressing the soil, to account for slippage and/or separation. h = the distance of the (effective sidewall) centroid from the ground surface.
- J_{w_s} and $I_{w_{ce}}$, which refer to rocking and correspond to A_{w_s} and $A_{w_{ce}}$, respectively. J_{w_s} = the sum of the polar moments of inertia about the (off-plane) axis of rotation of all surfaces actually shearing the soil; $I_{w_{ce}}$ = the sum of moments of inertia of all surfaces actually compressing the soil about their base axes parallel to the axis of rotation (x or y).
- $J_{w_{ce}}$, which pertains to torsion and is equal to the sum of the moments of inertia of all surfaces actually compressing the soil about the projection of the axis of rotation (z) onto their plane.

It is pointed out that most of the formulas of Table 15.2 are valid for symmetric *and* nonsymmetric contact along the perimeter of the vertical sidewalls and the surrounding soil. Note also that Table 15.2 compares the dynamic stiffnesses and dashpot coefficients of an embedded foundation, $\bar{K}_{emb} = K_{emb} \cdot k_{emb}$ and C_{emb}, with those of the corresponding surface foundation, $\bar{K}_{sur} = K_{sur} \cdot k_{sur}$ and C_{sur}.

Numerical Example A numerical example illustrates the use of Table 15.2 and the attached graphs. Figure 15.14 sketches the foundation, whose basemat is identical to the mat of the previous example (Fig. 15.14), but which is now placed at a depth $D = 6$ m. The sidewalls are in contact with the soil throughout the height D, but the engineer believes that the quality of contact at the top 2 m, will be too poor to be trusted.

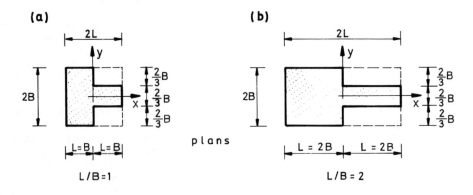

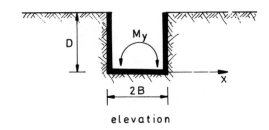

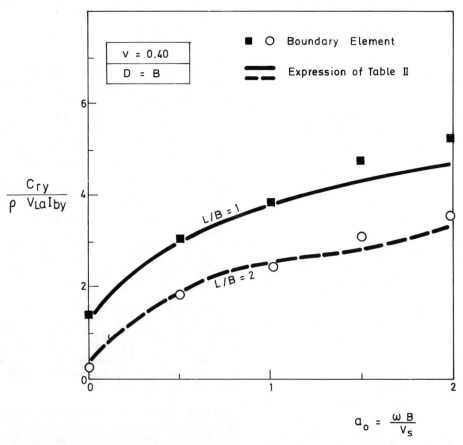

Fig. 15.16 Comparison of the results derived using the simple expressions given in this chapter with the results of rigorous (boundary-element) formulations for two T-shaped foundations ($L/B = 1$ and 2) embedded in a homogeneous halfspace.

He thus decides that the effective height of sidewall–soil contant is $d = 4$ m. The computations that follow make use of the results of the previous example for $\bar{K}_{sur}$ and C_{sur}.

VERTICAL MODE

$$A_w = 2 \times (14.9 + 4.3) \times 4.0 = 153.6 \text{ m}^2$$

$$D/B = 6/2.15 = 2.79$$

$$A_w/A_b = 153.6/57.6 = 2.67$$

Static and dynamic stiffness ("spring"):

$$K_{z,\text{emb}} = 5.5 \times 10^6 \times \left[1 + \frac{2.79}{21}(1 + 1.33 \times 0.26) \right]$$

$$\times [1 + 0.2 \times (2.67)^{2/3}]$$

$$= 5.5 \times 10^6 \times 1.18 \times 1.38 \cong 9 \times 10^6 \text{ kN/m}$$

The dynamic stiffness coefficient is obtained by linear interpolation between the "fully-embedded" value,

$$k_{z,d=6\text{m}} \cong 0.93[1 - 0.09(2.79)^{3/4}(1.3)^2] \cong 0.62$$

and the value for the foundation placed in an open trench, without sidewalls,

$$k_{z,d=0} \cong 0.93[1 + 0.09(2.79)^{3/4}(1.3)^2] \cong 1.24$$

Thus,

$$k_{z,\text{emb}} = k_{z,d=4\text{m}} \cong (4 \times 0.62 + 2 \times 1.24)/6 \cong 0.83$$

and

$$\bar{K}_{z,\text{emb}} = 9 \times 0.83 \times 10^6 = 7.4 \times 10^6 \text{ kN/m}$$

"Dashpot":

$$C_{z,\text{emb}} = 5.8 \times 10^4 + 2.0 \times 310 \times 153.6$$

$$\cong 15.0 \times 10^4 \text{ kN} \cdot \text{s} \cdot \text{m}^{-1}$$

$$\text{Total } C_{z,\text{emb}} = 15.0 \times 10^4 + \frac{2 \times 7.4 \times 10^6}{188.5} \times 0.05$$

$$= 15.4 \times 10^4 \text{ kN} \cdot \text{s} \cdot \text{m}^{-1}$$

LATERAL HORIZONTAL MODE (y)

$$h = 4 \text{ m} \qquad h/B = 4/2.15 = 1.86$$

$$A_w/L^2 = 153.6/7.45^2 \cong 2.77$$

Static and dynamic stiffness ("spring"):

$$K_{y,\text{emb}} = 4.79 \times 10^6 (1 + 0.15\sqrt{2.79})[1 + 0.52(1.86 \times 2.77)^{0.4}]$$

$$\cong 4.79 \times 10^6 \times 1.25 \times 2.0$$

$$\cong 12.0 \times 10^6 \text{ kN/m}$$

The dynamic stiffness coefficient is obtained with the help of the graphs that accompany Table 15.2. The "fully-embedded" value is

$$k_{y,d=6\text{m}} \cong 0.3$$

obtained by interpolating between the $L/B = 2$ and $L/B = 6$ plots, for $D/B \simeq 2.79$. Then for the partial embedment:

$$k_{y,\text{emb}} = k_{y,d=4\text{m}} \cong k_{y,d=6\text{m}} \cdot \frac{k(2/3)}{k(1)}$$

$$\cong 0.3 \times 1.3 \cong 0.4$$

$$\bar{K}_{y,\text{emb}} \cong 12.0 \times 0.4 \cong 4.8 \times 10^6 \text{ kN/m}$$

"Dashpot":

$$A_{w_s} = 2 \times 4.3 \times 4.0 = 34.4 \text{ m}^2$$

$$A_{w_{ce}} = 2 \times 14.9 \times 4.0 = 119.2 \text{ m}^2$$

$$C_{y,\text{emb}} = 3.6 \times 10^4 + 2.0 \times 310 \times 34.4 + 2.0 \times 500 \times 119.2$$

$$= 3.6 \times 10^4 + 2.1 \times 10^4 + 11.9 \times 10^4$$

$$\cong 17.6 \times 10^4 \text{ kN} \cdot \text{s} \cdot \text{m}^{-1}$$

$$\text{Total } C_{y,\text{emb}} = 17.6 \times 10^4 + \frac{2 \times 4.8 \times 10^6}{188.5} \times 0.05$$

$$\cong 17.9 \times 10^4 \text{ kN} \cdot \text{s} \cdot \text{m}^{-1}$$

ROCKING MODE rx (AROUND THE LONGITUDINAL AXIS)

$$d/B = 4/2.15 = 1.86$$

$$d/D = 4/6 = 0.67$$

$$B/L = 2.15/7.45 \cong 0.29$$

Static and dynamic stiffness ("spring"):

$$K_{rx,\text{emb}} = 2.7 \times 10^7 \{ 1 + 1.26 \times 1.86$$

$$\times [1 + 1.86 \times (0.67)^{-0.2} \times \sqrt{0.29}] \}$$

$$\cong 2.7 \times 5.89 \cong 15.9 \times 10^7 \text{ kN} \cdot \text{m}$$

$$k_{rx,\text{emb}} \cong k_{rx,\text{sur}} = 0.74$$

$$\bar{K}_{rx,\text{emb}} = 15.9 \times 10^7 \times 0.74 \cong 11.8 \times 10^7 \text{ kN} \cdot \text{m}$$

"Dashpot":

$$J_{w_s} = 2 \times \tfrac{1}{3} \times 4.3 \times 4.0 \times (2.15^2 + 4^2) \cong 236.5 \text{ m}^4$$

$$I_{w_{ce}} = 2 \times \tfrac{1}{3} \times 14.9 \times 4^3 \cong 635.7 \text{ m}^4$$

$$\sum_i (A_{w_{cei}}\Delta_i^2) = 2 \times (12.2 \times 4.0 \times 2.15^2 + 2.7 \times 4.0 \times 0.95^2)$$

$$\cong 235.3 \text{ m}^4$$

$$\tilde{c}_1 = 0.25 + 0.65\sqrt{1.3} \times (0.67)^{-0.65}(2.79)^{-0.25} \cong 1.0$$

$$C_{rx,\text{emb}} = 4.1 \times 10^4 + 2.0 \times 500 \times 635.7 \times 1.0 + 2.0 \times 310$$

$$\times (236.5 + 235.3) \times 1.0$$

$$\cong 4.1 \times 10^4 + 63.5 \times 10^4 + 29.2 \times 10^4$$

$$\cong 96.8 \times 10^4 \text{ kN} \cdot \text{s} \cdot \text{m}$$

$$\text{Total } C_{rx,\text{emb}} = 96.8 \times 10^4 + \frac{2 \times 11.8 \times 10^7}{188.5} \times 0.05$$

$$\cong 103 \times 10^4 \text{ kN} \cdot \text{s} \cdot \text{m}$$

TORSIONAL MODE (AROUND z, PASSING THROUGH BASE CENTROID)

"Spring":

$$K_{t,\text{emb}} = 1.5 \times 10^8 \times \{ 1 + 1.4 \times (1 + 0.29) \times (1.86)^{0.9} \}$$

$$\cong 6.2 \times 10^8 \text{ kN} \cdot \text{m}$$

$$k_{t,\text{emb}} \cong k_{t,\text{sur}} = 0.82$$

$$\bar{K}_{t,\text{emb}} = 6.2 \times 10^8 \times 0.82 \cong 5.1 \times 10^8 \text{ kN} \cdot \text{m}$$

TABLE 15.2 DYNAMIC STIFFNESSES AND DASHPOT COEFFICIENTS FOR ARBITRARILY SHAPED PARTIALLY OR FULLY EMBEDDED IN A HOMOGENEOUS HALFSPACE.

Vibration Mode	Dynamic Stiffness $\bar{K}_{emb} = K_{emb} \cdot k_{emb}(\omega)$		Radiation Dashpot Coefficient $C_{emb}(\omega)$	
	Static Stiffness K_{emb} — For foundations with arbitrarily-shaped basemat A_b of circumscribed rectangle $2L$ by $2B$; total sidewall–soil contact area A_w (or constant sidewall–soil contact height d)	Dynamic Stiffness Coefficient $k_{emb}(\omega)$, $0 < a_0 \leq 2$	General Foundation Shape	Rectangular Foundation $2L$ by $2B$ by d
Vertical, z	$$K_{z,emb} = K_{z,sur}\left[1 + \frac{1}{21}\frac{D}{B}(1 + 1.3\chi)\right]$$ $$\times \left[1 + 0.2\left(\frac{A_w}{A_b}\right)^{2/3}\right]$$ $K_{z,sur}$ obtained from Table 15.1 A_w = actual sidewall–soil contact area; for constant effective contact height d along the perimeter $A_w = (d) \times (\text{Perimeter})$, $\chi = A_b/4L^2$	Fully embedded: $$k_{z,emb} \simeq k_{z,sur}\left[1 - 0.09\left(\frac{D}{B}\right)^{3/4} a_0^2\right]$$ In a trench: $$k_{z,tre} \simeq k_{z,sur}\left[1 + 0.09\left(\frac{D}{B}\right)^{3/4} a_0^2\right]$$ Partially embedded: estimate by interpolating between the two $\}$ $\nu \leq 0.40$ Fully embedded, $L/B = 1-2$ $$k_{z,emb} \simeq 1 - 0.09(D/B)^{3/4}a_0^2$$ Fully embedded, $L/B > 3$ $$k_{z,emb} \simeq 1 - 0.35(D/B)^{1/2}a_0^{3.5}$$ $\}$ $\nu \approx 0.50$	$$C_{z,emb} = C_{z,sur} + \rho V_s A_w$$ with $C_{z,sur}$ and $\bar{c}_z$ according to Table 15.1	$$C_{z,emb} = 4\rho V_{La}BL\bar{c}_z$$ $$+ 4\rho V_s(B + L)d$$
Horizontal, y or x	$$K_{y,emb} = K_{y,sur}\left(1 + 0.15\sqrt{\frac{D}{B}}\right)$$ $$\times \left[1 + 0.52\left(\frac{h A_w}{B L^2}\right)^{0.4}\right]$$ $K_{y,sur}$ obtained from Table 15.1 $K_{x,emb}$ similarly computed from $K_{x,sur}$	$k_{y,emb}$ and $k_{x,emb}$ can be estimated in terms of $L/B, D/B$, and d/B for each value of a_0 from the graphs accompanying this table	$$C_{y,emb} = C_{y,sur} + \rho V_s A_{ws} + \rho V_{La} A_{wce}$$ $A_{ws} = \sum(A_{wi}\sin\vartheta_i)$ = total effective sidewall area shearing the soil $A_{wce} = \sum(A_{wi}\cos\vartheta_i)$ = total effective sidewall area compressing the soil ϑ_i angle of inclination of surface A_{wi} from loading direction $C_{y,sur}$ obtained from Table 15.1 $C_{x,emb}$ similarly computed from $C_{x,sur}$	$$C_{y,emb} = 4\rho V_{La}BL\bar{c}_y + 4\rho V_s Bd$$ $$+ 4\rho V_{La}Ld$$ $\bar{c}_y$ according to Table 15.1

	Static stiffness K_{emb}	Dynamic coefficient k_{emb}	Radiation damping coefficient C_{emb}	
	Expressions valid for any basemat shape but constant effective contact height d along the perimeter			
Rocking, rx (around the longitudinal axis)	$K_{rx,emb} = K_{rx,sur}$ $\cdot \left\{ 1 + 1.26 \dfrac{d}{B} \left[1 + \dfrac{d}{B} \left(\dfrac{d}{D} \right)^{-0.2} \sqrt{\dfrac{B}{L}} \right] \right\}$	$k_{rx,emb} \simeq k_{rx,sur}$	$C_{rx,emb} = C_{rx,sur} + \rho V_{La} I_{wce} \tilde{c}_1$ $+ \rho V_s \left(J_{ws} + \sum_i [A_{wcei} \Delta_i^2] \right) \tilde{c}_1$ $\tilde{c}_1 = 0.25 + 0.65 \sqrt{a_0} \left(\dfrac{d}{D} \right)^{-a_0/2} \times \left(\dfrac{D}{B} \right)^{-1/4}$ I_{wce} = total moment of inertia about their base axes parallel to x of all sidewall surfaces effectively compressing the soil Δ_i = distance of surface A_{wcei} from the x axis J_{ws} = polar moment of inertia about their base axes parallel to x of all sidewall surfaces effectively shearing the soil	$C_{rx,emb} = \frac{4}{3} \rho V_{La} B^3 L \tilde{c}_{rx}$ $+ \frac{4}{3} \rho V_{La} d^3 L \tilde{c}_1$ $+ \frac{4}{3} \rho V_s Bd (B^2 + d^2) \tilde{c}_1$ $+ 4 \rho V_s B^2 dL \tilde{c}_1$ with $\tilde{c}_1$ as in the preceding column and $\tilde{c}_{rx}$ according to Table 15.1
Rocking, ry (around the lateral axis)	$K_{ry,emb} = K_{ry,sur}$ $\cdot \left\{ 1 + 0.92 \left(\dfrac{d}{L} \right)^{0.6} \left[1.5 + \left(\dfrac{d}{L} \right)^{1.9} \times \left(\dfrac{d}{D} \right)^{-0.6} \right] \right\}$	$k_{ry,emb} \simeq k_{ry,sur}$	$C_{ry,emb}$ is similarly evaluated from $C_{ry,sur}$ with y replacing x and, in the equation for c_1, L replacing B	
Coupling term Swaying–rocking (x, ry) Swaying–rocking (y, rx)	$K_{xry,emb} \simeq \frac{1}{3} d K_{x,emb}$ $K_{yrx,emb} \simeq \frac{1}{3} d K_{y,emb}$	$k_{xry,emb} \simeq k_{yrx,emb} \simeq 1$	$C_{xry,emb} \simeq \frac{1}{3} d C_{x,emb}$ $C_{yrx,emb} \simeq \frac{1}{3} d C_{y,emb}$	as in the previous column
Torsional	$K_{t,emb} = K_{t,sur}$ $\times \left[1 + 1.4 \left(1 + \dfrac{B}{L} \right) \left(\dfrac{d}{B} \right)^{0.9} \right]$	$k_{t,emb} \simeq k_{t,sur}$	$C_{t,emb} = C_{t,sur} + \rho V_{La} J_{wce} \tilde{c}_2$ $+ \rho V_s \sum [A_{wi} \Delta_{zi}^2] \tilde{c}_2$ $\tilde{c}_2 \simeq \left(\dfrac{d}{D} \right)^{-0.5} \cdot \dfrac{a_0^2}{a_0^2 + \frac{1}{2} (L/B)^{-1.5}}$ J_{wce} = total moment of inertia of all sidewall surfaces effectively compressing the soil about the projection of the z axis onto their plane Δ_{zi} = distance of surface A_{wi} from the z axis	$C_{t,emb} = \frac{4}{3} \rho V_s BL (B^2 + L^2) \tilde{c}_t$ $+ \frac{4}{3} \rho V_{La} d (L^3 + B^3) \tilde{c}_2$ $+ 4 \rho V_s dBL (B + L) \tilde{c}_2$ with $\tilde{c}_2$ as in the preceding column and $\tilde{c}_t$ according to Table 15.1

NOTE: $V_{La} = \dfrac{3.4}{\pi(1 - \nu)} V_s$ is the apparent propagation velocity of compression–extension waves.

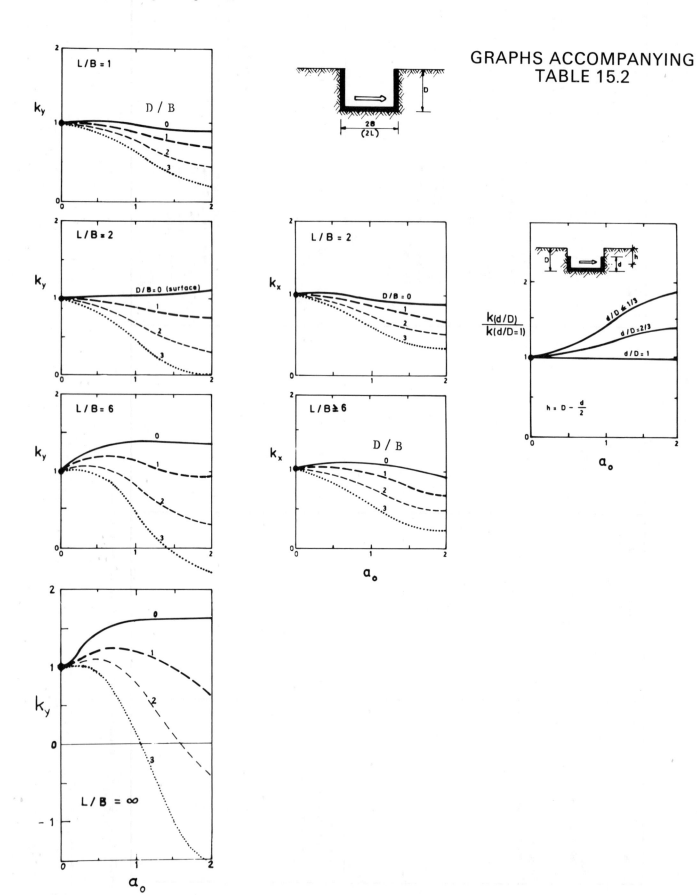

GRAPHS ACCOMPANYING
TABLE 15.2

"Dashpot":

$$\tfrac{1}{2} J_{w_{ce}} = 4.0 \times 14.9^3/12 + 4 \times 14.9 \times (8.14 - 7.45)^2$$

$$+ 4.0 \times 4.3^3/12$$

$$\cong 1157.5 \text{ m}^4$$

$$J_{w_{ce}} \cong 2315 \text{ m}^4$$

$$\sum_i (A_{w_i}\Delta_{z_i}^2) = 2 \times 12.2 \times 4.0 \times 2.15^2 + 4.3 \times 4.0 \times 6.76^2$$

$$+ 1.9 \times 4.0 \times 8.14^2 + 2 \times 2.7 \times 4.0 \times 0.95^2$$

$$+ 2 \times 1.2 \times 4.0 \times (8.14 - 2.7)^2$$

$$\cong 451 + 786 + 318 + 19 + 284$$

$$\cong 1858 \text{ m}^4$$

$$\tilde{c}_2 \cong (4/6)^{-0.5} \times \frac{1.3^2}{1.3^2 + (1/2) \times (3.5)^{-1.5}} \approx 1.17$$

$$C_{t,\text{emb}} = 5.5 \times 10^5 + 2.0 \times 500 \times 2315 \times 1.17 + 2.0 \times 310$$

$$\times 1858 \times 1.17$$

$$\cong 5.5 \times 10^5 + 27.1 \times 10^5 + 13.5 \times 10^5$$

$$\cong 46 \times 10^5 \text{ kN} \cdot \text{s} \cdot \text{m}$$

$$\text{Total } C_{t,\text{emb}} = 46 \times 10^5 + \frac{2 \times 5.1 \times 10^8}{188.5} \times 0.05$$

$$\cong 49 \times 10^5 \text{ kN} \cdot \text{s} \cdot \text{m}$$

It is apparent that embedment has produced very substantial increases for all "springs" (except the horizontal) and all "dashpots." We summarize these effects in terms of the dynamic stiffness ratio $\bar{K}_{\text{emb}}/\bar{K}_{\text{sur}}$ and the equivalent damping ratios ξ_{sur} (from the previous example) and ξ_{emb} (from the value of C_{emb} computed herein):

Mode	$\bar{K}_{\text{emb}}/\bar{K}_{\text{sur}}$	ξ_{sur}	ξ_{emb}
Vertical	1.45	38%	80%
Lateral	0.83	23%	>100%
Rocking	5.90	5%	42%
Torsion	4.25	15%	59%

Several conclusions of practical significance emerge from Table 15.2 and the illustrative example.

1. Increasing the embedment (in size and quality) may be a very effective way to reduce to acceptable levels the anticipated amplitudes of vibration, especially if these amplitudes arise due to rocking or torsion. Such an improvement would be effected mainly by the increase in radiation damping produced by waves emanating from the vertical sidewalls.
2. To rely on such a beneficial effect, however, the engineer must ensure that the quality of sidewall–soil contact is indeed high. In reality, unless special construction procedures are followed, separation ("gapping") and slippage are likely to occur near the ground surface where the initial confining pressures are small. Such effects may jeopardize the increase in damping and must be taken into account in the analysis. To this end, the areas and moments of inertia of the sidewalls on which damping and stiffness depend should be given suitably reduced values, rather than their nominal ones.

3. In view of the complexity of the problem of arbitrarily shaped partially embedded foundations, the formulas and graphs of Table 15.2 provide a very simple and complete solution, while allowing the engineer to use his experience and judgment. To give an idea as to how well the formulas of Table 15.2 may compare with rigorous theoretical solutions, Figure 15.16 refers to two foundations having T-shaped basemats and subjected to harmonic rocking oscillations. The circumscribed rectangles have $L/B = 1$ and 2 and each foundation is embedded at depth $D = B$. The rigorous results are from a dynamic boundary-element solution and are plotted as data points. The developed expressions for C_{rx}, given in Table 15.2, yield for each foundation the corresponding continuous lines. The agreement is indeed excellent and indicative of the capabilities of the simple methods utilized in this chapter. Also encouraging are comparisons of the presented solution with small-scale experimental measurements.

3. The Presence of Bedrock at Shallow Depth

Natural soil deposits are frequently underlain by very stiff material or even bedrock at a shallow depth, rather than extending to practically infinite depth as the homogeneous halfspace implies. The proximity of such a stiff formation to the oscillating surface modifies the static stiffnesses, K, the dynamic stiffness coefficients, $k(\omega)$, and the dashpot coefficients $C(\omega)$. Specifically, with reference to Table 15.3 and its graphs we see the following.

(1) The *static stiffnesses* in all modes increase with the relative depth to bedrock H/B. This is evident from the formulas of Table 15.3, which reduce to the corresponding halfspace stiffnesses when H/R approaches infinity.

Particularly sensitive to variations in the depth to rock are the vertical stiffnesses—the effect being far more pronounced with strip footings (factor 3.5 versus 1.3). Horizontal stiffnesses are also appreciably affected by H/R (factors of 2 for strip and 0.5 for circle), while the rotational stiffnesses (rocking and torsion) are the least affected. In fact, for $H/R > 1.5$ the response to torsional loads is essentially independent of the layer thickness.

An indication of the causes of this different behavior (between circular and strip footings and, in any footing, between the different types of loading) can be obtained by comparing the depths of the "zone of influence" (also called the "pressure bulb") in each case. Circular and square foundations on a homogeneous halfspace induce vertical normal stresses σ_z along the centerline that become practically negligible at depths exceeding $z_v \simeq 5R$; with strip foundations σ_z practically vanishes only below $z_v \simeq 15B$. The "depths of influence", z_h, for the horizontal shear stresses τ_{zx} due to lateral loading are of the order of $2R$ and $6B$ for circle and strip, respectively. On the other hand, for all foundation shapes (strip, rectangle, circle), moment loading is "felt" down to a depth, z_r, of about $2B$ or $2R$. For torsion, finally, $z_t \simeq 0.75R$ or $0.75B$.

Apparently, when a rigid formation "cuts" through the "pressure bulb" of a particular loading mode, it eliminates the corresponding deformations and thereby increases the stiffness.

(2) The variation of the *dynamic stiffness coefficient* with frequency reveals an equally strong dependence on H/B. On a stratum, $k(\omega)$ is not a smooth function, as with a halfspace, but exhibits undulations (peaks and valleys) associated with the natural frequencies (in shear and compression) of the stratum. In other words, the observed fluctuations are the

TABLE 15.3 DYNAMIC STIFFNESSES AND DASHPOT COEFFICIENTS FOR SURFACE FOUNDATIONS ON HOMOGENEOUS STRATUM OVER BEDROCK (sources are listed in the text).

Homogeneous Stratum G, ν, ρ — H — rigid formation

Foundation Shape		Circular Foundation of Radius $B = R$	Rectangular Foundation $2B$ by $2L$ $(L > B)$	Strip Foundation $2L \to \infty$
Static stiffnesses, K	Vertical, z	$K_z = \dfrac{4GR}{1-\nu}\left(1 + 1.3\dfrac{R}{H}\right)$	$K_z = \dfrac{2GL}{1-\nu}\left[0.73 + 1.54\left(\dfrac{B}{L}\right)^{3/4}\right]\left[1 + \dfrac{\frac{B}{H}}{0.5 + \frac{B}{L}}\right]$	$\dfrac{K_z}{2L} \simeq \dfrac{0.73G}{1-\nu}\left(1 + 3.5\dfrac{B}{H}\right)$
	Lateral, y	$K_y = \dfrac{8GR}{2-\nu}\left(1 + 0.5\dfrac{R}{H}\right)$	*	$\dfrac{K_y}{2L} \simeq \dfrac{2G}{2-\nu}\left(1 + 2\dfrac{B}{H}\right)$
	Lateral, x	$K_x = K_y$	*	—
	Rocking, rx	$K_{rx} = \dfrac{8GR}{3(1-\nu)}\left(1 + 0.17\dfrac{R}{H}\right)$	*	$\dfrac{K_{rx}}{2L} = \dfrac{\pi GB^2}{2(1-\nu)}\left(1 + 0.2\dfrac{B}{H}\right)$
	Rocking, ry	$K_{ry} = K_{rx}$	*	—
	Torsional, t	$K_t = \dfrac{16}{3}GR^3\left(1 + 0.10\dfrac{R}{H}\right)$	*	—
Dynamic stiffness coefficients, $k(\omega)$	Vertical, z	$k_z = k_z(H/R, a_0)$ is obtained from Graph III-1	$k_z = k_z(H/B, L/B, a_0)$ is plotted in Graph III-2 for rectangles and strip	
	Horizontal, y or x	$k_y = k_y(H/R, a_0)$ is obtained from Graph III-1	*	$k_y = k_y(H/B, a_0)$ is obtained from Graph III-3
	Rocking, rx or ry Torsional	$\begin{cases} k_\alpha(H/R) \simeq k_\alpha(\infty) \\ \alpha = rx, ry, t \end{cases}$	*	$k_{rx}(H/R) \simeq k_{rx}(\infty)$
Radiation dashpot coefficients, $C(\omega)$	Vertical, z	$C_z(H/B) \simeq 0$ at frequencies $f < f_c$, regardless of foundation shape $\quad C_z(H/B) \simeq 0.8C_z(\infty)$ at $f \geq 1.5f_c$ At intermediate frequencies: interpolate linearly.	$f_c = \dfrac{V_{La}}{4H}, \quad V_{La} = \dfrac{3.4V_s}{\pi(1-\nu)}$	
	Lateral, y or x	$C_y(H/B) \simeq 0$ at $f < \frac{3}{4}f_s; \quad C_y(H/B) \simeq C_y(\infty)$ at $f > \frac{4}{3}f_s$. At intermediate frequencies: interpolate linearly. $\quad f_s = V_s/4H$. Similarly for C_x		
	Rocking, rx or ry	$C_{rx}(H/B) \simeq 0$ at $f < f_c; \quad C_{rx}(H/B) \simeq C_{rx}(\infty)$ at $f > f_c$. Similarly for C_{ry}		
	Torsional, t	$C_t(H/B) \simeq C_t(\infty)$		

* Not available.

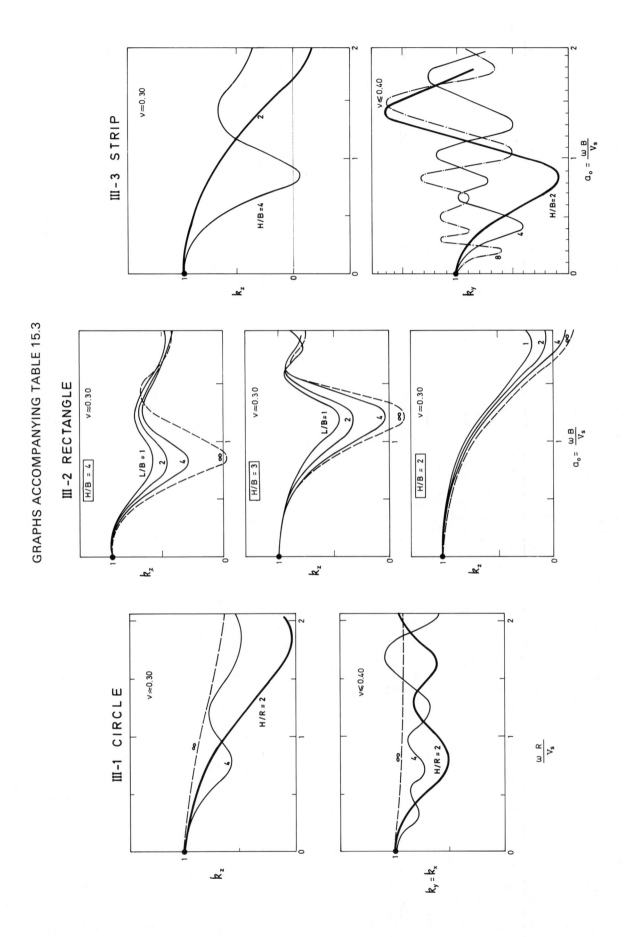

GRAPHS ACCOMPANYING TABLE 15.3

III-1 CIRCLE

III-2 RECTANGLE

III-3 STRIP

outcome of resonance phenomena: waves emanating from the oscillating foundation reflect at the soil–bedrock interface and return back to their source at the surface. As a result, the amplitude of foundation motion may significantly increase at frequencies near the natural frequencies of the deposit. Thus, the dynamic stiffness (being the inverse of displacement) exhibits troughs, which are very steep when the hysteretic damping in the soil is small (in fact, in certain cases, k would be exactly zero if the soil were ideally elastic).

For the "shearing" modes of vibration (swaying and torsion) the natural fundamental frequency of the stratum, which controls the behavior of $k(\omega)$, is

$$f_s = \frac{V_s}{4H} \tag{15.60}$$

while for the "compressing" modes (vertical, rocking) the corresponding frequency is

$$f_c \simeq \frac{V_{La}}{4H} = \frac{3.4}{\pi(1 - v)} f_s \tag{15.61}$$

(3) The variation of the *dashpot coefficients* with frequency reveals a twofold effect of the presence of a rigid base at relatively shallow depth. First, the $C(\omega)$ also exhibit undulations (crests and troughs) due to wave reflections at the rigid boundary. These fluctuations are more pronounced with strip than with circular foundations, but are *not* as significant as for the corresponding stiffnesses $k(\omega)$. Second, and far more important from a practical viewpoint, is that at low frequencies, below the first resonant ("cut-off") frequency of each mode of vibration, radiation damping is zero or negligible for all shapes of footings and all modes of vibration. This is due to the fact that no surface waves can exist in a soil stratum over bedrock at such low frequencies; and, since the bedrock also prevents waves from propagating downward, the overall radiation of wave energy from the footing is negligible or nonexistent.

Such an elimination of radiation damping may have severe consequences for heavy foundations oscillating vertically or horizontally, which would have enjoyed substantial amounts of damping in a very deep deposit (halfspace)—recall illustrative examples for Tables 15.1 and 15.2. On the other hand, since the low-frequency values of C in rocking and torsion are small even in a halfspace, operation below the cut-off frequencies may not be affected appreciably by the presence of bedrock.

Note that at operating frequencies f beyond f_s or f_c, as appropriate for each mode, the "stratum" damping $C(H/B)$ fluctuates about the "halfspace" damping $C(H/B = \infty)$. The "amplitude" of such fluctuations tends to decrease with increasing H/B; moreover, if some wave energy penetrates into bedrock (as does happen in real life thanks to some weathering of the upper mass of rock) the fluctuations tend to wither away—hence the recommendation of Table 15.3.

4. Foundation Embedded in Stratum

As can be seen from Table 15.4, embedding a foundation in a shallow stratum, rather than a halfspace, has one *additional* effect over those addressed in Table 15.2: the static stiffnesses tend to increase thanks to the decrease in the depth of the deforming zone underneath the foundation.

It is evident that the results summarized in Table 15.4 can follow from a proper combination of the pertinent results

of Table 15.2 (embedded foundation in halfspace) and of Table 15.3 (surface foundation on a stratum). No further explanation seems necessary.

5. Effect of G Increasing with Depth

Often, the assumption of homogeneous layer or halfspace may not be realistic, as the soil stiffness usually increases with depth, even in uniform soils. The prime cause is the increase of the confining pressure with depth, and the ensuing increase of G_{max} (for example, according to Equation 15.4). The effects of such an inhomogeneity could be assessed with the help of Table 15.5, which provides information for foundation on a number of inhomogeneous deep deposits, including (but not limited to):

- A deposit with (low-strain) shear modulus

$$G = G_0 \left(1 + \frac{z}{B} \right)^{1/2} \tag{15.62}$$

which is representative of many cohesionless soil deposits, in which the shear modulus is proportional to the square root of the confining pressure. $G_0 =$ the modulus at the ground surface, that is, at $z = 0$, and should not be confused with the low-strain modulus G_{max}. In this chapter we essentially always deal with very small strains and hence all G appearing in Tables 15.1 to 15.5 are low-strain ($\gamma < 10^{-5}$) values.

- A deposit with

$$G = G_0 \left(1 + \frac{z}{B} \right) \tag{15.63}$$

which is representative of deposits of saturated normally and slightly overconsolidated clays.

- A deposit with

$$G = G_0 \left(1 + \alpha \frac{z}{B} \right)^2 \tag{15.64}$$

which can simulate deposits with a relatively faster increase of G at large depths, α is a parameter determined by fitting the experimental test results.

The following trends are worthy of note in Table 15.5 and the accompanying graphs.

1. For the *static stiffnesses*, K, use of a single "effective" modulus, $G = G_{eff}$, in the formulas for a homogeneous halfspace (Table 15.1) would, at best, provide the correct stiffness in only one particular mode. This is because the "pressure bulb" of each mode reaches a different depth and is thus affected by different values of modulus. As one might expect, vertical loading (especially on a strip) penetrates the deepest—the "effective depth" z_{eff} is of the order of one to two times B, the foundation semiwidth or radius. Moment loading, on the other hand, is the least affected by inhomogeneity, with "effective depths" merely ranging from $\frac{1}{3}B$ to $\frac{1}{10}B$.
2. In strongly inhomogeneous soils, the *dynamic stiffness coefficients*, k, plotted as functions of $\omega B/V_{s0}$, are always smaller than those of a homogeneous halfspace with $V = V_{s0}$. The differences, however, are quite small and could be neglected in practical applications, as suggested in Table 15.5.

TABLE 15.4 DYNAMIC STIFFNESSES AND DASHPOT COEFFICIENTS FOR FOUNDATIONS EMBEDDED IN HOMOGENEOUS STRATUM OVER BEDROCK.*

Foundation Shape		Circular Foundation of Radius R	Strip Foundation
Static stiffnesses, K	Vertical	$K_{z,\mathrm{emb}} \simeq K_{z,\mathrm{sur}}\left(1+0.55\dfrac{d}{R}\right)\left[1+\left(0.85-0.28\dfrac{D}{R}\right)\dfrac{D}{H-D}\right]^{\dagger}$	$K_{z,\mathrm{emb}} \simeq K_{z,\mathrm{sur}}\left[1+0.2\left(\dfrac{d}{B}\right)^{2/3}\right]\left(1+3.5\dfrac{B}{H-D}\right)$
	Horizontal, y or x	$K_{y,\mathrm{emb}} \simeq K_{y,\mathrm{sur}}\left(1+\dfrac{d}{R}\right)\left(1+1.25\dfrac{D}{H}\right)^{\dagger}$	$K_{y,\mathrm{emb}} \simeq K_{y,\mathrm{sur}}\left(1+0.5\dfrac{d}{B}\right)\left(1+1.5\dfrac{D}{H}\right)$
	Rocking, rx or ry	$K_{rx,\mathrm{emb}} \simeq K_{rx,\mathrm{sur}}\left(1+2\dfrac{d}{R}\right)\left(1+0.65\dfrac{D}{H}\right)$	$K_{rx,\mathrm{emb}} \simeq K_{rx,\mathrm{sur}}\left(1+\dfrac{d}{B}\right)\left(1+0.65\dfrac{D}{H}\right)$
	Coupled swaying–rocking	$K_{yrx,\mathrm{emb}} \simeq \tfrac{1}{3}dK_{y,\mathrm{emb}}$	$K_{yrx,\mathrm{emb}} \simeq \tfrac{1}{3}dK_{y,\mathrm{emb}}$
	Torsional	$K_{t,\mathrm{emb}} \simeq K_{t,\mathrm{sur}}\left(1+2.67\dfrac{d}{R}\right)$	—
Dynamic stiffness coefficients, $k(\omega)$		The relationships between k_{emb} and k_{sur} follow approximately the same pattern as those between embedded and surface foundation on a homogeneous halfspace. Therefore, use the results of Table 15.2 as a first approximation.	
Radiation dashpot coefficients, $C(\omega)$		C_{emb} exceeds C_{sur} by an amount that depends on the geometry of the sidewall–soil contact surface and is practically independent of the presence or absence of a rigid base at shallow depths. Therefore, use the results of Table 15.2, but with C_{sur} corresponding to the layered profile and thus obtained according to Table 15.3 (approximate guideline).	

* Sources are listed in the text.

† $K_{z,\mathrm{sur}}, K_{y,\mathrm{sur}} \ldots$ are the stiffnesses for the corresponding surface foundations, and can be obtained from Table 15.3.

TABLE 15.5 DYNAMIC STIFFNESSES AND DASHPOT COEFFICIENTS FOR SURFACE FOUNDATIONS ON "DEEP" INHOMOGENEOUS DEPOSITS.

$$G(z) = G_0(1 + \alpha\zeta)^n$$

$$\nu, \rho = \text{constant}$$

$$\zeta = \frac{z}{B}$$

Examples of $G(z)$:

$G_0(1 + \alpha\zeta)^{1/m} \quad m > 1$

$G_0(1 + \alpha\zeta)$

$G_0(1 + \alpha\zeta)^2$

Vibration Mode	Static Stiffnesses K		"Equivalent" Depths $\bar{z}/B$	
	Square (2B by 2B)	Strip (2B by ∞)	Square	Strip
Vertical, z	$K_z = \dfrac{4.54}{1-\nu} G_0 B(1+\alpha)^n$	$K_z \simeq \dfrac{0.73}{1-\nu} G_0(1+2\alpha)^n$	1	2
Horizontal, x or y	$K_x \simeq \dfrac{9}{2-\nu} G_0 B\left(1+\tfrac{1}{2}\alpha\right)^n$	$K_y \simeq \dfrac{2}{2-\nu} G_0\left(1+\tfrac{2}{3}\alpha\right)^n$	$\tfrac{1}{2}$	$\tfrac{2}{3}$
Rocking, rx or ry	$K_{rx} \simeq \dfrac{3.6}{1-\nu} G_0 B^3\left(1+\tfrac{1}{3}\alpha\right)^n$	$K_{rx} \simeq \dfrac{\pi}{2(1-\nu)} G_0 B^2\left(1+\tfrac{1}{3}\alpha\right)^n$	$\tfrac{1}{3}$	$\tfrac{1}{3}$
Torsional, t	$K_t \simeq 7.93 G_0 B^3\left(1+\tfrac{1}{10}\alpha\right)^n$	—	$\tfrac{1}{10}$	—

The above expressions are only crude approximations based on limited information. They should be used judiciously. Also, one could utilize static computer codes and procedures. The strip stiffnesses are per unit length.

Dynamic Stiffness Coefficients $k(\omega)$

All Modes

k decreases only slightly faster with a_0 in a strongly inhomogeneous than in a homogeneous soil deposit. For practical applications one can make the approximation $k(\alpha) \approx k(\alpha = 0)$ with little error for realistically inhomogeneous deposits (i.e., if α is not too large)

Radiation Damping Coefficients $C(\omega)$

General Foundation Base Shape (area A_b, inertias I_{bx}, I_{by}, J_b)

Modes	General Expression	High-frequency Asymptotic Expression
Vertical, z	$C_z = \rho V_{s0} \dfrac{3.4}{\pi(1-\nu)} A_b \tilde{c}_z^*$	$\tilde{c}_z^* \simeq 1$ $(a_0 \geqslant 2)$
Horizontal, y	$C_y = \rho V_{s0} A_b \tilde{c}_y^*$	$\tilde{c}_y^* \simeq 1$ $(a_0 \geqslant 1.5)$
Horizontal, x	$C_s = \rho V_{s0} A_b \tilde{c}_x^*$	$\tilde{c}_x^* \simeq 1$ $(a_0 \geqslant 1.5)$
Rocking, rx	$C_{rx} = \rho V_{s0} \dfrac{3.4}{\pi(1-\nu)} I_{bx} \tilde{c}_{rx}^*$	$\tilde{c}_{rx}^* \simeq 1$ $(a_0 \geqslant 3.5)$
Rocking, ry	$C_{ry} = \rho V_{s0} \dfrac{3.4}{\pi(1-\nu)} I_{by} \tilde{c}_{ry}^*$	$\tilde{c}_{ry}^* \simeq 1$ $(a_0 \geqslant 2.5)$
Torsional, t	$C_t = \rho V_{s0} J_b \tilde{c}_t^*$	$\tilde{c}_t^* \simeq 1$ $(a_0 > 2.5)$

$$(V_{s0} = \sqrt{G_0/\rho})$$

All the dimensionless coefficients $\tilde{c}^*$ are invariably smaller than the corresponding coefficients $\tilde{c}$ for a homogeneous halfspace of shear modulus G_0 that were given in Table 15.1.
Graph V plots $\tilde{c}^*$ vs L/B and $a_0 = \omega B/V_{s0}$ for certain combinations of the inhomogeneity parameters α and n.

GRAPHS ACCOMPANYING TABLE 15.5

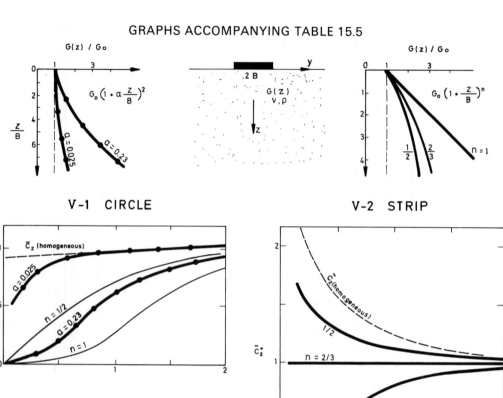

V-1 CIRCLE   V-2 STRIP

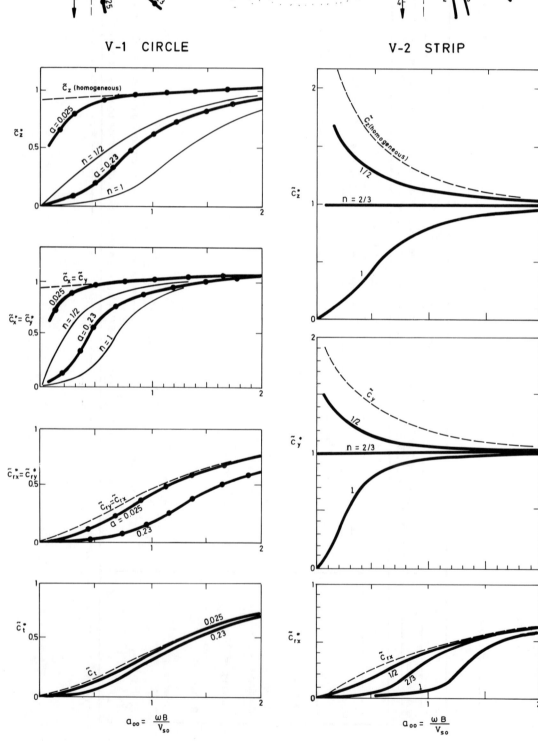

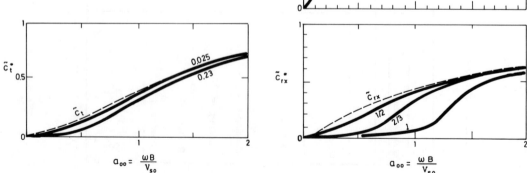

$$a_{oo} = \frac{\omega B}{V_{so}}$$

$$a_{oo} = \frac{\omega B}{V_{so}}$$

3. The strongest influence of modulus increasing with depth is on radiation damping, especially in the translational modes. *At low frequencies damping remains invariably lower than in a homogeneous halfspace of modulus G_0.* (Of course, if the homogeneous halfspace had a modulus $G = G_{eff}$, for which the static stiffnesses coincide, the differences would be even greater!) At higher frequencies the discrepancies between damping on the two media (inhomogeneous and homogeneous with $G = G_0$) tend to become vanishingly small. The cause and consequences of such a behavior are explained below.

A medium with continuously varying wave velocity can reflect propagating waves. In this case, even total reflection of the downward-propagating waves is possible owing to the increase of soil velocity with depth. A discontinuity in velocity (e.g., bedrock) is not necessary for such a reflection, since the rays in inhomogeneous media are not straight lines but curves (e.g., circular arcs for $G(z)$ as in Equation 15.63). Hence, wave energy cannot be radiated away as effectively as in homogeneous media—small radiation damping. In fact, as the degree of inhomogeneity increases, radiation damping decreases, in spite of the greater soil stiffness.

On the other hand, at "very high" frequencies the wavelengths of the emitted waves are "very small" and the source "sees" the transmitting medium as a homogeneous halfspace having modulus G_0 and velocities V_{s0} (in the shearing modes) and V_{La0} (in the compressing modes). It appears that "very small" may *not* be small in absolute terms, merely comparable with the dimensions of the source (the foundation). Consequently, "very high" frequencies are always within the range of frequencies of practical interest ($a_0 < 2$). At such "high" frequencies the dashpot coefficients for a surface foundation with arbitrary base shape become

$$C_z \approx \rho V_{La0} A_b \qquad (15.65a)$$

$$C_x = C_y \approx \rho V_{s0} A_b \qquad (15.65b)$$

$$C_{rx} \approx \rho V_{La0} I_{bx} \qquad (15.65c)$$

$$C_{ry} \approx \rho V_{La0} I_{by} \qquad (15.65d)$$

$$C_t \approx \rho V_{s0} J_b \qquad (15.65e)$$

The reader should not fail to observe (in Eqs. 15.65 as well as in Table 15.5) the significance of the near-surface soil modulus and wave velocity. Also, recall that capillary phenomena may, to a large extent, control such moduli in silty sands; since it is very difficult to simulate capillary effects in the laboratory, in-situ measurment of V_{s0} is of paramount importance. To this end, steady-state vibrating the free surface at very high frequencies, or using the emerging technique of spectral analysis of surface waves, may prove invaluable.

6. Foundations on Piles

The response of piles laterally loaded (by horizontal forces and moments) is independent of their length, in most practical situations. Only the uppermost part of the pile, of length l_c, experiences appreciable displacements. It is along this "*active*" length, l_c, that the imposed load is transmitted to the supporting soil. l_c is typically of the order of 5 to 10 pile diameters; for a given soil profile, l_c is a function of the pile with respect to the soil.

For three characteristic soil profiles, Table 15.6 presents simple algebraic expressions for estimating l_c of a circular solid pile with diameter d and Young's modulus E_p. For each profile, the only soil parameter that affects l_c is the reference Young's modulus, $\tilde{E}_s$, at a depth $z = d$.

For the three lateral impedances ("springs" and "dashpots"), $\mathcal{K}_{HH}, \mathcal{K}_{HM}$, and $\mathcal{K}_{MM}$, defined in Figure 15.17, Table 15.6 presents easy-to-use formulas, which, however, are valid only for piles with length

$$L > l_c \qquad (15.66)$$

Such piles are described as "flexible" piles in the literature. But note that a good majority of real-life piles, even some with large diameters, would fall into this category. Among the exceptions are short piers and caissons.

From a theoretical point of view, most of the formulas in Table 15.6 are reasonably accurate, as they are basically curve fits to rigorous numerical results. The real difficulty, however, is to select the proper profile and modulus for the supporting soil. Even with a uniform top layer, the secant soil modulus will change with the magnitude of induced strains, which decreases with depth. Other nonlinear phenomena, such as development of a gap between pile and soil near the ground surface, further complicate the problem. One solution might be to conduct a suitable full-scale or small-scale lateral pile load test in the field. By "suitable" we mean one that produces deformations comparable with those anticipated in the final design. Reference is made to Blaney and O'Neil (1986), and to Novak (1985), who have described such field tests, and to Richart and Chon (1977), who conducted a somewhat similar laboratory-scale test, among several others.

Another complication is the prediction of the impedances of pile groups, accounting for pile-to-pile interaction ("through-soil coupling," as it is often called). The surprisingly simple formulas for the complex-valued interaction factors, α_z and α_{HH}, have been developed recently by Dobry and Gazetas (1988) and Gazetas and Makris (1990), and yield results in excellent agreement with the rigorous solution of Kaynia and Kausel (1982). One uses such interaction factors exactly as the static interaction factors of Poulos (Poulos and Davis, 1980).

However, for the two inhomogeneous profiles, the rigorous results available in the published literature are scarce. The expressions presented are based on a simple physical model and are only tentative; they should be used with caution.

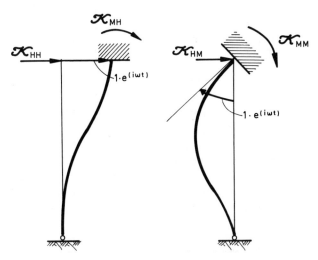

Fig. 15.17 Definition of pile head impedances.

TABLE 15.6 DYNAMIC STIFFNESSES AND DAMPING COEFFICIENTS FOR FLEXIBLE PILES ($L > l_c$).

	Linear Increase of Soil Modulus with Depth*	Parabolic Increase of Soil Modulus with Depth*	Constant Soil Modulus at All Depths
$L > l_c$ (pile diagram with H, L, l_c, d)	$E_s = \tilde{E}_s \dfrac{z}{d}$; $\tilde{V}_s = \left[\tilde{E}_s/2\rho(1+\nu)\right]^{\frac{1}{2}}$	$E_s = \tilde{E}_s \sqrt{\dfrac{z}{d}}$; $\tilde{V}_s = \left[\tilde{E}_s/2\rho(1+\nu)\right]^{\frac{1}{2}}$	$E_s = $ constant
"Active length"	$l_c \approx 2d(E_p/\tilde{E}_s)^{0.20}$	$l_c \approx 2d(E_p/\tilde{E}_s)^{0.22}$	$l_c \approx 2d(E_p/E_s)^{0.25}$
Natural shear frequency of deposit	$f_s = 0.19 V_{sH}/H$ where $V_{sH} = $ the S-wave velocity at depth $z = H$ (bottom of stratum)	$f_s = 0.223 V_{sH}/H$ where $V_{sH} = $ the S-wave velocity at depth $z = H$ (bottom of stratum)	$f_s = 0.25 V_s/H$
Static lateral (swaying) stiffness	$K_{HH} = 0.6d\tilde{E}_s(E_p/\tilde{E}_s)^{0.35}$	$K_{HH} = 0.8d\tilde{E}_s(E_p/\tilde{E}_s)^{0.28}$	$K_{HH} = dE_s(E_p/E_s)^{0.21}$
Lateral (swaying) stiffness coefficient	$k_{HH} \approx 1$	$k_{HH} \approx 1$	$k_{HH} \approx 1$
Lateral (swaying) coefficient: $C_{HH} = 2K_{HH}D_{HH}/\omega$	$\begin{cases} D_{HH} \approx 0.60\beta + 1.80fd\bar{V}_s^{-1}, \text{ for } f > f_s \\ D_{HH} \approx 0.60\beta, \text{ for } f \leqq f_s \end{cases}$	$\begin{cases} D_{HH} \approx 0.70\beta + 1.20fd(E_p/\tilde{E}_s)^{0.08}\bar{V}_s^{-1}, \\ \quad \text{for } f > f_s \\ D_{HH} \approx 0.70\beta, \text{ for } f \leqq f_s \end{cases}$	$\begin{cases} D_{HH} \approx 0.80\beta + 1.10fd(E_p/E_s)^{0.17}V_s^{-1}, \\ \quad \text{for } f > f_s \\ D_{HH} \approx 0.80\beta, \text{ for } f \leqq f_s \end{cases}$
Static rocking stiffness	$K_{MM} = 0.15d^3\tilde{E}_s(E_p/\tilde{E}_s)^{0.80}$	$K_{MM} = 0.15d^3\tilde{E}_s(E_p/\tilde{E}_s)^{0.77}$	$K_{MM} = 0.15d^3E_s(E_p/E_s)^{0.75}$
Rocking stiffness coefficient	$k_{MM} \approx 1$	$k_{MM} \approx 1$	$k_{MM} \approx 1$
Rocking dashpot coefficient: $C_{MM} = 2K_{MM}D_{MM}/\omega$	$\begin{cases} D_{MM} \approx 0.20\beta + 0.40fd\bar{V}_s^{-1}, \text{ for } f > f_s \\ D_{MM} \approx 0.20\beta, \text{ for } f \leqq f_s \end{cases}$	$\begin{cases} D_{MM} \approx 0.22\beta + 0.35fd(E_p/\tilde{E}_s)^{0.10}\bar{V}_s^{-1}, \\ \quad \text{for } f > f_s \\ D_{MM} \approx 0.22\beta, \text{ for } f \leqq f_s \end{cases}$	$\begin{cases} D_{MM} \approx 0.35\beta + 0.35fd(E_p/E_s)^{0.20}V_s^{-1}, \\ \quad \text{for } f > f_s \\ D_{MM} \approx 0.25\beta, \text{ for } f \leqq f_s \end{cases}$
Static swaying–rocking cross-stiffness	$K_{HM} = K_{MH} = -0.17d^2\tilde{E}_s(E_p/\tilde{E}_s)^{0.60}$	$K_{HM} = K_{MH} = -0.24d^2\tilde{E}_s(E_p/\tilde{E}_s)^{0.53}$	$K_{HM} = K_{MH} = -0.22d^2E_s(E_p/E_s)^{0.50}$
Swaying–rocking cross-stiffness coefficient	$k_{HM} \approx 1$	$k_{HM} \approx 1$	$k_{HM} \approx 1$
Swaying–rocking dashpot coefficient: $C_{HM} = 2K_{HM}D_{HM}/\omega$	$\begin{cases} D_{HM} \approx 0.30\beta + fd\bar{V}_s^{-1}, \text{ for } f > f_s \\ D_{HM} \approx 0.30\beta, \text{ for } f \leqq f_s \end{cases}$	$\begin{cases} D_{HM} \approx 0.60\beta + 0.70fd(E_p/\tilde{E}_s)^{0.05}\bar{V}_s^{-1}, \\ \quad \text{for } f > f_s \\ D_{HM} \approx 0.35\beta, \text{ for } f \leqq f_s \end{cases}$	$\begin{cases} D_{HM} \approx 0.80\beta + 0.85fd(E_p/E_s)^{0.18}V_s^{-1}, \\ \quad \text{for } f > f_s \\ D_{HM} \approx 0.50\beta, \text{ for } f \leqq f_s \end{cases}$

The axial stiffness of a pile depends not only on its relative compressibility (E_p/E_s) but also on the slenderness ratio L/d and the tip support conditions (end-bearing versus floating). See the pertinent geotechnical literature for a proper estimation of the static stiffness. The expressions given herein are *only* for estimates of the axial stiffness of floating piles in a homogeneous stratum of total thickness $H \approx 2L$.

	Column 1	Column 2	Column 3
Static axial stiffness	$K_z \approx 1.8 E_{sL} d \left(\dfrac{L}{d}\right)^{0.55} \left(\dfrac{E_p}{E_{sL}}\right)^{-(L/d)(E_p/E_{sL})}$ $E_{sL} = \tilde{E}_s \cdot (L/d)$	$K_z \approx 1.9 E_{sL} d \left(\dfrac{L}{d}\right)^{0.6} \left(\dfrac{E_p}{E_{sL}}\right)^{-(L/d)(E_p/E_{sL})}$ $E_{sL} = \tilde{E}_s \cdot \sqrt{(L/d)}$	$K_z \approx 1.9 E_s d \left(\dfrac{L}{d}\right)^{2/3} \left(\dfrac{E_p}{E_s}\right)^{-(L/d)(E_p/E_s)}$
Axial dynamic stiffness coefficient	$k_z \approx 1$ (for $a_0 = \omega d/V_{sL} < 0.5$; where V_{sL} is the S-wave velocity at depth L)	• $L/d < 20: k_z \approx 1$ • $L/d \geqslant 50: k_z \approx 1 + \tfrac{1}{3}\sqrt{a_0}$ interpolate in between (for $a_0 = \omega d/V_{sL} < 0.5$)	• $L/d < 15: k_z \approx 1$ • $L/d \geqslant 50: k_z \approx 1 + \sqrt{a_0}$ interpolate in between (for $a_0 = \omega d/V_s < 1$)

In all cases, k_z shows a narrow valley at the resonant frequency f_r of the soil stratum; as a first approximation, $f_r \approx f_c \approx \bar{V}_{La}/4H$ and $k_z(f_r) \approx 0.8$ for material soil damping $\beta = 0.05$. $\bar{V}_{La}$ is the average V_{La} over the whole stratum depth.

	Column 1	Column 2	Column 3
Axial radiation dashpot coefficient	$C_z \approx \tfrac{2}{3} a_0^{-1/3} \rho V_{sL} \pi d L r_d$ for $f > 1.5 f_r$ where: $r_d \approx 1 - e^{-2(E_p/E_{sL})(L/d)^{-2}}$ $C_z \approx 0$ for $f \leqslant f_r$ linearly interpolate for $f_r < f < 1.5 f_r$	$C_z \approx \tfrac{3}{4} a_0^{-1/4} \rho V_{sL} \pi d L r_d$ for $f > 1.5 f_r$ where: $r_d \approx 1 - e^{-1.5(E_p/E_{sL})(L/d)^{-2}}$ $C_z \approx 0$ for $f \leqslant f_r$ linearly interpolate for $f_r < f < 1.5 f_r$	$C_z \approx a_0^{-1/5} \rho V_s \pi d L r_d$ for $f > 1.5 f_r$ where: $r_d \approx 1 - e^{-(E_p/E_s)(L/d)^{-2}}$ $C_z \approx 0$ for $f \leqslant f_r$ linearly interpolate for $f_r < f < 1.5 f_r$

Pile-to-Pile Interaction Factors for Assessing the Response of Floating Pile Groups

	Column 1	Column 2	Column 3
Interaction factor α_z for axial in-phase oscillations of the two piles	$\alpha_z \approx \sqrt{2}\left(\dfrac{S}{d}\right)^{-3/4} \cdot e^{-0.5\beta\omega S/V_{sL}} \cdot e^{-i\omega\sqrt{2}S/V_{sL}}$ V_{sL} = the S-wave velocity at depth $z = L$; $\bar{V}_s = V_s$ at pile mid-length; S = axis-to-axis pile separation; β = soil hysteretic damping. Note: although α_z are complex numbers their use is identical to the familiar use of static interaction factors introduced by Poulos.	$\alpha_z \approx \sqrt{2}\left(\dfrac{S}{d}\right)^{-2/3} \cdot e^{-(2/3)\beta\omega S/V_{sL}} \cdot e^{-i\omega\sqrt{2}S/V_{sL}}$	$\alpha_z \approx \sqrt{2}\left(\dfrac{S}{d}\right)^{-1/2} \cdot e^{-\beta\omega S/V_s} \cdot e^{-i\omega S/V_s}$
Interaction factor α_{HH} for lateral in-phase oscillation	Very little information presently available	Very little information presently available	$\alpha_{HH}(90°) \approx (3/4)\alpha_z$ $\alpha_{HH}(0°) \approx 0.5\left(\dfrac{S}{d}\right)^{-1/2} \cdot e^{-\beta\omega S/V_{La}} \cdot e^{-i\omega S/V_{La}}$ $\alpha_{HH}(\theta°) \approx \alpha_{HH}(0°)\cos^2\theta + \alpha_{HH}(90°)\sin^2\theta$
Interaction factors: α_{MM} for in-phase rocking, and α_{MH} for swaying–rocking	$\alpha_{MM} \approx \alpha_{MH} \approx 0$	$\alpha_{MM} \approx \alpha_{MH} \approx 0$	$\alpha_{MM} \approx \alpha_{MH} \approx 0$

* $\tilde{E}_s$ and $\bar{V}_s$ (for the two inhomogeneous deposits) denote Young's modulus and S-wave velocity, respectively, at depth.

REFERENCES

Arya, S., O'Neil, M., and Pincus, G. (1979), *Design of Structures and Foundations for Vibrating Machines*, Gulf Publ. Co., Houston, Texas.

Barkan, D. D. (1962), *Dynamics of Bases and Foundations* (translated from Russian), McGraw-Hill Book Co., Inc., New York, N.Y.

Beredugo, Y. O. and Novak, M. (1972), Coupled horizontal and rocking vibrations of embedded footings, *Canadian Geotechnical Journal*, **9**, pp. 477–497.

Blaney, G. W. and O'Neil, M. W. (1986), Measured lateral response of mass on single pile in clay, *Journal of Geotechnical Engineering, ASCE*, **112**, pp. 443–457.

Booker, J. R., Balaam, N. P., and Davis, E. H. (1985), The behaviour of an elastic non-homogeneous halfspace, Parts I and II, *International Journal for Numerical and Analytical Methods in Geomechanics*, **9**, pp. 353–381.

Chow, Y. K. (1987), Vertical vibration of 3-D rigid foundation on layered media, *Earthquake Engineering and Structural Dynamics*, **15**, pp. 585–594.

Davies, T. G., Sen, R., and Banerjee, P. K. (1985), Dynamic behavior of pile groups in inhomogeneous soil, *Journal of Geotechnical Engineering, ASCE*, **111**, pp. 1635–1679.

Dobry, R. and Gazetas, G. (1985), Dynamic stiffness and damping of foundations by simple methods, *Vibration Problems in Geotechnical Engineering*, ed. G. Gazetas and E. T. Selig, ASCE, pp. 77–107.

Dobry, R. and Gazetas, G. (1988), Simple method for dynamic stiffness and damping of floating pile groups, *Géotechnique*, **38**, No. 4, pp. 557–574.

Dobry, R., Gazetas, G., and Stokoe, K. H., II (1986), Dynamic response of arbitrarily shaped foundations: Experimental verification, *Journal of Geotechnical Engineering, ASCE*, **112**, No. 2, pp. 136–149.

Drnevich, V. P. (1977), Resonant-column testing: Problems and solutions, *Dynamic Geotechnical Testing*, ASTM STP654, pp. 384–398.

Drnevich, V. P. (1985), Recent developments in resonant column testing, *Richart Commemorative Lectures*, ASCE, pp. 79–106.

Elsabee, F. and Morray, J. P. (1977), *Dynamic Behavior of Embedded Foundations*, Research Report R77-33, MIT.

Erden, S. M. (1974), Influence of shape and embedment of dynamic foundation response, Ph.D. Thesis, Dept. of Civil Engineering, University of Massachusetts, Amherst.

Fotopoulou, M. et al. (1989), Rocking damping of arbitrarily shaped embedded foundations, *Journal of Geotechnical Engineering, ASCE*, **115**, pp. 473–490.

Gazetas, G. (1982), Vibrational characteristics of soil deposits with variable wave velocity, *International Journal of Numerical and Analytical Methods in Geomechanics*, **6**, pp. 1–20.

Gazetas, G. (1983), Analysis of machine foundation vibrations: State of the art, *International Journal of Soil Dynamics and Earthquake Engineering*, **2**, pp. 2–42. (See also Errata in **5**, No. 3, 1987.)

Gazetas, G. (1984), Seismic response of end-bearing single piles, *International Journal of Soil Dynamics and Earthquake Engineering*, **3**, pp. 82–93.

Gazetas, G. (1987), Simple physical methods for foundation impedances, *Dynamic Behavior of Foundations and Buried Structures*, Elsevier Applied Science, New York, N.Y., chapter 2, pp. 45–94.

Gazetas, G. and Roesset, J. M. (1976), Forced vibrations of strip footings on layered soils, *Methods of Structural Analysis*, ASCE, Vol. 1, pp. 115–131.

Gazetas, G. and Roesset, J. M. (1979), Vertical vibrations of machine foundations, *Journal of the Geotechnical Engineering Division, ASCE*, **105**, pp. 1435–1454.

Gazetas, G. and Dobry, R. (1984), Horizontal response of piles in layered soils, *Journal of Geotechnical Engineering, ASCE*, **110**, pp. 20–40.

Gazetas, G. and Selig, E. T. (eds.) (1985), *Vibration Problems in Geotechnical Engineering*, ASCE.

Gazetas, G., Dobry, R., and Tassoulas, J. L. (1985), Vertical response of arbitrarily shaped embedded foundations, *Journal of Geotechnical Engineering, ASCE*, **111**, pp. 750–771.

Gazetas, G. and Tassoulas, J. L. (1987), Horizontal stiffness of arbitrarily shaped embedded foundations, *Journal of Geotechnical Engineering*, **113**, pp. 440–457.

Gazetas, G. and Makris, N. (1990), Dynamic pile–soil–pile interaction, I: Analysis of axial vibration, *Earthquake Engineering and Structural Dynamics*, **18** (in press).

Hadjian, A. H. and Luco, J. E. (1977), On the importance of layering on impedance functions, *Proceedings, 6th World Conference on Earthquake Engineering*, New Delhi.

Hall, J. R. (1985), Limits on dynamic measurements and instrumentation, *Richart Commemorative Lectures*, ASCE, pp. 108–119.

Hardin, B. O. (1978), The nature of stress–strain behavior of soils, *Earthquake Engineering and Soil Dynamics*, ASCE, Vol. I, pp. 3–90.

Hoar, R. J. and Stokoe, K. H. (1978), Generation and measurement of shear wave velocity, *Dynamic Geotechnical Testing*, ASTM STP654, pp. 3–29.

Jakub, M. and Roesset, J. M. (1977), *Dynamic Stiffness of Foundations: 2-D vs 3-D Solutions*, Research Report R77-36, MIT.

Johnson, G. R., Christiano, P., and Epstein, H. I. (1975), Stiffness coefficients for embedded footings, *Journal of the Geotechnical Engineering Division, ASCE*, **101**, pp. 789–800.

Kaynia, A. M. and Kausel, E. (1982), *Dynamic Stiffness and Seismic Response of Pile Groups*, Research Report R82-03, MIT.

Kausel, E. (1974), *Forced Vibrations of Circular Foundations on Layered Media*, Research Report R74-11, MIT.

Kausel, E. (1981), *An Explicit Solution for the Green Functions for Dynamic Loads in Layered Media*, Research Report R81-13, MIT.

Kausel, E. and Roesset, J. M. (1975), Dynamic stiffness of circular foundations, *Journal of the Engineering Mechanics Division, ASCE*, **101**, 771–785.

Kausel, E. and Ushijima, R. (1979), *Vertical and Torsional Stiffness of Cylindrical Footings*, Research Report R76-6, MIT.

Knox, D. P., Stokoe, K. H., and Kopperman, S. E. (1982), *Effect of State of Stress on Velocity of Low-Amplitude Shear Waves Along Principal Stress Directions in Dry Sand*, Research Report GR82-23, University of Texas, Austin.

Ladd, R. S. and Dutko, P. (1985), Small-strain measurements using triaxial apparatus, *Advances in the Art of Testing Soils Under Cyclic Conditions*, ed. V. Koshla, ASCE, pp. 148–166.

Lawrence, F. V. (1963), *Propagation Velocity of Ultrasonic Waves Through Sand*, Research Report R63-8, MIT.

Luco, J. E. (1974), Impedance functions for a rigid foundation on a layered medium, *Nuclear Engineering and Design*, **31**, pp. 204–217.

Luco, J. E. (1976), Vibrations of a rigid disc on a layered viscoelastic medium, *Nuclear Engineering and Design*, **36**, pp. 325–340.

Luco, J. E. (1982), Linear soil–structure interaction: A review, *Earthquake Ground Motion and its Effects on Structures*, ASME, AMD, Vol. 53, pp. 41–57.

Luco, J. E. and Westmann, R. A. (1971), Dynamic response of circular footings, *Journal of the Engineering Mechanics Division, ASCE*, **97**, pp. 1381–1395.

Lysmer, J. and Kuhlemeyer, R. L. (1969), Finite dynamic model for infinite media, *Journal of the Engineering Mechanics Division, ASCE*, **95**.

Lysmer, J., Udaka, T., Tsai, C.-F., and Seed, H. B. (1975), *FLUSH—A computer program for approximate 3-D Analysis of Soil–Structure Interaction Problems*, Report No. EERC 75-30, University of California, Berkeley.

Lysmer, J., Tabatabaie, R. M., Tajirian, F., Vahdani, S., and Ostadan, F. (1981), *SASSI—A System for Analysis of Soil–Structure Interaction*, Research Report GT81-02, University of California, Berkeley.

Major, A. (1980), *Dynamics in Civil Engineering*, Akadémical Kiadó, Budapest, Volumes I–IV.

Meek, J. W. and Veletsos, A. S. (1973), Simple models for foundations in lateral and rocking motions, *Proceedings, 5th World Conference on Earthquake Engineering*, Rome, paper 331.

Moore, P. J. (ed.) (1985), *Analysis and Design of Foundations for Vibrations*, A. A. Balkema, Rotterdam.

Nazarian, S. and Stokoe, K. H. (1983), Use of spectral analysis of surface waves for determination of moduli and thicknesses of pavement systems, *Transportation Research Record*, No. 954.

Nogami, T. and Konagai, K. (1985), Simple approach for evaluation of compliance matrix of pile groups, *Vibration Problems in Geotechnical Engineering*, ed. G. Gazetas and E. T. Selig, ASCE, pp. 47–74.

Nogami, T. (1979), Dynamic group effect of multiple piles under vertical vibration, *Proceedings, Engineering Mechanics Specialty Conference*, ASCE, pp. 750–754.

Novak, M. (1985), Experiments with shallow and deep foundations, *Vibration Problems in Geotechnical Engineering*, ed. G. Gazetas and E. Selig, ASCE, pp. 1–26.

Novak, M. (1987), State of the art in analysis and design of machine foundations, *Soil–Structure Interaction*, Elsevier/CML Publ., New York, N.Y., pp. 171–192.

Novak, M., Nogami, T., and Aboul-Ellas, F. (1978), Dynamic soil reactions for plane strain case, *Journal of the Engineering Mechanics Division, ASCE*, **104**, pp. 1024–1041.

Novak, M. and Sheta, M. (1980), Approximate approach to contact effects of piles, *Dynamic Response of Pile Foundation*, ed. M. O'Neil and R. Dobry, ASCE, pp. 53–79.

Pais, A. and Kausel, E. (1985), *Stochastic Response of Foundations*, Research Report R85-6, Dept. of Civil Engineering, MIT.

Ohsaki, Y. and Iwasaki, T. (1973), On dynamic shear moduli and Poisson's ratio of soil deposits, *Soils and Foundations*, **15**, No. 1.

Poulos, H. G. and Davis, E. H. (1980), *Pile Foundation Analysis and Design*, John Wiley and Sons, Inc., New York, N.Y.

Prakash, S. and Puri, V. K. (1988), *Foundations for Machines: Analysis and Design*, John Wiley and Sons, Inc., New York, N.Y.

Richart, F. E. (1975), Foundation vibrations, *Foundation Engineering Handbook*, 1st edn, Van Nostrand Reinhold, New York, N.Y., pp. 673–699.

Richart, F. E., Jr., Hall, J. R., Jr., and Woods, R. D. (1970), *Vibrations of Soils and Foundations*, Prentice-Hall, Inc., Englewood Cliffs, N.J.

Richart, F. E. and Chon, C. S. (1977), Notes on stiffness and damping of pile systems, *The Effect of Horizontal Loads on Piles*, Session 10, 9th ICSMFE, Tokyo, pp. 125–132.

Robertson, R. K., Campanella, R. G., Gillespie, D., and Rice, A. (1985), Seismic CPT to measure in-situ shear wave velocity, *Measurement and Use of S Wave Velocity*, ed. R. D. Woods, ASCE, pp. 35–49.

Roesset, J. M. (1980a), Stiffness and damping coefficients in foundations, *Dynamic Response of Pile Foundations*, ed. M. O'Neil and R. Dobry, ASCE, pp. 1–30.

Roesset, J. M. (1980b), The use of simple models in soil–structure interaction, *Civil Engineering and Nuclear Power*, ASCE, No. 1/3, pp. 1–25.

Sanchez-Salinero, I. (1982), *Static and Dynamic Stiffnesses of Single Piles*, Research Report GR82-31, University of Texas, Austin.

Seed, H. B. and Idriss, I. M. (1970), *Soil Moduli and Damping Factors for Dynamic Response Analyses*, Research Report EERC70-10, University of California, Berkeley.

Seed, H. B., Wong, R. T., and Idriss, I. M. (1986), Moduli and damping factors for dynamic analyses of cohesionless soils, *Journal of Geotechnical Engineering, ASCE*, **112**, pp. 1016–1032.

Stokoe, K. H. (1980), Field measurement of dynamic soil properties, *Civil Engineering and Nuclear Power*, **II**, pp. 7-1-1 to 7-1-31.

Stokoe, K. H. and Richart, F. E. (1974), Dynamic response of embedded machine foundations, *Journal of the Geotechnical Engineering Division, ASCE*, **100**, No. GT-4, pp. 427–447.

Tassoulas, J. L. (1981), *Elements for Numerical Analysis of Wave Motion in Layered Media*, Research Report R81-2, MIT.

Tassoulas, J. L. (1987), Dynamic soil–structure interaction, *Boundary Element Methods in Structural Mechanics*, ed. D. Beskos, ASCE.

Veletsos, A. S. and Wei, Y. T. (1971), Lateral and rocking vibrations of footings, *Journal of the Soil Mechanics and Foundations Division, ASCE*, **97**, No. SM-9, pp. 1227–1248.

Veletsos, A. S. and Verbic, B. (1973), Vibration of viscoelastic foundations, *Earthquake Engineering and Structural Dynamics*, **2**, pp. 87–102.

Velez, A., Gazetas, G., and Krishnan, R. (1983), Lateral dynamic response of constrained-head piles, *Journal of Geotechnical Engineering*, **109**, pp. 1063–1081.

Waas, G. (1972), Analysis method for footing vibrations through layered media, Ph.D. Thesis, University of California, Berkeley.

Werkle, H. and Waas, G. (1986), Dynamic stiffnesses of foundations on inhomogeneous soils, *Proceedings, 8th European Conference on Earthquake Engineering*, Lisbon, **2**, pp. 5.6/17–23.

Wolf, J. P. (1985), *Dynamic Soil–Structure Interaction*, Prentice-Hall, Englewood Cliffs, N.J.

Wolf, J. P. (1988), *Soil–Structure Interaction Analysis in Time Domain*, Prentice-Hall, Englewood Cliffs, N.J.

Wong, H. L. and Luco, J. E. (1976), Dynamic response of rigid foundations of arbitrary shape, *Earthquake Engineering and Structural Dynamics*, **4**, pp. 579–587.

Wong, H. L. and Luco, J. E. (1985), Tables of impedance functions for square foundations on layered media, *Soil Dynamics and Earthquake Engineering*, **4**, pp. 64–81.

Woods, R. D. (1978), Measurement of dynamic soil properties, *Proceedings, Earthquake Engineering and Soil Dynamics*, ASCE, **1**, pp. 91–178.

Woods, R. D., ed. (1985), *Measurement and Use of Shear Wave Velocity*, ASCE

Woods, R. D., and Stokoe, K. H. (1985), Shallow seismic exploration in soil dynamics, *Richart Commemorative Lectures*, ASCE, pp. 120–156.

16 EARTHQUAKE EFFECTS ON SOIL–FOUNDATION SYSTEMS

H. BOLTON SEED, Ph.D. (Deceased)
Professor of Civil Engineering
University of California at Berkeley

RONALD C. CHANEY, Ph.D., P.E.
Professor and Director
Fred Telonicher Marine Laboratory
Humboldt State University

SiBEL PAMUKCU, Ph.D.
Assistant Professor of Civil Engineering
Lehigh University

PART I: PRIOR TO 1975

H. BOLTON SEED, Ph.D. (Deceased)

16.1 INTRODUCTION

The damage resulting from earthquakes may be influenced in a number of ways by the characteristics of the soils in the affected area. Where the damage is related to a gross instability of the soil, resulting in large permanent movements of the ground surface, association of the damage with the local soil conditions is readily apparent. Thus, for example, deposits of loose granular soils may be compacted by the ground vibrations induced by the earthquake, resulting in large settlements and differential settlements of the ground surface. Typical examples of damage due to this cause are shown in Figures 16.1 and 16.2. Figure 16.1 shows an island near Valdivia, Chile, which was partially submerged as a result of the combined effects of tectonic land movements and ground settlement due to compaction in the Chilean earthquake of 1960. Figure 16.2 shows differential settlement of the backfill of a bridge in the Niigata earthquake of 1964.

In cases where the soil consists of loose granular materials, the tendency to compaction may result in the development of excess hydrostatic pressures of sufficient magnitude to cause liquefaction of the soil, resulting in settlements and tilting of structures as illustrated in Figure 16.3. Liquefaction of loose saturated sand deposits resulted in major damage to thousands of buildings in Niigata, Japan, in the earthquake of 1964 (Ohsaki, 1966).

Again, the combination of dynamic stresses and induced pore water pressures in deposits of soft clay and sands may result in major landslides such as that which developed in the Turnagain Heights area of Anchorage, Alaska, in the earthquake of March 27, 1964 (Seed and Wilson, 1967). An aerial view of the slide area is shown in Figure 16.4. The coastline in this area was marked by bluffs some 70 ft high sloping at about 1 on $1\frac{1}{2}$ down to the bay. The slide induced by the earthquake extended almost 2 miles along the coast and

extended inland an average distance of about 900 ft. The total area within the slide zone was thus about 130 acres. Within the slide area the original ground surface was completely devastated by displacements that broke up the ground into a complex system of ridges and depressions. In the depressed areas the ground dropped an average of 35 ft during the sliding. Houses in the area, some of which moved laterally as much as five or six hundred feet as the slide progressed, were completely destroyed. Major landslides of this type have been responsible for much damage and loss of life during earthquakes.

A somewhat less obvious effect of soil conditions on building damage is the influence they exert on the intensity of ground shaking and thereby on the structural damage that may develop even though the soils underlying a building may remain perfectly stable during an earthquake. A recent example of this effect is provided by the building damage in Caracas, Venezuela in the Caracas earthquake of 1967. In the east end of the city where the soils extended to depths ranging from 300 to 700 ft, four multistory apartment buildings collapsed and the structural damage intensity for buildings with more than nine stories was about 15 percent. However, in the west end of the city, which was slightly nearer the epicenter of the earthquake and where the soil had generally similar characteristics but was only about 60 to 280 ft deep, there were no collapses of multistory buildings and the structural damage intensity for buildings with more than nine stories was relatively low. The potential influence of local soil conditions on shaking and damage intensity in this way is one of their most far-reaching influences on earthquake damage and merits the most careful attention of soil engineers and geologists.

The following pages present a brief review of the current state-of-the-art concerning the engineering evaluation of the influence of local soil conditions on (1) ground response and shaking intensity; (2) soil settlement; (3) soil liquefaction; and (4) slope instability, during earthquakes.

Fig. 16.1 Partially submerged island near Valdivia, Chile (1960).

Fig. 16.2 Differential settlement between bridge abutment and backfill, Niigata (1964).

Fig. 16.3 Tilting of apartment buildings, Niigata (1964).

Fig. 16.4 Turnagain Heights landslide, Anchorage, Alaska (1964).

16.2 INFLUENCE OF SOIL CONDITIONS ON SHAKING INTENSITY AND ASSOCIATED STRUCTURAL DAMAGE

16.2.1 Ground Motions and Response Spectra

While the concept that the intensity of shaking and the incidence of damage during earthquakes is related in a general way to the local soil conditions has prevailed for many years, it seems likely that the first definitive study of the subject was that made by Wood (1908) following the 1906 San Francisco earthquake. From an investigation of damage locations and local geology Wood concluded:

> The investigation has clearly demonstrated that the amount of damage produced by the earthquake of April 18 in different parts of the city and county of San Francisco depended chiefly on the geological character of the ground.

Subsequent studies have shown different correlations between damage to houses and local soil conditions, but it is in comparatively recent years that instrumental data and analytical studies have provided the basis for an improved understanding of these effects.

Wood's concept of the variations in intensity of shaking at different locations in San Francisco, for example, were confirmed by recordings made at a number of locations during the 1957 San Francisco earthquake. The variations in maximum ground accelerations in relation to soil conditions are shown in Figure 16.5. It is apparent that although all the sites shown in Figure 16.5 were approximately the same distances from the zone of energy release, the ground accelerations at adjacent locations varied in some cases by as much as 100 percent, presumably as a result of the different soil conditions underlying the recording stations. Similar differences at adjacent sites have also been recorded in Osaka, Japan (Hisada et al., 1965). Variations in acceleration levels of this order of magnitude will inevitably lead to differences in structural damage intensity.

Important as the maximum ground acceleration at a building site may be, however, it does not alone determine the intensity of the shaking effects of a ground motion; these depend also on the frequency characteristics of the ground motion and its duration. For example, a very high acceleration developed for a very short period of time will cause little damage to many types of structures. A good example of this is provided by the ground motion recorded near Parkfield, California, in the earthquake of June 27, 1966. The maximum ground acceleration reached a value of $0.5g$ but probably because of its high frequency and the short duration of ground shaking, no significant damage to buildings was reported (Cloud, 1967). On the other hand, a motion with a relatively small amplitude that continues with a reasonably uniform frequency for a number of seconds can build up large accelerations and accompanying damage in certain types of structures. A good example of this effect is the damage to structures in Mexico City during the earthquake of July 28, 1957. The maximum acceleration in the central part of the city was estimated to be only about 0.05 to $0.1g$ (Merino y Coronado, 1957) but the frequency characteristics and duration were sufficient to cause the complete collapse of multi-story structures (Rosenblueth, 1960).

The combined influence of the amplitude of ground accelerations, their frequency components and, to some extent, the duration of the ground shaking on different structures is conveniently represented by means of a response spectrum (e.g., Housner, 1952; Hudson, 1956); that is, a plot showing the maximum response induced by the ground motion in single-degree-of-freedom oscillators of different fundamental periods, but having the same degree of internal damping. For example, the ground accelerations recorded in the El Centro earthquake of May 1940 are shown in the middle part of Figure 16.6. If the three simple structures shown in the upper part of Figure 16.6, having fundamental periods of 0.3, 0.5, and 1.0 sec and damping factors of 0.05 were subjected to this motion, the maximum accelerations developed in them would be $0.75g$, $1.02g$, and $0.48g$, respectively. It is apparent that the maximum acceleration induced in simple structures of this type varies with the fundamental periods of the structures. A graph showing the maximum accelerations induced in the entire range of such structures, with fundamental periods ranging from 0 to several seconds, is called an *acceleration response spectrum*. Such a graph for structures subjected to the ground motions recorded at El Centro is shown on the lower part of Figure 16.6. The maximum accelerations for the structures shown in the upper part of Figure 16.6, together with similar computations for structures with other fundamental periods, provide the means for plotting this response spectrum.

Clearly, similar computations could be made for structures with a similar range of fundamental periods but having different degrees of internal damping. Thus, it is customary to draw acceleration response spectra for a given ground motion for structures with several different degrees of internal damping, as shown in Figure 16.7.

Similarly, the computations could be made to determine not the maximum accelerations but either the maximum induced velocities or the maximum displacements. A plot showing the relationship between the maximum velocity induced by a given base motion in single-degree-of-freedom structures having a given degree of damping and the fundamental periods of the structures is termed a *velocity response spectrum*; such a spectrum for the El Centro ground motions is shown in Figure 16.8.

For any given ground motion, values of the spectral velocity S_v and the spectral acceleration S_a for a single-degree-of-freedom structure having a period T are related approximately by the equation

$$S_v \simeq \frac{T}{2\pi} \cdot S_a \qquad (16.1)$$

and it is therefore a simple matter to convert a velocity spectrum to an acceleration spectrum or vice versa.

It may be seen from the above discussion that the time history of the ground motions at a site is characterized by the corresponding response spectrum. Thus, differences in the time histories of motions at different sites may be conveniently evaluated by a comparison of their response spectra. More importantly, however, a response spectrum provides a convenient means of evaluating the maximum lateral forces developed in structures subjected to a given base motion. If the structure behaves as a single-degree-of-freedom system, the maximum acceleration and thus the maximum inertia force may be determined directly from the acceleration response spectrum from a knowledge of the fundamental period of the structure. If the structure behaves as a multi-degree-of-freedom system, the maximum responses can be determined for a number of modes and the overall maximum evaluated by some appropriate combination of the different modal effects. Normally the first mode has the greatest influence on the maximum response and thus the fundamental period, even for a multi-degree-of-freedom structure, has a dominant influence on the induced lateral forces.

Thus, the form of the response spectrum for a given ground motion is a major factor in determining the lateral forces induced on engineering structures. Of particular importance in the acceleration response spectrum, for example, is the

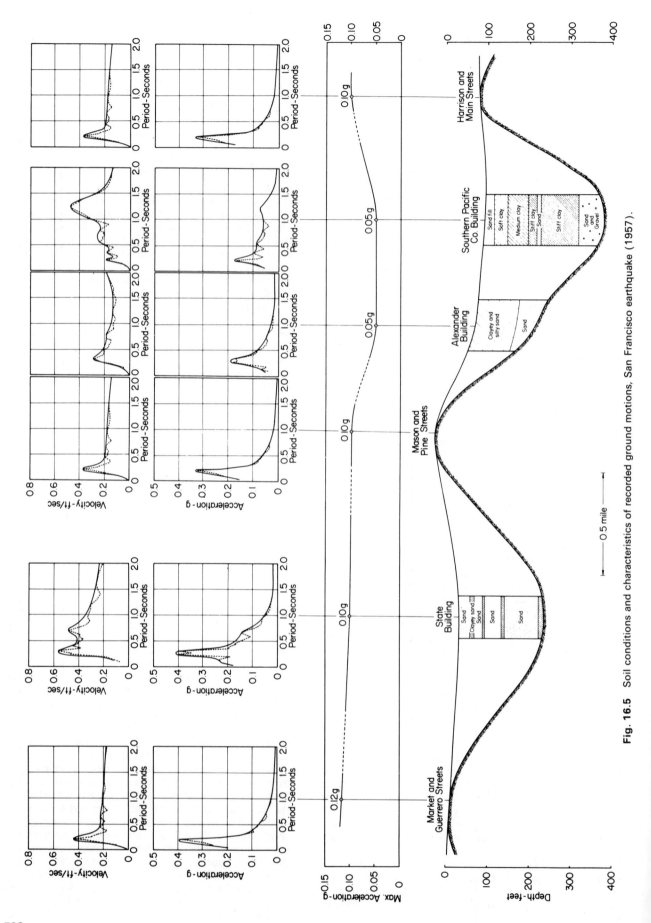

Fig. 16.5 Soil conditions and characteristics of recorded ground motions, San Francisco earthquake (1957).

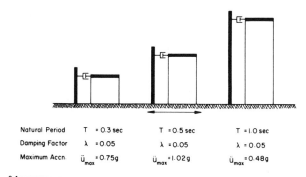

Natural Period	T = 0.3 sec	T = 0.5 sec	T = 1.0 sec
Damping Factor	λ = 0.05	λ = 0.05	λ = 0.05
Maximum Accn.	$\ddot{u}_{max}$ = 0.75g	$\ddot{u}_{max}$ = 1.02g	$\ddot{u}_{max}$ = 0.48g

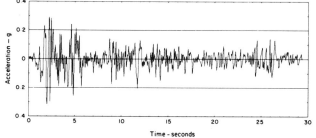

ACCELEROGRAM-EL CENTRO, CALIFORNIA EARTHQUAKE, MAY 18, 1940
(N-S COMPONENT)

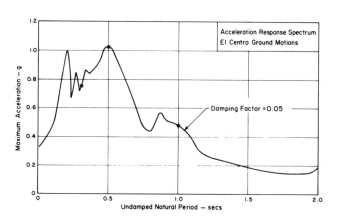

Fig. 16.6 Evaluation of acceleration response spectrum.

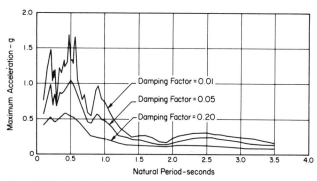

Fig. 16.7 Acceleration response spectra for El Centro (1940) earthquake.

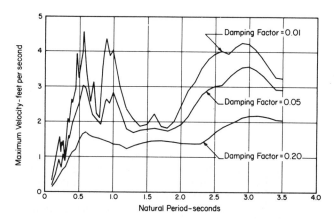

Fig. 16.8 Velocity response spectra for El Centro (1940) earthquake.

maximum ordinate and the fundamental period at which it occurs. This is readily illustrated by the data presented in Figure 16.9. In the lower part of the figure are shown two ground motion records, one for a site in San Francisco in the San Francisco earthquake of 1957 and one recorded in Pasadena during the Kern County, California, earthquake of 1952. Both records show about the same maximum ground acceleration. The response spectra for these ground motions are shown in the upper part of the figure. It is readily apparent from the spectra that responses of structures to the two ground motions will be radically different. For example, the maximum acceleration induced by the San Francisco ground motion in a one-degree-of-freedom structure having a fundamental period of about 0.9 sec would be only 0.04g; the maximum acceleration induced by the Pasadena ground motion in the same structure is seen to be 0.2g, an increase of about 400 percent. From the point of view of determining the maximum accelerations and lateral forces developed on structures during earthquakes, the establishment of the correct form of the response spectrum is clearly of primary importance.

It may be noted in Figure 16.9 that the ground motions compared were recorded at quite different epicentral distances.

Studies by Housner (1959) have shown that the frequency characteristics of the motions induced by any given earthquake change with increasing distance from the epicenter or zone of energy release. As the motions travel through the ground, the short period motions tend to be filtered out, with the result that the maximum ordinate of the response spectrum tends to develop at progressively higher values of the fundamental period.

However, it should also be recognized that even for sites in the same general area, the frequency characteristics of the ground motions, and thus the form of the response spectrum, may be profoundly influenced by the nature of the soil conditions underlying the sites. This is clearly shown by the spectra for ground motions at different locations in the 1957 San Francisco earthquake shown in Figure 16.5. For the sites shown in the figure, the highest values of spectral velocity for periods greater than about 0.5 sec were developed in the areas where ground accelerations had their lowest values.

The influence of soil conditions on the forms of response spectra is also illustrated by the series of six response spectra shown in Figure 16.10 (Seed and Idriss, 1969a), four of which were obtained from motions recorded in the same city in the same earthquake (Hisada et al., 1965) and all of which represent motions recorded at very considerable distances from the epicentral regions of earthquakes of similar magnitudes. Also shown in the figure are the soil conditions at the sites where the ground motions represented by the spectra were recorded. The spectra are arranged in sequence from A to F, corresponding to increasing degrees of "softness" of the soil conditions underlying the recording stations. To eliminate the influence of different amplitudes of surface accelerations, the ordinates of

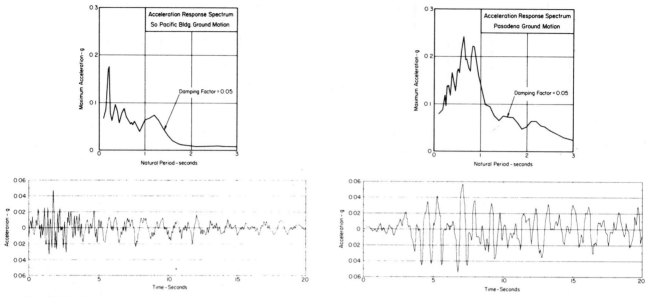

Fig. 16.9 Accelerograms and acceleration response spectra for two ground motions with comparable maximum accelerations.

the spectra have been normalized by dividing the spectral accelerations by the maximum ground acceleration at each site. Thus, the different forms of the normalized spectra reflect primarily the different frequency characteristics of the motions from which they were obtained.

It may be seen that for the recording made on rock at Site A, the peak ordinate of the response spectrum is developed at a period of about 0.3 sec, indicating a predominantly high frequency in the ground motion. However, for the still stiff (note high values of standard penetration resistance) but slightly softer soil deposit at site B, the peak ordinate of the spectrum occurs at 0.5 sec and as the ground conditions become progressively softer, as evidenced by the presence of increasingly greater depths of soft and medium stiff clays and silts, the periods at which the peak spectral accelerations are developed change as follows:

Sites *(arranged in increasing order of* *softness of soil conditions*	*Period (sec) at which* *maximum spectral acceleration* *is developed*
A	$\simeq 0.3$
B	$\simeq 0.5$
C	$\simeq 0.6$
D	$\simeq 0.8$
E	$\simeq 1.3$
F	$\simeq 2.5$

It is thus apparent that the frequency components of the motions at the different sites and the form of the response spectra change in a reasonably consistent fashion depending on the "softness" or "hardness" of the soil conditions. For the sites underlain by deposits of stiff soils, the peak ordinates of the acceleration response spectra tend to occur at a low value of the fundamental period, say 0.4 or 0.5 sec (see Fig. 16.10), indicating that at these locations the maximum accelerations would be induced in relatively stiff structures five or six stories in height. On the other hand, for the sites underlain by deep deposits of softer soils, the peak ordinates of the acceleration response spectra tend to occur at a rather high value of the fundamental period, say 1.5 to 2.5 sec (see Fig. 16.10), indicating that at these sites, the maximum accelerations would be induced in multistory structures, 20 to 30 stories in height. Thus, lateral forces on

structures and related building damage in the same general area may develop selectively; multistory structures may be severely affected where they rest on relatively soft soil deposits but adjacent stiffer structures on the same deposits may be hardly affected at all; conversely, multistory buildings on shallow, stiff soil deposits may be only slightly affected while adjacent stiff structures are subjected to large lateral forces.

16.2.2 Seismic Forces and Damage Potential

From the point of view of safety against building damage, the significant effects of earthquakes are the forces they induce on structures of all types and the effects of these forces on structural performance. If it is considered that the response of a structure to a given base motion is dominated by the influence of the first mode, then the maximum lateral forces would have the approximate distribution shown in Figure 16.11, decreasing from a maximum at the top of the structure to zero at the base. At the top of the structure the maximum acceleration would be equal to the spectral acceleration corresponding to the fundamental period, and since the participation factor for the first mode response is normally greater than 1, the maximum dynamic lateral force would be approximately equal to $W/g \cdot S_a$ where W is the weight of the structure. For a multistory building other modes besides the first will clearly influence the response but the response in the first mode provides a good approximation of the induced forces for preliminary analysis purposes. This force may be expressed as the product of the weight of the building and a maximum dynamic lateral force coefficient S_a/g.

Since the spectral acceleration value varies with the period of the structure for any one site and the forms of the spectra vary for different sites depending on the underlying soil conditions, it is apparent that buildings with different periods in different locations will be subjected to forces expressed by different maximum dynamic lateral force coefficients, the distribution depending on the variation of the quantity S_a/g in any given area. The variation of S_a/g along the section AB through San Francisco for the ground motions recorded in the 1957 earthquake, determined by reading off values from the acceleration spectra shown in Figure 16.5, is plotted in

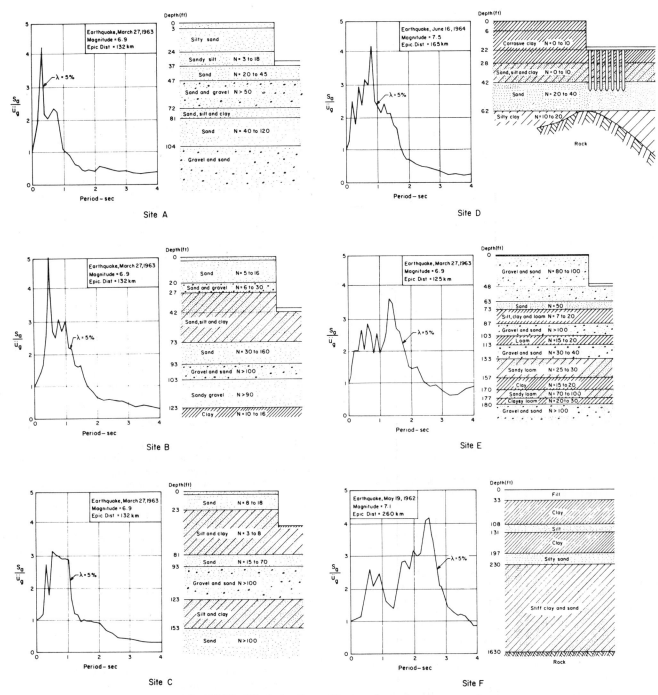

Fig. 16.10 Effect of soil conditions on form of response spectra.

Figure 16.12. Ranges are shown for building periods T in the ranges:

$$T = 0.3 \text{ to } 0.4 \text{ sec}$$
$$T = 0.6 \text{ to } 1.0 \text{ sec}$$
$$T = 1.2 \text{ to } 1.5 \text{ sec}$$

These periods are approximately related to the heights of different structures by the expression

$$T = \frac{N}{10} \tag{16.2}$$

where N is the number of stories.

It is readily apparent from the distribution of S_a/g shown in Figure 16.12 that for this earthquake, the lateral force expressed as a proportion of the weight of any building varied widely for different structures throughout a small section of the city. Maximum values were attained for buildings with periods ranging from 0.3 to 0.5 sec in the area of the State Building site and relatively smaller values developed for the same type of buildings in other areas. For very tall buildings, the maximum lateral forces were everywhere much less in proportion to the weights of the buildings, than for buildings in the three- to five-story range.

While spectral acceleration values provide a good index of the maximum lateral forces induced on buildings by any given

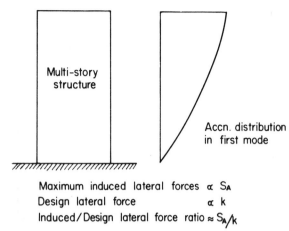

Maximum induced lateral forces $\propto S_A$

Design lateral force $\propto k$

Induced/Design lateral force ratio $\approx S_A/k$

Fig. 16.11 Schematic representation of first-mode forces on building.

ground motion, they do not necessarily provide the best index of the effects of these forces on a building. A large lateral force acting for a very short instant of time has little effect on a building. On the other hand, a somewhat smaller force acting for a substantially longer period of time may cause severe deformations. Thus, the potential damaging effect of a base motion might be considered to be approximately proportional to the product of the force developed and the period for which it sets; that is,

Potential damaging effect $\propto W \cdot S_a \cdot T$

$$\propto W \cdot S_v \quad \text{since } S_v \simeq \frac{T}{2\pi} \cdot S_a$$

Thus the spectral velocity S_v is probably a somewhat better measure of the potential damaging effect of a base motion than the spectral acceleration S_a.

The variations in S_v along the section AB are also shown in Figure 16.12. Again wide variations in the potential deforming effects of the motions are apparent; at any one site, the potential deforming effect of the earthquake, expressed by the spectral velocity S_v, varies widely depending on the period of the building, and for any one kind of building in a given period range, the value of S_v varies from site to site. For the 1957 earthquake, the potential deforming effects of the ground motion were apparently highest for buildings in the period range 0.3 to 0.4 sec located near the State Building and lowest for buildings in the period range 1.2 to 1.5 sec in the vicinity of the Alexander Building. However, in general, the variations in deformation potential expressed by the spectral velocity S_v across the profile are considerably less than the variations in the maximum dynamic lateral force coefficient expressed by the spectral acceleration, S_a/g.

In the previous paragraphs, the spectral acceleration and the spectral velocity have been shown to provide simple indices of the dynamic lateral forces developed on buildings and the damaging potential for buildings due to the ground motions produced by any given earthquake. If all buildings were designed to be equally resistant to earthquake effects, these indices would also provide a convenient means of expressing the damage potential of different buildings for any given earthquake. However, while most building codes used for earthquake-resistant design require that buildings of a given type be designed to withstand a static lateral force having a magnitude commensurate with the anticipated earthquake motions, and expressed as a lateral force coefficient k times the weight of the building, the magnitude of the coefficient usually varies with the fundamental period of the building, or the number of stories in the building. In general the lateral force coefficient decreases with increasing values of the fundamental period or increasing numbers of stories as illustrated in Figure 16.13. Thus, buildings are not generally designed to be equally resistant to the same induced forces.

The variations in design lateral forces should be taken into account in assessing the damage potential of any given earthquake for buildings designed in accordance with code requirements. In many cases, of course, buildings will be designed to withstand larger forces than the minimum values required by the codes, but in a general way it might be expected that the lateral force that a given type of building is designed to withstand would be proportional to the lateral force coefficient required by the local building code for that type of structure. In effect, then, the induced dynamic lateral force is approximately represented by $W \cdot S_a/g$ and the static lateral force an average building is designed to resist is equal to kW, where W is the weight of the building. Thus the ratio

$$F_r = \frac{\text{maximum induced dynamic lateral force}}{\text{static design lateral force}} \simeq \frac{S_a}{kg} \quad (16.3)$$

would provide a relative measure of the ability of different structures to withstand the destructive effects of an earthquake.

Alternatively, since the spectral velocity provides a better index of the damaging effects of a base motion, the relative abilities of different types of buildings to withstand different spectral velocities will be indicated by the ratio

$$D_r = \frac{\text{potential damaging effect of base motion}}{\text{design resistance}} \simeq \frac{S_v}{k} \quad (16.4)$$

Damage potential indices such as F_r and D_r (and possibly others such as S_v^2/k) provide a convenient, simple, and rational means of assessing, in a general way, the potentially damaging effect of earthquake motions on different structures. In particular the index D_r has been found to provide an extremely satisfactory basis for analyzing the damage resulting from the 1967 Caracas earthquake and might be expected to serve a similar purpose for other areas. However, their use in preventing damage depends on the ability of the engineer to predict the response spectra for the ground motions produced by earthquakes that may be expected to occur in any given area. Determination of the effects of local soil and foundation conditions on the characteristics of earthquake ground motions is therefore an essential part of damage prevention and analysis.

16.2.3 Methods of Determining the Effect of Soil Conditions on Ground Motion Characteristics

There are three main methods by which the effect of soil conditions on ground motions might be predicted:

1. By Accumulation of Strong Motion Records By accumulation of sufficient data on the ground response at a large number of sites with a wide range of soil conditions, due to different magnitudes of earthquakes at different epicentral distances, it would ultimately be possible to predict the probable motions at a new site by direct comparison of the appropriate conditions with previous data. While existing ground motion records provide a useful general guide for this purpose, there are insufficient records of strong ground motions and the associated soil conditions available to provide a basis for detailed analyses of specific sites.

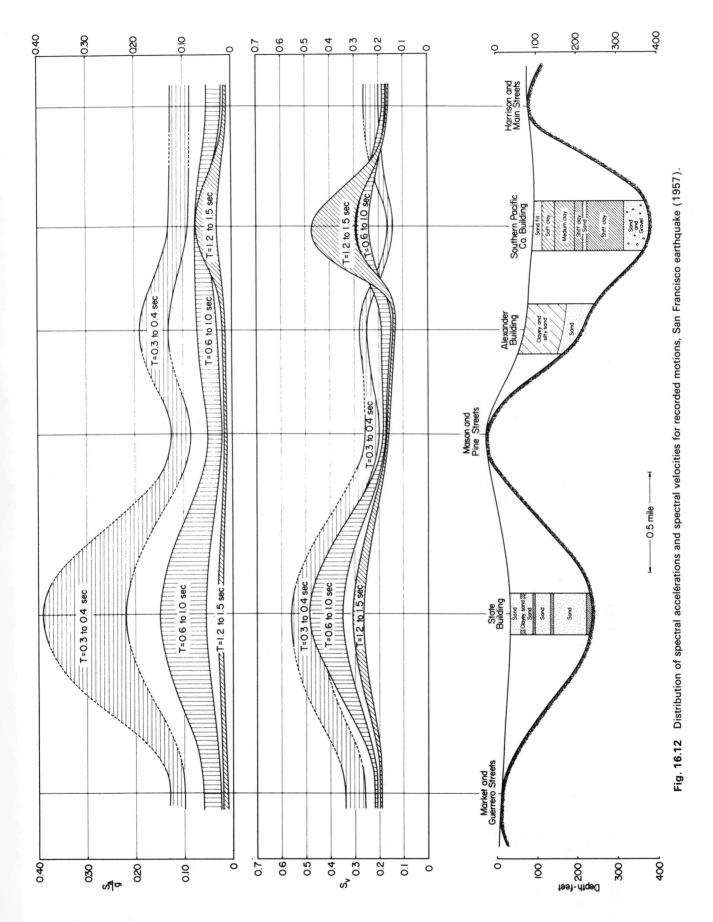

Fig. 16.12 Distribution of spectral accelerations and spectral velocities for recorded motions, San Francisco earthquake (1957).

603

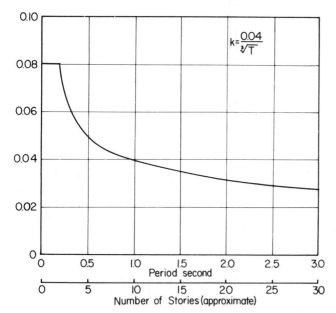

Fig. 16.13   Relationship between seismic coefficient and building period (SEAOC Code).

2. By Use of Microtremor Data   The influence of soil conditions on ground motion characteristics at any site could readily be determined if use could be made of small earthquakes and microtremors to provide a basis for evaluating site effects. In this event, mobile installations could be used to record the effects directly, since the frequency of occurrence of small earthquakes and microtremors would permit recordings to be made at frequent intervals. Unfortunately, because of the nonlinear stress–strain characteristics of soils, the behavior at small strain levels during very small earthquakes cannot be used as a direct basis for evaluating behavior at high strain levels during major earthquakes without the aid of an appropriate analytical procedure for extrapolating microtremor effects to strong motion conditions. Thus, while microtremor effects can serve an extremely useful purpose in establishing one bound on the range of possible behavior patterns and in checking the applicability of a proposed analytical procedure, they do not appear to provide, in themselves, a full predictive capability for engineering purposes.

3. By Use of Analytical Procedures   In many cases the ground motions developed near the surface of a soil deposit during an earthquake may be attributed primarily to the upward propagation of waves from an underlying rock formation and analytical procedures have been developed in recent years for determining ground response under these conditions. The methods of analysis depend on the configuration of the soil deposit.

(a) *Deposits with essentially horizontal boundaries.* For cases in which all boundaries of a stratified or homogeneous deposit are essentially horizontal, the soil may be treated as a series of semi-infinite layers as shown in Figure 16.14 and the analysis reduces to a one-dimensional problem. Two methods of approach have been used to analyze ground response under these conditions:

1. An analysis based on the use of the wave equation (Kanai, 1950, 1951; Zeevaert, 1963; Matthiesen et al., 1964; Herrera and Rosenblueth, 1965; Kobayashi and Kagami, 1966; Donovan and Matthiesen, 1968). In this approach the

Layer No. 1		G_1, λ_1, γ_1
2		G_2, λ_2, γ_2
3		G_3, λ_3, γ_3
4		G_4, λ_4, γ_4
n		G_n, λ_n, γ_n

Fig. 16.14   Semi-infinite soil deposit.

soil comprising each layer is considered to have uniform viscoelastic properties and the motion in the underlying bedrock to consist of a series of sinusoidal motions of different frequencies. The response at the surface of the deposit is then computed for a range of base rock frequencies providing a response amplification spectrum. The surface motions at a site resulting from a given base motion can then be evaluated by multiplying the Fourier spectrum of the base motion by the amplification spectrum to determine the motions at the ground surface.

2. An analysis in which the soil deposit is represented by a series of lumped masses connected by shear springs whose characteristics are determined by the stress–strain relationships of the soils in the various layers. Similarly the damping characteristics of the system are determined by the soil properties. The response of the system to a motion generated at the base can then be made by conventional dynamic analysis procedures (Penzien et al., 1964; Idriss and Seed, 1968a, 1969).

Whether the analysis is made by the wave propagation or the lumped-mass approach, meaningful results can only be obtained if the soil characteristics are correctly represented in the analytical procedure. In this respect it is important to recognize that soils have nonlinear stress–strain characteristics as shown in Figure 16.15a, which for analysis purposes may be represented by bilinear relationships as shown in Figure 16.15b (Penzien et al., 1964) or multilinear relationships (Valera, 1968). However, it has been found that essentially similar results can be obtained using an equivalent linear viscoelastic analysis in which the soil moduli and damping characteristics are selected to be compatible with the strains developed in the deposit as shown in Figure 16.16 (Idriss and Seed, 1968a). Thus, the equivalent shear moduli and damping characteristics of the soils

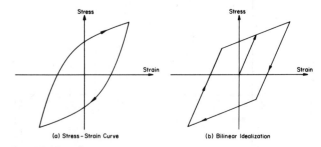

Fig. 16.15   Hysteric and equivalent bilinear stress–strain relationships for soil.

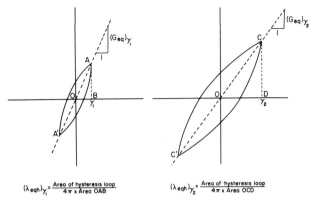

$$(\lambda_{eqh})_{\gamma_1} = \frac{\text{Area of hysteresis loop}}{4\pi \times \text{Area OAB}} \qquad (\lambda_{eqh})_{\gamma_2} = \frac{\text{Area of hysteresis loop}}{4\pi \times \text{Area OCD}}$$

Fig. 16.16 Determination of equivalent linear parameters for soil with nonlinear hysteretic characteristics.

are treated as strain-dependent properties and their values for high-intensity motions are significantly different from those applicable to low-intensity motions. Typical ranges for the variation of shear moduli and damping ratios with strain suggested by different investigators are shown in Figure 16.17. It should be emphasized that the results shown in this figure are of a general nature and are simply intended to illustrate the large variations in characteristics applicable for different strain amplitudes.

Analyses of simple soil profiles using the same soil characteristics have shown that wave propagation and lumped-mass analyses give the same results (Whitman, 1969). Furthermore both methods have been applied to soil profiles at locations for which records of the ground motions developed during earthquakes are available and shown to give results in reasonable agreement with the recorded values (Seed et al., 1968; Donovan and Matthiesen, 1968; Idriss and Seed, 1968b; Seed and Idriss, 1969a; Seed and Idriss, 1969b; Esteva et al., 1969). Apart from the use of strain-dependent soil characteristics, techniques for incorporating the nonlinear properties of soils in the wave propagation analysis procedures have not yet been developed, but since the use of equivalent linear techniques seems to be adequate for this purpose, meaningful results can be obtained by both methods of approach.

In making response analyses, it is of course important that the characteristics of the base rock motions be determined with reasonable accuracy, with regard to both amplitude and frequency characteristics. In this connection there is some question whether the motions developed in the rock at the base of a soil layer will be the same as those developed in an adjacent rock outcrop. However, such differences are likely to be small, and the self-compensating characteristics of a soil deposit subjected to a base excitation, substantial variations in amplitude of the estimated base motion will have only minor effects on the amplitude of the computed surface motions. Some typical examples for a number of different sites are shown in Table 16.1. It may be seen that significant variations in base motions of the order of ±50 percent lead to deviations varying between 5 and 25 percent from the mean value of the computed surface motions. Thus, while it is important to make reasonable assessments of base rock motions, extreme accuracy is often not required, especially in dealing with strong motions that are of major interest to the engineer, in order to make reasonably accurate assessments of surface motions. Similarly, variations in depth of a deposit, in excess of about two to three hundred feet, often have little influence on the characteristics of surface motions. On the other hand, good evaluations of soil properties

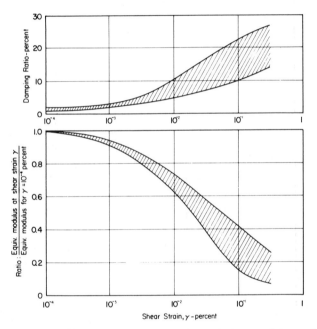

Fig. 16.17 Influence of shear strain on equivalent modulus and damping ratios.

TABLE 16.1 EFFECT OF AMPLITUDE OF ROCK MOTIONS ON GROUND RESPONSE.

Location	Range of Max. Rock Acceleration, g	Computed Max. Ground Surface Acceleration, g	Recorded Max. Ground Surface Acceleration, g
Stage Bldg., San Francisco	0.18 to 0.32	0.15 ± 10%	No record
Southern Pacific Bldg., San Francisco	0.18 to 0.32	0.21 ± 5%	No record
San Francisco Bay, California	0.17 to 0.44	0.16 ± 25%	No record
Niigata, Japan	0.10 to 0.25	0.14 ± 25%	0.155
Carabelleda, Venezuela	0.03 to 0.07	0.10 ± 12%	≃0.12[a]
Palos Grandes, Caracas	0.03 to 0.07	0.085 ± 20%	≃0.07[a]
West Caracas, Venezuela	0.03 to 0.07	0.10 ± 25%	No record

[a] Computed from structural damage observations.

in the upper two hundred feet or so of a deposit are often essential for good response evaluations, and it is to this end that the main efforts should be directed in studies of ground response. If the soil characteristics are correctly evaluated, and reasonably accurate assessments of base motions can be made, it should be possible to make reasonably good evaluations of the characteristics of ground surface motions using either of the two analysis procedures described above.

Several examples of comparisons between ground motion characteristics expressed in terms of response spectra, predicted by lumped-mass analyses, and those recorded at various sites during earthquakes are shown in Figures 16.18, 16.19, and 16.20. The degree of agreement is certainly indicative of the potential usefulness of these approaches for anticipating ground response for design purposes and for damage analysis studies. Their use for this latter purpose is discussed in a later section of this report.

(b) *Deposits with irregular or sloping boundaries.* If a deposit has irregular or sloping boundaries, it can no longer be treated as a semi-infinite layer and more complex analytical procedures, which take into account the two-dimensional aspects of the problem, are required. For this purpose the finite-element method of analysis, which is in effect the two-dimensional equivalent of the lumped-mass approach for semi-infinite layers, provides an appropriate method for response determination. The finite-element approach was first used to study the dynamic response of embankments by Clough and Chopra (1966); it has subsequently been applied to evaluate the response of earth banks and soil deposits underlain by sloping rock surfaces.

In this method a continuous medium is idealized as an assemblage of finite elements of appropriate sizes and shapes, connected at a finite number of nodal points as shown in Figures 16.21 and 16.22. The material properties of the prototype may be retained in the individual elements so that varying properties and geometric configurations can readily be handled.

The behavior of the idealized system is described by assigning an appropriate displacement field within each element. The force equilibrium of the system may then be expressed by a set of ordinary differential equations. These equations may be solved to determine the vibration mode shapes and frequencies of the system and the overall number of modes. Alternatively the solution may be obtained by the step-by-step method, which involves the direct integration of the equations of motion at discrete time intervals; this method is particularly well-suited for the solution of nonlinear problems (Valera, 1968; Wilson, 1968) for which the appropriate material characteristics may be introduced at each successive step of integration.

In applying the finite-element method of analysis to any given field problem, it is again necessary to take into account the nonlinear stress–deformation and damping characteristics of the soils comprising the deposit. This may be done either by

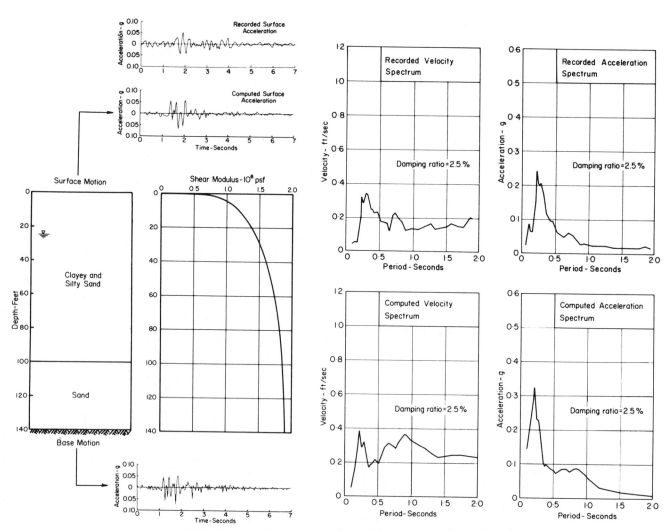

Fig. 16.18 Analysis of ground motions, Alexander Building, San Francisco.

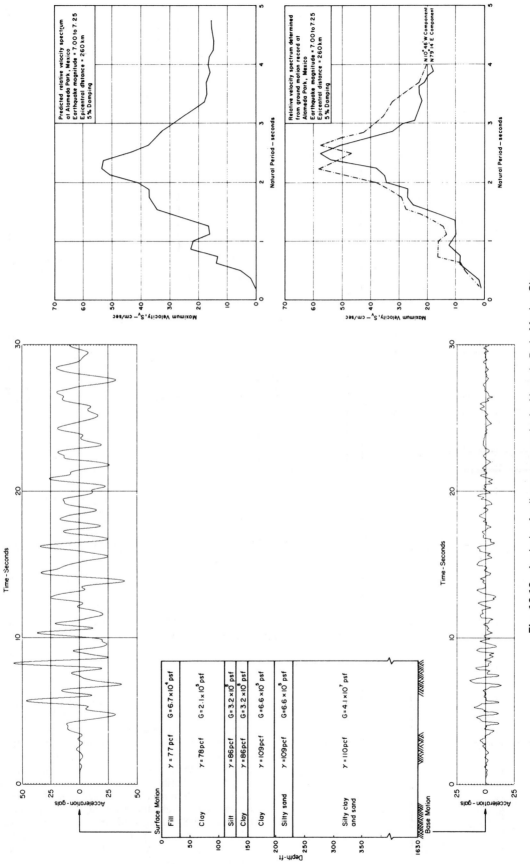

Fig. 16.19 Analysis of soil response at site in Alameda Park, Mexico City.

607

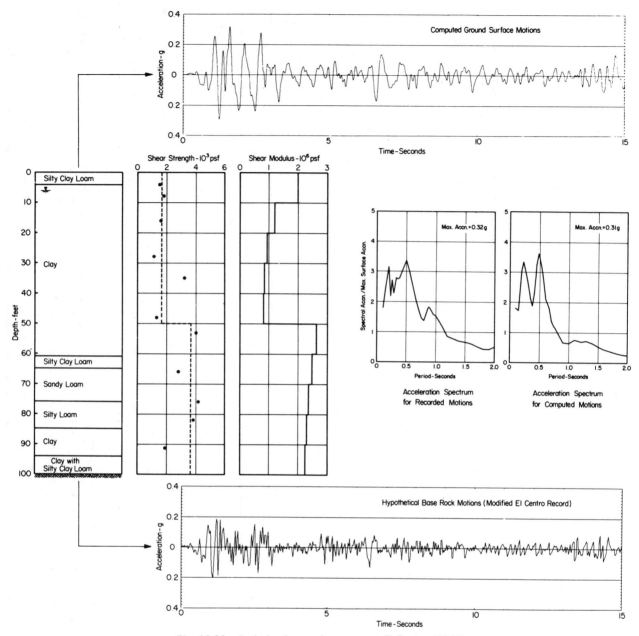

Fig. 16.20   Analysis of ground response at El Centro (1940).

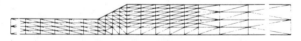

Fig. 16.21   Finite-element idealization of earth bank.

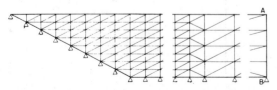

Fig. 16.22   Finite-element idealization of a soil deposit underlain by an inclined rock slope.

using multilinear representations of the actual stress–strain properties of the soil or by utilizing strain-dependent material characteristics in an equivalent linear analysis procedure.

The method has been used to investigate the response of embankments (Clough and Chopra, 1966), earth banks (Idriss and Seed, 1967; Idriss, 1968), and soil deposits underlain by sloping rock surfaces (Idriss et al., 1969; Dezfulian and Seed, 1969a, b); solutions have been developed for linear viscoelastic materials (Clough and Chopra, 1966), nonlinear materials (Valera, 1968; Dibaj and Penzien, 1969b), and nonlinear materials which may be represented by strain-compatible equivalent linear materials (Idriss et al., 1969); in addition studies have been conducted for rigid base motions with horizontal and vertical components and for travelling wave base motions (Dibaj and Penzien, 1969a; Dezfulian and Seed, 1969a, b).

Typical examples of the results obtained are shown in Figures 16.23, 16.24, and 16.25. The variations in amplitude of ground surface accelerations that may occur in a soil deposit underlain by a sloping rock surface, as determined by a finite-element analysis for motions normal to the crest of the slope and by semi-infinite layer analyses for motions parallel to the crest of the slope, are shown in Figure 16.23. It appears that sloping rock surfaces underlying soil deposits tend to reduce the motions below those developed in a similar deposit underlain by a horizontal rock surface in most cases.

Figure 16.24 shows the maximum ground surface accelerations developed in the vicinity of an earth bank 50 ft high composed of clay and subjected to motions in the underlying rock having a maximum acceleration of about 0.3*g*. It may be seen that the maximum accelerations developed just behind the crest of the slope are somewhat higher than those at locations well behind the crest of the slope, a result characteristic of a

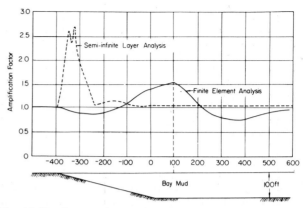

Fig. 16.23 Computed ground motions adjacent to sloping rock surface.

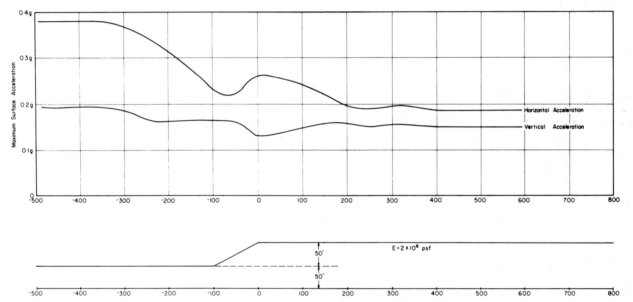

Fig. 16.24 Maximum ground surface accelerations in vicinity of earth bank.

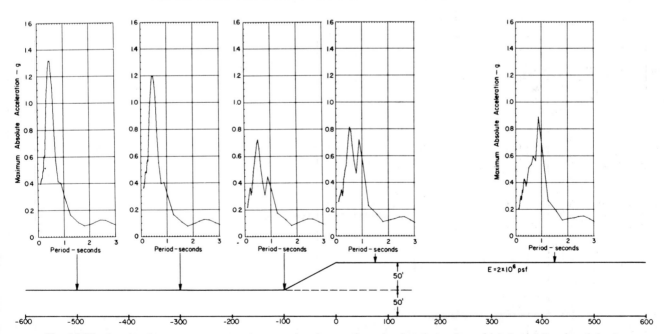

Fig. 16.25 Acceleration response spectra for ground surface motions developed at various points in vicinity of earth bank.

number of cases investigated and in accord with the higher intensities of damage that have sometimes been observed in these locations in Japan (Ohsaki, 1969) and in Anchorage, Alaska, as a result of the 1964 earthquake. Figure 16.25 shows an example of the marked changes in form of the response spectra that may occur in the vicinity of an earth bank.

Unfortunately there has been no opportunity to date to compare the response of soil deposits computed by the finite-element approach with those observed in the field. However, comparisons with the observed performance of small-scale embankments subjected to base motions on shaking tables show good agreement between computed and measured response (Kovacs et al., 1969). Furthermore, the method gives results in excellent accord with those computed by semi-infinite layer theories for deposits with horizontal boundaries and these have been shown to be in reasonably good agreement with observed ground motions. Thus, it seems likely that finite-element analyses can provide reasonably good determinations of two-dimensional problems of ground response.

16.2.4 Relationships Between Soil Conditions, Ground Motions, and Building Damage

It has long been recognized that the ground accelerations developed on the surface of soil deposits during earthquakes are usually greater than those recorded on adjacent rock outcrops. Inferential evidence that the soil conditions under-lying a site can cause a substantial increase in the intensity of the ground surface motions was presented by Wood (1908) in

his study of the distribution of damage and apparent intensity of shaking in the San Francisco Bay area during the earthquake of 1906, and by Duke (1958). In recent years a number of investigators—for example, Gutenberg (1957) in the United States and Kanai et al. (1954, 1959) in Japan—have obtained instrumental data showing that during small earthquakes and microtremors, the ground accelerations on soil deposits are usually considerably higher than those occurring on adjacent rock exposures. Wiggins (1964) and Blume (1965) drew a similar conclusion from studies of strong motion records, and the amplifying effects of soil deposits in transmitting underlying rock motions have been shown by analytical studies (see, for example, Figures 16.19 and 16.20). Because of these observations, the ratio of the maximum ground surface acceleration on soil to that developed in adjacent rock has been termed the ground acceleration amplification factor. Measured values for this factor range from values between 1 and 2 for strong motions to values in excess of 10 for microtremors, and it has often been suggested that amplification factors are higher for soft deposits than in firm deposits.

It is important to recognize, however, that soil deposits can also attenuate rock motions; that is, values of the amplification factor may sometimes be less than 1. For example, Figure 16.26 shows the ground accelerations recorded at 3 sites in San Francisco during the earthquake of 1957 (magnitude 5.7). It is apparent that the amplification factors for the two sites underlain by soil deposits were about 0.5 and 1.0 in this case. Furthermore, the accelerations recorded on soft soil deposits may sometimes be substantially less than those on firm deposits. Housner (1954) describes an example of this effect during the

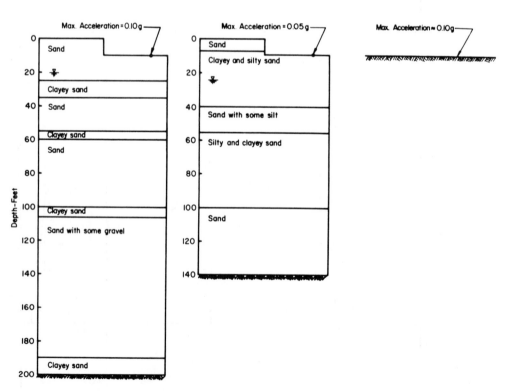

San Francisco Earthquake of March 22, 1957
Magnitude = 5.4
Epicentral Distance = 9.5 mi.

Fig. 16.26 Effects of soil conditions on maximum ground surface accelerations.

Seattle, Washington, earthquake of April 13, 1949:

> This shock was recorded at both Seattle and at Olympia, and both cities were approximately the same distance from the epicenter, so that it would be expected that the record of the ground motion would be of approximately the same intensity at both places. This, however, was not the case.... The Seattle record shows motion much less intense and with most of the high-frequency components missing. This is attributed to the fact that the instrument was adjacent to a sea wall... etc.

The maximum ground acceleration recorded at Seattle was only 0.08*g* compared with a value of about 0.3*g* at Olympia.

Whether or not a given soil deposit or a soil layer within a deposit will amplify or attenuate the motions in the underlying material, and the magnitude of these effects, seems to depend on the thickness and dynamic characteristics of the soil and the amplitude and frequency characteristics of the base motions. This is illustrated by recordings made at three levels in a soil deposit in Union Bay, Seattle (Seed and Idriss, 1969b). The soil conditions at the site consist of about 60 ft of peat, 40 ft of clay, with an underlying layer of glacial till. Accelerometers are located 10 ft below the ground surface in the peat layer, at the top of the clay layer, and at the top of the glacial till layer. The maximum accelerations recorded at these locations during two shocks, one a small local earthquake and the other an event occurring several hundred miles away are shown in Figure 16.27. During both events, the motions at the surface of the glacial till were amplified in the clay layer; however, while the peat deposit also amplified the motions from the distant seismic shock, it attenuated very considerably the stronger motions induced by the nearby earthquake. This was probably due in large measure to the higher damping characteristics of the peat under the larger strain amplitudes.

The influence of the amplitude of the base motions on the computed amplification factor for the soil deposits underlying the Alexander Building in San Francisco (see Fig. 16.5) is shown in Figure 16.28 (Idriss and Seed, 1968b). There is a marked reduction in amplification factor with increasing levels of base acceleration, a result also shown by field observations of soil response. It is for this reason that amplification factors measured during small earthquakes cannot be applied directly to predict performance during large earthquakes.

Finally, the influence of the frequency characteristics of the base motion on the computed response of a soil deposit is shown in Figure 16.29. The figure shows the relationship between the maximum surface acceleration, computed by a lumped-mass analysis, and the depth of a sand deposit for base motions having the same maximum acceleration (0.06*g*) but different predominant periods. It may be seen that the amplifying effects of different thicknesses of sand vary

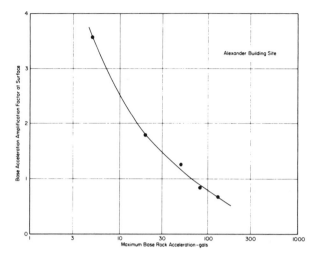

Fig. 16.28 Variations of amplification factors with maximum base rock accelerations.

considerably as the predominant period of the base motion is changed from about 0.3 sec (corresponding to an earthquake or magnitude 6½ at an epicentral distance of about 25 miles) to about 0.7 sec (corresponding to the motions produced by a very strong distant earthquake).

It is also important to note that, in general, the amplifying effects of a soil deposit on ground accelerations are greatest

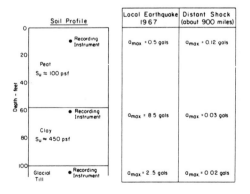

Fig. 16.27 Maximum ground accelerations recorded in Union Bay, Seattle.

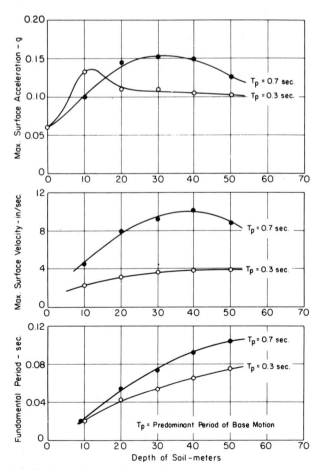

Fig. 16.29 Influence of depth of sand on response to base motion with maximum acceleration of 0.06*g*.

when the fundamental period of the deposit is close to the predominant period of the base motions. In Figure 16.29, the maximum surface acceleration is developed for a deposit having a thickness of 12 m and a fundamental period of about 0.3 sec when the predominant period of the base motion is 0.3 sec. However, the maximum surface acceleration is developed for a deposit having a thickness of about 30 m and a fundamental period of about 0.75 sec when the predominant period of the base motion is 0.7 sec. Thus the fundamental period of a soil deposit is an extremely important factor in assessing its probable response. However, it should be noted in the lower part of Figure 16.29 that the period of a given deposit will vary with the characteristics of the base motions (both frequency and amplitude) owing to the nonlinear characteristics of the soils comprising it. Such variations have been observed in the field and in the laboratory (Seed and Idriss, 1969a).

Of particular importance from a soil engineering point of view are the potential damaging effects of the ground motions. As discussed previously, these depend on the characteristics of the structure in addition to the various factors discussed above. For example, damage to wooden buildings of limited story heights might be expected to depend primarily on the maximum ground surface acceleration or velocity induced by the earthquake and to vary in any local area with the changes in surface acceleration or velocity levels resulting from different soil conditions. Thus, variations in potential damage patterns can readily be investigated. Suppose an area of alluvial sandy deposits of varying depths is shaken by a magnitude $8\frac{1}{4}$ earthquake, the zone of energy release being at a distance of about 125 km. The resulting motions in bedrock might be expected to have a maximum acceleration of about 0.06g and a predominant period of about 0.7 sec. The maximum ground surface accelerations and velocities developed at the surfaces of different depths of sandy alluvium due to such a base motion, computed by the lumped-mass method of analysis, are shown in Figure 16.30. The maximum ground surface motions increase with increasing depths of soil up to about 35 to 45 m and then decrease to some extent. It is interesting to compare these results with the damage to wooden buildings at Nagoya and along the River Kiku resulting from the Tonankai earthquake in Japan in 1944. The magnitude and epicentral distances were similar to those used to obtain the response data in Figure 16.30. Observed damage intensities at Nagoya and along the River Kiku, for different depths of soil, are also shown in Figure 16.30. The similarity in the form of the observed damage intensity relationships and the anticipated ground velocities is readily apparent.

For buildings with more than one or two stories designed in accordance with building code requirements for earthquake-resistant design, the damage intensities are more likely to be related to variations in a damage potential index such as $D_r = S_v/k$ as previously described. An example of the damage patterns that might be anticipated for such buildings is shown in Figures 16.31 and 16.32, for a firm deposit of sand and gravel with depths varying up to 150 m and subjected to base accelerations in the underlying rock having a maximum amplitude of 0.05g and a predominant period of about 0.35 sec. Values of the maximum ground surface acceleration for motions normal to the plane of the cross section shown in the figure, together with the associated acceleration and velocity response spectra, computed by the lumped-mass analysis procedure, are shown in Figure 16.31. Also shown are spectra of S_v/k, determined from the spectral velocities indicated by the velocity response spectra and values of the seismic coefficient k required by the Uniform Building Code (U.S.A.) for structures in seismic zone 2.

With the aid of this information it is a simple matter to read off values of S_v/k at different locations in the profile, for buildings with any given fundamental period. The distributions

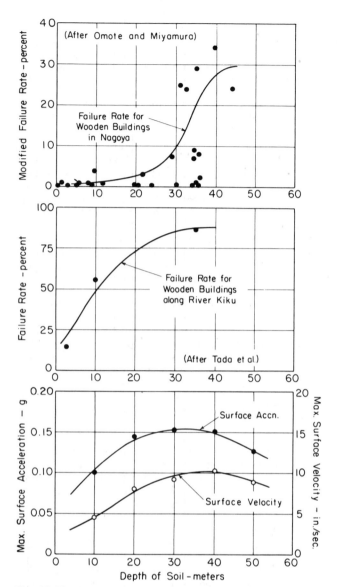

Fig. 16.30 Comparison of computed response of sand deposits with failure rates of wooden buildings in Tonankai earthquake (1944).

of S_v/k, determined from the values shown in Figure 16.31, for structures with fundamental periods of 0.2 sec, 0.4 sec, and 1.0 sec are plotted in relation to soil depth in Figure 16.32. It may be seen that the damage potential coefficient varies considerably with building characteristics and soil depths. For two- or three-story structures with a period of about 0.2 sec the damage intensity would appear to be essentially the same regardless of soil depth and everywhere substantially less than for structures with longer periods. For four- or five-story structures with a period of about 0.4 sec, the maximum damage intensity might be expected to develop where the depth of soil is about 20 to 50 m for this particular earthquake and soil type, and to be somewhat lower for greater or shallower depths. For ten- to twelve-story structures with a period of about 1 sec, the damage intensity would be expected to increase with soil depths up to 150 m or so, and where the soil reaches these depths to be considerably higher than that developed anywhere in the section for two- to five-story structures. Finally, for very high structures of 15 to 25 stories, the damage intensity would be expected to be even greater for soil depths of 150 to 250 m, though it would be relatively low for soil depths up to 80 m

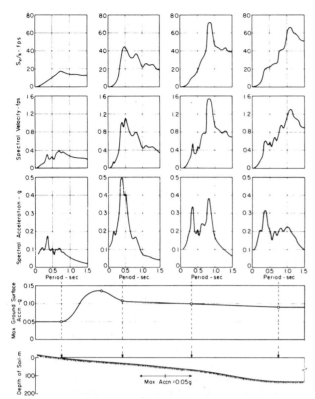

Fig. 16.31 Computed response spectra for sand and gravel deposit. (Maximum rock acceleration = 0.05*g*; predominant period of rock motions = 0.35 sec.)

or so. It may be noted that low intensities of damage are indicated for all types of structures on rock.

Results of this type may well serve to explain in a general way the damage patterns observed in several recent earthquakes. In Mexico City (1957), Anchorage, Alaska (1964), and Caracas (1967), for example, there was a notably higher incidence of damage for multistory buildings where soil depths ranged from 100 to 400 m than for similar structures on rock or shallow depths of soil. However, damage to lower structures was much less intense and essentially independent of soil depth. While details of damage patterns will of course depend on the particular soil characteristics in the area involved as well as the magnitude and location of the earthquake inducing the motions, damage analysis techniques seem to provide a basis for an improved understanding of these effects.

It is important to recognize that observed relationships between building damage and soil characteristics may be influenced by other factors besides ground shaking intensity. Excluding cases of major soil instability, these might include:

1. Variations in the quality of structural design; potential variations from this factor may well outweigh the effects of variations in soil conditions except where severe permanent soil movements are induced by the earthquake.
2. The effects of minor ground settlements and differential settlements resulting from the earthquake ground motions; these effects are likely to increase in a general way with increasing depths of soil.
3. The effects of previous settlements on the stresses existing in building structures before an earthquake; this is likely to be particularly important for buildings on soft ground and the effect will also increase with increasing depth of soil.

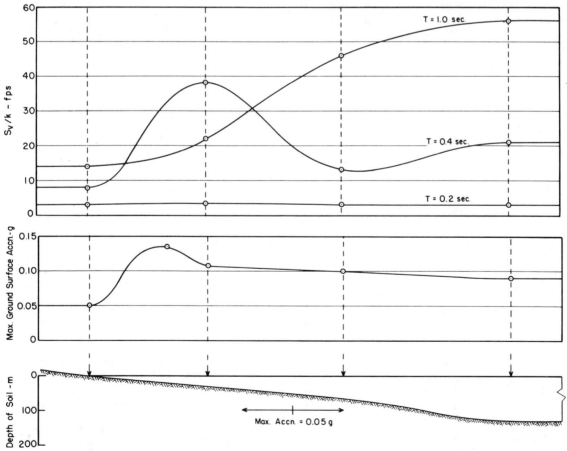

Fig. 16.32 Computed distribution of damage potential index S_v/k for sand and gravel deposit. (Maximum rock acceleration = 0.05*g*; predominant period of rock motions = 0.35 sec.)

In view of these factors, as well as those affecting shaking intensity and structural behavior, analysis of the effects of soil conditions on damage due primarily to the effects of ground shaking require an understanding of the complex interrelationships between the effects of soil types, soil depths, the amplitudes of ground motions, the frequency characteristics of ground motions, and the structural characteristics of buildings in order to analyze damage resulting from past earthquakes or prevent damage in future earthquakes.

16.3 INFLUENCE OF SOIL CONDITIONS ON GROUND SETTLEMENT

Since vibration has long been recognized as an effective method of compacting cohesionless soils, it is not surprising that the ground vibrations caused by earthquakes often lead to compaction of cohesionless soil deposits and associated settlement of the ground surface.

A quantitative measure of ground settlement of this type was provided by the behavior of a well casing at Homer during the Alaska earthquake of 1964 (Grantz et al., 1964). The casing had been installed to firm rock before the earthquake and projected about 1 ft above the ground surface. Following the earthquake the casing projected some $3\frac{1}{2}$ ft above the ground surface, indicating a decrease in thickness of the soil layer of about 2.5 ft. As shown in Figure 16.33 tectonic movements caused the rock surface to be lowered by 2 ft and this, together with the 2.5 ft of settlement caused by soil compaction, resulted in a total settlement of the ground surface of 4.5 ft.

A similar combination of effects in the Portage area of Alaska (1964) led to the two being inaccessible during periods of high tide. Here the combination of about 4 ft settlement of the rock due to tectonic movements together with about 4 ft of settlement due to compaction of the overlying soil, led to a ground surface settlement of about 8 ft (Grantz et al., 1964). As a result of the general flooding in this area during high-tide periods, the township had to move to a new location.

Similar problems of flooding and inundation of land due to settlement of soil by compaction or a combination of compaction and tectonic movements also occurred in the Chilean earthquake of 1960 (Retamal and Kausel, 1969) and

the Niigata, Japan, earthquake of 1964. In Valdivia, subsidence due to tectonic movements was about 6 ft and additional settlements due to soil compaction varied from 0 to about 3 ft.

Ground settlements due to compaction often lead to differential settlements of engineering structures—a phenomenon that is particularly well illustrated by the performance of bridge abutments. Often an abutment is supported on firm materials or on a pile foundation and undergoes relatively small settlements compared with the backfill material for the abutment, which rests directly on the ground surface and settles due to compaction of the soil on which it rests. Figure 16.2 shows a differential movement of several feet between a railroad bridge abutment and its backfill as a result of the Niigata earthquake of 1964.

In addition to the damage resulting from changes in elevation, differential settlements due to soil compaction and the resulting stresses induced in buildings may well have contributed significantly to the structural damage resulting from earthquakes in some locations. Tests on dry sands have shown that vertical accelerations in excess of $1g$ are required to cause any significant densification (Whitman and de Pablo, 1969). On the other hand, relatively small cyclic shear strains have been found to cause appreciable densification of loose sands under simple shear conditions. Thus, it seems likely that it is the horizontal motions induced by earthquakes that are primarily responsible for the settlements observed. Predictions of settlements based on this concept, using stresses computed by ground response analyses and settlement data obtained from cyclic simple shear tests have been found to be in reasonable agreement with the observed settlements of small sand layers in shaking table tests and seem to offer a reasonable basis for estimating potential settlements of sand deposits during earthquakes (Seed and Silver, 1972).

16.4 INFLUENCE OF SOIL CONDITIONS ON LIQUEFACTION POTENTIAL

One of the most dramatic causes of damage to engineering structures during earthquakes has been the development of liquefaction in saturated sand deposits, manifested either by the formation of boils and mudspouts at the ground surface, seepage of water through ground cracks, and, in some cases, by the development of quicksand-like conditions over substantial areas. Where the latter phenomenon occurs, buildings may sink substantially into the ground or lightweight buried structures may float upwards to the ground surface.

Nowhere has the phenomenon of liquefaction been more dramatically illustrated in recent years than in Niigata, Japan, during the earthquake of June 16, 1964. The epicenter of the earthquake (magnitude about 7.5) was located about 35 miles from Niigata but nevertheless the earthquake induced extensive liquefaction of the sand deposits in the low-lying areas of the town. Water began to flow out of cracks and boils during and immediately following the earthquake, as shown in Figure 16.34, causing liquefaction of the deposits and widespread damage. Many structures settled more than 3 ft in the liquefied soil and the settlement was often accompanied by severe tilting as shown in Figure 16.3. Thousands of buildings collapsed or suffered major damage as a result of those effects (Ohsaki, 1966).

Liquefaction has been reported in numerous other earthquakes (Seed, 1968; Ambraseys and Sarma, 1969), in some cases the upward flow of water from the ground continuing for as much as 30 min after the ground motions stopped.

The cause of liquefaction of sands has been understood, in a qualitative way, for many years. If a saturated sand is subjected to ground vibrations, it tends to compact and decrease in volume; if drainage is unable to occur, the tendency to decrease

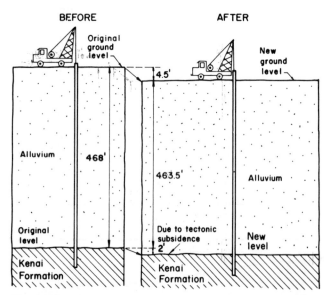

Fig. 16.33 Ground settlement around well casing at Homer during Alaska earthquake (1964).

with observed effects, these studies have shown that the resulting water movements can cause a considerable reduction in effective stresses in the upper layers and, under extreme conditions, can lead to a total loss of strength of the layer. Furthermore, even where liquefaction does not occur, the upward flow of water can lead to a marked reduction in bearing capacity of the upper layer.

16.4.1 Soil Conditions in Areas Where Liquefaction has Occurred

In spite of the large number of cases in which liquefaction has been reported to have occurred during earthquakes, there are relatively few details of the soil conditions in the affected areas. Several cases where detailed studies have been made are described below.

1. Niigata Earthquake In the Niigata earthquake the sand in the zone of liquefaction had a 10 percent size ranging from about 0.07 to 0.25 mm and a uniformity coefficient of about 10. Following the earthquake an extensive survey of the distribution of damaged structures was made. It was found that structures in the coastal dune area (designated zone A) suffered practically no damage. The major damage and evidence of liquefaction were concentrated in the lowland area, but even here two zones could be clearly recognized—one in which damage and lique-faction were extensive (designated zone C) and one in which damage was relatively light (zone B). Because all zones contained similar types of structures, the differences in extent of damage could be attributed to differences in the subsoil and foundation behavior. Studies were conducted by a number of Japanese engineers to determine the differences in soil conditions in the various zones.

The difference in behavior in zone A from that in zones B and C could readily be attributed to two major differences in soil characteristics. Although all zones were underlain by sandy soils to a depth of approximately 100 ft, in zone A the underlying sands were considerably denser than those in zones B and C, and, furthermore, the water table was at a much greater depth below the ground surface. In zones B and C, however, the general topography and depth of water table were essentially the same. It was therefore concluded that the difference in extent of damage in these two zones must be related in the characteristics of the underlying sands. Accordingly, considerable effort was made to determine any significant differences in the general soil conditions in these zones.

Because the soils involved are sands, efforts were concentrated on the determination of the relative density of the sands by means of standard penetration tests. Koizumi (1966) has presented the results of a number of borings made in zones B and C to show the variation of penetration resistance with depth in the two zones. There is a considerable scatter of the results in any one zone, but averaging the values obtained leads to the comparative values shown in Figure 16.35.

It may be seen that in zones B and C, the average penetration resistance of the sands is essentially the same in the top 15 ft. Below this the sands in zone B are somewhat denser than those in zone C. Below about 45 ft, the sands in both zones are relatively dense and are unlikely to be involved in liquefaction. It seems reasonable to conclude that the relatively small differences in penetration resistance of the sands in the depth range from 15 ft to 45 ft is responsible for the major difference in foundation and liquefaction behavior in the two zones.

In addition to comparing the soil conditions in the different damage zones, Japanese engineers have made a detailed study of the relationship between soil and foundation conditions

Fig. 16.34 Initial stages of water flow from ground, Niigata (1964).

in volume results in an increase in pore water pressure, and if the pore water pressure builds up to the point at which it is equal to the overburden pressure, the effective stress becomes zero, the sand loses its strength completely, and it develops a liquefied state.

Liquefaction of a sand in this way may develop in any zone of a deposit where the necessary combination of in-situ conditions and vibratory deformations may occur. Such a zone may be at the surface or at some depth below the ground surface, depending only on the state of the sand and the induced motions.

However, liquefaction of the upper layers of a deposit may also occur, not as a direct result of the ground motions to which they are subjected, but because of the development of liquefaction in an underlying zone of the deposit. Once liquefaction develops at some depth in a mass of sand, the excess hydrostatic pressures in the liquefied zone will dissipate by flow of water in an upward direction. If the hydraulic gradient becomes sufficiently large, the upward flow of water will induce a "quick" or liquefied condition in the surface layers of the deposit. Liquefaction of this type will depend on the extent to which the necessary hydraulic gradient can be developed and maintained; this, in turn will be determined by the compaction characteristics of the sand, the nature of ground deformations, the permeability of the sand, the boundary drainage conditions, the geometry of the particular situation, and the duration of the induced vibrations.

While most investigators have been concerned with the conditions inducing liquefaction, studies of the pore pressure distributions in sands during and following liquefaction have been presented by Maslov (1957), Housner (1958), Florin and Ivanov (1961), and Ambraseys and Sarma (1969). In accordance

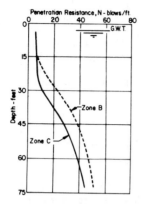

Fig. 16.35 Comparison of soil conditions in zone B (light damage) and zone C (heavy damage).

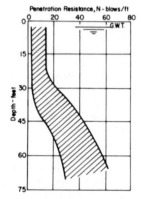

Fig. 16.36 Range of penetration resistance values in heavy damage zone.

and building performance in zone C. Here the variation of penetration resistance with depth falls within the shaded area shown in Figure 16.36, the standard penetration resistance in the top 25 ft generally being less than 15, but sometimes falling as low as 5. For each building in this zone the extent of damage caused by foundation failure was classified into one of four categories, ranging from the no-damage category I (buildings that settled up to 8 inches or tilted up to 20 minutes of angle) to the heavy damage category IV (buildings settling more than 3 ft or tilting more than 2.3°).

A study was made of the influence of foundation type on the settlement and tilting of reinforced concrete buildings in zone C (Kishida, 1965). Some of these buildings had shallow spread footing foundations; others were supported on short piles, typically extending to a depth of about 25 ft. For each foundation type, the proportion of buildings falling in light damage categories I and II was compared with the proportion in the heavy damage categories III and IV. The results of this comparison are shown in Table 16.2, from which it may be seen that the provision of short pile foundations apparently

TABLE 16.2 INFLUENCE OF TYPE OF FOUNDATION ON EXTENT OF DAMAGE.

Type of Foundation	No Damage and Slight Damage	Intermediate and Heavy Damage
Shallow (63 buildings)	36%	64%
Pile (122 buildings)	45%	55%

had little effect in reducing the damage caused by the earthquake ground motion.

For buildings with spread footing foundations, a study was made to determine the relationship between the penetration resistance of the sand at the base of the foundations and the extent of damage. The results of this study are shown in Figure 16.37. When the sand underlying the footings had a penetration resistance of less than 15, the buildings usually suffered heavy damage (categories III and IV). However, when the penetration resistance was between 20 and 25, the structures suffered only light damage, or none. Thus it appears that a penetration resistance of slightly more than 20 would be adequate to prevent foundation settlements exceeding about 6 inches in this earthquake.

The results of a similar study to determine the relationship between depth of piles, penetration resistance of the sand at the pile tip, and the extent of damage for pile-supported structures, are shown in Figure 16.38. From these data it may

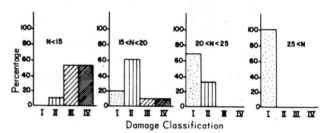

Fig. 16.37 Relationship between penetration resistance at base of foundation and extent of damage.

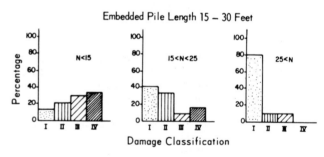

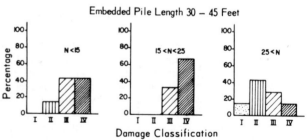

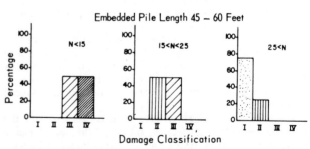

Fig. 16.38 Influence of penetration resistance at tip of piles on extent of damage.

be seen that for pile lengths varying from 15 to 60 ft, generally heavy damage (large settlements and tilting, or both) occurred when the penetration resistance of the sand at the pile tip was less than 15. However, for the same range of pile lengths, settlements and tilting were generally small when the penetration resistance of the sand at the pile tip exceeded 25. Thus, the penetration resistance values providing satisfactory performance for pile foundations are quite similar to those providing satisfactory performance for footing foundations.

Finally, for a wide range of buildings in the heavy damage zone, a study of the relationship between depth of foundation, penetration resistance, and extent of damage showed that (1) for foundations in the depth range 0 to 15 ft, a penetration resistance of $N = 14$ at the base of the foundation was apparently adequate to prevent major damage caused by settlement and tilting; (2) for foundations in the depth range 15 ft to 25 ft, a penetration resistance of the sand between 14 and 28 at the base of the foundation was required to prevent major damage; and (3) for foundations in the depth range 25 ft to 50 ft, a penetration resistance of $N = 28$ at the base of the foundation was required to prevent major damage.

These results provide a valuable guide in assessing the liquefaction of other sand deposits subjected to ground motions similar to those at Niigata.

2. Mino-Owari, Tonankai, and Fukui Earthquakes Kishida (1969) has made a detailed study of the soil conditions at places where sand volcanoes and eruption of water and soil were observed during the Mino-Owari (1981), Tonankai (1944), and Fukui (1948) earthquakes in Japan. At six sites investigated, the 10 percent size of the soil particles ranged from 0.05 to 0.25 mm and the uniformity coefficient was less than 5. Values of the standard penetration resistance in the upper 30 ft were typically less than 20 and often less than 10.

3. Jaltipan Earthquake In the Jaltipan (Mexico) earthquake of August 26, 1959, sudden settlements of about 1 m occurred in the foundations of the Naval shipyard at Coatzacoalcos as well as relative horizontal displacements of sections of an adjacent quay. Studies by Marsal (1961) led to the conclusion that these failures were due to partial liquefaction of a sandy silt and silty sand deposit for which D_{10} ranged from 0.01 to 0.1 mm with uniformity coefficients of the order of 2 to 10.

4. Alaska Earthquake The 1964 Alaska earthquake (magnitude $\simeq 8.3$) caused extensive damage to a wide variety of bridge foundations located at distances of 50 to 80 miles from the zone of major energy release (Ross et al., 1969). Damage included horizontal movement of abutment foundations toward stream channels, spreading and settlement of abutment fills, horizontal displacements and tilting of piers, and severe differential settlements of abutments and piers.

The greatest concentrations of severe damage occurred in regions characterized by thick deposits of saturated cohesionless soils. Ample evidence exists of liquefaction of these materials during the earthquake and this phenomenon probably played a major role in the development of foundation displacements and bridge damage. Typical foundation conditions in these areas consisted of piles driven through saturated sands and silts of low to medium relative density (standard penetration resistance less than about 20 to 25); of approximately 60 samples investigated from the heavy damage area, two-thirds of the samples had a 10 percent size ranging from about 0.01 to 0.1 mm and a uniformity coefficient of 2 to 4. On the other hand, bridges supported on gravels and gravelly sands, regardless of their penetration resistance values, generally showed small or no displacements, indicating no significant liquefaction of these materials under comparable conditions.

In the four case studies described above it appears that liquefaction has usually occurred in relatively uniform cohesionless soils for which the 10 percent size is between 0.01 and 0.25 mm and the uniformity coefficient between 2 and 10. Laboratory tests conducted to determine the susceptibility of soils to liquefaction under cyclic loading conditions also indicate that uniformly graded soils for which the 50 percent size lies in the range 0.02 to 0.4 mm are considerably more vulnerable to liquefaction than coarser or finer materials (Lee and Fitton, 1968). The liquefaction soils in the cases discussed typically had standard penetration resistance less than 25 blows per foot, but the available data are insufficient to permit general conclusions to be drawn for the range of ground motions likely to be encountered in practice.

16.4.2 Laboratory Investigations of Soil Liquefaction

A number of studies have been conducted to investigate the liquefaction characteristics of soils under laboratory test conditions. Several investigators (e.g., Maslov, 1957; Prakash and Mathur, 1965; Nunnally, 1966; Yoshimi, 1967) have attempted to establish the conditions producing liquefaction in terms of the acceleration at which the phenomenon can be observed to develop. Usually this is done by placing saturated sand in a box on a shaking table and observing the table accelerations at which liquefaction occurs. However, such results are inevitably influenced by the duration and frequency of the table motions to which the sand is subjected and possibly also by the geometry and deformation characteristics of the container in which the saturated sand is placed. Thus, it is difficult to extrapolate the results to field conditions. Furthermore Ambraseys and Sarma (1969) point out that the incidence of liquefaction in a deposit is more likely to be determined by the ground velocity rather than the ground accelerations, as observed by Puchkov (1962) in the field.

The difficulties in this type of approach have led other investigators to study the liquefaction characteristics of saturated sands under undrained cyclic loading conditions, in an attempt to simulate as closely as possible the stress conditions induced under field conditions (Lee and Seed, 1967; Peacock and Seed, 1968). Cyclic loading triaxial compression tests and simple shear tests have been used for this purpose. It has been shown that under these conditions liquefaction can readily be induced in loose to medium dense sands and its development is determined by (1) the magnitude of the cyclic shear stress or strain; (2) the number of stress or strain cycles; (3) the initial density; (4) the confining pressure; and (5) the initial shear stresses acting on the sand.

Figure 16.39 shows a typical relationship between the magnitude of the cyclic stress and the number of stress cycles required to induce failure for samples of saturated sand subjected to a confining pressure of 1 kg/cm^2. The larger the magnitude of the applied cyclic stresses, the fewer is the number of cycles required to induce liquefaction. However, the magnitude of the cyclic stresses required to induce liquefaction of a saturated sand increases rapidly with increase in the density of the sand.

Other factors being equal, the higher the confining pressure on a sand the greater is the cyclic shear stress required to induce liquefaction; thus, the presence of a surcharge will reduce the tendency of a deposit to liquefy. A similar conclusion has been drawn by Maslov (1957) and Ambraseys and Sarma (1969), based on other types of tests and anlytical considerations.

Finally, laboratory tests show that the magnitude of the cylic stress required to cause liquefaction is significantly

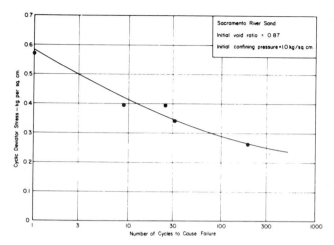

Fig. 16.39 Relationship between cyclic stress and number of cycles required to cause failure.

influenced by the initial stress conditions in the test specimen, a factor that would markedly change the liquefaction potential of the same sand under level-ground or sloping-ground conditions.

While these tests have thrown considerable light on the factors affecting liquefaction, it is necessary to use considerable judgment in applying the test data to analyzing field problems. The cyclic load triaxial test procedure suffers from the limitations that:

1. The initial ambient stress conditions are different from those on a sand element in situ.
2. The principal stress directions cannot rotate during the test as they do in the ground (only a 90° rotation of principal stress direction is possible in a triaxial compression test).
3. Under some conditions, and particularly for medium dense sands or dense sands, there is a possibility of water migration in the test specimen affecting the results.
4. It is difficult to interpret the test data if necking occurs in the test specimen.

Although the cyclic simple shear test eliminates these deficiencies, it has the following limitations:

1. It is difficult to prepare saturated samples of sand for testing under cyclic simple shear conditions.
2. Stress concentrations in the sample are likely to lead to premature failure of the test specimen.
3. Since the shear strain amplitude is normally limited to about 20 percent, it is not possible to determine whether the test specimen would dilate and stabilize at higher strain levels, in which case the mobility of the test specimens would not be indicative of the flow characteristics normally associated with liquefied soils.

These limitations in test procedures have led to the suggestion (Seed and Peacock, 1971) that the cyclic stresses causing failure under field conditions are likely to be about 35 to 45 percent lower than those causing failure in cyclic loading triaxial compression tests conducted under comparable conditions or about 15 to 20 percent higher than those causing failure in comparable simple shear tests. However, there is considerable need for refinement of these estimates, by either the development of new test procedures, an improved understanding of the significance of the limitations in current test procedures, or comparison of laboratory test data with conditions known to have caused failure in the field.

An alternative approach now under study is based on the concept that a soil cannot liquefy unless it can deform continuously without tending to dilate and thereby stabilize itself, and involves studies of the conditions under which dilatant and compressive volume-change tendencies are observed in undrained laboratory triaxial compression tests. The potential uses of this approach are likely to be presented in the near future.

16.4.3 Prediction of the Effects of Soil Conditions on Liquefaction Potential

From a practical point of view, the evaluation of the liquefaction potential of soil deposits at proposed construction sites is one of the most challenging problems facing the soil engineer working in seismically active regions of the world. Procedures used for making such evaluations are as follows.

1. Use of Past Experience Past experience of the conditions under which liquefaction has occurred in previous earthquakes will always be one of the most useful guides to the probable performance of other soil deposits. However, the limited extent of this experience at the present time and the limited number of well-defined case studies makes it extremely desirable to supplement this experience by appropriate analyses and test procedures whenever possible.

2. Use of Standard Blasting Tests A procedure used in the U.S.S.R. (Florin and Ivanov, 1961) involves the use of a standard blasting test to evaluate the liquefaction potential of sands in the field. For a sand deposit 25 to 35 ft thick, a charge of 5 kg of ammonite is exploded in the ground at a depth of 4.5 m and the resulting settlements of the ground surface are determined within a radius of 5 m from the explosion. Where the average settlement in this zone is less than 8 to 10 cm and the ratio of settlements from successive shots is less than about 0.6, it is considered that there is no need to provide measures against liquefaction of the soil.

3. Use of Ground Response Analyses and Laboratory Test Procedures A procedure that has been applied with some success to analyzing the liquefaction of soils in the Niigata earthquake (Seed and Idriss, 1967) and the failure of the Sheffield Dam in the Santa Barbara earthquake of 1925 (Seed et al., 1969), involves the use of ground response analyses to determine the stresses induced in a soil deposit during any given earthquake and the comparison of these stresses with those observed to cause failure in laboratory cyclic load tests. Such an approach involves the following steps:

1. Assess the magnitude of the ground motions likely to be developed in the base rock at the site under investigation; this assessment should involve the entire time history of the base motion throughout the period of the earthquake.
2. Determine the response of the overlying soils to the base motion, assuming that the deformations of the soils are caused primarily by the vertical propagation of shear waves as a result of the base motions. Such an analysis would permit the computation of the shear stresses, and their variation with time, at different depths in the soil deposits.
3. Idealize the shear stress history at the various depths to determine the significant number of stress cycles N and the equivalent uniform cyclic shear stress developed, τ_{dN}, at each level.
4. Determine, by means of cyclic load tests on representative samples of sand from the site, the cyclic shear stress τ_{lN} required to cause liquefaction of the sand in the significant number of stress cycles.

5. Compare the magnitude of equivalent cyclic stress developed at any depth (τ_{d_N}) with the cyclic stress causing liquefaction (τ_{l_N}) for the conditions existing at that depth to determine whether or not liquefaction will occur.

It is of course necessary to make appropriate corrections to the laboratory test data before using it in step (5) of the analyses.

This procedure provides a means for considering the effects of the amplitude and time history of the earthquake ground motions, the in-situ characteristics of the soils, the variation of overburden pressure with depth and the position of the water table. Different ground response analyses or laboratory test data may readily be incorporated in the procedure depending on the judgment and experience of the soil engineer involved.

16.5 INFLUENCE OF SOIL CONDITIONS ON LANDSLIDES

Earthquakes have been responsible for some of the largest landslides in recorded history and these, in turn, have caused enormous losses of both property and life. Possibly the earliest known landslide caused by an earthquake was the flow slide that carried the town of Helice into the sea during an earthquake in 373 B.C. in ancient Greece, with the complete loss of the town and all its inhabitants (Marinatos, 1960; Seed, 1968). Similar slides due to soil liquefaction or to the inducement of some degree of mobility in soil deposits as a result of the ground vibrations have been reported in many earthquakes since that time. A list of a number of such events is presented in Table 16.3 (Seed, 1968).

One of the most dramatic series of flow slides was that which occurred during the Kansu earthquake of December 16, 1920. Close and McCormick (1922) describe the events in the following terms:

> Of that most remarkable series of seismic disturbances which occurred throughout the world in November and December, 1920, the most phenomenal was undoubtedly the great Kansu earthquake of the late evening of December 16.... Landslides that eddied like waterfalls, crevasses that swallowed houses and camel trains, and villages that were swept away under a rising sea of loose earth, were a few of the subsidiary occurrences that made the earthquake in Kansu one of the most appalling catastrophes in history.
>
> The area of greatest destruction, 100 miles by 300 miles in extent, contains ten large cities besides numerous villages. In it is the heart of the loess country... where the loose earth cascaded down the valleys and buried every object in its path.
>
> It is in the loess area that the immense slides out of the terraced hills occurred, burying or carrying away villages... damming stream-beds and turning valleys into lakes, and accomplishing those hardly believable freaks which the natives name the "footsteps of the gods." The loss of nearly two hundred thousand lives and the total destruction of hundreds of towns and cities calls for reconstruction work on a staggering scale.

Similar slides in loess deposits have been reported as a result of earthquakes in the U.S.S.R. (Gubin, 1960).

Space does not permit an extensive review of the various types of slides caused by earthquakes, the damage they have caused, and methods of analyzing slope stability during earthquakes. Such reviews have been presented elsewhere in the past few years (Seed, 1967, 1968). However, it is pertinent to note that almost all of the slides listed in Table 16.3 have been associated with the liquefaction or mobilization of cohesionless soils and they can be grouped in the following categories.

(1) Flow slides caused by liquefaction of cohesionless soils usually involving sands, gravelly sands with sand seams, silty sands or loess. Typical examples are the slides in Kansu Province (1920), Chait (1949), Valdez, Alaska (1964), and Seward, Alaska (1964).

(2) Slides caused by liquefaction or water content redistribution in relatively thin seams or layers of sand, such as the 4th Avenue, L-Street, and Government Hill slides in Anchorage, Alaska, in 1964 and the slides near Lake Rinihue in the Chilean earthquake of 1960. A cross section showing the soil conditions in the L-Street slide area in Anchorage, where a block of soil 5000 ft long and 1200 ft wide moved laterally 14 ft, is shown in Figure 16.40. The sliding surface was located near the surface of the thin sand layer at about elevation 45. Block slides of this type have caused extensive damage to structures located in the grabens which form at the back end of the slides. However, structures located on the sliding block itself may be undamaged by the movements.

The soil conditions in another large landslide, extending over about 130 acres, which occurred near Lake Rinihue in the Chilean earthquake of 1960, are shown in Figure 16.41, together with cross sections through the slide area before and after sliding. In this case some parts of the slide mass moved laterally about 1200 ft. The soil conditions consisted of a surface deposit of sand and gravel varying from zero to about 150 ft thick, underlain by a 250-ft-thick deposit of lacustrine clay and a deeper bed of cemented sand and gravel. Field studies indicated that the surface of sliding was probably essentially horizontal and at a depth of about 150 ft in the lacustrine clay deposit, which was highly stratified, with alternating layers of silt and clay and frequent seams of fine sand. The failure was attributed to liquefaction of the silt and fine sand seams within the clay deposit (Davis and Karzulovich, 1961).

(3) Slides in clay deposits facilitated by liquefaction or water content redistribution in sand lenses. A good example of this type of slide is that which occurred at Turnagain Heights, Anchorage (Seed and Wilson, 1967; Seed, 1968), during the Alaska earthquake of 1964 (see Fig. 16.4). A cross section through the slide area is shown in Figure 16.42. In general the area is covered by a surface layer of sand and gravel varying in thickness between 5 and 20 ft, below which is a deep bed of clay, about 100 to 150 ft thick. This soil is a sensitive marine deposit of silty clay, with a shear strength decreasing from about 1 ton/ft^2 at its surface to about 0.45 ton/ft^2 at E1.0 and then increasing to about 0.6 ton/ft^2 at E1.−30; its sensitivity varies between about 5 and 30. The clay deposit contains numerous lenses of silt and fine sand particularly near the surface on which sliding occurred. These lenses varied in thickness from a fraction of an inch to several feet. Below the sliding surface, sand lenses were very thin and were only occasionally encountered. It is believed that liquefaction of the sand lenses or the formation of water films along the tops of the lenses in the vicinity of the slide surface played a major role in the development of this extensive slide.

(4) Slumping and collapse of fills due to liquefaction or failure of loose saturated silt and sand foundation soils. Typical examples are provided by embankment failures in Alaska (1964), Chile (1960), and Niigata (1964).

The slides listed in Table 16.3 were caused by earthquakes varying in magnitude from about $5\frac{1}{2}$ to $8\frac{1}{2}$ and they occurred at epicentral distances varying from several miles to hundreds of miles.

However, there are very few reported cases of slides developing in relatively homogeneous clay soils during earthquakes and there is a great need for information concerning this possibility.

In closing, it should be noted that it has not been possible in the course of this brief review to cover all of the types of

TABLE 16.3 LANDSLIDES DURING EARTHQUAKES DUE TO SOIL LIQUEFACTION.

Date	Earthquake	Magnitude	Location of Slide	Epicentral Distance, Miles	Type of Structure	Soil Type	Reference
373 B.C.	Helice	—	Helice	—	Coastal delta	—	Marinatos (1960)
1755	Lisbon	≃8.7	Fez	≃430	—	—	Lyell (1822), Richter (1958)
1783	Calabrian	—	Soriano	35	River banks	Fluvial sediments, clays with sand seams	Lyell (1822)
			Laureau		Hillsides	Volcanic sediments	
			Terranuova	5	River banks	Fluvial sediments	
					River banks	Fluvial sediments	
1811	New Madrid	—	Mississippi River Valley, Mo., Ark., Tenn., Ky., Ill., Ind.	Major slides 30 Minor slides 140	River valley banks and islands	Fluvial sediments, sands to muds	Fuller (1912)
1869	Cachar	—	Vicksburg, Miss.	290	Island	Fluvial sediments, sands to muds	Oldham (1882), Oldham and Mallet (1872)
			Barak River at Silchar	40 to 80	River banks	Fluvial—sand to clay	
1886	Charleston	—	15 miles SW of Ashley River	5 to 20	Railway fill	Fluvial and deltaic sands and silts	Dutton (1889)
			Ashley River at Greggs	5 to 10	River bank		
1897	India (Assam)	≃8.7	Shillong and Tutra regions	0 to 100	Canal banks Road embankments	Founded on alluvial plains	Oldham (1899), Richter (1958)
1899	Alaska (Yakutat)	—	Valdez		Submarine deposit	Deltaic and marine sediments—mainly silty sand and gravel	Coulter and Migliaccio (1966)
1902	St. Vincent	—	St. Vincent		Coastal delta	—	Hovey (1902)
1906	San Francisco	8.2	San Francisco area	10 to 30	Hillsides		Lawson et al. (1908)
1907	Karatag	—		—	Loess slopes	Loess	Gubin (1960)
1907	Chuyanchinsk	—		—	Loess slopes	Loess	Gubin (1960)
1908	Alaska	—	Valdez	25	Submarine deposit	Deltaic and marine sediments—mainly silty sand and gravel	Coulter and Migliaccio (1966)
1911	Alaska	6.9	Valdez	40	Submarine deposit	Deltaic and marine sediments—mainly silty sand and gravel	Tarr and Martin (1912)
1912	Alaska	7.25	Valdez		Submarine deposit	Deltaic and marine sediments—mainly silty sand and gravel	Coulter and Migliaccio (1966)
1920	Kansu Province	—	Kansu Province		Loess slopes	Loess	Close and McCormick (1922)
1923	Kwanto (Tokyo)	8.2	Yokohama area	40	Coastal hillsides		Wakimizu (1924)
			Tokyo area	60	Coastal hillsides	—	
1925	Santa Barbara	6.3	Santa Barbara	7	Earth dam	Silty sand	Dobry and Alvarez (1967)
1928	Chile	8.3	El Teniente	100	Tailings dam	Mining waste	Dobry and Alvarez (1967)
1933	Long Beach	6.3	Long Beach	20	Highway fills	Fills over marshland on shore roads	Wood (1933), Richter (1958)
			Newport Beach	3	Highway fills		
1934	Bihar Nepal	8.4	Sitamarhi to Purneau	0 to 80	Road and railway fills	Fluvial sediments, including sands	Roy (1939), Richter (1958)
1935	India (now West Pakistan)	7.6	Quetta	20 to 40	River banks	Alluvium—sand lenses Alluvium—uncertain gradation	West (1936), Richter (1958)
1940	El Centro	7.0	All-Americal Canal	6	Canal banks	Levees and foundations of deltaic sands	Ross (personal communication)
			Alamo Canal	7 to 255	Canal banks	Levees and foundations of deltaic sands	

	Location		Site	Structure		Soil conditions	Reference
1941	Garm	—	Solfatara Canal	Canal banks	25 to 30	Levees and foundations of deltaic sands	Gubin (1960)
			Brawley	Road and railway fills	20	Deltaic and fluvial sands	
1943	Faizabad	—	—	Loess slopes	—	Loess	Gubin (1960)
			—	Loess slopes	—	Loess	
1948	Fukui	7.2	Fukui plain	Levees, river banks, road and railway fills	0 to 15	Aeolian sands, beach sands, fluvial sands and silts	Tsuya (1950), Collins and Foster (1949), Butler et al. (1949)
1949	Chait	7.5	Surchob & Yasman River valleys	Loess slopes	5 to 25	Loess	Gubin (1960)
1950	Imperial Valley	5.4	Calipatria area	Canal banks	1 to 5	Deltaic and aeolian sands	Wood and Heck (1966)
1954	Anchorage	6.7	Rabbit Creek	Embankment	20 to 40	Fill on sand	Hansen (1965)
1957	San Francisco	5.3	Lake Merced	Lake banks	8	Aeolian and beach sands	
1959	Jaltipan	6.5	Coatzacoalcos	River banks, waterfront fill	20	Fine sandy silt, uniform and loose	de Cossio (1960), Marsal (1961)
			Minatitlou–Coatzacoalcos highway	Road and bridge approach fills	20 to 30	Fill over marshland	
1960	Chile	8.4	Rinihue	River banks	140	Fluvial and glacial sands	Duke and Leeds (1963), Lee (personal communication)
				Highway and railway fills		Foundations of fluvial and glacial gravels, sands, silts	
			Puerto Mount	Coastal terraces	240	Glacio-fluvial deposits	
				Sea walls and quay walls		Fill mainly sands to silty sands, loose	
			Valdivia	River banks	125	Fluvial sediments	
1964	Alaska	8.3	Anchorage	Coastal bluffs	70	Sand layers and lenses in clay deposit	Shannon and Wilson (1964)
			Valdez	Coastal data	40	Silty sands and gravel ($N \simeq 15$)	Coulter and Migliaccio (1966)
			Seward	Coastal data	90		
			Kenai Lake	Lake deltas	80	Deltaic sandy gravels, some sand lenses	McCulloch (1966)
1964	Niigata	7.3	Niigata area	Earth banks	35	Fluvial sand ($N < 15$)	Yamada (1966), Yokomura (1966), Kawakami and Asada (1966)
1965	Chile	7.2	El Cobre	Tailings dam	$\simeq 25$	Mining waste	Dobry and Alvarez (1967)
			La Patagua	Tailings dam	$\simeq 9$	Mining waste	
			Hierro Viejo	Tailings dam	$\simeq 16$	Mining waste	
			Los Maquis	Tailings dam	$\simeq 8$	Mining waste	
			El Cerrado	Tailings dam	$\simeq 18$	Mining waste	
1965	Seattle	6.7	Capital Lake Blvd., Olympia	Road causeway	38	Sand/gravel fill over lake and tidal sediments	Ross (personal communication)
			Union Pacific at Tumwater	Railway on benched slope	38	Cut/fill slope in outwash sands	
			Suquamish	Coastal bluff	26	Till over fine sand and silt strata	
			Port Orchard	Waterfront fill	18	Sand over beach sand and bay mud	
			E. Mercer Way, Mercer Island	Roadways on benched slopes	10	Sand on tills and outwash sands	
			Edmonds	Dumped fill on slope	29	Sandy till and refuse on till slope	
			Foster golf course, Duwamish	River terrace	10	Fluvial sands and silts	
			Victor	Highway fill	24	Sand fill at toe of coastal bluff	
1966	Parkfield	5.5	Cholame Creek north of Cholame	Stream banks	17	Fluvial sediments, sand strata or lenses	Ross (personal communication)

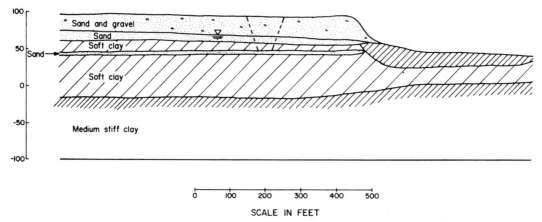

Fig. 16.40 Cross section through south end of L-Street slide area, Anchorage, Alaska. (*After Shannon and Wilson, 1964.*)

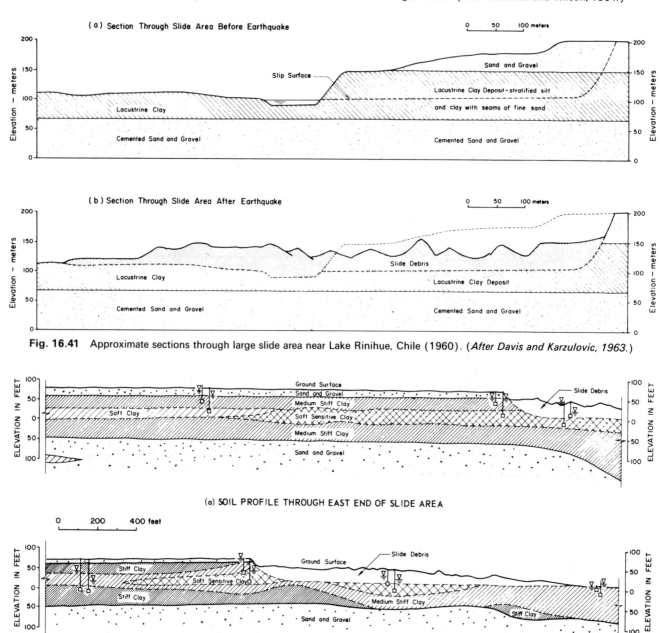

(a) Section Through Slide Area Before Earthquake

(b) Section Through Slide Area After Earthquake

Fig. 16.41 Approximate sections through large slide area near Lake Rinihue, Chile (1960). (*After Davis and Karzulovic, 1963.*)

(a) SOIL PROFILE THROUGH EAST END OF SLIDE AREA

(b) SOIL PROFILE THROUGH WEST END OF SLIDE AREA

Fig. 16.42 Cross sections through Turnagain Heights slide area, Anchorage, Alaska (1964).

problems in which soil conditions may affect the damage resulting from earthquakes. Omitted, for example, are discussions of the effects of soil conditions on the stability of slopes in relatively dry cohesionless soils and fills, the stability of retaining walls and waterfront bulkheads, or the deformations of bridge abutments. Each of these has been responsible for major damage to engineered structures during earthquakes, with the characteristics of the soils involved being a determining factor in their performance. However, damage due to these causes has usually been on a much smaller scale than that due to ground shaking, ground settlement, liquefaction, and slope stability, and concentration on these latter causes of earthquake damage was considered desirable for this reason. Even so, it has not been possible to mention all of the significant contributions on these subjects and some important studies may regrettably have gone unmentioned because they have not come to the attention of the author. Nevertheless it is hoped that the preceding review, which has attempted to emphasize techniques for developing a better understanding of the relationships between soil conditions and damage during earthquakes, will provide a useful guide to the current state of knowledge concerning these important aspects of soil behavior.

ACKNOWLEDGMENTS

Grateful acknowledgment should be expressed to many colleagues, especially I. M. Idriss, R. V. Whitman, R. W. Clough, K. L. Lee, H. Dezfulian, W. H. Peacock, and G. A. Ross, whose helpful suggestions and cooperation have contributed enormously to the preparation of this chapter.

PART II: FROM 1975 TO 1989

RONALD C. CHANEY, Ph.D., P.E. and **SIBEL PAMUKCU, Ph.D.**

16.6 INTRODUCTION

The following sections present a brief summary of a considerable body of work that has led to new insights and developments on earthquake effects on soil–foundation systems over the period 1975–1989 since Professor H. Bolton Seed wrote Part I of this chapter in 1975. In the following sections the engineering evaluation of the influence of local soil conditions on (1) soil settlement, (2) soil liquefaction, (3) slope instability, and (4) behavior of soft clays and silts, during earthquakes will be discussed.

To estimate the seismic risk for a given location and structure is a task that may involve the combined efforts of seismologists, geologists, geotechnical engineers, and structural engineers, among others. Therefore, the final result is bound to incorporate uncertainties caused by a variety of reasons. These include the size and location of the design earthquake, and the attenuation and modification of the seismic waves as they travel from the source to the site. In this chapter the focus is on the components of the risk that are associated with the specific subsoil and geologic conditions existing at and around the site, as shown schematically in Figure 16.43. For a situation such as shown in Figure 16.43, site conditions may affect the seismic threat to a structure on soil in the following two ways:

- The earthquake ground motions can be modified (amplified or dampened) by the presence of the soil.
- The soil can fail owing to the imposed seismic stresses.

Two of the most potentially serious ground failures as a result of earthquake motion are loss of bearing and frictional capacity owing to liquefaction and strength degradation and/or large deformation of soils. Liquefaction is a phenomenon observed with loose saturated granular soils when the shear strength goes to zero and the material behaves like a viscous liquid. As a result of liquefaction, lateral spreads, flow slides, floating of lightweight embedded structures, and increased lateral pressures on retaining structures may occur. The methods of evaluating the liquefaction hazard and potential of a site are given in a number of publications, which include Seed and Idriss (1982), Seed et al. (1983), National Research Council (1985), Koester and Franklin (1985), Earthquake Engineering Research Institute (1986), and Casagrande (1976). Evaluation of liquefaction potential can be conducted either through laboratory testing on undisturbed specimens or using in-situ test data.

In addition to granular soils, soft clays also undergo large deformations during earthquake loading. Strength loss and large deformations of such soils can result in failures such as flow slides, slope instabilities, and bending forces on piles and other embedded structures. Strength degradation of soft marine clays have also been shown to occur as a result of cyclic loading of waves. A phenomenon called "mud flow" is the flow of soft, mostly clay material in gullies of up to 0.5° of slope. These flows seem to be triggered when the material reaches a critical state of stress with progressive build-up of excess pore water pressures owing to the cyclic loading of waves (Pamukcu et al., 1983).

Earthquakes also cause strength loss in sensitive clays that give way to ground failures similar to liquefaction. Clays having sensitivities greater than 8 may be susceptible to such failures. Materials mostly susceptible to earthquake-induced natural slope failure are liquefiable or sensitive soils, cemented soils, loess, weakly cemented rocks, and rocks with prominent discontinuities. Natural slope failures occur either by earthquake-induced shear stresses exceeding the slope resistance or by loss of strength of brittle or sensitive materials. Man-made embankments and fills that are made of loose, saturated, fine-grained and cohesionless materials are most susceptible to failure. The type of failures observed are dam slide on its foundation, slope failure, loss of freeboard, displacement and failure of outlet works, and rupturing. The performance of earth dams and embankments depends on the soil type, degree of saturation, density, and strength and duration of the ground shaking.

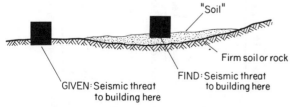

Fig. 16.43 Nature of the problem. (*After Roesset and Whitman, 1969.*)

16.7 INFLUENCE OF SOIL CONDITIONS ON GROUND SETTLEMENT

16.7.1 General Problems

A. Case Histories

It has long been observed that soils tend to settle and densify when they are subjected to earthquake shaking. Since even small settlements may sometimes have a significant effect on the performance of structures during earthquakes, it is important to evaluate settlement. An example of this behavior as a function of time has been reported by Zeevaert (1975) and is discussed in Chapter 17 of Zeevaert (1972) for a building located in Mexico City. The building was constructed in October 1952 about 5 years before the Mexico City earthquake of July 28, 1957. The building was constructed on an under-compensated foundation located at a depth of 6.5 m from the ground surface. The use of the undercompensated foundation resulted in an increment of stress on the order of 7 tonne/m^2 over the effective overburden stress at the foundation grade level. By 1957 the building was practically stabilized as shown by settlement records (refer to Figure 17.3). After the earthquake, the building started to settle, showing an immediate settlement on the order of 4.0 cm, which increased the following year by an additional 4.0 cm. The building was subsequently under-pinned in 1964 and is currently performing satisfactorily. Additional field observations of earthquake-induced settlements in saturated sands have been summarized by Tokimatsu and Seed (1987) as shown in Table 16.4. The mechanism behind this behavior is that if the soil is saturated and there is no possibility for drainage, so that constant volume conditions are maintained, the primary effect of the shaking is the generation of excess pore water pressures. Settlement then occurs as the excess pore pressures dissipate.

Depending on the characteristics of the soil and the length of the drainage path, the time required for all settlement to develop can vary considerably. For a saturated sand material the time for settlement due to the cylic loading to take place can be from almost immediately to about a day (Tokimatsu and Seed, 1987). In contrast, silty clay material may take years (Zeevaert, 1975). In dry sands the settlement occurs during the earthquake shaking under conditions of constant effective vertical stress.

Methods for evaluating the settlement of sand deposits under earthquake shaking for both saturated and dry conditions have been presented by Silver and Seed (1971), Lee and Albaisa (1974), and Tokimatsu and Seed (1987). In these procedures, either laboratory (cyclic triaxial, cyclic simple shear) or in-situ (standard penetration test) generated soil properties are utilized. In the following, the settlement of saturated sands and the settlement of dry sands will be discussed.

16.7.2 Settlement of Saturated Sands

The settlement of saturated sands under earthquake loading depends upon whether the sand experienced liquefaction or not. In the following both the nonliquefaction and postliquefaction responses will be discussed.

A. Nonliquefaction Response

The settlement of saturated sands resulting from the dissipation of excess pore water pressures for nonliquefaction conditions developed during cyclic loading was studied by Lee and Albaisa

TABLE 16.4 FIELD OBSERVATIONS OF EARTHQUAKE-INDUCED SETTLEMENTS IN SATURATED SANDS[a].

Earthquake	Year	Magnitude	Site	Maximum Acceleration	Thickness of Layer Causing Major Settlement, ft	Observed Settlement, in	Volumetric Strain, %	Average Stress	Average SPT $(N_1)_{60}$	Fines Content, %	Reference
Tokachioki	1968	7.9	Hachinohe P6	0.2	16	15 to 20	9	0.19	1.4	5	Ohsaki (1970)
Tokachioki	1968	7.9	Hachinohe P1	0.23	7	0.5 to 0.7	0.7	0.2	17.5	5	Ohsaki (1970)
Niigata	1964	7.5	Niigata C	0.16	30	8	2	0.16	11	2	Building Research Institute (1965)
Niigata	1964	7.5	Niigata A	0.18	30	0	0	0.17	22	2	
Miyagiken-Oki	1968	7.4	Arahama	0.2	30	8	2	0.22	14	0	Tohno and Yasuda (1981)

[a] After Tokimatsu and Seed (1987).

(1974). They concluded the following:

> the amount of reconsolidation volumetric strain for nonlique-faction conditions increases with increasing grain size of soil, decreasing relative density, and increasing excess pore pressure generated during the undrained cyclic loading, but is almost independent of how this excess pore pressure was generated.

Even though liquefaction may not have occurred, some excess pore pressure may be generated in sand deposits by earthquake shaking, and the dissipation of this pore pressure may result in small amounts of settlement. This condition, pointed out by Tokimatsu and Seed (1987), corresponds to the zone below the boundary line for liquefaction shown in Figure 16.44 and below the solid line marked volumetric strain = 1 percent in Figure 16.45. The pore pressure generated under such conditions may be expressed in terms of the normalized stress ratio (i.e., the ratio of the actual shear stress ratio to the shear stress ratio just causing liquefaction). The relationship between the shear stress ratio and the normalized stress ratio generally falls within the shaded area, and for most sands it may be represented by the broken line (Tokimatsu and Yoshimi, 1984) in Figure 16.46a.

Lee and Albaisa (1974) plotted volumetric strain as a function of induced pore pressure ratio (u/σ'_0). Combining the

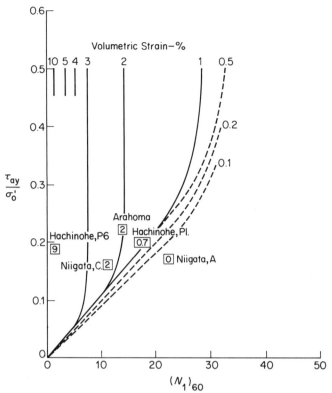

Fig. 16.45 Cyclic stress ratio versus SPT blow count as a function of volumetric strain, $M = 7.5$. (*After Tokimatsu and Seed, 1987.*)

graphs by Tokimatsu and Yoshimi and Lee and Albaisa, a graph of volumetric strain versus the normalized stress ratio was developed by Tokimatsu and Seed (1987) as shown in Figure 16.46b. A review of this figure shows that if the normalized stress ratio is less than 0.7, the induced volumetric strain is likely to be less than 0.1 percent and hence the effect of pore pressure generation on settlement is likely to be insignificant for most structures.

B. Postliquefaction Response

Tatsuoka et al. (1984) studied the volumetric strain after initial liquefaction (pore pressure ratio = 100 percent) and found that the amount of sett' ment can be significantly influenced by the maximum shear strain developed in the soil as well as the soil density, but that it is relatively insensitive to the effective overburden pressure. Thus, the maximum shear strain is an important factor influencing the probable settlements after liquefaction has developed, even though there are no significant changes in the maximum pore pressure beyond this stage.

The relationships between relative density (D_r) or the normalized SPT N-value ($N_1)_{60}$ and volumetric strain (ε_c) as a function of the shearing strain level (γ) is shown in Figure 16.47 (Tokimatsu and Seed, 1987). A flow chart showing the determination of ($N_1)_{60}$ is presented in Figure 16.55. The data in this figure incorporates the results from three different studies using three different sands. A review of Figure 16.47 shows that volumetric strain increases significantly with decreasing relative density or ($N_1)_{60}$ and increasing shearing strain (γ) levels in the soil. In addition, the figure also shows that even though three different sands were used the results show good general consistency.

Using Figure 16.44, which shows tentative values of the shear strain potential for any combination of cyclic shear stress ratio (τ_{av}/σ'_0) and normalized SPT N-value for a magnitude

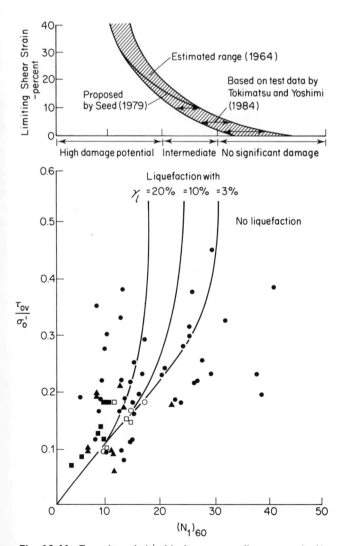

Fig. 16.44 Tentative relationship between cyclic stress ratio N_1 values and limiting strains for natural deposits of clean sand. (*After Seed et al., 1984.*)

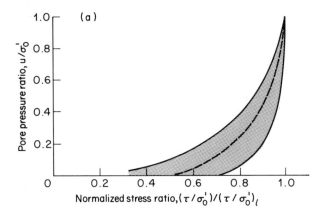

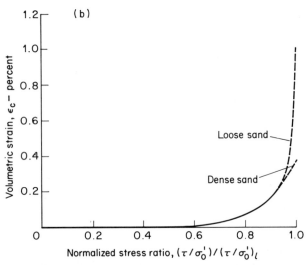

Fig. 16.46 Relationship between normalized stress ratio. (a) Pore pressure ratio. (b) Volumetric strain for sand. (*After Tokimatsu and Seed, 1987.*)

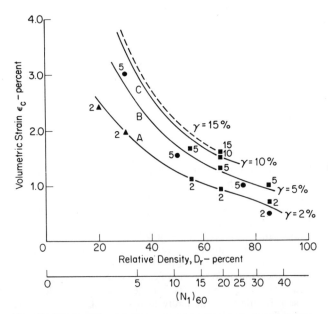

Fig. 16.47 Relationship between volumetric strain, induced strain, and relative density for sands. ●, Lee and Albaisa (1974); ■, Tatsuoka et al. (1984); ▲, Yoshimi et al. (1975). Numbers beside data points indicate maximum shear strain developed in the soil. (*After Tokimatsu and Seed, 1987.*)

TABLE 16.5 SCALING FACTORS FOR EFFECT OF EARTHQUAKE MAGNITUDE ON CYCLIC STRESS RATIO CAUSING LIQUEFACTION.

Earthquake Magnitude, M (1)	Number of Representative Cycles at 0.65_{max} (2)	Scaling Factor for Stress Ratio, r_m (3)
$8\frac{1}{2}$	26	0.89
$7\frac{1}{2}$	15	1.0
$6\frac{3}{4}$	10	1.13
6	5	1.32
$5\frac{1}{4}$	2 to 3	1.5

7.5 earthquake, the shear strain in-situ can be estimated (Seed et al., 1984; Kovacs et al., 1984). Combining the results shown in Figures 16.44 and 16.47, a graph as shown in Figure 16.45 can be developed that plots cyclic shear stress ratio (τ_{av}/σ_0') versus $(N_1)_{60}$ as a function of volumetric strain (ε_c) for a magnitude 7.5 earthquake. Using this figure and knowing the cyclic shear stress ratio (τ_{av}/σ_0'), the resulting volumetric strain (ε_c) for a site can be determined. A review of Figure 16.44 shows that the resulting volumetric strains after liquefaction may be as high as 2 to 3 percent for loose to medium sands and even higher for very loose sands. Figure 16.45 can be extended to events of other magnitude by noting that the main difference between different-magnitude earthquakes is the difference in number of cycles of stress they produce (Seed and Idriss, 1983). Table 16.5 summarizes the relative values of the stress ratio required to cause liquefaction for earthquakes of different magnitudes to the stress ratio required to cause liquefaction for a magnitude 7.5 event, together with the approximate numbers of cycles induced by the earthquakes. Thus, by multiplying the values of the ordinate for each equivolumetric strain in Figure 16.45 by the scaling factors shown in column 3 of Table 16.5, volumetric strain charts can be obtained for earthquakes with different magnitudes:

$$\left[\frac{\tau_{av}}{\sigma_0'}\right]_{m=7.5} = \left[\frac{\tau_{av}}{\sigma_0'}\right]_{m=m}\left[\frac{1}{r_m}\right] \qquad (16.5)$$

16.7.3 Settlement of Dry Sands

The settlement of dry sand due to cylic loading has been shown to be a function of the following: (1) relative density of the soil, (2) the magnitude of the cyclic shear strain, (3) the number of strain cycles, (4) multidirectional shaking, and not a function of vertical stress (Silver and Seed, 1971; Seed and Silver, 1972; Pyke et al., 1975).

Seed and Silver (1972) suggested a procedure for estimating the probable settlement of dry sand that involves a response analysis for the sand deposit to determine the induced shear strains developed at different depths in the soil deposit. This procedure is outlined below.

a. Determine a representative history of bedrock acceleration.
b. Divide the soil layer in n layers. They do not have to be of equal thickness.
c. Calculate the average value of the vertical effective stress for each layer.
d. Determine the relative density for each layer.
e. Using the relative density for each layer estimate its damping ratio and shear modulus.
f. Calculate the history of shear strains at the middle of all n layers.
g. Convert the irregular strain histories obtained in step (f) for each layer into average shear strains and equivalent number of uniform cycles.

h. Conduct laboratory tests with cyclic simple shear apparatus on representative soil specimens from each layer to obtain the vertical strains (ε_v) for the equivalent number of strain cycles calculated in step (g).

i. Calculate the total settlement as follows:

$$\Delta H = \varepsilon_{v1} H_1 + \varepsilon_{v2} H_2 + \cdots + \varepsilon_{vn} H_n \qquad (16.6)$$

where

$\varepsilon_{v1}, \ldots$ = average vertical strains determined in step (h)
$H_1, H_2, \ldots$ = layer thicknesses
ΔH = total settlement

Tests by Pyke et al. (1975) using multidirectional shear as well as unidirectional shear led to the conclusion that "the settlements caused by combined horizontal motions are about equal to the sum of the settlements caused by the components acting alone." Thus, the volumetric strain estimated from Figure 16.48 should be multiplied by a factor of 2 to take into account the effect of multidirectional horizontal shaking. The effect of the vertical acceleration is to also increase the settlement. The effect of the vertical acceleration on settlement of a sand layer with an initial relative density of 45 percent is shown in Figure 16.49 and the related computations are given in Table 16.6.

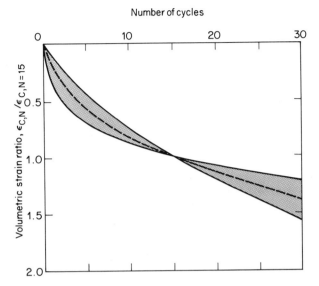

Fig. 16.48 Relationship between volumetric strain ratio and number of cycles for dry sands. (*After Tokimatsu and Seed, 1987.*)

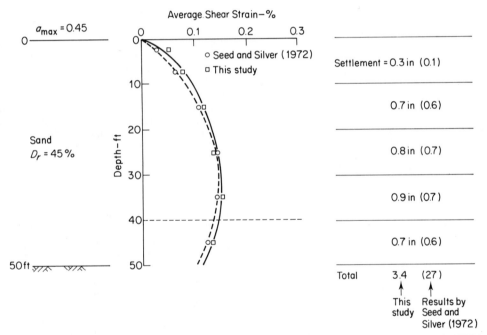

Fig. 16.49 Computation of settlement for 50-ft deep sand layer. (*After Tokimatsu and Seed, 1987.*)

TABLE 16.6 COMPUTATION OF SETTLEMENT FOR DEPOSIT OF DRY SAND[a].

Layer Number (1)	Thickness, ft (2)	$\sigma_0 = \sigma_0'$, psf (3)	D_r, % (4)	N_1 (5)	G_{max},[b] ksf (6)	$\gamma_{eff}(G_{eff}/G_{max})$ (7)	γ_{eff} (8)	$\varepsilon_{C,M=7.5}$, % (9)	$\varepsilon_{C,M=6.6}$,[c] % (10)	$2\varepsilon_{C,M=6.6}$,[d] % (11)	Settlement, in (12)
1	5	240	45	9	520	1.3×10^{-4}	5×10^{-4}	0.14	0.11	0.22	0.13
2	5	714	45	9	900	2.3	8	0.23	0.18	0.36	0.22
3	10	1425	45	9	1270	3.2	12	0.35	0.28	0.56	0.67
4	10	2375	45	9	1630	4.0	14	0.40	0.32	0.64	0.77
5	10	3325	45	9	1930	4.5	15	0.45	0.36	0.72	0.86
6	10	4275	45	9	2190	4.6	13	0.38	0.30	0.60	0.72
Total											3.37

[a] After Tokimatsu and Seed (1987).
[b] $G_{max} = K_2 \cdot 1000 (\sigma_m')^{1/2} \simeq 20 N_1^{1/3} (\sigma_m')^{1/2} \times 1000$.
[c] $\varepsilon_{C,M=6.6}/\varepsilon_{C,M=7.5} = 0.80$.
[d] Multidirectional effect.

16.8 INFLUENCE OF SOIL CONDITIONS ON LIQUEFACTION POTENTIAL

16.8.1 General Problem

The basic cause of liquefaction in saturated cohesionless soils is believed to be the build-up of excess hydrostatic pressure owing to the application of either a shock or cyclic loading. An evaluation of the effect of saturation on liquefaction behavior has been presented by Sherif et al. (1977) and Chaney (1978, 1987). As a consequence of the applied stresses, the structure of the cohesionless soil tends to become more compact, with a resulting transfer of stress to the pore water and a reduction in stress on the soil grains. The soil grain structure, in response, rebounds to the extent required to keep the volume constant, and this interplay of volume reduction and soil structure rebound determines the magnitude of increase in pore water pressure in the soil (Martin et al., 1975). Pore pressure build-up has been shown to be a function of the cyclic shear strain, and the number of cycles (Ladd et al., 1989; Dobry et al., 1982).

The basic phenomenon is illustrated schematically in Figure 16.50. This behavior is slightly modified by the effect of stress concentrations (refer to Chaney et al., 1980). As the pore water pressure approaches a value equal to the applied confining pressure, the sand begins to undergo deformations. If the sand is loose, the pore pressure will increase suddenly to a value equal to the applied confining pressure, and the sand will rapidly begin to undergo large deformations. If the sand will undergo virtually unlimited deformations without mobilizing significant resistance to deformation, it can be said to be liquefied.

In contrast, if the sand is dense, it may develop a residual pore water pressure after completion of one full cycle of loading which is equal to the confining pressure. On application of the next cycle of loading, or if the sand is subject to monotonic loading, the soil will tend to dilate. The corresponding pore water pressure will drop if the sand is undrained, and the soil will develop enough resistance to withstand the applied load.

In order for this to occur, the soil will have to undergo some deformation to develop the resistance.

If the cyclic loading continues, the amount of deformation required to produce a stable condition may increase. Ultimately, there will be a level of deformation at which the soil can withstand any load application. This type of behavior is termed cyclic mobility and is considerably less serious than liquefaction. In both cases there is a generation of excess pore water pressure that must be dissipated. The dissipation of the pore pressure can affect the behavior of lowerlying sediments by the formation of sand boils.

The movement of pore water due to excess pressure results in the erosion and movement of sediments from the liquefied zones to the soil surface. This migration of sediments leaves voids in the underlying soil stratum, which ultimately collapses owing to the overburden weight and thus causes the distortion of the surface.

16.8.2 Methods for Evaluating Liquefaction Potential

Liquefaction analyses share some of the same uncertainties as the calculation of ground motions. Of these uncertainties, the ones associated with the duration characteristics of the calculated seismic shear stresses in the soil become especially critical, as liquefaction and other modes of soil failure are fatigue-type phenomena (with damage to the soil structure increasing as the number of cycles of shaking increases). One important additional source of uncertainty for liquefaction analyses lies in the estimate of the cyclic shear strength of the soil.

A number of methods presently exist for estimating the liquefaction potential for a given site (Valera and Donovan, 1977). These methods should consider both the cyclic resistance of the soil and the cyclic shear stresses and strains caused by the earthquake. They can be separated into three methodologies: (1) simplified procedures (Fig. 16.51), (2) ground response analysis, and (3) empirical methods based on past performance. The first two methods are basically analytical procedures that depend on dynamic analysis and laboratory tests. The third procedure is an empirical procedure that depends primarily on field data (Seed, 1984). A summary of these three methodologies is shown in Figure 16.52. In the following sections each of these three methodologies will be discussed.

Simplified and semiempirical charts and procedures are widely used for estimating liquefaction potential of level, saturated sand sites. All of these methods are based on the few dozen case histories that have been compiled of sites that have been subjected to earthquakes, and did (or did not) experience liquefaction. Most of the procedures assume that the maximum horizontal ground surface acceleration and its duration characteristics are known (Ohsaki, 1970; Kishida, 1969; Seed and Idriss, 1971; Castro, 1975; Yegian and Oweis, 1976; Christian and Swiger, 1975; Seed, 1976), although Yegian and Whitman (1978) proposed a correlation based directly on the magnitude of the earthquake and its distance to the site. These methods without exception are based on the use of the standard penetration test (SPT) data as a measure of the cyclic shear strength of the soil at the site. Therefore, these methods are affected by the large uncertainty associated with this field test. Some of the authors mentioned above have included ranges or "gray" areas in their proposed correlations, to define in an approximate way the uncertainty of the method, while others have used probabilistic procedures.

In recent years additional in-situ devices (such as the flat plate dialatometer, cone penetration test, electrical measurements, shear wave measurements) have been employed by

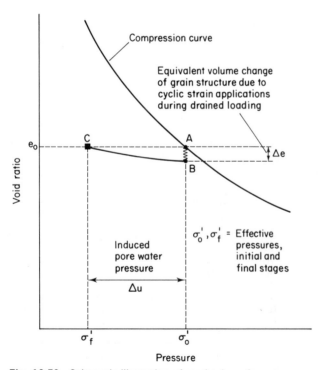

Fig. 16.50 Schematic illustration of mechanism of pore pressure generation during cyclic loading. (*After Seed and Idriss, 1982.*)

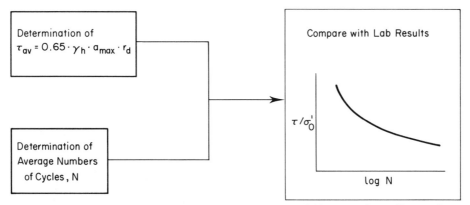

Fig. 16.51 Simplified procedures of determining liquefaction potential incorporating uniform cycle concepts.

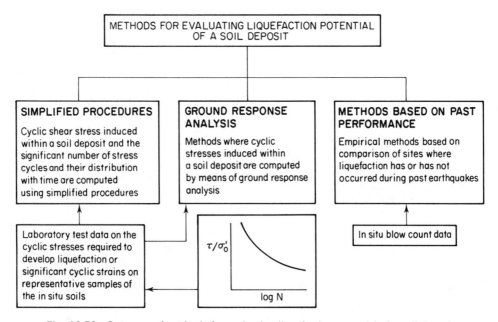

Fig. 16.52 Summary of methods for evaluating liquefaction potential of a soil deposit.

various investigators (Robertson and Campanella, 1985, 1986; Baldi et al., 1985; Arulanandan, 1977; Arulmoli et al., 1985; Arulanandan and Muraleetharan, 1988a,b; Bierschwale and Stokoe, 1984).

A. Methods Based on SPT

1. Methods Based only on N Ohsaki (1970) and Kishida (1969) were the first to propose procedures for evaluating liquefaction potential of a site based on the blow count N from the standard penetration test (SPT). In both of these methods the magnitude of the earthquake-induced stresses was not considered, but only the resistance of the soil. The Ohsaki procedure involved the criterion that if the blow count N exceeds twice the depth in meters, liquefaction will not occur. In contrast the relationship proposed by Kishida (1969) considered the standard penetration value versus depth as a function of a boundary line between liquefaction and no liquefaction as shown in Figure 16.53.

2. Methods Based on N Value, Acceleration, and Depth to the Water Table The dynamic stress ratio can be estimated from the maximum horizontal ground surface acceleration and the depth to the water table as shown in Figure 16.54. Then

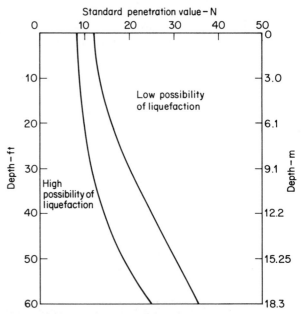

Fig. 16.53 Relationship between the possibility of liquefaction and N values at various depths. (*After Kishida, 1969.*)

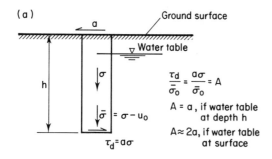

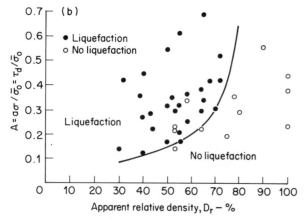

Fig. 16.54 (a) Relationship between ground surface acceleration, *a*, and stress factor, *A*. (b) Liquefaction data from field studies by Christian and Swiger (1975). (*After Byrne, 1976.*)

estimating the relative density of the sand deposit from the blow count (N), a graph of the dynamic stress ratio versus the relative density as a function of sites that have or have not experienced liquefaction can be constructed. Seed and Peacock (1971) and Whitman (1971) first presented data in this form. Subsequently Castro (1975), Christian and Swiger (1975), and Seed et al. (1975) extended the data as additional information became available. A graph developed by Christian and Swiger (1975) based on this methodology is shown in Figure 16.54b. Seed et al. (1983) have subsequently developed a relationship between a standardized blow count N_1 (normalized to 1 ton/ft^2) and the cyclic stress ratio τ_h/σ'_0. The cyclic stress ratio can be

determined using Equation 16.7:

$$\frac{(\tau_h)_{av}}{\sigma'_0} = 0.65 \frac{a_{max}}{g} \frac{\sigma_0}{\sigma'_0} r_d \qquad (16.7)$$

where

a_{max} = maximum acceleration at the ground surface
σ_0 = total overburden pressure on sand layer under consideration
σ'_0 = initial effective overburden pressure on sand layer under consideration
r_d = stress reduction factor varying from a value of 1 at the ground surface to a value of 0.9 at a depth of about 9.6 m (30 ft)
g = acceleration due to gravity
$(\tau_h)_{av}$ = average horizontal shear stress

A procedure to correct for the various equipment and procedural variations in the conduction of the SPT test using a standard size sampler (ASTM D1586, 1984) has been proposed by Seed et al. (1985). A summary of energy ratios for various SPT procedures corrected to a standardized 60 percent rod energy is presented in Table 16.7. The blow count from using non-standard size samplers in penetration tests, N'_m, may be converted to standard blow count, N_m, by the following approximate correlation (Lacroix and Horn, 1973). When possible, empirical correlations should be developed by direct testing for the particular nonstandard equipment or methods being used, and for the particular soil deposit being investigated.

$$N_m = N'_m \left(\frac{2}{D_1}\right)\left(\frac{12}{L_1}\right)\left(\frac{W_1}{140}\right)\left(\frac{H_1}{30}\right) = \frac{2N_1 W_1 H_1}{175 D_1^2 L_1} \qquad (16.8)$$

where

D_1 = outside diameter of the split spoon or conical point in inches
L_1 = depth of penetration in inches
W_1 = weight of hammer in pounds
H_1 = height of free fall of hammer in inches

In the Seed et al. procedure, a blow count N_m is corrected for four variables: (1) energy effect, (2) length of drill rod, (3) sampling barrel effect, and (4) overburden. A flow chart presenting these corrections based on Seed et al. (1985b) is presented in Figure 16.55. The result of this operation is a standardized blow count N_1.

A flow chart presenting methods of determining liquefaction potential based on past performance during earthquake is

TABLE 16.7 SUMMARY OF ENERGY RATIOS FOR SPT PROCEDURES[a].

Country (1)	Hammer Type (2)	Hammer Release (3)	Estimated Rod Energy, % (4)	Correction Factor for 60% Rod Energy (5)
Japan[b]	Donut	Free-fall	78	78/60 = 1.30
	Donut[c]	Rope and pulley with special throw release	67	67/60 = 1.12
United States	Safety[c]	Rope and pulley	60	60/60 = 1.00
	Donut	Rope and pulley	45	45/60 = 0.75
Argentina	Donut[c]	Rope and pulley	45	45/60 = 0.75
China	Donut[c]	Free-fall[d]	60	60/60 = 1.00
	Donut	Rope and pulley	50	50/60 = 0.83

[a] After Seed et al. (1985b).
[b] Japanese SPT results have additional corrections for borehole diameter and frequency effects.
[c] Prevalent method in this country today.
[d] Pilcon type hammers develop an energy ratio of about 60%.

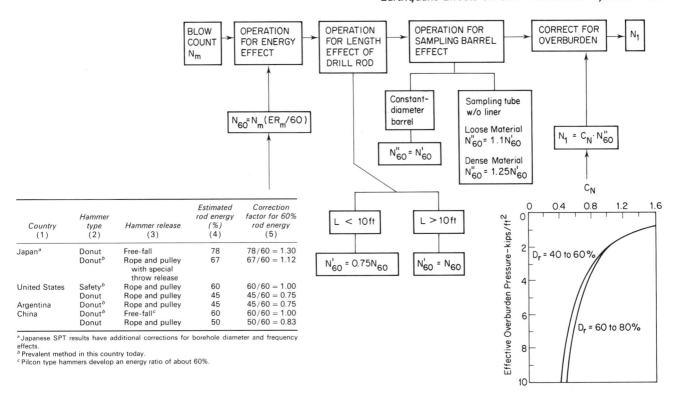

Fig. 16.55 Correction of SPT blow counts for use in dynamic studies.

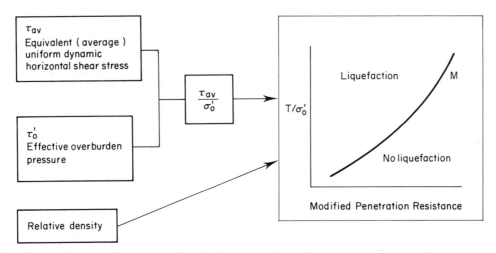

Fig. 16.56 Methods of determining liquefaction potential based on past performance during earthquakes.

shown in Figure 16.56. The result of this operation is a point on a graph of τ_{av}/σ_0' versus the modified penetration resistance $(N_1)_{60}$ as a function of earthquake magnitude for the occurrence or nonoccurrence of liquefaction. The location of this point on the graph determines the susceptibility of the deposit to liquefaction. A plot of the relationship between the cyclic stress ratio causing liquefaction and $(N_1)_{60}$ values for sands containing $\leqslant 5$ or $\geqslant 5$ percent fines are presented in Figures 16.57 and 16.58 for magnitude 7.5 earthquakes.

B. Methods Based on CPT

The cone penetration test (CPT) is believed to be the most reliable tool for delineation of stratigraphy through the continuous rapid measurement of penetration resistance, q_c,

and sleeve friction, f_s. The electric cone has also allowed the addition of pore pressure measurements during penetration. The procedure and equipment of the quasi-static electric cone penetration test have been standardized (ASTM D3441, 1979). A number of investigators have developed correlations between earthquake shaking conditions causing liquefaction or cyclic mobility and the cone penetration resistance of sands (Zhou, 1981; Douglas et al., 1981; Seed et al., 1983; Robertson and Campanella, 1985).

In the methodology developed by Zhou (1981), the critical value of cone penetration resistance, q_{crit}, separating liquefiable from nonliquefiable conditions to a depth of 15 m, is determined by an empirical equation that is a function of (1) depth of water level below ground surface, (2) depth to top of sand layer under consideration, and (3) a factor for shaking intensity. The

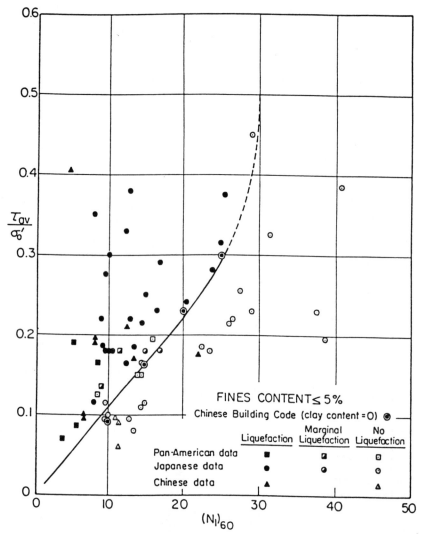

Fig. 16.57 Relationship between cyclic stress ratios causing liquefaction and $(N_1)_{60}$ values for sands with $\leqslant 5$ percent fines content for magnitude 7.5 earthquakes. (*After Seed et al., 1984.*)

equation was developed from field test data for the Tangshan earthquake area. The sand in this area was primarily a clean sand with little fine content ($D_{50} = 0.25$ mm). In the method developed by Douglas et al. (1981), CPT data is converted to equivalent STP N values by conducting preliminary studies at each new site to establish the correlation between q_c and N for the particular sand at the site. The equivalent SPT N values can then be used in SPT-based liquefaction resistance correlations. The Seed et al. (1983) method uses the available correlations between SPT and CPT test data, and applies them to the critical boundaries separating liquefiable from non-liquefiable conditions based on SPT data. A summary of the Seed et al. (1983) CPT method is shown in Figure 16.59.

The method proposed by Robertson and Campanella (1985) is developed along the same lines as the SPT method. This is accomplished by modifying the cone bearing, q_c, to an over-burden stress level of 1 kg/cm (1 tsf) using Equation 16.9:

$$Q_c = C_Q q_c \qquad (16.9)$$

where

Q_c = modified cone penetration resistance
C_Q = correction factor (refer to Figure 16.60)
q_c = cone bearing

Then to develop a CPT method for liquefaction evaluation involves two steps. These steps are (1) determination of the relative density corresponding to a given modified cone penetration resistance using information from Baldi et al. (1982), and (2) combination of this information with field liquefaction resistance date from Christian and Swiger (1975). This process results in Figure 16.61, which presents a correlation between liquefaction resistance and modified cone penetration resistance for sands based on relative density correlation.

C. Methods Based on Flat-Plate Dilatometer

The flat-plate dilatometer test (DMT) uses a flat plate 14 mm thick, 95 mm wide by 220 mm length. A flexible stainless steel membrane 60 mm in diameter is located on one face of the blade. The dilatometer is pushed into the ground at a constant rate of penetration of 2 cm/s. Readings are made every 20 cm in depth by inflating the membrane using high-pressure nitrogen gas supplied by a tube prethreaded through the rods. As the membrane is inflated, the pressures required to just lift the membrane off a sensing disk and to cause a 1 mm deflection at the center of the membrane are recorded. Before and after each penetration the dilatometer is calibrated for membrane stiffness. Using this procedure, three index parameters proposed

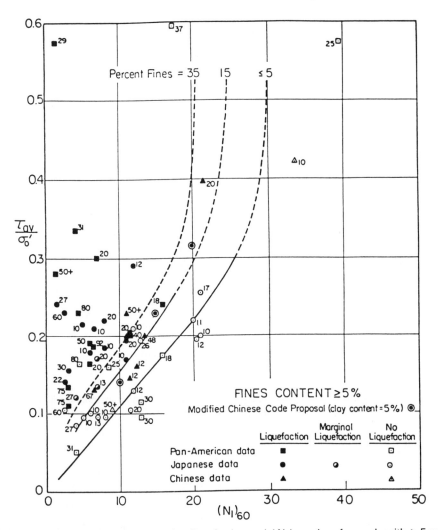

Fig. 16.58 Relationship between cyclic stress ratios causing liquefaction and $(N_1)_{60}$ values for sands with $\geqslant 5$ percent fines content for magnitude 7.5 earthquakes. (*After Seed et al., 1984.*)

by Marchetti (1982) are calculated. These parameters are (1) K_d, horizontal stress index, (2) I_d, material index, and (3) E_d, dilatometer modulus.

The horizontal stress index, K_d, was suggested by Marchetti (1982) as a parameter for assessing the liquefaction resistance under level-ground conditions of sands under cyclic loading. The K_d parameter appears to reflect the combination of all the following soil variables:

a. Relative density, D_r
b. In-situ stresses, K_0
c. Stress history and prestressing
d. Aging
e. Cementation

When the K_d parameter is low, none of the above variables is high, that is, the sand is loose, uncemented, in a low horizontal stress environment, and has little stress history. A sand under these conditions may liquefy or develop large strains under cyclic loading, using liquefaction as defined by Seed et al. (1983). A liquefaction correlation has been developed by Robertson and Campanella (1986) as shown in Figure 16.62. The DMT data shown in Figure 16.62 do not require modification for in-situ effective overburden pressure since this is accounted for

in the K_d parameter.

$$\frac{\tau_l}{\sigma'_{v0}} = \frac{K_d}{10} \qquad (16.10)$$

Comparison of Figure 16.62 with Equation 16.10 for estimating liquefaction as given by Marchetti (1982) shows that the figure predicts cyclic stress ratios significantly lower than those predicted using the formula. The correlation shown in Figure 16.62 is only applicable for testing in sands where penetration and expansion occur under drained conditions. Testing in silty sands or silts may generate significant pore pressures, which would influence the measured K_d values (Campanella and Robertson, 1983).

D. Comparison of Field Methods

A field and laboratory study was performed by Robertson and Campanella (1986) that included CPT, SPT, and undisturbed sampling using an 86-mm inside diameter (ID) thin-walled, fixed-piston sample tube. Laboratory cyclic triaxial tests were performed on the undisturbed samples. A comparison between the cyclic stress ratios to cause liquefaction predicted from the

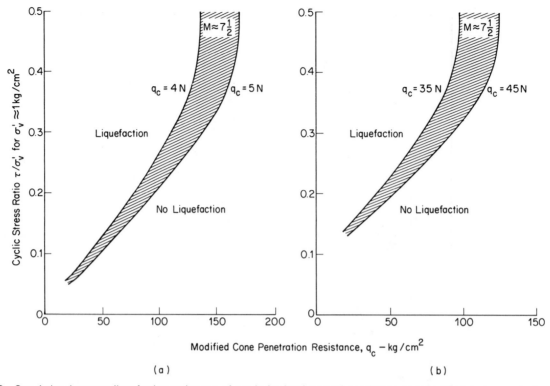

Fig. 16.59 Correlation between liquefaction resistance of sands for level-ground conditions and modified cone penetration resistance. (a) Clean sands: $D_{50} > 0.25$ mm. Based on $q_c/N = 4$ to 5 kg/cm^2. (b) Silty sands: $D_{50} < 0.15$ mm. Based on $q_c/N = 3.5$ to 4.5 kg/cm^2. (*After Seed et al., 1983.*)

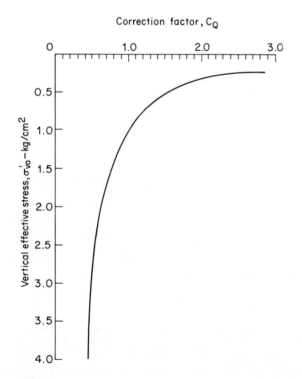

Fig. 16.60 Variation of correction factor, C_Q, with effective overburden pressure. (*After Robertson and Campanella, 1985.*)

DMT (Robertson and Campanella, 1986), SPT (Seed et al., 1983), and CPT (Robertson and Campanella, 1985) based methods and laboratory derived cyclic stress ratios are compared in Figure 16.63 for a sand site. A good agreement between the

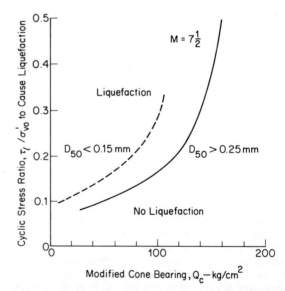

Fig. 16.61 Correlation between liquefaction resistance under level-ground conditions and modified cone penetration resistance for sands and silty sands. (*After Robertson and Campanella, 1985.*)

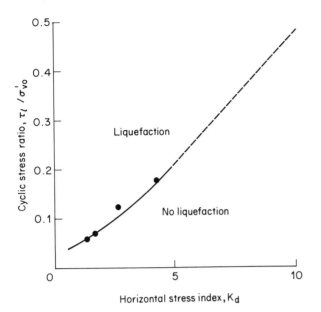

Fig. 16.62 Proposed correlation between liquefaction resistance under level-ground conditions and dilatometer horizontal stress index $(K_d = (P_0 - u_0)/\sigma'_{vo})$ for sands. (*After Robertson and Campenella, 1986.*)

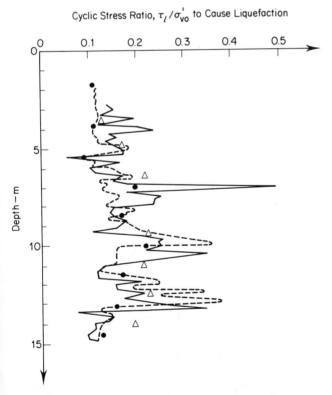

Fig. 16.63 Comparison of predicted cyclic stress ratio to cause liquefaction from DMT, SPT, CPT, and laboratory testing at McDonald's Farm Site, British Columbia. SPT, ●, Seed et al. (1983); CPT, ----, Robertson and Campanella (1985); DMT, ——, proposed correlation; △, laboratory. (*After Robertson and Campanella, 1986.*)

DMT and laboratory, SPT, and CPT derived liquefaction resistance is shown.

16.9 INFLUENCE OF SOIL CONDITIONS ON LANDSLIDES

16.9.1 General Problem

In seismically active areas, numerous cases are reported in which failure or severe damage to earth dams, embankments, and natural slopes have resulted from earthquake ground shaking. For this reason, earth dams, embankments, and natural slopes in seismically active areas require an evaluation of the degree of stability or potential instability which the embankment will experience during an earthquake.

In the coastal environment, the behavior shown in Figures 16.64 and 16.65 is similar to the behavior of a sand in a flow slide in which the sand suffers such a substantial reduction of its shear strength that the mass of soil seems to flow like a liquid. The soil spreads out until the shear stresses acting within the mass become so small that they are compatible with its reduced shear strength, at which time the slope of the soil surface may be only a few degrees. The fact that such failures resemble the flow of a heavy liquid is because the large loss in shear strength affects an important part of the mass rather than only the soil along a sliding surface. The magnitude of the deformation of the sand in a flow slide is governed by the initial shear stresses and the residual strength of the soil, which in turn is only a function of the void ratio of the sand.

A liquefaction failure, that is, flow of the sands, will occur only when there is a driving force that applies shear stresses to the soil, such as under sloping ground, under the foundations of buildings, or surrounding a buried structure that is lighter than the soil displaced. For a given void ratio, the susceptibility to liquefaction increases with confining pressure and with the shear stresses acting on the sand. Cyclic mobility can develop whether or not there are static driving forces inducing shear stresses in the soil mass (Castro, 1975; Castro and Poulus, 1977). In this case it was concluded that relative density is not a useful parameter for natural soil deposits and its use should be limited to compacted fills or laboratory research where homogeneous sand masses are involved (Castro, 1975).

To evaluate the stability or performance of the earthdam or embankment, there are at present three procedures available: (1) the factor of safety approach, (2) the strain potential approach, and (3) the permanent deformation approach. The factor of safety approach is based on the principle of limiting equilibrium, in which the strength available is compared with the applied stress. The difficulty with this procedure is that the stability or instability of the slope is depicted by a single number that does not convey to the analyst any feeling of the amount of movement involved. In response to this drawback, Seed (1966) proposed using a "strain potential" procedure in which the strain of an individual element or slice comprising the slope could be estimated. The drawback with this procedure arises in the basic assumption that each element or slice comprising the slope acts independently of every other slice or element. The third procedure, the "permanent deformation method" originally developed by Lee (1974), built upon the "strain potential" procedure. In this refinement the various elements comprising the slope being investigated are combined to act as an aggregate rather than as individual parts.

In the following sections, analytical methods of modeling the soil response along with a review of methods of evaluating slope performance will be discussed.

Type	Location	Type of Material Involved	Slope Angle (α)	Lateral Movement[a] L$_1$	L$_2$	Range of Volume of Material, yd^3	H	R	Δ	Comments
A. 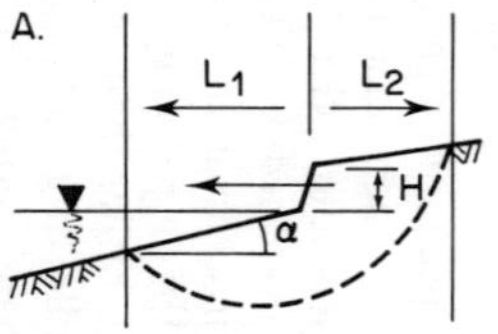	Along coast in loose to dense deposits	Sand and gravel	10° to 32°	3500 ft (1063 m)	1200 ft (366 m)	4 to 98 × 10^6 yd^3 (3 to 75 × 10^6 m^3)	20 ft–50 ft (6 m–15 m)	—	—	Largest slide reported to extend 8500 ft (2600 m) along coast. Depth of slides 120 to 150 t (37 to 46 m). Settlement ranged from 9 to 35 ft (2.7 to 10.7 m)
B. 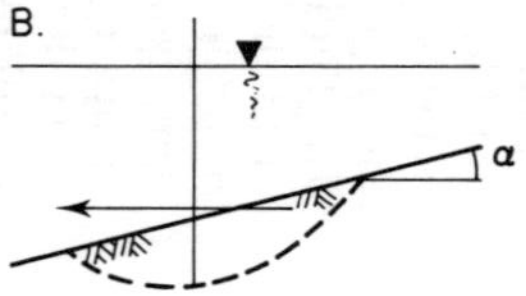	Unconsolidated sediments in deltaic deposits	Fine sand to medium sand with pebbles	3° to 32°	600 ft to 450 mi (183 m to 724 km)	—	4 to 98 × 10^6 yd^3 (3 to 75 × 10^6 m^3)	—	—	—	
C. 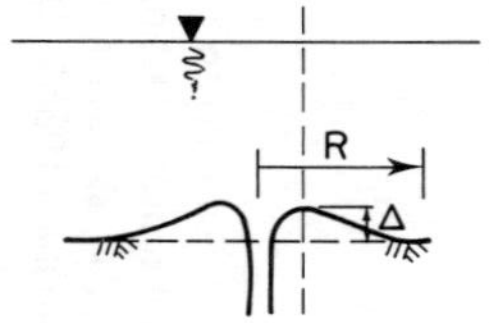	Loose sand $N < 10$ overlain by fine grain sediments	Fine to medium sand involved. Material eroded during movement to surface range from cobbles to clay	Occurs on both slopes or on horizontal	—	—	360 yd^{3a} (277 m^3)	—	Max, 30 ft[a] (9 m for circular boil)	3 ft[a] (0.9 m)	Venting occurred along induced fracturing or zone of weakness. Sand boils surface expression may be either elongated or approximately circular in shape. Sand boil typically starts 2 to 3 minutes after end of earthquake and will last up to 1 hr. Can occur under water. Maximum depth of material for sand boils originated at 66 ft (20 m)

[a] Maximum values.

Fig. 16.64 Typical earthquake liquefaction phenomena characteristics in coastal environment. (*After Chaney and Fang, 1985a.*)

Type of Loading	Typical Location	Slope Angle, α	Material Involved	Volume of Material Moved	Lateral Movement
Cyclic Loading					
A.  SUBSIDENCE OF JETTY	Wave loading along coast Note: Sand boils can occur in slide zone	3° to 4°	Fine to medium sands	75 to 3 × 10⁶ yd³ (58 to 2.3 × 10⁶ m³)	≤600 m
B. 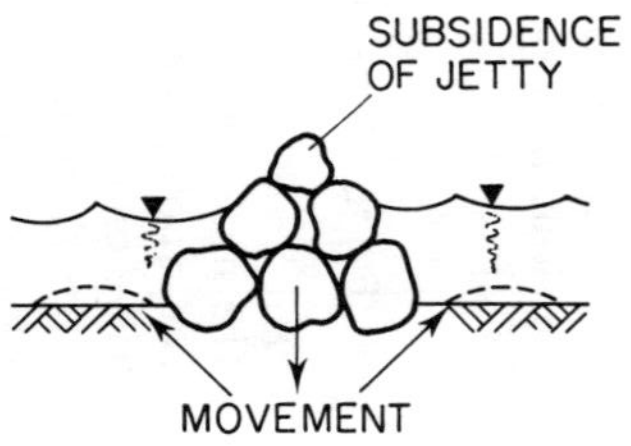 MOVEMENT	Fluvial deposit at mouth of river with rubble mound jetties		Fine sand		
Spontaneous Loading					
A. 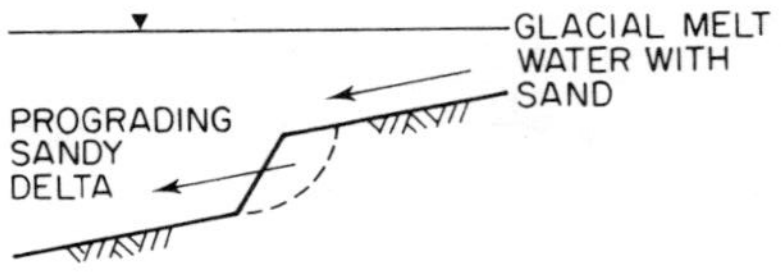 GLACIAL MELT WATER WITH SAND / PROGRADING SANDY DELTA	Fjord environment with active prograding sandy deltas fed by glacial melt water		Sands		
B. 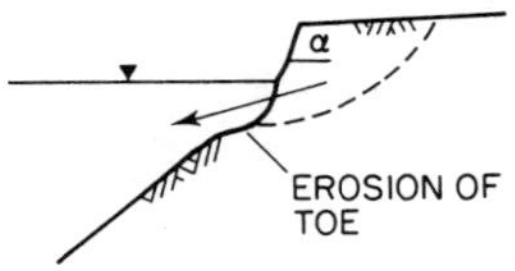 EROSION OF TOE	Fluvial deposits of sandy silts and clays along rivers or coasts	22° to 27°	Fine sand at 110 to 120 ft (33 m to 37 m)	4 × 10⁶ yd³ (3 × 10⁶ m³)[a]	800 ft (244 m)[a]

[a] Maximum values.

Fig. 16.65 Typical nonearthquake liquefaction phenomena characteristics in coastal environment. (*After Chaney and Fang, 1985a.*)

16.9.2 Methods of Modeling Soil Behavior

The effects of cyclic loading are extremely dependent upon the soil and the nature of cyclic deformation. The potential for pore water pressure generation in either clays or sands governs their behavior under cylic loading. Two analytical approaches have been developed to model this behavior:

1. Total stress–strain approaches, in which the cyclic strain or stress conditions of interest are simulated on laboratory specimens and behavior inferred directly
2. Effective stress–strain approaches, in which one attempts to determine analytically the pore water pressures generated and their influence on the soil behavior

Laboratory tests on the soils are utilized at a more fundamental level to determine pore water pressure generation characteristics. The effective stress approaches are more intuitively satisfactory from a fundamental standpoint. However, in present geotechnical practice it is not common that enough information is known about the in-situ stress condition or the stress–strain condition or the stress–strain pore water pressure characteristics of the material under cyclic loading conditions to enable effective implementation of this approach. Research and development of field, laboratory, and analytical tools will change this position in the future.

The total stress–strain approaches for cyclic loading have evolved into two methodologies. The first method involves the

TABLE 16.8 CHARACTERISTICS OF TOTAL AND EFFECTIVE STRESS ANALYSIS[a]

Total Stress Analysis

Input data requirements
1. Bulk unit weight
2. Stress–strain total stress parameter
3. Slope geometry
4. Stratification

Advantages
1. Concept is simple
2. Does not require knowledge of in-situ initial pore pressure distribution

Disadvantages
1. Requires accurate simulation of loading in laboratory
2. Does not give a fundamental understanding of phenomena
3. Requires extensive laboratory test program

Effective Stress Analysis

Input data requirements
1. Initial pore water pressure distribution
2. Bulk unit weights
3. Stress–strain effective stress parameters
4. Slope geometry
5. Stratification

Advantages
1. Gives a fundamental understanding of phenomena
2. Requires only limited laboratory test program to determine characteristics of pore water pressure generation. This assumes availability of effective stress–strain relation

Disadvantages
1. Requires knowledge of in-situ initial pore water pressure distribution
2. No universally accepted effective stress–strain relation is at present available

[a] After Chaney (1984).

development of parameters for input into a stress–strain matrix such as employed in a typical finite-element simulation. In general, the basic model of the soil consists of the following elements: (a) initial backbone curve (frequently that of Ramberg and Osgood, 1943); (b) criteria to describe the hysteretic behavior (normally Masing criteria are employed;) (c) formulation to degrade the backbone curve under either constant-strain or constant-stress cyclic loading.

The second methodology is a graphical procedure based on experimental results. This technique is used to extrapolate the resulting strain experienced by a sample in the laboratory under combined static and cyclic loads to the behavior of an element in the field. Typically, the cyclic triaxial or cyclic simple shear apparatus is employed. A summary of the input data requirements and the subsequent advantages/disadvantages of both the effective and total stress methodologies are presented in Table 16.8.

16.9.3 Evaluation of Slope Performance

Whenever a mass of soil has an inclined surface, the potential exists for part of the soil mass to slide from a higher location to a lower location. The initiation of sliding will occur if driving forces developed in the soil exceed the resisting strength. In theory the above problem is simple but certain practical considerations make precise stability analysis of slopes difficult in practice. The primary difficulties are: (a) identification of the failure surface; (b) variation of shear strength in time and space; and (c) identification and quantification of the loading in time and space.

A. Limit Equilibrium Methodology

There are a number of techniques available for performing limit equilibrium analysis (Chapter 10). The various techniques differ primarily in the manner in which they handle the slope geometry, the geometry of potential failure surfaces and utilization of either total or effective stress, and satisfy static equilibrium conditions. The seismic loading is modeled as a constant horizontal force. The primary methods are (1) infinite-slope analysis, (2) the Swedish circle method, and (3) the method of slices. The final evaluation of the stability or instability of the slope is made normally by the use of appropriate safety factors.

B. Finite-Element Method

The finite-element method has been used extensively for the evaluation of slope stability. The basic methodology utilized is based on the original work by Wilson (1963). The method itself is based on the determination of the resulting deformations of a frame under an imposed load. The deformations are based on the minimum potential energy.

In the use of the finite-element method, three distinct ways of quantifying the initial stability of slopes have been developed. The first two methods utilize element calculations in which the slope is modeled as a series of individual elements (Seed, 1966). The third method models the slope as an aggregate of elements.

The first method uses a calculated factor of safety for each element. In contrast, the second method is based on an evaluation of the potential strain experienced by each element. The third method involves predicting the overall resulting permanent deformations.

1. Factor of Safety Approach In this method the factor of safety is defined as the ratio between the cyclic strength and the induced dynamic stress in each element of the slope. One of the difficulties in this approach is that a different factor of safety is calculated for each element and no single overall definition of the factor of safety for the entire slope is readily determined. This difficulty is overcome by applying judgment or an averaging process to the results. Consideration of the mechanism of potential failure and the existence of a large number of safe elements may indicate a safe slope, whereas a large number of failure elements may suggest that the slope is unsafe (Fig. 16.66b).

2. Potential Strain Approach Knowing the factor of safety at each element based on failure defined by a specified strain in a laboratory test, it is possible to evaluate the equivalent cyclic strain at each element. These strains are strains that would develop in a laboratory test specimen that was subjected to the same static and seismic stresses imposed on a field element. However, an element in the field cannot strain like the test specimen, because it is constrained by the deformations of adjacent elements. Thus, with reference to the field loading case, the cyclic strains are referred to as potential strains. Judgment is again required to interpret the meaning of strain potential in terms of the overall dynamic stability of the slope. An example of this method is shown in Figure 16.67 for an earth embankment.

3. Permanent Deformation Approach Displacement-type analysis can be an important method in evaluating seismic loading effects on slopes. Lee (1974), Makdisi et al. (1978), and Chaney (1979, 1980, 1984, 1985) have developed more rigorous methods of estimating the permanent deformations experienced by earth structures during a seismic loading. These methods follow the basic Seed approach for calculating the factor of safety and strain potential for individual elements (Seed, 1966). Differences between the two methods depend on procedures used to calculate deformations from the strain potential values. The procedure initially developed by Lee (1974) has been extended by Chaney (1979, 1980, 1985) for partially saturated soils, and for marine soils (Chaney, 1984).

The Lee procedure involves the concept that permanent seismic deformations of a slope may be computed by evaluating dynamically induced softened slope stiffness values for soil elements with the resultant settling of the slope to a new condition being compatible with pseudo or apparent stress–strain properties of the soil comprising the slope. An example of a permanent deformation method for an earth embankment is shown in Figure 16.66a.

In summary, prediction of the performance of a slope knowing the cyclic loading depends on (a) quantifying the soil response to cyclic and static loading (total or effective stress analysis), and (b) selection of an appropriate slope stability analysis technique (limit equilibrium method or finite-element method). The selection of a slope stability analysis technique

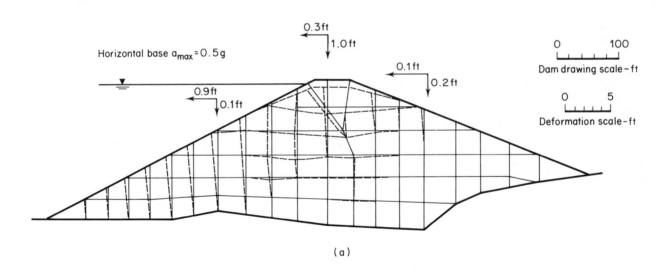

(a)

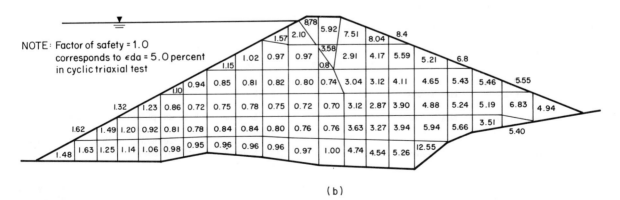

(b)

Fig. 16.66 Predicted performance of Lake Arrowhead new dam under Seed–Idriss simulated 8+ magnitude record, scaled to 0.5*g*. (a) Predicted permanent deformations, 2-D analysis. (b) Element factor of safety. (*After Chaney, 1979.*)

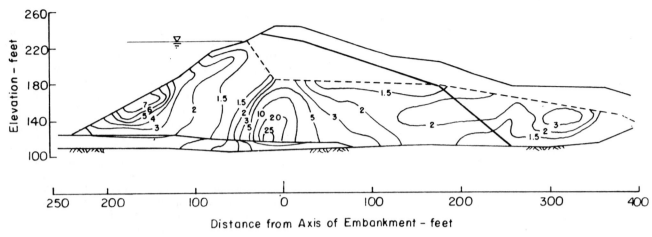

Fig. 16.67 Strain potential approach utilizing finite-element method. (*After Makdisi et al., 1978.*)

is dependent upon (1) the configuration of the failure surface; (2) the nature of cyclic loading; and (3) the required method of evaluating failure (factory of safety, potential strain, or permanent deformation). A summary of the various methods for evaluating instability is presented in Table 16.9.

16.9.4 Stability of Coastal Slope—An Analysis of Case Histories

A literature review was conducted to gain insight into the observed phenomena of liquefaction in the coastal environment (Chaney and Fang, 1985a). Because of limited data available, the review was expanded to include cases involving dams and streams/rivers. The rationale was that these additional cases still involved either a slope adjacent to a water surface or a submerged soil deposit. The results are presented in the form of five tables (Table 16.10(a) to (e)). Information contained in the tables includes, where appropriate, (1) event, (2) date, (3) site, (4) site characteristics, (5) observations, and (6) references.

A summary of the observed phenomena is presented in Figures 16.64 and 16.65. The following conclusions can be reached based on the analysis of case histories.

a. Liquefaction due to either cyclic or static loadings is present in the coastal environment.
b. Slides, either along coast or totally submerged, occur in fine to medium sand.
c. Liquefaction due to earthquake loading tends to involve a larger volume of material than nonearthquake loading.
d. Movement of slopes due to earthquake loading occurs over a range of slope angles from 3° to 32°. In contrast, movement of slopes due to wave loading occurs on slopes ranging from 3° to 4°.
e. Spontaneous liquefaction occurs on slopes ranging from 22° to 27°.
f. Sand boil surface expression can be either elongated or approximately circular in shape. In addition, sand boil can be either submerged or above the level of the ocean. The appearance of sand boil will typically occur 2 to 3 min after the end of the cyclic loading and will last up to 1 hr.

TABLE 16.9 SUMMARY OF METHODS FOR EVALUATION OF INITIATION OF INSTABILITY[a].

Method	Approach	Type of Failure Surface	Type of Stress Analysis Handled	Type of External Cyclic Loading Handled	Method of Evaluating Failure	Comment
Limit equilibrium	Plane failure surface	Plane	(a) Total (b) Effective	Seismic	Factor of safety	Morgenstern (1967)
	Circular failure surface	Circular	(a) Total (b) Effective	(a) Seismic (b) Wave	Factor of safety	Henkel (1970); Finn and Lee (1979)
Finite element	Factor of safety	Not required	Total	(a) Seismic (b) Wave	Factor of safety	Seed (1966)
	Potential strain	Not required	Total	(a) Seismic (b) Wave	Strain potential of individual elements	Seed (1966)
	Permanent deformation	Not required	(a) Total (b) Effective	(a) Seismic (b) Wave	Deformation	Lee (1974); Wright (1976); Serff (1976); Chaney (1979, 1980)

[a] After Chaney (1984).

TABLE 16.10(a) ALASKAN EARTHQUAKES LIQUEFACTION PHENOMENA.[a]

No.	Event	Seismicity			Site	g, %	Site Characteristics	Observations	Reference
		Date	Mag.	Dur, sec					
1.	Alaska	1964			Kenal		Submerged sloping (10° to 25°) stratified deltaic deposits of sandy pebbles and loose sand	Large underwater slides occurred where the deltas project into Kenal Lake. The slides were reported to have moved from 600 to 4100 ft carrying from 4 to 200 thousand yd³ of material	Holish and Hendron (1975)
2.	Alaska	1964			Kenal Lake		Saturated lacustrine deposits of sand and sandy gravel	The four largest slides (Lakeview, Lawing, Ship Creek, and Rocky Creek) were described. Details include: damage area, volume, lateral movement, depth of sliding pre- and postearthquake slopes. In general the unconsolidated sediments in deltas experienced lateral spreading. Piles (driven to 30 ft) in one case were observed to move laterally without any tilting	McCulloch (1966)
3.	Alaska	1964			Kenal Lake		Lake surrounded by steeply sloping rock walls with river deltas occurring at four locations: Lakeview, Lawing, Ship Creek and Rocky Creek. Submerged slopes varied between 10° and 20° and sloping heights of 220 to 520 ft	Submarine slides (between 66 000 and 227 000 yd³) occurred with the largest area being approximately 1 million ft²	Seed (1968)
4.	Alaska	1964			Knik River		Saturated gravels to silts	Longitudinal shift (≃ 2 ft) in location of piers supported on piles driven into cohesionless soils. No depths given	Ross et al. (1969)
5.	Alaska	1964			Martin-Bearing Rivers Area		Submerged river soils	Venting occurred along earthquake-induced fracturing. Sand boils up to 3 ft thick were observed to contain material sizes ranging from cobbles to clay. Some data provided on before and after slopes	Tuthill and Laird (1964)
6.	Alaska	1964			Seward		Sloping (15° to 20°) loose to medium dense deltaic sand and gravel	Submarine flow slide initiated landslides on shore and turbidity flows traveling several thousands of feet	Holish and Hendron (1975)
7.	Alaska	1964			Seward		Sloping saturated marine deposits of medium ($N \simeq 24$) clay and silt ($C_u \simeq 48$) and medium dense ($N = 31$) sand ($C_u \simeq 22$)	Largest slide (≃ 4000 ft long and extending 50 to 500 ft inland) on land followed submarine slides. Ground cracking was noted up to 800 ft beyond slides. Maximum depth of slide ≃ 150 ft. Slopes were observed to increase ≃ 5° following earthquake. Numerous sand boils (10 to 30 ft in diameter, sometimes coalescing in ridges) were observed along fractures depositing between 4 and 12 inches of material from depths of 25 to 75 ft. Settlements up to 3.5 ft were noted	Lemke (1967)
8.	Alaska	1964			Seward Highway Resurrection River Bridges		Saturated granular river bed soils overlying medium to dense sands ($N \simeq 28$ to 88 at −40 ft)	Bridge piers rotated owing to high lateral loads. Total liquefaction not likely but there may have been local cases in lenses of lower density	Ross et al. (1969)

TABLE 16.10(a) (Continued.)

No.	Event	Date	Seismicity Mag.	Dur, sec	g. %	Site	Site Characteristics	Observations	Reference
9.	Alaska	1964				Seward	Sloping (15 to 20°) coastline deposits of loose to medium dense sand and gravel	Approximately 4000 ft of coastline experienced slope failures from 50 to 500 ft inland. Maximum displaced depth ≃ 120 feet	Seed (1968)
10.	Alaska	1964				Seward		Fissures observed to occur on terraces parallel to stream beds. Sand, silt, and gravel forced to the surface formed ridges extending many feet. Sand was believed to have come from lenses about 12 ft deep	Wilson (1967)
11.	Alaska	1964				Seward	Sloping (10° to 32°) loose to dense sands (SP ≃ SM; C_u ≃ 5; D_{50} ≃ 0.15 mm) interfingering with silts and clayey silts (ML). Mean lowest low-water level at time of earthquake	Earthquake resulted in a sudden drawdown of water levels along the shore in conjunction with extensive submarine failures (500 to 600 ft by 1 mile long). 4000-ft long scarp on waterfront extended from 50 to 400 ft inland. Maximum depth of slide believed to be 150 ft. Ground cracking was observed 400 to 800 ft inland. Two large mud boils were reported approximately $\frac{1}{2}$ and $1\frac{1}{2}$ miles offshore. Failure progressed in sections throughout the strong shaking	Wilson (1967)
12.	Alaska	1964				Snow River	River sands, silty sands, and silts with organics. SPT blow counts vary between 5 and 10 in upper 60 ft and 15 to 80 at pile tip	Liquefaction resulted in bridge piling settling up to 10 ft when the piles terminated in loose sands and silts. Piling (driven to 90 to 100 ft) also experienced lateral displacement (up to 8 ft) and tilting (15°)	Ross et al. (1969)
13.	Alaska	1964				Turnagain		Ridges of sand 2 to 3 ft high and 3 to 6 ft wide by 100 ft long were formed by sand boils within the slide area. One such boil covering a 3200 ft² area was reported	Hansen (1965)
14.	Alaska	1964				Turnagain	Gently sloping, with 5- to 20-ft high bluffs (1.5:1) of sand and gravel overlying 100 to 150 ft of Bootlegger cove clay with numerous sand layers and lenses	Slide extended ≃ 8500 ft along coastline and between 600 and 1200 ft inland. Damage area covered ≃ 130 acres and had a maximum lateral movement of 2000 ft toward the bay. Greatest damage occurred in area where clay layer contained numerous sand lenses. Average drop in ground level ≃ 35 ft	Seed (1968)
15.	Alaska	1964				Turnagain	Site consists of relative horizontal surface with 50-ft embankment. Soils included surface layer of sand to depth of 15 to 20 ft underlain by Bootlegger Cove Clay 100 to 150 ft. Clay contains numerous discontinuous seams of silt and sand (up to 3 ft thick). Most seams about 40 to 50 ft below ground surface. Cyclic tests conducted on samples	Large slide occurred along bluff. Size was 8500 ft long and extended back 600 to 1200 ft. Ground dropped about 35 ft. Lenses in clay believed to have liquefied. As bank failed, the slide progressed back. Sliding surface defined at about 60 ft below the surface. Slope of slide was about 4 percent. Sensitivity of the clay contributed to the scale of occurrence. Ridges of sand boils noted. One boil spread over 3200 ft² area	Seed and Wilson (1967)

No.	Location	Year		Site	Description	Reference
16.	Alaska	1964		Turnagain	Slightly sloping buff with increased slope at shoreline. Sand ($D_{50} \simeq 0.1$) and gravel to 20 ft followed by Bootlegger clay having sand lenses up to 3 ft in thickness. Water table between 10 and 20 ft. Cyclic strength testing performed on undisturbed clays and remolded sands	Wilson (1967)
17.	Alaska	1964		Valdez	20 to 30 ft of loose to medium dense ($N = 1$ to 25) sandy silty gravel with cobbles underlain by medium to dense ($N \simeq 15$ to 50) gravelly sand with thin lenses of silt. Groundwater at approximately 65 ft below surface. Sand generally subangular; gravel subrounded. Numerous grain size analysis provided	Coulter and Migliaccio (1966)
18.	Alaska	1964		Valdez	Gentle sloping (15 to 20°) deltaic loose to medium dense ($N \simeq 7$ to 35) deposits of silt, sand and gravel. Submerged	Submerged landslide moving several thousand feet and carrying approximately 98 million yd³ of material	Holish and Hendron (1975)
19.	Alaska	1964		Valdez	Gentle sloping (4° to 10°) loose to medium dense ($N = 7$ to 25) silt, fine sand and gravel	Submarine landslide ($\simeq 98$ million yd³) extending inland up to 500 ft along 6500 ft of coastline with an estimated depth of 200 ft, and distance traveled of several thousand feet. Lateral cracking reported up to 3600 ft behind coastline. Settlements up to 9 feet observed in dock area. Pumping observed in many basements	Seed (1968)
20.	Alaska	1964		General		Settlements from less than 1 ft to several feet occurred in areas subjected to slumping and sliding as excess pore pressures dissipated. Nearly all major valleys in south central Alaska showed evidence of fissures and flows. Numerous sand boils occurred depositing material in cones, and (2 to 3 ft high and 100 ft long) ridges up to 6 feet wide and in one case a 1-acre area 3 ft deep. The sand in one cone was reportedly pumped from a depth of 12 ft. Many subaqueous slides reported in deep fjords of the Prince William Sound, and in several lakes. No slides were observed in less than 7 ft of water. Most severe slides normally occurred near an active stream channel	Scott (1973)
21.	Alaska	1964	0.15	General	Fluvial deposits with average relative density $\simeq 80$ percent ($N = 16$)		Whitman (1971)
22.	Alaska	1964		General	Coastal area	Revised hydrographic charts shows significant changes in elevation or seafloor. 20-fathom contour shifted 600 ft southward. $\frac{3}{4}$-fathom contour increased to 14 fathoms, $5\frac{1}{2}$-fathom contour became 23 fathoms, suggesting that large underwater flow slide occurred	U.S. Department of Commerce (1966)

[a] After Chaney and Fang (1985a).

643

TABLE 16.10(b) CALIFORNIAN EARTHQUAKES LIQUEFACTION PHENOMENA[a].

No.	Event	Date	Mag.	Dur, sec	g, %	Site	Site Characteristics	Observations	Reference
			Seismicity						
1.	San Francisco	4/18/06	8.3			Downtown (Market St.)	From 20 to 30 ft of slightly sloping (0.3°), artifically placed loose dune sand ($N=9$), over marsh deposits. Water table 8 to 13 ft below surface, were the most susceptible	Most prevalent type of slide involved lateral spreading (up to 2 ft). Alluvial deposits along streams and rivers (deltaic deposits) and loosely placed sandy fills	Youd and Hoose (1975)
						(South of Market St.)	Slightly sloping (0.8°) very loose to medium dense ($N=4$ to 12) artifically placed marsh deposits. Water table a few feet below surface	Lateral displacement up to 6 ft. Walks and streets settled up to 3 ft. Buildings on deep foundations not affected	
						(Mission Creek)	Slightly sloping (0.5°) very loose to medium dense ($N=2$ to 12) artifically placed fill as above between .3 and 30 ft. Water table between 5 and 20 ft	Settlements and sliding reported up to 6 ft	
						(Western San Francisco)		Qualitative description of settlement, sliding and sand boils that occurred in the zoological gardens	
2.	San Francisco	1906				Half Moon Bay	Water table near surface	Several examples of shallow (<10 ft) flow slides on slopes less than 20° were cited. Material was reported to flow large horizontal distances	Carnegie Institution of Washington (1908)
3.	San Francisco	1906				Half Moon Bay	Sloping (5°) saturated coarse alluvium with rock fragments	Referenced large flow landslide of 7000 to 10 000 yd^3, moving 300 ft	Holish and Hendron (1975)
4.	San Francisco	1906				Half Moon Bay		Large flow slides reported	Youd and Hoose (1975)
5.	Kern County	1952				Kern Island Canal		Water soaked banks slid into Kern Island Canal. Noted numerous cracks and offsets	Brodek (1952)
6.	San Fernando	3/22/57	5.3	5	0.12	Lake Merced	Soils around lake consist of fluvial and lacustrine deposits of sands, silts and clays. Sand layers, which were as much as 60 ft deep, were typically fine in grain size ($D_{50}=0.1$ to 0.3), low in density above the water table ($N<20$) and dense below the water table ($N>60$)	Several large landslides observed. Two of the slides appeared to be flow slides. Elevation of the flow surface thought to be below water table. Both flows extended about 80 ft back of embankment and were about 10 ft below the ground surface. N-values for failure zone were between 4 and 21	Ross et al. (1969)
7.	Parkfield	6/27/66	5.3 to 5.8	10	0.5	Stream banks of Cholame Valley	Surficial soils at the site consist of sands and silty sands in the upper 3 ft. Water table was at a depth of about 3 ft. Fluvial soils ranging from silts to gravel occur within the valley	Along a two-mile reach of Cholame Creek many small block slides developed on slopes of 5° to 10°. Appears that saturated uniform granular soils liquefied	Ross et al. (1969)
8.	San Fernando	1971				Juvenile Hall Area	Slightly sloping (1.5°) saturated loose sandy silts and silty sands	Settlement (0.5 ft) was observed in conjunction with sliding on the order of 6 ft. Numerous lateral spreading landslides at Van Norman Lakes. Failure occurred 10 to 25 ft below ground surface	Holish and Hendron (1975)
9.	San Fernando	1971				Van Norman Dam		Summary of Seed's (1975) and Lee's (1975) work	Cortright (1975)

#	Event	Year	Location		Site characteristics	Remarks	Reference
10.	San Fernando	1971	San Fernando Reservoir and Lower Juvenile Hall Area			Qualitative description of liquefaction cases given. Sliding in San Fernando Reservoir extended 650 ft northward, covering approximately 6000 ft^2 (2900 yd^3). Scarp on west shore is 0.5 miles long and has been associated with a linear array of sand boils. Lower Juvenile Hall area reported sliding (40 000 ft × 900 ft area) as much as 5 ft	Yerkes (1973)
11.	San Fernando	1971	Van Norman (lower)		140 ft medium dense hydraulic fill ($LD_{50} =$ 0.7 to 0.05 mm) overlying 35 ft of medium dense silty, sandy, gravel ($D_{50} \simeq$ 0.03 mm) alluvium. Dam utilizes a clay core bounded by coarse silty sand on edges. Cyclic strength test data available	Upper and lower dams located within 2 miles of one another. Postearthquake soil investigation included numerous subsurface borings, trenches, and laboratory testing. Relative density varied between 51 and 58 percent. Relative compaction varied from 83 to 100 percent depending upon the method of testing. Laboratory testing did not reveal any significant densification of loose zones during earthquakes	Lee et al. (1975)
			(upper)		80 ft of loose to medium dense hydraulic fill overlying 40 to 60 ft of medium dense alluvium (as described above). Cyclic strength test data available		
12.	San Fernando	1971	Van Norman (upper)	0.5 to 0.6	Site characteristics similar to those given by Lee et al. (1975). Slopes of embankments at 2½ to 1	Crest of dam moved downstream about 5 ft and settled about 3 ft. Entire dam moved. Believed part of movement due to liquefaction or partial liquefaction. No well-defined slip surface and several sand boils at downstream face. Slide believed to have occurred near end or after strong shaking	Seed et al. (1973)
			(lower)		Hydraulic filled dam with properties defined by Lee et al. Upstream slope at 2½ to 1 and downstream at 4½ to 1. Water table was 35 ft below crest	Upstream slope failed. Blocks moved 30 to 150 ft. Multiple shear zones found similar to deep-seated liquefaction in Anchorage. No pressure ridge formed. Trenched upstream face to 60 ft. Strong evidence of liquefaction near bottom of trench (mixing of soils at 70 ft). N-value typically less than 25 at 70 ft. Believed 20-ft layer near base liquefied	
13.	San Fernando	1971	Van Norman Dam		Slightly sloping (1.5°) soils consisting of firm surface layer followed by a soft saturated zone underlain by firmer soils	Lateral spreading landslides (4000 ft × 900 to 1300 ft) reported. Sand boils noted near disturbed zones	
14.	San Fernando	1971	Van Norman Dam			Contents of this report have been more recently summarized in Seed et al. (1975) and Lee et al. (1975)	Seed et al. (1973)
15.	San Fernando	1971	Van Norman Dam			Discusses landsliding at Van Norman Lakes and damage to constructed works associated with soil movements and foundation failures. Sand boils commonly formed below scarps (gives location of 25 such features); slides are attributed by partial or complete liquefaction	
16.	San Fernando	1971	General			General review and discussion of failures at the upper and lower San Fernando dams. In general, a combination of fine-grained soil and a high water table resulted in formation of sand boils and reported submerged sand boils in zones up to 200 ft wide. Materials in boils identified with the soils 4 to 6 ft below the upper fine-grained confining layer	

[a] After Chaney and Fang (1985a).

TABLE 16.10(c) JAPANESE EARTHQUAKES LIQUEFACTION PHENOMENA[a].

No.	Event	Date	Seismicity Mag.	Seismicity Dur, sec	Site	g, %	Site Characteristics	Observations	Reference
1.	Mino-Owari	10/28/91	8.4					Widespread and violent liquefaction in Nobi plain and Fukui plain. More than 50 cases noted. Limited to soft alluvial sites along various rivers. Water and sand ejected as high as 6 ft from wells. Over 1000 sand boils found	Kuribayashi and Tatsuoka (1975)
2.	Shonai	10/22/94	7.3					Liquefaction observed at 6 sites along Mogami River. Sand boils observed in Sakota clay (up to 2 ft high and 10 ft in diameter). Several tens of sand volcanoes observed in Hokurika region (up to 30 ft in diameter). Area liquefied again during Niigata earthquake	Kishida (1969)
3.	Gono (Anegawa)	8/14/09	6.9					Many sites liquefied that had previously liquefied during Mino-Owari earthquake	Kishida (1969)
4.	Fukui	1943	7.2		Kurugo River		Gently sloping to level, loose to medium dense, alluvial sand. Relative density varying between 50 to 70 percent	Sand boils, fissures and quick conditions encountered	Holish and Hendron (1975)
5.	Tohnankai	12/7/44	8.0		General			Caused liquefaction in coastal regions where soft alluvial deposits occur. In east seven different sites involved. Settling and tilting of wooden houses recorded. Large quantities of sand and water ejected from sand boils	Kuribayashi and Tatsuoka (1975)
6.	Niigata	6/16/64	7.3		Agano River		Site adjacent to Agano River. Soils are loose, saturated sands. Swedish soundings conducted	Sands liquefied to depth of 10 to 25 ft. Large subsidence occurred over 450 ft^2 area. Up to 10 ft of settlement. No damage observed when foundations located on deeper materials	Yokomura (1966)
7.	Niigata	1964			Niigata Port Area	0.16 to 0.18	Port facility with sand in upper 120 ft underlain by clay. N values for sand typically less than 30; D_{50} between 0.1 and 0.6. Up to 20% silt. Statis triaxial test data available	Damage to facility was aggravated by liquefaction. Loose material densified and dense material loosened. Density increase occurred in 0 to 45 ft range. Little damage occurred if elevation of structure was below 45 ft or when $N > 25$	Hayashi et al. (1966)
8.	Niigata	1964			Niigata		Alluvial deposits and artificial fills composed of loose ($N < 4$) saturated sands to 16 ft	Damage included slumping of heavy structures, floating of submerged structures (manholes) and sliding of roadway subgrades	Yamada (1966)
9.	Niigata	1964			Niigata		Loose to dense ($N = 5$ to 50 at 100 ft) fine to coarse sand with water table approximately 3 ft deep	Liquefaction widespread in lowland areas. Most damage associated with liquefaction of loose sand. Buildings not deeply embedded were most affected. Sand and mud volcanoes ejected sand 2 to 3 min after end of strong shaking	Ishihara (1974)

646

No.	Location	Year	Site/Structure	Soil Conditions	Field Observations	Reference
10.	Niigata	1964	Shinanogana (or Shinano Bridge)	Medium to fine sand varying from loose ($N < 5$) sand at surface to dense ($N = 50$) sand at 85 ft	Bridge piers rotated and tilted $\simeq 8°$ at a depth of 10 ft. Lateral movement reported up to 6 ft. Liquefaction occurred to depth of 42 ft	Yamada (1966)
11.	Niigata	1964	Shinano River	Level loose ($N = 5$) fine grained alluvial sand. Relative density $\simeq 50$ percent	Sand boils, fissures and quick conditions noted	Holish and Hendron (1975)
12.	Niigata	1964	Shinano River	Site next to river. Soils are loose sands ($N < 5$ to about 15 ft) over denser sands ($N \simeq 20$ at 36 ft). A very loose layer ($N \simeq 1$) was found at 45 to 50 ft. Water table located at surface. Note blow counts measured after earthquake	Large slide occurred next to river. Area of slide was approximately 750 ft by 450 ft. Distance of sliding was about 20 ft. Buildings located on sliding mass were not damaged	Kawakami and Asada (1966)
13.	Niigata	1964	Shonai	Soft saturated sand layer ($N < 10$) in upper 30 ft	Noted little damage in area where sand dunes occurred but heavy damage along river. Sand boils observed	Fukuoka (1966)
14.	Niigata	1964	Showa and Yachigo Bridge	Bridges across rivers. Medium or coarse sand in the upper 30 ft underlain by dense fine sand. Water table near surface; N values between 4 and 10. D_{60} ranged from 0.16 to 1.20; D_{10} were from 0.03 to 0.24	Bridges damaged as river banks pushed toward river. Piling bent; 2-ft concrete piles cracked. Attributed to horizontal sliding (depth at about 20 ft). Up to 2 ft of sliding noted	
15.	Niigata	1964	General		Ground and buildings began to crack 2 to 3 minutes after strong shaking. Sand boils and volcanoes ejected water and sand up to 60 inches into air. Material was correlated with zones approximately 15 to 20 ft deep. Extensive sinking, tilting and floating of structures and pipes observed	Ambraseys and Sarma (1969)
16.	Niigata	1964	General		Good summary of effects of Niigata earthquake including the areas in which soils liquefied, subsurface data, and description of damage	Japan National Committee on Earthquake Engineering (1965)
17.	Niigata	1964	General	Generally saturated sand with gravel to fine sand and silty sand. SPT blow count usually less than 10 in failure zone	Sand boils did not start until strong shaking had stopped and then continued for 2 to 3 min. Material accumulating around boils was correlated with soil 49 to 66 ft	Kawakami and Asada (1966)
18.	Niigata	1964	(Sakae Bridge over Nikko River) / General	Saturated loose ($N \simeq 5$) medium sands to silty sands ($D_{50} \simeq 0.25$ to 0.40) with less than 10% fines	Piles, 36 ft long, with the lower third embedded into dense material were observed to settle as much as 11 ft. Pipes embedded as much as 13 ft were also reported to have moved. Widespread liquefaction occurred. Zones limited to original riverbeds. National Highway settled more than 3 ft and moved laterally 12 ft. Movement caused by liquefaction of sandy banks	Kuribayashi and Tatsuoka (1975)

TABLE 16.10(c) (*Continued.*)

No.	Event	Seismicity Date	Mag.	Dur, sec	Site	g, %	Site Characteristics	Observations	Reference
19.	Niigata	1964			General	0.16	Loose to medium dense ($N = 5$ to 20) fine to medium ($D_{50} = 0.2$ mm) sand from 0 to 30 ft. Medium to very dense sands ($N = 25$ to 55) at 60 ft; Water table at 3 ft	Liquefaction to 20 ft resulted in settlements up to 40 inches, numerous sand boils, submerged structures were noted floating to the surface. Map of damaged area included	Seed and Idriss (1967)
20.	Tokachioki	5/16/68	7.8 to 7.9		Coastal		Dock facilities in coastal areas. Soils typically sandy and loose. Water table near surface	Damage to harbor structures included tilting of caissons supported at 20 ft and tilting of sheet-pile walls due to loss of anchor resistance. Quaywalls moved out nearly 2 ft over a distance of nearly 1000 ft. Traces of sand volcanoes found	Hayashi and Katagama (1970)
21.	Tokachioki	1968			Dikes and Roads		Dikes and road embankments composed of saturated fine loose	Found dike material with greater percentage of fines less damaged. Believes some road embankments composed of volcanic ash (Shirasu) ($D_r \simeq 100\%$) liquefied. Material transported too far for classical base-type failure. Found case where lateral resistance of pile was inadequate because of liquefaction of surficial material	Public Works Research Institute (1970)
22.	Tokachioki	1968			Earth dams		Small earth-filled dams made of volcanic ash and soft sand (volcanic loam). SPT values typically less than 6; D_{50} were approximately 0.4 mm with 30% passing the #200 sieve	Numerous dam failures occurred. Typical slopes were 2.5 to 3.5:1 for greater than 30-ft dams and 1.5 to 2.5:1 for less than 30-ft dams. No evidence of base flows; however, embankments appeared to have flowed	Moriya and Kawaguchi (1970)
23.	Tokachioki	1968			General		Level hydraulic backfill composed of loose alluvial sand (average N-value $\simeq 6$)	Formation of sand volcanoes, sewage pipes buoyed up	Holish and Hendron (1975)
24.	Tokachioki	1968			Nanaehama Beach		Site consists of approximately 15 ft or less going shoreward of hydraulic fill. Fill is typically coarse sand with $D_{50} = 2$ mm and D_r less than 75% ($N < 10$ in upper 15 ft). Material below 40 ft generally fine-grained	Complete liquefaction observed. Water spouts noted for 1 hr after earthquake. Material from spout from 3 to 40 ft range	*Soils and Foundations* (1968)

[a] After Chaney and Fang (1985a).

TABLE 16.10(d) MISCELLANEOUS EARTHQUAKES LIQUEFACTION PHENOMENA[a].

No.	Event	Date	Seismicity Mag.	Dur, sec	g, %	Site	Site Characteristics	Observations	Reference
1.	New Madrid	1811		7.4		Central Mississippi Valley	Loose alluvial sand in valley and silt and clayey silt on uplands. Alluvial materials saturated. Some gentle slopes	Major subsidence and sliding observed within 30 to 40 miles of epicenter. Caving of river banks and flowslides in bluffs adjacent to river recorded. Sand boils, fissures and subsidence occurred at numerous points	Holish and Hendron (1975)
2.	New Madrid	1811				Mississippi	Saturated sandy shoreline and islands	Large landslides reported along river banks. Largest slide involved a section one mile in depth along 6 miles of shoreline. Many sand bars and islands in the river channel completely disappeared. Numerous sand boils were also observed.	Holish and Hendron (1975)
3.	Charleston	1886		7		Atlantic Coast Plain	Loose alluvial silt and sand overlain by denser soil. Some gentle slopes	Sand boils, fissures and spouts of sands and silts are observed. Lateral spreading landslides observed at river banks (Ashely River) and railroad tracks (Rantoules)	Holish and Hendron (1975)
4.	Hawkes	2/3/31		7.9		Hawkes Bay Embankments	Earth embankments formed from cohesionless soil. Silt and silty sand overlain by dense gravel	Embankments spread even when formed from *dry* cohesionless soil. Embankments flowed toward river. Gravels slid on gentle slopes when silts and sand liquefied. Fine silts ejected into air. Chimneys, fireplaces and heavy machinery sank (as much as 2 ft without damage)	Ambraseys and Sarma (1969)
5.	Hawkes Bay	1931					Sandy embankment	Embankment generally experienced damage from lateral spreading upon liquefied bases	Seed (1968)
6.	Grand Banks	1935		7.5		Northern Atlantic		Metastable structure liquefied as result of the Grand Banks Earthquake. Material flowed over 450 miles in a 10-hr period. Rate and distance of travel determined by breaks in communication cables	Terzaghi (1956)
7.	USSR	1949		7.5		Chait	Loess soils deposited on steep mountain slopes along rivers. Soils saturated by rains prior to earthquake	Extensive flow slides along banks of 2 rivers were reported (largest damage area = 2.15 miles × 12 miles). Similar slides have been observed during 5 other earthquakes in U.S.S.R.	Seed (1968)
8.	Chile	5/22/60				Lake Vanquihue Concepcion	Saturated alluvium along lake front	Liquefaction believed to be caused by extensive damage to structures along lake front Sand boils up to 2 ft in diameter were observed along beach	Duke and Leeds (1963)

TABLE 16.10(d) (Continued.)

No.	Event	Date	Seismicity Mag.	Dur, sec	g, %	Site	Site Characteristics	Observations	Reference
9.	Chile	1960				Lake Lianquibue	Relatively flat site in vicinity of lake Lianquibue	Site involved in a lateral spreading landslide. Slumping and fissuring of ground accompanied by some movement toward lake. Horizontal movement only several feet	Holish and Hendron (1975)
10.	Chile	1960				Lake Rinihue	Lacustrine silts and clays overlain by glacial fluvial deposits	Three large flow slides occurred on the valley sides of the San Pedro River. Volumes of slide material were 40×10^6, 8×10^6 and 3×10^6 yd^3. Sliding distance was as much as 1000 ft	Seed (1968)
11.	Chile	1960				San Pedro River near Lake Rinihue	0 to 150 ft of sand and gravel overlying stratified layers of silt and clay with seams of fine sand	Three large landslides with damage areas varying from 35 to 300 acres. The 300-acre slide was reported to have moved up to 1200 ft laterally and 150 ft vertically. Volumes were estimated between 2 and 30 million yd^3. The slides were believed to be the result of fine sand layers within the clay	
12.	Chile	1960	8.4			Puerto Montt	Waterfront area consisted of approximately 15 ft of loose ($N = 5$) hydraulic fill (fine sand with silt and clay) overlying 5 ft of medium dense ($N = 30$) fine to coarse sand; 15 ft of gravel, followed by 16 ft of consolidated soft material with coarse sand	Numerous cases of ground cracking, subsidence (up to 3 ft), and lateral sliding reported in harbor area where structures were founded on or utilized the hydraulic fill as backfill. Approximately 2600 ft of retaining structures (up to $18\frac{1}{2}$ ft in height) failed as a result of increased pressures following liquefaction of the backfill material. Mud pumping was reported in conjunction with tilting, sliding and rotation of footings. Landslides in the sands and gravels were observed to leave scars 240 ft long × 16 ft high	Duke and Leeds (1963)
13.	Chile	1960	8.4			Puerto Montt	Waterfront area consisted of approximately 15 ft of loose ($N = 5$) hydraulic fill (fine sand with silt and clay) overlying 5 ft of medium dense ($N = 30$) fine to coarse sand; 15 ft of gravel, followed by 16 ft of consolidated soft material with coarse sand	Numerous cases of ground cracking, subsidence (up to 3 ft), and lateral sliding reported in harbor area where structures were founded on or utilized the hydraulic fill as backfill. Approximately 2600 ft of retaining structures (up to $18\frac{1}{2}$ ft in height) failed as a result of increased pressures following liquefaction of the backfill material. Mud pumping was reported in conjunction with tilting, sliding and rotation of footings. Landslides in the sands and gravels were observed to leave scars 240 ft long × 16 ft.	Duke and Leeds (1963)

No.	Location	Date	Magnitude	Distance	Acceleration	Soil Conditions	Field Evidence	Reference
14.	Chile Puerto Montt	1960				Gravelly sand and silt fill with relative densities between 50 and 55% ($N < 10$). Very gentle slopes	Underwent several feet of lateral movement. Attributed to partial or total liquefaction of a subsurface layer	Holish and Hendron (1975)
15.	Chile Puerto Montt	1960					Liquefaction damage reported to harbor retaining structures using sandy soils as backfill. Soil disruption was observed along approximately 2000(+) ft of shoreline	Seed (1968)
16.	Chile Rio San Pedro	1960				Embankment on bank of San Pedro River composed of lake clays, and a soil and ash layer. Embankment about 150 ft high	Large slide occurred when saturated material at the elevation of the toe liquefied (back from the toe about 1500 ft). Total slide area involved nearly 1 square mile	Davis and Karzulovic (1963)
17.	Chile Valdiva	1960				Unconsolidated water deposited alluvium and artificial fill composed of fine sand, silt, clay and glacial moraine fragments. Water table located approximately 6 ft below ground surface	Buildings founded upon unconsolidated fill experienced severe damage due to slumping and lurching	Duke and Leeds (1963)
18.	Chile Valdiva	1960				Loose saturated fine sand, silt, and clay fill underlain by fine sand and silt. Ground surface was flat	Ground settled and spread laterally toward river	Holish and Hendron (1975)
19.	Iran Lar Dam	1960				Silty fine sands with relative densities between 20 and 30%	Spherical voids (believed to be result of liquefaction) were found extensively in undisturbed samples to 80 ft. Isolated cases were encountered in a single borehole to 1000 ft	
20.	Seattle Baker River	1965				Hillside slope (3:1) consisting of clays and silty clays with lenses of fine clean or slightly silty sand	Slope failure was believed to be result of a failure in the clay following (days) an increase in stresses following liquefaction of sand	Seed (1968)
21.	Lima, Peru Lima Area	10/3/74	7.6	116–140	0.2 to 0.25	Subsurface soils underlying Lima consist of a conglomerate made up of boulders, cobbles and gravels in matrix of silty sand. Material is very hard. Where materials are looser, water table is deep, 180 ft	No cases of liquefaction reported in clay. Some light poles on beach were leaning after the earthquake thus suggesting liquefaction of surficial soils	EERI (1975)
22.	Montenegro, Yugoslavia General	4/15/79	7.0	15	0.35 to 0.44		Liquefaction along coastal areas resulting in uniform settlement of structures, landslides into ocean, and sand boils	Manojlovic (1980)
23.	Trinidad, California Big Lagoon Spit	11/8/80	7.1			Subsurface soils consisted of layering fine to medium well-graded to uniform sands	Sand boils consisting of medium coarse sand from 5 to 9 ft depth observed in addition to extensive lateral spreading	Jordan (1984)

[a] After Chaney and Fang (1985a).

TABLE 16.10(e) NONEARTHQUAKE LIQUEFACTION PHENOMENA[a].

No.	Event	Seismicity Date	Mag.	Dur, sec	Site	g, %	Site Characteristics	Observations	Reference
1.	General						Subsurface deposits of loose silts and sands deposited in a marine environment. Slopes were relatively gentle	Nonearthquake-initiated subaqueous flow slides occurred in the Trondheim Harbor (1888) and the Helsinki Harbor (1936). A very large earthquake-initiated flow slide occurred on the Grand Banks. These slides generally started at one point and spread in all directions	Anderson and Bjerrum (1967)
2.	Dutch Flow Slides						Banks in the Dutch province of Zeeland. Materials usually fine to medium sands (diams from 0.2 to 0.07 mm). Slopes at sites varied from 3° to 4°	During the period 1881 to 1946, over 200 flow slides recorded. The masses of material involved ranged from 75 to 3×10^6 yd^3. Slides caused by partial liquefaction of sands and silts	Koppejan (1948)
3.	Free Nigger Point (Mississippi River)						Bank adjacent to Mississippi River. Slopes on banks at 1:2 to 1:25; channel depth equal to about 170 ft. Materials include clays, silts, sands and gravels with fine sands from -60 to -100 ft. Typical $D_{50} = 0.12$ to 1.5 mm with up to 20% passing the No. 200 sieve. Blow counts for some sands less than 10. Portions of point were once occupied by river	Massive bank failure (4 million yd^3) when 170 ft of bank moved about 800 ft. Scour occurred at toe of bank. Thought to be liquefaction of sands at depth of 110 ft. Some evidence of failure between 1892 and 1895 at same location	U.S. Corps of Engineers (1950)
4.	Free Nigger Point (Mississippi River)						Data similar to that presented in U.S. Corps of Engineers (1950)	Liquefaction believed to have occurred at a depth of 120 ft due to oversteepening of bank by erosion	U.S. Corps of Engineers (1952)
5.	Menior Point (Mississippi River)						Fluvial deposits of sands, silts and clays along banks of Mississippi River. Grain size and density data similar to Free Nigger Point	General bank failure occurred in 1935 to 1937 of size comparable to Free Nigger Point slide. Soil conditions similar to Free Nigger Point conditions; speculation that liquefaction failure occurred at depth of about 110 ft	U.S. Corps of Engineers (1952)
6.	Grijalva River, Mexico						Fluvial deposit of fine sand 5 m thick at mouth of river	Subsidence of rubble mound jetties due to spontaneous liquefaction of fine sand foundation	Zeevaert (1983)
7.	Spitsbergen						Fjord environment with active prograding sandy deltas fed by glacial melt water	Delta-front chutes and irregularities in the outer slopes of the delta bar sands may be the result of liquefaction	Prior and Coleman (1983)
8.	Bay of Fundy							Wave-induced liquefaction	Dalrymple (1979, 1980)

[a] After Chaney and Fang (1985a).

16.10 BEHAVIOR OF CLAYS AND SILTS DURING CYCLIC LOADING

Clays and silts subjected to earthquake loading may undergo the following conditions: (1) the deterioration of undrained shear strength, (2) the degradation of stiffness, (3) the generation of excess pore pressures and their subsequent dissipation, (4) the accumulation of permanent strains. Some of the factors that affect cyclic strength of clays and silts are cyclic stress level and its frequency, number of cycles of loading, initial shear stress, stiffness, effective confining pressure, and overconsolidation ratio (OCR).

16.10.1 Strength Determination

During cyclic loading, excess pore pressures can build up, causing large cyclic strains. These strains may become so large that the soil would be considered "failed." The cyclic stress to cause failure can be related to the number of cycles of loading, which often shows a "fatigue" type behavior (Lee and Focht, 1976). It has also been shown that when clayey soils are loaded with "initial shear stress" or a fraction of their static failure stress, this may increase their cyclic shear strength significantly (Ishihara˙ and Yasuda, 1980). Figure 16.68 illustrates this phenomenon. The effect of strength deterioration due to cyclic loading is more pronounced when the stress reversal causes the soil to experience negative stresses at each cycle of loading. The effect of strength deterioration is correlated with plasticity index as shown in Figure 16.69, in which cyclic strength at zero initial shear stress normalized by static strength increases with plasticity index.

Experimental and theoretical correlations that relate precyclic to postcyclic undrained shear strength have been developed by a number of investigators (Thiers and Seed, 1969; Sangrey and France, 1980; Van Eekelen and Potts, 1978; Singh et al., 1978). Thiers and Seed demonstrated that the deterioration of undrained shear strength will be minimized when the cyclic strain is kept below one-half the precyclic undrained shear strain at failure. Van Eekelen and Potts (1978) derived a theoretical

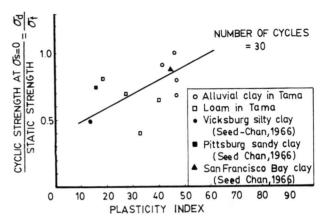

Fig. 16.69 Variation of cyclic strength ratio with plasticity index. (*After Ishihara and Yasuda, 1980.*)

expression with critical state parameters, κ and λ:

$$\frac{s_{uc}}{s_u} = \left(\frac{1 - u_e}{\sigma'_c}\right)^{\kappa/\lambda} \qquad (16.11)$$

where

s_{uc} = postcyclic undrained shear strength
s_u = precyclic undrained shear strength
u_e = excess pore pressure due to cyclic loading
σ_c = initial effective confining pressure
κ = rebound or recompression index expressed on the natural logarithmic scale ($= C_r/2.3$)
λ = compression index expressed on the natural logarithm scale ($= C_c/2.3$)

Sangrey et al. (1978) suggested that contractive clays, silts, and sands behave similarly under repeated loading. Figure 16.70 shows the correlation between the recompression index and normalized critical level of repeated loading (CLRL) for clays, silts, and sands, where e_0 is the initial void ratio and CLRL is the cyclic stress level at which soil passes into failure from equilibrium state. This correlation indicates that larger levels of cyclic stress are sustained by clays owing to their larger value of recompression index.

Meimon and Hicher (1980) correlated the postcyclic undrained shear strength of two laboratory-prepared clays (with OCR = 1 and OCR = 4), with maximum permanent axial strain. They found that if the maximum permanent strain did not exceed

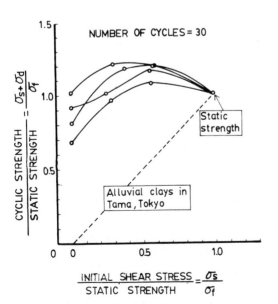

Fig. 16.68 Variation of cyclic strength ratio with initial stress for alluvial clays in Tokyo. (*After Ishihara and Yasuda, 1980.*)

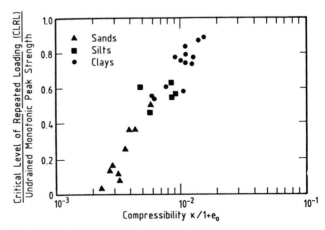

Fig. 16.70 Critical level of repeated loading (CLRL) from undrained tests on contractive soils. (*After Sangrey et al., 1978.*)

5 percent, the reduction in undrained shear strength was lower than 8 percent. With increasing strain level, the reduction could reach up to 40 percent. Experimental results also showed that the reduction did not necessarily depend upon OCR.

Loading rate has been shown to influence cyclic behavior of clays. The effect of loading rate on pre- and postcyclic undrained shear strength can be approximated by the following expression (Poulos, 1988):

$$\frac{s_u}{s_{ur}} = 1 + F_R \log_{10}\left(\frac{r}{r_r}\right) \qquad (16.12)$$

where

s_u = undrained shear strength
s_{ur} = reference value of s_u
r = actual loading rate
r_r = reference loading rate
F_R = rate factor (typically between 0.05 and 0.2)

This expression indicates that the undrained strength increases with increasing loading rate. An analytical model proposed by Prevost (1977) shows the effect of loading rate on the cyclic behavior of a clay soil in Figure 16.71. The model parameters were determined through slow monotonic and rapid cyclic simple shear tests of Drammen clay. Also shown in this figure are typical hysteresis loops obtained at constant strain amplitudes. The rapid cyclic stress–strain curve lies above the static curve, which indicates increase in stiffness and strength with increasing loading rate. The hysteresis loops develop an S shape which becomes more marked as the number of cycles of loading increases. It was reported that the gradient of the hysteresis loops at the peak shear stress remained constant and was approximately equal to the gradient of the static curve at the corresponding strain.

The effect of number of cycles of loading and frequency of loading on cyclic stress ratio ($= \tau/c_u$, where τ = cyclic stress; c_u = undrained shear strength) were studied by Procter and Khaffaf (1984). Figures 16.72a and 16.72b illustrate these effects. In Figure 16.72a the data indicate that, irrespective of strain amplitude, there is a minimum stress ratio achieved after a certain number of cycles of loading. A similar phenomenon was observed with shear modulus of soft marine clays, where the data indicated a minimum value of shear modulus after a

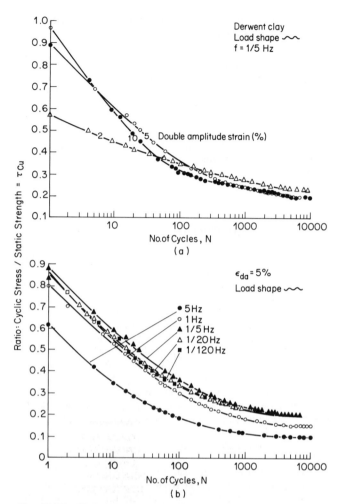

Fig. 16.72 Variation of cyclic stress ratio with number of cycles of loading, *N*, in reversed displacement controlled tests of remolded clay. (a) Constant frequency. (b) Constant strain. (*After Procter and Khaffaf, 1984.*)

critical value of number of cycles of loading (Pamukcu et al., 1983). In both cases this value of *N* (= number of cycles of loading) was around 10 000 and the data were obtained from strain or displacement controlled cyclic tests. Figure 16.72b shows effect of loading frequency on the cyclic stress ratio, where the ratio appears to be constant for frequencies below 0.1 Hz and decreases above this value.

16.10.2 Pore Pressure Build-up

Generation of excess pore water pressures under cyclic loading has been shown to cause marked reduction in undrained strength and stiffness of clay soils. Theoretical and empirical expressions have been developed that relate excess or residual pore pressures with factors such as cyclic stress and strain level, number of cycles of loading, and OCR. An empirical expression has been developed by Van Eekelen and Potts (1978) for the rate of generation of excess pore pressures. Another has been developed by Matsui et al. (1980) for the residual pore pressures:

$$\frac{u_r}{\sigma_c} = \beta\left[\log_{10}\left(\frac{\gamma_{c,\max}}{A_1(OCR-1)} + B_1\right)\right] \qquad (16.13)$$

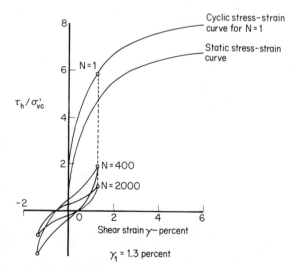

Fig. 16.71 Rapid cyclic and slow monotic stress–strain curves obtained from tests of Drammen clay. (*After Prevost, 1977.*)

where

u_r = residual pore pressure
σ_c = effective confining pressure
$\gamma_{c,\max}$ = single amplitude maximum cyclic shear strain
OCR = overconsolidation ratio
$\beta = 0.45$ (found experimentally)

and

PI	A_1	B_1
20	0.4×10^{-3}	0.6×10^{-3}
40	1.1×10^{-3}	1.2×10^{-3}
55	2.5×10^{-3}	1.2×10^{-3}

again, found experimentally.

Togrol and Guler (1984) suggested an empirical relation for normally consolidated clay that related the deviatoric stress at failure to excess pore pressure developed during cyclic loading:

$$q_f = 0.63p_c - 0.39u_{\text{din}} \qquad (16.14)$$

where

q_f = deviatoric stress at failure
p_c = consolidation pressure
u_{din} = excess pore water pressure

Using experimental values of maximum excess pore pressure developed and Equation 16.14, they found that the maximum reduction in undrained shear strength of the soil would be on the order of 35 percent under repeated load application.

Singh et al. (1978) indicated that build-up of excess pore pressure during cyclic loading of fine-grained soils produces an effect similar to increasing the soil's OCR in the unloading portion of consolidation. A similar conclusion was drawn by Anderson et al. (1980), where it was shown that the stress path of a statically sheared clay following cyclic loading is similar to that of an overconsolidated clay. This behavior was modeled to predict the undrained cyclic response of slightly over-consolidated clays using results of tests on normally consolidated clays (Azzouz et al., 1989). The model was based on AOCR, apparent overconsolidation hypothesis. The method was shown to provide good estimates of number of cycles to failure and development of excess pore pressures with respect to N.

Theoretical expressions derived by Egan and Sangrey (1978) relate residual excess pore pressure (due to plastic strain) and also maximum excess pore pressure developed in fine-grained soils under cyclic loading to critical-state soil parameters.

$$du_r = \left[1 - \exp\left(\frac{-\pi}{\kappa}\right)\right]p_0 \qquad (16.15)$$

and

$$du_{\max} = \left[1 - \exp\left(\frac{-\pi}{\kappa}\right)\left(1 - \frac{M}{3}\right)\right]p_0 \qquad (16.16)$$

where

κ = rebound or recompression index on the natural logarithmic scale ($= C_r/2.3$)
π = volume change potential ($= \kappa \cdot \ln(p_0/p_u)$), where p_u = mean effective stress at critical state
p_0 = initial mean effective stress
M = critical-state soil parameter (slope of critical state line in p–q space)

Ansal and Erken (1989) developed an empirical model that relates cyclic stress ratio and pore water pressure ratio for different values of N. Figure 16.73 illustrates this relation. A threshold cyclic stress ratio is defined that is similar to observations reported by Matsui et al. (1980) and Dobry et al. (1982), the latter being in sands and in terms of cyclic strain.

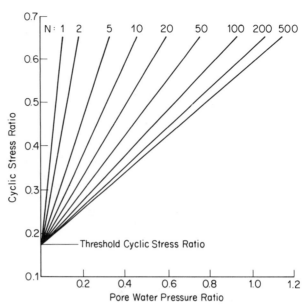

Fig. 16.73 Cyclic stress ratio–pore pressure relationship for different numbers of cycles. (*After Ansal and Erken, 1989.*)

The relationships reported by Ansal and Erken were established through a series of cyclic simple shear testing of normally consolidated kaolinite clay. In this model the pore pressure build-up is expressed as

$$u = \left[\left(\frac{\tau}{\tau_f}\right) - (\text{S.R.})_t\right]m \qquad (16.17)$$

where

u = pore press build-up
τ/τ_f = cyclic shear stress ratio, where τ_f = cyclic shear stress at failure
$(\text{S.R.})_t$ = threshold cyclic stress ratio
m = slope of the pore pressure line ($= \Delta u/\Delta(\tau/\tau_f)$) determined experimentally

The effect of frequency of loading on pore pressure build-up was also investigated. It was found that decrease in loading rate lead to an increase in the accumulated pore pressures with number of cycles and the effect of loading rate diminished after the initial cycles of loading. An indirectly related observation was made by Pamukcu and Suhayda (1987) in which the stiffness degradation of soft saturated clays with induced initial pore pressure ratios was more marked in slow monotonic loading than high-frequency dynamic loading for low strain amplitude range. This indicated larger pore pressure build-up in the slow monotonic loading case.

16.10.3 Reduction and Degradation of Stiffness

During cyclic loading, the stress–strain behavior of clays and silts is nonlinear and hysteretic. A hysteresis loop consists of three stages during one cycle of loading: the initial loading, unloading, and reloading stages. The initial loading consitutes the "backbone" curve of the loop. An idealized stress–strain hysteresis loop obtained for a soil specimen subjected to a symmetric cyclic shearing load along a plane free of initial shear stress is given in Figure 16.74. The backbone curve characterizes the nonlinear stress–strain behavior of clays. $G_{\max}$ is the maximum shear modulus defined as the slope of the initial tangent to the backbone curve. G_s is the secant modulus equal to the $\tau_{\max}/\lambda_{\max}$ ratio, where $\tau_{\max}$ and $\lambda_{\max}$ are the maximum cyclic shear stress and maximum shear strain, respectively.

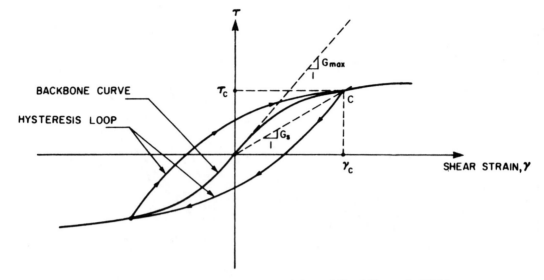

Fig. 16.74 Typical stress–strain hysteresis loop. (*After Idriss et al., 1978.*)

The Masing criterion is the most widely accepted rule for generating hysteresis loops from backbone curve (Masing, 1926). It simply states that the unloading and reloading branches of the loop are the same backbone curve with both stress and strain scales expanded by a factor of 2 and the origin translated. After the stress reversal, the tangent modulus at the tips of the loop is equal to G_{max}. The backbone curve is expressed in several mathematical formulations, which include bilinear (Thiers and Seed, 1969), multilinear, hyperbolic (Hardin and Drnevich, 1972), and Ramberg and Osgood (1943) formulations.

Damping ratio D is defined as the ratio of the energy dissipated to energy input during one cycle of loading. D is computed on the basis of the area contained within the hysteresis loop, and the equivalent secant modulus. Systems that satisfy the Masing criterion behave as though they have an equivalent viscous damping ratio independent of the frequency of vibration at a given strain amplitude.

Reduction of moduli with increasing strain amplitude is a major characteristic displayed by the nonlinear nature of the stress–strain relationship of soils. An idealized shear modulus reduction curve is given in Figure 16.75, whereby extrapolating the curve to zero strain, the maximum shear modulus, G_{max}, can be estimated at the intercept. Hardin and Drnevich (1972) suggested the use of the following empirical equation for calculation of G_{max} for many undisturbed cohesive soils as well

as sands:

$$G_{max} = 1230 \frac{(2.973 - e)^2}{(1 + e)} (OCR)^k \sigma_0'^2 \qquad (16.18)$$

where

e = void ratio
OCR = overconsolidation ratio
σ_0' = mean principal effective stress (psi)
k = a constant value which depends on plasticity index, PI, as given below
G_{max} = maximum shear modulus (psi)

PI	k
0	0.0
20	0.18
40	0.30
60	0.41
80	0.48
$\geqslant 100$	0.50

This equation, however, is said to produce low G_{max} values for void ratios in excess of 2.

Hardin and Drnevich (1972) related G/G_{max} to γ_h, hyperbolic strain, through the following equation.

$$\frac{G}{G_{max}} = \frac{1}{1 + \gamma_h} \qquad (16.19)$$

where

$$\gamma_h = \left(\frac{\gamma}{\gamma_r}\right) \cdot \left[1 + a \exp\left(\frac{-b \cdot \gamma}{\gamma_r}\right)\right] \qquad (16.20)$$

where

γ_r = reference strain
a, b = soil constants

Figure 16.76 shows the charts for variation of reference strain with respect to plasticity index, OCR, and void ratio. The suggested values of a and b are given in Table 16.11.

A plot of actual experimental results of shear modulus along with shear stress and damping ratio versus shearing strain is presented in Figure 16.77 for Shanghai soft clay. Represented on a semilogarithmic scale, normalized curves from various

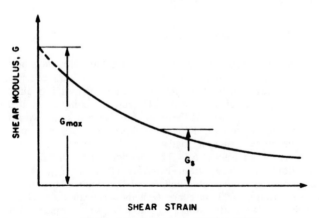

Fig. 16.75 Idealized shear modulus curve.

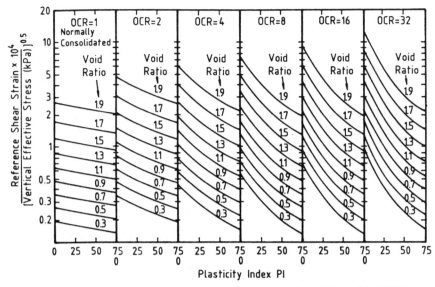

Fig. 16.76 Reference strain for clays. (*After Hardin and Drnevich, 1972.*)

investigators show the variation of shear modulus with shearing strain for different soils in Figure 16.78 (Fang et al., 1981). A hyperbolic curve was fitted to the normalized shear modulus versus shear strain variation data of Gulf of Mexico clay, where the agreement was good for the lower ranges of strain amplitude normalized by yield strain (Figure 16.79) (Pamukcu et al., 1983). For San Francisco Bay silty marine clay, consecutive hysteresis loops obtained for the first cycle of dynamic loading obtained at different controlled strain levels, Figure 16.80 illustrates the stiffness reduction as the loops deform and tilt with increasing strain amplitude (Idriss et al., 1978).

In clays, reduction of moduli is generally accompanied by degradation of the backbone curve. Progressive degradation of soil stiffness with increasing number of cycles of loading can be defined as progressive "softening" of the soil. Degradation is known to be mainly a function of the number of cycles of loading. An example of the effect is shown in Figure 16.81, where the first and the tenth cycle hysteresis loops at controlled strain are plotted. Degradation effects are formulated using degradation index δ, which is the ratio of the secant modulus in the Nth cycle to the initial secant modulus (Idriss et al., 1978). δ is a function of the number of cycles of load, N, and is defined as follows:

$$\delta = N^{-t} \qquad (16.21)$$

where t is the degradation parameter. It is defined as the slope of the semilogarithmic plot of secant moduli versus N. While it is strongly dependent on strain amplitude, it is essentially independent of confining stress and water conent (Idriss et al.,

1978; Stokoe, 1980). The degradation effect on the backbone curve constructed using the Ramberg–Osgood model on San Francisco Bay clay is shown in Figure 16.82. A review of Figure 16.82 shows that the backbone curve shifts down and flattens progressively with increasing values of δ. Test data superimposed on analytical backbone curves, degraded with number of cycles of loading is shown in Figure 16.83 for Gulf of Mexico clay (Pamukcu and Suhayda, 1984). Cyclic degradation of Drammen clay under sustained shear loading is illustrated in Figure 16.84 (Goulois et al., 1985).

The degradation effect diminishes with increasing number of cycles of load application. Therefore, one can predict a reasonable value of number of cycles at which the material can be assumed to have reached a steady-state condition with insignificant degradation of stiffness. Degradation and reduction of stiffness are often observed to become significant after a threshold strain amplitude. This threshold of shear strain is generally given between 10^{-3} percent and 10^{-2} percent for clays (Isenhower and Stokoe, 1981; Pamukcu and Suhayda, 1987). Vucetic (1988) has presented information that shows that clays exhibiting static normalized behavior with respect to the vertical consolidation stress, that is, along the lines of the SHANSEP method, also exhibit a similar cyclic normalized behavior.

Damping of soils under cyclic loading, like moduli, is strongly dependent on strain amplitude. The tilting of the hysteresis loops in Figure 16.80 is accompanied by the enlargement of the area they enclose as the strain amplitude increases. Since damping, by definition, is the ratio of the area enclosed by the loop to the area under secant modulus line, it can be observed that damping increases with increasing strain. Once again, damping ratio stays independent of strain amplitude up to a threshold value of the strain, which is between 10^{-3} percent to 10^{-2} percent for most clays. This value of the damping ratio is called the minimum damping, D_{min}.

Factors that Affect Measurement of Dynamic Properties of Clays

Dynamic properties of soils are dependent on a number of factors, such as overconsolidation ratio, effective stress, void ratio, and saturation (Athanasopoulos and Richart, 1983a, b;

TABLE 16.11 VALUES OF a AND b FOR SATURATED COHESIVE SOILS[a].

Application	a	b
Modulus	$1 + 0.25 \log N$	1.3
Damping	$1 + 0.2 f^{1/2}$	$0.2 f \exp(-\sigma_0') + 2.25\sigma_0' + 0.3 \log N$

[a] After Hardin and Drnevich (1972).
Frequency f is in cycles per second.
Mean effective stress σ_0' is in $\mathrm{kg \cdot cm^{-2}}$ ($1\ \mathrm{kg \cdot cm^{-2}} = 98\ \mathrm{kN \cdot m^{-2}}$).
N is number of cycles.

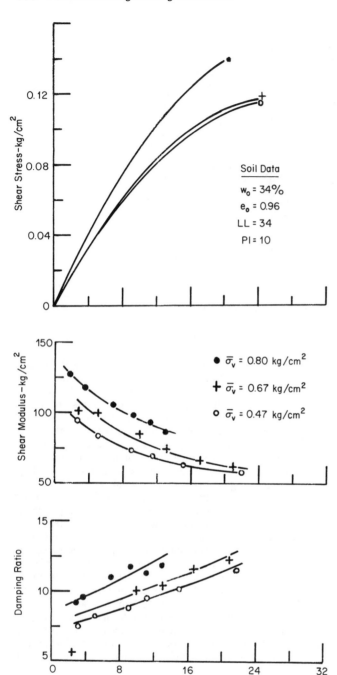

Fig. 16.77 Correlations of shear stress, shear modulus, damping ratio, and shearing strain of Shanghai soft clay. (*After Fang et al., 1981.*)

strength deterioration accounts for part of the influences in laboratory measurements.

Lefebvre and LeBoeuf (1987) studied the effects of strain rate and cyclic loading on sensitive marine clays, in which they showed that reducing the strain rate or cycling the load appears to weaken the clay by a fatigue phenomenon due to reduction of effective stress and weakening of brittle bonds in the clay skeleton. They concluded that strain rate effects in saturated clays cause the cyclic strength mobilized at high frequencies to be higher than the monotonic strength measured at standard strain rates. Pamukcu (1989) has shown that for normally consolidated soft kaolinite clay, the ratio of the static to dynamic shear modulus, measured at a strain amplitude of 10^{-4} percent, is on the order of 0.85. Athanasopoulos and Richart (1983a) showed that at rates of deformation characterizing initial stages of creep, clays exhibited significant loss of stiffness. However, they regained this loss and exhibited increase over the precreep value at later stages of creep with slower rates of deformation. This was not the case for overconsolidated clays, in which the recovery and increase of stiffness did not occur with extended creep.

Macky and Saada (1984) developed empirical models relating shear modulus, shear strain, and consolidation pressure for cross-anisotropic clays. They reported that under slow cyclic loading the G/G_{max} ratio decreased to about 50 percent of its original value at around 0.05 percent shear strain amplitude for the clays tested. Koutsoftas (1978) showed that the stiffness reduction was more significant than undrained strength reduction of two marine clays following cyclic loading. Athanosopoulos and Richart (1983b) showed that temporary release of confinement of cohesive soils caused a reduction in shear modulus. However, the initial value was regained when the confinement was reapplied over an interval of time. This modulus-regain time increased with the age of the cohesive soil. Temporary high-amplitude cyclic loading had a similar effect on soil moduli. These findings were suggested to explain some of the discrepancies between field and laboratory measurements of soil moduli.

Ray and Woods (1988) showed that, with cyclic loading, the soil skeleton can change even without generation of pore pressures. They conducted cyclic tests on dry sand and silt specimens and showed that silty soils exhibit reduction in shear modulus and damping ratios with the number of cycles of loading. Wu et al. (1984) studied the effect of saturation on sandy and silty soils, in which they found that capillary effects significantly increased the shear modulus of silty soils. For the specimens tested, the maximum increase in shear modulus (about two times the dry or fully saturated specimen) occurred between degrees of saturation of 5 and 20 percent. An empirical equation was developed to correct the shear modulus values measured at a particular degree of saturation for silty and sandy soils.

Analytical Methods to Predict Cyclic Response of Clays

There are a few analytical procedures that are used to estimate permanent strains resulting from cyclic loading of soils. Some of these approaches utilize hysteretic stress–strain relationships, and some use empirical relations (Marr and Christian, 1981; Bouckovalas et al., 1986). Prediction of permanent strains due to cyclic loading may be complicated in clays owing to the dependency of those strains on other factors, such as generation and dissipation of pore pressures and time involved in consolidation also. Hyde and Brown (1976) suggested that permanent strain behavior of clay and silt soils under repeated loading and creep loading may have similarities. They

Wu et al., 1984). The variation of shear modulus with pore pressure for Gulf of Mexico clay is shown in Figure 16.85 (Dyvik et al., 1983). The measurement of dynamic properties is often influenced by strain rate effects (Isenhower and Stokoe, 1981). The strain rate effect of the measured variation of shear modulus with shear strain amplitude is shown in Figure 16.86. Laboratory measurement of dynamic properties of soft soils are also often complicated by sample disturbance effects. Cuny and Fry (1973) reported $+50$ percent variation between laboratory and field measured values of soil moduli. Pore pressure build-up and

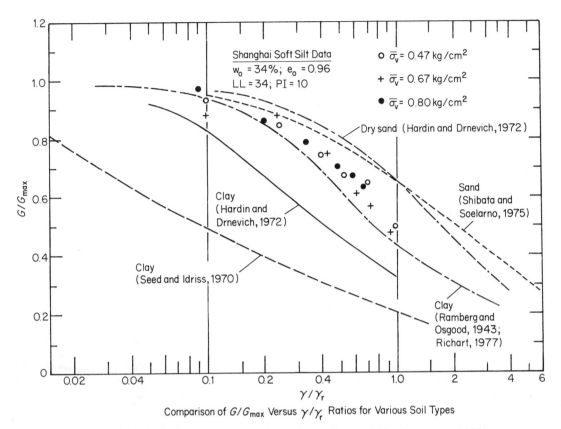

Comparison of G/G_{max} Versus γ/γ_r Ratios for Various Soil Types

Fig. 16.78 Modulus reduction curves for various soil types. (*After Fang et al., 1981.*)

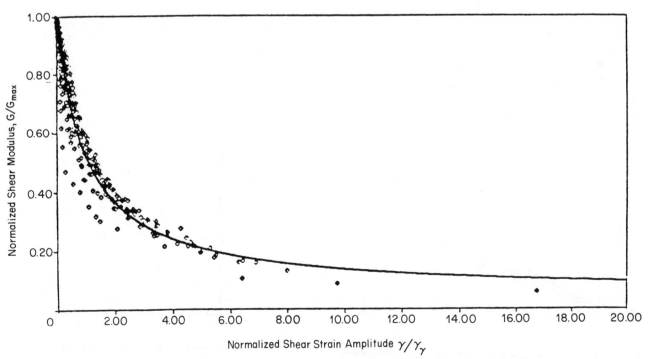

Normalized Shear Strain Amplitude γ/γ_γ

Fig. 16.79 Variation of shear modulus with strain amplitude. (*After Pamukcu et al., 1983.*)

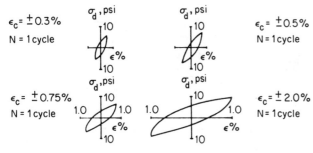

Fig. 16.80 Hysteresis loops measured during first cycle. (*After Idriss et al., 1978.*)

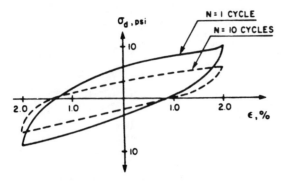

Fig. 16.81 Hysteresis loops measured during first and tenth cycles. (*After Idriss et al., 1978.*)

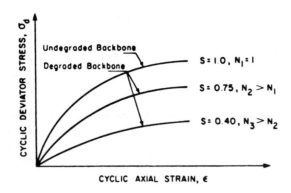

Fig. 16.82 Schematic illustration of construction of degraded backbone curves. (*After Idriss et al., 1978.*)

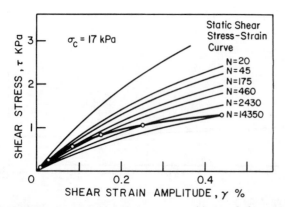

Fig. 16.83 Degraded backbone curves with actual shear stress–strain data points and the fitted hyperbolic curve. (*After Pamukcu and Suhayda, 1984.*)

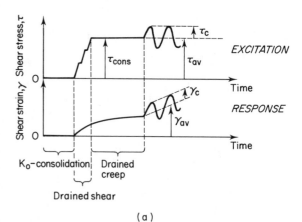

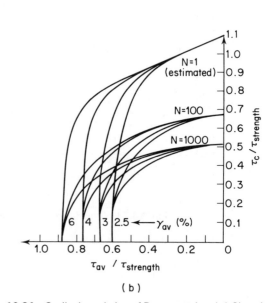

Fig. 16.84 Cyclic degradation of Drammen clay. (a) Shear loading and resulting shear strains. (b) Average shear strain network of curves for $N = 1$, 100, and 1000 cycles of loading. (*After Goulois et al., 1985.*)

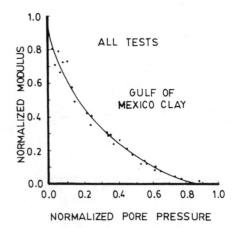

Fig. 16.85 Cyclic stress versus normalized pore pressure variation. (*After Dyvik et al., 1983.*)

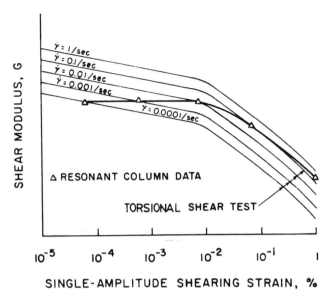

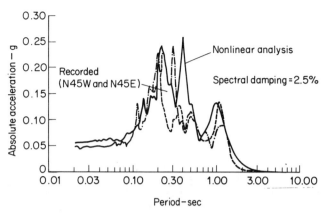

Fig. 16.88 Computed and recorded acceleration spectra, Southern Pacific Building. (*After Singh et al., 1981.*)

Fig. 16.86 Combined strain amplitude and strain rate effects on shear modulus. (*After Isenhower and Stokoe, 1981.*)

demonstrated that, as first approximation, permanent strains accumulated during one-way cyclic loading can be estimated from static creep load test data.

In evaluation of seismic response of soft clay sites it is more appropriate to use nonlinear analyses than equivalent linear

methods. A nonlinear analysis incorporates the nonlinear stress–strain response and modulus degradation effects where an equivalent linear method utilizes strain-dependent stiffness and damping. Some of the nonlinear models of stress–strain behavior are bilinear, multilinear, hyperbolic and Ramberg–Osgood idealizations. Singh et al. (1981) used a nonlinear model that incorporated modulus degradation developed by Idriss et al. (1978), on selected sites for seismic response analysis. The acceleration and velocity spectra results of nonlinear analysis, equivalent linear analysis, and recorded values are shown in Figures 16.87 and 16.88. As observed from these figures, a good agreement was obtained between ground response and the nonlinear analysis response. The study also showed the significance of parametric studies involving possible variations in assumed rock motions and soil parameters when using response calculations. It was concluded that, at the levels of earthquake intensities where modulus degradation of soft clay sites becomes significant, the nonlinear analysis would yield more realistic results than equivalent linear methods.

Another example of evaluation of seismic response of cohesive soils was given by Tsai et al. (1980). It was concluded that nonlinear deformation, failure, and degradation behavior of the soil profile can have significant influence on seismic response under strong levels of earthquake shaking.

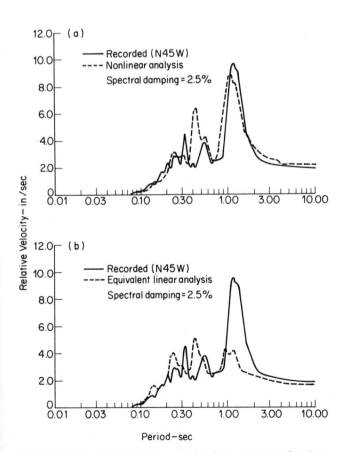

Fig. 16.87 Computed and recorded velocity spectra, Southern Pacific Building. (*After Singh et al., 1981.*)

16.11 REMEDY OF EARTHQUAKE DAMAGE ON SOIL–FOUNDATION SYSTEMS

Various methods of remedy for liquefiable soils are given in Table 16.12. Also, Figure 16.89 shows the grain size ranges of soil for the applicability of some of these methods.

The methods range from mechanical types such as blasting to chemical such as grouting (Chapter 9). Details on methods such as drain systems (Chapters 1 and 7), lime columns (Chapter 23), and deep compaction (Chapter 24) are presented elsewhere in this handbook.

Since earthquake effects on soil–foundation systems require various interdisciplinary knowledge to analyze, utilization of knowledge-based expert systems may provide effective results. Yu and Mikroudis (1987) and Fang et al. (1989) proposed the use of expert systems to aid architects and designers in seismic design. Also, the system can be used as a tool to collect, reorganize, and store systematic data from case studies, which may help researchers in related areas.

TABLE 16.12 IMPROVEMENT OF LIQUEFIABLE SOIL FOUNDATION CONDITIONS[a].

Method	Principle	Most Suitable Soil Conditions/Types	Maximum Effective Treatment Depth	Economic Size of Treated Area	Ideal Properties of Treated Material[b]	Applications[c]	Case[d]	Relative Costs[e]
				In-Situ Deep Compaction				
(1) Blasting	Shock waves and vibrations cause limited liquefaction, displacement, remolding, and settlement to higher density	Saturated, clean sands; partly saturated sands and silts after flooding	>40 m	Any size	Can obtain relative densities to 70 to 80%; may get variable density; time-dependent strength gain	Induce liquefaction in controlled and limited stages and increase relative density to potentially nonliquefiable range	2 3	Low ($2.00 to $4.00/$m^3$)
(2) Vibratory probe (a) Terraprobe (b) Vibrorods (c) Vibrowing	Densification by vibration; liquefaction-induced settlement and settlement in dry soil under overburden to produce a higher density	Saturated or dry clean sand; sand	20 m routinely (ineffective above 3 to 4 m depth); >30 m sometimes; Vibrowing, 40 m	>1000 m^2	Can obtain relative densities of 80% or more. Ineffective in some sands	Induce liquefaction in controlled and limited stages and increase relative density to potentially nonliquefiable range. Has been shown effective in preventing liquefaction	2 3	Moderate ($6.00 to $13.00/$m^3$)
(3) Vibrocompaction (a) Vibroflot (b) Vibro-composer system (c) Soil Vibratory stabilizing	Densification by vibration and compaction of backfill material of sand or gravel	Cohesionless soils with less than 20% fines	>30 m	>1000 m^2	Can obtain high relative densities (over 85%), good uniformity	Induce liquefaction in controlled and limited stages and increase relative densities to nonliquefiable condition. Is used extensively to prevent liquefaction. The dense columns of backfill provides (a) vertical support, (b) drains to relieve pore water pressure, and (c) shear resistance in horizontal and inclined directions. Used to stabilize slopes and strengthen potential failure surfaces or slip circles	1 2 Δ[f]	Low to moderate ($6.00 to $9.00/$m^3$)
(4) Compaction piles	Densification by displacement of pile volume and by vibration during driving, increase in lateral effective earth pressure	Loose sand soils; partly saturated clayey soils; loess	>20 m	>1000 m^2	Can obtain high densities, good uniformity. Relative densities of more than 80%	Useful in soils with fines. Increases relative densities to nonliquefiable range. Is used to prevent liquefaction. Provides shear resistance in horizontal and inclined directions. Useful to stabilize slopes and strengthen potential failure surfaces or slip circles	1 2 3	Moderate to high

662

	Principle	Most suitable soil conditions	Maximum effective treatment depth	Maximum size of treated area	Properties of treated material	Function / comment		Relative costs
(5) Heavy tamping (dynamic compaction)	Repeated application of high-intensity impacts at surface	Cohesionless soils best, other types can also be improved	30 m (possibly deeper)	>3300 m²	Can obtain high relative densities, reasonable uniformity. Relative densities of 80% or more	Suitable for some soils with fines; usable above and below water. In cohesionless soils, induces liquefaction in controlled and limited stages and increases relative density to potentially non-liquefiable range. Is used to prevent liquefaction	2 3	Low ($0.40 to $6.00/m³)
(6) Displacement/compaction grout	Highly viscous grout acts as radial hydraulic jack when pumped in under high pressure	All soils	Unlimited	Small	Grout bulbs within compressed soil matrix. Soil mass as a whole is strengthened	Increase in soil relative density and horizontal effective stress. Reduce liquefaction potential. Stabilize the ground against movement	1 2 3	Low to moderate ($3.00 to $15.00/m³)
(7) Surcharge/buttress	The weight of a surcharge/buttress increases the liquefaction resistance by increasing the effective confining pressures in the foundation	Can be placed on any soil surface	—	*Compression* >1000 m²	Increase strength and reduce compressibility	Increase the effective confining pressure in a liquefiable layer. Can be used in conjunction with vertical and horizontal drains to relieve pore water pressure. Reduce liquefaction potential. Useful to prevent movements of a structure and for slope stability	2 3	Moderate if vertical drains used
(8) Drains (a) Gravel (b) Sand (c) Wick (d) Wells (for permanent dewatering)	Relief of excess pore water pressure to prevent liquefaction. (Wick drains have comparable permeability to sand drains.) Primarily gravel drains; sand/wick may supplement gravel drain or relieve existing excess pore water pressure. Permanent dewatering with pumps	Sand, silt, clay	Gravel and sand >30 m; depth limited by vibratory equipment; wick, >45 m	*Pore Water Pressure Relief* >1500 m², any size for wick	Pore water pressure relief will prevent liquefaction	Prevent liquefaction by gravel drains. Sand and gravel drains are installed vertically; however, wick drains can be installed at any angle. Dewatering will prevent liquefaction but not seismically induced settlements	Gravel and sand 2 Δ′ Wick 1 2 3	Sand and gravel 0.3 m dia. ($11.50 to $21.50/m³); wick ($2.00 to $4.00/m³); dewatering very expensive
(9) Particulate grouting	Penetration grouting—fill soil pores with soil, cement, and/or clay	Medium to coarse sand and gravel	Unlimited	*Injection and Grouting* Small	Impervious, high strength with cement grout. Voids filled so they cannot collapse under cyclic loading	Eliminate liquefaction danger. Slope stabilization. Could potentially be used to confine an area of liquefiable soil so that liquefied soil could not flow out of the area	1 2 3	Lowest of grout methods ($3.00 to $30.00/m³)

TABLE 16.12 (*Continued.*)

Method	Principle	Most Suitable Soil Conditions/Types	Maximum Effective Treatment Depth	Economic Size of Treated Area	Ideal Properties of Treated Material[b]	Applications[c]	Case[d]	Relative Costs[e]
(10) Chemical grouting	Solutions of two or more chemicals react in soil pores to form a gel or a solid precipitate	Medium silts and coarser	Unlimited	Small	Impervious, low to high strength. Voids filled so they cannot collapse under cyclic loading	Eliminate liquefaction danger. Slope stabilization. Could potentially be used to confine an area of liquefiable soil so that liquefied soil could not flow out of the area. Good water shutoff	1 2 3	High ($75.00 to $250.00/m³)
(11) Pressure-injected lime	Penetration grouting— fill soil pores with lime	Medium to coarse sand and gravel	Unlimited	Small	Impervious to some degree. No significant strength increase. Collapse of voids under cyclic loading reduced	Reduce liquefaction potential	1 2 3	Low ($10.00/m³)
(12) Electrokinetic injection	Stabilizing chemicals moved into and fills soil pores by electroosmosis or colloids into pores by electrophoresis	Saturated sands, silts, silty clays	Unknown	Small	Increased strength, reduced compressibility, voids filled so they cannot collapse under cylic loading	Reduce liquefaction potential	1 2 3	Expensive
(13) Jet grouting	High-speed jets at depth excavate, inject, and mix a stabilizer with soil to form columns or panels	Sands, silts, clays	Unknown	Small	Solidified columns and walls	Slope stabilization by providing shear resistance in horizontal and inclined directions, which strengthens potential failure surfaces or slip circles. A wall could be used to confine an area of liquefiable soil so that liquefied soil could not flow out of the area	1 2 3	High ($250.00 to $650.00/m³)
Admixture Stabilization								
(14) Mix-in-place piles and walls	Lime, cement, or asphalt introduced through rotating auger or special inplace mixer	Sand, silts, clays, all soft or loose inorganic soils	>20 m (60 m obtained in Japan)	Small	Solidified soil piles or walls of relatively high strength	Slope stabilization by providing shear resistance in horizontal and inclined directions, which strengthens potential failure surfaces or slip circles. A wall could be used to confine an area of liquefiable soil so that liquefied soil could not flow out of the area	1 2 3	High ($250.00 to $650.00/m³)

(15) In-situ vitrification	Melts soil in place to create an obsidian-like vitreous material	All soils and rock	> 30 m	*Thermal Stabilization* Unknown	Solidified soil piles or walls of high strength. Impervious; more durable than granite or marble; compressive strength, 9 to 11 ksi; splitting tensile strength, 1 to 2 ksi	Slope stabilization by providing shear resistance in horizontal and inclined directions, which strengthens potential failure surfaces or slip circles. A wall could be used to confine an area of liquefiable soil so that liquefied soil could not flow out of the area	1 2 3	Moderate ($53.00 to $70.00/m³)
(16) Vibro-replacement stone and sand columns (a) Grouted (b) Not grouted	Hole jetted into fine-grained soil and backfilled with densely compacted gravel or sand hole formed in cohesionless soils by vibro techniques and compaction of backfilled gravel or sand. For grouted columns, voids filled with a grout	Sands, silts, clays	> 30 m (limited by vibratory equipment)	*Soil Reinforcement* > 1500 m²; fine-grained soils, > 1000 m²	Increased vertical and horizontal load-carrying capacity. Density increase in cohesionless soils. Shorter drainage paths	Provides (a) vertical support, (b) drains to relieve pore water pressure, and (c) shear resistance in horizontal and inclined directions. Used to stabilize slopes and strengthen potential failure surfaces or slip circles. For grouted columns no drainage provided but increased shear resistance. In cohesionless soil, density increase reduces liquefaction potential	1 2 Δ[f]	Moderate ($11.00 to $70.00/m³)
(17) Root piles, soil nailing	Small-diameter inclusions used to carry tension, shear, compression	All soils	Unknown	Unknown	Reinforced zone of soil behaves as a coherent mass	Slope stability by providing shear resistance in horizontal and inclined directions to strengthen potential failure surfaces or slip circles. Both vertical and angled placement of the piles and nails	1 2 3	Moderate to high

[a] After Ledbetter (1985).

[b] SP, SW, or SM soils that have average relative density equal to or greater than 85 percent and the minimum relative density not less than 80 percent are in general not susceptible to liquefaction (TM 5-818-1). D'Appolonia (1970) stated that for soil within the zone of influence and confinement of the structure foundation, the relative density should not be less than 70 percent. Therefore, a criteria may be used that relative density increase into the 70–90 percent range is in general considered to prevent liquefaction. These properties of treated materials and applications occur *only under ideal conditions* of soil, moisture, and method application. The methods and properties achieved are not applicable and will not occur in all soils.

[c] Applications and results of the improvement methods are dependent on: (a) soil profiles, types, and conditions, (b) site conditions, (c) earthquake loading, (d) structure type and condition, and (e) material and equipment availability. Combinations of the methods will most likely provide the best and most stable solution.

[d] Site conditions have been classified into three cases. Case 1 is for beneath structures, Case 2 is for the not-underwater free field adjacent to a structure, and Case 3 is for the underwater free field adjacent to a structure.

[e] The costs will vary depending on: (a) site working conditions, location, and environment, (b) the location, area, depth, and volume of soil involved, (c) soil type and properties, (d) materials (sand, gravel, admixtures, etc.), equipment, and skills available, and (e) environmental impact factors. The costs are average values based on: (a) verbal communication from companies providing the service, (b) current literature, and (c) literature reported costs updated for inflation.

[f] Δ means the method has potential use for Case 3 with special techniques required that would increase the cost.

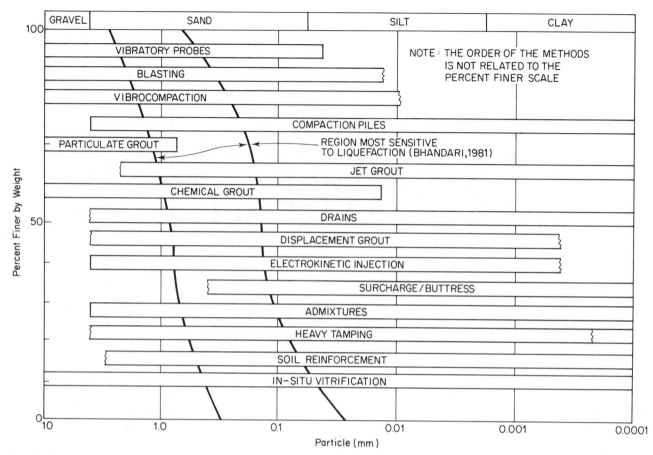

Fig. 16.89 Applicable grain size ranges for soil improvement methods. Jagged lines at the ends of a bar indicate the uncertainty of applicability of the method. (*After Ledbetter, 1985.*)

ACKNOWLEDGMENT

This summary section is respectfully dedicated to the late professors Drs. K. L. Lee and H. B. Seed, for their pioneering work in this field.

REFERENCES TO PART I

Ambraseys, N. and Sarma, S. (1969), Liquefaction of soils induced by earthquakes, *Bulletin of the Seismological Society of America*, **59**, No. 2, pp. 651–664.

Blume, J. A. (1965), Earthquake ground motion and engineering procedures for important installations near active faults, *Proceedings, 3rd World Conference on Earthquake Engineering*, New Zealand, **III**.

Butler, D. W., Muto, K., and Minami, K. (1949), Engineering report on the Fukui earthquake, Office of the Engineer, General Headquarters, Far East Command, Tokyo.

Close, U. and McCormick, E. (1922), Where the mountains walked, *The National Geographic Magazine*, **XLI**, No. 5.

Cloud, W. K. (1967), Intensity map and structural damage, Parkfield, California, earthquake of June 27, 1966, *Bulletin of the Seismological Society of America*, **57**, No. 6, pp. 1161–1179.

Clough, R. W. and Chopra, A. A. (1966), Earthquake stress analysis in earth dams, *Journal of the Engineering Mechanics Division, ASCE*, **92**, No. EM-2, Proc. Paper 4793, pp. 197–212.

Collins, J. J. and Foster, H. L. (1949), The Fukui earthquake, Hokuriku Region, Japan, 28 June 1948, *Geology*, **1**, Office of the Engineer, General Headquarters, Far East Command, February.

Coulter, H. W. and Migliaccio, R. R. (1966), Effects of the earthquake of March 27, 1964 at Valdez, Alaska, *Geological Survey Professional Paper 542-C*, U.S. Department of the Interior.

Davis, S. and Karzulovich, J. K. (1961), Deslizamientos en el valle del rio San Pedro Provincia de Valdivia Chiles, *Publication No. 20, Anales de la Faculted de Ciencieas Fiscal y Mathematicos*, University of Chile, Institute of Geology, Santiago.

de Cossio, R. D. (1960), Foundation failures during the Coatzacoalcos (Mexico) earthquake of 26 August 1959, *Proceedings, Second World Conference on Earthquake Engineering*, Tokyo, Japan.

Dezfulian, Houshang and Seed, H. Bolton (1969a), *Response of Nonuniform Soil Deposits to Traveling Seismic Waves*, Research Report, Geotechnical Engineering, University of California, Berkeley.

Dezfulian, Houshang and Seed, H. Bolton (1969b), *Seismic Response of Soil Deposits Underlain by Sloping Rock Boundaries*, Research Report, Geotechnical Engineering, University of California, Berkeley, June.

Dibaj, M. and Penzien, J. (1969a), Responses of earth dams to traveling seismic waves, *Journal of the Soil Mechanics and Foundations Division, ASCE*, **95**, No. SM-2, Proc. Paper 6453, pp. 541–560.

Dibaj, M. and Penzien, J. (1969b), *Nonlinear Seismic Response of Earth Structures*, Report No. EERC 69-2, University of California, Berkeley.

Dobry, R. and Alvarez, L. (1967), Seismic failures of Chilean tailing dams, *Journal of the Soil Mechanics and Foundations Division, ASCE*, **93**, No. SM-6, Proc. Paper 5582, pp. 237–260.

Donovan, N. C. and Matthiesen, R. B. (1968), Effects of site conditions on ground motions during earthquakes, State-of-the Art Symposium, EERI, San Francisco, California, February.

Duke, C. M. (1958), Effects of ground on destructiveness of large earthquakes, *Journal of the Soil Mechanics and Foundations Division, ASCE*, **84**, No. SM-3.

Duke, C. M. and Leeds, D. J. (1963), Response of soils, foundations and earth structures to the Chilean earthquakes of 1960, *Bulletin of the Seismological Society of America*, **53**, No. 2.

Dutton, C. E. (1889), *The Charleston Earthquake of August 31 1886*, Ninth Annual Report of the U.S. Geological Survey 1887–88, Washington.

Esteva, L., Rascon, O. A., and Gutieviez, A. (1969), Lessons from some recent earthquakes in Latin America, *Proceedings, Fourth World Conference on Earthquake Engineering*, Santiago, Chile.

Florin, V. A. and Ivanov, R. L. (1961), Liquefaction of saturated sandy soils, *Proceedings, 5th International Conference on Soil Mechanics and Foundation Engineering*, Paris, France.

Fuller, M. L. (1912), The new Madrid earthquake, *Bulletin 494*, U.S. Geological Survey, Department of the Interior, Washington, D.C.

Grantz, A., Plafker, G., and Kachedoorian, R. (1964), *Alaska's Good Friday Earthquake, March 27, 1964*, Geol. Survey, U.S. Dept of the Interior, Circular 491, Washington, D.C.

Gubin, I. E. (1960), *Regime of Seismicity on the Territory of Tadijikistan*, Academy of Sciences Press, U.S.S.R.

Gutenberg, B. (1957), The effects of ground on earthquake motion, *Bulletin of the Seismological Society of America*, **47**, No. 3, pp. 221–251.

Hansen, W. R. (1965), Effects of earthquake of March 27, 1964 at Anchorage, Alaska, *Geological Survey Professional Paper 542-A*, U.S. Department of the Interior, Washington, D.C.

Herrera, I. and Rosenblueth, E. (1965), Response spectra on stratified soil, *Proceedings, 3rd World Conference on Earthquake Engineering*, New Zealand.

Hisada, T., Nakagawa, K., and Izumi, M. (1965), Normalized acceleration spectra for earthquakes recorded by strong motions accelerographs and their characteristics related with subsoil conditions, *BRI Occasional Report No. 23*, Building Research Institute, Ministry of Construction, Tokyo, Japan.

Housner, G. W. (1952), Spectrum intensities of strong-motion earthquakes, *Proceedings, Symposium on Earthquake and Blast Effects on Structures*, Earthquake Engineering Research Institute, pp. 20–36.

Housner, G. W. (1954), Geotechnical problems of destructive earthquakes, *Geotechnique*, December.

Housner, G. W. (1958), The mechanism of sand blows, *Bulletin of the Seismological Society of America*, **48**, No. 2, pp. 155–161.

Housner, G. W. (1959), Behavior of structures during earthquakes, *Journal of the Engineering Mechanics Division, ASCE*, **85**, No. EM-4.

Hovey, E. O. (1902), Martinique and St. Vincent, a preliminary report upon the eruptions of 1902, *American Museum of Natural History Bulletin*, **16**, p. 340.

Hudson, D. E. (1956), Response spectrum techniques in engineering seismology, *Proceedings, First World Conference on Earthquake Engineering*, Berkeley, Calif.

Idriss, I. M. (1968), Finite element analysis for the seismic response of earth dams, *Journal of the Soil Mechanics and Foundations Division, ASCE*, **94**, No. SM-3, Proc. Paper 5929, pp. 617–636.

Idriss, I. M. and Seed, H. B. (1967), Response of earth banks during earthquakes, *Journal of the Soil Mechanics and Foundations Division, ASCE*, **93**, No. SM-3, pp. 61–82.

Idriss, I. M. and Seed, H. Bolton (1968a), Seismic response of horizontal soil layers, *Journal of the Soil Mechanics and Foundations Division, ASCE*, **94**, No. SM-4, pp. 1003–1031.

Idriss, I. M. and Seed, H. Bolton (1968b), An analysis of ground motions during the 1957 San Francisco earthquake, *Bulletin of the Seismological Society of America*, **58**, No. 6, pp. 2013–2032.

Idriss, I. M., Dezfulian, H., and Seed, H. B. (1969), *Computer Programs for Evaluating the Seismic Response of Soil Deposits with Nonlinear Characteristics using Equivalent Linear Procedures*, Research Report, Geotechnical Engineering, University of California, Berkeley.

Idriss, I. M. and Seed, H. Bolton (1969a), A variable damping solution for the evaluation of earthquake response of horizontal soil deposits, University of California, Berkeley.

Idriss, I. M. and Seed, H. Bolton (1969b), Influence of geometry and material properties on the seismic response of soil deposits, *Proceedings, 4th World Conference on Earthquake Engineering*, Santiago, Chile.

Kanai, K. (1950), The effect of solid viscosity of surface layer on the earthquake movements, *Bulletin of the Earthquake Research Institute*, **28**, pp. 31.

Kanai, K. (1951), Relation between the nature of surface layer and the amplitudes of earthquake motions, *Bulletin of the Earthquake Research Institute*, **29**.

Kanai, K., Tanaka, T., and Osada, K. (1954), Measurement of the microtremor, I, *Bulletin of the Earthquake Research Institute*, **32**, Part 2, pp. 199–209.

Kanai, K., Tanaka, T., and Yoshizawa, S. (1959), Comparative studies of earthquake motions on the ground and underground (Multiple reflection problem), *Bulletin of the Earthquake Research Institute*, **37**, Part 1, pp. 53–88.

Kawakami, F. and Asada, A. (1966), Damage to the ground and earth structures by the Niigata earthquake of June 16, 1964, *Soils and Foundations*, **6**, No. 1, pp. 14–30.

Kishida, H. (1965), Damage of reinforced concrete buildings in Niigata City with special reference to foundation engineering, *Soil and Foundation Engineering*, **6**, No. 1, Tokyo, Japan.

Kishida, H. (1969), Characteristics of liquefied sands during Mino-Owari, Tohnankai, and Fukui earthquakes, *Soils and Foundations*, **IX**, No. 1, pp. 75–92.

Kobayashi, H. and Kagami, H. (1966), A numerical analysis of the propagation of shear waves in multilayered ground, *Proceedings, Japan Earthquake Engineering Symposium*, Tokyo, Japan, pp. 15–20.

Koizumi, Yasunori (1966), Changes in density of sand subsoil caused by the Niigata earthquake, *Soils and Foundations*, **VI**, No. 2, pp. 38–44.

Kovacs, William, D., Seed, H. Bolton, and Idriss, Izzat M. (1969), *An Experimental Study of the Response of Clay Banks*, Research Report, Geotechnical Engineering, University of California, Berkeley.

Lawson, A. C. et al. (1908), *The California Earthquake of April 18, 1906*, Publication 87, Carnegie Institute, Washington.

Lee, K. L. and Seed, H. B. (1967), Cyclic stress conditions causing liquefaction of sand, *Journal of the Soil Mechanics and Foundations Division, ASCE*, **93**, No. SM-1, pp. 47–70.

Lee, Kenneth L. and Fitton, John A. (1968), Factors affecting the dynamic strength of soil, A Symposium presented at the 71st Annual Meeting, ASTM Vibration Effects of Earthquakes on Soils and Foundations, *ASTM STP 450*, San Francisco, June 23–28, 1968.

Lyell, C. (1822), *Principles of Geology*, Vol. 2, London.

Lyell, Charles (1849), *A Second Visit to the United States of North America*, Vol. 2, London, pp. 228–229.

Marinatos, S. N. (1960), Helice–Submerged town of classical Greece, *Archaeology*, **13**, No. 3.

Marsal, R. J. (1961), Behavior of a sandy uniform soil during the Jaltipan Earthquake, Mexico, *Proceedings, Fifth International Conference on Soil Mechanics and Foundation Engineering*, Paris, France.

Maslov, N. N. (1957), Questions of seismic stability of submerged sandy foundations and structures, *Proceedings, 4th International Conference on Soil Mechanics and Foundation Engineering*, London, England.

Matthiesen, R. B., Duke, C. M., Leeds, D. J., and Fraser, J. C. (1964), Site characteristics of southern California strong-motion earthquake stations, Part two, *Report No. 64-15*, Dept. of Engineering, University of California, Los Angeles.

McCulloch, D. S. (1966), Slide induced waves, seiching, and ground fracturing caused by the earthquake of March 27, 1964, at Kenai Lake, Alaska, *Geological Survey Professional Paper, 543-A*.

Merino y Coronado, J. (1957), El temblor del 28 de julio de 1957, *Anales Inst. de Geofis*, University of Mexico, **3**.

Nunnally, S. W. (1966), Development of a liquefaction index for cohesionless soils, Ph.D. Thesis, Northwestern University.

Ohsaki, Yorihiko (1966), Niigata earthquakes, 1964 building damage and soil condition, *Soils and Foundations*, **VI**, No. 2, pp. 14–37.

Ohsaki, Y. (1969), The effects of local soil conditions upon earthquake damage, Soil Dynamics Specialty Conference, Foundation Engineering, Preprints of State of the Art Lectures and Research Briefs, Mexico City.

Oldham, T. (1882), The Cachar earthquake of 10 January 1869, *Memoirs of the Geological Survey of India*, **19**.

Oldham, R. D. (1899), Report of the great earthquake of 12 June 1897, *Memoirs of the Geological Survey of India*, **29**.

Oldham, T. and Mallett, R. (1872), Secondary effects of the earthquake of 10 January 1869 in Cachar, Geologic Survey (London) *Quarterly Journal*, **28**.

Peacock, W. H. and Seed, H. B. (1968), Sand liquefaction under cyclic loading simple shear conditions, *Journal of the Soil Mechanics and*

Foundations Division, ASCE, **94**, No. SM-3, Proc. Paper 5957, pp. 689–708.

Penzien, J., Scheffey, C. F., and Parmelee, R. A. (1964), Seismic analysis of bridges on long piles, *Journal of the Engineering Mechanics Division*, ASCE, **90**, No. EM-3, pp. 223–254.

Prakash, S. and Mathur, J. N. (1965), Liquefaction of fine sand under dynamic loads, *Proceedings, 5th Symposium of the Civil and Hydraulic Engineering Dept.*, Indian Inst. of Science, Bangalore, India.

Puchkov, S. (1962), Correlation between the velocity of seismic oscillations of particles and the liquefaction phenomenon of water saturated sand, *Prob. Inz, Seism., Trudi No. 21*, Inst. Fiziki Zemli, Moscow (also *Problems in Engineering Seismology*, translated by Consultants Bureau, N.Y., p. 92).

Retamal, E. and Kausel, E. (1969), Vibratory compaction of the soil and tectonic subsidence during the 1960 earthquake in Valdivia, Chile, *Proceedings, Fourth World Conference on Earthquake Engineering*, Santiago, Chile.

Richter, C. F. (1958), *Elementary Seismology*, W. H. Freeman and Co., San Francisco.

Rosenblueth, E. (1960), The earthquake of 28 July 1957 in Mexico City, *Proceedings, 2nd World Conference on Earthquake Engineering*, Japan, I.

Ross, G. A., Seed, H. B., and Migliaccio, Ralph R. (1969), Bridge foundation behavior in Alaska earthquake, *Journal of the Soil Mechanics and Foundations Division*, ASCE, **94**, No. SM-4, Proc. Paper 6664, pp. 1007–1036.

Roy, S. C. (1939), The Bihar-Nepal earthquake of 1934: Seismometric study, *Memoirs of the Geological Survey of India*, **73**.

Seed, H. B. (1967), Slope stability during earthquakes, *Journal of the Soil Mechanics and Foundations Division*, ASCE, **93**, No. SM-4, Proc. Paper 5319, pp. 299–323.

Seed, H. B. (1968), Landslides during earthquakes due to liquefaction, *Journal of the Soil Mechanics and Foundations Division*, ASCE, **94**, No. SM-5, Proc. Paper 6110, pp. 1053–1122.

Seed, H. B. and Idriss, I. M. (1967), Analysis of soil liquefaction: Niigata earthquake, *Journal of the Soil Mechanics and Foundations Division*, ASCE, **93**, No. SM-3, pp. 83–108.

Seed, H. B. and Wilson, S. D. (1967), The Turnagain Heights landslide, Anchorage, Alaska, *Journal of the Soil Mechanics and Foundations Division*, ASCE, **93**, No. SM-4, Paper 5320, pp. 325–353.

Seed, H. B., Idriss, I. M., and Kiefer, L. W. (1968a), Characteristics of rock motions during earthquakes, *EERC 68–5*, University of California, Berkeley, September.

Seed, H. B., Lee, K. L., and Idriss, I. M. (1968b), An analysis of the Sheffield Dam failure. Report No. TE 68–2 to State of California Department of Water Resources, University of California, Berkeley.

Seed, H. B. and Idriss, I. M. (1969a), Influence of soil conditions on ground motions during earthquakes, *Journal of the Soil Mechanics and Foundations Division*, ASCE, **95**, No. SM-1, pp. 99–137.

Seed, H. B. and Idriss, I. M. (1969b), Analyses of ground motions at Union Bay, Seattle, during earthquakes and distant nuclear blasts, University of California, Berkeley.

Seed, H. B., Lee, K. L., and Idriss, I. M. (1969), An analysis of the Sheffield Dam failure, *Journal of the Soil Mechanics and Foundations Division*, ASCE, **95**, No. SM-6, Proc. Paper 6906, pp. 1453–1490.

Seed, H. B. and Peacock, W. H. (1971), Test procedures for measuring soil liquefaction characteristics, *Journal of the Soil Mechanics and Foundations Division*, ASCE, **97**, No. SM-8, Paper 8330, pp. 1099–1119.

Seed, H. B. and Silver, M. L. (1972), Settlement of dry sands during earthquakes, *Journal of the Soil Mechanics and Foundations Division*, ASCE, **98**, No. SM-4, Paper 8844, pp. 381–397.

Shannon and Wilson (1964), *Report on Anchorage Area Soil Studies*, Alaska, Report to the U.S. Army Engineer District, Anchorage, Alaska.

Tarr, R. S. and Martin, L. (1912), The earthquakes at Yakutats Bay, Alaska, in September 1899, *Geologic Survey Professional Paper 69*, U.S. Department of the Interior, Washington, D.C.

Tsuya, H. (1950), *The Fujui Earthquake of June 28, 1948*, Report of the Special Committee for the Study of the Fukui Earthquake, Tokyo.

Valera, Julio E. (1968), Seismic interaction of granular soils and rigid retaining structures, Ph.D. Thesis, University of California, Berkeley.

Wakimizu, T. (1924), *Kwaguku-Tisiki (Scientific Knowledge)*, **IV**, No. IV, Tokyo, April (in Japanese).

Whitman, Robert V. (1969), The current status of soil dynamics, *Applied Mechanics Review*, pp. 1–8.

Whitman, R. V. and de Pablo, P. O. (1969), Densification of sand by vertical vibrations, *Proceedings, Fourth World Conference on Earthquake Engineering*, Santiago, Chile.

Wiggins, J. H. (1964), Effects of site conditions on earthquake intensity, *Journal of the Structural Division*, ASCE, **90**, No. ST-2, pp. 279–313.

Wilson, E. L. (1968), A computer program for the dynamic stress analysis of underground structures, *Report No. 68-1*, Structural Engineering Laboratory, University of California, Berkeley.

Wood, H. O. (1908), Distribution of apparent intensity in San Francisco, *The California Earthquake of April 18, 1906*, Report of the State Earthquake Investigation Commission, Carnegie Institution of Washington, Washington, D.C., pp. 220–245.

Wood, H. O. (1933), Preliminary report on the Long Beach earthquake, *Bulletin of the Seismological Society of America*, **23**.

Wood, H. O. and Heck, N. H. (1966), *Earthquake History of the United States*, U.S. Department of Commerce, Washington, D.C.

Yamada, G. (1966), Damage to earth structures and foundations by the Niigata earthquake, June 16, 1964, *Soils and Foundations*, **6**, No. 1, pp. 1–13.

Yokomura, S. (1966), The damage to river dykes and related structures caused by the Niigata earthquake, *Soils and Foundations*, **6**, No. 1, pp. 38–53.

Yoshimi, Y. (1967), An experimental study of liquefaction of saturated sands, *Soils and Foundations*, **7**, No. 2.

Zeevaert, L. (1963), The effect of earthquakes in soft subsoil conditions, *Proceedings, 30th Annual Convention, Structural Engineers Association of California*, pp. 74–82.

REFERENCES TO PART II

Ambraseys, N. and Sarma, S. (1969), Liquefaction of soils induced by earthquakes, *Bulletin of the Seismological Society of America*, **59**, No. 2, April, pp. 651–664.

American Society for Testing and Materials—ASTM (1979), *D3441—Standard Method for Deep Quasi-Static, Cone and Friction Cone Penetration Tests of Soils*, ASTM, Philadelphia, Pa.

American Society for Testing and Materials—ASTM (1984), *D1586—Standard Method for Penetration Test and Split-Barrel Sampling of Soils*, ASTM, Philadelphia, Pa.

Anderson, A. and Bjerrum, L. (1967), Slides in subaqueous slopes in loose sand and silt, *Marine Geotechnique*, ed. A. F. Richards, University of Illinois Press, Urbana, Ill., pp. 221–239.

Anderson, K. H., Pool, J. H., Brown, S. F., and Rosenbrand, W. F. (1980), Cyclic and statis laboratory tests on drammen clay, *Journal of the Geotechnical Engineering Division*, ASCE, **106**, GT-5, pp. 499–529.

Ansal, A. M. and Erken, A. (1989), Undrained behavior of clay under cyclic shear stresses, *Journal of the Geotechnical Engineering Division*, ASCE, **115**, No. 7, pp. 968–983.

Arulanandan, K. (1977), Method and Apparatus for Measuring In-Situ Density and Fabric of Soils, patent application, Regents of the University of California, Berkeley, Calif.

Arulanandan, K. and Muraleetharan, K. K. (1988a), Level ground soil-liquefaction analysis using in situ properties: I, *Journal of the Geotechnical Engineering Division*, ASCE, **114**, No. 7, pp. 753–770.

Arulanandan, K. and Muraleetharan, K. K. (1988b), Level ground soil-liquefaction analysis using in situ properties: II, *Journal of the Geotechnical Engineering Division*, ASCE, **114**, No. 7, pp. 771–790.

Arulmoli, K., Arulanandan, K., and Seed, H. B. (1985), New method for evaluating liquefaction potential, *Journal of Geotechnical Engineering*, ASCE, **111**, No. 1, pp. 95–114.

Athanasopoulos, G. A. and Richart, F. E. (1983a), Effect of creep on shear modulus of clays, *Journal of Geotechnical Engineering*, ASCE, **109**, No. 10, pp. 1217–1232.

Athanasopoulos, G. A. and Richart, F. E. (1983b), Effect of stress release on shear modulus of clays, *Journal of Geotechnical Engineering*, ASCE, **109**, No. 10, pp. 1233–1245.

Azzouz, A. S., Malek, A. M., and Baligh, M. M. (1989), Cyclic behavior of clays under undrained simple shear, *Journal of the Geotechnical Engineering Division*, ASCE, **112**, No. 6, pp. 579–593.

Baldi, G., Bellotti, R., Ghionna, V., Jamiolkowski, M., and Pasqualini, E.

(1982). Design parameters for sands from CPT, *Proceedings of the Second European Symposium on Penetration Testing, ESOPT 11*, Amsterdam, The Netherlands, **2**, pp. 425–438.

Baldi, G., Bellotti, R., Ghionna, V., Jamiolkowski, M., and Pasqualini, E. (1985), Penetration resistance and liquefaction of sands, *Proceedings of the Eleventh International Conference on Soil Mechanics and Foundation Engineering*, A. A. Balkema, Rotterdam.

Bhandari, R. K. M. (1981), Dynamic consolidation of liquefiable sands, *Proceedings of the International Conference on Recent Advances in Geotechnical Earthquake Engineering and Soil Dynamics*, University of Missouri, Rolla, Miss., **2**, pp. 857–860.

Bierschwale, J. G. and Stokoe II, K. K. (1984), *Analytical Evaluation of Liquefaction Potential of Sands Subjected to the 1981 Westmoreland Earthquake*, Geotechnical Engineering Report GR-84-15, Civil Engineering Department, University of Texas, Austin, Texas.

Bouckovalas, G., Marr, W. A., and Christian, J. T. (1986), Analyzing permanent drift due to cyclic loads, *Journal of the Geotechnical Engineering Division, ASCE*, **112**, No. 6, pp. 579–593.

Brodek, R. (1952), *Record of Activities Following Earthquake of July 21, 1952*: Report to Kern County Land Co. (unpublished manuscript).

Building Research Institute (1965), *Niigata Earthquake and Damage To Reinforced Concrete Buildings in Niigata City*, Report, Building Research Institute, Ministry of Construction, Vol. 42 (in Japanese).

Byrne, P. M. (1976), An evaluation of the liquefaction potential of the Frazer Delta, *Canadian Geotechnical Journal*, **15**, No. 1, pp. 32–46.

Campanella, R. G. and Robertson, P. K. (1983), Flat plate dilatometer testing: Research and development at UBC, *First International Conference on Flat Plate Dilatometer*, Edmonton, Alberta, Canada.

Carnegie Institution of Washington (1908), Minor Geological Effects of the Earthquake, *The California Earthquake of April 18, 1906*, Report of the State Earthquake Investigation Commission, Vol. 1, Part II.

Casagrande, A. (1976), *Liquefaction and Cyclic Deformation of Sands— A Critical Review*, Harvard Soil Mechanics Series No. 88.

Castro, G. (1975), Liquefaction and cyclic mobility of saturated sands, *Journal of the Geotechnical Engineering Division, ASCE*, **101**, GT-6, pp. 551–569.

Castro, G. and Poulos, S. J. (1977), Factors affecting liquefaction and cyclic mobility, *Journal of the Geotechnical Engineering Division, ASCE*, **103**, GT-6, pp. 501–516.

Chaney, R. C. (1978), Saturation effects on the cyclic strengths of sand, *Proceedings, Speciality Conference on Earthquake Engineering and Soil Dynamics, ASCE*, Pasadena, Calif., **1**, pp. 342–358.

Chaney, R. C. (1979), Earthquake induced deformations of earth dams, *Proceedings, U.S. National Conference on Earthquake Engineering, EERI*, pp. 632–642.

Chaney, R. C. (1980), Seismically induced deformations in earthdams, *7th World Conference on Earthquake Engineering*, Istanbul, Turkey, **31**, pp. 483–486.

Chaney, R. C. (1984), Methods of predicting the deformation of the seabed due to cyclic loading, *Proceedings of Symposium on Seabed Mechanics*, International Union of Theoretical and Applied Mechanics, University of Newcastle, pp. 159–167.

Chaney, R. C. (1985), Methods of evaluating the performance of slopes under seismic loading, *Proceedings*, Discussion Session 7B—Strength Evaluation For Stability Analysis, *XI International Conference on Soil Mechanics and Foundation Engineering*, San Francisco, Calif.

Chaney, R. C. (1987), Cyclic response of soils to varying saturation, *Proceedings, International Symposium on Environmental Geotechnology*, Envo Publishing Co., Inc., Bethlehem, Pa., **2**, pp. 181–205.

Chaney, R. C., Stevens, E., Sheth, N., and Hencey, E. (1980), Effect of stress concentration on the liquefaction behavior of sand, *Geotechnical Testing Journal, ASTM*, **3**, pp. 97–104.

Chaney, R. C. and Fang, H. Y. (1985a), Liquefaction in the coastal environment: An analysis of case histories, *2nd Shanghai Symposium on Marine Geotechnology and Nearshore/Offshore Structures*, Shanghai, China, Tongji University Press, Shanghai, pp. 32–64.

Chaney, R. C. and Fang, H. Y. (1985b), Static and dynamic properties of marine sediments, *Proceedings of Symposium on Marine Geotechnology and Nearshore/Offshore Structures*, Shanghai, China, *ASTM STP 923*, pp. 74–111.

Christian, J. T. and Swiger, W. F. (1975), Statistics of liquefaction and SPT results, *Journal of the Geotechnical Engineering Division, ASCE*, **101**, No. GT-11, pp. 1135–1150.

Cortright, C. J. (1975), Effect of the San Fernando earthquake on the Van Norman Reservoir Complex, *Bulletin 196*, California Division of Mines and Geology, Sacramento, Calif., pp. 395–406.

Coulter, H. W. and Migliaccio, R. F. (1966), Effects of the earthquake of March 27, 1964 at Valdez, Alaska, *Geological Survey Professional Paper 542-C*, U.S. Department of the Interior, Washington, D.C.

Cuny, R. W. and Fry, Z. B. (1973), Vibratory in-situ and laboratory soil moduli compared, *Journal of the Geotechnical Engineering Division, ASCE*, **99**, No. 12, pp. 1055–1076.

Dalrymple, R. W. (1979), Wave-induced liquefaction: A modern example from the Bay of Fundy, *Sedimentology*, **26**, pp. 835–844.

Dalrymple, R. W. (1980), Wave induced liquefaction: An addendum, *Sedimentology*, **27**, pp. 461.

D'Appolonia, E. (1970), Dynamic loading, *Journal of the Soil Mechanics and Foundations Division, ASCE*, **96**, No. SM-1, pp. 49–72.

Davis, S. N. and Karzulovic, J. (1963), Landslides at Lago Rinihue, Chile, *Bulletin of the Seismological Society of America*, **53**, No. 6, pp. 1403–1414.

Dobry, R. et al. (1982), *Prediction of Porewater Pressure Buildup and Liquefaction of Sands During Earthquakes by the Cyclic Strain Method*, NBS Building Science Series 138, U.S. Dept. of Commerce, National Bureau of Standards.

Douglas, B. J., Olsden, R. S., and Martin, G. R. (1981), Evaluation of the cone penetrometer test for SPT liquefaction assessment, *Geotechnical Engineering Division, ASCE National Convention*, St. Louis, Mo., Session No. 24.

Duke, C. M. and Leeds, D. J. (1963), Response of soils, foundations and earth structures to the Chilean earthquake of 1960, *Bulletin of the Seismological Society of America*, **53**, No. 2, pp. 309–357.

Dyvik, R., Zimmie, T. F., and Schimelfenyg, P. (1983), *Cyclic Simple Shear Behavior of Fine Grained Soils*, Norwegian Geotechnical Institute, Publication No. 149.

Earthquake Engineering Research Institute (1975), *Engineering Aspects of the Lima, Peru Earthquake of October 3, 1974*, Report prepared by EERI Reconnaissance Team, May.

Earthquake Engineering Research Institute (1986), *Reducing Earthquake Hazards: Lessions Learned From Earthquakes*, Publication 86–02, El Cerrito, Calif.

Egan, J. A. and Sangrey, D. A. (1978), A critical state model for cyclic load pore pressure, *Proceedings, ASCE Special Conference on Earthquake Engineering and Soil Dynamics*, Pasadena, Calif., **1**, pp. 411–424.

Fang, H. Y., Chaney, R. C., and Pandit, N. S. (1981), Dynamic shear modulus of soft silt, *International Conference on Recent Advances in Geotechnical Earthquake Engineering and Soil Dynamics*, University of Missouri, Rolla, Mo., pp. 575–580.

Fang, H. Y., Mikroudis, G. K., and Zheng, H. (1989), Use of multi-domain knowledge-based expert system for planning, analysis and design of highrise buildings, *Proceedings, International Conference on Highrise Buildings*, Nanjing, China, **1**, pp. 169–174.

Finn, W. D. L. and Lee, M. K. W. (1979), Seafloor stability under seismic and wave loading, *Proceedings, Soil Dynamics in the Marine Environment, ASCE*, Reprint 3604.

Fukuoka, M. (1966), Damage to civil engineering structures, *Soils and Foundations*, **6**, No. 2, pp. 45–52.

Goulois, A. M., Whitman, R. V., and Hoeg, K. (1985), Effects of sustained shear stresses on the cyclic degradation of clay, *Proceedings of Symposium on Strength Testing of Marine Sediments, ASTM STP 883*, pp. 336–351.

Hansen, W. R. (1965), Effects of the earthquake of March 27, 1964 at Anchorage, Alaska, *Geological Survey Paper 542-A*, U.S. Department of the Interior, Washington, D.C.

Hardin, B. O. and Drnevich, V. P. (1972), Shear modulus and damping in soils: Design equations and curves, *Journal of the Soil Mechanics and Foundations Division, ASCE*, **98**, No. 7, pp. 667–692.

Hayashi, S. and Katagama, T. (1970), Damage to harbor structures by the Tokachioki earthquake, *Soils and Foundations*, **10**, No. 2, pp. 83–102.

Hayashi, S., Kubo, K., and Nakase, A. (1966), Damage to harbor structures by the Niigata earthquake, *Soils and Foundations*, **6**, No. 1, pp. 89–112.

Henkel, D. J. (1970), The role of waves in causing submarine landslides, *Geotechnique*, **20**, No. 1, pp. 75–80.

Holish, L. L. and Hendron, D. H. (1975), Liquefaction considerations for two submerged essential service cooling ponds, *1975 Structural*

Design of Nuclear Power Plant Facilities, Vol. 1-B, ASCE, New York, N.Y., pp. 887–931.

Hyde, A. F. L. and Brown, S. F. (1976), The plastic deformation of silty clay under creep and repeated loading, *Geotechnique*, **26**, No. 1, pp. 173–184.

Idriss, I. M., Dobry, R., and Singh, R. D. (1978), Nonlinear behavior of soft clays during cyclic loading, *Journal of the Geotechnical Engineering Division*, ASCE, **104**, No. 12, pp. 1427–1447.

Isenhower, W. M. and Stokoe II, K. H. (1981), Strain-rate dependent shear modulus of San Francisco Bay mud, *International Conference on Recent Advances in Geotechnical Earthquake Engineering and Soil Dynamics*, University of Missouri, Rolla, Mo., pp. 597–602.

Ishihara, K. (1974), Liquefaction of subsurface soils during earthquakes, *Technocrat*, **7**, No. 3, pp. 1–31.

Ishihara, K. and Yasuda, S. (1980), Cyclic strengths of undisturbed cohesive soils of western Tokyo, *International Symposium on Soils under Cyclic and Transient Loading*, Swansea, A. A. Balkema, Rotterdam, pp. 57–66.

Japan National Committee on Earthquake Engineering (1965), Niigata earthquake of 1964, *Proceedings, The Third World Conference on Earthquake Engineering*, Auckland and Wellington, New Zealand, **3**, pp. S-78–S-108.

Jordan, K. (1984), Modelling sand boils observed during the November 1980 Humboldt earthquake, Thesis presented in partial fulfillment of the requirements for degree of Bachelor of Science in Environmental Resources Engineering, Humboldt State University, Arcata, Calif.

Kawakami, F. and Asada, A. (1966), Damage to the ground and earth structures by the Niigata earthquake of June 16, 1964, *Soils and Foundations*, **6**, No. 1, pp. 14–30.

Kishida, H. (1969), Characteristics of liquefied sands during Mino-Owari, Tohnankai and Fukui earthquakes, *Soils and Foundations*, **IX**, No. 1, pp. 75–92.

Koester, J. P. and Franklin, A. G. (1985), *Current Methodologies for Assessing the Potential for Earthquake Induced Liquefaction in Soils*, Report NUREG/CR-4430 RA, U.S. Nuclear Regulatory Commission, Washington, D.C.

Koppejan, A. (1948), Slides in the Dutch Province of Zeeland, *Proceedings, Second International Conference on Soil Mechanics and Foundation Engineering*, Rotterdam, The Netherlands, **5**, pp. 89–96.

Koutsoftas, D. (1978), Effect of cyclic loads on undrained strength of two marine clays, *Journal of Geotechnical Engineering*, ASCE, **104**, No. GT-5, pp. 609–620.

Kovacs, W. D., Yokel, F. Y., Salomone, L. A., and Holtz, R. D. (1984), Liquefaction potential and the international SPT, *Proceedings of The Eight World Conference on Earthquake Engineering*, Prentice-Hall, Inc., Englewood Cliffs, New Jersey, **3**, pp. 243–268.

Kramer, S. L. and Seed, H. B. (1988), Initiation of soil liquefaction under static loading conditions, *Journal of Geotechnical Engineering Division*, ASCE, **114**, No. 4, pp. 412–430.

Kuribayashi, E. and Tatsuoka, F. (1975), Brief review of liquefaction during earthquakes in Japan, *Soils and Foundations*, **15**, No. 4, pp. 81–92.

Lacroix, Y. and Horn, H. M. (1973), Direct determination and indirect evaluation of relative density and its use on earthwork construction projects, *Evaluation of Relative Density and Its Role in Geotechnical Projects Involving Cohesionless Soils*, ASTM STP 523, pp. 251–280.

Ladd, R. S., Dobry, R., Dutko, P., Yokel, F. Y., and Chung, R. M. (1989), Pore-water pressure buildup in clean sands because of cyclic straining, *Geotechnical Testing Journal*, ASTM, **12**, No. 1, pp. 77–86.

Ledbetter, R. H. (1985), *Improvement of Liquefiable Foundation Conditions Beneath Existing Structures*, Technical Report REMR-GT-2, U.S. Army Corps of Engineers, Washington, D.C.

Lee, K. L. (1974), *Earthquake Induced Permanent Deformations of Embankments*, UCLA-ENG-7498, University of California, Los Angeles.

Lee, K. L. and Albaisa, A. (1974), Earthquake induced settlements in saturated sands, *Journal of the Soil Mechanics and Foundations Division*, ASCE, **100**, No. 4, pp. 387–400.

Lee, K. L., Seed, H. B., Idriss, I. M., and Makdisi, F. T. (1975), Properties of soil in the San Fernando hydraulic fill dams, *Journal of the Geotechnical Engineering Division*, ASCE, **101**, No. GT-8, pp. 801–821.

Lee, K. L. and Focht, J. A. (1976), Strength of clay subjected to cyclic loading, *Marine Geotechnology*, **1**, pp. 305–326.

Lefebvre, G. and LeBoeuf, D. (1987), Rate effects and cyclic loading of sensitive clays, *Journal of Geotechnical Engineering*, ASCE, **113**, No. 5, pp. 476–489.

Lemke, R. W. (1967), Effects of the earthquake of March 27, 1964 at Seward, Alaska, *U.S. Geological Survey Professional Paper 542-E*.

McCulloch, D. S. (1966), Slide-induced waves, seiching, and ground fracturing caused by the earthquake of March 27, 1964 at Kenai Lake, Alaska, *U.S. Geological Survey Professional Paper 543-A*.

Macky, T. A. and Saada, A. S. (1984), Dynamics of anisotropic clays under large strains, *Journal of Geotechnical Engineering*, ASCE, **110**, No. 4, pp. 487–504.

Makdisi, F. I., Seed, H. B., and Idriss, I. M. (1978), Analysis of Chabot Dam during the 1906 earthquake, *Proceedings, Earthquake Engineering and Soil Dynamics*, Pasadena, Calif., ASCE, **2**, pp. 569–587.

Manojlovic, M. (1980), On soil and rock instabilities during the earthquake in the Montenegro Coastal Region, *Proceedings of the International Symposium on Soils Under Cyclic and Transient Loading*, Swansea, A. A. Balkema, Rotterdam, pp. 555–560.

Marchetti, S. (1982), Detection of liquefiable sand layers by means of quasi-static penetration tests, *Proceedings of the 2nd European Symposium on Penetration Testing, ESOPT II*, Amsterdam, The Netherlands, **2**, pp. 689–695.

Martin, G. R., Finn, W. D. L., and Seed, H. B. (1975), Fundamentals of liquefaction under cyclic loading, *Journal of the Geotechnical Engineering Division*, ASCE, **101**, No. GT-5, pp. 423–438.

Marr, W. A. and Christian, J. T. (1981), Permanent displacements due to cyclic wave loading, *Journal of the Geotechnical Engineering Division*, ASCE, **107**, No. GT-8, pp. 1129–1149.

Masing, G. (1926), Eigenspannungen und Verfestigung beim Messing, *Proceedings of the Second International Congress of Applied Mechanics*.

Matsui, T., Ohara, H., and Ito, T. (1980), Cyclic stress–strain history and shear characteristics of clay, *Journal of the Geotechnical Engineering Division*, ASCE, **106**, No. 10, pp. 1011–1020.

Meimon, Y. and Hicher, P. Y. (1980), Mechanical behavior of clays under cyclic loading, *Proceedings of the International Symposium on Soils Under Cyclic and Transient Loading*, Swansea, A. A. Balkema, Rotterdam, pp. 77–88.

Morgenstern, N. R. (1967), Submarine slumping and the initiation of turbidity currents, *Marine Geotechnique*, ed. A. F. Richards, University of Illinois Press, Urbana, Ill., pp. 189–220.

Moriya, M. and Kawaguchi, M. (1970), Damage to small earthfill irrigation dams in Aomori Prefecture during the Tokachioki earthquake, *Soils and Foundations*, **10**, No. 2, pp. 72–82.

National Research Council, Committee on Earthquake Engineering (1985), *Liquefaction of Soils During Earthquakes*, National Academy Press, Washington, D.C., p. 240.

Ohsaki, Y. (1970), Effects of sand compaction on liquefaction during the Tokachioki earthquake, *Soils and Foundations*, **10**, No. 2, pp. 112–128.

Pamukcu, S. (1989), Shear modulus of soft marine clays, *Journal of Offshore Mechanics and Arctic Engineering*, ASME, **111**, No. 4, pp. 265–272.

Pamukcu, S., Poplin, J. K., Suhayda, J. N., and Tumay, M. T. (1983), Dynamic sediment properties, Mississippi Delta, *Proceedings of the Geotechnical Conference in Offshore Engineering*, ASCE, pp. 111–132.

Pamukcu, S. and Suhayda, J. N. (1984), Evaluation of shear modulus for soft marine clays, Mississippi Delta, *Strength Testing of Marine Sediments: Laboratory and In-Situ Measurements*, ASTM STP 883, ed. R. C. Chaney and K. R. Demars, pp. 352–362.

Pamukcu, S. and Suhayda, J. N. (1987), High resolution measurement of shear modulus of clay using triaxial vane device, *Soil Dynamics and Liquefaction*, ed. A. S. Carmak, Developments in Geotechnical Engineering No. 42, Elsevier and Computational Mechanics Publications, pp. 307–321.

Poulos, H. G. (1988), *Marine Geotechnics*, Chap. 3, Unwin Hyman, London.

Poulos, S. J., Castro, G., and France, J. W. (1985), Liquefaction evaluation procedure, *Journal of the Geotechnical Engineering Division*, ASCE, **111**, No. 6, pp. 772–792.

Prevost, J. H. (1977), Mathematical modelling of monotonic and cyclic undrained clay behavior, *International Journal of Numerical and Analytical Methods in Geotechnical Engineering*, **1**, No. 2, pp. 195–216.

Prior, D. B. and Coleman, J. (1983), Lateral movements of sediments, *Ocean Science and Engineering*, **8**, No. 2, pp. 113–155.

Procter, D. C. and Khaffaf, J. H. (1984), Cyclic triaxial tests on remoulded clay, *Journal of Geotechnical Engineering*, ASCE, **110**, No. 10, pp. 1431–1445.

Public Works Research Institute, Ministry of Construction (1970), Damage to roads, dikes, and highway bridges during the Tokachioki earthquake, *Soils and Foundations*, **10**, No. 2, pp. 15–38.

Pyke, R., Seed, H. B., and Chan, C. K. (1975), Settlement of sands under multidirectional shaking, *Journal of the Geotechnical Engineering Division*, ASCE, **101**, GT-4, pp. 379–398.

Ramberg, W. and Osgood, W. T. (1943), *Description of Stress–Strain Curves by Three Parameters*, Technical Note 902, NASA.

Ray, R. P. and Woods, R. D. (1988), Modulus and damping due to uniform and variable cyclic loading, *Journal of Geotechnical Engineering*, ASCE, **114**, No. 8, pp. 861–876.

Robertson, P. K. and Campanella, R. G. (1985), *Liquefaction Potential of Sands Using the CPT*, Research Report R69-15, Dept. of Civil Engineering, Massachusetts Institute of Technology, Cambridge, Mass.

Robertson, P. K. and Campanella, R. G. (1986), Estimating liquefaction potential of sands using the flat plate dilatometer, *Geotechnical Testing Journal*, **9**, No. 1, pp. 38–40.

Roesset, J. M. and Whitman, R. C. (1976), *Theoretical Background for Simplification Studies*, Research Report R69-15, Dept. of Civil Engineering, Massachusetts Institute of Technology, Cambridge.

Ross, G. A., Seed, H. B., and Migliaccio, R. R. (1969), Bridge foundation behavior in Alaska earthquake, *Journal of the Soil Mechanics and Foundations Division*, ASCE, **95**, No. SM-4, pp. 1007–1036.

Sangrey, D. A., Castro, G., Poulos, S. J., and France, J. W. (1978), Cyclic loading of sands, silts and clays, *Proceedings, ASCE Special Conference on Earthquake Engineering and Soil Dynamics*, Pasadena, Calif., **2**, pp. 836–851.

Sangrey, D. A. and France, J. W. (1980), Peak strength of clay soils after a repeated loading history, *International Symposium on Soils under Cyclic and Transient Loading*, Swansea, A. A. Balkema, Rotterdam, **1**, pp. 421–430.

Scott, R. F. (1973), Behavior of soils during the earthquake, *The Great Alaska Earthquake of 1964*, National Academy of Sciences, Washington, D.C., pp. 49–72.

Seed, H. B. (1966), A method for earthquake resistant design of earth dams, *Journal of the Soil Mechanics and Foundations Division*, ASCE, **92**, No. SM-1, pp. 13–41.

Seed, H. B. (1968), Landslides during earthquakes due to soil liquefaction, *Journal of the Soil Mechanics and Foundations Division*, ASCE, **94**, No. SM-5, pp. 1053–1122.

Seed, H. B. (1976), Evaluation of soil liquefaction effects on level ground during earthquakes, *Liquefaction Problems in Geotechnical Engineering*, American Society of Civil Engineers, Preprint 2725, presented at the ASCE National Convention, Philadelphia, Pa., pp. 1–104.

Seed, H. B. (1984), The role of case studies in the evaluation of soil liquefaction potential, *Proceedings, International Conference on Case Histories in Geotechnical Engineering*, **IV**, St. Louis, Mo.

Seed, H. B. and Idriss, I. M. (1967), Analysis of soil liquefaction: Niigata earthquake, *Journal of the Soil Mechanics and Foundations Division*, ASCE, **93**, No. SM-3, pp. 83–108.

Seed, H. B. and Wilson, S. D. (1967), The Turnagain Heights landslide, Anchorage, Alaska, *Journal of the Soil Mechanics and Foundations Division*, ASCE, **93**, No. SM-4, pp. 325–353.

Seed, H. B. and Idriss, I. M. (1971), Simplified procedure for evaluating soil liquefaction potential, *Journal of the Soil Mechanics and Foundations Division*, ASCE, **97**, No. SM-9, Proc. Paper 8371, pp. 1249–1273.

Seed, H. B. and Peacock, W. H. (1971), The procedure for measuring soil liquefaction characteristics, *Journal of the Soil Mechanics and Foundations Division*, ASCE, **97**, No. SM-8, pp. 1099–1119.

Seed, H. B. and Silver, M. L. (1972), Settlement of dry sands during earthquakes, *Journal of the Soil Mechanics and Foundations Division*, ASCE, **98**, No. SM-4, pp. 381–397.

Seed, H. B., Lee, K. L., Idriss, I. M., and Makdisi, F. I. (1973), *Analysis of the Slides in the San Fernando Dams During the Earthquake of February 9, 1971*, Report No. EERL 73-2, Earthquake Engineering Research Center, University of California, Berkeley, Calif. (Condensed version appears in the *Journal of the Geotechnical Engineering Division*, ASCE, **101**, No. GT-7, 1975, pp. 651–688 and **101**, No. GT-8, 1975, pp. 801–821).

Seed, H. B., Arango, I., and Chan, C. K. (1975), *Evaluation of Soil Liquefaction Potential During Earthquakes*, Report No. EERC 75-28, Earthquake Engineering Research Center, University of California, Berkeley, Calif.

Seed, H. B. and Idriss, I. M. (1982), *Ground Motions and Soil Liquefaction During Earthquakes*, Monograph series, Earthquake Engineering Research Institute, Berkeley, Calif.

Seed, H. B. and Idriss, I. M. (1983), Evaluation of liquefaction potential using field performance data, *Journal of Geotechnical Engineering*, ASCE, **109**, No. 3, pp. 458–482.

Seed, H., Idriss, I. M., and Arango, I. (1983), Evaluation of liquefaction potential using field performance data, *Journal of the Geotechnical Engineering Division*, ASCE, **111**, No. 12, pp. 458–482.

Seed, H. B., Tokimatsu, K., Harder, L. F., and Chung, R. M. (1984), *The Influence of SPT Procedures in Soil Liquefaction Resistance Evaluations*, Report No. UBC/EERC-84/15, Earthquake Engineering Research Center, University of California, Berkeley, Calif.

Seed, H., Tokimatsu, K., Harder, L. F., and Chung, R. M. (1985), The influence of SPT procedures in soil liquefaction resistance evaluations, *Journal of the Geotechnical Engineering Division*, ASCE, **105**, No. 5, pp. 201–255.

Serff, N. (1976), Earthquake-induced deformations of earth dams, Ph.D. Thesis, University of California, Berkeley.

Sherif, M. A., Isibashi, I., and Tsuchiya, C. (1977), Saturation effects on initial soil liquefaction, *Journal of the Geotechnical Engineering Division*, ASCE, **103**, No. GT-8, pp. 914–917.

Silver, M. L. and Seed, H. B. (1971), Volume changes in sands during cyclic loading, *Journal of the Soil Mechanics and Foundations Division*, ASCE, **97**, No. 9, pp. 1171–1182.

Singh, R. D., Kim, J. H., and Caldwell, S. R. (1978), Properties of clays under cyclic loading, *Proceedings, 6th Symposium on Earthquake Engineering*, University of Roorkee, **1**, pp. 107–112.

Singh, R. D., Ricardo, D., Doyle, E. H., and Idriss, I. M. (1981), Non-linear seismic response of soft clay sites, *Journal of the Geotechnical Engineering Division*, ASCE, **107**, No. GT-9, pp. 1201–1218.

Soils and Foundations (1968), Tokachioki Earthquake of 1968, **8**, No. 3, pp. 87–95.

Stokoe II, K. H. (1980), Dynamic properties of offshore silty samples, *12th Annual Offshore Technology Conference*, OTC 3771, Houston, Texas.

Taksuoka, F., Sasaki, T., and Yamada, S. (1984), Settlement in saturated sand induced by cyclic undrained simple shear, *Proceedings, 8th World Conference on Earthquake Engineering*, San Francisco, Calif., pp. 95–102.

Terzaghi, K. (1956), Varieties of submarine slope failures, *Proceedings, Eighth Texas Conference on Soil Mechanics and Foundation Engineering*, Texas, pp. 1–41.

Thiers, R. G. and Seed, H. B. (1969), Strength and stress–strain characteristics of clays subjected to seismic loading conditions, *Vibration Effects of Earthquakes on Soils and Foundations*, ASTM STP 450.

Togrol, E. and Guler, E. (1984), Effect of repeated loading on the strength of clay, *Soil Dynamics and Earthquake Engineering*, **3**, No. 4, pp. 184–190.

Tohno, I. and Yasuda, S. (1981), Liquefaction of the ground during the 1978 Miyagiken-Oki earthquake, *Soils and Foundations*, **21**, No. 3, pp. 18–34.

Tokimatsu, K. and Yoshimi, Y. (1984), Criteria of soil liquefaction with SPT and fines content, *Proceedings of the 8th World Conference on Earthquake Engineering*, San Francisco, Calif., **3**, pp. 255–262.

Tokimatsu, K. and Seed, H. B. (1987), Evaluation of settlements in sands due to earthquake shaking, *Journal of Geotechnical Engineering*, **113**, No. 8, pp. 861–878.

Tsai, C. F., Lam, I., and Martin, G. R. (1980), Seismic response of cohesive marine soils, *Journal of the Geotechnical Engineering Division*, ASCE, **106**, No. GT-9, pp. 997–1012.

Tuthill, S. J. and Laird, W. M. (1964), Geomorphic effects of the earthquake of March 27, 1964 in the Martin Bearing Rivers Area, Alaska, *U.S. Geological Survey Professional Paper 543-B*.

U.S. Corps of Engineers (1950), *Investigation of Free Nigger Point Crevasse, Mississippi River*, Waterways Experiment Station, Vicksburg, Miss.

U.S. Corps of Engineers (1952), *Bank Caving Investigations, Free Nigger Point and Point Menoir, Mississippi River*, Waterways Experimental Station, Report No. 15-1, Vicksburg, Miss.

U.S. Department of Commerce Report (1966), *The Prince William Sound, Alaska Earthquake of 1964 and Aftershocks, Vol. 1*, Publication 10-3, Coast and Geodetic Survey, Washington, D.C., pp. 230–238.

Valera, J. E. and Donovan, N. C. (1977), Soil liquefaction procedures—A review, *Journal of the Geotechnical Engineering Division*, ASCE, pp. 607–626.

Van Eekelen, H. A. M. and Potts, D. M. (1978), The behavior of drammen clay under cyclic loading, *Geotechnique*, **28**, No. 2, pp. 173–196.

Vucetic, M. (1988), Normalized behavior of offshore clay under uniform cyclic loading, *Canadian Geotechnical Journal*, **25**, No. 1, pp. 33–41.

Whitman, R. V. (1971), Resistance of soil to liquefaction and settlement, *Soils and Foundations*, **11**, No. 4, pp. 59–68.

Wilson, E. L. (1963), Finite element analysis of two-dimensional structures, Ph.D. dissertation, University of California, Berkeley.

Wilson, S. D. (1967), Landslides in the city of Anchorage, *The Prince William Sound, Alaska, Earthquake of 1964 and Aftershocks*, Publication 10-3, Vol. 2, Part A, U.S. Department of Commerce, Coast and Geodetic Survey, Washington, D.C., pp. 253–297.

Wright, S. G. (1976), Analysis for wave induced sea-floor movements, *Proceedings, 8th Annual Offshore Technology Conference*, Houston, Texas, **1**, pp. 41–52.

Wu, S., Gray, D., and Richart Jr, F. E. (1984), Capillary effects on dynamic modulus of sands and silts, *Journal of the Geotechnical Engineering Division, ASCE*, **110**, No. 9, pp. 1188–1203.

Yamada, G. (1966), Damage to earth structures and foundations by the Niigata earthquake, June 16, 1964, *Soils and Foundations*, **6**, No. 1, pp. 1–13.

Yegian, M. K. and Oweis, I. S. (1976), Discussion of Castro (1975), *Journal of Soil Mechanics and Foundations Division, ASCE*, **102**, No. GT-3, pp. 265–268.

Yegian, M. and Whitman, R. V. (1978), Risk analysis for ground failure by liquefaction, *Journal of the Geotechnical Engineering Division, ASCE*, **104**, No. GT-7, pp. 921–938.

Yerkes, R. F. (1973), Effects of the San Fernando earthquake as related to geology, *San Fernando, California, Earthquake of February 9, 1971*, Vol. 3, U.S. Department of Commerce Report, Washington, D.C., pp. 137–154.

Yokomura, S. (1966), The damage to river dykes and related structures caused by the Niigata earthquake, *Soils and Foundations*, **6**, No. 1, pp. 38–53.

Yoshimi, Y., Kuwabara, F., and Tokimatsu, K. (1975), One-dimensional volume change characteristics of sands under very low confining stresses, *Soils and Foundations*, **15**, No. 3, pp. 51–60.

Youd, T. L. and Hoose, S. N. (1975), Liquefaction during 1906 San Francisco earthquake, *Meeting Preprint 2517*, ASCE National Convention, Nov., pp. 1–25.

Yu, W. T. and Mikroudis, G. K. (1987), Development of SEICO— A knowledge-based expert system for building configuration in seismic areas, *Proceedings, International Symposium on Environmental Geotechnology*, Envo Publishing Co., Inc., Bethlehem, Pa., **2**, pp. 170–180.

Zeevaert, L. (1972), Introduction to earthquake problems in building foundations, *Foundation Engineering for Difficult Subsoil Conditions*, Van Nostrand Reinhold, New York, N.Y.

Zeevaert, L. (1975), Foundation problems in earthquake regions, *Analysis and Design of Building Foundations*, ed. H. Y. Fang, Envo Publishing Co., Inc., Bethlehem, Pa., pp. 753–770.

Zeevaert, L. (1983), Liquefaction of fine sand due to wave action, *Shore and Beach*, **51**, No. 2, pp. 32–36.

Zhou, S. G. (1981), Influence of fines on evaluating liquefaction of sand by CPT, *Proceedings of the 1981 International Conference on Recent Advances in Geotechnical Earthquake Engineering and Soil Dynamics*, St Louis, Mo., **1**, pp. 167–172.

17 FOUNDATION PROBLEMS IN EARTHQUAKE REGIONS

LEONARDO ZEEVAERT, Ph.D., C.E., P.E.
Professor of Civil Engineering, Emeritus
Universidad Nacional Autónoma de México

17.1 INTRODUCTION

A major earthquake produces a strong ground motion in the subsoil; consequently, underground and surface structures supported on the soil mass will be induced to move and take dynamic forces. The magnitude of the inertia forces are proportional to the acceleration at the depth at which the foundation structure is placed. Their action in the foundation structure may be estimated knowing the subsoil behavior. For this purpose, the maximum displacements, stresses, and accelerations should be determined in the soil mass.

The general geology of the affected area is important and the stratigraphy of the upper part of the subsoil comprising the soft sediments where the strong ground motion takes place should be determined. The present discussion will be confined to earthquake regions where unconsolidated sediments are underlain by firm deposits of stiff or very stiff materials.

When the seismic waves coming from the zone of generation hit the interphase of the firm ground with the unconsolidated sediments, two types of body waves are originated, namely irrotational and pure shear waves. These two types of waves propagate through the unconsolidated sediments with different velocities and produce different dynamical effects. The irrotational waves travel with higher velocity than the shear waves, and for translation require changes in volume of the soil. Therefore, in saturated soils they produce high pore pressures and, eventually, if the acceleration is high in noncohesive soils, liquefaction may take place at the ground surface. The displacements, however, may be small in saturated soils. On the other hand, the shear waves do not produce volume changes in the soil during their propagation; however, high shear distortions may be induced and shear stresses may develop running over the shear strength of the soil (Zeevaert, 1988).

From the above considerations, two types of foundation problems may be considered.

a. Problems induced by irrotational waves
b. Problems induced by transverse or shear waves

17.2 IRROTATIONAL SEISMIC WAVES

The stability of foundations in cohesionless fine sediments with very low cohesion may be importantly affected by the dilatation or irrotational waves. The shear strength of this type of soil is given by

$$S_{\text{sis}} = (\sigma - u_{\text{sis}}) \tan \phi_d \qquad (17.1)$$

in which σ is the in-situ effective stress, u_{sis} is the seismic pore water pressure induced by the seismic action and ϕ_d is the true angle of internal friction of the sediment.

Therefore, in a saturated soil the pore water pressure originated during a strong ground motion may be high and the shear strength may be completely lost, and liquefaction of the soil may take place. This phenomenon is detected at the ground surface by boils appearing after the earthquake. Nevertheless, even with smaller accelerations the shear strength is reduced to a point that the bearing capacity is affected and a foundation supported on this type of soil may suffer partial or total failure (see Fig. 17.1).

Fig. 17.1 Fuel tank tilted owing to reduction of the bearing capacity.

To illustrate this problem, consider the case of old fuel tanks that failed during a strong earthquake for which the ground acceleration may be estimated on the order of 0.2g. The site is located on an old river bed at the mouth of the Coatzacoalcos River, State of Veracruz, Mexico.

The author visited the area shortly after the earthquake and noticed the foundation failures due to loss in the soil shear strength and boils at the ground surface because of liquefaction produced during the strong ground motion in fine sand strata close to the ground surface. Figure 17.1 shows one of the tanks damaged, others suffered total collapse. Notice the light dwellings pushed out of the ground surface in a vertical direction because of wave action.

The subsoil at the site is a silty fine sand in a loose state to a depth of 2.5 m, underlain by a slightly organic silty fine sand to a depth of 9 m. From 9 to 17 m depth, clayey silty sand is encountered. At 17 m depth, firm ground is found (Fig. 17.2). The water table at the site of the tanks was found at a depth of 0.5 m from the ground surface. The geometrical dimensions of a fuel tank are shown in Figure 17.2 and it is supported on a concrete slab. The surface soil under the foundation slab has an angle of internal friction $\phi_d = 26°$ and relative density $D_r = 0.35$. The confining stress at the edge of the foundation is approximately 1.1 tonne/m². Under these conditions, the ultimate bearing capacity of the soil is (Zeevaert, 1983, pp. 207–209)

$$q_d = 1.2(\sigma_d N_q + 0.6\bar{\gamma} N_\gamma)(D_r + 0.1) \quad (17.2)$$

where $N_q = 13$, $N_\gamma = 10$ and the unit weight is 1.83 tonne/m². Hence

$$q_d \cong 10.4 \text{ tonne/m}^2 \quad (17.3)$$

On the other hand, the average soil reaction because of the tank weight when fully loaded and the concrete foundation slab, is 3.7 tonne/m². Hence, the factor of safety is on the order of 2.8. At the edges of the foundation slab the local ultimate bearing capacity of the sand is approximately 7.72 tonne/m². Hence,

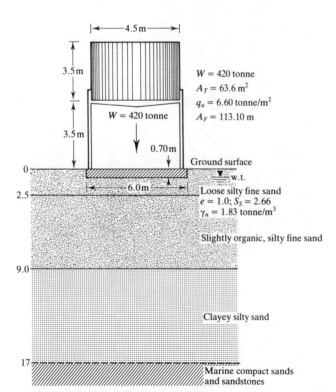

Fig. 17.2 Soil profile at site of fuel tank.

we will assume practically a uniform soil reaction under the foundation slab.

The reduction in bearing capacity of the soil may be approximately investigated by estimating the pore water pressure during the earthquake induced by the irrotational waves, propagating from firm ground to the surface, according to

$$v_d^2 \cdot \frac{\partial^2 w}{\partial z^2} = \frac{\partial^2 w}{\partial t^2} \quad (17.4)$$

in which w is the vertical displacement and v_d is the irrotational wave velocity. A particular solution of Equation 17.4, assuming v_d constant throughout the deposit, may be obtained by

$$w = w_0 \cos\left(\frac{\pi}{2D} z\right) \cdot \sin\left(\frac{2\pi}{T} t\right) \quad (17.5)$$

in which T is the resonant period of the soft soil deposit and w_0 is the maximum amplitude at the ground surface, $z = 0$. The maximum pressure is obtained at $t = \frac{1}{4}T, \frac{3}{4}T$, etc... Hence,

$$\sigma_z = E_c \cdot \frac{\partial w}{\partial z} \quad (17.6)$$

$$\sigma_z = -E_c w_0 \frac{\pi}{2D} \sin\left(\frac{\pi}{2} \cdot \frac{z}{D}\right) \quad (17.7)$$

at the ground surface

$$w_0 = \frac{G_{av}}{(2\pi)^2} T^2 \quad (17.8)$$

in which G_{av} is the maximum vertical ground surface acceleration. The ground dominant period $T = 4 \ D/v_d$ and $v_d^2 = E_c/\rho$, therefore the vertical displacement amplitude is

$$w_0 = \frac{4D^2}{\pi^2} \cdot \frac{\rho}{E_c} G_{av} \quad (17.9)$$

and substituting in (Eq. 17.7), the pressure configuration is obtained as

$$\sigma_z = -\left(\frac{2}{\pi} G_{av} D \rho\right) \sin\left(\frac{\pi}{2} \cdot \frac{z}{D}\right) \quad (17.10)$$

But during the earthquake in the saturated sediment $\sigma_z = -u_z$, hence

$$(u_z)_{sis} = \left(\frac{2}{\pi} G_{av} \cdot D \rho\right) \sin\left(\frac{\pi}{2} \cdot \frac{z}{D}\right) \quad (17.11)$$

Therefore, the seismic pore water pressure amplitude is

$$(u_0)_{sis} = \frac{2}{\pi} G_{av} \cdot D \rho \quad (17.12)$$

In this case the soil has a void ratio of 1.0, specific gravity of 2.66, and hence, $\gamma = 1.83$ tonne/m³.

Concerning the foundation of the tank, one can estimate the seismic pore water pressure with depth that reduces the confining stress under the foundation slab. Assume for this purpose a vertical ground surface acceleration of 0.2g. Using Equation 17.11 for $z = 0.70$ and $D = 17$ m, we get

$$(u_z)_{sis} = \frac{2}{\pi}(0.2)17(1.83) \sin\left(\frac{0.7\pi}{34}\right) \quad u_z = 0.26 \text{ tonne/m}^2$$

hence at any depth z,

$$(u_z)_{sis} = 3.96 \sin\left(\frac{\pi}{2} \cdot \frac{z}{H}\right) \quad (\text{tonne/m}^2) \quad (17.13)$$

Furthermore, we have to investigate the seismic angle of internal friction because ϕ_{sis} of the induced pore water pressure (Zeevaert, 1983, pp. 529–532)

$$\phi_{sis} = \sin^{-1}\left[\left(1 - \frac{u}{\sigma_c}\right)\sin\phi_d\right] \qquad (17.14)$$

in which σ_c is the octahedral or volumetric stress under the soil foundation.

The calculation for the average value of ϕ_{sis} with depth is presented in Table 17.1. It may be seen that the friction angle is reduced to approximately 13.2°. Therefore, the soil bearing capacity factors for the soil foundation take the values $N_q = 3.8$ and $N_\gamma = 1.9$, and the seismic bearing capacity reads, from Equation 17.2,

$$(q_d)_{sis} = 1.2[1.28(3.8) + 0.6(0.83)1.9] \cong 6.97 \text{ tonne/m}^2$$

In this case it is assumed that failure takes place at practically constant volume; therefore, we assume $(D_r + 0.1) = 1$. Under this seismic condition, the ultimate soil bearing capacity is on the order of 7.0 tonne/m^2.

The seismic overturning moment induces at the edge an excess soil pressure on the order of

For full tank conditions   14.1 tonne/m^2
For $\frac{1}{2}$ full tank conditions   7.51 tonne/m^2

From this analysis we can explain why tanks fully loaded at the time of the strong ground motion collapsed completely because of strong sinking on one side of the foundation slab, and others that were partially loaded suffered only different magnitudes of tilting like the one shown in the photograph of Figure 17.1.

This simple example illustrates the importance of analyzing the pore water pressures in the soil induced by earthquake irrotational waves. The high pore pressures induced by a strong

TABLE 17.1 COMPUTATION OF SEISMIC PORE WATER PRESSURE AND SEISMIC ANGLE OF FRICTION.

z	σ_{0z}	σ_{cz}	$\Delta\sigma_{cB}$	σ_c	$(u_z)_{sis}$	$(\phi_{sis})_1$	$(\phi_{sis})_2$
0.7	1.28	0.85	3.08	3.93	0.26	17.7	24.3
1.7	2.11	1.41	1.83	3.34	0.62	14.3	21.0
2.7	2.94	1.96	1.02	2.93	0.98	12.7	17.0
3.7	3.77	2.51	0.62	3.13	1.33	12.0	14.7
4.7	4.60	3.07	0.39	3.45	1.67	11.7	13.1
5.7	5.43	3.62	0.27	3.89	1.99	11.4	12.4
6.7	6.26	4.18	0.20	4.38	2.30	11.4	12.1
7.7	7.09	4.73	0.15	4.88	2.59	11.4	12.1
8.7	7.92	5.28	0.11	5.39	2.85	11.7	12.0

Average friction angle $\phi_{sis} = 13.2°$
Stresses in tonne/m^2 (1 tonne/m$^2 \simeq 10$ kN/m^2)

ALGORITHMS USED

$$\sigma_{cz} = 0.67\sigma_{0z}, \quad \phi_{sis} = \sin^{-1}\left[\left(1 - \frac{u_z}{\sigma_{cz}}\right)\sin 26°\right]$$

$$u_z = 3.96\sin\left(\frac{\pi}{2}\cdot\frac{z}{H}\right), \quad \Delta\sigma_{cB} = q_a\frac{2}{3}(1 + v)(1 - \cos\psi)$$

(Zeevaert, 1983, pp. 164–166)

$(\phi_{sis})_1$ = seismic angle of friction outside foundation
$(\phi_{sis})_2$ = seismic angle of friction at centre of foundation
$\Delta\sigma_{cB}$ = increment of confining stress due to weight of fully loaded tank, $q_u = 3.7$ tonne/m^2
v = Poisson's ratio = 0.25
$\cos^2\psi = 1/(1 + R^2/z^2)$
R = radius of foundation slab

earthquake may cause the failure of structures founded on fine soil that from the point of view of statics appeared to be stable over long time (Zeevaert, 1987).

17.3 SHEAR SEISMIC WAVES

The transverse or pure shear waves propagate from the interface at the firm ground into the unconsolidated sediments, producing important shear distortions in the soil mass. By the same token, the shear stresses may be important when added to the static shear stresses already in the ground supporting the foundations. Deep foundations and underground installations may be strongly stressed when large displacements are originated by the shear

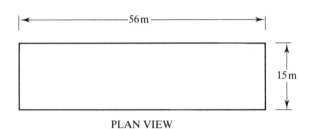

PLAN VIEW

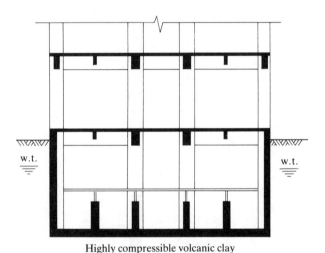

Highly compressible volcanic clay
Fig. 17.3 Cross section of building.

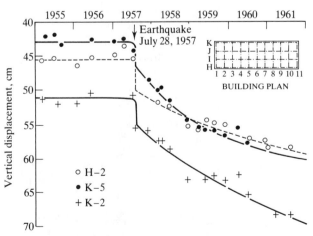

Fig. 17.4 Settlement of building after the earthquake.

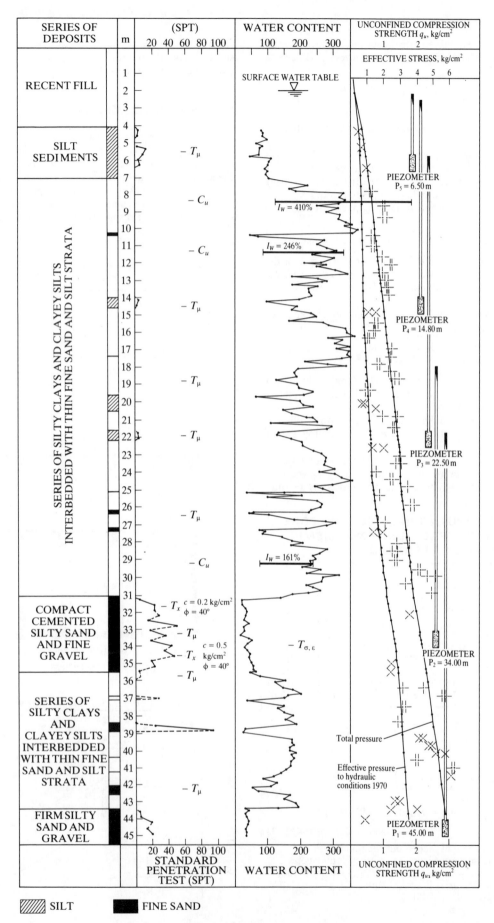

Fig. 17.5 Typical soil profile, Mexico City center.

waves. The behavior of the soil mass may be estimated with knowledge of the dynamic shear modulus of elasticity and shear wave velocity. From the dynamic stress–strain engineering characteristics of the soil, the interaction between the foundation structure and the soil mass may be investigated and the forces acting on the foundation structure and structural frame estimated (Zeevaert, 1988).

Building foundations working satisfactorily under static loading may be affected when the shear stresses in the subsoil induced by the earthquake added to the already existing static shear stresses reach the shear strength of the soil. In case of cohesionless fine sediments, the shear strength properties may be strongly affected during the transient seismic forces mainly when the irrotational waves act at the time of the shear waves. The irrotational waves inducing high pore water pressures in the soil will reduce the shear strength of the soil. Therefore, in cohesionless soils their action cannot be overlooked.

17.4 CASE HISTORY

A case history will illustrate the influence of the shear waves in the behavior of a large building foundation (Fig. 17.3). The building was constructed in October 1952, about 5 years before the major earthquake of Mexico City on July 28, 1957. The building was constructed on an undercompensated foundation at a depth of 6.5 m from the ground surface, permitting an increment of stress on the order of 7.0 tonne/m² over the effective overburden stress at the foundation grade elevation (Zeevaert, 1983, p. 523). By 1957 the building settled strongly; however, it stabilized, as shown by settlement records (Fig. 17.4). During the earthquake the building showed an immediate settlement on the order of 4.0 cm, which increased in the following year to another 4.0 cm. In 1964 the building was underpinned with friction piles, making the foundation work as a compensated friction pile foundation, without further important settlement.

The phenomenon observed during the earthquake and afterwards may be analyzed approximately by estimating the static and dynamic shear stresses in the subsoil and comparing them with the shear strength of the soil sediments determined by means of unconfined compression tests.

The subsoil behavior may be calculated from soil profile at the site and the dynamic shear modulus of elasticity (Fig. 17.5). A soil column may be analyzed using expressions for the shear wave component:

$$v_s^2 \cdot \frac{\partial^2 u}{\partial z^2} = \frac{\partial^2 u}{\partial t^2} \tag{17.15}$$

in which $v_s = \sqrt{\mu/\rho}$ is the shear wave velocity, where μ is the dynamic shear modulus of elasticity and ρ is the unit mass of the soil. Since the values of μ, ρ and therefore v_s change for the different strata forming the subsoil at the site, the integration of Equation 17.15 is performed step by step (Zeevaert, 1983, pp. 519–529). The results of the integration are reported in Table 17.2, where the maximum displacements of the ground and maximum shear stresses induced by a ground surface acceleration of 50 cm/s² are shown. The acceleration mentioned is the maximum ground surface acceleration at the site zone estimated during the earthquake of July 28, 1957 in Mexico City. The results of the calculation have been plotted in Figure 17.6.

The maximum shear stresses in the ground induced by the static excess load of $\Delta q = 7.0$ tonne/m² on the soil, may be estimated approximately at the center cross section of the

TABLE 17.2 ENGINEERING PHYSICAL PROPERTIES OF SUBSOIL MATERIALS AND PARAMETERS IN ALGORITHMS[a].

Stratum	Classification	Overburden stress σ_{oi}	Overburden vol. stress σ_{ci}	Unit weight γ_n	Soil rigidity μ	Unit mass ρ	Thickness stratum d	Shear wave velocity v_s	Period $\Sigma\Delta t$	$\delta_{i+1} = A_i\delta_i - B_i\tau_i$ $\tau_i = C_i(\delta_i + \delta_{i+1}) + \tau_i$ N_i	A_i	B_i	C_i	Horiz. soil displacement δ_i	Shear stress τ_i
a_1	Silty sand		0.92	1.38	1400	0.14	2.00	99.64	0.08	0.87	1.00	1.43	1.21	5.80	0.14
a_2	Silty sand	3.50	2.33	1.54	1930	0.16	2.10	110.87	0.16	0.77	1.00	1.09	1.42	5.78	0.31
b	Silty sand	5.60	4.67	1.43	2400	0.15	3.00	128.65	0.25	1.17	1.00	1.25	1.87	5.73	0.52
c	Clay	6.50	5.42	1.16	280	0.12	3.30	48.51	0.52	9.97	0.98	11.67	1.66	5.01	0.70
d	Clay	7.50	6.25	1.18	360	0.12	3.60	54.77	0.78	9.30	0.98	9.91	1.86	4.22	0.87
e	Sand	8.00	5.33	1.40	920	0.14	0.60	80.49	0.81	0.12	1.00	0.65	0.37	4.17	0.90
f	Clay	9.00	7.50	1.16	360	0.12	3.40	55.23	1.06	8.16	0.98	9.37	1.73	3.25	1.03
g	Clay	10.50	8.75	1.26	740	0.13	1.60	76.04	1.15	0.95	1.00	2.16	0.88	3.02	1.09
h	Sand	11.50	7.66	1.23	1060	0.13	0.90	92.09	1.18	0.21	1.00	0.85	0.48	2.93	1.12
i	Clay	12.00	10.00	1.21	760	0.12	1.10	78.29	1.24	0.43	1.00	1.45	0.59	2.77	1.15
j	Clay	12.50	8.33	1.36	740	0.14	0.60	73.23	1.27	0.15	1.00	0.82	0.36	2.67	1.17
k	Clay	14.00	11.67	1.19	470	0.12	5.00	62.85	1.59	13.74	0.97	10.50	2.58	1.38	1.27
l	Clay	20.00	16.67	1.22	720	0.12	4.00	76.20	1.80	5.93	0.99	5.52	2.14	0.66	1.32
m	Sand	28.00	18.65	1.78	5450	0.18	4.30	173.05	1.90	1.33	1.00	0.79	3.37	0.55	1.36
n	Clay	31.00	25.83	1.31	1120	0.13	2.50	91.77	2.01	1.60	1.00	2.23	1.43	0.25	1.37
o	Clay	33.00	27.50	1.28	2500	0.13	3.40	138.67	2.11	1.29	1.00	1.36	1.90	0.01	1.37
p	Clay	34.00	28.33	1.29	2600	0.13	1.20	140.88	2.14	1.56	1.00	0.46	0.68	= 0	1.38

[a] Algorithms taken from Zeevaert (1983, p. 517).

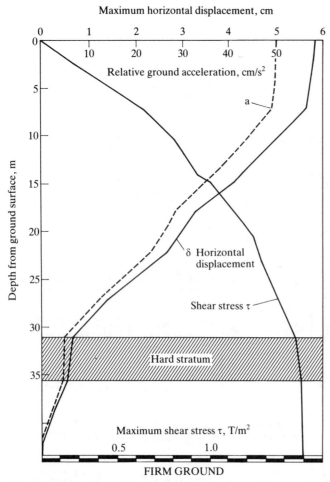

Fig. 17.6 Depth versus maximum horizontal displacement.

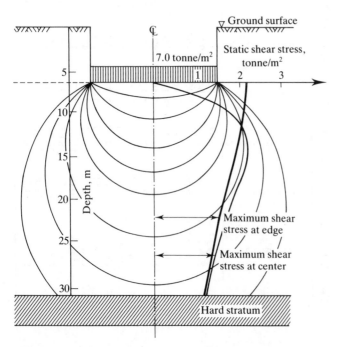

Fig. 17.7 Maximum static shear stress distribution under foundation structure.

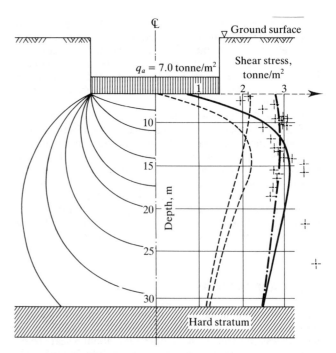

Fig. 17.8 Maximum shear stress including seismic effect: ———, at foundation center; —·—, at foundation edge; —¦—, $q_u/2$ (soil consistency).

foundation by means of (Zeevaert, 1983, p. 311)

$$\tau_{max} = \frac{\Delta q}{\pi} \sin \alpha_0 \qquad (17.16)$$

The shear distribution is calculated at the center of the section and at the edge based on a uniform soil reaction (Fig. 17.7).

The shear stresses induced in the soil mass by the strong ground motion have been added to the static maximum shear stresses (Fig. 17.8). The resulting maximum shear stresses taking place during the earthquake may be compared with the shear strength of the clay equal to $\frac{1}{2} q_u$. Notice that the shear stresses, static and dynamic, run up the shear strength of the soil in the upper part of the soil mass. This condition stressed the soil structure to a point that partial damage was produced and the compressibility increased, originating the consolidation process observed by the settlement records.

The experience just described shows the importance of analyzing the dynamic shear stresses when designing building foundations supported on high-compressibility and low-shear-strength clay sediments.

REFERENCES

Zeevaert, L. (1983), *Foundation Engineering for Difficult Subsoil Conditions*, 2d ed., Van Nostrand Reinhold Co., New York, N.Y.
  Chapter III.4, pp. 207–209
  Chapter XII.3, pp. 529–532
  Chapter XII.3, p. 523
  Chapter XII.3, pp. 519–529
  Chapter VII.6, p. 311
  Chapter III.2, pp. 164–166
Zeevaert, L. (1987), Seismo-Soil Dynamics Response of the Ground Surface and Building Foundations in Mexico City Earthquake, September 19, 1985, *ASCE Terzaghi Lecture*, Anaheim, October.
Zeevaert, L. (1988), *Seismo-Geodynamics of the Ground Surface*, Editora e Impresora International, S.A. de C.V., Mexico, D.F.

18 OFFSHORE STRUCTURE FOUNDATIONS

RONALD C. CHANEY, Ph.D., P.E.
Professor and Director
Fred Telonicher Marine Laboratory
Humboldt State University

KENNETH R. DEMARS, Ph.D.
Associate Professor of Civil Engineering
University of Connecticut

18.1 INTRODUCTION

18.1.1 Foundation Types

Marine foundations are used to transmit structural design loadings to the subsoil. The type of foundation element to be employed will depend on (1) the nature of loading, (2) the stiffness and strength of the surface sediments, and (3) the desires of the builder. A summary of the common platform types is shown in Figure 18.1. The two major foundation types are those that employ a surface loading mechanism (shallow foundations) and those that extend down through the surface sediments to a lower layer (deep foundations). An example of a foundation system for surface loading is the mat used on a gravity platform. The deep pile that is used on a jacket platform is an example of a deep foundation system. Examples of various marine foundation types are presented in Figures 18.2a and 18.2b.

This chapter is organized into an introductory section that discusses both the geologic and soil conditions of the marine environment. This introduction is followed by sections on loadings, pile structures, gravity platforms, anchor uplift capacity, pipelines, jack-up platforms, and hydraulic filled islands.

18.1.2 Geologic Conditions

The seafloor of the oceans is a complex environment consisting broadly of a continental shelf, continental slope, and abyssal plain areas. The majority of man-made construction takes place in the relatively shallow area of the continental shelf and to a lesser degree on the continental slopes. The complexity of the shelf areas is due to its geologic history and the action of the various environmental elements. These shelves vary in width depending on whether the continental margin is rising or slowly subsiding. Thus, the east coast of the United States has a very wide continental shelf owing to accretion, whereas that on the Pacific coast of South America is very narrow as a result of subduction of oceanic plates. Beyond the continental shelves are the slopes, averaging 4° down to the abyssal plain. Submarine canyons, which cut through both the shelf and slope, may have side slopes as great as 30°. They usually terminate in a fan on the deep seafloor.

The Pleistocene had a very dramatic influence on the continental shelf areas. For example, when the Wisconsin ice age was at its peak about 15 000 to 20 000 years ago, large quantities of water were stored at the poles, lowering the sea level by as much as 100 m. This lowering of the sea level resulted in the partial exposure of the continental shelves. On these coastal shelves, rivers discharging from the dry land cut channels into the sediments and land erosion processes took place. Since the rivers were steeper and velocities higher, sedimentary deposits were coarser. As the oceans have risen, the velocities have been reduced, and finer sediments—sands and silts—have been deposited on the shelves adjacent to large rivers.

During the ice ages, glaciers extended into the ocean basins, carving deep trenches such as the Norwegian Trench, Cook Inlet, and the Straits of San Juan de Fuca. With the subsequent warming period, the sea level has been rising, slowly but

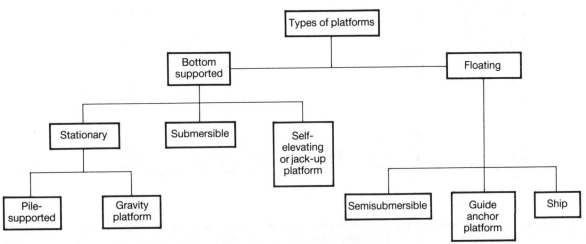

Fig. 18.1 Summary of platform types.

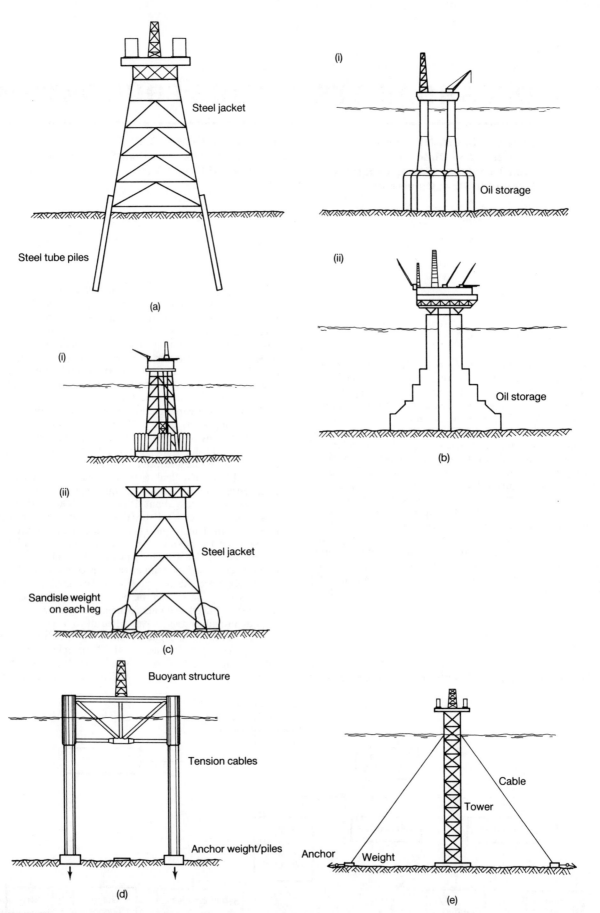

Fig. 18.2 Examples of marine foundation types. (a) Typical piled steel jacket structure. (b) Typical concrete gravity structures: (i) Condeep; (ii) Howard-Doris. (c) Hybrid steel gravity structures: (i) RDL hybrid; (ii) Sandpod. (d) Typical tension leg structure. (e) Guyed tower structure. (*After Gibson and Dowse, 1981.*)

inexorably flooding coastal areas, changing drainage patterns and creating new shoreline features. Sand dunes are dynamic features in this environment that form, migrate shoreward, and are then eroded and covered with the rising water. Major sand dunes in the southern North Sea and along the coast of The Netherlands have been submerged by the sea and now move back and forth as underwater sand dunes known as "megadunes." In a similar fashion, coastal silts and clays off New Jersey and Delaware have become overconsolidated by dune migration during the Pleistocene.

The shallow water deposits have been acted upon by water reworking and densifying the sands and silts. In the Arctic regions, nearshore silts have been subjected to cyclic freezing and thawing and to scouring by the keels of sea ice ridges. Off Greenland and Labrador, icebergs have reached down hundreds of meters to scour both sediments and rocks. Many such deep scours in rock can be seen on detailed underwater photos and acoustic images of the Straits of Belle Isle between Labrador and Newfoundland. Fault scars have been identified with 5-m high scarps in numerous offshore areas, and many more may exist, partially covered over by subsequent sediment deposition.

Many boulders are found on the floor of the North Sea and Gulf of Alaska. In the Cook Inlet, Alaska steep fjordlike walls of Turnagain Arm have rock falls that deposit large boulders on the tidal flats. In winter, the low tide freezes so that cakes of ice form around the boulders. On a subsequent high tide, the ice raft floats away, carrying the boulder with it. As the raft moves into the southern portion of Cook Inlet, the increased salinity and warmer water melts the ice, dropping the boulder.

18.1.3 Soil Conditions

A. General Background

The behavior of fine-grained sediment is governed to a large extent by its structure or fabric and mineralogy. The development of flocculated clay structures are encouraged in a seawater environment (Casagrande, 1932). The double-layer theory has been generalized to indicate that, for a given mineralogy, the electrochemical environment at the time of deposition is the most important variable in determining clay structure. If there are no environmental changes, the structure of fine-grained sediments acquired at the time of deposition is usually not destroyed or altered. Under such conditions, most sediments can be defined in terms of original structure and gross geologic events that produced changes in overburden pressure.

Marine sediments are derived from five main sources: weathering and erosion of land, volcanic ejecta, extraterrestrial matter, marine organism skeletons, and hydrothermal reactions. Some occur as clastic particles of terrestrial, volcanic, or cosmic origin; others are precipitates such as hydrogenous sediments (precipitated out of the sea or interstitial water), and many are biogenous sediments (from organisms that lived in the water or on the seabed). Based on their composition and origin, marine sediments can be divided into three major groups: lithogenous, biogenous, and hydrogenous sediments (Noorany, 1985). Geotechnical properties determined either in situ or in the laboratory of these marine materials can be found in the following references: George and Wood (1976), Demars and Chaney (1982), Briaud and Audibert (1985), Chaney and Demars (1985), Chaney et al. (1985), Chaney and Fang (1986a, b), Richards (1988).

B. Lithogenous Sediments

Lithogenous sediments are products of rock, volcanic ejecta, or meteorites. They can be subdivided into two groups: terrigenous sediments and pelagic clays. Terrigenous sediments are transported to the sea in the form of particles of gravel, sand, silt, and clay or a mixture of these components. Mixtures of terrigenous silt and clay are sometimes referred to as "mud." The terrigenous muds cover much of the continental margin and differ from deep-sea pelagic clays in that they have a considerable amount of silt with little or no authigenic materials such as zeolites.

Glacial marine sediments consist of gravel, sand, and other terrigenous materials that have been transported to deep water by floating icebergs (ice rafting) that gradually melt upon reaching warmer regions and drop their sediment loads. Another process, which has also been found in subtropical areas, is that of erosion, occurring when the sea was at a lower level than at present. The weaker deposits eroded away, dropping the boulders down and concentrating them. A third process occurs in granitic soils such as those of the east coast of Brazil and west coast of Africa, as well as in Hong Kong. As these deposits have weathered into residual soils, resistant cores have remained firm, thus becoming "boulders" formed in place (Gerwick, 1986).

Boulders also exist in clay deposits. Some of these arose as morainal deposits from glaciers, discharging their bed load into shallow water muds, which have since been overconsolidated by subsequent advances of the glaciers. These are the boulder clays of the North Sea. *Glacial till* is a term used to describe these unstratified conglomerate deposits of clay, gravel, cobbles, and boulders found in many Arctic and subarctic regions. The term is very nonspecific because some glacial tills have little binder and may be largely composed of gravel, whereas others may contain large boulders.

In addition to terrigenous sediments and pelagic clays, there are lithogenous hemipelagic silts and clays, which are terrigenous in origin but which have settled in the water and deposited on the floors of the continental slope and continental rise. They usually have more silt and less authigenic materials than deep-sea pelagic clays. In addition, their rate of sedimentation is over 10 times faster than that of typical deep-sea pelagic clays. They may be considered a transition between terrigenous muds and pelagic clays.

In many offshore areas there are very extensive accumulations of sands, in some cases due to longshore sediment transport of sand discharged from rivers, in other cases due to ancient sand dunes, such as found in the southern North Sea. Sand deposits in the North Sea and off Newfoundland have been subjected to the action of storm waves. The passage of storm waves results in the pore pressure being alternately raised and lowered along with the occurrence of drainage. Pore pressure variations of 35 kPa (5 psi) have been measured (Dunlap et al., 1978). After millions of cycles, the sand becomes extremely dense. Friction angles in excess of 40° have been reported. When these materials are sampled by conventional techniques, the sands are unavoidably disturbed; hence, laboratory tests will usually report lower densities and strengths than actually exist in the field.

C. Biogenous Sediments

Biogenous sediments consist of the remains of either marine plant or animal skeletons. They cover about one-half of the continental shelves, more than one-half of the deep sea abyssal plains, and parts of the continental slopes and rises. On a global scale, they cover more than 55 percent of the sea floor. Most of the biogenous sediments in shallow waters of the coastal zones are coarse-grained (sand sizes or larger); in deeper waters they are mostly fine-grained (silt and sand size). These finer biogenous sediments are usually referred to as oozes. In terms of composition of the mineral matter, they are calcareous, siliceous, or phosphatic.

1. Calcareous Biogenous Sediments

The biogenic component in these sediments consists chiefly of calcium carbonate ($CaCO_3$) (Chaney et al., 1982; Demars and Chaney, 1982). In the coastal zone and shallow waters of the continental shelf, they are primarily composed of shells and skeletal remains of benthic (bottom-dwelling) organisms such as corals and calcareous algae. As water increases, the proportion of plankton (drifter) organisms in the sediment increases. Thus, in the deep sea, biogenous sediments are made almost entirely of plankton remains, although some remains of the benthic foraminifera and clay minerals are also found.

In many tropical regions, the beach sand consists entirely of calcareous remains of corals, molluscs, and algae. Because coral sands are derived from mechanical erosion and breakdown of coral reefs and atolls by wave action (a mechanism similar to the process of erosion of clastic rocks), they are referred to as bioclastic sediments (Shepard, 1963; Berger, 1976).

Bioclastic Sediments The individual grains of bioclastic sediments are porous or hollow and have a rough texture. Such sands tend to crush more readily under stress than terrigenous sands. Calcareous sands may be found in cemented or uncemented states. It has been suggested by McClelland (1974), Murff (1987), Poulos and Lee (1988), Aggarwal et al. (1979), and Angemeer et al. (1975) that calcareous sands offer inferior support for pile foundations in comparison to non-calcareous sands on account of the following:

1. Crushing of sand particles during pile driving and the subsequent movement of crushed material into existing voids in the surrounding sand
2. Cementation, which prevents the development of lateral pressures on the pile wall

The compressibility of a calcareous sand is the result of four mechanisms:

1. Elastic deformation of the material
2. Rearrangement of the grains
3. Crushing of the grains
4. Breaking of cementation bonds when they exist

The first two mechanisms are common to all granular materials. The last two are very important for calcareous sands.

Grain Crushing The brittleness of bioclastic sand is characterized by the changing of its grain-size distribution under stress. Crushing magnitude can be expressed in terms of a crushing coefficient, CC, defined by Datta et al. (1979) as

$$CC = \frac{\text{percentage of particles of the sand, after subjection to stress, finer than D10 of the original sand}}{\text{percentage of particles of the original sand finer than D10 of the original sand}} \quad (18.1)$$

For skeletal sands, gain crushing appears progressively, beginning with low stresses and increases with increasing intraparticle porosity and with the grain angularity as shown in Table 18.1 and Figures 18.3a and 18.3b. Intraparticle porosity is determined typically by a vacuum saturation method. The intraparticle porosity is 0 percent for detrital sands and rises to 20 percent for some algal sands (Nauroy and LeTirant, 1983).

Cementation The existence of cementation bonds between grains may give calcareous sands a "cohesion" that is manifest as an apparent overconsolidation of the sand. Cementation is difficult to quantify because it is far from being homogeneous. It may be nodular. Weak cementation may be destroyed by core drilling, thus hiding its existence, although its influence on the installation and behavior of a pile may be real. In-situ measurement will certainly have to be developed.

For cemented deposits, a laboratory investigation is not totally relevant. Further it has not been possible so far to identify a parameter that quantitatively expresses the degree of cementation and usually the degree of cementation in any deposit is not uniform, which makes it difficult to interpret

TABLE 18.1 PHYSICAL PROPERTIES OF SANDS IN FIGURE 18.3[a].

Sand	Carbonate Content (percent)	Particle Characteristics	Particle Size	c_u[b]
A	92.2	Plate-like shell fragments, angular to subrounded particles with large intra-particle voids	Coarse	1.50
B	93.7	Plate-like shell fragments, subangular to subrounded particles with small intraparticle voids	Coarse to medium	2.11
C	91.3	Same as Sand B	Medium to coarse	2.12
D	95.0	Subrounded coralline debris particles with small intraparticle voids	Medium to fine	1.53
F	90.1	Rounded and oval nonskeletal particles	Medium to fine	1.90
G	46.9	Rounded and angular nonskeletal particles	Fine to medium	2.28
H	95.4	Subangular coralline debris particles	Medium to coarse	2.10
I	85.2	Angular to subrounded skeletal particles	Coarse to fine	3.15
E	—	Quartz, rounded particles	Medium to coarse	1.33

[a] Datta et al. (1980).
[b] Coefficient of uniformity

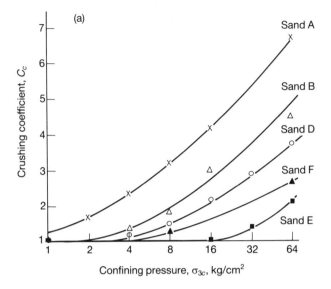

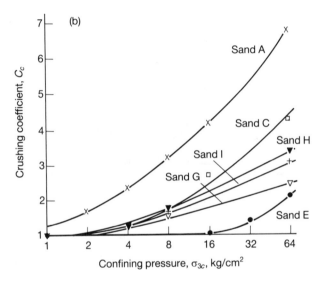

Fig. 18.3 Variation of crushing coefficient with confining pressure. (*Datta et al., 1980.*)

test data and conclusively identify trends (Saxena and Lastric, 1978).

Compressibility Compressibility is an essential parameter for characterizing the varying nature and behavior of calcareous materials. It integrates different effects such as (1) the physical characteristics of the grains (angularity, intraparticle porosity, brittleness), and (2) the way the grains are assembled (grain-size distribution, density, cementation). Compressibility of calcareous materials has been determined experimentally using an oedometer as shown in Figure 18.4 by Nauroy and LeTirant (1983). In Figure 18.4 the compressibility index C_p is defined as the slope of the $e-\log p$ curve. A review of Figure 18.4 shows that C_p is extremely variable and is dependent on the origin of the calcareous sand. In addition, the compressibility index C_p tends toward a limiting value for high confining pressures. Detrital calcareous sands have a compressibility that is comparable to that of ordinary siliceous sands (S). Skeletal sands with increasing intraparticle voids, which are generally more brittle, may be 10 to 100 times more compressible.

2. Siliceous Biogenous Sediments

The siliceous biogenic component in these sediments is made of opaline (hydrated silica dioxide: $SiO_2 + nH_2O$) skeletons of benthic organisms (sponges) on the continental shelves, and planktonic organisms on the slope and in the deep sea. The most common siliceous materials in the deep sea are frustules (hard siliceous shells) of diatoms, phytoplanktonic siliceous algae that range in size from a few micrometers (μm) to about 2 millimeters (mm), and radiolarians, a group of tiny protozoans that have various types of complex ornate skeletons with sizes commonly ranging from a few tens to a few hundreds of micrometers. In general, the particles sizes of most siliceous ooze microfossils fall in the silt size range of 0.03 to 0.07 mm.

D. Hydrogenous Sediments

These sediments are precipitates from seawater or sediment and rock interstitial water. They are also products of diagenetic alterations due to chemical reactions after deposition. Hydrogenous sediments include evaporites, nonskeletal carbonate sediments (calcareous sediments not made of skeletons of marine organisms), phosphorites, manganese nodules, and zeolites. Evaporites and nonskeletal carbonate deposits are only found in shallow waters, phosphorites are found on shelf areas and the top of seamounts, and ferromanganese nodules and zeolites are found primarily in the deep sea (Seibold and Berger, 1982).

1. Evaporites

These sediments are produced as a result of evaporation of seawater in semi-enclosed basins such as coastal lagoons and salt seas on the continental shelves. A significant basin where evaporites have formed is the Mediterranean, which was partially isolated from the Atlantic about 6 million years ago (Shepard, 1963; Seibold and Berger, 1982). Common evaporite sediments are sodium chloride (halite), calcium carbonate (argonite) and calcium sulfate (anhydrite and gypsum).

2. Nonskeletal Carbonate Deposits

These sediments are formed when high temperature and low CO_2 content in the seawater creates a favorable condition for the dissolved calcium carbonate to precipitate. One of the most ideal locations for this phenomenon to occur is the Bahamas, where the water is warm and highly alkaline, and carbonate deposits precipitate there at a rapid rate. The presence of shallow-water algae contributes to this process (Berger, 1976; Shepard, 1963; Seibold and Berger, 1982) by creating the delicate chemical balance that promotes precipitation of carbonates as crusts on the algae. In the Bahamas, aragonite needles (a few micrometers long) are believed to have formed with certain algae.

Another deposit of this type, which is very common on the Bahamian coast, is oolite. Oolites are sand-sized (about 0.3 mm in diameter), spherical particles formed of concentric layers of aragonite needles and organic matter.

3. Phosphorites

These sediments are commonly found on continental shelves, slopes, and tops of seamounts, and occur as sand to gravel-size nodules and cakes. They typically contain 20 to 30 percent P_2O_5. They are considered both hydrogenous (because they are deposited as a result of a chemical process in the seawater) and, at the same time, biogenous. They are also biogenous because some are composed of phosphate hard deposits such as fish debris and crustacean carapaces (Seibold and Berger, 1982).

	Sand	Location	CaCO$_3$ %	Origin	n_{intra}, %
o	CA	Carribean	98	Algal	20
▽	F	Florida Bay, USA	100	Shelly	15
+	CH	English Channel	90	Shelly	8
—	C2	Brittany, France	83	Shelly and coralline	8
□	AG2	Arabian Gulf	96	Coralline	—
△	C1	English Channel	90	Shelly	6
x	AG1	Arabian Gulf	90	Detrital	0
●	S	Fontainebleau, France	0	Terrigenous	0

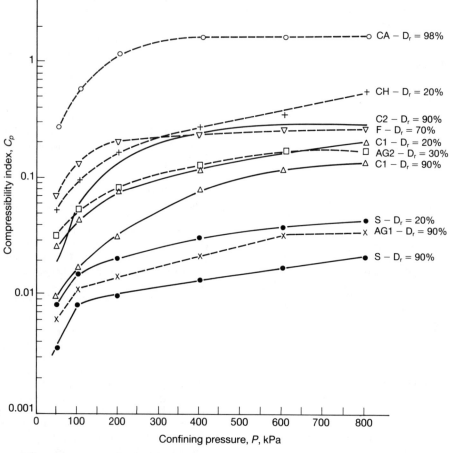

Fig. 18.4 Effect of confining pressure on compressibility index for some uncemented calcareous sands. (*Nauroy and Le Tirant, 1983.*)

18.2 LOADING ON FOUNDATIONS

Loadings that must be considered in the design of offshore structures include forces from waves, currents, ice, earthquakes, wind, and ship impact. These various forms of loadings produce compressive, lateral, and uplift forces that may strongly influence the design and construction of the structure. In addition, the cost of the structure will be directly proportional to the magnitude of the design load. Therefore, an understanding of the nature of these loadings is essential to achieve a proper design. Guidance for estimating environmental loadings can be found in Gaythwaite (1981), Dawson (1983), Fang and Chaney (1985, 1986), API Code (API, 1980), DNV Code (Det Norske Veritas, 1977), Hoeg and Tang (1977).

18.3 PILE STRUCTURES IN MARINE ENVIRONMENT

Piles are used to transmit structural loadings to the subsoil when the surface sediments cannot support the load. They may be single or in groups, and may be fully or partially embedded. Recent advancement in installation equipment has allowed these members to be driven in any desired environment and up to 1500 ft of water depth. The versatility of pile foundations has made them adaptable to most types of conditions and requirements. A detailed discussion of classical methods employed on land has been presented by Kézdi (1975) and Chapter 13. Methods employed offshore have been discussed by McClelland et al. (1967, 1969), McClelland (1974), and

McClelland and Cox (1976). In the following sections, methods and techniques applicable to the marine environment will be discussed.

The design of pile foundations for offshore structures must be able to satisfy the following criteria:

a. The piles and pile groups must have sufficient capacity to resist maximum anticipated loads with adequate factors of safety.
b. The piles and pile groups must have acceptable load deflection characteristics in all modes for the anticipated loading conditions.
c. Pile steel and grout stresses must remain within allowable limits under extreme loading conditions.
d. The available pile driving plant must be able to install the piles to their required penetration without overstressing.

18.3.1 Pile Design

A. Design Loading and Factor of Safety

The working load on the pile is taken as the capacity divided by an appropriate factor of safety. Generally, an appropriate factor of safety is 2.5 for a single pile. A value down to 2.0 may be used when a sufficient number of loading tests have been carried out or where there is a large body of local experience. Where there is less certainty of the value of the capacity, a value up to 3.0 should be used. A larger factor of safety should also be applied for cases involving large impact or vibratory loads or where soil properties may deteriorate with time.

B. Axial Capacity

The capacity (Q) of an axially loaded pile is dependent upon the components of resisting force developed by the end-bearing (i.e., toe or point resistance) (Q_p) and the frictional resistance (i.e., skin friction or shaft resistance) developed between the soil and the surface of the pile (Q_s), as shown in Figure 18.5.

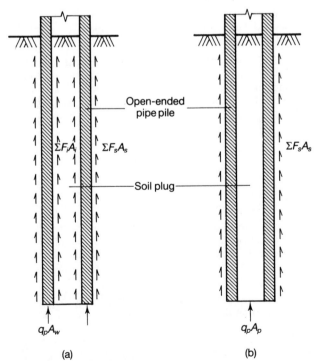

(a)   (b)

Fig. 18.5 Capacity of open-ended pipe piles: (a) unplugged; (b) plugged.

There are a number of computational techniques available to determine the axial capacity of a pile (Bea, 1975; Cox et al., 1979; Poulos, 1981b). These techniques vary from simple empirical procedures to more refined methods that look at the components of pile strength. The methods available to predict the axial capacity of offshore piles can lead to calculated capacities that vary by ± 25 percent from the mean of the predictions (Kraft et al., 1981).

The methods to calculate the end-bearing resistance (Q_p) and the skin friction (Q_s) depend upon whether the soil is a sand or clay. The relationship for capacity (Q_u) of a pile minus the weight of the pile (W) is given by Equation 18.2:

$$Q_u = Q_p + Q_s - W \qquad (18.2)$$

The capacity in compression for a pipe pile is based on two criteria, representing two extreme conditions: (1) the pile is considered as closed-ended, or (2) the pile behaves purely open-ended.

1. Closed-Ended or Plugged Pile
For a closed-ended or plugged pile, the compressive capacity results from

$$Q_c = \sum f_o A_o + q_p A_p - W \qquad (18.3)$$

where

Q_c = compressive capacity plugged or closed-ended pile
f_o = outer unit skin friction
A_o = outer pile shaft area
q_p = unit pile point end bearing
A_p = pile gross end-bearing area ($\frac{1}{4}\pi d_o^2$)
d_o = outer pile diameter

2. Open-Ended or Unplugged Pile
In contrast, the compressive capacity for an open-ended or unplugged pile can be computed using

$$Q_u = \sum f_o A_o + \sum f_i A_i + \sum q_w A_w - W \qquad (18.4)$$

where

Q_u = compressive capacity unplugged pile
f_o, f_i = outer and inner unit skin friction
A_o, A_i = outer and inner pile shaft area
q_w = unit pile wall end bearing
A_w = cross-sectional area of pile point ($t(d_o - t)$)
t = pile wall thickness
d_o = outer pile diameter

For any given penetration, Q_p and Q_u can be computed. The final assigned capacity is governed by the lower value of the two. Hence, comparing Equations 18.3 and 18.4, the pile will be plugged, that is closed-ended behavior will result, if

$$Q_o > Q_c \quad \text{or when} \quad \sum f_i A_i > q_p A_p'$$

where

A_p' = end-bearing area of pile plug ($\frac{1}{4}\pi d_i^2$)
d_i = inner pile diameter

This condition occurs in relatively low end-bearing soils, such as clays and silts. In dense sands the reverse condition often governs:

$$Q_c > Q_o \quad \text{hence} \quad q_p A_p' > \sum f_i A_i$$

In that case, piles are unplugged. Artificial plugs (concrete or grout) can be placed in the pile in order to develop the full end-bearing capacity, if required.

3. Tension Pile
The pull-out capacity of tension piles is derived only from the outer skin friction (Reese and Cox, 1976; Puesch, 1982). The

plugged or unplugged condition of the pile is not significant for pull-out capacity. This relationship is given by Equation 18.5:

$$F_t = \sum f_t A_o \qquad (18.5)$$

where

F_t = pull-out pile capacity
f_t = outer unit skin friction
A_o = outer pile shaft area

Equations 18.2 through 18.5 present the fundamental basis of any axial capacity determination, irrespective of the method to establish the soil parameters f and q. Several computation techniques are available to derive unit skin friction and unit end bearing either from laboratory and sampling data or from in-situ test data. The in-situ test most utilized is the cone penetration test (CPT). Details on the performance of the CPT can be found in Andresen et al. (1979) and de Ruiter (1971, 1975).

The method to be employed to determine f and q depends on whether the sediment is a sand or a clay.

4. Determination of End Bearing and Skin Friction

End Bearing Determination of the end-bearing component of axial capacity can be determined from bearing capacity theory as given in Equation 18.6:

$$q_p = c_{ub} N_c + \sigma'_v N_q + \frac{\gamma d N_\gamma}{2} \qquad (18.6)$$

where

N_c, N_q, N_γ = bearing capacity factors
γ = unit weight of soil
d = pile diameter
c_{ub} = cohesion of soil beneath base
σ'_v = vertical stress at level of pile base

In the following, the case of clay and sand will be presented.

(1) *Saturated clays.* In saturated clays, the undrained load capacity of the pile is normally considered the most critical value. Therefore, employing a total stress undrained analysis and assuming $\phi = 0$:

$$q_p = c_{ub} N_c + \sigma'_v \qquad (18.7)$$

where

c_{ub} = cohesion of soil beneath pile base
N_c = bearing capacity factor, usually taken as 9.0

(2) *Sands.* In contrast, for sands end bearing is computed using Equation 18.8:

$$q_c = \sigma'_v N_q \qquad (18.8)$$

where

σ'_v = vertical effective stress at pile tip
N_q = bearing capacity factor

Based on the assumption that the soil is perfectly plastic material, a wide scatter of N_q variations have been proposed by various authors, as shown in Figure 18.6. Notably, recommendations from American Petroleum Institute (API, 1980) and Det Norske Veritas (1977) are very conservative. Of the values shown in Figure 18.6, that of Berezantsev et al. (1961) is considered to be the most reliable (Nordlund, 1963; Vesić, 1965; Tomlinson, 1977; Canadian Foundation Engineering Manual (CFEM), 1978).

(3) *Calcareous sediments.* Nauroy and LeTirant (1983) have found that end bearing of calcareous sediments is influenced

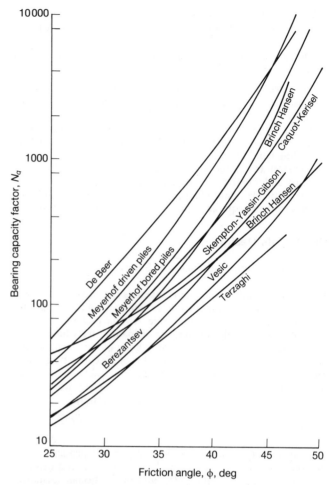

Fig. 18.6 Bearing capacity factors for deep circular foundations. (*After Vesić, 1967.*)

greatly by the compressibility of the calcareous sand material. This has also been shown by Datta et al. (1980) in Figure 18.7 using SPT N values.

(4) *CPT determination of end bearing for sands and clays.* The CPT method can be used to determine the unit point resistance for either sand or clay using the following relations (de Ruiter, 1975; Toolan and Fox, 1977; Ruiter and Beringen, 1979).

$$\text{Sand} \quad q_p = q_c$$
$$\text{Clays} \quad q_p = N_c S_u \qquad (18.9)$$

where

q_c = unit cone resistance
S_u = undrained shear strength

The computational methodology used to calculate the ultimate unit end bearing (q_u) in cohesionless soils is obtained from a CPT profile as shown in Figure 18.8. The calculated unit end bearing (q_u) by the CPT is then corrected as a function of a precompression ratio and grading.

(5) *Limiting values of end bearing.* According to Meyerhof (1976), Equation 18.9 is applicable only for piles with $L/D < L_c/D$, where L_c is the critical depth of penetration of pile, D is width or diameter of pile, and L is the depth of pile penetration. Beyond the critical depth ratio, q_p may be taken as q_c, which is constant with depth and equal to the value of q_p at L_c. The critical depth ratio is function of and may be estimated from

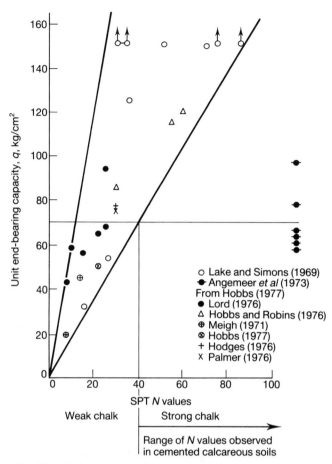

o Lake and Simons (1969)
•-Angemeer *et al* (1973)
From Hobbs (1977)
• Lord (1976)
△ Hobbs and Robins (1976)
⊕ Meigh (1971)
⊗ Hobbs (1977)
+ Hodges (1976)
× Palmer (1976)

SPT *N* values

Weak chalk   Strong chalk

Range of *N* values observed
in cemented calcareous soils

Fig. 18.7 Relation between unit end-bearing capacity and standard penetration resistance in chalk. (*Datta et al., 1980.*)

Figure 18.9. Both API (1980) and DNV (1977) have not recommended the critical depth for limiting values. But both methods have recommended limiting values for end bearing. DNV (1977) recommends a maximum end bearing value of 100 kPa for all soils. In contrast, the limiting values of end bearing recommended by API (1984) are based on density and gradation as shown in Table 18.2.

Skin Friction

(1) *Clay.* Available load test data on piles in clay show that the average frictional resistance expressed as a percentage of the average undrained shear strength or average effective overburden pressure decreases with increasing pile penetration. Three different procedures have been developed based on this data base. The first two, alpha method and lambda method, are based on total-stress concepts. The third, beta method, is based on effective stress concepts.

The lambda concept presented by Vijayvergiya and Focht (1972) and subsequently by others (Janbu, 1976; Meyerhof 1976; Flaate and Selnes, 1977) revealed that the available load test data on piles of clay show that the average frictional resistance expressed as a percentage of the average undrained shear strength or average effective overburden pressure decreases with increasing pile penetration. The percentage is termed the *mobilized friction ratio*, and the decreasing trend is called the *length effect*. The procedure is a quasieffective-stress approach because the correlation indirectly recognized the existing state of stress as represented by the undrained shear strength and the effective overburden pressure. The lambda empirical correlation shown in Figure 18.10 relied on data only from pipe

piles but included all types of clay profiles. For shallower depths, data from overconsolidated formations predominated. In addition, some of the data is from (1) piles with short setup time after driving, (2) piles redriven to deeper penetration after initial testing, and (3) piles with oversize closure plates. Questions have been raised concerning two areas: (1) greater pile penetration is required than obtained with the API RP2A $f = Su$, and (2) blind application of the procedure to short piles in soft clay will imply shaft friction considerably greater than the undrained shear strength.

In this procedure the total friction capacity Q_s can be obtained from

$$Q_s = \lambda(\sigma'_m + 2C_m)A_s \qquad (18.10)$$

where

λ = friction capacity coefficient
σ'_m = mean effective vertical pressure for depth of pile embedment
C_m = mean undrained shear strength for depth of pile embankment
A_s = surface area of the pile

In response to possible conservatism in lambda, a critical review of the lambda procedure was made by Focht and Kraft (1981) with the result that a correlation of lambda versus pile compressibility was developed (π_3). Pile compressibility is defined by Murff (1980) as follows:

$$\pi_3 = \frac{\text{ultimate capacity}}{\text{capacity if peak transfer is mobilized}} \qquad (18.11)$$

$$\pi_3 = \frac{\pi D L^2 f_{max}}{A E u^*} \qquad (18.12)$$

where

D = diameter of pile
L = length of pile
f_{max}—refer to Figure 18.11
A = cross-sectional area of pile
E = Young's Modulus
u^*—refer to Figure 18.11

Most offshore piles in normally consolidated clay have a π_3 value in the range of 10 to 100. A plot of data for various pile types versus π_3 on a log scale is presented in Figure 18.12. Focht and Kraft (1981) believe that lambda versus $\ln \pi_3$ is the best single correlation applicable to all clay profiles; however, for normally consolidated clays both lambda and beta with $\ln \pi_3$ provide good correlations.

In the beta method, Burland (1973) argued that the skin friction should be expressed in terms of effective stress as presented in Equation 18.13:

$$f_s = \beta \sigma'_v \qquad (18.13)$$

where

σ'_v = effective vertical overburden pressure at point
β = skin friction coefficient

For normally consolidated soils, Burland demonstrated that a lower limit for β is β_{nc} as given in Equation 18.14:

$$\beta_{nc} = (1 - \sin \phi') \tan \phi' \qquad (18.14)$$

where ϕ' = effective friction angle.

For overconsolidated soils, Meyerhof (1976) suggested the following:

$$\beta = \beta_{nc}(OCR)^{0.5} \qquad (18.15)$$

where OCR = overconsolidation ratio.

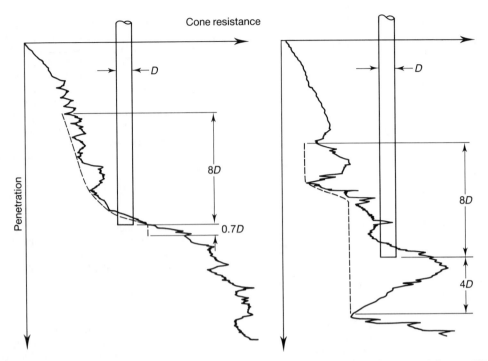

Fig. 18.8 Computation of the ultimate unit end bearing of a pile from a cone penetration test. (*Toolan and Coutts, 1980.*) The ultimate unit end bearing "q_u" is calculated from:

$$q_u = \frac{\dfrac{I + II}{2} + III}{2}$$

in which:

I   = average cone resistance below the foundation over a depth which may vary between 0.7D and 4D, selected in such a manner that the most unfavorable condition is obtained with respect to the computed bearing stress

II = minimum cone resistance recorded below the foundation over the same depth of 0.7D or 4D

III = average of the envelope of minimum cone resistance recorded above the foundation level over a height of 8D. In determining this average, values above the minimum selected under II are to be disregarded

D   = Outside diameter of pile.

Note: For an open-ended pile, the ultimate unit end bearing of the annulus may be assessed directly from the cone resistance (q_c). It must not exceed the limit value.

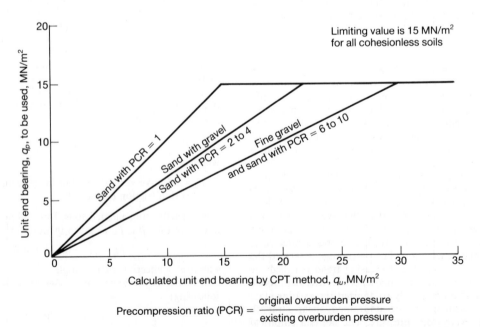

Fig 18.9 Correction factors for the point capacity of driven piles. (*After te Kemp; from Toolan and Coutts, 1980.*)

TABLE 18.2   CURRENT API (1984) RECOMMENDATIONS[a].

Density	Soil Description	Soil–Pile Friction Angle, δ, degrees	Limiting Skin Friction Values, kip/ft^2 (kPa)	N_q	Limiting End-Bearing Values, $kips/ft^2$ (MPa)
Very loose Loose Medium	Sand Sand–silt[b] Silt	15	1.0 (47.8)	8	40 (1.9)
Loose Medium Dense	Sand Sand–silt[b] Silt	20	1.4 (67.0)	12	60 (2.9)
Medium Dense	Sand Sand–silt[b]	25	1.7 (81.3)	20	100 (4.8)
Dense Very dense	Sand Sand–silt[b]	30	2.0 (95.7)	40	200 (9.6)
Dense Very dense	Gravel Sand	35	2.4 (114.8)	50	250 (12.0)

[a] The parameters listed in this table are intended as guidelines only. Where detailed information such as in-situ cone tests, strength tests on high-quality samples, model tests, or pile-driving performance is available, other values may be justified.

[b] Sand–silt includes those soils with significant fractions of both sand and silt. Strength values generally increase with increasing sand fractions and decrease with increasing silt fractions.

The alpha procedure was derived by Drewry, Weidler, and Hwong (1977) from the same data set used for the original lambda correlation. For pile-capacity calculations the procedure utilizes a multiplier α on the undrained shear strength (S_u) of the clay. The American Petroleum Institute (API) code has traditionally recommended a value of 1.0 for Gulf of Mexico type clays or a value that varies with strength for other types of clays. The API (1984) code varies the α value with the ratio of undrained soil strength to effective soil overburden (ψ) as given in Equation 18.16:

$$f_s = \alpha S_u \qquad (18.16)$$

where

$\alpha = 0.5\,\psi^{-0.5}$ for $\psi \leqslant 1.0$
$\alpha = 0.5\,\psi^{-0.25}$ for $\psi > 1.0$
f_s = pile friction
S_u = undrained shear strength
σ'_v = vertical effective stress
$\psi = S_u/\sigma'_v$

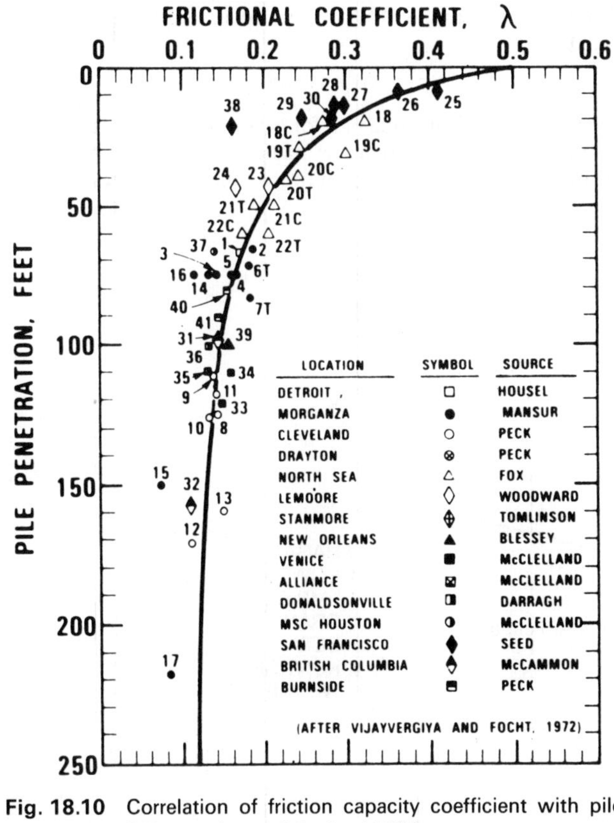

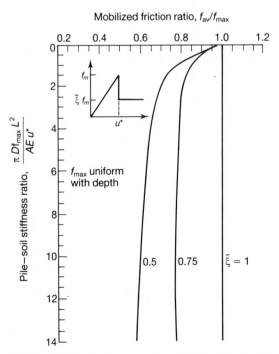

Fig. 18.10 Correlation of friction capacity coefficient with pile penetration. (*Vijayvergiya and Focht, 1972.*)

Fig. 18.11 Mobilized pile capacity with changes in pile–soil stiffness. (*Focht and Kraft, 1981.*)

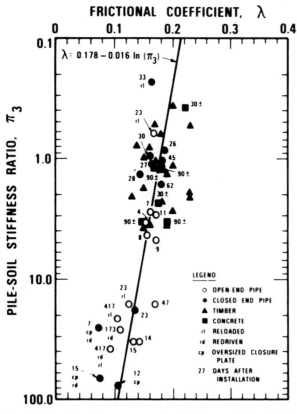

Fig. 18.12 λ Profile on terms of π_3. (*Focht and Kraft, 1981.*)

For a normally consolidated clay where ψ is constant and generally less than 0.4, the range of α lies between 0.8 and 1.0. For overconsolidated clays the ψ ratio varies and therefore α will also vary.

The α parameter can be equated to the β parameter as in Equation 18.17:

$$\beta = \alpha\,\psi \qquad (18.17)$$

The advantage of using the β parameter rather than α for the determination of skin friction is that σ_v' used in Equation 18.13

is an easier and more reliable parameter to measure in the field. The α parameters used in API (1984) can be rewritten as shown in Equation 18.18 (Toolan and Ims, 1988):

$$\beta = 0.5\,\psi^{0.5} \quad \text{for } \psi \leqslant 1$$

$$\beta = 0.5\,\psi^{0.75} \quad \text{for } \psi > 1 \qquad (18.18)$$

The capacities given by Equations 18.13 and 18.16 as compared to pile load test data is presented in Figures 18.13 and 18.14. A review of these figures shows that these equations tend to overpredict capacity for normally consolidated clays and underpredict capacity for heavily overconsolidated clays.

(2) *Sand.* Because of the fast rate of excess pore water stress dissipation, piles driven in sand are dominated by drained conditions and drained parameters are used in the analysis.

The average shaft resistance per unit area can be computed from the following equation based on the principles of Coulomb friction:

$$f_s = K\,\sigma_v'\tan\delta \qquad (18.19)$$

where

f_s = skin friction (lb/ft^2, kPa)
σ_v' = effective vertical stress (lb/ft^2, kPa)
K = coefficient of lateral earth pressure
δ = coefficient of friction between the soil and the pile shaft

The coefficient of lateral earth pressure (K) values depend significantly on many factors, and API (1980) guidelines suggested that K lay between 0.5 and 1.0 for compression loading and was 0.5 for tension loading. The API (1984) and subsequent editions have stated that for pipe piles, "it is usually appropriate to assume K as 0.8 for both tension and compression loading." The coefficient of friction, $\tan\delta$, between the soil and the pile shaft according to DNV (1977) recommendations may be taken small or equal to $\tan\phi$. Prior to the API (1984) edition, recommended practice for a medium dense to dense sand was for δ to be taken as given in Equation 18.20:

$$\delta = \phi - 5° \qquad (18.20)$$

Therefore, a typical value of δ for clean sand is 30°. In the API (1984) edition no reference is made to the angle of internal friction for the soil material but recommended values for δ are given in Table 18.1.

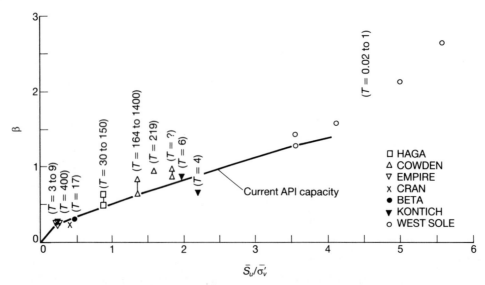

Fig. 18.13 Pile load test data. (*Toolan and Ims, 1988.*)

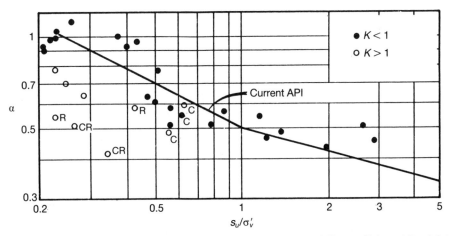

Fig. 18.14 Load test data from API data base. R = redriven; C = oversize shoe; K = flexibility coefficient. (*Randolph and Murphy, 1985.*)

(3) *Calcareous sediments.* Pile installation disturbs the surrounding soil and changes the stress state characterizing this soil in its virgin state. The normal stress on the pile wall evolves from σ_{e0} to a value of σ_e at the end of insertion. A graph showing the variations of lateral pressure in sand during the insertion of a model pile is shown in Figure 18.15. A review of Figure 18.15 shows that the variation of lateral pressure is a function of the type of material as well as whether the pile was open-ended or closed-ended. An open-ended tubular pile disturbs the soil much less than a closed-ended pile, although the existence of a plug may create conditions similar to those for closed-ended piles.

When the pile penetrates into the soil, the soil is compressed, deformed, and pushed laterally aside. The amplitude of the lateral push depends on the compressibility of the sand. The variation of σ_e is linked to the pushing back of the sand, and

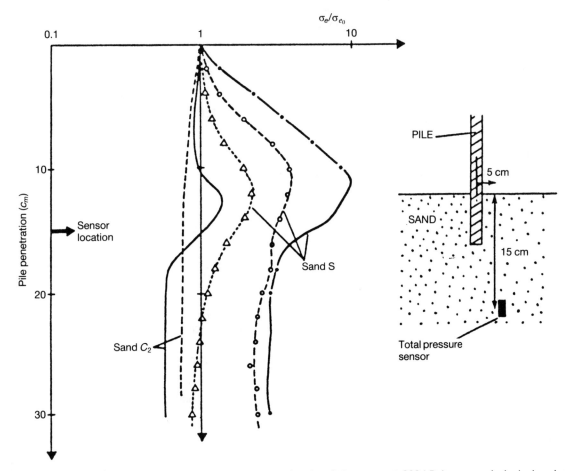

Fig. 18.15 Variations of lateral pressure in sand during pile penetration (confining pressure 200 kPa). ———, Jacked, closed-ended pile. — — —, Driven, closed-ended pile. ······, Driven, open-ended pile. S, terrigenous sand; C, shelly and coralline sand. (*Nauroy and Le Tirant, 1983.*)

hence to the compressibility. For the calcareous sands, the crushing of the grains is a major phenomenon. Under the great stresses imparted by driving, this crushing is quite extensive. Such grain crushing is thus responsible for a drop in σ_e along the pile and therefore low skin friction values.

Skin friction was shown by Nauroy and Le Tirant (1983) to be less for calcareous sands as compared to siliceous sands. The skin friction available for calcareous materials decreased as the compressibility of the materials increased as shown in Figure 18.16. In addition, the skin friction of closed-ended model piles was also shown not to be greatly affected by increasing confining pressure (Nauroy and Le Tirant, 1983).

The influence of sand cementation and the role of the plug in open-ended tubular piles cannot be clearly evaluated by laboratory tests and thus require recourse to in-situ pile experiments. The presence of in-situ cementation, even to a slight degree, may help prevent lateral pressure from developing (Angemeer et al., 1975) or may even cancel it out, thus reducing skin friction.

(4) *CPT determination of skin friction for sands and clays.* The cone penetration test can be used to determine the skin friction in both clays and sands as follows:

$$\text{Clays} \quad f_s = K_1 S_u$$
$$\text{Sand} \quad f_s = K_2 q_c \tag{18.21}$$

where

K_1 = empirical factor (1.0 in N.C. clays, and 0.5 in O.C. clays)
K_2 = empirical coefficient (1/300 compression, 1/400 tension)
$S_u = q_c/N_k$, N_k is cone factor varying between 15 and 20 with an average value of 16

(5) *Limiting values of skin friction.* Experimental work involving piles of sand has shown that unit skin friction between the pile and soil does not continue to increase unbounded as suggested by Equation 18.19. Limiting values have been used to establish upward bounds. Traditionally, values presented by McClelland et al. (1969) have been used. These recommendations provided a maximum unit skin friction of about 100 kPa. The API (1984) provides specific guidelines with limit values ranging from approximately 50 to 115 kPa depending on soil gradation and density. DNV (1977) recommends a maximum of 10 MPa for all soils. Similarly, de Ruiter and Beringen (1979) have also proposed limiting values of skin friction (0.12 MN/m²) for CPT-based offshore pile design.

5. Ultimate Capacity of Pile Groups

Determination of the load-carrying capacity of a pile group requires an estimation of single pile capacity, together with a judgment of the effect of interaction between piles in a group. Typical offshore pile groups are shown in Figure 18.17.

The ultimate capacity of a group of piles that derive most of their load-carrying capacities through point resistance can generally be taken as the sum of the ultimate load capacities of individual piles. However, for piles that transmit most of their load through shaft resistance, group effects must be taken into consideration. For piles in which shaft and point resistance are equally important, it is recommended that group effects be considered only for shaft resistance (Chellis, 1961; Vesić, 1969).

For piles that derive most of their capacities from shaft resistance, the load-carrying capacity of the group is generally less than that of a single pile multiplied by the total number of piles in a group. This reduction is often expressed in terms of

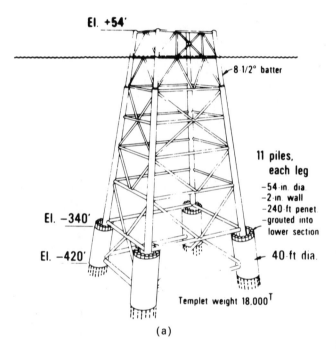

(a)

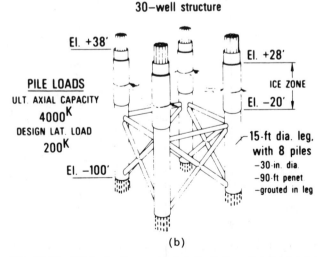

(b)

Fig. 18.17 Offshore pile groups. (a) North Sea; (b) Cook Inlet. (*O'Neill, 1983.*)

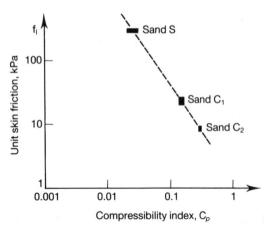

Fig. 18.16 Limiting unit skin friction versus limiting compressibility index. (*Nauroy and Le Tirant, 1983.*)

an efficiency factor (η) as given in Equation 18.22:

$$\eta = \frac{\text{ultimate load capacity of group}}{\text{sum of ultimate load capacities of individual piles}}$$

(18.22)

Groups in Clay Several techniques are available for computing pile group capacities in clay (Kézdi, 1975).

(1) Converse–Labarre formula (Chellis, 1961)

$$\eta = 1 - \phi\left(\frac{(n-1)m + (m-1)n}{90mn}\right) \quad (18.23)$$

where

$\phi = \arctan{(D/S)}$
$D =$ diameter of pile
$S =$ center-to-center spacing of piles
$n =$ number of rows of piles
$m =$ number piles in row

(2) Terzaghi and Peck (1967)

These authors suggest that the group capacity is the lesser of the sum of the ultimate capacities of individual piles or the ultimate capacity of the block as given in Equation 18.24:

$$Q = B_g L_g S_u N_c + 2L(B_g + L_g)S_u \quad (18.24)$$

where

$B_g =$ plan width of pile group
$L_g =$ plan length of pile group
$L =$ length of pile
$S_u =$ undrained shear strength
$N_c =$ empirical bearing capacity factor

Groups in Sand Unlike pile groups in clays, pile groups in sands may have an efficiency factor less than or exceeding unity, depending on the relative density of the sand mass, pile roughness, and spacing of piles in the group.

When groups of piles are driven into loose sand, the soil around the pile becomes highly compacted. If the pile spacing is close enough, this will increase the frictional resistance along the pile shafts and hence the ultimate group capacity may exceed the sum of ultimate capacities of individual piles.

If sand is so dense that the sand particles will dilate instead of undergo compaction, group efficiency may then be less than unity.

Thus, for dense sand, efficiency may be less than 1.0. For loose sand maximum efficiency is at a spacing of approximately two pile diameters. This maximum efficiency ranges between 1.8 and 2.2 (Lo, 1967).

Studies by Vesić (1969) on the behavior of pile groups in sand indicated that grouping has practically no effect on the ultimate point resistance. The point resistance efficiency remains constant at unity, independent of spacing. In contrast, Vesić found a significant increase in ultimate shaft resistance when piles are placed in a group. CFEM (1978) recommends that in sand, group efficiency of 1.0 may be taken unless actual load tests are performed.

Stresses on Underlying Strata Due to Pile Load A stress analysis is necessary to compute immediate settlements on loose granular deposits or consolidation settlements in clay deposits. The pile group transmits the load through the soil mass for friction piles or at the soil below the pile tip in end-bearing piles. The soil at or below the tip of piles must carry the load without excessive deformation. The stresses on strata or in the soil underlying a group of piles are typically calculated analytically.

Analytical method: Analytical solution of the stresses in the strata underlying a pile group is an extension of the Mindlin solution of a point load at the interior of an elastic solid. Geddes (1969) used the Mindlin solution for subsurface loading and given the influence factor the stresses at any location can be computed. Poulos and Davis (1968) also used the Mindlin solution to predict settlement. They presented a chart of settlement influence factor. Either the Geddes or the Poulos and Davis solution should provide the same deflection if properly used, since both are based on the Mindlin solution. The Geddes solution has an advantage that one can easily compute deflection from stresses, but stresses are not easily back-computed.

C. Pile Settlement

The settlement of a pile is equal to the displacement of the pile point and the elastic shortening of the pile between cap and point. For point-bearing piles the point displacement is relatively small and the principal displacement is the elastic shortening of the pile. For friction piles the point displacement will cause significant settlement. The settlement of a pile depends on the soil properties and the pile geometry and dimensions. In addition, the settlement also depends upon the type of pile and the method of construction. The settlement of a single pile can be computed by applying the theory of elasticity, step integration, or empirical rules and using data from load tests, finite-element analysis, or empirical methods. In routine analysis, solutions from elasticity are generally employed for cohesive soils and empirical methods in sandy soils.

1. Settlement in Cohesive Soils

Initial Elastic Settlement The elastic solutions are based on Mindlin's solution (Mindlin, 1936). As discussed previously, Geddes (1969) influence values can be used for calculating the settlement of a pile if the distribution of the total load between skin friction and point resistance is known. It should be emphasized that the amount of displacement required to mobilize shaft resitance is small (5 to 10 mm) and is practically independent of the soil type as well as the diameter and the length of the pile. In constrast, the movement required to mobilize the base resistance is large and generally increases with the diameter of the pile (Coyle and Reese 1966; Moretto, 1971). Taking this into account in settlement calculations requires a sophisticated method of analysis (Vijayvergiya, 1977).

On the basis of Mindlin's equations, Poulos and Davis (1980) have developed a design chart for rigid, cylindrical piles of diameter d and length L, embedded in a uniform elastic layer of thickness h that rests on a rigid base. The settlement (ρ) can then be calculated from Equation 18.25:

$$\rho = \frac{P}{LE} I_p \quad (18.25)$$

where

$L =$ length of pile
$P =$ load
$E =$ modulus of elasticity
$I_p =$ displacement influence factor, refer to Figure 18.18
$\rho =$ settlement

The most difficult step in this calculation is the selection of a suitable value of E.

Layered Soils The analysis can be used with reasonable confidence for cases in which the pile extends through an upper softer layer into a stiff material. The modulus of the equivalent

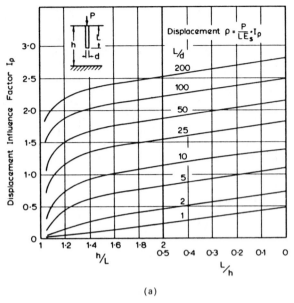

(a)

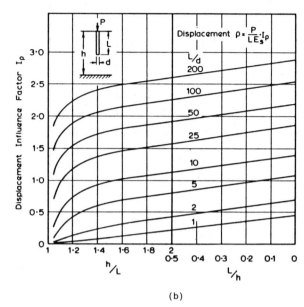

(b)

Fig. 18.18 Displacement of incompressible pile in finite layer. (a) $V_s = 0.4$; (b) $V_s = 0.5$. (*Poulos and Davis, 1968.*)

uniform soil layer can be taken as the weighted mean as given in Equation 18.26:

$$E = \sum \frac{E_k \delta_k}{L} \qquad (18.26)$$

where

L = length of pile
E_k = modulus of elasticity of kth layer
δ_k = thickness of kth layer

Pile Groups The settlement of a group of loaded piles is always greater than that of a single pile carrying the same load as that on each pile within the group (O'Neill, 1983). This occurs because the zone of soil stresses created by a pile group has a much greater dimension than that of a single pile.

The increase in settlement of each pile due to interaction of neighboring piles may be given by an interaction factor α as presented in Equation 18.27:

$$\alpha = \frac{\text{additional settlement due to adjacent pile}}{\text{settlement of pile under its own load}} \qquad (18.27)$$

The values of α for piles in semi-infinite soil mass and on a finite soil layer are shown in Figures 18.19 and 18.20, respectively. Thus, for a group of three piles, the value of α is twice the value for a group of two piles at the same spacing, while for a group of four piles at the spacing it is given by the following

$$\alpha = 2\alpha_1 + \alpha_2 \qquad (18.28)$$

where

α_1 = the value of α for two piles at a spacing S
α_2 = the value of α for two piles at a spacing of $\sqrt{2}S$

2. Consolidation Settlement
If a compressible layer is located below the pile tips, the settlement of a pile group can be computed as in the case of a spread footing. The stresses in the underlying layers may be estimated from empirical load distribution curves or Geddes

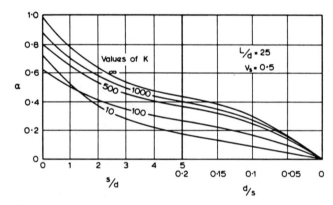

Fig. 18.19 Interaction factors for two floating piles in a semi-infinite mass. Note: (1) V_s = Poisson's ratio of soil mass. (2) $K = (E_p/E_s)R_A$; E_p = Young's modulus of pile; E_s = Young's modulus of soil mass; R_A = area of pile/$(\pi d^2/4)$; K = pile stiffness factor. (*Poulos and Davis, 1974.*)

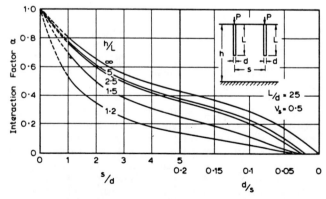

Fig. 18.20 Interaction factors for two incompressible piles in a finite layer. (*Poulos and Davis, 1974.*)

(1969) theory as discussed previously. The total settlement would be then the sum of the elastic settlement and the consolidation settlement.

Empirical Methods Empirical methods are generally used for computation of settlement in sand. In sandy material the total settlement is given by

$$\rho = \rho_s + \rho_{pp} + \rho_{ps} \qquad (18.29)$$

where

ρ_s = settlement due to axial deformation of pile shaft
ρ_{pp} = settlement of pile point caused by load transmitted at the point.
ρ_{ps} = settlement of pile point caused by load transmitted along the pile shaft

$$\rho = (Q_c + \alpha_s Q_A)\frac{L}{AE_p} \qquad (18.30)$$

where

Q_s = actual point load transmitted by the pile
Q_A = actual skin friction load transmitted by the pile
L = length of pile
E = modulus of elasticity of pile
A = area of pile
α_s = number that depends on the distribution of skin friction along the pile shaft

Pile Group A number of semiempirical methods have been proposed to calculate the settlements of pile groups in sand (O'Neill, 1983). Large deviations can be expected when the conditions at a particular site deviate from those at which the method was derived. Comparison with available test data indicates that often calculated settlement of a pile group will be too large when a modulus of elasticity that is constant with depth is used in the analysis.

(1) *Analytical semiempirical methods.* Three semiempirical methods are given below. The settlement of a pile group in sand is a function of the width of the pile group (Skempton, 1951) and can be expressed by Equation 18.31:

$$\frac{\rho_{group}}{\rho_{single}} = \left(\frac{4B+3}{B+4}\right)^2 \qquad (18.31)$$

where

B = width (least dimension) of pile group in meters
ρ_{group} = settlement of pile group
ρ_{single} = settlement of single group

Meyerhof (1959) modified this relationship to take account of the spacing of the piles and gave the following relationship:

$$\frac{\rho_{group}}{\rho_{single}} = \frac{(S/D)(5-S/3D)}{(1+1/R)^2} \qquad (18.32)$$

where

S = spacing of the piles
D = pile diameter
R = number of rows in the pile group

Vesić (1977) has proposed a simple relationship as given in Equation 18.33:

$$\rho_{group} = \rho_{single}\sqrt{B} \qquad (18.33)$$

(2) *Settlement from static cone penetration test (CPT).* Settlement in cohesionless soils can also be calculated from the results of static cone penetration test results. Typically in this method the pile group is assumed to be equivalent to a raft located at the lower third point of the pile penetration depth. The soil below the equivalent shaft is then divided into a number of layers and the compression of each layer is calculated separately. The reader is referred to empirical relationships of Skempton (1953) and Meyerhof (1959), and analytical methods of Poulos and Davis (1980).

(3) *Settlement from standard penetration test (SPT).* The settlement of a pile group in sand can be estimated conservatively from the results of SPT tests using the following relationship (Meyerhof, 1976).

$$\rho_{group} = \frac{0.09\, q\, I \sqrt{B}}{N} \qquad (18.34)$$

where

ρ_{group} = settlement of group (mm)
B = width of group (m)
q = net foundation pressure (kPa)
N = corrected SPT resistance (blows/30 cm) down to a depth that is equal to the width of the pile group below the bottom of the pile
I = influence factor, which can be evaluated by

$$I = \left(1 - \frac{D'}{8B}\right) \geqslant 0.5$$

D' = effective depth

D. Design of Piles for Lateral Loads

Analytical methods used to predict lateral deflections, rotations, and bending moments in the pile need to consider: (1) boundary conditions imposed on the pile by the superstructure, (2) varying moment of inertia of the pile, (3) soil layers of different stiffness, (4) nonlinear stress–strain behavior of the soil, and (5) group effects (Poulos, 1981a; Clausen et al., 1982). While it is feasible to formulate two- and three-dimensional nonlinear finite-element analyses (Smith, 1979), most design procedures use the semiempirical "p–y method". The p–y curves are analytical tools and not fundamental soil properties. The method is applied by treating the pile as a beam supported on discrete springs that act independently of each other (i.e., the Winkler foundation concept). A relationship is assumed between the lateral soil pressure, p, and the lateral pile displacement, y, at different positions along the pile (Eq. 18.35). This relationship is then solved for pile response:

$$EI\frac{d^4y}{d^4x} + P_x\frac{d^2y}{d^2x} - p - w = 0 \qquad (18.35)$$

where

P_x = axial load on pile
y = lateral deflection of the pile at a point x along the length of the pile
p = soil reaction per unit length
EI = flexural rigidity
w = distributed load along the length of the pile

The procedure that is followed by the computer to solve the nonlinear soil–structure interaction problem is shown in Figure 18.21 (Reese and Wang 1986). Figure 18.21a shows the pile and lateral load. Figure 18.21b shows a family of p–y curves with curves depicting the soil resistance opposite in direction to pile deflection. Also in Figure 18.21b are dashed lines showing the deflection of the pile for the first and second trial solutions. Figure 18.21c shows the upper p–y curve enlarged with the pile deflection at the depth represented by vertical dashed lines. The secant soil modulus (E_s) for the first and second trials is also

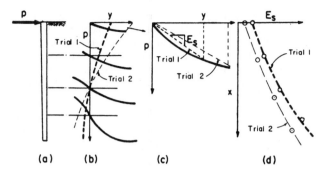

(a) **(b)** **(c)** **(d)**

Fig. 18.21 $p-y$ Curves iteration procedure. (*Reese and Wang, 1986.*)

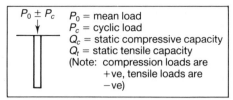

Zone A: cyclically stable. No reduction of load capacity after N cycles

Zone B: cyclically metastable. Some reduction of load capacity after N cycles

Zone C: cyclically unstable. Failure within N cycles or fewer

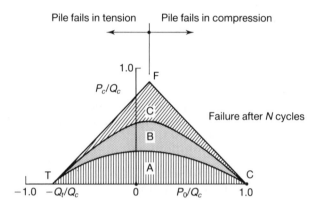

Fig. 18.22 Main features of cyclic stability diagram. (*Poulos, 1988.*)

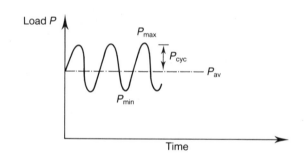

shown. Figure 18.21d shows possible values of soil modulus (E_s) plotted as a function of depth x. Interaction is continued until the difference in the deflections for the last two computations are less than a specified tolerance. The method is based on the back-analyses of a limited amount of actual pile test data using a "beam column" program (Reese, 1977). Typical methods for determining $p-y$ curves for various materials are given in the following references: clays, Matlock (1970); overconsolidated clays, Reese et al. (1975); sand, Reese et al. (1974). A detailed discussion of the limitations, advantages, and practical experiences has been presented by Sullivan et al. (1979) and Reese and Wang (1986).

Results using the "$p-y$ method" indicate that the lateral loads at mudline affect only the upper pile length of 10–15 diameters. The magnitudes of the computed bending moments in the pile are relatively insensitive to variations in the assumed $p-y$ curves along the pile, but the lateral displacements and rotations are clearly affected.

A pile in a group deflects much more than the same pile isolated when subjected to the same lateral load (Cooke et al., 1979; Matlock et al., 1980). The analysis of pile group behavior is by means of an incremental superposition of the nonlinear single-pile behavior based on a $p-y$ analysis and the interaction from the other piles in the group based on an elastic continuum analysis (Poulos, 1971).

E. Effect of Cyclic Loading

The design of piles to resist axial and lateral loading also requires an analysis of the influence of cyclic wave loading on the performance of a pile. The phenomena that need to be considered for axial response are (1) degradation in pile capacity due to cyclic loading, and (2) possible increase in pile axial capacity due to the rate of loading. Analysis of test results incorporating one or both of these effects have been presented by Matlock and Foo (1979), Bea (1980), Bea et al. (1980), Boulon et al. (1980), Karlsrud and Haugen (1985), Karlsrud et al. (1986), Nauroy et al. (1985), and Poulos (1979, 1981c, 1982, 1983). Recently Poulos (1988) has proposed a cyclic stability diagram as shown in Figure 18.22 to explain loading in both compression and tension.

Consider a pile subjected to a sustained axial tensile load P_{av} with a uniform cyclic axial load P_{cy} superimposed as shown in Figure 18.23. In a real storm the cyclic load magnitude and frequency will vary. After pile installation, but before pile loading, a soil element close to the pile shaft as shown in Figure 18.23 is subjected to effective stresses that change during reconsolidation after installation of the pile.

The application of the sustained tensile load P_{av} introduces an additional shear stress τ_{av} to the soil element, which may or may not have time to consolidate under the shear stress before the $\pm \tau_{cy}$ is introduced (Fig. 18.23). The results of laboratory

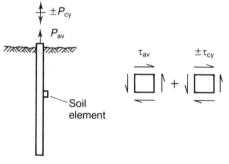

Fig. 18.23 Pile subjected to sustained and cyclic tensile loads. (*Hoeg, 1986.*)

studies can be presented schematically in the normalized diagram illustrated in Figure 18.24, where S_u denotes the undrained shear strength of the soil element when consolidated to the in-situ stresses and sheared to failure by monotonically increasing the shear stress. The figure shows a boundary line between a "stable" and an "unstable" region for a given number

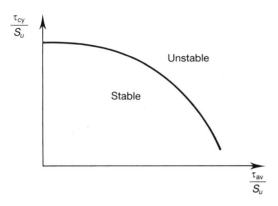

Fig. 18.24 Behavior of soil element shown in Fig. 18.23. (*Hoeg, 1986.*)

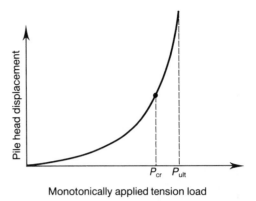

Fig. 18.25 Pile subjected to monotonically increasing load. (*Hoeg, 1986.*)

of cycles applied. Outside the boundary, the imposed stress conditions lead to a progressive accumulation of permanent (average) strains while undergoing cyclic loading. Inside the boundary envelope, permanent strains reach a constant value after the first few cycles. The position of the boundary between the "stable" and "unstable" regions depends on the number of cycles (or storm duration) the soil element is called upon to endure.

The same reasoning as used for the individual soil element may be applied to the pile in Figure 18.25. Assume that the pile, when subjected to a monotonically increasing load, pulls out of the ground when the load is equal to P_{ult}, while the somewhat smaller load P_{cr} denotes the load level at which the rate of creep deformation exceeds a certain acceptable value (Figure 18.25). The load P_{ult} or Q'_{us} will be used to normalize the average and cyclic pile loads in an interaction diagram corresponding to the one shown in Figure 18.26.

In the analyses of cyclic pile response it is necessary to consider the influence of load magnitude, rate, and number of cycles on the following factors:

1. Ultimate skin friction
2. Ultimate base resistance
3. Soil modulus

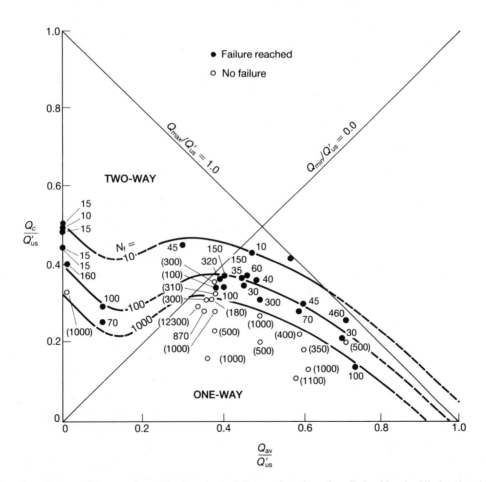

Fig. 18.26 Interaction diagram giving number of load cycles to failure as function of cyclic load levels. (*Karlsrud and Haugen, 1985.*)

To quantify cyclic loading effects it is convenient to define a degradation factor as follows (Poulos, 1983):

$$D = \frac{\text{property after cyclic loading}}{\text{property for static loading}} \quad (18.36)$$

The degradation factors for ultimate skin friction, ultimate base resistance, and soil modulus are D_τ, D_b, D_E, respectively.

1. Characterization of Cyclic Response

Degradation of Skin Friction (D_τ) Cyclic loading in general causes a reduction in the ultimate skin friction. This reduction is a function of the number of cycles of load application and the amount of cyclic displacement ($\pm \rho_c$). This conclusion is based on the results of small-scale model pile tests on both clays and sands. The relationship between the skin friction degradation factor (D_τ) as a function of cyclic displacement and number of cycles of loading for remolded clay is shown in Figure 18.27. The cyclic displacement ρ_c has been normalized both with respect to pile diameter (d) and also to the displacement required for full slip in a static test (ρ_{st}). A review of Figure 18.27 shows that a normalized cyclic displacement ratio (ρ_c/ρ_{st}) of 0.4 to 0.5 is required before degradation begins and D_τ becomes less than unity. For very large cyclic displacements ($\rho_c/\rho_{st} > 2$) reductions greater than 40 percent are shown to occur in skin friction after 1000 cycles. The model of degradation involving relationships such as shown in Figure 18.27 is referred to as the cyclic displacement model.

Holmquist and Matlock (1976) showed in model tests that two-way (compression/tension) caused a much more severe reduction in skin friction than one-way (compression) cycling (Fig. 18.28). Subsequently, Matlock and Foo (1979) developed an alternative model in which cyclic degradation at a point on the pile only occurs if there is plastic reversal of stress at that point.

Degradation of Ultimate Base Resistance There is very limited data of the degradation of base resistance under cyclic loading. One limited study by Van Weele (1979) found that in sand the degradation of base resistance could be more severe than for skin friction. Poulos (1983) suggests that because the data is so limited it can be assumed that the cyclic displacement

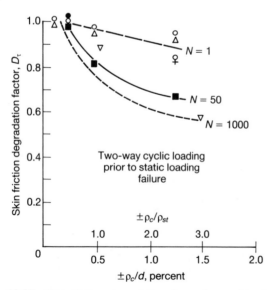

Fig. 18.27 Skin friction degradation factor for model piles in Hurstville clay. ρ_{st} = Pile displacement to mobilize full friction in static test. (*After Poulos, 1981c.*)

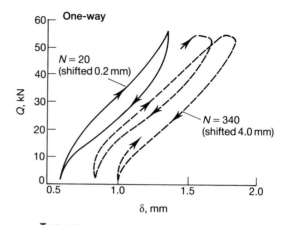

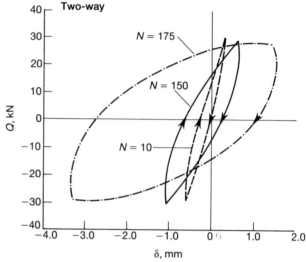

Fig. 18.28 Load–displacement curves during one-way and two-way cyclic loading. (*Karlsrud and Haugen, 1985.*)

degradation model can also apply to the ultimate base resistance with small modifications. The modification required is that in normalizing the cyclic displacement as shown in Figure 18.27 the value of ρ_{st} is the displacement to cause full mobilization of base resistance.

Degradation of Soil Modulus Model pile tests suggest that an expression similar to that developed by Idriss et al. (1978) from cyclic triaxial tests on clay may be applied to describe the degradation phenomena of soil modulus due to cycling. Therefore, for the cyclic displacement model of degradation, the modulus degradation factor D_E may be approximated by the expression

$$D_E = N^{-t} \quad (18.37)$$

where

   N = number of cycles
   t = a degradation parameter that is a function of normalized cyclic displacement

2. Lateral Response to Cyclic Loading

As a pile is cycled by lateral loadings, soil near the mudline deforms and forms a gap behind the pile, and water is drawn into this gap. As the pile is pushed in the other direction, water and some soil are expelled, extending the gap to some depth behind the pile. The development of the gap continues until the overburden pressures are sufficient to close any gaps. The upper zone of soil in which lateral stresses are not sufficient to close such gaps is termed the *zone of unconfined response*. The zone at depth, where the overburden pressures are sufficient to close any gaps, is termed the *zone of confined resistance* (Matlock, 1970). Experimentally derived lateral load (p) versus deformation (y) results for soil resistance in unconfined and confined response are shown in Figures 18.29 and 18.30, respectively. The gaping processes that are developed in the upper unconfined zone are shown in Figure 18.29 as dog-bone shaped hysteresis curves. In contrast, the hysteretic response of the soil in contact with the pile of the zone of confined response (Fig. 18.30) is similar to that derived from cyclic laboratory soil tests (i.e., cyclic triaxial, cyclic simple shear). In both of these figures, continued cyclic loading is shown to reduce peak

strength and stiffness, and increases the area of the hysteresis loops (i.e., damping).

The semiempirical $p-y$ curves can be adjusted to take into account the effects of repetitive cyclic loading (Scott et al., 1982). Superimposed cyclic axial loads reduce the lateral stiffness for flexible piles, but very little data exist to quantify these effects. In addition, the $p-y$ curves differ for piles in tension and compression. The lateral stiffness for a pile in tension is less than for one in compression because of the direction of the shaft friction along the pile, and hence the smaller normal stresses in the surrounding soil. It is difficult to incorporate present understanding of soil element behavior during cyclic loading using the $p-y$ method, and alternative methods of approach are being suggested (Richardson and Chaney, 1986). The soil element behavior, strain accumulation, and build-up and dissipation of excess pore pressures around the pile must be analyzed by means of finite-element or finite-difference continuum analyses.

The lateral deflections of the upper portion of the pile will lead to reduced axial skin friction and thus affect axial pile behavior. For an anchorage pile in tension, the lateral effects may prove significant. The cyclic axial loads on a pile in sustained tension are initially carried mainly by the upper portion of the pile. With time the cyclic effects will gradually extend deeper and will progress even further down because of lateral cycling, until a stable condition is reached.

In stiff, overconsolidated clays, cyclic lateral loads may open a gap (slot) between the pile and soil. The slot may gradually

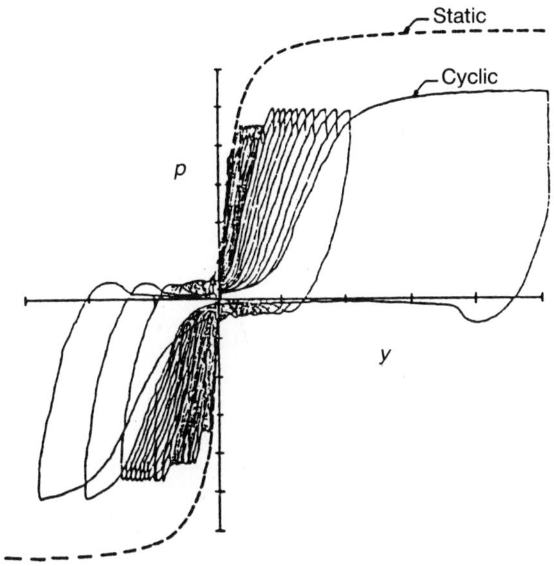

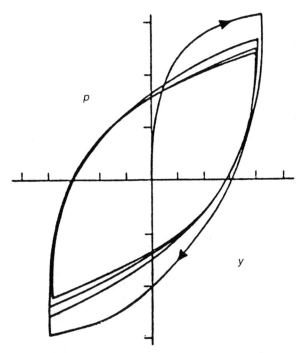

Fig. 18.29 Cyclic lateral load–displacement characteristics of soil in zone of unconfined response. (*Bea, 1980.*)

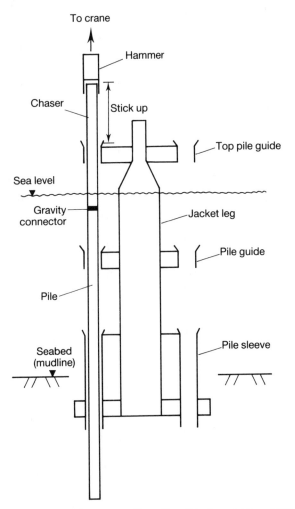

Fig. 18.31 Method of installing piles. (*Toolan and Fox, 1977.*)

Fig. 18.30 Cyclic lateral load–displacement characteristics of soil in zone of confined response. (*Bea, 1980.*)

grow wider and deeper because of erosion of soil caused by the squeezing of water out of the gap. This erosion, in addition to the seafloor surface erosion and scour, may significantly affect the pile design computations.

F. Construction and Installation of Pile Structures

Current practice to install piles has been by using a system of chasers and an above-water hammer (Fig. 18.31). A chaser is a pipe of the same diameter and wall thickness as the pile with a heavy torpedo-shaped weight at its lower end. The weight, known as a gravity connector, fits into the top of the pile and keeps the chaser in contact with the pile during driving. Each pile passes through a series of guides and a sleeve attached to the leg.

Piles are connected to the structure by grouting the annular space between the pile and its sleeve. A pile in its final position is shown in Figure 18.32. To increase the strength of the bond between the pile and the sleeve, shear connectors are placed over the length of the pile expected to be in the sleeve at final penetration. Packers may be placed at the bottom of the sleeves to prevent loss of grout.

Lateral loading on the jacket platform can produce high bending stresses in the piles as they emerge from the sleeve. To handle these stresses the wall thickness over a length of pile expected to be above and below the grout packer is increased to resist these stresses. This length is commonly called the "heavy wall section."

The installation of such piles must be carefully planned, since any delays may be costly.

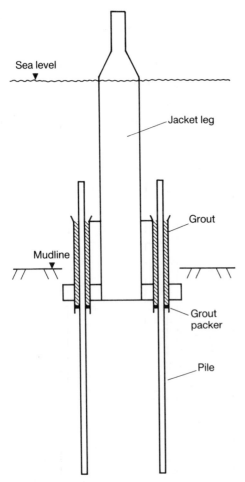

Fig. 18.32 Piles installed. (*Toolan and Fox, 1977.*)

1. Drivability Analysis

The ability of the available pile-driving equipment (hammers and chasers) to install the piles to the required penetration must be assessed. To understand the characteristics of offshore pile driving, consider the measurements in Figure 18.33. The pile systems involved in these tests represent typical situations for land (Fig. 18.33a) and offshore piles (Fig. 18.33b). In the land

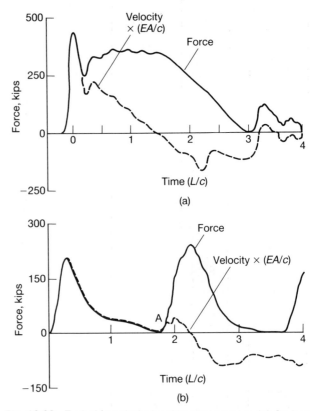

Fig. 18.33 Typical force/velocity–time measurement. (a) Onshore; (b) offshore. (*After Goble, 1983.*)

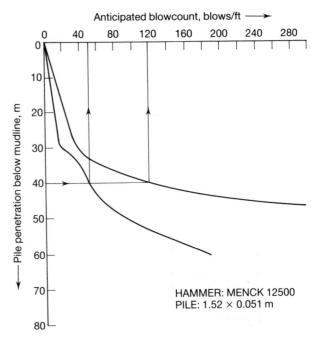

Fig. 18.34 Blow-count versus penetration depth. (*After de Ruiter and Beringen, 1979.*)

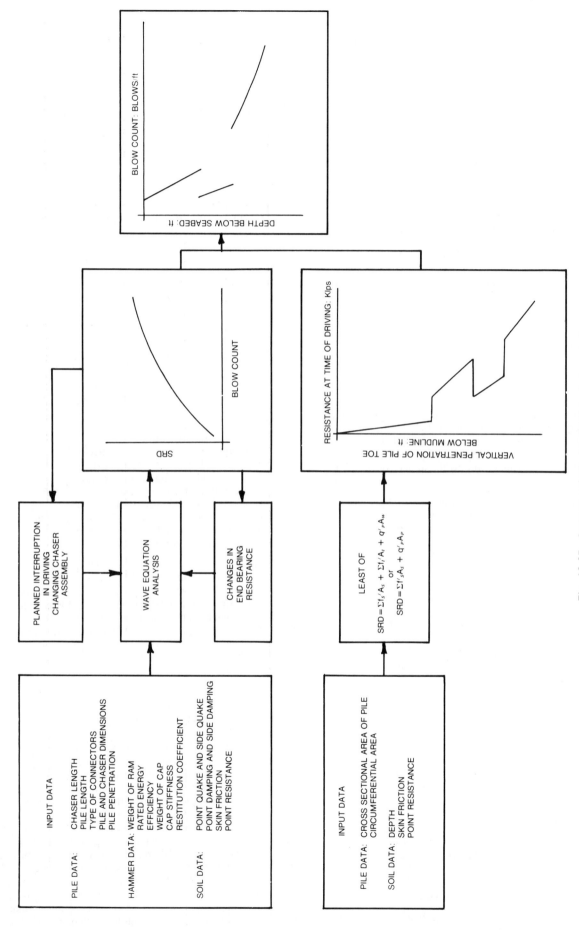

Fig. 18.35 Summary of pile drivability analysis procedure.

701

case, the presence of side friction forces near the pile top induces reflected compression forces that cause the force and velocity records to begin to deviate from each other. In contrast, the record from the offshore pile exhibits force–velocity proportionality, indicating no skin friction resistance, until almost all of the hammer energy has been transferred to the pile. The input and reflected forces in pile driving are commonly separated in the offshore case primarily because all offshore hammers are "light." In land applications, the ratio of ram weight to pile weight will typically be on the order of 2 for steel piles. In contrast, offshore ram weight ratios will commonly be less than 1/8. Thus, in the offshore case the hammer input force will decay much more rapidly relative to the time of pile tip reflection than is normal for land piles.

A drivability analysis consists of two primary phases. These are (1) a wave equation analysis to generate a curve of the static soil resistance at time of driving (SRD) versus blow count, and (2) an analysis to produce a curve of SRD versus depth. These two curves are then combined to generate a curve of blow count versus depth as shown in Figure 18.34. A schematic summary of the analysis process is shown in Figure 18.35. In the following section the methods employed to perform the wave equation analysis and to determine the soil resistance during driving will be discussed (Heerema, 1978, 1979).

Wave Equation Analysis　The first step in a typical drivability analysis is to perform wave equation analysis using a computer. The theory and application of the wave equation to pile driving problems has been described in a number of papers (Smith, 1962; Lowery et al., 1969; and Hirsch et al., 1975). In the typical wave equation computer program, the pile, soil, and hammer system are modeled as a series of masses, springs, and dashpots as shown in Figure 18.36. Owing to the impact of the ram, a

force wave starts traveling through the pile at a velocity of approximately 5000 m/s. The wave equation analysis then calculates for all elements in the system the corresponding velocities, displacements, and forces generated by the impact per time increment. The process is continued until the permanent set of the pile tip is achieved. This information then provides the expected blow count for a specified combination of soil resistance, pile, and hammer characteristics. The results from this analysis are normally presented as curves of blow count values versus soil resistance at time of driving (SRD).

Two different programs provide the basis of most of the current computer codes used in the analysis of offshore pile driving problems. The first is a program developed at Texas A & M University (Hirsch et al., 1976; Samson et al., 1963). The second is a program called WEAP developed by Goble and Associates under contract with the Federal Highway Administration (Goble and Rausche, 1980).

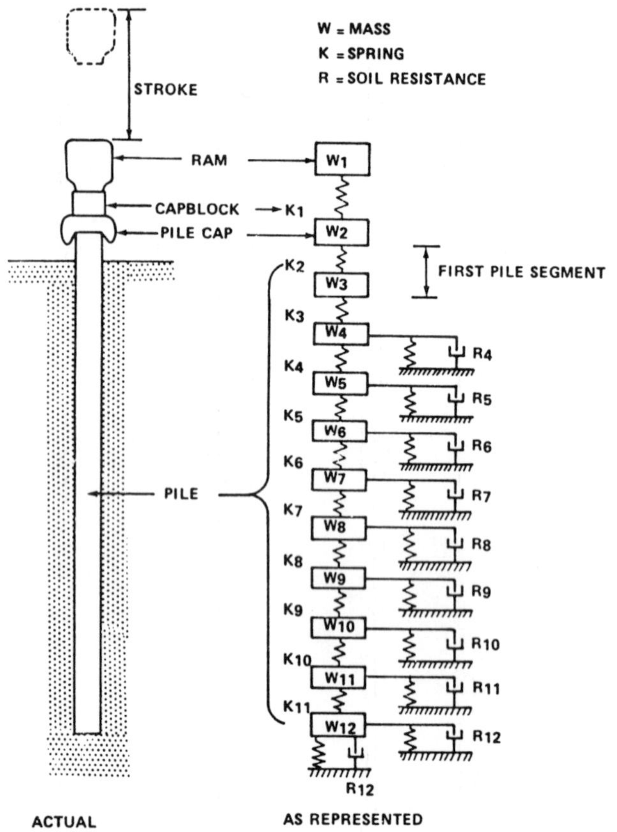

Determination of Soil Resistance During Driving (SRD)　There are two primary methods available to determine SRD. These methods are (1) CPT method (de Ruiter and Beringen, 1979), and (2) utilization of soil strength data from soil investigations (Toolan and Fox, 1977).

In the CPT method the SRD is computed using Equation 18.38, which assumes that fixed plugs during driving do not normally occur (de Ruiter and Beringen, 1979):

$$\text{SRD} = FD_i + FD_o + WD = \varepsilon F_s + A_p q_c \qquad (18.38)$$

where

$$\begin{aligned}
\text{SRD} &= \text{soil resistance during driving} \\
FD_i &= \text{inner skin friction during driving} \\
FD_o &= \text{outer skin friction during driving} \\
WD &= \text{wall end bearing during driving} \\
\varepsilon &= \text{empirical factor correlating hindcast } FD\text{s to computed outer skin friction} \\
A_p &= \text{cross-sectional area of pile point (annulus)} \\
q_c &= \text{cone resistance below the pile tip} \\
F_s &= \text{force due to skin friction}
\end{aligned}$$

In this relationship the ε factor has been computed based on combining driving results with CPT results from several North Sea locations. The typical range of values for ε is 1.6 at the surface to 1.0 at a depth of 50 m (de Ruiter and Beringen, 1979).

In the second method, soil strength data from soil investigations are used to compute the SRD. In this procedure, remolded friction values are used directly in sands and clays, while for pile tip resistances the undisturbed values are used (Toolan and Fox, 1977).

18.4　GRAVITY PLATFORMS

At a number of locations in the North Sea, large concrete gravity platforms have been used as alternatives to the more conventional pile-supported steel structures (Fig. 18.37). Among the many considerations that enter into the decision process between the pile-supported structures and gravity structures are the weather conditions during the platform installation period and the soil conditions at the site. Construction of pile-supported structures requires long duration at a fixed location in deep water offshore, whereas gravity structures can be constructed in deep, quiet water close to the shore and then towed out and placed on the desired site. The gravity structure also has a wide deck area and oil can be stored inside the platform. The unique aspects of the structure's foundation consists of dowels for positioning and concrete/steel skirts as shown in Figure 18.38. The gravity structure may be moved

Fig. 18.36　Pile driving model in wave equation. (*After de Ruiter and Beringen, 1979.*)

(a)

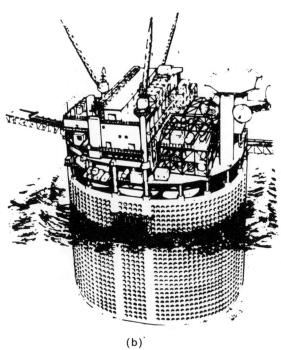

(b)

Fig. 18.37 (a) Statfjord A, Condeep. (*After FIP, 1978.*) (b) Ekofisk tank, Doris. (*After FIP, 1978.*)

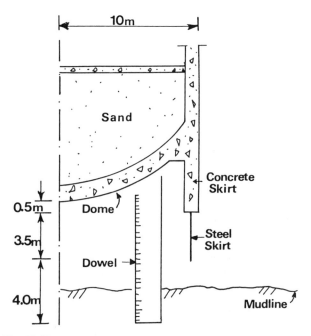

Fig. 18.38 Detail of the base of a typical gravity platform structure. (*Clausen, 1976.*)

and possibly reused by destabilizing the platform to zero submerged weight and by applying an excess water pressure under the base to pull the foundation skirts out of the seafloor.

The relatively flat seafloor topography and firm near-surface soil conditions are well suited to gravity type platforms. The principal features of offshore concrete gravity structures are their large size, their construction afloat and deployment reasonably complete, and their installation on a usually unprepared site.

Two types of structures have been constructed for the North Sea. One type involves a series of submerged cells comprising the concrete base supporting one or more concrete shafts (caissons) that in turn support the deck. The other type of structure consists of a surface-piercing tank more suitable for shallow waters because it gives generally higher obliquity and eccentricity of load than the first type. A few common types of gravity structures are shown in Figure 18.39. The caisson volume is designed to float the deck loads, and the foundation

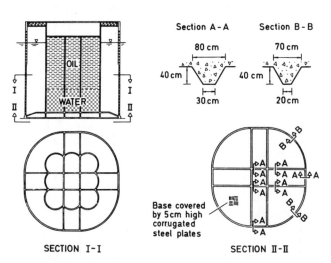

Fig. 18.39 A typical North Sea gravity platform design with concrete skirts. (*Schjetne et al., 1979.*)

base area has to be designed for the specific soil conditions at the site. Generally the structure's ballasting system provides weight by sluicing water into caisson compartments. During installation it is important to see that no overstressing of any part of the structure takes place owing to reaction forces from the ground and that the installation should not affect the foundation soil in an unfavorable manner, for instance, by washing out soil below the base.

18.4.1 Design Requirements

Gravity structures exert enormous forces on the seabed owing to their dimensions. These forces come from a combination of the weight of the structure and from the waves. As an example, the weight of Ekofisk tank in the North Sea is close to 200 000 tons when placed at the site. The 100-year wave expected in the North Sea varies from 24 m to 30 m depending on the location and depth of water. The primary force on large-displacement gravity structures is provided by the vertical component of the wave loading. The drag forces exerted by currents are typically neglected. There is a phase shift of 90° between the inertial force component and the drag force component. Consequently the peak horizontal force occurs when the wave surface elevation is near the water level. The resultant vertical load on the foundation changes during the passage of a wave. The effects of currents (velocity without acceleration) are usually unimportant for fluid loadings on the structure but they may be important for smaller objects such as conductors. In North Sea situations other loadings, such as those resulting from floating ice or seismic events, may provide total loadings that are greater than the 100-year wave (30.5 m). Loading is also provided by the storage of crude oil. This may have a specific gravity less than seawater so that an uplift is induced. Extra weight on the foundation will be provided if the specific gravity of the crude oil is greater than that of the seawater. If the crude oil is not evenly distributed in the storage compartments the uplift or extra weight will be eccentric, providing extra moments on the foundation.

The design of structures of this type involves some very unusual geotechnical problems. The requirements that gravity structures should be designed to meet are the following.

1. When subjected to maximum wave and wind loads the structure shall have an acceptable factor of safety against horizontal sliding and against a shear failure in the foundation soil.
2. The displacements experienced by the gravity structure should be acceptable when subjected to maximum wave and wind forces.
3. The factor of safety against a shear failure and the displacements should also be acceptable after the structure has been subjected to periods of repeated wave forces in severe storms.
4. The natural frequencies of the structure should differ from those of wave forces, ensuring that there will be no resonance.
5. The structure should be designed so that undermining by erosion is prevented and a good contact will be obtained between the base slab and the seafloor, and will be maintained during wave loading.
6. When required, precautions should be taken to prevent the foundation from becoming endangered by scouring caused by currents or wave action.

The overall foundation stability is checked for the situation of a long storm followed by the "100-year wave." The large forces from the platform have to be carried safely by the foundation, which now has reduced strength and stiffness owing to the previous cyclic loading. The stability analysis is performed statically owing to the long wave period, but the forces are multiplied by a dynamic amplification factor. During a storm there may be significant dissipation of excess pore pressure in sand, but for clay foundation soils no significant drainage can occur even during a series of succeeding storms. The cyclic displacements will decrease and stability will improve with time for platforms resting on sand or normally consolidated clay. However, for a structure on overconsolidated dilating clay, the long-term effect may increase cyclic displacements and reduce stability. So the behavior of clay subjected to cyclic loading is an important design problem for offshore gravity platforms. Anderson (1976) dealt with this problem rigorously and gave the following recommendations.

1. Behavior of clay subjected to cyclic loading may be explained in terms of effective stresses. For solution of practical problems, however, the stress-path method based on representative cyclic tests interpreted on total stress basis may be convenient.
2. The effect of cyclic loading increases with increasing cyclic shear stress levels and the number of cycles. It appears that beneath a certain cyclic shear stress level, cyclic loading has negligible effect.
3. At the same cyclic shear stress level τ/S_w, normally consolidated clay will be more resistant to cyclic loading than overconsolidated clay. This means that on a total stress basis a foundation on normally consolidated clay may be designed with lower safety factor for cyclic loading than one with overconsolidated clay.

Shear modulus and failure shear stress in undrained static loading are reduced by undrained cyclic loading. Effective shear stress parameters C' and ϕ' are not influenced by undrained cyclic loading for overconsolidated clay. For normally consolidated clay C' seems to increase. For both normally consolidated and overconsolidated clays, cyclic loading will influence the pore pressure response to undrained static loading.

18.4.2 Design Elements

The base geometry of a typical gravity structure is shown in Figure 18.40b. This example is of a Condeep structure construction in the North Sea. In addition to raft or mat base there is the addition of either concrete or steel skirts and dowels.

In Condeep structures three dowels of diameter 2.5 m and length 4 to 5 m below the tip of the steel skirts touch the seafloor first. These dowels are provided in order to prevent the structure from moving laterally when the steel skirts themselves reach the seafloor. When the steel skirts have penetrated 3.5 m into the soil, the concrete skirts come into contact, and after a further

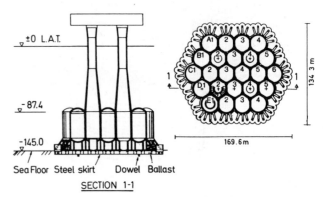

Fig. 18.40 A typical North Sea gravity platform design with steel skirts. (*Schjetne et al., 1979.*)

0.5 m penetration the undersides of the concrete domes touch the seafloor. The skirts serve a number of functions in the design:

1. Resistance to lateral loads
2. Transfer of loads through a soft mantle to harder underlying materials.
3. Scour prevention
4. Prevention of hydraulic instability
5. Avoidance of high contact pressure on base slabs
6. Containment of grout flow
7. Compartmenting of foundation

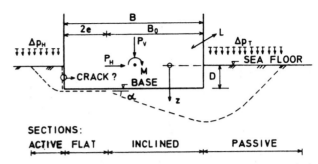

Fig. 18.42 Limiting equilibrium foundation stability analysis. (*After Lauritzsen and Schjetne, 1976.*)

18.4.3 Geotechnical Design of Foundation System

The following analyses are to be carried out for the soundness and safety of the platform and its foundation:

1. Platform installation, including skirt penetration analysis, the use of underpressures and grouting procedures
2. Analysis of load transfer from the structure to the soil
3. Overall foundation stability with respect to sliding and overturning of the platform
4. Initial and long-term settlements
5. Short-term cyclic displacements
6. Dynamic analysis of the soil structure system for wave loads and earthquakes
7. Stresses in conductor and risers due to short- and long-term displacements and settlements
8. Problems related to the removal of the platform

Several of the above analyses as shown in Figure 18.41 have to be carried out for different load combinations at different times in order to find the critical safety factor. For example, local contact stresses between the base and soil may be predominant in installation phase before the voids have been grouted. Similarly, the shear strength of the soil reduces after several years of production owing to cyclic loading to lower than it was shortly after the installation. A typical limiting equilibrium foundation stability analysis is shown in Figure 18.42.

Generally, a gravity platform is designed against a 100-year wave (defined such that over a long period of time a wave of this magnitude or greater will on the average occur once every 100 years). The data for a 100-year wave varies from site to site. For a typical North Sea site, the height will be between 24 m and 30 m, the wave length will be 600 m, and the wave period will be about 14 to 18 seconds. The resulting horizontal forces acting on the structure are thus very large compared with what is normally encountered on land and become the design criteria. To prevent sliding on the seafloor or shear failure in the sand beneath the foundation slab, the structure must exert on the foundation soil a vertical pressure that must be a certain proportion of the horizontal force.

One of the first structures built in the North Sea, Ekofisk, settled over 3.5 m owing to reservoir depletion (Fact Sheet, 1986; Clausen et al., 1975). Valhalla is settling at the same rate as Ekofisk (400 mm/yr) and Frigg is experiencing problems related to erosion (*Offshore Engineer*, 1986).

A. Bearing Capacity

The ultimate bearing capacity of spread foundations on uniform soils can be estimated with sufficient accuracy on the basis of plasticity theory, and the theoretical relationship can be represented by Equation 18.39 (Chapter 4):

$$q = \frac{Q}{BL} = C N_c + \gamma D N_q + 0.5 \gamma B N_\gamma \qquad (18.39)$$

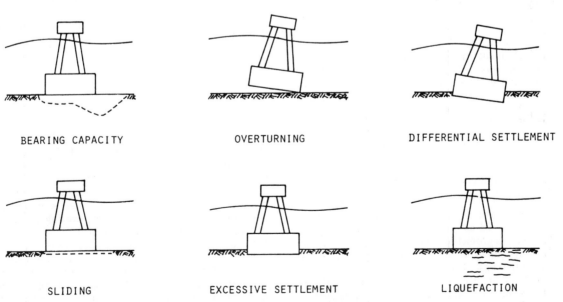

Fig. 18.41 Modes of foundation failure for gravity structures.

where

B = width of footing
D = depth of foundation
L = length of foundation
C = apparent cohesion of soil
γ = effective unit weight of soil
N_c, N_q, N_γ = bearing capacity factors, which are a function of the angle of internal friction

For footings like gravity structures for which the ratio of length to breadth is finite, the ultimate bearing capacity is given by

$$q = C N_c S_c d_c i_c + \gamma D N_q S_q d_q i_q + 0.5 \gamma B' N_\gamma S_\gamma d_\gamma i_\gamma$$

$$(18.40)$$

where

S_c, S_q, S_γ = shape factors
d_c, d_q, d_γ = depth factors
i_c, i_q, i_γ = inclination factors

For the special case that $\Phi = 0$ (undrained failure in clay) additive constants are used instead of factors and the following equation results:

$$q = C N_c (1 + S_c + d_c - i_c) \qquad (18.41)$$

The triaxial compression test should be used to determine shear strength parameters for circular and square footings, as the shear strength parameters obtained in plane strain compression tests give misleading results. For rectangular footings the friction angles Φ can be interpolated between plane strain and triaxial values in proportion to the side ratio B/L of the foundation. Since the angle of internal friction in plane strain compression tests is roughly 10 percent greater than in triaxial compression tests, the friction angles for rectangular foundations are roughly given by the following equation (Lade and Lee, 1976):

$$\Phi_{ps} = 1.5 \Phi_{tr} - 1.7 \ (\Phi_{tr} > 34°) \qquad (18.42)$$

where

Φ_{ps} = angle of internal friction for rectangular foundations (i.e., plain strain case for footings)
Φ_{tr} = angle of internal friction obtained from triaxial compression

In order to allow the eccentricity e of the resultant load R on the base of the foundation of width B, an effective foundation width B' given by $B' = B - 2e$ is used. For a double eccentricity of the load, an effective contact area can be determined in such a way that its center of gravity coincides with the resultant of the load. Both effective length and width are calculated from the following relationships:

$$B' = B - 2e_B \qquad L' = L - 2e_L$$

where

e_B = eccentricity in direction of the width
e_L = eccentricity in direction of the length

For shapes other than rectangular, the effective foundation area may be determined as that of the equivalent rectangle, constructed so that its geometric center coincides with the load center and that it follows as closely as possible the adjacent contour of the actual base area.

The influence of moment in addition to horizontal and vertical loads can be taken into account by using inclination factor given by DNV (1977) and Bowles (1988).

1. Inclination Factors

$$i_q = \left(1 - \frac{0.5 F_{Hd}}{F_{Vd} + A_f C_a \cot \phi}\right)^5$$

$$i_\gamma = 1 - \left(\frac{0.7 F_{Hd}}{F_{Vd} + A_f C_a \cot \phi}\right)^5$$

$$i_c = 0.5 + 0.5 \sqrt{1 - \frac{F_{Hd}}{A_f C_a}} \qquad (18.43)$$

where

F_H = horizontal load
F_V = vertical load
F_{Hd} = design horizontal load = $1.2 F_H$
F_{Vd} = design vertical load = $1.2 F_V$
A_f = effective foundation area ($B' \times L'$)
C_a = adhesion to base

A factor of 1.2 is recommended by DNV (1977) for possible unfavorable deviation of the loads from actual loads, thus allowing for unforeseen and abnormal actions.

The results of calculations should be used with care when the ratio F_{Hd}/F_{Vd} approaches or becomes less than 0.4.

2. Shape Factors

$$S_q = 1 + \frac{B}{L} \tan \phi$$

$$S_\gamma = 1 - 0.4 \frac{B}{L} \tan \phi$$

$$S_c = 1 + \frac{N_q}{N_c} \frac{B}{L} \qquad (18.44)$$

3. Depth Factors

For shallow foundations, especially those of offshore gravity structures, the depth factor has almost negligible effect on the calculated bearing capacity. In this connection, use:

$$d_c = d_q = d_r = 1$$

In special cases, values $d_q > 1.0$ and $d_c > 1.0$ may still be used, provided that the foundation installation procedure and other critical aspects allow for the mobilization of resisting shear stresses in the soil above the foundation level. In such cases the following expressions for d_q, valid for $d < b'$ define an upper limit of this contribution:

$$d_q = 1 + 2 \frac{D}{B} \tan \phi (1 - \sin \phi)^2$$

$$d_\gamma = 1.0$$

$$d_c = 1 + 0.4 \frac{D}{B} \qquad (18.45)$$

which approaches a limit value $d_c = 0.47$ for large depths.

For fairly homogeneous soil conditions the semiempirical bearing capacity formula suggested by Meyerhof (1965) can be used for preliminary estimates. To account for the influence on bearing capacity of a layered soil system, several investigators have proposed modifications to the existing classical bearing capacity theory (Reddy and Srinivasan, 1967; Brown and Meyerhof, 1969; Mitchell et al., 1972; Meyerhof, 1974). For a soft layer sandwiched between two stronger materials, the correction factors are proposed by Jurgenson (1934). Similarly, Davis and Booker (1973) have proposed correction factors for soils with increasing undrained strength with penetration. When the soil profile contains several layers of sand, silt, and clay, the above methods are not applicable. The bearing capacity formulas and correction factors given by various investigators are based on the assumed shapes of critical shear surfaces. To take the various factors into account, such as various configurations of soil properties, platform geometries, and applied loads, the slip surface method had been developed as shown in Figure 18.43.

The cross section over the entire length L is constant and the sliding body is cut off by vertical planes at both ends of the foundation. The sliding body is divided into four sections, an active section, a flat section, an inclined section, and a passive section (Schjetne et al., 1979); refer to Figure 18.42. The

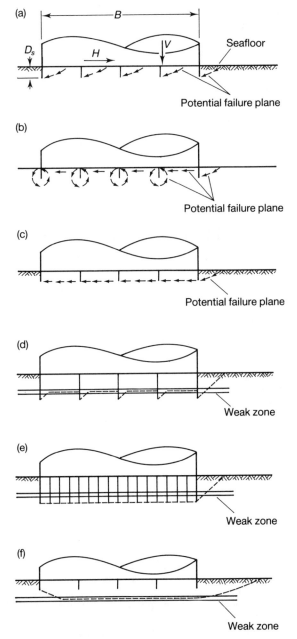

Fig. 18.43 Schematic illustration of some possible failure modes for sliding resistance. (a) Passive wedge failure. (b) Deep passive failure. (c) Sliding base failure. (d) Sliding failure in shallow weak zone with widely spaced skirts. (e) Sliding failure in shallow weak zone avoided with closely spaced skirts. (f) Sliding failure in deep weak zone. (*Young et al., 1975.*)

resistance to sliding for all surfaces of a sliding body are evaluated under force equilibrium conditions. The safety factor is found by overall horizontal force equilibrium. The depth z to the lowest point on the slip surface is varied in steps and the critical depth giving the minimum safety factor is established. The undrained shear strength values to be used along the different parts of the potential failure surface are determined from the results of triaxial compression, extension, and simple shear tests combined with the results from in-situ cone penetration tests.

The strength profile is determined for two different consolidation levels corresponding to no drainage and full drainage under the submerged weight of the platform. The effect of cyclic load on the reduction of shear strength has also to be taken into account.

B. Lateral Resistance

Owing to shear stresses induced in the foundation soils by the vertical eccentric foundation load, the magnitude of horizontal load that the foundation soils can sustain may decrease with an increase in vertical load. This fact must be recognized when computing and evaluating sliding resistance by the simplified procedures described below.

When the foundation is subjected to a horizontal load, there are at least three potential modes of sliding failure. The failure mode that develops depends primarily on the skirt height, the spacing and orientation of skirt elements, the net vertical foundation load, and the profile of soil strength characteristics. Following types of failure can occur as shown in Figure 18.43.

1. Types of Failure

Passive Wedge Failure When a horizontal load is applied to a mat with fully embedded skirts, resistance will be developed along the soil–structure interface and passive resistance on skirt elements. If a passive wedge develops under the mat, the mat moves upward with the wedge, which may destroy the sliding resistance on a portion of the soil–mat interface, and the vertical load is transferred to the smaller contact area on top of the failure wedge as shown in Figure 18.43. This transfer of load results in an increase in the passive resistance acting on the vertical skirt elements. Equations for estimating the magnitude of passive resistance can be obtained from theories of soil mechanics.

If the skirts are fully penetrated, the failure mode will be a passive wedge failure with some additional resistance due to the friction or adhesion acting on the end of the area of the skirts.

Deep Passive Failure Once the net vertical load exceeds some critical value, which varies with the skirt and soil conditions, the upward movement of the failure wedge beneath the mat is not possible. When this condition occurs the failure may extend below each skirt, as shown schematically in Figure 18.43.

When the passive failure develops without an upward mat movement, the passive resistance of the skirts may be computed using the concept of lateral resistance of piles (Matlock, 1970; API, 1980).

Sliding Base Failure Another possible sliding failure mode may be a plane passing through the skirt tips, as illustrated in Figure 18.43. In this case, sliding resistance along a horizontal plane is equated to the shear strength along this plane.

C. Foundation Deformations

Cyclic displacements and rotations, total settlement and tilting or differential settlement as shown in Figure 18.44 must be

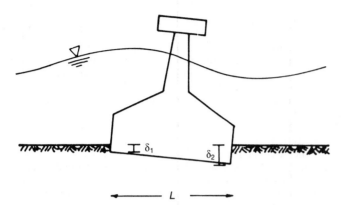

Fig. 18.44 Illustration of total and differential settlement for rigid foundation. Note: (1) δ_1, δ_2 = settlement; (2) differential settlement = $\delta_2 - \delta_1$.

within limits acceptable for operation of the platform. The settlement experienced by the Ekofisk Gravity Platform is shown in Figure 18.45. The foundation displacements may be separated into the following components: (1) initial settlement, (2) consolidation settlement, (3) secondary settlements, and (4) storm-induced settlements.

1. Initial Settlement

Common procedures have been to calculate initial settlements based on closed-form elasticity solutions and/or finite-element solutions.

At the feasibility stage, foundation movements are often estimated with available elastic solutions. The vertical (ρ_v), horizontal (ρ_h) and rotational (α) movements of a rigid circular foundation with radius (R) and supported on elastic half-space can be computed from the following equations (Young et al., 1975):

$$\rho_v = \frac{(1 - \mu^2)V}{2\,ER}$$

$$\rho_h = \frac{(7 - 8\mu)(1 + \mu)H}{16(1 - \mu)ER}$$

$$\alpha = \frac{3\,M(1 - \mu^2)}{4\,ER^3} \tag{18.46}$$

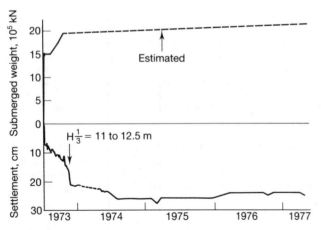

Fig. 18.45 Load–settlement history, Ekofisk tank. (*Clausen and Lunne, 1980.*)

where

$$V = \text{vertical load}$$
$$H = \text{horizontal load}$$
$$M = \text{overturning moment}$$
$$E \text{ and } \mu = \text{elastic properties}$$

Other solutions, which include the influence of layered soil system, soil anisotropy and foundation shape, are summarized by Poulos and Davis (1974). Carrier and Christian (1973) have presented a solution for vertical movement of a rigid, concentrically loaded circular foundation supported on a half-space with E increasing with penetration.

The reliability of movement predictions using elastic solutions is limited most by our ability to adequately select elastic parameters. Factors complicating the selection of elastic parameters include soil sampling and testing, disturbance, nonlinear stress–strain behavior, the influence of confining pressure, and time-dependent factors such as creep. The alteration of soil stress–strain properties and build-up of pore water pressures by cyclic foundation loading are further complications that may have a significant affect on the foundation movements during the cyclic loading. Not only may the weakening of the soil affect the movements during the storm, but the subsequent dissipation of the induced pore water pressures may result in foundation movements after the stress. The permanent deformation method, which was originally developed for estimating the movement of earthdams during earthquakes, may be applied to gravity structure problems (Lee, 1974; Chaney, 1984).

For preliminary estimation, Young's modulus (E) may be calculated from the following relationship.

$$E = \beta\,C_u \tag{18.47}$$

where

β = an empirical factor that is a function of soil type, stress level and overconsolidation ratio
C_u = undrained shear strength found from triaxial compression tests

The recommended values of β are 500 for soft sensitive clays, 1000 for firm to stiff clay, and 1500 for very stiff clay. For the highly overconsolidated clays a value of $\beta = 400$ can be used. Poisson's ratio is taken as 0.5 for undrained condition.

2. Consolidation Settlement

The consolidation settlement and its time history depend upon the permeability and compressibility of the soil and the drainage conditions. Cyclic loadings during storms induce settlements due to both shear strains during the storm and volumetric strains during and after the storm. The shear strains may be separated into two components. One is caused by the excess pore pressure from the cyclic loading (Eide et al., 1979b; Janbu et al., 1976). This excess pore pressure causes a reduction in effective stresses and hence in soil shear stiffness. There are permanent shear stresses in the soil due to initial in-situ stresses and submerged platform weight; and when the soil stiffness decreases, shear strains develop. The other shear strain component is caused by the cyclic shear stresses from the wave forces. These repetitive shear stresses cycle around the permanent shear stresses due to the initial in-situ stresses and stresses due to platform weight. The soil elements are thus subjected to unsymmetrical cyclic shear stresses and this will cause a permanent shear strain increase in the direction of permanent shear stress even if there is no excess pore pressure.

Parameters for calculation of consolidation settlement of overconsolidated clay are currently being found by two different procedures.

a. Use of constrained deformation modulus, *m*, as determined from oedometer or K_0-triaxial tests.
b. Use of drained Young's modulus (E) and Poisson's ratio (v) from empirical correlations.

For the first method it has been found most realistic to use the constrained deformation modulus (m) as determined from the reloading curve in the consolidation test. In the second method the sum of initial and consolidation settlements are calculated using $E = 150\,C_u$ and $v = 0.1$. The relationship is obtained from Butler (1975) by the back-calculation of actually measured values. The consolidation settlement is then found by subtracting the initial settlement.

The initial settlements must therefore be calculated separately. Using a value of Young's modulus given by Equation 18.48 and $v = 0.1$ the sum of the initial and consolidation settlements may be computed.

$$E = m\,\frac{(1 + v)(1 - 2v)}{(1 - v)} \qquad (18.48)$$

A relationship of the form $E = \alpha\,q_c$, where q_c is the cone tip resistance, has also been suggested. For very dense sand, α is of the order of 3 to 5. The time history of consolidation settlements is computed using the coefficient of consolidation, c_v, as determined from oedometer or triaxial tests. For normally consolidated sand it is of the order of 1.5 to 2.0.

3. Secondary and Storm-Induced Settlement

At present there is no well-established computational procedure for calculating secondary and storm-induced settlements for offshore gravity platforms, but analytical procedures have been proposed (Andersen et al., 1978; Prevost and Hughes, 1978; Mroz et al., 1978). For the gravity platforms in the North Sea, these settlements are estimated on the basis of experience from buildings on land subjected to repetitive loading, and available settlement readings of the first platform in the North Sea. Experience on land shows that cyclic load variations cause increased long-term settlement and that the secondary settlements, instead of decreasing with time, continue with unchanged rate over several decades (Bjerrum, 1973).

For the North Sea Gravity platforms, with a diameter of about 100 m and foundations on overconsolidated clay or dense sand, the estimate has been an average long-term settlement in addition to the consolidation settlement of about 1 cm/yr. This long-term settlement is intended to include both the ordinary secondary settlements and wave-induced settlements. As exemplified by the Ekofisk tank, the rate may be much higher in periods of severe storm loading, but in quiet periods the settlement rate will be smaller than the average suggested above.

D. Dowel and Skirt Penetration Resistance

Their are two primary purposes of the dowels: (1) to help position the platform during touchdown, and (2) to prevent lateral platform motion during penetration of the foundation skirts. Cross sections through the caisson base for a number of North Sea gravity structures are shown in Figure 18.46.

The rate of loading controls the ultimate soil resistance against skirts and dowels during the touchdown phase. Observed soil resistance to skirt and dowel penetration are shown in Figures 18.47 and 18.48, respectively. It is important to have reliable estimates of total penetration resistance as well as probable eccentricity in the penetration reaction.

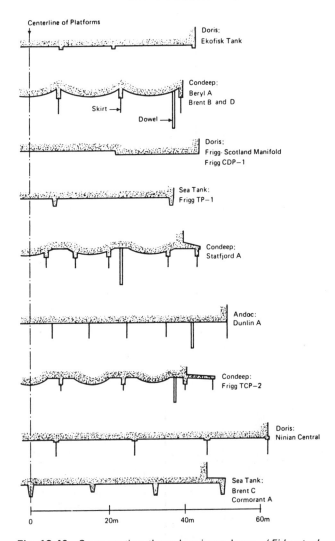

Fig. 18.46 Cross section through caisson bases. (*Eide et al., 1979a.*)

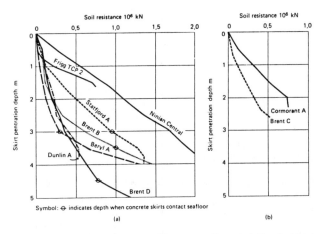

Fig. 18.47 Soil resistance to skirt penetration. (a) Steel skirts. (b) Concrete skirts used on sea tank structures. (*Eide et al., 1979a.*)

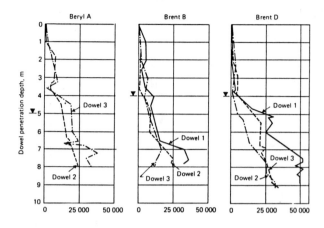

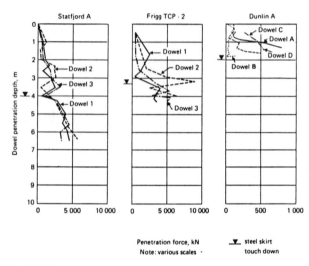

Penetration force, kN
Note: various scales ·

▼ steel skirt touch down

Fig. 18.48 Soil resistance to dowel penetration for structures equipped with dowels. (*Eide et al., 1979a.*)

Calculation of penetration resistance of solid dowels and thick-walled (concrete) skirts are performed by bearing capacity formulas. Penetration resistance for thin-walled (steel) skirts and dowels is based on recent experience (Lunne and St. John, 1979). The skirt tip resistance and skirt skin friction are given in Equation 18.49:

$$q_t(d) = K_t q_c(d) = \text{skirt tip resistance}$$
$$f(z) = K_f q_c(z) = \text{skirt skin friction} \qquad (18.49)$$

where

K_t and K_f = empirical coefficients varying with soil type
$q_c(z)$ = Dutch cone penetration (cpt) at depth z
d = penetration depth of skirt tip

Based on North Sea experience (Schjetne et al., 1979) the empirical coefficients are given by the following:

1. Dense to very dense silty fine sand $K_t = 0.3$ to 0.6
   $K_f = 0.001$ to 0.003
2. Very stiff, silty clay $K_t = 0.4$ to 0.6
   $K_f = 0.03$ to 0.045
3. Intermixed dense sand/clay layers $K_t = 0.5$
   $K_f = 0.006$ to 0.014

The low K_t and K_f values apply to the upper few meters. The upper range of the values may be used to predict maximum expected penetration resistance.

E. Base Contact Stress

The base contact stresses depend upon the geometry of the base, the seabed topography, and the soil properties. The base of a gravity structure may consist of a number of spherical domes as in the Condeep platform or it may have a flat base. The seabed topography must be mapped, and either the base must be made strong enough to withstand maximum expected base contact stress as shown in Figure 18.49 or the installation procedure must secure against high local contact stresses by the process of grouting. The maximum contact stress against the base is limited by the bearing capacity of the soil. The bearing capacity of the dense sand is considerably higher than for stiff clays. A layer of dense sand on top of stiff clays may also increase the local base contact stress. The maximum base contact stress against a dome penetrating into sand can be calculated by bearing-capacity theory. For domes penetrating into clay, the interpretation is somewhat more uncertain, mainly owing to uncertainties regarding the undrained shear strength of the upper clay layers. An example of the measured and back-calculated contact stresses at the tip of one of the Condeep Frigg TCP2 domes is presented in Figure 18.50.

F. Foundation Tilt

Foundation tilt may develop during installation owing to uneven skirt penetration resistance or sloping seafloor. It is desirable to keep the platform level, and for most platforms this has been attempted by applying eccentric ballasting during installation.

G. Piping and Erosion

During skirt penetration, the water entrapped within the skirt compartments flows out through outlet valves. The acceptable rate of penetration is therefore governed by the capacity of the outlet valves. If the penetration is too fast, an excess water pressure will build up within the skirt compartments, and piping may occur, causing the soil beneath the skirts to be washed away. If piping or erosion occur during installation, it will be

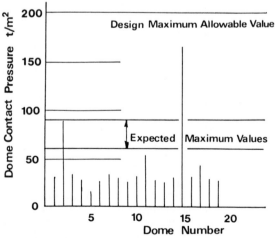

Fig. 18.49 Maximum dome contact pressure observed. (*Clausen, 1976.*)

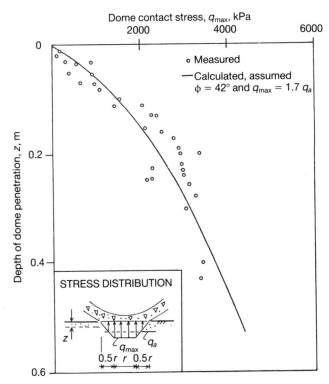

Fig. 18.50 Comparison between measured and calculated effective maximum base contact stresses. (*Kjekstad and Stub, 1978.*)

difficult to grout beneath the platform afterwards because the grout may flow out through the eroded channels. Unless successful grouting is obtained and all channels are filled with grout, water may flow in and out through the channels during a storm and cause further erosion.

H. Dynamic Analysis

In addition to static analysis, the response of the soil–structure system must be studied for cyclic loadings. Because of the size and relatively large gross bearing pressure of gravity structure foundations, their vertical movement can be several feet. Knowledge of the potential immediate and long-term foundation movements is required for planning and designing connections of pipelines and conductors to the platform as well as other structure appendages, such as a boat dock.

The natural frequency of structures tends to increase with increasing water depths and may coincide with the frequency of the larger waves. If the natural frequency of the soil–structure system coincides with the frequency of the exciting forces, large foundation movements may result. Therefore, reasonable response of the soil–structure system must also be available for evaluating the dynamic response of the structure. The contribution and importance of the soil response is increasing as structures are built for deeper waters.

18.4.4 Construction and Installation

Construction of the gravity platform structure has been described in detail by Eide et al. (1976, 1979a) and Gerwick (1986). When the construction phase is complete, the structure is towed to the site. After arrival at the site, towing lines are arranged in a star pattern to aid in maneuvering to the final target area. Positioning of the platform is controlled by various means: (1) acoustic transponders placed in advance in an approximate 1 to 1.5 km square on the seafloor, or (2) electronic, laser or visual measurement. For free-standing platforms the usual requirements have been that the center of the platform should be within a 30 to 50 m diameter circle, orientation within ± 2° to ± 5°.

The actual installation consists of four phases: touchdown, skirt penetration, base seating, and underbase grouting. In the following, each of these phases will be discussed.

A. Touchdown

Touchdown is when the structure's base first makes contact with the seafloor (Figure 18.51). The position and orientation of the platform is then held fixed so that any adjustments in position or orientation can be made before the skirts come into contact with the seabed. Touchdown can be recorded by instrumented dowels or any other appropriate system.

B. Skirt Penetration

As ballasting is increased, the skirts start to penetrate into the seabed (Figures 18.51b and c). The primary concern during the phase is (1) to prevent piping and the resulting erosion from occurring in the sediments below the base and (2) to maintain verticality. A slow penetration rate is usually employed to achieve these requirements. Typical penetration rates are as low as 0.1 to 0.2 m/hr.

At this stage in the installation procedure it is important to monitor the tendency of the platform to tilt owing to eccentric

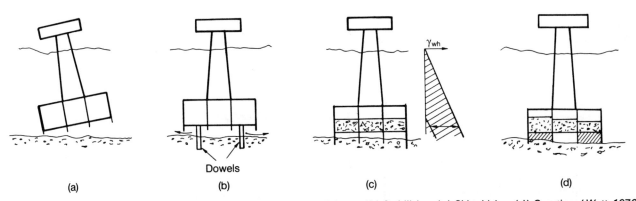

Fig. 18.51 Installation sequence for Condeep type platform. (a) Touchdown. (b) Stabilizing. (c) Skirt driving. (d) Grouting. (*Watt, 1976.*)

resistance of the seafloor materials to the skirt penetration. This should preferably be controlled by eccentric ballasting to keep the platform vertical. It may be difficult to correct tilt at a later stage if most of the ballast weight has been utilized.

C. Base Seating

The purpose of the control of base seating is to provide a gap between the base of the structure and the potentially uneven seafloor to prevent localized stress concentrations. The resulting gap will subsequently be filled with grout to allow for a uniform transfer of loads. For platforms such as Condeeps with spherical-shaped bottom domes, the depth of penetration of the domes in the seafloor materials has been up to 0.6 m.

D. Underbase Grouting

After the base of the gravity structure has been seated, the gap between the seafloor and the base is grouted (Figure 18.51d). The primary purposes for grouting are (1) to avoid further penetration and maintain verticality, (2) to develop uniform stresses on the platform and prevent localized stress concentrations, and (3) to avoid piping from water pockets below the base during environmental loading.

Grouting is especially important for uneven or sloping sites. The density of the grout is typically 12 to 15 kN/m³. The evacuation of displaced water and excess grout has either been directly into the sea or through a check valve in the base. In the latter case, the grout quality can be monitored.

18.5 ANCHOR UPLIFT CAPACITY

Until the 1960s, the anchors developed for marine applications were of the uplift pile, dead weight, or drag types. The specified uplift capacities of these last two anchor types were independent of seabed conditions or sediment properties and, as a result, the anchors were extremely large and cumbersome to handle at sea. Uplift piles require heavy equipment for installation. Drag anchors required a very large scope of line, which made it difficult to position the anchor with any precision. Deadweight anchors tend to drift under load and are unstable on a sloping seabed.

Industrial and military needs have placed greater emphasis on improved anchor systems as operations are performed in deeper water and with some positioning accuracy. Additionally, anchor requirements include high anchor capacity/weight ratio, reliability, efficiency, and ease of handling and deployment. These demands have led to the development of a large number of new anchor types and an assortment of theories to predict the uplift capacity of these anchors in different sediment types. This section will contain a review of these new anchor types and some anchor capacity theories. While the uplift pile remains an important anchor type, it was discussed in some detail in Section 18.3.1.B3.

18.5.1 Embedment Mechanisms

Taylor et al. (1975) cite several new anchor types, which are defined according to the mechanism of embedment or operation:

a. Propellant-actuated (explosive) anchor
b. Vibration embedment anchor

c. Screw-in anchor
d. Driven anchor
e. Drilled anchor
f. Jetted anchor
g. Free-fall anchor
h. Hydrostatic anchor

The objective of these embedment mechanisms or combinations of mechanisms (a–g) is to embed an anchor plate (fluke) or anchor pile (Fig. 18.52) to its design depth, D. The hydrostatic anchor is, by comparison, a surface anchor that relies on suction to attach it to the seabed (Wang et al., 1975, 1977) and to resist pullout. The hydrostatic anchor will not be discussed in further detail because of its special nature and limited use to date.

A. Propellant-actuated Anchor

The propellant-actuated or explosive embedment anchor uses an explosive charge and gun (Figure 18.53) to embed an anchor fluke in the seabed at a high velocity. The gun assembly includes a reaction vessel that minimizes recoil by virtue of its effective mass and hydrodynamic drag at detonation. While many types of flukes have been developed for soft or hard seabeds, the typical sediment fluke is either folded or rotated to configure a sleek geometry of low drag during penetration. The fluke is then keyed or rotated by the pull of the cable to a position where its maximum pullout resistance is mobilized. The gun and reaction vessel are often retrieved and reused but this approach necessitates the use of three cables—the anchor line, a gun retrieval line, and a detonation line—which can complicate operations at sea. A number of safety features are built into the system to eliminate premature detonation.

B. Vibration Embedment Anchor

The U.S. Navy has developed a vibration embedment anchor for use in shallow or deep water (Figure 18.54). The vibrator drive unit uses two counter-rotating eccentric masses to develop the vibrational force applied to the anchor fluke. The vibration unit can be either electrically or hydraulically powered and the vibrator and support frame can be recovered after the anchor is installed. The present system uses a Y-fluke that penetrates the sediment in a low-profile configuration and is keyed by pulling back on the anchor line to develop maximum breakout resistance. The support platform uses remote sensing to determine the attitude (plumbness) of the drive shaft and a displacement monitor to observe penetration depth and rate, which may be used to adjust vibration frequency.

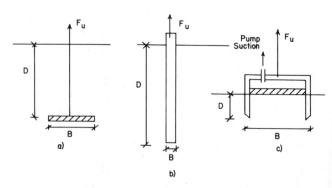

Fig. 18.52 Schematic diagram of (a) plate anchor, (b) pile anchor, and (c) hydrostatic anchor.

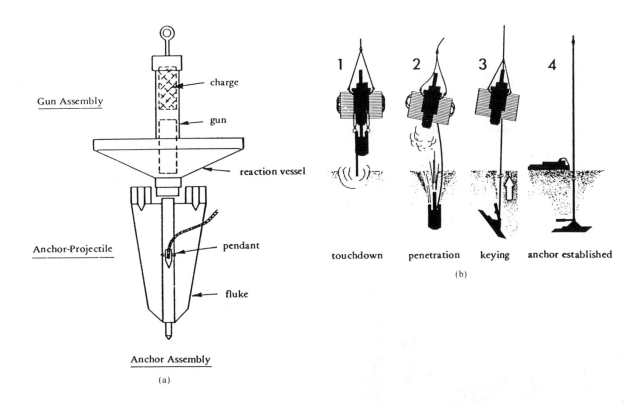

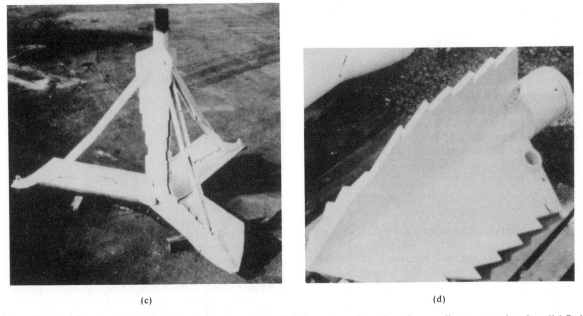

Fig. 18.53 Propellant-actuated or explosive embedment anchors. (a) General configuration of a propellant-actuated anchor. (b) Embedment and keying of a propellant-actuated anchor. (c) Anchor-projectile with hinged flukes extended. (d) Three-finned anchor-projectile for coral seafloor. (*Taylor et al., 1975.*)

Fig. 18.54 Deep-water vibrated anchor. (*Taylor et al., 1975.*)

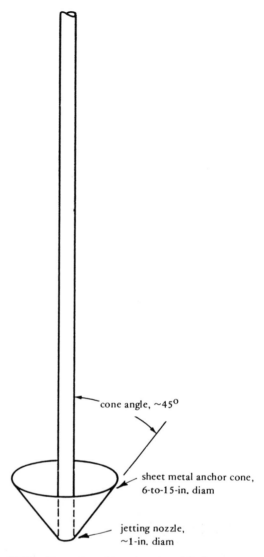

cone angle, ~45°

sheet metal anchor cone, 6-to-15-in. diam

jetting nozzle, ~1-in. diam

Fig. 18.55 Illustration of jetted anchor. (*Taylor et al., 1975.*)

C. Screw-in Anchor

The screw-in, auger, or helical anchor uses one or more single-turn helical plates attached to a rigid shaft to screw itself into the soil. This anchor system was developed for use on land to anchor guy wires for towers and poles. It has had limited use on the seabed to anchor pipelines and cables. Anchor penetration is dependent upon reaction force and torque from the drive unit in addition to soil strength.

D. Driven Anchor

A driven anchor is generally pounded into the soil to a design depth by repeated blows with a hammer. While this method is usually associated with pile anchors, it has been used to install plate anchors with the aid of a mandrel and follower.

E. Jetted Anchors

Jetted anchors use a jet or pulsating supply of water at the anchor tip to erode soil and allow anchor penetration. A typical small diver-operated unit as shown in Figure 18.55 may be capable of sustaining loads of 2 to 10 kips in a granular soil. When used in cohesive soil, there may not be sufficient density of the backfill material to develop a reasonable breakout capacity. Jetting techniques have been applied to other anchor types such as pile, deadweight, or mushroom anchors, but must

be used judiciously so as to develop sufficient soil density around the anchor after placement.

18.6.2 Anchor Holding Capacity

Many different anchor capacity theories have been developed to determine the vertical anchor capacity. These theories are based on soil plasticity analyses following observations of soil rupture surface geometry in the laboratory or field. Figure 18.56 summarizes some of the failure theories proposed by different researches, which suggest that each theory will provide a different estimate of anchor capacity. A comparison of measured and predicted anchor capacities for plate anchors in sand (Shaheen et al., 1987) shows that most theories tend to underpredict static capacity by a wide margin, and the theories of Matsuo (1967) and Meyerhof and Adams (1968) tend to provide slightly conservative results. While anchor capacity theories will be expected to provide better results in clay, there are several factors that are difficult to incorporate into anchor capacity predictions for all soil types. These factors include the effects of soil density and strength after anchor placement, anchor penetration depth, and loading conditions. As a result,

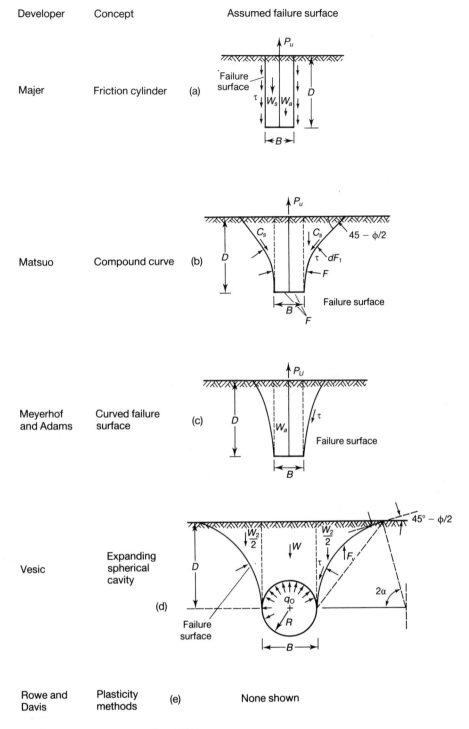

Developer	Concept		Assumed failure surface
Majer	Friction cylinder	(a)	
Matsuo	Compound curve	(b)	
Meyerhof and Adams	Curved failure surface	(c)	
Vesic	Expanding spherical cavity	(d)	
Rowe and Davis	Plasticity methods	(e)	None shown

Fig. 18.56 Anchor-capacity theories.

a liberal safety factor of 2 or 3 is recommended for capacity predictions.

The holding capacity equation for an anchor is similar to the bearing capacity equation for a deep foundation. The general uplift capacity, F_u, of an anchor is often expressed as the following:

$$F_u = A(c\, N_c' + \gamma'\, D N_q')\left(0.84 + 0.16\frac{B}{L}\right) \qquad (18.50)$$

where

A = projected fluke area
c = soil cohesion or undrained strength
γ' = buoyant unit weight
D = fluke embedment depth
$N_c',\ N_q'$ = holding capacity factors
B = fluke width or diameter
L = fluke length

This equation is sufficiently general that it can be applied to any anchor loading condition or soil type. However, it is usual to consider a soil as either cohesive or granular and as a result the following loading conditions must be considered:

1. Cohesive
   a. Short-term static loading
   b. Long-term static loading
   c. Short-term cyclic loading
2. Granular
   a. Short- or long-term static loading
   b. Long-term cyclic loading

For a given soil type, cohesive or granular, all of the loading conditions must be considered to evaluate long-term stability. Each load condition requires specific soil strength information for input to Equation 18.50 and soil strength and anchor geometry data are needed to select the proper holding capacity factors N_c', N_q'.

A. Cohesive Soil—Short-Term Static Loading

For short-term or undrained loading analysis ($\phi = 0$) the factor N_q' is zero and the factor N_c' is selected from Figure 18.57 for the anticipated relative embedment depth D/B and soil cohesion (a distance B above the anchor) in the soft rupture zone. The soil cohesion can be measured in place with a field vane or cone penetrometer or determined from laboratory tests on core samples such as the unconfined compression or triaxial CU test (Chapter 3). If soil strength data are not available, the strength profile can be estimated using $c/\sigma_{v0}' = 0.30$, where the vertical effective stress σ_{v0}' is estimated using a buoyant unit weight of 25 pcf.

B. Cohesive Soil—Long-Term Static Loading

During long-term loading, it is assumed that the soil around the anchor will drain under the pore-water pressure gradients introduced during anchor installation and loading. Drained

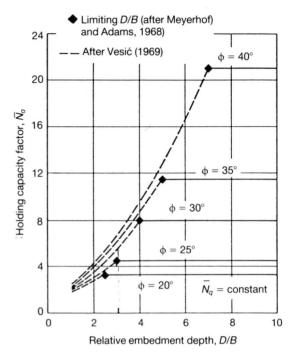

Fig. 18.58 Holding capacity factor N_q versus relative depth for cohesionless soil $c = 0$.

loading implies that the frictional properties ϕ of the clay will control anchor capacity and cohesion affects are destroyed ($c = 0$). A measurement of the clay friction angle can be made with a triaxial CD test or CIU test, which can be used with the relative embedment depth (Figure 18.58) to determine the factor N_q' for use in Equation 18.50.

C. Cohesive Soil—Short-Term Cyclic Loading

Cyclic loading during storms tends to remold a cohesive soil and reduce the cohesion or undrained strength. The remolded strength should therefore be used in Equation 18.50 to determine cyclic loading conditions.

D. Granular Soil—Short- or Long-Term Drained Loading

Since a granular soil will drain under static loading conditions, the measured friction angle can be used in Figure 18.58 to determine the factor N_q' for use in Equation 18.50 with $c = 0$.

E. Granular Soil—Long-Term Cyclic Loading

Long-term cyclic loading leads to a progressive failure whereby a few soil grains move around the fluke during each cycle of load. High safety factors of 8 to 10 are recommended for this loading condition.

18.6 PIPELINES

18.6.1 Installation

Four general methods are used to construct and place pipelines on the seabed (Miltz and Broussard, 1972). These methods include the reel barge, lay barge, flotation, and bottom pull. The reel barge method is limited to small-diameter pipe

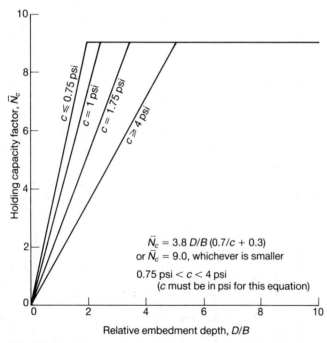

Fig. 18.57 Design curves of holding capacity factor $\bar{N}_c$ versus relative embedment depth (D/B). (*Taylor et al., 1975.*)

$\bar{N}_c = 3.8\, D/B\,(0.7/c + 0.3)$
or $\bar{N}_c = 9.0$, whichever is smaller

0.75 psi $< c < 4$ psi
(c must be in psi for this equation)

(less than 6-inch) that is fabricated in a yard, spooled onto a reel, proof tested, and barged to a site. The pipe is passed through a straightener as it is unreeled along the right of way. Horizontal tension is applied by tugs and anchors to control bending in the pipe as it is laid on the seabed at speeds up to 2 miles per hour.

The remaining three methods can be used for large pipelines up to 4 ft or more in diameter. With a lay barge, pipe sections are welded together on the barge in a conveyor-belt fashion as it moves along the right of way (Fig. 18.59). A stinger pontoon at the back of the barge is used to control the pipe curvature or stress level with the aid of a tensioning device at midships and an elaborate anchoring system at the corners of the barge. The anchors are used to position and propel the barge. With the flotation and bottom pull methods, the pipe is generally fabricated in long sections onshore at the right of way. With the flotation method, the pipe is towed along the right of way at the ocean surface with the aid of buoyancy units that are flooded or deflated to lower the pipeline to the seabed. A cable and anchor system is used to position the pipeline on the seabed. The bottom drag method is similar to the flotation method and uses a cable attached to the end of the pipeline to pull it into position along the seabed. A buoyancy system may be used to reduce bottom friction in cases where a heavy protective coating is applied to the pipeline.

Once a pipeline is in place on the seabed, a pipe trenching barge may be used to bury the pipeline as shown in Figure 18.60. Both water-jetting and plowing techniques have been used to create the pipe trench at great expense. However, the success of achieving complete burial has always been questioned because of the technical complexity of this operation and the natural forces involved.

Most pipelines experience a maximum state of stress during installation, from the combined flexural (curvature) and tensioning requirements. A buckling failure is quite common during installation and, in many instances, the protective coatings of concrete and mastic are damaged. Geotechnical/environmental concerns regarding pipeline loading and stability are equally important. To assess the value of the seabed as a foundation material, a thorough survey is needed to select a final route and to identify the conditions for which the pipeline must be designed.

18.6.2 Site Survey

In many cases the initial and final positions of a pipeline are dictated by the end structures, such as a production platform or deepwater terminal offshore and a processing plant onshore. Alternatively, the most desirable pipeline route may impact the positioning of one or more end structures. A thorough site survey should, therefore, be performed in steps and include a review of available information, a preliminary field survey, and final detailed survey.

For most coastal areas of the world there is a wealth of public information that can be used for this first step. This information includes detailed bathymetric maps, geological profiles and isopac maps, and navigation charts that define current velocity and direction. Highly developed areas around oil fields and large harbors or bays have likely generated a significant amount of site data that is of public or sometimes private record. Private data can often be purchased from the owner at a price significantly less than the acquisition price.

After two or three potential routes have been defined, a preliminary survey of each route by ship is necessary to define the best route. A typical survey would consist of continuous precision bathymetry, high-resolution subbottom profiling, and

shallow gravity core samples. A precision depth recorder with a frequency of 60 to 200 kHz would define the bottom topography to ± 1 or 2 ft. It may also help to identify important microtopographic features such as boulders, fault scarps, or foreign objects on the seabed.

High-resolution subbottom profiling with a 3.5 kHz sound source has become an extremely important tool for identifying geologic hazards or man-made features, as shown in Figure 18.61. The interpretation of a subbottom profile is very difficult and requires the services of a marine geologist/geophysicist. It is usually necessary to take borings along a proposed pipeline route every 2000 to 5000 ft to aid the interpretation of the subbottom profile records and to obtain sediment samples for measurement of physical and engineering properties such as gradation, Atterberg limits, and shear strength. Generally, gravity core samples 10 to 20 ft long are adequate for calibration of bathymetric profiles and laboratory properties determination, but deeper core samples with a length of 50 to 100 ft are always welcomed for defining deep stratigraphic features. Deeper core samples are often available at the site of an end structure.

When a final pipeline route is selected, a final detailed field survey may be necessary to fill in gaps in the data from the preliminary field survey. This may include additional high-quality sediment sampling and in-situ measurements to define geotechnical properties and geophysical survey lines for detailed mapping of hazards such as slumps, faults, buried Pleistocene river channels, old pipelines, or sunken vessels. The geophysical survey would consist of precision bathymetry, subbottom profiling, a magnetometer to locate metal objects, and side-scan sonar to evaluate topographic features. This phase of the survey should result in the marking of the pipeline route and structural hazards by placing a series of taut-wire buoys.

18.6.3 Geotechnical Design Considerations

Once the route survey has defined the potential hazards along the pipeline route, it is useful to prepare a map (Fig. 18.62) that shows the area over which the hazard may be a problem. This map will be the basis for preparing a logical plan to evaluate foundation design requirements.

A. Protection From Dragged Objects

Pipeline damage from anchors in active construction sites and from trawler boards and weights in active fishing areas is quite common. When damage is anticipated, it is necessary to design a pipe-protection system that will cause an anchor or fishing apparatus to deflect up and over the pipeline without inflicting damage (Brown, 1973). Figure 18.63 shows the common types of protection systems that have been used with limited success. Of the systems shown in Figure 18.63, pipeline burial is frequently performed with the aid of a trenching barge (Fig. 18.60) or by adding extra concrete protection to promote natural burial in soft sediments. At points where pipelines cross, it is usually necessary to place the pipeline on the sediment surface with an armor covering for protection.

B. Pipeline Flotation

A buried pipeline may be susceptible to flotation when buried in a soft cohesive soil with the natural water content greater than the liquid limit. Bonar and Ghazally (1973) performed laboratory tests with several pipelines and four cohesive soils

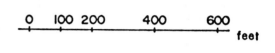

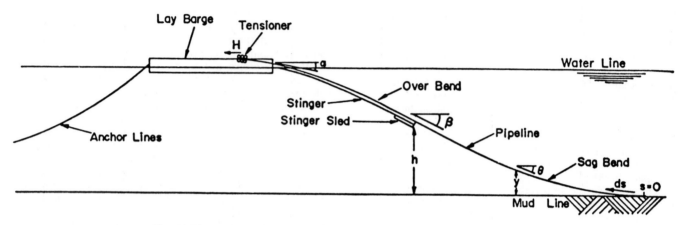

Fig. 18.59 A schematic of the laybarge and stinger configuration. (*Daley, 1973.*)

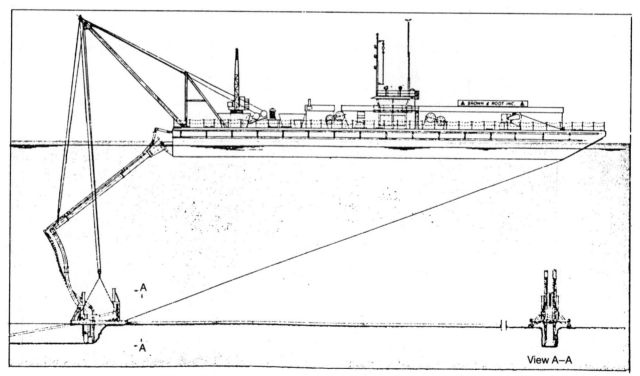

Fig. 18.60 A typical arrangement of a pipe trenching barge. (*Minor, 1966.*)

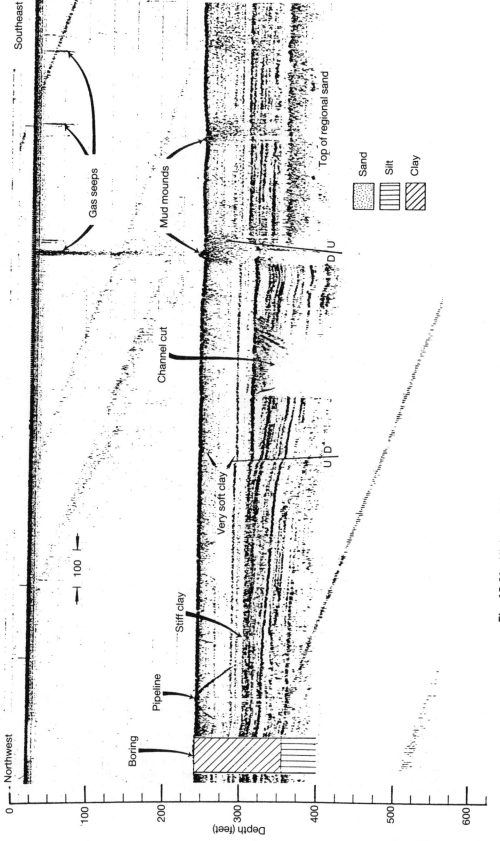

Fig. 18.61 High-resolution profiler section. (*Antoine and Trabant, 1976.*)

719

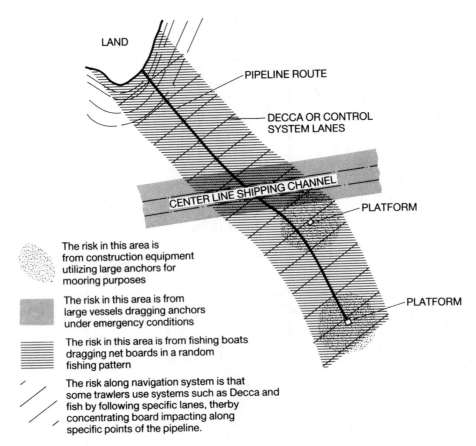

Fig. 18.62 Hazards along pipeline route. (*Brown, 1973.*)

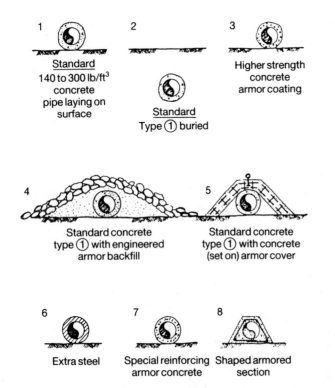

Fig. 18.63 Types of protection for offshore pipelines. (*Brown, 1973.*)

to examine the interaction of the flotation force and the resisting force defined as

$$F_{\text{flotation}} = (\gamma - \gamma_{\text{pipe}}) \frac{\pi D^2}{4} \qquad (18.51)$$

where

γ = unit weight of soil (lb/ft^3)
γ_{pipe} = unit weight of pipe and water contents at flotation (lb/ft^3)
D = pipe diameter (ft)
$F_{\text{flotation}}$ = flotation force (lb/ft)

$$F_{\text{resisting}} = 7.2\, c^{0.5}\, D \qquad (18.52)$$

where

c = soil cohesive shear strength (lb/ft^2)
$F_{\text{resisting}}$ = resisting force (lb/ft)

The constant value of 7.2 in Equation 18.52 for the resisting force includes any shape factor and units adjustment needed. No attempt was made to separate the units coefficient or other constant, and the shape factor.

At the start of flotation, these two forces are equal. Equating these two equations and solving for the unit weight of pipe and contents gives the following:

$$\gamma_{\text{pipe}} = \frac{9.17\, c^{0.5}}{D} - \gamma_{\text{soil}} \qquad (18.53)$$

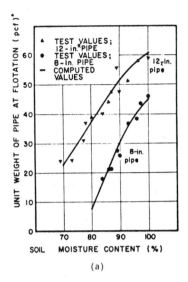

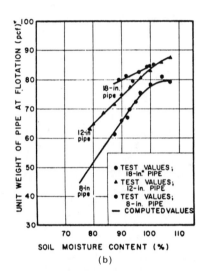

Fig. 18.64 Flotation of submarine pipe. Unit weight of pipe plus contents at flotation versus moisture for (a) soil 1, (b) soil 2. (*Bonar and Ghazally, 1973.*)

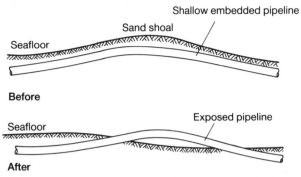

Fig. 18.66 Erosion of coastal sands can be hastened by wave-induced seepage stresses. (*Demars et al., 1977.*)

By substituting known values of c, D, and γ_{soil}, which is related to soil moisture content, it is possible to compute the critical γ_{pipe} at flotation. As shown in Figure 18.64, the computed and experimental values of γ_{pipe} at flotation compare very favorably for several pipe sizes.

In sandy soils, pipeline flotation is not likely to be a problem. An experimental study by Demars et al. (1977) shows that pipeline flotation can only occur at very low γ_{pipe} (< 50 pcf) when the sediment experiences cyclic liquefaction as shown in Figure 18.65. Buried pipelines may be exposed to current forces as a result of sand migration (Fig. 18.66) during storms or earthquake-induced liquefaction (Morgenstern, 1967; Henkel, 1970; Audibert and Nyman, 1977; Audibert et al., 1978; Thomas, 1978; Demars, 1979).

C. Fault Displacement

Many areas of the seabed experience fault displacement that results from sediment deposition or fluid withdrawal. Arnold (1967) and Trautman et al. (1985) have shown that a buried pipe that traverses a fault (Fig. 18.67) can be modeled using a finite-element analysis to determine moments, shears, and axial loads. They propose that the total uplift resistance in the soil be given by the following equation:

$$F_m = \left[1 - \frac{\pi D}{8 H} + K \tan \phi \frac{H}{D} \right] \gamma H D L \qquad (18.54)$$

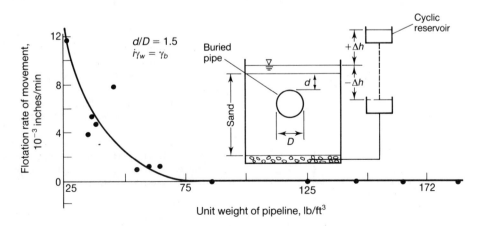

Fig. 18.65 Pipeline flotation curve for condition of cyclic sediment liquefaction. Schematic drawing of test apparatus is shown above. (*Demars et al., 1977.*)

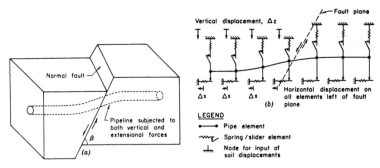

Fig. 18.67 View of pipeline affected by vertical soil displacement and numerical model. (a) Fault displacement. (b) Finite-element representation. (*Trautman et al., 1985.*)

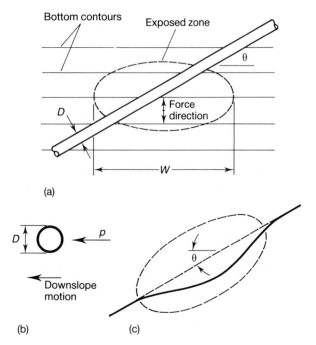

Fig. 18.68 Force–displacement conditions for pipeline subjected to external loading from sediment. (a) Initial condition. (b) Force per unit length. (c) Deformation after loading. (*Demars, 1978.*)

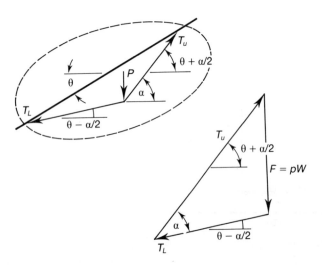

Fig. 18.69 Idealized pipeline displacement and force polygon. (*Demars, 1978.*)

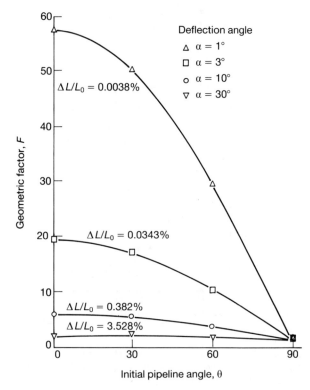

Fig. 18.70 Influence of initial pipeline orientation and allowable deflection angle on the geometric tension factor. (*Demars, 1978.*)

where

F = maximum vertical pipe force
γ = soil density
H = depth to pipe center below seabed
L = pipe length
ϕ = internal angle of friction of soil
K = coefficient of lateral earth pressure

This force equation assumes vertical failure surfaces above the pipe where shearing resistance to uplift is developed. The coefficient K depends on the soil type and pipe burial method and must be determined empirically.

D. Slope Instability

Many sloping areas of the seabed are soft as a result of rapid sedimentation rates or the presence of gas-charged sediments.

During earthquakes or storm-induced loading, a mass sediment instability (slump) may result and subbottom profiling is likely to reveal the presence and size of slump scars. Demars (1978) evaluated the pipeline tension when a slump moves through a buried pipeline (Fig. 18.68). The slump force P was idealized to act as a concentrated force (Fig. 18.69) for purposes of evaluating pipeline geometry effects such as pipeline orientation to the bottom contours θ and allowable deflection $(\Delta L/L_0)$ that results by laying a pipe in a sinuous fashion on the seabed. The analysis shows (Fig. 18.70) that the pipeline load (geometric tension factor) is minimized by orienting the pipeline perpendicularly to the bottom contours and by increasing the allowable deflection $(\Delta L/L_0)$ in areas of unstable sediment.

18.7 JACK-UP PLATFORMS

Oil and gas production in continental shelf areas are currently employing several hundred mobile jack-up platforms (McClelland et al., 1983). A plan and elevation of a typical jack-up rig supported on three legs is shown in Figure 18.71. Modern rigs use three or four legs, while earlier rigs had up to 12 legs. The length of the legs can be up to 120 to 140 m depending on the design. The maximum water depth where jack-up rigs can be used is approximately 100 m.

When a rig is towed onto the site, the legs are lowered onto the seabed to slow the platform's movement (Fig. 18.72). After the rig is stationary, the legs are then jacked further down, causing each leg to penetrate the seafloor until the

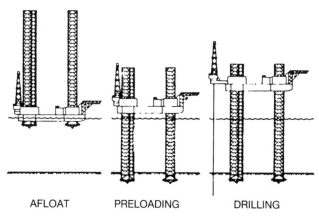

AFLOAT   PRELOADING   DRILLING

Fig. 18.72 Jack-up rig operational modes. (*McClelland et al., 1983.*)

bearing capacity of the soil can support the submerged weight of the spud can and leg. This corresponds to point A′ (Figs 18.73a and b). Figures 18.73a and b represent the cases of normally consolidated clay and a soft layer overlaying a hard layer, respectively. As the jacking operation is continued and the hull begins to lift out of the water, forces on the legs

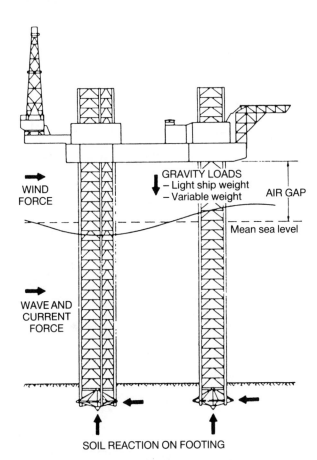

Fig. 18.71 Design loads on a jack-up rig. (*After McClelland et al., 1983.*)

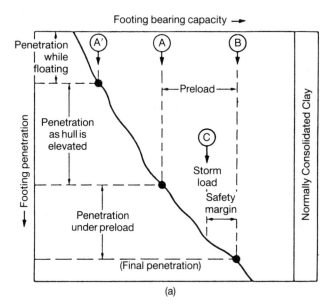

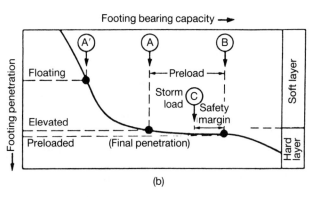

Fig. 18.73 Idealized installation and preloading of footings. (*McClelland et al., 1983.*)

increase owing to the loss of buoyancy, causing the legs to penetrate to point A. The rig mover continues to jack the hull out of the water at 10 to 30 m per hour until an air gap of approximately 1.5 m is achieved (McClelland et al., 1983). The required air gap depends on the wave height. The rig mover then levels the platform before beginning preloading to achieve a final penetration of the legs. The legs are preloaded in steps by filling tanks in the platform with sea water. The maximum preload typically corresponds to the load the platform might experience at the site during the drilling operation. The maximum load is maintained for several hours. The purpose of the preload is to force additional penetration of the footings to a level where the total bearing capacity exceeds an acceptable margin of safety as shown by point B in Figures 18.73a and b. After the preloading, the platform is raised to its final height, 10 to 20 m above the sea level, again depending mainly on the estimated maximum wave height.

The primary hazard with the presently used jack-up rigs is the risk of punch-through in layered soils, when one of the legs suddenly penetrates a relatively stiff layer as shown in Figures 18.74 and 18.75. The resulting sudden settlement of the structure may severely damage the legs and lead to the platform

collapsing. A number of failures caused by punch-through have been described by McClelland et al. (1983), Young et al. (1984), and Werno et al. (1987). Most of these failures have occurred when the hull is lifted out of the water, during the preloading of the legs, or during a severe storm. The geologic conditions that can cause a footing to experience a punch-through failure are generally recognized as (1) a sand or rock layer of limited thickness supported by a weaker underlying clay layer, or (2) a very strong clay layer underlain by a weaker clay. McClelland et al. (1983) have described a variety of geologic processes that can produce the hazardous subbottom sequence of a hard layer overlying a weaker layer.

18.7.1 Support Methods

From the foundation point of view, these platforms can be divided into two major categories (Werno et al. 1987):

1. *Footing type*, with three or more footings (i.e., spud cans) each supporting an individual leg. The spud cans used in this application are characterized by their different sizes and shapes as shown in Figure 18.76.
2. *Mat type*, with three or more legs supported by a common mat.

A summary of the support methods is presented in Figure 18.77. Mats, which generally cover the total area of the platform, are usually A-shaped. The thickness is about 3 m. The performance of mat-supported rigs has been described by Hirst et al. (1976), and by Young et al. (1981). They have the disadvantages that they require a relatively level bottom

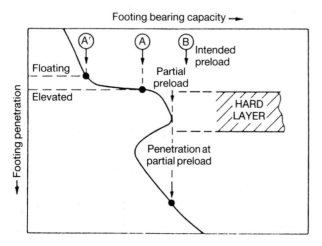

Fig. 18.74 Punch-through failure during preload. (*McClelland et al., 1983.*)

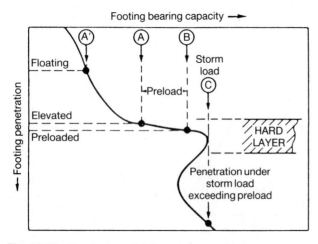

Fig. 18.75 Punch-through failure due to storm overload. (*McClelland et al., 1983.*)

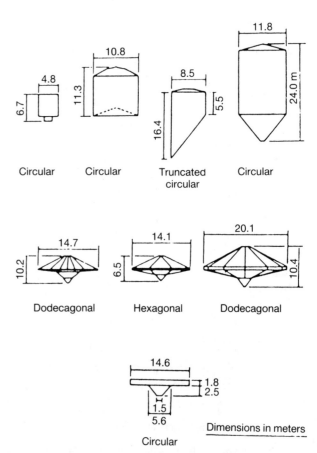

Fig. 18.76 Configuration of spud cans. (*After Werno et al., 1987.*)

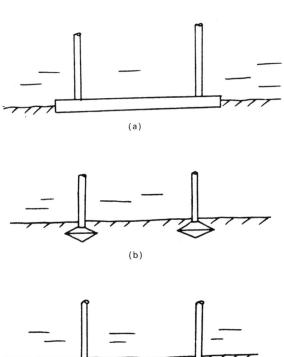

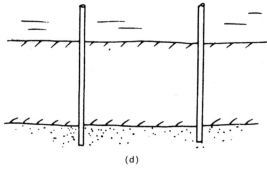

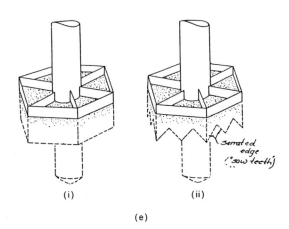

Fig. 18.77 Support methods. (a) Mat foundation. (b) Pad foundation. (c) Spud can foundation. (d) Pile foundation. (e) Grids: (i) straight edge; (ii) serrated edge. (*After Broms and James, 1985.*)

and that they are affected by differential settlements caused by eccentric or cyclic loading and by scour. However, the advantage is that a mat foundation can be used in very soft soil because of the large surface area.

Most jack-up rigs are supported on individual pads or spud cans with 12 to 16 m diameter as shown in Figure 18.76. The base is often enlarged in order to increase the bearing capacity. The typical load in each leg can be up to 45 MN or more. The main limitation of this type of foundation is the low bearing capacity in very soft soil and the risk of punch-through in layered or stratified soils. In cohesionless soils they are also affected by scour.

As the pads or spud cans are pushed down through the soil, a hole is formed. These holes may collapse suddenly, especially in sand, silt, and soft clay, or they may gradually fill up. The bearing capacity of the legs is thereby reduced especially in clay and silt. Gemeinardt and Focht (1970) have developed calculation methods that take this reduction of the bearing capacity into account. In the calculation of the ultimate bearing capacity, especially for clay, it is advisable to assume that the holes are closed. An additional factor is the remolding of the soil that occurs during penetration. The resulting reduction of the bearing capacity can be large for sensitive soils with a high sensitivity ratio (S_t).

Piles can be used where there is a layer with a high bearing capacity close to the bottom. Very long piles will be required when the depth to the bearing layer is large.

18.7.2 Prediction of Leg Penetration During Installation

The bearing capacity analysis of individual footings should fulfill two functions. The first is to predict the penetration below the seafloor to determine if the combined air gap, water depth, and footing penetration are less than or equal to the maximum available leg length. The second purpose is to evaluate the risk of a foundation failure caused by a punch-through.

A. Footings Supported on Uniform Soil

Several procedures are available for computing the bearing capacity of a foundation in cohesive soils provided the shear strength is fairly constant (i.e., Skempton, Meyerhof, and Hansen). In general, all procedures can be represented by the general form of the bearing-capacity equation given below (Young et al., 1984):

$$Q = A N_c S_u + \gamma_b V \qquad (18.55)$$

where

Q = footing capacity
A = cross-sectional area of embedded footing
N_c = dimensionless bearing-capacity factor
S_u = undrained shear strength
γ_b = buoyant unit weight of soil displaced by footing
V = embedded volume of footing

Young et al. (1984) indicated that accurate prediction of footing penetration is more dependent upon the selection of undrained shear strength than the analytical method.

The use of Skempton's equation was first suggested by Gemeinhardt and Focht (1970). Their study gave results comparing observed jack-up rig penetrations with those predicted by the available procedures in clay mentioned earlier.

They found that the most satisfactory results with the Skempton equation are obtained when: (1) D is defined as the depth to the footing tip; and (2) S_u is averaged over a depth equal to the footing width below the widest cross section. Later work by Young et al. (1981) indicated that better results are achieved by averaging the strength over a zone equal to $B/2$ instead of B as originally proposed, and defining D as the depth to the widest embedded cross section.

B. Footings Supported on Layered Soils

A number of analytical procedures ranging from a simplified projected area method to more complex analytical methods are available to evaluate the penetration depth and the bearing capacity for footings in stratified soil systems (Vesić, 1970; Meyerhof, 1974; Meyerhof and Hanna, 1978; Hanna and Meyerhof, 1980; Endley et al., 1981; Jacobsen et al., 1977). Young et al. (1984) have compared three of these methods (projected area method, Hanna and Meyerhof (1980), Jacobsen et al. (1977)) and found that the Hanna and Meyerhof (1980) method gives the closest fit between model test results and predicted capacities. The Hanna and Meyerhof equation for circular footings is

$$Q = A\left[6S_{uB} + 2\gamma_{b1}H^2\left(1 + \frac{2D}{H}\right)K_s\frac{\tan\phi}{B} \right] + \gamma_b V$$

$$(18.56)$$

where

S_{uB} = undrained shear strength of the lower soil stratum
H = thickness of upper layer in two layer system
D = depth of widest cross-sectional area
K_s = coefficient of punching shear
ϕ = internal friction angle of granular stratum
γ_b = buoyant unit weight of granular stratum

The parameter K_s can be determined in a two-part operation. The first part involves determining the punching shear parameter δ/ϕ_1 using Figure 18.78 and knowing the ratio q_2/q_1, where

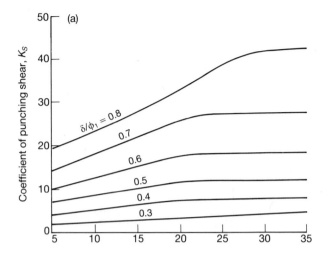

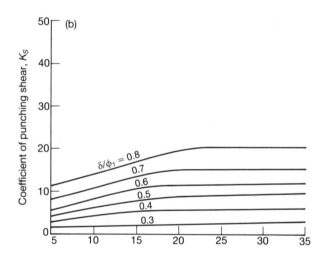

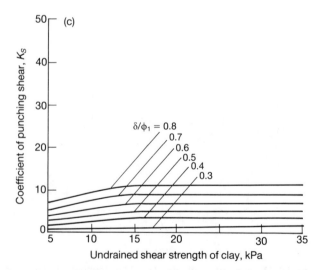

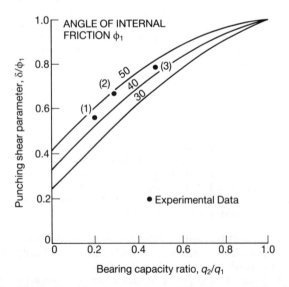

Fig. 18.78 Punching shear parameter. (*After Hanna and Meyerhof, 1980.*)

Fig. 18.79 Coefficients of punching shear: (a) $\phi_1 = 50°$; (b) $\phi_1 = 45°$; (c) $\phi_1 = 40°$. (*After Hanna and Meyerhof, 1980.*)

$q = 0.5\,\gamma_1\,B\,N_\gamma$ (for the homogeneous upper sand layer) and $q_2 = C_2\,N_c$ (for the homogeneous lower soft clayey layer). Then using Figure 18.79, K_s can be determined.

A procedure has been proposed by Werno et al. (1987) for the case of a multilayered soil. The method of computing the penetration is based on the interaction between the spud can (Fig. 18.76) and soil for the state of ultimate equilibrium of the soil. In this procedure the following assumptions are made:

1. The spud can consists of two independent parts: the slab foundation and the conical tip.
2. The conical tip is regarded as a set of rings (Fig. 18.80).
3. The bearing capacity of a ring is calculated as the difference between the bearing capacities of two subsequent circular foundations.
4. The ultimate bearing capacity of the spud can is the sum of the bearing capacities of foundation parts.
5. Summation of the bearing capacities is done as shown in Figure 18.81.

Five potential cases for the development of bearing capacity (shaded parts only) are shown in Figure 18.81. If the penetration depth is less than the height of the conical tip, the slab is not in contact with the seabed. The first case occurs when the conical tip sinks partially into the seabed and develops the bearing capacity (Fig. 18.81a). If the spud can is fully in contact with the homogeneous subsoil, it is assumed that the bearing capacity is created only by the slab (Figs. 18.81b and c). If the spud can has penetrated the first layer and the conical tip is partially in the next layer, the bearing capacity is created by both the slab and part of the tip when the lower layer is stronger (Fig. 18.81d). When the upper layer is stronger, the bearing capacity is created by the slab only (Fig. 18.81c).

As a first step, to obtain the load–penetration curve, the bearing capacity is computed according to the above procedure using the Hansen (1961) general formula. If the bearing capacity increases with depth, the computed values in the first-step

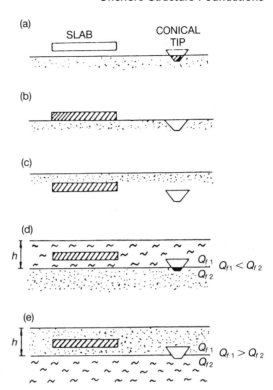

Fig. 18.81 First-step computation procedure. Q_f = bearing capacity at the depth h for the soil characteristics Q_{f1} of the upper layer and Q_{f2} of the lower layer. (*Werno et al., 1987.*)

penetrations are slightly overestimated, but this discrepancy has no influence on platform stability.

The second step of computation takes into account the influence of the weaker layer on the bearing capacity of the hard layer. The computations are based on assumptions (1) to (4) using Meyerhof and Hanna's methods as described in Hanna and Meyerhof (1980) or Meyerhof and Hanna (1978), depending on the soil characteristics. Examples of prediction results for a jack-up platform operating in the Baltic Sea are presented in Figure 18.82.

18.8 HYDRAULIC FILLED ISLANDS

In the Beaufort Sea, over 30 artificial islands have been built for exploration purposes and hence have a relatively short life in comparison with future production platforms or the present production platforms in the North Sea. A number of these artificial island deployments in the Beaufort Sea have included extensive site investigations, analyses, and instrumentation (Shinde et al., 1986). Of these 30 artificial islands, it is probable that about 20 to 30 percent have experienced unpredicted geotechnical problems, although only a few case histories have been published (Clark and Jordaan, 1987). Exceptions are those of the Nerlerk berm failure in 1984 (Mitchell, 1984; Sladen et al., 1985; Been et al., 1987); Adgo Island, which was a sand-bag-retained island that experienced instability (Shinde et al., 1986); Issugnak Island, where the above-water portion disappeared twice during construction (Shinde et al., 1986).

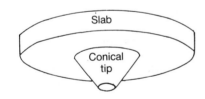

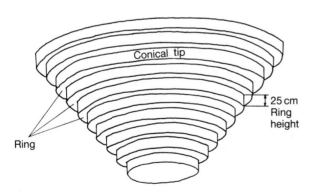

Fig. 18.80 Discrete scheme of the conical tip. (*Werno et al., 1987.*)

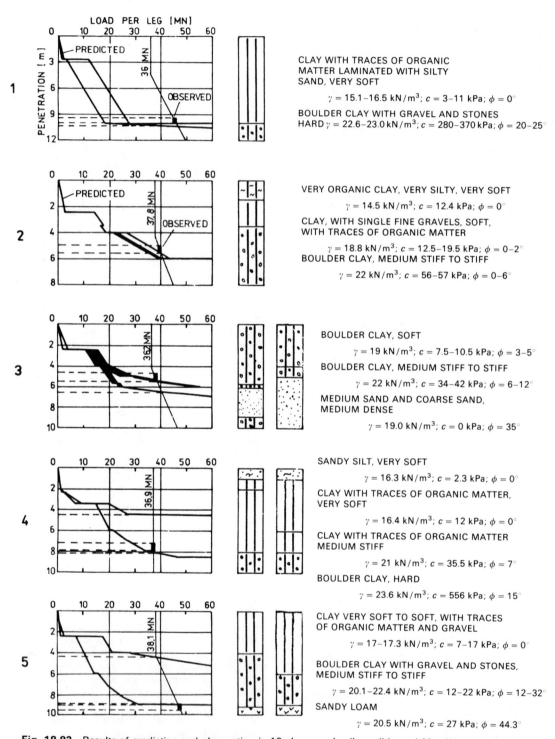

Fig. 18.82 Results of prediction and observation in 10 chosen subsoil conditions. (*After Werno et al., 1987.*)

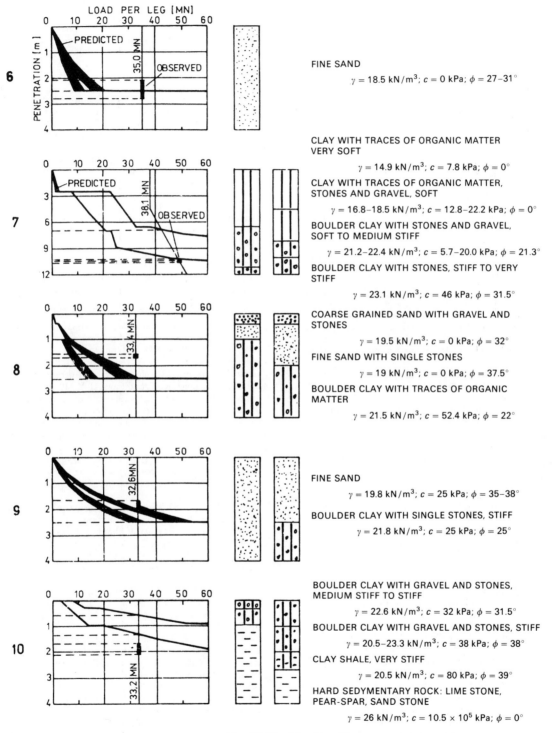

6

FINE SAND

$\gamma = 18.5$ kN/m³; $c = 0$ kPa; $\phi = 27-31°$

7

CLAY WITH TRACES OF ORGANIC MATTER
VERY SOFT

$\gamma = 14.9$ kN/m³; $c = 7.8$ kPa; $\phi = 0°$

CLAY WITH TRACES OF ORGANIC MATTER,
STONES AND GRAVEL, SOFT

$\gamma = 16.8-18.5$ kN/m³; $c = 12.8-22.2$ kPa; $\phi = 0°$

BOULDER CLAY WITH STONES AND GRAVEL,
SOFT TO MEDIUM STIFF

$\gamma = 21.2-22.4$ kN/m³; $c = 5.7-20.0$ kPa; $\phi = 21.3°$

BOULDER CLAY WITH STONES, STIFF TO VERY
STIFF

$\gamma = 23.1$ kN/m³; $c = 46$ kPa; $\phi = 31.5°$

8

COARSE GRAINED SAND WITH GRAVEL AND
STONES

$\gamma = 19.5$ kN/m³; $c = 0$ kPa; $\phi = 32°$

FINE SAND WITH SINGLE STONES

$\gamma = 19$ kN/m³; $c = 0$ kPa; $\phi = 37.5°$

BOULDER CLAY WITH TRACES OF ORGANIC
MATTER

$\gamma = 21.5$ kN/m³; $c = 52.4$ kPa; $\phi = 22°$

9

FINE SAND

$\gamma = 19.8$ kN/m³; $c = 25$ kPa; $\phi = 35-38°$

BOULDER CLAY WITH SINGLE STONES, STIFF

$\gamma = 21.8$ kN/m³; $c = 25$ kPa; $\phi = 25°$

10

BOULDER CLAY WITH GRAVEL AND STONES,
MEDIUM STIFF TO STIFF

$\gamma = 22.6$ kN/m³; $c = 32$ kPa; $\phi = 31.5°$

BOULDER CLAY WITH GRAVEL AND STONES, STIFF

$\gamma = 20.5-23.3$ kN/m³; $c = 38$ kPa; $\phi = 38°$

CLAY SHALE, VERY STIFF

$\gamma = 20.5$ kN/m³; $c = 80$ kPa; $\phi = 39°$

HARD SEDYMENTARY ROCK: LIME STONE,
PEAR-SPAR, SAND STONE

$\gamma = 26$ kN/m³; $c = 10.5 \times 10^5$ kPa; $\phi = 0°$

Fig. 18.82 (*Continued.*)

REFERENCES

Aggarwal, S. L., Malhotra, A. K., and Banerjee, R. (1979), Engineering properties of calcareous soils affecting the design of deep penetration piles for offshore structures, *Proceedings 9th Annual Offshore Technology Conference*, Houston, **3**, pp. 503–512.

Anderson, K. H. (1976), Behavior of clay subjected to undrained cyclic loading, Conference on the Behavior of Offshore Structures, BOSS 76, Trondheim, Norwegian Geotechnical Institute.

Anderson, K. H., Hansteen, O. L., Hoeg, K., and Prevost, J. H. (1978), *Soil Deformations Due to Cyclic Loads on Offshore Structures*, Norwegian Geotechnical Institute, No. 16.

Andresen, A., Berre, T., Kleven, A., and Lunne, T. (1979), Procedures used to obtain soil parameters for foundation engineering in the North Sea, *Marine Geotechnology*, **3**, pp. 201–266.

Angemeer, J., Carlson, E. D., and Klick, J. H. (1973) Techniques and results of offshore pile load testing in calcareous soils, Offshore Technology Conference, Houston, Paper 1894.

Angemeer, J., Carlson, E. D., Stroud, S., and Kurzeme, M. (1975), Pile load tests in calcareous soils conducted in 400 feet of water from a semisubmersible exploration rig, Offshore Technology Conference, Houston, Paper 2311.

Antoine, J. and Trabant, P. (1976), Geological features of shallow gas, *Proceedings, Houston Geophysical Society*, Houston, Texas.

API (1984), *Recommended Practice for Planning, Designing and Constructing Fixed Offshore Platforms*, American Petroleum Institute Publication RP-2A, Dallas, Texas.

Arnold, K. E. (1967), Soil movements and their effects on pipelines in the Mississippi Delta Region, M.S. Thesis, Tulane University, New Orleans.

Audibert, J. M. E. and Nyman, K. J. (1977), Soil restraint against horizontal motion of pipes, *Journal of the Geotechnical Engineering Division, ASCE*, **103**, No. GT-10, pp. 1119–1142.

Audibert, J. M. E., Lai, N. W., and Bea, R. G. (1978), Design of pipelines to resist seafloor instabilities and hydrodynamic forces, presented at the Energy Technology Conference and Exhibition, Houston, Texas.

Bea, R. G. (1975), Parameters affecting axial capacity of piles in clays, *Proceedings 7th Annual Offshore Technology Conference*, OTC 2307, pp. 611–623.

Bea, R. G. (1980), Dynamic response of piles in offshore platforms, *Proceedings, Dynamic Response of Pile Foundations: Analytical Aspects*, ASCE, pp. 80–109.

Bea, R. G., Audibert, J. M. E., and Dover, A. R. (1980), Dynamic response of laterally loaded and axially loaded piles, *Proceedings, 12th Offshore Technology Conference*, Houston, Texas, Paper OTC 3749, pp. 129–139.

Been, K., Jefferies, M. G., Crooks, J. H. A., and Rotherburg, L. (1987), The cone penetration test in sands: Part II, General inference of state, *Geotechnique*, **37**, pp. 285–299.

Berezantsev, V. G., Kristoforov, V. S., and Golubkov, V. N. (1961), Load bearing capacity and deformation of piled foundations, *Proceedings, 5th International Conference on Soil Mechanics and Foundation Engineering*, Paris, **2**, pp. 11–12.

Berger, W. H. (1976), Biogenous deep sea sediments: Production, preservation and interpretation, *Chemical Oceanography*, Vol. 5, 2d ed., ed. J. P. Riley and R. Chester, Academic Press, London, Chapter 29, pp. 265–388.

Bjerrum, L. (1973), Geotechnical problems involved in foundations of structures in the North Sea, *Geotechnique*, **23**, pp. 319–358.

Bonar, A. J. and Ghazzali, O. I. (1973), Research on pipeline flotation, *Journal of the Transportation Engineering Division, ASCE*, **99**, No. TE-2, pp. 211–233.

Boulon, M., Desrues, J., Foray, P., and Forque, M. (1980), Numerical model for foundation under cyclic loading, Application to piles, *International Symposium on Soils Under Cyclic and Transient Loading*, Swansea, A. A. Balkema, Rotterdam, pp. 681–694.

Bowles, J. E. (1988), *Foundation Analysis and Design*, McGraw-Hill Book Co., Inc., New York, N.Y.

Briaud, J. L. and Audibert, J. M. E. (eds.) (1985), The pressuremeter and its marine applications, Second International Symposium, *ASTM STP 950*, American Society for Testing and Materials.

Broms, B. and James, D. J. E. (1985), Foundation problems with jack-up rigs in East China Sea, *Proceedings, 2nd Shanghai Symposium on Marine Geotechnology and Nearshore/Offshore Structures*, Tongji University Press, Shanghai, pp. 3–31.

Brown, R. J. (1973), Pipeline design to reduce anchor and fishing board damage, *Journal of the Transportation Engineering Division, ASCE*, **99**, No. TE-2, pp. 199–210.

Brown, J. D. and Meyerhof, G. G. (1969), Experimental study of bearing capacity in layered clays, *Proceedings 7th International Conference on Soil and Foundation Engineering*, Mexico City, **2**.

Burland, J. B. (1973), Shaft friction of piles in clay—A simple fundamental approach, *Ground Engineering*, **6**, No. 3, pp. 30–42.

Butler, F. G. (1975), Heavily over consolidated clays, Review paper: Session III, Settlement of Structures, *Proceedings* of a conference organized by the British Geotechnical Society, Cambridge, Pentech Press, London, pp. 531–572.

CFEM (1978), *Canadian Foundation Engineering Manual*, Canadian Geotechnical Society, Montreal, Quebec, Canada.

Carrier, W. D. III, and Christian, J. T. (1973), Rigid circular plate resting on a non-homogeneous elastic half-space, *Geotechnique*, **23**, No. 1, pp. 67–84.

Casagrande, A. (1932), The structure of clay and its importance in foundation engineering, *Journal of the Boston Society of Civil Engineers*, **19**, No. 4.

Chaney, R. C. (1984), Methods of predicting the deformation of the seabed due to cyclic loading, *Seabed Mechanics*, ed. B. Denness, Graham and Trotman, London, pp. 159–167.

Chaney, R. C., Slonim, S. S., and Slonim, S. M. (1982), Determination of calcium content in soils, *Proceedings—Symposium on Performance and Behavior of Calcareous Soils*, ed. K. Demars and R. C. Chaney, ASTM STP 777, pp. 3–15.

Chaney, R. C. and Demars, K. R. (eds.) (1985), *Proceedings—Strength Testing of Marine Sediments: Laboratory and In Situ Measurements*, ASTM STP 883.

Chaney, R. C. and Fang, H. Y. (1986a), Static and dynamic properties of marine sediments, *Proceedings of Symposium on Marine Geotechnology and Nearshore/Offshore Structures*, ed. K. Demars and R. C. Chaney, ASTM STP 923, pp. 74–111.

Chaney, R. C. and Fang, H. Y. (eds.) (1986b), *Proceedings—Marine Geotechnology and Nearshore/Offshore Structures*, ASTM STP 923.

Chaney, R. C., Demars, K. R., and Fang, H. Y. (1985), Toward a unified approach to soil property characterization, *Proceedings—Strength Testing of Marine Sediments: Laboratory and In Situ Measurements*, ed. K. Demars and R. C. Chaney, ASTM STP 883, pp. 425–439.

Chellis, R. D. (1961), *Pile Foundations*, 2d ed., McGraw-Hill Book Co., Inc., New York, N.Y.

Clark, J. I. and Jordaan, I. J. (1987), Geotechnical predictions in ice affected marine environments, *Proceedings, International Symposium on Prediction and Performance in Geotechnical Engineering*, Calgary, Alberta, pp. 15–25.

Clausen, C. J. F. (1976), The Condeep story, *Offshore Soil Mechanics*, ed. P. George and D. Wood, Cambridge University Engineering Dept, and Lloyd's Register of Shipping, London, pp. 256–270.

Clausen, C. J. F., DiBiagio, E., Duncan, J. M., and Andersen, K. H. (1975), Observed behavior of the Ekofisk oil storage tank foundation, *Proceedings 7th Annual Offshore Technology Conference*, Houston, Texas, **3**, pp. 399–413.

Clausen, C. J. and Lunne, T. (1980), The application of soil investigation data to the design of offshore gravity platforms, *Offshore Site Investigation*, ed. E. A. Ardus, Graham and Trotman, London, pp. 247–256.

Clausen, C. J., Aas, P. M., and Almeland, I. B. (1982), Analysis of pile foundation system for a North Sea drilling platform, *Proceedings of the International Conference on the Behavior of Offshore Structures*, BOSS 82, Cambridge, Mass., **1**, pp. 141–157.

Cooke, R. W., Price, G., and Tarr, K. (1979), Jacked piles in London clay: A study of load transfer and settlement under working conditions, *Geotechnique*, **29**, No. 2, pp. 113–147.

Cox, W. R., Kraft, L. M., and Verner, E. A. (1979), Axial load tests on 14 inch pipe piles in clay, *Proceedings, Eleventh Annual Offshore Technology Conference*, Houston, Texas, pp. 1147–1151.

Coyle, H. W. and Reese, L. C. (1966), Load transfer for axially loaded piles in clay, *Journal of the Soil Mechanics and Foundation Division, ASCE*, pp. 1–26.

Daley, G. C. (1973), Optimization of tension level and stinger length for offshore pipeline installation, *Proceedings 5th Annual Offshore*

Technology Conference, Houston, Texas, Paper OTC 1875, pp. 473–478.

Datta, M., Gulhati, S. K., and Rao, G. V. (1979), Crushing of calcareous sands during shear, *Offshore Technology Conference,* Houston, Texas, OTC Paper No. 3525, pp. 1459–1467.

Datta, M., Gulhati, S. K., and Rao, G. V. (1980), An appraisal of the existing practice of determining the axial load capacity of deep penetration piles in calcareous sands, *12th Annual Offshore Technology Conference,* Paper No. OTC 3867, pp. 119–130.

Davis, E. H. and Booker, J. R. (1973), The effect of increasing strength with depth on the bearing capacity of clays, *Geotechnique,* **23**, No. 4, pp. 551–563.

Dawson, T. H. (1983), *Offshore Structural Engineering,* Prentice-Hall, Englewood Cliffs, N.J.

de Ruiter, J. (1971), Electric penetrometer for site investigations, *Journal of the Soil Mechanics and Foundation Division, ASCE,* **97**, No. SM-2, pp. 457–472.

de Ruiter, J. (1975), The use of in situ testing for North Sea soil studies, Offshore Europe Conference, Aberdeen.

de Ruiter, J. and Beringen, F. L. (1979), Pile foundations for large North Sea structures, *Marine Geotechnology,* **3**, No. 3, pp. 267–314.

Demars, K. R. (1978), Design of marine pipelines for areas of unstable sediment, *Journal of the Transportation Engineering Division, ASCE, TEI,* Proc. Paper 13455, pp. 107–112.

Demars, K. R. (1979), Design consideration for pipelines interacting with travelling waves, *Proceedings ASCE Coastal Structures 79,* Alexandria, Va., pp. 100–114.

Demars, K. R., Nacci, V. A., and Wang, M. C. (1977), Pipeline failures: A need for improved analyses and site surveys, *Proceedings Offshore Technology Conference,* Houston, Texas, Paper OTC 2966.

Demars, K. R. and Chaney, R. C. (eds.) (1982), *Proceedings—Symposium on Geotechnical Properties, Behavior and Performance of Calcareous Soils,* ASTM STP 777.

Det Norske Veritas (DNV) (1977), *Rules for Design, Construction and Inspection of Offshore Structures,* Hovik, Norway.

Drewry, J. M., Weidler, J. B., and Hwong, S. T. (1977), Predicting axial pile capacities for offshore platforms, *Petroleum Engineer,* **41**.

Dunlap, W. A., Bryant, W. R., Bennett, R. H., and Richards, A. F. (1978), Pore pressure measurements in unconsolidated sediments, *10th Annual Offshore Technology Conference,* Houston, Texas.

Eide, O. T., Larsen, L. G., and Mo, O. (1976), Installation of the Shell/Esso Brent B Condeep production platform, *Proceedings 8th Annual Offshore Technology Conference,* Houston, Texas, **1**, pp. 101–114.

Eide, O., Kjekstad, O., and Brylawski, E. (1979a), Installation of concrete gravity structures in the North Sea, *Marine Geotechnology,* **3**, pp. 315–368.

Eide, O. T., Andersen, K. H., and Lunne, T. (1979b), Observed foundation behaviour of concrete gravity platforms installed in the North Sea 1973–1978, *Proceedings, 2nd International Conference on the Behaviour of Offshore Structures,* BOSS 79, pp. 435–456.

Endley, S. N., Rapoport, V., Thompson, P. J., and Baglioni, V. P. (1981), Prediction of jack-up rig footing penetration, *Proceedings, 13th Offshore Technology Conference,* Houston, Texas, **4**, pp. 285–296.

Fact Sheet (1986), *The Norwegian Continental Shelf,* Royal Ministry of Petroleum and Energy, Norway.

Fang, H. Y. and Chaney, R. C. (1985), Causes of foundation instability of nearshore/offshore structures and improvement techniques, *Proceedings of Shanghai Symposium on Marine Geotechnology and Nearshore/Offshore Structures,* Shanghai, pp. 575–590.

Fang, H. Y. and Chaney, R. C. (1986), Geo-environmental and climatological conditions related to marine structural design along the China coastline, *Proceedings of Symposium on Marine Geotechnology and Nearshore/Offshore Structures,* ASTM STP 923, pp. 149–160.

FIP (1978), Federation International de la Precontrainte (FIP), Commission on Concrete Sea Structures. Working Group on Foundations, *Foundations of Concrete Gravity Structures in the North Sea,* SOA Draft.

Flaate, K. and Selnes, P. (1977), Side friction of piles in clay, *Proceedings, 9th International Conference on Soil Mechanics and Foundation Engineering,* Tokyo, **1**, pp. 517–522.

Focht Jr., J. A. and Kraft Jr., L. M. (1981), Prediction of capacity of long piles in clay: A status report, *Symposium on Geotechnical*

Aspects of Coastal and Offshore Structures, Bangkok, pp. 95–113.

Gaythwaite, J. (1981), *The Marine Environment and Structural Design,* Van Nostrand-Reinhold Co., New York, N.Y.

Geddes, J. D. (1969), Boussinesq based approximations to the vertical stress caused by pile type subsurface loadings, *Geotechnique,* **19**, No. 4, pp. 509–514.

Gemeinhardt, J. B. and Focht, J. A. (1970), Theoretical and observed performance of mobile rig footings on clay, *Proceedings, 2nd Offshore Technology Conference,* Houston, Texas, **1**, pp. 549–558.

George, P. and Wood, D. (eds.) (1976), *Offshore Soil Mechanics,* Cambridge University Engineering Department and Lloyd's Register of Shipping, London.

Gerwick Jr., B. C. (1986), *Construction of Offshore Structures,* John Wiley and Sons, Inc., New York, N.Y.

Gibson, R. E. and Dowse, B. E. W. (1981), The influence of geotechnical engineering on the evolution of offshore structures in the North Sea, *Canadian Geotechnical Journal,* **18**, No. 2, pp. 171–178.

Goble, G. G. (1983), Analysis of offshore pile driving—A review, *Proceedings of Conference on Geotechnical Practice in Offshore Engineering,* ed. S. G. Wright, ASCE, pp. 596–603.

Goble, G. G. and Rausche, F. (1980), Wave equation analysis of pile driving—WEAP program, Volumes I, II, and III, *FHWA Report No. FHWA-IP-76-14.1,* Goble and Associates, Warrensville Height, Ohio 44128.

Hanna, A. M. and Meyerhof, G. G. (1980), Design charts for ultimate bearing capacity of foundations on sand overlying soft clay, *Canadian Geotechnical Journal,* **17**, pp. 300–303.

Hansen, J. B. (1961), A general formula for bearing capacity, *Danish Geotechnical Institute Bulletin No. 28,* Copenhagen.

Heerema, E. P. (1978), Predicting pile driveability: Heather as an illustration of the 'friction fatigue' theory, European Offshore Petroleum Conference, London, Paper No. 50.

Heerema, E. P. (1979), Relationships between wall friction, displacement, velocity and horizontal stress in clay and sand for pile drivability, *Ground Engineering,* **12**, No. 1, pp. 55–65.

Henkel, D. S. (1970), The role of waves in causing submarine slides, *Geotechnique,* **20**, No. 1, pp. 75–80.

Hirsch, T. J., Koehler, A. M., and Sutton, V. J. R. (1975), Selection of pile driving equipment and field evaluation of pile bearing capacity during driving for the North Sea Forties field, *Proceedings, 7th Annual Offshore Technology Conference,* Paper No. 2247, pp. 37–49.

Hirsch, T. J., Carr, L., and Lowry, L. L. (1976), Pile driving analysis—Wave equation use manual TTI program, Vols I, II, and III, *FHWA Report No. FHWA-IP-76-13.1,* Texas Transportation Institute, Texas A & M University, College Station, Texas 77840.

Hirst, T. J., Steele, J. F., Remy, N. D., and Scales, R. E. (1976), Performance of mat-supported jack-up rigs, *Proceedings, 8th Annual Offshore Technology Conference,* Houston, Texas, **1**, pp. 821–830.

Hobbs, N. B. (1977), Behavior and design of piles in chalk—an introduction to the discussion of the papers on chalk, *Proceedings, Symposium on Piles in Weak Rock,* London, pp. 149–175.

Hoeg, K. (1986), Geotechnical issues in offshore engineering, *Marine Geotechnology and Nearshore/Offshore Structures,* ed. R. C. Chaney and H. Y. Fang, ASTM STP 923, pp. 7–50.

Hoeg, K. and Tang, W. H. (1977), Probabilistic considerations in the foundation engineering for offshore structures, *Proceedings of the Second International Conference on Structural Safety and Reliability,* Munich, pp. 267–296.

Holmquist, D. V. and Matlock, H. (1976), Resistance–displacement relationships for axially loaded piles in soft clay, *Proceedings 8th Offshore Technology Conference,* Houston, Texas, Paper OTC 2474, pp. 554–569.

Idriss, I. M., Dobry, R., and Singh, R. D. (1978), Nonlinear behavior of soft clays during cyclic loading, *Journal of the Geotechnical Engineering Division, ASCE,* **104**, No. GT-12, pp. 1427–1447.

Jacobsen, M., Christensen, K. V., and Sorensen, C. S. (1977), Gennemlokning aftynde sandlag, *Vag-och Vattenbuggaren, Sevenska Vag-och Vattenbuggares Riksforbund,* Stockholm, pp. 23–25.

Janbu, N. (1976), Static bearing capacity of friction piles, *Proceedings, European Conference on Soil Mechanics and Foundation Engineering,* **1.2**, pp. 479–488.

Janbu, N., Grande, L. and Eggereide, K. (1976), Effective stress stability analysis for gravity structures, *Proceedings, Behavior of Offshore Structure BOSS 76,* pp. 449–466.

Jurgenson, L. (1934), The application of theories of elasticity and plasticity of foundation problems, *Boston Society of Civil Engineers, Contributions to Soil Mechanics 1925–1940*, pp. 148–183.

Karlsrud, K. and Haugen, T. (1985), Behaviour of piles in clay under cyclic axial loading—Results of field model tests, *Behavior of Offshore Structures, BOSS 85*, Elsevier Science Publishers B.V., Amsterdam, pp. 589–600.

Karlsrud, K., Nadim, R., and Haugen, T. (1986), Piles in clay under cyclic axial loading—Field tests and computational modeling, *Proceedings 3d. International Conference on Numerical Methods in Offshore Piling*, Nantes, France, pp. 165–190.

Kézdi, A. (1975), Pile foundations, *Foundation Engineering Handbook*, eds. H. F. Winterkorn and H. Y. Fang, Van Nostrand Reinhold Co., New York, N.Y., pp. 556–600.

Kjekstad, O. and Stub, F. (1978), Installation of the ELF TCP-2 Condeep Platform at the Frigg Field, *Proceedings of the European Offshore Petroleum Conference*, London, **1**, pp. 121–130.

Kraft, L. M., Jr., Focht, J. A., Jr., and Amerasinghe, S. F. (1981), Friction capacity of piles driven into clay, *Journal of the Geotechnical Engineering Division*, ASCE, **107**, No. GT-11, pp. 1521–1541.

Lade, P. V. and Lee, K. L. (1976), *Engineering Properties of Soils*, Engineering Report, UCLA-ENG-7652.

Lauritzsen, R. and Schjetne, K. (1976), Stability calculations for offshore gravity structures, *Proceedings 6th Annual Offshore Technology Conference*, **1**, OTC 2431, pp. 75–82.

Lee, K. L. (1974), *Earthquake Induced Permanent Deformations of Embankments*, Engineering Report 7498, University of California, Los Angeles.

Lo, M. B. (1967), Discussion to paper by Y. O. Beredugo, *Canadian Geotechnical Journal*, **4**, No. 3, pp. 353–354.

Lord, J. A. (1976), A comparison of three types of driven cast in situ piles in chalk, *Geotechnique*, **26**, No. 1, pp. 73–93.

Lowery, L. et al. (1969), *Pile Driving Analysis State of the Art*, Research Report 33-13 (final), Texas Transportation Institute, College Station, Texas.

Lunne, T. and St. John, H. (1979), The use of cone penetration tests to compute penetration resistance of steel skirts underneath North Sea gravity platforms, *Proceedings of the European Conference on Soil Mechanics and Foundation Engineering*.

Matlock, H. (1970), Correlations for design of laterally loaded piles in soft clay, *Proceedings, Second Annual Offshore Technology Conference*, pp. 577–587.

Matlock, H. and Foo, S. C. (1979), Axial analysis of pile using a hysteretic and degrading soil model, *Proceedings Conference Numerical Methods in Offshore Piling*, ICE, London, pp. 165–185.

Matlock, H., Ingram, W. B., Kelley, A. E., and Bogard, D. (1980), Field tests of the lateral-load behavior of pile groups in soft clay, Offshore Technology Conference, Houston, Texas, OTC 3871, pp. 163–174.

Matsuo, M. (1967), Bearing capacity of anchor foundations, *Soils and Foundations*, **8**, No. 1, pp. 18–48.

McClelland, B. (1974), Design of deep penetration piles for ocean structures, *Journal of the Geotechnical Engineering Division*, ASCE, **100**, No. GT-7, pp. 709–747.

McClelland, B., Focht Jr., J. A., and Emrich, W. J. (1967), Problems in design and installation of heavily loaded pipe piles, *Proceedings, Conference Civil Engineering in the Oceans*, ASCE, pp. 601–634.

McClelland, B., Focht Jr., J. A., and Emrich, W. J. (1969), Problems in design and installation of offshore piles, *Journal of the Soil Mechanics and Foundations Division*, ASCE, **95**, No. SM-6, pp. 1491–1513.

McClelland, B. and Cox, W. R. (1976), Performance of pile foundations for offshore structures, *Proceedings, First International Conference, Behavior of Offshore Structures*, Trondheim, Norway, **1**, pp. 528–544.

McClelland, B., Young, A. G., and Remmes, B. D. (1983), Avoiding jack-up rig foundation failure, *Symposium on Geotechnical Aspects of Offshore and Nearshore Structures*, Bangkok, A.A. Balkema, Rotterdam, Netherlands, pp. 137–157.

Meyerhof, G. G. (1959), Compaction of sand and bearing capacity of piles, *Journal of the Soil Mechanics and Foundations Division*, ASCE, **85**, No. SM-6, pp. 1–30.

Meyerhof, G. G. (1965), Shallow foundations, *Journal of Soil Mechanics and Foundation Engineering*, ASCE, **91**, No. SM-2, pp. 21–31.

Meyerhof, G. G. (1974), Ultimate bearing capacity of footings on sand overlying clay, *Canadian Geotechnical Journal*, **11**, No. 2, pp. 223–229.

Meyerhof, G. G. (1976), Bearing capacity and settlement of pile foundations, *Journal of the Geotechnical Engineering Division*, ASCE, **102**, No. GT-3, pp. 197–228.

Meyerhof, G. G. and Adams, J. I. (1968), The ultimate uplift capacity of foundations, *Canadian Geotechnical Journal*, **5**, No. 4, pp. 225–244.

Meyerhof, G. G. and Hanna, A. M. (1978), Ultimate bearing capacity of foundations on layered soils under inclined load, *Canadian Geotechnical Journal*, **15**, pp. 565–572.

Milz, E. A. and Broussard, D. E. (1972), Technical capabilities in offshore pipeline operations to maximize safety, *Proceedings, Offshore Technology Conference*, Houston, Texas, Paper OTC 1711, pp. 122–133.

Mindlin, R. D. (1936), Force at a point in the interior of a semi-infinite solid, *Journal of Applied Physics*, **7**, No. 5, pp. 195–202.

Minor, L. E. (1966), Improving deep sea pipeline techniques, *Offshore*, June, pp. 54–57.

Mitchell, D. E. (1984), Liquefaction slides in hydraulically placed sands, *Proceedings, Fourth International Symposium on Landslides*, Toronto, Ontario.

Mitchell, R. J., Sangrey, D. A., and Webb, G. S. (1972), Foundations in the crust of sensitive clay deposits, *Proceedings on Performance of Earth and Earth Supported Structures*, Purdue University, Indiana, ASCE, **1**, No. 2, pp. 1051–1072.

Moretto, O. (1971), Cimientos Profundos; Sintesis esscogida del estado actual del conocimiento sobre La interaction con el suelo, *Raavista Latinoamericana de Geotecnica*, **1**, No. 2, pp. 96–141.

Morgenstern, N. R. (1967), Submarine slumping and the initiation of turbidity currents, *Marine Geotechnique*, ed. A. F. Richards, University of Illinois Press, pp. 189–220.

Mroz, Z., Norns, V. A., and Zienkiewcz, O. C. (1978), An anisotropic model for soils and its applications to cyclic loading, *International Journal for Numerical and Analytical Methods in Geomechanics*, **2**, pp. 203–221.

Murff, J. D. (1980), Pile capacity in a softening soil, *Numerical and Analytical Methods in Geomechanics*, **4**, No. 2, pp. 185–189.

Murff, J. D. (1987), Pile capacity in calcareous sands: State of the art, *Journal of Geotechnical Engineering*, ASCE, **113**, No. 5, Paper No. 21509, pp. 490–507.

Nauroy, J. F. and Le Tirant, P. (1983), Model tests of piles in calcareous sands, *Offshore Engineering Practice*, ASCE, pp. 356–369.

Nauroy, J. F., Brucy, F., and Le Tirant, P. (1985), Static and cyclic load tests on a drilled and grouted pile in calcareous sand, *BOSS 85*, pp. 577–587.

Noorany, I. (1985), Classification of marine sediments, *Proceedings, 2d Shanghai Symposium on Marine Geotechnology and Nearshore/Offshore Structures*, Tongji University Press, Shanghai, pp. 168–195.

Nordlund, R. L. (1963), Bearing capacity of piles in cohesionless soils, *Journal of the Soil Mechanics and Foundations Division*, ASCE, **89**, No. SM-3, pp. 1–35.

O'Neill, M. W. (1983), Group action in offshore piles, *Proceedings, Conference on Geotechnical Practice in Offshore Engineering*, ASCE, pp. 25–64.

Offshore Engineer (1986), Valhalla is sinking too, December 5.

Poulos, H. G. (1971), Behavior of laterally loaded piles: II—Pile groups, *Journal of the Soil Mechanics and Foundations Division*, ASCE, **97**, No. SM-5, pp. 733–751.

Poulos, H. G. (1979), Development of an analysis for cyclic axial loading of piles, *Proceedings, 3d. International Conference Numerical Methods in Geomechanics*, Aachen, **4**, pp. 1513–1530.

Poulos, H. G. (1981a), Pile foundations subjected to lateral loading, *Symposium on Geotechnical Aspects of Coastal and Offshore Structures*, Bangkok, pp. 79–93.

Poulos, H. G. (1981b), Pile foundations subjected to vertical loading, *Symposium on Geotechnical Aspects of Coastal and Offshore Structures*, Bangkok, pp. 61–78.

Poulos, H. G. (1981c), Cyclic axial response of single pile, *Journal of the Geotechnical Engineering Division*, ASCE, **107**, No. GT-7, pp. 41–58.

Poulos, H. G. (1982), Influence of cyclic loading on axial pile response, *Proceedings 2d Conference on Numerical Methods in Offshore Piling*, Austin, Texas.

Poulos, H. G. (1983), Cyclic axial response—alternative analyses, *Proceedings, Geotechnical Practice in Offshore Engineering*, ed. S. G. Wright, ASCE, pp. 403–421.

Poulos, H. G. (1988), Cyclic stability diagram for axially loaded piles, *Journal of the Geotechnical Engineering Division, ASCE,* **114,** No. 8, pp. 877–895.

Poulos, H. G. and Davis, E. H. (1968), The settlement behavior of single axially loaded incompressible piles and piers, *Geotechnique,* **XVIII,** No. 3, pp. 351–371.

Poulos, H. G. and Davis, E. H. (1974), *Elastic Solutions for Soil and Rock Mechanics,* John Wiley and Sons, Inc., New York, N.Y.

Poulos, H. G. and Davis, E. H. (1980), *Pile Foundation Analysis and Design,* John Wiley and Sons, Inc., New York, N.Y.

Poulos, H. G. and Lee, C. Y. (1988), Model test on grouted piles in calcareous sediment, *Proceedings, International Conference on Calcareous Sediments,* Perth, Australia, pp. 255–261.

Prevost, J. H. and Hughes, T. J. R. (1978), Mathematical modelling of cyclic soil behavior, *Proceedings of the Specialty Conference on Earthquake Engineering and Soil Dynamics,* ASCE, Pasadena, California, **2,** pp. 746–761.

Puesch, A. A. (1982), Basic data for the design of tension piles in silty soils, *Proceedings, 3d. BOSS Conference, Massachusetts,* **1,** pp. 147–157.

Randolph, M. F. and Murphy, B. S. (1985), Shaft capacity of driven piles in clay, *Proceedings 17th Offshore Technology Conference,* Houston, Texas, OTC 4883, pp. 371–378.

Reddy, A. S. and Srinivasan, R. J. (1967), Bearing capacity of footings on layered clays, *Journal of the Soil Mechanics and Foundations Division, ASCE,* **93,** No. SM-2, pp. 83–99.

Reese, L. C. (1977), Laterally loaded piles: Program documentation, *Journal of the Geotechnical Engineering Division, ASCE,* **103,** No. GT-4, pp. 287–305.

Reese, L. C., Cox, W. R., and Koop, F. D. (1974), Analysis of laterally loaded piles in sand, *Proceedings, Sixth Annual Offshore Technology Conference,* OTC paper No. 2080, pp. 473–483.

Reese, L. C., Cox, W. R., and Koop, F. D. (1975), Field testing and analysis of laterally loaded piles in stiff clay, *Proceedings, Seventh Annual Offshore Technology Conference,* OTC paper No. 2312, pp. 671–675.

Reese, L. C. and Cox, W. R. (1976), Pullout tests of piles in sand, *Proceedings, Eighth Annual Offshore Technology Conference,* pp. 527–538.

Reese, L. C. and Wang, S. T. (1986), Method of analysis of piles under lateral loading, *Marine Geotechnology and Nearshore/Offshore Structures,* eds. R. C. Chaney and H. Y. Fang, ASTM STP 923, pp. 199–211.

Richards, A. F. (ed.) (1988), *Vane Shear Strength Testing in Soils: Field and Laboratory Studies,* ASTM STP 1014.

Richardson, G. N. and Chaney, R. C. (1986), Evaluation of seismic lateral pile capacity, Mark Clark expressway, *Third U.S. National Conference on Earthquake Engineering,* Charleston, S.C.

Rowe, R. K. and Davis, E. H. (1982), The behavior of anchor plates in sand, *Geotechnique,* **32,** No. 1, pp. 25–41.

Samson, C. H., Hirsch, T. J., and Lowry, L. L. (1963), Computer study of the dynamic behavior of piling, *Journal of the Structural Division, ASCE,* **89,** No. ST-4, pp. 413–449.

Saxena, S. K. and Lastric, R. M. (1978), Static properties of lightly cemented sand, *Journal of the Geotechnical Engineering Division, ASCE,* **104,** No. GT-12, pp. 1449–1464.

Schjetne, K., Andersen, K. H., Lauritzsen, R., and Hansteen, O. E. (1979), Foundation engineering for offshore gravity structures, *Marine Geotechnology,* **3,** No. 4, pp. 369–421.

Scott, R. F., Tsai, C.-F., Steussy, D., and Ting, J. M. (1982), Full-scale dynamic lateral pile tests, *14th Annual Offshore Technology Conference,* Houston, Texas, **1,** pp. 435–450.

Seibold, E. and Berger, W. H. (1982), *The Sea Floor,* Springer-Verlag, New York, N.Y.

Shaheen, W. A., Chang, C. S., and Demars, K. R. (1987), Field evaluation of plate anchor theories in sand, *Proceedings, Offshore Technology Conference,* Houston, Texas, Paper No. OTC 5419, pp. 521–530.

Shepard, F. P. (1963), *Submarine Geology,* 2d ed., Harper and Row, New York, N.Y.

Shinde, S. B., Crooks, J. H. A., James, D. A., and Williams Fitzpatrick, S. (1986), Geotechnical design for Beaufort Sea structures, *Proceedings, Third Canadian Conference on Marine Geotechnical Engineering,* St. John's Nfld., **1,** pp. 347–362.

Skempton, A. W. (1951), The bearing capacity of clays, *Proceedings, Building Research Congress,* **1,** pp. 180–189.

Skempton, A. W. (1953), Discussion: Piles and pile foundations, settlement of pile foundations, *Proceedings 3d International Conference on Soil Mechanics and Foundation Engineering,* Zurich, **3,** p. 172.

Sladen, J. A., D'Hollander, R. D., Krahn, J., and Mitchell, D. E. (1985), Back analysis of the Nerlerk Berm liquefaction slides, *Canadian Geotechnical Journal,* **22,** pp. 579–588.

Smith, E. A. L. (1962), Pile driving analysis by the wave equation, *Transactions of the American Society of Civil Engineers,* **127,** Part 1, pp. 1145–1193.

Smith, I. M. (1979), A survey of numerical methods in offshore piling, *Proceedings of the Conference on Numerical Methods in Offshore Piling,* Institution of Civil Engineers, London, pp. 1–8.

Sullivan, W. R., Reese, L. C., and Fenske, C. W. (1979), Unified method for analysis of laterally loaded piles in clay, *Proceedings of the Conference on Numerical Methods in Offshore Piling,* Institution of Civil Engineers, London, pp. 135–146.

Taylor, R. J., Jones, D., and Beard, R. M. (1975), *Handbook for Uplift Resisting Anchors,* U.S. Navy, Civil Engineering Laboratory, Port Hueneme, Calif.

Terzaghi, K. and Peck, R. B. (1967), *Soil Mechanics in Engineering Practice,* 2d ed., John Wiley and Sons Inc., New York, N.Y.

Thomas, H. G. (1978), Discussion of "Soil restraint against horizontal motion of pipes," by J. M. E. Audibert and K. J. Nyman, *Journal of the Geotechnical Engineering Division, ASCE,* **10,** No. GT-9, pp. 1214–1216.

Tomlinson, M. J. (1977), *Pipe Design and Construction Practice,* Viewpoint Publications, London.

Toolan, F. E. and Fox, D. A. (1977), Geotechnical planning of piled foundations for offshore platforms, *Proceedings of the Institution of Civil Engineers,* London, Part I, **62,** pp. 221–230.

Toolan, F. E. and Coutts, J. S. (1980), The application of laboratory and in situ data to the design of deep foundations, *Offshore Site Investigations,* ed. D. A. Ardus, Graham and Trotman, London, pp. 231–246.

Toolan, F. E. and Ims, B. W. (1988), Impact of recent changes in the API recommended practice for offshore piles in sand and clays, *Underwater Technology,* **14,** No. 1, pp. 9–29.

Trautman, C. H., O'Rourke, T. D., and Kulhawy, F. H. (1985), Uplift force–displacement response of buried pipe, *Journal of the Geotechnical Engineering Division, ASCE,* **111,** No. GT-9, pp. 1061–1076.

Van Weele, A. F. (1979), Pile bearing capacity under cyclic loading compared with that under static loading, *Proceedings 2d Behavior of Offshore Structures Symposium (BOSS),* London, pp. 475–488.

Vesić, A. S. (1965), Ultimate loads and settlement of deep foundations in sand, *Proceedings of Symposium on Bearing Capacity and Settlement of Foundations,* Duke University, Durham, N.C., pp. 53–68.

Vesić, A. S. (1967), *A Study of Bearing Capacity of Deep Foundations,* Final Report Project B-189, Georgia Institute of Technology, Atlanta, Ga. pp. 231–236.

Vesić, A. S. (1969), Experiments with instrumented pile groups in sand, *Proceedings of Symposium on Performance of Deep Foundations,* ASTM STP 444, pp. 177–222.

Vesić, A. S. (1970), Load transfer in pile–soil system, *Design and Installation of Pile Foundation and Cellular Structures,* eds. H. Y. Fang and T. D. Dismuke, Envo Publishing Co., Lehigh Valley, Pa., pp. 47–74.

Vesić, A. S. (1977), *Design of Pile Foundations,* National Cooperative Highway Research Program Synthesis of Practice No. 42, Transportation Research Board, Washington, D.C.

Vijayvergiya, V. N. (1977), Soil–pile interaction for offshore structures, *Proceedings 14th Annual Meeting of the Society of Engineering Science, Inc.,* Bethlehem, Pa.

Vijayvergiya, V. A. and Focht Jr., J. A. (1972), A new way to predict capacity of piles in clay, *Proceedings, 4th Offshore Technology Conference,* Houston, Texas, **2,** pp. 856–874.

Wang, M. C., Nacci, V. A., and Demars, K. R. (1975), Behavior of the underwater suction anchor in soil, *Journal of Ocean Engineering,* **3,** No. 1, pp. 47–62.

Wang, M. C., Nacci, V. A., and Demars, K. R. (1977), Breakout capacity of model suction anchors in soil, *Canadian Geotechnical Journal,* **14,** No. 2, pp. 246–257.

Watt, B. J. (1976), Gravity structures—installation and other problems,

Offshore Soil Mechanics, eds. P. George and D. Wood, Cambridge University Engineering Department and Lloyd's Register of Shipping, London, pp. 285–305.

Werno, M., Juszkiewicz, and Inerowicz, M. (1987), Penetration of jack-up platform footings into the seabed, *Marine Geotechnology*, **7**, No. 2, pp. 65–78.

Winterkorn, H. F. and Fang, H. Y. (eds.) (1975), *Foundation Engineering Handbook*, Van Nostrand and Reinhold Co., New York, N.Y.

Young, A. G., Kraft, L. M., and Focht, J. A. (1975), Geotechnical considerations in foundation design of offshore gravity structures, *Offshore Technology Conference*, **III**, pp. 367–386.

Young, A. G., House, H. F., Herlfrich, S. C., and Thurner, D. (1981), Foundation performance of mat-supported jack-up rigs in soft clays, *Proceedings, 13th Annual Offshore Technology Conference*, Houston, Texas, **4**, pp. 273–284.

Young, A. G., Remmes, B. D., and Meyer, B. J. (1984), Foundation performance of offshore jack-up drilling rigs, *Journal of the Geotechnical Engineering Division, ASCE*, **110**, No. 7, Paper No. 18996, pp. 841–859.

19 FOUNDATIONS IN COLD REGIONS

ARVIND PHUKAN, Ph.D., P.E.
Professor of Civil Engineering
University of Alaska at Anchorage

19.1 INTRODUCTION

The design of foundations in cold regions differs significantly from that in temperate regions. Cold regions include both those areas with seasonal frost and perennially frozen ground (permafrost). On the basis of air temperature, snow depth, ice covers and permafrost, Bates and Bilello (1966) reported that about 48 percent of the northern hemisphere's land mass is categorized as cold regions and the southern most reaches of discontinuous permafrost over land masses approximately follow the 40°N latitude line, as illustrated in Figure 19.1. In cold regions, the upper soil layer, or active layer, experiences cycles of winter freezing and summer thawing.

In this region, the depth of the frost and thaw penetration varies widely depending on the air temperature, surface characteristics, and soil type and composition, including water content and thermal properties. These cyclical phase changes cause phenomena such as frost heave, downdrag force, and thaw settlement, which may cause foundation movement and seriously damage superstructures. Thus, the designer of foundations in cold regions must understand the physical, mechanical, and thermal properties of frozen soils. In addition, an evaluation of ambient temperature changes, soil deformation under freezing and thawing conditions, and long-term strength of foundation soils under different ground temperatures must be considered for the design life of a structure. The designer must ensure that a foundation meets the basic requirements for adequate depth, acceptable settlements and stability against failure. The basic foundation design concepts, such as allowable soil pressure and limiting short- and long-term settlement, that are applied to structure stability on unfrozen soils are also applicable to frozen soils. However, the design approach to foundations in frozon soils is more complex. Factors such as the ground's thermal regime, ice composition in frozen soils, load-deformation characteristics and creep behavior of foundation frozen soils, whether the structure is heated or unheated, construction schedule, and environmental constraints will govern the design and construction of foundations in frozen soils. The fundamentals of frozen-ground engineering are found in Phukan (1985).

The main purpose of this chapter is to present the fundamental aspects of foundation design in cold regions and the chapter is devoted to a discussion of frozen soils and temperature profile, design approach and considerations, thermal analysis, bearing capacity, and long-term settlement of shallow foundations and pile foundations.

19.2 FROZEN SOILS: PHASES AND TEMPERATURE PROFILE

Frozen soils consist of four phases—solid particles, unfrozen water, ice, and gas. As illustrated in Figure 19.2, the interphase relationships between soil particles and ice, ice and unfrozen water, unfrozen water and gas, unfrozen water and solid particles, and ice and gas are of prime importance and each constituent depends on the properties of each phase as well as external influences such as stress and temperature. Many soil mechanics texts (Lambe and Whitman, 1979; Holtz and Kovacs, 1981; and Chapter 3) have presented the physical relationships among the mineral, unfrozen water, and air components. The unfrozen water and ice phase composition, interface characteristics, and applicable thermodynamic relationship are more complex and are discussed by Anderson and Morgenstern (1973). The physical and mechanical properties of frozen soils are highly dependent on temperature and typical values of these properties are presented in Table 19.1. As the temperature is

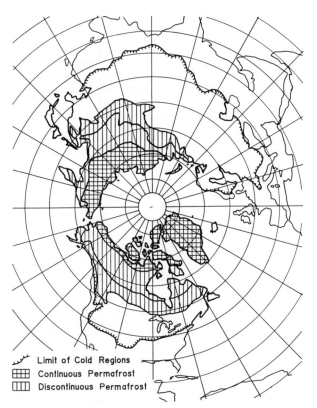

Limit of Cold Regions
Continuous Permafrost
Discontinuous Permafrost

Fig. 19.1 Boundaries of cold regions.

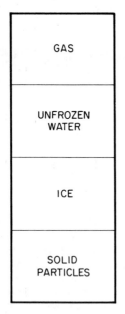

Fig. 19.2 Four-phase frozen soils.

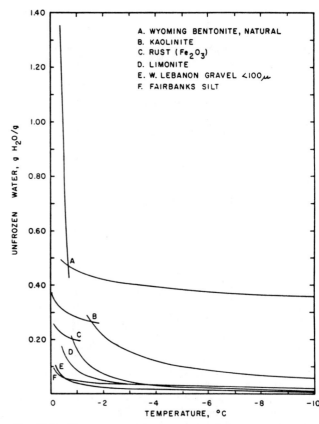

Fig. 19.3 Phase composition curves for six representative soils and soil constituents. (*After Anderson and Morgenstern, 1973.*)

lowered, the frozen soils become stronger and they become a three-phase material, of solid particles, ice, and gas. The relationship between the amount of unfrozen water and temperature is illustrated in Figure 19.3. Anderson and Tice (1973) have found that

$$W_{uw} = \alpha\theta^{\beta} \qquad (19.1)$$

where

W_{uw} = unfrozen water content
α, β = characteristics of soil parameter (Table 19.2)
θ = degree below freezing, °C

The presence of salt solutions will change the general trend presented in Figure 19.3 and this shift may be compared with the freezing point depression of the soil solution.

The behavior of frozen soils under stress is highly dependent on the composition of ice and its temperature. Ice is a highly plastic material and its properties depend strongly on the internal below-freezing temperature, and the duration and magnitude of load applied. As such, the ice composition in frozen soils, which is described in terms of nonvisible, visible,

and silt ice, must be known for effective foundation design (Phukan, 1985). Illustrations of ice descriptions are presented in Figure 19.4.

The thickness of frozen soils is generally obtained by plotting the annual temperature variation in the ground. Ground temperature fluctuates with air temperature as shown in Figure 19.5. The temperature tends to vary seasonally with depth to 30 to 50 ft (10 to 15 m). Below this depth of zero amplitude, the temperature generally increases under the geothermal gradient (average value of approximately 1.0°F per 100 ft (1°C in 30 m)) and depends on climate, geology, and geographic factors.

TABLE 19.1 MECHANICAL PROPERTIES OF VARIOUS FROZEN SOILS (Short-Term Condition).

Temperature	31°F (−55°C)	30°F (−1.11°C)	28°F (−2.22°C)	25°F (−3.88°C)	20°F (−6.66°C)
Fairbanks Silt Density 1.5 g/cm³					
Compressive strength, MN/m²	0.9	1.2	1.7	2.5	3.2
Tensile strength, MN/m²	0.2	0.8	1.2	1.8	2.3
Clay Density 1.70 g/cm²					
Compressive strength, MN/m²		1.4	2.0	2.8	3.5
Fine Sand Density 1.90 g/cm²					
Compressive strength, MN/m²		1.8	3.5	6.0	8.0
Silty Sand Density 1.80 g/m²					
Compressive strength, MN/m²		0.9	1.8	3.0	5.0

TABLE 19.2 SOIL PARAMETERS α AND β.

Soil Type	$\alpha \times 10^{-2}$	β
Limonite	8.81	−0.33
Kaolinite	23.80	−0.36
Wyomina bentonite	55.90	−0.29
Fairbanks silt	4.8	−0.33

a. NON-VISIBLE ICE

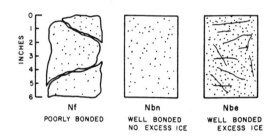

Nf	Nbn	Nbe
POORLY BONDED	WELL BONDED NO EXCESS ICE	WELL BONDED EXCESS ICE

b. VISIBLE ICE

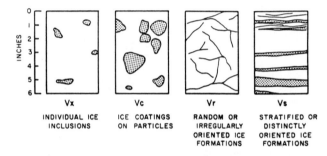

Vx	Vc	Vr	Vs
INDIVIDUAL ICE INCLUSIONS	ICE COATINGS ON PARTICLES	RANDOM OR IRREGULARLY ORIENTED ICE FORMATIONS	STRATIFIED OR DISTINCTLY ORIENTED ICE FORMATIONS

c. VISIBLE ICE – GREATER THAN 1 INCH THICK

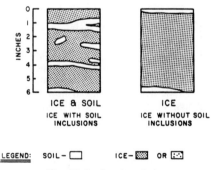

ICE & SOIL	ICE
ICE WITH SOIL INCLUSIONS	ICE WITHOUT SOIL INCLUSIONS

LEGEND: SOIL – ☐   ICE – ▨ OR ⬚

Fig. 19.4 Ice description.

As illustrated in Figure 19.5 (which is generally described as the "whiplash" curve), the thickness of frozen soils, whether seasonal or perennial, is determined from the temperature distribution. Seasonally frozen soils undergo cycles of winter freezing and summer thawing.

The terms perennially frozen soils or permafrost both describe ground thermal conditions that remain below 32°F (0°C) continuously for more than 2 years. Permafrost distribution in the world's cold regions is well known (Pewe, 1966; Brown, 1967; Baranov and Kudryavtsev, 1966). Factors such as climate, geology, hydrology, topography, and biology determine the permafrost distribution. The permafrost regions are further divided into two zones (Figure 19.1): a "continuous

zone" in the north, followed by a "discontinuous" zone further south.

In continuous permafrost zones, perennially frozen soils exist everywhere beneath the ground surface except in newly deposited, unconsolidated sediments where the climate has just begun to impose its influence on the ground surface. The thickness of continuous permafrost varies from about 200 to 300 ft (60 to 90 m) at the southern limit of the continuous zone, possibly increasing to 3000 ft (1000 m) in the northern part of the zone in Alaska and Canada. The temperature of the permafrost in the continuous zone at the depth of zero amplitude varies from 23°F (−5°C) in the south to about 5°F (−15°C) in the extreme north.

In discontinuous zones, some areas have perennially frozen soils beneath the surface and other areas are permafrost free. The temperature at the level of zero amplitude in the discontinuous zone varies from a few tenths of a degree below 32°F (0°C) at the southern limit to 23°F (−5°C) at the boundary with the continuous zone. Permafrost is also known to occur in many northern arctic coasts and the permafrost thickness under subsea conditions may vary widely. Offshore permafrost is beyond the scope of this chapter.

In continuous or discontinuous frozen soils, foundations will be within the depth of zero amplitude, and the ground temperature fluctuations must be determined. The temperature regime of frozen soils depends on present and past climate, terrain factors, and the complex energy exchange between the ground and the atmosphere. All aspects of foundations in cold regions are sensitive to heat transfer effects. Either a theoretical approach or actual ground temperature monitoring devices may be used to determine the temperature profile needed for the foundation design.

The warmest frozen ground temperature is given by

$$T_d = 32 - A_0 + A_0 \exp(-d\sqrt{\pi/aP}) \qquad (19.2)$$

where

T_d = temperature (°F) at depth d
A_0 = amplitude of temperature difference between top of permafrost (32°F) and depth at which there is no temperature fluctuation
a = thermal diffusity of frozen soils
P = a period of 365 days

Because there is a time lag between the air temperature and the ground temperature, the temperature profile calculated by Equation 19.2 is conservative. A time lag of approximately 3 months generally exists between the air temperature and the ground temperatures.

19.3 DESIGN APPROACH

In areas where there is no potential for thermal degradation, frozen soils with temperature well below freezing provide an excellent foundation support. In extreme cases, ice-rich frozen soils with near below-freezing temperatures may undergo extreme deformation even at low design load. Therefore, the design approach in frozen soils depends on the nature and composition of the frozen ground, ground temperature, type of structure to be founded and whether the structure will be heated or unheated, the magnitude of loads, etc. The following are the most common design approaches applied to the design of foundations in cold regions:

a. Maintaining frozen ground in its frozen state
b. Allowing frozen soils to undergo thawing caused by construction
c. Improving the site condition before construction
d. Using conventional designs appropriate to unfrozen soils

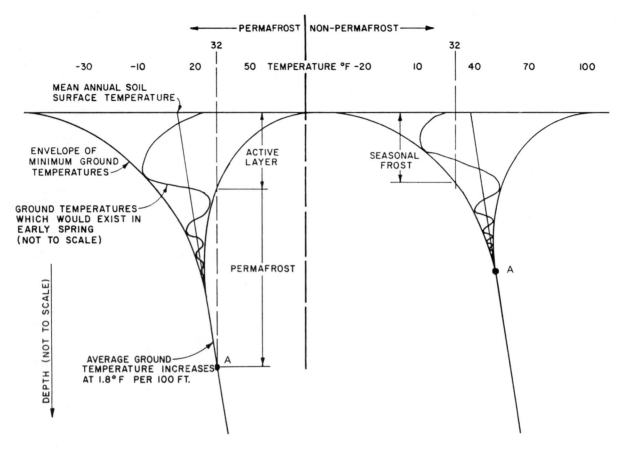

Fig. 19.5 Whiplash curve.

Method (a) is a passive method commonly applied in thaw-unstable materials. The method is particularly applicable in ice-rich frozen ground where significant potential for deformation as a result of thermal degradation exists. Self-refrigerated systems or two-phase heat tubes, mechanical refrigeration systems, and a combination of a gravel pad with insulation and a thermal heat removal device may be used to retain frozen soils in their frozen state.

Method (b) is suitable for thaw-stable material where the anticipated total settlement or differential settlement due to change in thermal regime will not exceed the acceptable limits of the structure. This active method is used mostly in competent material. Extensive site investigation, thaw-settlement tests, and predictions are needed for the application of this method.

Method (c) is applied primarily to seasonal frost or discontinuous permafrost areas where the thickness of the frozen soil layer is shallow enough to allow excavation of unsuitable material and replacement with competent thaw-stable material. Other site soil improvement techniques may involve prethawing the existing frozen ground and never allowing it to refreeze. Prethawing techniques, such as injection of steam or cold water, stripping of insulating surficial soils, and use of ground cover to increase solar heat input into frozen soils, may be applied to preclude the anticipated thaw settlement. However, problems such as metastability of soils, unworkability of ground surface, excessive pore pressure, or poor drainage may arise during the prethawing process and the designer must evaluate such detrimental effects before using this method.

Method (d) is only used when the supporting material consists of either ice-free competent bedrock or dense, thaw-stable sands and gravels. Depending on the magnitude of

structural loads, either shallow or deep foundations are installed using conventional allowable soil pressure theory.

19.4 DESIGN CONSIDERATIONS

Two basic considerations for foundation design in cold regions are stress effects on the supporting material and thermal phase effects between the structure and the foundation material. Stress effects include deformation and flow of material under the applied stress. Thermal phase effects encompass the heat flow between and thermal analysis of the structure and the supporting soils.

The designer must understand the behavior of frozen soils under different temperatures, confining pressures, and strain rates. Many publications (Tsytovich, 1975; Andersland and Anderson, 1978; Phukan, 1985) have presented the details of physical and mechanical properties of frozen soils. The designer must take into consideration the time-dependent as well as temperature-dependent behavior of frozen soils.

19.4.1 Creep Parameter and Long-Term Strength

Generally, the design of foundations in frozen soils is governed by the long-term strength. Creep parameters are determined from laboratory tests to define the long-term strength. As shown in Figure 19.6, the creep or time-dependent deformation behavior of frozen soils may result from (a) pressure melting of ice at points of soil–grain contact, (b) readjustment of the particles, (c) plastic deformation of pore ice, or (d) breakdown

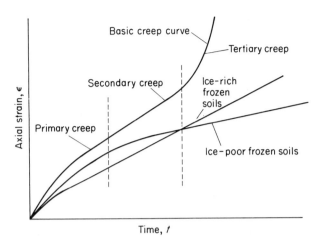

Fig. 19.6 Creep behavior.

of structural bonds between ice and soil grain. The deformation rate goes through three stages—primary, secondary, and tertiary—before the failure of frozen soils. Deformation during the primary stage is calculated from the following relationship (Vyalov, 1962; Assure, 1963):

$$\varepsilon = \left[\frac{\sigma t^{\Lambda}}{\omega(\theta + 1)^K} \right]^{1/m} \tag{19.3}$$

where

ε = primary strain
σ = applied stress
t = elapsed time of application of load
ω, K, Λ, m = creep parameters
θ = degree below freezing, °C

Typical values of creep parameters are given in Table 19.3.

The creep equation in the secondary state (Hult, 1966) may be written as

$$\varepsilon = \frac{\sigma}{E} + \varepsilon_k \left(\frac{\sigma}{\sigma_K} \right)^K + t\, \dot{\varepsilon}_c \left(\frac{\sigma}{\sigma_c} \right)^n \tag{19.4}$$

where

E = the Young's modulus
ε_k = arbitrary small strain for normalization
$\dot{\varepsilon}_c$ = arbitrary small strain for normalization
σ_k, σ_c = temperature-dependent creep parameters

Equations 19.3 and 19.4 are used to determine the long-term deformation of frozen soils. Generally, ice-poor frozen soils show distinct primary creep behavior, whereas ice-rich frozen soil behavior is dominated by secondary creep. The long-term strength (σ_{ult}) of frozen soils is given by

$$\sigma_{\text{ult}} = \frac{\beta}{\log_e(t/B)} \tag{19.5}$$

TABLE 19.3 CREEP PARAMETERS.

Soil Type	m	Λ	K	ω $psi(hr)^{\Lambda}/°C^K$	$psi(hr)^{\Lambda}/°F^K$
Ottawa sand	0.78	0.35	0.97	44.72	5500
Manchester fine sand	0.38	0.24	0.97	2.20	285
Suffield clay	0.42	0.14	1.00	0.73	93
Sandy silt	0.27	0.10	0.89	0.88	90

where β and B are parameters depending on soil type, its properties, and temperature and are determined from creep tests.

19.4.2 Thermal Analysis

The designer must perform an analysis of thermal behavior of the structure, ground and their interface. The basic heat transfer due to conduction, convection, and radiation must be understood through heat-flow calculations (Carslaw and Jaeger, 1959; Lunardini, 1981). For most engineering problems, conduction is the dominant heat-transfer process and its relationship to dry unit weight (γ_d) and water content is presented in Figures 19.7 and 19.8. The governing equation for steady-state conditions with no phase change is given by

$$\frac{\partial T}{\partial t} = a \frac{\partial^2 T}{\partial z^2} \tag{19.6}$$

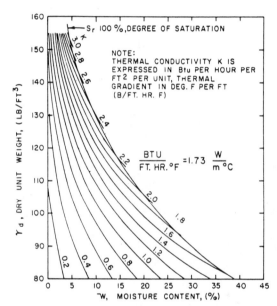

(A) COARSE GRAINED SOILS—FROZEN

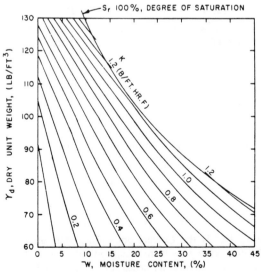

(B) FINE GRAINED SOILS—FROZEN

Fig. 19.7 Thermal conductivity of frozen soils. (A) Coarse-grained soils. (B) Fine-grained soils. (*After Kersten, 1949.*)

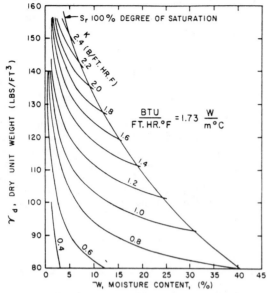

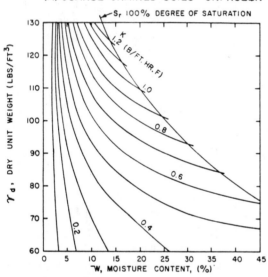

Fig. 19.8 Thermal conductivity of unfrozen soils. (A) Coarse-grained soils. (B) Fine-grained soils. (*After Kersten, 1949.*)

where

> T = temperature
> t = time
> z = depth
> a = thermal diffusivity = k/c
>   (k and c are thermal conductivity and volumetric heat capacity, respectively)

Equation 19.6 is rewritten for two dimensions as

$$\frac{\partial T}{\partial t} = a\left(\frac{\partial^2 T}{\partial z^2} + \frac{\partial^2 T}{\partial x^2}\right) \qquad (19.7)$$

Equation 19.7 is the Laplace equation and is used to solve various engineering problems using boundary conditions.

The steady temperature beneath heated or cooled areas on the ground surface is given by

$$T - T_g = \frac{T_s - T_g}{\pi}\tan^{-1} z/x \qquad (19.8)$$

where

> T = desired temperature
> T_g = mean ground surface temperature outside heated or cooled area
> z = depth
> x = horizontal direction

Equation 19.8 may be rewritten to determine the temperature regime under a long strip (such as a road of width $2a$) on the ground surface as

$$T - T_g = T_s - T_g \tan^{-1}\frac{2az}{x^2 + z^2 - a^2} \qquad (19.9)$$

19.4.3 Frost Penetration and Depth of Thaw

Determination of frost depth or thaw penetration is one of the most important thermal estimations in cold regions. Such calculations are required to estimate change and depth of the active layer, frost or uplift force, and downdrag force. The placement of shallow foundations or determination of effective pile depth is governed by such estimations. The frost or thaw depth (x) is given by

$$x = \Lambda\sqrt{\frac{48\,KnI_f\;(I_t)}{L}} \qquad (19.10)$$

for U.S. customary units, and

$$x = \Lambda\sqrt{\frac{7200\,K_a nI_f\;(I_t)}{L}} \qquad (19.11)$$

for S.I., where

> Λ = thermal constant, a function of thermal ratio (a) and fusion parameter (μ) (Fig. 19.9)
> K = average thermal conductivity for unfrozen and frozen soils in BTU/ft-hr-°F
> K_a = average thermal conductivity for unfrozen and frozen soils in J/S·m·°C
> n = the ratio between the air freezing or thawing index and surface freezing or thawing index (see Table 19.4 for typical values)
> I_f = air freezing index, degree-days.
> I_t = air thawing index, degree-days.
> L = latent heat of fusion of soil, BTU/ft³ or J/m³

EXAMPLE 19.1

Determine the depth of thaw penetration into a homogeneous frozen silt for the following conditions:

- Mean annual air temperature (MAT) = 36°F
- Air thawing index, I_t = 2500 degree-days
- Surface covered with snow
- Length of thawing season, t = 150 days
- Soil properties: dry unit weight density, (γ_d) = 85 pcf
  water content (ω) = 30 percent

Solution

Volumetric latent heat and fusion: $L = 144\gamma_d\omega$
$L = 144\,(85)\,(0.3) = 3672$ BTU/ft³
Average volumetric head, $C_a = \gamma_d(0.17 + 0.75\,\omega)$
$C_a = 100[0.17 + (0.75)(0.30)] = 39.5$ BTU/ft³-°F
Average thermal conductivity, $K_a = \frac{1}{2}\,(K_u + k_f)$, where
K_u = unfrozen thermal conductivity = 0.66 BTU/ft-hr-°F
(Fig. 19.8)

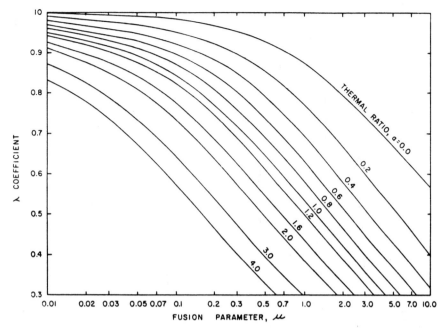

Fig. 19.9 Λ-values.

TABLE 19.4 *n* FACTORS.

Surface	Freezing Condition	Thawing Condition
Snow	1.0	—
Turf	0.5	1.0
Sand and gravel	0.9	1.5
Spruce	0.73	0.44
Willows	—	0.82
Weeds	—	0.86
Concrete road	0.8	2.03
Asphalt road	0.8	1.61
White painted surface	—	1.12

K_f = frozen thermal conductivity = 0.94 BTU/ft-hr-°F
$K_a = \frac{1}{2}(0.66 + 1.05) = 0.85$ BTU/ft-hr-°F

Assuming $n = 1$:

Average surface temperature, $V_s = \dfrac{nI_t}{t} = \dfrac{(1)(2500)}{150}$
$$= 16.66°F$$

Mean annual soil temperature, $V_0 = \text{MAT} - 32°F$
$$= 36 - 32 = 4°F$$

Thermal ratio, $a = \dfrac{V_0}{V} = \dfrac{4}{16.66} = 0.24$

Fusion parameter, $\mu = V_s Ca/L$
$$= \frac{(16.66)(39.5)}{3672} = 0.18$$

From Figure 19.9, $\Lambda = 0.92$.

Estimated depth of thaw, $x = \Lambda\sqrt{\dfrac{48K_a nI_t}{L}}$

$$= (0.92)\sqrt{\frac{(48)(85)(1)(2500)}{3672}}$$
$$= 4.8 \text{ ft}$$

EXAMPLE 19.2

Determine the depth of frost penetration into a homogeneous deposit for the following conditions:

- Mean annual air temperature (MAT) = 2°C
- Surface freezing index, $nI_f = 1300$ degree-days
- Length of freezing season, $t = 140$ days
- Soil conditions: silty gravelly soils
    unit weight, $\gamma = 2000$ kg/m³
    water content = 8 percent

Dry unit weight, $\gamma_d = \dfrac{\gamma}{1 + \omega} = \dfrac{2000}{1 + 0.08} = 1851.85$ kg/m³

Solution

$L = (334)(1851.85)(0.08) = 49.5$ MJ/m³

$C_{ave} = \dfrac{1851.85}{1000}\ \{0.71 + [(3.14)(0.08)]\} = 1.78$ MJ/m³·°C

$K_a = \dfrac{(2.04 + 2.16)}{2} = 2.10$ W/m·°C

$V_s = \dfrac{nI_f}{t} = \dfrac{1300}{140} = 9.28$°C

$V_0 = 2 - 0 = 2$°C
$$= \frac{2}{9.28} = 0.22$$
$$= \frac{(9.28)(1.78)}{49.5} = 0.33$$

From Figure 19.9, $\Lambda = 0.88$.
Estimated depth of frost penetration:

$$x = \Lambda\left(\frac{(7200\,K_a n I_f)}{L}\right)^{1/2}$$
$$= (0.88)\left[\frac{(7200)(24)(2.10)(1300)}{49.5 \times 10^6}\right]^{1/2}$$
$$= 2.7 \text{ m}$$

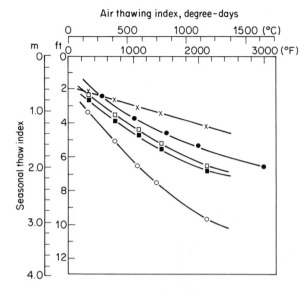

Fig. 19.10 Frost penetration under different conditions. ○, Gravel; □, sand; △, 24-inch (61-cm) gravel, sand; ×, 24-inch (61-cm) gravel, 1-inch (2.54-cm) insulation, sand. ●, White-painted mat, sand.

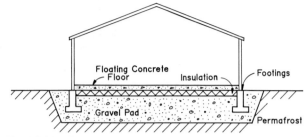

Fig. 19.11 Footings on an insulated gravel pad (not to scale.)

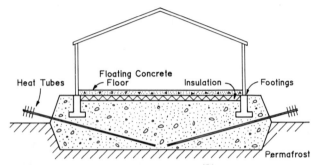

Fig. 19.12 Footings on a self-refrigerated gravel pad with heat tubes (not to scale.)

Phukan (1985) presented the determination of the depth of frost or thaw penetration for nonhomogeneous soil conditions, and some typical values of frost penetrations under different boundary conditions are illustrated in Figure 19.10.

19.5 SHALLOW FOUNDATIONS

Shallow foundations are usually chosen when competent subsoil conditions are encountered at shallow depths or when the superstructure's design load is not sufficient to require deep foundations. In cold regions, shallow foundations are generally constructed so that frost or thaw penetration will not affect the supporting foundation material. In addition to basic requirements regarding bearing capacity and long-term settlement, shallow foundations must be designed to withstand the adverse effect of thermal balance between the structure and the ground. Depending on the nature of the structure and the type and composition of supporting soils, including their temperature, the design approaches presented in Section 19.3 are used to maintain the thermal stability of foundation soils. Typical shallow foundation designs are illustrated in Figures 19.11 to 19.16. These foundations may be grouped into the following six major categories:

1. Footings on gravel pad with or without insulation
2. Raft on self-refrigerated gravel pad
3. Footings or raft on ventilated gravel pad
4. Footings or raft on mechanical refrigeration
5. Post and pad foundations
6. Ground sills on original ground or on gravel pad

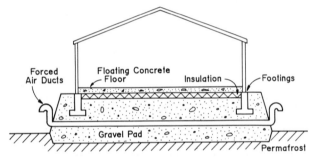

Fig. 19.13 Footings on a self-refrigerated gravel pad with forced air circulation (not to scale.)

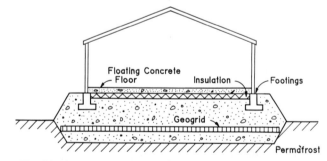

Fig. 19.14 Footings on a reinforced gravel pad (not to scale.)

19.5.1 Bearing Capacity of Frozen Ground

The classic theory of unfrozen soil-bearing capacity (presented in Chapter 4) may also be applied to frozen soils. The major difference is the use of long-term frozen soil strength based on the creep parameters presented in Equation 19.5. Also, the

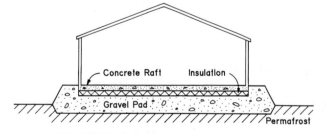

Fig. 19.15 Raft on grade (not to scale.)

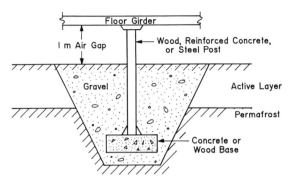

Fig. 19.16 Post and pad (not to scale.)

frozen soil shear strength is dominated by the cohesion (c) factor only. The bearing capacity of frozen soils is given by

$$q_{df} = C N_c \xi_c \tag{19.12}$$

where

c = cohesion = $\sigma_{ult}/2$
σ_{ult} = long-term strength of foundation frozen soil, given by Equation 19.5
N_c = bearing capacity factor = 5.14
ξ_c = geometry of foundation

Other investigators (Vyalov, 1959; Tsytovich, 1975) have published bearing-capacity values for typical soils and some of the recommended bearing capacity values are presented in Table 19.5. A few published data sources are available for ice-rich frozen soils at 23°F to 32°F (-5°C to 0°C) and an empirical relationship can be written as:

$$C \, (\text{KPa}) = 35 + 28 \, T \tag{19.13}$$

where T is the absolute value below freezing in degrees Celsius.

EXAMPLE 19.3

A square footing 6 ft by 6 ft is placed at a depth of 5 ft (below the active layer) in a deep stratum of frozen silt. Assume that there is air space between the floor and the natural ground to maintain thermal equilibrium conditions. Determine the bearing capacity of the frozen soils.

Solution

Assume that the long-term strength of frozen silt is determined from creep tests and is given by

$$\sigma_{ult} = \frac{1816}{\ln t_f + 8.15}$$

Assume the design life of structure t_f is

$$30 \text{ years} = (30)(365)(24) = 2.628 \times 10^6 \text{ hr}$$

$$\sigma_{ult} = \frac{1816}{\ln(t_f) + 8.15}$$

Bearing capacity = $C N_c \xi_c$:

$$N_c = 5.14; \quad \xi_c = 1 + \frac{N_q}{N_c} = 1 + \frac{1}{5.14} = 1.195$$

Bearing capacity = $(88.03/2)(5.14)(1.195)$
$\qquad\qquad\qquad = 270.35$ psi

19.5.2 Long-Term Settlement

Based on the bearing capacity shallow foundations may be designed in competent frozen soils with colder temperatures. However, long-term settlement generally governs the design, as frozen soil behavior is time-dependent as well as temperature-dependent. Ice-rich frozen soils or frozen soils with temperature close to freezing are especially susceptible to long-term settlement even at low loads. The long-term settlement in frozen soils is calculated from Equations 19.3 and 19.4.

Long-term settlement may also occur if the foundation frozen soils are subjected to gradual thermal degradation. If, under the design conditions, frozen soils are allowed to thaw, the total settlement(s) must be calculated by

$$S = \sum_{i=1}^{i=n} \varepsilon_i h_i + S_{ct} \tag{19.14}$$

where

ε_i = thaw strain
h_i = thickness of frozen soil layer
S_{ct} = consolidation settlement

The determination of consolidation settlement is complex, and one-dimensional linear theory (Morgenstern and Nixon, 1971) may be used for predicting settlement.

Typical thaw strain values for different soils are illustrated in Figure 19.17.

EXAMPLE 19.4

Determine the long-term settlement for the square footing given in Example 19.3. The temperature at the depth of zero amplitude is 25°F. Assume the following frozen soil properties:

- Frozen dry unit of soils, $\gamma_{df} = 80 \text{ lb/ft}^3$
- Thermal conductivity of frozen soils, $k_f = 1.0 \text{ BTU/ft-hr-}°\text{F}$
- Volumetric heat capacity of frozen soils, $C_f = 28.0 \text{ BTU/ft-hr-}°\text{F}$

TABLE 19.5 TYPICAL DESIGN BEARING CAPACITY VALUES (Ksf).

Temperature	Soil Type					
	Pure Ice & Organic Soils	Clayey Soils	Silty Soils	Sandy Soils	Gravelly Sand	Gravel
31°F (-55°C)	—	1.5	1.8	2.4	4.0	5.8
30°F (-1.11°C)	—	1.9	2.2	4.3	4.8	6.9
28°F (-2.22°C)	0.8	3.0	3.7	5.5	6.3	9.8
25°F (-3.88°C)	1.9	4.2	5.0	6.8	8.0	12.0
20°F (-6.66°C)	2.1	5.0	6.5	7.5	9.0	13.5

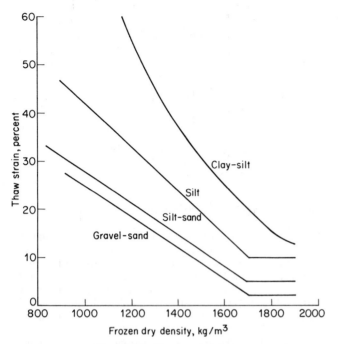

Fig. 19.17 Thaw strain values.

TABLE 19.6 TEMPERATURE CALCULATIONS AT DEPTH.

Depth from Ground surface, ft	Depth from Base of Footing, d, ft	T_d, °F
5	0	32.0
6	1	31.5
7	2	31.0
8	3	30.7
9	4	30.4
10	5	30.0
12	7	29.5
15	10	28.8
20	15	28.1
25	20	27.7

duct system beneath the floor to maintain the existing frozen ground condition at the site. Assume that the building site has the following soil properties and design data:

- Frozen silt, dry weight = 1450 kg/m³; water content, $\omega = 25\%$
- Air freezing index at the site = 4000 degree-days
- Freezing season = 215 days

Solution

Step 1. *Calculate resistance*

Resistance of concrete,

$$R_c = \frac{d}{k} = \frac{0.15 \text{ m}}{1.73 \text{W/m} \cdot {}^\circ\text{C}} = 0.09 \text{ m}^2 \cdot {}^\circ\text{C/W}$$

Resistance of insulation,

$$R_i = \frac{d}{k} = \frac{0.15}{0.057} = 2.63 \text{ m}^2 \cdot {}^\circ\text{C/W}$$

Total resistance,

$$R = R_c + R_i = 2.72 \text{ m}^2 \cdot {}^\circ\text{C/W}$$

Step 2. *Calculate depth of thaw*

Thawing season = 365 − freezing season   (days)
$$= 365 - 215 = 150 \text{ days}$$

Floor thawing index,

$$I_{tf} = (21 - 0) 150 = 3150 \text{ degree-days}$$

Latent heat of gravel,

$$L = (333.7)(1920)(0.05)$$
$$= 32.9 \text{ MJ/m}^3$$

Depth of thaw,

$$x = KR_f\left[\left(1 + \frac{7200\Lambda^2 I_{tf}}{kLR_f}\right)^{1/2} - 1\right]$$
$$= (1)(2.72)\left[1 + \left(\frac{7200(24)(97)^2(3150)}{(1)(32.9 \times 10^6)(2.72)^2}\right)^{1/2} - 1\right]$$
$$= 2 \text{ m}$$

Step 3. *Determine heat to be removed from gravel pad*

Heat to be removed from gravel including sensible heat

$$= (2)(32.87) + 1 \text{ percent}$$
$$= 70 \text{ MJ/m}^2$$

Solution

$$a = \frac{k_f}{c_f} = \frac{(1.0)(24)}{28.0} = 0.857 \text{ ft}^2/\text{day}$$

The warmest temperature distribution in the frozen soils is obtained by Equation 19.2.

$$T_d = 32 - A_0 + A_0 \exp(-d\sqrt{\pi/aP})$$
$$= 32 - 5 + 5 \exp(-d\sqrt{\pi/(0.857)(365)})$$
$$= 32 - 5 + 5 \exp(-0.1d)$$

The temperature calculations at depth are given in Table 19.6.

The allowable soil pressure, $q_{af} = q_{df}/FS$

Load on 6 ft × 6 ft footing = $(6 \times 6) q_{af}$
$$= (6)(6)\left(\frac{270.35}{2}\right)(144)$$
$$= 700\,747 \text{ lb} \approx 350 \text{ tons}$$

Creep parameters assumed are

$$m = 0.27, \Lambda = 1.10, k = 0.89$$
$$\omega = 90 \text{ psi} \cdot \text{hr/}^\circ\text{F}^K$$

Design period, $t = 340 \text{ yr} = 2.628 \times 10^6 \text{ hr}$

The long-term settlement calculation is presented in Table 19.7.

The following example illustrates the calculation of thermal flux between the structure and the foundation frozen soils and the determination of gravel pad thicknesses commonly used for shallow foundations in frozen soils (Figures 19.11 to 19.16).

EXAMPLE 19.5

A building (20 m × 100 m) is to be supported on a slab-on-grade with an average floor temperature of 21°C. Design a

TABLE 19.7 LONG-TERM SETTLEMENT CALCULATION.

Zone Thickness h, ft	Temperature Below Freezing, deg F	Vertical Stress psi	Strain $\varepsilon = \left[\dfrac{\sigma t^{\Lambda}}{\omega(\theta+1)^K} \right]^{1/m}$	Deformation $h' = (\varepsilon)(h)$
2	0.5	99.21	$\left[\dfrac{(99.21)(2.628 \times 10^6)^{0.10}}{(90)(0.5+1)^{0.89}} \right]^{1/0.27}$ $= 0.89$	1.78
2	0.8	60.00	$= 0.76$	0.15
3	1.25	36.75	$= 0.006$	0.018
3	1.60	23.12	$= 0.0006$	0.001
				1.949 ft (Excessive)

During freezing season, this heat must be removed over a period of 215 days:

$$\text{Average rate of flow} = \frac{70 \text{ MJ/m}^2}{(215)(24) \text{ hr}}$$

$$= 0.014 \text{ MJ/m}^2 \cdot \text{hr}$$

$$\approx 4 \text{ W/m}^2$$

Step 4. *Average thawing index at the surface*

$$I_t = \frac{Lx^2}{7200\Lambda^2 K} = \frac{(32.87)(2)^2}{(7200)(24)(0.97)^2(1)}$$

$$= 892 \text{ degree-days}$$

This thawing index must be compensated by an equal freezing index at the duct outlet to assure freeze-back.

Step 5. *Average temperatures*

Average pad surface temperature at the outlet of duct

$$= \frac{\text{freezing index}}{\text{length of freezing}} = \frac{-892}{215} = -4.2°C$$

Inlet air duct temperature

$$= \frac{\text{air freezing index}}{\text{length of freezing}}$$

$$= \frac{-4000}{215} = -18.6°C$$

Average temperature rise in duct $= -4.2 - (-18.6) = 14.4°C$

Step 6. *Heat flow*

Average heat flow between floor and inlet duct air,

$$= \frac{21 - (-18.6)}{2.72} = 14.5 \text{ W/m}^2$$

Average heat flow between floor and outlet duct

$$= \frac{21 - (4.2)}{2.72} = 9.6 \text{ W/m}^2$$

Average heat flow between total floor and duct

$$= \frac{14.5 + 9.6}{2} = 12 \text{ W/m}^2$$

Total heat flow $= 12 + 4 = 16 \text{ W/m}^2$

Step 7. *Calculation of the air velocity in the duct*

$$QLD = V A_d \rho C_p T_R$$

where

Q = total heat flow, W/m^2
L = length of duct, m
D = duct spacing, m
V = Velocity of duct air, m/s
A_d = cross-sectional area of duct, m^2
ρ = air density (assume 1.33 kg/m^3)
C_p = specific heat of air at constant pressure (assume 1000 J/kg·°C)
T_R = temperature rise in duct air, °C

$$V = \frac{QLD}{A_d \rho C_p T_R}$$

$$= \frac{(16)(100)(1.5)}{(0.073)(1.33)(1000)(14.4)}$$
(using duct dia = 1 ft = 305 mm)

$$= 1.7 \text{ m/s}$$

19.6 PILE FOUNDATIONS

Pile foundations are widely used in continuous and discontinuous permafrost regions to provide greater stability of the thermal region. In addition, they provide an effective support to transfer both heavy vertical and lateral loads to a depth where volume change, as well as reduction of shear strength due to temperature changes are minimal. In comparison to shallow foundations, pile foundations provide a more effective way to isolate structures from existing ground conditions and maintain thermal equilibrium. Seasonal frost heave and downdrag force due to thawing are eliminated by using an effective pile embedment length. The construction of foundations in poor foundation material, such as frozen, ice-rich, fine-grained soils, deep seasonal frost areas or frost-susceptible soils, is very difficult and pile foundations are the only alternative for such sensitive soil conditions. Frozen soil behavior depends on ice-matrix, temperature, rate of loading, etc. These factors must be considered in designing pile foundations in frozen soils (Phukan, 1977).

Some of the basic philosophy used in designing pile foundations in unfrozen soils (Chapter 13) may be applied in frozen ground. However, load-transfer mechanisms, installation techniques, and field pile loading tests are quite different. The load-carrying capacity of piles in cold regions is dependent on the adfreeze between the pile surface and the surrounding frozen natural soil or backfill material placed between the pile surface and the natural frozen ground. Figure 19.18 presents generalized load-transfer conditions at different times.

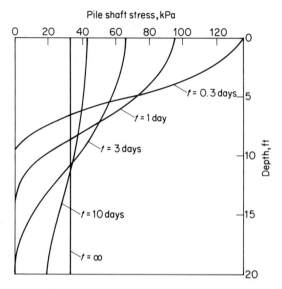

Fig. 19.18 Load transfer along pile shaft.

The state of the art of pile design has dramatically changed in the last decade owing to improvement in pile installation techniques and the use of a thermal device called the *heat tube*. Piles with thermal devices are called *thermal piles* and various configurations of heat tubes in thermal piles are illustrated in Figure 19.19. Depending on installation technique, piles are divided into three groups such as thermal piles, slurried piles and drilled-in-place piles. As explained earlier, the thermal piles may have different heat tube configurations as shown in Figure 19.19. The slurried piles do not have any heat tube and the annulus between the drilled hole and the pile surface is filled with slurry as illustrated in Figure 19.20. The drilled-in-place piles are generally driven into the soils and are most commonly installed in cases where a thawed soil profile is underlain by frozen soils.

In cold permafrost where frozen soils with colder temperature exist, the pile is placed in an oversized hole (generally by more than 0.5 ft or 0.152 m) and the annulus between the pile and the hole is backfilled with a sand slurry. This pile is called a slurried pile (Fig. 19.20) and typical sand slurry gradation is given by

U.S. Standard sieve size	Percent finer by weight
9.52 mm	100
No. 4	93 to 100
No. 10	70 to 100
No. 40	15 to 57
No. 200	0 to 17

Thermal piles may be installed similarly to slurried piles, but heat tubes are added to offer greater long-term thermal balance. Thermal piles may also be placed by (a) direct driving into natural frozen ground, (b) driving into undersized holes or steamed holes of diameter less than the pile diameter, (c) use of prethawed holes of diameter greater than the pile diameter. Thermal devices should be installed in piles placed in thaw-unstable frozen soils with temperatures close to freezing (i.e., warmer frozen soils).

In North America, prefabricated timber or steel piles are most commonly used in cold regions, whereas in the U.S.S.R., precast, reinforced-concrete piles are installed. The major problem with concrete piles is their weakness against frost heave force, which produces tensile forces in the pile. Concrete has

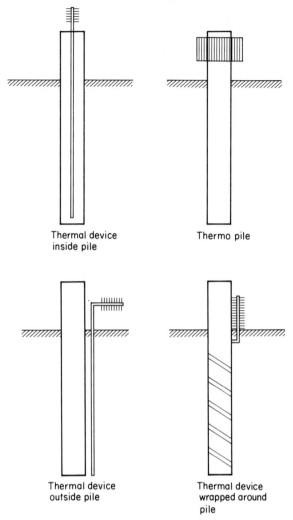

Fig. 19.19 Heat tube configuration.

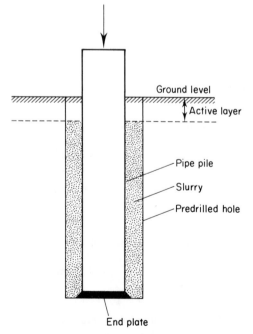

Fig. 19.20 Slurried pile.

low tensile strength. Moreover, economic considerations dictate the use of either steel or timber piles.

19.6.1 Single Pile Load Capacity

Pile design for frozen soils must satisfy both thermal and stress–deformation (rheological) requirements as outlined in Section 19.4. In addition, the design must comply with the following.

a. It should have an adequate factor of safety (at least 2) against gross failure.
b. The settlement (both short-term and long-term) should be within the acceptable limits of the superstructure.
c. There will be no unacceptable changes in the ground thermal regime due to the construction of the superstructure. Heat transmission by conduction down the pile must be prevented.
d. Freeze-back of placed slurry must be achieved before loading of piles.

Pile design is generally based on the long-term adfreeze strength between the pile surface and the surrounding frozen material. Typical values of adfreeze strength at temperatures below freezing are presented in Table 19.8 and the vertical load capacity of typical circular pile is presented in Figure 19.21.

Generally, only end bearing is considered when piles are embedded in dense thaw-stable granular material or ice-free bedrock.

The designer must consider both summer and winter loading conditions to determine pile load capacity. The pile load capacity in summer conditions is given by

$$Q + W_p + P_d = P_s + P_E \qquad (19.15)$$

Assuming $P_E = 0$,

$$Q + W_p + \pi d L_a \tau_a = \pi d L_e \tau_f \qquad (19.16)$$

where

Q = structural load on pile
W_p = weight of pile
P_d = downdrag force
P_s = shaft resistance
P_E = end bearing
d = diameter of pile
L_a = active layer thickness
L_e = effective embedment length of pile
τ_a = shear strength of soil in the active layer (limited to 500 psf or 400 MPa)
τ_f = adfreeze strength of pile surface and soils

Under winter conditions, the pile load capacity is given by

$$P_u - Q - W_p = P_s$$

or

$$P_u - Q - W_p = \pi d L_e \tau_f \qquad (19.17)$$

where P_u = uplift or frost heave force.

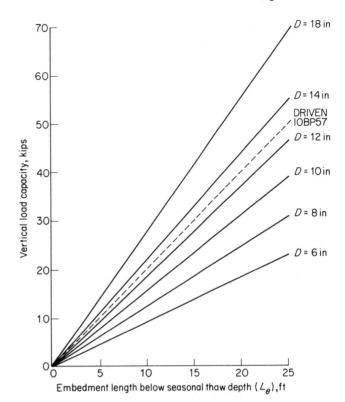

Fig. 19.21 Pile load capacity.

If, during the winter, when active uplift force exists, the structure has not been built, Equation 19.17 becomes

$$P_u - W_p = \pi d L_e \tau_f \qquad (19.18)$$

The uplift force (P_u) is the most critical term in Equation 19.18 because of the absence of structural load on the pile. Some typical values of uplift force are:

Soil Type	P_u
Silty (most frost-susceptible) soils	40 psi (270 kPa)
Organic soils	10 psi (70 kPa)
Silty granular soils	20 psi (140 kPa)

Using Equations 19.16 and 19.18, the most critical effective embedment length of the pile against the structural load is determined. When the required effective embedment length is significant, because of uplift force, the techniques that should be used to mitigate frost heave force are:

a. Isolation of the pile surface from the surrounding soils by wrapping with a suitable material such as visqueen
b. The use of heat tubes in the active layer to reduce frost heave force (it is theorized that the radial heat flow causes ice lenses to grow parallel to the pile surface, resulting in no upward pressure on the pile surface)
c. Replacement of frost-susceptible soils in the active layer with non-frost-susceptible soils

TABLE 19.8 TYPICAL VALUES OF ADFREEZE STRENGTH.

Temperature	31°F (−0.55°C)	30°F (−1.11°C)	28°F (−2.22°C)	25°F (−3.88°C)	20°F (−6.66°C)
Adfreeze strength, psi (kPa)	5 (34.5)	10 (68.9)	18 (124)	25 (172.4)	28 (193)

19.6.2 Single Pile Freeze-Back

The designer must calculate the freeze-back time required before allowing any load placement on pile installed in frozen ground. Either natural or artificial freeze-back may be used so that the installed pile regains adequate load capacity. The natural freeze-back method is used when sufficient cold reserves are available in the natural frozen ground. The best time for natural freeze-back is late spring when the ground temperature is lowest. In artificial freeze-back methods, the total heat that must be removed from the slurry is calculated and artificial refrigeration is used to remove the heat. Example 19.6 shows the step-by-step procedure for calculating refrigeration capacity requirements for artificial freeze-back.

EXAMPLE 19.6

A 12-inch diameter pile is placed in an 18-inch pre-augered, frozen, ice-rich silt deposit that is frozen well below the depth explored by drilling. A pile backfilled with slurry is installed to a depth of 35 ft. The following site conditions should be used to determine the pile load capacity and freeze-back time required.

PERMAFROST SILT:

- Average warmest temperature $= 30°F$
- Dry unit weight $= 85$ pcf
- Water content $= 25$ percent
- Uplift force $= 20$ psi
- Active layer $= 4$ ft

SLURRY BACKFILL SAND-WATER

- Placement temperature $= 40°F$
- Dry unit weight $= 100$ pcf
- Water content $= 15$ percent
- Slurry must be refrozen and cooled to $25°F$

Solution

(a) SUMMER CONDITIONS
Assuming shear strength in active layer $= 500$ psf,

Downdrag force, $P_d = \pi(1)(4)(500) = 6280$ lb

Shaft resistance, $P_s = \pi(1)(35 - 4)(144)(5 \text{ psi})$
$$= 70\,085 \text{ lb}$$
$$\text{(assumed } \tau_f \text{ at } 30°F = 5 \text{ psi)}$$

Neglecting the weight of the pile from Equation 19.16

$$Q + P_d = P_s$$

or

$$Q = P_s - P_d$$
$$= 70\,085 - 6280 = 63\,805 \text{ lb}$$

Allowable pile load capacity in summer condition

$$= \frac{63\,805}{2} = 31\,902 \text{ lb}$$

(b) WINTER CONDITIONS
Uplift force, $P_u = (\pi d L_a)(20 \text{ psi})$
$$= \pi(1)(4)(144)(20)$$
$$= 36\,173 \text{ lb}$$
Neglecting the weight of the pile, from Equation 19.18,
$$P_s = \pi d L_e \tau_f$$
$$= \pi(1)(35 - 4)(144)(5)$$
$$= 70\,085 \text{ lb}$$

The shaft resistance will resist the uplift force, therefore

$$\text{allowable pile load} = \frac{70\,085}{2}$$
$$= 35\,043 \text{ lb}$$

Summer conditions govern the design.

(c) FREEZE-BACK CALCULATION
Unfrozen volumetric heat capacity of the slurry is

$$C_u = \gamma_d(0.17 + (1)\omega)$$
$$= 100[0.17 + (1)(0.15)] = 32 \text{ BTU/ft}^3\text{-}°F$$

Frozen heat capacity of the slurry is

$$C_f = \gamma_d[0.17 + (0.5)(\omega)]$$
$$= 100[0.17 + (0.5)(0.15)] = 24.50 \text{ BTU/ft}^3\text{-}°F$$

Latent heat of slurry is

$$L = (144)(\gamma_d)(\omega)$$
$$= (144)(100)(0.15)$$
$$= 2160 \text{ BTU/ft}^3$$

Volume of slurry placed is

$$\pi(0.75 - 0.5)^2(35) = 6.869 \text{ ft}^3$$

Heat required to depress the slurry temperature to freezing point is

$$(C_u) \times (\text{volume of slurry})$$
$$\times (\text{temperature above freezing of slurry})$$
$$= (32)(6.869)(40 - 32)$$
$$= 1758 \text{ BTU}$$

Heat required to freeze the slurry is

$$(L) \times (\text{volume of slurry}) = (2160)(6.869) = 14\,837 \text{ BTU}$$

Heat required to depress the slurry temperature from the freezing point to $25°F$ is

$$(c_f) \times (\text{volume of slurry})$$
$$\times (\text{required temperature below freezing})$$
$$= (24.50)(6.689)(32 - 25)$$
$$= 1178 \text{ BTU}$$

Hence, total heat to be abstracted from the slurry is

$$1758 + 14\,837 + 1178 = 17\,773 \text{ BTU}$$

Assuming the artificial freeze-back capacity to be $17\,000$ BTU/day, freeze-back time by artificial refrigeration is

$$17\,773 = 1.045 \text{ day or } 25 \text{ hr}$$

19.6.3 Long-Term Pile Displacement

In ice-poor frozen soils, the pile settlement may be limited by calculating the allowable adfreeze strength and then determining the depth of pile embedment needed to carry the designed load with a suitable factor of safety. These soils have grain-to-grain contact and the behavior is dominated by primary creep, as illustrated in Figure 19.6. If the design load is limited within the primary creep range, the long-term settlement may be limited. On the other hand, the load–deformation behavior of ice-rich soils is largely dominated by secondary creep (Fig. 19.6) and the properties of ice and the soil temperature are responsible for the long-term deformation. Accordingly, a creep flow load should be introduced to mathematically describe the pile behavior. The creep rate of ice-rich frozen soils may be written as

$$\dot{\varepsilon} = B\tau^n \qquad (19.19)$$

where

$\dot\varepsilon$ = creep rate
τ = level of shear stress
B, n = creep parameters

Typical values of B and n for ice (Morgenstern et al., 1980) are:

Temperature, °C	B, kPa^{-n}yr^{-1}	n
-1	4.5×10^{-8}	3
-2	2.0×10^{-8}	3
-5	1.0×10^{-8}	3
-10	5.6×10^{-8}	3

If the constitutive behavior of material is characterized by Equation 19.19, the pile displacement rate (u_a) under constant tangial shear stress (τ) and uniform ground temperature is given by (Morgenstern et al., 1980)

$$\dot U_a = \frac{3^{(n+2)/2} a B \gamma^n}{n-1} \qquad (19.20)$$

where a = pile radius.
The long-term pile displacement is calculated by

$$U_a = \frac{3^{(n+2)/2} a B \gamma^n t}{n-1}$$

where t = design life of the structure.

EXAMPLE 19.7

Determine the anticipated pile settlement in 30 years for a load of 17.5 tons in Example 19.6.

Solution

Tangential shear stress is

$$\frac{(17.5)(0.00996)}{(\pi)(1)(0.305)(35)(0.305)} \text{ MN/m}^2 = 17.04 \text{ kPa}$$

Assuming the creep parameters

$n = 3$
$B(25°\text{F or } -3.88°\text{C}) = 1.5 \times 10^{-8} \text{ kPa}^{-n}\text{yr}^{-1}$

then from Equation 19.6,

$$U_a = \frac{3^{(3+1)/2}(0.5)(0.305)(1.5 \times 10^{-8})(17.04)^3(30)}{(3-1)}$$

$$= 0.0015 \text{ m} \quad \text{(negligible in this case)}$$

REFERENCES

Andersland, O. B. and Anderson, D. M. (eds.) (1978), *Geotechnical Engineering for Cold Regions*, McGraw-Hill Book Co., Inc., New York, N.Y.

Anderson, D. M. and Morgenstern, N. R. (1973), Physics, chemistry, and mechanics of frozen ground, North American Contribution, *Proceedings of the Second International Conference on Permafrost*, Yakutsk, USSR, National Academy of Sciences, Washington, D.C., pp. 257–288.

Anderson, D. M. and Tice, A. R. (1973), The unfrozen interfacial phase in frozen soil systems, *Ecological Studies*, **4**, pp. 197–125.

Assure, A. (1963), Discussion on creep of frozen soils, North American Contribution, *Proceedings of the First International Conference on Permafrost*, Lafayette, Indiana, National Academy of Sciences Publication 1287, pp. 339–340.

Baranov, I. J. and Kudryavtsev, V. A. (1966), Permafrost in eurasia, *Proceedings of the First International Conference on Permafrost*, Lafayette, Indiana, National Academy of Sciences Publication 1287, pp. 98–102.

Bates, R. E. and Billello, M. A. (1966), *Defining the Cold Regions of the Northern Hemisphere*, U.S. Army CRREL Technical Report 178.

Brown, R. J. E. (1967), *Permafrost in Canada*, National Research Council, Canada, Map 1246A.

Carslaw, H. S. and Jaeger, J. C. (1959), *Conduction of Heat in Solids*, 2d ed., Oxford University Press, London.

Di Pasquale, L., Gerlek, S., and Phukan, A. (1983), Design and construction of pile foundations in Yukon-Kuskokwim delta, Alaska, *Fourth International Conference on Permafrost*, Fairbanks.

Holtz, R. D. and Kovacs, W. D. (1981), *An Introduction to Geotechnical Engineering*, Prentice-Hall, Inc., Englewood Cliffs, N.J., p. 733.

Hult, J. A. H. (1966), *Creep in Engineering Structures*, Blaisdell, Waltham, Mass.

Johnston, G. H. (1981), *Permafrost: Engineering Design and Construction*, John Wiley and Sons, Inc., New York, N.Y.

Kersten, M. S. (1949), *Thermal Properties of Soils*, University of Minnesota, Engineering Experiment Station, Bulletin 28.

Lambe, T. W. and Whitman, R. V. (1979), *Soil Mechanics*, John Wiley and Sons, Inc., New York, N.Y.

Lunardini, V. J. (1981), *Heat Transfer in Cold Climate*, Van Nostrand Reinhold Co., New York, N.Y.

Morgenstern, N. R. and Nixon, J. F. (1971), One-dimensional consolidation of thawing soils, *Canadian Geotechnical Journal*, **8**, No. 4, pp. 448–565.

Morgenstern, N. R., Roggensack, W. D., and Weaver, J. S. (1980), The behavior of friction piles in ice and ice-rich soils, *Canadian Geotechnical Journal*, **17**, pp. 405–415.

Pewe, T. L. (1966), Ice-wedges in Alaska: Classification, distribution, and climatic significance, *Proceedings of the First International Conference on Permafrost*, Lafayette, Indiana, National Academy of Sciences Publication 1287, pp. 76–81.

Phukan, A. (1977), Pile foundation in frozen soils, *ASME Energy Technology Conference*, Houston, Texas.

Phukan, A. (1980), Design of deep foundations in discontinuous permafrost, *ASCE Spring Convention*, Technical Session on Deep Foundations, Portland, Oregon.

Phukan, A. (1981), *Design Guide for Roadways on Permafrost*, State of Alaska, Department of Transportation and Public Facilities, Division of Planning and Program Research Section, Fairbanks.

Phukan, A. (1985), *Frozen Ground Engineering*, Prentice-Hall, Inc., Englewood Cliffs, N.J.

Tsytovich, N. A. (1975), *The Mechanics of Frozen Ground*, McGraw-Hill Book Co., Inc., New York, N.Y.

Vyalov, S. S. (1959), *Rheological Properties and Bearings Capacity of Soils*, U.S. Army CRREL, Translation 74 (1965).

Vyalov, S. S. (1962), The strength and creep of frozen soils and calculations for ice–soil retaining structures, U.S. Army CRREL, Translation 76 (1965).

20 GEOTECHNICS OF HAZARDOUS WASTE CONTROL SYSTEMS

JEFFREY C. EVANS, Ph.D., P.E.
Associate Professor of Civil Engineering
Bucknell University

20.1 INTRODUCTION

Contamination of the subsurface environment with hazardous and toxic wastes has become the number one environmental problem. In the United States, a national cleanup effort has begun under C.E.R.C.L.A., the Comprehensive Environmental Response, Compensation and Liability Act (42 USC 9601 et seq., 1980), commonly known as Superfund, and its subsequent reauthorization under S.A.R.A., the Superfund Amendments Reauthorization Act (P.L. 99-499, October 17, 1986). Over 30 000 sites have been identified, each having the potential for the introduction of contaminants into the subsurface environment. Over 1000 sites have been specifically identified under the Superfund Program as requiring cleanup or remediation; these constitute the National Priorities List (N.P.L.). The geotechnical engineer will continue to play a key role in the assessment and cleanup of hazardous waste sites. More often than not, hazardous waste site conditions include contamination of groundwater and soil, necessitating control and remediation systems interacting with the subsurface environment.

The geotechnical engineer's role in solving these problems of society includes (1) the planning and implementation of site assessments to characterize the type, distribution, and migration of contaminants in the subsurface, and (2) the development of remedial alternatives in order to control contaminant migration and to protect public health and the environment. Principles of geotechnical engineering employed in conventional practice are frequently applied to these hazardous waste control problems. However, there are specific aspects of the art and practice of geotechnical engineering that must be considered differently from those for more conventional problems. The unique aspects of the application of geotechnology to hazardous waste control systems are presented in this chapter. These same principles may also be applied to environmental control systems for nonhazardous waste, such as systems for municipal solid and residual wastes.

It is the purpose of this chapter to describe the waste control systems typically employed to control the migration of groundwater and contaminants in the subsurface. Components of these systems may find application at new facilities or, more frequently, as an integral part of the overall remedial system for existing uncontrolled hazardous waste sites. The interactions between soils and contaminants are first examined. The measurement of hydraulic conductivity in both the laboratory and field is then discussed along with the compatibility between soils and the intruding contaminated pore fluid. Specifically, the variables in the laboratory and in the field that influence the hydraulic conductivity test results are presented. The basics of clay mineralogy and hydraulic conductivity testing form the basis of the soil–waste interaction studies described. With these fundamentals, geotechnically based hazardous waste control systems are examined in detail. Topics included are the mechanisms by which contaminants may migrate into the environment and an examination of the control systems that minimize the rate of transport into the environment. Particular emphasis is placed on widely-employed liners and slurry trench cutoff walls.

The chapter is organized consistently with the concept of a handbook. Specific engineering guidance is provided where possible and citations are presented to enable the reader to further investigate the underlying scientific principles. Topics are emphasized that represent those most frequently encountered in geotechnical engineering practice.

20.2 SOIL–WASTE INTERACTIONS

Clays are composed of atoms and molecules that are systematically organized into a crystalline structure. Unit cells of silica tetrahedra and octahedra (gibbsite and brucite) are combined to form a wide variety of clay minerals. The unit cell arrangement and the nature of the bonding between these basic building blocks control the behavior of the clay soils.

The most common clay mineral is kaolinite, in which an octahedral gibbsite (with a central atom of magnesium) layer is joined to a silica tetrahedral sheet to form a one-to-one structure. Intralayer bonding within the sheet is covalent and quite stable. The interlayer bonding (between adjacent layers) results from hydrogen bonds and is likewise quite stable. Since both the intralayer and interlayer bonding is strong, swelling and shrinkage due to changes in moisture content are typically relatively small. The cation exchange capacity (a measure of the propensity of the given clay mineral to exchange ions) is also quite small for kaolinitic minerals.

The two most commonly encountered minerals with a two-to-one structure are illites and smectites (montmorillonites). Smectites have two silica tetrahedral sheets on either side of a gibbsite sheet. The interlayer region is populated with water molecules and randomly located anions and cations; thus, the interlayer bonding is primarily from van der Waals forces. The interlayer bonds are relatively weak and easily separated by imposed stresses, including those from the adsorption of water or other polar fluids. A common form of smectite is bentonite, used in several geotechnical engineering applications to be subsequently discussed.

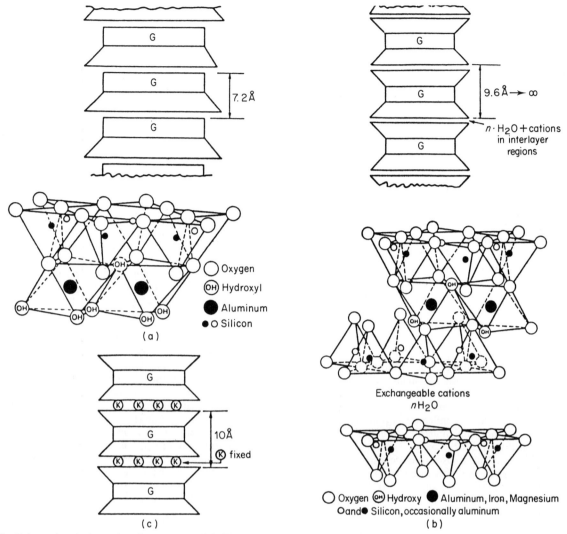

Fig. 20.1 Schematic of clay mineral structures. (a) Diagrammatic sketch of kaolinite. (b) Diagrammatic sketch of montmorillonite. (c) Diagrammatic sketch of illite. (*After Grim, 1967.*)

Although similar to smectite in its two-to-one structure, the interlayer region of illite is populated with potassium ions and a relatively strong bond results. The structures of kaolinite, illite, and montmorillonite are shown schematically in Figure 20.1 (Grim, 1967).

Isomorphous substitution occurs within the crystalline structure and a net negative charge results on the surface of the clay platelet. The negative charge is balanced by cations on the surface of the clay particle. In the presence of an electrolyte, the distribution of the ions adjacent to the clay's surface can be described using a diffuse ion layer model as developed by Gouy–Chapman (Van Olphen, 1977). This distribution of anions and cations adjacent to the clay's surface is a result of two force systems in opposition to each other. On one hand, the cations are attracted to the clay's surface owing to the net negative charge within the clay structure. Conversely, the cations in an electrolyte tend to be uniformly distributed owing to the repulsion forces between them. The resulting distribution as described by the Gouy–Chapman model is shown in Figure 20.2. Expansion of this diffuse ion layer, particularly in a material such as the montmorillonite with weak interlayer bonding, can result in swelling of the soil particles. Therefore, the changes in this diffuse ion layer impact the swelling and shrinkage of these clay particles.

Mechanisms from colloid chemistry have been used to formulate an expression for the "thickness" of this diffuse ion layer. The relationship shows that the diffuse ion layer is sensitive to variations in the dielectric constant of the fluid, the temperature, the electrolyte concentration in the pore water, and the cation valance. In addition, it has been shown that other variables such as the size of the cations in the diffuse ion

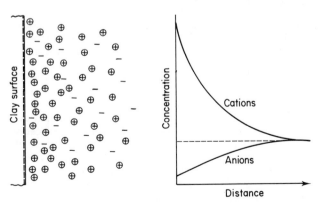

Fig. 20.2 Distribution of ions adjacent to clay surface.

TABLE 20.1 EFFECT OF PORE FLUID PARAMETERS ON DOUBLE-LAYER THICKNESS AND SOIL STRUCTURE[a].

Pore Fluid Parameter	Change in Pore Fluid Parameter	Change in Double-Layer Thickness	Change in Soil Structure Tendency
Electrolyte concentration	Increase Decrease	Decrease Increase	Flocculated Dispersed
Ion valence	Increase Decrease	Decrease Increase	Flocculated Dispersed
Dielectric constant	Increase Decrease	Increase Decrease	Dispersed Flocculated
Temperature	Increase Decrease	Increase Decrease	Dispersed Flocculated
Size of hydrated ion	Increase Decrease	Increase Decrease	Dispersed Flocculated
pH	Increase Decrease	Increase Decrease	Dispersed Flocculated
Anion adsorption	Increase Decrease	Increase Decrease	Dispersed Flocculated

[a] After Evans et al. (1985d).

layer, the pH of the fluid, and the anion absorption on the clay particles will also influence the diffuse ion layer thickness. These impacts have been summarized and are shown in Table 20.1 (Evans et al., 1985a). For additional detail, a parametric study was conducted to elucidate the influence of pore fluid parameters on the surface potential and the diffuse double layer (Jayadeva and Sridharan, 1982).

The repulsion of soil particles during deposition results in a fabric that would tend to be more dispersed (i.e., there would be attraction between particles) (Lambe, 1953). Thus, as a first approximation, the chemistry of the contaminated pore fluid in the soil can be examined to provide some guidance as to the change in the diffuse ion layer and the resulting interparticle forces.

Just as the soil structure at the mineralogical level is important, it is necessary to enlarge the perspective and examine the soil fabric at the micro and macro fabric levels in order to develop a basis for the understanding and analysis of environmental control systems. The effect of soil fabric was shown to significantly influence the engineering properties of compacted soil (Lambe, 1953). The compaction water content and compaction energy influence the soil behavior and are explained in terms of fabric descriptions, diffuse ion layers, and resulting interparticle forces. Studies of scanning electron micrographs have identified particle aggregations and interparticle soil fabric (Collins and McGown, 1974). Thus, the soil fabric, with its arrangement of particle assemblages and voids, will influence the soil behavior, particularly the hydraulic conductivity and the rate of contaminant transport. In addition to the microfabric, the macrofabric must also be considered.

For example, the microfabric may be such that the soil is well compacted with a uniform distribution of assemblages and particle voids. Testing of small samples would indicate the soil to be very impermeable. However, if large cracks should exist in the field in this material owing to, for example, desiccation, the field behavior would be significantly different from the laboratory behavior. In order to predict the field performance from laboratory samples, it is not only necessary to reproduce the density, moisture content, degree of compaction, type of clay, and type of pore fluid in the laboratory, it is also necessary to reproduce the fabric.

The foregoing discussion has focused primarily upon clayey materials. It is recognized that while granular materials are widespread in the environment, they are less subject to changes in behavior in response to permeation with contaminated pore fluids. Further, in most hazardous waste control systems it is the fine-grained fraction of the soil that controls the behavior of the mass.

20.3 PERMEABILITY AND COMPATIBILITY TESTING

The hydraulic conductivity of geotechnical materials is probably the most important parameter involved in the assessment of contaminant migration in the subsurface and the design of barriers for hazardous waste control. Permeability tests which measure the hydraulic conductivity of soils can be conducted in the laboratory or in the field. Permeability tests (typically in the laboratory) can also be used to assess the compatibility between contaminated groundwater/leachate and soil barrier material. In the following paragraphs, guidance is provided for the determination of hydraulic conductivity in both the laboratory and the field. The soil–waste interactions are summarized, indicating that, in general, dilute solutions of contaminants have little effect on the hydraulic conductivity of clayey soils, whereas concentrated organics may cause significant increases in the hydraulic conductivity.

It is important to differentiate between a permeability test and a compatibility test. A permeability test is conducted to measure the hydraulic conductivity. A compatibility test is conducted to study changes in hydraulic conductivity due to the contaminated environment.

Laboratory studies are useful for the determination of hydraulic conductivity. However, without some field testing, laboratory tests may be inadequate as a result of differences in soil structure and stress conditions between the laboratory and the field. Further, one can partially account for macro structural details in field tests that cannot be accounted for in the laboratory. If a representative soil sample is taken from a field-compacted clay liner and brought into the laboratory and tested, that test result is probably not representative of that material in place. In most cases, the laboratory test result will predict a lower hydraulic conductivity than field testing or performance data would indicate. To better assess the hydraulic conductivity, particularly of clay liners and covers, field testing is recommended.

20.3.1 Laboratory Permeability Testing

Laboratory tests for saturated clayey soils can be conducted on "undisturbed" samples obtained from the field or on laboratory-prepared samples. Laboratory samples may be prepared by compaction or by sedimentation and consolidation. The objective is to simulate field structure in the laboratory as closely as possible. Several types of permeameters (Figure 20.3) are used for laboratory permeability testing, including consolidation cell permeameters, compaction mold permeameters, and triaxial cell (flexible wall) permeameters.

Consolidation Cell Permeameter In a consolidation cell permeameter, a standard fixed-ring consolidometer is modified to permit the measurement of hydraulic conductivity. The soil sample is one-dimensionally consolidated to an applied vertical effective stress and a differential seepage pressure is applied across the sample. The head may be either a falling head or constant head. If the apparatus is so equipped, a back-pressure

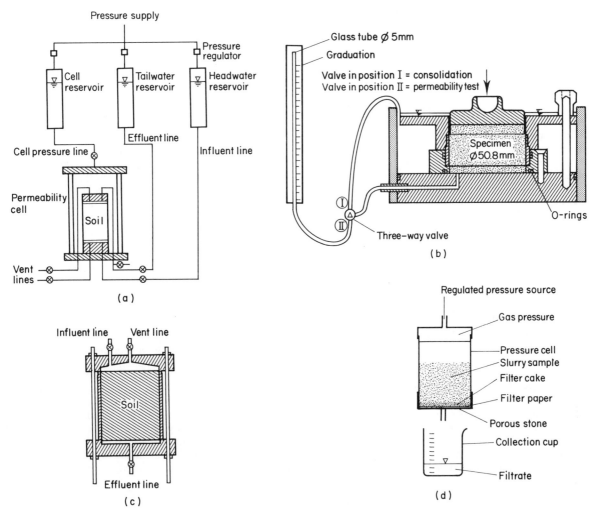

Fig. 20.3 Types of permeameters. (a) Hydraulic system employed for flexible-wall permeability. (b) Oedometer cell adapted for falling-head permeability tests. (c) Compaction-mold permeameter. (d) Filter-press test apparatus.

may be employed to aid in saturation. The consolidation cell permeameter is classified as a fixed-wall permeameter since the sample is laterally constrained by the fixed walls of the consolidation cell ring. Located on the top and bottom of the sample are porous stones that provide for uniform distribution of the water pressure. In addition to the direct measurement of hydraulic conductivity in a consolidation cell permeameter, the hydraulic conductivity may be calculated from the coefficients of consolidation, compressibility, and void ratio during a consolidation test. This method, however, may introduce greater variability into the coefficient of hydraulic conductivity than would normally be desired for environmental studies (Tavenas et al., 1983). Further, this method of calculation does not employ a direct measurement of the coefficient of hydraulic conductivity; instead, it is a calculation based upon Terzaghi's theory of consolidation. As with all of the test methods, the coefficient of permeability is dependent upon the stress level, as will subsequently be discussed. In a study of fixed-wall versus flexible-wall permeameters, it was found that permeameter type has little influence on the hydraulic conductivity of laboratory-compacted clays permeated with water, but may significantly influence test results with concentrated organic liquid permeants (Daniel et al., 1985).

Compaction Mold Permeameter The hydraulic conductivity of soils is commonly determined in a compaction mold per-

meameter. This test method is typically employed and recommended for the study of the hydraulic conductivity of compacted clay liners (Bowders et al., 1986). Using this technique, the soil is compacted directly into a specially modified compaction mold (which also serves as a permeameter). Porous stones are located on the bottom and the cell is equipped with gaskets to provide a positive seal. The top compaction collar then becomes a reservoir for the permeating fluid. The hydraulic gradient can be applied using either a falling-head or constant-head method. Permeation is downward. As with a consolidation cell permeameter, this permeameter is classified as a fixed-wall permeameter. The vertical hydraulic conductivity is computed using sample dimensions, measured flow rates, and Darcy's law. Depending upon the design of the control panel, back-pressure may be employed, although most frequently it is not. This technique offers a rapid test method, particularly for laboratory-prepared samples.

Triaxial Cell Permeameters The triaxial cell permeameter is commonly used to measure hydraulic conductivity (Zimmie et al., 1981; Carpenter and Stephenson, 1986). In this method, a cylindrical soil sample is encapsulated in a membrane, giving rise to its classification as a "flexible-wall" permeameter. Consolidating stresses may be applied either isotropically or anisotropically. Upon completion of consolidation, the sample may be back-pressured to dissolve the air and provide for saturation nearly equal to 100 percent. A differential head between the top cap and bottom pedestal may then be applied. Flow may be either up or down through the sample. Vertical hydraulic conductivity is measured. This technique lends itself to testing of undisturbed samples from Shelby tubes taken in natural soil deposits or in compacted clay liners. The primary benefit of this technique is the relatively easy application of back-pressure to aid saturation and the ability to control the state of stress of the sample. The other techniques are strain controlled in the radial direction. They have a fixed wall, which permits essentially no lateral deformation. In the triaxial cell permeameter, the sample is free to shrink or swell in a stress-controlled manner.

Filter Press Permeameter Hydraulic conductivity of bentonites and soil–bentonite materials has been measured using an API Filter press as a fixed-wall permeameter (D'Appolonia, 1980; Alther et al., 1985). A constant head of 100 psi is applied in a fixed-wall filter press designed to measure filtrate loss of clay slurries. The test can be run in the field with inexpensive equipment and relatively quickly. The results are often correlated to triaxial cell permeability tests and the field test is used as a quality control test.

Compatibility Testing Each of the above laboratory techniques may be used to determine the hydraulic conductivity of soils employed in hazardous waste control systems. However, it is of equal importance to assess potential changes in hydraulic conductivity in response to the contaminants being controlled.

First, it is necessary to select representative fluids for compatibility testing. For example, for a slurry trench cutoff wall used to control the flow of contaminated groundwater, representative samples of the groundwater would be used in the laboratory compatibility studies. For a clay liner employed to control leachate migration beneath a solid waste landfill, samples of the leachate would be utilized. Thus, the fluid employed in the laboratory testing must be representative of field conditions. Further, it is necessary to maintain the chemistry of the fluid throughout the testing period. This is difficult, and techniques used to maintain constant pore fluid chemistry will subsequently be discussed.

Finally, it is necessary to consider worst-case fluid conditions, as well as average conditions. For example, it has been demonstrated that solubilized organics have little effect upon clay materials, whereas concentrated organics may significantly alter the hydraulic conductivity. This, coupled with the relatively low solubility limits of organics, may dictate the necessity to test the soil with concentrated organics. Although concentrated organics may not be widely found in the subsurface, they often exist as nonaqueous-phase liquids (NAPL). Thus, over limited areas of the subsurface, the barrier may be in direct contact with concentrated organic fluids. It is therefore necessary to ascertain both the type and distribution of contaminants in contact with the barrier material and assess the associated compatibility.

Sources of Variability A number of sources of variability in laboratory testing of saturated cohesive soils have been identified. Some potential errors are representative of short-term testing to determine the hydraulic conductivity of soils with water. Others are unique to long-term compatibility testing. The following discusses the testing of variables as summarized in Table 20.2.

Common to both compatibility testing (long-term testing) and short-term hydraulic conductivity testing are a number of sources of error; the most important is the testing of nonrepresentative samples. In the laboratory, a relatively small sample is utilized and typically a "good" sample is trimmed for laboratory testing. This sample may not be representative of field conditions. The laboratory sample does not typically include desiccation cracks, fissures, and associated macrofabric defects.

Previous studies have shown that molding water content, compaction effort, and compaction method influence the hydraulic conductivity (Lambe, 1958). A recent study was conducted to investigate the type of permeameter, flow direction, sample size, storage time, and desiccation cracking (Boynton and Daniel, 1985). Differences in test parameters are presented for each type of permeameter in Table 20.3. It was found that the hydraulic conductivity of laboratory-compacted clay was essentially independent of permeameter type. In the laboratory, the hydraulic conductivity was independent of flow direction (isotropic), although field conditions may differ. Desiccation cracks formed in a matter of a few hours in the laboratory and required significant confining stress to close.

Sample disturbance that affects the test results may take the form of voids that exist owing to difficulty in sample trimming, or zones of clay that are smeared (changing the soil fabric). Smear of a clayey sample will result in a zone of lower permeability.

Lack of complete saturation has also been shown to reduce the measured value of hydraulic conductivity (because water flows around air bubbles in the sample). A back-pressure should be employed to provide improved saturation for short-term testing (Bishop and Henkel, 1962).

The use of distilled water as a permeant may alter the permeability. For short-term testing, it is important to utilize a permeant with a chemistry as close as possible to the original pore water. For an undisturbed specimen, a groundwater sample from the vicinity of the undisturbed sample is recommended as a permeant in the laboratory test. For samples of soil–bentonite backfill to be mixed in the field, samples of the mixing water should be obtained for hydraulic conductivity testing.

An important aspect of hydraulic conductivity testing is modeling of the state of stress. Since soils consolidate under increasing stresses, the application of confining stresses in excess of those that exist in the field can cause an unrealistically low void ratio and measured values of hydraulic conductivity lower than those that would actually exist in the field (Fig. 20.4). Based upon data gathered from a number of sites, confining pressures ranging from 0 to 8000 psf (0 to 400 kPa) for hazardous waste landfills and 0 to 6000 psf (0 to 300 kPa) for sanitary landfills are recommended (Pierce et al., 1986b). Further complicating the state of stress is the fact that there is a stress distribution in the sample in a triaxial cell permeameter. Shown in Figure 20.5 is this distribution of the stress for upward permeation of a triaxial specimen. In very soft soils, such as soil–bentonite backfill, differential consolidation will occur.

It has been shown (Olson and Daniel, 1979) that small changes in temperature may not significantly affect the measured values of hydraulic conductivity. Temperature corrections are, however, easily made.

Gradients representative of field conditions should be employed for the determination of hydraulic conductivity. Some studies have found that, as the gradient increases, the measured value of hydraulic conductivity also increases (Edil and Erickson, 1985). Other studies (Mitchell and Younger, 1967; Carpenter and Stephenson, 1986) have found decreasing values of hydraulic conductivity as the gradient increased. The decreasing values of hydraulic conductivity are apparently associated with particle migration, which causes subsequent clogging of the sample or porous stone. It is recommended that, in hydraulic conductivity testing, gradients similar to those encountered in the field be used. As will subsequently be discussed, this is not possible for compatibility testing.

TABLE 20.2 TESTING VARIABLES FOR LONG-TERM PERMEABILITY TESTS[a].

Variable	Typical "Cause"	Potential "Errors"	Potential "Solutions"
Gradient	Typically high (100 ±) to achieve adequate pore volume displacement in a reasonable testing period	Does not represent field conditions May cause migration of fines	Conduct replicate tests at "low" gradients to determine hydraulic conductivity and "high" gradients to determine compatibility
Effective stress	High enough to maintain positive effective stress at influent end of the sample	May cause differential consolidation May not represent field stress conditions May result in hydraulic conductivity measurements lower than "actual"	See above
Particle migration / reorientation	Induced by high gradients	May result in hydraulic conductivity measurements lower than actual	See above Perform grain size distribution analysis after testing to check for migration
Differential consolidation	See above	Reduced cross-sectional flow area; impeding flow	See above Use shorter samples
Maintaining saturation	Consolidation and backpressure typically applied using compressed air; tests may run for many weeks causing air in solution	Steady decrease in hydraulic conductivity observed due to entrapped air	Refresh permeant frequently Eliminate air / permeant interface Other equipment modifications
Influent equilibrium	Volatilization of organics from influent	Permeant chemistry may not represent field conditions	Refresh permeant frequently Eliminate air / permeant interface
Diffusion through membrane	Molecular diffusion of contaminants through membrane under chemical diffusion gradient	See above	Eliminate diffusion gradient by using contaminated cell water Use alternative membrane systems Use silicone oil as cell fluid
Biological activity	Biological growth due to long-term nature of tests	May cause decreases in flow rate owing to biological plugging	Use biocides
Type of "water"	Distilled water, tap water 0.01 N $CaSO_4$, 0.005 N $CaSO_4$ and 1.0 g / litre $MgSO_4$ are all "standard"	Water may alter hydraulic conductivity owing to pore fluid – soil interactions	Avoid distilled water Use "original pore fluid" where possible 0.01 N $CaSO_4$ or 0.005 N $CaSO_4$ impact
Low gradients	Used to "simulate" field conditions	Not practical for compatible testing Equipment precision may be inadequate	Procedures and equipment needs are different for short-term–low-gradient testing than for long-term–high-gradient compatibility testing

[a] Based on Evans and Fang (1988).

Considering the difficulties associated with conventional falling-head and constant-head testing, low-gradient permeability testing with the flow pump method warrants consideration (Olsen et al., 1985). Using this technique, a constant flow rate results in an induced hydraulic head across the sample measured with a sensitive differential pore pressure transducer. The flow-pump method may also be employed for compatibility testing (Olsen et al., 1988).

20.3.2 Compatibility Testing

Hydraulic conductivity testing is typically conducted in a relatively short period of time. However, long-term compatibility testing may require many weeks or months to fully assess the compatibility of the contaminated fluid with the barrier material. Testing time alone may be an inadequate parameter for measuring the compatibility of soil with the contaminated pore fluid. Pore volume displacement is a valuable indicator in assessing the adequacy of the test duration. To establish the compatibility of a soil barrier material with a leachate or other contaminated permeant fluid, it is necessary to continue the test until chemical equilibrium is established. Equilibrium may be defined as no change in the permeant fluid chemistry (influent versus effluent chemistry) and no change in hydraulic conductivity versus time, provided a pore volume displacement of at least 2 has been achieved. Thus, not only is time important, but the quantity of permeant that enters the sample must be adequate to interact with the sample to establish chemical equilibrium. For a sample diameter of 7 cm and length of 14 cm (at a void ratio of 0.50) the time required for one pore volume displacement can be estimated from Figure 20.6. For example, a soil with a permeability of 5×10^{-8} cm/s requires about 2 weeks per unit of pore volume displacement at a gradient equal to 100.

For changes in hydraulic conductivity that are due to changes in the diffuse ion layer system, generally a displacement of two pore volumes has been found to be adequate. It is also recommended that identically prepared and concurrently run water control tests be conducted to aid in distinguishing between equilibration and chemical effects (Bryant and Bodocsi, 1987).

TABLE 20.3 PERMEAMETER DEPENDENT VARIABLES FOR HYDRAULIC CONDUCTIVITY TESTS[a].

Test Parameter	Type of Permeameter		
	Compaction Mold	Consolidation Cell	Flexible Wall
Side-wall leakage	Leakage is possible	Applied vertical stress makes leakage unlikely	Leakage is unlikely
Void ratio, e	Relatively high e because applied vertical stress is zero	Relatively low e because a vertical stress is applied	Relatively low e because an all-around confining pressure is applied
Degree of saturation	Specimen may be unsaturated	Specimen may be unsaturated	Application of back-pressure is likely to cause essentially full saturation
Voids formed during trimming	Impossible; soil is tested in the compaction mold and is not trimmed	Voids may have formed, but application of a vertical stress should help in closing any voids	Voids are not relevant; the flexible membrane tracks the irregular surface of the soil specimen
Portion of sample tested	All of the compacted specimen is tested, including the relatively dense lower portion and the relatively loose upper portion; the dense lower portion may lead to measurement of relatively low k	Only the central portion of the specimen is tested; the upper and lower third of the specimen are trimmed away	1 cm of soil is trimmed off both ends of the compacted sample

[a] After Daniel et al. (1985).

Alternately, if dissolution or precipitation occurs within the soil structure, additional permeation may be necessary. Altering the clay structure by dissolution, causes changes in hydraulic conductivity that may continue as long as fresh permeant is introduced. To ascertain the nature of chemical changes, it is necessary to monitor the effluent chemistry. In this way, the rate of "breakthrough" of selected chemical constituents can be determined. It is most informative to continue the compatibility test until equilibrium of the effluent has been achieved (Bowders et al., 1986).

It is also necessary to continue compatibility tests to equilibrium, even if large changes in hydraulic conductivity occur initially. A close examination of the data by Brown and Anderson (1983) indicates that, after dramatic increases in hydraulic conductivity permeated with concentrated organic permeants, there is a new equilibrium hydraulic conductivity established. The same "leveling-off" after dramatic hydraulic conductivity increases was also observed for inorganic permeants (Alther et al., 1985). The new equilibrium hydraulic conductivity could be utilized in design, if so desired.

There are a number of variables unique to long-term compatibility testing that are not generally considered for short-term hydraulic conductivity. These variables are summarized in Table 20.2 and discussed below.

For long-term permeability testing, it is also important to examine the effect of gradient. Unlike the case of short-term testing to determine hydraulic conductivity, it is not possible to simulate field gradients in laboratory compatibility studies. This is because it is necessary to run the test long enough to establish adequate pore volume displacement. To aid in design of testing programs, Figure 20.6 presents the time required to achieve a pore volume displacement of 1 for a typical triaxial sample (2.8 inch diameter by 5.6 inches long, $e = 0.5$) as a function of gradient and hydraulic conductivity.

If realistic field gradients (typically less than 1) are used, the test times will be unreasonably long, that is, potentially many years. By increasing the gradient, the test duration for compati-

bility decreases to a manageable time. However, one may also alter the hydraulic conductivity as previously discussed. Further, a higher gradient may require the application of higher effective stresses, causing consolidation of the sample. The higher consolidating stresses have also been shown to diminish the effect of contaminants upon the soil (Acar et al., 1985) as shown in Figure 20.7. Changes in hydraulic conductivity are subdued because high confining stresses overcome the electrochemical stresses that might otherwise cause cracks in the sample. In addition, higher stresses may close microstructural defects in the sample that might otherwise permit rapid flow.

High gradients may also result in a misleading assessment of compatibility. At high gradients, the test time is decreased and most of the flow is through macropores with inadequate time for diffusion of contaminants into disconnected pores of the soil. Thus, a false sense of equilibrium is rapidly achieved. At low gradients, diffusion into the entire soil sample occurs, initiating the accompanying changes and requiring a long time to establish equilibrium. Caution must be exercised when testing for compatibility in recognizing the limitations of present test methods.

Since high hydraulic gradients are required for long-term permeability or compatibility testing, the effect of particle migration must be reconsidered. It is apparent that, when tests are first initiated at the high hydraulic gradients, the particles move, resulting in plugging and a lowering of the hydraulic conductivity. The effect of this particle reorientation upon the compatibility test results is not clear, but it is likely that it would subdue the changes in permeability.

It should be noted that most studies of compatibility employ a clay soil that has been prepared with water and usually first permeated with water. The results may be dramatically different if the soil is first mixed with concentrated contaminant organic liquids (Fernandez and Quigley, 1985). Water-insoluble hydrocarbon liquids were found to flow through "microchannels" or "macropores" and displace less than 10 percent of the original pore water when permeating water wet samples. In contrast,

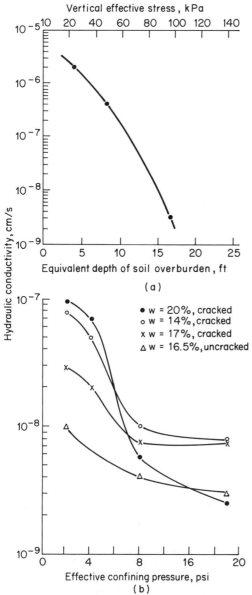

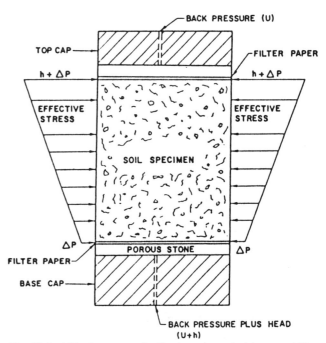

Fig. 20.5 Effective stress distribution on a triaxial permeability specimen. (*From Carpenter and Stephenson, 1986.*)

Fig. 20.4 Effect of consolidation stress upon hydraulic conductivity. ((a) *From Bowders et al., 1986*; (b) *from Boynton and Daniel, 1985.*)

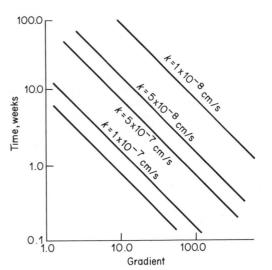

Fig. 20.6 Design chart for estimating test duration. (*From Evans and Fang, 1988.*)

water-soluble alcohols displaced the original pore water and caused 10-fold increases in hydraulic conductivity. Subsequent permeation with low-dielectric aromatics caused 1000-fold increases in hydraulic conductivity. Thus, just as the stress history of a soil influences the soil response under newly applied stresses, the sequence of permeation (i.e., its permeation history) influences the soil response to permeation.

Given the long-term nature of the compatibility test, one must also be concerned with biologic activity in the sample. This may not be a problem when using toxic permeants, but when testing with municipal landfill leachates, biologic growth may occur, resulting in a steady decrease in hydraulic conductivity to this biologic growth. This decrease in hydraulic conductivity may or may not exist in the field and may only be the product of laboratory testing.

In flexible-wall tests, diffusion through the membrane must be considered (Acar et al., 1985). For long-term compatibility testing, it is necessary to maintain a consistent chemistry of the

influent permeant. This can be accomplished in several different ways, depending upon the details of the testing apparatus (Evans and Fang, 1989). Options include daily replacement of the influent permeant, or isolation of the reservoir to maintain constant influent chemistry. The use of an air-over-liquid interface in devices termed accumulators has also been successfully employed (Daniel et al., 1984).

Given the difficulties associated with laboratory testing, it is nonetheless necessary to run compatibility tests in the laboratory. Given our environmental awareness, it is unacceptable to contaminate the ground in order to study changes in hydraulic conductivity and the test duration would be unacceptably long. Further, in the laboratory, dispersion characteristics of the clay–permeant system can be simultaneously evaluated (Korfiatis et al., 1986). Thus, to evaluate compatibility, laboratory studies are required.

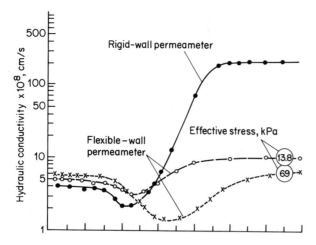

Fig. 20.7 Effective stress influence on compatibility test results. (*From Acar et al., 1985.*)

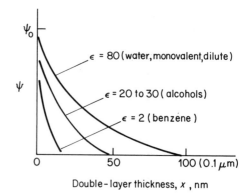

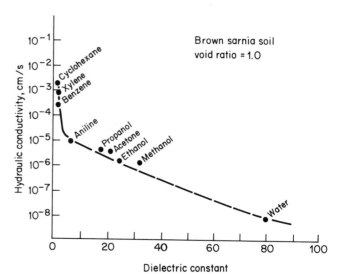

Fig. 20.8 Dielectric effect on diffuse double-layer thickness and hydraulic conductivity. (*From Fernandez and Quigley, 1985.*)

20.3.3 Compatibility Test Results

A number of studies have been undertaken to investigate specific soil–waste interactions, that is, the interaction of specific clays with specific chemicals. One such study (Brown and Anderson, 1983) employed four clayey soils including kaolinitic, illitic, and smectitic clay soils. These materials were evaluated in conventional permeability tests by permeation with concentrated (i.e., pure) organic fluids. The fluids included acidic, basic, neutral polar, and neutral nonpolar concentrated organic fluids. When permeated with water in a compaction mold permeameter, all four soils were found to be suitable clay liner materials having a hydraulic conductivity lower than 1×10^{-7} cm/s. These same soils were found to undergo large permeability increases when permeated with concentrated organic fluids. Mechanisms attributed to the large permeability increases included dissolution and piping. Selected concentrated acidic and basic organics were found to dissolve the alumina, iron, and silica, eroding the structure of the clay, resulting in the release of undissolved fragments which migrate with the percolating leachate. In addition, piping (the erosion of soil from the samples) was found, particularly along the sidewalls of the permeameter, where shrinkage associated with diffuse ion layer changes caused the bulk transport of permeant between the smooth wall of the compaction mold permeameter and the sample. This and other studies have documented significant increases in hydraulic conductivity owing to permeation with concentrated organic chemicals.

The findings of various researchers were reviewed and the changes in clay behavior were examined for compatibility with those results predicted by the Gouy–Chapman model. In most cases, the clay behavior due to changes in pore fluid composition were consistent with the changes predicted by the Gouy–Chapman model. In particular, the mechanism by which organic liquids increase hydraulic conductivity of compacted clays has been shown to be a result of changes in the surface electrical properties of the clay (due to the low dielectric of concentrated organics) which cause shrinkage and cracking (Brown and Thomas, 1987). This phenomenon is illustrated in Figure 20.8.

A number of compatibility studies on specific clay barriers and with specific leachates have been conducted to meet project requirements. For the range of leachates used and clay barrier materials studied, no significant changes in hydraulic conductivity were observed (Daniel and Liljestrand, 1984; Gipson, 1985; Finno and Schubert, 1986; Quigley et al., 1987; Fang and Evans, 1988). Further, clay barrier materials appear resistant to large permeability increases observed with concentrated organic leachates when stabilized by lime or cement (Daniel and Broderick, 1985). Likewise, clay liner materials permeated with dilute organic chemicals showed little change in hydraulic conductivity (Bowders and Daniel, 1987).

Studies from a variety of other investigators were recently reviewed (Madsen and Mitchell, 1987). The reviewers found that, in almost all the cases that they examined, concentrated organic fluids interacted adversely with the clays, resulting in shrinkage cracks and an associated increase in hydraulic conductivity. In contrast, however, dilute solutions of the organic fluids, that is, those dissolved in the water, had essentially no effect on the hydraulic conductivity. Further, they found that the effects of inorganic chemicals upon the hydraulic conductivity of soils were consistent with the effects of the particle's surface diffuse ion layers and surface charges, as indicated by the Gouy–Chapman model.

The behavior of clays, particularly smectites, in the presence of organics may be significantly modified if the clay is first organically activated. Organophilic clays can be prepared from dilute solutions of sodium bentonite and alkylammonium salts (Lagaly et al., 1984). These organophilic clays swell in the presence of certain organics and, conversely, are hydrophobic. Because organophilic clays do not swell in the presence of water, and because of their cost, they have not yet found extensive use in hazardous waste control systems.

20.3.4 Indicator Tests

In addition to direct measurement of compatibility in hydraulic conductivity tests, several indicator tests may be employed to assess compatibility, including Atterberg Limits, sedimentation volume, and cracking pattern tests. These tests are useful in quickly assessing the potential interaction between leachates or other contaminated fluids and clayey barrier materials.

Sedimentation Tests The sedimentation test is conducted by mixing a fixed dry mass (10 g) of soil with 100 ml of the fluid under study in a 100-ml graduated cylinder. The propensity of the fluid to cause the clay to shrink or swell is reflected in the sedimentation volume. These results tend to correlate with hydraulic conductivity test results (Brown and Anderson, 1983; Hettiaratchi and Hrudey, 1987; Ryan, 1987).

Standard Cracking Pattern Tests Standard cracking pattern tests are conducted by mixing 25 ml of stabilized clay slurry (5 percent clay with 95 percent demineralized water by weight) with an equal volume of the fluid under study. The resulting slurry is then poured onto a glass plate and allowed to air-dry. The resulting cracking pattern is evaluated with respect to crack density. Typical results are presented in Figure 20.9. This test has likewise shown reasonable correlation with compatibility test results (Alther et al., 1985; Ryan, 1987).

Atterberg Limits The Atterberg Limits have long been shown to be influenced by changes in pore fluid chemistry. Atterberg Limits can be determined after rewetting with contaminated pore fluid after the soil is air-dried. Figure 20.10 presents the effects of landfill leachate on a clayey liner material enriched with bentonite (Fang and Evans, 1988). At low bentonite contents, the liquid limit is not significantly affected, but as bentonite content increases, so does the impact of the leachate.

20.3.5 Field Permeability Testing

There are a number of field test methods available for evaluating hydraulic conductivity. The selection of the appropriate test method depends upon the desired precision, geologic conditions, and material being tested. For example, to test the hydraulic conductivity of a naturally occurring clay confining bed at a depth of 50 ft requires different equipment and procedures from that for testing a compacted clay cover over a landfill. The following paragraphs describe test methods and their uses and limitations.

In-situ determination of hydraulic conductivity of a naturally occurring clay stratum of low permeability typically involves a downhole permeability test. This may be accomplished with either constant or variable head, using either inflow or outflow. The test may be conducted in an open borehole or through a well screen and casing with a defined geometry. Depending upon the configuration, a variety of equations have been developed to permit analysis as shown in Figure 1.17 (Chapter 1). It is clear from an examination of the geometries of downhole permeability testing shown in Figure 1.17 that the horizontal permeability may have a significant influence upon the measurement of hydraulic conductivity. Thus, downhole permeability tests may be particularly useful in evaluating the anisotropy associated with natural deposits acting as barrier layers.

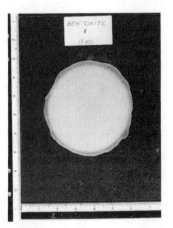

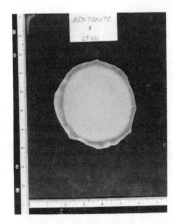
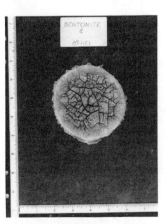

Fig. 20.9 Cracking pattern test results. (*From Alther et al., 1985.*)

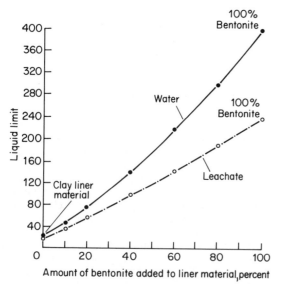

Fig. 20.10 Leachate effects upon the liquid limit. (*From Fang and Evans, 1988.*)

Recently, considerable progress has been made in the measurement of hydraulic conductivity in compacted clay liners (Daniel et al., 1984; Day and Daniel, 1985; Daniel and Trautwein, 1986). Based upon these studies, it has been found that field infiltrometers, such as those shown in Figure 20.11, may be useful in evaluating the hydraulic conductivity of compacted clay liners. In particular, double-ring infiltrometers minimize lateral flow that infiltrates from the inner ring. It should be noted, however, that the precision of such equipment is limited and the time associated with testing increases as hydraulic conductivity decreases.

There are several advantages to using field infiltrometers for the measurement of the hydraulic conductivity of clayey barrier

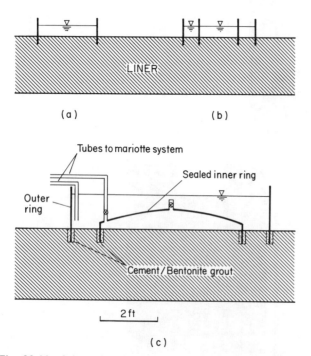

Fig. 20.11 Schematics of ring infiltrometers. (a) Single-ring infiltrometer. (b) Double-ring infiltrometer. (c) Sealed double-ring infiltrometer (SDRI). (*From Daniel and Trautwein, 1986.*)

materials. Most importantly, the larger ring encompasses a large volume of soil, resulting in a sample more "representative" of the system. In addition, the low gradients and stress levels may better represent actual field conditions. To improve the precision of field infiltrometer tests, sealed double-ring infiltrometers have been employed, which control evaporative losses, minimize temperature effects, and employ essentially one-dimensional flow (Daniel and Trautwein, 1986). Studies of clay liners with field infiltrometers have been useful in demonstrating several factors associated with the macrostructure of compacted clay liners. These factors will be discussed later in this chapter.

In-situ testing of vertical barrier walls, such as slurry trench cutoff walls, may be conducted through the use of downhole permeability testing. The hydraulic pressures must be low enough to avoid hydrofracturing of the materials (Bjerrum et al., 1972). Alternatively, pumping tests across the wall can be used to evaluate the completed hydraulic conductivity of vertical cutoff walls. Finally, field samples can be taken back to the laboratory for laboratory analysis of the hydraulic conductivity.

20.4 HAZARDOUS WASTE CONTROL SYSTEMS

The U.S.E.P.A. has prepared a handbook for remedial action at waste disposal sites (U.S.E.P.A., 1985). The handbook identifies and describes a variety of waste control technologies. Presented in Table 20.4 is a summary of remedial action alternatives. From this "menu," alternatives are selected as components for site-specific hazardous waste control systems. The art and science of geotechnical engineering is most frequently employed in the design, analysis, and construction of three major components of hazardous waste control systems. These are surface water controls, in particular, capping; groundwater controls, in particular, subsurface barriers; and leachate control—specifically, liners. The remainder of the chapter will discuss the geotechnical aspects of these hazardous waste control system components. Additional information on other aspects of leachate plume management can be found in Repa and Kofs (1985).

In designing the geotechnical components of hazardous waste control systems, one must consider the hydrologic pathways for contaminant migration which are presented in Figure 20.12 for a typical waste disposal site. As shown, precipitation on a waste disposal site can be returned to the environment through evapotranspiration or run-off of the site, or it may infiltrate the waste. Run-off may become contaminated and carry contaminants into surface water systems. Infiltration will result in the generation of leachate that may pass downward through the site into the groundwater or emerge as leachate seeps to surface water systems. Leachate reaching the groundwater is then either transported through the groundwater regime further into the groundwater system or discharged into surface water bodies. Thus, geotechnical engineering principles need to be applied to the control of the hydrologic pathways of contaminant migration.

Control systems must recognize both routes of exposure and transport processes. Geotechnical engineers are typically associated with groundwater control systems. The transport processes associated with groundwater contamination include advection, dispersion (both molecular diffusion and mechanical mixing), sorption and retardation, and transformation (both biologic and chemical). The impact of the hazardous waste control system on the environment must consider these processes.

The control of surface water is necessary to minimize the contamination of surface water and to prevent surface water infiltration that results in off-site transport of contaminated surface waters through groundwater and surface water transport

TABLE 20.4 SUMMARY OF REMEDIAL ACTION ALTERNATIVES[a].

Site Problem	Surface Water Controls	Air Pollution Controls	Leachate and Groundwater Controls	Gas Migration Control	Waste and Soil Excavation and Removal and Land Disposal	Contaminated Sediments Removal and Containment	In-Situ Treatment	Direct Waste Treatment	Contaminated Water Supply and Sewer Line Controls
Volatilization of chemicals into air		●							
Hazardous particulates released to atmosphere		●							
Dust generation by heavy construction or other site activities		●							
Contaminated site run-off	●								
Erosion of surface due to wind or water	●								
Surface seepage of leachate	●								
Flood hazard or contact of surface water body with wastes	●								
Leachate migrating vertically or horizontally			●				●		
High water table which may result in groundwater contamination or interfere with other remedial technologies			●						
Precipitation infiltrating into site to form leachate	●		●						
Evidence of methane or toxic gases migrating laterally underground				●					
On-site waste materials in non-disposed form: drums, lagooned waste, wastepiles					●		●	●	
Contaminated surface water, groundwater or other aqueous or liquid waste					●		●	●	
Contaminated soils					●		●	●	
Toxic and/or hazardous gases which have been collected								●	
Contaminated stream banks and sediments					●	●		●	
Drinking water distribution system contamination									●
Contaminated sewer lines									●

[a] From U.S.E.P.A. (1985).

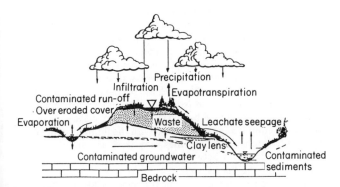

Fig. 20.12 Hydrologic pathways for contaminant migration. (*After U.S.E.P.A., 1985.*)

routes. Surface water controls can be constructed to prevent run-on and intercept run-off, to prevent or minimize infiltration, to control/minimize erosion, to collect and transfer surface waters for ultimate storage and discharge, and to protect from flooding. Conventional civil engineering technologies including the construction of dikes, berms and terraces, sedimentation ponds, levees and flood walls are used to control or prevent surface run-off and run-on. Capping of the site is used to prevent or minimize infiltration and to reduce erosion. Caps are typically multilayered systems with a barrier layer consisting of a compacted clay, a soil admixture, or a geosynthetic material. In a similar manner, liner systems are developed beneath facilities to minimize the rate at which leachate enters the subsurface and to permit collection by the leachate-collection system. The barrier materials for these liners are typically compacted clays, soil admixtures, or synthetic liners. Thus, the following discussion will focus on the geotechnical properties of the barrier material and identify the additional components associated with both cover and lining systems.

The rate of flow of contaminants through soil barriers is typically governed by dissolved constituents migrating with the water (advection) in response to a hydraulic gradient. Transport of contaminants through a membrane (assuming no physical defects such as rips, bad seams, or pinholes) depends on diffusion through the membrane. Thus, all barriers, including geomembranes, are permeable and the hydraulic and contaminant transport rates through barriers can be evaluated (Haxo et al., 1984; Giroud, 1984).

20.5 COVERS AND LINERS

20.5.1 Covers

A cover or cap system is typically multilayered. Using the E.P.A. regulations and guidelines under R.C.R.A, the Resource Conservation and Recovery Act, a multilayer system is recommended as shown in Figure 20.13. This multilayered cap is also occasionally termed a multimedia cap. The upper layer is a layer of sandy topsoil that supports vegetation. Vegetation serves to minimize erosion, maximize run-off, and maximize evapotranspiration. Underneath the topsoil layer is the drainage layer. This layer is typically a sand or gravelly sand and may also include drainage pipes. The function of this layer is to permit the ready drainage of infiltration off of the underlying barrier layer. Beneath the drainage layer is the low-permeability barrier layer. For hazardous waste landfills, this is typically a combination of a synthetic geomembrane and a soil barrier. For solid waste and other nonhazardous waste control systems, the barrier may consist of either synthetic, compacted clay or soil admix materials.

The fate of precipitation hitting a cover system is shown in Figure 20.14. The precipitation may run off the cover and be returned to surface water or be returned to the environment by evapotranspiration. Should the precipitation penetrate the vegetative layer, the drainage layer enhances lateral drainage to further minimize infiltration into the waste. Inevitably some precipitation will penetrate the barrier layer. The interaction with the covered materials results in leachate requiring treatment or control.

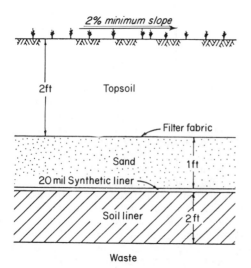

Fig. 20.13 Multilayered cap schematic. (*From U.S.E.P.A., 1985.*)

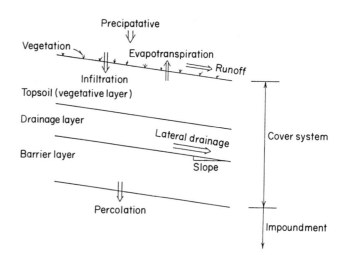

Fig. 20.14 Fate of precipitation landing on cover system.

Several considerations are involved in the selection of whether the barrier material should be a clay liner or a geosynthetic. Geosynthetics are generally more expensive and more labor-intensive in terms of joint sealing to ensure barrier integrity. However, if the covered area is underlain by a synthetic liner, it is necessary to use a cover at least as impermeable as the underlying synthetic liner to eliminate the "bathtub effect." Alternately, a clay barrier material may be utilized as long as there is adequate cover by the topsoil and sand to prevent degradation of the clay liner by freeze–thaw and wet–dry cycles. The overlying soils must also protect the cap from drying and cracking. In addition, filter fabric is often placed over the drainage layer to prevent clogging from the overlying vegetation and topsoil layer. Additional guidance for cap design is provided by the U.S.E.P.A. (E.P.A. Handbook 625-6-85/006 and 540/2-85-002). Design guidance for the drainage system beneath the cap is also provided by the E.P.A. (E.P.A. 600/2-86/058).

20.5.2 Liners

Compacted clay, clay admixtures, and synthetic barrier materials may also be used as liners in waste control systems. Liners are currently designed as multilayered systems. They frequently consist of a combination of both geosynthetic materials and clayey materials as shown in Figure 20.15. The uppermost material shown in the lining system of Figure 20.15 is the waste. Directly underlying the waste is a drainage system, which is called the leachate-collection system. This is the primary leachate-collection system to capture any liquids that result from consolidation of the waste or the transport of precipitation (rainfall and snowmelt) through the waste. Underlying the leachate-collection system is the primary liner. Beneath the primary liner is a secondary drainage system frequently identified as the leak-detection system. This leak-detection system acts as a secondary collection system and serves to pick up any materials that might pass through the primary liner. Underlying the leak-detection system is a secondary liner. This may be a combination system of both synthetic and clayey materials. Increased complexity is associated with the lining systems when geotextiles are used. Geotextiles may serve as part of the filter layer, as a separator between layers, and as protection against puncture for the liner system.

It is necessary for geotechnical engineers to determine the compatibility of the materials used in the system with the overlying waste; that is, the compatibility of the waste with the

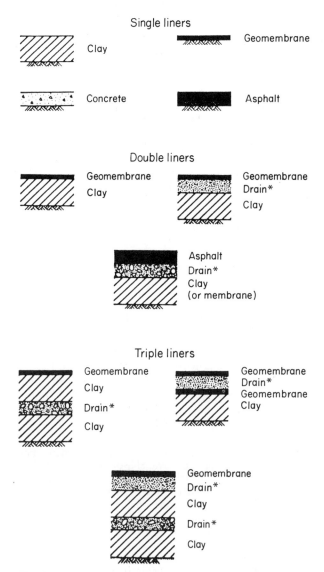

Single liners

Clay Geomembrane

Concrete Asphalt

Double liners

Geomembrane
Clay

Geomembrane
Drain*
Clay

Asphalt
Drain*
Clay
(or membrane)

Triple liners

Geomembrane
Clay
Drain*
Clay

Geomembrane
Drain*
Geomembrane
Clay

Geomembrane
Drain*
Clay
Drain*
Clay

Fig. 20.15 Typical liner systems. (*From Leach et al., 1988.*)

geomembranes, geotextiles, clay, and sand materials. In addition to degradation due to incompatibility, one frequently overlooked problem is that of chemical precipitation. This may cause plugging of the drainage materials and leachate-collection system. As precipitation (rainfall and snowmelt) passes through the waste, leachate is generated and includes dissolved constituents. The pressure and temperature conditions change as these liquids enter the drainage layer and leachate-collection system. Under this new pressure and temperature, precipitate may form. Precipitate tends to plug the drainage layer and plug the drain lines. Hence, the design should incorporate the future potential to repair or clean these drainage systems.

20.5.3 Barrier Materials for Covers and Liners

A. Compacted Clay Barriers

The use of compacted clay to minimize hydraulic transport is well established. Clay dikes, clay liners, and clay cores in dams have long been used to impede water flow. Owing to the critical nature of hazardous waste control systems, recent attention has been focused on the ability of clay barrier materials to minimize hydraulic transport. In a recent study (Day and Daniel, 1985) it was found that a prototype clay liner was significantly more permeable than laboratory data would indicate. In another study (Daniel, 1984), a number of case studies were evaluated and it was found that hydraulic conductivity in the field varied from 10 to 10 000 times that measured in the laboratory. The differences between the laboratory and field values have been primarily associated with macrofabric defects. Factors that have been found to be important include thickness of the clay lift, size of the clay "pods," moisture content, and plasticity of the material. Early studies demonstrated that compaction density alone was insufficient to control hydraulic conductivity (Lambe, 1953) and yet, historically, most clay liners are constructed without significant control of moisture (Pierce et al., 1986a). Given the same soil, at the same density, with the same compactive effort, the hydraulic conductivity of the sample may be an order of magnitude or more higher for a sample compacted on the dry side of optimum than a sample compacted on the wet side of optimum moisture content.

The hydraulic conductivity for a given soil deposit or compacted liner is not a unique value, but rather a statistical distribution of values about a mean. It appears that this distribution reasonably approximates a log–normal probability distribution function (Bergstrom and Kunkle, 1985). Thus, it is necessary to obtain adequate data to achieve an appropriate confidence level. Despite the potential discrepancy between field and laboratory values of hydraulic conductivity, clay liners and covers are frequently used as components in hazardous waste control systems. Their specific advantages include low cost and the ability to attenuate, that is, adsorb, contaminants that may migrate through the barrier. Further, calculations (Daniel and Schackelford, 1987) have shown that clay barriers, if properly constructed, may be more efficient than geomembranes, owing to mass transport through synthetic liners by diffusion.

In general, the most cost-effective covers can usually be completed utilizing native clay materials from local borrow sources. In many areas, local sources are available to provide compactible clays of relatively low hydraulic conductivity. The recommended minimum thickness of a clay barrier layer is generally 2 ft. The top few inches of the clay cap is not as well compacted as the remainder owing to lack of confining pressure at the surface. Further, in the long term, it is difficult to maintain the clay density in the top few inches, owing to potential desiccation, cracking, and frost action. The bottom of the clay barrier layer may be mixed with the subgrade material during installation. Therefore, the "effective thickness" of a nominal 2 ft cap is somewhat less.

Compacted clay covers must be protected from erosion due to rain water, cracking due to drying, differential subgrade movement, penetration by deep tap roots of vegetation, and rutting due to traffic. Conventional hydrologic studies are required to determine erosion potential and prevent erosion of the cover materials.

A clay cover can generally be placed on a slope that is 2 horizontal to 1 vertical or flatter. Flatter slopes more readily permit compaction and reduce the risk of local instability. A geotechnical analysis of slope stability is required to assess the potential for instability. The thickness of the clay on side slopes may be slightly less than the thickness on top of the area, since the larger side slope gradients enhance run-off, greatly reducing the time available for precipitation to percolate downward into the waste containment area.

B. Geomembrane Barriers

Barrier layers for liners and covers may also be constructed of geosynthetics. These polymeric membranes or geomembranes may be constructed from a wide variety of synthetic polymer membrane materials as listed below (also see Chapter 22). They have a relatively low hydraulic conductivity (on the order of 1×10^{-12} cm/s).

- Butyl rubber
- Chlorinated polyethylene (CPE)
- Chlorosulfonated polyethylene (CSPE–Hypalon)
- Epichlorohydrin rubber (ECO)
- Ethylene propylene rubber (EPDM)
- Low-density polyethylene (LDPE)
- High-density polyethylene (HDPE)
- Ethylene propylene terpolymer (EPT)
- Neoprene (chloroprene rubber)
- Polyvinyl chloride (PVC)
- Thermoplastic elastomers

Further, geomembranes are relatively easily installed. The major difficulty is prevention of membrane puncture during placement and provision of adequate connections so that the hydraulic conductivity of the seam is as low as the hydraulic conductivity of the entire sheet.

A recent study of the performance of synthetic membranes at waste sites indicated that failures were due to physical tears or punctures, chemical attack, field seaming, and gas pressure below the membrane. Further, success appears to be associated with designers/owners who demonstrate a thorough understanding of the importance of the liner and the inherent complexities of the system. By recognizing potential failure modes and their consequences, appropriate measures (in design and quality control) can be taken.

From a geotechnical standpoint, the important aspects of utilizing a geomembrane are the subgrade material, subgrade preparation, slope, and final cover. Equally important as these geotechnical aspects are the membrane design, placement procedures, field joining of seams, and field testing of the membrane and field welds, all of which are beyond the scope of this chapter. The subgrade soils must be free of materials that could puncture the membrane top seal, including sticks, large stones, and miscellaneous debris. Further, the subgrade should be graded and compacted to provide for run-off and to prevent liquid ponding. Careful consideration of side slopes is required to preclude sliding of cover materials along the membrane. As precipitation infiltrates and flows along the

membrane, a saturated and weakened zone in the cover material may develop, causing a slump of the cover material.

Just as the testing of soil materials is complex and difficult, evaluation of geomembrane barriers requires considerable expertise. Shown in Table 20.5 are test procedures for evaluating geomembranes for strength, toughness, durability, chemical resistance, stress cracking resistance, and physical properties. Data on the resistance of flexible membrane liners to chemicals has been compiled to assist in material selection prior to project-specific testing.

An experimental procedure has been developed for evaluating the compatibility of a geosynthetic with the site leachate. E.P.A. Method 9090 is used to determine the compatibility of a liner material exposed to leachate for a period of 120 days at one elevated temperature. Physical properties of the liner are measured before and after exposure, including:

- Tear resistance
- Puncture resistance
- Hardness
- Elongation at break
- Thickness
- Mass
- Length
- Width

A more rigorous test environment has been proposed (Lemmon et al., 1985), to include complete immersion, more frequent testing, and several test temperatures.

C. Soil–Bentonite Barriers

Barrier layers for covers and liners may be prepared by admixing bentonite into soil materials available on site. The high-swelling bentonite acts in the matrix of more permeable soils to reduce the hydraulic conductivity of the barrier layer. Soil–bentonite covers and liners also provide significant adsorbing capacity owing to the high cation exchange capacity of the bentonite.

The construction of a soil–bentonite barrier requires: (1) the application of the bentonite at a controlled rate (e.g., 2 to $4 \, \text{lb/ft}^2$); (2) thorough mixing with in-place soils into a predetermined loose lift thickness; (3) compaction of the material; (4) hydration of the bentonite. A 4- to 6-inch layer of clayey soil

TABLE 20.5 GEOMEMBRANE EVALUATION PROCEDURES.

Category	Test Property	Test Method	Typically Good Value
Strength	Tensile properties	ASTM D638, type-IV	2900 psi (20 N/mm^2)
	tensile strength at yield	dumb-bell at 2 ipm	4800 psi (33 N/mm^2)
	tensile strength at break		15 percent
	elongation at yield		800 percent
	elongation at break		
Toughness	Tear resistance initiation	ASTM D1004, die C	53 lb (236 N)
	Puncture resistance	FTMS 101B, method 2031	270 lb (1201 N)
	Low-temperature brittleness	ASTM D746, procedure B	−188°F (−83°C)
Durability	Percent carbon black	ASTM D1603	2.5 per cent
	Carbon black dispersion	ASTM D3015	A-1
	Accelerated heat aging	ASTM D573, D1349	Negligible strength change after 1 month at 110°C
Chemical resistance	Resistance to chemical waste mixtures	E.P.A. method 9090	10 percent tensile strength change over 120 days
	Resistance to pure chemical reagents	ASTM D543	10 percent tensile strength change over 7 days
Stress cracking resistance	Environmental stress crack resistance	ASTM D1693, condition C	2000 hr

with a hydraulic conductivity of 1×10^{-7} cm/s or less is typically constructed.

The principal advantage of a soil–bentonite barrier is the relatively low hydraulic conductivity that can be achieved. The principal disadvantage relates to the difficulty of controlling a uniform application rate. Other disadvantages may be cost and the thin nature of the barrier layer.

The effectiveness of a soil–bentonite barrier requires design and construction control of the subgrade materials, subgrade preparation, application rate, application uniformity, mixing, compaction, hydration, and cover. Deficiencies in any one of these variables can result in reduced effectiveness.

The base soil should be uniform and free of roots, sticks, cobbles, or other miscellaneous debris that would preclude a homogeneous blend for the specified barrier layer thickness. Further, the base soil must have water content and material characteristics that will not impede the uniform blending of the processed clay throughout the soil matrix. The moisture content should be near optimum for the soil–bentonite mixture to aid in the subsequent compaction. Compaction studies in a geotechnical laboratory are required to establish the moisture-density relationship for the soil–bentonite mixture. The subgrade must be sloped in such a way as to provide positive drainage characteristics and preclude ponding of precipitation. Slope stability of the processed clay barrier must also be considered. Slopes steeper than 3 horizontal to 1 vertical are not generally recommended, owing to the increased risk of slope instability.

The application rate is generally determined by consideration of the desired hydraulic conductivity and the available subgrade materials. For a given hydraulic conductivity, the application rate will vary with the base soil type. Laboratory studies should be undertaken to investigate the relationship between hydraulic conductivity, application rate, soil type, compaction water content, and degree of compaction. The design can then be optimized with respect to these variables.

Field control of the application rate is essential for the barrier to perform as predicted by the engineering studies. Three methods are generally employed for applying the powdered processed clay. The first involves the utilization of an agricultural-type spreader. The second involves the use of a pressurized container and distributor. The third method uses hand-spreading. With either of the first two methods the application rate can be checked by placing a relatively flat container (such as a trimmed cardboard or a tarpaulin) beneath the spreader as it passes. The weight of the material deposited in the container is then determined. The application rate is then computed as a weight per unit area. An additional check can be made by determining the total bentonite used for the total area treated. This results in the average application rate. Application rates should be checked frequently when mechanical applicators are employed.

The bentonite can also be applied by hand-spreading the material from bags in premarked grid squares. The bags are broken open and the material is raked into a "uniform" thickness across the grid. Hand-spreading may be more costly, but tighter control can generally be maintained. The probability of zones with inadequate application rates can be reduced with hand-spreading as compared with application by mechanical equipment.

After the application of the bentonite, the material is thoroughly mixed with the subgrade. Mixing should result in a uniform blend of processed clay and subgrade soils for a barrier layer of specified thickness. Adjustable rotary tillers appear to provide a reliable means for controlling barrier layer thickness and homogeneity. Tillers can blend the soil and processed clay with depth control devices, and can result in a fairly uniform layer thickness. Agricultural disks, graders, and other equipment have also been used, but quality control is more difficult.

Immediately after blending, the soil–bentonite barrier is compacted to the density determined during the design studies. Compaction is generally with smooth-drummed or pneumatic rollers. Sheepsfoot or padfoot rollers are not typically employed as the "feet" penetrate the relatively thin admixed layer. Hydration of the soil–bentonite may result from infiltrating precipitation for a cover or by ponding for a liner.

Final cover of soil–bentonite barriers to prevent desiccation and erosion is required. The cover thickness required for soil–bentonite is generally greater than that required for compacted clay caps, because there is little margin for disturbance without jeopardizing the barrier layer's integrity.

20.6 VERTICAL BARRIER SYSTEMS AND CUTOFF WALLS

20.6.1 Vertical Barrier Systems

Subsurface vertical barriers are frequently used to contain, capture, and redirect groundwater flow. These subsurface barriers to horizontal groundwater flow were originally developed for conventional geotechnical engineering applications of dewatering and control of groundwater beneath dams and levees. The most commonly employed vertical barriers to horizontal groundwater flow are slurry trench cutoff walls. Slurry trench cutoff walls are commonly constructed of soil–bentonite, cement–bentonite, plastic concrete, and reinforced concrete. The most common of these is soil–bentonite. In addition, grout curtains are employed for environmental applications. Sheet piling may be considered but, because of leakage through the joints, is infrequently used.

Vertical barriers to horizontal groundwater flow may be used in a variety of applications. By far the most common configuration is a circumferential cutoff wall that completely surrounds the wastes (Figure 20.16). The barrier may or may not be keyed into a low-permeability material beneath the site (Figure 20.17). A circumferential wall may be installed for several purposes. The first and the most common is in conjunction with a groundwater extraction and treatment system. In this application, there is a groundwater pumping and treatment system to "clean up" contaminated groundwater at the site. Without the circumferential cutoff wall, clean regional groundwater is drawn by pumping into the vicinity of the contaminants. This regional groundwater then becomes contaminated and requires the pumping and treatment of larger quantities of groundwater. With the installation of a circumferential cutoff wall, clean groundwater from the surrounding area is precluded from entering the pumping, extraction, and treatment system. Alternatively, in areas of low levels of contaminants and/or at the completion of cleanup, a circumferential cutoff wall serves as a passive barrier to reduce the rate of migration of any remaining contaminants into the environment. Thirdly, cutoff walls can provide for economical dewatering should excavation into the subsurface be required.

Short of a circumferential wall, cutoffs have been used in both up-gradient and down-gradient configurations (Fig. 20.16). On the up-gradient side, the cutoff wall is used to divert clean groundwater around the site. This diversion eliminates the contamination of clean groundwater as it passes beneath the contaminated site. An up-gradient vertical cutoff wall permits the groundwater withdrawal system to withdraw contaminated groundwater from offsite down-gradient locations. Down-gradient placement can also be used to contain and capture floating contaminants.

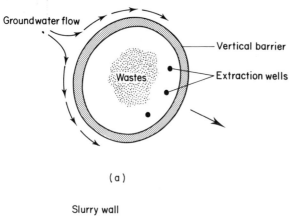

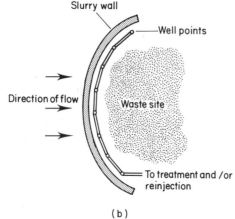

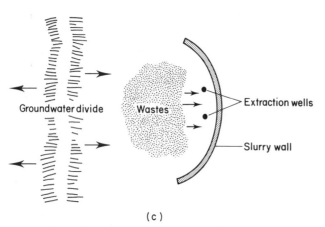

Fig. 20.16 Typical vertical barrier placement. (a) Plan of circumferential barrier wall placement. (b) Wellpoints located behind an up-gradient slurry wall, plan view. (c) Plan of down-gradient placement. (*After Spooner, 1984.*)

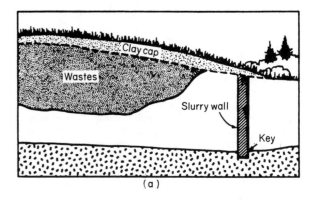

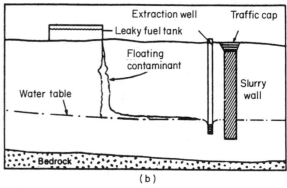

Fig. 20.17 Vertical barrier sections: keyed-in and hanging. (a) Keyed-in slurry wall. (b) Hanging slurry wall. (*After Spooner, 1984.*)

The applications shown in Figures 20.16 and 20.17 are independent of the type of vertical barrier chosen. However, site and subsurface conditions will frequently dictate the type of vertical barrier utilized. For example, to install a vertical barrier to horizontal groundwater flow in fractured rock, slurry trench cutoff wall techniques are not appropriate. In this geologic setting, a grout curtain is typically employed. Conversely, in cohesive lenticular soil conditions, slurry trench cutoff walls are typically most appropriate. The use of soil–bentonite in hazardous waste site remediation appears more common than other types (Morrison, 1983; Ward, 1984).

Feasibility studies and an alternatives analysis of the type of wall to select and/or whether to use a wall at all are required.

Feasibility studies must consider cutoff wall objectives, hydrogeologic considerations, constructability, and short- and long-term performance. In the feasibility study, costs must also be considered.

Vertical barrier wall objectives include considerations of whether the wall is a passive barrier to flow or, in conjunction with other pumping and treatment, part of an active system. The critical nature of the cutoff wall must be considered; that is, what is a permissible rate of contaminant transport across the wall. A study of cutoff wall objectives must assess strength requirements of the wall, if any. Hydrogeologic considerations include the availability of a low-permeability layer to key in the cutoff wall, changes in the groundwater regime due to the presence of the cutoff wall, and the aquifer characteristics through and into which the wall will be placed. Groundwater modeling is a useful tool to assess alternative configurations and design. Investigations of the influence of wall configuration, depth, and hydraulic conductivity upon the groundwater regime should be made. For example, a recent study of a proposed slurry wall employing groundwater modeling of alternatives demonstrated that the presence of a cutoff wall was ineffective in reducing contaminant loadings. This led the designers to conclude that the vertical barrier wall was ineffectual for the specific site and subsurface conditions.

Constructability considerations must also be made during the feasibility studies including: the reuse of materials; man-made interferences, such as debris, pipes, and fill; the depth to suitable aquatard or aquaclude; site preparation requirements; and final cover of the cutoff wall. During this time, input from cutoff wall contractors is useful. It is also necessary to consider cleanup costs and methods for disposal of excavated contaminated soils and slurry.

During the feasibility study, short-term and long-term performance must be considered. Short-term properties include strength, permeability, and compressibility of the cutoff wall; long-term properties include global incompatibility, local degradation, hydrofracturing, and potential settlement.

At completion of the feasibility studies, and with proper consideration of cutoff wall objectives, hydrogeologic conditions, constructability, and performance conditions, an alternatives analysis can be made that includes costs. Costs include construction, health and safety, design, specifications, and inspection. The alternatives most frequently considered for vertical barriers are soil–bentonite slurry trench cutoff walls, cement–bentonite slurry trench cutoff walls, plastic concrete slurry trench cutoff walls, other cutoff wall techniques, or the use of no cutoff. In some cases, a combination of more than one type of wall may turn out to be the optimal solution.

It is necessary to provide a cap for the cutoff wall. The cap is required to provide protection against desiccation of the wall and freeze–thaw degradation, and to support postconstruction traffic both parallel and transverse to the wall. Covers are often placed within a week of cutoff wall construction so as to avoid desiccation. Longer time periods may pass if the top foot or so of the wall is to be removed for construction of the final cap. If the wall is subject to postconstruction traffic, a cap with additional structural stability is required that includes geotextile layers to provide tensile support over the weak compressible cutoff wall and to minimize the thickness of roadbed material.

Slurry trench cutoff wall technology originated in Europe in two applications; for the construction of structural walls, and for groundwater control. According to D'Appolonia (1980), slurry walls first appeared in the United States in the 1940s and their widespread use began in the late 1960s and early 1970s. Slurry trench cutoff walls in environmental applications began in the early 1980s. In fact, a slurry trench cutoff wall was employed in the first Superfund site remediation implemented for coal tar contamination near Stroudsburg, Pennsylvania.

As mentioned before, slurry trench cutoff wall types include soil–bentonite, cement–bentonite, plastic concrete, and diaphragm (structural). Soil–bentonite walls employ a bentonite–water slurry for trench stability during excavation and a mixture of soil, bentonite, and water to create a material that is like high-slump concrete to act as the permanent wall backfill. Cement–bentonite is excavated under bentonite–water–cement slurry that is left to harden in place. Plastic concrete is excavated typically in a panel method using bentonite–water slurry to maintain trench stability. The panel is then backfilled (tremied) with a lean concrete that incorporates bentonite for added plasticity. Diaphragm or structural walls are excavated in a panel fashion with bentonite–water slurry for trench stability. The slurry is then replaced with structurally reinforced concrete. This technique has not been employed in environmental applications owing to the high cost. Further, the high structural rigidity associated with diaphragm walls is not normally required for environmental applications.

Finally, there are some additional vertical barrier walls that could be considered, including a composite wall, a biopolymer wall, vibrating-beam cutoff walls, and deep soil mixed walls. One such composite wall, termed an "EnviroWall" and developed by ICOS, a slurry-trench cutoff wall contractor, incorporates a polymeric membrane into the cutoff wall. Other techniques have also been employed to arrive at the placement of a geomembrane in the completed wall. A biopolymer wall employs a biodegradable drilling mud for excavation; the trench is backfilled with gravel to result in a continuous drain. The vibrating-beam cutoff wall is not a slurry trench cutoff wall, in that a trench is not evacuated. Rather, the wall is installed employing a grouting technique. In this technology a heavy beam is vibrated into the ground and cement–bentonite slurry/grout is injected into the hole created by the beam as the beam is withdrawn. Likewise, deep soil mixed walls can be formed by mixing the existing soil with bentonite–water slurry in place to form a vertical barrier.

20.6.2 Slurry Trench Cutoff Walls

A. Soil–Bentonite Slurry Walls

Soil–bentonite slurry trench cutoff walls are constructed in a two-step process. As shown in the schematic of slurry trench excavation and backfill in Figure 20.18, a trench is excavated in the subsurface, employing a slurry to maintain trench stability. The slurry is typically 5 percent bentonite (sodium montmorillonite) and 95 percent water. This viscous fluid has a unit weight of approximately 64 to 70 lb/ft^3 when first placed in the trench. Trench collapse is resisted by the hydrostatic force system, which counteracts the active earth pressures. As a result of the positive fluid pressure inside the trench, slurry will tend to "leak" out of the trench. In doing so, filter cake is formed as the filtrate enters into the formation. The filter cake is a very thin membrane of hydrated bentonite against which earth pressures and fluid pressures act. In this way, a stable trench can be excavated to depths in excess of 100 ft below the surface.

The methods of analysis of trench stability are well documented (Xanthakos, 1979). To analyze the trench stability, it is necessary to characterize both the slurry and the adjacent soil conditions.

Typically, the trench excavation is completed with a specially modified backhoe to depths of 55 to 60 ft. Below this depth, alternative equipment such as drag lines and clamshells would be utilized. Table 20.6 summarizes the excavation equipment used for slurry trench cutoff walls.

As shown in Figure 20.18, as the excavation progresses, the bentonite–water slurry is displaced with a backfill. The backfill is placed from the surface and moves in the direction of excavation. This backfill is the low-permeability material for the soil–bentonite slurry trench cutoff wall. Shown in Figure 20.19 are typical sections through a soil–bentonite slurry trench cutoff wall. Figure 20.19a shows the trench during the excavation phase of this two-step process. Providing the slurry level is maintained and the filter cake is formed, a stable trench

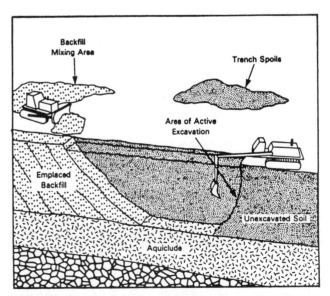

Fig. 20.18 Schematic of soil–bentonite slurry trench cutoff wall construction. (*From Spooner, 1984.*)

TABLE 20.6 EXCAVATION EQUIPMENT FOR SLURRY TRENCH CUTOFF WALLS[a].

Type	Trench Width, ft	Trench Depth, ft	Comments
Standard backhoe	1 to 5	50	Most rapid and least costly excavation method
Modified backhoe	2 to 5	80	Uses an extended dipper stick, modified engine and counterweighted frame; is also rapid and relatively low cost
Clamshell	1 to 5	>150	Attached to a kelly bar or crane; needs ≥ 18 ton crane; can be mechanical or hydraulic
Dragline	4 to 10	>120	Primarily used for wide, deep SB trenches
Rotary drill, percussion drill or large chisel	—	—	Used to break up boulders and to key into hard rock aquacludes. Can slow construction and result in irregular trench walls

[a] From Spooner (1984).

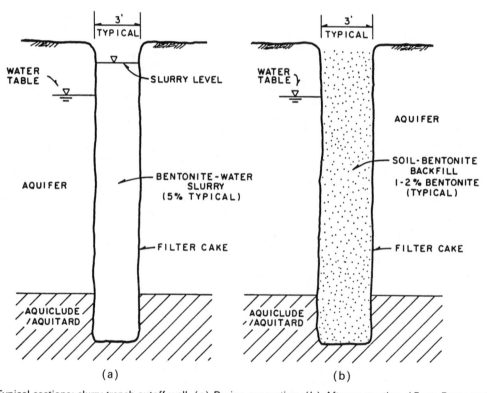

Fig. 20.19 Typical sections: slurry trench cutoff wall. (a) During excavation. (b) After excavation. (*From Evans and Fang, 1982.*)

results. The trench width varies from 1.5 to 5 ft with 3 ft being typical. In Figure 20.19b, the completed trench is shown with soil–bentonite backfill in place. The bentonite content may be as low as 1 percent or as high as 3 or 4 percent. The hydraulic conductivity measured in the laboratory on field-mixed samples of soil–bentonite backfill may vary from as high as 1×10^{-5} cm/s to as low as 5×10^{-9} cm/s, but generally range from 1×10^{-7} cm/s to 1×10^{-8} cm/s as shown in Figure 20.20 (Ryan, 1987). It has been suggested that the filter cake is the primary factor in the low permeability of the completed cutoff (Glover, 1982).

Since the integrity of the filter cake after backfill placement cannot be insured, it is recommended that the added benefit of the cake be considered as additional safety factor.

Backfill Design Considering the variability in backfill permeability, it is necessary to identify important parameters in the design of a soil–bentonite slurry trench cutoff wall. The most important parameters associated with performance of the soil–bentonite backfill are the grain-size distribution of the base soil and the water content. Soil–bentonite backfills have been

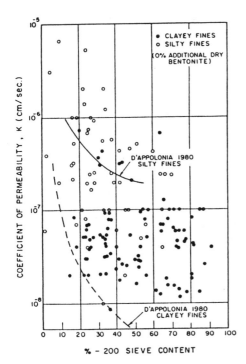

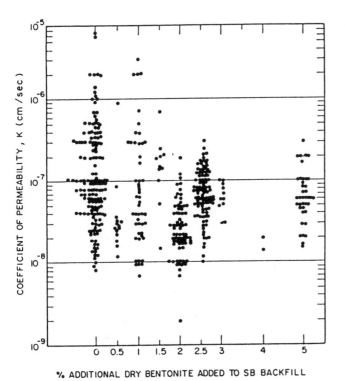

Fig. 20.20 Hydraulic conductivity of soil–bentonite walls. (*From Ryan, 1987.*)

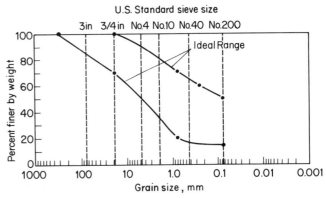

Fig. 20.21 Ideal grain-size distribution for soil–bentonite slurry trench cutoff wall backfill.

successfully prepared utilizing materials that varied from highly plastic clay to clean beach sand.

For environmental applications there are additional considerations not typically required for conventional dewatering applications. Most important is the need to minimize changes in hydraulic conductivity over time due to contact with contaminated water and leachate. Thus, the backfill should be designed to minimize potential changes in hydraulic conductivity. Further,

a lower permeability is frequently required for environmental applications than for dewatering applications.

This combination of permanence and low permeability can best be achieved with a well-graded base soil blended with soil–bentonite slurry. The well-graded base soil should contain particle size ranges varying from coarse to find gravel, through coarse to fine sands, silts and clays (Figure 20.21). A well-graded material utilized as the base soil, including sand- and gravel-size materials, will result in a backfill that is relatively incompressible. This is due to the grain-to-grain contacts of the coarser fraction of the material and the relative incompressibility of coarse-grained soils over fine-grained soils.

Low permeability is achieved owing to the well-graded nature of the material and the presence of both plastic and nonplastic fines in the matrix. The voids are progressively filled with smaller and smaller materials, resulting in a very tight, well-graded material (as exemplified in nature by many glacial tills). The addition of the small quantity of bentonite (1 percent or less) via the bentonite–water slurry contributes to the low permeability of the mix. The slurry provides viscosity to the backfill to allow it to behave as high-slump concrete and flow into the trench. In this manner, the backfill placement characteristics and long-term performance characteristics are optimized.

Low permeability can also be achieved by the addition of larger quantities of bentonite. Since the bentonite swells in the presence of water, the low permeability is achieved by the high swelling capacity of the bentonite. This swelling, however, is reversible. In the presence of contaminants the bentonite may shrink and the permeability of the wall may increase. This response is greater for bentonite-rich backfill than for backfill consisting of inactive fines, silts, and low-plasticity clays.

Backfill is most frequently mixed along the trench alignment, using materials excavated from the trench as the base soil. For added control over mixing, a remote mixing area may be employed with the mixed backfill trucked to the trench for placement. On some occasions, the backfill is mixed at a high enough water content to permit pumping rather than trucking (Ward, 1984). The higher the placement water content, however, the greater the subsequent consolidation settlement. It is suggested that pumping of backfill not generally be permitted.

Potential Defects As with any engineered and constructed system, it is necessary to consider potential failure mechanisms and control these through proper design and quality construction. Failure mechanisms in soil–bentonite slurry trench cutoff walls can be classified as two major types: construction defects and postconstruction property changes.

Summarized in Figure 20.22 are the defects that may develop during soil–bentonite slurry trench cutoff wall construction.

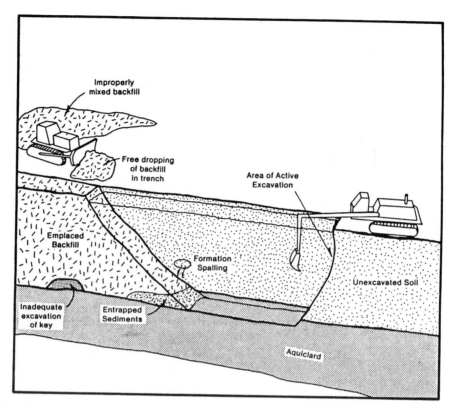

Fig. 20.22 Potential construction defects in soil–bentonite slurry trench cutoff walls. (*From Evans et al., 1985c.*)

Backfill inhomogeneity results from improperly mixed backfill, including lumps of unmixed soil and pockets of free slurry not fully blended with the base soil. Backfill inhomogeneity may also result from slurry entrapment during placement as a result of improper slump, free dropping of backfill into the trench, and the fundamental mechanisms of backfill movement beneath the surface (Evans et al., 1985b). Finally, backfill inhomogeneity may occur at the bottom of the trench owing to the presence of trench sediments during backfill placement. Trench sediments result from settlement of coarser-grained materials out of the slurry as the excavation proceeds; cave-in of the trench sides visible from the surface; and formation spalling—materials falling off the side walls of the trench that are not visible from the surface.

In addition to backfill inhomogeneity, defects in the completed cutoff wall result from inadequate excavation of the key. In all of the above defects, permeable materials remain in the completed cutoff wall resulting in a "window." The window permits increased rates of groundwater flow and contaminant transport through the wall.

Recent analyses of backfill placement have called into question the conventional viewpoint that the backfill sloughs forward as shown in Figure 20.18. It is conventionally considered that the backfill sloughs forward and results in a homogeneous blend of backfill in the trench. Soundings of the backfill generally show a smooth surface. This question was examined (Evans et al., 1985b) using both static equilibrium and viscous flow techniques. In addition, laboratory model studies were conducted and field observations were made. As a result of these studies, it has been demonstrated that the backfill does not slough forward, but slides on top of the backfill already in the trench as shown in Figure 20.23. Potential for entrapment therefore increases with increasing wall depth. Additional studies have confirmed these findings (McCandless and Bodocsi, 1988).

In addition to defects that may occur during construction, time-dependent property changes may also occur. Freeze–thaw

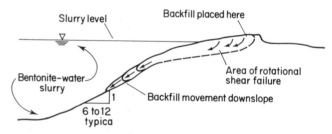

Fig. 20.23 Schematic of soil–bentonite backfilling. (*From Evans et al., 1985b.*)

cycles may cause ice lenses to form in the completed cutoff wall. Proper design of final cover for the cutoff wall will prevent freeze–thaw degradation of the completed wall. Desiccation of the backfill will occur both immediately after construction and in the long term, provided the completed cutoff wall is not covered on a timely basis or the depth of the final cover is inadequate. It is likely that the cutoff wall will also be subjected to wet–dry cycles owing to fluctuation of the water table. Although the wall is somewhat protected from drying by the moist subsurface environment, moisture content varies below and above the water table. The impact of a fluctuating water table on the properties of the completed cutoff wall has not been studied and is unknown at this time.

As a result of the fluid nature of the backfill (it is placed as high-slump concrete), consolidation settlement is inevitable. After consolidation settlement there is an unknown state of stress in the wall. The state of stress in the wall is shown in Figure 20.24. The relatively low shear strength associated with low confining pressures has been confirmed using cone penetration testing (Engemoen and Hensley, 1986). In particular, with large differential gradients across the wall, there is the potential for hydraulic fracturing. The U.S.E.P.A. guidelines

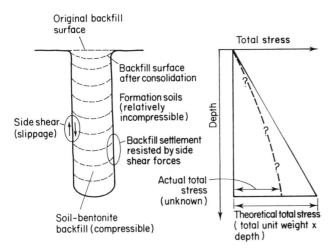

Fig. 20.24 Backfill settlement and stress state in soil–bentonite slurry trench cutoff walls. (*From Evans et al., 1985.*)

indicate that hydraulic fracturing will not occur if the differential head is maintained at less than 1 psi per foot of wall thickness.

Time-dependent property changes may occur owing to chemical incompatibility. This chemical incompatibility may be local, as a result of floating or sinking concentrated organics as shown in Figure 20.25, or global, where detrimental contaminants dissolved in the groundwater or leachate are exposed to the entire wall.

Design Studies Design studies for cutoff walls must be comprehensive and include a number of parameters. The design studies should consider wall configuration, including depth, thickness, layout, grade, and preparation of the working surface. The wall depth may be dictated by hydrogeologic calculations or the presence of an impermeable material beneath the site. If the wall is keyed into an impermeable material, a 3-ft minimum key is typically required. This key provides an adequate safety factor to account for variability between soundings and the inability to precisely identify the depth of the key.

The cutoff wall thickness is dictated by construction methods and hydrofracturing (the guidance of 1 psi per foot of wall thickness). Typical guidance from dam design requires 1 ft of thickness for each 20 ft differential in head; that is, a 100-ft head differential would require a 5-ft thick cutoff wall. The minimum wall thickness of a soil–bentonite wall is typically 30 inches and the maximum wall thickness is typically 5 ft, owing to constructability considerations. The width of the cutoff wall is

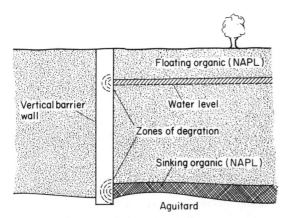

Fig. 20.25 Chemical degradation of vertical barrier walls.

confirmed by measuring and recording the width of the bucket that makes a continuous path through the trench.

The plan view layout of the cutoff wall configuration must consider the presence of discontinuities in layers in the subsurface as well as the presence of interferences such as pipelines, leachate-collection systems, overhead wires, and the like.

Verticality is typically specified to be within 3 percent of maximum. Since the slurry trench cutoff wall is constructed using a bentonite–water slurry, the excavating surface must be nearly level at grade, since the slurry surface will be level. In general, depending upon depth, excavations can be done on grades 2 percent or less.

Finally, it is necessary to provide a working surface (generally about 30 ft wide) for the backhoe and other excavating equipment. The mixing of backfill is either along the trench or in a remote area. For remote mixing, excavated materials are removed from the trench area for reuse or disposal and the completed backfill is trucked into the trench area from the remote mixing area.

Design studies must consider properties of the slurry. Slurry viscosity is measured using a Marsh viscometer and a minimum of 35 to 40 Marsh seconds slurry is typically specified. Additives are usually permitted with the approval of the engineer. Filtrate loss less than about 25 cm³ for a half-hour test at 100 psi in an API filter press is typically specified. The bentonite content of the slurry may be determined by the contractor. Water of adequate quality for slurry mixing must be provided. The following are typical water quality requirements for cutoff walls (Spooner, 1984):

- Hardness of <50 ppm
- Total dissolved solids content of <500 ppm
- Organics content of <50 ppm
- pH of about 7.0

The design should consider the minimum and maximum distance between the excavation face and the toe of the backfill being placed. Typical specifications require a minimum of 10 ft and a maximum of 50 to 100 ft between the toe of the backfill and the excavation face.

The design must consider excavation stability. Calculation procedures are provided by Xanthakos (1979) and others. These procedures do not account for cracking that may occur with stiff, cohesive materials overlying softer soils. These conditions are typically encountered with fill over marsh materials or fill over old trash and debris. Fluid levels are considered in the design calculations and typically must be kept to within 2 to 3 ft of the top of the trench. Most cases of instability experienced in the field are related to either the presence of stiff cohesive materials over a softer layer at some depth and/or the lowering of the fluid level in the trench.

The properties of the slurry in the trench are important both in the design calculations and in field quality control. Specifically, the slurry density must be high enough to maintain trench stability, but low enough to permit adequate placement of the backfill (which must displace this fluid in the trench). It is not unusual to require the slurry density to be at least 15 lb/ft³ more than the density of the backfill. Alternatively, some specifications include a maximum slurry density in the range from 80 to 85 lb/ft³.

Design studies must consider the nature of the backfill. Mix design studies investigate the interrelated variables of water content, bentonite, slump, and hydraulic conductivity. The objective for environmental applications is to minimize the bentonite content. A natural fines content in the range of 20 to 50 percent appears to be optimum when coupled with a well-graded material to reduce the compressibility of the backfill and produce a low permeability. The slump is typically specified to be in the range from 4 to 6 inches and the slope of the backfill in the field in the range of 5H : 1 to 10H : 1. The permeability of the material as measured in laboratory permeability tests can be 1×10^{-7} cm/s or lower.

The design of the backfill must consider permanence (compatibility). In some cases, the use of contaminated soils is recommended. Since the wall is being placed in a contaminated environment, the use of contaminated soils may result in the least amount of long-term property changes.

B. Cement–Bentonite Slurry Walls

Cement–bentonite slurry trench cutoff walls differ from soil–bentonite slurry trench cutoff walls in two ways. Firstly, the excavation slurry includes bentonite, water, and cement. Secondly, the slurry is left to harden in place and is not replaced with a subsequent backfill. The result is a low-solids-content barrier with higher strength (due to the cement). It has been clearly demonstrated that the hydraulic conductivity of cement–bentonite slurry trench cutoff walls is higher than that of soil–bentonite slurry trench cutoff walls. Further, the prediction of the hydraulic conductivity is more difficult, although trends have been shown to exist in specific studies. Widespread analysis has indicated that there is little correlation between cement content and permeability. Presented in Figure 20.26 is the data compiled for a number of cement–bentonite projects over the past 10 years. As can be seen, there is little relationship between cement content and permeability of the cutoff wall. It should be noted that, in laboratory studies, the three ingredients are typically cement, bentonite, and water. However, in the trench, there is an additional ingredient—soil from the excavated formations tends to be mixed in the slurry. This soil remains in suspension and hardens in place and may add up to 20 percent more solids to the mix. The data presented in Figure 20.26 include both laboratory-mixed samples and field-mixed samples tested in the laboratory. Little data is available regarding field-measured values of hydraulic conductivity of slurry trench cutoff walls.

The principal advantages of cement–bentonite over soil–bentonite are the added strength and the reduced potential for construction defects. Design considerations previously discussed are essentially the same for cement–bentonite as for soil–

bentonite. Since cement–bentonite hardens in place, it is occasionally possible to use steeper grades with cement–bentonite since they harden from one day to the next. Properties of soil–bentonite and cement–bentonite backfills are compared in Table 20.7.

A cement–bentonite cutoff wall is constructed of cement, bentonite, and water. The as-mixed bentonite–water slurry should have a stabilized Marsh viscosity of 35 seconds and a minimum unit weight of 64 pcf (1.03 g/cm^3). The cement–water ratio is typically between 0.10 and 0.30. For example, a cement–water ratio of 0.17 equates to 282 pounds of cement to one cubic yard of slurry (i.e., a typical 3-bag mix). The bentonite–water slurry is first mixed and then the cement is blended to achieve a homogeneous cement–bentonite–water slurry.

C. Plastic Concrete Slurry Walls

Plastic concrete is a mixture of cement, bentonite, aggregate, and water, which is placed in a trench excavated under a head of bentonite–water slurry, typically using a panel method of construction (Fig. 20.27). Studies reveal that plastic concrete can be prepared having a coefficient of hydraulic conductivity that is less than 1×10^{-7} cm/s. In addition, the shear strength of the material is significantly greater than that of soil–bentonite or cement–bentonite. As a result, under high differential head conditions, the potential for hydraulic fracturing is greatly reduced. Further, based upon limited data available to date, plastic concrete appears to offer greater resistance to degradation in response to permeation with contaminated pore fluids (Evans et al., 1987).

20.6.3 Other Control Systems

A. Grout Curtains

Vertical barriers to horizontal groundwater flow may also be required in subsurface materials that are not readily excavatable (i.e., rock). A grout curtain in rock is installed through the pressure injection of a viscous liquid (grout) into the rock fractures. The grout then sets or gels, resulting in a much lower hydraulic conductivity for the grouted rock mass. Grouts may be cement-based, pozzolan-based, or chemical. Grouting technology for hazardous waste control systems is derived from experience in conventional geotechnical applications. For additional detail, the reader is referred to publications dealing with grouting for groundwater control beneath dams and into foundation excavations. This section focuses on design and construction considerations unique to hazardous waste control systems.

It is first necessary to indicate a possibly detrimental environmental aspect of grouting, particularly with chemical grouts. As noted, hazardous waste control systems are frequently employed where the concentrations of organics are at low levels (i.e., parts per million or less) yet pose a threat to public health and the environment. However, the use of certain chemical grouts may further degrade the subsurface environment. For example, poisoning in Japan was found to be caused by acrylamide monomer, a main component in earlier chemical grouts (Ando and Mikita, 1977).

Studies of grout are typically conducted to measure strength and hydraulic conductivity. For hazardous waste control systems, one must also consider the nature and fate of both unreacted components and reaction products. Grout studies should also investigate the "setting" characteristics of the grout in the contaminated environment.

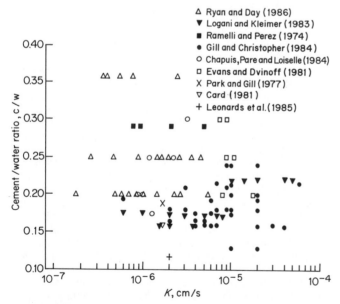

Fig. 20.26 Hydraulic conductivity of cement–bentonite walls. (*From Ryan, 1987.*)

Legend in figure:
△ Ryan and Day (1986)
▼ Logani and Kleimer (1983)
■ Ramelli and Perez (1974)
● Gill and Christopher (1984)
○ Chapuis, Pare and Loiselle (1984)
□ Evans and Dvinoff (1981)
X Park and Gill (1977)
▽ Card (1981)
+ Leonards et al. (1985)

Axis labels: Cement/water ratio, c/w (vertical); K, cm/s (horizontal)

TABLE 20.7 PROPERTIES OF SOIL–BENTONITE AND CEMENT–BENTONITE BACKFILL[a]**.**

Parameter	Soil–Bentonite Backfill	Cement–Bentonite Backfill
Density	Typically 105 to 120 p.c.f 1680 to 1920 kg/m³	Maximum likely 1300 kg/m³
Water content, percent by weight	25 to 35	55 to 70
Bentonite content, percent by weight	0.5 to 2	6
Other ingredients, percent by weight	Fines 10 to 20 Fines 20 to 40	Cement 18 Solids 30 to 45
Strength	Plastic. Very little strength; normally around 20 psf unconfined	Ultimate strength range: 5 to 55 psi Normal strength 20 to 45 psi
Permeability, cm/s	Minimum reported 5.0×10^{-9} Maximum reported $\sim 1 \times 10^{-5}$	$(1 \text{ to } 5) \times 10^{-6}$

[a] From Spooner (1984).

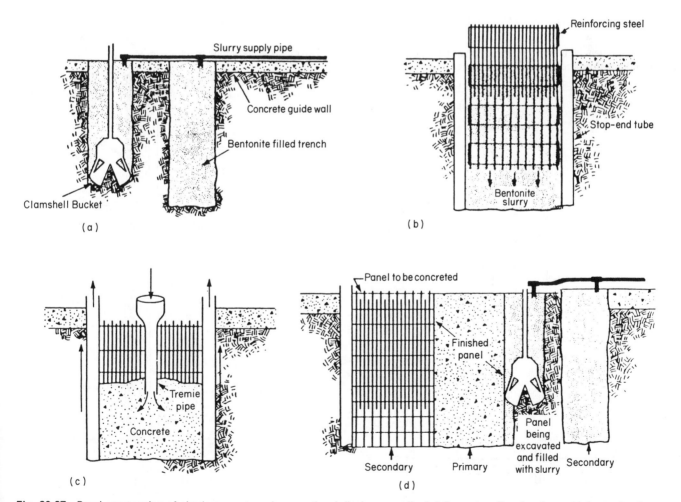

Fig. 20.27 Panel construction of plastic concrete and conventional diaphragm walls. (a) Excavate soil and replace with bentonite slurry. (b) Place stop-end tubes and reinforcing steel into fully excavated panel (for diaphragm wall only). (c) Pour tremie concrete to displace slurry. Remove stop-end tubes. (d) Different construction phases.

B. Vibrating-Beam Cutoff Walls

Barriers to horizontal groundwater flow and contaminant migration have been designed and constructed using the vibrating-beam method of construction, as shown in Figure 20.28. A vibratory pile driver is used to cause the penetration of a beam of specified dimensions into the subsurface to the design depth. Slurry is added through injection nozzles at the tip, both as the beam penetrates the subsurface, to aid in lubrication and penetration, and as the beam is withdrawn, to fill the void left by the beam. Subsequent beams are overlapped to provide a continuous barrier, typically with a maximum thickness of about 2 to 3 inches. Slurry is generally cement–bentonite, although bituminous grout has also been used.

Advantages of this system include the ability to construct a barrier wall under sites that are badly contaminated without exposing large volumes of excavated soil to the surface. This improves worker health and safety considerations, along with considerations of disposal of contaminated material. As a tradeoff, however, the control of the beam tip cannot be guaranteed, particularly with deep penetrations. In addition, penetration through cohesionless soils may be difficult. Even sands that are initially loose may be difficult to penetrate, because, as the beam is vibrated, soils adjacent to the beam become densified. Subsequent penetrations become more and more difficult. This may lead to the need to pre-auger to achieve beam penetration (McLay, 1987).

C. Deep Soil Mixing

Vertical barriers in the subsurface may be constructed by mixing bentonite with soil in place. The technique, developed in Japan

and termed a soil-mixed wall, was recently employed in the United States (Ryan, 1987). A special auger mixing shaft is inserted into the ground simultaneously with injection of a bentonite–water slurry. The installation sequence shown in Figure 20.29 results in a column of treated soil when multiple mixing shafts are employed. Continuity is achieved by overlapping penetrations. Since the slurry is added during mixing, the quantity of bentonite added may be limited. Since open

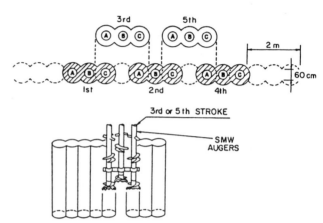

Fig. 20.29 Installation of soil-mixed wall barrier. (*From Ryan, 1987.*)

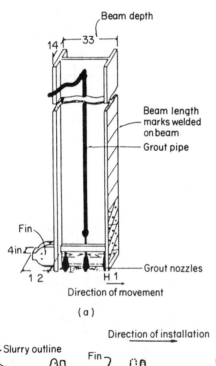

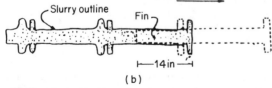

Fig. 20.28 Vertical barrier using the vibrating-beam method. (a) Typical beam configuration. (b) Plan view of slurry wall installation.

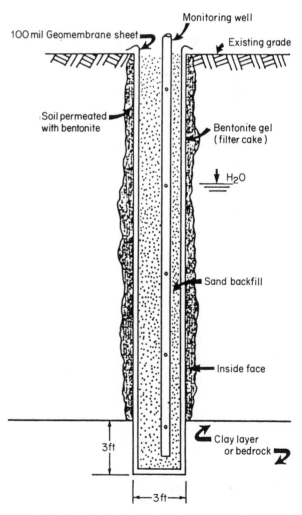

Fig. 20.30 Vertical barrier with a geomembrane.

excavation is not required, health and safety risks and costs are reduced when compared with conventional slurry trench techniques on contaminated sites.

D. Composite Walls

A composite vertical subsurface barrier is constructed by inserting a barrier material into a slurry trench cutoff wall. For example, steel sheet piling has been installed in cement–bentonite to construct a low-permeability barrier with significant resistance to hydraulic fracturing (Evans and Dvinoff, 1981). For hazardous waste control systems, synthetic membranes have been inserted in soil–bentonite cutoff walls (Fig. 20.30). The widespread application of these techniques has been limited by cost and inherent construction difficulties associated with membrane placement and joining of adjacent sheets.

E. Vertical Drainage Barriers

In addition to the vertical barriers discussed above, which act as passive cutoffs, horizontal flow of groundwater and leachate can be cut off with continuous drains. Drains, or leachate-collection systems, may be constructed with conventional construction techniques (U.S.E.P.A., 1985). Alternately, trenches have been excavated with biodegradable slurries and then backfilled with free-draining materials to form a subsurface drainage barrier. Recently, for example, a trench was excavated using "vegetables," that is, a vegetable-based fluid to maintain trench stability (*Civil Engineering Magazine*, 1988). Innovative techniques will continue to develop, employing the fundamental geotechnical engineering principles in new and interesting ways.

20.7 GENERAL GUIDANCE

In summary, hazardous waste control systems frequently employ geotechnical engineering concepts and techniques. The implementation of a geotechnical engineering technology to an environmental control system requires considerations that would not normally be employed. The primary considerations are the interaction between the contaminants and the soil and the need to have systems with fewer defects than would normally be required for geotechnical engineering applications. In environmental control systems, even small defects, allowing small amounts of contaminants to migrate through may be detrimental. In addition, one must consider mechanisms of contaminant transport that go beyond hydraulic transport mechanisms. Thus the consideration of diffusion through the system must be considered. The application of imperfect technologies to the complex technical conditions associated with hazardous waste control will not result in alternatives that are risk-free. Evaluation of the risks may result in conflicts between technical feasibility and social acceptability (LaGrega and Evans, 1987). The application of hazardous waste control technologies must recognize the limitations of the design, analysis, and construction of these technologies.

ACKNOWLEDGMENTS

Special thanks are due my wife (and secretary), Laurel Evans, for her patience and assistance through the typing and editing of the many drafts of this chapter. It could not have been done without her.

REFERENCES AND BIBLIOGRAPHY

Acar, Y. B., Hamidon, A., Field, S. D., and Scott, L. (1985), The effect of organic fluids on hydraulic conductivity of compacted kaolinite, *Hydraulic Barriers in Soil and Rock*, ed. A. I. Johnson et al., ASTM STP 874, pp. 171–187.

Acar, Y. B. and Ghosn, A. A. (1986), Role of activity in hydraulic conductivity of compacted soils permeated with acetone, *Proceedings of the International Symposium on Environmental Geotechnology*, ed. H. Y. Fang, **I**, pp. 403–412.

Adaska, W. S. and Cavalli, N. J. (1984), Cement barriers, *Proceedings of the 5th National Conference on the Management of Uncontrolled Hazardous Waste Sites*, Washington, D. C., pp. 126–130.

Alther, G. R. (1982), The role of bentonite in soil sealing applications, *Bulletin of the Association of Engineering Geologists*, **19**, No. 4, pp. 401–409.

Alther, G. R. (1986), The effect of the exchangeable cations on the physico-chemical properties of Wyoming bentonites, *Applied Clay Science*, **1**, pp. 273–284.

Alther, G. R., Evans, J. C., Witmer, K. A., and Fang, H. Y. (1985), Inorganic permeant effects upon bentonite, *Hydraulic Barriers in Soil and Rock*, ASTM STP 874, pp. 64–74.

American Petroleum Institute (1980), *Recommended Practice for Standard Procedure for Testing Drilling Fluids*, Specification RP 13B, 8th ed., Dallas.

Ando, S. and Mikita, M. (1977), Environmental impacts on groundwater by chemical grouting, *Proceedings of the Ninth International Conference on Soil Mechanics and Foundation Engineering*, Tokyo, Japan, **I/2**, pp. 17–31.

Bergstrom, W. R. and Kunkle, G. R. (1985), Statistical evaluation of hydraulic conductivity data for waste disposal sites, *Management of Toxic and Hazardous Wastes*, ed. H. G. Bhatt, R. M. Sykes, and T. L. Sweeney, Lewis Publishers, Chelse, Mich., Chapter 9, pp. 81–104.

Bishop, A. W. and Henkel, F. J. (1962), *The Measurement of Soil Properties in the Triaxial Test*, Edward Arnold, London.

Bjerrum, L. and Huder, J. (1957), Measurement of the permeability of compacted clays, *Proceedings of the Fourth International Conference on Soil Mechanics and Foundation Engineering*, London, England, **1**, pp. 6–10.

Bjerrum, L., Nash, J. K. T. L., Kennard, R. M., and Gibson, R. E. (1972), Hydraulic fracturing in field permeability testing, *Geotechnique*, **22**, No. 2, pp. 319–332.

Bowders, J. J., Daniel, D. E., Broderick, G. P., and Liljestrand, H. M. (1986), Methods for testing the compatibility of clay liners with landfill leachate, *Hazardous and Industrial Solid Waste Testing: Fourth Symposium*, ed. J. D. Petros, Jr., W. J. Lacey, and R. A. Conway, ASTM STP 886, pp. 233–250.

Bowders, J. J. and Daniel, D. E. (1987), Hydraulic conductivity of compacted clay to dilute organic chemicals, *Journal of Geotechnical Engineering*, **113**, No. 12, pp. 1432–1448.

Boynton, S. S. and Daniel, D. E. (1985), Hydraulic conductivity tests on compacted clay, *Journal of Geotechnical Engineering*, **111**, No. 4, pp. 465–477.

Brown, K. W. and Anderson, D. C. (1983), *Effects of Organic Solvents on the Permeability of Clay Soils*, EPA-600/2-83-016.

Brown, K. W. and Thomas, J. C. (1984), Conductivity of three commercially available clays to petroleum products and organic solvents, *Hazardous Waste*, **1**, No. 4, pp. 545–553.

Brown, K. W. and Thomas, J. C. (1987), A mechanism by which organic liquids increase the hydraulic conductivity of compacted clay materials, *Soil Science Society of America Journal*, **51**, pp. 1451–1459.

Bryant, J. and Bodocsi, J. (1987), *Precision and Reliability of Laboratory Permeability Measurements*, EPA/600/52-86/097.

Carpenter, G. W. and Stephenson, R. W. (1986), Permeability testing in the triaxial cell, *Geotechnical Testing Journal*, **9**, No. 1, pp. 3–9.

Chen, S., Low, P. F., Cushman, J. H., and Roth, C. B. (1987a), The effect of organic compounds on the physical properties of clays: I. Swelling, *Journal Paper No. 11001*, Purdue University Agricultural Experiment Station, West Lafayette, Ind.

Chen, S., Low, P. F., Cushman, J. H., and Roth, C. B. (1987b), The effect of organic compounds on the physical properties of clays: II. Flocculation, *Journal Paper No. 11002*, Purdue University Agricultural Experiment Station, West Lafayette, Ind.

Chen, S., Low, P. F., Cushman, J. H., and Roth, C. B. (1987c), Organic compound effects on swelling and flocculation of Upton Montmorillonite, *Soil Science Society of America Journal,* **51**, pp. 1444–1450.

Collins, K. and McGown, A. (1974), The form and function of microfabric features in a variety of natural soils, *Geotechnique,* **24**, No. 2, pp. 223–254.

Comprehensive Environmental Response Compensation and Liability Act of 1980 (P. L. 96-510).

Daniel, D. E. (1984), Predicting hydraulic conductivity of clay liners, *Journal of Geotechnical Engineering,* **110**, No. 2, pp. 285–300.

Daniel, D. E. and Liljestrand, H. M. (1984), Effects of landfill leachate on natural liner systems, report to Chemical Manufacturers Association, Geotechnical Engineering Center, University of Texas, Austin.

Daniel, D. E., Trautwein, S. J., Boynton, S. S., and Foreman, D. E. (1984), Permeability testing with flexible wall permeameters, *Geotechnical Testing Journal,* **7**, No. 3, pp. 113–122.

Daniel, D. E., Anderson, D. C., and Boynton, S. S. (1985), Fixed-wall versus flexible-wall permeameters, *Hydraulic Barriers in Soil and Rock,* ed. A. I. Johnson, R. K. Frobel, N. J. Cavalli, and C. B. Pettersson, ASTM STP 874, pp. 107–126.

Daniel, D. E. and Broderick, G. P. (1985), Stabilization of compacted clay against attack by concentrated organic chemicals, report to Chemical Manufacturers Association, *Geotechnical Engineering Report GR85-18,* Geotechnical Engineering Center, University of Texas, Austin.

Daniel, D. E. and Trautwein, S. J. (1986), Field permeability test for earthen liners, *Proceedings of the ASCE Specialty Conference on the Use of In-Situ Tests in Geotechnical Engineering,* ed. S. Clemence, pp. 146–160.

D'Appolonia, D. J. (1980), Soil–bentonite slurry trench cutoffs. *Journal of Geotechnical Engineering, ASCE,* **106**(4), pp. 399–417.

Day, S. R. and Daniel, D. E. (1985), Hydraulic conductivity of two prototype clay liners, *Journal of Geotechnical Engineering, ASCE,* **III**, No. 8, pp. 957–970.

Dunn, R. J. (1986), Clay liners and barriers—Considerations of compacted clay structure, *Proceedings of the International Symposium on Environmental Geotechnology,* ed. H. Y. Fang, **1**, pp. 293–302.

Dunn, R. J. and Mitchell, J. K. (1984), Fluid conductivity testing of fine-grained soils, *Journal of Geotechnical Engineering,* **110**, No. 11, pp. 1648–1665.

Edil, T. B. and Erickson, A. E. (1985), Procedure and equipment factors affecting permeability testing of a bentonite–sand liner material, *Hydraulic Barriers in Soil and Rock,* ed., A. I. Johnson, R. K. Frobel, N. J. Cavalli, and C. B. Pettersson, ASTM STP 874, pp. 155–170.

Engemoen, W. O. and Hensley, P. J. (1986), *ECPT Investigation of a Slurry Trench Cutoff Wall, Use of In Situ Tests in Geotechnical Engineering,* ASCE, New York.

Evans, J. C. and Dvinoff, A. H. (1981), *Geotechnical Investigation—Proposed Cement–Bentonite Slurry Trench Wall, Perry Nuclear Plant,* Woodward and Clyde Consultants.

Evans, J. C. and Fang, H. Y. (1982), Geotechnical aspects of the design and construction of waste containment systems, *Proceedings of the 3d National Conference on the Management of Uncontrolled Hazardous Waste Sites,* Washington, D.C., pp. 175–182.

Evans, J. C., Kugelman, I. J., and Fang, H. Y. (1983), Influence of industrial wastes on the geotechnical properties of soils, *Proceedings of the Fifteenth Mid-Atlantic Industrial Waste Conference,* Bucknell University, Lewisburg, Pa., pp. 557–568.

Evans, J. C. and Manuel, E. N. (1985), Geotechnical property testing of hazardous materials and contaminated soils, *Proceedings of the 6th National Conference on the Management of Uncontrolled Hazardous Waste Sites,* Washington, D.C., pp. 369–373.

Evans, J. C., Fang, H. Y., and Kugelman, I. J. (1985a), Influence of hazardous and toxic wastes on the engineering behavior of soils, *Management of Toxic and Hazardous Wastes,* ed. H. G. Bhatt, R. M. Sykes, and T. L. Sweeney, Lewis Publishers, Inc., Chelsea, Mich., Chapter 21, pp. 237–264.

Evans, J. C., Lennon, G. P., and Witmer, K. A. (1985b), Analysis of soil–bentonite backfill placement in slurry walls, *Proceedings of the 6th National Conference on the Management of Uncontrolled Hazardous Waste Sites,* Washington, D.C., pp. 357–361.

Evans, J. C., Fang, H. Y., and Kugelman, I. J. (1985c), Containment of hazardous materials with soil–bentonite slurry walls, *Proceedings of the 6th National Conference on the Management of Uncontrolled Hazardous Waste Sites,* Washington, D.C., pp. 249–252.

Evans, J. C., Fang, H. Y., and Kugelman, I. J. (1985d), Organic fluid

effects on the permeability of soil–bentonite slurry walls, *Proceedings of the National Conference on Hazardous Wastes and Environmental Emergencies,* Cincinnati, Ohio, pp. 267–271.

Evans, J. C. and Fang, H. Y. (1986), Triaxial equipment for permeability testing with hazardous and toxic permeants, *Geotechnical Testing Journal, ASTM,* **9**, No. 3, pp. 126–132.

Evans, J. C. and LaGrega, M. L. (1987), Remediation of Superfund sites: Any feasible solutions?, *Toxic and Hazardous Wastes: Proceedings of the Nineteenth Mid-Atlantic Industrial Waste Conference,* Bucknell Univeristy, pp. 310–329.

Evans, J. C., Stahl, E. D., and Droof, E. (1987), Plastic concrete cutoff walls, *Geotechnical Practice for Waste Disposal '87,* ASCE Geotechnical Special Publication No. 13, pp. 462–472.

Evans, J. C. and Fang, H. Y. (1988), Triaxial permeability and strength testing of contaminated soils, *Advanced Triaxial Testing of Soil and Rock,* ed. R. T. Donaghe, R. C. Chaney, and M. L. Silver, ASTM STP 977, pp. 387–404.

Fang, H. Y., Slutter, R. G., and Koerner, R. M. (1977), Load bearing capacity of compacted waste disposal materials, *Proceedings of the Ninth International Conference on Soil Mechanics and Foundation Engineering,* Tokyo, Japan, **IV/2,** 265–278.

Fang, H. Y. and Evans, J. C. (1988), Long term permeability tests using leachate on a compacted clayey liner material, *Field Methods for Ground Water Contamination and their Standardization,* ASTM STP 963, pp. 397–404.

Fernandez, F. and Quigley, R. M. (1985), Hydraulic conductivity of natural clays permeated with simple liquid hydrocarbons, *Canadian Geotechnical Journal,* **22**, pp. 205–214.

Fernandez, F. and Quigley, R. M. (1987), Effect of viscosity on the hydraulic conductivity of clayey soils permeated with water-soluble organics, *Proceedings of the 40th Canadian Geotechnical Engineering Conference,* Regina, Saskatchewan, pp. 313–319.

Finno, R. J. and Schubert, W. R. (1986), Clay liner compatibility in waste disposal practice, *Journal of Environmental Engineering, ASCE,* **112**, No. 6, pp. 1070–1084.

Gipson, A. H., Jr. (1985), Permeability testing on clayey soil and silty sand–bentonite mixture using acid liquor, *Hydraulic Barriers in Soil and Rock,* ed. A. I. Johnson, R. K. Frobel, N. J. Cavalli, and C. B. Pettersson, ASTM STP 874, pp. 140–154.

Giroud, J. P. (1984), Impermeability: The myth and a rational approach, *Proceedings of the International Conference on Geomembranes,* Denver, Colo., pp. 157–162.

Glover, E. W. (1982), Containment of contaminated ground water: An overview, *Proceedings of the Second National Symposium on Acquifer Restoration and Ground Water Monitoring,* National Water Well Association, Columbus, Ohio, pp. 17–22.

Grim, R. E. (1967), *Clay Mineralogy,* 2d ed., McGraw-Hill Book Company, Inc., New York, N.Y.

Haxo, H. E., Miedema, J. A., and Nelson, N. A. (1984), Permeability of polymeric membrane lining materials, *Proceedings of the International Conference on Geomembranes,* Denver, Colo., pp. 151–156.

Hettiaratchi, J. P. A. and Hrudey, S. E. (1987), Influence of contaminant organic–water mixtures on shrinkage of impermeable clay soils with regard to hazardous waste landfill liners, *Hazardous Waste & Hazardous Materials,* **4**, No. 4, pp. 377–388.

Jayadeva, M. S. and Sridharan, A. (1982), A study on potential–distance relationship of clays, *Indian Geotechnical Journal,* **12**, No. 1, pp. 83–97.

Koerner, R. R. (1986), *Designing with Geosynthetics,* Prentice-Hall, Inc., Englewood Cliffs, N.J.

Koerner, R. M., Martin, J. P., and Lord, Jr., A. E. (1986), Geomembranes in solid waste disposal, *Proceedings, International Symposium on Environmental Geotechnology,* ed. H. Y. Fang, **I**, pp. 285–292.

Korfiatis, G. P., Demetracopoulos, A. D., and Schuring, J. R. (1986), Laboratory testing for permeability and dispersivity of cohesive soils, *Proceedings of the International Symposium on Environmental Geotechnology,* ed. H. Y. Fang, **I**, pp. 363–369.

Lagaly, G., Tributh, H., Sander, H., and Cracuim, C. (1984), Vorgange bei der Sada-aktivierung von Bentoniten am Beispiel eines Bentonites von Neuseeland, *Keram. Z.* **33**, No. 5, pp. 278–283.

Lambe, T. W. (1953), *The Permeability of Fine-Grained Soils,* ASTM STP 163, pp. 456–467.

Lambe, T. W. (1958), The engineering behavior of compacted clay, *Journal of the Soil Mechanics and Foundations Division, ASCE,* **84**, No. SM-2, pp. 1654-1 to 1654-34.

Leach, A., Harper, T., and Tape, R. (1988), Current practice in the use of geosynthetics in heap leaching, *Geotechnical Fabrics Report,* **6,** No. 1, pp. 14–16.

Lemmon, Jr., A. W., Craig, C. G., Grotta, H., Thomas, R., Pfau, J., and Sharpe, R. (1985), Compatibility of hazardous waste landfill and impoundment liners, *Proceedings of the National Conference on Hazardous Wastes and Environmental Emergencies,* Cincinnati, Ohio, pp. 272–276.

Lentz, R. W., Horst, W. D., and Uppot, J. O. (1985), The permeability of clay to acidic and caustic permeants, *Hydraulic Barriers in Soil and Rock,* ed. A. I. Johnson, R. K. Frobel, N. J. Cavalli, and C. B. Pettersson, ASTM STP 874, pp. 127–139.

Low, P. F. (1980), The swelling of clay: II, *Soil Science Society of America Journal,* **44,** pp. 667–676.

Low, P. F. (1981), The swelling of clay, III, *Soil Science Society of America Journal,* **45,** pp. 1074–1078.

Madsen, F. T. and Mitchell, J. K. (1987), *Chemical Effects on Clay Hydraulic Conductivity and their Determination,* Open File Report No. 13, Environmental Institute for Waste Management Studies, The University of Alabama, Tuscaloosa.

Manuel, E. M., Evans, J. C., and Singh, R. D. (1987), Discussion of hydraulic conductivity of two prototype clay liners, by S. R. Day and D. E. Daniel, *Journal of the Geotechnical Engineering Division, ASCE,* **113,** No. GT-7, pp. 804–806.

McCandless, R. M. and Bodocsi, A. (1988), *Final Report: Investigation of Slurry Cutoff Wall Design and Construction Methods for Containing Hazardous Wastes,* Office of R & D, Hazardous Waste Engineering Research Laboratory, U.S.E.P.A., Cincinnati, Ohio.

McGown, A., Marsland, A., Radwan, A. M., and Gabr, A. W. A. (1980), Recording and interpreting soil macrofabric data, *Geotechnique,* **30,** No. 4, pp. 417–447.

McLay, D. S. (1987), Installation of a cement–bentonite slurry wall using the vibrated beam method—A case history, *Proceedings of the 19th Mid-Atlantic Industrial Waste Conference,* ed. J. C. Evans, Technomic Publishing Company, Inc., Lancaster, Pa., pp. 272–282.

Mitchell, J. L., Hooper, D. R., and Campanell, R. G. (1965), Permeability of compacted clay, *Journal of the Soil Mechanics and Foundation Division, ASCE,* **92,** No. SM-4, pp. 41–66.

Mitchell, J. K. and Younger, J. S. (1967), *Abnormalities in Hydraulic Flow through Fine-Grained Soil,* ASTM STP 417, pp. 106–139.

Morrison, A. (1983), Arresting a toxic plume, *Civil Engineering,* American Society of Civil Engineers, August, pp. 35–37.

Murray, R. S. and Quirk, J. P. (1982), The physical swelling of clays in solvents, *Soil Science Society of America Journal,* **46,** pp. 865–868.

Newton, J. P. (1987), Advanced chemical fixation for organic content wastes, Draft, from International Wastes Technologies, Wichita, Kans.

Olsen, H. W. (1985), Osmosis: A cause of apparent deviations from Darcy's law, *Geotechnique,* **22,** pp. 238–241.

Olsen, H. W., Nichols, R. W., and Rice, T. L. (1985), Low gradient permeability measurements in a triaxial system, *Geotechnique,* **35,** No. 2, pp. 145–157.

Olsen, H. W., Rice, T. L., and Nichols, R. W. (1988), Measuring effects of permeant composition on pore-fluid movement in soil, *Proceedings Symposium on Field Methods for Groundwater Contamination Studies,* ASTM STP 963, pp. 331–342.

Olsen, H. W., Morin, R. H., and Nichols, R. W. (1988), Flow pump applications in triaxial testing, *Proceedings of Symposium on Advanced Triaxial Testing for Soil and Rock,* ASTM STP 977, pp. 68–81.

Olson, R. E. and Daniel, D. E. (1979), Field and laboratory measurements of the permeability of saturated and partially saturated fine-grained soils, *Symposium on Permeability and Groundwater Contaminant Transport,* ASTM, Philadelphia, Pennsylvania.

Olson, R. E and Daniel, D. E. (1981), *Measurement of the Hydraulic Conductivity of Fine-Grained Soils,* ASTM STP 746, pp. 18–64.

Pierce, J. J., Sallfors, G., and Peterson, E. (1986a), Clay liner construction and quality control, *Journal of Environmental Engineering, ASCE,* **112,** No. 2, pp. 13–24.

Pierce, J. J., Sallfors, G., and Murray, L. (1986b), Overburden pressures exerted on clay liners, *Journal of Environmental Engineering, ASCE,* **112,** No. 2, pp. 280–291.

Quigley, R. M., Fernandez, F., Yanful, E., Helgason, T., Margaritis, A., and Whitby, J. L. (1987), Hydraulic conductivity of contaminated natural clay directly below a domestic landfill, *Canadian Geotechnical Journal,* **24,** pp. 377–383.

Rad, N. S. and Acar, Y. B. (1984), A study of membrane-permeant compatibility, *Geotechnical Testing Journal,* **7,** No. 2, pp. 104–106.

Repa, E. and Kofs, C. (1985), *Leachate Plume Management,* EPA/540/ 2-85/004, U.S.E.P.A., Cincinnati, Ohio.

Ryan, C. R. (1987), Vertical barriers in soil for pollution containment, *Geotechnical Practice for Waste Disposal,* ASCE, New York.

Shackleford, C. E. (1989), *Diffusion of Contaminants Through Waste Containment Barriers,* Transportation Research Record 1219, Geotechnical Engineering 1989, Transportation Research Board, National Research Council, pp. 169–182.

Sparks, D. L. (1986), *Soil Physical Chemistry,* CRC Press, Inc., Boca Raton, Fla.

Spooner, P. (1984), *Slurry Trench Construction for Pollution Migration Control,* EPA-540/2-84-001, U.S.E.P.A., Cincinnati, Ohio.

Tavenas, F., Leblond, P., Jean, P., and Leroveil, S. (1983), The permeability of natural soft clays. Part I: Methods of laboratory measurement, *Canadian Geotechnical Journal,* **20,** pp. 629–644.

U.S.E.P.A. (1985), *Handbook: Remedial Action at Waste Disposal Sites (Revised),* Hazardous Waste Engineering Research Laboratory, Office of Emergency & Remedial Response, Washington D.C., Office of R & D, Hazardous Waste Engineering Research Laboratory, U.S.E.P.A., Cincinnati, Ohio.

U.S.E.P.A. (1986), *Method 9100: Saturated Hydraulic Conductivity Saturated Leachate Conductivity and Intrinsic Permeability,* SW-846.

Van Olphen, H. (1977), *An Introduction to Clay Colloidal Chemistry for Clay Technologists, Geologists and Soil Scientists,* 2d ed., John Wiley & Sons, Inc., New York, N.Y.

Waller, F. S. and Evans, J. C. (1985), Geotechnics of lagoon closures, *Proceedings of the Third Annual Hazardous Materials Management Conference,* Philadelphia, Pa.

Ward, L. M. (1984), Closeup on cleanup at Lipari waste site, *Hazardous Materials and Waste Management,* May–June, pp. 40–42.

Weeks, O. L. (1986), *Liner Systems Used for the Containment of Solvents and Solvent-Contaminated Hazardous Wastes,* Open File Report No. 12, Environmental Institute for Waste Management Studies, The University of Alabama, Tuscaloosa.

Xanthakos, P. (1979), *Slurry Walls,* McGraw-Hill Book Co., Inc., New York, N.Y.

Zimmie, T. F., Doynow, J. S., and Wardell, J. T. (1981), Permeability testing of soils for hazardous waste disposal sites, *Proceedings of the Xth International Conference on Soil Mechanics and Foundation Engineering,* Stockholm, pp. 403–568.

21 REINFORCED EARTH

F. SCHLOSSER
Ecole National des Ponts et Chaussées
Paris, France, and
Terrasol, Paris, France

M. BASTICK
Terre Armée International,
Paris, France

21.1 INTRODUCTION

Reinforced Earth was invented in 1963, by the French architect–engineer Henri Vidal. It is a construction material made of a frictional backfill material reinforced by linear flexible strips generally placed horizontally (Fig. 21.1). Since its invention, Reinforced Earth has found a wide use in many different areas of civil engineering, notably in retaining walls, seawalls, dams, bridge abutments, and foundation slabs. This technique has been adopted worldwide and the total number of Reinforced Earth structures built each year has been continuously increasing as indicated in Figure 21.2.

This important development over the past 20 years and the concept of reinforced soil introduced by Vidal contributed to the birth of a new area of soil improvement: soil reinforcement. This area now includes several reinforcement techniques; most are based on frictional interaction between inclusions and soil, but others are based on the mobilization of passive earth pressure by means of anchors and grids.

The number, type, and arrangement of the inclusions may be quite variable. Depending on the type of the inclusion, two extreme cases should be considered (Schlosser et al., 1983):

1. *Uniform inclusion*, for which the soil–reinforcement interaction develops at any point along the inclusion. In this case, a relatively high and uniform density of reinforcements will result in a new composite material called "Reinforced Soil."
2. *Composite inclusion*, for which the shape of the inclusion is such that the soil–reinforcement interaction is concentrated

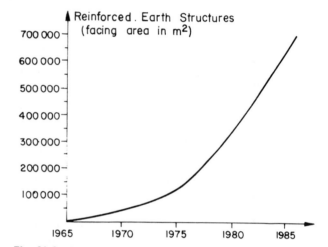

Fig. 21.2 Evolution of the total amount of Reinforced Earth built per year worldwide.

TABLE 21.1 CLASSIFICATION OF SOIL-REINFORCEMENT SYSTEMS.

Density of Reinforcement	Type of Reinforcement	
	Uniform	*Composite*
Periodical Isolated	Reinforced soil Membrane piles	Multianchorage systems Anchorages

at particular locations. Generally, as for multianchorage systems, these points are located at the ends of the inclusions.

These considerations lead to the classification of soil reinforcement systems presented in Table 21.1.

21.2 PRINCIPLE AND ADVANTAGES OF REINFORCED EARTH

21.2.1 Principle

The basic concept of Reinforced Earth material is presented in Figure 21.3. This material results from the association of two components having different moduli of elasticity. A stress

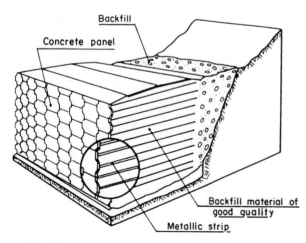

Fig. 21.1 Perspective view of a Reinforced Earth wall.

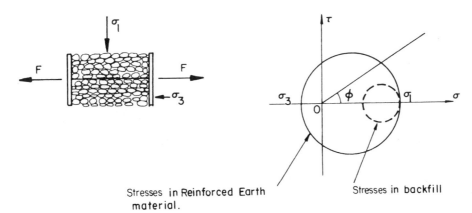

Stresses in Reinforced Earth material.

Stresses in backfill

Fig. 21.3   Basic concept of Reinforced Earth.

applied to the mass will cause strain in the soil that will transmit the tensile load to the strips. The displacements are restrained in the direction of the strips, causing the reinforced mass to behave like a cohesive anisotropic material. Hence, the concept of Reinforced Earth is based on a frictional earth–reinforcement interaction.

As indicated in Figure 21.4, this mechanism of interaction results in shear stresses exerted by the backfill material at the reinforcement interface. The tensile force along a reinforcement varies continuously. Its variation is proportional to the above shear stress according to the following formula:

$$\tau = \frac{1}{2b}\frac{dT}{dl} \qquad (21.1)$$

where

τ = shear stress at the interface
T = tensile force in the reinforcement
l = location along the strip
b = width of reinforcement

The mobilization of the shear stress τ requires a relative displacement of the reinforcement with respect to the earth. This shows that the deformability of the inclusions takes part in the distribution of the forces along the reinforcements. On the other hand, the shear stress τ is limited with respect to the

normal stress exerted on the reinforcement by the value of the soil–reinforcement coefficient of friction, the value of which will be discussed further.

21.2.2   Advantages

Apart from the savings that could be realized over a conventional reinforced concrete or masonry retaining wall, several advantages contributed to the wide and swift development of Reinforced Earth:

1. *Ease and rapidity of construction* is achieved by the systematic use of easily handled standard prefabricated elements.
2. *Adaptability to various slopes and soil conditions*: only mid-size construction equipment is necessary to erect a structure.
3. *Flexibility of the resulting structure*, which allows construction even on relatively soft soils: the only limitation on differential settlements concerns the facing. In order to prevent architectural damage, differential settlements should be limited to 1 or 2 percent.
4. *Aesthetic of the structure and architectural finish of the facing*: the adaptability of the panel technique enables the designer to best fit the shape of the structure to the environment, and to select an appropriate finish (relief, texture, color).

21.3   HISTORY AND DEVELOPMENT

The first Reinforced Earth structure built was the retaining wall of Pragnières, France in 1965. Soon after that, in 1968–69 a large project including 10 retaining walls on unstable slopes near Nice, France, gave the impulse for large research programs and technological developments. This important effort explains the worldwide commercial success of the technique.

Research programs included, laboratory tests on reinforced sand performed at the LCPC in France from 1969–77, and reduced-scale models, both two- and three-dimensional. About 20 full-scale experiments, conducted in various countries since 1968, provided enough data to propose the first design methods and validate the finite-element models that are now used to improve our understanding of the behavior of standard structures and to develop new design methods, particularly for specific applications.

Three events marked the technological development of Reinforced Earth. First, the choice of galvanized steel for strips and facing, after unsuccessful tentative experiments with polyester-coated fiberglass, stainless steel, and aluminum. The

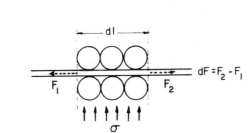

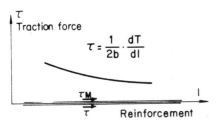

Fig. 21.4   Variation of the tensile forces in the reinforcement and shear stresses exerted on the reinforcement. (*After Mitchell and Schlosser, 1979.*)

second event was the development in 1971 of a typical cruciform panel for the facing in replacement of the original U-shaped metallic facing elements. This type of prefabricated element offers the possibility of architectural finishes and curved facings. It is now representative worldwide of Reinforced Earth and its development. In 1975, the Reinforced Earth Company patented the ribbed strip. This new technological improvement issued directly from research on the soil reinforcement frictional interaction. Owing to the restrained dilatancy effect, the presence of the ribs on a reinforcement leads to a much higher apparent friction coefficient.

21.4 BEHAVIOR OF REINFORCED EARTH

21.4.1 Reinforced Earth Material

Considering Reinforced Earth essentially as a material having its own properties, and not as an anchoring system, the "Laboratoire Centre des Ponts et Chaussées" undertook, in 1969, a series of triaxial tests on samples of sand reinforced by horizontal disks of aluminum foil. These tests showed the effects of various parameters (sand density, reinforcement spacing, tensile strength of the aluminum foil) on the properties of a small volume of Reinforced Earth. These tests first demonstrated the two possible failure modes for a Reinforced Earth mass, that is, failure by breakage of the reinforcement and failure by slippage or lack of adherence between soil and reinforcements.

The samples were reinforced in a two-dimensional way that differed from the technology of Reinforced Earth structures using strip reinforcements. This difference is of minor importance with regard to the reinforcement-breakage, but it turns out to be significant in the case of reinforcement-slippage.

Following this study, similar tests were carried out by Young (1972), Laréal and Bacot (1973), Haussmann (1976), and Romstad et al. (1978). The essential finding is that the presence of reinforcements improves noticeably the mechanical properties of the sand.

Depending on the confining pressure (σ_3), two modes of failure may be observed, as described below (Fig. 21.5)

- *Low values of* σ_3. Failure of the sample occurs by slippage of the reinforcements. The failure curve in the principal stress

axis can be approximated by a straight line passing through the origin (Haussmann, 1976). The strength of the reinforced sand can be represented by a global apparent friction angle ϕ_r greater than ϕ, the internal friction angle of the sand. ϕ_r is directly dependent on the reinforcement density.

Higher values of σ_3. Failure of the sample occurs by breakage of the reinforcements. The failure curve, in the principal stress axis, is a straight line parallel to the failure line of the sand alone. Hence the strength of the reinforced sand can be represented by an internal angle of friction ϕ and an anisotropic cohesion intercept C. The value of C is directly proportional to the density of reinforcements and to their tensile strength, according to the formula:

$$C = \sqrt{K_p} \cdot \frac{RT}{\Delta H} \qquad (21.2)$$

where

$$K_p = \tan^2\left(\frac{\pi}{4} + \frac{\phi}{2}\right)$$

RT = unit tensile strength of reinforcements
ΔH = vertical spacing of reinforcement layers

The rupture in the sample occurs along the classical Coulomb failure plane inclined at an angle of $[(\pi/4) + (\phi/2)]$ on the horizontal. However, cracks appear in the reinforcements near the peak deviator stress. These cracks are approximately located on a cylinder which radius is $\frac{2}{3}R$ (Fig. 21.6a).

Calculations using the finite-element method, as well as other triaxial tests involving stress measurements in the reinforcements, have shown that the distribution of the tensile stresses exhibits two peaks corresponding to the location of the cracks (Fig. 21.6b). Hence, the reinforcements provide links between the core of the sample (resisting zone) and an outer shell where the sand tends to slip out (active zone). These links result in the apparent cohesion of the sample.

21.4.2 Behavior and Mechanism of Reinforced Earth Structures

As observed during triaxial tests on reinforced sand samples, the tensile force along a reinforcement is not constant but

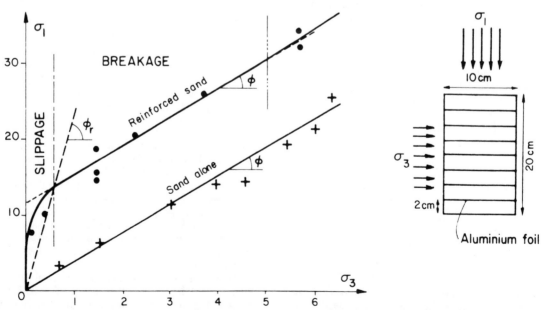

Fig. 21.5 Failure curves for triaxial reinforced and unreinforced samples. (*After Schlosser et al., 1972.*)

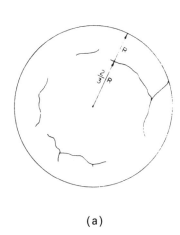

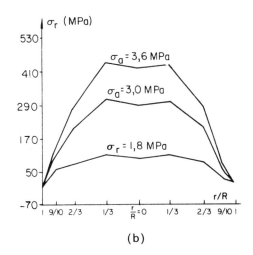

(a)   (b)

Fig. 21.6 Distribution of tensile stresses in disks of a reinforced sand triaxial sample. (a) Location of cracks observed in the aluminum reinforcing disks. (Triaxial sample stressed to peak value but not failed.) (b) Tensile stresses along reinforcements in triaxial tests. (*After Schlosser et al., 1972; Madani 1979.*)

exhibits peaks. For a given load pattern the position of the maximum for different reinforcement layers defines a line of maximum tensile force. Generally, the maximum tensile force line defines two zones in a Reinforced Earth mass (Fig. 21.7):

- An "active zone," where the soil tends to slip out of the structure but is maintained by the friction along the strips. The shear stresses exerted are directed outwards and lead to a decrease of the tensile force in the reinforcements (Zone 1, Fig. 21.7).
- A "resistant zone," where the shear stresses mobilized to prevent the sliding of the reinforcements are directed inwards, toward the free end of the reinforcements (Zone 2, Fig. 21.7).

The link between these two zones, provided by the reinforcements, results in an apparent cohesion of the Reinforced Earth material.

The boundary line between the two zones (line of maximum tension) represents a potential failure surface for the structure. Its position depends on various features such as geometry, applied loads, and dynamic effects. It may also depend on the extensibility of the strips. In intricate situations several of these lines (relative maxima) may appear.

In full-scale experiments on Reinforced Earth walls with vertical facing, the maximum tensile force line was found to be vertical in the upper part of the structure and very different from the classical Coulomb's plane observed behind retaining walls (Fig. 21.7). This is due to the presence of horizontal and practically inextensible metallic strips used for reinforcement, which, by restraining lateral displacements, completely change the pattern of strains and stresses in the soil. This was demonstrated by Bassett and Last (1978) using the Cambridge strain failure criterion. According to this theory, the potential failure lines coincide with the zero-extension lines.

In the case of a Reinforced Earth mass, the first family of zero extension lines (α) coincides with the direction of the horizontal strips, while the second family (β) is vertical and corresponds to the direction of the potential failure plane. This is quite different from what exists in the case of a conventional retaining wall, where the zero extension lines are inclined at $\pm[(\pi/4 + \phi/2)]$ with respect to the horizontal plane (Fig. 21.8).

Recent experimentations of walls reinforced with polyester strips (John et al., 1983) and recent finite-element studies seem to indicate that the line of maximum tension tends to move

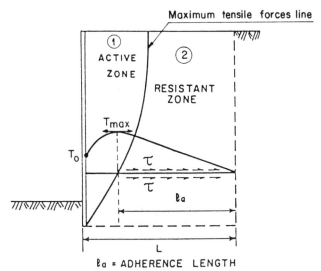

Fig. 21.7 Line of maximum tensile forces, and active and resisting zones in a Reinforced Earth wall.

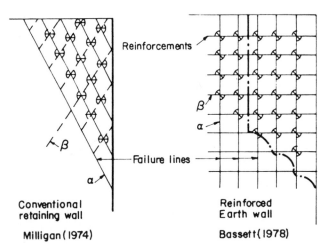

Fig. 21.8 Zero-extension lines and potential failure planes in the cases of conventional retaining wall and Reinforced Earth wall. (*After Bassett and Last, 1978.*)

back toward the Coulomb failure wedge when more extensible strips are used, which confirms the theory outlined above.

21.4.3 Friction Between Soil and Inclusions

The mechanism of friction along linear inclusions such as Reinforced Earth strips involves a three-dimensional mechanism that cannot take place in the case of a two-dimensional reinforcement.

Figure 21.9a illustrates the mechanism of soil inclusion interaction in a dilatant soil. The pull-out of an inclusion induces shear displacements in a zone of the surrounding soil. The volume of this zone is significantly increased by the presence of ribs on the surface of the strips. In a compacted granular soil this sheared zone tends to dilate. However, the fact that the corresponding volume change is restrained by the surrounding soil results in an increase $\Delta\sigma_v$ in the normal stresses applied to the strip.

The importance of the dilatancy effect led to the consideration of an apparent friction coefficient μ^* (sometimes denoted f^*), which is defined as the ratio of the maximum shear stress along the inclusion to the initial normal stress σ_0, which, for design purposes, can be approximated by the overburden pressure γz ($\gamma =$ unit weight of the fill, $z =$ depth of strip). This apparent coefficient is highly dependent on the dilatancy

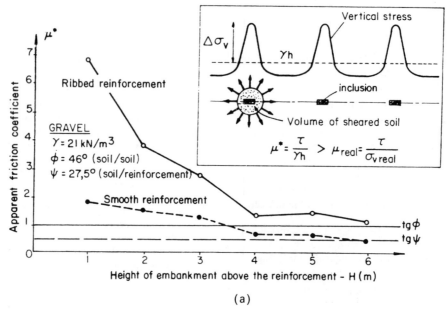

(a)

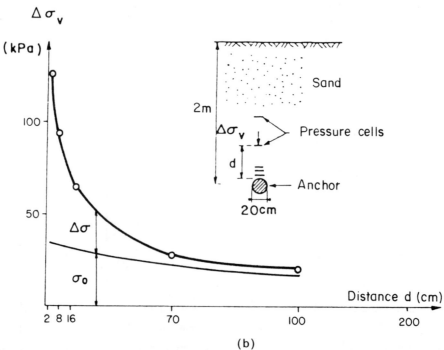

(b)

Fig. 21.9 Restrained dilatancy effect on linear inclusion soil friction. (a) Restrained dilatancy effect mechanism. (*After Schlosser and Elias, 1978.*) (b) Normal pressure measurements around a tensioned inclusion. (*After Plumelle, 1984.*)

behavior of the soil. It can reach values much larger than the soil–metal friction coefficient tan ψ, or even than the soil–soil friction tan ϕ (ϕ: internal angle of friction of the fill).

The apparent coefficient of friction μ^* can be written as

$$\mu^* = \mu \frac{\sigma_0 + \Delta\sigma_v}{\sigma_0} \qquad (21.3)$$

where μ is the true coefficient of friction between soil and reinforcement (no dilatancy effect) and $\Delta\sigma_v$ is the increase of normal stress on the reinforcement due to the restrained dilatancy.

Figure 21.9b presents the results obtained from measurements of $\Delta\sigma_v$ around a linear inclusion during pull-out.

Various parameters significantly affect the value of the apparent coefficient of friction μ^*. Available information on these factors has been reviewed by Schlosser and Elias (1978), McKittrick (1979), Mitchell and Schlosser (1979), and Schlosser and Guilloux (1981). We provide here a list and short discussion of the most influential parameters.

(1) *Density of the backfill.* The restrained dilatancy effect mentioned above can only take place in dense granular soil. The denser the backfill the greater the μ^* value; hence the importance of a good compaction.

(2) *Surface characteristics of the strips.* The dilatancy effect in a given soil can be enhanced by forcing more material to be sheared during the pull-out of the strip. In the case of high adherence (H.A.) strips this is achieved by the ribs. Another very interesting effect of shearing a larger quantity of soil lies in the fact that the peak, which occurs for a slightly larger displacement, is much flatter. This justifies the use of the maximum value of μ^* for ribbed reinforcements and the residual value for smooth reinforcements in the design of reinforced earth structures (Fig. 21.10).

(3) *Overburden pressure.* For a given density, a soil will be less dilatant if the confining pressure is higher. This explains why the favorable effect of dilatancy on the apparent coefficient of friction decreases when the average overburden stress (γH) increases (see Guilloux et al., 1979; Schlosser and Guilloux, 1981). In Figure 21.11, presenting the results of a large number of pull-out tests performed on actual structures, the importance of this parameter is clearly shown. Practical values used for design may be found in Section 21.5.

(4) *Backfill type.* The importance of the type of backfill used, and especially of the grain-size distribution, was studied at the Laboratoire Central des Ponts et Chaussées, Paris (1970–74). Two types of soil were tested: an artificial soil made from a mixture of glass balls and powdered clay, and mixtures of natural soils. Saturated soil samples with varying amounts of fines were tested in a direct shear box at a deformation rate of 1.2 mm/min. The results indicate that the most important parameter is the relative volume of the fine-grained portion to the granular portion. The internal friction angle decreases when the fine-grained portion increases (Fig. 21.12). This can be explained as follows.

1. As long as the fine-grained portion is small, the number of contacts between the grains of the granular skeleton is not significantly affected. Therefore, the value of the internal friction angle remains constant.
2. From a critical value of the relative volume of fines (characterized by the granular void ratio, e_g), the number of contacts between the grains of the skeleton and consequently the internal friction angle decreases.
3. When there is no more contact between the grains of the granular skeleton, the value of the internal friction angle tends to zero and the soil becomes purely cohesive.

Results have further shown that the grain size differentiating the fine-grained portion from the granular portion is 15 μm. Specifications resulting from these studies are presented in Section 21.6.

(5) *Water content of the backfill.* While the water content is only of minor importance when considering the friction characteristics of a coarse granular backfill, its influence may become essential when the portion of fines increases. Figure 21.12 shows the influence of an increase of the water content from OPN (optimum water content) to saturation on the pull-out resistance of a strip in very clayey gravel. The maximum pull-out strength for the saturated soil is reduced to approximately one-third of that determined at OPN.

21.4.4 Durability and Strip Material Choice

The long-term behavior of Reinforced Earth structures is governed by the durability of the strips. Based on mechanical criteria, the obvious choice for the strip material is steel.

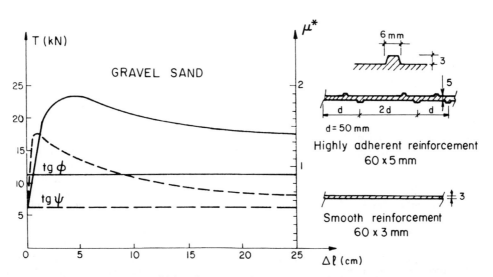

Fig. 21.10 Pull-out tests on high-adherence and smooth strips. (*After Schlosser and Elias, 1978.*)

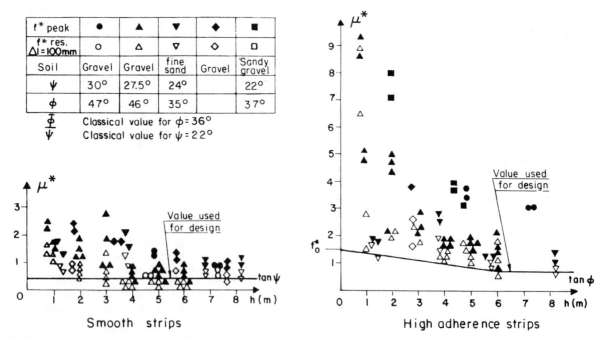

Fig. 21.11 Variation of the apparent coefficient of friction μ^* with respect to the overburden pressure. (*After Schlosser and Elias, 1978.*)

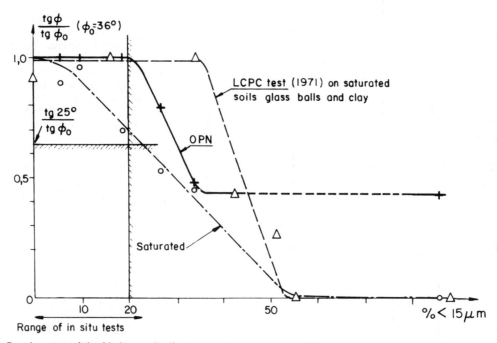

Fig. 21.12 Development of the friction angle of clayey sand compacted at OPN and saturated. (*After Schlosser and Long, 1975.*)

In order to avoid the disadvantage that steel corrodes, other materials have been experimented with since the invention of Reinforced Earth.

In 1966, the first experimental Reinforced Earth wall using fiberglass-reinforced plastic strips and facing units was built. Unfortunately, the plastic material was attacked by bacteria and the wall was destroyed within 10 months. This revealing experience demonstrated the necessity of carefully studying the long-term behavior of the strips and was the starting point of research programs on the durability of various materials under buried conditions. We present below a summary of the results of these research programs.

Organic Materials Although they may maintain a good visual aspect these materials do degrade with time. Many different chemical and physical phenomena, each with its own kinetics, contribute to the degradation. The decay of the mechanical properties of the structural element is the net result of these multiple aging phenomena, which take place concurrently. This superposition of phenomena makes it difficult to predict the long-term properties of a plastic material, since they are very dependent not only on the type of organic material but on the physical and chemical environment and on the stress history. While one particular phenomenon, with its specific time rate, may be critical during laboratory tests, a completely different one may govern over the long term. The usefulness of accelerated tests in such a context is very uncertain.

During the degradation process, physical characteristics may vary greatly; for example, ductile members tend to become brittle, which for Reinforced Earth application is not acceptable. Yet another difficulty lies in the fact that mechanical properties are not simply dependent on physicochemical parameters, but that threshold effects may confound any attempt to extrapolate the evaluation of mechanical properties with time.

Hence, while organic material will probably become a good choice in the future, our knowledge is at present totally inadequate for their use with a good long-term safety.

Metals: Generalities on Corrosion   Corrosion is an electrochemical reaction in which metallic atoms lose one or several electrons at the anode and change into ions that may or may

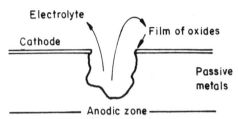

Fig. 21.13   Pitting corrosion in passive metals.

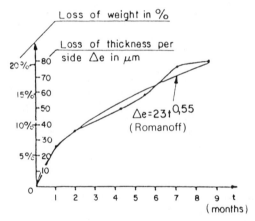

Fig. 21.14   Observed and theoretical corrosion during a full-scale test. (*After Guilloux and Jailloux, 1979.*)

not go into solution. The free electrons then flow in the metal to the cathodic zone where they can react either with water to form negative ions or with positive ions, depending on the pH and on the environment. Many different ions can be involved in the process, but there will always be a flow of electrons in the metal and a change from metal to oxide at the anode.

Corrosion under buried conditions may take various forms depending on the soil characteristics and on the metal (see Romanoff, 1957; Haviland et al., 1968; Gourmelon et al., 1975). The governing properties of the fill are as follows: (1) A free-draining material is usually not very corrosive, and is preferred for use in earth reinforcement in conjunction with metal strips; (2) pH controls the type of cathodic reaction and significantly affects the corrosion rate. Also, the ion content of the soil (SO_4^{2-} and Cl^- ions, especially) has been found to greatly influence the rate of corrosion.

Passive Metals   Metals in the passive state, such as aluminum and stainless steel, exhibit a rather high initial susceptibility to corrosion but are protected from further attack by a thin layer of oxide. This layer, under a given environment, will be very stable. However, this will remain true only for a given range of pH values that depends on the nature of the metal. Some ions, like Cl^-, are also notorious for destabilizing the protective layer of oxides. When this happens at a localized point, corrosion occurs on a very restricted zone, resulting in the swift formation of pits or crevices (Fig. 21.13). The exact conditions for pitting to occur are difficult to determine with accuracy. What is certain is that this form of corrosion, once started, is very destructive, since, in the case of a reinforcing strip, only a few pits can reduce the tensile resistance to nearly zero. Consequently, it is advisable not to use passive metals.

Black Steel   With black steel, corrosion will almost certainly occur. However, owing to the large area affected, it will proceed in a rather homogeneous manner, with no or only shallow pitting. In this case, the mechanical strength of the strip decreases regularly, and the decrease is approximately proportional to the weight of the corroded steel. Extensive studies have been carried out to study the corrosion rate of black steel in soils of various compositions. The results show that there is always a protective effect of the corrosion by-products, leading to a progressive reduction in the speed of corrosion (Romanoff, 1957).

Figure 21.14 shows the results of a corrosion test conducted on a full-scale wall compared with a corrosion law first proposed

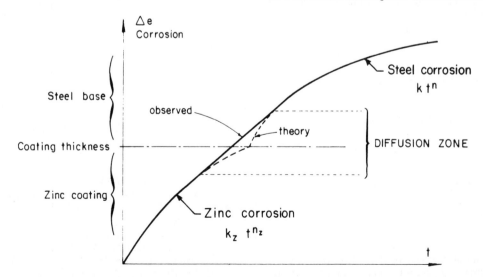

Fig. 21.15   Corrosion with time of a galvanized steel strip under buried conditions. (*After Schlosser and Bastick, 1985.*)

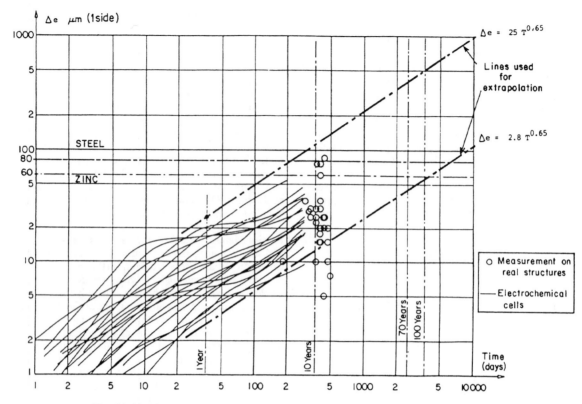

Fig. 21.16 Observed corrosion from various tests. (*After Schlosser and Bastick, 1985.*)

by Romanoff after a large series of tests performed under buried conditions.

Galvanized Steel Galvanized steel is protected by a layer of zinc, which has a low rate of corrosion. The superiority of this coating derives from the fact that zinc, being more electronegative than iron, will always be attacked first. If a scratch exposes an area of bare steel, an electrochemical cell will form between the zinc and the iron in which the zinc is the anode and is corroded while the iron acts as the cathode. This phenomenon is called *cathodic protection* and prevents any pitting. Studies have shown that even after the zinc has been completely corroded, the zinc oxide still plays a protective role by inhibiting some of the chemical reactions, thereby decreasing the corrosion rate. Figure 21.15 shows the corroded thickness as a function of time for a galvanized steel strip under buried conditions.

Test results from the Terre Armée research program, which has been going on for over a decade, are presented in Figure 21.16. This program includes both analysis of strip samples extracted from actual structures and laboratory tests performed in electrochemical cells.

As can be seen, the general trend observed is more or less a straight line, in the log–log scale, which confirms the damping of the corrosion rate with time.

21.5 DESIGN METHODS

21.5.1 Generalities

The design of a Reinforced Earth structure is usually divided in two phases: external or overall stability and internal design.

External design first takes into account a few preliminary dimensioning rules, specific to Reinforced Earth, such as the length of strip to height ratio (L/H). Then, the safety against three classical modes of failure is checked: sliding on the base, overturning, and bearing capacity. These verifications are conducted along the same lines as for other types of structures (e.g., concrete gravity walls). It should be noted, however, that thanks to the flexibility of the Reinforced Earth mass, the safety factor with regard to bearing capacity failure can be taken as 2 instead of the 3 required for more rigid structures.

Internal design is usually performed using a local equilibrium method, where each strip layer is verified independently. The two possible modes of failure, breakage of the reinforcement and lack of adherence are verified. This will be detailed in the following paragraphs.

In some difficult cases, internal and external stabilities may largely interfere. This is often encountered, for example, in the case of superimposed or stepped walls and walls founded on unstable slopes. In such cases, global limit equilibrium procedures derived from Bishop's modified method and adapted to take into account the presence of reinforcements are used.

21.5.2 Walls

In order to ensure that the behavior of the structure conforms to the representation on which the design method is based, it is necessary to satisfy some geometrical conditions. The most important of these conditions is the L/H ratios, for which a minimum of 0.5 has been set. This value results from the large experience now available and on full-scale test results. Other conditions should also be satisfied. However, space does not allow us to detail them.

The embedment depth at the toe of the wall should be in any case at least equal to 0.4 m. In order to prevent bearing capacity failures, the embedment depth should also be greater than $H/20$, where H is the height of the wall (wall on a flat foundation).

Safety Against Strip Breakage At each level, the value of T_{max}, the maximum tensile force in the reinforcing layer, is calculated according to the following expression:

$$T_{max} = K\sigma_v \Delta H \quad \text{(per linear meter of facing)} \quad (21.4)$$

The vertical stress, σ_v, is calculated by considering the equilibrium of the Reinforced Earth mass located above the considered level and subjected to the pressure of the random backfill located behind the wall. The Meyerhof distribution is used to evaluate an average value of σ_v around the point of maximum tension T_{max}.

The coefficient K is similar to a coefficient of lateral earth pressure. In a wall it decreases from top to toe, as indicated in Figure 21.17, which presents results from several full-scale experiments. For design purposes, the value of K is taken to be equal to K_0 at the top and decreases linearly to K_a at a depth of 6 meters (solid line on Fig. 21.17).

Once the tension is calculated for each layer, the minimum number of strips per linear meter of facing (n), is computed according to the following formula:

$$T_{max} = \frac{1}{FS_b} n R_T \quad (21.5)$$

where R_T is the tensile resistance of a unit strip, and FS_b the factor of safety against a breakage failure. Depending on the sensitivity of the structure and on the country code, FS_b may range from 1.5 to 1.65 or more.

Safety Against Lack of Adherence The line of maximum tensile force presented in Section 21.4 (Fig. 21.7), can be schematically represented as shown in Figure 21.18a. This shape is a conservative envelope of the actual positions observed in numerous experiments, both reduced and full-scale, and confirmed by finite-element studies. Similarly, the variation of

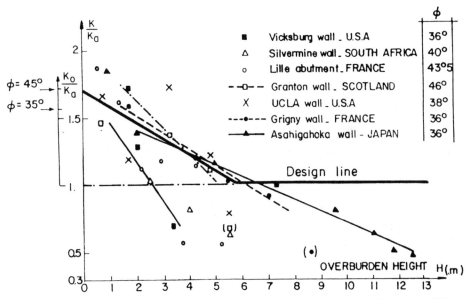

Fig. 21.17 Variation of K for several instrumented walls. (*After Mitchell and Schlosser, 1979.*)

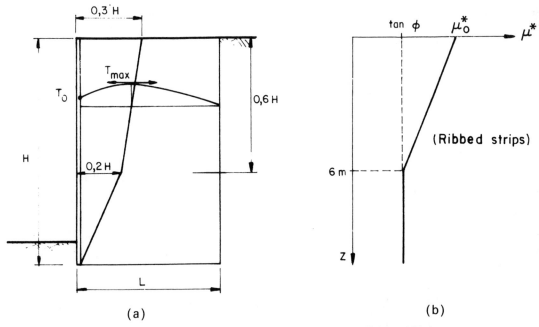

(a) (b)

Fig. 21.18 Schematic tensile force line and apparent coefficient of friction.

the apparent coefficient of friction μ^* (discussed in Section 21.4) as a function of depth may be approximated as shown in Figure 21.18b.

The resistance R_F of a unit strip against pull-out may be calculated as:

$$R_F = \int_d^L 2b\mu^* \gamma z \, dx \qquad (21.6)$$

where d and L are as in Figure 21.18, b = width of a strip, and γz = overburden pressure. The safety against pull-out failure is then ensured if the following formula is satisfied:

$$T_{max} < \frac{1}{FS_a} n R_F \qquad (21.7)$$

where FS_a is the factor of safety against a failure by lack of adherence. Depending on the sensitivity of the structure and on the country code, FS_a may range from 1.35 to 1.5 or more.

21.5.3 Abutments

In the case of an abutment, or of a heavily surcharged wall, the Reinforced Earth mass fulfills two functions for which it must be designed: retaining function and load-bearing function. As in elasticity theory, it is assumed for design purposes that the effects of the two functions can be evaluated separately and then superimposed, as shown in Figure 21.19.

All the aspects relating to the retaining function are identical to those developed in Section 21.5.

Safety Against Strip Breakage At each level, the maximum value of the total tensile force may be evaluated with the same formula as in Section 21.5, in which σ_v is taken as the sum of σ_{v1} due to the retaining function and σ_{v2} due to the load-bearing functions:

$$T_{max} = (T_1 + T_2)_{max} = K(\sigma_{v1} + \sigma_{v2})_{max} \Delta H \qquad (21.8)$$

σ_{v2} is obtained by using the Boussinesq diffusion formula on the applied load distribution. In order to take into account the limiting of the diffusion by the facing, the superimposition principle is once again used, as shown in Figure 21.20. The facing is assumed to undergo very little movement and can therefore be considered as an axis of symmetry for the problem, including the mirror image of the wall and the surcharge.

Safety Against Lack of Adherence Depending on the load intensity, the line of maximum tensile force may either correspond to the retaining behavior scheme (as shown in Fig. 21.21) or it may be located just below the center of the surcharge. In practice, the safety would be checked for both cases.

In the same manner as in Section 21.5 the resistance of a unit strip against pull-out may be computed as:

$$R_F = \int_d^L 2b\mu^*(\gamma z + \sigma_{v2}) \, dx \qquad (21.9)$$

where d = distance between the maximum tensile force line and the facing, L = strip length, b = strip width, γz = overburden pressure, σ_{v2} = vertical stress due to surcharge diffusion.

21.5.4 Sloped Walls

The general scheme for designing a wall with an inclined facing is similar to that of a vertical wall. When the angle of the facing with the horizontal decreases from vertical, the line of maximum tensile forces moves towards the facing and the forces in the strips decrease. At an angle corresponding to ϕ, the angle of internal friction, the active zone is reduced to a line along the facing and the forces in the strips vanish.

Safety Against Strip Breakage The maximum tensile force in a strip layer is computed as in Section 21.5 with the equation:

$$T_{max} = K\lambda\gamma z \, \Delta H \qquad (21.10)$$

where K is computed as shown in Figure 21.22b, λ is a coefficient close to unity taking into account overturning, γ is the unit weight of the fill, z is the depth from the very top of the structure, and ΔH is the vertical spacing between strip layers.

K_α corresponds to the active earth pressure on a sloped plane. K_α and K_α^0 may be computed with the following formulas:

$$K_\alpha = \frac{\sin^2(\alpha - \phi)}{\sin\alpha(\sin\alpha + \sin\phi)^2}$$

$$K_\alpha^0 = \frac{\sin^2(\alpha - \phi)}{\sin\alpha(\sin\alpha + \sin\alpha)} \qquad (21.11)$$

where ϕ = angle of internal friction and α = inclination of the facing on horizontal.

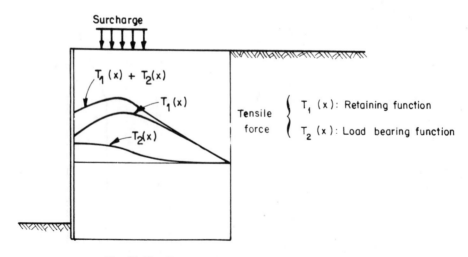

Fig. 21.19 Abutment design: superimposition principle.

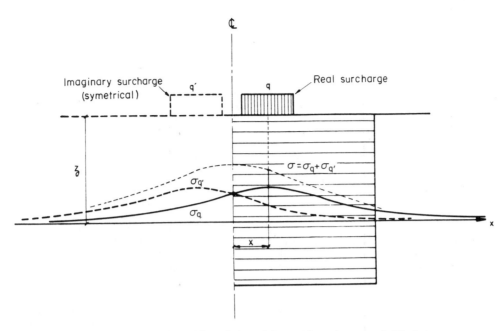

Fig. 21.20 Image load for solution of the problem of truncated diffusion.

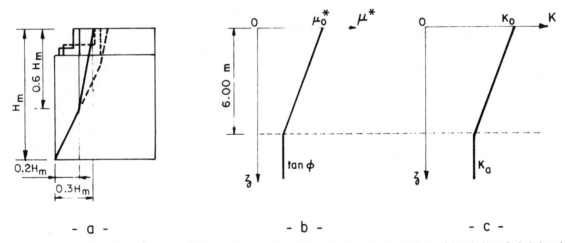

- a -   - b -   - c -

Fig. 21.21 Position of the line of maximum tensile forces in an abutment for (a) various load positions, (b) variation of μ^*, (c) variation of K.

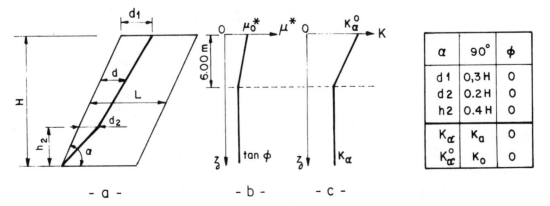

- a -   - b -   - c -

Fig. 21.22 Sloped wall design: line of maximum tensile forces, μ^*, and K values.

Safety Against Lack of Adherence In practice, the line of maximum tensile force is linearly interpolated between the two extreme cases $\alpha = 90°$ and $\alpha = \phi$ as indicated in Figure 21.22(a). The resistance against pull-out for a unit strip is computed with the equation:

$$R_F = \int_d^L 2b\mu^*\sigma_v(x)\,dx \qquad (21.12)$$

where d and L are as shown in Figure 21.22, $b =$ width of strip, and $\sigma_v(x) =$ vertical pressure at a distance x from the facing resulting from an elastoplastic finite element. Analysis charts have been developed by the Reinforced Earth Company.

21.5.5 Rafts

Reinforced Earth rafts may be used in two different circumstances. First, to bridge a sinkhole, where the function of the structure relies mainly on resistance characteristics; second to reduce differential settlements under a construction or a heavy load, where the behavior is controlled by the relative extensibility of the strips and of the soils involved (backfill material and foundation).

In the first case, elastic thin-shell theory has been used to develop practical design methods. In the second instance, only sophisticated elastoplastic finite-element studies, taking in account the parameters of the materials used, lead to a reliable design.

21.6 MATERIAL SPECIFICATIONS

Three components are necessary to build a Reinforced Earth structure: facing, reinforcements, and backfill material.

21.6.1 Facing

The facing has only a limited part in the mechanical behavior of a Reinforced Earth structure, and its primary function is to restrain the soil between the strips and prevent local erosion. However, Reinforced Earth being a flexible material, very often used on relatively soft soil undergoing subsidence, the facing must be very flexible.

The wall should also be architecturally pleasing. The facing is the outer part of the structure and is therefore visible. It is most efficiently made of small prefabricated elements, to ease the construction and reduce the cost.

Two types of facings are used:

1. *Steel or metallic facing.* The basic element is a metallic cylinder of semielliptic section, which is very flexible and stable with respect to the thrust exerted by the backfill. Each individual skin element is very light, which makes it very useful in remote areas where transportation, access, and handling are difficult (Fig. 21.23a).
2. *Concrete facing.* This type, developed later (in 1971), consists of cross-shaped interlocking concrete panels. The joints are designed to allow overall deformations to take place without cracking in concrete. This type of facing has been more widely used because of its versatility; standard panels allow curved walls, and can easily be fitted to ascending tops. Special concrete panels have been developed for sloped or inclined walls and for grassed walls. Concrete facing also allows a wide variety of architectural finishes (Fig. 21.23b).

21.6.2 Reinforcing Strips

The reinforcing strips are the key elements that transmit the load from the active to the passive zone. They should have good adherence to the backfill material (ribbed strips), small deformability (no material undergoing creep), relatively large elongation at break (ductile metal), and good durability. All these considerations led to the choice of ribbed galvanized-steel strips. As a result of the research program conducted on corrosion, standard sacrificial corrosion thicknesses have been defined for a number of environments and service lives. These are presented in Table 21.2. Structures not corresponding to the

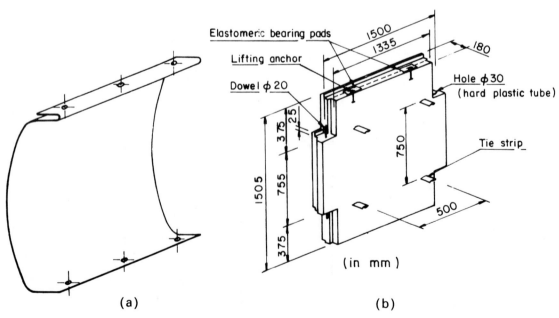

Fig. 21.23   (a) Metallic facing. (b) Concrete facing.

cases presented in Table 21.2 are designed after a specific corrosion study.

21.6.3 Backfill

Two criteria should be taken into account in choosing the backfill material: mechanical (related mainly to adherence) and chemical (related to durability). Numerous tests carried out to study the mechanical criteria, that is, the importance of the grain-size distribution, have led to the following conclusions.

1. The percentage of grains smaller than 15 μm seems to be a very good criterion, since it correlates very well with the measured value of μ^* determined from a large number of experiments.

TABLE 21.2 SACRIFICIAL THICKNESSES TAKEN FOR DESIGN.

	Sacrifice Thickness, mm							
	Black Steel				Galvanized Steel			
Minimum service life, years	5	30	70	100	5	30	70	100
Out of water	0.5	1.5	3.0	4.0	0	0.5	1.0	1.5
Fresh water	0.5	2.0	4.0	5.0	0	1.0	1.5	2.0
Coastal structures	1.0	3.0	5.0	7.0	0	—^{a}	—	—

^{a} — not applicable.

TABLE 21.3 MECHANICAL CRITERIA FOR THE CHOICE OF THE BACKFILL MATERIAL.

Percentage by Weight < 80 μm	
< 15%	Accepted
> 15%	
· and	
Percentage by Weight < 15 μm	
< 10%	Accepted
10% to 20%	
H.A. reinforcements with internal friction angle $\phi > 25°$	Accepted
H.A. reinforcements with internal friction angle $\phi < 25°$	Rejected
Smooth reinforcements with soil–reinforcement friction angle $\psi > 22°$	Accepted
Smooth reinforcements with soil–reinforcement friction angle $\psi < 22°$	Rejected
> 20%	Rejected

TABLE 21.4 ELECTROCHEMICAL CRITERIA FOR REINFORCED EARTH FILL.

	Structures Out of water	Structures in Fresh Water
Resistivity, $\Omega \cdot$cm	$\rho > 1000$	$\rho > 3000$
pH	$5 < \text{pH} < 10$	$5 < \text{pH} < 10$
Cl^-	< 200 ppm	< 100 ppm
SO_4^{2-}	< 1000 ppm	< 500 ppm

2. The grains larger than 15 μm constitute what can be called a frictional skeleton, which is responsible for the internal friction angle of the soil. This value decreases with the fine-grained portion as explained in Section 21.4.
3. The soil–strip adherence is directly dependent on the internal friction angle of the soil.

Practical rules have evolved from these observations, along with quantitative results of numerous measurements. These rules, applicable for Reinforced Earth structures are presented in Table 21.3.

Following the research performed by Terre Armée, guidelines have been set for choosing the backfill as far as corrosion is concerned. These practical rules are described in Table 21.4.

21.7 EFFECTS OF WATER AND DYNAMIC LOADING

21.7.1 Effects of Water

The action of water, or more generally the actions related to water, can result in multiple effects, amongst which the principal factors relating to Reinforced Earth structures are:

- *The buoyancy phenomenon and hydrostatic pressure.* Since the facing is nearly free-draining, very little difference in water elevation should be observed on both sides of the facing, unless the drawdown is very quick. Choice of a backfill with good permeability is highly recommended in partly submersible structures.
- *Seepage forces.* The choice of a coarse backfill will reduce these forces.
- *Variation of* μ^* (the apparent coefficient of friction). Good quality backfill will also provide a high value for μ^* even in wet conditions.

21.7.2 Seismic Behavior

The seismic behavior of Reinforced Earth has been studied since the invention of this technique. Research programs on this topic have included reduced-scale and half-scale models on shaking tables, finite-element studies, and observations on full-scale structures after seismic events. A general conclusion is that, thanks to its flexibility, Reinforced Earth may undergo severe shaking without damage. The distribution of the periodic reinforcement in the mass of the structure leads to a dispersion and a dissipation of the seismic energy, preventing the build-up of concentrated forces.

The effects of an earthquake on the external stability of a Reinforced Earth structure can be evaluated along the same lines as for any other flexible gravity retaining system. Internally, the seismic event will lead to dynamic incremental loads in the strips. The distribution of the corresponding incremental tensile forces is different from that occurring in the static case. However, as long as the seismic intensity is not very high this will have very little influence on the position of the resultant line of maximum tensile forces (see Fig. 21.24). Under strong ground movements the line will tend to move away from the wall facing. The distance from the facing in the upper part of the wall can be represented as:

$$d = (0.3 + 0.5a/g)H \tag{21.13}$$

where a = ground acceleration, g = acceleration due to gravity, and H = wall height.

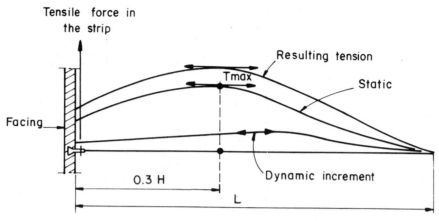

Fig. 21.24 Variation of the tensile force in a strip due to seismic effects.

21.8 APPLICATIONS

21.8.1 Advantages and Main Areas of Application

Since its invention, Reinforced Earth has found wide use in many different areas of civil engineering, owing to:

- The mechanical flexibility of the material
- The use of prefabricated elements (strips, facing) to speed up construction
- The generally lower cost of the method as compared with traditional techniques

Reinforced Earth is able to withstand large deformations with no structural damage. This high flexibility, close to that of an earth embankment, allows its use on unstable slopes or on highly compressible foundation soils.

Since 1968, when the first important project was built, Reinforced Earth has found many different applications. The following examples will give an idea of the versatility of this technique.

- Retaining walls for mountain roads, where poor foundation or unstable slopes exclude any other techniques (first example, Pragnières wall, France, 1965).
- Groups of walls, tier walls or stepped walls for large highway projects (example, Roquebrune–Menton highway, France, 1968–1969, 10 retaining walls on unstable slopes with a total facing area of 5500 m²).

- Retaining walls for express roads, where low cost and fast construction are required (many examples of circular roads, including Madrid (Spain), Atlanta (U.S.A.)).
- Retaining walls for building house supporting terraces in difficult slopes.
- Bridge abutments, where Reinforced Earth flexibility makes it preferred for poor soils, when the schedule is tight, and for low cost (first example, 14 m high bridge abutment of Thionville, France, 1972).
- Various structures for railways, where a good resistance to vibration is required.
- Sea walls, marine walls, and dams, owing to the good resistance of Reinforced Earth against wave action and water erosion.
- Industrial structures—glory-holes or slot storages, protection dykes for LNG storage or fuel reservoirs (first slot storage facility, Reno Junction, Black Thunder Mine, U.S.A., 1977).
- Foundation slabs for protection against soil subsidence due to collapse of underground cavities (example, slab over sinkholes on highway 202, Pennsylvania, U.S.A., 1974).

21.8.2 Structures in Mountainous Areas

Reinforced Earth found wide use in mountainous areas, where the presence of slopes, very often unstable, requires the use of a combination of various techniques. A good recent example of this type of situation is the construction of the A40 freeway,

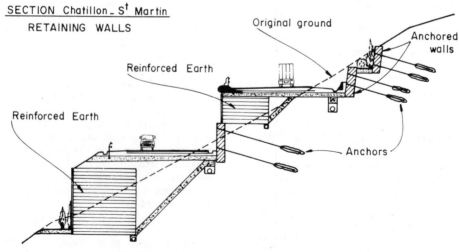

Fig. 21.25 A40 French highway, typical cross section.

which constitutes the last section of the Rome to Paris highway at the point where it crosses the pre-Alps near Geneva. Several sections of the project were to be built on steep colluvium slopes (30° to 40°). Reinforced Earth, soil nailing, and prestressed anchors were used in conjunction. A typical cross section representative of the 50 000 m² of Reinforced Earth is given in Figure 21.25.

21.8.3 Structures on Compressible Soils

Owing to its flexibility and to the relatively uniform pressure at the foundation level, Reinforced Earth may be used on relatively compressible soil where other techniques would require the use of costly deep foundations (Fig. 21.26).

In the case of very soft soils, Reinforced Earth can be used in conjunction with other soil-improvement techniques such as preloading and stone columns (Fig. 21.27).

21.8.4 Other Specific Applications

Although few dams have been built with Reinforced Earth, more and more examples are showing the large savings that can be achieved through the use of this technique. The two most

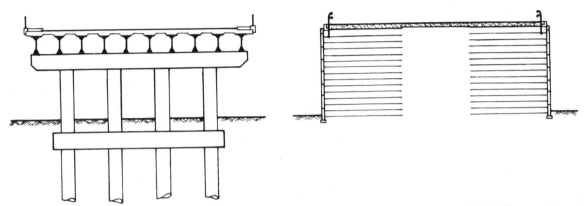

Fig. 21.26 Replacement of a viaduct on piles by a double-faced Reinforced Earth wall on soft soil (Val d'Esnoms, France).

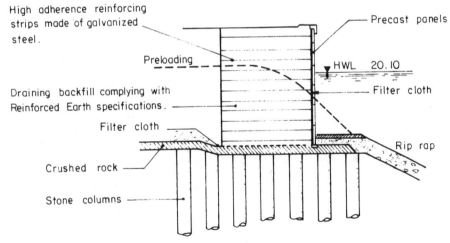

Fig. 21.27 Reinforced Earth used in conjunction with stone columns (Idaho Clark Fork highway, cross section, station 334 + 00).

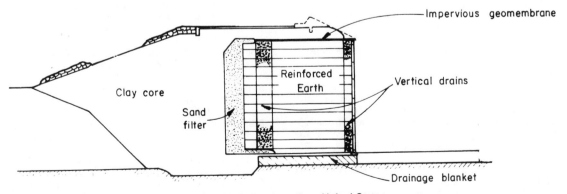

Fig. 21.28 Tailor draw dam, United States.

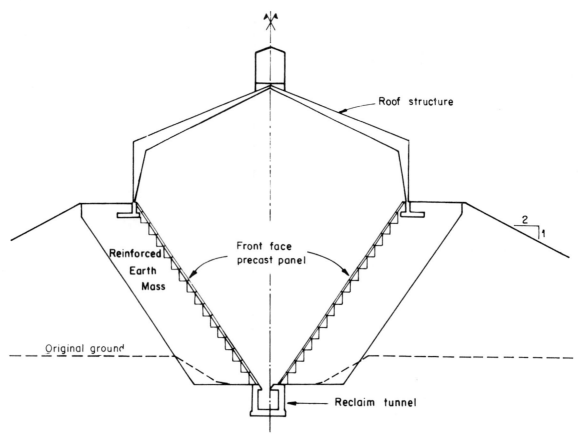

Fig. 21.29 Typical cross section of a slot storage facility.

spectacular advantages are:

1. Savings in earth work volume, which is most important when dams need to be raised
2. Suppression of the spillway, which is always a very expensive and delicate structure (Fig. 21.28)

Among other recent applications of the technique, one of the most spectacular is in slot storage. In these structures, two inclined faced walls create a V-shaped storage space. The bottom of the V usually includes a tunnel for reclaiming the stored material either by conveyor belt of by trains (Fig. 21.29).

REFERENCES

Bassett, R. H. and Last, H. C. (1978), Reinforced earth below footings and embankments, *Symposium on Earth Reinforcement*, ASCE, Pittsburgh, pp. 202–231.

Gourmelon, J. P., Preynat, J. P., and Raharinaivo, A. (1975), Durabilité des ouvrages, *Dimensionnement des Ouvrages en Terre Armée Murs et Culées de Ponts*, Collection organized by the Association Amicale des Ingénieurs Anciens Elèves de l'Ecole Nationale des Ponts et Chaussées, pp. 97–105.

Guilloux, A. and Jailloux, J. M. (1979), Essai de rupture en vraie grandeur d'un mur en terre armée par corrosion accélérée, *International Conference on Soil Reinforcement: Reinforced Earth and Other Techniques*, ENPC–LCPC, Paris, **2**, pp. 503–508.

Guilloux, A., Schlosser, F., and Long, N. T. (1979), Etude du frottement sable-armature en laboratoire, *International Conference on Soil Reinforcement: Reinforced Earth and Other Techniques*, ENPC–LCPC, Paris, **1**, pp. 35–40.

Haussmann, M. R. (1976), Strength of reinforced soil, *Australian Road Research Board*, Session 13, **8**, pp. 1–8.

Haviland, J. E., Bellair, P. J., and Morrel, V. (1968), Durability of corrugated metal, *Highway Research Record* No. 242, Washington D.C.

John, N. W. M., Ritson, R., Johnson, P. B., and Petley, D. J. (1983), Instrumentation of reinforced soil walls, *Proceedings, 8th ECSMFE*, Helsinki, **2**, pp. 509–512.

Laréal, P. and Bacot, J. (1973), Etude sur modèles réduits bi-dimensionnels de la rupture de massifs en terre armée, *Travaux*, No. 463.

Madani, C. (1979), Etude du mécanisme interne et du comportement dynamique de la terre armée à l'appareil triaxial, Doctorate Thesis, Ecole Nationale des Ponts et Chaussées, Paris.

McKittrick, D. P. (1979), Reinforced earth: application of theory and research to practice, *Ground Engineering*, **12**, No. 1, pp. 19–31.

Mitchell, J. K. and Schlosser, F. (1979), Mechanism, behavior and design methods for earth reinforcement, General Report, *International Conference on Soil Reinforcement: Reinforced Earth and Other Techniques*, ENPC–LCPC, Paris, **3**, pp. 23–74.

Plumelle, C. (1984), Improvement of the bearing capacity of soil by inserts of group and reticulated micropiles, *International Conference on In-situ Soil and Rock Improvement*, Paris, pp. 83–89.

Romanoff, M. (1957), *Underground Corrosion*, Circular No. 579, N.B.S., Washington, D.C.

Romstad, K. M., Al-Yassin, Z., Herrman, L. R., and Shen, C. K. (1978), Stability analysis of Reinforced Earth retaining structures, *Symposium on Earth Reinforcement*, ASCE, Pittsburgh, pp. 685–713.

Schlosser, F., Long, N. T., Guegan, Y., and Legeay, G. (1972), Etude de la Terre Armée à l'appareil triaxial, *Rapport de recherche*, No. 17, LCPC, Paris.

Schlosser, F. and Long, N. T. (1975), Choix de matériau de remblai, *Dimensionnement des Ouvrages en Terre Armée Murs et Culées de Ponts*, Collection organized by the Association Amicale des Ingénieurs Anciens Elèves de l'Ecole Nationale des Ponts et Chaussées, pp. 141–148.

Schlosser, F. and Elias, V. (1978), Friction in reinforced earth, *Symposium on Earth Reinforcement*, ASCE, Pittsburgh, pp. 735–763.

Schlosser, F. and Guilloux, A. (1981), Le frottement dans le reinforcement des sols, *Révue Française de Géotechnique*, No. 16, pp. 66–77.

Schlosser, F., Jacobsen, H. M., and Juran, I. (1983), Soil reinforcement, General Report, *Proceedings, 8th ECSMFE*, Helsinki, **3**, pp. 83–104.

Schlosser, F. and Bastick, M. (1985), Reinforced Earth: new aspects and new applications, *3rd International Geotechnical Seminar on Soil Improvement Methods*, Singapore, pp. 273–284.

Schlosser, F. and Delage, P. (1987), Reinforced soil retaining structures and polymeric materials, *NATO Advanced Research Workshop on the Application of Polymeric Reinforcement in Soil Retaining Structures*, Kingston, Canada, NATO ASI Series E147, published in 1988, pp. 3–65.

Yang, Z. (1972), Strength and deformation characteristics of reinforced sand, Ph.D. Thesis, University of California, Los Angeles.

22 GEOSYNTHETICS IN GEOTECHNICAL ENGINEERING

ROBERT M. KOERNER, Ph.D., P.E.
H. L. Bowman Professor of Civil Engineering,
and Director, Geosynthetic Research Institute,
Drexel University

22.1 INTRODUCTION

Geosynthetics are a rapidly emerging family of materials used in geotechnical engineering in a wide variety of applications. They are almost exclusively polymeric and consist of the following major types (Koerner, 1990):

- Geotextiles
- Geogrids
- Geonets
- Geomembranes
- Geocomposites

When following the concept of "design-by-function" (Koerner, 1984) one must decide on a primary function for the specific application considered and select the appropriate type of geosynthetic; see Table 22.1 for the various options available. It should be noted that within each type of geosynthetic listed in Table 22.1 there exists a tremendous variety of product styles and configurations, which will be described in the sections to follow. Since the literature is abundant on product applications, design concepts will be emphasized throughout.

22.2 GEOTEXTILES

22.2.1 Overview

Geotextiles are porous, flexible polymeric fabrics made to serve one or more of the functions listed in Table 22.1. Most are made from polypropylene or polyester, but specialty situations sometime require other polymers, for example, polyethylene or polyaramide. The basic resins are usually augmented by anti-degradants (such as carbon black) and other fillers and/or additives and made into fibers. These fibers take the shape of monofilaments, monofilament yarns (multifilaments), staple yarns, and slit or split films (or tapes). The fibers are then made into fabrics of which woven and nonwoven styles dominate; see Figure 22.1. Very few are knitted fabrics. The resulting series of options available to a geotextile manufacturer leads to a tremendous variety of available products; see Table 22.2 for a listing of commercially available products. The possibility always exists for developing specialty products as well.

Consideration of the above range of available geotextiles should dispel any notion of a design or specification based on a "product x or equal" concept. No two geotextiles are truly "equal" and design and selection must be based on a rational

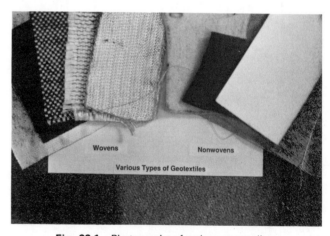

Fig. 22.1 Photographs of various geotextiles.

TABLE 22.1 GEOSYNTHETIC TYPE VERSUS AVAILABLE FUNCTION.

Type	Available Functions				
	Separation	Reinforcement	Filtration	Drainage	Moisture Barrier
Geotextile	P or S	P or S	P or S	P or S	n/a
Geogrid	S	P	n/a	n/a	n/a
Geonet	S	n/a	n/a	P	n/a
Geomembrane	S	n/a	n/a	n/a	P
Geocomposite	P or S	P or S	P or S	P or S	P or S

Note: P = primary function; S = secondary function; n/a = not applicable.

TABLE 22.2 APPROXIMATE NUMBER OF COMMERCIALLY AVAILABLE GEOTEXTILES.

Region	Manufacturers	Number of Types of Available Geotextiles
U.S.A.	50	450 +
Canada	10	70
Europe	30	150 +
Australia/New Zealand	10	40
South and Central America	10	50
Asia	15	70
Africa	5	30
Total	130	860 +

approach. Such an approach is embodied in the design-by-function concept. At the heart of this concept is the formulation of a factor of safety in the traditional engineering manner, that is,

$$F.S. = \frac{\text{allowable (or test) value}}{\text{required (or design) value}} \qquad (22.1)$$

where F.S. must be greater than 1, the actual magnitude depending upon the implication of failure, which is always site-specific.

Regarding the allowable (or test) value for the various properties of geotextiles, there is a large amount of worldwide activity. At least 30 organizations are working on geotextile test methods and standards. In the U.S.A., the American Society of Testing and Materials (ASTM) is the lead organization, which has grouped their activity into physical, mechanical, hydraulic, endurance, and durability categories. Included in each group are index (or comparison and quality control oriented) tests and performance (or design oriented) tests, the latter being preferred for engineering design. Rather than describe all of the available tests, only those relevant to the designs presented in this chapter will be described and referenced. Furthermore, they will be described and explained when they are needed and not as a separate section.

Regarding the required (or design) value in the factor of safety equation, geotechnical engineering procedures will generally be required. In reinforcement problems this will require an analysis of stress and strain, while in hydraulic problems this will require estimates of flow and soil retention considerations. In some cases, altogether new concepts will be required.

The subsections to follow in this geotextile section are written to conform to the major functions that geotextiles can perform; recall Table 22.1. Each will be treated spearately with a descriptive problem illustrating the type of application involved in the utilization of the particular function.

22.2.2 Geotextiles in Separation

While there are many applications where geotextiles can be used to separate two dissimilar materials, their use beneath pavement stone-base courses and above the underlying soil subgrade is very common. The objective is to keep the stone from penetrating into the soil and the soil from intruding into the stone. By so doing, the drainage capability of the stone base is preserved for the lifetime of the pavement. This drainage preservation is a significant feature in pavement lifetime, particularly with fine-grained soil subgrades or in areas of cold weather where freeze–thaw cycling occurs.

Presented in Koerner (1990) are four separate design situations, but only one will be illustrated here. As shown in Figure 22.2, the pressure exerted from the heaviest loaded vehicle will induce stress in the geotextile, causing the underlying soil to protude up into the void created by adjacent stone particles. Thus, there is a tendency for the geotextile to burst in an out-of-plane manner. The following factor-of-safety equation approximates the situation:

$$F.S. = \frac{p_t d_t}{p_a d_v} \qquad (22.2)$$

where

p_t = burst test pressure of candidate geotextile
d_t = diameter of test device
p_a = maximum applied pressure
d_v = diameter of stone aggregate void

Using a Mullen burst test method (ASTM D774) where $d_t = 30.5$ mm and $d_v = 0.4\ d_a$ (where d_a = diameter of the

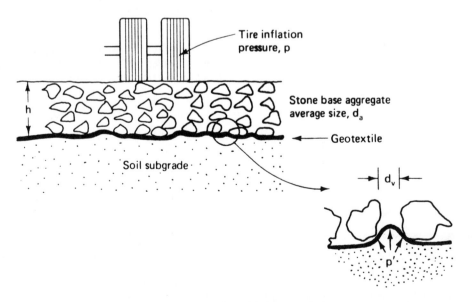

Fig. 22.2 Detail used in burst analysis.

aggregate), a more general equation can be developed:

$$\text{F.S.} = \frac{p_t(30.5)}{p_a(0.4\,d_a)} \tag{22.3}$$

$$\text{F.S.} = \frac{76.2\,p_t}{d_a p_a} \tag{22.4}$$

where d_a must be in millimeters and the units of p_t and p_a must be the same. Thus for 75-mm aggregate, a maximum pressure of 550 kPa and a candidate geotextile having an allowable burst strength of 2000 kPa, the resulting factor of safety is 3.7. Other problems involving grab strength, puncture resistance, and impact resistance can also be formulated. They all illustrate the concept of design by function.

22.2.3 Geotextiles in Reinforcement

Geotextiles having varying degrees of tensile strength can obviously be used to reinforce soil, which is notoriously weak in tension. Geotextile reinforcement of unpaved roads on very weak soil subgrades (e.g., CBR < 2) has nicely illustrated this feature (see Hausmann, 1986, for a review of various analytic techniques). Other areas of considerable activity are geotextile-reinforced walls (Yako and Christopher, 1987) and stabilization of existing slopes (Koerner and Robins, 1986). Nowhere, however, is there greater activity than in the construction of embankments over extremely soft soils. This work, pioneered by the U.S. Army Corps of Engineers, has produced remarkable results (Fowler and Koerner, 1987). River-transported fine-grained soils of near zero shear strength have been used as embankment foundation material when supported by a high-strength geotextile. Often, these embankments are used for subsequently dredged soil containment dikes or for building directly thereon. A recent conference has been directed at this activity (Koerner, 1988).

At the heart of the analysis is a limit equilibrium method modified for the inclusion of a geotextile. As illustrated in Figure 22.3, a traditional slope stability procedure using undrained shear strengths can be used. For moment equilibrium considerations,

$$\text{F.S.} = \sum \frac{\text{resisting moments}}{\text{driving moments}} \tag{22.5}$$

$$\text{F.S.} = \frac{(\tau_e L_{ab} + \tau_f L_{bc})R + T_a Y}{wX} \tag{22.6}$$

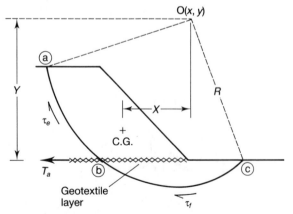

Fig. 22.3 General configuration used to modify slope stability analysis to include a geotextile reinforcement layer.

where

τ_e = shear strength of embankment soil (often neglected owing to lack of confinement or low strength)
L_{ab} = arc length a–b
τ_f = shear strength of foundation soil (usually very low)
L_{bc} = arc length b–c
R = radius from critical center to failure arc
T_a = allowable tensile strength of geotextile
Y = moment arm of geotextile (sometimes taken as R)
w = weight of soil in failure zone
X = moment arm from center of gravity to center of failure arc

It is easily seen in the above equation that the geotextile's strength can be increased as necessary to drive the factor of safety up to an acceptable value. Strengths of up to 500 kN/m have been used to date. Note that these strengths must be tested in a plane strain condition, the closest simulated test being the wide-width tensile test (ASTM D4595-86). This test uses a test specimen 200 mm wide by 100 mm in height and loads it at a constant strain rate until failure. Also, note that fabric creep must be allowed for; thus, the ultimate strength must be somewhat reduced. Considering that a strength gain in the underlying foundation soil will generally occur, a creep factor of safety of 1.5 to 2.5 should generally be adequate. Other considerations that are important to consider in design are sewn-seam requirements, the effect of holes (e.g., when vertical strip drains are installed), manner of fill placement, required fabric modulus, friction between fabric and embankment soil, and anchorage length (Fowler and Koerner, 1987; Koerner, 1988). The point that should be emphasized, however, is that with the advent and use of high-strength geotextiles we can almost "build on water."

22.2.4 Geotextiles in Filtration

There are myriad applications for geotextiles used as filters, for example, underdrains, behind retaining walls, as capillary breaks, etc. Geotextiles used in soil as filters for liquids must fulfill two mutually contradicting requirements. The first is that the fabric voids must be sufficiently open to allow the liquid to pass through without building excess pore water pressure, while the second is that the fabric voids must be sufficiently tight to prevent excess loss of upstream soil particles. Superimposed upon both is the requirement that the soil must not clog the fabric, thereby blocking flow. The first requirement of adequate flow is handled by forming a factor of safety in the form of a permittivity comparison, that is,

$$\text{F.S.} = \frac{\psi_{\text{allow}}}{\psi_{\text{req}}} \tag{22.7}$$

where permittivity ψ is defined as

$$\psi = \frac{k_n}{t} \tag{22.8}$$

and k_n is coefficient of permeability normal to the fabric, and t is the fabric thickness. This term is necessary owing to the sensitivity of fabric thickness, which varies under applied normal load, hydraulic gradient, etc. The fabric's allowable permittivity value is obtained from a laboratory permeability test (ASTM D4491), either with the fabric unloaded or, better, loaded; see Table 22.3 for typical values. The required permittivity value is estimated or designed using a form of Darcy's equation. The latter approach is generally preferred where flow net techniques are often required; see Koerner (1990) for examples.

TABLE 22.3 TYPICAL PERMITTIVITY AND PERMEABILITY VALUES OF GEOTEXTILES.

Fabric Type	Permittivity (s^{-1})	Permeability (cm/s)
Woven, monofilament	1000 to 0.1	10 to 0.01
Nonwoven, needle-punched	50 to 0.1	1 to 0.01
Nonwoven, heat-set	10 to 0.1	0.1 to 0.005
Nonwoven, resin-bonded	1 to 0.005	0.05 to 0.001
Woven, slit film	1 to 0.01	0.01 to 0.001

The second mechanism of soil retention is handled by comparing the fabric's opening size to the size of the soil to be retained. Some form of the following relationship is usually used:

$$O_f = \lambda d_s \qquad (22.9)$$

where

O_f = opening size of the fabric (often taken as the 95 percent opening size)

d_s = diameter of soil to be retained (often taken as the 85 percent finer size)

λ = function of soil gradation, soil density, liquid type, hydraulic gradient, etc.

In its simplest form, Carroll (1983) has suggested,

$$O_{95} < (2 \text{ or } 3) \times d_{85} \qquad (22.10)$$

However, numerous other more detailed approaches are also available; see Bertacchi and Cazzuffi (1985) in this regard.

The third consideration is that undue clogging (short-term or long-term) of the geotextile must not occur. In discussing clogging one must consider how a geotextile filter works. Essentially, the geotextile represents a catalyst that is intended to force the upstream soil to do its own filtering. Obviously, some of the fine soil particles directly against the fabric will be lost through, or within, the fabric, but the amount must not be excessive. During this action the soil should be "tuning" itself to come into equilibrium with the applied flow regime. Many postulated mechanisms have been presented (McGown, 1978). As a check, laboratory testing may also be performed. Two options are currently available. One is the gradient ratio test (Haliburton and Wood, 1982), which measures the hydraulic gradient of flow through the fabric plus 25 mm of soil and compares this value to the hydraulic gradient through 50 mm of soil by itself. If this ratio is greater than 3.0, the candidate fabric–soil combination is not compatible. If it is less than 3.0, clogging should not occur. The test was originally developed to evaluate cohesionless soils and woven monofilament fabrics. For other soils and/or fabrics, the test is not well behaved (Halse et al., 1987). This leads to the second option for evaluating soil–fabric compatibility, which is the long-term flow test (Koerner and Ko, 1982). Here the site situation is simulated as closely as possible, that is, in terms of soil type, fabric, hydraulic conditions, etc., and long-term flow is measured in constant-head column tests. Typical response curves are shown in Figure 22.4. The initial decrease in flow is due to soil densification and is not meaningful. From the transition time on, however, the response is very significant. If the flow rate continues to decrease, clogging is occurring and the situation is not acceptable. If, however, the flow rate stabilizes, the soil–fabric combination has accommodated the flow regime and hydraulic equilibrium has been achieved. Intermediate situations call for continued testing, which can take up to 10 000 hours (approximately one year).

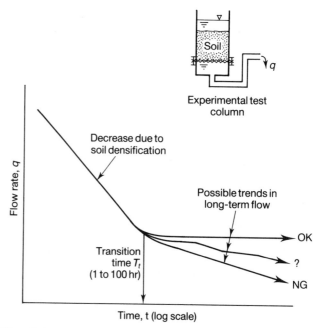

Fig. 22.4 Flow rate response curves for assessing soil–fabric clogging potential.

22.2.5 Geotextiles in Drainage

When geotextiles are used as drains as in chimney drains, fin drains, embankment drains, etc., the flowing liquid moves within the plane of the fabric. All fabrics possess this capability but to widely varying degrees. See Table 22.4, where the term *transmissivity* (as described below) will be used to describe in-plane flow. For reference purposes, the traditional permeability coefficient is also given in Table 22.4.

$$\theta = k_p t \qquad (22.11)$$

where

θ = transmissivity

k_p = planar coefficient of permeability

t = fabric thickness

The design of geotextile drains follow along similar lines to that for geotextile filters. The specific elements are adequate flow, soil retention, and soil–fabric compatibility. The only difference is in the first part, adequate flow, which is determined on the basis of the previously defined transmissivity, and is used as follows:

$$\text{F.S.} = \frac{\theta_{\text{allow}}}{\theta_{\text{req}}} \qquad (22.12)$$

where

θ_{allow} = allowable (or test) value of the geotextile

θ_{req} = required (or design) value

The allowable transmissivity value is determined from a laboratory test of either radial or planar in-plane flow configuration (Koerner and Bove, 1983). The results are strongly dependent on the applied normal pressure and can be made to simulate in-situ conditions quite closely (ASTM D4716). As suggested from Table 22.4, considerable data is available. The value of required transmissivity (the denominator of Equation 22.12) is determined by geotechnical design methods that usually require use of Darcy's formula, either directly or

TABLE 22.4   APPROXIMATE RANGE OF VALUES FOR GEOTEXTILE TRANSMISSIVITY.

Fabric Type	Transmissivity $(m^3/s \cdot m)$	Permeability Coefficient (cm/s)
Woven, slit film	1.5×10^{-9}	0.001
Nonwoven, heat-set	3.0×10^{-9}	0.002
Woven, monofilament	2.0×10^{-8}	0.015
Nonwoven, resin-bonded	7.0×10^{-8}	0.02
Nonwoven, needled (thin)	2.0×10^{-6}	0.3
Nonwoven, needled (medium)	10.0×10^{-6}	0.8
Nonwoven, needled (heavy)	20.0×10^{-6}	1.0

by means of a flow net. Other design guides may also be used (Koerner, 1990).

22.2.6.   Geotextiles as Moisture Barriers

By infiltrating the voids of a geotextile with bitumen, polymer, or other filter, a moisture barrier can be created. This procedure has advantages in certain applications but rarely creates a complete geomembrane of the type to be discussed in the geomembrane section. In applications where the properties of a geotextile are important (e.g., tensile strength, puncture resistance, impact resistance, etc.) and yet some moisture release from the system is permitted, the technique has been utilized. These include water-reservoir liners, where some leakage is tolerable, and membrane-encapsulated soil layers (MESLs). This latter concept is used where a moisture-sensitive subgrade or subbase soil is fully encapsulated (bottom, sides, and top) within a bitumen-infilled geotextile. The preservation of the as-placed moisture content of the encapsulated soil is maintained, thereby providing temporary stability (Koerner, 1990).

Other than for these relatively limited applications, however, the function of a geosynthetic moisture barrier is best provided by a geomembrane that will be discussed separately later in the chapter.

22.3   GEOGRIDS

22.3.1   Overview

Geogrids are deformed or nondeformed netlike polymeric materials used in geotechnical engineering-related construction activities for reinforcement; see Figure 22.5. Their primary function is reinforcement; however, they sometimes can be used for separation of large-sized particles as well. It should be noted that geogrids are *not* geonets, which are used exclusively as drainage cores and will be treated separately in the next section.

There are a wide variety of manufacturing approaches used to make geogrids and hence their final shapes vary considerably. Furthermore, the area is quite active and new products are appearing regularly. The original geogrids utilize a geomembrane sheet with holes punched into it at regular spacings. The sheet is then cold-worked over a series of rollers that are successively moving faster, stretching the product as it travels along. The original holes become ellipses when the final product, with approximately an 8-to-1 draw ratio, is completed. Thus, the polymer, which is high-density polyethylene, is mechanically drawn well beyond its yield point, thereby providing enhanced modulus, strength, and creep resistance. A related product is biaxially deformed, providing a geogrid with balanced strength

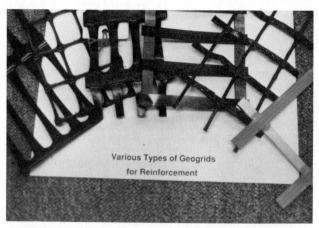

Fig. 22.5 Photographs of commercially available geogrids.

properties in two perpendicular directions. A similarly manufactured product is made by mechanically drawing the punched sheets by gripping the transverse ribs and elongating the longitudinal ribs.

Geogrids are also available from bonding of mutually perpendicular high-strength polymer strips together at their cross-over points, or nodes. One company uses high-tenacity polyester strips, which are ultrasonically bonded at their nodes, while a different firm's are melt-bonded at their nodes. In this latter case it is the polypropylene sheathing covering the high-tenacity polyester fibers that is melt-bonded.

A number of different geogrids are also manufactured by entangling high-strength polyester yarns at their intersections (or nodes), thus providing a gridlike material. A surface coating is used to maintain the gridlike shape. The actual processing can be done in a variety of ways.

22.3.2   Properties of Geogrids

Since the primary function of geogrids is in soil reinforcement, their tensile strength plays a critical role. This strength is usually assessed in a wide-width test or on the basis of individual rib strength if the ribs are spaced widely apart. The result is best expressed in force per unit width dimensions, which is the necessary value for plane-strain-related problem solutions. Note, however, that this ultimate value must be reduced for long-term creep considerations by a suitable factor of safety. This value is between 2.0 and 4.0 depending upon the type of polymer, manufacturing style, design lifetime, and criticality of structure. Installation damage and long-term degradation should also be considered.

Fig. 22.6 Geogrids with stone embedded within structure.

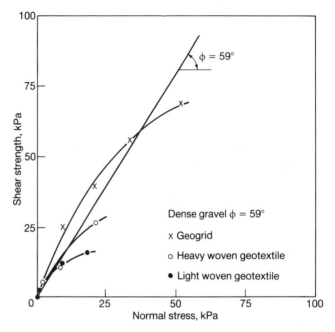

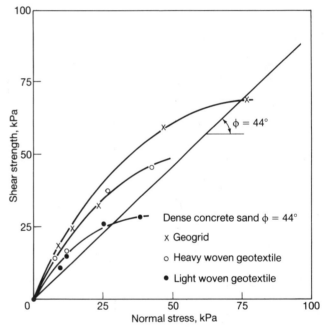

Fig. 22.7 Anchorage (pullout) behavior of geogrids vs. granular soils compared to several geotextiles.

Also of significance is junction, or node, strength. This is due to the functioning of the geogrid when it is being stressed in a soil system. Since the soil has complete strike-through of the geogrid's apertures (see Fig. 22.6), a portion of the resisting mechanism is bearing resistance against the geogrid's transverse ribs. (There is also friction resistance along the surface of the geogrid, but this is greatly product-dependent.) The stress in the transverse ribs is transferred to the longitudinal ribs, where it resists the imposed stresses, for example, at a wall facing panel. This stress must be transferred through the junctions or nodes, hence their importance.

Since bearing capacity of soil against the transverse ribs is a major resisting mechanism when using geogrids for reinforcement, this feature should be evident in anchorage or pullout tests. Figure 22.7 shows the performance of reinforcement geogrids relative to geotextiles and soil by itself. The enhanced behavior over the soil by itself and several geotextiles is readily seen. Note, however, that this behavior is for granular soils. When dealing with fine-grained silts and clays, particularly at high water contents, the response is not as beneficial.

22.3.3 Reinforcement Design with Geogrids

Since geogrids are directly competitive with geotextiles in soil reinforcement applications, the designs are essentially identical. The slope stabilization design presented earlier with geotextiles is of this type. Here the illustration will be in the form of reinforced soil for the formation of a vertical wall. The wall front can be of the wraparound style, or the geogrids can be attached to facing panels of various materials and styles. Design procedures follow along Rankine-type plastic equilibrium concepts using active earth pressure conditions for dead loads and Boussinesq concepts for live loads. The entire process is illustrated in Koerner (1990). Note that the allowable geogrid strength considering at least installation damage and creep must be included in the design. The entire process has been made simpler by use of design charts for some geogrid products. For example, Schmertmann et al. (1987) have developed a series of design charts to be used in spacing and length determination; see Figure 22.8, which is used in the following problem.

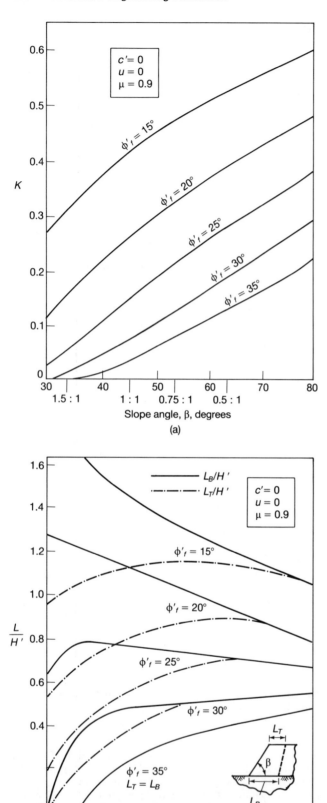

Fig. 22.8 Design guide for geogrid reinforced walls. (a) Reinforcement force coefficient. (b) Reinforcement length ratio. (*After Schmertmann et al., 1987.*)

EXAMPLE PROBLEM

Given: All assumptions valid. Required F.S. = 1.5.
Soil properties: $\phi'_p = 36°$; $c = 0$; $\gamma = 20\,\text{kN/m}^3$
Slope parameters: $H = 12\,\text{m}$; $\beta = 45°$; $q = 40\,\text{kPa}$

Design Steps:
1. Calculated modified slope height:
   $H' = 12\,\text{m} + 40\,\text{kPa}/(20\,\text{kN/m}^3) = 14\,\text{m}$
2. Calculated factored soil friction angle:
   $\phi'_f = \tan^{-1}(\tan 36°/1.5) = 25.8°$
3. Obtain K and calculate total geogrid force:
   $K = 0.15$ (from Fig. 22.8a)
   $T = \frac{1}{2}(0.15)(20\,\text{kN/m}^3)(14\,\text{m})^2 = 294\,\text{kN/m}$
4. Select geogrid design strength and calculate number of geogrid layers:
   $\alpha_d = 30\,\text{kN/m}$ (from laboratory testing)
   $N = (294\,\text{kN/m})/(30\,\text{kN/m}) = 9.8$; use 10 layers

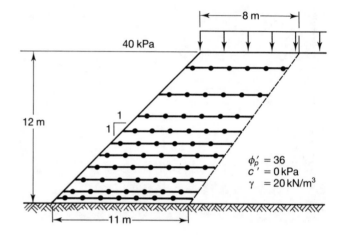

5. Obtain length ratios and calculate geogrid lengths:
   $L_T/H' = 0.55$ $L_B/H' = 0.76$ (from Fig. 22.8b)
   $L_T = 0.55(14\,\text{m}) = 7.7\,\text{m}$; use 8 m
   $L_B = 0.76(14\,\text{m}) = 10.6\,\text{m}$; use 11 m
   Space geogrid layers inversely proportional to depth.

22.3.4 Other Geogrid Reinforcement Situations

Geogrids have been successfully used to support unpaved roads on very weak subgrades. Use and design parallels geotextile work and the review by Hausmann (1986) is equally applicable to geogrids as to geotextiles. Other possibilities using geogrids are also available. Since interlocking of soil, in this case stone aggregate, is possible, the idea of reinforcing the stone base course in highways and railroads has been successful. Here the effective modulus of the stone base is increased owing to the lateral confinement afforded by the geogrid. Thinner base course thicknesses or improved lifetime should result. Attempts have also been made in reinforcing asphalt pavement by sandwiching the geogrid within the structural section itself. Work by Haas (1984) has been significant in this regard and field attempts are ongoing.

22.4 GEONETS

22.4.1 Overview

Geonets are deformed or nondeformed netlike polymeric materials used in geotechnical engineering-related construction

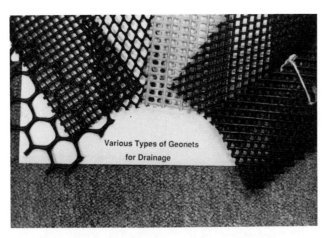

Fig. 22.9 Photographs of various types of geonets.

activities for in-plane drainage; see Figure 22.9. They are rarely, if ever, used for other functions; recall Table 22.1.

As with most geosynthetic material categories, there are a wide variety of possible manufacturing approaches used in making geonets. Perhaps the most common manufacturing technique is to extrude the molten polymer through slits in counter-rotating dies, which forms a tight net of closely spaced ribs. This net is then opened up by forcing it over a tapered

mandrel until it reaches its final configurations, when it is cooled, rolled and shipped. The resulting geonet has intersecting sets of ribs at 60° to 75° apart with the crossover points being integrally bonded to one another. The ribs can be square or slightly rectangular in cross section. A number of competing products are available; see Koerner (1990). A slight variation of the above technique is to add a foaming agent to the polymer mix and then process it as just described. The foaming agent is released and forms micrometer-sized gas-filled spheres within the rib cross sections. Geonets formed in this manner can have very high ribs (resulting in increased flow capability) in comparison to the solid formed ribs. Still other variations are possible wherein a flat substrate has a built-up section of ribs superimposed on it and a completely extruded shape.

22.4.2 Properties of Geonets

Since the primary function of geonets is in-plane drainage, their flow capability in this mode is the most important property to determine. The laboratory test utilizes parallel flow on square or rectangular-shaped specimens. Obviously, the size of the test specimen must be sufficiently large to eliminate scale effects; usually 150 mm by 150 mm, or larger test specimens are used. The geonet is placed in a leakproof membrane or container and measured under a prescribed normal pressure and hydraulic gradient for its resulting flow rate; see Figure 22.10a for a solid-rib product and Figure 22.10b for a foamed-rib product.

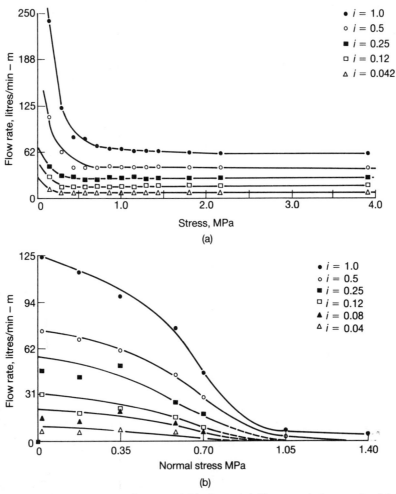

Fig. 22.10 Flow rate behavior of drainage geonets between rigid plates. (a) Flow rate behavior of solid-rib geonet (0.15 in thick). (b) Flow rate behavior of foamed-rib geonet (0.30 in thick).

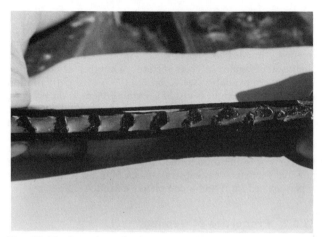

Fig. 22.11 Intrusion of adjacent materials into flow space of geonets.

It should be noted that these results are for geonets between solid plate surfaces; thus, the results represent the maximum flow that the core is capable of transmitting versus a geotextile or geomembrane on the surface that could intrude into the core space. See Figure 22.11 for this type of intrusion. Obviously, the performance test should simulate this behavior as closely as possible. Another feature about the flow curves of Figure 22.10 is that the consideration of duration of load is largely absent. According to current ASTM testing procedures (ASTM D4716), the dwell time for normal pressure application is 15 minutes and the time to take the flow measurements is also 15 minutes. For considering long-term behavior of the geonets, such testing times are usually very short. Creep of the geonet, and thereby reduction of flow rates, must generally be considered. Thus, long-duration tests are usually warranted.

22.4.3 Drainage Design Using Geonets

As with all geosynthetic design, the approach should be the formation of a factor of safety comparing the allowable property of the candidate geonet with the required, or design, value for the situation considered. With geonets, either transmissivity or flow rates can be the compared values, that is,

$$\text{F.S.} = \frac{\theta_{\text{allow}}}{\theta_{\text{req}}} \qquad (22.13)$$

or

$$\text{F.S.} = \frac{q_{\text{allow}}}{q_{\text{req}}} \qquad (22.14)$$

This selection is based upon consideration of whether the geonet is saturated or not. If the geonet is saturated (along with existence of laminar flow), then the transmissivity relationship can be used. If it is not saturated (or turbulent flow conditions exist), then flow rate at a specified hydraulic gradient should be used. The decision is site-specific. In either case, the flow rate value must be determined at the maximum applied normal pressure that the geonet will be subjected to.

Design must also consider creep of the geonet and intrusion of adjacent geosynthetics into the core space. Creep of the geonet itself is usually handled by having tests conducted at 2 or 3 times the maximum anticipated pressure to see whether flow is maintained. (Geonets sometimes fail by a "lay-down" of intersecting ribs against one another.) The intrusion problem

is much more formidable. Not only is elastic intrusion a problem (as seen in Fig. 22.11), but also long-term creep intrusion. Note that this latter condition is out-of-plane creep of the geotextile or geomembranes into the geonet core space. Long-term tests are indeed warranted if the situation is of critical concern.

22.5 GEOMEMBRANES

22.5.1 Overview

Geomembranes are essentially impermeable sheets of polymeric materials used as liquid barriers. Known also as pond liners or flexible membrane liners, they are used to contain all types of liquids, solids, and vapors. While nothing is truly impermeable, the "equivalent permeability coefficient" of geomembranes is in the range of 10^{-11} to 10^{-14} cm/sec. This being 2 to 5 orders of magnitude lower than clay liner materials classifies them as impermeable, at least in an engineering sense. We say "equivalent permeability coefficient" to recognize that liquid does not flow through voids in a geomembrane like it does a soil. Liquid eventually passes through a geomembrane by vapor diffusion which is governed by Fick's Law of Diffusion. Furthermore, it is concentration driven rather than hydraulic gradient driven and is therefore a very slow and complicated phenomenon involving liquid/vapor/liquid phase changes. The vapor diffusion test and its calculations to arrive at an equivalent permeability coefficient is available in Koerner (1990). This discussion presumes, of course, that the geomembranes can be adequately selected, designed, installed, and maintained in such a manner that holes do not occur.

The polymers that are used to manufacture geomembranes are either thermoplastic (reversible upon repeated melting cycles), thermoset (nonreversible once cured), or combinations of both. At present, however, thermoplastic geomembranes prevail, the major types being the following:

- Polyethylene; very low-density (VLDPE), medium-density (MDPE), and high-density (HDPE)
- Polyvinyl chloride (PVC)
- Chlorinated polyethylene (CPE)
- Chlorosulfonated polyethylene (CSPE)
- Ethylene interpolymer alloy (EIA)

The manufacturing of these sheet materials follows three possible routes: extrusion, clandering, or spread-coating. These are indicated in Figure 22.12, where the term *reinforced membrane* is introduced. This reinforcement is an open woven fabric that is sandwiched within the laminated sheets of a calendered geomembrane, or as the dense nonwoven substrate when the spread-coating method is used. It (the fabric), however, does not reinforce the soil subgrade, its purpose is only to provide improved tear and impact resistance to the geomembrane and make it dimensionally stable. As with all other geosynthetics, a wide variety of geomembranes are commercially available.

22.5.2 Geomembrane Properties

Laboratory test methods for the evaluation of geomembrane properties are handled by a number of organizations, including ASTM, NSF, EPA, and Bu Rec. The specific properties are often grouped into categories of which the following are typical:

- Physical tests (specific gravity, thickness, mass per unit area, water and solvent vapor transmission)
- Mechanical tests (tensile strength, puncture resistance, impact resistance, hydrostatic resistance, friction behavior, pullout behavior)

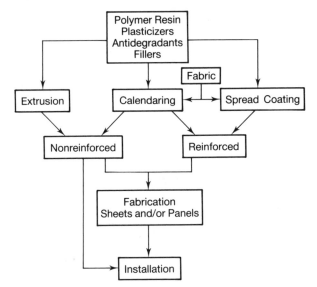

Fig. 22.12 Various manufacturing routes in formation of a geomembrane.

- Chemical tests (liquid compatibility, ozone resistance, ultraviolet resistance)
- Biological tests (microorganism compatibility)
- Endurance tests (creep behavior, abrasion resistance, soil burial resistance, durability/aging behavior)

Space precludes a complete discussion of each of the above properties, but some comments are in order on the more significant tests as far as design is concerned.

The tensile strength of a specific geomembrane is extremely important during design by function. Shown in Figure 22.13 are a series of stress-vs.-strain curves for various types of geomembranes. These curves are obtained by testing "dogbone" specimens of 6 mm width at their throat or uniform 25 mm width samples. Note that the vertical axis is not a true stress unit but rather force per unit width. In order to obtain the proper dimensions it is necessary to divide by the geomembrane thickness. Although dimensionally correct, this introduces significant inaccuracy since the thickness varies greatly during

the test. While the curves of Figure 22.13 are indeed the design curves that will be used subsequently, the results (particularly the strain) might be quite different when using wide-width or out-of-plane tension tests. Work of this type is ongoing in a number of organizations.

The second test method to be discussed in some detail is the frictional behavior between geomembranes and soil. This test is modeled directly after the direct shear test common to geotechnical engineering testing. Instead of soil-to-soil friction, however, the test requires soil-to-geomembrane friction. The data of Table 22.5 was obtained in a 100-mm by 100-mm shear box with the geomembrane fixed in the lower portion of the shear box and the soil placed above it. The table lists various geomembranes to granular soils (where only friction is present) and various geomembranes to cohesive soils (where both cohesion and friction are present). Easily seen is that the frictional resistance of a specific geomembrane to a specific soil (in a specific condition) is indeed site-specific, that is, it must be individually evaluated. Also to be noted are the relatively low frictional characteristics of smooth HDPE geomembranes.

The third test method to be discussed in some detail is chemical compatibility. Indeed, all is lost if the geomembrane material is incompatible with the liquid it is meant to contain. For liquid reservoirs containing a single known liquid, the selection is not too difficult. There exists a large database of geomembrane compatibility to various liquids. These liquids include oils, sludges, chemicals, liquids of various pH values, etc. For solid waste liners, however, the situation is much more formidable. Here the waste liquid (called "leachate") is largely unknown. As a result, a worst-case scenerio of leachate selection is used. Quite often, organic solvents and phenols are selected.

Nevertheless, the leachate having been selected, geomembrane compatibility tests often are done in accordance with the EPA 9090 test procedure (U.S.E.P.A., 1984). Here candidate geomembrane test samples are incubated in the leachate for times up to 120 days. Incubation is often done at elevated temperatures. Periodically, the samples are removed, cut into test specimens and tested for tensile strength, puncture, tear, thickness, and dimensional stability. Results are graphically compared to the as-received material in terms of relative changes. While some geomembranes react to leachate and are easily identified, many do not have well-defined reactions. In such cases one must consider experimental variations and test inaccuracies that make a decision on proper selection very

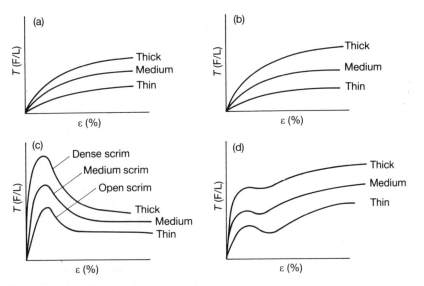

Fig. 22.13 Stress-vs.-strain response curves of various geomembranes in index tension testing. (a) PVC response curves. (b) CPE response curves. (c) CSPE-R response curves. (d) HDPE response curves.

TABLE 22.5   FRICTIONAL CHARACTERISTICS OF GEOMEMBRANES TO VARIOUS SOILS.

(a) Granular Soils (Koerner et al., 1986a)

Geomembrane	Concrete Sand ($\phi = 30°$)	Ottawa Sand ($\phi = 28°$)	Mica Schist Sand ($\phi = 26°$)
EPDM-R	24° (0.80)	20° (0.71)	24° (0.92)
PVC			
Rough	27° (0.90)	—	25° (0.96)
Smooth	25° (0.83)	—	21° (0.81)
CSPE-R	25° (0.83)	21° (0.75)	23° (0.88)
HDPE	18° (0.60)	18° (0.64)	17° (0.65)

(b) Cohesive Soils (Koerner et al., 1986a)

	ML-CL				CL-ML				CL			
	c	$E_c(\%)$	$\phi(°)$	$E_\phi(\%)$	c	$E_c(\%)$	$\phi(°)$	$E_\phi(\%)$	c	$E_c(\%)$	$\phi(°)$	$E_\phi(\%)$
Soil itself	9·0	100	38	100	12·0	100	34	100	20	100	30	100
	c_a	$E_c(\%)$	$\delta(°)$	$E_\phi(\%)$	c_a	$E_c(\%)$	$\delta(°)$	$E_\phi(\%)$	c_a	$E_c(\%)$	$\delta(°)$	$E_\phi(\%)$
Geomembrane												
PVC	8·5	94	39	100	3·7	31	23	69	14·0	70	16	53
CPE	8·0	89	40	100	3·2	27	24	71	13·0	65	16	57
EPDM-R	5·0	55	33	87	5·0	42	23	67	8·0	40	23	77
HDPE	5·0	88	26	68	2·0	17	23	67	14·0	70	15	50

difficult. Often, in the case of aggressive leachates, only HDPE (with crystallinity greater than 50 percent) is sufficiently resistant for proper geomembrane composition.

22.5.3   Liquid-Containment Liners

The design of a liquid-containment liner follows along clearly defined and sequential steps. These are as follows.

- Site selection
- Geometric layout
- Cross-section selection
- Geomembrane material selection
- Thickness design
- Side slope design
- Cover soil considerations
- Anchor trench details
- Final details and miscellaneous items

Each step in the above list uses information gained from preceding steps. Most have an analytic basis, which must be tuned to the actual situation and to its criticality.

To illustrate the situation, a thickness problem will be outlined. Note that a completely unyielding geomembrane for liquid-barrier purposes can be very thin and in the limit even a molecular thickness could suffice. Two situations prohibit this, however; one is the installation stresses, the other is localized bending. The following design is for bending, from which a thickness calculation can be made and then compared to a minimum thickness for installation survivability. Using a free body as shown below, one can take the sum of forces in a horizontal direction:

$$\Sigma F_x = 0; \qquad T \cos \beta = T_U + T_L$$

$$(\sigma_a t) \cos \beta = (\sigma'_U \tan \delta_U + \sigma'_L \tan \delta_L) X$$

$$t = \frac{(\sigma'_U \tan \delta_U + \sigma'_L \tan \delta_L) X}{\sigma_a \cos \beta} \qquad (22.15)$$

where

t = required geomembrane thickness
σ'_U = effective normal stress of cover soil (negligible in most cases)
σ'_L = effective normal stress beneath geomembrane
    = γH
γ = unit weight of contained liquid
H = height (depth) of contained liquid
δ_U = friction angle of geomembrane to cover soil
δ_L = friction angle of geomembrane to subsoil
X = embankment depth to mobilize σ_a
σ_a = allowable or yield stress of geomembrane
β = subsidence angle

All values in the equation are known or can be measured with the exception of the embedment depth value X. While not a standardized test, this value can be evaluated in the laboratory. The test involves sandwiching the geomembrane between parallel plates at the desired normal stress. The anchorage depth required to mobilize the allowable or yield stress is the desired value; see GRI (1987) for the procedure and details. For the value σ_a, the curves of Figure 22.13 should be used. For well-defined breaks or yield points, as in (c) or (d), the choice is obvious. For geomembranes without a well-defined target, as in (a) and (b), one must select a maximum allowable strain, for example, 50 percent, and use the corresponding value of allowable stress.

In following other aspects of the design procedure it will be seen that cover soil stability is very troublesome. Generally, very flat side slopes are required (3 to 1 or flatter) unless special precautions such as geogrid or geotextile reinforcement are taken. Obviously, seams are very significant and these will be discussed later in the section.

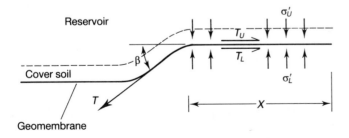

TABLE 22.6   WASTE GENERATED IN U.S.A.[a].

	Millions of Tons				Average Annual Growth (%)	
	1977	*1988*	*1993*	*2000*	*1977–8*	*1988–93*
By type of waste						
Heavy metals	51	114	149	196	7.6	5.5
Organic chemicals	42	100	132	180	8.2	5.7
Petroleum derived	16	33	44	60	6.8	5.9
Inorganic chemicals	17	35	43	55	6.8	4.2
Other hazardous waste	5	9	13	19	5.5	7.6
Total	131	291	380	510	7.5	5.5
By method of disposal						
Landfill/surface impound	12	200	225	165	29.1	2.4
Treatment/stabilization	2	13	50	150	18.5	30.9
Incineration	neg[b]	15	35	95	—	18.5
Resource recovery	2	12	30	75	17.7	20.1
Deep-well injection	5	14	15	10	9.8	1.4
Illegal disposal	110	35	20	5	– 9.9	– 10.6
Other methods	neg[b]	2	5	10	—	20.1

[a] Hanson (1989). Source: Freedonia Group.
[b] neg = negligible.

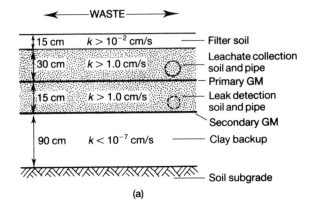

(a)

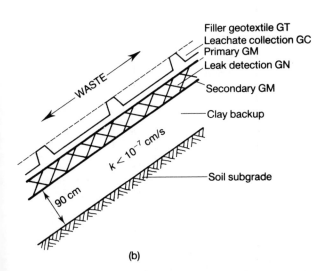

(b)

Fig. 22.14 Bottom and side liner schemes for hazardous waste disposal cells. (a) Regulated cross section. (b) Geosynthetic alternate cross section.

22.5.4   Solid-Waste-Containment Liners

Geomembranes used in the containment of solid waste pose much harsher conditions than do liquid-containment liners. This comes about owing to solids mobilizing shear stress on side slopes, the necessity of collecting and removing leachate, the necessity of providing redundancy if the primary liner leaks, and the general negative emotions that landfills of all types provoke. It should be clearly recognized that the amount of waste generated in the U.S.A. for treatment, storage and/or disposal is enormous; see Table 22.6. Clearly, the hazardous waste group is the target of most concern and, appropriately, of the most severe legislation and regulations. Because of its significance, it will be the focus of this section. Bear in mind, however, that a geomembrane design for a hazardous-waste landfill can always be modified to a less-critical situation.

The current regulated cross section for landfills, surface impoundments, and waste piles is shown in Figure 22.14a. Considering, however, that HDPE is the liner material most often resulting from leachate immersion testing, the alternative cross section of Figure 22.14b is seeing greater use, particularly on side slopes. In the design of these systems there are many individual elements. Invariably they consist of a design model, laboratory test input, and a resulting factor of safety based upon the yield stress of the geomembrane; recall Figure 22.13d. Tables 22.7 and 22.8 summarize the design elements for the geomembrane and the drainage geosynthetics, respectively (Koerner and Richardson, 1987). Each of these models is worked out in detail with illustrative problems in an E.P.A. design manual by Richardson and Koerner (1987).

22.5.5   Other Geomembrane Applications

Other geomembrane applications, all of which can be approached on a rational design basis, are as follows (Koerner, 1990):

• Reservoir covers
• Canal liners

TABLE 22.7 VARIOUS DESIGN MODELS FOR GEOMEMBRANES IN WASTE DISPOSAL SITUATIONS[a].

Problem	Liner Stress	Free-Body Diagram	Required Properties Geomembrane	Required Properties Liner	Typical Factor of Safety
1. Liner self-weight	Tensile		G, t, σ_y, δ_L	β, H	10 to 100
2. Weight of filling	Tensile		$t, \sigma_y, \delta_U, \delta_L$	β, h, γ, H	0.5 to 10
3. Impact during construction	Impact		I	d, w	0.1 to 5
4. Weight of landfill	Compression	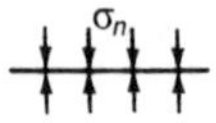	σ_y	γ, H	10 to 50
5. Puncture	Puncture		σ_p	γ, H, P, A_p	0.5 to 10
6. Anchorage	Tensile	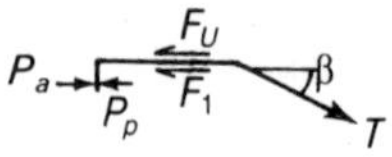	$t, \sigma_y, \delta_U, \delta_L$	β, γ, ϕ	0.7 to 5
7. Settlement of landfill	Shear	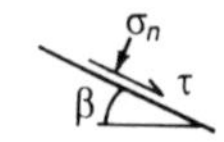	τ, δ_U	β, γ, H	10 to 100
8. Subsidence under landfill	Tensile	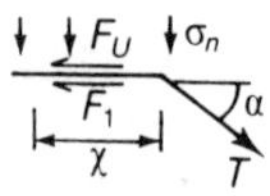	$t, \sigma_y, \delta_U, \delta_L, \chi$	α, γ, H	0.3 to 10

[a] After Koerner and Richardson (1987).
Notes:

- *Geomembrane Properties*
  - G = specific gravity
  - t = thickness
  - σ_y = yield stress (or allowable stress)
  - τ = shear stress
  - I = impact resistance
  - σ_p = puncture stress
  - δ_U = friction with material above
  - δ_L = friction with material below
  - χ = mobilization distance

- *Landfill Properties*
  - β = slope angle
  - H = height
  - γ = unit weight
  - h = lift height
  - α = subsidence angle
  - ϕ = friction angle
  - d = drop height
  - W = weight
  - p = puncture force
  - A_p = puncture area

TABLE 22.8 VARIOUS DESIGN CONSIDERATIONS FOR DRAINAGE GEOCOMPOSITES IN WASTE DISPOSAL SITUATIONS[a].

Problem	Reason	Approach	Required Properties Geocomposite	Required Properties Landfill	Severity of Problem
1. Strength of core	Avoid crushing	F.S. = $\sigma_{ult}/\sigma_{max}$	σ_{ult}	γ, H	Minor
2. Flow in core	First approximation	F.S. = $\theta_{allow}/\theta_{act}$	θ_{allow}	$\gamma, H, i, q_{act}, \theta_{act}$	Minor
3. Creep of core	First reduction	F.S. = $\theta_{allow}/\theta_{act}$	θ_{allow}	$\gamma, H, q_{act}, \theta_{act}$	Nil to major
4a. Elastic intrusion of geomembrane	Second reduction	Elastic plate theory	E, μ, x, y	$\gamma, H, q_{act}, \theta_{act}$	Major
4b. Elastic intrusion of geotextile	Second reduction	Elastic plate theory	E, μ, x, y	$\gamma, H, q_{act}, \theta_{act}$	Major
5a. Creep intrusion of geomembrane	Third reduction	Creep theory	$\dot{\varepsilon}(\sigma, t), x, y$	γ, H, t	Unknown
5b. Creep intrusion of geotextile	Third reduction	Creep theory	$\dot{\varepsilon}(\sigma, t), x, y$	γ, H, t	Unknown

[a] After Koerner and Richardson (1987).
Notes:

- *Geocomposite Properties*
  - σ_{ult} = ultimate compression strength
  - σ_{max} = maximum stress
  - σ = applied stress
  - θ_{allow} = transmissivity
  - t = time
  - E = modulus of elasticity
  - μ = Poisson's ratio
  - x, y = core dimensions
  - $\dot{\varepsilon}(\sigma, t)$ = strain rate

- *Landfill Properties*
  - γ = unit weight
  - H = height
  - i = hydraulic gradient
  - q_{act} = actual (design) flow rate
  - θ_{act} = actual (design) transmissivity
  - t = time

TABLE 22.9 TYPES OF GEOMEMBRANE SEAMING METHODS[a].

(a) Adhesive and Tapes

Base Polymer of Common Geomembrane Systems	Solvent		Bodied Solvent		Solvent Adhesive		Contact Adhesive		Vulcanizing tape/ Adhesive		Tape, F
	M[b]	F[b]	M	F	M	F	M	F	M	F	F
Thermoplastics											
Polyvinyl chloride (PVC)	X	X			X	X	X	X			X
Nitrile-PVC (TN-PVC)	X	X			X	X	X	X			X
Ethylene interpolymer alloy (EIA)											
Crystalline thermoplastics											
Low-density polyethylene (LDPE)							X	X			X
High-density polyethylene (HDPE)							X	X			X
Elastomers											
Butyl rubber (IIR)							X	X	X	X	
Ethylene propylene diene monomer (EPDM)							X	X			X
Neoprene (polychloroprene)							X	X			
Epichlorohydrin rubber (CO)							X	X			
Thermoplastic elastomers											
Chlorinated polyethylene (CPE)	X	X	X	X	X	X	X	X			X
Hypalon (chlorosulfonated polyethylene) (CSPE)	X	X	X	X	X	X	X	X			X
Thermoplastic EPDM (T-EPDM)							X	X			

(b) Thermal and Mechanical

Base Polymer of Common Geomembrane Systems	Thermal methods					Extrusion (Fusion) Welding, F	Mechanical, F
	Hot Air		Hot Wedge		Dielectric, M		
	M	F	M	F			
Thermoplastics							
Polyvinyl chloride (PVC)	X	X			X		X
Nitrile-PVC (TN-PVC)	X	X			X		X
Ethylene interpolymer alloy (EIA)	X	X					
Crystalline thermoplastics							
Low-density polyethylene (LDPE)	X	X	X	X			X
High-density polyethylene (HDPE)		X	X	X		X	X
Elastomers							
Butyl rubber (IIR)							
Ethylene propylene diene monomer (EPDM)							
Neoprene (polychloroprene)							
Epichlorohydrin rubber (CO)							
Thermoplastic elastomers							
Chlorinated polyethylene (CPE)	X	X			X		
Hypalon (chlorosulfonated polyethylene (CSPE)	X	X			X		
Thermoplastic EPDM (T-EPDM)	X	X					

[a] After Frobel (1984).
[b] M, manufactured or factory seams; F, field fabrication.

TABLE 22.10 OVERVIEW AND CRITIQUE OF NONDESTRUCTIVE GEOMEMBRANE SEAM TESTS[a].

Nondestructive Test Method	Primary User			General Comments					
	Contractor	Design Engr. Insp.	Third Party Insp.	Cost of Equipment, $	Speed of Tests	Cost of Tests	Type of Result	Recording Method	Operator Dependency
1. Air lance	Yes	n/a	n/a	$200	Fast	Nil	Yes–No	Manual	Very high
2. Mechanical point (pick) stress	Yes	n/a	n/a	Nil	Fast	Nil	Yes–No	Manual	Very high
3. Vacuum chamber (negative pressure)	Yes	Yes	n/a	$1000	Slow	V. high	Yes–No	Manual	High
4. Dual seam (positive pressure)	Yes	Yes	n/a	$200	Fast	Mod.	Yes–No	Manual	Low
5. Ultrasonic pulse echo	n/a	Yes	yes	$5000	Moderate	High	Yes–No	Automatic	Moderate
6. Ultrasonic impedance	n/a	Yes	yes	$7000	Moderate	High	Qualitative	Automatic	Unknown
7. Ultrasonic shadow	n/a	Yes	yes	$5000	Moderate	High	Qualitative	Automatic	Low

[a] After Koerner and Richardson (1987).

- Caps and closures of landfills
- Earth dam retrofit liners
- Concrete dam retrofit liners
- Vertical cutoff barriers
- Secondary underground storage tank liners
- Ground vapor barriers (against moisture, methane, radon, etc.)
- Heap leaching pads in the mining industry
- Thermal extraction from salt ponds

22.5.6 Geomembrane Details

Details are important in the construction of any system, but nowhere are they more important than in geomembrane systems. One leak can defeat the purpose of the entire liner system. Thus, worksmanship takes on an extremely high priority. Even third-party construction quality assurance (CQA) consulting is required on many installations. While every location of the liner is a potential problem, the field seams are rightfully the focus of most attention. The seaming methods are related to the type of polymer; see Table 22.9 for an overview (Frobel, 1984).

Regarding inspection of the seams, the choices are between destructive and nondestructive testing. Destructive tests require the cutting of a coupon from the geomembrane and then testing it in shear or in peel (tension). Such tests are good indicators of the manner and technique of seaming, but do require patching and tell nothing of the continuity between individual tests. Thus, the need for nondestructive tests, of which there are many; see Table 22.10. These methods are elaborated upon in various references. Of them, the air-lance and pick tests are really contractors' methods for investigating missed seam areas. Current specifications often call for vacuum-box testing, but when required for 100 percent of the seams this is very time-consuming, tedious, boring, and costly. Future trends may be toward the ultrasonic techniques and, in particular, the ultrasonic shadow method (Lord et al., 1986; Koerner et al., 1987).

Details around pipes, connections, and fittings are very difficult to design and construct. As much as possible, modulus mismatches should be kept to a minimum; that is, a liner should not be anchored firmly to steel or concrete. Batten strips, rounded or tapered corners, slack zones, etc., must all be considered in the design of such details. Most manufacturers have excellent experience in this regard and their promotional literature has a wealth of good information.

In the final analysis, however, it is the field installation contractor who holds the key to a successful and trouble-free project. Experience, workmanship, and attention to detail cannot be overemphasized.

22.6 GEOCOMPOSITES

22.6.1 Overview

The relatively ill-defined area called geocomposites consists of manufactured products using combinations of geotextiles, geogrids, geonets, and/or geomembranes in laminated or composite form. Generally, but certainly not always, the end-product is completely polymeric. Other options include using fiberglass or steel for tensile reinforcement, sand in compression or as a filler, dried clay for subsequent expansion as a liner, or bitumen as a waterproofing agent. Since a geocomposite can be made for any specific function (recall Table 22.1) this section will be subdivided to include all of the primary geosynthetic functions. Owing to the widespread use of drainage geocomposites, however, this particular area will be emphasized.

22.6.2 Geocomposites in Separation

Many erosion control systems are being made from a continuous mat of polymer (nylon and PVC have been used successfully) held together with an open geotextile substrate. Drainage ditches and swales used to intercept runoff in highway and railroad soil slopes regularly use such systems in place of concrete ditches. By using a geosynthetic, a number of advantages are gained:

- Conformability to irregular surfaces
- Flexibility for subsequent soil subgrade movements
- Promotion of vegetative growth, and thus good asthetics
- Low runoff velocities
- Elimination of need for downstream energy dissipation
- Significantly lower expense than concrete or rip-rap

Many State DOTs are specifying and using such systems. While exposed to ultraviolet degradation, the systems appear to have lifetimes adequate for their use. Vegetative growth is certainly helpful in this regard.

22.6.3 Geocomposites in Reinforcement

Many attempts at increasing the tensile properties of modulus and strength and/or decreasing creep by using materials with strength greater than polymers have been tried. Steel stands have been used effectively to make heavy filter mattresses laid directly on the seafloor to support concrete piers. Fiberglass has been used by a number of companies, which is usually woven or knitted in continuous filaments for enhanced strength properties. Both of these materials (steel and fiberglass) have drawbacks of long-term corrosion that must be contemplated in the design.

Quite a different approach is to use continuous-filament polymer yarn along with soil to form reinforced slopes. The technique uses a soil spray with many fibers included in it to actually construct the entire slope (Leflaive, 1986). Almost-vertical slopes that support relatively large surcharge loads have been constructed.

A number of specific polymers have been used in conjunction with one another with excellent results. One of these types uses high-tenacity polyester yarns that are grouped together and held in a polypropylene sheath to form rods, strips, grids, links, and webs (Koerner, 1986b). Another approach has used polyester yarn in the warp direction and polypropylene yarn in the fill direction to make anisotropic high-strength fabrics. The possibilities are enormous and the geosynthetic manufacturing industry appears ready to develop new products as the demand warrants.

22.6.4 Geocomposites in Filtration

The most outstanding example of a geocomposite filter is the Dutch soil–geotextile filter mattress (Visser and Mouw, 1982). Here a composite is formed consisting of the following layers from bottom to top:

- High-strength steel-reinforced geotextile
- 110 mm of sand
- Geotextile separator
- 110 mm of sandy gravel
- Geotextile separator

- 140 mm of gravel
- Geotextile upper layer

The system was made into 200-m long by 42-m wide mattresses and placed on the ocean floor to support concrete piers for a storm surge barrier. The Eastern Scheldt River Barrier is probably the most significant project incorporating geosynthetics to date, particularly considering that the envisioned lifetime is 200 years (Visser and Mouw, 1982).

The use of two geotextile filters needled together, each designed for the particular soil placed adjacent to them, has seen common use and has been quite successful. Other options are also available.

22.6.5 Geocomposites in Drainage

This area is perhaps the fastest-moving of all of the geocomposite options available to the user. It can be subdivided into three topics, since each area is very application-specific.

"Sheet drains" consist of rolls or panels of drainage cores protected by a geotextile filter on one side (to allow for liquid to enter) and a geotextile, geomembrane, or structure on the other side; see Figure 22.15a. The cores vary greatly in their polymer type, shape, and configuration. This being the case, it should come as no surprise that the flow rate capability should vary greatly as well (Koerner et al., 1986a). The test for evaluating these products is identical to that described in the geonet section—recall Section 22.4. It should be noted that geonets protected as described above are indeed geocomposites and all of the details described there, namely, intrusion, creep, etc., apply here as well. At least two features distinguish geocomposite drains from geonets; first, geocomposite drains usually have greater flow rate capability and, second, they usually have a pronounced collapse strength. Figure 22.16 illustrates these points for selected products, where the sensitivity to normal stress and hydraulic gradient is evident. Thus, each product must have this type of information available for design purposes.

Regarding the design (or required) flow-rate values for use in a factor-of-safety calculation, the intended application is most significant. Some applications (in approximate order of increasing flow-rate demand) are:

- Fine-grained soil drainage
- Roof-garden drainage
- Plaza-deck drainage
- Sport-field drainage
- Capillary breaks
- Seeping-slope drainage (soil or rock)
- Leachate collection systems
- Surface-water drains
- Retaining-wall drains

The actual calculation of required flow-rate capability is obtained by appropriate theory, such as Darcy's formula, flow nets, empirical guides and charts, etc. (Koerner et al., 1986b).

A final comment on sheet-drain specifications is in order, since the products available differ so widely in their performance. There is no "or equal" in this area. A proper specification should require a specified flow rate, at a given normal stress, at a given hydraulic gradient. If desirable, further details as to normal stress dwell time, soil adjacent to the candidate product, etc., can also be included.

"Strip drains" (or prefabricated vertical drains, also incorrectly called wick drains) are used to rapidly consolidate fine-grained saturated soil. They are true drainage geocomposites, consisting of a drainage core completely surrounded by a geotextile filter; see Figure 22.15b. As with sheet drains,

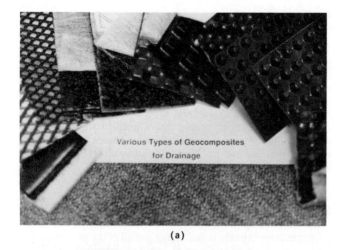

(a)

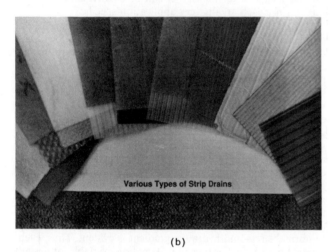

(b)

Fig. 22.15 Photographs of geocomposite drainage systems. (a) Various types of sheet drains. (b) Various types of strip drains (prefabricated vertical drains).

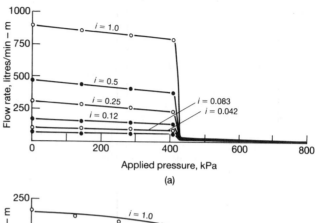

(a)

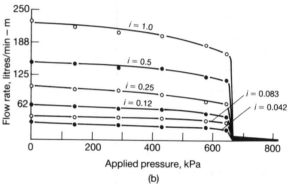

(b)

Fig. 22.16 Flow rate capability for selected geocomposite drainage systems. (a) Flow rate behavior of product *a*. (b) Flow rate behavior of product *b*.

there are many competing styles of strip drains currently available. Their design can be based on volumetric flow rate considerations, but sufficient analytic work has been done such that an equation for time for consolidation is currently available (Hansbo, 1979):

$$t = \frac{D^2}{8c_h}\left(\ln\frac{D}{d} - 0.75\right)\left(1 - \frac{1}{\bar{U}}\right) \qquad (22.16)$$

where

t = time for U percent consolidation
$\bar{U}$ = consolidation expressed as a ratio
c_h = horizontal (radial) coefficient of consolidation
D = strip drain spacing (for triangular pattern use 1.05 spacing; for rectangular pattern use 1.13 spacing)
d = equivalent diameter of strip drain ($\simeq$ circumference/π)

A completely worked example with a design chart using this formula is available in Koerner (1990). Further modifications and a comprehensive review of the subject are available in Kraemer and Smith (1986).

One point of concern, however, that should be considered in designing strip drains is their flexibility. Since the axial shortening of strip drains during consolidation of the surrounding soil could be 20 to 30 percent of their original installed length, one must ask how this shortening is accomplished. If a

very localized area is affected (as one would expect, since soils are invariably nonhomogeneous), the stiff drains could kink, thereby stopping flow. Thus, strip drains should be categorized as stiff, intermediate, or flexible and viewed in light of their anticipated shortening. A laboratory test to evaluate this phenomenon is available (Suits et al., 1985, 1987).

Lastly, it is interesting to consider what a large number of strip drains penetrating through a weak soil stratum would contribute toward soil stability. Since soft soil foundation failures as surcharge is increased are common in rapid consolidation projects, the strip drains could indeed provide a reinforcing effect to the site. Typically, the tensile strength of strip drains is 1 to 2 kN per drain. Analysis is ongoing in this area as well as others involving use of strip drains.

"Edge drains" have recently emerged as geocomposite drains in their own right. Originated in 1985 there are today from 10 to 15 competing products. They are generally 300 or 450 mm high, from 25 to 37 mm thick and hundreds of meters long. They are installed in a vertical position immediately adjacent to the edge of a highway pavement for the purpose of subsurface drainage. At spacings of 100 and 200 m they are intercepted and the drainage is diverted to a ditch or swale away from the pavement area.

Both design and testing of highway edge drains are still in a formative stage but are being actively pursued by a number of organizations (see Koerner, 1990, for some of these details).

22.6.6 Geocomposites as Moisture Barriers

Various combinations of geosynthetics, along with asphalt, clay, elastomers, etc., can be combined to make effective moisture barriers. The spread-coating method described in Figure 22.12 is of this type. Many more complex situations are also available. Some of the larger types have asphalt layers within geotextiles that are rolled onto a site and have heat-sealing for their seaming technique. Dry bentonite clay has been effectively sandwiched between geotextiles, placed, and moistened allowing the clay to swell, thereby forming the required barrier. Fiberglass fabric has been used between bitumen sheets for preventing crack reflection in highway repairs. The list of possibilities is essentially endless.

This last comment is perhaps fitting to close this chapter on geosynthetics. Indeed, the entire area is new, exciting, and full of possibilities to effectively solve a wide range of geotechnical engineering-related problems.

REFERENCES

ASTM D4491, Standard Test Method, *Water Permeability of Geotextiles by Permittivity* (Permittivity Under Load is in ASTM D-35 Committee).

ASTM D4595, Standard Test Method, *Tensile Properties of Geotextiles by the Wide-Width Strip Method.*

ASTM D774, Standard Test Method for Mullen Burst Testing of Textiles.

ASTM D4716, Standard Test Method for Testing Constant Head Hydraulic Transmissivity (In-Plane Flow) of Geotextile and Geotextile Related Products.

Bertacchi, P. and Cazzuffi, D. (1985), Geotextile filters for embankment dams, *Water Power and Dam Construction*, **36**, No. 14, pp. 14–22.

Carroll, R. G., Jr. (1983), Geotextile filter criteria, *Engineering Fabrics in Transportation Construction*, TRB 916, Washington, D.C., pp. 46–53.

Fowler, J. and Koerner, R. M. (1987), Stabilization of very soft soils using geosynthetics, *Proceedings, Geosynthetics '87*, New Orleans, La., IFAI, pp. 289–300.

Frobel, R. K. (1984), Method of constructing and evaluating geomembrane seams, *Proceedings of the International Conference on Geomembranes*, Denver, Colo., IFAI, pp. 359–364.

GRI (1987), *Standard Test Method for Embedment Depth for Anchorage Mobilization, GRI Test Method GM 2-87*, Geosynthetic Research Institute, Philadelphia, Pa.

Haas, R. (1984), Structural behavior of Tensar reinforced pavements and some field applications, *Proceedings of a Symposium on Polymer Grid Reinforcement in Civil Engineering*, ICE, London, Paper 5.1.

Haliburton, T. A. and Wood, P. D. (1982), Evaluation of U.S. Army Corps of Engineers gradient ratio test for geotextile performance, *Proceedings of the 2d International Conference on Geotextiles*, Las Vegas, Nev., IFAI, pp. 97–101.

Halse, Y. H., Koerner, R. M., and Lord, A. E. Jr. (1987), Filtration properties of geotextiles under long term testing, *Proceedings, Penn DOT/ASCE Conference*, Harrisburg, Pa., pp. 1–12.

Hansbo, S. (1979), Consolidation of clay by band shaped perforated drains, *Ground Engineering*, July, pp. 16–25.

Hanson, D. J. (1989), Hazardous waste management: planning to avoid future problems, *Chemical and Engineering News*, July 31, pp. 9–18.

Hausmann, M. R. (1986), Fabric reinforced unpaved road design methods—Parametric studies, *Proceedings of the Third International Conference on Geotextiles*, Vienna, Austria, IFAI, pp. 19–24.

Koerner, R. M. (1984), A note on geotextile design methods, *Geotechnical Fabrics Report*, **2**, No. 2, pp. 28–29.

Koerner, R. M. (ed.) (1988), *Proceedings of Use of High Strength Geotextiles to Stabilize Very Soft Soils*, Geosynthetic Research Institute, Drexel University, Philadelphia, Pa., Published in *Journal of Geotextiles and Geomembranes*. **6**, Nos. 1–3, 252 pp.

Koerner, R. M. (1990), *Designing with Geosynthetics*, 2nd Edition, Prentice-Hall, Inc., Englewood Cliffs, N.J.

Koerner, R. M. and Ko, F. (1982), Laboratory studies on long-term drainage capability of geotextiles, *Proceedings of the 2d International Conference on Geotextiles*, Las Vegas, Nev., 1982, IFAI, pp. 91–95.

Koerner, R. M. and Bove, J. A. (1983), In-plane hydraulic properties of geotextiles, *Geotechnical Testing Journal, ASTM*, **6**, No. 4, pp. 190–195.

Koerner, R. M. and Robins, J. C. (1986), In-situ stabilization of soil slopes using nailed geosynthetics, *Proceedings of the Third International Conference on Geotextiles*, Vienna, Austria, IFAI, pp. 395–400.

Koerner, R. M., Martin, J. P., and Koerner, G. R. (1986a), Shear strength parameters between geomembranes and cohesive soils, *Journal of Geotextiles and Geomembranes*, **4**, pp. 21–30.

Koerner, R. M., Luciani, V. A., Freese, J. S., and Carroll, R. G., Jr. (1986b), Prefabricated drainage composites: Evaluation and design guidelines, *Proceedings of the Third International Conference on Geotextiles*, Vienna, Austria, IFAI, pp. 551–556.

Koerner, R. M. and Richardson, G. N. (1987), Design of geosynthetic systems for waste disposal, *Proceedings of Geotechnical Practice for Waste Disposal '87*, Ann Arbor, Mich., ASCE, pp. 65–85.

Koerner, R. M. et al. (1987), Geomembrane seam inspection using the ultrasonic shadow method, *Proceedings, Geosynthetics '87*, New Orleans, La., IFAI, pp. 493–504.

Kraemer, S. R. and Smith, A. D. (1986), *Geocomposite Drains*, FHwA Contract No. DTFH 61-83-C-00101, Washington, D.C.

Leflaive, E. (1986), The reinforcement of soils by continuous threads, *Proceedings of the Third International Conference on Geotextiles*, Vienna, Austria, IFAI, pp. 523–528.

Lord, A. E. Jr., Koerner, R. M., and Crawford, R. B. (1986), NDT techniques to assess geomembrane seam quality, *Proceedings of Management of Uncontaminated and Hazardous Waste*, Washington, D.C., pp. 272–276.

McGown, A. (1978), The properties of nonwoven fabrics presently identified as being important in public works applications, Index 78 Programme, University of Strathclyde, Glasgow, Scotland.

Richardson, G. N. and Koerner, R. M. (1987), *Geosynthetic Design Guidance for Hazardous Waste Landfill Cells and Surface Impoundments*, Geosynthetic Research Institute, Philadelphia, Pa.

Schmertmann, G. R. et al. (1987), Design charts for geogrid reinforced soil slopes, *Proceedings, Geosynthetics '87*, New Orleans, La., IFAI, pp. 108–120.

Suits, L. D., Gemme, R. L., and Masi, J. J. (1985), The effectiveness of prefabricated drains on the laboratory consolidation of remolded soils, *ASTM Symposium on Soils Laboratory Testing*, Ft. Lauderdale, Fla.

Suits, L. D., Gemme, R. L., and Masi, J. J. (1987), *Standard Test Method for Strip Drain Kinking Efficiency, GRI Test Method GC 6-87*, Geosynthetic Research Institute, Philadelphia, Pa.

U.S.E.P.A. (1984), *EPA-Method 9090—Compatibility Tests for Waste and Membrane Liners*, Office of Solid Waste, Washington, D.C.

Visser, T. and Mouw, K. A. G. (1982), The development and application of geotextiles on the Oosterschelde Project, *Proceedings of the 2d International Conference on Geotextiles*, Las Vegas, Nev., IFAI, pp. 265–270.

Yako, M. A. and Christopher, B. R. (1987), Polymerically reinforced retaining walls and slopes in North America, *Workshop on Applications of Polymeric Reinforcement in Soil Retaining Structures*, ed. P. Jarrett, Royal Military College, Kingston, Ontario, Canada, pp. 213–226.

23 DEEP COMPACTION OF GRANULAR SOILS

BENGT B. BROMS, Ph.D.
Professor,
Nanyang Technological Institute, Singapore

23.1 INTRODUCTION

Several new or improved methods for deep compaction of granular soils have been developed during the last few years:

- to control settlements
- to increase the bearing capacity
- to prevent or to reduce the risk of liquefaction

Significant advances have been made with respect to the efficiency of the methods, the prediction of the improvement that can be achieved, and the checking of the results using, for example, penetration, pressuremeter and dilatometer tests, crosshole measurements, and other in-situ methods. In this chapter, the principles, the applications, and the design methods that can be used in the different compaction methods are reviewed and compared.

Deep compaction methods can be broadly classified into vibration, displacement, and loading methods depending on the mechanism of the compaction shown below.

1. *Vibration methods*
   (Vibrocompaction)
   1.1 Vibroflotation
   1.2 Vibrocompaction
       Vibro-Wing
       Foster Terraprobe
       Franki Y-Probe
   1.3 Blasting
2. *Displacement methods*
   (Vibrodisplacement)
   2.1 Stone and gravel columns and sand compaction piles
   2.2 Compaction piles
   2.3 Dynamic consolidation (heavy tamping)
   2.4 Pressure grouting
3. *Loading methods*
   3.1 Preloading

The vibration methods, vibroflotation, vibrocompaction, and blasting, are mainly effective in clean sand and gravel below the groundwater table where the vibrations will cause local liquefaction around the probe. Displacement methods and preloading can also be used in silty or clayey soils above the groundwater table where the soil is only partially saturated. Dynamic consolidation (heavy tamping) and blasting generate high pore water pressures that reduce the shear strength of the soil. This means that the soil particles can be displaced by the shear waves (S-waves) that follow after the initial compression waves (P-waves) caused by the impact. The volume of soil affected by dynamic consolidation and blasting is usually relatively large (up to 10 m or more from the point of impact or blasting). At vibroflotation and vibrocompaction the soil is compacted locally up to a few meters from the source. It is also possible to replace the compressible layers with a compacted granular fill.

The selection of a particular compaction method for a project depends on the required depth and degree of compaction, gradation of the soil, content of fines, degree of saturation, location of groundwater table, risks involved, available equipment and time, experience of the contractor, and the costs. It is often much easier to compact a granular material to a specified relative density when the soil initially is loose compared with the case when the initial relative density is high. The homogeneity of the soil is also improved by the compaction. Sometimes a combination of methods can be useful when the depth is large (Johnson et al., 1983; Keller et al., 1987; Mitchell and Welsh, 1989), the soil conditions are variable (Cognon et al., 1983; Schmertmann et al., 1986), or when a high relative density is required (Solymar and Reed, 1986).

A conventional foundation method (driven or bored piles) must often be used if sufficient time is not available for the required field and laboratory tests or for the compaction.

The most difficult part in geotechnical investigations and in evaluating alternative soil improvement methods is to determine the improvement that is really necessary, the required depth of the compaction and the compressibility of the compacted soil. For example, the compressibility of granular soils depends on a large number of factors that are not related to the relative density of the soil. It should be noted that the properties of the even clean sand continue to improve for several weeks or months after the initial compaction without any volume changes, as has been observed, for example, by Mitchell (1986) and by Solymar and Reed (1986).

As discussed by Engelhardt and Golding (1975), Engelhardt and Kirsch (1977), and Seed and Booker (1977) among others, deep compaction down to 30 m or more may be required, for example, for earth and rockfill dams, nuclear plants, airports, and harbors in order to reduce the liquefaction potential. In many other cases compaction down to 5 or 10 m may be sufficient.

Deep compaction is often necessary for land reclamation where land is scarce, for example in Singapore, Hong Kong, and Japan. In Japan alone more than 5 percent of the flat land available in the country has been reclaimed during the last 20 years (Aboshi, 1984). For reclamation projects it is important to estimate accurately the decrease of the volume caused by the compaction. The reduction is usually between 5 and 15 percent of the thickness, depending on the initial relative density of the compressible layers.

Deep compaction methods for granular soils have been discussed at several recent conferences and symposia (London, 1975, 1983; Brighton, 1979; Paris, 1980, 1984; Stockholm, 1981; Helsinki, 1983; Bangkok, 1982; Singapore, 1985). Different soil improvement methods used in the United States have been reviewed by the ASCE Committee on Placement and Improvement of Soils (ASCE, 1977). State-of-the-art reports have been published by Mitchell (1981), Mitchell and Katti (1981), and Greenwood and Kirsch (1983).

Extensive field and laboratory testing is generally required to evaluate the effectiveness of different compaction methods in terms of the relative compaction or relative density that can be achieved. Penetration tests are normally used to determine the thickness, location, and lateral extent of the different soil layers. The compressibility and the bearing capacity of the different strata are usually estimated from cone (CPT) and standard penetration (SPT) tests or from weight soundings (WST). Disturbed but representative samples are usually sufficient to determine the grain-size distribution and the content of fines. Pressuremeter and plate load tests are often useful to establish in situ the shear strength and the compressibility of the different strata. Dilatometer tests can sometimes be useful to evaluate the compressibility and the coefficient of lateral earth pressure at rest, both of which are of interest, for example, in finite-element analyses (FEM). Pore pressure conditions can be determined with a piezocone, often in combination with cone penetration tests. Relatively thin sand or silt seams can be detected by pore pressure soundings, which can also be used to evaluate the coefficient of consolidation (Robertson, 1986).

The required improvement is normally expressed in terms of a relative density defined by the relationship

$$I_D = \frac{e - e_{\min}}{e_{\max} - e_{\min}} \tag{23.1}$$

where $e_{\max}$ and $e_{\min}$ are the maximum and the minimum void ratios of the sand, respectively. The maximum and minimum void ratios are generally determined in accordance with the procedure described in ASTM D2049.

As pointed out, for example, by Green and Padfield (1983), the relative density is difficult to determine in the field owing to the disturbance caused by the sampling. Furthermore, the relative density is often not a very satisfactory index of the compressibility, the shear strength, or the liquefaction potential of granular soils because these properties are affected not only by the relative density but also by such factors as preloading, time, content of fines, and the effective overburden pressure. Another difficulty is the changes of particle size, particle-size distribution, and of maximum and minimum void ratios that occur in many deep compaction methods owing to the crushing of the soil particles during the compaction. It is therefore often preferable to relate the required compaction to a minimum penetration resistance determined by penetration tests, CPT or STP, rather than to a certain relative density. However, the results from the penetration tests should be corrected with respect to the effective overburden pressure and thus to the depth and the location of the groundwater table.

Often a relatively high relative density is specified for earth and rockfill dams to reduce the risks of liquefaction. A minimum relative density of 80 percent has, for example, been prescribed by Iyengar (1983) for an earth dam in India. For the Jebba Dam in Nigeria, a minimum relative density of 60 percent was required down to 20 m depth. At the Changi Airport in Singapore a relative density of 75 percent was specified (Choa et al., 1979). For vibratory machinery such as generators, a minimum relative density of 80 percent is frequently required.

Local variations of the compaction are normally not significant. Such variations can be caused by lenses of organic material or by bad workmanship. This can be checked with penetration tests—CPT, SPT or WST.

23.2 VIBRATION METHODS

23.2.1 Vibroflotation

The oldest of the deep compaction methods is vibroflotation, which was developed in Germany in the 1930s. In this method, 0.30-m to 0.45-m diameter rod vibrators (vibroflots) are used, as illustrated in Figure 23.1. The lateral vibrations of the vibrator are caused by rotating eccentric weights. The vibrations can be transmitted to the soil along the whole length of the vibrator (Keller system) or at the tip (Bauer system).

The efficiency of the vibroflotation method has gradually been improved. Vibrators with up to 300 kN centrifugal force and a frequency of 30 to 50 Hz are available (as at 1990). The maximum amplitude is 5 to 23 mm. The total weight of the vibrators is 4 to 8 tonnes and the length is 2.0 to 4.5 m. There is a trend to increase the amplitude and the weight further and to reduce the frequency.

The vibrators are provided with water jets top and bottom to improve the penetration rate and to remove some of the fines in the soil. Air can be injected to improve the penetration at large depths. Compressed air is used to stabilize the borehole in the dry vibro-replacement method.

The vibrators can also be provided with fins to increase the efficiency and to reduce the twisting during compaction. A cutting ring is used to widen the hole when there are cemented layers or seams in the soil.

The vibroflot is vibrated and jetted down to the required depth of the compaction. Typical penetration rates are 1.0 to 2.0 m/min. At the required depth the bottom water jets are turned off and the side jets are turned on. The purpose of the side jets is to loosen the soil above the vibrator. The volume of water should be sufficient to stabilize the borehole and to compensate seepage looses. The flow rate is up to 3 m³/min and the maximum water pressure is 0.8 MPa. The water level in the borehole should always be kept higher than the groundwater level, to prevent caving of the sides of the hole.

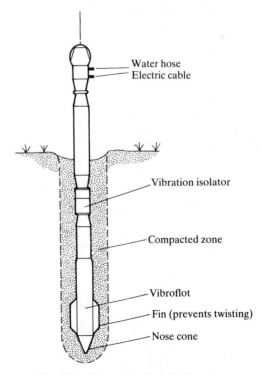

Fig. 23.1 Compaction by vibroflotation.

The vibroflot is then slowly withdrawn in 0.3-m to 0.5-m intervals in order to compact the loose sand to about 1.5 to 3.0 m around the probe. The withdrawal rate is normally 0.3 m/min. About 0.5 to 1 min of compaction is required at each level. The additional improvement by extending the duration of compaction to more than 1 to 3 min at each level is generally small. Silty soils, calcareous sand, and coral debris may even be weakened by the additional vibration.

The crater that develops at the surface around the probe during compaction is filled with granular material. Up to 1.5 m³/m length might be required. The most suitable backfill material is coarse sand or gravel with little or no fines. The preferred particle size is 10 to 40 mm. Coarse gravel can, however, cause arching around the vibrator during withdrawal. The granular material can also be added through the vibroflot at the bottom of the vibrator (bottom feeding), thereby reducing the risk of arching. Compressed air is often used to force the material down through the extension tube (Brown, 1977).

The maximum depth of the compaction is usually 25 m. However, at the Jebba Dam in Nigeria the method was used down to 30 m depth. Vibroflotation has also been used in Canada down to the same depth (30 m) in order to reduce the liquefaction potential (Solymar and Reed, 1986). The production rate was high: 300 to 600 m per probe per 10-hour shift.

The power consumption required to operate the vibrator increases as the relative density of the soil increases. This effect can be used to monitor the compaction during the withdrawal (d'Appolonia, 1953; Brown, 1977). However, there are also several other factors besides relative density and degree of compaction that affect the power consumption (peak demand). Morgan and Thomson (1983) suggest that it is advisable to check also the amplitude and the frequency during the compaction using an accelerometer attached to the tip of the vibrator, since the amplitude decreases in general with increasing compaction. However, it is not well known how the relative density affects the amplitude. Calibration is therefore required at each site, with for example, CPT or SPT.

The compaction is generally highest next to the vibrator. The relative density is usually about 100 percent up to 0.30–0.55 m from the probe. The compaction may be reduced just around the hole left after the vibroflot. The compaction decreases gradually with increasing distance from the probe. The lowest relative density is usually obtained halfway between the compaction points. It has been observed that the penetration resistance can be reduced temporarily some distance (3 m) away from the vibrator just after the compaction. The variation of the relative density can be large depending on the soil conditions and the spacing. An average relative density of 75 percent can normally be obtained for clean sand (Fig. 23.2) when the spacing is 1.5 to 2.0 m.

The largest uncertainty with vibroflotation is the evaluation of the compressibility and the thickness of the compacted zone. It is often assumed that the compaction and the reduction of the compressibility correspond to the volume of the added material and the surface settlements during the compaction. It is therefore important to measure both the volume of the added material and the settlements around the compaction point. The results of the compaction are often expressed in terms of a soil improvement factor (n), the ratio of the estimated settlements before and after the compaction.

Vibroflotation has mainly been applied to reducing the settlements of oil tanks, industrial buildings, earth and rockfill dams (e.g., West, 1976; Solymar, 1984), bridges, machine foundations, and other relatively flexible structures, and to reducing the liquefaction potential of silt and fine sand (Engelhardt and Golding, 1975; Dobson, 1987). Applications in the U.K. have been described by West (1976) and by Greenwood (1970, 1975). The method has also been used in

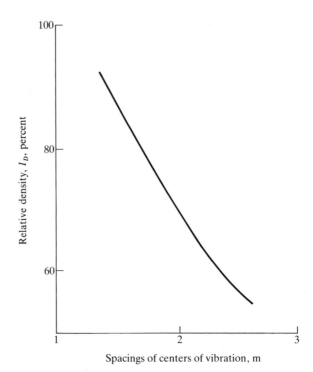

Fig. 23.2 Relative density (I_D) of clean sand midway between centers of vibration.

Canada (Solymar and Reed, 1986) and China (Wang et al., 1988) to reduce the liquefaction potential and in the United States (d'Appolonia et al., 1955).

Mainly sandy soils can be treated. The maximum content of fines in soils that can be compacted successfully by vibroflotation is about 20 to 25 percent, as illustrated in Figure 23.3 (Mitchell and Katti, 1981). The clay content should not exceed 3 percent. The effectiveness of vibroflotation is also reduced in gravelly soils when the coefficient of permeability exceeds about 0.01 m/s. The effectiveness is also reduced in cemented soils.

Additional compaction using a 3- to 5-tonne vibratory roller is generally required at the surface to densify the soil down to 1.0–1.5 m depth where the compaction with vibroflotation or vibrocompaction is poor.

Brown (1977) has proposed a suitability number (β) for the classification of soils with respect to the effectiveness of the compaction by the vibroflotation method:

$$\beta = 1.7 \sqrt{\frac{3}{(d_{50})^2} + \frac{1}{(d_{20})^2} + \frac{1}{(d_{10})^2}} \qquad (23.2)$$

This suitability number depends on the particle size d_{50}, d_{20}, and d_{10}, in millimeters, at 50, 20, and 10 percent passing by weight, respectively. The rating varies from excellent when the suitability number is less than 10 to poor and unsuitable when it exceeds 30. The time required for compaction decreases with decreasing suitability number and thus with increasing particle size.

Vibroflotation can in some cases be combined with other deep compaction methods to increase the effectiveness. For example, Johnson et al. (1983) used vibroflotation in combination with heavy tamping in order to compact a 25-m thick fill. The lower part of the fill was compacted by vibroflotation while tamping was used for the upper part. Andreu et al. (1983) and Basore and Boitano (1969) used piles in combination with vibroflotation because of the variation of the compaction obtained using vibroflotation alone. Solymar and Reed (1986)

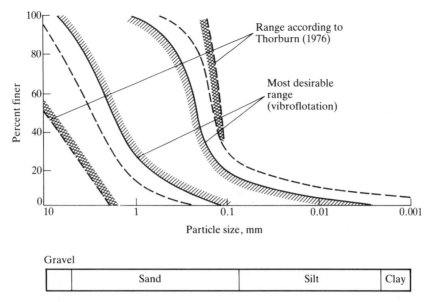

Fig. 23.3 Required grain-size distribution in vibroflotation. (*After Mitchell and Katti, 1981.*)

have described a case where vibroflotation was combined with blasting in order to reduce the liquefaction potential.

The most reliable method for checking the compaction by vibroflotation is with cone penetration tests (CPT). Standard penetration tests (SPT) are often unreliable, as pointed out by Green and Padfield (1983), and the interpretation of the results is uncertain. It should be noted that the penetration resistance increases in general with time. It is not unusual that the penetration resistance after one month can be 50 to 100 percent higher than that immediately after the compaction.

It is often necessary to calibrate the vibroflot before the start of the compaction at a particular site to determine the thickness of the compacted zone, the required spacing, and the relative density that can be reached. The amplitude and frequency of the vibrations can be checked with an accelerometer mounted at the tip of the vibrator.

23.2.2 Vibrocompaction

Vibrocompaction can be used to compact granular soils down to about 40 m depth. In this deep compaction method, a vibratory hammer is used, which is attached to a pipe or a probe that is vibrated down into the soil. Examples are the Vibro-Wing, the Foster Terraprobe, and the Franki Y-Probe. Other shapes of the probe have also been found to be effective (Saito, 1977).

In contrast to vibroflotation, the vibrations are now in the vertical direction, and the amplitude is 10 to 25 mm. The production rate is high at vibrocompaction, 4 to 5 times higher than that of vibroflotation, but the increase in the relative density of the soil around the probe and the affected volume are less (Brown and Glenn, 1976).

In the *Vibro-Wing* method, shown in Figure 23.4, a steel rod that is provided with 0.8-m long "wings" spaced 0.5 m apart in the vertical direction is driven down by a vibratory hammer to the required depth of the compaction. The shaft is pulled out slowly as the soil is vibrated. The withdrawal rate depends on the time required to reach the required relative density. The rate is mainly governed by the permeability of the soil. A higher withdrawal rate can be used for coarse sand than for fine sand.

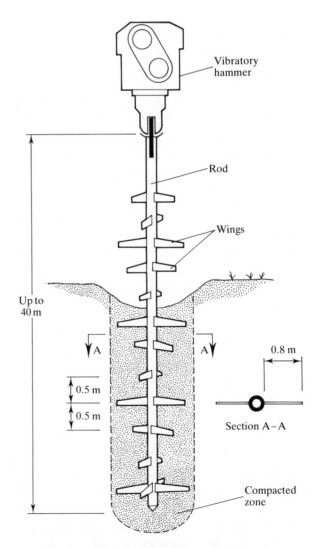

Fig. 23.4 The Vibro-Wing method.

The frequency of vibration is about 20 Hz (Massarsch and Broms, 1983; Massarsch, 1985).

The Vibro-Wing method has been used in Rostock Harbor, G.D.R. The compaction of the hydraulic fill, in this case a fine to medium sand, was done in a triangular pattern and the spacing was 2.5 m. A plate vibrator was used at the surface, where the compaction was poor. The compaction was checked by cone penetration and seismic crosshole tests. Figure 23.5 shows the results from CPT before and after the compaction (Massarsch and Broms, 1983).

In the *Terraprobe* method developed by Foster Engineering in the United States, an open-ended 0.76-m diameter pipe is utilized as illustrated in Figure 23.6a. The frequency of the vibratory hammer that is used to drive the probe is 15 Hz. The length of the probe is generally 3 to 5 m longer than the required depth of the compaction. The probe is vibrated down and extracted after the compaction of the soil by the hammer. Sand or gravel is added after the extraction to compensate for the settlements around the probe.

Several cases where the method has been used successfully in the United States have been described by the ASCE Committee on Placement and Improvement of Soils (ASCE, 1977), by Janes (1973), and by Leycure and Schroeder (1987).

The *Franki Y-Probe* is star-shaped as it consists of three long steel plates that are joined together at angles of 120° as shown in Figure 23.6b. The plates are 0.5 m wide and 20 mm thick (Massarsch, 1985). Ribs, which are spaced 2 m apart, are welded on both sides of the plates in order to increase the efficiency of the probe. The frequency of the vibrations can be varied to correspond to the natural frequency of the ground (10 to 15 Hz) to reduce the time required for the compaction.

Field tests are normally required for major projects to establish the optimum spacing, which is usually 1.5 to 3.0 m depending on the soil conditions. The results are generally checked with cone (CPT) or standard penetration tests (SPT). The method has been used successfully, for example, to compact a coarse sand fill below water in the harbor in Zeebrugge in Belgium (de Wolf et al., 1980, 1983; van Impe, 1985).

Wallays (1982a, b), Holeyman and Wallays (1984), and Holeyman and Broms (1986) have related the degree of compaction that can be achieved, and thus the efficiency of the method, to the silt and clay content of the soil and thus to the sleeve friction ratio (FR) as determined by cone penetration test (CPT). The relative improvement has been found to decrease with increasing friction ratio and with increasing initial relative density of the soil and thus with increasing cone resistance (q_c). Saito (1977) has suggested that at least part of the increase of the penetration resistance is caused by an increase of the lateral earth pressure from the vibratory compaction.

The best results with deep vibratory compaction are usually obtained with clean sand and gravel below the groundwater level where the soil is saturated. The method is effective only when the content of fines (silt- and clay-size particles) is less than 20 to 25 percent (Greenwood and Kirsch, 1983). A relative density of 70 to 80 percent can usually be obtained below 25 m depth and higher values at shallow depths (Solymar and Reed, 1986). The efficiency decreases rapidly with increasing silt and clay contents. The clay content should preferably be less than 3 percent. The particle size, gradation, relative density, and depth are also important. The compaction is generally poor close to the surface down to 1.5–2.0 m depth. Additional compaction using a 3- to 5-tonne vibratory roller is usually required. A relatively large area (> 1500 m²) has to be treated before the method becomes economical because of the relatively high costs for the mobilization of the equipment.

The required spacing of the compaction points depends on the permeability of the soil and on the depth. The spacing is normally 1.0 to 2.0 m. Because of the vertical vibrations of the

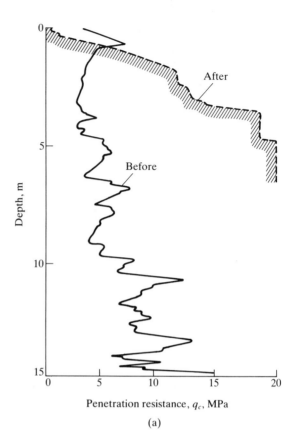

Penetration resistance, q_c, MPa

(a)

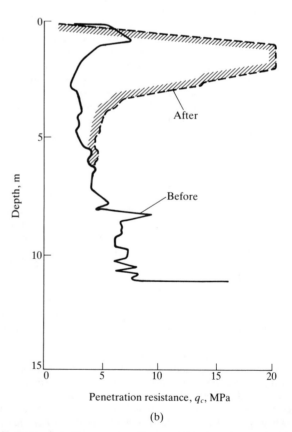

Penetration resistance, q_c, MPa

(b)

Fig. 23.5 (a) Cone penetration resistance before and after compaction by the Vibro-Wing method. (b) Cone penetration resistance before and after surface compaction (2.5 × 3 m plate.)

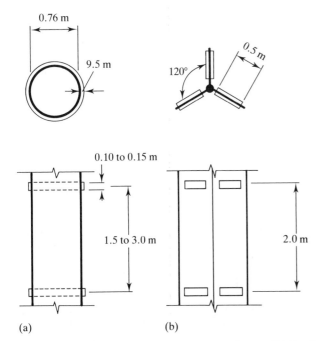

Fig. 23.6 Vibrocompaction methods. (a) L. B. Foster. (b) Franki Y-Probe.

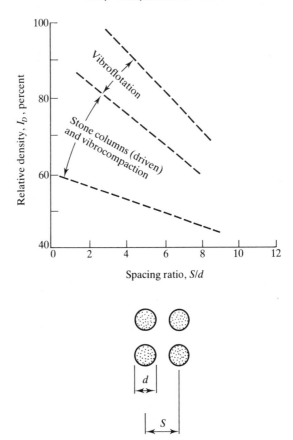

Fig. 23.7 Comparison of different compaction methods.

probe (Fig. 23.7), a smaller spacing is generally required for vibrocompaction compared with vibroflotation. A higher average relative density can in general be obtained at the same spacing with vibroflotation than with vibrocompaction. The spacing must often be reduced in coarse sand compared with that required for fine sand with a relatively low permeability because of the reduced width of the compacted zone around the probe. The spacing must also be reduced in silty soils because of the reduced efficiency with increasing silt and clay content. Even a few percent of silt- or clay-size particles can considerably reduce the effect of the compaction. The effect of the content of fines ($< 74\ \mu m$) on the compaction that can be achieved is shown in Figure 23.8. It can be seen that even a small amount of silt- and clay-size particles has a large effect of the maximum penetration resistance that can be obtained.

Most of the compaction is obtained during the first 2 to 5 minutes. Therefore it is generally more economical to reduce the spacing in order to reach the required degree of compaction than to increase the compaction time. A higher relative density can be obtained with coarse sand than with fine sand. The time required for the compaction depends mainly on the permeability and thus on the content of fines. The effect will be low when the permeability is less than 10^{-5} m/s. The maximum capacity is 10 to 15 probings/hour.

The densification varies with depth. Leycure and Schroeder (1987) observed that the highest relative density was obtained at 5 m depth and that the density decreased with increasing overburden pressure. The low density next to the surface was attributed to "overvibration" which occurs when the particle acceleration is high and the effective overburden pressure is low. Similar observations have been made by Janes (1973). Also the slope of the ground surface affects the compaction. Leycure and Schroeder (1987) report that the relative density that could be achieved was reduced by about 10 percent when the ground was sloping.

The size of the liquefied zone around the probe and the compaction will be limited in coarse sand and gravel when $k > 10^{-2}$ m/s. Gravel and coarse sand are more difficult to compact by vibrocompaction than medium to fine sand.

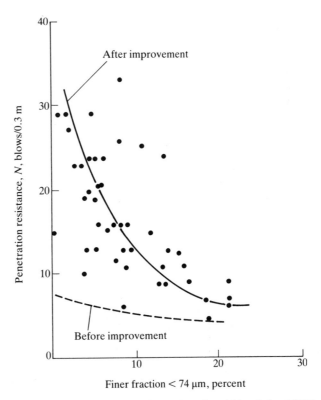

Fig. 23.8 Effect of fines in vibrocompaction. (*After Saito, 1977.*)

Vibroflotation is preferred in these soils because of the lateral displacements of the vibrator during compaction.

23.2.3 Blasting

Granular soils have been compacted successfully by blasting down to about a depth of 40 m in order to reduce the liquefaction potential and the settlements. This method is usually very economical when it can be used, even when the area to be treated is relatively small. The largest obstacle in many countries is often obtaining permission to use and to store the explosives required for the blasting.

Blasting is mainly used below the groundwater level, since the compaction is caused by partial liquefaction of the soil. The method can also be used when the silt and clay contents of the soil are relatively high because of the disturbance of the soil caused by the blasting. It has also been used successfully to compact loess after prewetting (Litvinov, 1966; Abelev, 1976).

Blasting is not a new method. In the U.S.S.R., it has mainly been used for harbors, roads, and airfields, and for earth and rockfill dams to reduce the risk of liquefaction and the settlements (Litvinov 1973; Denchev 1980; Ivanov, 1980, 1983). In the United States, the method has been used, for example, to compact a 6-m deep layer with fine to medium sand at the Franklin Dam (Lyman, 1942). Blasting has also been used in Canada for the compaction of sand tailings (Klohn et al., 1981) and to reduce the liquefaction potential of sand (Solymar and Reed, 1986).

Barendsen and Kok (1983) applied the method in Amsterdam Harbor, where the cost for the blasting was less than half of the estimated cost for other soil improvement methods. The method was also used in the harbor of Zeebrugge in Belgium (Carpentier et al., 1985; van Impe, 1985) and in the Gdansk Harbor (Dembicki et al., 1980b; Dembicki and Kisielowa, 1983). There the blasting reduced the volume by 6 percent. The relative density was increased from 35 percent to over 80 percent.

The main limitations of the method are the risks involved with the explosives and the difficulties in predicting the results. Interval blasting can be used to increase the effectiveness of the method and to reduce the risks (Dowding and Hryciw, 1986). However, adjacent structures may be damaged by the vibrations generated by the blasting and landslides can be triggered.

Explosive charges are typically placed 3 to 6 m apart in jetted or drilled boreholes at elevations that correspond to 50–75 percent of the desired depth of the compaction.

Individual charges are usually 1 to 12 kg or 10 to 30 g/m^3 of the soil to be compacted. The spacing of the boreholes is usually 5 to 15 m (for example, Prugh, 1963). The charges can also be placed under water just above the mudline (Dembicki et al., 1980a; Ivanov, 1980). Compaction can generally not be increased further by increasing the quantity of explosives or by reducing the spacing of the boreholes.

Blasting is often repeated not more than two to three times since the improvement after three rounds is usually small. It should be noted that the density of the soil just around the detonation point will be low and that layers that are initially very dense ($D_r > 0.7$) may be loosened by the blasting. The results therefore vary within the compacted soil mass.

The increase of the relative density is generally 15 to 30 percentage points. The largest improvement is obtained when the soil is initially loose. A relative density of 65 to 75 percent can normally be reached below 10 to 12 m depth. At shallower depths the maximum relative density is 75 to 85 percent (Solymar and Reed, 1986). The surface settlement is usually 2 to 10 percent of the total thickness of the compacted layer.

Loose sand is usually compacted down to a depth that corresponds to 1.5 times the depth of the charge. For medium dense sand the depth of compaction is 1.2 to 1.3 times the depth. The maximum depth that can be compacted is usually 15 to 20 m.

The efficiency of the method depends mainly on the pore water pressures that are generated by the blasting, and the size of the liquefied zone around the detonation point. The compaction by the blasting depends mainly on the factor $\sqrt[3]{W}/R$, where W is the size (in kilograms) of the charge and R is the radius (in meters) of the compacted zone. Figure 23.9 shows that the increase of the pore water pressure due to the blasting will be small if $\sqrt[3]{W}/R$ is less than 0.09 to 0.15 (Barendsen and Kok, 1983). The content of fines should be less than 25 percent.

The degree of saturation of the soil is also important. Even a small amount of gas will reduce the effects of the blasting significantly. Loose saturated fine sand is compacted up to 15 to $25\sqrt{W}$ (m) while for dense fine sand the compaction is less than $7\sqrt{W}$ (m), where W is the charge in kilograms. The relative density of the soil can be increased up to 10 m or more from the detonation point. Thus, the volume of the compacted soil is much larger than with vibroflotation or vibrocompaction.

High pressures are generated in water and in saturated sand by blasting. The peak pressure from a 1-kg charge may exceed 14 MPa at a distance of 4 m (Langefors and Kihlstrom, 1978).

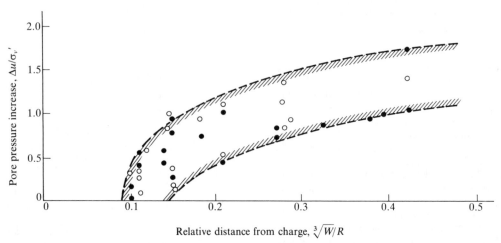

Fig. 23.9 Relationship between charge and pore pressure increase. (*After Barendsen and Kok, 1983.*)

A few large charges are often more effective than a large number of small charges. However, the required embedment depth has to be increased with increasing size of the charge so as to prevent craters from forming at the surface.

The gas from the explosions escapes through fissures and cracks in the soil, which can reduce the efficiency of the compaction. The surface settles after the gas has escaped and the excess pore water pressures have dissipated. Additional compaction is generally required at the ground surface because of the low relative density of the soil down to about 2 to 3 m depth. Pilot et al. (1981) have pointed out that the thickness of the poorly compacted surface zone increases with increasing size of the charges.

The effect of the blasting depends mainly on its ability to break down the initial soil structure. It also depends on the increase of the pore water pressure (Δu) and the resulting reduction of the effective overburden pressure (σ'_{vo}). Barendsen and Kok (1983) have shown that the ratio ($\Delta u/\sigma'_{vo}$) should be at least 0.8. The peak pore water pressure from the blasting is about the same in water and in saturated sand, but is greatly reduced when the soil is partially saturated. Even a small air content (> 1 percent) will reduce the effectiveness of the blasting.

There is at present (1990) no generally accepted method available for predicting the compaction caused by blasting (Mitchell, 1981). The increase of the settlement decreases for each charge with increasing number of charges (Prugh, 1963; Kummeneje and Eide, 1961). The first charge is normally responsible for 50 to 60 percent of the total settlement. The additional settlement by the second charge is about 20 percent. Figure 23.10 shows the settlement of mine tailings as a function of the distance (R) from the buried charge (Ivanov, 1983). As expected, the settlements were the largest just above the charge.

Test blasting is recommended for large jobs in order to determine the optimum spacing of the boreholes, the size of the charges, and the intervals and the sequence of the blasting (Carpentier et al., 1985).

Compaction caused by blasting is generally checked by measuring the surface settlements. The compaction can also be determined with cone (CPT) or standard penetration (SPT) tests and or with weight soundings (WST). The initial increase of the penetration resistance just after the blasting has in some cases been low despite relatively large surface settlements. However, the penetration resistance of clean sand often increases with time, up to several weeks or months after the blasting,

owing to "aging" (Mitchell and Solymar, 1984; Mitchell, 1986). In order to determine whether and when the blasting should be repeated it is often advisable to measure the excess pore water pressures that are generated and also the gradual dissipation of the excess pore water pressures with time.

23.3 DISPLACEMENT METHODS

23.3.1 Stone, Gravel, and Sand Columns

Stone and gravel columns and sand compaction piles (vibro-replacement) can also be used to compact granular soils down to about 25 m depth. Different methods are used to install the stone or gravel columns and to compact the granular material in the columns. For example, the casing used for the installation can be redriven several times during the withdrawal. The stone and gravel columns can also be grouted with cement mortar in order to increase the bearing capacity and to reduce the settlements.

In the *Franki method* an open thick-walled steel casing with 0.5 to 0.7 m diameter is used (Fig. 23.11a). The casing, which is initially closed at the bottom by a gravel plug, is driven down to the required depth by an internal 4- to 9-tonne drop hammer. At the desired depth of compaction, the gravel plug is extruded by further driving so that a foot or a bulb is formed below the bottom of the casing. Additional material is added as the casing is withdrawn. The sand around the column, and the stones or gravel in the column shaft are subsequently compacted by the hammer during the withdrawal of the casing.

The diameter of the stone or gravel columns depends mainly on the shear strength and the compressibility of the surrounding soil. It can be estimated from the volume of the material (stones or gravel) added and the settlements of the ground around the columns during the compaction. In loose sand the diameter of the stone or gravel columns is about 0.8 m. The spacing is usually 1.5 to 3.0 m, depending on the initial relative density of the deposit and required degree of compaction. In general the required spacing decreases with decreasing particle size. The maximum length of the columns is about 35 m.

Wallays (1982b) has proposed a method of estimating the improvement of the relative compaction in terms of a factor (f) that depends on the volume of material added and on the silt and clay contents of the soil which reduce the efficiency of the method.

Vibroflotation can also be used for the installation of stone or gravel columns (e.g., Glover, 1982; Munfakh, 1984) as illustrated in Figure 23.11b. Granular material (stone or gravel) is added during the compaction as the vibrator (vibroflot) is being withdrawn. The soil is liquefied locally around the vibrator during compaction and new material can be added either at the surface (top feeding) or through the extension tube and the vibrator (bottom feeding). The thickness of the liquefied zone is 0.30 to 0.55 m (Thorburn, 1976). The thickness of the liquefied zone decreases with increasing permeability and thus with increasing particle size. The improvement of the compaction beyond 2.5 m from the vibrator is usually insignificant regardless of the duration of the compaction. The diameter of the columns will vary with the relative density of the surrounding soil. The diameter is generally larger at the surface than at the base. Also, soils with a relatively high silt content (up to 30 percent) can be compacted (Thorburn, 1976) provided the spacing of the columns is small. The time required for constructing a single stone column is 10 to 30 min. The main disadvantage of the method is often the difficulty of disposing of the excess water from the installation of the stone columns.

Stone columns were used successfully by, for example, Thorburn and MacVicar (1974) to improve the foundation for

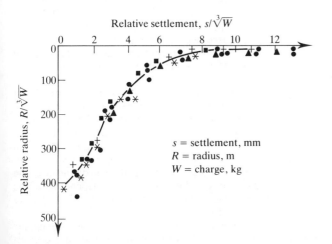

Fig. 23.10 Relationship between settlement and size of charge. (*After Ivanov, 1983.*)

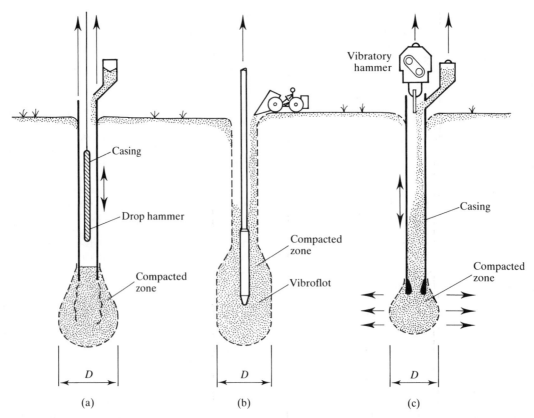

Fig. 23.11 Compaction methods. (a) Franki stone columns. (b) Vibroflotation. (c) Sand compaction piles.

a three-story apartment block in Glasgow, Scotland. The reduction of the costs was substantial, up to 50 percent compared with alternative foundation methods (Thorburn, 1976). The method has also been used by Andreu et al. (1983) to reduce the liquefaction potential at a thermal power plant in Spain. There the relative density was increased by more than 20 percent. The required minimum penetration resistance as given by CPT after the compaction was 10 MPa.

In India (Datye and Nagaraju, 1981; Datye, 1982; Ranjan and Rao, 1983; Datye and Madhav, 1988) the equipment used for the installation of bored piles has been utilized for the manufacture of stone and gravel columns. Well-graded granular material is compacted inside the casing by a simple drop hammer during the withdrawal of the casing. The diameter of the resulting stone or gravel columns is relatively small, 250 to 350 mm.

Also, a casing with a trap door (flap valve) at the bottom has been used at times for the manufacture of stone and gravel columns. When the casing is withdrawn, the hole is filled with stone, gravel, or sand through the flap valve. This method has been used by Solymar (1986) to reduce the liquefaction potential of loose silty fine sand in Java, Indonesia.

Air can be used above the groundwater table to keep the hole open. The method has been used in Hong Kong for oil-storage tanks on reclaimed land. In Canada, the method has been used to reduce the liquefaction potential. There the length of the columns was 7.5 to 12 m (Solymar and Reed, 1986). In this case, the sand consumption was 0.51 m³/m column. This corresponds to an average diameter of the columns of 0.76 m.

To compact the granular material in the shaft the casing can be redriven several times during the withdrawal. The density of the upper part of the stone or gravel columns can also be improved by confining the soil around the columns at the surface with a skirt, or a large-diameter concrete pipe placed around

the columns (Ranjan and Rao, 1983; Rao and Ranjan, 1985). It is possible to reduce further the settlements by grouting the columns (Smoltczyk, 1983; Jebe and Bartels, 1983). Grouting is especially effective close to the ground surface, since the applied load is transferred to the columns within a depth that corresponds to about three to four column diameters.

Sand compaction piles using the Vibrocompozer method are common in Japan as described by Aboshi et al. (1979) (Fig. 23.11c). Since the method was developed in the late 1950s more than 80 million meters of sand compaction piles have been installed in Japan mainly in cohesive soils (Murayama and Ichimoto, 1982) and in hydraulic fills. The sand compaction piles increase the consolidation rate in clayey and silty soils because they also function as drains. They have also been used outside Japan, for example, in Taiwan (Moh et al., 1981).

In the Vibrocompozer method a heavy 0.4-m to 1.5-m diameter casing is used, which is driven down to the required depth of the compaction by a vibratory hammer. The vertical force is 0.4 to 0.6 MN, the amplitude 15 to 18 mm, and the frequency 10 Hz.

During the driving, the casing is closed at the bottom by a sandplug. The inside diameter of the casing is reduced at the bottom to prevent the displacement of the plug. After the casing has been driven to the required depth, it is partially filled with sand or gravel, which flows out when the casing is withdrawn. The casing is redriven several times during the withdrawal to compact the sand in and around the shaft. The diameter of the resulting sand columns is about 0.7 to 2.0 m, depending on the diameter of the casing and on the soil conditions. The spacing is normally 1.7 to 2.5 m.

Stone or gravel columns are usually placed in a triangular or rectangular pattern. The compacted zones should overlap in order to reduce the settlements when the stone columns are used to support footings. The bearing capacity and the settlements depend mainly on the compressibility of the compacted material. Design charts have been proposed by d'Appolonia (1953) and by Thorburn (1976). An allowable bearing pressure of 0.5 MPa is often possible for clean sand.

The depth of treatment and the power consumption are generally recorded automatically in order to check the effectiveness of the compaction. The penetration resistance of the sand columns as determined by STP is usually about 30 blows/0.3 m. However, the resistance is affected by the properties of the surrounding soil as well as by the particle size of the added material.

Stone and gravel columns have been used to reduce the settlement of oil-storage tanks, grain silos, ore or coal storage yards, roads, airfields, industrial buildings, and low-cost housing. Stone columns have also been used to increase the bearing capacity along the edges of oil tanks and of embankments where the bearing capacity is low because of the low confining pressure. However, the settlement of buildings will often be excessive when they are supported on stone or gravel columns that have not been compacted by redriving the casing during the installation or have not been grouted. Gravel and stone columns have also been used successfully to reduce the liquefaction potential (Seed and Booker, 1977). The gravel should be well graded to prevent clogging of the columns and loss of drainage efficiency.

One large advantage with stone and gravel columns is their flexibility and high ductility compared with concrete piles. The required length of the columns will be less than that of end-bearing piles because of their high skin friction resistance and large surface area. Often only the upper parts of the columns have to be compacted because of the reduced compressibility of the soil with increasing depth. It is rarely necessary to compact more than the upper 8 m (Greenwood and Kirsch, 1983).

The initial spacing of stone or gravel columns is generally relatively large. A square or a triangular grid pattern can often be used. Additional columns can be installed between the initial columns when the improvement halfway between the compaction points has not been sufficient. The final spacing is usually 1.5 to 3.5 m, depending on the particle size and content of fines.

The maximum improvement is governed by the content of fines. Hussin and Ali (1987) have reported that no appreciable improvement was obtained when the content of fines exceeded 12 percent.

Stone columns can be placed in clusters or rows when they are used as foundation for buildings. A 0.3-m thick sand blanket is usually constructed at the surface as a working pad when stone columns are used in soft clays. In cohesive soils the columns function both as large-diameter drains in the soil and as load-carrying members that reduce the settlement. The compaction of the soil can be checked by, for example, cone penetration tests (CPT) or standard penetration tests (SPT).

Wallays (1982b) has related the improvement that can be obtained by vibrocompaction or with stone columns to a factor f that is defined as the ratio of the average cone resistance after and before compaction. The factor f is a function of the volume of sand or gravel added (s) and the volume of the zone of influence (S):

$$f = 1 + \frac{s}{S} \tan \alpha$$

The term $\tan \alpha$ is a coefficient that depends on the gradation index of the soil and on the initial cone resistance. The average expected increase of the penetration resistance is thus 60 percent at $\tan \alpha = 6$ and $s/S = 0.1$. At $\tan \alpha = 10$ the expected increase is 100 percent.

Wallays (1982b) proposed the relationship shown in Figure 23.12, which relates the maximum ($\tan \alpha_m$) and the minimum

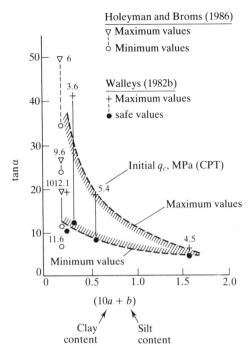

Fig. 23.12 Effect of the clay and silt content on compaction. (*After Wallays, 1982b.*)

value ($\tan \alpha_s$) with the gradation index at an initial cone resistance of 5 MPa. The gradation index is a dimensionless number ($10a + b$), where a and b are the fractions finer than 2 μm and 74 μm, respectively. Thus the reduction of the compaction caused by 1 percent clay is equivalent to that by 10 percent silt. The gradation index takes into account the observation that clay-size particles have a much larger effect on the maximum compaction that can be obtained compared with silt. Test data indicate that the value of $\tan \alpha$ will increase when the initial cone resistance is less than 5 MPa. It is reduced when the resistance exceeds 5 MPa.

The friction ratio as determined by electrical cone penetration tests is an indication of the silt and clay contents of the soil and thus of the increase of the relative compaction that can be obtained with stone columns. The friction ratio therefore corresponds to the term ($10a + b$) proposed by Wallays (1982b). In Figure 23.13 the factor $\tan \alpha_m$ has been plotted at $s/S = 0.10$

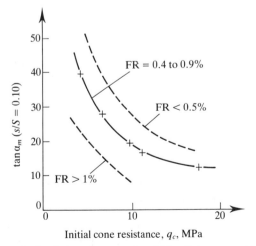

Fig. 23.13 Prediction of compaction. (*After Holeyman and Broms, 1986.*)

as a function of the initial cone resistance (q_c) at different values of the friction ratio (FR). It can be seen that the relative improvement that can be reached decreases with increasing friction ratio (FR) and with increasing initial penetration resistance of the soil. The improvement is small when FR > 3 percent. It is thus important to monitor closely the volume of the material added so that the increase of the relative density of the compacted zone and the factor s/S can be estimated.

Settlements can generally be estimated from the following assumptions:

- The change of the average void ratio corresponds to the volume of the displaced soil and the settlement of the ground surface during the compaction
- The stone or gravel columns and the surrounding granular soil behave as a composite material
- The stress distribution in the ground corresponds to the confined modulus of the stone or gravel columns and that of the compacted granular material between the columns as discussed by, for example, Aboshi et al. (1979) and Goughnour and Bayuk (1979). Thus, the area ratio of the stone columns and the surrounding compacted soil has a large effect on the estimated settlements. It should be noted that the confined modulus of both the stone columns and of the surrounding compacted soil increases with increasing settlement.

The settlement can also be estimated by the finite-element method (FEM) (Balaam et al., 1977) when the spacing of the columns is small. It is then assumed that the settlement and the bearing capacity will correspond to that of dense sand.

The main uncertainty with the stone column method is the evaluation of the compressibility of the compacted material in the columns. A stress ratio of 4 to 12, the ratio of the axial effective stress in the columns and the vertical effective stress in the soil around the columns, has been reported by Morgenthaler et al. (1977) for a sandy silt with gravels. The confined modulus is generally in the range 40 to 70 MPa.

The compaction can be checked with load tests, penetration tests (CPT or SPT), or pressuremeter tests. Penetration tests should be carried out at different distances from the vibrator, for example, at 0.5, 1.0, 1.5, and 2 m, so that the extent of the compacted zone can be determined. A test installation is generally required for large jobs. However, the height of the test fill has to be large because of the usually high bearing capacity of the compacted material. Large-diameter plate load tests can also be used.

If the Franki method is used the machine itself which is required for the installation of the stone columns can be utilized to determine the average confined modulus of the compacted soil and hence the improvement by the compaction. The machine is relatively heavy and it is equipped with four pads so that it can be moved. The settlement of the centre pad when the machine is lifted from the ground is measured before and after the installation of the columns. The reduction of the settlements after the compaction can then be determined directly, and thus the improvement factor n. Stone or gravel columns will normally reduce the settlements 40 to 60 percent.

23.3.2 Compaction Piles

Pile driving is one of the most effective methods for compacting loose sand and gravel. The method is also effective in silty soils above the groundwater table because of the displacement caused by the pile driving. Compaction piles can therefore be used in finer-grained soils than can vibroflotation or vibrocompaction. The best effect is usually obtained below the groundwater table where the soil is saturated. The compaction is partly caused

by the vibrations from driving the piles and partly by the displacement of the soil caused by the piles. The diameter of the compacted zone around each pile is $7D$ to $12D$, where D is the pile diameter (Robinsky and Morrison, 1964; Kishida, 1967). The size of the compacted zone increases in general with increasing initial relative density of the soil. The soil is also compacted below the piles down to a depth that corresponds to about one pile diameter. The maximum economic depth is about 20 m. It is usually possible to compact the soil to a relative density of 75 to 80 percent (Solymar and Reed, 1986). The method is economical for relatively small areas compared with other soil improvement methods.

The increase of the relative density of the soil can be estimated from the total volume of the inserted piles and from the settlement of the ground surface observed during the installation. The improvement of the compaction can also be checked with CPT, SPT, and WST. Test piles are generally required for determining the maximum depth to be compacted and the required spacing of the piles.

Building settlements can be estimated as if the structures were supported on dense sand. The reduction of the settlement by the compaction piles is generally large because the piles increase the horizontal pressures in the ground and thus the confinement. The relative improvement is less when the initial relative density of the soil is high.

Timber piles are often used for compaction piles because of their low cost and favorable shape. Timber piles should be treated to resist rotting. The high bearing capacity of tapered piles has been pointed out by Lindqvist and Petaja (1981). The use of tapered concrete piles is also common, especially if they will extend above the groundwater table.

The required spacing of the piles is generally 1.2 to 1.5 m. A relatively large spacing is normally chosen for the first few piles. Additional piles are then driven if the required penetration resistance and relative density have not been reached. In Finland (Jarvio and Petaja, 1983), 4.5-m long tapered precast concrete piles have been used to compact loose sand below a bridge pier. The piles were driven from the outside towards the centre of the treated area.

23.3.3 Dynamic Consolidation (Heavy Tamping)

Another method for compacting granular soils down to about 30 m depth is dynamic consolidation. Then a heavy steel or a concrete weight (6 to 172 tonnes has been reported in the literature) is dropped repeatedly 3 to 8 times at the same spot from 10 to 40 m height (Fig. 23.14). With a 6- to 30-tonne weight the soil is compacted down to 3 to 12 m depth, depending on the drop height and the mass of the weight.

In the U.S.S.R., 4- to 7-tonne weights dropped from a height of 5 to 7 m are used to compact loess. The weights are pear-shaped in order to increase penetration into the soil. The resulting holes in the ground surface are filled with sand or gravel. The soil can be compacted down to 2.0 to 3.5 m depth by this method (Abelev, 1976).

Compaction by dynamic consolidation is usually done systematically, in a rectangular or triangular pattern. The spacing between the impact points depends mainly on the depth of the compressible layer, the permeability, and the location of the groundwater level. Deep (1.0 to 2.0 m) craters are formed in the soil by the heavy weight when the weight of fall is large. The craters are filled with sand after each pass. The heave around the craters is generally small.

The repeated blows by the weights are believed to cause partial liquefaction of granular soils below the groundwater table and radial cracks around the point of impact (Menard

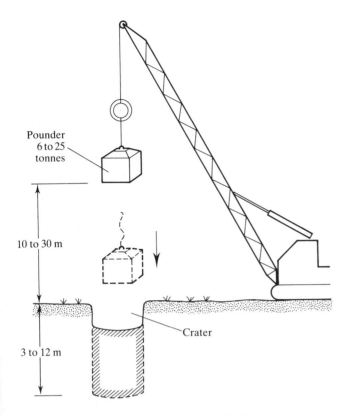

Pounder
6 to 25
tonnes

10 to 30 m

3 to 12 m

Crater

Fig. 23.14 Dynamic consolidation.

and Broise, 1975). The relative density of the soil is increased when water is squeezed out of the soil through the fissures. Reconsolidation of the soil can occur rapidly. Geysers and small wash-out cones have even been observed at the surface. The depth of the improvement is normally larger below the ground-water table where the soil is saturated compared with dry sand (Smits and DeQuelerij, 1989).

The initial spacing of the impact points is usually 5 to 10 m and corresponds to the thickness of the compressible layer. This is done in order to compact first the lower part of the layer. It is often advantageous to use a high height of fall for the first few blows to extend the compaction as deep as possible. The additional compaction obtained after the first five blows is often small. The spacing can then be reduced for the subsequent passes; thereby, the granular material located closer to the surface will also be compacted. A 1-m thick layer with free-draining material is normally placed over the area before the compaction to improve the transfer of energy to the soil. Menard and Broise (1975) suggested that a network of vertical cracks that improves the drainage is created by the regular systematic tamping. Irregular tamping can disrupt the continuity of this network.

Standard cranes are normally used to lift the weight ("pounder") when the required depth of the compaction is less than 10 m. The maximum capacity of many cranes is 25 tonnes. The maximum drop height is about 30 m. Special equipment is required when the soil below 10 m depth is to be compacted. A specially built tripod crane, the Giga machine with 1.7 MN lifting capacity, was used at Nice Airport (Gambin, 1983). Precautions have to be taken to prevent injuries to personnel and damage to the cranes when standard cranes are used.

The cross section of the weight is usually square, circular, or octagonal. Circular or octagonal weights are preferred for the initial tamping, while the square shape is of advantage

during the final "ironing" phase. Hollow weights have been used for compaction of granular soils under water (Hanzawa, 1981).

Dynamic consolidation is particularly useful for compacting rockfills below water and for bouldery soils where other methods cannot be used or are difficult to apply. The method has been used in Sweden (Hansbo, 1977) and in Norway (Bolgerud and Haug, 1983) for the compaction of rockfills. Soil layers containing up to 1-m diameter boulder have been compacted successfully by this method (Mori, 1977).

In the United States, Leonards et al. (1980) used dynamic consolidation to compact a 5-m thick layer with fine to medium sand. A 6-tonne weight was dropped from 12 m height in a square pattern. Successful applications of the method in the United States have also been reported by Keller et al. (1987), Kummerie and Dumas (1988), and by Song and Gambin (1988). Dynamic compaction has also been used successfully in Karlstad, Sweden to compact a hydraulic fill down to 10 m depth. The applied energy was 2.4 MNm/m^2. The compaction increased the pressuremeter modulus by more than 500 percent for the first 7 m. At 15 m depth the increase was more than 200 percent (Hansbo et al., 1974). In Finland, the method has been used to compact a silty sand for a bridge abutment (Hartikainen and Valtonen, 1983) and in Helsinki Harbor (Koponen, 1983). At Nice Airport in France a granular fill was compacted down to 10 m depth using dynamic consolidation. Dynamic compaction is commonly used worldwide (see, e.g., Mayne et al., 1983). The method has also been utilized in Lagos to compact a silty sand for a drydock (Gambin, 1983) as well as Saudi Arabia (Ghosh and Tabba, 1988). The method has also been tried in Singapore at the Changi airport (Choa et al., 1979) in combination with vertical drains and at Paya Lebar airport (Ramaswamy et al., 1982). In Canada, a 15-tonne weight with a height of fall of 15 m was used to reduce the liquefaction potential. The average increase of the cone penetration resistance was from 6 MPa to 10–15 MPa (Solymar and Reed, 1986). In China, the method has mainly been used to reduce the risk of liquefaction (Fan et al., 1988).

Dynamic consolidation has mainly been used to compact granular fills when there are no buildings nearby that can be damaged by the vibrations generated by the impact of the falling weight. Waste dumps, sanitary landfills, and mine waste have also been compacted successfully by heavy tamping (e.g., West and Slocombe, 1973; West, 1976; Welsh, 1983; Coupe, 1986). The method has also been used to compact silty soils such as loess, as discussed by Abelev (1976) and by Minkov et al. (1980). However, the effectiveness of the method decreases rapidly with increasing clay content. Keller et al. (1987) have found that the method was not effective when the clay content exceeded 10 percent. Similar results have also been reported by Hussin and Ali (1987). Reconsolidation of the soil generally occurs rapidly owing to the radial cracks that form in the soil during the impact (in French "claquage"). The compaction process is thus similar in many ways to that of blasting and vibro-compaction. Menard and Broise (1975) have suggested that the gas dissolved in the pore water plays an important role during the compaction and the following consolidation. The method has also been tried in clay but the results have been uncertain.

A single pass may be sufficient for coarse-grained soils, while for silty or clayey sands several passes might be required. The time interval between each pass is generally 1 to 3 weeks, depending on the dissipation rate of the excess pore water pressures that are generated by the pounding and thus on the permeability of the soil. Up to 3 to 4 weeks might be required for cohesive soils. It is usually preferable to compact granular soils with several passes and a relatively small number of blows at each pass. Because of the high cost of the mobilization of the equipment relatively large areas have to be treated

($> 10\,000\,\text{m}^2$) before the method becomes economic. The production rate can be as high as $10\,000\,\text{m}^2/\text{month}$.

The main limitations of dynamic consolidation are the limited depth of the improvement and the vibrations generated by the impact. Surrounding buildings or other structures may be damaged by the vibrations caused by the pounding. The frequency range of the vibrations is generally between 2 and 12 Hz. Dominating frequencies are 3 to 4 Hz (Menard and Broise, 1975). The frequency of the surface waves (Rayleigh waves) that are generated by the impact generally decreases with increasing distance from the point of impact.

In Figure 23.15 is shown the attenuation of the peak particle velocity as a function of the distance from the impact point. The peak velocity decreases approximately with D^2, where D is the distance from the impact. It should be noted that the peak velocity generally increases with increasing degree of compaction.

According to the German Standard DIN 4150 a maximum particle velocity less than 6 mm/s will probably not likely damage adjacent structures supported on spread footings owing to settlement of the underlying granular material. In saturated silt, even a small velocity (3 mm/s) can cause excessive settlements when the groundwater table is close to the ground surface. A much higher velocity (50 to 60 mm/s) is normally required to damage the structure directly. The velocity at a distance of 30 m from the point of impact is usually less than 50 mm/s (Menard and Broise, 1975). The height of fall has only a small effect on the amplitude of the vibrations. See also Mitchell (1981).

The depth (d) of the compaction can generally be estimated from the following empirical expression (Leonards et al., 1980):

$$d(m) = 0.5\sqrt{WH} \qquad (23.3)$$

where W is the mass of the pounder in tonnes and H is the height of fall in meters. The average depth varies in general between $0.3\sqrt{WH}$ and $0.8\sqrt{WH}$ (Mayne et al., 1984). The particle size and the impact velocity both affect the depth of the compaction. The depth generally decreases with decreasing particle size.

The increase of the relative density is generally highest close to the pounder; between one-third to one-half the compaction depth. It decreases with increasing distance below the ground

surface. The reported depth of the improvement varies between different investigators, depending on how the compacted zone has been defined. The surface settlement is generally 5 to 15 percent of the depth of the improved layer, depending on the initial relative density of the soil.

High lateral pressures develop in the soil during compaction and may exceed the effective overburden pressure. Thus settlements of buildings supported on spread footings or rafts will often be smaller than those estimated from penetration tests (CPT or SPT).

The maximum improvement generally increases with increasing applied energy. The required energy typically ranges from $1\,\text{MNm/m}^2$ to $6\,\text{MNm/m}^2$ in order to reach the desired compaction. The mass of the pounder and the height of fall depend mainly on the thickness of the layer to be compacted. The energy per blow (WH) usually varies between 1.5 and 5.0 MNm but can be as high as 10 to 20 MNm. It should be noted that the wire attached to the weight can reduce the energy by up to 20 percent (Lukas, 1986).

In Figure 23.16 is shown for clean to silty sand, the increase due to compaction of the cone penetration resistance (CPT) above a critical depth $0.5\sqrt{WH}$ after the compaction (Mayne et al., 1984). The penetration resistance after compaction is up to 40 to 50 blows/0.3 m at SPT. The limit pressure at pressuremeter tests is 2.0 to 2.5 MPa (Lukas, 1986). Normally the improvement of the penetration resistance for loose sand is 3 to 5 times. It can be seen that the results fall within a relatively narrow band. This figure can be used to estimate the maximum depth that can be obtained from a proposed compaction program.

It should be noted that the improvement at the edge of the compacted area will be less than at the center. The width of this poorly compacted zone is approximately twice the thickness of the compacted layer. It is therefore important that the treatment extends beyond the area that will be loaded.

The improvement is low down to a depth of about 1.5 to 2.0 m. The compaction is generally terminated by "ironing" the area with the height of fall of the weight reduced to 1–3 m. The spacing is also reduced so that the compacted areas overlap. The surface can also be compacted by a heavy vibratory roller (3 to 5 tonne).

Cone penetration tests (CPT), standard penetration tests (SPT) (e.g. Cleaud et al., 1983; Song and Gambin, 1988), weight soundings (WST) (Hartikainen and Valtonen, 1983), and pressuremeter tests (PMT) (e.g. Kummerie and Dumas, 1988)

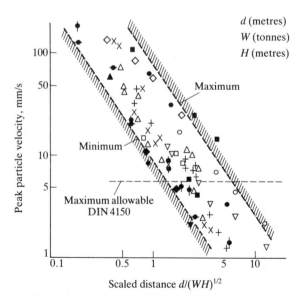

Fig. 23.15 Attenuation of ground vibration. (*After Mayne et al., 1984.*)

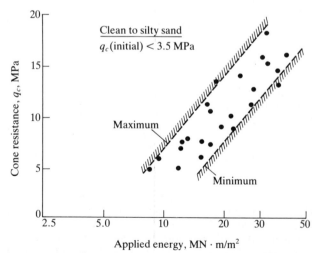

Fig. 23.16 Relationship between q_c and applied energy. (*After Mayne et al., 1984.*)

can be used to check the degree of compaction. About one penetration test is required every 2000 to 3000 m² after each pass. An increase of the penetration resistance of 300 to 400 percent can be expected in sand and gravel. In silt the increase is usually about 200 percent.

It is often desirable to carry out a compaction test before the mass (*W*) of the pounder, the height of fall (*H*), and the number of passes are selected. It is also desirable to measure the increase of the pore pressures in the soil and the vibration level (peak particle velocity) in adjacent buildings during the testing. Plate load tests (Hansbo, 1977, 1978) and dilatometer tests can also be useful. The depth of the crater and the settlement of the ground surface should also be determined after each blow during the test compaction. These measurements are helpful in selecting of the optimum number of blows per pass. However, test compaction is expensive because the equipment is very heavy and difficult to transport.

The improvement of rockfills is difficult to check. The most reliable method is to measure the settlements caused by the compaction. Surface settlements are typically 5 to 10 percent of the thickness of the compressible layers. In rockfills, the surface settlement can sometimes exceed 10 percent. Large-size plate load tests have been used at times. The wave velocity and the amplitude of the surface Rayleigh waves have also been taken as an indication of the effectiveness of the compaction. Another method for checking the compaction is measurement of the retardation when the weight strikes the fill. This can be done with an accelerometer attached to the weight (Hansbo, 1977; Jessberger and Beine, 1981). The reaction force of the underlying rock fill can be calculated from the known mass of the pounder and the measured retardation.

23.3.4 Squeeze and Compaction Grouting

Granular soils can also be improved by squeeze and compaction grouting. At *squeeze* grouting a relatively thin cement slurry is used, where the grout either penetrates into the soil or causes hydraulic fracturing ("claquage"). The orientation of the cracks and the fissures in the soil depends mainly on the initial stress conditions in the ground. Cracks normally develop in the compacted material perpendicularly to the direction of the minor principal stress. In normally loaded soils (where the current stress state has never been exceeded in the past) the cracks are generally vertical. In preloaded or over-consolidated soils the cracks are predominantly horizontal.

A relatively thick slurry is used in *compaction or consolidation grouting*, which is a commonly used method in the United States (e.g., Shilley, 1982; Warner, 1982; Baker et al., 1983; Schmertmann et al., 1986). The method has also been used in France (Plumelle, 1989; Robert, 1989). Compaction grouting is mainly applied in soft compressible silts and sandy silts below the groundwater table. The effectiveness of the method generally increases with increasing depth. The grout pressure is relatively high, 2 to 4 MPa. The diameter of the boreholes required for the grouting should be at least 50 mm. The spacing of the boreholes is generally 2 to 4 m for the primary and secondary boreholes. A closer spacing may be required when a high relative density is required. The maximum depth is about 30 m (Brown and Warner, 1973). Inclined boreholes reduce the effectiveness of the compaction grouting. The inclination should not exceed 20 degrees (Salley et al., 1987). The holes are generally grouted from the top down. The method is usually more expensive than vibrocompaction or dynamic compaction.

In loose sand the grouting may initially cause some settlements owing to the compaction. The increase of the lateral pressure in the soil by the grouting reverses the direction of the shear stresses in the soil, as pointed out by Escario (1983). The

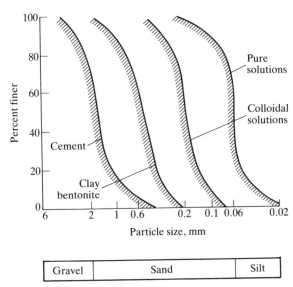

Fig. 23.17 Grouting of granular soils.

required volume of grout can be estimated from the required relative density of the soil. The results are generally checked by SPT or CPT. An improvement of the penetration resistance of SPT of 3 to 5 times is often obtained. At CPT the expected improvement is about 100 percent.

The main disadvantage with sqeeze and compaction grouting is the difficulty in predicting the results. The volume of the injected grout, the grout pressures, surface settlements and the heave should be measured and recorded during the grouting. Inclinometer pipes installed along the perimeter of the treated area are often useful in order to check the lateral displacements and the effectiveness of the grouting (Salley et al., 1987). Test grouting is often required in order to determine the improvement and the settlements caused by the grouting and the required spacing of the grout holes and the maximum grout pressure to avoid surface heave. Salley et al. (1987) recommended that the maximum surface heave should not exceed 19 mm (0.75 inch). Compaction grouting has also been used to strengthen the foundation for light structures constructed on silty soils.

Penetration grouting using cement, bentonite, silica, or different chemicals has also been tried for stabilizing granular soils. The main purpose of the grouting is usually to reduce the permeability of the soil below, for example, earth or rockfill dams rather than to reduce the compressibility or to increase the shear strength and the bearing capacity of the foundation soils.

The application of different grouts is shown in Figure 23.17. Normal cement grouts can only be used in coarse sand. Colloidal (silica fume) and pure solutions are required in fine sand. Chemical grouts are generally expensive and therefore seldom used. Sometimes there are problems with toxicity and with pollution of the groundwater. There are also some uncertainties about the long-term strength of some chemical grouts.

23.4 LOADING METHODS

23.4.1 Preloading

Preloading followed by complete or partial unloading can also be used to reduce the settlements of granular soils. The method has the advantages compared with other deep compaction

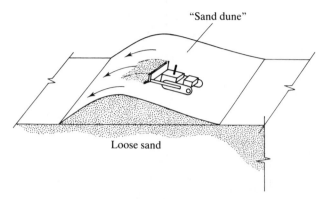

Fig. 23.18 Traveling dune. (*After Escario, 1983.*)

TABLE 23.1 COMPARISON BETWEEN DIFFERENT PENETRATION TESTING METHODS.

	Very Loose	Loose	Medium Dense	Dense	Very Dense
SPT, N_{30} blows/0.3 m	< 4	4 to 10	10 to 30	30 to 50	> 50
CPT, q_c, MPa	< 5	5 to 10	10 to 15	15 to 20	> 20
I_D, %	< 15	15 to 35	35 to 65	65 to 85	> 85
ϕ', degrees	< 30	30 to 32	32 to 35	35 to 38	> 38

[a] After Mitchell and Katti (1981).

methods that the bonding between soil particles is not destroyed during the loading (Escario, 1983), and that the compressibility can be greatly reduced with only minor changes of the unit weight and small settlements (e.g., Lee and Focht, 1975; Ladd et al., 1977; Ishihara and Okada, 1978a, b). An additional advantage is that compressible clay or silt layers in the granular material can be detected by measuring the total settlements and the settlement rates caused by the preloading.

The preloading can be done in the form of a traveling sand dune (Escario, 1983), as illustrated in Figure 23.18, when a heavy tractor is used to move the soil. The soil below the fill is subjected to high normal and shear stresses because the preloading increases both the vertical and the lateral pressures in the underlying soil mass.

The improvement of the compaction can be checked by measuring the settlement of the surface, with penetration tests and weight soundings (CPT, SPT, and WST). Plate load test and pressuremeter tests can also be useful.

23.5 CONTROL METHODS

Different methods, such as penetration tests, sampling, pressuremeter tests, or crosshole measurements, can be used to check the effectiveness of the compaction. It should be noted, that local variations of the compaction can occur. However, localized pockets with loose material are generally not significant unless they are extensive (Greenwood and Kirsch, 1983) or located close to the ground surface.

The most simple method to evaluate the compaction is by leveling using surface markers. The required accuracy is usually ± 2 mm. It is, however, possible to measure settlements with an accuracy of ± 0.1 mm if required. Such high precision is normally not warranted. The estimated increase of the relative density can be indirectly correlated with improvements of the stiffness and of shear strength of the soil.

Sampling is often difficult or misleading in granular soils because of the change of the density that takes place during the drilling and when a piston or an open drive sampler is pushed into the soil. Freezing of the soil before sampling has also been tried to reduce this disturbance. However, the volume of the samples changes during the freezing because the volume of ice is 9 percent larger than that of the water in the soil before the freezing.

The most commonly used method for checking the effectiveness of the compaction is with different types of penetrometers; standard penetration tests (SPT), cone penetration tests (CPT), weight soundings (WST), or ram soundings (DP). Reference testing procedures have been proposed for these methods and are available in the literature. The liquefaction potential of granular soils is still commonly evaluated by SPT (Seed et al.,

1985). In Germany and Bulgaria, light penetrometers are used. In Southeast Asia, it is common to check the degree of compaction with the so-called Mackintosh probe. However, it is preferable to use CPT (Smoltczyk, 1983) or WST for the checking, since the results from SPT can be misleading (Green and Padfield, 1983). Vane tests have also been tried but the results could be misleading in medium dense to dense sands and silts. A comparison between different penetration testing methods is shown in Table 23.1.

It should be noted that the penetration resistance is affected not only by the relative density but that other factors such as the time after the compaction are also important. An increase of the penetration resistance several months after the compaction has been reported by Mitchell and Solymar (1984) and by Solymar and Reed (1986). The effect is generally more pronounced in clean sands compacted by blasting than in other soil types. The increase of the penetration resistance with time is generally small for silty soils.

It is also common to check the improvement of the compaction using the Menard pressuremeter, where a cylindrical probe is used that is lowered down into a borehole. The pressure and volume of water required to expand the center cell of the pressuremeter and thus the wall of the borehole are measured. From the test data a limit pressure (p_l) and a pressuremeter modulus (E_M) can be determined. (The limit pressure is an indication of the shear strength of the compacted soil, while the pressuremeter modulus is a measure of the compressibility.) One limitation of the method is the disturbance of the soil caused by the drilling of the boreholes required for the test. This disturbance can be reduced by careful augering. It should be noted that the pressuremeter modulus corresponds to the shear modulus of the soil in the horizontal direction, while settlements are governed by the confined modulus in the vertical direction.

Test pits can be used above the groundwater level to investigate the effect of the compaction down to about 5 m depth. Different methods are available for evaluating the unit weight of the compacted soil in test pits at or close to the ground surface. Sand, oil, or rubber balloons can be used to estimate the relative compaction or the relative density of the compacted material. However, the results are often misleading because of the disturbance of the soil caused by the sampling (Green and Padfield, 1983). It is also common to use nuclear density meters, where both the density and the water content of the compacted soil can be determined down to about 6 m depth. However, the probe has to be carefully calibrated at each site for the different soils.

Test fills are also useful for large projects in order to check the settlements and the improvement of the compaction. Test fills have been used to determine the relative density required for oil-storage tanks and for structures founded on spread footings and rafts. However, the method is time-consuming and therefore costly. The height of the fill can be increased gradually during the testing or in steps. The settlements should preferably be measured when the load is increased as well as some time after the increase of the applied load. The time interval between successive load increments should be kept constant—at a few hours to several days. It is desirable to measure the settlements at different depths so that the effect of the overburden pressures on the compaction can be evaluated.

Plate load tests can be carried out at the surface or in pits above the groundwater level (West, 1976). Screw plate tests are useful in loose to medium dense sand below the groundwater table. In this testing method, a screw-shaped plate is used, which is loaded by a hydraulic jack located just above the plate. However, it is usually difficult to screw the plate down into dense sand.

Shear wave velocity and the shear modulus determined by in-situ crosshole seismic measurements have also been used as an indication of the degree of compaction. The shear modulus can also be determined from the velocity of the surface Rayleigh waves (for example, Stokoe and Nazarian, 1983). This method is particularly useful for checking the compaction of rockfills, since boreholes are not required. Seismic crosshole tests were used by, for example, Massarsch and Broms (1983) and by Hoenig (1984) to determine the compaction of hydraulic fills.

The compressibility, the bearing capacity and the liquefaction potential of the compacted soil are often of greater interest than the relative density. There are no direct correlations of these properties with either the relative compaction or the relative density. It is well known, for example, that the liquefaction potential is affected by the stress history of the soil and by the depth below the ground surface.

Careful planning of the checking of the compaction is required, since the cost for the control tests is usually high. A combination of control methods is often useful, for example, cone penetration tests (CPT) in combination with crosshole measurements.

There is a trend to instrument the equipment used for the compaction so that the efficiency of the method can be checked. There is also a need to monitor the pore water pressures, the particle velocity at different distances from the compaction point, and the settlements.

23.6 COST COMPARISONS

A large number of different methods are available for deep compaction of granular soils, such as vibroflotation; stone, gravel, or sand columns; vibrocompaction; dynamic consolidation; blasting; compaction piles; grouting; and preloading. With these soil-improvement methods the foundation costs can often be reduced compared with driven or bored piles.

The relative cost for different compaction methods varies with local conditions (for example, accessibility, remoteness of the site), soil conditions (depth and thickness of the layers to be compacted), available equipment and materials, experience

TABLE 23.2 RELATIVE COST OF DIFFERENT DEEP COMPACTION METHODS COMPARED WITH BLASTING.

Compaction Method	Relative Cost, percent
Blasting	100
Deep vibratory compaction	200
Dynamic consolidation (heavy tamping)	300
Vibroflotation	400
Stone, gravel and sand compaction piles, compaction grouting	500

of the contractor, and the competitive situation. In northern Europe the relative cost per cubic meter of the treated soil compared with blasting is approximately as shown in Table 23.2 under favorable conditions.

Blasting is usually the most economic method where it can be used, compared with stone or gravel columns or sand compaction piles. It should be noted that the relative cost can vary considerably, depending on the local conditions, available equipment, and the competitive situation. The choice of method is often governed by the limitations of alternative methods.

REFERENCES

Abelev, M. V. (1976), Compacting loess soils in the USSR. *Proceedings, Ground Treatment by Deep Compaction*, Institution of Civil Engineers, London, pp. 79–82.

Aboshi, H. (1984), Soil improvement techniques in Japan, *Seminar on Soil Improvement and Construction Techniques in Soft Ground*, Singapore, pp. 3–16.

Aboshi, H., Ichitomo, E., Enoki, M., and Haroda, K. (1979), The Compozer, a method to improve characteristics of soft clays by inclusion of large diameter sand columns, *Proceedings of the International Conference on Reinforcement*, ENPC, Paris, France, **1**.

Anagnosti, P. (1985), Grouting of soils, *Proceedings, 3d International Geotechnical Seminar, Soil Improvement Methods*, Singapore, pp. 33–44.

Andreu, J., Arcones, A., and Soriano, A. (1983), Ground improvement and pile driving at Los Barrios (Spain), *Proceedings of the European Conference on Soil Mechanics and Foundation Engineering*, Helsinki, **1**, pp. 193–198.

d'Appolonia, E. (1953), Loose sands—Their compaction by vibroflotation, *Symposium on Dynamic Testing of Soils*, ASTM STP No. 156, pp. 138–154.

d'Appolonia, E., Miller, C. E., and Ware, T. M. (1955), Sand compaction by vibroflotation. *Transactions of ASCE*, **120**, pp. 154–168.

ASCE (1977), *Soil Improvement, History, Capabilities, and Outlook*, ASCE Committee on Placement and Improvement of Soils, Geotechnical Engineering Division, New York.

Baker, W. H., Cording, E. J., and MacPherson, H. H. (1983), *Compaction Grouting to Control Ground Movements During Tunneling Under Ground Space*, Pergamon Press, Oxford, **7**, pp. 205–212.

Barendsen, D. A. and Kok, L. (1983), Prevention and repair of flow-slides by explosion densification, *Proceedings of the 8th European Conference on Soil Mechanics and Foundation Engineering*, Helsinki, **1**, pp. 205–212.

Basore, C. E. and Boitano, J. P. (1969), Sand densification by piles and vibroflotation, *Journal of the Soil Mechanics and Foundations Division, ASCE*, **95**, No. SM-6, pp. 1303–1323.

Bolgerud, O. and Haug, A. K. (1983), Dynamic consolidation of rockfill at an oil refinery site, *Proceedings of the 8th European Conference on Soil Mechanics and Foundation Engineering*, Helsinki, **1**, pp. 213–218.

Brown, D. R. and Warner, J. (1973), Compaction grouting, *Journal of the Soil Mechanics and Foundations Division, ASCE*, **99**, No. SM-8, Paper 9908, pp. 589–601.

Brown, R. E. (1977), Vibroflotation compaction of cohesionless soils, *Journal of the Geotechnical Engineering Division, ASCE*, **103**, No. GT-12, pp. 1437–1451.

Brown, R. E. and Glenn, A. J. (1976), Vibroflotation and terra-probe comparison, *Journal of the Geotechnical Engineering Division, ASCE*, **102**, No. GT-10, pp. 1059–1072.

Carpentier, R., de Wolf, P., van Damme, L., de Rouck, J., and Bernard, A. (1985), Compaction by blasting in offshore harbour construction, *Proceedings of the 11th International Conference on Soil Mechanics and Foundation Engineering*, San Francisco, **3**, pp. 1687–1692.

Choa, V., Karunaratne, G. P., Ramaswamy, S. D., Vijiaratnam, A., and Lee, S. L. (1979), Compaction of sand fill at Changi Airport, *Proceedings of the 6th Asian Regional Conference on Soil Mechanics and Foundation Engineering*, Singapore, **1**, pp. 137–140.

Cleaud, J. J., Bourbon, L., and Karaki, P. (1983), Analysis of results obtained on a dynamic compaction site, *Proceedings of the 8th*

European Conference on Soil Mechanics and Foundation Engineering, Helsinki, **1**, pp. 19–22.

Cognon, J. M., Liausu, P., and Vialard, R. (1983), Combination of the drain and surcharge method with dynamic compaction. *Proceedings of the 8th European Conference on Soil Mechanics and Foundation Engineering*, Helsinki, **1**, pp. 291–222.

Coupe, P. S. (1986), An extension of dynamic consolidation, *Ground Engineering*, **19**, No. 3, pp. 14–21.

Datye, K. R. (1982), Settlement and bearing capacity of foundation system with stone columns, *Symposium on Soil and Rock Improvement Techniques Including Geotextile, Reinforced Earth and Modern Piling Methods*, AIT, Bangkok, pp. A.1.1–A.1.27.

Datye, K. R. and Nagaraju, S. S. (1981), Design approach and field control for stone columns, *Proceedings of the 10th International Conference on Soil Mechanics and Foundation Engineering*, Stockholm, **3**, pp. 637–644.

Datye, K. R. and Madhav, M. R. (1988), Case histories of foundations with stone columns, *Proceedings, 2d International Conference on Case Histories in Geotechnical Engineering*, St. Louis, **2**, pp. 1075–1086.

Dembicki, E., Kisielowa, N., Nowakowski, H., and Novakowski, Z. (1980a), Consolidation dynamic des vases a l'explosif (Dynamic consolidation of mud soils by means of blasting charges). *Proceedings, International Conference on Compaction*, Paris, **1**, pp. 295–300.

Dembicki, E., Kisielowa, N., Nowakowski, H., and Osiecimski, R. (1980b), Compactage des fonds marines sableux a l'exposif (Compaction of sandy marine subsoils by means of blasting charges), *Proceedings, International Conference on Compaction*, Paris, **1**, pp. 301–305.

Dembicki, E. and Kisielowa, N. (1983), Technology of soil compaction by means of explosion, *Proceedings of the 8th European Conference on Soil Mechanics and Foundation Engineering*, Helsinki, **1**, pp. 229–230.

Denchev, P. (1980), Compaction of loess by saturation and explosion, *Proceedings, International Conference on Compaction*, Paris, **1**, pp. 313–317.

Dobsen, T. (1987), Case histories of the vibro system to minimize the risk of liquefaction, *Soil Improvement—A Ten Year Update*, ASCE, Geotechnical Special Publication, No. 12, pp. 167–183.

Dowding, C. H. and Hryciw, R. D. (1986), A laboratory study of blast densification of saturated sand, *Journal of Geotechnical Engineering*, ASCE, **112**, No. 2, pp. 187–199.

Engelhardt, K. and Golding, H. C. (1975), Field testing to evaluate stone column performance in seismic areas, *Geotechnique*, **25**, No. 1, pp. 61–69.

Englehardt, K. and Kirsch, K. (1977), Soil improvement by deep vibratory techniques, *Proceedings of the 5th Southeast Asian Conference on Soil Mechanics and Foundation Engineering*, Bangkok.

Escario, V. (1983), Ground improvement related with soil liquefaction. Squeeze and compaction grouting, *Proceedings of the 8th European Conference on Soil Mechanics and Foundation Engineering*, Helsinki, **3**, pp. 1027–1036.

Fan, W., Shi, M., and Qui, Y. (1988), Ten years of dynamic consolidation in China, *Proceedings, 2d International Conference on Case Histories in Geotechnical Engineering*, St. Louis, **2**, pp. 1047–1054.

Gambin, M. P. (1983), The Menard dynamic consolidation method at Nice Airport, *Proceedings of the European Regional Conference on Soil Mechanics and Foundation Engineering*, Helsinki, **1**, pp. 231–234.

Gambin, M. and Bolle, G. (1983), Sea bed improvement for Lagos dry dock, *Proceedings of the 8th European Regional Conference on Soil Mechanics and Foundation Engineering*, Helsinki, **3**, pp. 837–840.

Glover, J. C. (1982), Sand compaction and stone columns by the vibroflotation process, *Symposium on Soil and Rock Improvement Techniques Including Geotextiles, Reinforced Earth and Modern Piling Methods*, AIT, Bangkok, pp. A.7.1–A.7.18.

Gosh, N. and Tabba, M. M. (1988), Experience in ground improvement by dynamic compaction and preloading at Half Moon Bay—Saudi Arabia, *Proceedings, 2d International Conference on Case Histories in Geotechnical Engineering*, St. Louis, **2**, pp. 1055–1061.

Goughnour, R. R. and Bayuk, A. A. (1979), Analysis of stone columns. Soil matrix interaction under vertical load, *Proceedings, International Conference on Soil Reinforcement*, ENPC, **1**, pp. 271–277.

Green, P. A. and Padfield, C. J. (1983), A field study of ground improvement using vibroflotation, *Proceedings of the 8th European Conference on Soil Mechanics and Foundation Engineering*, Helsinki, **1**, pp. 241–248.

Greenwood, D. A. (1970), Mechanical improvement of soils below ground surface, *Proceedings, Symposium on Ground Engineering*, Institution of Civil Engineers, London, pp. 11–22.

Greenwood, D. A. (1975), Vibroflotation, design and operation, *Methods of Treatment of Unstable Ground*, ed. F. G. Bell, Butterworths, London, pp. 189–210.

Greenwood, D. A. and Kirsch, K. (1983), Specialist ground treatment by vibratory and dynamic methods, *Proceedings of the International Conference on Advances in Piling and Ground Treatment for Foundations*, The Institution of Civil Engineers, London, pp. 17–45.

Guilloux, A. and Blondeau, F. (1989), Le traitement des sols fins par injection solid, *Proceedings of the 12th International Conference on Soil Mechanics and Foundation Engineering*, Rio de Janeiro, **2**, pp. 1367–1368.

Hanzawa, H. (1981), Improvement of a quick sand, *Proceedings of the 10th International Conference on Soil Mechanics and Foundation Engineering*, Stockholm, **3**, pp. 683–686.

Hansbo, S. (1977), Dynamic consolidation of rockfill at Uddevalla shipyard, *Proceedings of the 9th International Conference on Soil Mechanics and Foundation Engineering*, Tokyo, **2**, pp. 241–246.

Hansbo, S. (1978), Dynamic consolidation of soils by a falling weight, *Ground Engineering*, **11**, No. 5, pp. 27–36.

Hansbo, S., Pramborg, B., and Nordin, P. O. (1974), The Vanern Terminal—An illustrative example of dynamic consolidation of hydraulically placed fills of organic silt and sand, *Sols-Soils*, Paris, No. 25, pp. 5–11.

Hartikainen, J. and Valtonen, M. (1983), Heavy tamping of ground of Aimarautio Bridge, *Proceedings of the 8th European Regional Conference on Soil Mechanics and Foundation Engineering*, Helsinki, **1**, pp. 249–252.

Hoenig, H. (1984), In-situ quality control of vibrator compaction, *Proceedings of the International Conference on In-Situ Soil and Rock Reinforcement*, Paris, pp. 373–382.

Holeyman, A. and Wallays, M. (1984), Deep compaction by ramming, *Proceedings of the International Conference on In-Situ Soil and Rock Reinforcement*, Paris, pp. 367–372.

Holeyman, A. and Broms, B. B. (1986), Sand compaction piles, *Proceedings of the International Conference on Deep Foundations*, Beijing, China, **2**, pp. 2.26–2.31.

Hussin, J. D. and Ali, S. (1987), Soil improvement at the Trident submarine facility, *Soil Improvement—A Ten Year Update*, ASCE, Geotechnical Special Publication, No. 12, pp. 215–231.

van Impe, W. F. (1985), Soil improvement in Belgium, *Proceedings, 3d International Seminar, Soil Improvement Methods*, Singapore, pp. 201–228.

Ishihara, K. and Okada, S. (1978a), Yielding of overconsolidated sand and liquefaction model under cyclic stress, *Soils and Foundations*, **18**, No. 1, pp. 57–72.

Ishihara, K. and Okada, S. (1978b), Effects of stress history on cyclic behaviour of sand, *Soils and Foundations*, **18**, No. 4, pp. 31–45.

Ivanov, P. L. (1980), Consolidation of saturated soils by explosions, *Proceedings of the International Conference on Compaction*, Paris, **1**, pp. 331–337.

Ivanov, P. L. (1983), Prediction and control techniques to compact loose soils by explosions, *Proceedings of the 8th European Conference on Soil Mechanics and Foundation Engineering*, Helsinki, **1**, pp. 253–254.

Iyengar, M. (1983), Improvement of a cohesionless deposit to support a DMT process building in a seismic area, *Proceedings of the 8th European Conference on Soil Mechanics and Foundation Engineering*, Helsinki, **1**, pp. 255–258.

Janes, H. W. (1973), Densification of sand for drydock by Terra-probe, *Journal of the Soil Mechanics and Foundations Division*, ASCE, No. SM-6, pp. 451–470.

Jarvio, E. and Petaja, J. (1983), Improvement of the bearing capacity of underwater marine fine sand strata by compaction piling, *Proceedings of the 8th European Conference on Soil Mechanics and Foundation Engineering*, Helsinki, **2**, pp. 851–856.

Jebe, W. and Bartels, K. (1983), The development of compaction methods with vibrators from 1976 to 1982, *Proceedings of the 8th European Conference on Soil Mechanics and Foundation Engineering*, Helsinki, **1**, pp. 259–266.

Jessberger, H. L. and Beine, R. A. (1981), Heavy tamping, theoretical and practical aspects, *Proceedings of the 10th International Conference on Soil Mechanics and Foundation Engineering*, Stockholm, **3**, pp. 695–699.

Johnson, D., Nicholls, R., and Thomson, G. H. (1983), An evaluation of ground improvement at Belawan Port, North Sumatra, *Proceedings of the 8th European Conference on Soil Mechanics and Foundation Engineering*, Helsinki, **1**, pp. 45–52.

Keller, T. O., Castro, G., and Rogers, J. H. (1987), Steel Creek Dam foundation densification, *Soil Improvement—A Ten Year Update*, ASCE, Geotechnical Special Publication, No. 12, pp. 136–166.

Kishida, H. (1967), Ultimate bearing capacity of piles driven into loose sand, *Soils and Foundations*, **7**, No. 3, pp. 20–29.

Klohn, E. J., Garga, V. K., and Shukin, W. (1981), Densification of sand tailings by blasting, *Proceedings of the 10th International Conference on Soil Mechanics and Foundation Engineering*, Stockholm, **3**, pp. 725–730.

Koponen, H. (1983), Soil improvement by deep compaction at the site of a harbour storage, *Proceedings of the 8th European Conference on Soil Mechanics and Foundation Engineering*, Helsinki, **1**, pp. 267–269.

Kummeneje, O. and Eide, O. (1961), Investigation of loose sand deposits by blasting, *Proceedings of the 5th International Conference on Soil Mechanics and Foundation Engineering*, Paris, **1**, pp. 491–497.

Kummerie, R. P. and Dumas, J. C. (1988), Soil improvement using dynamic compaction for Bristol Resource Recovery Facility, *Proceedings, 2d International Conference on Case Histories in Geotechnical Engineering*, St. Louis, **2**, pp. 921–927.

Ladd, C. C., Foott, R., Ishihara, K., Poulos, H. G., and Schlosser, F. (1977), Stress deformation and strength characteristics, *Proceedings of the 9th International Conference on Soil Mechanics and Foundation Engineering*, Tokyo, **2**, pp. 421–494.

Langefors, V. and Kihlstrom, B. (1978), *The Modern Technique of Rock Blasting*, John Wiley and Sons, Inc., New York, N.Y.

Lee, K. L. and Focht, J. A. (1975), Liquefaction potential at Ecofisk Tank in the North Sea, *Journal of the Geotechnical Division, ASCE*, **101**, No. GT-1, pp. 1–18.

Leonards, G. A., Cutter, W. A., and Holtz, R. D. (1980), Dynamic compaction of granular soils, *Journal of the Geotechnical Engineering Division, ASCE*, **106**, No. GT-1, pp. 35–44.

Leycure, P. and Schroeder, W. L. (1987), Slope effects on probe densification of sands, *Soil Improvement—A Ten Year Update*, ASCE, Geotechnical Special Publication, No. 12, pp. 167–183.

Lindqvist, L. and Petaja, J. (1981), Experience in the evaluation of the bearing capacity of tapered friction piles in postglacial sand and silt strata, *Proceedings of the 10th International Conference on Soil Mechanics and Foundation Engineering*, Stockholm, **2**, pp. 759–766.

Litvinov, I. M. (1966), Accelerated method of deep compaction of slumping loess soil—by the II type with preliminary flooding and directed explosives, *Soil Mechanics and Foundation Engineering*, **2**, pp. 116–121.

Litvinov, I. M. (1973), Deep compaction of soils with the aim of considerably increasing their carrying capacity, *Proceedings of the 8th International Conference on Soil Mechanics and Foundation Engineering*, Moscow, **43**, pp. 392–394.

Lukas, R. G. (1986), *Dynamic Compaction for Highway Construction. Vol. 1, Design and Construction Guidelines*, Report No. FHWA/RD—86/133, Federal Highway Administration Office of Research and Development, Washington, D.C.

Lyman, A. K. B. (1942), Compaction of cohesionless foundation soil by explosives, *Transactions of ASCE*, **107**, Paper 2160.

Massarsch, K. R. (1985), Deep compaction of sand using vibratory probes, *Proceedings of the Third International Geotechnical Seminar, Soil Improvement Methods*, Singapore, pp. 9–17.

Massarsch, K. R. and Broms, B. B. (1983), Soil compaction by Vibro Wing method, *Proceedings of the 8th European Conference on Soil Mechanics and Foundation Engineering*, Helsinki, **1**, pp. 275–278.

Mayne, P. W., Jones, J. S., and Dumas, J. C. (1984), Ground response to dynamic compaction, *Journal of Geotechnical Engineering*, **110**, No. 6, pp. 757–774.

Menard, L. and Broise, Y. (1975), Theoretical and practical aspects of dynamic consolidation, *Geotechnique*, **25**, No. 1, pp. 3–18.

Minkov, M., Evstatiev, D., and Denchev, P. (1980), Dynamic compaction of loess, *Proceedings of the International Conference on Compaction*, Paris, **1**, pp. 345–349.

Mitchell, J. K. (1981), Soil improvement. State-of-the-art report, *Proceedings of the 10th International Conference on Soil Mechanics and Foundation Engineering*, Stockholm, **4**, pp. 509–565.

Mitchell, J. K. (1986), Practical problems from surprising soil behaviour, *Journal of Geotechnical Engineering, ASCE*, **112**, No. 3, pp. 259–289.

Mitchell, J. K. and Katti, R. K. (1981), Soil improvement—General report, *Proceedings of the 10th International Conference on Soil Mechanics and Foundation Engineering*, Stockholm, **4**, pp. 567–575.

Mitchell, J. K. and Solymar, Z. V. (1984), Time-dependent strength gain in freshly deposited or densified sand, *Journal of Geotechnical Engineering, ASCE*, **110**, No. 11, pp. 1559–1576.

Mitchell, J. K. and Welsh, J. P. (1989), Soil improvement by combining methods, *Proceedings of the 12th International Conference on Soil Mechanics and Foundation Engineering*, Rio de Janeiro, **2**, pp. 1393–1396.

Moh, Z. C., Ou, C. D., Woo, S. M., and Yu, K. (1981), Compaction sound piles for soil improvement, *Proceedings of the 10th International Conference on Soil Mechanics and Foundation Engineering*, Stockholm, **3**, pp. 749–752.

Morgan, J. G. D. and Thomson, G. H. (1983), Instrumentation methods for control of ground density in deep vibrocompaction, *Proceedings of the 8th European Conference on Soil Mechanics and Foundation Engineering*, Helsinki, **1**, pp. 59–72.

Morgenthaler, M., Cambou, B., and Sanglerat, G. (1977), Colonnes ballastees, essais de chargement et calculs par la method des elements finis, *Revue Française Geotechnique*, **5**.

Mori, H. (1977), Compaction of the deep fill of boulder soils by impact-force, *Proceedings of the 5th Southeast Asia Conference on Soil Engineering*, Bangkok, pp. 389–399.

Munfakh, G. A. (1984), Soil reinforcement by stone columns—varied case applications, *International Conference on In-Situ Soil and Rock Reinforcement*, Paris, pp. 157–162.

Murayama, S. and Ichimoto, E. (1982), Sand compaction pile method, *Symposium on Soil and Rock Improvement Techniques Including Geotextiles, Reinforced Earth and Modern Piling Methods*, AIT, Bangkok, pp. A.5.1–A.5.1.3.

Pilot, G., Colas de Francs, E., Puntous, R., and Queyroi, D. (1981), Compactage par explosif d'un ramblai hydraulic, *Proceedings of the 10th International Conference on Soil Mechanics and Foundation Engineering*, Stockholm, **3**, pp. 757–760.

Plumelle, C. (1989), Compactage statique horizontal par injection solide (Compaction grouting), *Proceedings of the 12th International Conference on Soil Mechanics and Foundation Engineering*, Rio de Janeiro, **2**, pp. 1401–1402.

Prugh, B. J. (1963), Densification of soils by explosive vibrations, *Journal of the Construction Division, ASCE*, **89**, No. CO-1, pp. 79–100.

Ramaswamy, S. D., Aziz, M. A., and Lee, S. L. (1982), In-depth stabilization of soils in Singapore, *Proceedings of the Regional Conference on Tall Buildings and Urban Habitat*, Kuala Lumpur, pp. 5.28–5.34.

Ranjan, G. and Rao, B. G. (1983), Skirted granular piles for ground improvement, *Proceedings of the 8th European Conference on Soil Mechanics and Foundation Engineering*, Helsinki, **1**, pp. 297–300.

Rao, B. G. and Ranjan, G. (1985), Settlement analysis of skirted granular piles, *Journal of Geotechnical Engineering*, **III**, No. II, pp. 1264–1283.

Robert, J. (1989), Amélioration des sols par intrusion de mortier, *Proceedings of the 10th International Conference on Soil Mechanics and Foundation Engineering*, Stockholm, **2**, pp. 1407–1408.

Robertson, P. K. and Campanella, R. G. (1983), Interpretation of cone penetration tests. Part I, Sand, *Canadian Geotechnical Journal*, **20**, No. 4, pp. 718–733.

Robinsky, E. I. and Morrison, D. E. (1964), Sand displacement and compaction around model friction piles, *Canadian Geotechnical Journal*, **1**, No. 2, p. 81.

Saito, A. (1977), Characteristics of penetration resistance of a reclaimed sandy deposit and their change through vibratory compaction, *Soils and Foundations*, **17**, No. 4.

Salley, J. R., Foreman, B., Baker, W. H., and Henry, J. F. (1987), Compaction grouting test programme Pinopolis West Dam, *Soil Improvement—A Ten Year Update*, ASCE, Geotechnical Special Publication, No. 12, pp. 245–269.

Schmertmann, J., Baker, W., Gupta, R., and Kessler, K. (1986), CPT/DMT QC of ground modification at a power plant, *Proceedings of the ASCE Specialty Conference, In Situ 86*, VPI, Blacksburg, Virginia.

Seed, H. B. (1979), Soil liquefaction and cyclic mobility evaluation for level ground during earthquakes, *Journal of the Geotechnical Engineering Division, ASCE*, **105**, No. GT-2, pp. 201–255.

Seed, H. B. and Booker, J. R. (1977), Stabilization of potentially liquefiable sand deposits using gravel drains, *Journal of the*

Geotechnical Engineering Division, ASCE, **103**, No. GT-7, pp. 757–768.

Seed, H. B., Tokimotsu, K., Harder, L. F., and Chung, R. M. (1985), Influence of SPT procedures in soil liquefaction resistance evaluation, *Journal of the Geotechnical Engineering Division, ASCE,* **111**, No. 12, pp. 1435–1445.

Shibazaki, M. and Ohta, S. (1982), A unique underpinning of soil solidification utilizing super-high pressure liquid jet, *ASCE Specialty Conference on Grouting in Geotechnical Engineering,* New Orleans, pp. 680–693.

Shilley, A. N. (1982), Compaction grouting for foundation stabilization, *ASCE Specialty Conference on Grouting in Geotechnical Engineering,* New Orleans, pp. 923–937.

Smits, M. T. J. H. and DeQuelerij, L. (1989), The effect of dynamic compaction on dry granular soils, *Proceedings of the 12th International Conference on Soil Mechanics and Foundation Engineering,* Rio de Janeiro, **2**, pp. 1419–1422.

Smoltczyk, U. (1983), Deep compaction, *Proceedings of the 8th European Conference on Soil Mechanics and Foundation Engineering,* Helsinki, **3**, pp. 1105–1116.

Solymar, Z. V. (1984), Compaction of alluvial sands by deep blasting, *Canadian Geotechnical Journal,* **21**, pp. 305–321.

Solymar, Z. V. (1986), Ground improvement by compaction piling, *Journal of Geotechnical Engineering, ASCE,* **112**, No. 12, pp. 1069–1083.

Solymar, Z. V., Iloabachie, B. C., Gupta, R. C., and Williams, L. R. (1984), Earth foundation treatment at Jebba Dam site, *Journal of Geotechnical Engineering, ASCE,* **110**, No. 10, pp. 1415–1445.

Solymar, Z. V. and Reed, D. J. (1986), A comparison of foundation compaction techniques, *Canadian Geotechnical Journal,* **23**, No. 3, pp. 271–280.

Song, B. and Gambin, M. (1988), Dynamic compaction—an unusual application, *Proceedings, 2d International Conference on Case Histories in Geotechnical Engineering,* St. Louis, **2**, pp. 969–975.

Stokoe, K. H. and Nazarian, S. (1983), Effectiveness of ground improvement from spectral analysis of surface waves, *Proceedings of the 8th European Regional Conference on Soil Mechanics and Foundation Engineering,* Helsinki, **1**, pp. 91–94.

Thorburn, S. (1976), Building structures supported by stabilized ground, *Proceedings, Ground Treatment by Deep Compaction,* Institution of Civil Engineers, London, pp. 83–94.

Thorburn, S. and MacVicar, R. S. L. (1974), The performance of buildings founded on river alluvium, *Proceedings of the Cambridge Conference on Settlement of Structures,* Paper V/12.

Wallays, M. (1982a), Deep compaction by vertical and horizontal vibration, *Symposium on Soil and Rock Improvement Techniques Including Geotextiles, Reinforced Earth and Modern Piling Methods,* AIT, Bangkok, pp. A.4.1–A.4.22.

Wallays, M. (1982b), Deep compaction by casing driving, *Symposium on Soil and Rock Improvement Techniques Including Geotextiles, Reinforced Earth and Modern Piling Methods,* AIT, Bangkok, pp. A.6.1–A.6.20.

Wang, Y., Zhang, W., and Qiao, T. (1988), Evaluation of the effect of saturated silty and fine sand foundation improved by vibroflotation in seismic area, *Proceedings, 2d International Conference on Case Histories in Geotechnical Engineering,* St. Louis, **2**, pp. 963–968.

Warner, J. (1982), Compaction grouting—The first thirty years, *ASCE Specialty Conference, Grouting in Geotechnical Engineering,* New Orleans.

West, J. M. (1976), The role of ground improvement in foundation engineering, *Proceedings, Ground Treatment by Deep Compaction.* Institution of Civil Engineers, London, pp. 71–78.

West, J. M. and Slocombe, B. (1973), Dynamic consolidation as an alternative foundation, *Ground Engineering,* **6**, No. 6, pp. 52–54.

Welsh, J. P. (1983), Dynamic compaction of sanitary landfill to support superhighway, *Proceedings of the 8th European Conference on Soil Mechanics and Foundation Engineering,* Helsinki, **1**, pp. 319–321.

de Wolf, P., de Rouck, J., Allaert, J., and Delapierrre, J. (1980), Zeebrugge—Front harbour problems caused by the compaction of the foundation ground of the dyke, *Proceedings of the International Conference on Compaction,* Paris, **1**, pp. 307–311.

de Wolf, P., Carpentier, R., Alaert, J., and de Rouck, J. (1983), Ground improvement for the construction of the new outer harbour at Zeebrugge, Belgium, *Proceedings of the 8th European Conference on Soil Mechanics and Foundation Engineering,* Helsinki, **3**, pp. 827–832.

24 STABILIZATION OF SOIL WITH LIME COLUMNS

BENGT B. BROMS, Ph.D.
Professor,
Nanyang Technological Institute, Singapore

24.1 INTRODUCTION

24.1.1 General

The behavior of very soft clay or silt can be improved with lime or cement columns. In this soil stabilization method, the soft soil is mixed in situ either with unslaked lime (CaO) or with cement using a tool shaped like a giant dough mixer, as illustrated in Figure 24.1. Other materials can be used, such as gypsum (Holm et al., 1983a), fly-ash and furnace slag (Nieminen, 1978), hydroxyaluminum (Bryhn et al., 1983), and potassium chloride (Eggestad and Sem, 1976).

The mixing tool is first screwed into the soil down to a depth that corresponds to the desired length of the columns. The maximum length is at present 15 m (50 ft). The tool is then slowly withdrawn ($\leqslant 2.5$ cm/revolution) as unslaked lime or cement is forced down into the soil through holes located just above the horizontal blades of the mixing tool using compressed air. Since the blades are inclined (tilted), the stabilized soil will be compacted during the withdrawal. The resulting columns have the same diameter as the mixing tool (Fig. 24.2) (Broms and Boman, 1975a, b, 1976, 1977a, b; Holm et al., 1981, 1983a, b; Broms, 1983, 1984a, b, 1985).

The equipment for the manufacture of the lime columns (mast, rotary table, and kelly) is mounted on either a standard front-wheel loader or on a special carrier as shown in Figures 24.3a and 24.3b, respectively. Behind the loader or carrier is a

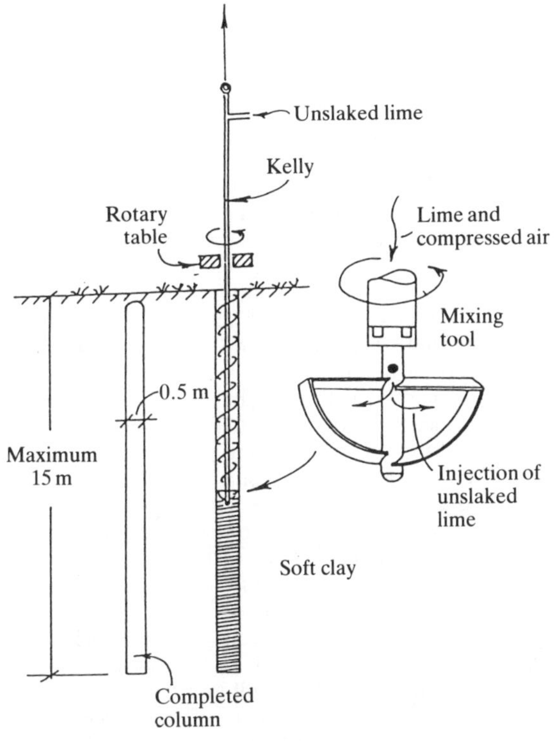

Fig. 24.1 Manufacture of lime columns.

Fig. 24.2 Excavated lime columns.

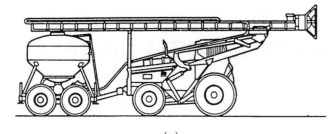

(a)

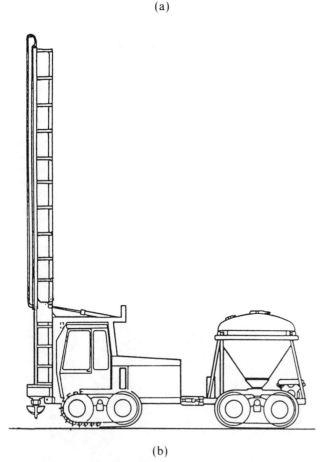

(b)

Fig. 24.3 (a) Volvo BM LM 641. (b) Lime column machine Linden-Alimak LPS4.

container with a capacity of 2.5 m³ (90 ft³) for the storage of the lime or the cement.

A similar technique, the so-called deep mixing method, using lime or cement (DLM and DCM), has been developed in Japan (e.g., Okumura and Terashi, 1975; Kawasaki et al., 1981, 1984; Terashi and Tanaka, 1981, 1983; Aboshi, 1984). In this method, up to eight mixing units, which are usually mounted on a large barge, are used for the mixing of lime or the cement with the soft soil. The blades of the mixing units overlap in order to increase the efficiency of the mixing. Thorough mixing is required, especially when cement is used. The area of the lime or cement columns varies between 1.5 and 9.5 m², depending on the number and the diameter of the mixing tool. The maximum length is 40 m. The method has mainly been applied in Japan for ports and harbors (Okumura and Terashi, 1975; Terashi and Tanaka, 1981, 1983).

Another type of lime column is used in India, Japan, and Taiwan, where unslaked lime is placed and compacted in predrilled holes (Chiu and Chin, 1963; Holeyman et al., 1983).

The method has also been used in Singapore and Malaysia. The expansion that takes place during the slaking of the lime is then utilized to consolidate the soil between the boreholes. (The volume increase of the quicklime is about 85 percent when water is added). The resulting reduction of the water content of the soil around the boreholes corresponds to the increase of the volume of the lime in the boreholes. This method is mainly effective in silty soils, where a relatively small reduction of the water content will effectively increase the shear strength and reduce the compressibility of the soil and thus the settlements.

Lime slurry has also been injected into boreholes in the soil to reduce the swelling potential and to reduce the heave. Primarily, stiff fissured swelling clays have been treated. This method was first tried in the United States in about 1960. Field tests have indicated, however, that the diffusion rate of the calcium ions in uniform soft clay is very low. Only a few millimeters around the boreholes will be affected, even after several years. Lime slurry has also been used to stabilize in-situ soft Bangkok clay (Petchgate and Tungboonterm, 1990).

24.1.2 Stabilization of Soft Clay or Silt with Lime or Cement

Part of the immediate increase of the shear strength of soft clay or silt when mixed with lime or cement is caused by flocculation of the clay and part by a reduction of the water content. For example, Ekström and Tränk (1979) found that the average water content had been reduced by 15 percent when 10 percent CaO was mixed with a soft clay with an initial water content of about 60 percent. The reduction of the water content in percent is generally somewhat larger than the added quicklime in percent with respect to the dry weight of the soil (Ahnberg and Holm, 1986).

The shear strength of clay stabilized with lime will normally be higher than that of the undisturbed clay about 1 to 2 hours after the mixing. This strength increase is often obtained even when the sensitivity of the clay is high and a large part of the initial shear strength is lost due to the remolding of the clay during the mixing.

The shear strength of the stabilized soil gradually increases with time through pozzolanic reactions when the lime reacts with the silicates and aluminates in the clay. The clay content of the soil should therefore not be less than 20 percent when lime is used. The sum of the silt and clay fractions should preferably exceed 35 percent, which is normally the case when the plasticity index of the soil exceeds 10. Fly-ash can be added when the clay content is not sufficient. Cement can also be used.

The pozzolanic reactions take place over many months and years. A high ground temperature and a high pH value (pH $\geqslant$ 12) accelerates the chemical reactions because the solubility of the silicates and the aluminates increases with increasing pH value. The pH value will normally exceed 12 with even a few percent of lime.

For inorganic clays with medium to low plasticity ($I_p \langle 50$) the maximum increase of the shear strength is generally obtained with 6 to 8 percent quicklime with respect to the dry weight of the soil. With cement, a higher content of 15 to 25 percent is generally required. The optimum lime and cement content increases with increasing water content and with increasing plasticity index of the soil.

The stabilized soil is normally firm to hard and the texture is grainy. The behavior is essentially that of an over-consolidated clay (Balasubramaniam and Buencuceso, 1989; Balasubramaniam et al., 1990). It has a low compressibility compared with the unstabilized soil and the preconsolidation pressure is increased. The sensitivity of the stabilized clay is low, typically 1 to 3. The plastic limit of the stabilized soil is increased with lime or cement, while the plasticity index is reduced as reported by, e.g., Bell and Tyrer (1989). The plastic limit usually increases with time. The liquid limit can also be affected, but the increase is normally small in comparison with the change of the plastic limit. A reduction of the liquid limit may occur if the clay is organic.

The permeability of the soil is generally increased by the lime. It should be noted that clay stabilized with lime may become frost-sensitive, as pointed out by Brandl (1981). The frost sensitivity and the permeability are reduced if cement is used instead.

The undrained shear strength of the stabilized clay can under favorable conditions be as high as 0.5 to 1.0 MPa (5 to 10 tsf) after 1 year. Normally an increase of 10 to 50 times can be expected if the initial shear strength is low—10 to 15 kPa. Approximately one-third of the final shear strength is usually obtained after 1 month, and approximately three-quarters after three months. The relative increase of the shear strength (in percent) decreases in general with increasing liquid limit. The shear strength increases in general with increasing compaction of the stabilized soil.

The shear strength of the soil in the columns can be less than that of samples stabilized in the laboratory. However, in most cases the undrained shear strength of the stabilized soil in the columns will exceed that determined by unconfined compression or undrained triaxial tests on laboratory samples.

The shear strength of the stabilized soil in the columns is not uniform even when the mixing of the lime or the cement with the clay has been done very carefully. Because of aggregation, the measured shear strength will vary with the testing method and with the size of the tested samples.

The shear strength of the stabilized soil as determined by unconfined compression tests increases with time as indicated in Figure 24.4. The largest relative increase of the shear strength has been observed for inorganic varved silty clays. However, the effect of the lime decreases in general with increasing water content, as shown by Holm et al. (1983b) and with increasing organic content. For organic soils with a high water content and liquid limit the increase is often very low. Even a relatively small amount of organic material can reduce the strength increase.

Gypsum can be used together with unslaked lime to stabilize organic soils with a water content of up to about 120 percent when lime alone is not effective. With gypsum, needlelike particles (ettringite) are formed that bind together the individual clay particles when the gypsum reacts with the clay. The best effect has been obtained with about one-third gypsum and two-thirds unslaked lime.

The gypsum also speeds up the chemical reactions. After 10 to 100 days the shear strength of the stabilized soil mixed with gypsum and lime can be two to four times higher than that when only unslaked lime is used (Holm et al., 1983a). Gypsum

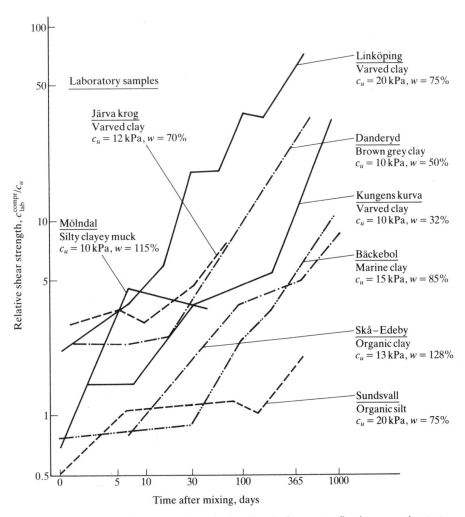

Fig. 24.4 Increase of shear strength with time. Results from unconfined compression tests.

and slaked lime has been used to stabilize soils with a water content up to 300 to 400 percent (Kuno et al., 1989). It should be noted that the behavior of the soil stabilized with gypsum and lime is relatively brittle and that the ductility is low (Kujala, 1983).

Lime and fly-ash have also been used to stabilize organic soils with a water content of up to 100 percent. About 10 to 30 percent is required with respect to the dry weight of the soil. Cement has also been effective and soils with a water content up to about 200 percent have been treated successfully (Kuno et al., 1989) irrespective of the organic content. A combination of gypsum and cement has been used to improve significantly the shear strength of soil with a water content up to 300 percent. A relatively high cement content is usually required (15 to 25 percent).

The chemical reactions and the increase of the shear strength with time are affected by the ground temperature. The rate increases with increasing temperature (Ahnberg and Holm, 1986). At temperatures below about 4°C the increase of the shear strength is very slow. In countries with a cold climate it is therefore important to prevent the ground from freezing by isolating the ground surface with straw, rockwool, etc., during the winter months.

Sufficient heat is released during the slaking of the lime that the ground temperature is increased to 30 to 50°C within the stabilized soil mass after a few days when the spacing of the columns is 2 m or less. At the center the temperature could be as high as 100°C immediately after the installation (Ahnberg and Holm, 1986; Ahnberg et al., 1989). With cement the increase is less than half of that of quicklime. This increase of the ground temperature speeds up the chemical reactions in the soil, which can appreciably affect the strength increase with time. This effect should be considered when the time required to reach a certain shear strength and bearing capacity is critical.

One other possible limitation of the lime column method is that certain plants are affected when the lime content and pH value of the soil are very high. With gypsum, adjacent concrete structures may deteriorate if sulfate-resistant cement is not used.

24.1.3 Bearing Capacity of Lime Columns

The bearing capacity of excavated columns and the shear strength are affected by joints and fissures in the stabilized soil. The shear strength as determined from load tests on excavated columns is generally lower than the shear strength obtained from fall-cone, laboratory vane, or undrained direct shear tests on small samples cut from the columns.

The shear strength as determined by fall-cone or vane tests or by a pocket penetrometer varies along the diameter of a lime or cement column. The shear strength of the stabilized soil at the center of a column is low because of the hole left after the kelly. The hole is filled with almost pure lime. The average shear strength of a column can therefore be underestimated if, for example, penetration tests (CPT or SPT) are used to check the shear strength of the lime columns, since the penetrometers have a tendency to follow the weak hole at the center of the column during a test.

The columns are often surrounded by a thin layer of very soft clay, which can affect the bearing capacity. The shear strength of this layer may be lower than the initial shear strength of the clay (Bryhn et al., 1983).

Undrained triaxial tests on whole column segments and on small-sized samples cut from the columns, as well as direct shear tests in the laboratory and in the field, have shown that the shear strength increases with increasing confining pressure until a limiting value has been reached that corresponds

approximately to the shear strength of the clay matrix. A relatively high angle of internal friction (30 to 45°) has been observed by, for example, Brandl (1981) at triaxial tests or direct shear tests when the cell pressure or the normal pressure is low. Kujala (1983) has reported an angle of 23° from triaxial tests when lime was used and an angle of 40° for gypsum and lime. Kujala and Nieminen (1983) found that the angle was about 10° higher for gypsum and lime compared with lime only. This increase of the shear strength reflects the dilatancy of the stabilized soil when the confining pressure or the normal pressure is low, as pointed out by Serra et al. (1983). Brandl (1981) observed that the friction angle increased with increasing lime content. A similar behavior has been observed for stiff fissured clays.

Results from triaxial and direct shear tests in situ indicate that the ultimate bearing capacity of a single lime column can be estimated conservatively as though the angle of internal friction of the column material is at least 30° when the shear strength is less than the shear strength of the clay matrix as determined from small samples. When this limiting shear strength has been reached, the angle of internal friction of the stabilized soil is reduced to zero and the column material behaves as an ideal plastic material.

The ultimate strength of excavated columns has been compared with the shear strength as determined from laboratory-prepared samples. Bryhn et al. (1983) have reported that the shear strength of small samples of stabilized clay cut from columns in the field has been 50 to 150 percent higher than that of laboratory samples. The results from laboratory tests can therefore normally be used to estimate the long-term ultimate strength and the creep load of the columns. It should be noted that the size of the columns has to be considered in the evaluation. There have been cases, however, where the in-situ shear strength has been lower than that determined by unconfined compression tests on samples prepared in the laboratory.

The in-situ confining pressure affects the bearing capacity of the lime or cement columns. The shear strength of the column material determined by undrained triaxial or direct shear tests at high normal pressures will normally govern the bearing capacity and the settlements, except close to the ground surface where the confining pressure is low. The in-situ lateral confining pressure on the columns increases when the quicklime is mixed with the soil and during the compaction. Measurements indicate that the lateral confining pressure after the mixing will normally be at least equal to the total overburden pressure.

The increase of the lateral pressure caused by the lateral deformations of the columns when they are loaded could possibly be considered in the design. The maximum confining pressure at the depth z corresponds to the limit pressure ($\rho z + k c_u$) where ρz is the total overburden pressure and c_u is the undrained shear strength of the surrounding unstabilized clay. The coefficient k is a factor that depends on the modulus of elasticity of the unstabilized clay. For soft clays a value of 5.0 is often used. In the case that the undrained shear strength of the clay around the columns is 10 kPa, the maximum confining pressure will be 76 kPa at the bottom of a 1-m (3-ft) thick crust. The corresponding increase of the ultimate bearing capacity is 220 kPa or 45 kN for a 0.5-m diameter column. However, it is not certain that this increase of the bearing capacity can be utilized, owing to the relatively large deformation required to mobilize the limit pressure.

The shear strength of the stabilized clay and the creep limit of the lime columns can be much higher than those determined from unconfined compression tests on laboratory-prepared samples (Ekström and Tränk, 1979). Test results indicate that the creep limit of a column at the same age as the samples is normally 3 to 5 times larger than the shear strength determined by unconfined compression tests. This relationship is uncertain and can only be used as a guide. A better method is to estimate the creep strength from plate load tests or from screw-plate tests in situ.

It should be noted that the bearing capacity of the lime columns may decrease with time when the pH value of the groundwater is very low (acid) or the content of carbon dioxide (CO_2) is high.

24.2 PRINCIPLE OF THE LIME COLUMN METHOD

24.2.1 General

Lime or cement columns have been used in Sweden, Norway, and Finland for the following purposes.

- To reduce the total and the differential settlements of light structures (one- to two-storey buildings)
- To increase the settlement rate and to control the settlements of relatively heavy structures
- To improve the stability of embankments, slopes, trenches, and deep cuts
- To reduce the vibrations from, for example, traffic, blasting, pile driving, etc.

In Sweden and Finland alone, about 25 000 lime columns are manufactured per year. Almost 2 million meters have been installed since 1975.

In soft clay, lime or cement columns can often be an economical alternative to conventional foundation methods, depending on such factors as the size, weight, and flexibility of the structure, the depth and shear strength of the different compressible strata, the risks and the consequences of a failure, and the effects of a lowering of the ground level.

Both the total and the differential settlements (Figure 24.5) can be controlled with lime columns as discussed by Holm et al. (1981). The settlements will be small as long as the load in the columns is less than the creep limit. With concrete or steel piles the total weight of the building has to be supported by the piles; with lime or cement columns only a part of the weight

has to be considered in order to control the settlements. The lime column method is therefore often an economical alternative to piles and to a floating foundation (raft) even when the thickness of the compressible strata below the structure is large.

The permeability of the lime column is often high compared with the surrounding untreated soft clay. Test data indicate that the lime normally increases the permeability 100 to 1000 times for soft clays and that the lime columns function as drains. A further increase can be obtained with gypsum. The permeability of the stabilized soil is generally reduced with cement.

For road embankments and dykes the lime column method is often an economical alternative to sand or band drains, as illustrated in Figure 24.6. The spacing of the lime columns can be relatively large because of the large diameter (0.5 to 0.6 m) of the columns and the relatively small disturbance of the soil caused by the installation of the columns. The consolidation time is reduced further by the reduction of the compressibility of the soil by the columns.

The bearing capacity of soft clay is also increased by lime columns and a relatively high overload can be used to preload and to consolidate the soil without exceeding the bearing capacity. The lime column method can thus be an economical alternative to other soil-improvement methods such as light-weight fills, pressure berms, excavation and replacement, and embankment piles.

Lime or cement columns can also be used to stabilize trenches and deep excavations instead of sheet piles, as shown in Figure 24.7. The columns should partly overlap so that they form a solid wall. The weight of the wall is utilized to resist the lateral earth pressure and any water pressures acting on the wall in the same way as a gravity retaining wall. The required thickness of the column wall increases with increasing depth of the excavation. The effectiveness of the walls can be increased by inclining the columns backwards away from the excavation or by arranging the columns in the form of buttresses. If the water pressure behind the wall can be reduced, then the lime column method can be economical even when the excavation depth is relatively large (4 to 6 m). Ekström and Tränk (1979) have estimated that the cost of a lime column wall is about 25 percent less than that of a conventional sheet-pile wall.

Lime or cement columns can also be used to protect structures supported on piles. The lime columns will reduce the negative skin friction on the structural piles caused, for example, by a lowering of the groundwater table or by a fill placed around

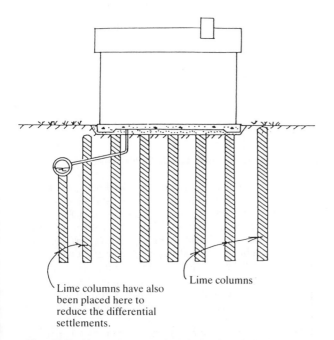

Lime columns have also been placed here to reduce the differential settlements.

Lime columns

Fig. 24.5 Lime columns as foundation for a light building.

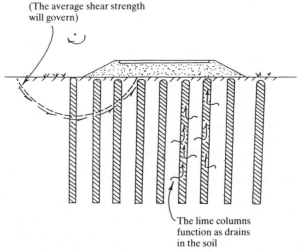

Potential failure surface
(The average shear strength will govern)

The lime columns function as drains in the soil

Fig. 24.6 Lime columns installed for a road embankment.

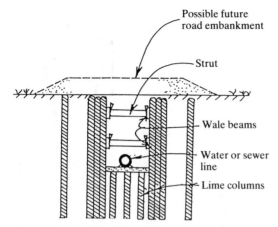

Fig. 24.7 Stabilization of trenches with lime columns.

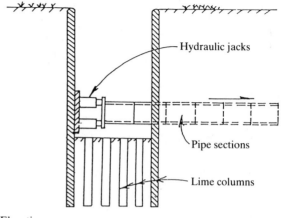

Elevation

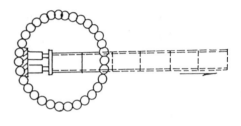

Plan

Fig. 24.8 Use of lime columns to stabilize a deep shaft in soft clay.

or behind the structure. Furthermore, the lime columns will decrease the settlements so that adjacent or connecting water and sewer lines will not be damaged. Holm et al. (1983b) have described a case where lime columns have been used behind a pile-supported bridge abutment for a motorway (E4) in Sweden constructed on a 7.5-m thick layer of very soft clay with an undrained shear strength of only 6 to 9 kPa.

Lime and cement columns can also be used to reduce the creep of the soft clay beneath buildings supported on piles. The lateral displacement of the clay may fail the structural piles below the exterior walls of the building. This will occur when the total overburden pressure at the level of the basement floor exceeds five to six times the undrained shear strength of the soft clay. Several cases have been reported in Sweden where piles have failed owing to lateral creep. In this case, the lime column method is an alternative to a lightweight fill or to embankment piles that will carry part of the weight of the fill around the structure.

Lime columns have also been applied in the construction of deep-lying sewer lines, where the concrete segments are jacked through the soil (Sahlberg, 1979; Sahlberg and Boman, 1979). The lime columns are used to stabilize the walls and the bottom of the deep shafts required with this construction method, as illustrated in Figure 24.8. The concrete pipe elements can be jacked through the lime column wall, which should be designed to resist a lateral earth pressure that corresponds to that for a braced excavation. The shear strength and the bearing capacity of the lime column wall have to be sufficiently high to resist the circumferential force in the wall caused by the lateral earth pressure. The lime column wall can be used as reaction when the pipe segments are pushed through the lime column wall and through the unstabilized soil. Lime columns should also be installed below the bottom of the shaft to prevent failure by bottom heave.

Another application is illustrated in Figure 24.9 where lime or cement columns are used to increase the stability of an anchored sheet-pile wall. The columns have a dual purpose in front of the wall. During the construction they will increase the passive earth pressure at the toe of the wall, but the lime or cement columns will also reduce the settlements of footings constructed on the stabilized soil inside the excavation. Lime columns can be placed behind the wall to reduce the anchor force and the lateral earth pressure acting on the sheet piles. The lateral earth pressure is reduced, since the lime columns increase the average shear strength of the clay behind the wall. However, it may be difficult or impossible to withdraw the sheet piles if they have been driven through the lime columns, as pointed out by Bredenberg (1983) because of the high shear strength of the stabilized soil.

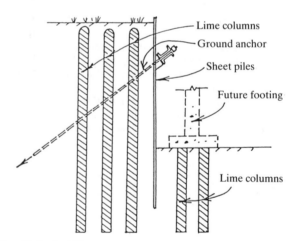

Fig. 24.9 Stabilization of an anchored sheet-pile wall with lime columns.

24.2.2 Results from a Full-Scale Test

Results from a full-scale load test at Skå-Edeby, the test field of the Swedish Geotechnical Institute (SGI) located about 30 km west of Stockholm, Sweden, are shown in Figure 24.10. Two (8 m × 15 m) areas were loaded by a 0.6-m (2-ft) thick gravel fill. The applied load, 10 kPa (0.33 ksf), corresponded to the total weight of a one-storey family house. One area was stabilized by 120 lime columns about 6-m long and spaced 1.4 m (4.6 ft) apart. The total volume of the columns is approximately equal to 10 percent of the total volume of soil enclosed by the columns. There were no lime columns below the second test area.

The soil below the approximately 1.0-m thick surface crust consisted of soft postglacial clay with a relatively high sulfide

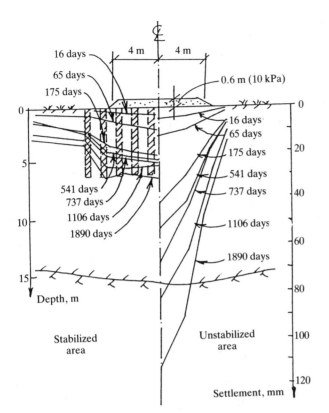

Fig. 24.10 Settlement of two test fills at Skå-Edeby, Sweden from 0 to 6 m depth.

clay at a depth of 3 to 4 m was about 100 percent, as shown in Figure 24.11.

The settlements of the two test fills increased gradually with time owing to consolidation of the underlying very soft clay. The maximum settlement was almost 120 mm below the center of the reference area after 5 years, compared with 30 mm for the stabilized area. The compression of the bottom layers below 6 m depth was 32 mm for the stabilized soil, compared with 17 mm for the unstabilized area because of the difference in the consolidation rate. The settlement rate has gradually decreased with time. Today (1990), the settlement rate of the stabilized area is very low, indicating that the degree of consolidation is almost 100 percent, while the settlement rate of the reference area, almost 20 years after the construction, is still high.

For the stabilized area the maximum differential settlement, that is, the difference in settlement between the edge and the center of the fill, is very small. The maximum angular rotation below the loaded part was only 1/833 for the stabilized area 1.5 years after the installation of the lime columns. It later decreased. The maximum angular rotation of the reference area is 1/37. The reduction of the angular rotation by the lime columns has thus been more than 95 percent.

It can be seen from Figure 24.10 that the settlements outside the loaded part are larger for the stabilized area than for the reference area. This difference indicates that a large part of the weight of the fill for the stabilized area has been transferred to the surrounding soil along the periphery of the reinforced soil block.

24.2.3 Bearing Capacity of Single Lime or Cement Columns

The ultimate strength and the creep limit of single columns are only important when the applied load is so high that the creep limit of the stabilized soil will be exceeded. Load tests in-situ and triaxial tests indicate that the columns can undergo very large deformations when confined without reduction of the bearing

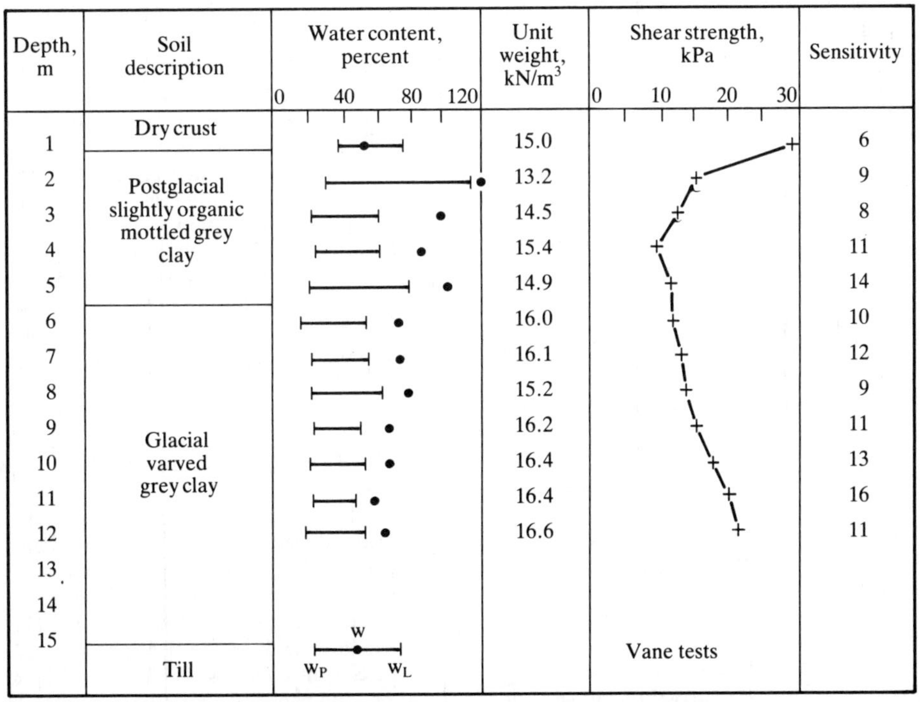

content down to a depth of approximately 15 m. The clay was normally consolidated to slightly overconsolidated. The minimum shear strength determined by vane tests was about 8 kPa at a depth of 4 m. The shear strength increased approximately linearly with depth. The water content of the

Depth, m	Soil description	Water content, percent 0 40 80 120	Unit weight, kN/m³	Shear strength, kPa 0 10 20 30	Sensitivity
1	Dry crust		15.0		6
2	Postglacial slightly organic mottled grey clay		13.2		9
3			14.5		8
4			15.4		11
5			14.9		14
6			16.0		10
7			16.1		12
8			15.2		9
9	Glacial varved grey clay		16.2		11
10			16.4		13
11			16.4		16
12			16.6		11
13					
14					
15		w w_P w_L		Vane tests	
	Till				

Fig. 24.11 Soil conditions at Skå-Edeby, Sweden.

capacity, except close to the ground surface where the confining pressure is low.

The ultimate strength of a single lime or cement column is governed either by the shear strength of the surrounding soft clay (soil failure) or by the shear strength of the column material (column failure). At soil failure the ultimate bearing capacity of a single column depends on both the skin friction resistance along the surface of the columns and on the point resistance. The short-term ultimate bearing capacity of a single column in soft clay can be estimated from the expression

$$Q^{col} = (\pi d H_{col} + 2.25 \pi d^2) c_u \qquad (24.1)$$

where d is the diameter of the column ($d = 0.5$ or 0.6 m), H_{col} is the column length and c_u is the average undrained shear strength of the surrounding soft clay determined, for example, by fall-cone or field vane tests. It has been assumed that the skin friction resistance is equal to the undrained shear strength of the clay (c_u) and that the point resistance corresponds to $9c_u$. When the shear strength exceeds about 30 kPa, a reduced shear strength should be used to calculate the skin resistance. Experience with driven piles indicates that the skin friction resistance of a single column corresponds at least to the undrained shear strength of the untreated soil as determined, for example, by field vane tests when the shear strength is less than 30 kPa (0.31 tsf). It is possible that the skin friction resistance can be less than c_u owing to the low shear strength of the clay next to the columns. Normally the shear strength of the clay is increased up to 30 to 50 mm around the columns. With potassium chloride (KCl) the affected zone can be relatively large (200 mm). The ultimate bearing capacity can thus be slightly lower than that calculated by Equation 24.1.

The point resistance of floating lime or cement columns that do not penetrate the compressible strata is generally low compared with the skin friction resistance, while the point resistance of columns that extend down to an underlying firm layer with a high bearing capacity (e.g., gravel or till) can be high. A large part of the applied load will then be transferred to the underlying soil through the bottom of the columns.

The bearing capacity of a single column at column failure depends on the shear strength of the column material. Load tests on single excavated columns indicate that failure takes place along the relatively weak joint planes. The shear strength along the joints of the fissures will then govern the compressive strength rather than the shear strength of the clay matrix. The shear strength of the columns may be only 20 percent of that determined in the laboratory for small samples cut from the columns (Aas, 1983). The behavior of the column material is thus similar to that of a stiff fissured clay.

The bearing capacity of the columns is also affected by the confining pressure from the surrounding unstabilized soil. Triaxial, direct shear, and in-situ load tests indicate that the ultimate strength of the columns increases with increasing confining pressure and that the increase corresponds to an angle of internal friction of at least 30° (triaxial and direct shear tests have shown that the angle of internal friction of the column material at low confining pressures varies between 30° and 45°). In a few cases a lower friction angle has been reported (Kujala, 1983). However, the shear strength of the clay matrix as determined by fall-cone or vane tests on small-sized samples cut from the columns represents an upper limit. This shear strength is about two to four times the shear strength along the joints as determined by unconfined compression tests. The measured cohesion of the column material (c_{col}) varies between 0 and 150 kPa (Holm, 1979; Kujala, 1983).

The corresponding failure envelope curve, which is mainly based on results from direct shear tests carried out in situ on excavated columns, is shown in Figure 24.12. The short-term

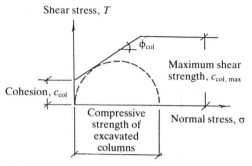

Fig. 24.12 Assumed failure diagram of lime-stabilized soil (Coulomb–Mohr rupture curve).

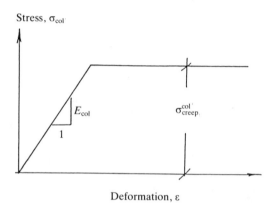

Fig. 24.13 Assumed stress–strain relationship of lime-stabilized soil.

ultimate bearing capacity Q_{ult}^{col} at the depth z can be estimated from the relationship

$$Q_{ult}^{col} = A_{col}(3.5 c_{col} + 3\sigma_h) \qquad (24.2)$$

where c_{col} is the cohesion of the column material and σ_h is the total lateral pressure. It has been assumed that the angle of internal friction of the soil is 30°. The factor 3 corresponds to the coefficient of passive earth pressure at $\phi_u^{col} = 30°$. The lowest bearing capacity is normally obtained at the bottom of the dry crust, where the undrained shear strength of the surrounding unstabilized soil is low.

The long-term ultimate strength of the columns can be lower than the short-term strength owing to creep. The long-term creep strength of a column (Q_{creep}^{col}) is usually 0.65 to 0.80 Q_{ult}^{col}. The load–deformation relationship shown in Figure 24.13 can normally be used to calculate the load distribution by assuming that the load–distribution relationship is linear up to the creep strength σ_{creep}^{col} and that the slope corresponds to the confined modulus of the column material (M_{col}). When the creep strength is exceeded, the load in the columns is assumed to be constant.

24.2.4 Bearing Capacity of Column Groups

The ultimate bearing capacity of a lime or cement column foundation depends on both the shear strength of the untreated soil between the columns and the shear strength of the column material. Possible failure modes of a group with lime or cement columns are indicated in Figure 24.14. The ultimate capacity is either governed by the bearing capacity of the block with lime or cement columns, as shown in Figure 24.14a, or by the local bearing capacity along the edge of the block, as indicated in Figure 24.14b, when the spacing of the columns is large.

column group will then be

$$Q_{\text{ult}}^{\text{group}} = 2c_u H(B + L) + (6 \text{ to } 9)c_u BL \qquad (24.3)$$

The factor 6 refers to a foundation where the length L is large compared with the width $B(L > B)$. For a square foundation the factor 9 can be used. However, a relatively large deformation, 5 to 10 percent of the width of the loaded area, is required to mobilize the maximum end bearing resistance. It is therefore proposed that the end bearing resistance should not be utilized in the design.

The ultimate bearing capacity with respect to a local bearing capacity failure along the edge of the column block is governed by the average shear strength of the soil along the approximately circular failure surface shown in Figure 24.14b. This average shear strength can be calculated in the same way as the stability of a slope.

The ultimate bearing capacity with respect to local failure can be estimated from the expression

$$q_{\text{ult}} = 5.5\, c_{\text{av}}(1 + 0.2\, b/l) \qquad (24.4)$$

where b and l are the width and the length of the locally loaded area, respectively, and c_{av} is the average shear strength along the assumed failure surface.

The average shear strength of the stabilized area is affected by the relative column area a (NA_{col}/BL) and by the shear strength of the column material. If the shear strength of the dry surface crust is considered in the calculations, then a shear strength equal to about one-third of that determined by fall-cone or vane tests should be used. It is proposed to use a global factor of safety of 2.5.

It is only in exceptional cases that the ultimate bearing capacity of the lime or cement columns will govern the design. The shear strength of the unstabilized soil alone without the lime columns is often sufficient that the total weight of the structure can be carried with adequate safety. Usually the total or the differential settlement governs.

24.2.5 Total and Differential Settlements

The lime or the cement columns will reduce both the total and the differential settlements. Bredenberg (1983) has reported that lime columns have reduced the settlements of a 15-m thick layer with soft clay from about 590 mm to 130 mm, or by almost 80 percent. Holm et al. (1983b) observed a 60 percent reduction for an embankment, while for a single-storey building the reduction was 75 percent (Holm et al., 1981). The settlements of a 2 m and 4 m high embankment stabilized by 10 m long cement columns were reduced by 58 and 46 percent, respectively, as reported by Wei et al. (1990). The water content of the underlying soft organic clay and soft marine clay was 146 and 80 percent, respectively. With lime columns the reduction was by 49 and 25 percent of a 2 m and 4 m high embankment. The cement and lime contents of the columns were 25 and 10 percent, respectively.

Test data (e.g., Soyez et al., 1983) indicate that the lime or cement columns and the untreated soil between the columns deform as a unit and that the axial shortening of the columns corresponds to the settlement of the untreated soil between the columns. Holm et al. (1983b) have, however, reported that the vertical deformation of the unstabilized soil between the columns close to the ground surface could be somewhat larger than the axial deformation of the columns. The maximum difference is normally only a few percent of the total settlement even when the spacing of the columns is 3 to 4 diameters (1.5 to 2 m), since the relative displacement to mobilize the maximum skin friction resistance along a single column is small (a few millimeters).

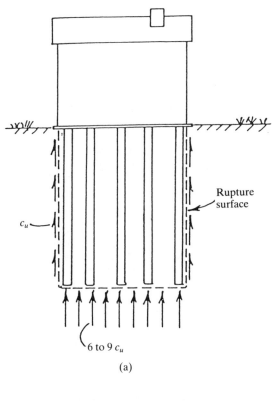

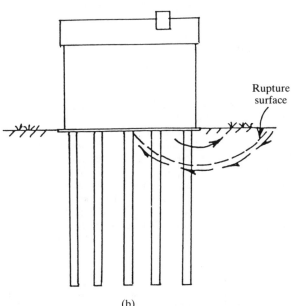

Fig. 24.14 (a) Block failure. (b) Local shear failure.

It is also possible that the shear resistance along a failure surface that cuts through the whole block may govern the bearing capacity in certain cases.

It is proposed to evaluate the ultimate bearing capacity at block failure as indicated in Figure 24.14a when the spacing of the lime columns is small. The ultimate bearing capacity will in this case correspond to the sum of the skin resistance along the perimeter of the lime column group ($2c_u H(B + L)$, where B, L and H are the width, length and height of the column group, respectively) and the end bearing capacity (base resistance) of the block, which is 6 to 9 times the undrained shear strength of the soil (c_u). The total bearing capacity of the lime

Total Settlements

When the axial shortening of the columns corresponds to the settlement of the untreated soil between the columns, the load distribution will depend on the relative stiffness of the columns with respect to the surrounding untreated soil as long as the axial stress in the columns (σ_{col}) is less than the creep limit of the column material. The load distribution will then depend on the confined modulus of the column material (M_{col}) and of the unstabilized soil (M_{soil}). At the same relative deformation,

$$\sigma_{col} = \frac{Q_{col}}{A_{col}} = \frac{q}{a + (1-a)M_{soil}/M_{col}} \quad (24.5)$$

where $q = W/BL$ is the average unit weight of the building; a is the relative column area (NA_{col}/BL), the ratio of the total area of the columns (NA_{col}) and the stabilized area (BL); and M_{soil} and M_{col} are the confined moduli of the untreated soil and of the column material, respectively. If, for example, the relative column area (a) is 0.1 to 0.2, which corresponds to a column spacing of 1.4 to 1.0 m (4.6 to 3.3 ft) and $M_{col} = 10M_{soil}$, then more than half the total load will be carried by the columns. The settlements will then be reduced by more than 50 percent.

When the creep limit of the column material is exceeded, the plastic deformations of the columns can be large. The load carried by the unstabilized soil between the columns ($W - NQ_{creep}^{col}$) will then correspond to the difference between the total applied load (W) and the sum of the creep loads of the individual columns (NQ_{creep}^{col}). This net load will cause a settlement and consolidation of the untreated soil between the columns. The settlement of the unstabilized soil caused by this net load can be calculated from the compression index C_c as determined by, for example, consolidation tests.

The lime columns and the soil enclosed by the columns will act as a rigid block when the spacing between the individual lime columns does not exceed 1.5 to 2.0 m. The columns will then function as vertical reinforcement in the soil. The interaction of the reinforced block and the surrounding soil is primarily governed by the relative dimensions of the block, the magnitude of the applied loads, and the shear strength of the surrounding untreated soil. The applied load is transferred to the surrounding untreated soil mainly through the periphery of the block (Figure 24.15), since a very small relative displacement is required to mobilize the shear strength of the soil. The part of the load that is transferred through the bottom of the reinforced block is initially small. It increases, however, with increasing load level and with time. Approximate calculations using the finite-element method (FEM) indicate that immediately after construction and before any appreciable consolidation of the untreated soil around or below the reinforced block has taken place, 70 to 80 percent of the total weight from a light single-storey structure will be transferred to the surrounding soil along the perimeter of the reinforced block and about 20 to 30 percent through the bottom of the block.

A redistribution of the load transfer will take place with time because of the consolidation of the surrounding soil and a larger part of the load will be transferred through the bottom of the block. It can be seen from Figure 24.10 that the settlements outside the fill are large for the stabilized area in comparison with the reference area. This increase of the settlements indicates that the load from the fill is mainly transferred to the surrounding soil through the periphery of the block.

The maximum angular rotation (α) will be small when the columns are long and the average shear stress along the perimeter of the block is less than the average shear strength (c_u) of the surrounding soil. The average shear stress along the perimeter of the reinforced block (τ_{per}) can be estimated from

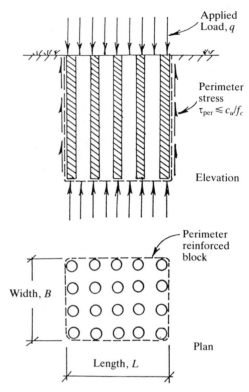

Fig. 24.15 Perimeter shear stress.

the expression

$$\tau_{per} = \frac{0.8\,W}{2(B+L)H} < \frac{c_u}{f_c} \quad (24.6)$$

where B, L, and H are the width, length, and height, respectively, of the block, W is the total weight of the structure, and f_c is a partial safety factor that should be at least 1.3 to 1.5. It has been assumed that 20 percent of the total load is transferred to the surrounding soil through the bottom of the block immediately after the application of load.

The average shear stress along the perimeter of the block should be less than the average shear strength c_u of the surrounding clay in order to keep the differential settlements small. At $f_c = 1.5$, Equation 24.6 can be rewritten as

$$\frac{H}{B} \geqslant \frac{0.6\,q/c_u}{(B/L + 1.0)} \quad (24.7)$$

where $q = W/BL$. For example, at $q/c_u = 2.0$ and $B/L = 1.0$, then $H/B > 0.60$. The required length of the columns at $B = 8$ m (26 ft) is thus 4.8 m (14.5 ft).

If the load transferred from a structure to the reinforced block is not uniformly distributed, the columns should be concentrated in the heaviest-loaded parts so that the stress increase in the unstabilized clay will be as uniform as possible. The stress distribution is affected by the relative stiffness of the structure with respect to the reinforced block. The unit load is often higher below the outside walls of the structure than at the center. In that case it is advisable to concentrate the columns at the exterior walls. However, the maximum distance between two adjacent columns should not exceed half the length of the columns.

The total settlement of a structure supported on lime columns can be estimated as illustrated in Figure 24.16. The maximum total settlement is equal to the sum of the local settlement of the reinforced block (Δh_1) and the local settlement

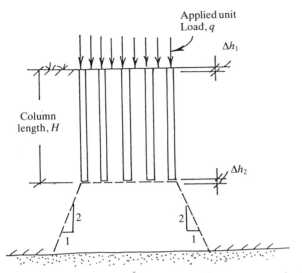

Fig. 24.16 Calculation of settlement when the creep strength of the lime columns is not exceeded.

of the unstabilized soil below the block (Δh_2). The settlement (Δh_1) is affected by the stress distribution in the block and thus by the interaction of the columns with the surrounding unstabilized soil. Test data indicate that the relative deformation of the columns will be about the same as that of the unstabilized soil between the columns when the distance between the columns is less than about 1.5 to 2.0 m, except possibly close to the ground surface if the surface crust is thin (< 1 m).

Two cases have to be investigated. In the first case (Case A) the applied load is relatively low so that the creep limit of the columns, Q_{creep}^{col}, will not be exceeded. In the second case (Case B) the applied load is relatively high. The axial load in the columns will then correspond to the creep limit.

Case A In this case the relative stiffness of the columns with respect to the unstabilized soil between the columns will govern the load distribution. The equivalent confined modulus of the column material increases in general with increasing creep strength. The relationship $M_{col} = (50 \text{ to } 100)c_{col}$ can often be used to estimate the settlements. Normally the modulus is 15 to 25 MPa. It is preferable, however, to evaluate M_{col} directly from oedometer tests. The increase of the confined modulus of the stabilized soil is usually larger than the increase of the shear strength, as pointed out by Brandl (1981).

The lime columns will reduce the settlements down to the bottom of the reinforced block. The weight q of a structure or a fill is carried partly (q_1) by the columns and partly (q_2) by the unstabilized soil between the columns. The resulting settlement (Δh_1) can be estimated conservatively from the expression

$$\Delta h_1 = \frac{qH}{a M_{col} + (1 - a)M_{soil}} \qquad (24.8)$$

where a is the relative column area ($N A_{col}/BL$).

For example, at $q = 20 \text{ kPa}$ (0.22 tsf), $M_{col} = 10M_{soil}$, $a = 0.2$, $H = 5 \text{ m}$ (16 ft), $M_{soil} = 250 c_u$, and $c_u = 20 \text{ kPa}$ (0.21 tsf), the estimated settlement Δh_1 from Equation 24.8 is 7.1 mm. The results from oedometer tests should be used in the calculations in the case that the preconsolidation pressure is exceeded.

The settlement reduction ratio (β), that is, the ratio of the total settlements down to the bottom of the reinforced block with and without lime columns, can be estimated from the

relationship

$$\beta = \frac{M_{soil}}{a M_{col} + (1 - a)M_{soil}} \qquad (24.9)$$

It can be seen from Equation 24.9 that the local settlement of the reinforced block decreases with increasing relative area of the columns (a), as can be expected, and with increasing value on the equivalent confined modulus of the column material. At $a = 0.2$ and $M_{col}/M_{soil} = 50$ the reduction ratio β is 0.093. The lime columns will thus reduce the local settlements by more than 90 percent down to the bottom of the reinforced block.

The average stress increase q in the reinforced block depends on the load transferred to the surrounding soil through the perimeter of the block and thus on the relative stiffness of the block with respect to the surrounding soil and on the thickness of the compressible soil layers below the block. In order to simplify the calculations it is proposed to calculate the average stress increase q in the block from the following equation:

$$q = \frac{0.8 W}{BL} \qquad (24.10)$$

It is thus assumed that the stress increase is constant throughout the height of the block. This is a conservative assumption. If more accurate evaluation is required, then the stress increase should be evaluated from the assumption that the average shear stress along the perimeter corresponds to the shear strength of the soil.

An average confined modulus (M_{av}) can also be calculated from Equation 24.9:

$$M_{av} = \frac{q}{h_1/H} = a M_{col} + (1 - a)M_{soil} \qquad (24.11)$$

At $a = 0.2$, $M_{col} = 10M_{soil}$, then $M_{av} = 2.8M_{soil}$. The average confined modulus is thus considerably larger than the confined modulus of the untreated soil even when the spacing of the columns is relatively large.

It has been assumed in the derivation of Equations 24.8 and 24.9 that the stress–strain relationship of the column material is linear up to the creep limit. When the creep limit is exceeded, the calculation method described in Case B should be used.

The local settlement (Δh_2) below the block can be estimated by dividing the soil below the block into layers and by calculating the compression of each layer separately. The estimated total settlement corresponds to the sum of the settlements of the individual layers. The stress increase in each layer can be determined, for example, by the 2:1 method. Lahtinen and Vepsalainen (1983) found a good agreement between this method and the settlements calculated by the finite-element method (FEM).

Case B The total applied load is so high in this case that the axial load in the individual columns corresponds to the creep limit. The settlement of the reinforced block (Δh_1) can then be estimated as shown in Figure 24.17, where the applied load has been divided into one part q_1 that is carried by the columns (Fig. 24.17a) and a second part q_2 that is carried by the unstabilized soil between the columns (Fig. 24.17b). The unit load q_1 is governed by the creep strength of the columns:

$$q_1 = \frac{N Q_{creep}^{col}}{BL} \qquad (24.12)$$

where N is the total number of columns, and B and L are the width and the length of the reinforced block, respectively.

The part q_2 ($q - q_1$), which is carried by the unstabilized soil between the columns, governs the settlement (Δh_1) down to the bottom of the reinforced block. This settlement can be

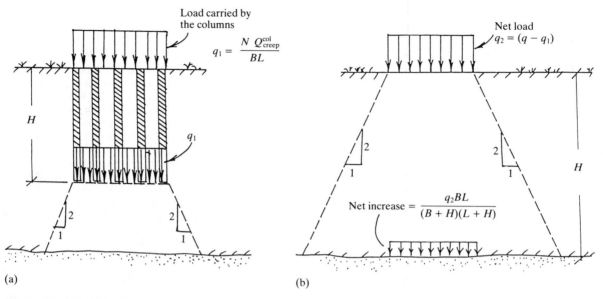

Fig. 24.17 Calculation of settlement when the creep strength of the lime columns is exceeded. (a) Load carried by the lime columns. (b) Load carried by the unstabilized clay between columns.

determined by dividing the block into layers and by calculating the settlement of each layer separately.

The settlement Δh_2 below the reinforced block can be estimated from the assumption that the load q_1 is transferred down to the bottom of the reinforced block while the load q_2 is applied at the ground surface. The total settlement is determined by first dividing the soil below the block into layers. Then the settlement of each layer is evaluated separately in the same way as for Case A. This is a conservative assumption since the stress increase for the unstabilized soil decreases with depth and the resulting compression of the soil may be less than the creep limit of the columns. Holm (1979) has used this method to estimate the settlement of a road embankment with satisfactory results. The compression modulus was evaluated from oedometer tests. The estimated creep load of the columns was 65 percent of the calculated ultimate strength.

Differential Settlements

Damage of structures is mainly caused by large differential settlements, which for a block reinforced with lime or cement columns are mainly governed by the shear deformations in the unstabilized soil between the columns (Fig. 24.18).

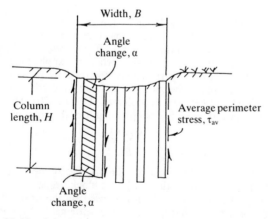

Fig. 24.18 Calculation of differential settlement (angle change α).

The differential settlements will be small as long as the average shear stress along the perimeter of the reinforced block is less than the average shear strength of the unstabilized clay. Holm et al. (1981) reported, for example, a maximum angular rotation of 1/700 for a single storey building constructed on lime columns.

In the following derivation it is assumed that the columns are stiff in comparison with the unstabilized soil between the columns and that the axial shortening of the columns is small and can be neglected. The maximum angular rotation (α) between two column rows will then be proportional to the average shear stress τ_{av} along the perimeter of the reinforced block and the average shear modulus G_{soil} as expressed by the relationship

$$\alpha = \frac{\tau_{av}}{G_{soil}} \qquad (24.13)$$

In this case the soil between two adjacent column rows will behave in a similar way as in a direct shear test. The average shear stress in the soil then corresponds to the average shear stress along the perimeter of the reinforced block. At $\tau_{av} = c_u/f_c$ and $G_{soil} = 100c_u$ the maximum angular rotation (α) will be $(1/100)f_c$. At $f_c = 2.0$ the angular rotation is 1/200.

Test results (Fig. 24.10) indicate that the largest differential settlement will occur outside the loaded area and that the increase of the differential settlements with time will be small. The decrease of the shear modulus that takes place with time due to consolidation of the untreated soil will be compensated at least partly by the increase of the load transferred through the bottom of the block and by the resulting decrease of the average shear stress along the perimeter of the block.

Settlement–Time Relationships

The pore water pressure in the unstabilized soil between the lime columns is increased owing to the weight of the structure even when the load is less than the creep strength of the columns. The lime column block settles as the excess pore water pressures dissipate and the effective pressures in the soil increase. Measurements indicate that the pore pressures are increased also in the unstabilized clay around the columns during the installation owing to the expansion of the lime during the slaking. The pore pressures may even approach the total overburden pressure.

The addition of unslaked lime usually increases the permeability of soft clay, as pointed out, for example, by Brandl (1981), Evans and Bell (1981), Serra et al. (1983), and others. However, for marine clays, Terashi and Tanaka (1983) have observed a reduction of the permeability with increasing lime content. The reduction was large when cement was used to stabilize the soil. The observed increase of the consolidation rate can at least partly be attributed to the increase of the coefficient of consolidation caused by the reduced compressibility of the stabilized soil.

The lime columns usually act as drains in soft clay. The settlement rate is increased even when the load in the lime columns is below the apparent preconsolidation pressure of the stabilized soil. Test results from Sweden and Finland indicate that the effective drain radius of 0.5-m diameter lime columns is 0.25 m and that the disturbance of the surrounding clay caused by the installation of the columns is small in contrast to that by displacement-type sand drains, where the permeability of the surrounding soil is reduced. The coefficient of consolidation (c_{vh}) with respect to horizontal flow is often three to four times higher than c_{vv}.

Holm (1979) has described a case where lime columns were used to stabilize a road embankment constructed on very soft clay. When the spacing of the lime columns was 1.0 m and 1.2 m, the consolidation was essentially completed after 3.5 and 6.5 months, respectively. For a single-storey building constructed on a 5-m thick clay layer the increase of the settlements after the first 3 months after the installation of the lime columns was small (Holm et al., 1981).

The following relationship can be used according to Barron (1948) to calculate the degree of consolidation (U) as a function of the time (t) when the apparent preconsolidation pressure of the lime columns is not exceeded:

$$U = 1 - \exp\left(-\frac{2 c_{vh} t}{R^2 f(n)}\right) \qquad (24.14)$$

where

U = degree of consolidation
c_{vh} = coefficient of consolidation ($k_h/m_v \rho_w$)
t = consolidation time
R = radius of influence of a single lime column
$n = R/r$, where r is the column radius (0.25 or 0.30 m)

The term $f(n)$ is a function of n:

$$f(n) = \frac{n^2}{n^2 - 1} \ln n - \frac{3}{4} + \frac{1}{4 n^2} \qquad (24.15)$$

It has been assumed in the derivation of this equation that the apparent preconsolidation pressure of the columns will not be exceeded, that the lime columns do not disturb the surrounding soil during the installation, and that the thickness of the smear zone around the columns is small and can be neglected. When the preconsolidation pressure will be exceeded, the net load on the unstabilized soil, the difference between the total load and the load carried by the lime columns, should be used to calculate the settlement rate.

The settlement rate is thus a function of the permeability of the soil in the horizontal direction, k_h, and the average confined modulus, M_{av}, in the vertical direction when the creep limit of the columns is not exceeded. Holm (1979) found a good agreement between measured and calculated settlement rates by this method.

The coefficient of consolidation of the unstabilized soil can be determined from pore pressure soundings (Torstensson,

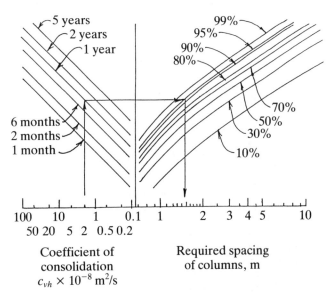

Fig. 24.19 Calculation of required spacing of lime columns when used as drains. (Example: $c_{vh} = 2.10^{-8} \, \text{m}^2/\text{s}$. At $t = 1$ year and $U = 95$ percent, the spacing should be 1.5 m.)

1975) or from triaxial tests. For normally consolidated clays the coefficient c_{vh} ($= k_h/m_v \rho_w$) is generally 1 to 5 times the coefficient of consolidation c_{vv} ($= k_v/m_v \rho_w$) determined from conventional oedometer tests. For varved clays the ratio c_{vh}/c_{vv} can be very large because the permeability is high in the horizontal direction, parallel with the stratifications. The permeability in the horizontal direction is governed by the layers with the highest permeability, while the layers with the lowest permeability determine the permeability of the soil in the vertical direction.

The diagram shown in Figure 24.19, which is based on Equation 24.14, can be used to determine the required spacing of the columns in order to reach a certain degree of consolidation within a given time. The spacing has been calculated for the example shown in the figure for the case when the coefficient of consolidation (c_{vh}) is $2 \times 10^{-8} \, \text{m}^2/\text{s}$ and the required degree of consolidation is 95 percent after 1 year.

24.2.6 Slope Stability

The stability of slopes in soft clay can be improved with lime or cement columns. This method is thus an alternative to concrete, timber, or steel piles. The driving of the piles has in some cases initiated slides because of the resulting high pore water pressures. The reduction of the shear strength caused by the installation of lime columns is small, since the disturbance of the soil around the columns is small during the mixing of the lime or the cement with the soil. The dissipation of any excess pore pressures caused by the lime columns occurs rapidly, since the columns function as vertical drains in the soil. With cement the permeability is reduced. The risk of initiating a slide during the installation of the lime columns is therefore small in comparison with the case of timber or precast concrete piles. The method has been used successfully, for example, in Sweden and Norway. Lime columns have been used at Tuve in Göteborg, Sweden, to stabilize parts of an area that have been affected by a recent landslide.

The average shear strength along a potential failure or rupture surface in the soil can be estimated from the equation

$$c_{av} = c_u(1 - a) + s_{col}a \qquad (24.16)$$

where c_u is the undrained shear strength of the soil as determined by, for example, field vane tests, s_{col} is the average shear strength of the stabilized clay within the columns, and a is the relative column area. The shear strength s_{col} of the stabilized soil depends on the normal pressure and thus on the relative stiffness of the columns with respect to the surrounding unstabilized soil. It is proposed to use the total overburden pressure in the calculation of the shear strength and a friction angle (ϕ_u^{col}) of 30°. The calculated shear strength of the stabilized soil should not exceed $c_{col,max}$, the shear strength of the clay matrix in the columns. The spacing of the lime columns and the volume of the soft clay requiring stabilization can be evaluated by analyzing the stability of the slope along different potential failure surfaces. The safety factor should be at least 1.3 to 1.5.

The lime column method can also be used with advantage when high pore water pressures in thin pervious silt or sand layers have reduced the stability of a slope, since the lime columns function as vertical drains. The increase of the average shear strength of the soil by the columns will also contribute to the stability of the slope.

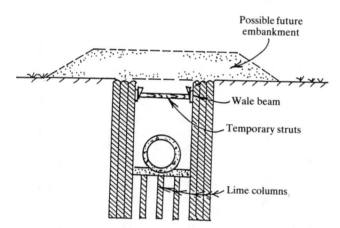

Fig. 24.20 Trench stabilized with lime columns.

24.2.7 Stability of Excavations and Trenches

Lime or cement columns can be used instead of sheet piles in trenches and excavations as illustrated in Figure 24.20. The method has mainly been applied when the depth is less than 4 to 5 m. Failure by overturning or by bottom heave should be considered as well as shear failure of the lime or cement column wall with respect to a circular or a plain failure surface through the columns.

Overturning When the lime columns are used to increase the stability of deep excavations, their behavior is partly governed by the thin layer of lime with a low tensile strength that surrounds each column. Tensile cracks can develop along this weak layer during the excavation of the trench. Failure by overturning can occur of the individual columns when the cracks become filled with water (Broms and Boman, 1975b). This type of failure can be prevented by placing the columns so closely that they overlap. Thereby the weak layer between the individual columns is eliminated. The water and earth pressures causing failure of the column wall by overturning are resisted by the weight of the wall in the same way as for a gravity retaining wall (Fig. 24.21a). The center of rotation is located at the bottom of the excavation close to the surface of the wall.

In the case that the depth of the excavation exceeds a critical value $2c_u/\gamma$, where γ is the unit weight of the soil, the resulting lateral earth pressure will also affect the stability, as illustrated in Figure 24.21b. The actual lateral earth pressure will be less than that calculated by the Rankine earth pressure theory because that method does not take into account the friction or adhesion along the lime column wall.

The lateral earth pressure on the lime column wall can be reduced with lime columns placed behind the wall. The resulting lateral earth pressure corresponds to the average shear strength of the soil, taking into account the bearing capacity of the columns. The effectiveness of the lime columns can be increased by inclining the columns. Thereby the internal moment arm is increased.

Shear Failure The shear strength of the lime column wall should also be considered. The stability can be estimated by assuming that failure takes place along an inclined failure surface, as illustrated in Figure 24.22, or along a horizontal plane through the bottom of the excavation. In order to simplify

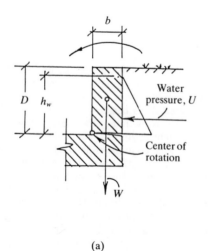

(a)

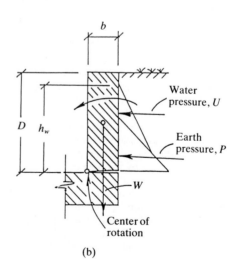

(b)

Fig. 24.21 Failure by overturning. (a) $D \leqslant \dfrac{2c_u}{\gamma_{soil}}$. (b) $D > \dfrac{2c_u}{\gamma_{soil}}$.

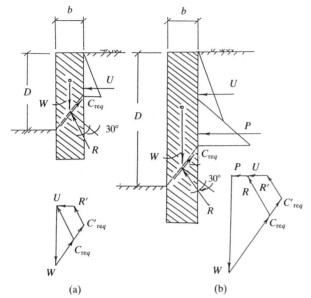

Fig. 24.22 Shear failure. (a) $D \leqslant \dfrac{2c_u}{\gamma_{\text{soil}}} + b\sqrt{3}$. (b) $D > \dfrac{2c_u}{\gamma_{\text{soil}}} + b\sqrt{3}$.

the calculations it can be assumed that the failure surface is inclined at 60°.

Bottom Heave Failure by bottom heave can occur when the depth of the excavation exceeds $6c_u/\gamma$ and the length of the excavation is large compared with the depth. The bottom can be stabilized with lime columns as illustrated in Figure 24.23. When the columns overlap, the length of the columns (H_d) that is required to prevent failure by bottom heave can be estimated from the relationship

$$H_d > \frac{BD\gamma_{\text{soil}}}{2\,s_{\text{col}}} - \frac{3\,c_u B}{s_{\text{col}}} \tag{24.17}$$

It should be noted that the shear strength s_{col} is affected by the normal pressure along the failure surface. If the columns do not

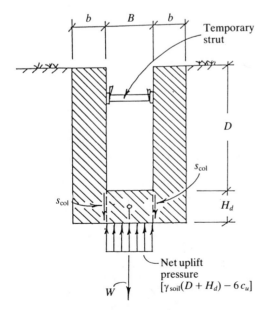

Fig. 24.23 Bottom heave.

overlap then the initial shear strength of the soft clay (c_u) between the columns should be used in the analysis instead of the shear strength of the column material (s_{col}).

It has been assumed in the derivation of Equation 24.17 that the length of the excavation is large compared with the width and that the thickness of the clay layer below the bottom of the excavation exceeds the width of the excavation. The risk of failure by bottom heave can be reduced by excavating the trench in short sections.

24.3 APPLICATIONS OF THE LIME COLUMN METHOD

24.3.1 General

The equipment for the lime columns (mast, rotary table, and kelly) can be mounted on a standard front-wheel loader (Fig. 24.3a) or a tracked carrier (Fig. 24.3b) in order to improve the mobility. Both units are provided with automatic recorders that indicate the amount of lime or cement injected into the soil as a function of the depth and the length of the columns.

The unslaked lime stored in the 2.5-m³ container unit pulled by the loader or by the carrier (Figs. 24.3a and b) is normally sufficient for the manufacture of 15 approximately 10-m long columns. The lime or cement should be stored at the site in silos or in pressure tanks. Compressed air is used to transport the lime from the pressure tanks to the container. On small jobs, bags are used.

The unslaked lime should be finely ground so that the maximum particle size is less than 0.2 mm. The gradation is also important to prevent clogging of the lime during the mixing. It is also important that the lime should be as pure as possible. However, moderate amounts of magnesium oxide normally will not affect the strength increase.

A surface fill might be required in order to improve the bearing capacity sufficiently that the lime columns can be installed. Sand or gravel can be used. The maximum particle size should not exceed 50 mm, since larger particles may interfere with the installation of the columns.

The maximum capacity of the tracked carriers (LPS 4) shown in Figure 24.3b is 50 to 60 columns 10 m long in about 8 to 10 hours. Normally 40 to 50 columns (500 m) can be manufactured under favorable conditions. The maximum capacity of a front-wheel loader is about 30 to 40 columns in 8 hours.

The quality of individual lime columns can vary even if the soil appears to be uniform. Even a relatively small change of, for example, the organic content of the soil can affect the shear strength, the bearing capacity, and the creep limit. Load tests indicate that the creep limit of excavated single columns is normally distributed and that the standard deviation is usually less than 20 percent. When the spacing of the lime columns in a group does not exceed 1.5 to 2 m, then the load in a weak column can be transferred to the adjacent columns with a higher bearing capacity. The difference in settlement between two adjacent columns is usually small even if some of the columns are loaded to the creep limit.

The time can be shortened before an area can be loaded by preloading the soil after the installation of the columns when the bearing capacity of the columns is still low, as illustrated in Figure 24.24. Then a large part of the applied load will initially be carried by the unstabilized clay between the columns, as shown in Figure 24.24a. The increase of the pore pressure will approximately correspond to the increase of the total stress in the unstabilized clay when the clay is normally consolidated. Since the lime columns also act as drains, the consolidation of

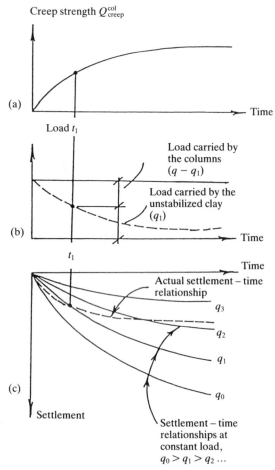

Fig. 24.24 Effect of preloading.

the clay and the dissipation of the excess pore water pressures will occur rapidly.

The bearing capacity of the lime columns will gradually increase with time and with increasing settlement, since the axial deformation of the columns will be about the same as that of the soil between the columns. The load on the unstabilized soil between the columns will decrease with increasing time. For example, at the time t_1 in Figure 24.24a the load on the unstabilized clay between the columns is about 50 percent of the total applied load. The consolidation rate will then correspond to curve q_0 in Figure 24.24c. (The time–settlement relationships shown in Figure 24.24c correspond to the constant loads q_0, q_1, etc., where $q_0 > q_1 > q_2$.) Some time later the load carried by the columns has increased. The consolidation rate will then correspond to curve q_2. The real time–settlement relationship will therefore correspond to the dotted curve shown in Figure 24.24c, which can be determined from the creep limit of the lime columns and the consolidation rate of the undisturbed clay at different stress levels from oedometer tests, for example.

24.3.2 Embankments and Fills

Fills or embankments can normally be constructed to the full height immediately after the installation of the lime or cement columns when the bearing capacity of the columns is still low because the stability is generally satisfactory even without the columns. A large part of the settlements will occur within in a few months, since the lime columns also act as drains in the soft clay.

The spacing and the length of the lime columns are governed by the available construction time and by the allowable total and differential settlements. It should be noted that the spacing of the lime columns should not exceed the thickness of the fill or of the embankment in the case when the dry surface crust is thin or has been removed. Otherwise the settlements of the unstabilized soil between the columns can be large and the columns may protrude into the embankment. This risk can be reduced by placing a layer of fabric over the columns.

It is normally not economic to carry the total weight of the embankment by the lime columns. It is usually possible also to utilize the bearing capacity of the unstabilized soil between the columns, especially if the soil has been preloaded. The bearing of the columns may not be sufficient immediately after the installation. It may then be necessary to combine the lime columns with some other soil-stabilization method such as pressure beams or soil reinforcement, where fabric or other types of reinforcement are placed in the soil in order to increase the bearing capacity especially along the edge of the embankment or the fill. The pressure berms can be removed after a few months, when the bearing capacity of the lime or cement columns is sufficient to carry the total weight of the fill. The global factor of safety should be at least 1.3 to 1.5.

If rock fill is used, it is often desirable to place a layer of fabric or gravel before the fill in order to prevent the rock fill from penetrating into the underlying soft clay (Holm, 1979). It should be noted that the coefficient of consolidation will decrease when the creep limit of the columns has been exceeded because of the reduction of the average confined modulus of the stabilized soil.

24.3.3 Foundation of Structures

Lime or cement columns can be used as foundation for light structures instead of, for example, precast concrete or steel piles. The lime column method is often economic when the maximum bearing capacity of traditional foundation methods, such as piles, cannot be utilized fully. With lime or cement columns a relatively thin floor slab can be used because the slab will be supported at a large number of points compared with the case when only a few concrete or steel piles are used.

The total settlement that can be allowed is normally relatively large provided the settlements are even. Connecting water and sewer lines often rupture when the maximum total settlement of the structure exceeds 150 to 200 mm (6 to 8 in). The maximum total settlement that can be tolerated depends on such factors as the material used for the sewer or water lines, the diameter and length of the pipe segments, as well as the flexibility of the joints. It is often advisable also to install lime columns outside the structure, as illustrated in Figure 24.25, so that the transition from the structure to the unstabilized area is as smooth as possible.

Structures are primarily damaged by differential settlements. Partition walls often crack when the angular distortion is 1/200 to 1/300 and the height of the wall is larger than about half the length of the wall. If the height is less than one-quarter of the length, cracks normally do not appear until the angle change is about 1/150. Also, the settlement rate affects the cracking. A relatively large angular distortion can often be accepted owing to creep in the construction material, for example, reinforced concrete, when the settlements increase gradually, as is normally the case for a soft normally consolidated clay.

Brick structures or buildings constructed of lightweight concrete blocks as well as tiled walls are very sensitive to differential settlements. The maximum angular rotation is usually 1/400 to 1/500. The strength of the mortar used in brick buildings has also a large influence on the differential

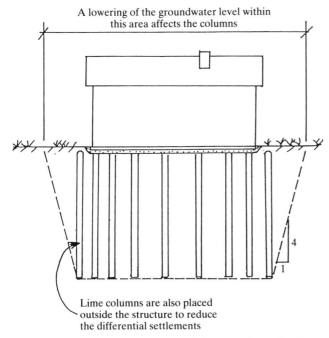

A lowering of the groundwater level within this area affects the columns

Lime columns are also placed outside the structure to reduce the differential settlements

Fig. 24.25 Effect of a lowering of the groundwater level.

settlements that can be allowed without significant cracking. The differential settlement that can cause excessive cracking of a brick building with cement mortar is small in comparison with that in older structures where lime mortar has been used. Doors and windows start to jam when the angular rotation is about 1/150. Prefabricated wooden houses can usually undergo a relatively large angular rotation without damage, possibly as large as 1/200. Buildings constructed of lightweight building blocks start to crack when the maximum angular rotation is about 1/400. The corresponding maximum angular rotation for a brick building is about 1/350.

The lime or cement columns should preferably penetrate all compressible layers when the depth to firm bottom is less than 15 m (50 ft), so that the settlements below the reinforced block will be as small as possible. This is especially important when the thickness varies. Otherwise the differential settlements caused, for example, by a lowering of the groundwater level can be excessive. It should be noted that the lime content at the bottom of the columns (~ 0.3 m) can be low, because the unslaked lime is forced into the soil just above the mixing tool. It is proposed that an additional compression of 5 to 10 mm should be allowed for in the settlement calculations.

A lowering of the groundwater level may increase the load in the columns owing to negative skin friction. It is, therefore, advisable in this case also to place a few lime columns outside the structure, since the largest angular rotation caused by a lowering of the groundwater level will occur between the two outside rows. It is also important that the load distribution is as even as possible.

The bottom slab can be designed as if it is supported on a series of elastic springs (Winkler foundation). The stiffness of the springs depends on the compressibility and creep strength of the columns, the width of the loaded area, and the stiffness of the slab. In the calculations of the equivalent spring constant for a point or a line load it is proposed to use an effective depth (H_e) below the loaded area that corresponds to six to seven times the thickness of the concrete slab. The spring constant can be estimated from

$$k = \frac{M_{av}}{H_e} \qquad (24.18)$$

where H_e is the effective thickness of the compressible layer and M_{av} is the average compression modulus (Eq. 24.11) of the stabilized soil. The ratio of the confined modulus of the unstabilized clay between the lime or cement columns and that of the column material varies normally between 10 and 50. The compression modulus of the unstabilized clay in the dry crust is usually $250c_u$ to $500c_u$ where c_u is the undrained shear strength of the clay.

Lime columns can be an economic foundation method for light industrial buildings. Heavy buildings, structures with large free spans, or statically indeterminate structures are normally supported on steel or concrete piles. Usually the ground floor can be supported on lime or cement columns even when the applied load is relatively high. The spacing and length of the columns are governed either by the maximum total settlement or by the maximum differential settlement that can be tolerated. The possibility of reducing potentially excessive settlements by preloading should always be investigated. The thickness of the floor slabs and the required reinforcement can be estimated in the same way as for a light structure by assuming that the slab is supported on a series of elastic springs (Winkler foundation).

Fills and embankments required for roads, loading, and parking areas around industrial buildings can cause large settlements, especially in combination with a lowering of the groundwater level. The structural piles below the exterior walls of the buildings can then be affected by the resulting negative skin friction. The piles will in that case also carry part of the weight of the embankment or of the fill. The settlement of the structural piles can in extreme cases be so large that the supported structure is damaged and connecting sewer and water lines may fail. This can be prevented by placing lime or cement columns also outside the structure below the fill or the embankment.

Lime or cement columns can in general not compete with concrete or steel piles as foundations for heavy structures, since the bearing capacity of the concrete or steel piles can usually be utilized fully. Lime columns can, however, be used in combination with a floating foundation (raft) in order to reduce the risks of bottom heave and the differential settlements caused, for example, by a lowering of the groundwater level.

24.3.4 Excavations and Trenches

It may also be possible to use the lime or cement columns as walls in deep excavations in soft clay instead of sheet-pile and diaphragm walls. The basement walls can be cast directly against the lime columns when they overlap or against rockwool or fiberglass mats. The mats will function both as drains and as insulation. Lime or cement columns can be used as lateral support in trenches. The columns should overlap so that they form continuous walls at both sides of the excavation. The dimensions of the wall are in general governed by overturning when the creep strength of the columns is larger than 20 to 30 kN (2.2 to 3.3 tons) and vertical lime columns are used. The trenches can generally be excavated about 1 month after the installation of the columns.

When the bottom of the trench has been stabilized with lime columns to prevent failure by bottom heave, it is often sufficient to place a sand or gravel layer at the bottom in order to support even large-diameter sewer pipes when the spacing of the lime columns is relatively small (< 1 m).

The stability of a lime column wall with respect to overturning can be increased by inclining the wall backwards, as illustrated in Figure 24.26, or by placing the columns so that they function as buttresses, as shown in Figure 24.27. The unstabilized soil between the buttresses will then also contribute to the stability of the wall with respect to overturning. If the lime or cement

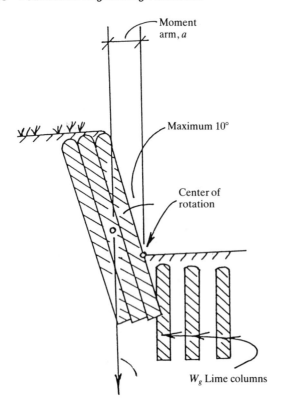

Fig. 24.26 The use of inclined columns to increase the stability of a deep cut.

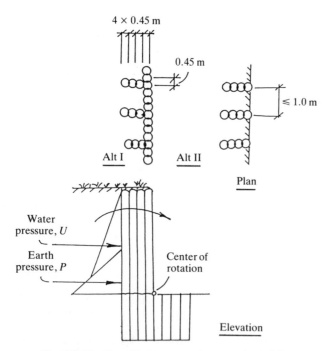

Fig. 24.27 Use of buttresses to increase the stability.

column wall is tilted backwards, the moment arm of the weight of the wall is increased because the center of rotation is located close to the front face of the lime column wall, as shown in Figure 24.26. Even a relatively small inclination of the wall will have a substantial effect. The stability of the wall will then be governed by the average shear strength of the soil along a circular or plane failure surface through or below the wall.

Sahlberg (1979) has described a case where inclined lime columns were used to stabilize a deep cut in a very soft clay ($c_u = 10\,\text{kPa}$). It was estimated that the lime columns reduced the construction costs by as much as 30 to 40 percent compared with a conventional sheet-pile wall. Also, Lahtinen and Vepsalainen (1983) have described a case where lime columns were used successfully to stabilize a deep cut.

The column wall for a deep cut or trench is normally designed by first calculating the active Rankine earth pressure acting on the wall using a reduced shear strength. The shear strength of the soil as measured, for example, by field vane tests is usually reduced to take into account the variation of the shear strength of the soft clay and the effect of the loading rate. The required width of the lime column wall can then be determined with respect to overturning around a point located at the bottom of the excavation (Fig. 24.26). The number of lime columns is governed by the required factor of safety with respect to a shear failure along a horizontal, an inclined, or a circular failure surface through the wall.

The length of the columns is governed by the required passive earth pressure acting on the lime or cement column wall below the bottom of the excavation. The passive earth pressure should exceed the active earth pressure and the water pressure behind the wall. Stability of the excavation with respect to bottom heave should also be checked.

The bracing should be designed to resist at least the potential water pressure behind the wall, since the cracks behind the wall may be filled with water during, for example, a heavy rainstorm. For deep excavations or when the bearing capacity of the columns is uncertain the wall should be supported at least at two levels. When the columns are inclined and the water pressure behind the wall can be controlled, the bracing can be omitted.

24.4 LABORATORY AND FIELD INVESTIGATIONS

24.4.1 General

The creep limit of the lime columns and the shear strength of the stabilized soil will normally govern the design and the economy of the method. The possible increase of the shear strength with lime or cement should be investigated as early as possible, since it takes up to 4–6 months to carry out the necessary laboratory and field tests.

The costs can be reduced if the limitations of the method are considered early in the design. During the preliminary design stage it is necessary to know the magnitude and the distribution of the applied loads, the total and differential settlements that can be allowed, the strength and deformation properties of the lime or cement columns during the construction, the initial soil conditions, the need for predrilling, etc. The cost can sometimes be reduced by combining the lime column method with other soil-improvement methods such as preloading, pressure berms, or embankment piles. Relatively extensive field and laboratory investigations are normally required in comparison with other foundation methods.

24.4.2 Evaluation of the Lime Column Method

During the preliminary stage of a project one normally has some knowledge from previous investigations about the depth and thickness of the different compressible layers. A preliminary assessment of the lime column method and of the costs can often be done on the basis of this information. The largest increase of the shear strength and of the bearing capacity using lime can generally be expected for soft clays with a low plasticity index and a low water content. The effect of quicklime on the shear strength of organic soils, peat or muck, is generally low.

The increase of the shear strength with unslaked lime, gypsum, cement, or fly-ash can be estimated from samples prepared in the laboratory. The shear strength of the stabilized soil has to be at least three to five times the initial shear strength before the lime column method becomes economic. Bryhn et al. (1983) have, for example, reported that the shear strength of small samples cut from columns stabilized in the field is 50 to 150 percent higher than that of samples prepared in the laboratory. Holm et al. (1981) found that the shear strength determined in situ on actual columns by pressuremeter tests was one to two times higher than the shear strength determined by unconfined compression tests on laboratory-prepared samples.

The prepared samples should be stored in a moist room or under water after the mixing. The temperature during the storage should preferably correspond to the ground temperature. A lime content of 6 to 8 percent with respect to the dry weight of the soil is normally sufficient for inorganic clays with a medium to low plasticity ($w_L < 50$) in order to reach a shear strength of 50 to 200 kPa after 1 year as determined by unconfined compression tests. The required lime content generally increases with increasing water content and with increasing organic content of the soil. The maximum is about 12 percent. The shear strength is often reduced when a higher percentage of lime is used.

Gypsum ($CaSO_4$) can sometimes be effective in combination with lime in order to stabilize organic soils with a water content between about 90 and 180 percent because the gypsum speeds up the chemical reactions in the soil. The highest strength (after 10 to 100 days) is usually obtained with 75 percent lime and 25 percent gypsum.

At water contents larger than about 180 percent, the increase of the shear strength with lime or cement is normally too small to be economic. For organic soils, the shear strength usually increases with increasing percentage of lime and gypsum up to a limiting value. The maximum is about 16 percent (Kujala, 1983). About 80 percent of the 1-year strength is normally obtained after 2 months. Cement or fly-ash may be beneficial in soils with a very low plasticity index ($I_p < 10$).

The unconfined compressive strength of samples prepared in the laboratory is normally determined 1, 7, 28, 90, 180, and 360 days after the mixing. After 1 to 3 months it is often possible to get an indication from the laboratory samples of the long-term creep strength. For large jobs (> 1000 columns) it is advisable to check the bearing capacity and the creep limit with test columns in the field.

The creep limit is affected by the confining pressure in situ. This increase can be estimated from the angle of internal friction of the stabilized soil ($\phi_u^{col} \geqslant 30°$) and from the total overburden pressure ($z\gamma$).

The shear strength determined by fall-cone or laboratory vane tests is often two to three times higher than the shear strength determined by unconfined compression tests owing to cracks and fissures in the stabilized clay. The shear strength as determined by fall-cone or vane tests corresponds to the shear strength of the clay matrix, while the shear strength as determined by unconfined compression tests is affected by cracks and fissures in the stabilized soil.

The bearing capacity of the lime columns can be determined in the field by pressuremeter tests, in-situ screw plate tests, or load tests on excavated columns. These testing methods were primarily used during the development of the lime column method. To check the quality of the lime columns after the

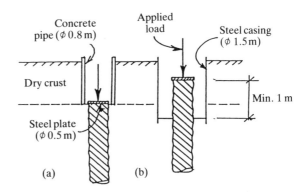

Fig. 24.28 Load tests of lime columns in situ.

installation, the lime column penetrometer can be used as described below.

The equipment used for the screw plate load tests consist of a 160-mm diameter screw-shaped plate and a hydraulic jack located just above the plate. The plate and the hydraulic jack are screwed together to the required depth. The displacement of the plate is measured by the jack at each load increment 15, 45, 105, and 165 seconds after the load has been increased, so that the creep strength of the columns can be determined.

The creep strength of the lime columns can also be determined from in-situ plate load tests as shown in Figure 24.28a or from load tests on excavated columns (Fig. 24.28b). In the in-situ plate load tests the columns are first excavated down to the bottom of the surface crust, where the ultimate strength and the bearing capacity of the columns will be the lowest. This bearing capacity also reflects the increase of the strength that is caused by the surrounding soil below the structure or the fill. When the bottom of the dry crust has been reached, a concrete block or steel plate is placed on the leveled surface of the excavated column. The diameter of the circular steel plate or concrete block should be the same as the diameter of the lime or cement column (0.5 or 0.6 m). A concrete pipe can be used to stabilize the sides of the excavation required for the test. This method can only be used when the thickness of the dry crust is relatively thin, less than about 1.5 to 2.0 m.

The applied load and the axial deformation of the lime or cement columns down to a depth that corresponds to about two column diameters (1.0 m) should preferably be measured as well as the total deformations of the columns. It is also desirable to measure the axial deformations of the columns at a depth that correspond to one column diameter (0.5 or 0.6 m) so that the compression modulus of the column material can be calculated.

The deformations of the column should be measured at each load increment 15 and 165 seconds after the increase of the load in the same way as in a plate load test, when the applied load is kept constant. The failure load is generally defined as the load when the axial deformation of the column is 50 or 60 mm (10 percent of the diameter of the column).

Excavated columns can also be tested as illustrated in Figure 24.28b when the dry crust is relatively thick. The diameter of the casing supporting the sides of the excavation should be relatively large so as to provide sufficient space around the column for the test. The free length of the excavated column should be at least 1.0 to 1.2 m, twice the diameter of the columns.

In the evaluation of the allowable axial load a partial factor of safety (f_a) of about 1.2 to 1.3 may be used, or the load that corresponds to 95 percent probability. Test results from screw plate tests indicate that the creep limit is normally distributed and that the standard deviation is about 20 percent.

24.5 CONTROL METHODS

24.5.1 Control Program

The control program for a lime or cement column installation depends primarily on the size and the complexity of the work, and on the consequences of a failure. It is important to check the length and the continuity of the columns as well as the average shear strength of the stabilized soil in the columns.

Standard penetration tests (SPT) or cone penetration tests (CPT) should not be used to check the quality of the lime columns, since the bearing capacity will be underestimated. The scatter of the results is often large because the penetrometer generally follows the relatively weak hole left after the kelly at the center of the columns.

24.5.2 Lime Column Penetrometer

A modified penetrometer has therefore been developed that is provided with two wings or blades to check the quality of the lime columns as described by Holm (1979), Holm et al. (1981) and Ahnberg et al. (1989) (Fig. 24.29). The lower edge of the wings is enlarged in order to reduce the skin friction resistance. Test data indicate that the undrained shear strength of the stabilized soil corresponds to $p_u/11$, where p_u is the unit penetration resistance with respect to the enlarged lower part of the wings. It is desirable to check 1 to 3 percent of all columns on an actual job with this. The number of tests should be large enough that the results can be analyzed statistically.

Pressuremeter tests can also be used to check the quality of the lime or cement columns and to determine the shear strength of the stabilized clay. The undrained shear strength is usually taken as 18 percent (1/5.5) of the limit pressure p_l (Baguelin et al., 1978; Holm et al., 1981). The compression modulus can be estimated from the measured volume–pressure relationship.

24.5.3 Plate Load and Screw Plate Tests

On large jobs (> 1000 to 2000 columns) the bearing capacity of the columns and the creep limit should also be checked in situ with plate load tests or screw plate tests. About one column in every 1000 should be tested at the depth where the creep limit is expected to be the lowest. This critical section is usually located close to the groundwater table just below the stiff surface crust. With this testing method the confining pressure provided by the surrounding unstabilized soil will be taken into account. The load tests can be combined with column penetrometer tests in the same column so that a comparison between the different testing methods can be made.

24.5.4 Sampling of Lime Columns

Complete columns can be checked using the square 11-m long sampler shown in Figure 24.30. The sampler is driven over an existing column using an air or vibratory hammer suspended from a crane (Broms et al., 1978). The dimensions of the square sampler (0.6 m) are slightly larger than the diameter of the columns so that whole columns can be recovered. The sampler is provided with trap doors at the bottom to prevent the column from slipping out when the sampler is withdrawn. The sampler can be opened along the diagonal after it has been recovered.

The recovered lime columns are usually cut along the axis so that the variation of the shear strength can be determined

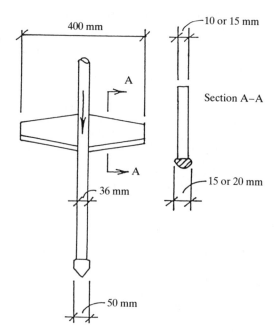

Fig. 24.29 Lime column penetrometer (Model II).

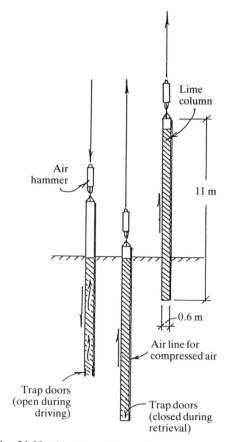

Fig. 24.30 Sampling of a complete lime column.

with, for example, a pocket penetrometer or with fall-cone or laboratory vane tests. Also, the variation of the water and lime content should be determined as described by Broms et al. (1978).

The ultimate strength of the column sections can be determined as indicated in Figure 24.31. A steel frame is placed

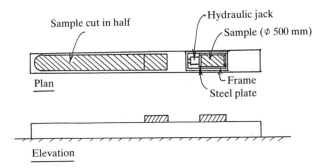

Fig. 24.31 Testing of column segments.

over the 1-m long column segments so that the segments can be loaded axially by a hydraulic jack.

24.5.5 Control of Lime Content

The lime content and the length of the columns should be recorded for every column. The lime content can also be checked from samples taken from the columns using a helical auger (50 mm diameter) or an open drive sampler.

24.6 SCHEDULE FOR DESIGN AND CONSTRUCTION

General

The design of a line column foundation should start very early, preferably 6 to 12 months before the installation of the lime columns since it takes time to obtain the necessary information about the long-term bearing capacity, the creep limit, the compressibility of the stabilized clay, and the effectiveness of different additives from the field and laboratory investigations. The following five steps are usually followed.

Step 1: Evaluation of the Lime Column Method First the creep limit of the lime columns and the compression modulus of the stabilized soil are estimated. Normally some information is available from previous investigations about the soil conditions, such as depth and thickness of the different compressible layers and of the shear strength and the compressibility of the soft soil. From the estimated creep limit and the compression modulus a rough estimate can be made of the number of lime columns and the required length as well as of the cost. It is important to take into account the accessibility of the site, the size of the job (the total number of lime columns), as well as the time of the year. If the lime columns are to be installed during the winter the ground may be frozen, which will require predrilling.

Step 2: Field and Laboratory Investigations Samples from representative layers should be tested in the laboratory in order to determine the creep limit of the columns and the compressibility at different times after the mixing. The soil from two to three representative layers is mixed with, for example, 6 to 8 percent unslaked lime when the soil is inorganic and the liquid limit and the water content are low (< 80 percent). The optimum lime content generally increases with increasing organic content. About 8 to 10 percent lime may be required for organic soils with a water content of up to 90 percent. Gypsum can be beneficial if the water content exceeds 90 percent.

The stabilized soil is compacted in plastic cylinders of, for example, 36 mm diameter and 72 mm length using a Harvard Miniature Compaction Device or by a wooden tamper. The compaction can also be done by hand if the initial shear strength of the stabilized soil is low. The cylinders should be sealed after compaction or wrapped in plastic bags. The samples should be stored under water or in a moist room until they are tested. The shear strength of the samples is normally determined by unconfined compression or fall-cone tests 1, 7, 28, 90, 180, and 365 days after the mixing. Also, a pocket penetrometer can be useful. Samples with 50, 75, or 100 mm diameter should be prepared as well so that the compressibility of the stabilized soil can be determined by oedometer tests at different times after the mixing.

On large jobs (> 1000 to 2000 columns) it is desirable to carry out in situ two to four plate load tests or screw plate tests on test columns installed at the actual site in order to check the results from the laboratory tests. The columns should be tested to failure approximately 3 months after manufacture.

Step 3: Design of Lime Column Foundations After 1 to 3 months, when the results from the laboratory investigations are available, the detailed design of the lime column foundation can begin. The creep limit (σ_{creep}^{sol}) of the columns can be estimated about 3 months after the mixing from the shear strength of the samples prepared in the laboratory. Also, the confined modulus of the column material can then be estimated; it normally varies between 50 and 100 times c_{col}.

The maximum total settlement normally governs the design of a lime or cement column foundation when the depth of the compressible layers is less than about 10 m and the lime columns extend down to firm bottom.

For buildings it is important to check that the factor of safety with respect to local failure along the edge of the lime column group is sufficient. The bearing capacity of the unstabilized soil is usually high enough to carry the applied load.

The stability of slopes and deep excavations should be checked for different arrangements and spacing of the lime or cement columns from Equation 24.16. The stability of deep excavations or cuts should be determined when the cracks behind the lime column wall are filled with water. Also, the stability of the lime column wall with respect to overturning should be investigated.

It is difficult to evaluate the confined modulus of the stabilized soil. It is preferable to use the results from oedometer tests on compacted samples or the results from plate load tests in situ. The load q_2 carried by the unstabilized soil between the columns can be estimated from the relative compression of the columns ($\Delta h_1 / H_{col}$). The difference ($q - q_2$) will be carried by the lime columns. It is often assumed in the analysis that the stress increase is constant for the full height of the columns. Since the axial stress σ_{col} in the lime columns is known, the required number of lime columns can be calculated in order to reduce the settlements to an acceptable level.

When the total thickness of the compressible strata exceeds 10 to 15 m, then the settlements of the compressible strata below the columns has to be considered. The settlements (Δh_2) of the compressible strata below the columns can be estimated conservatively as illustrated in Figure 24.17. It is generally advantageous to make the columns as long as possible.

The maximum differential settlement can be determined from Equation 24.13. The required factor of safety f_c in order to limit the angular rotation can be evaluated from the equation

$$f_c = \frac{1}{100\,\alpha} \qquad (24.19)$$

At a maximum angular rotation (α) of $1/300$ then $f_c = 3.0$. If the average shear strength of the soil is 15 kPa, the perimeter shear stress should not exceed 5 kPa (Eq. 24.6).

Step 4: Field Instrumentation It is desirable to install the settlement gauges and the surface markers 1 to 2 months before the lime columns in order to check the settlements of the different strata and to ensure that the columns are working properly, especially if the area with the lime columns will be preloaded. Precautions have to be taken so that the gauges will not be damaged during the installation of the columns.

Pore pressure gauges (open pipes) should be installed in the different pervious strata to check the pore water pressures within or below the compressible clay or silt layer. The pore pressure gauges should, if possible, be placed at least 1 year before the start of the construction in order to get some indication of the seasonal changes to the groundwater level. The groundwater level in the area can be affected by adjacent construction, for example.

If the area is to be preloaded it is desirable to place the fill required for the preloading as soon as possible after the installation of the lime columns. It may be necessary to construct the fill in stages if the bearing capacity of the underlying soil is not sufficient to carry the weight of the fill.

Step 5: Control Program The surface settlement should be measured every month for at least 6 to 12 months after the columns have been installed and loaded. It is also desirable to measure the settlements at the bottom of the lime columns when they do not extend down to firm bottom.

During the installation of the lime columns the following points should be checked:

- Quality of the lime (particle size and water content)
- Location and inclination of the individual lime columns
- Date of manufacture, amount of lime used, and length of individual columns
- Withdrawal rate (mm/revolution)
- Total amount of lime used per shift (the volume should correspond to the number of columns manufactured and the average lime content)

After the installation it is important to check:

- The record of each lime column, to determine the length and the amount of lime used
- The shear strength of some of the columns (1 to 3 percent) with the lime column penetrometer
- The bearing capacity of a few columns for large jobs with in-situ plate load tests or in-situ screw plate tests
- The uniformity and quality of the columns (whole columns may be retrieved)

Disturbed samples obtained with a helical auger can generally be used to check the lime content and the uniformity of the columns while open drive or piston samplers will be required to obtain undisturbed samples so that the undrained shear strength of the stabilized clay can be determined. The shear strength of the recovered samples can be evaluated by fall-cone tests, with a pocket penetrometer, or with unconfined compression tests. Owing to the high shear strength of the stabilized clay, it is frequently difficult to push a Shelby tube or a piston sampler into a lime column. Predrilling may be required. It should be noted that the shear strength can be low at the center of the columns.

The stability of excavations and deep cuts can be checked by measuring during the excavation the lateral displacement and the tilt of the lime column walls with surface markers, tilt meters, or inclinometers. Slopes can be checked by measuring the lateral displacement at the ground surface and the pore water pressures. For large installations, the horizontal ground movements should be determined as well using inclinometers.

REFERENCES

Aas, G. (1983), Stabilization of clay with hydroxy aluminium. In-situ determination of strength properties of stabilized columns, Nordic Seminar on Soil Stabilization, Esbo, Finland.

Aboshi, H. (1984), Soil improvement techniques in Japan, *Proceedings, Seminar on Soil Improvement and Construction Techniques in Soft Ground*, Singapore, pp. 3–16.

Ahnberg, H. and Holm, G. (1986), *Kalkpelarmetoden (The lime column method)*, Swedish Geotechnical Institute, Report No. 31, Linköping, Sweden (in Swedish).

Ahnberg, H., Bengtsson, P.-E., and Holm, G. (1989), Prediction of strength of lime columns, *Proceedings of the 10th International Conference on Soil Mechanics and Foundation Engineering*, Stockholm, **2**, pp. 1327–1330.

Baguelin, F., Jezequel, J. F., and Shields, D. H. (1978), *The Pressuremeter and Foundation Engineering*. Trans Tech Publications, Clausthal, Germany.

Balasubramaniam, A. S. and Buensuceso, B. R. (1989), On the overconsolidated behavior of lime treated soft clay, *Proceedings of the 12th International Conference on Soil Mechanics and Foundation Engineering*, Rio de Janeiro, **2**, 1335–1338.

Balasubramaniam, A. S., Buensuceso, B. R., Phien-Wej, N., and Bergado, D. T. (1990), Engineering behavior of lime stabilized soft Bangkok clay, *Proceedings of the 10th Southeast Asian Geotechnical Conference*, Taipei, **1**, pp. 23–28.

Barron, R. A. (1948), Consolidation of fine-grained soils by drain wells, *Transactions of ASCE*, **113**, No. 2346, pp. 718–754.

Bell, F. G. and Tyrer, M. J. (1989), The enhancement of the properties of clay soils by the addition of cement or lime, *Proceedings of the 12th International Conference on Soil Mechanics and Foundation Engineering*, Rio de Janeiro, **2**, pp. 1339–1343.

Brandl, H. (1981), Alteration of soil parameters by stabilization with lime, *Proceedings of the 10th International Conference on Soil Mechanics and Foundation Engineering*, Stockholm, **3**, pp. 587–594.

Bredenberg, H. (1983), Lime columns for ground improvement at new cargo terminal in Stockholm, *Proceedings of the 8th European Conference on Soil Mechanics and Foundation Engineering*, Helsinki, Finland, **2**, pp. 881–884.

Bredenberg, H. and Broms, B. (1983), Lime columns as foundations for buildings, *International Conference on Advances in Piling and Ground Treatment for Foundations*, The Institution of Civil Engineers, London, England.

Broms, B. (1983), Stabilization of soft clay with lime columns, *Proceedings, Seminar on Soil Improvement and Construction Techniques in Soft Ground*, Singapore, pp. 120–133.

Broms, B. (1984a), *Stabilization of Soil with Lime Columns. Design Handbook*, Lime Column AB, Kungsbacka Sweden, third edition.

Broms, B. (1984b), Stabilization of soft clay with lime columns, *Proceedings, Seminar on Soil Improvement and Construction Techniques in Soft Ground*, Singapore, pp. 120–133.

Broms, B. B. (1985), Stabilization of very soft clay in waste ponds at Tampines, Singapore, *Proceedings, 3d International Geotechnical Seminar on Soil Improvement Methods*, Singapore, pp. 147–152.

Broms, B. and Boman, P. (1975a), Lime stabilized columns—a new construction method, *XVth World Road Congress, Question 1. Planning and Construction of Highways*, Mexico City, pp. 22–28.

Broms, B. and Boman, P. (1975b), Lime stabilized columns, *Proceedings of the 5th Asian Regional Conference*, Bangladore, India, **1**, pp. 227–234.

Broms, B. and Boman, P. (1976), Stabilization of deep cuts with lime columns, *Proceedings of the 5th European Conference on Soil Mechanics and Foundation Engineering*, Tokyo, Japan, **1**, pp. 207–210.

Broms, B. B. and Boman, P. (1977a), Lime columns—A new type of vertical drain, *Proceedings of the 9th International Conference on Soil Mechanics and Foundation Engineering*, Tokyo, **1**, pp. 427–432.

Broms, B. B. and Boman, P. (1977b), Stabilization of soil with lime, *Ground Engineering*, **12**, No. 4, pp. 23–32.

Broms, B., Boman, P., and Ingelsson, I. (1978), *Investigation of Lime*

Columns at Smistavägen, Huddinge, Sweden, Report, Department of Soil and Rock Mechanics, Royal Institute of Technology, Stockholm, Sweden.

Bryhn, O. R., Løken, T., and Aas, G. (1983), Stabilization of sensitive clays with hydroxy-aluminium compared with unslaked lime, *Proceedings of the 8th European Conference on Soil Mechanics and Foundation Engineering*, Helsinki, **2**, pp. 885–896.

Chiu, K. H. and Chin, K. Y. (1963), The study of improving bearing capacity of Taipei silt using quicklime piles, *Proceedings of the 2d Asian Regional Conference on Soil Mechanics and Foundation Engineering*, **1**, pp. 367–393.

Eggestad, A. and Sem, H. (1976), Stability of excavations improved by salt diffusion from deep wells, *Proceedings of the 6th European Conference on Soil Mechanics and Foundation Engineering*, Vienna, Austria, **1.1**, pp. 211–216.

Ekström, A. and Tränk, R. (1979), Kalkpelarmetoden. Tillämpningar för stabilisering av brostöd och rörgrav (The lime column method. Stabilization of bridge abutments and trenches), *Proceedings of the Nordic Geotechnical Meeting*, Helsinki, Finland, pp. 258–268 (in Swedish).

Evans, G. L. and Bell, D. H. (1981), Chemical stabilization of loess, New Zealand, *Proceedings of the 10th International Conference on Soil Mechanics and Foundation Engineering*, Stockholm, Sweden, **3**, pp. 649–658.

Holeyman, A., Frank, S. A., and Mitchell, J. K. (1983), Assessment of quicklime pile behaviour, *Proceedings of the 8th European Conference on Soil Mechanics and Foundation Engineering*, Helsinki, **2**, pp. 897–902.

Holm, G. (1979), Kalkpelarförstärkning for urgrävning och vägbank vid Stenungsund (Stabilization of an excavation and a road embankment with lime columns at Stenungsund, Sweden), *Proceedings of the Nordic Geotechnical Meeting*, Helsinki, Finland, pp. 269–284 (in Swedish).

Holm, G., Bredenberg, H., and Broms, B. B. (1981), Lime columns as foundations for light structures, *Proceedings of the 10th International Conference on Soil Mechanics and Foundation Engineering*, Stockholm, **3**, pp. 687–694.

Holm, G., Tränk, R., and Ekström, A. (1983a), Improving lime column strength with gypsum, *Proceedings of the 8th European Conference on Soil Mechanics and Foundation Engineering*, Helsinki, Finland, **2**, pp. 903–907.

Holm, G., Tränk, R., Ekström, A., and Torstensson, B.-A. (1983b), Lime columns under embankments—A full scale test, *Proceedings of the 8th European Conference on Soil Mechanics and Foundation Engineering*, Helsinki, Finland, **2**, pp. 909–912.

Kawasaki, T., Niina, A., Saitoh, S., Suzuki, Y., and Honjyo, Y. (1981), Deep mixing method using cement hardening agent, *Proceedings of the 10th International Conference on Soil Mechanics and Foundation Engineering*, Stockholm, **3**, pp. 721–724.

Kawasaki, T., Saitoh, S., Suzuki, Y., and Babasaki, R. (1984), Deep mixing method using cement slurry, *Proceedings, Seminar on Soil Improvement and Construction Techniques in Soft Ground*, Singapore, pp. 17–38.

Kujala, K. (1983), The use of gypsum in deep stabilization, *Proceedings of the 8th European Conference on Soil Mechanics and Foundation Engineering*, Helsinki, Finland, **2**, pp. 925–928.

Kujala, K. and Nieminen, P. (1983), On the reaction of clays stabilized with gypsum lime, *Proceedings of the 8th European Conference on Soil Mechanics and Foundation Engineering*, Helsinki, Finland, **2**, pp. 929–932.

Kuno, G., Kutara, K., and Miki, M. (1989), Chemical stabilization of soft soils containing humic acid, *Proceedings of the 12th International Conference on Soil Mechanics and Foundation Engineering*, Rio de Janeiro, **2**, pp. 1381–1384.

Lahtinen, P. O. and Vepsalainen, P. E. (1983), Dimensioning deep-stabilization using the finite element method, *Proceedings of the 8th European Conference on Soil Mechanics and Foundation Engineering*, Helsinki, Finland, **2**, pp. 933–936.

Nieminen, P. (1978), Soil stabilization with gypsum lime, *Proceedings of the International Conference on the Use of By-products and Waste in Civil Engineering*, Paris, France, **1**, pp. 229–235.

Okumura, T. and Terashi, M. (1975), Deep-lime-mixing method of stabilization for marine clays, *Proceedings of the 5th Asian Regional Conference on Soil Mechanics and Foundation Engineering*, Bangalore, India, **1**, pp. 69–75.

Petchgate, K. and Tungboonterm, P. (1990), Installation of lime columns and their performance, *Proceedings of the 10th Southeast Asian Geotechnical Conference*, Taipei, **1**, pp. 121–124.

Sahlberg, O. (1979), Kalkpelarstabilisering för djup ledningsschakt i lös lera (Stabilization of an excavation in soft clay with lime columns), *Proceedings of the Nordic Geotechnical Meeting*, Helsinki, Finland, pp. 295–296 (in Swedish).

Sahlberg, O. and Boman, P. (1979), *Kalkpelarmetoden—Uppföljning av en rörgravs-stabilisering (The Lime Column Method—Evaluation of a Stabilized Cut)*, Swedish Council for Building Research, Project 780.

Serra, M., Robinet, J. C., Mohkam, M., and Daonh, T. (1983), Soil improvement of dykes by liming, *Proceedings of the 8th European Conference on Soil Mechanics and Foundation Engineering*, **2**, Helsinki, pp. 947–950.

Soyez, B., Magnan, J. P., and Delfaut, A. (1983), Loading tests on a hydraulic fill stabilized by limetreated soil columns, *Proceedings of the 8th European Conference on Soil Mechanics and Foundation Engineering*, Helsinki, **2**, pp. 951–954.

Terashi, M. and Tanaka, H. (1981), Ground improvement by deep mixing method, *Proceedings of the 11th International Conference on Soil Mechanics and Foundation Engineering*, Stockholm, Sweden, **3**, pp. 777–780.

Terashi, M. and Tanaka, H. (1983), Settlement analysis for deep mixing method, *Proceedings of the 8th International Conference on Soil Mechanics and Foundation Engineering*, Tokyo, **1**, pp. 191–194.

Torstensson, B. A. (1975), The pressure sounding instrument, *ASCE Specialty Conference. In-situ Measurement of Soil Properties*, Raleigh, **2**, pp. 48–54.

Wei, J., Ho, S. K., Chua, C. H., Chong, M. K., and Broms, B. B. (1990), Cement and lime columns for ground improvement—a field study at Pasir Ris, *Proceedings of the 10th Southeast Asian Geotechnical Conference*, Taipei, **1**, pp. 139–144.

25 DURABILITY AND PROTECTION OF FOUNDATIONS

THOMAS D. DISMUKE, M.S.C.E., P.E.
Consulting Engineer
Bethlehem Steel Corporation

25.1 INTRODUCTION

Failure of structures due to deterioration of the foundations occurs infrequently in some applications and far too frequently in others. Waterfront facilities (particularly in marine and tidal exposures) have a much higher incidence of material deterioration than land-based facilities. Protection of materials in aggressive environments and operations (or use) has long been recognized as a necessity. Satisfactory protection systems are available, but are, in many instances, misapplied or ignored. Foundation deterioration is usually thought of in terms of electrochemical and chemical phenomena; however, causes of deterioration also include heat, abrasion, and inadequate time-dependent strength properties. Deterioration due to low-temperature environments and temperature cycling occurs rarely and is usually related, in the case of concrete, to severe exposure while curing. Foundations are not subjected to the temperature extremes that the supported structures are. This is a decided benefit.

A list of environmental influences on piles in marine structures is given in Table 25.1 (Dismuke et al., 1981).

Similar tables could be developed for foundations of land-based structures, but since the land environment is generally not as aggressive, the list would be shorter for comparable severity. Further discussions on chemical and biological causes are presented by Dillon (1986). Gauffreau (1987) has presented a review of pile protection methods.

The following is a list of those items that may have to be considered in the practical application of material selection for new construction or rehabilitation of underground or submerged foundations:

- Facility function and failure criteria
- Properties of construction material, soil, and water
- Site data
- Behavior of materials in soil and water
- Evaluation of site data

Data on the durability of copolymers, fabrics and the conventional constructional materials (metal, wood and concrete) when exposed to hazardous soils is limited. As a result, the studies made to date—primarily in uncontaminated soils—

TABLE 25.1 ENVIRONMENTAL INFLUENCE ON PILES IN MARINE STRUCTURES[a].

Agent	Mechanism
Physical	
Tide	Thermal cycles
Wind	Fatigue and overstressing
Current	Erosion by sand
Wave action	Fatigue and overstressing
Ice	Overstressing (freeze–thaw cycles)
Ship impact	Overstressing
Chemical	
Brackish and seawater (submerged, tidal and splash zones)	Corrosion
Polluted water	Corrosion and direct attack
Fire	Burning
Fresh water (submerged, tidal and splash zones)	Corrosion
Biological	
Fouling organisms	Chemical by-products
Aerobic bacteria	Chemical by-products
Anaerobic bacteria	Chemical by-products
Marine borers	Ingestion

[a] After Dismuke et al. (1981).

are used as guides in the selection and protection of materials in geostructures.

Foundations, as herein considered, include piles, pile caps, piers, footings, grade beams, pavements, retaining structures, fills, subbases, and base courses. The last three items are included because they are composed of man-made materials or are naturally occurring processed materials.

25.2 FACILITY FUNCTION AND FAILURE CRITERIA

Failure occurs when the facility does not perform its function satisfactorily over a given time period. Material failure in structures constructed in soil or water is, in most cases, very costly to remedy. This fact should spur the facility designer to utilize adequate materials and construction procedures.

Failure is usually due to out-of-tolerance structure or component movement. Some facilities have very low tolerance for foundation movement, while others can tolerate considerable movement. Table 25.2 (Sowers, 1962) shows the maximum allowable settlement for a number of structures and some equipment. The table is not universally accepted but does give an order of magnitude of movement. As a general rule, the foundations for rotating or reciprocating equipment are designed to much stricter movement tolerances than are those for structures. Exceptions are notable—such as foundations for radiotelescopes and colliders, which are intolerant of movement. Most large movements are due to soil or rock settlement, negative pile loading, etc., where the source of the movement is external to the structure and material deterioration is not a factor. However, there have been instances where fills and/or subbases and base courses were constructed of material that did deteriorate. Although the seat of the structure movement

was external to the footings, it must be considered that the foundation system was not durable.

The severity of material deterioration should be based on the function of the specific member involved. For instance, if the member is used to carry loads the average loss of material or strength and the location of the loss are important. If the member is a liquid or gas storage container the structural condition may be adequate, but the containment function may not be satisfactory.

One of the most common failures in construction of buildings is associated with the vapor barrier. A number of synthetics are improperly used and cause concrete curing problems when concrete is placed directly on the barrier. If the wrong synthetic is used and the concrete contacts the barrier, it may become brittle and crack.

25.3 PROPERTIES OF CONSTRUCTION MATERIALS, SOIL, AND WATER

25.3.1 Construction Materials

Table 25.3 shows the general application of existing materials commonly used in underground construction. The table listing reflects the importance of some fairly long-used materials such as soil–cement, asphalt, fiberglass, and polyvinyl chloride.

Since concrete, steel, and wood are the principal materials used for foundation construction, some properties of these materials are given in Table 25.4. The effects of temperature on steel and concrete are shown in Figures 25.1 through 25.4 (deJesus, 1980). As noted in Section 25.1, the foundation in-situ temperature effects are minor. However, the effects of temperature on materials during construction can be disastrous.

TABLE 25.2 ALLOWABLE STRUCTURE MOVEMENT[a].

Type of Movement	Limiting Factor	Maximum Settlement[b]
Total settlement	Drainage	6 to 12 in
	Access	12 to 24 in
	Probability of nonuniform settlement:	
	Masonry walled structure	1 to 2 in
	Framed structures	2 to 4 in
	Smokestacks, silos, mats	3 to 12 in
Tilting	Stability against overturning	Depends on height and width
	Tilting of smokestacks, towers	0.004L
	Rolling of trucks, etc.	0.01L
	Stacking of goods	0.01L
	Machine operation—cotton loom	0.003L
	Machine operation—turbogenerator	0.0002L
	Crane rails	0.003L
	Drainage of floors	0.01 to 0.02L
Differential movement	High continuous brick walls	0.0005 to 0.001L
	One-story brick mill building, wall cracking	0.001 to 0.002L
	Plaster cracking (gypsum)	0.001L
	Reinforced-concrete building frame	0.0025 to 0.004L
	Reinforced-concrete building curtain walls	0.003L
	Steel frame, continuous	0.002L
	Simple steel frame	0.005L

[a] After Sowers (1962).
[b] L = distance between adjacent columns that settle different amounts, or between any two points that settle differently. Higher values are for regular settlements and more tolerant structures. Lower values are for irregular settlements and critical structures.

TABLE 25.3 USES OF VARIOUS MATERIALS.

Material	Structural					Nonstructural (Containment)					
	Subbase	Foundation	Pavement	Piles	Pipe	Vapor Barrier	Pipe	Drain	Liner	Caps	Tanks
Metals											
Iron					X		X	X			
Steel		X		X	X		X	X			X
Aluminum					X		X	X			
Copper							X				
Wood											
Hardwood				X							
Softwood		X		X							
Concrete											
Plain		X	X								
Reinforced	X	X	X	X							X
Prestressed	X	X	X	X			X	X			X
Soil–cement	X										
Soil											
Fine-grained (clays)									X		
Coarse-grained	X										
Asphalt			X						X		
Fiberglass							X	X			
PVC							X	X	X		
Flexible polymers						X		X	X	X	

TABLE 25.4 SOME PROPERTIES OF COMMON MATERIALS OF CONSTRUCTION.

Material	Steel	Ordinary Concrete	Wood[a]
Young's modulus, E, psi	29×10^6	2 to 5×10^6	1.2 to 2.0×10^6
Compressive strength, psi	36 to 200×10^3	3000 to 5000	0.9 to 1.25×10^{2b}
			0.2 to 0.70×10^{2c}
Tensile strength, psi	36 to 200×10^3	300 to 500	1.9 to 2.5×10^{3d}
Freezing–thawing resistance, cycles	N/A	500 to 700	
Permeability	Very low	Medium	N/A
Alkaline resistance	Fair	Fair	Poor
Sulfate resistance	Poor	Fair	Poor
Unit weight, pcf	480 to 490 pcf	140 to 150 pcf	30 to 60 pcf

[a] Includes Red Pine, Red Oak, Southern Pine and Douglas Fir.
[b] Parallel to grain.
[c] Perpendicular to grain.
[d] Extreme fiber in bending.

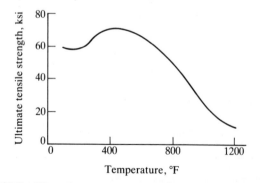

Fig. 25.1 Effect of temperature on tensile strength of low-carbon steel.

Fang and Mehta (1976) have described polymer-impregnated concrete and its resistance to deterioration. In special cases it may become very useful.

Preservative treatment can reduce the strength of wood (AITC, 1985). Retention of preservative within a 1-inch depth is shown in Table 25.5 (NCEL, Tech Data Sheet, 1979). A comparison of the strengths of treated and untreated wood piles is given in Table 25.6 (NCEL, Tech Data Sheet, 1979).

25.3.2 Soil

The widely varying properties of soil and rock make it difficult to characterize them in contrast to steel and concrete, or even wood.

Most of the items listed below are important for determining behavior in soil—but not all items are important for all materials.

- Soil moisture
- Soil resistivity
- Redox potential
- Physical properties
- Soil disturbance
- Chemical constituents
- Presence of sulfate-reducing bacteria
- pH

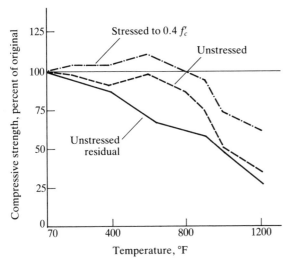

Fig. 25.2 Effect of temperature on compressive strength of siliceous aggregate concrete.

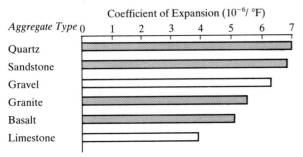

Fig. 25.3 Thermal coefficient of expansion of concrete constituents.

Table 25.7 is a compilation of typical soil data.

Fills of all varieties constitute an important part of the soils that are of concern. A generalized fill classification as given below is used to define the various placement methods and constituents (Dismuke, 1987):

- Controlled (CF)—an engineered fill made of specified soils and constructed in a specified manner
- Controlled Contaminated (CCF)—same as above except that the soil contains one or more chemically undesirable elements
- Uncontrolled (UF)—a nonengineered fill made of unspecified material placed in a random manner. No compaction requirements specified
- Uncontrolled Contaminated (UCF)—same as for UF except that the soil contains one or more chemically undesirable elements

In the above classification it should be noted that the word contaminated is used in reference to the effect on the materials of construction rather than in the biologic sense.

25.3.3 Water

Properties of water of a given classification (fresh, salt, etc.) do not greatly vary. However, there is considerable variation between fresh water (as used for human and animal consumption), salt (marine) water, and the mixture as found at the land–marine interface. Four of the most important properties of water that affect steel are composition, salinity, pH and resistivity (or conductivity). Resistivity reflects the salinity content. Table 25.8 gives the averaged composition of seawater. If seawater is polluted, ions other than those given may be present and greatly affect the seawater's aggressiveness. Table 25.9 shows the variation in salinity of seawater at various locations. pH is a measure of a solution's alkaline vs. acid strength. Table 25.10 gives pH values for various solutions. (Values less than 7 are acidic and values greater than 7 are alkaline.) Resistivity (electrical), expressed in ohm-centimeters, is the resistance between opposite faces of any one-centimeter cube of material (water, soil, etc.). It is an indirect measure of the relative corrosiveness of water to steel and steel–concrete combinations. The lower the resistivity of water, the more aggressive it is, but large resistivity differences adjacent to each other, even of high-resistivity waters, may cause some aggressive action. Table 25.11 gives some electrical resistivity values for water. Thermoclines often found—particularly in estuarial waters—can have large gradients that can influence deterioration rates.

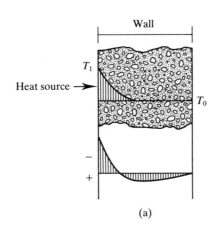
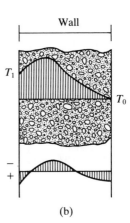

Fig. 25.4 Nonlinear temperature distribution and thermal stress in concrete. (a) Heat source impinging. (b) Heat source removed.

TABLE 25.5 PRESERVATIVE RETENTION OF MARINE PILES WITHIN A 1-INCH DEPTH.

Treatment[a, b]	Specific Gravity	Average Preservative Retention of Five Piles per Treatment, lb/ft³				
		Creosote	Total Metallic Salts	Chromium	Copper	Arsenic
Southern Pine						
Untreated	0.54	0	0	0	0	0
Creosote	0.54	28.9	0	0	0	0
2.5 ACA	0.54	0	6.76	0	3.43	3.33
1.0 ACA, kiln, creosote	0.53	29.8	2.04	0	1.05	0.99
1.0 ACA, air, creosote	0.55	29.5	1.70	0	0.81	0.89
1.0 CCA, kiln, creosote	0.54	27.6	1.58	0.68	0.23	0.67
1.0 CCA, air, creosote	0.55	31.1	1.39	0.63	0.18	0.57
2.5 CCA	0.56	0	5.18	2.46	0.83	1.89
Douglas Fir						
Untreated	0.47	0	0	0	0	0
Creosote	0.44	21.6	0	0	0	0
2.5 ACA	0.44	0	4.67	0	2.37	2.30
1.0 ACA, kiln, creosote	0.46	30.6	1.12	0	0.53	0.59
1.0 ACA, air, creosote	0.46	30.5	1.05	0	0.51	0.54
1.0 CCA, kiln, creosote	0.44	18.7	0.79	0.36	0.12	0.31
1.0 CCA, air, creosote	0.46	16.1	0.55	0.27	0.07	0.22

[a] ACA = ammoniacal copper arsenite.
CCA = chromated copper arsenate.

[b] Number represents pounds of chemical per cubic foot of sapwood.

TABLE 25.6 AVERAGE MECHANICAL PROPERTIES OF PILES.

Type of Treatment	No. of Test Piles	Flexural Properties			Compressive Strength, F_c, psi
		Modulus of Rupture, psi	Modulus of Elasticity in Flexure, 10^6 psi	Average Absorbed Energy in Flexure in-lb/in³	
Fir					
Untreated	5	8,394	1.922	6.338	3,346
Creosote	5	6,862	1.584	4.202	a
ACA dual[b]	10	6,111	1.537	3.059	2,714
CCA dual[b]	10	3,844	1.171	3.364	2,333
ACA	5	5,620	1.416	2.078	2,462
Pine					
Untreated	5	8,007	1.942	5.240	a
Creosote	5	5,950	a	a	a
ACA dual[b]	10	4,725	1.568	2.829	a
CCA dual[b]	10	4,167	1.441	2.413	a
ACA	5	5,534	1.538	a	a
CCA	5	5,410	a	a	a

[a] No value is provided because of the large spread in measured values for a small number of samples.
[b] Includes both air-dried and kiln-dried specimens (five each).

TABLE 25.7 TYPICAL SOIL PROPERTIES.

Property	Clays	Silts	Sand	Gravel	Rock
Specific gravity	2.7	2.7	2.65	2.65	2.65
Weight, pcf[a]	89 to 129	81 to 136	118 to 135	120 to 190	125 to 210
Poissons Ratio (μ)	0.1 to 0.5	0.3 to 0.35	0.2 to 0.4	0.1 to 0.4	0.1 to 0.4
Modulus of elasticity, psi	0.5 to 2.5×10^3		1.5 to 12.0×10^3	14.0 to 18×10^3	7×10^5 to 156×10^5
Permeability, cm/sec	< 0.0000001	10^{-3} to 10^{-7}	10^{-3} to 10^{-5}	10^{-1}	10^{-3} to 10^{-9}
Particle size, mm	0.08	0.006 to 0.08	0.08 to 4.0	4.0 to 70.0	N/A

[a] Unit weight in the saturated state.
For additional data see Tables 3.8 and 3.12

TABLE 25.8 COMPOSITION OF SEAWATER[a].

Ions	Grams per Kilogram[b]
Chloride	18.8
Sodium	10.56
Sulfate	2.649
Magnesium	1.294
Calcium	0.410
Potassium	0.380
Bicarbonate	0.142

[a] From *Handbook of Corrosion Protection for Steel Pile Structures in Marine Environments* (Dismuke et al., 1981).
[b] Seawater in the open varies somewhat in its salinity, which is defined as the number of grams of dissolved matter per 1000 grams of water.

TABLE 25.9 SALINITY OF SEAWATERS[a].

Water Body	Dissolved Solids, ppm
Baltic Sea	8 000
Black Sea	22 000
Irish Sea	32 500
Atlantic Ocean	37 000
Mediterranean Sea	41 000
Caspian Sea	130 000
Dead Sea	260 000

[a] Source same as in Table 25.8.

TABLE 25.10 pH OF VARIOUS SOLUTIONS[a].

Water	pH
Orange juice	4 to 3
Tap water (CO_2 saturated)	6 to 5.5
Pure water	7.0
Clean seawater	7.2 to 8.2
Lime water	10 to 11
Concrete	12.5

[a] Source same as Table 25.8.

TABLE 25.11 ELECTRICAL RESISTIVITY OF NATURAL WATERS[a,b].

Type of Water	Resistivity (ohm-cm)
Pure	20 000 000
Distilled	500 000
Rain	20 000
Tap	1 000 to 5 000
Brackish river	200
Arctic Sea	35
Open sea	20 to 25
Tropical sea	16

[a] Approximate values for indicating relative order of magnitude for respective waters.
[b] Source same as in Table 25.8.

The effect on foundations of the various forms of sea life ranges from benign to drastic. Table 25.1 shows how some of the more commonly encountered marine and freshwater life forms affect structures.

Many harbor waters are fouled with oils and other substances. The oils, in particular, have protected waterfront structure foundations, whether constructed of steel, concrete, or wood. However, serious deterioration of all three materials generally occurs if they are located near raw-sewage outfalls. The continuing effort to reduce water pollution has resulted in less pollution and, therefore, less protection for many marine structures. The opposite is true for structures that are located on estuarial or inland waterways with industrial and municipal waste loads.

25.4 SITE DATA

25.4.1 Soil

The site investigation should include:

- Surface and subsurface profiling
- Soil borings
- Extraction of soil and water samples at specified elevations
- Drainage characteristics
- Site history
- Stray current survey
- Temperature and moisture data

For ferrous materials the effect of the many ions in the soil is not well defined. The interaction between the ions, oxygen, etc., is not well understood. For durability studies the ionic content should be determined. The most common ions that affect durability include:

- Sulfate
- Chlorine
- Nitrate
- Phosphate
- Sodium
- Calcium
- Sulfides (hydrogen sulfide)

The site data that is generally obtained for design purposes is also applicable for use in planning the rehabilitation of structures when emphasis is usually placed on the condition of any existing structure. Unsuitable information is often obtained as a result of not defining what is to be measured. An example is when the strength of a deteriorated steel member must be determined, but the site investigation only reports pit depths instead of average remaining metal thickness and the location of the remaining metal.

Figure 25.5 shows an example of the type of informative investigation that has been made. The data was obtained in connection with a study of the condition of steel piles that had been in situ for a number of years.

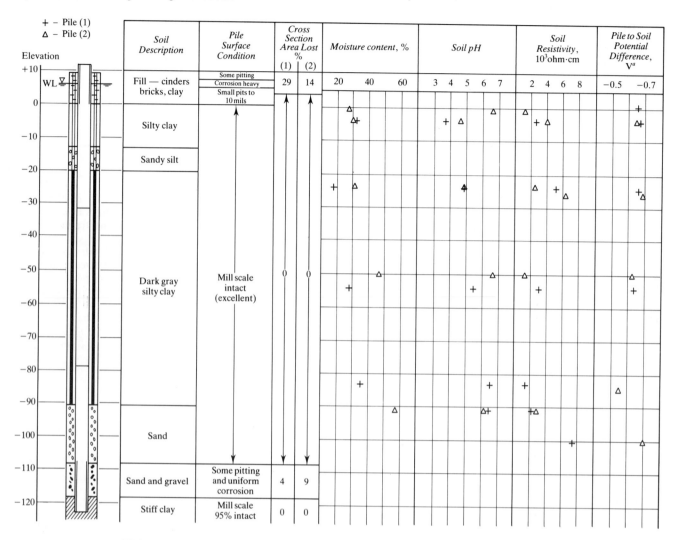

Fig. 25.5 Comparison of condition of steel H-piles with soil properties.

As with all site investigations, what is determined from an analysis of site data affects the project justification. If the soil is identified as a clay, but has been exposed to certain environments, it may exhibit siltlike characteristics. That would affect the soil's strength, permeability, and possibly its aggressiveness towards materials. The effect on the supporting soil and foundations of chemicals that have leaked from tanks and pipelines is often economically severe. Identification of the chemicals and the effect on the soil and/or foundation can be difficult—particularly if site investigation is not thorough.

25.4.2 Water

The desired site data for water sites includes the following:

- Type of water (fresh, saline)
- Tidal range
- Wind directions
- Wave characteristics
- Storm frequency
- Water current
- Bottom profile
- Type of soil
- Ice formation
- Water characteristics (pH, resistivity, constituents, etc.)
- Presence of pollutants
- Organisms

Deterioration of all three common construction materials can occur by erosion, albeit in different ways. Concrete and wood are usually subject to the direct mechanical removal of material by sands. (Armored concrete and wood have been used to reduce the rate of erosion.) Steel is usually deteriorated by the corrosion–erosion cycle. Protective corrosion products are removed by erosion and then the corrosion rate increases. Protective coatings or shielding must be able to withstand ice movement, debris impingement, and erosion caused by sand during storms. Natural currents have caused harmful build-up of municipal and industrial wastes.

25.5 BEHAVIOR OF MATERIALS IN SOIL AND WATER

The most useful information concerning foundation material behavior is obtained by determining:

1. The condition of existing foundations (if available).
2. The constituents and properties of the surrounding soil and/or water that existed during the exposure period of the foundation.
3. Site conditions such as site drainage, existence of stray currents, aggressive wastes, and structure use.

Various organizations such as the National Institute of Standards and Technology, the National Association of Corrosion Engineers, and most materials producers and their respective trade organizations have either conducted or supported investigations, or the collection and dissemination of data on the durability of materials in soil and water environments. The American Society For Testing and Materials, through Subcommittee G01.10, has published an excellent group of papers on underground corrosion of metals (Escalante, 1985).

25.5.1 Soil

Dry soils, such as those found in many high plains areas, are usually not aggressive to steel and concrete or treated wood. The lack of moisture is responsible for this benign condition. However, many dry areas contain high-alkaline-content soils that may be quite destructive to various types of cement

(ACI, 1980). Types II and V cements are resistant to the destructive effect of alkaline soils. Alkaline aggregate can also be destructive to concrete. Soils with high (> 2.0 percent) water soluble sulfate can cause severe deterioration if Type V cement is not used. Lean concretes are rapidly deteriorated by sulfates.

Steel reinforcing bars and piling are, in general, not affected by alkaline soils. As will be noted later, passivation of steel is due to the alkalinity of concrete. Acid soils can deteriorate concrete structures—sometimes rapidly. Unless free oxygen is present, steel is not seriously affected by acidic soils of natural deposition. Figure 25.6 shows the surface condition, in the fill area, of a steel H-pile that was driven through a fill classified as UCF. The aggressive materials included fresh cinder. The pile had over 90 percent of the cross-sectional area remaining after 20 years of exposure.

Fills and base courses for buildings, roadways and airport runways have been made using steel slag. A number of structures have been affected by swelling of the slag (Crawford and Burn, 1969). Figure 25.7 shows the cracks in a wire-mesh-reinforced

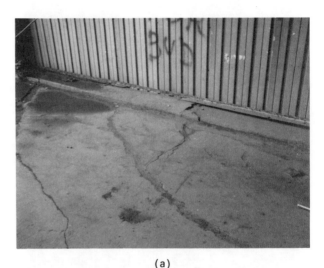

(a)

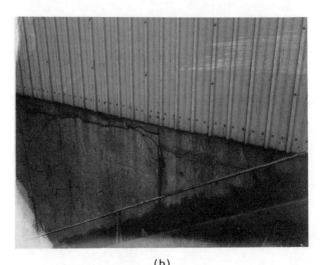

(b)

Fig. 25.7 Cracking of concrete floor and wall footing due to expansion of uncured steel slag. (a) Crack in 6-in thick concrete floor with wire mesh reinforcing. Floor heaved a maximum of 3 in. (b) Foundation cracks at end of floor cracks.

Fig. 25.6 Appearance of a steel H-pile embedded in fill composed of cinders, clay, organic material, and steel slag.

floor and the wall footing. Blast-furnace slag does not swell and has been used for many years for slag cement, concrete, and asphalt aggregate and base courses.

Adequately treated wood can be used in most soils. Untreated wood can also be used in most soils, but should be limited to below-groundwater elevation.

25.5.2 Water

Fresh water is not injurious to concretes and wood. Steel will react with water at a water–air interface and slowly develop surface corrosion products that protect the steel and further deterioration is almost nil.

Fig. 25.8 Deteriorated prestressed concrete piles in seawater.

Fig. 25.9 Repair of deteriorated steel H-pile (in brackish water).

Fig. 25.10 Appearance of concrete above and below the water–air interface. (Pile was located in an estuary.)

The initial corrosion rate of steel in seawater and fresh water is similar, but a protective film forms on the steel in fresh water and greatly mitigates the corrosion rate.

Figure 25.8 shows deterioration of prestressed concrete piles supporting a railroad structure. Figure 25.9 shows the deterioration (and ongoing repairs) of steel piles supporting a pier surrounded by brackish water. The piles had been in place for 25 years. Figure 25.10 shows the appearance of concrete partially immersed in estuarial water for 27 years. The immersed area was subjected to salt wedge intrusions and industrial wastes (Dismuke, 1988). The industrial wastes were apparently somewhat acidic.

25.6 EVALUATION OF SITE DATA

The conventional site data should be evaluated, with respect to material durability, as to:

- Facility life
- Facility use
- Facility function
- Ease of rehabilitation
- Site restriction
- Possible use change

For waste disposal structures the lack of more precise data on material durability appears to have been an important influence in the development of regulations. This is shown by the requirements for long periods of site monitoring and number of types of monitoring locations. It is difficult to determine how much of the tendency to gold plate waste disposal facilities because of biologic and durability duress is due to lack of information.

An example of how wrong site data and improper evaluation of the data can affect conclusions is shown in Figure 25.11. The figure shows the surface condition of a steel pile section after exposure. There has been extensive pitting and overall section loss. Since this is a structural member, the location of, and amount of, cross-sectional loss is of far more importance than knowledge of pit depths and pitting rate of corrosion. In this case, the rate of erosion had been reported with regard to pit depths. Since the pitting is extensive on all surfaces it is difficult to obtain an accurate thickness. The specimen shown was ground smooth over most of one flange and ultrasonic and micrometer measurements were made to compare the results.

Fig. 25.11 Surface condition of a deteriorated steel section.

There was a 25 percent difference between the two methods of measurement. This much difference could determine whether replacement or repair was necessary.

The correct type of cement and aggregate must be chosen to develop durable concrete for given conditions. However, of the three common construction materials, concrete fails most often because of poor quality control at the construction stage.

25.7 PROTECTION OF FOUNDATIONS

25.7.1 Methods of Preventing or Mitigating Deterioration

Plain and reinforced concrete are the most common foundations used. Concrete is inherently stable in the majority of soils if proper materials are used and care is used during placement. As previously discussed, alkaline soils are the most frequently encountered naturally occurring soils that may deteriorate concrete. Types I and II cements are normally used for foundations in those areas.

Coatings are frequently used on concrete walls to eliminate or mitigate water intrusion. However, the coatings to prevent concrete deterioration are usually found in marine environments.

The coatings are routinely applied to concrete piles, pile caps, and wharf structures of all types. As with other foundation materials, fresh water exposure is not deleterious to concrete. In industrial and highway construction, protective coatings are frequently used for protection from acids and salts spilled or placed on the concrete bridge decks, and other highway slabs are usually reinforced with epoxy-coated reinforcing bars (rebar) in the areas where chlorides are used to control snow and ice. Coated rebar is occasionally being used in waterfront structures.

Concrete used in prestressed components is an excellent material whose compressive strength is generally higher than that used in reinforced concrete. Because of the high quality of the concrete the cover over the prestressed strands may be less than that when lower-strength concrete is used in reinforced-concrete construction. Also, the prestressed wire keeps the concrete in compression, which reduces cracking of the concrete. However, the wire is susceptible to stress corrosion and the strands must be adequately covered by the concrete. Currently, a minimum 4 inches of cover is used when reinforced concrete is exposed to marine environments and 3 inches when exposed to soil. These covers provide adequate protection to the reinforcing bars when the in-situ concrete quality is suitable.

Usually, the protection of the reinforcing bar increases with increasing concrete strength.

Epoxy coating and galvanizing of reinforcing bar in concrete are commonly used. Epoxy coatings are used in the majority of concrete highway pavements and bridge decks in the northern United States in the attempt to control deterioration caused by the application of chlorides. Cathodic protection systems are used in bridge decks and have, in some cases, been used for some slab-on-grade installations. Steel bearing piles driven in undisturbed soil do not require protection. If the piles extend above the ground or dredge line, the exposure conditions at and above the soil–water–air interface will indicate whether protection at the interface is required. When piles are installed through fills, protection may be required if the fill is of recent deposit and contains acids that have not been leached out. Fly-ash and cinder fills may not be aggressive after several years of exposure. The rate of reduction of aggressive elements in fills depends on the rainfall, permeability of the soil, and drainage characteristics.

Several large, long-term tests in marine environments have been made at three sites along the eastern seaboard to evaluate the performance of coatings and cathodic protection on steel piles. An interim report of results at the NBS test location (Escalante et al., 1977) showed that after 8 years' exposure a coal-tar epoxy with an inorganic zinc-rich primer performed best, as shown in Figure 25.12. Later data, after 13 years' exposure, showed that coal-tar epoxy with an organic zinc-rich primer and a fiberglass-filled polyester were somewhat better. Bare piles with cathodically protected (galvanic) erosion zones had low corrosion rates in those zones. Coal-tar epoxy coatings are used for protecting steel piles at the waterline in some locations where the piles are fully embedded and in marine environments. For driven sheet-pile bulkheads the side in contact with the soil is rarely coated, nor is that area always necessary to include when determining the required current for a cathodic protection system.

Untreated wood piles, as indicated before, last many years if located below groundwater line. Treated piles (see Table 25.5) are used for exposure to and above the soil–water–air interface and to resist fungi and marine borers in marine environments. The choice of treatment depends not only on the exposure conditions, but also on the effect of the treatment on the mechanical properties. As Table 25.6 indicates, there are some differences in properties. Tropical hardwoods (Basralocus, Bongossi, etc.), which resist fire, marine borers, and fungi, are not frequently used because of cost, but do have low maintenance costs and long life.

Cathodic protection of steel and reinforced and prestressed concrete piles used in waterfront construction has been used for many years. One problem with cathodic protection is the high probability of current interruption due to environmental (ice, storms) and operational (propeller wash, debris, etc.) action. Maintenance costs can be high compared with that of a pipeline protection system. Cathodic protection of steel or concrete piling in soil has never been proven to be necessary. In some instances, cathodic protection has been applied, but the circumstances were technically or otherwise unusual. The application of pipeline cathodic protection methods has not been proven to be applicable to piles.

25.7.2 Monitoring and Inspection of Foundations For Deterioration

Of the major causes of foundation failures, two-material deterioration and undercutting (erosion) are usually monitored by physical inspection on a regular basis. For most structures, excessive movement triggers an inspection. For land-based

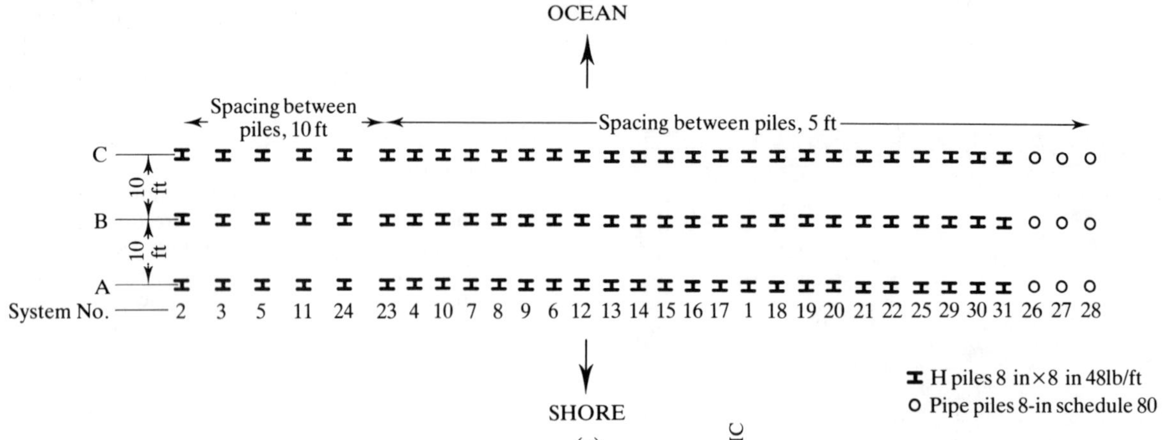

OCEAN

↑

Spacing between piles, 10 ft ← → ← Spacing between piles, 5 ft →

C — I I I I I IIIIIIIIIIIIIIIIIIIIIIIIII o o o

B — I I I I IIIIIIIIIIIIIIIIIIIIIIIIII o o o

A — I I I I I IIIIIIIIIIIIIIIIIIIIIIIIIIII o o o

10 ft (between A–B), 10 ft (between B–C)

System No. — 2 3 5 11 24 23 4 10 7 8 9 6 12 13 14 15 16 17 1 18 19 20 21 22 25 29 30 31 26 27 28

↓

SHORE

(a)

I H piles 8 in × 8 in 48 lb/ft
o Pipe piles 8-in schedule 80

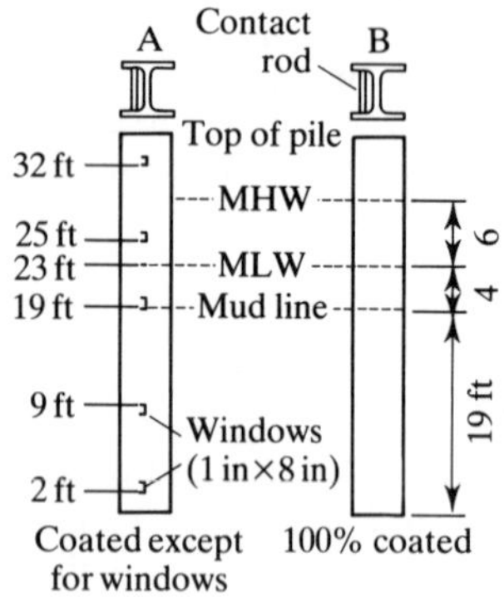

A   Contact rod   B   C

Top of pile

32 ft — MHW
25 ft — MLW (23 ft)
23 ft
19 ft — Mud line

9 ft
Windows (1 in × 8 in)
2 ft

Coated except for windows

100% coated

Coated (35 ft) / Uncoated (12 ft)

6 ft / 4 ft / 19 ft

(b)

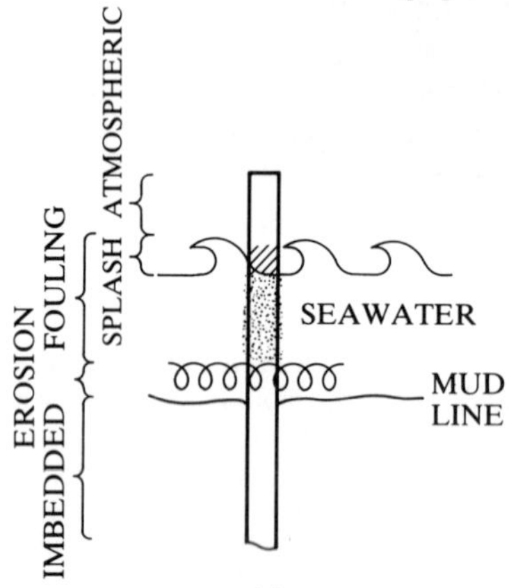

ATMOSPHERIC
SPLASH
FOULING
EROSION
IMBEDDED

SEAWATER

MUD LINE

(c)

System	Coating Description	Overall Average Corrosion Rate, mpy[a]	Average Corrosion Rate Within Zone, mpy[a]			
			Imbedded Zone 0 to 15 ft[b]	Erosion Zone 15 to 21 ft[b]	Immersed Zone 21 to 29 ft[b]	Atmosphere Zone 29 to 35 ft[b]
21	Coal-tar epoxy/zinc-rich inorganic	< 0.01	< 0.01	< 0.01	< 0.01	< 0.01
13	Vinyl/flame-sprayed aluminum	0.03	0.01	0.17	0.07	0
20	Epoxy polyamide/zinc-rich inorganic	0.05	0.02	0.22	0.10	0
8	Aluminum-pigmented coal-tar epoxy	0.07	0.07	0.06	0.08	0.03
29	Polyester glass flake	< 0.10	< 0.10	< 0.10	< 0.10	< 0.10
15	Polyvinylidene chloride/flame-sprayed zinc	0.10	0	0.14	0.12	0.29
10	Galvanized	0.14	0	0.67	0.32	0.06
17	Phenolic mastic	0.14	0.11	0.11	0.15	0.21
14	Flame-sprayed aluminum	0.16	0.19	0.39	0.19	0.03
9	Aluminum-pigmented coal-tar epoxy	0.18	0.18	0.08	0.21	0.04
18	Coal-tar epoxy/zinc-rich organic	0.19	0.17	0.15	0.21	0.24
19	Vinyl/zinc-rich inorganic	0.20	0.19	0.22	0.18	0.31
22	Vinyl mastic/zinc-rich inorganic	0.24	0.02	1.4	0.61	0
4	Coal-tar epoxy	0.53	0.17	0.21	0.27	2.1
25	Coal-tar epoxy on mariner steel	0.53	0.18	0.44	0.45	1.6
7	Coal-tar epoxy plus armor	0.55	0.13	0.07	0.07	2.7
6	Coal-tar epoxy	0.80	0.27	0.72	0.46	2.9
16	Vinyl red lead/flame-sprayed zinc	1.4	0.08	3.2	1.8	2.3
12	Polyvinylidene chloride	2.4	0.81	4.9	3.6	3.5
30	Bare carbon steel	4.9	0.9	8.9	6.7	10.5
1	Bare carbon steel	6.0	1.8	9.7	7.9	12.2
31	Bare carbon steel	7.3	2.8	10.5	9.0	13.9
26	Bare carbon steel	Pipe[c]	—	—	—	—
27	Coal-tar epoxy	Pipe[c]	—	—	—	—
28	Coal-tar epoxy plus armor	Pipe[c]	—	—	—	—

[a] For Iron: mils per year × 5.48 = milligrams per square decimeter per day.
[b] Distance from bottom of pile.
[c] Flange thickness measurements not made on pipe piles.

(d)

866

structures where there is low deterioration potential the need for inspection of foundations is minimal. A major exception to this is the case of bridges. For inland bridges, erosion of the piling material and soil around the piling or around and under the pier footings must be considered as possible to probable. For those structures located in estuarial areas and on the waterfront, regular inspections are necessary.

Most inspections are carried out by viewing the foundations and/or making measurements of thickness, width, etc. Direct and ultrasonic measurements usually suffice, but electrical and acoustic methods are increasingly being used for evaluating the condition and rate of deterioration of foundations. For steel piles, electrochemical polarization techniques are being used (Schwerdtfeger and Romanoff, 1972) to monitor performance. Acoustic methods can be used to determine continuity and in some instances cross-sectional variations in all materials.

REFERENCES

ACI Committee 201 (1980), *Guide to Durable Concrete*, ACI Manual of Concrete Practice, Part 1, American Concrete Institute.

AICT (1985), American Institute of Timber Construction, *Timber Construction Manual*, 3d ed.

Crawford, C. B. and Burn, K. M. (1969), Building damage from expansive steel slag backfill, *Journal of the Soil Mechanics and Foundations Division*, ASCE, pp. 1325–1334.

deJesus, J. M. (1980), Concrete Structures Exposed to Heat, presented at American Iron and Steel Engineers, Philadelphia Section Meeting.

Dillon, C. F. (1986), *Corrosion Control in the Chemical Process Industries*, McGraw-Hill Book Co., New York, N.Y.

Dismuke, T. D. (1987), Durability and protection of geostructural members in hazardous ground, *Proceedings of the 1st International Symposium on Environmental Geotechnology*, Envo Publishing Co., Inc., Bethlehem, Pa., **2**, pp. 251–257.

Dismuke, T. D. (1988), *The Chickasawbogue Bridge Failure*, American Iron and Steel Institute.

Dismuke, T. D., Coburn, S. K., and Hirsch, C. M. (eds.) (1981), *Handbook of Corrosion Protection for Steel Pile Structures in Marine Environments*, American Iron and Steel Institute.

Escalante, E. (ed.) (1985), *Underground Corrosion of Metals*, ASTM STP 741, American Society for Testing Materials.

Escalante, E., Iverson, W. P., Gerhold, W. F., Sanderson, B. T., and Alumbaugh, R. L. (1977), *Corrosion and Protection of Steel Piles in a Natural Seawater Environment*, NBS Monograph 158, National Bureau of Standards (currently National Institute for Technology and Standards).

Fang, H. Y. and Mehta, H. C. (1976), Corrosion prevention in geostructures with special emphasis on polymer-concrete systems, *Analysis and Design of Building Foundations*, Envo Publishing Co., Inc., Bethlehem, Pa., pp. 611–635.

Gauffreau, P. E. (1987) A review of pile protection methods in a corrosive environment, *Proceedings of the 1st International Symposium on Environmental Geotechnology*, Envo Publishing Co., Inc., Bethlehem, Pa., **2**, pp. 372–378.

NCEL Tech Data Sheet (1979), *Mechanical Properties of Preservative Treated Marine Piles*, and *Preservative Retention of Marine Piles Within a 1-Inch Depth*, Civil Engineering Laboratory, Naval Construction Battalion Center, Port Hueneme, Calif.

Schwerdtfeger, W. J. and Romanoff, M. (1972), *NBS Papers On Underground Corrosion of Steel Piling*, Monograph 127, National Bureau of Standards.

Sowers, G. F. (1962), Shallow foundations, *Foundation Engineering*, ed. G. A. Leonards, McGraw-Hill Book Co., New York, N.Y., pp. 525–632.

Fig. 25.12 (opposite) Layout, exposure zones and corrosion rates of steel piles at the National Bureau of Standards Test Site in Dams Neck, Virginia. (a) Plan of steel pile test specimens. (b) Coating arrangement on piles. (c) Zones of exposure. (d) Corrosion rates of H-piles based on flange thickness.

26 GROUND ANCHORS AND SOIL NAILS IN RETAINING STRUCTURES

ILAN JURAN, D.Sc.
Professor and Head
Department of Civil and Environmental Engineering
Brooklyn Polytechnic University

VICTOR ELIAS, P.E.
V. Elias & Associates, P.A.
Consulting Engineers

26.1 INTRODUCTION

Ground anchor and soil nail retaining systems are designed to stabilize and support natural and engineered structures and to restrain their movement using tension-resisting elements. The basic design concept consists of transferring the resisting tensile forces generated in the inclusions into the ground through the friction (or adhesion) mobilized at the interfaces. These systems allow the engineer to efficiently use the in-situ ground in providing vertical or lateral structural support. They present significant technical advantages over conventional rigid gravity retaining walls or external bracing systems that result in substantial cost savings and reduced construction period. Therefore, during the past few decades, ground anchors, and more recently soil nails, have been increasingly used in civil engineering projects.

The use of these systems in permanent structures requires careful evaluation of the durability of the structural elements and assessment of the long-term system performance. A variety of inclusions, corrosion-protection systems, and installation techniques have been progressively developed by specialty contractors. This chapter briefly describes the construction process and the main structural elements. It presents the main aspects of ground–inclusion interaction, illustrates the observed behavior of instrumented structures, and outlines durability considerations, performance criteria, and design approaches that have been developed to ensure the internal and external stability of these composite retaining systems.

26.2 PRINCIPLES, HISTORICAL DEVELOPMENT, AND FIELDS OF APPLICATION

26.2.1 Permanent Ground Anchors

Permanent ground anchors are prestressed cement-grouted tendons used in soils or rock to restrain and control the displacements of structural elements such as walls or slabs. They have been developed mainly by specialty contractors involved in temporary excavation support systems and in some cases are proprietary. The anchors are installed in drilled holes and prestressed to the design load in order to mobilize and transfer the required resisting force from the ground to the structural element. Temporary ground anchors are used for a specified construction period and their service life is generally less than 2 years. Permanent ground anchors are corrosion-protected to insure their long-term performance throughout the design service life of the structure.

Figure 26.1 shows a schematic diagram of a permanent ground anchor. The basic components of the ground anchor are:

• *The tendon* is made of prestressing steel wires, strands, or bars and includes:
  a. The anchor bond length—where the tendon is fixed in the primary grout bulb and transfers the tension force to the surrounding ground. The anchor bond length is designed to provide the required load pull-out capacity of the anchor.
  b. The unbonded length—where the tendon is free to elongate elastically transferring the resisting force from the anchor bond length to the structural element (i.e., wall face, slab, etc.). It is designed to reach the underlying substratum or, in

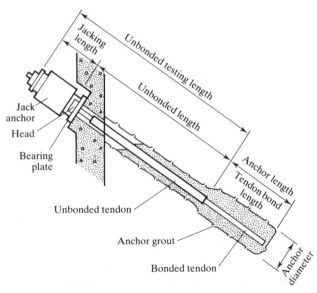

Fig. 26.1 Permanent ground anchors.

homogeneous soils, to locate the anchor bond length beyond the potentially unstable soil mass adjacent to the structural element.

- *The anchor grout*, also called primary grout, is generally a portland cement-based mixture or a polymer resin and is used to transfer the anchor force to the ground. Secondary grout can be injected into the drilled hole after stressing to provide corrosion protection for unsheeted tendons.
- *The anchorage* is a device attached to the tendon that consists of a plate and an anchor head (or threaded nut) and permits stressing and lock-off of the prestressing steel.

During the past 50 years, permanent ground anchors have been extensively used by contractors to provide vertical and lateral support for natural and engineered structures. Typical applications of ground anchors are illustrated in Figure 26.2. They have found widespread acceptance in a variety of civil engineering projects including cut slope retaining systems, tied-back diaphragm or soldier pile walls, bridge abutments, stabilization of natural slopes and cliffs, tunnel portals, underpinning, repair or reconstruction of quay walls, dam spillways, loading ramps, hangars, etc. They have also been frequently used as tiedown supports for dams, transmission towers, and waterfront structures, primarily to resist uplift water pressures and rotational loadings.

Tiebacks were first used to anchor structures in rock. The earliest permanent rock tiedowns were installed by the French engineer Coyne for anchoring the Jument lighthouse (1930) and raising the Cheurfas Dam, Algeria (1934). By the late 1950s, use of permanent rock tiedowns had become common practice in renovation and construction of dams (Evans, 1955; Morris, 1956; Middleton, 1961) and towers (Weatherby, 1982). In the 1950s contractors began to use tiebacks for temporary supports

of deep excavations. The first permanent soil tiebacks in the United States were installed in 1961 in a very stiff silty clay for the construction of retaining walls for the Michigan expressway (Jones and Kerkhoff, 1961). However, in spite of long-term European experience, permanent ground anchors had not been in common use in the United States until the late 1970s, mainly because of engineering concerns with regard to long-term performance, potential time-dependent (creep) movement, corrosion protection of the tendon, and the need to establish reliable quality control testing procedures to verify the short- and long-term holding capacity. Technological efforts have been continuously invested by specialty contractors to overcome these limitations, develop efficient corrosion-protection systems, improve grouting methods and installation procedures, and increase the tension capacity of the prestressed tendons.

The rapid acceptance and growing use of ground anchors can be attributed mainly to significant technical advantages resulting in substantial cost savings and reduced construction period. Specifically, in urban areas the use of ground anchors often allows significant reduction in right-of-way acquisition and permits the elimination of temporary support systems, external bracings, or the need for underpinning existing structures near to excavation sites. The increasing confidence in ground anchor use for permanent structures is primarily due to reliable quality control procedures that involve routine performance and proof testing of all production anchors under loads exceeding the design load. Performance specifications and codes of practice, based on experience and long-term observations of permanent anchor installations, have been developed in European countries (French Recommendations, Bureau Securitas, 1977; FIP Rules, 1974; German Standards, DIN, 1972, 1976; PTI Recommendations, 1980) and more recently in the United States (FHWA; see Cheney, 1984) to specify design, construction, and monitoring procedures.

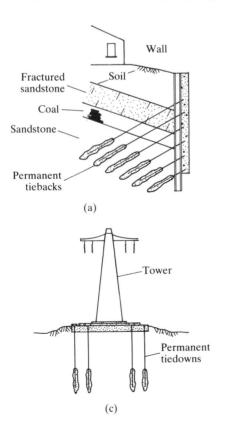

(a)

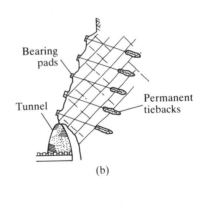

(b)

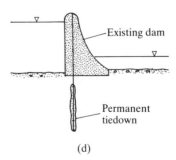

(c)   (d)

Fig. 26.2 Typical applications of permanent ground anchors. (a) Concrete wall. (b) Landslide and tunnel portal. (c) Permanent tower tiedown. (d) Dams. (*After Weatherby, 1982.*)

26.2.2 Soil Nailing

Soil nailing is an in-situ soil reinforcement technique that has been used during the past two decades, mainly in France and Germany, in cut slope retaining systems and slope stabilization. The fundamental concept of soil nailing consists of reinforcing the ground by passive inclusions, closely spaced, to create in situ a coherent gravity structure and thereby to increase the overall shear strength of the in-situ soil and restrain its displacements. This technique has emerged essentially as an extension of the "New Austrian Tunneling Method" (Rabcewicz, 1964-65), which combines reinforced shotcrete and rockbolting to provide a flexible support system for the construction of underground excavations.

Although soil nailing technology is relatively new, it has been used in a variety of civil engineering projects including stabilization of railroad and highway cut slopes (Rabejac and Toudic, 1974; Hovart and Rami, 1975; Stocker et al., 1979; Cartier and Gigan, 1983; Schlosser, 1983); construction of excavation retaining structures in urban areas, for high-rise buildings and underground facilities (Louis, 1981; Gassler and Gudehus, 1981; Shen et al., 1981); landslide stabilization (Guilloux et al., 1983; Blondeau et al., 1984); tunnel portals in steep and unstable stratified slopes (Louis, 1981); and other civil and industrial projects. Typical applications of soil nailing are illustrated in Figure 26.3. Several nailed soil-retaining structures have been instrumented to establish a data base for evaluation of structure performance and development of reliable design methods (Stocker et al., 1979; Gassler and Gudehus, 1981; Schlosser, 1983; Plumelle, 1986). In North America, the system was initially used in Vancouver, B.C., in the late 1960s in temporary excavation supports for industrial and residential buildings (Shen et al., 1981). Presently, soil nailing systems can be considered for any temporary or permanent application where conventional cut retaining systems, such as cast-in-place reinforced-concrete walls or tied-back walls, are applicable. As demonstrated by Gassler and Gudehus (1981), soil-nailed retaining structures can withstand both static and dynamic vertical loads at their upper surface without undergoing excessive displacements. Therefore, they can be effectively used in the construction of bridge abutments. Soil nailing also appears to provide an efficient and economical technique for repair and reconstruction of existing structures, particularly tie-back walls and reinforced soil retaining structures.

In soil-nailed retaining structures, the inclusions are generally steel bars or other metallic elements that can resist tensile stresses, shear stresses, and bending moments. They are either placed in drilled boreholes and grouted along their total length or driven into the ground. The nails are not prestressed but are closely spaced (e.g., one driven nail per 2.5 ft^2, one grouted nail per 10–50 ft^2) to provide an anisotropic apparent cohesion to the native ground. The facing of the soil-nailed structure is not a major structural load-carrying element but rather ensures local stability of the soil between reinforcement layers and protects the ground from surface erosion and weathering effects. It generally consists of a thin layer of reinforced shotcrete (4- to 6-in thick), constructed incrementally from the top down. The facing and the nails are placed immediately after each excavation stage to restrain ground decompression and therefore to prevent deterioration of the original mechanical properties and shear strength characteristics of the native ground. Prefabricated or cast-in-place concrete panels have increasingly been used in the construction of permanent structures to satisfy specific aesthetic and durability design criteria.

As with ground anchors, soil nailing has been primarily used for temporary retaining structures. This is mainly due to the

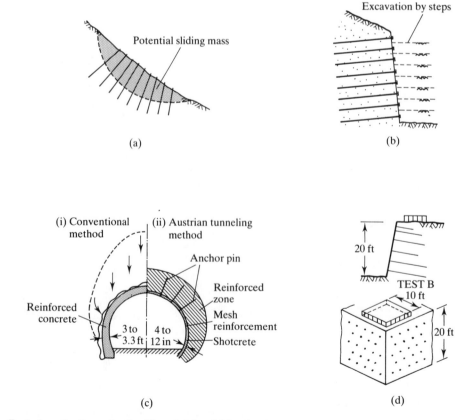

Fig. 26.3 Typical applications of soil nailing. (a) Landslide. (b) Retaining structures. (c) Tunnel portal. (d) Abutments.

engineering concerns with regard to durability of metallic inclusions in the ground and shortcomings of facing technology. In recent years, technological developments have been mostly focused on producing low-cost corrosion-protected nails and prefabricated concrete or steel panels to overcome these limitations. Soil nailing has now become a common construction method in European countries and over 100 temporary and permanent retaining structures have been constructed. In the United States, the engineering use of this technology for permanent structures is currently growing with increasing local experience followed by development of standard specifications for design, construction, quality control, and monitoring of soil-nailed structures.

26.3 TECHNOLOGY, CONSTRUCTION PROCESS, AND STRUCTURAL ELEMENTS

The main components of an anchored (or nailed) ground retaining system are the in-situ ground, the tension-resisting inclusions (anchors or nails), and the facing or the structural retaining element. The economy of the system is predominantly dependent upon the technology used (i.e., structural elements and installation process of inclusions) and the construction rate achieved.

26.3.1 Inclusions and Installation Techniques

A. Ground Anchors

A variety of anchors have been developed by specialty contractors using different anchor tendons, drilling methods, grout control procedures, and corrosion-protection systems. Selection of anchors and installation process for a specific application will generally depend upon the subsurface soil (or rock) type, groundwater conditions, site restrictions, and availability of equipment. Figure 26.4 illustrates schematically the four basic types of commonly used ground anchors.

Low-pressure-grouted straight-shafted ground anchors are used in most types of soils. They are commonly installed using hollow-stem augers or tremie grouting under low (or no) grout pressure.

Hollow-stem augers are extensively used in the United States in both granular and cohesive soils. The anchor installation process includes three basic stages: insertion of the tendon into the auger, drilling, and grouting through the auger while it is being extracted.

Low-pressure-grouted anchors are usually tremie-grouted (grout pressures lower than 150 psi) in drillholes that are cased in cohesionless soils but can be open in competent rocks or clayey soils. In granular soils, drilling is generally performed by driving a casing (4 or 6 inches in diameter) and using air or water flushing to remove the soil.

Cohesive soils are mostly rotary drilled without casing. Bentonite and cement slurries may be used to prevent caving of the drill holes.

Pressure-injected anchors are installed in sandy or gravelly soils under grout pressures exceeding 150 psi. Drilled or driven casing (3.5 or 3 inches in diameter) with a solid closure point may be used to seal the hole and to allow the bond zone to be grouted under high pressure during the extraction of the casing.

Postgrouting technique is generally used in both granular and cohesive soils to enlarge the primary grout bulb by multiple stages of controlled high-pressure grouting. An inflatable bag or packer is used to isolate the bond length and allow the grout to be pressurized through single or multiple postgrouting phases. A special grout tube (called tube-à-manchette) using a double packer is designed to allow repetitive grouting through

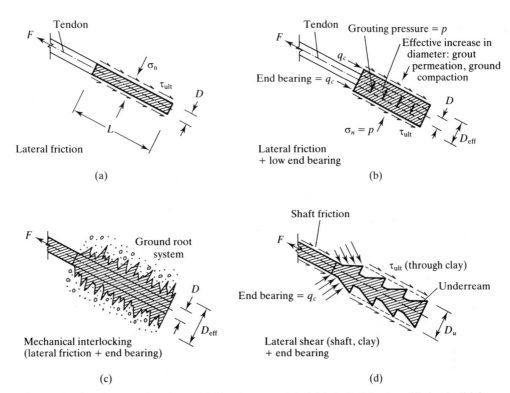

Fig. 26.4 Interaction mechanisms in ground anchors. (a) Tremie-grouted straight-shafted (rocks, stiff clays). (b) Low-pressure-grouted anchors. (c) High-pressure-grouted anchors. (d) Underreamed anchors (stiff to hard cohesive clays).

valves located along the bond zone. Typical examples of regroutable anchors are the Soletanche IRP anchor (Jorge, 1969), the TMD–Bachy anchor (Clement and Navarro, 1972), and the TPT anchor (Mastrantuono and Tomiolo, 1977). The Soletanche IRP anchor and the tube-à-manchette packer are schematically illustrated in Figure 26.5.

Single and multi-underreamed anchors are used in stiff cohesive soils and weak rock. Their installation process involves the use of an underreaming device that is basically a cutting tool. First, the cylindrical anchor shaft is drilled, usually using a continuous flight auger. Then, the cutting tool is mechanically expanded at a controlled rate to the design size of the underream. The soil is removed by water flushing; neat cement grout is tremie-grouted into the drill hole; and the tendon is inserted. Depending on the underreaming device, several underreams

can be cut simultaneously. The spacing between the underreams is selected to induce a shear failure along a cylinder passing through the tips of the underreams.

The main structural element of each ground anchor is the steel tendon, which may consist of bars, wires, or strands. Strands and wires have advantages with respect to tensile strength (ultimate tensile strength: 270 ksi for strands and 240 ksi for wires), and ease of transportation, storage, and fabrication. However, bars (ultimate tensile strength of 150 to 160 ksi) are more readily protected against corrosion and, in the case of shallow, low-capacity anchors, are usually easier and cheaper to install. Often, availability and cost will be the determining factors. In the United States, bars and seven-wire strands are the most commonly used tendons. High-capacity tendons made of 18 strands with a diameter of 0.50 or 0.60 inch are also available for high-capacity tiedown applications.

B. Soil Nailing

The steel reinforcing elements used for soil nailing can be classified as (a) driven nails, (b) grouted nails, (c) jet-grouted nails, and (d) corrosion-protected nails.

Driven nails, commonly used in France and Germany, are small-diameter (15 to 46 mm) rods or bars, or metallic sections, made of mild steel with a yield strength of 350 MPa (50 ksi). They are closely spaced (2 to 4 bars per square meter) and create a rather homogeneous composite reinforced soil mass.

The nails are driven into the ground at the designed inclination using a vibropercussion pneumatic or hydraulic hammer with no preliminary drilling. Special nails with an axial channel can be used to allow for grout sealing of the nail to the surrounding soil after its complete penetration. This installation technique is rapid and economical (4 to 6 per hour). However, it is limited by the length of the bars (maximum length about 20 m) and by the heterogeneity of the ground (e.g., presence of boulders).

Grouted nails are generally steel bars (15 to 46 mm in diameter) with a yield strength of 60 ksi. They are placed in boreholes (10 to 15 cm in diameter) with a vertical and horizontal spacing varying typically from 1 to 3 m depending on the type of the in-situ soil. The nails are usually cement-grouted by gravity or under low pressure. Ribbed bars can be used to improve the nail–grout adherence, and special perforated tubes have been developed to allow injection of the grout through the inclusion.

Jet-grouted nails are composite inclusions made of a grouted soil with a central steel rod, which can be as thick as 30 to 40 cm. A technique that combines the vibropercussion driving and high-pressure (greater than 20 MPa) jet grouting has been developed recently by Louis (1986). The nails are installed (Fig. 26.6) using a high-frequency (up to 70 Hz) vibropercussion hammer, and cement grouting is performed during installation. The grout is injected through a small-diameter (few millimeters) longitudinal channel in the reinforcing rod under a pressure that is sufficiently high to cause hydraulic fracturing of the surrounding ground. However, nailing with a significantly lower grouting pressure (about 4 MPa) has been used successfully, particularly in granular soils. The jet-grouting installation technique provides recompaction and improvement of the surrounding ground and increases significantly the pull-out resistance of the composite inclusion.

Corrosion-protected nails generally use double protection schemes similar to those commonly used in ground anchor practice. Proprietary nails have recently been developed by specialty contractors (Intrafor-Cofor; Solrenfor) to be used in permanent structures. These corrosion-protection schemes are described in a later section of this chapter.

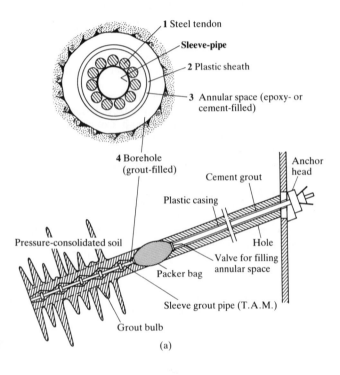

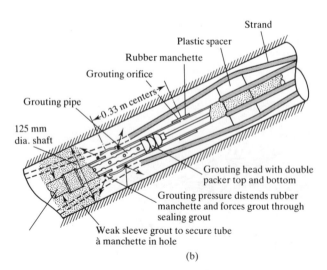

Fig. 26.5 (a) Schematic section of an IRP anchor. (b) Detail of tube-à-manchette for pressure grouting control. (*After Pfister et al., 1982.*)

Fig. 26.6a Construction process of a soil-nailed wall illustrating excavation, shotcreting, and nailing.

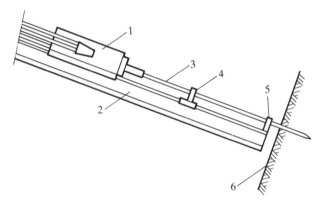

Fig. 26.6b Jet bolting: installation of reinforcing elements. 1. Vibropercussion hammer. 2. Sliding support. 3. Reinforcement to be inserted. 4. Sliding guide. 5. Fixed guide. 6. Soil to be treated.

26.3.2 Facing and Structural Retaining Elements

In an anchored retaining system the wall has a major structural role. It has to resist the tensile forces transferred by the anchors, the lateral pressure applied by the retained soil, and bending moments. The wall has to be stiff enough to restrain the ground displacement induced by the excavation process. The facing in a nailed soil-retaining system has only a minor mechanical role. The maximum tensile forces generated in the nails are significantly greater than those transferred to the facing. The main function of the facing is to ensure local stability of the ground between the nails and to limit its decompression. Hence, the facing has to be continuous, fit the irregularities of the cut slope surface, and be flexible enough to withstand ground displacement during excavation. The structural elements used to build the anchored wall are therefore basically different from those used to construct the facing of a nailed soil-retaining structure.

A. Structural Elements of an Anchored Wall

An anchored wall can be constructed with a wide variety of structural elements, using different installation techniques. Selection of the structural element for a specific application will generally depend on the subsurface soil (or rock) type, groundwater conditions, local construction practice, availability of material and equipment, and performance requirements. The structural elements can be evaluated in terms of their stiffness, ease of handling and installation, durability, water-tightness or continuity, and ease of removal. The elements commonly used can be broadly classified into four major categories: driven sheet piles, soldier piles and lagging walls, cylinder walls, and concrete diaphragm or slurry walls. Typical properties of each system are indicated in Table 26.1.

Sheet-pile walls usually consist of interlocking steel sheets driven into the ground prior to excavation. They are fairly impervious and easy to handle and install in soft clays, cohesionless silts, or loose sands. However, they are difficult to use in compact granular soils containing cobbles or boulders. As compared with other elements, they are relatively flexible and the wall displacement will in general be larger. They are commonly used for marine bulkhead construction (see Chapter 12).

Soldier piles and lagging (Figure 26.7a) usually consist of steel H-beams that are either driven into the ground or placed in predrilled boreholes prior to excavation. Concrete bored piles with reinforcement or permanent casing have also been used. As excavation proceeds, the ground between these piles is retained by lagging of wood planks, cast-in-place, or precast concrete elements. H-beam soldier piles and lagging walls are probably the excavation support system most widely used in the United States for temporary supports. They are easy to install in most types of soils, and present a significant advantage specifically in compact or irregular strata that would obstruct sheet piling. They can be readily adapted to different site conditions and irregular wall alignments. The main disadvantage of this retaining system is that the wall is rather pervious and subsurface water flow may cause local instabilities. A properly lagged wall should permit drainage, drawdown, and fluctuation of water level without flow of the retained soil.

Cylinder walls consist of an array of cylindrical caissons that are usually constructed of reinforced concrete or mixed-in-place soil–cement and are closely spaced to form a continuous wall. They can be cast-in-place and installed using several techniques such as hollow-stem augers, rotary drilling equipment, deep mixing methods, or jet-grouting. Depending on the stiffness of the individual cylinders, such a wall may be rigid enough to support lateral loads with limited deflection. To achieve water-tightness and properly retain the soil, shotcrete or lagging in the space between the cylinders may be required. Alternatively, the cylinders can overlap to produce a continuous, impervious wall. In addition to their rigidity, cylinder walls offer the advantage of adaptability to irregular site alignments and can be used in a variety of ground conditions.

Slurry walls or concrete diaphragm walls are generally formed in a trench supported by viscous mud slurry (see Chapter 20). Concrete is tremied into the trench, displacing the mud slurry upward. Reinforcement of the wall is made by vertical steel sections, precast reinforced-concrete members, or cages of reinforcing steel. Recent developments include the use of precast concrete panels. These walls can be designed to achieve a specified degree of stiffness and water-tightness, and can be integrated in the permanent structure. They are often used where lowering of the water table would adversely affect adjoining structures. Their main disadvantage is the relatively high cost and the need for specialized construction equipment and experienced contractors. They also may present environmental problems pertaining to slurry disposal.

Cast-in-place reinforced-concrete panels have been used in the construction of multitied-back walls (Kerisel et al., 1981). Figure 26.7b shows a schematic cross section of a 30-m deep open excavation retained by ten layers of prestressed anchors. This anchored wall was constructed from the top down with successive stages of (1) excavation, (2) in-place casting of the reinforced-concrete panel (2-ft thick, 9-ft high), and (3) anchoring.

TABLE 26.1 PROPERTIES OF STRUCTURAL WALL ELEMENTS.

System	Properties				Technical Features
	$EI\text{-}KSF/F$ $\times 10^3$	Moment KF/F	Depth Range, ft	Water-tight	
VERTICAL STEEL SHEETING	3 to 50	10 to 125	15 to 70	Fair	Readily available Effective in soft ground Low cost
SOLDIER BEAM AND CONCRETE/WOOD LAGGING (8WF to 14WF at 6ft to 8ft c–c)	3 to 40	7 to 70	15 to 60	Controlled by lagging	Ease of installation in competent ground Readily available Low cost
STEEL BEAM OR BAR REINFORCED CYLINDER PILES — TANGENT ($d = $ 18in to 36in)	70 to 800	100 to 400	20 to 60	Poor to fair	Common technique Can be stiffened by adding core Can be widely spaced Water-tightness can be improved by overlapping
SLURRY WALL — STEEL BEAM REINFORCEMENT ($t = $ 24in to 36in)	350 to 1600	30 to 400	20 to 100	Good	High strength Durable Can be permanent wall High cost
SLURRY WALL — REINFORCING BARS ($t = $ 24in to 36in)	300 to 1000	15 to 260	20 to 100	Good	High strength Durable Can be permanent Higher cost

For deep excavations this multianchored wall offers some technical advantages as compared with a diaphragm wall: (1) precasting of concrete prior to excavation is not required; (2) as each panel is anchored, wall thickness is significantly smaller; (3) adaptability to variation in ground conditions; and (4) effective control of the wall performance during excavation by properly adjusting the individual panels.

B. Facing in Nailed Retaining Systems

Depending on the application and soil (or rock) type, four types of facing are presently used.

Shotcrete facing (10 to 25 cm thick) is currently used for most temporary retaining structures in soils. This facing technology provides a continuous, flexible surface layer that can fill voids and cracks in the surrounding ground. It is generally reinforced with a welded wire mesh (Fig. 26.7c) and its required thickness is obtained by successive layers of shotcrete (each 9 to 12 cm thick). This technique is relatively simple and inexpensive, but it may not provide the technical quality and aesthetics

required for permanent structures. In particular the durability of the shotcrete facing can be affected by groundwater, seepage, and environmental factors such as climatic changes, e.g. freezing, which may induce cracking. In addition, with shotcrete facings, provision of efficient drainage at the concrete–soil interface is difficult.

Welded wire mesh is generally used to provide a facing in fragmented rocks or intermediate soils (chalk, marl, shales).

Concrete and steel facings: Cast-in-place reinforced-concrete is frequently used for permanent structures.

Prefabricated concrete or steel panels have also been adapted for permanent structures. These panels can be designed to meet a variety of aesthetic, environmental, and durability criteria. They also provide efficient means of integrating continuous drainage behind the facing. Figure 26.8 shows Solrenfor metallic panels for inclined facing (Louis, 1986). The rectangular steel panels are bolted together and soil nails are installed through their common corners. Also shown in Figure 26.8 is a prefabricated concrete panel with continuous geotextile drainage. Concrete panels have also been combined with prefabricated steel panels and cast-in-place concrete.

(a)

(b)

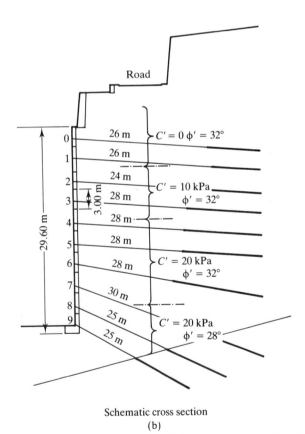

Schematic cross section

(b)

Fig. 26.7 (a) Soldier piles and logging anchored wall. (*Courtesy Nicholson Construction Co.*) (b) Las Palmeras open excavation in Monte Carlo. (*Kerisel et al., 1981.*)

cont.

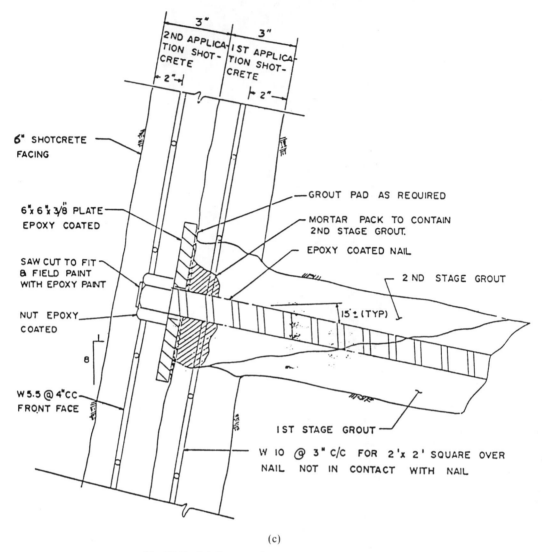

3"

2ND APPLICA-TION SHOT-CRETE

3"

1ST APPLICA-TION SHOT-CRETE

2"

2"

6" SHOTCRETE FACING

6"x 6"x 3/8" PLATE EPOXY COATED

SAW CUT TO FIT & FIELD PAINT WITH EPOXY PAINT

NUT EPOXY COATED

8

W 5.5 @ 4"CC FRONT FACE

GROUT PAD AS REQUIRED

MORTAR PACK TO CONTAIN 2ND STAGE GROUT.

EPOXY COATED NAIL

2 ND STAGE GROUT

15'± (TYP)

1ST STAGE GROUT

W 10 @ 3" C/C FOR 2'x 2' SQUARE OVER NAIL NOT IN CONTACT WITH NAIL

(c)

Fig. 26.7 (c) Construction detail of a shotcrete facing.

(a)

(b)

Fig. 26.8 (a) Prefabricated steel panels. (b) Prefabricated concrete panels and nail connections. (*Courtesy Solrenfor.*)

26.3.3 Drainage

Groundwater (see Chapters 1 and 7) is a major engineering concern relative to construction of anchored and nailed soil-retaining structures. An appropriate drainage system must be provided to (a) prevent generation of excessive hydrostatic pressures on the facing (or the structural wall element), (b) protect the facing element and particularly shotcrete facing from deterioration induced by water contact, (c) prevent saturation of the nailed ground, which can significantly affect the structure displacement and may cause instability during and after excavation. In anchored walls, prefabricated vertical drains, porous engineering fabrics, or subhorizontal drains can be used for drainage of the subsurface flow. In soil nailing, shallow drainage (plastic pipes, 10 cm in diameter, 30 to 40 cm long) is usually used to protect the facing, while subhorizontal slotted plastic tubes are used for deep drainage of the nailed ground. In the case of permanent structures with prefabricated panels, a continuous drain such as a geotextile can be placed behind the facing.

26.4 SOIL–INCLUSION INTERACTION: PULL-OUT CAPACITY ESTIMATES

The load-transfer mechanisms between a grouted anchor (or nail) and the subsurface soil (or rock) as well as the ultimate pull-out capacity depend upon several parameters, including installation technique, drilling and grouting method, grouting pressure, size and shape of the grouted inclusion, engineering properties of the in-situ soil and specifically its relative density (or overconsolidation ratio), permeability, and shear strength characteristics (see Chapter 3).

The grain size and porosity of the in-situ soil govern the grout conductivity. In sands, gravels and weathered rock, with hydraulic conductivities of 10^{-1} to 10^{-2} cm/sec, grout will permeate through the pores or natural fractures of the ground. In fine-grained cohesionless soils (silts and fine sands), with hydraulic conductivity smaller than 10^{-3} cm/sec, the grout cannot penetrate the small pores but rather compacts locally, under pressure, the surrounding ground. Increasing the grout pressure will induce a greater grout permeation into the ground and/or a more effective ground densification. Consequently, under high-pressure grouting, high radial stresses are locked into the soil surrounding the anchor, increasing its pull-out capacity.

26.4.1 Load Transfer in Ground Anchors

Figure 26.4 illustrates the basic soil–inclusion interaction mechanisms for the main types of ground anchors.

Tremie-grouted straight-shafted anchors, which are more commonly used in rock and very stiff to hard cohesive soils, generate their pull-out resistance through the lateral shear mobilized at the grout–ground interface. The pull-out capacity of these anchors is often estimated by

$$P = \pi \cdot D \cdot L \cdot \tau_{ult} \qquad (26.1)$$

where τ_{ult} is the ultimate lateral shear stress at the ground–grout interface (also called shaft friction), D and L are, respectively, the effective diameter and bond length of the grouted anchor.

It should be noted that the effective anchor diameter D is difficult to estimate, since it is highly dependent upon ground porosity and grout permeability. It is commonly assumed that in competent rocks (Littlejohn and Bruce, 1975)

$$\tau_{ult} = 10\% \cdot S_a \qquad \text{for } S_a < 600 \text{ psi} \qquad (26.1a)$$

where S_a is the uniaxial compressive strength. In cohesive soils,

$$\tau_{ult} = \alpha \cdot S_u \qquad (26.1b)$$

where α is an adhesion factor, and S_u is the average undrained shear strength of the soil.

The adhesion factor (α) generally varies (Tomlinson, 1957; Peck, 1958; Woodward et al., 1961) within the range of 0.3 to 0.75, with the lower values obtained for stiffer and harder clays.

Low-pressure grouted anchors are installed under an effective grout pressure lower than 150 psi (or, in cohesive soils, under pressure that would not fracture the ground) most commonly using hollow-stem augers or tremie grouting, an open hole in cohesive soils or cored rotary-drilled holes in cohesionless soils. The grouting pressure will induce an increase of the effective diameter of the grout bulb by permeation or local compaction of the ground. Therefore, the pull-out resistance of these anchors is highly dependent upon the grout pressure. It is primarily derived from the ultimate interface shear stress but an end-bearing resistance can be mobilized owing to an effective increase of the grout bulb diameter. The pull-out resistance is commonly estimated using Equation 26.1.

For cohesionless soils,

$$\tau_{ult} = p \cdot A \cdot \tan \phi \qquad (26.1c)$$

where p is the effective grout pressure, ϕ is the internal friction angle of the soil, and A is a dimensionless empirical coefficient smaller than 1 (Hanna, 1982). For practical applications p is generally limited to less than 50 psi or 2 psi per foot of overburden (Littlejohn, 1970).

High-pressure grouted anchors are installed under effective grout pressures exceeding 150 psi often using postgrouting techniques or pressure injection (feasible only in cohesionless soils). Figure 26.9, by Jorge (1969), illustrates the significant effect of grouting pressure on the ultimate load-transfer rate (or ultimate lateral shear stress) of multiphase postgrouted anchors in different types of soils. The high-pressure grouting

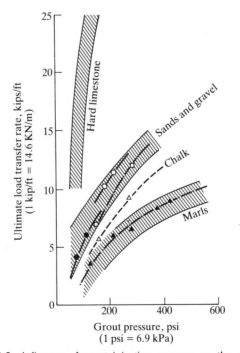

Fig. 26.9 Influence of grout injection pressure on the ultimate bearing capacity of anchors. ○, Medium sand. □, Alluvium and marl. ●, Alluvium sand. △, Soft chalk. ▲, Gypsum marl. (*After Jorge, 1969.*)

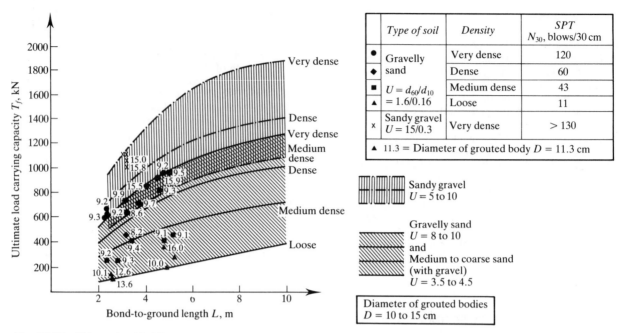

	Type of soil	Density	SPT N_{30}, blows/30 cm
●	Gravelly	Very dense	120
◆	sand	Dense	60
■	$U = d_{60}/d_{10}$	Medium dense	43
▲	$= 1.6/0.16$	Loose	11
×	Sandy gravel $U = 15/0.3$	Very dense	> 130

▲ 11.3 = Diameter of grouted body $D = 11.3$ cm

Sandy gravel
$U = 5$ to 10

Gravelly sand
$U = 8$ to 10
and
Medium to coarse sand
(with gravel)
$U = 3.5$ to 4.5

Diameter of grouted bodies
$D = 10$ to 15 cm

Fig. 26.10 Ultimate load holding capacity of anchors in sandy-gravel and gravelly sand. (*After Ostermayer and Scheele, 1977.*)

results in a grout root (or fissure) system that mechanically interlocks with the surrounding ground, increasing substantially the pull-out capacity of the anchor. Particularly, in dense granular soils this interlocking phenomena generates high tendency for the soil to dilate, which in turn results in a normal stress concentration at the grout–ground interface. The effect of pressure injection on the soil–anchor interaction is difficult to evaluate. Empirical relationships were provided by Ostermayer (1974) for estimating the ultimate lateral shear stress for high-pressure-grouted anchors, with and without postgrouting, in fine-grained soils (sandy silts to highly plastic clays). Ostermayer and Sheele (1977) developed empirical curves, reproduced in Figure 26.10, to estimate the ultimate pull-out capacity of pressure-injected anchors in granular soils as a function of anchor length, soil type, density, and uniformity. These curves, derived from 30 pull-out tests on anchors installed under grout pressures of about 70 psi, illustrate that the ultimate capacity of the anchor is not proportional to its length.

Underreamed anchors, which are mainly used in stiff to hard cohesive soils derive their pull-out capacity from adhesion along their shaft above the underreams, end bearing of the first underream, and lateral shear along a cylinder established by the tips of the underreams. For the cylinder to be effectively established, the spacing between the underreams should not exceed 1.5 times their diameter (Bassett, 1977). Estimate of the pull-out capacity of these anchors (Littlejohn, 1970) is based on empirical formulas that are conventionally used for pile design in cohesive soils.

The load transfer along pressure-injected and high pressure postgrouted anchors has been investigated by several authors (Bustamente, 1975, 1976; Ostermayer and Sheele, 1977; Shields et al., 1978; Bustamente, 1980; Davis and Plumelle, 1982). Figure 26.11 illustrates, for ultimate pull-out loads, the distributions of the lateral interface shear stress along pressure-injected anchors in gravelly sands of different densities (Ostermayer and Sheele, 1977). Similar results were reported for postgrouted anchors (Bustamente, 1972) in river sands (Fig. 26.12a) and for straight-shafted anchors (Feddersen, 1974) in highly over-consolidated, stiff, plastic clays. Figure 26.12b shows the results

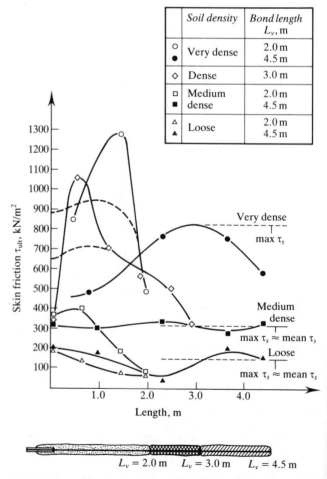

	Soil density	Bond length L_v, m
○ ●	Very dense	2.0 m 4.5 m
◇	Dense	3.0 m
□ ■	Medium dense	2.0 m 4.5 m
△ ▲	Loose	2.0 m 4.5 m

$L_v = 2.0$ m $L_v = 3.0$ m $L_v = 4.5$ m

Fig. 26.11 Distribution of the lateral interface shear stress along pressure-injected anchors at the ultimate load. (*After Ostermayer and Scheele, 1977.*)

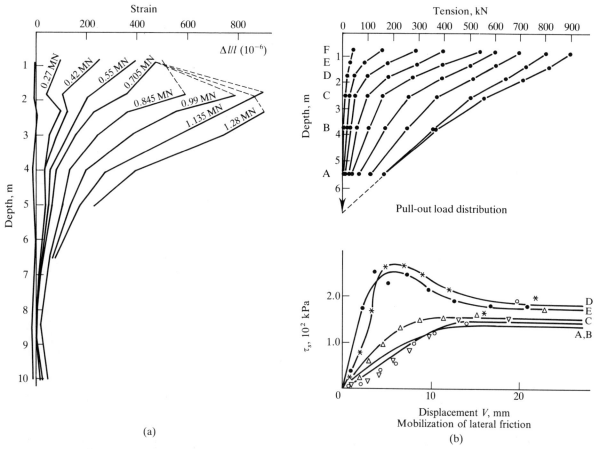

(a)

Fig. 26.12 (a) Distribution of deformation along the length of an IRP anchor. (*After Bustamante, 1972*). (b) Mobilization of the lateral friction along an anchor in a plastic clay. (Winnezeele, *Bustamente, 1980.*)

of a pull-out test on an instrumented anchor in a plastic clay (Bustamente, 1980). The slope of the tension force distribution along the anchor corresponds to the lateral interface shear stress mobilized at a specific depth under the applied pull-out force. As shown in Figure 26.12, the shear stress–upward anchor displacement curves obtained for different depths indicate overconsolidation of the subsurface soil layer and illustrate that the anchor displacement required to fully mobilize the ultimate shear stress is about 5 to 10 mm. The results of these studies demonstrate that in dense granular soils and highly overconsolidated clays the load-transfer rate along the anchor is not constant and the pull-out capacity is therefore not proportional to the anchor length.

The variation of the load-transfer rate along the anchor is mainly the result of its extensibility during pull-out testing. It is primarily dependent upon the relative rigidity (or elastic modulus ratio) of the anchor and the grout–soil interface and is particularly pronounced in highly dilatant stiff soils. Wernick (1977), Schlosser and Elias (1978), and Plumelle and Gasnier (1984) have shown that the restrained tendency of the soil to dilate during shearing results in a normal stress concentration at soil–inclusion interfaces that affects significantly the load transfer rate. As shown in Figure 26.13, vertical stresses as high as four times the overburden pressure were measured in a medium dense river sand at the anchor interface. The higher the (anchor-to-soil) elastic modulus ratio the more uniform is the load-transfer rate. Figure 26.11 shows that in loose to medium dense granular soils the interface lateral shear stress is approximately constant along the anchor. It is of interest to

indicate that these results are consistent with those obtained for ribbed metallic strips in Reinforced Earth structures (Schlosser and Elias, 1978).

Several authors have attempted to analyze the load transfer along the anchor using the "*t–z*" method (Coyle and Reese, 1966; Davis and Plumelle, 1982), which is commonly applied in design of friction piles, or more complex interface soil models (Zaman et al., 1984; Frank et al., 1982). However, rational anlysis of the ground–anchor interaction requires appropriate interface properties that are difficult to estimate.

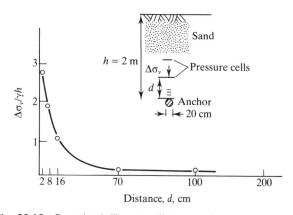

Fig. 26.13 Restrained dilatancy effect around a ground anchor. (*After Plumelle and Gasnier, 1984.*)

26.4.2 Soil–Nail Interaction

In soil nailing, similarly to ground anchors, the load transfer mechanism and the ultimate pull-out resistance of the nails depend primarily upon soil type and installation technique.

The pull-out resistance of *driven nails* in a dense granular soil was correlated by Cartier and Gigan (1983) with design recommendations for Reinforced Earth walls (Schlosser and Segrestin, 1979). These recommendations use the concept of an "apparent friction coefficient" that is derived from Equation 26.1 assuming

$$\tau_{ult} = \gamma \cdot h \cdot \mu^* \qquad (26.1d)$$

where

γ is the unit weight of the soil
h is the overburden height above the nail
μ^* is the apparent friction coefficient

As shown in Figure 26.14, the apparent friction coefficient (μ^*) obtained from pull-out tests in a nailed granular wall corresponds to the design value generally used for the ribbed metallic strips in Reinforced Earth walls. At relatively low depth, owing to the restrained dilatancy effect, the value of μ^* is significantly greater than 1 and it decreases with depth to $\tan(\phi)$.

Laboratory-scale pull-out tests were conducted by Elias and Juran (1988) in a medium dense sand to evaluate the effect of the nail installation process on the apparent friction coefficient. Figure 26.15 shows that the construction process for Reinforced Earth (i.e., placing the nail during the construction and compacting the sand around the nail) produces a substantially higher apparent friction coefficient than nailing by driving the nail into the compacted sand embankment. In the latter case, nail driving will significantly reduce the restrained dilatancy effect on the pull-out resistance. Therefore, design guidelines for Reinforced Earth walls cannot be extrapolated to soil-nailed structures.

Grouted nails are generally gravity-grouted. Their pull-out resistance is therefore expected to be approximately the same as that of an equivalent straight-shafted anchor installed under low (or no) grout pressure. The drilling of the borehole for the grouted nail produces an unloading of the disturbed surrounding soil that can significantly affect its mechanical properties. The soil–nail interaction is primarily dependent upon soil recompaction due to grouting. In cohesionless soils, grouting pressures of 50 to 100 psi are commonly used to prevent caving as the casing is withdrawn. This grouting pressure will induce ground

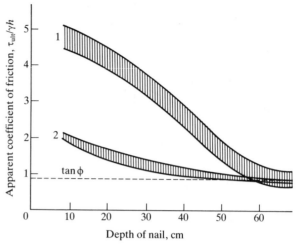

Fig. 26.15 Laboratory pull-out test results. (1) Nails placed during backfilling. (2) Nails inserted during excavation. (*After Elias and Juran, 1988.*)

recompaction associated with grout penetration into permeable gravelly seams, thereby increasing substantially the pull-out resistance of the nail. Apparent friction coefficient values as high as 3 to 6 have been reported (Elias and Juran, 1988). Figure 26.16a shows a cross-sectional view of an excavated

(a)

(b)

Fig. 26.16 (a) Cross-sectional view of an excavated grouted nail in a granular soil, illustrating effect of grout permeation. (b) Sectional view of an excavated nail in silty clay soil.

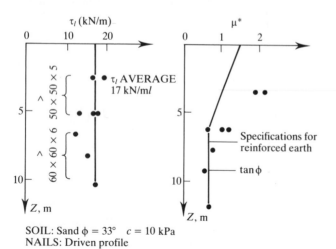

SOIL: Sand $\phi = 33°$ $c = 10$ kPa
NAILS: Driven profile

Fig. 26.14 Soil–reinforcement friction between a driven nail and a granular soil. (*After Cartier and Gigan, 1983.*)

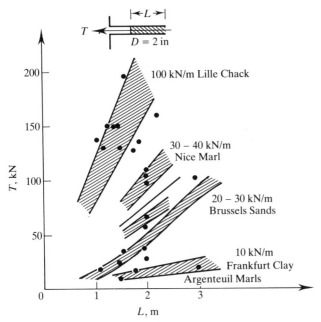

Fig. 26.17 Variation of pull-out resistance of grouted nails with embedment length. (*After Louis, 1986.*)

nailed soil, illustrating grout permeation into an alluvial soil grouted under pressure of less than 70 psi. Figure 26.16b shows that in a fine-grained cohesive soil the tremie grouting results in a rather smooth soil–inclusion interface. The presence of water at the interfaces, specifically in plastic soils, will generate a lubrification effect, decreasing substantially the pull-out resistance of the nail. Figure 26.17 (Louis, 1986) shows a

summary of pull-out test results obtained with low-pressure-grouted nails in different types of soils.

Jet-grouted nails are installed under a grout pressure that can exceed 20 MPa and is sufficiently high to cause hydraulic fracturing of the surrounding ground (Louis, 1981). Similarly to high-pressure grouting of anchors, the jet-grouting installation technique produces a mechanical interlocking between the penetrating grout and the surrounding ground that results in a substantial increase of the effective nail diameter. It also provides recompaction of the surrounding ground that significantly improves the pull-out resistance of the composite nailed-soil inclusion. Field pull-out tests on jet-grouted nails (Louis, 1986) yielded ultimate lateral shear stress values as high as 400 kPa in sands and 1000 kPa in sandy gravels.

26.4.3 Estimates of Pull-Out Cpacity from In-situ Tests

To date, estimates of the pull-out resistance of anchors and nails are mainly based upon empirical formulas (or ultimate lateral interface shear stress values) derived from field experience. These formulas are useful for feasibility evaluation and preliminary design. Table 26.2 provides a summary of estimated ultimate interface lateral shear stress (or ultimate load-transfer rate) values for soil nails and ground anchors as a function of soil (or rock) type and installation technique. Recently, increasing attempts have been made to develop field correlations between the ultimate lateral shear stress (τ_{ult}) and the engineering properties of soils obtained from commonly used in-situ tests such as the Standard Penetration Test (Fujita et al., 1977) or the self-boring pressuremeter test (Bustamente, 1975, 1976). Recognizing apparent similitude between the soil response to high-pressure anchor grouting and to the expansion of a

TABLE 26.2 ESTIMATED ULTIMATE INTERFACE LATERAL SHEAR STRESS VALUES FOR GROUND ANCHORS AND SOIL NAILS.

Grouted Nails Construction Method	Soil Type	Ultimate Lateral Shear Stress, kips/ft	
		Soil Nailing (Elias and Juran, 1988)	Permanent Ground Anchors (Cheney, 1984)[a]
Rotary drilled	Silty sand	2 to 4	5 to 9
	Silt	1.2 to 1.6	
	Piedmont residual	1.5 to 2.5	
Driven casing	Sand	6	7 to 13
	Dense sand/gravel	8	10 to 20
	Dense moraine	8 to 12	
	Sandy colluvium	2 to 4	
	Clayey colluvium	1 to 2	
Jet grouted	Fine sand (medium dense)		3.5 to 4.5[b]
	Sand	8	4.5 to 8.5[b]
	Sand/gravel	20	8.5 to 11.5[b]
Augered	Soft clay	0.4 to 0.6	
	Stiff to hard clay	0.8 to 1.2	2 to 4
	Clayey silt	1 to 2	1.5[c]
	Calcareous sandy clay	4 to 6	
	Silty sand fill	0.4 to 0.6	

[a] Cheney recommends a safety factor of 2.5 with respect to the ultimate lateral shear stress values indicated in this table.
[b] Values obtained for pressure-injected anchors by Jorge (1969).
[c] Design value proposed by Weatherby (1982) for hollow-stem augered anchor (assuming a diameter of 6 inches) in both sandy and clayey soils.

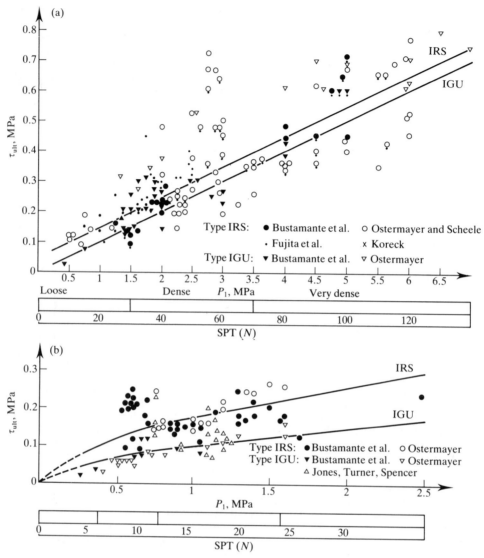

Fig. 26.18 Empirical relationships for the determination of the lateral interface shear stress. (a) Lateral interface shear stress for sand and gravel. (b) Lateral interface shear stress for silty clay soils. (*After Bustamante and Doix, 1985.*)

pressuremeter cell, the French Central Laboratory of Bridges and Roads (L.C.P.C.) has conducted an extensive research program including 94 pull-out tests in 34 sites to provide a data base for field correlations.

Figure 26.18 shows the empirical relationships derived by Bustamente and Doix (1985) to estimate the ultimate lateral shear stress values (τ_{ult}) in different types of soils and rocks as a function of the limit pressure p_l obtained from the pressuremeter test or the SPT N value. These guidelines take into account the improvement of the soil surrounding the anchors by different modes of injection, considering single-stage pressure-grouted (IGU) anchors (grout pressure of $0.5p_l < p < p_l$) and multi-stage postgrouted (IRS) anchors (grout pressure > p_l). Also shown in Figure 26.18 is the wide scatter of the field data obtained by the L.C.P.C. and other investigators (Ostermayer, 1974; Fujita et al., 1977; Ostermayer and Sheele, 1977; Koreck, 1978; Jones and Turner, 1980; Jones and Spencer, 1984) that have been compiled by Bustamente and Doix to establish these empirical relationships. The pull-out capacity of the anchor is estimated using Equation 26.1. The effective anchor diameter is estimated using a correction factor (a) to allow for diametral expansion due to high-pressure grouting. The a values for IGU

type anchors range from 1.1 in weathered rocks, silty clays, and fine sands, to 1.4 in highly dilatant granular soils, while the a values for IRS type anchors range from 1.4 in granular soils and weathered rock, to 1.8 in stiff clays and marls.

The available field data pertaining to the pull-out capacity of nails is presently still too limited to substantiate development of reliable correlations. An attempt has been made by Guilloux and Schlosser (1984) and Louis (1986) to correlate the measured pull-out capacity of both driven and grouted nails with the French recommendations (L.C.P.C. and S.E.T.R.A., 1985) for the determination of lateral shaft friction on bored and driven concrete piles from pressuremeter test results. Figure 26.19 shows that in fine-grained soils (i.e., fine sands, silts, nonplastic clays) predicted τ_{ult} values correlate reasonably well with pull-out test results, while in dilatant gravelly soils, compacted moraine, or fissured rocks they may significantly underestimate the measured ultimate lateral shear stress.

It appears that further research and field testing could significantly improve the database for estimating the pull-out capacity of ground anchors and soil nails. The pressuremeter test appears also to provide valuable data for grouting procedure, such as the maximum injection pressure that can be used

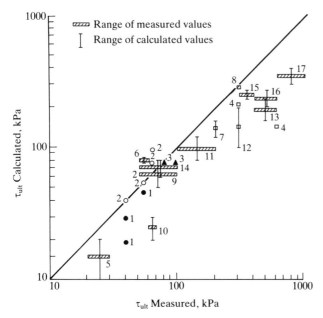

Fig. 26.19 Comparison between measured and estimated values for ultimate lateral shear stress (*Tests results reported by Guilloux and Schlosser, 1984; Louis 1986.*)
1. Driven bars in fine-grained soil.
2. Grouted bars in fine-grained soil.
3. Driven bars in granular soil.
4. Grouted bars in weathered rock.
5. Grouted bars in soft clay.
6. Grouted bars in stiff clay.
7. Grouted bars in marl.
8. Grouted bars in still marl.
9. Grouted bars in clayey silt.
10. Drilled and grouted bars in silt.
11. Drilled and grouted bars in silty sand.
12. Driven casing grouted bars in sand.
13. Driven casing grouted bars in moraine.
14. Driven casing grouted bars in colluvium clay.
15. Drilled and grouted bars in marl–limestone.
16. Drilled and grouted bars in soft rock.
17. Drilled and grouted bars in fissured rock.

without fracturing the ground. However, in light of the large variability of parameters affecting the load-transfer mechanism, specifically in heterogeneous soils, empirical correlations can only be used for preliminary design. The load-carrying capacity of each production anchor should be tested according to established standard testing procedures to assess its performance over the anticipated service life of the structure. In soil nailing, pull-out tests are required to provide reliable data for final design and to verify on site the design assumptions by testing to failure of nonservice witness nails.

26.5 APPLICATION CRITERIA: ADVANTAGES AND LIMITATIONS

Ground anchor and soil-nail systems are designed to stabilize and support natural and engineered structures and to restrain their movement using tension-resisting inclusions. The basic design concept consists of transferring the resisting tensile forces generated in the inclusions into the ground through the friction mobilized at their interfaces. These systems allow the engineer to make use efficiently of the in-situ ground in providing vertical or lateral structural support. They present significant technical advantages over bracing systems that provide external supports or conventional gravity retaining structures (e.g., reinforced-concrete walls, uplift slabs, etc.), which often require substantial weight to maintain stability. For cut slope retaining structures, which currently constitute the major application of these systems, the main technical advantages can be summarized as follows.

1. Incorporation of the temporary excavation support system in the permanent facility.
2. Reduction in the amount of excavation and the concrete work required for footing.
3. Elimination of backfilling behind the wall.
4. Elimination of foundation piles to support the structure in mountainous areas with unstable slopes or sites underlain by compressible soils.
5. Reduction in quantities of the reinforced concrete required for the construction of the retaining wall.
6. Reduction in construction disturbance and right-of-way acquisition, which in urban sites may eliminate the need for underpinning nearby structures.
7. Adaptability to different site conditions and soil profiles, allowing for cost-effective use in repair and reconstruction of existing structures.

In addition, soil nailing appears to offer some unique advantages over tie-back systems that significantly affect the cost and the construction rate of the structure, including:

1. Relatively rapid and flexible installation process of the shotcrete facing and the unstressed nails.
2. Use of light and relatively inexpensive structural elements (i.e., nails and relatively thin shotcrete or concrete facing) that may be substantially cheaper than ground anchors and structural elements (i.e., cast-in-place slurry wall, sheet piles, soldier piles, etc.) used in permanent tie-back walls.
3. Greater flexibility in adapting on site the technology (i.e., structural elements and installation technique) and structure geometry to site conditions and soil profile.
4. Light construction equipment can be used for drilling or vibropercussion and gravity grouting, which is of particular interest on sites with difficult access.
5. Soil nailing uses a large number of nails; therefore, failure of any nail may not detrimentally affect the stability of the system as would be the case for a conventional tied-back system.
6. In heterogeneous ground with boulders, dense gravelly soils, or weathered or hard rock, small-diameter drilling for soil nailing is more feasible and cost-effective than installation of soldier piles.
7. Structural flexibility: Nailed-soil structures are more flexible than conventional reinforced-concrete retaining walls and can therefore withstand larger total and differential settlements.
8. Resistance to seismic loading: As a coherent yet flexible mass, nailed-soil structures provide a relatively high degree of structural damping and therefore appear to be well adapted for construction in seismically active regions.

The main limitations of ground anchors and soil nailing technology are:

1. Permanent underground easements are required.
2. In fine-grained soils, effective groundwater drainage systems may be difficult to construct and to maintain.
3. In plastic clayey soils, creep can significantly affect long-term performance and structure displacements.
4. In soft cohesive soils, pull-out capacity of inclusions cannot be economically mobilized.

5. Durability considerations may impose severe limitations on the use of metallic inclusions in aggressive environments.

In addition, it should be pointed out that in a nailed-soil retaining structure a certain soil-to-reinforcement relative displacement is required to mobilize their interaction and generate the required resisting nail forces. Therefore, in urban areas the potential use of this technique can be limited by the requirement to prevent movement of structures in the immediate vicinity of excavation sites. Monitoring of the structure displacements can be implemented to verify that they are compatible with the required performance.

26.6 FEASIBILITY EVALUATION

To evaluate the feasibility and engineering use of permanent ground anchors or soil-nailing systems, soil conditions and existing physical constraints have to be considered.

The presence of utilities such as subways or other underground facilities and the need to obtain underground easements may preclude installation of anchors (or nails) and can significantly affect the project cost.

Durability considerations require an evaluation of the aggressiveness of the ground and the pore water, particularly when field observations indicate corrosion of existing structures. Ground anchors and nailing should not be used for permanent structures in corrosive soils (e.g., soils with high contents of cinder, ash or slag fills, rubble fills, industrial or acid mine wastes, etc.). The soil tests most commonly used to evaluate ground aggressiveness are electrical resistivity, pH, and sulfate concentration. The critical values for ground aggressiveness commonly associated with ASTM standards are outlined in FHWA DP-68-IR, Permanent Ground Anchors, and are summarized in Table 26.3.

Assessment of the suitability of the subsurface soil (or rock) to provide short- and long-term pull-out capacity of the anchor (or nail) requires a determination of its engineering properties, specifically, shear strength and creep characteristics.

In rock, the overall strength is controlled by the existing joints or discontinuity system. Highly fractured rocks with open joints or cavernous limestone are difficult to grout and therefore potential use of ground anchors or soil nails should be carefully assessed.

Permanent ground anchors and soil nails generally cannot be cost-effectively installed in loose granular soils with SPT blow count number (N) lower than 10 or relative density lower than 0.30. Nailing becomes practically unfeasible in poorly graded cohesionless soils with a uniformity coefficient of less than 2. In such soils, nailing would require stabilization of the cut face prior to excavation by grouting or slurry wall construction.

In fine-grained cohesive soils, long-term pull-out performance of the anchors (or nails) is a critical design criterion. Permanent ground anchors and soil nails are, generally, not feasible in soft

cohesive soils with undrained shear strength smaller than 0.5 tsf, or soils susceptible to creep. A number of national codes (German Standards and French Recommendations) index the creep susceptibility to the Atterberg limits and natural moisture content of the soil. They preclude the use of permanent ground anchors in organic soils, and plastic clayey soils with liquid limit (LL) greater than 50 and liquidity index (LI) greater than 0.2 (or consistency index (I_c) less than 0.9). Soils with a plasticity index (PI) greater than 20 must also be carefully assessed for creep. In light of the limited experience with soil nailing in clayey soils, the applicability criteria developed for ground anchors are recommended for feasibility evaluation of soil-nailed structures.

26.7 SHORT- AND LONG-TERM PERFORMANCE OF ANCHORS AND NAILS

The effective load transfer from the anchor to the surrounding ground requires a relative displacement between these two components of the retaining system. For ground anchors, this relative displacement is generated by prestressing the anchor immediately after installation. In the passive soil nails, resisting forces are generated owing to ground displacement during the construction. Evaluation of the short- and long-term performance of ground anchors and nails requires determination of their load–displacement–time behavior for the specific application and site conditions. Short-term performance is defined by a time-independent load–displacement relationship, while an assessment of the long-term performance should account for the effect of time-dependent phenomena such as creep and relaxation.

26.7.1 Short-term Performance

A static loading of anchors or nails can cause several "short-term" failure mechanisms:

a. Failure of the steel tendon or nails.
b. Shear failure of the soil mass owing to insufficient depth of anchor embedment.
c. Failure of the grout–tendon or nail bond.
d. Failure of the soil–grout bond.

The engineering design of the anchored (or nailed) retaining system for specific application and site conditions should provide a proper selection of the inclusion (i.e., mechanical properties, length, inclination, spacings, and corrosion protection) to prevent any of these failure modes.

(a) *Selection of tendon or nail* section should insure that the working stress in the inclusion does not exceed its ultimate tensile strength with an acceptable factor of safety. The Post Tensioning Institute (PTI) recommends limiting the working tensile stress in prestressed steel to 60 percent of the ultimate tensile strength for permanent structures and 80 percent for temporary applications.

(b) To prevent a *shear failure* of the shallow soil mass overlying the upper anchors, the bond zone should be located at a minimum depth of embedment that is generally of the order of 15 ft (4.6 m). This embedment length should also permit high-pressure grouting without damage to existing facilities.

(c) To insure that the strength of the ground is fully mobilized *the grout–tendon* (or nail) *bond* should not be exceeded. The mechanism of grout–tendon bond involves three components: adhesion, friction, and mechanical interlock. The neat cement

TABLE 26.3 FEASIBILITY CRITERIA WITH REGARD TO GROUND AGGRESSIVENESS.

Test	ASTM Standard	Critical Values
Resistivity	G-57-78 (ASTM)	Below 2000 ohm/cm
pH	G-51-77 (ASTM)	Below 4.5
Sulfate	California DOT test 407	Above 500 ppm
Chlorides	California DOT test 422	Above 100 ppm

grout generally used provides steel–cement bonding values of the order of 1 to 2 MPa, mainly owing to the mechanical grout interlocking against irregularities in the tendon surface (i.e., ribbing of the bar, threading of rebars, or stranding of wires).

(d) Failure of the *ground–grout bond* will result in sliding of the anchor or nail. The bonded anchor or nail length should be designed to ensure that the force mobilized in the inclusion does not exceed its pull-out capacity with an acceptable factor of safety. The empirical relations currently employed to estimate the pull-out capacity (or the ultimate lateral shear stress) of anchors or nails can only be used for a preliminary design. Pull-out tests on soil nails are required to provide reliable data for final design, and for anchored walls each production anchor should be tested to ensure its load-carrying capacity throughout the anticipated service life of the structure. For practical reasons, a minimum bond length of 15 ft is generally required for ground anchors in soils. Experience has shown that bond lengths exceeding 40 ft do not efficiently increase the anchor capacity.

Anchor inclination should be as small as possible. However, steep inclinations may be dictated by practical considerations, such as right-of way constraints, buried utilities, and soil profile. A minimum inclination of about 10° to 15° is generally required to facilitate and insure effective grouting, particularly under low pressure. Higher inclination of tieback anchors will result in a transfer of significant vertical forces to the structural element (i.e., concrete wall or soldier pile) and is generally used only to reach deep bearing strata or to avoid existing structures.

26.7.2 Long-Term Performance

Long-term performance of anchors or nails depends primarily upon the potential of the ground–inclusion system to creep. Theoretically, creep can develop in the three basic components of the system: the ground surrounding the bond zone, the grout, and the steel (i.e., tendon and/or connections). However, in practice, creep deformations of the cement grout and the steel are found to be insignificant, while fine-grained clayey soils may undergo large creep deformation that will result in time-dependent anchor displacement. Large creep displacements have been reported for multi-underreamed anchors (Ostermayer, 1974) and pressure-injected anchors (Bustamente et al., 1978; Bustamente, 1980) in plastic clayey soils. Relaxation of the steel tendon (i.e., stress decrease under constant strain) can also affect the long-term performance. However, for a stress level lower than the elastic limit of the steel the stress loss will generally not exceed 5 percent of the lock-off stress and its effect on the displacement will be negligible.

Creep is a time-dependent deformation of the soil structure under a sustained loading owing to a continuous fabric reorientation. The creep potential of a clayey soil is highly dependent upon the composition and structure of its minerals, its depositional (preconsolidation) history, and its natural moisture content (or consistency index). Several investigators (Murayama and Shibata, 1958; Bishop, 1966; Singh and Mitchell, 1968; Edgers et al., 1973) have shown that, as illustrated in Figure 26.20a, for most soils, under a sustained deviatoric stress, the log of strain rate is linearly decreasing with the log of time. Singh and Mitchell (1968) reported that the slope m of this linear relationship appears to be a soil property and is independent of the deviatoric stress level. The m parameter, which can be obtained from laboratory creep tests, can be used to assess the creep potential of the soil. Values of m smaller than 1 indicate a relatively high potential for accelerated creep associated with a strength loss that will induce a creep rupture. Bustamente (1980) showed that Singh and Mitchell's creep theory appears

to consistently describe the observed time-dependent anchor displacement under a constant load. He therefore suggested that the creep displacement under a sustained load can be esitmated using Singh and Mitchell's type equation

$$\Delta l = \Delta l_0 + \frac{A e^{\alpha T}}{1 - m} (t^{1-m} - 1) \tag{26.2}$$

where T and Δl_0 are respectively the applied pull-out force and the initial displacement prior to creep; A, α, and m are interface creep parameters that are obtained from the experimental $\log \dot{\Delta l}$–$\log t$ and $\log \dot{\Delta l} - T$ curves, and $\dot{\Delta l}$ is the displacement rate.

Figures 26.20b and c (Bustamente, 1980) illustrate the creep behavior of an anchor in a plastic clay and the determination of the relevant interface creep parameters. The test results indicate a steady increase of the creep displacement almost up to failure, which is consistent with the $m = 1$ value derived from the experimental $\log \dot{\Delta l}$–t curves.

In spite of the apparent similarity between the laboratory creep test results and the soil–anchor interface creep behavior observed in situ, more fundamental studies are required in order to develop a rational creep model for anchors in plastic fine-grained soils.

In practice, the critical creep load of an anchor or nail is obtained from a load-controlled pull-out test following a standard testing and interpretation procedure (DIN 4125, 1972, 1974; Bureau Securitas, 1977; Cheney, 1984). The French standard testing procedure is schematically illustrated in Figure 26.21a. Figure 26.21b shows actual results of a load-controlled pull-out test on an anchor in a plastic clay (Bustamente, 1980). It consists of 1-hour sustained load increments of $0.1F_g$ (where F_g is the elastic limit strength of the steel tendon at which permanent elongation is 0.1 percent). For each load increment the anchor displacement (s) is plotted versus log time (T). An upward concavity of the creep curve indicates an accelerated creep inducing failure. The slope of the s vs. log T line is plotted against the applied pull-out load to determine the critical creep load F_c. The allowable anchor working load F_w is the smaller of either $0.9F_c$ or $0.6F_g$. The loading increment period can significantly affect the test result. Therefore, a second test is conducted that includes a 72-hour sustained loading stage at $0.9F_w$ to verify the long-term anchor performance.

26.7.3 Repetitive Loading

Anchored structures are often subjected to repetitive (or fluctuating) live loads such as tidal variations, wind or sea wave loadings, etc. Permanent ground anchors must be designed to withstand such repetitive loadings throughout the service period of the structure, which may include millions of cycles. Documented technical data on the long-term performance of anchors under repetitive loadings are still very limited. Repetitive loading tests on anchors for a seawall in France showed (Pfister et al., 1982) that for peak cyclic load levels smaller than 63 percent of P_u (where P_u is the ultimate static pull-out capacity) anchor displacement became negligible after five cycles. However, for larger cyclic loads anchor displacement continued to increase at a constant or increasing rate. Begemann (1973) reported that repetitive uplift loads on steel H-piles in sand under cyclic load amplitude as low as 35 percent of P_u generated progressive pull-out of the piles. Laboratory model studies of repetitive loading on plate anchors and friction piles have been conducted by several investigators (Hanna et al., 1978; Andreadis et al., 1978; Hanna, 1982) and suggest some trends in the anticipated anchor response to cyclic loading. Specifically, Al-Mosawe (1979) and Hanna (1982) showed that displacement rate (per

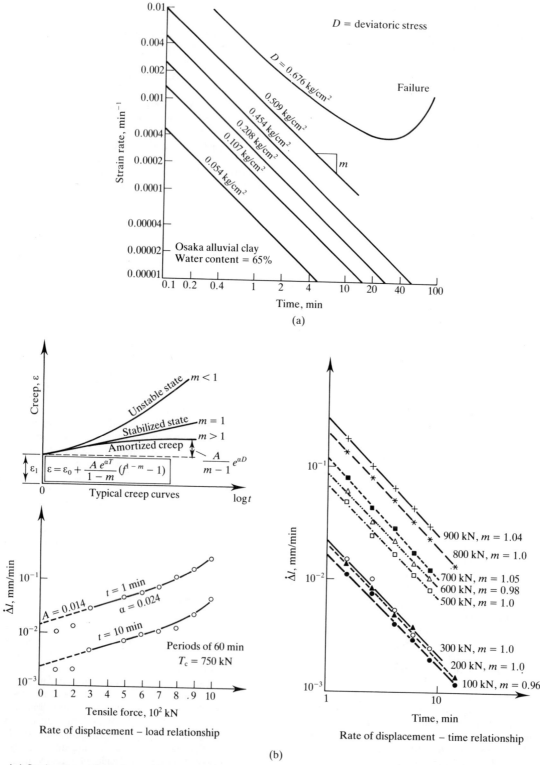

Fig. 26.20 (a) Strain rate vs. time relationship during undrained creep of alluvial clay. (*After Murayama and Shibata, 1958.*) (b) Modelling creep of anchors in clays. (Winnezeele, *Bustamante, 1980.*)

cycle) of plate anchors (Fig. 26.22a) and friction piles under repeated tensile loads gradually decreased with increasing number of cycles but did not cease. For a large number of cycles (Fig. 26.22b), large strains occurred under a cyclic stress amplitude as low as 25 percent of P_u. The higher the cyclic load amplitude, the smaller is the number of cycles required to

generate large strains. Alternating cyclic loading (tension to compression) accelerates the degradation of the anchor resistance. Prestressing the anchor increases its resistance to repeated loading.

In granular soils the effect of repeated loading on the anchor capacity appears to be mainly related to soil densification due

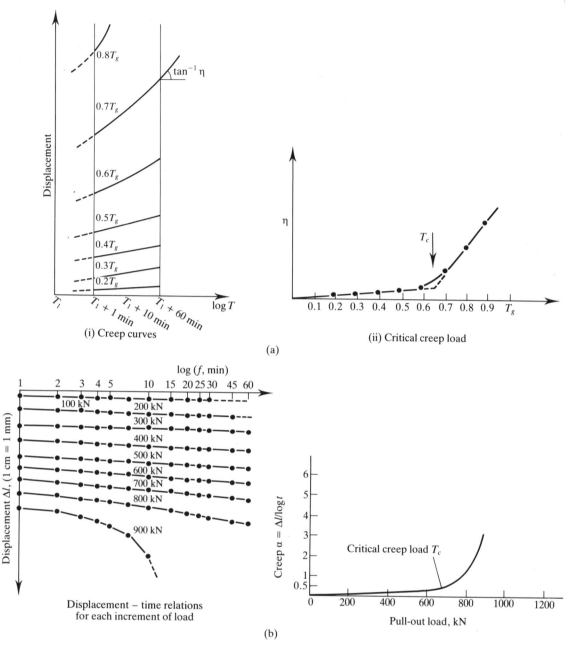

Fig. 26.21 (a) Anchor tension test for determination of critical creep load. (*Bureau Securitas, 1977*.) (b) Load-controlled pull-out test on an anchor in plastic clay. (*After Bustamente, 1980*.)

to particle reorientation, which results in a decrease of the normal stress at the soil–inclusion interface. In fine-grained soils of low permeability, the cyclic shear stress may result in a gradual increase of pore water pressure, decreasing the effective normal stress at the interfaces. However, further research and field testing are required in order to develop a database for a rational evaluation of anchor performance in clayey soils under low-frequency repetitive loading.

26.7.4 Anchor Testing and Acceptance Criteria

Anchor load tests are conducted following standard testing procedures (e.g., German DIN 4125, British Standard on Ground Anchors, French TA 77, or FHWA recommendations

for Permanent Ground Anchors by Cheney, 1984) and are conventionally classified in three categories:

1. Preproduction tests to evaluate pull-out capacity or critical creep load and to establish the design load. The testing procedure is illustrated in Figure 26.21.
2. Performance tests to verify on a limited number of production anchors that: (a) design load may be safely carried, (b) effective free length corresponds to design requirements, and (c) residual movement is within tolerable range.
3. Proof tests to verify that the load–deflection behavior of each production anchor is consistent with the specified acceptance criteria.

The *performance test* consists of incrementally applying cycles of anchor loading and unloading until the reference test

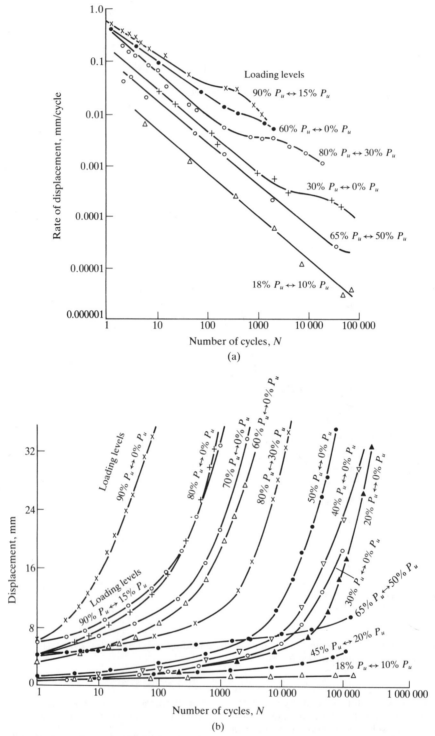

Fig. 26.22 (a) Effect of number of cycles on the rate of anchor displacement. (b) Effect of number of load cycles on anchor displacement. P_u = ultimate pull-out load. (*After Al-Mosawe, 1979.*)

load is attained. In order to determine long-term creep potential, each load increment is maintained until measured deflection is negligible (i.e., displacement rate is smaller than a specified displacement increment per log cycle of time) and a 1-hour creep test is conducted under the reference test load. Cheney (1984) recommends that the reference test load should be 1.5 times the design working load in cohesionless soils and 1.25 times the design working load in cohesive soils. The performance

tests are conducted on the first anchors (minimum of two anchors) to verify the selected installation procedure and provide reference data for the proof tests.

The *proof test* consists of a single cycle of incremental loading to the reference test loads specified above followed by unloading. Each load increment is maintained until measured deflection is negligible. The test results are compared with performance test results on an adjacent anchor.

Three acceptance criteria have been established (Cheney, 1984):

1. To ensure that the load transfer reaches the anchor bond length, the deflection of the anchor head should exceed 80 percent of the calculated elastic elongation of the unbonded tendon length.
2. The total anchor deflection measured at the maximum test load should not exceed the calculated elastic elongation of the tendon length measured from the anchor head to the center of the bond length. This criterion (not valid for anchors in layered soils or for underreamed anchors) ensures that the center of gravity of the bond stress distribution has not been transferred beyond the midpoint of the bond length.
3. Creep displacement should not exceed 0.08 inches during the final log cycle of time.

26.8 DURABILITY CONSIDERATIONS

The long-term resistance of permanent ground anchors (or nails) depends primarily upon their resistivity to corrosion, ground aggressiveness, and groundwater compositions. Underground steel corrosion is induced when a difference in potential exists between two points that are electrically connected and immersed in an electrolyte. Electrochemical cells may develop between the steel tendon and an external metal element or in local regions of inhomogeneities within the metal surface of the same tendon. In either case the chemical reaction between the groundwater and anodic regions in the exposed steel tendon results in time-dependent metal loss. For the corrosion process to occur, oxygen has to be supplied to the metal and therefore air–moisture solutions, specifically in industrial areas, and soil layers containing high oxygen content are highly corrosive. The major variables that affect the corrosion rate are:

1. Ground aggressiveness: organic soils, and acidic or highly alkaline soils that contain large concentrations of soluble salts such as sulfates, chlorides or bicarbonates, are highly corrosive.
2. Groundwater composition: acidic, alkaline, or salt solutions have high electrical conductivity, inducing high corrosion rate.
3. Differential aeration: high oxygen concentration (e.g., in fill or near the ground surface) results in a cathodic environment, its local variation in the ground generates electrochemical cells and thereby accelerates the corrosion rate.
4. High stresses or cyclic stresses in the steel tendon accelerate corrosion and may generate, in anodic environments, brittle stress-corrosion cracking.
5. Environmental hazards including bimetallic action, large temperature changes, anerobic bacteria, and stray currents in the ground (i.e., currents caused by a mass transit facility, electrical transmission, or transport systems) will generate a highly corrosive environment.

The corrosion process can develop through different mechanisms, such as uniform surface corrosion, localized pit corrosion, stress corrosion, corrosion fatigue, and hydrogen embrittlement. The type of corrosion will significantly affect the degradation rate of the steel tendon and the efficiency of potential protection systems.

Documented technical data on the long-term corrosion performance of ground anchors are very limited since only few permanent installations have been in service more than 25 years (Weatherby, 1982). However, performance trends can be anticipated on the basis of extensive research that has been conducted by the National Bureau of Standards (Romanoff, 1957) on underground corrosion.

A detailed review of corrosion-induced anchor failures by Weatherby (1982) yielded pertinent conclusions, specifically in demonstrating that:

1. Quenched and tempered prestressing steels have been involved in a significant number of tieback failures.
2. The unprotected portion of the tendon just behind the anchor head is highly susceptible to corrosion.
3. All reported failures occurred in the unbonded length of the tendon, mostly near the anchor head, where poor corrosion protection (or none) was provided.

Based on these conclusions, FHWA recommendations for permanent ground anchors (Cheney, 1984) require that all anchors used for permanent applications be corrosion-protected in the unbonded length and at the anchor head. For routine applications, only a single degree of corrosion protection is required, which may consist of a grease-filled sheath along the free stressing length and grout cover (minimum 0.5-inch thick) in the bond zone.

A variety of corrosion-protection systems have been developed. They mostly rely on the following basic principles (Weatherby, 1982; Hanna, 1982).

Simple corrosion protection relies upon cement grout to generate a noncorrosive high-pH environment and protect the tendon in the bond zone. Plastic sheaths filled with anticorrosion grease, special epoxy pitch, or cement mix, and heat-shrinkable sleeves are commonly used for corrosion protection along the free stressing length.

Coating with electrostatically applied resin-bonded epoxy can be applied to increase the corrosion protection in the bond zone. Intact resin-bonded coatings, being dielectric, will preclude the formation of galvanic cells in areas affected by microcracking.

Complete encapsulation of the steel tendon is accomplished by grouting it into a uniformly corrugated plastic or steel tube to provide double protection. The annular space between the tube and the tendon is usually filled with neat cement grout containing admixtures to control bleeding of water from the grout.

Compression steel tubes are used by European contractors to protect the tendon in the bond zone. The tube, which is high-pressure grouted into the ground, maintains the pressure-injected grout under compression, preventing microcracking. The unbonded length is generally protected using a grease-filled PVC tube.

Secondary grouting is commonly used to protect the unbonded length of the tendon. First, the anchor (primary) grout is tremied into the bond zone and the tendon is tested and locked-off. Then, the secondary (antibleed) grout is tremied around the unbonded length of the anchor, bonding it to the surrounding ground. Cheney (1984) recommends that secondary grouting be used only for semipermanent or low-risk applications. Postgrouting technique can be effectively used to provide repeated high-pressure grouting in the bond zone with corrosion protection of the tendon.

For permanent applications of *soil nailing*, based on current experience, it is recommended (Elias and Juran, 1988) that a minimum grout cover of 1.5 inches be achieved along the total length of the nail. Secondary protection should be provided by electrostatically applied resin-bonded epoxy on the bars with a minimum thickness of about 14 mils. In aggressive environments, full encapsulation is recommended. It may be achieved, as for anchors, by encapsulating the nail in corrugated plastic or steel tube grouted into the ground. For driven nails, a preassembled encapsulated nail, shown in Figure 26.23, has been developed by the French contractor Solrenfor (Louis, 1986).

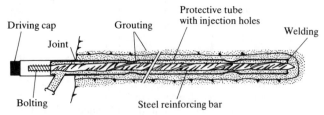

Fig. 26.23 "TBHA" nail patented and developed by Solrenfor for permanent structures.

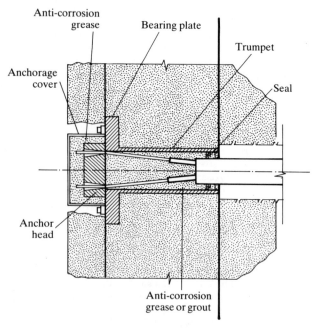

Fig. 26.24 Anchor head corrosion protection. (*After Weatherby, 1982.*)

The *anchor head* is highly susceptible to corrosion, particularly below the bearing plate. It is usually protected by encapsulation within a plastic or steel cap filled with anti-corrosion grease, or cement grout. This encapsulation should permit prestressing of the tendon and accommodate load changes in the anchor during its service life. Figure 26.24 shows anchor head details for multistrand and bar tendons.

26.9 DESIGN OF ANCHORED WALLS AND NAILED SOIL-RETAINING STRUCTURES

26.9.1 Basic Behavior and Design Concepts

The basic design concept of an anchored or soil-nailed retaining structure relies upon the transfer of resisting tensile forces generated in the inclusions (anchors or nails) into the ground through friction (or adhesion) mobilized at the interfaces. As illustrated in Figure 26.25, the ground exerts the driving forces (i.e., lateral earth pressure on the wall or weight of a potentially sliding soil mass) while providing the anchor bond resistance. The frictional interaction between the ground and the quasi "nonextensible" steel inclusions restrain the ground movement during and after excavation. The resisting tensile forces mobilized in the inclusions induce an apparent increase of normal stresses along potential sliding surfaces (or rock joints), increasing the

overall shear resistance of the native ground. The main engineering concern in the design of these retaining systems is to ensure that ground–inclusion interaction is effectively mobilized to restrain ground displacements and can secure the structure stability with an appropriate factor of safety.

In an anchored wall, the resisting tension force is mobilized by prestressing the anchor to the design working load immediately after its installation. As excavation proceeds, wall deflection is mainly controlled by the bending stiffness of the wall, the prestress load in the anchors, and the anchor longitudinal stiffness (or elastic modulus).

The effect of prestress anchor loads on wall movement measured in tied-back walls constructed both in sands and in clayey soils is illustrated in Figure 26.26 (Clough, 1975). In spite of the large scatter in the field data, which is mainly due to the differences in the construction process, structural wall components, and subsurface soil type, these results demonstrate that an increase in the prestress level results in a significant decrease of ground movement. For the sake of comparison, prestress loads based on earth pressure design values proposed by Terzaghi and Peck (1967) for braced excavations, as discussed later in this chapter, are reported.

Figure 26.27 (Goldberg et al., 1976) illustrates for braced excavations the effect of the wall bending stiffness, defined as EI/S_v^4 (where E and I are, respectively, the elastic modulus and moment of inertia of the wall, S_v is the vertical spacing between braces), on its lateral deflection in clayey soils characterized by the stability number $\gamma H/S_u$ (where H is the total structure height and S_u is the undrained shear strength of the soil). Clough and Tsui (1974) have reported that wall movement and ground settlement in anchored walls are generally smaller than those observed in braced excavations.

The design procedure of anchored walls should include the following steps.

1. Select structural wall element, and for the specified wall type, estimate the design working prestress loads in the anchors required to limit ground movement to allowable displacement values.
2. Select anchor type, corrosion protection system, length, and spacings, and verify that the anchor resistance (tensile strength and pull-out capacity) is sufficient to withstand the design working load and testing overloads to ensure long-term performance.
3. Verify that the anchor bond length is located beyond the potential sliding surface.
4. Verify that the global stability of the retaining system (ground–anchors–wall) and the surrounding ground with respect to general sliding along a potential failure is maintained with an acceptable factor of safety.
5. Structural design of the wall element with respect to the applied system of forces and bending moments.
6. Evaluation of basal stability of the wall elements (i.e., required soldier pile penetration, bearing capacity of the foundation soil below a diaphragm wall, required embedment depth of a sheet pile, etc.)
7. Select drainage system.

In soil-nailed retaining structures, the reinforcement by closely spaced passive inclusions results in a composite coherent material and, as schematically illustrated in Figure 26.25b, the maximum tensile forces generated in the nails are significantly greater than those transferred to the facing. The locus of maximum tensile forces separates the soil-nailed mass into two zones: an active zone (or potential sliding soil or rock wedge) where lateral shear stresses are mobilized at the interfaces to restrain the outward ground movement, and a resistant (or stable) zone where the generated nail forces are transferred into

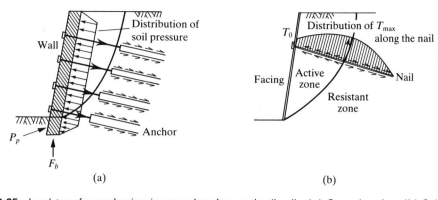

Fig. 26.25 Load-transfer mechanism in ground anchors and soil nails. (a) Ground anchor. (b) Soil nailing.

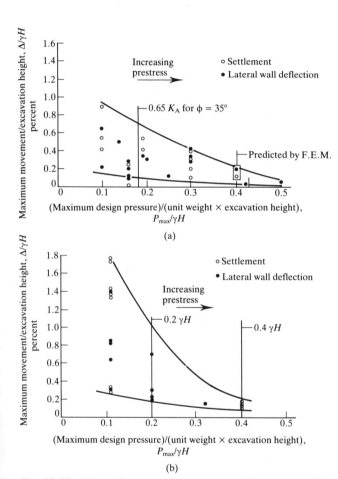

(a)

(b)

Fig. 26.26 Effect of prestress pressure on wall movement. (a) Sands. (b) Clays. (*After Clough, 1975.*)

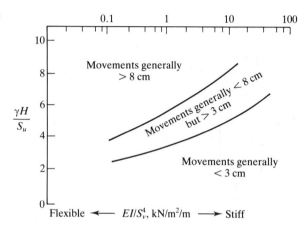

Fig. 26.27 Effect of wall stiffness on lateral wall movement. (*After Goldberg et al., 1976.*)

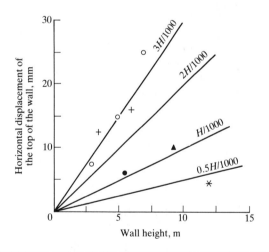

Soil	Nail	Reference
+ Medium sand	Driven	Gassler et al. (1981)
▲ Silty sand (SM)	Grouted	Shen et al. (1981)
● Fine sand (SP) to clayey sand (SC)	Driven	Cartier and Gigan (1983)
✻ Residual clayey silt weathered shale, sandstone	Grouted	Juran and Elias (1986)
○ Fontainbleau Sand (SP)	Grouted	Plumelle (1986)

Fig. 26.28 Horizontal displacement ot soil-nailed walls. (*After Juran and Elias, 1987.*)

the ground. Laboratory model tests (Juran et al., 1984) have demonstrated that this maximum tensile force line coincides with the potential sliding surface in the soil.

The soil–nail interaction is mobilized during construction and structure displacement occurs as the resisting forces are progressively generated in the nails. Therefore, it has been essential to monitor actual structures, to measure the facing displacements in different types of soils and to verify that they are compatible with performance criteria. Figure 26.28 shows field measurements of horizontal facing displacements in several soil-nailed structures (Gassler and Gudehus, 1981; Shen et al., 1981; Cartier and Gigan, 1983; Juran and Elias, 1987; Plumelle, 1986). In spite of the differences in the types of inclusions, installation techniques, and soil profiles, these data illustrate that in nonplastic soils maximum facing displacement does not exceed 0.3 percent of the structure height. This ground movement is comparable to that observed in braced and anchored retaining systems.

The design procedure of a soil-nailed retaining structure should include the following steps.

1. For the specified structure geometry (depth and cut slope inclination), ground profile, and boundary (surcharge) loadings, estimate working nail forces and location of the potential sliding surface.
2. Select the reinforcement (type, cross-sectional area, length, inclination, and spacing) and verify local stability at each reinforcement level, that is, verify that nail resistance (strength and pull-out capacity) is sufficient to withstand the estimated working forces with an acceptable factor of safety.
3. Verify that the global stability of the nailed-soil structure and the surrounding ground is maintained during and after excavation with an acceptable factor of safety.
4. Estimate the system of forces acting on the facing (i.e., lateral earth pressure and nail forces at the connections) and design the facing for specified architectural and durability criteria.
5. For permanent structures, select corrosion protection relevant to site conditions.
6. Select the drainage system for groundwater piezometric levels.

The following section outlines the design procedures currently used for development of earth pressure diagrams, selection of inclusions, and evaluation of the stability of multianchored and soil-nailed retaining structures.

26.9.2 Estimate Working Loads and Structure Displacements

Several approaches have been developed to estimate the prestress anchor loads required to limit wall displacements to a tolerable range. Similar approaches have been proposed for the design of nailed soil-retaining structures. They can be broadly classified into four main categories:

a. Empirical design earth pressure diagrams
b. "$p-y$" lateral soil reaction method
c. Finite-element analyses
d. Kinematical limit analysis method

A. Empirical Earth Pressure Diagrams

Selection of an appropriate earth pressure diagram for the determination of anchor prestress or nail loads depends upon the tolerable level of wall and ground movements. Generally, the allowable displacements of anchored walls are limited to the range of displacements observed in braced excavations. Therefore, it is common practice to estimate the prestress loads using the design diagrams proposed by Terzaghi and Peck (1948, 1967) and Tschebotarioff (1951) to estimate bracing loads (Hanna, 1982; Pfister et al., 1982; Cheney, 1984). These diagrams are schematically illustrated in Figure 26.29 (also see Chapter 12). For a multianchor system with uniform anchor spacings, the design anchor load F_w can be expressed as a normalized, nondimensional parameter:

$$TN = \frac{F_w}{(\gamma H \cdot S_h \cdot S_v)} \qquad (26.3)$$

at the relative depth of z/H. For sands,

$$(c/\gamma H < 0.05): \quad TN = 0.65 K_a \qquad (26.3a)$$

For a cohesive soil with both cohesion (c) and friction angle (ϕ):

$$TN = K_a\left(1 - \frac{4c}{\gamma H} \cdot \frac{1}{\sqrt{K_a}}\right) \qquad (26.3b)$$

where $K_a = \tan^2(\pi/4 - \phi/2)$ is the Rankine active earth pressure coefficient, H is the total excavation depth, S_h and S_v are, respectively, the horizontal and vertical anchor spacings.

Juran and Elias (1987) have shown through analysis of field measurements obtained on soil-nailed retaining structures that these earth pressure design diagrams provide a rational estimate of working tensile forces generated in the nails. On the basis of the reviewed field data, Terzaghi and Peck's design diagram for sands has been slightly modified (Fig. 26.29) in order to calculate nail forces. Figure 26.30 shows nail forces and structure displacements measured in two instrumented soil-nailed structures (field data reported by Shen et al., 1981; Plumelle, 1986). Table 26.4 summarizes the main characteristics of these structures. The measured nail forces were found to agree fairly well with the assumed earth pressure design diagram. These results illustrate that the observed behavior of nailed cut slopes is similar to that of braced excavations.

The use of the empirical earth pressure diagrams in the design of anchored and soil-nailed retaining walls presents some

Notes:
– Vertical cut slope
– Horizontal upper surface

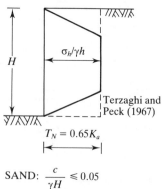

SAND: $\dfrac{c}{\gamma H} \leq 0.05$

$K_a = \tan^2(\pi/4 - \phi/2)$

σ_h = lateral earth pressure
γ_h = overburden pressure

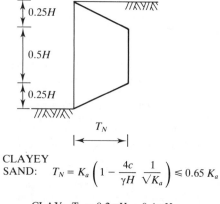

CLAYEY
SAND: $T_N = K_a\left(1 - \dfrac{4c}{\gamma H}\dfrac{1}{\sqrt{K_a}}\right) \leq 0.65\,K_a$

CLAY: $T_N = 0.2\,\gamma H \rightarrow 0.4\,\gamma H$

Fig. 26.29 Empirical earth pressure design diagram. (*After Juran and Elias, 1987.*)

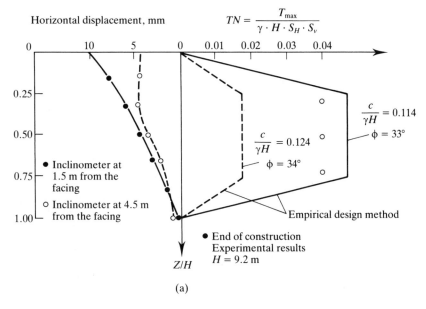

(a)

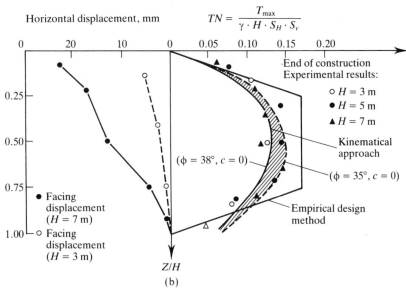

(b)

Fig. 26.30 (a) Davis wall—Experimental data and theoretical predictions of tension forces. (b) Full-scale experiment CEBTP—Experimental data and theoretical predictions of tension forces.

TABLE 26.4 CHARACTERISTICS OF ANALYZED SOIL-NAILED STRUCTURES[a].

Structure	H, m	Soil Classification	ϕ, degrees	c, kPa	γ, kN/m³	Type of Nail	L, m	S_h/S_v, m	Inclination,[b] degrees	Installation Technique[c]	F_l, kN/m
Davis wall (Shen et al., 1981)	9.2	Heterogeneous SM	36.5	18.5	16.3	#8 rebars	6	1.85 1.85	20	A	—
CEBTP wall (Plumelle, 1986)	7.5	SP	38	0 to 3	15	Al. tubes 40 × 1 mm or 30 × 2 mm	6 to 8	1.15 1.00	10	B	4.5 to 5.5

[a] After Juran and Elias (1987).
[b] *Facing:* reinforced shotcrete.
[c] *Installation technique:* A—grouted nails in 10-cm boreholes.
    B—grouted nails in 6.3-cm boreholes.

limitations. In particular, these diagrams have been developed for the conventional cases of bracing supports with the simple geometry of a vertical wall, horizontal ground surface, and lateral braces. Therefore, they cannot be used to assess the effect of varying design parameters such as inclination of the facing or the wall slope surcharge, wall stiffness, inclination, and rigidity of the inclusions, etc., on the working forces in the inclusions and structure displacements. In addition, as shown in Figure 26.30a, in cohesive soils the empirical earth pressure diagram is highly sensitive to small variations in soil properties and is, therefore, difficult to use confidently in design.

B. "p–y" Lateral Soil Reaction Model

This model is commonly used in design of laterally loaded structures (piles, foundations, or walls) to assess the lateral displacement and earth pressure distribution generated by a given system of external loads (or boundary displacements). It is based on Winkler's (1867) solution for the elastic bending of beams assuming that the soil can be represented by a series of independent elastic springs. At any point of the soil–structure interface the lateral soil stress p is assumed to be proportional to the wall deflection y, that is,

$$p = K_h \cdot y \qquad (26.4)$$

where K_h is the modulus of horizontal subgrade reaction.

Nonlinear, elastoplastic "p–y" relationships are generally used to more adequately represent the soil response. The elastoplastic reaction model proposed by Terzaghi (1948) for laterally loaded retaining walls has been adapted (Pfister et al., 1982). This model, which relates the earth pressure coefficient K to the relative wall displacement y/H, is schematically illustrated in Figure 26.31. For an assumed "p–y" relationship, the differential equation of the elastic bending of the wall is analytically or numerically integrated, yielding wall deflection, bending moments, and earth pressure distribution.

This design approach permits an evaluation of the effect of actual wall stiffness and anchor elasticity on the wall movement. As reported by Pfister et al. (1982), it generally predicts fairly well earth pressure distributions and wall deflections measured in actual structures. An iterative procedure is commonly used to assess the effect of the main parameters and search for design optimization.

The main drawback of this design approach lies in the difficulty of determining an appropriate, characteristic "p–y" relationship (or lateral reaction modulus) for the soil. Several

investigators (Baguelin et al., 1978; Briaud et al., 1983) have proposed semiempirical methods for deriving K_h values from pressuremeter test results. Pfister et al. (1982) provided useful charts, based on field experience, relating the K_h value to the shear strength parameters of the soil.

C. Finite-Element Analysis

The finite-element method has been used by several investigators to analyze the behavior of anchored walls (Clough et al., 1974; Clough and Tsui, 1974; Simpson et al., 1979; Barla and Mascardi, 1974; Egger, 1972) and soil-nailed retaining structures (Shen et al., 1981; Juran et al., 1985; Shafiee, 1986). These analyses involve different constitutive equations for the soil and interface elements to simulate soil–wall and soil–inclusion interaction. Attempts have been made by several investigators (Clough et al., 1974; Simpson et al., 1979; Shen et al., 1981) to compare finite-element predictions with observed behavior of instrumented structures. However, the use of finite-element methods in design is currently limited by the relatively high cost and raises significant difficulties with regard to the following.

1. The actual construction stages and installation process of the inclusions are difficult, if not practically impossible, to simulate.
2. The complex soil–inclusion and soil–wall interaction is difficult to model. Several interface models have been developed (Zaman et al., 1984; Frank et al., 1982), but their implementation in design requires relevant interface properties that are difficult to determine properly.
3. Various elastoplastic soil models can presently be used to predict soil behavior during excavation. However, determination of soil model parameters generally requires specific and rather elaborated testing procedures, which limits the practical use of these models.

The finite-element method has therefore been used primarily as a research tool to evaluate the effect of the main design parameters on the engineering behavior of the structure, ground movement, and working forces in the inclusions.

Clough and Tsui (1974) have shown through finite-element simulations of anchored sheet pile and slurry walls in a medium clayey soil, illustrated in Figure 26.32a, that an increase in flexibility of the wall and/or tiebacks results in larger structure displacements. With respect to soil-nail walls, Figure 26.32b shows the results of a parametric finited-element study (Juran et al., 1985; Shafiee, 1986) to evaluate the effect of bending stiffness and nail inclination on facing displacement in vertical nailed cut slopes. These results illustrate that, for nail inclinations used in practice (10° to 15°), the greater the nail bending stiffness is, the smaller is the facing displacement. As shown in Figure 26.33a, for inclined nails an increase in the bending stiffness results in a decrease of the maximum tensile forces. The behavior of the inclined nails is substantially different from that of horizontally placed nails. During construction, the relatively flexible inclined nails tend to undergo a local deformation, approaching the horizontal direction of maximum extension strain in the soil. This local deformation, which is controlled by the bending stiffness of the nails, results in an increase of the structure/facing displacements. For horizontal nails, as illustrated by both reduced-scale model tests and numerical test simulations (Figs. 26.32b and 26.33b), the bending stiffness has practically no effect on the mobilized nail forces and structure displacements. Although the finite-element results are rather qualitative, they provide a significant insight into the fundamental understanding of the system behavior and relevant input into the selection of the main design parameters.

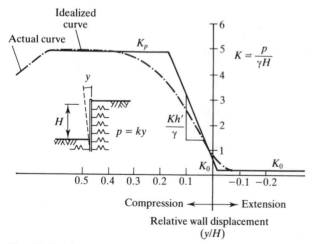

Fig. 26.31 Terzaghi's idealized relationship used in *p–y* analysis.

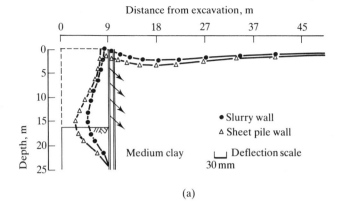

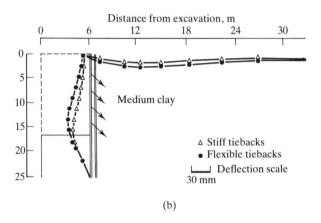

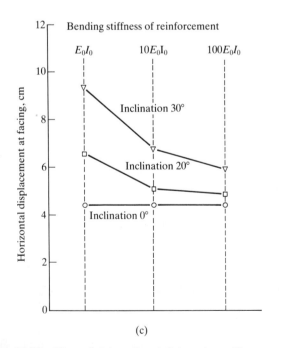

Fig. 26.32 Effect of (a) wall and (b) anchor stiffness on the movements of an anchored structure in clay; finite-element simulations. (*After Clough and Tsui, 1974.*) (c) Effect of the bending stiffness and the inclination of reinforcement on the facing displacements. (*After Shafiee, 1986.*)

D. Kinematical Limit Analysis Design Method

This limit analysis approach was developed (Juran and Beech, 1984; Juran et al., 1988) for the design of nailed soil-retaining structures. It allows for the evaluation of the effect of the main design parameters (i.e., structure geometry, inclination, spacing, and bending stiffness of nails) on the tension and shear forces generated in the nails during construction. The main design assumptions, shown in Figure 26.34, are:

a. Failure occurs by a quasi-rigid body rotation of the active zone, which is limited by a log-spiral failure surface.
b. The locus of the maximum tension and shear forces at failure coincides with the failure surface developed in the soil.
c. The shearing resistance of the soil, defined by Coulomb's failure criterion, is entirely mobilized along the sliding surface.
d. The shearing resistance of stiff inclusions, defined by Tresca's failure criterion, is mobilized in the direction of the sliding surface in the soil.
e. The horizontal components of the interslice forces E_h (Fig. 26.33) are equal.
f. The effect of a slope (or horizontal surcharge F_h), at the upper surface of the nailed soil mass, on the tension forces in the inclusions is linearly decreasing along the failure surface.

The effect of the bending stiffness is analyzed using a conventional "$p-y$" analysis procedure, assimilating the relatively flexible nail to a laterally loaded infinitely long pile. This solution implies that at the failure surface the bending moment in the nail is zero, whereas the tension and shear forces are maximum. It involves a nondimensional bending stiffness parameter, defined as

$$N = \left(\frac{K_h D}{\gamma H}\right) \frac{L_0^2}{S_h \cdot S_v} \qquad (26.5)$$

where $L_0 = [(4EI)/(K_h D)]^{1/4}$ is the transfer length, which characterizes the relative stiffness of the inclusion to the soil, E, I and D are the elastic modulus moment of inertia and diameter of the nail, respectively.

Generally, the length of the inclusion L is substantially greater than three times the transfer length L_0 and it can therefore be considered as infinitely long.

A unique, kinematically admissible, failure surface that verifies all the equilibrium conditions of the active zone can be defined. In order to establish the geometry of this failure surface, it is necessary to determine its inclination A_0 with respect to the upper ground surface. Observations on both full-scale structures (Schlosser, 1983; Juran and Elias, 1987) and laboratory model walls (Juran et al., 1984) show that for relatively flexible nails the failure surface is practically vertical at the upper part of the structure ($A_0 = 0$).

The normal soil stress along this failure surface is calculated using Kotter's equation. The maximum tension force (T_{max}) in each inclusion is calculated from the horizontal force equilibrium of the slice comprising the inclusion. Following assumption (d), analysis of the state of stress in the inclusion yields the ratio of the mobilized shear (T_c) to tension (T_{max}) forces as a function of the inclination of the inclusion with respect to the failure surface.

In order to implement the kinematical analysis approach in a detailed design of soil-nailed structures, a computer code has been developed (Juran et al., 1988). It provides for each reinforcement level at the relative depth of Z/H (where H is the total excavation depth) the nondimensional design parameters corresponding to the normalized maximum tension force

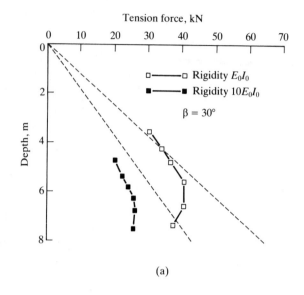

(a)

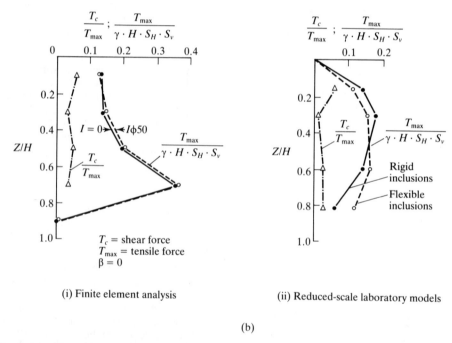

(i) Finite element analysis   (ii) Reduced-scale laboratory models

(b)

Fig. 26.33 (a) Effect of bending stiffness of the inclusions on the maximum tensile forces in inclined nails. (*After Shafiee, 1986.*) (b) Effect of the bending stiffness of the inclusions on the maximum tensile forces in horizontal nails. (*After Juran et al., 1985.*)

($TN = T_{max}/\gamma \cdot H \cdot S_h \cdot S_v$), the normalized shear force ($TS = T_c/\gamma \cdot H \cdot S_h \cdot S_v$), and the normalized distance of the locus of the maximum tension force from the facing ($SH = S/H$). Figure 26.35a illustrates these design parameters for a typical soil-nailed vertical cut slope, 12 m deep, in a silty soil ($\phi = 35°$, $c/\gamma H = 0.05$) using #8 rebar nails ($N = 0.33$) at an inclination of $\beta = 15°$. For preliminary design in homogeneous soils, simplified yet conservative, design charts have been prepared (Juran et al., 1988). Figure 26.35b provides design charts established for the common geometry of vertical facing and horizontal ground surface considering perfectly flexible nails with 15° inclination.

Figure 26.30 shows a comparison between predicted and measured values of maximum tension forces in nailed soil-retaining structures. It illustrates that the kinematical design

approach provides a reasonable estimate of tension forces mobilized in the inclusions. Specifically, the results of the full-scale experiment conducted in France on a 7-m deep granular soil-nailed wall (field data reported by Plumelle, 1986), which are reported in Figure 26.30c for several excavation depths, illustrate that the total excavation depth has only a negligible effect on both the normalized tension forces in the nails and the geometry of the active zone. Therefore, at any relative depth (Z/H) the maximum nail tension force is approximately proportional to the total excavation depth. The predicted distribution of the maximum tension forces agrees fairly well with the earth pressure design diagrams proposed for braced excavations. It can be concluded that the kinematical design approach can also be used to estimate working forces in bracing supports and anchor prestress loads.

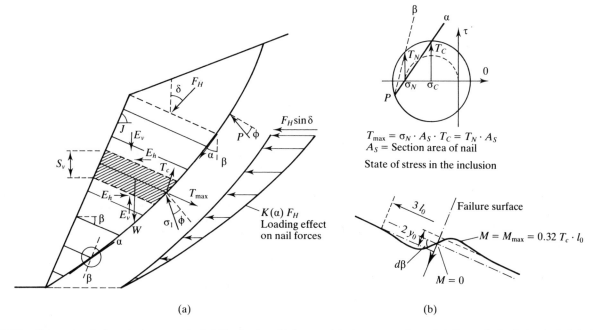

Fig. 26.34 Kinematical limit analysis approach. (a) Mechanics of failure and design assumptions. (b) Theoretical solution for infinitely long bar adopted for design purposes. (*After Juran et al., 1988.*)

26.9.3 Stability Analysis of Anchored and Soil-Nailed Retaining Structures

The design of anchored or soil-nailed retaining structures should verify:

a. The local stability at the level of each inclusion.
b. The global stability of the structure and the surrounding ground with respect to a rotational or translational failure along potential sliding surfaces.
c. For anchored walls, the basal stability must be ensured by sufficient penetration of the wall element (i.e., soldier pile, diaphragm wall, or sheet pile).

A. Local Stability Analysis (soil-nailed and multianchored walls)

At the level of each inclusion the design should satisfy the following internal failure criteria.

(1) *Pull-out failure of the inclusion:*

$$\frac{\tau_{ult}}{F_l} \geqslant \frac{T_{max}}{\pi D L_a} \tag{26.6}$$

where τ_{ult} is the limit interface lateral shear stress, T_{max} is the maximum tensile force in the nail or the design prestress load F_w in the anchor, L_a is the adherence (or bond) length, and F_l is the safety factor with respect to pull-out. This design criterion implies that for a multianchored wall or a soil-nailed cut slope, the structure geometry defined by the L/H ratio (where L is the total inclusion length) should satisfy

$$\left[\frac{L}{H}\right] \geqslant \left[\frac{S}{H}\right] + F_l\left[\frac{TN}{(\pi \cdot \mu)}\right] \tag{26.7}$$

where

$$TN = \frac{T_{max}}{\gamma \cdot H \cdot S_h \cdot S_v}$$

$$\mu = \frac{\tau_{ult} \cdot D}{\gamma \cdot S_h \cdot S_v}$$

and S is the inclusion length in the active (or unbonded) zone.

(2) *Breakage failure of the inclusion:* For *flexible nails* and anchors that withstand only tension forces,

$$\frac{f_{all} \cdot A_s}{\gamma \cdot H \cdot S_v \cdot S_h} \geqslant TN \tag{26.8}$$

where f_{all} and A_s are the allowable tension stress and cross-sectional area of the inclusion, respectively.

For *rigid nails* that can withstand both tension and shear forces,

$$\frac{f_{all} \cdot A_s}{\gamma \cdot H \cdot S_v \cdot S_h} \geqslant K_{eq} \tag{26.9}$$

where

$$K_{eq} = [(TN)^2 + 4 \cdot (TS)^2]^{1/2}$$

$$TS = \frac{T_c}{\gamma \cdot H \cdot S_h \cdot S_v}$$

and T_c is the maximum shear force in the inclusion.

(3) *Failure by excessive bending of a stiff inclusion:*

$$M_p > F_m \cdot M_{max} \tag{26.10}$$

where M_p is the plastic bending moment of the inclusion and F_m is a factor of safety with respect to plastic bending (usually, allowable tension stress is used to calculate M_p with $F_m = 1$). The bending moment M_{max} is derived from the "$p-y$" analysis:

$$M_{max} = 0.32 T_c \cdot L_0$$

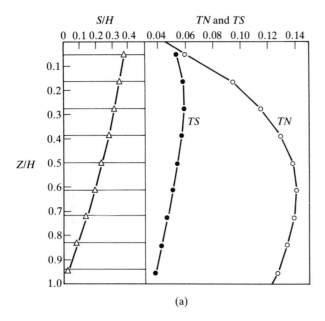

(a)

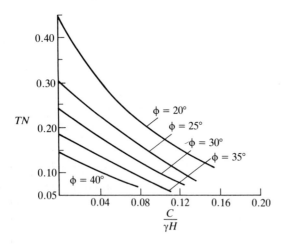

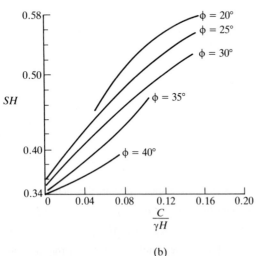

(b)

Fig. 26.35 (a) Typical example of design output provided by the kinematical limit analysis approach. (b) Design charts for perfectly flexible nails. Notes: 1, Nail inclination 15°. 2, Vertical force. 3, Horizontal backfill. 4, No surcharge. (*After Juran et al., 1988.*)

hence,

$$\frac{M_p/L_0}{\gamma \cdot H \cdot S_v \cdot S_h} > 0.32 F_m \cdot TS \tag{26.11}$$

B. Global Stability Analysis

This analysis consists of evaluating a global safety factor of the anchored or soil-nailed retaining structure and the surrounding ground with respect to a rotational or translational failure along potential sliding surfaces. It requires determination of the critical sliding surface, which may be dictated by the stratification of the subsurface soil or, in rock, by an existing system of joints and discontinuities. The potential sliding surface can be located inside or outside the anchored or soil-nailed mass. Evaluation of the global safety factor is generally based on limit equilibrium approaches. Slope stability analysis procedures have been developed to account for the available limit pull-out, tension, and shearing resistance of the inclusions crossing the potential sliding surfaces.

Limit equilibrium methods commonly used in the design of anchored walls (Kranz, 1953; Broms, 1968; Bureau Securitas, 1977) and soil-nailed retaining structures (Stocker et al., 1979; Shen et al., 1981; Schlosser, 1983) involve different definitions of the safety factors, and a variety of assumptions with regard to the shape of the failure surface, the type of soil–inclusion interaction, and the resisting forces in the inclusions.

1. Anchored Walls For anchored walls it is common practice to assume a planar failure surface passing through the toe of the wall at an inclination of ϕ or $(\pi/4 - \phi/2)$. Rankine's failure surface has been recommended by Cheney (1984). The shearing resistance of the soil, as defined by Mohr–Coulomb's failure criterion, is assumed to be entirely mobilized along the potential failure surface. The global safety factor is defined as the ratio of the sum of the available resisting limit forces (R_L) in the inclusions to the total force (R_m) required to maintain limit equilibrium, that is,

$$FS = \frac{R_L}{R_m} \tag{26.12}$$

As shown in Figure 26.36 the total force R_m required to maintain limit equilibrium is readily obtained using the polygon of forces acting on the rigid soil wedge. The resisting forces (R_L) are provided by the pull-out capacity of the anchors. Cheney recommends that the anchor bond length be located at a distance of at least $H/5$ beyond the assumed failure surface or a minimum distance of 15 ft from the facing, whichever is greater.

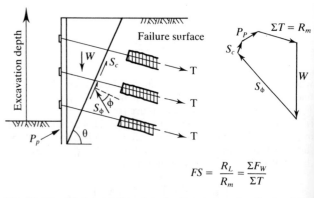

$$FS = \frac{R_L}{R_m} = \frac{\Sigma F_W}{\Sigma T}$$

Fig. 26.36 Global stability analysis of an anchored wall using force equilibrium method with a plane failure surface.

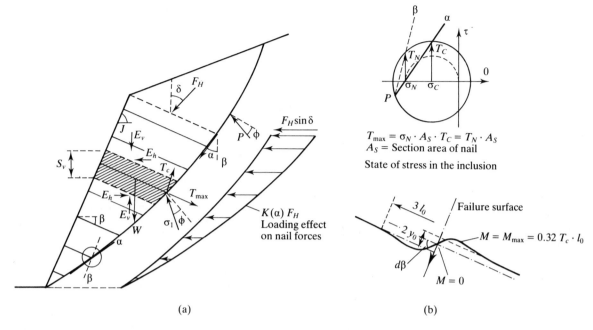

Fig. 26.34 Kinematical limit analysis approach. (a) Mechanics of failure and design assumptions. (b) Theoretical solution for infinitely long bar adopted for design purposes. (*After Juran et al., 1988.*)

26.9.3 Stability Analysis of Anchored and Soil-Nailed Retaining Structures

The design of anchored or soil-nailed retaining structures should verify:

a. The local stability at the level of each inclusion.
b. The global stability of the structure and the surrounding ground with respect to a rotational or translational failure along potential sliding surfaces.
c. For anchored walls, the basal stability must be ensured by sufficient penetration of the wall element (i.e., soldier pile, diaphragm wall, or sheet pile).

A. Local Stability Analysis (soil-nailed and multianchored walls)

At the level of each inclusion the design should satisfy the following internal failure criteria.

(1) *Pull-out failure of the inclusion:*

$$\frac{\tau_{\text{ult}}}{F_l} \geqslant \frac{T_{\max}}{\pi D L_a} \qquad (26.6)$$

where τ_{ult} is the limit interface lateral shear stress, $T_{\max}$ is the maximum tensile force in the nail or the design prestress load F_w in the anchor, L_a is the adherence (or bond) length, and F_l is the safety factor with respect to pull-out. This design criterion implies that for a multianchored wall or a soil-nailed cut slope, the structure geometry defined by the L/H ratio (where L is the total inclusion length) should satisfy

$$\left[\frac{L}{H}\right] \geqslant \left[\frac{S}{H}\right] + F_l\left[\frac{TN}{(\pi \cdot \mu)}\right] \qquad (26.7)$$

where

$$TN = \frac{T_{\max}}{\gamma \cdot H \cdot S_h \cdot S_v}$$

$$\mu = \frac{\tau_{\text{ult}} \cdot D}{\gamma \cdot S_h \cdot S_v}$$

and S is the inclusion length in the active (or unbonded) zone.

(2) *Breakage failure of the inclusion:* For *flexible nails* and anchors that withstand only tension forces,

$$\frac{f_{\text{all}} \cdot A_s}{\gamma \cdot H \cdot S_v \cdot S_h} \geqslant TN \qquad (26.8)$$

where f_{all} and A_s are the allowable tension stress and cross-sectional area of the inclusion, respectively.

For *rigid nails* that can withstand both tension and shear forces,

$$\frac{f_{\text{all}} \cdot A_s}{\gamma \cdot H \cdot S_v \cdot S_h} \geqslant K_{\text{eq}} \qquad (26.9)$$

where

$$K_{\text{eq}} = [(TN)^2 + 4 \cdot (TS)^2]^{1/2}$$

$$TS = \frac{T_c}{\gamma \cdot H \cdot S_h \cdot S_v}$$

and T_c is the maximum shear force in the inclusion.

(3) *Failure by excessive bending of a stiff inclusion:*

$$M_p > F_m \cdot M_{\max} \qquad (26.10)$$

where M_p is the plastic bending moment of the inclusion and F_m is a factor of safety with respect to plastic bending (usually, allowable tension stress is used to calculate M_p with $F_m = 1$). The bending moment $M_{\max}$ is derived from the "$p-y$" analysis:

$$M_{\max} = 0.32 T_c \cdot L_0$$

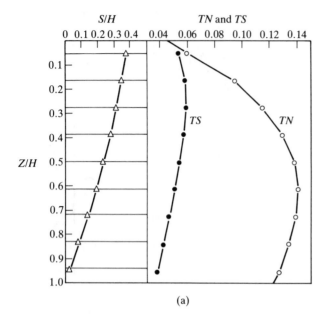

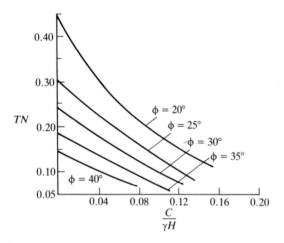

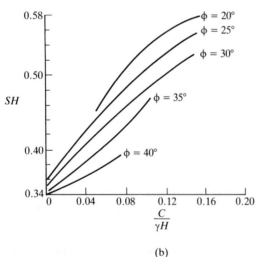

(b)

Fig. 26.35 (a) Typical example of design output provided by the kinematical limit analysis approach. (b) Design charts for perfectly flexible nails. Notes: 1, Nail inclination 15°. 2, Vertical force. 3, Horizontal backfill. 4, No surcharge. (*After Juran et al., 1988.*)

hence,

$$\frac{M_p/L_0}{\gamma \cdot H \cdot S_v \cdot S_h} > 0.32 F_m \cdot TS \qquad (26.11)$$

B. Global Stability Analysis

This analysis consists of evaluating a global safety factor of the anchored or soil-nailed retaining structure and the surrounding ground with respect to a rotational or translational failure along potential sliding surfaces. It requires determination of the critical sliding surface, which may be dictated by the stratification of the subsurface soil or, in rock, by an existing system of joints and discontinuities. The potential sliding surface can be located inside or outside the anchored or soil-nailed mass. Evaluation of the global safety factor is generally based on limit equilibrium approaches. Slope stability analysis procedures have been developed to account for the available limit pull-out, tension, and shearing resistance of the inclusions crossing the potential sliding surfaces.

Limit equilibrium methods commonly used in the design of anchored walls (Kranz, 1953; Broms, 1968; Bureau Securitas, 1977) and soil-nailed retaining structures (Stocker et al., 1979; Shen et al., 1981; Schlosser, 1983) involve different definitions of the safety factors, and a variety of assumptions with regard to the shape of the failure surface, the type of soil–inclusion interaction, and the resisting forces in the inclusions.

1. Anchored Walls For anchored walls it is common practice to assume a planar failure surface passing through the toe of the wall at an inclination of ϕ or $(\pi/4 - \phi/2)$. Rankine's failure surface has been recommended by Cheney (1984). The shearing resistance of the soil, as defined by Mohr–Coulomb's failure criterion, is assumed to be entirely mobilized along the potential failure surface. The global safety factor is defined as the ratio of the sum of the available resisting limit forces (R_L) in the inclusions to the total force (R_m) required to maintain limit equilibrium, that is,

$$FS = \frac{R_L}{R_m} \qquad (26.12)$$

As shown in Figure 26.36 the total force R_m required to maintain limit equilibrium is readily obtained using the polygon of forces acting on the rigid soil wedge. The resisting forces (R_L) are provided by the pull-out capacity of the anchors. Cheney recommends that the anchor bond length be located at a distance of at least $H/5$ beyond the assumed failure surface or a minimum distance of 15 ft from the facing, whichever is greater.

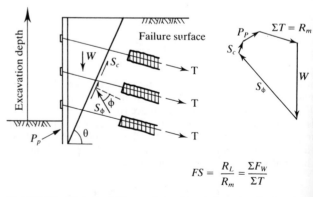

$$FS = \frac{R_L}{R_m} = \frac{\Sigma F_W}{\Sigma T}$$

Fig. 26.36 Global stability analysis of an anchored wall using force equilibrium method with a plane failure surface.

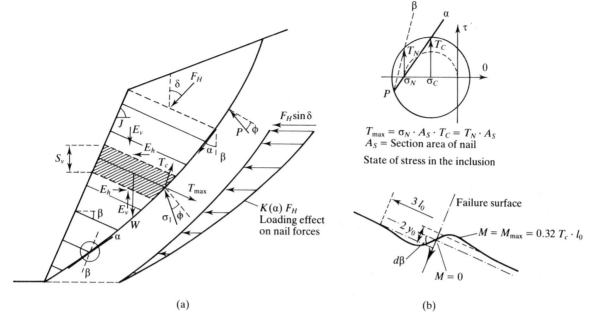

Fig. 26.34 Kinematical limit analysis approach. (a) Mechanics of failure and design assumptions. (b) Theoretical solution for infinitely long bar adopted for design purposes. (*After Juran et al., 1988.*)

26.9.3 Stability Analysis of Anchored and Soil-Nailed Retaining Structures

The design of anchored or soil-nailed retaining structures should verify:

a. The local stability at the level of each inclusion.
b. The global stability of the structure and the surrounding ground with respect to a rotational or translational failure along potential sliding surfaces.
c. For anchored walls, the basal stability must be ensured by sufficient penetration of the wall element (i.e., soldier pile, diaphragm wall, or sheet pile).

A. Local Stability Analysis (soil-nailed and multianchored walls)

At the level of each inclusion the design should satisfy the following internal failure criteria.

(1) *Pull-out failure of the inclusion:*

$$\frac{\tau_{ult}}{F_l} \geqslant \frac{T_{max}}{\pi D L_a} \tag{26.6}$$

where τ_{ult} is the limit interface lateral shear stress, T_{max} is the maximum tensile force in the nail or the design prestress load F_w in the anchor, L_a is the adherence (or bond) length, and F_l is the safety factor with respect to pull-out. This design criterion implies that for a multianchored wall or a soil-nailed cut slope, the structure geometry defined by the L/H ratio (where L is the total inclusion length) should satisfy

$$\left[\frac{L}{H}\right] \geqslant \left[\frac{S}{H}\right] + F_l\left[\frac{TN}{(\pi \cdot \mu)}\right] \tag{26.7}$$

where

$$TN = \frac{T_{max}}{\gamma \cdot H \cdot S_h \cdot S_v}$$

$$\mu = \frac{\tau_{ult} \cdot D}{\gamma \cdot S_h \cdot S_v}$$

and S is the inclusion length in the active (or unbonded) zone.

(2) *Breakage failure of the inclusion:* For *flexible nails* and anchors that withstand only tension forces,

$$\frac{f_{all} \cdot A_s}{\gamma \cdot H \cdot S_v \cdot S_h} \geqslant TN \tag{26.8}$$

where f_{all} and A_s are the allowable tension stress and cross-sectional area of the inclusion, respectively.

For *rigid nails* that can withstand both tension and shear forces,

$$\frac{f_{all} \cdot A_s}{\gamma \cdot H \cdot S_v \cdot S_h} \geqslant K_{eq} \tag{26.9}$$

where

$$K_{eq} = [(TN)^2 + 4 \cdot (TS)^2]^{1/2}$$

$$TS = \frac{T_c}{\gamma \cdot H \cdot S_h \cdot S_v}$$

and T_c is the maximum shear force in the inclusion.

(3) *Failure by excessive bending of a stiff inclusion:*

$$M_p > F_m \cdot M_{max} \tag{26.10}$$

where M_p is the plastic bending moment of the inclusion and F_m is a factor of safety with respect to plastic bending (usually, allowable tension stress is used to calculate M_p with $F_m = 1$). The bending moment M_{max} is derived from the "$p-y$" analysis:

$$M_{max} = 0.32 T_c \cdot L_0$$

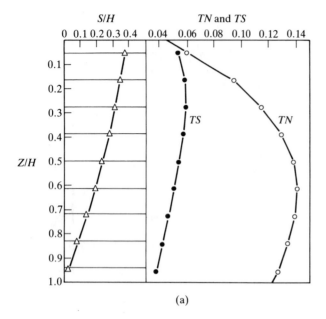

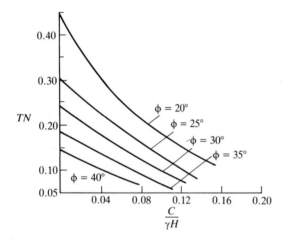

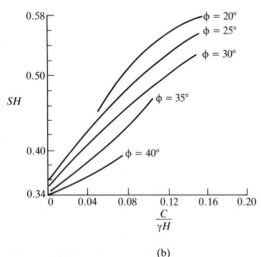

(b)

Fig. 26.35 (a) Typical example of design output provided by the kinematical limit analysis approach. (b) Design charts for perfectly flexible nails. Notes: 1, Nail inclination 15°. 2, Vertical force. 3, Horizontal backfill. 4, No surcharge. (*After Juran et al., 1988.*)

hence,

$$\frac{M_p/L_0}{\gamma \cdot H \cdot S_v \cdot S_h} > 0.32 F_m \cdot TS \qquad (26.11)$$

B. Global Stability Analysis

This analysis consists of evaluating a global safety factor of the anchored or soil-nailed retaining structure and the surrounding ground with respect to a rotational or translational failure along potential sliding surfaces. It requires determination of the critical sliding surface, which may be dictated by the stratification of the subsurface soil or, in rock, by an existing system of joints and discontinuities. The potential sliding surface can be located inside or outside the anchored or soil-nailed mass. Evaluation of the global safety factor is generally based on limit equilibrium approaches. Slope stability analysis procedures have been developed to account for the available limit pull-out, tension, and shearing resistance of the inclusions crossing the potential sliding surfaces.

Limit equilibrium methods commonly used in the design of anchored walls (Kranz, 1953; Broms, 1968; Bureau Securitas, 1977) and soil-nailed retaining structures (Stocker et al., 1979; Shen et al., 1981; Schlosser, 1983) involve different definitions of the safety factors, and a variety of assumptions with regard to the shape of the failure surface, the type of soil–inclusion interaction, and the resisting forces in the inclusions.

1. Anchored Walls For anchored walls it is common practice to assume a planar failure surface passing through the toe of the wall at an inclination of ϕ or $(\pi/4 - \phi/2)$. Rankine's failure surface has been recommended by Cheney (1984). The shearing resistance of the soil, as defined by Mohr–Coulomb's failure criterion, is assumed to be entirely mobilized along the potential failure surface. The global safety factor is defined as the ratio of the sum of the available resisting limit forces (R_L) in the inclusions to the total force (R_m) required to maintain limit equilibrium, that is,

$$FS = \frac{R_L}{R_m} \qquad (26.12)$$

As shown in Figure 26.36 the total force R_m required to maintain limit equilibrium is readily obtained using the polygon of forces acting on the rigid soil wedge. The resisting forces (R_L) are provided by the pull-out capacity of the anchors. Cheney recommends that the anchor bond length be located at a distance of at least $H/5$ beyond the assumed failure surface or a minimum distance of 15 ft from the facing, whichever is greater.

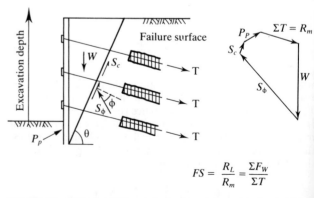

$$FS = \frac{R_L}{R_m} = \frac{\Sigma F_W}{\Sigma T}$$

Fig. 26.36 Global stability analysis of an anchored wall using force equilibrium method with a plane failure surface.

The early work of Terzaghi (1948) demonstrated that the shape of the failure surface behind rigid retaining walls is highly dependent upon the displacement mode and associated earth pressure distribution. In braced excavations or anchored walls, restraining the lateral soil displacement results in a nonplanar failure surface that is likely to be vertical in the upper part of the wall. Therefore, the assumption of a Rankine failure surface is inconsistent with the actual behavior of these structures. Cheney's recommendations provide an initial assessment for the location of the anchor bond zone that has to be verified by global stability analysis.

Kranz's (1953) limit equilibrium method generalized by Broms (1968) considers (Fig. 26.37) a bilinear sliding surface passing through the midpoint of the anchor shear force (or hinge) in the wall. The soil resistance to shearing (i.e., cohesion component S_c and frictional component S_ϕ) along the lower segment of the potential failure surface is assumed to be entirely mobilized. The retained soil is assumed to apply an active lateral earth pressure P_a on the vertical segment of the potential failure surface calculated using Rankine's theory (i.e., $P_a = \frac{1}{2} K_a \cdot \gamma \cdot Z_0^2$, where Z_0 is the depth of the midpoint of the anchor bond length).

The polygon of forces acting on the soil wedge at limit equilibrium involves the active earth pressure of the retained soil P_a, the shearing resistance of the soil (i.e., S_c and S_ϕ), and the wall reaction P_A. The polygon of forces acting on the structural wall element involves the soil pressure P_A, the resisting prestress anchor load required to maintain equilibrium, the passive resistance of the foundation soil P_p and the basal soil reaction F_b.

Kranz (1953) considered the limit equilibrium of the soil wedge neglecting the passive (P_p) and basal (F_b) reaction forces applied by the foundation soil. Equation 26.12 is used to define the factor of safety, that is, $FS = R_L / R_m$, with $R_L = F_w$ and $R_m = T$.

Broms (1968) extended Kranz's method, taking into account the passive soil resistance P_p and the basal reaction F_b in the equilibrium of forces acting on the anchored wall–soil wedge system. The factor of safety was defined as $FS = P_p^{\lim} / P_p$, where $P_p^{\lim}$ is the limit passive resistance of the foundation soil that can be estimated using Rankine's theory and P_p is the passive soil reaction required to maintain the limit equilibrium of the anchored wall–soil wedge system.

In its generalized form (Fig. 26.37), Kranz's (1953) and Brom's (1968) limit equilibrium method requires evaluation of the passive soil resistance P_p and of the basal reaction F_b. The passive soil resistance on the wall element is estimated using Rankine's theory or Coulomb's limit equilibrium method, which are commonly used in design of retaining walls. The basal reaction F_b depends upon the interaction between the wall element and the foundation soil.

For soldier piles and sheet piles, the basal reaction is mobilized through lateral shaft friction and can be estimated using bearing-capacity formulas that are commonly used in design of friction piles. For cast-in-place diaphragm walls, the basal reaction can be estimated using Meyerhof's (1953) bearing-capacity formula for shallow foundations under inclined eccentric loads.

In the global stability analysis of the anchored wall–soil wedge system, the safety factors with respect to the passive soil resistance P_p and basal reaction F_b should be consistent with current design procedures commonly used for retaining walls, friction piles, or shallow foundations. The prestress anchor load required to maintain the system equilibrium can be estimated from the polygon of forces. The global safety factor should be evaluated with respect to pull-out failure of the anchor following Kranz's original definition given by Equation 26.12.

The major advantage of Brom's method is its simplicity. However, it is noted that this method is based on a restrictive assumption concerning the shape of the potential sliding surface and under its extended form requires an evaluation of the soil–wall element interaction, which is difficult to assess. Furthermore, the assumed definition of the failure surface is difficult to generalize for the case of multianchored walls. A generalized Kranz's method is used in the French code of practice (Bureau Securitas, 1977) to allow for the design of multianchored walls; however, it leads to significant over-estimates of the required prestress anchor loads. Kranz's and Brom's methods provide a safety factor with respect to failure along a potential sliding surface passing through the anchors. However, the global stability of the anchored wall system and the surrounding ground with respect to general sliding along potential failure surfaces that do not cross the anchors should be investigated using conventional slope stability analysis methods.

Slope stability analysis procedures are increasingly used in design of anchored wall systems for the evaluation of the global safety factor with respect to general sliding along a potential failure surface that may pass within or outside the anchored wall system. These procedures (Pfister et al., 1982; Blondeau et al., 1984) mostly use conventional limit equilibrium slice methods (see Chapter 10) that have been modified to account for the prestress load and pull-out capacity of the anchors crossing the potential sliding surface.

2. Soil-Nailed Retaining Structures The three limit equilibrium methods currently most used for nailed soil-retaining structures are the so-called German method (Stocker et al., 1979), Davis method (Shen et al., 1981), and French method (Schlosser, 1983).

(a) THE GERMAN METHOD. Stocker and coworkers (1979) proposed a limit force equilibrium method (Fig. 26.38) that, following the principles of Kranz's method, assumes a bilinear sliding surface and adapts the same definition for the global factor of safety (Eq. 26.12). The inclination (θ_1) of the failure surface is iteratively determined to yield the minimum factor of safety. Gassler and Gudehus (1981) have shown, through stability analyses, that the minimum safety factor value is usually obtained for $\theta_2 = (\pi/4 - \phi/2)$ and $\theta_s = 90°$. In this analysis, the resisting limit nail force is provided by the pull-out capacity of the nails (i.e., the pull-out capacity of the portion of the nail located beyond the potential failure surface).

The assumed bilinear failure surface is mainly based on a limited number of model tests where failure was generated by substantial surcharge loading. However, it does not appear to be consistent with the observed behavior of nailed soil-retaining structures that are primarily subjected to their self-weight. In particular, stability analyses show that this bilinear failure

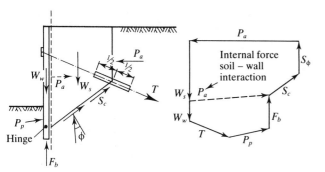

Fig. 26.37 Generalized Kranz's force equilibrium method for stability analysis of an anchored wall. (*After Broms, 1968.*)

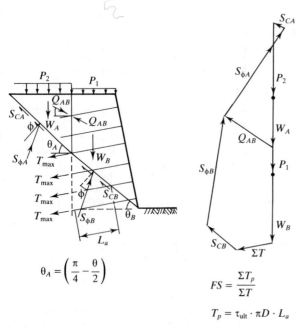

$$\theta_A = \left(\frac{\pi}{4} - \frac{\theta}{2}\right)$$

$$FS = \frac{\Sigma T_p}{\Sigma T}$$

$$T_p = \tau_{\text{ult}} \cdot \pi D \cdot L_a$$

Fig. 26.38 Force equilibrium method of global stability analysis of nailed soil-retaining structure. (*After Stocker et al., 1979.*)

surface is generally not entirely contained within the nailed-soil mass and therefore yields an active zone (or potential failure wedge) that is substantially larger than that observed on actual structures.

(b) THE DAVIS METHOD. Shen et al. (1981) developed a similar force equilibrium method assuming a parabolic failure surface, passing either entirely or partially within the nailed-soil mass. Their assumption is based on the contours of factor of safety derived from finite-element simulations, as shown in Figure 26.39a. Nails are assumed to withstand only tension forces and their failure can therefore be generated by either breakage or pull-out. In this analysis it is implicitly assumed that the safety factors with respect to the shear strength of the soil (i.e., $F_c = c/c_m$ and $F_\phi = \tan \phi/\tan \phi_m$, where c_m and ϕ_m are, respectively, the soil cohesion and internal friction angle mobilized along the potential sliding surface) and the ultimate interface lateral shear stress (i.e., $F_L = \tau_{\text{ult}}/\tau_m$, where τ_m is the lateral shear stress mobilized at the soil–inclusion interfaces) are equal to the global safety factor FS, that is,

$$FS = F_c = F_\phi = F_L \tag{26.13}$$

A slope stability analysis procedure, using a modified method-of-slices, has been implemented to iteratively determine the critical sliding surface and the minimum factor of safety.

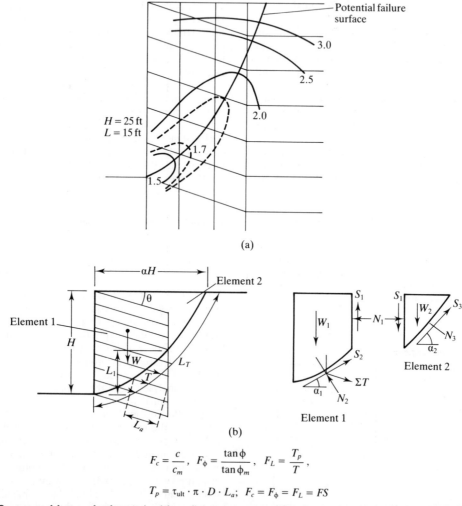

$$F_c = \frac{c}{c_m}, \quad F_\phi = \frac{\tan \phi}{\tan \phi_m}, \quad F_L = \frac{T_p}{T},$$

$$T_p = \tau_{\text{ult}} \cdot \pi \cdot D \cdot L_a; \quad F_c = F_\phi = F_L = FS$$

Fig. 26.39 (a) Contours of factor of safety derived from finite-element analysis. (b) Limit force equilibrium method for stability analysis of nailed soil-retaining structures. (*After Shen et al., 1981.*)

To calculate the interslice forces, a stress ratio parameter K (i.e., ratio of the lateral to vertical stresses at the interslice) is input with K values of 0.4 for frictional soils and 0.5 for cohesive soils.

Shen et al. (1981) have evaluated their design procedure through analysis of observed failure surfaces and failure heights of centrifugal soil-nailed model walls. As shown in Figure 26.39b the method's predictions agree fairly well with the experimental results.

(c) THE FRENCH METHOD. Common to all the limit equilibrium methods specified above is the assumption that the inclusions withstand only tension forces. A more general approach for the stability analysis of soil-nailed retaining structures has been developed by Schlosser (1983), considering the two fundamental mechanisms of soil–inclusion interaction (i.e., lateral friction and passive normal soil reaction). This method takes into account both the tension and shearing resistance of the inclusions as well as the effect of their bending stiffness. For an inclusion that withstands both tension (T_{max}) and shear (T_c) forces, the mobilized limit forces are calculated according to the principle of maximum plastic work considering Tresca's failure criterium. The T_c/T_{max} ratio is a function of the inclination of the inclusion with respect to the potential failure surface. The tensile strength of the inclusion is defined by the elastic limit $f_{all} = f_y$, and the shear resistance by $R_c = f_y/2$. The limit shear force that can be generated in the nail depends upon the mobilized passive lateral soil pressure on the inclusion. In order to prevent plastic flow (or creep) of the soil between the inclusions, the maximum lateral soil pressure p_{lim} should not exceed half of the ultimate lateral pressure of the characteristic "$p–y$" curve. In French practice, this lateral soil pressure is limited to the creep pressure obtained from a pressuremeter test. The shear force in the inclusion should therefore not exceed

$$T_c = p_{lim} \cdot L_0 \cdot D/2 \qquad (26.14)$$

A multicriteria analysis, illustrated in Figure 26.40, is conducted to evaluate the global stability of the nailed-soil system with respect to four potential failure modes: shear failure of the soil along the critical sliding surface, pull-out failure of the nail, nail breakage by either excessive bending or combined effect of tension and shear forces, and creep or plastic flow of the soil between the nails. The global factor of safety is defined by Equation 26.13 (i.e., $F_c = F_\phi = F_l = F_s$) and a minimum safety factor of 1.5 is generally required. This multicriteria analysis procedure uses a slices method (see Chapter 10) that is modified to take into account the resisting nail forces in the equilibrium of each slice. This procedure permits an evaluation

of the effect of soil stratification, groundwater flow, and seismic loading on the global structure stability. It can also be used for the design of mixed structures associating ground anchors and soil nailing.

Postfailure analyses of several nailed soil-retaining structures reported by Blondeau et al. (1984) have illustrated that with an appropriate input design value of the ultimate lateral shear stress this design procedure could predict fairly well the pull-out failure of the structures.

3. Evaluation of Global Stability Analysis Procedures The Davis and the French design procedures have been evaluated (Juran and Elias, 1987) through the analysis of field data from the full-scale structures. These analyses suggest that:

1. In soil-nailed cut slopes the factors of safety with respect to soil strength are fairly close to 1.
2. The Davis method generally yields safety factor values that are about 15 percent lower than those predicted using the French method.
3. Observed loci of the maximum tension forces in these structures agree fairly well with predicted location of the potential failure surface (Fig. 26.41).

These results further suggest that in soil-nailed retaining structures, owing to the staged construction process the soil resistance to shearing along the potential failure surface is

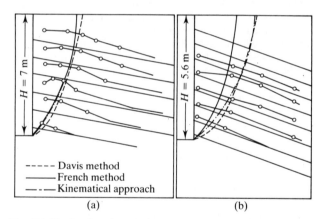

- - - - - Davis method
———— French method
—·—·— Kinematical approach

(a)   (b)

Fig. 26.41 Predicted and observed locus of maximum tension forces in nails. (a) Full-scale experiment CEBTP. (*Experimental results, Plumelle, 1986.*) (b) Parisian wall. (*Experimental results, Cartier and Gigan, 1983.*) (*After Juran and Elias, 1987.*)

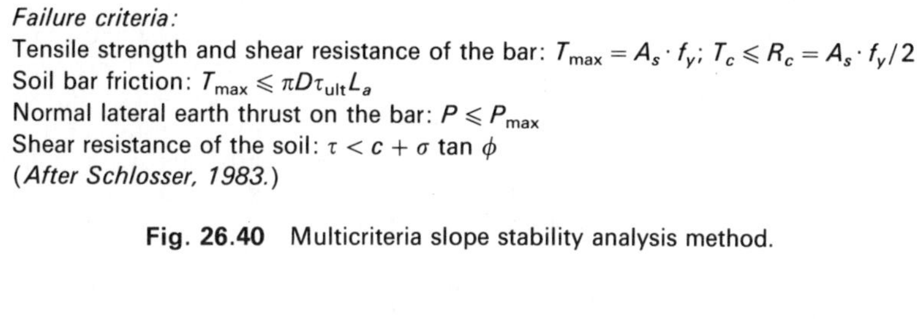

Forces in the bar

Failure criteria:
Tensile strength and shear resistance of the bar: $T_{max} = A_s \cdot f_y$; $T_c \leqslant R_c = A_s \cdot f_y/2$
Soil bar friction: $T_{max} \leqslant \pi D \tau_{ult} L_a$
Normal lateral earth thrust on the bar: $P \leqslant P_{max}$
Shear resistance of the soil: $\tau < c + \sigma \tan \phi$
(*After Schlosser, 1983.*)

Fig. 26.40 Multicriteria slope stability analysis method.

practically mobilized at the early stages of excavation. As the excavation proceeds, the load increments are entirely transferred to the inclusions. Therefore, for internal stability analysis of these structures, the definition of the factor of safety adapted by Kranz for anchored walls and Stocker et al. for soil-nailed structures appears to be more consistent with the observed structure behavior. It implies that the global factor of safety should be evaluated with respect to the pull-out resistance of the inclusions considering a safety factor of 1 with respect to shear resistance of the soil. For permanent anchored walls Cheney (1984) recommends a minimum safety factor of 1.5, whereas for soil-nailed walls Elias and Juran (1988) recommend a minimum safety factor of 2. Gassler and Gudehus (1981) recommend for soil-nailed structures the use of residual soil-strength parameters (c and ϕ) factored by 1.25 to comply with statistical evaluation criteria concerning the probability of failure.

The major limitation of the slope stability analysis procedures currently used in design of soil-nailed retaining structures lies in the basic definition of a global factor of safety. Observations on both full-scale structures and reduced-scale laboratory models have illustrated that pull-out failure is a progressive phenomenon that is generally induced by the sliding of the upper inclusions. Therefore, this internal failure mechanism cannot be adequately defined using a "global" value of a unique safety factor for all the inclusions. The local stability (or safety factor) at the level of the sliding inclusion can be significantly more critical than the "global" stability with respect to general sliding in the retaining system or the surrounding ground. For a reliable design of these composite structures, it is therefore essential that the engineer should attempt to rationally evaluate both the local internal stability at the level of each inclusion and the "global" structure stability.

C. Structural Design of Anchored Walls or Facing Elements and Stability Analysis of the Wall Base

The design of the structural elements of an anchored wall or of the facing of a soil-nailed retaining structure involves three main steps:

1. Estimation of the system of applied forces and bending moments.
2. Structural analysis (i.e., calculation of internal stresses and moments) and selection of appropriate wall or facing elements.
3. Stability analysis of the structural element base.

The basic differences in construction process of a soil-nailed facing and anchored wall elements result in a substantially different interaction of these structural elements with the foundation soil. As illustrated in Figure 26.25, the flexible facing of a soil-nailed structure does not transfer any load to the foundation soil while the anchored wall is designed to mobilize the passive soil resistance and transfer the vertical component of the anchor load to the foundation soil. Therefore, in the case of inclined anchors, the basal stability of the structural wall element has to be carefully evaluated.

The system of forces acting on the facing or wall elements includes the resisting forces in the inclusions (anchors or nails), the lateral soil pressure and the reaction of the foundation soil on an anchored wall. The forces in the inclusions are the prestress anchor loads or the forces in nail connections, which are assumed to be equal to the maximum tension nail forces. The lateral soil reaction on the anchored wall is governed by the passive resistance of the foundation soil, which can be estimated using conventional limit equilibrium approaches.

The design of the structural element (i.e., soldier pile, concrete wall, sheet pile, reinforced shotcrete, etc.) is generally done by assimilating this element to a beam or raft (width equal to the vertical or horizontal spacing between the inclusions) on simple supports formed by the inclusions. For the design loads in the supports and the estimated lateral passive soil reaction, the moments in the structural element can be calculated and the selection of the element follows usual structural design procedures. "$p-y$" type analyses are commonly used in design of diaphragm walls (Pfister et al., 1982) to assess the effect of wall stiffness and anchor longitudinal rigidity on the structure displacements.

Evaluation of the basal stability of anchored walls requires a rational estimate of the vertical load transferred to the foundation soil. This vertical load is highly dependent upon the frictional interaction between the structural element, the foundation soil, and the retained ground, which is difficult to assess. A conservative working assumption neglecting the effect of soil–structure interaction above the foundation level has been recommended by Cheney (1984). This design assumption governs the required wall penetration and may significantly affect construction cost. Therefore, additional research and site monitoring are needed to provide a relevant database for the development of a more rational design assumption.

For the estimated vertical load, the basal stability of the structural wall element can be analyzed using conventional bearing-capacity limit equilibrium methods. Required penetration depth of soldier sheet piles can be estimated using bearing-capacity formulas that are commonly used in design of friction piles. The basal stability of a diaphragm wall can be evaluated using Meyerhof's (1963) formula for the bearing capacity of shallow foundations under inclined eccentric loads.

26.10 TECHNOLOGICAL DEVELOPMENT AND RESEARCH NEEDS

The increasing use of ground anchors and soil nails in permanent structures is a key parameter in current technological developments. Durability of the inclusions, long-term performance in fine-grained soil, and environmental/architectural requirements for the anchored wall or soil-nailed facing have become major design considerations.

The potential application of the technology in more aggressive environments has stimulated specialty contractors to continuously invest in the development of more reliable and cost-effective corrosion-protection schemes. In particular it is noted that driven soil nails, which are commonly used in Europe, are not corrosion-protected. Their implementation in the construction of permanent structures requires innovative improvements in the manufacturing and/or installation process to provide adequate protective coatings that may resist construction damage. Attempts have been made by French contractors to provide corrosion protection of driven nails using a variety of techniques such as coupling nail driving with jet grouting, driving encapsulated nails, or driving prefabricated nails that consist of prestressed bars in compression tubes.

For the construction of permanent soil-nailed retaining structures, shortcomings of the shotcrete facing technology have generated development of prefabricated facing elements (e.g., concrete or steel panels) that may efficiently accommodate adequate facing drainage and economically comply with durability and architectural requirements. In addition, technological efforts have been invested by European contractors to develop cost-effective driving–installation processes, such as jet nailing, that may efficiently improve the pull-out capacity while substantially increasing construction rate and thereby decreasing the project cost.

As it is often the case in geotechnical engineering, technology and construction practice with ground anchors and soil nailing have always preceded any fundamental research on the behavior or long-term performance of the anchored or nailed ground system and any experimental or theoretical basis for development of appropriate design methods. However, implementation of newly developed corrosion-protection systems has generated a substantial requirement for innovative quality control procedures to properly assess the performance of the proposed protection scheme. Furthermore, the growing use of the technology in permanent structures and potential economical applications in fine-grained soils have raised significant research needs pertaining to the following.

(1) Long-term performance in fine-grained soils. The creep response of the soil is highly dependent upon the strain path and the induced shear strain rate, which in turn are mainly governed by the mode of soil–inclusion interaction (i.e., lateral shaft friction or passive soil resistance on transversal underreams) and the applied stress level. Therefore, basic research on the effect of the installation process (i.e., regrouting procedure, grouting pressure) and inclusion shape on the long-term inclusion performance may provide a relevant database for extending the range of fine-grained soils in which ground anchors and soil nails can be confidently used.

(2) Estimate of the pull-out capacity of grouted nails. The current design procedure relies upon the estimate of an ultimate interface lateral shear stress. However, the load transfer along a grouted inclusion is nonlinear and therefore the interface ultimate lateral shear stress derived from pull-out test results is length-dependent. This scale effect is difficult to assess and the use of short testing nails may result in an overestimate of the pull-out capacity of production nails. Development of a more rational load-transfer model and relevant testing interpretation procedure may significantly improve the estimate of the pull-out capacity of grouted nails.

(3) Evaluation of soil–structure interaction in the anchored wall–retained ground–foundation soil system. This complex soil–structure interaction may significantly affect the load transferred to the foundation soil and thereby the required penetration depth of the structural wall element.

(4) Performance of ground anchors and soil nails under repetitive cyclic loadings.

It is anticipated that monitoring of the long-term performance of structures, development of reliable quality control procedures, testing of witness inclusions for corrosion studies, and basic research on soil–inclusion and soil–structure interactions under both monotonic and cyclic loadings, will significantly contribute to improving the state of design and practice. This may substantially enhance technological innovations and cost-effective use in more aggressive environments and in a wider range of fine-grained soils.

REFERENCES

Al-Mosawe, M. M. (1979), The effect of repeated and alternating loads on the behavior of dead and prestressed anchors in sand, Thesis, University of Sheffield, England.

Andreadis, A., Harvey, R. C., and Burley, E. (1978), Embedment anchors subjected to repeated and alternating loads, *Ground Engineering*, **11**, No. 3.

Baguelin, F., Jezequel, J. F., and Shields, D. H. (1978), *The Pressuremeter and Foundation Engineering*, Trans Tech Publications, Clausthal, Germany.

Barla, G. and Mascardi, C. (1974), High anchored wall in Genoa, *Conference on Diaphragm Walls and Anchorages*, Institute of Civil Engineers, London, pp. 123–128.

Bassett, R. H. (1977), Underreamed ground anchors, Specialty session No. 4, *Proceeding of the 9th International Conference on Soil Mechanics and Foundation Engineering*, Tokyo, pp. 11–17.

Begemann (1973), Alternating loads and pulling tests on steel I-beam piles, *Proceedings of the 8th International Conference on Soil Mechanics and Foundation Engineering*, **2.1**, pp. 13–17.

Bishop, A. W. (1966), The strength of soils as engineering materials, *Geotechnique*, **16**, pp. 91–128.

Blondeau, F., Christiansen, M., Guilloux, A., and Schlosser, F. (1984), TALREN: methode de calcul des ouvrages en tere renforcee, *Proceedings of the International Conference on In-situ Soil and Rock Reinforcements*, Paris, pp. 219–224.

Briaud, J. L., Smith, T., and Meyer, B. (1983), Pressuremeter gives elementary model for laterally loaded piles, *International Symposium on In-situ Testing*, Paris, **2**, pp. 217–221.

British Standards Institute (1980), October 1980 Draft British Code of Practice for Ground Anchors, DSB 22 Committee.

Broms, B. B. (1968), Swedish tieback system for sheet pile walls, *Proceedings of the 3d Budapest Conference on Soil Mechanics and Foundation Engineering*, pp. 391–403.

Bureau Securitas (1977), *Recommendations Concerning the Concepts, the Calculation, the Execution, and the Control of Ground Anchors*, T.A. 77, Editions Eyrolles, 61, Boulevard Saint-Germain 75005, Paris.

Bustamante, M. (1972), Essais pre'alables de tirants precontraints definitifs pour la rive gauche de la Seine, Pont de Saint-Cloud–Pont de Sevres, *Travaux*, No. 450.

Bustamante, M. (1975), Mesure des elongations dans les pieux et tirants a l'aide d'extensometres amovibles, *Travaux*, No. 489.

Bustamante, M. (1976), Essais de pieux de haute capacite scelles par injection sous haute pression, *Proceedings of the 6th European Conference on Soil Mechanics and Foundation Engineering*, Vienna.

Bustamante, M. (1980), Capacité d'ancrage et comportement des tirants injectes, scelles dans une argile plastique, Thèse docteur-ingenieur ENPC, Paris.

Bustamante, M., Delmas, F., and Lacour, J. (1977), Behavior of prestressed anchors in plastic clay, *Proceedings of the 9th International Conference on Soil Mechanics and Foundation Engineering*, Special Session No. 4, Tokyo.

Bustamante, M., Delmas, F., and Lacour, J. (1978), Comportement de tirants preconstraints dans une argile plastique, *Revue Française de Geotechnique*, **3**.

Bustamante, M. and Doix, B. (1985), Une methode pour le calcul des tirants et des micropieux injectes, *Bulletin de liaison des Laboratoires des Ponts et Chaussees*, No. 140.

Cartier, G. and Gigan, J. P. (1983), Experiments and observations on soil nailing structures, *Proceedings of the 7th European Conference on Soil Mechanics and Foundation Engineering*, Helsinki, Finland.

Cheney, R. S. (1984), Permanent Ground Anchors, Federal Highway Administration Report No. FHWA-DP-68-1R.

Clement, P. and Navarro, M. (1972), Les tirants en terrain meuble type TM, *Travaux*, No. 450.

Clough, G. W. (1975), Deep excavations and retaining structures, *Proceedings of the Conference on Foundations of Tall Buildings*, Lehigh University.

Clough, G. W. and Tsui, Y. (1974), Performance of tied-back walls in clay, *Journal of the Geotechnical Engineering Division, ASCE*, **100**, No. GT-12, pp. 1259–1273.

Clough, G. W., Weber, P. R., and Lamont, J. (1974), Design and observations of a tied back wall, *Proceedings of the Specialty Conference on Performance of Earth and Earth Supported Structures*, ASCE, Purdue, **2**, pp. 1367–1389.

Coyle, M. M. and Reese, L. C. (1966), Load transfer for axially loaded piles in clay, *Journal of the Soil Mechanics and Foundations Division, ASCE*, **92**, No. SM-2, pp. 1–26.

Darbin, M., Jailloux, J. M., and Montuelle, J. (1978), Performance and research on the durability of reinforced earth reinforcing strips, *ASCE Symposium on Earth Reinforcement*, Pittsburgh, Pa., pp. 305–333.

Davis, A. and Plumelle, C. (1982), Identification et étude des parametres controlant le comportement des tirants d'ancrage dans un sable fin, *Annales de ITBTP*, No. 401.

DIN (1972, 1976), Deutsche Industrie Norm, *Soil and Rock Anchors: Temporary Soil Anchors, Analysis, Structural Design and Testing*, DIN 4125, Part 1, pp. 1–9, 1972; Part 2, pp. 1–9, 1976.

Edgers, L., Ladd, C. C., and Christian, J. T. (1973), *Undrained Creep of Atchafalaya Levee Foundation Clays*, Vol. 1, Report R73-16, Soils Publication 319, Department of Civil Engineering, MIT.

Egger, P. (1972), Influence of wall stiffness and anchor prestressing on earth pressure distributions, *Proceedings of the 5th European Conference on Soil Mechanics and Foundation Engineering*, Madrid, **1**, pp. 259–264.

Elias, V. and Juran, I. (1988), *Draft Manual of Practice for Soil Nailing*, prepared for U.S. Department of Transportation, FHWA, Contract DTFH-61-85-C-00142.

Evans, R. H. (1955), Application of prestressed concrete to water supply and drainage, *Public Health Engineering Division Meeting*, Public Health Paper No. 12, London.

Feddersen, I. (1974), Verpessanker in Lockergestein (Grouted Anchors in Soils), *Bauingenieur*, **49**, No. 8, pp. 302–310.

FIP (1974), Federation Internationale de la Précontrainte, Ground Anchors, *Proceedings of the 7th Congress of the FIP*, New York, pp. 33–42.

Frank, R., Guenot, A., and Humbert, P. (1982), Numerical analysis of contact in geomechanics, *Proceedings of the 4th International Conference on Numerical Methods in Geomechanics*, Edmonton.

Fujita, K., Ueda, K., and Kusabuka, M. (1977), A method to predict the load–displacement relationship of ground anchors, Specialty Session No. 4, *Proceedings of the 9th International Conference on Soil Mechanics and Foundation Engineering*, Tokyo.

Gassler, G. and Gudehus, G. (1981), Soil nailing: Some mechanical aspects of in situ reinforced earth, *Proceedings of the 10th International Conference on Soil Mechanics and Foundation Engineering*, Stockholm, Sweden, **3**, pp. 665–670.

Goldberg, D. T., Jawouski, W. E., and Gordon, M. D. (1976), *Lateral Support Systems and Underpinning, Construction Method*, Final Report No. FHWA-RD-75-130.

Guilloux, Notte and Gonin (1983), Experience on a retaining structure by nailing, *Proceedings of the 7th European Conference on Soil Mechanics and Foundation Engineering*, Helsinki, Finland.

Guilloux, A. and Schlosser, F. (1984), Soil nailing: Practical applications, *Proceedings of the Symposium on Soil and Rock Improvement Techniques*, AIT, Bangkok.

Hanna, T. H. (1982), *Foundations in Tension*, Trans Tech Publications, Series on Rock and Soil Mechanics, Vol. 6.

Hanna, T. H., Sivapalon, E., and Senturk, A. (1978), The behavior of dead anchors subjected to repeated and alternating loads, *Ground Engineering*, **11**, No. 3.

Hovart, C. and Rami, R. (1975), Elargissement de l'emprise SNCF pour la desserte de Saint-Quentin-en-Yvelines, *Revue Travaux*.

Jones, D. A. and Turner, M. J. (1980), Load tests on post-grouted micropiles in London clay, *Ground Engineering*, **6**, No. 13.

Jones, D. A. and Spencer, I. M. (1984), Clay anchors: A Caribbean case history, *Ground Engineering*, **17**, No. 1.

Jones, N. C. and Kerkhoff, G. O. (1961), Beleld caissons anchor walls as Michigan remolds an expressway, *Engineering News Record*, pp. 28–31, 195–197.

Jorge, G. R. (1969), The regroutable IRP anchorage for soft soils, low capacity or karstic rocks, *Proceedings of the 7th International Conference on Soil Mechanics and Foundation Engineering*, Specialty Session No. 14 and 15, pp. 159–163.

Jorge, J. (1970), Le tirant IRP reinjectable pour terrains meubles, karstiques, ou a faibles caracteristiques geotechniques, *Proceedings of the 2d International Congress on Rock Mechanics*, Mexico.

Juran, I., Beech, J., and Delaure, E. (1984), Experimental study of the behavior of nailed soil retaining structures on reduced scale models, *Proceedings of the International Conference on In-situ Soil and Rock Reinforcements*, Paris.

Juran, I. and Beech, J. (1984), Analyse Théorique du Comportement d'un Soutenement en Sol Cloue, *Proceedings of the International Conference on In-Situ Reinforcement of Soils and Rock Reinforcements*, Paris, pp. 301–307.

Juran, I., Shafiee, S., and Schlosser, F. (1985), Numerical study of nailed soil retaining structures, *Proceedings of the 11th International Conference on Soil Mechanics and Foundation Engineering*, San Francisco, **4**, pp. 1713–1717.

Juran, I. and Elias, V. (1987), Soil nailed retaining structures: Analysis of case histories, *ASCE Special Geotechnical Publication* No. 12, pp. 232–245.

Juran, I., Baudrand, G., Farrag, K., and Elias, V. (1988), Kinematical limit analysis approach for the design of nailed soil retaining structures, *Proceedings of the International Geotechnical Symposium on Theory and Practice of Earth Reinforcement*, Fukuoka Kyushu, Japan.

Kerisel, J., Robert, J., Schlosser, F., Juran, I., Causse, G., and Romon, C. (1981), Experimentation d'un Mur d'Ancrages Multiples (Experiments on a multi-anchored wall), *Proceedings of the 10th International Conference on Soil Mechanics and Foundation Engineering*, Stockholm, **2**, pp. 157–161.

Koreck, W. (1978), Small diameter bored injection piles, *Ground Engineering*, May.

Kranz, E. (1953), *Über die Verankerung von Spundwänden*, Wilh. Ernst & Sohn, 2 Aufl., Berlin.

Littlejohn, G. S. (1970), Soil anchors, *Proceedings of the Conference on Ground Engineering*, Institution of Civil Engineers, London, pp. 33–44.

Littlejohn, G. S. and Bruce, D. A. (1975), Rock anchors state-of-the-art, Part I: Design and Part II: Construction, *Ground Engineering*, May.

Louis, C. (1981), Nouvelle methode de soutenement des sols en deblais, *Revue Travaux* No. 533.

Louis, C. (1986), Theory and practice in soil nailing temporary or permanent works, *ASCE Annual Conference*, Boston.

L.C.P.C.-S.E.T.R.A. (1985), Regles de justification des fondations sur pieux.

Mastrantuono, C. and Tomiolo, A. (1977), First application of a totally protected anchorage, *Proceedings of the 9th International Conference on Soil Mechanics and Foundation Engineering*, Specialty Session, Tokyo, pp. 107–112.

McKittrick, D. P. (1979), Reinforced earth: Application of theory and research practice, *Ground Engineering*, **12**, No. 1, pp. 19–31.

Meyerhof, G. G. (1953), The bearing capacity of foundations under eccentric and inclined loads, *Proceedings of the 3d International Conference on Soil Mechanics and Foundation Engineering*, Zurich, **1**, p. 440.

Meyerhof, G. G. (1963), Some recent research on the bearing capacity of foundations, *Canadian Geotechnical Journal*, **1**, pp. 16–26.

Middleton, H. (1961), Raising the Argal dam, *The Consulting Engineer*, **II**.

Mitchell, J. K. et al. (1987), Reinforcement of earth slopes and embankments, *National Cooperative Highway Research Program Report* No. 290, Transportation Research Board, June.

Morris, S. S. (1956), Steenbras dam strengthened by post tensioning cables, *Civil Engineering*, **2**.

Murayama, S. and Shibata, T. (1958), *On the Rheological Characteristics of Clays*, Part I, Bulletin No. 26, Disaster Prevention Research Institute, Kyoto, Japan.

Ostermayer, M. (1974), Construction, carrying behavior and creep characteristics of ground anchors, *Conference on Diagram Walls and Anchorages*, Institute of Civil Engineers, London.

Ostermayer, M. and Sheele, F. (1977), Research on ground anchors in non-cohesive soils, *Proceedings of the 9th International Conference on Soil Mechanics and Foundation Engineering*, Tokyo.

Peck, R. B. (1958), A study of the comparative behavior of friction piles, *Highway Research Board Special Report* No. 36, p. 72.

Pfister, P., Evers, G., Guilland, M., and Davidson, R. (1982), *Permanent Ground Anchors: Soletanche Design Criteria*, Federal Highway Administration Report No. FHWA-RD-81-150.

PTI (1980), Post-Tensioning Institute, *Recommendations for Prestressed Rock and Soil Anchors*, PTI, 301 W. Osborne, Suite 3500, Phoenix, Ariz., 85013, p. 57.

Plumelle, C. (1986), Full scale experimental nailed-soil retaining structures, *Revue Française de Geotechnique*, No. 40, pp. 45–50.

Plumelle, C. (1987), Experimentation en vraie grandeur d'une paroi clouée, *Revue Française de Geotechnique*, No. 40, pp. 45–50.

Plumelle, C. and Gasnier, R. (1984), Etude Experimentale en Vraie Grandeur de Tirants d'Ancrage (Full scale tests on ground anchors), *Proceedings of the International Symposium on In-Situ Reinforcement in Soils and Rocks*, Paris.

Rabcewicz, L. V. (1964-65), The new Austrian tunnelling method, Parts I to III, *Water Power*, London, Dec. 1964 and Jan. 1965.